Capilano College Library

WITHDRAWN

Capilano
College

LIBRARY DONATION

This book is

a Gift of

Allen Clapp

WITHDRAWN

Environmental Engineers' Handbook

Second Edition

Capilano College Library

Environmental Engineers' Handbook

Second Edition

David H.F. Liu
Second Edition Editor

Béla G. Lipták
Handbook Editor

Paul A. Bouis
Special Consultant

LEWIS PUBLISHERS

Boca Raton   New York

54617297
.61949548
.01045441

Copyright © 1997, 1974 by Béla G. Lipták
Second Edition   All Rights Reserved
Published by CRC Press LLC, 2000 Corporate Blvd., N.W., Boca Raton,
Florida 33431

No part of this book may be reproduced, transmitted or stored
in any form or by any means, electronic or mechanical,
without prior written permission from the publisher

Designed by Arlene Putterman
Illustrations by General Graphic Services, Inc.
Manufactured in the United States of America

Library of Congress Cataloging-in-Publication Data

Environmental engineers' handbook / David H.F. Liu, editor
  2nd ed.
    p.   cm.
  Rev. ed. of: Environmental engineers' handbook / Béla G. Lipták.
  1st ed. 1973–74.
    Includes bibliographical references and index.
    ISBN 0-8493-9971-8
    1. Environmental engineering.   I. Liu, David.
  TD145.E574   1997
  628—dc21   96-46781
                                                                  CIP

1 2 3 4 5 6 7 8 9 0   6 5 4 3 2 1 0 9 8 7

Acknowledgments

On behalf of my late husband, David Liu, I would like to convey his sincere gratitude and respect for all the coauthors who helped, directly or indirectly, currently or in the past, in this book's development. With your help, he accomplished his goal: a comprehensive, authoritative, and current reference. The valuable expertise, strong support, and dedication of all the coauthors will make the *Environmental Engineers' Handbook* an unqualified success.

Special appreciation is extended to Béla Lipták and Paul Bouis, who did the final technical review of manuscript, art and page proofs, sharing their valuable time and advice to complete David's work.

Thanks also to the publishing staff under the leadership of Chris Kuppig, General Manager, with special mention of the key role played by Kathryn Conover, a devoted editor.

Irene Liu
Princeton, New Jersey
November, 1996

Contents

6   Noise Pollution   449

7   Wastewater Treatment   507

8   Removing Specific Water Contaminants   927

11 Hazardous Waste   1249

Contributors

Irving M. Abrams
BCh, PhD; Manager, Technical Development,
Diamond Shamrock Chemical Company

Carl E. Adams, Jr.
BSCE, MSSE, PhDCE, PE; Technical Director,
Associated Water & Air Resources Engineers, Inc.

Elmar R. Altwicker
BS, PhD; Professor, Department of Chemical Engineering,
Rensselaer Polytechnic Institute

Donald B. Aulenbach
BSCh, MS, PhDS; Associate Professor,
Bio-Environmental Engineering,
Rensselaer Polytechnic Institute

Richard C. Bailie
BSChE, MSChE, PhDChE;
Professor of Chemical Engineering,
West Virginia University

Edward C. Bingham
BSCh, MBA; Technical Assistant to General Manager,
Farmers Chemical Association, Inc.

L. Joseph Bollyky
PhD; President,
Pollution Control Industries Ozone Corp.

David R. Bookchin, Esq.
BA, JD, MSL; private practice, Montpelier, Vermont

Paul A. Bouis
BSCh, PhDCh; Assistant Director, Research &
Development, Mallinckrodt Baker, Inc.

Jerry L. Boyd
BSChE; Chief Process Application Engineer, Eimco
Corp.

Thomas F. Brown, Jr.
BSAE, EIT; Assistant Director, Environmental Engineering,
Commercial Solvents Corp.

Barrett Bruch
BSME, BSIE; Oil Spill Control Project Leader,
Lockheed Missiles & Space Company

Robert D. Buchanan
BSCE, MSCE, PE; Chief Sanitary Engineer,
Bureau of Indian Affairs

Don E. Burns
BSCE, MSCE, PhD-SanE; Senior Research Engineer,
Eimco Corp.

Larry W. Canter
BE, MS, PhD, PE;
Sun Company Chair of Ground Water Hydrology,
University of Oklahoma

Paul J. Cardinal, Jr.
BSME; Manager, Sales Development, Envirotech Corp.

Charles A. Caswell
BS Geology, PE; Vice President,
University Science Center, Inc.

Samuel Shih-hsien Cha
BS, MS; Consulting Chemist, TRC Environmental Corp.

Yong S. Chae
AB, MS, PhD, PE; Professor and Chairman,
Civil and Environmental Engineering, Rutgers University

Karl T. Chuang
PhDChE; Professor, Department of Chemical
Engineering, University of Alberta

Richard A. Conway
BS, MSSE, PE; Group Leader, Research & Development,
Union Carbide Corp.

George J. Crits
BSChE, MSChE, PE; Technical Director,
Cochrane Division, Crane Company

Donald Dahlstrom
PhDChE; Vice President and Director of Research &
Development, Eimco Corp.

Stacy L. Daniels
BSChE, MSSE, MSChE, PhD; Development Engineer,
The Dow Chemical Company

Ernest W.J. Diaper
BSc, MSc; Manager,
Municipal Water and Waste Treatment,
Cochrane Division, Crane Company

Frank W. Dittman
BSChE, MSChE, PhD, PE;
Professor of Chemical Engineering, Rutgers University

Wayne F. Echelberger, Jr.
BSCE, MSE, MPH, PhD; Associate Professor of Civil
Engineering, University of Notre Dame

Mary Anna Evans
BS, MS, PE; Senior Engineer, Water and Air Research,
Inc.

Jess W. Everett
BSE, MS, PhD, PE; Assistant Professor, School of Civil
Engineering and Environmental Engineering, University
of Oklahoma

David C. Farnsworth, Esq.
BA, MA, JD, MSL; Vermont Public Service Board

J.W. Todd Ferretti
President, The Bionomic Systems Corp.

Ronald G. Gantz
BSChE; Senior Process Engineer,
Continental Oil Company

William C. Gardiner
BA, MA, PhD, PE; Director, Electrochemical Development,
Crawford & Russell, Inc.

Louis C. Gilde, Jr.
BSSE; Director, Environmental Engineering,
Campbell Soup Company

Brian L. Goodman
BS, MS, PhD; Director, Technical Services,
Smith & Loveless Division, Ecodyne Corp.

Ahmed Hamidi
PhD, PE, PH, CGWP; Vice President,
Sadat Associates, Inc.

Negib Harfouche
PhD; President, NH Environmental Consultants

R. David Holbrook
BSCE, MSCE; Senior Process Engineer, I. Krüger, Inc.

Sun-Nan Hong
BSChE, MSChE, PhD; Vice President, Engineering,
I. Krüger, Inc.

Derk T.A. Huibers
BSChE, MSChE, PhDChE, FAIChE; Manager,
Chemical Processes Group, Union Camp Corp.

Frederick W. Keith, Jr.
BSChE, PhDChE, PE; Manager, Applications Research,
Pennwalt Corp.

Edward G. Kominek
BS, MBA, PE; Manager, Industrial Water & Waste Sales,
Eimco Processing Machinery Division, Envirotech Corp.

Lloyd H. Ketchum, Jr.
BSCE, MSE, MPH, PhD, PE; Associate Professor,
Civil Engineering and Geological Sciences,
University of Notre Dame

Mark K. Lee
BSChE, MEChE; Project Manager,
Westlake Polymers Corp.

David H.F. Liu
PhD, ChE; Principal Scientist, J.T. Baker, Inc. a division
of Procter & Gamble

Béla G. Lipták
ME, MME, PE; Process Control and Safety Consultant,
President, Liptak Associates, P.C.

János Lipták
CE, PE; Senior Partner, Janos Liptak & Associates

Andrew F. McClure, Jr.
BSChE; Manager, Industrial Concept Design Division,
Betzon Environmental Engineers

George W. McDonald
PhD, ChE; Pulping Group Leader, Research and
Development Division, Union Camp Corp.

Francis X. McGarvey
BSChE, MSChE; Manager, Technical Center,
Sybron Chemical Company

Kent Keqiang Mao
BSCE, MSCE, PhDCE, PE; President,
North America Industrial Investment Co., Ltd.

Thomas J. Myron, Jr.
BSChE; Senior Systems Design Engineer,
The Foxboro Company

Van T. Nguyen
BSE, MSE, PhD; Department of Civil Engineering,
California State University, Long Beach

Frank L. Parker
BA, MS, PhD, PE;
Professor of Environmental and Water Resources
Engineering, Vanderbilt University

Joseph G. Rabosky
BSChE, MSE, PE; Senior Project Engineer, Calgon Corp.

Gurumurthy Ramachandran
BSEE, PhD; Assistant Professor,
Division of Environmental and Occupational Health,
University of Minnesota

Roger K. Raufer
BSChE, MSCE, MA, PhD, PE; Associate Director,
Environmental Studies,
Center for Energy and the Environment,
University of Pennsylvania

Parker C. Reist
ScD, PE; Professor of Air and Industrial Hygiene
Engineering, University of North Carolina

LeRoy H. Reuter
MS, PhD, PE; Consultant

Bernardo Rico-Ortega
BSCh, MSSE; Product Specialist,
Pollution Control Department,
Nalco Chemical Company

Howard C. Roberts
BAEE, PE; Professor of Engineering (retired)

Reed S. Robertson
BSChE, MSEnvE, PE; Senior Group Leader, Nalco
Chemical Company

David M. Rock
BSChE, MSChE, PE; Staff Engineer,
Environmental Control, American Enka Company

F. Mack Rugg
BA, MSES, JD, Environmental Scientist,
Project Manager, Camp Dresser & McKee Inc.

Alan R. Sanger
BSc, MSc, DPhil; Consultant and Professor,
Department of Chemical Engineering,
University of Alberta

Chakra J. Santhanam
BSChE, MSChE, ChE, PE;
Senior Environmental Engineer, Crawford & Russell, Inc.

E. Stuart Savage
BSChE, PE; Manager, Research and Development,
Water & Waste Treatment, Dravco Corp.

Letitia S. Savage
BS Biology; North Park Naturalist,
Latodami Farm Nature Center,
Allegheny County Department of Conservation

Frank P. Sebastian
MBA, BSME; Senior Vice President, Envirotech Corp.

Gerald L. Shell
MSCE, PE; Director of Sanitary Engineering,
Eimco Corp.

Wen K. Shieh
PhD; Department of Systems Engineering,
University of Pennsylvania

Stuart E. Smith
BChE, MSChE, MSSE, PE; Manager,
Industrial Wastewater Operation, Environment/One Corp.

John R. Snell
BECE, MSSE, DSSE, PE; President,
John R. Snell Engineers

Paul L. Stavenger
BSChE, MSChE; Director of Technology,
Process Equipment Division, Dorr-Oliver, Inc.

Michael S. Switzenbaum
BA, MS, PhD; Professor,
Environmental Engineering Program,
Department of Civil and Environmental Engineering,
University of Massachusetts, Amherst

Floyd B. Taylor
BSSE, MPH, PE, DEE; Environmental Engineer,
Consultant

Amos Turk
BS, MA, PhD; Professor Emeritus,
Department of Chemistry,
The City College of New York

Curtis P. Wagner
BA, MS; Senior Project Manager, TRC Environmental,
Inc.

Cecil C. Walden
BA, MA, PhD; Associate Director, B.C. Research, Canada

Roger H. Zanitsch
BSCE, MSSE; Senior Project Engineer, Calgon Corp.

William C. Zegel
ScD, PE, DEE; President, Water and Air Research, Inc.

Preface

Dr. David H.F. Liu passed away during the preparation of this revised edition. He will be long remembered by his coworkers, and the readers of this handbook will carry his memory into the 21st Century

Engineers respond to the needs of society with technical innovations. Their tools are the basic sciences. Some engineers might end up working *on* these tools instead of working *with* them. Environmental engineers are in a privileged and challenging position, because their tools are the totality of man's scientific knowledge, and their target is nothing less than human survival through making man's peace with nature.

When, in 1974, I wrote the preface to the three-volume first edition of this handbook, we were in the middle of an energy crisis and the future looked bleak, I was worried and gloomy. Today, I look forward to the 21st Century with hope and confidence. I am optimistic because we have made progress in the last 22 years and I am also proud, because I know that this handbook made a small contribution to that progress. I am optimistic because we are beginning to understand that nature should not be conquered, but protected, that science and technology should not be allowed to evolve as "value-free" forces, but should be subordinated to serve human values and goals.

This second edition of the *Environmental Engineers' Handbook* contains most of the technical know-how needed to clean up the environment. Because the environment is a complex web, the straining of some of the strands affects the entire web. The single-volume presentation of this handbook recognizes this integrated nature of our environment, where the various forms of pollution are interrelated symptoms and therefore can not be treated separately. Consequently, each handbook section is built upon and is supported by the others through extensive cross-referencing and subject indexes.

The contributors to this handbook came from all continents and their backgrounds cover not only engineering, but also legal, medical, agricultural, meteorological, biological and other fields of training. In addition to discussing the causes, effects, and remedies of pollution, this handbook also emphasizes reuse, recycling, and recovery. Nature does not cause pollution; by total recycling, *nature makes resources out of all wastes*. Our goal should be to learn from nature in this respect.

The Condition of the Environment

To the best of our knowledge today, life in the universe exists only in a ten-mile-thick layer on the 200-million-square-mile surface of this planet. During the 5 million years of human existence, we lived in this thin crust of earth, air, and water. Initially man relied only on inexhaustible resources. The planet appeared to be without limits and the laws of nature directed our evolution. Later we started to supplement our muscle power with exhaustible energy sources (coal, oil, uranium) and to substitute the routine functions of our brains by machines. As a result, in some respects we have "conquered nature" and today we are directing our own evolution. Today, our children grow up in man-made environments; virtual reality or cyberspace is more familiar to them than the open spaces of meadows.

While our role and power have changed, our consciousness did not. Subconsciously we still consider the planet inexhaustible and we are still incapable of thinking in time-frames which exceed a few lifetimes. These human limitations hold risks, not only for the planet, nor even for life on this planet, but for our species. Therefore, it is necessary to pay attention not only to our physical environment but also to our cultural and spiritual environment.

It is absolutely necessary to bring up a new generation which no longer shares our deeply rooted subconscious belief in continuous growth: A new generation which no longer desires the forever increasing consumption of space, raw materials, and energy.

It is also necessary to realize that, while as individuals we might not be able to think in longer terms than centuries, as a society we must. This can and must be achieved by developing rules and regulations which are appropriate to the time-frame of the processes which we control or influence. The half-life of plutonium is 24,000 years, the replacement of the water in the deep oceans takes 1000 years. For us it is difficult to be concerned about the consequences of our actions, if those consequences will take centuries or millennia to evolve. Therefore, it is essential that we develop both an educational system and a body of law which would protect our descendants from our own shortsightedness.

Protecting life on this planet will give the coming generations a unifying common purpose. The healing of environmental ills will necessitate changes in our subconscious and in our value system. Once these changes have occurred, they will not only guarantee human survival, but will also help in overcoming human divisions and thereby change human history.

The Condition of the Waters

In the natural life cycle of the water bodies (Figure 1), the sun provides the energy source for plant life (algae), which produces oxygen while converting the inorganic molecules into larger organic ones. The animal life obtains its muscle energy (heat) by consuming these molecules and by also consuming the dissolved oxygen content of the water.

When a town or industry discharges additional organic material into the waters (which nature intended to be disposed of as fertilizer on land), the natural balance is upset. The organic effluent acts as a fertilizer, therefore the algae overpopulates and eventually blocks the transparency of the water. When the water becomes opaque, the ultraviolet rays of the sun can no longer penetrate it. This cuts off the algae from its energy source and it dies. The bacteria try to protect the life cycle in the water by attempting to break down the excess organic material (including the dead body cells of the algae), but the bacteria require oxygen for the digestion process. As the algae is no longer producing fresh oxygen, the dissolved oxygen content of the water drops, and when it reaches zero, all animals suffocate. At that point the living water body has been converted into an open sewer.

In the United States, the setting of water quality standards and the regulation of discharges have been based on the "assimilative capacity" of the receiving waters (a kind of pollution dilution approach), which allows discharges into as yet unpolluted waterways. The Water Pollution Act of 1972 would have temporarily required industry to apply the "best practicable" and "best available" treatments of waste emissions and aimed for zero discharge by 1985. While this last goal has not been reached, the condition of American waterways generally improved during the last decades, while on the global scale water quality has deteriorated.

Water availability has worsened since the first edition of this handbook. In the United States the daily withdrawal rate is about 2,000 gallons per person, which represents roughly one-third of the total daily runoff. The bulk of this water is used by agriculture and industry. The average daily water consumption per household is about 1000 gallons and, on the East Coast, the daily cost of that water is $2–$3. As some 60% of the discharged pollutants (sewage, industrial waste, fertilizers, pesticides, leachings from landfills and mines) reenter the water supplies, there is a direct relationship between the quality and cost of supply water and the degree of waste treatment in the upstream regions.

There seems to be some evidence that the residual chlorine from an upstream wastewater treatment plant can combine in the receiving waters with industrial wastes to form carcinogenic chlorinated hydrocarbons, which can enter the drinking water supplies downstream. Toxic chemicals from the water can be further concentrated through the food chain. Some believe that the gradual poisoning of the environment is responsible for cancer, AIDS, and other forms of immune deficiency and self-destructive diseases.

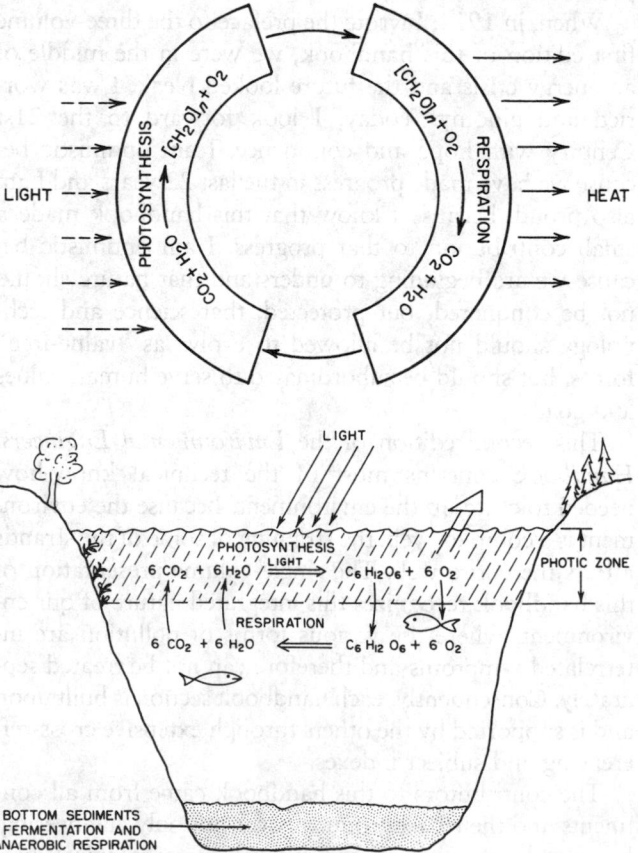

FIG. 1 The natural life cycle.

While the overall quality of the waterways has improved in the United States, worldwide the opposite occurred. This is caused not only by overpopulation, but also by ocean dumping of sludge, toxins, and nuclear waste, as well as by oil leaks from off-shore oil platforms. We do not yet fully understand the likely consequences, but we can be certain that the ability of the oceans to withstand and absorb pollutants is not unlimited and, therefore, international regulation of these discharges is essential. In terms of international regulations, we are just beginning to develop the required new body of law. The very first

case before the International Court of Justice (IJC) wherein it might be argued that rivers are not the property of nation states, and that the interests of nations must be balanced against the interests of mankind, will be heard by IJC in 1997 in connection with the Danube, just about coincident with the publication of this handbook.

The Condition of the Air

There is little question about the harmful effects of ozone depletion, acid rain, or the greenhouse effect. One might

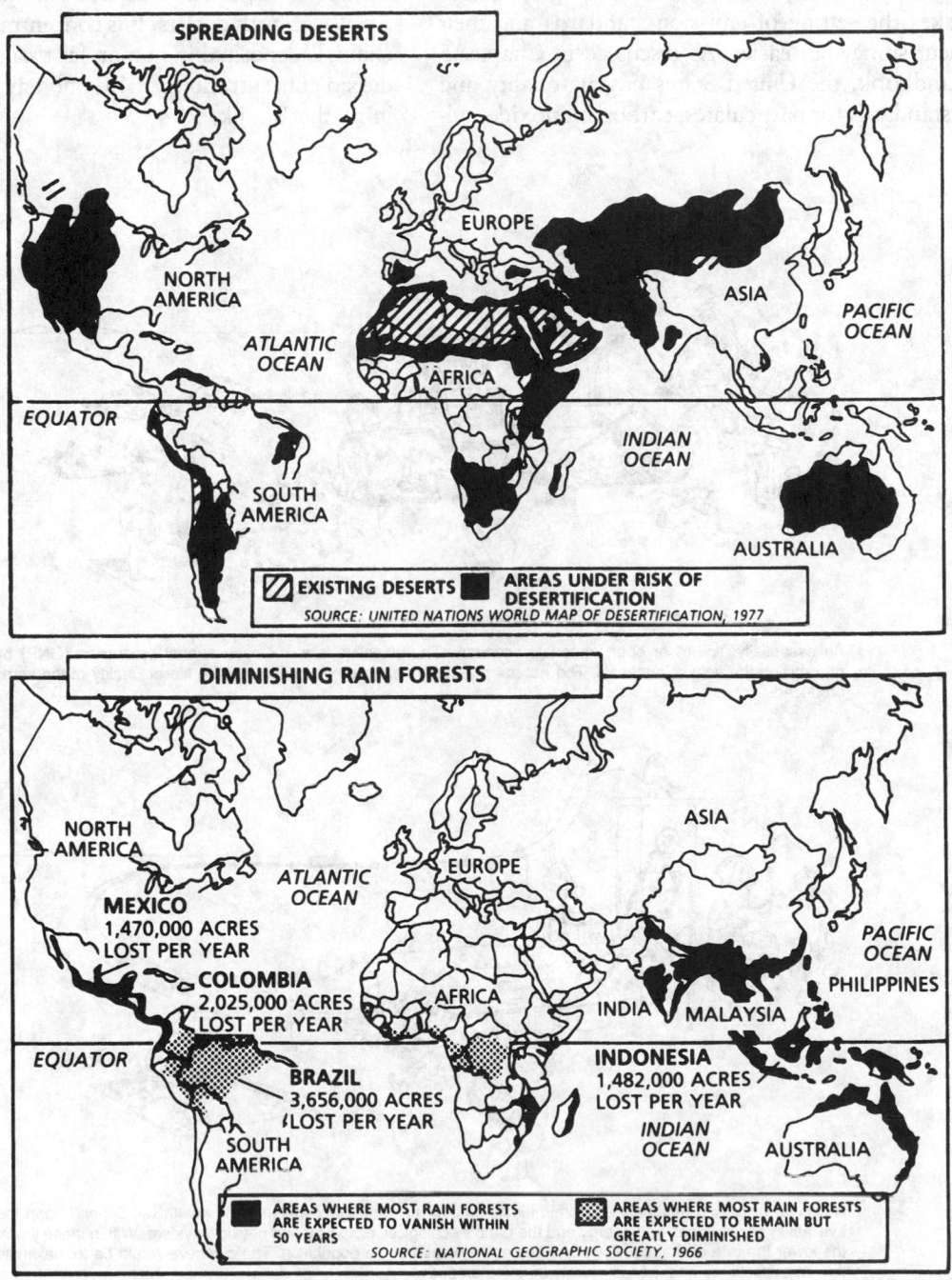

FIG. 2 Areas of diminishing rain forests and spreading deserts.

debate if the prime cause of desertification is acid rain, excessive lumbering, soil erosion, or changes in the weather, but it is a fact that the rain forests are diminishing and the deserts are spreading (Figure 2). We do not know what quantity of acid fumes, fluorinated hydrocarbons, or carbon dioxide gases can be released before climatic changes become irreversible. But we know that the carbon dioxide content of the atmosphere has substantially increased, that each automobile releases 5 tons of carbon dioxide every year, and that the number of gas-burning oil platforms in the oceans is approaching 10,000.

Conditions on the land and in the waters are determined by complex biosystems. The nonbiological nature of air makes the setting of emission standards and their enforcement somewhat easier. As discussed in Chapter 5 of this handbook, the United States has air quality and emission standards for particulates, carbon monoxide, sulfur and nitrogen oxides, hydrocarbons, photochemical oxidants, asbestos, beryllium, and mercury.

For other materials, such as the "possible human carcinogens," the furans and dioxins (PCDD and PCDF), there are no firm emission or air quality standards yet. These materials are the byproducts of paper bleaching, wood preservative and pesticide manufacturing, and the incineration of plastics. Because typical municipal solid waste (MSW) in the U.S. contains some 8% plastics, incineration is probably the prime source of dioxin emissions. Dioxins are formed on incinerator fly ash and end up either in landfills or are released into the atmosphere. Dioxin is suspected to be not only a carcinogen but also a cause of birth defects. It is concentrated through the food chain, is deposited in human fat tissues, and in some cases dioxin concentrations of 1.0 ppb have already been found in mother's milk.

An essentially "linear" or open materials economy. The objective is to increase annual production (GNP) by maximizing the flow of materials. The natural pressure, therefore, is to decrease the life or quality of the items produced.

A circular or closed materials economy. Limits on the total amount of materials or wealth will depend upon the availability of resources and energy and the earth's ecological, biological and physical system. Within these limits, the lower the rate of material flow, the greater the wealth of the population. The objective would be to maximize the life expectancy and, hence, quality of items produced.

FIG. 3 The "open" and "closed" material-flow economies.

Although in the last decades the air quality in the U.S. improved and the newer standards (such as the Clean Air Act of 1990) became stricter, lately we have seen misguided attempts to reverse this progress. Regulations protecting wetlands, forbidding clear-cutting of forests, and mandating use of electric cars have all been relaxed or reversed. In the rest of the world, the overall trend is continued deterioration of air quality. In the U.S., part of the improvement in air quality is due not to pollution abatement but to the exporting of manufacturing industries; part of the improvement is made possible by relatively low population density, not the result of conservation efforts.

On a per capita basis the American contribution to worldwide pollutant emissions is high. For example, the yearly per capita generation of carbon dioxide in the U.S. is about 20 tons. This is twentyfold the per capita CO_2 generation of India. Therefore, even if the emission levels in the West are stabilized or reduced, the global generation of pollutants is likely to continue to rise as worldwide living standards slowly equalize.

The Condition of the Land

Nature never produces anything that it can not decompose and return into the pool of fresh resources. Man does. Nature returns organic wastes to the soil as fertilizer. Man often dumps such wastes in the oceans, buries them in landfills, or burns them in incinerators. Man's deeply rooted belief in continuous growth treats nature as a commodity, the land, oceans, and atmosphere as free dumps. There is a subconscious assumption that the planet is inexhaustible. In fact the dimensions of the biosphere are fixed and the planet's resources are exhaustible.

The gross national product (GNP) is an indicator based on the expectation of continuous growth. We consider the economy healthy when the GNP and, therefore, the quantity of goods produced increases. The present economic model is like an open pipeline which takes in resources at one end and spills out wastes at the other. The GNP in this model is simply a measure of the rate at which resources are being converted to wastes. The higher the GNP, the faster the resources are exhausted (Figure 3). According to this model, cutting down a forest to build a parking lot increases the GNP and is therefore good for the economy. Similarly, this open-loop model might suggest that it is cheaper to make paper from trees than from waste paper, because the environmental costs of paper manufacturing and disposal are not included in the cost of the paper, but are borne separately by the whole community.

In contrast, the economic model of the future will have to be a closed-loop pipeline (Closed-GNP). This will be achieved when it becomes more profitable to reuse raw materials than to purchase fresh supplies. This is a function of economic policy. For example, in those cities where only newspapers printed on recycled paper are allowed to be sold, there is a healthy market for used paper and the volume of municipal waste is reduced. Similarly, in countries where environmental and disposal costs are incorporated into the total cost of the products (in the form of taxes), it is more profitable to increase quality and durability than to increase the production quantity (Figure 3).

In addition to resource depletion and the disposal of toxic, radioactive, and municipal wastes, the natural environment is also under attack from strip mining, clear cutting, noise, and a variety of other human activities. In short, there is a danger of transforming the diverse and stable ecosystem into an unstable one which consists only of man and his chemically sustained food factory.

Energy

When man started to supplement his muscle energy with outside sources, these sources were all renewable and inexhaustible. The muscle power of animals, the burning of wood, the use of hydraulic energy were man's external energy sources for millions of years. Only during the last couple of centuries have we started to use exhaustible energy sources, such as coal, oil, gas, and nuclear. This change in energy sources not only resulted in pollution but has also caused uncertainty about our future because we can not be certain if the transition from an exhausted energy source to the next one can be achieved without major disruptions.

The total energy content of all fossil deposits and uranium 235 (the energy source of "conventional" nuclear plants) on the planet is estimated to be 100×10^{18} BTUs. Our present yearly energy consumption is about 0.25×10^{18} BTUs. This would give us 400 years to convert to an inexhaustible energy source, if our population and energy demand were stable and if some energy sources (oil and gas) were not depleted much sooner than others.

Breeder reactors have not been considered in this evaluation because the plutonium they produce is too dangerous to even contemplate a plutonium-based future. This is not to say that conventional nuclear power is safe. Man has not lived long enough with radiation to know if millions of cubic feet of nuclear wastes can be stored safely.

We receive about 100 Watts of solar energy on each square meter of the Earth's surface, or a yearly total of about 25×10^{18} BTUs. Therefore, 1% of the solar energy received on the surface of the planet could supply our total energy needs. If collected on artificial islands or in desert areas around the Equator, where the solar radiation intensity is much higher than average, a fraction of 1% of the globe's surface could permanently supply our total energy needs. If the collected solar power were used to obtain hydrogen from water and if the compressed hydrogen were used as our electric, heat, and transportation energy source, burning this fuel would result in the emission of only clean, nonpolluting steam. Also, if the combustion took place in fuel cells, we could nearly double the present efficiency of electric power generation (about 33%) or the efficiency of the internal combustion engine

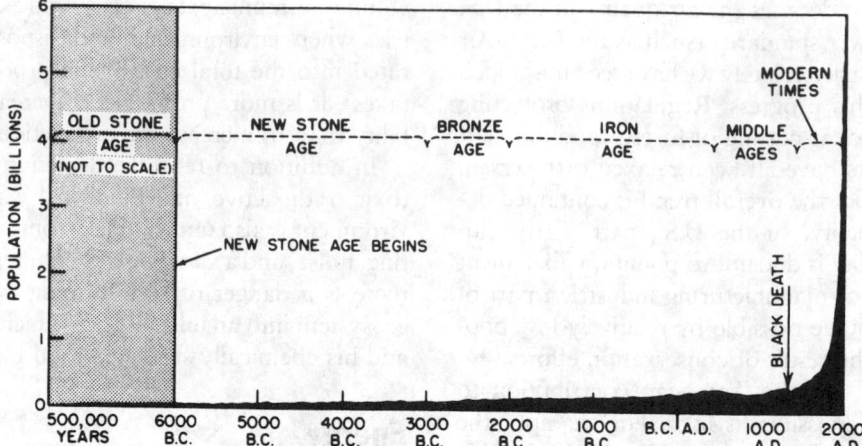

FIG. 4 Growth of human population.

(about 25%) and thereby substantially reduce thermal pollution.

Today, as conventional energy use increases, pollution tends to rise exponentially. As the population of the U.S. has increased 50% and our per capita energy consumption has risen 25%, the emission of pollutants has soared by 2000%. While the population of the world doubles in about 50 years, energy consumption doubles in about 20 and electric energy use even faster. In addition to chemical pollution, thermal pollution also rises with fossil energy consumption, because for each unit of electricity generated, two units of heat energy are discharged into the environment.

It is time to redirect our resources from the military—whose job it is to protect dwindling oil resources—and from deep sea drilling—which might cause irreversible harm to the ocean's environment—and use these resources to develop the new, permanent, and inexhaustible energy supplies of the future.

Population

Probably the most serious cause of environmental degradation is overpopulation. More people live on Earth today than all the people who died since Creation (or, if you prefer, the "accidental" beginning of "evolution"). Three hundred years ago the world's population doubled every 250 years. Today it doubles in less than a life span. When I was editing the first edition of this handbook, the population of the planet was under 4 billion; today it is nearing 6 billion (Figure 4). During that same time period, the population of the Third World increased by more than the total population of the developed countries.

The choice is clear: we either take the steps needed to control our numbers or nature will do it for us through famine, plague, and loss of fertility. We must realize that the teaching which was valid for a small tribe in the desert ("Conquer nature and multiply") is no longer valid for the overpopulated planet of today. We must realize that, even if we immediately take all the steps required to stabilize the population of the planet, the total number will still reach some 15 billion before it can be stabilized.

To date, food production has kept pace with population growth, but only at a drastic price: increases in pesticide (300%) and fertilizer (150%) use, which in turn further pollutes the environment.

The total amount of land suitable for agriculture is about 8 billion acres. Of that, 3.8 billion acres are under cultivation and, with the growth of the road systems and cities, the availability of land for agricultural uses is shrinking. The amount of water available for irrigation is also dropping. Without excessive fertilization, one acre of land is needed to feed one person: therefore, the human population has already exceeded the number supportable without chemical fertilizers. As chemical fertilizer manufacturing is based on the use of crude oil, models simulating world trends predict serious shortages in the next century (Figure 5).

While all these trends are ominous, the situation is not hopeless. The populations of the more developed countries seem to have stabilized, the new communication technologies and improved mass transit are helping to stop or even reverse the further concentration of urban masses. Environmental education and recycling have been successful in several nations. New technologies are emerging to serve conservation and to provide nonpolluting and inexhaustible energy sources.

When Copernicus discarded the concept of an earth-centered and stationary Universe, the Earth continued to travel undisturbed in its orbit around the Sun, yet the consequences of this discovery were revolutionary. Copernicus' discovery changed nothing in the Universe, but it changed our subconscious view of ourselves as the "centerpiece of creation." Today, our concept of our immediate universe, the Earth, is once again changing and this change is even more fundamental. We are realizing that the planet is exhaustible and that our future depends

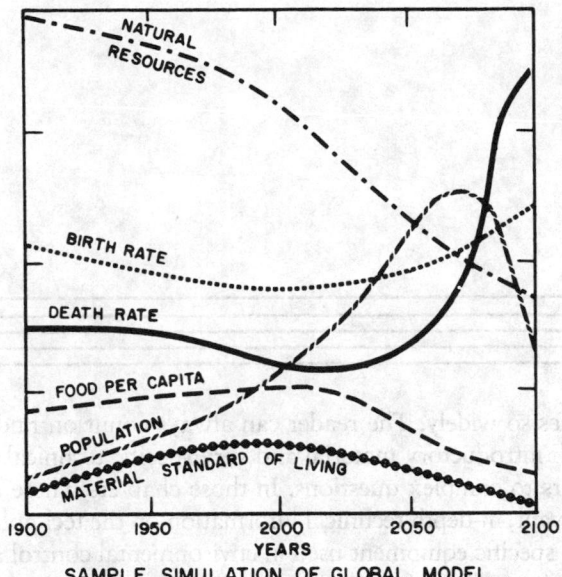

NATURAL RESOURCES

BIRTH RATE

DEATH RATE

FOOD PER CAPITA

POPULATION

MATERIAL STANDARD OF LIVING

1900   1950   2000   2050   2100
YEARS
SAMPLE SIMULATION OF GLOBAL MODEL

FIG. 5 Computer simulation of world trends.

on our own behavior. It took several centuries for Copernicus' discovery to penetrate our subconscious. Therefore, we should not get impatient if this new understanding does not immediately change our mentality and life style. On the other hand, we must not be complacent. Human ingenuity and the combined talent of people, such as the contributors and readers of this handbook, can solve the problems we face, but this concentrated effort must not take centuries. We do not have that much time.

Protecting the global environment, protecting life on this planet, must become a single-minded, unifying goal for all of us. The struggle will overshadow our differences, will give meaning and purpose to our lives and, if we succeed, it will mean survival for our children and the generations to come.

Béla G. Lipták
Stamford, Connecticut
November, 1996

Foreword

The revised, expanded, and updated edition of the *Environmental Engineers' Handbook* covers in depth the interrelated factors and principles which affect our environment and how we have dealt with them in the past, how we are dealing with them today, and how we might deal with them in the future. Although the book is clearly aimed at the environmental professional, it is written and structured in a way that will allow others outside the field to educate themselves about our environment, and what can and must be done to continue to improve the quality of life on spaceship earth. The book covers in detail the ongoing global transition among the cleanup of the remains of abandoned technology, the prevention of pollution from existing technology, and the design of future zero emission technology. The relationship of cost to benefit is examined and emphasized throughout the book.

The Preface will remind the reader of Charles Dickens' famous *A Christmas Carol*, and we should reflect on its implications carefully as we try to decide the cost-to-benefit ratio of environmental control technology. The book begins with a thorough review of environmental law and regulations that are then further detailed in individual chapters. The chapter on environmental impact assessment is the bridge between the release of pollutants and the technology necessary to reduce the impact of these emissions on the global ecosystem. Chapters on the source control and/or prevention of formation of specific pollutants in air, water, land, and our personal environment follow these introductory chapters. A chapter on solid waste is followed by the final chapter on hazardous waste, which tries to strike a balance between the danger of hazardous wastes and the low probability that a dangerous environmental event will occur because of these wastes.

The type of information contained in every chapter is designed to be uniform, although there is no unified format that each chapter follows, because subject matter varies so widely. The reader can always count on finding both introductory material and very specific technical answers to complex questions. In those chapters where it is relevant, in-depth technical information on the technology and specific equipment used in environmental control and clean up will be found. Since analytical results are an intricate part of any environmental study, the reader will find ample sections covering the wide variety of analytical methods and equipment used in environmental analysis. Several chapters have extensive sections where the derivation of the mathematical equations used are included. Textual explanations usually also accompany these mathematical-based sections.

A great deal of effort has gone into providing as much information as possible in easy-to-use tables and figures. We have chosen to use schematic diagrams rather than actual pictures of equipment, devices, or landscapes to explain or illustrate technology and techniques used in various areas. The length of this book is testimony to the level of detail that has been included in order to make the book a single-source handbook. The reader will also find ample references if additional information is required. The author of a section is given at the end of each section and we encourage readers to contact the author directly with any questions or comments. Although extensive review and proofreading of the manuscript was done, we ask readers who find errors or omissions to bring them to our attention.

Finally, we wish to acknowledge the numerous individuals and organizations who either directly or indirectly have contributed to this work, yet have not been mentioned by name.

Paul A. Bouis
Bethlehem, Pennsylvania
January, 1997

O = No measurable impact is expected to occur as a result of considering the project action relative to the environmental factor

M = Some type of mitigation measure can be used to reduce or avoid a small adverse, adverse, or significant adverse impact

NA = The environmental factor is not applicable or relevant for the proposed project

Development of a Simple Matrix

Developing a specific matrix for the project, plan, program, or policy being analyzed is better than using a generic matrix. The following steps can be used in preparing a simple interaction matrix:

1. List all anticipated project actions and group via temporal phases
   a. Construction
   b. Operation
   c. Postoperation
2. List pertinent environmental factors from environmental setting and group according to physical–chemical, biological, cultural, and socioeconomic categories; and spatial considerations such as site and region or upstream, site, and downstream
3. Discuss preliminary matrix with study team members and advisors to team or study manager
4. Decide on impact rating scheme (such as numbers, letters, or colors) to be used
5. Talk through the matrix as a team and make ratings and notes to identify impacts and summarize impacts (documentation)

Summary Observations on Simple Matrices

The following observations are based upon numerous examples of matrices and over twenty years of experience in using such matrices:

1. In using a simple interaction matrix, one must carefully define the spatial boundaries associated with environmental factors, as well as each environmental factor; the temporal phases and specific actions associated with the proposed project; and the impact rating or summarization scales used in the matrix.
2. The most important concept in using an interaction matrix is to consider the matrix as a tool for purposes of analysis, with the key need being to clearly state the rationale used for the impact ratings assigned to a given temporal phase (or project action) and a given spatial boundary (or environmental factor).
3. The development of one or more interaction matrices can be a useful technique in discussing a proposed action and its potential environmental impacts. This development can be helpful in the early stages of a study

to assist each team member in understanding the implications of the project and developing detailed plans for more extensive studies on particular factors.

4. The interpretation of information on resultant impact rating scales should be carefully considered, particularly when large differences in spatial boundaries, as well as temporal phases, for a proposed project may exist.
5. Interaction matrices are useful for delineating the impacts of the first and second or multiple phases of a two-phase or multiphase project; the cumulative impacts of a project when considered relative to other past, present, and reasonably foreseeable future actions in the area; and the potential positive effects of mitigation measures. Creative codes can be used in the matrix to delineate this information.
6. If interaction matrices are used to display comparisons between different alternatives, the same basic matrix must be used in terms of spatial boundaries and temporal phases for each alternative being analyzed. Completion of such matrices for each alternative can provide a basis for tradeoff analysis.
7. Impact quantification can provide a valuable basis for the assignment of impact ratings for different project actions and different environmental factors.
8. Color codes can be used to display or communicate information on anticipated impacts. Beneficial impacts could be displayed with green or shades of green; whereas, detrimental or adverse effects could be displayed with red or shades of red. Impact matrices can be used without incorporating number rating scales. For example, circles of varying size can be used to denote range of impacts.
9. One concern related to interaction matrices is that project actions or environmental factors are artificially separated when they should be considered together. Footnotes can be used in a matrix to identify groups of impacts which should be considered together. Footnote use allows the delineation of primary and secondary effects of projects.
10. The development of a preliminary interaction matrix does not mean that it must be included in a subsequent EIA report. The preliminary matrix can be used as an internal working document or tool in study planning and development.
11. Importance weighting for environmental factors and project actions can be used in a simple interaction matrix. If this approach is chosen, the rationale upon which differential importance weights are assigned must be carefully delineated.
12. One important advantage of a simple interaction matrix is that its use forces consideration of actions and impacts related to a proposed project within the context of other related actions and impacts. In other words, the matrix aids in preventing overriding attention being given to one action or environmental factor.

TABLE 2.3.2 USDA CHECKLIST FOR ADDRESSING AND SUMMARIZING ENVIRONMENTAL IMPACTS (U.S. DEPARTMENT OF AGRICULTURE 1990)

Topical Issue	Yes	Maybe	No	Comments

Land Form

Will the project result in:

Unstable slopes or embankments?

Extensive disruption to or displacement of the soil?

Impact to land classified as prime or unique farmland?

Changes in ground contours, shorelines, stream channels, or river banks?

Destruction, covering, or modification of unique physical features?

Increased wind or water erosion of soils?

Foreclosure on future uses of site on a long-term basis?

Air/Climatology

Will the project result in:

Air pollutant emissions which will exceed federal or state standards or cause deterioration of ambient air quality (e.g., radon gas)?

Objectionable odors?

Alteration of air movements, humidity, or temperature?

Emissions of hazardous air pollutants regulated under the Clean Air Act?

Water

Will the project result in:

Discharge to a public water system?

Changes in currents or water movements in marine or fresh water?

Changes in absorption rates, drainage patterns, or the rate and amount of surface water runoff?

Alterations to the course or flow of flood waters?

Impoundment, control, or modifications of any body of water equal to or greater than ten acres in surface area?

Discharges into surface waters or alteration of surface water quality including, but not limited to, temperature or turbidity?

Alteration of the direction or rate of flow of groundwaters?

Alterations in groundwater quality?

Contamination of public water supplies?

Violation of State Stream Quality Standards, if applicable?

Location in a riverine or coastal floodplain?

Exposure of people or property to water-related hazards such as flooding?

Location in a state's coastal zone and subject to consistency with the State Coastal Zone Management Plan?

Impact on or construction in a wetland or inland floodplain?

Solid Waste

Will the project:

Generate significant solid waste or litter?

Noise

Will the project:

Increase existing noise levels?

Expose people to excessive noise?

Plant Life

Will the project:

Change the diversity or productivity of species or number of any species of plants (including trees, shrubs, grass, crops, microflora, and aquatic plants)?

Continued on next page

1

Environmental Laws and Regulations

David Bookchin | David Farnsworth

Introduction

Environmental law consists of all legal guidelines that are intended to protect our environment. Much of the environmental legislation in the United States is initiated at the federal level. Various regulatory agencies may then prepare regulations, which define how activity must be conducted to comply with the law. In practice, the terms *law, statute,* and *regulation* are often used interchangeably. Regulations are generally more volatile than laws (statutes), of more applicability in determining compliance.

However, to obtain copies of laws or regulations, one must differentiate between statutes (laws) and regulations. Laws can be accessed through their public law number from the U.S. Printing Office and are compiled under the *United States Code (USC).* Regulations are printed in the *Federal Register (FR)* and are compiled annually in the *Code of Federal Regulations (CFR).*

Regulatory compliance is a significant aspect of conducting business today. The scheme of obligations posed by environmental legislation represents two costs: the effort and expenditure required to achieve compliance and the fines, penalties, and liabilities that may be incurred as a result of noncompliance. Whether preparing for environmental audits, developing an emergency response plan, or participating in an environmental impact study, environmental engineers must be conversant in environmental law and environmental policy. Ignorance of regulatory requirements is viewed by federal, state, and local governments as no excuse for noncompliance.

An overview of federal environmental laws is provided in this chapter. The chapter is divided into four sections and an appendix.

Government Agencies and Administrative Law. This section outlines some of the procedures under which laws are developed and applied. It is a "broadbrush" characterization of administrative law which focuses on the practice of government agencies.

Information Laws. This section includes statutes used to gather and disseminate information as a central part of their regulatory schemes. This section includes the National Environmental Policy Act and the Emergency Planning and Community Right-to-Know Act.

Natural Resource Laws. This section includes statutes such as the Endangered Species Act and the Coastal Zone Management Act which protect habitat and regulate land use.

Pollution Control Laws. Statutes discussed in this section, such as the Clean Air Act, Clean Water Act, Resource Conservation and Recovery Act, and Toxic Substances Control Act, generally focus on regulating the pollutants which create risk to human health and the environment.

Federal Environmental Protection Agencies. The organization of the Environmental Protection Agency and the addresses and telephone numbers of the headquarters and regional offices and state and territorial agencies are presented in the appendix.

This chapter provides an overview and a general understanding of the key features of the major environmental statutes. The discussions of statutes should pave a way for further, in-depth study into the environmental laws.

It should be noted that environmental laws are dynamic and subject to change, interpretation, and negotiation. Although the following discussions of these federal laws provide important information, the reader is advised to determine if any updates or revisions of these laws are in effect. The information provided on these statutes is no substitute for up-to-date advice from licensed practitioners.

1.1
ADMINISTRATIVE LAW

This section provides an overview of government agencies and their characteristics, limitations on agencies, and the judicial review of agency actions.

Government Agencies

The government can be divided into the executive, legislative, and judicial branches. Agencies within the executive branch perform a large part of the day-to-day business on environmental protection. This branch is comprised of many agencies including the Environmental Protection Agency (EPA) and other cabinet-level agencies, such as the Department of Interior and the Department of Commerce.

Administrative agencies have the essential attributes of the three branches of our government. They generally have legislative, executive, and judicial powers.[1] As organizations, agencies possess many of the same powers and limits as the three government branches do.

LEGISLATIVE

Agencies regulate according to the statutes developed by Congress. In addition, agencies are responsible for developing and promulgating regulations. Regulations generally are more specific statements of the rules found in statutes. For example, in response to Congressional mandates in the Clean Water Act, the EPA has promulgated specific regulations for storm water permits.

Agencies often develop regulations through an informal rulemaking that involves input from the EPA's technical and policy specialists and from interest groups which expect to be affected by those regulations.[2] Agencies initially develop proposed regulations. The EPA then publishes the proposed regulations and allows a period for public comment. (See Section 4.6). This process allows interested parties, such as industries and nongovernmental organizations, to review the proposed regulations and provide the EPA with their comments.

If enough interest exists, hearings may be scheduled to discuss and clarify the proposed regulations. The input of various parties during the comment and hearing period, like the input of legislators, all goes into what finally becomes the regulation or rule. Once all the comments are reviewed, the agency publishes a final rule or regulation.

EXECUTIVE

After an agency promulgates regulations, the rules are implemented or applied. Usually, the agency which develops the regulations also applies them. Under the Clean Water Act, for example, the EPA has the authority not only to promulgate regulations, but also to implement them.[3]

JUDICIAL

Agencies are also adjudicatory. In other words, they work like courts and hand down judgments regarding issues which arise in the context of their programs. When an agency adjudicates, it performs trial-type procedures which are similar to civil trials performed by the judicial branch of government.[4] Parties participate in hearings, present evidence and testimony, conduct cross-examinations, and develop a written record. Hearings take place before a neutral administrative law judge. Finally, agency adjudications may be appealed within an agency as well as to state or federal courts.

Limitations on Agencies

The three branches of government exercise numerous controls over agencies. For example, Congress is responsible for creating and empowering agencies as well as defining an agency's role.[5] Congress has also developed the Administrative Procedures Act (APA) (5 USC §§501–506) which sets forth various standards for all agency actions.

The executive branch controls the nomination of agency directors and administrators. However, these upper-level

---

1. The discussion here focuses on executive branch agencies. However, the term *executive* is used as an adjective to describe, in general, the executive functions of agencies.
2. Usually, when an agency legislates or develops regulations, it follows procedures commonly known as notice and comment or informal rulemaking. Informal rulemaking requires the agency to notify the public that it is considering developing a rule, commonly referred to as a proposed rule. The agency must publish a draft of the proposed rule and invite comments from the public in response. Other kinds of rulemakings include formal, hybrid, and negotiated rulemaking. However, the scope of this discussion does not go beyond the informal rulemaking regulations.

3. The EPA can also delegate the authority to implement regulations to a state environmental agency. Many states, for instance, have their own water discharge permit programs which they implement themselves. Others do not. This delegation, however, does not change the executive function which agencies—state or federal—possess.
4. Some significant differences between agency adjudications and standard civil bench trials include relaxed rules of evidence. Pretrial discovery (information-gathering) rules may also be different.
5. Enabling legislation is typically the law that creates an agency, gives it authority, and defines its role.

appointments are subject to confirmation by the Senate. Congress and the executive branch also control an agency's budget. These provisions translate into a large amount of control over an agency. Finally, courts define and limit agency action. They review agency decisions within the judicial framework of statutory and common law.

Due process is one of the most fundamental legal principals which courts apply to agencies when reviewing their relationship to and treatment of citizens. The term is found in the fifth and fourteenth amendments to the U.S. Constitution. The fifth amendment states that "No person [shall] be deprived of life, liberty or property without due process of law."[6]

Due process generally implies sufficient notice and a right to a hearing. It involves the application of certain procedures which seek to assure fairness, participation, accuracy, and checks on the concentration of power in government's hands.

Judicial Review of Agency Actions

Several observations can be made about the court system's review of agency decisions. First, parties must initially use or *exhaust* all the avenues of agency review before they take their complaints to the court system. Second, several U.S. Supreme Court decisions define a court's role in reviewing agency actions. Generally, the Supreme Court has held that courts should acknowledge and accommodate agency expertise, rely upon the controlling statutory authority in making their judgements, and avoid imposing further rulemaking procedures on an agency without showing extraordinary circumstances.[7]

EXHAUSTION

Parties that disagree with the results of an agency adjudication are typically required to *exhaust administrative remedies* within that agency before going to the court system. This requirement means that if an agency has established appeal procedures, the party must follow those procedures before entering an appeal in court. Unless a party fully exhausts agency review, it cannot take the next step and get review in the court system.

STANDARDS OF REVIEW

The APA (5 *USC* §§501–706) provides a statutory basis for the review of agency actions, with two exceptions.[8] The APA (5 *USC* §701[a]) does not apply "to the extent

that (1) statutes preclude judicial review; or (2) agency action is committed to agency discretion by law."[9] The first exception applies, for example, where a statute explicitly precludes review. The second exception has been clarified by judicial interpretation.

The *Citizens to Preserve Overton Park, Inc., v. Volpe* (401 U.S. 402, 411 [1971]) case involved the second exception. In this case, the Court reviewed the secretary of transportation's authorization of funds to build a highway through a public park. The statute at issue allowed the secretary to use funds for highways except in situations where a feasible and prudent alternative was available. Environmentalists successfully argued that the secretary of transportation did not have the discretion to authorize the funds, as he maintained, and that he had not considered alternatives to the highway construction.

The *Overton Park* case emphasizes the arbitrary and capricious standard for nonadjudicative agency actions. This test establishes a minimum standard which agencies must meet to justify their decisions. In reviewing the record upon which an agency bases its decision, a court must find some basis for the agency's decision. If no basis exists for the agency's decision within the record, a court can hold that the agency was arbitrary and capricious, i.e., that it failed to meet the minimum standard for justifying its decision. In the *Overton Park* case, the Supreme Court found a sufficient basis for overturning the lower court's decision that upheld the original agency action.

DEFERENCE TO THE AGENCY

Although many cases deal with administrative law and the role of agencies, the *Chevron U.S.A. Inc., v. Natural Resources Defense Council, Inc.* (467 U.S. 837 [1984]) case readily demonstrates how courts should review an appeal from an agency action.

Because courts frequently lack the expertise to make technical decisions associated with environmental issues, they often show deference to agencies. If an agency presents a justifiable basis for its decisions, a court frequently relies on the agency's expertise. In the *Chevron* case, the Supreme Court reviewed the EPA's interpretation and administration of the Clean Air Act. The Court was faced with the issue of what rules of interpretation to apply in considering whether the EPA was justified in defining a Clean Air Act term: *stationary source*.

The *Chevron* case establishes the procedures for a court to follow in reviewing an agency's interpretation of the statutes it administers. First, a court must ask: "has Congress spoken to the issue explicitly (*Chevron U.S.A.*

6. The Fourteenth Amendment to the U.S. Constitution contains similar language: "[N]or shall any State deprive any person of life, liberty, or property, without due process of law. . . ."
7. *See Baltimore Gas and Electric Co. v. Natural Resources Defense Council (NRDC)*, 462 U.S. 87 (1983); *Chevron U.S.A., Inc. v. NRDC*, 467 U.S. 837 (1984); *Vermont Yankee Nuclear Power Corp. v. NRDC*, 435 U.S. 519 (1978).

8. The standard of review of factual issues in adjudications is the substantial evidence test. This standard requires a reviewing court to uphold the decision of a lower court unless the reviewing court can find no substantial evidence in the record to support the holding.
9. *See also* Levin. 1990. *Understanding unreviewability in administrative law.* Minn. L.R. 74:689.

Inc., v. Natural Resources Defense Council, Inc., 842). In other words, does the language in the statute discuss the issue? If it does not, but rather "the statute is silent or ambiguous with respect to the specific issue, the question for the court is whether the agency's [interpretation] is based on a permissible construction of the statute" (*Id.*). The court noted that the agency's interpretation did not have to be the only interpretation, or even one which the court would have adopted. Rather, an agency only has to provide a "permissible construction of the statute." (*Id. at* 843).

Finding Regulations

The *Code of Federal Regulations (CFR)* is the primary source for information on government regulations. The *CFR* is a government publication which contains nearly all federal regulations and is compiled annually in July. It is organized by title and updated quarterly. New regulations which are not yet in the *CFR* can often be found in the *Federal Register (FR)*.

Each volume of the CFR provides guidelines on how to use it. The volume cover lists the number, parts included, and revision date. An *Explanation* section at the beginning of each volume lists information such as issue dates, legal status, and how to use the *CFR*. More detailed information on using the *CFR* is included at the end of the volume. The *Finding Aids* section is composed of the following subsections:

1. Materials approved for incorporation by reference
2. Table of *CFR* titles and chapters
3. Appendix to List of *CFR* sections affected
4. List of *CFR* sections affected.

CFR Title 29 contains regulations mandated by the Occupation Safety, Health, and Safety Administration (OSHA); Title 40 contains EPA regulations; the Department of Transportation (DOT) regulations are found in Title 49. Regulatory actions are codified in numbered parts and sections. These parts designate general subject areas, and sections within each part are numbered consecutively. Thus, 40 *CFR* 141.11 is interpreted as an EPA regulation in which 141 identifies the regulation as the National Primary Drinking Water Regulations, and 11 specifies maximum contamination levels for inorganic chemicals in drinking water supplies.

The *FR* is a weekly and daily, official newspaper of the regulatory side of the federal government, published by the Government Printing Office. Much of the material in the *FR* eventually is incorporated into the *CFR*. The *FR* typically contains notice of repealed regulations and proposed regulations. The contents are organized alphabetically by issuing agency, such as, the National Labor Relations Board and National Mediation Board.

While the *FR* is the most up-to-date source of federal regulations, going through each *FR* published subsequent to the newest *CFR* available is time-consuming. Rather than going through each *FR* to establish any changes in regulation, a researcher can consult a monthly companion to the *CFR* entitled the *List of CFR Sections Affected (LSA)*.

The *LSA* can be used once a researcher has established the date at which the *CFR* coverage ends. The most recent *LSA* should then be consulted. A researcher can refer to the regulation by title and number. The *LSA* indicates whether the regulation has been revised or amended. If changes have been made, the *FR* which contains the altered regulations is referenced.

—*David Bookchin*
David Farnsworth

Reference

Administrative Procedures Act. 1988. *U.S. Code.* Vol. 5, secs. 501–706.

1.2 INFORMATION LAWS

This section provides an overview of the information laws including:

- The National Environmental Policy Act
- The Freedom of Information Act
- The Occupational Safety and Health Act
- The Emergency Planning and Community Right-to-Know Act

National Environmental Policy Act

The National Environmental Policy Act (42 USC §§4321–4370; 40 CFR Parts 1500–1508).

STATUTORY ROADMAP

§4321	Congressional declaration of purpose
SUBCHAPTER 1.	POLICIES AND GOALS
§4331	Congressional declaration of national environmental policy
§4332	Cooperation of agencies, reports, availability of information, recommendations, international and national coordination of efforts
§4333	Conformity of administrative procedures to national environmental policy
§4334	Other statutory obligations of agencies
SUBCHAPTER 2.	COUNCIL ON ENVIRONMENTAL QUALITY
§4341	Reports to Congress; recommendations for legislation
§4342	Establishment, membership, chairman, appointments
§4343	Employment of personnel, experts, and consultants
§4344	Duties and functions
SUBCHAPTER 3.	MISCELLANEOUS PROVISIONS

PURPOSE

The National Environmental Policy Act of 1969, 1992 (42 USC §4321 et seq.), commonly referred to as NEPA, is a procedural statute created to insure that certain federal projects are analyzed for their environmental impacts before they are implemented. The NEPA was the first major environmental law enacted in the 1970s. It was signed into law by President Nixon on January 1, 1970.

NEPA's purposes are far-reaching. They serve as a foundation for environmental goals in the United States and for many policies set forth in other environmental statutes. First, the NEPA (§2, 42 USC §4321) sets forth a national policy to "encourage productive and enjoyable harmony between man and his environment [and] to promote efforts which will prevent or eliminate damage to the environment and biosphere and stimulate the health and welfare of man." In addition, the NEPA (§101[a], 42 USC §4331[a]) establishes a continuing federal government policy "to use all practicable means and measures ... to create and maintain conditions in which man and nature can exist in productive harmony, and fulfill the social, economic, and other requirements of present and future generations of Americans."

SPECIFIC PROVISIONS

The heart of NEPA 42 USC §4332 is based in section 102. In accordance with this section, federal agencies must comply with NEPA's procedural mandates if these agencies are conducting a federal action that significantly affects the quality of the human environment. The procedural requirements are meant to further the policies of the NEPA.

Council of Environmental Quality

The NEPA (§202, 42 USC §4342) created the Council of Environmental Quality (CEQ), composed of three members appointed by the president. The CEQ's functions include:

Assisting the president in preparing an annual environmental quality report to Congress

Gathering, analyzing, and interpreting information about current and prospective trends in environmental quality

Reviewing federal programs in light of NEPA's environmental policy and making subsequent recommendations to the president

Recommending other national policies to the president which improve environmental quality

Conducting studies to make recommendations to the president on matters of policy and legislation (NEPA §204, 42 USC §4344).[1]

The CEQ issued the initial guidelines to meet the NEPA's procedural requirements. After seven years, the CEQ replaced the guidelines with official regulations pur-

1. See also, Whitney. 1991. The role of the president's Council on Environmental Quality in the 1990's and Beyond. J. Envtl. L. 6:81.

suant to Executive Order 11991. The new regulations apply to all federal agencies and seek to improve implementing the NEPA's procedural mandates (40 *CFR* §1500–1508).

Environmental Impact Statements

The NEPA achieves its policies and objectives by requiring federal agencies to consider the environmental effects of their activities. In accordance with NEPA section 102 (42 *USC* §4332[c]), every federal agency's recommendation or report on proposals for legislation and other *federal actions significantly affecting the quality of the human environment* must include a detailed statement by the responsible official on

1. The environmental impact of the proposed action
2. Any adverse environmental effects which cannot be avoided if the proposal is implemented
3. Alternatives to the proposed action
4. The relationship between local short-term uses of man's environment and the maintenance and enhancement of long-term productivity
5. Any irreversible and irretrievable commitments of resources involved in the proposed action if it is implemented.[2]

This detailed statement, known as an environmental impact statement or EIS, is not intended to be a simple disclosure document. Rather, federal agencies are required to make thorough inquiries into federal projects before the projects are undertaken. The purpose of the EIS is to insure that NEPA's policies and goals are incorporated into the actions of the federal government. The EIS must include an assessment of the environmental impacts of a project and propose reasonable alternatives to minimize the adverse impacts of the project. Environmental impact statements should be clear, concise, and supported by evidence showing that the agency made the necessary analysis (40 *CFR* §1502.1).

Section 102 contains key statutory language which has resulted in significant judicial and administrative interpretation. These interpretations have typically served to broaden the NEPA's jurisdiction. For example, a "major federal action" is not limited to projects funded or carried out by the federal government. Instead, courts have interpreted "major federal actions" to include projects which merely require federal approval or are potentially subject to federal control.[3] Courts have also addressed questions involving the scope of an EIS as well as what triggers the EIS mandate.[4]

2. *National Environmental Policy Act.* Section 102(c). *U.S. Code.* Vol. 42, sec. 4332(c). Emphasis added.
3. *See, e.g., Minnesota Public Interest Group v. Butz,* 498 F.2d 1314 (8th Cir. 1974).
4. *See* Battle, J.B. 1986. *Environmental decisionmaking and NEPA.* Cincinnati: Anderson Publishing Co.)

The CEQ regulations also serve to interpret the jurisdiction of the NEPA. For example, they propose that federal actions typically fall within one of four categories: the adoption of official policy, formal plans or programs, the approval of specific construction projects, or management activities in a defined geographic area (40 *CFR* §1508). Thus, the courts define the NEPA's procedural mandates, i.e., jurisdiction, and EIS scope and content, through statutory and regulatory interpretation.

Environmental Assessments

The NEPA requires an agency to prepare an environmental assessment (EA) when the need for an EIS is unclear. EAs create a reviewable record to assess if an EIS is required. Both federal agencies and courts need a reviewable environmental record to determine whether a major federal action is significantly affecting the environment.

The EA should contain evidence and analysis sufficient to determine if the agency should prepare an EIS or make a finding of no significant impact (FONSI) (40 *CFR* §1508.9). The EA is basically a mini-EIS. It is a brief document which includes a discussion of the need for the proposed action, alternatives to the proposed action, environmental impacts, and a list of agencies and persons consulted.

Categorical Exclusion

Federal agencies must make an initial inquiry to determine if an EIS is needed for a proposal or if the proposal falls under categorical exclusion. The NEPA provides for "a category of actions which do not individually or cumulatively have a significant effect on the human environment . . . and [for] which, therefore, neither an environmental assessment nor an environmental impact statement is required" (40 *CFR* §1508.4). Thus, under limited circumstances, neither an EIS nor an EA is required.

SUMMARY

The NEPA establishes a broad, protective national environmental policy as a goal to be furthered by the procedural mandates of it and other environmental statutes (*NEPA* §101).

The NEPA requires all federal agencies to prepare an environmental impact statement for major federal actions significantly affecting the quality of the environment (*NEPA* §102).

The NEPA requires the president to submit an annual environmental quality report to Congress (*NEPA* §201).

The NEPA creates the CEQ to assist the president in preparing the environmental quality report, to develop national environmental policies, and to create rules for implementing the procedural requirements of the NEPA (*NEPA* §§202–204).

Freedom of Information Act

The Freedom of Information Act (5 *USC* §552).

STATUTORY ROADMAP

§552   Public information; agency rules, opinions, orders, records, and proceedings

PURPOSE

The Freedom of Information Act (FOIA) (1988, 5 *USC* §552) was enacted in 1966 to assure public access to certain federal agency records. The United States Supreme Court has stated that FOIA's purpose is "to ensure an informed citizenry, vital to the functioning of a democratic society, needed to check against corruption and to hold the governors accountable to the governed." (*National Labor Relations Board v. Robbins Tire & Rubber Co.*, 437 U.S. 214, 242 [1978]).

SPECIFIC PROVISIONS

The FOIA includes provisions for disseminating available information, defining key terms, procedural requirements, statutory exemptions and exclusions, and using a reverse FOIA.

Available Information

The FOIA requires federal agencies to publish information related to agency business in the *FR*. This information includes descriptions of agency organization, functions, procedures, and substantive rules and statements of general policy (*FOIA*, 5 *USC* §552[a][1]).

Agencies are also required to provide public access to "reading-room" materials. These materials include adjudicatory opinions, policy statements, and administrative staff manuals. Agencies must index the materials to facilitate public inspection (*FOIA*, 5 *USC* §552[a][1]). They must also provide an opportunity to review and copy the materials (*FOIA*, 5 *USC* §552[a][2]).

An FOIA request can be made for any reason regardless of relevancy. However, the act has nine exceptions to this disclosure requirement in which a record may fall (*FOIA*, 5 *USC* §552[b]), along with three law enforcement exclusions (*FOIA*, 5 *USC* §552[c]). The exclusions and exemptions balance the needs of an informed public against the security and confidentiality required of certain government information.

Definitions

The FOIA applies only to records maintained by federal agencies as defined by the act (*FOIA*, 5 *USC* §552[f]).

Agencies include any executive or military department or establishment, government or government-controlled corporation, or any independent regulatory agency. The FOIA does not require disclosure of records from state agencies, municipalities, courts, Congress, or private citizens. Nor does it require disclosure from the executive office or any presidential staff whose sole purpose is to counsel the president. However, states may have a functional equivalent of this federal act.

The FOIA does not explicitly define the term *record*. Nevertheless, the Supreme Court (*Department of Justice v. Tax Analysts*, 492 U.S. 136, 144–145 [1989]) has established a two-part test for determining an agency record. An agency record must be (1) created or obtained by an agency and (2) under the agency's control at the time of the request.

Any *person* can make an FOIA request. Under the act (*FOIA*, 5 *USC* §551[2]), a person includes United States citizens, foreign citizens, partnerships, corporations, associations, and foreign and domestic governments. However, no person can make an FOIA request in violation of the law.

Procedural Requirements

An information request under section (a)(3) must follow procedural requirements including a fee payment to cover governmental costs. Every federal agency must publish its own specific procedural regulations in the *FR* (*FOIA*, 5 *USC* §§552[a][3], [a][4][A]). The regulations include the types of records maintained by the agency, a description of how to access such records, fees and fee waivers, and the agency's administrative appeal procedures. Generally, any person can access agency records provided that the agency's procedures are followed (*FOIA*, 5 *USC* §552[a][3][B]) and the request reasonably describes the records sought (*FOIA*, 5 *USC* §552[a][3][A]).

Once an agency receives an FOIA request, the agency must inform the applicant of its decision to grant or deny the request within ten working days (*FOIA*, 5 *USC* §552[a][6][C]). If access is granted, an agency typically releases the records after the ten day period (*FOIA*, §552[a][6][C]). Agencies can obtain time extensions if the request involves an extensive or voluminous search, or if the request requires consultation with other agencies (*FOIA*, 5 *USC* §552[a][6][B]).

Agencies which deny requests must provide the applicant with the reasons for denial, the right of appeal, and the names of the persons responsible for the denial (*FOIA*, 5 *USC* §552[a][6][A][1]). If the administrative appeal upholds the denial, the administrative opinion must also provide the appellee with the reasons for denial, the right for judicial review in the federal courts, and the name of the persons responsible for the denial (*FOIA*, 5 *USC* §552[a][6][A][ii]).

Statutory Exemptions and Exclusions

Agencies are required to provide FOIA applicants with the records they request unless the request falls within one of the statutory exemptions or exclusions. When one of the nine statutory exemptions applies, agencies can use discretion to disclose or withhold the information. The exemptions apply to the following nine types of documents (*FOIA, 5 USC* §552[b]):

1. Classified documents
2. Internal personnel rules and practices
3. Information exempt under other laws
4. Trade secrets and other privileged or confidential information
5. Internal agency letters and memoranda
6. Information relating to personal privacy
7. Certain records or information relating to law enforcement
8. Information relating to financial institutions
9. Geological information

In addition to exemptions, the FOIA lists three types of documents which are excluded from public access. The three FOIA exclusions were added to the act as part of the Freedom of Information Reform Act of 1986 and were designed to protect sensitive law enforcement matters.[5]

Reverse FOIA

A reverse FOIA prevents the disclosure of information. It is designed to protect businesses and corporations that submit information to an agency. This protection is allowed when a third party makes an FOIA request to obtain the agency records containing that business' information (*CNA Financial Corp. v. Donovan*, 830 F.2d 1132 [D.C. Cir. 1988]). Nevertheless, an agency can release the records if an exemption does not apply, or if one does apply, but, in the agency's discretion, the release is justified (*CNA Financial Corp. v. Donovan*).

SUMMARY

The FOIA ensures public access to certain information obtained, generated, and held by the government.

The FOIA contains nine exemptions which balance the public's interest in information against the government's interest in efficient operation and security.

Occupational Safety and Health Act

The Occupational Safety and Health Act (29 *USC* 651 *et seq.*; 29 *CFR* Parts 1910, 1915, 1918, 1926).

5. *See Freedom of Information Act. U.S. Code.* Vol. 5 sec. 552(c)(1)–(3).

STATUTORY ROADMAP

§651	Congressional statement of findings and declaration of purpose and policy
§654	Duties of employers and employees
§655	Standards
§656	Administration
§657	Inspections, investigations, and record-keeping
§659	Enforcement procedures
§660	Judicial review
§666	Civil and criminal penalties

PURPOSE

The Occupational Safety and Health Act of 1970 (OSH Act) (29 *USC* §651 *et seq.*) differs from the other federal laws examined in this overview because it is directed toward protecting the workplace and its environment rather than the more traditional ambient environment. The OSH Act's purpose (§2[b], 29 *USC* §651[b]) makes this direction evident in "assur[ing] so far as possible every working man and woman in the Nation safe and healthful working conditions and preserv[ing] our human resources. . . ." This discussion of the OSH Act concentrates on the act's focus towards controlling hazardous substances in the occupational environment.

SPECIFIC PROVISIONS

The act creates two general duties for employers to keep the workplace free from hazards. First, employers must provide employees with a place of employment "free from recognized hazards that are causing or likely to cause death or serious physical harm . . ." (*OSH Act* §5[a][1], 29 *USC* §654[a][1]). Secondly, and more directly related to controlling hazardous substances in the environment, employers must comply with the occupational safety and health standards promulgated under the act (*OSH Act* §5[a][2], 29 *USC* §654[a][2]). In addition, employees must comply with the act's standards as well as all other rules and regulations related to the act (*OSH Act* §5[b], 29 *USC* §654[b]).

Occupational Health and Safety Administration Standards

The Department of Labor's Occupational Health and Safety Administration (OSHA) is required to promulgate health and safety standards to protect workers at their places of employment (*OSH Act* §6, 29 *USC* §655). The original standards, sometimes referred to as source standards, have been in effect since April 28, 1971. These standards originated from private groups such as the National Fire Protection Association as well as from previously established federal safety standards. While some of the original source standards were revoked because they were un-related to health or safety,[6] most of the standards are in effect today. All other OSHA standards are adopted in accordance with the procedures in section 6(b) of the act (*OSH Act* §6[b], 29 *USC* §655[b]).[7]

Source standards generally apply to air contaminants in the workplace for which the act creates threshold limits which cannot be exceeded. Approximately 380 substances are currently subject to these limits.[8] The OSHA has adopted approximately twenty additional standards pursuant to section 6(b). These standards are largely based upon acute health effects, chronic health effects, and carcinogenicity.

The scope of OSHA's standards is divided into two principal areas, General Industry Standards (29 *CFR* pt. 1910) and Construction Industry Standards (29 *CFR* pt. 1926). Nevertheless, certain industries may be exempt from a standard when another federal agency "exercise[s] statutory authority to prescribe or enforce standards or regulations affecting occupational safety or health (*OSH Act* §4[b][1], 29 *USC* §653[b][1]).

The act also provides a temporary variance and a permanent variance which facilities can obtain to avoid the OSHA standards. A temporary variance can be granted for up to two years from the effective date of a standard provided that either the means for meeting the standard are not currently available or the controls cannot be installed by the standard's effective date (*OSH Act* §6[b][6], 29 *USC* §655[b][6]).[9] A permanent variance can be granted when the employer can demonstrate the workplace is "as safe and healthful as those which would prevail if he complied with the standard (*OSH Act* §6[d], 29 *USC* §655[d]).[10]

Hazard Communication Standard

In November of 1983, the OSHA published a hazard communication standard (HCS) which requires employers to inform employees of the hazards associated with the chemicals they are exposed to in the workplace (29 *CFR* §1910.1200). The HCS also requires employers to inform employees of how to protect themselves from health risks associated from such exposure (29 *CFR* §1910.1200). Finally, the HCS creates labeling standards for containers of hazardous substances in the workplace (29 *CFR* §1910.1200[f][1]). In effect, the HCS created an information dissemination system in which employers obtain information from manufacturers, importers, and distributers of chemicals and, in turn, employers inform and train employees regarding potential hazards.

The HCS requires chemical manufacturers and im

porters to prepare a material safety data sheet (MSDS) for every hazardous chemical produced or imported (29 *CFR* §1910.1200[g]). Limited exceptions exist for trade secrets.[11] The initial MSDS and all subsequent revisions must be provided to all current and future distributers and manufacturing purchasers. Some of the minimum MSDS requirements include identifying the name and hazardous characteristics of the chemical, the health hazards of the chemical, the permissible exposure limit, precautions for safe handling and use, and emergency and first aid measures.[12] Employers must maintain copies of all MSDSs and assure that employees have access to them during working hours.

PREEMPTION

OSHA's hazardous communication standard preempts, or takes precedence over, all state right-to-know legislation.[13] Notably, occupational safety is the only federal right-to-know legislation which explicitly preempts similar state legislation.

ENFORCEMENT

The OSHA inspects workplaces to insure compliance with its standards. If an employer refuses to allow an OSHA compliance officer onto the premises to conduct an inspection, the compliance officer must obtain a warrant based upon probable cause. The OSHA can then issue a citation if it believes that the act is being violated (*OSH Act* §9[a], 29 *USC* §658[a]). The citation references the alleged violation, fixes a reasonable time for abatement, and proposes a penalty (*OSH Act* §9[a], 29 *USC* §658[a]). Employers must contest the citation within fifteen days of receipt or the citation becomes final and enforceable (*OSH Act* §§10[a,b], 17[1], 29 *USC* §§659[a,b], 666[1]).

SUMMARY

The OSH Act assures safe and healthful working conditions in the nation's workplace and preserves the nation's human resources (§651).

The OSH Act requires employers to provide employees with a workplace free from recognized hazards that are likely to cause death or serious bodily harm (§654).

The OSH Act requires employers to comply with specific occupational safety and health standards promulgated pursuant to the act (§654).

The OSH Act creates the OSHA, which inspects and investigates conditions in the workplace, provides for citations and notices of proposed penalties for violations, and provides for both civil and criminal penalties (§§657–660).

6. See *Federal Register 43*, (1978):49726.
7. See also *Code of Federal Regulations*. Title 29, part 1911.
8. See *Code of Federal Regulations*, Title 29, sec. 1910.1000.
9. See also *Code of Federal Regulations*. Title 29, sec. 1905.
10. See also ibid.

11. See *Code of Federal Regulations*, Title 29, sec. 1910.1200(i).
12. See ibid., sec. 1910.1200(g)(2).
13. See *United Steelworkers of America v. Auchter,* 763 F.2d 728 (3rd Cir. 1985); see also *Gade v. National Solid Waste Management Ass'n,* 112 S.Ct. 2374 (1992).

Emergency Planning and Community Right-To-Know Act

The Emergency Planning and Community Right-To-Know Act of 1986 (42 *USC* §§11001–11050).

STATUTORY ROADMAP

SUBCHAPTER I.   EMERGENCY PLANNING AND NOTIFICA-
                  TION
SUBCHAPTER II.   REPORTING REQUIREMENTS
SUBCHAPTER III.   GENERAL PROVISIONS

PURPOSE

The Emergency Planning and Community Right-to-Know Act (EPCRA) was enacted in 1986 as Title III of the Superfund Amendments and Reauthorization Act (*EPCRA* 001, 42 *USC* §§11001–11050). Despite its origin, the EPCRA is not a part of the Comprehensive Environmental Compensation and Liability Act but rather is an individual federal statute. The EPCRA provides for the gathering and dissemination of information on local industries' use of hazardous substances. It also provides for local community planning to deal with potential chemical-related emergencies, such as the accidental release of methyl isocyanate in Bhopal, India in 1984.

SPECIFIC PROVISIONS

The EPCRA provides for emergency planning and notification and specifies reporting requirements and its relationship to other laws.

Emergency Planning and Notification

The EPCRA (§301, 42 *USC* §11001) requires states to establish a state-level emergency response commission and local emergency planning districts to prepare and implement emergency plans. Each planning district designates a local emergency planning committee comprised of impact groups in the community (*EPCRA* §301[c], 42 *USC* §11001[c]).[14] Each local committee establishes its own procedures and rules for handling public requests for information.

The planning and notification requirements of the EPCRA are triggered by certain extremely hazardous substances. The EPA lists over 350 chemicals which it considers extremely hazardous. The list is published in Appendix A of the *Chemical Emergency Preparedness*

Program Interim Guidance.[15] Any facility which has a threshold amount of a listed substance must notify the state emergency response commission. Threshold amounts vary depending upon the toxicity of the substance.

Any release of a regulated substance triggers the statute's emergency notification procedures.[16] Generally, any facility must report the release of a reportable quantity of any listed substance to the local emergency planning committee and the state emergency response commission (40 *CFR* §355.40). The EPCRA divides releases into four categories (*EPCRA* §304[a], 42 *USC* §11004[a]); all trigger the notification requirement by the facility owner or operator (*EPCRA* §304[b], 42 *USC* §11004[b]). The reporting requirements have several statutory exceptions. These exceptions include releases resulting in exposure solely within the facility's boundaries and any federally permitted release pursuant to *CERCLA* §101(10) (*EPCRA* §304[b], 42 *USC* §11004[b]).

Reporting Requirements

The EPCRA requires facility owners or operators to complete forms providing information about chemicals found within or released from a facility. With the exception of the limited provisions for securing trade secrets (*EPCRA* §322, 42 *USC* §11042), the information is generally made available to the public (*EPCRA* §324[a], 42 *USC* §11044[a]).

Facility owners and operators who prepare an OSH Act (29 *USC* §§651–658) MSDS for hazardous chemicals must submit an MSDS for each applicable chemical to the state emergency response commission, the local emergency planning committee, and the local fire department (*EPCRA* §311[a], 42 *USC* §11021[a]). An MSDS contains the name and hazardous characteristics of each applicable chemical, the related health hazards, permissible exposure limit, precautions for safe handling and use, and emergency and first aid measures.

Alternatively, the facility can submit a list of all chemicals for which an OSH Act MSDS is required (*EPCRA* §311[a][2][A], 42 *USC* §11021[a][2][A]). The facility must identify the hazardous components of such chemicals, and the EPA may require additional information (*EPCRA* §§311[a][2][A–B], 42 *USC* §§11021[a][2][A–B]).

Facilities must also submit emergency and hazardous chemical inventory forms to provide information on the types, location, and quantities of hazardous chemicals at the facilities (*EPCRA* §312, 42 *USC* §11022). The inventory forms are divided into tier I information, which is

---

14. For example, representatives from the state, local officials, fire departments, community groups, and owners and operators of facilities subject to the EPCRA.

15. *See Federal Register 51*, (17 November 1986):41570.
16. Release is defined as "any spilling, leaking, pumping, pouring, emitting, discharging, injecting, escaping, leaching, dumping, or disposing into the environment (including the abandonment or discarding of barrels, containers and other closed receptacles) . . ." (*Emergency Planning and Community Right-to-Know Act*. sec. 329[5]. *U.S. Code*. Vol. 42, sec. 11049[5].)

provided in all instances, and tier II information, which is provided upon special request (*EPCRA* §§312[a][2], [e], 42 *USC* §§11022[a][2], [e]). The EPA can also request information on individual hazardous chemicals (*EPCRA* §312[d][1][C], 42 *USC* §11022[d][1][C]). As with the MSDS, this inventory form must be filed with the state emergency response commission, the local emergency planning committee, and the local fire department (*EPCRA* §312[a][1], 42 *USC* §11022[a][1]).

Finally, facility owners and operators must submit an annual toxic chemical release form to provide information about toxic chemicals released from a facility during its normal business operations (*EPCRA* §313[a], 42 *USC* §11023[a]). This requirement applies to facilities which employ ten or more full-time employees; which are categorized in Standard Industrial Classification (SIC) Codes 20 through 39; and which manufacture, use, or process a toxic chemical above the stated threshold quantity (*EPCRA* §313[a], 42 *USC* §11023[a]).[17]

The EPA also has authority to require individual facilities to complete the form although such facilities are not under the appropriate SIC code (*EPCRA* §313[b][2], 42 *USC* §11023[b][2]). The release form should provide information to the federal, state, and local governments, as well as to the general public (*EPCRA* §313[h], 42 *USC* §11023[h]).

Relationship to Other Laws

The EPCRA (§321[a], 42 *USC* §11041[a]) does not explicitly preempt any state or local law or interfere with any obligations or liabilities under any other federal law. This relationship differs from the OSH Act requirements related to hazardous information disclosure in the workplace which explicitly preempts any related state laws. Nevertheless, any state or local law which requires facilities to file a MSDS must at least comply with the format and content requirements under the EPCRA (§321[b], 42 *USC* §11041[b]).

SUMMARY

Facilities must complete an MSDS containing the name and hazardous characteristics of each applicable chemical, the related health hazards, permissible exposure limit, precautions for safe handling and use, and emergency and first aid measures (*EPCRA* §311).

Facilities must complete annual emergency and hazardous chemical inventory forms which are sent to the EPA and an appointed state official (*EPCRA* §312).

Facilities must complete toxic chemical release forms to report on regular operational releases of hazardous substances from them. The forms are filed annually with the EPA and an appointed state official (*EPCRA* §313).

The EPCRA provides for administrative, civil, and criminal penalties for noncomplying owners and operators of facilities (*EPCRA* §325).

The EPCRA provides for citizen suits against facility owners and operators, the EPA, a state governor, or a state emergency response commission for inaction under the act (*EPCRA* §326[a]).

—David Bookchin
David Farnsworth

References

Code of Federal Regulations. Title 29, secs. 1905, 1910, 1915, 1918, 1926; Title 40, sec. 355.40, parts 1500–1508.

Emergency Planning and Community Right-to-Know Act. 1986. Secs. 301, 304, 311–313, 321–322, 324, 329. *U.S. Code.* Vol. 42, secs. 11001–11050.

Freedom of Information Act. 1988. *U.S. Code.* Vol. 5, secs. 551–552.

Freedom of Information Reform Act. 1986. U.S. Public Law 99–570, secs. 1801–1804, Stat. 3207, 3207–48.

National Environmental Policy Act. Secs. 2, 101–102, 202, 204. *U.S. Code.* Vol. 42, secs. 4321–4370.

Occupational Safety and Health Act. 1970. Secs. 2, 4–6, 9–10, 17. *U.S. Code.* Vol. 29, sec. 651 *et seq.*

Superfund Amendments and Reauthorization Act, Title III. 1986. U.S. Public Law 99–499, 100 Stat. 1613. *U.S. Code.* Vol. 42, secs. 11001–11050.

17. *See also Emergency Planning and Community Right-to-Know Act.* Sec. 313(f). *U.S. Code.* Vol. 42, sec. 11023(f) (threshold for reporting).

1.3
NATURAL RESOURCE LAWS

This section discusses laws enacted to protect natural resources including:

- The Endangered Species Act
- The Coastal Zone Management Act

Endangered Species Act

The Endangered Species Act of 1973 (16 USC §1531 et seq., 50 CFR §17.3).

STATUTORY ROADMAP

§1531	Purposes and policy
§1532	Definitions
§1533	Determinations
§1534	Land acquisition
§1535	Cooperation with states
§1536	Interagency cooperation
§1537	International cooperation
§1537(a)	Convention implementation
§1538	Prohibited acts
§1539	Exceptions
§1540	Enforcement
§1541	Endangered plants
§1542	Appropriations
§1543	Construction with the Marine Mammal Protection Act
§1544	Annual cost analysis

PURPOSE

Enacted in 1973, the Endangered Species Act (ESA) (16 USC §§1531–1543) is a relatively simple statute which is sweeping in its scope. The ESA seeks to protect species of fish, wildlife, and plants, and the habitat associated with those species. Congress has declared the purposes of the ESA (§2[b], 16 USC §1531[b]) are "to provide a means whereby the ecosystems upon which endangered and threatened species depend may be conserved, to provide a program for the conservation of such endangered species and threatened species . . . ," and to meet the United States' duties under other fish and wildlife protection treaties and conventions. The ESA regulates mainly by prohibiting persons from taking listed species and by protecting habitat.

SPECIFIC PROVISIONS

This overview of the ESA examines its regulatory structure in conjunction with key terms defined under the act.

Species

The terms *species* and *fish and wildlife* are broadly defined under the act. Species includes not only true species, but also subspecies and *distinct populations* of fish, wildlife, or plants (*ESA* §3[16], 16 USC §1532[6]). Fish and wildlife is defined as any member of the animal kingdom, "including without limitation any mammal, fish, bird (including any migratory, nonmigratory, or endangered bird for which protection is also afforded by treaty or other international agreement), amphibian, reptile, mollusk, crustacean, arthropod, or other invertebrate, and includes any part, product, egg, or offspring thereof, or the dead body parts thereof" (*ESA* §3[8], 16 USC §1531[8]).

Taking

The concept of *taking* an endangered species is much more than simply, for example, shooting a bald eagle. Taking is defined broadly to include actions like "harass, harm, pursue, hunt, shoot, wound, kill, trap, capture, or collect, or attempt to engage in any such conduct" (*ESA* §3[19], 16 USC §1532[19]).

The term *harm* is even more broadly defined in Fish and Wildlife Service regulations promulgated pursuant to ESA section 3(19), and later in the court case which first construed those regulations.[1] Harm is defined as an act which kills or injures. "Such an act may include significant habitat modification or degradation where it actually kills or injures wildlife by significantly impairing essential behavioral patterns, including breeding, feeding, or sheltering" (*50 CFR* §17.3 [1990]).

The concepts of harm and taking were applied in the *Palila v. Hawaii Department of Land and Natural Resources* (649 F.Supp. 1070 [1986], aff'd, 852 F.2d 1106 [9th Cir. 1988]) case which involved a state-maintained flock of sheep which was destroying the habitat of the palila, an endangered species of bird. The palila relied on the vegetation on which the sheep were browsing. The court found that the actions of the sheep constituted harm, and therefore taking under the Fish and Wildlife Service's regulations which had been promulgated pursuant to the ESA.

---

1. The Fish and Wildlife Service eventually settled on the current definition of *harm* which is found in *Federal Register 46*, (4 November 1981): 54748, 54750.

Listing Endangered Species

To receive protection under the act, species must first be listed. The listing process is an assessment of the relative vulnerability of certain species.[2] While it has been changed several times, the listing process can be initiated by private individuals, who petition the Secretary of Interior.[3] To determine if a species should be listed, the Secretary of the Interior considers the best scientific and commercial data available and also takes into account the efforts being made by any state or foreign nation, or political subdivision of a state or foreign nation to protect such species (*ESA* §4[b][1][A], 16 *USC* §1533[b][1][A]).

Classifying Endangered Species

Once listed, species are classified as either endangered or threatened. Those species "in danger of extinction throughout all or a significant portion of their range" are classified as endangered (*ESA* §3[6], 16 *USC* §1532[6]). For a species to be classified as threatened, the petitioner must show that the species "are likely to become an endangered species within the foreseeable future throughout all or a significant portion of their range" (*ESA* §3[6], 16 *USC* §1532[6]).

Critical Habitat

Once the secretary determines that a species should be listed as either endangered or threatened, he must also establish a critical habitat for that species (*ESA* §3[5][A][i], 16 *USC* §1532[5][A][i]). Critical habitats are specific areas within the geographical areas occupied by the species at the time the species are listed. Critical habitats also have physical and biological features considered essential to conserving the species and which may require special management considerations or protection (*ESA* §3[5][A][i], 16 *USC* §1532[5][A][i]).[4]

Persons

Finally, a *person* for purposes of the ESA, is also defined broadly. A person is "an individual, corporation, partner-

ship, trust, association, or any private entity, or an officer, employee agent, department, or instrumentality of the Federal Government, of any state, municipality, or political subdivision of the state; or any other entity subject to the jurisdiction of the United States" (*ESA* §3[13], 16 *USC* §1532[13]).

SUMMARY

The ESA provides broad definitions of taking, species, and person (*ESA* §3).

The ESA specifies listing procedures and classification on the basis of species' vulnerability (*ESA* §4).

The ESA provides for listed species' critical habitat (*ESA* §3[5]).

Coastal Zone Management Act

The Coastal Zone Management Act (16 *USC* §1451 *et seq.*; 15 *CFR* Parts 931.1 *et seq.*, 930.1 *et seq.*, 923.1 *et seq.*, 926.1 *et seq.*).

STATUTORY ROADMAP

§1453	Definitions
§1454(b)	Program requirements
§1455	Administrative grants
§1455b	Protection of coastal waters

PURPOSE

The Coastal Zone Management Act (CZMA) (16 *USC* §§1451–1464), originally enacted in 1972 and reauthorized in 1990 (*Coastal Zone Act Reauthorization Amendments* §6201–17), controls land use along the nation's coastal zone. This area is particularly susceptible to pressure from population and development. Consequently, Congress developed the CZMA to protect and enhance the nation's coastal zone (*CZMA* §302[a], 16 *USC* §1451[a]). The act seeks to achieve this goal by implementing four national policies and calling for states to implement programs which meet minimum federal standards. The four national policies under the CZMA are:

1. To preserve, protect and develop, and where possible to restore and enhance the coastal zone (*CZMA* §303[1], 16 *USC* §1452[1])
2. To assist states in developing their own coastal management programs (*CZMA* §303[2], 16 *USC* 1452[2])
3. To encourage the preparation of management plans for special areas to protect natural resources and allow for reasonable coastal-dependent economic growth (*CZMA* §303[3], 16 *USC* §1452[3])

---

2. The term is used by Michael Bean. *See* Bean, M. *The evolution of national wildlife law.* 1983. New York. For provisions regarding the listing and identification of endangered and threatened wildlife and plant life, *see Code of Federal Regulations,* Title 50, secs. 7.1, 217.1 *et seq.,* 424.01 *et seq.,* and Title 60, sec. 227.1 *et seq.*

3. The Secretary of Commerce may, in certain cases, be the authority to petition. The Department of the Interior, through its Secretary, can also initiate the listing process. Apart from the petition process, the Secretary of Agriculture is authorized to enforce the ESA with respect to plants. *See Endangered Species Act,* sec. 3(15), *U.S. Code* Vol. 16, sec. 1532(15). These sections also apply to the changing of a species' status and the removal of species from the list, i.e., delisting.

4. Specific areas outside the geographical areas occupied by the species are considered as critical habitat under certain circumstances.

4. To encourage the participation of federal, state, regional, and local government bodies in achieving the purposes of the act (*CZMA* §303[4], 16 *USC* §1452[4])

SPECIFIC PROVISIONS

The CZMA provides for coastal zone management programs and specifies its applicability.

State Coastal Zone Management Programs

To promote its policies, the CZMA requires states to develop coastal management programs to meet the performance standards prescribed in it. An acceptable state coastal program must, at a minimum, provide for:

1. The protection of wetlands, floodplains, estuaries, beaches, dunes, barrier islands, coral reefs, and fish and wildlife and their habitat within the coastal zone
2. The management of coastal development in hazardous areas to minimize loss to life and property
3. Priority consideration given to coastal-dependant uses and an orderly process for situating major facilities related to defense, energy, fisheries development, recreation, ports and transportation, and the location, to the maximum extent practicable, of new commercial and industrial developments in or adjacent to areas where such development exists
4. Public access to the coast for recreational purposes
5. Assistance in redevelopment of waterfronts and other aesthetic, cultural, and historic coastal features
6. Coordination of government decision-making regarding the coastal zone, and coordination with federal agencies
7. Public participation in coastal management decision-making
8. Comprehensive planning, conservation, and management for living resources (*CZMA* §303[2][A–K], 16 *USC* §1452[2][A–K])

Many states with coastal management programs have implemented coastal area permit programs which regulate development in the coastal zone. Typically, states establish criteria which an applicant must meet. For example, in California and North Carolina, permits are required for activities that affect designated areas such as wetlands, estuaries, and shorelines. Permit requirements vary from state to state. However, the permitting process generally serves as a review of projects which may create detrimental effects to a state's coastal zone. Permits are enforced through the surveillance and monitoring of permitted projects.

Applicability

The act applies only to areas designated as the coastal zone. This area varies from state to state and, consequently, from program to program. The CZMA adopts the territorial sea as its seaward or outermost limit. This limit extends three nautical miles from the shore and, according to federal law, all of the submerged lands and resources within that area are owned by the states. The inland boundary, however, is the portion of the coastal zone which varies. The act calls for the coastal zone to extend "only to the extent necessary to control shorelands, the uses of which have a direct and significant impact on the coastal waters" (*CZMA* §304[1], 16 *USC* §1453[1]). Thus, the coastal zone can include an entire state, such as Florida or Hawaii, or it can be a much smaller portion of land.

SUMMARY

The CZMA seeks to protect coastal resources, manage those resources, and prevent conflicts in their use.

The CZMA mandates the development of state programs, meeting federal minimum requirements, which permit and regulate activity in their own coastal zones. While state programs vary, the federal government has authority to require state programs to promote the four coastal management goals and to meet the federal minimum performance standards. Coastal programs are typically implemented around a permitting application and review scheme.

—*David Bookchin*
David Farnsworth

References

Coastal Zone Act Reauthorization Amendments of 1990. U.S. Public Law 101–508, sec. 6201–17. U.S. Code, Congressional Administrative News.

Coastal Zone Management Act. Secs. 302–304. U.S. Code. Vol. 16, secs. 1451 *et seq.*

Code of Federal Regulations. Title 15, parts 923, 926, 930–931, Title 50, sec. 17.3.

Endangered Species Act. Secs. 2–4, 7. U.S. Code. Vol. 16, secs. 1531 *et seq.*

1.4
POLLUTION CONTROL LAWS

This section discusses the pollution control laws including:

- The Clean Air Act
- The Resource Conservation and Recovery Act
- The Comprehensive Environmental Response, Compensation, and Liability Act
- The Noise Control Act
- The Safe Drinking Water Act
- The Federal Water Pollution Control Act
- The Toxic Substance Control Act
- The Federal Insecticide, Fungicide, and Rodenticide Act
- The Pollution Prevention Act

Clean Air Act

The Clean Air Act (42 *USC* §§7401–7671q; 40 *CFR* Part 50).

STATUTORY ROADMAP

SUBCHAPTER I.	PROGRAMS AND ACTIVITIES
Part A.	*Air quality and emission limitations*
§7401	Findings and purpose
§7409	National primary and secondary ambient air quality standards
§7410	State implementation plans
§7411	Standards of performance for new stationary sources
§7412	Hazardous air pollutants
§7413	Federal enforcement procedures
Part C.	*Prevention of significant deterioration*
§7470	Purpose
§7472	Initial classifications
§7473	Increments and ceilings
§7474	Area redesignation
§7475	Preconstruction requirements
§7479	Definitions
Part D.	*Plan requirements for nonattainment areas in general*
Subpart 1.	Nonattainment areas in general
Subpart 2.	Additional provisions for ozone nonattainment areas
Subpart 3.	Additional provisions for carbon monoxide nonattainment areas
Subpart 4.	Additional provisions for particulate matter nonattainment areas
Subpart 5.	Additional provisions for areas designated nonattainment for sulfur oxides, nitrogen dioxide, and lead
SUBCHAPTER II.	EMISSION STANDARDS FOR MOVING SOURCES
Part A.	*Motor vehicle emission and fuel standards*
Part C.	*Clean fuel vehicles*
SUBCHAPTER III.	GENERAL PROVISIONS
§7603	Emergency powers
§7604	Citizen suits
§7607	Administrative proceedings and judicial review
§7619	Air quality monitoring
SUBCHAPTER IV.	ACID DEPOSITION CONTROL
SUBCHAPTER V.	PERMITS
SUBCHAPTER VI.	STRATOSPHERIC OZONE PROTECTION

PURPOSE

The goal of the Clean Air Act (CAA) (§§101–618, 42 *USC* §§7401–7671q) is to prevent and control the discharge of pollutants into the air which can harm human health and natural resources. It regulates pollution by establishing ambient air quality standards at which pollutants can be safely tolerated. The act also regulates emission sources through a system of limitations on specified pollutants and a permit program for major sources.

The current act is comprised of several amendments that address air pollution problems over the last twenty years. Most of the major provisions were developed in the 1970, 1977, and 1990 amendments (*CAA* [1990], *CAA* §§101–618, 42 *USC* §§7401–7671q). The central part of the act is its provisions for National Ambient Air Quality Standards for criteria pollutants. Other features of the act include standards for areas which meet the ambient air quality standards, attainment areas, and more stringent standards for nonattainment areas. In addition, the act provides for prevention of significant deterioration, air toxics, state implementation plans, permits, and the control of mobile sources.

SPECIFIC PROVISIONS

The CAA has specific provisions for the National Ambient Air Quality Standards, attainment and nonattainment, ozone, prevention of significant deterioration, air toxics, state implementation plans, permits, and mobile sources. These provisions are summarized next.

National Ambient Air Quality Standards

National Ambient Air Quality Standards (NAAQS) are health-based standards used to measure and protect the air around us.[1] Section 108 directs the EPA to identify criteria pollutants which may reasonably endanger public health and welfare. The list of criteria pollutants includes:

- Sulfur dioxide (SO_2)
- Carbon monoxide
- Ozone (smog)
- Nitrogen oxides (NO)
- Lead
- Particulate matter

The act directs the EPA to establish two levels of NAAQS for the criteria pollutants. First, the EPA developed primary air quality standards to protect human health and provide a margin of safety to protect sensitive members of the population, e.g., pregnant woman, children, and the elderly. The EPA can also establish stricter secondary standards to avoid adverse impacts on the environment and protect public welfare by preventing harm to agricultural crops and livestock.

Attainment and Nonattainment

The attainment of ambient standards is a central purpose of the CAA. Areas are in attainment if they meet the NAAQS. Pollution sources in attainment areas are subject to the prevention of significant deterioration (PSD) requirements discussed next.

An area is designated in nonattainment if it exceeds the NAAQS for a given pollutant more than once a year. This determination is made for each criteria pollutant. Consequently, an area can be in nonattainment for one pollutant, while at the same time be in attainment and subject to PSD provisions for another.

Though similar to the classification scheme in the 1977 amendments, the 1990 amendments divide nonattainment areas into categories depending on the severity of each area's problem. It also allows boundaries to be adjusted to accommodate areas that have come into attainment. Amended section 107(d) requires states to designate regions as:

"Nonattainment"—any area that does not meet (or contributes to ambient air quality in a nearby area that does not meet) national primary or secondary ambient air quality standards for the pollutant;

"Attainment"—any area that does meet the standard for the pollutant, or

"Unclassifiable"—any area that cannot be classified attainment or nonattainment on the basis of available information.

Any state not meeting implementation deadlines is subject to a moratorium on the construction of new major stationary sources of pollution or on any major modifications of existing major sources in nonattainment areas. A major source is one with emissions or potential to emit 100 tons or more per year of a pollutant subject to regulation under the act.[2] A major modification is any physical change in the operation of a major stationary source that would result in a significant net increase in a pollutant subject to regulation under the act.[3]

Section 173 requires states to adopt permit programs for the construction or modification of major stationary sources in nonattainment areas. It also imposes a more demanding technology requirement, the lowest achievable emission rate (LAER) on those sources. Finally, permits for modified sources in nonattainment areas can be issued only if emissions from existing sources in the area decrease enough to offset the increase in emissions from the new or modified source and continue reasonable further progress.

Ozone

Ozone, commonly known as smog, is the most serious and common nonattainment problem. The 1990 amendments contain requirements for ozone nonattainment areas (*CAA* §181, 42 *USC* 7511). Section 181 divides ozone nonattainment areas into five categories depending on their degree of nonattainment:

- Marginal
- Moderate
- Serious
- Severe
- Extreme

Each category is assigned a design value, which is a measure of its ambient air quality expressed in parts per million. Attainment deadlines vary among areas and are measured from November 15, 1990, the enactment date of the 1990 amendments. The attainment deadline for each category follows:

- Marginal areas—three years
- Moderate areas—six years
- Serious areas—nine years
- Severe areas—fifteen years
- Extreme areas—twenty years

On the basis of its category and design value, each area is responsible for its own category requirements as well as

1. *See U.S. Code*, Vol. 42, secs. 7408–9; *Federal Register 36*, 1971, 8186; *Code of Federal Regulations.* 1991, Title 40, part 50.

2. *See Code of Federal Regulations*, 1984, Title 40, sec. 52.24(f)(4)(i)(a).
3. *See Clean Air Act*, sec. 110(a)(2)(I); *U.S. Code,* Vol. 42, sec. 7410(a)(2)(I); *Code of Federal Regulations*, Title 40, sec. 52.24(f)(5)(i).

the requirements for categories of lesser nonattainment. Thus, moderate areas are responsible for their own requirements as well as those applicable to marginal areas. For example, a state with a marginal ozone area must submit:

1. An inventory of all actual emissions to the EPA by November 1992
2. Revisions to its state implementation plans (SIP) designed to meet the three-year deadline for attainment, including the implementation of reasonably available control technology (RACT)
3. Permit programs for new and modified sources
4. An increase in the offset requirement for new sources and modifications, from 1 to 1 to 1.1 to 1
5. The retention of the vehicle inspection and maintenance program previously required for the area

A moderate area must submit requirements (1) through (5) listed for marginal areas, as well as all requirements assigned to moderate areas which include:

1. A revision of its SIP to provide for reasonable further progress through a 15% reduction in emissions within six years for volatile organic compounds (VOCs)
2. Any other annual reductions of VOCs and NO necessary to reach attainment by the deadline
3. The implementation of RACT for each category of VOC source covered by existing control techniques guidance.

Prevention of Significant Deterioration

Areas in attainment for NAAQS are subject to provisions for PSD (*CAA* §§160–169[B], 42 *USC* §§7470–7479[B]). PSD provisions seek to protect areas that already enjoy cleaner air than that required by the ambient standards, and to keep that air from deteriorating in quality.[4] PSD provisions also require new major facilities to apply for a preconstruction permit and to use the best available control technology (BACT) for each pollutant regulated under this section.[5]

The BACT is determined on a case-by-case basis and reflects the most effective controls currently in use. The Act defines the BACT as the maximum degree of reduction which considers energy, environmental, economic, and other impacts.

PSD regulations divide the country into three classes on the basis of air quality:

Class I (large national parks and wilderness areas)
Class II (very clean areas allowing some industrial growth)
Class III (Class II areas established for industrial development)

For each class, the act establishes maximum allowable increases over baseline concentrations (increments) of pollutants (SO_2, NO, and PM–10). Baseline concentrations are established based on the ambient concentrations measured at permit application. Allowable increases cannot create concentrations greater than those sanctioned by the NAAQS.

Section 165 of the act controls PSD permits (*CAA* §165, 42 *USC* §7475. This section requires new and modified sources to undergo a preconstruction review. The review includes a hearing with public comment to assess a project's potential impact and to consider alternatives to the proposed project. After completing air quality monitoring, the owner or operator must also demonstrate that the expected emissions from the operation or construction of the project will not exceed the limit on the increment of clean air in that area.

Air Toxics

The issue of air toxics is treated separately from attainment and nonattainment of ambient air quality and the prevention of significant deterioration (*CAA* §112, 42 *USC* §7412.[6] The 1970 act authorized the EPA to set health-based national emission standards for hazardous air pollutants (NESHAPS). However, the EPA acted on only a limited number. The 1990 amendments, under section 112, mandates the establishment of technology-based standards for 189 hazardous substances. Section 112 also calls for the use of maximum achievable control technology (MACT) to regulate the emission of these substances.

The 1990 amendments create a two-step approach to controlling air toxics. First, new sources must use MACT to achieve a reduction level equal to levels reached by the least-polluting existing sources within given categories. All categories of sources are supposed to be promulgated by November 2000: 25% of all standards are to be promulgated by November 1994, and 50% of all standards are to be completed by November 1997. The second step establishes residual risk standards which are more stringent than MACT. Residual risk standards should be used to protect the public health with an ample margin of safety.

The EPA has already issued some MACT standards. One example is the hazardous organic NESHAP (HON) rule, which sets standards for reducing the emissions of 149 toxic chemicals from synthetic–organic chemical manufacturing (SOCMI) processes.[7] The rule applies to processes that develop chemicals considered primary products, which are distinct from other by-products created in these processes. The HON applies to any plant that is con-

---

4. *See Code of Federal Regulations,* Title, 40 sec. 52.51 for PSD requirements and PSD provisions in SIPs.
5. *See Federal Register 53* at 40656.

6. *See also Code of Federal Regulations,* Title 40, part 61.
7. National Emission Standards for Hazardous Air Pollutants for Source Categories; Organic Hazardous Air Pollutants from the Synthetic Organic Chemical Manufacturing Industry and Seven Other Processes, *Federal Register 57,* (31 December 1992):62608.

sidered a major source and is a model for future MACT standards developed at the federal level.

The EPA has also developed a final rule on the toxics early reductions program for hazardous air pollutants (57 FR 61970 et seq.).[8] This program encourages industrial facilities to pursue early reductions before the final MACT standards are established. Facilities that qualify for the program can defer their compliance with MACT standards for six years if they agree to and can demonstrate reductions of their emissions by 90 and 95% before the MACT standards take effect (CAA §112[d], 42 USC §7412[d], 40 CFR §63.70 et seq.). Because industries under this program can choose how to reduce their toxics emissions, in theory they have the flexibility to make the most economically viable choices for themselves.

The EPA has published an initial list of categories of major sources and area sources of air toxics (57 FR [16 July 1992] 31576). An area source is defined as a stationary source that is not a major source (CAA §112[a][2], 42 USC §7412[a][2]). The regulation of area sources is meant to include small diverse sources, such as dry cleaners and service stations, which substantially contribute to the emission of hazardous pollutants. Instead of being subject to MACT, areas sources are subject to less stringent standards called generally available control technology (GACT).

State Implementation Plans

The act requires each state to submit a plan for implementation and enforcement of the national standards within its jurisdiction (CAA §110, 42 USC §7410, 40 CFR §§51.01 et seq., 52.01 et seq., 52.2370 et seq.). SIPs are based on emission inventories and computer models that predict whether violations of the NAAQS will occur. States can decide to meet the NAAQS as long as the regulatory requirements they choose enable them to attain the national standards. Once an SIP is approved, it becomes an element of state and federal law and is enforceable under either. In attainment areas, SIP planning must also account for PSD issues.

In nonattainment areas, SIPs provide for attainment of NAAQS and include:

1. Provisions for RACT
2. Reasonable progress in attaining the required reductions
3. A current emissions inventory
4. Permits for new and modified major stationary sources
5. Emission offset requirements

6. A contingency plan if unable to meet NAAQS by the specified date

SIPs must also contain procedures for the review and permitting of new or modified stationary sources and their impact on the attainment and maintenance of NAAQS. SIPs regulate these sources on the basis of their general provisions. SIPs also contain provisions that apply to individual sources. To obtain a variance or be exempted from SIP provisions, however, involves a lengthy process which includes revising the SIP and review by the state authorities and the EPA.

Some SIPs contain provisions for alternative compliance, which streamline review and allow for in-state revisions of the requirements for individual facilities. In some contexts, the use of "bubbles" provides another means for facilities to make changes within their plant without triggering an SIP revision.[9] Put simply, bubbling is the consolidation of multiple emission sources. Bubbling creates one source (for regulatory purposes), which allows certain limited changes in emissions from the sources within the bubble.

If a state fails to gain approval of its SIP, the EPA can promulgate a plan for that state. This plan is called a federal implementation plan (FIP). An implementation plan includes the following requirements:[10]

1. Enforceable emission limits
2. Necessary monitoring
3. Program for enforcement including a permit program
4. Interstate and international requirements
5. Adequate personal funding and authority
6. Self-monitoring requirements
7. Emergency authority
8. Revision authority
9. Nonattainment plan
10. PSD plans
11. Air quality modelling
12. Federal requirements
13. Consultation and participation

Permits

Generally, the federal government delegates the responsibility for issuing permits to the states through the SIPs. Prior to implementing the new operating permit program under the 1990 amendments, new source review was the means for issuing permits to new or modified industrial facilities. Permit requirements vary, depending on whether the project is in a PSD or nonattainment area.[11] Apart from PSD

---

8. This program is implemented through the Title V operating permits program. The EPA has developed a guidance document outlining procedures for facilities to follow in applying for this program, Enabling Document for Regulations Governing Compliance Extensions for Early Reductions of Hazardous Air Pollutants (EPA-450/3-91-013 [December 1992]).

9. The EPA first developed a bubble policy in 1979 but revised it to include not only existing sources but also emissions trading and new sources. Federal Register 51, (1986):43814.

10. For SIP minimum criteria, see Federal Register 56, (26 August 1991):42216.

11. See discussions of Prevention of Significant Deterioration and Nonattainment supra.

and nonattainment concerns, permitting is a two-step process which requires initial authorization to construct a project and then authorization to operate a source.

The permit program under the 1990 amendments brings together the requirements for individual facilities and places them in one document. It also provides a financial basis for the program through permit fees. New permits contain emissions standards, monitoring, record-keeping, and reporting requirements. Major sources, sources subject to air toxics regulation, and all sources subject to new source performance standards are required to obtain a permit.

Mobile Sources

The 1970 act regulated emissions from automobiles. It called for a 90% reduction of emissions below the then-current levels. Subsequently, the 1977 amendments adjusted the NO emissions to allow for a 75% reduction. These standards are applied on a national basis because automobiles move from state to state.

The 1990 amendments go further in reducing emissions from mobile sources (*CAA* §§202–250, 42 *USC* §§7521–7590). They establish more stringent tailpipe standards and establish clean fuel requirements.

To implement the new emission standards, the 1990 amendments establish a tier system which phases in standards for NO between 1994 and 1995, and for nonmethane hydrocarbons and carbon monoxide between 1994 and 1998. A second tier, if the EPA finds it necessary, will provide for even stricter standards. These standards will be developed between 2003 and 2006. The EPA must also develop controls for emissions resulting from fuel evaporation.

The clean fuel program promotes the use of some fuels like methanol, ethanol, reformulated gasoline, and other fuels such as electricity and natural gas. Reformulated gas, for areas with high ozone levels, has lower levels of oxygen, aromatic hydrocarbons, and benzene. For carbon monoxide nonattainment areas, the oxygenated fuel program will require a minimum oxygen content in fuels.

The 1990 amendments also target fleets of cars and trucks. Fleets of ten or more which can be centrally fueled are subject to the clean fuel requirements. Lesser standards apply to heavy-duty trucks. These standards apply to all cars and trucks in serious, severe, and extreme ozone nonattainment areas.

OTHER FEATURES OF THE CAA AMENDMENTS OF 1990

The 1990 amendments to the CAA establish a new nonattainment program and contain new deadlines and divide nonattainment areas into several categories.

Deadlines range from 1993 for marginal areas to 2010 for Los Angeles.

Section 112 on air toxics lists 189 regulated hazardous air pollutants and directs the EPA to establish technology-based standards based on MACT. The amendments also provide for a program to prevent accidental releases.

Modeled on the Clean Water Act's National Pollution Discharge Elimination System, sections 501 to 507 of title V establish a federal permit program for existing stationary sources. The EPA can veto permits not in compliance with the act, and citizens have certain rights to challenge them.

The new acid rain program seeks to reduce sulphur dioxide emissions nationwide to 8.9 million tons per year. It also mandates a reduction of annual NO emissions by approximately two million tons. These reduction provisions are part of a market-based emission allowance program.

SUMMARY

The CAA sets primary and secondary standards for major pollutants (§7409).

The CAA requires states to establish air quality control regions and develop SIPs to achieve the nation's ambient air standards within reasonably statutorily defined time periods (§§7404, 7406).

The CAA requires the EPA to establish NSPSs to limit emissions from new or expanding sources to the fullest extent possible (§7411).

The CAA requires the EPA to establish emission standards for hazardous air pollutants (§7412).

The CAA mandates the EPA to follow specific strategies for limiting emissions from stationary sources in both attainment and nonattainment areas (§§7470 *et seq.*, 7501 *et seq.*).

The CAA mandates the EPA to follow a specific strategy for controlling mobile air emission sources (§7521 *et seq.*).

The EPA can commence a civil or criminal action for a violation of any requirement of an applicable implementation plan. Additionally, the CAA provides for citizen suits to insure that the EPA is performing all of its nondiscretionary duties (§§7413, 7604).

The CAA amendments of 1990 add substantial content to the act including provisions focusing on urban air pollution, mobile sources, toxic pollutants, permits, and the problems associated with acid rain.

Resource Conservation and Recovery Act

The Resource Conservation and Recovery Act (42 *USC* §6901 *et seq.*, 40 *CFR* Parts 240–271).

STATUTORY ROADMAP

PURPOSE

Enacted in 1976, the Resource Conservation and Recovery Act (RCRA) (1992, 42 *USC* §§6901–6991i) regulates the transportation, handling, storage, and disposal of solid and hazardous waste. The RCRA was amended in 1984 by the Hazardous and Solid Waste Amendments. Originally conceived as a law to control the disposal of solid waste and encourage recycling, the RCRA's emphasis is the regulation of hazardous waste.

The central mechanism in the RCRA is the manifest system which creates documentation on hazardous waste and monitors the movement of hazardous waste. The RCRA also establishes standards for treatment, storage and disposal facilities, and state hazardous waste programs.

The general objective of the RCRA (§1003[a], 42 *USC* §6902[a]) is "to promote the protection of health and the environment and [the conservation of] valuable material and energy resources...." This objective should be achieved through the development of solid waste management plans and hazardous waste management practices. The RCRA also sets forth a national waste management policy: "wherever feasible, the generation of hazardous waste is to be reduced or eliminated as expeditiously as possible. Waste that is nevertheless generated should be treated, stored, or disposed of so as to minimize the present and future threat to human health and the environment" (*RCRA* §1001[b], 42 *USC* §6902[b]).

SPECIFIC PROVISIONS

The RCRA provides for a manifest system, regulated wastes, regulating hazardous wastes, releases and ground water monitoring, closure, and underground storage tanks.

The Manifest System

The RCRA requires waste to be monitored from its generation to its ultimate disposal. Thus, the RCRA is commonly referred to as a cradle-to-grave statute. A manifest is a written description of a container's contents, source, and destination. The manifest must accompany the waste at all times.[12] This information must also be made available to the regulatory authorities, enabling them to monitor waste generation, shipment, and disposal.

Regulated Wastes

A person's status under the RCRA depends upon the type of waste he is handling. The RCRA regulates solid waste and hazardous waste, both of which are defined by the statute and its regulations.

SOLID WASTE

The RCRA defines solid waste as "any garbage, refuse, sludge from a waste treatment plant, water supply treatment plant, or air pollution control facility and other discarded material, including solid, liquid, semisolid, or contained gaseous material resulting from industrial, commercial, mining and agricultural operations, and from community activities..." (*RCPA* §1004, 42 *USC* §6903[27]). The remainder of the definition excludes certain materials from solid waste, thus omitting them from solid waste regulation under the RCRA.

States are encouraged to develop solid waste management plans which enable waste to be controlled on a local level. While the RCRA does not directly regulate state solid waste programs, it directs the EPA to develop guidelines for states to follow. These guidelines are developed under subtitle D of the act. In addition, the RCRA establishes minimum requirements upon which states can model solid waste plans. The requirements prohibit open dumping of solid waste and mandate the closing or the upgrading of operational open dumps.

HAZARDOUS WASTE

Under the RCRA (§1004[5], 42 *USC* §6903[5]), hazardous waste is defined as any solid waste which, because of its quantity, concentration, or physical, chemical, or infectious characteristics may—

(A) cause, or significantly contribute to an increase in mortality or an increase in serious irreversible, or incapacitating reversible, illness; or

(B) pose a substantial present or potential hazard to human health or the environment when improperly treated, stored, transported, or disposed of, or otherwise managed (*RCRA* §1004[5], 42 *USC* 6903[5]).

---

12. *Code of Federal Regulations*. Title 40, parts 264, 265. Manifest requirements and the reporting requirements are in Subparts E of these parts.

Solid waste is considered hazardous if it is listed as hazardous or if it has hazardous characteristics.[13] If the EPA designates a waste hazardous through rulemaking, then it is a listed hazardous waste. Alternatively, the EPA can find a waste to be hazardous based upon the waste's characteristics.

The EPA has developed four criteria for measuring the characteristics of substances: ignitability, corrosivity, reactivity, and toxicity (*RCRA* §1022[a], 42 *USC* §6921[a]). Depending on their classification, wastes are assigned a hazardous waste classification number: ignitable (D001), corrosive (D002), and reactive (D003).

A liquid solid waste is determined ignitable if it has a flash point lower than 60 degrees centigrade.[14] A solid material is ignitable if it is "capable at a temperature of 25 degrees centigrade and a pressure of one atmosphere, of causing fire through friction, absorption of moisture or spontaneous chemical changes, and when ignited, burns so vigorously and persistently that it creates a hazard" (40 *CFR* §261.21).

A corrosive waste is a solid waste with a pH of less than 2 or greater than 12.5 (40 *CFR* §261.22). Reactive wastes are solid wastes which are normally unstable, react violently or generate toxic fumes when they come in contact with water, contain cyanide or sulfide, (when exposed to pH conditions between 2 and 12.5) generate toxic fumes, or are explosive.

The final criterion for determining a hazardous waste is the toxicity characteristic. This characteristic is determined by the toxicity characteristic leachate procedure (TCLP) test.[15] Waste which contains concentrations greater than those listed in the TCLP tables are considered toxicity characteristic hazardous wastes.[16]

Specific rules apply to mixtures of a hazardous waste with nonhazardous waste, as well as to their treatment. The mixture of toxicity characteristic hazardous waste with nonhazardous waste is considered hazardous if it retains a hazardous characteristic (40 *CFR* §261.3[a][2][iii]). The mixture of listed hazardous waste with nonhazardous waste is considered hazardous unless other criteria are met.[17] Finally, any residue from the treatment of listed or toxicity characteristic hazardous wastes is also considered hazardous waste.[18]

13. Certain solid wastes including household wastes are excluded from hazardous waste designation depending on conditions. *Code of Federal Regulations.* Title 40, sec. 261.4.
14. 140 degrees Fahrenheit. *Code of Federal Regulations.* Title 40, sec. 261.21.
15. In the past, the EPA used the extraction procedure (EP) test which was designed to test for inorganic compounds only. The current TCLP test can test for both inorganic and organic compounds.
16. *See Code of Federal Regulations.* Title 40, sec. 261.24, Table No. 1.
17. *See ibid.*, sec. 261.3(a)(2)(iv).
18. This rule is known as the derived-from rule. Ibid., sec. 261.3(c)(2).

Regulating Hazardous Wastes

Hazardous wastes are regulated under the RCRA's subtitle C which incorporates the manifest system to track waste from its point of generation to its point of disposal. The regulations vary according to where the hazardous waste is located and how long the waste will remain at the site. Generally, the RCRA regulates generators; transporters; and treatment, storage, and disposal facilities (TSDFs).

GENERATORS

Individuals who generate hazardous waste are subject to licensing and other regulatory requirements under the RCRA. A generator is any person whose act or process produces hazardous waste or whose actions cause hazardous waste to become subject to regulation (40 *CFR* §262.10). Generators are subject to a number of requirements, including:

1. Testing materials they generate to determine if they are hazardous
2. Keeping accurate records which identify quantities and constituents of the hazardous waste they generate
3. Labeling storage, transport, and disposal containers
4. Providing information on the wastes they are transporting, storing, or disposing
5. Complying with manifest system and notification rules (40 *CFR* §261.10)

Any person who generates hazardous waste must obtain a generator identification number (40 *CFR* §262.12). That person must also notify the EPA at least thirty days prior to generating the waste (40 *CFR* §262.12). A generator that transports or provides for the transport of hazardous waste for treatment, storage, or disposal offsite must fully prepare a manifest before the waste leaves the site (40 *CFR* §§262.20–262.23). The manifest must include the generator's name, address, telephone number, EPA identification number, and a description of the waste's constituents and quantity. It must also contain the name and EPA identification number of each transporter, the designated receiving facility, and an alternative receiving facility (40 *CFR* §262.20–262.23).

Typically, the generator must sign the manifest, obtain the signature of the first transporter, and enter the date on which the waste is accepted. The generator must also submit a copy of the manifest to the state agency in charge of administering that state's hazardous waste program. If the destination of the waste is out of state, the generator typically must use the receiving state's manifest form and send a copy of it to that state's agency in charge. A generator must also retain a copy of the manifest for a minimum of three years.

The TSDF receiving the waste must return a copy of the manifest, signed by the TSDF operator, to the generator. If the generator does not receive notice from the TSDF within a specified time, the generator must contact the

TSDF. If the manifest does not come, the generator must notify the state agency in charge of the hazardous waste program. The notice provided to the state agency is referred to as an exception report. It includes a copy of the original manifest and a letter stating the efforts made by the generator to locate the material which it shipped offsite (40 *CFR* §262.42).

The RCRA provides an exemption for small quantity generators, those which create less than 100 kg of hazardous waste per month (40 *CFR* §261.5). However, acute hazardous waste in excess of 100 kg per month, or with a total of 100 kg or greater of residue, contaminated soil, or debris from a cleanup of a spill of acute hazardous waste, subjects the generator to full regulation. A conditionally exempt small quantity generator can treat hazardous waste onsite or send the waste to an authorized TSDF.

TRANSPORTERS

Transporters of hazardous waste are subject to many of the same regulatory requirements as generators. For instance, a transporter must adhere to strict recordkeeping and labeling requirements and must comply with the manifest system (*RCRA* §3003, 42 *USC* §6923). Additional requirements are found at 40 *CFR* part 263.

TSDFs

The types of TSDFs vary as much as the technologies for handling hazardous waste. For instance, a TSDF can be an incinerator; chemical, physical, or biological treatment plant; or a container or tank system. They also include generators that store hazardous waste onsite for more than ninety days as well as those generators that treat or dispose of wastes themselves.

The RCRA (§3004, 42 *USC* §6924; 40 *CFR* pts. 264, 265, 267) places controls on TSDFs. All TSDF owners and operators must obtain an RCRA permit or be designated as interim status to be legally operating (*RCRA* §3005, 42 *USC* §6925; 40 *CFR* pt. 270). The permit must last the life of the unit, including the closure period. Interim status applies to facilities which:

1. Are in existence on or prior to November 19, 1980, or the effective date of statutory or regulatory changes which make the facility subject to the RCRA's permit requirements
2. Have complied with the general notification requirements of section 6930
3. Have applied for an RCRA permit

Part A of the RCRA permit requires applicants to provide their name, address, and facility location. Also required are the operator's name and address; a general description of the business; a scale drawing of the facility; a description of the processes used to treat, store, and dispose of hazardous waste; as well as a list and an estimate of the quantity of hazardous waste onsite. In addition, Part A requires a list of the permits which the facility holds under other environmental programs.

Part B of the permit calls for detailed technical information explaining the type of TSDF involved (e.g., an incinerator) and a thorough chemical and physical analysis of the facility's hazardous waste. It also requires a description of security procedures; the facility inspection schedule; as well as procedures for preventing spills, runoff, contamination, and exposure. Applicants must also provide an outline of training programs, a copy of the facility closure plan, and proof of insurance coverage. Finally, the permit must be signed by a responsible corporate officer.

TSDFs have numerous other requirements under the RCRA. A TSDF, for instance, must obtain an identification number from the EPA. They must also comply with notice requirements, analyze all wastes that their facility produces, and develop a waste analysis plan.

Releases and Groundwater Monitoring

The RCRA regulations (40 *CFR* pts. 264–265, subpts. F) establish procedures to follow in the event of a release from a TSDF. To assure detection, compliance, and corrective action, TSDFs must practice groundwater monitoring by installing monitoring wells.

Facilities must initially perform detection monitoring for hazardous constituents listed in their RCRA permit. When groundwater protection standards are exceeded for the facility's hazardous constituents, more stringent compliance monitoring is also required.

Certain facilities are excluded from groundwater monitoring requirements. Facilities are excluded if they:

1. Receive and contain no liquid
2. Are designed and operate without liquid
3. Contain inner and outer layers of containment enclosing
4. Have a leak detection system
5. Have continual operation and maintenance of a leak system
6. Do not allow migration of hazardous constituents beyond the outer containment layer.

Closure

TSDFs must assure their financial ability to cover any liability involved in closure before they are granted a permit to open (40 *CFR* pts. 264–265, subpts. G). TSDF permits contain closure plans which assure a regulated closing. When a TSDF plans to close, it must meet requirements which assure a clean closure (i.e., no contaminated equipment or contamination in the soil or groundwater on the facility site remains).

Underground Storage Tanks

The RCRA sought to regulate underground storage tanks after its 1984 amendments (*RCRA* §§9001–9010, 42 *USC* §6991–6991i), but final regulations for the underground storage tank program were not promulgated until 1988 (*53 FR* [23 September 1988] 37082). The regulations require release detection and prevention, as well as corrective action.

Underground storage tanks are defined as "any one or combination of tanks (including underground pipes connected thereto) which is used to contain an accumulation of regulated substances, and the volume of which (including the underground pipes connected thereto) is 10 per centum or more beneath the surface of the ground (*RCRA* §9001[1], 42 *USC* §6991[1]).

Owners or operators of underground storage tanks must notify the EPA of their existence. Tanks installed after December 22, 1988, must comply with a national code of practice. This code establishes requirements for new tanks, such as corrosion protection and leak detection devices. All underground storage tank owners must comply with the regulations for releases of regulated substances. The regulations also impose requirements for closure of tanks as well as various aspects of financial responsibilities.

SUMMARY

The RCRA provides statutory goals and objectives and declares the national solid waste and hazardous waste disposal policy (*RCRA* §§1002–1003).

The EPA establishes standards for generators of hazardous waste. These standards include requirements for recordkeeping, labeling of storage, transport, and disposal containers, and the use of a manifest system to track hazardous wastes (*RCRA* §3002).

The EPA establishes standards for transporters of hazardous wastes, including requirements for recordkeeping, labeling, and the use of a manifest system (*RCRA* §3003).

The EPA establishes standards for TSDFs. These requirements ensure the safe handling of hazardous wastes by TSDF owners and operators (*RCRA* §3004).

TSDFs must obtain a permit from the EPA or an authorized state (*RCRA* §3005).

The EPA develops guidelines for state and regional solid waste plans. Plans must meet minimum federal requirements including provisions for prohibiting new open dumps and the closing or upgrading of all existing open dumps (*RCRA* §§4002–4003).

The RCRA provides for federal enforcement, citizens suits, penalties for imminent and substantial endangerment, and judicial review (*RCRA* §§3008, 7002, 7003, 7006).

Owners of underground storage tanks must provide notice to a designated state agency of their existence and specifications. States must maintain inventories of all underground storage tanks containing regulated substances (*RCRA* §9002).

The EPA establishes regulations for underground storage tanks to control releases. The minimum requirements include maintaining a leak detection system, controlling inventory, reporting and recordkeeping, and maintaining evidence of financial responsibility (*RCRA* §9003).

The EPA establishes regulations for a medical waste tracking program (demonstration program). The medical wastes subject to the program are listed and published by the EPA (*RCRA* §§11001–11003).

Comprehensive Environmental Response, Compensation, and Liability Act

The Comprehensive Environmental Response, Compensation, and Liability Act (42 *USC* §§9601–9675; 40 *CFR* Part 300).

STATUTORY ROADMAP

SUBCHAPTER I. RELEASES, LIABILITY, COMPENSATION

§9603	Notification requirements respecting released quantitites, regulations
§9604	Response authorities
§9605	National contingency plan
§9606	Abatement actions
§9607	Liability
§9608	Financial responsibility
§9609	Civil penalties and awards
§9611	Use of fund
§9616	Schedules
§9621	Cleanup standards
§9622	Settlements

SUBCHAPTER III. MISCELLANEOUS PROVISIONS

§9659	Citizens suits

SUBCHAPTER IV. POLLUTION INSURANCE

PURPOSE

The Comprehensive Environmental Response, Compensation, and Liability Act (CERCLA) (1992, 42 *USC* §§9601–9675), commonly referred to as the superfund, was enacted in 1980 in response to problems associated with hazardous waste disposal.[19] Together with the Superfund Amendments and Reauthorization Act (SARA) of 1986, the statute provides a multibillion dollar fund for

---

19. *See* Grad. 1982. A legislative history of the Comprehensive Environmental Response Compensation and Liability (Superfund) Act of 1980. *J. Envtl. L.* 8 Colum.: 1.

the remediation or cleaning up of hazardous waste sites. The superfund's resources, however, are limited and are not intended to be the primary financial source for covering the costs of cleanup. Rather, CERCLA's far-reaching liability scheme serves to impose the costs of cleanup on those responsible for the waste problem. It also serves as an incentive to prevent hazardous waste sites from becoming unmanageable.

By the late 1970s, the hazardous waste disposal problem was a great concern in the United States. Thousands of abandoned waste sites posed serious risks to public health and the environment. Several well-known environmental disasters—Love Canal and Valley of Drums—swayed public opinion and had a major influence on CERCLA's enactment.[20]

SPECIFIC PROVISIONS

An understanding of the basic structure of the law and definitions of its key terms is a prerequisite to understanding the specific provisions of the CERCLA.

Basic Structure of the Law

The EPA is responsible for carrying out CERCLA's mandates. Generally, the EPA can invest the superfund's monies toward cleanup whenever a hazardous substance is released or threatened to be released into the environment. The act authorizes response actions or hazardous waste cleanups which include both remedial and removal activities. However, CERCLA's structure goes beyond this general authority. To understand the details of the statute, the CERCLA can be broken down into four areas to simplify its framework:

- Reporting provisions for releases of hazardous substances
- Investigation, prioritization, and response provisions
- Liability, enforcement, and settlement provisions
- Superfund financing provisions

Definition of Key Terms

The following definitions are necessary for a basic understanding of the CERCLA.

1. A *hazardous substance* under the CERCLA (§101[14], 42 *USC* §9601[14]) is defined largely on the basis of

its meaning under other environmental statutes.[21] The definition includes:

Any hazardous waste under the RCRA
Any hazardous air pollutant under section 112 of the CAA
Any hazardous substance or toxic pollutant under sections 307 or 311 of the CWA
Any hazardous chemical substances or mixtures regulated pursuant to section 7 of the Toxic Substances Control Act
Any other materials designated as hazardous under CERCLA section 102
Petroleum products, natural gas, and synthetic gas used as fuel are excluded from regulation as hazardous substances.

2. A *release* is broadly defined as "any spilling, leaking, pumping, pouring, emitting, emptying, discharging, injecting, escaping, leaching, dumping, or disposing into the environment (including the abandonment or discarding of barrels, containers, and other closed receptacles containing any hazardous substance or pollutant or contaminant). . . ." This list is followed by a number of exclusions (*CERCLA* §101[22], 42 *USC* §9601[22]) including releases in the workplace, which only cause exposure to that confined area, and releases of source, byproduct, or special nuclear materials relating to nuclear incidents, which are covered under the Atomic Energy Act of 1954 (42 *USC* §2011 *et seq.*).

3. The *environment* is broadly defined to include anything except the inside of a building or any other manmade enclosed areas. The environment includes all land surface and subsurface strata, ambient air, and all waters within the United States or its jurisdiction (*CERCLA* §101[8], 42 *USC* §9601[8]).

4. A *response* is a federally funded activity which includes both removal and remedial actions and any enforcement activities related to those actions (*CERCLA* §101[25], 42 *USC* §9601[25]).

5. A *removal action* is a cleanup action aimed at containing or minimizing a release and mitigating damage to the public health without detailed preliminary studies. Removal actions are also referred to as site stabilization actions and generally consist of short-term cleanup measures, such as the installation of fencing around a site, provisions for alternative water supplies, and temporary housing for displaced individuals (*CERCLA* §101[23], 42 *USC* §9601[23]).

6. A *remedial action* is a long-term or permanent cleanup measure to prevent or minimize the release of a hazardous substance into the environment. A remedial action should also prevent the migration of hazardous substances which may endanger public health or wel-

---

20. Two of the most notorious sites were New York's Love Canal and Kentucky's Valley of the Drums. At the New York site, an estimated 141 pounds of dioxin were buried underground at what became a housing development. The Kentucky site involved nearly 20,000 drums of hazardous waste containing approximately 200 organic chemicals and thirty metals. Roughly one-third of the drums were rusted through and leaking their contents into the ground.

21. CERCLA hazardous substances are listed in *Code of Federal Regulations*, Title 40, part 302.

fare or the environment. The statute provides a nonexhaustive list of remedial actions, which includes storage, confinement, and perimeter protection of hazardous substances as well as onsite treatment and monitoring to assure that the actions are reaching their goals (*CERCLA* §101[24], 42 *USC* §9601[24]).

Reporting Provisions

CERCLA's notification provisions help the EPA and state agencies identify where response actions are needed. The CERCLA mandates two general reporting requirements: spill reporting and facility notification requirements.

Owners of facilities or vessels which spill hazardous substances into the environment must report those releases to the National Response Center (NRC) at the U.S. Coast Guard in Washington, DC (*CERCLA* §103[a], 42 *USC* §9603[a]). Owners must also notify the appropriate state agency because states typically have similar reporting requirements. To fall under this reporting requirement, the spill must be greater than or equal to a reportable quantity for the hazardous substance as established under CERCLA section 102(a), and it must not be a federally permitted release. Persons who fail to properly notify the NRC may be subject to sanctions and imprisonment (*CERCLA* §103[b], 42 *USC* §9603[b]).

In addition, the CERCLA (§103[c], 42 *USC* §9603[c]) requires facilities to notify the EPA when they treat, store, or dispose of hazardous wastes unless they are within the RCRA notification requirement. This notification requirement was due by June 9, 1981. However, the EPA recommends reports of subsequent findings if the facility does not have an RCRA permit or is without RCRA interim status. The notification should specify the type and amount of hazardous substance found at the facility, any suspected releases, and any other information which the EPA requires. Failure to comply may result in sanctions.

Investigating, Prioritizing, and Cleaning Up Hazardous Waste Sites

Investigating, prioritizing, and cleaning up hazardous waste sites is accomplished through CERCLA's provisions for the National Contingency Plan (NCP), national priorities list (NPL), investigating sites, and selecting a remedy for site cleanups.

NCP

The EPA's NCP details guidelines for cleaning up hazardous waste sites.[22] The minimum requirements of the NCP call for:

Developing methods and procedures for discovering, investigating, evaluating, and remedying any releases or

22. The NCP is published in the *Code of Federal Regulations*, Title 40, part 300.

threats of releases of hazardous substances from facilities

Explaining the roles and responsibilities of governmental and nongovernmental entities in carrying out the plan

Establishing an NPL and a hazardous ranking system (HRS)

Setting forth requirements for removal and remedial response actions (*CERCLA* §105, 42 *USC* §9605)

The NCP also provides for national and regional teams to oversee hazardous waste emergencies and other details for responding to hazardous waste releases.

NPL

The EPA prepares the NPL to determine which superfund sites should get primary cleanup attention. Remedial actions can be conducted only on listed sites. The EPA ranks the sites using the site risk factors of the HRS, which is located in Appendix A of the NCP (40 *CFR* pt. 300 app. A).

States initially develop lists which prioritize sites for remedial action. States revise the lists annually based on some of the risk factors in the HRS (*CERCLA* §105[8][A], 42 *USC* §9605[8][A]). The lists are submitted to the EPA and contribute towards the NPL's annual revision. The NPL is published in Appendix B of the NCP (40 *CFR* pt. 300 app. B).

INVESTIGATING SITES

The CERCLA (§116, 42 *USC* §9616) imposes deadlines for inspections of listed sites. The EPA or private parties conducting cleanup first perform a remedial investigation and feasibility study (RI/FS) to evaluate the release and cleanup alternatives before proceeding with further action. The EPA has broad authority to enter and investigate sites and gather information, provided it has a reasonable basis to believe that a release of a hazardous substance may occur.

SELECTING A REMEDY FOR SITE CLEANUPS

As previously defined, a removal action is a short-term control measure implemented to contain the release of a hazardous substance before a remedial action is implemented. A removal action begins with a preliminary assessment and, if necessary, a site inspection, all based on readily available information. According to the NCP, removal actions may also require compliance with any legally applicable or relevant and appropriate standard, requirement, criteria, or limitation (ARARs), depending on the urgency and the scope of the situation.

Remedial actions also begin with a preliminary assessment and a site inspection. The NCP then requires the EPA to develop a plan assessing the remedial alternatives for site cleanup. The CERCLA (§121, 42 *USC* §9621) establishes cleanup standards containing criteria applicable to remedial actions. The standards typically require that remedial actions comply with ARARs, are cost effective, and use permanent solutions and treatment technologies to the maximum extent practicable.

Following the remedy selection process, the EPA publishes its selected remedy in a record of decision. The selection must be supported by site-specific facts, analysis, and policy determinations. Finally, the EPA develops a remedial design used to implement the remedial action.

Liability, Enforcement, and Settlement

Most superfund cases are resolved through settlements between the potentially responsible parties (PRPs) and the government. Settlements are common because CERCLA's liability scheme imposes substantial penalties on such a broad group of persons. The group of PRPs who can be liable includes:

Both current and past owners of a site
Generators of waste who arranged for treatment or disposal of waste at a site
Persons who transported waste to a site (*CERCLA* §107[a], 42 *USC* 9607[a])

In addition, courts have interpreted the CERCLA to impose strict liability on any and all PRPs. This interpretation means that persons are liable for releases of hazardous substances irrespective of fault. CERCLA liability is also both joint and several. This concept means that any or all PRPs can be accountable for the entire cost of cleanup. Generally, a PRP's liability is unlimited for response costs (*CERCLA* §107[c], 42 *USC* 9607[c]).

The CERCLA provides several narrow defenses to its liability scheme. A PRP can avoid liability upon a showing that the release and resulting damages were solely caused by:

An act of God
An act of war
An act or omission of a third party who had no contractual relation to the PRP, and the PRP exercised due care and took adequate precautions against any foreseeable acts or omissions (*CERCLA* §107[b], 42 *USC* 9607[b]).

The limited defenses provided for PRPs combined with the CERCLA's stringent liability scheme make it difficult for PRPs to avoid liability. Therefore, a PRP, once identified, usually enters negotiations for settlement with the EPA.

Settlements are authorized under CERCLA section 122. The most common types of CERCLA settlements involve response actions and result in either the PRP performing the removal or remedial action (*CERCLA* §104[a], 42 *USC* 9604[a]) or covering a portion of the cleanup costs. When a PRP negotiates to perform the response action, its duties are explicitly stated in a consent order or decree (*CERCLA* §122[d], 42 *USC* 9622[d]). The EPA can also

authorize a PRP to perform investigations or studies related to a response action (*CERCLA* §104[b], 42 *USC* 9604[b]). These duties are also described in a consent order or decree.

A PRP is bound by any settlement agreement into which it enters. If a PRP fails to comply with a settlement agreement, administrative order, or consent decree, it is subject to civil penalties of up to $25,000 per day for each violation for as long as the violation continues (*CERCLA* §122[1], 42 *USC* 9622[1]).

Superfund Use and Financing

The SARA amendments provided for $8.5 billion to be used over a five year period. The Omnibus Budget Reconciliation Act of 1990 appropriated an additional $5.1 billion for the period commencing October 1, 1991, and ending September 30, 1994. These resources can finance site investigation and cleanup costs for hazardous substance releases but are generally not used without attempting to cover the costs through the liabilities of private parties. If insufficient funds are derived from private parties, the EPA uses the superfund monies.

The CERCLA designates several uses for the fund which include:

Covering the cost of government response costs
Covering the cost of injury, destruction, or loss of natural resources including damage assessment
Providing grants for technical assistance (*CERCLA* §111, 42 *USC* §9611)

Use of the fund for natural resource damage,[23] however, is contingent upon exhausting all administrative and judicial claims to recover the damage cost from persons who may be liable (*CERCLA* §111[2], 42 *USC* §9611[2]).

The superfund generates its monies largely from a network of environmental taxes. These taxes include an excise tax on crude oil and petroleum products, an excise tax on feedstock chemicals, and an environmental tax on corporate profits. Superfund revenues are also generated through appropriations from general revenues, interest, cost recovery actions, and penalties obtained from lawsuits.

SUMMARY

- The CERCLA includes reporting provisions.
- The CERCLA has provisions for investigation, prioritization, and response.
- The CERCLA provides for liability, enforcement, and settlement.
- Superfund financing provisions are outlined in the act.

23. The term refers to "land, fish, wildlife, biota, air, water, groundwater, drinking water supplies, and other such resources ..." *Comprehensive Environmental Response, Compensation, and Liability Act.* Sec. 101(16). *U.S. Code,* Vol. 42, sec. 9601(16).

Noise Control Act

The Noise Control Act (42 *USC* §4901 *et seq.*; 40 *CFR* Parts 204, 211).

STATUTORY ROADMAP

§4901	Congressional findings and statement of policy
§4902	Definitions
§4903	Federal programs
§4904	Identification of major noise sources
§4905	Noise emission standards for products distributed in commerce
§4907	Labeling
§4908	Imports
§4909	Prohibited acts
§4910	Enforcement
§4911	Citizens suits
§4912	Records, reports, and information
§4913	Quiet communities, research, and public information
§4914	Development of low noise emission products
§4915	Judicial review
§4916	Railroad noise emission standards
§4917	Motor carrier noise emission standards
§4918	Authorization of appropriations

PURPOSE

In 1972 Congress enacted the Noise Control Act (1992, 42 *USC* §§4901–4918) in response to the growing concern of the effects of noise on public health and welfare. Congress recognized that noise had to be adequately controlled, particularly in urban areas; that this would best be accomplished at state and local levels; and that national uniformity of treatment would best serve this purpose. Consistent with these congressional findings, the purpose of the act is "to establish a means for effective coordination of federal research and activities in noise control, to authorize the establishment of federal noise emission standards for products distributed in commerce, and to provide information to the public respecting the noise emission and noise reduction characteristics of such products" (42 *USC* §4901[b]).

SPECIFIC PROVISIONS

The Noise Control Act provides noise emission standards and other requirements to control noise.

Noise Emission Standards

The Noise Control Act requires the EPA to establish noise emission standards for various products for which such determinations are feasible (42 *USC* §4905). The EPA must first, however, develop and publish criteria relating to noise (42 *USC* §4904). These criteria reflect the scientific knowledge available to show the effects of differing quantities of noise. The EPA must also publish at least one report containing products which are major sources of noise and an assessment of techniques available for controlling the noise emitted from those products. The noise control techniques should include technical data, costs, and alternative methods of noise control. The criteria and reports should be reviewed and revised from time to time.

Products identified in the report and which fall within a specified category must have regulations established for them. The specified categories of products are:

1. Construction equipment
2. Transportation equipment
3. Any motor or engine
4. Electrical or electronic equipment (42 *USC* §4905)

In addition, the EPA can promulgate regulations for other products where such regulations are feasible and necessary to protect the public from harmful noise effects. Regulations should include a noise emission standard which considers "the magnitude and conditions of use of such product (alone or in combination with other noise sources), the degree of noise reduction available through the use of the best available technology, and the cost of compliance" (42 *USC* §4905[c]).

Other Requirements

Products which emit excessive noise under the act must be labeled to warn its prospective user (42 *USC* §4907). The distribution of products which violate standards and labelling requirements is prohibited (42 *USC* §4909). Moreover, the act provides for enforcement through criminal penalties and allows any person to commence a civil action under its citizens suits provision (42 *USC* §§4910, 4911).

The 1978 amendments to the Noise Control Act added provisions relating to the Quiet Communities Program (42 *USC* §4913). The purpose of the amendments was to increase state and local noise control programs. The amendments provide for federal grants and contracts to research noise pollution and inform the public about noise pollution issues and concerns. They also develop a national noise environmental assessment program to identify human trends in noise exposure, identify ambient noise levels and compliance data, and determine the effectiveness of noise abatement actions. Finally, the amendments establish regional technical assistance centers and provide for technical assistance to state and local governments to promote local noise control.

In 1981 Congress implemented the administration's proposal to cease funding the EPA Office of Noise Abatement and Control (ONAC). The act, however, was not repealed. The standards promulgated under both the Noise Control Act and Quiet Communities Act remain in effect. Because the noise emission and labeling standards remain in effect, they preempt state and local governments from adopting other standards.[24]

SUMMARY

The EPA establishes noise emission standards for products for which such standards can feasibly be developed (§4905).

Manufacturers of products which emit excessive noise must sufficiently label and warn consumers of the products' loud tendencies (§4907).

States are encouraged to develop noise control programs to deal with local noise-related issues (§4913).

Safe Drinking Water Act

The Safe Drinking Water Act (16 *USC* §§300f–300j–11, 40 *CFR* Parts 140–149).

STATUTORY ROADMAP

Part A.	*Definitions*
Part B.	*Public water systems*
Part C.	*Protection of underground sources of drinking water*
Part D.	*Emergency powers*
Part E.	*General provisions*
Part F.	*Additional requirements to regulate the safety of drinking water*

PURPOSE

Congress first enacted the Safe Drinking Water Act (SDWA) in 1974 (§§1401–1451, 42 *USC* §§300f–300j–11) to protect the nation's drinking water supplies from contaminants.[25] To achieve this goal, the act establishes federal standards for drinking water quality and regulates public water systems and the underground injection of waste. The act's current regulatory scheme, based upon its 1986 amendments (SDWA U.S. Public Law 99–339), still follows this general regulatory approach.

SPECIFIC PROVISIONS

The SDWA has specific provisions for public water systems and underground injections.

Public Water Systems

The SDWA requires states to enforce various drinking water standards established by the EPA. Specifically, the standards apply to public drinking water systems which provide piped water to the public for various uses including consumption. The act, however, limits the definition of public water systems to those systems which have a minimum of fifteen service connections or provide water supply to at least twenty-five individuals (*SDWA* §1401[4], 42 *USC* §300f[4]).

The EPA promulgates national primary drinking water regulations for public water systems (*SDWA* §1412, 42 *USC* §300g–1). Under these regulations, the EPA must establish both primary and secondary drinking water standards. Notably, the primary standards are enforceable by the states, while the secondary standards are guidelines or goals to be achieved.

PRIMARY STANDARDS

The EPA establishes primary standards for constituents which may effect human health.[26] Initially, the EPA has set maximum contaminant level goals for contaminants found to have adverse effects on human health and which are known or anticipated to occur in public water systems (*SDWA* §§1412[a][3], [b][3][A], 42 *USC* §§300g–1[a][3], [b][3][A]). These health-based goals are set "at a level at which no known or anticipated adverse effects on the health of persons occur and which allows an adequate margin of safety" (*SDWA* §1412[b][4], 42 *USC* §300g–1[b][4]).

The EPA also establishes maximum contaminant levels which specify the levels to which water contamination must be reduced.[27] These primary standards are set as close to the maximum contaminant level goals as is feasible (*SDWA* §1412g–1[b][4], 42 *USC* §300g–1[b][4]).[28] If the maximum contaminant levels are not economically or technologically feasible, the EPA must specify any known treatment techniques which will sufficiently reduce the levels of the contaminants.

---

24. The Federal Aviation Administration has authority to abate airport noise under the Noise Control Act and the Aviation Noise and Capacity Act (1990), U.S. Public Law 101–508, secs. 9301–09. The DOT is responsible for the enforcement of the EPA's railroad and motor carrier emission standards. While DOT is funded for this purpose, they do not have the authority to promulgate regulations under the act.

25. *Contaminant* is defined as any physical, chemical, biological, or radiological substance or matter in water. *Safe Drinking Water Act.* Sec. 1401(6). *U.S. Code.* Vol. 42, sec. 300f(6).

26. *See Code of Federal Regulations.* Title 40, parts 141–142.

27. The act defines *maximum contaminant level* to mean "the maximum permissible level of a contaminant in water which is delivered to any user of a public water system." *Safe Drinking Water Act.* Sec. 1401(3). *U.S. Code.* Vol. 42, sec. 300f(3).

28. *Feasible* is defined under the act to mean "feasible with the use of the best technology, treatment techniques and other means which [the EPA] finds, after examination for efficacy under field conditions and not solely under laboratory conditions, are available (taking cost into consideration)." *Safe Drinking Water Act.* Sec. 1412g–1(b)(5). *U.S. Code.* Vol. 42, sec. 300g–1(b)(5).

SECONDARY STANDARDS

The EPA also establishes secondary standards on drinking water issues related to public welfare.[29] For example, secondary standards specify maximum contaminant levels for constituents affecting the odor, color, cloudiness, acidity, and taste of water in public water systems. States are not required to enforce secondary standards.

Underground Injections

The SDWA regulates underground injections[30] through a permitting scheme which is implemented and administered by individual states. The EPA promulgates regulations for state programs, but the states must adopt their own underground injection control programs (*SDWA* §§1421–1422, 42 *USC* §§300h–h–1). These programs control and authorize underground injections through the issuance of state permits. To be granted a permit, an applicant must demonstrate that the injection will not endanger drinking water sources. Under the act, an underground injection endangers drinking water sources:

> if such injection may result in the presence in underground water which supplies or can reasonably be expected to supply any public water system of any contaminant, and if the presence of such contaminant may result in such system's not complying with any national primary drinking water regulation or may otherwise adversely affect the health of persons (*SDWA* §1421[d][2], 42 *USC* §300h[d][2]).

SUMMARY

The EPA promulgates national primary drinking water regulations for public water systems. These regulations include maximum contamination level goals and maximum contamination levels for each contaminant which may have an adverse effect on human health and may exist in a public water system (*SDWA* §1412).

The EPA delegates the states with primary enforcement responsibility for public water systems provided the state can satisfactorily meet the minimum federal requirements (*SDWA* §1413).

The EPA has ultimate authority to enforce the national primary drinking water regulations where it determines that a state has not commenced appropriate enforcement actions. Moreover, owners and operators of public water systems must notify consumers of violations of the act including the failure to meet applicable maximum contaminant levels, treatment technique requirements, or testing procedures under national primary drinking water regulations (*SDWA* §1414).

A state with primary enforcement responsibility can grant variances and exemptions from national primary drinking water regulations to public water systems within its jurisdiction. For instance, a maximum contaminant level requirement or treatment technique requirement in the regulations can be exempted provided the applicable conditions are satisfied (*SDWA* §§1415–1416).

The act prohibits the use of lead pipes, solder, or flux in the installation or repair of any public water system or in the plumbing of a facility connected to a public water system which provides water for human consumption (*SDWA* §1417).

The state underground injection programs contain minimum requirements to assure that underground injections do not endanger drinking water sources. An approved state program gives the state primary enforcement responsibility for underground water sources. State programs regulate underground injections through permit system (*SDWA* §§1421–1422).

The EPA can exercise its emergency powers and take any action necessary to protect human health when it receives information that a contaminant is present or is likely to enter a public water system or an underground source of drinking water (*SDWA* §1431).

The EPA promulgates regulations requiring public water systems to implement monitoring programs for unregulated contaminants. The regulations include a list of unregulated contaminants which require monitoring. The results of the monitoring go to the primary enforcement authority (*SDWA* §1445).

The act specifically provides for judicial review and includes a citizens suit provision (*SDWA* §§1448–1449).

The act contains other provisions to regulate the safety of drinking water, including a recall of all drinking water coolers with lead-lined tanks, and federal assistance programs for states to remedy lead contamination in school drinking water (*SDWA* §1461–1466).

Federal Water Pollution Control Act

The Federal Water Pollution Control Act (33 *USC* §§1251–1387; 40 *CFR* Parts 100–140, 40 *CFR* Parts 400–470).

STATUTORY ROADMAP

TITLE I.	RESEARCH AND RELATED PROGRAMS
§1251	Declaration of goals and policies
TITLE II.	GRANTS FOR CONSTRUCTION OF TREATMENT WORKS
TITLE III.	STANDARDS AND ENFORCEMENT
§1311	Effluent limitations
§1313	Water quality standards and implementation plans
§1314	Information and guidelines
§1319	Federal enforcement

---

29. *See Code of Federal Regulations*. Title 40, part 143.
30. *Underground injection* is defined in the act as "the subsurface emplacement of fluids by well injection [and does not include] the underground injection of natural gas for purposes of storage." *Safe Drinking Water Act.* Sec. 1421(d)(1). *U.S. Code.* Vol. 42, sec. 300h(d)(1).

PURPOSE

In 1972, Congress enacted the Federal Water Pollution Control Act, commonly known as the Clean Water Act (CWA) (1990, 33 *USC* §§1251–1387). The primary purpose of the act is "to restore and maintain the chemical, physical, and biological integrity of the nation's waters (*CWA* §101[a], 33 *USC* §1251[a]). Additionally, the act set forth numerous goals and policies related to protecting the nation's waters including (*CWA* §101[a][1], 33 *USC* §1251[a][1]):

To eliminate discharge into navigable waters
To protect fish, shellfish, and wildlife
To provide for recreation in and on the U.S. waters
To prohibit the discharge of toxic pollutants in toxic amounts
To develop and implement programs to control nonpoint sources of pollution

Specifically, the 1972 act targeted the elimination of pollutant discharges into navigable waters by 1985. This original deadline was not achieved. The act's subsequent amendments, however, extended the deadlines and attempted to adopt more realistic goals and solutions to the water pollution problem.

SPECIFIC PROVISIONS

The CWA has specific provisions for the National Pollutant Discharge Elimination System, categorizing pollutants, technology-based controls and effluent limitations, water quality standards, variances, nonpoint source controls, and dredge and fill permits. These provisions are discussed next.

National Pollutant Discharge Elimination System

The CWA controls water pollution primarily through its National Pollutant Discharge Elimination System (NPDES). Section 301 of the act states that "the discharge of any pollutant by any person shall be unlawful" except as provided by other sections within the act (33 *USC* §1311). Section 402 of the act establishes the main ex-

ception (33 *USC* §1342), which provides for issuing NPDES permits to control direct discharges of pollutants into U.S. waters. Direct dischargers include both industrial sources and publicly owned treatment works (POTWs). NPDES permits are issued through either the EPA or a delegated state (*CWA* §402[b], 33 *USC* §1342[b]). The EPA can delegate NPDES authority to a state if the state adopts a permit program which meets minimum federal criteria.

Permits must be obtained by any point source which discharges pollutants into waters of the United States. This regulation generally includes the two types of direct dischargers mentioned previously. A *point source* is defined in the CWA (40 *CFR* §122.2 [1990]) regulations as "any discernible, confined, and discrete conveyance, including but not limited to, any pipe, ditch, channel, tunnel, conduit, well, discrete fissure, container, rolling stock, concentrated animal feeding operation, landfill leachate collection system, vessel or other floating craft from which pollutants are or may be discharged." *Pollutants* and *waters of the United States* are also specifically defined in the regulations.

An applicant must obtain a permit from either the EPA or an authorized state agency. Point sources are regulated based on numerous factors including the type of pollutants discharged and the character of the receiving water. The permitting authority establishes effluent limits which are written into the permit. Before a permit is issued, both public notice and an opportunity to be heard are required. Permits typically have a five-year duration period and can be renewed.

The CWA generally uses technology-based effluent standards and water quality standards (also known as ambient water standards) to control water pollution. Technology-based effluent standards apply to all direct dischargers and require treatment of pollutants prior to discharge. The goal of technology-based effluent standards is to minimize actual pollutants going into the nation's waters by installing pollution control technology.

Water quality standards, on the other hand, are based on the receiving water's ability to absorb a given pollutant. They are not to be confused with ambient water quality-related effluent limits[31] but are site-specific standards which vary according to the characteristics and designated uses of the regulated water segment.

Categorizing Pollutants

The CWA divides pollutants into three categories and uses this framework to create effluent limits for dischargers. The three categories of pollutants are conventional, toxic, and nonconventional. Conventional pollutants include fecal coliform, biochemical oxygen demand, suspended solids, and pH. Toxic pollutants are listed in the CFR (40 *CFR* §401.15). They include chemicals which in small

---

31. See *Clean Water Act*, Sec. 302; *U.S. Code*. Vol. 33, sec. 1312.

amounts cause adverse effects on fish, wildlife, and humans. Finally, nonconventional pollutants include all nontoxic pollutants which are not categorized as conventional pollutants.

Technology-based Controls and Effluent Limitations

The regulating agency establishes technology-based effluent limits which can be incorporated into a facility's permit. These effluent limits are based on the SIC classification of the facility, the type of pollutant the facility discharges, and whether the facility is considered a new or an existing source.

The EPA publishes national effluent guidelines for categories and subcategories of industry types in the CFR. These guidelines establish technology-based effluent limits for standard pollutants discharged from the particular category or subcategory of an industry. The EPA has not yet created effluent guidelines for all categories and subcategories of industries.

At a minimum, existing sources must satisfy the best practicable control technology currently available (BPT) standard. BPT establishes a national floor for effluent treatment (CWA §304[b][1][A–B], 33 USC §1314[b][1][A–B]). Also existing sources must meet treatment standards according to the three categories of pollutants as follows:

Conventional pollutants are subject to the standard BCT standard (CWA §304[b][4][B], 33 USC §1314[b][4][B]).
Toxic pollutants are subject to the more stringent best available technology economically feasible (BAT) standard, together with additional health-based standards (CWA §307, 33 USC §1317).
Nonconventional pollutants are also subject to the BAT standard (CWA §301[b][2][F], 33 USC 1311[b][2][F]).

New sources are regulated pursuant to a different set of technology-based controls. The CWA (§306[a][2], 33 USC §1316[a][2]) defines a new source as one whose construction began after new source performance standards (NSPS) were proposed for it, provided those standards were adopted. The act declares that NSPS be established "through the best available demonstrated control technology, processes, operating methods, or other alternatives, including, where practicable, a standard permitting no discharge of pollutants" (CWA §306[a][1], 33 USC §1316[a][1]). The EPA is required to compile a list of categories of new sources for which it must propose and publish NSPSs (CWA §306[b][1][A], 33 USC §1316[b][1][A]). The EPA must also publish toxic effluent standards (CWA §307[a][2], 33 USC §1317[a][2]) and water quality-related effluent limits (CWA §302, 33 USC §1312).

The EPA has not established a national effluent guideline for every industry which requires a permit. In these cases, the EPA or authorized state permit writers establish a technology-based limit according to their best professional judgment (BPJ). This limit is known as a BPJ permit limit.

Water Quality Standards

Water quality standards are an additional control to protect waterbodies. Generally, water quality standards focus on the receiving water's ability to integrate, dilute, or absorb pollutants. A facility may have to install controls beyond those required to achieve technology-based standards to meet water quality standards.

States must adopt water quality standards and submit them to the EPA for all bodies of water within the state (CWA §303[a], 33 USC §1313[a]). The standards are based upon state designated uses for specific bodies of water together with water quality criteria for the designated use. The EPA reviews the state standards and replaces them if they fail to meet minimum federal requirements (CWA §§303[c][3–4], 33 USC §§1313[c][3–4]). States are further required to specify and prioritize the waters within their boundaries where technology-based standards are insufficient to meet the established water quality standards (CWA §303[d][1][A], 33 USC §1313[d][1][A]).

States must set a total maximum daily load (TMDL) for specific waterbodies. The TMDL is the maximum amount of a pollutant that a defined water segment can accept while still achieving the applicable water quality standard. In accordance with section 303(d) of the CWA (§303[d][1][C], 33 USC §1313[d][1][C]), a TMDL "shall be established at a level necessary to implement the applicable water quality standards with seasonal variations and a margin of safety which takes into account any lack of knowledge concerning the relationship between effluent limitations and water quality." States must submit a list of the prioritized waters and their applicable TMDLs to the EPA for approval (CWA §303[d][2], 33 USC §1313[d][2]). Nevertheless, states do have discretion in allocating the TMDLs among dischargers and establishing corresponding effluent limits on the facilities.

Variances

In some cases, point sources can obtain variances which provide alternatives to the national effluent controls in section 304. Among the variances in the act are the fundamentally different factors (FDF) variance (CWA §301[n], 33 USC §1311[n]; 40 CFR pt. 125 subpt. D), the §301(c) economic variance (CWA §301[c], 33 USC §1311[c]), the §301(g) water quality variance (CWA §301[g], 33 USC §1311[g]), and the net/gross limitation (40 CFR §122.45[g]). The most notable of these variances is the FDF which may apply when a discharger's characteristics are fundamentally different from the sources which the EPA considered to determine the applicable technology-based standard.

Nonpoint Source Controls

Nonpoint sources are a significant contributor to water pollution. Nonpoint sources are diffuse sources of pollution created generally by land use. Nonpoint source pollution is caused, for example, by the rainfall and runoff from urban areas, mining, and agricultural sites.

Under the 1972 amendments to the act, the section 208 planning process delegates control of nonpoint source pollution to the states. This section called on state governors to designate areas with substantial water quality control problems and to select a regional planning agency to help develop section 208 plans to address nonpoint issues. This approach is not considered successful in controlling this type of pollution.

The 1987 amendments to the act, however, developed section 319 to address the nonpoint source pollution problem which was ineffectual in section 208 (*CWA* §319, 33 *USC* §1329). States are required to designate water bodies that are unable to meet water quality standards without a nonpoint source program. States must then establish nonpoint source plans for these waterbodies. The plans require best management practices regulations for different categories of sources along the waters, such as construction sites or manure storage sites. Plans must include implementation schedules and other regulatory measures and are subject to EPA approval.

Dredge and Fill Permits

Section 404 of the act requires permits for discharging dredge and fill materials into waters of the United States. The Army Corps of Engineers, generally, administers the program. However, the EPA has the authority to review section 404 permits and can veto certain permit conditions or an entire permit (*CWA* §334, 33 *USC* §1344[c]). This section is broadly applied based upon the liberal interpretation of key terms such as *waters of the United States* and *dredge and fill materials.*

Tributaries of navigable waters and wetlands are protected under section 404. Regulated activities extend well beyond dredge and fill operations. For instance, section 404 controls activities such as bridge and dam building, flood control construction activities, and dike and drainage systems.

The Army Corps of Engineers can consider several factors when making its decision to issue a section 404 permit. The corps can consider individually or cumulatively the following twelve issues:

... conservation, economics, aesthetics, general environmental concerns, wetlands, historic properties, fish and wildlife values, flood hazards, floodplain values, land use, navigation, shore erosion and accretion, recreation, water supply and conservation, water quality, energy needs, safety, food and fiber production, mineral needs, considerations of property ownership and, in general, the needs and welfare of the people (33 *CFR* §320.4).

Section 404 also includes a review of the requirements developed under other federal environmental acts. A 404 permit application is subject to review under the NEPA, Fish and Wildlife Coordination Act,[32] CZMA, Wild and Scenic Rivers Act,[33] and the ESA.

SUMMARY

The CWA declares several national goals and policies to restore and maintain the nation's waters (*CWA* §101).

The CWA makes the discharge of any pollutant by any person unlawful unless the discharge is in compliance with specified provisions in the act (*CWA* §301).

States are required to adopt water quality standards for identified waters within their boundaries (*CWA* §303).

The EPA must develop information and guidelines for the implementation and administration of programs under the act. (*CWA* §304).

The EPA is required to establish federal NSPSs for categories of new sources (*CWA* §306).

The CWA establishes procedures for inspection, monitoring, and entry. Owners and operators of point sources are required to maintain records and install and use monitoring equipment or methods. The EPA or its authorized representative can enter and inspect point sources and their records (*CWA* §308).

The CWA's federal enforcement provision authorizes compliance schedules and orders, administrative orders and penalties, as well as criminal and civil penalties (*CWA* §309).

States must adopt nonpoint source management programs to improve water quality within the state (*CWA* §319).

The NPDES establishes a permit program to control the discharge of pollutants from point sources. The program is implemented by the EPA or a delegated state (*CWA* §402).

The Army Corps of Engineers is authorized to issue permits for the discharge of dredged and fill material into waters of the United States. This provision has a significant impact on wetland protection (*CWA* §404).

The EPA can bring suit on behalf of the United States pursuant to its emergency powers. Such powers are used to protect the health and welfare of persons against imminent and substantial endangerment from a pollution source (*CWA* §504).

The CWA has a citizen's suit provision allowing any citizen to commence a civil action against any person who

---

32. This law, at a minimum, requires consultation with the U.S. Fish and Wildlife Service to assure protection of habitat and species under their protection.

33. This act and the ESA are administered by the Department of Interior (except in the case of rivers in national forests which are subject to the Department of Agriculture).

violates the act. This provision includes suits against the EPA for failure to perform a nondiscretionary duty (*CWA* §505).

Toxic Substances Control Act

The Toxic Substances Control Act (15 *USC* §2601 *et seq.*; 40 *CFR* §§700–799).

STATUTORY ROADMAP

SUBCHAPTER I.   CONTROL OF TOXIC SUBSTANCES

§2603   Testing of chemical substances and mixtures

§2604   Manufacturing and processing notices

§2605   Regulation of hazardous chemical substances and mixtures

§2606   Imminent hazards

§2607   Reporting and retention of information

SUBCHAPTER II.   ASBESTOS HAZARD EMERGENCY RESPONSE

SUBCHAPTER III. INDOOR RADON ABATEMENT

SUBCHAPTER IV.   LEAD EXPOSURE REDUCTION

PURPOSE

The Toxic Substances Control Act (TSCA) (1992, 15 *USC* §§2601–2654) was enacted in 1976 as the first major federal law governing toxic substances. The purpose of the TSCA is to understand the health risks of certain chemical substances by developing production and health risk data from the manufacturers of the chemicals and chemical products.

In developing the TSCA, Congress had the following three policy objectives:

1. To require industry to develop data on the effect of chemical substances and mixtures on health and the environment
2. To assure that adequate authority exists to regulate chemical substances
3. To assure that this authority not "impede unduly or create unnecessary economic barriers to technological innovation" in the manufacture of chemicals (*TSCA* §2[b], 15 *USC* §2601[b])

Thus, the TSCA is described as a risk/benefit-balancing statute.

SPECIFIC PROVISIONS

The TSCA requires manufacturers to notify the EPA at least ninety days prior to manufacturing a new chemical substance or placing it in the market for a new use. The notice must contain health and safety data and an assessment of the impact of that chemical on human health and the environment. The TSCA applies to chemical manufacturers and processors. A processor is any person who prepares a chemical substance or mixture after its manufacture for distribution into commerce (*TSCA* §§3[10–11], 15 *USC* §§2602[10–11]).

Testing

The TSCA (§4, 15 *USC* §2603) authorizes the EPA to require testing of chemical substances. The EPA has published testing guidelines pursuant to this authority. Moreover, the EPA can compel manufacturers to test chemical substances for which insufficient data exist to determine their effects on human health and the environment (40 *CFR* pts. 796, 797, 798). To determine if such chemicals present unreasonable risks, this requirement applies to both new chemical substances, as well as those on the market. The EPA can require toxicity tests on a substance "when there is [a] more-than-theoretical basis for suspecting that some amount of exposure occurs and that the substance is sufficiently toxic at that exposure level to present 'unreasonable risk of injury to health'" (*Chemical Manufacturers Assn. v. EPA*, 859 F.2d 977, 984 (D.C. Cir. 1988).

The act creates the Interagency Testing Committee (ITC) to recommend chemicals for testing (*TSCA* §4[e][2], 15 *USC* §2603[e][2]). The ITC consists of members from eight separate federal agencies who recommend testing. The ITC gives priority to those chemical substances likely to cause or contribute "to cancer, gene mutations, or birth defects."

Upon listing a chemical substance, the EPA has one year to either initiate a rulemaking proceeding requiring testing or publish in the *FR* the reason why a proceeding was not initiated. The EPA initiates action if testing data or other information provide a basis to conclude that a chemical substance presents a significant risk of human cancer, genetic mutation, or birth defects. Where such significant risk exists, the EPA must act within 180 days of receipt of the information. This period can be extended up to ninety days for good cause.

Premanufacture Notice Requirements

Chemical manufacturers must provide the EPA with premanufacture notice before any new chemical is manufactured, processed, or put on the market (*TSCA* §5, 15 *USC* §2604). The manufacturer must also provide notice if an existing chemical substance is being used in a new manner. The EPA has regulations to assess the safety of potentially marketable chemicals (40 *CFR* pts. 720–723).

The manufacturer gives the premanufacture notice at least ninety days prior to the intended manufacturing or processing. This notice includes test data showing that the chemical does not present an unreasonable risk. The EPA can prohibit or limit the manufacturing or processing of the chemical if it receives insufficient information from the manufacturer to make a proper evaluation.

Reporting

The EPA requires manufacturers, processors, distributors, and importers of chemical substances to report information on chemicals so that the EPA can effectively administer and enforce the TSCA (§8[a], 15 USC §2607[a]). The EPA promulgated the Preliminary Assessment Information Rule (40 CFR pt. 712, subpt. B) and the Comprehensive Assessment Information Rule (40 CFR pt. 704) pursuant to this authority. These rules designate chemical substances which, together with the ITC list, must be reported by manufacturers and processors. Both rules provide forms and instructions for reporting and request information on the qualities of chemicals and worker exposure.

Manufacturers and processors must also maintain records of any allegations of harmful effects caused by chemicals (TSCA §8[c], 15 USC §2607[c]; 40 CFR pt. 717). The EPA can inspect these records upon request. Sources of allegations include lawsuits, workers compensations claims, and employee or customer complaints. A company must retain records of adverse reactions of employees for thirty years from the time an incident is reported. All other records of adverse reactions must be retained for a five-year period.

The EPA can also request and obtain health and safety studies on chemical substances from persons dealing in chemicals (TSCA §8[d], 15 USC §2607[d]). The EPA defines *health and safety studies* as "any study of any effect of a chemical substance or mixture on health or the environment" (40 CFR §716.3). The studies encompass those conducted or initiated by the person, known to the person, or reasonably ascertainable to the person.

Finally, the EPA requires persons dealing in chemicals to report any information which suggests a chemical substance may substantially risk injury to human health or the environment (TSCA §8[e], 15 USC 2607[e]). The EPA does not have the authority to publish regulations for this requirement, but it has published several guidance documents (43 FR [16 March 1978]; TSCA §8[e]).

Authority to Regulate Hazardous Chemical Substances

The EPA has broad authority to regulate the manufacture, processing, distribution, use, or disposal of chemical substances (TSCA §6, 15 USC §2605). The EPA can limit or prohibit the manufacture or distribution of a chemical substance or may simply require public warnings. This regulatory authority is based on a finding that a reasonable basis exists that such activities "present or will present an unreasonable risk of injury to health or the environment." The EPA must also consider the benefits of the chemical

substance, any economic effects of the regulatory action, and the health risks created by the chemical's manufacture. When the EPA regulates a chemical substance, it must apply the least burdensome means of providing adequate protection.

TSCA's LIMITED REGULATORY PRACTICE

The EPA has promulgated regulations directed at polychlorinated biphenyls (PCBs), chlorofluorocarbons (CFCs), asbestos, and 2,3,7,8-tetrachlorodibenzo-p-dioxin (TCDD) waste. The controls on the use of PCBs are a good example of this regulation.

PCBs were developed for use in industrial equipment such as electrical transformers and capacitors. PCBs are regulated under section 6 of the TSCA (§6[e], 15 USC 2605[e]) and under the rule promulgated pursuant to section 6.[34] Section 6 bans the manufacturing, processing, or distribution of PCBs with a few exceptions. The rule also requires a tracking and manifest system with notification and recordkeeping requirements similar to the RCRA. Moreover, the EPA has a PCB Penalty Policy (August 1990) which subjects violators to penalties of up to $25,000 per day.

SUMMARY

The EPA requires testing of chemical substances to insure that they do not present an unreasonable risk of injury to health or the environment (TSCA §4).

Manufacturers must provide the EPA with notice if they intend to manufacture a new chemical substance or use an existing chemical substance in a new way (TSCA §5).

The EPA has broad regulatory powers over the manufacture, processing, use, or disposal of chemical substances if those substances are found to potentially present an unreasonable risk of injury to health or the environment (TSCA §6).

The EPA can commence a civil action to seize a chemical substance or for other relief to protect against an imminently hazardous chemical substance (TSCA §7).

The EPA may require recordkeeping and the submission of reports for inspection from manufacturers of chemical substances (TSCA §8).

Federal Insecticide, Fungicide, and Rodenticide Act

The Federal Insecticide, Fungicide, and Rodenticide Act (7 USC §§136–136y; 40 CFR Parts 162–180).

---

34. In addition to regulating the disposal of PCB wastes, the TSCA regulates the manufacture and use of products which use PCBs. *Code of Federal Regulations*. Title 40, sec. 761.20.

STATUTORY ROADMAP

PURPOSE

The Federal Insecticide, Fungicide, and Rodenticide Act (FIFRA) is the primary statutory tool regulating pesticides.[35] The FIFRA does not have a specific section designating its purpose. However, it implicitly seeks to prevent adverse effects on human health and the environment by controlling the sale, distribution, and use of pesticides within the United States (7 *USC* §136a[c][5]).

The FIFRA is a risk/benefit statute which calls for an analysis of the risks created by a pesticide compared to the benefits of the product if it enters the market. The 1972 FIFRA amendments (*Federal Environmental Pesticide Control Act [FEPCA]*, 1972) establish the current structure of the act which concentrates on the regulation of pesticides through their registration, classification, and labeling.

SPECIFIC PROVISIONS

The FIFRA requires all pesticides to be registered with the EPA. Pesticides are then classified according to general or

35. The *FIFRA* was enacted in 1972 and is codified in the *U.S. Code*, Vol. 7, sec. 136 *et seq.* It was amended in 1975, 1978, 1980, 1983, 1984, 1988 (major amendments, U.S. Public Law 100–532, 102 Stat. 2654) and 1990. The *FIFRA* regulations are found in the *Code of Federal Regulations*, Title 40, secs. 162–180.

restricted use (7 *USC* §136a). The EPA registers a pesticide if it determines that the pesticide does not create unreasonable adverse effects on human health and the environment. The registration process requires applicants to provide information on the product's chemistry, residue chemistry, toxicology, field reentry protection, aerial drift evaluation, effect on nontarget organisms, and product performance (40 *CFR* §158.20[c]).

Defining Pesticides

Since the FIFRA regulates pesticide sale, distribution, and use, the scope of the act largely depends on the definition of the term *pesticide*. The FIFRA (§2[u], 7 *USC* §136[u]) defines a pesticide as "any substance or mixture of substances intended for preventing, destroying, repelling, or mitigating any pest, and any substance or mixture of substances intended for use as a plant regulator, defoliant, or desiccant. . . ."

The FIFRA contains other definitions including explanations of the terms within the definition of pesticide. For instance, an understanding of the term *pest* is needed to apply the definition of pesticide. A pest is defined as "any insect, rodent, nematode, fungus, weed, or any other form of terrestrial aquatic plant or animal life or virus, bacteria, or other micro-organism (except [those] on or in living man or other living animals) which [the EPA] declares to be a pest . . ." (*FIFRA* §2[t], 7 *USC* §136[t]). The EPA has promulgated regulations which clarify what is and is not a pesticide for the purposes of this statute.[36]

Registering Pesticides

Section 3 of the FIFRA (7 *USC* §136a) requires persons to register pesticides with the EPA before they are sold or distributed within the United States. FIFRA's registration process allows the EPA to evaluate the risks of the pesticide being proposed. This evaluation is accomplished by examining application information.

Applicants must file a statement which includes:

The name and address of the applicant and the name of any other person who appears on the label of the pesticide

The name of the pesticide

A complete copy of the pesticide's label, instructions for its use, and a statement of all claims made of the pesticide

The complete formula of the pesticide

A request that the pesticide be classified for general use, restricted use, or both (*FIFRA* §3[c][1], 7 *USC* §136a[c][1])

36. *See Code of Federal Regulations.* Title 40, part 162.

Additionally, the EPA can request the applicant to include a full description of tests, test results, and other data associated with the pesticide. The EPA has guidelines stating the types of information which an applicant must submit to register a pesticide (*FIFRA* §3[c][2], 7 *USC* §136a[c][2]).[37]

The EPA generally registers a pesticide provided that:

The pesticide is what it is claimed to be in the application

The applicant complies with labeling and other FIFRA application requirements

The pesticide performs its function without unreasonable adverse effects to the environment

The pesticide, when used "in accordance with widespread and commonly recognized practice," does not cause unreasonable adverse effects to the environment (*FIFRA* §3[c][5], 7 *USC* 136a[c][5]).

The EPA must weigh the economic, social, and environmental costs and benefits to determine whether a pesticide causes unreasonable adverse effects to the environment (*FIFRA* §2[bb], 7 *USC* §136[bb]). The applicant has the burden of providing all information needed by the EPA to determine if the pesticide meets the registration standard.

The FIFRA (§3[c][7], 7 *USC* §136a[c][7]) also provides for conditionally registered pesticides under limited conditions. The EPA allows conditional registration if the proposed pesticide is substantially similar to an existing registered pesticide and no unreasonable risks on the environment exist. The conditional registration requires the applicant to submit additional data on the pesticide.

Finally, the act provides for experimental use permits which can be issued by the EPA or an authorized state (*FIFRA* §5, 7 *USC* §136c; 40 *CFR* pt. 172). These permits allow the applicant to sell, distribute, and use the pesticide under restricted conditions until the pesticide is properly registered.

Classification of Pesticides

As part of the registration procedure, the EPA classifies a pesticide for general use, restricted use, or both (*FIFRA* §3[d], 7 *USC* 136a[d]). A pesticide is classified for general use if it does not cause unreasonable adverse effects on the environment when applied according to its instructions and applicable warnings.

A pesticide is classified for restricted use if its application likely causes unreasonable adverse effects on the environment without additional regulatory restrictions. If the EPA classifies a pesticide for restricted use based on a finding that "the acute dermal or inhalation toxicity of the pesticide presents a hazard to the applicant or to other persons," the pesticide can be applied only by or under the supervision of a certified applicator. The FIFRA (§11, 7 *USC* §136i) has procedures and requirements to become

37. The guidelines are published in the *Code of Federal Regulations*, Title 40, part 158.

a certified applicator which are implemented by the EPA or through a federally approved state certification program.

Suspension and Cancellation of Registration

The EPA is required to cancel a registered pesticide five years after its registration unless the registrant requests to have the pesticide reregistered (*FIFRA* §6[a], 7 *USC* §136d[a]). Additionally, the EPA can issue an immediate suspension order to a registrant to prevent an imminent hazard (*FIFRA* §6[c], 7 *USC* §136d[c]).

Reregistration of Pesticides

Many pesticides were registered under the FIFRA prior to the 1972 amendments. However, those pesticides were not subject to any risk/benefit analysis prior to registration. Moreover, many registered pesticides also had insufficient warnings and labels. Section 4 of the FIFRA (7 *USC* §136a–l) requires that these pesticides and any pesticide issued before November 1, 1984, reregister with the EPA. The implementation of this process has been slow.

In 1978, Congress sought to facilitate the reregistration process by giving the EPA authority to obtain data from holders of existing registered pesticides (*FIFRA* §3[c][2][B], 7 *USC* §136a[c][2][B]). Registrants then have ninety days to provide evidence that they are taking steps to obtain the data. The EPA can issue a notice of intent to suspend the registration if the registrant fails to comply. This notice can lead to a full suspension of the pesticide which could last indefinitely or until the registrant complies with the request for data.

Congress amended the FIFRA in 1988 to improve the reregistration process. Congress adopted a new section 4 to the FIFRA which created the following five phases for reregistration:

Phase 1. Listing and prioritizing active ingredients of pesticides for reregistration

Phase 2. Submission of notices of intent to reregister, and identification of and commitments to replace missing or inadequate data

Phase 3. Submission of studies on the active ingredients of the pesticides

Phase 4. Independent initial review by the EPA of submissions under phases 2 and 3, identification of outstanding data requirements, and request for additional data if necessary

Phase 5. Review of data submitted, and appropriate regulatory action implemented (*FIFRA* §4, 7 *USC* §136a–l)

Additionally, Congress adopted a fee program to cover the expected costs of the new reregistration process (*FIFRA* §4[i], 7 *USC* §136a–l[i]).

SUMMARY

Pesticides must be registered with the EPA before being manufactured, distributed, or used in the United States (*FIFRA* §3).

Pesticides issued prior to November 1, 1984, must be reregistered (*FIFRA* §4).

The EPA can cancel or suspend a registered pesticide based upon failure to seek reregistration after five years, an imminent hazard, or a showing of unreasonable adverse effects on the environment (*FIFRA* §6).

The FIFRA makes it unlawful to sell or distribute any pesticide which is unregistered or is otherwise improperly accounted for according to the act (*FIFRA* §12).

The FIFRA provides for civil and criminal penalties for violating the act (*FIFRA* §14).

The FIFRA provides for judicial review of nondiscretionary actions by the EPA under the act (*FIFRA* §16).

Pollution Prevention Act

The Pollution Prevention Act of 1990 (42 *USC* 13101 *et seq.*).

STATUTORY ROADMAP

§13101	Policy
§13102	Definitions
§13103	EPA's role
§13104	Grants
§13105	Source reduction clearinghouse
§13106	Source reduction and recycling data collection
§13107	EPA report
§13108	Savings provisions
§13109	Authorization of appropriations

PURPOSE

The Pollution Prevention Act (PPA) of 1990 (§§6601–6610, 42 *USC* §§13101–13113 [Supp. II 1990]) is an holistic approach to environmental protection which seeks to reduce or eliminate pollutants before they are generated. This prevention is done by altering products, procedures, or raw materials in the manufacturing process. The EPA defines pollution prevention as "the use of materials, processes or practices that reduce or eliminate the creation of pollutants or wastes at the source. [It] includes practices that reduce the use of hazardous materials, energy, water or other resources, and practices that protect natural resources through conservation and more efficient use" (U.S. EPA. 1990).

SPECIFIC PROVISIONS

The PPA (§6602[b], 42 *USC* §1310[b]) establishes a hierarchy of pollution control methods. The fundamental idea behind the hierarchy is to prevent pollution before it occurs. Pollution which cannot be avoided should be reduced. In turn, that which cannot be reduced should be recycled, then treated. Only after all those methods have been applied should pollution be released into the environment through disposal. This hierarchy is stated as the central policy of the act as follows:

> pollution should be prevented or reduced at the source whenever feasible; pollution that cannot be prevented should be recycled in an environmentally safe manner whenever feasible; pollution that cannot be prevented or recycled should be treated in an environmentally safe manner whenever feasible; and disposal or other release into the environment should be employed only as a last resort and should be conducted in an environmentally safe manner (*PPA* §6602[b], 42 *USC* §13101[b]).

The hierarchy reflects a preference to eliminate pollution before it occurs, thereby preventing it from crossing into various media, or even jurisdictions, where its more difficult to trace, recapture, and regulate.

Reporting

The PPA (§6607, 42 *USC* §13106) requires certain facilities to complete a toxic chemical source reduction and recycling report.[38] Under the PPA, each owner or operator of a facility which must file an annual toxic chemical release form under section 313 of the SARA of 1986 must also include a source reduction and recycling report for the prior calendar year. The report includes the following sections:

1. The quantity of each chemical entering the wastestream, or otherwise released into the environment, prior to recycling, treatment, or disposal, and the percentage change from the previous year
2. The amount of the chemical from the facility which is recycled during the calendar year, the recycling process used, and the percentage change from the previous year.
3. The source reduction practices used with respect to that chemical, reported in accordance with the following categories:
   a. Equipment, technology, process, or procedure modifications
   b. Reformulation or redesign of products

---

38. The EPA has also called for pollution prevention plans in storm water permits under the CWA. *See* U.S. EPA. *Proposed rule, national pollutant discharge elimination system general permits and reporting requirements for storm water discharges associated with industrial activity. Federal Register 56,* (1991):4094 (proposed 16 August 1991). The EPA has requested comments on the use of incentives under the RCRA to reduce or eliminate the generation of hazardous waste. *See* U.S. EPA. *Waste minimization incentives: Notice and request for comment on desirable and feasible incentive to reduce or eliminate the generation of hazardous waste. Federal Register 55,* (1990):40881.

   c. Substitution of raw materials

   d. Improvement in management, training, inventory control, materials handling, or other general operational phases of industrial facilities

4. The amount expected to be reported under sections 1 and 2 for the two calendar years immediately following the calendar year for which the report is being filed

5. A ratio of production in the reporting year to that in the previous year

6. Techniques used to identify source reduction opportunities including employee recommendations, external and internal audits, and material balance audits

7. The amount of toxic chemical released into the environment resulting from a catastrophic event, remedial action, or other one-time event and which is not associated with production processes during the reporting year

8. The amount of chemical from the facility which is treated during the calendar year and the percentage change from the previous year

The EPA has several other approaches to encourage pollution prevention, including an information clearinghouse, technical and financial assistance to industry and local government, and waste audits. In addition, the PPA calls for the EPA to "coordinate source reduction activities in each agency office and coordinate with appropriate offices to promote source reduction practices in other federal agencies ... (PPA §6604[b][3], 42 USC §13103[b][3]).

The practical implications of the PPA appear in the opportunity for facilities to meet emission or discharge limits under other programs, e.g., the CAA and CWA, by reducing at the source rather than removing at the end of the pipe. The informational aspect of the PPA also requires facilities to reassess where they might be wasting raw materials and where they can reuse materials which otherwise would be discharged into various media.

SUMMARY

- The PPA creates a hierarchy of pollution control methods (§6602).
- The PPA establishes reporting requirements for the use of toxic substances (§6607).

—*David Bookchin*
David Farnsworth

References

Clean Air Act (CAA). Secs. 101–618. *U.S. Code.* Vol. 42, 7401–7671q.

Clean Air Act of 1990. U.S. Public Law 101–549, 104 Stat. 2399.

Clean Water Act (CWA). Secs. 101, 301, 304, 306–307, 319, 334, 402. *U.S. Code.* Vol. 33, secs. 1251–1387.

Code of Federal Regulations. Title 33, pt. 320; Title 40, pts. 50, 63, 122, 125, 158, 162–180, 204, 211, 240–271, 300, 400–407, 700–799.

Comprehensive Environmental Response, Compensation, and Liability Act (CERCLA). Secs. 101, 103–107, 111, 116, 121–122. *U.S. Code.* Vol. 42, secs. 9601–9675.

Federal Environmental Pesticide Control Act (FEPCA). 1972. U.S. Public Law 92–516, 86 Stat. 973.

Federal Insecticide, Fungicide, and Rodenticide Act (FIFRA). Secs. 2–7, 11. *U.S. Code.* Vol. 7, sec. 136.

Federal Register 43, (16 March 1978):11110; EPA Office of Toxic Substances, *TSCA* sec. 8(e) Reporting Guide (June 1991).

—— *53,* (23 September 1988):37082.

—— *57,* (16 July 1992):31576.

——, 61970 *et seq.*

Hazardous and Solid Waste Amendments. 1984. U.S. Public Law 98–616, 98 Stat. 3221.

Noise Control Act. U.S. Code. Vol. 42, secs. 4901–4918.

Omnibus Budget Reconcilation Act of 1990. U.S. Public Law 101–508, 104 Stat. 1388–501.

Pollution Prevention Act (PPA). Secs. 6601–6610. *U.S. Code.* Vol. 42, secs. 13101–13113.

Polychlorinated Biphenyls (PCB) Penalty Policy. Envtl. L. Rep. 20 (August 1990):35235.

Resource Conservation and Recovery Act (RCRA). Secs. 1001, 1003–1004, 1022, 3003–3005, 9001–9010. *U.S. Code.* Vol. 42, 6901 *et seq.*

Safe Drinking Water Act (SDWA). Secs. 1401–1451. *U.S. Code.* Vol. 42, sec. 300.

Safe Drinking Water Act 1986 Amendments. U.S. Public Law 99–339, *The Weekly Compilation of Presidential Documents* 22, (19 June 1986):831.

Superfund Amendments and Reauthorization Act (SARA) of 1986. Public Law 99–499, 100 Stat. 1613 (Codified at *U.S. Code.* Vol. 42, secs. 9601–9675 [1988]).

Toxic Substances Control Act (TSCA). Secs. 2–6, 8. *U.S. Code.* Vol. 15, secs. 2601–2654.

U.S. Environmental Protection Agency (EPA). 1990. *Environmental Protection Agency pollution prevention directive.* (Draft) 13 May. Cited in U.S. EPA, *Pollution prevention 1991: Progress on reducing industrial pollutants.* October 1991 at 4.

Bibliography

Arbuckle, J.G., G.W. Frick, R.M. Hall, Jr., M.L. Miller, T.F.P. Sullivan, and T.A. Vanderver, Jr. 1983. *Environmental law handbook.* 7th ed., Rockville, Md.: Government Institutes, Inc.

2

Environmental Impact Assessment

Larry W. Canter

Introduction

The National Environmental Policy Act (NEPA) of 1969 (PL91–190) has been referred to as the Magna Carta for the environment in the United States (Council on Environmental Quality 1993). The thrust of this act, as well as that of subsequent Council on Environmental Quality (CEQ) guidelines and regulations, is to ensure that balanced decision making occurs in the total public interest. Project planning and decision making should consider technical, economic, environmental, social, and other factors. Environmental impact assessment (EIA) can be defined as the systematic identification and evaluation of the potential impacts (effects) of proposed projects, plans, programs, or legislative actions, relative to the physical–chemical, biological, cultural, and socioeconomic components of the environment. The primary purpose of the EIA process, also called the NEPA process, is to encourage the consideration of the environment in planning and decision making and to ultimately arrive at actions which are more environmentally compatible.

This chapter focuses on NEPA and the EIA process. Following an initial section related to background conceptual and administrative considerations, the chapter's emphasis is on practical methods and approaches used for impact identification, prediction, and assessment (interpretation), and on comparative evaluations of alternatives. Writing considerations and follow-on environmental monitoring are also addressed. The final sections relate to emerging issues and international activities.

Reference

Council on Environmental Quality. 1993. *Environmental quality.* Twenty-third Annual Report. January:151–172. Washington, D.C.: U.S. Government Printing Office.

2.1
BACKGROUND CONCEPTUAL AND ADMINISTRATIVE INFORMATION

Section 102 of the NEPA has three primary parts related to the EIA process. Part A specifies that all federal government agencies will use a systematic, interdisciplinary approach, which ensures the integrated use of the natural and social sciences and environmental design arts in planning and decision making that may impact the human environment. Part B requires agencies to identify and develop methods and procedures to ensure that presently unquantified environmental amenities and values are considered in decision making along with economic and technical considerations. This part has provided impetus for the development of environmental assessment methods. Part C states the necessity for preparing environmental statements (called environmental impact statements or EISs) and identifies basic items to be included.

To aid the implementation of the EIS requirement, the NEPA also created the Council on Environmental Quality (CEQ) within the Executive Office of the President of the United States. This council has the role of providing overall coordination to the EIA process in the United States. CEQ issued guidelines in 1971 and 1973 for federal agencies to follow in conjunction with EISs. In 1978, the CEQ issued regulations which became effective in mid-1979 for responding to the requirements of the NEPA (Council on Environmental Quality 1978).

Key Definitions

A key feature of the CEQ regulations is the concept of three levels of analysis; Level 1 relates to a categorical exclusion determination, Level 2 to the preparation of an environmental assessment (EA) and finding of no significant impact, and Level 3 to the preparation of an EIS (U.S. Environmental Protection Agency 1989). Figure 2.1.1 depicts the interrelationships between these three levels. Key definitions from the CEQ regulations related to Figure 2.1.1 include federal action, categorical exclusion, EA, finding of no significant impact, significant impact, and EIS.

Federal actions include the adoption of official policy (rules, regulations, legislation, and treaties) which result in or alter agency programs; adoption of formal plans; adoption of programs; and approval of specific projects, such as construction or management activities located in a defined geographic area, and actions approved by permit or other regulatory decision as well as federal and federally assisted activities. The EIA process is typically applied to proposed projects. Key information needed in applying the EIA process to a proposed project includes items such as:

1. A description of the type of project and how it functions or operates in a technical context

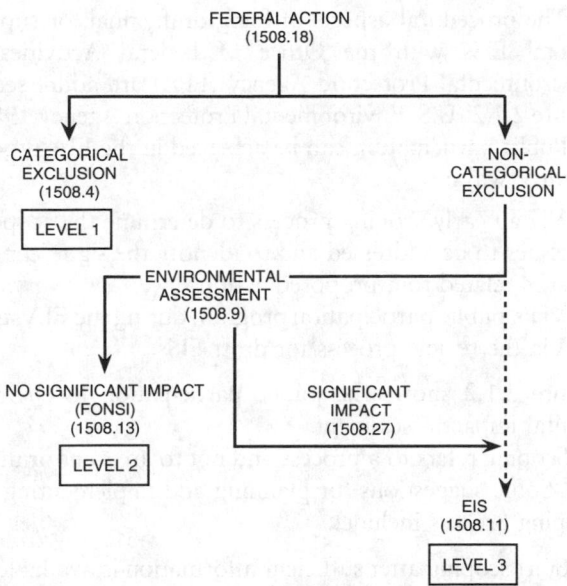

FIG. 2.1.1 Three levels of analysis in the EIA process. Number in parentheses denotes paragraph in CEQ regulations which contains definition (Council on Environmental Quality 1987).

2. The proposed location for the project and why it was chosen
3. The time period required for project construction
4. The potential environmental requirements or outputs (stresses) from the project during its operational phase, including land requirements, air pollution emissions, water use and water pollutant emissions, and waste generation and disposal needs
5. The identified current need for the proposed project in the location where it is proposed (this need could be related to housing, flood control, industrial development, economic development, and many other requirements); project need must be addressed as part of the environmental documentation
6. Any alternatives which have been considered, with generic alternatives for projects including site location, project size, project design features and pollution control measures, and project timing relative to construction and operational issues; project need in relation to the proposed project size should be clearly delineated; the range of alternatives may be limited due to the individual preferences of project sponsors, primary focus on traditional engineering solutions, and time pressures for decision making (Bacow 1980)

A categorical exclusion refers to a category of actions which do not individually or cumulatively have a significant effect on the human environment and have no such effect in procedures adopted by a federal agency in implementation of the CEQ regulations. Neither an EA nor an EIS is required for categorical exclusions.

An EA is a concise public document that serves to provide sufficient evidence and analysis for determining whether to prepare an EIS or a finding of no significant impact (FONSI), aid an agency's compliance with the NEPA when no EIS is necessary, or facilitate preparation of an EIS when one is necessary. A FONSI is a document written by a federal agency briefly presenting the reasons why an action, not otherwise excluded, will not have a significant effect on the human environment and for which an EIS will not be prepared. A mitigated FONSI refers to a proposed action that has incorporated mitigation measures to reduce any significant negative effects to insignificant ones.

The key definition in the EIA process is *significantly* or *significant impact* since a proposed action which significantly affects the human environment requires an EIS. *Significantly* as used in the NEPA requires considerations of both context and intensity. Context means that significance must be analyzed relative to society as a whole (human, national), the affected region, the affected interests, the locality, and whether the effects are short- or long-term. Intensity refers to the severity of impact. The following should be considered in evaluating intensity:

1. Impacts that may be both beneficial and adverse (A significant effect may exist even if the federal agency believes that on balance the effect will be beneficial)
2. The degree to which the proposed action affects public health or safety
3. Unique characteristics of the geographic area, such as proximity to historic or cultural resources, park lands, prime farmlands, wetlands, wild and scenic rivers, or ecologically critical areas
4. The degree to which the effects on the quality of the human environment are likely to be controversial
5. The degree to which the possible effects on the human environment are uncertain or involve unique or unknown risks
6. The degree to which the action may establish a precedent for future actions with significant effects or represents a decision in principle about a future consideration
7. Whether the action is related to other actions with individually insignificant but cumulatively significant impacts (Significance exists if a cumulatively significant impact on the environment is anticipated. Significance cannot be avoided by terming an action temporary or by breaking it down into component parts)
8. The degree to which the action may adversely affect districts, sites, highways, structures, or objects listed in or eligible for listing in the *National Register of Historic Places* or may cause loss or destruction of significant scientific, cultural, or historical resources
9. The degree to which the action may adversely affect an endangered or threatened species or its habitat that has been determined to be critical under the Endangered Species Act of 1973

10. Whether the action threatens a violation of federal, state, or local law or requirements imposed for the protection of the environment

An EIS is a detailed written statement that serves as an action-forcing device to ensure that the policies and goals defined in the NEPA are infused into the ongoing programs and actions of the federal government. It provides a full and fair discussion of significant environmental impacts and informs decision makers and the public of the reasonable alternatives which would avoid or minimize adverse impacts or enhance the quality of the human environment. An EIS is more than a disclosure document; it is used by federal officials in conjunction with other relevant material to plan actions and make decisions.

The definition of *significantly* from the CEQ regulations considers both context and intensity. Context is primarily related to the "when and where" of the impacts. The preceding ten points for intensity can be divided into two groups as follows: those related to environmental laws, regulations, policies, and executive orders (points 2, 3, 8, 9, and 10 in the list); and those related to other considerations and how they in turn may implicate environmental laws, regulations, policies, and executive orders (points 1, 4, 5, 6, and 7 in the list).

Three types of EISs are pertinent (draft, final, or supplemental to either draft or final). In addition to the earlier generic definition of an EIS, the following information from the CEQ guidelines or regulations is germane (Council on Environmental Quality 1987, 1973):

1. Draft EIS: The draft EIS is the document prepared by the lead agency proposing an action; it is circulated for review and comment to other federal agencies, state and local agencies, and public and private interest groups. Specific requirements with regard to timing of review are identified in the CEQ regulations. In the draft statement the agency will make every effort to disclose and discuss all major points of view on the environmental impacts of the alternatives, including the proposed action.
2. Final EIS: The final EIS is the draft EIS modified to include a discussion of problems and objections raised by reviewers. The final statement must be on file with EPA for at least a thirty-day period prior to initiation of construction on the project. The format for an EIS is delineated in Sections 1502.10 through 1502.18 of the CEQ regulations.
3. Supplemental EIS: Lead agencies will prepare supplements to either draft or final EISs if the agency makes substantial changes in the proposed action that are relevant to environmental concerns; or significant new circumstances or information relevant to environmental concerns and bearing on the proposed action or its impacts exists. Lead agencies may also prepare supplements when the agency determines that the purposes of the Act will be furthered by doing so.

The procedural aspects of filing draft, final, or supplemental EISs with the Office of Federal Activities of Environmental Protection Agency (EPA) are addressed in Figure 2.1.2 (U.S. Environmental Protection Agency 1989).

Public participation can be achieved in the EIA process in three ways:

1. Via an early scoping process to determine the scope of issues to be addressed and to identify the significant issues related to a proposed action
2. Via a public participation program during the EIA study
3. Via the review process for draft EISs

Figure 2.1.2 shows the public participation in environmental impact assessment.

Scoping refers to a process and not to an event or meeting. Some suggestions for planning and implementing the scoping process include:

1. Start scoping after sufficient information is available on the proposed action.
2. Prepare an information packet.
3. Design a unique scoping process for each project, plan, program, and policy.
4. Issue a public notice.
5. Carefully plan and conduct all public meetings.
6. Develop a plan for using received comments.
7. Allocate EIS work assignments and establish a completion schedule (Council on Environmental Quality 1981).

Strategic environmental assessment (SEA) refers to the EIA process applied to policies, plans, or programs (Lee and Walsh 1992). A programmatic EIS is analogous to an SEA. Programmatic EISs (PEISs) can be used in the United States to address the environmental implications of the policies and programs of federal agencies. PEISs can be used to address the impacts of actions that are similar in nature or broad in scope, including cumulative impacts (Sigal and Webb 1989). Site-specific or local-action EAs and EISs can be tiered from the PEIS.

While the NEPA includes the concept of applying the EIA process to policies, plans, and programs, the majority of EISs prepared in the first twenty-five years have been on projects. However, the NEPA should be used in the formulation of policies and the planning of programs (Bear 1993). EAs and EISs can be prepared for policies (including rules, regulations, and new legislation), plans, and programs. Plans and programs can include operational considerations for extant single or multiple projects and repair, evaluation, maintenance, and rehabilitation of extant projects. Decommissioning of existing facilities also requires the application of the EIA process.

Impact Significance Determination

Impacts resulting from proposed actions can be considered in one or more of the following categories:

- Beneficial or detrimental
- Naturally reversible or irreversible

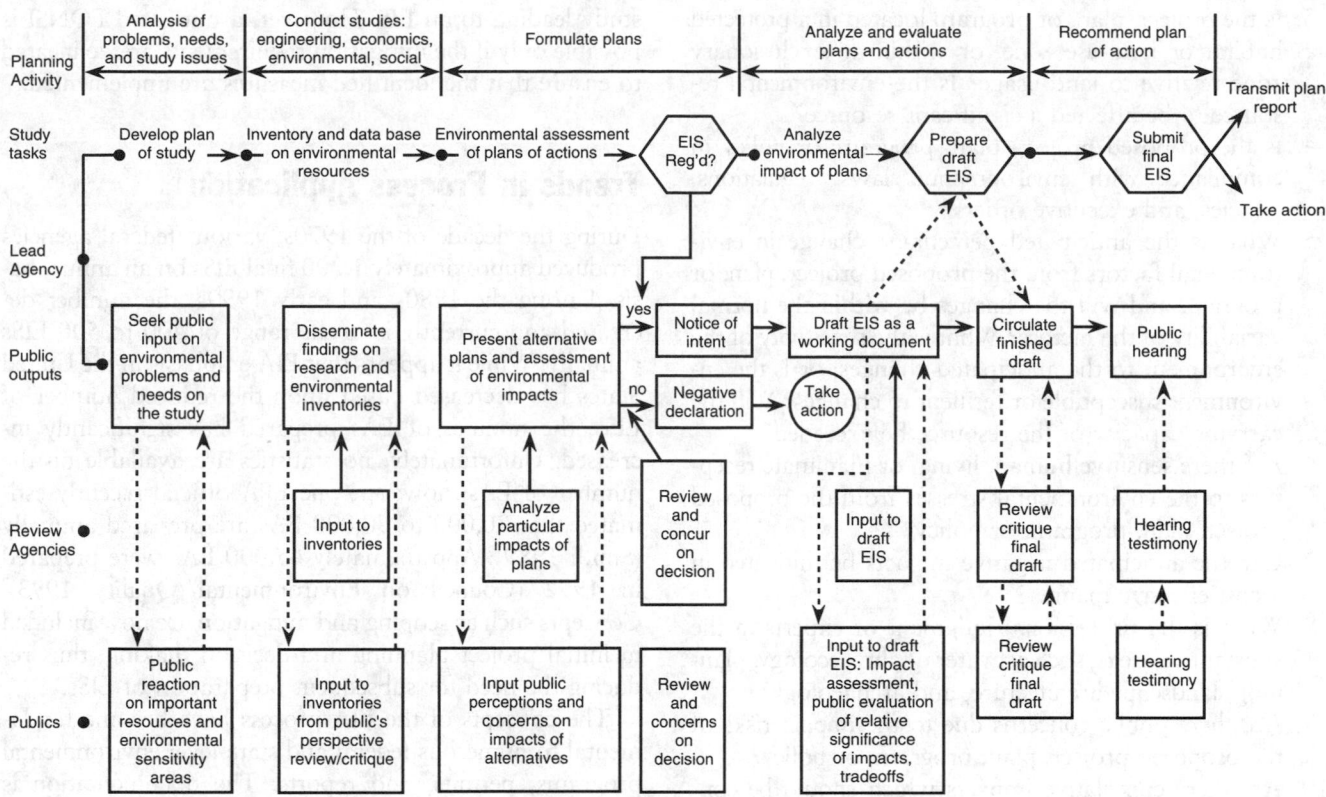

FIG. 2.1.2 Public participation in environmental impact assessment.

- Reparable via management practices or irreparable
- Short term or long term
- Temporary or continuous
- Construction or operational phase
- Local, regional, national, or global
- Accidental or planned (recognized beforehand)
- Direct or primary, or indirect or secondary
- Cumulative or single

Two key terms from the categories are direct or indirect, and cumulative; their definitions are as follows (Council on Environmental Quality 1987):

Effects (or impacts): These terms can be considered as synonymous. Two broad categories of effects are direct and indirect. Direct effects are caused by the action and occur at the same time and place. Indirect effects are caused by the action and occur later or farther removed in distance, but are still reasonably foreseeable. Indirect effects may include growth-inducing effects and other effects related to induced changes in the pattern of land use, population density or growth rate, and related effects on air and water and other natural systems, including ecosystems. Effects include ecological (such as the effects on natural resources and on the components, structures, and functioning of affected ecosystems), aesthetic, historic, cultural, economic, social, or health, whether direct, indirect, or cumulative. Effects also include those resulting from actions which may have both beneficial and detrimental effects, even if on balance the agency believes that the effect will be beneficial.

Cumulative impact: The impact on the environment which results from the incremental impact of the action when added to other past, present, and reasonably foreseeable future actions regardless of what agency (federal or nonfederal) or person undertakes such other actions. Cumulative impacts can result from individually minor, but collectively significant, actions taking place over a period of time.

Based on these categories of impacts, several simple to more structured options can be used to determine impact significance. At a minimum, the definition of *significantly* in the CEQ regulations could be applied as described in the previous section.

A sequenced approach for impact significance determination is appropriate. A sequenced approach considers several levels in determining the potential significance of impacts from a proposed federal action. A sequenced approach is achieved by applying the following questions in the order shown (the answers to any question can be used to determine if an EIS should be prepared):

1. Does the proposed project, plan, program, or policy cause impacts that exceed the definition of significant impacts as contained in pertinent laws, regulations, or executive orders?
2. Is a quantitative threshold criterion exceeded in terms of project, plan, or program type, size, or cost?

3. Is the project, plan, or program located in a protected habitat or land-use zone, or within an exclusionary zone relative to land usage? Is the environmental resource to be affected a significant resource?

4. Is the proposed project, plan, program, or policy in compliance with environmental laws, regulations, policies, and executive orders?

5. What is the anticipated percentage change in environmental factors from the proposed project, plan, or program, and will the changes be within the normal variability of the factors? What is the sensitivity of the environment to the anticipated changes; or is the environment susceptible or resilient to changes? Will the carrying capacity of the resource be exceeded?

6. Are there sensitive human, living, or inanimate receptors to the environmental stresses from the proposed project, plan, program, or policy?

7. Can the anticipated negative impacts be mitigated in a cost-effective manner?

8. What is the professional judgment of experts in the substantive areas, such as water quality, ecology, planning, landscape architecture, and archaeology?

9. Are there public concerns due to the impact risks of the proposed project, plan, program, or policy?

10. Are there cumulative impacts which should be considered, or impacts related to future phases of the proposed action and associated cumulative impacts?

Detailed specific questions related to these ten groups of questions can be developed.

One thing that can be done in conjunction with identified significant negative impacts is to consider appropriate mitigation measures to reduce negative impacts within reasonable environmental and economic constraints. Relative to practice in the United States, mitigation includes (Council on Environmental Quality 1987):

1. Avoiding the impact altogether by not taking a certain action or parts of an action

2. Minimizing impacts by limiting the degree or magnitude of the action and its implementation

3. Rectifying the impact by repairing, rehabilitating, or restoring the affected environment

4. Reducing or eliminating the impact over time by preservation and maintenance operations during the life of the action

5. Compensating for the impact by replacing or providing substitute resources or environments

These measures should be used in sequence or ease of application, beginning with avoiding the impact.

Negative impacts fall into three categories: insignificant, significant but mitigable, or significant but not mitigable. When potentially significant negative impacts are identified, and if they can be reduced via mitigation to something of lesser concern, a mitigated FONSI can be prepared following an EA and without doing a comprehensive study leading to an EIS. However, a mitigated FONSI is possible only if the mitigation requirements are delineated to ensure that the identified measures are implemented.

Trends in Process Application

During the decade of the 1970s, various federal agencies produced approximately 1,200 final EISs on an annual basis. During the 1980s and early 1990s, the number decreased and currently is in the range of 400 to 500 EISs annually. While it appears that EIA emphasis in the United States has decreased based upon the reduced number of EISs, the number of EAs prepared has significantly increased. Unfortunately, no statistics are available on the number of EAs; however, one EPA official recently estimated that 30,000 to 50,000 EAs are prepared annually (Smith 1989). Approximately 45,000 EAs were prepared in 1992 (Council on Environmental Quality 1993). Concepts such as scoping and mitigation are now included in initial project planning and decision making, thus reducing the need for subsequent preparation of EISs.

The concepts of the EIA process have become fundamental to numerous federal and state-level environmental programs, permits, and reports. The documentation is analogous to a "targeted" EA or EIS. Examples of relevant permits and reports include:

1. Air quality permits and related reporting required by the Clean Air Act Amendments of 1990

2. Point source wastewater discharge permits and related reporting required by the National Pollutant Discharge Elimination System (NPDES) program of the Federal Water Pollution Control Act Amendments (also known as Clean Water Act) and its subsequent amendments

3. Industrial area storm water discharge permits and related reporting required by the NPDES program of the Clean Water Act of 1987

4. Permits for dredging and filling activities in navigable waters as required by Section 404 of the Clean Water Act of 1972

5. Remedial investigations, feasibility studies, and records of decision on uncontrolled hazardous waste sites identified under the auspices of the Comprehensive Environmental Response, Compensation, and Liability Act (CERCLA or Superfund) of 1981 and the Superfund Amendments and Reauthorization Act (SARA) of 1986 (Sharples and Smith 1989)

6. Replacements, permits, and reports on underground storage tanks regulated by the Resource Conservation and Recovery Act Amendments of 1984

7. Operating permits and closure plans for sanitary landfills or hazardous waste landfills required by the Resource Conservation and Recovery Act Amendments (also called the Hazardous and Solid Waste Act) of 1984 (Sharples and Smith 1989)

8. Reports prepared on site (property transfer) assessments to establish owner, buyer, and lender liability for contamination
9. Reports prepared on regulatory audits
10. Environmental reporting requirements related to new chemical and new product licensing

—Larry W. Canter

References

Bacow, L.S. 1980. The technical and judgmental dimensions of impact assessment. *Environmental Impact Assessment Review* 1, no. 2:109–124.

Bear, D. 1993. NEPA: substance or merely process. *Forum for Applied Research and Public Policy* 8, no. 2:(Summer):85–88.

Council on Environmental Quality. 1973. Preparation of environmental impact statements: Guidelines. *Federal Register* 38, no. 147 (1 August):20550–20562.

———. 1978. National Environmental Policy Act—Regulations. *Federal Register* 43, no. 230 (29 November):55978–56007.

———. 1981. *Memorandum: Scoping guidance.* 30 April. Washington, D.C.

———. 1987. *40 Code of federal regulations.* Chap. 5, 1 July:929–971. Washington, D.C.: U.S. Government Printing Office.

———. 1993. *Environmental quality.* Twenty-third Annual Report. January:151–172. Washington, D.C.: U.S. Government Printing Office.

Lee, N., and F. Walsh. 1992. Strategic environmental assessment: An overview. *Project Appraisal* 7, no. 3 (September):126–136.

Sharples, F.E., and E.D. Smith. 1989. *NEPA/CERCLA/RCRA Integration.* CONF-891098-9. Oak Ridge, Tennessee: Oak Ridge National Laboratory.

Smith, E.D. 1989. *Future challenges of NEPA: a panel discussion.* CONF-891098-10. Oak Ridge, Tennessee: Oak Ridge National Laboratory.

Sigal, L.L., and J.W. Webb. 1989. The programmatic environmental impact statement: Its purpose and use. *The Environmental Professional* 11, no. 1:14–24.

U.S. Environmental Protection Agency. 1989. *Facts about the National Environmental Policy Act.* LE-133. September. Washington, D.C.

2.2
EIA METHODS: THE BROAD PERSPECTIVE

Several activities are required to conduct an environmental impact study, including impact identification, preparation of a description of the affected environment, impact prediction and assessment, and selection of the proposed action from a set of alternatives being evaluated to meet identified needs. The objectives of the various activities differ, as do the methods for accomplishing the activities (Lee 1988). The term *method* refers to structured scientific or policy-based approaches for achieving one or more of the basic activities. Table 2.2.1 contains a delineation of eighteen types of methods arrayed against seven activities that are typically associated with an EIA study. An *x* in the table denotes that the listed method type is or may be directly useful for accomplishing an activity. However, the absence of an *x* for any given type of method does not mean that it has no usefulness for the activity; it merely suggests that it may be indirectly related to the activity.

Based on the information in Table 2.2.1, the following observations can be made:

1. Each type of method has potential usefulness in more than one EIA study activity.
2. Each EIA activity has three or more method types which are potentially useful.
3. In a given EIA study, several types of methods will probably be used even though the study may not completely document all of the methods used. Several reviews of actual method adoption in the EIA process have suggested lack of widespread usage; however, this usage probably reflects a focus on a few of the types of methods (such as matrices or checklists), and not the more inclusive list of methods contained in Table 2.2.1.
4. Each of the types of methods have advantages and limitations; examples of these for checklists, decision-focused checklists, matrices, and networks are described in subsequent sections.
5. While numerous types of methods have been developed, and additional methods are being developed and tested, no universal method can be applied to all project types in all environmental settings. An all-purpose method is unlikely to be developed due to lack of technical information as well as the need for exercising subjective judgment about predicted impacts in the environmental setting where the potential project may occur. Accordingly, the most appropriate perspective is to consider methods as tools which can be used to aid the impact assessment process. In that sense, every method should be project- and location-specific, with the basic concepts derived from existing methods. These methods can be called ad hoc methods.
6. Methods do not provide complete answers to all questions related to the impacts of a potential project or set of alternatives. Methods are not "cookbooks" in which a successful study is achieved by meeting the requirements of them. Methods must be selected based on appropriate evaluation and professional judgment, and

TABLE 2.2.1 SYNOPSIS OF EIA METHODS AND STUDY ACTIVITIES

Types of Methods in EIA	Define Issues (Scoping)	Impact Identification	Describe Affected Environment	Impact Prediction	Impact Assessment	Decision Making	Communication of Results
Analogs (look-alikes) (case studies)	X	X		X	X		
Checklists (simple, descriptive, questionnaire)		X	X				X
Decision-focused checklists (MCDM; MAUM; DA; scaling, rating, or ranking: weighting)					X	X	X
Expert opinion (professional judgment, Delphi, adaptive environmental assessment, simulation modeling)		X		X	X		
Expert systems (impact identification, prediction, assessment, decision making)	X	X	X	X	X	X	
Laboratory testing and scale models		X		X			
Literature reviews		X		X	X		
Matrices (simple, stepped, scoring)	X	X		X	X	X	X
Monitoring (baseline)			X	X			
Monitoring (field studies of analogs)				X	X		
Networks (impact trees and chains)		X	X	X			
Overlay mapping (GIS)			X	X	X		X
Photographs and photomontages			X	X			X
Qualitative modeling (conceptual)				X	X		
Quantitative modeling (media, ecosystem, visual, archaeological, systems analysis)				X	X		
Risk assessment	X	X	X	X	X		
Scenarios				X		X	
Trend extrapolation				X	X		

X: Potential for direct usage of method for listed activity

MCDM = multicriteria decision making; MAUM = multiattribute utility measurement; DA = decision analysis; GIS = geographical information system.

they must be used with the continuous application of judgment relative to data input as well as analysis and interpretation of results.

7. Methods which are simpler in terms of data and personnel resources requirements, and in technical complexity, are probably more useful in the EIA process.

Examples of selected types of methods are presented in subsequent sections, including: matrices and simple checklists for impact identification and describing the affected environment; impact prediction based on a range of methods, including analogs and qualitative and quantitative modeling; and decision making based on decision-focused checklists.

—*Larry W. Canter*

Reference

Lee, N. 1988. An overview of methods of environmental impact assessment. Environmental Impact Assessment Workshop, November. Seville, Spain.

2.3
INTERACTION MATRIX AND SIMPLE CHECKLIST METHODS

Interaction matrices were one of the earliest methods used. The simple matrix refers to a display of project actions or activities along one axis, with appropriate environmental factors listed along the other axis of the matrix. Many variations of the simple interaction matrix have been used in environmental impact studies, including stepped matrices (Economic and Social Commission for Asia and the Pacific 1990, Lohani and Halim 1990, and International Institute for Applied Systems Analysis 1979).

The matrix method developed by Leopold et al. 1971, is an example. The method involves the use of a matrix with one hundred specified actions and eighty-eight environmental items. Figure 2.3.1 illustrates the concept of the Leopold matrix; in its usage, each action and its potential for creating an impact on each environmental item is considered. Where an impact is anticipated, the matrix is marked with a diagonal line in the interaction box. The second step in using the Leopold matrix is to describe the interaction in terms of its magnitude and importance.

The magnitude of an interaction is the extensiveness or scale and is described by assigning a numerical value from one to ten, with ten representing a large magnitude and one a small magnitude. Values near five on the magnitude scale represent impacts of intermediate extensiveness. Assigning a numerical value for the magnitude of an interaction is based on an objective evaluation of facts.

The importance of an interaction is related to the significance, or assessment of the consequences, of the anticipated interaction. The scale of importance also ranges from one to ten, with ten representing an interaction of high importance and one an interaction of low importance. Assignment of an importance numerical value is based on the subjective judgment of the interdisciplinary team working on the environmental assessment study.

A simpler approach than the Leopold matrix can be used in an environmental impact study. Using the matrix entails considering the potential impacts, either beneficial or detrimental, of each project action relative to each environmental factor. Each interaction is delineated in terms of a predefined code denoting the characteristics of the impacts and whether certain undesirable features could be mitigated. Table 2.3.1 displays the concept of this type of an interaction matrix for a proposed wastewater collection, treatment, and disposal project in Barbados (Canter 1991). For this analysis, the following definitions are used for the codes:

SB = Significant beneficial impact (represents a highly desirable outcome in terms of either improving the existing quality of the environmental factor or enhancing that factor from an environmental perspective)

SA = Significant adverse impact (represents a highly undesirable outcome in terms of either degrading the existing quality of the environmental factor or disrupting that factor from an environmental perspective)

B = Beneficial impact (represents a positive outcome in terms of either improving the existing quality of the environmental factor or enhancing that factor from an environmental perspective)

A = Adverse impact (represents a negative outcome in terms of either degrading the existing quality of the environmental factor or disrupting that factor from an environmental perspective)

b = Small beneficial impact (represents a minor improvement in the existing quality of the environmental factor or a minor enhancement in that factor from an environmental perspective)

a = Small adverse impact (represents a minor degradation in the existing quality of the environmental factor or a minor disruption in that factor from an environmental perspective)

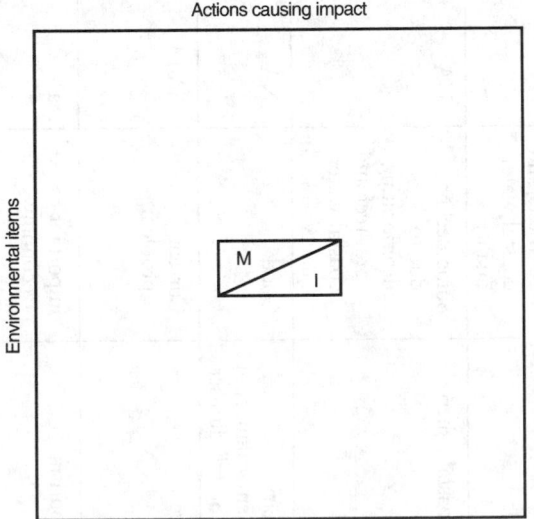

FIG. 2.3.1 Leopold Interaction Matrix (Leopold et al. 1971). M = magnitude; I = importance.

TABLE 2.3.1 INTERACTION MATRIX FOR SOUTH COAST SEWERAGE PROJECT (CANTER 1991)

Environmental Factor/Resource	Existing Quality	Construction Phase				Operation Phase			
		Collection System	Treatment Plant	Outfall Line	Resultant Quality	Collection System	Treatment Plant	Outfall Line	Resultant Quality
Air quality	In compliance with air quality standards	A/M	A/M	a	Dusts, CO	a (odor at lift station sites)	A/M	O	localized odor
Noise	Typical of urban residential areas	A/M	A/M	a	increase in local noise	a (pumps)	a	a (pumps)	small increase in noise
Ground water	Satisfactory for area	O	O	O	same as existing	b	b	b	better quality due to less sheet water discharge
Graeme Hall	"Natural" biological resource	NA	a/M (no encroachment)	NA	some disturbance, recovery expected	NA	O	NA	same as existing
Beach erosion/coral reef/coastal water quality	Erosion of 0.1 to 0.3 m/yr, deteriorating coral reef and coastal water quality	NA	NA	a (water quality)	turbidity increase	b	SB	NA	improve quality
Coastal fisheries	Some decline due to deteriorating coral reef and coastal water quality	NA	NA	a	local turbidity	b	SB	NA	improve quality
Marine environment at outfall diffuser	Good	NA	NA	a	some local disturbance	NA	NA	a	small decrease in quality
Traffic	Current problem	SA/M	a	a	increase in congestion	a	a	a	continued problem due to tourism increase
Tourism	Important to economy	a	NA	a	traffic congestion might cause decrease	B	B	B	increase in economy

A = adverse impact; M = mitigation measure planned for adverse impact; a = small adverse impact; O = no anticipated impact; NA = environmental factor not applicable; SA = significant adverse impact; b = small beneficial impact; B = beneficial impact; SB = significant beneficial impact

TABLE 2.3.2 *Continued*

Topical Issue	Yes	Maybe	No	Comments

Reduce the numbers or affect the habitat of any state or
    federally designated unique, rare, or endangered
    species of plants? (Check state and federal lists of
    endangered species.)

Introduce new species of plant into the area or create a
    barrier to the normal replenishment of existing species?

Reduce acreage or create damage to any agricultural crop?

Animal Life

Will the project:

Reduce the habitat or numbers of any state or federally
    designated unique, rare, or endangered species of
    animals? (Check state and federal lists and the
    Migratory Bird Treaty Act.)

Introduce new species of animals into an area or create
    a barrier to the migration or movement of animals
    or fish?

Cause attraction, entrapment, or impingement of animal life?

Harm existing fish and wildlife habitats?

Cause emigration resulting in human–wildlife interaction
    problems?

Land Use

Will the project:

Substantially alter the present or planned use of an area?

Impact a component of the National Park system, the
    National Wildlife Refuge system, the National Wild
    and Scenic River system, the National Wilderness
    system, or National Forest land?

Natural Resources

Will the project:

Increase the use rate of any natural resource?

Substantially deplete any nonreusable natural resource?

Be located in an area designated as or being considered
    for wilderness, wild and scenic river, national park,
    or ecological preserve?

Energy

Will the project?

Use substantial amounts of fuel or energy?

Substantially increase the demand on existing sources of
    energy?

Transportation and Traffic Circulation

Will the project result in:

Movement of additional vehicles?

Effects on existing parking facilities or demands for new
    parking?

Substantial impact on existing transportation system(s)?

Alterations to present patterns of circulation or movement
    of people or goods?

Increased traffic hazards to motor vehicles, bicyclists, or
    pedestrians?

Construction of new roads?

Public Service

Will the project have an effect on, or result in, a need for
    new or altered governmental services in any of the
    following areas:

Fire protection?

Schools?

Other governmental services?

Continued on next page

TABLE 2.3.2 *Continued*

Topical Issue	Yes	Maybe	No	Comments

Utilities

Will the project result in a need for new systems or
alterations to the following utilities:

Power and natural gas?

Communications systems?

Water?

Sewer or septic tanks?

Storm sewers?

Population

Will the project:

Alter the location or distribution of human population
in the area?

Accident Risk

Does the project:

Involve the risk of explosion or release of potentially
hazardous substances including oil, pesticides,
chemicals, radiation, or other toxic substances in the
event of an accident or "upset" condition?

Human Health

Will the project:

Create any health hazard or potential health hazard?

Expose people to potential health hazards?

Economic

Will the project:

Have any adverse effect on local or regional economic
conditions, e.g., tourism, local income levels, land values,
or employment?

Community Reaction

Is the project:

Potentially controversial?

In conflict with locally adopted environmental plans and
goals?

Aesthetics

Will the project:

Change any scenic vista or view open to the public?

Create an aesthetically offensive site open to the public
view (e.g., out of place with character or design of
surrounding area)?

Significantly change the visual scale or character of the
vicinity?

Archaeological, Cultural, and Historical

Will the project:

Alter archaeological, cultural, or historical sites, structures,
objects, or buildings, either in or eligible for inclusion
in the National Register (e.g., be subject to the Historic
Preservation Act of 1974)?

Hazardous Waste

Will the project:

Involve the generation, transport, storage or disposal of
any regulated hazardous waste (e.g., asbestos, if
demolition or building alterations is involved)?

Simple Checklists

Checklist methods range from listings of environmental factors to highly structured approaches. Structured approaches involve importance weightings for factors and the application of scaling techniques for the impact of each alternative on each factor. Simple checklists represent lists of environmental factors (or impacts) which should be addressed; however, no information is provided on specific data needs, methods for measurement, or impact prediction and assessment. Simple checklists were extensively used in the initial years of EIA studies, and they still represent a valid approach for providing systemization to an environmental impact study.

Table 2.3.2 shows a simple questionnaire checklist developed by the Cooperative Research Service of the U.S. Department of Agriculture (1990) for use related to projects that might impact agricultural lands. This extensive checklist can be used in both planning and summarizing an environmental impact study. It can also be used to identify environmental factors to be addressed in preparing a description of the affected (baseline) environment. Another example of a simple checklist is that developed by the Asian Development Bank (1987) for use on major dam, reservoir, and hydropower projects. This checklist also includes mitigation information. Other checklists have been developed: some focus on categories of impacts, such as health impacts (U.S. Agency for International Development 1980, World Bank 1982, and World Health Organization Regional Office for Europe 1983).

Simple checklists of environmental factors and impacts to consider are helpful in planning and conducting an environmental impact study, particularly if one or more checklists for the project type are used. The following summary comments are pertinent for simple checklists:

1. Because published checklists represent the collective professional knowledge and judgment of their developers, they have a certain level of professional credibility and useability.

2. Checklists provide a structured approach for identifying key impacts and pertinent environmental factors to consider in environmental impact studies.
3. Checklists stimulate and facilitate interdisciplinary team discussions during the planning, conduction, and summarization of environmental impact studies.
4. Checklists can be modified (items added or deleted) to make them more pertinent for particular project types in given locations.

—*Larry W. Canter*

References

Asian Development Bank. 1987. *Environmental guidelines for selected industrial and power development projects.* Manila, Philippines.

Canter, L.W. 1991. *Environmental impact assessment of south coast sewerage project.* Report submitted to Inter-American Development Bank, July. Washington, D.C.

Economic and Social Commission for Asia and the Pacific. 1990. *Environmental impact assessment—Guidelines for water resources development.* ST/ESCAP/786:19–48. New York: United Nations.

International Institute for Applied Systems Analysis. 1979. *Expect the unexpected—An adaptive approach to environmental management.* Executive Report 1. Laxenburg, Austria.

Leopold, L.B., et al. 1971. *A Procedure for evaluating environmental impact. U.S. Geological Survey,* Circular 645. Washington, D.C.

Lohani, B.N., and N. Halim. 1990. *Environmental impact identification and prediction: Methodologies and resource requirements.* Background papers for course on environmental impact assessment of hydropower and irrigation projects, 13–31 August:152–182. Bangkok, Thailand: International Center for Water Resources Management and Training (CEFIGRE).

U.S. Agency for International Development. 1980. *Environmental design considerations for rural development projects.* Washington, D.C.

U.S. Department of Agriculture. 1990. *Checklist for summarizing the environmental impacts of proposed projects.* Stillwater, Oklahoma: Cooperative State Research Service.

World Bank. 1982. *The environment, public health, and human ecology: Considerations for economic development.* Washington, D.C.

World Health Organization Regional Office for Europe. 1983. *Environmental health impact assessment of irrigated agricultural development projects, guidelines and recommendations: Final report.* Copenhagen, Denmark.

2.4
TECHNIQUES FOR IMPACT PREDICTION

A key technical element in the EIA process is the prediction of impacts (effects) for both the without-project and with-project conditions. Numerous technical approaches can be used. As an example, the principles and guidelines of the Water Resources Council (1983) delineate several approaches which can be used in the EIA process for water resources projects. These approaches include:

1. Adoption of forecasts made by other agencies or groups
2. Use of scenarios based on differing assumptions regarding resources and plans
3. Use of expert group judgment via the conduction of formalized Delphi studies or the use of the nominal group process
4. Extrapolation approaches based upon the use of trend analysis and simple models of environmental components
5. Analogy and comparative analyses which involve the use of look-alike resources and projects and the application of information from such look-alike conditions to the planning effort.

A criticism of many early EISs is that the impact predictions were not based on formalized and repeatable methods with predefined relationships, such as mathematical equations, physical models, and other structured approaches. Accordingly, many environmental impact studies have been criticized based on their lack of scientific approach and technical validity. This criticism is diminishing as more knowledge is gained based on the use of quantitatively based prediction techniques in environmental impact studies and the development of additional techniques through routine scientific research projects.

Classification of Prediction Techniques

A study prepared for the Ministry of Environment in the Netherlands involved the examination of 140 case studies of EIAs and related studies (Environmental Resources Limited 1982). The objectives were to identify predictive techniques used in the practice of EIAs, prepare descriptions of the techniques, and classify the techniques in terms of the effect predicted and the method used. A total of 280 predictive techniques were identified and broadly classified for use in determining effects on the atmospheric environment, effects on the surface aquatic environment, effects on the subsurface environment (groundwater and soils), effects on the acoustic environment, direct effects on

plants and animals, direct effects on landscape, and accidental effects. Table 2.4.1 contains a summary of the different types of methods used in predictive techniques for each of the seven main effect groups (Environmental Resources Limited 1982). The identified methods can be divided into experimental methods, mathematical models, and survey techniques. Table 2.4.2 displays a systematic grouping of prediction techniques (Environmental Resources Limited 1982).

Experimental methods used for prediction include physical models, field experiments, and laboratory experiments. Physical models include scaled-down representations, in two or three dimensions, of the study area after an activity has been implemented. Field experiments refer to in situ tracer experiments where tracers are used to predict the behavior of releases to surface waters (usually marine) or to groundwater. Laboratory experiments refer to bioassay methods to determine the effect of pollution on a particular species. Standard toxicological methods are used to determine the effect of a pollutant (or mix of pollutants) in water on a species, usually fish. Laboratory experiments are useful where no data exist on the effect of a pollutant on plants or animals. An advantage is that these experiments can be set up to represent the environment in which the effect may occur by using, for example, water from the river to which a pollutant is discharged.

Mathematical models refer to predictive techniques which use mathematical relationships between system variables to describe the way an environmental system reacts to an external influence. Mathematical models can be divided into those models which are empirical or "black-box" models, where the relationships between inputs and outputs are established from analysis of observations in the environment; and those models which are internally descriptive, that is, where the mathematical relationships within the model are based on some understanding of the mechanisms of the processes occurring in the environment.

Survey techniques are based on the identification and quantification of existing or future aspects of the environment that might be affected, in terms of their sensitivity to change or of the importance of their loss or disturbance. The three main groups of survey techniques include inventory techniques, evaluation techniques, and visibility techniques.

Inventory techniques involve determining the distribution of things which may be affected by an activity (receptors) usually because of their proximity to an activity. Evaluation techniques refer to surveys to determine the

TABLE 2.4.1   METHODS USED IN PREDICTIVE TECHNIQUES FOR DIFFERENT EFFECTS (ENVIRONMENTAL RESOURCES LIMITED 1982)

Atmospheric Effects
SOURCES
  Emission factor techniques
  Empirical techniques
EFFECTS ON AIR QUALITY
  Experimental methods
    wind tunnels
    water channels
  Mathematical models
    roll-back models
    dispersion models
      simple box
      Gaussian plume
      K-theory
      long-range transport
      long-term prediction
    empirical models
HIGHER-ORDER EFFECTS
  Mathematical models
    simple dilution models
      for soils and water
    pathway models for
      human exposure
    empirical models
      dose effect
  survey techniques
    inventory technique

Aquatic Effects
SOURCES
  Simple steady-state run-off
    models
  Complex dynamic run-off
    models
  Accidental spills
HYDRAULIC EFFECTS
  Experimental models
    hydraulic models
  Mathematical models
    dynamic models
EFFECTS ON WATER QUALITY
  Experimental methods
    hydraulic models
    in situ tracer experiments
  Mathematical models for
    rivers

Aquatic Effects (Cont'd)
    estuaries
    coastal waters
    lakes
    river-reservoir systems
  simple mixing models
  dissolved oxygen models
  steady-state estuary models
  complex coastal waters
    dispersion models
HIGHER-ORDER EFFECTS
  Experimental models
    bioassay
  Mathematical models
    population, productivity, and
      nutrient cycling models
    partition models
    empirical models
  Survey techniques
    inventory techniques

Subsurface Effects
SOURCES
  Simple leachate flow models
    (Darcy's Law)
  Simple leachate quality models
    Darcy's Law
    empirical
HYDRAULIC EFFECTS
  Experimental models
    field tests
  Mathematical models
    steady-state dispersion models
    complex models
EFFECTS ON GROUNDWATER QUALITY
  Experimental methods
    in situ tracer experiments
  Mathematical models
    steady-state dispersion
    complex models
  Evaluation techniques
EFFECTS ON SOILS
  Mathematical models
    mixing models
    simple steady-state models

    complex models
    empirical models

Effects on Plants and Animals
  Mathematical models
    population, productivity, and
      nutrient cycling models
  Survey techniques
    evaluation techniques
    inventory techniques

Effects on Landscape
  Experimental methods
    still 2-D models
    moving 2-D models
    3-D models
  Mathematical models
    empirical models
  Survey techniques
    evaluation methods
    visibility techniques
    inventory techniques

Acoustic Effects
ACTIVITY
  Mobile sources
    roads
    railways
    airports
  Stationary sources
SOURCES
  Emission models
ACOUSTIC EFFECTS
  Experimental methods
    physical models
  Mathematical models
    steady-state ambient sound
      and noise models
HIGHER-ORDER EFFECTS
  Mathematical models
    empirical annoyance models
  Survey techniques
    inventory techniques
ACCIDENTAL EFFECTS
  Hazard and operability studies
  Event and fault tree analysis
  Consequence modelling

value of the environmental aspect that will be lost or disturbed as the result of an activity, and where possible its change in value after the activity is undertaken. The principle of evaluation techniques is that they place a value on an environmental aspect in some location where it will be affected by the activity. The evaluation is usually made in terms of an index based on several characteristics of the environment which are considered to contribute to or detract from the value of the environmental aspect in question. Visibility techniques form a group which cannot easily be categorized in either of the other types. These techniques are used in landscape assessment for determining the zone from which an activity will be visible.

Current Use of Techniques

Currently the range of impact prediction techniques in the EIA process is broad and encompasses the use of analogies through sophisticated mathematical models. In a specific environmental impact study, several prediction tech-

TABLE 2.4.2 SYSTEMATIC GROUPING OF
PREDICTION TECHNIQUES
(ENVIRONMENTAL RESOURCES
LIMITED 1982)

Experimental Methods
Physical models
 illustrative models
 working models
Field experiments
Laboratory experiments

Mathematical Models
Empirical models
 site-specific empirical models
 generalized empirical models
Internally descriptive models
 emission factor models
 roll-back models
 simple mixing models
 steady-state dispersion models
 complex mathematical models

Survey Techniques
Inventory techniques
Evaluation techniques
Visibility techniques

niques may be required due to the availability of data and specific mathematical models (Stakhiv et al. 1981 and Stakhiv 1986). In addition, as greater attention is given to the global environment and to potential global consequences of large projects or activities, mesoscale environmental consequences must be considered.

SIMPLE TECHNIQUES

Perhaps the simplest approach for impact prediction is to use analogies or comparisons to the experienced effects of existing projects or activities. This approach can be termed a "look-alike" approach in that information gathered from similar types of projects in similar environmental settings can be used to descriptively address the anticipated impacts of a proposed project or activity. Professional judgment is necessary when analogies are used for specific impacts on the environment.

Another approach is the inventory technique. In this approach, the environmental engineer compiles an inventory of environmental resources by assembling existing data or conducting baseline monitoring. With this approach, a presumption is that the resources in the existing environment, or portions of it, will be lost as a result of the proposed project or activity. This technique can be perceived as a worst-case prediction, and for certain types of resources it represents a reasonable approach for use in environmental impact studies. Again, professional judgment is necessary to interpret information related to the existing environment and the potential consequences of a proposed

project or activity. This approach can also be aided by the analogy approach.

An often used approach for impact prediction is to incorporate checklists or interaction matrices as a part of the impact study. Several types of checklists have been developed, ranging from simple listings of anticipated impacts by project type, to questionnaire checklists which incorporate a series of detailed questions and provide structure to the impact prediction activity. Some checklists include the use of scaling, rating, or ranking the impact of alternatives and incorporate relative-importance weights to the environmental factors. These checklists can aggregate the impacts of a project into a final index or score which can be used to compare alternatives. In this context, these checklists are similar to multicriteria decision-making techniques which are used in environmental planning and management.

Interaction matrices include simple x–y matrices that identify impacts and provide a basis for categorization of impact magnitude and importance. In stepped matrices, secondary and tertiary consequences of project actions are delineated. The most sophisticated types of matrices are networks or impact trees in which systematic approaches trace the consequences of a given project or activity. A key point relative to both checklist and matrix methods is that they tend to be qualitative in terms of the predicted impacts; however, they do represent useful tools for impact prediction.

INDICES AND EXPERIMENTAL METHODS

A fourth category of impact prediction approaches in environmental impact studies is environmental indices (Ott 1978). An environmental index is a mathematical or descriptive presentation of information factors used for classifying environmental quality and sensitivity and predicting the impacts of a proposed project or activity. The basic concept for impact prediction is to anticipate and quantify the change in the environmental index as a result of the project or activity, and then to consider the difference in the index from the with- and without-project conditions as a measure of impact. Numerous environmental indices have been developed for air quality, water quality, noise, visual quality, and quality of life (a socioeconomic index which can include a large number of specific factors). One widely used type of index is based on habitat considerations and uses the Habitat Evaluation Procedures developed by the U.S. Fish and Wildlife Service (1980) or a Habitat Evaluation System developed by the U.S. Army Corps of Engineers (1980). This system is based on the development of a numerical index to describe habitat quality and size. A key advantage of index approaches for impact prediction is that these approaches can be related to available information and they provide a systematic basis for considering the consequences of a project or activity.

The fifth category of impact prediction approaches is experimental methods. These methods encompass the conduction of laboratory experiments to develop factors or coefficients for mathematical models, and the conduction of large-scale field experiments to measure changes in environmental features as a result of system perturbations. In addition, physical models are used to examine impacts related to hydrodynamics and ecological changes within microcosms of environmental settings. Experimental methods are primarily useful in dealing with physical–chemical components and biological features of the environmental setting.

MATHEMATICAL MODELS

The most sophisticated approach for impact prediction involves the use of mathematical models. Numerous types of mathematical models account for pollutant transport and fate within the environmental setting. Other models describe environmental features and the functioning of ecosystems. This review does not delineate the state-of-the-art of mathematical modeling with regard to environmental impact studies but discusses the availability of types of models which can be used in studies.

Stakhiv (1986) discusses several types of models used for forecasting in water resources planning. He notes the availability of predictive deterministic models for forecasting, including models addressing demographic, socioeconomic, and economic changes; and models for ecological, water quality, energy, hydraulics, hydrology, and land-use changes.

With regard to air quality dispersion, numerous models address point, line, and area sources of air pollution and the results of dispersion from these sources. In addition, within recent years, models are available for long-range transport of pollution and for atmospheric reactions leading to photochemical smog formation and acid rain. Many air quality models are available in software form for personal computers; they represent a usable technology for many studies.

Anderson and Burt (1985) note the following about hydrologic modeling:

> All models seek to simplify the complexity of the real world by selectively exaggerating the fundamental aspects of a system at the expense of incidental detail. In presenting an approximate view of reality, a model must remain simple enough to understand and use, yet complex enough to be representative of the system being studied.

Anderson and Burt (1985) classify hydrological models into three types:

1. Black-box models: These models contain no physically based transfer function to relate input to output; they depend upon establishing a statistical correspondence between input and output.

2. Conceptual models: These models occupy an intermediate position between the deterministic approach and empirical black-box analysis. They are formulated on the basis of a simple arrangement of a small number of components, each of which is a simplified representation of one process element in the system being modeled; each element of the model consists of a nonlinear reservoir in which the relationship between outflow (Q) and storage (S) is given by

$$S = K \cdot Q^n \qquad 2.4(1)$$

where K and n represent constants.

3. Deterministic models: These models are based on complex physical theory. However, despite the simplifying assumptions to solve the flow equations, these models have huge demands in terms of computational time and data requirements and are therefore costly to develop and operate.

Figure 2.4.1 shows a generic method for selecting a mathematical hydrological model (Anderson and Burt 1985). The method emphasizes the dependence of the

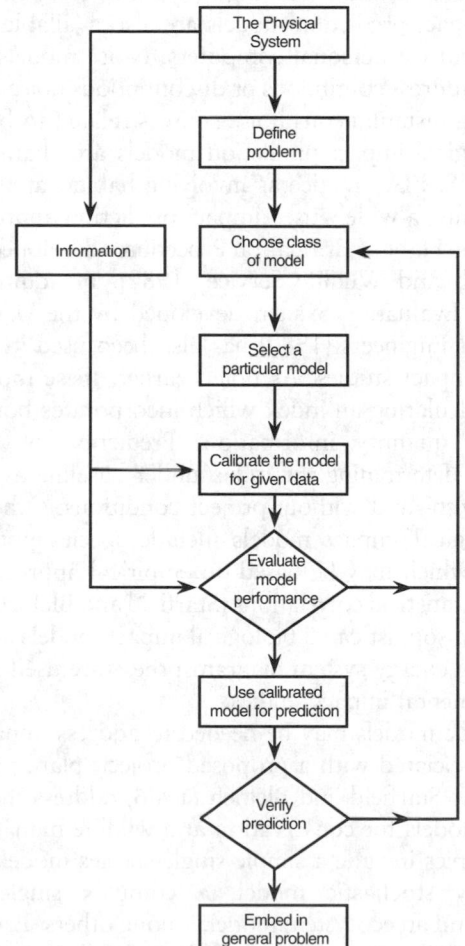

FIG. 2.4.1 Method for selecting a mathematical model (Anderson and Burt 1985).

modeling upon a clear definition of the problem to be solved, and upon the data base to describe the physical system.

Surface and groundwater quality and quantity models are also plentiful, with major research developments within the last decade occurring in solute transport in subsurface systems. Surface water quality and quantity models range from one-dimensional steady-state models to three-dimensional dynamic models which can be used for rivers, lakes, and estuarine systems (Anderson and Burt 1985). Groundwater flow models now include subsurface processes, such as adsorption and biological decomposition. The International Ground Water Modeling Center, along with the National Water Well Association, has hundreds of groundwater models in software form that address specific environmental consequences of projects or activities.

Noise impact prediction models are available for point, line, and area sources of noise generation. These models range in complexity from simple calculations involving the use of nomographs, to sophisticated computer modeling for airport operations. The technology for noise impact prediction is well developed as a result of numerous research studies related to highways and airports. Many noise impact prediction models are also available in software form for personal computers. Noise models can be used to address continuous or discontinuous noise sources, including instantaneous noise sources related to blasting.

Biological impact prediction models are characterized in the U.S. EIA practice as involving habitat approaches. Specifically, a widely used impact prediction approach involves the Habitat Evaluation Procedures developed by the U.S. Fish and Wildlife Service (1980). In addition, the Habitat Evaluation System developed by the U.S. Army Corps of Engineers (1980) has also been used in a number of impact studies. As noted earlier, these models involve calculating an index which incorporates both quality and quantity information. Prediction of impacts involves determining the index under baseline as well as future with- and without-project conditions. Other types of biological impact models include species population models which may be based on empirical approaches involving statistical correlations (Starfield and Bleloch 1986). The most sophisticated biological impact models are those involving energy system diagrams; these are used in some environmental impact studies.

Ad hoc models may be needed to address impact concerns associated with a proposed project, plan, program, or policy. Starfield and Bleloch (1986) address the building of models for conservation and wildlife management. Their topics include a simple single-species model, an exploratory stochastic model, a complex single-species model, and an ecosystem model, among others. Each topic begins with a management problem and describes how a model can be constructed to address that problem. They also describe modifications to the model in light of the

available (or unavailable) data, how the model can be exercised, and what can be learned from it. They presume that the problem that needs to be modeled is often poorly defined; the processes and mechanisms are not well understood, and the data are often scant and difficult to obtain. Their work also includes information on the development of expert systems, as well as for resource management.

Model building for biological systems is also addressed by Armour and Williamson (1988) in their procedure for organizing and simplifying complex information into a cause-and-effect model through an interdisciplinary exercise. Information includes prerequisites to help ensure model completion, applications of model information, diagnosing and correcting modeling problems when users encounter difficulties, and technical limitations of the approach.

Predictive modeling is also possible for ascertaining the potential for archaeological resources in geographical study areas. Such modeling is primarily based upon evaluating factors that indicate the likelihood of archaeological resources being found, relating the factors to existing information, evaluating the likelihood for early occupations in the area, and other environmental and sociological factors. This type of modeling is often used to determine the need for planning and conducting archaeological field surveys.

Visual quality is also a subject in selected impact studies. Visual impact modeling approaches have been developed by several federal agencies, including the U.S. Forest Service, U.S. Bureau of Land Management, U.S. Soil Conservation Service, and U.S. Army Corps of Engineers (Smardon, Palmer, and Felleman 1986). These visual impact models typically involve evaluating a series of factors, in some cases quantitatively and in other cases descriptively, and assembling the information into an overall visual quality index for the study area. In this context, these models are similar to the environmental indices approach described earlier.

Impact prediction related to the socioeconomic environment is often associated with the use of human population and econometric models. Population forecasting can range from simple projections of historical trends to complicated cohort analysis models. Econometric models relate the population and economic characteristics of study areas so that interrelationships can be depicted between population changes and changes in economic features within given study areas. In addition to the econometric models and models related to population change, other impact predictions for the socioeconomic environment are addressed by the use of multiplier factors applied to population changes. In this regard, several input-output models can be used in environmental impact studies.

Table 2.4.3 provides a noninclusive listing of state-of-the-art books or reports related to quantitative models useful for impact forecasting. A common ingredient in the ap-

TABLE 2.4.3 SELECTED KEY REFERENCES RELATED TO IMPACT FORECASTING

Topical Area of Impact	Key Reference(s)
Air quality	Zanetti 1990
Surface water quantity	Anderson and Burt 1985
Surface water quality	Henderson-Sellers 1991
	James 1993
	U.S. Army Corps of Engineers 1987
Groundwater quantity	Domenico and Schwartz 1990
Groundwater quality	Water Science and Technology Board 1990
Noise	Magrab 1975
	World Health Organization 1986
Terrestrial or aquatic habitat	U.S. Army Corps of Engineers 1980
Terrestrial or aquatic species	U.S. Fish and Wildlife Service 1980
Cultural resources	King 1978
Visual quality	Smardon, Palmer, and Felleman 1986
Socioeconomic impacts	Canter, Atkinson, and Leistritz 1985
Health impacts	Turnbull 1992

proaches for impact prediction is that decisions for a project or activity must use the best available predictive technology in view of the location, size, and type of project or activity, as well as the budget available for the environmental impact study. In that regard, sophisticated mathematical models may not be used due to the need for extensive data input and model calibration. Accordingly, a range of approaches is usually necessary in conducting an environmental impact study for a project or activity.

—Larry W. Canter

References

Anderson, M.G., and T.P. Burt. 1985. Modelling strategies. Chap. 1 in *Hydrological Forecasting*. Edited by M.G. Anderson and T.P. Burt. New York: John Wiley and Sons, Ltd. 1–13.

Armour, C.L., and S.C. Williamson. 1988. Guidance for modeling causes and effects in environmental problem solving. *Biological—89(4)* October. Ft. Collins, Colorado: National Ecology Research Center, U.S. Fish and Wildlife Service.

Canter, L.W., S.F. Atkinson, and F.L. Leistritz. 1985. *Impact of growth.* Chelsea, Michigan: Lewis Publishers, Inc.

Domenico, P.A., and F.W. Schwartz. 1990. *Physical and Chemical Hydrogeology,* New York: John Wiley and Sons, Inc.

Environmental Resources Limited. 1982. *Environmental impact assessment—Techniques for predicting effects in EIA.* Vol. 2, February. London, England.

Henderson-Sellers, B. 1991. *Decision support techniques for lakes and reservoirs.* Vol. 4 of *Water quality modeling.* Boca Raton, Florida: CRC Press.

James, A., ed. 1993. *An introduction to water quality modeling,* West Sussex, England: John Wiley and Sons, Ltd.

King, T.F. 1978. *The archaeological survey: methods and uses.* Washington, D.C.: Heritage Conservation and Recreation Service, U.S. Department of the Interior.

Magrab, E.B. 1975. *Environmental Noise Control.* New York: John Wiley and Sons.

Ott, W.R. 1978. *Environmental indices: Theory and practice.* Ann Arbor, Michigan: Ann Arbor Science Publishers, Inc.

Smardon, R.C., J.F. Palmer, and J.P. Felleman. 1986. *Foundations for visual project analysis.* New York: John Wiley and Sons, Inc.

Stakhiv, E., et al. 1981. Executive summary. Pt. 1 of *Appraisal of selected Corps preauthorization reports for environmental quality planning and evaluation.* Report No. CERL-TR-N-118, September. Champaign, Illinois: U.S. Army Construction Engineering Research Laboratory.

Stakhiv, E.Z. 1986. Achieving social and environmental objectives in water resources planning: Theory and practice. In *Proceedings of an engineering foundation conference on social and environmental objectives in water resources planning and management,* May:107–125. New York: American Society of Civil Engineers.

Starfield, A.M., and A.L. Bleloch. 1986. *Building models for conservation and wildlife management.* New York: Macmillan Publishing Company.

Turnbull, R.G., ed. 1992. *Environmental and health impact assessment of development projects.* London: Elsevier Science Publishers, Ltd.

U.S. Army Corps of Engineers. 1980. *A habitat evaluation system for water resources planning.* August. Vicksburg, Mississippi: Lower Mississippi Valley Division.

———. 1987. *Water quality models used by the Corps of Engineers.* Information Exchange Bulletin, vol. E-87-1, March. Vicksburg, Mississippi: Waterways Experiment Station.

U.S. Fish and Wildlife Service. 1980. *Habitat evaluation procedures (HEP).* ESM 102, March. Washington, D.C.

Water Resources Council. 1983. Economic and environmental guidelines for water and related land resources implementation studies. Chap. 3 in *Environmental quality (EQ) procedures.* 10 March. Washington, D.C.

Water Science and Technology Board. 1990. *Ground water models—Scientific and regulatory applications.* Washington, D.C.: National Academy Press.

World Health Organization. 1986. *Assessment of noise impact on the urban environment.* Environmental Health Series no. 9. Copenhagen, Denmark: Regional Office for Europe.

Zanetti, P. 1990. *Air pollution modeling—Theories, computational methods, and available software.* New York: Van Nostrand Reinhold.

2.5
DECISION-FOCUSED CHECKLISTS

Environmental impact studies typically address a minimum of two alternatives, and sometimes as many as ten, but usually three to five alternatives. The minimum number typically represents a choice between construction and operation of a project versus project nonapproval. The alternatives can encompass a range of considerations. Typical categories of alternatives, expressed generically, include:

1. Site location alternatives
2. Design alternatives for a site
3. Construction, operation, and decommissioning alternatives for a design
4. Project size alternatives
5. Phasing alternatives for size groupings
6. No-project or no-action alternatives
7. Timing alternatives relative to project construction, operation, and decommissioning

Decision-focused checklists are systematic methods for comparing and evaluating alternatives. Scaling, rating, or ranking-weighting checklists can be used in such comparisons and evaluations. In scaling checklists, an algebraic scale or letter scale is assigned to the impact of each alternative on each environmental factor. In ranking checklists, alternatives are ranked from best to worst in terms of their potential impacts on identified environmental factors, while rating uses a predefined rating approach. These checklists are useful for comparative evaluations of alternatives, thus they provide a basis for selecting the preferred alternative.

In weighting-scaling checklists, relative importance weights are assigned to environmental factors and impact scales are determined for each alternative relative to each factor. Weighting-ranking checklists involve importance weight assignments and the relative ranking of the alternatives from best to worst in terms of their impacts on each environmental factor. Numerous weighting-scaling and weighting-ranking checklists are available for environmental impact studies. These methods represent adaptations of routinely used multicriteria or multiattribute decision-making techniques; such techniques are also called decision-analysis techniques.

Conceptual Basis for Tradeoff Analysis

To achieve systematic decision making among alternatives, tradeoff analysis should be used. Tradeoff analysis involves the comparison of a set of alternatives relative to a series of decision factors. Petersen (1984) notes that in a tradeoff analysis, the contributions of alternative plans are compared to determine what is gained or foregone in choosing one alternative over another. Table 2.5.1 displays a tradeoff matrix for systematically comparing the groups of alternatives or specific alternatives within a group relative to a series of decision factors (Canter, Atkinson, and Leistritz 1985).

The following approaches can be used to complete the tradeoff matrix in Table 2.5.1:

1. Qualitative approach: Descriptive, synthesized, and integrated information on each alternative relative to each decision factor is presented in the matrix.
2. Quantitative approach: Quantitative, synthesized, and integrated information on each alternative relative to each decision factor is displayed in the matrix.
3. Ranking, rating, or scaling approach: The qualitative or quantitative information on each alternative is summarized via the assignment of a rank, rating, or scale value relative to each decision factor (the rank or rating or scale value is presented in the matrix).
4. Weighting approach: The importance weight of each decision factor relative to each other decision factor is considered, with the resultant discussion of the information on each alternative (qualitative; quantitative; or ranking, rating, or scaling) being presented in view of the relative importance of the decision factors.
5. Weighting-ranking, -rating, or -scaling approach: The importance weight for each decision factor is multiplied

TABLE 2.5.1   TRADEOFF ANALYSIS FOR DECISION MAKING (CANTER, ATKINSON, AND LEISTRITZ 1985)

	Alternatives				
Decision Factors	*1*	*2*	*3*	*4*	*5*
Degree of Meeting Needs and Objectives					
Economic Efficiency					
Social Concerns (public preference)					
Environmental Impacts Biophysical Cultural Socioeconomic (includes health)					

by the ranking, rating, or scaling of each alternative, then the resulting products for each alternative are summed to develop an overall composite index or score for each alternative; the index may take the form of:

$$\text{Index } j = \sum_i^n (IW)_i \, (R)_{ij} \qquad 2.5(1)$$

where:

$\text{Index } j$ = the composited index for the jth alternative
n = number of decision factors
IW_i = importance weight of ith decision factor
R_{ij} = ranking, rating, or scaling of jth alternative for ith decision factor

Decision making which involves the comparison of a set of alternatives relative to a series of decision factors is not unique to considering environmental impacts. This decision-making problem is classic and is often referred to as multiattribute or multicriteria decision making, or decision analysis.

Importance Weighting for Decision Factors

If the qualitative or quantitative approach is used to complete the matrix as shown in Table 2.5.1, information for this approach related to the environmental impacts can be based on impact prediction. This information is also needed for the approaches involving importance weighting and ranking, rating, or scaling. If the importance weighting approach is used, the assignment of importance weights to decision factors, or at least the arrangement of them in a rank ordering of importance, is critical. Table 2.5.2 lists some structured importance weighting or ranking techniques which can be used to achieve this step. These techniques have been used in numerous environmental decision-making projects. Brief descriptions of several techniques from Table 2.5.2 are illustrated. In addition to the structured techniques, less formal approaches, such as reliance on scoping, can be used as a basis for importance weighting.

Ranking techniques for importance weighting basically involve the rank ordering of decision factors in their rela-

TABLE 2.5.2 EXAMPLES OF IMPORTANCE WEIGHTING TECHNIQUES USED IN ENVIRONMENTAL STUDIES

Ranking
Nominal group process
Rating
Predefined importance scale
Multiattribute utility measurement
Unranked paired comparison
Ranked paired comparison
Delphi

tive order of importance. If n decision factors exist, rank ordering involves assigning 1 to the most important factor, 2 to the second-most important factor, and so forth, until n is assigned to the least important factor. The rank order numbers can be reversed; that is, n can be assigned to the most important factor, $n - 1$ to the second-most important factor, and so forth, until 1 is assigned to the least important factor. The Nominal Group Process (NGP) technique (Voelker 1977) illustrates a ranking technique and is described next.

The NGP technique is an interactive group technique and was developed in 1968 (Voelker 1977). It was derived from social-psychological studies of decision conferences, management-science studies of aggregating group judgments, and social work studies of problems surrounding citizen participation in program planning. The NGP technique is widely used in health, social service, education, industry, and government organizations. For example, Voelker (1977) describes use of the NGP technique to rank decision factors in siting nuclear power plants. The following four steps are involved in the NGP technique for importance weighting:

1. Nominal (silent and independent) generation of ideas in writing by a panel of participants
2. Round-robin listing of ideas generated by participants on a flip chart in a serial discussion
3. Group discussion of each recorded idea for clarification and evaluation
4. Independent voting on priority ideas, with mathematical rank ordering determining the group decision

Rating techniques for importance weighting basically involve assigning importance numbers to decision factors and sometimes their subsequent normalization via a mathematical procedure. Two examples of rating techniques are the use of a predefined importance scale (Linstone and Turoff 1978) and the use of the multi-attribute utility measurement (MAUM) technique (Edwards 1976). The NGP technique can also be used for rating the importance of decision factors. Decision factors can be assigned numerical values based on predefined importance scales. Table 2.5.3 delineates five scales with definitions to consider in assigning numerical values to decision factors (Linstone and Turoff 1978). Use of the predefined scales can aid in systematizing importance weight assignments.

Paired comparison techniques (unranked and ranked) for importance weighting involve comparisons between decision factors and a systematic tabulation of the numerical results of the comparisons. These techniques are used extensively in decision-making efforts, including numerous examples related to environmental impact studies. One of the most useful techniques is the unranked paired comparison technique developed by Dean and Nishry (1965). This technique, which can be used by an individual or group, compares each decision factor to each other decision factor in a systematic manner.

TABLE 2.5.3 EXAMPLE OF IMPORTANCE SCALE (LINSTONE AND TUROFF 1978)

Scale Reference	Definitions
1. Very important	A most relevant point First order priority Has direct bearing on major issues Must be resolved, dealt with, or treated
2. Important	Is relevant to the issue Second order priority Significant impact, but not until other items are treated Does not have to be fully resolved
3. Moderately important	May be relevant to the issue Third order priority May have impact May be a determining factor to major issue
4. Unimportant	Insignificantly relevant Low priority Has little impact Not a determining factor to major issue
5. Most unimportant	No priority No relevance No measurable effect Should be dropped as an item to consider

Scaling, Rating, or Ranking of Alternatives

Scaling, rating, or ranking of each alternative for each decision factor is the second major aspect in using the multicriteria decision-making approach. Rating and ranking concepts are described in the previous subsection on importance weighting. Several techniques can be used for this evaluation of alternatives in a decision. Examples of techniques include the use of the alternative profile concept, a reference alternative, linear scaling based on the maximum change, letter or number assignments designating impact categories, evaluation guidelines, unranked paired comparison techniques, and functional curves.

Bishop et al. (1970) discuss the alternative profile concept for impact scaling. This concept is represented by a graphic presentation of the effects of each alternative relative to each decision factor. Each profile scale is expressed on a percentage basis, ranging from a negative to a positive 100%, with 100% being the maximum absolute value of the impact measure adopted for each decision factor. The impact measure represents the maximum change, either plus or minus, associated with an alternative being evaluated. If the decision factors are displayed along with the impact scale from +100% to −100%, a dotted line can be used to connect the plotted points for each alternative and thus describe its profile. The alternative profile concept is useful for visually displaying the relative impacts of a series of alternatives.

Salomon (1974) describes a scaling technique for evaluating cooling system alternatives for nuclear power plants. To determine scale values, his technique uses a reference cooling system, and each alternative system is compared with it. He assigned the following scale values to the alternatives based on the reference alternative: very superior (+8), superior (+4), moderately superior (+2), marginally superior (+1), no difference (0), marginally inferior (−1), moderately inferior (−2), inferior (−4), and very inferior (−8).

Odum et al. (1971) discuss a scaling technique in which the actual measures of the decision factor for each alternative plan are normalized and expressed as a decimal of the largest measure for that factor. This technique represents linear scaling based on the maximum change.

A letter scaling system is described by Voorhees and Associates (1975). This method incorporates eighty environmental factors oriented to the types of projects conducted by the U.S. Department of Housing and Urban Development. The scaling system assigns a letter grade from A+ to C− for the impacts, with A+ representing a major beneficial impact and C− an undesirable detrimental change.

Duke et al. (1977) describe a scaling checklist for the environmental quality (EQ) account for water resource projects. Scaling follows establishing an evaluation guideline for each environmental factor. An evaluation guideline is defined as the smallest change in the highest existing quality in the region that is considered significant. For example, assuming that the highest existing quality for dissolved oxygen in a region is 8 mg/l, if a reduction of 1.5 mg/l is considered as significant, then the evaluation guideline is 1.5 mg/l, irrespective of the existing quality in a given regional stream. Scaling is accomplished by quantifying the impact of each alternative relative to each environmental factor, and if the net change is less than the evaluation guideline, it is insignificant. If the net change is greater and moves the environmental factor toward its highest quality, then it is considered a beneficial impact; the reverse exists for those impacts that move the measure of the environmental factor away from its highest existing quality.

One of the most useful techniques for scaling, rating, or ranking alternatives relative to each decision factor is the unranked paired comparison technique described by Dean and Nishry (1965). This technique was mentioned earlier relative to its use for importance weighting of decision factors. Again, this technique can be used by an individual or group for the scaling, rating, or ranking of alternatives.

Functional curves, also called value functions and parameter function graphs or curves, can also be used in environmental impact studies for scaling, rating, or ranking the impacts of alternatives relative to decision factors. Figure 2.5.1 shows an example of a functional curve for species diversity (Dee et al. 1972). Dee et al. (1972) describe the following seven steps used in developing a functional curve (relationship) for an environmental parameter (decision factor):

Step 1. Obtain scientific information on the relationship between the parameter and the quality of the environment. Also, obtain experts in the field to develop the value functions.

Step 2. Order the parameter scale so that the lowest value of the parameter is zero and it increases in the positive direction—no negative values.

Step 3. Divide the quality scale (0–1) into equal intervals, and express the relationship between an interval and the parameter. Continue this procedure until a curve exists.

Step 4. Average the curves over all experts in the experiment to obtain a group curve. (For parameters based solely on judgment, determine value functions by a representative population cross section.)

Step 5. Indicate to the experts estimating the value function the group curve and expected results of using the curves. Decide if a modification is needed; if needed, go to Step 3; if not, continue.

Step 6. Do Steps 1 through 5 until a curve exists for all parameters.

Step 7. Repeat the experiment with the same group or another group to increase the reliability of the functions.

Development of a Decision Matrix

The final step in multicriteria decision making is to develop a decision matrix displaying the products of the importance weights (or ranks) and the alternative scales (or ranks). Table 2.5.4 shows a simple weighting-rating checklist used in an environmental impact study of sites for a wastewater treatment plant (Wilson 1980). Two groups of importance weights are used, with factors assigned an importance weight of 2 being more important than factors assigned an importance weight of 1. Each of the three sites in the study is rated on a scale from 1 to 3, with 1 denoting an undesirable site and 3 a desirable site relative to thirteen decision factors. The best overall site based on a composite evaluation is Site C.

Examples of Decision-Focused Checklists

The Battelle Environmental Evaluation System (EES) is a weighting-scaling checklist developed for the U.S. Bureau of Reclamation. It contains seventy-eight environmental factors organized into seventeen components and four categories as shown in Figure 2.5.2 (Dee et al. 1972). An interdisciplinary team assigns importance weights to each of the categories, components, and factors based on use of the ranked paired comparison technique. Impact scaling in the Battelle EES uses functional relationships for each of the seventy-eight factors (Dee et al. 1972).

The basic concept of the Battelle EES is that an index expressed in environmental impact units (EIUs) can be developed for each alternative and baseline environmental condition. Mathematical formulation of this index is as follows:

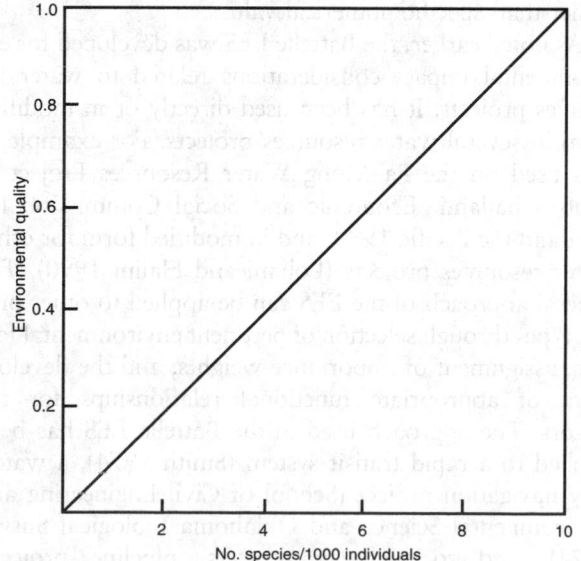

FIG. 2.5.1 Functional curve for species diversity (Dee et al. 1972).

$$EIU_j = \sum_{i=1}^{n} (EQ)_{ij} (PIU)_i \qquad 2.5(2)$$

TABLE 2.5.4 WEIGHTING-RATING CHECKLIST FOR WASTEWATER TREATMENT PLANT SITE EVALUATION (WILSON 1980)

Factors to be considered	Significance of factor	Importance to decision (2 = most)	Site A	Site B	Site C
				Rating scale: 1 = worst; 3 = best	
1. Construction cost	One-time cost with federal share	1	2	2.5	3
2. Operating cost	Ongoing cost, includes energy costs (all local share)	2	1	2	3
3. Nonpotable reuse	Safe, economic benefit; key is proximity to users	2	1	2	3
4. Potable reuse	More long-range than item 3 (above); best sites are near well field or water plant; industrial sewage should not be present	1	3	2.5	1
5. Odor potentials	Assumes good plant design and operation	2	2	1	3
6. Other land use conflicts	Potential to interfere with agricultural/residential land	1	2	1	3
7. Availability of site area	Future expansion capability, flexibility	2	3	2	1
8. Relationship to growth area	Assumes growth to state line (in fifty-year time frame)	1	3	2	1
9. Construction impacts	Reworking of lines	1	1	2	3
10. Health of workers	Air pollution	1	3	3	2
11. Implementation capability	Land acquisition problems	1	3	1	2
12. Operability	One plant is better than two	1	1	2	3
13. Performance reliability	Assumes equal treatment plants	1	2	2	2
14. TOTAL	Rating times Importance, totaled Highest number = best site		34	32	40

where:

EIU_j = environmental impact units for jth alternative
EQ_{ij} = environmental quality scale value for ith factor and jth alternative
PIU_i = parameter importance units for ith factor

Use of the Battelle EES consists of obtaining baseline data on the seventy-eight environmental factors and, through use of their functional relationships, converting the data into EQ scale values. These scale values are then multiplied by the appropriate PIUs and aggregated to obtain a composite EIU score for the baseline setting. For each alternative being evaluated, the anticipated changes in the seventy-eight factors must be predicted. The predicted factor measurements are converted into EQ scale values using the appropriate functional relationships. The EQ scale values are then multiplied by the PIUs and aggregated to arrive at a composite EIU score for each alternative. This numerical system displays tradeoffs between the alternatives in terms of specific environmental factors, intermediate components, and categories. Profes-

sional judgment is necessary in the interpretation of the numerical results, with the focus on comparative analyses, rather than specific numerical values.

As noted earlier, the Battelle EES was developed for environmental impact considerations related to water resources projects. It has been used directly or in modified form in several water resources projects. For example, it was used on the Pa Mong Water Resources Project in South Thailand (Economic and Social Commission for Asia and the Pacific 1990) and in modified form for other water resources projects (Lohani and Halim 1990). The general approach of the EES can be applied to other project types through selection of pertinent environmental factors, assignment of importance weights, and the development of appropriate functional relationships for the factors. The approach used in the Battelle EES has been applied to a rapid transit system (Smith 1974), a waterway navigation project (School of Civil Engineering and Environmental Science and Oklahoma Biological Survey 1974), and to highway projects, pipeline projects, channel improvement projects, and wastewater treatment plants (Dee et al. 1973).

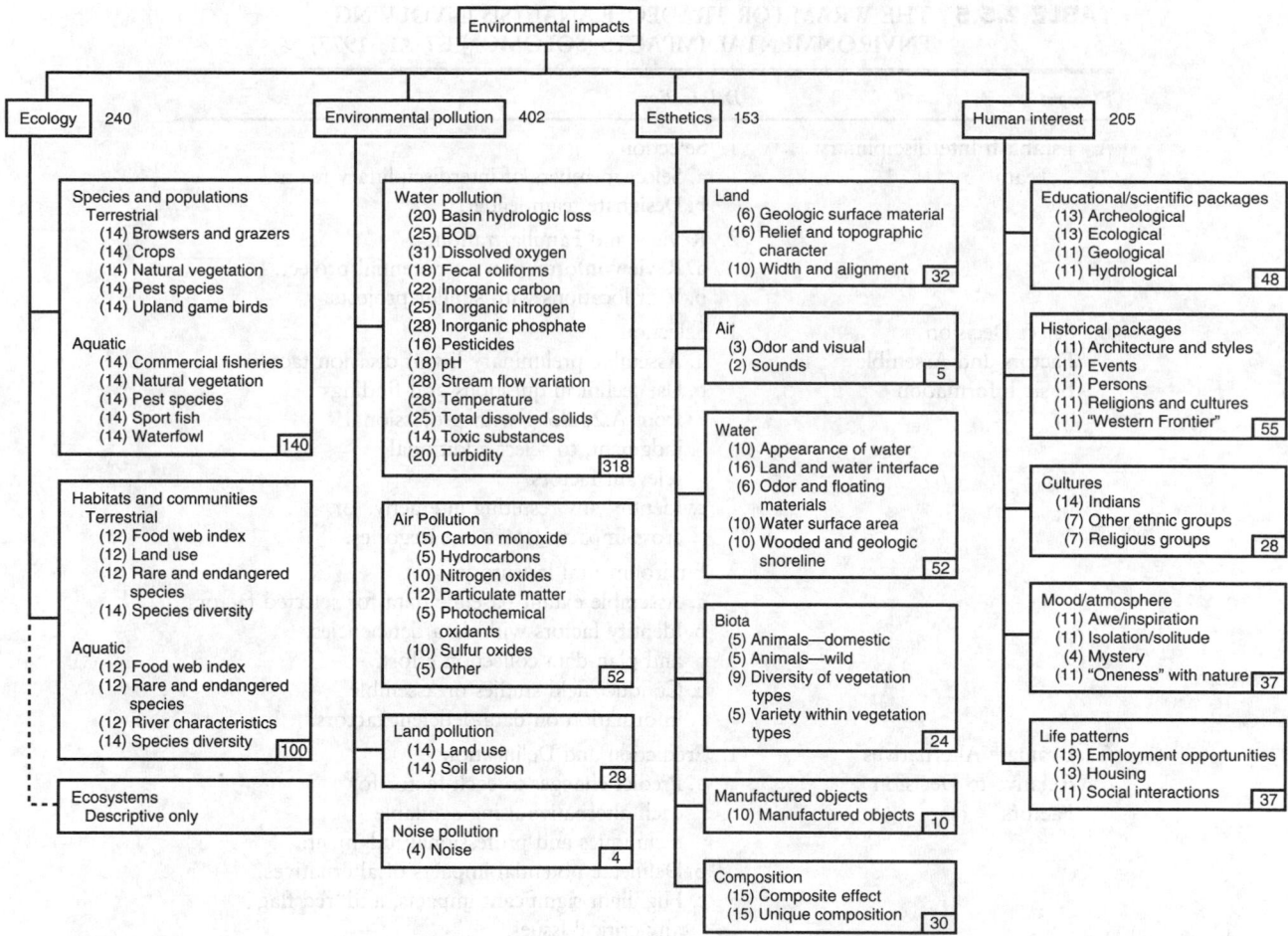

FIG. 2.5.2 Battelle environmental evaluation system. Numbers in parentheses are parameter importance units. Numbers enclosed in boxes represent the total (Dee et al. 1972).

Another example of a weighting-scaling checklist for water resources projects is the Water Resources Assessment Methodology (WRAM) developed by the U.S. Army Corps of Engineers (Solomon et al. 1977). Table 2.5.5 lists the details of the methodology. Key elements include the selection of an interdisciplinary team; selection and inventory of assessment variables (environmental factors); impact prediction, assessment, and evaluation; and documentation of the results. Impact prediction, assessment, and evaluation is the element that includes weighting and scaling. The weighted ranking technique (unranked paired comparison technique) is used to determine the relative factor importance coefficient (FIC) for each assessment variable. Importance weight assignments are required for each study and should reflect the importance of the variables in the given geographical location. Impact scaling in the WRAM uses functional graphs (relationships), linear proportioning, or the development of alternative choice coefficients (ACCs).

In the late 1970s and early 1980s the concept of commensuration was introduced into the water resources planning vocabulary. Commensuration refers to measuring different things by a single standard or measure (Lord, Deane,

and Waterstone 1979). In essence, commensuration develops common units of measurement of various plans, with these units serving as the basis for tradeoff analysis among the plans. Lord, Deane, and Waterstone noted that the four components of commensuration are:

1. To identify the factors which are commensurable
2. To determine whose value judgments about those factors are to be considered
3. To discover what those value judgments are
4. To combine the judgments of all selected individuals into a single collective set of judgments

As a result of the emphasis on commensuration, several water resources methods were developed; the methods are basically weighting-scaling, -rating, or -ranking checklists. Four examples of these methods are discussed. Brown, Quinn, and Hammond (1980) address impact scaling for alternative water development plans, with particular emphasis on environmental and social impacts. Mumpower and Bollacker (1981) developed the Evaluation and Sensitivity Analysis Program (ESAP), which is a computerized environmental planning technique to evaluate alternative water resource management plans.

TABLE 2.5.5 THE WRAM FOR TRADEOFF ANALYSIS INVOLVING
ENVIRONMENTAL IMPACTS (SOLOMON ET AL. 1977)

Element	*Delineation*

A. Establish Interdisciplinary Team	1. Selection
	  a. Select members of interdisciplinary team.
	  b. Designate team leader.
	2. Review and Familiarization
	  a. Review information on potential project.
	  b. Visit locations with similar projects.

B. Select Decision Factors and Assemble Basic Information	1. Selection
	  a. Assemble preliminary list of decision factors.
	  b. Use technical questions and findings from A.2, along with professional judgment, to select additional relevant factors.
	  c. Identify any resulting interactive or cross-impact factors or categories.
	2. Environmental Inventory
	  a. Assemble extant baseline data for selected factors.
	  b. Identify factors with data deficiencies, and plan data collection effort.
	  c. Conduct field studies or assemble information on data-deficient factors.

C. Evaluate Alternatives Relative to Decision Factors	1. Prediction and Delineation
	  a. Predict changes in each factor for each alternative using available techniques and professional judgment.
	  b. Delineate potential impacts of alternatives.
	  c. Highlight significant impacts, and "red flag" any critical issues.
	2. Weighting and Scaling
	  a. Use unranked paired comparison technique, or some other importance weighting technique, to determine importance coefficients for each factor (FIC).
	  b. Scale, rate, or rank predicted impacts through development of alternative choice coefficients, or use of some other technique for evaluation of alternatives relative to decision factors (ACC).
	3. Evaluation and Interpretation of Results
	  a. Multiply FIC by ACC to obtain final coefficient matrix. Sum coefficient values for each alternative.
	  b. Use values in final coefficient matrix as basis for description of impacts of alternatives and tradeoffs between alternatives.
	  c. Discuss any critical issues and predicted impacts.

D. Document Results	1. Rationale
	  a. Describe rationale for selection of decision factors.
	  b. Describe procedure for impact identification and prediction, and rationale for weighting; scaling, rating, or ranking; and interpreting results.
	2. Reference Sources of Information

Anderson (1981) also developed a multiple-objective, multiple-publics method for evaluating alternatives in water resources planning called the cascaded tradeoffs method. This method prepares an overall ranking of planning alternatives on the basis of public values. The key feature of the method is that it provides for tradeoffs across both issue dimensions (decision factors) and publics. Finally in terms of commensuration approaches, Brown and Valenti (1983) developed the Multiattribute Tradeoff System (MATS). MATS is a computer program that helps planners evaluate multiattribute alternatives to determine each alternative's relative worth or desirability. MATS leads the user through a series of questions (a tradeoff analysis) which focuses on the relative importance of various characteristics of the alternatives. The program documents the judgments which lead to developing a policy for evaluating alternatives. Importance weighting in MATS involves use of the unranked paired comparison technique, described earlier. Impact scaling is based on functional curves, also described earlier.

Current Trends in Decision-Making Tools

A recent trend in decision making in environmental studies is the use of computer software. For example, Torno et al. (1988) developed a training manual to evaluate the environmental impacts of large-scale water resources development projects. Enough information is provided to enable the knowledgeable user to evaluate any water body of interest. The training primarily uses a multiobjective, multicriteria decision analysis approach. An interactive computer program simplifies application of the method described in the training manual and serves as a valuable learning aid.

A decision-support system computer model, called d-SSYS, can help determine the relative weights of evaluation parameters used to evaluate projects and the utility function for each of the attributes (Klee 1988). A unique feature of this model is that it incorporates three types of uncertainties: (1) those dealing with the factor (parameter) weights; (2) those dealing with the worth of each project with regard to each factor; and (3) those dealing with the utilities of the attributes. This model is applicable to any problem of competing alternatives.

Summary Observations on Decision-Focused Checklists

Weighting-scaling, weighting-ranking, or weighting-rating checklists are valuable for displaying tradeoffs between alternatives and their associated environmental impacts; thus they are useful in selecting a proposed action. Computerization of these decision-focused checklists is a current trend. The following observations pertain to this category of EIA methods:

1. Several approaches are available for assigning importance weights and for achieving scaling, rating, or ranking. These approaches have relative features and can be considered in choosing the approaches to be used in a study.
2. The process used for importance weighting of the individual decision factors and the rationale used in determining the relative importance weights of individual factors must be described. The rationale used for the scaling, rating, or ranking of individual alternatives relative to each decision factor must also be described. In fact, the description of the rationale is probably more important than the final numerical scores or classifications which result from them.
3. The most debatable point relative to weighting-scaling, -rating, or -ranking checklists is the assignment of importance weights to individual decision factors. Several approaches for importance weighting involve the use of public participation. Where this participation can be incorporated, it should legitimize the overall decision-making process.
4. The use of these checklists can structure the decision process in comparing alternatives and selecting the one to become the proposed action. The process of using these checklists can structure the decision process and provide a tradeoff basis for comparisons and evaluations of alternatives.
5. Due to the similarities between these decision-making approaches for environmental studies and other types of decision-making approaches, a wide range of computer software has become available within recent years to aid the process. This computer software is typically user friendly and can guide the assignment of importance weights and the scaling, rating, or ranking of alternatives. The software can then be used to calculate final index scores for each alternative. In addition, due to the availability of computer software, sensitivity analyses of the overall decision process can be easily conducted by examining the influence of changes in importance weights as well as impact scaling, rating, or ranking assignments to individual alternatives. Use of software for sensitivity analysis can indicate the relative sensitivity of the scores to individual changes.
6. The weighting-scaling, -rating, or -ranking checklist must be kept simple to facilitate the decision-making process. Additional alternatives and decision factors do not necessarily indicate that a better overall decision will be made.
7. These types of checklists can be used at several points in overall project planning and decision making. For example, they can be used early in a study to reduce the number of alternatives to allow a more detailed analysis of a smaller number of alternatives. In addi-

tion, using this process can reduce the number of decision factors so that in the final selection, a smaller number of alternatives is compared in relation to the key decision factors.

8. Use of these checklists forces decision making in context; that is, it keeps the decision maker from giving too much attention to a single issue in the decision-making process. It forces the decision maker to consider each issue and impact in relation to other issues and impacts.

9. These checklists can be used for several types of decision making, based on other considerations than environmental. They can also be used for decision making that evaluates and compares the environmental impacts as well as economic characteristics of different alternatives. Finally, such approaches can be used for systematic decision making considering environmental impacts, economic evaluations, and engineering feasibility. In other words, their use can range from considering only the environment to considering a composite of the three "Es" of decision making (environment, economics, and engineering).

—Larry W. Canter

References

Anderson, B.F. 1981. *Cascaded tradeoffs: A multiple objective, multiple publics method for alternatives evaluation in water resources planning.* August. Denver Colorado: U.S. Bureau of Reclamation.

Bishop, A.B., et al. 1970. *Socio-economic and community factors in planning urban freeways.* (September). Menlo Park, California: Department of Civil Engineering, Stanford University.

Brown, C.A., R.J. Quinn, and K.R. Hammond. 1980. *Scaling impacts of alternative plans.* June. Denver, Colorado: U.S. Bureau of Reclamation.

Brown, C.A. and T. Valenti. 1983. *Multi-attribute tradeoff system: User's and programmer's manual.* March. Denver, Colorado: U.S. Bureau of Reclamation.

Canter, L.W., S.F. Atkinson, and F.L. Leistritz. 1985. *Impacts of growth.* Chelsea, Michigan: Lewis Publishers, Inc.

Dean, B.V., and J.J. Nishry. 1965. Scoring and profitability models for evaluating and selecting engineering products. *Journal Operations Research Society of America* 13, no. 4 (July–August):550–569.

Dee, N., et al. 1972. *Environmental evaluation system for water resources planning.* Final report. Columbus, Ohio: Battelle-Columbus Laboratories.

———. 1973. *Planning methodology for water quality management: Environmental evaluation system.* July. Columbus, Ohio: Battelle-Columbus Laboratories.

Duke, K.M., et al. 1977. *Environmental quality assessment in multi-objective planning.* Final report to U.S. Bureau of Reclamation, November. Denver, Colorado.

Economic and Social Commission for Asia and the Pacific. 1990. *Environmental impact assessment—Guidelines for water resources development.* ST/ESCAP/786:19–48. New York: United Nations.

Edwards, W. 1976. *How to use multi-attribute utility measurement for social decision making.* SSRI Research Report, 76–3 (August). Los Angeles: Social Science Research Institute, University of Southern California.

Klee, K.J. 1988. *d-SYSS: A computer model for the evaluation of competing alternatives.* EPA 600/S2-88/038, July. Cincinnati, Ohio: U.S. Environmental Protection Agency.

Linstone, H.A., and M. Turoff. 1978. *The Delphi Method—Techniques and applications.* Reading, Massachusetts: Addison-Wesley Publishing Company.

Lohani, B.N. and N. Halim. 1990. *Environmental impact identification and prediction: Methodologies and resource requirements.* Background papers for course on environmental impact assessment of hydropower and irrigation projects, 13–31 August:152–182. Bangkok, Thailand: International Center for Water Resources Management and Training (CEFIGRE).

Lord, W.B., D.H. Deane, and M. Waterstone. 1979. *Commensuration in federal water resources planning: Problem analysis and research appraisal.* Report no. 79-2, April. Denver, Colorado: U.S. Bureau of Reclamation.

Mumpower, J. and L. Bollacker. 1981. *User's manual for the evaluation and sensitivity analysis program.* Technical report E-81-4, March. Vicksburg, Mississippi: U.S. Army Engineer Waterways Experiment Station.

Odum, E.P., et al. 1971. *Optimum pathway matrix analysis approach to the environmental decision making process—test case: Relative impact of proposed highway alternates.* Athens, Georgia: Institute of Ecology, University of Georgia.

Petersen, M.S. 1984. *Water resource planning and development.* 84–85. Englewood Cliffs, New Jersey: Prentice-Hall, Inc.

Salomon, S.N. 1974. Cost-benefit methodology for the selection of a nuclear power plant cooling system. Paper presented at the Energy Forum. Spring Meeting of the American Physics Society, Washington, D.C., 22 April.

School of Civil Engineering and Environmental Science and Oklahoma Biological Survey. 1974. *Mid-Arkansas river basin study—Effects assessment of alternative navigation routes from Tulsa, Oklahoma to vicinity of Wichita, Kansas.* June. Norman, Oklahoma: University of Oklahoma.

Smith, M.A. 1974. *Field test of an environmental impact assessment methodology.* Report ERC-1574, August. Atlanta, Georgia: Environmental Resources Center, Georgia Institute of Technology.

Solomon, R.C., et al. 1977. *Water resources assessment methodology (WRAM): Impact assessment and alternatives evaluation.* Technical report Y-77-1, February. Vicksburg, Mississippi: U.S. Army Engineer Waterways Experiment Station.

Torno, H.C., et al. 1988. *Training guidance for the integrated environmental evaluation of water resources development projects.* Paris, France: UNESCO.

Voelker, A.H. 1977. *Power plant siting, An application of the nominal group process technique.* ORNL/NUREG/TM-81, February. Oak Ridge, Tennessee: Oak Ridge National Laboratory.

Voorhees, A.M. and Associates. 1975. *Interim guide for environmental assessment: HUD field office edition.* June. Washington, D.C.

Wilson, L.A. 1980. *Personal communication.*

2.6
PREPARATION OF WRITTEN DOCUMENTATION

Perhaps the most important activity in the EIA process is preparing written reports which document the impact study findings. The resultant document or documents include EAs, EISs, environmental impact reports, environmental impact declarations, and FONSIs. To illustrate the importance of written documentation in the EIA process, the CEQ regulations (1987) state in paragraph 1502.8:

> Environmental impact statements shall be written in plain language and may use appropriate graphics so that decision makers and the public can readily understand them. Agencies should employ writers of clear prose or editors to write, review, or edit statements, which will be based upon the analysis and supporting data from the natural and social sciences and the environmental design arts.

The broad topics included in an EA prepared to meet the requirements of the CEQ (1987) are:

- Need for the proposal
- Description of alternatives
- Environmental impacts of proposed action and alternatives
- List of agencies and persons consulted

Table 2.6.1 shows a topical outline for an EIS (Council on Environmental Quality 1987).

Basic principles of technical writing must be applied in written documentation. These principles can be used in both planning the document and preparing written materials. Five such principles are delineated by Mills and Walter (1978):

1. Always have in mind a specific reader, and assume that this reader is intelligent, but uninformed.
2. Before you start to write, decide the purpose of your report; make sure that every paragraph, sentence, and word makes a clear contribution to that purpose, and makes it at the right time.
3. Use language that is simple, concrete, and familiar.
4. At the beginning and end of every section, check your writing according to this principle: First, tell your readers what you are going to tell them, then tell them, and then tell them what you have told them.
5. Make your report visually attractive.

The target audience of impact study documentation typically consists of two groups: (1) a nontechnical audience represented by decision makers and interested members of the public and (2) a technical audience represented by professional colleagues in government agencies and specific public groups who have interest in the study. Accordingly, written documentation on impact studies must address the information needs of both nontechnical and technical audiences.

—Larry W. Canter

References

Council on Environmental Quality. 1987. *40 Code of federal regulations.* Chap. 5, 1 July:929–971. Washington, D.C.: U.S. Government Printing Office.

Mills, G.H. and J.A. Walter. 1978. *Technical writing.* Dallas, Texas: Holt, Rinehart and Winston.

TABLE 2.6.1 ENVIRONMENTAL IMPACT STATEMENT OUTLINE (COUNCIL ON ENVIRONMENTAL QUALITY 1987)

Section	Comments
Cover Sheet	The cover sheet shall not exceed one page. It shall include a list of the responsible agencies, including the lead agency and any cooperating agencies; the title of the proposed action; the name, address, and telephone number of the person at the agency who can supply further information; a designation of the statement as a draft, final, or draft or final supplement; a one-paragraph abstract of the statement; and the date by which comments must be received.
Summary	Each environmental impact statement shall contain a summary which adequately and accurately summarizes the statement. The summary shall stress the major conclusions, areas of controversy (including issues raised by agencies and the public), and the issues to be resolved (including the choice among alternatives). The summary will normally not exceed fifteen pages.
Purpose and Need	The statement shall briefly specify the underlying purpose and need to which the agency is responding in proposing the alternatives, including the proposed action.
Alternatives Including the Proposed Action	This section is the heart of the environmental impact statement. Based on the information and analysis presented in the sections on the Affected Environment and the Environmental Consequences, it should present the environmental impacts of the proposal and the alternatives in comparative form, thus sharply defining the issues and providing a clear basis for choice among options by the decision maker and the public.
Affected Environment	The environmental impact statement shall succinctly describe the environment of the area(s) to be affected or created by the alternatives under consideration. The descriptions shall be no longer than is necessary to understand the effects of the alternatives. Data and analyses in a statement shall be commensurate with the importance of the impact, with less important material summarized, consolidated, or simply referenced. Agencies shall avoid useless bulk in statements and shall concentrate effort and attention on important issues. Verbose descriptions of the affected environment are themselves no measure of the adequacy of an environmental impact statement.
Environmental Consequences	This section forms the scientific and analytic basis for the comparisons of alternatives. The discussion will include the environmental impacts of the alternatives, including the proposed action, any adverse environmental effects which cannot be avoided should the proposal be implemented, the relationship between short-term uses of man's environment and the maintenance and enhancement of long-term productivity, and any irreversible or irretrievable commitments of resources which would be involved in the proposal should it be implemented.
List of Preparers	The environmental impact statement shall list the names, together with their qualifications (expertise, experience, professional disciplines), of the persons who were primarily responsible for preparing the environmental impact statement or significant background papers, including basic components of the statement. Where possible, the persons who are responsible for a particular analysis, including analyses in background papers, shall be identified. Normally the list will not exceed two pages.
Appendices	If an agency prepares an appendix to an environmental impact statement, the appendix shall: (a) consist of material prepared in connection with the EIS; (b) normally consist of material which substantiates any analysis fundamental to the impact statement; (c) normally be analytic and relevant to the decision to be made; and (d) be circulated with the environmental impact statement or be readily available on request.

2.7
ENVIRONMENTAL MONITORING

The CEQ regulations (1987) enunciate the principle of post-EIS environmental monitoring in sections 1505.3 and 1505.2(c). The CEQ regulations focus on monitoring in conjunction with implementing mitigation measures. Monitoring can also be used to determine the effectiveness of each of the types of mitigation measures.

Sadler and Davies (1988) delineate three types of environmental monitoring associated with the life cycle of a project, plan, or program; these include baseline monitoring, effects or impact monitoring, and compliance monitoring. Baseline monitoring refers to the measurement of environmental variables during a representative preproject period to determine existing conditions, ranges of variation, and processes of change. Effects or impact monitoring involves the measurement of environmental variables during project construction and operation to determine changes which may have resulted from the project. Finally, compliance monitoring is periodic sampling or continuous measurement of levels of waste discharge, noise, or similar emissions, to ensure that conditions are observed and standards are met. Pre-EIS monitoring includes baseline monitoring, while post-EIS monitoring encompasses effects or impact monitoring, and compliance monitoring.

Numerous purposes and implied benefits can be delineated for pre- and post-EIS environmental monitoring. For example, Marcus (1979) identifies the following six general purposes or uses of information from post-EIS monitoring:

1. Provides information for documentation of the impacts that result from a proposed federal action, with this information enabling more accurate prediction of impacts associated with similar federal actions.
2. Warns agencies of unanticipated adverse impacts or sudden changes in impact trends
3. Provides an immediate warning whenever a preselected impact indicator approaches a pre-selected critical level.
4. Provides information for agencies to control the timing, location, and level of impacts of a project. Control measures involve preliminary planning as well as the possible implementation of regulation and enforcement measures.
5. Provides information for evaluating the effectiveness of implemented mitigation measures.
6. Provides information to verify predicted impacts and thus validate impact prediction techniques. Based on these findings, techniques, such as mathematical models, can be modified or adjusted.

Environmental monitoring can serve as a basic component of a periodic environmental regulatory auditing program for a project (Allison 1988). In this context, auditing can be defined as a systematic, documented, periodic, and objective review by regulated entities of facility operations and practices related to environmental requirements (U.S. Environmental Protection Agency 1986). The purposes of environmental auditing are to verify compliance with environmental requirements, evaluate the effectiveness of in-place, environmental management systems, and assess risks from regulated and unregulated substances and practices. Some direct results of an auditing program include an increased environmental awareness by project employees, early detection and correction of problems and thus avoidance of environmental agency enforcement actions, and improved management control of environmental programs (Allison 1988). Several references are available describing protocols and experiences in auditing related to the EIA process (Canter 1985a; Munro, Bryant, and Matte-Baker 1986; PADC Environmental Impact Assessment and Planning Unit 1982; Sadler 1987; United Nations Environment Program, 1990).

Careful planning and implementation of an environmental monitoring program is necessary to meet the stated purposes of monitoring. Three premises relative to monitoring programs in the United States are:

1. An abundance of environmental monitoring data is routinely collected by various governmental agencies and the private sector. These data typically need to be identified, aggregated, and interpreted.
2. Environmental monitoring programs are expensive to plan and implement; therefore, every effort should be made to use extant monitoring programs or modify extant programs.
3. Due to overlapping environmental management and monitoring responsibilities of many local, state, and federal government agencies, environmental monitoring planning must be coordinated with several agencies.

Several fundamental books and articles are useful in the detailed planning and implementation of a monitoring program. References are available for air-quality monitoring (Noll and Miller 1977 and Lodge 1989), surface water quality monitoring (Canter 1985b and Loftis et al. 1989), groundwater quality monitoring (Aller et al. 1989), noise monitoring (Lipscomb and Taylor 1978), species and habitat monitoring (Brown and Dycus 1986; Gray 1988; Horner, Richey, and Thomas 1986; Ontario Ministry of the Environment 1989; Roberts and Roberts 1984; and Spellerberg 1991), social impact assessment monitoring (Krawetz, MacDonald, and Nichols 1987), and health effects monitoring (Burtan 1991 and Schweitzer 1981). General references which encompass several types of environmental monitoring (Cheremisinoff and Manganiello 1990, Gilbert 1987, and Keith 1991) are also available.

—Larry W. Canter

References

Aller, L., et al. 1989. *Handbook of suggested practices for the design and installation of ground-water monitoring wells.* EPA 600/4-89/034. Dublin, Ohio: National Water Well Association.

Allison, R.C. 1988. Some perspectives on environmental auditing. *The Environmental Professional* 10:185–188.

Brown, R.T. and D.L. Dycus. 1986. Characterizing the influence of natural variables during environmental impact analysis. *Rationale for Sampling and Interpretation of Ecological Data in the Assessment of Freshwater Ecosystems,* 60–75. Philadelphia: American Society for Testing and Materials.

Burtan, R.C. 1991. Medical monitoring's expanding role. *Environmental Protection* 2, no. 6 (September):16–18.

Canter, L.W. 1985a. Impact prediction auditing. *The Environmental Professional* 7, no. 3:255–264.

———. 1985b. *River water quality monitoring.* Chelsea, Michigan: Lewis Publishers, Inc.

Cheremisinoff, P.N. and B.T. Manganiello. 1990. *Environmental field sampling manual.* Northbrook, Illinois: Pudvan Publishing Company.

Council on Environmental Quality. 1987. *40 Code of federal regulations.* Chap. 5, 1 July:929–971. Washington, D.C.: U.S. Government Printing Office.

Gilbert, R.O. 1987. *Statistical methods for environmental pollution.* New York: Van Nostrand Reinhold.

Gray, R.H. 1988. *Overview of a comprehensive environmental monitoring and surveillance program: The role of fish and wildlife.* PNL-SA-15922, May. Richland, Washington: Battelle Pacific Northwest Labs.

Horner, R.R., J.S. Richey, and G.L. Thomas. 1986. Conceptual framework to guide aquatic monitoring program design for thermal electric power plants. *Rationale for Sampling and Interpretation of Ecological Data in the Assessment of Freshwater Ecosystems,* 86–100. Philadelphia: American Society for Testing and Materials.

Keith, L.H. 1991. *Environmental sampling and analysis: A practical guide.* Chelsea, Michigan: Lewis Publishers, Inc.

Krawetz, N.M., W.R. MacDonald, and P. Nichols. 1987. *A framework for effective monitoring.* Hull, Quebec: Canadian Environmental Assessment Research Council.

Lipscomb, D.M. and A.C. Taylor, eds. 1978. *Noise control handbook of principles and practices.* New York: Van Nostrand Reinhold Company.

Lodge, J.P., ed. 1989. *Methods of air sampling and analysis.* 3d ed. Chelsea, Michigan: Lewis Publishers, Inc.

Loftis, J.C., R.C. Ward, R.D. Phillips, and C.H. Taylor. 1989. *An evaluation of trend detection techniques for use in water quality monitoring programs.* EPA/600/S3-89/037, September. Cincinnati, Ohio: U.S. Environmental Protection Agency, Center for Environmental Research Information.

Marcus, L.G. 1979. *A methodology for post-EIS (environmental impact statement) monitoring.* Geological Survey Circular 782. Washington, D.C.: U.S. Geological Survey.

Munro, D.A., T.J. Bryant, and A. Matte-Baker. 1986. *Learning from experience: A state-of-the-art review and evaluation of environmental impact assessment audits.* Hull, Quebec: Canadian Environmental Assessment Research Council.

Noll, K.E., and T.L. Miller. 1977. *Air monitoring survey design,* Ann Arbor, Michigan: Ann Arbor Science Publishers, Inc.

Ontario Ministry of the Environment. 1989. *Investigation, evaluation, and recommendations of biomonitoring organisms for procedures development for environmental monitoring.* MIC-90-00873/WEP. Toronto.

PADC Environmental Impact Assessment and Planning Unit. 1982. *Post-developing audits to test the effectiveness of environmental impact prediction methods and techniques.* Aberdeen, Scotland: University of Aberdeen.

Roberts, R.D. and T.M. Roberts. 1984. *Planning and ecology.* New York: Chapman and Hall.

Sadler, B., ed. 1987. *Audit and evaluation in environmental assessment and management: Canadian and international experience.* 2 vols. Hull, Quebec: Environment Canada.

Sadler, B. and M. Davies. 1988. *Environmental monitoring and audit: Guidelines for post-project analysis of development impacts and assessment methodology.* August:3–6, 11–14. Aberdeen, Scotland: Centre for Environmental Management and Planning, Aberdeen University.

Schweitzer, G.E. 1981. Risk assessment near uncontrolled hazardous waste sites: Role of monitoring data. *Proceedings of National Conference on Management of Uncontrolled Hazardous Waste Sites,* October:238–247. Silver Spring, Maryland: Hazardous Materials Control Research Institute.

Spellerberg, I. F. 1991. *Monitoring ecological change.* 181–182. Cambridge, England: Cambridge Unviersity Press.

United Nations Environment Program. 1990. *Environmental auditing.* Technical Report Series no. 2. Paris, Cedex, France.

U.S. Environmental Protection Agency. 1986. Environmental auditing policy statement. *Federal Register* 51, no. 131, 9 July:25004.

2.8
EMERGING ISSUES IN THE EIA PROCESS

After twenty-five years of experience in applying the EIA process in the United States, the process is maturing and the resultant procedures and documentation are becoming more technically sound and appropriate in including the environment in project planning and decision making. However, issues continue to emerge related to the process. These issues include new topical items and the application of new tools and techniques. Emerging items are related to cumulative impacts, focused activities related to the EIA process, the inclusion of risk assessment, the need to address impacts on biodiversity (Council on Environmental Quality 1993) and global issues, the importance of monitoring and environmental auditing, particularly as related to major projects, the need for environmentally responsible project management, and the emerging use of market-based approaches in project planning and evaluation.

Cumulative impact considerations in the EIA process are becoming increasingly important as projects are recognized as not being constructed and operated in isolation, but in the context of existing and other planned projects in the environs. Methods for identifying key cumulative impact concerns need to be developed and addressed in a responsible manner in the EIA process.

Risk assessment is a tool developed in the 1970s primarily for evaluating environmental regulatory strategies. In recent years, interest in the application of risk assessment within the EIA process (Canter 1993a) has increased. Risk assessment provides not only a focus on human health concerns, but can also be applied in ecological analyses. Risk assessment is anticipated to become more effectively integrated in the EIA process in the coming years.

As was noted earlier, the EIA process is being applied in a more focused manner to environmental permits and specific environmental remediation projects. In contrast, interest is increasing in considering regional and broader issues in the EIA process. One example is the delineated need for addressing the impacts of projects on biodiversity (Council on Environmental Quality 1993). This need results from the realization that projects can have implications on biodiversity, and this item should be considered, particularly for larger scale projects. A related item is the need to address project impacts in a transboundary context. This issue is particularly important as related to the impacts of a country's projects on other countries. Finally, the implications for some large-scale projects, in terms of acid rain as well as global warming, may have to be considered.

The need to consider the EIA process in relation to project management and follow-on is another topic on the EIA agenda. Follow-on monitoring to document experienced impacts, as well as auditing, was mentioned earlier. Monitoring and auditing can be used for project management decisions to minimize detrimental impacts (Canter 1993b).

Market-based approaches in environmental management include topics such as the use of mitigation banking for wetland losses and the application of emissions trading relative to both air pollutant and water pollutant emissions. The economic evaluation of potential environmental impacts is also of interest. This topic should receive more attention in the coming years, and soon such economic analyses may possibly be included with traditional cost-benefit analyses in project planning and decision making.

A number of new tools and techniques can be effectively used within the EIA process. The GIS represents an emerging technology that can facilitate resource identification and evaluation. GIS technology is already being used in larger scale projects.

Expert systems also represent an emerging tool which could find applicability within the EIA process. To date, most expert systems developed in environmental engineering focus on hazardous waste management and groundwater pollution evaluation, along with pollution control facilities. The development of expert systems for use in the EIA process is needed.

Public participation in the EIA process often leads to conflict. Several techniques are being developed in environmental mediation which have potential applicability in the EIA process. Examples include alternative dispute resolution and environmental dispute resolution techniques. A key feature of these techniques is the incorporation of a third-party intervener. The intervener negotiates between the project proponent and those interests who have raised opposition to the project on environmental grounds.

As increased information becomes available on environmental systems and processes, the associated development of mathematical models can be used for project impact quantification and evaluation. Such modeling is anticipated to increase, particularly for major projects that may have significant environmental impact issues.

In summary, the EIA process represents a young field within environmental engineering, while at the same time signs that it is a maturing field of practice are evident. Although many policy and social implications are associated with the EIA process, the process is fundamentally a technical or scientific process. Accordingly, the application of scientific approaches is fundamental to the effective implementation of this process in project planning and decision making.

—*Larry W. Canter*

2.9
INTERNATIONAL ACTIVITIES IN ENVIRONMENTAL IMPACT ASSESSMENT

The 1970 effective date of the NEPA in the United States signaled the beginning for many countries to adopt laws analogous to the NEPA. Over seventy-five countries now have laws requiring impact studies on proposed development projects. While the resultant reports are not always referred to as EISs, they do represent the documentation of studies similar in concept to those conducted in the United States. Terms other than EIS used by other countries include environmental impact assessment reports, environmental assessment reports, environmental impact documents, and environmental impact reports.

From a global perspective, the following observations relate to countries or portions of the world that have adopted EIA legislation. In North America, Canada, the United States, and Mexico have EIA laws. Some Central American countries have also adopted EIA legislation, and the majority of countries in South America have done similarly. Some larger countries with major development projects, such as Brazil, Argentina, and Venezuela, have active programs related to EIA.

In the European context, the European Community (EC) has a directive on EIA which must be met by all member countries. In addition, these countries also have their own EIA legislation which can be more stringent than the EC directive. The Scandinavian countries such as Sweden, Denmark, and Finland have EIA legislation and are active in the application of the EIA process. Many of the east European countries which were in the former Soviet Union are in the initial stages of adopting EIA legislation.

With regard to the African Continent, several countries in the northern tier, most notably Morocco and Egypt, have EIA legislation. South Africa also has such legislation. Other countries in the African Continent are in various stages of development of EIA legislation. In the Middle East, several countries have legislation or expressed interest in the EIA process; examples include Kuwait and Saudi Arabia.

In Asia and Southeast Asia, most countries have adopted EIA legislation; examples include India, Sri Lanka, Thailand, Malaysia, and Indonesia. Both Taiwan and the People's Republic of China have EIA legislation, as does Japan. Finally, Australia and New Zealand have active programs in planning and conducting impact studies.

For those countries that have not adopted their own EIA legislation, most are involved in some fashion in impact studies through the auspices of international lending agencies, such as the World Bank or Regional Development Banks, or through requirements of bilateral donor agencies providing funding for development projects. While the specific EIA requirements of donor countries may vary, the general concepts are applied throughout the developing world.

Both substantive and emphasis differences exist between the EIA processes of other countries and that in the United States. For example, in the United States, analyzing alternatives and choosing the one that balances environmental impacts and economic efficiency is emphasized. Most other countries emphasize alternatives analysis less than the EIA process in the United States does. In other words, detailed analysis in most other countries is for the proposed project.

A second difference is that many countries have highly structured land-use planning systems. In this context, the EIA process is often incorporated within the overall planning system. This incorporation causes procedural differences compared to the practice in the United States where an overall land-use planning system is absent.

A major difference between EIA practice in the United States and that in lesser developed countries is the limited availability of environmental data in the latter. Numerous

References

Canter, L.W. 1993a. Pragmatic suggestions for incorporating risk assessment principles in EIA studies. Invited paper, *The Environmental Professional* 15, no. 1:125–138.

———. 1993b. The role of environmental monitoring in responsible environmental management. Invited paper, *The Environmental Professional* 15, no. 1:76–87.

Council on Environmental Quality. 1993. *Incorporating biodiversity considerations into environmental impact analysis under the National Environmental Policy Act.* January. Washington, D.C.

extant environmental monitoring systems exist in the United States, and this information can be used to describe the affected environment or baseline conditions. In many developing countries, such environmental data are either absent or only minimally available. Accordingly, monitoring baseline conditions in the EIA practice is emphasized in many countries.

EIA practice in the United States has typically focused on the preparation of an EA or an EIS. Follow-on activities are only minimally addressed. In contrast, post-EIS activities are emphasized in many countries. These activities include collecting monitoring data and using this information in project management to minimize negative environmental consequences.

As a final point of comparison, the United States practice is often characterized by litigation, wherein opponents to projects can file a lawsuit against project sponsors on the basis of not satisfying the spirit and intent of the NEPA. This litigative concept is essentially absent in most other countries.

In summary, the international EIA practice throughout the world is largely patterned on the principles used in the United States. Future activities related to EIA suggest further coordination and integration of EIA requirements on a worldwide basis, with emphasis on both procedures and methodology. In addition, transboundary impacts, in which the impacts of development projects in one country may be manifested in other nearby countries, are being emphasized. These concerns will facilitate greater coordination between countries regarding the EIA process.

—*Larry W. Canter*

Pollution Prevention in Chemical Manufacturing

David H.F. Liu

3.1
REGULATIONS AND DEFINITIONS

Pollution prevention, as defined under the Pollution Prevention Act of 1990, means source reduction and other practices that reduce or eliminate the creation of pollutants through (1) increased efficiency in the use of raw materials, energy, water, or other resources or (2) protection of natural resources by conservation. Under the Pollution Prevention Act, recycling, energy recovery, treatment, and disposal are not included within the definition of pollution prevention. Practices commonly described as in-process recycling may qualify as pollution prevention. Recycling conducted in an environmentally sound manner shares many of the advantages of pollution prevention—it can reduce the need for treatment or disposal and conserve energy and resources.

Pollution prevention (or source reduction) is an agency's first priority in the environmental management hierarchy for reducing risks to human health and the environment from pollution. This hierarchy includes (1) prevention, (2) recycling, (3) treatment, and (4) disposal or release. The second priority in the hierarchy is the responsible recycling of any waste that cannot be reduced at the source. Waste that cannot feasibly be recycled should be treated according to environmental standards that are designed to reduce both the hazard and volume of waste streams. Finally, any residues remaining from the treatment of waste should be disposed of safely to minimize their potential release into the environment. Pollution and related terms are defined in Table 3.1.1.

Regulatory Background

Three key federal programs have been implemented to address pollution production: the Pollution Prevention Act of 1990, the Environmental Protection Agency's (EPA's) 33/50 Voluntary Reduction Program, and the Clean Air Act Amendments' (CAAA's) Early Reduction Program for Maximum Achievable Control Technology (MACT). Table 3.1.2 compares the features of these programs, from which the following key points are noted:

Air toxics are used as a starting point for multimedia pollution prevention (that is consistent with two-thirds of the reported 3.6 billion lb released into the air).

Reductions in hazardous air pollutants will occur incrementally during different years (1992, 1994, 1995, and beyond).

Flexibility or variability in the definition of the base year, the definition of the source, and credits for reductions are possible.

The Pollution Prevention Strategy focuses on cooperative effort between the EPA, industry, and state and local governments as well as other departments and agencies to forge initiatives which address key environmental threats. Initially, the strategy focused on the manufacturing sector and the 33/50 program (formerly called the Industrial Toxics Project), under which the EPA sought substantial voluntary reduction of seventeen targeted high-risk industrial chemicals (see Table 3.1.3).

Hazardous and Toxic Chemicals

The following five key laws specifically address hazardous and toxic chemicals.

National Emission Standards for Hazardous Air Pollutants (NESHAP), Hazardous Air Emissions—This law addresses six specific chemicals (asbestos, beryllium, mercury, vinyl chloride, benzene, and arsenic) and one generic category (radionuclides) released into the air.

Clear Water Act, Priority Pollutants—This act addresses 189 chemicals released into water including volatile substances such as benzene, chloroform, and vinyl chloride; acid compounds such as phenols and their derivatives; pesticides such as chlordane, dichlorodiphenyl trichloroethane (DDT), and toxaphene; heavy metals such as lead and mercury; polychlorinated biphenyls (PCBs); and other organic and inorganic compounds.

Resource Conservation and Recovery Act (RCRA), Hazardous Wastes—This act addresses more than 400 discarded commercial chemical products and specific chemical constituents of industrial chemical streams destined for disposal on land.

Superfund Amendments and Reauthorization Act (SARA) Title III, Section 313: Toxic Substances—This act addresses more than 320 chemicals and chemical categories released into all three environmental media. Under specified conditions, facilities must report releases of these chemicals to the EPA's annual Toxic Release Inventory (TRI).

SARA Section 302: Extremely Hazardous Substances—This act addresses more than 360 chemicals for which facilities must prepare emergency action plans if these chemicals are above certain threshold quantities. A release of these chemicals to air, land, or water requires a facility to report the release to the state emergency response committee (SERC) and the local emergency planning committee (LEPC) under SARA Section 304.

TABLE 3.1.1 DEFINITIONS OF POLLUTION PREVENTION TERMS

Waste

In theory, waste applies to nonproduct output of processes and discarded products, irrespective of the environmental medium affected. In practice, since passage of the RCRA, most uses of waste refer exclusively to the hazardous and solid wastes regulated under RCRA and do not include air emissions or water discharges regulated by the Clean Air Act or the Clean Water Act.

Pollution/Pollutants

Pollution and pollutants refer to all nonproduct output, irrespective of any recycling or treatment that may prevent or mitigate releases to the environment (includes all media).

Waste Minimization

Waste minimization initially included both treating waste to minimize its volume or toxicity and preventing the generation of waste at the source. The distinction between treatment and prevention became important because some advocates of decreased waste generation believed that an emphasis on waste minimization would deflect resources away from prevention towards treatment. In the current RCRA biennial report, waste minimization refers to source reduction and recycling activities and now excludes treatment and energy recovery.

Source Reduction

Source reduction is defined in the *Pollution Prevention Act of 1990* as "any practice which (1) reduces the amount of any hazardous substance, pollutant, or contaminant entering any waste stream or otherwise released into the environment (including fugitive emissions) prior to recycling, treatment, and disposal; and (2) reduces the hazards to public health and the environment associated with the release of such substances, pollutants, or contaminants. The term includes equipment or technology modifications, process or procedure modifications, reformulations or design of products, substitution of raw materials, and improvements in housekeeping, maintenance, training, or inventory control." Source reduction does not entail any form of waste management (e.g., recycling and treatment). The act excludes from the definition of source reduction "any practice which alters the physical, chemical, or biological characteristics or the volume of a hazardous substance, pollutant, or contaminant through a process or activity which itself is not integral to and necessary for the production of a product or the providing of a service."

Waste Reduction

This term is used by the Congressional Office of Technology Assessment synonymously with source reduction. However, many groups use the term to refer to waste minimization. Therefore, determining the use of waste reduction is important when it is encountered.

Toxic Chemical Use Substitution

Toxic chemical use substitution or material substitution describes replacing toxic chemical with less harmful chemicals even though relative toxicities may not be fully known. Examples include substituting a toxic solvent in an industrial process with a less toxic chemical and reformulating a product to decrease the use of toxic raw materials or the generation of toxic by-products. This term also refers to efforts to reduce or eliminate the commercial use of chemicals associated with health or environmental risks, including substitution of less hazardous chemicals for comparable uses and the elimination of a particular process or product from the market without direct substitution.

Toxics Use Reduction

Toxics use reduction refers to the activities grouped under source reduction where the intent is to reduce, avoid, or eliminate the use of toxics in processes and products so that the overall risks to the health of workers, consumers, and the environment are reduced without shifting risks between workers, consumers, or parts of the environment.

Pollution Prevention

Pollution prevention refers to activities to reduce or eliminate pollution or waste at its source or to reduce its toxicity. It involves the use of processes, practices, or products that reduce or eliminate the generation of pollutants and waste or that protect natural resources through conservation or more efficient utilization. Pollution prevention does not include recycling, energy recovery, treatment, and disposal. Some practices commonly described as in-process recycling may qualify as pollution prevention.

Resource Protection

In the context of pollution prevention, resource protection refers to protecting natural resources by avoiding excessive levels of waste and residues, minimizing the depletion of resources, and assuring that the environment's capacity to absorb pollutants is not exceeded.

Cleaner Products

Cleaner products or clean products refers to consumer and industrial products that are less polluting and less harmful to the environment and less toxic and less harmful to human health.

Environmentally Safe Products, Environmentally Preferable Products, or Green Products

The terms environmentally safe products, environmentally preferable products, or green products refer to products that are less toxic and less harmful to human health and the environment when their polluting effects during their entire life cycle are considered.

Life Cycle Analysis

Life cycle analysis is a study of the pollution generation characteristics and the opportunities for pollution prevention associated with the entire life cycle of a product or process. Any change in the product or process has implications for upstream stages (extraction and processing of raw materials, production and distribution of process inputs) and for downstream stages (including the components of a product, its use, and its ultimate disposal).

Source: U.S. Environmental Protection Agency, 1992, *Pollution prevention 1991: Research program*, EPA/600/R-92/189 (September). (Washington, D.C.: Office of Research and Development).

TABLE 3.1.2 SUMMARY OF POLLUTION PREVENTION REGULATORY INITIATIVES

	Pollution Prevention Act of 1990	*CAAA Early Reduction Program*	*EPA 33/50 Voluntary Reduction Program*
Goals	Reporting requirements: Collect and disseminate information on pollution to all media and provide financial aid to states	For air only, reduction for source by 90% for gaseous hazardous air pollutants (HAPs) and 95% for particulate HAPs; uses hazard index for weighting reductions of highly toxic pollutants	Voluntary reduction of pollutants to all media by 33% by the end of 1992 and by 50% by the end of 1995
Number and Type of Chemicals	All SARA 313 chemicals	All 189 HAPs listed in the CAAAs of which 35 are considered high-risk HAPs	17 chemicals, all of which are listed HAPs
Affected Sources	Facilities with ten or more employees, within standard industrial classification (SIC) 20–39, handling amounts greater than specified threshold limits for reporting	Facility-specific sources emitting more than 10 tn/yr of one HAP or more than 25 tn/yr of combined HAPs; flexible definition of source; credits for other reductions, including regulatory reductions, 33/50 reductions, or production shutdown or curtailment	Any SARA reporting companies; source can be all facilities operated by a company
Reporting Requirements	Annual, via new EPA Form R; report amounts of waste, recycle, and treated materials, amounts treated or disposed onsite and offsite, and treatment methods; project next two years	Six-year extension for implementing MACT; must enter into an enforceable commitment prior to EPA defining MACT in regulations; four submittal requirements: source identification, base-year HAP emissions, reduction plan, and statement of commitment	EPA Form R
Compliance Measurement or Baseline	For production throughput baseline production from prior year	Emissions in 1987 or later	Measured by annual EPA Form R relative to 1988 baseline year
Deadline(s)	7/1/92 for calendar year 1991 and every year thereafter	Achieve early reduction prior to MACT for the source or achieve reduction by 1/1/94 for sources with MACT prior to 1994	End of years 1992 and 1995
Enforcement	Penalties up to $25,000 per day	The company may rescind prior to 12/1/93 without penalty; voluntary but enforceable once committed	None
For More Information	42 USCS § 13.01	Public Law 101-549, 11/15/90, 104 Stat. 2399-2712	The 33/50 program, U.S. EPA Office of Toxic Substances, Washington, DC, July 1991

Source: William W. Doerr, 1993, Plan for future with pollution prevention, *Chemical Engineering Progress* (May).

TABLE 3.1.3 PRIORITY CHEMICALS TARGETED IN THE 33/50 PROJECT FOR THE INDUSTRIAL SECTOR POLLUTION PREVENTION STRATEGY

Target Chemicals	Million Pounds Released in 1988
Benzene	33.1
Cadmium	2.0
Carbon Tetrachloride	5.0
Chloroform	26.9
Chromium	56.9
Cyanide	13.8
Dichloromethane	153.4
Lead	58.7
Mercury	0.3
Methyl Ethyl Ketone	159.1
Methyl Isobutyl Ketone	43.7
Nickel	19.4
Tetrachloroethylene	37.5
Toluene	344.6
1,1,1-Trichloroethane	190.5
Trichloroethylene	55.4
Xylene	201.6

Source: U.S. Environmental Protection Agency, 1992, *Pollution prevention 1991: Research program*, EPA/600/R-92/189 (September). (Washington, D.C.: Office of Research and Development).

Source Reduction versus Discharge Reduction

The EPA has taken a strong position on pollution prevention by regarding source reduction as the only true pollution prevention activity and treating recycling as an option. Industry's position prior to the act (and effectively unchanged since) was to reduce the discharge of pollutant waste into the environment in the most cost-effective manner. This objective is achieved in some cases by source reduction, in others by recycling, in others by treatment and disposal, and usually in a combination of these methods. For this reason, this handbook examines all options in the pollution prevention hierarchy.

Traditionally, regulations change, with more stringent controls enacted over time. Therefore, source reduction and perhaps recycling and reuse (instead of treatment or disposal) may become more economically attractive in the future.

State Programs

Many states have enacted legislation that is not voluntary, particularly those states with an aggressive ecological presence. Facilities should consult the pollution prevention legislation in their states on (1) goals, (2) affected chemicals, (3) affected sources, (4) reporting requirements, (5) exemptions, (6) performance measurement basis, (7) deadlines, and (8) other unique features.

Any company responding to the pollution prevention legislation in its state should consider a coordinated approach to satisfy the requirements of the federal programs as follows:

EPA Form R data and state emission data should be carefully reviewed, compared, and reported consistently. Scheduling activities for compliance should be integrated with the EPA's 33/50 program and the CAAA's Early Reduction Program prior to MACT for source reduction to be effective.

The Pollution Prevention Act contains new tracking and reporting provisions. These provisions require companies to file a toxic chemical source reduction and resource recycling report file for each used chemical listed under SARA 313 for TRI reporting under the Federal Emergency Planning and Community Right-to-Know Act (EPCRA). These reports, which do not replace SARA Form R, cover information for each reporting year including:

- The amount of the chemical entering the waste stream before recycling, treatment, or disposal
- The amount of the chemical that is recycled, the recycling method used, and the percentage change from the previous year
- The source reduction practice used for the chemical
- The amount of the chemical that the company expects to report for the two following calendar years
- A ratio of the current to the previous year's chemical production
- Techniques used to identify source reduction opportunities
- Any catastrophic releases
- The amount of the chemical that is treated onsite or offsite
- Optional information about source reduction, recycling, and other pollution control methods used in previous years

In addition, the appropriate state environmental protection agency should be contacted for detailed information on reporting requirements, including the pollution prevention plan (PPP) and PPP summary.

—*David H.F. Liu*

3.2
POLLUTION PREVENTION METHODOLOGY

In recent years, several waste reduction methodologies have been developed in government, industry, and academe. These methodologies prescribe a logical sequence of tasks at all organization levels, from the executive to the process area. Despite differences in emphasis and perspective, most stepwise methodologies share the following four common elements:

A *chartering phase,* in which an organization affirms its commitment to a waste reduction program; articulates policies, goals, and plans; and identifies program participants

An *assessment phase,* in which teams collect data, generate and evaluate options for waste reduction, and select options for implementation

An *implementation phase,* in which waste reduction projects are approved, funded, and initiated

An *ongoing auditing function,* in which waste reduction programs are monitored and reductions are measured. Usually feedback from the auditing function triggers a new iteration of the program.

Model Methodologies

The EPA and the Chemical Manufacturers' Association have published their pollution prevention methodologies. These methodologies provide a model for companies to use in developing methodologies.

EPA METHODOLOGY

The recent publication of the U.S. EPA's *Facility pollution prevention guide* (1992) represents a major upgrade to their methodology (see Figure 3.2.1). It places additional emphasis on the management of a continuous waste reduction program. For example, the single chartering step prescribed in the previous manual (U.S. EPA, 1988) was expanded to four iteration steps in the new guide. Also, where auditing was a constituent task of implementation in the previous manual, the new guide presents it as a discrete, ongoing step. The guide's inclusion of "maintain a pollution prevention program" as part of the methodology is also new.

The methodology prescribed in the new guide is a major step forward. The previous manual correctly assumed that assessments are the basis of a waste reduction program. However, the new methodology increases the likelihood that assessment is performed because it prescribes waste reduction roles at all levels of the organization.

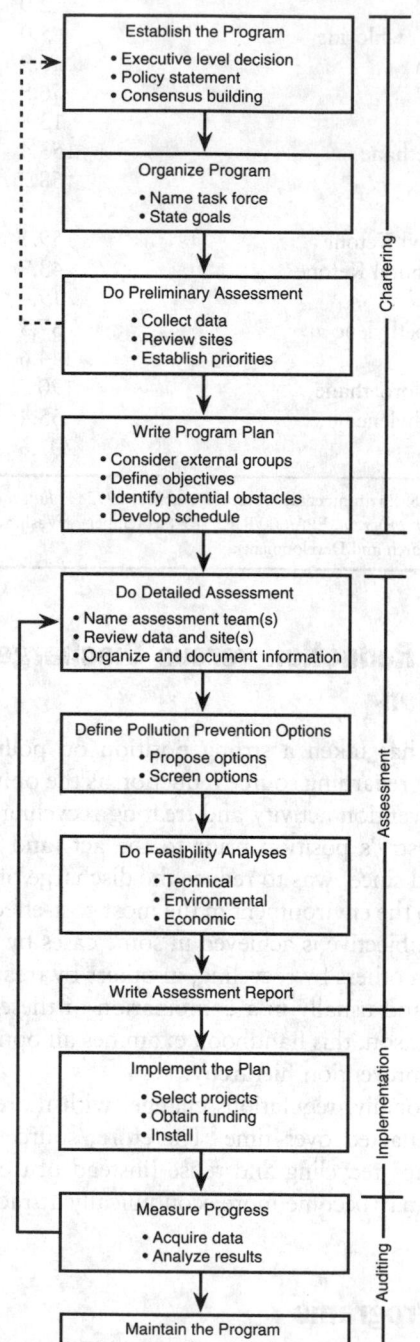

FIG. 3.2.1 EPA pollution prevention methodology. Chartering, assessment, implementation, and auditing elements are common to most methodologies.

RESPONSIBLE CARE

The Chemical Manufacturers' Association (CMA) (1991) has published its *Responsible Care Code,* to which all member organizations have committed. The codes aim to improve the chemical industry's management of chemicals, safety, health, and environmental performance.

Figure 3.2.2 presents the responsible care codes for pollution prevention. The codes do not constitute a methodology in that they do not prescribe how any organization implements them. Rather, they describe hallmarks that successful pollution prevention programs share. The codes also provide a series of checkpoints for an organization to incorporate into its methodology.

Determinants of Success

Today most corporations are committed to pollution prevention programs. Any lack of progress that exists represents the failure of a methodology to transfer corporate commitment into implementation at the production area. Area managers must meet multiple demands with limited amounts of time, people, and capital. Pollution prevention often competes for priority with ongoing demands of production, safety, maintenance, and employee relations. These competing demands for the area manager's attention present barriers to pollution prevention. A pollution prevention methodology can overcome these barriers in two ways:

By providing corporate enablers for the production areas
By providing production areas with a set of tools to simplify and shorten the assessment phase

Pollution prevention policies are effective when they are developed to mesh with the firm's overall programs (Hamner 1993). Total quality management (TQM) complements and aids pollution prevention. In many aspects, the goals of safety and pollution prevention are compatible. However, some aspects, such as lengthened operating cycles to reduce waste generation, increase the likelihood of accidents. The optimal pollution prevention program requires balancing these two potentially contradictory requirements.

CORPORATE ENABLERS

The output of the chartering step performed at the executive level can be viewed as a set of enablers designed to assist waste reduction at the process level. Enablers consist of both positive and negative inducements to reduce waste. They take a variety of forms, including the following:

- Policy statements and goals
- Capital for waste reduction projects
- People resources
- Training

Code 1
A clear commitment by senior management through policy, communications, and resources to ongoing reductions at each of the company's facilities in releases to air, water, and land.

Code 2
A quantitative inventory at each facility of wastes generated and released to the air, water, and land measured or estimated at the point of generation or release.

Code 3
Evaluation, sufficient to assist in establishing reduction priorities, of the potential impact of releases on the environment and the health and safety of employees and the public.

Code 4
Education of and dialog with employees and members of the public about the inventory, impact evaluation, and risks to the community.

Code 5
Establishment of priorities, goals, and plans for waste and release reduction, taking into account both community concerns and the potential safety, health, and environmental impacts as determined under Codes 3 and 4.

Code 6
Ongoing reduction of wastes and releases, giving preference first to source reduction, second to recycling and reuse, and third to treatment.

Code 7
Measure progress at each facility in reducing the generation of wastes and in reducing releases to the air, water, and land by updating the quantitative inventory at least annually.

Code 8
Ongoing dialog with employees and members of the public regarding waste and release information, progress in achieving reductions, and future plans. This dialog should be at a personal, face-to-face level, where possible, and should emphasize listening to others and discussing their concerns and ideas.

Code 9
Inclusion of waste and release prevention objectives in research and in the design of new or modified facilities, processes, or products.

Code 10
An ongoing program for promotion and support of waste and release reduction by others.

Code 11
Periodic evaluation of waste management practices associated with operations and equipment at each member company facility, taking into account community concerns and health, safety, and environmental impacts, and implement ongoing improvements.

Code 12
Implementation of a process for selecting, retaining, and reviewing contractors and toll manufacturers, that takes into account sound waste management practices that protect the environment and the health and safety of employees and the public.

Code 13
Implementation of engineering and operating controls at each member company facility to improve prevention of and early detection of releases that may contaminate groundwater.

Code 14
Implementation of an ongoing program for addressing past operating and waste management practices and for working with others to resolve identified problems at each active or inactive facility owned by a member company taking into account community concerns and health, safety, and environmental impacts.

FIG. 3.2.2 Responsible care codes for pollution prevention.

- Project accounting methods that favor waste reduction
- Awards and other forms of recognition
- Newsletters and other forms of communication
- Personnel evaluations based in part on progress in meeting waste reduction goals
- Requirements for incorporating waste reduction goals into business plans

Corporate managers can choose enablers to overcome barriers at the plant level.

ASSESSMENT TOOLS

The procedures that a methodology recommends for performing assessment activities are assessment tools. For example, the weighted-sum method of rating is a tool for prioritizing a list of waste reduction implementations. Alternative tools include simple voting or assigning options to each category as do-now or do-later. An effective methodology avoids presenting a single tool for performing an assessment activity. Providing multiple tools from which a production area can choose imparts flexibility to a methodology and makes it suitable for a variety of processes and waste streams.

Project Methodology

Proactive area managers need not wait for direction from the top to begin reducing waste. Each area can make its own commitment to waste reduction and develop its own vision of a waste-free process. Thus, chartering can occur at the area level. Establishing an area waste reduction program provides a degree of independence that can help bridge the differences between corporate commitment and implementation at the process area. Figure 3.2.3 is an example of what such a program may look like.

Some suggestions for enhancing the effectiveness of the program follow (Trebilcock, Finkle, and DiJulia 1993; Rittmeyer 1991).

Chartering Activities

Selecting the waste streams for assessment is the first step in chartering a waste reduction program. This step is sometimes done at a high organizational level. Program planners should gather the minimum amount of data required to make their selections and use the fastest method possible to prioritize them. Methods such as weighted-sum ranking and weighting are not necessary for streams produced by a single area.

Other tools for prioritizing a waste stream can be considered. For example, Pareto diagrams are a simple way to rank waste streams by volume. Smaller waste volumes can be given high priority if they are toxic or if regulatory imperatives are anticipated. A Pareto analysis of a typical

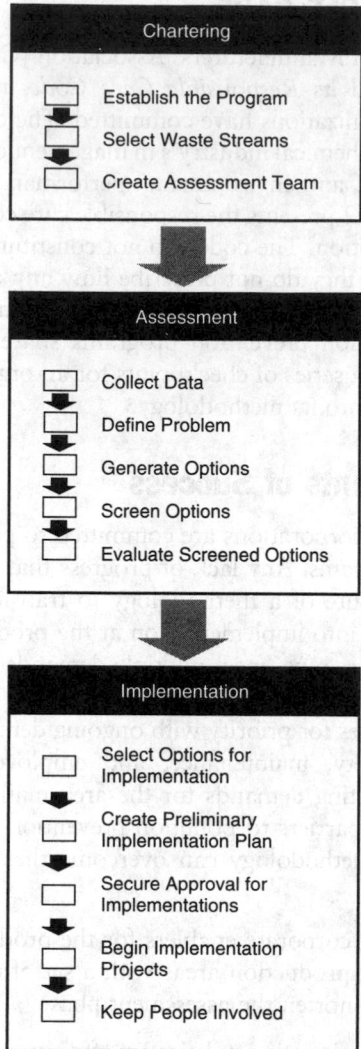

FIG. 3.2.3 A pollution prevention methodology for the production area.

chemical plant is likely to show that the top 20% of the waste stream accounts for more than 80% of the total waste volume.

In addition to selecting the major waste streams, planners should select a few small, easily reduced streams to reinforce the program with quick success.

Assessment Phase

Some general observations from the assessment phase follow.

An assessment should be quick, uncomplicated, and structured to suit local conditions. Otherwise, it is viewed as an annoyance intruding on the day-to-day concern of running a production process.

Assessment teams should be small, about six to eight people, to encourage open discussion when options are generated.

Including at least one line worker on an assessment team provides insight into how the process operates.

Including at least one person from outside the process on an assessment team provides a fresh perspective.

Area inspections and brainstorming meetings are valuable tools during the assessment phase.

Determining the source of the waste stream, as opposed to the equipment that emits it, is important before the option generation step.

Overly structured methods of screening options do not overcome group biases and are regarded as time-wasters by most teams.

Particularly helpful is the inclusion of people from outside the process on each assessment team. Outsiders provide an objective view. Their presence promotes creative thinking because they do not know the process well enough to be bound by conventions. Appointing outsiders as the assessment team leaders can capitalize on the fresh prospectives they provide.

The following is a task-by-task analysis of the assessment phase of a project (Trebilcock, Finkle, and DiJulia 1993).

DATA COLLECTION

Assessment teams should not collect exhaustive documentation, most of which is marginally useful. Material balances and process diagrams are minimum requirements, but many assessments require little more than that.

For each assessment, some combination of the following information is useful during the assessment phase:

- Operating procedures
- Flow rates
- Batch sizes
- Waste concentrations within streams
- Raw materials and finished product specifications
- Information about laboratory experiments or plant trials.

The project team may want to obtain or generate a material balance before the area inspection. The material balance is the most useful piece of documentation. In most cases, having sufficient data to compile a material balance is all that is required for an assessment. Table 3.2.1 lists the potential sources of material balance information.

Energy balances are not considered useful because of their bias in the waste stream selection. Energy consumption is rated low as a criterion for selecting streams, and few of the options generated during an assessment have a significant impact on energy consumption. However, energy costs are included in the calculations for economic feasibility. Similarly, water balances are not considered useful, but water costs are included in the calculations for economic feasibility.

TABLE 3.2.1 SOURCES OF MATERIAL BALANCE INFORMATION

Samples, analyses, and flow measurements of feed stocks, products, and waste streams
Raw material purchase records
Material inventories
Emission inventories
Equipment cleaning and validation procedures
Batch make-up records
Product specifications
Design material balances
Production records
Operating logs
Standard operating procedures and operating manuals
Waste manifests

AREA INSPECTION

An area inspection is a useful team-building exercise and provides team members with a common ground in the process. Without an inspection, outside participants may have trouble understanding discussions during subsequent brainstorming.

PROBLEM DEFINITION

The sources and causes of waste generation should be well understood before option generation begins. A preassessment area inspection helps an assessment team understand the processes that generate pollution. Table 3.2.2 presents guidelines for such a site inspection. The assessment team should follow the process from the point where raw material enters the area to the point where the products and waste leave the area.

Determining the true source of the waste stream before the option generation part of the assessment phase is important. Impurities from an upstream process, poor process control, and other factors may combine to contribute to waste. Unless these sources are identified and their relative importance established, option generation can focus on a piece of equipment that emits the waste stream and may only produce a small part of the waste. As Figure 3.2.4 shows, the waste stream has four sources. Two of these sources are responsible for about 97% of the waste. However, because these sources were not identified beforehand, roughly equal numbers of options address all four sources. Fortunately, the causes of the waste stream were understood before the assessment was complete. But knowing the major sources of the waste beforehand would have saved time by allowing members to concentrate on them.

Several tools can help identify the source of the waste. A material balance is a good starting point. A cause-and-effect fishbone diagram, such as shown in Figure 3.2.4, can identify the sources of the waste and indicate where to look for reductions. Sampling to identify components

TABLE 3.2.2 GUIDELINES FOR SITE INSPECTION

Prepare an agenda in advance that covers all points that require clarification. Provide staff contacts in the area being assessed with the agenda several days before the inspection.

Schedule the inspection to coincide with the operation of interest (e.g., make-up chemical addition, bath sampling, bath dumping, start up, and shutdown

Monitor the operation at different times during the shift, and, if needed, during all three shifts, especially when waste generation highly depends on human involvement (e.g., in painting or parts cleaning operations).

Interview the operators, shift supervisors, and foremen in the assessed area. Do not hesitate to question more than one person if an answer is not forthcoming. Assess the operators' and their supervisors' awareness of the waste generation aspects of the operation. Note their familiarity (or lack of) with the impacts their operation may have on other operations.

Photograph the area of interest, if warranted. Photographs are valuable in the absence of plant layout drawings. Many details are captured in photographs that otherwise may be forgotten or inaccurately recalled.

Observe the housekeeping aspects of the operation. Check for signs of spills or leaks. Visit the maintenance shop and ask about any problems in keeping the equipment leak-free. Assess the overall cleanliness of the site. Pay attention to odors and fumes.

Assess the organizational structure and level of coordination of environmental activities between various departments.

Assess administrative controls, such as cost accounting procedures, material purchasing procedures, and waste collection procedures.

of the waste stream can provide clues to their sources. Control charts, histograms, and scatter diagrams can depict fluctuations in waste stream components and thus provide more clues.

OPTIONS GENERATION

For all but the most obvious waste problems, brainstorming is the best tool for generating waste reduction options. The best format for these meetings is to freely col-

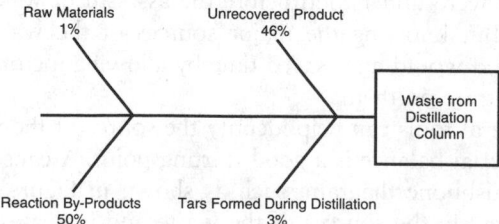

FIG. 3.2.4 Sources of waste.

lect ideas and avoid discussing them beyond what is necessary to understand them. Team members are encouraged to suggest ideas regardless of their practicality. Scribes capture suggestions and record them on cause-and-effect fishbone charts. The fishbone charts enable grouping options into categories such as chemistry, equipment modification, and new technology.

Identifying potential options relies on both the expertise and creativity of the team members. Much of the requisite knowledge comes from members' education and on-the-job experience. However, the use of technical literature, contacts, and other information sources is helpful. Table 3.2.3 lists some sources of background information for waste minimization techniques.

OPTIONS SCREENING

The EPA methodology offers several tools for screening options which vary in complexity from simple voting by the assessment team to more rigorous weighted-sum ranking and weighting.

TABLE 3.2.3 SOURCES OF BACKGROUND INFORMATION ON WASTE MINIMIZATION OPTIONS

Trade associations
As part of their overall function to assist companies within their industry, trade associations generally provide assistance and information about environmental regulations and various available techniques for complying with these regulations. The information provided is especially valuable since it is industry-specific.

Plant engineers and operators
The employees that are intimately familiar with a facility's operations are often the best source of suggestions for potential waste minimization options.

Published literature
Technical magazines, trade journals, government reports, and research briefs often contain information that can be used as waste minimization options.

State and local environmental agencies
A number of state and local agencies have or are developing programs that include technical assistance, information on industry-specific waste minimization techniques, and compiled bibliographies.

Equipment vendors
Meetings with equipment vendors, as well as vendor literature, are useful in identifying potential equipment-oriented options. Vendors are eager to assist companies in implementing projects. However, this information may be biased since the vendor's job is to sell equipment.

Consultants
Consultants can provide information about waste minimization techniques. A consultant with waste minimization experience in a particular industry is valuable.

In assessments using the weighted-sum method, follow-up meetings are held after brainstorming sessions. The meetings begin with an open discussion of the options. Sometimes, a team concludes that an option does not really reduce waste and removes it from the list. At other times, the team combines interdependent options into a single option or subdivides general options into more specific options.

After the team agrees on the final option list, they generate a set of criteria to evaluate the options. When the criteria are adopted, the team assigns each one a weight, usually between 0 and 10, to signify its relative importance. If the team feels that a criterion is not an important process or is adequately covered by another criterion, they can assign it a value of 0, essentially removing the criterion from the list.

After the weights are established, the team rates each option with a number from 0 to 10 according to how well it fulfills each criterion. Multiplying the weight by the rating provides a score for that criterion; the sum of all scores for all criteria yields the option's overall score.

The weighted-sum method has some potential pitfalls. An option can rank near the top of the list because it scores high in every criteria except probability of success or safety. However, an unsatisfactory score of these two criteria is enough to reject an option regardless of its other merits. High scores achieved by some impractical options probably indicate that the assessment team has used too many weighted criteria.

Another problem with ranking and weighting is that many options cannot be evaluated quickly. Some options must be better defined or require laboratory analysis, making ranking them at a meeting difficult.

Weighting and ranking meetings are not entirely fruitless. Often discussions about an option provide a basis for determining its technical and environmental feasibility.

One of the simpler tools offered by the EPA is to classify options into three categories: implement immediately, marginal or impractical, and more study required.

Other tools can be used to quickly screen options. These include cost–benefits analysis, simple voting, and listing options' pros and cons.

FEASIBILITY ANALYSIS OR OPTION EVALUATION

The most difficult part of the feasibility evaluation is the economic analysis. This analysis requires estimating equipment costs, installation costs, the amount of waste reduction, cost saving to the process, and economic return.

For projects with significant capital costs, a more detailed profitability analysis is necessary. The three standard profitability measures are:

- Payback period
- Net present value (NPV)
- Internal rate of return (IRR)

The payback period is the amount of time needed to recover the initial cash outlay on the project. Payback periods in the range of three to four years are usually acceptable for a low-risk investment. This method is recommended for quick assessment of profitability.

The NPV and IRR are both discounted cash flow techniques for determining profitability. Many companies use these methods to rank capital projects that are competing for funds. Capital funding for a project may hinge on the ability of the project to generate positive cash flows well beyond the payback period and realize an acceptable return on investment. Both the NPV and IRR methods recognize the time value of money by discounting future net cash flows. For an investment with a low-risk level, an aftertax IRR of 12 to 15% is typically acceptable.

Most spreadsheet programs for personal computers automatically calculate the IRR and NPV for a series of cash flows. More information on determining the IRR or NPV is available in any financial management, cost accounting, or engineering economics text.

When the NPV is calculated, the waste reduction benefits are not the only benefits. Most good options offer other benefits such as improved quality, reduced cycle times, increased productivity, and reduced compliance costs (see Table 3.2.4). The value of these additional benefits is often more than the value derived from reducing waste.

Implementation Phase

Waste reduction options that involve operational, procedural, or material changes (without additions or modifications to equipment) should be implemented as soon as the potential savings have been determined.

Some implementations consist of stepwise changes to the process, each incrementally reducing the amount of waste. Such changes can often be made without large capital expenditures and can be accomplished quickly. This approach is common in waste reduction. When expenditures are small, facilities are willing to make the changes without extensive study and testing. Several iterations of incremental improvement are often sufficient to eliminate the waste stream. Other implementations require large capital expenditures, laboratory testing, piloting, allocating resources, capital, installation, and testing.

Implementation resources should be selected that are as close to the process as possible. Engineers should not do what empowered personnel can do. External resources should not be solicited for a job that an area person can handle. A well-motivated facility can be self-reliant.

AUDITING

Measuring the success of each implementation is important feedback for future iterations of the pollution prevention program. Waste streams are eliminated not by a

TABLE 3.2.4 OPERATING COSTS AND SAVINGS ASSOCIATED WITH WASTE MINIMIZATION PROJECTS

Reduced waste management costs

This reduction includes reductions in costs for:

Offsite treatment, storage, and disposal fees

State fees and taxes on hazardous waste generators

Transportation costs

Onsite treatment, storage, and handling costs

Permitting, reporting, and recordkeeping costs

Input material cost savings

An option that reduces waste usually decreases the demand for input materials.

Insurance and liability savings

A waste minimization option can be significant enough to reduce a company's insurance payments. It can also lower a company's potential liability associated with remedial clean-up of treatment, storage, and disposal facilities (TSDFs) and workplace safety. (The magnitude of liability savings is difficult to determine).

Changes in costs associated with quality

A waste minimization option may have a positive or negative effect on product quality. This effect can result in higher (or lower) costs for rework, scrap, or quality control functions.

Changes in utility costs

Utility costs may increase or decrease. This cost includes steam, electricity, process and cooling water, plant air, refrigeration, or inert gas.

Changes in operating and maintenance labor, burden, and benefits

An option can either increase or decrease labor requirements. This change may be reflected in changes in overtime hours or in changes in the number of employees. When direct labor costs change, the burden and benefit costs also change. In large projects, supervision costs also change.

Changes in operating and maintenance supplies

An option can increase or decrease the use of operating and maintenance supplies.

Changes in overhead costs

Large waste minimization projects can affect a facility's overhead costs.

Changes in revenues from increased (or decreased) production

An option can result in an increase in the productivity of a unit. This increase results in a change in revenues. (Note that operating costs can also change accordingly.)

Increased revenues from by-products

A waste minimization option may produce a by-product that can be sold to a recycler or sold to another company as a raw material. This sale increases the company's revenues.

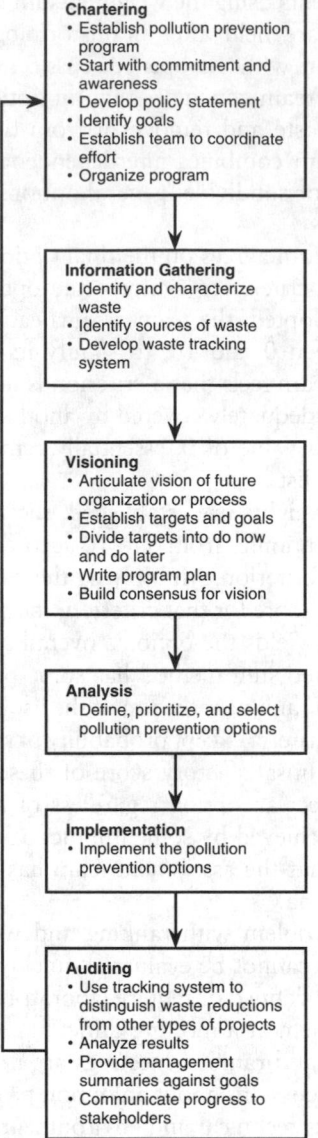

FIG. 3.2.5 Upgraded methodology.

Meeting minutes and worksheets used for analyses can be structured in such a way that merely collecting them in a folder is enough documentation.

METHODOLOGY UPGRADE

The EPA methodology has evolved from a method for conducting assessments to a comprehensive pollution prevention program. It will probably evolve again as experience with its application grows. Joint projects between the EPA and industry, such as the Chambers Works Project (U.S. EPA 1993), provide input to future iterations. The EPA is well-placed to develop an industry standard for pollution prevention methodologies.

An important strength of the current methodology is its recognition that pollution prevention requires participation from all levels of an organization. It contains well-articulated prescriptions about management commitment.

single, dramatic implementation, but by a series of small improvements implemented over time. Therefore, the last step is to renew the program.

Waste assessment should be documented as simply as possible. Capturing waste reduction ideas that were proposed and rejected may be useful in future iterations of the program. However, writing reports is not necessary.

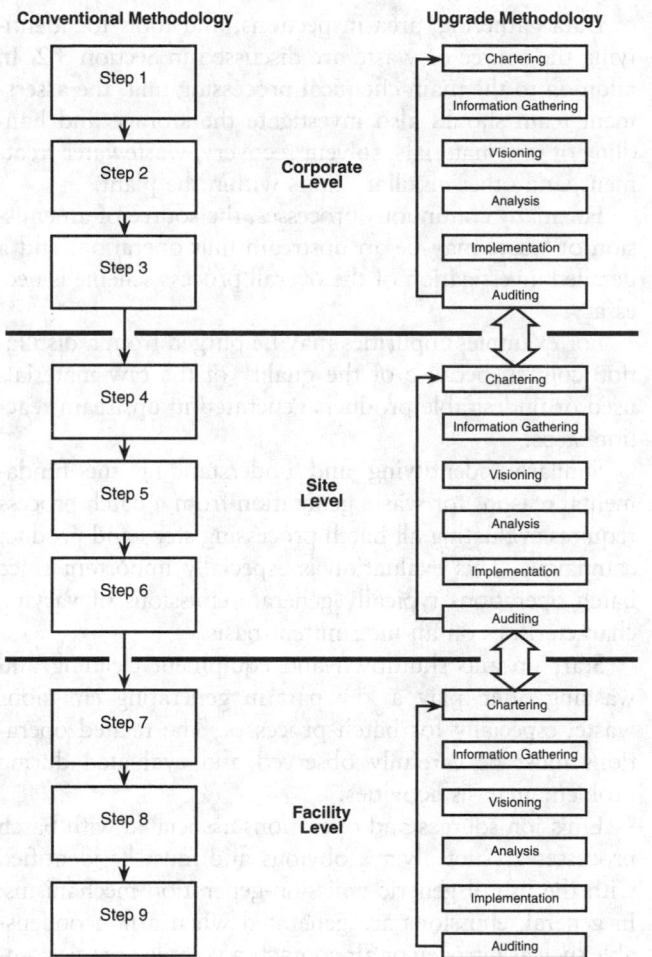

Conventional Methodology

Step 1

Step 2

Step 3

Step 4

Step 5

Step 6

Step 7

Step 8

Step 9

Upgrade Methodology

Corporate Level
- Chartering
- Information Gathering
- Visioning
- Analysis
- Implementation
- Auditing

Site Level
- Chartering
- Information Gathering
- Visioning
- Analysis
- Implementation
- Auditing

Facility Level
- Chartering
- Information Gathering
- Visioning
- Analysis
- Implementation
- Auditing

FIG. 3.2.6 Comparison of conventional and upgraded methodologics.

Figure 3.2.5 shows a suggested methodology update (U.S. EPA 1993). One unique feature is that all steps must be performed at all organization levels. This concept is illustrated in Figure 3.2.6. Most methodologies consist of a series of steps: the first few of which are performed at the highest organization levels, and the last of which are performed at the line organization. However, the new methodology prescribes that each step of the plan be performed at each level of the organization.

The activities recommended for each step consider the limited time and resources available for pollution prevention. Instead of prescribing "how-tos", the methodology provides a variety of tools from which local sites can choose. The hope is that waste reduction opportunities can be identified quickly, leaving more time for people to perform the implementations that actually reduce waste.

—David H.F. Liu

References

Hamner, Burton. 1993. Industrial pollution prevention planning in Washington state: First wave results. Paper presented at AIChE 1993 National Meeting, Seattle, Washington, August 1993.

Rittmeyer, Robert W. 1991. Prepare an effective pollution-prevention program. *Chem. Eng. Progress* (May).

Trebilcock, Robert W., Joyce T. Finkle, and Thomas DiJulia. 1993. A methodology for reducing wastes from chemical processes. Paper presented at AIChE 1993 National Meeting, Seattle, Washington, August 1993.

U.S. Environmental Protection Agency (EPA). 1988. *Waste minimization opportunity assessment manual.* Washington, D.C.

———. 1992. *Facility pollution prevention guide.* EPA/600/R-92/088. Washington, D.C.

———. 1993. *DuPont Chambers Works waste minimization project.* EPA/600/R-93/203 (November). Washington, D.C.: Office of Research and Development.

3.3
POLLUTION PREVENTION TECHNIQUES

In the current working definition used by the EPA, source reduction and recycling are considered the most viable pollution prevention techniques, preceding treatment and disposal. A detailed flow diagram, providing an in-depth approach to pollution prevention, is shown in Figure 3.3.1.

Of the two approaches, source reduction is usually preferable to recycling from an environmental perspective. Source reduction and recycling are comprised of a number of practices and approaches which are shown in Figure 3.3.2.

A pollution prevention assessment involves three main steps as shown in Figure 3.3.3. This section focuses on defining the problem and developing pollution prevention strategies.

Defining the Problem

Unlike other field assessments, the pollution prevention assessment focuses on determining the reasons for releases and discharges to all environmental media. These reasons

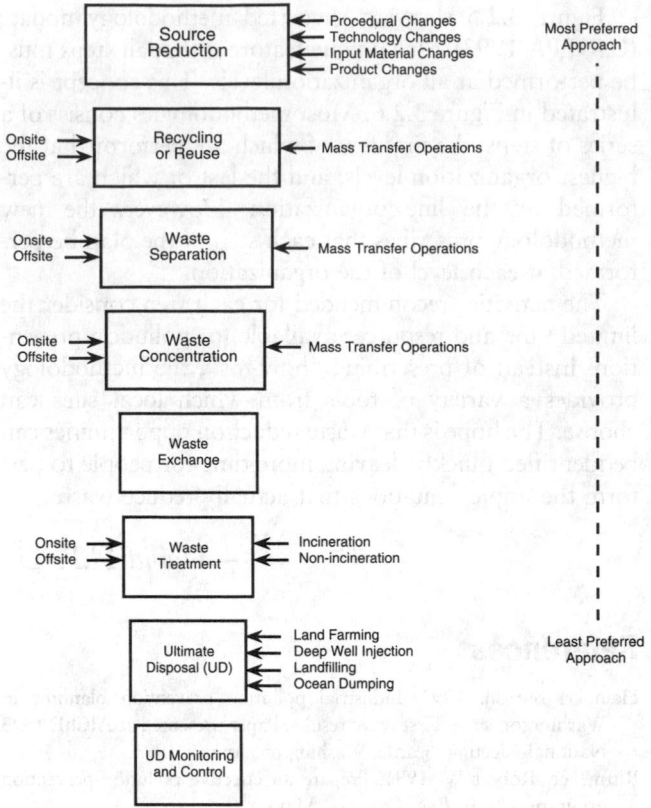

FIG. 3.3.1 Pollution prevention hierarchy.

can be identified based on the premise that the generation of emissions and waste follow recurring patterns independent of the manufacturing process (Chadha and Parmele, 1993).

Emissions and waste are generated due to process chemistry, engineering design, operating practices, or maintenance procedures. Classifying the causes into these four generic categories provides a simple but structured framework for developing pollution prevention solutions.

Data gathering, area inspections, and tools for identifying the source of waste are discussed in Section 3.2. In addition to the main chemical processing unit, the assessment team should also investigate the storage and handling of raw materials, solvent recovery, wastewater treatment, and other auxiliary units within the plant.

For many continuous processes, the source of an emission or waste may be an upstream unit operation, and a detailed investigation of the overall process scheme is necessary.

For example, impurities may be purged from a distillation column because of the quality of the raw materials used or undesirable products generated in upstream reaction steps.

Similarly, identifying and understanding the fundamental reasons for waste generation from a batch process requires evaluating all batch processing steps and product campaigns. This evaluation is especially important since batch operations typically generate emissions of varying characteristics on an intermittent basis.

Start up and shutdown and equipment cleaning and washing often play a key part in generating emissions waste, especially for batch processes. The related operations must be carefully observed and evaluated during problem analysis activities.

Emission sources and operations associated with batch processes are not always obvious and must be identified with the use of generic emission-generation mechanisms. In general, emissions are generated when a noncondensable such as nitrogen or air contacts a volatile organic compound (VOC) or when uncondensed material leaves a process.

Thus, for batch processes involving VOCs, processing steps such as charging the raw material powders, pressure transfer of the vessel's contents with nitrogen, solvent cleaning of the vessel's contents with nitrogen, and solvent cleaning of the vessels between batches should be closely

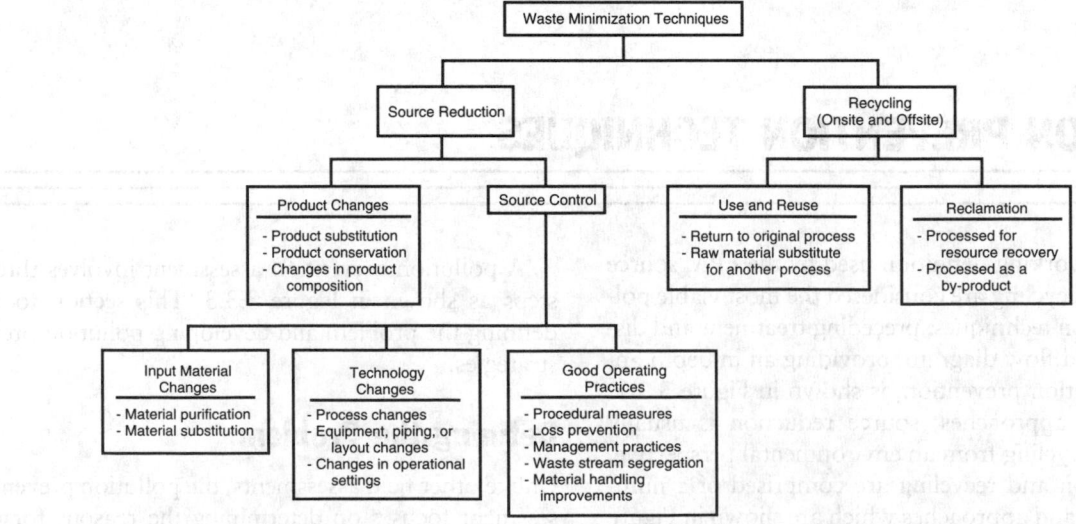

FIG. 3.3.2 Waste minimization techniques.

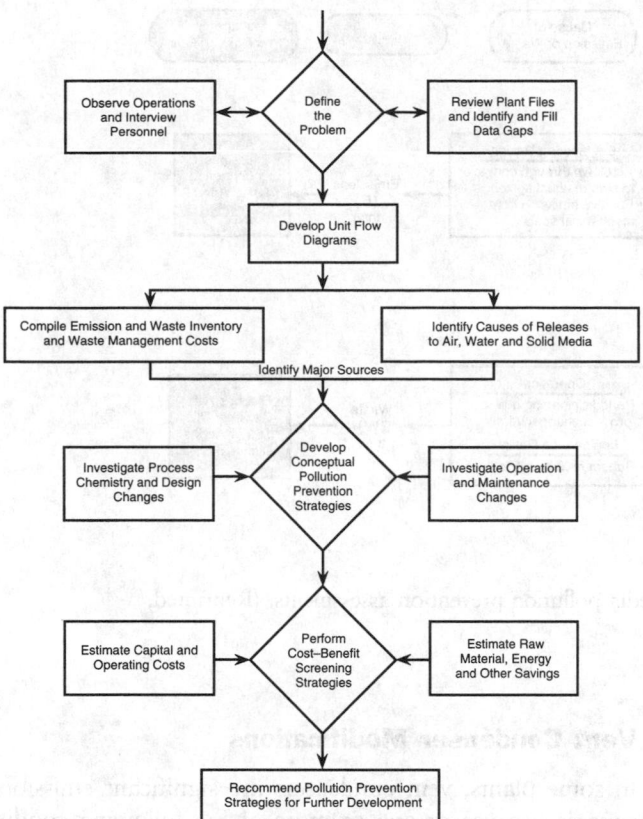

FIG. 3.3.3 Methodology for multimedia pollution prevention assessments. (Reprinted, with permission, from N. Chadha, 1994, Develop multimedia pollution prevention strategies, *Chem. Eng. Progress* [November].)

observed. The operator may leave charging manholes open for a long period or use vessel cleaning procedures different from written procedures (if any), which can increase the generation of emissions and waste. The field inspection may also reveal in-plant modifications such as piping bypasses that are not reflected in the site drawings and should be assessed otherwise.

The unit flow diagram (UFD) shown in Figure 3.3.4 is a convenient way to represent the material conversion relationships between raw materials, solvents, products, by-products, and all environmental discharges. The UFD is a tool that systematically performs a unit-by-unit assessment of an entire production process from the perspective of discharges to sewers and vents. This visual summary focuses on major releases and discharges and prioritizes a facility's subsequent pollution prevention activities.

Developing Conceptual Strategies

The next step is to develop conceptual strategies that specifically match the causes of emissions and waste generation. Addressing the fundamental causes helps to develop long-term solutions rather than simply addressing the symptoms.

A simple tool for brainstorming ideas and developing options is to use checklists based on practical experience. Tables 3.3.1 to 3.3.4 list 100 pollution prevention strategies based on changes in engineering design, process chemistry, operating procedures, and maintenance practices. These tables are based on the experiences of Chadha (1994), Chadha and Parmele (1993), Freeman (1989), Nelson (1989), and the U.S. EPA (1992) and are not comprehensive. The variety of technical areas covered by these checklists emphasizes the importance of a multimedia, multidisciplinary approach to pollution prevention.

Source Reduction

Source reduction techniques include process chemistry modifications, engineering design modifications, vent condenser modifications, reducing nitrogen usage, additional automation, and operational modifications.

PROCESS CHEMISTRY MODIFICATIONS

In some cases, the reasons for emissions are related to process chemistry, such as the reaction stoichiometry, kinetics, conversion, or yields. Emission generation is minimized by strategies varying from simply adjusting the order in which reactants are added to major changes that require significant process development work and capital expenditures.

Changing the Order of Reactant Additions

A pharmaceutical plant made process chemistry modifications to minimize the emissions of an undesirable by-product, isobutylene, from a mature synthesis process. The process consisted of four batch operations (see Figure 3.3.5). Emissions of isobutylene were reduced when the process conditions that led to its formation in the third step of the process were identified.

In the first reaction of the process, tertiary butyl alcohol (TBA) was used to temporarily block a reactive site on the primary molecule. After the second reaction was complete, TBA was removed as tertiary butyl chloride (TBC) by hydrolysis with hydrochloric acid. To improve process economics, the final step involved the recovery of TBA by reacting TBC with sodium hydroxide. However, TBA recovery was incomplete because isobutylene was inadvertently formed during the TBA recovery step.

An investigation indicated that the addition of excess NaOH caused alkaline conditions in the reactor that favored the formation of isobutylene over TBA. When the order of adding the NaOH and TBC was reversed and the NaOH addition rate was controlled to maintain the pH between 1 and 2, the isobutylene formation was almost completely eliminated. Therefore, installing add-on emission controls was unnecessary, and the only capital expense was the installation of a pH control loop.

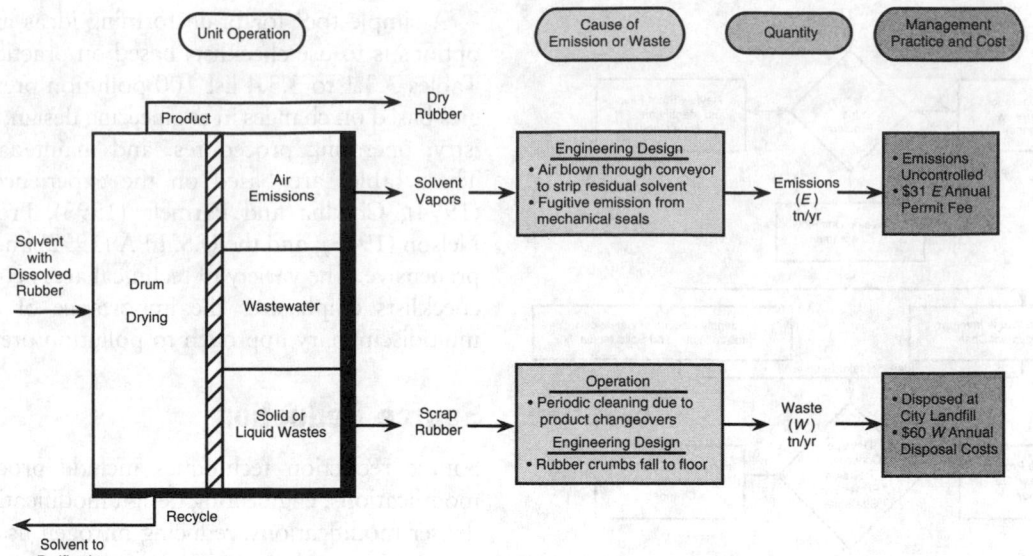

FIG. 3.3.4 Typical unit flow diagram for multimedia pollution prevention assessments. (Reprinted, with permission, from Chadha 1994.)

Changing the Chemistry

In one plant, odorous emissions were observed for several years near a drum dryer line used for volatilizing an organic solvent from a reaction mixture. Although two dryer–product lines existed, the odors were observed only near one line.

The analysis and field testing indicated that the chemical compounds causing the odors were produced in upstream unit operations due to the hydrolysis of a chemical additive used in the process. The hydrolysis products were stripped out of the solution by the process solvent and appeared as odorous fumes at the dryer. Conditions for hydrolysis were favorable at upstream locations because of temperature and acidity conditions and the residence time available in the process. Also, the water for the hydrolysis was provided by another water-based chemical additive used in the dryer line that had the odor problem.

Because the cause of the odorous emission was the process chemistry, the plant had to evaluate ways to minimize hydrolysis and the resulting formation of odorous products. Ventilation modifications to mitigate the odor levels would not be a long-term solution to the odor problem.

ENGINEERING DESIGN MODIFICATIONS

Emissions can be caused by equipment operating above its design capacity, pressure and temperature conditions, improper process controls, or faulty instrumentation. Strategies vary from troubleshooting and clearing obstructed equipment to designing and installing new hardware.

Vent Condenser Modifications

In some plants, vent condensers are significant emission sources because of one or more of the following conditions:

Field modifications bypass vent condensers, but the associated changes are not documented in the engineering drawings.

The vent stream is too dilute to condense because of changes in process conditions.

The condenser is overloaded (e.g., the heat-transfer area is inadequate) due to gradual increases in production capacity over time.

The overall heat-transfer coefficient is much lower than design because of fouling by dirty components or condenser flooding with large quantities of noncondensable nitrogen gas.

The condenser's cooling capacity is limited by improper control schemes. In one case, only the coolant return temperature was controlled.

In each case, design modifications are needed to reduce emissions.

REDUCING NITROGEN USAGE

Identifying ways to reduce nitrogen usage helps to minimize solvent emissions from a process. For example, every 1000 cu ft of nitrogen vents approximately 970 lb of methylene chloride with it at 20°C and 132 lb of methylene chloride with it at −10°C. The problem is aggravated if fine mists or aerosols are created due to pressure transfer or entrainment and the nitrogen becomes supersaturated with the solvent.

TABLE 3.3.1 ENGINEERING DESIGN-BASED POLLUTION PREVENTION STRATEGIES

Storage and Handling Systems
Install geodesic domes for external floating-roof tanks.
Store VOCs in floating-roof tanks instead of fixed-roof tanks.
Store VOCs in low-pressure vessels instead of atmospheric storage tanks.
Use onsite boilers instead of wet scrubbers for air pollution control.
Select vessels with smooth internals for batch tanks requiring frequent cleaning.
Install curbs around tank truck unloading racks and other equipment located outdoors.
Load VOC-containing vessels via dip pipes instead of splash loading.
Install closed-loop vapor recycling systems for loading and unloading operations.

Process Equipment
Use rotary-vane vacuum pumps instead of steam ejectors.
Use explosion-proof pumps for transferring VOCs instead of nitrogen or air pressure transfer.
Install canned or magnetic-drive sealless pumps.
Install hard-faced double or tandem mechanical seals or flexible face seals.
Use shell-and-tube heat exchangers instead of barometric condensers.
Install welded piping instead of flanges and screwed connections.
Install lining in pipes or use different materials of construction.
Install removable or reusable insulation instead of fixed insulation.
Select new design valves that minimize fugitive emissions.
Use reboilers instead of live steam for providing heat in distillation columns.
Cool VOC-containing vessels via external jackets instead of direct-contact liquid nitrogen.
Install high-pressure rotary nozzles inside tanks that require frequent washing.

Process Controls and Instrumentation
Install variable-speed electric motors for agitators and pumps.
Install automatic high-level shutoffs on storage and process tanks.
Install advanced process control schemes for key process parameters.
Install programmable logic controllers to automate batch processes.
Install instrumentation for inline sampling and analysis.
Install alarms and other instrumentation to help avoid runaway reactions, trips, and shutdowns.
Install timers to automatically shut off nitrogen used for blowing VOC-containing lines.

Recycle and Recovery Equipment
Install inplant distillation stills for recycling and reusing solvent.
Install thin-film evaporators to recover additional product from distillation bottoms and residues.
Recover volatile organics in steam strippers upstream of wastewater treatment lagoons.
Selectively recover by-products from waste using solvent extraction, membrane separation, or other operations.
Install equipment and piping to reuse noncontact cooling water.
Install new oil–water separation equipment with improved designs.
Install static mixers upstream of reactor vessels to improve mixing characteristics.
Use a high-pressure filter press or sludge dryer for reducing the volume of hazardous sludge.
Use reusable bag filters instead of cartridge filters for liquid streams.

Source: N. Chadha, 1994, Develop multimedia pollution prevention strategies, *Chem. Eng. Progress* (November).

Some plants can monitor and reduce nitrogen consumption by installing flow rotameters in the nitrogen supply lines to each building. Within each building, simple engineering changes such as installing rotameters, programmable timers, and automatic shutoff valves can minimize solvent emissions.

ADDITIONAL AUTOMATION

Sometimes simply adding advanced process control can produce dramatic results. For example, an ion-exchange resin manufacturer improved the particle size uniformity

of resin beads by installing a computerized process control. This improvement reduced the waste of off-spec resins by 40%.

OPERATIONAL MODIFICATIONS

Operational factors that impact emissions include the operating rate, scheduling of product campaigns, and the plant's standard operating procedures. Implementing operational modifications often requires the least capital compared to other strategies.

TABLE 3.3.2 PROCESS CHEMISTRY AND
TECHNOLOGY-BASED STRATEGIES

Raw Materials

Use different types or physical forms of catalysts.

Use water-based coatings instead of VOC-based coatings.

Use pure oxygen instead of air for oxidation reactions.

Use pigments, fluxes, solders, and biocides without heavy metals or other hazardous components.

Use terpene or citric-acid-based solvents instead of chlorinated or flammable solvents.

Use supercritical carbon dioxide instead of chlorinated or flammable solvents.

Use plastic blasting media or dry ice pellets instead of sand blasting.

Use dry developers instead of wet developers for nondestructive testing.

Use hot air drying instead of solvent drying for components.

Use no-clean or low-solids fluxes for soldering applications.

Plant Unit Operations

Optimize the relative location of unit operations within a process.

Investigate consolidation of unit operations where feasible.

Optimize existing reactor design based on reaction kinetics, mixing characteristics, and other parameters.

Investigate reactor design alternatives to the continuously stirred tank reactor.

Investigate a separate reactor for processing recycling and waste streams.

Investigate different ways of adding reactants (*e.g.*, slurries versus solid powders).

Investigate changing the order of adding reaction raw materials.

Investigate chemical synthesis methods based on renewable resources rather than petrochemical feedstocks.

Investigate conversion of batch operations to continuous operations.

Change process conditions and avoid the hydrolysis of raw materials to unwanted by-products.

Use chemical additives to oxidize odorous compounds.

Use chemical emulsion breakers to improve organic–water separation in decanters.

Source: Chadha, 1994.

Market-driven product scheduling and inventory considerations often play an important part in the generation of waste and emissions. A computerized material inventory system and other administrative controls can address these constraints. Another common constraint for pollution prevention projects is conformance with product quality and other customer requirements (Chadha 1994).

An example of reducing emissions through operational modifications is a synthetic organic chemical manufacturing industry (SOCMI) plant that wanted to reduce emissions of a cyclohexane solvent from storage and loading and unloading operations. The tank farms had organic liquid storage tanks with both fixed-roof and floating-roof storage tanks. The major source of cyclohexane emissions was the liquid displacement due to periodic filling of fixed-roof storage tanks. Standard operating procedures were modified so that the fixed-roof storage tanks were always kept full and the cyclohexane liquid volume varied only in the floating-roof tanks. This simple operational modification reduced cyclohexane emissions from the tank farm by more than 20 tn/yr.

Another example is a pharmaceutical manufacturer who wanted to reduce emissions of a methylene chloride solvent from a process consisting of a batch reaction step followed by vacuum distillation to strip off the solvent. The batch distillation involved piping the reactor to a receiver vessel evacuated via a vacuum pump. The following changes were made in the operating procedures to minimize emissions:

The initial methylene chloride charge was added at a reactor temperature of −10°C rather than at room temperature. Providing cooling on the reactor jacket lowered the methylene chloride vapor pressure and minimized its losses when the reactor hatch was opened for charging solid reactants later in the batch cycle.

The nitrogen purge to the reactor was shut off during the vacuum distillation step. The continuous purge had been overloading the downstream vacuum pump system and was unnecessary because methylene chloride is not flammable. This change reduced losses due to the stripping of methylene chloride from the reaction mix.

The temperature of the evacuated receiving vessel was lowered during the vacuum distillation step. Providing maximum cooling on the receiving vessel minimized methylene chloride losses due to revaporization at the lower pressure of the receiving vessel.

Table 3.3.5 shows another checklist that can be integrated into an analysis structured like a hazard and operability (HAZOP) study but focuses on pollution prevention.

Recycling

Reuse and recycling (waste recovery) can provide a cost-effective waste management approach. This technique can help reduce costs for raw materials and waste disposal and possibly provide income from a salable waste. However, waste recovery should be considered in conjunction with source control options.

Waste reuse and recycling entail one or a combination of the following options:

- Use in a process
- Use in another process
- Processing for reuse
- Use as a fuel
- Exchange or sale

TABLE 3.3.3 OPERATIONS-BASED POLLUTION PREVENTION STRATEGIES

Inventory Management
Implement a computerized raw material inventory tracking system.
Maintain product inventory to minimize changeovers for batch operations.
Purchase raw materials in totes and other reusable containers.
Purchase raw materials with lower impurity levels.
Practice first-in/first-out inventory control.

Housekeeping Practices
Recycle and reuse wooden pallets used to store drums.
Implement procedures to segregate solid waste from aqueous discharges.
Implement procedures to segregate hazardous waste from nonhazardous waste.
Segregate and weigh waste generated by individual production areas.
Drain contents of unloading and loading hoses into collection sumps.

Operating Practices
Change filters based on pressure-drop measurements rather than operator preferences.
Increase relief valve set pressure to avoid premature lifting and loss of vessel contents.
Optimize reflux ratio for distillation columns to improve separation.
Optimize batch reaction operating procedures to minimize venting to process flares.
Optimize electrostatic spray booth coater stroke and processing line speed to conserve coating.
Implement a nitrogen conservation program for processes that commonly use VOCs.
Minimize the duration for which charging hatches are opened on VOC-containing vessels.
Use vent condensers to recover solvents when boiling solvents for vessel cleaning purposes.
Reduce the number or volume of samples collected for quality control purposes.
Develop and test new markets for off-spec products and other waste.
Blend small quantities of off-spec product into the salable product.

Cleaning Procedures
Use mechanical cleaning methods instead of organic solvents.
Operate solvent baths at lower temperatures and cover when not in use.
Reduce the depth of the solvent layer used in immersion baths.
Reduce the frequency of the solvent bath change-out.
Use deionized water to prepare cleaning and washing solutions.
Develop written operating procedures for cleaning and washing operations.

Source: Chadha, 1994.

The metal finishing industry uses a variety of physical, chemical, and electrochemical processes to clean, etch, and plate metallic and nonmetallic substrates. Chemical and electrochemical processes are performed in numerous chemical baths, which are following by a rinsing operation.

Various techniques for recovering metals and metal salts, such as electrolysis, electrodialysis, and ion exchange, can be used to recycle rinse water in a closed-loop or open-loop system. In a closed-loop system, the treated effluent is returned to the rinse system. In an open-loop, the treated effluent is reused in the rinse system, but the final rinse is accomplished with fresh water. An example of a closed-loop system is shown in Figure 3.3.6.

Due to the cost associated with purchasing virgin solvents and the subsequent disposal of solvent waste, onsite recycling is a favorable option. Recycling back to the generating process is favored for solvents used in large volumes in one or more processes.

Some companies have developed ingenious techniques for recycling waste streams that greatly reduced water consumption and waste regeneration. At a refinery, hydrocarbon-contaminated wastewater and steam condensate are first reused as washwater in compressor aftercoolers to prevent salt buildup. The washwater is then pumped to a fluid catalytic cracker column to absorb ammonium salts from the vapor. The washwater, now laden with phenol, hydrogen sulfide, and ammonia, is pumped to a crude column vapor line, where organics extract the phenol from the wastewater. This step reduces the organic load to the downstream end-of-pipe wastewater treatment process which includes steam stripping and a biological system (Yen 1994).

A general pollution prevention option in the paper and pulp industry is to use closed-cycle mill processes. An example of a closed-cycle bleached kraft pulp mill is shown in Figure 3.3.7. This system is completely closed, and water is added only to the bleached pulp decker or to the last

TABLE 3.3.4 MAINTENANCE-BASED STRATEGIES

Existing Preventive Maintenance (PM) Program

Include centrifuges, dryers, and other process equipment in the PM program.

Include conveyors and other material handling equipment in the PM program.

Minimize pipe and connector stresses caused by vibration of pumps and compressors.

Minimize air leaks into VOC-containing equipment operating under vacuum.

Minimize steam leaks into process equipment.

Adjust burners to optimize the air-to-fuel ratio.

Implement a computerized inventory tracking system for maintenance chemicals.

Use terpene or citric-acid-based maintenance chemicals instead of chlorinated solvents.

Proactive PM Strategies

Monitor fugitive emissions from pumps, valves, agitators, and instrument connections.

Monitor fouling and leaks in heat exchangers and other process equipment.

Monitor vibration in rotating machinery.

Inspect and test interlocks, trips, and alarms.

Inspect and calibrate pH, flow, temperature, and other process control instruments.

Inspect and test relief valves and rupture disks for leaks.

Inspect and periodically replace seals and gaskets.

Source: Chadha, 1994.

---

dioxide stage washer of the bleach plant. The bleach plant is countercurrent, and a major portion of the filtrate from this plant is recycled to the stock washers, after which it flows to the black liquor evaporators and then to the recovery furnace. The evaporator condensate is steam stripped and used as a major water source at various points in the pulp mill. A white liquor evaporator is used to sep-

TABLE 3.3.5 EXAMPLE CHECKLIST OF POLLUTION REDUCTION METHODS

Material Handling

Recycling, in-process or external

Reuse or alternative use of the waste or chemical

Change in sources from batch operations (for example, heel reuse, change in bottom design of vessel, vapor space controls, dead-space controls)

Installation of isolation or containment systems

Installation of rework systems for treating off-spec materials

Change in practices for managing residuals (consolidation, recirculation, packaged amounts, reuse and purification)

Use of practices or equipment leading to segregated material streams

Recovery or rework of waste streams generated by maintenance or inspection activities

Chemical or Process Changes

Treatment or conversion of the chemical

Chemical substitution

Process change via change in thermodynamic parameters (temperature, pressure, chemical concentration, or phase) or installation of phase-separation equipment (such as vapor suppression systems, vessels with reduced vapor spaces, and filtration or extraction equipment)

Altering line or vessel length or diameter to make changes in the amount of product contained in lines or equipment that are purged

Installation of recirculation systems for process, water, gas inerting, or discharge streams as a substitute for single-pass streams

Time-Related Issues

Change in frequency of operation, cleaning, release, or use

Change in sequence of batch operations

Source: W.W. Doerr, 1993, Plan for the future with pollution prevention, *Chem. Eng. Progress* (January).

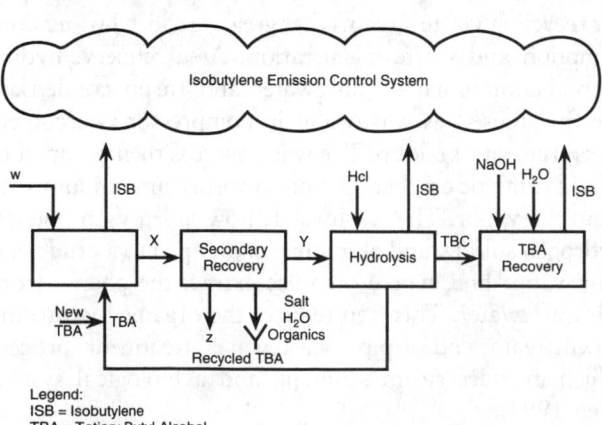

Legend:
ISB = Isobutylene
TBA = Tetiary Butyl Alcohol
TBC = Tetiary Butyl Chloride

FIG. 3.3.5 Process chemistry changes to reduce emissions. (Reprinted, with permission, from N. Chadha and C.S. Parmele, 1993, minimize emissions of toxics via process changes, *Chem. Eng. Progress* [January].)

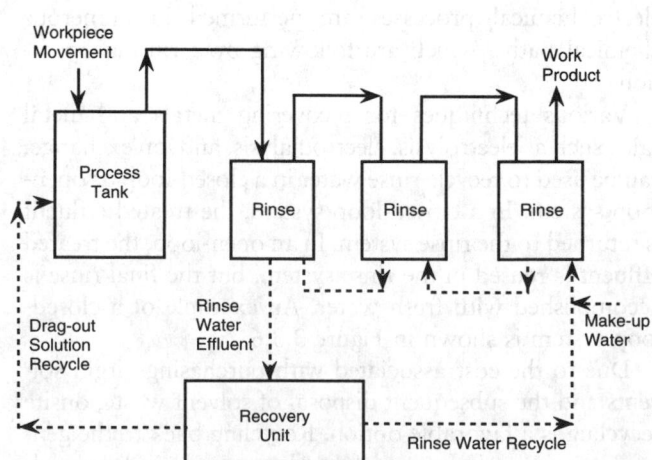

FIG. 3.3.6 Closed-loop rinse water recovery system.

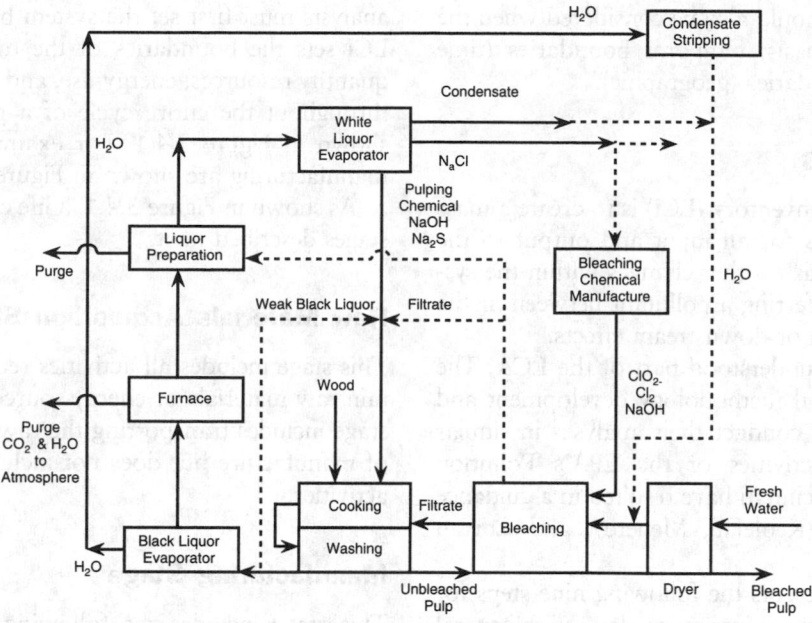

FIG. 3.3.7 Closed-cycle mill.

arate NaCl since the inlet stream to the water liquor evaporator contains a large amount of NaCl due to the recycling of bleach liquors to the recovery furnace (Theodore and McGuinn 1992).

—*David H.F. Liu*

References

Chadha, N. 1994. Develop multimedia pollution prevention strategies. *Chem. Eng. Progress* (November).

Chadha, N. and C.S. Parmele. 1993. Minimize emissions of toxics via process changes. *Chem. Eng. Progress* (January).

Doerr, W.W. 1993. Plan for the future with pollution prevention. *Chem. Eng. Progress* (January).

Freeman, H.W., ed. 1989. *Hazardous waste minimization: Industrial overview.* JAPCA Reprint Series, Aior and Waste Management Series. Pittsburgh, Pa.

Nelson, K.E. 1989. Examples of process modifications that reduce waste. Paper presented at AIChE Conference on Pollution Prevention for the 1990s: A Chemical Engineering Challenge, Washington, D.C., 1989.

Theodore, L. and Y.C. McGuinn. 1992. *Pollution prevention.* New York: Van Nostrand Reinhold.

U.S. Environmental Protection Agency (EPA). 1992. *Pollution protection case studies compendium.* EPA/600/R-92/046 (April). Washington, D.C.: EPA Office of Research and Development.

Yen, A.F. 1994. Industrial waste minimization techniques. *Environment '94*, a supplement to *Chemical Processing, 1994.*

3.4
LIFE CYCLE ASSESSMENT (LCA)

Life cycle refers to the cradle-to-grave stages associated with the production, use, and disposal of any product. A complete life cycle assessment (LCA), or ecobalance, consists of three complementary components:

Inventory analysis, which is a technical, data-based process of quantifying energy and resource use, atmospheric emissions, waterborne emissions, and solid waste

Impact analysis, which is a technical, quantitative, and qualitative process to characterize and assess the effects of the resource use and environmental loadings identified in the inventory state

Improvement analysis, which is the evaluation and implementation of opportunities to effect environmental improvement

Scoping is one of the first activities in any LCA and is considered by some as a fourth component. The scoping process links the goal of the analysis with the extent, or scope, of the study (i.e., that will or will not be included).

The following factors should also be considered when the scope is determined: basis, temporal boundaries (time scale), and spatial boundaries (geographic).

Inventory Analysis

The goal of a life cycle inventory (LCI) is to create a mass balance which accounts for all input and output to the overall system. It emphasizes that changes within the system may result in transferring a pollutant between media or may create upstream or downstream effects.

The LCI is the best understood part of the LCA. The LCA has had substantial methodology development and now most practitioners conduct their analyses in similar ways. The research activities of the EPA's Pollution Research Branch at Cincinnati have resulted in a guidance manual for the LCA (Keoleian, Menerey, and Curran 1993).

The EPA manual presents the following nine steps for performing a comprehensive inventory along with general issues to be addressed:

- Define the purpose
- Define the system boundaries
- Devise a checklist
- Gather data
- Develop stand-alone data
- Construct a model
- Present the results
- Conduct a peer review
- Interpret the results

DEFINING THE PURPOSE

The decision to perform an LCI is usually based on one or more of the following objectives:

To establish a baseline of information on a system's overall resource use, energy consumption, and environmental loading

To identify the stages within the life cycle of a product or process where a reduction in resource use and emissions can be achieved

To compare the system's input and output associated with alternative products, processes, or activities

To guide the development of new products, processes, or activities toward a net reduction of resource requirements and emissions

To identify areas to be addressed during life cycle impact analysis.

SYSTEM BOUNDARIES

Once the purposes for preparing an LCI are determined, the analyst should specifically define the system. (A *system* is a collection of operations that together perform some clearly defined functions.) In defining the system, the analysts must first set the system boundaries. A complete LCI sets the boundaries of the total system broadly to quantify resources, energy use, and environmental releases throughout the entire cycle of a product or process, as shown in Figure 3.4.1. For example, the three steps of manufacturing are shown in Figure 3.4.2.

As shown in Figure 3.4.1, a life cycle comprises the four stages described next.

Raw Materials Acquisition Stage

This stage includes all activities required to gather or obtain raw materials or energy sources from the earth. This stage includes transporting the raw materials to the point of manufacture but does not include material processing activities.

Manufacturing Stage

This stage includes the following three steps shown in Figure 3.4.2:

Materials manufacture—The activities required to process a raw material into a form that can be used to fabricate a product or package. Normally, the production of many intermediate chemicals or materials is included in this category. The transport of intermediate materials is also included.

Product fabrication—the process step that uses raw or manufactured materials to fabricate a product ready to be filled or packaged. This step often involves a consumer product that is distributed for use by other industries.

Filling, packaging, and distribution—processes that prepare the final products for shipment and transport the

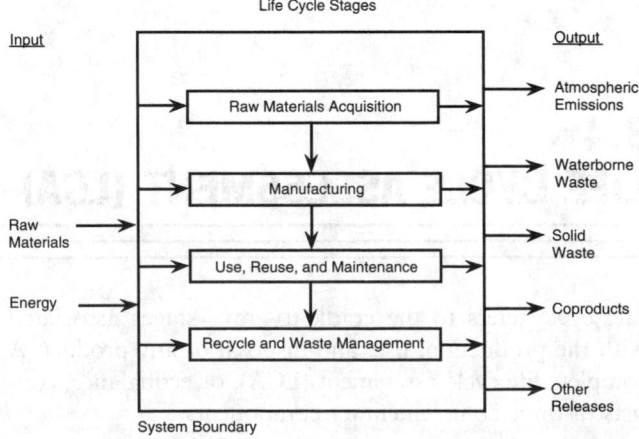

FIG. 3.4.1 Defining system boundaries. (Reprinted from G.A. Keoleian, Dan Menerey, and M.A. Curran, 1993, *Life cycle design guidance manual*, EPA/600/R-92/226 [January], Cincinnatti, Ohio: U.S. EPA, Risk Reduction Engineering Laboratory, Office of Research and Development.)

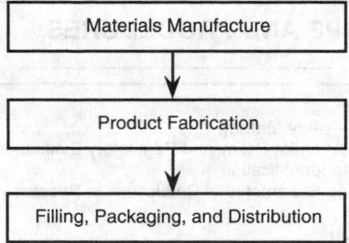

FIG. 3.4.2 Steps in the manufacturing stage. (Reprinted from Keoleian, Menerey, and Curran, 1993.)

products to retail outlets. In addition to primary packaging, some products require secondary and tertiary packaging and refrigeration to keep a product fresh, all of which should be accounted for in the inventory.

Use, Reuse, and Maintenance Stage

This stage begins after the product or material is distributed for use and includes any activity in which the product or package is reconditioned, maintained, or serviced to extend its useful life.

Recycling and Waste Management Stage

This stage begins after the product, package, or material has served its intended purpose and either enters a new system through recycling or enters the environment through the waste management system.

Examples of System Boundaries

Figure 3.4.3 shows an example of setting system boundaries for a product baseline analysis of a bar soap system. Tallow is the major material in soap production, and its primary raw material source is the grain fed to cattle. The production of paper for packaging the soap is also included. The fate of both the soap and its packaging end the life cycle of this system. Minor input could include the energy required to fabricate the tires on the combine that plants and harvests the grain.

The following analysis compares the life cycles of bar soap made from tallow and liquid hand soap made from synthetic ingredients. Because the two products have different raw material sources (cattle and petroleum), the analysis begins with the raw material acquisition steps. Because the two products are packaged differently and have different formulas, the materials manufacture and packaging steps must be included. Consumer use and waste management options should also be examined because the different formulas can result in varying usage patterns. Thus, for this comparative analysis, an analyst would have to inventory the entire life cycle of the two products.

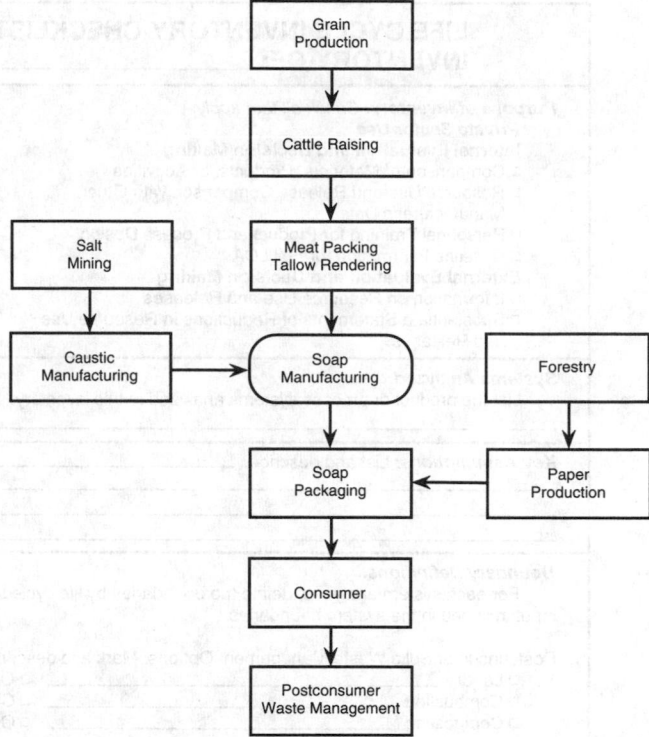

FIG. 3.4.3 Example system flow diagram for bar soap. (Reprinted from Keoleian, Menerey, and Curran, 1993.)

Again, the analyst must determine the basis of comparison between the systems. Because one soap is a solid and the other is a liquid, each with different densities and cleaning abilities per unit amount, comparing them on equal weights or volumes does not make sense. The key factor is how much of each is used in one hand-washing to provide an equal level of function or service.

A company comparing alternative processes for producing one petrochemical product may not need to consider the use and disposal of the product if the final composition is identical.

A company interested in using alternative material for its bottles while maintaining the same size and shape may not need filling the bottle as part of its inventory system. However, if the original bottles are compared to boxes of a different size and shape, the filling step must be included.

After the boundaries of each system are determined, a flow diagram as shown in Figure 3.4.3 can be developed to depict the system. Each system should be represented individually in the diagram, including production steps for ancillary input or output such as chemicals and packaging.

INVENTORY CHECKLIST

After inventory purposes and boundaries are defined, the analyst can prepare an inventory checklist to guide data collection and validation and to enable the computational model. Figure 3.4.4 shows a generic example of an in-

LIFE CYCLE INVENTORY CHECKLIST PART I—SCOPE AND PROCEDURES
INVENTORY OF:_____

Purpose of Inventory: *Check all that apply.*

Private Sector Use

Internal Evaluation and Decision Making
☐ Comparison of Materials, Products, or Activities
☐ Resource Use and Release Comparison with Other Manufacturer's Data
☐ Personnel Training for Product and Process Design
☐ Baseline Information for Full LCA

External Evaluation and Decision Making
☐ Information on Resource Use and Releases
☐ Substantiate Statements of Reductions in Resource Use and Releases

Public Sector Use

Evaluation and Policy Making
☐ Support Information for Policy and Regulatory Evaluation
☐ Information Gap Identification
☐ Aid in Evaluating Statements of Reductions in Resources Use and Releases

Public Education
☐ Support Materials for Public Education Development
☐ Curriculum Design Assistance

Systems Analyzed:
List the product or process systems analyzed in this inventory: _____
_____

Key Assumptions: List and describe. _____
_____
_____
_____

Boundary Definitions:
For each system analyzed, define the boundaries by life cycle stage, geographic scope, primary processes, and ancillary input included in the system boundaries.

Postconsumer Solid Waste Management Options: Mark and describe the options analyzed for each system.
☐ Landfill _____   ☐ Open-Loop Recycling _____
☐ Combustion _____   ☐ Closed-Loop Recycling _____
☐ Composting_____   ☐ Other _____

Basis for Comparison:
☐ This is not a comparative study.   ☐ This is a comparative study.
State basis for comparison between systems: *(Example: 1000 units, 1000 uses)* _____

If products or processes are not normally used on a one-to-one basis, state how the equivalent function was established.
_____

Computational Model Construction:
☐ System calculations are made using computer spreadsheets that relate each system component to the total system.
☐ System calculations are made using another technique. Describe: _____
_____

Descibe how input to and output from postconsumer solid waste management are handled. _____
_____
_____

Quality Assurance: State specific activities and initials of reviewer.
Review performed on: ☐ Data Gathering Techniques _____   ☐ Input Data _____
                                  ☐ Coproduct Allocation _____   ☐ Model Calculations and Formulas _____
                                                                                               ☐ Results and Reporting _____

Peer Review: State specific activities and initials of reviewer.
Review performed on: ☐ Scope and Boundary_____   ☐ Input Data _____
                                  ☐ Data Gathering Techniques _____   ☐ Model Calculations and Formulas _____
                                  ☐ Coproduct Allocation_____   ☐ Results and Reporting _____

Results Presentation:
☐ Methodology is fully described.
☐ Individual pollutants are reported.
☐ Emissions are reported as aggregated totals only.
   Explain why: _____
   _____
☐ Report is sufficiently detailed for its defined purpose.

☐ Report may need more detail for additional use beyond defined purpose.
☐ Sensitivity analyses are included in the report.
   List: _____
☐ Sensitivity analyses have been performed but are not included in the report. List: _____
_____

FIG. 3.4.4   A typical checklist of criteria with worksheet for performing an LCI. (Reprinted from Keoleian, Menerey, and Curran, 1993.)

LIFE CYCLE INVENTORY CHECKLIST PART II—MODULE WORKSHEET

Inventory of: _____   Preparer: _____

Life Cycle Stage Description: _____

Date: _____   Quality Assurance Approval: _____

MODULE DESCRIPTION: _____

	Data Value [a]	Type [b]	Data [c] Age/Scope	Quality Measures [d]
MODULE INPUT				
Materials				
Process				
Other [e]				
Energy				
Process				
Precombustion				
Water Usage				
Process				
Fuel-related				
MODULE OUTPUT				
Product				
Coproducts [f]				
Air Emissions				
Process				
Fuel-related				
Water Effluents				
Process				
Fuel-related				
Solid Waste				
Process				
Fuel-related				
Capital replacement				
Transportation				
Personnel				

(a) Include units.

(b) Indicate whether data are actual measurements, engineering estimates, or theoretical or published values and whether the numbers are from a specific manufacturer or facility or whether they represent industry-average values. List a specific source if pertinent, e.g., obtained from Atlanta facility wastewater permit monitoring data.

(c) Indicate whether emissions are all available, regulated only, or selected. Designate data as to geographic specificity, e.g., North America, and indicate the period covered, e.g., average of monthly for 1991.

(d) List measures of data quality available for the data item, e.g., accuracy, precision, representativeness, consistency-checked, other, or none.

(e) Include nontraditional input, e.g., land use, when appropriate and necessary.

(f) If coproduct allocation method was applied, indicate basis in quality measures column, e.g., weight.

FIG. 3.4.4 *Continued*

ventory checklist and an accompanying data worksheet. The LCA analyst may tailor this checklist for a given product or material.

PEER REVIEW PROCESS

Overall a peer review process addresses the four following areas:

- Scope and boundaries methodology
- Data acquisition and compilation
- Validity of key assumptions and results
- Communication of results

This peer review panel could participate at several points in the study: reviewing the purpose, system boundaries, assumptions, and data collection approach; reviewing the compiled data and the associated quality measures; and reviewing the draft inventory report, including the intended communication strategy.

GATHER DATA

Data for a process at a specific facility are often the most useful for analysis. Development teams may be able to generate their own data for in-house activities, but detailed information from outside sources is necessary for other life cycle stages. Sources of data for inventory analysis include:

Predominately In-House Data:

- Purchasing records
- Utility bills
- Regulatory records
- Accident reports
- Test data and material or product specifications

Public Data:

- Industry statistics
- Government reports including statistical summaries and regulatory reports and summaries
- Material, product, or industry studies
- Publicly available LCAs
- Material and product specifications
- Test data from public laboratories

Analysts must be careful in gathering data. The data presented in government reports may be outdated. Also, data in such reports are often presented as an average. Broad averages may not be suitable for accurate analysis. Journal articles, textbooks, and proceedings from technical conferences are other sources of information for an inventory analysis but may also be too general or outdated. Other useful sources include trade associations and testing laboratories. Many public laboratories publish their results. These reports cover such issues as consumer product safety, occupational health issues, or aspects of material performance and specifications.

Develop Stand-Alone Data

Stand-alone data is a term that describes the set of information developed to standardize or normalize the subsystem module input and output for the product, process, or activity being analyzed. (A *subsystem* is an individual step or process that is part of the defined system.) Stand-alone data must be developed for each subsystem to fit the subsystems into a single system. Two goals are necessary to achieve in this step:

Presenting data for each subsystem consistently by reporting the same product output from each subsystem
Developing the data in terms of the life cycle of only the product being examined in the inventory

A standard unit of output must be determined for each subsystem. All data could be reported in terms of producing a certain number of pounds, kilograms, or tons of a subsystem product.

Once the data are at a consistent reporting level, the analyst must determine the energy and material requirements

and the environmental releases attributed to the production of each coproduct using a technique called coproduct allocation. One commonly used allocation method is based on relative weight. Figure 3.4.5 illustrates this technique.

Once the input and output of each subsystem are allocated, the analyst can establish the numerical relationships of the subsystems within the entire system flow diagram. This process starts at the finished product of the system and works backward; it uses the relationships of the material input and product output of each subsystem to compute the input requirements from each of the preceding subsystems.

CONSTRUCT A COMPUTATION MODEL

The next step in an LCI is model construction. This step consists of incorporating the normalized data and material flows into a computational framework using a computer spreadsheet or other accounting technique. The sys-

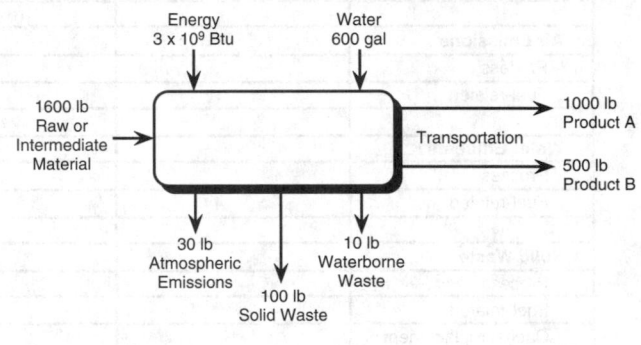

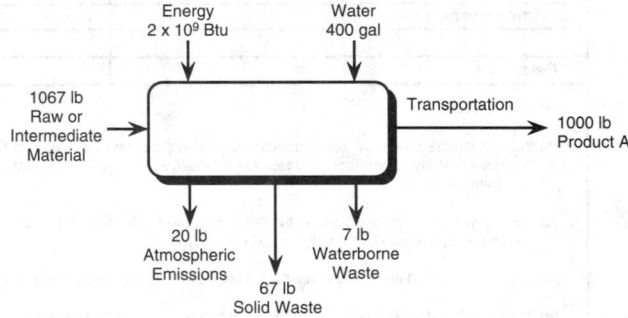

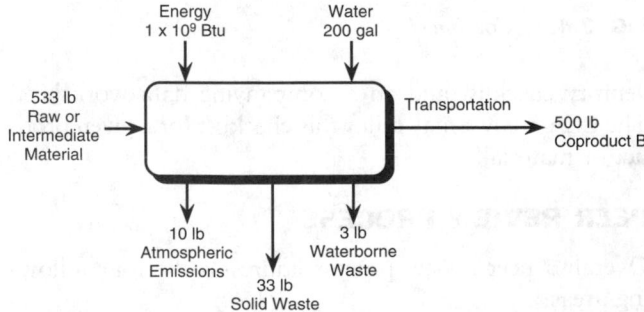

FIG. 3.4.5 Example coproduct allocation based on relative weight. (Reprinted from Keoleian, Menerey, and Curran, 1993.)

tem accounting data that result from the model computations give the total results for energy and resource use and environmental releases from the overall system.

The overall system flow diagram, derived in the previous step, is important in constructing the computational model because it numerically defines the relationships of the individual subsystems to each other in the production of the final product. These numerical relationships become the source of *proportionality factors,* which are quantitative relationships that reflect the relative subsystem contributions to the total system. The computational model can also be used to perform sensitivity analysis calculations.

PRESENT THE RESULTS

The results of the LCI should be presented in a report that explicitly defines the systems analyzed and the boundaries that were set. The report should explain all assumptions made, give the basis for comparison among the systems, and explain the equivalent usage ratio used. Using a checklist or worksheet as shown in Figure 3.4.4 provides a process for communicating this information.

A graphic presentation of information augments tabular data and aids interpretation. Both bar charts (either individual bars or stacked bars) and pie charts help the reader to visualize and assimilate the information from the perspective of gaining ownership or participation in the LCA.

For internal industrial use by product manufacturers, pie charts showing a breakout by raw materials, process, and use or disposal have been useful in identifying waste reduction opportunities.

Interpret and Communicate the Results

The interpretation of the results of the LCI depends on the purpose for which the analysis was performed. Before any statements regarding the results of the analysis are published, the analyst should review how the assumptions and boundaries were defined, the quality of the data used, and the representativeness of the data (e.g., whether the data were specific to one facility or representative of the entire industry).

The assumptions in analysis should be clearly documented. The significance of these assumptions should also be tested. For LCIs, sensitivity analysis can reveal how large the uncertainty in the input data can be before the results can no longer be used for the intended purpose.

The boundaries and data for many internal LCAs require that the results be interpreted for use within a particular corporation. The data used may be specific to a company and may not represent any typical or particular product on the market. However, because the data used in this type of analysis are frequently highly specific, ana-

lysts can assume a fairly high degree of accuracy in interpreting the results. Product design and process development groups often benefit from this level of interpretation.

The analyst should present the results of externally published studies comparing products, practices, or materials cautiously and consider the assumptions, boundaries, and data quality in drawing and presenting conclusions. Studies with different boundary conditions can have different results, yet both can be accurate. These limitations should be communicated to the reader along with all other results. Final conclusions about results from LCIs can involve value judgments about the relative importance of air and water quality, solid waste issues, resource depletion, and energy use. Based on the locale, background, and life style, different analysts make different value judgments.

LIMITATIONS AND TRENDS

Data quality is an ongoing concern in LCA due in part to the newness of the field. Additional difficulties include:

- Lack of data or inaccessible data
- Time and cost constraints for compiling data

Performing an LCA is complex, but the time and expense required for this task may be reduced in the future. The methodology has advanced furthest in Europe where it is becoming part of public policy-making and environmental initiatives (C&E News 1994).

The discipline has produced the two following organizations dedicated to the methodology:

The Society of Environmental Toxicology and Chemistry (SETAC), founded in 1979 and currently based in Pensacola, Florida and in Brussels. Its members are individuals working to develop LCA into a rigorous science.

The Society for the Promotion of LCA Development (SPOLD), founded in 1992 and based in Brussels. Its members are companies who support LCA as a decision making tool.

SPOLD is conducting a feasibility study on creating a database of lifetime inventories for commodities such as basic chemical feedstocks, electricity, packaging, water, and services.

Another public information source is the Norwegian database on LCA and clean production technology, which is operated by the World Industries Committee for the Environment (WICE) in Frederickstad, Norway. Although it does not inventory data, the database lists LCAs with information on product type, functional units, and system boundaries. The database already contains fifty LCAs and can be accessed by computer modem (telephone: 47 69 186618). According to project coordinator Ole Hanssen (1993), WICE's long term objective is to integrate LCA with pollution prevention and process innovation.

Impact Analysis

The impact analysis component of the LCA is a technical, quantitative, and qualitative process to characterize and assess the effects of the resource requirements and environmental loading (atmospheric and waterborne emissions and solid waste) identified in the inventory stage. Methods for impact analysis under development follow those presented at a SETAC workshop in 1992. The EPA's Office of Air Quality Planning has two documents which address life cycle impact analysis. (See also Chapter 2.)

The key concept in the impact analysis component is that of stressors. The stressor concept links the inventory and impact analysis by associating resource consumption and the releases documented in the inventory with potential impact. Thus, a *stressor* is a set of conditions that may lead to an impact. For example, a typical inventory quantifies the amount of SO_2 releases per product unit, which may then produce acid rain and then in turn affect the acidification of a lake. The resultant acidification might change the species composition and eventually create a loss of biodiversity.

Impact analysis is one of the most challenging aspects of LCA. Current methods for evaluating environmental impact are incomplete. Even when models exist, they can be based on many assumptions or require considerable data. The following sections describe several aspects of impact assessment and their limitations when applied to each of the major categories of environmental impact.

RESOURCE DEPLETION

The quantity of resources extracted and eventually consumed can be measured fairly accurately. However, the environmental and social costs of resource depletion are more difficult to assess. Depletion of nonrenewable resources limits their availability to future generations. Also, renewable resources used faster than they can be replaced are actually nonrenewable.

Another aspect of resource depletion important for impact assessment is resource quality. Resource quality is a measurement of the concentration of a primary material in a resource. In general, as resources become depleted, their quality declines. Using low-quality resources requires more energy and other input while producing more waste.

ECOLOGICAL EFFECTS

Ecological risk assessment is patterned after human health risk assessment but is more complex. As a first step in the analysis, the ecological stressors are identified; then the ecosystem potentially impacted is determined. Ecological stressors can be categorized as chemical (e.g., toxic chemicals released into the atmosphere), physical (e.g., habitat destruction through logging), or biological (e.g., the introduction of an exotic species).

The Ecology and Welfare Subcommittee of the U.S. EPA Science Advisory Board has developed a method for ranking ecological problems (Science Advisory Board 1990). The subcommittee's approach is based on a matrix of ecological stressors and ecosystem types (Harwell and Kelly 1986). Risks are classified according to the following:

- Type of ecological response
- Intensity of the potential effect
- Time scale for recovery following stress removal
- Spatial scale (local or regional biosphere)
- Transport media (air, water, or terrestrial)

The recovery rate of an ecosystem to a stressor is a critical part of risk assessment. In an extreme case, an ecological stress leads to permanent changes in the community structure or species extinction. The subcommittee classifies ecosystem responses to stressors by changes in the following:

Biotic community structure (alteration in the food chain and species diversity)
Ecosystem function (changes in the rate of production and nutrient cycling)
Species population of aesthetic or economic value
Potential for the ecosystem to act as a route of exposure to humans (bioaccumulation)

Determining potential risks and their likely effects is the first step in ecological assessment. Many stressors can be cumulative, finally resulting in large-scale problems. Both habitat degradation and atmospheric change are examples of ecological impact that gain attention.

Habitat Degradation

Human activities affect many ecosystems by destroying the habitat. When a habitat is degraded, the survival of many interrelated species is threatened. The most drastic effect is species extinction. Habitat degradation is measured by losses in biodiversity, decreased population size and range, and decreased productivity and biomass accumulation.

Standard methods of assessing habitat degradation focus on those species of direct human interest: game fish and animals, songbirds, or valuable crops (Suter 1990).

Ecological degradation does not result from industrial activity alone. Rapid human growth creates larger residential areas and converts natural areas to agriculture. Both are major sources of habitat degradation.

Atmospheric Change

A full impact assessment includes all scales of ecological impact. Impact can occur in local, regional, or global scales. Regional and local effects of pollution on atmosphere include acid rain and smog. Large-scale effects include global climate change caused by releases of greenhouse gases and increased ultraviolet (UV) radiation from ozone-depletion gases.

A relative scale is a useful method for characterizing the impact of emissions that deplete ozone or lead to global warming. For example, the heat-trapping ability of many gases can be compared to carbon dioxide, which is the main greenhouse gas. Similarly, the ozone-depleting effects of emissions can be compared to chlorofluorocarbons such as CFC-12. Using this common scale makes interpreting the results easier.

Environmental Fate Modeling

The specific ecological impact caused by pollution depends on its toxicity, degradation rate, and mobility in air, water, or land. Atmospheric, surface water, and groundwater transport models help to predict the fate of chemical releases, but these models can be complex. Although crude, equilibrium partitioning models offer a simple approach for predicting the environmental fate of releases. Factors useful for predicting the environmental fate include:

- Bioconcentration factor (BCF)—the chemical concentration in fish divided by the chemical concentration in water
- Vapor pressure
- Water solubility
- Octanol/water partition coefficient—the equilibrium concentration in octanol divided by the equilibrium chemical concentration in the aqueous phase
- Soil/water partition coefficient—the chemical concentration in soil divided by the chemical concentration in the aqueous phase

Once pathways through the environment and final fate are determined, impact assessment focuses on the effects. For example, impact depends on the persistence of releases and whether these pollutants degrade into further hazardous by-products.

HUMAN HEALTH AND SAFETY EFFECTS

Impact can be assessed for individuals and small populations or whole systems. The analyst usually uses the following steps to determine the impact on human health and safety: (1) hazard identification, (2) risk assessment, (3) exposure assessment, and (4) risk characterization. (See Section 11.8.)

Determining health risks from many design activities can be difficult. Experts, including toxicologists, industrial hygienists, and physicians, should be consulted in this process. Data sources for health risk assessment include biological monitoring reports, epidemiological studies, and bioassays. Morbidity and mortality data are available from sources such as the National Institute of Health, the Center for Disease Control, and the National Institute of Occupational Safety and Health.

The following ways are available to assess health impact: the threshold limited value–time-weighted average (TLV–TWA), the medium lethal dose (LD), the medium lethal concentration (LC), the no observed effect level (NOEL), and the no observed adverse effect level (NOAEL). (See Section 11.8.)

Other methods are used to compare the health impact of residuals. One approach divides emissions by regulatory standards to arrive at a simple index (Assies 1991). This normalized value can be added and compared when the emission standard for each pollutant is based on the same level of risk. However, this situation is rare. In addition, such an index reveals neither the severity nor whether the effects are acute or chronic. Properly assessing the impact of various releases on human health usually requires more sophistication than a simple index.

Impact on humans also includes safety. Unsafe activities cause particular types of health problems. Safety usually refers to physical injury caused by a chemical or mechanical force. Sources of safety-related accidents include malfunctioning equipment or products, explosions, fires, and spills. Safety statistics are compiled on incidences of accidents, including hours of lost work and types of injuries. Accident data are available from industry and insurance companies.

Health and safety risks to workers or users also depend on ergonomic factors. For tools and similar products, biomechanical features, such as grip, weight, and field of movement influence user safety and health.

ASSESSING SYSTEM RISK

Human error, poor maintenance, and interactions of products or systems with the environment produce consequences that should not be overlooked. Although useful for determining human health and safety effects, system risk assessment applies to all other categories of impact. For example, breakdowns or accidents waste resources and produce pollution that can lead to ecological damage. Large, catastrophic releases have a different impact than continual, smaller releases of pollutants.

In risk assessment, predicting how something can be misused is often as important as determining how it is supposed to function. Methods of risk assessment can be either relatively simple or quite complex. The most rigorous methods are usually employed to predict the potential for high-risk events in complex systems. Risk assessment models can be used in design to achieve inherently safe products. Inherently safe designs result from identifying and removing potential dangers rather than just reducing possible risks (Greenberg and Cramer 1991). A brief outline of popular risk assessment methods follows.

Simple Risk Assessment Procedures

These procedures include the following:

- Preliminary hazard analysis
- Checklists
- What-if analysis

A preliminary hazard analysis is suited for the earliest phases of design. This procedure identifies possible hazardous processes or substances during the conceptual stage of design and seeks to eliminate them, thereby avoiding the costly and time-consuming delays caused by later design changes.

Checklists ensure that the requirements addressing risks have not been overlooked or neglected. Design verification should be performed by a multidisciplinary team with expertise in appropriate areas.

A what-if analysis predicts the likelihood of possible events and determines their consequences through simple, qualitative means. Members of the development team prepare a list of questions that are answered and summarized in a table (Doerr 1991).

Mid-Level Risk Assessment Procedures

These procedures include the following:

- Failure mode and effects analysis (FEMA)
- HAZOP study

The FEMA is also a qualitative method. It is usually applied to individual components to assess the effect of their failure on the system. The level of detail is greater than in a what-if analysis (O'Mara 1991). HAZOPs systematically examine designs to determine where potential hazards exist and assign priorities. HAZOPs usually focus on process design.

Relatively Complex Risk Assessment Procedures

These procedures include the following:

- Faulty tree analysis (FTA)
- Event tree analysis (ETA)
- Human reliability analysis (HRA)

FTA is a structured, logical modeling tool that examines risks and hazards to precisely determine unwanted consequences. FTA graphically represents the actions leading to each event. Analysis is generally confined to a single system and produces a single number for the probability of that system's failure. FTA does not have to be used to generate numbers; it can also be used qualitatively to improve the understanding of how a system works and fails (Stoop 1990).

ETA studies the interaction of multiple systems or multiple events. ETA is frequently used with FTA to provide quantitative risk assessment. Event trees are also used to assess the probability of human errors occurring in a system.

HRA can be a key factor in determining risks and hazards and in evaluating the ergonomics of a design. HRA can take a variety of forms to provide proactive design recommendations.

LIMITATIONS

LCA analysts face other fundamental dilemmas. How to examine a comprehensive range of effects to reach a decision? How to compare different categories of impact? Assessment across categories is highly subjective and value laden. Thus, impact analysis must account for both scientific judgment and societal values. Decision theory and other approaches can help LCA practitioners make these complex and value-laden decisions.

Impact assessment inherits all the problems of inventory analysis. These problems include lack of data and time and cost constraints. Although many impact assessment models are available, their ability to predict environmental effects varies. Fundamental knowledge in some areas of this field is still lacking.

In addition to basic inventory data, impact analysis requires more information. The often complex and time-consuming task of making further measurements also creates barriers for impact analysis.

Even so, impact analysis is an important part of life cycle design. For now, development teams must rely on simplified methods. LCA analysts should keep abreast of developments in impact analysis so that they can apply the best available tools that meet time and cost constraints.

Improvement Analysis

The improvement analysis component of LCA is a systematic evaluation of the need and opportunities to reduce the environmental burden associated with energy and raw material use and waste emissions throughout the life cycle of a product, process, or activity. Improvement analysis has not received the immediate attention of the LCA methodology development community. Improvement analysis is usually conducted informally throughout an LCA evaluation as a series of what-if questions and discussions. To date, no rigorous or even conceptual framework of this component exists. Ironically, this component of the LCA is the reason to perform these analyses in the first place. SETAC has tentative plans to convene a workshop in 1994 (Consoil 1993).

—*David H.F. Liu*

References

Assies, J.A. 1991. Introduction paper. *SETAC-Europe Workshop on Environmental Life Cycle Analysis of Products, Leiden, Netherlands: Center for Environmental Science (CML), 2 December 1991.*

Battelle and Franklin Associates. 1992. *Life cycle assessment: Inventory guidelines and principles.* EPA/600/R-92/086. Cincinnati, Ohio: U.S. EPA, Risk Reduction Engineering Laboratory, Office of Research and Development.

Consoil, F.J. 1993. Life-cycle assessments—current perspectives. *4th Pollution Prevention Topical Conference, AIChE 1993 Summer National Meeting, Seattle, Washington, August, 1993.*

Doerr, W.W. 1991. WHAT-IF analysis. In *Risk assessment and risk management for the chemical process industry.* Edited by H.R. Greenberg and J.J. Cramer. New York: Van Nostrand Reinhold.

Greenberg, H.R. and J.J. Cramer. 1991. *Risk assessment and risk management for the chemical process industry.* New York: Van Nostrand Reinhold.

Harwell, M.A. and J.R. Kelly. 1986. *Workshop on ecological effects from environmental stresses.* Ithaca, N.Y.: Ecosystems Research Center, Cornell University.

O'Mara, R.L. 1991. Failure modes and effects analysis. In *Risk assessment and risk management for the chemical process industry.* Edited by H.R. Greenberg and J.J. Cramer. New York: Van Nostrand Reinhold.

Science Advisory Board. 1990. *The report of Ecology and Welfare Subcommittee, Relative Risk Reduction Project.* SAB-EC-90-021A. Washington, D.C.: U.S. EPA.

Stoop, J. 1990. Scenarios in the design process. *Applied Ergonomics* 21, no. 4.

Suter, Glenn W.I. 1990. Endpoints for regional ecological risk assessment. *Environmental Management* 14, no. 1.

3.5
SUSTAINABLE MANUFACTURING (SM)

In the report, *Our Common Future, sustainable development* is defined as "meets the needs of the current generation without compromising the needs of future generations" (United Nations World Commission on the Environment and Development 1987). The concept of sustainability is illustrated by natural ecosystems, such as the hydrologic cycle and the food cycle involving plants and animals. These systems function as semi-closed loops that change slowly, at a rate that allows time for natural adaptation.

In contrast to nature, material flows through our economy in one direction only—from raw material toward eventual disposal as industrial or municipal waste (see part (a) in Figure 3.5.1). Sustainable development demands change. When a product's design and manufacturing process are changed, the overall environmental impact can be reduced. Green design emphasizes the efficient use of materials and energy, reduction of waste toxicity, and reuse and recycling of materials (see part (b) in Figure 3.5.1).

SM seeks to meet consumer demands for products without compromising the resource and energy supply of future generations. SM is a comprehensive business strategy that maximizes the economic and environmental returns on a variety of innovative pollution prevention techniques (Kennedy 1993). These techniques including the following:

Design for environment (DFE) directs research and development (R&D) teams to develop products that are environmentally responsible. This effort revolves on product design.

Toxics use reduction (TUR) considers the internal chemical risks and potential external pollution risks at the process and worker level.

LCA defines the material usage and environmental impact over the life of a product.

SM embeds corporate environmental responsibility into material selection, process and facility design, marketing, strategic planning, cost accounting, and waste disposal.

Product Design and Material Selection

By following the design strategies described next, designers can meet environmental requirements.

PRODUCT SYSTEM LIFE EXTENSION

Extending the life of a product can directly reduce environmental impact. In many cases, longer-lived products save resources and generate less waste because fewer units are needed to satisfy the same need. Doubling the life of a product translates into a pollution prevention of 50% in process transportation and distribution and a waste reduction of 50% at the end of the product's life.

Understanding why products are retired helps designers to extend the product system life. Reasons why products are no longer in use include:

- Technical obsolescence
- Fashion obsolescence
- Degraded performance or structural fatigue caused by normal wear over repeated use
- Environmental or chemical degradation
- Damage caused by accident or inappropriate use

To achieve a longer service life, designers must address issues beyond simple wear. A discussion of specific strategies for product life extension follows.

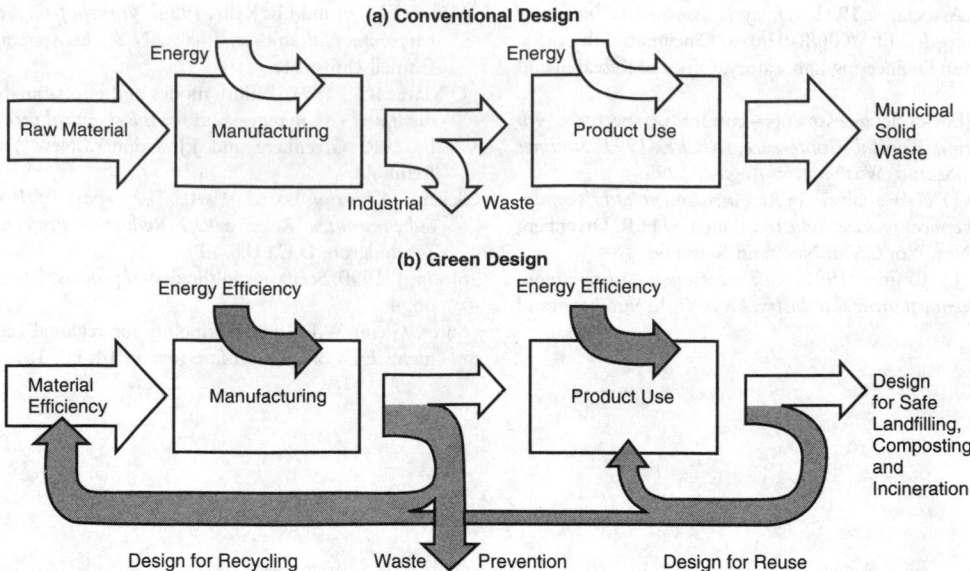

FIG. 3.5.1 How product design affects material flows. Making changes in a product's design reduces overall environmental impact. The green design emphasizes the efficient use of material and energy, reduction of waste toxicity, and reuse and recycling of materials. (Reprinted from U.S. Congress Office of Technology Assessment 1992, *Green products by design: Choices for a cleaner environment* [U.S. Government Printing Office].)

Appropriate Durability

Durable items can withstand wear, stress, and environmental degradation over a long useful life. Development teams should enhance durability only when appropriate. Designs that allow a product or component to last beyond its expected useful life are usually wasteful.

Enhanced durability can be part of a broader strategy focused on marketing and sales. Durability is an integral part of all profitable leasing. Original equipment manufacturers who lease their products usually gain the most from durable design.

For example, a European company leases all the photocopiers it manufactures. The company designs drums and other key components of their photocopiers for maximum durability to reduce the need for replacement or repair. Because the company maintains control of the machines, they select materials to reduce the cost and impact of disposal.

Adaptability

Adaptability can extend the useful life of a product that quickly becomes obsolete. To reduce the overall environmental impact, designers should design a product so that a sufficient portion of it remains after obsolete parts are replaced.

Adaptable designs rely on interchangeable components. For example, an adaptable strategy for a new razor blade design ensures that the new blade mounts on the old handle so that the handle does not become part of the waste stream.

A large American company designed a telecommunication control center using a modular work station approach. Consumers can upgrade components as needed to maintain state-of-the-art performance. Some system components change rapidly, while others stay in service for ten years or more.

Reliability

Reliability is often expressed as a probability. It measures the ability of a system to accomplish its design mission in the intended environment for a certain period of time.

The number of components, the individual reliability of each component, and the configuration are important aspects of reliability. Parts reduction and simplified design can increase both reliability and manufacturability. A simple design may also be easier to service. All these factors can reduce resource use and waste.

Designers cannot always achieve reliability by reducing parts or making designs simple. In some cases, they must add redundant systems to provide backup. When a reliable product system requires parallel systems or fail-safe components, the cost can rise significantly. Reliable designs must also meet all other project requirements.

Reliability should be designed into products rather than achieved through later inspection. Screening out potentially unreliable products after they are made is wasteful because such products must be repaired or discarded. Both environmental impact and cost increase.

For example, a large American electronics firm discovered that the plug-in boards on the digital scopes it designs failed in use. However, when the boards were returned for testing, 30% showed no defects and were sent back to customers. Some boards were returned repeatedly, only to pass tests every time. Finally, the company discovered that a bit of insulation on each of the problem boards' capacitors was missing, producing a short when they were installed in the scope. The cause was insufficient clearance between the board and the chassis of the scope; each time the board was installed it scraped against the side of the instrument. Finding the problem was difficult and expensive. Preventing it during design with a more thorough examination of fit and clearance would have been simpler and less costly.

Remanufacturability

Remanufacturing is an industrial process that restores worn products to like-new condition. In a factory, a retired product is first completely disassembled. Its usable parts are then cleaned, refurbished, and put into inventory. Finally, a new product is reassembled from both old and new parts, creating a unit equal in performance and expected life to the original or currently available alternative. In contrast, a repaired or rebuilt product usually retains its identity, and only those parts that have failed or are badly worn are replaced.

Industrial equipment or other expensive products not subject to rapid change are the best candidates for remanufacturing.

Designs must be easy to take apart if they are to be remanufactured. Adhesives, welding, and some fasteners can make this process impossible. Critical parts must be designed to survive normal wear. Extra material should be present on used parts to allow refinishing. Care in selecting materials and arranging parts also helps to reduce excessive damage during use. Design continuity increases the number of interchangeable parts between different models in the same product line. Common parts make remanufacturing products easier.

For example, a midwestern manufacturer could not afford to replace its thirteen aging plastic molding machines with new models, so it chose to remanufacture eight molders for one-third the cost of new machines. The company also bought one new machine at the same time. The remanufactured machines increased efficiency by 10 to 20% and decreased scrap output by 9% compared to the old equipment; performance was equal to the new molder. Even with updated controls, operator familiarity with the remanufactured machines and use of existing foundations and plumbing further reduced the cost of the remanufactured molders.

Reusability

Reuse is the additional use of an item after it is retired from a defined duty. Reformulation is not reuse. However, repair, cleaning, or refurbishing to maintain integrity can be done in the transition from one use to the next. When applied to products, reuse is a purely comparative term. Products with no single-use analogs are considered to be in service until discarded.

For example, a large supplier of industrial solvents designed a back-flush filter that could be reused many times. The new design replaced the single-use filters for some of their onsite equipment. Installing the back-flush filter caused an immediate reduction in waste generation, but further information about the environmental impact associated with the entire multiuse filter system is necessary to compare it to the impact of the single-use filters (Kusz 1990).

MATERIAL LIFE EXTENSION

Recycling is the reformation or reprocessing of a recovered material. The EPA defines recycling as "the series of activities, including collection, separation, and processing, by which products or other materials are recovered from or otherwise diverted from [the] solid waste stream for use in the form of raw materials in the manufacture of new products other than fuel" (U.S. EPA 1991a).

Recycled material can follow two major pathways: closed loop and open loop. In closed-loop systems, recovered material and products are suitable substitutes for virgin material. In theory a closed-loop model can operate for an extended period of time without virgin material. Of course, energy, and in some cases process material, is required for each recycling. Solvents and other industrial process ingredients are the most common materials recycled in a closed loop.

Open-loop recycling occurs when the recovered material is recycled one or more times before disposal. Most postconsumer material is recycled in an open loop. The slight variations or unknown composition of such material usually cause it to be downgraded to a less demanding use.

Some material also enters a cascade open-loop model in which it is degraded several times before the final discard. For example, used white paper can be recycled into additional ledger or computer paper. If this product is then dyed and not de-inked, it can be recycled as mixed grade after use. In this form, it can be used for paper board or packing, such as trays in produce boxes. Currently, the fiber in these products is not valuable enough to recover. Ledger paper also enters an open-loop system when it is recycled into facial tissue or other products that are disposed of after use.

Recycling can be an effective resource management tool. Under ideal circumstances, most material can be recovered many times until it becomes too degraded for further use. Even so, designing for recyclability is not the strategy for meeting all environmental requirements. As an example, studies show that refillable glass bottles use less life cycle energy than single-use recycled glass to deliver the same amount of beverages (Sellers and Sellers 1989).

When a suitable infrastructure is in place, recycling is enhanced by:

- Ease of disassembly
- Ease of material identification
- Simplification and parts consolidation
- Material selection and compatibility

In most projects, the material selection is not coordinated with environmental strategies. For instance, a passenger car currently uses 50 to 150 different materials. Separating this mixture from a used car is impossible. Designers can aid recycling by reducing the number of incompatible materials in a product. For example, a component containing parts of different materials could be designed with parts made from the same material.

Some polymers and other materials are broadly incompatible. If such materials are to be recycled for similar use, they must be meticulously separated for high purity.

Some new models in a personal system/2 product line are specifically designed with the environment in mind. These models use a single polymer for all plastic parts. The polymer has a molded-in finish, eliminating the need for additional finishes, and molded-in identification symbols. In addition, the parts snap together, avoiding the use of metal pieces such as hinges and brackets. These design features facilitate recycling, principally through easy disassembly, the elimination of costly plastic parts sorting, and the easy identification of polymer composition (Dillon 1993).

MATERIAL SELECTION

Because material selection is a fundamental part of design, it offers many opportunities for reducing environmental impact. In life cycle design, designers begin material selection by identifying the nature and source of raw materials. Then, they estimate the environmental impact caused by resource acquisition, processing, use, and retirement. The depth of the analysis and the number of life cycle stages varies with the project scope. Finally, they compare the proposed materials to determine the best choices.

Minimizing the use of virgin material means maximizing the incorporation of recycled material. Sources of recycled feedstock include in-house process scrap, waste material from another industry, or reclaimed postconsumer material.

The quality of incoming material determines the amount of unusable feedstock and the amount of time required to prepare the material. Therefore, product design dimensions should closely match incoming feedstock dimensions to minimize machining, milling, and scrap generation.

Material Substitution

Material substitutions can be made for product as well as process materials, such as solvents and catalysts. For example, water-based solvents or coatings can sometimes be substituted for high-VOC alternatives during processing. Also, materials that do not require coating, such as some metals or polymers, can be substituted in the product.

For example, an American company replaced its five-layer finish on some products with a new three-layer substitute. The original finish contained nickel (first layer), cadmium, copper, nickel, and black organic paint (final layer). The new finish contains nickel, a zinc–nickel alloy, and black organic paint. This substitution eliminates cadmium, a toxic heavy metal, and the use of a cyanide bath solution for plating the cadmium. The new finish is equally corrosion resistant. It is also cheaper to produce, saving the company 25% in operating costs (U.S. EPA 1991b).

A large textile dye house in Chelsea, Massachusetts, complied with local sewer limits by working with its imported fabric suppliers and clients to select only those fabrics with the lowest zinc content. The company thus avoided installing a $150,000 treatment plant. (Kennedy 1993).

Finally, reducing the use of toxic chemicals results in fewer regulatory concerns associated with handling and disposing hazardous material and less exposure to corporate liability and worker health risks. For example, a water-based machining coolant can reduce the quantity of petroleum oils generated onsite and allow parts to be cleaned more effectively using a non-chlorinated or water-based solvent.

Reformulation

Reformulation is an appropriate strategy when a high degree of continuity must be maintained with the original product. Rather than replacing one material with another, the designer alters the percentages to achieve the same result. Some material may be added or deleted if the original product characteristics are preserved.

REDUCED MATERIAL INTENSIVENESS

Resource conservation can reduce waste and directly lower environmental impact. A less material-intensive product may also be lighter, thus saving energy in distribution or use. When reduction is simple, benefits can be determined with a vigorous LCA.

For example, a fast-food franchise reduced material input and solid waste generation by decreasing the paper napkin weight by 21%. Two store tests revealed no change in the number of new napkins used compared to the old design. Attempts to reduce the gage of plastic straws, however, caused customer complaints. The redesigned straws were too flimsy and did not draw well with milkshakes

(Environmental Defense Fund and McDonalds' Corporation 1991).

ENERGY-EFFICIENT PRODUCTS

Energy-efficient products reduce energy consumption and greenhouse gas emissions. For example, the EPA's Energy Star Program initiates a voluntary program to reduce the power consumption of laser printers when inactive. The EPA's Green Lights Program is aimed at persuading companies to upgrade their lighting systems to be more efficient.

Process Management

Although process design is an integral part of product development, process improvement can be pursued outside of product development.

PROCESS SUBSTITUTION

Processes that create major environmental impact should be replaced with more benign ones. This simple approach to impact reduction can be effective. For example, copper sheeting for electronic products was previously cleaned with ammonium persulfate, phosphoric acid, and sulfuric acid at one large American company's facility. The solvent system was replaced by a mechanical process that cleaned the sheeting with rotating brushes and pumice. The new process produces a nonhazardous residue that is disposed in a municipal solid waste landfill.

A large American chemical and consumer products company switched from organic solvent-based systems for coating pharmaceutical pills to a water-based system. The substitution was motivated by the need to comply with regulations limiting emissions of VOCs. To prevent the pills from becoming soggy, a new sprayer system was designed to precisely control the amount of coating dispensed. A dryer was installed as an additional process step. The heating requirements increased when the water-based coatings were used. However, for a total cost of $60,000, the new system saved $15,000 in solvent costs annually and avoided the expense of $180,000 in end-of-the-pipe emission controls that would have been required if the old solvent system had been retained (Binger 1988).

Process redesign directed toward plant employees can also yield health and safety benefits, as well as reduce cost. In addition, through certain process changes, a facility can reduce its resource demands to a range where closing the loop or completely eliminating waste discharges from the facility is economically feasible. Unless a company fine tunes each process first, however, the waste volume may overwhelm the equipment's capacity to recycle or reuse it. For example, an electroplating process that does not have an optimized rinsing operation must purchase metal recovery equipment with a capacity of five to ten times that needed under optimal rinsing conditions.

The EPA has published several pollution prevention manuals for specific industries. Each manual reviews strategies for waste reduction and provides a checklist.

PROCESS ENERGY EFFICIENCY

Process designers should always consider energy conservation including:

Using waste heat to preheat process streams or do other useful work

Reducing the energy requirement for pumping by using larger diameter pipes or cutting down frictional losses

Reducing the energy use in buildings through more efficient heating, cooling, ventilation, and lighting systems

Saving energy by using more efficient equipment. Both electric motors and refrigeration systems can be improved through modernization and optimized control technology.

Conserving process energy through the insulation of process tanks, monitoring, and regulating temperatures to reduce energy cost and resource use in energy generation

Using high-efficiency motors and adjustable-speed drives for pumps and fans to reduce energy consumption

Reducing energy use through proper maintenance and sizing of motors

Renewable energy sources such as the sun, wind, and water offer electricity for the cost of the generating equipment. Surplus electricity can often be sold back to the utilities to offset electrical demand. A decrease in the demand for electricity resulting from the use of renewable resources increases the environmental quality.

PROCESS MATERIAL EFFICIENCY

A process designed to use material in the most efficient manner reduces both material input and waste. For example, new paint equipment can reduce overspray, which contains VOCs.

Environmental strategies for product design are also applicable to facilities and equipment. Designers can extend the useful life of facilities and processes by making them appropriately durable. Flexible manufacturing can be an effective life extension for facilities. Through its Green Light Program, the EPA educates companies about new lighting techniques and helps them conserve energy.

For example, a large American electronics company designed a flux dispensing machine for use on printed circuit boards. This low solid flux (LSF) produces virtually no excess residue when it is applied, thus eliminating a cleaning step with CFCs and simplifying operations. Performance of the boards produced with the new LSF was maintained, and the LSF helped this manufacturer reduce CFC emissions by 50% (Guth 1990).

INVENTORY CONTROL AND MATERIAL HANDLING

Improved inventory control and material handling reduces waste from oversupply, spills, or deterioration of old stock. This reduction increases efficiency and prevents pollution. Proper inventory control also ensures that materials with limited shelf lives have not degraded. Processes can thus run at peak efficiency while directly reducing the waste caused by reprocessing.

On-demand generation of hazardous materials needed for certain processes is an example of innovative material handling that can reduce impact.

Storage facilities are also an important element of inventory and handling systems. These facilities must be properly designed to ensure safe containment of material. They should be adequately sized for current and projected needs.

A large American electronics firm developed an on-demand generation system for producing essentially toxic chemicals that had no substitutes. Less harmful precursors were reacted to form toxic chemicals for immediate consumption. The company now produces arsine, an acutely toxic chemical essential for semiconductor production, as it is needed. This system avoids transporting arsine to manufacturing sites in compressed cylinders and using specially designed containment facilities to store the arsine. The company no longer must own three special storage facilities which cost $1 million each to build and maintain (Ember 1991).

Efficient Distribution

Efficient distribution includes improving transportation and packaging.

TRANSPORTATION

The environmental impact caused by transportation can be reduced by several means including:

- Choosing an energy-efficient route
- Reducing air pollutant emissions from transportation
- Using the maximum vehicle capacity where appropriate
- Backhauling materials
- Ensuring proper containment of hazardous material
- Choosing routes carefully to reduce potential exposure from spills and explosions

Table 3.5.1 shows transportation efficiencies. Time and cost considerations, as well as convenience and access, determine the best choice for transportation. When selecting a transportation system, designers should also consider infrastructure requirements and their potential impacts.

PACKAGING

As a first step, products should be designed to withstand both shock and vibration.

Designers can use the strategies that follow to design packaging for efficient distribution.

Packaging Reduction

Packaging reduction includes elimination, reusable packaging, product modifications, and material reduction.

In elimination, appropriate products are distributed unpackaged. In the past, many consumer goods such as screw drivers, fasteners, and other items were offered unpackaged. Wholesale packaging can be eliminated. For example, furniture manufacturers commonly ship furniture uncartoned.

With reusable packaging, wholesale items that require packaging are commonly shipped in reusable containers. Tanks of all sizes, wire baskets, plastic boxes, and wooden hooks are frequently used for this purpose.

Even when products require primary or secondary packaging to ensure their integrity during delivery, product modifications can decrease packaging needs. Designers can further reduce the amount of packaging by avoiding unusual product features or shapes that are difficult to protect.

In material reduction, products that contain an ingredient in dilute form can be distributed as concentrates. In some cases, customers can simply use concentrates in reduced quantities. A larger, reusable container can also be sold in conjunction with concentrates. This method allows customers to dilute the products if appropriate. Examples of product concentrates include frozen juice concentrates and concentrated versions of liquid and powdered detergent. Material reduction can also be pursued in packaging design. Many packaging designers have reduced material use while maintaining performance. Reduced thickness of corrugated containers (board grade reduction) is one example. In addition, aluminum, glass, plastic, and steel containers have continually been redesigned to require less material to deliver the same volume.

TABLE 3.5.1   TRANSPORTATION EFFICIENCIES

Mode	Btu/tn-mi
Waterborne	365
Class 1 Railroad	465
All Pipelines[1]	886
Crude oil pipeline	259
Truck	2671–3460
Air[2]	18,809

[1]Average figure; ranges from 236 Btu/tn-mi for petroleum to approximately 2550 Btu/tn-mi for coal slurry and natural gas.

[2]All-cargo aircraft only. Belly freight carried on passenger airlines is considered free because the energy used to transport it is credited to the passengers. Thus, the efficiency figure for all air freight is a misleading 9548 Btu/tn-mi.

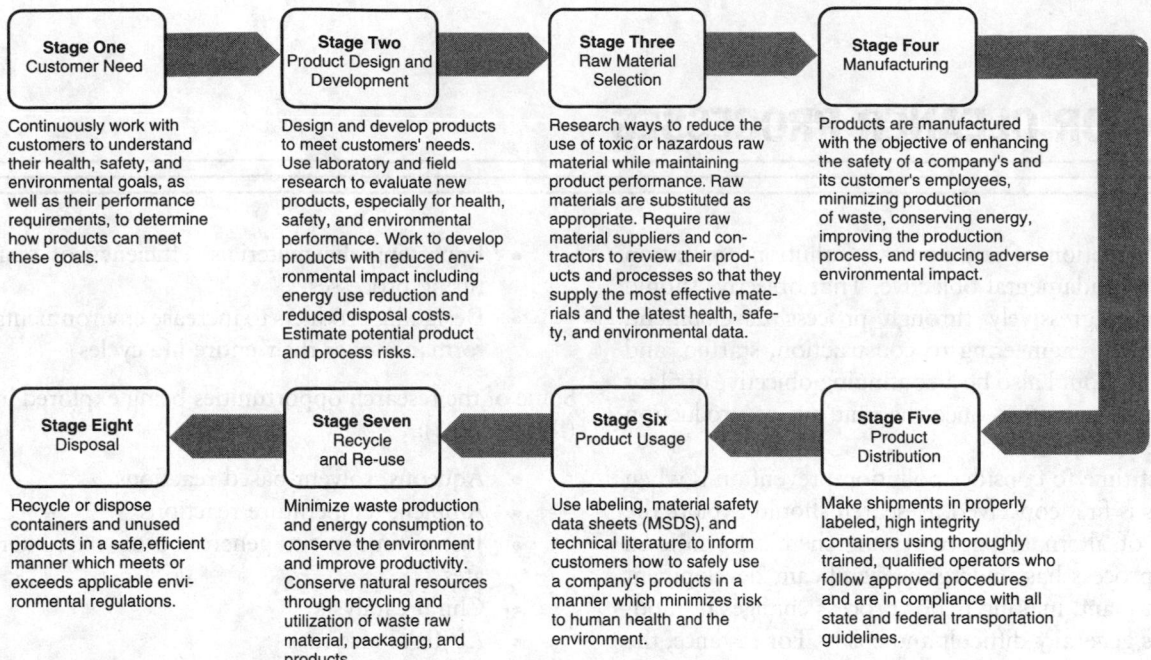

FIG. 3.5.2 XYZ Product stewardship.

Material Substitution

One common example of this strategy is to substitute more benign printing inks and pigments for those containing toxic heavy metals or solvents. Also, whenever possible, designers can create packaging with a high recycled content. The necessary design elements for most reusable packaging systems include:

- a collection or return infrastructure
- procedures for inspecting items for defects or contamination
- repair, cleaning, and refurbishing capabilities
- storage and handling systems

Degradable Materials

Degradable materials can be broken down by biological or chemical processes or exposure to sunlight. Degradability is a desirable trait for litter deposited in aesthetically pleasing natural areas. However, a number of challenging problems must be resolved before the use of degradable packaging becomes a commonly accepted strategy.

Improved Management Practices

Designing new business procedures and improving existing methods also play a role in reducing environmental impact. Business management strategies apply to both manufacturing and service activities. For example, forcing aircraft to use a plug-in system at an airport rather than using their own auxiliary power systems results in a reduction of air pollution, especially in countries with clean hydroelectricity.

Figure 3.5.2 shows a product steward's program covering each part of the product's stages, including design, manufacturing, marketing, distribution, use, recycling, and disposal.

—*David H.F. Liu*

References

Binger, R.P. 1988. Pollution prevention plus. *Pollution Engineering* 20.

Dillon, Patricia S. 1993. From design to disposal: Strategies for reducing the environmental impact of products. Paper presented at the 1993 AIChE Summer National Meeting, August 1993.

Ember, L.R. 1991. Strategies for reducing pollution at the source are gaining ground. *C&E News* 69, no. 27.

Environmental Defense Fund and McDonald's Corporation. 1991. *Waste Reduction Task Force, final report.*

Guth, L.A. 1990. *Applicability of low solids flux.* Princeton, N.J.: AT&T Bell Labs.

Kennedy, Mitchell L. 1993. Sustainable manufacturing: Staying competitive and protecting the environment. *Pollution Prevention Review* (Spring).

Kusz, J.P. 1990. Environmental integrity and economic viability. *Journal of Industrial Design Society of America* (Summer).

Sellers, V.R. and J.D. Sellers. 1989. *Comparative energy and environmental impacts for soft drink delivery systems.* Prairie Village, Kans.: Franklin Associates.

United Nations World Commission on the Environment and Development. 1987. *Our common future.* England: Oxford University Press.

U.S. Environmental Protection Agency (EPA). 1991a. Guidance for the use of the terms "recycled" and "recyclable" and the recycling emblem in environmental marketing claims. 49992-50000.

———. 1991b. *Pollution prevention, 1991: Progress on reducing industrial pollutants.* EPA 21P-3003. Washington, D.C.: Office of Pollution Prevention.

3.6
R & D FOR CLEANER PROCESSES

From the inception of any process, pollution prevention should be a fundamental objective. That objective should be pursued aggressively through process development, process design, engineering to construction, startup, and operation. It should also be a continuing objective of plant engineers and operators once the unit begins production (see Figure 3.6.1).

The best time to consider pollution prevention is when the process is first conceived. Research should explore the possibility of alternate pathways for chemical synthesis. Once the process has undergone significant development at the pilot plant, making major process changes or modifications is generally difficult and costly. For instance, the pharmaceutical industry is restricted from process modifications once the clinical efficacy of the drug is established.

An international consensus is growing on the need to use pollution prevention and clean production principles for the following:

- Changing industrial raw materials to less toxic chemicals

- Improving the materials' efficiency of manufacturing processes
- Designing products to increase environmental performance over their entire life cycles

Some of the research opportunities being explored include (Illman 1993):

- Aqueous, solvent-based reactions
- Ambient-temperature reactions
- Just-in-time in situ generation of toxic intermediates
- Chiral catalysts
- Artificial enzymes
- Built-in recyclability

Environmental Load Indicator

ICI uses a rough indicator of environmental impact called the environmental load factor (ELF) to choose the reaction route that is best for the environment. The ELF equals the net weight of raw materials, solvents, catalysts, and

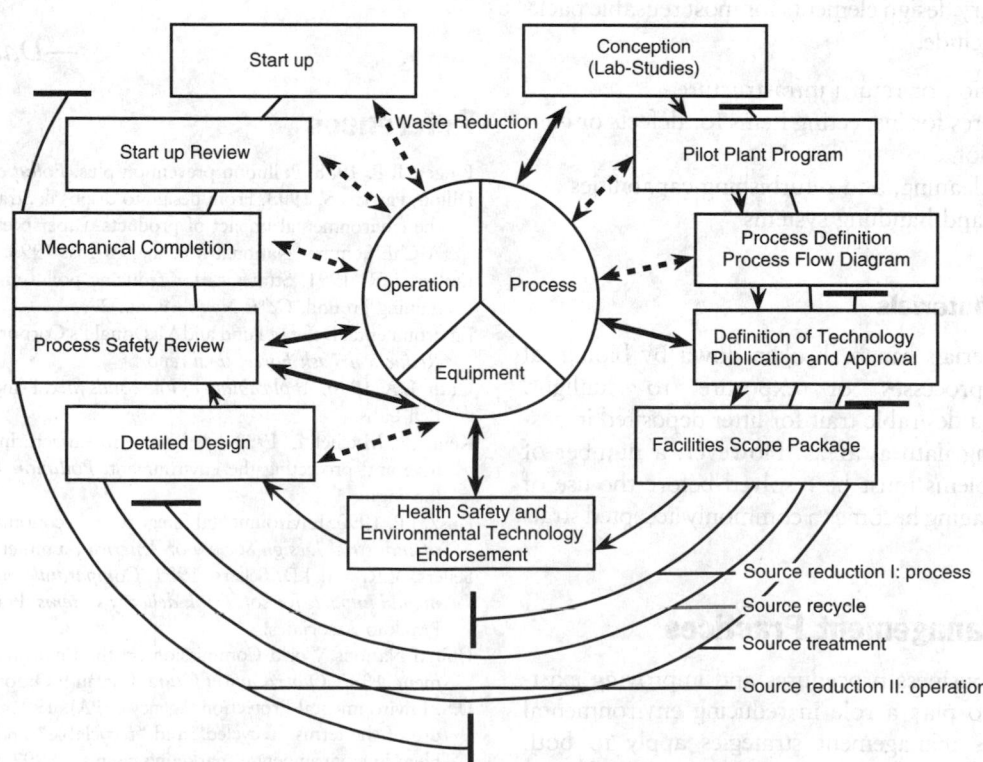

FIG. 3.6.1 Waste reduction and new technology development. (Reprinted, with permission, from Ronald L. Berglund and Glenn E. Snyder, 1990, Waste minimization: The sooner the better, *Chemtech* [December].)

other chemicals used to make a unit weight of product. Subtracting the weight of the finished product from the weight of all material fed to the process and dividing that difference by the weight of the finished product calculates the ELF. Therefore, pollution prevention dictates that researchers minimize the use of additives. Additives must be separated from a product, and at some point, they too become waste. In addition, installations designed to protect the environment are also invariably sources of waste (Hileman 1992).

The waste ratio is an indicator used at the 3M Company to measure the progress of the waste-reduction strategies. This ratio is defined rather simply as:

$$\text{Waste ratio}(\%) = (\text{Waste/total output})(100) \quad \textbf{3.6(1)}$$

The quantities are measured by weight. Total output includes good output plus waste. Good output includes finished goods, semifinished goods, and by-products (however, by-product material that is beneficially burned for fuel counts as waste). Waste is the residual from the manufacturing site before it is subjected to any treatment process. The material that is recycled is not included as waste (Benforado, Riddlehover, and Gores 1991).

At 3M, the waste ratio varies from 10 to 20% for batch polymerizations to 99% for products that are not favored by reaction kinetics or that require multistep purification operations. New products undergo special screening if the initial waste ratio exceeds 50%. This screening is important not only for meeting waste-reduction targets but also for assessing the economic viability of the product as treatment and disposal costs escalate.

Process Chemistry

Preventing pollution in process chemistry includes the choice of the reaction route, catalyst technology, the choice of the reagents, and the choice of the solvents.

CHOICE OF REACTION ROUTE

One major way to reduce waste in the manufacture of complex organic substances is to reduce the number of steps required from raw materials to the final product. Every intermediate must usually be purified after each step, and nearly all purification processes produce waste. For example, a new ICI route to a fungicide process reduces the number of reaction steps from six to three while using fewer solvents and other chemicals. The total effect decreases the amount of waste to only 10% of that generated in the previous process (Hileman 1992). In addition, reducing steps means decreasing capital and operating costs.

Figure 3.6.2 compares the six-step ibuprofen process with the three-step process. Both processes start with the Friedel–Crafts conversion of sec-butylbenzene to p-sec-butylacetophenone. The six-step process also uses alu-

minum chloride as a catalyst with potential problems of water quench, emulsions, difficult extractions or filtration, and a large volume of water-borne waste. The three-step process uses the cleaner hydrogen fluoride as a catalyst and reduces the need for treatment of the water effluent and eliminates the formation of HCl.

Molecular Design in Basel, Switzerland has reaction retrieval systems that can be used to design improved processes to clients' target compounds (Stinson 1993). These reaction retrieval systems allow the design of a sequence that has only four steps, begins from more easily accessible substituted phthalimides to spirosuccinimides, and eliminates the generation of toxic by-products (see Figure 3.6.3). One of the steps uses trimethylsilyl cyanide which could be added to the company's catalog for sales to others.

In various forms, such computer software has been under development around the country for about twenty-five years. Its purpose is to help chemists identify new syntheses for target molecules from the myriad of potential routes and to suggest novel chemical reactions that might be investigated.

Most of these programs are retrosynthetic—that is they generate syntheses for target molecules by working back-

FIG. 3.6.2 Hoechst route to ibuprofen versus conventional route.

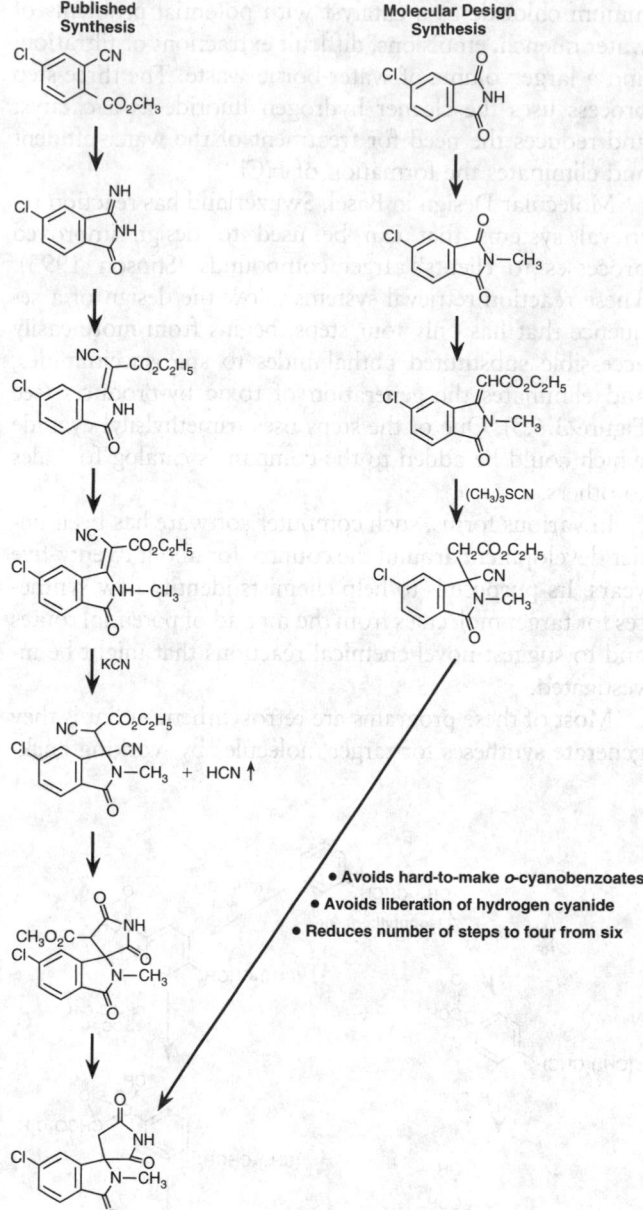

FIG. 3.6.3 Databases to improve synthetic schemes. (Reprinted, with permission, from Molecular Design.)

ward from the target to the candidate starting materials. Other programs are synthetic—that is they identify side reactions, by-products, and the effect of various conditions on reaction outcomes. However, none of these programs are built with environmentally benign synthesis in mind.

Out of about twenty software tools examined by J. Dirk Nies (Chemical Information Services, Oakville, Maryland) and colleagues, three programs appear to be useful for providing theoretical alternative synthesis pathways in support of EPA pollution prevention initiatives: Cameo, which operates synthetically, and Syngen and Lhasa, which both operate retrosynthetically. The user, however, must decide which pathways are environmentally safer by considering

the health and environmental hazards of the starting reagents (Illman 1993).

CATALYST TECHNOLOGY

Catalytic technologies offer great potential for reducing waste and energy consumption, minimizing the use and transportation storage of hazardous materials, and developing products that are safer for the environment. Some examples of catalyst-based products and processes that reduce pollutant emissions follow.

Production of Environmentally Safer Products

This example describes a process with negligible side reactions in which 1,1-dichloro-1-fluoroethane (HCFC-141b) is a replacement for stratospheric-ozone-depleting fluorotrichloromethane (CFC-11).

Low yields are obtained when 1,1 dichloroethylene ($CH_2{=}CCl_2$) is reacted with hydrogen fluoride using conventional catalysts because the reaction favors trichloroethane (HFC-143a) as follows:

$$CH_2{=}CCl_2 + HF \longrightarrow CH_3CFCl_2 + CH_3CF_2Cl + CH_3CF_3$$
$$\text{HCFC-141b} \quad \text{HCFC-142b} \quad \text{HFC-143a}$$
$$3.6(2)$$

However, when a specially prepared aluminum fluoride catalyst is used, nearly all the reactant is converted to HCFC-141b, and less than 500 ppm of the reactant remain in the product. The HCFC-141b does not even need to be purified (see Figure 3.6.4).

Management of Hazardous and Toxic Materials

The DuPont company has a catalytic process for making methylisocyanate (MIC), starting from materials far less hazardous than the traditional phosgene. The MIC is produced only moments before it is converted to a pesticide so that only small quantities of MIC exist at any one time. This pathway is safer for both the environment and worker health and safety. The catalytic process does not produce hydrochloric acid.

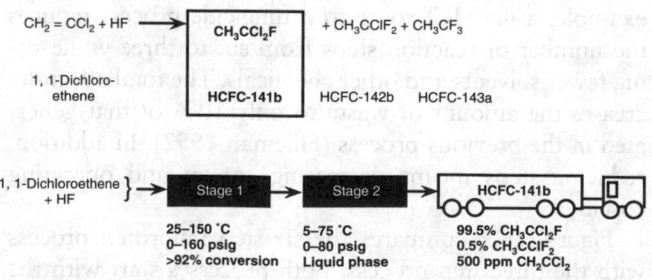

FIG. 3.6.4 High-yield catalytic hydrochlorofluorocarbon process.

In the traditional process, the following reaction occurs:

$$CH_3NH_2 + COCl_2 \longrightarrow CH_3NCO + 2HCl \quad 3.6(3)$$
$$(\text{Methylamine} + \text{Phosgene} \longrightarrow \text{MIC} + \text{Hydrochloric Acid})$$

In the DuPont catalytic process, the following reactions occur:

$$CH_3NH_2 + CO \longrightarrow CH_3NHCHO \quad 3.6(4)$$

$$CH_3NHCHO + 1/2\ O_2 \longrightarrow CH_3NCO + H_2O \quad 3.6(5)$$
$$\text{in situ}$$

Environmentally Benign Reagent for Carbonylation or Methylation

Dimethylcarbonate (DMC) is an alternative to toxic and dangerous phosgene, dimethyl sulphate, or methyl chloride. DMC is easy to handle, offers economic advantages, and is a versatile derivatives performer. It uses readily available, low-cost raw material, a clean technology, and a clean product. The result is a route based on the oxidation of carbon monoxide with oxygen in methanol as follows:

$$CO + 2\ CH_3OH + 1/2\ O_2 \longrightarrow (CH_3O)_2CO + H_2O \quad 3.6(6)$$

This process operates at medium pressure using a copper chloride catalyst (see Figure 3.6.5).

Environmentally Safer Route to Aromatic Amines

The conventional industrial reaction involves activating the C—H bond by chloride oxidation. A variety of commercial processes use the resulting chlorobenzenes to produce substituted aromatic amines. Since neither chlorine atom ultimately resides in the final product, the ratio of the by-product produced per pound of product generated in these processes is unfavorable.

In addition, these processes typically generate waste streams that contain high levels of inorganic salts which are expensive and difficult to treat.

Michael K. Stern, of Monsanto, uses nucleophilic substitution for hydrogen to generate intermediates for manufacturing 4-amino-diphenylamine, eliminating the need for halogen oxidation (see Figure 3.6.6). Stern and coworkers synthesize p-nitroaniline (PNA) and p-phenylenediamine (PPD) using nucleophilic aromatic substitution for

hydrogen instead of a pathway using chlorine oxidation. Benzamide and nitrobenzene react in the presence of a base under aerobic conditions to give 4-nitrobenzanilide in high yield. Further treatment with methanolic ammonia gives PNA and regenerates benzamide.

Minimizing Waste Disposal Using Solid Catalysts

This example includes processes for producing ethylbenzene, methyl tert-butyl ether (MTBE), and ethyl tert-butyl ether (ETBE), and many other processes.

Historically, the $AlCl_3$ catalyst system has been the technology for ethylbenzene synthesis. However, the disposal of the waste stream presented an environmental problem. The Mobil ZSM-5 catalyst permits a solid-catalyst, vapor-phase system which gives yields comparable to the $AlCl_3$, catalyzed system but without the environmental problems associated with $AlCl_3$.

A new, strongly acidic, ion-exchange catalyst is replacing more acid catalyst applications for producing MTBE, ETBE, olefin hydration, and esterification. The high-temperature catalyst, Amberlyst 36, can tolerate temperatures up to 150°C. Compared to sulfuric acid, the selectivity of the acidic resins is higher, product purity is better, and unlike acid, separating the product from the catalyst is not a problem.

Petrochemicals from Renewable Resources

This example describes making alpha-olefins from carboxylic acids.

A catalytic technique (the Henkel process) offers a green alternative for making alpha-olefins because it can produce them from fatty acids instead of from petroleum. In the reaction, an equimolar mixture of a carboxylic acid and acetic anhydride is heated to 250°C in the presence of a palladium or rhodium catalyst as follows:

$$RCH_2CH_2CO_2H + (CH_3CO)_2O \longrightarrow$$

$$RCH=CH_2 + CO + 2\ CH_3COOH \quad 3.6(7)$$

This reaction causes the carboxylic acid to undergo decarbonylation and dehydration to a 1-alkene having one

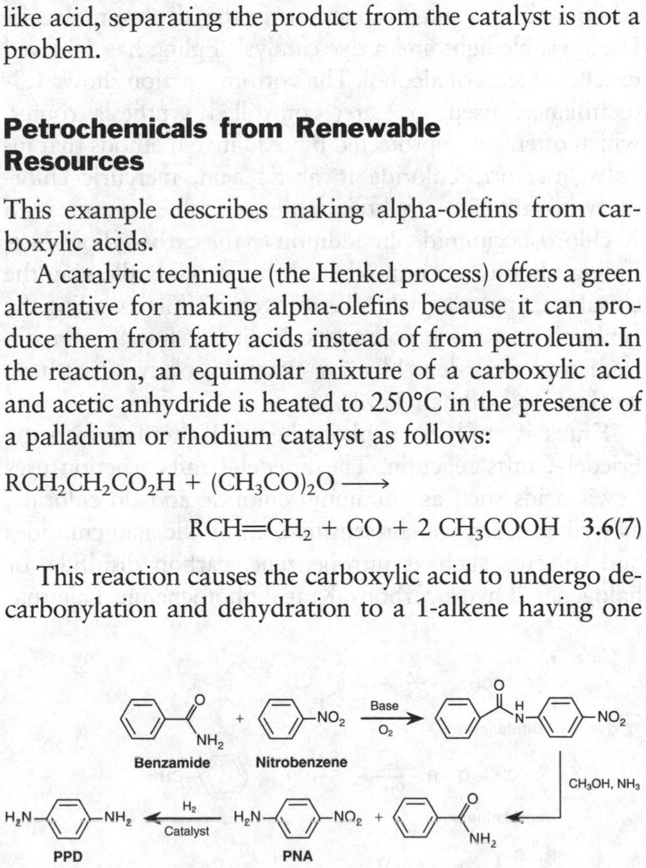

FIG. 3.6.5 Block diagram of the EniChem Synthesis process for DMC.

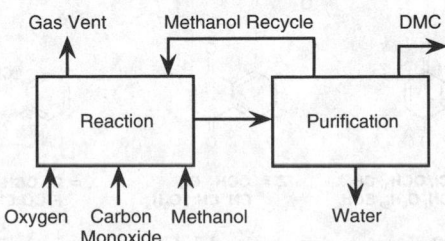

FIG. 3.6.6 Environmentally safer route to aromatic amines. (Reprinted, with permission, from D.L. Illman, 1993, Green technology presents challenge to chemists, *C&E News* [6 September].)

less carbon atom. The catalyst lives are long. The spent catalyst can be recovered for reuse (Borman 1993). Only minute quantities of by-products are generated.

Avoiding Toxic Catalysts, Toxic Acids, and Solvents

These examples include using a dye and a light in green oxidations and an alternate to the Friedel–Crafts reaction.

The use of nontoxic dyes as catalysts in oxidation reactions that were previously carried out with toxic compounds of metals such as cadmium, lead, mercury, nickel, and chromium are being explored by Epling. His strategy uses a dye to absorb visible light and then transfer an electron efficiently and selectively to cause a reaction. Some of these reactions are shown in Figure 3.6.7.

In this figure, with light as a reagent and dye as a catalyst, deprotection of organic functional groups proceeds under neutral conditions without the use of heavy metals or chemical oxidants. The reaction at the top of the figure shows deprotection of dithianes. Shown in the middle, the benzyl ether protecting group is often used to protect an alcohol during organic synthesis. The usual ways to remove this blocking group—catalytic hydrogenation or alkalimetal reduction—involve conditions that can result in additional, unwanted reductions in the alcohol molecule. Using visible light and a dye catalyst, Epling has achieved excellent yields of alcohol. The bottom reaction shows 1,3-oxathianes, used in stereo-controlled synthesis routes, which often are deprotected by oxidative methods that involve mercuric chloride in acetic acid, mercuric chloride with alkaline ethanolic water, or silver nitrate with N-chlorosuccinimide. In addition to the carbonyl product, Epling obtains the nonoxidized thioalcohol, allowing the chiral starting material to be recovered while avoiding the generation of toxic pollutants. Eosins (yellow), erthrosin (red), and methylene blue are examples of dyes that have worked well (Illman 1993).

Kraus is studying a photochemical alternative to the Friedel–Crafts reaction. The Friedel–Crafts reaction uses Lewis acids such as aluminum chloride and tin chloride, as well as corrosive, air-sensitive, and toxic acid chlorides and solvents such as nitrobenzene, carbon disulfide, or halogenated hydrocarbons. Kraus' photochemical alterna-

tive exploits the reaction between a quinone and an aldehyde and is initiated by a simple lamp. Some of the products produced with this alternative are shown in Figure 3.6.8.

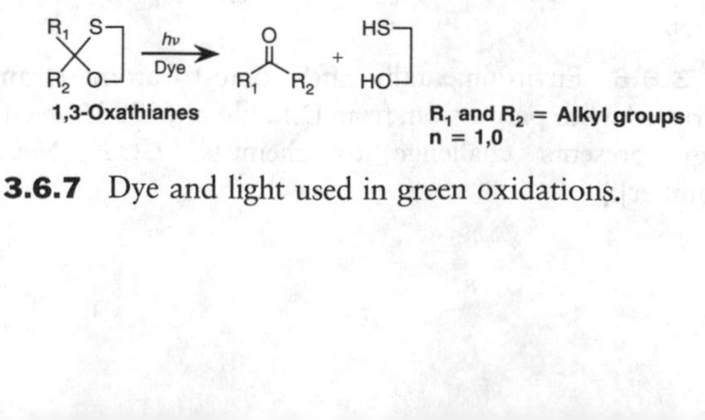

This figure show that no restrictions appear to exist for functional groups meta and para to the formyl group of the benzaldehyde. Ortho groups that are compatible with the reaction conditions include alkoxy groups, alkyl groups, esters, and halogens. Many substituted benzoquinones react with aromatic and aliphatic aldehydes according to this scheme.

Organic Chemicals from Renewable Resources

This example discusses processes that use genetically engineered organisms as synthetic catalysts.

D-glucose can be derived from numerous agricultural products as well as waste streams from processing food products. Frost has developed a technology using genetically engineered microbes as synthetic catalysts to convert D-glucose to hydroquinone, benzoquinone, catechol, and adipic acid used in nylon production (see Figure 3.6.9).

The technology shown in this figure presents two challenges: directing the largest possible percentage of the consumed D-glucose into the common pathway of aromatic amino acid synthesis and assembling new biosynthetic pathways inside the organism to siphon carbon flow away from those amino acids and into the synthesis of the industrial chemicals. To synthesize hydroquinone and benzoquinone, 3-dehydroquinate (DHQ) is siphoned from the common pathway by the action of quinic acid dehydrogenase. Catechol and adipic acid synthesis rely on siphoning off 3-dehydroshikimate (DHS).

Until recently, enzymes have been used largely for degradative processes in the food and detergent industries.

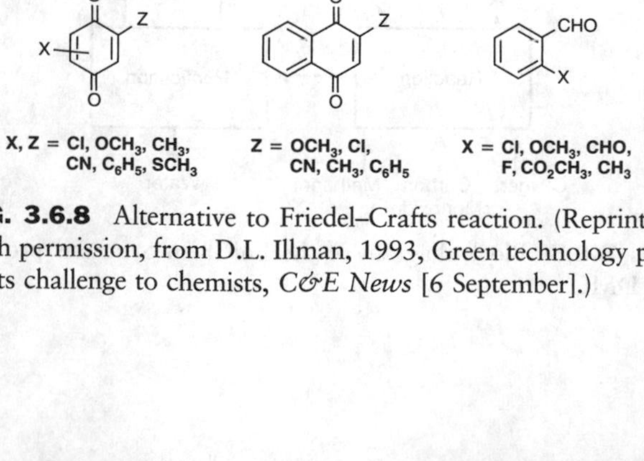

FIG. 3.6.8 Alternative to Friedel–Crafts reaction. (Reprinted, with permission, from D.L. Illman, 1993, Green technology presents challenge to chemists, *C&E News* [6 September].)

FIG. 3.6.7 Dye and light used in green oxidations.

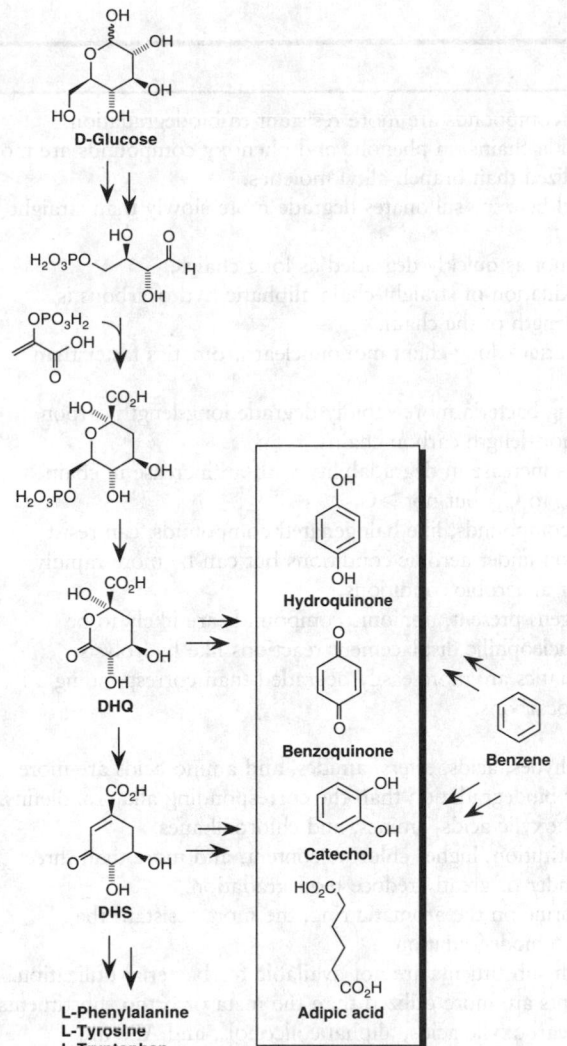

FIG. 3.6.9 Industrial chemicals produced from D-glucose by engineered bacteria.

However, in the future, they will be used increasingly for synthesis. The advantages are that enzymes are selective, they are nonhazardous, and operating conditions are moderate. For example, *S. aureus,* a naturally occurring protease (enzyme), is used in a route to convert aspartic acid and phenylalanine methylester into aspartame. This use avoids the pretreatment needed to block a side reaction that forms a nonsweet product (Parkinson and Johnson 1989).

CHOICE OF REAGENTS

In oxidations, potassium dichromate or permanganate, or lead tetraacetate, is used in the laboratory. However, these stoichiometric reactions involve mole-for-mole amounts of high-molecular-weight oxidants. Therefore, the oxidants must be used in large, absolute quantities. Such reactions are costly and generate large volumes of an effluent, possibly containing toxic metal salts that must be treated in the scaled-up process (Stinson 1993).

Hydrogen peroxide, peracetic acid, or tert-butyl hydroperoxide should be investigated as alternatives in oxidations catalyzed by trace amounts of transition metals. Such reactions could be easily worked up and result in only water, acetic acid, or tert-butanol as by-products. (Stinson 1993).

DMC can replace phosgene in carbonylation reactions and dimethyl sulfate and methyl chloride in methylation reactions. Oxalyl chloride is another alternative to phosgene.

Phosgene is a valuable agent for converting acids to acid chlorides, amides to nitriles, primary amines to isocyanates, and alcohols to chloroformates. The more tractable disphosgene (trichloromethyl chloroformate, $ClCO_2Cl_3$) can be used as a phosgene substitute in small-scale reactions (Stinson 1993).

An alternative to the highly toxic and potentially explosive fluorinating reagents (such as HF, $FClO_3$, and CF_3OF) and the reagent called Selectfluor (or F-TEDA-BF4) is 1-chloromethyl-4-fluoro-1,4-diazonia[2,2,2] bicyclooctane bis(tetrafluoroborate). Selectfluor provides a tool for developing and producing high-performance fluorine-containing drugs (Illman 1993).

Carbon dioxide has advantages over traditional pH-control chemicals, such as sulfuric acid. The savings come from both the favorable cost of carbon dioxide and the elimination of the costs of handling, storing, and disposing sulfuric acid. Charging a sodium nitrile solution with carbon dioxide under high pressure drops the pH low enough to diazotize p-anisidine. Releasing the pressure afterward raises the pH so that the coupled product precipitates out.

CHOICE OF SOLVENTS

The principal functions of a solvent are to (1) provide a practical homogeneous reaction mass, (2) act as a heat-transfer agent, and (3) cause products or by-products to precipitate out, thereby improving the yield.

Chemicals under the 33/50 program started by the EPA in February 1991 should be avoided. The solvent should be chosen based on environmental grounds (ease of workup and solvent recovery) and then on optimizing in that solvent.

Less-hazardous organic solvents, such as ethyl acetate and isopropylacetate, should be used in place of more toxic (or more rigidly controlled by the EPA) solvents, such as methylene chloride and benzene. The current trend is to replace chlorinated solvents with nonchlorinated solvents.

Table 3.6.1 provides a set of general rules for the biodegradability of some organic molecules.

The use of polar solvents such as dimethyl sulfoxide (DMSO) or dimethylformamide (DMF) should be minimized. These solvents speed up many reaction rates, but

TABLE 3.6.1   GENERAL RULES FOR BIODEGRADABILITY

Chemical Structure	Factors
Branched structures	Highly branched compounds are more resistant to biodegradation. 1. Unbranched side chains on phenolic and phenoxy compounds are more easily metabolized than branch alkyl moieties. 2. Branched alkyl benzene sulfonates degrade more slowly than straight chains.
Chain length	Short chains are not as quickly degraded as long chains. 1. The rate of oxidation of straight-chain aliphatic hydrocarbons is correlated to length of the chain. 2. Soil microbes attack long-chain mononuclear aromatics faster than short chains. 3. Sulfate-reducing bacteria more rapidly degrade long-length carbon chains than short-length carbon chains. 4. ABS detergents increase in degradability with an increase in chain length from C_6 to C_{12} but not $>C_{12}$.
Oxidized compounds	Highly oxidized compounds, like halogenated compounds, can resist further oxidation under aerobic conditions but can be more rapidly degraded under anaerobic conditions.
Nonionic compounds	With active halogens present, nonionic compounds are likely to be degraded by nucleophilic displacement reactions like hydrolysis.
Saturated and unsaturated compounds	Unsaturated aliphatics are more easily degraded than corresponding saturated hydrocarbons.
Substituents on simple organic molecules	1. Alcohols, aldehydes, acids, esters, amides, and amino acids are more susceptible for biodegradation than the corresponding alkanes, olefins, ketones, dicarboxylic acids, amines, and chloroalkanes. 2. Increased substitution, higher chlorine content, and more than three cyclic rings hinder or greatly reduce biodegradation. 3. The more chlorine on the aromatic ring, the more resistant the compound is to biodegradation. 4. Aromatics with substituents are not available for bacterial utilization. Para substituents are more utilized than the meta or ortho substituents. 5. Mono- and dicarboxylic acids, aliphatic alcohols, and ABS are decreasingly degraded when H is replaced by the CH_3 group. 6. Ether functions are sometimes resistant to biodegradation.

the reaction mixtures then need dilution with water, extraction, and evaporation and cause difficulties in solvent recovery and generate contaminated effluent water for treatment. However, using a 10% solution of DMF or DMSO in toluene might be worthwhile because the increased reaction rates can be combined with easier solvent recovery.

DeSimone and colleagues demonstrated the use of supercritical CO_2 as a medium for dispersion polymerizations. They used CO_2 and a specially engineered, free-radical initiator and polymeric stabilizer to effect the polymerization of methyl methyacrylare.

The stabilizer contains a carbon-dioxide-phobic background, which attaches to the growing polymer particle, and a carbon-dioxide-philic side-chain, which is soluble in the supercritical medium and stabilizes the polymer colloid as the reaction proceeds. Use of the stabilizer allows high degrees of polymerization, leading to micrometer-size particles with a narrow size distribution.

Carbon dioxide can be easily separated from the reaction mixture since the pressure can be released and the gas vented to the atmosphere. This scheme offers the possibility of avoiding the generation of hazardous waste streams, including aqueous streams contaminated with leftover monomer and initiator, which is one of the plastic industry's greatest cleanup problems.

Physical Factors

In addition to the preceding chemical factors, a number of physical factors have an important bearing on the chemical reaction. The most important of these are:

- Reaction pressure
- Reaction temperature
- Ratio of reactants
- Product workup

Most reactions occur under atmospheric pressure; while oxidation, hydrogenation, and some polymerization occur at higher pressures. The effects of pressure and temperature on equilibrium and the reaction rate are discussed in Section 3.7.

For capacity, feeding the reactants in their stoichiometric ratio and at a maximum concentration is advantageous. However, if one of the reactants is relatively expensive, using the other, less expensive, reactant in excess produces a higher yield with respect to the former reactant. This method is especially effective when recovering the more expensive reactant from the product stream is difficult.

Process Optimization

In process optimization, the chemical engineer uses statistical, factorial experiments while simultaneously varying the reaction parameters such as temperature and stoichiometry. Computer programs are then used to generate contour maps of yield versus temperature and stoichiometry, revealing the global maximum yield. Finally, the engineer optimizes for the overall process cost but not necessarily for yields. Although certain stoichiometry gives the highest yields, the costs of indicated molar excesses of reagents may be higher for it than other stoichiometries, raising the process cost overall.

The workup is a separate phase of optimization because the workup of a reaction mixture differs depending on the ratio of product to other substances. The workup is different in a reaction that gives a 60% yield than in another that gives a 95% yield.

The yields to be maximized in a workup should be chromatographically determined yields and not isolated ones. Thus, with the yield maximized, the workup can be optimized.

Whenever feasible, distillation provides the most convenient and cheapest workup procedure. On a workup of reaction mixtures, quenching with water to precipitate an organic product or extract the aqueous layer with organic solvents should be avoided. When the organic product is precipitated by water, it is usually not in the best crystalline form. Thus, the solid is hard to filter and takes a long time to dry, which results in a long production time. Water precipitation also results in the water being contaminated with organics, which must be treated (Benforado, Riddlehover, and Gores 1991).

The organic solvent extraction of a water-quenched reaction mixture is also messy. The extraction is slow and inefficient. Things tend to be extracted nonselectively. Also, several water washings of organic phases produce a large amount of water effluent.

Instead of water quenching, chemical engineers can usually induce a solid product to crystallize by adjusting the solvent concentration and temperature (Stinson 1993). These adjustments result in a purer product, easier solvent

recovery by distillation of the filtrate, and less effluent to treat.

Process Development

Once the process has been defined in the laboratory, expected yields are known, and the waste streams that are most likely to be produced are quantified, the chemical engineer's next step is to verify these data on a scale that can be used to design the commercial manufacturing process. This step is done in a pilot plant.

PILOT PLANT STUDIES

Obviously, some waste reduction work must be done. In addition to verifying the chemistry proven in the laboratory, an effective waste elimination strategy dictates that the pilot plant be used to quantify the nonproductive activities which include:

- Start up and shutdown losses
- Reactor washings between operations
- Sampling and analytical losses
- Catalyst usage and losses
- Incidental losses from spills and equipment cleanings
- Packaging requirements for raw materials and products

The key parameters to evaluate during a pilot plant study include:

- Flexibility in the selection of raw materials to minimize waste volume and toxicity
- Methods of improving process reliability to minimize spills and off-spec production
- The ability to track and control all waste streams
- The potential impacts of the process on the public, including odor generation, visible emissions, fear generated by the handling of toxic materials, emergency considerations, and so on

Process reliability is generally considered from the health and safety perspective. Reliability also affects waste generation by preventing situations that might result in releases or off-spec products. Sequencing operations to reduce equipment cleaning and reactor washing between steps also reduces the amount of nonproduction waste associated with the process.

A major problem that interferes with the ability of many operating plants to minimize their waste is a lack of adequate process measurement. Thus, the installation of point-of-generation measurement systems should be incorporated into the process and plant design.

INTEGRATED PROCESS DEVELOPMENT

The aim of pollution prevention R&D is to modify processes or test alternate routes to minimize waste streams in the first place. For waste that is unavoidable, this goal involves integrating recovery or waste destruction into the process. In this procedure, the chemical engineer must evaluate the type and quality of the starting materials, the optimal recycling options, and improved treatment or elimination of process waste as a whole at the development stage. Thus for each process, optimum chemical and physical conditions must first be established in the laboratory and then at the pilot plant. The engineer must establish process balances and develop technologies for the optimal treatment of waste streams.

The decision tree approach (see Figure 3.6.10) relates a production plant and its waste streams, with emphasis on process safety. Integrated process development allows a dynamic search for the optimum process conditions by iterative use of the established tools of process chemists and engineers. The described approach for an environmentally sound chemical process applies to both individual process steps and complete, complex, multistage processes.

Figure 3.6.11 shows the process balances of letter acid production before and after the development work (Laing 1992). The upgraded process requires about sixty man-years' work for chemists, chemical engineers, and other engineers. The cost of the new plant is several hundred millions of dollars. Large production volumes, similarities in products, and the possibility of building a new plant at a new site facilitated a near perfect solution.

The following benefits were achieved:

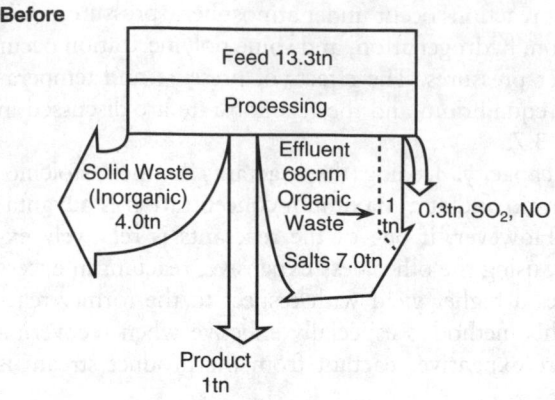

Before

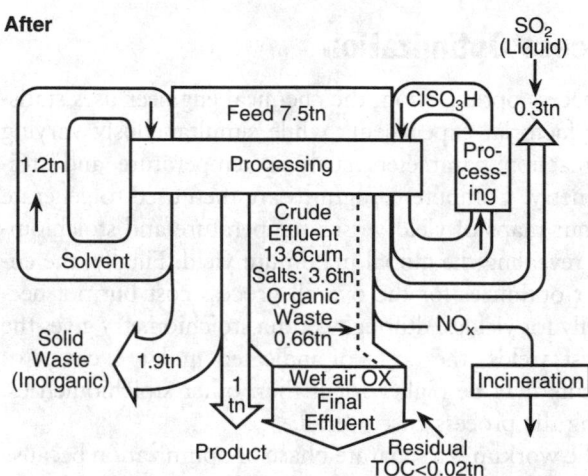

After

FIG. 3.6.11 Letter acid production. (Reprinted, with permission, from Ian G. Laing, 1992, Waste minimisation: The role of process development, *Chemistry & Industry* [21 September].)

Raw material consumption reduced from 13.3 tn to 7.5 tn per ton of product—a decrease of 44%.

The solvents were recycled, and the hydrochloric acid (gas) was converted back to chlorosulphonic acid and recycled in the process.

Sulfur dioxide gas was purified, liquified, and sold for external recycling (no sulfur dioxide emissions).

All waste was reduced by about half and the aqueous effluent by about 80%.

The organic load in the aqueous effluent was reduced by more than 97% through the integration of wet air oxidation, and waste gas was reduced to zero through the inclusion of waste air incineration.

As in another plant, the waste was drastically reduced but not totally eliminated. Even with ever increasing automation and processing technologies, the optimization between the input and output streams of a production process is never complete. The potential for improving efficiency always exists.

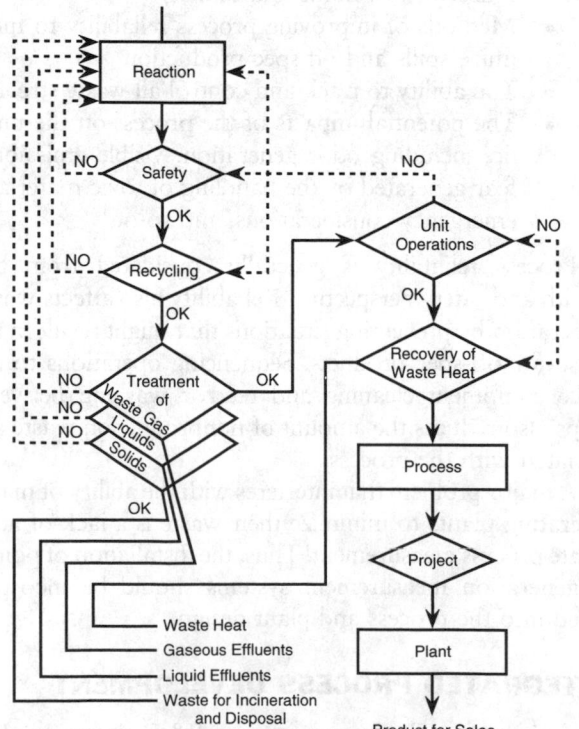

FIG. 3.6.10 Integrated process development.

—*David H.F. Liu*

References

Benforado, D.M., G. Riddlehover, and M.D. Gores. 1991. Pollution prevention: One firm's experience. *Chem Eng* (September).

Borman, Stu. 1993. Process makes alpha-olefins from carboxylic acid. *C&E News* (11 January).

Hileman, Bette. 1992. CHEMRAWN in Moscow provides practical approaches to industry goals. *C&E News* (9 November).

Illman, D.L. 1993. Green technology presents challenge to chemists. *C&E News* (6 September).

Parkinson, G. and E. Johnson. 1989. Designer catalysts are all the rage. *Chem. Eng.* (September)

Stinson, Stephen C. 1993. Customer chemicals. *C&E News* (8 February).

3.7
REACTION ENGINEERING

A pollution prevention strategy is to begin at the reactor, which is the heart of the process. The reactor is where raw materials are converted into products and waste by products. Reactor design is therefore a vital step in the overall design of a process. Here, the chemical engineer must regard two things as being fixed beforehand: the scale of operation and the thermodynamics and kinetics of the given reaction.

Batch and Continuous Operations

For batch operations, the physical properties such as temperature, concentration, pressure, and reaction rate change at any point within the reactor as the reaction proceeds. In continuous operations, these properties are subject to only small, if any, local fluctuations. Two key differences on waste regeneration between batch and continuous processes are:

Waste streams from batch processes are generally intermittent, whereas those from continuous processes are continuous.

The composition and flow rates of waste streams leaving batch processes typically vary, whereas those of continuous processes are fairly constant.

The variability of waste from batch processes creates more difficult waste management problems. For example, if the total volume of waste must be handled, the instantaneous maximum flow is higher in a batch plant, and larger equipment is required to handle this waste. Also, waste generation rates are often high during start up and shutdown periods, and these periods occur most frequently in batch units. Therefore, waste reduction factors generally favor continuous rather than batch processing (Rossiter, Spriggs, and Klee 1993).

The main factors and heuristics usually considered in a decision between batch and continuous operations are (Rossiter, Spriggs, and Klee 1993):

Production rate: Under 500,000 lb/yr, batch processing is invariably used; between 500,000 and 5,000,000 lb/yr, batch processing is common; at higher rates, continuous processing is preferred.

Product life: Batch plants are better suited to products with short life spans where a rapid response to the market is required.

Multiproduct capabilities: If the unit must make several similar products using the same equipment, batch processing is usually preferred.

Process reasons: A number of process-related factors can lead to batch processing being preferred; for example, cleaning requirements that need frequent shutdowns, difficulties in scaling up laboratory data, or complicated process recipes.

If potentially serious environmental problems are anticipated with a process, the selection of a continuous unit is favored. Batch operations are preferred for reactions where rapid fouling occurs or contamination is feared.

In practice, some of the other factors mentioned previously can dictate that a batch operation is preferred. Then, the chemical engineer should consider smoothing intermittent or variable flow streams (for example, by adding buffer storage capacity) to simplify processing and recovery of waste material.

The capital cost for a batch operation is often less than for a corresponding continuous process. Therefore, it is frequently favored for new and untried processes which will be changed to a continuous operation at a more advanced stage of development.

As a final observation on the use of heuristics, Haseltine (1992) notes that the inherent flexibility of batch plants often makes raw material and product substitution simpler in these processes.

Waste Production in Reactors

Under normal operating conditions, five sources of waste production exit reactors (Smith and Petela 1991):

If unreacted feed material cannot be recycled back to the reactor, then low conversion in the reactor leads to waste for that unreacted feed.

The primary reaction can produce waste by products; for example:

$$\text{FEED 1 + FEED 2} \longrightarrow \text{PRODUCT + WASTE PRODUCT} \quad 3.7(1)$$

Secondary reactions can produce waste by products; for example:

$$\text{FEED 1 + FEED 2} \longrightarrow \text{PRODUCT} \longrightarrow \text{WASTE PRODUCT} \quad 3.7(2)$$

Impurities in the feed material become waste or can react to produce additional waste by products.

The catalyst is either degraded and requires changing or is lost and cannot be recycled.

REDUCING WASTE FROM SINGLE REACTIONS

If the reaction forms a waste by-product, as in Equation 3.7(1), the chemical engineer can only avoid the waste by product by using a different reaction path, e.g., a change in feedstock, different reaction chemistry, and ultimately a different process.

Increasing Conversion for Single Irreversible Reactions

If separating and recycling unreacted feed material is difficult, a high conversion in the reactor is necessary. For an irreversible reaction, a low conversion can be forced to a higher conversion by a longer residence time in the reactor, a higher temperature, or higher pressure. A longer residence time is usually the most effective means.

For continuous reactors, this increase in residence time means adding extra volume to the reactor. For batch reactors, higher conversion can mean a longer cycle time, a bigger reactor, or a new reactor in parallel. Sometimes, the chemical engineer can increase the residence time in the existing reactor without increasing the cycle time simply by rescheduling other operations in the process.

Increasing Conversion for Single Reversible Reactions

Maximum conversion is the equilibrium conversion, which cannot be exceeded even with a long residence time. However, options are available for increasing conversion.

EXCESS REACTANTS

Using an excess of one of the reactants is a well-known technique (see part (a) in Figure 3.7.1).

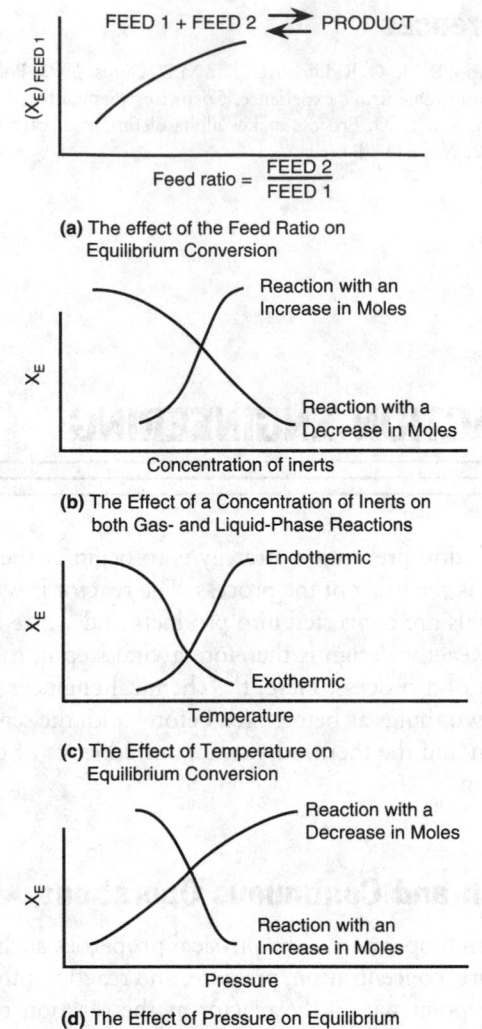

(a) The effect of the Feed Ratio on Equilibrium Conversion

(b) The Effect of a Concentration of Inerts on both Gas- and Liquid-Phase Reactions

(c) The Effect of Temperature on Equilibrium Conversion

(d) The Effect of Pressure on Equilibrium Conversion of Gas-Phase Reactions

FIG. 3.7.1 How reactor conditions affect equilibrium conversion for reversible reactions.

PRODUCT REMOVAL DURING REACTION

Sometimes the product (or one of the products) can be removed continuously from the reactor as the reaction progresses; for example, the product can be allowed to vaporize from a liquid-phase reaction mixture. Another way is to carry out the reaction in stages with intermediate separation of the products between each stage.

INERT CONCENTRATION

Sometimes an inert is present in the reactor. This inert might be a solvent in a liquid-phase reaction or an inert gas in a gas-phase reaction. If the total number of moles increases as the reaction proceeds, adding inert material increases the equilibrium conversion (see part (b) in Figure 3.7.1). If the number of moles decreases, decreasing the concentration of the inert increases the equilibrium conversion. If the number of moles remains the same, the inert material has no effect on the equilibrium conversion.

REACTION TEMPERATURE

Sometimes the chemical engineer can change the temperature to force a higher equilibrium conversion (see part (c) in Figure 3.7.1). For endothermic reactions, the temperature should be as high as possible without exceeding the limitations on construction materials, catalyst life, and safety. For exothermic reactions, the ideal temperature decreases continuously as conversion increases.

REACTION PRESSURE

Changing the reactor pressure can also force a higher equilibrium conversion in gas-phase reactions (see part (d) in Figure 3.7.1). Reactions involving a decrease in the number of moles should operate at the highest possible pressure. For reactions involving an increase in the number of moles, the ideal pressure should decrease continuously as conversion increases. The chemical engineer can reduce the pressure by either operating at a lower absolute pressure or adding an inert diluent.

Note that forcing a high conversion reduces the load on the separation system and may allow it to operate more effectively. Thus changes to the reactor reduce waste from the separation system.

REDUCING WASTE FROM MULTIPLE REACTION SYSTEMS

All of the arguments presented for a single reaction apply to the primary reaction in a multiple reaction system. Besides suffering the losses described for single reactions, multiple reaction systems also form waste by-products in secondary reactions.

Reactor Type

The correct type of reactor must be selected. The CPI uses a variety of reactor types, but most emulate one of three ideal models used in reaction kinetic design theory: the ideal-batch, continuous well-mixed, and plug-flow models (see Figure 3.7.2). In ideal-batch and plug-flow reactors, material spends the same time in the reactor. By contrast, in the continuous well-mixed reactor, the residence time is widely distributed. A series of continuous well-mixed reactors approaches the plug-flow reactor in behavior.

The differences in mixing characteristics between ideal-batch and plug-flow reactors and ideal-batch and continuous well-mixed reactors can significantly effect waste minimization in multiple reaction systems.

In the continuous well-mixed reactor, the incoming feed is instantly diluted by the product which has been formed. Thus, an ideal-batch or plug-flow reactor maintains a higher average concentration of feed than a continuous well-mixed reactor.

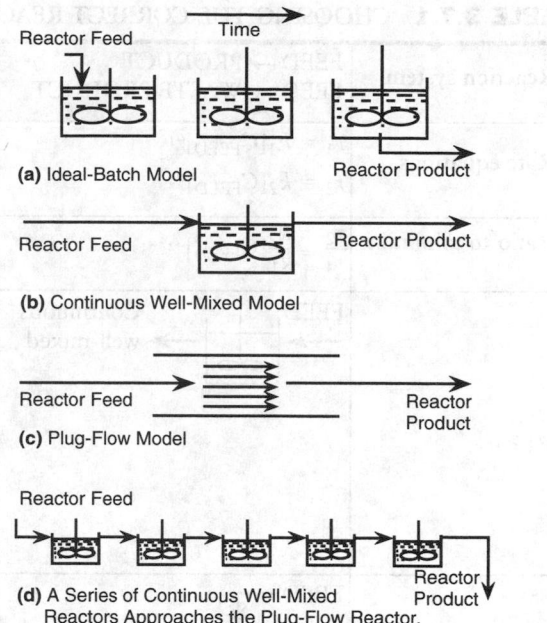

(a) Ideal-Batch Model

(b) Continuous Well-Mixed Model

(c) Plug-Flow Model

(d) A Series of Continuous Well-Mixed Reactors Approaches the Plug-Flow Reactor.

FIG. 3.7.2 Models for reactor design. (Reprinted, with permission, from Robin Smith and E. Petela, 1991, Waste minimisation in the process industries, Part 2: Reactors, *The Chemical Engineer* [12 December].)

Reactor Selection

As shown in the two sets of parallel reactions in Table 3.7.1, the feed material can react either to the PRODUCT or in parallel to the WASTE BY-PRODUCT. By looking at the ratio of the rates of the secondary and primary reactions in Table 3.7.1, the chemical engineer can choose conditions to minimize that ratio.

For some two-feed reaction systems (as shown in Table 3.7.1), semibatch and semiplug-flow processes can be used. In a semibatch process, the reactor is charged with one of the feeds at the start of the reaction, and the other feed is added gradually. The semiplug-flow scheme uses a series of well-mixed reactors, and one of the feeds is charged gradually as the reaction progresses.

Instead of the parallel reactions shown in Table 3.7.1, reactions can also be in series. This reaction system with its corresponding rate equations is as follows:

$$FEED \longrightarrow PRODUCT \qquad r = k \, [C(FEED)]$$
$$3.7(3)$$

$$PRODUCT \longrightarrow WASTE \; BY\text{-}PRODUCT \quad r = k \, [C(PRODUCT)]$$
$$3.7(4)$$

In this reaction system, the FEED reacts to the PRODUCT without any parallel reactions, but the PRODUCT continues to react in series to the WASTE BY-PRODUCT. If the FEED's residence time in the reactor is too short, insufficient PRODUCT is formed. However, if the FEED remains in the reactor too long, this excess time increases its chances of becoming WASTE BY-PRODUCT. Thus, the FEED should ideally have a fixed, well-defined residence

TABLE 3.7.1 CHOOSING THE CORRECT REACTOR TYPE TO MINIMIZE WASTE FOR PARALLEL REACTIONS

Reaction system	FEED → PRODUCT FEED → WASTE PRODUCT		FEED 1 + FEED 2 → PRODUCT FEED 1 + FEED 2 → WASTE BY-PRODUCT	
Rate equations	$r_1 = k_1[C_{FEED}]^{a_1}$ $r_2 = k_2[C_{FEED}]^{a_2}$		$r_1 = k_1[C_{FEED\ 1}]^{a_1}[C_{FEED\ 2}]^{b_1}$ $r_2 = k_2[C_{FEED\ 1}]^{a_2}[C_{FEED\ 2}]^{b_2}$	
Ratio to minimize	$\dfrac{r_2}{r_1} = \dfrac{k_2}{k_1}[C_{FEED}]^{a_2-a_1}$		$\dfrac{r_2}{r_1} = \dfrac{k_2}{k_1}[C_{FEED,\,1}]^{a_2-a_1}[C_{FEED,\,2}]^{b_2-b_1}$	
$a_2 > a_1$	FEED → [reactor] → Continuous well-mixed	$b_2 > b_1$	FEED 1, FEED 2 → [reactor] →	Continuous well-mixed
		$b_2 < b_1$	FEED 1 (top) — [reactor] — FEED 2	Semibatch
			FEED 2 — [reactors in series] ← FEED 1	Semiplug flow
$a_2 < a_1$	FEED → [reactor] Batch	$b_2 > b_1$	FEED 2 (top) — [reactor] — FEED 1	Semibatch
			FEED 1 — [reactors in series] ← FEED 2	Semiplug flow
	FEED → → Plug-flow	$b_2 < b_1$	FEED 1 — [reactor] — FEED 2	Batch
			FEED 1, FEED 2 → →	Plug-flow

Source: Smith and Petela, 1991.

time in the reactor. This requirement implies that to minimize waste from multiple reactions in series, an ideal-batch or plug-flow reactor performs better than a continuous well-mixed reactor.

Reactor Conversion

For the series reaction of Equations 3.7(3) and 3.7(4), a low concentration of PRODUCT in the reactor minimizes the formation of WASTE BY-PRODUCT. This reduction can be achieved by low conversion in the reactor.

If the reaction involves more than one feed, operating with the same low conversion on all feeds is not necessary. By using an excess of one feed, the chemical engineer can operate with relatively high conversion of other feed material and still inhibit series waste by-product formation.

Unfortunately, low conversion in the reactor (whether on one or all of the feeds) can increase both energy use and the cost of separation and recycling. However, if an existing separation system has spare capacity, reducing conversion to reduce waste by-product formation via series reaction is still worth considering.

Whereas increasing conversion always increases the formation of waste by-products via series reactions, the same is not always true of parallel reactions. Whether waste by-product formation via parallel reactions increases or decreases with increasing conversion depends on the order of the primary and secondary reactions.

Reactor Concentration

One or more of the following actions improves selectivity:

For a given reactor design, use an excess of one of the feeds when more than one feed is involved.

If the by-product reaction is reversible and involves a decrease in the number of moles, increase the concentration of inerts.

If the by-product reaction is reversible and involves an increase in the number of moles, decrease the concentration of inerts.

Separate the product during the reaction before continuing further reaction and separation.

Recycle waste by-products to the reactor. Where this recycling is possible, the waste by-products should be recovered and recycled to extinction.

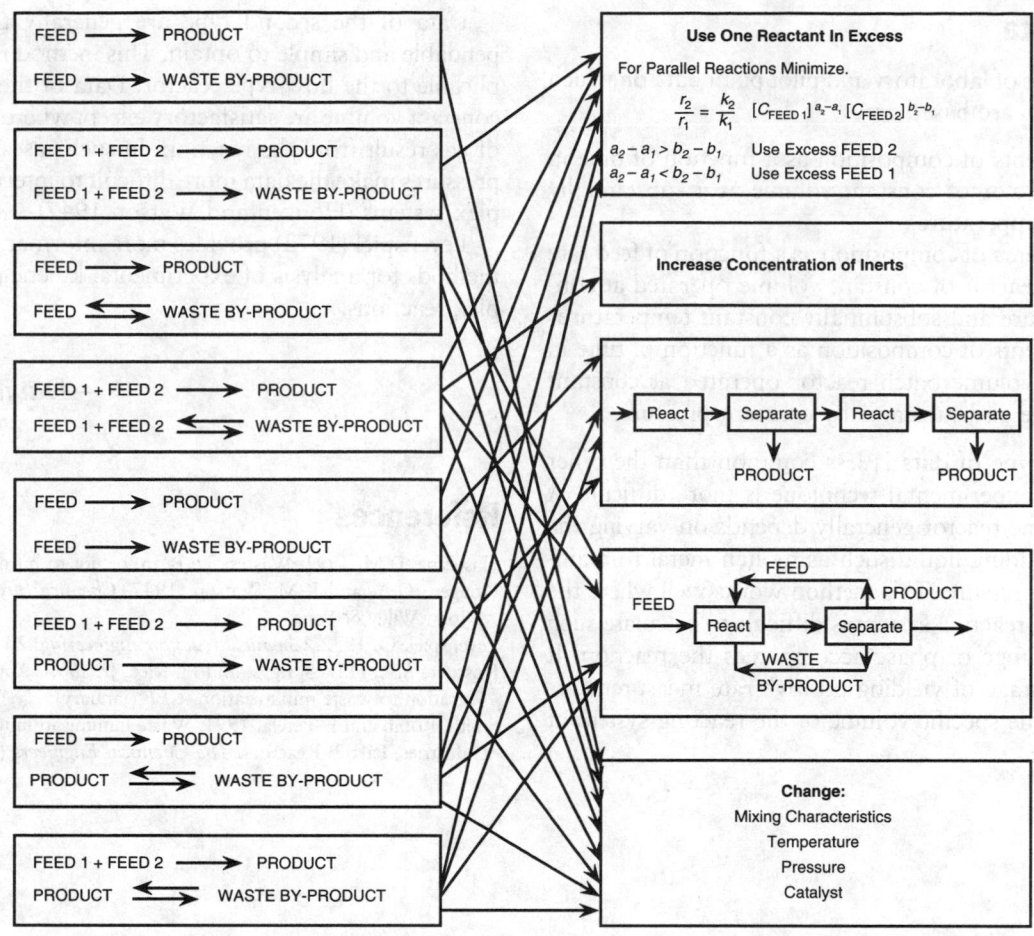

FIG. 3.7.3 Overall strategy for minimizing waste from secondary reactions. (Reprinted, with permission, from Smith and Petela, 1991.)

Each of these measures, in appropriate circumstances, minimizes waste. The effectiveness of these techniques depends on the reaction system. Figure 3.7.3 shows where each technique is appropriate for a number of reaction systems.

Temperature, Pressure, and Catalysts

In a system of multiple reactions, a significant difference can exist between the primary and secondary reactions in the way they are affected by changes in temperatures or pressure. Temperature and pressure should be manipulated to minimize waste (see Figure 3.7.3).

Catalysts also have a significant influence on the waste production with multiple reactions. Changing a catalyst is a complex problem which can ultimately mean a new process.

Impurities and Catalyst Loss

When feed impurities react, this reaction wastes feed material, products, or both. Avoiding such waste is usually only possible by purifying the feed. Thus, higher feed pu-

rification costs must be evaluated against lower cost for raw materials, product separation, and waste disposal.

Using heterogeneous rather than homogeneous catalysts can also reduce waste from catalyst loss. Homogeneous catalysts can be difficult to separate and recycle, and this difficulty leads to waste. Heterogeneous catalysts are more common, but they degrade and need replacement. If contaminants in the feed material or recycling shortens the catalyst life, extra separation to remove those contaminants before they enter the reactor might be justified. If the catalyst is sensitive to extreme conditions such as high temperature, the following measures can help avoid local hot spots and extend the catalyst life:

- Better flow distribution
- Better heat transfer
- A catalyst diluent
- Better instrumentation and control

Fluid-bed catalytic reactors tend to lose the catalyst through attrition of the solid particles generating fines which are then lost. More effective separation and recycling of fines reduce catalyst waste to a point. Improving the mechanical strength of the catalyst is probably the best solution in the long run.

Kinetic Data

The three types of laboratory and pilot plant data on which reactor designs are based are:

1. Measurements of composition as a function of time in a batch reactor of constant volume at a substantially constant temperature
2. Measurements of composition as a function of feed rate to a flow reactor of constant volume operated at constant pressure and substantially constant temperature
3. Measurements of composition as a function of time in a variable-volume batch reactor operated at constant temperature and substantially constant pressure.

The third type of data is less common than the other two, and the experimental technique is more difficult. A variable-volume reactor generally depends on varying the level of a confining liquid such as molten metal to maintain constant pressure. This method works well where the volume of the reacting system is difficult to calculate such as when a change of phase accompanies the reaction. It has the advantage of yielding positive rate measurements and data on the specific volume of the reacting system at the same time.

Data of the second type are generally the most dependable and simple to obtain. This method is directly applicable to the flow-type reactor. Data of the first type at constant volume are satisfactory except where added moles of gas result from the reaction. In such cases, the varying pressures make the data more difficult to interpret for complex systems (Hougen and Watson 1947).

Levenspiel (1972) provides more information about the methods for analysis of experimental kinetic data of complex reactions.

—*David H.F. Liu*

References

Haseltine, D.M. 1992. Wastes: To Burn, or not to burn? *CEP* (July).

Hougen, O.A. and K.M. Watson. 1947. *Chemical process principles.* John Wiley & Sons.

Levenspiel, O. 1972. *Chemical reaction engineering.* 2d ed. John Wiley.

Rossiter, A.P., H.D. Spriggs, and H. Klee, Jr. 1993. Apply process integration to waste minimization. *CEP* (January).

Smith, Robin and E. Petela. 1991. Waste minimisation in the process industries, Part 2: Reactors. *The Chemical Engineers* (12 December).

3.8
SEPARATION AND RECYCLING SYSTEMS

Waste minimization within industry is synonymous with increasing the efficiency of separation systems. Efficiency means both the sharpness of the separation (i.e., how well the components are separated from each other) and the amount of energy required to effect the separation. If separation systems can be made more efficient so that the reactant and the intermediate in the reactor effluent can be separated and recycled more effectively, this efficiency can reduce waste.

This section discusses the use of schemes that minimize waste from separation and recyling systems and the use of new separation technologies for waste reduction.

Minimizing Waste

Figure 3.8.1 illustrates the basic approach for reducing waste from separation and recycling systems. The best sequence to consider the four actions depends on the process. The magnitude of effect that each action has on waste minimization varies for different processes.

RECYCLING WASTE STREAMS DIRECTLY

If waste streams can be recycled directly, this way is clearly the simplest to reduce waste and should be considered first. Most often, the waste streams that can be recycled directly are aqueous streams which, although contaminated, can be substituted for part of the freshwater feed to the process.

Figure 3.8.2 is a flowsheet for the production of isopropyl alcohol by the direct hydration of propylene. Propylene containing propane as an impurity is reacted with water to give a mixture which contains propylene, propane, water, and isopropyl alcohol. A small amount of by-products, principally di-isopropyl ether, is formed. Unreacted propylene is recycled to the reactor, and a purge is taken so that the propane does not build-up. The first distillation column (C1) removes the light ends (including di-isopropyl ether). The second distillation column (C2) removes as much as water as possible to approach the azeotropic composition of the isopropyl alcohol–water mixture. The final column (C3) performs an azeotropic distillation using di-isopropyl ether as an entrainer.

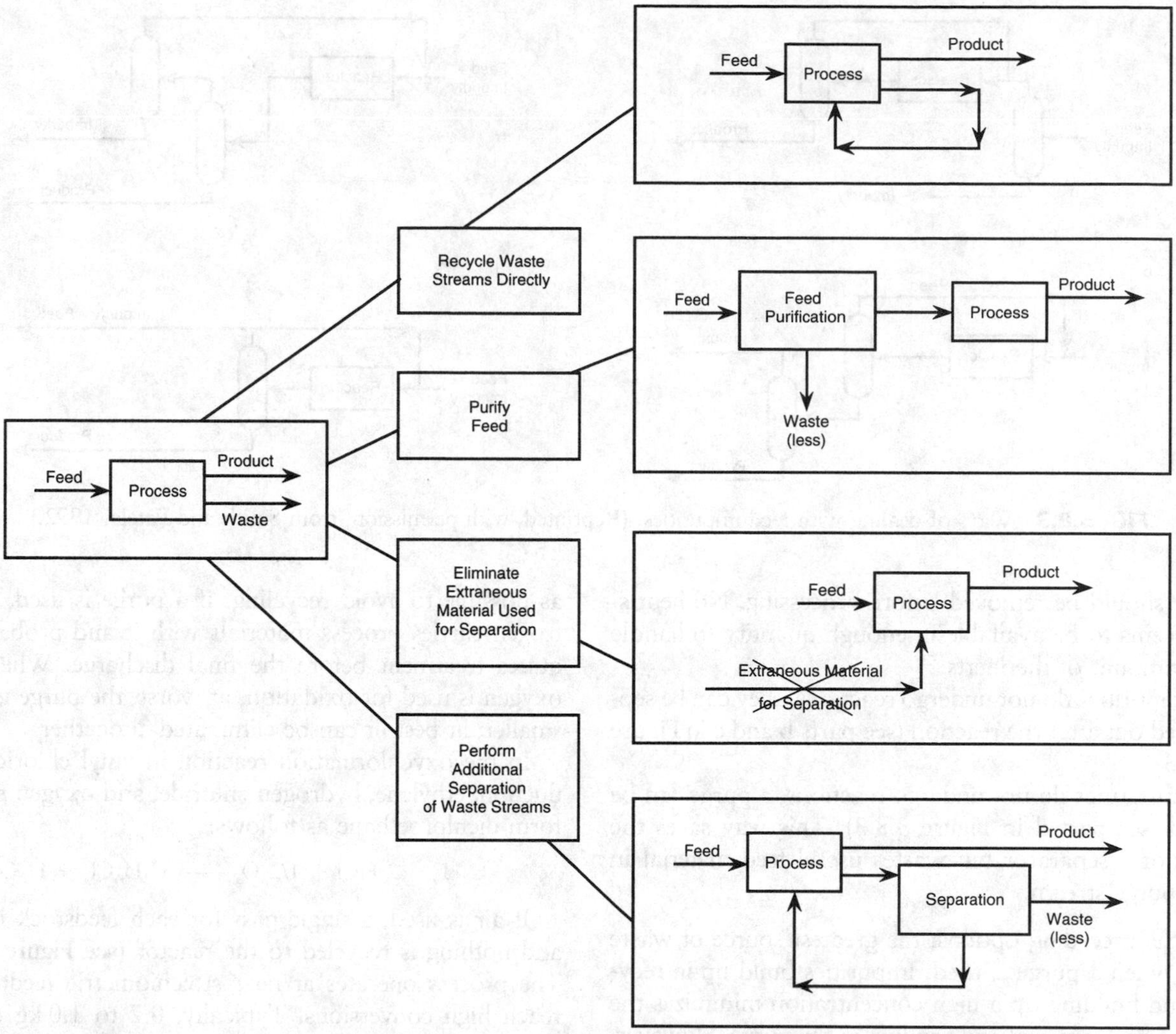

FIG. 3.8.1 Four general ways in which waste from the separation and recycling systems can be minimized.

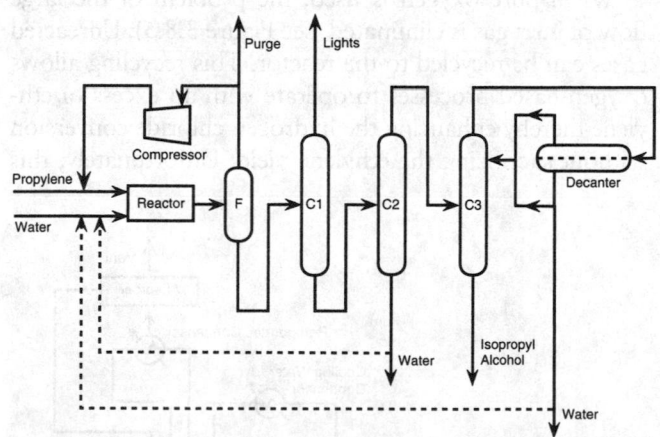

FIG. 3.8.2 Flowsheet for the production of isopropyl alcohol by direct hydration of propylene. (Reprinted, with permission, from R. Smith and Eric Patela, 1992, Waste minimization in the process industries, Part 3: Separation and recycle systems, *The Chemical Engineer* [13 February].)

Waste water leaves the process from the bottom of the second column (C2) and the decanter of the azeotropic distillation column (C3). These streams contain small quantities of organics which must be treated before the final discharge. The chemical engineer can avoid this treatment by recycling the wastewater to the reactor inlet and substituting it for part of the fresh water feed (dotted line in Figure 3.8.2).

Sometimes waste streams can be recycled directly but between different processes. The waste streams from one process may become the feedstock for another.

FEED PURIFICATION

Impurities that enter with the feed inevitably cause waste.

Figure 3.8.3 shows a number of ways to deal with feed impurities including:

If the impurities react, they should be removed before they enter the process (see part a in Figure 3.8.3).

If the impurities are inert, are present in fairly large amounts, and can be easily separated by distillation,

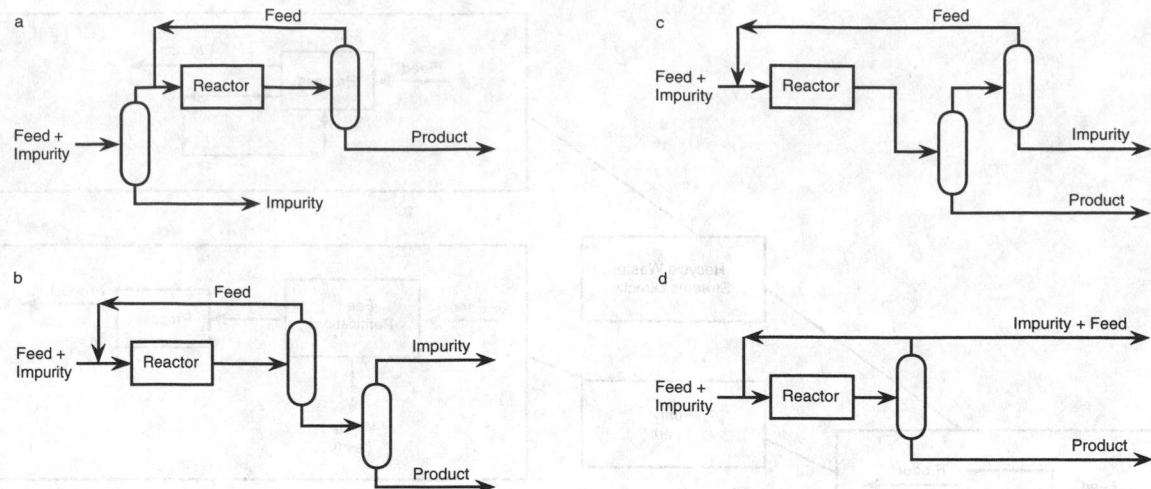

FIG. 3.8.3 Ways of dealing with feed impurities. (Reprinted, with permission, from Smith and Patela, 1992.)

they should be removed before processing. No heuristic seems to be available in enough quantity to handle the amount of the inerts.

If the impurities do not undergo reactions, they can be separated out after the reaction (see parts b and c in Figure 3.8.3).

If the impurities do not undergo reactions, a purge can be used (see part d in Figure 3.8.3). This way saves the cost of a separator but wastes useful feed material in the purge stream.

Of the preceding options, the greatest source of waste occurs when a purge is used. Impurities build up in recycling and building up a high concentration minimizes the waste of feed material and product in the purge. However, two factors limit the extent to which feed impurities can be allowed to build up:

High concentrations of inert material can have an adverse effect on the reactor performance.

As more and more feed impurities are recycled, the recycling cost increases (e.g., through increased recycling gas compression costs) to the point where that increase outweighs the savings in raw material lost in the purge.

In general, the best way to deal with a feed impurity is to purify the feed before it enters the process. In the isopropyl alcohol process (see Figure 3.8.2), the propane, an impurity in propylene, is removed from the process via a purge. This removal wastes some propylene together with a small amount of isopropyl alcohol. The purge can be virtually eliminated if the propylene is purified by distillation before entering the process.

Many processes are based on an oxidation step for which air is the first obvious source of oxygen. Clearly, because the nitrogen in air is not required by the reaction, it must be separated at some point. Because gaseous separations are difficult, nitrogen is normally separated using a purge, or the reactor is forced to as high a conversion

as possible to avoid recycling. If a purge is used, the nitrogen carries process materials with it and probably requires treatment before the final discharge. When pure oxygen is used for oxidation, at worst, the purge is much smaller; at best, it can be eliminated altogether.

In the oxychlorination reaction in vinyl chloride production, ethylene, hydrogen chloride, and oxygen react to form dichloroethane as follows:

$$C_2H_4 + 2\ HCl + 1/2\ O_2 \longrightarrow C_2H_4Cl_2 + H_2O \quad \textbf{3.8(1)}$$

If air is used, a single pass for each feedstock is used, and nothing is recycled to the reactor (see Figure 3.8.4). The process operates at near stoichiometric feedrates to reach high conversions. Typically, 0.7 to 1.0 kg of vent gases are emitted per kilogram of dichloroethane produced.

When pure oxygen is used, the problem of the large flow of inert gas is eliminated (see Figure 3.8.5). Unreacted gases can be recycled to the reactor. This recycling allows oxygen-based processes to operate with an excess of ethylene thereby enhancing the hydrogen chloride conversion without sacrificing the ethylene yield. Unfortunately, this

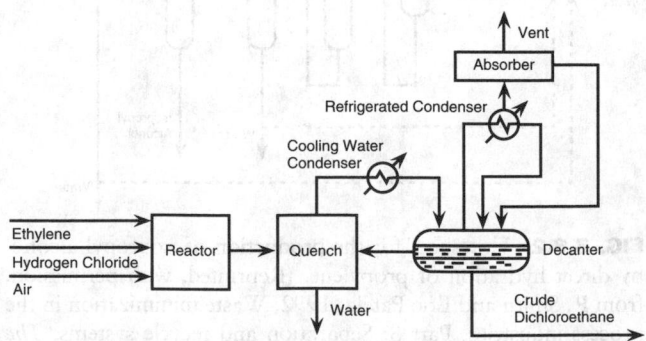

FIG. 3.8.4 The oxychlorination step of the vinyl chloride process with air feed. (Reprinted, with permission, from Smith and Patela, 1992.)

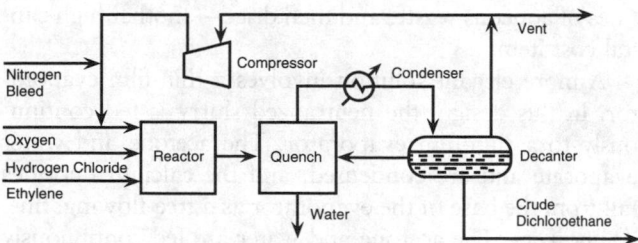

FIG. 3.8.5 The oxychlorination step of the vinyl chloride process with oxygen feed. (Reprinted, with permission, from Smith and Patela, 1992.)

processing introduces a safety problem downstream of the reactor; unconverted ethylene can create explosive mixtures with oxygen.

This problem can be avoided when a small bleed of nitrogen is introduced. Since nitrogen is drastically reduced in the feed and essentially all ethylene is recycled, only a small purge must be vented. This method results in a 20-to-100-fold reduction in the size of the purge compared to a process that uses air as the oxidant (Reich 1976).

ELIMINATION OF EXTRANEOUS SEPARATION MATERIALS

The most obvious example of an extraneous material used for separation is a solvent, either aqueous or organic. An acid or alkali can be used to precipitate other materials from a solution. When these extraneous materials used for separation can be recycled with high efficiency, no major problem exists. Sometimes, however, they cannot, and the discharge of that material creates waste. Reducing this waste involves using an alternative method of separation, such as evaporation instead of precipitation.

A flowsheet for a liquid-phase, vinyl chloride process is shown in Figure 3.8.6. The reactants, ethylene and chlorine dissolved in recirculating dichloroethane, are reacted to form more dichloroethane. The temperature is maintained between 45 and 65°C, and a small amount of ferric chloride is present to catalyze the reaction. The reaction generates considerable heat.

In early designs, the reaction heat was typically removed by means of heat transfer to cooling water. Crude dichloroethane was withdrawn from the reactor as a liquid, acid-washed to remove ferric chloride, then neutralized with a dilute caustic, and purified by distillation. The material used for separating the ferric chloride could be recycled to a point, but a purge had to be taken. This process created waste streams contaminated with chlorinated hydrocarbons which had to be treated before disposal.

The problem with the process shown in Figure 3.8.6 is that the ferric chloride is carried from the reactor with the product and must be separated by washing. A reactor design that prevents the ferric chloride from leaving the reactor would avoid the effluent problems created by the washing and neutralization. Because the ferric chloride is nonvolatile, one way to prevent ferric chloride from leaving the reactor is to allow the heat of the reaction to rise to the boiling point and remove the product as a vapor, leaving the ferric chloride in the reactor. Unfortunately, if the reaction is allowed to boil, two problems result:

Ethylene and chlorine are stripped from the liquid phase, giving a low conversion.
Excessive by-product formation occurs.

This problem is solved in the reactor shown in Figure 3.8.7. Ethylene and chlorine are introduced into circulating liquid dichloroethane. They dissolve and react to form more dichloroethane. No boiling takes place in the zone where the reactants are introduced or in the zone of reaction. As shown in Figure 3.8.7, the reactor has a U-leg in which dichloroethane circulates as a result of the gas lift and thermosiphon effects. Ethylene and chlorine are introduced at the bottom of the up-leg, which is under sufficient hydrostatic head to prevent boiling.

The reactants dissolve and immediately begin to react to form further dichloroethane. The reaction is essentially complete at a point two-thirds of the way up the rising leg. As the liquid continues to rise, boiling begins, and finally the vapor–liquid mixture enters the disengagement

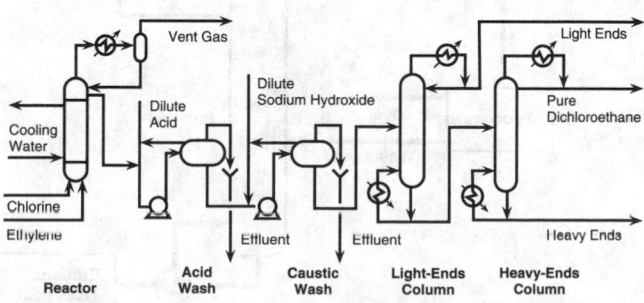

FIG. 3.8.6 The direct chlorination step of the vinyl chloride process using a liquid-phase reactor. (Reprinted, with permission, from Smith and Patela, 1992.)

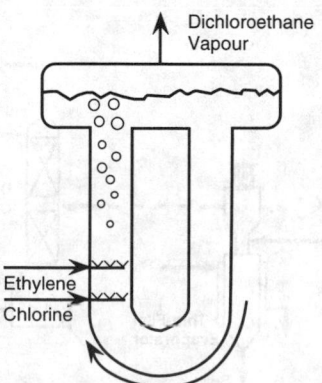

FIG. 3.8.7 A boiling reactor used to separate the dichloroethane from the ferric chloride catalyst. (Reprinted, with permission, from Smith and Patela, 1992.)

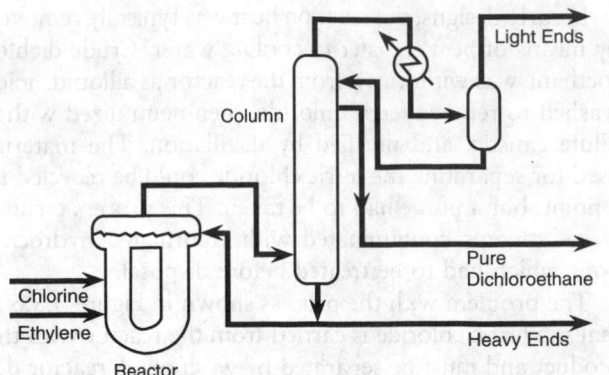

FIG. 3.8.8 The direct chlorination step of the vinyl chloride process using a boiling reactor. This design eliminates the washing and neutralization steps and the resulting effluents. (Reprinted, with permission, from Smith and Patela, 1992.)

drum. A slight excess of ethylene ensures essentially 100% conversion of the chlorine.

As shown in Figure 3.8.8, the vapor from the reactor flows into the bottom of a distillation column, and high-purity dichloroethane is withdrawn as a sidestream, several trays from the column top. The design shown in Figure 3.8.8 is elegant in that the reaction heat is conserved to run the separation, and no washing of the reactor products is required. This design eliminates two aqueous streams which inevitably carry organics with them, require treatment, and cause loss of materials.

With improved heat recovery, using the energy system inherent in the process can often drive the separation system and operate at little or no increase in operating costs.

Figure 3.8.9 is a schematic for the recovery of acetone in a liquid waste containing hydrofluoric acid. This stream originates in the manufacture of an antibiotic precursor. It is pumped to a holding tank. Next, the hydrofluoric acid is neutralized with calcium hydroxide in a stirred reactor. The calcium fluorides that precipitate are difficult to filter. A totally enclosed, high-press filter press was tested. The filter cake had to be washed, generating large quan-

tities of aqueous waste, and then dried—another high capital cost item.

A more elegant solution involves a thin-film evaporator. In this design, the neutralized slurry is fed continuously to a thin-film evaporator. The acetone and water evaporate and are condensed, and the calcium fluorides fall from the base of the evaporator as a free-flowing, fine, dry powder. The acetone and water are fed continuously to a distillation column. The water leaves the bottom of the column and is sent to a water treatment plant. The acetone is upgraded to a sufficiently high purity to be reused directly in the production.

ADDITIONAL SEPARATION AND RECYCLING

Perhaps the most extreme examples of separation and recycling are purge streams. Purges deal with both feed impurities and the by-products of the reaction. Purifying the feed can reduce the size of some purges. However, if purification is not practical or the purge must remove a by-product of the reaction, additional separation is necessary.

Figure 3.8.10 shows the recovery of acetone from an aqueous waste stream by distillation. As the fractional recovery of acetone increases when the reflux ratio is fixed, the cost of column and auxiliary equipment increases. Alternately, fixing the number of plates in the column eliminates additional column cost, and increasing recovery by increasing the reflux ratio increases the energy consumption for separation.

For each fractional recovery, a tradeoff exists between the capital and the energy required to obtain the optimum reflux ratio. The result is that the cost of separation (capital and energy) increases with increasing recovery (see Figure 3.8.11). On the other hand, increasing recovery saves the cost of some of the lost acetone. Adding the cost of raw materials to the cost of separation and recycling

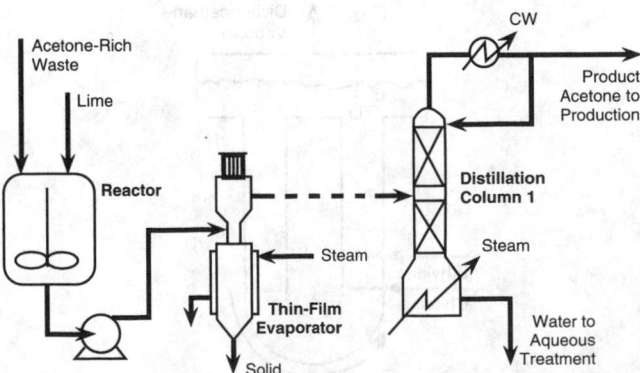

FIG. 3.8.9 Acetone recovery. The thin-film evaporator provides a solution for hard-to-filter calcium fluorides. (Reprinted, with permission, from Smith and Patela, 1992.)

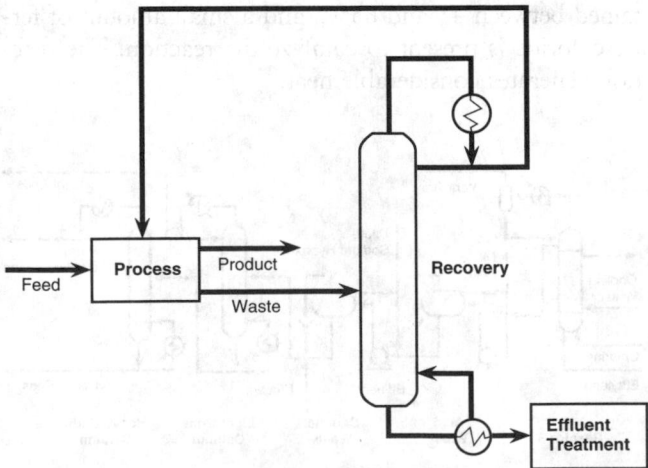

FIG. 3.8.10 Process improved by recycling the excess reactant and solvent used in the reaction.

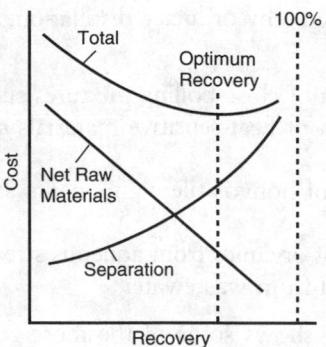

FIG. 3.8.11 Optimum recovery determined by trading-off the effluent treatment costs and raw materials costs against the costs of separation.

gives a curve which shows the minimum cost at a particular acetone recovery.

Again, the unrecovered material from the separation becomes an effluent that requires treatment before it can be discharged to the environment. Thus, adjusting the raw material cost to the net value involves adding the cost of waste treatment of the unrecovered material. In fact with some separations, such as that between acetone and water, separation is possible to a level that is low enough for discharge without biological treatment.

Figure 3.8.12 shows an example of improving a process by recycling the excess reactant and solvent used in the reaction. Initially, the total waste generated from this process amounts to 0.8 lb/lb of product produced. After the changes, this figure drops to 0.1 lb/lb of product produced, and manufacturing costs drop by more than 20%.

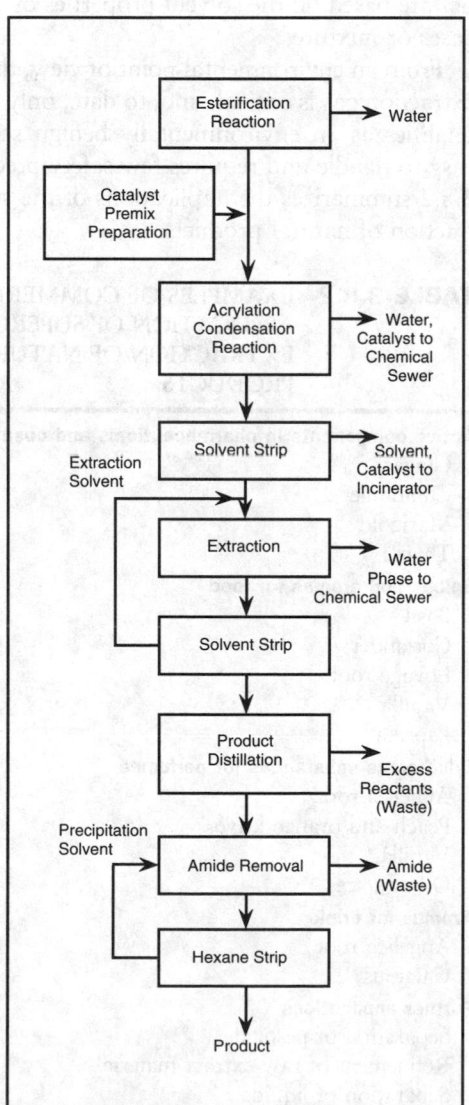

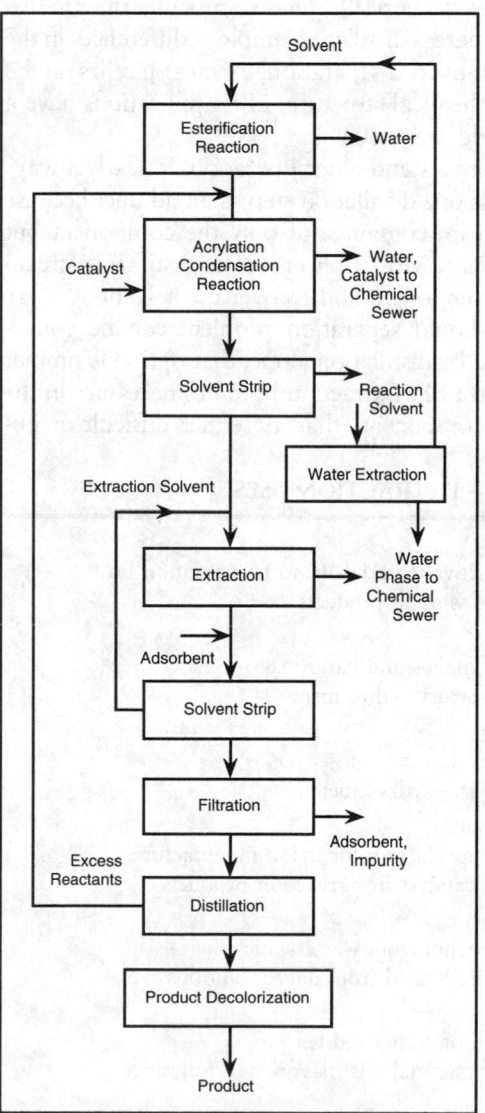

FIG. 3.8.12 Changing a process to recycle excess reactants. The change from the process on the left to the process on the right not only slashed the outlet of waste, but also lowered the manufacturing costs significantly.

Separation Technology

In the CPI, separation processes account for a large part of the investment and a significant portion of the total energy consumption. The dominant separation process in the chemical industry (for liquids) is distillation. In terms of cleaner engineering, a goal is to find methods that provide a sharper separation than distillation, thus reducing the amount of contaminated product streams (i.e., waste), improving the use of raw materials, and yielding better energy economy. For this purpose, this section discusses some unconventional techniques that offer the potential for high separation efficiency and selectively.

EXTRACTION

Distillation is used predominately in the process industries for separating the components of a liquid mixture. Liquid–liquid extraction (LLE) has some similarities to distillation; but whereas distillation employs differences in the boiling point to make a separation, extraction relies on the differences in chemical structure. LLE applications have a broad range and are growing.

LLE is seldom a stand-alone operation. It nearly always requires at least one distillation step as an adjunct because the extract stream contains not only the component but also some solvent. Most LLE processes distill this stream to purify the component and recover the solvent.

If a liquid–liquid separation problem can be conveniently handled by distillation alone, that option is simpler and better than LLE. Extraction becomes necessary in the separation of components that are either difficult or impossible to handle by ordinary distillation. Key examples are:

The separation of close-boiling mixtures such as isomers

The separation of heat-sensitive materials such as antibiotics

The recovery of nonvolatile components as in extractive metallurgy

The removal of organics from aqueous streams, e.g., phenol removal from wastewater

Table 3.8.1 shows some of the more common industrial extraction processes.

SUPERCRITICAL EXTRACTION

Supercritical extraction is essentially a liquid extraction process employing compressed gases under supercritical conditions instead of solvents. The extraction characteristics are based on the solvent properties of the compressed gases or mixtures.

From an environmental point of view, the choice of the extraction gas is critical, and, to date, only the use of CO_2 qualifies as an environmentally benign solution. CO_2 is easy to handle and requires few safety precautions. Table 3.8.2 summarizes the applications of the supercritical extraction of natural products.

TABLE 3.8.1   EXTRACTION USES

Pharmaceuticals
  Recovery of active materials from fermentation broths
  Purification of vitamin products
Chemicals
  Separation of olefins and paraffins
  Separation of structured isomers
Metals Industry
  Copper production
  Recovery of rare-earth elements
Polymer Processing
  Recovery of caprolactam for nylon manufacture
  Separation of catalyst from reaction products
Effluent Treatment
  Removal of phenol from waste water
  Recovery of acetic acid from dilute solutions
Foods
  Decaffeination of coffee and tea
  Separation of essential oils (flavors and fragrances)
Petroleum
  Lube oil quality improvement
  Separation of aromatics and aliphatics (e.g., benzene, toluene, xylenes)

TABLE 3.8.2   EXAMPLES OF COMMERICAL APPLICATION OF SUPERCRITICAL EXTRACTION OF NATURAL PRODUCTS

Active components in pharmaceuticals and cosmetics

Ginger	Calmus
Camomile	Carrots
Marigold	Rosemary
Thyme	Salvia

Spices and aromas for food

Basil	Cardamom
Coriander	Ginger
Lovage root	Marjoram
Vanilla	Myristica
Paprika	Pepper

Odoriferous substances for perfumes

Angelica root	Ginger
Peach and orange leaves	Parsley seed
Vanilla	Vetiver
Oil of spices	

Aromas for drink

Angelica root	Ginger
Calamus	Juniper

Further applications

Separation of pesticides
Refinement of raw extract material
Separation of liquids
Extraction of cholesterol

Source: M. Saari, 1987. Prosessiteollisuuden Erotusmenetelmät, *VTT Res. Note,* and reference 730.

The advantage offered by supercritical extraction is that it combines the positive properties of both gases and liquids, i.e., low viscosity with high density, which results in good transport properties and high solvent capacity. Also, under critical conditions, changing the pressure and temperature varies the solvent characteristics over a range.

Figure 3.8.13 is a flow diagram of a typical supercritical extraction process using supercritical CO_2. Mixing organic components into CO_2 generally enhances their solvent power while inert gases (Ar, N_2) reduce the solvent power. Supercritical extraction is developing rapidly and may become an alternative worth considering not only for fine chemical separation but also for bulk processes.

MEMBRANES

Membrane processes constitute a well-established branch of separation techniques (see Table 3.8.3). They work on continuous flows, are easily automated, and can be adapted to work on several physical parameters such as:

- Molecular size
- Ionic character of compounds
- Polarity
- Hydrophilic or hydrophobic character of components

Microfiltration, ultrafiltration, and reverse osmosis differ mainly in the size of the particles that the membrane can separate as follows:

Microfiltration uses membranes having pore diameters of 0.1 to 10 μm for filtering suspended particles, bacteria, or large colloids from solutions.

Ultrafiltration uses membranes having pore diameters in the range of 22,000 Å for filtering dissolved macromolecules, such as proteins from solutions.

Reverse osmosis membranes have pores so small that they are in the range of the thermal motion of polymer chains, e.g., 5 to 20 Å.

Electrodialysis membranes separate ions from an aqueous solution under the driving force of an electrostatic potential difference.

The membrane is also used in pervaporization that permits the fractionation of liquid mixtures by partial vaporization through a membrane, one side of which is under reduced pressure or flushed by a gas stream. Currently, the only industrial application of pervaporation is the dehydration of organic solvents, particularly dehydration of 90% plus ethanol solutions.

LIQUID MEMBRANES

Liquid membrane technology offers a novel membrane separation method in which the separation is affected by the solubility of the component to be separated rather than by its permeation through pores, as in conventional membrane processes. The component to be separated is extracted from the continuous phase to the surface of the liquid membrane, through which it diffuses into the interior liquid phase.

The liquid membrane can be created in an emulsion or on a stabilizing surface of a permeable support (e.g., polymer, glass, or clay) as shown in Figure 3.8.14. The advantage of an emulsion is the large specific surface.

Figure 3.8.15 is a simplified diagram of a continuous emulsion liquid extraction process. The emulsion is prepared in the first stage of the process (water in oil [W/O] emulsion) by the emulgation of the inner liquid phase I in an organic phase II. In the permeation stage, the emulsion is dispersed into the continuous phase to be treated to form a water/oil/water (W/O/W) emulsion in which the mater-

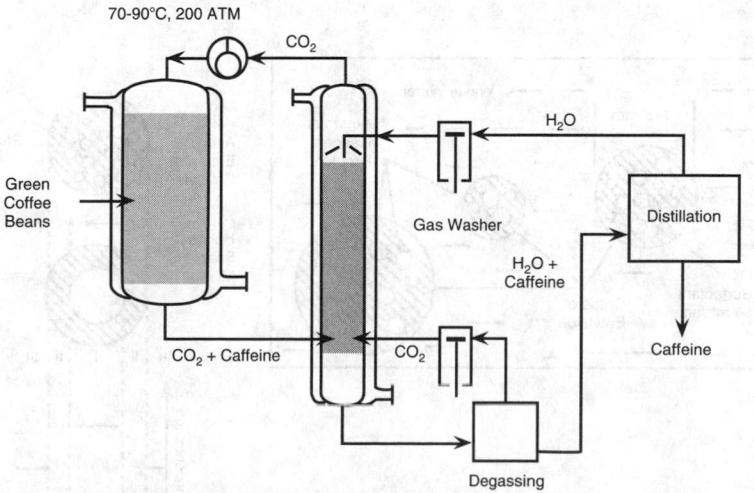

FIG. 3.8.13 A typical supercritical extraction process for coffee decaffeination.

TABLE 3.8.3   MAIN MEMBRANE SEPARATION PROCESSES: OPERATING PRINCIPLES AND APPLICATION

Separation process	Membrane type	Driving force	Method of separation	Range of application
Microfiltration	Symmetric microporous membrane, 0.1 to 10 μA pore radius	Hydrostatic pressure difference, 0.1 to 1 bar	Sieving mechanism due to pore radius and adsorption	Sterile filtration clarification
Ultrafiltration	Asymmetric microporous membrane, 1 to 10 μA pore radius	Hydrostatic pressure difference, 0.5 to 5 bar	Sieving mechanism	Separation of macromolecular solutions
Reverse osmosis	Asymmetric skin-type membrane	Hydrostatic pressure, 20 to 100 bar	Solution–diffusion mechanism	Separation of salt and microsolutes from solutions
Dialysis	Symmetric micro-porous membrane, 0.1 to 10 μA pore size	Concentration gradient	Diffusion in convection-free layer	Separation of salts and microsolutes from macromolecular solutions
Electrodialysis	Cation and anion exchange membranes	Electrical potential gradient	Electrical charge of particle and size	Desalting of ionic solution
Gas separation	Homogeneous or porous polymer	Hydrostatic pressure concentration gradient	Solubility, diffusion	Separation from gas mixture
Supported liquid membranes	Symmetric microporous membrane with adsorbed organic liquid	Chemical gradient	Solution diffusion via carrier	Separation
Membrane distillation	Microporous membrane	Vapor-pressure	Vapor transport into hydrophobic membrane	Ultrapure water concentration of solutions

Source: E. Orioli, R. Molinari, V. Calabrio, and A.B. Gasile, 1989, Membrane technology for production—integrated pollution control systems, Seminar on the Role of the Chemical Industry in Environmental Protection, CHEM/SEM. 18/R. 19, Geneva.

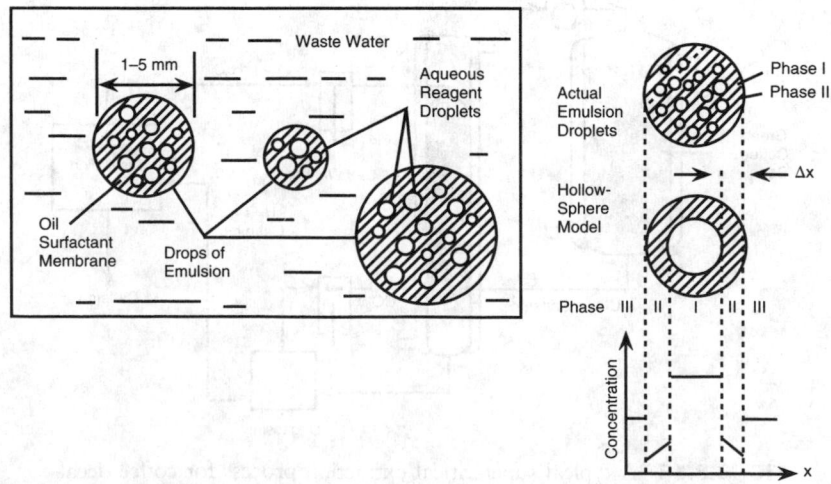

FIG. 3.8.14   Liquid membrane drops. (Reprinted, with permission, from M. Saari, *Prosessiteollisuuden Erotus Menetelmat. VTT Res. Note,* 730.)

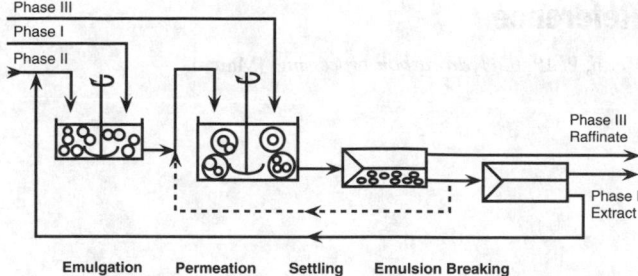

FIG. 3.8.15 Continuous emulsion liquid membrane extraction. (Reprinted, with permission from R. Marr, H. Lackner, and J. Draxler, 1989, *VTT Symposium 102*. Vol. 1, 345.)

ial transfer takes place. In the sedimentation stage, the continuous phase is separated from the emulsion by gravity. In the emulsion-breaking stage, the W/O emulsion is broken, and the organic inner phase is separated.

Compared to LLE, liquid membrane extraction, has the following advantages:

- Requires less solvent
- Is more compact because the extraction and stripping are performed simultaneously

Currently, the main areas of development are in increasing the emulsion stability and in the possibility of including catalyst reactions in the inner phase. Although liquid membrane extraction is not yet widely available, promising results have been reported for a variety of applications, and it appears to offer distinct advantages over alternative methods.

BIOSORBERS

A number of selected, nonliving, inactivated materials of biological origin, such as algae, bacteria, and their products, have been screened for their ability to adsorb metals from a solution. Different biomass types exhibited different performance for the metal species tested. The pH for the test solutions also influenced biomass adsorption performance (see Table 3.8.4).

The biosorptive uptake of chromium and gold was rather specific. It was not affected by the presence of other cations such as Cu^{2+} UO^{2+}, Ca^{2+}, and Ni^{2+} or anions

TABLE 3.8.4   NEW BIOSORBENTS FOR NONWASTE TECHNOLOGY

Application	Biosorbent	pH
Chromium and gold	*Halimeda opuntia* and *Sargassum natans*	4 to 6 for chromium; <3 for gold
Cobalt	*Ascophyllum nodosum*	4.5
Silver	*Chondris crispus*	2 to 6
Arsenic	*Saccaromyces cerevisiae*	4 to 9
Platinum	*Palmaria palmata*	<3

Source: N. Kuyucak and B. Volensky, 1989, New biosorbers for non-waste technology, *VTT Symposium 102*.

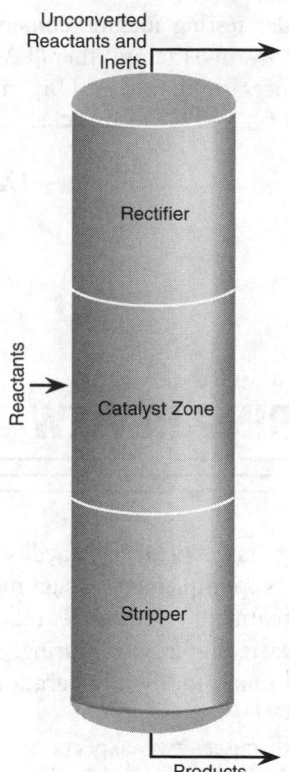

FIG. 3.8.16 In reactive distillation, conversion increased by continuously removing the product from the reactants.

such as NO^-, SO^{2-}, and PO^{3-}. The kinetics of metal uptake, except for gold, was generally rapid.

Sequestered metals could be eluted from the biomass, and the biosorbent material could be reused many times. This work demonstrates the potential of a new group of biosorbent materials of microbial origin which can be effectively used in novel processes of metal recovery from dilute solutions.

REACTIVE DISTILLATION

Also known as catalytical distillation, this technique involves the use of a catalyst within a distillation column (see Figure 3.8.16). When the reaction and distillation occur in one step, a separate reaction step is eliminated.

The chemical reactions best suited for reactive distillation are those characterized by unfavorable reaction equilibria, high reaction heat, and significant reaction rates at distillation temperatures. For example, reactions with unfavorable equilibria are those in which the reaction products contain high concentrations of unconverted reactants. For these reactions, continuously removing one or more products from the reacting mixture substantially increases product conversion. A recently introduced structured packing that incorporates a catalyst may have significant benefits with this application.

A commercial process that uses catalytic distillation is the production of methyl-*tert*-butyl ether (MTBE), an octane enhancer made from methanol and isobytlene. Other applications under testing include cumene from benzene and propylene; *tert*-amyl methyl ether (TAME); and ETBE, an octane enhancer similar to MTBE, made by reactive distillation from C_4 and C_5 feedstocks.

—*David H.F. Liu*

3.9
ENGINEERING REVIEW

During the early stages of a new facility's design, ample opportunity exists to implement design modifications that reduce waste treatment via source reduction or reuse. Incorporating waste elimination during process design is less complicated than modifying operations at an existing plant (Jacobs 1991).

This section discusses two aspects of pollution prevention from the early stage of engineering; these aspects are plant configuration and layout (Berglund and Lawson 1991). Then, it details a ten-step procedure and checklist that identifies and analyzes all environmental issues and pollution prevention opportunities during the plant design stage (Kraft 1992).

Plant Configuration

In the context of pollution prevention, two aspects of plant configuration and layout are especially important: process integration and safety.

PROCESS INTEGRATION

An integrated plant uses all by-products and co-products within the plant itself. This use minimizes the interplant transportation of raw materials, by-products, and wastes. Such a facility is likely to be large and complex. Thus, pollution prevention can supplement or replace the economics of scale for planning large, integrated facilities.

The importance of considering pollution prevention from the early design stage cannot be overemphasized. This consideration is critical for avoiding intrinsic wastes including unreacted raw materials, impurities in the reactants, unwanted by-products, and spent auxiliary materials (catalysts, oils, solvents, and others).

THE SAFETY LINK

Chemical plants today are designed and operated to minimize chances of the worst possible outcome, such as ex-

Reference

Reich, P. 1976. *Hydrocarbon processing.* (March).

plosions, fires, and operator exposure to toxic material. On the other hand, the pollution prevention perspective tries to maximize chances of the best possible outcome, namely zero emissions.

Reducing storage, monitoring for fugitive leaks, and using more integrated facilities are compatible objectives of both those operating goals. However, in other aspects, such as lengthened operating cycles to reduce waste generation during maintenance, efforts to achieve zero emissions can increase the likelihood of unsafe incidents, such as process upsets or the episodic (as opposed to continual) release of a toxic material. The optimal pollution prevention program balances the objectives between these potentially contradictory goals.

Ten-Step Procedure

The ten-step procedure is summarized in Table 3.9.1. Completing these steps ensures that all environmental issues are addressed and all opportunities to reduce waste are effectively defined and analyzed. This technique has been successfully applied in the design of a grassroots plant (Kraft 1992).

TABLE 3.9.1   ENVIRONMENTAL REVIEW PROCEDURE TO EVALUATE NEW PLANT DESIGNS

1. Conduct initial screening and predesign assessments.
2. Assign project environmental leadership responsibility.
3. Define the project's environmental objectives.
4. Identify the need for any permits.
5. Determine the environmental compliance requirements.
6. Perform an overall waste minimization analysis.
7. Apply best environmental practices for emission-free and discharge-free facilities.
8. Determine waste treatment and disposal requirements.
9. Perform engineering evaluations of waste management options.
10. Complete project environmental overview.

STEP 1—PERFORM INITIAL ASSESSMENTS

The initial screening of the project is performed to:

- See if any environmental issues exist
- Perform an environmental site assessment evaluation
- Define and evaluate environmental baseline information

The initial screening answers the following questions:

Does the project involve the use of chemical ingredients?

Does the project involve equipment containing fuels, lubricants, or greases?

Does the potential exist for reducing or eliminating waste, internally recycling materials, or reusing by-products?

Do potential problems exist with the existing site conditions, such as the presence of contaminated soil or groundwater?

Does the project have the potential to contaminate or impair groundwater or soil?

Does the project involve the storage or transport of secondary waste?

If the responses to all of these questions are negative, then the responses are documented, and no further review is required. However, if the answer to any of these questions is yes, then environmental leadership responsibility for the project is assigned (Step 2), and the remaining steps are followed.

Many projects suffer delays and unforecasted expense due to site contamination. Therefore, the site should be checked for potential contamination as soon as possible. The site assessment should do the following:

Determine whether site remediation is needed before construction

Define the proper health and safety plans for construction activities

Determine the appropriate disposal options for any excavated soil

Identify the regulatory requirements that apply

Environmental site assessments include a review of the files about past site operations, an examination of aerial photographs, tests for potential soil and groundwater contamination, and the identification of the environmental constraints that can delay or prevent construction. The time and expense for the assessment should be incorporated into the project time-line and cost estimates.

Environmental baseline information usually includes:

Background air quality prior to project start up

Current emissions at existing sites and potential impact of these emissions on a new project

Monitoring equipment needed to verify environmental compliance after start up

Impact of the construction and operation of a new facility on existing waste treatment facilities and current air, land, and water permits

Determination of whether an environmental impact analysis (EIA) should be performed. EIAs are common at greenfield sites, especially in Europe, and are generally performed by outside consultants.

STEP 2—ASSIGN LEADERSHIP RESPONSIBILITY

The leader's role is to identify and coordinate all resources and ensure that the environmental analysis steps outlined in this procedure are followed. This role should be assigned as early as possible.

STEP 3—DEFINE ENVIRONMENTAL OBJECTIVES

Environmental objectives include a statement supporting government regulations and company policy, a list of specific goals for emissions and discharges or reduction of emissions and discharges, and other project-specific objectives. These objectives focus preferentially on source reduction and recycling rather than waste treatment.

Table 3.9.2 is an example of an environmental charter. Table 3.9.3 presents a hierarchy (prioritized list) of emissions and discharges, and Table 3.9.4 lists the types of emissions and discharges. The charter and these lists serve as a starting point and should be modified to suit the project. The hierarchy of emissions and discharges varies depending on the geographical location (for example, CO_2 can rank higher in the hierarchy in Europe than in the United States).

STEP 4—IDENTIFY PERMIT NEEDS

Obtaining permits to construct and operate a new facility is often the most critical and time-limiting step in a project schedule. Therefore, this step should be started as early as possible in the project.

Permit requirements or limits are not always clearly defined and can often be negotiated with government regulatory agencies. The types of permits required depend on the process involved, the location of the facility, the types of existing permits at an existing facility, and whether new permits or modifications to the existing permits are needed. Typically, permits are required for any part of a process that impacts the environment, such as:

Any treatment, storage, or disposal system for solid or hazardous waste

The exhaust of anything other than air, nitrogen, oxygen, water, or carbon dioxide (Carbon dioxide may require a permit in the future.)

The use of pesticides or herbicides

Incineration or burning

TABLE 3.9.2 ENVIRONMENTAL CHARTER

TO
Design facilities that operate as close to emission- or discharge-free as technically and economically feasible

IN A WAY THAT
Complies with existing and anticipated regulations as well as the established standards, policies, and company practices. Emission- and discharge-reduction priorities are based on a hierarchy of emissions and discharges (see Table 3.9.3) and include various types of emissions and discharges (see Table 3.9.4)

Develops investment options to reduce and eliminate all liquid, gaseous, and solid discharges based on best environmental practices. These options are implemented if they yield returns greater than capital costs. Failing to meet this standard, options may still be implemented subject to nonobjection of the business, production, and research and development functions.

Considers waste management options in the following priority order:
1. Process modifications to prevent waste generation
2. Process modifications to be able to
   a. Recycle
   b. Sell as co-product
   c. Return to vendor for reclamation or reuse. Where materials are sold or returned to the vendor, the project team should ensure that customers and vendors operate in an environmentally acceptable manner.
3. Treatment to generate material with no impact on the environment.

Considers all potential continuous and fugitive emissions in the basic design data as well as all noncontinuous events such as maintenance and clean-out, start-up, and routine or emergency shutdowns.

Allows no hazardous waste to be permanently retained onsite unless the site has a regulated hazardous landfill.

Documents all emissions prior to and after waste minimization efforts.

Interacts with other internal or external processes or facilities to generate a combined net reduction in emissions. Interaction that leads to a net decrease in emissions is considered in compliance with this charter, while that leading to a net increase in emissions is considered not in compliance.

Where decisions are made to delay the installation of emission reduction facilities or to not eliminate specific emissions, considers providing for the future addition of such facilities at minimal cost and operating disruption.

Lists, where possible, specific goals for emissions and discharges or emission and discharge reductions, especially with regard to hazardous and toxic substances.

SO THAT
New facilities provide a competitive advantage in the marketplace based on their environmental performance.

TABLE 3.9.3 HIERARCHY OF EMISSIONS AND DISCHARGES

Carcinogens
Hazardous and Toxic Substances
  Air emissions
    Locally regulated pollutants
    Nonregulated pollutants
  Liquid and solid discharges
    Heavy metals
    Locally regulated pollutants
    Nonregulated pollutants
Nonhazardous Substances
  Low concentrations of toxic materials
  Air emissions
  Liquid and solid emissions
  Inorganic salts
Nonhazardous Waste
  Packaging, sanitary, biotreatment wastewater and sludge
Carbon dioxide
Water and Air (oxygen and nitrogen)
  Drinking and breathing standards

Note: Odor, visible plumes, thermal pollution, and noise should also be considered and may rank high on the hierarchy, depending on the project.

TABLE 3.9.4 TYPES OF EMISSIONS AND DISCHARGES

Direct process streams (including after treatment)
Fugitive emissions
Oils, lubricants, fuels, chlorofluorocarbons, heat transfer fluids
Noncontact process water (e.g., cooling tower water and steam)
Batch process waste (e.g., dirty filters, fly ash, and water washes)
Packaging materials
Old equipment disposal
Office and cafeteria waste
Contaminated soil
Contaminated groundwater
Sediment and erosion control
Stormwater runoff discharges
Construction debris

TABLE 3.9.5 STREAM-BY-STREAM INVENTORY (CHECKLIST A)

1. Name of project (process step, production unit, plant)
2. Operating unit
3. Person completing this analysis
4. List each raw material and its major constituents or contaminants used in this process step, production unit, or plant
5. List each stream by type (feed, intermediate, recycle, nonuseful)

Stream Type	Stream Name and Number	State (Vapor, Liquid, Solid)	Quantity (Volume)	Potential Environmental Issue(s)

Dredging in a water body or any activity that impacts wetlands
Erosion and sedimentation control
Monitoring or dewatering wells
Any action that constructs or alters landfills or land-treatment sites
Any system that constructs or alters water systems
Any system that constructs or alters sanitary wastewater collection or treatment systems
Stormwater runoff

STEP 5—DETERMINE COMPLIANCE REQUIREMENTS

Compliance is determined from the emission and discharge limits specified in the application permit. Going beyond the regulatory requirements and company goals can im-
prove goodwill or image, proactively address possible future regulations, and enhance the company's competitive advantage. If a company decides to go beyond the regulatory requirements, it should do so via waste reduction or reuse rather than waste treatment.

STEP 6—ANALYZE WASTE MINIMIZATION OVERALL

An accurate flow sheet that identifies all major process streams and their composition is important for meaningful waste minimization results.

First, all process streams should be classified into one of the four categories—nonuseful (waste), feed, intermediate, and recyclable—with potential environmental issues noted. Checklist A shown in Table 3.9.5 can be used for this analysis.

TABLE 3.9.6 STREAM-BY-STREAM WASTE MINIMIZATION ANALYSIS (CHECKLIST B)

1. Name of project (process step, production unit, plant)
2. Operating unit
3. Person completing this analysis
4. Stream information (see Nonuseful Streams on Checklist A, Question #5)
   Stream Number
   Stream Name
   State (Vapor, Liquid, Solid)
5. What technologies, operating conditions, and process changes are being evaluated to:
   a. Minimize this stream at the source?
   b. Reuse, recover, or recycle this stream further than originally planned?
   c. Process this stream into a useful product not originally planned?
   d. Reduce the pollution potential, toxicity, or hazardous nature of this stream?
6. What other technologies, operating conditions, and process changes must be evaluated to:
   a. Totally eliminate this stream at the source?
   b. Allow reuse, recovery, or recycling of the stream?
   c. Develop a useful product?
   d. Reduce the pollution potential, toxicity, and hazardous nature of this stream?
7. How could raw material changes eliminate or reduce this nonuseful stream (See Checklist A, Question #4)?
8. What considerations are being given to combining or segregating streams to enhance recycling and reuse or optimize treatment?
9. Does this stream have fuel value? If so, what is being done to recover this fuel value?
10. What consideration is being given to using this stream as a raw material in other company production lines?
11. How does the way in which this stream is handled meet or exceed corporate, operating unit, and site waste elimination or minimization goals?
12. What is the regulatory inventory status (if any) of the chemical components of this stream? Do any premanufacture regulatory requirements exist for these components?

Fill out a separate form for each Nonuseful Stream (prior to treatment or abatement) that is not recycled internally in the manufacturing process via hard piping.

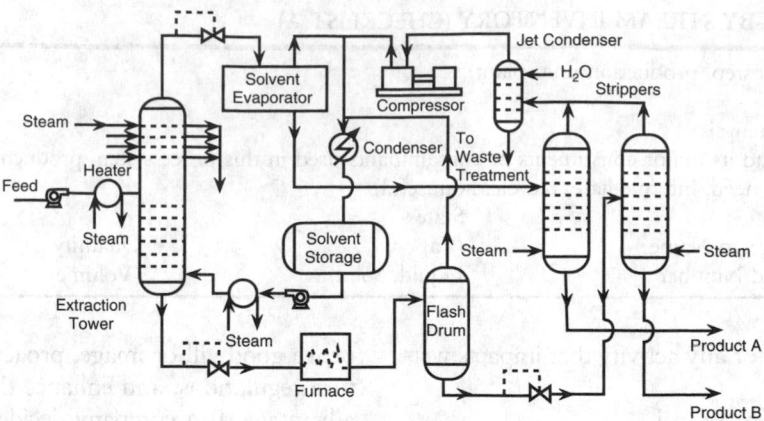

FIG. 3.9.1 Two-phase separators in a chemical process setup. (Reprinted, with permission, from S.G. Woinsky, 1994, Help cut pollution with vapor/liquid and liquid/liquid separators, *CEP* [October].)

Next, the focus is on the nonuseful streams. Checklist B (see Table 3.9.6) should be completed for each nonuseful stream. Each nonuseful stream should be analyzed as follows:

Can it be eliminated or minimized at the source?

If not, can the need for waste treatment be avoided or minimized via reuse, recycling, or co-product sale?

If not, the stream must be treated or rendered as nonhazardous to the environment. (Waste treatment is discussed in Step 8.)

The first two routes often result in attractive economic returns in addition to the environmental benefits. Treatment, while having environmental benefits, seldom has an economic return. For example, separating a gaseous raw material from a reactor purge stream for recycling or reuse may be advantageous. Removing VOCs from the air leaving the dryer involves passing the air through a carbon bed and recirculating it to the heater and dryer.

Reactors can be redesigned to minimize the buildup of latex with a polished interior surface.

Vapor–liquid and liquid–liquid separators can reduce pollution in a plant. Figure 3.9.1 shows a process flowsheet before the detailed engineering phase. Table 3.9.7 summarizes the points in Figure 3.9.1 where separation devices need to be considered. Since the quantities of the recovered material are small, high-value products are the most likely targets for separators, especially from liquid–liquid separators. However, for vapor–liquid separators, the cost is usually low since only the cost of a separator pad is normally involved, and moderately valued products can be targets for these separators (Woinsky 1994).

A major problem that interferes with the ability of operating plants to minimize their waste is a lack of adequate process measurements, such as flows and pH. Installing measurement devices where the waste is generated allows individual streams with waste-reduction potential to be identified and controlled (Jacobs 1991).

TABLE 3.9.7 USE OF LIQUID–LIQUID AND VAPOR–LIQUID SEPARATORS IN FIGURE 3.9.1

Type of Device	Location	Reason
1. Liquid–liquid	In bottom of extraction tower.	Reduces utilities and size of furnace and flash drum.
2. Liquid–liquid	Above feed in extraction tower.	Avoids contamination of Product A with Product B and loss of Product B.
3. Vapor–liquid	In solvent evaporator.	Reduces utilities, avoids fouling of heat exchange equipment, and reduces size of heat exchangers.
4. Vapor–liquid	In flash drum.	Same as above.
5. Vapor–liquid	In top of jet condenser.	Protects compressor.
6. Vapor–liquid	In top of Product A stripper.	Avoids loss of Product A and reduces wastewater hydrocarbon load.
7. Vapor–liquid	In top of Product B stripper	Avoids loss of Product B and reduces wastewater hydrocarbons load.

Source: Woinsky, 1994.

TABLE 3.9.8 BEST PRACTICES FOR EMISSION- AND DISCHARGE-FREE FACILITIES (CHECKLIST C)

1. Name of project (process step, production unit, plant)
2. Operating unit
3. Person completing this analysis
4. What waste or emission is being generated from nonroutine operations (such as changeovers, clean-outs, start ups and shutdowns, spills, and sampling)? List the waste, its major constituents, frequency, and quantity.
5. What techniques or hardware are being used to eliminate or minimize waste from nonroutine operations?
6. What design practices are being used to eliminate fugitive emissions (e.g., from valves, pumps, flanges, and maintenance practices)?
7. Where is closed-loop piping used to eliminate emissions (e.g., interconnections of vents)?
8. What provisions are being made to contain process materials while other steps are in process or are taken offline (e.g., keeping materials within pipes or reactors or going to total recycling)?
9. Where is emission-free equipment specified (e.g., pumps and compressors)?
10. Where does the slope of lines change to eliminate waste via proper draining back to the process?
11. Where are special materials of construction selected to eliminate waste (e.g., so conservation vents do not rust and stick open, cooling coils do not leak, metals do not leach from tanks, and sewers do not leak)?
12. Where are equipment or unit operations close-coupled to eliminate waste?
13. What is being done to eliminate or minimize putting waste in drums?
14. What is being done to eliminate or minimize the amount of excess empty drums to dispose (e.g., changes to receive raw materials in returnable packages or bulk shipment)?
15. What design practices are being used to prevent, through containment and early detection, soil and groundwater contamination due to leaks and spills (e.g., aboveground piping, secondary containment, and no inground facilities)?
16. Where is secondary containment not provided and why?
17. Where are chemicals stored or transported inground and why?
18. What measures are being taken to ensure the control, recovery, and proper disposal of any accidental release of raw materials, intermediates, products, or waste?

TABLE 3.9.9 STREAM-BY-STREAM TREATMENT DISPOSAL ANALYSIS (CHECKLIST D)

1. Name of project (process step, production unit, plant)
2. Operating unit
3. Person completing this analysis
4. Stream information (see Nonuseful Streams on Checklist A,
   Question #5)
   Stream number
   Stream name
   State (Vapor, Liquid, Solid)
5. Does this stream:
   a. Contain toxic chemicals on any regulatory list? If yes, which ones?
   b. Become a hazardous waste under any regulations? If yes,
      is it a listed or a characteristic waste?
6. What permitting requirements are triggered if this stream is to enter the environment?
   Water:
   Land:
   Air:
   Local:
   Other:
7. Is treatment of this stream required before release to the
   environment? If no, what is the basis for this decision?
8. How is this stream, or waste derived from treating it, to be disposed?
9. If flaring of this stream is proposed, what alternatives to flaring were considered?
10. If landfilling of this stream, or waste derived from it, is proposed:
    a. What can be done to eliminate the landfilling?
    b. Must this stream be stabilized before landfilling?
    c. Must this waste be disposed in a secure Class I hazardous waste landfill?
11. Is offsite treatment, storage, or disposal of this waste proposed? If yes, what could be done to dispose of this waste onsite?

Fill out a separate form for each Nonuseful Stream listed in Checklist A and being discharged or emitted into the environment from this process step, production unit, or plant.

TABLE 3.9.10 PROJECT DESIGN ENVIRONMENTAL SUMMARY (CHECKLIST E)

Date

1. Project name   Project number
   Authorization date   Proof year
   Location   Operating unit
2. Project purpose
3. Project environmental leader
4. Environmental project objectives developed and attached?
5. List permits required to construct or operate this facility
   (use attachments if necesssary)

	Date	On	Person
Permit	Required	Critical Path?	Responsible

6. How is this project environmentally proactive (i.e., goes beyond regulatory and corporate requirements)?
7. Describe the method used to develop overall waste minimization and treatment options.
8. Describe the method used to ensure that best environmental practices have been incorporated.
9. How is the project going to impact environmental goals and plans?
   Corporate goals
   Operating unit goals
   Site goals
10. Competitive benchmarking; best competitors emissions or discharges:

	More	Same	Less	Don't Know	Quality of Competitive Information*
Air					
Water					
Land					

*High quality (well-known), Poor quality (guess), or Don't know

11. What has been done to assess the impact of this project on the community, and who has been informed?

12. Environmental results:

Emission/ Discharge	Before[1] (thousand lb/yr)	After (thousand lb/yr)	Waste Mgmt Hierarchy Method[2]	Reason for Not Zero Discharge[3]	Included in Company Env. Plan?
Air					
Carcinogens					
Regulated[4]					
All other[5]					
Greenhouse/ Ozone Depletors[6]					
Water					
Carcinogens					
Regulated[4]					
Hazardous					
Other[7]					
Land					
Carcinogens					
Regulated[4]					
Hazardous					
Other[7]					

Comments:

13. Total environmental investment (thousands of dollars)
    Percent of project cost
    Source reduction
       Investment _____ IRR (%) _____
    Recycling or reuse
       Investment _____ IRR (%) _____
    Waste treatment
       Investment _____ IRR (%) _____
    For company compliance
    For regulatory compliance

Explanations:

1. For stand-alone projects, before is the amount of waste generated and after is the amount emitted or discharged to the environment. For other projects, before and after refer to the amounts emitted or discharged to the environment before and after the project.
2. Indicate A, B, C, or D; A = Source reduction; B = Recycling or Reuse; C = Treatment, D = Disposal.
3. Indicate A, B, C, or D; A = Costs, B = Technology not available; C = Time schedule precluded; D = Other (explain).
4. Use SARA 313 for U.S. and countries that have no equivalent regulation; use equivalent for countries that have a SARA 313 equivalent.
5. Includes NO, SO, CO, NH_3, nonSARA or equivalent VOCs, and particulates.
6. Indicates CO_2, N_2O, CCl_4, methyl chloroform, and nonSARA CFCs.
7. Can include (for example) TOD, BOD, TSS, and pH.

Finally, the operating conditions (e.g., temperature and pressure) and procedures, equipment selection and design, and process control schemes must be evaluated. Some minor alterations to operating conditions, equipment design, and process control may afford significant opportunities for waste minimization at the source.

STEP 7—APPLY BEST ENVIRONMENTAL PRACTICES

The designer should review the entire process to minimize or eliminate unplanned releases, spills, and fugitive emis-sions. (For example, recycle loops are used to sample batches. The use of reusable tote tanks should be promoted, and small refrigerated condensers should be installed on selected units to cut the emissions of costly volatile materials). This review should include all equipment pieces, seals, operating procedures, and so on.

The following hierarchy lists ways to eliminate or minimize fugitive emissions:

Prevent or minimize leaks at the source by eliminating equipment pieces or connections where possible and upgrading or replacing standard equipment with equipment that leaks less or does not leak at all.

Capture and recycle or reuse to prevent or minimize the need for abatement.

Abate emissions to have no impact on the environment.

Use the checklist shown in Table 3.9.8 in this analysis.

STEP 8—DETERMINE TREATMENT AND DISPOSAL OPTIONS

This step defines the waste treatment for nonuseful streams that cannot be reused or eliminated at the source. The goal here is to define the most cost-effective treatment method to render emissions and discharges nonharmful to the environment.

Waste treatment seldom has attractive economics. Waste treatment is used only as the last resort after all options to eliminate waste at the source or reuse waste are exhausted. The checklist shown in Table 3.9.9 can be used to analyze waste treatment.

STEP 9—EVALUATE OPTIONS

Performing engineering evaluations for Steps 6, 7, and 8 is the next step and is especially important when more than one option is available to achieve the same result. Choosing between options based on economic considerations re-quires information such as capital investment, operating costs, and the cost of capital. The net present value calculations and internal rate of return can economically justify one alternative over another.

STEP 10—SUMMARIZE RESULTS

Checklist E shown in Table 3.9.10 is a suggested form for this report. If projects follow a formal approval procedure, this form shows that appropriate environmental reviews were conducted.

Finally, auditing the construction and start up ensures that the environmental recommendations are implemented.

—*David H.F. Liu*

References

Berglund, R.L. and C.T. Lawson. 1991. Preventing pollution in the CPI. *Chem. Eng.* (September).

Jacobs, Richard A. 1991. Waste minimization—Part 2: Design your process for waste minimization. *CEP* (June).

Kraft, Robert L. 1992. Incorporating environmental reviews into facility design. *CEP* (August).

Woinsky, S.G. 1994. Help cut pollution with vapor/liquid and liquid/liquid separators. *CEP* (October).

3.10
PROCESS MODIFICATIONS

The ideal way to reduce or eliminate waste products is to avoid making them in the first place. Almost every part of a process presents an opportunity for waste reduction. Pollutant generation follows repeating patterns that are independent of an industry. Specific improvements involve raw materials, reactors, distillation columns, heat exchangers, pumps, piping, solid processors, and process equipment cleaning.

This section describes several practical ideas and options for preventing polluting generation. When these options are not practical, a second technique is used: recycling waste products back to the process. This section is not an exhaustive compilation of all possibilities. This information is intended to serve as a basis for discussion and brainstorming. Each organization must evaluate the suitability of these and other options for their own needs and circumstances.

Understanding the sources and causes of pollutant or waste generation is a prerequisite to brainstorming for the equipment and process improvement options. For example, when confronted with a tar stream leaving a distillation column, an investigator can formulate the problem as reducing the tar stream. This formulation could lead the investigator to consider measures for optimizing the column. However, further investigation of the problem may show that the distillation is responsible for only a small portion of the tar and that variable raw material quality is responsible for some percentage of the by-product formation. Therefore, the most effective route for reducing the tar stream may not involve the distillation column at all (U.S. EPA 1993).

Raw Materials

Raw materials are usually purchased from an outside source or transferred from an onsite plant. Studying each raw material determines how it affects the amount of waste produced. Also, the specification of each raw material entering the plant should be closely examined. The options described next can prevent pollution generated from raw materials.

Improving Feed Quality

Although the percentage of impurities in a feed stream may be low, it can be a major contributor to the total waste produced by a plant. Reducing the level of impurities involves working with the supplier of a purchased raw material, working with onsite plants that supply feed streams, or installing new purification equipment. Sometimes the effects are indirect (e.g., water gradually kills the reactor catalyst causing formation of by-product, so a drying bed is added).

Using Off-Spec Materials

Occasionally, a process uses off-spec materials (that would otherwise be burned or landfilled) because the quality that makes a material off-spec is not important to the process.

Improving Product Quality

Impurities in a company's products can create waste in their customers' plants. Not only is this waste costly, it can cause some customers to look elsewhere for higher quality raw materials. A company should take the initiative to discuss the effects of impurities with their customers.

Using Inhibitors

Inhibitors prevent unwanted side reactions or polymer formation. A variety of inhibitors are commercially available. If inhibitors are already being used, a company should check with suppliers for improved formulations and new products.

Changing Shipping Containers

If the containers for raw materials being received cannot be reused and must be burned or landfilled, they should be changed to reusable containers or bulk shipments. Similarly, a company should discuss the use of alternative containers for shipping plant products with customers.

Reexamining the Need for Each Raw Material

Sometimes a company can reduce or eliminate the need for a raw material (especially one which ends up as waste) by modifying the process or improving control. For example, one company cut in half the need for algae inhibitors in a cross-flow cooling tower by shielding the water distribution decks from sunlight.

Reactors

The reactor is the center of a process and can be a primary source of waste. The quality of mixing in a reactor is crucial. Unexpected flow patterns and mixing limitations cause problems in commercial reactors designed from bench-scale research data.

The three classes of mixing are defined as follows:

Macro scale mixing refers to blending the feed so that every gallon in a reactor has the same average composition. Mixing on the macro scale is controlled through the use of agitators, educators, and chargers.

Micro scale mixing refers to interdispersing the feed to give uniform composition to the 10–100 mμ scale. Mixing on the micro scale is controlled by eddies and is not affected much by agitation.

Molecular scale mixing is complete when every molecule in the reactor is surrounded by exactly the same molecules at least on a time-averaged basis. This mixing is almost totally driven by diffusion.

A completely well-stirred-tank reactor implies good mixing in all these levels. Many agitated vessels are well mixed on the macro scale, however, no commercial reactor is perfectly well mixed or thoroughly plug-flow. This incomplete mixing may be unimportant if the reaction is dominant or all reactions are slow. In many applications, the reaction and mixing times are often similar and must be carefully studied.

Sometimes one level of mixing controls the reaction rates in a bench reactor, and another level controls them in the plant. For example, in the following commercial reaction:

$$A + B \rightleftharpoons C \qquad 3.10(1)$$
$$C + D \longrightarrow P + Q \qquad 3.10(2)$$
$$A + A \longrightarrow X \qquad 3.10(3)$$

where A, B, and D are reactants; P is the product; and Q and X are by-products; a major discrepancy in yields between laboratory and plant reactors was found. The scale of mixing regimes was determined to be the source of the difference. The reaction in Equation 3.10(1) was fast, the reaction in Equation 3.10(2) was the slow-rate determining step and the reaction in Equation 3.10(3) was intermediate in speed. In the laboratory reactor, A reacted with B before it could form X. In the plant, it took a few seconds to get A and B together. During this mixing time, enough A was lost to X to seriously reduce the yield. A 1-mm-diameter stream disperses much faster than a 50-mm one.

The vessel flow patterns, feed introduction methods, and mixing at all levels can create waste and operational problems. Companies should consider the following options to prevent pollution generated in reactors.

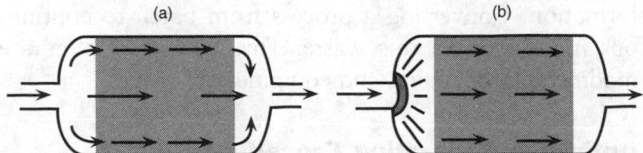

FIG. 3.10.1   Feed distribution systems. (a) Poor feed distribution in a fixed-catalyst bed causes poor conversion and poor yield. (b) Uniform feed flow improves both yield and conversion.

Improving Physical Mixing in the Reactor

Modifications to the reactor such as adding or improving baffles, installing a higher revolutions-per-minute motor on the agitator, or using a different mixer blade design (or multiple impellers) improves mixing. Pumped recirculation can be added or increased. Two fluids going through a pump, however, do not necessarily mix well, and an inline static mixer may be needed to ensure good contacting.

Distributing Feed Better

Part (a) in Figure 3.10.1 shows the importance of distributing feeds. The reactants enter at the top of a fixed-catalyst bed. Part of the feed short-circuits through the center of the reactor having inadequate time to convert to the product. Conversely, the feed closer to the walls remains in the reactor too long and overreacts creating by-products that become waste. Although the average residence time in the reactor is correct, inadequate feed distribution causes poor conversion and poor yield.

One solution is to add a distributor that causes the feed to move uniformly through all parts of the reactor (see part (b) in Figure 3.10.1). Some form of collector may also be necessary at the bottom to prevent the flow from necking down to the outlet.

Improving Methods of Adding Reactants

The purpose of this option is to make the reactant concentrations closer to the ideal before the feed enters the reactor. This change helps avoid the secondary reactions which form unwanted by-products. Part (a) in Figure 3.10.2 shows the wrong way to add reactants. The ideal concentration probably does not exist anywhere in this reactor. A consumable catalyst should be diluted in one of the feed streams (one which does not react in the presence of the catalyst). Part (b) in Figure 3.10.2 shows one approach of improving the addition of reactants using three inline static mixers.

Improve Catalysts

Searching for better catalysts should be an ongoing activity because of the significant effect a catalyst has on the reactor conversion and product mix. Changes in the chemical makeup of a catalyst, the method of preparation, or its physical characteristics (such as size, shape, and porosity) can lead to substantial improvements in the catalyst life and effectiveness (Nelson 1990).

Providing Separate Reactors for the Recycling Stream

Recycling by-product and waste streams is an excellent way of reducing waste, but often the ideal reactor conditions for converting recycling streams to usable products are different from conditions in the primary reactors. One solution is to provide a separate, smaller reactor for handling recycling and waste streams (see Figure 3.10.3). The temperatures, pressures, and concentrations can then be optimized in both reactors to take maximum advantage of reaction kinetics and equilibrium.

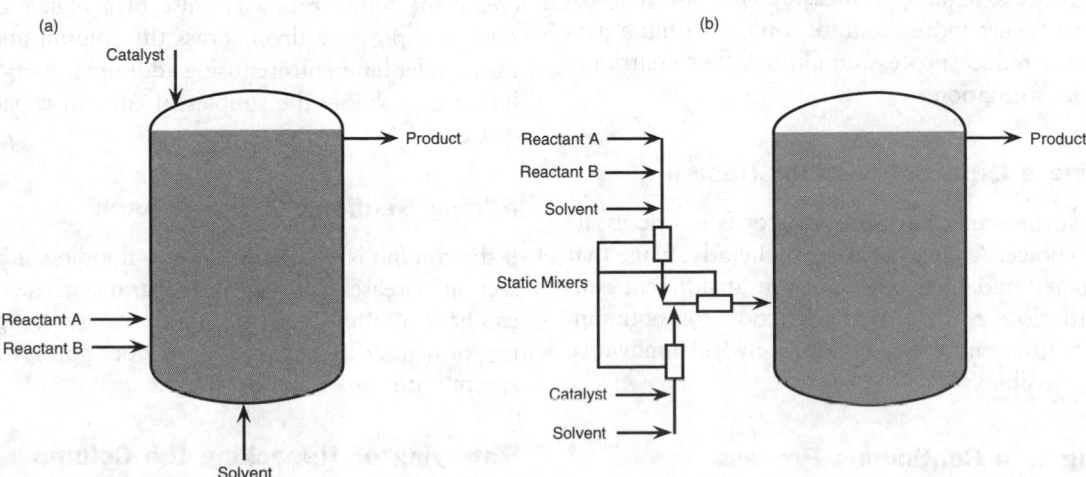

FIG. 3.10.2   Adding feed to a reactor. (a) This method results in poor mixing. (b) Inline static mixers improve reactor performance.

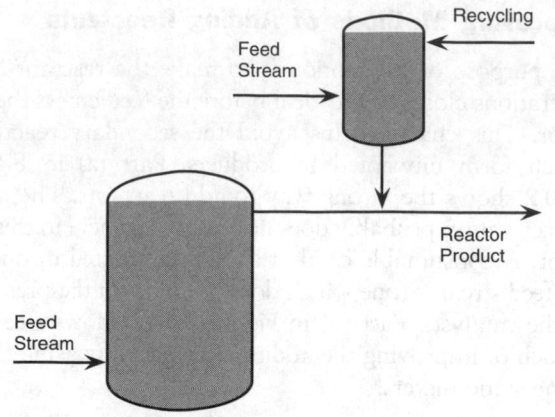

FIG. 3.10.3 A separate, small reactor for recycling waste streams.

Examining Heating and Cooling Techniques

A company should examine the techniques for heating and cooling the reactor to avoid hot or cold spots in a fixed-bed reactor or overheated feed streams, both of which usually give unwanted by-products.

Providing Online Analysis

Online analysis and control of process parameters, raw material feed rates, or reaction conversion rates can significantly reduce by-products and waste.

Implementing Routine Calibration

Routine calibration of process measurement and control equipment can minimize inaccurate parameter set points and faulty control.

Upgrading Process Controls

Upgrading process parameter measurement and control equipment to ensure more accurate control within a narrow range can reduce process conditions that contribute to by-product formation.

Considering a Different Reactor Design

The classic stirred-tank, back-mix reactor is not necessarily the best choice. A plug reactor has the advantage that it can be staged and each stage can run at different conditions, with close control of the reaction for optimum product mix (minimum waste). Many hybrid innovative designs are possible.

Converting to a Continuous Process

The start ups and shutdowns associated with batch processes are a common source of waste and by-product

formation. Converting a process from batch to continuous mode reduces this waste. This option may require modification of piping and equipment.

Optimizing Operating Procedures

This option includes investigating different ways of adding the reactant (e.g., slurry or solid powders), changing process conditions and avoiding the hydrolysis of raw materials to unwanted by-products, and using chemical emulsion breakers to improve organic–water separation in decanters. A common cause of by-product formation is a reaction time that is either too short or too long. In such cases, increasing or decreasing the feed rate can reduce by-products. Optimization of the reactant ratio can reduce excess constituents that may be involved in side, by-product-forming reactions.

Distillation Columns

Distillation columns typical produce waste as follows:

By allowing impurities that ultimately become waste to remain in a product. The solution is better separation. In some cases, normal product specifications must be exceeded.

By forming waste within the column itself usually because of high reboiler temperatures which cause polymerization. The solution is lower column temperatures.

By having inadequate condensation, which results in vented or flared products. The solution is improved condensing.

Some column and process modifications that reduce waste by attacking one or more of these three problems are outlined next.

Increasing the Reflux Ratio

The most common way of improving separation is to increase the reflux ratio. The use of a higher reflux ratio raises the pressure drop across the column and increases the reboiler temperature (using additional energy). This solution is probably the simplest if column capacity is adequate.

Adding Sections to the Column

If the column is operating close to flooding, adding a new section increases capacity and separation. The new section can have a different diameter and use trays, regular packing, or high-efficiency packing. It does not have to be consistent with the original column.

Retraying or Repacking the Column

Another method of increasing separation is to retray or repack part or all of a column. Both regular packing and high-efficiency packing lower the pressure drop through a column, decreasing the reboiler temperature. Packing is no longer limited to a small column; large-diameter columns have been successfully packed.

Changing the Feed Tray

Many columns are built with multiple feed trays, but valving is seldom changed. In general, the closer the feed conditions are to the top of the column (high concentration of lights and low temperature), the higher the feed tray; the closer the feed conditions are to the bottom of the column (high concentration of heavies and high temperature), the lower the feed tray. Experimentation is easy if the valving exists.

Insulating the Column or Reboiler

Insulation is necessary to prevent heat loss. Poor insulation requires higher reboiler temperatures and also allows column conditions to fluctuate with weather conditions (Nelson 1990).

Improving the Feed Distribution

A company should analyze the effectiveness of feed distributors (particularly in packed columns) to be sure that distribution anomalies are not lowering the overall column efficiency (Nelson 1990).

Preheating the Column Feed

Preheating the feed should improve column efficiency. Supplying heat in the feed requires lower temperatures than supplying the same amount of heat to the reboiler, and it reduces the reboiler load. Often, the feed is preheated by a cross-exchange with other process streams.

Removing Overhead Products

If the overhead contains light impurities, obtaining a higher purity product may be possible from one of the trays close to the top of the column. A bleed stream from the overhead accumulator can be recycled back to the process to purge the column lights. Another solution is to install a second column to remove small amounts of lights from the overheads.

Increasing the Size of the Vapor Line

In a low-pressure or vacuum column, the pressure drop is critical. A larger vapor line reduces the pressure drop and decreases the reboiler temperature (Nelson 1990).

Modifying Reboiler Design

A conventional thermosiphon reboiler is not always the best choice, especially for heat-sensitive fluids. A falling-film reboiler, a pumped recirculation reboiler, or high-heat flux tubes may be preferable to minimize product degradation.

Using Spare Reboilers

Shutting down the column because of reboiler fouling can generate waste, e.g., the material in the column (the hold up) can become an off-spec product. A company should evaluate the economics of using a spare reboiler.

Reducing Reboiler Temperature

Temperature reduction techniques such as using lower pressure steam, desuperheating steam, installing a thermocompressor, or using an intermediate heat-transfer fluid also apply to the reboiler of a distillation column.

Lowering the Column Pressure

Reducing the column pressure also decreases the reboiler temperature and can favorably load the trays or packing as long as the column stays below the flood level. The overhead temperature, however, is also reduced, which may create a condensing problem. If the overhead stream is lost because of an undersized condenser, a company can consider retubing, replacing the condenser, or adding a supplementary vent condenser to minimize losses. The vent can also be rerouted back to the process if the process pressure is stable. If a refrigerated condenser is used, the tubes must be kept above 32°F if any moisture is in the stream.

Forwarding Vapor Overhead Streams

If the overhead stream is sent to another column for further separation, using partial condensers and introducing the vapor to the downstream column may be possible.

Upgrading Stabilizers or Inhibitors

Many distillation processes use stabilizers that reduce the formation of tars as well as minimize unfavorable or side reactions. However, the stabilizers not only become large components of the tar waste stream but also make the waste more viscous. The more viscous the waste stream, the more salable product the waste stream carries with it. Upgrading the stabilizer addition system requires less stabilizer in the process.

The upgrade can include the continuous versus batch addition of a stabilizer or the continuous or more frequent analysis of a stabilizer's presence coupled with the automatic addition or enhanced manual addition of the stabilizer. Another option is to optimize the point of addition, the column versus the reboiler, along with the method of addition.

A stabilizer typically consists of a solid material slurried in a solvent used as a carrier. Options for waste reduction also focus on the selection of one of these two components. The addition of a stabilizer in powder form eliminates the solvent. The use of the product as a carrier component is one of the best options.

Improving the Tar Purge Rate

Continuous distillation processes require a means of removing tar waste from the column bottoms. Optimizing the rate at which tars are purged can reduce waste. An automatic purge that controls the lowest possible purge rate is probably best. If an automatic purge is not possible, other ways exist to improve a manually controlled or batch-operated tar purge. If a batch purge is used, more frequent purges of smaller quantities can reduce overall waste (EPA 1993).

Some processes that purge continuously are purged at excessively high rates to prevent valve plugging. More frequent cleaning or installing a new purge system (perhaps with antisticking interior surfaces) permits lower purge rates.

Treating the Column Bottoms to Further Concentrate Tars

Treating the tar stream from the bottom of a distillation column for further removal of the product by a wipe-film evaporator may be a viable option.

Automating Column Control

For a distillation process, one set of operating conditions is optimum at any given time. Automated control systems respond to process fluctuations and product changes swiftly and smoothly, minimizing waste production.

Converting to a Continuous Process

The start ups and shutdowns associated with batch processes are a common source of waste and by product formation. Converting a process from batch to continuous mode reduces this waste. This option may require modifications to piping and equipment. Before any equipment modification is undertaken, a company should do a computer simulation and examine a variety of conditions. If the column temperature and pressure change, equipment ratings should also be reexamined.

Heat Exchangers

A heat exchanger can be a source of waste, especially with products that are temperature-sensitive. A number of techniques can minimize the formation of waste products in heat exchangers. Most techniques are associated with reducing tube-wall temperatures.

Using a Lower Pressure Steam

When plant steam is at fixed-pressure levels, a quick option is to switch to steam at a lower pressure, reducing the tube-wall temperatures.

Desuperheating Plant Steam

High-pressure plant steam can contain several hundred degrees of superheat. Desuperheating steam when it enters a process (or just upstream of an exchanger) reduces tube-wall temperatures and increases the effective area of heat transfer because the heat transfer coefficient of condensing steam is about ten times greater than that of superheated steam.

Installing a Thermocompressor

Another way of reducing the tube-wall temperature is to install a thermocompressor. These relatively inexpensive units work on an ejector principle, combining high- and low-pressure steams to produce an intermediate-pressure steam.

Figure 3.10.4 illustrates the principle of thermocompression. The plant steam at 235 psig upgrades 30-psig steam to 50 psig. Before a thermocompressor was installed, only 235-psig steam was used to supply the required heat.

Using Staged Heating

If a heat-sensitive fluid must be heated, staged heating minimizes degradation. For example, the process can begin with waste heat, then use low-pressure steam, and finally use superheated, high-pressure steam (see Figure 3.10.5).

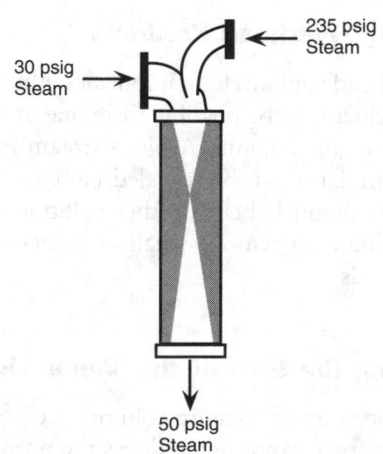

FIG. 3.10.4 Thermocompressor to upgrade 30-psig steam.

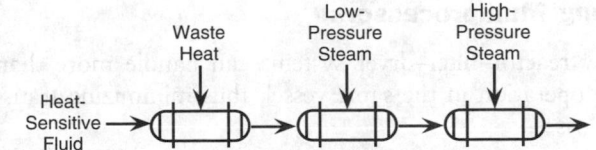

FIG. 3.10.5 Staged heating to reduce product degradation.

Using Air-Fin Coolers

This option reduces the use of cooling water.

Using Online Cleaning Techniques

Online cleaning devices such as recirculated sponge balls or reversing brushes keep tube surfaces clean so that lower temperature heat sources can be used.

Using Scraped-Wall Exchangers

A scraped-wall exchanger consists of a set of rotating blades inside a vertical, cylindrical, jacketed column. They can be used to recover salable products from viscous streams. A typical application is to recover a monomer from polymer tars.

Monitoring Exchanger Fouling

Exchanger fouling does not always occur steadily. Sometimes an exchanger fouls rapidly when plant operating conditions change too fast or when a process upset occurs. Other actions, such as switching pumps, unloading tank cars, adding new catalysts, or any routine action, can influence fouling. However, estimating the effect is possible.

The first step in reducing or eliminating the fouling causes is for the company to identify the causes by continuously monitoring the exchanger and correlating any rapid changes with plant events.

Using Noncorroding Tubes

Corroded tube surfaces foul more quickly than noncorroded tube surfaces. Changing to noncorroding tubes can significantly reduce fouling.

Controlling the Cooling Water Temperature

Excessive cooling water temperature can cause scale on the cooling water side.

Pumps

Two options can prevent the pollution generated by pumps.

Recovering Seal Flushes and Purges

A company should examine each seal flush and purge for a possible source of waste. Most can be recycled to the process with little difficulty.

Using Seal-less Pumps

Leaking pump seals lose product and create environmental problems. Using can-type, sealless pumps or magnetically driven sealless pumps eliminates these losses.

Piping

Even plant piping can cause waste, and simple piping changes can result in major reductions. The process changes described next are options for preventing pollution from piping.

Recovering Individual Waste Streams

In many plants, various streams are combined and sent to a waste treatment facility as shown in Figure 3.10.6. A company should consider each waste stream individually. The nature of the impurities may make recycling or otherwise reusing a stream possible before it is mixed with other waste streams and becomes unrecoverable. Stripping, filtration, drying, or some type of treatment may be necessary before the stream can be reused.

Avoiding Overheating Lines

If a process stream contains temperature-sensitive materials, a company should review both the amount and temperature level of line and vessel tracing and jacketing. If plant steam levels are too hot, a recirculated warm fluid can prevent the process stream from freezing. A company should choose a fluid that does not freeze if the system is shut down in winter. Electric tracing is also an option.

Avoiding Sending Hot Material to Storage

Before a temperature-sensitive material is sent to storage, it should be cooled. If this coding is uneconomical because the stream from the storage must be heated when used, hot steam can be piped directly into the suction line of the storage tank pump as shown in Figure 3.10.7. The storage tank pump must be able to handle hot material without cavitating.

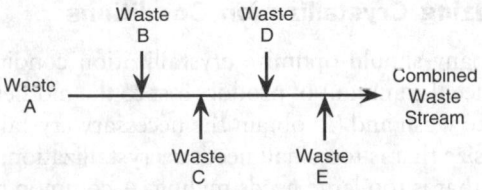

FIG. 3.10.6 Typical plant mixing of waste streams.

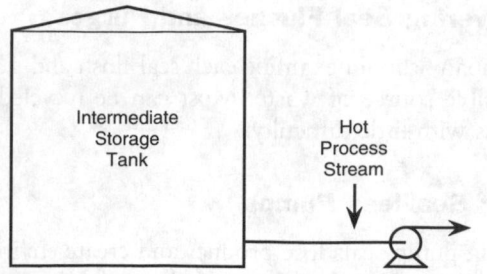

FIG. 3.10.7 Hot stream piped directly into the suction line of the storage tank pump.

Eliminating Leaks

Leaks are a major contributor to a plant's overall waste, especially if the products cannot be seen or smelled. A good way to document leaks is to measure the quantity of raw materials that must be purchased to replace lost stream (e.g., the amount of refrigerant purchased).

Changing the Material of Construction

The type of metal used for vessels or piping can cause color problems or act as a catalyst in the formation of by-products. If this problem occurs, an option is to change to more inert metals. Using lined pipes or vessels is often a less expensive alternative to complex metallurgy. Several coatings are available for different applications.

Monitoring Major Vents and Flare Systems

Flow measurements need not be highly accurate but should give a reasonable estimate of how much product is lost and when those losses occur. Intermittent losses, such as equipment purges, can be particularly elusive. Corrective action depends on the situation. Frequently, a company can reduce or eliminate venting or flaring by installing piping to recover products that are vented or flared and reuse them in the process. Storage tanks, tank cars, and tank trucks are common sources of a vented product. A condenser or small vent compressor may be all that is needed in these sources. Additional purification may be required before the recovered streams can be reused.

Solid Processing

The options described next can prevent pollution generated in solid processing.

Optimizing Crystallization Conditions

A company should optimize crystallization conditions to (1) reduce the amount of product lost to the mother liquor and cake wash and (2) obtain the necessary crystal size. A crystal size that is too small needs recrystallization; a crystal size that is too large needs milling. A common practice is to add small crystals known as seeds into the solution immediately before incipient crystallization.

Using Multiprocessors

New reactor–filter–dryer systems can handle more than one operation in the same vessel, thus minimizing transfer losses.

Coupling the Centrifuge and Dryer

With this option, the solid is separated, washed, and dropped into the dryer to minimize transfer losses. A company can also use an easily emptied, bottom-discharge centrifuge.

Using Bag Filters

A company should consider the use of bag filters instead of cartridge filters. Products can be recovered, and the bags can be washed and reused, while used filter cartridges must be disposed.

Installing a Dedicated Vacuum System

This option can be used to clean spilled powders for companies producing dry formulations.

Minimizing Wetting Losses

To minimize wetting losses, a company can modify the tank and vessel dimensions to reduce contact (e.g., the use of conical vessels).

Process Equipment Cleaning

Equipment cleaning is one of the most common areas of waste generation. The reduction of solvent wash waste from metal cleaning and degreasing operations as well as various applications in the paint industries is well documented. This section focuses on the options used in the chemical industry's reduction of solvent wash waste (i.e., vessels and associated piping required in the clean-out).

Cleaning Equipment Manually

Manual cleaning reduces the amount of solvent used because: manual washing can be more efficient than an automated wash system; and personnel can vary the amount of solvent needed from wash to wash depending on the condition of the equipment (cleanliness).

A variation of this option involves personnel entering the equipment and wiping the product residue off the equipment interior walls with hand-held wipers or spatulas which would minimize or eliminate the need for a subsequent solvent wash. A company should thoroughly review the safety aspect of this option, particularly the nature and extent of personnel exposure, before implementation.

Draining Equipment Between Campaigns

Better draining can reduce the amount of product residue on equipment walls and thereby minimizes or eliminates the solvent needed in a subsequent wash. A company can improve its draining simply by lengthening the time between the end of a production batch or cycle and the start of the washout procedure. For a packed distillation column, maintaining a slight positive pressure (with nitrogen) on the column for twenty-four to forty-eight hours facilitates draining. The residue product is thereby swept off the packing and accumulates in the bottom of the column.

Prewashing Equipment with a Detergent and Water Solution

Prewashing contaminated equipment with a soap and water solution minimizes or eliminates the solvent needed in a subsequent wash step.

Flushing the Equipment with the Product and Recycling It Back to the Process

This option applies when more than one product is produced with the same equipment. Prior to processing another product, a company can withhold a small reserve of a product from a previous similar process and then use it to flush the equipment. The contaminated product (used as a flush) can then be reworked or reprocessed to make it acceptable for use.

Flushing with Waste Solvent from Another Process

Instead of using fresh solvent, a company can use the waste solvent from another process in the plant for the equipment flush. This option reduces the plant's total waste load.

Minimizing the Amount of Solvent Used to Wash Equipment

Often, a company can minimize the amount of solvent used for a flush without changing the resulting cleanliness of the equipment.

Increasing Campaign Lengths

With careful scheduling and planning, a company can increase product campaign lengths and thereby reduce the number of equipment washings needed.

Optimizing the Order of Product Changeovers

Often the specifications for products produced in the same equipment are different. One set of specifications may be more stringent than another. Through careful planning and inventory control, a company can make product changeovers from products with tighter specifications to those with looser specifications.

Washing Vessels Immediately to Avoid Solidification

Often, product residue dries, thickens, and hardens in the equipment between solvent washouts. Immediately washing out vessels between campaigns makes the residue easier to remove when it does not set on the equipment interior walls.

Replacing the Solvent with a Nonhazardous Waste

The solvent wash can be replaced with a less hazardous or nonhazardous (i.e., water) flush material. Another variation is to replace the solvent with a less volatile solvent thus reducing fugitive emissions. The solvent can then be recovered and recycled.

Using a High-Pressure Water Jet

A new cleaning system uses a special nozzle and lance assembly which is connected to a high-pressure water source and inserted through a flange at the vessel bottom (see Figure 3.10.8).

As shown in the figure, a chain-drive moves the lance up and down the carriage as needed. A swivel joint at the base of the lance permits free rotation. The nozzle at the tip of the spinning lance has two apertures, which emit

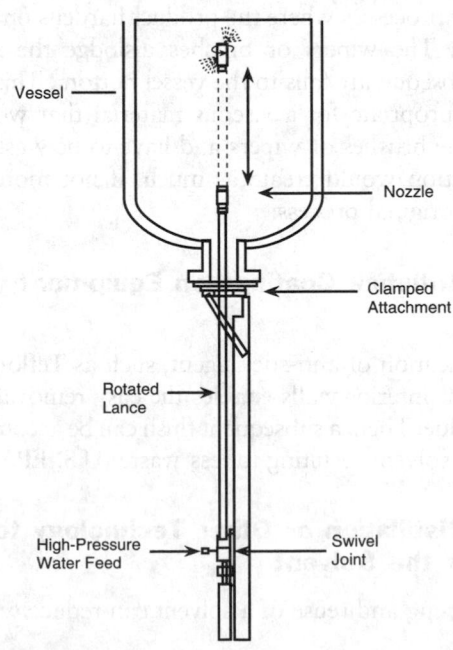

FIG. 3.10.8 High-pressure water system.

cone-shaped sprays of water at 10,000 psi with a combined flow rate of 16 gpm. The operation of the lance is controlled from a panel well removed from the vessel. The process is designed so that no high-pressure spray leaves the interior of the vessel. These precautions assure operator safety during vessel washout.

All solvent waste is eliminated. The product removed from the equipment walls can be separated from the water and recovered for further waste reduction. Even in processes where water cannot be introduced into a vessel, an alternative exists. Vessels can be cleaned with solid carbon dioxide (dry ice) particles suspended in a nitrogen gas carrier. The solid CO_2 cleans in a manner similar to that of sand blasting, leaving only the material removed from the equipment (U.S. EPA 1993).

Using a Rotating Spray Head

A rotating spray head can be used to clean vessel interiors. This system minimizes solvent use by allowing the solvent to contact all contaminated surfaces in an efficient manner (U.S. EPA 1993).

Using Pipe-Cleaning Pigs

Pigs are pipe-cleaning mechanisms made of various materials. They are actuated by high-pressure water, product, or air. Pigs remove the residual build-up on pipe walls thereby minimizing or eliminating subsequent washing.

Using a Wiping or Brushing System

This option uses a system of wipers or brushes that cleans off residual product. (This system is somewhat analogous to a car wash except that it washes the interior vessel walls as opposed to the outside of a car.) This system is appropriate for processes where the product hardens on the vessel walls. The wipers or brushes dislodge the material which subsequently falls to the vessel bottom. This system is not appropriate for a viscous material that would adhere to the brushes or wipers and have to be washed out; this situation would create as much, if not more, waste than the original process.

Using Antistick Coatings on Equipment Walls

The application of anti-stick agent, such as Teflon, to the equipment interior walls enables the easy removal of leftover residue. Then, a subsequent flush can be accomplished with less solvent resulting in less waste (U.S. EPA 1993).

Using Distillation or Other Technology to Recover the Solvent

The recycling and reuse of a solvent can reduce waste significantly.

Using Dedicated Equipment to Make Products

This option eliminates the necessity of washing out equipment between production campaigns thus eliminating the flush solvent stream.

Installing Better Draining Equipment

During the design of a new process, a company can minimize flush solvent waste by designing equipment to facilitate draining. This equipment includes vessels with sloping interior bottoms and piping arrangements with valve low points or valves that drain back to the main vessels. After a product campaign, the residue is drained from each equipment section into a movable, insulated collection vessel. The collected material is then reintroduced into the process during the next campaign.

Other Improvements

A company can make a number of other improvements to reduce waste. These options are described next.

Avoiding Unexpected Trips and Shutdowns

A good preventive maintenance program and adequate spare equipment are two ways to minimize trips and unplanned shutdowns. Another way is to provide early warning systems for critical equipment (e.g., vibration monitors). When plant operators report unusual conditions, minor maintenance problems are corrected before they become major and cause a plant trip.

Reducing the Number and Quantity of Samples

Taking frequent and large samples can generate a large amount of waste. The quantity and frequency of sampling should be reduced, and the samples returned to the process after analysis.

Recovering Products from Tank Cars and Trucks

The product drained from tank cars and trucks (especially those dedicated to a single service) can often be recovered and reused.

Reclaiming Waste Products

Sometimes waste products—not all of which are chemical streams—can be reclaimed. Rather than sending waste products to a burner or landfill, some companies have found ways to reuse them. This reuse can involve physical cleaning, special treating, filtering, or other reclamation techniques. Also, converting a waste product into a salable product may require additional processing or creative salesmanship, but it can be an effective means of reducing waste.

Installing Reusable Insulation

When conventional insulation is removed from equipment, it is typically scrapped and sent to a landfill. A number of companies manufacture reusable insulation which is particularly effective on equipment where the insulation is removed regularly for maintenance (e.g., head exchanger heads, manways, valves, and transmitters).

Maintaining External Painted Surfaces

Even in plants handling highly corrosive material, external corrosion can cause pipe deterioration. Piping and valves should be painted before being insulated, and all painted surfaces should be well maintained.

—*David H.F. Liu*

References

Nelson, Kenneth E. 1990. Use these ideas to cut waste. *Hydrocarbon Processing* (March).

U.S. Environmental Protection Agency (EPA). 1993. *DuPont Chambers Works waste minimization project.* EPA/600/R-93/203 (November). Washington, D.C.: Office of Research and Development.

3.11
PROCESS INTEGRATION

Process integration is defined as the act of putting together (or integrating) the various chemical reactors, physical separations, and heating and cooling operations that constitute a manufacturing process in such a way that the net production cost is minimized. *Pinch technology* is the term used for the series of principles and design rules developed around the concept of a process pinch within the general framework of process integration. Pinch technology is a methodology for the systematic application of the first and second laws of thermodynamics to process and utility systems.

Pinch technology is a versatile tool for process design. Originally pioneered as a technique for reducing the capital and energy costs of a new plant, pinch technology is readily adaptable to identifying the potential for energy savings in an existing plant. Most recently, it has become established as a tool for debottlenecking, yield improvement, capital cost reduction, and enhanced flexibility. With the concern for the environment, design engineers can use the power of pinch technology to solve environmental problems.

This section addresses the following three areas in which pinch technology has been identified as having an important role (Spriggs, Smith, and Petela 1990):

- Flue gas emissions
- Waste minimization
- Evaluation of waste treatment options

Before these areas are described, a brief review of pinch technology is necessary.

Pinch Technology

Pinch technology provides a clear picture of the energy flows in a process. It identifies the most constrained part of the process—the process pinch. By correctly constructing composite heating and cooling curves, a design engineer can quantitatively determine the minimum hot and cold utility requirements. This tool is called *targeting*. Another tool, the *grand composite curve*, determines the correct types, levels, and quantities of all utilities needed to drive a process.

Once targets are set, the design engineer can design the equipment configuration that accomplishes the targeted minimum utilities. A key feature of pinch technology is that energy and capital targets for the process are established before the design of the energy recovery network and utility systems begins.

FUNDAMENTALS

Part a in Figure 3.11.1 shows two streams plotted in the temperature–enthalphy (T/H) diagram, one hot (i.e., requiring cooling) and one cool. The hot stream is represented by the line with an arrow pointing to the left, and the cold stream by the line with an arrow pointing to the right.

For feasible heat exchange between the two streams, the hot stream must be hotter than the cool stream at all points. However, because of the relative temperatures of the two streams, the construction of the heating and cooling curves shown in part a of Figure 3.11.1 represents a limiting case illustrated by the flow diagram shown in part b of the figure. The heat exchange between the hot stream countercurrent to the cold stream cannot be increased because the temperature difference between the hot and cold streams at the cold end of the exchanger is zero. This difference means that the heat available in the hot stream below 100°C must be rejected to the cooling water, and the balance of the heat required by the cold stream must be made up from steam heating.

In part c of Figure 3.11.1, the cold stream is shifted on the H-axis relative to the hot stream so that the minimum temperature difference, ΔT_{min}, is no longer zero but positive and finite. The effect of this shift increases utility heating and cooling by equal amounts and reduces the load on the heat exchanger by the same amount. The arrangement, which is now practical because ΔT_{min} is nonzero, is shown in the flow diagram in part d of Figure 3.11.1. Clearly, further shifting implies larger ΔT_{min} values and larger utility consumption.

COMPOSITE CURVES

A design engineer can analyze the heat exchanges between several hot and cold streams in the same way as in the preceding two-stream, heat exchange example. A single composite of all hot and a single composite of all cold streams

can be produced in the T/H diagram and handled in the same way as the two-stream problem.

The first step in constructing composite curves is to correctly identify the streams that undergo enthalpy changes as hot or cold. A *hot stream* is defined as one that requires cooling; a *cold stream* is defined as one that requires heating. The objective is to determine the minimum amount of residual heating or cooling necessary after the heat interchange between the process streams has been fully exploited.

The design engineer extracts stream data from the process flowsheet which contains heat and material balance information. The items of interest are mass flow rates, specific heat capacity (CP), and supply and target temperatures. This procedure is called *data extraction*.

Starting from the individual streams, the design engineer can construct one composite curve of all hot streams in the process and another of all cold streams by simply adding the heat contents over the temperature range.

In part a of Figure 3.11.2, three hot streams are plotted separately, with their supply and target temperatures defining a series of interval temperatures T_1 to T_5. Between T_1 and T_2, only stream B exists, so the heat available in this interval is given by $CP_B (T_1 - T_2)$. However, between T_2 and T_3, all three streams exist, so the heat available in this interval is $(CP_A + CP_B + CP_C) \times (T_2 - T_3)$. A series

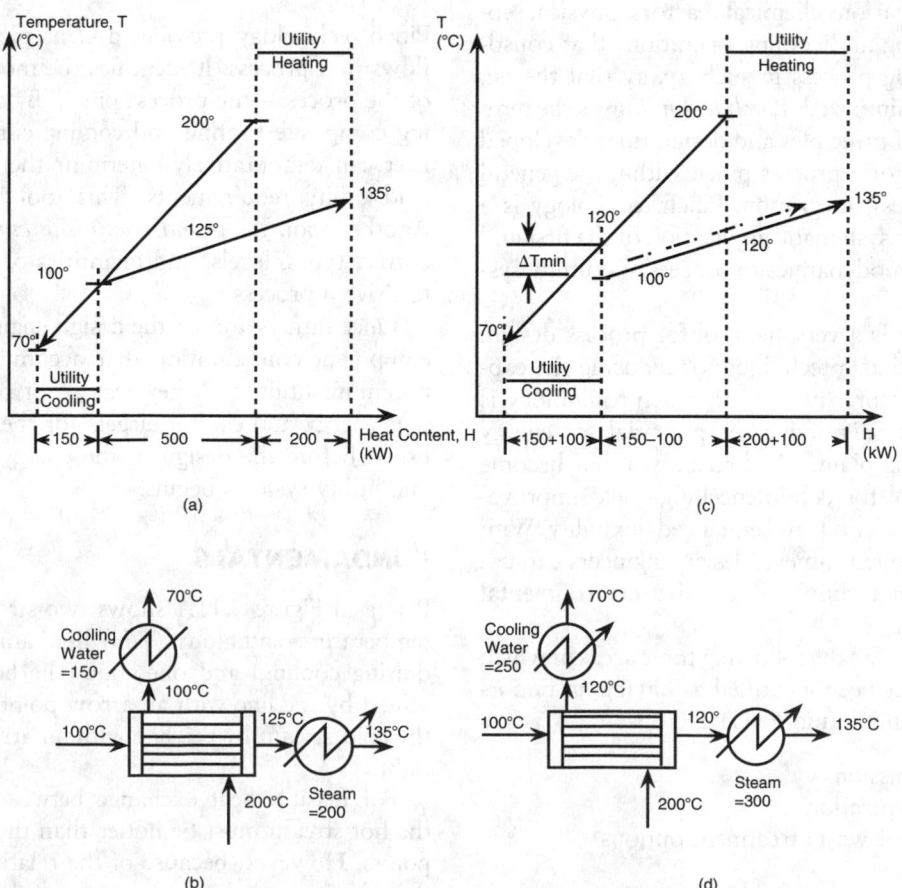

FIG. 3.11.1 Two-stream heat exchange in the T/H diagram.

of values of ΔH for each interval can be obtained in this way, and the result can be replotted against the interval temperatures as shown in part b of Figure 3.11.2. The resulting T/H plot is a single curve representing all hot streams. A similar procedures gives a composite of all cold streams in the problem.

Figure 3.11.3 shows a typical pair of composite curves. Shifting the curves leads to behavior similar to that shown in the two-stream problem. However, the kinked nature of the composites means that ΔT_{min} can occur anywhere in the interchange region and not just at one end. For a given value of ΔT_{min}, the utility quantities predicted are the minimum required to solve the heat recovery problem. Although many streams are in the problem, in general, ΔT_{min} occurs at only one point termed the pinch. Therefore, a network can be designed which uses the minimum utility requirement, where only the heat exchangers at the pinch must operate at ΔT values down to ΔT_{min}.

Figure 3.11.4 shows that the pinch divides the overall system into two thermodynamically separate systems, each of which is in enthalpy balance with its utility target. This example shows that utility targets can only be achieved if no heat transfers across the pinch. To guarantee minimum energy consumption, the design engineer must ensure that heat is not transferred across the pinch in developing a structure. The following design rules must be followed:

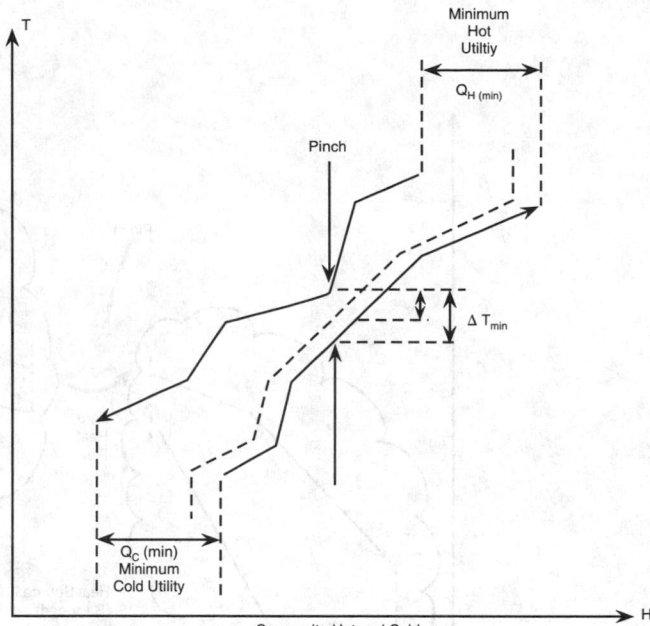

FIG. 3.11.3 Energy targets and the pinch with composite curves.

- Heat must not be transferred from hot streams above the pinch to cold streams below the pinch.
- Utility cooling cannot be used above the pinch.
- Utility heating cannot be used below the pinch.

GRAND COMPOSITE CURVE

Composite curves show the scope for energy recovery and the hot and cold utility targets. Generally, several utilities at different temperature levels and of different costs are available to a design engineer. Another pinch technology tool, the grand composite curve, helps the design engineer to select the best individual utility or utility mix.

The grand composite curve presents the profile of the horizontal (enthalpy) separation between the composite curves with a built-in allowance for ΔT_{min}. As shown in Figure 3.11.5, its construction involves bringing the composite curves together vertically (to allow for ΔT_{min}) and then plotting the horizontal separation (α in Figure 3.11.5).

Figure 3.11.6 shows how the grand composite curve reveals where heat is transferred between utilities and process and where the process can satisfy its own heat demand (Linnhoff, Polley, and Sahdev 1988).

Applications in Pollution Prevention

This section describes the application of pinch technology in pollution prevention in flue gas emissions and waste minimization.

FLUE GAS EMISSIONS

The relationship between energy efficiency and flue gas emissions is clear. The more inefficient the use of energy,

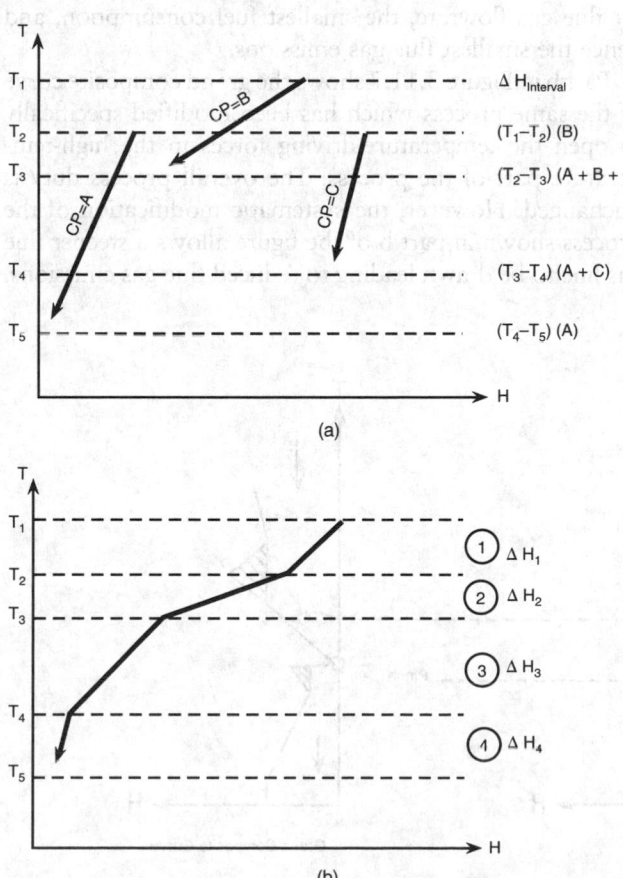

FIG. 3.11.2 Construction of composite curves.

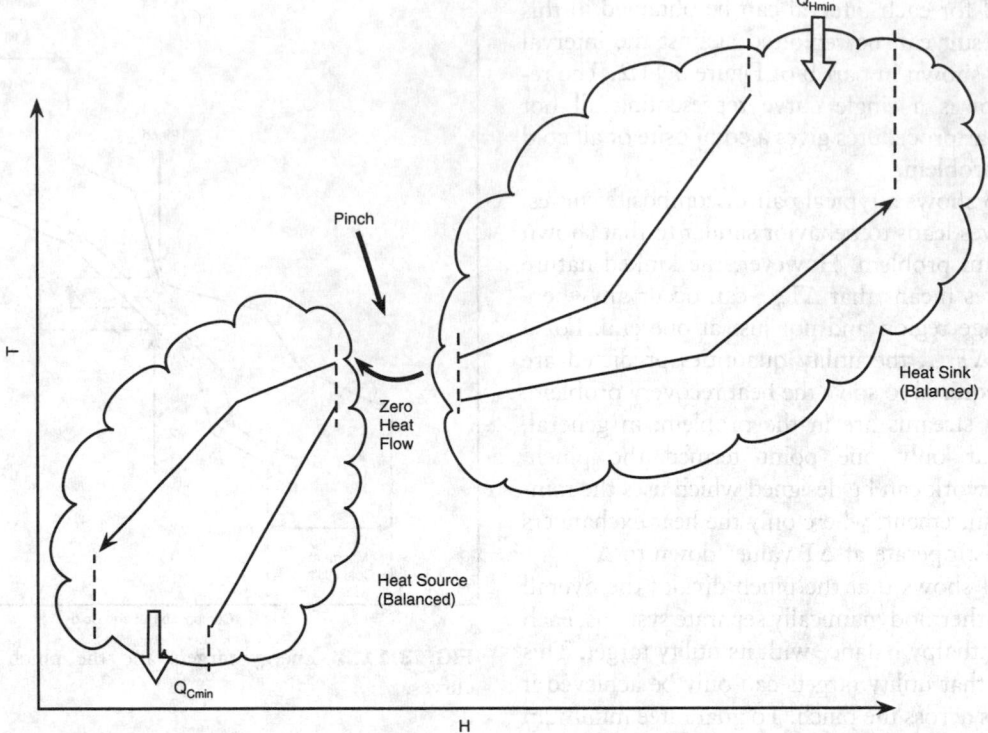

FIG. 3.11.4 Division of a process into two thermodynamically separate systems.

the more fuel burned and the greater the flue gas emissions. Pinch technology can be used to improve energy efficiency through better integration and thus reduce flue gas emissions.

In addition, a design engineer can systematically direct basic modifications to a process to reduce flue gas emissions. For example, part a of Figure 3.11.7 shows a process grand composite. Because the process requires a high temperature, a furnace is required. Part a in Figure 3.11.7 shows the steepest flue gas line which can be drawn against

the existing process. This line corresponds with the smallest flue gas flowrate, the smallest fuel consumption, and hence the smallest flue gas emissions.

Part b in Figure 3.11.7 shows the grand composite curve of the same process which has been modified specifically to open the temperature-driving forces in the high-temperature part of the process. The overall process duty is unchanged. However, the systematic modification of the process shown in part b of the figure allows a steeper flue gas line to be drawn leading to reduced flue gas emissions.

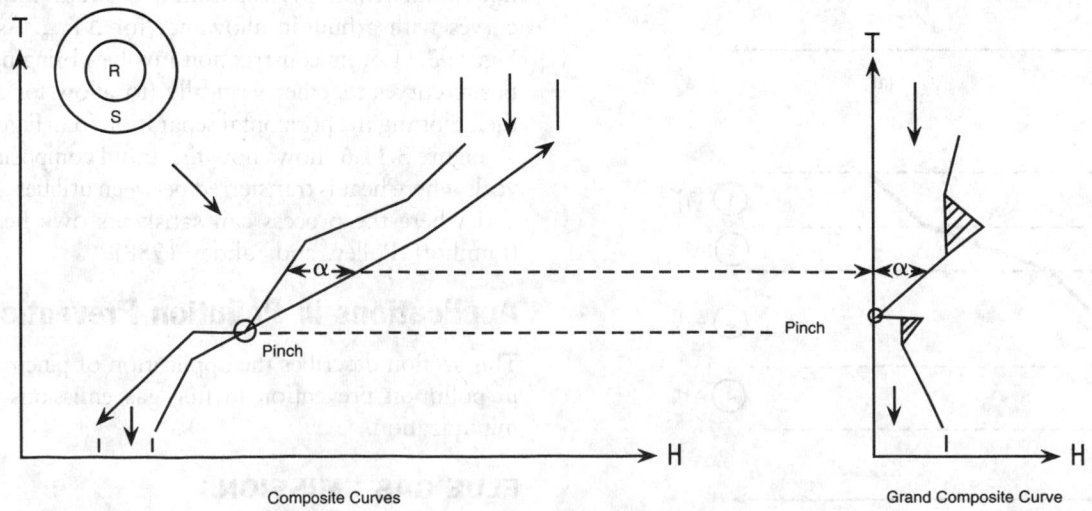

FIG. 3.11.5 Grand composite curve presenting the profile of the horizontal separation between the composite curves with a built-in allowance for ΔT_{min}.

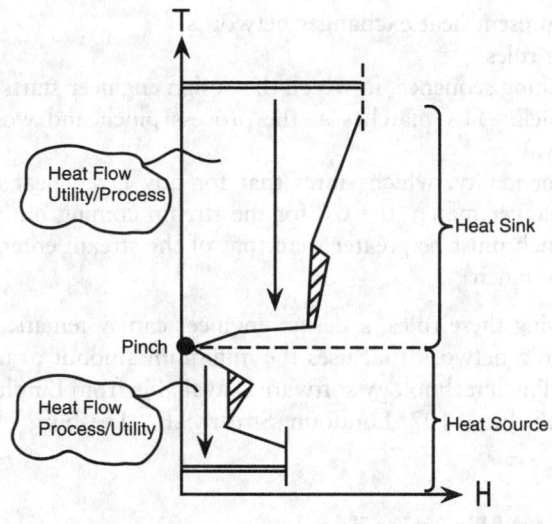

Grand Composite Curve

FIG. 3.11.6 Grand composite curve revealing where heat is transferred between utilities and the process.

WASTE MINIMIZATION

In waste minimization, the objective is to make a process more efficient in its use of raw materials. This objective is achieved through improvements in the reaction and separation systems within the process. Because these systems often require the addition or removal of heat, a design engineer can use pinch technology to identify the cost-effective process modifications. The following example from a fine chemical plant illustrates this use (Rossiter, Rutkowski, and McMullen 1991).

Figure 3.11.8 shows the process, which is a batch operation. The stirred-tank reactor is filled with two feeds (F1 and F2) and is heated to the reaction temperature with the steam in the reactor jacket. The temperature is then maintained at this level to allow the reaction to proceed to the required extent, after which the vessel is cooled by water passing through the reactor jacket. This cooling causes the product to crystallize out of the solution. The solid product is then separated from the liquid by filtration, and the filtrate is rejected as an effluent.

Because of the high effluent treatment costs, the company sought means to improve product recovery and thereby reduce the effluent treatment requirements. They considered the following two options:

Cooling of the effluent with refrigeration to reduce the product solubility, thereby increasing product recovery

Evaporating the effluent to reduce the volume of the liquid and thus increase the amount of solid products formed

Both of these options required capital expenditure and additional energy to reduce the effluent and improve the raw material efficiency. However, pinch analysis high-

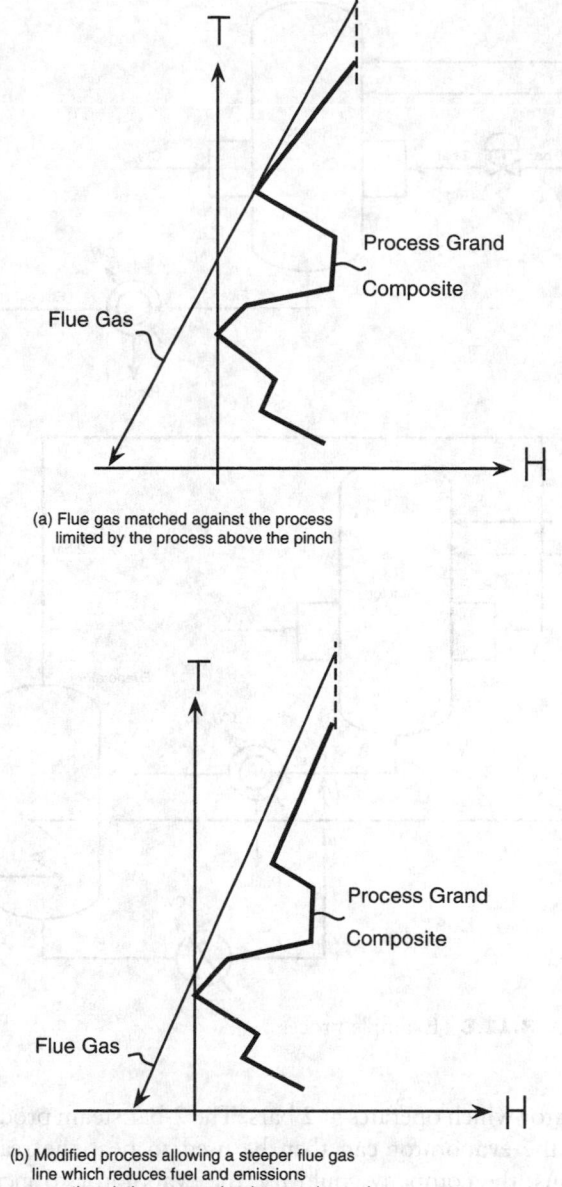

(a) Flue gas matched against the process limited by the process above the pinch

(b) Modified process allowing a steeper flue gas line which reduces fuel and emissions even though the process duty has not changed.

FIG. 3.11.7 Grand composite curve allowing the minimum flue gas to be drawn leading to reduced flue gas emissions. (a) Flue gas matched against the process limited by the process above the pinch. (b) Modified process allowing a steeper flue gas line which reduces fuel and emissions even though the process duty has not changed.

lighted an opportunity to totally eliminate the effluent without increasing the energy costs.

Steam to the original process shown in part a of Figure 3.11.8 was supplied at 10 bars. However, the grand composite curve for the process (see Figure 3.11.9), which plots the net heat flow against the actual, required processing temperatures rather than the existing practice, clearly shows that steam at only 2 bars is hot enough for the reactor requirements.

This observation led to the modified design (see part b in Figure 3.11.8), in which 10-bar steam drives an evap-

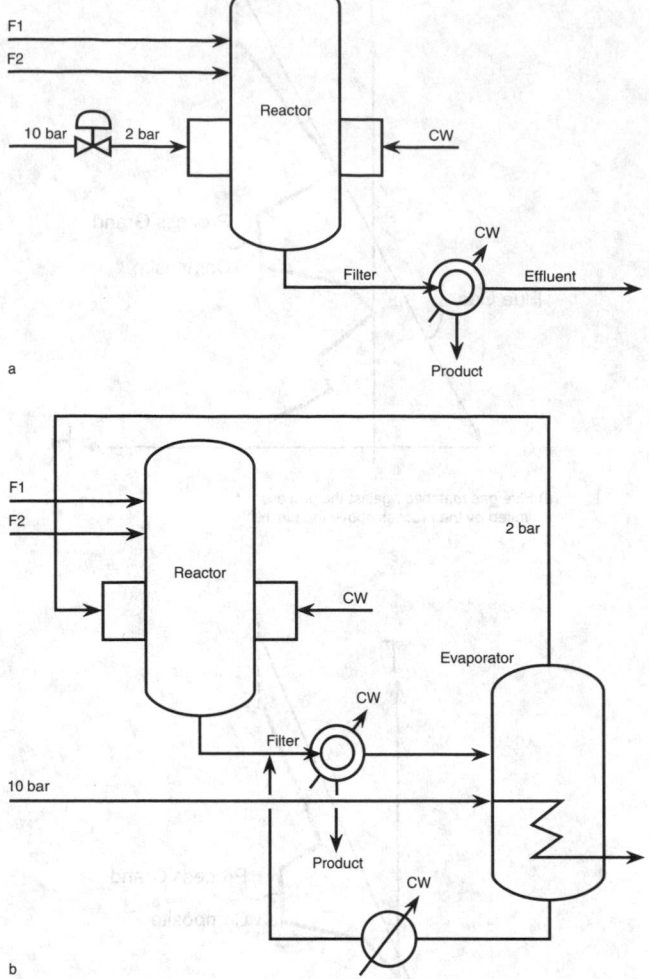

FIG. 3.11.8 Example process.

orator which operates at 2 bars. The 2-bar steam produced in the evaporator can then be used to heat the reactor. Thus, the company could use the evaporator to increase product recovery and eliminate the effluent without any increase in energy costs.

Of course, to make this design work in a batch plant the operations must be sequenced in such a way that the evaporator is running at the same time that the reactor is being heated. In practice, this sequencing is easily achieved. The resulting process modification is a more cost-effective means of waste minimization than those developed without pinch technology and has the added benefit of enhanced product recovery.

Designing a Heat Exchange Network

After targeting is complete, the remaining task is to design a heat exchange network that meets energy and capital targets. The following design aids are available to accomplish this task (Linnhoff et al. 1982):

Grid representations, which provide a convenient way to

represent heat exchanger networks

Pinch rules

Matching sequence, in which the design engineer starts by placing HX matches at the process pinch and works away

CP inequality, which states that for any given heat exchanger match, the CP for the stream coming out the pinch must be greater than that of the stream entering the pinch

Using these rules, a design engineer can systematically design a network that uses the minimum amount of utilities. Pinch technology software is available from Linnhoff March Inc., 107 Loudoun Street S.E., Leesburg, VA 22075.

Waste Minimization

Figure 3.11.10 summarizes a suggested approach for applying pinch technology to environmental problems (Spriggs, Smith, and Petela 1990). Waste minimization is clearly the place to start. Solving environmental problems at the source is not always the simplest solution, but it is usually the most satisfactory solution in the long term. Reducing the problem at the source by modifications to the process reaction and separation technology has the dual benefit of reducing raw material and effluent treatment costs.

Once the design engineer has exhausted the possibilities for waste minimization by process modifications, the

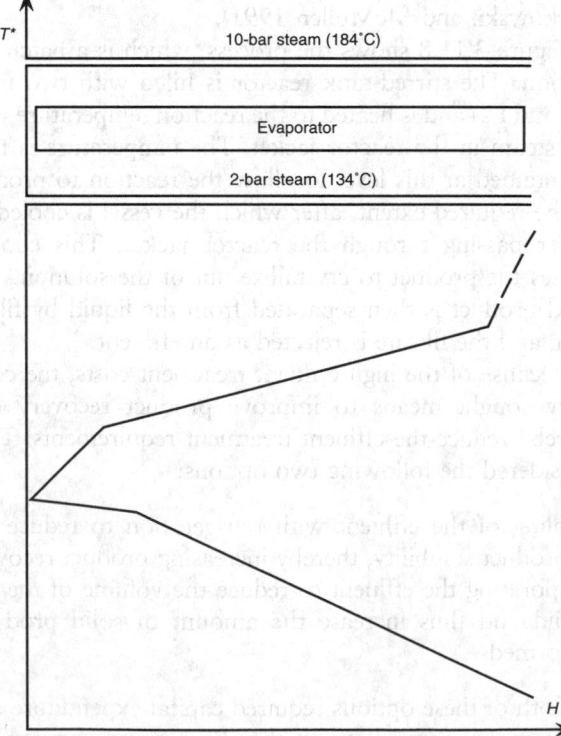

FIG. 3.11.9 Grand composite of example process.

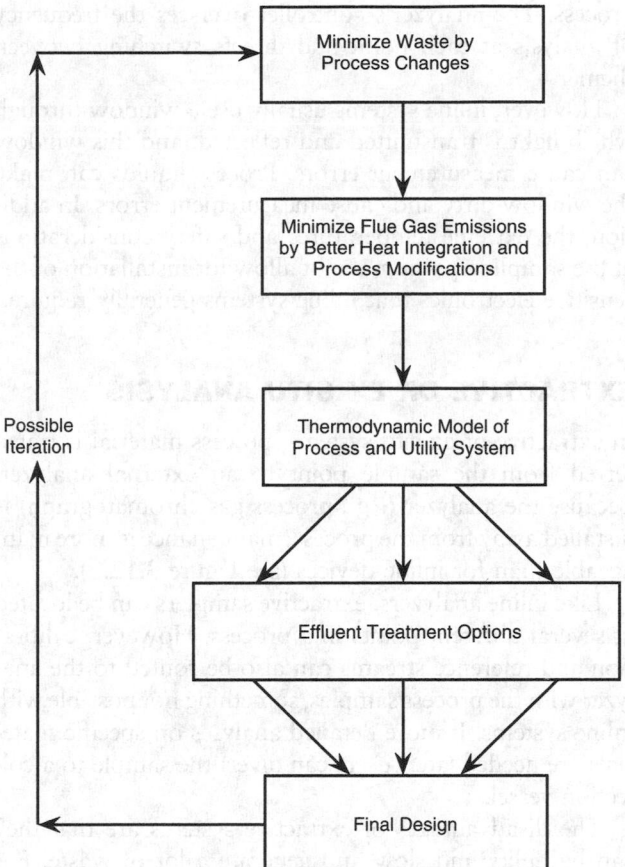

FIG. 3.11.10 A suggested approach for applying pinch technology to environmental problems.

minimum flue gas emissions can be established (see Figure 3.11.10). This stage establishes the thermodynamic model for the process and utility system.

Next, the design engineer assesses the alternative waste treatment options (see Figure 3.11.10). The design engineer must assess these options by considering the process and its waste treatment systems together and any possibilities for integration between them.

Figure 3.11.10 also shows that some earlier decisions may need to be readdressed after the waste treatment options and considered. At each stage, pinch technology establishes the economic tradeoffs.

—*David H.F. Liu*

References

Linnhoff, B. et al. 1982. User guide on process integration for the efficient use of energy. London: Institute of Chemical Engineers. (Available in the United States through Pergamon Press, Elmsford, N.Y.)

Linnhoff, B., G.T. Polley, and V. Sahdev. 1988. General process improvement through pinch technology. *Chem. Eng. Prog.* (June): 51–58.

Rossiter, A.P., M.A. Rutkowski, and A.S. McMullen. 1991. Pinch technology identifies process improvements. *Hydrocarbon Processing* (January).

Spriggs, H.D., R. Smith, and E.A. Petela. 1990. Pinch technology: Evaluate the energy/environmental economic trade-offs in industrial processes. Paper presented at Energy and Environment in the 21st Century Conference. Massachusetts Institute of Technology, Cambridge, MA, March 1990.

3.12
PROCESS ANALYSIS

Online analysis of the physical properties or chemical composition in dynamic processes allows for realtime control. Thus, a company can detect potentially harmful by-products in process streams immediately, especially in a continuous stream, to prevent the production of large quantities of off-spec products. In addition, online analyzers cut down product variation and raw material waste and help plants minimize energy use.

Onsite data gathering is becoming increasingly important as waste streams become more complex. A waste treatment facility benefits from the ability to identify a change in the waste profile. Multiple sensor and instrumentation systems serve this need in generating realtime data. On-demand interrogation coupled with limit alarms announce changing conditions and facilitate a response action (Breen and Dellarco 1992).

Several analytical methods, including gas chromatography, liquid chromatography, infrared and near-infrared (NIR) spectroscopy, and wet chemistry analyzer, have successfully transferred from the laboratory to the process line. Each method has its own price, accuracy, complexity, and maintenance requirements. A thorough knowledge of this information is required to install an economic and effective system.

The major parts of an online process analyzer are the sampling apparatus; the analyzer; and the methods used for data correlation, reporting, and communication.

Sampling

In online analysis, as in all analytical chemistry, sampling is the most critical and least accurate step. In addition, 80 to 90% of all maintenance problems experienced by an online analyzer occur during sampling. Filtration and dilution may be required when the sample arrives at the analyzer; the more equipment used, the greater the potential for malfunctions.

All sampling techniques for online analysis fall into one of the three following general categories:

Direct insertion of the sampler into the process (*inline or in situ analysis*)

Continuous extraction of process material for delivery to the analyzer via transfer lines (*ex situ or extractive analysis*)

Discrete or grab sampling (*atline or nearline analysis*)

INLINE OR IN SITU ANALYSIS

In inline or in situ systems, the sample is not transported from the sampling point to the analyzer because the analyzer is at the sampling point. Probe-type analyzers have been used for a long time. One of the newer probes is the fiber-optic probe (FOP), which uses fiber-optic waveguides to return process-modified light from the probe to the spectrum analyzer usually located in a control room (see Figure 3.12.1). No transfer time or sample waste occurs because the measurement is made on the moving process material.

Another advantage of inline analyzers is that they can be multiplexed; that is, installed at many points in the process. The analyzer's controller oversees the frequency of analysis at each point and directs switching between them.

However, inline systems usually use a window through which light is transmitted and reflected, and this window can cause measurement errors. Process liquids can make the window dirty and cause measurement errors. In addition, the extreme temperatures and safety considerations at the sampling point may not allow for installation of the sensitive electronics that inline systems generally require.

EXTRACTIVE OR EX SITU ANALYSIS

In extractive or ex situ systems, process material is transferred from the sample point to an external analyzer. Because the analyzer (e.g., process gas chromatograph) is installed away from the process, maintenance is more manageable than for inline devices (see Figure 3.12.2).

Like inline analyzers, extractive samplers can be located at several different points in a process. However, calibration and reference streams can also be routed to the analyzer with the process samples, something not possible with inline systems. If more detailed analyses on specific materials are needed later, users can divert the sample to a collection vessel.

The disadvantages of extractive systems are that they can be bulky and slow and generate a lot of waste. For

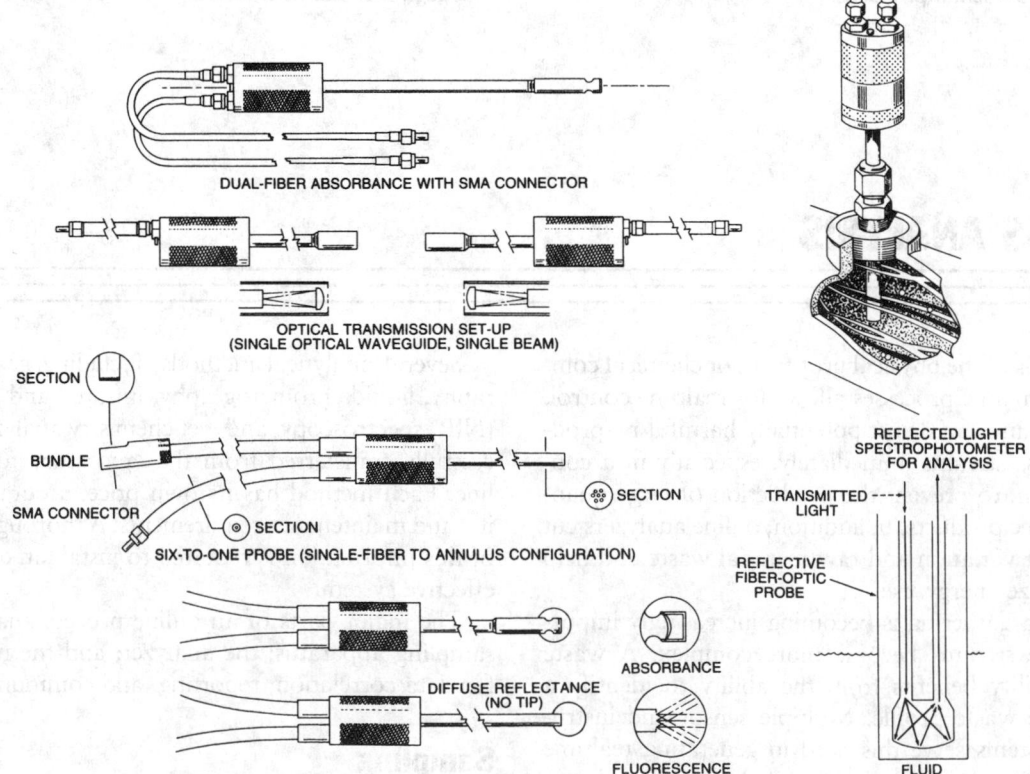

FIG. 3.12.1 FOPs providing data on absorbance, diffuse reflectance, fluorescence, and scattering. (Reprinted, with permission, from Guided Wave Inc.)

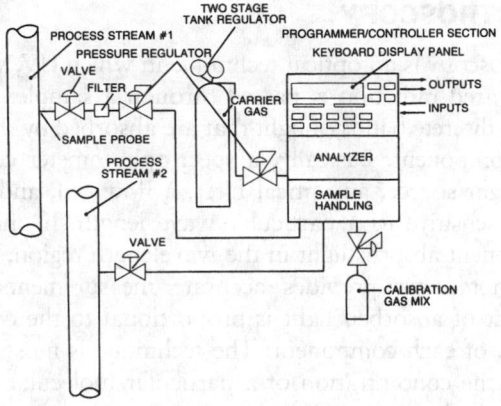

FIG. 3.12.2 Basic elements of a multistream, process gas chromatograph system.

example, an analyzer sampling six different streams requires an enclosure at the process to house the related machinery. Depending on the distance that the sample must travel to the analyzer, times ranging from 20 to 60 sec are common. Filtration, dilution, or concentration may also be required when the sample arrives at the analyzer. When heated samplers and transfer lines are required to keep a sample at a particular temperature, installation and maintenance costs can also be significant.

DISCRETE OR GRAB SAMPLING

In discrete or grab sampling, aliquots of process material are simply collected by hand and delivered to the analyzer, which is located at the process (atline) or in a lab (offline). While this type of sampling can be sufficient for some analytical requirements, it is not a true online analysis method.

Analyzers

Analyzers range from property- or compound-specific sensors, such as pH probes and oxygen detectors, to chemical and optical analyzers, chromatographs, and spectrophotometers.

SPECIFIC SENSORS

Specific sensors are the simplest type of online analyzers. They are generally used to measure either the physical parameters of a gas or liquid stream, such as pH, temperature, turbidity, or oxidation-reduction potential, or easily detected compounds, such as oxygen, cyanide, or chlorine. In most cases, these sensors have continuous output.

Sensors are relatively inexpensive and easy to install and require little or no maintenance. While they can be selective and sensitive, specific sensors can become fouled by constant contact with process materials, particularly those with high particle concentration.

While the output from specific sensors is considered to be in real time, all process analyzers, including sensors, experience lag times and stabilization times. The lag time is the time required for the process material to pass through the sensor's sampling element. The stabilization time, called T90, is the time required for the sensor to reach 90% of its final output. Typical T90 times are about 20 to 60 sec.

GAS CHROMATOGRAPHY (GC)

GC is one of the most widely used analyzers in the petrochemical and refining industry. It offers flexibility of applications, high sensitivity analysis, and multicomponent analysis. The widespread use of GC is a result of its versatility. If a sample can be vaporized, an effective separation is often possible.

Process GC sampling is extractive. A small sample of process material is obtained with a sample valve and vaporized in a preheater. The sample is pushed by an inert carrier through a packed solid capillary tube. Components within the sample have varying degrees of affinity for the column packing. The more the component is attracted to the column packing, the slower it moves. As the components come out of the column (that is, are eluted), they are recorded as a function of time by a detector. Depending upon the sensitivity of the detector (see Figure 3.12.3) and separation quality of the instrument, GCs routinely achieve accuracy within 0.25 to 2%. The accuracy of most analyzers is a fixed percentage of their full-scale range.

GC offers continuous results, but the retention time of the sample (i.e., the time taken for the separated components to pass down the column) must be considered. Typical retention times range between 1 and 20 min, depending on the number of components and their vaporization temperatures. A retention longer than 20 min is impractical and generally unacceptable for online analysis.

A big drawback of GC is maintenance. Instruments generally have many electrical and mechanical components. Preventive maintenance is a must. Repairs are not always fast or easy. Repairing a chromatograph can take hours or even days.

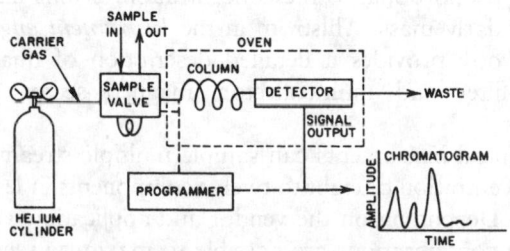

FIG. 3.12.3 Basic schematic of a chromatograph.

LIQUID CHROMATOGRAPHY (HPLC)

HPLC is similar to gas chromatography. HPLCs also use carriers (mobile phase), columns, and detectors. The sample is passed through the column under high pressure. However, because the HPLC column is much smaller in diameter than the process GC column, plugging results if the liquid sample is not free of particulates. Not many process streams can be conditioned to provide the necessary clean sample. To date, HPLCs have not been used with great success in liquid component process analysis.

WET CHEMISTRY ANALYZERS

Wet chemistry analyzers all work similarly. A sample-handling system extracts a clean sample from the process (to prevent plugging). The sample is injected into a chemically treated solution, and a chemical reaction takes place. The reaction can be a change in color, pH, or conductivity. The change is proportional to the concentration of a single component of interest in the process stream (Lang 1991).

The fastest wet chemistry technique is process flow injection analysis (PFIA). Here, a clean process sample flows continuously through a sample injection valve. At user-selected intervals, a fixed volume of sample, usually in microliters, is injected into a constantly flowing liquid carrier stream. Precise mixing generates a specific chemical reaction, and an appropriate detector (UV or visible, pH, or conductivity) gives a signal proportional to the concentration in the sample of the component of interest. The response time is usually fast so that one or two determinations can be made per minute.

MASS SPECTROMETERS

Mass spectrometers offer performance, versatility, and flexibility, sometimes exceeding that of GC. They are applicable for quantitative analysis of organic and inorganic compounds. A small vapor sample is drawn into the instrument through an inlet leak. A mass spectrometer works by ionizing a sample and then propelling the ions into a magnetic field. This field deflects each ion in proportion to its charge/mass ratio and causes it to strike one of a number of collectors. The signal from each collector is directly proportional to the concentration of ions having a particularly mass. Ahlstrom in the *Instrument engineers' handbook* provides a detailed description of quadruple mass-filter and multicollector, magnetic-sector instruments.

Typical instruments can sample multiple streams with a concentration of eight to twelve components in less than 5 sec. Depending on the vendor and applications, a variety of configurations are possible to maximize sensitivity and utility.

SPECTROSCOPY

Spectroscopy is an optical technique in which UV, visible, or infrared radiation is passed through a sample. Filters isolate discrete bands of light that are absorbed by the specific component. Basically, a spectrophotometer consists of a light source, an optical filter, a flow cell, and a detector sensitive to a particular wave length. If only one component absorbs light in the wavelength region, a simple photometer provides accurate measurements. The amount of absorbed light is proportional to the concentration of each component. The technique is most useful where the concentration of a particular molecular group such as hydroxyls, paraffins, olefins, naphthenes, and aromatics must be determined (see Figure 3.12.4).

Spectroscopy is faster and mechanically simpler than chromatography, and either direct-insertion or extractive sampling can be used. Spectroscopy is not precise when multicomponent mixtures are measured because a chemical's absorbances at individual wavelengths often interfere with each other. Multiple filters or a scanning instrument must be used in these solutions.

System computers are now equipped with data reduction programs called chemometrics. These programs compare sample spectra to known spectra stored in a database as a learning or training set. Establishing such a database requires examining and storing the spectral properties and reference methods for a significant number of calibration samples before the unit goes online.

For GCs controlled by external computers, the calibration requires only one sample, or six at most, to establish linearity in the expected concentration range. However, the chemometric models used by online spectrophotometers become more reliable as more samples are added to the learning set (Crandall 1993).

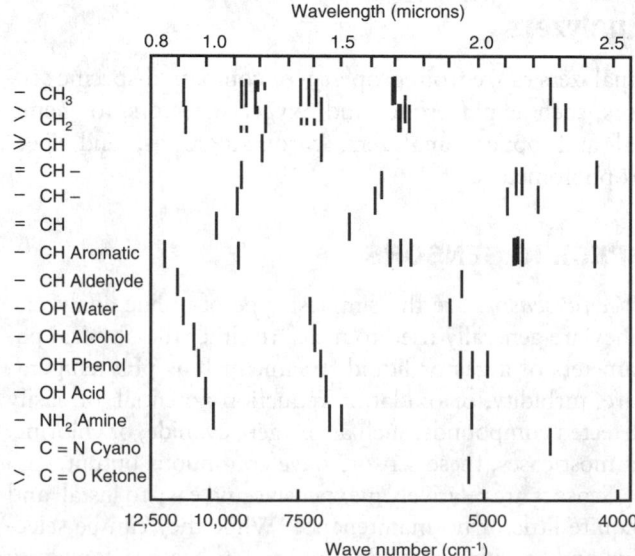

FIG. 3.12.4 An example of how the location of absorption bands shows functional groups in the NIR.

The precision of spectroscopic methods can be 0.1% of the full-scale reading. A system's accuracy is a reflection of its learning set. The accuracy of an online spectrophotometer is only as good as the accuracy of the reference method used to calibrate it.

NEAR INFRARED ANALYSIS

Near infrared (NIR) analysis is a new process liquid measurement technology that is growing. NIR spectroscopy is an optical scanning technique that operates in a range of wavelengths between 800 and 2400 nm. The primary advantage of NIR analysis is that because the sample probe is placed directly in the process stream, an extractive sample handling system is not needed. More importantly, process NIR analysis addresses applications that have not been tried by other technologies.

Traditionally, the measurement of liquid components using NIR analysis has been done by analyzers that operate at only one wavelength. Recently, analyzers using gratings, filter wheels, and other moving parts have been developed to vary the infrared wavelengths, making measurements of multiple liquid components possible. However, when these moving parts are in the process environment, they require frequent maintenance.

Traditionally, NIR analysis has been used to measure moisture in the process industry. However, its ability to perform scanning and the advent of fast computers with sophisticated computer software (e.g., chemometric) have expanded its applications, particularly in polymers.

A summary of proven and potential applications of spectroscopy follows. Table 3.12.1 summarizes a proven, closed-loop control application for NIR spectroscopy. Any molecule containing a carbon–hydrogen, hydroxyl (O—H), carboxyl (C═O), or amine (N—H) bond and many inorganic species adsorbs NIR radiation. Water allows NIR radiation to pass through. Thus, NIR analysis can perform water analysis or determine the concentration of the materials that are mixed with it. Table 3.12.2 lists online NIR projects that are in development for processes which have not been amenable to the technique in the past. The NIR analyzer can monitor alkylation, reforming, blending, and isomerization in addition to distillation.

With the advent of fiber optics, online spectroscopy has become safer and more flexible. Fiber-optic cable allows the analyzer and online probes to be separated up to 1000 m. The incident light travels along the cable to the probe, where sample absorption occurs. The reflected signal travels back to the detector through the cable, where it is analyzed (see Figure 3.12.1).

If process safety is a high priority, fiber optics may be the best technique to use because only the probe is in contact with the process material; the analyzer is in a safe location. The tradeoff is that the spectral quality of such systems can suffer and they can be expensive.

The maintenance of online spectrophotometers is equipment- and application-oriented. The maintenance of the equipment is fairly simple. Application maintenance consists of the scientific and engineering work required to validate the system's results and generate the learning set as process formulations and the associated feedstocks change. Application maintenance is often overlooked by users, but it is critical to the reliability of online analysis.

Even today's simplest analyzer has a microprocessor that refines raw data, does calculations, and displays results. Analyzers can have graphic interfaces and can be networked with other analyzers and data systems.

Every analyzer should be linked to a plant's distributed control system (DCS). Thus, process analytical data can

TABLE 3.12.1 PROVEN CLOSED-LOOP CONTROL APPLICATIONS FOR NIR SPECTROSCOPY

General applications	Measurement of product purity
	Detection of known impurities or contaminants
	Moisture determination down to ppm levels
	Measurement of indicators and outliers
Petroleum processing	Distillation of aromatics
	Reformation and fluid–catalytic cracking of olefins
	Gasoline blending
	Determination of octane number
Plastics processing	Measurement of OH in polyethylene glycol
	Measurement of epoxy in prepreg
	Condensation in and polymerization of urethanes
	Polymer ratios in batch blending of adhesives
	Degree of curing in resin coatings
Food processing	Alcohol and water measurement in beer blending
	Water and oil measurement in cheese and other foods

Source: Robert Classon, 1993, Expanding the range of online process analysis. *Chemical Engineering* (April).

TABLE 3.12.2 ONLINE NIR PROJECTS IN DEVELOPMENT ON PROCESSES NOT PREVIOUSLY AMENABLE TO THE TECHNIQUE

Industry	Process	Parameter(s) Measured
Plastics processing	Ethoxylation and propoxylation esterification	Hydroxide (O—H) Carboxyl group (C=O)
	Polymerization	Purity, molecular weight
	Blending	Polymer ratios
Pharmaceuticals	Blending	Alcohol and various other parameters
Fine chemicals	Addition and condensation reactions	Various functional-group analyses
Pesticides and herbicides	Lethal reactions	Active ingredients
	Endpoint indicator	Acid concentration
Food processing	Batch blending	Sugars, brix, carbohydrates, edible oils, amino acids, iodine levels
Petroleum processing	Blending	Octane, alcohol, MTBE

Source: Classon, 1993.

be shared along a network just like data from input–output devices of the DCS. Users along the network with either a DCS interface or engineering work can access the process data. Many suppliers of chromatography systems offer such networks (see Figure 3.12.5).

System Design and Support

The objective of system design and support is to install an online analyzer that measures physical properties and chemical compositions in dynamic processes. This objective means selecting the right equipment, making sure it is

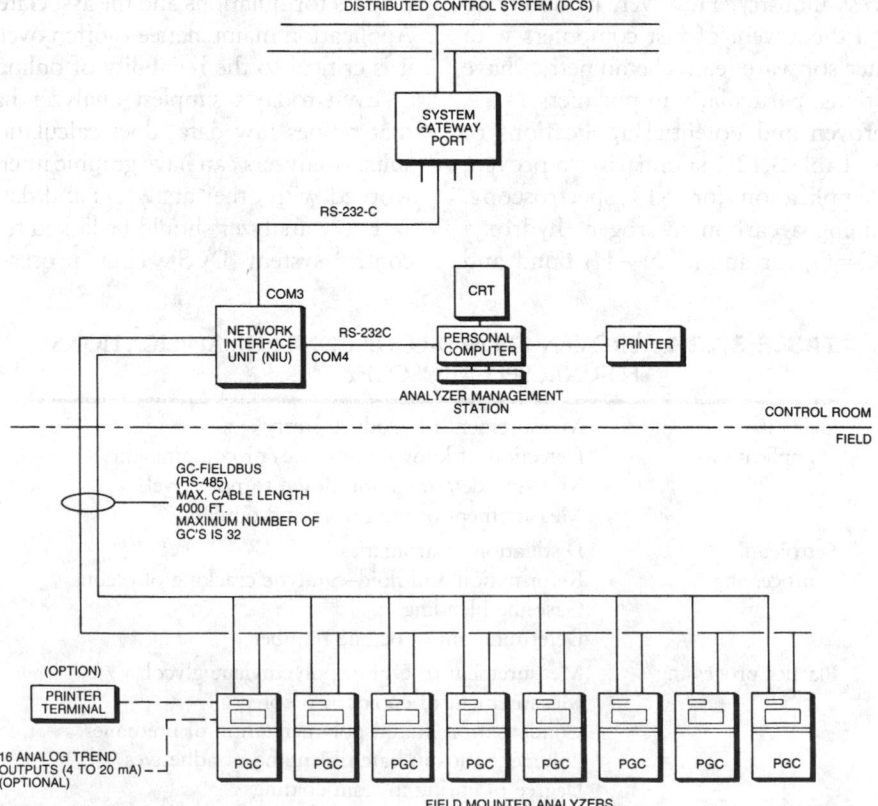

FIG. 3.12.5 A typical multiprocess GC interface to the DCS. In this configuration, only process GC data, validation, system alarms, and stream sequence control are available to the DCS operator. All control and data communication is available to the AMS operator. (Reprinted, with permission, from The Foxboro Co.)

reliable, and making sure it provides the proper analytical data to the users.

The goals of a project are determined by the analytical requirements of the plant. These requirements include the chemical components or properties to be determined, as well as the ranges, precision, accuracies, and response times. The analyst must define the minimum analytical requirements, the optimum requirements, and the degree of flexibility between them.

Technologies compete. One method of analysis can overlap another in capability which means that more than one technology can give satisfactory results and the analyst must choose between technologies.

The analyst should test equipment before installation to verify its reliability and define the probable maintenance requirements. Regardless of how simple or complex the

online systems is, every system requires constant validation and maintenance.

—*David H.F. Liu*

References

Ahlstrom, R.C. 1995. Mass spectrometers. In *Instrument engineers' handbook* 3d ed., edited by B.G. Liptak. Radnor, Pa.: Chilton Book Company.

Breen, Joseph J. and Michael J. Dellarco, eds. 1992. *Pollution prevention in industrial process: The role of process analytical chemistry.* American Chemical Society.

Crandall, J. 1993. How to specify, design and maintain online analyzers. *Chemical Engineering* (April).

Lang, Gary 1991. New on-line process analyzers expand NIR capabilities. *I&CS* (April).

3.13
PROCESS CONTROL

Benefits in Waste Reduction

Modern technology allows the installation of sophisticated computer control systems that respond more quickly and accurately than human beings. That capability can be used to reduce waste as follows.

IMPROVING ONLINE CONTROL

Good process control reduces waste by minimizing cycling and improving a plant's ability to handle normal changes in flow, flow temperatures, pressures, and composition. Statistical quality control techniques help analyze process variations and document improvements. Additional instrumentation or online monitors are necessary, but good control optimizes process conditions and reduces a plant's trips, a major source of waste (Nelson 1990).

OPTIMIZING DAILY OPERATIONS

If a computer is incorporated into the control scheme, it can be programmed to analyze the process continuously and optimize operating conditions. If a computer is not an integral part of the control scheme, a company can per-

form offline analysis and use it as a guide for setting process conditions (Nelson 1990).

AUTOMATING START UPS, SHUTDOWNS, AND PRODUCT CHANGEOVERS

Large quantities of waste are produced during plant start ups, shutdowns, and product changeovers, even when these events are well planned. Programming a computer to control these events brings the plant to stable operating conditions quickly and minimizes the time spent generating off-spec products. In addition, since minimum time is spent in unwanted running modes, equipment fouling and damage are reduced.

UNEXPECTED UPSETS AND TRIPS

Even with the best control systems, upsets and trips occur. Not all upsets and trips can be anticipated, but operators who have years of plant experience probably remember the important ones and know the best ways to respond. With computer control, optimum responses can be preprogrammed. Then, when upsets and trips occur, the computer takes over, minimizing downtime, spills, equipment damage, product loss, and waste generation.

Distributed Control Systems

The DCS is the dominant form of instrumentation used for process control (Liptak 1994). The equipment in a DCS is separated by function and is installed in two different working areas of a process installation. The equipment that an operator uses to monitor process conditions and manipulate the set points of the process operation is located in a central control room. From this location, the operator can (1) view the information transmitted from the processing areas on a CRT and (2) change control conditions from a keyboard. The controlling portions of the system, which are distributed at various locations throughout the processing area, perform the following two functions at each location:

- Measure analog variable and discrete input
- Generate output signals to actuators that change process conditions

Input and output signals can be analog or digital. By means of electric transmission, the system communicates information between the central location and the remote controller locations.

Figure 3.13.1 shows a generic arrangement for the components in a DCS. The operator's console in the control room is called the high-level operator's interface (HLOI). It can be connected through a shared communication facility (data highway) to several distributed system components. These components can be located either in rooms adjacent to the control room or in the field. Such distributed local control units (LCUs) can also have a limited amount of display capability and are called the low-level operator's interface (LLOI).

The console or HLOI is the work center for an operator. In this area, the operator follows a process and uses the fast and accurate translation of raw data into useful trends and patterns to decide the required actions. One CRT is usually dedicated to each section of a plant; each of these CRTs requires an operator's continuous attention. The HLOI also includes keyboards, usually one for each CRT, which allow the operator to enter set points or other parameters or to closely examine particular portions of the process for further information. The HLOI peripherals include disks, tapes or other recorders, and printer units.

The DCS operator depends on the CRT displays for plant information. Three principal types of displays are the group display, the overview display, and the detailed display. A graphic display capacity shows a picture on the screen so that the operator can look at a portion of the process more realistically than by watching a row of bar graphs. Figure 3.13.2 is a graphic display representation of a fractionation column. The display includes process and operation information, and it can be interactive, dynamically changing as real-time information changes.

Trend displays are the DCS equivalents of chart records. They are a profile of the values of process variables showing the changes that occur over a period of time. Some detail displays (see Figure 3.13.3) include a real-time trend graph of the process variable values during a selected period. In some displays, several trend graphs can be displayed at the same time, allowing a comparison of the history of several variables. Trends over longer periods (up to a week) can be saved on floppy-disk memory and displayed on command.

A single best distributed control solution does not exist. The right control system for an application is a function of the process to be controlled (Funk and McAllister 1989).

In the broadest sense, manufacturing processes are either continuous (e.g., a petroleum refinery or an ethylene plant), or batch (e.g., specialty chemicals or pharmaceuti-

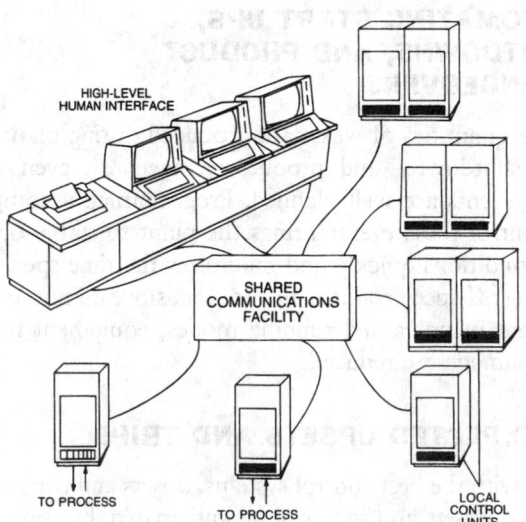

FIG. 3.13.1 A typical DCS. The panel boards and consoles are eliminated, and the communications are over a shared data highway, which minimizes the quantity of wiring while allowing unlimited reconfiguration flexibility. (Reprinted, with permission, from M.P. Lukas, 1986, *Distributed control systems,* Van Nostrand Reinhold Co.)

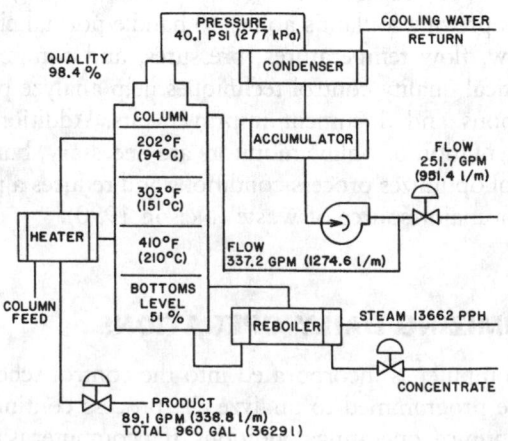

FIG. 3.13.2 Graphic display.

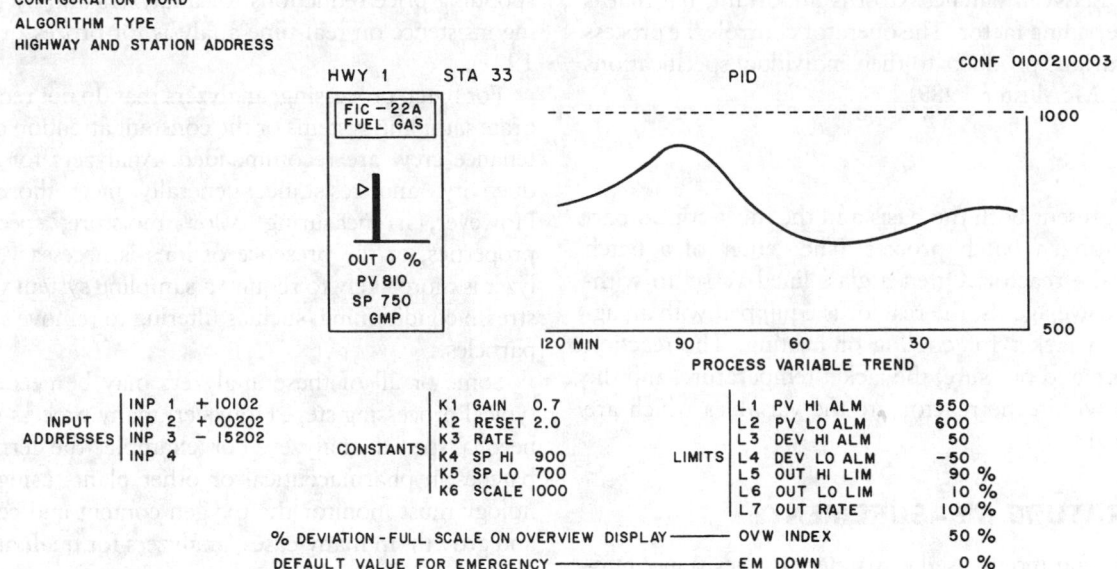

CONFIGURATION WORD
ALGORITHM TYPE
HIGHWAY AND STATION ADDRESS

HWY 1   STA 33   PID   CONF 0100210003

FIC - 22A
FUEL GAS

OUT 0 %
PV 810
SP 750
GMP

120 MIN   90   60   30
PROCESS VARIABLE TREND

1000

500

INPUT ADDRESSES			CONSTANTS			LIMITS		
INP 1	+10102		K1	GAIN	0.7	L1	PV HI ALM	550
INP 2	+00202		K2	RESET	2.0	L2	PV LO ALM	600
INP 3	-15202		K3	RATE		L3	DEV HI ALM	50
INP 4			K4	SP HI	900	L4	DEV LO ALM	-50
			K5	SP LO	700	L5	OUT HI LIM	90 %
			K6	SCALE	1000	L6	OUT LO LIM	10 %
						L7	OUT RATE	100 %

% DEVIATION - FULL SCALE ON OVERVIEW DISPLAY ——— OVW INDEX   50 %
DEFAULT VALUE FOR EMERGENCY ———————————— EM DOWN   0 %

FIG. 3.13.3 Detail display.

cals). In reality, few plants are purely batch or continuous. Instead, they often fall in between and blend some aspects of both.

MASS FLOW

In batch flow, some physical mass of materials is processed together as a unit. Recipe management is essential. Predetermined quantities of raw materials are added to the vessel and blended and reacted with the entire batch being completed at one time. A plant can keep records by batch and, if necessary, troubleshoot by batch.

In a continuous process, the product is comingled continuously. The process is either on or off; material flow never stops while the process is running. Raw materials enter at the beginning of the process, and the final product comes out continuously.

Control Information Requirements

In a batch process, 5-min averages of process values are seldom meaningful. Chemicals can be added to the batch at various times, at which point the temperature may drop. Then, the batch can be heated for a period of time, held at a particular temperature for a while, and then allowed to cool. In this process, the average temperature for the whole cycle offers no real information about the batch properties. A different type of data gathering and archiving is needed that is more event-oriented.

Continuous processes rely heavily on regulatory controls. In these processes, the control system can be set to read a temperature data point every 5 sec, average it for 5 min, and present the resulting data as a representative sample of the process temperature.

In batch systems, linkage to product specifications and the daily shipping schedule are commonly needed and can be provided by DCS. In continuous processes, recipe management is less of a concern, and shipments are typically in bulk.

CONTROL HARDWARE

Batch systems use a greater number of digital devices. Ranges of 65 to 85% digital and only 15 to 35% analog signals are typical. On–off valves and limit switches comprise the majority of digital devices although alarms, interlocks, and emergency shutdown equipment are also important (Procyk 1991).

In general, the analog PID controller is less prominent in batch operations because a batch process operates on a changing state instead of the steady-state environment maintained by the proportional integral and derivative (PID)-loop set point.

Batch processes are usually operated with more modularity and flexibility than are continuous processes. Thus, design engineers must anticipate and plan for what will be needed for future recipes and batches.

SAFETY SYSTEMS

Safety systems on continuous processes aim to minimize shutdowns due to the expense associated with a shutdown or an interruption in production. This issue may have little relevance for batch processes.

BATCH AUTOMATION

The justification for a DCS to control a batch process focuses on maintaining on-spec quality or minimizing the dead time between batches. Cost is important, but that is not the overriding factor. The operator controls the process to make various products to their individual specifications (Funk and McAllister 1989).

Sensors

Sensors represent both the basis and the most critical part of automating a batch process. The center of a batch process is the reactor. Often a glass-lined vessel to withstand corrosive agents, the reactor is equipped with an agitator and a jacket for cooling or heating. The reaction temperature and pressure, the jacket temperature, and the fluid level within the reactor are the variables which are often sensed.

TEMPERATURE MEASUREMENTS

The mostly commonly used instrument is a resistance temperature detector (RTD) with an electronic temperature transmitter. The RTD provides high accuracy.

LEVEL MEASUREMENTS

Level measurements can present challenging problems. First, the properties from which the level can be inferred usually vary. Further, mechanical and electrical complications exist, such as swirling, swishing, and frothing at the top layer. No single level sensor fits all applications. Every level measurement situation must be evaluated separately. The design engineer must consider the metallurgy and configuration of the vessel, the temperature and pressure ranges, the chemical and physical characteristics of the liquid, the nature of the agitation, the electrical area classification, maintenance practices, and perhaps other factors.

PRESSURE AND VACUUM MEASUREMENTS

A range of conventional sensors can be used for pressure and vacuum measurements. The presence of corrosive vapors may require the use of diaphragms and pancake flange designs linked to the main body of the sensor by a capillary.

FLOW MEASUREMENTS

An array of flow sensors is suitable for batch processes, ranging from a rotameter to a mass flowmeter. However, careful evaluation is required of the phase (solid, liquid, or powder), viscosity, flow range, corrosivity, required accuracy, and other parameters in special cases.

ANALYZERS

The online analyzer is becoming more prevalent. This trend is stimulated by the technological advances and corre-

sponding price reductions for analyzers and by the growing insistence on real-time analysis for processors (Procyk 1991).

For batch processing, analyzers that do not require elaborate sampling systems or the constant attention of a maintenance crew are recommended. Analyzers for pH, conductivity, and resistance generally meet those criteria. However, if measuring color, moisture, spectroscopic properties, or the presence of ions is necessary, the analyzer is more likely to require a sampling system with some stream conditioning, such as filtering to remove suspended particles.

Some or all of these analyzers may be needed in any typical processing step. However, many process industries need a special analyzer. For example, the fermentation batches in pharmaceutical or other plants using biotechnology must monitor the oxygen content and cell density and growth. In many cases, analyzers for qualitative properties do not exist, and a plant must try to correlate those properties with ones that can be measured using DCS.

Step-by-Step Batch DCS

Table 3.13.1 shows the control activities of an entire batch in a hierarchial manner from the sensors and elements to the business planning level. Level 1 activities involve process and product management and production management; Level 2 activities involve a batch and unit management; and Level 3 activities involve sequential, regulatory, and discrete control and safety interlocking.

PROCESS AND PRODUCT MANAGEMENT

Production management consists of three control activities: recipe management, production scheduling, and batch history management.

Recipe Management

A recipe is the complete set of data and operations that defines the control requirements of a type or grade of product. A recipe is composed of (1) the header, (2) equipment requirements, (3) the formula, and (4) the procedure.

The procedure defines the generic strategy for producing a batch product. A procedure consists of subprocedures, subprocedures of operations, operations of phases, phases of control steps, and control steps of control instructions. Figure 3.13.4 diagrams this relationship.

The recipe management function maintains a database of site recipes for various products, formulas, and procedures. The control recipe contains specific information on the batch equipment and units that can run each operation within the procedure. A master recipe is constructed from the site recipes using the formulas, procedures, and equipment-specific information. The master recipe is se-

TABLE 3.13.1 CONTROL ACTIVITY MODEL

Level	Function	Activity
Level 1	Process/product management	Production planning, inventory planning, and general recipe management
	Production management	Recipe management, production scheduling, and batch history management
Level 2	Batch management	Recipe generation and selection, batch execution supervision, unit activities coordination, and log and report generation
	Unit management	Unit supervision, allocation management, and unit coordination
Level 3	Sequential/regulatory/ discrete control	Device, loop, and equipment module control, predictive control, model-based control, and process interlocking
	Safety interlocking	—

lected and accessed by the batch management activity which converts it to a control recipe. The control recipe is the batch-specific recipe that is ready to run. Figure 3.13.5 shows the recipe hierarchy.

Using process and product knowledge, the recipe management function analyzes a process and determines the basic phases. These basic phases, along with product knowledge from the laboratory chemist, are used to construct the general (corporate-wide) recipe. Plant knowledge (for example, raw material availability) from the plant site engineer is used to transform the general recipe into a site-specific recipe. Equipment knowledge (for example, what vessels and piping are available in the plant) is used to transform this site-specific recipe into a master recipe. This master recipe is used as the basis for a control recipe when a batch is ready to be produced. Figure 3.13.6 summarizes the activities involved during recipe management.

Production Scheduling

Schedules serve as a guide for the production requirements in terms of the availability of equipment, personnel, raw materials, facilities, equipment, and process capacity. The schedule should have many of the following objectives:

- To minimize the processing time
- To minimize the deviation from a master plan
- To optimize the production of the product within quality guidelines
- To minimize energy costs
- To minimize the use of raw materials

The responsibility of the production scheduler is to develop a detailed, time-based plan of the activities necessary to achieve the production targets set by the production plan. The production scheduler must be able to dynamically allocate a new schedule at any time. Reallocating or creating a schedule automatically via some algorithm or manually via user intervention should be feasible.

Schedulers can be implemented in several ways. Linear programs, expert systems, or other multivariable techniques have been used successfully. The scheduler must provide a procedure and method for batch sizing and is the logical place where lot assignments are made. Figure 3.13.7 presents the production scheduling model currently defined by ISA standards committee (Jensen 1994).

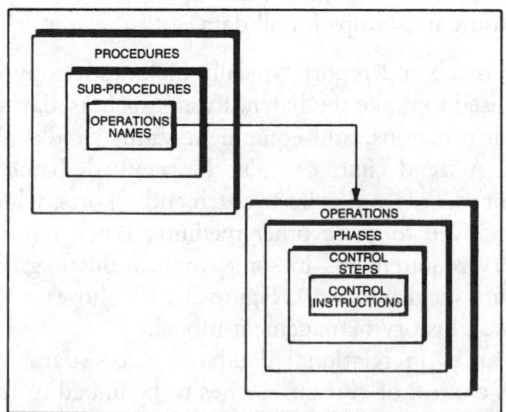

FIG. 3.13.4 Procedure model.

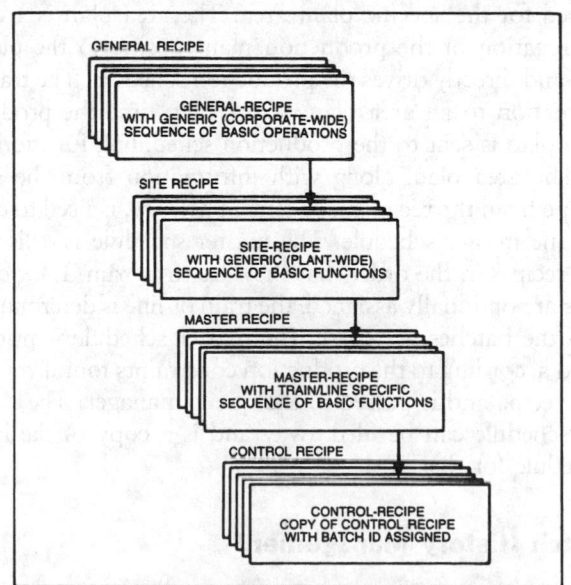

FIG. 3.13.5 Recipe model.

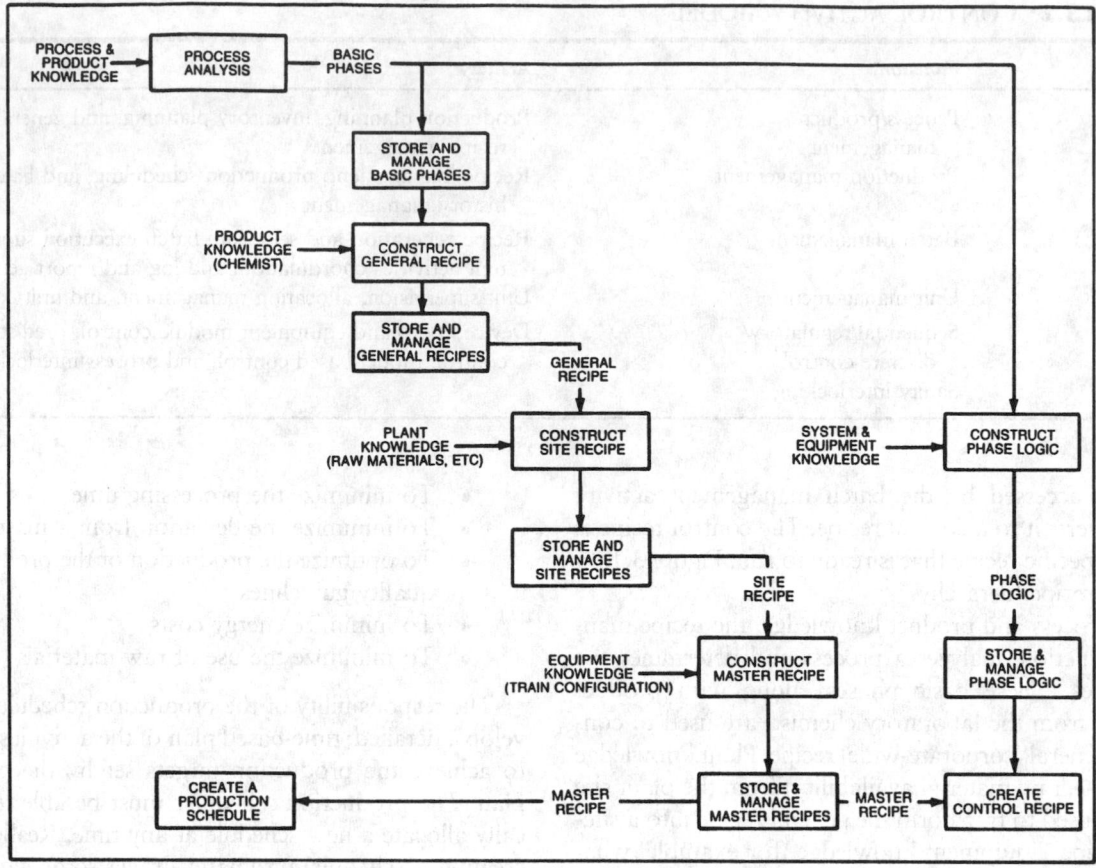

FIG. 3.13.6 Recipe activities.

As shown by the model, the production plan is input to the scheduling model. The plan is first transformed to an area plan. Knowledge about the process equipment is required at this time. The area plan is a listing of the end items which are to be produced, how many of each item are to be produced, and when the items are to be produced for the specific plant area. The area plan is a disaggregation of the production plan specific to the plant site and directly drives the production schedule. The transformation to an area plan occurs each time the production plan is sent to the production scheduling function.

The area plan, along with information from the site recipe from the recipe management activity, is used to create the master schedule. The master schedule is a list of the recipes in the order that they are to be run. Lot numbers are optionally assigned, the train or line is determined, and the batches are sized. The master schedule is prioritized according to the production constraints found via the site recipe and is passed to the queue manager. The master schedule can be filed away and is a copy of the best schedule for that area.

Batch History Management

Batch history management involves collecting and maintaining integrated, identifiable sets of dissimilar data. Batch tracking is the collection of this data. It is generally event triggered and contains the following related data:

Continuous process data (flow, temperatures, and pressures)

Event data (operator actions, alarms, and notes)

Recipe formula data (set points and times)

Calculated data (totalization, material usage, and accounting data)

Manual entries with an audit trail (location of change and operator of record)

Stage, batch, and lot identification

Time and date stamps on all data

The batch end report typically includes a copy of the recipe used to make the batch. Events such as alarms, operator instructions, and equipment status should also be logged. A trend chart can also be retained. Batch management records and collects batch end reports, which are then archived to some other medium. Batch reports are statutory requirements in some applications (e.g., in the pharmaceutical industry). Figure 3.13.8 shows a simplified batch history management model.

Advances in relational databases allow data for the process control of current batches to be linked to the histories of previous batches. Using standard query language (SQL) calls to access batch history provides new ways to

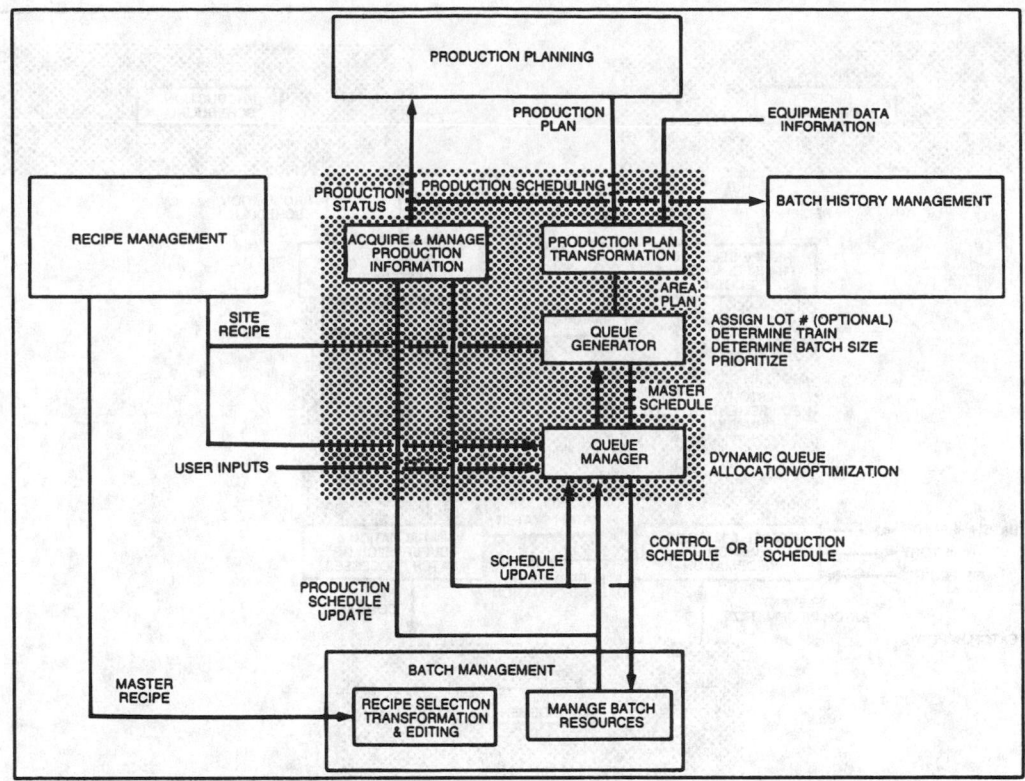

FIG. 3.13.7 Production scheduling model.

analyze and report batch histories. Other analysis techniques, such as statistical process control (SPC) and statistical quality control (SQC), can be applied at this level.

MANAGEMENT INTERFACES

Batch management interfaces with the user, recipe management, production scheduling, unit management, sequential control, and regulatory and discrete control. Its functions include (1) recipe selection, transformation, and editing; (2) initiation and supervision of batch processes; (3) management of batch resources; and (4) acquisition and management of batch information. Figure 3.13.9 shows the batch management model currently defined by the ISA standards committee.

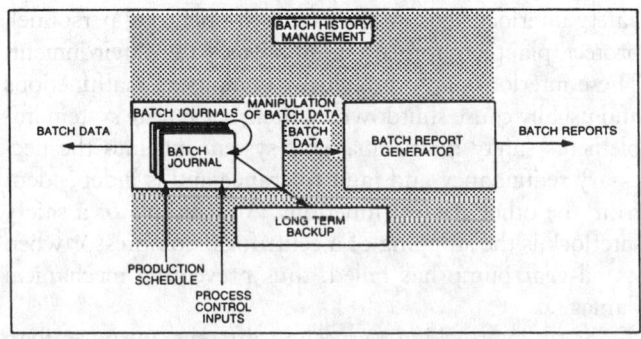

FIG. 3.13.8 Batch history management.

UNIT MANAGEMENT

A unit is a physical grouping of equipment and the unit control functions required to execute a batch. A process unit is a group of mechanical equipment, with each piece performing, somewhat independently, a portion of the chemical process. Examples of process units are filters, batch reactors, heat exchangers, and distillation columns. Control consists of the process states, known as phases, required to perform the unit operations. Examples are charging, heating, cooling, agitation, reacting, discharging, and washing.

Unit management interfaces with the user, batch management, sequential control, regulatory control, and discrete control in performing its function of (1) communicating with other unit equipment modules, loops, devices, and elements; (2) acquiring resources; (3) executing phases; and (4) handling phase exceptions. (In a batch program, the basic action is called a statement; several statements make up a step; a number of steps comprise a phase; phases can be combined into an operation; and a sequence of operations makes up a batch.) Figure 3.13.10 shows the unit management model defined by the ISA.

CONTROL FUNCTIONS

Sequential, regulatory, and discrete control functions interface directly with elements and actuators to change the process. These control functions are defined as:

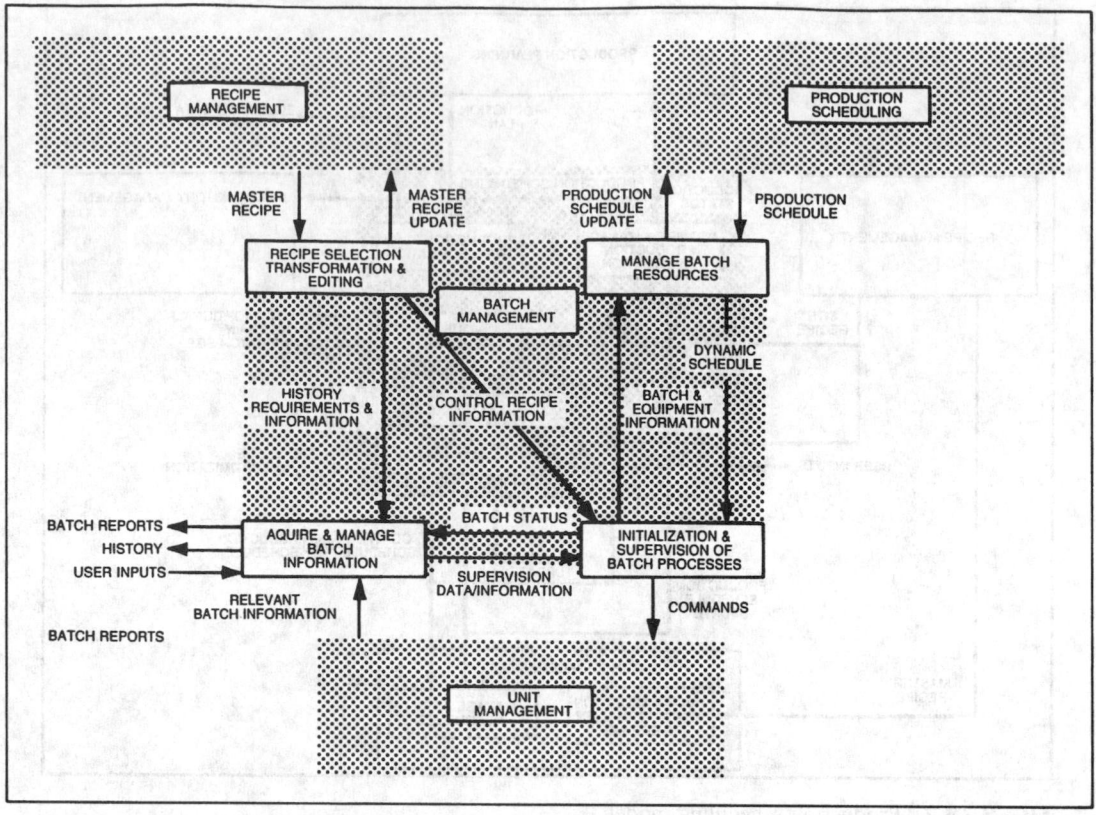

FIG. 3.13.9 Batch management model.

Discrete control maintains the process states at target values chosen from a set of stable states.

Regulatory control maintains the measurements of a process as close as possible to their set-point values during all events, including set-point changes and disturbances.

Sequential control sequences the process through a series of states as a function of time.

The user implements these control functions using devices, loops, and equipment modules, which are defined as:

A *device* is an item of process equipment that is operated as a single entity and can have multiple states or values. The user initiates discrete states (using hardware and software) to control discrete devices such as solenoid valves, pumps, and agitators.

A *loop* is a combination of elements and control functions which is arranged so that signals pass between the elements to measure and control a process variable. A PID control algorithm is a common control loop function.

An example of an *equipment module* is the sequential control of dehydrogenator bed control valves which put one bed online while the other bed is being regenerated according to a time schedule.

Process interlocking and advanced control, in the form of feedforward, predictive, or model-based control, are additional control functions which provide a higher level of automation for additional benefits.

Compared to continuous processing, additional control algorithms and control methodology are normally used in batch processing. Functions such as time-based PID (heat-soaked ramp), sequencers, and timers are required. Batch processing commonly uses techniques such as enabling and disabling control functions based on a phase state, enabling and disabling alarms on devices and loops, and employing antireset windup protection on PI or PID loops. Batch processes are device-oriented, while continuous processes are loop-oriented.

SAFETY INTERLOCKING

Safety interlocks ensure the safety of operating personnel, protect plant equipment, and protect the environment. These interlocks are initiated by equipment malfunctions and usually cause shutdowns. Often, a separate system implements safety interlocks. This system includes the necessary redundancy and fault tolerance and is independent from the other control functions. An example of a safety interlock is the stopping of a centrifugal compressor when its oil gear pump has failed, thus preventing mechanical damage.

Safety interlocking serves a different purpose than process interlocking and permissive interlocking. Process

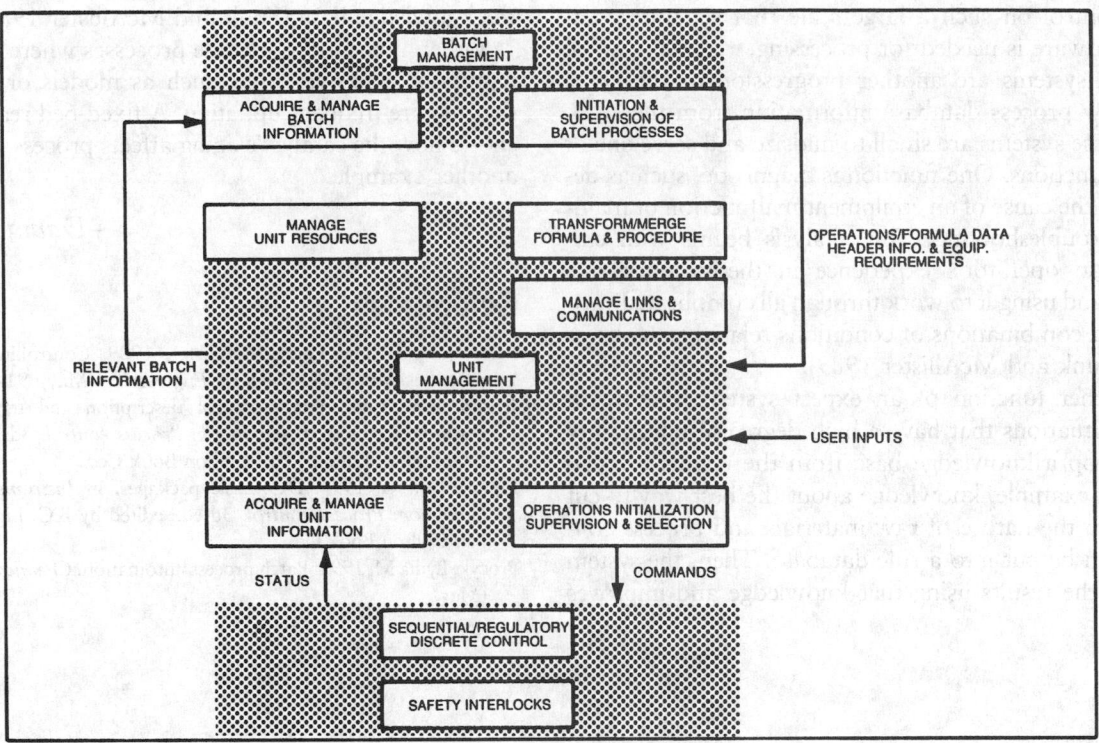

FIG. 3.13.10 Unit management model.

interlocking can be safety-related, but it is primarily associated with the process. An example of a process interlock is to stop charging a material if the agitator is not running. A permissive interlock establishes an orderly progression of sequences. An example of a permissive interlock is to not allow the feeding of an extruder before the barrel temperature has reached a minimum value.

Continuous Process Automation

Continuous processes require a control system that can minimize cost and optimize grade changes. Multiple products and grades are often made from the same feedstock in a continuous plant. However, since a plant cannot be shutdown between grades, a change-over period occurs during which the product is not on-spec. Thus, shortening the change-over period can minimize the amount of off-spec material being made during grade changes.

Integrated DCSs are essential to control complex manufacturing situations. A single control loop can be handled with traditional instruments alone. However, controlling complex situations requires a broader span of information than is available through discrete instrumentation.

Continuous processes rely heavily on regulatory controls and inferred variables. The DCS must also interface with highly sophisticated instruments and analyzers that have some control capabilities of their own and offer significant inference capabilities. For example, the Btu value of fuel cannot be measured unless it is burned in a calorimeter, but then it is not a useful process control variable.

However, the heating value of the fuel can be inferred from the thermal conductivity, temperature, and pressure of the incoming nature gas.

DCSs achieve high degrees of process accuracy because they incorporate process data archives which allow a better understanding of the process. These data archives allow people other than the operator to view and understand what happens in the process and be able to work on improving it. Optimum operation no longer depends on the operator monitoring the process every minute. Instead, the control system watches the process every second.

Many projects can be identified as being beneficial to the business. These projects frequently require the timely, accurate, and comprehensive information provided by DCSs as input. Such projects encompass a variety of areas including process optimizers, expert systems, quality lab interfaces, statistical process control, and plantwide maintenance programs.

Process optimization is a natural progression of the DCS. A company can use the information available through the DCS archives to optimize process conditions to conserve energy or save raw materials. They can also use the information to modify product mixes and product specifications to optimize costs on a global, rather than individual unit, basis (Funk and McAllister 1989)

Process optimizers provide advanced supervisory control. They monitor the current operating conditions, run advanced algorithms, and return recommended set-point changes to the control systems. However, process optimizers control on such a large scale that separate computer hardware is needed for processing.

Expert systems are another progression of the DCS. They draw process database information from the DCS. Often, these systems are small to midsize and serve one of the two functions. One function is diagnostic, such as determining the cause of an equipment malfunction or maintenance troubleshooting. This analysis begins with embedding an operator's experience in the system's rule database and using it to work through all complicated variations and combinations of conditions related to faults or alarms (Funk and McAllister 1989).

The other function of an expert system is to work through situations that have a high degree of uncertainty and develop a knowledge base from the results of decisions. For example, knowledge about the best way to cut costs given the nature of raw materials and process conditions can be put into a rule database. Then, the system monitors the results using that knowledge and improves the knowledge base (Funk and McAllister 1989).

Other applications include processes where process dynamics change over time, such as models or controllers that require frequent updating. A fixed-bed reactor which shows how the catalyst's aging affects process dynamics is another example.

—David H.F. Liu

References

Funk, John C. and Larry McAllister. 1989. Controlling continuous processes with DCS. *Chemical Engineering* (May): 91–96.

Jensen, B.A. 1994. Batch control description and terminology. In *Instrument engineers' handbook: Process control.* 3d ed., edited by B.G. Liptak. Radnor, Pa.: Chilton Book Co.

Liptak, Bela G. 1994. DCS-basic packages. In *Instrument engineers' handbook: Process control.* 3d ed., edited by B.G. Liptak. Radnor, Pa.: Chilton Book Co.

Procyk, Lydia M. 1991. Batch process automation. *Chemical Engineering* (May).

3.14
PUBLIC SECTOR ACTIVITIES

EPA Pollution Prevention Strategy

Pollution prevention, while not new to the EPA, has emerged as a priority in the 1990s. This prioritization represents a fundamental change from the historical interpretation of the agency's mission as protecting human and environmental health through pollution control. The EPA's pollution control emphasis was to eliminate the options of releasing and transferring industrial pollution in the environment and to increase the cost of the remaining options of treatment and disposal. The net effect has been to encourage industry to limit their pollution through source reduction.

The formal shift in policies and priorities for the EPA is reflected in the 1990 passage of the CAAA and the Pollution Prevention Act. The EPA issued a pollution prevention strategy in 1991 to articulate its position and objectives. This policy serves the following two purposes:

To guide and direct incorporating pollution prevention into the EPA's existing regulatory and nonregulatory program

To specify a program with stated goals and a time for their accomplishment

The EPA's goal is to incorporate pollution prevention into every facet of its operations including enforcement actions, regulations, permits, and research.

This strategy confronts the institutional barriers that exist within the EPA which is divided along single environmental medium lines. The agency has accomplished the following:

Established an Office of Pollution Prevention and Toxics which coordinates the agencywide pollution prevention policy

Created a Waste Minimization Branch in the Office of Solid Waste to coordinate waste minimization and pollution prevention under the RCRA

Charged the EPA Risk Reduction Engineering Laboratory with conducting research on industrial pollution prevention and waste minimization technologies

Developed a Pollution Prevention Advisory Committee to ensure that pollution prevention is incorporated throughout the EPA's programs

All areas of the EPA are developing initiatives to promote a pollution prevention ethic across the agency. These initiatives are characterized by the use of a range of tools including market incentives, public education and information, technical assistance, research and technology applications, and the traditional regulatory and enforcement actions. Examples include:

Establishing cash rewards for EPA facilities and individuals who devise policies and actions to promote pollution prevention

Public commending and publicizing of industrial facility pollution prevention success stories

Coordinating the development and implementation of regulatory programs to promote pollution prevention

Clustering rules to evaluate the cumulative impact of standards in industry, which encourage early investment in prevention technologies and approaches

The EPA is further implementing the 33/50 program which calls for the involuntary cooperation of industry in developing pollution prevention strategies to reduce environmental releases of seventeen selected chemicals by the year 1995 (see Section 3.1).

The EPA's pollution prevention program is multifaced and expansive. The Pollution Prevention Clearing-House (PPIC) provides current news and information on recent developments in this rapidly changing arena. The PPIC Technical Support Hotline is (703) 821-4800.

Three programs of interest are the Green Lights Program, the Golden Carrot Program, and the Energy Star Computers Program. These programs are described next.

GREEN LIGHTS PROGRAM

The Green Lights Program, launched by the EPA in January 1991, is designed to prevent pollution by encouraging the use of energy-efficient lighting in offices, stores, factories, and other facilities across the country. Lighting consumes about 25% of the nation's electricity, and more than half of the electricity used for lighting is wasted by inefficient technology and design practices. Under the Green Lights Program, the EPA has asked businesses, governments, and other institutions to install energy-efficient lighting over a five-year period, but only where it is profitable and lighting quality is maintained or improved.

Over 600 companies have made voluntary commitments to participate in this program, representing 2.5 billion sq ft of business space. Given the commitments of current participants, over the next five years, they will reduce their electric bills by an estimated $760 million. In addition, the program will prevent the generation of 7.4 million tn of CO, 59,500 tn of SO, and 25,300 tn of NO emissions.

GOLDEN CARROT PROGRAM

The Golden Carrot Program, promoted by the EPA's Office of Air and Radiation, encourages manufacturers to design superefficient refrigerators that use no CFCs for cooling or insulation. Refrigerators and freezers use about 20% of the nation's electricity and vary in efficiency. To stimulate manufacturers to develop more efficient CFC-free units, twenty-three electric utilities have pooled their resources to offer a $30 million incentive (the golden carrot) to the winner of a product design competition in superefficient refrigeration.

The EPA estimates that the Golden Carrot Program is capable of reducing electric power consumption by 3 to 6 billion kWh, saving $240 to 480 million annually in consumer electric bills.

ENERGY STAR COMPUTERS PROGRAM

The EPA's Energy Star Computers Program is a voluntary, market-based partnership with computer manufacturers to promote energy-efficient personal computers in an effort to reduce the air pollution caused by the generation of electricity. Office equipment is the fastest growing electrical load in the commercial sector. Computer systems alone account for approximately 5% of the commercial electricity consumption—a figure which could reach 10% by the year 2000. Dramatic, cost-effective, efficiency improvements are available for both hardware power consumption and the control of operation hours, offering up to 90% energy savings for many computer applications.

To date eight computer manufacturers—Apple, IBM, Hewlett-Packard, Digital, Compaq, NCR, Smith Corona, and Zenith Data Systems—have signed partnership agreements with the EPA to participate in the program. By the year 2000, the EPA's Energy Star Computers Program and other campaigns to promote energy-efficient computer equipment will probably save 25 billion kWh of electricity annually—down from an estimated consumption of 70 billion kWh per year. These savings will reduce CO emissions by 20 million metric tn, SO emissions by 140,000 metric tn, and NO emissions by 75,000 metric tn.

CROSS-CUTTING RESEARCH

Three major research areas are targeted in the cross-cutting research component of pollution prevention research. Cross-cutting issues have been selected because of their importance in furthering the science of pollution prevention and the agency's ability to promote and implement pollution prevention as the preferred approach to environmental protection. Cross-cutting issues include (1) tool development, (2) application of tools, and (3) measurement of progress.

A balanced cross-cutting research program addresses the development of innovative tools for pollution prevention including technological, informational, and evaluative tools. The cross-cutting research strategy for tools development focuses on performing industry-specific pollution prevention assessments, incorporating pollution prevention factors into process simulation models, developing and testing LCA methodology, and improving the agency's understanding of how individuals and corporations make decisions and the factors that affect their behavior.

Demonstrating the effectiveness of pollution prevention approaches is critical to increasing reliance on this preferred approach to environmental management. The application of the tool research area focuses on incorporating pollution prevention considerations into the EPA's rule-making process, developing and demonstrating innovative pollution prevention technologies, transferring timely information on pollution prevention approaches, and determining the most effective ways to use incentives and education to promote prevention.

Continued environmental progress depends upon knowing what has worked and how well and what has been less successful and why. Environmental engineers can then use this information to identify areas for additional research, improve approaches, and develop new approaches. The development of techniques to measure and evaluate the effectiveness of pollution prevention approaches is critical to determine which approaches effectively prevent pollution and which approaches fail. These techniques are useful for measuring progress and establishing priorities for research and other activities.

Industrial Programs and Activities

Industries, working harder to be good neighbors and responsible stewards of their products and processes, are aggressively engaged in pollution prevention activities both as trade associations and as individual companies (U.S. EPA 1991). An extended description of successful trade association and company programs is available from the EPA's Office of Pollution Prevention.

TRADE ASSOCIATION PROGRAMS

The CMA, American Petroleum Institute (API), and National Paints and Coating Association (NPCA) are examples of trade associations committed to pollution prevention. Their programs are described next.

CMA

The CMA started its Responsible Care Program in 1988 to improve the chemical industry's management of chemicals. All CMA members are required to participate and adhere to the ten guiding principles of the program. The principles speak of protecting health, safety, and the environment but do not address pollution prevention specifically. The program outlines the framework for the reduction of waste and releases to the environment (see Section 3.2). To evaluate progress, the CMA requires its companies to submit an annual report that identifies progress in implementation and quantifies facility-specific releases and wastes (CMA 1990).

The program is being implemented in all parts by CMA's 175 member companies. These companies have about 2000 facilities and produce about 90% of the U.S. chemicals. In May 1993, the Synthetic Organic Manufacturing Association (COMA) voted to become a Responsible Care partner association.

American Petroleum Institute (API)

The API also has a prescribed set of guiding environmental principles its members are encouraged to follow. The API's eleven principles generically promote action to protect health, safety, and the environment. One of the API's principles addresses pollution prevention by requiring its members to reduce overall emissions and waste generation (Chevron Corporation 1990). The API articulates the eleven principles as goals to which members should aspire.

National Paints and Coating Association (NPCA)

The NCPA has their Paint Pollution Prevention Program (April 1990). The goal of the program is "the promotion of pollution prevention in our environment through effective material utilization, toxics use, and emissions reduction and product stewardship in the paint industry" (NPCA 1990). The statement recommends that each NCPA member company establishes a waste reduction program that includes setting priorities, goals, and plans for waste reduction with preference first to source reduction, second to recycling and reuse, and third to treatment.

COMPANY PROGRAMS

The scope of company programs varies considerably. Some are limited to one environmental medium, while others are multimedia. Some focus on certain types of pollutants, such as toxic release inventory (TRI) chemicals; others are more wide ranging. All include some forms of pollution prevention but vary in their emphasis. Most adopt the EPA's environmental management hierarchy: source reduction first, followed by recycling, treatment, and disposal.

A review of company pollution prevention activities reveals that some companies have programs they are willing to share with the public and other companies consider their efforts internal and proprietary. The more accessible programs are usually with large multifacility companies. Some programs such as Dow Chemical's Waste Reduction Always Pays (WRAP) and 3M's Pollution Prevention Pays (3P program) are well known programs.

Public interest groups have questioned the accountability and reliability of the accomplishments claimed. Legitimate questions remain on whether the cited reductions are real and result from pollution prevention methods or whether they are artifacts of changes in reporting requirements or analytical methods or from waste transfer between sites or between media (U.S. Printing Office 1990).

The concerns notwithstanding, a major change in the industrial perspective on the way business is done has occurred. The programs initiated by industry on pollution prevention are important because they raise the expectation for future progress. If the successes are real and include financial gains, other firms will likely follow the leaders into this new era of environmental protection.

State and Local Programs

By 1991, close to fifty state laws were in place (U.S. EPA 1991). More than half the states have passed pollution prevention laws, some states passing more than one. Other states have legislation pending or on their agenda.

The state pollution prevention laws vary in their provisions. Some have detailed requirements. They target specific source reduction goals and provide measures to meet them. Other states have general statutes, dealing with pollution prevention as state policy to be the preferred method of handling hazardous waste. Some states have no formal laws but have operationally included pollution prevention into their programs.

FACILITY PLANNING REQUIREMENTS

A new trend in state pollution prevention requirements is facility planning. These statutes require industrial facilities to submit pollution prevention plans and update them periodically. Most plans cover facilities obliged to report federal TRI data.

Many of the facility planning laws require industry to consider only pollution prevention options. Others are broader in scope but consider pollution prevention as the preferred approach when technically and economically practical. Facilities that are required to prepare plans must either prepare and submit progress reports or annual reports. Facilities failing to complete adequate plans or submit progress reports may be subject to enforcement actions or negative local publicity.

STATE POLLUTION PREVENTION PROGRAMS

State programs are at best a barometer of activity in pollution prevention. Programs vary along with their enabling statutes. Some programs are mature, well-established, and independent. Others consist of little more than a coordinator, who pulls together the pollution prevention aspects from other state environmental programs. Some states delegate their pollution prevention to third-party groups at universities and research centers and provide state funding for their operation.

Program elements of state programs include raising the general awareness of benefits from pollution prevention, reducing information and technological barriers, and creating economic and regulatory incentives for pollution prevention. Some also attempt to foster changes in the use of toxic materials and the generation and release of toxic by-products.

LOCAL PROGRAMS

The effects of hazardous waste production are felt first at the local level. Rather than relying on state and federal efforts, local governments are often in a better position to identify the needs and limitations of local facilities.

Local governments can also be flexible in dealing with specific problems. One example is the potential offered by publicly owned treatment works (POTWs). POTWs receive and process domestic, commercial, and industrial sewage. Under delegated federal authority, they can restrict industrial and commercial pollution from the wastewater they receive. Pollution prevention is becoming a recurrent theme in the operation of POTWs.

Nongovernmental Incentives

Colleges and universities play a vital role in developing pollution prevention ethics among scientists, business people, and consumers. The efforts of academia assure environmental awareness among students who will design and manage society's institutions and develop ties between industry and the campus (U.S. EPA 1991).

ACADEMIA

University professors have identified a range of research topics in pollution prevention. Under cooperative programs with state agencies, the EPA has sponsored research on product substitutes and innovative waste stream reduction processes. An increasing number of industries are also beginning to support university research. The evolution of the pollution prevention perspective is reflected in academic environmental programs.

The progression starts with an industry's initial control efforts of good housekeeping, inventory control, and minor operating changes. In the waste minimization stage, an industry uses technologies to modify processes and reduce effluents. The 1990s have introduced highly selective separation and reaction technologies predicated on the precepts of design for technology and toxic use reduction.

The American Institute of Chemical Engineers (AIChE) founded the American Institute of Pollution Prevention (AIPP) to assist the EPA in developing and implementing pollution prevention. The AIChE aggressively encourages industry sponsorship of university research. Targeted research areas include identification and prioritization of waste streams, source reduction and material substitution, process synthesis and control, and separation and recovery technology (through its Center for Waste Reduction Technologies).

The American Chemical Society's effort has been more modest. Clearly contributions are needed from synthetic and organic and inorganic chemists to build more environmentally friendly molecules, molecules designed for the environment, while still fulfilling their intended function and use. The Center for Process Analytical Chemistry provides an important role in pollution prevention.

COMMUNITY ACTION

The public, as consumers and disposers of toxic-chemical-containing products, is a major source of toxic pollution. It must and has become involved in toxic pollution prevention.

Public involvement has resulted from state-wide initiatives, the action of interest groups, and individual initiatives. Environmentalists concerned with pollution control advocate source reduction over waste treatment as the preferred environmental option. However, the lack of public information about industrial releases to the environment has effectively blocked community action groups from addressing the toxic issue. The TRI and the right-to-know laws have changed that barrier irreversibly.

—David H.F. Liu

References

Chemical Manufacturers' Association (CMA). 1990. *Improving performance in the chemical industry.* (September).

Chevron Corporation. 1990. *1990 report on the environment: A commitment to excellence.*

National Paints and Coating Association (NCPA). 1990. Paint pollution prevention policy statement. *Pollution Prevention Bulletin* (April).

U.S. Environmental Protection Agency (EPA). 1991. Pollution prevention 1991: Progress on reducing industrial pollutants. EPA 21P-3003.

U.S. Printing Office. 1990. *Toxics in the community: National and local perspective.*

Bibliography

Borman, Stu. 1993. Chemical engineering focuses increasingly on the biological. *C&E News* (11 January).

Cusack, Roger W., P. Fremeaux, and Don Glatz. 1991. A fresh look at liquid-liquid extraction. *Chem. Eng.* (February).

Enichem systheses unpacks DMC derivatives potential. 1992 *The Chemical Engineer* (9 April).

Johansson, Allan. 1992. *Clean technology.* Lewis Publishers.

Keoleian, G.A., D. Menerey, and M.A. Curran. 1993. *A life cycle approach to product system design.* Pollution Prevention Review (Summer).

———. 1993. *Life cycle design manual.* EPA 600/R-92/226, Cincinnati, Ohio: U.S. EPA, Pollution Prevention Research Branch.

Laing, Ian G. 1992. Waste minimisation: The role of process development. *Chemistry & Industry* (21 September).

Lipták, Béla G. 1994. *Instrument engineer's handbook: Process control.* 3d ed., edited by B.G. Liptak. Radnor, Pa.: Chilton Book Co.

Martel, R.A. and W.W. Doerr. 1993. Comparison of state pollution prevention legislation in the Mid-Atlantic and New England states. Paper presented at AIChE Seattle Summer National Meeting, August 1993.

Modell, Donald J. 1989. DCS for batch process control. *Chemical Engineering* (May).

New linear alkylbenzene process. 1991. *The Chemical Engineer* (17 January).

Rittmeyer, R.W. 1991. Prepare effective pollution-prevention program. *Chem. Eng. Progress* (May).

Theodore, L. and Y.C. McGuinn. 1992. Pollution prevention. New York: Van Nostrand Reinhold.

U.S. Environmental Protection Agency (EPA). 1989. *EPA manual for the assessment of reduction and recycling opportunities for hazardous waste (Arrow project).* Cincinnati, Ohio: Alternative Technologies Division, Hazardous Waste Engineering Research Laboratory.

4

Standards

William C. Zegel

Air Quality Standards

4.1
SETTING STANDARDS

Introduction

Today's air quality standards have emerged from sections 109 and 112 of the 1970 Clean Air Act (CAA) Amendments and Title III of the 1990 CAA Amendments.

SECTION 109, 1970 CAA AMENDMENTS

The 1970 CAA Amendments define two primary types of air pollutants for regulation: criteria air pollutants and hazardous air pollutants. Under section 108, criteria pollutants are defined as those that "cause or contribute to air pollution that may reasonably be anticipated to endanger public health or welfare . . . the presence of which in the ambient air results from numerous or diverse mobile or stationary sources." Under section 109, the EPA identifies pollutants that meet this definition and prescribes national primary air quality standards, "the attainment and maintenance of which . . . allowing an adequate margin of safety, are requisite to protect the public health."

National secondary air quality standards are also prescribed, "the attainment and maintenance of which . . . is requisite to protect the public welfare from any known or anticipated effects associated with the presence of the air pollutant." Welfare effects include injury to agricultural crops and livestock, damage to and the deterioration of property, and hazards to air and ground transportation. The National Ambient Air Quality Standards (NAAQS) are to be attained and maintained by regulating stationary and mobile sources of the pollutants or their precursors.

SECTION 112, HAZARDOUS AIR POLLUTANTS

Under section 112, the 1970 amendments also require regulation of hazardous air pollutants. A hazardous air pollutant is defined as one "to which no ambient air standard is applicable and that . . . causes, or contributes to, air pollution which may reasonably be anticipated to result in an increase in serious irreversible, or incapacitating reversible, illness." The EPA must list substances that meet the definition of hazardous air pollutants and publish national emission standards for these pollutants providing "an ample margin of safety to protect the public health from such hazardous air pollutant[s]." Congress has provided little additional guidance, but identified mercury, beryllium, and asbestos as pollutants of concern.

TITLE III, 1990 CAA AMENDMENTS

Although the control of criteria air pollutants is generally considered a success, the program for hazardous air pollutants was not. By 1990, the EPA regulated only seven of the hundreds of compounds believed to meet the definition of hazardous air pollutants.

Title III of the 1990 CAA Amendments completely restructured section 112 to establish an aggressive new program to regulate hazardous air pollution. Specific programs have been established to control major-source and area-source emissions. Title III establishes a statutory list of 189 substances that are designated as hazardous air pollutants. The EPA must list all categories of major sources and area sources for each listed pollutant, promulgate standards requiring installation of the maximum achievable control technology (MACT) at all new and existing major sources in accordance with a statutory schedule, and establish standards to protect the public health with an ample margin of safety from any residual risks remaining after MACT technology is applied.

Ambient Concentration Limits

Air pollution control strategies for toxic air pollutants are frequently based on ambient concentration limits (ACLs). ACLs are also referred to as acceptable ambient limits (AALs) and acceptable ambient concentrations (AACs). A regulatory agency sets an ACL as the maximum allowable ambient air concentration to which people can be exposed. ACLs generally are derived from criteria developed from human and animal studies and usually are presented as weight-based concentrations in air, possibly associated with an averaging time.

The EPA uses this approach for criteria air pollutants but not for toxic air pollutants. The CAA Amendments of 1970 require the EPA to regulate toxic air pollutants through the use of national emission standards. The 1990 amendments continue and strengthen this requirement. However, state and local agencies make extensive use of ACLs for regulatory purposes. This extensive use is because, for most air pollutants, ACLs can be derived easily and economically from readily available health effects information. Also, the maximum emission rate for a source that corresponds to the selected ACL can be determined easily through mathematical modeling. Thus, the regulator can determine compliance or noncompliance. Lastly, the use of ACLs relieves regulators from identifying and specifying acceptable process or control technologies.

ACLs are frequently derived from occupational health criteria. However, ACLs are susceptible to challenge because no technique is widely accepted for translating standards for healthy workers exposed for forty hours a week to apply to the general population exposed for twenty-four hours a day. Another disadvantage of ACLs is that both animal and occupational exposures, from which health criteria are developed, are typically at concentrations greater than normal community exposures. This difference requires extrapolation from higher to lower dosages and often from animals to humans.

DERIVATION OF AMBIENT CONCENTRATION LIMITS

ACLs are typically derived from health criteria for the substance in question. They are usually expressed as concentrations such as micrograms per cubic meter (μg/cu m). Health criteria are generally expressed in terms of dose—the weight of the pollutant taken into the body divided by the weight of the body. To convert a dose into a concentration, assumptions must be made about average breathing rates, average consumption of food and water, and the amount of each that is available to the body (adsorption factors). The EPA has a generally accepted procedure for this process (U.S. EPA 1988, 1989).

Other methods of deriving ACLs are based upon an absolute threshold (CMA 1988). These methods set ACLs at some fraction of an observed threshold or established guideline. A margin of safety is generally added depending on the type and severity of the effect on the body, the quality of the data, and other factors. Still other methods depend upon extrapolation from higher limits established for other similar purposes.

The health criteria felt most appropriate for deriving ACLs is the risk reference dose (RfD) established by the EPA (Patrick 1994). The EPA has developed RfDs for both inhalation and ingestion pathways (U.S. EPA 1986). They require much effort to establish and are generally designed for long-term health effects.

USE OF THE RfD

RfDs are developed for ingestion and inhalation exposure routes. If a relevant inhalation RfD is available, regulatory agencies should use it as the basis for deriving an ACL for an air pollutant. The EPA is currently deriving reference values for inhalation health effects in terms of micrograms per cubic meter. These risk reference concentrations (RfCs) provide a direct link with ACLs. Without more specific information on inhalation rates for the target population, regulators frequently assume the volume of air breathed by an average member of a typical population to be 20 cubic meters per day, which is considered a conservative value.

When an inhalation RfD is not available, regulators must derive an ACL from another source. One approach is to use an ingestion RfD to estimate an RfC. However, this technique can be inaccurate because absorption through the digestive system is different from absorption through the respiratory system.

RfDs and RfCs are available through the EPA's Integrated Risk Information System (IRIS). Many state and local regulatory agencies use the EPA-derived RfDs and RfCs to establish ACLs. These reference values are available through the EPA's National Air Toxics Information Clearing House (NATICH). Because of the large number of state and local agencies, NATICH does not always have the latest information. Therefore, the practicing engineer should get the latest information directly from the local agency.

USE OF OCCUPATIONAL EXPOSURE LIMITS

In some cases, neither RfDs nor RfCs are available, and regulators must use another source of information to derive ACLs. Occupational limits, usually in the form of threshold limit values (TLVs) and permissible exposure limits (PELs), are often used to establish ACLs. Both establish allowable concentrations and times that a worker can be exposed to a pollutant in the work place. TLVs and PELs are particularly useful in establishing acute exposure ACLs.

The American Conference of Governmental Industrial Hygienists (ACGIH) develops TLVs. Three types of TLVs are the time-weighted average (TLV-TWA), the short-term exposure limit (TLV-STEL), and the ceiling limit (TLV-C). The TLV-TWA is the time-weighted average concentration for a normal eight-hour work day and forty-hour work week to which almost all workers can be repeatedly exposed without adverse effects. TLV-STELs are fifteen-minute time-weighted average concentrations that should not be exceeded during the normal eight-hour work day, even if the TLV-TWA is met. TLV-Cs are concentrations that should never be exceeded.

PELs are established by the U.S. Occupational Safety and Health Administration (OSHA) and are defined in much the same way as the TLVs. OSHA adopted the ACGIH's TLVs when federal occupational standards were originally published in 1974. Since that time, many of the values have been revised and published as PELs.

These occupational levels were developed for relatively healthy workers exposed only eight hours a day, forty hours a week. They do not apply to the general population, which includes the young, the old, and the sick and which is exposed twenty-four hours a day, seven days a week. However, using safety factors, regulators can use occupational levels as a basis for extrapolation to community levels. Different regulatory agencies use different safety factors.

USE OF OTHER APPROACHES

When no RfD has been derived, regulators can use the level at which no observed adverse effects have been found (NOAEL) or the lowest level at which adverse effects have been observed (LOAEL), with appropriate safety factors. These levels are similar in nature and use to the RfDs. Related levels are the no observed effect level (NOEL) and the lowest observed effect level (LOEL), respectively. Other sources of information are the minimal risk level (MRL), the level that is immediately dangerous to life and health (IDLH), emergency response planning guidelines (ERPG), and emergency exposure guideline levels (EEGL) for specific pollutants. These last four levels are for special situations; for these levels to be useful in assessing danger to the general public, regulators must severely attenuate them by safety factors. However, in the absence of other data, these levels can be useful in establishing an ACL or standard.

A pollutant's NOAEL is the highest tested experimental exposure level at which no adverse effects are observed. The NOEL is the highest exposure level at which no effects, adverse or other, are observed. The NOEL is generally less useful since factors other than toxicity can produce effects.

A pollutant's LOAEL is the lowest tested experimental exposure level at which an adverse health effect is observed. Since the LOAEL does not convey information on the no-effect level, it is less useful than the NOAEL, but it can still be useful. The LOEL is the lowest level at which any effect is observed, adverse or not. As a result, it is generally less useful than the NOEL.

MRLs are derived by the Agency for the Toxic Substances and Disease Registry (ATSDR), which was formed under the Comprehensive Environmental Response, Compensation and Liability Act (CERCLA) of 1980. The CERCLA requires ATSDR to prepare and update toxicological profiles for the hazardous substances commonly found at superfund sites (those sites on the National Priority List) that pose the greatest potential risk to human health. As part of the profiles, ATSDR derives MRLs for both inhalation and ingestion exposures.

The National Institute for Occupational Safety and Health (NIOSH) developed IDLHs primarily to select the most effective respirators to use in the work place. IDLHs are the maximum pollutant concentration in the air from which healthy male workers can escape without loss of life or suffering irreversible health effects during a maximum thirty-minute exposure. Another way of thinking of IDLHs is that if levels are above these standards, respirators must be used to escape the area of contamination.

The American Industrial Hygiene Association (AIHA) has derived ERPGs at three levels for several substances. Level 1 is the lowest level; it represents the maximum pollutant concentration in the air at which exposure for one hour results in mild, transient, adverse health effects. Level 2 is the concentration below which one hour of exposure does not result in irreversible or serious health effects or

TABLE 4.1.1 SUMMARY OF NAAQSs

| Pollutant | Averaging Time | Standard (@ 25°C and 760 mm Hg) | |
		Primary	Secondary
Particulate matter, 10 micrometers (PM_{10})	Annual arithmetic mean	50 $\mu g/m^3$	Same as primary
	24-hour	150 $\mu g/m^3$	Same as primary
Sulfur dioxide (SO_2)	Annual arithmetic mean	0.03 ppm (80 $\mu g/m^3$)	Same as primary
	24-hour	0.14 ppm (365 $\mu g/m^3$)	Same as primary
	3-hour	None	0.5 ppm (1300 $\mu g/m^3$)
Carbon monoxide (CO)	8-hour	9 ppm (10 mg/m^3)	Same as primary
	1-hour	35 ppm (40 mg/m^3)	Same as primary
Ozone (O_3)	1-hour per day	0.12 ppm (235 $\mu g/m^3$)	Same as primary
Nitrogen dioxide (NO_2)	Annual arithmetic mean	0.053 ppm (100 $\mu g/m^3$)	Same as primary
Lead (Pb)	Quarterly arithmetic mean	1.5 $\mu g/m^3$	Same as primary

Source: CFR Title 40, Part 50. Environmental Protection Agency. U.S. Government Printing Office, 1993.

Notes: All standards with averaging times of 24 hours or less, and all gaseous fluoride standards, are not to have more than one actual or expected exceedance per year.

$\mu g/m^3$ or mg/m^3 = microgram or milligram per cubic meter

TABLE 4.1.2 STATE AND LOCAL AGENCY USE OF AMBIENT CONCENTRATION LIMITS

State	Derivation of ACL	State	Derivation of ACL
Alabama	TLV/40 (one-hour), TLV/420 (annual)	New Hampshire	TLV/100 (twenty-four-hour) low toxicity
Alaska	Case-by-case analysis		TLV/300 (twenty-four-hour) medium
Arizona	0.0075 × Lower of TLV or TWA		toxicity
Arkansas	TLV/100 (twenty-four-hour), LD_{50}/10,000		TLV/420 (twenty-four-hour) high
California	Risk assessment used		toxicity
Colorado	Generally uses risk assessment	New Jersey	Case-by-case analysis
Connecticut	TLV/50 low toxicity	New Mexico	TLV/100 (eight-hour)
	TLV/100 medium toxicity	New York	TLV/50 (eight-hour) low toxicity
	TLV/200 high toxicity		TLV/300 (eight-hour) high toxicity
Delaware	TLV/100	North Carolina	TLV/10 (one-hour) acute toxicity
Florida	Ranges from TLV/50 to TLV/420		TLV/20 (one-hour) systemic toxicity
	depending upon the situation		TLV/160 (twenty-four-hour) chronic
Georgia	TLV/100 (eight-hour), noncarcinogens		toxicity
	TLV/300 (eight-hour), carcinogens	North Dakota	TLV/100 (eight-hour)
Hawaii	TLV/200	Ohio	TLV/42
Idaho	Case-by-case analysis	Oklahoma	TLV/10, TLV/50, TLV/100
	BACT can be required	Oregon	TLV/50, TLV/300
Illinois	Case-by-case analysis	Pennsylvania	TLV/42, TLV/420, TLV/4200 (one-week)
Indiana	Case-by-case analysis	Rhode Island	Case-by-case analysis
Iowa	Case-by-case analysis	South Carolina	TLV/40 (eight-hour) low toxicity
Kansas	TLV/100 (twenty-four-hour), irritants		TLV/100 (eight-hour) medium toxicity
	TLV/420 (annual), serious effects		TLV/200 (eight-hour) high toxicity
Kentucky	Case-by-case analysis	South Dakota	Case-by-case analysis
Louisiana	TLV/42 (one-hour) screening level	Tennessee	TLV/25, screening
Maine	Case-by-case analysis	Texas	TLV/100 (thirty-minute)
Maryland	Varies, TLV/100 (eight-hour)		TLV/1000 (annual)
Massachusetts	Health-based program	Utah	TLV/100 (twenty-four-hour)
Michigan	TLV/100 (eight-hour)	Vermont	TLV/420 (eight-hour)
Minnesota	TLV/100 (eight-hour)	Virginia	TLV/60 (eight-hour), TLV/100
Mississippi	TLV/100 (ten-minute)	Washington	TLV/420
Missouri	TLV/75 to TLV/7500 (eight-hour)	West Virginia	Case-by-case analysis
Montana	TLV/42	Wisconsin	TLV/42 (twenty-four-hour), screening
Nebraska	Case-by-case analysis	Wyoming	TLV/4
Nevada	TLV/42 (eight-hour) and case-by-case analysis		

Source: David R. Patrick, ed, 1994, *Toxic air pollution handbook* (New York: Van Nostrand Reinhold).

in symptoms that could impair the ability to take protective action. Level 3 is the concentration below which most individuals could be exposed for one hour without experiencing or developing life-threatening health effects.

The National Research Council for the Department of Defense has developed EEGLs. These levels may be unhealthy, but the effects are not serious enough to prevent proper response to emergency conditions to prevent greater risks, such as fire or explosion. These peak levels of exposure are considered acceptable in rare situations, but they are not acceptable for constant exposure.

Compliance with ACLs

ACLs are useful tools for reducing pollution levels. They also establish a framework to prioritize actions in reducing pollution. Generally, ACLs require sources to reduce their pollutant emissions to a level that assures that the ACL is not exceeded at the property boundary or other nearby public point. If a monitoring method is established for a pollutant, a regulator can demonstrate compliance using mathematical dispersion modeling techniques of measured emissions or ambient monitoring.

SOURCE AND AMBIENT SAMPLING

Regulators can sample emissions at the source by withdrawing a sample of gases being released into the atmosphere. The sample can be analyzed by direct measurement or by extraction and analysis in the field or in a laboratory. Flow rate measurements also are needed to establish the rate of a pollutant's release by the source. In a similar manner, the ambient air can be sampled and analyzed by extraction and analysis or by direct measurement.

AIR DISPERSION MODELING

The regulating agency can estimate the concentrations of pollutants from a source to which a community is exposed by performing mathematical dispersion modeling if they know the rate at which the pollutants are being released. They can also model the ACL backwards to establish the maximum allowable rate of release at the pollutant source.

The EPA has guidelines for using the most popular models (U.S. EPA 1986). Models are available for various meteorological conditions, terrains, and sources. Meteorological data are often difficult to obtain but crucial for accurate results from mathematical models.

CURRENT USE OF ACLs

The NAAQSs in Table 4.1.1 are ACLs derived from the best available data. State and local regulators have also used an array of ACLs for regulating toxic air pollutants. Examples are shown in Table 4.1.2.

—*William C. Zegel*

References

Chemical Manufacturers Association (CMA). 1988. *Chemicals in the community: Methods to evaluate airborne chemical levels.* Washington, D.C.

Patrick, D. R. ed. 1994. *Toxic air pollution handbook.* New York: Van Nostrand Reinhold.

U.S. Environmental Protection Agency (EPA). 1986. *Integrated Risk Information System (IRIS) database.* Appendix A, *Reference dose (RfD): Description and use in health risk assessments.* Washington, D.C.: Office of Health and Environmental Assessment.

———. 1989. *Exposure factors handbook.* EPA 600/8-89-043. Washington, D.C.: Office of Health and Environmental Assessment.

———. 1988. *Superfund exposure assessment manual.* EPA 540/1-88-001, OSWER Directive 9285.5-1. Washington, D.C.: Office of Emergency and Remedial Response.

4.2
TECHNOLOGY STANDARDS

Technology standards, used to control point and area sources of air pollutants, are based upon knowledge of the processes generating the pollutants, the equipment available to control pollutant emissions, and the costs of applying the control techniques. Technology standards are not related to ACLs but rather to the technology that is available to reduce pollution emissions. In the extreme, a technology standard could be to ban a process, product, or raw material.

Standards Development Process

In response to the requirements of the 1970 CAA Amendments, the EPA established a model process to develop technology standards. Because of their strong technological basis, technology standards are based on rigorous engineering and economic investigations. The EPA process consisted of three phases:

- Screening and evaluating information availability
- Gathering and analyzing data
- Making decisions

In the first phase, the regulating agency reviews the affected source category or subcategory, gathers available information, and plans the next phase. In the second phase, the processes, pollutants, and emission control systems used by facilities in this category are evaluated. This phase includes measuring the performance of emission control systems; developing costs of the control systems; and evaluating the environmental, energy, and economic effects associated with the control systems. Several regulatory alternatives are also selected and evaluated. In the third phase, regulators select one of the regulatory alternatives as the basis for the standard and initiate the procedures for rule making.

Elements of an Emission Standard

Emission standards must clearly define what sources are subject to it and what it requires. Standards should contain four main elements: applicability; emission limits; compliance procedures and requirements; and monitoring, reporting, and record-keeping requirements.

APPLICABILITY

The applicability provision defines who and what are subject to the emission standard requirements. This provision includes a definition of the affected source category or subcategory, the process or equipment included, and any size limitations or exemptions. Any distinction among classes, types, and sizes of equipment within the affected source category is part of the applicability.

EMISSION LIMITS

Emission limits specify the pollutant being regulated and the maximum permissible emission of that pollutant. In developing emission limits, regulators evaluate the performance, cost, energy, and environmental effects of alternate control systems. As a result of this evaluation, a control system is selected as the basis for the standard.

COMPLIANCE REQUIREMENTS

This part of the standard specifies the conditions under which the facility is operated for the duration of the compliance test. Generally, a facility is required to operate under normal conditions. Operation under conditions greater than or much less than design levels is avoided unless it represents normal operation.

This part of the standard also specifies the test methods to be used and the averaging time for the test. The test method is usually either reference, equivalent, or alternative. The reference method is widely known and is usually published as part of the regulations. An equivalent method is one that has been demonstrated to have a known, consistent relationship with a reference method. An alternative method is needed when the characteristics of individual sources do not lend themselves to the use of a reference or equivalent method. An alternative method must be demonstrated to produce consistent and useable results. Averaging time for an emission standard is important if the source is variable in its emissions. A short averaging time is more variable and more likely to exceed a standard than a long averaging time.

MONITORING, REPORTING, AND RECORD KEEPING

Monitoring, reporting, and record-keeping requirements ensure that the facility is operating within normal limits and that control equipment is being properly operated and maintained. Data are generally kept at the facility for review at any time, but regular reporting of critical data to the regulatory agency may be required.

Ambient Air Quality Standards

In accordance with the CAA, as amended, the EPA has established the NAAQS for criteria pollutants. The NAAQS is based on background studies, including information on health effects, control technology, costs, energy requirements, emission benefits, and environmental impacts.

The pollutants selected as criteria pollutants are sulfur dioxide, particulate matter (now PM_{10} and previously TSP or total suspended particulates), nitrogen oxides, carbon monoxide, photochemical oxidants (ozone), volatile organic compounds, and lead. The NAAQS represents the maximum allowable concentration of pollutants allowed in the ambient air at reference conditions of 25°C and 760 mm Hg. Table 4.1.1 shows the pollutant levels of the national primary and secondary ambient air quality standards.

States are responsible for ensuring that the NAAQS is met. They can establish statewide or regional ambient air quality standards that are more stringent than the national standards. To achieve and maintain the NAAQS, states develop state implementation plans (SIPs) containing emission standards for specific sources. When an area fails to meet an NAAQS, it is considered a nonattainment area. More stringent control requirements, designed to achieve attainment, must be applied to nonattainment areas.

The 1990 amendments to the CAA (1) require states to submit revised SIPs for nonattainment areas, (2) accelerate attainment timetables, and (3) require federally imposed controls if state nonattainment plans fail to achieve attainment. In addition, the amendments expand the number and types of facilities that are regulated under SIPs.

Hazardous Air Pollution Standards

The 1990 amendments to the CAA totally revise section 112 with regard to hazardous air pollutants, including national emission standards for hazardous air pollutants (NESHAP). They also direct the EPA administrator to establish standards that require the installation of MACT.

NESHAP

Although section 112 of the 1970 CAA granted the EPA broad authority to adopt stringent emission standards for hazardous air pollutants, as of this writing only seven pollutants are listed as hazardous air pollutants. These pollutants are beryllium, mercury, vinyl chloride, asbestos, benzene, radionuclides, and arsenic. Table 4.2.1 shows the NESHAP. Almost all these standards are technology standards.

MACT/GACT

A hazardous air pollutant is now defined as "any air pollutant listed pursuant to" section 112(b). In section 112(b), Congress established an initial list of 189 hazardous air pollutants. These listed chemicals are initial candidates for regulation under section 112, and the EPA can add other chemicals to the list.

The control of these substances is to be achieved through the initial promulgation of technology-based emission standards. These standards require major sources to install MACT and area sources to install generally available control technologies (GACT). Major sources are defined as those emitting more than 10 tons per year of any one hazardous air pollutant or more than 25 tons per year of all hazardous air pollutants. MACT/GACT standards

TABLE 4.2.1 NATIONAL EMISSION STANDARDS FOR HAZARDOUS AIR POLLUTANTS

Affected Facility	Emission Level	Monitoring
Asbestos		
Asbestos mills	No visible emissions or meet equipment standards	No requirement
Roadway surfacing	Contain no asbestos, except temporary use	No requirement
Manufacturing	No visible emissions or meet equipment standards	No requirement
Demolition/renovation	Wet friable asbestos or equipment standards and no visible emissions	No requirement
Spraying friable asbestos		
Equipment and machinery	No visible emissions or meet equipment standards	No requirement
Buildings, structures, etc.	<1 percent asbestos dry weight	No requirement
Fabricating products	No visible emissions or meet equipment standards	No requirement
Friable insulation	No asbestos	No requirement
Waste disposal	No visible emissions or meet equipment and work practice requirements	No requirement
Waste disposal sites	No visible emissions; design and work practice requirements	No requirement
Beryllium		
Extraction plants	1. 10 g/hour, or	1. Source test
Ceramic plants	2. 0.01 μ/m^3 (thirty-day)	2. Three years CEM[a]
Foundries		
Incinerators		
Propellant plants		
Machine shops (Alloy >5 percent by weight beryllium)		
Rocket motor test sites		
Closed tank collection of combustion products	75 μg min/m^3 of air within 10 to 60 minutes during two consecutive weeks	Ambient concentration during and after test
	2 g/hour, maximum 10 g/day	Continuous sampling during release
Mercury		
Ore processing	2300 g/24 hour	Source test
Chlor-alkali plants	2300 g/24 hour	Source test or use approved design, maintenance and housekeeping
Sludge dryers and incinerators	3200 g/24 hour	Source test or sludge test
Vinyl Chloride (VC)		
Ethylene dichloride (EDC) manufacturing	1. EDC purification: 10 ppm[b]	Source test/CEM[a]
	2. Oxychlorination: 0.2 g/kg of EDC product	Source test
VC manufacturing	10 ppm[b]	Source test/CEM[a]
Polyvinyl chloride (PVC) manufacturing		
Equipment	10 ppm[b]	Source test/CEM[a]
Reactor opening loss	0.02 g/kg	Source test
Reactor manual vent valve	No emission except emergency	

Continued on next page

TABLE 4.2.1 *Continued*

Affected Facility	Emission Level	Monitoring
Sources after stripper	Each calendar day: 1. Strippers—2000 ppm (PVC disposal resins excluding latex); 400 ppm other	Source test
	2. Others—2 g/kg (PVC disposal resins excluding latex); 0.4 g/kg other	Source test
EDC/VC/PVC manufacturing		
Relief valve discharge	None, except emergency	
Loading/unloading	0.0038 m³ after load/unload or 10 ppm when controlled	Source test
Slip gauge	Emission to control	
Equipment seals	Dual seals required	
Relief valve leaks	Rupture disc required	
Manual venting	Emissions to control	
Equipment opening	Reduce to 2.0 percent VC or 25 gallon	
Sampling (>10 percent by weight VC)	Return to process	
LDAR[d]	Approved program required	Approved program
In-process wastewater	10 ppm VC before discharge	Source test
Inorganic Arsenic		
Glass melting furnace	Existing: <2.5 Mg/year[c] or 85 percent control	Method 108
	New or modified: <0.4 Mg/year or 85 percent control	Continuous opacity and temperature monitor for control
Copper converter	Secondary hooding system Particle limit 11.6 mg/dscm[d]	Methods 5 and 108A Continuous opacity for control
	Approved operating plan	Airflow monitor for secondary hood
Arsenic trioxide and metallic arsenic plants using roasting/ condensation process	Approved plan for control of emissions	Opacity monitor for control Ambient air monitoring
Benzene		
Equipment leaks (Serving liquid or gas ≥10 percent by weight benzene; facilities handling 1000 Mg/ year and coke oven by-product exempt)	Leak is 10,000 ppm using Method 21; no detectable emissions (NDE) is 500 ppm using Method 21	
Pumps	Monthly LDAR,[e] dual seals, 95 percent control or NDE[f]	Test of NDE[f]
Compressors	Seal with barrier fluid, 95 percent control or NDE[f]	Test for NDE[f]
Pressure relief valves	NDE[f] or 95 percent control	Test for NDE[f]
Sampling connection systems	Closed purge or closed vent	
Open-end valves/lines	Cap, plug, or second valve	

Continued on next page

TABLE 4.2.1 *Continued*

Affected Facility	Emission Level	Monitoring
Valves	Monthly LDAR[e] (quarterly if not leaking for two consecutive months) or NDE[f]	Test for NDE[g]
Pressure relief equipment	LDAR[e]	
Product accumulators	95 percent control	
Closed-vent systems and control devices	NDE or 95 percent control	Monitor annually
Coke by-product plants		
Equipment and tanks	Enclose source, recover, or destroy. Carbon adsorber or incinerator alternate	Semiannual LDAR,[e] annual maintenance
Light-oil sumps	Cover, no venting to sump	Semiannual LDAR[e]
Napthalene equipment	Zero emissions	
Equipment leaks (serving ≥10 percent by weight)	See 40 CFR 61, subpart J.	
Exhauster (≥1 percent by weight)	Quarterly LDAR[e] or 95 percent control or NDE[f]	Test for NDE[f]
Benzene storage vessels		
Vessels with capacity >10,000 gallon	Equipped with:	
	1. Fixed roof with internal floating roof-seals, or	Periodic inspection
	2. External floating roof with seals, or	Periodic inspection
	3. Closed vent and 95 percent control	Maintenance plant and monitoring
Benzene transfer		
Producers and terminals (loading >1 300 000/year)	Vapor collection and 95 percent control	Annual recertification
Loading racks (marine rail, truck)	Load vapor-tight vessels only	Yes
Exemptions:		
Facilities loading <70 percent benzene		
Facilities loading less than required of >70 percent benzene		
Both of above subject to record-keeping		
Waste Operations	1. Facilities ≥10 Mg/year in aqueous wastes must control streams ≥10 ppm. Control to 99 percent or <10 ppm	Monitor control and treatment. Also, periodically monitor certain equipment for emissions >500 ppm and inspect equipment
Chemical manufacturing plants		
Petroleum refineries		
Coke by-product plants		
TSDF[g] treating wastes from the three preceding	2. If >10 ppm in wastewater treatment system: Wastes in <10 ppm Total in <1 Mg/year	
	3. >1 Mg/year to <10 Mg/year	Report annually
	4. <1 Mg facilities	One-time report
Radionuclides		
DOE facilities (radon not included)	10 mrem/year[h] radionuclides (any member of the public)	Approved EPA computer model and Method 114 or direct monitoring (ANSIN13.1-1969)

Continued on next page

TABLE 4.2.1 *Continued*

Affected Facility	Emission Level	Monitoring
NRC licensed facilities and facilities not covered by subpart H	10 mrem/year[h] radionuclides (any member of the public)	Approved EPA computer model or Appendix E
	3 mrem/year iodine (any member of the public)	Emissions determined by Method 114 or direct monitoring (ANSIN13.1-1969)
Calciners and nodulizing kilns at elemental phosphorus plants	2 curies per year (polonium-210)	Method 111
Storage and disposal facilities for radium-containing material, owned/operated by DOE	20 pCi/m² per second[i] (radon-222)	None specified
Phosphogypsum stacks (waste from phosphorus fertilizer production)	20 pCi/m² per second[i] (radon-222)	Method 115
Disposal of uranium mill tailings (operational)	20 pCi/m² per second[i] (radon-222)	Method 115

Source: Adapted from David R. Patrick, ed, 1994, *Toxic air pollution handbook* (New York: Van Nostrand Reinhold).

[a]CEM = continuous emission monitor.
[b]Before opening equipment, VC must be reduced to 2.0 percent (volume) or 25 gallons, whichever is larger.
[c]Mg/year = megagrams per year.
[d]mg/dscm = milligrams per dry standard cubic meter.
[e]LDAR = leak detection and repair.
[f]NDE = no detectable emissions.
[g]TSDF = treatment, storage, and disposal facilities.
[h]mrem/year = millirems per year (the rem is the unit of effective dose equivalent for radiation exposure).
[i]pCi/m² per second = picocuries per square meter per second.

are developed to control hazardous air pollutant emissions from both new and existing sources.

The 1990 amendments establish priorities for promulgating standards. The EPA, in prioritizing its efforts, is to consider the following:

The known or anticipated adverse effects of pollutants on public health and the environment

The quality and location of emissions or anticipated emissions of hazardous air pollutants that each category or subcategory emits

The efficiency of grouping categories or subcategories according to the pollutants emitted or the processes or technologies used

The EPA is to promulgate standards as expeditiously as practicable, but the 1990 amendments also established a minimum number of sources that must be regulated pursuant to a schedule. At this writing, standards for forty categories and subcategories are to be promulgated. The following standards are among those that have been promulgated:

On September 22, 1993, the EPA issued national emission standards for perchloroethylene (PCE) dry cleaning facilities

On October 27, 1993, coke oven battery standards were promulgated

On April 22, 1994, the EPA announced its final decisions on the hazardous organic NESHAP rule (HON), which requires sources to achieve emission limits reflecting the application of the MACT

By November 15, 1994, emission standards for 25 percent of the listed categories and subcategories were promulgated. Another 25 percent must be promulgated by November 15, 1997. All emission standards must be promulgated by November 15, 2000. Generally, existing sources must meet promulgated standards as expeditiously as practicable, but no later than three years after promulgation.

Other Technology Standards

Other technology standards include new source performance standards, best available control technology (BACT) and lowest achievable emission rate (LAER) standards, best available control technology for toxics (T-BACT) standards, and reasonably available control technology (RACT) standards. These standards are discussed next.

TABLE 4.2.2 NEW SOURCE PERFORMANCE STANDARDS FOR SOME SOURCES POTENTIALLY EMITTING TOXIC AIR POLLUTANTS

Source Category	Citation[a]	Pollutants Regulated[b]
Incinerators (>50 tons/day)	Subpart E	PM
Municipal waste combusters (>250 tons/day)	Subpart Ea	PM, organics, NO_x, acid gases
Portland cement	Subpart F	PM
Asphalt plants	Subpart I	PM
Petroleum refineries	Subpart J	PM, CO, SO_2, VOC
Petroleum storage vessels (>40,000 gallon)	Subpart K, Ka	VOC
Secondary lead smelters	Subpart L	PM
Secondary brass/bronze	Subpart M	PM
Basic oxygen furnaces	Subpart N, Na	PM
Sewage treatment plants	Subpart O	PM
Primary copper smelters	Subpart P	PM, SO_2
Primary zinc smelters	Subpart Q	PM, SO_2
Primary lead smelters	Subpart R	PM, SO_2
Primary aluminum reduction	Subpart S	Fluorides
Phosphoric acid plants	Subpart T	Fluorides
Superphosphate acid plants	Subpart U	Fluorides
Diammonium phosphate plants	Subpart V	Fluorides
Triple superphosphate plants	Subpart W	Fluorides
Triple superphosphate storage	Subpart X	Fluorides
Coal preparation plants	Subpart Y	PM
Ferroalloy production	Subpart Z	PM
Electric arc furnaces	Subpart AA, AAa	PM
Kraft pulp mills	Subpart BB	PM, TRS
Glass manufacturing	Subpart CC	PM
Surface coating—metal furniture	Subpart EE	VOC
Lime manufacturing	Subpart HH	PM
Lead-acid battery manufacturing	Subpart KK	Lead
Metallic minerals	Subpart LL	PM
Surface coating—automobiles and light-duty trucks	Subpart MM	VOC
Phosphate rock plants	Subpart NN	PM
Ammonium sulfate manufacture	Subpart PP	PM
Graphic arts and printing	Subpart QQ	VOC
Surface coating—tapes and labels	Subpart RR	VOC
Surface coating—large appliances	Subpart SS	VOC
Surface coating—metal coils	Subpart TT	VOC
Asphalt processing/roofing	Subpart UU	PM
Equipment leaks—organic chemical manufacturing industry	Subpart VV	VOC
Surface coating—beverage cans	Subpart WW	VOC
Bulk gasoline terminals	Subpart XX	VOC
Residential wood heaters	Subpart AAA	PM
Rubber tire manufacturing	Subpart BBB	VOC
Polymer manufacturing	Subpart DDD	VOC
Flexible vinyl and urethane coating and printing	Subpart FFF	VOC
Equipment leaks in petroleum refineries	Subpart GGG	VOC
Synthetic fiber production	Subpart HHH	VOC
Air oxidation processes—organic chemical manufacturing	Subpart III	VOC
Petroleum dry cleaners (dryer capacity ≥38 kg)	Subpart JJJ	VOC
Onshore natural gas processing		
Equipment leaks	Subpart KKK	VOC
SO_2 emissions	Subpart LLL	SO_2
Distillation processes—organic chemical manufacturing	Subpart NNN	VOC
Nonmetallic minerals	Subpart OOO	PM
Wool fiberglass insulation manufacturing	Subpart PPP	PM
Petroleum refinery wastewater	Subpart QQQ	VOC
Magnetic tape manufacturing	Subpart SSS	VOC
Surface coating—plastic parts for business machines	Subpart TTT	VOC

Source: David R. Patrick, ed, 1994, *Toxic air pollution handbook* (New York: Van Nostrand Reinhold).

[a]All citations are in the *Code of Federal Regulations*, Title 40, part 60.
[b]PM = particulate matter; CO = carbon monoxide; SO_2 = sulfur dioxide; NO_x = nitrogen oxides; VOC = volatile organic compounds; TRS = total reduced sulfur.

NEW SOURCE PERFORMANCE STANDARDS

Section 111 of the 1990 CAA Amendments authorizes the EPA to establish new source performance standards for any new stationary air pollution source category that causes, or significantly contributes to, air pollution that may endanger public health or welfare. The new source performance standards should reflect the degree of emission limitation achieved by applying the best demonstrated system of emission reduction. In considering the best, the EPA must balance the level of reduction against cost, other environmental and health impacts, and energy requirements. Table 4.2.2 presents a list of new source performance standards.

BACT/LAER

The CAA, as amended, provides for the prevention of significant deterioration (PSD) program. This program ensures that sources of air pollutants in relatively unpolluted areas do not cause an unacceptable decline in air quality. Under this program, no major source can be constructed or modified without meeting specific requirements, including demonstrating that the proposed facility is subject to the BACT for each regulated pollutant. A major source is one that emits more than 100 tons per year of regulated pollutants.

In nonattainment areas, proposed sources undergo a new source review. This review includes permits for the construction and operation of new or modified major sources that require the LAER. In nonattainment areas for ozone, a major source is one that emits as little as 10 tons of pollutant per year. The precise definition of a major source varies with the severity of ozone exceedances in the area.

Because BACT and LAER standards are determined on a case-by-case basis, no standards are published. The EPA has established the BACT/LAER Clearinghouse to assist in the consistent selection of BACT and LAER standards. This clearinghouse is designed to assist local and state regulatory agencies rather than industries.

T-BACT

Before enactment of the 1990 amendments to the CAA, many states developed programs for toxic air pollutants. Some states developed regulations that required new and modified sources of toxic air pollutants to minimize emissions by using T-BACT. These programs can be modified with EPA guidance from the 1990 amendments.

RACT/CTG

States with NAAQS exceedances have adopted and submitted SIPs to the EPA detailing how they plan to meet the NAAQS within a reasonable time. These SIPs require the installation of RACT for selected stationary sources. Regulating agencies determine RACT on a case-by-case basis within each industry, considering the technological and economic circumstances of the individual source. The EPA has issued a control techniques guideline (CTG) document to provide guidance on RACT for the control of volatile organic compound (VOC) emissions in nonattainment areas. The 1990 amendments require the EPA to issue CTGs within three years for eleven categories of stationary sources for which CTGs have not been issued.

—William C. Zegel

4.3
OTHER AIR STANDARDS

This section discusses other air standards including state and local air toxic programs and air toxics control in Japan and some European countries.

State and Local Air Toxics Programs

In 1989, the State and Territorial Air Pollution Program Administrators (STAPPA) and the Association of Local Air Pollution Control Officials (ALAPCO) conducted a comprehensive survey of state and local agency toxic air pol-

lution activities. This survey showed that every state had an air toxics program. The approaches used by states varied but generally fell into three categories:

- Formal regulatory programs
- Comprehensive policies
- Informal programs

The approaches used by local agencies are as diverse as the state programs, but they can be categorized similarly.

TABLE 4.3.1 HARMFUL SUBSTANCES IN JAPAN (JUNE 22, 1971)

Substance	Facility	Standard Value (mg/Nm³)
Cadmium and its compounds	Baking furnace and smelting furnace for manufacturing glass using cadmium sulfide or cadmium carbonate as raw materials.	1.0
	Calcination furnace, sintering furnace, smelting furnace, converter and drying furnace for refining copper, lead, or cadmium.	
	Drying facility for manufacturing cadmium pigment or cadmium carbonate.	
Chlorine and hydrogen chloride	Chlorine quick cooling facility for manufacturing chlorinated ethylene.	30 (chlorine)
	Dissolving tank for manufacturing ferric chloride.	80 (HCl)
	Reaction furnace for manufacturing activated carbon using zinc chloride.	
	Reaction facility and absorbing facility for manufacturing chemical products.	
	Waste incinerator (HCl)	700
Fluorine, hydrogen fluoride and silicon fluoride	Electrolytic furnace for smelting aluminium (harmful substances are emitted from discharge outlet).	3.0
	Electrolytic furnace for smelting aluminum (harmful substances are emitted from top).	1.0
	Baking furnace and smelting furnace for manufacturing glass using fluorite or sodium silicofluoride as raw material.	10
	Reaction facility, concentrating facility, and smelting furnace for manufacturing phosphoric acid.	
	Condensing facility, absorbing facility, and distilling facility for manufacturing phosphoric acid.	
	Reaction facility, drying facility, and baking furnace for manufacturing sodium triple-phosphate.	
	Reaction furnace for manufacturing superphosphate of lime.	15
	Baking furnace and open-hearth furnace for manufacturing phosphoric acid fertilizer.	20

Source: David L. Patrick, ed, 1994, *Toxic air pollution handbook* (New York: Van Nostrand Reinhold).

State and local programs are growing as the federal air toxics program, under Title III of the 1990 amendments, and the new federal and state operating permit program, established under Title V of the 1990 amendments, are fully implemented.

Air Toxics Control in Japan

Japan has taken strong steps to control what are known in the United States as criteria pollutants from both stationary and mobile sources, with the exception of lead. Lead is included in a group of special particulate pollutants. These special particulates include lead and its compounds; cadmium and its compounds; chlorine and hydrogen chloride; fluorine, hydrogen fluoride, and silicon fluoride. The emission standards for these four classes of pollutants are associated with categories of sources and are shown in Table 4.3.1.

Investigations of emission rates and the environmental effects of potentially toxic air pollutants are ongoing. Some substances have been found to have a long-term impact

on the environment, although present levels are not considered toxic. Japan has established regulations to control releases of asbestos and is examining other toxic materials for possible regulation, including various chlorinated volatile organics and formaldehyde.

Air Toxics Control in Some European Countries

Most western countries have some control program for U.S. criteria pollutants. Inter-country transport of air pollutants is a subject of study and concern. However, in most European countries, control of toxic air pollutants is not yet the subject of a regulatory program. Sweden has an action program to reduce or ban the use of harmful chemicals. The Swedes have identified thirteen compounds or categories of compounds for this program, including methylene chloride, trichloroethylene, tetrachloroethylene, lead and lead compounds, organotin compounds, chloroparaffins, phthalates, arsenic and its compounds, creosote, cad-

TABLE 4.3.2 DUTCH TARGET AND LIMIT VALUES FOR PRIORITY SUBSTANCE (QUANTITIES IN µg/m³ FOR AIR OR µg/l FOR WATER)

Substances	Target Value	Limit Value
Trichloroethane	50	50
Surface water	0.1	
Tetrachloroethane	25	2,000
Surface water	0.1	
Benzene	1	10
Phenol	1	100
Styrene	8	100
Acrylonitrile	0.1	10
Toluene		3 mg/m(3)
1,2-Dichlorethane	1	
Ethylene oxide	0.3	
Methylbromide	1 (year)	
	100 (hour)	
Vinyl chloride	1	
Propylene oxide	1	
Dichloromethane	20	
Trichloromethane	1	
Tetrachloroethane	1	
Epichlorohydrin	2	

Source: David R. Patrick, ed, 1994, *Toxic air pollution handbook* (New York: Van Nostrand Reinhold).

TABLE 4.3.3 EXAMPLE TOXIC AIR POLLUTION REGULATIONS IN SOME EUROPEAN COUNTRIES

Country	Air Toxic	Comments
France	Carcinogens	See Note 1
Germany	Volatile halogenated hydro carbons; 20 metals; various inorganics; organics	Specific industrial processes regulated by Technical Instructions on Air Quality Control
United Kingdom	Any pollutant	See Note 2
	Metals; metalloids; asbestos; halogens; phosphorous; and compounds	Local control required

Source: Private communication, Water and Air Research, 1994.

Note 1: Based on the 1982 Seveso Directive, a 28 December 1983 circular defines use of risk assessment; over 300 installations are subject to risk assessment studies.

Note 2: Integrated national control of processes with a potential for pollution to air, land, or water. Authorization is required; operator must install best practical means of control.

mium and its compounds, and mercury and its compounds.

The Netherlands has a national strategy to control toxic air pollutants. This strategy includes sustainable development through reducing to acceptable or negligible levels the risks posed to humans and the environment by one or more toxic substances. The Netherlands developed environmental standards based on all relevant information compiled in a basic document and established a no-effects level for human and ecosystem exposure. Table 4.3.2 shows the target value, which is generally below the no-effect level, and the limit value for the priority substances selected by the Dutch.

Table 4.3.3 summarizes other toxic air pollutant regulations in European countries.

—William C. Zegel

Noise Standards

4.4
NOISE STANDARDS

Sound is transmitted through the air as a series of compression waves. The energy of the noise source causes air molecules to oscillate radially away from the source. This oscillation results in a train of high-pressure regions following one another, travelling at a speed of approximately 760 miles per hour in sea-level air.

Noise can be described in terms of its loudness and its pitch, or frequency. Loudness is measured in decibels (dB). The dB scale, shown in Table 4.4.1, is a logarithmic scale—a 20 dB sound is ten times louder than a 10 dB sound. Pitch is a measure of how high or low a sound is. Pitch is measured in cycles per second (cps), or hertz (Hz). This measurement is the number of compression waves passing a point each second. The human ear is sensitive to sounds in the range of 20 to 20,000 Hz, but the ear is not as sensitive to low- and high-frequency sounds as it is to medium-frequency sounds (Figure 4.4.1).

Human Response to Noise

The ability of humans to hear decreases with age and exposure to noise. As we age, the organ that translates sound into nerve impulses slowly degenerates. Continuous exposure to loud noises can result in a permanent loss of hearing. Generally, the louder the noise, the less time it takes to induce a permanent hearing loss. Lower-frequency noise does less damage than higher-frequency sounds at the same level of loudness. However, even a partial hearing loss can severely impact an individual's ability to comprehend speech, negatively impacting that person's comfort level at social gatherings or when interacting with strangers. In children, hearing is important for learning language, and hearing loss can limit development.

Noise also affects sleep and stress levels, albeit more subtly than it affects hearing loss. Sleep disturbance can take the form of preventing sleep, making it difficult to fall asleep, causing a person to wake after falling asleep, or altering the quality of sleep. A high level of background noise, particularly if it is of variable levels, can change the stress and comfort levels of entire neighborhoods.

Wildlife Response to Noise

The effects of noise on wildlife are similar to its effects on humans. Additionally, noise can affect a creature's ability to obtain food or to breed. Some species that depend on detecting sounds and subtle differences in sound to locate food, to avoid becoming food, or to locate a mate may experience difficulties in high-noise environments. Short, loud noises that do not permanently affect a creature's hearing seem to have much less impact than steady background noise.

Occupational Noise Standards

The 1970 Occupational Safety and Health Act sets permissible limits on noise exposure for most commercial and industrial settings. Table 4.4.2 presents these limits. Exposure to impulsive or impact noise should not exceed a peak sound pressure of 140 dB. When a worker is exposed daily to more than one period of noise at different levels, these noise exposures can be compared to the standards in Table 4.4.2 by adding the ratio of the time allowed at the noise level to the time of exposure at that level for each period. If that sum is greater than one, then the mixed exposure exceeds the standards.

Land Use and Average Noise Level Compatibility

Noise conditions are characterized in terms of A-weighted decibels (dBA), using the following common descriptors: (1) equivalent sound level for twenty-four-hour periods, $L_{eq}(24)$, and (2) day–night sound level, L_{dn}. The former is a time-weighted average; the latter is weighted more heavily for noise during the night (for more detail, refer to Chapter 6).

In general, local ordinances regulate noise outside the workplace, usually as a nuisance, and normally do not have applicable standards. Similarly in most situations, no federal or state noise standards apply. Regulatory agencies have developed guidelines to assist in land use planning and in situating major facilities that generate significant

TABLE 4.4.1 A DECIBEL SCALE

Sound Intensity Factor	Sound Level (dB)	Sound Sources	Perceived Loudness	Effects Damage to Hearing	Community Reaction to Outdoor Noise
1,000,000,000,000,000,000	180—	• Rocket engine			
100,000,000,000,000,000	170—				
10,000,000,000,000,000	160—				
1,000,000,000,000,000	150—	• Jet plane at takeoff	Painful	Traumatic injury	
100,000,000,000,000	140—			Injurious range; irreversible damage	
10,000,000,000,000	130—	• Maximum recorded rock music			
1,000,000,000,000	120—	• Thunderclap • Textile loom • Auto horn, 1 meter away	Uncomfortably loud		
100,000,000,000	110—	• Riveter • Jet flying over at 300 meters			
10,000,000,000	100—	• Newspaper press		Danger zone; progressive loss of hearing	
1,000,000,000	90—	• Motorcycle, 8 meters away • Food blender • Diesel truck, 80 km/hr, at 15 meters away	Very loud	Damage begins after long exposure	Vigorous action
100,000,000	80—	• Garbage disposal			
10,000,000	70—	• Vacuum cleaner • Ordinary conversation	Moderately loud		Threats Widespread complaints
1,000,000	60—	• Air conditioning unit, 6 meters • Light traffic noise, 30 meters			Occasional complaints
100,000	50—	• Average living room			
10,000	40—	• Bedroom • Library	Quiet		No action
1,000	30—	• Soft whisper			
100	20—	• Broadcasting studio	Very quiet		
10	10—	• Rustling leaf	Barely audible		
1	0—	• Threshold of hearing			

Source: Turk et al, 1978, Environmental Science (Philadelphia: Saunders), 523.

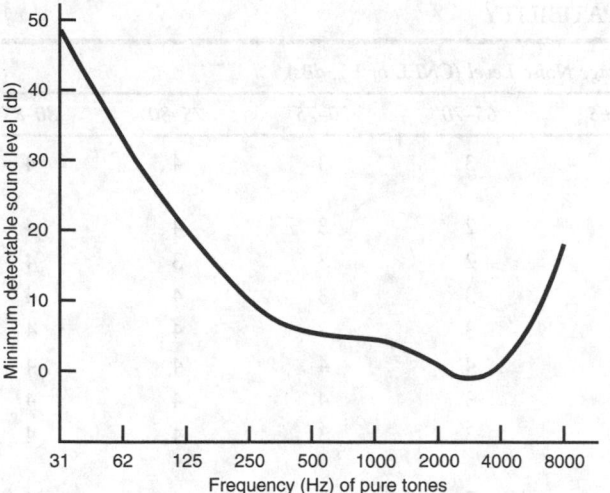

FIG. 4.4.1 Sensitivity of the human ear to various frequencies. Reprinted by permission from Daniel D. Chiras, 1985, *Environmental science*, Menlo Park, CA: Benjamin/Cummings Publishing Co.

TABLE 4.4.2 DAMAGE RISK CRITERIA FOR STEADY NOISE

Duration[a] Daily Exposure	White Noise (dBA)	1 Octave Bandwidth (dBA)	Pure Tone (dBA)
8 hr	90	85	80
4 hr	90	85	80
2 hr	92	87	82
1 hr	95	90	85
30 min	98	93	88
15 min	102	97	92
7 min	108	103	98
3 min	115	110	105
1½ min	125	120	115

Level (dB re 0.0002 dynes/cm²)

Source: B.G. Liptak, ed, 1974, *Environmental engineers' handbook*, Vol. 3 (Radnor, Penna.: Chilton Book Company).

[a]If ear protectors are not worn, even the shortest exposure is considered hazardous at levels above 135 dBA. If ear protectors *are* worn, no exposure to levels above 150 dB, however short, is considered safe. These criteria assume that hearing loss will be within acceptable limits if, after 10 years, it is no greater than 10 dB below 1000 Hz, 15 dB up to 2000 Hz, or 20 dB up to 3000 Hz.

levels of noise. The EPA guidelines were developed to protect public health and welfare (U.S. EPA 1974). These guidelines are summarized in Table 4.4.3.

The Department of Housing and Urban Development (HUD) has established guidelines for noise levels in residential areas. They define categories of acceptability as follows: acceptable if the L_{dn} is less than 65 dBA, normally unacceptable if the L_{dn} is greater than 65 dBA but less than 75 dBA, and unacceptable if the L_{dn} is greater than 75 dBA (HUD 1979). According to EPA studies, the majority of complaints occur when the L_{dn} exceeds 65 dBA (U.S. EPA 1973).

When land uses are noise sensitive, as with hospitals, parks, outdoor recreation areas, music shells, nursing homes, concert halls, schools, libraries, and churches, more restrictive guidelines are used. Conversely, less restrictive guidelines are used for commercial and agricultural land uses. For example, Table 4.4.4 shows a set of U.S. Navy noise guidelines for various land uses.

Traffic Noise Abatement

In America, a common source of community noise is automobile and truck traffic; yet by their nature, roads must be continuous and connected. Therefore, noise abatement strategies must be applied when the actual or projected noise from a highway exceeds its guidelines. This strategy can be considered a technology standard in that barriers, traffic management, alignment modifications, and landscaping have limited ability to reduce noise levels.

Community Exposure to Airport Noise

Aircraft can directly affect the noise levels of wide areas since no natural or manmade barriers are present. Usually, aircraft noise is infrequent and, when averaged over a twenty-four-hour period, is below guideline levels. However, near airports some areas have high noise levels, measured as $L_{eq}(24)$ or L_{dn}. In these areas, regulators can

TABLE 4.4.3 SUMMARY OF NOISE LEVELS IDENTIFIED AS REQUISITE TO PROTECT PUBLIC HEALTH AND WELFARE WITH AN ADEQUATE MARGIN OF SAFETY

Effect	Level	Area
Hearing loss	$L_{eq}(24) = 70$ dBA*	All areas.
Outdoor activity interference and annoyance	$L_{dn} = 55$ dBA	Outdoors in residential areas where people spend widely varying amounts of time and other places in which quiet is a basis for use.
	$L_{eq}(24) = 55$ dBA	Outdoor areas where people spend limited amounts of time, such as school yards and playgrounds.
Indoor activity interference and annoyance	$L_{dn} = 45$ dBA	Indoor residential areas.
	$L_{eq}(24) = 45$ dBA	Other indoor areas with human activities such as schools, etc.

Source: U.S. Environmental Protection Agency, 1974, *Information on levels of environmental noise requisite to protect public health and welfare with an adequate margin of safety*, EPA/550-9-74-004 (U.S. Environmental Protection Agency).
*Based on annual averages of the daily level over a period of 40 years.

TABLE 4.4.4 LAND USE AND AVERAGE NOISE LEVEL COMPATIBILITY

Land Use	Average Noise Level (CNEL or L_{dn} dBs)						
	50–55	55–60	60–65	65–70	70–75	75–80	80–85
Residential, single family, duplex mobile homes	1	1	2	3	3	4	4
Residential—multiple family	1	1	1	2	3	4	4
Transient lodging	1	1	1	2	3	3	4
Schools, libraries, churches	1	1	2	3	3	4	4
Hospitals, nursing homes	1	1	2	3	3	4	4
Music shells	2	2	3	4	4	4	4
Auditoriums, concert halls	1	2	3	3	4	4	4
Sport arenas, outdoor spectator sports	1	1	2	3	3	4	4
Parks, playgrounds	1	2	2	3	3	4	4
Natural recreation areas	1	1	2	2	2	4	4
Golf courses, riding stables, water recreation, cemeteries	1	1	2	2	3	3	4
Office buildings, personal, business and professional	1	1	1	2	2	3	4
Commercial, retail, movie theaters, restaurants	1	1	1	2	2	3	4
Commercial—wholesale, some retail, industrial, manufacturing	1	1	1	1	2	2	3
Livestock farming, animal breeding	1	1	1	1	2	3	4
Agriculture (except livestock), mining, fishing	1	1	1	1	2	3	4

Source: U.S. Navy, 1979.

Key:

1 = *Clearly Compatible*—The average noise level is such that indoor and outdoor activities associated with the land use can be carried out with essentially no interference from noise.

2 = *Normally Compatible*—The average noise level is great enough to be of some concern, but common building construction should make the indoor environment compatible with the usual indoor activities, including sleeping.

3 = *Normally Incompatible*—The average noise level is significantly severe so that unusual and costly building construction may be necessary to ensure an adequate environment for indoor activities. Barriers must be erected between the site and prominent noise sources to make the outdoor environment tolerable.

4 = *Clearly Incompatible*—The average noise level is so severe that construction costs to make the indoor environment acceptable for activities would probably be prohibitive. The outdoor environment would be intolerable for outdoor activities associated with the land use.

apply a type of technology standard by changing flight patterns and flight times to reduce noise impacts. New jet aircraft are required to use low-noise engines. The abatement strategy of insulating impacted structures against the intrusion of noise can also reduce the influence of aircraft noise.

Railroad Noise Abatement

Railroad traffic has noise characteristics that create special abatement problems. Safety horns and whistles are loud and designed to be heard; further, they must be sounded at specific locations. Trains can be long and can maintain a noise level for ten to twenty minutes. Because trains move twenty-four hours a day, night noise events are possible. The tracks are well established and cannot be easily moved.

All of these factors reduce abatement standards to the use of barriers, possibly with landscaping, and traffic management.

—*William C. Zegel*

References

Department of Housing and Urban Development (HUD). 1979. *Residential area noise level guidelines.* Department of Housing and Urban Development.

U.S. Environmental Protection Agency (EPA). 1973. *Public health and welfare criteria for noise.* EPA 550/9-73-002. Washington, D.C.: U.S. Environmental Protection Agency.

U.S. Environmental Protection Agency (EPA). 1974. *Information on levels of environmental noise requisite to protect public health and welfare with an adequate margin of safety.* EPA/550-9-74-004. U.S. Environmental Protection Agency.

Water Standards

4.5
WATER QUALITY STANDARDS

This section discusses the legislative activity, ACLs, technology standards, water quality goals, and toxic pollutants related to water quality standards.

Legislative Activity

The first substantive water pollution legislation in the United States, the Water Pollution Control Act, was passed in 1948. In 1956, the Federal Water Pollution Control Act, commonly called the Clean Water Act (CWA), provided the first long-term control of water pollution. The act has been amended several times. A key amendment in 1972 establishes a national goal of *zero discharge* by 1985. This concept refers to the complete elimination of all water pollutants from navigable waters of the United States. This amendment also called upon the EPA to establish effluent limitations for industries and make money available to construct sewage treatment plants. The amendments in 1977 direct the EPA to examine less common water pollutants, notably toxic organic compounds.

This legislation has resulted in the development of a complex series of water quality standards. These standards define the levels of specific pollutants in water that protect the public health and welfare and define the levels of treatment that must be achieved before contaminated water is released. The most notable of these standards are the water quality criteria set by the EPA. These criteria describe the levels of specific pollutants that ambient water can contain and still be acceptable for one of the following categories:

- Class A—water contact recreation, including swimming
- Class B—able to support fish and wildlife
- Class C—public water supply
- Class D—agricultural and industrial use

Water quality standards are set by the states and are subject to approval by the EPA. These standards define the conditions necessary to maintain the quality of water for its intended use. Per a provision of the CWA, existing uses of a body of water must be maintained (i.e., uses that downgrade water quality resulting in a downgraded use category are not allowed).

The primary enforcement mechanism established by the CWA, as amended in 1977, is the National Pollution Discharge Elimination System (NPDES). The NPDES is administered by the states with EPA oversight. Facilities that discharge directly into waters of the United States must obtain NPDES permits.

Under NPDES, permits for constructing and operating new sources and existing sources are subject to different standards. Discharge permits are issued with limits on the quantity and quality of effluents. These limits are based on a case-by-case evaluation of potential environmental impacts. Discharge permits are designed as an enforcement tool, with the ultimate goal of meeting ambient water quality standards.

Most states have assumed primary authority for the enforcement and permit activities regulated under the CWA. In those states that have not assumed primacy, discharges to surface waters require two permits, one from the EPA under CWA and one from the state under its regulations. In addition, such discharges are frequently regulated by local governments.

The EPA does not have permit responsibility under section 404 of the CWA, nor does it have responsibility for discharges associated with marine interests. Consequently, other federal water programs affect water quality. Table 4.5.1 summarizes selected regulations promulgated by the U.S. Army Corps of Engineers, the Coast Guard, and the EPA.

Figure 4.5.1 shows the relationship of water quality criteria, water quality standards, effluent guidelines, effluent limitations and permit conditions.

ACLs

Water quality standards are frequently expressed in terms of ambient concentration. The regulatory agency determines ACLs, which are the maximum concentration of a contaminant in water to which people are exposed. The degree of human exposure depends upon the use of the water body. Thus, different ACLs apply to different water bodies. Generally, regulators derive ACLs from health effects information. Using ACLs, regulators can mathematically determine the maximum contribution that an ef-

TABLE 4.5.1 CLEAN WATER REGULATIONS

Agency/Reference	Topic
CG: 33CFR	
153–157	Oil Spills
159	Marine Sanitation Devices
COE: 33CFR	
209	Navigable Waters
320–330	Permit Programs
EPA: 40CFR	
109	Criteria for State, Local, and Regional Oil Removal Contingency Plans
110	Discharge of Oil
112	Oil Pollution Prevention
113	Liability Limits for Small Onshore Oil Storage Facilities
114	Civil Penalties for Violations of Oil Pollution Prevention Regulations
116	Designation of Hazardous Substances
117	Determination of Reportable Quantities for Hazardous Substances
121	State Certification of Activities Requiring a Federal License or Permit
122	NPDES Permit
123	State NPDES Permit Program Requirements
125	Criteria and Standards for the National Pollutant Discharge Elimination System
130	Water Quality Planning and Management
131	Approving State Water Quality Standards
133	Secondary Treatment Information
136	Test Procedures for the Analysis of Pollutants
140	Performance Standards for Marine Sanitation Devices
141	National Primary Drinking Water Regulations
142	Primary Drinking Water Implementation Regulations
143	National Secondary Drinking Water Regulations
220–225, 227–229	Ocean Dumping Regulations and Criteria
230	Discharge of Dredge or Fill Material into Navigable Waters
231	Disposal Site Determination Under the CWA
233	State Dredge or Fill (404) Permit Program Requirements
403	Pretreatment Standards

Source: Compiled from Code of Federal Regulations.
Abbreviations: CG = Coast Guard; COE = Corps of Engineers; EPA = Environmental Protection Agency.

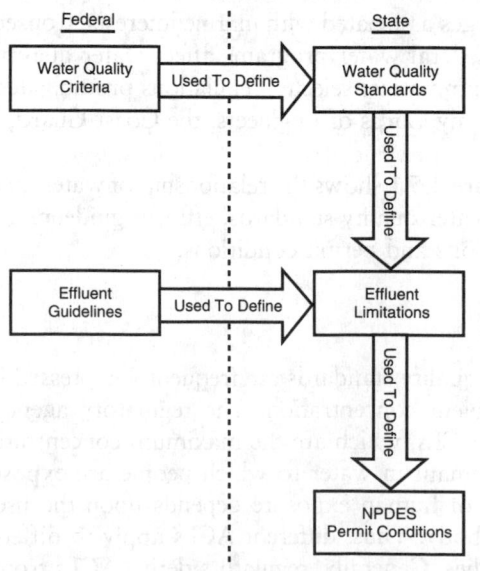

FIG. 4.5.1 Relationship of elements used in defining NPDES permit conditions.

fluent source makes to a water body without violating relevant ACL(s).

Technology Standards

Permits for discharges of pollutants into the waters of the United States are subject to the NPDES effluent limits and permit requirements. Such source standards are generally based upon the best available technology (BAT). These technology standards can be effluent guidelines, effluent limits, and the definition of BAT.

Effluent guidelines define uniform national guidelines for specific pollutant discharges for each type of industry regulated. These are not, in fact, guidelines but are regulatory requirements. Federal effluents guidelines and standards cover more than fifty industrial categories, as shown in Table 4.5.2.

Effluent limits are specific control requirements that apply to a specific point–source discharge. They are based

TABLE 4.5.2   CATEGORICAL INDUSTRIAL
EFFLUENT GUIDELINES AND
STANDARDS

40 CFR Part	Source
405	Dairy Products
406	Grain Mills
407	Canned and Preserved Fruits and Vegetables
408	Canned and Preserved Seafood
409	Sugar Processing
410	Textiles
411	Cement Manufacturing
412	Feedlots
413	Electroplating
414	Organic Chemicals
415	Inorganic Chemicals
417	Soaps and Detergents
418	Fertilizer Manufacturing
419	Petroleum Refining
420	Iron and Steel Manufacturing
421	Nonferrous Metals
422	Phosphate Manufacturing
423	Steam Electric Power Generating
424	Ferroalloy Manufacturing
425	Leather Tanning and Finishing
426	Glass Manufacturing
427	Asbestos Manufacturing
428	Rubber Processing
429	Timber Products
430	Pulp, Paper, and Paper Board
431	Builders Paper and Board Mills
432	Meat Products
433	Metal Finishing
435	Offshore Oil and Gas Extraction
436	Mineral Mining and Processing
439	Pharmaceutical Manufacturing
440	Ore Mining and Dressing
443	Paving and Roofing Materials
446	Paint Formulating
447	Ink Formulating
454	Gum and Wood Chemicals Manufacturing
455	Pesticides Chemicals Manufacturing
457	Explosives Manufacturing
458	Carbon Black Manufacturing
459	Photographic Processing
460	Hospital
461	Battery Manufacturing Point Source Category
463	Plastics Molding and Forming
464	Metal Molding and Casting
465	Coil Coating
466	Porcelain Enameling
467	Aluminum Forming
468	Copper Forming
469	Electrical and Electronic Components
471	Nonferrous Metals Forming and Metal Powders

Source: Compiled from the Code of Federal Regulations.

on both national effluent guidelines and state water quality standards, as shown in Figure 4.5.1.

In some cases, a standard consists of a treatment technology that the regulatory agency accepts as the BAT for that type of source or a unique combination of source and receiving water body.

Water Quality Goals

The discharge standards under the NPDES refer to specific potential contaminants. A series of water quality goals are associated with each potential contaminant. Goals for a specific situation depend on the established use of the water body. The regulated contaminants vary from state to state. Schultz has classified more than fifty water quality parameters into four groups based on the frequency of their use in state ACLs and associated NPDES water quality standards (Schultz 1972).

All state water quality standards classify the following nine parameters: dissolved oxygen (D), pH, coliform, temperature, floating solids (oil–grease), settleable solids, turbidity–color, taste–odors, and toxic substances. In 50 to 99 percent of the state standards, three groups of parameters are categorized. In most regions, these frequently sampled parameter groups (radioactivity, total dissolved solids, and U.S. Public Health Service Drinking Water Standards) are sampled less frequently than the first nine. Sixteen parameters, eleven of which are heavy metals and other toxic substances, are found in the 20 to 49 percent of the state standards. Eighteen parameters appear in less than 20 percent of the state standards.

Table 4.5.3 shows the optimum and maximum values of water quality characteristics related to type of use published by California as an example of water quality goals.

Effluent Standards

NPDES permits are issued to municipal and industrial discharge sources to ensure that they do not violate water quality standards. In addition, state and federal monitoring, inspection, and enforcement ensures compliance with standards and permits.

MUNICIPAL EFFLUENT LIMITS

Municipal effluent limits are less complex than industrial limits. All publicly owned treatment works must meet a secondary treatment level (Table 4.5.4). This treatment level implies the following technologies: mechanical removal of solids by screening and settling, removal of additional organic wastes and solids by treating the waste with air or oxygen and allowing bacteria to consume the organic chemicals, and chlorination. Table 4.5.5 summarizes the requirements of this program.

TABLE 4.5.3 OPTIMUM AND MAXIMUM VALUES OF WATER QUALITY CHARACTERISTICS IN RELATION TO TYPE OF BENEFICIAL USE

Characteristics	Domestic Water Supply	Recreation			Wildlife Propagation				Irrigation			Industrial				Aesthetic Enjoyment
		Bathing and Swimming		Boating and Fishing	Fish		Foul Refuge	Shellfish Culture	Truck Garden	Citrus Fruits	Other Crops	Food Processing		Cooling and Other		
		Fresh Water	Salt Water		Fresh Water	Salt Water			Vegetables			Fresh Water	Salt Water	Fresh Water	Salt Water	
1. Bacterial—per ml.																
Coliform (opt.)	1.0	none	1.0	10	10	10	100	1.0	1.0	10	100	0.1	1.0	1.0	10	
Coliform (max.)	50	1.0	10	100	100	100	1,000	5	10	100	100	1.0	3.0	10	100	
2. Organic—ppm.																
B.O.D. (opt.)	none	5	5	10	10	10	10	5				none	1	5	5	20
B.O.D. (max.)	0.5	10	10	30	30	30	50	20				5	10	10	20	100
D.O. (opt.)	5	5	5	5	5	5	5	5				5	5	3.0	3.0	5.0
D.O. (min.)	2	2	2	2	3	2	2	2				1	1	1.0	1.0	1.0
Oil (opt.)	none	none	none	none	none	none	none	none	none	none	none	none	none	5	5	none
Oil (max.)	2	2	2	5	5	5	5	2	5	5	5	2	5	10	10	10
3. Reaction																
pH (opt.)	6.8–7.2	6.8–7.2	6.8–7.2		6.5–8.5	6.5–8.5	6.5–8.5	6.8–7.2	6.5–8.5	6.5–8.5	6.5–8.5	6.5–8.5	6.5–8.5	4.0–10.0	4.0–10.0	
pH (critical)	6.6–8.0	6.5–8.6	6.5–8.6		6.5–8.5	6.5–8.5	6.5–8.5	6.6–8.0	6.0–9.0	6.0–9.0	6.0–9.0	6.0–9.0	6.0–9.0	4.0–10.0	4.0–10.0	
4. Physical—ppm.																
Turbid. (opt.)	5	5	5	10	5	5	10	5	5			5	5			50
Turbid. (max.)	20	20	30	50	10	20	100	50	20			20	50			
Color (opt.)	10	10	10	10	5	10	10	10	10			10	10			20
Color (max.)	30	30	30	50	10	20	100	50	30			30	50			100
Susp. solids (opt.)	10	50	50		10	10	50	10				10	10	50	50	
Susp. solids (max.)	100	100	100		20	50	250	100				50	100	150	150	
Float. solids (opt.)	none	none	none	none	none	none	slight	none				none	none	none	none	slight
Float. solids (max.)	gross	gross	gross	gross	gross	gross	gross	gross				slight	slight	slight	slight	gross

5. Chemical—ppm.

Total solids (opt.)	500	1000	500	500	500	500	500
Total solids (max.)	1500	5000	1500	1500	2000	1500	1500
Cl (opt.)	250	1000	100	250	250	100	500
Cl (max.)	750	2500	500	750	750	500	1000
F (opt.)	0.5–1.0			0.5–1.0			
F (max.)	1.5			5			
Toxic metals (opt.)	none	0.5	0.1	5	0.1	none	none
Toxic metals (max.)	0.05	10	5	2.5	10	0.1	0.1
Phenol (opt.)	1*	0.1	1*	5*	1*	1*	0.5
Phenol (max.)	5*	1	50*	20*	50*	10*	5*
Boron (opt.)			0.5	0.5	0.5		
Boron (max.)			1.0	1.0	5		
Na ratio† (opt.)		35–50†	35–50†	35–50†	35–50†	90†	90†
Na ratio† (max.)		80†	75†	80†	80†	90†	90†
Hardness (opt.)	100					100	100
Hardness (max.)	250					500	500
6. Temp.—°F. (max.)	65	60	60	60	60	70	O
7. Odor‡ (max.)	N	N	M	M	O	N	O
8. Taste‡ (max.)	M	M	M	M	O	N	O

Source: California State Water Pollution Control Board, 1952.
*Parts per billion.
†Percent.
‡Key: D—disagreeable; M—marked; N—noticeable; O—obnoxious.

TABLE 4.5.4 NATIONAL PRETREATMENT STANDARDS

1. General Prohibitions: A user may not introduce into a publicly owned treatment works, POTW, any pollutant(s) which cause "pass through" without a change in nature or "interference" with treatment processes, including sludge use or disposal.
2. Specific Prohibitions: The following pollutants shall not be introduced into a POTW:
   a. Pollutants which create a fire or explosion hazard in the POTW;
   b. Pollutants which will cause corrosive structural damage to the POTW, but in no case discharges with pH lower than 5.0, unless the works is specifically designed to accommodate such discharges;
   c. Solid or viscous pollutants in amounts which will cause obstruction to the flow in the POTW resulting in interference;
   d. Any pollutant, including oxygen, demanding pollutants (BOD, etc.) released in a discharge at a flow rate and/or pollutant concentration which will cause interference with the POTW;
   e. Heat in amounts which will inhibit biological activity in the POTW resulting in interference, but in no case heat in such quantities that the temperature at the POTW exceeds 40° C (104° F) unless the approval authority, upon request of the POTW, approves alternate temperature limits.
3. Categorical Requirements: Standards specifying quantities or concentrations of pollutants or pollutant properties which may be discharged to a POTW by existing or new industrial users in specific industrial subcategories will be established as separate regulations under the appropriate subpart of 40 CFR Chapter 1, Subchapter N. These standards, unless specifically noted otherwise, shall be in addition to the general prohibitions established above (40 CFR 403.5).

Source: U.S. Environmental Protection Agency, 1973–1985, Secondary treatment regulation, *Code of Federal Regulations,* Title 40, part 133 (Washington, D.C.: U.S. Government Printing Office).

INDUSTRIAL EFFLUENTS

Industrial effluents are subject to state and federal requirements (see Table 4.5.2). States must tailor effluents discharge programs to water quality goals. Regulating agencies must monitor each segment of a water body at specified intervals and evaluate the impact of each discharger. Based on the evaluation, each water body is put into one of two groups:

Effluent limited—Water bodies for which the standards are met or will be met if the nationwide limits are implemented

Water quality limited—Water bodies that cannot meet the quality standards even when nationwide limits are applied. Effluent requirements must be tailor-made for each discharger along this group of water.

The NPDES requires all dischargers to file a standardized report with the EPA and their state agency. This report indicates the quality and quantity of their discharges. Where appropriate, the state and the EPA can respond to the report with a set of effluent limits, an abatement schedule to meet the effluent limits, and a monitoring schedule. A permit is issued upon agreement by the discharger, interested citizens, the state regulatory agency, and the EPA.

STORM WATER DISCHARGE

Storm water is defined as storm water runoff, surface runoff, and drainage. The regulations apply only to point-source discharges. A permit is required for both direct discharges into U.S. waters and for indirect discharges into a facility's drainage system or a separate city storm water system. However, an industry that separates its nonindustrial storm water, such as drainage from office buildings and parking lots, from the plant's industrial storm water does not have to include the nonindustrial areas in their permit.

These industrial discharges must meet effluent standards under sections 301 and 402 of the CWA, including the use of BAT and best practicable control technology (PCT). Industrial discharges must apply best management practices (BMPs) to control and reduce storm water discharges. These standards are technological and are measures and controls designed to eliminate or minimize pollutant loadings in storm water discharges.

General BMPs include good housekeeping, preventative maintenance, visual inspection, spill prevention and response, sediment erosion prevention, management of runoff, employee training, and record keeping and re-

TABLE 4.5.5 SECONDARY TREATMENT REQUIREMENTS

Pollutant		Effluent Limitations*
BOD₅†	30 (45) mg/L	Maximum 30-day average
	45 (65) mg/L	Maximum 7-day average
	85% (65) removal	Minimum 30-day average
Suspended Solids	30 (45) mg/L	Maximum 30-day average
	45 (65) mg/L	Maximum 7-day average
	85% (65) removal	Minimum 30-day average
pH	6.0–9.0	Range

Source: U.S. Environmental Protection Agency, 1973–1985, Secondary treatment regulation, *Code of Federal Regulations,* Title 40, part 133 (Washington, D.C.: U.S. Government Printing Office).

*( ) denotes value applicable to treatment equivalent to secondary treatment. Adjustment available for effluents from trickling filter facilities and waste stabilization pond facilities.

†If CBOD₅ is approved substitute for BOD₅, the CBOD₅ limitations are:
40 mg/L   Maximum 30-day average
60 mg/L   Maximum 7-day average
65% removal Minimum 30-day average

TABLE 4.5.6 PRIORITY POLLUTANTS*

#	Pollutant	#	Pollutant	#	Pollutant
1	*Acenaphthene	40	4-chlorophenyl phenyl ether	85	*Tetrachloroethylene
2	*Acrolein	41	4-bromophenyl phenyl ether	86	*Toluene
3	*Acrylonitrile	42	Bis(2-chloroisopropyl) ether	87	*Trichloroethylene
4	*Benzene	43	Bis(2-chloroethoxy) methane *halomethanes (other than those listed elsewhere	88	*Vinyl chloride (chloro-ethylene) pesticides and metabolites
5	*Benzidine	44	Methylene chloride (dichloromethane)	89	*Aldrin
6	*Carbon tetrachloride (tetrachloromethane) *chlorinated benzenes (other than dichlorobenzenes)	45	Methyl chloride (chloromethane)	90	*Dieldrin
7	Chlorobenzene	46	Methyl bromide (bromomethane)	91	*Chlordane (technical mixture and metabolite DDT and metabolites)
8	1,2,4 trichlorobenzene	47	Bromoform (tribromomethane)	92	4,4'-DDT
9	Hexachlorobenzene *chlorinated ethanes (including 1,2-dichloroethane, 1,1,1-trichloroethane and hexachloroethane	48	Dichlorobromomethane	93	4,4'-DDE (p,p'-DDX)
10	1,2-dichloroethane	49	Trichlorofluoromethane**	94	4,4'-DDD (p,p'-TDE) *endosulfan & metabolite
11	1,1,1-trichloroethane	50	Dichlorodifluoromethane**	95	A-endosulfan-Alpha
12	Hexachloroethane	51	Chlorodibromomethane	96	B-endosulfan-Beta
13	1,1-dichloroethane	52	*Hexachlorobutadiene	97	Endosulfan sulfate *endrin and metabolites
14	1,1,2-trichloroethane	53	*Hexachlorocyclopentadiene	98	Endrin
15	1,1,2,2-tetrachloroethane	54	*Isophorone	99	Endrin aldehyde *heptachlor and metabolites
16	Chloroethane *chloroalkyl ethers (chloromethyl, chlorecethyl & mixed ethers)	55	*Naphthalene	100	Heptachlor
17	bis(chloromethyl) ether**	56	*Nitrobenzene *nitrophenols (including 2,4-dinitrophenol and dinitrocresol)	101	Heptachlor epoxide *hexchlorocyclohexane (all isomers)
18	Bis(2-choloroethyl) ether	57	2-Nitrophenol	102	a-BHC-Alpha
19	2-chloroethyl vinyl ether (mixed) *chlorinated naphthalene	58	4-Nitrophenol	103	b-BHC-Beta
20	2-chloronaphthalene *chlorinated phenols (other than those listed elsewhere; includes trichlorophenols and chlorinated cresols)	59	*2,4-dinitrophenol	104	r-BHC-(lindane)-Gamma
21	2,4,6-trichlorophenol	60	4,6-dinitro-o-cresol *nitrosamines	105	g-BCH-Delta *polychlorinated biphenyls (PCB's)
22	Parachlorometa cresol	61	N-nitrosodimethylamine	106	PCB-1242 (Arochlor 1242)
23	*Chloroform (trichloromethane)	62	N-nitrosodiphenylamine	107	PCB-1254 (Arochlor 1254)
24	*2-chlorophenol *dichlorobenzenes	63	N-nitrosodi-n-propylamine	108	PCB-1221 (Arochlor 1221)
25	1,2-diclorobenzene	64	*Pentachlorophenol	109	PCB-1232 (Arochlor 1232)
26	1,3-dichlorobenzene	65	*Phenol *phthalate esters	110	PCB-1248 (Arochlor 1248)
27	1,4-cichlorobenzene *dichlorobenzidine	66	Bis(2-ethylhexyl) phthalate	111	PCB-1260 (Arochlor 1260)
28	3,3'-dichlorobenzidine *dichloroethylenes (1,1-dichloroethylene and 1,2-dichloroethylene)	67	Butyl benzyl phthalate	112	PCB-1016 (Arochlor 1016)
29	1,1-dichloroethylene	68	Di-n-butyl phthalate	113	*Toxaphene
30	1,2-trans-dichloroethylene	69	Di-n-octyl phthalate	114	*Antimony (total)
31	*2,4-dichlorophenol *dichloropropane and dichloropropene	70	Diethyl phthalate	115	*Arsenic (total)
32	1,2-dichloropropane	71	Dimethyl phthalate *polynuclear aromatic hydrocarbons	116	*Asbestos (fibrous)
33	1,2-dichloropropylene (1,3-dichloropropene)	72	Benzo(a)anthracene (1,2-benzanthracene)	117	*Beryllium (total)
34	*2,4-dimethylphenol *dinitrotoluene	73	Benzo(a)pyrene (3,4-benzopyrene)	118	*Cadmium (total)
35	2,4-dinitrotoluene	74	3,4-benzofluoranthene	119	*Chromium (total)
36	2,6-dinitrotoluene	75	Benzo(k)fluoranthene (11,12-benzofluoranthene)	120	Copper (total)
37	*1,2-diphenylhydrazine	76	Chrysene	121	*Cyanide (total)
38	*Ethylbenzene	77	Acenaphthylene	122	*Lead (total)
39	*Fluoranthene *haloethers (other than those listed elsewhere	78	Anthracene	123	*Mercury (total)
		79	Benzo(ghi)perylene (1,1-benzoperylene)	124	*Nickel (total)
		80	Fluorene	125	*Selenium (total)
		81	Phenanthrene	126	*Silver (total)
		82	Dibenzo(a,h)anthracene (1,2,5,6-dibenzanthracene)	127	*Thallium (total)
		83	Indeno (1,2,3-cd)pyrene (2,3-o-phenylenepyrene)	128	*Zinc (total)
		84	Pyrene	129	2,3,7,8-tetrachlorodibenzo-p-dioxin (TCDD)

Source: E.J. Shields, 1985, *Pollution Control Engineers' Handbook* (Northbrook, Ill.: Pudvan).

*The original list consisted of 65 specific compounds and chemical classes, indicated in bold type. When various forms of certain categories were broken out, the original list included 13 metals, 114 organic chemicals, asbestos, and cyanide. Three organic chemicals (**) were deleted in 1981.

211

TABLE 4.5.7 TOXIC POLLUTANT EFFLUENT STANDARDS

Toxic Pollutant	Source	Effluent Limitation	
Aldrin/Dieldrin	Manufacturer	Prohibited	
	Formulator	Prohibited	
DDT	Manufacturer	Prohibited	
	Formulator	Prohibited	
Endrin	Manufacturer Existing	1.5 μg/L	(A)
		0.0006 kg/kkg	(B)
		7.5 μg/L	(C)
	New	0.1 μg/L	(A)
		0.00004 kg/kkg	(B)
		0.5 μg/L	(C)
	Formulator	Prohibited	
Toxaphene	Manufacturer Existing	1.5 μg/L	(A)
		0.00003 kg/kkg	(B)
		7.5 μg/L	(C)
	New	0.1 μg/L	(A)
		0.000002 kg/kkg	(B)
		0.5 μg/L	(C)
	Formulator	Prohibited	
Benzidine	Manufacturer	10 μg/L	(A)
		0.130 kg/kkg	(B)
		50 μg/L	(C)
	Applicator	10 μg/L	(A)
		25 μg/L	(C)

Source: U.S. Environmental Protection Agency, 1977, Toxic pollutant effluent standards, *Code of Federal Regulations,* Title 40, part 129 (Washington, D.C.: U.S. Government Printing Office).

(A) An average per working day, calculated over any calendar month.
(B) Monthly average daily loading per quantity of pollutant produced.
(C) A sample(s) representing any working day.

porting. These BMPs can be applied to a variety of industrial facilities and are considered to be *source reduction* practices. These BMPs eliminate or reduce the pollutant generation at the source.

The next best BMP options include reuse or recycling of storm water in the industrial processes, followed by runoff management controls, such as vegetative swales and infiltration devices. Finally, if none of these applications control the storm water pollutants, regulators should explore treatment options. Treatment options include engineered systems, such as oil–water separators, sedimentation tanks, and metal precipitation systems, depending on the pollutants to be removed.

Toxic Pollutants

The CWA regulates many toxic pollutants through NPDES permits. Table 4.5.6 lists priority pollutants that the EPA has designated as toxic. Effluent standards have been promulgated for some of these pollutants, listed in Table 4.5.7.

The water environment can be monitored for toxic pollutants in the following three ways:

By biological and chemical analyses of the water
By a bioassay in which organisms are placed in water samples and their reaction is compared with control. This method is receiving attention for monitoring toxic discharges.
By plants and animals that are reactive to types and degrees of pollution. The absence of organisms known to be intolerant of pollution also serves as an indication of pollution; the presence and expansion of pollution-tolerant organisms is a related index of pollution.

—*William C. Zegel*

References

Schultz, S. 1972. *Design of USAF water quality monitoring program.* AD-756-504, Springfield, Va.: National Technical Information Service.

4.6
DRINKING WATER STANDARDS

The Safe Drinking Water Act (SDWA) is designed to achieve uniform safety and quality of drinking water in the United States. To achieve this goal, the SDWA identifies contaminants and establishes maximum acceptable levels for those contaminants. The major provisions of the act with respect to establishing drinking water quality standards are (1) the establishment of primary regulations to protect public health and (2) the establishment of secondary regulations that are related to the taste, odor, and appearance of drinking water.

Drinking Water Regulation

The EPA's philosophy in setting drinking water regulation is to initially assess the potential for harm and determine the feasibility of attainment (57 *FR* pt. 219 [13 November 1985] 46936). Regulatory actions must be scientifically, legally, defensibly, technically, and economically feasible. This feasibility requires careful study and analysis and extensive communication with those affected by the regulations (i.e., state agencies, public water supplies, and the scientific community), as well as other interested parties in the public sector.

Terms such as *regulations, standards, goals, guidelines, criteria, advisories, limits, levels,* and *objectives* describe numerical or narrative qualities of drinking water that protect public health. Distinctions between these terms generally fall into either (1) legally enforceable concentrations (regulations or standards) or (2) concentrations that represent desirable water quality but are not enforceable (goals or criteria). In addition, both regulations and goals usually represent either (1) a health-related NOAEL or (2) a level representing a balance between health risks and the feasibility of achieving these levels.

MAXIMUM CONTAMINANT LEVEL GOALS

Maximum contaminant level goals (MCLGs) must be set at a level at which "no known or anticipated adverse effects on the health of persons occur and which (sic) allows an adequate margin of safety" (Cotruvo 1987, 1988). First, the highest NOAEL is based upon an assessment of human or animal data (usually from animal experiments). To determine the RfD for regulatory purposes, the NOAEL is divided by an uncertainty factor (UF). This process corrects for the extrapolation of animal data to human data, the existence of weak or insufficient data, and individual differences in human sensitivity to toxic agents, among other factors.

For MCLG purposes, the NOAEL must be measurable in terms of concentration in drinking water (e.g., milligrams per liter). An adjustment of the RfD, which is reported in milligrams per kilograms (body weight) per day to milligrams per liter, is necessary. This adjustment is made by factoring in a reference amount of drinking water consumed per day and a reference weight for the consumer. The NOAEL, in milligrams per liter, is called the drinking water equivalent level (DWEL).

DWELs are calculated as follows:

$$DWEL = \frac{[NOAEL \text{ in } mg/kg \cdot day](70 \text{ kg})}{(UF)(2 \text{ L/day})} \qquad 4.6(1)$$

where:

NOAEL = no observed adverse effect level
70 kg   = assumed weight of an adult
2 L/day = assumed amount of water consumed by an adult per day
UF   = uncertainty factor (usually 10, 100, or 1000)

(Note: An uncertainty factor of 10 is used when good acute or chronic human exposure data are available and supported by acute and chronic data in other species; an uncertainty factor of 1000 is used when acute or chronic data in all species are limited or incomplete [National Academy of Science, 1977]).

To determine the MCLG, regulators account for the contribution from other sources of exposure, including air and food. Comprehensive data are usually not available on exposures from air and food. In this case, the MCLG is determined by

MCLG = DWEL
$\qquad$ × (percentage of the drinking water contribution) **4.6(2)**

When specific data are not available, regulators often use a 20 percent drinking water contribution.

Maximum contaminant levels (MCLs) are set "as close to" the MCLGs "as is feasible." The term *feasible* means feasible with the use of the BAT, which is a technology standard rather than an ACL, taking costs into consideration (Cotruvo 1987, 1988). The general approach to setting MCLs is to determine the feasibility of controlling contaminants. This approach requires (1) evaluating the availability and cost of analytical methods, (2) evaluating the availability and performance of technologies and other factors related to feasibility and identifying those that are best, and (3) assessing the costs of applying the technologies to achieve various concentrations.

EPA PROCESS FOR SETTING STANDARDS

EPA regulation development involves interpreting the mandates and directives of the SDWA, performing technical and scientific assessments to meet SDWA requirements, preparing regulations that blend technical and scientific aspects with policy considerations, reviewing draft regulations within the agency, and facilitating public review of the draft and proposed regulations.

Figure 4.6.1 shows the development of draft EPA regulations. The EPA presents analytical methods and monitoring techniques in the methods and monitoring support document. The following factors are among those they consider in specifying which analytical methods should be approved:

- Reliability (precision and accuracy) of the analytical results
- Specificity in the presence of interferences
- Availability and performance of laboratories
- Rapidity of analysis to permit routine use
- Costs of analysis

Guidance to implement the monitoring requirements is also in the document. States have an active role in determining appropriate monitoring requirements. The EPA generally specifies a minimum frequency and provides guidance on factors to consider when assessing a system's vulnerability to contamination.

The treatment technologies and costs document summarizes the availability and performance of the treatment technologies that can reduce contaminants in drinking water. Costs of treatment are determined for each technology for many sizes of water systems. The EPA determines the BAT based upon a number of factors, some of which include technologies that

- Have the highest efficiency of removal
- Are compatible with other types of water treatment processes
- Are available as manufactured items or components
- Are not limited to application in a particular geographic region
- Have integrity for a reasonable service life as a public work
- Are reasonably affordable by large metropolitan or regional systems
- Can be mass-produced and put into operation in time for implementation of the regulations

PUBLIC PARTICIPATION

Figure 4.6.2 shows the public participation part of the process. Public workshops, meetings, and hearings are conducted from early in the process until completion. These meetings allow informal and formal discussions of issues

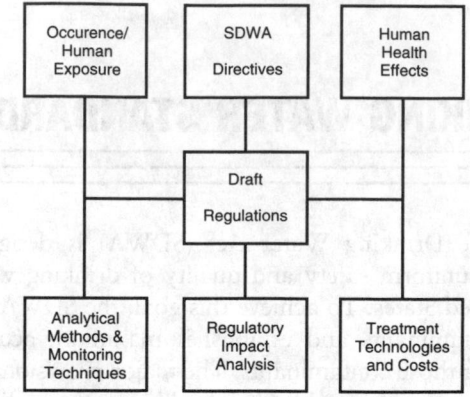

FIG. 4.6.1 Development of draft U.S. EPA regulations (Adapted from the *Federal Register*).

and the opportunity for the public to provide data and information to the EPA. The advance notice of proposed rule making (ANPRM) is an optional step that provides an extra opportunity for public comment. Public comment periods generally last from 30 to 45 days and up to 120 days. The EPA reviews each submitted comment and uses it to prepare the proposed and final regulations. They also prepare a detailed document that presents each comment and the EPA's responses.

EPA Drinking Water and Raw Water Standards

Table 4.6.1 presents the national primary drinking water standards. These standards are enforceable MCLs until they are revised. Associated with these concentration standards are the surface water treatment standards (filtration

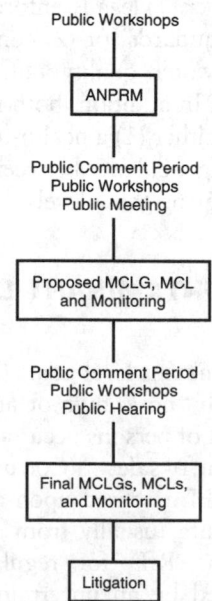

FIG. 4.6.2 Public participation in regulation development (Adapted from the *Federal Register*).

TABLE 4.6.1 NATIONAL PRIMARY DRINKING WATER STANDARDS

Contaminants	MCLG (mg/L)	MCL (mg/L)	Potential Health Effects from Ingestion of Water	Sources of Contaminant in Drinking Water
Fluoride	4.0	4.0	Skeletal and dental fluorosis	Natural deposits; fertilizer, aluminum industries; water additive
Volatile Organics				
Benzene	zero	0.005	Cancer	Some foods; gas, drugs, pesticide, paint, plastic industries
Carbon Tetrachloride	zero	0.005	Cancer	Solvents and their degradation products
p-Dichlorobenzene	0.075	0.075	Cancer	Room and water deodorants, and mothballs
1,2-Dichloroethane	zero	0.005	Cancer	Leaded gas, fumigants, paints
1,1-Dichloroethylene	0.007	0.007	Cancer, liver and kidney effects	Plastics, dyes, perfumes, paints
Trichloroethylene	zero	0.005	Cancer	Textiles, adhesives and metal degreasers
1,1,1-Trichloroethane	0.2	0.2	Liver, nervous system effects	Adhesives, aerosols, textiles, paints, inks, metal degreasers
Vinyl Chloride	zero	0.002	Cancer	May leach from PVC pipe; formed by solvent breakdown
Coliform and Surface Water Treatment				
Giardia lambia	zero	TT	Gastroenteric disease	Human and animal fecal waste
Legionella	zero	TT	Legionnaire's disease	Indigenous to natural waters; can grow in water heating systems
Standard Plate Count	N/A	TT	Indicates water quality, effectiveness of treatment	
Total Coliform*	zero	<5%+	Indicates gastroenteric pathogens	Human and animal fecal waste
Turbidity*	N/A	TT	Interferes with disinfection, filtration	Soil runoff
Viruses	zero	TT	Gastroenteric disease	Human and animal fecal waste
Phase II: Inorganics				
Asbestos (>10 μm)	7MFL	7MFL	Cancer	Natural deposits; asbestos cement in water systems
Barium*	2	2	Circulatory system effects	Natural deposits; pigments, epoxy sealants, spent coal
Cadmium*	0.005	0.005	Kidney effects	Galvanized pipe corrosion; natural deposits; batteries, paints
Chromium* (total)	0.1	0.1	Liver, kidney, circulatory disorders	Natural deposits; mining, electroplating, pigments
Mercury* (inorganic)	0.002	0.002	Kidney, nervous system disorders	Crop runoff; natural deposits; batteries, electrical switches
Nitrate*	10	10	Methemoglobinemia	Animal waste, fertilizer, natural deposits, septic tanks, sewage
Nitrite	1	1	Methemoglobinemia	Same as nitrate; rapidly converted to nitrate
Selenium*	0.05	0.05	Liver damage	Natural deposits; mining, smelting, coal/oil combustion
Phase II: Organics				
Acrylamide	zero	TT	Cancer, nervous system effects	Polymers used in sewage/wastewater treatment
Alachlor	zero	0.002	Cancer	Runoff from herbicide on corn, soybeans, other crops
Aldicarb*	0.001	0.003	Nervous system effects	Insecticide on cotton, potatoes, others; widely restricted
Aldicarb sulfone*	0.001	0.002	Nervous system effects	Biodegradation of aldicarb
Aldicarb sulfoxide*	0.001	0.004	Nervous system effects	Biodegradation of aldicarb
Atrazine	0.003	0.003	Mammary gland tumors	Runoff from use as herbicide on corn and noncropland
Carbofuran	0.04	0.04	Nervous, reproductive system effects	Soil fumigant on corn and cotton; restricted in some areas
Chlordane*	zero	0.002	Cancer	Leaching from soil treatment for termites

Continued on next page

TABLE 4.6.1 *Continued*

Contaminants	MCLG (mg/L)	MCL (mg/L)	Potential Health Effects from Ingestion of Water	Sources of Contaminant in Drinking Water
Chlorobenzene	0.1	0.1	Nervous system and liver effects	Waste solvent from metal degreasing processes
2,4-D*	0.07	0.07	Liver and kidney damage	Runoff from herbicide on wheat, corn, rangelands, lawns
o-Dichlorobenzene	0.6	0.6	Liver, kidney, blood cell damage	Paints, engine cleaning compounds, dyes, chemical wastes
cis-1,2-Dichloroethylene	0.07	0.07	Liver, kidney, nervous, circulatory	Waste industrial extraction solvents
trans-1,2-Dichloroethylene	0.1	0.1	Liver, kidney, nervous, circulatory	Waste industrial extraction solvents
Dibromochloropropane	zero	0.0002	Cancer	Soil fumigant on soybeans, cotton, pineapple, orchards
1,2-Dichloropropane	zero	0.005	Liver, kidney effects; cancer	Soil fumigant; waste industrial solvents
Epichlorohydrin	zero	TT	Cancer	Water treatment chemicals; waste epoxy resins, coatings
Ethylbenzene	0.7	0.7	Liver, kidney, nervous system	Gasoline; insecticides; chemical manufacturing wastes
Ethylene dibromide	zero	0.00005	Cancer	Leaded gas additives; leaching of soil fumigant
Heptachlor	zero	0.0004	Cancer	Leaching of insecticide for termites, very few crops
Heptachlor epoxide	zero	0.0002	Cancer	Biodegradation of heptachlor
Lindane	0.0002	0.0002	Liver, kidney, nerve, immune, circulatory	Insecticide on cattle, lumber, gardens; restricted 1983
Methoxychlor	0.04	0.04	Growth, liver, kidney, nerve effects	Insecticide for fruits, vegetables, alfalfa, livestock, pets
Pentachlorophenol	zero	0.001	Cancer; liver and kidney effects	Wood preservatives, herbicide, cooling tower wastes
PCBs	zero	0.0005	Cancer	Coolant oils from electrical transformers; plasticizers
Styrene	0.1	0.1	Liver, nervous system damage	Plastics, rubber, resin, drug industries; leachate from city landfills
Tetrachloroethylene	zero	0.005	Cancer	Improper disposal of dry cleaning and other solvents
Toluene	1	1	Liver, kidney, nervous, circulatory	Gasoline additive; manufacturing and solvent operations
Toxaphene	zero	0.003	Cancer	Insecticide on cattle, cotton, soybeans; cancelled 1982
2,4,5-TP	0.05	0.05	Liver and kidney damage	Herbicide on crops, right-of-way, golf courses; cancelled 1983
Xylenes (total)	10	10	Liver, kidney; nervous system	By-product of gasoline refining; paints, inks, detergents
Lead and Copper				
Lead*	zero	TT†	Kidney, nervous system damage	Natural/industrial deposits; plumbing, solder, brass alloy faucets
Copper	1.3	TT‡	Gastrointestinal irritation	Natural/industrial deposits; wood preservatives, plumbing
Phase V: Inorganics				
Antimony	0.006	0.006	Cancer	Fire retardants, ceramics, electronics, fireworks, solder
Beryllium	0.004	0.004	Bone, lung damage	Electrical, aerospace, defense industries
Cyanide	0.2	0.2	Thyroid, nervous system damage	Electroplating, steel, plastics, mining, fertilizer
Nickel	0.1	0.1	Heart, liver damage	Metal alloys, electroplating, batteries, chemical production
Thallium	0.0005	0.002	Kidney, liver, brain, intestinal	Electronics, drugs, alloys, glass
Phase V: Organics				
Adipate, (di(2-ethylhexyl))	0.4	0.4	Decreased body weight; liver and testes damage	Synthetic rubber, food packaging, cosmetics
Dalapon	0.2	0.2	Liver, kidney	Herbicide on orchards, beans, coffee, lawns, road/railways
Dichloromethane	zero	0.005	Cancer	Paint stripper, metal degreaser, propellant, extraction

Contaminant	MCLG	MCL	Health effects	Sources
Dinoseb	0.007	0.007	Thyroid, reproductive organ damage	Runoff of herbicide from crop and noncrop applications
Diquat	0.02	0.02	Liver, kidney, eye effects	Runoff of herbicide onland & aquatic weeds
Dioxin	zero	0.00000003	Cancer	Chemical production by-product; impurity in herbicides
Endothall	0.1	0.1	Liver, kidney, gastrointestinal	Herbicide on crops, land/aquatic weeds; rapidly degraded
Endrin	0.002	0.002	Liver, kidney, heart damage	Pesticide on insects, rodents, birds; restricted since 1980
Glyphosate	0.7	0.7	Liver, kidney damage	Herbicide on grasses, weeds, brush
Hexachlorobenzene	zero	0.001	Cancer	Pesticide production waste by-product
Hexachlorocyclopentadiene	0.05	0.05	Kidney, stomach damage	Pesticide production intermediate
Oxamyl (Vydate)	0.2	0.2	Kidney damage	Insecticide on apples, potatoes, tomatoes
PAHs (benzo(a)pyrene)	zero	0.0002	Cancer	Coal tar coatings; burning organic matter; volcanoes, fossil fuels
Phthalate, (di(2-ethylhexyl))	zero	0.006	Cancer	PVC and other plastics
Picloram	0.5	0.5	Kidney, liver damage	Herbicide on broadleaf and woody plants
Simazine	0.004	0.004	Cancer	Herbicide on grass sod, some crops, aquatic algae
1,2,4-Trichlorobenzene	0.07	0.07	Liver, kidney damage	Herbicide production; dye carrier
1,1,2-Trichloroethane	0.003	0.005	Kidney, liver, nervous system	Solvent in rubber, other organic products; chemical production wastes
Other Proposed (P) and Interim (I) Standards				
Beta/photon emitters (I) and (P)	zero	4 mrem/yr	Cancer	Decay of radionuclides in natural and manmade deposits
Alpha emitters (I) and (P)	zero	15 pCi/L	Cancer	Decay of radionuclides in natural deposits
Combined Radium 226/228 (I)	zero	5 pCi/L	Bone cancer	Natural deposits
Radium 226* (P)	zero	20 pCi/L	Bone cancer	Natural deposits
Radium 228* (P)	zero	20 pCi/L	Bone cancer	Natural deposits
Radon (P)	zero	300 pCi/L	Cancer	Decay of radionuclides in natural deposits
Uranium (P)	zero	0.02	Cancer	Natural deposits
Sulfate (P)	400/500	400/500	Diarrhea	Natural deposits
Arsenic* (I)	0.05	0.05	Skin, nervous system toxicity	Natural deposits; smelters, glass, electronics wastes; orchards
Total Trihalomethanes (I)	zero	0.10	Cancer	Drinking water chlorination by-products

Source: U.S. Environmental Protection Agency, 1994, EPA 810-F-94-001A (February) (Washington, D.C.: U.S. EPA Office of Water).

*Indicates original contaminants with interim standards which have been revised.

†Action level = 0.015 mg/L
‡Action level = 1.3 mg/L
TT = Treatment technique requirement
MFL = Million fibers per liter
pCi = picocurie—a measure of radioactivity
mrem = millirems—a measure of radiation absorbed by the body

TABLE 4.6.2 NATIONAL SECONDARY DRINKING WATER STANDARDS

Constituent	SMCL Level (mg/L)	Constituent	SMCL Level (mg/L)
Chloride (Cl)	250	Manganese (Mn)	0.05
Color, color units	15	Odor, threshold odor number	3
Copper (Cu)	1	pH, pH units	6.5–8.5
Corrosivity	Noncorrosive	Sulfate (SO₄)	250
Fluoride	2.0	Total dissolved solids (TDS)	500
Surfactants (MBAS)	0.5	Zinc (Zn)	5.0
Iron (Fe)	0.3		

Health Advisory

Constituent	Level (mg/L)
Sodium	20

Source: U.S. Environmental Protection Agency, *Federal Register* 54, part 124 (29 June 1989):27544.

and disinfection) and total coliform standards promulgated on June 29, 1989. (54 *FR* pt. 124 [29 June 1989]: 27486, 27544; U.S. EPA 1989)

Table 4.6.2 provides the EPA's secondary standards that set levels of drinking water contaminants that affect the aesthetic value of drinking water, such as taste, odor, color, and appearance. These standards are not enforceable by the federal government, but states are encouraged to adopt them. States can establish higher or lower levels based on local conditions, such as availability of alternative source waters, provided that public health and welfare are not adversely affected.

In some areas of the country, raw surface water is used directly as a domestic water supply. California has issued standards for such water used as a source of drinking water. Table 4.6.3 presents these California standards as an example.

Canadian Drinking Water Guidelines

In Canada, drinking water is a shared federal–provincial responsibility. In general, provincial governments are responsible for an adequate, safe supply, whereas the Federal Department of National Health and Welfare develops

TABLE 4.6.3 STANDARDS FOR RAW WATER USED AS SOURCES OF DOMESTIC WATER SUPPLY

Constituents		Excellent Source of Water Supply, Requiring Disinfection Only, as Treatment	Good Source of Water Supply, Requiring Usual Treatment Such as Filtration and Disinfection	Poor Source of Water Supply, Requiring Special or Auxiliary Treatment and Disinfection
B.O.D. (5-day) ppm	Monthly Average:	0.75	1.5–2.5	2.0–5.5
	Maximum Day, or sample:	1.0	3.0–3.5	4.0–7.5
Coliform MPN per 100 ml	Monthly Average:	50–100	240–5,000	10,000–20,000
	Maximum Day, or sample:	....	<20%>5,000 <5%>20,000	
Dissolved oxygen	ppm. average	4.0–7.5	2.5–7.0	2.5–6.5
	% saturation	50–75	25–75	....
pH	Average	6.0–8.5	5.0–9.0	3.8–10.5
Chlorides, max.	ppm.	50	250	500
Iron and manganese together	Max. ppm.	0.3	1.0	15
Fluorides	ppm.	1.0	1.0	1.0
Phenolic compounds	Max. ppm.	none	.005	.025
Color	ppm.	0–20	20–70	150
Turbidity	ppm.	0–10	40–250	....

Source: California State Water Pollution Control Board, 1952.

TABLE 4.6.4 CANADIAN GUIDELINES FOR DRINKING WATER QUALITY (1987)

Parameter	Type[a]	MAC[b]	IMAC[b]	AO[b]	Notes
Alachlor	P	—	—	—	1
Aldicarb	P	0.009	—	—	2(A)
Aldrin and dieldrin	P	0.0007	—	—	1
Antimony	I	c	—	—	1
Arsenic	I	0.05	—	—	1
Asbestos	I	d	—	—	3
Atrazine	P	—	0.06	—	2(A)
Azinphos-methyl	P	0.02	—	—	2(A)
Barium	I	1.0	—	—	1
Bendiocarb	P	0.04	—	—	2(A)
Benzene	O	0.005	—	—	2(A)
Benzo(a)pyrene	O	0.00001	—	—	2(A)
Boron	I	5.0	—	—	1
Bromoxynil	P	—	0.005	—	2(A)
Cadmium	I	0.005	—	—	3
Carbaryl	P	0.09	—	—	2(R)
Carbofuran	P	0.09	—	—	2(A)
Carbon tetrachloride	O	0.005	—	—	2(A)
Cesium 137	R	50 Bq/L			
Chlordane	P	0.007	—	—	1
Chloride	I	—	—	<250	5
Chlorobenzene; 1,2-di-	O	0.2	—	0.003	2(A)
Chlorobenzene; 1,4-di-	O	0.005	—	0.001	2(A)
Chlorophenol; 2,4-di-	O	0.9	—	0.0003	2(A)
Chlorophenol; penta-	O	0.06	—	0.03	2(A)
Chlorophenol; 2,3,4,6-tetra-	O	0.1	—	0.001	2(A)
Chlorophenol; 2,4,6-tri-	O	0.005	—	0.002	2(A)
Chlorpyrifos	P	0.09	—	—	2(A)
Chromium	I	0.05	—	—	3
Coliform organisms	—	e			
Color	—	—	—	<15 TCU	5
Copper	I	—	—	<1.0	5
Cyanazine	P	—	0.01	—	2(A)
Cyanide	I	0.2	—	—	4
2,4-D	P	0.1	—	—	1
DDT and metabolites	P	0.03	—	—	1
Diazinon	P	0.02	—	—	4
Dicamba	P	0.12	—	—	2(A)
1,2-Dichloroethane	O	—	—	—	1
1,1-Dichloroethylene	O	—	—	—	1
Dichloromethane	O	0.05	—	—	2(A)
Diclofop-methyl	P	0.009	—	—	2(A)
Dieldrin and aldrin	P	0.0007	—	—	1
Dimethoate	P	—	0.02	—	2(A)
Dinoseb	P	—	—	—	1
Dioxins	O	—	—	—	1
Diquat	P	0.07	—	—	2(A)
Diuron	P	0.15	—	—	2(A)
Endrin	P	—	—	—	2(D)
Ethylbenzene	O	—	—	<0.0024	2(A)
Fluoride	I	1.5[f]	—	—	4
Furans	O	—	—	—	1
Gasoline	O	g	—	—	2(A)
Glyphosate	P	—	0.28	—	2(A)
Hardness	I	—	—	h	4
Heptachlor and heptachlor epoxide	P	0.003	—	—	1
Iodine 131	R	10 Bq/L			
Iron	I	—	—	<0.3	5
Lead	I	0.05	—	—	1
Lindane	P	0.004	—	—	1
Linuron	P	—	—	—	1
Malathion	P	0.19	—	—	2(A)
Manganese	I	—	—	<0.05	5
MCPA[i]	P	—	—	—	1
Mercury	I	0.001	—	—	3
Methoxychlor	P	0.9	—	—	2(R)
Methyl-parathion	P	0.007	—	—	1
Metolachlor	P	—	0.05	—	2(A)

Continued on nex page

TABLE 4.6.4 *Continued*

Parameter	Type[a]	MAC[b]	IMAC[b]	AO[b]	Notes
Metribuzin	P	0.08	—	—	2(A)
Nitrate	I	10.0[j]	—	—	1
Nitrilotriacetic acid (NTA)	O	0.05	—	—	4
Nitrite	I	1.0[j]	—	—	1
Odor	—	—	—	Inoffensive	5
Paraquat	P	—	0.01	—	2(A)
Parathion	P	0.05	—	—	2(R)
PCBs	O	—	—	—	1
Pesticides (total)	P	0.1	—	—	1
pH	—	—	—	6.5–8.5[k]	5
Phenols	O	—	—	—	2(D)
Phorate	P	—	0.002	—	2(A)
Picloram	P	—	—	—	1
Radium 226	R	1 Bq/L			
Selenium	I	0.01	—	—	3
Silver	I	—	—	—	2(D)
Simazine	P	—	0.01	—	2(A)
Sodium	I	[l]	—	—	3
Strontium 90	R	10 Bq/L			
Sulfate	I	500	00	<150	1
Sulfide (as H$_2$S)	I	—	—	<0.05	5
2,4,5-T	P	0.28	—	<0.02	2(A)
2,4,5-TP	P	—	—	—	2(D)
Taste	—	—	—	Inoffensive	5
TCA[m]	P	—	—	—	1
Temephos	P	—	0.28	—	2(A)
Temperature	—	—	—	<15°C	5
Terbufos	P	—	0.001	—	2(A)
Tetrachloroethylene	O	—	—	—	1
Toluene	O	—	—	<0.024	2(A)
Total dissolved solids	I	—	—	<500	5
Toxaphene	P	—	—	—	2(D)
Triallate	P	0.23	—	—	2(A)
1,1,1-Trichloroethane	O	—	—	—	1
Trichloroethylene	O	—	—	—	1
Trifluralin	P	—	—	—	1
Trihalomethanes	O	0.35	—	—	1
Tritium	R	40,000 Bq/L			
Turbidity	—	1 NTU[n]	—	<5 NTU[o]	1
Uranium	I	0.1	—	—	2(R)
Xylenes	O	—	—	<0.3	2(A)
Zinc	I	—	—	<5.0	5

Source: Health and Welfare Canada, 1987, *Guidelines for Canadian drinking water quality.*

[a]I—Inorganic constituent; O—Organic constituent; P—Pesticide; R—Radionuclide.

[b]Unless otherwise specified, units are mg/L; limits apply to the sum of all forms of each substance present. MAC = maximum acceptable concentration; IMAC = interim maximum acceptable concentration; AO = aesthetic objectives.

[c]An objective concentration only was set in 1978, based on health considerations.

[d]Assessment of data indicates no need to set numerical guideline.

[e]No sample should contain more than 10 total coliform organisms per 100 mL; not more than 10 percent of the samples taken in a 30-day period should show the presence of coliform organisms; not more than two consecutive samples from the same site should show the presence of coliform organisms; and none of the coliform organisms detected should be fecal coliform.

[f]*It is recommended, however, that the concentration of fluoride be adjusted to 1.0 mg/L, which is the optimum level for the control of dental caries. Where the annual mean daily temperature is less than 10°C, a concentration of 1.2 mg/L should be maintained.*

[g]Assessment of data indicates no need to set a numerical guideline.

[h]Public acceptance of hardness varies considerably. Generally hardness levels between 80 and 100 mg/L (as CaCO$_2$) are considered acceptable; levels greater than 200 mg/L are considered poor but can be tolerated; those in excess of 500 mg/L are normally considered unacceptable. Where water is softened by sodium-ion exchange, it is recommended that a separate unsoftened supply be retained for culinary and drinking purposes.

[i]2-Methyl-4-chlorophenoxyacetic acid.

[j]As nitrate- or nitrite-nitrogen concentration.

[k]Dimensionless.

[l]It is recommended that sodium be included in routine monitoring programs since levels may be of interest to authorities who wish to prescribe sodium-restricted diets for their patients.

[m]Trichloroacetic acid.

[n]For water entering a distribution system. *A maximum of 5 NTU may be permitted if it can be demonstrated that disinfection is not compromised by the use of this less stringent value.*

[o]At the point of consumption.

Notes:

1. Under review for possible revision, deletion from, or addition to the guidelines.

2. It is proposed that a guideline be added for this parameter for the first time (A); a change be made to the previous guideline (R); or the guideline be deleted (D). If after 1 year, no evidence comes to light that questions the appropriateness of the proposal, it will be adopted as the guideline.

3. Reassessment of data indicates no need to change 1978 recommendation.

4. Adapted from *Guidelines for Canadian Drinking Water Quality: 1978*; reassessment considered unnecessary at this time.

5. Previously listed in *Guidelines for Canadian Drinking Water Quality: 1978* as a maximum acceptable concentration based on aesthetic considerations.

TABLE 4.6.5 EUROPEAN ECONOMIC COMMUNITY STANDARDS

Parameters	Expression of the Results	Guide Level	Maximum Admissible Concentration
For Parameters Concerning Toxic Substances			
Arsenic	As μg/L		50
Beryllium	Be μg/L		
Cadmium	Cd μg/L		5
Cyanides	CN μg/L		50
Chromium	Cr μg/L		50
Mercury	Hg μg/L		1
Nickel	Ni μg/L		50
Lead*	Pb μg/L		50 (in running water)
Antimony	Sb μg/L		10
Selenium	Se μg/L		10
Vanadium	V μg/L		
Pesticides and related products*	μg/L		0.1
Substances considered separately			
Total			0.5
PAH[a]	μg/L		0.2
For Organoleptic Parameters			
Color	mg/L Pt/Co scale	1	20
Turbidity*	mg/L SiO_2	1	10
	Jackson units	0.4	4
Odor*	Dilution number	0	2 at 12°C
			3 at 25°C
Taste*	Dilution number	0	2 at 12°C
			3 at 25°C
For Physicochemical Parameters (in Relation to the Water's Natural Structure)			
Temperature	°C	12	25
Hydrogen ion concentration*	pH unit	$6.5 \leq pH \leq 8.5$	
Conductivity*	μS/cm at 20°C	400	
Chlorides*	Cl mg/L	25	
Sulfates	SO_4 mg/L	25	500
Silica*	SiO_2 mg/L		
Calcium	Ca mg/L	100	
Magnesium	Mg mg/L	30	50
Sodium*	Na mg/L	20	150–175
Potassium	K mg/L	10	12
Aluminum	Al mg/L	0.05	0.2
Total hardness*			
Dry residues	mg/L after drying at 180°C	1500	
Dissolved oxygen*	% O_2 saturation		
Free carbon dioxide*	CO_2 mg/L		
For Parameters Concerning Substances Undesirable in Excessive Amounts			
Nitrates	NO_3 mg/L	25	50
Nitrites	NO_2 mg/L		0.1
Ammonium	NH_4 mg/L	0.05	0.5
Kjeldahl nitrogen*	N mg/L		1
$KMnO_4$ oxidizability*	O_2 mg/L	2	5

Continued on next page

TABLE 4.6.5 *Continued*

Parameters	Expression of the Results	Guide Level	Maximum Admissible Concentration
Total organic carbon (TOC)*	C mg/L		
Hydrogen sulfide	S μg/L		Undetectable organoleptically
Substances extractable in chloroform	mg/L dry residue	0.1	
Dissolved or emulsified hydrocarbons*	μg/L		10
Phenols (phenol index)*	C_6H_5OH μg/L		0.5
Boron	B μg/L	1000	
Surfactants	μg/L (lauryl sulfate)		200
Other organochlorine compounds*	μg/L	1	
Iron	Fe μg/L	50	200
Manganese	Mn μg/L	20	50
Copper*	Cu μg/L	100, 3000	
Zinc*	Zn μg/L	100, 5000	
Phosphorus	P_2O_5 μg/L	400	5000
Fluoride*	F μg/L		
	8–12°C		1500
	25–30°C		700
Cobalt	Co μg/L		
Suspended solids		None	
Residual chlorine*	Cl μg/L		
Barium	Ba μg/L	100	
Silver*	Ag μg/L		10

Parameters*	Expression of the Results	Minimum Required Concentration (Softened Water)
For Minimum Required Concentration for Softened Water Intended for Human Consumption		
Total hardness	Ca mg/L	60
Hydrogen ion concentration	pH	
Alkalinity	HCO_3 mg/L	30
Dissolved oxygen		

Source: Official Journal of the European Communities 23, Official Directive no. L229/11-L229/23 (30 August 1980).

Note: Certain of these substances may even be toxic when present in very substantial quantities.

*Refer to EEC standards for comments.

quality guidelines and conducts research. Guidelines for Canadian drinking water quality are developed through a joint federal–provincial mechanism and are not legally enforceable unless promulgated as regulations by the appropriate provincial agency. Table 4.6.4 lists the current guidelines for Canadian drinking water quality.

European Economic Community Drinking Water Directives

The European Economic Community (EEC), established by a treaty of the Council of the European Communities, issued a council directive relating to the quality of water

TABLE 4.6.6 RECOMMENDED WATER TESTS FOR EXISTING HOME WELLS (NONPUBLIC WATER SYSTEMS)

Test Name	MCL or SMCL
Recommended:	
Bacteria (Total Coliform)[1]	None detected
Nitrate[1]	10 mg/l NO_3^-
Lead[1]	0.05 mg/l
Consider:	
Volatile organic[1] chemical scan	If positive retest for specific chemicals
Hardness (Total)	150 mg/l
Iron	0.3 mg/l
Manganese	0.05 mg/l
Sodium	50 mg/l
pH	6.5–8.5
Corrosivity	Langelier Index +−1.0
Radioactivity (Gross Alpha)[2]	5 pico curies/l
Mercury[2]	.002 mg/l

Source: S.D. Faust, 1974, *Water from home wells—Problems and treatment,* *Circular 594-B* (Cook College, Rutgers University: New Jersey Experiment Station).

[1]Denotes an MCL-based on health standard. If these levels are exceeded consult with the local health department for interpretation and guidance.

[2]In wells between 50–150 feet deep in South Jersey, the Department of Environmental Protection also recommends that the homeowner consider these tests. Consult with your local health officer for the applicability of these tests to your municipality.

for human consumption. Specifically, the EEC standards provide for both setting standards for toxic chemicals and bacteria that present a health hazard and defining the physical, chemical, and biological parameters for different water uses, specifically for human consumption. The member states are directed to bring laws, regulations, and administrative provisions into force to comply with the directive on water standards (Table 4.6.5).

Home Wells

State health agencies generally regulate the water quality from home wells. As an example, Table 4.6.6 presents a list of recommended water tests for home wells in New Jersey. The list of tests is designed to ensure maximum water supply safety while keeping the cost of testing at a minimum.

As an absolute minimum, testing should include the coliform test for bacteriological safety. Additional chemical testing is warranted:

1. if the home is located in a heavily industrialized area; near service stations, machine shops, or dry cleaners
2. if the home is near a hazardous waste source or a landfill
3. if the home is near houses that have reported problems
4. if the water has an unusual chemical taste, odor, or color

Testing for specific organic chemicals is usually expensive. For accurate and reliable results, tests should be performed in a state-certified laboratory.

Bottled Water

The United States Food and Drug Administration (FDA) regulates bottled water on a national level. Some states have promulgated their own standards for bottled water (Shelton 1994). The FDA has quality standards for bottled drinking water and Good Manufacturing Practice Regulations for processing and bottling all bottled water. These standards and regulations outline in detail the sanitary conditions under which the water is to be obtained, processed, bottled, and tested. They require that water be obtained from sources free from pollution and be of good sanitary quality when judged by the results of bacteriological and chemical analysis. Water bottlers must list the addition of salt and carbon dioxide on their labels.

—*William C. Zegel*

References

Cotruvo, J.A. 1987. Risk assessment and control decisions for protecting drinking water quality. In I.H. Suffet and M. Malaiyandi, eds., *Advances in Chemistry* (Washington, D.C.: American Chemical Society).

———. 1988. Drinking water standards and risk assessment. *Regulatory Toxicology and Pharmacology* 8, no. 3 (September): 288.

Federal Register 50, part 219 (13 November 1985): 46936.

———. 54, part 124 (29 June 1989): 27486.

———: 27544.

Shelton, T.B. 1994. *Interpreting drinking water analysis.* Cook College, Rutgers University: Publication Distribution Center.

U.S. Environmental Protection Agency (U.S. EPA). 1989. *Guidance manual for compliance with filtration and disinfection requirements for public water supplies using surface waters.* PB 90 148016/AS (October) National Technical Information Service.

4.7
GROUNDWATER STANDARDS

Vast reserves of water are in the ground in many areas of the world. Little is known about the quality of this groundwater, except in areas where aquifers are being exploited. In Europe and the United States, where groundwater represents a significant source of fresh water, between 5 and 10 percent of all investigated wells have nitrate levels higher than the maximum recommended value of 45 mg/L[A]. Many organic pollutants find their way into groundwater as seepage from dumps, leakage from sewers and fuel tanks, and runoff from agricultural land or paved surfaces.

Because groundwater is cut off from the atmosphere's oxygen supply, its capacity for self-purification is low. The microbes that perform this function in surface waters need oxygen to function. Microbes that do not use oxygen are in groundwater, but their destruction of pollutants is slow. Thus, although the pollution of rivers and lakes can be rapidly reversed, pollution of groundwater is not easily reversed. Generally, the only practical control for groundwater pollution is to eliminate sources of contamination, particularly in areas of rapid aquifer recharge from the surface.

For regulatory purposes, groundwaters are classified according to use, generally potable and nonpotable. In some areas, such as Florida, almost all drinking water comes from groundwater. They have four classes of groundwater, with the additional discrimination based on levels of dissolved solids. Because of Florida's strong dependence on groundwater, it has developed some of the most sophisticated regulations and standards to manage the resource. For this reason, this section focuses on the current regulation of groundwater in Florida as a model that may become more common in the next century.

Groundwater Classifications

Florida's groundwaters are classified as follows:

Class G-I: Potable groundwaters in single source aquifers that have a total dissolved solids content less than 3,000 mg/L

Class G-II: Potable groundwaters in single source aquifers that have a total dissolved solids content less than 10,000 mg/L

Class G-III: Nonpotable groundwaters in unconfined aquifers that have a total dissolved solids content of 10,000 mg/L or greater, or with a content of 3,000–10,000 mg/L that have been reclassified as having no

potential as a future drinking water source, or have been designated as an exempt aquifer

Class G-IV: Nonpotable groundwaters in confined aquifers that have a total dissolved solids content of 10,000 mg/L or greater

Other areas usually classify potable and nonpotable groundwaters in some manner.

Groundwater Standards

In Florida, any discharge into Class G-I and G-II groundwaters must comply with the water quality criteria of each classification and with minimum criteria. Discharges into Class G-III groundwaters must comply only with minimum criteria, and discharges into Class G-IV groundwaters must comply with minimum criteria only when the state regulatory agency determines a danger to public health, safety, or welfare.

Minimum criteria include all substances in concentrations that are harmful to plants, animals, or organisms native to the soil and responsible for treatment or stabilization of the discharge. The minimum criteria also include substances in concentrations that:

- Are carcinogenic, mutagenic, teratogenic, or toxic to humans
- Are acutely toxic to indigenous species of significance to the aquative community
- Pose a serious danger to public health, safety, or welfare
- Create or constitute a nuisance
- Impair the reasonable and beneficial use of adjacent waters
- Waters classified as Classes G-I and G-II must also meet the primary and secondary drinking water standards

Wellhead Protection

A developing area of regulation is the control of land use in the vicinity of drinking water wells or in the recharge area(s) for wells. This control restricts uses that could release contaminants and adversely affect the quality of groundwaters. It imposes standards upon the contemplated use of land in such an area, called the wellhead protection (WHP) area. The WHP area is the surface and subsurface area surrounding a public water well or wellfield through which contaminants could pass and eventually reach the groundwater supply.

Implementation of a WHP plan can cover a range of actions. The minimum action is to develop plans for alternate sources of drinking water in the event of well contamination, to inventory potential sources of contamina-

tion in the area, and to educate the population of the area as to the potential danger. Stronger actions include control of nonpoint sources of pollution, banning certain land uses and facilities, and a strong enforcement agency.

—William C. Zegel

International Standards

4.8
ISO 14000 Environmental Standards

ISO 14000 is a different kind of environmental standard than others discussed in this chapter. It is a series of process standards developed by the International Organization for Standardization (ISO). They consist of a family of voluntary environmental management standards and guidelines. The purpose of the standards is to establish an organizational environmental ethic and enhance an organization's ability to measure and attain standards of environmental performance. As such, it has the potential to help companies act as responsible environmental citizens by providing a commonly accepted basis for corporate commitment to the environment and to provide a platform from which environmental professionals may take their companies in new directions.

ISO 14000 is also an international system for the certification of users. This extends the standards into the domains of international policy and trade, which is consistent with the mission of ISO to facilitate the international exchange of goods and services. The system consists of:

1. Environmental Management System Standard
2. Environmental Auditing Standard
3. Environmental Labeling Standard
4. Environmental Performance Evaluation Standard
5. Life Cycle Analysis Standard
6. Product Standards
7. Terms & Definitions

The standards deal with management systems, *not* with performance; they are voluntary and require no public reporting; and they are designed for organizations in the developed world. As such, they will provide a company with internal benefits, but will not fully address a company's concerns about international trade requirements, regulatory compliance, or public image as environmentally responsible corporate citizens. It is up to the company implementing the environmental management system at the heart of ISO 14000 to integrate it into the business in a manner that will realize financial and environmental performance improvements.

The Environmental Management System Standard is the centerpiece of ISO 14000. The essential elements of this standard are summarized below. Please refer to the final standards and their associated documents for further guidance and clarification.

1. Top management defines the organization's environmental policy.
2. The organization establishes and maintains:
   a. a procedure or process to identify the environmental issues pertaining to its activities, products, and services that it can control and over which it has an influence, in order to determine those aspects of operations that have or can have a significant impact upon the environment;
   b. a procedure or process to identify and have access to legal and other requirements that are directly applicable to the environmental aspects of its activities, products, and services;
   c. documented environmental objectives and targets for each relevant function and at each level within the organization;
   d. a program for setting and achieving its objectives and targets.
3. The plan is implemented by:
   a. management defining a structure and providing resources to effectively manage environmental issues;
   b. identifying the training, education, and skills needed for all personnel whose work may significantly affect the environment;

c. establishing and maintaining internal and external communication procedures regarding environmental aspects of the organization's activities and its environmental management system;

d. identifying and maintaining information on operations, plans, and procedures related to the environmental management system;

e. establishing and maintaining procedures for controlling all the documents required for effective implementation of the environmental management system;

f. developing and maintaining documented procedures to facilitate implementation of the organization's environmental policies, objectives, targets, and programs;

g. establishing and maintaining procedures for prevention of—and response to—accidents and emergency situations, and for prevention or mitigation of the environmental impacts that may be associated with them.

4. The plan is monitored and corrective actions taken by establishing and maintaining:

a. documented procedures to monitor and measure the key characteristics of those processes that can have a significant impact on the environment;

b. documented procedures both for handling and investigating non-conformance with the environmental management system and for initiating corrective and preventive action;

c. procedures for the identification, maintenance, and disposition of the environmental records needed to implement and operate the environmental management system;

d. a program and procedures for periodically conducting environmental management system audits.

5. To ensure the continuing suitability and effectiveness of the environmental management system, a process must be established and maintained—and implemented at defined intervals—for management to review and evaluate the efficacy of the environmental management system.

—*William C. Zegel*

Bibliography

Schumacher, A. 1988. *A Guide to Hazardous Materials Management.* 62–69. Quorum Books.

U.S. Environmental Protection Agency (EPA). 1977. Toxic pollutant effluent standards. *Code of Federal Regulations.* Title 40, part 129. Washington, D.C.: U.S. Government Printing Office.

5

Air Pollution

Elmar R. Altwicker | Larry W. Canter | Samuel S. Cha | Karl T. Chuang | David H.F. Liu | Gurumurthy Ramachandran | Roger K. Raufer | Parker C. Reist | Alan R. Sanger | Amos Turk | Curtis P. Wagner

Pollutants: Sources, Effects, and Dispersion Modeling

5.1
SOURCES, EFFECTS, AND FATE OF POLLUTANTS

Air pollution is defined as the presence in the outdoor atmosphere of one or more contaminants (pollutants) in quantities and duration that can injure human, plant, or animal life or property (materials) or which unreasonably interferes with the enjoyment of life or the conduct of business. Examples of traditional contaminants include sulfur dioxide, nitrogen oxides, carbon monoxide, hydrocarbons, volatile organic compounds (VOCs), hydrogen sulfide, particulate matter, smoke, and haze. This list of air pollutants can be subdivided into pollutants that are gases or particulates. Gases, such as sulfur dioxide and nitrogen oxides exhibit diffusion properties and are normally formless fluids that change to the liquid or solid state only by a combined effect of increased pressure and decreased temperature. Particulates represent any dispersed matter, solid or liquid, in which the individual aggregates are larger than single small molecules (about 0.0002 μm in diameter) but smaller than about 500 micrometers (μm). Of recent attention is particulate matter equal to or less than 10 μm in size, with this size range of concern relative to potential human health effects. (One μm is 10^{-4} cm).

Currently the focus is on air toxics (or hazardous air pollutants [HAPs]). Air toxics refer to compounds that are present in the atmosphere and exhibit potentially toxic effects not only to humans but also to the overall ecosystem. In the 1990 Clean Air Act Amendments (CAAAs), the air toxics category includes 189 specific chemicals. These chemicals represent typical compounds of concern in the industrial air environment adjusted from workplace standards and associated quality standards to outdoor atmospheric conditions.

The preceding definition includes the quantity or concentration of the contaminant in the atmosphere and its associated duration or time period of occurrence. This concept is important in that pollutants that are present at low concentrations for short time periods can be insignificant in terms of ambient air quality concerns.

Additional air pollutants or atmospheric effects that have become of concern include photochemical smog, acid rain, and global warming. Photochemical smog refers to the formation of oxidizing constituents such as ozone in the atmosphere as a result of the photo-induced reaction of hydrocarbons (or VOCs) and nitrogen oxides. This phenomenon was first recognized in Los Angeles, California, following World War II, and ozone has become a major air pollutant of concern throughout the United States.

Acid rain refers to atmospheric reactions that lead to precipitation which exhibits a pH value less than the normal pH of rainfall (the normal pH is approximately 5.7 when the carbon dioxide equilibrium is considered). Recently, researchers in central Europe, several Scandinavian countries, Canada, and the northeastern United States, have directed their attention to the potential environmental consequences of acid precipitation. Causative agents in acid rain formation are typically associated with sulfur dioxide emissions and nitrogen oxide emissions, along with gaseous hydrogen chloride. From a worldwide perspective, sulfur dioxide emissions are the dominant precursor of acid rain formation.

Another global issue is the influence of air pollution on atmospheric heat balances and associated absorption or reflection of incoming solar radiation. As a result of increasing levels of carbon dioxide and other carbon-containing compounds in the atmosphere, concern is growing that the earth's surface is exhibiting increased temperature levels, and this increase has major implications in shifting climatic conditions throughout the world.

Sources of Air Pollution

Air pollutant sources can be categorized according to the type of source, their number and spatial distribution, and the type of emissions. Categorization by type includes natural and manmade sources. Natural air pollutant sources include plant pollens, wind-blown dust, volcanic eruptions, and lightning-generated forest fires. Manmade sources include transportation vehicles, industrial processes, power plants, municipal incinerators, and others.

POINT, AREA, AND LINE SOURCES

Source categorization according to number and spatial distribution includes single or point sources (stationary), area or multiple sources (stationary or mobile), and line sources. Point sources characterize pollutant emissions from industrial process stacks and fuel combustion facility stacks. Area sources include vehicular traffic in a geographical area as well as fugitive dust emissions from open-air stock piles of resource materials at industrial plants. Figure 5.1.1 shows point and area sources of air pollution. Included in these categories are transportation sources, fuel combustion in stationary sources, industrial process losses, solid waste disposal, and miscellaneous items. This organization of source categories is basic to the development of emission inventories. Line sources include heavily travelled highway facilities and the leading edges of uncontrolled forest fires.

GASEOUS AND PARTICULATE EMISSIONS

As stated earlier, air pollution sources can also be categorized according to whether the emissions are gaseous or particulates. Examples of gaseous pollutant emissions include carbon monoxide, hydrocarbons, sulfur dioxide, and nitrogen oxides. Examples of particulate emissions include smoke and dust emissions from a variety of sources. Often, an air pollution source emits both gases and particulates into the ambient air.

PRIMARY AND SECONDARY AIR POLLUTANTS

An additional source concept is that of primary and secondary air pollutants. This terminology does not refer to the National Ambient Air Quality Standards (NAAQSs), nor is it related to primary and secondary impacts on air quality that result from project construction and operation. Primary air pollutants are pollutants in the atmosphere that exist in the same form as in source emissions. Examples of primary air pollutants include carbon monoxide, sulfur dioxide, and total suspended particulates. Secondary air pollutants are pollutants formed in the atmosphere as a result of reactions such as hydrolysis, oxidation, and photochemical oxidation. Secondary air pollutants include acidic mists and photochemical oxidants. In terms of air quality management, the main strategies are directed toward source control of primary air pollutants. The most effective means of controlling secondary air pollutants is to achieve source control of the primary air pollutant; primary pollutants react in the atmosphere to form secondary pollutants.

EMISSION FACTORS

In evaluating air quality levels in a geographical area, an environmental engineer must have accurate information on the quantity and characteristics of the emissions from numerous sources contributing pollutant emissions into the ambient air. One approach for identifying the types and estimating the quantities of emissions is to use emission factors. An emission factor is the average rate at which a pollutant is released into the atmosphere as a result of an activity, such as combustion or industrial production, divided by the level of that activity. Emission factors relate the types and quantities of pollutants emitted to an indicator such as production capacity, quantity of fuel burned, or miles traveled by an automobile.

EMISSION INVENTORIES

An emission inventory is a compilation of all air pollution quantities entering the atmosphere from all sources in a geographical area for a time period. The emission inventory is an important planning tool in air quality management. A properly developed inventory provides information concerning source emissions and defines the location,

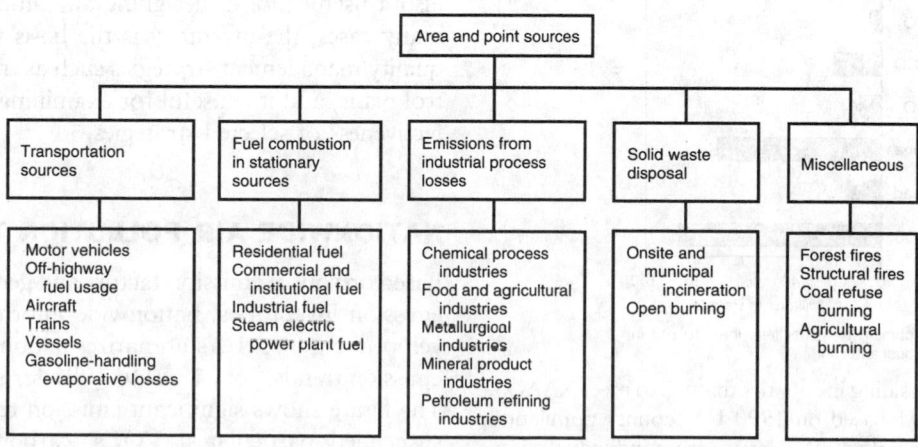

FIG. 5.1.1 Source categories for emission inventories.

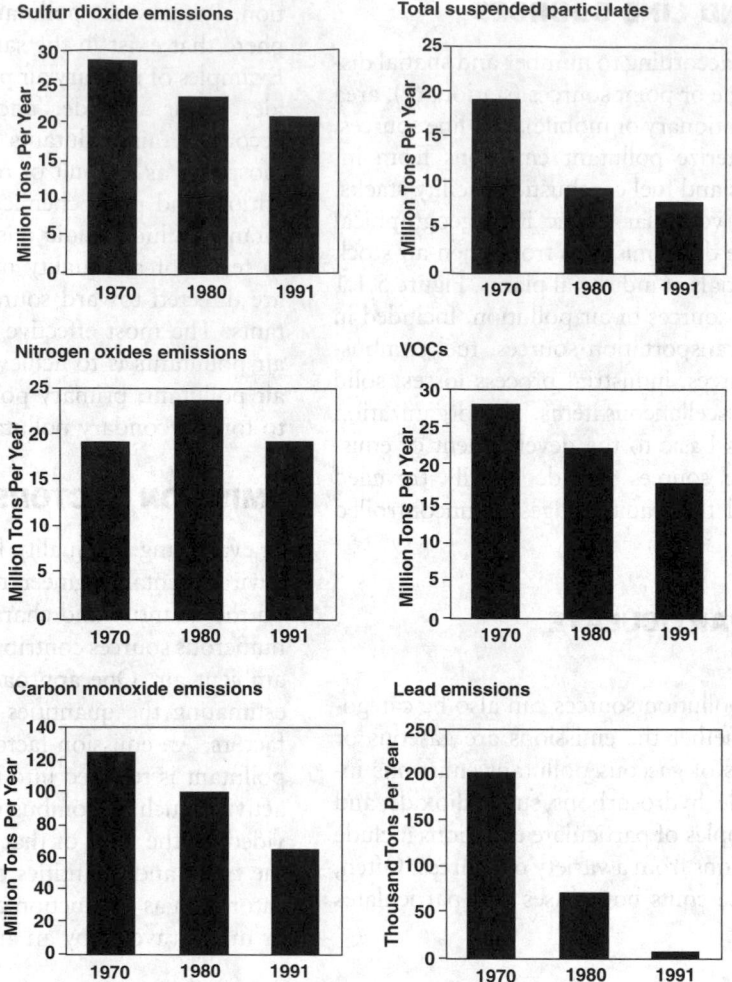

FIG. 5.1.2 Air pollution emission trends in the United States, 1970–1991. (Reprinted from Council on Environmental Quality, 1993, *Environmental quality,* 23rd Annual Report, Washington, D.C.: U.S. Government Printing Office [January].)

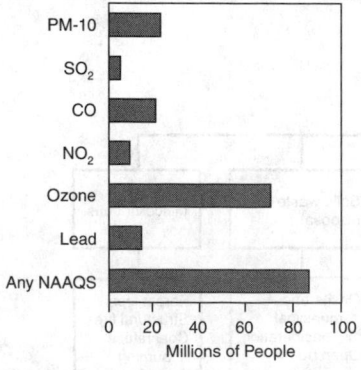

PM-10 = particulate matter less than 10 μm in diameter (dust and soot).

FIG. 5.1.3 People residing in counties that fail to meet NAAQS. Numbers are for 1991 based on 1990 U.S. county population data. Sensitivity to air pollutants can vary from individual to individual. (Reprinted from Council on Environmental Quality 1993.)

magnitude, frequency, duration, and relative contribution of these emissions. It can be used to measure past successes and anticipate future problems. The emission inventory is also a useful tool in designing air sampling networks. In many cases, the inventory is the basis for identifying air quality management strategies such as transportation control plans, and it is useful for examining the long-term effectiveness of selected strategies.

NATIONWIDE AIR POLLUTION TRENDS

Based on source emission factors and geographically based emission inventories, nationwide information can be developed. Figure 5.1.2 summarizes nationwide air pollution emission trends from 1970 to 1991 for six key pollutants. The figure shows significant emission reductions for total suspended particulates, VOCs, carbon monoxide, and lead. The greatest reduction from 1982–1991 was an 89% reduction in lead levels in the air resulting primarily from

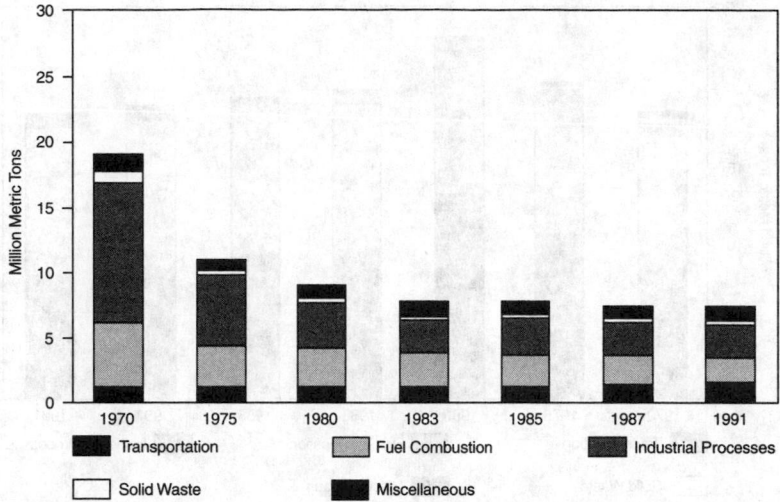

FIG. 5.1.4 U.S. emissions of particulates by source, 1970–1991. (Reprinted from Council on Environmental Quality 1993.)

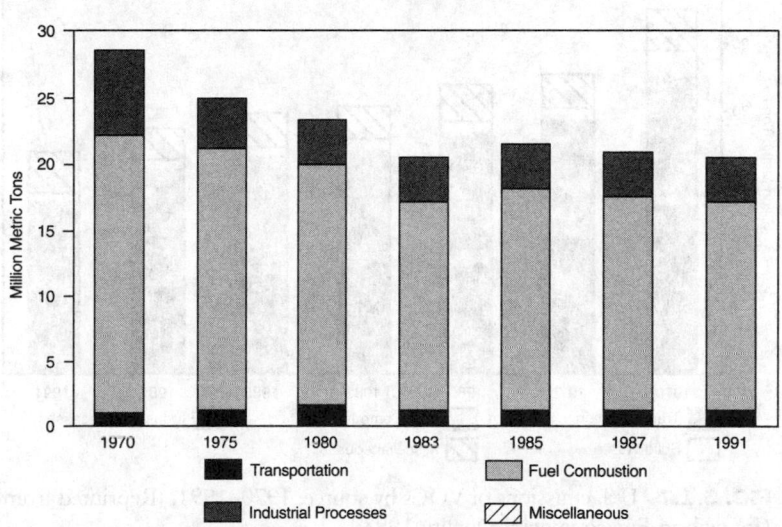

FIG. 5.1.5 U.S. emissions of sulfur oxides by source, 1970–1991. (Reprinted from Council on Environmental Quality 1993.)

the removal of lead from most gasoline. In addition, the gradual phase in of cleaner automobiles and powerplants reduced atmospheric levels of carbon monoxide by 30%, nitrogen oxides by 6%, ozone by 8%, and sulfur dioxide by 20%. Levels of fine particulate matter (PM-10, otherwise known as dust and soot) dropped 10% since the PM-10 standard was set in 1987 (Council on Environmental Quality 1993).

Despite this progress, 86 million people live in U.S. counties where the pollution levels in 1991 exceeded at least one national air quality standard, based on data for a single year. Figure 5.1.3 shows this data. Urban smog continues to be the most prevalent problem; 70 million people live in U.S. counties where the 1991 pollution levels exceeded the standard for ozone.

Many areas release toxic pollutants into the air. The latest EPA toxics release inventory shows a total of 2.2 billion lb of air toxics released nationwide in 1990 (Council on Environmental Quality 1993).

The primary sources of major air pollutants in the United States are transportation, fuel combustion, industrial processes, and solid waste disposal. Figures 5.1.4 through 5.1.9 show the relative contribution of these sources on a nationwide basis for particulates, sulfur oxides, nitrogen oxides, VOCs, carbon monoxide, and lead. Table 5.1.1 contains statistics on the emissions from key sources of these six major pollutants.

Figure 5.1.10 shows anthropogenic sources of carbon dioxide emissions, mainly fuel combustion, from 1950–1990. Table 5.1.2 contains information on the source contributions. Solid and liquid fuel combustion have been the major contributors.

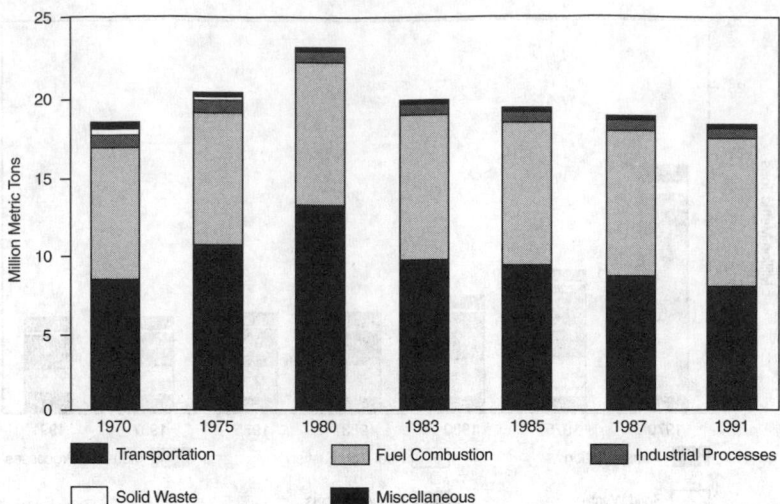

FIG. 5.1.6 U.S. emissions of nitrogen oxides by source, 1970–1991. (Reprinted from Council on Environmental Quality 1993.)

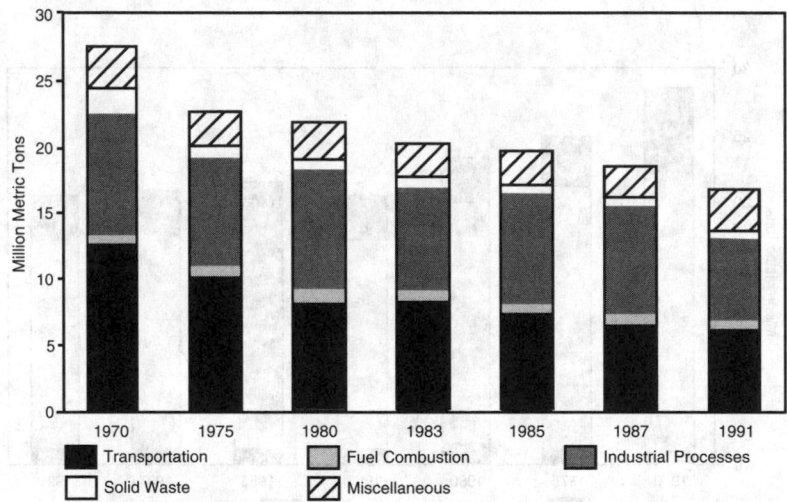

FIG. 5.1.7 U.S. emissions of VOCs by source, 1970–1991. (Reprinted from Council on Environmental Quality 1993.)

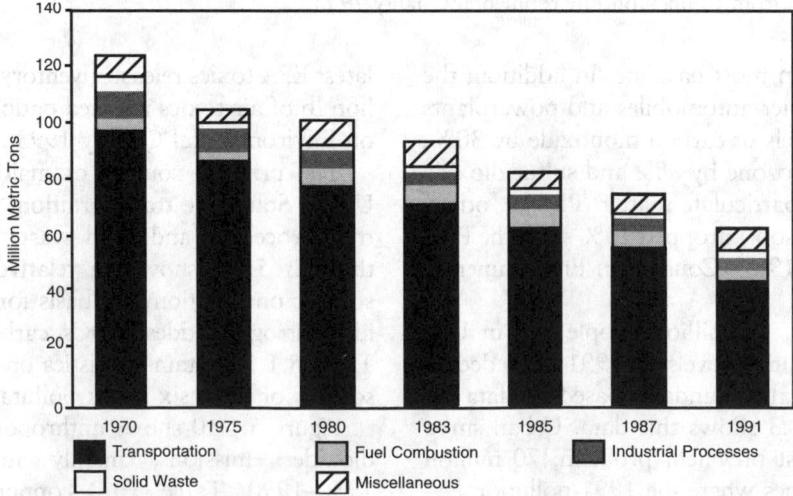

FIG. 5.1.8 U.S. emissions of carbon monoxide by source, 1970–1991. (Reprinted from Council on Environmental Quality 1993.)

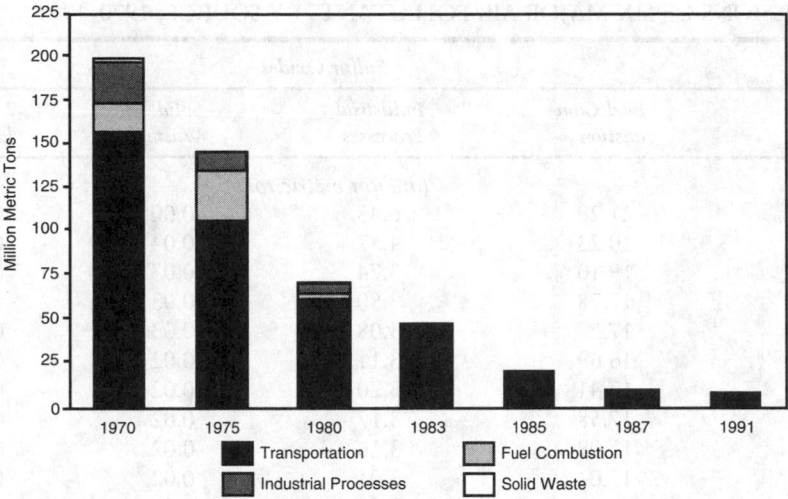

FIG. 5.1.9 U.S. emissions of lead by source, 1970–1991. (Reprinted from Council on Environmental Quality 1993.)

Effects of Air Pollution

Manifold potential effects result from air pollution in an area. These effects are manifested in humans, animals, plants, materials, or climatological variations.

The potential effects of air pollution can be categorized in many ways. One approach is to consider the type of effect and identify the potential air pollutants causing that effect. Another approach is to select an air pollutant such as sulfur dioxide and list all potential effects caused by sulfur dioxide. The types of potential air pollutant effects include aesthetic losses, economic losses, safety hazards, personal discomfort, and health effects. Aesthetic effects include loss of clarity of the atmosphere as well as the presence of objectionable odors. Atmospheric clarity loss can be caused by particulates and smog as well as by visibility reductions due to nitrate and sulfate particles. Objectionable odors encompass a range of potential air pollutants; the majority are associated with the gaseous form. Examples of odorous air pollutants include hydrogen sulfide, ammonia, and mercaptans. Mercaptans are thio alcohols which are characterized by strong odors often associated with sulfur.

ECONOMIC LOSSES

Economic losses resulting from air pollutants include soiling, damage to vegetation, damage to livestock, and deterioration of exposed materials. Soiling represents the general dirtiness of the environment that necessitates more frequent cleaning. Examples include more frequent cleaning of clothes, washing of automobiles, and repainting of structures. Soiling is typically due to particulate matter being deposited, with the key component being settleable particulates or dustfall.

Examples of damage to vegetation are numerous and include both commercial crops and vegetation in scenic areas. Most vegetation damage is due to excessive exposure to gaseous air pollutants, including sulfur dioxide and nitrogen oxides. Oxidants formed in the atmosphere due to photochemically induced reactions also cause damage to vegetation. Some studies indicate that settleable particulates also disrupt normal functional processes within vegetation and thus undesirable effects take place. An example is the deposit of settleable particulates around a cement plant.

VISIBLE AND QUANTIFIABLE EFFECTS

The visible and quantifiable effects of air pollution include tree injury and crop damage, with examples occurring nationwide (Mackenzie and El-Ashry 1989). Many influences shape the overall health and growth of trees and crops. Some of these influences are natural: competition among species, changes in precipitation, temperature fluctuations, insects, and disease. Others result from air pollution, use of pesticides and herbicides, logging, land-use practices, and other human activities. With so many possible stresses, determining which are responsible when trees or crops are damaged is difficult. Crop failures are usually easier to diagnose than widespread tree declines. By nature, agricultural systems are highly managed and ecologically simpler than forests. Also, larger resources have been devoted to developing and understanding agricultural systems than natural forests. Figure 5.1.11 shows the states in the contiguous United States where air pollution can affect trees or crops (Mackenzie and El-Ashry 1989).

The air pollutants of greatest national concern to agriculture are ozone (O_3), sulfur dioxide (SO_2), nitrogen dioxide (NO_2), sulfates, and nitrates. Of these, ozone is of greatest concern; the potential role of acid deposition at ambient levels has not been determined. At present deposition rates, most studies indicate that acid deposition does no identifiable harm to foliage. However, at lower-than-ambient

TABLE 5.1.1 U.S. EMISSIONS OF SIX MAJOR AIR POLLUTANTS BY SOURCE, 1970–1991

Sulfur Oxides

Year	Transportation	Fuel Combustion	Industrial Processes	Solid Waste	Miscellaneous	Total
			(million metric tons)			
1970	0.61	21.29	6.43	0.00	0.10	28.42
1975	0.64	20.23	4.57	0.04	0.02	25.51
1980	0.90	19.10	3.74	0.03	0.01	23.78
1981	0.89	17.78	3.80	0.03	0.01	22.51
1982	0.83	17.27	3.08	0.03	0.01	21.21
1983	0.79	16.69	3.11	0.02	0.01	20.62
1984	0.82	17.41	3.20	0.02	0.01	21.47
1985	0.88	17.58	3.17	0.02	0.01	21.67
1986	0.87	17.09	3.16	0.02	0.01	21.15
1987	0.89	17.04	3.01	0.02	0.01	20.97
1988	0.94	17.25	3.08	0.02	0.01	21.30
1989	0.96	17.42	3.10	0.02	0.01	21.51
1990	0.99	16.98	3.05	0.02	0.01	21.05
1991	0.99	16.55	3.16	0.02	0.01	20.73

Nitrogen Oxides

Year	Transportation	Fuel Combustion	Industrial Processes	Solid Waste	Miscellaneous	Total
			(million metric tons)			
1970	8.45	9.11	0.70	0.40	0.3	18.96
1975	10.02	9.33	0.68	0.14	0.15	20.33
1980	12.46	10.10	0.68	0.10	0.23	23.56
1981	10.42	10.01	0.64	0.10	0.19	21.35
1982	9.74	9.84	0.55	0.09	0.15	20.37
1983	9.35	9.60	0.55	0.08	0.23	19.80
1984	9.10	10.16	0.58	0.08	0.19	20.11
1985	9.15	9.38	0.56	0.08	0.21	19.39
1986	8.49	9.55	0.56	0.08	0.16	18.83
1987	8.14	10.05	0.56	0.08	0.19	19.03
1988	8.19	10.52	0.58	0.08	0.28	19.65
1989	7.85	10.59	0.59	0.08	0.19	19.29
1990	7.83	10.63	0.59	0.08	0.26	19.38
1991	7.26	10.59	0.60	0.01	0.21	18.76

Reactive VOCs

Year	Transportation	Fuel Combustion	Industrial Processes	Solid Waste	Miscellaneous	Total
			(million metric tons)			
1970	12.76	0.61	8.93	1.80	3.30	27.40
1975	10.32	0.60	8.19	0.88	2.54	22.53
1980	8.10	0.95	9.13	0.67	2.90	21.75
1981	8.94	0.95	8.24	0.65	2.44	21.22
1982	8.32	1.01	7.41	0.63	2.13	19.50
1983	8.19	1.00	7.80	0.60	2.65	20.26
1984	8.07	1.01	8.68	0.60	2.64	20.99
1985	7.47	0.90	8.35	0.60	2.49	19.80
1986	6.88	0.89	7.92	0.58	2.19	18.45
1987	6.59	0.90	8.17	0.58	2.40	18.64
1988	6.26	0.89	8.00	0.58	2.88	18.61
1989	5.45	0.91	7.97	0.58	2.44	17.35
1990	5.54	0.62	8.02	0.58	2.82	17.58
1991	5.08	0.67	7.86	0.69	2.59	16.88

Continued on next page

TABLE 5.1.1 *Continued*

Year	Transportation	Fuel Combustion	Industrial Processes	Solid Waste	Miscellaneous	Total
			Carbon Monoxide			
			(million metric tons)			
1970	96.85	4.21	8.95	6.40	7.20	123.61
1975	86.15	4.03	6.88	2.93	4.77	104.76
1980	77.38	6.59	6.34	2.09	7.57	99.97
1981	77.08	6.65	5.87	2.01	6.43	98.04
1982	72.26	7.07	4.35	1.94	4.91	90.53
1983	71.40	6.97	4.34	1.84	7.76	92.31
1984	67.68	7.05	4.66	1.84	6.36	87.60
1985	63.52	6.29	4.38	1.85	7.09	83.12
1986	58.71	6.27	4.20	1.70	5.15	76.03
1987	56.24	6.34	4.33	1.70	6.44	75.05
1988	53.45	6.27	4.60	1.70	9.51	75.53
1989	49.30	6.40	4.58	1.70	6.34	68.32
1990	48.48	4.30	4.64	1.70	8.62	67.74
1991	43.49	4.68	4.69	2.06	7.18	62.10

Year	Transportation	Fuel Combustion	Industrial Processes	Solid Waste	Miscellaneous	Total
			National Total Suspended Particulates			
			(million metric tons)			
1970	1.18	5.07	10.54	1.10	1.10	18.99
1975	1.30	3.28	5.19	0.44	0.75	10.96
1980	1.31	3.04	3.31	0.33	1.08	9.06
1981	1.33	2.96	3.03	0.32	0.94	8.58
1982	1.30	2.75	2.57	0.31	0.75	7.67
1983	1.28	2.72	2.39	0.29	1.09	7.77
1984	1.31	2.76	2.80	0.29	0.93	8.08
1985	1.38	2.47	2.70	0.29	1.01	7.85
1986	1.36	2.46	2.43	0.28	0.78	7.31
1987	1.39	2.44	2.38	0.28	0.93	7.42
1988	1.48	2.40	2.48	0.28	1.30	7.94
1989	1.52	2.41	2.46	0.27	0.92	7.57
1990	1.54	1.87	2.53	0.28	1.19	7.40
1991	1.57	1.94	2.55	0.34	1.01	7.41

Year	Transportation	Fuel Combustion	Industrial Processes	Solid Waste	Miscellaneous	Total
			National PM-10 Particulates			
			(million metric tons)			
1985	1.32	1.46	1.90	0.21	0.73	5.61
1986	1.31	1.48	1.74	0.20	0.54	5.27
1987	1.35	1.49	1.70	0.20	0.66	5.40
1988	1.43	1.45	1.73	0.20	0.96	5.76
1989	1.47	1.49	1.77	0.20	0.65	5.59
1990	1.48	1.04	1.81	0.20	0.87	5.42
1991	1.51	1.10	1.84	0.26	0.73	5.45

Year	Agricultural Tilling	Construction	Mining & Quarrying	Paved Roads	Unpaved Roads	Wind Erosion
			National PM-10 Fugitive Particulates			
			(million metric tons)			
1985	6.20	11.49	0.31	5.95	13.34	3.23
1986	6.26	10.73	0.28	6.18	13.30	8.52

Continued on next page

TABLE 5.1.1 *Continued*

	National PM-10 Fugitive Particulates					
Year	Agricultural Tilling	Con-struction	Mining & Quarrying	Paved Roads	Unpaved Roads	Wind Erosion
1987	6.36	11.00	0.34	6.47	12.65	1.32
1988	6.43	10.58	0.31	6.91	14.17	15.88
1989	6.29	10.22	0.35	6.72	13.91	10.73
1990	6.35	9.11	0.34	6.83	14.20	3.80
1991	6.32	8.77	0.36	7.39	14.36	9.19

	Lead					
Year	Transpor-tation	Fuel Com-bustion	Industrial Processes	Solid Waste	Miscel-laneous	Total
			(thousand metric tons)			
1970	163.60	9.60	23.86	2.00	0.00	199.06
1975	122.67	9.39	10.32	1.45	0.00	143.83
1980	59.43	3.90	3.57	1.10	0.00	68.00
1981	46.46	2.81	3.05	1.10	0.00	53.42
1982	46.96	1.70	2.71	0.94	0.00	52.31
1983	40.80	0.60	2.44	0.82	0.00	44.66
1984	34.69	0.49	2.30	0.82	0.00	38.30
1985	14.70	0.47	2.30	0.79	0.00	18.26
1986	3.45	0.47	1.93	0.77	0.00	6.62
1987	3.03	0.46	1.94	0.77	0.00	6.21
1988	2.64	0.46	2.02	0.74	0.00	5.86
1989	2.15	0.46	2.23	0.69	0.00	5.53
1990	1.71	0.46	2.23	0.73	0.00	5.13
1991	1.62	0.45	2.21	0.69	0.00	4.97

Source: Council on Environmental Quality, 1993.

Notes: Estimates of emissions from transportation sources have been recalculated using a revised model. These estimates supersede those reported in 1992's report and are not directly comparable to historical estimates calculated using different models. PM-10 refers to particulates with an aerodynamic diameter smaller than 10 μm. These smaller particles are likely responsible for most adverse health effects of particulates because of their ability to reach the thoracic or lower regions of the respiratory tract. Detail may not agree with totals because of independent rounding.

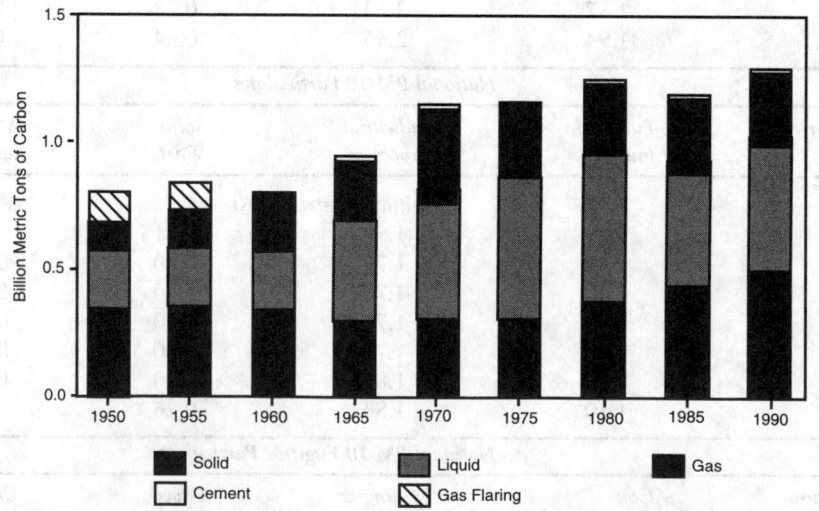

FIG. 5.1.10 U.S. emissions of carbon dioxide from anthropogenic sources, 1950–1990. (Reprinted from Council on Environmental Quality, 1993.)

TABLE 5.1.2 U.S. EMISSIONS OF CARBON DIOXIDE FROM ANTHROPOGENIC SOURCES 1950–1990

Year	Solid	Liquid	Gas	Cement	Gas Flaring	Total	Per Capita
			(million metric tons of carbon)				*(metric tons)*
1950	347.1	244.8	87.1	5.3	11.8	696.1	4.57
1951	334.5	262.2	102.7	5.7	11.7	716.7	4.63
1952	296.6	273.2	109.9	5.8	12.5	697.9	4.44
1953	294.3	286.6	115.5	6.1	11.9	714.5	4.46
1954	252.2	290.2	121.2	6.3	10.6	680.5	4.18
1955	283.3	313.3	130.8	7.2	11.4	746.0	4.50
1956	295.0	328.5	138.1	7.6	12.7	781.9	4.63
1957	282.7	325.8	147.6	7.1	1.9	775.1	4.51
1958	245.3	333.0	155.8	7.5	9.3	750.8	4.29
1959	251.5	343.5	169.9	8.1	8.4	781.4	4.40
1960	253.4	349.8	180.4	7.6	8.3	799.5	4.43
1961	245.0	354.1	187.4	7.7	7.7	801.9	4.37
1962	254.2	364.3	198.7	8.0	6.3	831.5	4.46
1963	272.5	378.8	210.3	8.4	5.6	875.6	4.63
1964	289.7	389.7	219.8	8.8	5.0	912.9	4.76
1965	301.1	405.6	228.0	8.9	4.7	948.3	4.88
1966	312.7	425.9	246.4	9.1	5.5	990.7	0.08
1967	321.1	443.6	258.5	8.8	7.2	1039.2	5.23
1968	314.8	471.9	277.4	9.4	7.6	1081.0	5.38
1969	319.7	497.4	297.8	9.5	7.7	1132.0	5.58
1970	322.4	514.8	312.1	9.0	7.2	1165.5	5.68
1971	305.7	530.5	323.3	9.7	4.2	1173.2	5.66
1972	310.4	575.5	327.6	10.2	3.6	1227.3	5.86
1973	334.0	605.4	321.7	10.6	3.6	1275.4	6.03
1974	330.1	580.7	307.9	10.0	2.4	1231.1	5.76
1975	317.6	565.1	286.0	8.4	1.9	1179.0	5.46
1976	351.6	608.1	291.3	9.0	2.0	1262.0	5.78
1977	355.6	641.9	260.5	9.7	2.0	1269.7	5.76
1978	361.2	655.0	264.7	10.4	2.2	1293.4	5.80
1979	378.7	634.6	274.8	10.4	2.4	1300.9	5.77
1980	394.6	581.0	272.5	9.3	1.8	1259.3	5.53
1981	403.0	533.1	264.2	8.8	1.4	1210.6	5.26
1982	390.1	502.2	245.4	7.8	1.4	1146.9	4.93
1983	405.5	500.1	233.8	8.7	1.4	1149.4	4.89
1984	427.8	507.1	241.5	9.6	1.6	1187.5	5.01
1985	448.0	505.6	236.7	9.6	1.4	1201.3	5.02
1986	439.7	531.1	222.6	9.7	1.4	1204.5	4.99
1987	463.3	545.3	233.8	9.6	1.8	1253.8	5.15
1988	491.4	566.3	244.6	9.5	2.1	1313.8	5.35
1989	498.4	566.5	252.4	9.5	2.1	1328.9	5.37
1990	508.1	542.9	247.9	9.5	1.9	1310.3	5.26

Source: Council on Environmental Quality, 1993.

pH levels, various impacts include leaf spotting, acceleration of epicuticular wax weathering, and changes in foliar leaching rates. When applied simultaneously with ozone, acid deposition also reduces a plant's dry weight (Mackenzie and El-Ashry 1989).

BIODIVERSITY

Air pollution can effect biodiversity. For example, prolonged exposure of the vegetation in the San Bernardino Mountains in southern California to photochemical oxidants has shifted the vegetation dominance from ozone-sensitive pines to ozone-tolerant oaks and deciduous shrubs (Barker and Tingey 1992). The fundamental influencing factors include the pollutant's environmental partitioning, exposure pattern, and toxicity and the sensitivity of the affected species. Biodiversity impacts occur on local, regional, and global scales. Local plume effects reduce vegetation cover, diversity, and ecosystem stability. Regional impacts occur via exposure to photochemical ox-

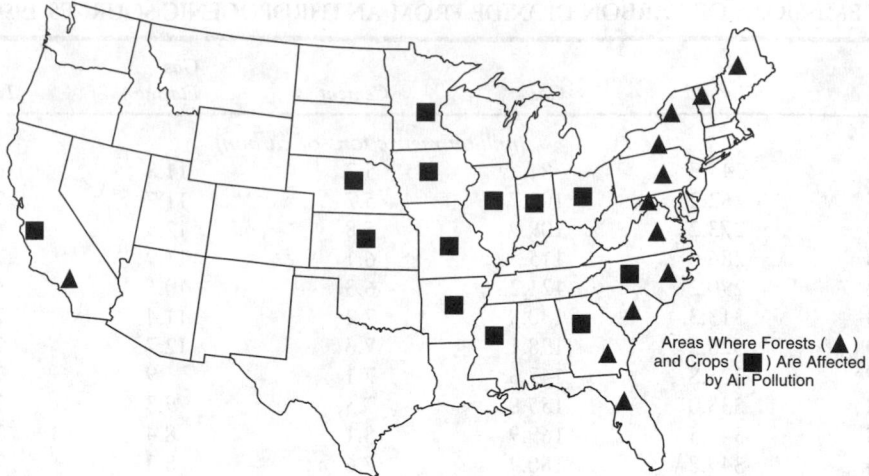

FIG. 5.1.11 Areas where air pollution affects forest trees and agricultural crops. (Reprinted, with permission, from J.J. Mackenzie and M.T. El-Ashry, 1989, Tree and crop injury: A summary of the evidence, chap. 1 in *Air pollution's toll on forests and crops*, edited by J.J. Mackenzie and M.T. El-Ashry, New Haven, Conn.: Yale University Press.)

idants, wet or dry acid or metal deposition, and the long-range transport of toxic chemicals.

Air pollution effects on biodiversity are difficult to document. Unlike habitat destruction, which results in a pronounced and rapid environmental change, the effects of air pollution on biota are usually subtle and elusive because of their interactions with natural stressors. Years can be required before the ecological changes or damage within ecosystems become evident due to continuous or episodic exposure to toxic airborne contaminants or global climate changes (Barker and Tingey 1992).

A number of domestic animals are subject to air pollutant effects. The most frequently cited example is the effects of fluoride on cattle. Other air pollutants also affect animals, including ammonia, carbon monoxide, dust, hydrogen sulfide, sulfur dioxide, and nitrogen oxides.

DETERIORATION OF EXPOSED MATERIALS

The deterioration of exposed materials includes the corrosion of metals, weathering of stone, darkening of lead-based white paint, accelerated cracking of rubber, and deterioration of various manmade fabrics. Sulfur dioxide accelerates the corrosion of metals, necessitating more frequent repainting of metal structures and bridges. The weathering of stone is attributed to the effects of acidic mists formed in the atmosphere as a result of oxidative processes combined with water vapor. Some types of acidic mists include sulfuric acid, carbonic acid, and nitric acid.

HEALTH EFFECTS

The category of health effects ranges from personal discomfort to actual health hazards. Personal discomfort is characterized by eye irritation and irritation to individuals with respiratory difficulties. Eye irritation is associated with oxidants and the components within the oxidant pool such as ozone, proxyacetylnitrate, and others. The burning sensation experienced routinely in many large urban areas is due to high oxidant concentrations. Individuals with respiratory difficulties associated with asthma, bronchitis, and sinusitis experience increased discomfort as a result of oxidants, nitrogen oxides, and particulates.

Health effects result from either acute or chronic exposures. Acute exposures result from accidental releases of pollutants or air pollution episodes. Episodes with documented illness or death are typically caused by persistent (three to six days) thermal inversions with poor atmospheric dispersion and high air pollutant concentrations (Godish 1991). Exposures to lower concentrations for extended periods of time have resulted in chronic respiratory and cardiovascular disease; alterations of body functions such as lung ventilation and oxygen transport; impairment of performance of work and athletic activities; sensory irritation of the eyes, nose, and throat; and aggravation of existing respiratory conditions such as asthma (Godish 1991).

An overview of ambient air quality indicates the potential health effects. Table 5.1.3 shows ambient air quality trends in major urban areas in the United States. The table uses the pollutants standard index (PSI) to depict trends for fifteen of the largest urban areas.

Table 5.1.4 summarizes the effects attributed to specific air pollutants. Many of these effects are described in previous examples, thus this table is a composite of the range of effects of these air pollutants. Table 5.1.5 contains information on the effects of sulfur dioxide. The effects are arranged in terms of health, visibility, materials, and veg-

TABLE 5.1.3 AIR QUALITY TRENDS IN MAJOR URBAN AREAS, 1980–1991

PMSA	1980	1981	1982	1983	1984	1985	1986	1987	1988	1989	1990	1991
					(number of PSI days greater than 100)							
Atlanta	7	9	5	23	8	9	17	19	15	3	16	5
Boston	8	2	5	16	6	2	0	5	11	1	1	5
Chicago	na	3	31	14	8	4	5	9	18	2	3	8
Dallas	10	12	11	17	10	12	5	6	3	3	5	0
Denver	35	51	52	67	59	37	43	34	18	11	7	7
Detroit	na	18	19	18	7	2	6	9	17	12	3	7
Houston	10	32	25	43	30	30	28	31	31	19	35	39
Kansas City	13	7	0	4	12	4	8	5	3	2	2	1
Los Angeles	220	228	195	184	208	196	210	187	226	212	163	156
New York	119	100	69	65	53	21	16	16	35	9	10	16
Philadelphia	52	29	44	56	31	25	21	36	34	19	11	24
Pittsburgh	20	17	14	36	24	6	9	15	31	11	12	3
San Francisco	2	1	2	4	2	5	4	1	1	0	1	0
Seattle	33	42	19	19	4	26	18	13	8	4	2	0
Washington	38	23	25	53	30	15	11	23	34	7	5	16
Total	567	576	488	619	492	396	400	409	484	315	276	285

Source: Council on Environmental Quality, 1993, *Environmental quality,* 23rd Annual Report (Washington, D.C.: U.S. Government Printing Office [January]).

Notes: PMSA = Primary Metropolitan Statistical Area. PSI = Pollutant Standards Index. na = not applicable. The PSI index integrates information from many pollutants across an entire monitoring network into a single number which represents the worst daily air quality experienced in the urban area. Only carbon monoxide and ozone monitoring sites with adequate historical data are included in the PSI trend analysis above, except for Pittsburgh, where sulfur dioxide contributes a significant number of days in the PSI high range. PSI index ranges and health effect descriptor words are as follows: 0 to 50 (good); 51 to 100 (moderate); 101 to 199 (unhealthful); 200 to 299 (very unhealthful); and 300 and above (hazardous). The table shows the number of days when the PSI was greater than 100 (= unhealthy or worse days).

TABLE 5.1.4 QUALITATIVE SUMMARY OF THE EFFECTS ATTRIBUTED TO SPECIFIC POLLUTANTS

Air Pollutant	Effects
Particulates	Speeds chemical reactions; obscures vision; corrodes metals; causes grime on belongings and buildings; aggravates lung illness
Sulfur oxides	Causes acute and chronic leaf injury; attacks a wide variety of trees; irritates upper respiratory tract; destroys paint pigments; erodes statuary; corrodes metals; ruins hosiery; harms textiles; disintegrates book pages and leather
Hydrocarbons (in solid and gaseous states)	May be cancer-producing (carcinogenic); retards plant growth; causes abnormal leaf and bud development
Carbon monoxide	Causes headaches, dizziness, and nausea; absorbs into blood; reduces oxygen content; impairs mental processes
Nitrogen oxides	Causes visible leaf damage; irritates eyes and nose; stunts plant growth even when not causing visible damage; creates brown haze; corrodes metals
Oxidants:	
Ozone	Discolors the upper surface of leaves of many crops, trees, and shrubs; damages and fades textiles; reduces athletic performance; hastens cracking of rubber; disturbs lung function; irritates eyes, nose, and throat; induces coughing
Peroxyacetyl nitrate (PAN)	Discolors the lower leaf surface; irritates eyes; disturbs lung function

etation. Many health effects and visibility are related to the combination of sulfur dioxide and particulates in the atmosphere.

Numerous acute air pollution episodes have caused dramatic health effects to the human population. One of the first occurred in the Meuse Valley in Belgium in 1930 and was characterized by sixty deaths and thousands of ill people. In Donoro, Pennsylvania in 1948, seventeen people died, and 6000 of the population of 14,000 were reported ill. In Poza Rica, Mexico in 1950, twenty-two people died, and 320 people were hospitalized as a result of an episode. Several episodes with excess deaths have been recorded in London, England, with the most famous being in 1952 when 3500 to 4000 excess deaths occurred over a one-

TABLE 5.1.5 EFFECTS ATTRIBUTED TO SULFUR DIOXIDE

Category of Effect	Comments
Health	a. At concentrations of about 1500 $\mu g/m^3$ (0.52 ppm) of sulfur dioxide (24-hr average) and suspended particulate matter measured as a soiling index of 6 cohs or greater, mortality can increase.
	b. At concentrations of about 715 $\mu g/m^3$ (0.25 ppm) of sulfur dioxide and higher (24-hr mean), accompanied by smoke at a concentration of 750 $\mu g/m^3$, the daily death rate can increase.
	c. At concentrations of about 500 $\mu g/m^3$ (0.19 ppm) of sulfur dioxide (24-hr mean), with low particulate levels, mortality rates can increase.
	d. At concentrations ranging from 300 to 500 $\mu g/m^3$ (0.11 to 0.19 ppm) of sulfur dioxide (24-hr mean) with low particulate levels, increase hospital admissions of older people for respiratory disease can increase; absenteeism from work, particularly with older people, can also occur.
	e. At concentrations of about 715 $\mu g/m^3$ (0.25 ppm) of sulfur dioxide (24-hr mean) accompanied by particulate matter, illness rates for patients over age 54 with severe bronchitis can rise sharply.
	f. At concentrations of about 600 $\mu g/m^3$ (about 0.21 ppm) of sulfur dioxide (24-hr mean) with smoke concentrations of about 300 $\mu g/m^3$, patients with chronic lung disease can experience accentuation of symptoms.
	g. At concentrations ranging from 105 to 265 $\mu g/m^3$ (0.037 to 0.092 ppm) of sulfur dioxide (annual mean) accompanied by smoke concentrations of about 185 $\mu g/m^3$, the frequency of respiratory symptoms and lung disease can increase.
	h. At concentrations of about 120 $\mu g/m^3$ (0.046 ppm) of sulfur dioxide (annual mean) accompanied by smoke concentrations of about 100 $\mu g/m^3$, the frequency and severity of respiratory diseases in school children can increase.
	i. At concentrations of about 115 $\mu g/m^3$ (0.040 ppm) of sulfur dioxide (annual mean) accompanied by smoke concentrations of about 160 $\mu g/m^3$, mortality from bronchitis and lung cancer can increase.
Visibility	At a concentration of 285 $\mu g/m^3$ (0.10 ppm) of sulfur dioxide with a comparable concentration of particulate matter and relative humidity of 50%, visibility can be reduced to about 5 mi.
Materials	At a mean sulfur dioxide level of 345 $\mu g/m^3$ (0.12 ppm) accompanied by high particulate levels, the corrosion rate for steel panels can increase by 50%.
Vegetation	a. At a concentration of about 85 $\mu g/m^3$ (0.03 ppm) of sulfur dioxide (annual mean), chronic plant injury and excessive leaf drop can occur.
	b. After exposure to about 860 $\mu g/m^3$ (0.3 ppm) of sulfur dioxide for 8 hr, some species of trees and shrubs show injury.
	c. At concentrations of about 145 to 715 $\mu g/m^3$ (0.05 to 0.25 ppm), sulfur dioxide can react synergistically with either ozone or nitrogen dioxide in short-term exposures (e.g., 4 hr) to produce moderate to severe injury to sensitive plants.

Source: National Air Pollution Control Administration, 1969, *Air quality criteria for sulfur oxides,* Pub. No. AP-50 (Washington, D.C. [January]: 161–162).

week time period. Other episodes occurred recently in locations throughout the United States, and others are anticipated in subsequent years. Generally, the individuals most affected by these episodes are older people already experiencing difficulties with their respiratory systems. Common characteristics of these episodes include pollutant releases from many sources, including industry, and limiting atmospheric dispersion conditions.

ATMOSPHERIC EFFECTS

Air pollution causes atmospheric effects including reductions in visibility, changes in urban climatological characteristics, increased frequency of rainfall and attendant meteorological phenomena, changes in the chemical characteristics of precipitation, reductions in stratospheric ozone levels, and global warming (Godish 1991). The latter three effects can be considered from a macro (large-scale) perspective and are addressed in Section 5.5.

Particulate matter can reduce visibility and increase atmospheric turbidity. Visibility is defined as the greatest distance in any direction at which a person can see and identify with the unaided eye (1) a prominent dark object against the sky at the horizon in the daytime, and (2) a known, preferably unfocused, moderately intense light source at night. In general, visibility decreases as the concentration of particulate matter in the atmosphere increases. Particle size is important in terms of visibility reduction, with sizes in the micron and submicron range of greatest importance. Turbidity in ambient air describes the phenomena of back scattering of direct sunlight by particles in the air, thus reducing the amount of direct sunlight

TABLE 5.1.6 QUALITY FACTORS OF URBAN AIR IN RATIO TO THOSE OF RURAL AIR EXPRESSED AS 1

Urban	Quality Factor
10	Dust particles
5	Sulfur dioxide
10	Carbon dioxide
25	Carbon monoxide
0.8	Total sunshine
0.7	Ultraviolet, winter
0.95	Ultraviolet, summer
1.1	Cloudiness
2	Fog, winter
1.3	Fog, summer

reaching the earth. As an illustration of the effect of turbidity increases in the atmosphere, the total sunshine in urban areas is approximately 80% of that in nearby rural areas. The ultraviolet (UV) component of sunlight in the winter in urban areas is only 70% of that in nearby rural areas; in the summer the UV component in urban areas is 95% of the rural areas' value.

Table 5.1.6 summarizes the quality factors of urban air in ratio to those of rural air when rural air is a factor of 1. The quantity of urban air pollutants and some of the results of the effects of cloudiness and fog are evident in urban areas more than rural areas. Urban areas and the associated air pollutants also influence certain climatological features such as temperature, relative humidity, cloudiness, windspeed and precipitation.

RAINFALL QUALITY

One issue related to the general effects of air pollution is the physical and chemical quality of rainfall. Air pollution can cause the pH of rainfall to decrease, while the suspended dissolved solids and total solids in rainfall increase. Nitrogen and phosphorus concentrations in rainfall can also increase as a result of the atmospheric releases of pollutants containing these nutrients. Finally, increases in lead and cadmium in rainfall are also a result of air pollutant emissions.

An important issue related to air pollution effects is acid rainfall and the resultant effects on aquatic ecosystems. Acid rainfall is any rainfall with a pH less than 5.7. The natural pH of rainfall is 5.7 and reflects the presence of weak carbonic acid (H_2CO_3) resulting from the reaction of water and carbon dioxide from green plants. Rainfall becomes more acid as a result of acidic mists such as H_2SO_4 and HNO_3. Atmospheric emissions of carbon monoxide also add to the carbonic acid mist in the atmosphere and cause the pH of rainfall to be less than 5.7. Numerous locations in the United States have rainfall with the pH values around 4.0. Some of the lowest recorded pH values of rainfall are 2.0 to 3.0.

The chief concerns related to acid rainfall are the potential adverse effects. For example, the pH of the soil can be changed and this change can have unfavorable implications. Changes in the pH in soil can cause changes in adsorption and desorption patterns and lead to differences in nonpoint source water pollution as well as changes in nutrients in both surface runoff as well as from infiltration to groundwater. Acid rain can decrease plant growth, crop growth, and growth in forested areas. Acid rainfall can accelerate the weathering and erosion of metals, stone buildings, and monuments. One concern is related to changes in the quality of surface water and the resultant potential toxicity to aquatic species.

Tropospheric Ozone—A Special Problem

The most widespread air quality problem in the United States is exceedances of the ozone standard (0.12 ppm for 1 hr per year) in urban areas. The ozone standard is based on protecting public health. Ozone is produced when its precursors, VOCs and nitrogen oxides (NO_x), combine in the presence of sunlight (Office of Technology Assessment 1989). VOCs, a broad class of pollutants encompassing hundreds of specific compounds, come from manmade sources including automobile and truck exhaust, evaporation of solvents and gasoline, chemical manufacturing, and petroleum refining. In most urban areas, such manmade sources account for the majority of VOC emissions, but in the summer in some regions, natural vegetation produces an almost equal quantity. NO_x arises primarily from fossil fuel combustion. Major sources include highway vehicles and utility and industrial boilers.

About 100 nonattainment areas dot the country from coast to coast, with *design values* (a measure of peak ozone concentrations) ranging from 0.13 ppm to as high as 0.36 ppm. Figure 5.1.12 summarizes the data for the 3-year period 1983–85 (Office of Technology Assessment 1989). Generally, the higher the design value, the stricter the emission controls needed to meet the standard.

From one-third to one-half of all Americans live in areas that exceed the standard at least once a year. As shown in Figure 5.1.13, 130 of the 317 urban and rural areas exceeded 0.12 ppm for at least 1 hr between 1983 and 1985 (Office of Technology Assessment 1989). Sixty had concentrations that high for at least 6 hr per year. A number of areas topped the standard for 20 or more hr, with the worst, Los Angeles, averaging 275 hr per year.

Ozone's most perceptible short-term effects on human health are respiratory symptoms such as coughing and painful deep breathing (Office of Technology Assessment 1989). It also reduces people's ability to inhale and exhale normally, affecting the most commonly used measures of lung function (e.g., the maximum amount of air a person can exhale in 1 sec or the maximum a person can exhale

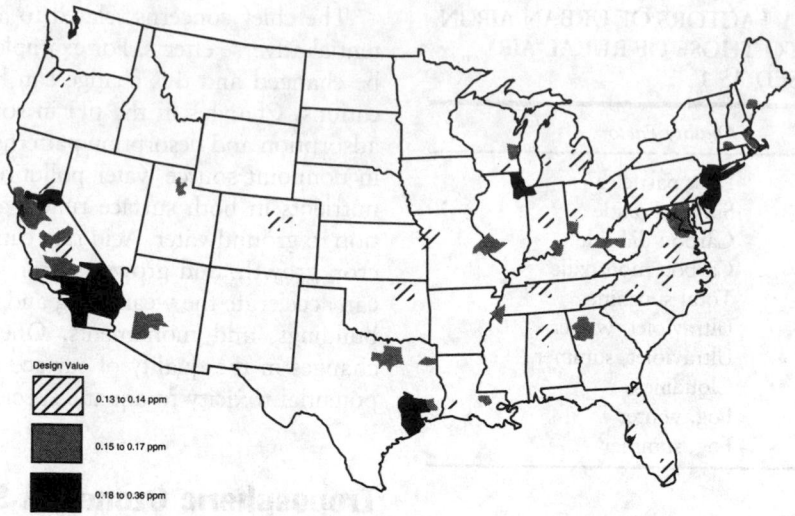

FIG. 5.1.12 Areas classified as nonattainment for ozone based on 1983–85 data. The shading indicates the fourth highest daily maximum one-hour average ozone concentration, or design value, for each area. (Reprinted from Office of Technology Assessment, 1989, *Catching our breath—Next steps for reducing urban ozone*, OTA-0-412, Washington, D.C.: U.S. Congress [July].)

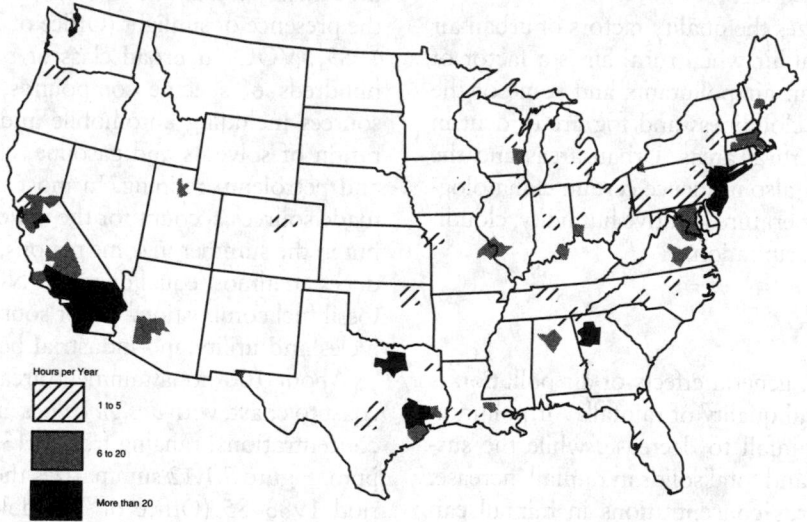

FIG. 5.1.13 Areas where ozone concentrations exceeded 0.12 ppm at least one hour per year on average, from 1983–85. Data from all monitors located in each area were averaged in the map construction. The shading indicates the number of hours that a concentration of 0.12 ppm was exceeded. The areas shown have 130 million residents. (Reprinted from Office of Technology Assessment, 1989.)

after taking a deep breath). As the intensity of exercise rises so does the amount of air drawn into the lungs and thus the dose of ozone. The more heavily a person exercises at a level of ozone concentration and the longer the exercise lasts, the larger the potential effect on lung function.

The U.S. Environmental Protection Agency (EPA) has identified two subgroups of people who may be at special risk for adverse effects: athletes and workers who exercise heavily outdoors and people with preexisting respiratory

problems (Office of Technology Assessment 1989). Also problematic are children, who appear to be less susceptible to (or at least less aware of) acute symptoms and thus spend more time outdoors in high ozone concentrations. Most laboratory studies show no special effects in asthmatics, but epidemiologic evidence suggests that they suffer more frequent attacks, respiratory symptoms, and hospital admissions during periods of high ozone. In addition, about 5 to 20% of the healthy adult population appear to be *responders,* who for no apparent reason are more sensitive than average to a dose of ozone.

At the summertime ozone levels in many cities, some people who engage in moderate exercise for extended periods can experience adverse effects. For example, as shown in Figure 5.1.14, on a summer day when ozone concentrations average 0.14 ppm, a construction worker on an 8-hr shift can experience a temporary decrease in lung function that most scientists consider harmful (Office of Technology Assessment 1989). On those same summer days, children playing outdoors for half the day also risk the effects on lung function that some scientists consider adverse. And some heavy exercisers, such as runners and bicyclists, notice adverse effects in about 2 hr. Even higher levels of ozone, which prevail in a number of areas, have swifter and more severe impacts on health.

Brief Synopsis of Fate of Air Pollutants

Atmospheric dispersion of air pollutants from point or area sources is influenced by wind speed and direction, atmospheric turbulence, and atmospheric stability (Godish 1991).

EFFECTS OF WIND SPEED AND DIRECTION

Horizontal winds play a significant role in the transport and dilution of pollutants. As wind speed increases, the volume of air moving by a source in a period of time also increases. If the emission rate is relatively constant, a doubling of the wind speed halves the pollutant concentration, as the concentration is an inverse function of the wind speed.

Pollutant dispersion is also affected by the variability in wind direction (Godish 1991). If the wind direction is relatively constant, the same area is continuously exposed to high pollutant levels. If the wind direction is constantly shifting, pollutants are dispersed over a larger area, and concentrations over any exposed area are lower. Large changes in wind direction can occur over short periods of time.

EFFECTS OF ATMOSPHERIC TURBULENCE

Air does not flow smoothly near the earth's surface; rather, it follows patterns of three-dimensional movement which are called turbulence. Turbulent eddies are produced by two specific processes: (1) thermal turbulence, resulting from atmospheric heating, and (2) mechanical turbulence caused by the movement of air past an obstruction in a windstream. Usually both types of turbulence occur in any atmospheric situation, although sometimes one prevails. Thermal turbulence is dominant on clear, sunny days with light winds. Although mechanical turbulence occurs under a variety of atmospheric conditions, it is dominant on windy nights with neutral atmospheric stability. Turbulence enhances the dispersion process although in mechanical turbulence, downwash from the pollution source can result in high pollution levels immediately downstream (Godish 1991).

EFFECTS OF ATMOSPHERIC STABILITY

In the troposphere, temperature decreases with height to an elevation of approximately 10 km. This decrease is due to reduced heating processes with height and radiative

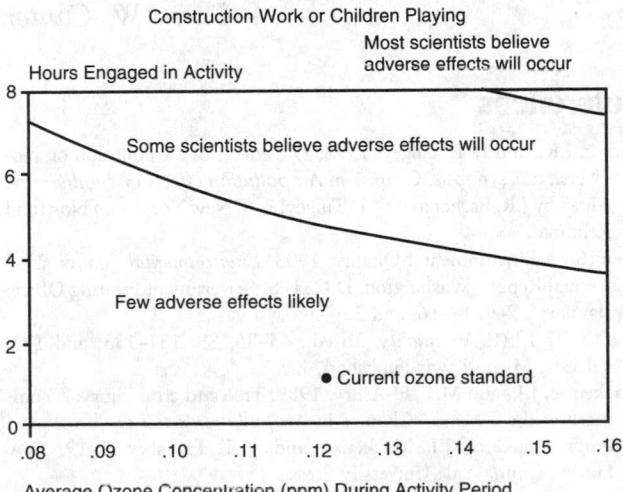
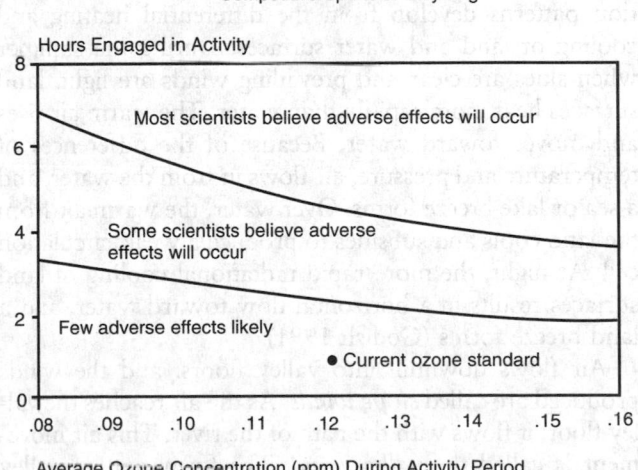

FIG. 5.1.14 Likelihood of adverse effects from ozone while exercising. The likelihood of experiencing adverse effects depends on 1) the ozone concentration, 2) the vigorousness of the activity, and 3) the number of hours engaged in that activity. The figure on the left shows the number of hours to reach an adverse effect under moderate exercise conditions (e.g., construction work or children playing). The figure on the right shows that fewer hours are needed under heavy exercise (e.g., competitive sports or bicycling). The current 1-hr ozone standard is shown for comparison. (Reprinted from Office of Technology Assessment 1989.)

cooling of air and reaches its maximum in the upper levels of the troposphere. Temperature decrease with height is described by the lapse rate. On the average, temperature decreases $-0.65°C/100$ m or $-6.5°C/km$. This decrease is the normal lapse rate. If warm dry air is lifted in a dry environment, it undergoes adiabatic expansion and cooling. This adiabatic cooling results in a lapse rate of $-1°C/100$ m or $-10°C/km$, the dry adiabatic lapse rate.

Individual vertical temperature measurements vary from either the normal or dry adiabatic lapse rate. This change of temperature with height for measurement is the *environmental lapse rate*. Values for the environmental lapse rates characterize the stability of the atmosphere and profoundly affect vertical air motion and the dispersion of pollutants (Godish 1991).

If the environmental lapse rate is greater than the dry adiabatic lapse rate, dispersion characteristics are good to excellent. The greater the difference, the more unstable the atmosphere and the more enhanced the dispersion. If the environmental lapse rate is less than the dry adiabatic lapse rate, the atmosphere becomes stable, and dispersion becomes more limited. The greater the difference from the adiabatic lapse rate, the more stable the atmosphere and the poorer the dispersion potential (Godish 1991).

EFFECTS OF TOPOGRAPHY ON AIR MOTION

Topography can affect micro- and mesoscale air motion near point and area sources. Most large urban centers in this country are located along sea (New York City and Los Angeles) and lake (Chicago and Detroit) coastal areas, and heavy industry is often located in river valleys, e.g., the Ohio River Valley. Local air flow patterns in these regions have a significant impact on pollution dispersion processes. For example, land–water mesoscale air circulation patterns develop from the differential heating and cooling of land and water surfaces. During the summer when skies are clear and prevailing winds are light, land surfaces heat more rapidly than water. The warm air rises and moves toward water. Because of the differences of temperature and pressure, air flows in from the water, and a sea or lake breeze forms. Over water, the warm air from the land cools and subsides to produce a weak circulation cell. At night, the more rapid radiational cooling of land surfaces results in a horizontal flow toward water, and a land breeze forms (Godish 1991).

Air flows downhill into valley floors, and the winds produced are called *slope winds*. As the air reaches the valley floor, it flows with the path of the river. This air movement is called the *valley wind*. The formation of valley wind lags several hours after slope winds. Because of a smaller vertical gradient, downriver valley winds are lighter

and because of the large volume, cool dense air accumulates, flooding the valley floor and intensifying the surface inversion that is normally produced by radiative cooling (Godish 1991). The inversion deepens over the course of the night and reaches its maximum depth just before sunrise. The height of the inversion layer depends on the depth of the valley and the intensity of the radiative cooling process.

Mountains affect local air flow by increasing surface roughness and thereby decreasing wind speed. In addition, mountains and hills form physical barriers to air movement.

In summary, the atmospheric dispersion of air pollution emissions depends on the interplay of a number of factors which include (1) the physical and chemical nature of the pollutants, (2) meteorological parameters, (3) the location of the source relative to obstructions, and (4) downwind topography (Godish 1991).

OTHER FACTORS

In addition to dispersion, wet and dry removal processes as well as atmospheric reactions affect the concentrations of air pollutants in the atmosphere. Atmospheric reactions include ozone or acid rain formation. In dry removal, particles are removed by gravity or impaction, and gases diffuse to surfaces where they are absorbed or adsorbed. Wet removal is the major removal process for most particles and can be a factor in the removal of gaseous contaminants as well. Wet removal can involve the in-cloud capture of gases or particles (rainout) or the below-cloud capture (washout). In washout, raindrops or snowflakes strike particles and carry them to the surface; gases are removed by absorption (Godish 1991).

—*Larry W. Canter*

References

Barker, J.R., and D.T. Tingey. 1992. The effects of air pollution on biodiversity: A synopsis. Chap. 1 in *Air pollution effects on biodiversity*, edited by J.R. Barker and D.T. Tingey, 3–8. New York: Van Nostrand Reinhold.

Council on Environmental Quality. 1993. *Environmental quality*. 23rd Annual Report. Washington, D.C.: U.S. Government Printing Office. (January): 7–9, 14–16, and 326–340.

Godish, T. 1991. *Air quality*, 2d ed., 65–85, 89, 131–133, and 173. Chelsea, Mich.: Lewis Publishers, Inc.

Mackenzie, J.J., and M.T. El-Ashry. 1989. Tree and crop injury: A summary of the evidence. Chap. 1 in *Air pollution's toll on forests and crops*, edited by J.J. Mackenzie and M.T. El-Ashry, 1–19. New Haven, Conn.: Yale University Press.

Office of Technology Assessment. 1989. *Catching our breath—Next steps for reducing urban ozone*. OTA-0-412. Washington, D.C.: U.S. Congress. (July): 4–9.

5.2
VOCs AND HAPs EMISSION FROM CHEMICAL PLANTS

Emission Points

Emission sources (or points) of volatile organic chemicals (VOCs) and hazardous air pollutants (HAPs) in a chemical plant can be classified into three groups: (1) process point sources, (2) process fugitive sources, and (3) area fugitive sources (U.S. EPA 1991). VOCs refer to compounds which produce vapors at room temperature and pressure; whereas, HAPs include VOCs as well as nonvolatile organics and inorganics present as vapors or particulates.

PROCESS POINT SOURCES

Process point sources of VOCs and HAPs can be individually defined for a chemical plant. Chemical reactors, distillation columns, catalytic cracking units, condensers, strippers, furnaces, and boilers are examples of point sources that discharge both air toxics and criteria pollutants through vent pipes or stacks. Emission reductions or control are achieved through process changes focused on pollution prevention and the use of add-on control devices such as adsorbers, absorbers, thermal or catalytic incinerators, fabric filters, or electrostatic precipitators (ESPs).

PROCESS FUGITIVE SOURCES

Although typically more numerous than process point sources, process fugitive sources can also be individually defined for a chemical plant. Inadvertent emissions from or through pumps, valves, compressors, access ports, storage tank vents, and feed or discharge openings to a process classify such units or equipment as process fugitive sources. Vent fans from rooms or enclosures containing an emissions source can also be classified this way (U.S. EPA 1991). Once process fugitive emissions are captured by hooding, enclosures, or closed-vent systems, they can often be controlled by add-on devices used for process point sources.

AREA FUGITIVE SOURCES

Large surface areas characterize area fugitive sources. Examples of such sources include waste storage ponds and raw material storage piles at many chemical plants. VOC and HAP control measures for area fugitive sources typically focus on release prevention measures such as the use of covers or chemical adjustments in terms of the pH and oxidation state for liquid wastes.

Classification of VOCs and HAPs

The HAPs described in this manual are not limited to the specific compounds listed in current laws such as the CAAAs of 1990, the Resource Conservation and Recovery Act (RCRA), or the Toxic Substances Control Act. HAPs can be classified relative to the type of compounds (i.e., organic or inorganic) and the form in which they are emitted from process point, process fugitive, or area fugitive sources (i.e., vapor or particulate).

This section discusses two examples of VOC and HAP emissions from chemical plant classes. Table 5.2.1 summarizes emissions from the inorganic chemical manufacturing industry. This industry produces basic inorganic chemicals for either direct use or use in manufacturing other chemical products. Although the potential for emissions is high, in many cases they are recovered due to economic reasons. As shown in Table 5.2.1, the chemical types of inorganic emissions depend on the source category, while the emission sources vary with the processes used to produce the inorganic chemical.

The second example is from petroleum-related industries, including the oil and gas production industry, the petroleum refining industry, and the basic petrochemicals industry. Table 5.2.2 summarizes the emission sources within these three categories. Sources of emissions from the oil and gas production industry include blowouts during drilling operations; storage tank breathing and filling losses; wastewater treatment processes; and fugitive leaks in valves, pumps, pipes, and vessels. In the petroleum refining industry, emission sources include distillation and fractionating columns, catalytic cracking units, sulfur recovery processes, storage tanks, fugitives, and combustion units (e.g., process heaters). Fugitive emissions are a major source in this industry. Emission sources in the basic petrochemicals industry are similar to those from the petroleum refining segments (U.S. EPA 1991).

Table 5.2.3 summarizes the potential HAP emissions from the petroleum refining segment of the petroleum industries. A large proportion of the emissions occur as organic vapors; for example, benzene, toluene, and xylenes are the principal organic vapor emissions. These organic vapors are due to the chemical composition of the two starting materials used in these industries: crude oil and natural gas. Crude oil is composed chiefly of hydrocarbons (paraffins, napthalenes, and aromatics) with small amounts of trace elements and organic compounds containing sulfur, nitrogen, and oxygen. Natural gas is largely saturated hydrocarbons (mainly methane). The remainder

TABLE 5.2.1 POTENTIAL HAPS FOR INORGANIC CHEMICAL MANUFACTURING INDUSTRY

| Source Category | Potential HAPs | | Potential Emission Sources | | |
| | Inorganic | | | | |
	Vapor	Particulate	Process Point	Process Fugitive	Area Fugitive
Aluminum chloride	4,10		X	X	
Aluminum fluoride	17		X	X	
Ammonia	1		B,D,E	K	J,S
Ammonium acetate	1		X	X	
Ammonium-nitrate, sulfate, thiocyanate, formate, tartrate	1		C,F,I,L	Q	
Ammonium phosphate	1,17		X	X	
Antimony oxide	5		X	X	
Arsenic-disulfide, iodide, pentafluoride, thioarsenate, tribromide, trichloride, trifluoride, trioxide, orthoarsenic acid	2	2	H,U	K,Q,T	J,S
Barium-carbonate, chloride, hydroxide, sulfate, sulfide		6	C,E,G,I,L,U	N,P,Q,T	
Beryllium-oxide, hydroxide		7	X	X	
Boric acid and borax		9	X	X	
Bromine	8,10		X	X	
Cadmium (pigment)—sulfide, sulfoselenide, lithopone		15	X	X	
Calcium-carbide, arsenate, phosphate	3,17	2	H	K,P	
Chlorine	10	25	H,C	K,R	J
Chlorosulfonic acid	19,34		X	X	
Chromic acid	12	11,12	H	K,N,O,Q	J,S
Chromium-acetate, borides, halides, etc.		11	X	X	
Chromium (pigment)—oxide		11	X	X	
Cobalt—acetate, carbonate, halides, etc.		13	X	X	
Copper sulfate	14		X	X	
Fluorine	17		X	X	
Hydrazine	1,39		X	X	
Hydrochloric acid	10,20	20	B		
Hydrofluoric acid	17		B,G	K,R	
Iodine (crude)	10	38	X	X	
Iron chloride	10,20	20	X	X	
Iron (pigment)—oxide	40		X	X	
Lead-arsenate, halides, hydroxides, dioxide, nitrate	3	2,21	G,L	P,Q	
Lead chromate	22		G,R	P,Q	
Lead (pigments)—oxide, carbonate, sulfate		21	G,R	P,Q	
Manganese dioxide (Potassium permanganate)	24	23	G,L	Q,P,T	
Manganese sulfate		23	G,L	Q,P,T	
Mercury-halides, nitrates, oxides		25	X	X	

Continued on next page

TABLE 5.2.1 *Continued*

Source Category	Potential HAPs		Potential Emission Sources		
	Inorganic		Process Point	Process Fugitive	Area Fugitive
	Vapor	*Particulate*			
Nickel-halides, nitrates, oxides		26		P,Q	
Nickel sulfate	27	26	L	Q,T	
Nitric acid	28	28	B,H	K,N,R	J,S
Phosphoric acid					
Wet process	10,17,18,30	30	H,C,W	K,N,P,T	J,S
Thermal process			B,G	K,N,R,T	J,S
Phosphorus	17		X	X	
Phosphorus oxychloride	10		X	X	
Phosphorus pentasulfide	29,31	29	X	X	
Phosphorus trichloride	32,10,29	29	X	X	
Potassium-bichromate, chromate	16	16	I		
Potassium hydroxide	10	25	X	X	
Sodium arsenate		2	H	K,P	
Sodium carbonate	1		I,L,V	P	
Sodium chlorate	10		X	X	
Sodium chromate-dichromate	16	16	G,I,L,M	P,Q	
Sodium hydrosulfide	18		X	X	
Sodium-siliconfluoride, fluoride	17	16	X	X	
Sulfuric acid	33,34	33	A,B,C,H	K,R	J,S
Sulfur monochloride-dichloride	10		X	X	
Zinc chloride	36,21	21	X	X	
Zinc chromate (pigment)	35		X	X	
Zinc oxide (pigment)	37		X	X	

Source: U.S. Environmental Protection Agency, 1991, *Handbook: Control technologies for hazardous air pollutants.* EPA/625/6-91/014 (Cincinnati, Ohio [June]).

Pollutant Key
1. ammonia
2. arsenic
3. arsenic trioxide
4. aluminum chloride
5. antimony trioxide
6. barium salts
7. beryllium
8. bromine
9. boron salts
10. chlorine
11. chromium salts
12. chromic acid mist
13. cobalt metal fumes
14. copper sulfate
15. cadmium salts
16. chromates (chromium)
17. fluorine
18. hydrogen sulfide
19. hydrogen chloride
20. hydrochloric acid
21. lead
22. lead chromate
23. manganese salts
24. manganese dioxide
25. mercury
26. nickel
27. nickel sulfate
28. nitric acid mist
29. phosphorus
30. phosphoric acid mist
31. phosphorus pentasulfide
32. phosphorus trichloride
33. sulfuric acid mist
34. sulfur trioxide
35. zinc chromate
36. zinc chloride fumes
37. zinc oxide fumes
38. iodine
39. hydrazine
40. iron oxide

Source Key
A. converter
B. absorption tower
C. concentrator
D. desulfurizer
E. reformer
F. neutralizer
G. kiln
H. reactor
I. crystallizer
J. compressor and pump seals
K. storage tank vents
L. dryer
M. leaching tanks
N. filter
O. flakers
P. milling, grinding, and crushing
Q. product handling and packaging
R. cooler (cooling tower and condenser)
S. pressure relief valves
T. raw material unloading
U. purification
V. calciner
W. hot well
X. no information

TABLE 5.2.2 EMISSION SOURCES FOR THE PETROLEUM-RELATED INDUSTRIES

Source Category	Potential HAP Emission Sources		
	Process Point	Process Fugitive	Area Fugitive
Oil and Gas Production			
Exploration, site preparation and drilling	A	C	D,E
Crude processing	G	F,H	
Natural gas processing	G,J,K	H	
Secondary and tertiary recovery techniques	G		I
Petroleum Refining Industry			
Crude separation	G,J,L	F,H,M,N	I
Light hydrocarbon processing	O,G	F,H	Q
Middle and heavy distillate processing	G,O,P,R	F,H	I
Residual hydrocarbon processing	B,G,K,O,R	H	I
Auxiliary processes	G	F,H	I
Basic Petrochemicals Industry			
Olefins production	G,K,O	F,H	I
Butadiene production	G,J,L,O,R	F,H,N	I
Benzene/toluene/xylene (BTX) production	G,K,O,R	F,Q	I
Naphthalene production	G,L,O	F,H	I
Cresol/cresylic acids production	G,L	F,H	I
Normal paraffin production	G,O	F,H	I

Source: U.S. EPA, 1991.

Source Key

A. blowout during drilling
B. visbreaker furnace
C. cuttings
D. drilling fluid
E. pipe leaks (due to corrosion)
F. wastewater disposal (process drain, blow-down, and cooling water)

G. flare, incinerator, process heater, and boiler
H. storage, transfer, and handling
I. pumps, valves, compressors, and fittings
J. absorber
K. process vent
L. distillation and fractionation
M. hotwells

N. steam ejectors
O. catalyst regeneration
P. evaporation
Q. catalytic cracker
R. stripper

TABLE 5.2.3 POTENTIAL HAPs FOR PETROLEUM REFINING INDUSTRIES (SPECIFIC LISTING FOR PETROLEUM REFINING SEGMENT)

Process	Potential HAPs			
	Organic		Inorganic	
	Vapor	Particulate	Vapor	Particulate
Crude separation	a,b,d,e,f,g,h,i,j,k,l,m,o, A,B,C,D,E,F,J	o	c,m,t,u,v,x,y,L	p,I,Q,R
Light hydrocarbon processing	g,h,i,n,N,O,P	R	t,v	G,H,Q
Middle and heavy distillate processing	a,d,e,f,g,h,i,j,k,l, F,J,K,O,P,S,T	o,R	m,t,u,v,x,y,L	p,q,G,H,I,Q,U
Residual hydrocarbon processing	a,d,e,f,g,h,i,j,k,l,n, F,J,M,N,P,S,T	o,R	m,s,t,u,v,x,y,L	p,q,G,H,I,Q,U
Auxiliary processes	a,b,d,e,f,g,h,i,j,k,l,n, A,B,C,D,J,K,M,T	o,R	c,m,s,u,y,L	p,q,r,z,I

Source: U.S. Environmental Protection Agency (EPA), 1991, *Handbook: Control technologies for hazardous air pollutants*, EPA/625/6-91/014, Cincinnati, Ohio, (June) 2-1 to 2-13.

Pollutant Key

a. maleic anhydride
b. benzoic acid
c. chlorides
d. ketones
e. aldehydes
f. heterocyclic compounds (e.g., pyridines)
g. benzene
h. toluene
i. xylene
j. phenols
k. organic compounds containing sulfur (sulfonates, sulfones)
l. cresols
m. inorganic sulfides
n. mercaptans
o. polynuclear compounds (benzopyrene, anthracene)

p. vanadium
q. nickel
r. lead
s. sulfuric acid
t. hydrogen sulfide
u. ammonia
v. carbon disulfide
x. carbonyl sulfide
y. cyanides
z. chromates
A. acetic acid
B. formic acid
C. methylethylamine
D. diethylamine
E. thiosulfide
F. methyl mercaptan

G. cobalt
H. molybdenum
I. zinc
J. cresylic acid
K. xylenols
L. thiophenes
M. thiophenol
N. nickel carbonyl
O. tetraethyl lead
P. cobalt carbonyl
Q. catalyst fines
R. coke fines
S. formaldehyde
T. aromatic amines
U. copper

can include nitrogen, carbon dioxide, hydrogen sulfide, and helium. Organic and inorganic particulate emissions, such as coke fires or catalyst fires, can be generated from some processes (U.S. EPA 1991).

—*Larry W. Canter*

Reference

U.S. Environmental Protection Agency (EPA). 1991. *Handbook: Control technologies for hazardous air pollutants.* EPA/625/6-91/014. Cincinnati, Ohio. (June) 2-1 to 2-13.

5.3
HAPs FROM SYNTHETIC ORGANIC CHEMICAL MANUFACTURING INDUSTRIES

The Synthetic Organic Chemical Manufacturing Industry (SOCMI), as a source category, emits a larger volume of a variety of HAPs compared to other source categories (see Table 5.3.1). In addition, individual SOCMI sources tend to be located close to the population. As such, components of SOCMI sources have been subject to various federal, state, and local air pollution control rules. However, the existing rules do not comprehensively regulate emissions for all organic HAPs emitted from all emissions points at both new and existing plants.

By describing hazardous organic national emission standards for air pollutants (NESHAP), or the HON, this section describes the emission points common to all SOCMI manufacturing processes and the maximum achievable control technology (MACT) required for reducing these emissions.

Hazardous Organic NESHAP

The HON is one of the most comprehensive rules issued by the EPA. It covers more processes and pollutants than

TABLE 5.3.1 EMISSION POTENTIAL ACCORDING TO BASIC MANUFACTURING CATEGORY

Emission Potential[a]	Category	% Total Industry Emissions (U.S.)
1	Chemical synthesis	64
2	Fermentation	19
3	Extraction[b]	7
4	Formulation	5
5	Other[c]	5

Notes:
[a]Decreasing order.
[b]Listed as botanicals.
[c]Includes research and development, animal sources, and biological products.

previous EPA air toxic programs (40 CFR Part 63). For example, one major portion of the rule applies to sources that produce any of the 396 SOCMI products (see Table 5.3.2) that use any of the 112 organic HAPs (see Table 5.3.3) either in a product or as an intermediate or reactant. An additional 37 HAPs are regulated under another part of the HON (40 CFR Part 63). The HON lists 189 HAPs regulated under the air toxic program.

The focus of this rule is the SOCMI. For purposes of the MACT standard, a SOCMI manufacturing plant is viewed as an assortment of equipment—process vents, storage tanks, transfer racks, and wastewater streams—all of which emit HAPs. The HON requires such plants to monitor and repair leaks to eliminate fugitive emissions and requires controls to reduce toxics coming from discrete emission points to minuscule concentrations. Table 5.3.4 summarizes the impacts of these emission sources.

PROCESS VENTS

A process vent is a gas stream that is continuously discharged during the unit operation from an air oxidation unit, reactor process unit, or distillation operation within a SOCMI chemical process. Process vents include gas streams discharged directly to the atmosphere after diversion through a product recovery device. The rule applies only to the process vents associated with continuous (nonbatch) processes and emitting vent streams containing more than 0.005 wt % HAP. The process vent provisions do not apply to vents from control devices installed to comply with wastewater provisions. Process vents exclude relief valve discharges and other fugitive leaks but include vents from product accumulation vessels.

Halogenated streams that use a combustion device to comply with 98% or 20 parts per million by volume (ppmv) HAP emissions must vent the emissions from the combustion device to an acid gas scrubber before venting to the atmosphere.

TABLE 5.3.2 SOCML

	Chemical Name[a]	
Acenaphthene	Chloroacetophenone (2-)	Dihydroxybenzoic acid (Resorcylic acid)
Acetal	Chloroaniline (p-)	Dilsodecyl phthalate
Acetaldehyde	Chlorobenzene	Dilsooctyl phthalate
Acetaldol	Chlorodifluoroethane	Dimethylbenzidine (3,3'-)
Acetamide	Chlorodifluoromethane	Dimethyl ether
Acetanilide	Chloroform	Dimethylformamide (N,N-)
Acetic acid	Chloronaphthalene	Dimethylhydrazine (1,1-)
Acetic anhydride	Chloronitrobenzene (1,3-)	Dimethyl phthalate
Acetoacetanilide	Chloronitrobenzene (o-)	Dimethyl sulfate
Acetone	Chloronitrobenzene (p-)	Dimethyl terephthalate
Acetone cyanohydrin	Chlorophenol (m-)	Dimethylamine
Acetonitrile	Chlorophenol (o-)	Dimethylaminoethanol (2-)
Acetophenone	Chlorophenol (p-)	Dimethylaniline (N,N)
Acrolein	Chloroprene	Dinitrobenzenes (NOS)
Acrylamide	Chlorotoluene (m-)	Dinitrophenol (2,4-)
Acrylic acid	Chlorotoluene (o-)	Dinitrotoluene (2,4-)
Acrylonitrile	Chlorotoluene (p-)	Dioxane
Adiponitrile	Chlorotrifluorourethane	Dioxolane (1,3-)
Alizarin	Chrysene	Diphenyl methane
Alkyl anthraquinones	Cresol and cresylic acid (m-)	Diphenyl oxide
Allyl alcohol	Cresol and cresylic acid (o-)	Diphenyl thiourea
Allyl chloride	Cresol and cresylic acid (p-)	Diphenylamine
Allyl cyanide	Cresols and cresylic acids (mixed)	Dipropylene glycol
Aminophenol sulfonic acid	Crotonaldehyde	Di(2-methoxyethyl)phthalate
Aminophenol (p-)	Cumene	Di-o-tolyguanidine
Aniline	Cumene hydroperoxide	Dodecyl benzene (branched)
Aniline hydrochloride	Cyanoacetic acid	Dodecyl phenol (branched)
Anisidine (o-)	Cyanoformamide	Dodecylaniline
Anthracene	Cyclohexane	Dodecylbenzene (n-)
Anthraquinone	Cyclohexanol	Dodecylphenol
Azobenzene	Cyclohexanone	Epichlorohydrin
Benzaldehyde	Cyclohexylamine	Ethane
Benzene	Cyclooctadienes	Ethanolamine
Benzenedisulfonic acid	Decahydronaphthalene	Ethyl acrylate
Benzenesulfonic acid	Diacetoxy-2-Butene (1,4-)	Ethylbenzene
Benzil	Dialyl phthalate	Ethyl chloride
Benzilic acid	Diaminophenol hydrochloride	Ethyl chloroacetate
Benzoic acid	Dibromomethane	Ethylamine
Benzoin	Dibutoxyethyl phthalate	Ethylaniline (n-)
Benzonitrile	Dichloroaniline (inbred isomers)	Ethylaniline (o-)
Benzophenone	Dichlorobenzene (p-)	Ethylcellulose
Benzotrichloride	Dichlorobenzene (m-)	Ethylcyanoacetate
Benzoyl chloride	Dichlorobenzene (o-)	Ethylene carbonate
Benzyl acetate	Dichlorobenzidine (3,5-)	Ethylene dibromide
Benzyl alcohol	Dichlorodifluoromethane	Ethylene glycol
Benzyl benzoate	Dichloroethane (1,2-) (Ethylene dichloride) (EDC)	Ethylene glycol diacetate
Benzyl chloride	Dichloroethyl ether	Ethylene glycol dibutyl ether
Benzyl dichloride	Dichloroethylene (1,2-)	Ethylene glycol diethyl ether (1,2-diethoxyethane)
Biphenyl	Dichlorophenol (2,4-)	Ethylene glycol dimethyl ether
Bisphenol A	Dichloropropene (1,3-)	Ethylene glycol monoacetate
Bis(Chloromethyl)Ether	Dichlorotetrafluoroethane	Ethylene glycol monobutyl ether acetate
Bromobenzene	Dichloro-1-butene (3,4-)	Ethylene glycol monobutyl ether
Bromoform	Dichloro-2-butene (1,4-)	Ethylene glycol monoethyl ether acetate
Bromonaphthalene	Diethanolamine	Ethylene glycol monoethyl ether
Butadiene (1,3-)	Diethyl phthalate	Ethylene glycol monohexyl ether
Butanediol (1,4-)	Diethyl sulfate	Ethylene glycol monomethyl ether acetate
Butyl acrylate (n-)	Diethylamine	Ethylene glycol monomethyl ether
Butylbenzyl phthalate	Diethylaniline (2,6-)	Ethylene glycol monooctyl ether
Butylene glycol (1,3-)	Diethylene glycol	Ethylene glycol monophenyl ether
Butyrolacetone	Diethylene glycol dibutyl ether	Ethylene glycol monopropyl ether
Caprolactam	Diethylene glycol diethyl ether	Ethylene oxide
Carbaryl	Diethylene glycol dimethyl ether	Ethylenediamine
Carbazole	Diethylene glycol monobutyl ether acetate	Ethylenediamine tetracetic acid
Carbon disulfide	Diethylene glycol monobutyl ether	Ethylenimine (Aziridine)
Carbon tetrabromide	Diethylene glycol monoethyl ether acetate	Ethylhexyl acrytate (2-isomer)
Carbon tetrachloride	Diethylene glycol monoethyl ether	Fluoranthene
Carbon tetrafluoride	Diethylene glycol monohexyl ether	Formaldehyde
Chloral	Diethylene glycol monomethyl ether acetate	Formamide
Chloroacetic acid	Diethylene glycol monomethyl ether	Formic acid

Continued on next page

TABLE 5.3.2 *Continued*

Fumaric acid	Naphthylamine sulfonic acid (2,1-)	Styrene
Glutaraldehyde	Naphthylamine (1-)	Succinic acid
Glyceraldehyde	Naphthylamine (2-)	Succinonitrile
Glycerol	Nitroaniline (m-)	Sulfanilic acid
Glycerol tri(polyoxypropylene)ether	Nitroaniline (o-)	Sulfolane
Glycine	Nitroanisole (o-)	Tartaric acid
Glyoxal	Nitroanisole (p-)	Terephthalic acid
Hexachlorobenzene	Nitrobenzene	Tetrabromophthalic anhydride
Hexachlorobutadiene	Nitronaphthalene (1-)	Tetrachlorobenzene (1,2,4,5-)
Hexachloroethane	Nitrophenol (p-)	Tetrachloroethane (1,1,2,2-)
Hexadiene (1,4-)	Nitrophenol (o-)	Tetrachlorophthalic anhydride
Hexamethylenetetramine	Nitropropane (2-)	Tetraethyl lead
Hexane	Nitrotoluene (all isomers)	Tetraethylene glycol
Hexanetriol (1,2,6-)	Nitrotoluene (o-)	Tetraethylenepentamine
Hydroquinone	Nitrotoluene (m-)	Tetrahydrofuran
Hydroxyadipaldehyde	Nitrotoluene (p-)	Tetrahydronapthalene
Iminodiethanol (2,2-)	Nitroxylene	Tetrahydrophthalic anhydride
Isobutyl acrylate	Nonylbenzene (branched)	Tetramethylenediamine
Isobutylene	Nonylphenol	Tetramethylethylenediamine
Isophorone	N-Vinyl-2-Pyrrolidine	Tetramethyllead
Isophorone nitrile	Octene-1	Thiocarbanilide
Isophthalic acid	Octylphenol	Toluene
Isopropylphenol	Paraformaldehyde	Toluene 2,4 diamine
Lead phthalate	Paraldehyde	Toluene 2,4 diisocyanate
Linear alkylbenzene	Pentachlorophenol	Toluene diisocyanates (mixture)
Maleic anhydride	Pentaerythritol	Toluene sulfonic acids
Maleic hydrazide	Peracetic acid	Toluenesulfonyl chloride
Malic acid	Perchloroethylene	Toluidine (o-)
Metanilic acid	Perchloromethyl mercaptan	Trichloroaniline (2,4,6-)
Methacrylic acid	Phenanthrene	Trichlorobenzene (1,2,3-)
Methanol	Phenetidine (p-)	Trichlorobenzene (1,2,4-)
Methionine	Phenol	Trichloroethane (TCA) (1,1,1-)
Methyl acetate	Phenolphthalein	TCA (1,1,2-)
Methyl acrylate	Phenolsulfonic acids (all isomers)	Trichloroethylene (TCE)
Methyl bromide	Phenyl anthranilic acid (all isomers)	Trichlorofluoromethane
Methyl chloride	Phenylenediamine (p-)	Trichlorophenol (2,4,5-)
Methyl ethyl ketone	Phloroglucinol	Trichlorotrifluoroethane (1,2,2-1,1,2)
Methyl formate	Phosgene	Triethanolamine
Methyl hydrazine	Phthalic acid	Triethylamine
Methyl isobutyl carbinol	Phthalic anhydride	Triethylene glycol
Methyl isocyanate	Phthalimide	Triethylene glycol dimethyl ether
Methyl mercaptan	Phthalonitrile	Triethylene glycol monoethyl ether
Methyl methacrylate	Picoline (b-)	Triethylene glycol monomethyl ether
Methyl phenyl carbinol	Piperazine	Trimethylamine
Methyl tert-butyl ether	Polyethylene glycol	Trimethylcyclohexanol
Methylamine	Polypropylene glycol	Trimethylcyclohexanone
Methylaniline (n-)	Propiolactone (beta-)	Trimethylcyclohexylamine
Methylcyclohexane	Propionaldehyde	Trimethylolpropane
Methylcyclohexanol	Propionic acid	Trimethylpentane (2,2,4-)
Methylcyclohexanone	Propylene carbonate	Tripropylene glycol
Methylene chloride	Propylene dichloride	Vinyl acetate
Methylene dianiline (4,4'-isomer)	Propylene glycol	Vinyl chloride
Methylene diphenyl diisocyanate (4,4'-) (MDI)	Propylene glycol monomethyl ether	Vinyl toluene
Methylionones (a-)	Propylene oxide	Vinylcyclohexane (4-)
Methylpentynol	Pyrene	Vinylidene chloride
Methylstyrene (a-)	Pyridine	Vinyl(N)-pyrrolidone (2-)
Naphthalene	p-tert-Butyl toluene	Xanthates
Naphthalene sulfonic acid (a-)	Quinone	Xylene sulfonic acid
Naphthalene sulfonic acid (b-)	Resorcinol	Xylenes (NOS)
Naphthol (a-)	Salicylic acid	Xylene (m-)
Naphthol (b-)	Sodium methoxide	Xylene (o-)
Naphtholsulfonic acid (1-)	Sodium phenate	Xylene (p-)
Naphthylamine sulfonic acid (1,4-)	Stilbene	Xylenol

Source: Code of Federal Regulations, Title 40, Part 63.104, *Federal Register 57,* (31 December 1992).

[a]Isomer means all structural arrangements for the same number of atoms of each element and does not mean salts, esters, or derivatives.

TABLE 5.3.3 ORGANIC HAPs

Chemical Name[a,b]

Acetaldehyde	Dimethylformamide	Methylene chloride (Dichloromethane)
Acetamide	1,1-Dimethylhydrazine	Methylene diphenyl diisocyanate (MDI)
Acetonitrile	Dimethyl phthalate	4,4'-Methylenedianiline
Acetophenone	Dimethyl sulfate	Naphthalene
Acrolein	2,4-Dinitrophenol	Nitrobenzene
Acrylamide	2,4-Dinitrotoluene	4-Nitrophenol
Acrylic acid	1,4-Dioxane (1,4-Diethyleneoxide)	2-Nitropropane
Acrylonitrile	1,2-Diphenylhydrazine	Phenol
Allyl chloride	Epichlorohydrin (1-Chloro-2,3-	p-Phenylenediamine
Aniline	epoxypropane)	Phosgene
o-Anisidine	Ethyl acrylate	Phthalic anhydride
Benzene	Ethylbenzene	Polycyclic organic matter[d]
Benzotrichloride	Ethyl chloride (Chloroethane)	Propiolactone (beta-isomer)
Benzyl chloride	Ethylene dibromide (Dibromoethane)	Propionaldehyde
Biphenyl	Ethylene dichloride (1,2-Dichloroethane)	Propylene dichloride (1,2-
Bis(chloromethyl)ether	Ethylene glycol	Dichloropropane)
Bromoform	Ethylene oxide	Propylene oxide
1,3-Butadiene	Ethylidene dichloride (1,1-	Quinone
Caprolactam	Dichloroethane)	Styrene
Carbon disulfide	Formaldehyde	1,1,2,2-Tetrachloroethane
Carbon tetrachloride	Glycol ethers[c]	Tetrachloroethylene (Perchloroethylene)
Chloroacetic acid	Hexachlorobenzene	Toluene
2-Chloroacetophenone	Hexachlorobutadiene	2,4-Toluene diamine
Chlorobenzene	Hexachloroethane	2,4-Toluene diisocyanate
Chloroform	Hexane	o-Toluidine
Chloroprene	Hydroquinone	1,2,4-Trichlorobenzene
Cresols and cresylic acids (mixed)	Isophorone	1,1,2-TCA
o-Cresol and o-cresylic acid	Maleic anhydride	TCB
m-Cresol and m-cresylic acid	Methanol	2,4,5-Trichlorophenol
p-Cresol and p-cresylic acid	Methyl bromide (Bromomethane)	Triethylamine
Cumene	Methyl chloride (Chloromethane)	2,2,4-Trimethylpentane
1,4-Dichlorobenzene(p-)	Methyl chloroform (1,1,1-	Vinyl acetate
3,3'-Dichlorobenzidine	Trichloroethane)	Vinyl chloride
Dichloroethyl ether (Bis(2-	Methyl ethyl ketone (2-Butanone)	Vinylidene chloride (1,1-
chloroethyl)ether)	Methyl hydrazine	Dichloroethylene)
1,3-Dichloropropene	Methyl isobutyl ketone (Hexone)	Xylenes (isomers and mixtures)
Diethanolamine	Methyl isocyanate	o-Xylene
N,N-Dimethylaniline	Methyl methacrylate	m-Xylene
Diethyl sulfate	Methyl tert-butyl ether	p-Xylene
3,3'-Dimethylbenzidine		

Source: 40 *CFR* Part 63.104.

Notes: [a]For all listings containing the word "Compounds" and for glycol ethers, the following applies: Unless otherwise specified, these listings include any unique chemical substance that contains the named chemical (i.e., antimony, arsenic) as part of that chemical's infrastructure.

[b]Isomer means all structural arrangements for the same number of atoms of each pigment and does not mean salts, esters, or derivatives.

[c]Includes mono- and di-ethers of ethylene glycol, diethylene glycol, and triethylene glycol R-(OCH$_2$CH$_2$)$_n$-OR where n = 1, 2, or 3; R = alkyl or aryl groups; and R' = R, H, or groups which, when removed, yield glycol ethers with the structure: R-(OCH$_2$CH$_2$)$_n$-OH Polymers are excluded from the glycol category.

[d]Includes organic compounds with more than one benzene ring, and which have a boiling point greater than or equal to 100°C.

STORAGE VESSELS

A storage vessel is a tank or vessel storing the feed or product of a SOCMI chemical manufacturing process when the liquid is on the list of HAPs (see Table 5.3.3). The storage vessel provisions require that one of the following control systems is applied to storage vessels:

- An internal floating roof with proper seals and fittings
- An external floating roof with proper seals and fittings
- An external floating roof converted to an internal floating roof with proper seals and fittings
- A closed-vent system with 95% efficient control

TABLE 5.3.4   NATIONAL PRIMARY AIR POLLUTION IMPACTS IN THE FIFTH YEAR[a]

| Emission Points | Baseline Emissions (Mg/yr) | | Emission Reductions | | | |
| | HAP | VOC[c] | (Mg/yr) | | (Percent) | |
			HAP	VOC[b]	HAP	VOC[b]
Equipment leaks	66,000	84,000	53,000	68,000	80	81
Process vents	317,000	551,000	292,000	460,000	92	83
Storage vessels	15,200	15,200	5,560	5,560	37	37
Wastewaster collection and treatment operations	198,000	728,000	124,000	452,000	63	62
Transfer loading operations	900	900	500	500	56	56
Total	597,000	1,380,000	475,000	986,000	80	71

Source: Code of Federal Regulations, Title 40, part 63; Clean Air Act Amendments, amended 1990, Section 112.

[a]These numbers represent estimated values for the fifth year. Existing emission points contribute 84% of the total. Emission points associated with chemical manufacturing process equipment built in the first 5 yr of the standard contribute 16% of the total.

[b]The VOC estimates consist of the sum of the HAP estimates and the nonHAP VOC estimates.

TRANSFER OPERATIONS

Transfer operations are the loading of liquid products on the list of HAPs from a transfer rack within the SOCMI chemical manufacturing process into a tank truck or railcar. The transfer rack includes the total loading arms, pumps, meters, shutoff valves, relief valves, and other piping and valves necessary to load trucks or railcars.

The proposed transfer provisions control transfer racks to achieve a 98% organic HAP reduction or an outlet concentration of 20 ppmv. Combustion devices or product recovery devices can be used. Again, halogenated streams that use combustion devices to comply with the 98% or 20 ppmv emission reduction must vent the emissions from the combustion device to an acid scrubber before venting to the atmosphere.

WASTEWATER

The wastewater to which the proposed standard applies is any organic HAP-containing water or process fluid discharged into an individual drain system. This wastewater includes process wastewater, maintenance-turnaround wastewater, and routine and routine-maintenance wastewater. Examples of process wastewater streams include those from process equipment, product or feed tank drawdown, cooling water blowdown, steam trap condensate, reflux, and fluid drained into and material recovered from waste management units. Examples of maintenance-turnaround wastewater streams are those generated by the descaling of heat exchanger tubing bundles, cleaning of distillation column traps, and draining of pumps into individual drain system. A HAP-containing wastewater stream is a wastewater stream that has a HAP concentration of 5 parts per million by weight (ppmw) or greater and a flow rate of 0.02 liters per minute (lpm) or greater.

The proposed process water provisions include equipment and work practice provisions for the transport and handling of wastewater streams between the point of generation and the wastewater treatment processes. These provisions include the use of covers, enclosures, and closed-vent systems to route organic HAP vapors from the transport and handling equipment. The provisions also require the reduction of volatile organic HAP (VOHAP) concentrations in wastewater streams.

SOLID PROCESSING

The product of synthetic organic processes can be in solid, liquid, or gas form. Emissions of solid particulates are also of concern. One reason is that particulate emissions occur with drying, packaging, and formulation operations. Additionally, these emissions can be in the respirable size range. Within this range, a significant fraction of the particulates can be inhaled directly into the lungs, thereby enhancing the likelihood of being absorbed into the body and damaging lung tissues.

Toxic Pollutants

Table 5.3.3 shows that halogenated aliphatics are the largest class of priority toxics. These chemicals can cause damage to the central nervous system and liver. Phenols are carcinogenic in mice; their toxicity increases with the

TABLE 5.3.5 HEALTH EFFECTS OF SELECTED HAPS

Pollutant	Major Health Effects
Acryronitrile (CH_2═CH—C═N)	Dermatitis; haematological changes; headaches; irritation of eyes, nose, and throat; lung cancer
Benzene (C_6H_6)	Leukemia; neurotoxic symptoms; bone marrow injury including anaemia, and chromosome aberrations
Carbon disulfide (CS_2)	Neurologic and psychiatric symptoms, including irritability and anger; gastrointestinal troubles; sexual interferences
1,2 Dichloroethane ($C_2H_2Cl_2$)	Damage to lungs, liver, and kidneys; heart rhythm disturbances; effects on central nervous systems, including dizziness; animal mutagen and carcinogen
Formaldehyde (HC HO)	Chromosome aberrations; irritation of eyes, nose, and throat; dermatitis; respiratory tract infections in children
Methylene chloride (CH_2Cl_2)	Nervous system disturbances
Polychlorinated bi-phenyls (PCB) (coplanar)	Spontaneous abortions; congenital birth defects; bioaccumulation in food chains
Polychlorinated dibenzo-dioxins and furans	Birth defects; skin disorders; liver damage; suppression of the immune system
Polycyclic organic matter (POM) [including benzo(a)pyrene (BaP)]	Respiratory tract and lung cancers; skin cancers
Styrene (C_6H_5—CH═CH_2)	Central nervous system depression; respiratory tract irritations; chromosome aberrations; cancers in the lymphatic and haematopoietic tissues
Tetrachloroethylene (C_2Cl_4)	Kidney and genital cancers; lymphosarcoma; lung, cervical, and skin cancers; liver dysfunction; effects on central nervous system
Toluene (C_6H_5—CH_3)	Dysfunction of the central nervous system; eye irritation
TCE (C_2HCl_3)	Impairment of psychomotoric functions; skin and eye irritation; injury to liver and kidneys; urinary tract tumors and lymphomas
Vinyl chloride (CH_2═CHCl)	Painful vasospastic disorders of the hands; dizziness and loss of consciousness; increased risk of malformations, particularly of the central nervous systems; severe liver disease; liver cancer; cancers of the brain and central nervous system; malignancies of the lymphatic and haematopoietic system

Source: OECD.

degree of chlorination of phenolic molecules. Maleic anhydride and phthalic anhydride are irritants to the skin, eyes, and mucous membranes. Methanol vapor is irritating to the eyes, nose, and throat; this vapor explodes if ignited in an enclosed area.

Table 5.3.5 lists the health effects of selected HAPs. Because of the large number of HAPs, enumerating the potential health effects of the category as a whole is not possible. However, material safety data sheets (MSDS) for the HAPs are available from chemical suppliers on request, and handbooks such as the *Hazardous chemical data book* (Weiss 1980) provide additional information.

—*David H.F. Liu*

References

Code of Federal Regulations. Title 40, Part 63. *Federal Register* 57, (31 December 1992).

Weiss, G., ed. 1980. *Hazardous chemicals data book.* Park Ridge, N.J.: Noyes Data Corp.

5.4 ATMOSPHERIC CHEMISTRY

Pollutants enter the atmosphere primarily from natural sources and human activity. This pollution is called *primary pollution,* in contrast to *secondary pollution,* which is caused by chemical changes in substances in the atmosphere. Sulfur dioxides, nitric oxides, and hydrocarbons are major primary gaseous pollutants, while ozone is a secondary pollutant, the result of atmospheric photochemistry between nitric oxide and hydrocarbons.

Pollutants do not remain unchanged in the atmosphere after release from a source. Physical changes occur, especially through dynamic phenomena, such as movement and scattering in space, turbulent diffusion, and changes in the concentration by dilution.

Changes also result from the chemistry of the atmosphere. These changes are often simple, rapid chemical reactions, such as oxidation and changes in temperature to condense some gases and vapors to yield mist and droplets. After a long residence of some gaseous pollutants in the atmosphere, these gases convert into solid, finely dispersed substances. Solar conditions cause chemical reactions in the atmosphere among various pollutants and their supporting media. Figure 5.4.1 shows simplified schemes of the main chemical changes of pollutants in the atmosphere.

Basic Chemical Processes

A basic chemical process in the atmosphere is the oxidation of substances by atmospheric oxygen. Thus, sulfur dioxide (SO_2) is oxidized to sulfur trioxide (SO_3), and nitric oxide to nitrogen dioxide. Similarly, many organic substances are oxidized, for example, aldehydes to organic acids and unsaturated hydrocarbons. While pollutant clouds are transported and dispersed to varying degrees, they also age. Pollutant cloud aging is a complex combination of homogeneous and heterogeneous reactions and physical processes (such as nucleation, coagulation, and the Brownian motion). Chemically unlike species can make contact and further branch the complex pattern (see Figure 5.4.1). Table 5.4.1 summarizes the major removal reactions and sinks. Most of these reactions are not understood in detail.

Sulfur oxides, in particular SO_2, have been studied with respect to atmospheric chemistry. However, an understanding of the chemistry of SO_2 in the atmosphere is still far from complete. Most evidence suggests that the eventual fate of atmospheric SO_2 is oxidation to sulfate. One problem that complicates understanding atmospheric SO_2 processes is that reaction paths can be homogeneous and heterogeneous. Two processes convert SO_2 to sulfate: catalytical and photochemical.

CATALYTIC OXIDATION OF SO$_2$

In clear air, SO_2 is slowly oxidized to SO_3 by homogeneous reactions. However, studies show that the rate of SO_2 oxidation in a power plant plume can be 10 to 100 times the clear-air photooxidation rate (Gartrell, Thomas, and Carpenter 1963). Such a rapid rate of reaction is similar to that of oxidation in solution in the presence of a catalyst.

SO_2 dissolves readily in water droplets and can be oxidized by dissolved oxygen in the presence of metal salts, such as iron and manganese. The overall reaction can be expressed as:

$$2\,SO_2 + 2\,H_2O + O_2 \xrightarrow{\text{catalyst}} 2\,H_2SO_4 \qquad \text{5.4(1)}$$

Catalysts for the reaction include sulfates and chlorides of manganese and iron which usually exists in air as sus-

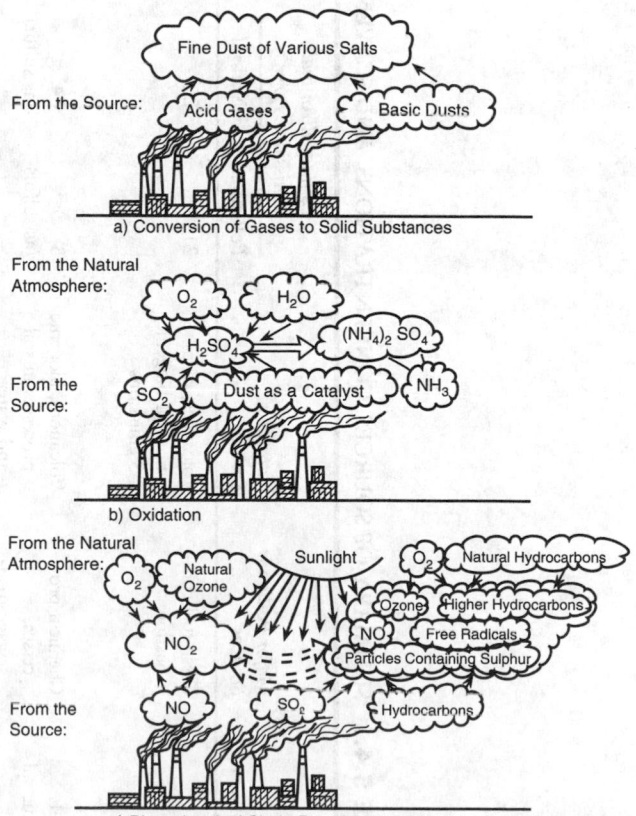

a) Conversion of Gases to Solid Substances

b) Oxidation

c) Photochemical Chain Reactions—Principle of Smog Formation

FIG. 5.4.1 Examples of chemical reactions in the atmosphere.

TABLE 5.4.1 SUMMARY OF SOURCES, CONCENTRATIONS, AND SCAVENGING PROCESSES OF ATMOSPHERIC TRACE GASES

Contaminant	Major Pollutant Sources	Natural Sources	Estimated Annual Emissions Tg/yr*		Atmospheric Background Concentrations	Estimated Atmospheric Residence Time	Removal Reactions and Sinks	Remarks
			Pollutants	Natural				
SO_2	Fossil fuel combustion	Volcanoes, reactions of biogenic S emissions	212[1]	20[2]	About 0.1 ppb[3]	1–4 days	Oxidation to sulfate by photochemical reactions or in liquid droplets	High reaction rates in summer due to photochemical processes
H_2S and organic sulfides[4]	Chemical processes, sewage treatment	Volcanoes, biogenic processes in soil and water	3 (as sulfur)	84[5] (as sulfur)	H_2S: 0.05–0.1 ppb; COS: 0.5 ppb[6]; CS_2: 0.05 ppb[6]	H_2S: 1–2 days; COS: 1–2 yr[6]	Oxidation to SO_2 and SO_4	Atmospheric data are incomplete; COS residence time can be 20 yr.[2]
CO	Auto exhaust, general combustion	Forest fires, photochemical reactions	700[7]	2100[7]	0.1–0.2 ppm (N. Hemisphere) 0.04–0.06 ppm (S. Hemisphere)	1–3 mon	Photochemical reactions with CH_4, and OH	No long-term changes in the atmosphere have been detected.
NO, NO_2	Combustion	Biogenic processes in soil, lightning	75[8] (as NO_2)	180[9] (as NO_2)	About 0.1 ppb[10]	2–5 days	Oxidation to nitrate	Natural processes mostly estimated; background concentrations are in doubt but may be as low as 0.01 ppb.
NH_3	Waste treatment, combustion	Biogenic processes in soil	6[11]	260[9]	About 10 ppm[9]	1–7 days	Reaction with SO_2 to form $(NH_4)_2SO_4$	Atmospheric measurements are sparse.
N_2O	Small amounts from combustion	Biogenic processes in soil	3[12]	340[13]	300 ppb	20–100 yr	Photochemical in stratosphere	Some estimates place natural source at 100 Tg or less.[12]
CH_4	Combustion, natural gas leakage	Biogenic processes in soil and water	160[14]	1050[14]	1.5 ppm	8 yr[15]	Reaction with OH to form CO	Pollutant source includes 60 Tg yr^{-1} from biomass burning.
Isoprene and terpenes	None	Biogenic plant emissions	None	830[7]	0	1–2 hr	Photochemical reactions with OH and O_3	Not found in ambient atmosphere away from source regions
Total nonCH$_4$ hydrocarbons	Combustion	Biogenic processes in soil and vegetation	40[17]	2×10^4 [16]	0–1 μg m^{-3} for C$_2$'s	Hours to a few days	Photochemical reactions with NO and O_3	Concentration given for C$_2$s in rural atmosphere

Species	Anthropogenic source	Natural source	Amount	Amount	Concentration	Residence time	Removal process	Remarks
CO_2	Combustion	Biological processes	22,000[16]	10^6 [13]	345 ppm (1981)	2–4 yr	Biogenic processes, photosynthesis, absorption in oceans	Forest destruction and changes in earth's biomass may add 20–30 × 10^3 Tg CO_2/yr to atmosphere[18]
CH_3Cl	Combustion	Oceanic biological processes	2[19]	4–6[16,19]	600 ppt[19,20]	1–2 yr[19]	Stratospheric reactions	Photochemical reactions in stratosphere may impact on O_3 layer
HCl, Cl_2	Combustion, Cl manufacturing	Atmospheric reactions of NaCl, volcanoes	4[21]	100–200[23]	About 0.5 ppb[23]	About 1 wk	Precipitation	Volcanoes can release 10–20 Tg Cl yr^{-1} [22]

Source: Elmer Robinson, (Pullman, Wash.: Washington State University).

Notes: *Tg/yr = 10^{12} gm/yr or 10^6 metric tn/yr

[1]Based on 1978 global fuel usage and estimated sulfur contents.

[2]Major reference is R.D. Cadle, 1980, *Rev. Geophys. Space Phys.* 18, 746–752.

[3]P.J. Maroulis, A.L. Torres, A.B. Goldberg, and A.R. Bandy, 1980, *J. Geophys. Res.* 85, 7345–7349.

[4]Includes COS, CS_2, $(CH_3)_2S$, $(CH_3)_2S_2$, CH_4, and SH.

[5]Adapted from D.F. Adams, S.O. Farwell, E. Robinson, and M.R. Pack, 1980, *Biogenic sulfur emissions in the SURE region.* Final report by Washington State University for Electric Power Research Institute, EPRI Report No. EA-1516.

[6]A.L. Torres, P.J. Maroulis, A.B. Goldberg, and A.R. Bandy, 1980, *J. Geophys. Res.* 85, 7357–7360.

[7]P.R. Zimmerman, R.B. Chatfield, J. Fishman, P.J. Crutzen, and P.L. Hanst, 1978, *Geophys. Res. Lett.* 5, 679–682.

[8]Based on 1978 global combustion estimates.

[9]I.E. Galbally, *Tellus* 27, 67–70.

[10]Approximate value combining values given in several references.

[11]R. Söderlund, and B.H. Svensson, 1976, The global nitrogen cycle, in *SCOPE Report 7*, Swedish National Science Research Council, Stockholm.

[12]1978 fuel usage figures apply to the following references: R.F. Weiss, and H. Craig, *Geophys. Res. Lett.* 3, 751–753; and D. Pierotti, and R.A. Rasmussen, 1976, *Geophys. Res. Lett.* 3, 265–267.

[13]E. Robinson, and R.C. Robbins, Emissions, concentrations, and fate of gaseous atmospheric pollutants, in *Air pollution control*, edited by W. Strauss, 1–93, Part 2 of New York: Wiley.

[14]J.C. Sheppard, H. Westberg, J.F. Hopper, and K. Ganesan, 1982, *J. Geophys. Res.* 87, 1305–1312.

[15]L.E. Heidt, J.P. Krasnec, R.A. Lueb, W.H. Pollock, B.E. Henry, and P.J. Crutzen, 1980, *J. Geophys. Res.* 85, 7329–7336.

[16]R.E. Graedel, 1979, *J. Geophys. Res.* 84, 273–286.

[17]Reference 13 tabulation updated to approximate 1978 emissions.

[18]G.M. Woodwell, R.H. Whittaker, W.A. Reiners, G.E. Likens, C.C. Delwiche, and D.B. Botkin, 1978, *Science* 199, 141–146.

[19]R.A. Rasmussen, L.E. Rasmussen, M.A.K. Khalil, and R.W. Dalluge, 1980, *J. Geophys. Res.* 85, 7350–7356.

[20]E. Robinson, R.A. Rasmussen, J. Krasnec, D. Pierotti, and M. Jakubovic, 1977, *Atm. Environ.* 11, 213–215.

[21]J.A. Ryan, and N.R. Mukherjee, 1975, *Rev. Geophys. Space Phys.* 13, 650–658.

[22]R.D. Cadle, 1980, *Rev. Geophys. Space Phys.* 18, 746–752.

[23]Based on estimated reaction of NaCl to form Cl_2.

pended particles. At high humidities, these particles act as condensation nuclei or undergo hydration to become solution droplets. The oxidation then proceeds by absorption of both SO_2 and O_2 by the liquid aerosols with subsequent chemical reactions in the liquid phase. The oxidation slows considerably when the droplets become highly acidic because of the decreased solubility of SO_2. However if sufficient ammonia is present, the oxidation process is not impeded by the accumulation of H_2SO_4. Measurements of particulate composition in urban air often show large concentrations of ammonium sulfate.

PHOTOCHEMICAL REACTIONS

In the presence of air, SO_2 is slowly oxidized to SO_3 when exposed to solar radiation. If water is present, the SO_2 rapidly converts to sulfuric acid. Since no radiation wavelengths shorter than 2900 Å reach the earth's surface and the dissociation of SO_2 to SO and O is possible only for wavelengths below 2180 Å, the primary photochemical processes in the lower atmosphere following absorption by SO_2 involve activated SO_2 molecules and not direct dissociation. Thus, the conversion of SO_2 to SO_3 in clear air is a result of a several-step reaction sequence involving excited SO_2 molecules, oxygen, and oxides of sulfur other than SO_2. In the presence of reactive hydrocarbons and nitrogen oxides, the conversion rate of SO_2 to SO_3 increases markedly. In addition, oxidation of SO_2 in systems of this type is frequently accompanied by aerosol formation.

A survey of possible reactions by Bufalini (1971) and Sidebottom et al. (1972) concludes that the most important oxidation step for the triplet state $^3SO_2$ from among those involving radiation only is:

$$^3SO_2 + O_2 \xrightarrow{hr} SO_3 + O \quad (3400 \text{ to } 4000 \text{ Å}) \quad 5.4(2)$$

Other primary substances absorbing UV radiation include sulfur and nitrogen oxides and aldehydes. UV radiation excites the molecules of these substances, which then react with atmospheric molecular oxygen to yield atomic oxygen. Analogous to SO_2 oxidation, aldehydes react as follows:

$$HCHO + O_2 \xrightarrow{hr} HCOOH + O \quad 5.4(3)$$

Atomic oxygen can also be formed by the following reactions:

$$H_2S + O_2 \longrightarrow H_2O + S + O \quad 5.4(4)$$

$$NO + O_2 \longrightarrow NO_2 + O \quad 5.4(5)$$

$$CH_4 + O_2 \longrightarrow CH_3OH + O \quad 5.4(6)$$

$$C_2H_6 + O_2 \longrightarrow C_2H_4 + H_2O + O \quad 5.4(7)$$

$$CO + O_2 \longrightarrow CO_2 + O \quad 5.4(8)$$

SO_2 and aldehydes react irreversibly, whereby the amount of atomic oxygen formed by these processes is relatively small and corresponds to the amount of SO_2 and aldehydes in the atmosphere. In the reaction of nitrogen dioxide, however, the absorption of UV radiation leads to the destruction of one bond between the nitrogen and oxygen atoms and to the formation of atomic oxygen and nitrogen oxide. Further reactions lead to the formation of atomic oxygen and nitrogen oxide as follows:

$$NO_2 \xrightarrow{hr} NO + O \quad 5.4(9)$$

$$NO_2 + O \longrightarrow NO + O_2 \quad 5.4(10)$$

$$O + O_2 \longrightarrow O_3 \quad 5.4(11)$$

The regenerated nitrogen dioxide can reenter the reaction, and this process can repeat until the nitrogen dioxide converts into nitric acid or reacts with organic substances to form nitrocompounds. Therefore, a low concentration of nitrogen dioxide in the atmosphere can lead to the formation of a considerable amount of atomic oxygen and ozone. This nitrogen dioxide is significant in the formation of oxidation smog.

Olefins with a large number of double bonds also react photochemically to form free radicals. Inorganic substances in atomic form in the atmosphere also contribute to the formation of free radicals. On reacting with oxygen, some free radicals form peroxy compounds from which new peroxides or free radicals are produced that can cause polymerization of olefins or be a source of ozone. The photochemistry is described by the thirty-six reactions for the twenty-seven species in Table 5.4.2 which includes four reactive hydrocarbon groups: olefins, paraffins, aldehydes, and aromatics.

Particulates

Atmospheric reactions are strongly affected by the number of suspended solid particles and their properties. The particles supply the surfaces on which reactions can occur thus acting as catalysts. They can also affect the absorption spectrum through the adsorption of gases (i.e., in the wavelength range of adsorbed radiation) and thus affect the intensities of radiation absorption and photochemical reactions. Moreover, solid particles can react with industrially emitted gases in common chemical reactions.

Combustion, volcanic eruptions, dust storms, and sea spray are a few processes that emit particles. Many particulates in the air are metal compounds that can catalyze secondary reactions in the air or gas phase to produce aerosols as secondary products. Physical processes such as nucleation, condensation, absorption, adsorption, and coagulation are responsible for determining the physical properties (i.e., the number concentration, size distribution, optical properties, and settling properties) of the formed aerosols. Particles below 0.1 μ, (known as Aitken nuclei), although not significant by gravity, are capable of serving as condensation nuclei for clouds and fog. Secondary effects are the results of gas-phase chemistry and photochemistry that form aerosols.

TABLE 5.4.2 GENERALIZED CHEMICAL KINETIC MECHANISM IN PHOTOCHEMICAL BOX MODEL

1.	$NO_2 \xrightarrow{hv} NO + O$	19. $OLEF + O \longrightarrow RO_2 + ALD + HO_2$
2.	$O + O_2 + M \longrightarrow O_3 + M$	20. $OLEF + O_3 \longrightarrow RO_2 + ALD + HO_2$
3.	$O_3 + NO \longrightarrow NO_2 + O_2$	21. $OLEF + HO \longrightarrow RO_2 + ALD$
4.	$O_3 + NO_2 \longrightarrow NO_3 + O_2$	22. $PARAF + HO \longrightarrow RO_2$
5.	$NO_3 + NO \longrightarrow 2NO_2$	23. $ALD \xrightarrow{hv} 0.5RO_2 + 1.5HO_2 + 1.0CO$
6.	$NO_3 + NO_2 + H_2O \longrightarrow 2HONO_2$	24. $ALD + HO \longrightarrow 0.5RlO_2 + 0.5HO_2 + HO_2$
7.	$HONO \xrightarrow{hv} HO + NO$	25. $RO_2 + NO \longrightarrow RO + NO_2$
8.	$HO + NO \xrightarrow{(O_3)} HO_2 + CO_2$	26. $RO + O_2 \longrightarrow ALD + HO_2$
9.	$HO_2 + NO_2 \longrightarrow HONO + O_2$	27. $RlO_2 + NO_2 \longrightarrow PAN$
10.	$HO_2 + NO \longrightarrow HO + NO_2$	28. $RO + NO_2 \longrightarrow RONO_2$
11.	$HO_2 + NO_2 + M \longrightarrow HOONO_2 + M$	29. $RO_2 + O_3 \longrightarrow RO + 2O_2$
12.	$HOONO_2 \longrightarrow HO_2 + NO_2$	30. $RlO_2 + NO \xrightarrow{(O_3)} RO_2 + NO_2$
13.	$HO + HONO \longrightarrow NO_2 + H_2O$	31. $PAN \longrightarrow RlO_2 + NO_2$
14.	$HO + NO_2 + M \longrightarrow HONO_2 + M$	32. $AROM + HO \xrightarrow{(O_3)} R_2O_2 + 2ALD + CO$
15.	$HO + NO + M \longrightarrow HONO + M$	33. $R_2O_2 + NO \longrightarrow R_2O + NO_2$
16.	$HO_2 + O_3 \longrightarrow HO + 2O_2$	34. $R_2O + O_2 \longrightarrow ALD + HO_2 + 2CO$
17.	$HO + O_3 \longrightarrow HO_2 + O_2$	35. $R_2O_2 + O_3 \longrightarrow R_2O + 2O_2$
18.	$HO_2 + HO_2 \longrightarrow H_2O_2 + O_2$	36. $RlO_2 + O_3 \longrightarrow RO_2 + 2O_2$

Source: K.L. Demerjian and K.L. Schere, 1979, *Proceedings, Ozone/Oxidants: Interactions with Total Environment II* (Pittsburgh: Air Pollution Association).

Note: M stands for any available atom or molecule which by collision with the reaction product carries off the excess energy of the reaction and prevents the reaction product from flying apart as soon as it is formed.

The removal of particles (aerosols and dust) from the atmosphere involves dry deposition by sedimentation, washout by rainfalls and snowfalls, and dry deposition by impact on vegetation and rough surfaces.

A volcanic eruption is a point source which has local effects (settling of particles and fumes) and global effects since the emissions can circulate in the upper atmosphere (i.e., the stratosphere) and increase the atmospheric aerosol content.

From the point of view of atmospheric protection, some of these reactions are favorable as they quickly yield products that are less harmful to humans and the biosphere. However, the products of some reactions are even more toxic than the reactants, an example being peroxylacetyl nitrate.

The atmospheric chemical reactions of solid and gaseous substances in industrial emissions are complex. A deeper analysis and description is beyond the scope of this section.

Long-Range Planning

Other long-range problems caused by atmospheric chemical reactions occur in addition to those of sulfur and nitrogen compounds. States and provinces must formulate strategies to achieve oxidant air quality standards. They must assess both the transport of oxidants from outside local areas and the estimated influx of precursors that create additional oxidants. Lamb and Novak (1984) give the principal features of a four-layer regional oxidant model (see Figure 5.4.2) designed to simulate photochemical processes over time scales of several days and space scales

of 1000 km. Temporal resolution yields hourly concentrations from time steps of 30 min and spatial resolution of about 18 km. The model includes the following processes:

- Horizontal transport
- Photochemistry using thirty-five reactions of twenty-three species
- Nighttime chemistry
- Nighttime wind shear, thermal stratification, and turbulent episodes associated with nocturnal jet
- Cumulus cloud effects, including venting from the mixed layer and photochemical reactions caused by their shadows
- Mesoscale vertical motion induced by terrain and horizontal divergence
- Mesoscale eddy effects on trajectories and growth rates of urban plumes
- Terrain effects on flow and diffusion
- Subgrid-scale chemical processes due to subgrid-scale emissions
- Natural sources of hydrocarbons and nitrogen oxides
- Wet and dry removal processes

The model was initially applied to the northeastern quarter of the United States. A 1980 emissions inventory gathered data on nitrogen oxides, VOCs, carbon dioxide, sulfur oxides, and total suspended particulate matter. In the model, volatile organics are considered as four reactive classes: olefins, paraffins, aldehydes, and aromatics. Applying the model requires acquiring and preparing emission and meteorological information for an area and a

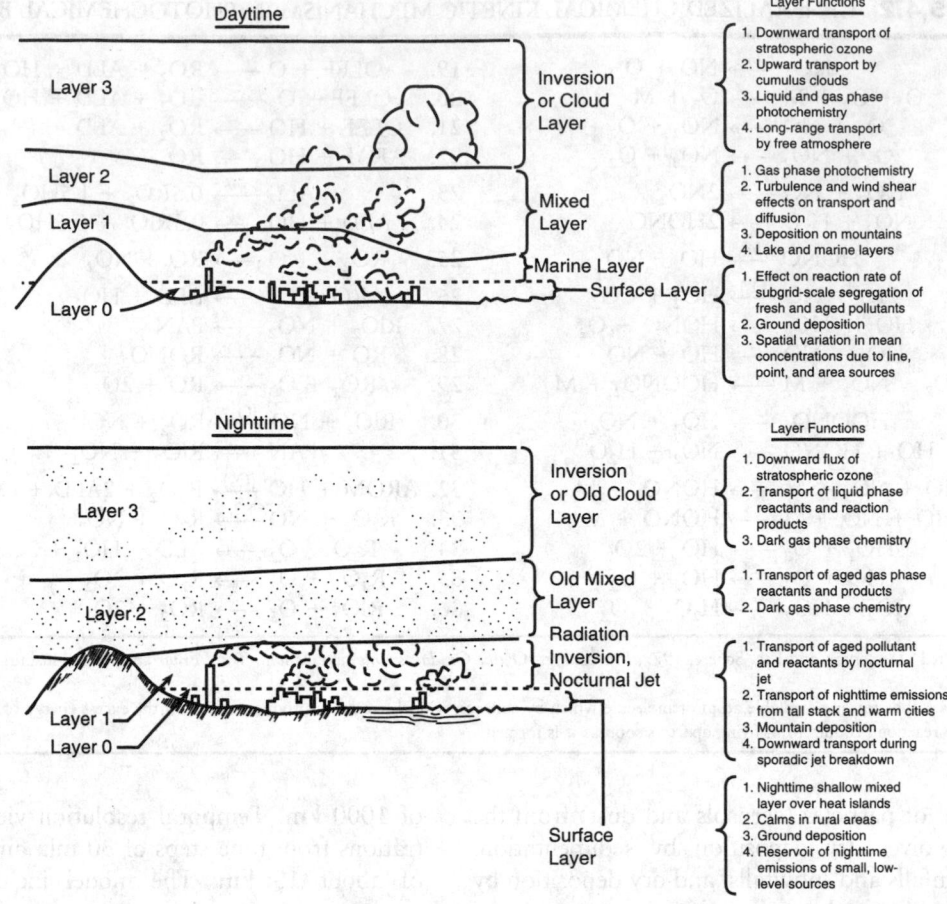

FIG. 5.4.2 Schematic diagram of the dynamic layer structure of the regional model. (Reprinted, with permission, from R.G. Lamb and J.H. Novak, 1984, *Proceedings of EPA–DECD International Conference on Long Range Transport Models for Photochemical Oxidants and Their Precursors,* EPA-600/9-84/006, Research Triangle Park, N.C.: U.S. EPA.)

three- to four-month commitment of a person with knowledge of the model (Turner 1986).

—*David H.F. Liu*

References

Bufalini, M. 1971. The oxidation of sulfur dioxide in polluted atmospheres: A review. *Environ. Sci. Technol. 5,* no. 685.

Gartrell, F.E., F.W. Thomas, and S.B. Carpenter. 1963. Atmospheric oxidation of SO_2 in coal burning power plant plumes. *Am. Ind. Hygiene Assoc. J.* 24, no. 113.

Lamb, R.G., and J.H. Novak. 1984. *Proceedings of EPA-OECD International Conference on Long Range Transport Models for Photochemical Oxidants and Their Precursors.* EPA-600/9-84/006. Research Triangle Park, N.C.: U.S. EPA.

Sidebottom, H.W., C.D. Badcock, G.E. Jackson, J.G. Calvert, G.W. Reinhardt, and E.K. Damon. 1972. Photooxidation of sulfur dioxide. *Environ. Sci. Technol.* 6, no. 72.

Turner, D. Bruce. 1986. The transport of pollutants. Vol. VI in *Air pollution,* edited by Arthur C. Stern. Academic Press, Inc.

5.5
MACRO AIR POLLUTION EFFECTS

Macro air pollution effects refer to those consequences of air pollution exhibited on a large geographical scale, with the scale ranging from regional to global. Examples of such effects include acid rain, losses in the stratospheric ozone layer, and global warming.

Acid Rain Effects

Acid precipitation causes multiple effects on both terrestrial and aquatic ecosystems. Also, acid precipitation and dry deposition can affect materials and even human health. Demonstrated effects on terrestrial ecosystems include necrotic lesions on foliage, nutrient loss from foliar organs, reduced resistance to pathogens, accelerated erosion of the waxes on leaf surfaces, reduced rates of decomposition of leaf litter, inhibited formation of terminal buds, increased seedling mortality, and heavy metal accumulation (Cowling and Davey 1981). Soil and vegetation and crop-related effects include soil acidification, calcium removal, aluminum and manganese solubilization, tree growth reduction, reduction of crop quality and quantity, elimination of useful soil microorganisms, and selective exchange of heavy metal elements for more beneficial mono- and divalent cations (Glass, Glass, and Rennie 1979). Soil microbiological processes such as nitrogen fixation, mineralization of forest litter, and nitrification of ammonium compounds can be inhibited, the degree depending on the amount of cultivation and soil buffering capacity (Cowling and Davey 1981).

EFFECTS ON FORESTS

Field studies of the effects of acid precipitation on forests have been conducted in the United States and Europe. Reports of decreased growth and increased mortality of forest trees in areas receiving high rates of atmospheric pollutants emphasize the need to understand and quantify both the mechanisms and kinetics of changes in forest productivity. The complex chemical nature of combined pollutant exposures and the fact that these changes can involve both direct effects to vegetation and indirect and possibly beneficial effects mediated by a variety of soil processes make quantification of such effects challenging. However, evidence is growing on the severity of forest problems in central Europe due to acid precipitation. For example, in West Germany, fully 560,000 hectares of forests have been damaged (Wetstone and Foster 1983).

EFFECTS ON SOIL

Acid precipitation can affect soil chemistry, leaching, and microbiological processes. In addition, various types of soils exhibit a range of sensitivities to the effects of acid rain; for example, some soils are more sensitive than others. Factors influencing soil sensitivity to acidification include the lime capacity, soil profile buffer capacity, and water–soil reactions (Bache 1980). Wiklander (1980) reviews the sensitivity of various soils, and Peterson (1980) identifies soil orders and classifications according to their response to acid precipitation.

Two important effects of acid precipitation on soil are associated with changes in the leaching patterns of soil constituents and with the potential removal and subsequent leaching of chemical constituents in the precipitation. For example, Cronan (1981) describes the results of an investigation of the effects of regional acid precipitation on forest soils and watershed biogeochemistry in New England. Key findings include the following:

1. Acid precipitation can cause increased aluminum mobilization and leaching from soils to sensitive aquatic systems
2. Acid deposition can shift the historic carbonic acid/organic acid leaching regime in forest soils to one dominated by atmospheric H_2SO_4
3. Acid precipitation can accelerate nutrient cation leaching from forest soils and can pose a threat to the potassium resources of northeastern forested ecosystems
4. Progressive acid dissolution of soils in the laboratory is an important tool for predicting the patterns of aluminum leaching from soils exposed to acid deposition.

Soil microorganisms and microbiological processes can be altered by acid precipitation. The effects of acid precipitation include changes in bacterial numbers and activity, alterations in nutrient and mineral cycling, and changes in the decomposition of organic matter.

EFFECTS ON GROUNDWATER

As groundwater quality is becoming increasingly important, a concern is growing related to the effects of acid precipitation on quality constituents. Direct precipitation in recharge areas is of particular concern. The most pronounced effects are associated with increased acidity causing accelerated weathering and chemical reactions as the precipitation passes through soil and rock in the process of recharging an aquifer. The net effect on groundwater is reduced water quality because of increased mineralization.

EFFECTS ON SURFACE WATER

Acid precipitation causes many observable, as well as nonobservable, effects on aquatic ecosystems. Included are changes in water chemistry and aquatic faunal and floral species. One reason for changes in surface water chemistry is the release of metals from stream or lake sediments. For example, Wright and Gjessing (1976) note that concentrations of aluminum, manganese, and other heavy metals are higher in acid lakes due to enhanced mobilization of these elements in acidified areas.

Due to the extant water chemistry and sediment characteristics, some surface water is more susceptible to changes in water chemistry than others. Several surface water sensitivity studies leading to classification schemes have been conducted. For example, Hendrey et al. (1980) analyzed bedrock geology maps of the eastern United States to determine the relationship between geological material and surface water pH and alkalinity. They verified map accuracy by examining the current alkalinity and pH of water in several test states, including Maine, New Hampshire, New York, Virginia, and North Carolina. In regions predicted to be highly sensitive, the alkalinity in upstream sites was generally low, less than 200 microequivalents per liter. They pinpoint many areas of the eastern United States in which some of the surface water, especially upstream reaches, are sensitive to acidification.

Acid precipitation affects microdecomposers, algae, aquatic macrophytes, zooplankton, benthos, and fish (Hendry et al. 1976). For example, many of the 2000 lakes in the Adirondack Region of New York are experiencing acidification and declines or loss of fish populations. Baker (1981) found that, on the average, aluminum complexed with organic ligands was the dominant aluminum form in the dilute acidified Adirondack surface water studied. In laboratory bioassays, speciation of aluminum had a substantial effect on aluminum and hydrogen ions, and these ions appeared to be important factors for fish survival in Adirondack surface water affected by acidification.

EFFECTS ON MATERIALS

Acid precipitation can damage manmade materials such as buildings, metals, paints, and statuary (Glass, Glass, and Rennie 1980). For example, Kucera (1976) has reported data on the corrosion rates of unprotected carbon steel, zinc and galvanized steel, nickel and nickel-plated steel, copper, aluminum, and antirust painted steel due to sulfur dioxide and acid precipitation in Sweden. Corrosion rates are higher in polluted urban atmospheres than in rural atmospheres because of the high concentrations of airborne sulfur pollutants in urbanized areas. Economic damage is significant in galvanized, nickel-plated, and painted steel and painted wood.

EFFECTS ON HEALTH

Acid precipitation affects water supplies which in turn affects their users. Taylor and Symons (1984) report the results of the first study concerning the impact of acid precipitation on drinking water; the results report health effects in humans as measured by U.S. EPA maximum contaminant levels. The study sampled surface water and groundwater supplies in the New England states, but it also included other sites in the northeast and the Appalachians. No adverse effects on human health were demonstrated, although the highly corrosive nature of New England water may be at least partly attributable to acidic deposition in poorly buffered watersheds and aquifers.

Losses in Stratospheric Ozone Layer

The stratospheric ozone layer occurs from 12 to 50 km above the earth; the actual ozone concentration in the layer is in the order of ppmv (Francis 1994). Ozone can be both formed and destroyed by reactions with NO_x; of recent concern is the enhanced destruction of stratospheric ozone by chlorofluorocarbons (CFCs) and other manmade oxidizing air pollutants. The natural ozone layer fulfills several functions related to absorbing a significant fraction of the ultraviolet (uv) component of sunlight and terrestrial infrared radiation, and it also emits infrared radiation.

Several potential deleterious effects result from decreasing the stratospheric ozone concentration. Of major concern is increased skin cancer in humans resulting from greater UV radiation reaching the earth's surface. Additional potential concerns include the effects on some marine or aquatic organisms, damage to some crops, and alterations in the climate (Francis 1994). While environmental engineers are uncertain about all seasonal and geographic characteristics of the natural ozone layer and quantifying these effects, the effects are recognized via precursor pollutant control measures included in the 1990 CAAAs.

Precursor pollutants that reduce stratospheric ozone concentrations via atmospheric reactions include CFCs and nitrous oxide. Principal CFCs include methylchloroform and carbon tetrachloride; these CFCs are emitted to the atmosphere as a result of their use as aerosol propellants, refrigerants, foam-blowing agents, and solvents. Example reactions for one CFC (CFC-12) and ozone follow (Francis 1994):

$$CCl_2F_2 + UV \longrightarrow Cl\cdot + CClF_2 \qquad 5.5(1)$$
$$Cl\cdot + O_3 \longrightarrow ClO + O_2 \qquad 5.5(2)$$

$Cl\cdot$ denotes atomic chlorine. ClO is also chemically reactive and combines with atomic oxygen as follows:

$$ClO + O\cdot \longrightarrow Cl\cdot + O_2 \qquad 5.5(3)$$

Because of the preceding cycling of $Cl\cdot$, environmental engineers estimate that a single Cl atom can destroy an average of 100,000 ozone molecules (Francis 1994).

Global Warming

The potential effects of global climate change can be considered in terms of ecological systems, sea-level rise, water resources, agriculture, electric demand, air quality, and health effects (Smith and Tirpak 1988). Since climate influences the location and composition of plants and animals in the natural environment, changes in climate have numerous consequences on ecological systems. One consequence includes shifts in forests in geographic range and composition; for example, the current southern boundary of hemlock and sugar maple in the eastern United States could move northward by about 400 mi (Smith and Tirpak 1988). An example of compositional change is that the mixed boreal and northern hardwood forest in northern Minnesota could become all northern hardwood.

ECOLOGICAL IMPACT

Biodiversity is also impacted by climate change, with *diversity* defined as the variety of species in ecosystems and the variability within each species. Examples of impact include an increased extinction of many species; growth or losses in freshwater fish populations depending on geographic location; and mixed effects on migratory birds, with some arctic-nesting herbivores benefiting and continental nesters and shorebirds suffering (Smith and Tirpak 1988).

Sea-level rises can cause increased losses of coastal wetlands, inundation of coastal lowlands, increased erosion of beaches, and increased salinity in estuaries. Coastal wetlands currently total 13,145 sq mi; with a 1-m rise in sea level, 26 to 66% of these wetlands can be lost, with the majority occurring in states bordering the Gulf of Mexico (Smith and Tirpak 1988).

IMPACT ON WATER RESOURCES

The main consequences of climatic changes to inland waters include the following (da Cunha 1988): (a) changes in the global amount of water resources and in the spatial and temporal distribution of these resources; (b) changes in soil moisture; (c) changes in extreme phenomena related to water resources, i.e., floods and droughts; (d) changes in water quality; (e) changes in sedimentation processes; and (f) changes in water demand.

The consequences of climatic change on water quality include possible changes in the precipitation regime and the occurrence of acid rain. Direct consequences of climatic changes on water quality occur. For example, temperature increases can decrease levels of dissolved oxygen in the water. Second, the biochemical oxygen demand (BOD) also increases with temperature. These two effects can decrease the dissolved oxygen concentration in a surface water system. Also, climatic changes can have indirect consequences on water quality since a decrease of river discharges, particularly during the dry season, can increase the concentration of pollutants in water bodies.

Climatic changes influence not only water availability but also water demand. For example, water demand for irrigation is largely affected by climatic change which conditions evapotranspiration. Water demands for domestic or industrial use are also affected by climatic change, for example, as a result of temperature increases that influence water consumption for cooling systems, bathing, washing, and gardening.

The simplest way to view the implications of global climate change on water resources is to consider the relationship between increasing atmospheric CO_2 and the hydrologic cycle; this relationship is shown in Figure 5.5.1. The following comments relate to the implications of Figure 5.5.1 (Waggoner and Revelle 1990):

1. When the atmosphere reaches a new and warmer equilibrium, more precipitation balances faster evaporation. This faster hydrologic cycle is predicted to raise the global averages of the up-and-down arrows of precipitation and evaporation in Figure 5.5.1 by 7 to 15%. However, the predicted change in precipitation is not a uniform increase. The net of precipitation and evaporation, that is, soil moisture, is predicted to fall in some places. Also, precipitation is predicted to increase in some seasons and decrease in others. Although global climatic models disagree on where precipitation will decrease and although they do not simulate the present seasonal change in precipitation correctly, disregarding the warning of a drier climate is unwise.

2. To be most usable for water resource considerations, frequency distributions of precipitation (and flood and drought projections) are needed. Models to develop such information are in their infancy.

3. A dimensionless elasticity for runoff and precipitation is:

[Percentage change in runoff/percentage

change in precipitation]   5.5(4)

If the elasticity for runoff and precipitation is greater than 1, the percentage change in runoff is greater than the percentage change in precipitation that causes it. Therefore, a general conclusion about the transformation of climate change into runoff change can be stated. Over diverse climates, the elasticities of percentage change in runoff to percentage change in precipitation and evaporation are 1 to 4. The percentage change in runoff is greater than the percentage change in the forcing factor.

Based on this brief review of water resource issues related to global climate change, the following summary comments can be made:

1. Global climate change caused by the greenhouse effect appears to be a reality although scientific opinion differs as to the rate of change.

2. Numerous inland and coastal water resource management issues are impacted when temperature and precipitation patterns change.

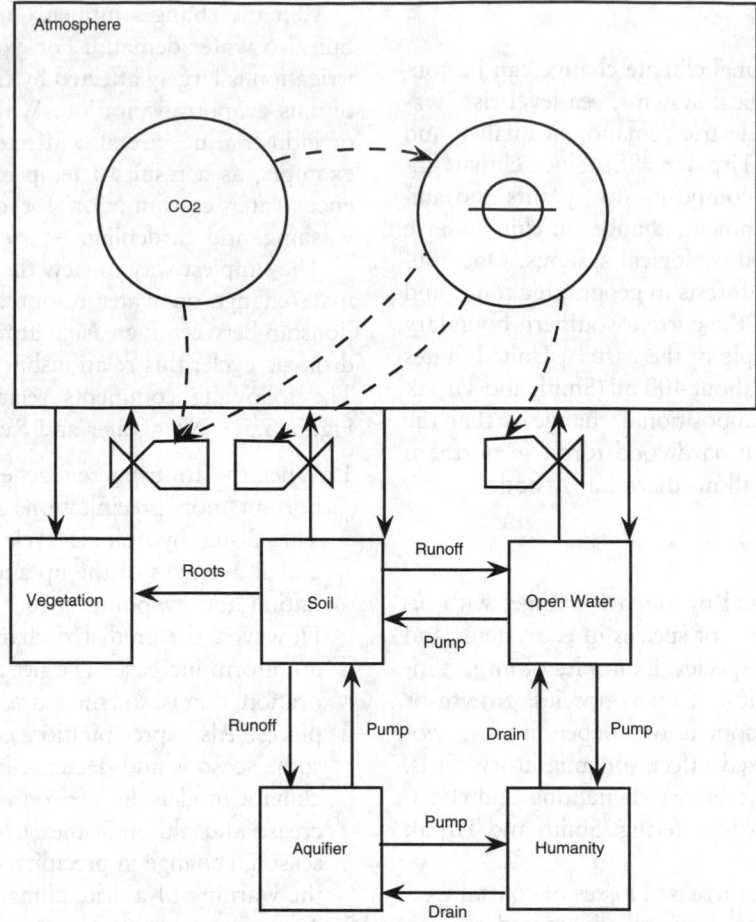

FIG. 5.5.1 The hydrologic cycle. The large rectangle at the top represents the water in the atmosphere. The rectangles beneath represent the water in vegetation, in the soil reached by roots, in the aquifers below, and in open bodies of water. The rectangle for humanity represents the water in people and pipes. Arrows represent fluxes. Valves are placed on three fluxes to show their control by CO_2. CO_2 directly affects transpiration from foliage by enlarging it through faster photosynthesis and by narrowing leaf pores. Indirectly, CO_2 warms the air temperature, speeding the evaporation from vegetation, soil, and open water. Faster evaporation decreases the water that runs off to bodies of water or into aquifers, increases pumping to irrigate soil, and even raises humanity's demands. The arrows from air down to vegetation, soil, and open water are the flux and addition to them by precipitation. Water resources are the pools signified by the rectangles for open water and aquifers. (Reprinted, with permission, from P.E. Waggoner and R.R. Revelle, 1990, Summary, Chap. 19 in *Climate change and U.S. water resources*, edited by P.E. Waggoner, New York: John Wiley and Sons.)

3. The global perspective presented in most studies is not specific enough to address the water resource implications in regional and local areas.

4. Global climate change might be controlled by effective programs to reduce the atmospheric emissions of greenhouse gases.

5. Global climate change must rise higher on national and international political agendas before effective societal measures can be implemented to control or manage the water resources implications.

IMPACT ON AGRICULTURE

Agricultural productivity in the United States is based on the temperate climate and rich soils. Global warming exhibits direct and indirect geographical effects on agricultural productivity. Direct effects occur through changes in the length of the growing season, the frequency of heat waves, and altered patterns of rainfall; while indirect effects result from changes in topsoil management practices. Dryland yields of corn, wheat, and soybeans could decrease in many regions as a result of higher temperatures

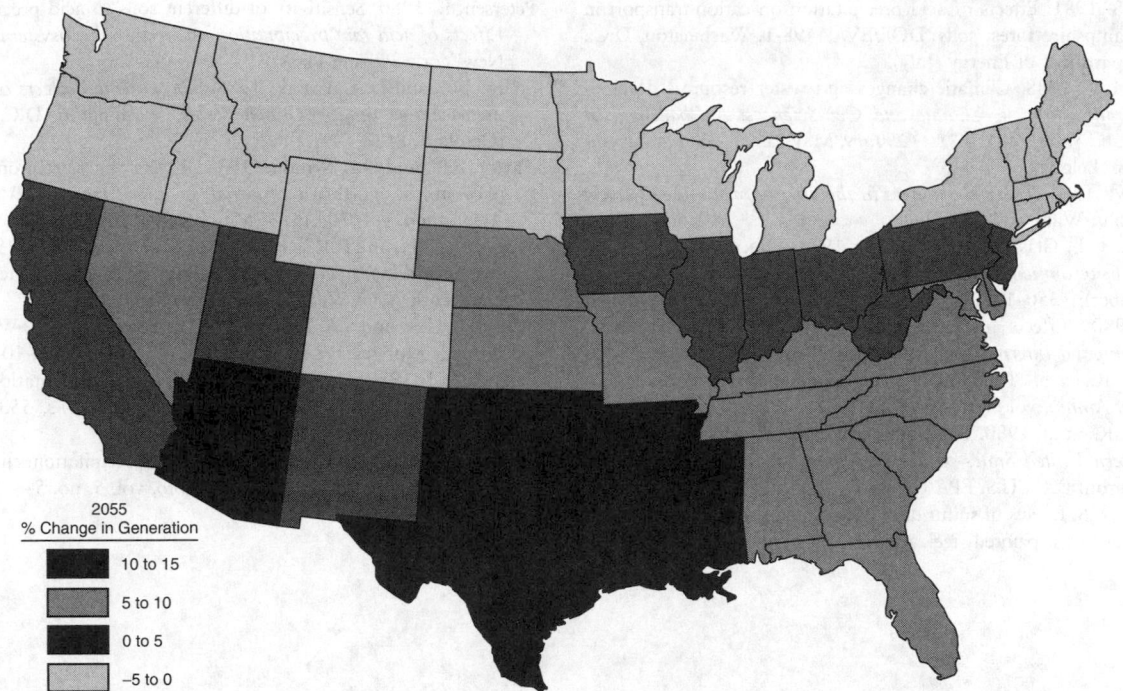

FIG. 5.5.2 Changes in electric generation by state induced by climate change to 2055. (Reprinted from J.B. Smith and D.A. Tirpak, 1988, *The potential effects of global climate change on the United States,* Washington, D.C.: U.S. EPA [October].)

which shorten a crop's life cycle (Smith and Tirpak 1988). In contrast, in northern areas such as Minnesota, dryland yields of corn and soybeans could double as warmer temperatures extend the frost-free growing season. Crop acreage could decrease by 5 to 25% in Appalachia, the southeast, and the southern Great Plains areas; conversely, acreage could increase by 5 to 17% in the northern Great Lakes states, the northern Great Plains, and the Pacific northwest areas (Smith and Tirpak 1988). Irrigation would probably increase in many areas because irrigated yields are more stable than dry-land yields under conditions of increased heat stress and reduced precipitation.

Climate-sensitive electric end uses include space heating and cooling and, to a lesser degree, water heating and refrigeration. Summer cooling electric demands would increase, while winter heating demands would decrease. As a result of climate change, annual electric demands are expected to increase by 4 to 6% by the year 2055. Figure 5.5.2 summarizes regional demand changes. Additional power plants will be required to meet peak demands.

IMPACT ON AIR QUALITY

Global climate change also has implications for ambient air quality. Increased emissions of SO_x, NO_x, and CO are associated with power plants meeting increased electric demands. Plume rises from stacks would decrease due to higher ambient air temperatures and reduced buoyancy effects; these decreases would be manifested in higher

ground-level concentrations of air pollutants located closer to their stacks. Air pollutant dispersion is also affected by weather variables such as windspeed and direction, temperature, precipitation patterns, cloud cover, atmospheric water vapor, and global circulation patterns (Smith and Tirpak 1988). Ozone pollution problems in many urban areas would worsen due to higher reaction rates at higher temperatures and lengthened summer seasons.

IMPACT ON HUMAN HEALTH

Human health effects are manifested by changes in morbidity and increases in mortality, particularly for the elderly during hotter and extended summer periods. Geographical patterns in relation to health effects are also expected.

—Larry W. Canter

References

Bache, B.W. 1980. The sensitivity of soils to acidification. In *Effects of acid rain precipitation on terrestrial ecosystems.* 569–572. New York: Plenum Press.

Baker, J.P. 1981. Aluminum toxicity to fish as related to acid precipitation and Adirondack surface water quality. Ph.D. diss. Cornell University, Ithaca, New York (January).

Cowling, E.B., and C.B. Davey. 1981. Acid precipitation: Basic principles and ecological consequences. *Pulp and Paper,* vol. 55, no. 8 (August): 182–185.

Cronan, C.S. 1981. Effects of acid precipitation on cation transport in New Hampshire forest soils. DOE/EV/04498-1. Washington, D.C.: U.S. Department of Energy (July).

da Cunha, L.V. 1988. Climatic changes and water resources development. *Symposium on Climate and Geo-Sciences: A Challenge for Science and Society in the 21st Century,* May 22–27, 1988, Louvain-la-Neuve, Belgium.

Francis, B.M. 1994. *Toxic substances in the environment.* 42–47. New York: John Wiley and Sons, Inc.

Glass, N.R., G.E. Glass, and P.J. Rennie. 1979. Effects of acid precipitation. *Environmental Science and Technology,* vol. 13, no. 11 (November): 1350–1352.

———. 1980. Effects of acid precipitation in North America. *Environmental International,* vol. 4, no. 5–6: 443–452.

Hendrey, G.R., et al. 1976. Acid precipitation: Some hydrobiological changes. *Ambio,* vol. 5, no. 5–6: 224–228.

Hendrey, G.R., et al. 1980. *Geological and hydrochemical sensitivity of the eastern United States to acid precipitation.* EPA-600/3-80-024. Washington, D.C.: U.S. EPA (January).

Kucera, V. 1976. Effects of sulfur dioxide and acid precipitation on metals and anti-rust painted steel. *Ambio,* vol. 5, no. 5–6: 248–254.

Petersen, L. 1980. Sensitivity of different soils to acid precipitation. In *Effects of acid rain precipitation on terrestrial ecosystems.* 573–577. New York: Plenum Press.

Smith, J.B., and D.A. Tirpak. 1988. *The potential effects of global climate change on the United States.* Washington, D.C.: U.S. EPA (October): 8–33.

Taylor, F.B., and G.E. Symons. 1984. Effects of acid rain on water supplies in the northeast. *Journal of the American Water Works Association,* vol. 76, no. 3 (March): 34–41.

Waggoner, P.E., and R.R. Revelle. 1990. Summary. Chap. 19 in *Climate change and U.S. water resources,* edited by P.E. Waggoner, 447–477. New York: John Wiley and Sons, Inc.

Wetstone, G.S., and S.A. Foster. 1983. Acid precipitation: What is it doing to our forests. *Environment,* vol. 25, no. 4 (May): 10–12, 38–40.

Wiklander, L. 1980. The sensitivity of soils to acid precipitation. In *Effects of acid rain precipitation on terrestrial ecosystems,* 553–567. New York: Plenum Press.

Wright, R.F., and E.T. Gjessing. 1976. Acid precipitation: changes in the chemical composition on lakes. *Ambio,* vol. 5, no. 5–6: 219–224.

5.6
METEOROLOGY

Clinging to the surface of the earth is a thin mantle of air known as the atmosphere (Figure 5.6.1). Calling the atmosphere thin may be confusing; however, 99% of the atmosphere mass lies within just 30 km (19 mi) of the earth's surface, and 90% of the atmosphere's mass lies within just 15 km (9 mi) of the surface.

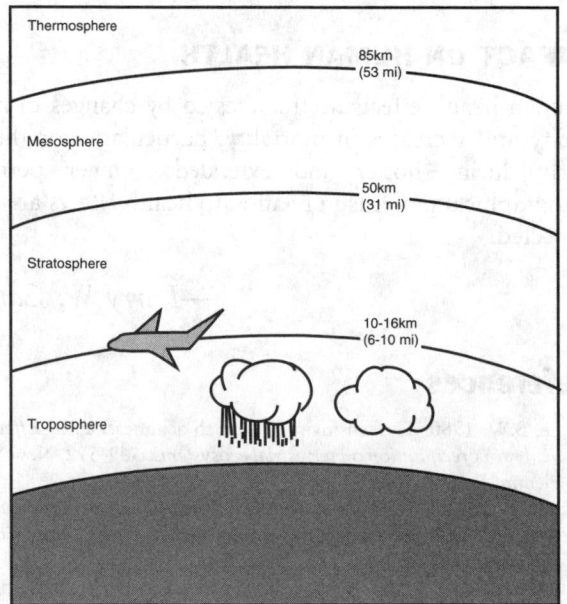

FIG. 5.6.1 Classification of atmosphere based on temperature. (This figure is not drawn to scale.)

The atmosphere is often classified in terms of temperature. Starting at the earth's surface and moving upward, temperature generally decreases with increasing altitude. This region, termed the *troposphere,* is of most interest to meteorologists because it is where weather and air pollution problems occur.

The *tropopause* is the boundary between the troposphere and the *stratosphere.* Below the tropopause, atmospheric processes are governed by turbulent mixing of air; but above it, they are not. In the stratosphere, temperature increases with height because of the high ozone concentration. Ozone absorbs radiation from the sun, resulting in an increase in stratospheric temperature.

The meteorological elements that have the most direct and significant effects on the distribution of air pollutants in the atmosphere are wind speed, wind direction, solar radiation, atmospheric stability, and precipitation.

Wind

The effects of wind on the distribution of air pollutants in the atmosphere involve understanding the scales of air motion, wind rose, and turbulence.

SCALES OF AIR MOTION

Wind is the motion of air relative to earth's surface. On the macroscale, the movement originates in the unequal distribution of atmospheric temperature and pressure over the earth's surface and is influenced by the earth's rotation. The direction of wind flow is characteristically from high pressure to low, but the Coriolis force deflects the air current out of these expected patterns (see Figure 5.6.2). These phenomena occur on scales of thousands of kilometers and are exemplified by the semipermanent high- and low-pressure areas over oceans and continents.

On the mesoscale and microscale, topographical features critically influence wind flow. Surface variations have an obvious effect on wind velocity and the direction of air flow. Monsoons, sea and land breezes, mountain–valley winds, coastal fogs, windward precipitation systems, and urban heat islands are all examples of the influence of regional and local topography on atmospheric conditions. Mesoscale phenomena occur over hundreds of kilometers; microscale phenomena, over areas less than 10 kilometers.

For an area, the total effect of these circulations establishes the hourly, daily, and seasonal variation in wind speed and direction. The frequency distribution of wind direction indicates the areas toward which pollutants are most frequently transported.

WIND ROSE

Wind speed determines the travel time of a pollutant from its source to a receptor and accounts for the amount of pollutant diffusion in the windward direction. Therefore, the concentration of pollutant at any receptor is inversely proportional to the wind speed. Wind direction determines in what direction a pollutant travels and what receptor is affected at a given time. Wind direction is normally defined by a *wind rose*, a graphic display of the distribution of wind direction at a location during a defined period. The characteristic patterns can be presented in either tabular or graphic forms.

Wind speed is usually measured by an anemometer, which consists of three or four hemispherical cups arranged around a vertical axis. The faster the rotation of the cups, the higher the speed of the wind. A wind vane indicates wind direction. Although wind is three-dimensional in its movement, generally only the horizontal component is denoted because the vertical component is much smaller.

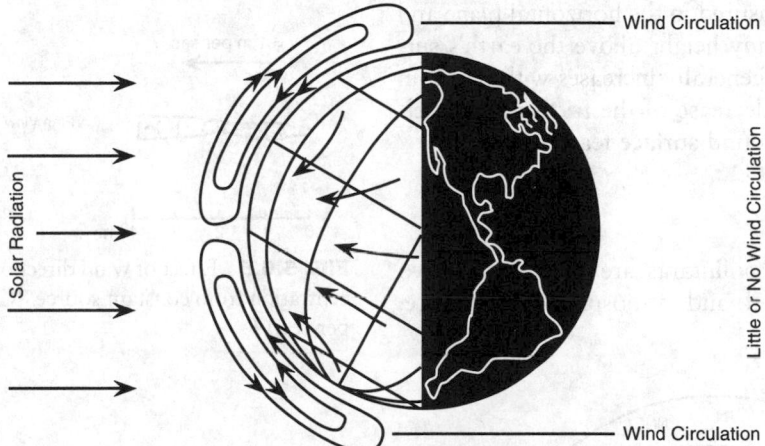

A. If the earth did not turn, air would circulate in a fixed pattern.

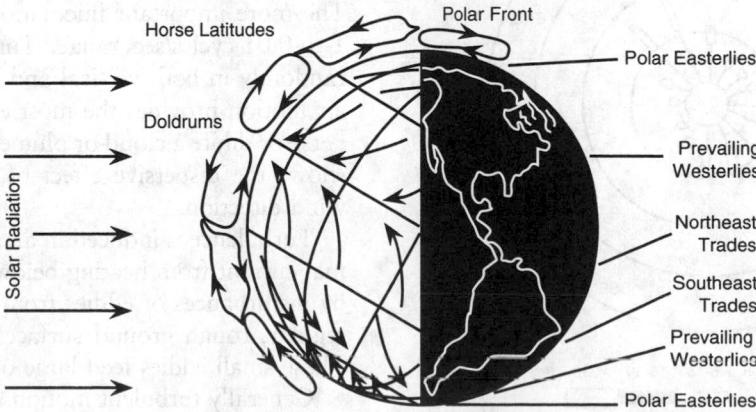

B. The earth turns, creating variable wind patterns.

FIG. 5.6.2 Global wind patterns. (Reprinted, with permission, from the American Lung Association.)

A wind rose is a set of wind statistics that describes the frequency, direction, force, and speed (see Figure 5.6.3). In this plot, the average wind direction is shown as one of the sixteen compass points, each separated by 22.5° measured from true north. The length of the bar for a direction indicates the percent of time the wind came from that direction. Since the direction is constantly changing, the time percentage for a compass point includes those times for wind direction at 11.25° on either side of the point. The percentage of time for a velocity is shown by the thickness of the direction bar. Figure 5.6.3 shows that the average wind direction from the southwest direction is 19% of the time and 7% of the time the southwesterly wind velocity is 16–30 mph.

Figure 5.6.4 shows the particulate fallout around an emission source and a wind rose based on the same time period.

The wind rose is imprecise in describing a point in a study region because the data are collected at one location in the region and not at each location. The data are often a seasonal or yearly average and therefore not accurate in describing any point in time in an ideal representation of atmospheric diffusion. A final limitation of wind rose is that the wind is only measured in the horizontal plane and is assumed identical at any height above the earth's surface. (Note: wind speed generally increases with height in lower levels due to the decrease of the frictional drag effect of the underlying ground surface features.)

TURBULENCE

In general, atmospheric pollutants are dispersed by two mechanisms: wind speed and atmospheric turbulence.

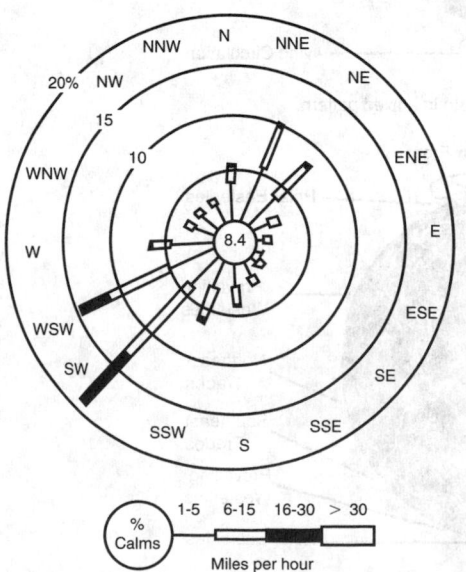

FIG. 5.6.3 Wind rose showing direction and velocity frequencies.

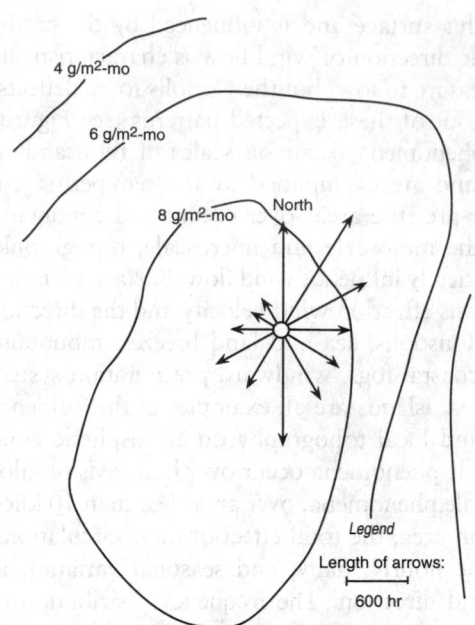

FIG. 5.6.4 Wind rose and corresponding particulate fallout pattern.

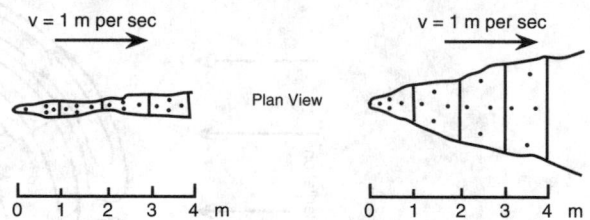

FIG. 5.6.5 Effect of wind direction variability or pollutant concentration from constant source. (Continuous emission of 4 units per sec.)

Atmospheric turbulence usually includes those wind flow fluctuations that have a frequency of more than 2 cycles/hr. The more important fluctuations have frequencies in the 1-to-0.01-cycles/sec range. Turbulent fluctuations occur randomly in both vertical and horizontal directions. This air motion provides the most effective mechanism to disperse or dilute a cloud or plume of pollutants. Figure 5.6.5 shows the dispersive effect of fluctuations in horizontal wind direction.

Turbulence is induced in air flow in two ways: by thermal current from heating below (*thermal turbulence*) and by disturbances or eddies from the passage of air over irregular, rough ground surfaces (*mechanical turbulence*). These small eddies feed large ones.

Generally turbulent motion and, in turn, the dispersive ability of the atmosphere, are enhanced during solar heating over rough terrain. Conversely, turbulence is suppressed during clear nights over smooth terrain.

Lapse Rates and Stability

Lapse rates, stability, and inversions also affect the dispersion of pollutants in the atmosphere.

LAPSE RATES

In the stratosphere, the temperature of the ambient air usually decreases with an increase in altitude. This rate of temperature change is called the *elapse rate.* Environmental engineers can determine this rate for a place and time by sending up a balloon equipped with a thermometer. The balloon moves through the air, not with it, and measures the temperature gradient of ambient air, called the *ambient lapse rate,* the *environmental lapse rate,* or the *prevailing lapse rate.*

Using the ideal-gas law and the law of conservation of energy, environmental engineers have established a mathematical ratio for expressing temperature change against altitude under adiabatic conditions (Petterssen 1968). This rate of decrease is termed the *adiabatic elapse rate,* which is independent of the prevailing atmospheric temperature.

Dry air, expanding adiabatically, cools at 9.8°C per km (or 5.4°F per 1000 ft), which is the dry adiabatic lapse rate (Smith 1973). In a wet as in a dry adiabatic process, a parcel of air rises and cools adiabatically, but a second factor affects its temperature. Latent heat is released as water vapor condenses within the saturated parcel of rising air. Temperature changes in the air are then due to the liberation of latent heat as well as the expansion of air. The wet adiabatic lapse rate (6°C/km) is thus less than the dry adiabatic lapse rate. Since a rising parcel of effluent gases is seldom completely saturated or completely dry, the adiabatic lapse rate generally falls somewhere between these two extremes.

STABILITY

Ambient and adiabatic lapse rates are a measure of atmospheric stability. Figure 5.6.6 shows these stability conditions. The atmosphere is *unstable* as long as a parcel of air moving upward cools at a slower rate than the surrounding air and is accelerated upward by buoyancy force. Moving downward, the parcel cools slower and is accelerated downward. Under these conditions, vertical air motions and turbulence are enhanced.

Conversely, when a rising parcel of air is cooler than the surrounding air, the parcel settles back to its original elevation. Downward movement produces a warmer parcel, which rises to its original elevation. Under these conditions, vertical movement is dampened out by adiabatic cooling or warming, and the atmosphere is *stable.*

Figure 5.6.7 shows that the boundary line between stability and instability is the dry adiabatic lapse line. When

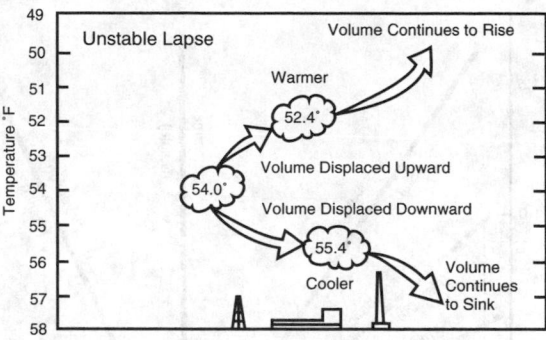

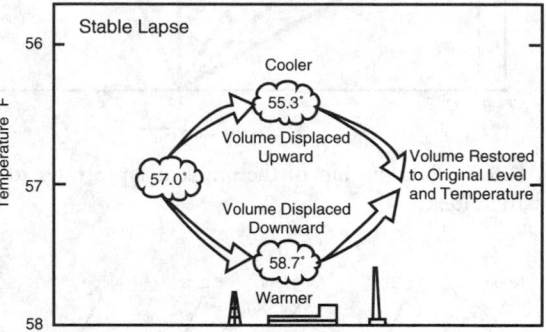

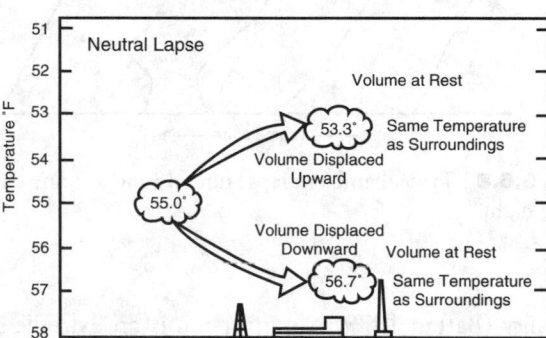

FIG. 5.6.6 The effect of lapse rate on vertical stability.

the ambient lapse rate exceeds the adiabatic lapse rate, the ambient lapse rate is termed *superadiabatic,* and the atmosphere is highly unstable. When the two lapse rates are equal, the atmosphere is *neutral.* When the ambient lapse rate is less than the dry adiabatic lapse rate, the ambient lapse rate is termed *subadiabatic,* and the atmosphere is stable (Figure 5.6.8). If the air temperature is constant throughout a layer of atmosphere, the ambient lapse rate is zero, the atmosphere is described as *isothermal,* and the atmosphere is stable (Battan 1979).

When the temperature of ambient air increases (rather than decreases) with altitude, the lapse rate is negative, or inverted, from the normal state. A negative lapse rate occurs under conditions referred to as an *inversion,* a state in which warmer air blankets colder air. Thermal or temperature inversions represent a high degree of atmospheric

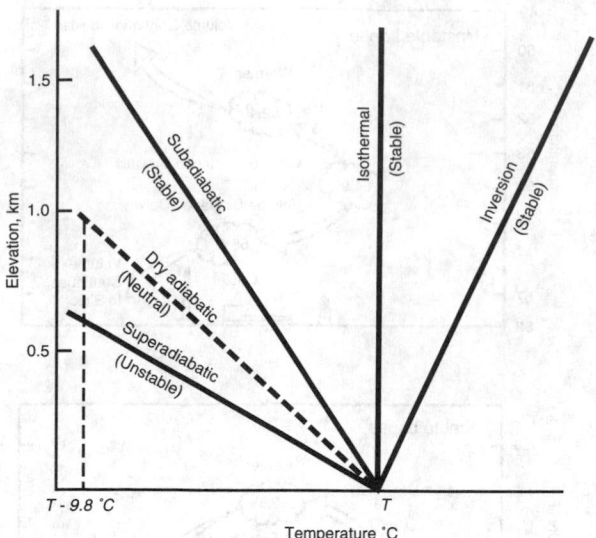

FIG. 5.6.7 Relationship of the ambient lapse rates to the dry adiabatic rate.

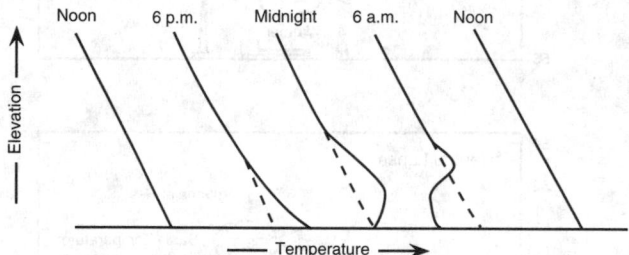

FIG. 5.6.8 Typical ambient lapse rates during a sunny day and clear night.

stability (Battan 1979). An inversion is an extreme subadiabatic condition, thus almost no vertical air movement occurs.

INVERSIONS

Three types of inversions develop in the atmosphere: radiational (surface), subsidence (aloft), and frontal (aloft).

Radiational Inversions

A radiational inversion occurs at low levels, seldom above a few hundred feet, and dissipates quickly. This type of inversion occurs during periods of clear weather and light to calm winds and is caused by rapid cooling of the ground by radiation. The inversion develops at dusk and continues until the surface warms again the following day. Initially, only the air close to the surface cools, but after several hours, the top of the inversion can extend to 500 ft (see Figure 5.6.9). Pollution emitted during the night is caught under this "inversion lid."

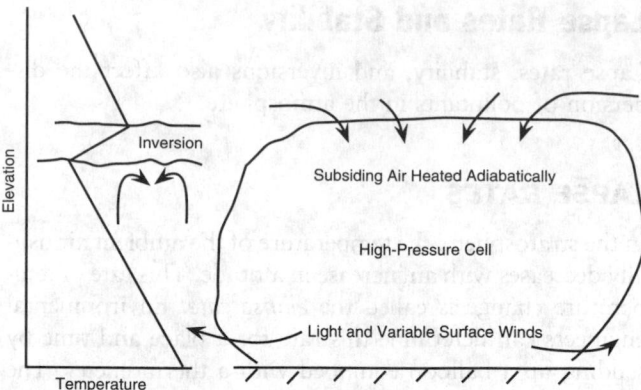

FIG. 5.6.9 Formation of subsidence inversion.

Subsidence Inversions

A subsidence inversion is important in pollution control because it can affect large areas for several days. A subsidence inversion is associated with either a stagnant high-pressure cell or a flow aloft of cold air from an ocean over land surrounded by mountains (Cooper and Alley 1986). Figure 5.6.9 shows the inversion mechanism.

A significant condition is the subsidence inversion that develops with a stagnating high-pressure system (generally associated with fair weather). Under these conditions, the pressure gradient becomes progressively weaker so that winds become light. These light winds greatly reduce the horizontal transport and dispersion of pollutants. At the same time, the subsidence inversion aloft continuously descends, acting as a barrier to the vertical dispersion of the pollutants. These conditions can persist for several days, and the resulting accumulation of pollutants can cause serious health hazards.

Fog almost always accompanies serious air pollution episodes. These tiny droplets of water are detrimental in two ways: (1) fogs makes the conversion of SO_3 to H_2SO_4 possible, and (2) fogs sits in valleys and prevents the sun from warming the valley floor to break inversions, thus prolonging air pollution episodes.

Figure 5.6.10 shows the frequency of stagnation periods of high-pressure cells over the eastern United States.

Frontal Inversions

A frontal inversion usually occurs at high altitudes and results when a warm air mass overruns a cold air mass below. This type of inversion is not important from a pollution control standpoint.

Precipitation

Precipitation serves an effective cleansing process of pollutants in the atmosphere as follows:

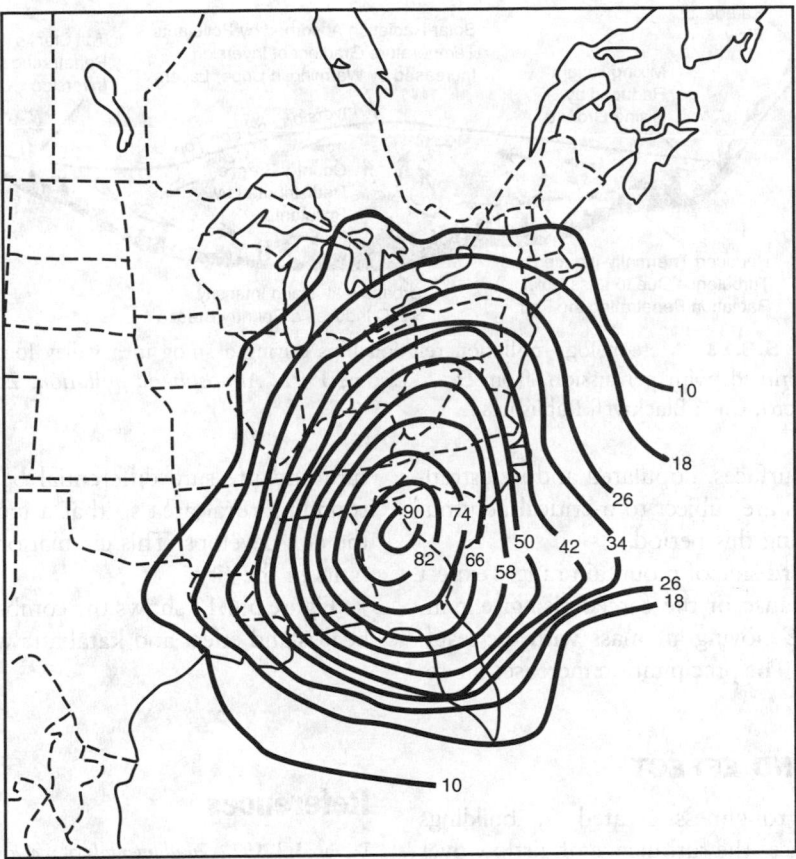

FIG. 5.6.10 Frequency of stagnating high-pressure cells over the eastern United States. The contours give the number of periods of four or more successive days between 1936–1965.

The washing out or scavenging of large particles by falling raindrops or snowflakes (washout)

The accumulation of small particles in the formation of raindrops or snowflakes in clouds (rainout)

The removal of gaseous pollutants by dissolution or absorption

The efficiencies of these processes depend on complex relationships between the properties of the pollutants and the characteristics of precipitation. The most effective and prevalent process is the washout of large particles in the lower layer of the atmosphere where most pollutants are released.

Topography

The topographic features of a region include both natural (hills, bridges, roads, canals, oceans, rivers, lakes, and foliages) and manmade (cities, bridges, roads, and canals) elements in a region. The prime significance of topography is its effects on meteorological elements, particularly the local or small-scale circulations that develop. These circulations contribute either favorably or unfavorably to the transport and dispersion of the pollutants.

LAND–SEA BREEZE

In the daytime, land heats rapidly, which heats the air above it. The water temperature remains relatively constant. The air over the heated land surface rises producing low pressure compared with the pressure over water. The resulting pressure gradient produces a surface flow off the water toward land. This circulation can extend to a considerable distance inland. Initially, the flow is onto the land, but as the breeze develops, the Coriolis force gradually shifts the direction so that the flow is more parallel to the land mass. After sunset and several hours of cooling by radiation, the land mass is cooler than the water temperature. Then, the reverse flow pattern develops, resulting in a wind off the land. During a stagnating high-pressure system when the transport and dispersion of pollutants are reduced, this short-period, afternoon increase in airflow can prevent the critical accumulation of pollutants.

MOUNTAIN–VALLEY WINDS

In the valley region, particularly in winter, intensive surface inversions develop from air cooled by the radiation-

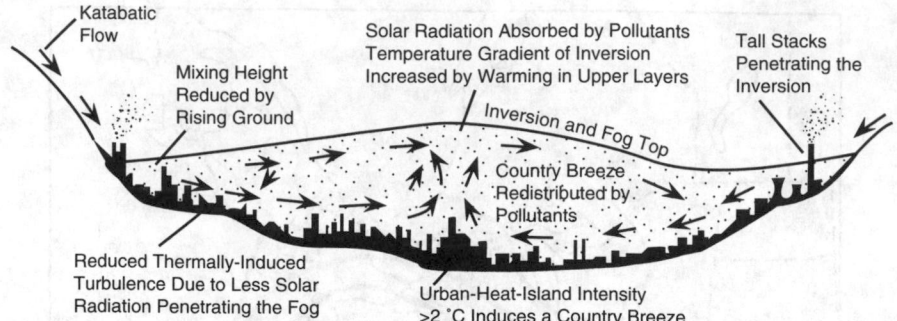

FIG. 5.6.11 Meteorology–pollution relationships during a smog in a valley location. (Reprinted, with permission, from D.M. Elsom, 1992, *Atmospheric pollution*, 2d ed., Oxford, U.K.: Blackwell Publishers.)

ally cooled valley wall surfaces. Populated and industrialized bottom valley areas are subject to a critical accumulation of pollutants during this period.

Areas on the windward side of mountain ranges expect added precipitation because of the forced rising, expansion, and cooling of the moving air mass with the resultant release of moisture. The precipitation increases the removal of pollutants.

URBAN-HEAT-ISLAND EFFECT

The increased surface roughness created by buildings throughout a city enhances the turbulence of airflow over the city, thus improving the dispersion of the pollutants emitted. However, at the same time, the city's buildings and asphalt streets act as a heat reservoir for the radiation received during the day. This heat plus the added heat from nighttime heating during cool months creates a tempera-

ture and pressure differential between the city and surrounding rural area so that a local circulation inward to the city develops. This circulation concentrates the pollutants in the city.

Figure 5.6.11 shows the combined effects of the urban-heat-island effect and katabatic winds.

—*David H.F. Liu*

References

Battan, L.J. 1979. *Fundamentals of meteorology.* Englewood Cliffs, N.J.: Prentice-Hall.

Cooper, C.D., and F.C. Alley. 1986. *Air pollution control—A design approach.* Prospect Heights, Ill.: Waveland Press, Inc.

Petterssen, S. 1968. *Air pollution.* 2d ed. New York: McGraw-Hill.

Smith, M.E., ed. 1973. *Recommended guide for the prediction of the dispersion of airborne effluents.* New York: ASME.

5.7
METEOROLOGIC APPLICATIONS IN AIR POLLUTION CONTROL

This section gives examples of meteorologic applications in air pollution control problems.

Air Pollution Surveys

Air pollution surveys are unique in their development and conduct. A common goal is to obtain a representative sample from an unconfined volume of air in the vicinity of one or more emission sources.

Depending on the objectives of an air pollution survey, a mobile or fixed sampler can be used. Other than the obvious considerations such as accessibility and the relationship to interfering pollutant sources, the principal factors in site selection are meteorology and topography. The controlling factor for site selection is wind movement. With some knowledge of the predominant wind direction, the environmental engineer can predict the path of pollution from the emission source to the point of ground-level impact and determine the most suitable location for an air monitoring site. The most convenient method for performing this analysis is to use the wind rose described in Section 5.5.

Besides wind direction and wind speed, other meteorological data necessary for sample correlations are temperature, cloud cover, and lapse rate where possible. The environmental engineer uses local temperatures to estimate the contribution of home heating to the total pollutant emission rates.

The simplest case is one where one wind direction predominates over a uniform topography for an isolated plant emitting a single pollutant that remains unchanged in the atmosphere. Two monitors are used: one monitors the effects of the source and the other is placed upwind to provide background concentrations. Where wind directions vary and other emission sources are operating nearby, the environmental engineer requires additional samples to identify the concentrations attributable to the source.

Environmental engineers often use a variation of the wind rose, called a *pollution rose,* to determine the source of a pollutant. Instead of plotting all winds on a radial graph, they use only those days when the concentration of a pollutant is above a minimum. Figure 5.7.1 is a plot of pollution roses. Only winds carrying SO_3 levels greater than 250 $\mu g/m$ are plotted. The fingers of the roses point to plant three. Pollution roses can be plotted for other pollutants and are useful for pinpointing sources of atmospheric contamination.

Selection of Plant Site

In selecting a plant site, planners should consider the air pollution climatology of the area. They should prepare seasonal wind roses to estimate pollution dispersion patterns. Wind roses based on average winds excluding frontal weather systems are especially helpful. Planners must con-

sider the frequency of stagnant weather periods and the effects of topography and local wind systems, such as land–sea breezes and mountain–valley winds, with respect to dispersion patterns and nearby residential and industrial areas.

The location of the plant within an area can depend on local wind speed and directions data. For example, residential areas may lie downwind of a proposed plant, in line with the prevailing wind direction. Considering a more suitable site would reduce the air pollution impact of the plant. Figure 5.7.2 illustrates this point.

Data on temperature, humidity, wind speed and direction, and precipitation are generally available through official weather agencies. Other potential sources of information are local airports, military installations, public utilities, and colleges and universities. The National Climatological Data Center, Asheville, North Carolina, is a major source of information. The center also contracts to prepare specific weather summaries and frequency.

Allowable Emission Rates

In plant planning, planners should consider local, state, and federal air pollution authorities, which can shutdown or curtail plant emission activities during times of air pollution emergency. Plants must have standby plans ready for reducing the emission of air contaminants into the atmosphere.

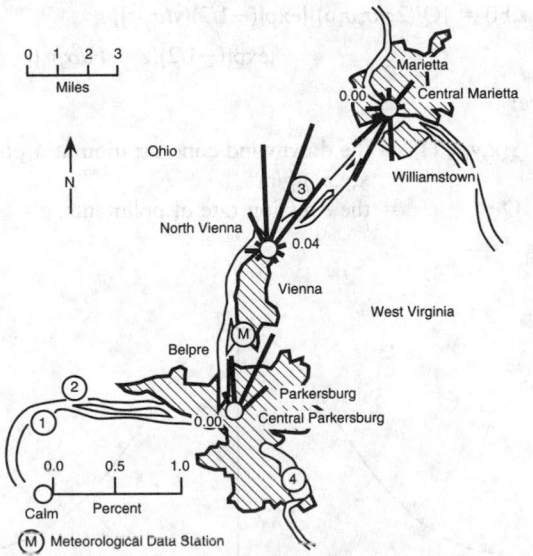

FIG. 5.7.1 Pollution roses, with SO_2 concentrations greater than 250 $\mu g/m$. The major suspected sources are the four chemical plants, but the data indicates that Plant Three is the primary culprit. (Reprinted, with permission, from P.A. Vesilind, 1983, *Environmental pollution and control,* Ann Arbor Science Publishers.)

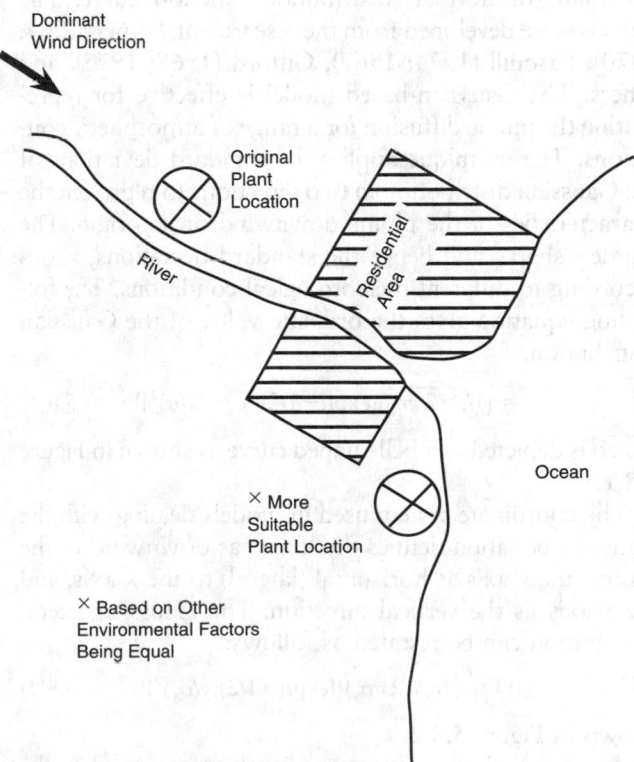

FIG. 5.7.2 Plant site selection within a localized region influenced by meteorological conditions.

A plant must control emission rates to ensure that problems do not occur even during poor dispersion conditions. This control requires full knowledge of the frequency of poor dispersion weather. In addition, weather conditions should be considered when plant start ups are scheduled or major repairs that may produce more emissions are undertaken.

Stack Design

Section 5.4 presents stack design procedures and lists the weather parameters required. The stack height design must consider the average height at stack elevation, average temperature, average mixing conditions (stability), and average lapse rate. The stack height design must also consider the average height and frequencies of inversions. For emission sources such as generating stations, the ideal stack height should exceed the most frequent inversion height. Also, planners should consider not only the averages of temperature, wind speed, and stability, but also the frequency with which worst-case combinations of these parameters occur.

—*David H.F. Liu*

5.8
ATMOSPHERIC DISPERSION MODELING

The Gaussian Model

The goal of air quality dispersion modeling is to estimate a pollutant's concentration at a point downwind of one or more emission sources. Since the early 1970s, the U.S. EPA has developed several computer models based on the Gaussian (or normal) distribution function curve. The models were developed from the research of Turner (1964; 1970); Pasquill (1974; 1967), Gifford (1968; 1975), and others. The Gaussian-based model is effective for representing the plume diffusion for a range of atmospheric conditions. The technique applies the standard deviations of the Gaussian distribution in two directions to represent the characteristics of the plume downwind of its origin. The plume's shape, and hence the standard deviations, varies according to different meteorological conditions. The following equation gives the ordinate value of the Gaussian distribution:

$$y = [1/(\sqrt{2\pi}\sigma)]\{\exp[(-1/2)(x - \bar{x}/\sigma)^2]\} \qquad 5.8(1)$$

which is depicted as a bell-shaped curve as shown in Figure 5.8.1.

The coordinate system used in models dealing with the Gaussian equation defines the x axis as downwind of the source, the y axis as horizontal (lateral) to the x axis, and the z axis as the vertical direction. The Gaussian lateral distribution can be restated as follows:

$$\chi(y) = [1/(\sqrt{2\pi}\sigma_y)]\{\exp[(-1/2)(y/\sigma_y)^2]\} \qquad 5.8(2)$$

Shown in Figure 5.8.2.

A second, similar Gaussian distribution describes the distribution of the plume in the vertical, or z, direction. The distribution of the plume around the centerline in both the y and z directions can be represented when the two single distributions in each of the two coordinate directions are multiplied to give a double Gaussian distribution. Projecting this distribution downwind through x gives the volume of space that contains the plume as shown in Figure 5.8.3.

Shifting the centerline upward a distance H corrects the equation for emissions at the effective stack height (stack height plus plume rise above stack) as follows:

$$\chi(x,y,z;H) = [Q/(2\pi\sigma_y\sigma_z u)]\{\exp[(-1/2)(y/\sigma_y)^2]\}$$
$$\{\exp[(-1/2)(z - H/\sigma_z)^2]\} \qquad 5.8(3)$$

where:

$\chi(x,y,z; H)$ = the downwind concentration at a point x,y,z, $\mu g/m^3$

Q = the emission rate of pollutants, g/s

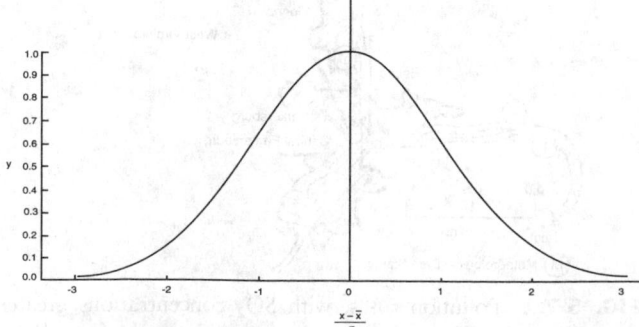

FIG. 5.8.1 The Gaussian distribution (or normal) curve. (Reprinted from D.B. Turner, 1970, *Workbook of atmospheric dispersion estimates (Revised)*, Office of Air Programs Pub. No. AP-26, Research Triangle Park, N.C.: U.S. EPA.)

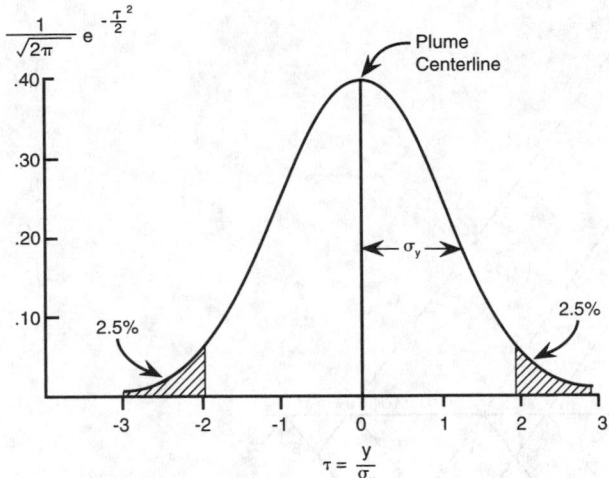

FIG. 5.8.2 Properties of the Gaussian distribution. Adapted, with permission, from H.A. Panofsky and J.A. Dutton, 1984. *Atmospheric Turbulence: Models and Methods for Engineering Applications,* John Wiley & Sons, New York.

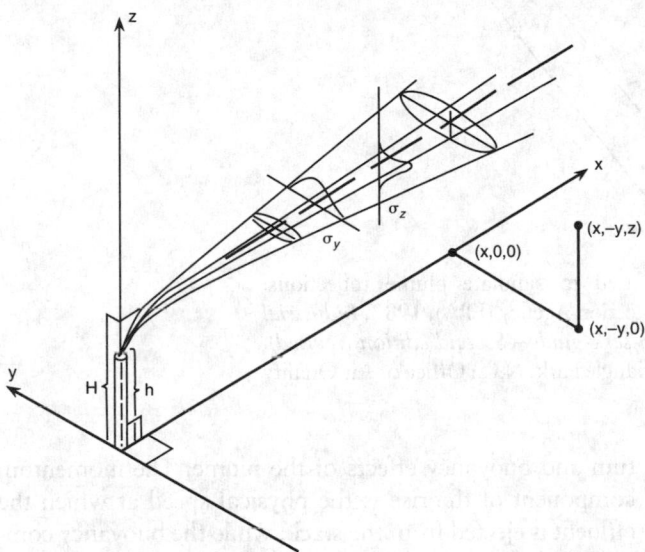

FIG. 5.8.3 Coordinate system showing Gaussian distribution in the horizontal and vertical. (Reprinted from D.B. Turner, 1970, *Workbook of atmospheric dispersion estimates (Revised),* Office of Air Programs Pub. No. AP-26, Research Triangle Park, N.C.: U.S. EPA.)

σ_y, σ_z	= the plume standard deviations, m
u	= the mean vertical wind speed across the plume height, m/s
y	= the lateral distance, m
z	= the vertical distance, m
H	= the effective stack height, m

As the plume propagates downwind, at some point the lowest edge of the plume strikes the ground. At that point, the portion of the plume impacting the ground is reflected upward since no absorption or deposition is assumed to occur on the ground (conservation of matter). This reflec-

tion causes the concentration of the plume to be greater in that area downwind and near the ground from the impact site. Functionally this effect can be mimicked, within the model with a virtual point source created identical to the original, emitting from a mirror image below the stack base as shown in Figure 5.8.4. Adding another term to the equation can account for this reflection of the pollutants as follows:

$$\chi(x,y,z; H) = [Q/(2\pi\sigma_y\sigma_z u)]\{\exp[(-1/2)(y/\sigma_y)^2]\}$$
$$\{\exp[(-1/2)(z - H/\sigma_z)^2] + \exp[(-1/2)(z + H/\sigma_z)^2]\} \quad 5.8(4)$$

When the plume reaches equilibrium (total mixing) in the layer, several more iterations of the last two terms can be added to the equation to represent the reflection of the plume at the mixing layer and the ground. Generally, no more than four additional terms are needed to approximate total mixing in the layer.

For the concentrations at ground level, z can be set equal to zero, and Equation 5.8(4) reduces as follows:

$$\chi(x,y,H) = [Q/(\pi\sigma_y\sigma_z u)]\{\exp[(-1/2)(y/\sigma_y)^2]\}$$
$$\{\exp[(-1/2)(H/\sigma_z)^2]\} \quad 5.8(5)$$

In addition, the plume centerline gives the maximum values. Therefore, setting y equal to zero gives the following equation:

$$\chi(x,H) = [Q/(\pi\sigma_y\sigma_z u)]\{\exp[(-1/2)(H/\sigma_z)^2]\} \quad 5.8(6)$$

which can estimate the concentration of pollutants for a distance x.

Finally, if the emission source is located at ground level with no effective plume rise, the equation can be reduced to its minimum as follows:

$$\chi(x) = Q/(\pi\sigma_y\sigma_z u) \quad 5.8(7)$$

A number of assumptions are typically used for Gaussian modeling. First, the analysis assumes a steady-state system (i.e., a source continuously emits at a constant strength; the wind speed, direction, and diffusion characteristics of the plume remain steady; and no chemical transformations take place in the plume). Second, diffusion in the x direction is ignored although transport in this direction is accounted for by wind speed. Third, the plume is reflected up at the ground rather than being deposited, according to the rules of conservation of matter (i.e., none of the pollutant is removed from the plume as it moves downwind). Fourth, the model applies to an ideal aerosol or an inert gas. Particles greater than 20 μm in diameter tend to settle out of the atmosphere at an appreciable rate. More sophisticated EPA models consider this deposition, as well as the decay or scavenging of gases. Finally, the calculations are only valid for wind speeds greater than or equal to 1 m per sec.

Application of the Gaussian models is limited to no more than 50 km due to extrapolation of the dispersion coefficients. Other factors that influence the Gaussian dis-

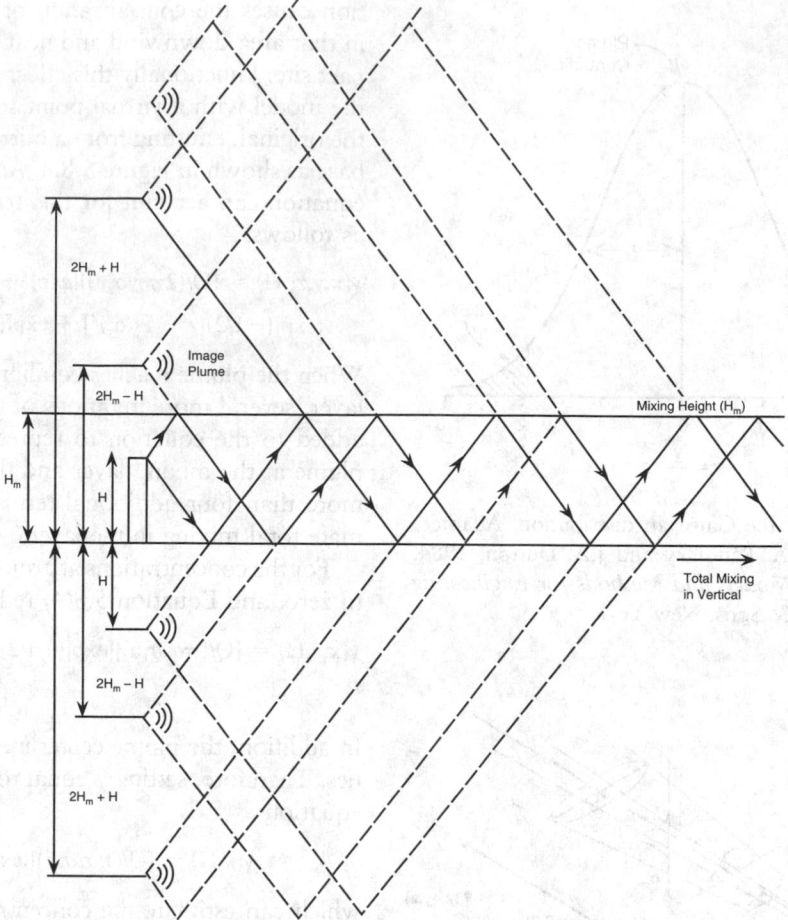

FIG. 5.8.4 Multiple plume images used to simulate plume reflections. (Reprinted from U.S. Environmental Protection Agency (EPA), 1987, *Industrial source complex (ISC) dispersion model user's guide—Second edition (revised)*, Vol. 1, EPA-450/4-88-002a, Research Triangle Park, N.C.: Office of Air Quality Planning and Standards, [December].)

persion equations are ground roughness, thermal characteristics, and meteorological conditions. Plume dispersion tends to increase as each of these factors increases, with the models most sensitive to atmospheric stability.

PLUME CHARACTERISTICS

Boilers or industrial furnaces that have a well-defined stack and emissions resulting from products of combustion are the most common sources of pollutants modeled. The hot plume emitted from the stack rises until it has expanded and cooled sufficiently to be in volumetric and thermal equilibrium with the surrounding atmosphere. The height at which the plume stabilizes is referred to as the effective plume height (H) and is defined as:

$$H = h_s + \Delta H \qquad 5.8(8)$$

where (h_s) is the physical stack height portion and (ΔH) is the plume rise portion as shown in Figure 5.8.5. The plume rise is the increase in height induced by both the momen-

tum and buoyancy effects of the plume. The momentum component of the rise is the physical speed at which the effluent is ejected from the stack, while the buoyancy component is due to the thermal characteristics of the plume in relation to ambient air. In the modeling, pollutants are assumed either to emit from a point directly above the stack at the effective plume height or to gradually rise over some distance downwind of the emission point until the effective plume height is reached.

The standard plume rise formula used in most EPA air dispersion models is based on a review of empirical data performed by Briggs (1969). (The next section includes a more detailed description.) The Briggs formula takes into account both momentum and buoyancy factors as well as meteorological conditions.

DISPERSION COEFFICIENTS

Sigma y (σ_y) and sigma z (σ_z) are the standard deviations of the Gaussian distribution functions. Since they are used

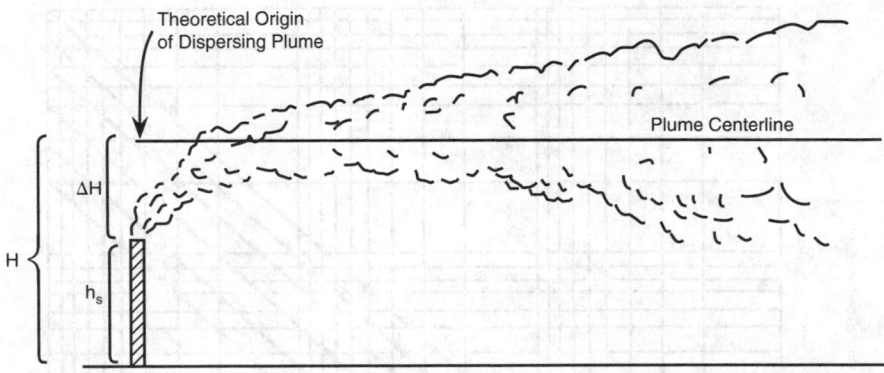

FIG. 5.8.5 Effective stack height (H), with dispersion beginning at a theoretical point above the stack. (Adapted with permission from American Society for Mechanical Engineers [ASME] Air Pollution Control Div., 1973, *Recommended guide for the prediction of the dispersion of airborne effluents*, 2d ed., New York: ASME.)

in describing a plume as it disperses, they increase with time and distance traveled. The rates of growth for σ_y and σ_z depend upon meteorological conditions. Sigma y is generally larger than σ_z, since no stratification obstacles are in the y (horizontal) direction. Sigma y and σ_z are strongly influenced by heat convection and mechanical turbulence, and, as these forces become more pronounced, the sigmas increase more quickly.

As a standard deviation, σ_y and σ_z characterize the broadness or sharpness of the normal distribution of pollutants within the plume. As both sigmas increase, the concentration value of a pollutant at the plume centerline decreases. However, the total amount of the pollutant in the plume remains the same; it is merely spread out over a wider range and thus, the concentration changes. Approximately two thirds of the plume is found between plus or minus one sigma, while 95% of the plume is found between plus or minus 2 sigma as shown in Figure 5.8.2. The plume edge is considered to be that concentration that is one-tenth the concentration of the centerline.

Sigma y and σ_z are generally determined from equations derived using empirical data obtained by Briggs (1969) and McElroy and Pooler (1968) and research performed by Bowne (1974). Figures 5.8.6 and 5.8.7 show plots of these curves. These plots were developed from the observed dispersion of a tracer gas over open, level terrain.

STABILITY CLASSES

In the late 1960s, Pasquill developed a method for classifying atmospheric conditions which was later modified by Gifford (1975), resulting in six stability classes, labeled A through F. The method was based on the amount of incoming solar radiation, cloud cover, and surface wind speed as shown in Table 5.8.1.

Stability greatly affects plume behavior as demonstrated by the dispersion curves discussed above. Classes E and F indicate stable air in which stratification strongly dampens mechanical turbulence, typically with strong winds in a constant wind direction. These conditions can produce a fanning plume that does not rise much and retains a narrow shape in the vertical dimension for a long distance downwind as shown in part (a) of Figure 5.8.8.

A situation where a plume in a stable layer is brought quickly to the surface by turbulence in a less stable layer is termed *fumigation* and is shown in part (b) of Figure 5.8.8. This can occur as the result of heat convection in the morning.

Class D stability is neutral, with moderate winds and even mixing properties. These conditions produce a coning plume as shown in part (c) of Figure 5.8.8. Classes A, B, and C represent unstable conditions which indicate various levels of extensive mixing. These conditions can produce a looping plume as shown in part (d). If the effective stack height exceeds the mixing height, the plume is assumed to remain above it, and no ground-level concentrations are calculated. This effect is known as lofting and is shown in part (e).

Rough terrain or heat islands from cities increase the amount of turbulence and change the classification of ambient conditions, usually upward one stability class.

In general, a plume under stability class A conditions affects areas immediately near the emission source with high concentrations. Class F stability causes the plume to reach ground level further away, with a lower concentration (unless terrain is a factor).

WIND SPEED

The wind speed is the mean wind speed over the vertical distribution of a plume. However, usually the only wind

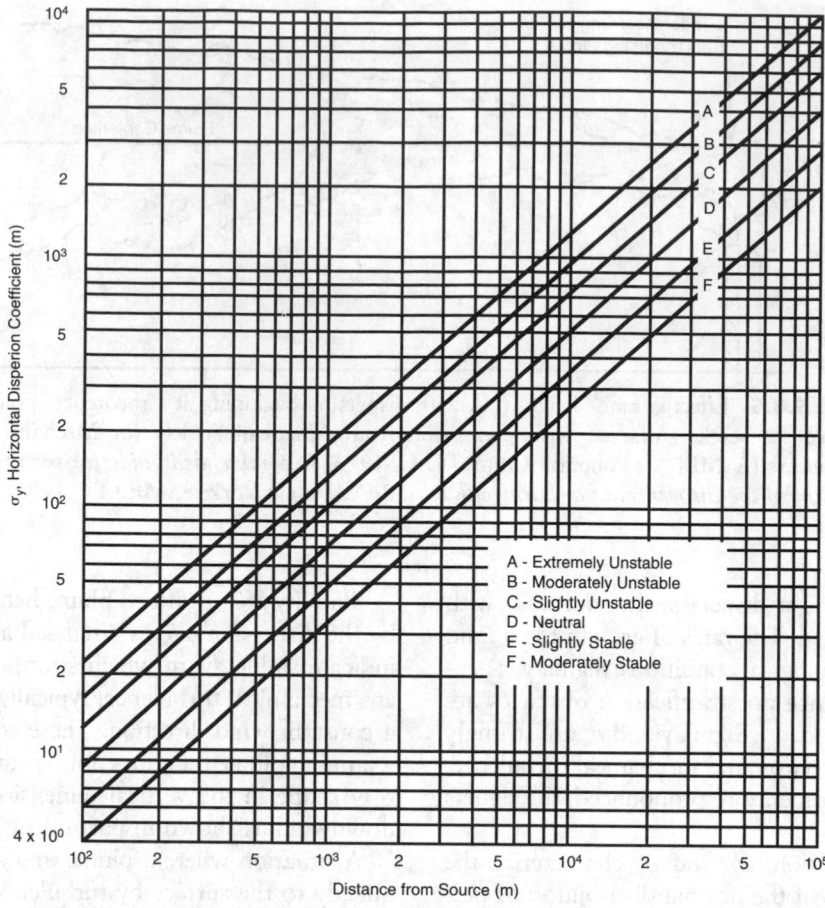

FIG. 5.8.6 Horizontal dispersion coefficient as a function of downwind distance from the source. (Reprinted from D.B. Turner, 1970, *Workbook of atmospheric dispersion estimates (Revised),* Office of Air Programs Pub. No. AP-26, Research Triangle Park, N.C.: U.S. EPA.)

speed available is that monitored at ground-level meteorological stations. These stations record ambient atmospheric characteristics, usually at the 10-m level, and typically with lower wind speeds than those affecting the plume. These lower speeds are due to the friction caused by the surface as shown in Figure 5.8.9. Therefore, the wind speed power law must be used to convert near-surface wind speed data into a wind speed representative of the conditions at the effective plume height. The wind speed power law equation is as follows:

$$u_2 = u_1 * (z_2/z_1)^p \qquad 5.8(9)$$

where u_1 and z_1 correspond to the wind speed and vertical height of the wind station, while u_2 and z_2 pertain to the characteristics at the upper elevation. This formula is empirical, with the exponent derived from observed data. The exponent (p) varies with the type of ambient weather conditions, generally increasing with stability and surface roughness (Irwin 1979). It can range from 0.1 for calm conditions to 0.4 for turbulent weather conditions. Table 5.8.2 shows exponents for various types of surface characteristics. Table 5.8.3 shows selected values for both ur-

ban and rural modes as used for the six stability categories in the industrial source complex (ISC3) models.

Plume Rise and Stack Height Considerations

In dispersion modeling, the plume height is critical to the basic equation for determining downwind concentrations at receptors. Several factors affect the initial dispersion of the plume emitted from a stack, including the plume rise, the presence of buildings or other features disturbing the wind stream flow, and the physical stack height. This section discusses these factors.

PLUME RISE

Various attempts have been made to estimate the plume rise from stationary sources. Two types of equations have resulted: theoretical and empirical. Theoretical models are generally derived from the laws of buoyancy and momentum. They are often adjusted for empirical data. Empirical models are developed from large amounts of ob-

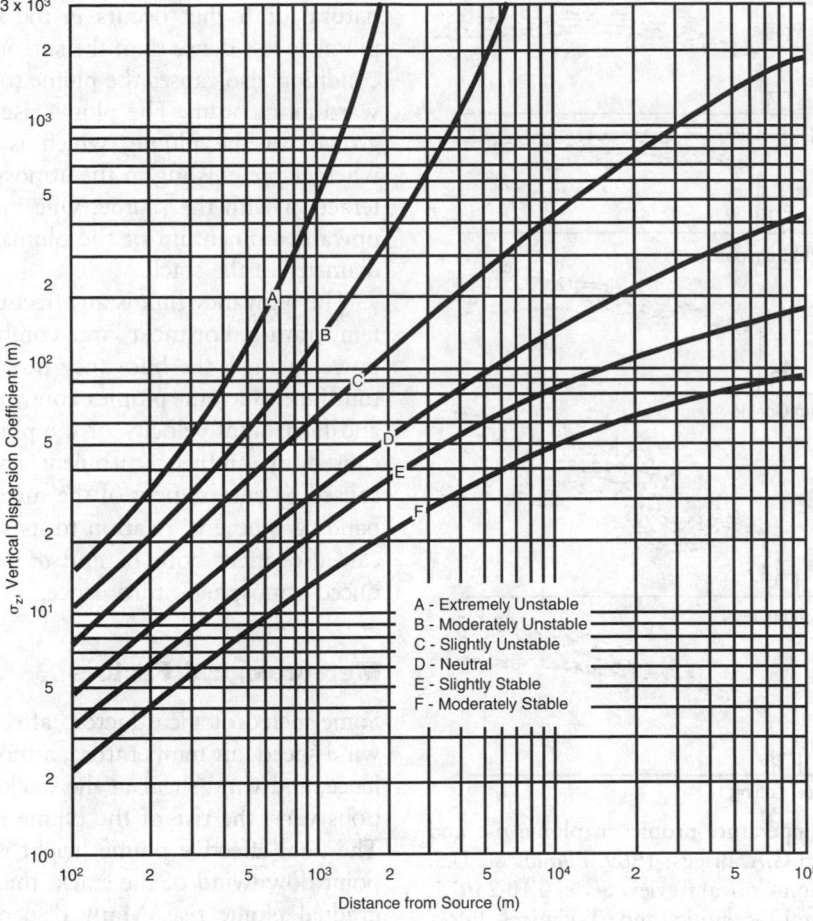

FIG. 5.8.7 Vertical dispersion coefficient as a function of downwind distance from the source. (Reprinted from Turner, 1970.)

served data such as tracer studies, wind tunnel experiments, and photographic evidence. Most plume rise equations apply to uniform or smoothly varying atmospheric conditions. An important consideration is that while they often predict the plume rise reasonably well under similar conditions, they can give wrong answers for other conditions.

Momentum and Buoyancy Factors

The plume rises mainly due to two factors: 1) the velocity of the exhaust gas, which imparts momentum to the plume, and 2) the temperature of the exhaust gas, which gives the plume buoyancy in ambient air. The momentum flux comes from mechanical fans in duct systems and the

TABLE 5.8.1   KEY TO PASQUILL STABILITY CATEGORIES

Surface Wind Speed (at 10 m) (m/sec)	Day			Night	
	Incoming Solar Radiation			Thinly Overcast or $\geq\frac{4}{8}$ Low Cloud	Clear or $\leq\frac{3}{8}$ Cloud
	Strong	Moderate	Slight		
<2	A	A–B	B		
2–3	A–B	B	C	E	F
3–5	B	B–C	C	D	E
5–6	C	C–D	D	D	D
>6	C	D	D	D	D

Source: D.B. Turner, 1970, *Workbook of atmospheric dispersion estimates* (Revised), Office of Air Programs Pub. No. AP-26 (Research Triangle Park, N.C.: U.S. EPA).
Note: The neutral class *D* should be assumed for overcast conditions during day or night.

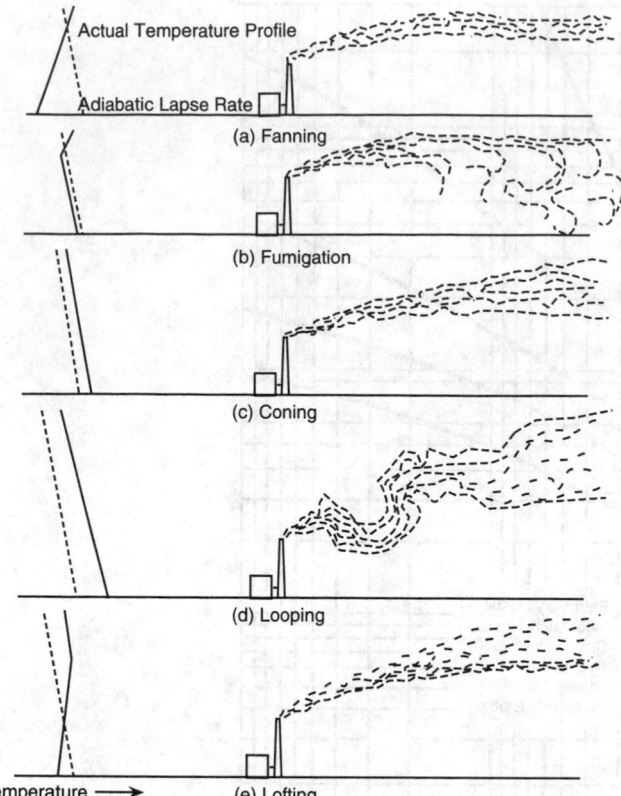

FIG. 5.8.8 Effect of temperature profile on plume rise and diffusion. (Reprinted from G.A. Briggs, 1969, *Plume rise*, U.S. Atomic Energy Commission Critical Review Series T10-25075. Clearinghouse for Federal Scientific and Technical Information.)

natural draft that occurs in the stack. If the gas in the plume is less dense than the surrounding atmosphere, this condition also causes the plume to rise and adds to its upward momentum. The plume rise also depends upon the growth of the plume, which is caused by turbulence, whether pre-existing in the atmosphere or induced by interaction with the plume. One method of increasing the upward momentum of the plume is to constrict the exit diameter of the stack.

The buoyancy flux is an effect of the plume's increased temperature. For most large combustion sources (such as power plants), the buoyancy flux dominates the momentum flux. Buoyant plumes contribute to both the vertical and horizontal velocity of the plume, in addition to that caused by ambient turbulent levels. This condition is caused by entrainment of the surrounding air into the expanding plume in relation to its surroundings. Buoyancy can also affect both σ_y and σ_z because of buoyancy-induced atmospheric turbulence.

Meteorological Factors

Some meteorological factors affecting plume rise include wind speed, air temperature, atmospheric stability, turbulence, and wind shear at the stack height. As these conditions vary, the rise of the plume is enhanced or reduced. The final effective plume height is reached only at some point downwind of the stack; this condition is known as gradual plume rise. Many dispersion models (especially where terrain is an important consideration) incorporate special algorithms to analyze ambient concentrations in

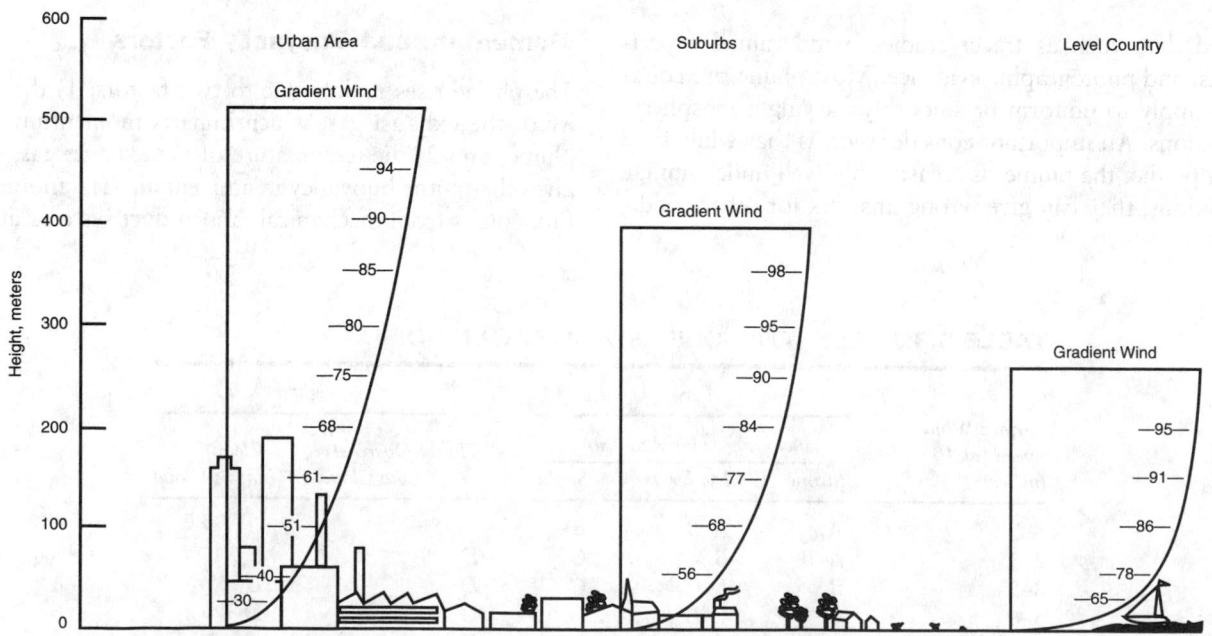

FIG. 5.8.9 Variation of wind with height for different roughness elements (figures are percentages of gradient wind). (Reprinted from D.B. Turner, 1970, *Workbook of atmospheric dispersion estimates (Revised)*, Office of Air Programs Pub. No. AP-26, Research Triangle Park, N.C.: U.S. EPA and based on A.G. Davenport, 1963, the relationship of wind structure to wind loading. Presented at Int. Conf. on the Wind Effects on Buildings and Structures, Nat. Physical Laboratory, Teddington, Middlesex, England, 26-28 June.)

TABLE 5.8.2   EXPONENTS FOR POWER LAW VELOCITY PROFILE EQUATION

Surface Configuration	Stability	p
Smooth open country	Unstable	0.11
	Neutral	0.14
	Moderate stability	0.20
	Large stability	0.33
Nonurban—varying roughness and terrain	Daytime—unstable and neutral	0.1–0.3
	Nighttime—stable including inversion	0.2–0.8
Urban (Liverpool)	Unstable, $\Delta\theta^a < 0$	0.20
	Neutral, $\Delta\theta = 0$	0.21
	Stable, $0 \le \Delta\theta < 0.75$	$0.21 + 0.33\,\Delta\theta$
Flat open country, $z_G{}^b = 274$ m	Neutral	0.16
Woodland forest, $z_G = 396$ m	Neutral	0.28
Urban area, $z_G = 518$ m	Neutral	0.40

Source: Gordon H. Strom, 1976, Transport and diffusion of stack effluents, Vol. 1, in *Air pollution,* 3d ed., edited by Arthur C. Stern, p. 412 (New York: Academic Press).

Notes: [a]$\Delta\theta$ = Potential temperature difference between 162- and 9-m elevations.
[b]z_G = Height of planetary boundary layer to gradient wind.

the region before the final rise is attained. However, models usually apply the final effective stack height at the point above the source to simplify calculations.

An early equation to predict plume rise was developed by Holland (1953) for the Atomic Energy Commission. This equation accounted for momentum and buoyancy by having a separate term for each. A major problem with the equation was that it did not account for meteorological effects. At the time of his work, Holland recognized that the formula was appropriate only for class D stability conditions.

BRIGGS PLUME RISE

Briggs (1969) developed a plume rise formula that is applicable to all stability cases. The Briggs plume rise formula is now the standard used by the EPA in its disper-

TABLE 5.8.3   WIND PROFILE EXPONENTS IN ISC3 DISPERSION MODELS

Stability Category	Rural Exponent	Urban Exponent
A	0.07	0.15
B	0.07	0.15
C	0.10	0.20
D	0.15	0.25
E	0.35	0.30
F	0.55	0.30

Source: U.S. Environmental Protection Agency (EPA), 1987, *Industrial source complex (ISC) dispersion model user's guide—Second edition (Revised),* Vol. 1, EPA-450/4-88-002a (Research Triangle Park, N.C.: Office of Air Quality Planning and Standards [December]).

sion models, but it is not as straightforward as Holland's approach. For the Briggs equation, determining the buoyancy flux, F_b, is usually necessary using the following equation:

$$F_b = gv_s d_s^2[(\Delta T)/4T_s]\qquad 5.8(10)$$

where:

g = the gravitational constant (9.8 m/s^2)
v_s = the stack gas exit velocity, m/s
d_s = the diameter of the stack, m
T_s = the stack gas temperature, deg. K
ΔT = the difference between T_s and T_a (the ambient temperature), deg. K

When the ambient temperature is less than the exhaust gas temperature, it must be determined whether momentum or buoyancy dominates. Briggs determines a crossover temperature difference $(\Delta T)_c$ for $F_b \ge 55$, and one for $F_b < 55$. If ΔT exceeds $(\Delta T)_c$, then a buoyant plume rise algorithm is used; if less, then a momentum plume rise equation is employed. The actual algorithms developed by Briggs to calculate the plume rise further depend upon other factors, such as the atmospheric stability, whether the plume has reached the distance to its final rise (i.e., gradual rise), and building downwash effects (U.S. EPA 1992d).

DOWNWASH

All large structures distort the atmosphere and interfere with wind flow to some extent. These atmospheric distortions usually take the form of a wake, which consists of a pocket of slower, more turbulent air. If a plume is emitted near a wake, it is usually pulled down because of the lower pressure in the wake region. This effect is termed *downwash*. When downwash occurs, the plume is brought down to the ground near the emission source more quickly.

A wake that causes downwash usually occurs as the result of one of three physical conditions: 1) the stack, referred to as stack-tip downwash, 2) local topography, or 3) nearby large structures or building downwash. Figure 5.8.10 shows examples of each of these conditions.

Stack-Tip Downwash

Stack-tip downwash occurs when the ambient wind speed is high enough relative to the exit velocity of the plume so that some or all of the plume is pulled into the wake directly downwind of the stack, as shown in Figure 5.8.11. This downwash has two effects on plume rise. First, the pollutants drawn into the stack wake leave the stack region at a lower height than that of the stack and with a lower upward velocity. Second, the downwash increases the plume cross section, which decreases the concentration.

To avoid stack-tip downwash, environmental engineers should consider the ratio of emission velocity (v_s) to wind speed velocity at the stack height (u_s) in the stack design. If $v_s < 1.5\ u_s$, then the physical stack height should be adjusted by the following equation:

$$h_{std} = 2d_s[(v_s/u_s) - 1.5] \qquad 5.8(11)$$

If $v_s/u_s > 1.5$, then stack-tip downwash is avoided since the exhaust gas is emitted from the stack at sufficient velocity to clear the downwash area on the downwind side of the stack. If $v_s/u_s < 1.0$, downwash will probably occur, possibly seriously. For intermediate values of v_s/u_s,

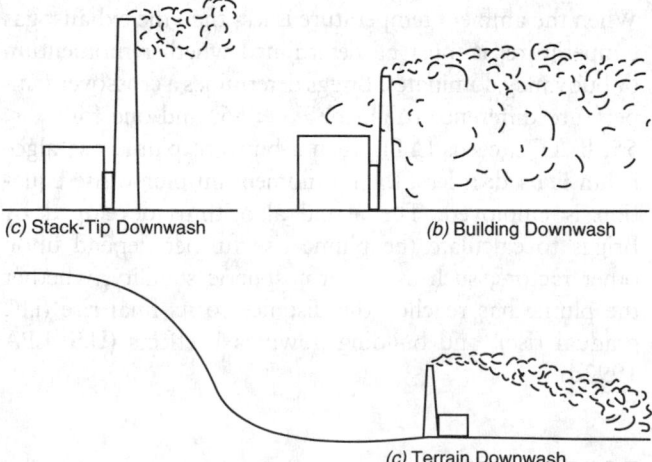

FIG. 5.8.10 Physical conditions that cause downwash. Reprinted from G.A. Briggs, 1969. *Plume rise,* U.S. Atomic Energy Commission Critical Review Series TID-25075, Clearinghouse for Federal Scientific and Technical Information.

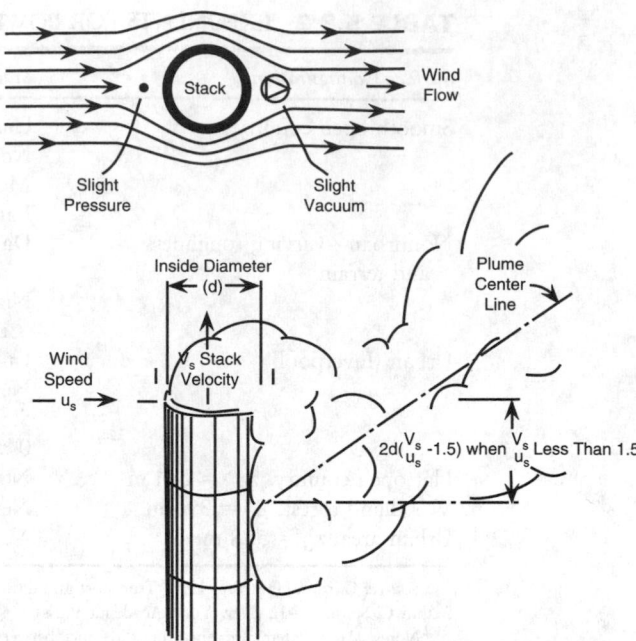

FIG. 5.8.11 Stack-tip downwash. (Reprinted, with permission, from Trinity Consultants, Inc., 1989, *Atmospheric diffusion notes,* Issue no. 13, Dallas, Tex. [June].)

downwash may occur depending on the ambient conditions at the time.

Downwash Caused by Topography

Downwash can also be caused by local topography. Large hills or mountains can change the normal wind patterns of an area. If the stack is located closely downwind of a hill above stack height, the air flowing off the hill can cause the plume to impact closer to the stack than normal as shown in part (c) of Figure 5.8.10. Modeling of these situations often employs physical models in wind tunnels. A recently developed model, the complex terrain dispersion model plus algorithms for unstable situations (CTDM-PLUS), employs a critical hill height calculation that determines if the plume impacts the hill or follows the uninterrupted laminar flow around it (U.S. EPA 1989).

Building Downwash

Large structures surrounding the stack also affect ambient wind conditions. The boundaries of the wake region resulting from surrounding structures are not sharply defined. They depend on the three-dimensional characteristics of the structure and are time dependent. The extent of distortion depends extensively upon building structure geometry and wind direction. Generally, a single cylindrical structure (e.g., a free-standing silo) has little influence on the wind flow compared to a rectangular structure.

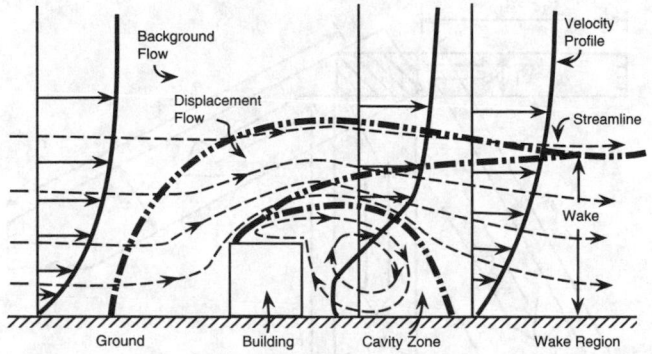

FIG. 5.8.12 Building structure downwash with cavity zone and wake effect. (Reprinted, with permission, from American Society of Mechanical Engineers [ASME].)

Part (b) in Figure 5.8.10 shows the building downwash that occurs when the plume is drawn into a wake from a nearby structure. Two zones exist within the downwash area of a structure. The first zone, which extends approximately three building heights downwind, is the cavity region where plume entrapment can occur. The second zone, which extends from the cavity region to about ten times the lesser dimension of the height or projected width, is the wake region where turbulent eddies exist as a result of structure disturbance to the wind flow. Figure 5.8.12 shows an example of these zones. Generally, the cavity region concentration is higher than the wake region concentration due to plume entrapment. Bittle and Borowsky

(1985) examined the impact of pollutants in cavity zones and found several calculations apply depending on the building and stack geometry. Beyond these zones, wind flow is unaffected by the structure.

The first downwash calculations were developed as the result of studies in a wind tunnel by Snyder and Lawson (1976) and Huber (1977). However, these studies were limited to a specific stability, structure shape, and orientation to the wind. Additional work by Hosker (1984), Schulman and Hanna (1986), and Schulman and Scire (1980) refined these calculations. Figure 5.8.13 shows the areas where the Huber–Snyder and Schulman–Scire downwash calculations apply. The following equation determines whether the Huber–Snyder or Schulman–Scire downwash calculations apply:

$$h_s = H + 0.5L \qquad 5.8(12)$$

where:

h_s = the physical stack height
H = the structure height
L = the lesser dimension of the height or projected width

The adjustments are made to the dispersion parameters.

To avoid building downwash, the EPA has developed a general method for designing the minimum stack height needed to prevent emissions from being entrained into any wake created by the surrounding buildings. In this way, emissions from a stack do not result in an excessive concentration of the pollutant close to the stack. This ap-

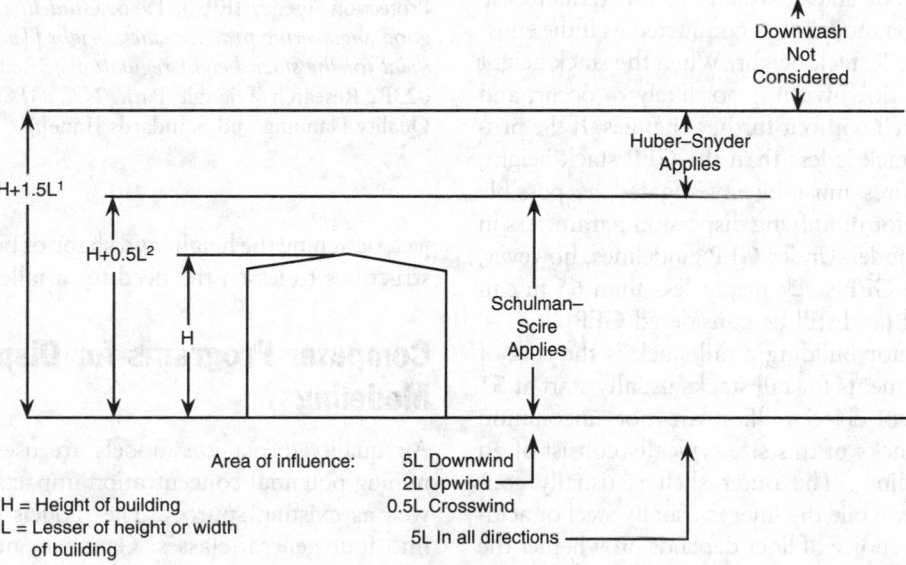

$H+1.5L^1$

$H+0.5L^2$

H

Downwash
Not
Considered

Huber–Snyder
Applies

Schulman–
Scire
Applies

Area of influence: 5L Downwind
2L Upwind
0.5L Crosswind

H = Height of building
L = Lesser of height or width
of building

5L In all directions

[1]Compare H+1.5L to plume elevation (stack height above grade + momentum plume rise 2L downwind-stack-tip downwash)

[2]Compare H+.5L to physical stack height above grade

FIG. 5.8.13 Selection of downwash algorithms in the ISCST model. (Reprinted, with permission, from Trinity Consultants, Inc., 1989, *Atmospheric diffusion notes,* Issue no. 13, Dallas, Tex. [June].)

proach is called the good engineering practice (GEP) of stack height design (U.S. EPA 1985) and retroactively covers all stacks built since December 31, 1970. The following equation determines the GEP stack height:

$$H_{GEP} = H + 1.5(L) \qquad 5.8(13)$$

where:

H_{GEP} = the GEP stack height
H = the maximum height of an adjacent or nearby structure
L = the lesser dimension of the height or projected width of an adjacent or nearby structure

The projected width of a structure is the exposed area perpendicular to the wind as shown in Figure 5.8.14. Environmental engineers should check all structures within 5L of the stack for their possible effect on the plume. The range of influence for a given structure is defined by Tickvart (1988) as 2L upwind of the structure, 5L downwind, and 0.5L on the sides parallel to the wind flow.

TALL STACKS

If concentration impacts are excessive, constructing a tall stack is one approach to reduce them. A tall stack dilutes ambient ground-level concentrations near the emission source. However, this approach does not reduce emission levels or total pollution loadings in a region; it merely provides greater initial dispersion at the source. The EPA regulates stack height under its tall stacks policy to encourage better control technology application.

Tall stacks describes stacks that are greater than the GEP stack height. For stacks which are taller than GEP guidelines, dispersion modeling is conducted as if the emission source has a GEP stack height. When the stack height is at the GEP level, downwash is not likely to occur, and modeling can proceed without further changes. If the proposed or existing stack is less than the GEP stack height, surrounding structures must be investigated as possible downwash sources for modifying dispersion parameters in the air dispersion model. Under GEP guidelines, however, a source that has a GEP stack height less than 65 m can raise it to that level (and still be considered GEP).

A consideration for building a tall stack is the cost of construction; investments for tall stacks usually start at $1 million, and costs of $4–5 million are not uncommon (Vatavuk 1990). Stacks of this size typically consist of an outer shell and a liner. The outer shell is usually constructed of concrete, while the liner is usually steel or acid-resistant brick. The choice of liner depends on whether the exhaust gas is above the acid dew point (steel above it, brick below). Given these costs, building a stack taller than GEP is rare; however, constructing a stack below GEP and conducting the additional downwash modeling required may be worthwhile. In some cases, if the environmental engineer is involved early in the design phase of the pro-

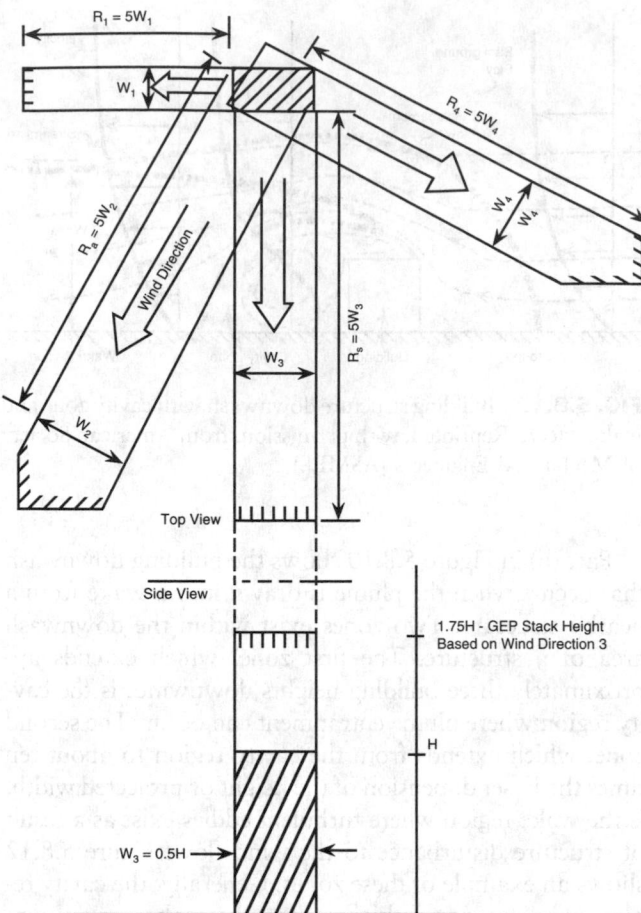

FIG. 5.8.14 GEP determination of projected structure width and associated region of adverse influence for a structure in four different wind directions. (Reprinted from U.S. Environmental Protection Agency (EPA), 1985, *Guideline for determination of good engineering practice stack height [Technical support document for the stack height regulations] [Revised]*, EPA-450/4-80-023R, Research Triangle Park, N.C.: U.S. EPA, Office of Air Quality Planning and Standards [June].)

ject, designing the height and shape of buildings and nearby structures to lessen the need for a taller stack is possible.

Computer Programs for Dispersion Modeling

Air quality dispersion models are useful tools for determining potential concentration impacts from proposed as well as existing sources. The models can be categorized into four general classes: Gaussian, numerical, statistical (empirical), and physical. The first three models are computer based, with numerical and Gaussian models dominating the field. This section focuses on Gaussian-based models since they are the most widely applied. This wide application is almost entirely due to their ease of application and the conservative estimates they provide, despite any of their shortcomings in precisely describing a plume's diffusion in the atmosphere.

Gaussian-based models generally require three types of input data: source emission data, receptor data, and meteorological data, though the latter two can be assumed in some cases. Source emission data provide the characteristics of the pollutant released to the atmosphere. Receptor data provide the location where a predicted concentration is desired. Meteorological data provide the conditions for the model to determine how the emissions are transported from the source to the receptor.

GUIDELINE MODELS

To promote consistency in applying models, the U.S. EPA has developed a document entitled *Guideline on Air Quality Models* (1978; 1986; 1987; 1993a; 1996). In this document, the EPA summarizes the performance of models in several comparative analyses and suggests the best applications of the model. As a result of the EPA's evaluations, those that perform well for a general set of conditions are classified as *Appendix A* models, and these models are listed in Table 5.8.4. Of the models listed, only the urban airshed model (UAM) and the emission and dispersion model (EDMS) are not Gaussian-based.

Models not classified as Appendix A but recognized as having potential application for a specific case can be used pending EPA approval. These models are designated as Appendix B models and are listed in Table 5.8.5. However, a performance demonstration may be required for an Appendix B model to demonstrate its suitability over a standard Appendix A model. This section addresses Appendix A models.

All Appendix A and B models and user's documentation are available from the U.S. Department of Commerce's National Technical Information Service (NTIS), in Springfield, Virginia 22161. In addition, model codes and selected abridged user's guides are available from the U.S. EPA's Support Center for Regulatory Air Models Bulletin Board System (SCRAM BBS), which can be accessed on the Internet (at http://ttnwww.rtpnc.epa.gov).

MODEL OPTIONS

Due to the range and combinations of physical conditions, dispersion models are often simplified and designed for specialized applications in limited situations. The EPA has designated Appendix A models for use under such specific applications and may require the selection of predetermined options for regulatory application. Some examples of specialized functions include simple screening versus refined modeling, terrain features, surrounding land use, pollutant averaging period, number and type of sources to be modeled, and additional influences to the release. These functions are discussed next.

SCREENING AND REFINED MODELS

Dispersion models have two levels of sophistication. The first, referred to as *screening modeling,* is a preliminary approach designed to simplify a source's emissions and provide conservative plume concentration impact estimates. The model user compares the results of screening modeling to the NAAQS, prevention of significant deterioration (PSD) increments, and/or ambient significance levels to determine if a second level of analysis, *refined modeling,* is required for a better estimate of the predicted concentrations. The purpose of screening is to identify if the potential for exceeding applicable air quality threshold levels exists, and thus the need for refined analysis. Screening eliminates the time and expense of refined modeling if the predicted concentrations do not approach the applicable levels.

A refined model provides a more detailed analysis of the parameters and thus gives a more accurate estimate of the pollutant concentration at receptors. However, a refined model demands more specific input data. The specific data can include topography, better receptor grid resolution, downwash or other plume adjustments, and pollutant decay or deposition.

SIMPLE AND COMPLEX TERRAIN

Air quality dispersion models can also be divided into three categories based on their application to terrain features (Wilson 1993) which address the relationship between receptor elevations and the top of the stack. Early models were developed without any terrain consideration (i.e., they were flat). In simple terrain models, receptors are located below the stack top, while in complex terrain models, receptors are located at or above the stack top.

More recently, model developers have focused attention on elevations which are located between the stack top and the final height of the plume rise. These elevations are classified as intermediate terrain (see Figure 5.8.15), and calculations within the intermediate region can be evaluated by either simple or complex models. The model predicting the highest concentration for a receptor is conservatively selected for that point. Thus, intermediate terrain is an overlapping region for model predictions.

A model, CTDMPLUS, is included in the EPA's Appendix A to address this concern. The model can predict concentrations at the stack top or above for stability conditions and has been approved for intermediate and complex terrain applications (U.S. EPA 1989).

URBAN AND RURAL CLASSIFICATION

EPA recommends models for either urban or rural applications, and most models contain an option for selecting urban or rural dispersion coefficients. These coefficients

TABLE 5.8.4 U.S. EPA PREFERRED AIR QUALITY DISPERSION MODELS

The following is a list of U.S. EPA approved Appendix A guideline models and their intended application. The user is referred to the *Guideline on Air Quality Models* (U.S. EPA 1978; 1986; 1987; 1993) and the appropriate user's guide (see the following references) to select and apply the appropriate model.

Terrain	Mode	Model	Reference
Screening			
Simple	Both	SCREEN3	U.S. EPA 1988; 1992a
Simple	Both	ISC3	Bowers, Bjorklund, and Cheney 1979; U.S. EPA 1987; 1992b; 1995
Simple	Both	TSCREEN	U.S. EPA 1990b
Simple	Urban	RAM	Turner and Novak 1978; Catalano, Turner, and Novak 1987
Complex	Rural	COMPLEXI	Chico and Catalano 1986; Source code.
Complex	Urban	SHORTZ	Bjorklund and Bowers 1982
Complex	Rural	RTDM3.2	Paine and Egan 1987
Complex	Rural	VALLEY	Burt 1977
Complex	Both	CTSCREEN	U.S. EPA 1989; Perry, Burns, and Cinnorelli 1990
Line	Both	BLP	Schulman and Scire 1980
Refined			
Simple	Urban	RAM	Turner and Novak 1978; Catalano, Turner, and Novak 1987
Simple	Both	ISC3	Bowers, Bjorklund, and Cheney 1979; U.S. EPA 1987; 1992b; 1995
Simple		EDMS	Segal 1991; Segal and Hamilton 1988; Segal 1988
Simple	Urban	CDM2.0	Irwin, Chico, and Catalano 1985
Complex	Both	CTDMPLUS	Paine et al. 1987; Perry et al. 1989; U.S. EPA 1990a
Line	Both	BLP	Schulman and Scire 1980
Line	Both	CALINE3	Benson 1979
Ozone	Urban	UAM-V	U.S. EPA 1990a
Coastal		OCD	DiCristofaro and Hanna 1989

References for Table 5.8.4

Benson, P.E. 1979. CALINE3—*A versatile dispersion model for predicting air pollutant levels near highways and arterial streets, Interim Report.* FHWA/CA/TL-79/23. Washington, D.C.: Federal Highway Administration.

Bjorklund, J.R., and J.F. Bowers. 1982. *User's instructions for the SHORTZ and LONGZ computer programs.* EPA 903/9-82-004a and b. Philadelphia: U.S. EPA Region 3.

Bowers, J.F., J.R. Bjorklund, and C.S. Cheney. 1979. *Industrial source complex (ISC) dispersion model user's guide.* EPA-450/4-79-030 and 031. Research Triangle Park, N.C.: U.S. EPA, Office of Air Quality Planning and Standards.

Burt, E.W. 1977. *VALLEY model user's guide.* EPA-450/2-77-018. Research Triangle Park, N.C.: U.S. EPA.

Catalano, J.A., D.B. Turner, and J.H. Novak. 1987. *User's guide for RAM.* 2d ed. EPA-600/3-87-046. Research Triangle Park, N.C.: U.S. EPA, Office of Air Quality Planning and Standards (October).

DiCristofaro, D.C., and S.R. Hanna. 1989. *OCD: The offshore and coastal dispersion model, version 4.* Vols. I & II. Westford, MA.: Sigma Research Corporation.

Irwin, J.S., T. Chico, and J. Catalano. 1985. *CDM 2.0 (climatological dispersion model) user's guide.* Research Triangle Park, N.C.: U.S. EPA.

Paine, R.J., and B.A. Egan. 1987. *User's guide to the rough terrain diffusion model (RTDM) (Rev. 3.20).*

Paine, R.J., D.G. Strimaitis, M.G. Dennis, R.J. Yamartino, M.T. Mills, and E.M. Insley. 1987. *User's guide to the complex terrain dispersion model.* Vol. 1. EPA-600/8-87-058a. Research Triangle Park, N.C.: U.S. Environmental Protection Agency.

Perry, S.G., D.J. Burns, and A.J. Cimorelli. 1990. *User's guide to CDTMPLUS: Volume 2. The screening mode (CTSCREEN).* EPA-600/8-90-087. Research Triangle Park, N.C.: U.S. EPA.

Perry, S.G., D.J. Burns, L.H. Adams, R.J. Paine, M.G. Dennis, M.T. Mills, D.G. Strimaitis, R.J. Yamartino, and E.M. Insley. 1989. *User's guide to the complex terrain dispersion model plus algorithms for unstable situations (CTDMPLUS).* Volume 1: Model Descriptions and User Instructions. EPA-600/8-89-041. Research Triangle Park, N.C.: U.S. EPA.

Segal, H.M. 1988. *A microcomputer pollution model for civilian airports and Air Force bases—Model application and background.* FAA-EE-88-5. Washington, D.C.: Federal Aviation Administration.

Segal, H.M. 1991. *EDMA-Microcomputer pollution model for civilian airports and Air Force bases: User's guide.* FAA-EE-91-3. Washington, D.C.: Federal Aviation Administration.

Segal, H.M., and P.L. Hamilton. 1988. *A microcomputer pollution model for civilian airports and Air Force bases—Model description.* FAA-EE-88-4. Washington, D.C.: Federal Aviation Administration.

Schulman, L.L., and J.S. Scire. 1980. *Buoyant line and point source (BLP) dispersion model user's guide.* Doc. P-7304B. Concord, Mass.: Environmental Research & Technology, Inc.

Turner, D.B., and J.H. Novak. 1978. *User's guide for RAM.* Vols. 1 & 2. EPA-600/8-78-016a and b. Research Triangle Park, N.C.: U.S. EPA.

U.S. Environmental Protection Agency (EPA). 1988. *Screening procedures for estimating the air quality impact of stationary sources, draft for public comment.* EPA-450/4-88-010. Research Triangle Park, N.C.: U.S. EPA, Office of Air Quality Planning and Standards (August).

————. 1989. *User's guide to the complex terrain dispersion model plus algorithms for unstable situations (CTDMPLUS).* Vol. 1. EPA-600/8-89-041. Research Triangle Park, N.C.: U.S. EPA, Atmospheric Research and Exposure Assessment Laboratory (March).

————. 1990a. *User's guide to the urban airshed model.* Vols. 1–8. EPA 450/4-90-007a-c, d(R), e-g, EPA-454/B-93-004. Research Triangle Park, N.C.: U.S. EPA.

————. 1990b. *User's guide to TSCREEN: A model for screening toxic air pollutant concentrations.* EPA-450/4-91-013. Research Triangle Park, N.C.: U.S. EPA.

————. 1992a. *The SCREEN2 model user's guide.* EPA-450/4-92-006. Research Triangle Park, N.C.: U.S. EPA.

————. 1992b. *User's guide for the industrial source complex (ISC2) dispersion models.* Vols. 1–3. EPA-450/4-92-008a, b, and c. Research Triangle Park, N.C.: U.S. EPA, Office of Air Quality Planning and Standards (March).

————. 1995. *User's guide for the industrial source complex (ISC3) dispersion models.* Vols 1 & 2. EPA-454/B-95-003a & b. Research Triangle Park, N.C.: U.S. EPA.

affect the plume's spread in the y and z directions and are determined by the characteristics of the area surrounding the site. In heavily developed areas such as cities, the urban heat island and structures affect the surrounding atmosphere, increasing turbulence and air temperature. In the country, the foliage and undeveloped land reduce longwave radiation and generate less turbulence. The amount of turbulence generated by each of these locations affects how the plume is dispersed in the atmosphere.

Irwin (1978) recommends two methods for categorizing the surrounding area as urban or rural. One technique relies on a methodology developed by Auer (1978), which characterizes the land use within a 3-km radial area. The second technique is based on a population density threshold (750 people/sq km) within a 3-km area. The Auer method is typically the preferred approach; it would identify a large industrial plant with storage yards (i.e., a steel mill) as an industrial (urban) site instead of an unpopulated rural site. Table 5.8.6 lists the Auer land use categories.

AVERAGING PERIODS

While several averaging periods are available for pollutants, the models are divided into two groups. The first group, referred to as short-term models, handles averaging periods from an hour to a year using hourly meteorological conditions. Note that most refined short term models calculate concentrations for block averaging periods rather than for running average periods within a day. Screening models fall into this category and predict concentrations for one averaging period only.

Long-term models use meteorological conditions ranging from a month or season to one or more years. The climatological data used in these models are generated from hourly data into joint frequency–distribution tables of wind speed, wind direction, and Pasquill–Gifford stability categories. These data are referred to as stability array (STAR) data sets.

SINGLE AND MULTIPLE SOURCES

To simplify calculations, some models predict concentrations for only one source at a time, while others can calculate concentrations for a complex combination of sources. Screening models are usually limited to one source. As a compromise to these two extremes, some models colocate numerous sources at the same point and add individual source concentrations together. This approach is usually applied for conservative screening estimates.

TYPE OF RELEASE

Four types of releases can be simulated: point, area, line, and volume sources. Some models can be limited to the types of releases they handle. For most applications, line sources can be treated as area or volume sources, as well as in individual line models (see the subsection on mobile and line source modeling). Point sources are the most common modeling application and are defined by height, temperature, diameter, and velocity parameters.

In some cases, small or insignificant point sources may be grouped into either point or area sources to simplify the calculation. Area sources are only defined by a release height and the length of a side (which represents a square area). Models calculate the emissions from area sources using a virtual point source located some distance upwind so that the plume dispersion is equal to the width of the box at the center of the area source (see Figure 5.8.16). The models typically ignore concentrations calculated upwind and over the area source.

Volume sources represent continuous releases over a length, such as a conveyor or monitor. They are defined in the models by a release height, length of a side, and initial dispersions in the y and z directions (σ_y and σ_z). Simulation of these sources is similar to the virtual point source used for area sources.

ADDITIONAL PLUME INFLUENCES

Some models handle various factors that affect the plume. These options include stack-tip downwash, buoyancy-induced dispersion, gradual plume rise, and building or topography downwash factors as described in previous sections. Other factors include pollutant decay (half-life) and deposition. Pollutant decay considers chemical changes or scavenging of the pollutant in the atmosphere, typically through the use of a half-life duration factor. Deposition accounts for the settling of the pollutant from the plume. This deposition can either occur through dry (gravitational) settling or wet (rainwash) removal. Dry deposition modeling usually requires particle size and settling velocities for the pollutant emitted; wet deposition also requires meteorological data about precipitation frequency and characteristics.

METEOROLOGY

As previously mentioned, the type of meteorology used in a model depends on the pollutant averaging period as well as the level of modeling and site characteristics. If the modeling is performed for a state implementation plan (SIP) or a PSD permit, five years of meteorological data are typically required.

Most screening models use a preselected set of meteorological conditions that provide the worst-case plume transport and ground-level concentrations. The basic data used by models include wind speed, wind direction, stability, temperature, and mixing height.

TABLE 5.8.5 ALTERNATIVE MODELS

The following is a list of U.S. EPA selected Appendix B models to the preferred guideline models. The user is referred to the *Guideline on Air Quality Models Appendix B* (U.S. EPA 1993), the appropriate user's guide (see the following references), and contact with the U.S. EPA Regional Modeler to select and apply the model.

Model	Reference
AVACTA-II	Zannetti, Carboni, and Lewis 1985
DEGADIS 2.1	U.S. EPA 1989
ERT Visibility	ENSR Consulting and Engineering 1990
HGSYSTEM	Post 1994a; Post 1994b
HOTMAC/RAPTAD	Mellor and Yamada 1974; 1982; Yamada and Bunker 1988
LONGZ	Bjorklund and Bowers 1982
MESOPUFF-II	Scire et al. 1984; U.S. EPA 1992b
MTDDIS	Wang and Waldron 1980
PANACHE	Transoft Group 1994
PLUVUE-II	U.S. EPA 1992a
PAL-DS	Petersen 1978; Rao and Snodgrass 1982
PPSP	Brower 1982; Weil and Brower 1982
RPM-IV	U.S. EPA 1993
SCSTER	Malik and Baldwin 1980
SHORTZ	Bjorklund and Bowers 1982
SDM	PEI Associates 1988
SLAB	Ermak 1990
Simple Line Source Model	Chock 1980
WYNDvalley	Harrison 1992

References for Table 5.8.5

Bjorklund, J.R., and J.F. Bowers. 1982. *User's instructions for the SHORTZ and LONGZ computer programs*. EPA 903/9-82-004a and b. Philadelphia: U.S. EPA, Region 3.

Brower, R. 1982. *The Maryland power plant siting program (PPSP) air quality model user's guide*. PPSP-MP-38. Baltimore: Maryland Department of Natural Resources.

Chock, D.P. 1980. *User's guide for the simple line-source model for vehicle exhaust dispersion near a road*. Warren, Mich.: Environmental Science Department, General Motors Research Laboratories.

ENSR Consulting and Engineering. 1990. *ERT visibility model: Version 4; Technical description and user's guide*. M2020-003. Acton, Mass.: ENSR Consulting and Engineering.

Ermak, D.L. 1990. *User's manual for SLAB: An atmospheric dispersion model for denser-than-air releases (UCRL-MA-105607)*. Lawrence Livermore National Laboratory.

Harrison, H. 1992. *A user's guide to WYNDvalley 3.11, An Eulerian-grid air quality dispersion model with versatile boundaries, sources, and winds*. Mercer Island, Wash.: WYNDsoft, Inc.

Malik, M.H., and B. Baldwin. 1980. *Program documentation for multi-source (SCSTER) model*. EN7408SS. Atlanta: Southern Company Services, Inc.

Mellor, G.L., and T. Yamada. 1974. A hierarchy of turbulence closure models for planetary boundary layers. *Journal of Atmospheric Sciences* 31:1791–1806.

Mellor, G.L., and T. Yamada. 1982. Development of a turbulence closure model for geophysical fludi problems. *Rev. Geophys. Space Phys.* 20:851–875.

PEI Associates. 1988. *User's guide to SDM—A shoreline dispersion model*. EPA-450/4-88-017. Research Triangle Park, N.C.: U.S. EPA.

Post, L. (ed.). 1994a. *HGSYSTEM 3.0 technical reference manual*. Chester, United Kingdom: Shell Research Limited, Thornton Research Centre.

Post, L. 1994b. *HGSYSTEM 3.0 user's manual*. Chester, United Kingdom: Shell Research Limited, Thornton Research Centre.

Rao, K.S., and H.F. Snodgrass. 1982. *PAL-DS model: The PAL model including deposition and sedimentation*. EPA-600/8-82-023. Research Triangle Park, N.C.: U.S. EPA, Office of Research and Development.

Scire, J.S., F.W. Lurmann, A. Bass, and S.R. Hanna. 1984. *User's guide to the Mesopuff II model and related processor programs*. EPA-600/8-84-013. Research Triangle Park, N.C.: U.S. EPA.

Transoft Group. 1994. *User's guide to fluidyn-PANACHE, a three-dimensional deterministic simulation of pollutants dispersion model for complex terrain*. Cary, N.C.: Transoft Group.

U.S. Environmental Protection Agency (EPA). 1989. *User's guide for the DEGADIS 2.1—Dense gas dispersion model*. EPA-450/4-89-019. Research Triangle Park, N.C.: U.S. EPA.

———. 1992a. *User's manual for the plume visibility model, PLUVUE II (Revised)*. EPA-454/B-92-008. Research Triangle Park, N.C.: U.S. EPA.

———. 1992b. *A modeling protocol for applying MESOPUFF II to long range transport problems*. EPA-454/R-92-021. Research Triangle Park, N.C.: U.S. EPA.

———. 1993. *Reactive Plume Model IV (RPM-IV) User's Guide*. EPA-454/B-93-012. Research Triangle Park, N.C.: U.S. EPA, (ESRL).

Wang, I.T., and T.L. Waldron. 1980. *User's guide to MTDDIS mesoscale transport, diffusion, and deposition model for industrial sources*. EMSC6062.1UR(R2). Newbury Park, Calif.: Combustion Engineering.

Weil, J.C., and R.P. Brower. 1982. *The Maryland PPSP dispersion model for tall stacks*. PPSP-MP-36. Baltimore, Md.: Maryland Department of Natural Resources.

Yamada, T., and S. Bunker, 1988. Development of a nested grid, second moment turbulence closure model and application to the 1982 ASCOT Brush Creek data simulation. *Journal of Applied Meteorology* 27:562–578.

Zannetti, P., G. Carboni, and R. Lewis. 1985. *AVACTA-II user's guide (Release 3)*. AV-OM-85/520. Monrovia, Calif.: AeroVironment, Inc.

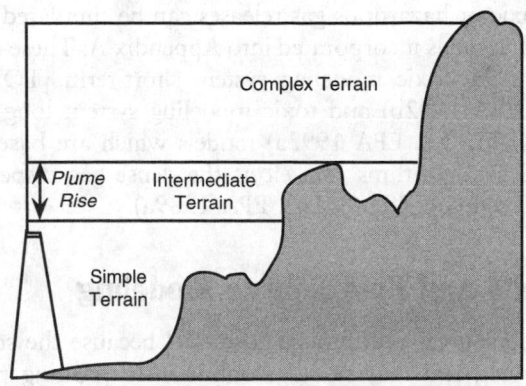

FIG. 5.8.15 Terrain categories. (Reprinted, with permission, from Trinity Consultants, Inc., 1993, *Air issues review*, Issue no. 5, Dallas, Tex. [September].)

A minimum of one year of site-specific data is preferred for refined and complex terrain models. However, these data often either are not available or require a year to collect. In most cases (i.e., noncomplex terrain), the National Weather Service (NWS) observations from a nearby station can be substituted in refined modeling applications. Five years of meteorological data from a representative station provide the best, reasonable representation of climatology at the station.

For refined, simple terrain modeling, representative data from both a NWS surface and upper air station are the required minimum. The model user can modify the data for input using standard EPA meteorological preprocessor programs. For complex terrain modeling, site-specific data are critical to represent the conditions within the local topographic regime. In most cases, site-specific data are incomplete for all the required parameters and must usually be supplemented by NWS preprocessed data. To retain

TABLE 5.8.6 IDENTIFICATION AND CLASSIFICATION OF LAND USE TYPES

Type	Use and Structures	Vegetation
I1	Heavy industrial	
	Major chemical, steel, and fabrication industries generally 3–5 story buildings, flat roofs	Grass and tree growth extremely rare; <5% vegetation
I2	Light–moderate industrial	
	Rail yards, truck depots, warehouse, industrial parks, minor fabrications; generally 1–3 story buildings, flat roofs	Very limited grass, trees almost totally absent; <5% vegetation
C1	Commercial	
	Office and apartment buildings, hotels; >10 story heights, flat roofs	Limited grass and trees; <15% vegetation
R1	Common residential	
	Single-family dwelling with normal easements; generally one story, pitched roof structures; frequent driveways	Abundant grass lawns and light–moderately wooded; >70% vegetation
R2	Compact residential	
	Single-, some multiple-, family dwellings with close spacing; generally <2 story height, pitched roof structures; garages (via alley), no driveways	Limited lawn sizes and shade trees; <30% vegetation
R3	Compact residential	
	Old multifamily dwellings with close (<2m) lateral separation; generally 2 story height, flat roof structures; garages (via alley) and ash pits, no driveways	Limited lawn sizes, old established shade trees; <35% vegetation
R4	Estate residential	
	Expansive family dwelling on multiacre tracts	Abundant grass lawns and light wooded; >80% vegetation
A1	Metropolitan natural	
	Major municipal, state, or federal parks, golf courses, cemeteries, campuses; occasional single-story structures	Nearly total grass and light wooded; >95% vegetation
A2	Agricultural rural	Local crops (e.g., corn, soybean); 95% vegetation
A3	Undeveloped	
	Uncultivated, wasteland	Mostly wild grasses and weeds, lightly wooded; >90% vegetation
A4	Undeveloped rural	Heavy wooded; 95% vegetation
A5	Water surfaces	
	Rivers, lakes	

Source: A.H. Auer, 1978, Correlation of land use and cover with meteorological anomalies, *Journal of Applied Meteorology* 17.

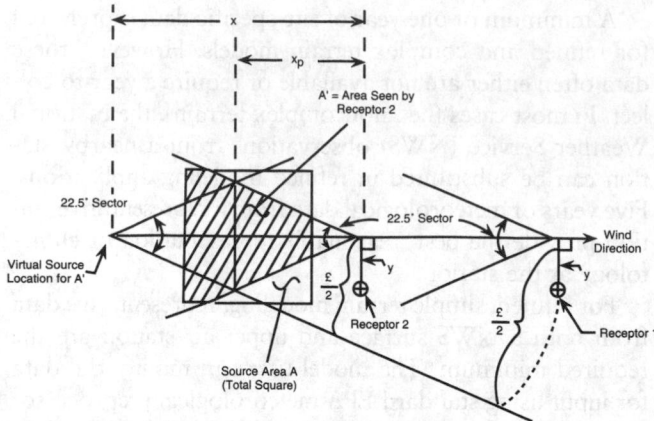

FIG. 5.8.16 Schematic of the virtual point source as projected from an area source. (Reprinted from TRW Systems Group, 1969, *Air quality display model*, Washington, D.C.: National Air Pollution Control Administration, DHEW, U.S. Public Health Service.)

consistency, the EPA has a protocol to follow when data are substituted for missing observations.

Both short- and long-term models also require mixing height data to define the upper limit of the area where effluent mixing occurs (the ground being the lower limit). Holzworth (1972) developed a set of figures and tables for seasonal and annual mixing heights, which are typically used in long-term modeling. For short-term applications, model users can interpolate hourly mixing height values (using the EPA's RAMMET preprocessor) based on twice-a-day upper air data collected by radiosonde measurements at numerous sites throughout the country and available through the National Climatic Center (NCC).

OTHER MODELS

In addition to the Gaussian-based models in Appendix A, computer programs have also been created for modeling other specific cases or pollutants. These other models include photochemical reactions or Eulerian or LaGrangian equations. Some, such as the chemical mass balance (CMB7) or fugitive dust model (FDM), are available for addressing specific pollutants, while others are implemented on a case-by-case basis for unique source or topographic conditions. The environmental engineer is directed to the appropriate U.S. EPA Region office to consult on selection and approval of the appropriate model for these cases.

The urban airshed model (UAM) is an Appendix A model applicable to most regions of the United States (USEPA 1990). The reactive plume model (RPM-IV) (USEPA 1993) accounts for ozone and other photochemically reactive pollutants over a smaller area than the urban airshed model.

Toxic or hazardous gas releases can be simulated with several models incorporated into Appendix A. These models are the toxic modeling system short-term (TOXST) (U.S. EPA 1992b) and toxic modeling system long-term (TOXLT) (U.S. EPA 1992a) models which are based on the ISC3 algorithms as well as the dense gas dispersion model (DEGADIS2.1) (U.S. EPA 1989a).

Mobile and Line Source Modeling

Mobile sources are difficult to model because the source is moving and may not be continuously emitting pollutants at a constant emission rate. Nonetheless, environmental engineers often model these sources using line source techniques that assign emissions at fixed points along a line. The U.S. EPA has developed a program called MOBILE5b to calculate emissions from mobile sources based on EPA tests of emissions from a variety of vehicle classes and types. In certain cases, environmental engineers can use the ISC3 models to estimate concentrations at receptors, but the EPA has identified the California line (CALINE3), CAL3QHC or CAL3QHCR, and the buoyant line and point source (BLP) models as the most appropriate for calculating emission concentrations from line sources.

CALINE3 MODEL

The CALINE3 model (Benson 1979) was designed to estimate highway traffic concentrations of nonreactive pollutants for the state of California. The model is a steady-state, Gaussian-based model capable of predicting 1-hr to 24-hr average concentrations from urban or rural roadway emission sources located at grade, in cut sections, in fill sections, and over bridges as shown in Figure 5.8.17. The program assumes all roadways have uncomplicated topography (i.e., simple terrain). Any wind direction and roadway orientation can be modeled. Primary pollutants can be modeled, including particulate matter, for which the model uses deposition and settling velocity factors. However, unlike the other models, CALINE3 does not account for any plume rise calculations.

CAL3QHC MODEL

While the CAL3QHC model (U.S. EPA 1992c) is not officially part of the Appendix A models, it is a recommended model for CO emissions in the *Guidelines on Air Quality Models* (U.S. EPA 1993). The model predicts carbon monoxide concentrations at signalized intersections. In effect, the model determines the increase in emissions and their resultant concentrations during queuing periods at stoplights. A version of the model, referred to as CAL3QHCR,

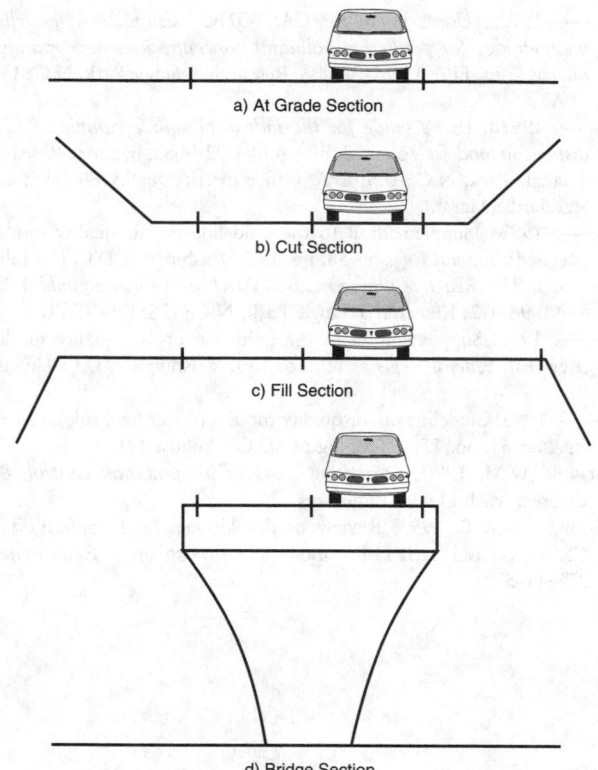

FIG. 5.8.17 Four roadway cross sections treated by the CA-LINE3 model.

has recently been created to process up to a year of meteorology, vehicle emissions, traffic volume, and signalization data.

BLP MODEL

The BLP model (Schulman and Scire 1980) is a Gaussian-based, plume dispersion model that deals with plume rise and downwash effects from stationary line sources. It was originally designed to simulate emissions from an aluminum reduction plant. The model can predict short-term concentrations in rural areas for buoyant, elevated line sources. However, it does not treat deposition or settling of particles.

—Roger K. Raufer
Curtis P. Wagner

References

Auer, A.H. 1978. Correlation of land use and cover with meteorological anomalies. *Journal of Applied Meteorology* 17:636–643.

Benson, P.E. 1979. CALINE3—A versatile dispersion model for predicting air pollutant levels near highways and arterial streets, Interim report. FHWA/CA/TL-79/23, Washington, D.C.: Federal Highway Administration.

Bittle, C.R., and A.R. Borowsky. 1985. A review of dispersion modeling methods for assessing toxic releases. Presented at the 78th Annual Meeting of the Air Pollution Control Association, June, 1985. 85–25B.2.

Bowne, N.E. 1974. Diffusion rates. *Journal of the Air Pollution Control Association* 24:832–835.

Briggs, G.A. 1969. *Plume rise*. U.S. Atomic Energy Commission Critical Review Series TID-25075. Clearinghouse for Federal Scientific and Technical Information.

Davenport, A.G. 1963. The relationship of wind structure to wind loading. Presented at Int. Conf. on the Wind Effects on Buildings and Structures, Nat. Physical Laboratory, Teddington, Middlesex, England, 26–28 June. Cited in Turner 1970.

Gifford, F.A. 1968. Meteorology and atomic energy—1968. Chap. 3 in *U.S. atomic energy report*. TID-24190. Oak Ridge, Tenn.

———. 1975. *Lectures on air pollution and environmental impact analyses*. Chap. 2. Boston, Mass.: American Meterological Society.

Holland, J.Z. 1953. *A meteorological survey of the Oak Ridge Area*. ORO-99, 540. Washington, D.C.: Atomic Energy Commission.

Holzworth, G.C. 1972. *Mixing heights, wind speeds, and potential for urban air pollution throughout the contiguous United States*. Office of Air Programs Pub. no. AP-101. Research Triangle Park, N.C.: U.S. EPA.

Hosker, R.P. 1984. Flow and diffusion near obstacles. In *Atmospheric science and power production*, edited by D. Randerson, DOE/TIC-27601. Washington, D.C.: U.S. Department of Energy.

Huber, A.H. 1977. Incorporating building/terrain wake effects on stack effluents. Preprint vol. *AMS-APCA Joint Conference on Applications of Air Pollution Meteorology, Salt Lake City, UT, Nov. 29–Dec. 2, 1977*.

Irwin, J.S. 1978. *Proposed criteria for urban versus rural dispersion coefficients (Draft staff report)*. Research Triangle Park, N.C.: U.S. EPA, Meteorology and Assessment Division.

———. 1979. A theoretical variation of the wind profile law exponent as a function of surface roughness and stability. *Atmospheric Environment* 13:191–194.

McElroy, J. and F. Pooler. 1968. St. Louis dispersion study. Vol. 2, *Analysis*. AP-53. Arlington, Va.: National Air Pollution Control Administration, U.S. Department of Health, Education, and Welfare.

Pasquill, F. 1974. *Atmospheric diffusion*. 2d ed. London: Van Nostrand.

———. 1976. *Atmospheric dispersion parameters in Gaussian-plume modeling, Part 2: Possible requirements for a change in the Turner workbook values*. EPA-600/4-76-030b. Research Triangle Park, N.C.: U.S. EPA, Office of Research and Development.

Schulman, L.L., and S.R. Hanna. 1986. Evaluation of downwash modifications to the industrial source complex model. *Journal of the Air Pollution Control Association* 24:258–264.

Schulman, L.L., and J.S. Scire. 1980. *Buoyant line and point source (BLP) dispersion model user's guide*. Doc. P-7304B. Concord, Mass.: Environmental Research & Technology, Inc.

Snyder, W.H., and R.E. Lawson. 1976. Determination of a necessary height for a stack close to a building—A wind tunnel study. *Atmospheric Environment* 10:683–691.

Stern, A.C., H.C. Wohlers, R.W. Boubel, and W.P. Lowry. 1973. *Fundamentals of air pollution*. New York: Academic Press.

Tickvart, J.A. 1988. *Memorandum on stack-structure relationships*. Research Triangle Park, N.C.: U.S. EPA, Office of Air Quality Planning and Standards (11 May).

Turner, D.B. 1964. A diffusion model for an urban area. *Journal of Applied Meteorology* 3:83–91.

———. 1970. *Workbook of atmospheric dispersion estimates (Revised)*. Office of Air Programs Publication No. AP-26. Research Triangle Park, N.C.: U.S. EPA.

U.S. Environmental Protection Agency (EPA). 1978. *Guideline on air quality models*. EPA-450/2-78-027. Research Triangle Park, N.C.: U.S. EPA, Office of Air Quality Planning and Standards (April).

———. 1985. *Guideline for determination of good engineering practice stack height (Technical support document for the stack height regulations) (Revised).* EPA-450/4-80-023R. Research Triangle Park, N.C.: U.S. EPA, Office of Air Quality Planning and Standards (June).

———. 1986. *Guideline on air quality models (Revised).* EPA-450/2-78-027R. Research Triangle Park, N.C.: U.S. EPA, Office of Air Quality Planning and Standards (July).

———. 1987. *Supplement A to the guideline on air quality models (Revised).* EPA-450/2-78-027R. Research Triangle Park, N.C.: U.S. EPA, Office of Air Quality Planning and Standards (July).

———. 1989a. *User's guide for the DEGADIS 2.1—Dense gas dispersion model.* EPA-450/4-89-019. Research Triangle Park, N.C.: U.S. EPA.

———. 1989b. *User's guide to the complex terrain dispersion model plus algorithms for unstable situations (CTDMPLUS).* Vol. 1, EPA-600/8-89-041. Research Triangle Park, N.C.: U.S. EPA Atmospheric Research and Exposure Assessment Laboratory (March).

———. 1990. *User's guide to the urban airshed model.* Vols. 1-8. EPA-450/4-90-007a to g, EPA-454/B-93-004. Research Triangle Park, N.C.: U.S. EPA.

———. 1992a. *Toxic modeling system long-term (TOXLT) user's guide.* EPA-450/4-92-003. Research Triangle Park, N.C.: U.S. EPA.

———. 1992b. *Toxic modeling system short-term (TOXST) user's guide.* EPA-450/4-92-002. Research Triangle Park, N.C.: U.S. EPA.

———. 1992c. *User's guide for CAL3QHC version 2: A modeling methodology for predicting pollutant concentrations near roadway intersections.* EPA-454/R-92-006. Research Triangle Park, N.C.: U.S. EPA.

———. 1992d. *User's guide for the industrial source complex (ISC2) dispersion models.* Vols. 1-3. EPA-450/4-92-008a, b, and c. Research Triangle Park, N.C.: U.S. EPA, Office of Air Quality Planning and Standards (March).

———. 1993a. Supplement B to the guideline on air quality models (Revised). *Federal Register* 58, no. 137. Washington, D.C. (20 July).

———. 1993b. *Reactive plume model IV (RPM-IV) user's guide.* EPA-454/B-93-012. Research Triangle Park, N.C.: U.S. EPA (ESRL).

———. 1995. Supplement C to the guideline on air quality models (Revised). *Federal Register* 60, no. 153. Washington, D.C. (August 9).

———. 1996. Guideline on air quality models (direct final rule). *Federal Register* 61, no. 156. Washington, D.C. (August 12).

Vatavuk, W.M. 1990. *Estimating costs of air pollution control.* 82. Chelsea, Mich.: Lewis Publishers.

Wilson, Robert B. 1993. Review of development and application of CRSTER and MPTER models. *Atmospheric Environment* 27B:41–57.

Air Quality

5.9
EMISSION MEASUREMENTS

Data from emission measurements benefit a company in many ways. Companies use these data for permits and compliance audits, for determining the effectiveness of control equipment, for the design of pollution control strategies, and for implementation of waste minimization and pollution prevention programs. The work involves sampling and testing procedures and physical and chemical measurements.

Planning an Emissions Testing Program

The first step in planning an emissions testing program is to define its objectives. Essentially, the testing objectives dictate the accuracy of the data needed, which dictate the following four conditions of a testing program:

Stream—This testing condition specifies whether sampling is performed directly at the point where the pollutant is generated (e.g., valves, pumps, or compressors) or on ambient air.

Frequency—This testing condition specifies whether samples are taken periodically or continuously.

Method—This testing condition specifies whether standardized reference methods are used to analyze for a particular compound or if a customized method must be developed.

Location—This testing condition specifies whether samples are taken from the field to the laboratory or are tested directly in the field. For stack sampling, environmental engineers must refine arrangements to obtain representative samples for analysis. Figure 5.9.1 shows the requirements for stack sampling.

Table 5.9.1 provides recommendations for the four preceding primary testing conditions based on the end uses of the data.

Analyzing Air Emissions

The technologies for sampling air are many and varied. The choice of the air sampling and analysis method de-

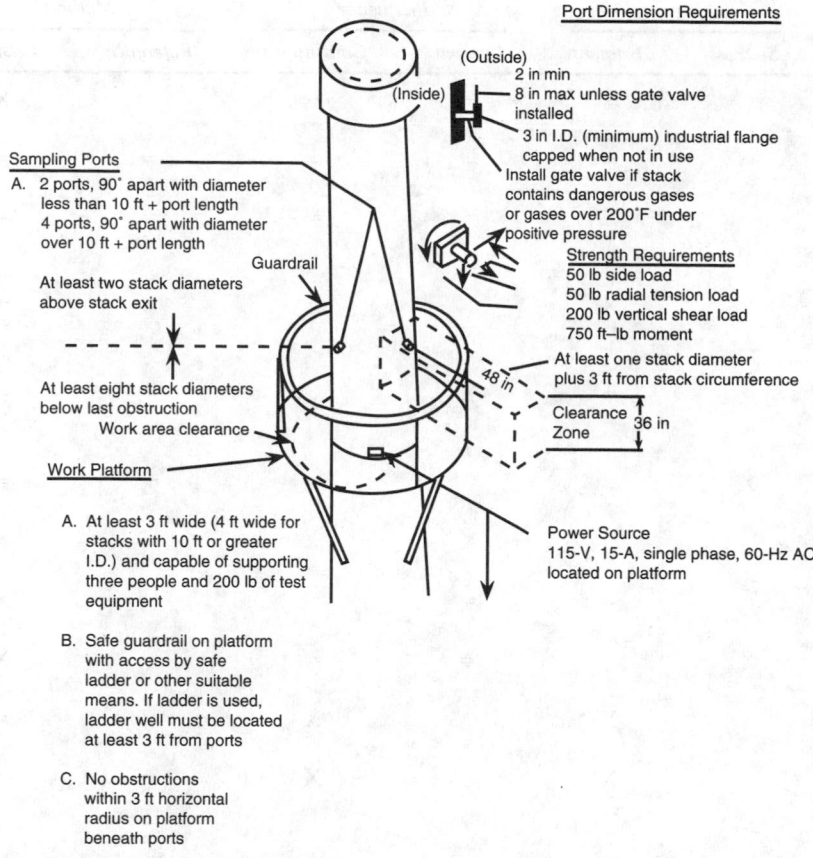

FIG. 5.9.1 Typical sampling point provisions. (Reprinted from U.S. Environmental Protection Agency [EPA].)

pends on the source. For example, emissions from stacks can be measured directly at release points. If the amounts and concentrations of contaminants are needed for a large area, then area or remote methods are preferable to source testing. To detect fugitive emissions from equipment, such as valves, flanges, pumps, or motors, screening or bagging tests are used, in which the device is enclosed to capture leaks.

STACK SAMPLING

The U.S. EPA has published standard sampling routines for stack gases. These reference methods give representative concentration data for compounds. These test methods are validated by laboratory and field studies, and the data obtained from them have predictable accuracies and reproducibilities. Table 5.9.2 lists the standard sampling routines for stack gases (see also, *Code of Federal Regulations*, Title 40, Part 60 App. A).

The sampling train described in reference methods 5 and 8 together with operating procedures are discussed in Section 5.14. The following sampling method numbers refer to the methods identified in EPA guidance documents and reports (U.S. EPA 1989; 1990; Harris et al. 1984):

Semivolatile Organics (Method 0010)

The sampling train in reference method 5 must be modified to include an adsorbent trap (see Figures 5.9.2 and 5.9.3) when organic compounds with a boiling point between 100 and 300°C are present in the gas. The adsorbent module is inserted between the filter and the first impinger. Gas chromatography–mass spectrometry (GC–MS) is typically used to analyze for organic compounds.

Volatile Organics (Method 0030)

The volatile organics sampling train (VOST) is used for organics that boil below 100°C. The VOST system involves drawing a stack gas sample through two adsorbent tubes in series. The first tube contains Tenax resins, and the second contains Tenax and activated carbon. The pollutants adsorbed on these tubes are then desorbed. GC–MS is a typical method for low-boiling-point organics.

HCl and Cl₂ (Method 0050)

This method is used to collect HCl and Cl_2 in stack gases. It collects emission samples isokinetically and can be combined with reference method 5 for particulate determination.

TABLE 5.9.1 TEST PLANNING MATRIX

Primary Objective	Stream		Frequency		Method		Location	
	Source	Ambient	Interval	Continuous	Reference	Custom	Lab	Field
Test compliance with regulations	√	=	=	=	√√	X	√	=
Improve overall emission inventory	=	=	=	=	=	=	=	=
Identity and characterize emission sources	√	X	=	=	=	=	=	=
Conduct performance testing of process controls	√	X	√	=	=	=	=	=
Detect intermittent or transient emissions	=	=	X	√	=	=	X	√
Gather personnel exposure data	X	√	=	=	=	=	=	=
Provide early warning of leaks	=	√	X	√	=	=	XX	√√
Develop defensible data	=	=	=	=	√√	X	=	=
Track effluent quality	X	√	=	=	√	X	=	=
Characterize water and wastes	√	X	√	X	√	=	√	=
Support waste minimization and pollution prevention	√	X	√	X	=	=	=	=

Source: Graham E. Harris, Michael R. Fuchs, and Larry J. Holcombe, 1992, A guide to environmental testing, *Chem. Eng.* (November): 98–108.

Notes: The construction of an environmental testing program begins with a list of the objectives that must be met. When that list is complete, the table indicates the proper elements to employ. Depending upon the end use of the data, testing can be performed directly in the field with portable instrumentation.

Key:
√√ Required or highly preferred
√ Preferred or likely to be used
= Neutral—Either is acceptable
X Discouraged or unlikely to be used
XX Not permitted or not available

TABLE 5.9.2 SELECTED EPA REFERENCE METHODS

Method 1—Sample and velocity traverses for stationary sources

Method 2—Determination of stack gas velocity and volumetric flow rate (type S pitot tube)

Method 2A—Direct measurement of gas volume through pipes and small ducts

Method 2B—Determination of exhaust gas volume flow rate from gasoline vapor incinerators

Method 3—Gas analysis for carbon dioxide, oxygen, excess air, and dry molecular weight

Method 3A—Determination of oxygen and carbon dioxide concentrations in emissions from stationary sources (instrumental analyzer procedure)

Method 4—Determination of moisture content in stack gases

Method 5—Determination of particulate emissions from stationary sources

Method 5A—Determination of particulate emissions from the asphalt processing and asphalt roofing industry

Method 5B—Determination of nonsulfuric acid particulate matter from stationary sources

Method 5D—Determination of particulate matter emissions from positive pressure fabric filters

Method 5E—Determination of particulate emissions from the wool fiberglass insulation manufacturing industry

Method 5F—Determination of nonsulfate particulate matter from stationary sources

Method 6—Determination of sulfur dioxide emissions from stationary sources

Method 6A—Determination of sulfur dioxide, moisture, and carbon dioxide emissions from fossil fuel combustion sources

Continued on next page

Method 6B—Determination of sulfur dioxide and carbon dioxide daily average emissions from fossil fuel combustion sources

Method 6C—Determination of sulfur dioxide emissions from stationary sources (instrumental analyzer procedure)

Method 7—Determination of nitrogen oxide emissions from stationary sources

Method 7A—Determination of nitrogen oxide emissions from stationary sources

Method 7B—Determination of nitrogen oxide emissions from stationary sources (UV spectrophotometry)

Method 7C—Determination of nitrogen oxide emissions from stationary sources

Method 7D—Determination of nitrogen oxide emissions from stationary sources

Method 7E—Determination of nitrogen oxide emissions from stationary sources (instrumental analyzer procedure)

Method 8—Determination of sulfuric acid mist and sulfur dioxide emissions from stationary sources

Method 9—Visual determination of the opacity of emissions from stationary sources

Method 10—Determination of carbon monoxide emissions from stationary sources

Method 10A—Determination of carbon monoxide emissions in certifying continuous emission monitoring systems at petroleum refineries

Method 11—Determination of hydrogen sulfide content of fuel gas streams in petroleum refineries

Method 12—Determination of inorganic lead emissions from stationary sources

Method 13A—Determination of total fluoride emissions from stationary sources (SPADNS zirconium lake method)

Method 13B—Determination of total fluoride emissions from stationary sources (specific ion electrode method)

Method 14—Determination of fluoride emissions from potroom roof monitors of primary aluminum plants

Method 15—Determination of hydrogen sulfide, carbonyl sulfide, and carbon disulfide emissions from stationary

Method 15A—Determination of total reduced sulfur emissions from sulfur recovery plants in petroleum refineries

Method 16—Semicontinuous determination of sulfur emissions from stationary sources

Method 16A—Determination of total reduced sulfur emissions from stationary sources (impinger technique)

Method 16B—Determination of total reduced sulfur emissions from stationary sources

Method 17—Determination of particulate emissions from stationary sources (instack filtration method)

Method 18—Measurement of gaseous organic compound emissions by GC

Method 19—Determination of sulfur dioxide removal efficiency and particulate matter, sulfur dioxide, and nitrogen oxide emission rates

Method 20—Determination of nitrogen oxides, sulfur dioxide, and oxygen emissions from stationary gas turbines

Method 21—Determination of VOC leaks

Method 22—Visual determination of fugitive emissions from material sources and smoke emissions from flares

Method 24—Determination of volatile matter content, water content, density, volume solids, and weight solids of surface coating

Method 24A—Determination of volatile matter content and density of printing inks and related coatings

Method 25—Determination of total gaseous nonmethane organic emissions as carbon

Method 25A—Determination of total gaseous organic concentration using a flame ionization analyzer

Method 25B—Determination of total gaseous organic concentration using a nondispersive infrared analyzer

Method 27—Determination of vapor tightness of gasoline delivery tank using pressure-vacuum test

Appendix B—Performance Specifications

Performance Specification 1—Performance specifications and specification test procedures for transmissometer systems for continuous measurement of the opacity of stack emissions

Performance Specification 2—Specifications and test procedures for SO_2 and NO_x continuous emission monitoring systems in stationary sources

Performance Specification 3—Specifications and test procedures for O_2 and CO_2 continuous emission monitoring systems in stationary sources

---

Source: Code of Federal Regulations, Title 40, part 60, App. A.

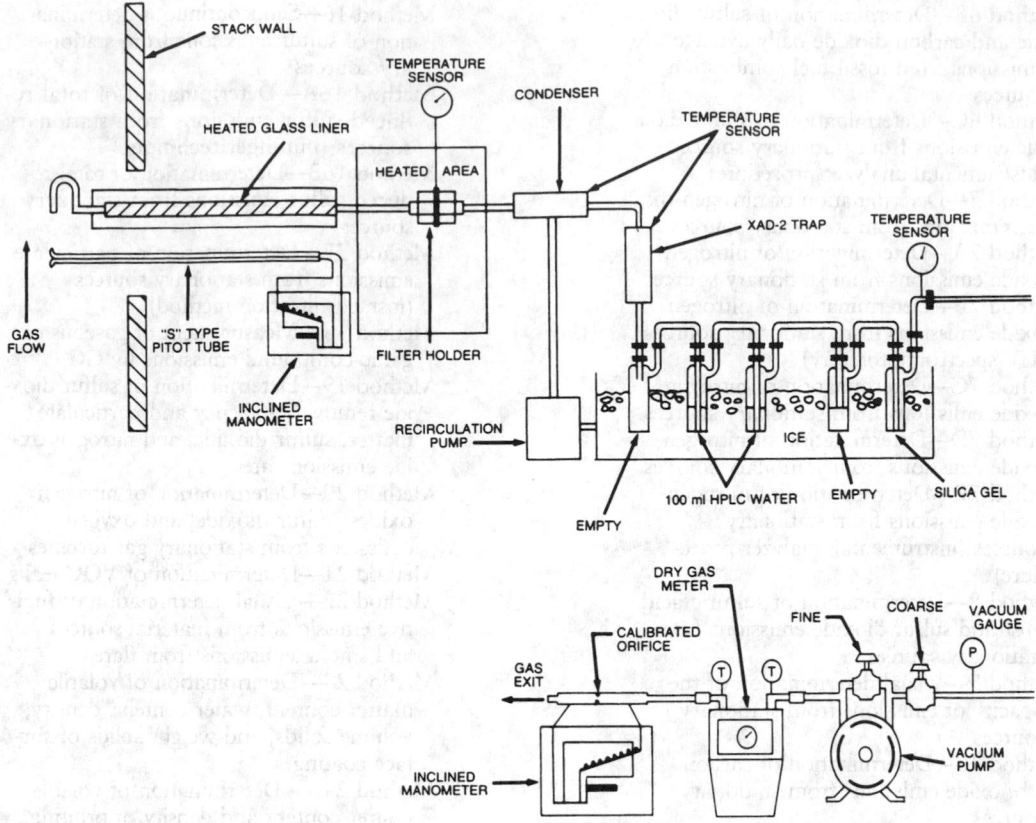

FIG. 5.9.2 Sampling train.

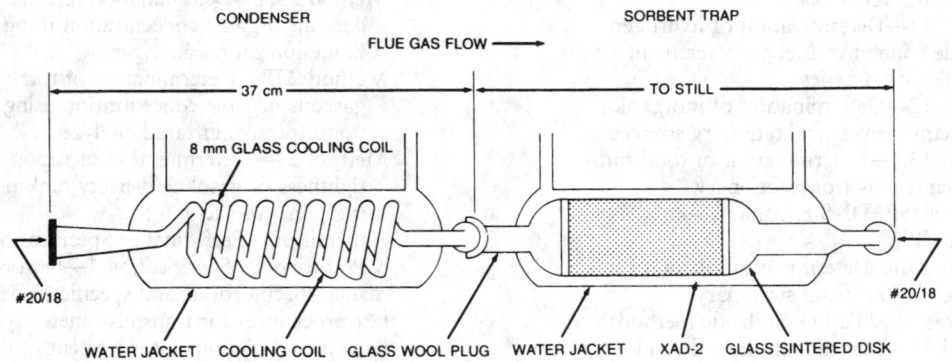

FIG. 5.9.3 Condenser and absorbent trap.

Trace Metals (Method 0012)

The metals sampling train is used to determine the total chromium, cadmium, arsenic, nickel, manganese, beryllium, copper, zinc, lead, selenium, phosphorous, thallium, silver, antimony, barium, and mercury in incinerator stack emissions. The stack gas is withdrawn isokinetically from the source with the particulate emissions collected in a probe and on heated filters and the gaseous emissions collected in a series of chilled impingers. These impingers contain a solution of dilute nitric acid combined with hydrogen peroxide in two impingers and an acidic potassium permanganate solution in two impingers. Analyzing the metals in particulates involves using either inductive coupled argon plasma (ICAP) spectroscopy or a graphite furnace atomic adsorption (GFAA) spectrometer.

Formaldehyde (Method 0011)

This method is used to collect formaldehyde in stack emissions.

Other Sampling Methods

Reference methods are continuously being developed to keep up with the demands of new regulations. Among the

TABLE 5.9.3 NESHAPS TEST METHODS

101	Particulate and gaseous mercury emissions from the air streams of chloralkali plants
101A	Particulate and gaseous mercury emissions from sewage sludge incinerators
102	Particulate and gaseous mercury emissions from the hydrogen streams of chloralkali plants
103	Beryllium screening
104	Beryllium emissions from stationary sources
105	Mercury in sewage sludges from wastewater treatment plants
106	Vinyl chloride emissions from stationary sources
107	Vinyl chloride in process wastewater, PVC resin, slurry, wet cake, and latex samples
107A	Vinyl chloride in solvents, resin–solvent solutions, PVC resin, resin slurries, wet resin, and latex
108	Particulate and gaseous arsenic emissions
111	Polonium-210 emissions from stationary sources

Source: Code of Federal Regulations, Title 40, Part 61, App. B.
Note: Analytical methods specified under NESHAPS cover toxic and radioactive pollutants not addressed by standard EPA methods.

most recent introductions are the aldehydes–ketones sampling train and the hexavalent chromium sampling train (U.S. EPA 1990).

AIR TOXICS IN AMBIENT AIR

In addition to the reference methods, the EPA has developed methods to detect toxic and radioactive pollutants under NESHAP. Table 5.9.3 lists these test methods as well as methods for the analysis of low concentrations of organics in ambient air.

The EPA's Atmospheric Research and Exposure Assessment Laboratory (AREAL) has also developed a compendium of methods for quantifying HAPs in ambient air. Table 5.9.4 identifies the methods, and Figure 5.9.4 shows the categories of HAPs in the compendium. Two popular methods are compendium method TO-13 for semivolatiles and TO-12 for VOCs. Method TO-13 describes a sampling and analysis procedure for semivolatiles, such as benzo(a)pyrene and polynuclear aromatic hydrocarbons (PAHs). Compendium method TO-14 involves the collection of VOHAPs in stainless canisters.

A copy of the compendium of methods can be obtained from the following:

U.S. EPA, Atmospheric Research and Exposure Assessment Laboratory (AREAL), MD-77, Research Triangle, NC 27711.
Noyes Publication, Mill Road at Grand Ave., Park Ridge, NJ 07656.

A copy of the statement of work (SOWs) for HAPs from superfund sites can be obtained from the U.S. EPA, Office of Solid Waste and Energy Response, Analytical Operation Branch, 401 M St., S.W., Washington, DC 20460.

Many of the sampling and analytic procedures recommended most likely need additional development and validation to improve accuracy and precision. A method that requires validation is not an inferior method; the method simply requires additional experimentation to define accuracy, precision, and bias. The environmental engineer begins experimental work with analyses of a known concentration of the target pollutant. Then, the engineer determines the potential interferences by repeating the tests

TABLE 5.9.4 COMPENDIUM METHODS

Number	Method
TO-1	Determination of VOCs in ambient air using Tenax adsorption and GC–MS.
TO-2	Determination of VOCs in ambient air by carbon molecular sieve adsorption and GC–MS.
TO-3	Determination of VOCs in ambient air using cryogenic preconcentration techniques and GC with flame ionization and electron capture detection.
TO-4	Determination of organochlorine pesticides and PCBs in ambient air.
TO-5	Determination of aldehydes and ketones in ambient air using high-performance liquid chromatography (HPLC).
TO-6	Determination of phosgene in ambient air using HPLC.
TO-7	Determination of N-nitrosodimethylamine in ambient air using GC.
TO-8	Determination of phenol and methylphenols (cresols) in ambient air using HPLC.
TO-9	Determination of polychlorinated dibenzo-p-dioxins (PCDDs) in ambient air using high-resolution GC–high-resolution mass spectrometry.
TO-10	Determination of organochlorine pesticides in ambient air using low-volume polyurethane foam (PUF) sampling with GC–electron capture detector (GC–ECD).
TO-11	Determination of formaldehyde in ambient air using adsorbent cartridge followed by HPLC.
TO-12	Determination of nonmethane organic compounds (NMOC) in ambient air using cryogenic preconcentration and direct flame ionization detection (PDFID).
TO-13	Determination of PAHs in ambient air using high-volume sampling with GC–MS and HPLC analysis.
TO-14	Determination of VOCs in ambient air using SUMMA polished canister sampling and GC analysis.

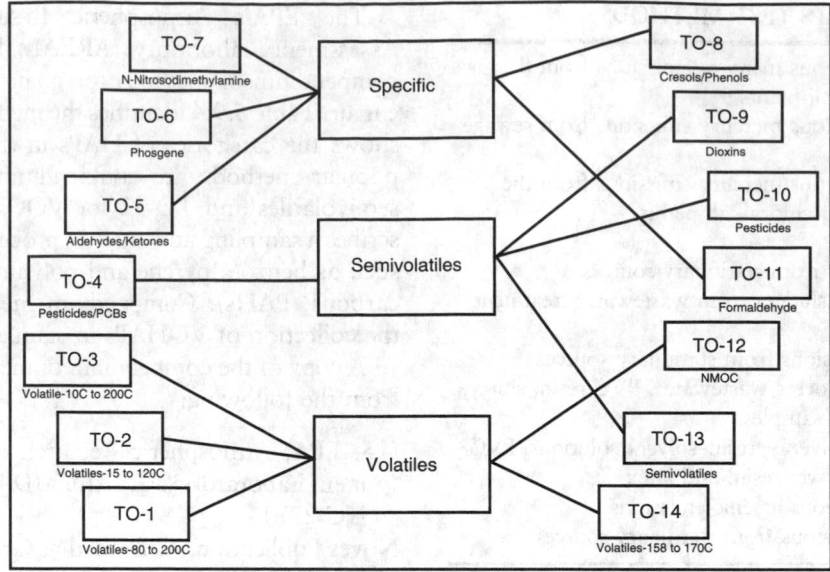

FIG. 5.9.4 Compendium of methods.

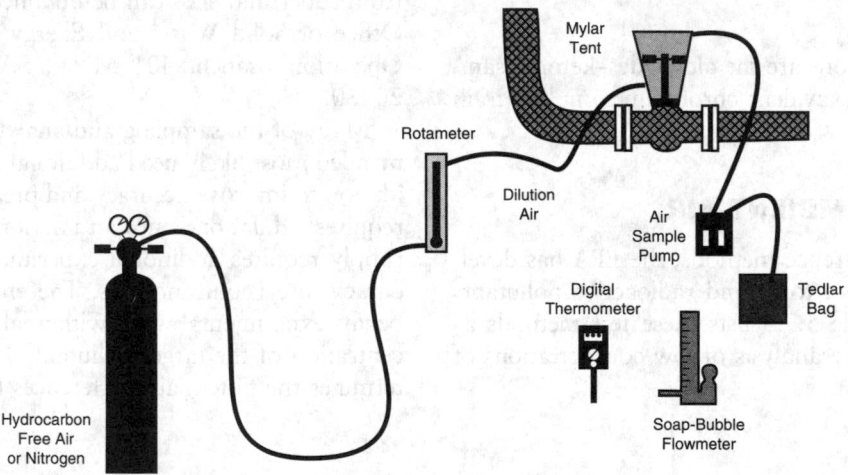

FIG. 5.9.5 Bagging test—Blow-through method.

with a gas that models the emission matrix. Validation is completed with field tests on an actual emission stream. The EPA has prepared standard protocols for validation of sampling and analysis methods (U.S. EPA 1991).

EQUIPMENT EMISSIONS

Equipment emission measurements are made by a portable hydrocarbon analyzer at potential leak points. The analyzer is a flame ionization detector (FID) that measures total hydrocarbons over a range of 1 to 10,000 ppm on a logarithmic scale. Since these measurements are basic for regulatory compliance, procedures are formalized in the EPA's reference method 21.

Environmental engineers can determine emission rates from stand-alone equipment such as valves and pumps by using bagging tests in which the source is enclosed and

leaks are captured in a known value of air or inert gas. Two different methods are commonly used: the vacuum method and the blow-through method. Figure 5.9.5 shows the blow-through method. Multiplying the concentration of the exhaust stream by the flow rate of air through the equipment calculates the emission rates. Samples are usually analyzed with a total-hydrocarbon analyzer. Toxics are detected by gas chromatography (GC).

Monitoring Area Emissions

Measuring the concentration of air pollutants in a large area presents different problems. The potential sources of these emissions are many. For example, VOCs are emitted from hazardous waste sites; organic and inorganic gases are emitted from landfills, lagoons, storage piles, and spill sites. Area emissions can be measured either directly

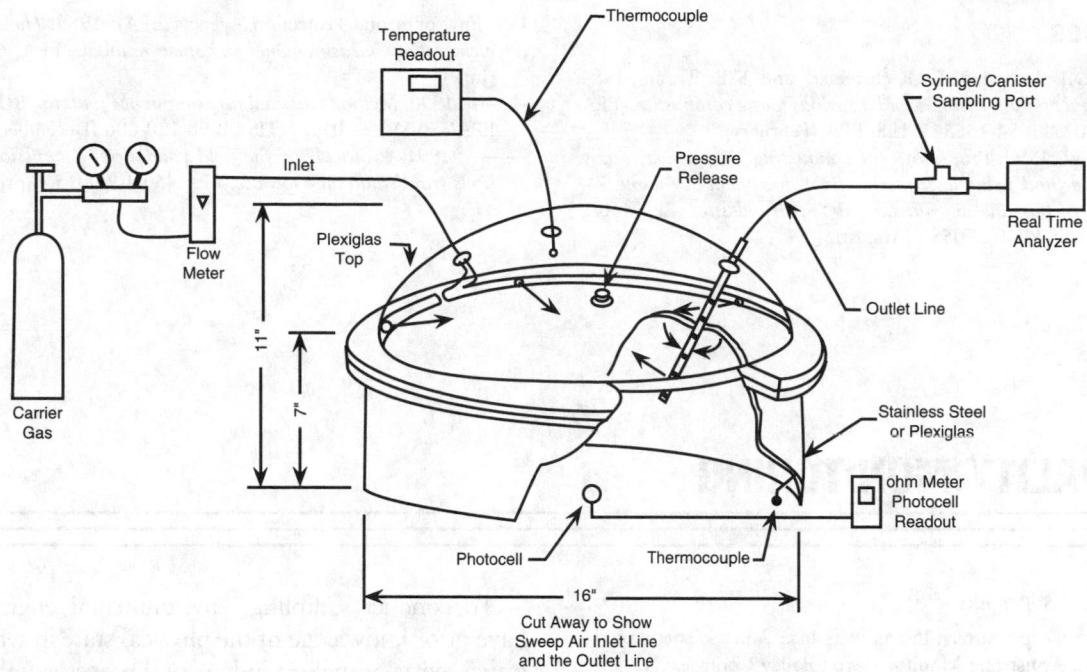

FIG. 5.9.6 Cutaway diagram of the emission isolation flux chamber and support equipment. (Reprinted, with permission, from J.C. Harris, C.E. Rechsteiner, and K.E. Thrun, 1984, *Sampling and analysis methods for hazardous waste combustion*, EPA-600/8-84/002, PB 84-155545, U.S. EPA [February].)

or indirectly. In direct measurements, ambient air is sampled at various points in the area of emission. Indirect methods sample ambient air at points upwind and downwind of the emission.

DIRECT MEASUREMENTS

Environmental engineers use an emission-flux chamber to make direct measurements of concentrations (see Figure 5.9.6). The atmospheric emissions in an area enter the chamber where they are mixed with clean dry air or nitrogen that is fed in at a fixed rate. The analyzer measures the pollutant concentration. The emission flux is the exit gas concentration multiplied by the flow rate divided by the surface area covered by the chamber.

When the area is several acres in size, measurements must be taken at various points to develop an overall emission rate. The number of measurements needed depends on the precision required and the size of the source.

Flux chambers are best suited to measure small to medium size areas (as large as a few acres) in which the pollutant concentration is fairly homogeneous. Because the flux chamber is isolated, measurements are independent from environmental influences such as wind; therefore, the measurement data are independent of the meteorological conditions at the site and are comparable from day to day and site to site.

INDIRECT METHODS

Indirect methods are best suited for the measurement of emission rates from large, heterogeneous sources. The environmental engineer measures the emission flux indirectly by collecting ambient concentrations upwind and downwind of the emission source.

The main disadvantage of indirect methods is that the analytical results rely on meteorological conditions, which can invalidate the data or preclude the collection altogether. In addition to weather patterns, buildings and hills also influence dispersion characteristics and limit sampling. The sensitivity of the analytical method is also critical since the ambient concentration of the emission can be low.

The downwind measurement determines the average concentration of contaminants in the plume. The upwind measurement monitors background readings. The volume of air passing over the monitors in a time period is recorded, and a computer model calculates an emission rate from the concentration data. The average emission rate is equal to the difference in mass measurements (downwind minus upward concentration) divided by the transit time across the source. Environmental engineers have tried various approaches for making these estimates using conventional ambient air monitoring methods (Schmidt et al. 1990).

In some cases, gaseous tracers are released at the emission source. These tracers mimic the dispersion of the emissions. When a tracer is released at a known rate, the environmental engineer can determine the emission rates of compounds by comparing the measured pollutant concentration to that of the tracer at the same location. Sections 5.12 and 5.13 describe other techniques such as continuous emission monitoring and remote sensing.

—David H.F. Liu

References

Harris, J.C., D.J. Larsen, C.E. Rechsteiner, and K.E. Thrun. 1984. *Sampling and analysis method for hazardous waste combustion.* EPA-600/8-84/002, PB 84-155545. U.S. EPA (February).

Schmidt, C. et al. 1990. *Procedures for conducting air–pathway analyses for Superfund activities, Interim final documents: Volume 2—Estimation of baseline air emission at Superfund sites.* EPA-450/1-89-002a, NTIS PB90-270588 (August).

U.S. Environmental Protection Agency (EPA). 1989. *Hazardous waste incineration measurement guidance manual.* EPA 625/6-89/021 (June).

———. 1990. *Methods manual for compliance with the BIF regulations.* EPA/530-SW-91-010, NTIS PB 90-120-006 (December).

———. 1991. *Protocol for the field validation of emission concentrations from stationary sources.* EPA 450/4-90-015 (April).

5.10
AIR QUALITY MONITORING

PARTIAL LIST OF SUPPLIERS

Air Monitor Corp.; Airtech Instruments Inc.; Ametek/Thermox; Anarad Inc.; Armstrong Monitor Corp.; Bailey Controls Co.; Barringer Research Ltd.; Bran & Luebbe Analyzing; Dasibi Environmental Corp.; Delphian Corp.; Environment One; Environmental Technology Group Inc.; Enviroplan Inc.; The Foxboro Co.; GasTech Inc.; G C Industries Inc.; GEC Canada Ltd.; General Metal Works Inc.; Horiba Instrument Inc.; International Sensor Technology; Lear Siegler Measurement Control Corp.; National Draeger Inc.; MSA Instrument Div.; Purafil Inc.; Research Appliance Co. Div. of Andersen Samplers; Scientific Industries Corp.; Sensidyne Inc.; Servomex Co.; Sieger Gasalarm; Siemens Energy & Automation; Sierra Monitor Corp.; Sigrist-Photometer Ltd.; SKC Inc.; Teledyne Analytical Instruments; VG Instruments; Yokogawa Corp. of America

This section describes ambient air sampling, air quality monitoring systems, and microprocessor-based portable ambient air analyzers.

Sampling of Ambient Air

All substances in ambient air exist as either particulate matter, gases, or vapors. In general, the distinction is easily made; gases and vapors consist of substances dispersed as molecules in the atmosphere, while particulate matter consists of aggregates of molecules large enough to behave like particles. Particulate matter, or particulates, are filterable, can be precipitated, and can settle out in still air. By contrast, gases and vapors do not behave in this way and are homogeneously mixed with the air molecules.

SAMPLING METHOD SELECTION

A substance, such as carbon monoxide, exists only as a gas; an inorganic compound like iron oxide exists only as a particle. Many substances exist as either particles or vapors, however; substances that are gases can be attached by some means to particulate matter in the air.

To conduct sampling, environmental engineers must have prior knowledge of the physical state in which a substance exists or make a judgment. Devices suitable for collecting particulate matter do not usually collect gases or vapors; hence, selection of an incorrect sampling method can lead to erroneous results. Fortunately, considerable knowledge concerning the more common pollutants is available, and in most cases, selecting a suitable sampling method is not difficult.

GENERAL AIR SAMPLING PROBLEMS

Certain general observations related to sampling ambient air must be recognized. For example, the quantity of a substance contained in a volume of air is often extremely small; therefore, the sample size for the analytical method must be adequate. Even heavily polluted air is not likely to contain more than a few milligrams per cubic meter of most contaminants; and frequently, the amount present is best measured in micrograms, or even nanograms, per cubic meter.

For example, the air quality standard for particulates is 75 $\mu g/M^3$. A cubic meter of air, or 35.3 cu ft, is a large volume for many sampling devices, and a considerable sampling period is required to draw such a quantity of air through the sampler. When atmospheric mercury analyses are made, the environmental engineer must realize that background levels are likely to be as low as several nanograms per cubic meter. In general, most substances are of concern at quite low levels in ambient air.

In addition to the problems from low concentrations of the substances being sampled, the reactivity of some substances causes other problems, resulting in changes after collection and necessitating special measures to minimize such changes.

Whenever a substance is removed from a volume of air by sampling procedures, the substance is altered, and the analysis can be less informative or even misleading. Ideally, environmental engineers should perform analyses of the unchanged atmosphere using direct-reading devices that give accurate information concerning the chemical and physical state of contaminants as well as concentration information. Such instruments exist for some substances, and many more are being developed. However, conventional air sampling methods are still used in many instances and will continue to be required for some time.

Gas and Vapor Sampling

The methods for gas and vapor sampling include collection in containers or bags, absorption, adsorption, and freeze-out sampling.

COLLECTION IN CONTAINERS OR BAGS

The simplest method of collecting a sample of air for analysis is to fill a bottle or other rigid container with it or to use a bag of a suitable material. Although sampling by this method is easy, the sample size is distinctly limited, and collecting a large enough sample for subsequent analysis may not be possible.

Bottles larger than several liters in capacity are awkward to transport; and while bags of any size are conveniently transported when empty, they can be difficult to handle when inflated. Nevertheless, collecting several samples in small bags can prove more convenient than taking more complex sampling apparatus to several sampling sites. If analyzing the contaminant is possible by GC procedures or gas-phase infrared spectroscopy, samples as small as a liter or less can be adequate and can be easily collected with bags.

Several methods exist for filling a rigid container such as a bottle. One method is to evacuate the bottle beforehand, then fill it at the sampling site by drawing air into the bottle and resealing it (see Figure 5.10.1). Alternatively, a bottle can be filled with water, which is then allowed to drain and fill with the air. A third method consists of passing a sufficient amount of air through the bottle using a pumping device until the original air is completely displaced by the air being sampled.

Plastic bags are frequently filled by use of a simple hand-operated squeeze bulb with valves on each end (see Figure 5.10.1) that are connected to a piece of tubing attached to the sampling inlet of the bag. In most cases, this procedure is satisfactory; but the environmental engineer must be careful to avoid contaminating the sampled air with the sampling bulb or losing the constituent on the walls of the sampling bulb. Problems of this kind can be avoided when the bag is placed in a rigid container, such as a box, and air is withdrawn from the box so that a negative pressure is created, resulting in air being drawn into the bag.

Selecting bag materials requires care; some bags permit losses of contaminants by diffusion through the walls, and others contribute contaminants to the air being sampled.

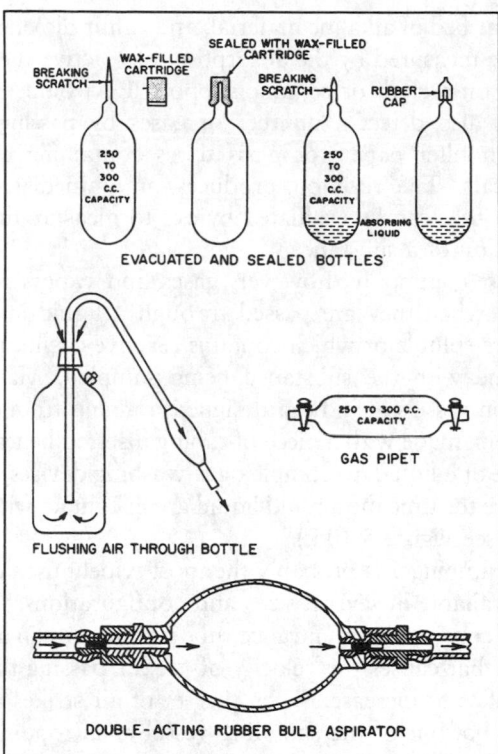

FIG. 5.10.1 Devices for obtaining grab samples.

A number of polymers have been studied, and several are suitable for many air sampling purposes. Materials suitable for use as sampling bags include Mylar, Saran, Scotchpak (a laminate of polyethylene, aluminum foil, and Teflon), and Teflon.

Even when a bag is made of inert materials, gas-phase chemical reactions are always possible, and after a period of time, the contents of the bag are not identical in composition to the air originally sampled. Thus, a reactive gas like sulfur dioxide or nitric oxide gradually oxidizes, depending on the storage temperature. Generally, analyses should be performed as soon as possible after the samples are collected. Losses by adsorption or diffusion are also greater with the passage of time and occur to some extent even when the best available bag materials are used.

The use of small bags permits the collection of samples to be analyzed for a relatively stable gas, such as carbon monoxide, at a number of locations throughout a community, thus permitting routine air quality measurements that might otherwise be inordinately expensive.

ABSORPTION

Most air sampling for gases and vapors involves absorbing the contaminant in a suitable sampling medium. Ordinarily, this medium is a liquid, but absorption can also take place in solid absorbents or on supporting materials such as filter papers impregnated with suitable absorbents. Carbon dioxide, for example, is absorbed in a granular bed of alkaline material, and sulfur dioxide is frequently measured by the absorption of reactive chemicals placed on a cloth or ceramic support. Environmental engineers also detect a number of gases by passing them through filter papers or glass tubes containing reactive chemicals. The reaction produces an immediate color change that can be evaluated by eye to measure the concentration of a substance.

Most commonly, however, gases and vapors are absorbed when they are passed through a liquid in which they are soluble or which contains reactive chemicals that combine with the substance being sampled. Many absorption vessels have been designed, ranging from simple bubblers, made with a piece of tubing inserted beneath the surface of a liquid, to complex gas-washing devices, which increase the time the air and liquid are in contact with each other (see Figure 5.10.2).

The impinger is probably the most widely used device. It is available in several sizes and configurations. An impinger consists of an entrance tube terminating in a small orifice that causes the velocity of the air passing through the orifice to increase. When this jet of air strikes a plate or the bottom of the sampling vessel at an optimal distance from the orifice, an intense impingement or bubbling action occurs. This impingement results in more efficient absorption of gases from the airstream than takes place if the air is simply bubbled through at low velocity. The two most frequently used impingement devices are the *standard impinger* and the *midget impinger* (see Figure 5.10.3). They are designed to operate at an airflow of 1 cubic foot per minute (cfm) and 0.1 cfm, or 28.3 and 2.8 lpm, respectively. Using such devices for sampling periods of 10 or 30 min results in a substantial amount of air passing through the devices, thus permitting low concentrations of trace substances to be determined with improved sensitivity and accuracy. Many relatively insoluble gases, such as nitrogen dioxide, are not quantitatively removed by passing through an impinger containing the usual sampling solutions.

The most useful sampling devices for absorbing trace gases from air are those in which a gas dispersion tube made of fritted or sintered glass, ceramic, or other material is immersed in a vessel containing the absorption liquid (see Figure 5.10.2). This device causes the gas stream to be broken into thousands of small bubbles, thus promoting contact between the gas and the liquid with resulting high-collection efficiencies. In general, fritted absorbers are more applicable to sampling gases and vapors than impingers and are not as dependent on flowrates as impingers. Fritted absorbers are available from scientific supply companies and come in various sizes suitable for many sampling tasks. Prefiltering the air prior to sampling with a fritted absorber is advisable to prevent the gradual accumulation of dirt within the pores of the frit.

The use of solid absorbents is not widely practiced in ambient air sampling because the quantity of absorbed gases is usually determined by gravimetric means. With the exception of carbon dioxide, few atmospheric gases lend themselves to this type of analysis.

ADSORPTION

Adsorption, by contrast with absorption, consists of the retention of gaseous substances by solid adsorbents which, in most cases, do not chemically combine with the gases. Instead, adsorptive forces hold the gases or vapors which can subsequently be removed unchanged. Any solid substance adsorbs a small amount of most gases; but to be useful as an adsorbent, a substance must have a large surface area and be able to concentrate a substantial amount of gas in a small volume of adsorbent.

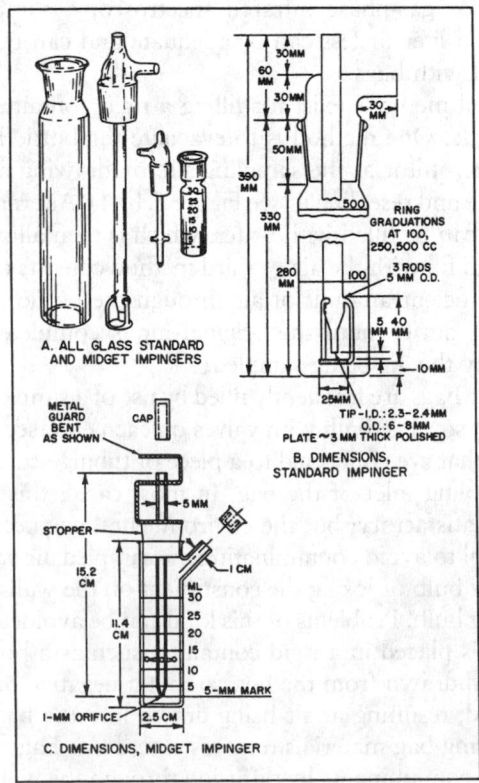

FIG. 5.10.3 Standard and midget impingers.

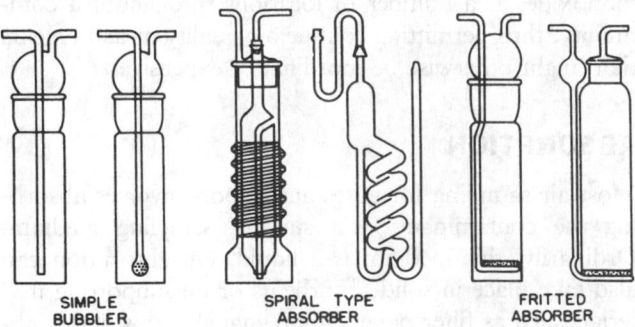

FIG. 5.10.2 Gas-absorbing vessels.

Activated carbon or charcoal and activated silica gel are most widely used for this purpose. A small quantity of either adsorbent placed in a U-tube or other container through which air is passed quantitatively removes many vapors and gases from a large volume of air. These gases can then be taken to the laboratory where desorption removes the collected substances for analysis. Desorption commonly involves heating the adsorbent and collecting the effluent gases or eluting the collected substances with a suitable organic liquid.

For most organic vapors, subsequent analysis by GC, infrared (IR), or UV spectroscopy is most convenient. For some purposes, either silica gel or activated carbon is used; but the use of silica gel is not recommended because it also adsorbs water vapor, and a short sampling period in humid air can saturate the silica gel before sufficient contaminant is adsorbed. Because charcoal does not adsorb water, it can be used in humid environments for days or even weeks if the concentration of the contaminant is low.

The ease of sampling using adsorbents is offset somewhat by the difficulty of quantitatively desorbing samples for analysis. When published data are not available to predict the behavior of a new substance, the environmental engineer should perform tests in the laboratory to determine both the collection efficiency and the success of desorption procedures after sample collection.

FREEZE-OUT SAMPLING

Vapors or gases that condense at a low temperature can be removed from the sampled airstream by passage through a vessel immersed in a refrigerating liquid. Table 5.10.1 lists liquids commonly used for this purpose. Usually forming a sampling train in which two or three coolant liquids progressively lower the air temperature in its passage through the system is recommended.

All freeze-out systems are hampered somewhat by the accumulation of ice from water vapor and can eventually become plugged with ice. Flow rates through a freeze-out train are also limited because a sufficient residence time in the system is necessary for the air to be cooled to the required degrees. For these reasons and because of the inconvenience of assembling freeze-out sampling trains, they

are not used for routine sampling purposes unless no other approach is feasible.

However, freeze-out sampling is an excellent means of collecting substances for research studies inasmuch as the low temperatures tend to arrest further chemical changes and ensure that the material being analyzed remains in the sampling container ready for analysis after warming. Analysis is usually conducted by GC, IR, or UV spectrophotometry or by mass spectrometry.

Particulate Matter Sampling

Particulate matter sampling methods include filtration, impingement and impaction, electrostatic precipitation, and thermal precipitation.

FILTRATION

Passing air through a filter is the most convenient method to remove particulate matter (see Figure 5.10.4). Before filtration is used to obtain a sample, however, the purpose of the sample should be considered. Many filters collect particulates efficiently, but thereafter removing the collected matter may be impossible except by chemical treatment.

If samples are collected for the purpose of examining the particles and measuring their size or noting morphological characteristics, many filters are not suitable because the particles are imbedded in the fibrous web of the filter and cannot readily be viewed or removed. If the sample is collected for the purpose of performing a chemical analysis, the filter must not contain significant quantities of the substance. If the purpose of sampling is to collect an amount of particulate matter for weighing, then a filter which can be weighed with the required precision must be selected. Many filtration materials are hygroscopic and change weight appreciably in response to changes in the relative humidity.

Filters can be made of many substances, and almost any solid substance could probably be made into a filter. In practice, however, fibrous substances, such as cellulose or paper, fabrics, and a number of plastics or polymerized materials, are generally used. The most readily available filters are those made of cellulose or paper and used in chemical laboratories for filtering liquids. These filter papers come in a variety of sizes and range in efficiency from loose filters that remove only larger particles to papers that remove fine particles with high efficiency. All filters display similar behavior, and ordinarily a high collection efficiency is accompanied by increased resistance to airflow.

Certain filtration media are more suited to air sampling than most paper or fibrous filters. Of these, membrane filters have the greatest utility, and commercially available membrane filters combine high-collection efficiencies with low resistance to flow. Such filters are not made up of a fibrous mat but are usually composed of gels of cellulose

TABLE 5.10.1 COOLANT SOLUTIONS FOR FREEZE-OUT SAMPLING

Coolant	Temperature (°C)
Ice water	0
Ice and salt (NaCl)	−21
Dry ice and isopropyl alcohol	−78.5
Liquid air	−147
Liquid oxygen	−183
Liquid nitrogen	−196

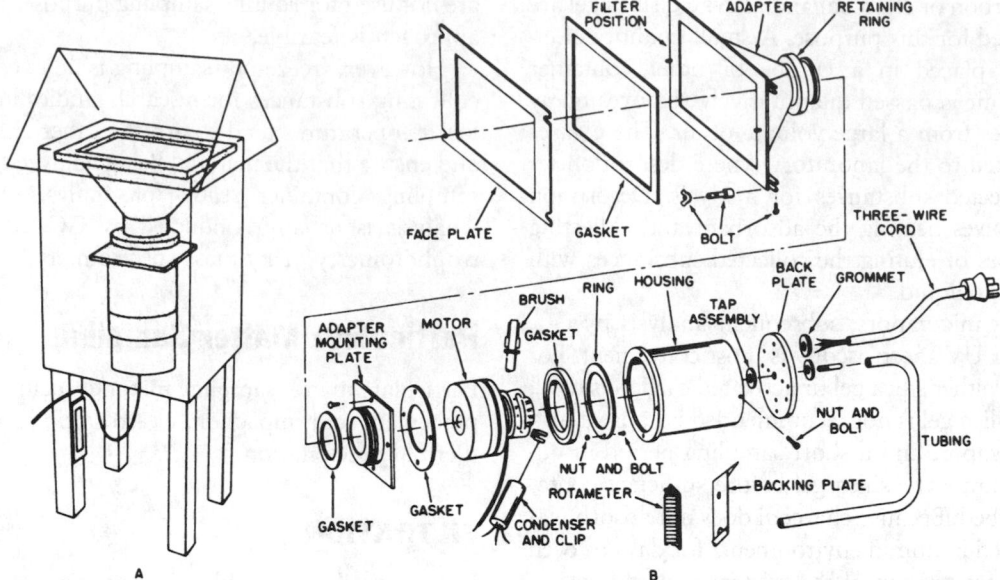

FIG. 5.10.4 High-volume sampler: (A) assembled sampler and shelter; (B) exploded view of typical high-volume air sampler.

esters or other polymeric substances that form a smooth surface of predictable characteristics.

Such filters contain many small holes, or pores, and can be made to exacting specifications so that their performance characteristics can be predicted. In addition, the filters usually have high chemical purity and are well suited to trace metal analyses. Some membrane filters can also be transparent and thus permit direct observation of collected particles with a microscope. Alternatively, the filters can be dissolved in an organic solvent, and the particles can be isolated and studied. Most membrane filters are not affected by relative humidity changes and can be weighed before and after use to obtain reliable gravimetric data.

Another kind of filter that is widely used in sampling ambient air is the fiberglass filter. These filters are originally made of glass fibers in an organic binder. Subsequently, the organic binder is removed by firing, leaving a web of glass that is efficient in collecting fine particles from the air. The principal advantages of using this type of filter are its low resistance to air flow and its unchanging weight regardless of relative humidity.

However, these filters are not well suited to particle size studies. In addition, they are not chemically pure, and the environmental engineer must ensure that the filter does not contribute the substance being analyzed in unknown quantities. In the United States, most data relating to suspended particulate matter in our cities have been obtained on filters made of fiberglass and used in conjunction with a sampling device referred to as a high-volume sampler (see Figure 5.10.4). Many other kinds of filters are available, but most sampling needs are well met by membrane or fiberglass filters.

IMPINGEMENT AND IMPACTION

The impingers previously described for sampling gases and vapors (see Figure 5.10.3) can also be used for the collection of particles and, in fact, were originally developed for that purpose. However, in ambient air sampling, they are not used because their collection efficiency is low and unpredictable for the fine particles present in ambient air. The low sampling rates also make them less attractive than filters for general air sampling, but instances do arise when impingers can be satisfactorily used. When impingers are used, the correct sampling rates must be maintained since the collection efficiency of impingers for particles varies when flow rates are not optimal.

Impactors are more widely used in ambient air sampling. In these devices, air is passed through small holes or orifices and made to impinge or impact against a solid surface. When these devices are constructed so that the air passing through one stage is subsequently directed onto another stage containing smaller holes, the resulting device is known as a *cascade impactor* and has the capability of separating particles according to sizes.

Various commercial devices are available. Figure 5.10.5 shows one impactor that is widely used and consists of several layers of perforated plates through which the air must pass. Each plate contains a constant number of holes, but the hole size progressively decreases so that the same air volume passing through each stage impinges at an increased velocity. The result is that coarse particles are deposited on the first stage and successively finer particles are removed at each subsequent stage. Although these instruments do not achieve exact particle size fractions, they

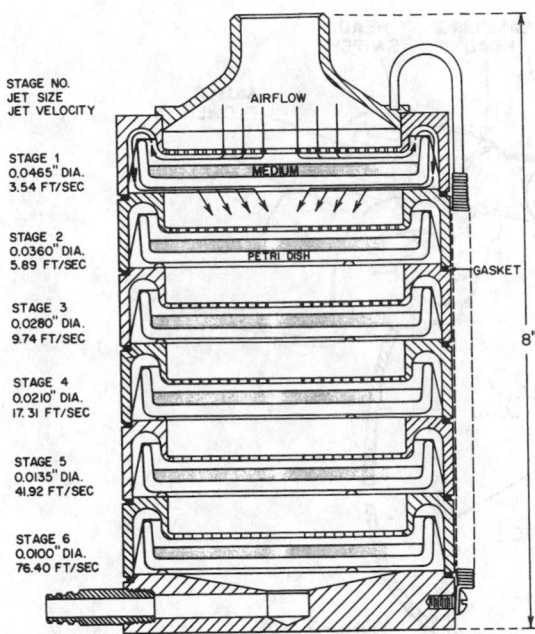

STAGE NO.
JET SIZE
JET VELOCITY

AIRFLOW

STAGE 1
0.0465" DIA.
3.54 FT/SEC

MEDIUM

STAGE 2
0.0360" DIA.
5.89 FT/SEC

PETRI DISH

GASKET

STAGE 3
0.0280" DIA.
9.74 FT/SEC

8"

STAGE 4
0.0210" DIA.
17.31 FT/SEC

STAGE 5
0.0135" DIA.
41.92 FT/SEC

STAGE 6
0.0100" DIA.
76.40 FT/SEC

FIG. 5.10.5 Cascade impactor (Andersen sampler).

do perform predictably when the characteristics of the aerosol being sampled are known.

In use, an environmental engineer assembles a cascade impactor after scrupulously cleaning each stage and applying, if necessary, a sticky substance or a removable surface on which the particles are deposited. After a period of sampling during which time the volume of air is metered, the stages can be removed, and the total weight of each fraction is determined, as well as its chemical composition. Such information is sometimes more useful than a single weight or chemical analysis of the total suspended particulate matter without regard to its particle size.

ELECTROSTATIC PRECIPITATION

Particulate matter can be quantitatively removed from air by ESPs. Devices that operate on the same principal but are much larger are frequently used to remove particulate matter from stack gases prior to discharging into the atmosphere. Several commercially available ESPs can be used for air sampling, and all operate on the same general principle of passing the air between charged surfaces, imparting a charge to particles in the air, and collecting the particles on an oppositely charged surface or plate.

In one of widely used commercial devices (see Figure 5.10.6), a high-voltage discharge occurs along a central wire; the collecting electrode is a metallic cylinder placed around the central wire while the air is passing through the tube. An intense corona discharge takes place on the central wire; the particles entering the tube are charged and are promptly swept to the walls of the tube where they remain firmly attached. With this method, collecting a sample for subsequent weighing or chemical analysis and

examining the particles and studying their size and shape is possible. However, the intense electrical forces can produce aggregates of particles that are different from those in the sampled air.

ESPs are not as widely used as filters for ambient air sampling because they are less convenient and tend to be heavy due to the power pack necessary to generate the high voltage. Nevertheless, they are excellent instruments for obtaining samples for subsequent analysis and sample, at high-flow rates and low resistance.

THERMAL PRECIPITATORS

Whenever a strong temperature gradient exists between two adjacent surfaces, particles tend to be deposited on the colder of these surfaces. Collecting aerosols by this means is termed *thermal precipitation,* and several commercial devices are available.

Because thermal forces are so weak, a large temperature difference must be maintained in a small area, and the air flow rate between the two surfaces must be low in order not to destroy the temperature gradient and to permit particles to be deposited before moving out of the collection area. As a result of these requirements, most devices use a heated wire as the source of the temperature differential and deposit a narrow ribbon of particles on the cold surface. Airflows are small, on the order of 10 to 25 ml per minute. At such low rates, the amount of material collected is normally insufficient for chemical analysis or weight determinations but is ample for examination by optical or electron microscopy.

Collection for particle size studies is the principal use for thermal precipitation units, and they are well suited to collecting samples for such investigations. Because the collecting forces are gentle, the particles are deposited unchanged. The microscopic examination gives information that can be translated into data concerning the number of particles and their morphological characteristics. The use of a small grid suitable for insertion into an electron microscope is also convenient as the collecting surface; this use eliminates additional manipulation of the sample prior to examination by electron microscopy.

Air Quality Monitoring Systems

Many options are available for the type of air quality information which can be collected, and the cost of air quality monitoring systems varies greatly. Only with a thorough understanding of the decisions which must be made based on the information received from the air quality monitoring system can an environmental engineer make an appropriate selection.

Regardless of the instruments used to measure air quality, the data are only as good as they are representative of the sampling site selected.

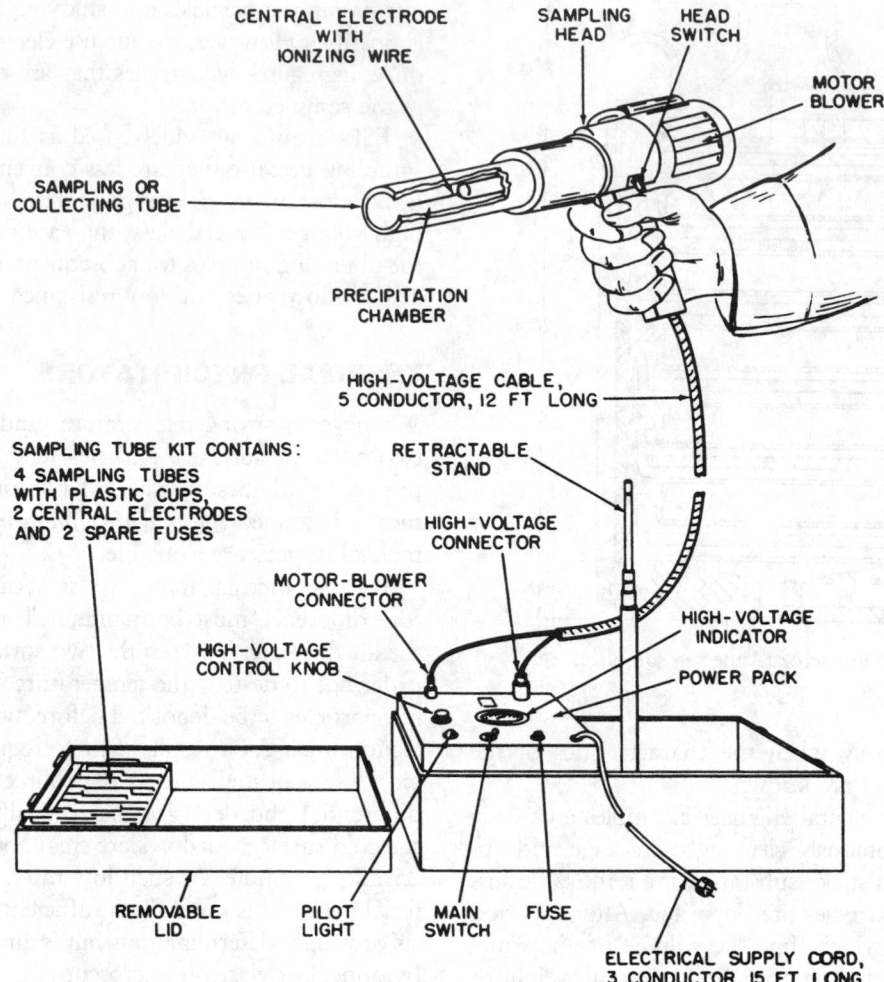

FIG. 5.10.6 Electrostatic precipitator. (Reprinted, with permission, from Mine Safety Appliances Company.)

The simplest air quality monitors are static sensors that are exposed for a given length of time and are later analyzed in the laboratory. In some cases, these devices provide all the required information. More commonly, a system of automatic instruments measuring several different air quality parameters is used. When more than a few instruments are used, the signals from these instruments can be retained on magnetic tape rather than on recorder charts.

The most common errors in the design of air quality monitoring systems are poor site location and the acquisition of more data than necessary for the purpose of the installation.

PURPOSES OF MONITORING

The principal purpose of air quality monitoring is to acquire data for comparison to regulated standards. In the United States, standards have been promulgated by the federal government and by many states. Where possible, these standards are based on the physiological effect of the air pollutant on human health. The averaging time over which various concentration standards must be maintained differs for each pollutant. (See Chapter 4 for tabulations of the ambient air quality standards promulgated by the federal government.)

Some air quality monitoring systems determine the impact of a single source or a concentrated group of sources of emissions on the surrounding area. In this case, determining the background level, the maximum ground-level concentration in the area, and the geographical extent of the air pollutant impact is important.

When the source is isolated, e.g., a single industrial plant in a rural area, the system design is straightforward. Utilizing the meteorological records available from nearby airports or government meteorological reporting stations, environmental engineers can prepare a wind rose to estimate the principal drift direction of the air pollutant from the source. Then they can perform dispersion calculations to estimate the location of the expected point of maximum ground-level concentration. As a general rule with stacks between 50 and 350 ft tall, this point of maximum concentration is approximately 10 stack heights downwind.

The air quality monitoring system should include at least one sensor at the point of expected maximum ground-level concentration. Additional sensors should be placed not less than 100 stack heights upwind (prevailing) to provide a background reading, and at least two sensors should be placed between 100 and 200 stack heights downwind to determine the extent of the travel of the pollutants from the source. If adequate resources are available, sampling at the intersection points of a rectilinear grid with its center at the source is recommended.

With such a system for an isolated source, environmental engineers can obtain adequate data in a one-year period to determine the impact of the source on the air quality of the area. Because of the variability in climatic conditions on an annual basis, few areas require less than one year of data collection to provide adequate information. If a study is performed to determine the effect of process changes on air quality, the study may have to continue for two to five years to develop information that is statistically reliable.

Some air quality monitoring is designed for the purpose of investigating complaints concerning an unidentified source. This situation usually happens in urban areas for odor complaints. In these cases, human observers use a triangulation technique to correlate the location of the observed odor with wind direction over several days. Plotting on a map can pinpoint the offending source in most cases. While this technique is not an air pollution monitoring system in an instrumental sense, it is a useful tool in certain situations.

Air pollution research calls for a completely different approach to air quality monitoring. Here, the purpose is to define some unknown variable or combination of variables. This variable can be either a new atmospheric phenomenon or the evaluation of a new air pollution sensor. In the former case, the most important consideration is the proper operation of the instrument used. In the latter case, the most important factor is the availability of a reference determination to compare the results of the new instrument against.

MONITORING IN URBAN AREAS

Urban areas are of major interest in air pollution monitoring in the United States since most of the population lives in these areas. The most sophisticated and expensive air quality monitoring systems in the United States are those for large cities (and one or two of the largest states) where data collection and analysis are centralized at a single location through the use of telemetry. Online computer facilities provide data reduction.

The three philosophies that can be used in the design of an urban air quality monitoring system are locating sensors on a uniform area basis (rectilinear grid), locating sensors at locations where pollutant concentrations are high, and locating sensors in proportion to the population distribution. The operation of these systems is nearly identical, but the interpretation of the results can be radically different.

The most easily designed systems are those where sensors are located uniformly on a geographical basis according to a rectilinear grid. Because adequate coverage of an urban area frequently requires at least 100 sensors, this concept is usually applied only with static or manual methods of air quality monitoring.

Locating air quality sensors at points of maximum concentration indicates the highest levels of air pollutants encountered throughout the area. Typically, these points include the central business district and the industrial areas on the periphery of the community. This type of data is useful when interpreted in the context of total system design. One or two sensors are usually placed in clean or background locations, so that the environmental engineer can estimate the average concentration of air pollutants over the entire area. The basis of this philosophy is that if the concentrations in the dirtiest areas are below air quality standards, certainly the cleaner areas will have no problems.

The design of air quality monitoring systems based on population distribution means placing air quality sensors in the most populous areas. While this philosophy may not include all high-pollutant concentration areas in the urban region, it generally encompasses the central business district and is a measure of the air pollution levels to which most of the population are exposed. The average concentrations from this type of sampling network are an adequate description from a public health standpoint. This system design can, however, miss some localized high concentration areas.

Before the system is designed, the system designer and those responsible for interpreting the data must agree on the purpose of air quality monitoring.

SAMPLING SITE SELECTION

Once the initial layout has been developed for an air quality monitoring system, specific sampling sites must be located as close as practical to the ideal locations. The major considerations are the lack of obstruction from local interferences and the adequacy of the site to represent the air mass, accessibility, and security.

Local interferences cause major disruptions to air quality sensor sites. A sampler inlet placed at a sheltered interior corner of a building is not recommended because of poor air motion. Tall buildings or trees immediately adjacent to the sampling site can also invalidate most readings.

The selection of sampling sites in urban areas is complicated by the canyon effect of streets and the high density of pollutants, both gaseous and particulate, at street level. In order for the data to represent the air mass sampled, environmental engineers must again review the purpose of the study. If the data are collected to determine areawide pollutant averages, the sampler inlet might best be located in a city park, vacant lot, or other open area. If this location is not possible, the sampler inlet could be at the roof level of a one- or two-story building so that street-level effects are minimized. On the other hand, if the physiological impact of air pollutants is a prime consideration, the samplers should be at or near the breathing level of the people exposed. As a general rule, an elevation of 3 to 6 m above the ground is an optimum elevation.

The sampling site location can be different for the same pollutant depending on the purpose of the sampling. Carbon monoxide sampling is an example. The federally promulgated air quality standards for carbon monoxide include both an 8-hr and a 1-hr concentration limit. Maximum 1-hr concentrations are likely to be found in a high-traffic density, center-city location. People are not ordinarily exposed to these concentrations for 8-hr periods. When sampling for comparison with the 1-hr standard, the environmental engineer should locate the sensor within about 20 ft of a major traffic intersection. When sampling for comparison with the 8-hr standard, the engineer should locate the sensor near a major thoroughfare in either the center-city area or in the suburban area with the sampler less than 50 ft from the intersection. The reason for two different sampling site locations is to be consistent with the physiological effects of carbon monoxide exposure and the living pattern of most of the population. If only one site can be selected, the location described for the 8-hr averaging time is recommended.

When the sampling instruments are located inside a building and an air sample is drawn in from the outside, using a sampling pipe with a small blower is advantageous to bring outside air to the instrument inlets. This technique improves sampler line response time. An air velocity of approximately 700 ft per min in the pipe balances problems of gravitational and inertial deposition of particulate matter when particulates are sampled.

The sampling site should be accessible to operation and maintenance personnel. Since most air pollution monitoring sites are unattended much of the time, sample site security is a real consideration; the risk of vandalism is high in many areas.

STATIC METHODS OF AIR MONITORING

Static sensors used to monitor air quality require the minimum capital cost. While averaging times are in terms of weeks and sensitivity is low, in many cases, static monitors provide the most information for the amount of investment. Although inexpensive, they should not be rejected but should be considered as useful adjuncts to more sophisticated systems.

Dustfall Jars

The simplest of all air quality monitoring devices is the dustfall jar (see Figure 5.10.7). This device measures the fallout rate of coarse particulate matter, generally above about 10 μm in size. Dustfall and odor are two major reasons for citizen complaints concerning air pollution. Dustfall is offensive because it builds up on porches and automobiles and is highly visible and gritty to walk upon.

Dustfall seldom carries for distances greater than $\frac{1}{2}$ mi because these large particles are subject to strong gravitational effects. For this reason, dustfall stations must be spaced more closely than other air pollution sensors for a detailed study of an area.

Dustfall measurements in large cities in the United States in the 1920s and 1930s commonly indicated dustfall rates in hundreds of tn per sq mi per mon. These levels are considered excessive today, as evidenced by the dustfall standards of 25 to 30 tn per sq mi per mon promulgated by many of the states. While the measurement of low or moderate values of dustfall does not indicate freedom from air pollution problems, measured dustfall values in excess of 50 to 100 tn per sq mi per mon are an indication of excessive air pollution.

The large size of the particulate matter found in dustfall jars makes it amenable to chemical or physical analysis by such techniques as microscopy. These analyses are useful to identify specific sources.

Lead Peroxide Candles

For many years, sulfur dioxide levels have been determined through the use of lead peroxide candles. These devices, known as candles because they are a mixture of lead peroxide paste spread on a porcelain cylinder about the size and shape of a candle, are normally exposed for periods of 1 mon. Sulfur gases in the air react with lead peroxide to form lead sulfate. Environmental engineers analyze the sulfate according to standard laboratory procedures to indicate the atmospheric levels of sulfur gases during the period of exposure.

A modification of this technique that simplifies the laboratory procedure is to use a fiber filter cemented to the inside of a plastic petri dish (a flat-bottom dish with shallow walls used for biological cultures). The filter is satu-

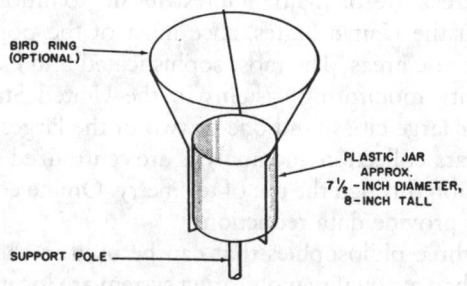

FIG. 5.10.7 Dustfall jar.

rated with an aqueous mixture of lead peroxide and a gel, commonly gum tragacanth, and is allowed to dry. These dishes or plates are exposed in an inverted position for periods of 1 to 4 wk.

The lead peroxide estimation of sulfur dioxide has inherent weaknesses. All sulfur gases, including reduced sulfur, react with lead peroxide to form lead sulfate. More importantly, the reactivity of lead peroxide depends on its particle size distribution. For this reason, the results from different investigators are not directly comparable. Nevertheless, a network of lead peroxide plates over an area provides a good indication of the exposure to sulfur gases during the exposure period. This technique is useful for determining the geographical extent of sulfur pollution.

Other Static Methods

Environmental engineers have modified the technique of using fiber filter cemented to a petri dish by using sodium carbonate rather than lead peroxide to measure sulfur gases. This method also indicates the concentration of other gases, including nitrogen oxides and chlorides. Engineers have measured relative levels of gaseous fluoride air pollution using larger filters, e.g., 3-in diameter, dipped in sodium carbonate, and placed in shelters to protect them from the rain. With all these static methods, the accuracy is low, and the data cannot be converted directly into ambient air concentrations. They do, however, provide a low-cost indicator of levels of pollution in an area.

Environmental engineers have evaluated the corrosive nature of the atmosphere using standardized steel exposure plates for extended periods to measure the corrosion rate. This method provides a gross indication of the corrosive nature of the atmosphere. As with other static samplers, the results are not directly related to the concentration of air pollutants.

Manual Analyses

Manual analyses for air quality measurements are those that require the sample first be collected and then analyzed in the laboratory. Manual instruments provide no automatic indication of pollution levels.

The manual air sampling instrument in widest use is the high-volume sampler. With this method, ambient air is drawn through a preweighed filter at a rate of approximately 50 atmospheric cubic feet per minute (acfm) for a period of 24 hr. The filter is then removed from the sampler, returned to the laboratory, and weighed. The weight gain, combined with the measured air volume through the sampler, allows the particulate mass concentration, expressed in micrograms per cubic meter, to be calculated.

Reference methods for nearly all gaseous air pollutants involve the use of a wet sampling train in which air is drawn through a collecting medium for a period of time. The exposed collecting medium is then returned to the laboratory for chemical analysis. Sampling trains have been developed that allow sampling of five or more gases simultaneously into separate bubblers. Sequential samplers, which automatically divert the airflow from one bubbler to another at preset time intervals, are also available.

These sampling methods can be accomplished with a modest initial investment; however, the manpower required to set out and pick up the samples, combined with the laboratory analysis, raises the total cost to a point where automated systems can be more economical for long-term studies.

Instrumental Analyses

As the need for accurate data that can be statistically reduced in a convenient manner increases, automated sampling systems become necessary. The elements of an automated system include the air-flow handling system, the sensors, the data transmission storage and display apparatus, and the data processing facility. The overall system is no more valuable than its weakest link.

SENSORS

The output reliability from an air quality sensor depends on its inherent accuracy, sensitivity, zero drift, and calibration. The inherent accuracy and sensitivity are a function of the the instrument's design and its operating principle. Zero drift can be either an electronic phenomenon or an indication of difficulties with the instrument. In instruments that use an optical path, lenses become dirty. In wet chemical analyzers, the flow rates of reagents can vary, changing both the zero and the span (range) of the instrument. Because of these potential problems, every instrument should have routine field calibration at an interval determined in field practice as reasonable for the sensor.

Environmental engineers calibrate an air quality sensor using either a standard gas mixture or a prepared, diluted gas mixture using permeation tubes. In some cases, they can currently sample the airstream entering the sensor by using a reference wet chemical technique.

The operator of air quality sensors should always have a supply of spare parts and tools to minimize downtime. At a minimum, operator training should include instruction to recognize the symptoms of equipment malfunction and vocabulary to describe the symptoms to the person responsible for instrument repair. Ideally, each operator should receive a short training session from the instrument manufacturer or someone trained in the use and maintenance of the instrument so that the operator can make repairs on site. Since this training is seldom possible in practice, recognition of the symptoms of malfunction becomes increasingly important.

DATA TRANSMISSION

The output signal from a continuous monitor in an air quality monitoring system is typically the input to a strip

TABLE 5.10.2 COMPOUNDS THAT CAN BE ANALYZED BY THE MICROPROCESSOR-CONTROLLED PORTABLE IR SPECTROMETER

Compound	Range of Calibration (ppm)	Compound	Range of Calibration (ppm)
Acetaldehyde	0 to 400	Ethylene oxide	0 to 10 and 0 to 100
Acetic acid	0 to 50	Ethyl ether	0 to 1000 and 0 to 2000
Acetone	0 to 2000	Fluorotrichloromethane (Freon 11)	0 to 2000
Acetonitrile	0 to 200	Formaldehyde	0 to 20
Acetophenone	0 to 100	Formic acid	0 to 20
Acetylene	0 to 200	Halothane	0 to 10 and 0 to 100
Acetylene tetrabromide	0 to 200	Heptane	0 to 1000
Acrylonitrile	0 to 20 and 0 to 100	Hexane	0 to 1000
Ammonia	0 to 100 and 0 to 500	Hydrazine	0 to 100
Aniline	0 to 20	Hydrogen cyanide	0 to 20
Benzaldehyde	0 to 500	Isoflurane	0 to 10 and 0 to 100
Benzene	0 to 50 and 0 to 200	Isopropyl alcohol	0 to 1000 and 0 to 2000
Benzyl chloride	0 to 100	Isopropyl ether	0 to 1000
Bromoform	0 to 10	Methane	0 to 100 and 0 to 1000
Butadiene	0 to 2000	Methoxyflurane	0 to 10 and 0 to 100
Butane	0 to 200 and 0 to 2000	Methyl acetate	0 to 500
2-Butanone (MEK)	0 to 250 and 0 to 1000	Methyl acetylene	0 to 1000 and 0 to 5000
Butyl acetate	0 to 300 and 0 to 600	Methyl acrylate	0 to 50
n-Butyl alcohol	0 to 200 and 0 to 1000	Methyl alcohol	0 to 500 and 0 to 1000
Carbon dioxide	0 to 2000	Methylamine	0 to 50
Carbon disulfide	0 to 50	Methyl bromide	0 to 50
Carbon monoxide	0 to 100 and 0 to 250	Methyl cellosolve	0 to 50
Carbon tetrachloride	0 to 20 and 0 to 200	Methyl chloride	0 to 200 and 0 to 1000
Chlorobenzene	0 to 150	Methyl chloroform	0 to 500
Chlorobromomethane	0 to 500	Methylene chloride	0 to 1000
Chlorodifluoromethane	0 to 1000	Methyl iodide	0 to 40
Chloroform	0 to 100 and 0 to 500	Methyl mercaptan	0 to 100
m-Cresol	0 to 20	Methyl methacrylate	0 to 250
Cumene	0 to 100	Morpholine	0 to 50
Cyclohexane	0 to 500	Nitrobenzene	0 to 20
Cyclopentane	0 to 500	Nitromethane	0 to 200
Diborane	0 to 10	Nitrous oxide	0 to 100 and 0 to 2000
m-Dichlorobenzene	0 to 150	Octane	0 to 100 and 0 to 1000
o-Dichlorobenzene	0 to 100	Pentane	0 to 1500
p-Dichlorobenzene	0 to 150	Perchloroethylene	0 to 200 and 0 to 500
Dichlorodifluoromethane (Freon 12)	0 to 5 and 0 to 800	Phosgene	0 to 5
1,1-Dichloroethane	0 to 200	Propane	0 to 2000
1,2-Dichloroethylene	0 to 500	n-Propyl alcohol	0 to 500
Dichloroethyl ether	0 to 50	Propylene oxide	0 to 200
Dichloromonofluoromethane (Freon 21)	0 to 1000	Pyridine	0 to 100
Dichlorotetrafluoroethane (Freon 114)	0 to 1000	Styrene	0 to 200 and 0 to 500
		Sulfur dioxide	0 to 100 and 0 to 250
Diethylamine	0 to 50	Sulfur hexafluoride	0 to 5 and 0 to 500
Dimethylacetamide	0 to 50	1,1,2,2-Tetrachloro 1,2-difluoroethane (Freon 112)	0 to 2000
Dimethylamine	0 to 50	1,1,2,2-Tetrachloroethane	0 to 50
Dimethylformamide	0 to 50	Tetrahydrofuran	0 to 500
Dioxane	0 to 100 and 0 to 500	Toluene	0 to 1000
Enflurane	0 to 10 and 0 to 100	Total hydrocarbons	0 to 1000
Ethane	0 to 1000	1,1,2-TCA	0 to 50
Ethanolamine	0 to 100	TCE	0 to 200 and 0 to 2000
2-Ethoxyethyl acetate	0 to 200	1,1,2-Trichloro 1,2,2-TCA (Freon 113)	0 to 2000
Ethyl acetate	0 to 400 and 0 to 1000	Trifluoromonobromomethane (Freon 13B1)	0 to 1000
Ethyl alcohol	0 to 1000 and 0 to 2000		
Ethylbenzene	0 to 200	Vinyl acetate	0 to 10
Ethyl chloride	0 to 1500	Vinyl chloride	0 to 20
Ethylene	0 to 100	Vinylidene chloride	0 to 20
Ethylene dibromide	0 to 10 and 0 to 50	Xylene (Xylol)	0 to 200 and 0 to 2000
Ethylene dichloride	0 to 100		

Source: Foxboro Co.

chart recorder, magnetic tape data storage, or an online data transmission system. The output of most air quality sensors is in analog form. This form is suitable for direct input to a strip chart recorder; but in automated systems, the analog signal is commonly converted to a digital signal. For those sensors that have linear output, the signal can go directly to the recorder or transmission system. For sensors with logarithmic output, it may be advantageous to convert this signal to a linear form.

Many early automated air quality monitoring systems in the United States had difficulty with the data transmission step in the system. In some cases, this difficulty was caused by attempts to overextend the lower range of the sensors so that the signal-to-noise ratio was unfavorable. In other cases, matching the sensor signal output to the data transmission system was poor. With developments and improvements in systems, these early difficulties were overcome.

Online systems, i.e., those that provide an instantaneous readout of the dynamic situation monitored by the system, add an additional step to air quality monitoring systems. Data from continuous monitors can be stored on magnetic tape for later processing and statistical reduction. In an online system, this processing is done instantaneously. The added expense of this sophistication must be evaluated in terms of the purposes of the air quality monitoring system. When decisions with substantial community impact must be made within a short time, this real-time capability may be necessary.

DATA PROCESSING

The concentration of many air pollutants follows a log normal rather than a normal distribution. In a log normal distribution, a plot of the logarithm of the measured values more closely approximates the bell-shaped Gaussian distribution curve than does a plot of the numerical data. Suspended particulate concentrations are a prime example of this type of distribution. In this case, the geometric mean is the statistical parameter that best describes the population of data. The arithmetic mean is of limited value because it is dominated by a few occurrences of high values. The geometric mean, combined with the geometric standard deviation, completely describes a frequency distribution for a log normally distributed pollutant.

Environmental engineers should consider the averaging time over which sample results are reported in processing and interpreting air quality data. For sulfur dioxide, various agencies have promulgated air quality standards based on annual arithmetic average, monthly arithmetic average, weekly arithmetic average, 24-hr arithmetic average, 3-hr arithmetic average, and 1-hr arithmetic average concentrations. The output of a continuous analyzer can be averaged over nearly any discrete time interval. To reduce the computation time, an environmental engineer must consider the time interval over which continuous analyzer output is averaged to obtain a discrete input for the calculations. If a 1-hr average concentration is the shortest

time interval of value in interpreting the study results, using a 1- or 2-min averaging time for input to the computation program is not economic.

Environmental engineers must exercise caution in using strip chart recorders to acquire air quality data. The experience of many organizations, both governmental and industrial, is that the reduction of data from strip charts is tedious. Many organizations decide that they do not really need all that data once they find a large backlog of unreduced strip charts. Two cautions are suggested by this experience. First, only data that is to be used should be collected. Second, magnetic tape data storage followed by computer processing has advantages.

The visual display of air quality data has considerable appeal to many nontechnical personnel. Long columns of numbers can be deceptive if only one or two important trends are shown. The use of bar charts or graphs is frequently advantageous even though they do not show the complete history of air quality over a time span.

Portable and Automatic Analyzers

Microprocessor-controlled spectrometers are available to measure concentrations of a variety of gases and vapors in ambient air. These units can be portable or permanently installed and can comply with environmental and occupational safety regulations. In the IR spectrometer design, an integral air pump draws ambient air into the test cell, operating at a flow rate of 0.88 acfm (25 l/min). The microprocessor selects the wavelengths for the components and the filter wheel in the analyzer allows the selected wavelengths to pass through the ambient air sample in the cell. The microprocessor automatically adjusts the path length through the cell to give the required sensitivity. Because of the folded-path-length design, the path length can be increased to 20 m (60 ft), and the resulting measurement sensitivity is better than 1 ppm in many cases.

As shown in Table 5.10.2, practically all organic and some inorganic vapors and gases can be monitored by these IR spectrometers. The advantage of the microprocessor-based operation is that the monitor is precalibrated for the analysis of over 100 Occupational Safety and Health Administration (OSHA)-cited compounds. The memory capacity of the microprocessor is sufficient to accommodate another ten user-selected and user-calibrated gases. Analysis time is minimized because the microprocessor automatically sets the measurement wavelengths and parameters for any compound in its memory. A general scan for a contaminant in the atmosphere takes about 5 min, while the analysis of a specific compound can be completed in just a few minutes. The portable units are battery-operated for 4 hr of continuous operation and are approved for use in hazardous areas.

—R.A. Herrick
R.G. Smith
Béla G. Lipták

5.11
STACK SAMPLING

Type of Sample
Gas containing particulates

Standard Design Pressure
Generally atmospheric or near-atmospheric

Standard Design Temperature
−25 to 1500°F (−32 to 815°C)

Sample Velocity
400 to 10,000 ft (120 to 3000 m) per minute

Materials of Construction
316 or 304 stainless steel for pitot tubes; 304 or 316 stainless, quartz, or Incoloy for sample probes

Partial List of Suppliers
Andersen Samplers, Inc.; Applied Automation/Hartmann & Braun Process Control Systems Div.; Bacharach, Inc.; Columbia Scientific Industries Corp.; Cosa Instruments; Gastech, Inc.; Mine Safety Appliance Co., Instrument Division; Sensidyne, Inc. Gas & Particulate Detection Systems; Scientific Glass & Instruments Inc.; Sierra Monitor Corp.; Teledyne Analytical Instrument, Teledyne Inc.

A complete EPA particulate sampling system (reference method 5) is comprised of the following four subsystems (U.S. EPA 1971; 1974; 1977; 1987; 1989; 1991; Morrow, Brief, and Bertrand 1972):

1. A pitot tube probe or pitobe assembly for temperature and velocity measurements and for sampling
2. A two-module sampling unit that consists of a separate heated compartment with provisions for a filter assembly and a separate ice-bath compartment for the impinger train and bubblers

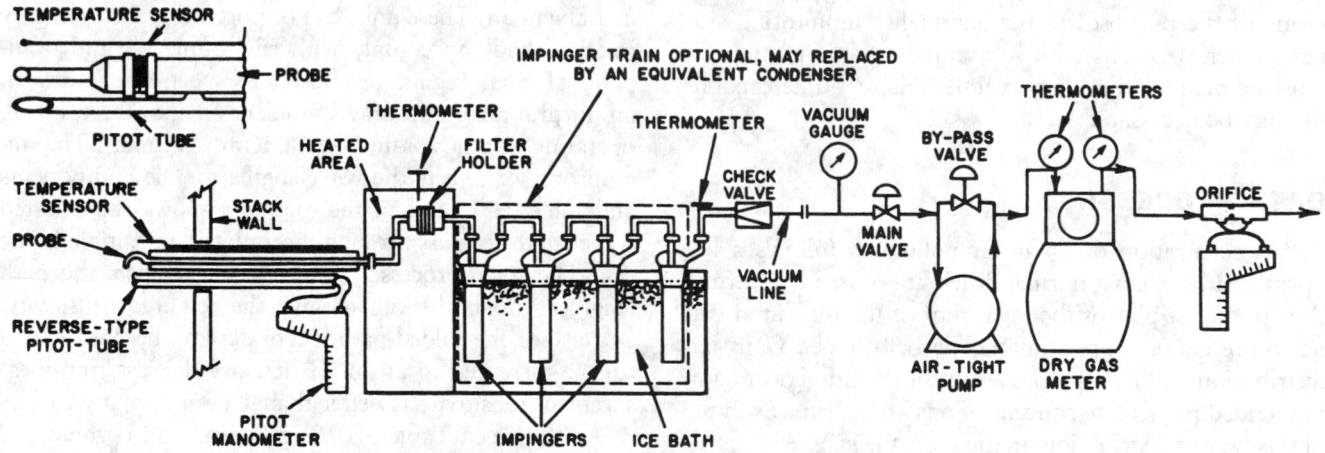

EPA PARTICULATE SAMPLING TRAIN (METHOD 5)
(FEDERAL REGISTER, VOL. 36, NOS. 234 AND 247)

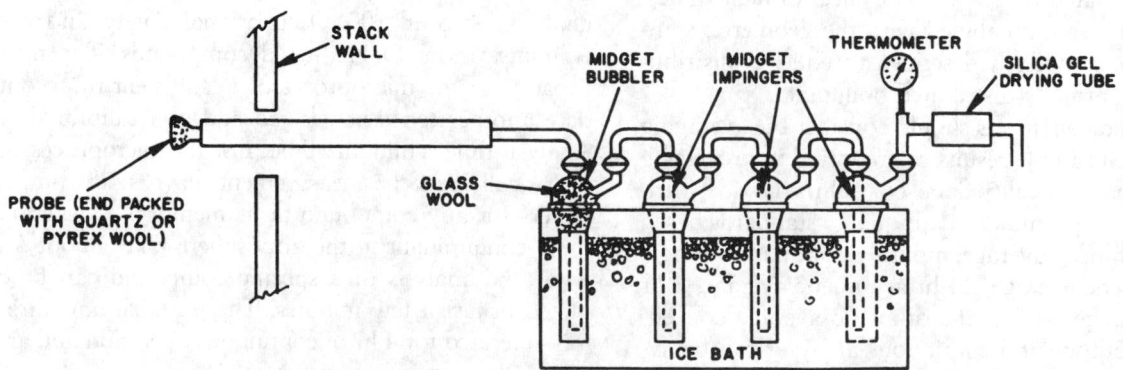

SAMPLING CASE FOR SO₂, SO₃ AND H₂SO₄ MIST (METHOD 8)

FIG. 5.11.1 (Top) EPA particulate sampling train (reference method 5); (Bottom) Sampling case for SO_2, SO_3, and H_2SO_4 mist (reference method 8). (Reprinted from *Federal Register* 36, nos. 234 and 247.)

3. An operating or control unit with a vacuum pump and a standard dry gas meter
4. An integrated, modular umbilical cord that connects the sample unit and pitobe to the control unit

Figure 5.11.1 is a schematic of a U.S. EPA particulate sampling train (reference method 5). As shown in the figure, the system can be readily adapted for sampling sulfur dioxide (SO_2), sulfur trioxide (SO_3), and sulfuric acid (H_2SO_4) mist (reference method 8) (U.S. EPA 1971; 1977).

This section gives a detailed description of each of the four subsystems.

Pitot Tube Assembly

Procuring representative samples of particulates suspended in gas streams demands that the velocity at the entrance to the sampling probe be equal to the stream velocity at that point. The environmental engineer can equalize the velocities by regulating the rate of sample withdrawal so that the static pressure within the probe is equal to the static pressure in the fluid stream at the point of sampling. A specially designed pitot tube with means for measuring the pertinent pressures is used for such purposes. The engineer can maintain the pressure difference at zero by automatically controlling the sample drawoff rate.

Figure 5.11.2 shows the pitot tube manometer assembly for measuring stack gas velocity. The type S (Stauscheibe or reverse) pitot tube consists of two opposing openings, one facing upstream and the other downstream during the measurement. The difference between the impact pressure (measured against the gas flow) and the static pressure gives the velocity head.

Figure 5.11.3 illustrates the construction of the type S pitot tube. The external tubing diameter is normally between $\frac{3}{16}$ and $\frac{3}{8}$ in (4.8 and 9.5 mm). As shown in the figure, the distance is equal from the base of each leg of the tube to its respective face-opening planes. This distance (P_A and P_B) is between 1.05 and 1.50 times the external tube diameter. The face openings of the pitot tube should be aligned as shown.

Figure 5.11.4 shows the pitot tube combined with the sampling probe. The relative placement of these components eliminates the major aerodynamic interference effects. The probe nozzle has a bottom-hook or elbow design. It is made of seamless 316 stainless steel or glass with a sharp, tapered leading edge. The angle of taper should be less than 30°, and the taper should be on the outside to preserve a constant internal diameter. For the probe lining, either borosilicate or quartz glass probe liners are used for stack temperatures to approximately 900°F (482°C); quartz liners are used for temperatures between 900 and 1650°F (482 and 899°C). Although borosilicate or quartz glass probe linings are generally recommended, 316 stainless steel, Incoloy, or other corrosion-resistant metal can also be used.

INSTALLATION

The environmental engineer selects the specific points of the stack for sampling to ensure that the samples represent the material being discharged or controlled. The en-

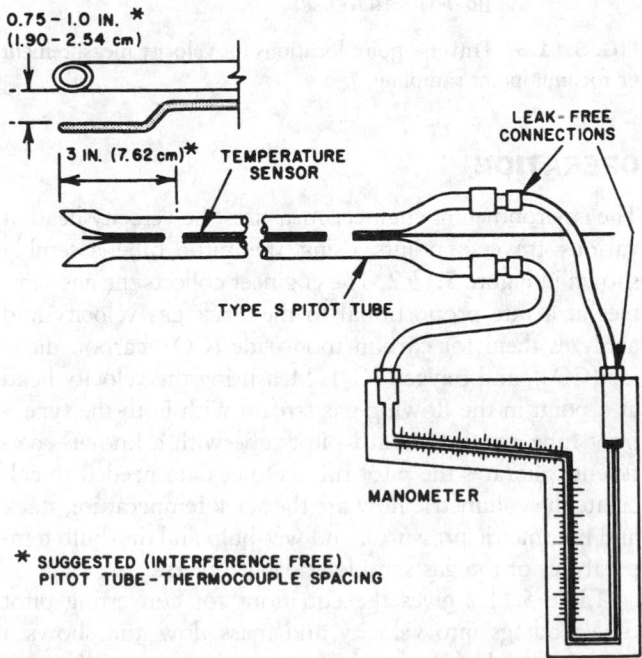

0.75 - 1.0 IN. *
(1.90 - 2.54 cm)

3 IN. (7.62 cm)*

TEMPERATURE SENSOR

LEAK-FREE CONNECTIONS

TYPE S PITOT TUBE

MANOMETER

* SUGGESTED (INTERFERENCE FREE) PITOT TUBE - THERMOCOUPLE SPACING

FIG. 5.11.2 Type S pitot tube manometer assembly.

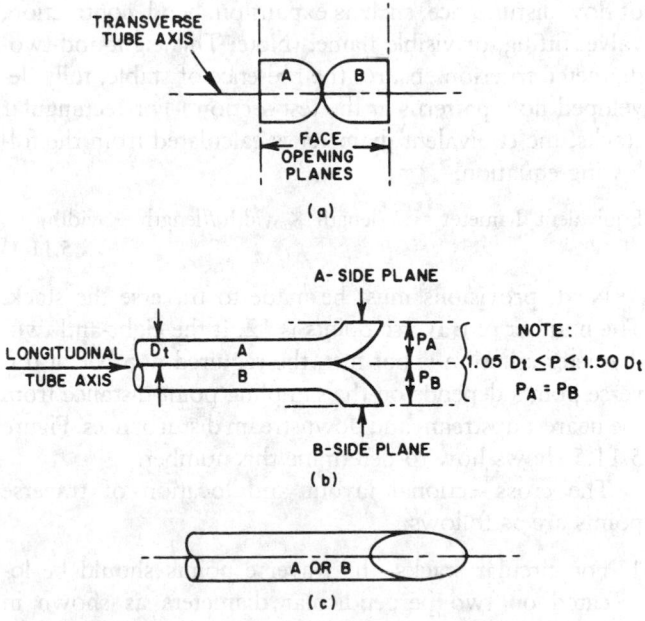

TRANSVERSE TUBE AXIS

A   B

FACE OPENING PLANES

(a)

LONGITUDINAL TUBE AXIS

A-SIDE PLANE

D_t A
        B

P_A
P_B

B-SIDE PLANE

NOTE:
$1.05\ D_t \leq P \leq 1.50\ D_t$
$P_A = P_B$

(b)

A OR B

(c)

FIG. 5.11.3 Properly constructed type S pitot tube. (a) End view: face-opening planes perpendicular to transverse axis, (b) Top view: face-opening planes parallel to longitudinal axis, (c) Side view: both legs of equal length and center lines coincident; when viewed from both sides, baseline coefficient values of 0.84 can be assigned to pitot tubes constructed this way.

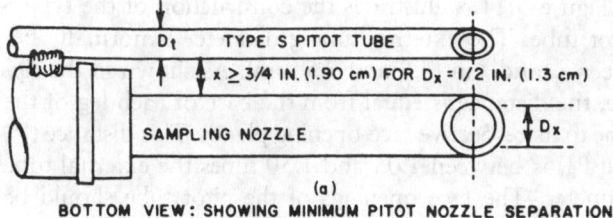

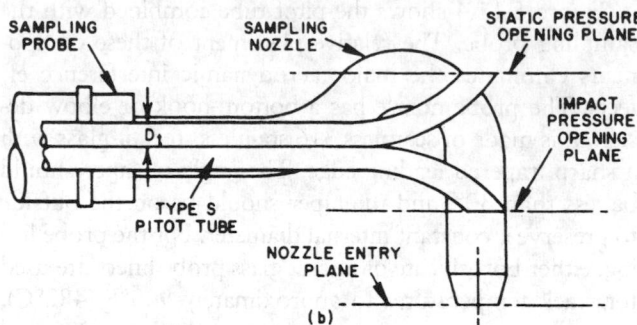

FIG. 5.11.4 Proper pitot tube with sampling probe nozzle configuration to prevent aerodynamic interference. (a) Bottom view: minimum pitot nozzle separation, (b) Side view: the impact pressure opening plane of the pitot tube located even with or above the nozzle entry plane so that the pitot tube does not interfere with gas flow streamlines approaching the nozzle.

gineer determines these points after examining the process or sources of emissions and their variation with time.

In general, the sampling point should be located at a distance equal to at least eight stack or duct diameters downstream and two diameters upstream from any source of flow disturbance, such as expansion, bend, contraction, valve, fitting, or visible flame. (Note: This eight-and-two-diameter criterion ensures the presence of stable, fully developed flow patterns at the test section.) For rectangular stacks, the equivalent diameter is calculated from the following equation:

$$\text{Equivalent diameter} = 2\,(\text{length} \times \text{width})/(\text{length} + \text{width})$$
$$\text{5.11(1)}$$

Next, provisions must be made to traverse the stack. The number of traverse points is 12. If the eight-and-two-diameter criterion is not met, the required number of traverse points depends on the sampling point distance from the nearest upstream and downstream disturbances. Figure 5.11.5 shows how to determine this number.

The cross-sectional layout and location of traverse points are as follows:

1. For circular stacks, the traverse points should be located on two perpendicular diameters as shown in Figure 5.11.6 and Table 5.11.1.
2. For rectangular stacks, the cross section is divided into as many equal rectangular areas as traverse points so that the length-to-width ratio of the elemental area is between one and two. The traverse points are located at the centroid of each equal area as shown in Figure 5.11.6.

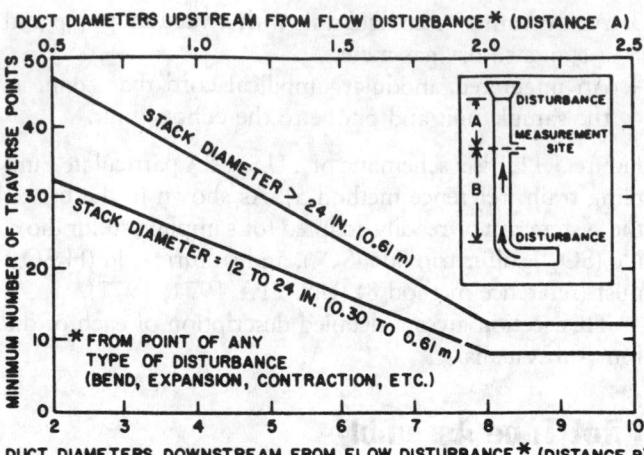

FIG. 5.11.5 Minimum number of traverse points for particulate traverses.

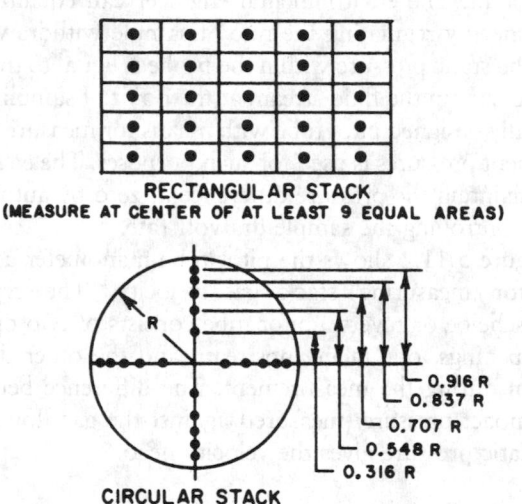

FIG. 5.11.6 Traverse point locations for velocity measurement or for multipoint sampling.

OPERATION

The environmental engineer measures the velocity head at various traverse points using the pitot tube assembly shown in Figure 5.11.2. The engineer collects the gas samples at a rate proportional to the stack gas velocity and analyzes them for carbon monoxide (CO), carbon dioxide (CO_2), and oxygen (O_2). Measuring the velocity head at a point in the flowing gas stream with both the type S pitot tube and a standard pitot tube with a known coefficient calibrates the pitot tube. Other data needed to calculate the volumetric flow are the stack temperature, stack and barometric pressures, and wet-bulb and dry-bulb temperatures of the gas sample at each traverse.

Table 5.11.2 gives the equations for converting pitot tube readings into velocity and mass flow and shows a typical data sheet for stack flow measurements (Morrow, Brief, and Bertrand 1972).

TABLE 5.11.1 LOCATION OF TRAVERSE POINTS IN CIRCULAR STACKS
(Percent of stack diameter from inside wall to traverse point)

Traverse Point Number on a Diameter	Number of Traverse Points on a Diameter											
	2	4	6	8	10	12	14	16	18	20	22	24
1	14.6	6.7	4.4	3.2	2.6	2.1	1.8	1.6	1.4	1.3	1.1	1.1
2	85.4	25.0	14.6	10.5	8.2	6.7	5.7	4.9	4.4	3.9	3.5	3.2
3		75.0	29.6	19.4	14.6	11.8	9.9	8.5	7.5	6.7	6.0	5.5
4		93.3	70.4	32.3	22.6	17.7	14.6	12.5	10.9	9.7	8.7	7.9
5			85.4	67.7	34.2	25.0	20.1	16.9	14.6	12.9	11.6	10.5
6			95.6	80.6	65.8	35.6	26.9	22.0	18.8	16.5	14.6	13.2
7				89.5	77.4	64.4	36.6	28.3	23.6	20.4	18.0	16.1
8				96.8	85.4	75.0	63.4	37.5	29.6	25.0	21.8	19.4
9					91.8	82.3	73.1	62.5	38.2	30.6	26.2	23.0
10					97.4	88.2	79.9	71.7	61.8	38.8	31.5	27.2
11						93.3	85.4	78.0	70.4	61.2	39.3	32.3
12						97.9	90.1	83.1	76.4	69.4	60.7	39.8
13							94.3	87.5	81.2	75.0	68.5	60.2
14							98.2	91.5	85.4	79.6	73.8	67.7
15								95.1	89.1	83.5	78.2	72.8
16								98.4	92.5	87.1	82.0	77.0
17									95.6	90.3	85.4	80.6
18									98.6	93.3	88.4	83.9
19										96.1	91.3	86.8
20										98.7	94.0	89.5
21											96.5	92.1
22											98.9	94.5
23												96.8
24												98.9

Based on the range of velocity heads, the environmental engineer selects a probe with a properly sized nozzle to maintain isokinetic sampling of particulate matter. As shown in Figure 5.11.7, a converging stream develops at the nozzle face if the sampling velocity is too high. Under this subisokinetic sampling condition, an excessive amount of lighter particles enters the probe. Because of the inertia effect, the heavier particles, especially those in the range of 3μ or greater, travel around the edge of the nozzle and are not collected. The result is a sample indicating an excessively high concentration of lighter particles, and the weight of the solid sample is in error on the low side.

Conversely, portions of the gas stream approaching at a higher velocity are deflected if the sampling velocity is below that of the flowing gas stream. Under this superisokinetic sampling condition, the lighter particles follow the deflected stream and are not collected, while the heavier particles, because of their inertia, continue into the probe. The result is a sample indicating a high concentration of heavier particles, and the weight of the solid sample is in error on the high side.

Isokinetic sampling requires precisely adjusting the sampling rate with the aid of the pitot tube manometer readings and nomographs such as APTD–0576 (Rom). If the pressure drop across the filter in the sampling unit becomes too high, making isokinetic sampling difficult to maintain, the filter can be replaced in the midst of a sample run.

Measuring the concentration of particulate matter requires a sampling time for each run of at least 60 min and

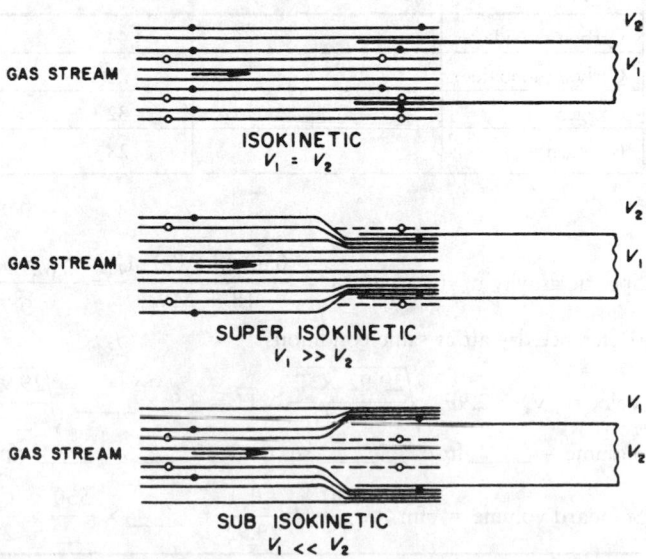

FIG. 5.11.7 Particle collection and sampling velocity.

TABLE 5.11.2 PITOT TUBE CALCULATION SHEET

Stack Volume Data

Stack No. .... Station .... Date .... Page ....

Name of Firm ....

Point	Position, in	Reading, H, in of H_2O	$\sqrt{H}$	Temperature t_b, °F	Velocity, V_b ft/sec
1					
2					
3					
4					
5					
6					
7					
8					
9					
10					
11					
12					
13					
14					
15					
16					
	Totals				
	Average				
	Absolute temperature, $T_s = t_s + 460 =$ °R.				

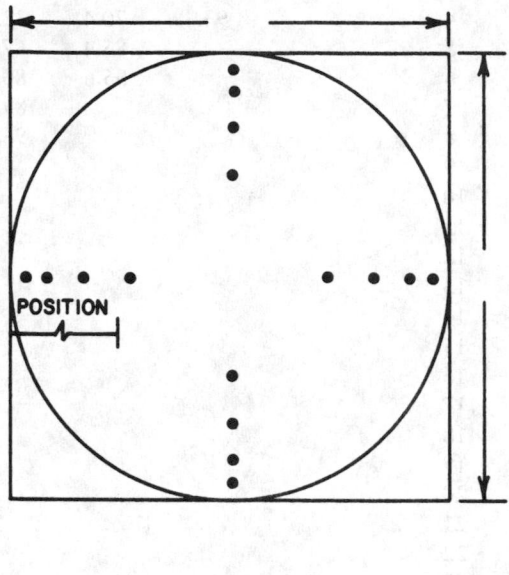

Dry-bulb temperature, $t_d =$ _____ °F

Wet-bulb temperature, $t_u =$ _____ °F

Absolute humidity, $W =$ _____ lb H_2O/lb dry gas

Stack area, $A_s =$ _____ sq ft

Barometer, $P_b =$ _____ in, Hg

Stack gage pressure = _____ in, H_2O

Stack absolute pressure, $P_s =$ _____ $\dfrac{\text{in, } H_2O}{13.6} \pm P_b =$ _____ in, Hg

Pitot correction factor, $F_s =$ _____

Component	Volume Fraction, Dry Basis x Molecular Weight	= Weight Fraction, Dry Basis
Carbon dioxide	44	=
Carbon monoxide	28	=
Oxygen	32	=
Nitrogen	28	=

Average dry gas molecular weight, $M =$ _____

Specific gravity of stack gas, $G_s = \dfrac{0.62\ M\ (W + 1)}{18 + MW} = \dfrac{0.62 \times ___ \times ___}{18 + ___} = $ _____

(Reference dry air at same conditions)

Velocity, $V_s = 2.9 F_s \sqrt{\dfrac{29.92 \times T_s}{P_s \times G_s}}\ \sqrt{H} = 2.9 \times ___ \sqrt{\dfrac{29.92 \times ___}{___ \times ___}}\ \sqrt{H} = $ _____ ft/sec

Volume = _____ ft/sec $\times$ _____ sq. ft. $\times$ 60 _____ = _____ cfm

Standard volume = cfm $\times \dfrac{530}{T_s} \times \dfrac{P_s}{29.92} = $ _____ $\times \dfrac{530}{___} \times \dfrac{___}{29.92} = $ _____ scfm

a minimum volume of 30 dry acfm (51 m³/hr) (U.S. EPA 1974).

Two-Module Sampling Unit

The two-module sampling unit has a separate heated compartment and ice-bath compartment.

HEATED COMPARTMENT

As shown in Figure 5.11.1, the probe is connected to the heated compartment that contains the filter holder and other particulate-collecting devices, such as the cyclone and flask. The filter holder is made of borosilicate glass, with a frit filter support and a silicone rubber gasket. The compartment is insulated and equipped with a heating system capable of maintaining the temperature around the filter holder during sampling at 248 ± 25°F (120 ± 14°C), or at another temperature as specified by the EPA. The thermometer should measure temperature to within 5.4°F (3°C). The compartment should have a circulating fan to minimize thermal gradients.

ICE-BATH COMPARTMENT

The ice-bath compartment contains the system's impingers and bubblers. The system for determining the stack gas moisture content consists of four impingers connected in series as shown in Figure 5.11.1. The first, third, and fourth impingers have the Greenburg–Smith design. The pressure drop is reduced when the tips are removed and replaced with a ½-in (12.5 mm) ID glass tube extending to ½ in (12.5 mm) from the bottom of the flask. The second impinger has the Greenburg–Smith design with a standard tip. During sampling for particulates, the first and second impingers are filled with 100 ml (3.4 oz) of distilled and deionized water. The third impinger is left dry to separate entrained water. The last impinger is filled with 200 to 300 g (7 to 10.5 oz) of precisely weighed silica gel (6 to 16 mesh) that has been dried at 350°F (177°C) for 2 hr to completely remove any remaining water. A thermometer, capable of measuring temperature to within 2°F (1.1°C), is placed at the outlet of the last impinger for monitoring purposes. Adding crushed ice during the run maintains the temperature of the gas leaving the last impinger at 60°F (16°C) or less.

Operating or Control Unit

As shown in Figure 5.11.1, the control unit consists of the system's vacuum pump, valves, switches, thermometers, and the totalizing dry gas meter and is connected by a vacuum line with the last Greenburg–Smith impinger. The pump intake vacuum is monitored with a vacuum gauge just after the quick disconnect. A bypass valve parallel with the vacuum pump provides fine control and permits re-circulation of gases at a low-sampling rate so that the pump motor is not overloaded.

Downstream from the pump and bypass valve are thermometers, a dry gas meter, and calibrated orifice and inclined or vertical manometers. The calibrated orifice and inclined manometer indicate the instantaneous sampling rate. The totalizing dry gas meter gives an integrated gas volume. The average of the two temperatures on each side of the dry gas meter gives the temperature at which the sample is collected. The addition of atmospheric pressure to orifice pressure gives meter pressure.

Precise measurements require that the thermometers are capable of measuring the temperature to within 5.4°F (3°C); the dry gas meter is inaccurate to within 2% of the volume; the barometer is inaccurate within 0.25 mmHg (torr) (0.035 kPa); and the manometer is inaccurate to within 0.25 mmHg (torr) (0.035 kPa).

The umbilical cord is an integrated multiconductor assembly containing both pneumatic and electrical conductors. It connects the two-module sampling unit to the control unit, as well as connecting the pitot tube stack velocity signals to the manometers or differential pressure gauges.

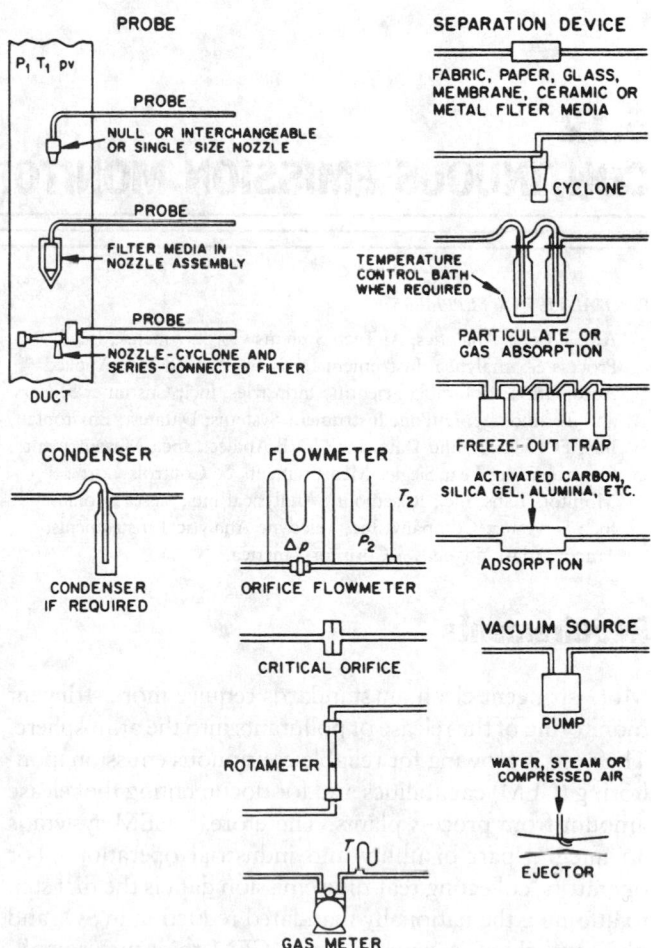

FIG. 5.11.8 Components of common sampling systems.

Sampling for Gases and Vapors

Some commonly used components in stack sampling systems are shown in Figure 5.11.8. If ball-and-socket joints and compression fittings are used, any arrangement of components is readily set up for field use. Environmental engineers select the stack sampling components on the basis of the source to be sampled, the substances involved, and the data needed.

Industrial hygienists developed a summary of sampling procedure outlines for specific substances (Vander Kolk 1980). After considering the complications that might arise from the presence of interfering substances in the gas samples, an environmental engineer should use the procedural outlines as a starting point in assembling a stack sampling system. The American Society for Testing and Materials (1971) provides other recommended sampling procedures for gases and vapors.

—*David H.F. Liu*

References

American Society for Testing and Materials. 1971. *Standards of methods for sampling and analysis of atmospheres.* Part 23.

Morrow, N.L., R.S. Brief, and R.R. Bertrand. 1972. Sampling and analyzing air pollution sources. *Chemical Engineering* 79, no. 2 (24 January): 84–98.

Rom, J.J. *Maintenance, calibration and operation of isokinetic source sampling equipment.* APTD-0576. U.S. EPA.

U.S. Environmental Protection Agency (EPA). 1971. Standards of performance for new stationary sources. *Federal Register* 36, no. 159 (17 August): 15,704–15,722.

———. 1974. Standards of performance for new stationary sources. *Federal Register* 30, no. 116 (14 June): 20,790–820,794.

———. 1977. Standards of performance for new stationary sources, Revision to reference method 1–8. *Federal Register* 42, no. 160 (18 August): 41,754–841,789.

———. 1987. Standards of performance for new stationary sources. *Federal Register* 52, no. 208 (28 October): 41,424–41,430.

———. 1989. Standards of performance for new stationary sources. *Federal Register* 54, no. 58 (28 March): 12,621–126,275.

———. 1990. Standards of performance for new stationary sources. *Federal Register* 55, no. 31 (14 February): 5,211–5,217.

———. 1991. Standards of performance for new stationary sources. *Federal Register* 56, no. 30 (13 February): 5,758–5,774.

Vander Kolk, A.L. 1980. *Michigan Department of Public Health, private communications.* (17 September).

5.12
CONTINUOUS EMISSION MONITORING

PARTIAL LIST OF SUPPLIERS

ABB Process Analytics; Al Tech Systems Corp.; Ametek, Inc., Process & Analytical Instrument Division; Anarad Inc.; Applied Automation; Columbia Scientific Industries, Inc.; Customer Sensors and Technology; DuPont Instrument Systems; Datatest; Enviroplan, Inc.; FCI Fluid; Fluid Data, Inc.; KVB Analect, Inc.; Measurement Control Corp; Lear Siegler Measurements & Controls Corp.; Monitors Labs, Inc.; Rosemount Analytical Inc.; Sierra Monitor Inc.; Servomex Company, Inc; Teledyne Analytical Instruments; Tracor Atlas; Yogogawa Corp. of America.

Requirements

More stringent clean air standards require more stringent monitoring of the release of pollutants into the atmosphere. The need is growing for reliable continuous emission monitoring (CEM) capabilities and for documenting the release amount from process plants. Therefore, a CEM system is an integral part of utility and industrial operations. For operators, collecting real-time emission data is the first step to attaining the nationally mandated reduction in SO_x and NO_x emissions. A company uses CEM to ensure compliance with the Acid Rain Program requirements of the CAAA.

A CEM system is defined by the U.S. EPA in Title 40, Part 60, Appendix B, Performance Specification 2 of the *Code of Federation Regulations,* as all equipment required to determine a gas concentration or emission rate. The regulation also defines a CEM system as consisting of subsystems that acquire, transport, and condition the sample, determine the concentration of the pollutant, and acquire and record the results. For the measurement of opacity, the specifics of the major subsystems are slightly different but basically the same.

The EPA has codified the standards of performance, equipment specifications, and installation and location specifications for the measurement of opacity, total reduced sulfur (TRS), sulfur dioxide, carbon dioxide, and hydrogen sulfide. These standards include requirements for the data recorder range, relative accuracy, calibration drift and frequency of calibration, test methods, and quarterly and yearly audits. The regulations also require opacity to be measured every 10 sec, the average to be recorded every 6 min, and pollutants to be measured a minimum of one cycle of sampling, analyzing, and data recording every 15 min. The readings from the gas analyzer must agree with a stack sampler to within 20% relative accuracy.

This section provides an overview of the CEM technology, the components of a proper analysis system, and some details in the reliable and accurate operation of CEM.

System Options

Figure 5.12.1 shows the various CEM options. All systems fall into one of two categories: in situ or extractive. Extractive units can be further classified into full-extractive (wet or dry basis) or dilution-extractive systems.

IN SITU SYSTEMS

An in situ system measures a gas as it passes by the analyzer in a stack. Figure 5.12.2 shows an in situ probe-type analyzer. The measurement cavity is placed directly into the sample flow system to measure the gas received on a wet basis.

The most commonly accepted in situ analyzer is the zirconium oxide (ZrO_2) oxygen analyzer. It is the most reliable method for measuring and controlling a combustion process. Since the introduction of ZrO_2 analyzers, many other gases have been measured in situ with light absorption instrumentation, such as UV and IR spectrometers. The gases that can be measured with such spectrometers include CO, CO_2, SO_2 and NO.

Most of the first analyzers designed to measure these gases have disappeared. Most noticeably, across-the-stack technology units are no longer used in this country due primarily to the promulgation of Title 40, Part 60,

Appendix F of the *Code of Federal Regulations* by the EPA in 1986. Appendix F requires that an analyzer be able to complete gas calibrations and that it be certified quarterly. An across-the-stack unit works like an opacity instrument in that it lacks the capability of using protocol gases for cylinder gas audits (CGA) as allowed by the EPA under Appendix F.

EXTRACTIVE SYSTEMS

Extractive CEM systems can be configured for either dry- or wet-basis measurements. Both configurations can achieve the CAAA-required 10% relative accuracy.

Dry-Extractive Systems

A standard extractive system (see Figure 5.12.3) extracts a gas sample from the stack and delivers it to an analyzer cabinet through heated sample lines. Filtration removes particulates at the sample probe. (However, uncontrolled condensation can block the sample line because it cannot trap and hold fines that pass through the stack filter.) The dry-extractive CEM system removes moisture through a combination of refrigeration, condensation, and permeation tube dryers that pass the sample through water-excluding membranes. This system helps keep the gas analyzer dry and removes any interferences caused by water.

To convert the volumetric concentration into a mass emission rate when using velocity-measuring flow moni-

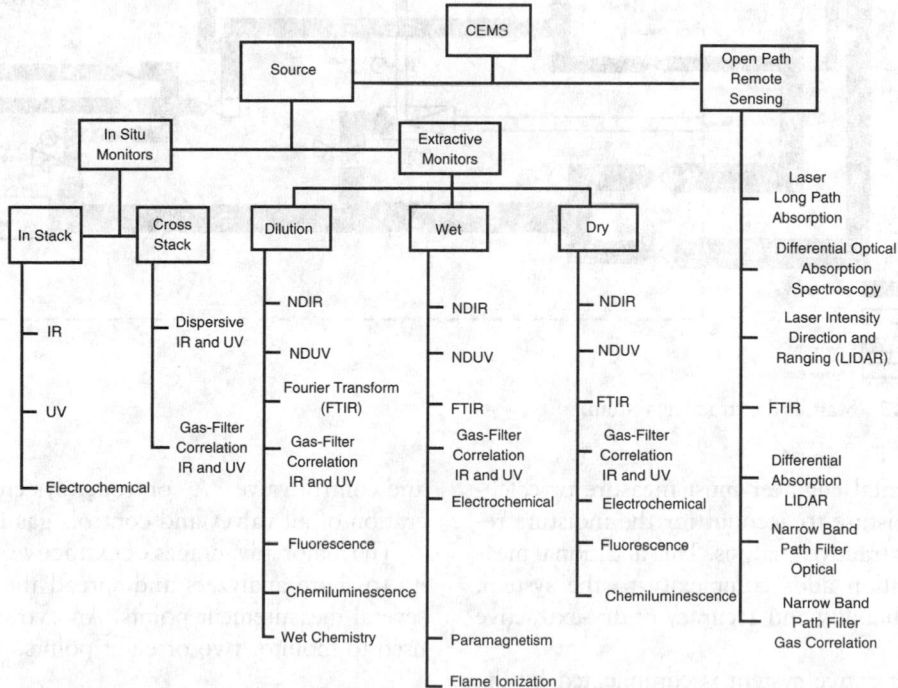

FIG. 5.12.1 CEM options. (Reprinted, with permission, from J. Schwartz, S. Sample, and R. McIlvaine, 1994, Continuous emission monitors—Issues and predictions, *Air & Waste*, Vol. 44 [January].)

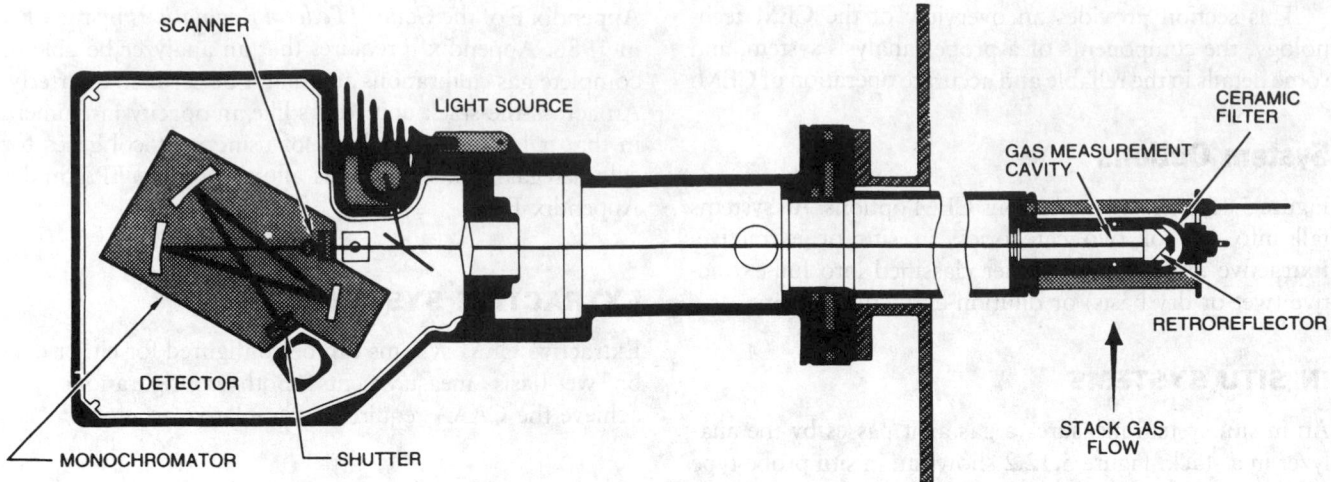

FIG. 5.12.2 Probe-type stack gas analyzer. (Reprinted, with permission, from Lear Siegler Inc.)

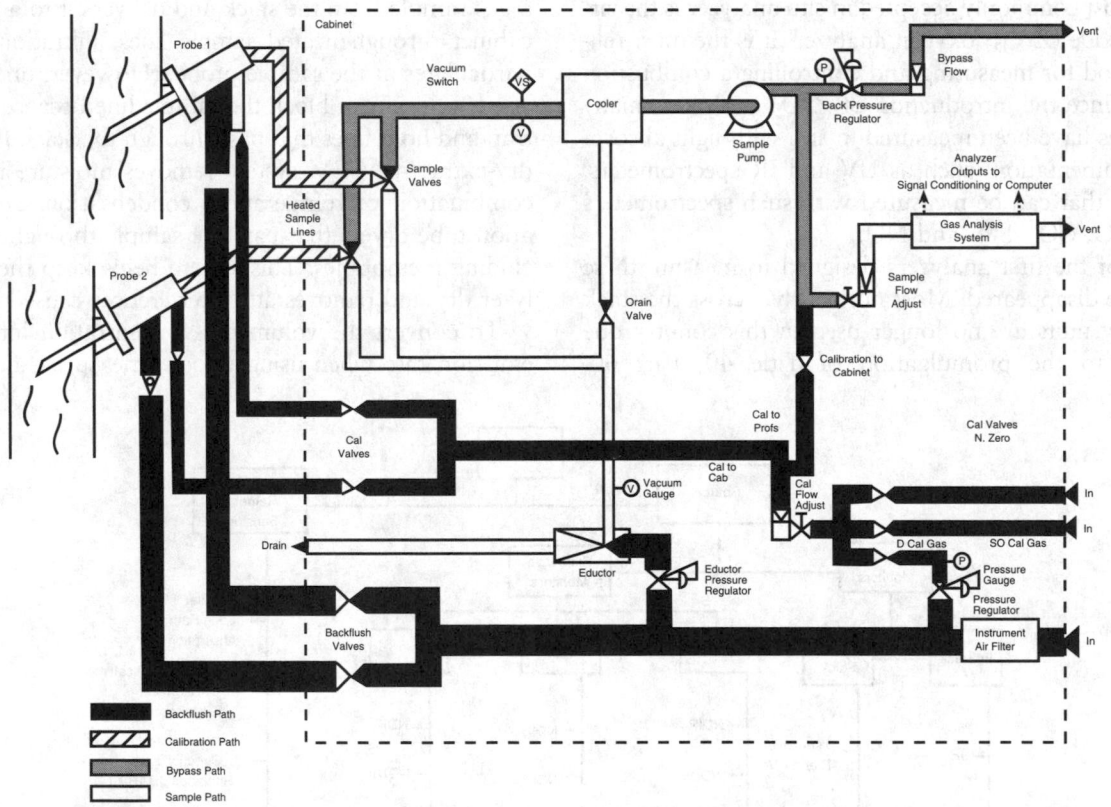

FIG. 5.12.3 Standard extractive system.

tors, the environmental engineer must measure or calculate the flue-gas moisture to account for the moisture removed before dry-extractive analysis. This additional measurement or calculation adds complexity to the system, which lowers the reliability and accuracy of dry-extractive CEM systems.

Operating an extractive system is complicated by the daily calibration, zero and span checks of the analyzers, and the need to backpurge the sample handling system to clear the probe and filters. These tasks are accomplished with additional lines and connections to the stack probe and control valves. A controller system sequences the operation of all valves and controls gas flows.

The major advantages of extractive systems is their ability to share analyzers and spread the analyzer cost over several measurement points. An extractive system can be used to monitor two or eight points.

Wet-Extraction Systems

Wet-extraction CEM systems are similar to dry-extraction ones except that the sample is maintained hot, and moisture in the flue gas is retained throughout. Heat tracing is the critical component of a wet-extractive system. Sample temperatures must be maintained between 360° and 480°F to prevent acid gases from condensing within the sampling lines and analyzer.

Since nothing is done to the sample before analysis, wet-extractive CEM has the potential of being the most accurate measurement technique available. However, only a limited number of vendors supply wet-extractive CEM systems. In addition, increased flexibility and accuracy makes the wet-extractive CEM systems more costly than dry-extractive ones.

Dilution-Extractive Systems

The newest, accepted measurement system is dilution-extractive CEM. By precisely diluting the sample system at stack temperature with clean, dry (lower than −40°F dew-point) instrument air, dilution-extractive systems eliminate the need for heat tracing and conditioning of extracted samples. Particulates are filtered out at the sample point. Thus, a dilution system measures all of the sample along with the water extracted with the sample on a wet basis. Figure 5.12.4 shows a schematic of a dilution-extractive system.

The dilution-extraction CEM system uses a stack dilution probe (see Figure 5.12.5) designed in Europe. A precisely metered quantity of flue gas is extracted through a critical orifice (or sonic orifice) mounted inside the probe. Dilution systems deliver the sample under pressure from the dilution air to the gas analyzers. Thus, the system protects the sample from any uncontrolled dilution from a leak in the sample line or system.

The environmental engineer selects the dilution ratios based on the expected water concentration in the flue gas and the lower limit of the ambient air temperature to avoid freezing. This kind of sample line is still considerably less expensive than the heated sample line for wet-extractive CEM.

When choosing the dilution ratio, the engineer must also consider the lowest pollutant concentration that can be detected by the monitoring device. These systems use a dilution ratio ranging from 12:1 to more than 700:1. Irrespective of the dilution ratio, the diluted sample must match the analyzer range.

SYSTEM SELECTION

The majority of the systems purchased for existing utility boilers are the dilution-extractive type. Beyond the utility industry, the dry-extractive, wet-extractive, and dilution-extractive systems are all viable choices. The accuracy, reliability, and cost differences must be considered in the selection of a system for a specific application. The CEM system selection process can also be limited by local regulatory agencies. These restrictions can include preapproval procedures that limit the type of system an agency accepts. For example, with some types of emission monitors, some states require that only dedicated, continuous measurement are used; that is no time-sharing. They can also define such terms as "continuous." This factor directly defines the requirements for a data acquisition system.

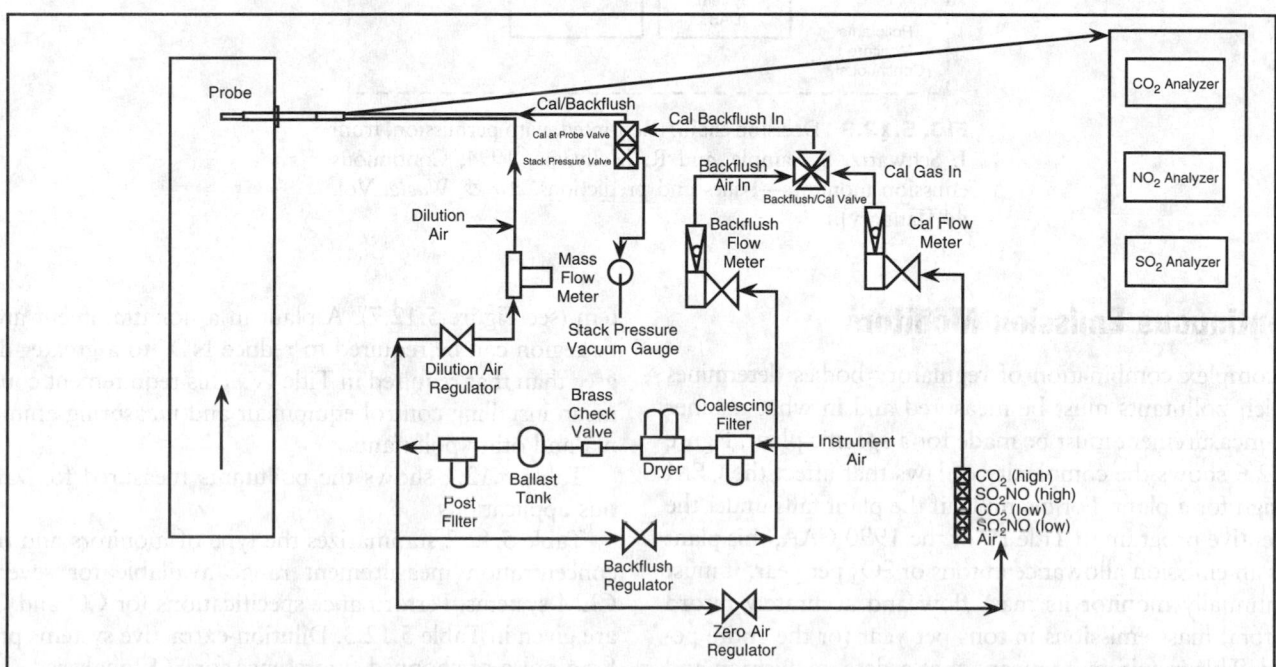

FIG. 5.12.4 Dilution-extractive system schematic. (Reprinted, with permission, from Lear Siegler.)

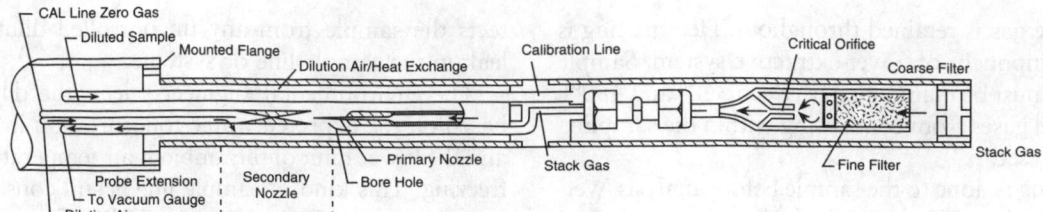

FIG. 5.12.5 Dilution probe.

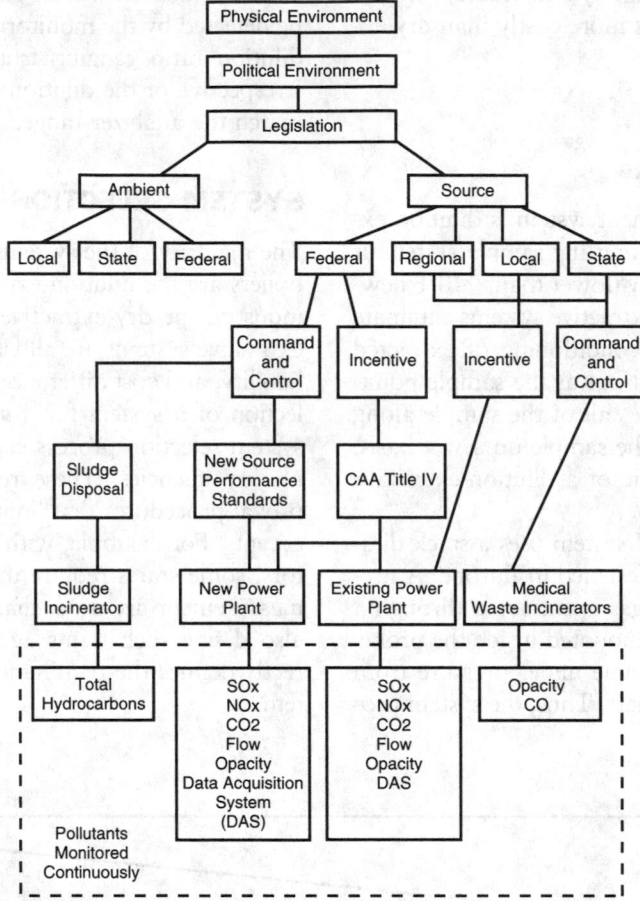

FIG. 5.12.6 Decision chain. (Reprinted, with permission, from J. Schwartz, S. Sample, and R. McIlvaine, 1994, Continuous emission monitors—Issues and predictions, *Air & Waste,* Vol. 44 [January].)

Continuous Emission Monitors

A complex combination of regulatory bodies determines which pollutants must be measured and in what manner the measurement must be made for a specific plant. Figure 5.12.6 shows the complexity of laws that affect the CEM design for a plant. For example, if the plant falls under the incentive program in Title IV of the 1990 CAA, this plant has an emission allowance in tons of SO_2 per year. It must continually monitor its mass flow and accurately record its total mass emissions in tons per year for the entire period. This requirement means that a data acquisition and reporting system must be incorporated into the CEM sys-

tem (see Figure 5.12.7). A plant in a nonattainment area or region can be required to reduce NO_x to a greater degree than that required in Title IV. This requirement could mean installing control equipment and measuring ammonia and other pollutants.

Table 5.12.1 shows the pollutants measured for various applications.

Table 5.12.2 summarizes the type of monitors and the concentration measurement range available for several CEM systems. Performance specifications for CO and O_2 are given in Table 5.12.3. Dilution-extractive systems prefer a pulse of chopped fluoresence for SO_2 analysis.

NO is generally measured using chemiluminescence (see

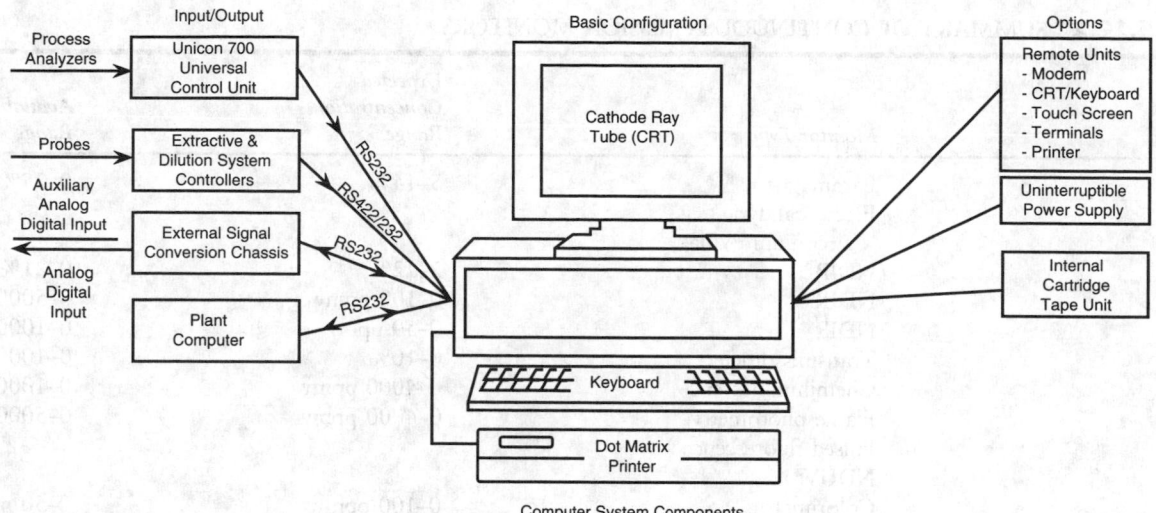

FIG. 5.12.7 Data acquisition and reporting computer system.

TABLE 5.12.1 MEASUREMENT REQUIREMENTS

	SO_2	TRS	NO_x	THC	HCl	CO	CO_2	O_2	NH_3	Flow	Opacity	DAS
Utility												
New gas			1			1		1	*			1
New coal	2		1			1	1	1	*	1	1	1
Part 75	1		1					1	*	1		1
Incinerators												
Municipal	2		1			2	1	2			1	1
Sewage				1								1
Hazardous	1		1	1	1	1		1			1	1
Pulp & Paper												
Recovery boiler		1						1			1	1/3
Lime kiln		1						1			1	1/3
Power boiler	1		1				†	1			1	1/3

Source: J. Schwartz, S. Sample, and R. McIlvaine, 1994, Continuous emission monitors—Issues and predictions, *Air & Waste*, Vol. 44 (January).
Notes: *Ammonia monitor required if selective catalytic reduction (SCR) or selective noncatalytic reduction (SNCR) is utilized.
†Can be substituted for O_2.

Figure 5.12.8). The term NO_x refers to all nitrogen oxides; however, in air pollution work, NO_x refers only to NO (about 95–98%) and NO_2. Because NO is essentially transparent in the visible and UV light region, it must be converted to NO_2 before it can be measured. NO can then react with ozone to form NO_2 with chemiluminescence. The light emitted can be measured photometrically as an indication of the reaction's extent. With excess ozone, the emission is in proportion to the amount of NO. Including NO_2 (which reacts slowly with ozone) involves bringing the sample gas in contact with a hot molybdenum catalyst. This technique converts all NO_2 to NO before it reacts with ozone.

LOCATION OF SYSTEMS

Title 40, Part 60 of the *Code of Federal Regulations* includes guidelines for instrument location within specific performance specifications. GEP dictates a location sepa-

rate from bends, exits, and entrances to prevent measurements in areas of high gas or material stratification. Environmental engineers can determine the degree of stratification of either the flowrate or pollution concentration by taking a series of preliminary measurements at a range of operating conditions. For gas measurements, certification testing confirms the proper location. Under the certification, readings from the gas analyzer must agree with the stack sampler measurements to within 20% relative accuracy (RA).

To convert the volumetric pollution concentration (ppmv) measured by the CEM system to the mass emission rate (lb/hr) required by the CAAA, an environmental engineer must measure the flowrate. Proper placement of the sensor in the flue-gas stream is a key requirement. The goal is to measure flow that represents the entire operation at various boiler loads. Measurement must compare favorably with the standard EPA reference methods to demonstrate the RA of the unit.

TABLE 5.12.2 SUMMARY OF CONTINUOUS EMISSION MONITORS

Pollutant	Monitor Type	Expected Concentration Range	Available Range[a]
O_2	Paramagnetic	5–14%	0–25%
	Electrocatalytic (e.g., zirconium oxide)		
CO_2	NDIR[b]	2–12%	0–21%
CO	NDIR	0–100 ppmv	0–5000 ppmv
HCl	NDIR	0–50 ppmv	0–10000 ppmv
Opacity	Transmissometer	0–10%	0–100%
NO_x	Chemiluminescence	0–4000 ppmv	0–10000 ppmv
SO_2	Flame photometry	0–4000 ppmv	0–5000 ppmv
	Pulsed fluorescence		
	NDUV[c]		
SO_3	Colorimetric	0–100 ppmv	0–50 ppmv
Organic compounds	GC (FID)[d]	0–50 ppmv	0–100 ppmv
	GC (ECD)[e]		
	GC (PID)[f]		
	IR absorption		
	UV absorption		
HC	FID	0–50 ppmv	0–100 ppmv

Source: E.P. Podlenski et al., 1984, *Feasibility study for adapting present combustion source continuous monitoring systems to hazardous waste incinerators,* EPA-600/8-84-011a, NTIS PB 84-187814, U.S. EPA (March).

Notes: [a]For available instruments only. Higher ranges are possible through dilution.
[b]Nondispersion infrared.
[c]Nondispersion ultraviolet.
[d]Flame ionization detector.
[e]Electron capture detector.
[f]Photo-ionization detector.

TABLE 5.12.3 PERFORMANCE SPECIFICATIONS FOR CO AND O_2 MONITORS

Parameter	CO Monitors		O_2 Monitors
	Low Range	High Range	
Calibration drift (CD) 24 hr	≤6 ppm[a]	≤90 ppm	≤0.5% O_2
Calibration error (CE)	≤10 ppm[a]	≤150 ppm	≤0.5% O_2
Response time RA[b]	≤2 min ([c])	≤2 min ([c])	≤2 min (incorporated in CO and RA calculation)

Source: U.S. Environmental Protection Agency (EPA), 1990, *Method manual for compliance with BIF regulations,* EPA/530-SW-91-010, NTIS PB 90-120-006.

Notes: [a]For Tier II, CD and CE are ≤3% and ≤5% of twice the permit limit, respectively.
[b]Expressed as the sum of the mean absolute value plus the 95% confidence interval of a series of measurements.
[c]The greater of 10% of the performance test method (PTM) or 10 ppm.

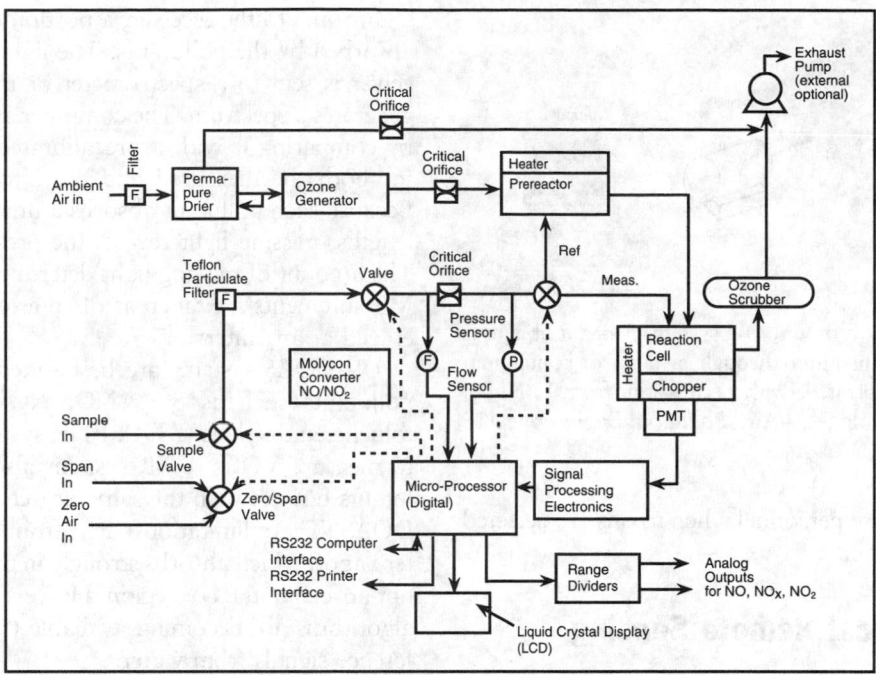

FIG. 5.12.8 System schematic of nitrogen oxides analyzer.

MAINTENANCE OF SYSTEMS

Because downtime can result in a fine, designing and engineering CEM systems to maximize reliability is essential. Certification follows start up and commissioning. Once a CEM system is running, users must devote appropriate resources to the ongoing maintenance and calibration requirements.

Proper maintenance procedures require periodically inspecting the probe and replacing the filter material. A good blowback system increases the time interval between these maintenance inspections.

Cochran, Ferguson, and Harris (1993) outline ways of enhancing the accuracy and reliability of CEM systems.

—David H.F. Liu

Reference

Cochran, J.R., A.W. Ferguson, and D.K. Harris. 1993. Pick the right continuous emissions monitor. *Chem. Eng.* (June).

5.13
REMOTE SENSING TECHNIQUES

PARTIAL LIST OF SUPPLIERS

Altech Systems Corp.; Ametek, Process and Analytical Instruments Div.; Extrel Corp.; The Foxboro., Environmental Monitoring Operations; KVB Analect, Inc.; MDA Scientific, Inc.; Nicolet Instrument Corp.; Radian Corp.; Rosemount Analytical, Inc.

The open-path, optical remote sensing techniques make new areas of monitoring possible. These techniques measure the interaction of light with pollutants in real time (see Figure 5.13.1). In an open-path configuration, path lengths—the distance between a light source and a reflecting device—can range to kilometers. Thus, these techniques are ideal for characterizing emission clouds wafting over process areas, storage tanks, and waste disposal sites. Open-path optical sensing systems can monitor gas concentrations at the fence lines of petrochemical plants and hazardous waste sites.

Wide-area systems are also being used indoors for emergency response and to demonstrate compliance with OSHA regulations. When used indoors, these systems can quickly detect routine or accidental releases of many chemical species. They are frequently fitted to control systems

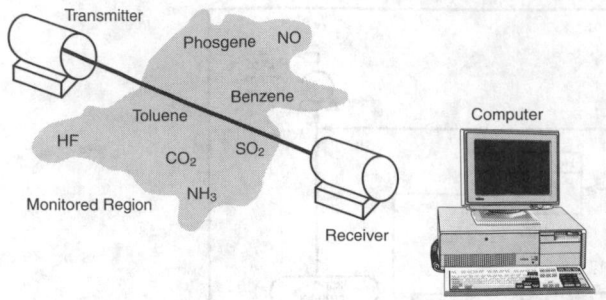

FIG. 5.13.1 Open-path monitoring system. A beam of light is transmitted from a transmitter through an area of contamination to a receiver. (Reprinted, with permission, from L. Nudo, 1992, Emerging technologies—Air, *Pollution Engineering* [15 April].)

and alarms to evacuate personnel when toxins are detected above a threshold limit.

Open-Path Optical Remote Sensing Systems

Table 5.13.1 describes several sensing techniques. Of the techniques listed, the FTIR and UV differential optical absorption spectroscopy (UV-DOAS) are the most widely publicized. They both measure multiple compounds in open areas in near real time.

These systems use a high-pressure xenon lamp in the UV-DOAS or a grey-body source in the FTIR to produce a broadband light that is collimated by a telescope into a narrow beam. This beam is then transmitted through the area of contamination to a receiver. The transmission path can range from several feet, in stacks, to 1 to 3 km, in ambient air over large areas (see Figure 5.13.1).

In route to the receiver, a portion of the light energy is absorbed by the pollutants. The light that reaches the receiver is sent to a spectrometer or interferometer, which generates a spectrum. The computer analyzes the spectrum by comparing it with a precalibrated reference spectrum for both the measured and the interfering components. Because each pollutant absorbs a unique pattern of wavelengths, missing light reveals the presence of a pollutant. The amount of missing light determines its concentration. Measurements are taken as often as every minute and averaged at any interval.

UV-DOAS systems are best suited to monitor criteria pollutants such as SO_2, NO, NO_2, O_3 and benzene, toluene, and xylene (BTX). FTIR systems are better suited to measure VOCs. FTIR systems also detect criteria pollutants but not with the same degree of sensitivity as UV-DOAS. These limitations stem from interferences of water vapor, which absorbs strongly in the IR region but does not absorb in the UV region. However, powerful software algorithms are becoming available to filter out the interference signals from water.

In general, more chemicals absorb light in the IR spectrum than in the UV spectrum. Thus, FTIR systems measure a broader range of compounds (more than fifty) than UV-DOAS systems, which detect about twenty-five (Nudo 1992). With the aid of a contractor, the EPA has developed reference IR spectra for approximately 100 of the 189 HAPs listed under Title III. The maximum number of HAPs to which FTIR might ultimately be applicable is about 130 (Schwartz, Sample, and McIlvaine 1994). The EPA plans to initiate its own investigation into this area. Additional investigation into the potential of FTIR in continuous applications is underway at Argonne National Labs.

TABLE 5.13.1 REMOTE SENSING TECHNIQUES FOR AIR ANALYSIS

Method	Operating Principle
FTIR	A broadband source allows collection and analysis of a full IR spectrum. This method compares standard spectra with field readings to determine constituents and concentrations
UV spectroscopy	UV spectra are collected over a limited absorption spectral region. This method uses the differential absorptions of the compounds in the air to determine the identity and concentrations of contaminants
Gas-filter correlation	A sample of the gas to be detected is used as a reference. This method compares the correlation between the spectrum of the sample gas to the gas in the measurement path to determine concentration
Filtered band-pass absorption	This method measures absorption of the gas in certain bands to detect composition and concentration
Laser absorption	This method uses one or more lasers to measure absorption at different wavelengths
Photoacoustic spectroscopy	The pressure change from the deactivation of excited molecules is measured in a closed acoustic chamber
LIDAR	This method measures molecular or aerosol backscatter by either differential absorption or Raman scattering to identify gases and determine their concentrations. Unlike the other methods listed, this method provides ranging information on measurements
Diode-laser spectroscopy	A developing technology for open-air use. This method scans a line feature of the gas spectrally to identify and quantify the compound

Source: G.E. Harris, M.R. Fuchs, and L.J. Holcombe, 1992, A guide to environmental testing, *Chem. Eng.* (November).

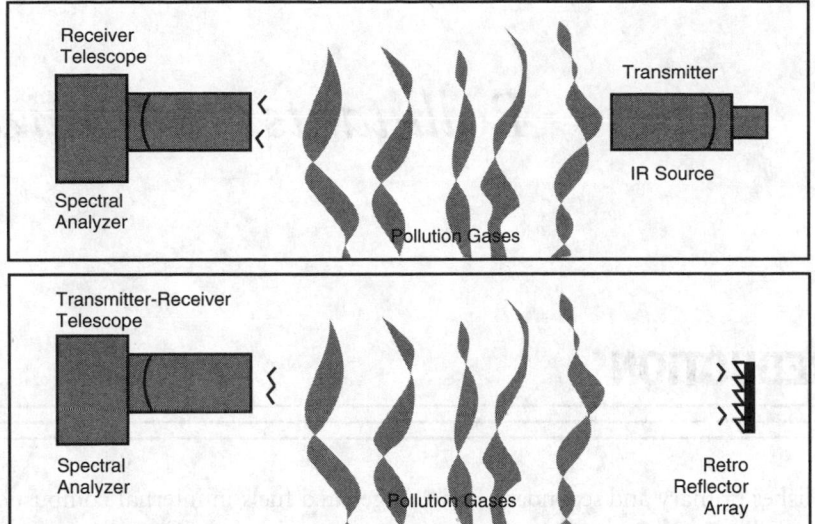

FIG. 5.13.2 Bistatic and unistatic systems. Bistatic systems (top), which place the IR source and receiver in different positions, require two power sources and precise alignment to record properly. Unistatic systems (bottom) are simpler to operate. Most installations use four separate bi- or unistatic source–receiver pairs—one for each side of a property. The next step may be to use one centrally located IR source on a tower, with retroreflectors around it.

UV-DOAS systems are more selective. They have a lower limit of detection and fewer problems with interferences from water vapor and carbon dioxide. For the best features of both systems, these complimentary systems can be integrated.

Instrumentation

The instrumentation used in remote sensing techniques includes bistatic systems and unistatic systems.

BISTATIC SYSTEMS

Early bistatic design (see the top of Figure 5.13.2) placed the light source and the receiver at opposite ends of the area to be monitored. Precise alignment, to assure that the receiver gets the full strength of the transmitted signal, is often a problem. Therefore, two separate electrical sources are usually required to operate these systems (Shelley 1991).

UNISTATIC SYSTEMS

A new unistatic design (see the bottom of Figure 5.13.2), developed by MDA Scientific, uses a single telescope that acts as both source and receiver. The IR beam traverses the area being monitored and returns to the receiver from an array of cubed mirrors, or retro-reflectors, on the opposite end of the field.

These clusters collect and refocus the beam and, unlike flat-faced mirrors, need not be precisely aligned to achieve near-perfect reflection along a light beam's original path.

This design makes the unistatic system easier to set up and reduces error during analysis.

LIMITATIONS

With open-path systems, sophisticated, meteorological modeling software is necessary to integrate information about regional topography and moment-to-moment changes in atmospheric conditions. Such modeling programs are still in the early stages. Therefore, discrete-point sampling is needed to confirm optical sensing results. Another shortcoming of using open-path systems is the assumption that the distribution of molecules in the beam is homogeneous. Path-averaging assumptions can underestimate a localized problem.

Despite some remaining problems, the use of perimeter monitoring is rising for documenting cumulative emissions over process facilities. The EPA is developing procedures to demonstrate equivalence with reference test methods for use with some toxic gases (Schwartz, Sample, and McIlvaine 1994). These procedures are required before regulatory agencies and industry can adopt optical techniques.

—*David H.F. Liu*

References

Nudo, L. 1992. Emerging technologies—Air. *Pollution Engineering* (15 April).

Schwartz, J., S. Sample, and R. McIlvaine. 1994. Continuous emission monitors—Issues and predictions. *Air & Waste,* Vol. 44 (January).

Shelley, S. 1991. On guard! *Chem. Eng.* (November): 31–39.

Pollutants: Minimization and Control

5.14
POLLUTION REDUCTION

The CAA of 1970 establishes primary and secondary standards for criteria pollutants. Primary standards protect human health, while secondary standards protect materials, crops, climate, visibility, and personal comfort. The standard for total suspended particulates (TSP) is an annual geometric mean of 75 $\mu g/m^3$ and a maximum 24 hr average of 260 $\mu g/m^3$. This standard is currently being reviewed by the U.S. EPA for possible change.

Prevention of air pollution from industrial operations starts within the factory or mill. Several alternatives are available to prevent the emission of a pollutant including:

- Selecting process inputs that do not contain the pollutant or its precursors
- Operating the process to minimize generation of the pollutant
- Replacing the process with one that does not generate the pollutant
- Using less of the product whose manufacture generates the pollutant
- Removing the pollutant from the process effluent.

Raw Material Substitution

Removing some pollutants involves simply substituting materials which perform equally well in the process but which discharge less harmful products to the environment. This method of air pollution reduction usually produces satisfactory control at a low cost.

Typical examples are the substitution of high-sulfur coal with low-sulfur coal in power plants. This substitution requires little technological change but results in a substantial pollution reduction. Changing to a fuel like natural gas or nuclear energy can eliminate all sulfur emissions as well as those of particulates and heavy metals. However, natural gas is more expensive and difficult to ship and store than coal, and many people prefer the known risks of coal pollution than the unknown risks of nuclear power. Coal gasification also greatly reduces sulfur emissions. Another example is substituting gasoline with ethanol or oxy-genated fuels in internal combustion engines. This substitution reduces the O_3 pollution in urban areas. Alternative energy sources, such as wind or solar power, are being explored and may become economically feasible in the future.

Process Modification

Chemical and petroleum industries have changed dramatically by implementing automated operations, computerized process control, and completely enclosed systems that minimize the release of materials to the outside environment.

A process modification example is industry reducing the oxidation of SO_2 to SO_3 by reducing excess air from ~20% to <1% when burning coal, resulting in reduced sulfuric acid emissions. However, this process change has increased fly ash production.

Powders and granulated solids are widely used in industry. Handling these materials at locations such as transfer points or bagging and dumping operations generates dust that can affect worker health. Plinke et al. (1991) found that the amount of dust generated by an industrial process depends on the size distribution of the dust, the ratio of impaction forces that disperse the dust during material handling, and the cohesion properties of the dust, such as moisture content. The implication is that by modifying both the material handling processes and the properties of the powders, the dust generated can be reduced.

Similar approaches are being used in municipal trash incinerators which emit carcinogenic dioxins. By adjusting the temperature of incineration, dioxin emissions can be prevented.

Marketing Pollution Rights

Some economists view absolute injunctions and rigid limits as counterproductive. They prefer to rely on market mechanisms to balance costs and effects and to reduce pollution. Corporations can be allowed to offset emissions by buying, selling, and banking pollution rights from other factories at an expected savings of $2 to 3 billion per year. This savings may make economic sense for industry and even protect the environment in some instances, but it may be disastrous in some localities.

Demand Modification

Demand modification and lifestyle changes are another way to reduce pollution. To combat smog-causing emissions, some districts in California have proposed substantial lifestyle changes. Aerosol hair sprays, deodorants, charcoal lighter fluid, gasoline-powered lawn mowers, and drive-through burger stands could be banned. Clean-burning, oxygenated fuels or electric motors would be required for all vehicles. Car-pooling would be encouraged, and parking lots would be restricted.

Gas Cleaning Equipment

Industry controls air pollution with equipment that removes contaminants at the end of the manufacturing process. Many such devices exist and are described in the next sections. These devices display two characteristics: the size and cost of the equipment increase with the volume of gas cleaned per unit of time, and the cost of removing a contaminant rises exponentially with the degree of removal. The degree of pollutant removal is described as collection efficiency, while the operating cost depends on the pressure drop across the unit and its flow capacity.

—*Gurumurthy Ramachandran*

Reference

Plinke, M.A.E., D. Leith, D.B. Holstein, and M.G. Boundy. 1991. Experimental examination of factors that affect dust generation. *Am. Ind. Hyg. Assoc. J.* 52, no. 12:521–528.

5.15
PARTICULATE CONTROLS

Control Equipment

If waste reduction is not possible, either by process or material change, waste particulate material must be removed from the process air stream. Selecting the type of control equipment depends largely on the characteristics of the particulate material to be removed. Factors such as the physical form of the particulates (whether solid or liquid), particle size and size distribution, density and porosity of the particulates, and particle shape (spheres, fibers, or plates) affect particle behavior.

Control equipment can be divided into three broad classes: gravitational and inertial collectors, electrostatic precipitators (ESPs), and filters. Included in the category of gravitational and inertial collectors are wet and dry scrubbers, gravity settling chambers, and cyclone collectors. ESPs include both single and two-stage units. Finally, filters encompass a variety of media ranging from woven and nonwoven fabric, fiber, and paper media. In addition to media, filters are often classified according to the method of cleaning.

Particulate Size

Although particulate material varies in both size and composition, certain particle types are usually associated with a range of particle sizes. For example, particulate material produced by crushing or grinding or resuspended from settled dusts is generally made up of particles with diameters larger than 1 μm. On the other hand, particulate material produced by condensation or gas-phase chemical reactions is comprised of many small particles, all much smaller than 1 μm in diameter. Figure 5.15.1 illustrates typical particle diameters for a variety of substances, including particulates such as beach sand and pollens. For comparison, this figure also includes the size range for types of electromagnetic radiation, estimates of gas molecule diameters, size ranges for fogs, mists and raindrops, and ranges for inspirable particles.

Other Factors

Generally, particle size is the most important factor in the selection of a collector since the range of particle sizes to be collected strongly affects the expected cost and efficiency. In some instances, other factors such as serviceability or the pressure drop across the collector are the most important factors. Table 5.15.1 lists the characteristics of various particulate removal equipment and compares some of these features. For the most part, the collection devices that are the least expensive to install and operate are also the least efficient.

Table 5.15.2 also compares the collection characteristics and total annual costs of various air cleaning equip-

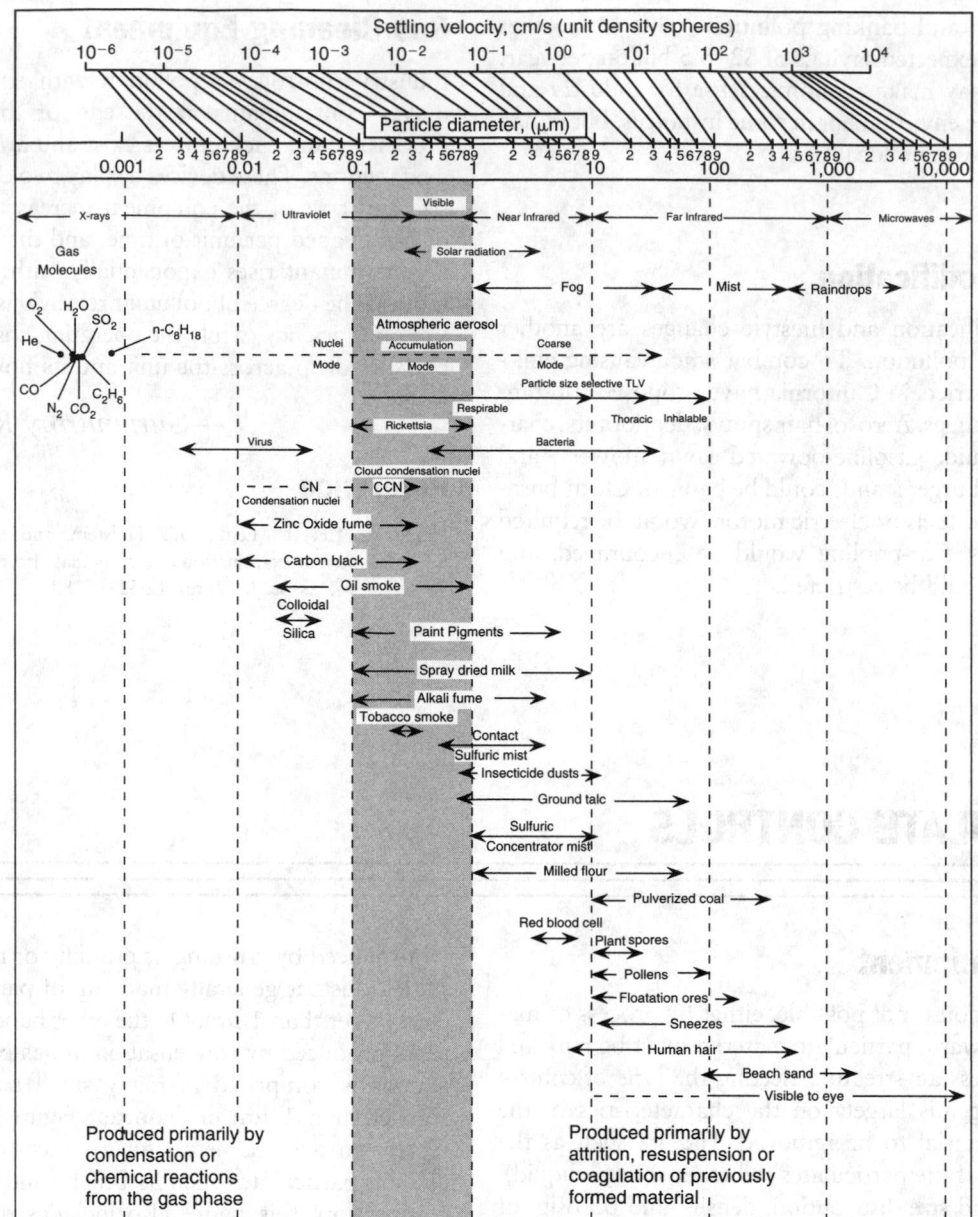

FIG. 5.15.1 Molecular and aerosol particle diameters, copyright © P.C. Reist. Molecular diameters calculated from viscosity data. See Bird, Stewart, and Lightfoot, 1960, *Transport phenomona,* Wiley. (Adapted from Lapple, 1961, *Stanford Research Institute Journal,* 3d quarter; and J.S. Eckert and R.F. Strigle, Jr., 1974, *JAPCA* 24:961–965.)

TABLE 5.15.1 CHARACTERISTICS OF AIR AND GAS CLEANING METHODS AND EQUIPMENT FOR COLLECTING AEROSOL PARTICLES

Characteristics	Gravitational and Inertial Collectors				ESPs		Filters			
	Dry		*Wet*		*2-Stage Low Voltage*	*1-Stage High Voltage*	*Fabric*		*Ventilation*	*Absolute*
	Settling Chamber	*Cyclone Chamber*	*Low Energy*	*High Energy*			*Inside Collector — Reverse Gas Cleaning*	*Outside Collector — Pulse-Jet Cleaning*		
Type			*Wetted Cyclone*	*Venturi*	*Electronic Air Cleaner*	*Wire and Plate*	*Shaker Cleaning*		*Pleated Paper Media*	*Paper*
			Impingement & Entrainment	*Disintegrator*	*Furnace Air Cleaner*	*Wire and Tube*				*Deep Bed*
Contaminants	Crushing, grinding, machining	Crushing, grinding, machining	Crushing, grinding, machining	Metallurgical fumes	Room air, oil mist	Fly ash acid mist	All dry dusts	All dry dusts	Room air	Precleaned room air
Loadings (g/m^3)	0.1–100	0.1–100	0.1–100	0.1–100	<0.1	0.1–10	0.1–20	0.1–20	<0.01	<0.001
Overall efficiency, %	High for >10 μm	High for >10 μm	High for >10 μm	High for >1 μm	High for >0.5 μm	High for >0.5 μm	High for all sizes	High for all sizes	High for >5 μm	High for all sizes
Pressure drop (mm Hg)	4–20	4–20	4–20	15–200	4–10	4–20	5–20	5–20	2–10	2–4
Initial cost	Low	Low	Moderate	Moderate	Moderate	High	Moderate to high	Moderate	Low	Moderate
Operating cost	Moderate	Moderate	Moderate	High	Low	Moderate	Moderate	Moderate	Low	High
Serviceability	Good (erosion)	Good (corrosion)	Good (corrosion)	Good (corrosion)	Poor (shorts)	Fair (shorts)	Fair to good	Fair to good (blinding)	Fair	Fair to poor
Limitations	Efficiency	Disposal	Operating cost	Operating cost	Loading	Resistivity	Temperature	Temperature	Loading	Loading

Source: M.W. First and D. Leith, 1994, Personal communication.

TABLE 5.15.2 COMPARISON OF PARTICULATE REMOVAL SYSTEMS

| Unit | Collection Characteristics | | | Space Required | Cost[a] $/yr/m³ |
	0.1–1 μm	1–10 μm	10–50 μm		
Standard cyclone	Poor	Poor	Good	Large	7
High-efficiency cyclone	Poor	Fair	Good	Moderate	11
Baghouse (cotton)	Fair	Good	Excellent	Large	14
Baghouse (dacron, nylon, orlon)	Fair	Good	Excellent	Large	17
Baghouse (glass fiber)	Fair	Good	Good	Large	21
ESP	Excellent	Excellent	Good	Large	21
Dry scrubber	Fair	Good	Good	Large	21
Baghouse (teflon)	Fair	Good	Excellent	Large	23
Impingement scrubber	Fair	Good	Good	Moderate	23
Spray tower	Fair	Good	Good	Large	25
Venturi scrubber	Good	Good	Excellent	Small	56

Source: A.C. Stern, R.W. Boubel, D.B. Turner, and D.L. Fox, 1984, *Fundamentals of air pollution*, 2d ed., 426.
[a]Includes water and power cost, maintenance cost, operating cost, capital, and insurance costs (1984 dollars).

ment. These cost comparisons, although representative of air cleaners as a whole, may not reflect actual cost differences for various types of air cleaning equipment when applied to specific applications. Thus, these tables should be used as guides only and should not be relied upon for accurate estimates in a specific situation.

—*Parker C. Reist*

5.16
PARTICULATE CONTROLS: DRY COLLECTORS

Gravity Settling Chambers

FEATURE SUMMARY

Pressure Drop
Less than 0.5 in of water

Operating Temperatures
Up to 1000°C

Applications
Precleaners for removing dry dust produced by grinding in cement and lime kilns, grain elevators, rock crushers, milling operations, and thermal coal dryers

Dust Particle Sizes
Greater than 50 μm

Settling chambers serve as preliminary screening devices for more efficient control devices. These chambers remove large particles from gas streams by gravity. The chambers slow the gas sufficiently to provide enough time for particles to collect (Leith, Dirgo, and Davis 1986). Figure 5.16.1 shows a chamber of length L, width W, and height H. A mass balance for an infinitesimal slice dL yields:

$$V_gHWC = V_gHW(C - dC) + WdL(-V_{ts})C \quad 5.16(1)$$

where:

V_g = the uniform gas velocity
V_{ts} = the particle terminal settling velocity
C = particle concentration in the slice dL

This equation assumes that the particles are well-mixed in every plane perpendicular to the direction of air flow due to lateral turbulence. This assumption is reasonable because even at low velocities, gas flow in industrial-scale settling chambers is turbulent. Rearranging Equation 5.16(1) and integrating between the two ends of the chamber yields the collection efficiency of the chamber as:

$$\eta = 1 - \exp\left[\frac{-V_{ts}L}{V_gH}\right] \quad 5.16(2)$$

The terminal settling velocity is as follows:

$$V_{ts} = \frac{\rho_p d^2 C_c g}{18\mu} \quad 5.16(3)$$

where:

ρ_p = the particle density
d = the particle diameter
μ = the viscosity of the medium
C_c = the Cunningham slip correction factor

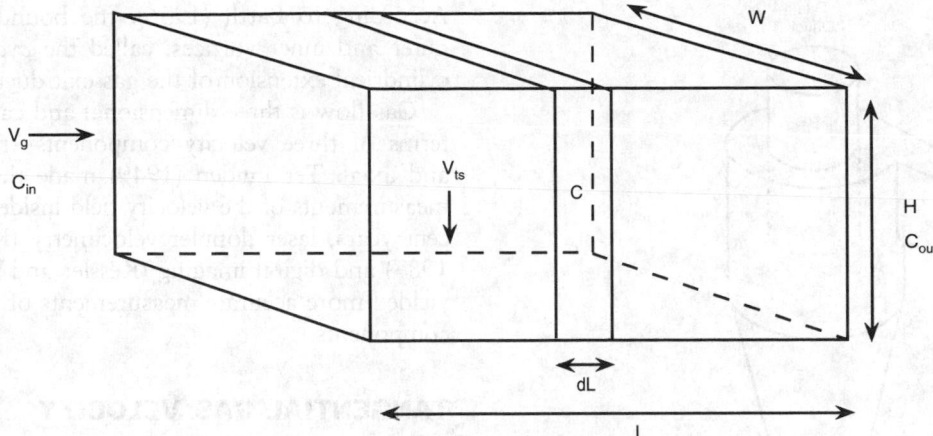

FIG. 5.16.1 Schematic diagram of settling chamber with complete lateral mixing due to turbulence.

Equation 5.16(2) shows that settling chambers have high collection efficiency for low gas velocities, high terminal settling velocities, and large ratios of chamber length to height. Thus, gas velocities are usually below 3 m/sec, which has the added benefit of preventing reentrainment. A large L/H ratio ensures a long residence time and a short vertical distance for the particle to travel to be collected. The equation yields high terminal settling velocities for large particle sizes, therefore, these devices efficiently remove particles greater than 50 μm in diameter.

Settling chambers are characterized by low capital costs and low pressure drop (~0.5 in water gauge [w.g.]). They can be used under temperature and pressure extremes (~1000°C and ~100 atm). Settling chambers are used as precleaners to remove dry dusts produced by grinding, e.g. in coal dryers, grain elevators, and rock crushers.

Cyclones

FEATURE SUMMARY

Types of Designs
Reverse flow, or straight through, with tangential, scroll, and swirl vane entries

Gas Flow Rates
50 to 50,000 m³/hr

Pressure Drop
Between 0.5 and 8.0 in of water

Operating Temperatures
Up to 1000°C

Applications
Cement and lime kilns, grain elevators, milling operations, thermal coal dryers, and detergent manufacturing

Dust Particle Sizes
Greater than 5 μm

Partial List of Suppliers
Advanced Combustion Systems Inc.; Alfa-Laval Separation Inc.; Bayliss-Trema Inc.; Beckert and Heister Inc.; Clean Gas Systems Inc.; Dresser Industries Inc.; Ducon Environmental Systems; Emtrol Corp.; Fisher-Klosterman Inc.; Hough International Inc.; HVAC Filters Inc.; Interel Environmental Technologies Inc.; Joy Technologies Inc.; Quality Solids Separation Co.; Thiel Air Technologies Corp.; United Air Specialists Inc.; Wheelabrator Inc.

Cyclones operate by accelerating particle-laden gas in a vortex from which particles are removed by centrifugal force. One of the most widely used dust collecting devices, cyclones are inexpensive to construct and easy to maintain because they do not have any moving parts. Although cyclones are inefficient for collecting particles smaller than 5 μm in diameter, they operate with low to moderate pressure drops (0.5 to 8.0 in w.g.).

Cyclones can be constructed to withstand dust concentrations as high as 2000 g/m³, gas temperatures as high as 1000°C, pressures up to 1000 atm, and corrosivity. Such conditions are encountered in the pressurized, fluidized-bed combustion of coal, where cyclones are an economic control option. In addition, cyclones are used to control emissions from cement and lime kilns, grain elevators, grain drying and milling operations, thermal coal dryers, and detergent manufacturing.

Various cyclone designs have been proposed; however the reverse-flow cyclone is the type most commonly used for industrial gas cleaning. Figure 5.16.2 shows the dimensions of a reverse-flow cyclone. Dust-laden gas enters the cyclone at the top; the tangential inlet causes the gas to spin. After entry, the gas forms a vortex with a high tangential velocity that gives particles in the gas a high centrifugal force, moving them to the walls for collection. Below the bottom of the gas exit duct, the spinning gas gradually migrates to the cyclone axis and moves up and out the gas exit. Thus, the cyclone has an outer vortex moving downward and an inner vortex flowing upward (Dirgo and Leith 1986). Collected dust descends to the duct outlet at the bottom of the cone.

Cyclone dimensions are usually expressed as multiples of the diameter D. The dimension ratios a/D, b/D, D_e/D,

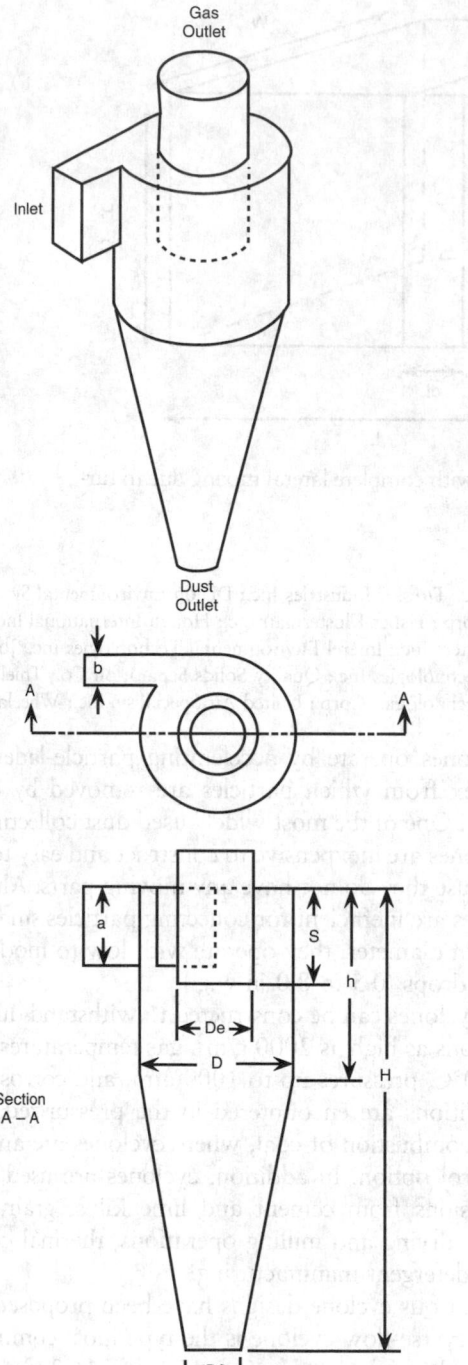

FIG. 5.16.2 Dimensions of a reverse-flow cyclone.

S/D, h/D, H/D, and B/D allow cyclones which differ in size to be compared. Table 5.16.1 gives some standard designs.

GAS FLOW PATTERNS

Understanding and characterizing the performance of a cyclone requires knowledge of the gas flow patterns within. The overall gas motion consists of two vortices: an outer vortex moving down and an inner vortex moving up.

According to Barth (1956), the boundary between the outer and inner vortices, called the *cyclone core,* is the cylindrical extension of the gas exit duct.

Gas flow is three dimensional and can be described in terms of three velocity components—tangential, radial, and axial. Ter Linden (1949) made the first systematic measurements of the velocity field inside a cyclone. In recent years, laser doppler velocimetry (Kirch and Loffler 1987) and digital imaging (Kessler and Leith 1991) have yielded more accurate measurements of the three velocity components.

TANGENTIAL GAS VELOCITY

Early theoretical treatments of cyclone flow patterns (Lissman 1930; Rosin, Rammler, and Intelman 1932) described the relationship between the tangential gas velocity V_t and the distance r from the cyclone axis as:

$$V_t r^n = \text{constant} \qquad 5.16(4)$$

The vortex exponent n is +1 for an ideal liquid and −1 for rotation as a solid body, while the usual range for n is from 0.5 to 0.9 (Ter Linden 1949; Shepherd and Lapple 1939; First 1950).

Equation 5.16(4) implies that tangential velocity increases from the walls to the cyclone axis where it reaches infinity. Actually, in the outer vortex, tangential velocity increases with decreasing radius to reach a maximum at the radius of the central core. In the inner vortex, the tangential velocity decreases with decreasing radius. Figure 5.16.3 shows Iozia and Leith's (1989) anemometer measurements of tangential velocity. Measurements by Ter Linden (1949) show similar velocity profiles.

RADIAL AND AXIAL GAS VELOCITY

Radial gas velocity is the most difficult velocity component to measure experimentally. Figure 5.16.4 shows Kessler and Leith's (1991) measurements of radial velocity profiles in a cyclone. Despite some uncertainty in the measurements, some trends can be seen. Radial flow velocity increases toward the center of the cyclone due to the conservation of mass principle. A tendency also exists for the maximum radial velocity in each cross section to decrease with decreasing cross-sectional height.

Figure 5.16.5 shows the axial gas velocity in a reverse-flow cyclone. The gas flows downward near the cyclone wall and upward near the cyclone axis. The downward velocity near the wall is largely responsible for transporting dust from the cyclone wall to the dust outlet.

MODELING GAS FLOWS

The fluid dynamics of a cyclone are complex, and modeling the detailed flow pattern involves solving the strongly

TABLE 5.16.1 STANDARD DESIGNS FOR REVERSE FLOW CYCLONES

Source	Duty	D	a/D	b/D	D_e/D	S/D	h/D	H/D	B/D	ΔH	Q/D² (m/hr)
Stairmand[a]	High efficiency	1	0.50	0.20	0.50	0.50	1.5	4.0	0.375	5.4	5500
Swift[b]	High efficiency	1	0.44	0.21	0.40	0.50	1.4	3.9	0.40	9.2	4940
Lapple[c]	General purpose	1	0.50	0.25	0.50	0.625	2.0	4.0	0.25	8.0	6860
Swift[b]	General purpose	1	0.50	0.25	0.50	0.60	1.75	3.75	0.40	7.6	6680
Stern et al.[d]	Consensus	1	0.45	0.20	0.50	0.625	0.75	2.0	—	—	—
Stairmand[a]	High through-put	1	0.75	0.375	0.75	0.875	1.5	4.0	0.375	7.2	16500
Swift[b]	High through-put	1	0.80	0.35	0.75	0.85	1.7	3.7	0.40	7.0	12500

Source: D. Leith, 1984, Cyclones, in *Handbook of powder science and technology,* edited by M.A. Fayed and L. Otten (Van Nostrand Reinhold Co.).
Notes: [a]Stairmand, C.J. 1951. The design and performance of cyclone separators. *Trans. Instn. Chem. Engrs.* 29:356.
[b]Swift, P. 1969. Dust controls in industry. *Steam and Heating Engineer* 38:453.
[c]Lapple, C. 1951. Processes use many collector types. *Chemical Engineering* 58:144.
[d]Stern, A.C., K.J. Kaplan, and P.D. Bush. 1956. *Cyclone dust collectors.* New York: American Petroleum Institute.

coupled, nonlinear, partial differential equations of the conservation of mass and momentum. Boysan, Ayers, and Swithenbank (1982) have developed a mathematical model of gas flow based on the continuity and momentum conservation principles, accounting for the anisotropic nature of turbulence and its dissipation rate. The velocity

and pressure profiles of their model agree remarkably with the experimental measurements of Ter Linden (1949). However, as they suggest, the information obtained from such modeling is more detailed than that required by a process design engineer.

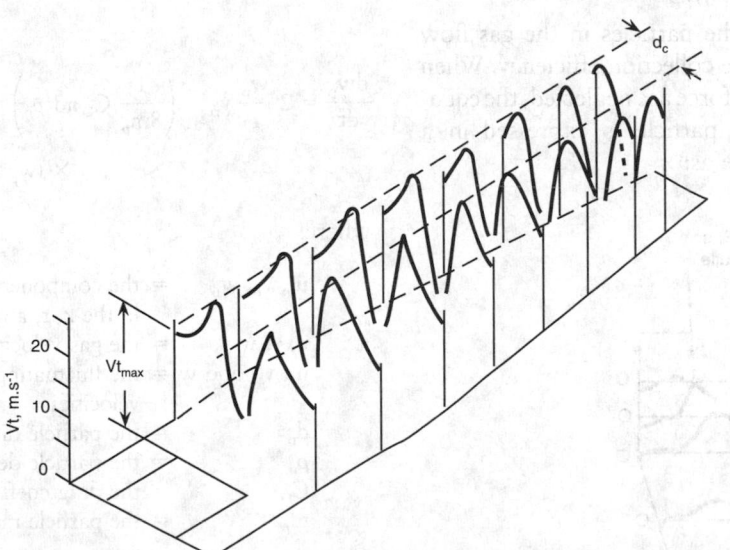

FIG. 5.16.3 Tangential gas velocity in a reverse flow cyclone. The figure is a vertical cross section of half the cyclone. Each of the nodes represents tangential gas velocity. (Reprinted, with permission, from D.L. Iozia and D. Leith, 1989, Effect of cyclone dimensions on gas flow pattern and collection efficiency, *Aerosol Sci. and Technol.* 10:491.)

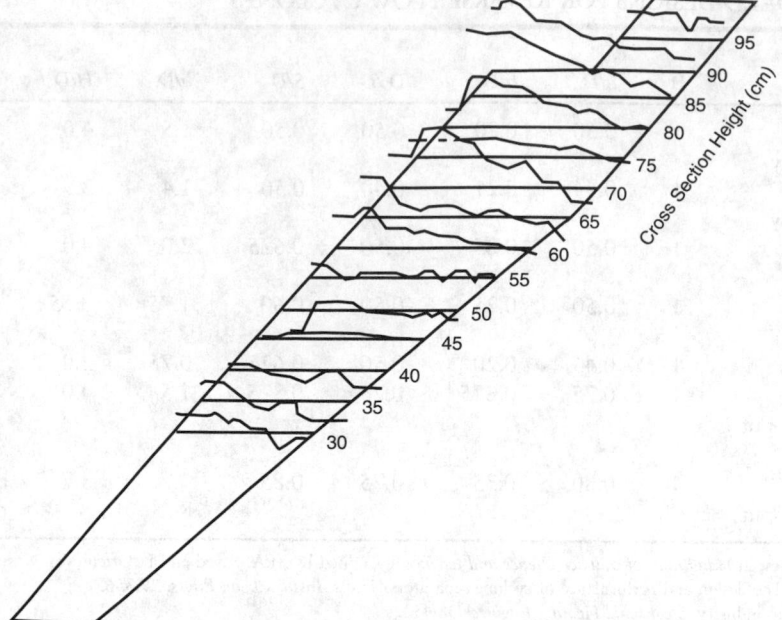

FIG. 5.16.4 Radial gas velocity in a reverse-flow cyclone. (Reprinted, with permission, from M. Kessler and D. Leith, 1991, Flow measurement and efficiency modeling of cyclones for particle collection, *Aerosol Sci. and Technol.* 15.)

COLLECTION EFFICIENCY

Collection efficiency η is defined as the fraction of particles of any size that is collected by the cyclone. The plot of collection efficiency against particle diameter is called *a fractional* or *grade efficiency curve*.

Determining the path of the particles in the gas flow field is essential to estimate the collection efficiency. When all external forces except drag force are neglected, the equation of motion for a small particle is expressed in a Lagrangian frame of reference as:

$$\frac{du_p}{dt} = -\left(\frac{1}{8m_p} C_D \pi d^2 \rho_p\right)(u_p - u - u')|u_p - u - u'|$$

$$\frac{dv_p}{dt} = \frac{w_p^2}{r} - \left(\frac{1}{8m_p} C_D \pi d^2 \rho_p\right)(v_p - v - v')|v_p - v - v'|$$

$$5.16(5)$$

$$\frac{dw_p}{dt} = 2\frac{w_p}{r} v_p - \left(\frac{1}{8m_p} C_D \pi d^2 \rho_p\right)$$

$$\times (w_p - w - w')|w_p - w - w'|$$

where:

u_p, v_p, w_p = the components of the particle velocities in the z, r, and θ directions
u, v, w = the gas velocities in the same directions
u', v', and w' = the fluctuating components of the gas velocities
d_p = the particle diameter
ρ_p = the particle density
C_D = the drag coefficient
m_p = the particle mass

In addition to Equations 5.16(5), the following equation applies:

$$\frac{dz}{dt} = u_p; \quad \frac{dr}{dt} = v_p; \quad \text{and} \quad \frac{d\theta}{dt} = \frac{w_p}{r} \quad 5.16(6)$$

Boysan, Ayers, and Swithenbank (1982) used a stochastic approach whereby the values of u', v', and w', which pre-

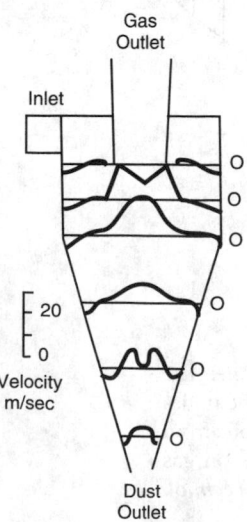

FIG. 5.16.5 Axial gas velocity in a reverse-flow cyclone.

vail during the lifetime of a fluid eddy in which the particle is traversing, are sampled assuming that these values possess a Gaussian probability distribution. They used the lifetime of the fluid eddy as a time interval over which the gas velocity remains constant. This assumption allowed the direct solution of the equations of motion to obtain local, closed-form solutions. They obtained the values of z, r, and θ by a simple stepwise integration of the equation of the trajectory. Many random trajectories were evaluated, and a grade efficiency curve was constructed based on these results. The results agreed with the experimental results of Stairmand (1951).

Kessler and Leith (1991) approximated the drag force term using Stokes' law drag for spheres. They solved the equations of motion numerically using backward difference formulas to solve the stiff system of differential equations. They modeled the collection efficiency for a single, arbitrarily sized particle by running several simulations for that particle. For each particle size, the ratio of collected particles to total particles described the mean of a binomial distribution for the probability distribution. Figure 5.16.6 shows this curve, three other grade efficiency models (Barth [1956], Dietz [1981], and Iozia and Leith [1990]), and the direct measurements of Iozia and Leith.

Early collection efficiency treatments balanced the centrifugal and drag forces in the vortex to calculate a *critical particle* size that was collected with 50 or 100% efficiency. The tangential velocity of the gas and the particle were assumed to be equal, and the tangential velocity was given by the vortex exponent law of Equation 5.16(4). As previously shown, the maximum tangential velocity, V_{max}

occurs at the edge of the central core. The average inward radial velocity at the core edge is as follows:

$$V_r = -\frac{Q}{2\pi r_{core}(H - S)} \qquad 5.16(7)$$

For particles of the critical diameter, the centrifugal and drag forces balance, and these static particles remain suspended at the edge of the core. Larger particles move to the wall and are collected, while smaller particles flow into the core and out of the cyclone. Barth (1956) and Stairmand (1951) used different assumptions about V_{max} to develop equations for the critical diameter. Collection efficiencies for other particle sizes were obtained from a separation curve—a plot that relates efficiency to the ratio of particle diameter to the critical diameter. However, theories of the critical particle type fail to account for turbulence in the cyclone (Leith, Dirgo, and Davis 1986); moreover, a single separation curve is not universally valid.

Later theories attempted to account for turbulence and predict the entire fractional efficiency curve (Dietz 1981; Leith and Licht 1972; Beeckmans 1973). Of these, the theory by Leith and Licht (1972) has been frequently used; however, some of the assumptions on which the theory is based have been invalidated. As with most efficiency theories, the interactions between particles is not taken into account. The assumption of complete turbulent mixing of aerosol particles in any lateral plane has been proven untrue.

Iozia and Leith (1989) used experimental data to develop an equation to predict the *cut diameter* d_{50}, which is the particle size collected with 50% efficiency. Their empirical equations are as follows:

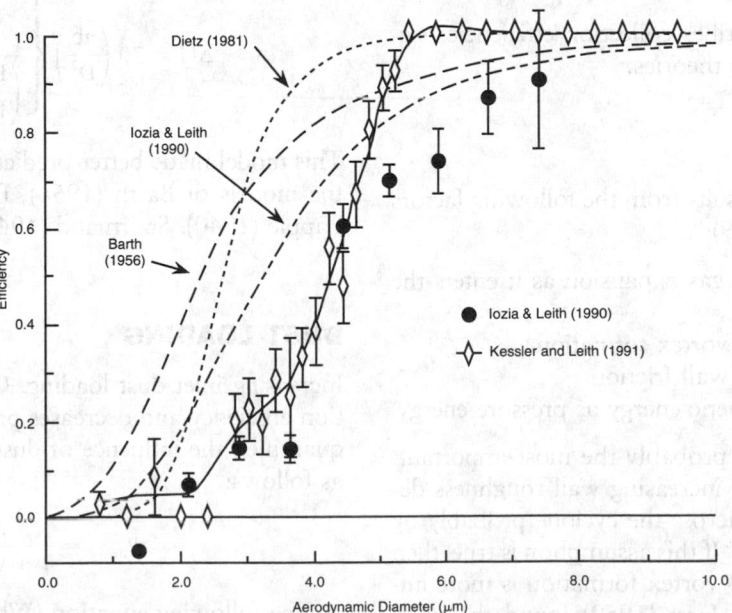

FIG. 5.16.6 Measured collection efficiency from studies by Barth, Dietz, and Iozia and Leith and modeled collection efficiency by Kessler and Leith.

$$d_{50} = \left(\frac{9\mu Q}{\pi z_c \rho_p V_{max}^2}\right)^{1/2} \qquad 5.16(8)$$

where:

$$V_{max} = 6.1 V_i \left(\frac{ab}{D^2}\right)^{0.61} \left(\frac{D_e}{D}\right)^{-0.74} \left(\frac{H}{D}\right)^{-0.33} \qquad 5.16(9)$$

where V_i is the inlet gas velocity, and z_c is the core length which depends on core diameter d_c as follows:

$$d_c = 0.52 D \left(\frac{ab}{D^2}\right)^{-0.25} \left(\frac{D_e}{D}\right)^{1.53} \qquad 5.16(10)$$

When $d_c < B$, the core intercepts the cyclone wall, and the following equation applies:

$$z_c = (H - S) - \left(\frac{H - h}{\frac{D}{B} - 1}\right)\left(\frac{d_c}{B} - 1\right) \qquad 5.16(11)$$

When $d_c < B$, the core extends to the bottom of the cyclone, and the following equation applies:

$$z_c = (H - S) \qquad 5.16(12)$$

Iozia and Leith (1989, 1990) also developed the following logistic equation for the fractional efficiency curve:

$$\eta = \frac{1}{1 + \left(\frac{d_{50}}{d}\right)^{\beta}} \qquad 5.16(13)$$

where the parameter β is estimated from the cut size d_{50}, and the geometry of the cyclone is as follows:

$$\ln(\beta) = 0.62 - 0.87 \ln(d_{50}(cm))$$
$$+ 5.21 \ln\left(\frac{ab}{D^2}\right) + 1.05 \left(\ln\left(\frac{ab}{d^2}\right)\right)^2 \qquad 5.16(14)$$

The preceding model describes collection efficiencies significantly better than other theories.

PRESSURE DROP

A cyclone pressure drop results from the following factors (Shepherd and Lapple 1939):

1. Loss of pressure due to gas expansion as it enters the cyclone
2. Loss of pressure due to vortex formation
3. Loss of pressure due to wall friction
4. Regain of rotational kinetic energy as pressure energy

Factors 1, 2, and 3 are probably the most important. Iinoya (1953) showed that increasing wall roughness decreases the pressure drop across the cyclone probably by inhibiting vortex formation. If this assumption is true, then energy consumption due to vortex formation is more important than wall friction. First (1950) found that wall friction is not an important contributor to pressure drop.

Devices such as an inlet vane, an inner wall extension of the tangential gas entry within the cyclone body to a position close to the gas exit duct, and a cross baffle in the gas outlet lower the pressure drop (Leith 1984). However, such devices work by suppressing vortex formation and decrease collection efficiency as well. Designing cyclones for low pressure drop without attaching devices that also lower collection efficiency is possible.

Cyclone pressure drop has traditionally been expressed as the number of inlet velocity (v_i) heads ΔH. The following equation converts velocity heads to pressure drop ΔP:

$$\Delta P = \Delta H \left(\frac{1}{2} \rho_g v_i^2\right) \qquad 5.16(15)$$

The value of ΔH is constant for a cyclone design (i.e., cyclone dimension ratios), while ΔP varies with operating conditions.

Many analytical expressions for determining ΔH from cyclone geometry are available (Barth 1956; First 1950; Shepherd and Lapple 1940; Stairmand 1949; Alexander 1949). None of these expressions predicts pressure drop accurately for a range of cyclone designs; predictions differ from measured values by more than a factor of two (Dirgo 1988). Further, evaluations of these models by different investigators produced conflicting conclusions as to which models work best.

Dirgo (1988) and Ramachandran et al. (1991) developed an empirical model for predicting pressure drop, which was developed through statistical analysis of pressure drop data for ninety-eight cyclone designs. They used stepwise and backward regression to develop the following expression for ΔH based on cyclone dimension ratios:

$$\Delta H = 20 \left(\frac{ab}{D_e^2}\right) \left\{\frac{\left(\frac{S}{D}\right)}{\left(\frac{H}{D}\right)\left(\frac{h}{D}\right)\left(\frac{B}{D}\right)}\right\}^{1/3} \qquad 5.16(16)$$

This model made better predictions of pressure drop than the models of Barth (1956), First (1950), Shepherd and Lapple (1940), Stairmand (1949), and Alexander (1949).

DUST LOADING

Increasing inlet dust loading, C_i (g/cm^3), increases collection efficiency and decreases pressure drop. Briggs (1946) quantified the influence of dust loading on pressure drop as follows:

$$\Delta P_{dusty} = \frac{\Delta P_{clean}}{1 + 0.0086 C_i^{1/2}} \qquad 5.16(17)$$

The following equation (Whiton 1932) gives the effect on efficiency of changing the inlet loading from C_{i1} to C_{i2}:

$$\frac{100 - \eta_1}{100 - \eta_2} = \left(\frac{C_{i2}}{C_{i1}}\right)^{0.182} \qquad 5.16(18)$$

CYCLONE DESIGN OPTIMIZATION

Cyclone design usually consists of choosing an accepted standard design or a manufacturer's proprietary model that meets cleanup requirements at a reasonable pressure drop. However, investigators have developed analytical procedures to optimize cyclone design, trading collection efficiency against pressure drop.

Leith and Mehta (1973) describe a procedure to find the dimensions of a cyclone with maximum efficiency for a given diameter, gas flow, and pressure drop. Dirgo and Leith (1985) developed an iterative procedure to improve cyclone design. While holding the cyclone diameter constant, their method alters one cyclone dimension and then searches for a second dimension to change to yield the greatest collection efficiency at the same pressure drop.

They selected the gas outlet diameter D_e as the primary dimension to vary. This variation changed the pressure drop. Then, they varied the inlet height a, inlet width b, and gas outlet duct length S, one at a time, to bring pressure drop back to the original value. They predicted the d_{50} for each new design from theory. The new design with the lowest d_{50} became the baseline for the next iteration. In the next iteration, they varied D_e again with the three other dimensions to find the second dimension change that most reduced d_{50}. They continued iterations until the predicted reduction in d_{50} from one iteration to the next was less than one nanometer.

Ramachandran et al. (1991) used this approach with the efficiency theory of Iozia and Leith, Equations 5.16(8)–5.16(14), and the pressure drop theory of Dirgo, Equation 5.16(16), to develop optimization curves. These curves predict the minimum d_{50} and the dimension ratios of the optimized cyclone for a given pressure drop (see Figures 5.16.7 and 5.16.8).

To design a cyclone, the design engineer must obtain the inlet dust concentration and size distribution and other design criteria such as gas flow rate, temperature, and particle density, preferably by stack sampling. When a cyclone is designed for a plant to be constructed, stack testing is impossible, and the design must be based on data obtained from similar plants. Once the size distribution of the dust is known, the design engineer chooses a value of d_{50}. For each size range, the engineer calculates the collection efficiency using Equation 5.16(13). The overall efficiency is calculated from the following equation:

$$\eta_{overall} = \sum_i f_i \eta_i \qquad 5.16(19)$$

where f_i is the fraction of particles in the ith size range. By trial and error, the engineer chooses a value of d_{50} to

obtain the required overall efficiency. This d_{50} is located on the optimization curve (e.g., for H = 5D) of Figure 5.16.7, and the pressure drop corresponding to this d_{50} is found. Figure 5.16.8 gives the cyclone dimension ratios. The cyclone diameter is determined from the following equation:

$$D_2 = D_1 \left(\frac{\rho_{p2} Q_2}{\rho_{p1} Q_1}\right)^{1/3} \qquad 5.16(20)$$

where D_1, ρ_{p1}, and Q_1 are the cyclone diameter (0.254 m), particle density (1000 kg/m³), and flow (0.094 m³/sec) of the cyclone optimized in Figure 5.16.8; and ρ_{p2} and Q_2 are the corresponding values for the system being designed. The following equation gives the design pressure drop:

$$\Delta P_2 = \Delta P_1 \left(\frac{Q_2 D_1^2}{Q_1 D_2^2}\right)^2 \qquad 5.16(21)$$

where ΔP_1 is the pressure drop from Figure 5.16.7 corresponding to the chosen d_{50}.

If the pressure drop ΔP_2 is too high, then the design engineer should explore other options, such as choosing a taller cyclone as the starting point or reducing the flow through the cyclone by installing additional cyclones in parallel. These options, however, increase capital costs. The design engineer has to balance these various factors.

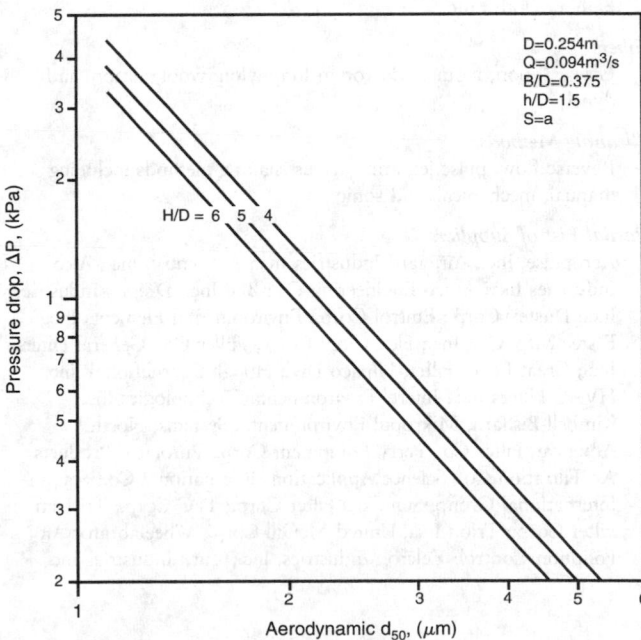

FIG. 5.16.7 Optimum pressure drops versus d_{50} for different cyclone heights. (Reprinted, with permission, from G. Ramachandran, D. Leith, J. Dirgo, and H. Feldman, 1991, Cyclone optimization based on a new empirical model for pressure drop, *Aerosol Sci. and Technol.* 15.)

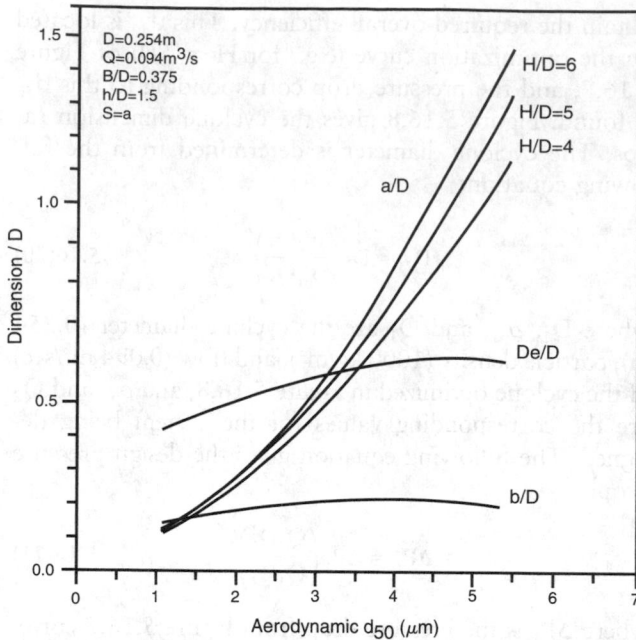

FIG. 5.16.8 Optimum dimension ratios versus d_{50} for different cyclone heights. (Reprinted, with permission, from Ramachandran et al. 1991.)

Filters

FEATURE SUMMARY

Types of Filter Media
Paper, woven fabric with a low air/cloth ratio, felted fabric with high air/cloth ratio

Fibers
Glass, Teflon, Nomex, dacron, orlon, nylon, wool, cotton, and dynel

Cleaning Methods
Reverse flow, pulse jet, and various shaking methods including manual, mechanical, and sonic

Partial List of Suppliers
Aeropulse, Inc.; Airguard Industries Inc.; Air Sentry, Inc.; Alco Industries Inc.; Amco Engineering Co.; BGI Inc.; Dresser Industries Inc.; Dustex Corp.; Emtrol Corp.; Environmental Elements Inc.; Esstee Mfg. Co., Inc.; Flex-Cleen Corp.; Fuller Co.; General Filters Inc.; Great Lakes Filter, Filpaco Div.; Hough International, Inc.; HVAC Filters Inc.; Interel Environmental Technologies Inc.; Kimbell-Bishard; Mikropul Environmental Systems; North American Filter Co.; Perry Equipment Corp.; Purolator Products Air Filtration Co.; Science Applications International Co.; Stamm International Group; Standard Filter Corp.; TFC Corp.; Tri-Dim Filter Corp.; Trion Inc.; United McGill Corp.; Wheelabrator Air Pollution Control; Zelcron Industries, Inc.; Zurn Industries Inc.

FIBER FILTERS

Fiber bed filters collect particles within the depth of the filter so that a dust cake does not build up on the filter surface. Fiber sizes range from less than 1 μm to several hundred μm. The materials used for fibrous filters include cellulose, glass, quartz, and plastic fibers. Fiber beds are used as furnace filters, air conditioning filters, and car air filters. They are also used for sampling in air pollution and industrial hygiene measurements. The collected particles cannot generally be cleaned from the bed, so these filters must be replaced when the pressure drop becomes too high.

Filtration Theory

The basis of this theory is the capture of particles by a single fiber. The single-fiber efficiency η_{fiber} is defined as the ratio of the number of particles striking the fiber to the number of particles that would strike the fiber if streamlines were not diverted around the fiber. If a fiber of diameter d_f collects all particles contained in a layer of thickness y, then the single-fiber efficiency is y/d_f (see Figure 5.16.9).

The general approach involves finding the velocity field around an isolated fiber, calculating the total collection efficiency of the isolated fiber due to the mechanisms of particle deposition, expressing the influence of neighboring fibers (interference effects) by means of empirical corrections, and finally obtaining the overall efficiency of a filter composed of many fibers. The following equation relates the overall efficiency of a filter composed of many fibers in a bed E to the single-fiber efficiency:

$$E = 1 - \exp\left[\frac{-4\eta_{fiber}\alpha L}{\pi d_f(1 - \alpha)}\right] \qquad 5.16(22)$$

where:

α = the solidity or packing density of the filter
L = the filter thickness
d_f = the fiber diameter

Note than even if $\eta_{fiber} = 1$, the total filter efficiency can be low if α is low or the filter thickness is small. The derivation of this equation assumes that particles are well-mixed in every plane perpendicular to the gas flow direction, no particles are reentrained into the gas stream, all fibers are perpendicular to the gas flow direction, and the gas has a uniform velocity.

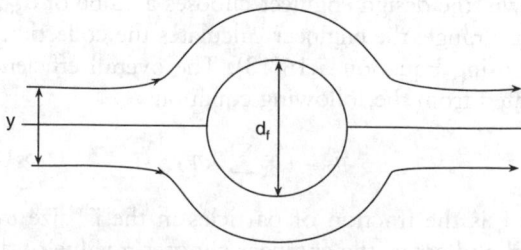

FIG. 5.16.9 Single-fiber collection efficiency.

Filtration Mechanisms

As air passes through a filter, the trajectories of particles deviate from the streamlines due to various mechanisms, most important of which are interception, inertial impaction, diffusion, gravity, and electrostatic forces. The radiometric phenomena such as thermophoresis and diffusiophoresis are usually negligible. The single-fiber efficiency η_{fiber} is approximated as the sum of the efficiencies due to each of the preceding mechanisms acting individually as follows:

$$\eta_{fiber} = \eta_{inter} + \eta_{impaction} + \eta_{diffusion} + \eta_{gravity} + \eta_{elec} \quad 5.16(23)$$

The maximum value of the single-fiber efficiency is 1.0; so if several of the mechanisms have large efficiencies, their sum cannot exceed unity.

Interception

Particles are collected when the streamline brings the particle center to within one particle radius from the fiber surface. This effect is described by a dimensionless interception parameter R, which is the ratio of particle diameter d_p to fiber diameter as follows:

$$R = \frac{d_p}{d_f} \quad 5.16(24)$$

Assuming Kuwabara flow, η_{inter} is given by the following equation:

$$\eta_{inter} = \frac{1 + R}{2Ku} \left[2\ln(1 + R) - 1 \right.$$
$$\left. + \alpha + \left(\frac{1}{1 + R}\right)^2\left(1 - \frac{\alpha}{2}\right) - \frac{\alpha}{2}(1 + R)^2 \right] \quad 5.16(25)$$

where Ku is the Kuwabara hydrodynamic factor given by the following equation:

$$Ku = -\frac{1}{2}\ln \alpha - \frac{3}{4} + \alpha - \frac{\alpha^2}{4} \quad 5.16(26)$$

Interception is independent of the flow velocity around a fiber.

Inertial Impaction

A particle, because of its inertia, may be unable to adjust to the rapidly changing curvature of streamlines in the vicinity of a fiber and may cross the streamlines to hit the fiber. The dimensionless number used to describe this effect is the Stokes number, which is the ratio of particle stopping distance to fiber diameter as follows:

$$Stk = \frac{\tau U}{d_f}, \quad \text{where } \tau = \frac{1}{18} d_p^2 \frac{(\rho_p - \rho_m)}{\mu} C_c \quad 5.16(27)$$

where:

C_c = the Cunningham slip correction factor
U = the free-stream velocity (cm/sec)
μ = the viscosity of the medium (poise)
ρ_p and ρ_m = the densities of the particle and medium, respectively (gm/cc)

The slip correction factor is given by the equation $C_c = 1 + [2\lambda(1.257)/d_p]$ where λ is the mean free path of gas molecules and is equal to 0.071 μm for air molecules at 25°C and $d_p > 2\lambda$.

Stechkina, Kirsch, and Fuch (1969) calculated the inertial impaction efficiency for particles using the Kuwabara flow field as follows:

$$\eta_{impaction} = \frac{1}{(2Ku)^2}$$
$$\times [(29.6 - 28\alpha^{0.62})R^2 - 27.5R^{2.8}]Stk \quad 5.16(28)$$

Inertial impaction is a significant collection mechanism for large particles, high gas velocities, and small-diameter fibers.

Brownian Diffusion

Very small particles generally do not follow streamlines but wobble randomly due to collisions with molecules of gas. Random Brownian motion causes incidental contact between a particle and the fiber. At low velocities, particles spend more time near fiber surfaces, thus enhancing diffusional collection. A dimensionless parameter called the Peclet number Pe is defined as a ratio of convective transport to diffusive transport as follows:

$$Pe = \frac{d_c U}{D} \quad 5.16(29)$$

where:

d_c = the characteristic length of the collecting medium
U = the average gas velocity
D = the diffusion coefficient of the particles as follows:

$$D = \frac{kT C_c}{3\pi\mu d_p} \quad 5.16(30)$$

where:

k = the Boltzman constant (1.38×10^{-16} erg/°Kelvin)
T = the absolute temperature
μ = the viscosity of the medium (poise)
d_p = the particle diameter

Lee and Liu (1982a,b) use a boundary layer model commonly used in heat and mass transfer analysis with a flow field around multiple cylinders that account for flow interference due to neighboring fibers as follows:

$$\eta_{diffusion} = 2.58 \frac{1 - \alpha}{Ku} Pe^{-2/3} \quad 5.16(31)$$

Gravitational Settling

Particles deviate from streamlines when the settling velocity due to gravity is sufficiently large. When flow is downward, gravity increases collection; conversely, when flow is upward, gravity causes particles to move away from the collector resulting in a negative contribution. Gravity is important for large particles at low flow velocities. The dimensionless parameter for this mechanism is as follows:

$$Gr = \frac{V_{ts}}{U} \qquad 5.16(32)$$

where U is the free stream velocity and V_{ts} is the settling velocity. The single-filtration efficiency due to gravity is as follows:

$$\eta_{gravity} = \frac{Gr}{1 + Gr} \qquad 5.16(33)$$

Electrostatic Forces

Aerosol particles can acquire an electrostatic charge either during their generation or during flow through a gas stream. Likewise, the filter fibers can acquire a charge due to the friction caused by a gas stream passing over them. The following three cases can occur:

1. Charged Particle–Charged Fiber

$$\eta_{elec} = \frac{4Qq}{3\mu dd_f V_g} \qquad 5.16(34)$$

where Q is the charge per unit length of the fiber and q is the charge on the particle.

2. Charged Fiber–Neutral Particle

$$\eta_{elec} = \frac{4}{3}\left(\frac{\varepsilon - 1}{\varepsilon + 1}\right)\left(\frac{d^2Q^2}{d_f^3 \mu V_g}\right) \qquad 5.16(35)$$

where ε is the dielectric constant of the particle.

3. Charged Particle–Neutral Fiber

$$\eta_{elec} = \sqrt{\frac{\varepsilon - 1}{\varepsilon - 1}\left(\frac{q^2}{3\pi\mu dd_f^2 V_g(2 - \ln Re)}\right)^{1/2}} \qquad 5.16(36)$$

where μ is the viscosity of the medium and Re is the Reynold's number given by $(d_f V_g \rho_g)/\mu$.

Pressure Drop of Fibrous Filters

The following equation gives the pressure drop for fibrous filters (Lee and Ramamurthi 1993):

$$\Delta P = \frac{16\eta\alpha UL}{Kud_f^2} \qquad 5.16(37)$$

The measured pressure drop across a filter is used with Equation 5.16(37) to determine the effective fiber diameter d_f for use in the filtration efficiency theory, previously described (Lee and Ramamurthi 1993). Predicting the pressure drop for real filters is not straightforward. Comparing the measured pressure drop and calculated pressure drop based on an ideal flow field indicates how uniformly the media structure elements, such as fibers or pores, are arranged.

FABRIC FILTERS (BAGHOUSES)

Baghouses separate fly ash from flue gas in separate compartments containing tube-shaped or pocket-shaped bags or fabric filters. Baghouses are effective in controlling both total and fine particulate matter. They can filter fly ash at collection efficiencies of 99.9% on pulverized, coal-fired utility boilers. Other baghouse applications include building material dust removal, grain processing, oil mist recovery in workplace environments, soap powders, dry chemical recovery, talc dust recovery, dry food processing, pneumatic conveying, and metal dust recovery.

The main parameters in baghouse design are the pressure drop and air-to-cloth ratio. Pressure drop is important because higher pressure drops imply that more energy is required to pull gas through the system. The *air-to-cloth ratio* determines the unit size and thus, capital cost. This ratio is the result of dividing the volume flow of gas received by a baghouse by the total area of the filtering cloth and is usually expressed as acfm/ft². This ratio is also referred to as the *face velocity*. Higher air-to-cloth ratios mean less fabric, therefore less capital cost. However, higher ratios can lead to high pressure drops forcing energy costs up. Also, more frequent bag cleanings may be required, increasing downtime. Fabric filters are classified by their cleaning method or the direction of gas flow and hence the location of the dust deposit.

Inside Collectors

During filtration, as shown in Figure 5.16.10, dusty gas passes upward into tubular or pocket-shaped bags that are closed at the top. Tubular bags are typically 10 m tall and 300 mm in diameter. A dust cake builds on the inside bag surface during filtration. Clean gas passes out through the filter housing. Filtration velocities are about 10 mm/sec (2 cfm/ft²). Many bags in a compartment act in parallel, and a fabric filter usually comprises several compartments.

Two cleaning techniques are used with inside collectors: reverse flow and shaking. In reverse flow, as shown in Figure 5.16.10, filtered gas from the outlets of other compartments is forced backward through the bags. Between the support rings, this reverse flow causes the bags to partially collapse causing the dust cake on the inside to deform, crack, and partially dislodge. A framework of rings keeps the bags open during cleaning. After a few minutes, the reverse gas stops, and filtration resumes. Cleaning is usually performed offline. Typical air-to-cloth ratios are 2 cfm/ft^2, and dust cake weights range from 0.5 to 1.5 lb/ft^2. The pressure drop across the fabric and dust cake with reverse flow is 0.5 to 1.0 in w.g. Some inside collectors used for oil mist recovery are not cleaned but are replaced when they become saturated with oil.

When equilibrium is established, the forces acting to remove the residual dust cake must equal those tending to retain it (Carr and Smith 1984). The removing forces are mechanical flexing and deformation, aerodynamic pressure, gravity, erosion, and acceleration by snapping the bags. Forces acting to retain the dust cake are adhesion and cohesion. *Adhesion* refers to the binding forces between particles and fibers when they contact each other; whereas *cohesion* refers to the bonding forces that exist between the collected particles. Forces acting to remove the residual dust cake increase with thickness or weight of the dust cake, while those tending to retain it decrease or remain the same. Thus, when equilibrium is achieved, a residual dust cake is established.

In shaking, the bag tops are connected to an oscillatory arm, causing the dust on the inside bag surfaces to separate from the fabric and fall into the hopper. Shaking is usually used in conjunction with reverse-gas cleaning. Although bags in shake and deflate units contain no anti-collapse rings, they retain a nominally circular cross section because the reverse-gas flow is low. The shaking force is applied to the tops of the bags causing them to sway and generating traveling waves in them. Deformation of the bags is significant in dislodging the dust cake.

Bags made from woven fabrics are generally used for these filters. Most large filters cleaned by reverse flow collect fly ash at coal-fired plants (Noll and Patel 1979). Bags are usually made of glass fiber to withstand the hot flue gases. Bag lifetime has exceeded 20,000 hr for woven bags. The long life is due to the low filtration velocity and infrequent cleaning (Humphries and Madden 1981). Filter problems in the utility industry are caused by improper bag specifications, installation, or tensioning. Other problems are related to *bag blinding*, a gradual, irreversible increase in pressure drop.

Outside Collectors

As shown in Figure 5.16.11, in outside collectors, dusty gas flows radially inward through cylindrical bags held open by a metal frame inside them. The bags are typically 3 m tall and 200 mm in diameter. Dust collects on the outside bag surfaces. Clean gas passes out of the top of each bag to a plenum.

Cleaning outside collectors usually involves injecting a pulse of compressed air at the outlet of each bag. This pulse snaps the bag open and drives the collected dust away from the bag surface into the hopper. Because pulse-jet cleaning takes a fraction of a second, it can be done online without interrupting the gas flow to a compartment.

Pulse-jet filters generally use thick felt fabrics to reduce dust penetration even when the dust cake is not thick. Filtration velocities through a pulse-jet filter are several times higher than through a reverse gas filter; therefore, pulse-jet filters can be smaller and less expensive. Operating at high filtration velocities can lead to an excessive pressure drop, dust penetration, and fabric wear. These filters are most commonly used to control industrial dust.

Pulse-jet cleaning can be ineffective when the dust is fine. The particles are driven beneath the fabric surface and are not easily removed. As fine dust accumulates below the surface, the pressure drop across the filter gradually increases. If the pressure drop becomes too high, the blinded bags must be replaced. With online cleaning of pulse-jet filters, as little as 1% of the dust on a bag can fall to the hopper after each cleaning pulse (Leith, First, and Feldman 1977).

Pressure Drop

The pressure drop is the sum of the pressure drop across the filter housing and across the dust-laden fabric (Leith and Allen 1986). The pressure drop across the housing is proportional to the square of the gas-flow rate due to turbulence. The pressure drop across the dust-laden fabric is

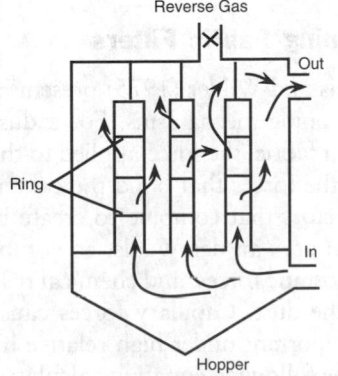
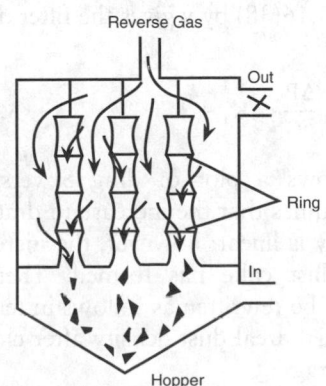

FIG. 5.16.10 Reverse-gas flow cleaning in inside collectors.

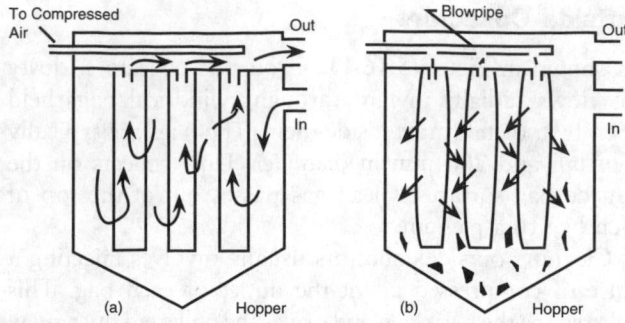

FIG. 5.16.11 Pulse-jet cleaning in outside collectors.

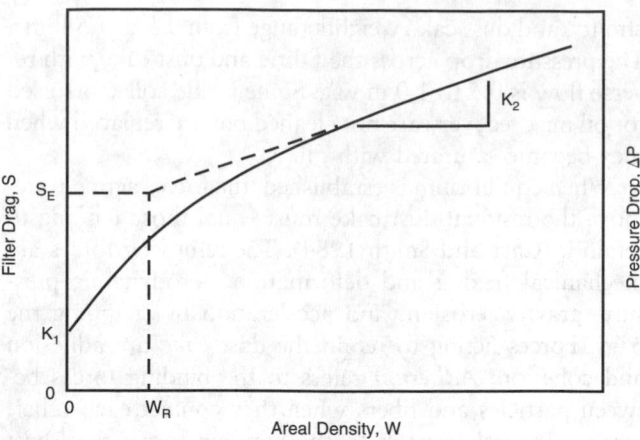

FIG. 5.16.12 Pressure drop and drag for a fabric filter versus areal density of the dust deposit.

the sum of the pressure drop across the clean fabric and the pressure drop across the dust cake as follows:

$$\Delta P = \Delta P_f + \Delta P_d = K_1 v + K_2 v w \qquad 5.16(38)$$

where:

v = the filtration velocity
K_1 = the flow resistance of the clean fabric
K_2 = the specific resistance of the dust deposit
w = the fabric dust areal density

K_1 is related to Frazier permeability, which is the flow through a fabric in cfm/ft^2 of fabric when the pressure drop across the fabric is 0.5 in w.g. as follows:

$$K_1(\text{Pa s m}^{-1}) = \frac{24590}{\text{Frazier Permeability(cfm/ft}^2 \text{ at 0.5 in w.g.)}}$$

$$5.16(39)$$

The following equation determines the dust loading adding to the fabric between cleanings w_0:

$$w_0 = C_{in} v t \qquad 5.16(40)$$

where:

C_{in} = the dust inlet concentration
t = the time between cleanings

Dividing Equation 5.16(38) by v gives the filter drag as follows:

$$S = \frac{\Delta P}{v} = K_1 + K_2 w \qquad 5.16(41)$$

Figure 5.16.12 shows a plot of drag S versus w. Equation 5.16(41) assumes that the increase in drag with increasing areal density is linear; however, the increase is linear only after a dust cake has formed. Therefore, Equation 5.16(41) can be rewritten as follows in terms of the extrapolated residual areal dust density after cleaning w_R and the effective drag S_E.

$$S = S_E + K_2(w - w_R) \qquad 5.16(42)$$

Evaluation of Specific Resistance K₂

The Kozeny–Carman relationship (Billings and Wilder 1970; Carman 1956) is often used to describe pressure drop across a dust deposit. Rudnick and First (1978) showed that the Happel (1958) cell model corrected for slip flow gives better K_2 estimates than the Kozeny–Carman relationship for a dust cake with no interaction between the fabric and cake. All theoretical models for K_2 are sensitive to dust deposit porosity and dust particle size distributions. Porosity cannot usually be estimated correctly and varies with filtration velocity, humidity, and other factors. Whenever possible, K_2 should be measured rather than calculated from theory. Leith and Allen (1986) suggest that the dust collected on a membrane filter and K_2 should be calculated from the increase in pressure drop $(\Delta P_2 - \Delta P_1)$ with filter weight gain $(M_2 - M_1)$ as follows:

$$K_2 = \left(\frac{A}{v}\right)\left[\frac{(\Delta P_2 - \Delta P_1)}{(M_2 - M_1)}\right] \qquad 5.16(43)$$

where A is the surface area of the membrane filter.

Cleaning Fabric Filters

Dennis and Wilder (1975) present a comprehensive study on cleaning mechanisms. For a dust cake to be removed from a fabric, the force applied to the cake must be greater than the forces that bond the cake to the fabric. The major factors that combine to create both adhesion and cohesion are van der Waals, coulombic and induced dipole electrostatic forces, and chemical reactions between the gas and the dust. Capillary forces caused by surface tension are important under high relative humidity conditions.

The following equation calculates the drag across a single bag:

$$S = \frac{1}{\sum\limits_{i=1}^{n}\left(\dfrac{a_i}{S_i}\right)} \qquad 5.16(44)$$

where a_i is the ith fraction of the total bag area and S_i is the drag through that area (Stephan, Walsh, and Herrick 1960). Substituting the values a_i and S_i for the cleaned and uncleaned portions in Equation 5.16(44) calculates the drag for the entire filter. When the local filtration velocity through each bag area (which depends on local drag) is used, the additional dust collected on each area over time can be determined. This collection in turn changes the local drag. This procedure allows an iterative prediction of pressure drop versus time.

Cleaning Inside Collectors

The factors responsible for cake removal in inside collectors have received little attention. Cake rupture due to fabric flexing is the primary cleaning mechanism. The other main mechanism is the normal stress to the cake from reverse pressurization.

Shaker filters use a motor and an eccentric cam to produce the cleaning action. Cleaning depends on factors such as bag shape and support, rigidity, bag tension, shaker frequency, and shaker amplitude. The following equation gives the acceleration a developed by the shaker arm:

$$a = 4\pi^2 f^2 l_d \qquad 5.16(45)$$

where:

f = the frequency of shaking
l_d = the amplitude of the sinusoidal motion

Modeling the transfer of motion to the bags is analogous to a vibrating string, whereby a wave travels down the bag and is reflected from its end. If the reflected wave reaches the arm at the same time that the next wave is produced, resonance occurs; this time is when the transfer of cleaning energy to the bag is optimum.

Pulse-Jet Cleaning

Leith and Ellenbecker (1980b) use Equation 5.16(38) as the basis of a model to predict pressure drop across a pulse-jet cleaned filter. Their model assumes that the fraction of the dust deposit removed by a cleaning pulse is proportional to the separation force applied and that impulse and conservation of momentum determine bag motion as follows:

$$\Delta P = K_v v^2 + \frac{P_s + K_1 v - \sqrt{(P_s - K_1 v) - 4w_o v \left(\dfrac{K_2}{K_3}\right)}}{2}$$

$$5.16(46)$$

where K_v depends on the pressure drop characteristics of the venturi at the top of each bag. Typically, $K_v = 57500$ Pa s^2/m^2. The value P_s is the pressure within the bag generated by the cleaning pulse, and it depends on venturi design as well as pulse pressure P as follows:

$$P_s(\text{Pa}) = 164[P(\text{kPa})]^{0.6} \qquad 5.16(47)$$

The values K_1 and w_o are from Equations 5.16(39) and 5.16(40), respectively. The term (K_2/K_3) is a constant that depends on the interaction of the dust and the fabric.

An environmental engineer can use Equation 5.16(46) to estimate the pressure drop of an existing filter when the operating conditions change. When the pressure drop under the initial operating conditions is known, the engineer can use Equation 5.16(46) to determine (K_2/K_3). Then, using this value of (K_2/K_3), the engineer can determine the pressure drop for other operating conditions.

Collection Efficiency

The collection efficiency of a fabric filter depends on the interactions between the fabric, the dust, and the cleaning method. However, a clear understanding of these factors is not available, and existing models cannot be generalized beyond the data sets used in their development.

Impaction, diffusion, and interception in a dust cake are effective. Essentially all incoming dust is collected through a pulse-jet filter, with or without a dust cake. Nevertheless, some dust penetrates a fabric filter. It does so because gas bypasses the filter by flowing through pinholes in the dust cake or because the filter fails to retain the dust previously collected (seepage).

Pinholes form in woven fabrics at yarn intersections. With dust loading, some pinholes are bridged, while the gas velocity increases through open pinholes. Incoming particles bounce through pinholes rather than collect. A disproportionately high fraction of gas flows through unbridged pinholes due to their lower resistance. As the fabric flexes during cleaning, some dust particles dislodge and recollect deeper in the fabric. After several cycles of dislodgment and recollection, the particles completely pass through the filter. Particle penetration by this mechanism is called *seepage*.

Dennis et al. (1977) and Dennis and Klemm (1979) developed a model for predicting effluent fly ash concentration with fabric loading for woven glass fabrics. However, their models contain empirical constants that may be inappropriate for other fabrics. Leith and Ellenbecker (1982) showed that seepage is the primary mechanism for penetration through a pulse-jet-cleaned felt fabric. They developed a model (Leith and Ellenbecker 1980a) for outlet flux N assuming that all incoming dust is collected by the filter and that seepage of previously collected particles through the fabric accounts for all the dust emitted once the filter is conditioned. Seepage occurs as the bag strikes its supporting cage at the end of a cleaning pulse. The impact dislodges particles from the filter; the particles are then carried into the outlet gas stream. The following equation calculates the outlet flux:

$$N = \frac{kw^2v}{t} \qquad 5.16(48)$$

where:

N = the outlet flux
w = the areal density of the dust deposit
v = the filtration velocity
t = the time between cleaning pulses to each bag
k = a constant that depends on factors such as dust characteristics, fabric types, and length of filter service

Figure 5.16.13 plots the outlet flux measured in laboratory experiments against the flux predicted using Equation 5.16(48) with k = 0.002 m − s/kg. These data are for different felt surfaces with different filtration velocities and different dusts. However, all data cluster about the same line.

Figure 5.16.14 plots the mass outlet flux against particle diameter for a pulse-jet filter. Comparatively little flux results from the seepage of small particles because of their small mass. The flux due to large particles is also little because they do not pass through the filter. Intermediate size particles contribute the most to the flux since they are large enough to have appreciable mass and small enough to seep through. However, no theory exists to predict outlet flux as a function of particle size.

However, Christopher, Leith, and Symons (1990) evaluate mass penetration as a function of particle size. They developed the following empirical equation:

$$N = k \left(\frac{w^2v}{t} \right)^n \qquad 5.16(49)$$

where the exponent n is given by:

$$n = 0.916(d_p)^{0.257} \qquad 5.16(50)$$

where d_p is the particle diameter in μm.
The following equation gives model constant k:

$$k = a(d_p)^{4.82} \qquad 5.16(51)$$

where a = 4.45 × 10^{-6} for Ptfe-laminated fabric and a = 2.28 × 10^{-5} for untreated polyester felt fabric.

Bag Design

Bags are chosen for temperature and chemical resistance, mechanical stability, and the ability to collect the dust cake and then allow it to be easily removed. In utility applications, fiberglass bags are used exclusively. For low temperature applications (<280°F), acrylic is an economical alternative.

Bag fabric research focuses on two broad objectives: improving performance and economics. Different weaves

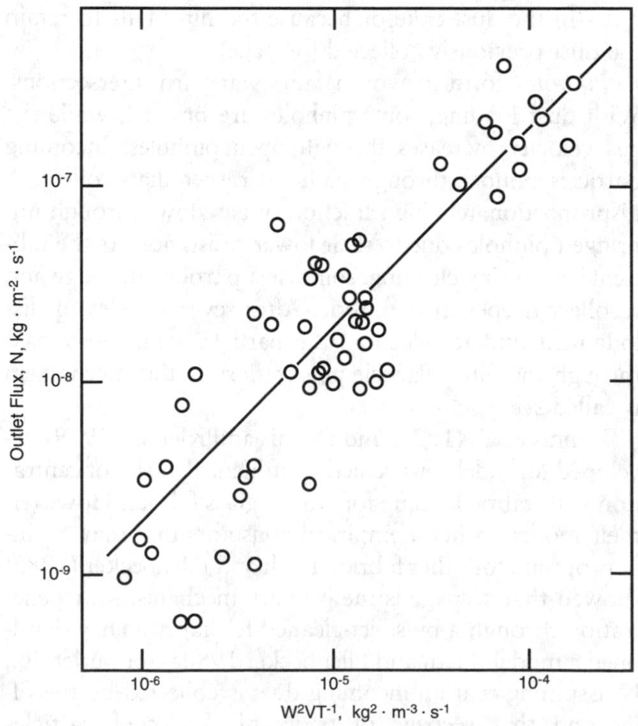

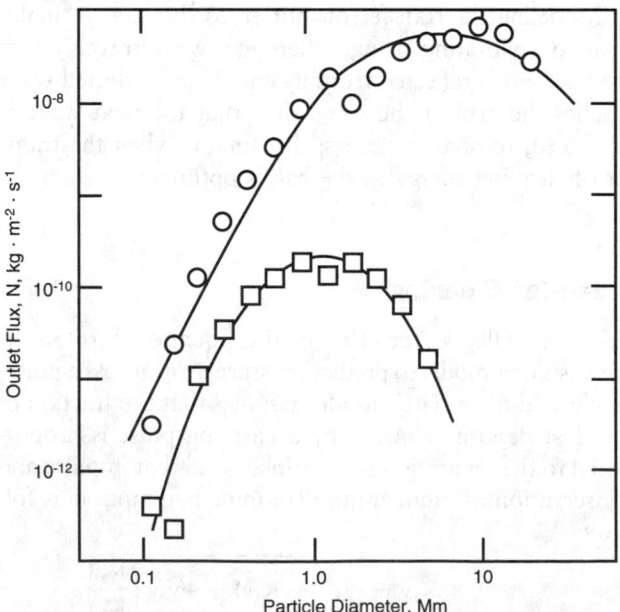

FIG. 5.16.13 Outlet flux versus w² v/t for laboratory experiments with a pulse-jet filter.

FIG. 5.16.14 Outlet flux versus particle diameter for a pulse-jet filter at 50 mm/sec velocity collecting granite dust on Ptfe-laminated bags (squares) and at 100 mm/sec velocity collecting fly ash on polyester felt with untreated surface (circles).

TABLE 5.16.2   RELATIVE PROPERTIES OF POPULAR FIBERS

Resistance to Dry Heat		Resistance to Moist Heat		Resistance to Abrasion		Relative Tensile Strength	Specific Gravity
Glass	550°	Glass	E	Nylon	E	Nylon	1.14
Teflon	400°	Teflon	E	Dacron	E	Polypro	0.90
Nomex	400°	Nomex*	E	Polypro	E	Dacron	1.38
Dacron	275°	Orlon	G	Nomex	E	Glass	2.54
Orlon	260°	Nylon*	G	Orlon	G	Polyeth	0.92
Rayon	200°	Rayon	G	Dynel	G	Rayon	1.52
Polypro	200°	Cotton	G	Rayon	G	Nomex	1.38
Nylon	200°	Wool	F	Cotton	G	Cotton	1.50
Wool	200°	Dacron*	F	Acetate	G	Dynel	1.30
Cotton	180°	Polypro	F	Polyeth	G	Orlon	1.14
Dynel	160°	Polyeth	F	Wool	F	Wool	1.32
		Dynel	F	Teflon	F	Teflon	2.10
		Acetate	F	Glass	P	Acetate	1.33

Resistance to Mineral Acids		Resistance to Alkalies		Relative Staple Fiber Cost		Relative Filament Fiber Cost	
Glass†	E	Teflon	E	Teflon	7.0	Teflon	54.
Teflon	E	Polypro	E	Nomex	3.2	Nomex	16.
Polypro	E	Nylon	G	Glass	6.	Acrylic	5.
Polyeth	G	Nomex	G	Orlon (Acrylic)	5.	Dacron	3.
Orlon	G	Cotton	G	Wool	5.	Nylon	3.
Dacron	G	Dynel	G	Nylon	5.	Polypro	2.8
Dynel	G	Polyeth	G	Dynel	4.	Polyeth	2.3
Nomex	F	Dacron	F	Dacron	3.5	Rayon	1.6
Nylon	P	Orlon	F	Acetate	2.0	Acetate	1.4
Rayon	P	Rayon	F	Rayon	1.5	Glass	1.0
Cotton	P	Acetate	P	Cotton	1.0		
Acetate	P	Glass	P				
Wool	G.	Wool	P				

Notes: *These fibers degrade in hot, moist atmospheres; Dacron is affected most, Nomex next, nylon the least.
†Glass is destroyed by gaseous HF at dew point temperatures.
These ratings are only a general guide: E = Excellent, G = Good, F = Fair, P = Poor.

and finishes for fiberglass are being studied, as well as alternative fabrics such as Teflon, Nomex, Ryton, and felted glass (see Table 5.16.2).

Bag weave is significant because different weaves produce different types of filter cakes. *Warp* is the system of yarns running lengthwise in a fabric, and *fill* is the system running crosswise. The style, weight, thickness, porosity, and strength are factors to consider in weave selection. Finishes are a lubricant against abrasion and protect against acid and alkali attack. Teflon, silicon, and graphite are the most popular finishes.

—Gurumurthy Ramachandran

References

Alexander, R. Mc K. 1949. Fundamentals of cyclone design and operation. *Proc. Australas Inst. Mining Metall.*, n.s., 203:152–153.

Barth, W. 1956. Design and layout of the cyclone separator on the basis of new investigations. *Brennstoff-Warme-Kraft* 8:1–9.

Beeckmans, J.M. 1973. A two-dimensional turbulent diffusion model of the reverse flow cyclone. *J. Aerosol Sci.* 4:329.

Billings, C.E., and J. Wilder. 1970. *Fabric filter cleaning studies*. ATD-0690, PB-200-648. Springfield, Va.: NTIS.

Boysan, F., W.H. Ayers, and J. Swithenbank. 1982. A fundamental mathematical modeling approach to cyclone design. *Trans. I. Chem. Eng.* 60:222–230.

Briggs, L.W. 1946. Effect of dust concentration on cyclone performance. *Trans. Amer. Inst. Chem. Eng.* 42:511.

Carman, P.C. 1956. *Flow of gases through porous media*. New York: Academic Press.

Carr, R.C., and W.B. Smith. 1984. Fabric filter technology for utility coal fired power plants. *JAPCA* 43, no. 1:79–89.

Christopher, P.C., D. Leith, and M.J. Symons. 1990. Outlet mass flux from a pulse-jet cleaned fabric filter: Testing a theoretical model. *Aerosol Sci. and Technol.* 13:426–433.

Dennis, R., R.W. Can, D.W. Cooper, R.R. Hall, V. Hampl, H.A. Klemm, J.E. Longley, and R.W. Stern. 1977. *Filtration model for coal fly ash with glass fabrics*. EPA report, EPA-600/7-77-084. Springfield, Va.: NTIS.

Dennis, R., and H.A. Klemm. 1979. A model for coal fly ash filtration. *J. Air Pollut. Control Assoc.* 29:230.

Dennis, R., and J. Wilder. 1975. *Fabric filter cleaning studies.* EPA-650/2-75-009. Springfield, Va.: NTIS.

Dietz, P.W. 1981. Collection efficiency of cyclone separators. *A.I.Ch.E. Journal* 27:888.

Dirgo, J. 1988. Relationships between cyclone dimensions and performance. Sc.D. thesis, Harvard University, Cambridge, Mass.

Dirgo, J.A., and D. Leith. 1985. Performance of theoretically optimized cyclones. *Filtr. and Sep.* 22:119–125.

Dirgo, J.A., and D. Leith. 1986. Cyclones. Vol. 4 in *Encyclopedia of fluid mechanics,* edited by N.P. Cheremisinoff. Houston: Gulf Publishing Co.

First, M.W. 1950. Fundamental factors in the design of cyclone dust collectors. Sc.D. thesis, Harvard Univ., Cambridge, Mass.

Happel, J. 1958. Viscous flow in multiparticle systems: Slow motion of fluids relative to beds of spherical particles. *A.I.Ch.E. Journal* 4:197.

Humphries, W., and J.J. Madden. 1981. Fabric filtration for coal fired boilers: Nature of fabric failures in pulse-jet filters. *Filtr. and Sep.* 18:503.

Iinoya, K. 1953. Study on the cyclone. Vol. 5 in *Memoirs of the faculty of engineering,* Nagoya Univ. (September).

Iozia, D.L., and D. Leith. 1989. Effect of cyclone dimensions on gas flow pattern and collection efficiency. *Aerosol Sci. and Technol.* 10:491.

Iozia, D.L., and D. Leith. 1990. The logistic function and cyclone fractional efficiency. *Aerosol Sci. and Technol.* 12:598.

Kessler, M., and D. Leith. 1991. Flow measurement and efficiency modeling of cyclones for particle collection. *Aerosol Sci. and Technol.* 15:8–18.

Kirch, R., and F. Loffler. 1987. Measurements of the flow field in a gas cyclone with the aid of a two-component laser doppler velocimeter. *ICALEO '87 Proc., Optical Methods in Flow and Particle Diagnostics, Laser Inst. of America* 63:28–36.

Lee, K.W., and B.Y.H. Liu. 1982a. Experimental study of aerosol filtration by fibrous filters. *Aerosol Sci. and Technol.* 1:35–46.

Lee, K.W., and B.Y.H. Liu. 1982b. Theoretical study of aerosol filtration by fibrous filters. *Aerosol Sci. and Technol.* 1:147–161.

Lee, K.W., and M. Ramamurthi. 1993. Filter collection. In *Aerosol measurement, principles, techniques, and applications,* edited by K. Willeke and P.A. Baron. Van Nostrand Reinhold Co.

Leith, D. 1984. Cyclones. In *Handbook of powder science and technology,* edited by M.A. Fayed and L. Otten. Van Nostrand Reinhold Co.

Leith, D., and R.W.K. Allen. 1986. Dust filtration by fabric filters. In *Progress in filtration and separation,* edited by R.J. Wakeman. Elsevier.

Leith, D., J.A. Dirgo, and W.T. Davis. 1986. Control devices: Application, centrifugal force and gravity, filtration and dry flue gas scrubbing. Vol. 7 in *Air pollution,* edited by A. Stern. New York: Academic Press, Inc.

Leith, D., and M.J. Ellenbecker. 1980a. Theory for penetration in a pulse jet cleaned fabric filter. *J. Air Pollut. Control Assoc.* 30:877.

———. 1980b. Theory for pressure drop in a pulse-jet cleaned fabric filter. *Atmos. Environ.* 14:845.

———. 1982. Dust emission characteristics of pulse jet cleaned fabric filters. *Aerosol Sci. and Technol.* 1:401.

Leith, D., M.W. First, and H. Feldman. 1977. Performance of a pulse-jet filter at high filtration velocity II. Filter cake redeposition. *J. Air Pollut. Control Assoc.* 27:636.

Leith, D., and W. Licht. 1972. The collection efficiency of cyclone type particle collectors—A new theoretical approach. *AIChE Symposium Ser.* 68:196–206.

Leith, D., and D. Mehta. 1973. Cyclone performance and design. *Atmos. Environ.* 7:527.

Lissman, M.A. 1930. An analysis of mechanical methods of dust collection. *Chem. Met. Eng.* 37:630.

Noll, K.E., and M. Patel. 1979. Evaluation of performance data from fabric filter collectors on coal-fired boilers. *Filtr. Sep.* 16:230.

Ramachandran, G., D. Leith, J. Dirgo, and H. Feldman. 1991. Cyclone optimization based on a new empirical model for pressure drop. *Aerosol Sci. and Technol.* 15:135–148.

Rosin, P., E. Rammler, and W. Intelman. 1932. Grundhagen und greuzen der zyklonentstaubung. *Zeit. Ver. Deutsch. Ing. Z.* 76:433.

Rudnick, S., and M.W. First. 1978. *Third Symposium on Fabric Filters for Particulate Collection.* Edited by N. Surprenant. EPA-600/7-78-087, 251. Research Triangle Park, N.C.: U.S. EPA.

Shepherd, C.B., and C.E. Lapple. 1939. Flow pattern and pressure drop in cyclone dust collectors. *Ind. Eng. Chem.* 31:972.

Shepherd, C.B., and C.E. Lapple. 1940. Flow pattern and pressure drop in cyclone dust collectors. *Ind. Eng. Chem.* 32:1246.

Stairmand, C.J. 1949. Pressure in cyclone separators. *Engineering* (London) 168:409.

Stairmand, C.J. 1951. The design and performance of cyclone separators. *Trans. Instn. Chem. Engrs.* 29:356.

Stechkina, I.B., A.A. Kirsch, and N.A. Fuch. 1969. Studies in fibrous aerosol filters—IV. Calculation of aerosol deposition in model filter in the range of maximum penetration. *Ann. Occup. Hyg.* 12:1–8.

Ter Linden, A.J. 1949. Investigations into cyclone dust collectors. *Proc. Inst. Mech. Engrs.* (London) 160:233–240.

Whiton, L.C. 1932. Performance characteristics of cyclone dust collectors. *Chem. Met. Eng.* 39:150.

5.17
PARTICULATE CONTROLS: ELECTROSTATIC PRECIPITATORS

FEATURE SUMMARY

Types of Designs
   Single- or two-stage

Applications
   Single-stage units are used in coal-fired power plants to remove fly ash, cement kiln dust, lead smelter fumes, tar, and pulp and paper alkali salts. Two-stage units are used for air conditioning applications.

Flue Gas Limitations
   Flow up to 4×10^6 acfm, temperature up to 800°C, gas velocity up to 10 ft/sec, pressure drop is under 1 in w.g.

Treatment Time
   2–10 sec

Power Requirement
   Up to 17.5 W per m³/min

Dust Particle Size Range
   0.01 μm to greater than 1000 μm

Collection Efficiency
   99.5 to 99.99%

Partial List of Suppliers
   Air Cleaning Specialists, Inc.; Air Pol Inc.; Air Quality Engineering, Inc.; ASEA Brown Boveri Inc.; Babcock and Wilcox, Power Generation Group; Belco Technologies Corp.; Beltran Associates, Inc.; Dresser Industries Inc.; Ducon Environmental Systems; GE Company; Joy Environmental Technologies, Inc.; North American Pollution Control Systems; Research-Cottrell Companies; Scientific Technologies, Inc.; United Air Specialists, Inc.; Universal Air Precipitator Corp.; Wheelabrator Air Pollution Control.

Electrostatic precipitation uses the forces of an electric field on electrically charged particles to separate solid or liquid aerosols from a gas stream. The aerosol is deliberately charged and passed through an electric field causing the particles to migrate toward an oppositely charged electrode which acts as a collection surface. Gravity or rapping the collector electrode removes the particles from the precipitator. Various physical configurations are used in the charging, collection, and removal processes.

ESPs are characterized by high efficiencies, even for small particles. They can handle large gas volumes with low pressure drops and can be designed for a range of temperatures. On the other hand, they involve high capital costs, take up a lot of space, and are not flexible after installation to changes in operating conditions. They may not work on particles with high electrical resistivity.

Commercial ESPs accomplish charging using a high-voltage, direct-current corona surrounding a highly charged electrode, such as a wire. The large potential gra-

dient near the electrode causes a corona discharge comprising electrons. The gas molecules become ionized with charges of the same sign as the wire electrode. These ions then collide with and attach to the aerosol particles, thereby charging the particles. Two electrodes charge the particles, and two electrodes collect the particles, with an electric field between each pair.

When the same set of electrodes is used for both charging and collecting, the precipitator is called a *single-stage precipitator*. Rapping cleans the collecting electrodes; thus, these precipitators have the advantage of continuous operation. The discharge electrode consists of a wire suspended from an insulator and held in position by a weight at the bottom. A power source supplies a large direct current (DC) voltage ($\sim$50 kV) which can be either steady or pulsed.

Parallel plate ESPs are more widely used in industry. Here, gas flows between two vertical parallel plates with several vertical wires suspended between them. The wires are held in place by weights attached at the bottom. These wires constitute the charging electrodes, and the plates are the collecting electrodes. Rapping the plate removes the collected dust; the dust gathers in a dust hopper at the bottom. Figure 5.17.1 shows a single-stage, parallel plate precipitator with accessories such as hoppers, rappers, wire weights, and distribution baffles for the gas.

If different sets of electrodes are used for charging and collecting, the precipitator is called a *two-stage precipitator* (see Figure 5.17.2). Cleaning involves removing the collecting plates and washing them or washing the collecting plates in place. In a two-stage ESP, the charging section is short, providing a short residence time, and the collection section is five or more times longer to provide sufficient time for collection (Crawford 1976). Two-stage ESPs are used in air conditioning applications, while single-stage ESPs are used in industrial applications where dust loadings are higher and space is available.

Corona Generation

When the potential difference between the wire and plate electrodes increases, a voltage is reached where an electrical breakdown of the gas occurs near the wire. When gas molecules get excited, one or more of the electrons can shift to a higher energy level. This state is transient; once the excitation has ceased, the molecule reverts to its ground state, thereby releasing energy. Part of this energy converts

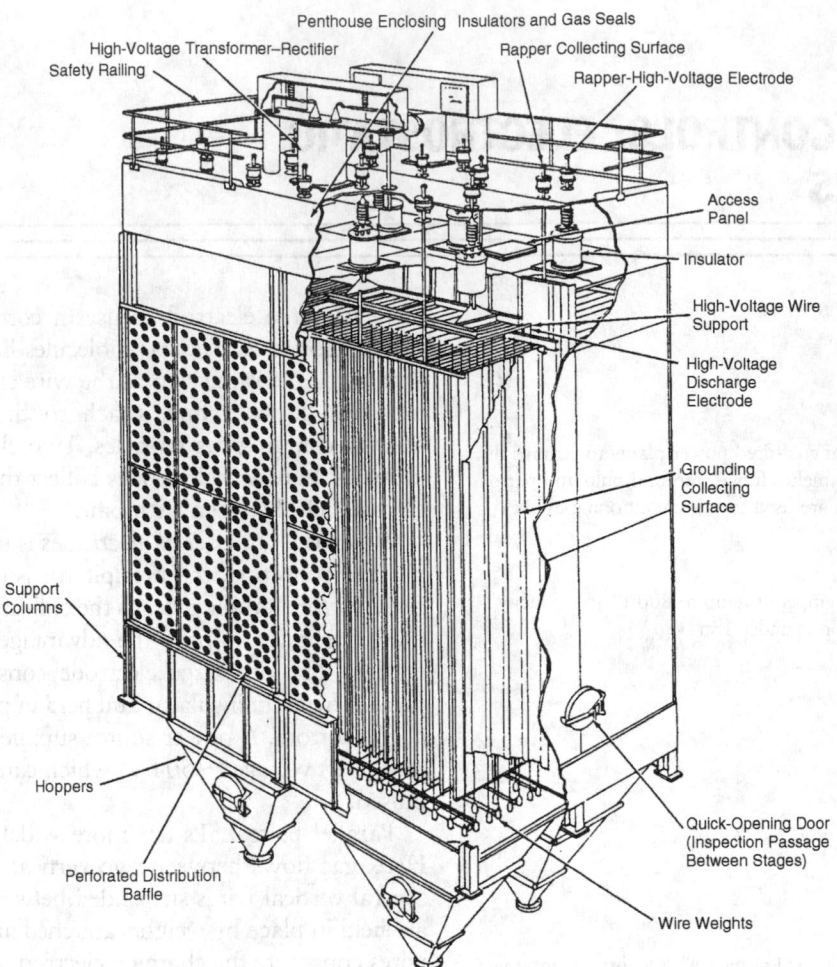

Penthouse Enclosing Insulators and Gas Seals
High-Voltage Transformer–Rectifier
Safety Railing
Rapper Collecting Surface
Rapper–High-Voltage Electrode
Access Panel
Insulator
High-Voltage Wire Support
High-Voltage Discharge Electrode
Grounding Collecting Surface
Support Columns
Hoppers
Quick-Opening Door (Inspection Passage Between Stages)
Perforated Distribution Baffle
Wire Weights

FIG. 5.17.1 Single-stage, parallel plate ESP with accessories.

to light. The bluish glow around the wire is the corona discharge. A different situation occurs when an electron or ion imparts additional energy on an excited molecule. This process causes a cascade or avalanche effect described next.

The space between the wire and the plate can be divided into an active and a passive zone (see Figure 5.17.3). In the active zone, defined by the corona glow discharge, electrons leave the wire electrode and impact gas molecules, thereby ionizing the molecules. The additional free electrons also accelerate and ionize more gas molecules. This avalanche process continues until the electric field decreases to the point that the released electrons do not acquire sufficient energy for ionization. The behavior of these charged particles depends on the polarity of the electrodes; a negative corona is formed if the discharge electrode is negative, and a positive corona is formed if the discharge electrode is positive.

In a negative corona, positive ions are attracted toward the negative wire electrode, and electrons are attracted toward the positive plate or cylinder electrode. Beyond the corona glow region, the electric field diminishes rapidly, and if electronegative gases are present, the gas molecules

become ionized by electron impact. The negative ions move toward the plate electrode. In the passive zone, these ions attach themselves to aerosol particles and serve as the principal means for charging the aerosol. The ion concentration is typically 10^7 to 10^9 ions/cm^3.

The corona current, and therefore the charge density in the space between the electrodes, depends on factors such as the ionic mobility, whether the gas is electropositive or electronegative, and whether the corona is positive or negative. If the gas is electropositive with low electron affinity like N_2, H_2, or an inert gas, its molecules absorb few electrons. Thus, the current is predominantly electronic. Due to the higher mobility of electrons, the corona current is high. Conversely, an electronegative gas like O_2 has high electron affinity and absorbs electrons easily. Here, the current is due to negative ions, and thus the corona current is low due to the lower mobility of the gas ions.

When a corona is negative, the free electrons leaving the active zone are transformed into negative ions with a substantially lower mobility on their way to the plates. The negative charge carriers thus cover the first part of their path as fast, free electrons and the second part as slower ions; their average mobility is lower than that of free elec-

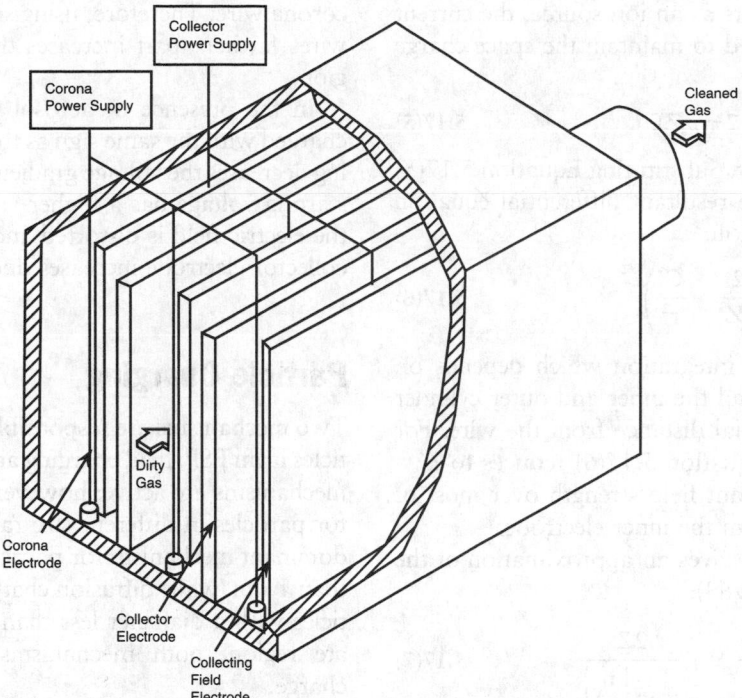

FIG. 5.17.2   Parallel plate, two-stage ESP.

trons but higher than that of the large ions. On the other hand, when a corona is positive, the positive charge carriers are large, slow ions by origin and retain this form throughout their motion. Consequently, a negative corona always has a higher corona current than a positive corona for an applied voltage.

Negative coronas are more commonly used in industrial applications, while positive coronas are used for cleaning air in inhabited spaces. A negative corona is accompanied by ozone generation and, therefore, is usually not used for cleaning air in inhabited spaces. However, most industrial gas-cleaning precipitators use a negative corona because of its superior electrical characteristics which increase efficiency at the temperatures at which they are used.

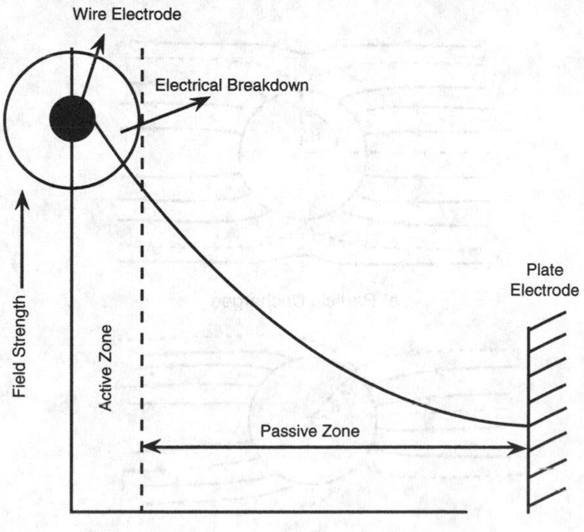

Current–Voltage Relationships

For a wire-in-cylinder configuration, the current–voltage relationship can be derived from Poisson's equation as follows:

$$\nabla^2 V = -\frac{4\pi}{\varepsilon_0}\rho_s \qquad 5.17(1)$$

where:

ρ_s = the space charge per unit volume
ε_0 = the permittivity of the medium.

The equations in this treatment follow the electrostatic system of units (ESU). The value V is the electric potential which is related to the electric field $\vec{E}$ as follows:

$$-\nabla V = \vec{E} \qquad 5.17(2)$$

Thus, the relationship can be expressed as follows:

$$\nabla^2 V = \frac{\partial^2 V}{\partial r^2} + \frac{1}{r}\frac{\partial V}{\partial r} \qquad 5.17(3)$$

In the presence of a charge sufficient to produce corona discharge, the preceding equations can be expressed as:

$$\frac{dE}{dr} + \frac{1}{r}E = -\frac{4\pi}{\varepsilon_0}\rho_s \qquad 5.17(4)$$

FIG. 5.17.3   Variation of field strength between wire and plate electrodes.

Assuming that the wire acts as an ion source, the current i applied to the wire is used to maintain the space charge as follows:

$$i = 2\pi r \rho_s Z E \qquad 5.17(5)$$

where Z is the ion mobility. Substituting Equation 5.17(5) in 5.17(4) and solving the resultant differential equation yields the following equation:

$$E = \left(\frac{2i}{Z} + \frac{C^2}{r^2}\right)^{1/2} \qquad 5.17(6)$$

where C is a constant of integration which depends on corona voltage, current, and the inner and outer cylinder diameters and r is the radial distance from the wire. For large values of i and r, Equation 5.17(6) reduces to $E = (2i/Z)^{1/2}$, implying a constant field strength over most of the cross section away from the inner electrode.

The following equation gives an approximation of the corona current i (White 1963):

$$i = V (V - V_0) \frac{2Z}{r_0^2 \ln\left[\dfrac{r_0}{r_i}\right]} \qquad 5.17(7)$$

where:

    V = the operating voltage
    V_0 = the corona starting voltage
    r_0 and r_i = the radii of the cylinder and wire, respectively.

The value V_0 can be estimated by use of the following equation:

$$V_0 = 100\delta f r_i \left(1 + \frac{0.3}{\sqrt{r_i}}\right) \ln\left[\frac{r_0}{r_i}\right] \qquad 5.17(8)$$

where δ is a correction factor for temperature and pressure as follows:

$$\delta = \frac{293}{T}\frac{P}{760} \qquad 5.17(9)$$

The factor f is a wire roughness factor, usually between 0.5 and 0.7 (White 1963).

Equation 5.17(8) shows that reducing the size of the corona wire decreases the applied voltage necessary to initiate corona V_0. Decreasing the corona starting voltage increases the corona current for an applied voltage.

An analysis similar to that shown for wire-in-cylinder configuration can be applied to the parallel plate and wire electrode configuration. The equations are not as simple because the symmetry of the cylindrical precipitator simplifies the mathematics. The plate-type precipitator also has an additional degree of freedom—corona wire spacing. Qualitatively, if the corona wires are spaced close together, the system approaches the field configuration of a parallel plate capacitor, which yields a constant field strength in the interelectrode space. For an applied voltage, this configuration reduces the electric field near the corona wire. Therefore, using smaller wires or spacing the wires farther apart increases the current density in a region.

In the presence of aerosol particles, the particles get charged with the same sign as the wire. This particle charging decreases the voltage gradient, and therefore the corona starting voltage has a higher value. Another effect is that the electric field is distorted and the electric field near the collector electrode increases due to image forces.

Particle Charging

Two mechanisms are responsible for charging aerosol particles in an ESP: field charging and diffusion charging. Both mechanisms are active; however, each becomes significant for particles in different size ranges. Field charging is the dominant mechanism for particles with a diameter greater than 1 μm, while diffusion charging predominates for particles with a diameter less than 0.2 μm. In the intermediate region, both mechanisms contribute a significant charge.

FIELD CHARGING

The presence of particles with a dielectric constant greater than unity causes a localized deformation in the electric field (see Figure 5.17.4). Gas ions travel along electric field lines, and because the lines intercept the particle matter, the ions collide with these particles and charge them. When the particle reaches a *saturation charge*, additional ions are repelled and charging stops. The amount of charge q on a particle is the product of the number of charges n and the electronic charge e (q = ne). The following equation gives the rate of particle charging:

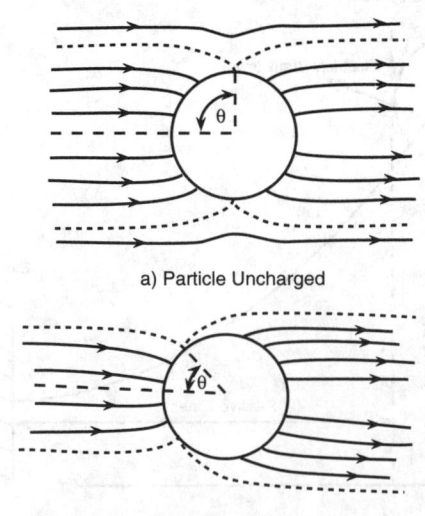

a) Particle Uncharged

b) Particle Partially Charged

FIG. 5.17.4 Distortion of an electric field around an aerosol particle.

$$\frac{d\left(\frac{n}{n_s}\right)}{dt} = \pi N_i eZ \left(1 - \frac{n}{n_s}\right)^2 \qquad 5.17(10)$$

On integration, this equation yields the number of charges on a particle as a function of time as follows:

$$n = n_s \left(\frac{t}{t + \tau}\right) \qquad 5.17(11)$$

where n_s is the saturation charge given by $(3\varepsilon/\varepsilon + 2)$ $(d^2 E/4e)$. The saturation charge is the maximum charge that can be placed on a particle of diameter d by a field strength E. The value ε is the dielectric constant of the particle. The value τ_e is a time constant for the rapidity of charging and is equal to $(1/\pi e Z_i N_i)$, where Z_i is the ion mobility and N_i is the ion concentration.

DIFFUSION CHARGING

Particles are charged by unipolar ions in the absence of an electric field. The collision of ions and particles occurs due to the random thermal motion of ions. If every ion that collides with a particle is retained, the rate of charging with respect to time is as follows:

$$\frac{dn}{dt} = \frac{\pi}{4} d^2 \bar{c} N_i \exp\left[-\frac{2ne^2}{dkT}\right] \qquad 5.17(12)$$

For an initially uncharged particle, integrating this equation gives the number of charges gained due to diffusion charging as follows:

$$n = \left(\frac{dkT}{2e^2}\right) \ln\left[1 + \frac{\pi d\bar{c}e^2 N_i}{2kT}t\right] \qquad 5.17(13)$$

where:

    k = the Boltzman constant
    T = the absolute temperature
    e = the charge of one electron

Here, $\bar{c}$ is the mean thermal speed of the ions and for a Maxwellian distribution is given by $\sqrt{8kT/\pi m}$, where m is the ionic mass.

When both field and diffusion charging are significant, adding the values of the n calculated with Equations 5.17(11) and 5.17(13) is not satisfactory; adding the charging rates due to field and diffusion charging is better and gives an overall rate. This process yields a nonlinear differential equation with no analytic solution and therefore has to be solved numerically. An overall theory of combined charging agrees with experimental values but is computationally cumbersome.

Migration Velocity

The following equation determines the velocity of a charged particle suspended in a gas and under the influence of an electric field by equating the electric and Stokes drag forces on the particle:

$$V_p = \frac{neEC_c}{3\pi\mu d} \qquad 5.17(14)$$

Here, V_p is the particle migration velocity toward the collector electrode, and E is the collector electric field. The value for n depends on whether field or diffusion charging is predominant. Particles charged by field charging reach a steady-state migration velocity, whereas the migration velocity for small particles charged by diffusion charging continues to increase throughout the time that the dust is retained in the precipitator.

ESP Efficiency

The most common efficiency theory is the Deutsch–Andersen equation. The theory assumes that particles are at their terminal electrical drift velocity and well-mixed in every plane perpendicular to the gas flow direction due to lateral turbulence. The theory also assumes plug flow through the ESP, no reentrainment of particles from the collector plates, and uniform gas velocity throughout the cross section. Collection efficiency is expressed as follows:

$$\eta = 1 - \exp\left[-\frac{V_p A}{Q_g}\right] \qquad 5.17(15)$$

where A is the total collecting area of the precipitator and Q_g is the volumetric gas flow. The derivation of this equation is identical to that for gravity settling chambers except that the terminal settling velocity under a gravitational field is replaced with terminal drift velocity in an electric field.

The precipitator performance can differ from theoretical predictions due to deviations from assumptions made in the theories. Several factors can change the collection efficiency of an idealized precipitator. These factors include particle agglomeration, back corona, uneven gas flow, and rapping reentrainment.

Cooperman (1984) states that the Deutsch–Andersen equation neglects the role of mixing (diffusional and large-scale eddy) forces in the precipitator. These forces can account for the differences between observed and theoretical migration velocities and the apparent increase in migration velocity with increasing gas velocity. He postulates that the difference in particle concentration along the precipitator length produces a mixing force that results in a particle velocity through the precipitator that is greater than the gas velocity. At low velocities, the effect is pronounced, but it is masked at higher velocities. He presents a more general theory for predicting collection efficiency by solving the mass balance equation as follows:

$$D_1 \frac{\partial^2 C}{\partial x^2} + D_2 \frac{\partial^2 C}{\partial y^2} - V_g \frac{\partial C}{\partial x} + V_p \frac{\partial C}{\partial x} = 0 \qquad 5.17(16)$$

where:

    C   = the particle concentration
    D_1 and D_2 = the longitudinal and transverse mixing co-
                 efficients, respectively

V_g = the gas velocity in the x direction
V_p = the migration velocity in the y direction
(see Figure 5.17.5)

The initial conditions chosen are $C(0,y) = C_0$, and $C(x,y) \to 0$ as $y \to 0$, where C_0 is the particle concentration at the inlet. The first condition implies a constant concentration across the inlet, while the second condition states that an infinitely long precipitator has zero penetration.

The boundary conditions are $D_2\,\partial C/\partial y + V_pC = 0$ at the center plane, and $D_2\,\partial C/\partial y - fV_pC = 0$ at the boundary layer. The first condition states that the diffusional flux balances the flux due to migration velocity, and no net particle flux occurs across the center plane. The second condition contains a reentrainment factor f. This factor is an empirical parameter indicating what fraction of the particles that enter the boundary layer are later reentrained back into the gas stream. The complete analytic solution is fairly complicated, and therefore only the first term of the solution is used; but it yields sufficient accuracy as follows:

$$\eta = 1 - \exp\left[\left(\frac{V_g}{2D_1}\right) - \sqrt{\left(\frac{V_g}{2D_1}\right)^2 + \left(\frac{D_2}{D_1}\right)\left(\frac{\lambda_1}{b}\right)^2 + \frac{V_p^2}{4D_1D_2}}\right]L$$

5.17(17)

$$\tan\lambda_1 = \frac{\beta}{\chi}\,\frac{2\lambda_1(1-f)}{\lambda_1^2 - (1-f)\dfrac{\beta^2}{\chi^2}}$$

$$\beta = \frac{bV_p}{2D_1}; \quad \chi = \frac{D_2}{D_1}$$

where b is the distance from the center line to the collector plate and λ_1 is the smallest positive solution to the preceding transcedental equation.

The Deutsch–Andersen equation and other efficiency theories are limiting cases of this general theory. To use this exact theory, an environmental engineer has to first decide on appropriate mixing coefficients. For a large D_2 (i.e., large transverse mixing), the equation for efficiency reduces to the following equation:

$$\eta = 1$$
$$- \exp\left[\left(\frac{V_gb}{2D_1}\right) - \sqrt{\left(\frac{V_gb}{2D_1}\right)^2 + 2(1-f)\frac{V_pb}{2D_1}}\right]\frac{L}{b}$$

5.17(18)

Expanding the square root term as a Taylor series up to the first three terms and rearranging gives the following equation:

$$\eta = 1 - \exp\left[-(1-f)\left(1 - \frac{V_p}{V_g^2}\right)\frac{D_1(1-f)}{b}\frac{V_pA_c}{Q_g}\right]$$

5.17(19)

This equation is analogous to the Deutsch–Andersen equation except that two modifying factors are added. The $(1-f)$ factor corrects for reentrainment, and the second factor is related to the diffusional transport in the gas flow direction that modifies the gas velocity. Cooperman (1984) suggests a value of 18,000 cm²/sec for D_1.

Thus, efficiency depends on longitudinal mixing as well as reentrainment.

DUST RESISTIVITY

A major factor affecting ESP performance is the electrical resistivity of dust. If the resistivity is low (high conductivity), the charge on a particle leaks away quickly as particles collect on the plate. Thus, reentrainment becomes possible. As van der Waals' forces may be insufficient to bind the particles, the particles hop or creep through the precipitator. Conversely, if the particle resistivity is high, a charge builds on the collected dust. This charge reduces field strength, reducing ionization and the migration velocity of particles through the gas.

A plot of resistivity versus temperature shows a maximum resistivity between 250 and 350°F (see Figure 5.17.6). This relationship is unfortunate because operators cannot reduce the ESP temperature below 250°F without risking condensation of H_2SO_4 on the plate surfaces.

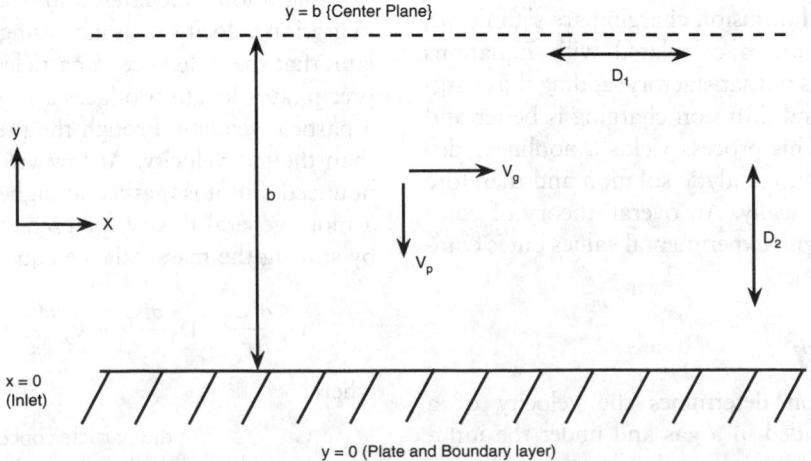

FIG. 5.17.5 Top view of wire-plate precipitator.

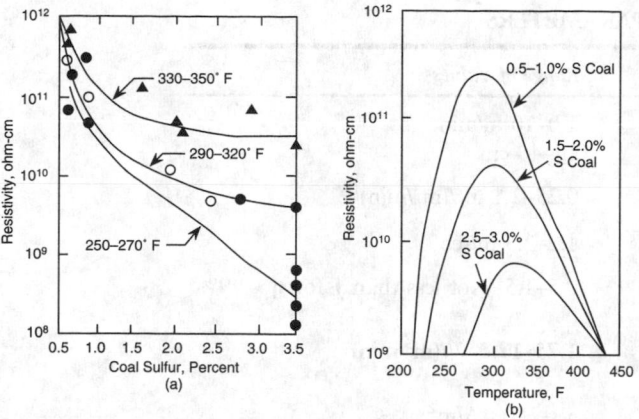

FIG. 5.17.6 Variation of fly ash resistivity with coal sulfur content and flue-gas temperature.

Increasing the temperature above 350°F results in excessive heat loss from the stack.

Resistivity decreases with increased sulfur content in coal because of the increased adsorption of conductive gases on the fly ash. Resistivity changes were responsible for the increased fly ash emissions when power plants switched from high-sulfur coal to low-sulfur coal to reduce SO_2 emissions in the United States.

In some cases of high resistivity in dust, adding a conditioning agent to the effluent gases substantially reduces resistivity and enhances particle collection. Examples are adding SO_3 to gas from power generator boilers and NH_3 to gas from catalytic cracking units used in petroleum refining. The overall efficiency can increase from 80% before injection to 99% after injection.

PRECIPITATOR DESIGN

Precipitator design involves determining the sizing and electrical parameters for an installation. The most important parameters are the precipitation rate (migration velocity), specific collecting area, and specific corona power (White 1984). In addition, the design includes ancillary factors such as rappers to shake the dust loose from the plates, automatic control systems, measures for insuring high-quality gas flow, dust removal systems, provisions for structural and heat insulation, and performance monitoring systems.

The design engineer should determine the size distribution of the dust to be collected. Based on this information, the engineer can calculate the migration velocity (also known as the precipitation rate) V_p using Equation 5.17(14) for each size fraction. The engineer calculates the number of charges on a particle n using Equation 5.17(11) or 5.17(13), depending on whether field or diffusion charging is predominant. Diffusion charging is the dominant charging mechanism for particles less than 0.2 μm, while field charging is predominant for particles greater than 1 μm. For particles of intermediate sizes, both mechanisms

are significant. The engineer can also calculate V_p empirically from pilot-scale or full-scale precipitator tests. The value V_p also varies with each installation depending on resistivity, gas flow quality, reentrainment losses, and sectionalization. Therefore, each precipitator manufacturer has a file of experience to aid design engineers in selecting a value of V_p. A high migration velocity value indicates high performance.

After selecting a precipitation rate, the design engineer uses the Deutsch–Andersen relationship, Equation 5.20(15) or 5.20(18), to determine the collecting surface area required to achieve a given efficiency when handling a given gas flow rate. If Equation 5.20(18) is used, the engineer can choose the value of the reentrainment factor f empirically from pilot-scale studies or previous experience or set it to zero as an initial guess. The quantity A/Q_g is called the specific collection area.

The corona power ratio is P_c/Q_g, where P_c is the useful corona power. The design engineer determines the power required for an application on an empirical basis. The power requirements are related to the collection efficiency and the gas volume handled. Figure 5.17.7 plots the collection efficiency versus the corona power ratio. At high efficiencies, large increments of corona power are required for small increments in efficiency. The precipitation rate (migration velocity) is related to corona power as follows:

$$V_p = \frac{kP_c}{A} \qquad 5.17(20)$$

where k is an empirical constant that depends on the application. The Deutsch–Andersen equation can therefore be expressed as follows:

$$\eta = 1 - \exp\left[-\frac{kP_c}{Q}\right] \qquad 5.17(21)$$

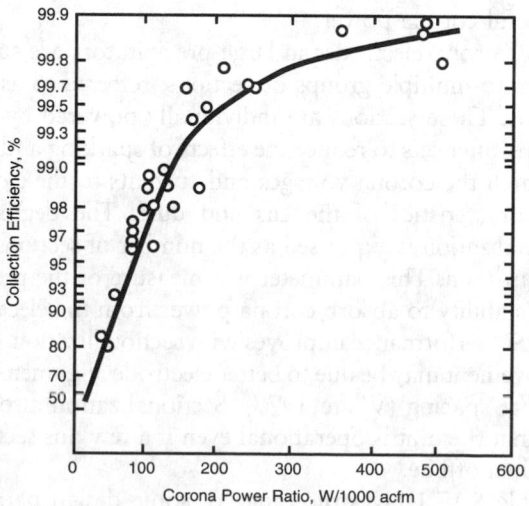

FIG. 5.17.7 Collection efficiency as a function of the corona power ratio.

TABLE 5.17.1 RANGES FOR ESP DESIGN PARAMETERS

Parameter	Range of Values
Precipitation rate V_p	1.0–10 m/min
Channel width D	15–40 cm
Specific collection area $\dfrac{\text{Plate Area}}{\text{Gas Flow}}$	0.25–2.1 m²/(m³/min)
Gas velocity u	1.2–2.5 m/sec
Aspect ratio R = $\dfrac{\text{Duct Length}}{\text{Height}}$	0.5–1.5 (Not less than 1 for $\eta > 99\%$)
Corona power ratio $\dfrac{\text{Corona Power}}{\text{Gas Flow}} = \dfrac{P_c}{Q}$	1.75–17.5 W/(m³/min)
Corona current ratio $\dfrac{\text{Corona Current}}{\text{Plate Area}}$	50–750 µA/m²
Plate area per electrical set A_s	460–7400 m²
Number of electrical sections N_s	
a. In the direction of gas flow	2–8
b. Total number of sections	1–10 bus sections/(1000 m³/min)
Spacing between sections	0.5–2 m
L_{en}, L_{ex}	2–3 m
Plate Height; Length	8–15 m; 1–3 m

Power Density versus Resistivity Ash Resistivity, ohm-cm	Power Density, W/m²
10^4–10^7	43
10^7–10^8	32
10^9–10^{11}	27
10^{11}	22
10^{12}	16
10^{13}	10.8

Source: C.D. Cooper and F.C. Alley, 1986, *Air pollution control: A design approach* (Boston: PWS Publishers).

This relationship gives the corona power necessary for a given precipitation efficiency, independent of the precipitator design. The maximum corona power capability of the electrical sets should be considerably higher than the useful corona power.

The corona electrodes in large precipitators are subdivided into multiple groups or sections, referred to as bus sections. These sections are individually powered by separate rectifier sets to reduce the effects of sparking and better match the corona voltages and currents to the electrical characteristics of the gas and dust. The degree of sectionalization is expressed as the number of sections per 1000 m³/min. This parameter is a measure of the precipitator's ability to absorb corona power from the electrical sets. ESP performance improves with sectionalization. This improvement may be due to better electrode alignment and accurate spacing (White 1977). Sectionalization also implies that the unit is operational even if a few bus sections are taken offline.

Table 5.17.1 gives the range of some design parameters. A design engineer can specify the basic geometry of an ESP using the information in the table.

The following equation determines the total number of channels in parallel or the number of electrical sections in a direction perpendicular to that of gas flow N_d:

$$N_d = \frac{Q}{uDH} \qquad 5.17(22)$$

where:

Q = the volumetric gas flow rate
u = the linear gas velocity
D = the channel length (plate separation)
H = the plate height

The overall width of the precipitator is N_dD. The overall length of the precipitator is as follows:

$$L_0 = N_sL_p + (N_s - 1)L_s + L_{en} + L_{ex} \qquad 5.17(23)$$

where:

N_s = the number of electrical sections in the direction of flow
L_p = the plate length

L_s = the spacing between the electrical sections

L_{en} and L_{ex} = the entrance and exit section lengths, respectively.

The following equation determines the number of electrical sections N_s:

$$N_s = \frac{RH}{L_p} \qquad 5.17(24)$$

where R is the aspect ratio.

When the number of ducts and sections have been specified, the design engineer can use the following equation to calculate the actual area:

$$A_a = 2HL_pN_sN_d \qquad 5.17(25)$$

The preceding procedure provides a rational basis for determining the plate area, total power, and degree of sectionalization required.

—*Gurumurthy Ramachandran*

References

Cooperman, G. 1984. A unified efficiency theory for electrostatic precipitators. *Atmos. Environ.* 18, no. 2:277–285.

Crawford, M. 1976. Air pollution control theory. McGraw Hill, Inc.

White, H.J. 1963. Industrial electrostatic precipitation. New York: Addison Wesley.

———. 1977. Electrostatic precipitation of fly ash—Parts I, II, III, and IV. *JAPCA* 27 (January–April).

———. 1984. The art and science of electrostatic precipitation. *JAPCA* 34, no. 11:1163–1167.

5.18
PARTICULATE CONTROLS: WET COLLECTORS

FEATURE SUMMARY

Types of Designs
Low or high energy

Limitations
Low efficiency and liquid waste disposal for low-energy designs; operating costs for high-energy designs

Loadings
0.1 to 100 g/m³

Pressure Drop
Low energy—0.5 to 2 cm H_2O; high energy—2 to 100 cm H_2O

Overall Efficiency
Low energy—high for >10 μm; high energy—high for >1 μm

Partial List of Suppliers
Aerodyne Development Corp; Air-Cure Environmental Inc./Ceilcote Air Pollution Control; Air Pol Inc.; Beco Engineering Co.; Ceilcote Co.; Croll Reynolds Co. Inc.; Fairchild International; Hild Floor Machine Co. Inc.; Joy Technologies Inc./Joy Environmental Equipment Co.; Lurgi Corp USA; Merck & Co. Inc./Calgon Corp; Safety Railway Service Corp/Entoleter Inc.; Sonic Environmental Systems; Spendrup & Associates Inc./Spendrup Fan Co.; Svedala Industries Inc./Allis Mineral Systems; Kennedy Van Saun; Wheelabrator Air Pollution Control; Zurn Industries Inc./Air Systems Div.

A particulate scrubber or wet collector is a device in which water or some other solvent is used in conjunction with inertial, diffusion, or other forces to remove particulate matter from the air or gases. The scrubbing process partially mimics natural processes where dust-laden air is cleaned by rain, snow, or fog. The first industrial scrubbers attempted to duplicate this natural cleaning, with dusty air ascending through a rain of liquid droplets in a large, vertical tube. Subsequent developments reduced the space requirements for scrubbers.

The first patent for a particle scrubber design was issued in Germany in 1892, and the first gas scrubber with rotating elements was patented about 1900. Venturi scrubbers were developed just after World War II, mainly in the United States, and represented a breakthrough in scrubber design (Batel 1976).

General Description

In the scrubbing process, gas containing dust particles contacts a scrubbing liquid (often water). Here, some of the dust particles are captured by the scrubbing liquid. The scrubbing liquid can also condense on the dust particles, forming liquid droplets which are more easily removed than the dry dust particles. Finally, the mist droplets, which may or may not contain dust particles, are removed in an entrainment separator. Figure 5.18.1 schematically shows the process. The primary collection mechanisms for particle removal are inertial impaction, interception, diffusion or diffusiophoresis, and electrical attraction. At times, other mechanisms, such as those listed in Table 5.18.1, are involved in collection. However, in all cases, the dominant mechanism is inertial impaction followed by diffusion.

Scrubbers can be divided into three categories. In the first category, the scrubbing liquid is a spray, and collection occurs when particles are embedded by impaction in the scrubbing liquid surface (see part a in Figure 5.18.2). In the second category, collection occurs by impingement

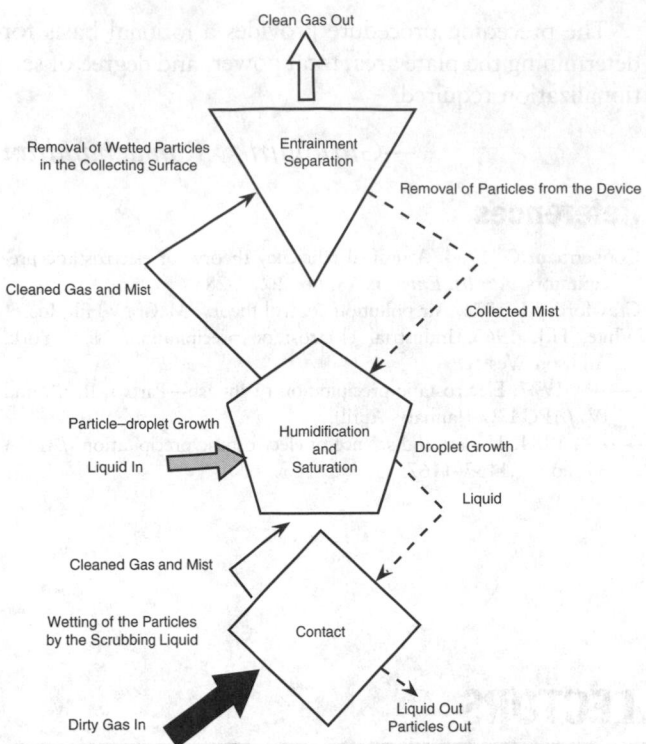

FIG. 5.18.1 Schematic diagram of scrubbing process.

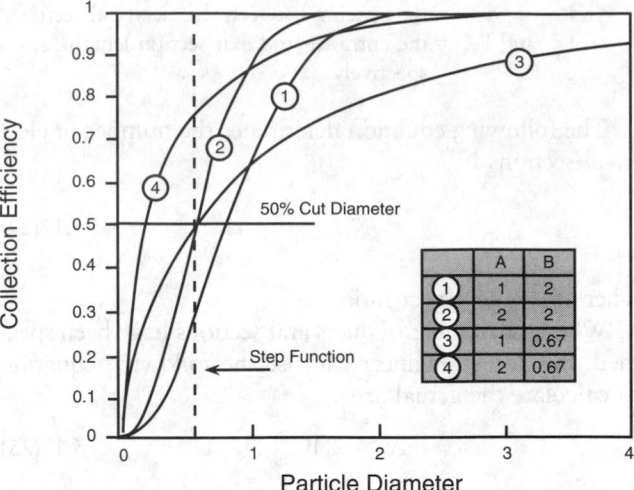

FIG. 5.18.3 Typical grade efficiency curves as computed from Efficiency = $\exp(Ad_{pa}^{B})$.

on a wetted surface (see part b of Figure 5.18.2). Finally, in the third category, particle-laden air is bubbled through the scrubbing liquid, and particles are removed by impingement (see part c of Figure 5.18.2).

The main factor influencing a scrubber's efficiency is the size of the dust particles being removed. Scrubber collection efficiency also varies with scrubber design and operating conditions. Figure 5.18.3 shows examples of typical grade efficiency curves. Because of the efficiency variation due to particle size, collection efficiency is often expressed as a function of one particle size, the *50% cut diameter,* that is, the particle diameter for which collection efficiency is 50% (Calvert et al. 1972; Calvert 1974, 1977). The rationale for this definition is that the grade efficiency curve for all scrubbing is steepest at the 50% point, and for estimating purposes this curve is considered to be a step function—all particles greater than this size are collected, all smaller are not (see dotted line on Figure

5.18.3). Because collection efficiency is also a function of particle density, the concept of the aerodynamic particle diameter must be considered. The aerodynamic diameter of an aerosol particle is the diameter of a unit density sphere (density = 1 g/cm^3) which has the same settling velocity as the particle. With this definition, the relationship of a spherical particle of diameter d_s and density ρ to its aerodynamic diameter d_a is expressed as:

$$d_a = d_s(\rho C_c)^{1/2} \qquad \textbf{5.18(1)}$$

For particles with diameters of several micrometers or less, a slip correction term C_c must be included in the aerodynamic diameter conversion as follows:

$$C_c = 1 + \frac{2\lambda}{d_s}\left[1.257 + 0.400\exp\left(-\frac{1.1d_s}{2\lambda}\right)\right] \qquad \textbf{5.18(2)}$$

The term d_s is the dust particle diameter, and λ is the mean free path of gas molecules. For air at 20°C, λ has a value of 0.0687 μm.

Scrubber Types

Over the years, many scrubber designs have been developed based on the three preceding scrubbing processes. Some representative examples are described next.

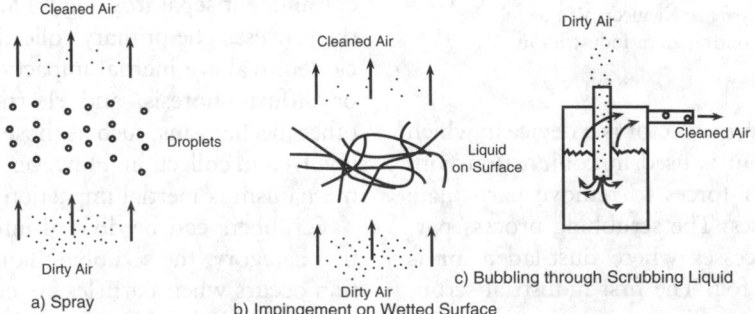

FIG. 5.18.2 Schematic illustration of various scrubbing mechanisms.

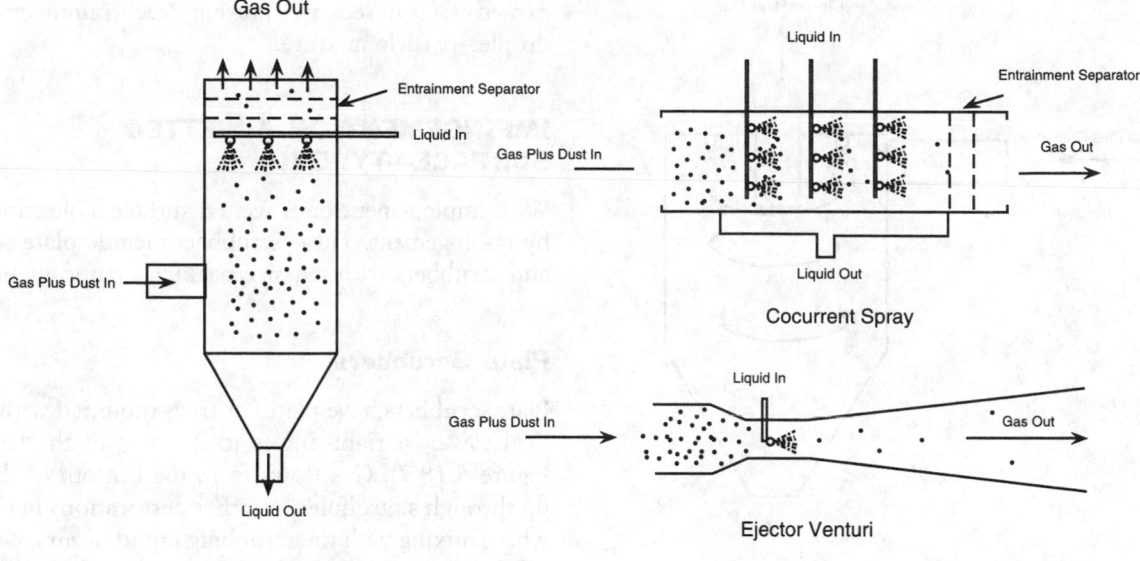

FIG. 5.18.4 Example of preformed spray unit.

SPRAY COLLECTORS—TYPE I

With spray-type units, particles are collected on liquid droplets within the scrubbing unit. The liquid droplets are formed either by atomization, where the flow rate of the scrubbing liquid and pressure in the atomizing nozzle control the droplet size and number, or by the moving gas stream, which atomizes and accelerates the resulting droplets.

Preformed Spray

With a preformed spray, droplets are formed by a nozzle and are then sprayed through the dust-laden air (see Figure 5.18.4). Removal of the dust is primarily by impaction, although diffusion and diffusiophoresis can also contribute to particle collection. Spray scrubbers, taking advantage of gravitational settling, achieve cut diameters around 2 μm, with high-velocity sprays capable of reducing this lower limit to about 0.7 μm (aerodynamic diameter) (Calvert et al. 1972).

Gas-Atomized Spray

With a gas-atomized spray, high-velocity, particle-laden gas atomizes liquid into drops, and the resulting turbulence and velocity difference between droplets and particles enhances particle–droplet collisions and hence particle removal (see Figure 5.18.5). Typical gas velocities range from 60 to 120 m/sec (200–400 f/sec), and the pressure drop in these units is relatively high. A typical example of this type of unit is the venturi scrubber. Venturi scrubbers have achieved cut diameters down to 0.2 μm (Calvert 1977).

Centrifugal Scrubbers

With centrifugal scrubbers, the scrubbing liquid is sprayed into the gas stream at the same time the unit is imparting a spinning motion to the mixture of particles and droplets (see Figure 5.18.6). The result is a centrifugal deposition of the particle–droplet mixture on the outer walls of the scrubber. Depending on the amount of scrubbing liquid used, the collection efficiency is good for particles down to about 1 to 2 μm diameter (aerodynamic) (Calvert et al. 1972). The tangential velocities in these units should not

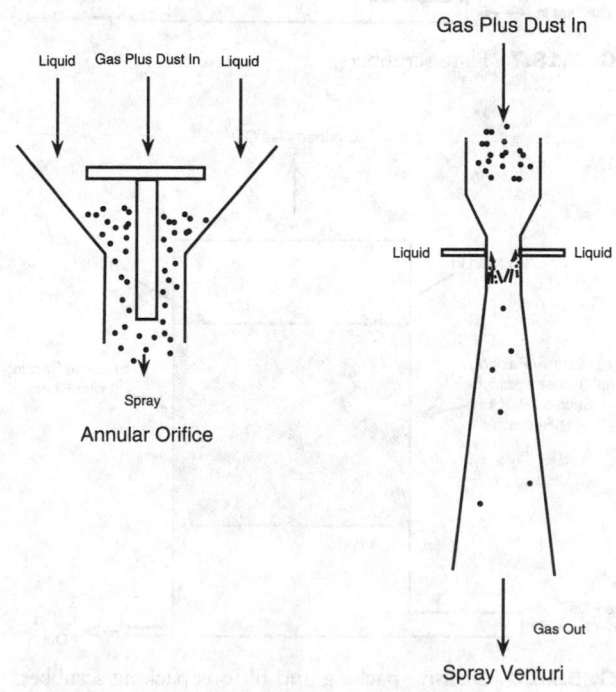

FIG. 5.18.5 Examples of gas-atomized spray unit.

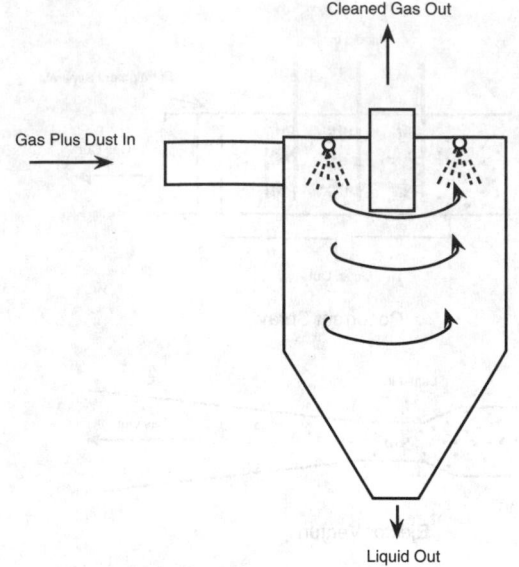

FIG. 5.18.6 Centrifugal scrubbers.

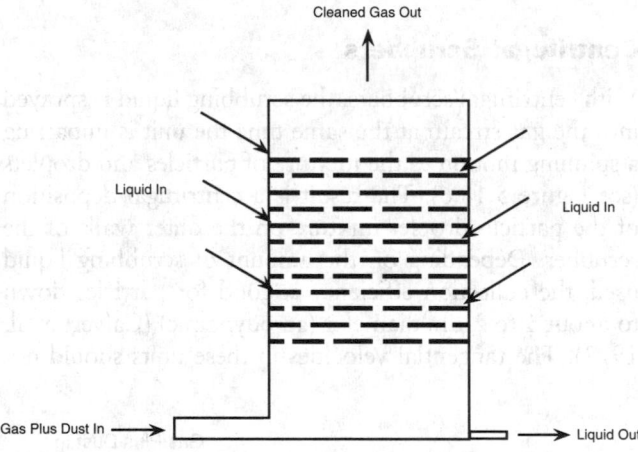

FIG. 5.18.7 Plate scrubbers.

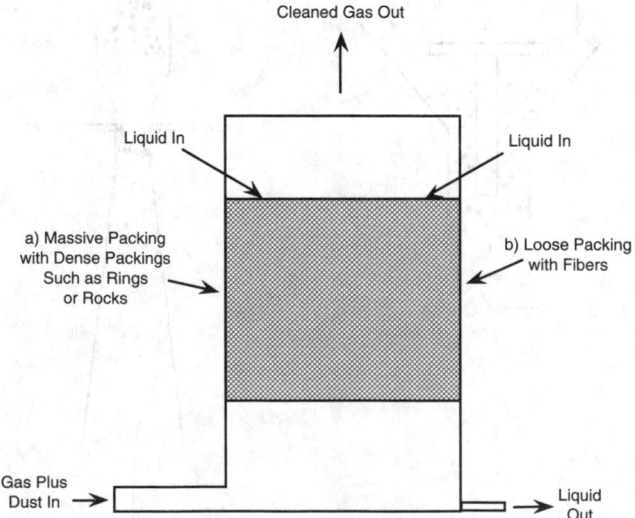

FIG. 5.18.8 Massive packing and fibrous packing scrubbers.

exceed 30 m/sec to prevent reentrainment of the droplet–particle mixture.

IMPINGEMENT ON A WETTED SURFACE—TYPE II

With impingement on a wetted surface, collection occurs by impingement. These scrubbers include plate scrubbers and scrubbers with massive packing or fibrous packing.

Plate Scrubbers

Plate scrubbers have plates or trays mounted within a vertical tower at right angles to the axis of the tower (see Figure 5.18.7). Gas flows from the bottom of the tower up through slots, holes, or other perforations in the plates where mixing with the scrubbing liquid occurs. Collection efficiency increases as the cut diameter decreases. A cut diameter of about 1 μm (aerodynamic diameter) is typical for $\frac{1}{8}$-in holes in a sieve plate (Calvert et al. 1972). Calvert (1977) points out that increasing the number of plates does not necessarily increase collection efficiency. Once particles around the size of the cut diameter are removed, adding more plates does little to increase collection efficiency.

Massive Packing

Packed-bed towers packed with crushed rock or various ring- or saddle-shaped packings are often used as gas scrubbers (see a in Figure 5.18.8). The gas–liquid contact can be crossflow, concurrent, or countercurrent. Collec-tion efficiency rises as packing size falls. According to Calvert (1977), a cut diameter of about 1.5 μm (aerodynamic) can be achieved with columns packed with 1-in Berl saddles or Raschig rings. Smaller packings give higher efficiencies; however, the packing shape appears to have little importance.

Fibrous Packing

Beds of fibers (see part b in Figure 5.18.8) are also efficient at removing particles in gas scrubbers. Fibers can be made from materials such as steel, plastic, or even spun glass. With small diameter fibers, efficient operation is achieved. Efficiency increases as the fiber diameter decreases and also as gas velocity increases. The collection is primarily by impaction and interception. Diffusion is important for small particles, although to increase diffusional collection, lower gas velocities are necessary. Cut diameters can be as low as 1.0 to 2.0 μm (aerodynamic) or in some cases as low as 0.5 μm (aerodynamic) (Calvert et al. 1972). A major difficulty with fibrous beds is that they are prone to plugging; even so, they are widely used.

BUBBLING THROUGH SCRUBBING LIQUID—TYPE III

In scrubbers that bubble gas through a scrubbing liquid, the particulate-containing gas is subjected to turbulent mixing with the scrubbing liquid. Droplets form by shear forces, or the dust particles are impacted directly onto the scrubbing liquid. Droplets are also formed by motor-driven impellers (that are either submerged or in the free air of the scrubber). These impellers serve not only to form small droplets but also to enhance impaction of the dust particles.

Baffle and Secondary-Flow Scrubbers

These units remove particulates from an air stream by continually changing the flow direction and velocity as the gas flows through the unit (see Figure 5.18.9). This motion results in intimate particle mixing in the gas and the spray droplets of scrubbing liquid. Zig-zag baffles or louvers are examples of how flow direction and velocity are altered internally. With these units, cut diameters as low as 5 to 10 μm can be achieved with low pressure drops. Plugging can be a problem with some heavy particle loadings.

Impingement and Entrainment Scrubbers

With this type of scrubber design, the particle–gas mixture is bubbled through or skimmed over the scrubbing liquid surface which atomizes some droplets and mixes the particles with the scrubbing liquid (see Figure 5.18.10). Both particle impaction on the liquid surface and on the atomized drops and some diffusion contribute to particle collection, which is effective with high-velocity entrainment for cut diameters down to 0.5 μm (aero-

dynamic diameter). The pressure drop for this class of device is high.

Mechanically Aided Scrubbers

With mechanically aided scrubbers, a motor-driven device in the scrubber produces spray droplets and mixes the incoming gas and scrubbing liquid more intimately (see Figure 5.18.11). The motor-driven device can be a fan with the scrubbing liquid introduced into the fan rotor, or it can be a paddle arrangement (disintegrator) in the scrubbing liquid that produces droplets. These units have cut diameters down to 2.0 μm (aerodynamic), or even 1.0 μm for some disintegrator designs (Calvert et al. 1972).

Fluidizied- (Moving) Bed Scrubbers

Fluidized-bed scrubbers are similar to packed-bed units except that the packing material is light enough to float in the gas stream (see Figure 5.18.12). The packing material expands to about twice its original depth. This expansion intimately mixes the particulates and scrubbing liquid and permits effective collection down to cut diameters of 1 μm (aerodynamic) (Calvert 1977).

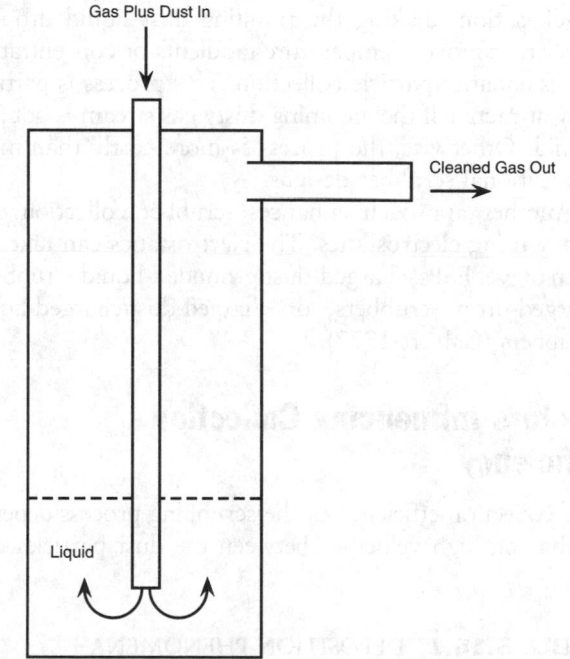

FIG. 5.18.10 Impingment scrubber.

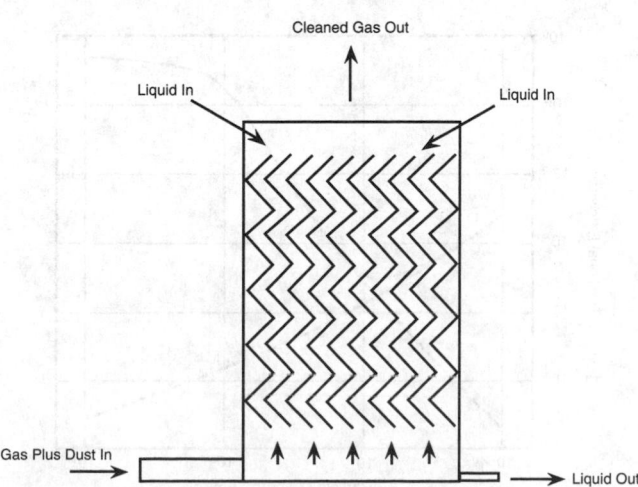

FIG. 5.18.9 Baffle-type scrubber.

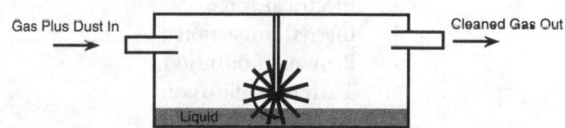

FIG. 5.18.11 Mechanically aided scrubber.

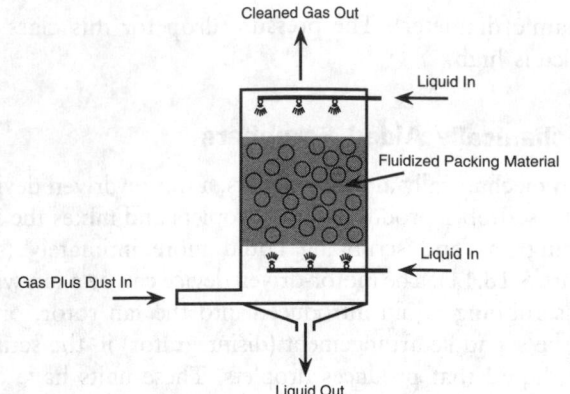

FIG. 5.18.12 Fluidized- or mobile-bed scrubber.

SCRUBBERS USING A COMBINATION OF DESIGNS

Besides the preceding basic scrubber types, a number of scrubber designs use two or more of the collection mechanisms in Table 5.18.1. With a process known as flux force/condensation scrubbing (Calvert and Jhaveri 1974), collection phenomena such as diffusiophoresis or thermophoresis combine with condensation to improve particle removal. This process uses steam to increase the deposition of scrubbing liquid on the dust particles by condensation, making the resulting dust–liquid droplets easier to remove. Temperature gradients or concentration fluxes enhance particle collection. This process is particularly attractive if the incoming dusty gas stream is hot and humid. Otherwise, the process is more costly than more conventional scrubber designs.

Another approach enhances scrubber collection efficiency using electrostatics. The electrostatics can take the form of wet ESPs, charged-dust/grounded-liquid scrubbers, charged-drop scrubbers, or charged-dust/charged-liquid scrubbers (Calvert 1977).

Factors Influencing Collection Efficiency

The collection efficiency of the scrubbing process depends on having high velocities between the dust particles and

TABLE 5.18.1   DEPOSITION PHENOMENA

1.	Interception
2.	Magnetic force
3.	Electrical force
4.	Inertial impaction
5.	Brownian diffusion
6.	Turbulent diffusion
7.	Gravitational force
8.	Photophoretic force
9.	Thermophoretic force
10.	Diffusiophoresis force

the collecting liquid (Overcamp and Bowen 1983). These high velocities enhance the impaction efficiency. Because fine particles rapidly attain the velocity of their surroundings, continually mixing the dirty air stream with the collecting liquid is necessary for good impaction, and short distances between the particles and the collecting surface are required for efficient diffusive collection (Nonhebel 1964).

Considering all collection mechanisms in an ideal scrubber, Figure 5.18.13 shows the shape of a typical efficiency versus particle diameter curve for impaction plus diffusion. In the figure a minimum collection efficiency is apparent that is typical of the low efficiency in the transition from impaction collection processes to diffusion-related processes. Above this minimum, efficiency varies roughly as a function of the dust particle diameter squared, and below it roughly as $(1/d^2)$ (Reist 1993). Efficiency is approximately proportional to particle density. As might be expected, because of the variety of scrubbing configurations available, no one collection mechanism is dominant, and no single approach determines all scrubber performance (Crawford 1976).

CONTACTING POWER RULE

Through experience, investigators have observed that scrubber efficiency for similar aerosols, regardless of scrubber design, is related to the power consumed by the scrubber in making the liquid–particle contact. Thus, scrubber grade efficiency appears to be a function of power input. Scrubbers are high-energy devices. That is, in general, the higher the energy input per unit volume of gas treated, the more efficient the scrubber is for smaller sized particle collection. Also, higher power inputs imply smaller scrubbing liquid drop sizes and more turbulence.

Semrau (1960) shows that the efficiency of any scrubber for similar aerosols is strongly affected by the power

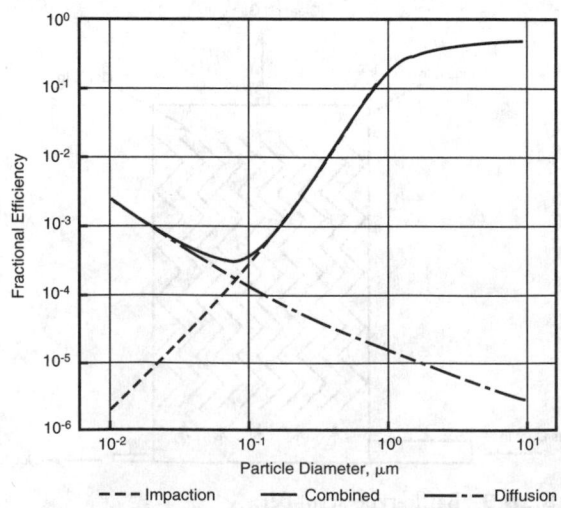

FIG. 5.18.13 Fractional efficiency for combined impaction and diffusion.

dissipated in operating the scrubber (termed *contacting power*) and little affected by scrubber size, geometry, or the manner in which the power is applied.

In this text, contacting power is the power per unit of the gas volumetric flow rate used in contacting the aerosol to be cleaned. This power is ultimately dissipated as heat. The power per unit of the gas volumetric flow rate is considered an effective friction loss and represents friction loss across the scrubbing unit, neglecting losses due to kinetic energy changes in the flowing gas stream or losses due to equipment operating dry (Semrau 1960, 1977).

Contacting power is stated in the following contacting power rule:

$$N_t = \alpha P_T^\gamma \qquad 5.18(3)$$

where N_t is the number of transfer units. Transfer units are related to the fractional collection efficiency as follows:

$$\varepsilon = 1 - e^{-N_t} \qquad 5.18(4)$$

In Equation 5.18(3), the terms α and γ are empirical constants that depend on the properties of the dust being collected. Table 5.18.2 lists the values for α and γ for various dust plus scrubber combinations. The contacting power P_T is calculated from the following equation (Semrau 1960):

$$P_T = P_G + P_L + P_M \qquad 5.18(5)$$

where:

P_G = the power input required to overcome the gas pressure drop across the collector

P_L = the power input required to produce droplets through the spray nozzles

P_M = the power input required to drive any rotor, if present.

Figure 5.18.14 plots N_T as a function of contacting power for various dusts. In the preceding equations, the power terms are expressed in kWhr/1000 m³. Table 5.18.2 gives the factors for converting power to either hp/1000 cfm, or effective friction loss (inches or centimeters of water).

Using the contacting power approach is best when some knowledge of the scrubber's performance for a specific aerosol is available. For example, when pilot plant designs are scaled up to full-size units for a dust control problem, contacting power is helpful in predicting the overall efficiency of the new, full-size unit. Or, if a scrubbing unit is to be replaced with a new and different design, the contacting power approach is helpful in estimating the performance of the new unit. However, in determining contacting power, a design engineer must insure that only the portion of energy input representing energy dissipated in scrubbing is used (Semrau 1977). Also, little data exists relating gas-phase and mechanical contacting power. Since the contacting power approach is an empirical approach, extrapolating the results into areas with little data requires caution.

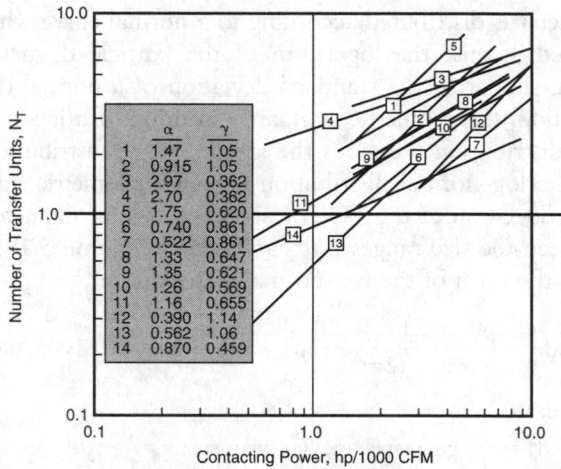

	α	γ
1	1.47	1.05
2	0.915	1.05
3	2.97	0.362
4	2.70	0.362
5	1.75	0.620
6	0.740	0.861
7	0.522	0.861
8	1.33	0.647
9	1.35	0.621
10	1.26	0.569
11	1.16	0.655
12	0.390	1.14
13	0.562	1.06
14	0.870	0.459

1. Raw lime kiln dust - venturi and cyclonic spray
2. Prewashed lime kiln dust - venturi, pipeline, and cyclonic spray
3. Talc dust - venturi
4. Talc dust - orifice and pipeline
5. Black liquor - venturi and cyclonic spray
6. Black liquor - venturi, pipeline, and cyclonic spray
7. Black liquor - venturi evaporator
8. Phosphoric acid mist - venturi
9. Foundry cupola dust - venturi
10. Open-hearth steel - venturi
11. Talc dust - cyclone
12. Copper sulfate - solivore, and mechanical spray generators
13. Copper sulfate - hydraulic nozzles
14. Ferrosilicon furnace fume - venturi and cyclonic spray

FIG. 5.18.14 Performance curve for scrubbing aerosols. (Adapted from K.T. Semrau, 1960, Correlation of dust scrubber efficiency, *J. APCA* 10:200.)

TABLE 5.18.2   FORMULAS FOR PREDICTING CONTACTING POWER

	Symbol	Metric Units[b]	Dimensional Formula
Effective friction loss	F_E	cm H_2O	Δp[a]
Gas phase contacting power	P_G	kWh/1000 m³	0.02724 F_E
Liquid phase contacting power	P_L	kWh/1000 m³	0.02815 ρ_t (Q_L/Q_G)[c]
Mechnical contacting power	P_M	kWh/1000 m³	16.67 (W_s/Q_G)[c]
Total contacting power	P_T	kWh/1000 m³	$P_G + P_L + P_M$

Notes: [a]The effective friction loss is approximately equal to the scrubber pressure loss Δp.

[b]1.0 kWh/1000 m³ = 2.278 hp/1000 ft³/min

[c]This quantity is actually power input and represents an estimate of contacting power; Q_L is in l/min, Q_G is in m³/min, and W_s is net mechanical power input in kW.

USE OF THE AERODYNAMIC CUT DIAMETER

Because of the variety of parameters involved, accurately designing a scrubber for an installation without prior knowledge of scrubber performance at that or a similar installation is difficult. Factors such as air flow rate, scrubbing liquid flow rate, inlet gas temperature, scrubbing liquid temperature, relative humidity, concentration, size and size distribution of the aerosol in the incoming gas, scrubber droplet size and size distribution, and the type of scrubber can confound the design engineer and cloud the design process. Even so, some equations are available to make initial estimates for scrubber design.

Dust size distributions are often represented by log–normal distributions; that is, the particle frequency for a given diameter is distributed according to a normal curve when plotted against the logarithm of the particle diameter. Then, similar to the standard deviation of a normal distribution, the geometric standard deviation of a log–normal distribution measures the spread of the distribution. With a log–normal distribution having a geometric standard deviation of σ_g, 67% of all particles have diameters between the size ranges d_g/σ_g and $\sigma_g d_g$. Equation 5.18(6) gives the form of the log–normal distribution:

$$\int_0^d f(x)dx = \int_0^d \frac{1}{x \ln \sigma_g (2\pi)^{0.5}} \exp\left[-\frac{(\ln x - \ln d_g)^2}{2 \ln^2 \sigma_g}\right] dx \quad \textbf{5.18(6)}$$

where:

d_g = the geometric median diameter
σ_g = the geometric standard deviation
x = any particle diameter of interest.

Thus, the right side of Equation 5.18(6) gives on integration the fraction of particles whose diameters are less than or equal to x. Integrating between limits of x = 0 to ∞ gives a result of 1, and between limits of x = 0 to d_g, a result of 0.5.

Two particle size distributions must be considered in scrubber design. Besides the size distribution of the dust to be collected, the grade efficiency must also be considered when the collector operates under specific conditions. As previously discussed, according to Calvert et al. (1972), the grade efficiency can be effectively represented by the *aerodynamic cut diameter* d_{ac}. If $d_{ac} = d_g$, a collection efficiency of 50% is estimated; if $d_{ad} = d_g/\sigma_g$, the estimated collection efficiency is 84%. Hence, for any level of efficiency and a log–normal aerosol, the use of Equation 5.18(6) determines the equivalent d_{ac}.

Figure 5.18.15 represents an integrated form of Equation 5.18(6) which gives any collection efficiency (50% or greater) as a function of the ratio d_{ac}/d_g with σ_g as a parameter. Thus, for example, if a collection efficiency of 99% is needed for a log–normal, unit-density aerosol with $d_g = 5.0\ \mu m$ and $\sigma_g = 2$, Figure 5.18.15 shows that a multiplication factor of about 0.2 is indicated. The required d_{ac} is then $5 \times 0.2 = 1\ \mu m$.

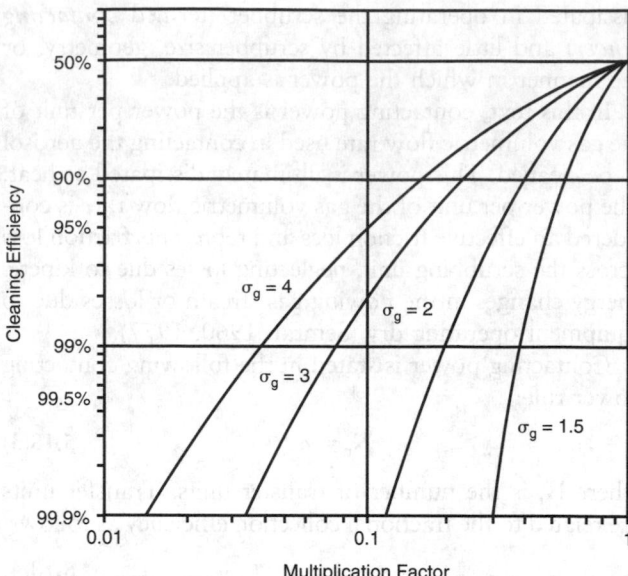

FIG. 5.18.15 Cleaning efficiency as a function of the multiplication factor for log–normal particle distributions of various skewness.

DETERMINATION OF d_{ac} AS A FUNCTION OF SCRUBBER OPERATING PARAMETERS

A number of theoretical equations estimate scrubber performance with reasonable accuracy for a specific scrubber type.

Venturi

Calvert (1970) and Calvert et al. (1972) considered the venturi scrubber and developed the following equation for collection efficiency ε:

$$\varepsilon = \exp\left[-\frac{2Q_L u_G \rho_L d_d}{55 Q_G \mu_G} F(K_{pt}, f)\right] \quad \textbf{5.18(7)}$$

where:

Q_L/Q_G = the volumetric flow ratio of liquid to gas
u_G = the gas velocity in cm/sec
ρ_L = the density of the liquid in gm/cm^3
μ_G = the viscosity of the liquid in poises
d_d = the drop diameter in cm
f   = an operational factor which ranges from 0.1 to 0.3, but is often taken as 0.25

The term f includes the influence of factors such as collection by means other than impaction, variation in drop sizes, loss of liquid to venturi walls, particle growth by condensation, and degree of utilization of the scrubbing liquid.

The following equation gives function $F(K_{pt}, f)$:

$$F(K_{pt}, f) =$$
$$\left[0.7 + K_{pt}f - 1.4 \ln\left(\frac{K_{pt}f + 0.7}{0.7}\right) - \frac{0.49}{K_{pt}f + 0.7}\right]\frac{1}{K_{pt}} \quad \textbf{5.18(8)}$$

where K_p is defined as:

$$K_p = \frac{v_r d_a^2}{9\mu_G d_d} \qquad 5.18(9)$$

The term v_r represents the relative velocity between the dust particle and the droplet, and d_a is the aerodynamic particle diameter as defined in Equation 5.18(1). The average liquid droplet diameter d_d is given by the empirical equation of Nukiyama and Tanasawa (1938) as:

$$d_d = \frac{50}{u_G} + 91.8 \left(\frac{Q_L}{Q_G}\right)^{1.5} \qquad 5.18(10)$$

The drop diameter d_d is expressed in units of centimeters (cm), and the gas velocity u_G in cm/sec.

When $\varepsilon = 0.5 = 50\%$, Equation 5.18(7) can be solved for $d_a = d_{ac}$ for a variety of Q_L/Q_G ratios. The solutions to these calculations are plotted as shown Figure 5.18.16. For these calculations, $k = 0.25$, $\mu_G = 1.82 \times 10^{-4}$ poises, and $u_G = v_r$ with u_G as a parameter ranging from 5 m/sec to 15 m/sec. Figure 5.18.16 can then be used to estimate d_{ac} for venturi scrubber operation at different Q_L/Q_G ratios and different throat velocities.

Rudnick et al. (1986) evaluated three widely used venturi scrubber equations (Calvert [1970], Yung et al. [1978], and Boll [1973]) and compared the equations with experimental data. They concluded that Yung's model, a modification of the Calvert model, predicts data better than the other two. However, the difference was not that great, and since the Yung model is complicated, the Calvert model remains useful for a first approximation of venturi scrubber performance.

The pressure drop across a venturi scrubber (in centimeters of water) can be estimated from the following equation (Leith, Cooper, and Rudnick 1985):

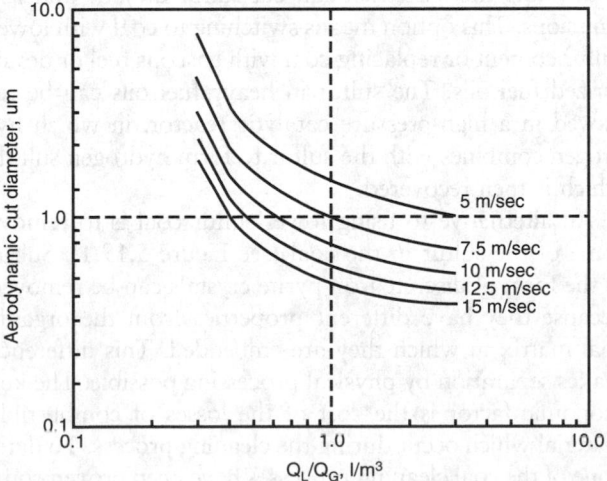

FIG. 5.18.16 Predicted venturi performance. (Adapted from S. Calvert, J. Goldshmid, D. Leith, and D. Mehta, 1972, *Scrubber handbook*, Riverside, Calif.: Ambient Purification Technology, Inc.)

$$\Delta p = 1.02\rho_L \frac{Q_L}{Q_G} u_{Gt}^2 \left[\beta\left(1 - \frac{u_{Gf}}{u_{Gt}}\right) + \left(\frac{u_{Gf}}{u_{Gt}}\right)^2\right] \qquad 5.18(11)$$

The term u_{Gf} is the gas velocity at the exit of the venturi, and u_{Gt} is the gas velocity in the throat. The term β can be estimated from the following equation:

$$\beta = 2(1 - X^2 + \sqrt{X^4 - X^2}) \qquad 5.18(12)$$

and X is defined as follows:

$$X = 1 + 3L_t C_D \rho_G/(16d_d\rho_L) \qquad 5.18(13)$$

where L_t is the throat length, and C_D is the drag coefficient, which can be estimated up to a flow Reynolds number of about 1000 by the following equation:

$$C_D = \frac{24}{Re}(1 + 0.15Re^{0.687}) \qquad 5.18(14)$$

Above this Reynolds number, the drag coefficient is estimated by $C_D = 0.44$. The Reynolds number is given by the following equation:

$$Re = \frac{u_G d \rho_g}{\mu_G} \qquad 5.18(15)$$

Countercurrent Spray Tower

For a countercurrent spray tower as shown in Figure 5.18.4, the following equation estimates the collection efficiency (Calvert 1968):

$$\varepsilon = 1 - \exp\left[-\frac{3Q_L v_{d/g} Z \eta_d}{2Q_G v_{d/w} d_d}\right] \qquad 5.18(16)$$

where:

$v_{d/g}$ = the velocity of the droplets relative to the gas
$v_{d/w}$ = the velocity of the droplets relative to the walls
$(v_{d/w} = v_{d/g} + v_{g/w})$
Z = the height or length of the scrubber in cm

The following equation gives η_d:

$$\eta_d = \left(\frac{Stk}{Stk + 0.7}\right)^2 \qquad 5.18(17)$$

The following equation gives the Stokes number Stk:

$$Stk = \frac{d_a^2 \rho_p v_{a/g}}{9\mu_G d_d} \qquad 5.18(18)$$

Impingement and Entrainment-Type Scrubber

For this type of unit (see Figure 5.18.10), the efficiency can be estimated from the following equation (Calvert et al. 1972):

$$\varepsilon = 1 - \exp\left[\frac{2\Delta p d_d}{55 u_G \mu_G} F(K_{pt}, f)\right] \qquad 5.18(19)$$

Usually a value of 0.25 is chosen for f. According to Schiffter and Hesketh (1983), the drop diameter d_d is about one-third that calculated from Equation 5.18(10). The pressure drop in these units is given by the following equation:

$$\Delta p = \frac{\rho_L Q_L u_G}{A} \qquad 5.18(20)$$

where A is the collecting area, defined as the total surface area of the collecting elements perpendicular to the gas flow direction.

—Parker C. Reist

References

Batel, W. 1976. *Dust extraction technology*. Stonehouse, England: Technicopy Ltd.

Boll, R.H. 1973. *Ind. Eng. Chem. Fundam.* 12:40–50.

Calvert, S. 1968. Vol. 3 in *Air pollution*. 2d ed. Edited by A.C. Stern. New York: Academic Press.

———. 1970. Venturi and other atomizing scrubbers: Efficiency and pressure drop. *AIChE J.* 16:392–396.

———. 1974. Engineering design of fine particle scrubbers. *J. APCA* 24:929–934.

———. 1977. How to choose a particulate scrubber. *Chemical Eng.* (29 August):55.

Calvert, S., J. Goldshmid, D. Leith, and D. Mehta. 1972. *Scrubber handbook*. Riverside, Calif.: Ambient Purification Technology, Inc.

Calvert, S. and N.C Jhaveri. 1974. Flux force/condensation scrubbing. *J. APCA* 24:946–951.

Crawford, M. 1976. *Air pollution control theory*. New York: McGraw-Hill, Inc.

Leith, D., D.W. Cooper, and S.N. Rudnick. 1985. Venturi scrubbers: Pressure loss and regain. *Aerosol Sci. and Tech.* 4:239–243.

Nonhebel, G. 1964. *Gas purification processes*. London: George Newnes, Ltd.

Nukiyama, S. and Y. Tanasawa. 1938. Experiment on atomization of liquids. *Trans. Soc. Mech. Eng. (Japan)* 4:86.

Overcamp, T.J., and S.R. Bowen. 1983. Effect of throat length and diffuser angle on pressure loss across a venturi scrubber. *J. APCA* 33:600–604.

Reist, P.C. 1993. *Aerosol science and technology*. New York: McGraw-Hill.

Rudnick, S.N, J.L.M. Koelher, K.P. Matrin, D. Leith, and D.W. Cooper. 1986. Particle collection efficiency in a venturi scrubber: Comparison of experiments with theory. *Env. Sci. and Tech.* 20:237–242.

Schifftner, K.C. and H. Hesketh. 1983. *Wet scrubbers*. Ann Arbor, Mich.: Ann Arbor Science.

Semrau, K.T. 1960. Correlation of dust scrubber efficiency. *J. APCA* 10:200–207.

———. 1977. Practical process design of particulate scrubbers. *Chemical Eng.* (26 September):87–91.

Yung, S., S. Calvert, H.F. Barbarika, and L.E. Sparks. 1978. *Env. Sci. and Tech.* 12:456–459.

5.19
GASEOUS EMISSION CONTROL

This section presents some options for reducing the emission of pollutants. It focuses on the complex series of processes and phenomena generally grouped under acid rain. Acid rain is associated with the release of sulphur and nitrogen oxides into atmosphere via the burning of fossil fuels. The deposition of sulfur and nitrogen compounds is one of the most pressing large-scale air pollution problems.

The options presented are based on general pollution prevention techniques. This section focuses on source control via technology changes. Source reduction measures aim at long-term reduction, while other options can limit pollution during unfavorable conditions.

Energy Source Substitution

This option includes the transition to solar energy, nuclear energy, and other methods of obtaining energy. Of the pollutants generated by burning fuels, the emission of sulfur oxides and ash are directly attributable to fuel composi-

tion. Using the right fuel can cut sulfur dioxide and ash emissions. This option means switching to coal with lower sulfur content or replacing coal with gaseous fuel or desulfurized fuel oils. The sulfur in heavy fuel oils can be removed in a high-pressure catalytic reactor, in which hydrogen combines with the sulfur to form hydrogen sulfide which is then recovered.

An alternative to using lower-sulfur coal is to remove some of the sulfur in the coal (see Figure 5.19.1). Sulfur in the form of discrete iron pyrite crystals can be removed because they have different properties from the organic coal matrix in which they are embedded. This difference makes separation by physical processing possible. The key economic factor is the cost of the losses of combustible material which occur during the cleaning process. To date, none of the coal cleaning processes have been proven commercially successful. Their attractiveness is also diminished by their inability to recover, on average, more than half the sulfur content (e.g., the pyritic fraction) of the fuel (Bradshaw, Southward, and Warner 1992).

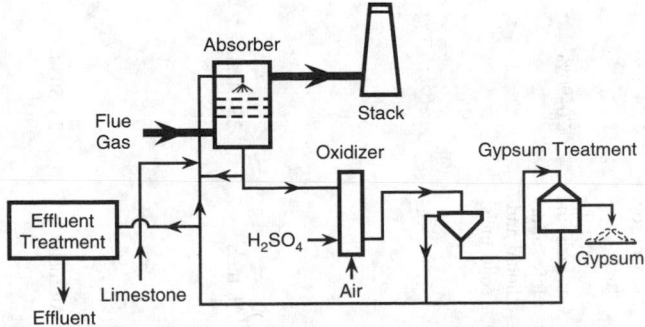

FIG. 5.19.1 Schematic diagram of a limestone–gypsum FGD plant.

Nitrogen oxides form at firebox temperatures by the reaction of the oxygen and nitrogen in the air and fuel. The thermal fixation of atmospheric nitrogen and oxygen in the combustion air produces *thermal NO$_x$*, while the conversion of chemically bound nitrogen in the fuel produces *fuel NO$_x$*.

For natural gas and light distillate oil firing, nearly all NO emissions result from thermal fixation. With residual fuel oil, the contribution of fuel-bound nitrogen can be significant and in some cases predominant. This contribution is because the nitrogen content of most U.S. coals ranges from 0.5 to 2% whereas that of fuel oil ranges from 0.1 to 0.5%. The conversion efficiencies of fuel mixture to NO$_x$ for coals and residual oils have been observed between 10 and 60% (U.S. EPA 1983). Figure 5.19.2 shows the possible fates of fuel nitrogen. One option of reducing NO$_x$ is to use low-nitrogen fuel.

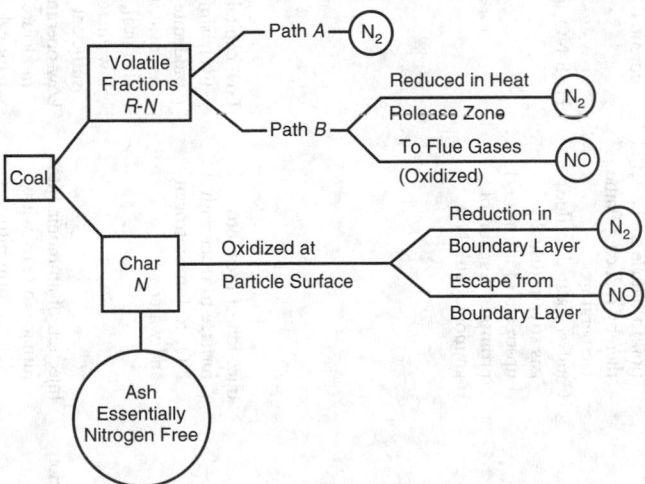

FIG. 5.19.2 Possible fates of nitrogen contained in coal. (Reprinted, with permission, from M.P. Heap et al., 1976, The optimization of burner design parameters to control NO formation in pulverized coal and heavy oil flames, *Proceedings of the Stationary Source Combustion Symposium—Vol. II: Fuels and Process Research Development*, EPA-600/2-76-152b [Washington, D.C.: U.S. EPA].

Process Modifications

The formation rates of both thermal NO$_x$ and fuel NO$_x$ are kinetically or aerodynamically limited, with the amount of NO$_x$ formed being less than the equilibrium value (Wark and Warner 1981). Combustion conditions dominate the formation rate of NO$_x$, and modifying the combustion process can suppress it. Rapidly mixing oxygen with the fuel promotes both thermal and fuel NO$_x$ formation.

MacKinnon (1974) developed a kinetic model from experimental studies of a heated mixture of N$_2$, O$_2$, and air as well as air. His kinetic equations provide insight to the strategies for controlling the formation of thermal NO$_x$ as follows:

The peak temperature should be reduced.
The gas residence time at the peak temperature should be reduced.
The oxygen concentration in the highest temperature zone should be reduced.

Regardless of the mechanisms, several general statements can be made about fuel NO$_x$. It depends highly on the air/fuel ratio. The percent conversion to fuel NO$_x$ declines rapidly with an increasing fuel equivalent ratio. (The *fuel equivalent ratio* is a multiple of the theoretical fuel/air ratio and is the inverse of the stoichiometric ratio. The stoichiometric ratio is unity when the actual air/fuel ratio equals the theoretical air to fuel needed for complete combustion.) The fuel equivalent ratio primarily affects the oxidation of the volatile R–N fraction (where R represents an organic fragment) rather than the nitrogen remaining in the char. The degree of fuel–air mixing also strongly affects the percent conversion of fuel nitrogen to NO$_x$, with greater mixing resulting in a greater percent conversion.

COMBUSTION CONTROL

Several process modifications can reduce NO$_x$ formation. Table 5.19.1 summarizes the commercially available NO$_x$ control technologies as well as their relative efficiencies, advantages and disadvantages, applicability, and impacts.

The simplest combustion control technology is the *low-excess-air* (LEA) operation. This technology reduces the excess air level to the point of some constraint, such as carbon monoxide formation, flame length, flame stability, and smoke. Unfortunately, the LEA operation only moderately reduces NO$_x$.

Off-Stoichiometric Combustion (OSC)

OSC, or staged combustion, combusts fuel in two or more stages. The primary flame zone is fuel-rich, and the secondary zones are fuel-lean. This combustion is achieved through the following techniques. These techniques are generally applicable only to large, multiple-burner, combustion devices.

TABLE 5.19.1 SCREENING POTENTIAL NO$_X$ CONTROL TECHNOLOGIES

Technique	Description	Advantages	Disadvantages	Impacts To Consider	Applicability	NO$_x$ Reduction
LEA	Reduces oxygen availability	Easy operational modification	Low NO$_x$ reduction potential	High carbon monoxide emissions, flame length, flame stability	All fuels	1–15%
OSC a. BOOS b. OFA c. Air Lances	Staged combustion, creating fuel-rich and fuel-lean zones	Low operating cost, no capital requirement for BOOS	a. Typically requires higher air flow to control carbon monoxide b. Relatively high capital cost c. Moderate capital cost	Flame length, forced draft fan capacity, burner header pressure	All fuels; Multiple-burner devices	30–60%
LNB	Provides internal staged combustion, thus reducing peak flame temperatures and oxygen availability	Low operating cost, compatible with FGR as a combination technology to maximize NO$_x$ reduction	Moderately high capital cost; applicability depends on combustion device and fuels, design characteristics, and waste streams	Forced-draft fan capacity, flame length, design compatibility, turndown flame stability	All fuels	30–50%
FGR	Up to 20–30% of the flue gas recirculated and mixed with the combustion air, thus decreasing peak flame temperatures	High NO$_x$ reduction potential for natural gas and low-nitrogen fuels	Moderately high capital cost, moderately high operating cost, affects heat transfer and system pressures	Forced-draft fan capacity, furnace pressure, burner pressure drop, turndown flame stability	Gas fuels and low-nitrogen fuels	40–80%
W/SI	Injection of steam or water at the burner, which decreases flame temperature	Moderate capital cost, NO$_x$ reductions similar to FGR.	Efficiency penalty due to additional water vapor loss and fan power requirements for increased mass flow	Flame stability, efficiency penalty	Gas fuels and low-nitrogen fuels	40–70%
RAPH	Air preheater modification to reduce preheat, thereby reducing flame temperature	High NO$_x$ reduction potential	Significant efficiency loss (1% per 40°F)	Forced-draft fan capacity, efficiency penalty	Gas fuels and low-nitrogen fuels	25–65%
SCRI	Catalyst located in flue gas stream (usually upstream of air heater) promotes reaction of ammonia with NO$_x$	High NO$_x$ removal	Very high capital cost, high operating cost, extensive ductwork to and from reactor required, large volume reactor must be sited, increased pressure drop may require induced-draft fan or larger forced-draft fan, reduced efficiency, ammonia sulfate removal equipment for air heater required, water treatment of air heater wash required	Space requirements, ammonia slip, hazardous waste disposal	Gas fuels and low-sulfur liquid and solid fuels	70–90%
SNCR—Urea Injection	Injection of urea into furnace to react with NO$_x$ to form nitrogen and water	Low capital cost, relatively simple system, moderate NO$_x$ removal, nontoxic chemical, typically low energy injection sufficient	Temperature dependent, design must consider boiler operating conditions and design, NO$_x$ reduction may decrease at lower loads	Furnace geometry and residence time, temperature profile	All fuels	25–50%
SNCR—Ammonia Injection	Injection of ammonia into furnace to react with NO$_x$ to form nitrogen	Low operating cost, moderate NO$_x$ removal	Moderately high capital cost; ammonia handling, storage, vaporization	Furnace geometry and residence time, temperature profile	All fuels	25–50%

The *burner-out-of-service* (BOOS) technique terminates the fuel flow to selected burners while leaving the air registers open. The remaining burners operate fuel rich, thereby limiting oxygen availability, lowering peak flame temperatures, and reducing NO_x formation. The unreacted products combine with the air from the terminated fuel burners to complete burnout before exiting the furnace.

Installing *air-only (OFA) ports* above the burner zone also achieves staged combustion. This technique redirects a portion of the air from the burners to the OFA ports. A variation of this concept, *lance air*, has air tubes installed around the periphery of each burner to supply staged air.

Combustion Temperature Reduction

Reducing the combustion temperature effectively reduces thermal NO_x but not fuel NO_x. One way to reduce the temperature further is to introduce a diluent, as in *flue gas recirculation* (FGR). FGR recirculates a portion of the flue gas back into the windbox. The recirculated flue gas, usually 10–20% of the combustion air, provides sufficient dilution to decrease NO_x emissions.

An advantage of FGR is that it can be used with most other combustion control methods. However, in retrofit applications, FGR can be expensive. In addition to requiring new large ducts, FGR may require major modifications to fans, dampers, and controls.

Water or steam injection (W/SI) is another method that works on the principle of combustion dilution, similar to FGR. In addition, W/SI reduces the combustion air temperature. In some cases, W/SI is a viable option when moderate NO_x reductions are required for compliance.

OTHER MODIFICATIONS

Reduction of the air preheat temperature (RAPH) is another viable technique for cutting NO_x emissions. This technique lowers the peak flame temperatures, thereby reducing NO_x formation. The thermal efficiency penalty, however, can be substantial.

Post-combustion control techniques such as SCR and SNCR by ammonia or urea injection are described in Section 5.24.

The techniques of OSC, FGR, and RAPH can be effectively combined. The OSC techniques, namely LEA and two-stage combustion, reduce the quantities of combustion gases reacting at maximum temperature, while FGR and RAPH directly influence the maximum level of combustion temperatures. The reductions obtained by combining individual techniques are not additive but multiplicative. However, the combined conditions necessary to achieve such low levels of oxides are not compatible with operational procedures.

Figure 5.19.3 summarizes the NO_x control technology choices. An environmental engineer can use this figure and Table 5.19.1 to identify the potential control technologies for boilers and process heaters. After identifying the applicable technologies, the engineer must conduct an economic analysis to rank the technologies according to their cost effectiveness. Management can then select the optimum NO_x control technology for a specific unit.

Design Feature Modifications

Several design feature modifications also reduce NO_x emissions including modified burners, burner location and spacing, tangential firing, steam temperature control, air and fuel flow patterns, and pressurized fluidized-bed combustion.

MODIFIED BURNERS

While SCR and SNCR control NO_x emissions by treating the NO_x after it has been formed in the combustion reaction, modifications to the combustion equipment or burners also significantly reduces NO_x formation. Using such modified burners has a number of advantages, the major ones being simplicity and low cost. At the same time, since burners are the primary component of a furnace, implementing new ones should be tried cautiously.

Stage-Air Burners

Staged-air burner systems divide incoming combustion air into primary and secondary paths. All fuel is injected into

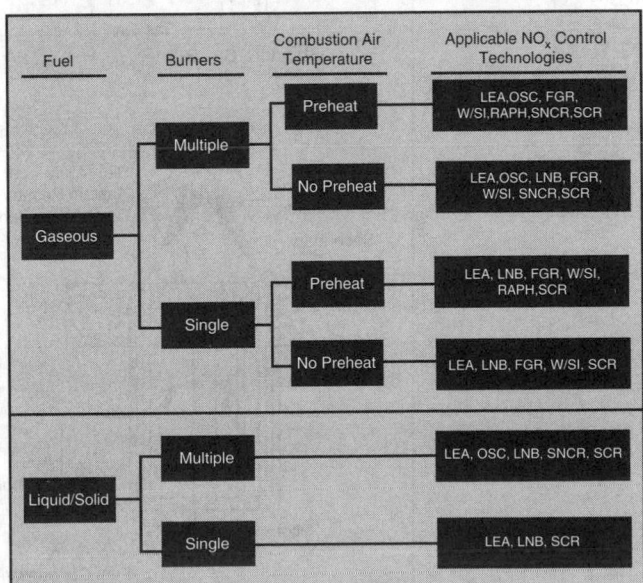

FIG. 15.9.3 Guidelines to identify potential NO_x control technologies. (Reprinted, with permission, from A. Garg, 1994, Specify better low-NO burners for furnaces, *Chem. Eng. Prog.* [November].)

the throat of the burner and is combined with the primary air, which flows through the venturi and burns (see the right side of Figure 5.19.4).

Staged-air burners are simple and inexpensive, and NO_x reductions as high as 20 to 35% have been demonstrated (Garg 1992). These burners are most suitable for forced-draft, liquid-fuel applications and lend themselves to external, flue gas circulations. The main disadvantage of these burners is the long flames, which must be controlled.

Staged-Fuel Burners

In staged-fuel burners, fuel is injected into the combustion zone in two stages, thus creating a fuel-lean zone and delaying completion of the combustion process. This lean combustion reduces peak flame temperatures and reduces thermal NO_x. The remainder of the fuel–gas is injected into the secondary zone through secondary combustion nozzles (see the left side of Figure 5.19.4).

The combustion products and inert gases from the primary zone reduce the peak temperatures and oxygen concentration in the secondary zone, further inhibiting thermal NO_x formation. Some of the NO_x formed in the first stage combustion zone is reduced by the hydrogen and carbon monoxide formed in the staged combustion.

Staged-fuel burners can reduce NO_x emissions by 40–50% (Garg 1994). The flame length of this type of burner is about 50% longer than that of a standard gas burner. Staged-fuel burners are ideal for gas-fired, natural-draft applications.

Ultra-Low-NO_x Burners

Several designs combine two NO_x reduction steps into one burner without any external equipment. These burners typically incorporate staged air with internal FGR or staged fuel with external FGR. In the staged-fuel burners with internal FGR, the fuel–gas pressure

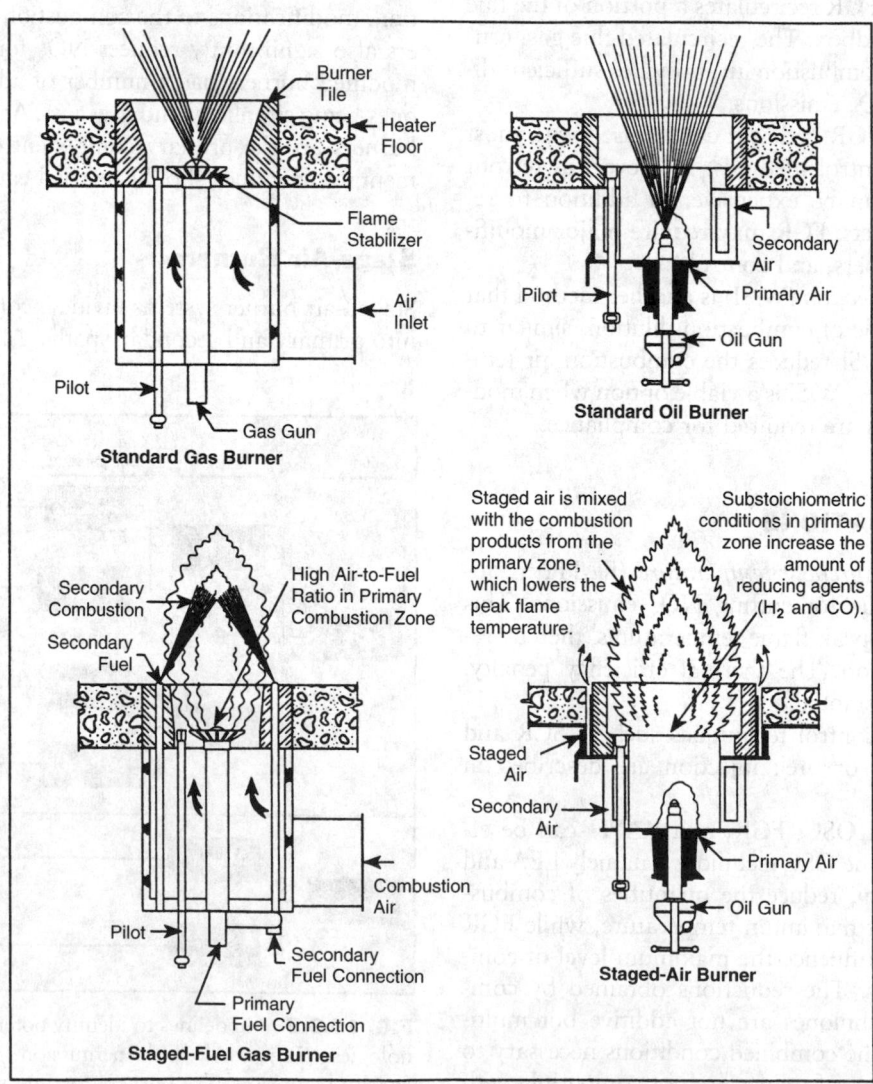

FIG. 5.19.4 OFC achieved by air staging or fuel staging. (Reprinted, with permission, from Garg, 1994.)

induces recirculation of the flue gas (see Figure 5.19.5), creating a fuel-lean zone and reducing oxygen partial pressure.

In staged-air burners with internal FGR, fuel mixes with part of the combustion air, creating a fuel-rich zone. The high-pressure atomization of liquid fuel or fuel gas creates FGR. Pipes or parts route the secondary air in the burner block to complete combustion and optimize the flame profile.

Air or fuel–gas staging with internal FGR can reduce NO_x emissions by 55 to 75%. The latter design can be used with liquid fuels, whereas the former is used mostly for fuel–gas applications.

Low-NO_x burners have been installed in a variety of applications in both new and revamped plants. Table 5.19.2 summarizes the performance of several of these installations.

BURNER LOCATIONS AND SPACING

NO_x concentrations vary with burner type, spacing, and location in coal-fired generation plants. Cyclone burners are known for their highly turbulent operations and result in high-level emissions of NO_x in these plants.

The amount of heat released in the burner zone seems to have a direct effect on NO_x concentrations (see Figure 5.19.6). The OFC reduction techniques are the most effective for larger generating units with larger burners. These techniques are effective since essentially all NO is formed in the primary combustion zone, and as the burner size increases, the primary zone becomes less efficient.

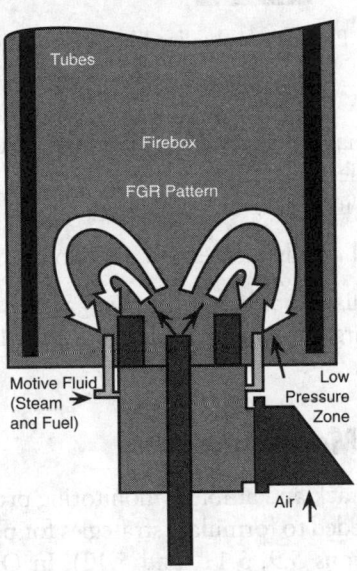

FIG. 5.19.5 Combining staged fuel burners with internal FGR. (Reprinted, with permission, from A. Garg, 1992, Trimming NO_x from furnaces, *Chem. Eng. Prog.* [November].)

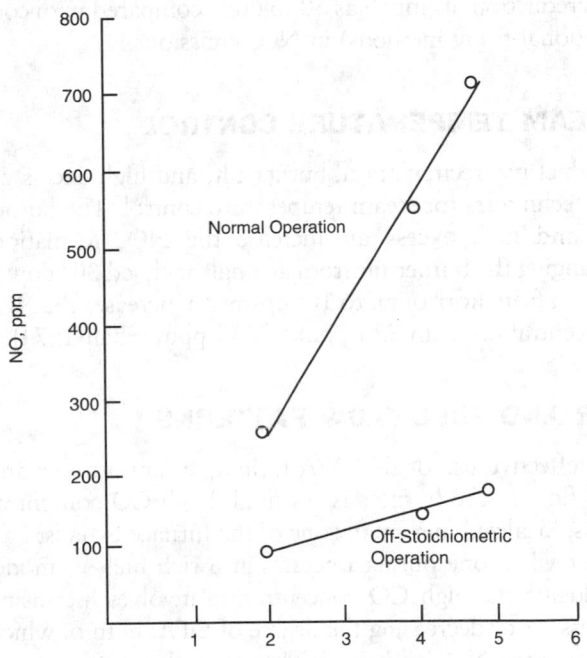

FIG. 5.19.6 Effect of nitric oxide concentrations with increasing burner zone heat release rate: 480 MW unit, natural gas fuel.

TABLE 5.19.2   APPLICATIONS OF LNB BURNERS

Heater Application	Burner Type	NO$_x$ Emission Level, ppm
Crude, vacuum, and coker heaters (cabin)	Forced-draft, staged fuel, preheated air	60
Vertical cylindrical refinery heaters	Natural-draft, staged-fuel, with internal FGR	25
Down-fired hydrogen reformer	Induced-draft, staged-fuel-gas	60*
Vertical cylindrical refinery heaters	Forced-draft, staged-fuel, with internal FGR, preheated air	60
Upfired ethane cracker	Natural-draft, staged-fuel	85

Source: A. Garg, 1994, Specify better low-NO_x burners for furnaces, *Chem. Eng. Prog.* (January).
Notes: *With steam injection.

Different burner spacing has essentially no effect on the NO_x concentration in boiler emissions when the boilers are at full-load operations. However, at a reduced load, closer burner spacing results in higher NO_x releases because close burner spacing inhibits effective bulk recirculation into the primary combustion zone.

The distribution of air, through the primary air ducts located immediately above and below the fuel nozzles and through the secondary air ducts located immediately above and below the primary ducts, can be used as a means to reduce NO_x emissions.

TANGENTIAL FIRING

In tangential firing, the furnace is used as a burner, resulting in a lower flame temperature and in a simultaneous reduction (as much as 50 to 60% compared with conventional firing methods) in NO_x emissions.

STEAM TEMPERATURE CONTROL

Product gas recirculation, burner tilt, and high excess air are techniques for steam temperature control. The burner tilt and high excess air increase the NO_x formation. Changing the burner tilt from an angle inclined 30° downstream from horizontal to 10° upstream increases the NO_x concentration from 225 ppm to 335 ppm (Shah 1974).

AIR AND FUEL FLOW PATTERNS

For effective use of the OFA technique, uniform air and fuel flow to the burners is essential. High CO concentrations, localized in a small zone of the furnace exhaust gas, occur when one burner operates at a rich fuel–air mode. Reducing the high CO concentration involves increasing excess air or decreasing the degree of OFA, both of which increase the NO_x emissions. Therefore, the effective solution is to reduce the fuel flow to the burner.

PRESSURIZED FLUIDIZED-BED COMBUSTION

Mixing an adsorbent (e.g, limestone) directly into the fluidized bed in which coal is burned achieves direct sulfur dioxide control (see Figure 5.19.7). Fluidization is achieved via the combustion air that enters the base of the bed (Halstead 1992). Environmental engineers can apply this basic concept in several ways to meet the needs of chemical industries as well as power generation plants.

The potential advantages of fluidized-bed combustion are as follows:

Lower combustion temperature resulting in less fouling and corrosion and reduced NO_x formations
Fuel versatility including range of low-grade fuels, such as char from synthetic fuel processing

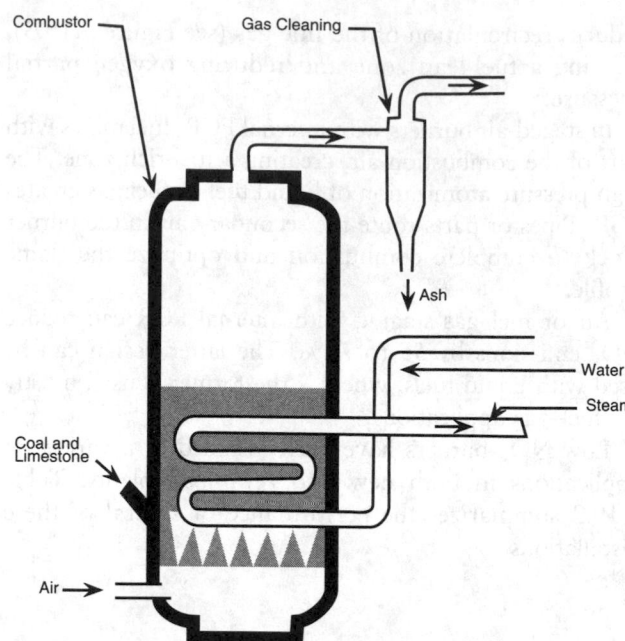

FIG. 5.19.7 Pressurized fluidized-bed combustion system.

Higher thermal efficiency including high heat release and heat transfer
Waste solids in dry form

The potential disadvantages are as follows:

Large particulate loading in the flue gas
Potentially large amounts of solid waste, which are SO_2 absorbent.

Pollution Monitoring

Continuous stack and ambient monitoring provides the information needed to formulate strategies for pollutant control (see Sections 5.9, 5.11, and 5.12). In Osaka, Japan, telemeters monitoring pollutant emissions from factories and stations sampling ambient quality perform environmental monitoring. An environmental monitoring system uses remote-sensing by earth-observing satellites and an environmental information system to forecast the environmental impact of pollution.

A procedure change can ease the effects of sulfur or nitrogen oxide pollution. Automatic changes can be made during high levels of ambient pollution. For example, the Tennessee Valley Authority (TVA) has been switching to low-sulfur fuels during adverse meteorological conditions.

—David H.F. Liu

References

Bradshaw, A.D., Sir Richard Southward, and Sir Frederick Warner, eds. 1992. *The treatment and handling of wastes.* London: Chapman & Hall.

Garg, A. 1992. Trimming NO$_x$ from furnaces. *Chem. Eng. Prog.* (November).

———. 1994. Specify better low-NO$_x$ burners for furnaces. *Chem. Eng. Prog.* (January).

MacKinnon, D.J. 1974. Nitric oxide formation at high temperatures. *Journal of the Air Pollution Control Association* 24, no. 3 (March).

Shah, I. 1974. Furnace modifications. In Vol. 2 of *Environmental engineers' handbook*, edited by B.G. Liptak. Radnor, Pa.: Chilton Book Company.

U.S. Environmental Protection Agency (EPA). 1983. *Control techniques for nitrogen oxide emissions from stationary sources.* Revised 2d ed. EPA-450/3-83-002. Research Triangle Park, N.C.: U.S. EPA.

Wark, K., and C.F. Warner. 1981. Air pollution, its origin and control. New York: Harper & Row Publishers.

5.20
GASEOUS EMISSION CONTROL: PHYSICAL AND CHEMICAL SEPARATION

Major air pollutants are gases such as carbon monoxide, nitrogen oxides, sulfur oxides, and VOCs. Generally, the pollutant concentrations in waste air streams are relatively low, but emissions can still exceed the regulatory limits. Removing air pollutants is achieved by the following methods:

Absorption by a liquid solution
Condensation of pollutants by cooling the gas stream
Adsorption on a porous adsorbent
Chemical conversion of pollutants into harmless compounds

Sometimes methods are combined to treat a feed stream. For example, the absorption of SO$_2$ can be performed in an absorber using an aqueous lime solution. The key step in the separation is absorption although reactions occur between the SO$_2$ and lime in the absorber. Therefore, this SO$_2$ separation is considered *absorption*.

Choosing the air pollutant removal method depends mostly on the physical and chemical properties of the pollutant and the conditions (i.e., temperature, pressure, volume, and concentration) under which the pollutant is treated. The methods chosen for reducing air pollution must not increase pollution in other sectors of the environment. For example, transferring the air pollutant into liquid or solid absorption agents that subsequently contaminate the environment is not a solution to the problem.

Absorption

Absorption is a basic chemical engineering operation and is probably the most well-established gas control technique. It is used extensively in the separation of corrosive, hazardous, or noxious pollutants from waste gases. The major advantage of absorption is its flexibility; an absorber can handle a range of feed rates.

Absorption, also called scrubbing, involves transferring pollutants from a gas phase to a contacting solvent. The transfer occurs when the pollutant partial pressure in the gas phase is higher than its vapor pressure in equilibrium with the solvent. To maximize the mass-transfer driving force (i.e., the difference in pollutant concentration between the gas and liquid phase), the absorber generally operates in a countercurrent fashion.

Absorption systems can be classified as physical absorption and absorption with a chemical reaction. In physical absorption, the pollutants are dissolved in a solvent and can be desorbed for recovery. The absorption of ammonia by water or the absorption of hydrocarbon by oil are typical examples.

If a solvent to absorb a significant quantity of the pollutant cannot be found, a reactant mixed with the solvent can be used. The pollutants must first be absorbed into the liquid phase for the reaction to occur. In this case, the pollutant concentration in the liquid is reduced to a low level. As a result, high absorption capacity is achieved.

The reaction can be reversible or irreversible. Typical reversible reactions are H$_2$S/ethanolamines, CO$_2$/alkali, carbonates, and some flue gas desulfurization (FGD) systems. The reversible reactions allow the pollutants to be recovered in a concentrated form and the solvent to be recycled to the absorber. If the reactions are irreversible, the reaction products must be disposed or marketed (e.g. ammonium sulfate). A few FGD systems are irreversible (vide infra).

ABSORPTION OPERATIONS

Absorption systems design involves selecting a solvent and the design of the absorber.

Solvent Selection

Solubility is the most important consideration in the selection of a solvent for absorption. The higher the solubility, the lower the amount of solvent required to remove a given amount of pollutants. The solvent should also be relatively nonvolatile to prevent an excessive carryover in the gas effluent. Other favorable properties include low flammability and viscosity, high chemical stability, acceptable corrosiveness, and low toxicity and pollution potential. The final selection criterion is an economic comparison with other control technologies.

Absorber Design

Any gas–liquid contactors that promote the mass transfer across the phase boundary can be used in absorption operations. The most popular devices are spray, packed, and tray columns as well as venturi scrubbers. For gas pollution control, the combination of high gas flowrate and low pollutant concentration suggests that the absorber should exhibit a low pressure drop. The mass-transfer efficiency of the absorber determines the height of the column, but it is not as important a consideration as the pressure drop. Clearly, spray and packed columns are the best devices to satisfy the preceding criteria. Spray columns are used where fouling and low pressure drops are encountered. The design of a spray column is straightforward and is detailed in other publications (Kohl 1987). The packed column is the device used most often.

Two types of packings are used for absorption: random packing and structured packing. Table 5.20.1 shows a list of packing suppliers. Structured packing tends to be proprietary, and the design procedure is normally obtained from suppliers. Random packings (e.g., Pall rings and Intalox saddles) can be purchased from several suppliers,

and the system design procedure is published. Table 5.20.2 shows the properties of typical random packings.

As a general rule, randomly packed columns have the following characteristics (Strigle 1987):

A pressure drop between 2 and 5 cm H_2O/m of packed depth

Air velocity between 1.7 and 2.4 m/sec for modern, high-capacity plastic packings

An inlet concentration of pollutants below 0.5% by volume

A superficial liquid rate in the range of 1.35 to 5.5 $L/m^2 \cdot sec$

Figure 5.20.1 is a schematic diagram of a typical packed column installation.

Design Procedure

In designing a packed column, the design engineer must consider the solvent rate, column diameter, and column height.

SOLVENT RATE

Once the feed and effluent specifications are established, the design engineer can calculate the minimum solvent rate. The minimum rate is the rate below which separation is impossible even with a column of infinite height. The engineer calculates flowrates as illustrated in Figure 5.20.2. The mole fraction of the pollutant in the solvent (X_T) and effluent gas (Y_T) at the top of the column as well as that in the feed gas (Y_B) are known. A straight line is drawn from (X_T, Y_T) to intercept the equilibrium line at Y_B. The slope of this line represents the ratio of the minimum solvent rate to the feed gas rate.

The normal operating solvent rate should be 30 to 70% above the minimum rate.

TABLE 5.20.1 ABSORPTION PACKING SUPPLIERS

Supplier	Address	Random Packing	Structured Packing
ACS Industries	Woonsocket, RI		x
Ceilcote Air Pollution Control	Berea, OH	x	
Chem-Pro Corp.	Fairfield, NJ	x	
Clean Gas Systems, Inc.	Farmingdale, NY	x	
Glitsch, Inc.	Dallas, TX	x	x
Jaeger Products, Inc.	Houston, TX	x	
Julius Montz Co.	Hilden, Germany		x
Koch Engineering Co., Inc.	Wichita, KS	x	x
Kuhni, Ltd.	Basel, Switzerland		x
Lantec Products, Inc.	Agoura Hills, CA	x	
Munters Corp.	Fort Myers, FL		x
Norton Chemical Process Product Corp.	Akron, OH	x	x
Nutter Engineering	Tulsa, OK	x	x
Sulzer Brothers Ltd.	Winterthur, Switzerland		x

TABLE 5.20.2 PROPERTIES OF RANDOM PACKINGS

Packing Type	Nominal Size (mm)	Elements (per m³)	Bed Weight (kg/m³)	Surface Area (m²/m³)	ε Void Fraction	F_p Packing Factor (m⁻¹)
Intalox saddles (ceramic)	13	730,000	720	625	0.78	660
	25	84,000	705	255	0.77	320
	38	25,000	670	195	0.80	170
	50	9400	670	118	0.79	130
	75	1870	590	92	0.80	70
Intalox saddles (metal)	25	168,400	350	n.a.	0.97	135
	40	50,100	230	n.a.	0.97	82
	50	14,700	181	n.a.	0.98	52
	70	4630	149	n.a.	0.98	43
Pall rings (metal)	16				0.92	230
	25	49,600	480	205	0.94	157
	38	13,000	415	130	0.95	92
	50	6040	385	115	0.96	66
	90	1170	270	92	0.97	53
Raschig rings (ceramic)	13	378,000	880	370	0.64	2000
	25	47,700	670	190	0.74	510
	38	13,500	740	120	0.68	310
	50	5800	660	92	0.74	215
	75	1700	590	62	0.75	120
Berl saddles (ceramic)	13	590,000	865	465	0.62	790
	25	77,000	720	250	0.68	360
	38	22,800	640	150	0.71	215
	50	8800	625	105	0.72	150
Intalox saddles (plastic)	25	55,800	76	206	0.91	105
	50	7760	64	108	0.93	69
	75	1520	60	88	0.94	50
Pall rings (plastic)	16	213,700	116	341	0.87	310
	25	50,150	88	207	0.90	170
	50	6360	72	100	0.92	82

Source: A.L. Kohl, 1987, Absorption and stripping, in *Handbook of separation process technology* (New York: John Wiley).
Note: n.a. = not applicable.

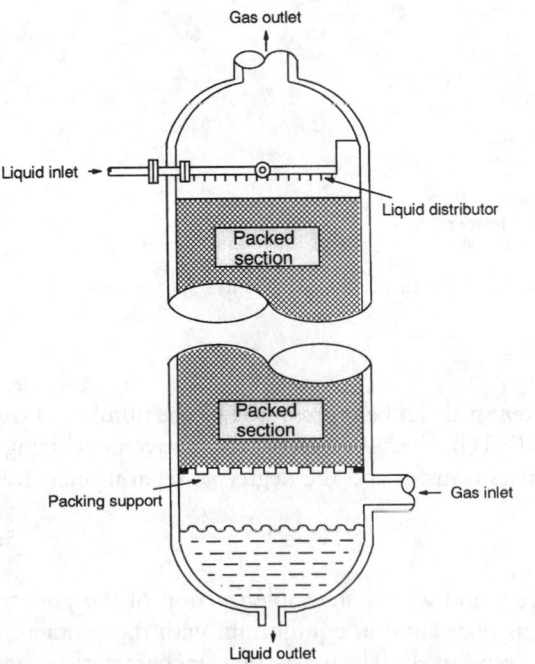

Gas outlet

Liquid inlet

Liquid distributor

Packed section

Packed section

Packing support

Gas inlet

Liquid outlet

FIG. 5.20.1 Schematic diagram of a typical packed column.

COLUMN DIAMETER

Since the pressure drop directly impacts the operating cost for the blower to deliver the feed gas, the design engineer should choose the tower diameter to meet the pressure drop specification. Figure 5.20.3 shows the value for a widely used correlation for the pressure drop calculation. The correlation includes a parameter, packing factor F_P, which is a constant for a given packing size and shape (Fair 1987). The value of F_p can be obtained from Table 5.20.2. With known physical properties and flowrates for the liquid and gas streams, the engineer first determines the specific gas rate G, kg/m² · sec in the column and then calculates the column cross-sectional area by dividing the total gas flowrate (G_T, kg/s) by G.

In cases where the pressure drop is not a prime consideration (e.g., treatment of a pressurized process stream), the column can be designed to operate at 70–80% of flooding rate. For packings below 2.5 cm size, the design engineer can use the flooding line shown in Figure 5.20.3 to calculate the flooding rate. If the packing is larger than 3.7 cm, the following empirical equation determines the pressure drop at flooding (McCabe 1993):

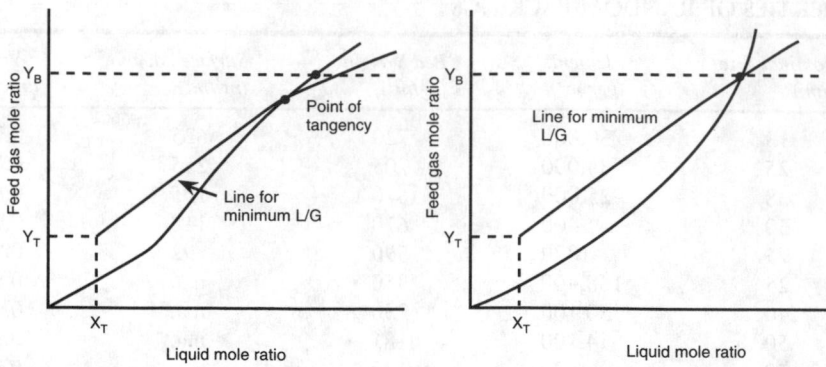

FIG. 5.20.2 Calculation of minimum solvent rates for two different shapes of equilibrium line.

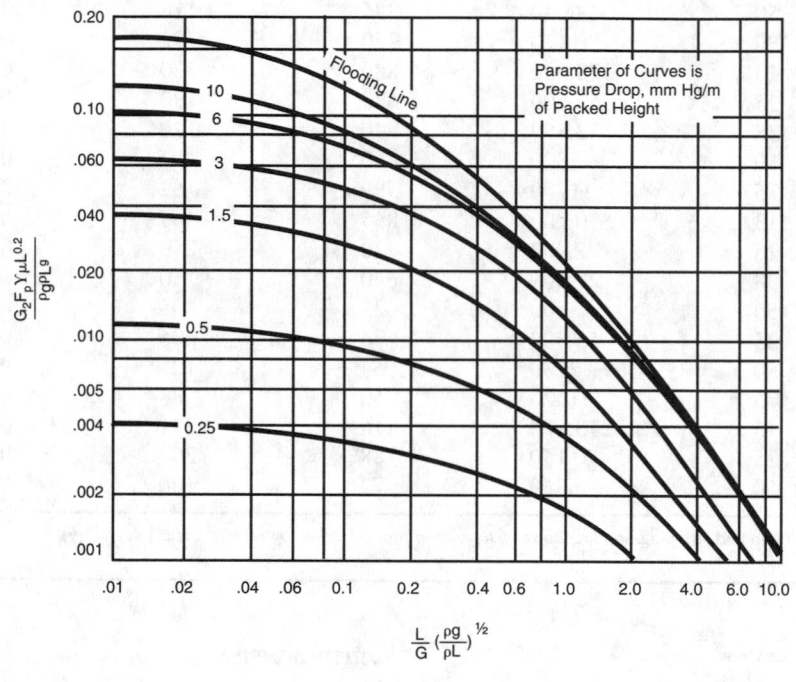

Key:
L = Liquid rate, kg/s m²
G = gas rate, kg/s m²
ρL = Liquid density, kg/m³
ρg = Gas density, kg/m³
Fp = Packing factor, m⁻¹
μL = Viscosity of Liquid, mPa s
Y = Ratio, (density of water)/(density of liquid)
g = Gravitational constant, 9.81 m/s²

FIG. 5.20.3 Generalized correlation for the calculation of pressure drops in packed columns.

$$\Delta P_{flood} = 0.307 F_P^{0.7} \qquad 5.20(1)$$

Using Figure 5.20.3, the engineer can then calculate the G value and hence the required column diameter.

COLUMN HEIGHT

The required height of an absorption column is determined primarily by the degree of pollutant removal and the mass-transfer characteristics of the system. The degree of pollu-

tant removal can be represented by the number of transfer units (NTU) which is a function of the average driving force for mass transfer and the degree of separation as follows:

$$NTU = \int_{y_T}^{y_B} \frac{dy}{y - y^*} \qquad 5.20(2)$$

where y and y^* are the mole fraction of the pollutant in the gas phase and in equilibrium with the contacting solvent, respectively. The mass-transfer characteristics include the fluid properties, operating conditions, and column internal parameters. The following equation expresses these effects:

$$\text{HTU} = \text{Height of a Transfer Unit} = \frac{G}{K_y a} \qquad 5.20(3)$$

where K_y is the overall mass-transfer coefficient and a is the effective interfacial area available for mass transfer. The required column height H can then be calculated from the following expression:

$$H = \text{HTU} \cdot \text{NTU} = \frac{G}{K_y a} \int_{y_T}^{y_B} \frac{dy}{y - y^*} \qquad 5.20(4)$$

McCabe (1993) provide the derivation of the preceding equations. To solve Equation 5.20(4), the design engineer must specify y_B and y_T and obtain G from the column diameter calculations, y^* from the equilibrium relationship, and $K_y a$ from engineering correlations. The values for y^* for physical absorption are available from Henry's law equation as follows:

$$y^* = \frac{H}{P_T} x \qquad 5.20(5)$$

where H is the Henry's law constant and P_T is the total pressure. The accuracy of estimating column height depends mostly on the $K_y a$ correlation chosen for the calculation. In mass-transfer operations, K_y can be broken down to the mass-transfer coefficient in the gas phase k_y and in the liquid phase k_x. This two-resistance theory is widely accepted (McCabe 1993) and leads to the following relationships:

$$\frac{1}{K_y} = \frac{1}{k_y} + \frac{m}{k_x} \quad \text{and} \quad y^* = mx \qquad 5.20(6)$$

These relationships assume that the resistances to mass transfer in the gas and liquid phases occur in series, and the gas–liquid interface does not contribute significant resistance to the mass transfer. If k_y is much greater than k_x/m, then $K_y \cong m/k_x$, and the rate of absorption is controlled by the liquid-phase transfer. On the other hand, if $k_y << k_x/m$, then the system is gas-phase controlled. To some extent, the value of m or H controls the phase control mechanism. For a soluble system (e.g., absorption of NH_3 by water), the value of m is small, and thus the mass transfer is gas-phase controlled. For a CO_2/water system, the value of m is large, and the absorption is liquid-phase controlled.

The calculations also require estimating the effective interfacial area a. The area a is generally smaller than the geometric area of the packing because some packing area is not wetted and some area is covered by a stagnant, liquid film that is already saturated with the pollutant (i.e., inactive for mass transfer).

Onda, Takeuchi, and Okumoto (1968) have expressions for the individual mass-transfer coefficients k_x and k_y as well as the effective interfacial area a, and the de-

rived values can be used in the absence of experimental data. The equations are as follows:

$$k_y \left(\frac{RT}{a_p D_G} \right) = 5.23 \left(\frac{G}{a_p \mu_G} \right)^{0.7} (Sc_G)^{1/3} (a_p d_p)^{-2} \qquad 5.20(7)$$

$$k_x \left(\frac{\rho L}{g \mu_L} \right)^{1/3} = 0.0051 \left(\frac{L}{a_p \mu_L} \right)^{2/3} (Sc_L)^{-1/2} (a_p d_p)^{0.4} \qquad 5.20(8)$$

where:

L = mass flowrate of liquid (kg/sec·m²)
G = mass flowrate of gas (kg/sec·m²)
D_G = diffusion coefficient (m²/sec)
Sc_L = Schmidt number, $(\mu_L/\rho_L D_L)$
k_y = gas-phase mass-transfer coefficient (kmol/N·s)
k_x = liquid-phase mass-transfer coefficient (m/s)
ρ_L = density of liquid (kg/m³)
μ_L = viscosity of liquid (N·s/m²)
d_p = effective packing diameter (the diameter of a sphere with equal surface area)
a_p = dry outside surface area of packing

In this model, a is assumed to equal a_w, the wetted area of the packing, and is calculated by the following equation:

$$a = a_w$$

$$= a_p \left\{ 1 - \exp \left[-1.45 (Re_L)^{-0.1} (Fr_L)^{-0.05} (We_L)^{0.2} \left(\frac{\sigma}{\sigma_c} \right)^{-0.75} \right] \right\}$$

$$5.20(9)$$

where:

a = the effective interfacial area for mass transfer (m²/m³)
a_w = wetted area of packing
σ_c = a critical surface tension = 61 dyn/cm for ceramic packing, 75 dyn/cm for carbon steel packing, and 33 dyn/cm for polyethylene packing
Re_L = $L/a_p \mu_L$
Fr_L = $a_p L^2/(g\rho_L)^2$
We_L = $L^2/a_p \rho_L$
Sc_L = $\dfrac{\mu_L}{\rho_L D_L}$
Sc_G = $\dfrac{\mu_G}{\rho_G D_G}$

Absorption with a chemical reaction in the liquid phase is involved in most applications for gas control. Reaction in the liquid phase reduces the equilibrium partial pressure of the pollutant over the solution, which increases the driving force for mass transfer. If the reaction is irreversible, then $y^* = 0$, and the NTU is calculated as follows:

$$\text{NTU} = \int_{y_T}^{y_B} \frac{dy}{y - y^*} = \ln \frac{y_B}{y_T} \qquad 5.20(10)$$

A further advantage of the reaction is the possible increase in the liquid-phase mass-transfer coefficient k_x and the effective interfacial area a. Perry and Green (1984) describe the design methods in detail.

COMMERCIAL APPLICATIONS

Most commercial applications of absorption for gas control fall into two categories: regenerative and nonregenerative systems. For regenerative systems, the solvent leaving the absorber is enriched in pollutant. This mixture is normally separated by distillation. The solvent is then recycled back to the absorber, and the pollutant can be recovered in a concentrated form. Regeneration is possible when the absorber uses a physical solvent or a solvent containing compounds that react reversibly with the pollutant. The recovery of H_2S from hydrocarbon processing using an amine as the solvent is a major application of regenerative absorption. Figure 5.20.4 shows a typical flowsheet.

For the capture of H_2S and volatile sulfur compounds (e.g., thiols, COS, and CS_2) released during the transportation of sour liquids, a portable system is required. Am-gas Scrubbing Systems Ltd. (Calgary, Canada) manufactures both portable and stationary units that use 26% aqueous ammonia. The spent solution oxidizes to form ammonium sulfate, which is recovered from the spent absorbent and can be marketed as fertilizer. Alternate absorbents for portable units include basic hypochlorite (bleach) solutions, but the stability of the solution decreases with consumption of the base as the H_2S is absorbed. A complicating factor in the use of basic fluid absorbent systems is that CO_2 is also a weak acid and is absorbed to sequentially form bicarbonate (HCO_3^-) and carbonate (CO_3^{2-}) ions, thereby consuming the absorbent.

The absorption of SO_2 by dimethylaniline from off-gases at copper, zinc, lead, and nickel smelters is another application. The desorbed SO_2 produces H_2SO_4. The SO_2 concentration in the off-gases is approximately 10%.

For nonregenerative or throwaway systems, irreversible reactions occur in the liquid phase. The reagents in the solvent are consumed and must be replenished for absorption efficiency. The resulting products are discarded. Typical examples are the absorption of acid gases by caustic solutions. Nonregenerative systems are economical only when the reagents are inexpensive or the volume involved is small.

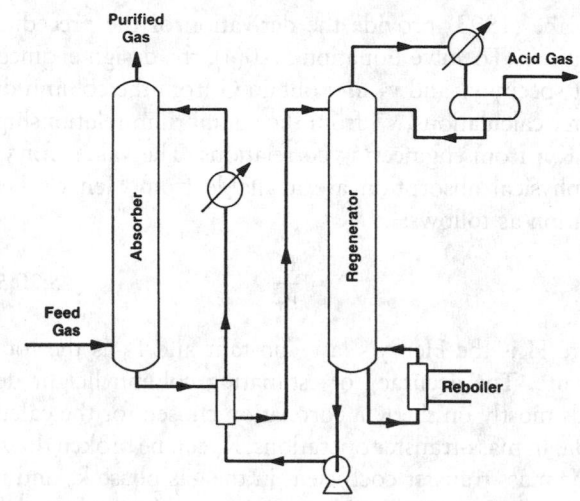

FIG. 5.20.4 Flowsheet for the recovery of H_2S using amine as a solvent.

TABLE 5.20.3 FLUE GAS DESULFURIZATION PROCESSES

Process Generics	Process Operations	Active Material	Key Sulfur Product
Throwaway processes			
1. Lime or limestone	Slurry scrubbing	CaO, $CaCO_3$	$CaSO_3/CaSO_4$
2. Sodium	Na_2SO_3 solution	Na_2CO_3	Na_2SO_4
3. Dual alkali	Na_2SO_3 solution, regenerated by CaO or $CaCO_3$	$CaCO_3/Na_2SO_3$ or CaO/NaOH	$CaSO_3/CaSO_4$
4. Magnesium promoted-lime or limestone	$MgSO_3$ solution, regenerated by CaO or $CaCO_3$	$MgO/MgSO_4$	$CaSO_3/CaSO_4$
Regenerative processes			
1. Magnesium oxide	$Mg(OH)_2$ slurry	MgO	15% SO_2
2. Sodium (Wellman–Lord)	Na_2SO_3 solution	Na_2SO_3	90% SO_2
3. Citrate	Sodium citrate solution	H_2S	Sulfur
4. Ammonia	Ammonia solution, conversion to SO_2	NH_4OH	Sulfur (99.9%)
Dry processes			
1. Carbon adsorption	Adsorption at 400°K, reaction with H_2S to S, reaction with H_2 to H_2S	Activated carbon/H_2	Sulfur
2. Spray dryer	Absorption by sodium carbonate or slaked lime solutions	$Na_2CO_3/Ca(OH)_2$	Na_2SO_3/Na_2SO_4 or $CaSO_3/CaSO_4$

The most important application of absorption for gas control technology is FGD. Because flue gas from power plants contributes about two-thirds of the U.S. emissions, large efforts have been spent on developing an effective control technology. Currently, several hundred FGD systems are in commercial operation. The absorption can be regenerative or nonregenerative as well as wet or dry; in effect, four categories of FGD processes exist. Table 5.20.3 lists the processes in use. The following sections describe the most important FGD processes.

Nonregenerative Systems

Most FGD systems are nonregenerative. The reagents used in the absorber are alkaline compounds that react with SO_2. Most well-established technology (see Figure 5.20.5) is based on limestone and lime. Other processes are based on $NaOH$, Na_2CO_3, and NH_4OH. Since lime and limestone are the most inexpensive and abundant reagents, they are used in about 75% of the installed FGD systems. In this process, the reaction products are $CaSO_3$ and $CaSO_4$ in a sludge form that must be disposed. These systems can remove 90% of the SO_2 in the flue gases. Although lime is more reactive than limestone, it is more expensive; therefore, lime is not used as widely as limestone.

The main drawback of lime- and limestone-based FGD systems is scaling and plugging of the column internals. This problem is eliminated in the dual alkali system (see Figure 5.20.6) by absorption with a Na_2SO_3/Na_2SO_4 solution which is then sent to a separate vessel where lime or limestone and some $NaOH$ are introduced. The lime or limestone precipitates the sulfite and sulfate and regenerates $NaOH$ for reuse in the absorption column. The major disadvantage of this process is the loss of soluble sodium salts into the sludge which may require further treatment. As a result, this process is not used as widely as the lime and limestone process.

Regenerative Systems

Regenerative processes have higher costs than throw-away processes. However, regenerative processes are chosen when disposal options are limited. Regenerative processes produce a reusable sulfur product. In Japan, where the government mandates FGD, regenerative processes are used almost exclusively.

The Wellman–Lord process is the most well-established regenerative process which uses an aqueous sodium sulfite solution as the solvent. Figure 5.20.7 is a schematic diagram of the Wellman–Lord process. This process consists of the four following subprocesses:

Flue gas pretreatment. In this subprocess, flue gas from an ESP is blown through a venturi prescrubber. The prescrubber removes most of the remaining particles and any existing SO_3 and HCl, which upset the SO_2 absorption chemistry. The prescrubber also cools and humidifies the flue gas. A liquid purge stream from the prescrubber removes the solids and chlorides.

SO_2 absorption by sodium sulfite solution. In the SO_2 absorber tower, the flue gas from the prescrubber contacts the aqueous sodium sulfite, and the SO_2 is absorbed and reacted with Na_2SO_3 in the liquid to form sodium bisulfite. Since excess O_2 is always present, some of the Na_2SO_3 oxidizes to Na_2SO_4; some of the Na_2SO_3 reacts with the residual SO_3 to form Na_2SO_4 and sodium bisulfite. The sodium sulfate does not further SO_2 absorption. A continuous purge from the bottom of the absorber prevents excessive sulfate buildup. Since the absorber bottom is rich in bisulfite, most of the stream is routed for further processing.

Purge treatment. Part of the liquid stream leaving the absorber is sent to the chiller and crystallizer, where the less soluble, sodium sulfate crystals are formed. The slurry is centrifuged, and the solids are dried and discarded. The bisulfite-rich, centrifugal material is returned to the process.

Sodium bisulfite regeneration. The remaining part of the liquid stream from the absorber is sent to a heated evap-

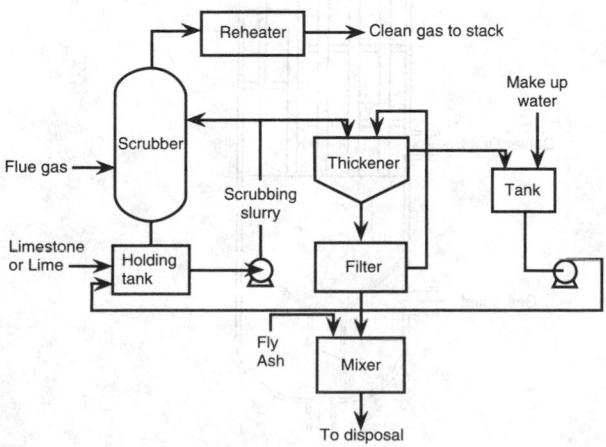

FIG. 5.20.5 Schematic flow diagram of a limestone-based FGD system.

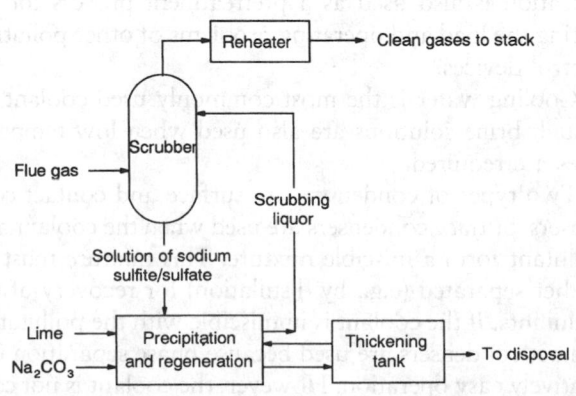

FIG. 5.20.6 Schematic diagram of a dual alkali FGD system.

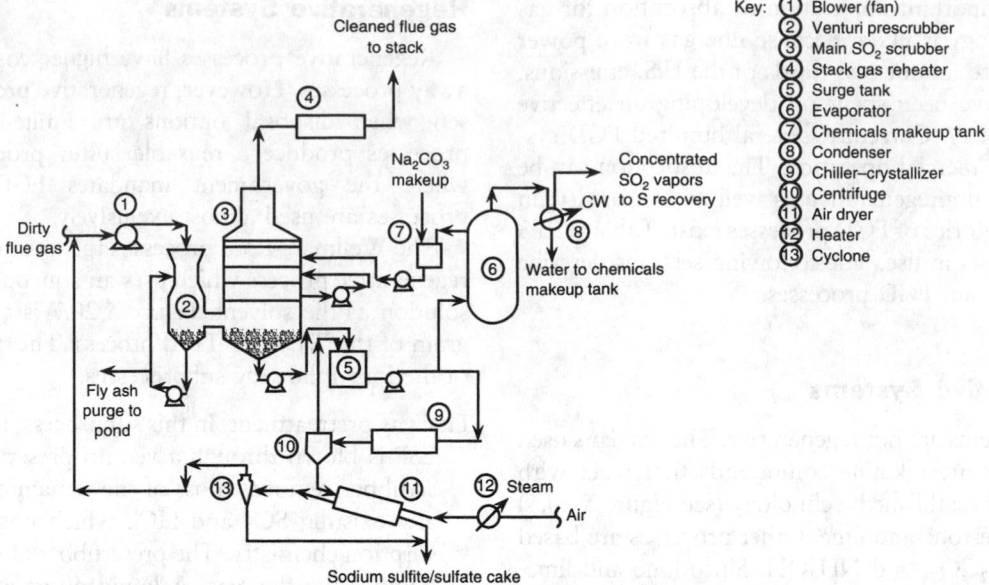

Key:
① Blower (fan)
② Venturi prescrubber
③ Main SO₂ scrubber
④ Stack gas reheater
⑤ Surge tank
⑥ Evaporator
⑦ Chemicals makeup tank
⑧ Condenser
⑨ Chiller–crystallizer
⑩ Centrifuge
⑪ Air dryer
⑫ Heater
⑬ Cyclone

FIG. 5.20.7 Schematic process flow diagram of the Wellman–Lord SO₂ scrubbing and recovery system. (Reprinted from U.S. Environmental Protection Agency (EPA), 1979.)

orator and crystallizer, where sodium bisulfite is decomposed to Na_2SO_3 and SO_2. The gas stream contains 85% SO_2 and 15% H_2O, thus the SO_2 can be used as feed stock for producing S or sulfuric acid. To replace the lost Na_2SO_3, this subprocess adds soda ash (Na_2CO_3) to make up sodium. The Na_2CO_3 reacts readily with SO_2 in the absorber tower to give sodium sulfite.

Condensation

In cases where pollutants have low vapor pressures, condensation is effective for removing a significant part of the vapor. The condenser works by cooling the feed gas to a temperature below the dew point of the feed gas. Although condensation can also occur by increasing the pressure without changing the temperature, this method is seldom used for the control of air pollution.

Because condensation is a simple process and the design techniques for condensers are well-established, condensation is also used as a pretreatment process for reducing the load and operating problems of other pollution control devices.

Cooling water is the most commonly used coolant although brine solutions are also used when low temperatures are required.

Two types of condensers are surface and contact condensers. Surface condensers are used when the coolant and pollutant form a miscible mixture. This mixture must be further separated (e.g., by distillation) for recovery of the pollutants. If the coolant is immiscible with the pollutants, contact condensers are used because phase separation is a relatively easy operation. However, the coolant is not contaminated to the extent of causing water pollution problems.

Surface condensers are normally shell and tube type and should be set vertically. Vapor should only condense inside the tubes. This arrangement prevents a stagnant zone of inert gas (air) that might blanket the heat transfer surfaces. Figure 5.20.8 shows a typical surface condenser. The feed gas enters the top of the condenser and flows concurrently downward with the condensate.

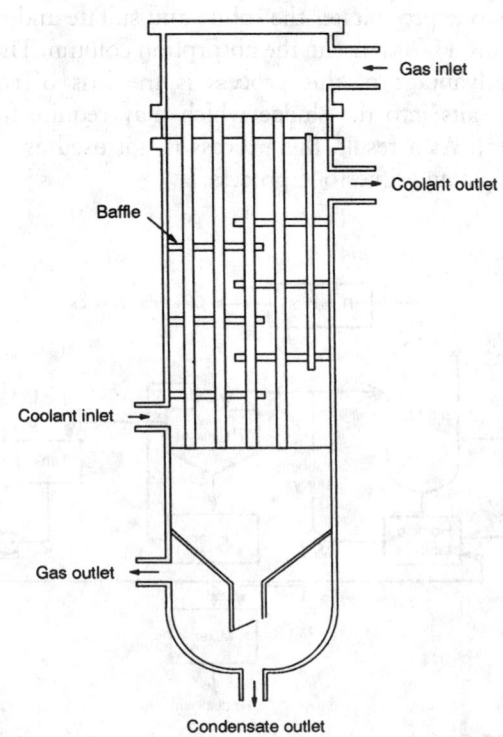

FIG. 5.20.8 Surface condenser.

Contact condensers are much smaller and less expensive than surface condensers. In the design shown in Figure 5.20.9, part of the coolant is sprayed into the gas space near the top of the condenser. This design is similar to the spray column used in absorption. The remainder of the coolant is directed into a discharge throat to complete the condensation. The pressure regain in the downstream cone of the venturi is often sufficient to eliminate the need for a pump or a barometric leg.

Because the coolant required to operate contact condensers is ten to twenty times that for surface condensers, the latter is the predominate device used in air pollution control applications. The design of a surface condenser is based on complete resistance to heat transfer on the condensing side in the layer of condensate. A mean condensing coefficient is calculated from appropriate correlations. Calculating the overall heat transfer coefficient and the log–mean temperature driving force estimates the required heat transfer area. Many engineering textbooks (Perry and Green 1984) provide additional design details.

Adsorption

Adsorption is a separation process based on the ability of adsorbents to remove gas or vapor pollutants preferentially from a waste gas stream (Yang 1987; Kohl and Riesenfeld 1985). The process is particularly suitable when pollutants are (1) noncombustible, (2) insoluble in liquids, or (3) present in dilute concentration.

The mechanism of adsorption can be classified as either physical adsorption or chemisorption. In physical adsorption, gas molecules adhere to a solid surface via van der Waals forces. The process is similar to the condensation of a vapor. It is a reversible process. Desorption occurs by lowering the pressure, increasing the temperature, or purging the adsorbent with an inert gas. In chemisorp-

tion, gas molecules are adsorbed by forming chemical bonds with solid surfaces. This process is sometimes irreversible. For example, oxygen chemisorbed on activated carbon can only be desorbed as CO or CO_2. Chemisorbed spent adsorbents cannot normally be regenerated under mild temperatures or vacuum.

ADSORPTION ISOTHERM

For a particular single gas–solid combination, one of five types of adsorption isotherm is found (see Figure 5.20.10). In separation processes, the favorable isotherms, types 1 and 2 are most frequently encountered.

From considering kinetics for a single gas impacting a uniform solid surface and adsorbing without chemical change, Langmuir (1921) deduced that the fraction of a surface covered by a monolayer varies with the partial pressure of the adsorbate P_a. The following equation gives the mass of adsorbate adsorbed per unit mass of adsorbent m_a:

$$m_a = \frac{k_1 P_a}{k_2 P_a + 1} \qquad 5.20(11)$$

where k_1 and k_2 are constants. Equation 5.20(11) is known as the Langmuir isotherm, and the shapes are types 1 and 2 in Figure 5.20.10.

Assuming that the number of sites of energy Q (N_Q) is related to a base value for Q, Q_0, as follows:

$$N_Q = ae^{-Q/Q_0} \qquad 5.20(12)$$

then Equation 5.21(11) reduces to the following approximation:

$$\Theta = \text{const}.P^{1/n} \qquad 5.20(13)$$

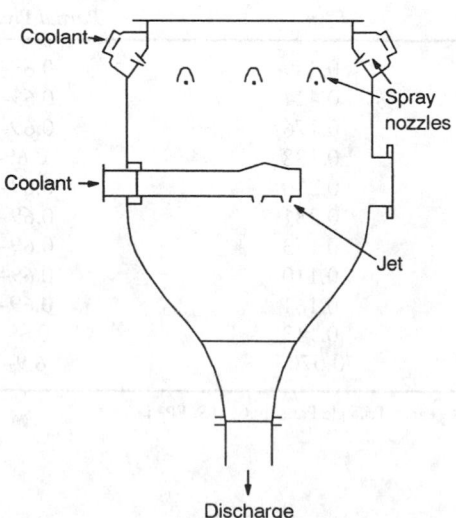

FIG. 5.20.9 Contact condenser.

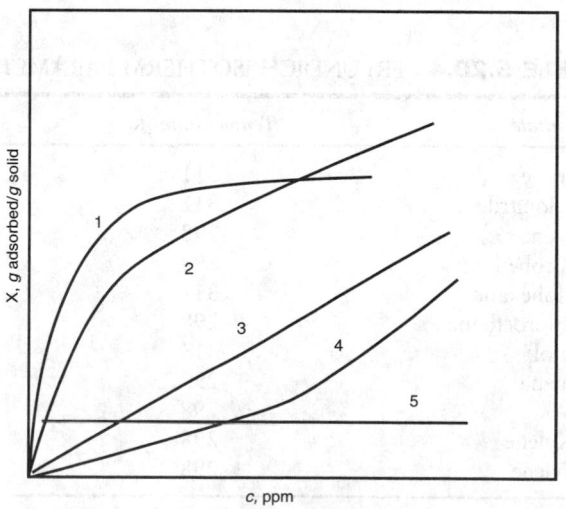

Key: 1 = High loading (highly favorable)
     2 = Favorable
     3 = Linear
     4 = Low loading (unfavorable)
     5 = Irreversible

FIG. 5.20.10 Adsorption loading profiles.

When expressed as the volume of adsorbate adsorbed per unit mass of adsorbent ν, as follows:

$$\nu = kP^{1/n} \qquad 5.20(14)$$

or expressed in terms of concentration in the gas phase c and at the absorbent surface w, as follows:

$$c = \alpha w^{\beta} \qquad 5.20(15)$$

the expression is known as the Freundich isotherm. For a binary, or higher, mixture in which the components compete for surface sites, the expression is more complex (Yang 1987). Table 5.20.4 gives the values for parameters k and n in the Freundich isotherm for selected adsorbates.

Ideally, an adsorbent adsorbs the bulk of a pollutant from air even when that material is in low concentration. The adsorption profile for such behavior is called *favorable* and is shown as curves 1 and 2 in Figure 5.20.10. If high concentrations must be present before significant quantities are adsorbed (curve 4), the profile is called *unfavorable adsorption*.

As a pollutant is adsorbed onto the adsorbent, the concentration in the air stream falls. The adsorbent continues to absorb the pollutant from the air stream until it is close to saturation. Thus, as the air stream passes through a bed of adsorbate, the pollutant concentration varies along the bed length. This adsorption wave progresses through the bed with time (see Figure 5.20.11). At time t_1, the bed is fresh, and essentially all pollutant is adsorbed close to the entrance of the bed. At time t_2, the early part of the bed is saturated, but the pollutant is still effectively adsorbed. At time t_3, a small concentration of pollutant remains in the air stream at the exit. When the pollutant concentration at the exit meets or exceeds the limiting value, the bed is spent and must be replaced or regenerated. For some pollutants, the breakthrough of detectable amounts of pollutants is unacceptable. To avoid exceeding acceptable or regulated limits for pollutant concentrations in exit gases, replacing or regenerating adsorbent beds well before the end of the bed's lifetime is essential.

The profiles shown in Figure 5.20.11 are for a single pollutant in an air stream. An adsorber normally operates in a vertical arrangement to avoid bypassing gases. For multiple pollutants, the process is more complex as each pollutant can exhibit different behavior or competition can occur between pollutants for adsorption sites (Yang 1987; Kast 1981). The bed design must be able to adsorb and retain each pollutant.

ADSORPTION EQUIPMENT

The major industrial applications for adsorption processes have both environmental and economic objectives. The abatement of air pollution includes removing noxious and odorous components. Additionally, adsorption is used for the recovery of solvents and reagents, such as carbon disulfide, acetone, alcohol, aliphatic and aromatic hydrocarbons, and other valuable materials, from effluent streams (Kast 1981).

Typically, a process in which the adsorbate is recovered uses temperature swing adsorption (TSA), pressure swing adsorption (PSA), or continuous or periodic removal of the adsorbent from the system (Kast 1981). The TSA and PSA technologies are normally used only for the removal of high concentrations of adsorbate. For the majority of environmental applications, small concentrations must be removed from air, and continuous or periodic removal technologies are more appropriate.

TABLE 5.20.4 FREUNDICH ISOTHERM PARAMETERS FOR SOME ADSORBATES

Adsorbate	Temperature (K)	k × 100	n	Partial Pressure (Pa)
Acetone	311	1.324	0.389	0.69–345
Acrylonitrile	311	2.205	0.424	0.69–103
Benzene	298	12.602	0.176	0.69–345
Chlorobenzene	298	19.934	0.188	0.69–69
Cyclohexane	311	7.940	0.210	0.69–345
Dichloroethane	298	8.145	0.281	0.69–276
Phenol	313	22.116	0.153	0.69–207
Toluene	298	20.842	0.110	0.69–345
TCA	298	25.547	0.161	0.69–276
m-Xylene	298	26.080	0.113	0.69–6.9
p-Xylene	298	28.313	0.0703	6.9–345

Source: U.S. Environmental Protection Agency (EPA), 1987, *EAB control cost manual*, 3d ed. (Research Triangle Park, N.C. U.S. EPA).
Notes: The amount adsorbed is expressed in kg adsorbate/kg adsorbent.
The equilibrium partial pressure is expressed in Pa.
Data are for the adsorption on Calgon-type BPL activated carbon (4 × 10 mesh).
Data should not be extrapolated outside of the partial pressure ranges shown.

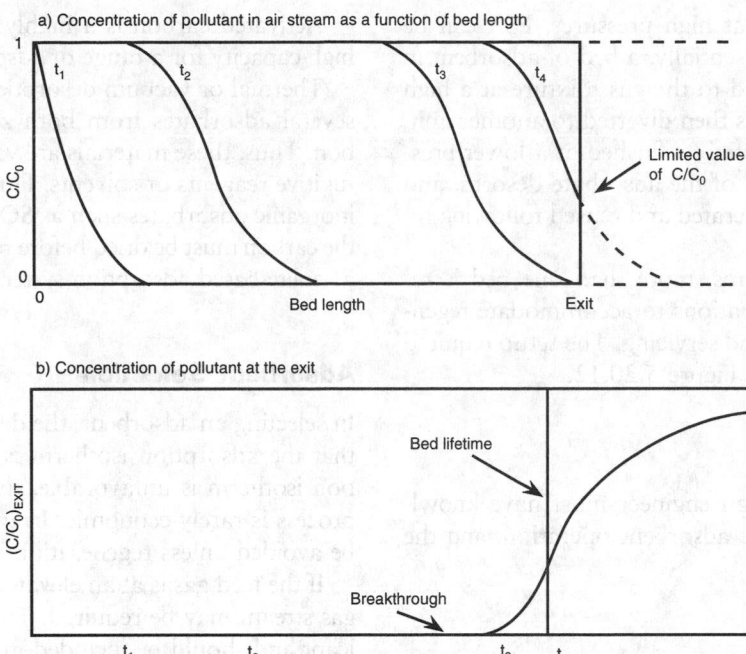

FIG. 5.20.11 Adsorption profiles for a single contaminant in an air stream.

Periodic Removal

This system is the simplest to design or build but is not necessarily the most efficient or cost-effective to operate. Essentially, the adsorbate is left in a fixed-bed or fluidized-bed adsorber until the adsorptive capacity of the bed is approached. Then, the adsorbate is removed for destruction, disposal, or regeneration. A new adsorbent is then charged to the system. For systems in which either the adsorbent or adsorbate is hazardous or contact with air is deleterious, this method presents issues requiring expensive solutions. Further, during changeover of the spent adsorbent, the adsorber is not used. Therefore, a minimum of two adsorber units is necessary for any plant in continuous operation.

Continuous Removal

By simultaneously introducing and removing adsorbent, this system achieves continuous operation. This system requires the bed, or the material in the bed, to be in continuous motion. A fixed-bed can rotate so that the fraction of the bed exposed to the feed stream is continuously charged. The spent adsorbent then rotates to a separate zone where the system regenerates it either by flushing with a carrier gas, contact with a reagent, or desorption of the adsorbate by sweeping with a hot gas. Because the adsorbent is necessarily porous, no significant reverse pressure differential exists between the feed and flushing systems.

A solid adsorbent cascading through the feed gas can be continuously removed from the base of the adsorber unit. Regenerated adsorbent can then be recycled to the unit.

TSA

When the bond between the adsorbent and adsorbate is strong (i.e., chemisorption), regenerating the adsorbent requires elevated temperatures and purging with a nonadsorbing gas. If this process occurs in the adsorber unit, a TSA process is used. For continuous processing, a minimum of three adsorber units are required for economic operation: at least one in operation, one undergoing regeneration, and one cooling down following regeneration. Typically, the gas flushing the unit that is cooling down is then passed through the bed undergoing regeneration. A heater ensures that the flush gas is at a temperature to effect desorption. The effluent stream is then cooled, and if the adsorbate is to be recovered, the flush gas and adsorbate are separated by physical methods (distillation, condensation, or decanting). Regeneration can also require additional steps, including physical or chemical treatment, washing, and drying.

PSA

When the adsorption capacity of an adsorbent is a function of pressure or one component in a gaseous mixture is preferentially adsorbed at high pressures, PSA can be used for gas separation. Essentially, a bed of adsorbent in one unit is initially exposed to the gas mixture at a high pressure. The feed stream is then diverted to another unit, while the initial unit is vented or flushed at a lower pressure. A significant fraction of the adsorbate desorbs, and the bed in this unit is regenerated and reused following repressurization.

For each of these systems, more than one bed is required for continuous operations to accommodate regeneration of spent material and servicing. The setup required is similar to that shown in Figure 5.20.12.

ADSORBER DESIGN

In adsorber design, a design engineer must have knowledge about adsorbents, the adsorbent operation, and the regeneration operation.

Adsorbents

Favorable properties of an adsorbent include good thermal stability; good mechanical integrity, especially for use in a continuous removal or fluidized-bed system; high surface area and activity; a large capacity for adsorption; and facile regeneration. In some cases, the spent adsorbent must be disposed of, in which case it must be inexpensive and environmentally benign or of commercial value following minor processing.

Activated carbon and the series of microporous aluminosilicates know as zeolites are important adsorbents. Table 5.20.5 presents the properties of the widely used zeolites. Because water vapor is not considered an air pollutant, the adsorption capacity of a zeolite for water is not a benefit.

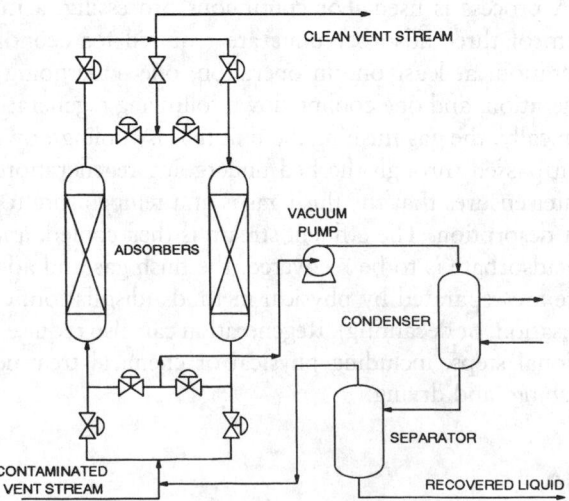

FIG. 5.20.12 Activated carbon system for solvent recovery using modified PSA. (Data from AWD Technologies, Inc., 1993, *Hydrocarbon processing* [August]:76.)

Activated carbon is a highly porous adsorbent with a high capacity for a range of adsorbates (see Table 5.20.6).

Thermal or vacuum desorption permits the recovery of several adsorbates from both zeolites and activated carbon. Thus, these materials are valuable for the recovery of fugitive reagents or solvents. Flushing with water removes inorganic adsorbates such as SO_2 adsorbed on carbon but the carbon must be dried before reuse. Figure 5.20.13 shows a zeolite-based adsorption system for solvent vapors.

Adsorbent Selection

In selecting an adsorbent, the design engineer must ensure that the adsorption isotherm is favorable. If the adsorption isotherm is unfavorable, the loading is low, and the process is rarely economic. Irreversible adsorption should be avoided unless regeneration is not required.

If the feed gas is at an elevated temperature, cooling the gas stream may be required. The cost of cooling is significant and should be included in cost estimates.

Some adsorbates decompose in adsorption or regeneration. Many decomposition products (e.g., from chlorinated hydrocarbons, ketones, and acetates) are acidic and can corrode the system. They also need an additional separation step if the recovered adsorbate is valuable.

Using an adsorbent system for continuous solvent recovery from a vent stream requires using multiple beds. Regeneration of the adsorbent and recovery of the solvent can be accomplished by several means, including PSA (see Figure 5.20.12). Alternatively, a system can recover the solvent by sequentially steam treating the spent adsorbate to remove the adsorbed solvent, drying the carbon with hot air, and cooling the bed to the operating temperature. The number of beds required depends on the effective time-on-stream of the adsorbent bed and the time required to regenerate the carbon. The determination of these parameters is described next. Table 5.20.7 shows typical operating parameters for adsorbers.

Adsorption Operation

The design and operation of an adsorber requires knowledge of the adsorption isotherm, mass transfer, and axial dispersion of adsorbates. Heat transfer is also significant at high pollutant concentrations. Generally, if the feed rate is less than 12 m^3/sec, purchasing an adsorber package from a reputable supplier is more economical than custom building a unit.

Figure 5.20.14 shows the operation of a typical adsorber. For an ideal breakthrough curve, all the pollutant fed in time t_s is adsorbed. During this time period, the concentration on the adsorbent surface increases from the initial value w_o to the saturation (or equilibrium) value w_s. Thus, the following equation applies for a feed gas with linear velocity u_o and concentration c_o:

TABLE 5.20.5 TYPICAL PROPERTIES OF UNION CARBIDE TYPE X MOLECULAR SIEVES

Basic Type	Nominal pore diameter Å	Available from	Bulk density, lb/ft³	Heat of adsorption (maximum), Btu/lb H_2O capacity, % wt	Equilibrium H_2O capacity % wt	Molecules Adsorbed	Molecules Excluded	Applications
3A	3	Powder $\frac{1}{16}$-in pellets $\frac{1}{8}$-in pellets	30 44 44	1800	23 20 20	Molecules with an effective diameter <3 Å, e.g., including H_2O and NH_3	Molecules with an effective diameter > 3 Å, e.g., ethane	The preferred molecular sieve adsorbent for the commercial dehydration of unsaturated hydrocarbon streams such as cracked gas, propylene, butadiene, and acetylene. Also used for drying polar liquids such as methanol and ethanol.
4A	4	Powder $\frac{1}{16}$-in pellets $\frac{1}{8}$-in pellets 8 × 12 beads 4 × 8 beads 14 × 30 mesh	30 45 45 45 45 44	1800	28.5 22 22 22 22 22	Molecules with an effective diameter < 4 Å, including ethanol, H_2S, CO_2, SO_2, C_2H_4, C_2H_6, and C_3H_6	Molecules with an effective diameter > 4 Å, e.g., propane	The preferred molecular sieve adsorbent for static dehydration in a closed-gas or liquid system. Used as a static desiccant in household refrigeration systems; in packaging of drugs, electronic components, and perishable chemicals; and as a water scavenger in paint and plastic systems. Also used commercially in drying saturated hydrocarbon streams.
5A	5	Powder $\frac{1}{16}$-in pellets $\frac{1}{8}$-in pellets	30 43 43	1800	28 21.5 21.5	Molecules with an effective diameter < 5 Å, including n-C_4H_9OH, n-C_4H_{10}, C_3H_7 to $C_{22}H_{46}$, R-12	Molecules with an effective diameter < 5 Å, e.g., iso compounds and all 4-carbon rings	Separates normal paraffins from branched-chain and cyclic hydrocarbons through a selective adsorption process.
10X	8	Powder $\frac{1}{16}$-in pellets $\frac{1}{8}$-in pellets	30 36 36	1800	36 28 28	Iso paraffins and olefins, C_6H_6, molecules with an effective diameter < 8 Å	Di-n-butylamine and larger	Aromatic hydrocarbon separation.
13X	10	Powder $\frac{1}{16}$-in pellets $\frac{1}{8}$-in pellets 8 × 12 beads 4 × 8 beads 14 × 30 mesh	30 38 38 42 42 38	1800	36 28.5 28.5 28.5 28.5 28.5	Molecules with an effective diameter > 10 Å	Molecules with an effective diameter > 10 Å, e.g., $(C_4F_9)_3N$	Used commercialy for general gas drying, air plant feed purification (simultaneous removal of H_2O and CO_2), and liquid hydrocarbon and natural gas sweetening (H_2S and mercaptan removal)

Source: J.L. Kovach, 1988, Gas-phase adsorption, in *Handbook of separation techniques for chemical engineers* 2d ed. (New York: McGraw-Hill).

TABLE 5.20.6 MAXIMUM CAPACITY OF ACTIVATED CARBON FOR VARIOUS SOLVENTS FROM AIR AT 20°C AND 1 ATM

Adsorbate	Maximum Capacity, kg/kg Carbon
Carbon tetrachloride, CCl_4	0.45
Butyric acid, $C_4H_8O_2$	0.35
Amyl acetate, $C_7H_{14}O_2$	0.34
Toluene, C_7H_8	0.29
Putrescene, $C_4H_{12}N_2$	0.25
Skatole, C_9H_9N	0.25
Ethyl mercaptan, C_2H_6S	0.23
Eucalyptole, $C_{10}H_{18}O$	0.23
Ethyl acetate, $C_4H_5O_2$	0.19
Sulfur dioxide, SO_2	0.10
Acetaldehyde, C_2H_4O	0.07
Methyl chloride, CH_3Cl	0.05
Formaldehyde, HCHO	0.03
Chlorine, Cl_2	0.022
Hydrogen sulfide, H_2S	0.014
Ammonia, NH_3	0.013
Ozone, O_3	decomposes to O_2

Source: P.C. Wankat, 1990, *Rate-controlled separations* (London: Elsevier).

$$t_s = \frac{LB_s(w_s - w_o)}{u_o c_o} \qquad 5.20(16)$$

where L and P_s are the bed length and the bulk density of the adsorber, respectively. For new and completely regenerated adsorbents, $w_o = 0$.

The actual breakpoint time t is always less than t_s. If the adsorption zone is small compared to the bed length L, most of the adsorbent is used.

In an ideal case of no mass-transfer resistance and no axial dispersion, the adsorption zone becomes zero, i.e., $t = t_s$. However, since mass-transfer resistance always exists, the practical operating capacity is normally 25–50% of the theoretical isotherm value.

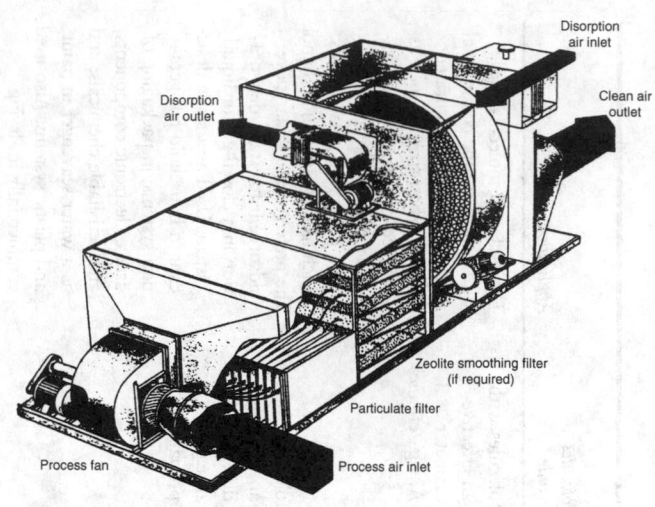

FIG. 5.20.13 Zeolite-based adsorption system for solvent vapors. The large wheel in the middle of the unit adsorbs vapors and is simultaneously regenerated.

In a fixed, cylindrical bed adsorber, the adsorption zone shown in Figure 5.20.14 moves down the column at a nearly constant pattern. The shape and velocity of the adsorption zone are influenced by the adsorption isotherm. For Freundich-type adsorption, in which $c_o = \alpha w_s^\beta$ as expressed in Equation 5.20(15), the following equation gives velocity of the adsorption zone:

$$u_{ad} = \frac{u_o}{\rho_s} (\alpha)^{1/\beta} (c_o)^{\beta - 1/\beta} \qquad 5.20(17)$$

and the following equation expresses the height of the adsorption zone:

$$z = \frac{u_o}{K} \int_0^1 \frac{d\gamma}{\gamma - \gamma^\beta} ; \qquad \gamma = c/c_o \qquad 5.20(18)$$

where K is the overall gas–solid mass-transfer coefficient (s^{-1}) and c is the breakthrough concentration set by the

TABLE 5.20.7 TYPICAL OPERATING PARAMETERS FOR ADSORBERS

Parameter	Range	Design
Superficial gas velocity	20 to 50 cm/sec (40 to 100 ft/min)	40 cm/sec (80 ft/min)
Adsorbent bed depth*	3 to 10 HAZ	5 HAZ
Adsorption time	0.5 to 8 hr	4 hr
Temperature	−200 to 50°C	
Inlet concentration	100 to 5000 vppm	
Adsorbent particle size	0.5 to 10 mm	4 to 8 mm
Adsorbent void volume	38 to 50%	45%
Steam regeneration temperature	105 to 110°C	
Inert gas regenerant temperature†	100 to 300°C	
Regeneration time	½ adsorption time	
Number of adsorbers	2 to 6	2 to 3

*HAZ is the height of the adsorption zone (see Figure 5.20.1).
†Maximum temperature is 900°C for carbon in nonoxidizing atmosphere and 475°C for molecular sieves.

a) Schematic of adsorption bed in operation.

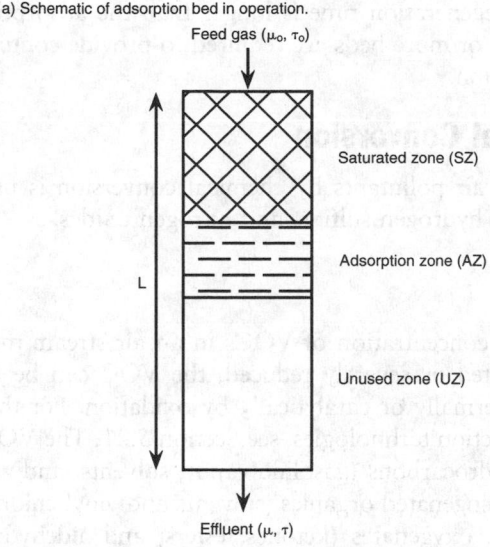

b) Profile of relative concentration of a pollutant through the bed.

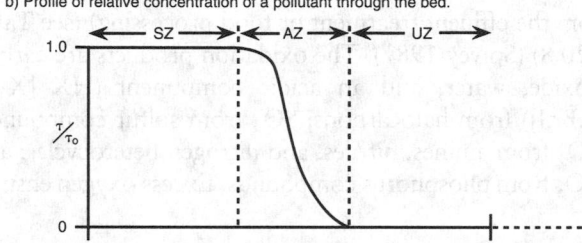

FIG. 5.20.14 Adsorber operation.

design engineer (kg/m³) to meet the environmental standards. The value of the integral on the right side of Equation 5.20(18) is undefined for limits of 0 and 1. However, taking limits close to these values (e.g., 0.01 and 0.99) defines the integral and solves the equation. The breakthrough time t_β is then as follows:

$$t_\beta = L - \frac{z}{u_{ad}} \qquad 5.20(19)$$

These design equations estimate operating time before breakthrough occurs. The accuracy of the calculations depends on the estimation of K, which is determined by the combination effect of diffusion external to the adsorbent, in the pores and on the pore surface, as well as axial dispersion. Predictions of K values from existing mass-transfer correlations are available but often are unreliable. Thus, adsorber designs are generally based on laboratory data. The scale-up then involves using the same adsorbent size and superficial gas velocity.

For longer beds, the column height of the adsorption zone is a small fraction of the bed length. A longer bed leads to better utilization of the adsorbent. The longer bed also results in a lower degree of backmixing (i.e., higher mass-transfer coefficient). However, proper design is required to avoid an excessive pressure drop. The energy consumption for a blower to overcome a pressure drop is a significant part of the overall operating cost. In the absence of experimental data, the Ergun equation or

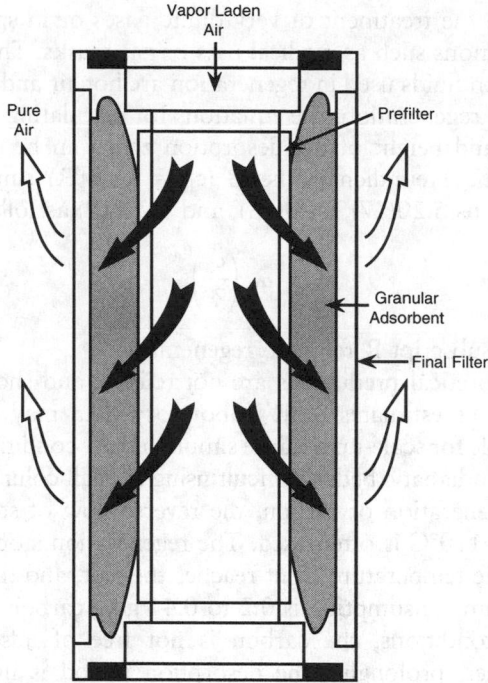

FIG. 5.20.15 Canister adsorption system used to remove trace contaminants from air. (Reprinted, with permission, from Balston, Inc.)

equations provided by adsorbent suppliers (e.g., UOP) can provide an estimate. The Ergun equation is as follows:

$$\frac{\Delta P}{L} = \frac{(1-\varepsilon)G^2}{g_c \varepsilon^3 d_p \rho_G}\left[1.75 + \frac{150(1-\varepsilon)\mu_G}{d_p G}\right] \qquad 5.20(20)$$

where:

ΔP = pressure drop, lb_f/ft^2
L = bed depth, ft
g_c = gravitational constant, 4.17×10^8 lb_m = ft/lb_f − h^2
ε = void fraction
ρ_G = gas density, lb_m/ft^3
G = superficial gas velocity, lb_m/ft^2 − h
μ_G = gas viscosity, lb_m/ft^2 − h
d_p = particle diameter, ft

For nonspherical particles, d_p is defined as the equivalent diameter of a sphere having the same specific surface (external area of the particle/bed volume) as the particle.

When the pressure drop across the adsorbent bed must be small, the shape of both the bed and, consequently, the adsorption wave are different from those for a cylindrical packed adsorber. Figure 5.20.15 shows an example of using a canister-type bed for low pressure drop operations.

Regeneration Operation

For continuous operation, when breakthrough occurs, the adsorbent has reached its operating capacity. The adsorption operation must then be performed in another, unsaturated bed. The spent adsorbent is regenerated unless it is used in the treatment of very dilute gases or in specialty applications such as medical uses or gas masks. The most common fluids used in regeneration are hot air and steam. During regeneration, the equations for calculating the velocity and height of the desorption zone can be derived from the Freundich isotherm ($c_R = \alpha_R w_s^\beta R$) similar to Equations 5.20(17), 5.20(18), and 5.20(19) as follows:

$$c_R = \alpha_R \left(\frac{c_o}{\alpha}\right)^{\beta_R/\beta} \qquad 5.20(21)$$

where subscript R refers to regeneration.

Theoretical predictions are not reliable and should be treated as estimates only. Laboratory data may not be available for scale-up because simulating the conditions for a large adiabatic bed is difficult using a small column. For the regeneration of carbon, the reverse flow of steam at 105 to 110°C is often used. The regeneration stops soon after the temperature front reaches the exit, and the typical steam consumption is 0.2 to 0.4 kg/kg carbon. Under these conditions, the carbon is not free of adsorbate. However, prolonging the desorption period is not economic because an excessive amount of steam is required to remove the remaining adsorbate due to the unfavorable shape of the tail of the desorption wave.

If the regeneration time is longer than the adsorption time, three or more beds are required to provide continuous operation.

Chemical Conversion

Removing air pollutants by chemical conversion is used for VOCs, hydrogen sulfide, and nitrogen oxides.

VOCs

When the concentration of VOCs in an air stream must be eliminated or severely reduced, the VOC can be destroyed thermally or catalytically by oxidation. For thermal destruction technologies, see Section 5.21. The VOCs include hydrocarbons (gasoline vapor, solvents, and aromatics), halogenated organics (solvents and vinyl chloride monomer), oxygenates (ketones, esters, and aldehydes), and odorous compounds (amines, mercaptans, and others from the effluent treatment or food processing) (see Table 5.20.8) (Spivey 1987). The oxidation products are carbon dioxide, water, and an acidic component (HX [X = Cl,Br,I]) from halocarbons; SO_x from sulfur compounds; NO_x from amines, nitriles, and nitrogen heterocycles; and P_2O_5 from phosphorus compounds. Excess oxygen ensures

TABLE 5.20.8 EXAMPLES OF VOC DESTRUCTION APPLICATIONS

Sources of Pollution	Types of Pollution	Oxidation Catalysts
Chemicals Manufacture	**Solvents**	**Metals/Supported Metals**
-Plant operations	-Hydrocarbons	Metal =Ag
-Petrochemicals	-Aromatics	Pt
-Storage	-Ketones	Pd
	-Esters	Ru
Chemicals Use	-Alcohols	Rh
-Coating processes		Ni
-Electronics industry		
-Furniture manufacture	**Odor Control**	
-Painting	-Amines	Support =SiO_2
-Dry cleaning	-Thiols/thioethers	Al_2O_3
-Paper industry	-Heterocyclics	Zeolite
-Wood coating	-Aldehydes	ThO_2
-Printing	-Acids	ZrO_2
	-Food by-products	Polymers
	-Smelter off-gases	
Environment Management		**Metal Oxides**
-Landfill	**Other Effluent**	MgO
-Hazardous waste handling	-Monomers	V_2O_5
	-Plasticizers	Cr_2O_3
Food/Beverage	-CO	MoO_3
-Breweries	-Formaldehyde	Pr_2O_3
-Bakeries		Fe_2O_3
-Food processing		MnOx
-Ovens		NiO
		TiO_2
Energy		Co_3O_4
-Power generation		La_2MO_4 (M = (Cu,Ni))
-Stationary engines		$CoMoO_4$
-Diesel trucks		Mixed oxides

Notes: For specific applications, one or more catalyst types can be effective. The selection of a particular catalyst system is based on the combination of the VOC to be destroyed and the technology selected.

TABLE 5.20.9. DESTRUCTIBILITY OF VOC BY CATALYTIC OXIDATION

VOC	Relative Destructibility	Catalytic Ignition Temperature[a] (0°C)
Formaldehyde	High	<30
Methanol	↑	<30
Acetaldehyde		100
Trimethylamine		100
Butanone (Methyl ethyl ketone)		100
n-Hexane		120
Phenol		150
Toluene		150–180
Acetic acid		200
Acetone		200
Propane	Low	250–280
(Chlorinated hydrocarbons)		(400)

Source: Data from F. Nakajima, 1991, Air pollution control with catalysis—Past, present, and future, *Catalysis Today* 10:1.

Note: [a]Catalyst: Pd/activated alumina.

that oxidation is complete and that partial oxidation products (which can be pollutants) are not formed. Scrubbing with a mild basic aqueous solution removes the acidic component in the tail streams.

The oxidation of a VOC is exothermic. However, because a VOC is present only as a dilute component, the reaction heat is insufficient to maintain the temperature required for oxidation. Therefore, additional heating is required, or a catalyst must be used that enables total oxidation at lower temperatures (see Table 5.20.8).

Most catalytic operations also require elevated temperatures to ensure a complete reaction at a suitably fast rate. Therefore, either the catalyst bed is heated, or the feed stream is preheated. Inlet gas streams are kept 50–150°C higher than the ignition temperatures (Nakajima 1991). When the preheating uses an internal flame, the process is complex, involving both the products of combustion in the flame and processes occurring at the catalyst. The operating parameters (temperature and space velocity) for an application depend on the destructability of the VOC (see Table 5.20.9).

In a typical application, more than one VOC is destroyed. Frequently, oxidation at a catalyst is competitive under stoichiometric or partial oxidation conditions, or one VOC inhibits catalytic oxidation of another. Using excess oxygen in air overcomes these problems. A range of parameters applies in catalytic oxidation for VOC destruction. These parameters depend on the application (Spivey 1987).

Mechanisms

The oxidation of a VOC (S) at a catalyst can involve a species at the surface and in the vapor phase. The Langmuir–Hinshelwood mechanism (Scheme 1) requires the adsorption of each species at nearby sites and subsequent reaction and desorption.

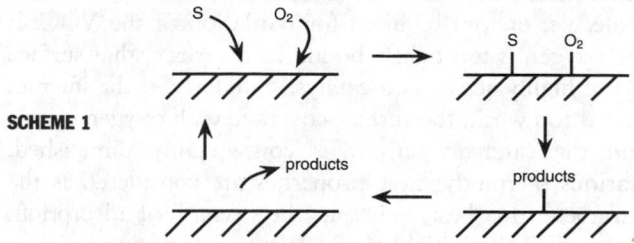

SCHEME 1

The Mars–van Krevelen mechanism of catalytic oxidations (Scheme 2) explicitly requires a redox process in which oxygen is consumed from the catalyst surface by reaction with the VOC and then is replenished by oxygen from the vapor phase.

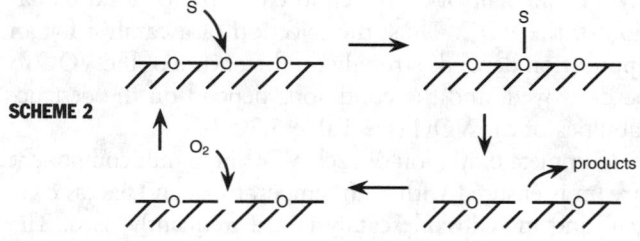

SCHEME 2

The Eley–Rideal mechanism (Scheme 3) is similar to the Mars–van Krevelen mechanism except that the products are formed from adsorbed oxygen and the VOC in the gas phase.

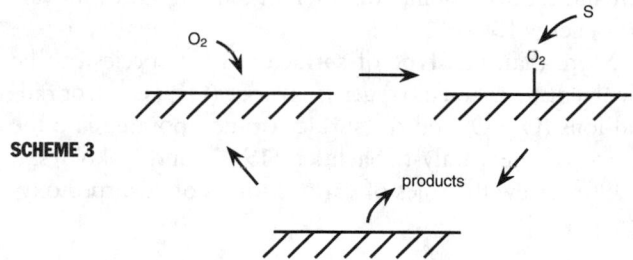

SCHEME 3

As these schemes show, for metal or nonreducible oxide catalysts, excess oxygen in the gas phase means that the catalyst surface is well-covered with oxygen and that little if any VOC is adsorbed. Thus, the Eley–Rideal mechanism is expected to be important. For metal oxide catalysts containing readily reducible metals, the Mars–van Krevelen mechanism is important.

Metal oxides that are n-type semiconductors are rich in electrons and are generally not highly active as oxidation catalysts. Vanadium pentoxide is the notable exception. In contrast, p-type semiconductors are conductive because of the electron flow into positive holes. The electron-deficient surfaces of such metal oxides readily adsorb oxygen, and if the adsorption is not too strong, they are active catalysts. Insulators, which are inexpensive and not friable or thermally unsuitable, have values as supports for more expensive, catalytically active metal oxides or noble metals.

For all mechanisms, a key factor is the strength of the interaction between the surface and the oxygen (atom, molecule, or ion) required for oxidation of the VOC. If the oxygen is too tightly bound to a surface, that surface is not highly active as a catalyst. Similarly, if the interaction is too weak, the surface coverage with oxygen is low, and the catalytic activity is consequently diminished. Various thermodynamic properties are considered as the parameter that best represents the strength of adsorption. For metals, the initial heat of oxygen adsorption is a reasonable choice (Bond 1987). For metal oxides, the reaction enthalpy for reoxidation of the used catalyst (Mars–van Krevelen mechanism) is considered the most representative (Satterfield 1991). The maximum rate for VOC oxidation over an oxide catalyst is estimated to occur when the reaction enthalpy for reoxidation of the catalyst is one-half of the reaction enthalpy for total oxidation of the VOC. Thus, the selection of a catalyst for an application depends strongly on the nature of the VOC to be destroyed, and the conditions depend on the destructabilities of the VOC (see Table 5.20.9).

Complete oxidation of each VOC in a multicomponent stream is ensured with high temperatures and excess oxygen, and mixed-oxide catalysts are frequently used. The mixed oxides, especially when promoted with alkali or alkaline earth metal oxides, frequently have activities that are different from the combination of properties of the components, and in general the activities are higher. This phenomenon probably arises from two factors: the availability and mobility of the different forms of available oxygen and the accessibility of different binding sites with various energy levels.

More than one type of surface oxygen species can be involved: adsorbed dioxygen (O_2), ions (O^{2-}; O_2^{2-}), or radical ions (O^-; O_2^-) on the surface or incorporated into the lattice of the catalyst. Sachtler (1970) and Sokolovskii (1990) review the roles of various forms of adsorbed oxygen.

Kinetics

The following equations summarize the steps required for destruction of a VOC by an Eley–Rideal mechanism:

$$O_2 + [\ ] \longrightarrow [O_2] \qquad 5.20(22)$$

$$[O_2] + [\ ] \longrightarrow 2[O] \qquad 5.20(23)$$

$$VOC + [O] \longrightarrow [S_a]_i + H_2O \qquad 5.20(24)$$

$$[S_a]_i + [O] \longrightarrow CO_2 + [\ ] \qquad 5.20(25)$$

where [ ] represents a surface site and $[S_a]_i$ represents the ith in a series of partially oxidized species S_a at the surface of the catalyst. The following rate expression is derived from the preceding mechanism:

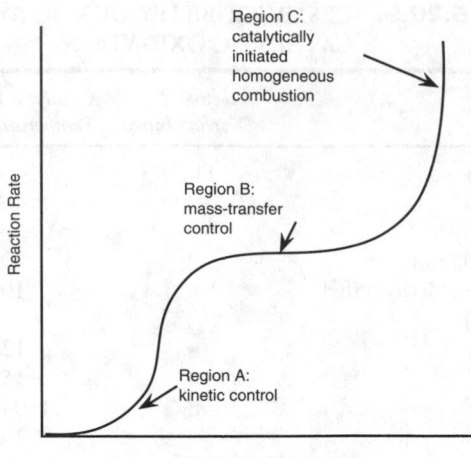

FIG. 5.20.16 Overall reaction rate. (Reprinted, with permission, from D.L. Trimm, 1991, Catalytic combustion, Chap. 3 in *Studies in inorganic chemistry 1991* 11:60.)

$$r = \frac{k_a P_{O_2} P_{VOC}}{k_b P_{O_2} + v k_c P_{VOC}} \qquad 5.20(26)$$

where k_a, k_b, and k_c are constants and v is the stoichiometric coefficient of oxygen in the overall oxidation reaction of the VOC. Under conditions of excess oxygen where $k_b P_{O_2} \gg v k_c P_{VOC}$, this equation reduces to the following approximate form:

$$r \cong k P_{VOC} \qquad 5.20(27)$$

For most applications, the kinetics are described by the following fractional power expression:

$$r = k_b P_{O_2}^a P_{VOC}^b \qquad 5.20(28)$$

in which a and b are fractional coefficients with values close to zero and unity, respectively.

The oxidation of a VOC can occur at the catalyst surface and in the gas phase. The overall reaction rate is the sum of these two components and is a strong function of temperature (see Figure 5.20.16) (Prasad 1984, Trimm 1991). The catalytic oxidation of hydrocarbons over supported metal catalysts is thought to occur via dissociative chemisorption of the VOC, followed by reaction with co-adsorbed oxygen at the surface and then desorption of the combustion products (Chu and Windawi 1996). The rate determining step in this Langmuir–Hinshelwood type of mechanism is hydrogen abstraction from the VOC. Thus the ease of oxidation of the VOC is directly related to the strength of the C–H bond. Methane is more difficult to oxidize than other paraffins, aromatics, or olefins, and oxygenates are relatively easier to oxidize.

Reactors

The reactor's heat requirement heat arises mainly from the need to preheat the inlet gases or to heat the catalyst bed. For efficient operation of a catalytic oxidation system, the exhaust heat must be recovered and used to preheat the feed, as shown in Figure 5.20.17 for a system manufac-

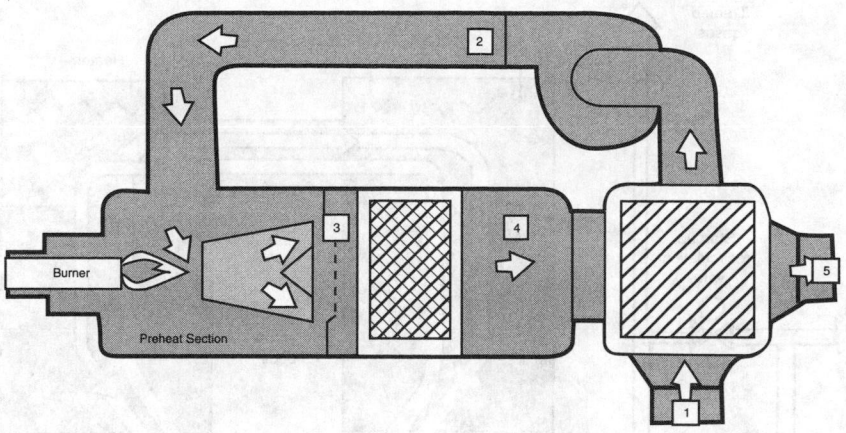

FIG. 5.20.17 Schematic diagram of catalytic process with burner. (Reprinted, with permission, from Salem Engelhard.)

tured by Salem Engelhard (South Lyon, Mich.). The residual heat is recovered by a secondary heat exchanger and used for area heating or other purposes requiring low-grade energy.

In some cases, destruction of all VOCs and intermediates requires a higher operating temperature than that required to combust a single VOC (see Figure 5.20.18).

The waste gases from several chemical, printing, or related industries contain mixtures of halogenated and non-halogenated VOCs. Converting each VOC requires a combination of catalysts. Further, scrubbing the effluent from the reactor is necessary to remove the acidic components generated. As an example of such a system, the catalytic solvent abatement (CSA) process designed and marketed by Tebodin V.B. is shown in Figure 5.20.19.

The heat for a catalytic incinerator can be applied directly to the catalyst bed, rather than from a burner. The swingtherm system (see Figure 5.20.20) manufactured by Mo-Do Chemetics, Ltd. (Vancouver, Canada and Ornskolsvik, Sweden) uses dual beds at a temperature of 300–350°C. The air stream is fed at 60°C and is heated by contacting a ceramic at 320°C. It then passes through a platinum-based catalyst. The exothermal VOC oxidation raises the temperature of the gases to 350°C. The gases then pass through the other bed to heat the ceramic rings; the effluent gases are cooled to 90°C. When the second reactor is warmed to 320°C, the inflow is reversed. The flow reversal occurs every 2–5 min. For VOC concentrations below 300 ppm, auxiliary heating is required.

HYDROGEN SULFIDE

Large sources of hydrogen sulfide (H_2S) are treated as a resource from which sulfur is recovered. The normal recovery is to oxidize a part of the H_2S to sulfur dioxide (SO_2) and then react H_2S and SO_2 over an alumina-based catalyst in the Claus process as follows:

$$2H_2S + SO_2 \longrightarrow 3S + 2H_2O \qquad 5.20(29)$$

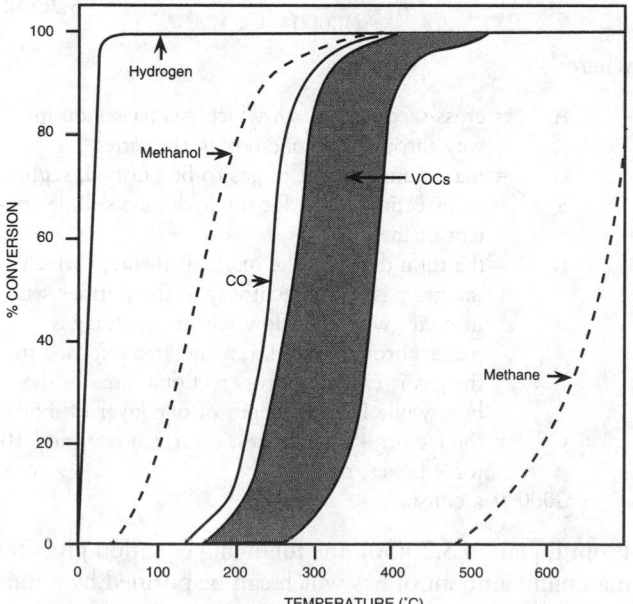

FIG. 5.20.18 Conversion curves for VOC destruction using a catalytic incinerator. (Data from Brown Engineering, Seattle, Wash; Johnson Matthey, Catalytic Systems Division, Wayne, Pa.)

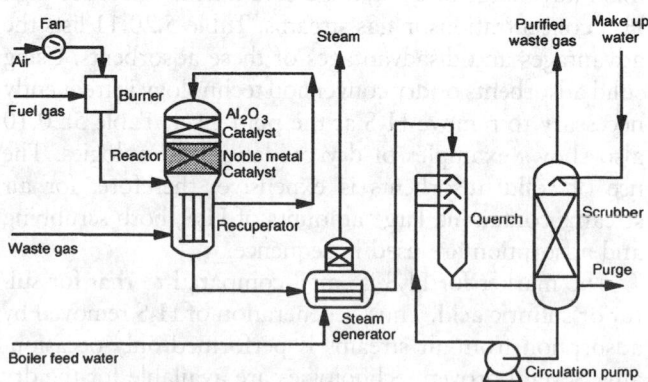

FIG. 5.20.19 CSA system. (Data from Tebodin B.V., The Hague, the Netherlands.)

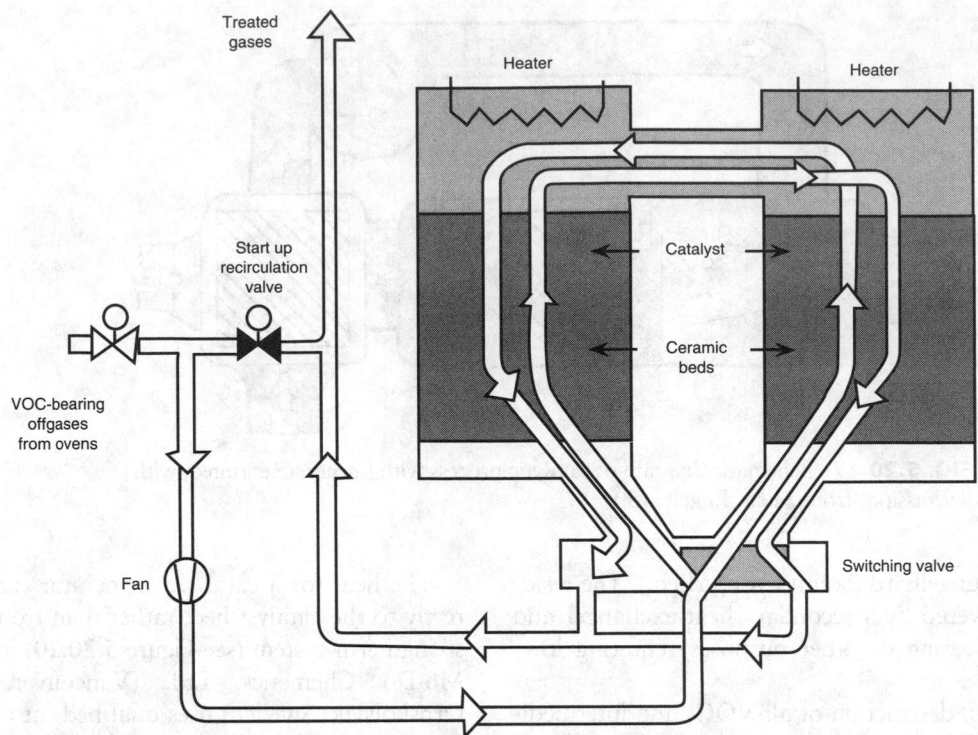

FIG. 5.20.20 Swingtherm process. (Reprinted, with permission, from *Chemical Engineering*, 1993 [October]:153.)

H_2S is initially removed from a source, such as sour natural gas, by dissolution in an alkanolamine solvent. However, for low concentrations of H_2S in air, this method is not sufficiently effective for total removal of H_2S (Kohl and Riesenfeld 1985). Instead, the H_2S is either oxidized completely to SO_2, and then the system removes the SO_2 by scrubbing or capture, or it is captured by reaction with solid or liquid-phase adsorbents.

For stationary units, any liquid or solid base reacts with H_2S. Magnetite (Fe_3O_4), limestone ($CaCO_3$), lime (CaO), zinc carbonate ($ZnCO_3$), and zinc oxide (ZnO) are each effective, and Fe_3O_4, CaO, and ZnO are used commercially.

Table 5.20.10 lists the major adsorbents for H_2S. The main advantage of using these solutions is to reduce high H_2S concentrations in gas streams. Table 5.20.11 lists the advantages and disadvantages of these adsorbents. Using solid adsorbents or dry conversion technology is frequently necessary to remove H_2S at the ppm level. Table 5.20.10 also shows examples of dry oxidation technologies. The use of solid adsorbents is expensive; therefore, for air streams containing large amounts of H_2S, both scrubbing and adsorption are used in sequence.

The market for H_2S is small compared to that for sulfur or sulfuric acid. Thus, regeneration of H_2S removed by adsorption from air streams is performed only occasionally. Several proven technologies are available for the dry oxidation of H_2S to sulfur.

Dry Oxidation

The dry oxidation process using iron oxide is inexpensive but not pleasant or simple to operate. The following equation empirically gives the suitable box sizes (Steere Engineering Co.):

$$A = \frac{GS}{3000(D + C)} \qquad 5.20(30)$$

where:

A = cross-section through which gas passes on its way through any one box in the series

G = maximum amount of gas to be purified, scf/hr

S = a correction factor for the hydrogen sulfide content of the inlet gas

D = the total depth of the oxide, ft through which the gas passes consecutively in the purifier set. In boxes with split flow where half the gas passes through each layer, the area exposed to the gas is twice the cross-sectional area of the box, while D is the depth of one layer of oxide.

C = the factor: 4 for 2 boxes, 8 for 3 boxes, and 10 for 4 boxes, respectively

3000 = a constant

From Equation 5.20(30), the following equation gives the maximum amount of gas which can be purified by a unit:

$$G = \frac{3000(D + C)A}{S} \qquad 5.20(31)$$

TABLE 5.20.10 ADSORBENTS FOR HYDROGEN SULFIDE

Liquid Adsorbents	Trade or Generic Names	Source	Comments
Alkanolamines:			
Monoethanolamine	MEA	Dow, U.S.A.	Moderately selective
Diethanolamine	DEA	Dow, U.S.A.	Moderately selective
Triethanolamine	TEA	Dow, U.S.A.	Selective
N-methydiethanolamine	MDEA	Dow, U.S.A.	Selective
Diisopropanolamine	DIPA	Dow, U.S.A.	Selective
Aqueous potassium carbonate	Hot Carbonate		Moderately selective
Aqueous tripotassium phosphate			Selective
Alkazid process			
Dialkylglycine sodium salt	DIK Solution		Selective
Sodium alanine solution	M Solution		Moderately selective
Sodium phenolate solution	S Solution		Not selective
Aqueous Ammonia			
+Air/hydroquinone	Perox		Recovers as sulfur
+Air	Am-gas	Am-gas, Canada	Recovers as $(NH_4)_2SO_4$
Aqueous Sodium Carbonate			
+Quinoline sulfonate salt and sodium metavanadate	Stretford	W.C. Holmes and Company, U.K.	Recovers as sulfur
+Naphthaquinone sulfonate salt	Takahax	Tokyo Gas Co. and Nittetsu Chemical Engineers Co.	Recovers as sulfur
Aqueous zinc acetate			Precipitates ZnS
Aqueous calcium acetate +Heat to decompose product			$Ca(SH)_2$ soluble; CaS precipitates
Iron Oxide +Controlled reoxidation	Dry Box		Sulfur recovery, for ppm H_2S streams
Zinc Oxide, ZnO			Useful for ppm H_2S; only mod. selective
Calcium oxide, CaO	Lime		Product releases H_2S on decomp. oxidizes to sulfate
Copper, Cu (or other metals)			Reagent, not adsorbent; forms CuS and H_2, but requires expensive regeneration

Table 5.20.12 tabulates the values of S. The values for the factor C are based on the number of boxes in the series being used at the time of operation and on the assumption that the flow is reversed during purification.

The designer must use the appropriate size of iron oxide particles for the required pressure drop across the bed. The empirical relationship was developed for one system (Prasad 1984) and can be used as an approximation for similar systems:

$$\Delta P = 3.14d^{-0.61} \qquad 5.20(32)$$

Perry (1984) recommends an allowable pressure drop of 1–2 psi/ft at 800–1000 psig and linear gas velocities of 5–10 ft/min. Typical operating conditions are shown in Table 5.20.13.

Activated Carbon Process

Activated carbon is an effective catalyst for the oxidative conversion of H_2S in hydrocarbon gas streams to elemental sulfur under mild conditions. I.G. Farbenindustrie developed the first process to exploit this capability. Such a unit may process up to 200,000 cu ft/hr of water–gas, with a pressure drop of 25 in of water.

Reactions of H_2S and SO_2

The Sulfreen process designed by Lurgi (Germany) and Elf Aquitaine (France) reduces residual sulfur compounds in tail gases from Claus plants for sulfur recovery from natural gas. Alumina or carbon catalysts

TABLE 5.20.11   TECHNICAL ADVANTAGES AND DISADVANTAGES FOR AMINES USED AS H_2S ADSORBENTS

Amine	Advantages	Disadvantages
MEA	Known technology Low cost (approximately $1.6/kg) Simple regeneration. 20–30% solution has low freezing point (to −50°C).	Some degradation by CO_2, COS, and CS_2 requires use of make-up. May need addition of foaming prevention agents. Some corrosion effects.
DEA	Known technology. Low cost. Simple regeneration. Better H_2S selectivity than MEA.	Some degradation. Slightly higher cost than MEA. May need foam prevention agent.
TEA	Known technology. Low cost (approximately $1.4/kg). Simple regeneration. Highly selective to H_2S. Less degradation. 40% solution has reasonable freezing point (−32°C).	Approximately twice as expensive as MEA, based on adsorption capacity. Minimum (fluid) operating temperature is higher than for MEA.
MDEA or DIPA	Highest selectivity to H_2S. Lowest regeneration temperature; hence energy savings. Lower heat of reaction Lower corrosion effects. Low vaporization losses. Not degraded by COS or CS_2. Higher initial costs are offset by reduced make-up costs.	Higher initial cost for amine. If an MDEA or DIPA regeneration plant is not nearby, a regeneration unit is required.

are used, and efficiencies as high as 90% can be attained for streams containing 1.50% H_2S and 0.75% SO_2. Similar technologies are used in the cold-bed adsorption process, developed by AMOCO Canada Petroleum Company Ltd., and the MCRC sulfur recovery process, licensed by Delta Engineering Corporation (USA). Sulfur recoveries up to 99% are attainable for each of these processes.

TABLE 5.20.12   CORRECTION FACTOR S FOR THE HYDROGEN SULFIDE CONTENT OF THE INLET GAS

Grains H_2S/100 scf of Unpurified Gas	Factor
1000 or more[a]	720
900	700
800	675
700	640
600	600
500	560
400	525
300	500
200 or less	480

[a]1000 grains/100 scf is 22.9 grams/cu m

Oxidation to Oxides of Sulfur

An alternative strategy in sulfur recovery is the oxidation to oxides of sulfur. For treating air with low concentrations of H_2S, this strategy does not require finding a market for small amounts of sulfur produced, and it can remove any adsorbed products from the catalyst bed thermally or by flushing with water.

The Katasulf process (Germany) for cleaning gas streams for domestic use can operate in various configurations. Depending on the pollutants and their concentrations in the feed gas, a prewasher may be necessary in addition to the main gas washer for SO_2 removal from gases exiting the catalyst chamber. Other applications use sidestream or split-stream arrangements. In each case, the catalyst is activated carbon, alumina, or a combination of two of three metals (iron, nickel, and copper) and operates close to 400°C. The Katasulf process is effective for H_2S and NH_3 removal and can be operated to reduce HCN and organic sulfur compounds.

NITROGEN OXIDES

The major stationary sources of nitrogen oxide (NO_x) emissions come from the combustion of fossil fuels. The

TABLE 5.20.13 TYPICAL OPERATING CONDITIONS OF IRON OXIDE PURIFIERS

Variable Condition	Conventional Boxes	Deep Boxes	High-Pressure	Tower Purifiers	Continuous
			Type of Purifier		
Gas volume treated, millions of cubic feet (mmcf)/day	6.0	4.3	15.0	24.0	2.0
Hydrogen sulfide content, gr/100 scf	1000	740	10	500–950	1000
Pressure, psig	low	40*	325	low	
Number of units in series (boxes or towers)	5	6	4†	6§	3
Cross-sectional area per unit, sq ft	960	1200	24	776	71
Number of layers per unit	4	1	1	28	1
Depth of layer, ft	2.25; 1.50	4	10	1.4	40
Temperature: °F:					
In	60	73		85	
Out	70	93		100	
Space velocity, cu ft/(hr) (cu ft)	7.15	6.66	37.4‡	5.38	9.4
R ratio	35.6	40	112‡	32	28

Source: Data from A.L. Kohl and F.C. Risenfeld 1985, *Gas purification,* 4th ed. (Houston: Gulf Publishing Company).
Notes: *Inches of water.
†Two series of four units, three units in operation in each series.
‡At 325 psig.
§Two series of six towers.

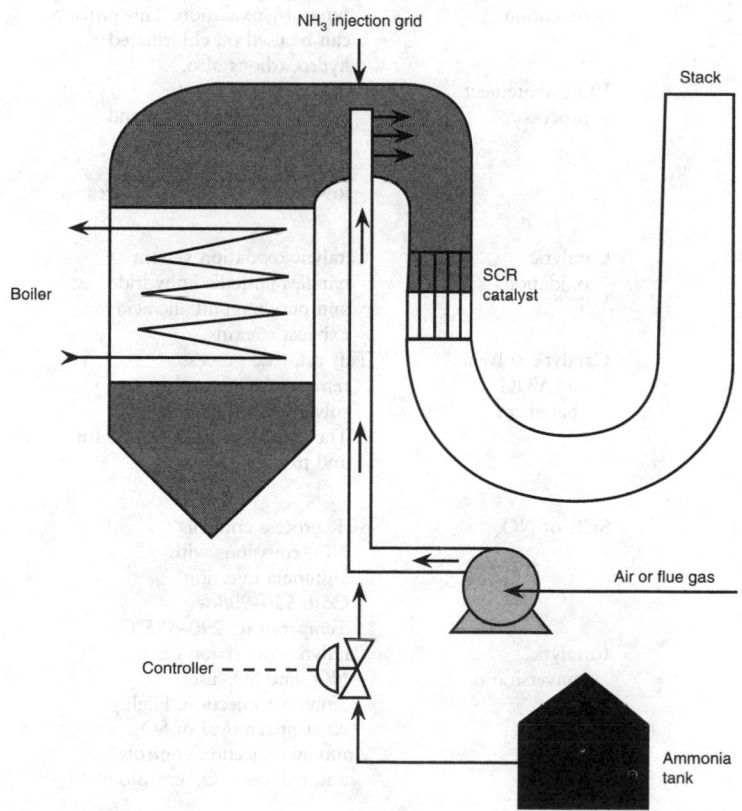

FIG. 5.20.21 SCR of NO_x.

TABLE 5.20.14 TECHNOLOGY MANUFACTURERS AND SUPPLIERS

Company	Process	Application	Number of Installations	Date of Last Installation
VOC Abatement				
AWD Technologies, Inc., subsidiary of Dow Chemical Co.	Modified PSA	Modified PSA process captures VOCs from variable concentration vent emission streams.	20	1993
Calgon Carbon Corp.	Granular activated-carbon adsorption	Granular activated-carbon adsorbs VOCs from air and other vapor streams.	100	NA
Callidus Technology, Inc.	Vacuum-regenerated activated-carbon vapor recovery	Vacuum-regenerated activated-carbon vapor recovery system controls VOC vapors from loading operations.	2	NA
Jaeger Products, Inc.	Vent gas adsorption	Process removes small amounts of acid, basic, or organic vapors into a liquid by absorption with and without chemical reaction.	100	1993
Membrane Technology and Research Inc.	Membrane vapor separation for VOC recovery	Membrane separation with proven condensation and compression techniques removes select organic compounds.	14	1993
Chlorinated Hydrocarbon Abatement				
Tebodin B.V., Consultants and Engineers	Catalytic solvent abatement process	Catalytic solvent abatement process treats exhaust gases containing chlorinated hydrocarbons and is suitable for PVC facility.	1	1991
Thermatrix	Flameless thermal oxidation	A packed-bed reactor destroys hazardous organic fumes by oxidation. This process can be used on chlorinated hydrocarbons also.	10	1992
Vara International, division of Calgon Carbon Corp.	VOC abatement process	Combination of fixed-carbon-bed adsorption and thermal oxidation concentrates and oxidizes low-VOC concentration streams.	6	NA
Catalytic VOC Oxidation				
CSM Environmental Systems, Inc.	Catalytic oxidation	Catalytic oxidation system handles phthalic anhydride and pure terephthalic acid exhaust streams.	20	NA
Haldor Topsøe Inc.	Catalytic solvent and VOC abatement	This catalytic process removes VOCs and solvents from exhaust air. The catalyst is poison-resistant and trouble-free.	134	1993
Catalytic NO$_x$ Reduction				
Engelhard Corp.	SCR of NO$_x$	SCR process controls NO$_x$ emissions with ammonia injection. Cost: \$20–90/kw. Temperature: 290–595°C.	25	NA
Lurgi AG	Catalytic conversion of NO$_x$	Catalytic conversion of NO$_x$ and SO$_x$ uses ammonia injection. Highly efficient removal of SO$_2$.	NA	NA
Research–Cottrell Co.	SCR	Ammonia injection controls and reduces NO$_x$ emissions.	NA	NA

Continued on next page

TABLE 5.20.14 *Continued*

Company	Process	Application	Number of Installations	Date of Last Installation
SNCR of NO$_x$				
Shell	Catalytic NO$_x$ reduction	Low temperature catalyst can achieve 90% reduction of NO$_x$.	4	1993
Exxon Research and Engineering Co.	Thermal NO$_x$ reduction	Reduces NO$_x$ emissions in flue gas streams with aqueous and anhydrous ammonia and has 80% efficiency. Temperature: 700–1100°C.	130	1993
Lurgi AG	SNCR of NO$_x$	Noncatalytic process reduces NO$_x$ compounds using aqueous ammonia or urea.	NA	NA
Nalco Fuel Tech	SNCR of NO$_x$	SNCR of NO$_x$ from stationary combustion sources uses stabilized aqueous urea. Cost: $500–1500/tn NO$_x$. Temperature: 815–1100°C.	70	NA
Research–Cottrell/Nalco Fuel Tech	SNCR of NO$_x$	Process uses controlled urea injection and chemicals to reduce NO$_x$ emissions.	70	NA
Thermal NO$_x$ Reduction				
ABB Stal	Dry low NO$_x$ combustion	System reduces NO$_x$ emissions for gas turbines.	4	NA
Catalytic SO$_x$ Reduction				
Haldor Topsøe A/S	Catalytic SO$_x$ and NO$_x$ removal	Catalytic process removes SO$_2$ and NO$_x$ from flue gases and offgases. NO$_x$ is reduced by ammonia to N$_2$ and H$_2$O.	29	1993
Noncatalytic SO$_x$ Reduction				
Exxon Research and Engineering Co.	Wet gas scrubbing process	Process removes particulates and SO$_x$ from FCC unit. Simple system has economic and operating advantages over other control systems.	14	1992
Lurgi AG	Wet gas scrubbing	Dust, noxious gases, and heavy metals are scrubbed from flue gas and recovered as a slurry form.	NA	NA
Incineration				
NAO Inc.	Thermal incineration of VOCs	Thermal oxidizer destroys VOC emissions with 99.9% efficiency and can recover valuable thermal energy.	NA	NA
Praxair, Inc.	Oxygen combustion	Process uses high-velocity oxygen jets to replace air that recirculates organics within the incinerator.	10	1993
Other				
Edwards Engineering Corp.	Hydrocarbon/solvent vapor recovery system	Hydrocarbon and solvent vapor recovery system, based on the Rankine refrigeration cycle, condenses vapors to liquids.	340	NA
Institut Français du Pétrole/Babcock Enterprises	Clean combustion of heavy fuel oil	AUDE boiler burns heavy fuel oil and petroleum residues cleanly and meets European directives on atmospheric discharges.	2	1993

Source: Data from *Hydrocarbon Processing,* 1993 (August):75, 102.
Notes: NA = not applicable.

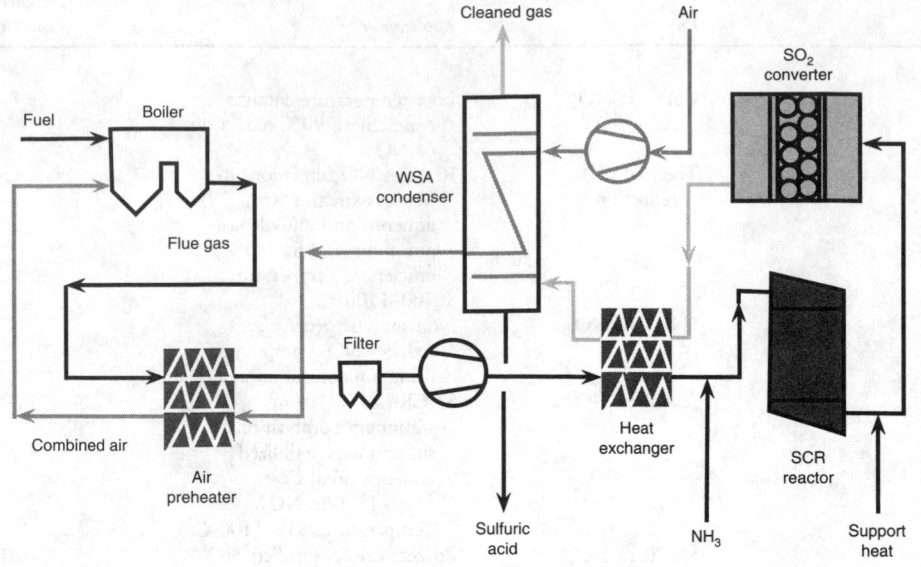

FIG. 5.20.22 SNOX process for combined treatment of NO$_x$ and SO$_x$.

flue gas usually contains 2–6% oxygen and a few hundred ppm of NO$_x$, which consists of 90–95% NO and 5–10% NO$_2$. The only viable postcombustion process for NO$_x$ emission control is chemical conversion to N$_2$ by a reducing agent. Several gases can be used for this purpose including methane, hydrogen, carbon monoxide, and ammonia. However, because the flue gas contains much more than O$_2$ than NO$_x$, the reducing agent must selectively react with NO$_x$ rather than O$_2$ to minimize its consumption (Environment Canada: Task Force Report 1989; Nakatsuji 1991). To date, many commercial installations based on the selective reduction of NO$_x$ by NH$_3$ can remove about

80% of NO$_x$ in the flue gas. The governing equations for the ammonia-based technology are as follows:

$$4NO + 4NH_3 + O_2 \longrightarrow 4N_2 + 6H_2O \qquad 5.20(33)$$

$$2NO_2 + 4NH_3 + O_2 \longrightarrow 3N_2 + 6H_2O \qquad 5.20(34)$$

$$4NH_3 + 5O_2 \longrightarrow 4NO + 6H_2O \qquad 5.20(35)$$

The first two reactions dominate when the flue gas is heated to about 1000°C. Above 1100°C, Equation 5.20(35) becomes significant leading to the unwanted formation of NO. If the temperature is below 800°C, the reaction rate is too low for practical use. Thus, the process

TABLE 5.20.15 CATALYST SUPPLIERS

Supplier	Materials (partial list)
Corning Corp. Corning, NY	Ceramic supports
Johnson Matthey, Catalytic Systems Div. Wayne, PA	Noble metals
W.R. Grace & Co., TEC Systems Div. DePere, WI.	Supports, metals/compounds, supported metals, other
W.R. Grace & Co., Davison Chemical Div. Baltimore, MD	Oxides, supported oxides, other
UOP Des Plaines, IL	Zeolites, adsorbents
Engelhard Corporation Iselin, NJ	Noble metals, supported metals/oxides
Norton Chemical Process Products Corp. Akron, OH	NO$_x$ reduction catalysts

Notes: This list is not comprehensive. The authors have attempted to provide current information but recognize that data may have changed. Inclusion of a company in this list is not an endorsement of that company's products and should not be construed as such.

is temperature sensitive. With the use of a suitable catalyst, the NO_x reduction can be carried out at 300–400°C, a temperature normally available in a flue gas system. This process is called Selective Catalytic Reduction (SCR). Figure 5.20.21 is a schematic diagram for the SCR process. In this process, at least 1% O_2 should be present in the flue gas, thus it is suitable for boiler and furnace applications. Several simple catalytic and noncatalytic systems are in commercial operation at several sites (see Table 5.20.14).

Under ideal reaction conditions, one mole of NH_3 is required to convert one mole of NO; however, in practice, one mole of NH_3 reduces about 0.8–0.97 moles of NO_x. The excess NH_3 is required because of the side reaction with O_2 and incomplete mixing of the ammonia with the flue gas. Much of the ammonia slip ends up in the fly ash, and the odor can become a problem when the ash is sent to a landfill or sold to cement plants.

The SCR processes are simple, requiring only a proper catalyst and an ammonia injection system. The catalysts currently in use are TiO_2-based which can be mixed with vanadium or molybdenum and tungsten oxides. These catalysts eliminate the formation of ammonium bisulfate which can plug the downstream equipment, a problem in earlier SCR systems.

A new Shell de-NO_x catalyst comprises vanadium and titanium, in high oxidation states, impregnated onto silica with a high surface area (300 m^2/g). The data for a commercial and semicommercial operation indicate consistent performance for periods up to one year (Groeneveld et al. 1988). The catalyst is deactivated by high concentrations of SO_2, but the poisoning is reversible with heating. A version of this process is installed to control dust-containing flue gas. This process employs a parallel passage system in which the flue gas permeates through the catalyst separating into passages for feed and purified streams.

The Shell technology is being developed for other applications. In particular, a lateral flow reactor is being developed which demonstrates a low pressure drop. The design of this reactor should allow for convenient installation and maintenance (Groeneveld et al. 1988).

Flue gases, especially from coal-burning boiler units or power generation, contain both NO_x and SO_x, with fly ash and metal-containing particulates. Processes have been developed that convert the NO_x to nitrogen, which is vented, and the SO_x to sulfuric acid, which is removed by scrubbing. Figure 5.20.22 is a schematic of the SNOX process (Haldor Topsøe A/S, Denmark). This process recovers up to 95% of the sulfur in the SO_x as sulfuric acid and reduces 95% of the NO_x to free nitrogen. All fly ash and metals are essentially captured.

Table 5.20.15 is a recent list of catalyst suppliers, and Table 5.20.14 lists the manufacturers of environmental technologies (Environmental Processes '93 1993).

—Karl T. Chuang
Alan R. Sanger

References

Bond, G.C. 1987. *Heterogeneous catalysis: Principles and application.* 2d ed. Oxford: Oxford University Press.

Chu, W., and H. Windawi. 1996. *Chem. Eng. Progress.* (March): 37.

Environment Canada: Task Force Report 1989. *Development of a national nitrogen oxide (NO_x) and volatile organic compounds (VOC) management plan for Canada.* (July).

Environmental processes '93. 1993. *Hydrocarbon Processing* 72, no. 8 (August):67.

Fair, J.R. 1987. Distillation. In *Handbook of separation process technology.* New York: John Wiley.

Groeneveld, M.J., G. Boxhoorn, H.P.C.E. Kuiper, P.F.A. van Grinsven, H. Gierman, and P.L. Zuideveld. 1988. Preparation, characterization and testing of new V/Ti/SiO_2 catalysts for DeNO$_x$ing and evaluation of shell catalyst S-995. *Proc. 9th Int. Congr. Catalysis*, edited by M. Ternan and M.J. Phillips, vol. 4: 1743. Ottawa: Chemical Institute of Canada.

Kast, W. 1981. Adsorption from the gas phase—Fundamentals and processes. *Ger. Chem. Eng.* 4:265.

Kohl, A.L. 1987. Absorption and stripping. In *Handbook of separation process technology.* New York: John Wiley.

Kohl, A.L., and F.C. Riesenfeld 1985. *Gas purification.* 4th ed. Houston, Gulf.

Langmuir, I. 1921. The mechanism of the catalytic action of platinum in the reactions $2CO + O_2 = 2CO_2$ and $2H_2 + O_2 = 2H_2O$. *Trans. Faraday Soc.* 17:621.

McCabe, W.L., J.C. Smith, and P. Harriott. 1993. Unit operations of chemical engineering. 5th ed. New York: McGraw-Hill.

Nakajima, F. 1991. Air pollution control with catalysis—Past, present and future. *Catalysis Today.* 10:1.

Nakatsuji, T., and A. Miyamoto. 1991. Removal technology for nitrogen oxides and sulfur oxides from exhaust gas. *Catalysis Today* 10:21.

Onda, K.H., H. Takeuchi, and Y. Okumoto. 1968. Mass transfer coefficients between gas and liquid phases in packed columns. *J. Chem. Eng.* (Japan) Vol. 1, no. 1:56.

Perry, R.H., and D.W. Green. 1984. *Perry's chemical engineers' handbook.* 6th ed. New York: McGraw-Hill.

Prasad, R., L.A. Kennedy, and E. Ruckenstein. 1984. *Catalytic combustion. Catal. Rev.—Sci. Eng.* 26:1.

Sachtler, W.H.M. 1970. The mechanism of catalytic oxidation of some organic molecules. *Catal. Rev.—Sci. Eng.* 4:27.

Satterfield, C.N. 1991. Heterogeneous catalysis in practice. 2d ed. New York: McGraw-Hill.

Sokolovskii, V.D. 1990. Principles of oxidative catalysis on solid oxides. *Catal. Rev.—Sci. Eng.* 32:1.

Spivey, J.J. 1987. Complete catalytic oxidation of volatile organics. *Ind. Eng. Chem. Res.* 26:2165.

Strigle, R.F. 1987. *Random packings and packed towers.* Houston: Gulf.

Trimm, D.L. 1991. Catalytic combustion. Chap. 3 in *Studies in inorganic chemistry 1991* 11:60.

Yang, R.T. 1987. *Gas separation by adsorption processes.* Boston: Butterworths.

5.21
GASEOUS EMISSION CONTROL: THERMAL DESTRUCTION

This section addresses the thermal destruction of gaseous wastes. These wastes are predominantly VOC-containing. The principal methods are thermal combustion and incineration and flaring, although a number of new technologies are emerging. First, this section provides an overview of these methods including major technological and some cost considerations. Then, it cites several examples of industries that employ these technologies to achieve destruction and removal efficiencies (DREs) in excess of 98% and low products of incomplete combustion (PIC) emissions.

Next this section reviews the thermodynamic and kinetic fundamentals that form the basis of thermal incineration and illustrates the principles with two sample calculations. Most of this section is devoted to the design considerations of thermal incinerators; the remaining section is allocated to flares and the emerging technologies.

Overview of Thermal Destruction

The principal methods for thermal destruction include thermal combustion and incineration, flaring, and other emerging technologies.

THERMAL COMBUSTION AND INCINERATION

Thermal combustion and incineration is the principal approach used for VOCs (usually expressed in concentration terms) but is equally applicable to liquids and solids with sufficient heat content. This technique is different from catalytic destruction, which is discussed in Section 5.20. From a global mass balance and energy balance point of view, the design and implementation of a thermal incinerator is straightforward. Thermal efficiencies can be estimated closely with CO_2 and H_2O as the principal products. For dilute streams, achieving the required temperature may require auxiliary fuel.

In the area of PIC-formation, global approaches become unworkable. Environmental engineers may have to use detailed measurement techniques to verify emission types and levels not only from the combustion process but also for the pollution control equipment.

Thermal destruction is simplest when (at least theoretically) CO_2 and H_2O are the expected products. The process becomes more complicated when hetero atoms (i.e., N, S, and Cl) or inorganics are involved, especially if corrosive products (i.e., HCl and SO_2) are formed. The dividing line between gaseous and liquid wastes is not always sharp because liquid wastes can contain many species of high-vapor pressure and gaseous wastes can carry liquid residues in droplet form.

The main variables controlling the efficiency of a combustion process are temperature, time, and turbulence; the three Ts of combustion. At a constant combustion chamber temperature, the DRE, defined in terms of the following equation:

$$DRE \% = \frac{VOC_{in} - VOC_{out}}{VOC_{in}} \times 100 \qquad 5.21(1)$$

increases with residence time; increasing the temperature increases the DRE at a constant time. At efficient combustion temperatures, the rate can become mixing limited.

Modern thermal oxidation systems can accomplish +99% DRE for capacities ranging from 1000–500,000 cfm and VOC concentrations of 100–2000 ppmv. Typical residence times are 1 sec or less at temperatures of 1300–1800°F. Inlet VOC concentrations above 25% of the lower explosion limit (LEL) are generally avoided due to the potential explosion hazards. Temperatures near 1800°F and long residence times can lead to elevated nitrogen oxide levels, which may have to be controlled separately (if lowering the combustion temperature is not feasible).

Thermal incinerators are usually coupled to two types of thermal energy recovery systems: regenerative and recuperative. Both methods transfer the heat content of the combustion exhaust gas stream to the incoming gas stream (Ruddy and Carroll 1993). In a regenerative system, an inert material (such as a dense ceramic) removes heat from the gases exiting the furnace. Such a ceramic storage bed eventually approaches the temperature in the combustor, consequently reducing the heat transfer. Therefore, the hot exhaust stream contacts a cooler bed, while the incoming gas stream passes through the hot bed.

As shown in Figure 5.21.1, the VOC-laden gas stream enters bed #1 which warms this gas stream by transferring heat from a previous cycle. Some VOCs are destroyed here, but most of them are oxidized in the combustion chamber. The flue gases from this combustion exit through bed #2 and transfer most of their enthalpy in the process. Within seconds of a heating and cooling cycle, the beds are switched, and the incoming stream now enters bed #2. Consequently, a near steady-state operation is approached,

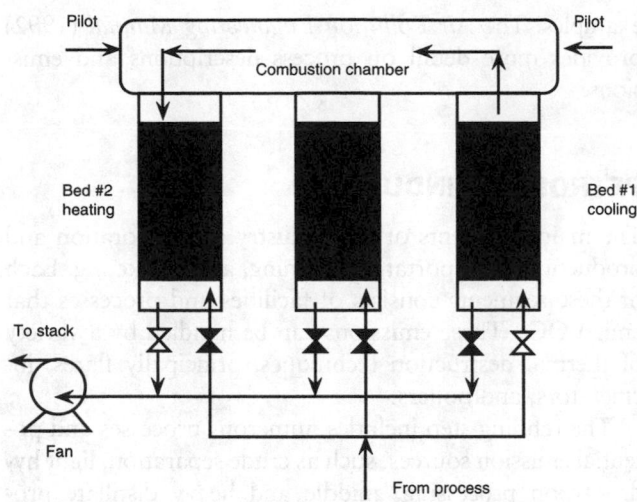

FIG. 5.21.1 Diagram of a regenerative thermal oxidizer.

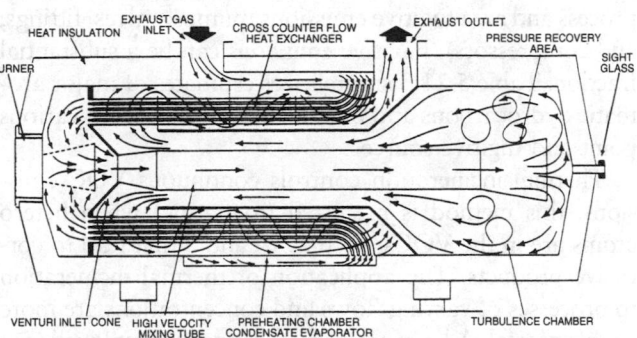

FIG. 5.21.2 Recuperation-type afterburner.

and more heat can be recovered from these systems than from a typical thermal incinerator (about 70%). Using multiple beds can lead to heat recoveries to 95%. The need for any auxiliary fuel depends on the potential thermal energy of the VOC-laden stream.

Recuperative thermal oxidation systems typically use a shell-and-tube design heat exchanger to recover heat from the flue gases for heating the incoming gases. Operating temperatures are reached quickly in these systems. Recuperative heat exchange is also more suited to cyclic operations and variations in VOC feed rates and concentrations. Figure 5.21.2 is a schematic of such a system.

Table 5.21.1 summarizes the emission sources, VOC categories, and typical operating parameters (flow rates and concentrations), as well as the cost aspects of thermal VOC control technologies.

FLARING

Flaring is the process of disposing of industrial and other combustible waste gas streams via a visible flame (flare). The flame can be enclosed in a chimney or stack.

Flares are widely used for the disposal of waste gases from several industrial processes, including process start ups, shutdowns, and emergencies. These processes are characterized by variable or intermittent flow. Flares are principally used where the heating value cannot be economically recovered. Applications include petroleum production, blast furnace and coke oven combustible gases, industrial chemical production, and landfills. Refinery

TABLE 5.21.1 THERMAL VOC CONTROL TECHNOLOGIES

Technology	Emission Source	VOC Category	Emission Rate Volumetric (scfm)	VOC (ppmv)
Thermal oxidation[a]	PV, ST TO, WW	AHC, HHC A, K	<20,000 w, w/o HR ≥20,000 w HR	20–1000 w/o HR ≥1,000 w HR
Flaring	PV, ST, TO, WW, F	AHC, A, K		

Source: E.C. Moretti and N. Mukhopadhyay, 1993, VOC control: Current practices and future trends, *Chem. Eng. Progress* 89, no. 7:20–26.

Notes: Users of VOC control technology project (1993) that about 43% of their capital expenditures will be in thermal oxidizers.

[a]Removal efficiencies   95–99+%

Capital Costs $10–200/cfm (recuperative)
    $30–450/cfm (regenerative)

Annual Operating Costs $15–90/cfm (recuperative)
    $20–150/cfm (regenerative)

Up to 95% energy recovery is possible

Not recommended for batch operations

Key:

PV = process vents
ST = storage tanks
TO = transfer operations
WW = wastewater operations
F = fugitive
AHC = aliphatic and aromatic HCs
HHC = halogenated hydrocarbons
A = alcohols, glycol ethers, ethers, epoxides, and phenols
K = ketones and aldehydes
HR = heat recovery

TABLE 5.21.2 EMERGING TECHNOLOGIES

Molten salt oxidation
Molten metal catalytic extraction
Molten glass
Plasma systems
Corona discharge

flares are typically elevated and steam-assisted. Flares can be open or enclosed and burn with a type of diffusion flame. Auxiliary fuel may have to be added to support flare combustion.

EMERGING TECHNOLOGIES

Numerous emerging technologies have been used or proposed for the thermal destruction of wastes, more so for liquid and solid wastes than for gaseous wastes. The reason is clear: most gaseous wastes that are candidates for thermal destruction carry large quantities of inert diluents (CO_2, H_2O, and N_2), whereas some of these emerging technologies are energy intensive because they operate at high temperatures.

Some of these technologies have been available for decades and will probably not find widespread application. However, some technologies combine thermal destruction with the formation of a useful product. Consequently, product formation enhances the advantages of thermal incineration, in which energy recovery is the chief economic benefit.

Preconcentration (adsorption and desorption or absorption and desorption) can be a suitable pretreatment to one of these techniques, though rarely practiced. Also, numerous emerging technologies are not thermal in nature. Table 5.21.2 lists several emerging technologies based primarily on the availability of practical and theoretical information but not restricted to gaseous wastes.

Source Examples

Hundreds of processes emitting VOCs are candidates for thermal destruction. This section presents several diverse examples. The *Air Pollution Engineering Manual* (1992) provides more detail on process descriptions and emissions.

PETROLEUM INDUSTRY

The major segments of this industry are exploration and production, transportation, refining, and marketing. Each of these segments consists of facilities and processes that emit VOCs. These emissions can be handled by a variety of thermal destruction techniques, principally flares, incinerators, and boilers.

The refining step includes numerous processes and potential emission sources, such as crude separation, light hydrocarbon processing, middle and heavy distillate processing, residual hydrocarbon processing, and auxiliary processes. Emissions are classified as process point and process and area fugitive emissions (pumps, valves, fittings, and compressors). Fugitive emissions can be a substantial fraction. Table 5.21.3 summarizes estimates of major aromatic hydrocarbons and butadiene emissions from various point and fugitive sources.

Thermal incineration controls continuous VOC emissions; this method is preferred (to flaring) when hetero atoms are in the VOCs, such as Cl and S that lead to corrosive products. The application of thermal incineration to processes of varying flows and concentrations are more problematical. A large safety factor must be employed (not to exceed 25% of the LEL). On the other hand, if the concentrations are too low, auxiliary fuel is required.

CHEMICAL WOOD PULPING

The major gaseous emissions in this industry are odorous, total reduced sulfur (TRS) compounds, characteristic of Kraft pulp mills. The principal components are hydrogen sulfide (H_2S), methyl mercaptan (CH_3SH), dimethyl sulfide ($(CH_3)_2S$), and dimethyl disulfide ($(CH_3)_2S_2$). In addition, emissions of noncondensible gases such as acetone and methanol are common. Emissions from uncontrolled sources, such as digester and evaporator, relief, and blow

TABLE 5.21.3 ESTIMATED EMISSIONS FROM PETROLEUM REFINERY

Chemical	Point, tn/yr	Fugitive, tn/yr	Total, tn/yr
Benzene	114	29	142
Toluene	437	111	548
Xylene (total)	31	1751	1782
Butadiene	3	1	4
Trimethyl benzene (1,2,4)	310	141	452

Source: Air Pollution Management Association, 1992, *Air pollution engineering manual*, edited by A.J. Buonicore and W.T. Davis (New York: Van Nostrand Reinhold).

gases, are expressed in lb/tn of air-dried pulp. Emissions also arise from black liquor oxidation tower vents.

Kraft pulp mills that began construction or modification after September 24, 1976 are subject to the new source performance standards (NSPS) for particulate matter and TRS emissions. Also, in 1979, the U.S. EPA issued retrofit emission guidelines to control TRS emissions at existing facilities not subject to the NSPS.

Today, combustion controls most major and minor sources of TRS emissions (National Council of the Paper Industry for Air and Stream Improvement 1985). Most commonly, existing combustors, such as power boilers and lime kilns, are used as well as specifically dedicated incinerators. The principal oxidation product is sulfur dioxide; a caustic scrubber is often installed after the incinerator to neutralize that gas. Two types of noncondensible gases are produced by the Kraft pulping process: low volume, high concentration (LVHC) and high volume, low concentration (HVLC). The latter can be burnt only in a boiler capable of accepting such large gas volumes without disrupting the unit's efficiency.

LANDFILL GAS EMISSIONS

The major component of landfill gas is methane; less than 1% (by volume) consists of nonmethane organic compounds. Air toxics detected in landfills include such compounds as benzene, chlorobenzene, chloroform, TSRs, tetrachloroethylene and toluene, and xylenes (Air and Waste Management Association 1992). Landfill gas is generated by chemical and biological processes on municipal solid waste (MSW). The gas generation rate is affected by parameters such as the type and composition of the waste, the fraction of biodegradable materials, the age of the waste, the moisture content and pH, and the temperature. Anaerobic decomposition can produce internal temperatures to 37°C (98.6°F); gas production rates are highest for moisture contents of 60–78%.

Control measures are usually based on containment combined with venting and collection systems. Low permeability solids for cover and slurry walls reduce the landfill gas movement. Capping is the process that uses a cover soil of low permeability and low porosity.

Gases can be vented or collected. Collection systems consist of several vertical and/or horizontal recovery wells that collect and convey gas within the landfill via a piping header system to a thermal oxidizer such as a flare. Blowers or compressors are used for this purpose. Gas collection efficiencies for active landfills with capping and gas collection can reach 90%.

The principal thermal destruction method for landfill gas is flaring. Emissions from landfill gas flares are a function of the gas flow rates, the concentration of the combustible component (which determines the temperature), and the residence time required. These emissions are estimated from combustion calculations; they are generally not measured.

RENDERING PLANTS

Thermal destruction methods are often applied to odorous emissions. Rendering plants are prime examples of an odorous emission. In these plants, animal and poultry byproducts are processed to produce fallow, grease, and protein meals. Batch and continuous processes are used. Other sources (National Council of the Paper Industry for Air and Stream Improvement 1985; Prokop 1985) describe these processes in detail.

In a batch cooker system (the basic rendering process), cookers are charged with raw material; a cook is made under controlled time, temperature, and pressure conditions; the cooked material is discharged; and the cycle is repeated. Under continuous conditions, the raw material is charged to the cooker.

The principal odorous emissions from these processes are N- and S-containing organic compounds, which are listed in Table 5.21.4. Additional compounds include higher molecular weight organic acids, pyrazines, alcohols, and ketones. These compounds are mostly noncondensibles and arise under the cooking conditions (~220°F). The type of raw material and its age have a significant effect on the odor intensity. Continuous systems tend to be enclosed and have a greater capability of confining odors. In the batch process, the steam rate varies between 450–900 ft^3 min^{-1}. Steam is subsequently condensed and cooled to below 120°F. Odorous noncondensibles range in odor intensity from 5000–10^6 odor units/scf; the volumetric emission rate of noncondensibles varies between 25–75 ft^3 min^{-1}.

Since rendering plants use boilers for steam generation and drying, odor control by incineration usually uses the existing boilers. The following factors should be considered when boiler incineration of odor-intense effluents is implemented:

Excess air at odor pickup points should be avoided
If possible, the odor-containing stream should be used as primary combustion air
Moisture and particle concentrations should be low
High-intensity odors must contact the furnace flame
Sufficient residence time at T > 1200°F must be provided in existing boilers.

In addition, the use of an existing boiler for odor incineration must conform to engineering, safety, and insurance requirements. Sections 5.26 and 5.27 provide more information on odor and its control.

Combustion Chemistry

Thermal incineration is based on combustion chemistry including stoichiometry and kinetics. These fundamentals and sample calculations are described next.

TABLE 5.21.4 ODOROUS COMPOUNDS IN RENDERING PLANT EMISSIONS

Compound Name	Formula	Molecular Weight	Detection Threshold (ppm, v/v)	Recognition (ppm, v/v)
Acetaldehyde	CH_3CHO	44	0.067	0.21
Ammonia	NH_3	17	17	37
Butyric acid	C_3H_7COOH	88	0.0005	0.001
Dimethyl amine	$(CH_3)_2NH$	45	0.34	—
Dimethyl sulfide	$(CH_3)_2S$	62	0.001	0.001
Dimethyl disulfide	CH_3SSCH_3	94	0.008	0.008
Ethyl amine	$C_2H_5NH_2$	45	0.27	1.7
Ethyl mercaptan	C_2H_5SH	62	0.0003	0.001
Hydrogen sulfide	H_2S	34	0.0005	0.0047
Indole	$C_6H_4(CH)_2NH$	117	0.0001	—
Methyl amine	CH_3NH_2	31	4.7	—
Methyl mercaptan	CH_3SH	48	0.0005	0.0010
Skatole	C_9H_9N	131	0.001	0.050
Trimethyl amine	$(CH_3)_3N$	59	0.0004	—

Source: Air Pollution Management Association, 1992.

STOICHIOMETRY

The starting basis of thermal incineration is the complete combustion of a hydrocarbon to carbon dioxide and water in air as follows:

$$C_xH_y + \left(x + \frac{y}{4}\right)O_2 + \left(x + \frac{y}{4}\right)3.78\,N_2 \longrightarrow$$
$$x\,CO_2 + \frac{y}{2}\,H_2O + \left(x + \frac{y}{4}\right)3.78\,N_2 \quad 5.21(2)$$

This equation accounts for all major atoms. If the VOC contains Cl, S, or N in appreciable amounts, these components must be accounted for in the stoichiometry; usually HCl, SO_2, and NO are combustion products. If the components of a mixture are known, equations can be written for each species; if the components are not known, an apparent chemical formula can be based on the weight percents of each combustible element (e.g., C, H, N, and S).

Any oxygen in VOCs (alcohols and ketones) is subtracted from the stoichiometric oxygen requirements. The formation of thermal NO (from nitrogen in the air) is not accounted for in stoichiometric combustion equations; only when nitrogen is present in the fuel–VOC mixture is NO formation (from fuel or VOC N) accounted for stoichiometrically.

For dilute gas streams, achieving high-DRE and low-PIC emissions (not accounted for in the combustion stoichiometry) requires auxiliary fuel to maintain a minimum temperature and residence time.

Using excess air is common practice. This use is expressed in terms of the air/fuel ratio (mass based) or the equivalence ratio, defined as ϕ as follows:

$$\phi = \frac{(VOC/oxygen)_{actual}}{(VOC/oxygen)_{stoichio.}} \quad 5.21(3)$$

SAMPLE CALCULATION #1

A hazardous chlorinated hydrocarbon is combusted in air according to the following stoichiometry:

$$C_xH_yCl_z + \left(x + \frac{y - z}{4}\right)O_2$$
$$\longrightarrow x\,CO_2 + \frac{y - z}{2}\,H_2O + z\,HCl$$

Calculate the stoichiometric HCl mole fraction for x = 6, y = 3, and z = 2. If the combustion is conducted in a way that leads to 4 oxygen mole % in the stack gas, what is the equivalence ratio; if, under these conditions, the DRE is 99.99%, what is the emitted $C_6H_3Cl_2$ concentration?

SOLUTION

$$C_6H_3Cl_2 + \left(6 + \frac{3 - 2}{4}\right)O_2 + \left(6 + \frac{3 - 2}{4}\right)3.78\,N_2 \longrightarrow$$
$$6CO_2 + \frac{3 - 2}{2}\,H_2O + 2HCl + 6.25 \times 3.78\,N_2$$

Total moles in stack gas: 32.125; mole fraction HCl = 0.062

If E = excess oxygen, then by stoichiometry, 6.25E oxygen is in the product gases. Therefore, the following equation applies:

$$0.04 = \frac{6.25E}{32.125 + 6.25E + 6.25 \times 3.78E}$$

Solving, E = 0.254; then

$$(oxygen)_{actual} = 6.25 + 6.25 \times 0.254 = 7.84$$

$$\phi = \frac{1/7.84}{1/6} = 0.765$$

1 mole $C_6H_3Cl_2$ in → 0.0001 mole $C_6H_3Cl_2$
Total moles out = 32.125 + 7.589 = 39.714

$$\frac{0.0001}{39.714} \times 10^6 = 2.52 \text{ ppm}$$

KINETICS

Stoichiometry cannot account for finite DRE and PIC concentrations. Global kinetics can address only the former. Finite DRE can be expressed as a fractional conversion as follows:

$$\ln \frac{C_{vocf}}{C_{voci}} = -kt = \ln(1 - DRE) \qquad 5.21(4)$$

where k is a (pseudo) first-order rate coefficient s^{-1}. The product of rate and residence time determines the conversion (i.e., a high DRE is achieved only by a sufficiently high rate [high temperature] and sufficient residence time).

This presentation implies a plug-flow reactor and irreversibility; the latter is usually a good assumption in combustion reactions. Some evidence exists that a modified model incorporating an ignition-delay time gives a better fit to experimental data (Lee et al. 1982).

In Equation 5.21(4), the dimensionless group $-kt$ presents the natural log of the unreacted fraction. Conversion increases rapidly with an increase in kt. This product implies a batch reactor. For a flow reactor (plug flow), t is replaced by $\bar{t}$, the mean residence time, defined as V/Q (reactor volume divided by the volumetric flow rate at a constant density) or $L/\bar{u}$ for tubular reactors (L = reactor/combustor length; $\bar{u}$ = average velocity).

Some reported data shows evidence that continuously stirred tank reactors (CSTR), or perfect mixer, behavior occurs at high conversion (Hemsath and Suhey 1974). The following equation expresses this behavior:

$$\frac{C_{vocf}}{C_{voci}} = \left(\frac{1}{1 + k\bar{t}}\right) \qquad 5.21(5)$$

Composite behavior is sometimes an explanation for the observed results, i.e., a model consisting of a plug-flow reactor followed by one or more CSTRs.

In all combustion reactions, some CO is always formed. The following two-step global model accounts for this formation:

$$VOC \xrightarrow{k_1} CO \xrightarrow{k_2} CO_2 \qquad 5.21(6)$$

where, assuming excess oxygen, the following equations apply:

$$r_{VOC} = -k_1 (VOC) \qquad 5.21(7)$$

$$r_{CO} = -k_1 (VOC) - k_2 (CO) \qquad 5.21(8)$$

$$r_{CO_2} = k_2 (CO) \qquad 5.21(9)$$

Qualitatively, the resulting concentration dependence is of the form shown in Figure 5.21.3. Under typical incinera-

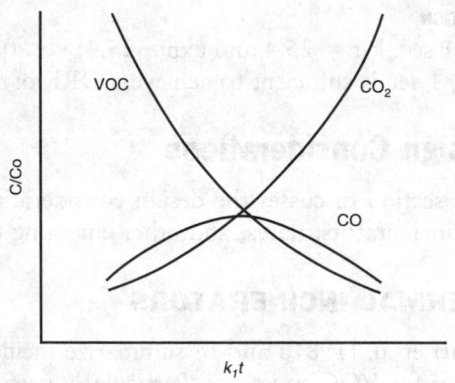

key:
c = Concentration at time t or exit concentration
C_0 = Initial concentration
k_1t = Product of rate coefficient x time, where t can be taken as the residence time at a given temperature

FIG. 5.21.3 Qualitative relationship between VOC, CO, and CO_2 concentration as a function of k_1t.

tion conditions, the preceding chemical sequence is irreversible. Using C_{VOC} (t = 0) = $C_{VOC,i}$ and $C_{CO,i} = C_{CO_2,i} = 0$, the solution of Equations 5.21(7) to 5.21(9) is as follows:

$$\frac{C_{VOC,t}}{C_{VOC,i}} = \exp(-k_1t) \qquad 5.21(10)$$

$$\frac{C_{CO,t}}{C_{VOC,i}} = \frac{1}{1 - \frac{k_2}{k_1}} (\exp(-k_2t) - \exp(-k_1t)) \qquad 5.21(11)$$

$$\frac{C_{CO_2,t}}{C_{VOC,i}} = 1 - \exp(-k_1t)$$

$$- \frac{1}{1 - \frac{k_2}{k_1}} (\exp(-k_2t) - \exp(-k_1t)) \qquad 5.21(12)$$

This scheme can account for variable levels of CO in the exit gases, depending on the VOCs, temperature, and residence time. This analysis does not account for other PICs; usually when CO is low, other PICs are low also. Since the rate coefficients k_1 and k_2 are the Arrhenius-type and are strongly temperature-dependent, temperature is the critical variable.

SAMPLE CALCULATION #2

The incineration of VOCs requires a minimum residence time at a specified temperature to achieve a certain DRE. Given the global rate expression of Equation 5.21(2) and a pseudo-first-order rate coefficient $k_1 = A \exp(-E/RT)$, where E = 45 Kcal $mole^{-1}$, A = 1.5 × 10^{11} s^{-1}, R = gas constant = 2 × 10^{-3} Kcal $mole^{-1}$ K^{-1}, and T = temperature K, would 1 sec at 1000 K (1341°F) be sufficient to achieve a DRE of 99.99%?

$$k_1 = 1.5 \times 10^{11} \exp\left(-\frac{45}{2 \times 10^{-3} \times 10^3}\right) = 25.4 \text{ s}^{-1}$$

SOLUTION

For 1 sec, $k_1t = 25.4$ and $\exp(-25.4) \ll 0.0001$; therefore, 1 sec is sufficient to achieve a DRE of 99.99%.

Design Considerations

This section discusses the design considerations for thermal incinerators, flares, and other emerging technologies.

THERMAL INCINERATORS

Katari et al. (1987a and b) summarize incineration techniques for VOC emissions. Particularly useful are a table on the categorization of waste gas streams (see Table 5.21.5) and a flow chart that determines the suitability of a waste gas stream for incineration and the need for auxiliaries. The categorization includes the % oxygen, VOC content vis-a-vis LEL, and heat content. Thus, a mixture of VOC and inert gas with zero or a negligible amount of oxygen (air) and a heat content >100 Btu scf^{-1} (3.7 MJ m^{-3}) can be used as a fuel mixed with sufficient oxygen for combustion (see category 5 in Table 5.21.5).

Figure 5.21.4 is a schematic of the incineration system. Waste gas from the process, auxiliary fuel (if needed), and combustion air (if needed) are combined in the combustion chamber under conditions (i.e., time, temperature, and turbulence) to achieve minimum conversion (DRE >99%). The temperature inside the combustion chamber should be well above the ignition temperature (1000–1400°F) of most VOCs; residence times of 0.3–1.0 sec may be sufficient. If higher DREs are required (>99.99%), both residence time and temperature may have to be increased depending on the VOC composition. This increase is also necessary the more nonuniform (in terms of VOC components) the waste gas stream is.

The majority of industrial waste gases for thermal destruction fall under category 1 of Table 5.21.5. The following parameters must be quantified (see Figure 5.21.4):

Heat content of the waste gas and Tp from process (Tp 5 Tw, if no heat exchanger is used)
The % LEL (VOC content and types)
The waste gas volumetric flow rate T_w
TE required based on the necessary DRE

The nature of the VOCs in the waste gas stream determine the temperature (T_E) at the exit of the combustor (not equal to the adiabatic flame temperature).

The composition of the waste gases determines the combustion air requirements; required temperatures and flow rates determine the auxiliary fuel requirements, furnace chamber size, and heat exchanger capacity. The suggested temperatures are 1800°F for 99% DREs at approximately 1 sec residence time. Applicable incinerator types include liquid injection, rotary kiln, fixed-hearth, and fluidized-beds (of which several variants exist). Dempsey and Oppelt (1993) describe these units in more detail. Although these authors principally address hazardous waste, their article

TABLE 5.21.5 CATEGORIZATION OF WASTE GAS STREAMS

Category	Waste Gas Composition	Auxiliaries and Other Requirements
1	Mixture of VOC, air, and inert gas with >16% O_2 and a VOC content <25% LEL (i.e., heat content <13 Btu/ft^3)	Auxiliary fuel is required. No auxiliary air is required.
2	Mixture of VOC, air, and inert gas with >16% O_2 and a VOC content between 25 and 50% LEL (i.e., heat content between 13 and 26 Btu/ft^3)	Dilution air is required to lower the heat content to <13 Btu/ft^3. (Alternative to dilution air is installation of LEL monitors.)
3	Mixture of VOC, air, and inert gas with <16% O_2	This waste stream requires the same treatment as categories 1 and 2 except the portions of the waste gas used for fuel burning must be augmented with outside air to bring its O_2 content to about 16%.
4	Mixture of VOC and inert gas with zero to negligible amount of O_2 (air) and <100 Btu/ft^3 heat content	This waste stream requires direct oxidation with a sufficient amount of air.
5	Mixture of VOC and inert gas with zero to negligible amount of O_2 (air) and >100 Btu/ft^3 heat	This waste stream requires premixing and use as a fuel.
6	Mixture of VOC and inert gas with zero to negligible amount of O_2 and heat content insufficient to raise the waste gas to the combustion temperature	Auxiliary fuel and combustion air for both the waste gas VOC and fuel are required.

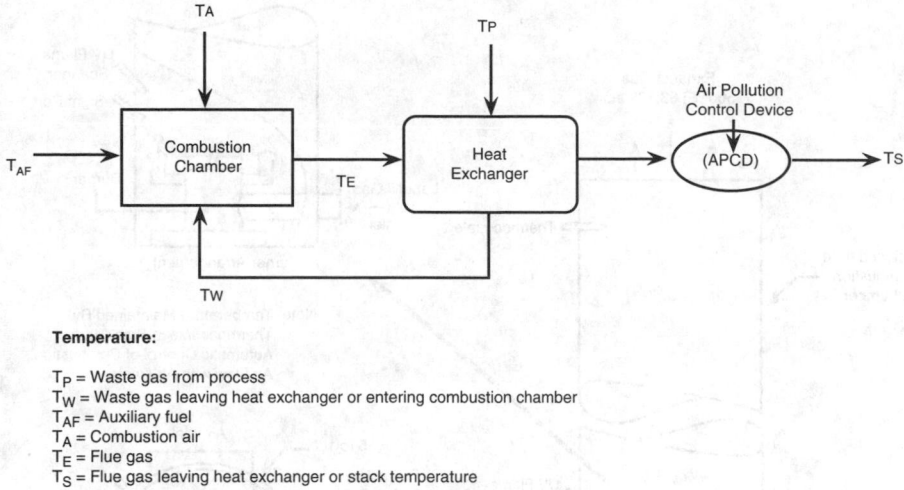

FIG. 5.21.4 Schematic diagram of incinerator system.

Temperature:

T_P = Waste gas from process
T_W = Waste gas leaving heat exchanger or entering combustion chamber
T_{AF} = Auxiliary fuel
T_A = Combustion air
T_E = Flue gas
T_S = Flue gas leaving heat exchanger or stack temperature

is a comprehensive review that includes regulatory aspects, current practice, technology, emissions and their measurements, control parameters, performance indicators, and risk assessment.

In thermal incinerators, T_w is usually below 1000–1100°F (to avoid preignition); as T_w increases, auxiliary fuel requirements can decrease. For a system with a recuperative heat exchanger, T_w can be calculated as follows:

$$T_w = T_p + \eta \, (T_E - T_p) \qquad 5.21(13)$$

where η represents the efficiency of the heat exchanger (see Figure 5.21.2). Standard heat transfer texts provide the equations for estimating η as well as the requisite material and energy balances.

If the wastes to be destroyed contain chlorine, higher temperatures may be required, and APCD needs increase, specifically the necessity to control HCl emissions. In addition, the PIC mix is more complex when organochlorine compounds are present and may require the application of Appendix VIII compound sampling techniques. Dempsey and Oppelt summarize these methods in a table.

CEMs are often used (or are required) for combustion gas components such as CO, CO_2, O_2, NO_x, and THC, one or more of which serves as a performance indicator. Again, if chlorinated compounds are combusted, continuous monitoring of HCl can be necessary. The methodology is the same as for CO and CO_2, namely NDIR. Nitrogen-containing compounds can form NO_x, S-containing compounds lead to SO_2 and perhaps some SO_3, P-containing compounds lead to P_2O_5 (a highly corrosive compound), and Br-, F-, and I-compounds form the corresponding acids.

Gaseous hazardous waste can lead to higher molecular weight PICs; for example, the combustion of methyl chloride (while yielding mainly CO_2 and HCl) can also lead to species such as chloroethanes and chlorobenzenes.

FLARES

Figure 5.21.5 shows an example of an enclosed flare, as used in landfill gas disposal. A multiple-head, gas burner is mounted at ground level inside a refractory-lined, combustion chamber that is open at the top. Enclosed flares have better combustion efficiencies than elevated flares, and emission testing is more readily performed. Typical minimum performance parameters include 0.3 sec residence time at 1000°F and CO < 100 ppm. Most refinery flares are elevated and steam assisted (see Figure 5.21.6). The steam promotes turbulence, and the induction of air into the flare improves combustion. The amount of steam required depends partly on the C/H ratio of the VOCs to be destroyed; a high ratio requires more steam to prevent a smoking flare.

U.S. EPA studies (Joseph et al. 1983; McCrillis 1988) identify several parameters important to flare design: flare head design, flare exit velocity, VOC heating value, and whether the flame is assisted by steam or air. They acquired the data from a specially constructed flare test facility. For a given flare head design, they found that the flame stability (the limit of flame stability is reached when the flame propagation speed is exceeded by the gas velocity) is a function of the fuel (VOC), gas velocity at the flare tip, and the lower heating value (LHV) of the fuel. For a given velocity, a minimum LHV is required for stable combustion. A combustion efficiency of >98% is maintained as long as the ratio of the LHV to the minimum LHV required for stable combustion is >1.2. Below that value, the DRE drops rapidly.

Gas (VOC) composition has a major influence on the stability limit. Figure 5.21.7 illustrates this influence with the results from the EPA flare test facility. Particularly interesting is the difference between methyl chloride (MeCl), propane (Prop), and butadiene (But) at a given exit velocity. The figure shows that with most gases, an increase in the flare exit velocity must be accompanied by an in-

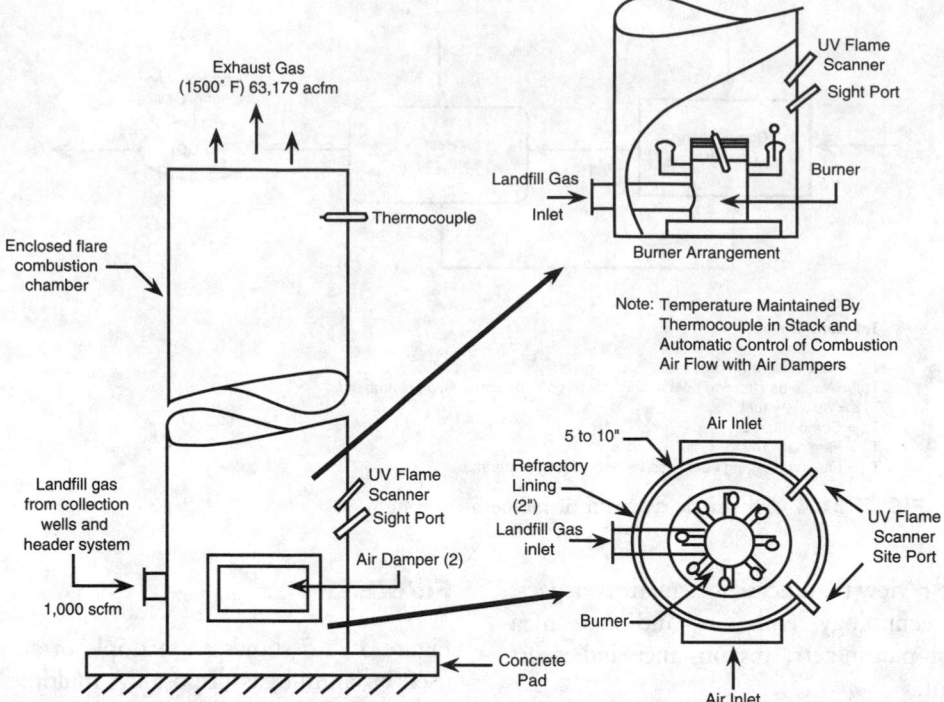

FIG. 5.21.5 Typical enclosed flare. (Adapted from John Zinc Inc., General arrangement drawing model ZTOF enclosed flare.)

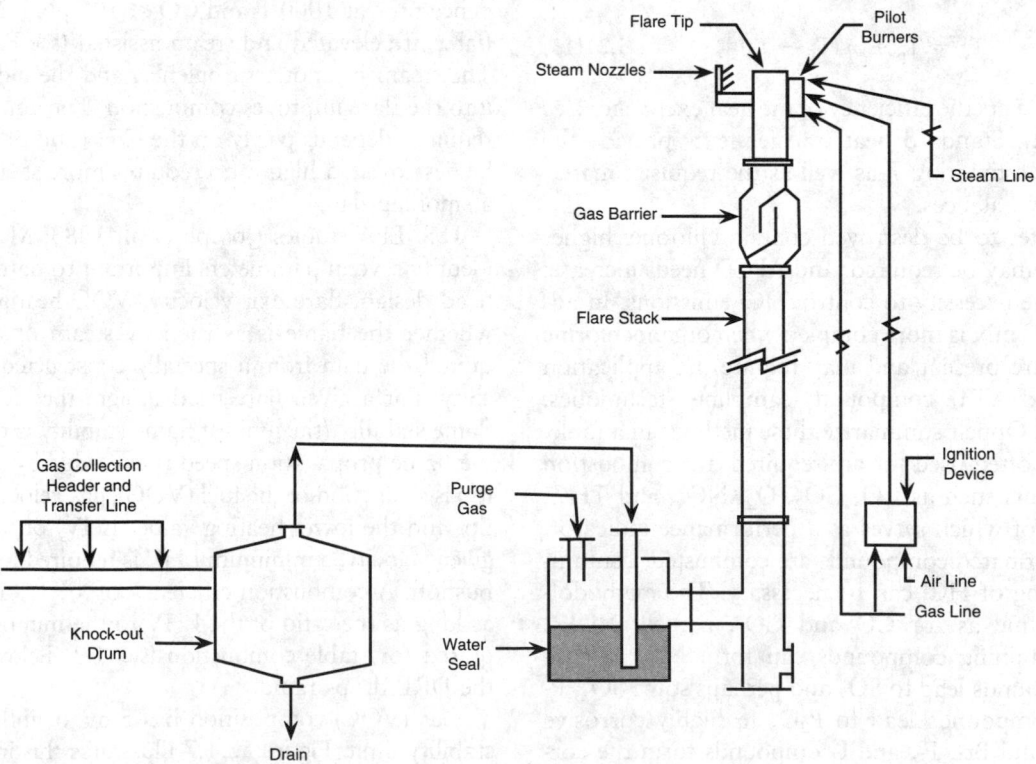

FIG. 5.21.6 Steam-assisted elevated flare system.

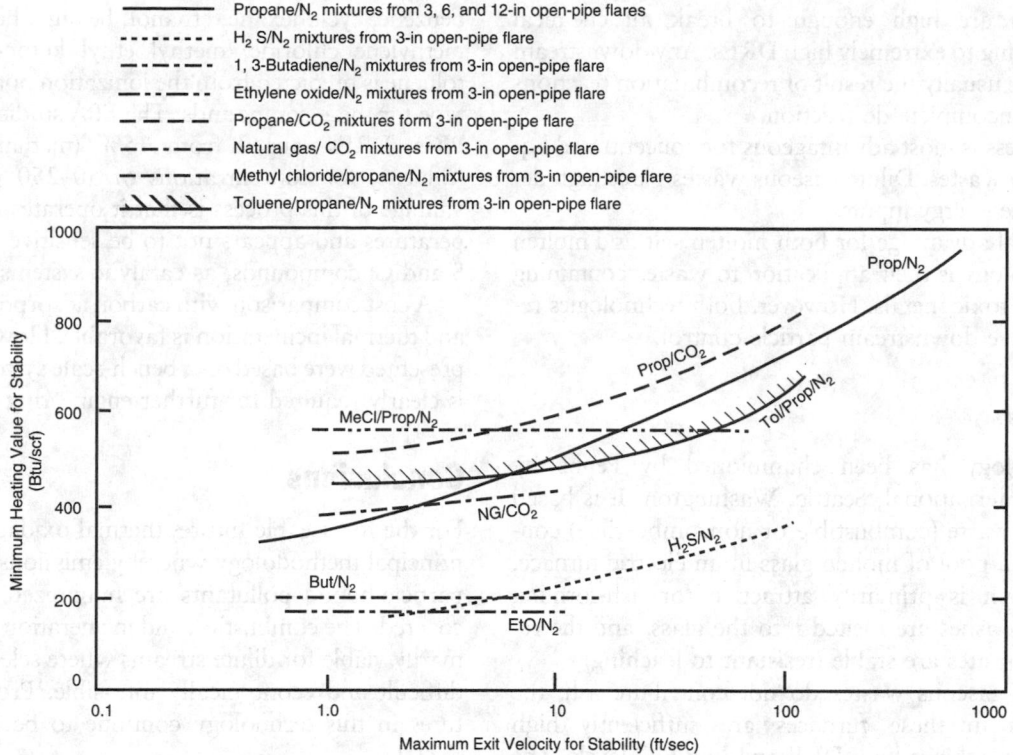

Propane/N₂ mixtures from 3, 6, and 12-in open-pipe flares
H₂ S/N₂ mixtures from 3-in open-pipe flare
1, 3-Butadiene/N₂ mixtures from 3-in open-pipe flare
Ethylene oxide/N₂ mixtures from 3-in open-pipe flare
Propane/CO₂ mixtures form 3-in open-pipe flare
Natural gas/ CO₂ mixtures from 3-in open-pipe flare
Methyl chloride/propane/N₂ mixtures from 3-in open-pipe flare
Toluene/propane/N₂ mixtures from 3-in open-pipe flare

FIG. 5.21.7 Flame stability limits for different gas mixtures flared without pilot assist from a 3-in open-pipe nozzle.

crease in the heating value of the waste to maintain a stable flame. For Cl-containing compounds, the flame stability should correlate with the H/Cl ratio.

Stone et al. (1992) outline design procedures for flares. The U.S. EPA requirements for flares are specified in 40 *CFR* Section 60.18. The requirements are for steam-assisted, air-assisted, and nonassisted flares. For steam-assisted, elevated flares, the following points must be addressed:

An exit velocity at the flare tip <60 ft/sec (for 300 Btu/scf gas streams) and <40 ft/sec (for >1000 Btu/scf gas streams). Between 300–1000 Btu/scf, the following equation applies:

$$\log V_{max} = \frac{NHV + 1214}{852} \qquad 5.21(14)$$

where V_{max} = the maximum permitted velocity and NHV = the net heating value (BTU/scf)
Absence of visible emissions (5 min exception period for any 2 consecutive hr)
Presence of a flame at all times when venting occurs
NHV not less than 300 Btu/scf.

EMERGING TECHNOLOGIES

This section discusses the design considerations for emerging technologies including molten salt oxidation, molten metal reactions, molten glass, plasma systems, and corona destruction.

Molten Salt Oxidation

Molten salt technology is an old technology. The process combines combustible wastes and air in a molten salt batch (which can be a single component such as sodium carbonate or a mixture) within a molten salt reactor, usually constructed of ceramic or steel. Typical temperatures are 1500–1900°F and several seconds of residence time for the gas phase.

An attractive feature is the heating value of the fuel which should be sufficient to maintain the salt medium in the molten state. Consequently, this technology is best applied to combustible liquid and solid wastes, rather than gaseous wastes, as the latter are usually too dilute. Another attractive aspect is the neutralization of acidic species, such as HCl and SO_2, which form from Cl- or S-containing wastes.

Molten Metal Reactions

Much higher temperatures are possible for a molten metal bath (primarily iron), up to 3000°F. This technology works best for wastes that are low in oxygen, leading to an off-gas composed primarily of CO and H_2. A molten metal bath requires large heat input by induction; the resulting temperatures are high enough to break all chemical bounds, leading to extremely high DREs. Any downstream emissions are usually the result of recombination reactions rather than incomplete destruction.

This process is most advantageous for concentrated liquid and solid wastes. Dilute gaseous wastes place high demands on the energy inputs.

A potential advantage for both molten salt and molten metal technology is their application to wastes containing quantities of toxic metals. However, both technologies require extensive downstream particle control.

Molten Glass

This technology has been championed by Penberthy Eletromelt International, Seattle, Washington. It is based on charging waste (combustible or noncombustible) continuously to a pool of molten glass in an electric furnace. This approach is primarily attractive for ash-forming wastes; these ashes are melted into the glass, and the resulting composites are stable (resistant to leaching).

Although gaseous wastes do not contribute ash, the temperatures in these furnaces are sufficiently high (>2300°F) to achieve high DRE and low PIC levels. The residence time for the molten glass phase is long (hours), but the residence time for the gas phase is much shorter.

Plasma Systems

A plasma incinerator burns waste in a pressurized stream of preheated oxygen. Because temperatures can reach 5000°F, applying this technology to dilute streams is prohibitive. Such a system consists of a refractory-lined, water-cooled preheater, where a fuel is burned to heat oxygen to about 1800°F. This preheater is followed by the combustion chamber, where the waste and oxygen are mixed and auto-ignition occurs. From here, the gases pass into a residence chamber, where the destruction is completed at maximum temperature, followed by a quench chamber, a scrubber to remove acid gases, and a stack.

At plasma temperatures, the degree of dissociation of molecules is high, consequently reactivity increases; reaction rates are much higher than at normal incinerator temperatures. The high combustion efficiencies that are achievable lead to a compact design. High heat recovery is also possible. Thus, these systems lend themselves to a portable design.

Corona Destruction

The EPA has been researching VOC and air toxic destruction since 1988 (Nunez et al. 1993). This work features a fixed-bed packed with high dielectric-constant pellets, such as barium titanate, that are energized by an AC voltage applied through stainless steel plates at each end of the bed. The destruction efficiencies for VOCs such as benzene, cyclohexane, ethanol, hexane, hexene, methane, methylene chloride, methyl ethyl ketone, styrene, and toluene is predicted from the ionization potential and bond type for these compounds. The EPA studies did not report PICs; DREs ranged from 15% (methane) to ~100% (toluene) for concentrations of 50–250 ppmv. One advantage of this process is that it operates at ambient temperatures and appears not to be sensitive to poisoning by S and Cl compounds, as catalytic systems are.

A cost comparison with carbon adsorption and catalytic and thermal incineration is favorable. However, the results presented were based on a bench-scale system only; scaleup is clearly required for further engineering evaluation.

Conclusions

For the foreseeable future, thermal oxidation remains the principal methodology whereby emissions of gaseous and particle-bound pollutants are minimized and heat is recovered. The combustion and incineration approach is primarily viable for dilute streams where selective recovery is difficult and economically infeasible. Proposed expenditures in this technology continue to be high (see Table 5.21.1).

Where wastes are well characterized and contain no chlorine, few problems are encountered. Thermal destruction is a mature technology. Future improvements include enhanced energy recovery, smaller sizes, and continuous monitors for compliance. The presence of chlorine and other hetero atoms that lead to corrosive products (e.g., HCl) impose constrains on the material choices, gas cleanup, and monitoring requirements. The chlorine level entering the unit should be closely controlled so that downstream cleanup is efficient. In some cases, monitoring for PICs, including polychlorinated dibenzo-p-dioxins and dibenzofurans (PCDD/F), is required.

Flares will come under increasing scrutiny in the future, and performance improvements will be required. No routine methods are available to measure emissions from flares. PIC inventories are inadequate. However, in regions where flaring is common, VOC emissions are expected to impact the photochemical smog potential of the atmosphere; hence, these emissions must be described in more detail.

—*Elmar R. Altwicker*

References

Air and Waste Management Association. 1992. *Air pollution engineering manual.* Edited by A.J. Buonicore and W.T. Davis. New York: Van Nostrand Reinhold.

Dempsey, C.R., and E.T. Oppelt. 1993. Incineration of hazardous waste: A critical review update. *J. Air Waste Mgt. Assoc.* 43, no. 1:25–73.

Hemsath, K.H., and P.E. Suhey. 1974. *Fume incineration kinetics and its applications.* Am. Inst. of Chemical Engineers Symp. Series No. 137, 70, 439.

Joseph, D., J. Lee, C. McKinnon, R. Payne, and J. Pohl. 1983. *Evaluation of the efficiency of industrial flares: Background—Experimental design facility.* EPA-600/2-83-070, NTIS No. PB83-263723. U.S. EPA.

Katari, V.S., W.M. Vatavuk, and A.H. Wehe. 1987a. Incineration techniques for control of volatile organic compound emissions, Part I, Fundamentals and process design considerations. *J. Air Pollut. Control Assoc.* 37, no. 1:91–99.

———. 1987b. Incineration techniques for control of volatile organic compound emissions, Part II, Capital and annual operating costs. *J. Air Pollut. Control Assoc.* 37, no. 1:100–104.

Lee, K.C., N. Morgan, J.L. Hanson, and G. Whipple. 1982. Revised predictive model for thermal destruction of dilute organic vapors and some theoretical explanations. Paper No. 82-5.3, *Air Pollut. Control Assoc. Annual Mtg.* New Orleans, La.

McCrillis, R.C. 1988. Flares as a means of destroying volatile organic and toxic compounds. Paper presented at EPA/STAPPA/ALAPCO Workshop on Hazardous and Toxic Air Pollution Control

Technologies and Permitting Issues, Raleigh, NC and San Francisco, CA.

National Council of the Paper Industry for Air and Stream Improvement. 1985. *Collection and burning of Kraft non-condensible gases—Current practices, operating experience and important aspects of design and operations.* Technical Bulletin No. 469.

Nunez, C.M., G.H. Ramsey, W.H. Ponder, J.H. Abbott, L.E. Hamel, and P.H. Kariher. 1993. Corona destruction: An innovative control technology for VOCs and air toxics. *J. Air Waste Mgt. Assoc.* 43, no. 2:242–247.

Prokop, W.H. 1985. Rendering systems for processing animal by-product materials. *J. Am. Oil Chem. Soc.* 62, no. 4:805–811.

Ruddy, E.N., and L.A. Carroll. 1993. Select the best VOC control strategy. *Chem. Eng. Progress* 89, no. 7:28–35.

Stone, D.K., S.K. Lynch, R.F. Pandullo, L.B. Evans, and W.M. Vatavuk. 1992. Flares. Part I: Flaring technologies for controlling VOC containing waste streams. *J. Air Waste Mgt. Assoc.* 42, no. 3:333–340.

5.22
GASEOUS EMISSION CONTROL: BIOFILTRATION

The biological treatment of VOCs and other pollutants has received increasing attention in recent years. Biofiltration involves the removal and oxidation of organic compounds from contaminated air by beds of compost, peat, or soil. This treatment often offers an inexpensive alternative to conventional air treatment technologies such as carbon adsorption and incineration.

The simplest biofiltration system is a soil bed, where a horizontal network of perforated pipe is placed about 2 to 3 ft below the ground (Bohn 1992) (see Figure 5.22.1). Air contaminants are pumped through the soil pores, adsorbed on the surface of the moist soil particles, and oxidized by microorganisms in the soil.

Mechanisms

Biofiltration combines the mechanism of adsorption, the washing effect of water (for scrubbing), and oxidation. Soils and compost have porosity and surface areas similar to those of activated carbon and other synthetic adsorbents. Soil and compost also have a microbial population of more than 1 billion antiomycetes (microorganism resembling bacteria and fungi) per gram (Alexander 1977). These microbes oxidize organic compounds to carbon dioxide and water. The oxidation continuously renews the soil beds adsorption capacity (see Figure 5.22.2).

Another distinction is that the moisture in the waste gas stream increases the adsorption capacity of water-soluble gases and is beneficial for the microbial oxidation on which the removal efficiency of biofilters depends. Conversely, the moisture adsorbed by synthetic adsorbents reduces

their air contaminant adsorption capacity and removal efficiency. In addition, biofilter beds also adsorb and oxidize volatile inorganic compounds (VICs) to form calcium salts.

Gases in air flowing through soil pores adsorb onto or, as in GC, partition out on the pore surfaces so that VOCs remain in the soil longer than the carrier air. Soil–gas partition coefficients indicate the relative strengths of retention. The coefficients increase with VOC molecular weight and the number of oxygen, nitrogen, and sulfur functional groups in the VOC molecules. In dry soils, the coefficients for VOCs have been reported from 1 for methane to 100,000 for octane. However, under moist conditions, because of the water-soluble nature, the soil–gas partition coefficient for octane is probably a thousand, and acetaldehyde is around several thousand (Bohn 1992).

The biofilter's capacity to control air contaminants depends on the simultaneous operation of both adsorption and regeneration processes. Thus, overloading the system through excessive air flow rates can affect the biofilter's removal efficiency so that adsorption rates are lower than the rates at which chemicals pass through the filter. Once all adsorption sites are occupied, removal efficiency diminishes rapidly.

A second limiting factor is the microbial regeneration rate of the adsorbed chemical, which must equal or exceed the adsorption rate. Toxic chemicals can interfere with microbial processes until a bacterial population develops that can metabolize the toxic chemical. Biofilter bed acidity also reduces the removal efficiency because the environment for soil bacterial is inhospitable. In most cases

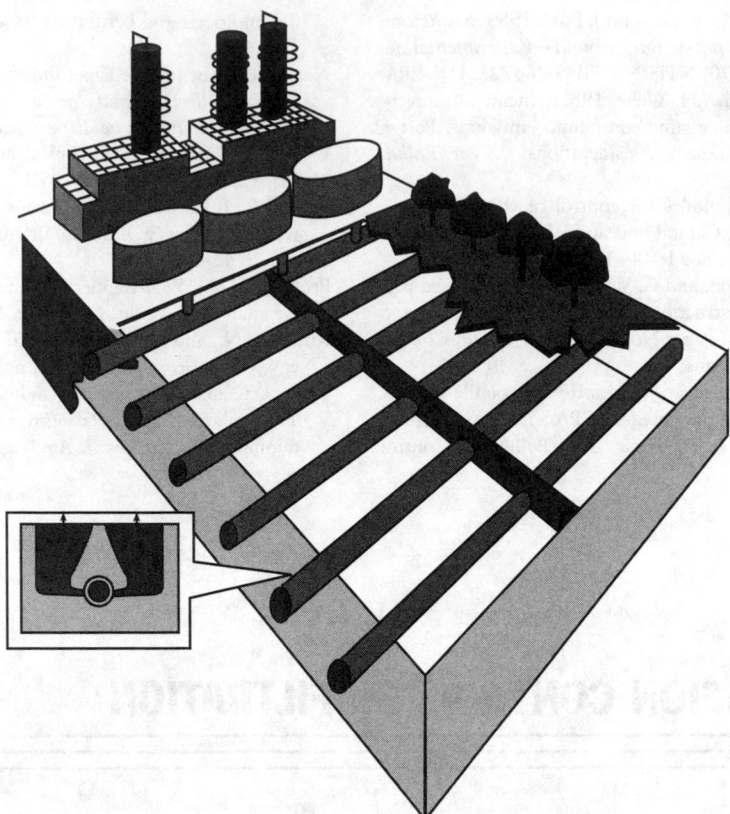

FIG. 5.22.1 Soil bed. A biofilter consists of a bed of soil or compost, beneath which is a network of perforated pipe. Contaminated air flows through the pipe and out the many holes in the sides of the pipe (enlarged detail), thereby being distributed throughout the bed.

Adsorption of ammonia (NH₃), hydrogen sulfide (H₂S), and other odors on the soil particle surface.

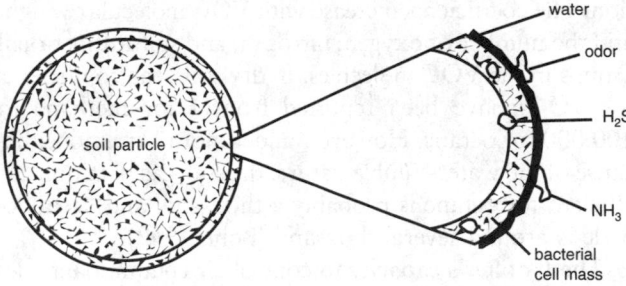

REGENERATION REACTIONS
Odorous hydrogen sulfide is oxidized to odorless sulfate
$$H_2S + 2O_2 \rightarrow SO_4^{2-} + 2H^+$$
Ammonia is dissolved in water and oxidized to odorless nitrate
$$NH_3 + H_2O \rightarrow NH_4^+ + OH_2^-$$
$$2NH_4^+ + 3O_2 \rightarrow 2 NO_3^- + 8H^+$$
Bacteria oxidize odorous volatile organics to odorless carbon dioxide and water
$$volatile\ organics + O_2 \rightarrow CO_2 + H_2O$$

FIG. 5.22.2 Adsorption of odors and regeneration of active sites.

of biofilter failure, the limiting factor is filter overloading rather than microbiological processes because of the great diversity and number of soil bacteria.

After start up, biofilter beds require an adaptation time for the microbes to adapt to a new air contaminant input and to reach steady state. For rapidly biodegradable compounds, the adaptation time is no more than several hours. As the biodegradability decreases, the adaptation time can take weeks. After start up, the bed is resistant to shock load effects. Table 5.22.1 summarizes the biodegradability of various gases.

Fixed-Film Biotreatment Systems

Most biological air treatment technologies are fixed-film systems that rely on the growth of a biofilm layer on an inert organic support such as compost or peat (biofilters) or on an inorganic support such as ceramic or plastic (biotrickling filters). For both systems, solid particles must be removed from waste gases before the gases enter the system; particulates plug the pores. Both systems are best suited for treating vapor streams containing one or two major compounds. When properly designed, biofilters are well suited for treating streams that vary in concentration from minute to minute.

TABLE 5.22.1   GASES CLASSIFIED ACCORDING TO DEGRADABILITY

Rapidly Degradable VOCs	Rapidly Reactive VICs	Slowly Degradable VOCs	Very Slowly Degradable VOCs
Alcohols	H_2S	Hydrocarbons*	Halogenated hydrocarbons†
Aldehydes	NO_x	Phenols	Polyaromatic hydrocarbons
Ketones	(but not N_2O)	Methyl chloride	CS_2
Ethers	SO_2		
Esters	HCl		
Organic acids	NH_3		
Amines	PH_3		
Thiols	SiH_4		
Other molecules containing O, N, or S functional groups	HF		

Source: H. Bohn, 1992, Consider biofiltration for decominating gases, *Chem. Eng. Prog.* (April).
Notes: *Aliphatics degrade faster than aromatics such as xylene, toluene, benzene, and styrene.
†Such as TCE, TCA, carbon tetrachloride, and pentachlorophenol.

Figure 5.22.3 is a schematic diagram of a biofiltration system. A biofiltration system uses microorganisms immobilized in the form of a biofilm layer on an adsorptive filter substrate such as compost, peat, or soil. As a contaminated vapor stream passes through the filtered bed, pollutants transfer from the vapor to the liquid biolayer and oxidize. More sophisticated enclosed units allow for the control of temperature, bed moisture content, and pH to optimize degradation efficiency. At an economically viable vapor residence time (1 to 1.5 min), biofilters can be used for treating vapor containing about 1500 $\mu g/l$ of biodegradable VOCs.

Figure 5.22.4 is a schematic diagram of a biotrickling filter for treating VOCs (Hartman and Tramper 1991). Biotrickling filters are similar to biofilters but contain conventional packing instead of compost and operate with recirculating liquid flowing over the packing. Only the recirculating liquid is initially inoculated with a microorganism, but a biofilm layer establishes itself after start up. The automatic addition of acid or base monitors and controls the pH of the recirculating liquid.

The pH within a biofilter is controlled only by the addition of a solid buffer agent to the packing material at the start of the operation. Once this buffering capacity is exhausted, the filtered bed is removed and replaced with fresh material. For the biodegradation of halogenated contaminants, biofilter bed replacement can be frequent. Therefore, biotrickling filters are more effective than biofilters for the treatment of readily biodegradable halogenated contaminants such as methylene chloride.

Biotrickling filters, possibly because of higher internal biomass concentrations, offer greater performance than biofilters at higher contaminant loadings. At a 0.5-min vapor residence time, the maximum concentration of styrene that can be degraded with 90% efficiency using biotrickling filters is two times greater than what can be degraded using biofilters.

Applicability and Limitations

Biofiltration has been used for many years for odor control at slaughter houses in Germany, the Netherlands, the United Kingdom, and Japan and to a limited extent in the United States. Recently, many wastewater and sludge treatment facilities have used biofilters for odor control purposes. The use of biofilters to degrade more complex air contaminants from chemical plants has occurred only

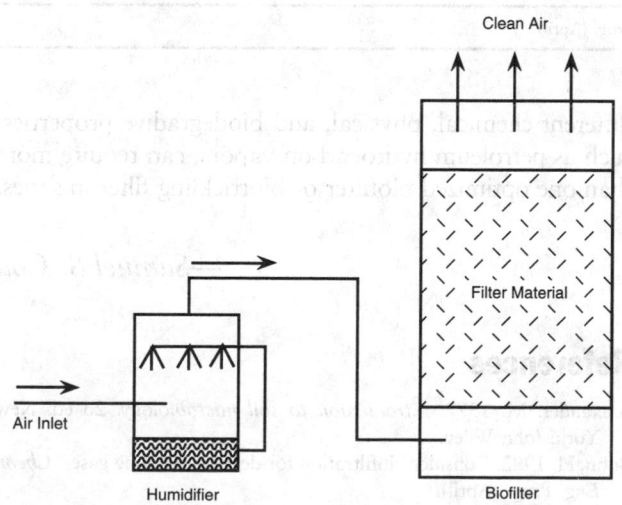

FIG. 5.22.3 Biofilter schematic diagram. (Reprinted, with permission, from A.P. Togna and B.R. Folsom, 1992, Removal of styrene from air using bench-scale biofilter and biotrickling filter reactors, Paper No. 92-116.04, *85th Annual Air & Waste Management Association Meeting and Exhibition, Kansas City, June 21–26.*)

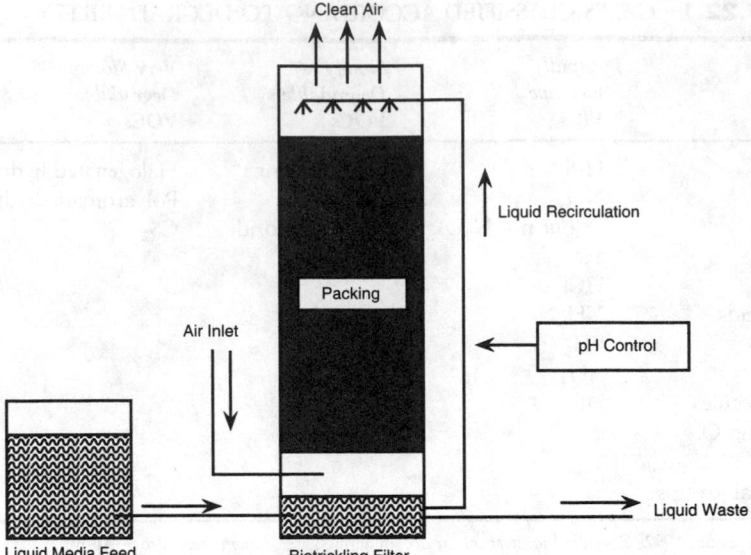

FIG. 5.22.4 Biotrickling filter schematic diagram. (Reprinted, with permission from Togna and Folsom, 1992.)

TABLE 5.22.2 CPI APPLICATIONS OF BIOFILTRATION

Company	Location	Application
S.C. Johnson & Son, Inc.	Racine, Wis.	Propane and butane removal from room air; 90% removal efficiency about 3000 cfm
Monsanto Chemical Co.	Springfield, Mass.	Ethanol and butyraldehyde removal from dryer air; 99% removal, 28,000 cfm; styrene removal from production gases
Dow Chemical Co.	Midland, Mich.	Chemical process gases
Hoechst Celanese Corp.	Coventry, R.I.	Process gases
Sandoz	Basel, Switzerland	Chemical process gases
Esso of Canada	Sarnia, Ontario	Hydrocarbon vapors from fuel storage tanks (proposed)
Mobil Chemical Co.	Canandaigua, N.Y.	Pentane from polystyrene–foam molding (proposed)
Upjohn Co.	Kalamazoo, Mich.	Pharmaceutical production odors; 60,000 cfm (proposed)

Source: H. Bohn, 1992, Consider biofiltration for decominating gases, *Chem. Eng. Prog.* (April).

within the last few years. Table 5.22.2 lists some chemical process industry CPI applications.

Biotrickling filters are more effective than biofilters for the treatment of readily biodegradable halogenated contaminants such as methylene chloride. Biofiltration can be unsuitable for highly halogenated compounds such as TCE, TCA, and carbon tetrachloride because they degrade slowly aerobically.

The limiting factors of soil-bed treatments are biodegradability of the waste and the permeability and chemistry of the soil. Because these factors vary, the design of soil beds is site-specific. For VIC removal, the lifetime of the bed depends on the soil's capacity to neutralize the acids produced. Any complex mixture with widely different chemical, physical, and biodegradive properties, such as petroleum hydrocarbon vapors, can require more than one optimized biofilter or biotrickling filter in series.

—*Samuel S. Cha*

References

Alexander, M. 1977. *Introduction to soil microbiology.* 2d ed. New York: John Wiley.

Bohn, H. 1992. Consider biofiltration for decontaminating gases. *Chem. Eng. Prog.* (April).

Hartman, S., and J. Tramper. 1991. Dichloromethane removal from waste gases with trickling-bed bioreactor. *Bioprocess Eng.* (June).

Fugitive Emissions: Sources and Controls

5.23
FUGITIVE INDUSTRIAL PARTICULATE EMISSIONS

This section is concerned with the dust generated in processing operations that is not collected through a primary exhaust or control system. Such emissions normally occur within buildings and are discharged to the atmosphere through forced- or natural-draft ventilation systems.

Industrial fugitive emissions contribute more than 50% of the total suspended and inhalable emissions (Cowherd and Kinsey 1986). In addition, these particulates frequently contain toxic or hazardous substances.

Sources

Most dry processing operations generate dust. Generation points include the following:

Dumping of materials
Filling of materials
Drying of materials
Feeding or weighing of materials
Mechanical conveying of materials
Mixing or blending of materials
Pneumatic conveying of materials
Screening or classifying of materials
Size reduction of materials
Bulk storage of materials
Emptying of bulk bags

Generally, these materials are finely divided products that are easily airborne, resulting in sanitation problems and possibly fire or explosions.

Emission Control Options

Fugitive particulates can be controlled by three basic techniques: process modification, preventive measures, and add-on capture and removal equipment.

PROCESS MODIFICATION

Changing a process or operation to reduce emissions can be more practical than trying to control the emissions. For example, a pneumatic conveyance system eliminates the emission problems of a conveyor belt, which can require any or all preventive procedures (i.e., enclosures, wet suppression, housekeeping of spilled materials, and stabilization of materials) and add-on equipment. Another solution is to use new intermediate bulk containers (IBCs), such as shown in Figure 5.23.1.

PREVENTIVE MEASURES

Housekeeping provides numerous beneficial results of good operation and maintenance (O&M). Good O&M includes the prevention of process upsets and defective equipment, prompt cleanup of spillage before dust becomes airborne, the partial or complete enclosure or shielding of dust sources.

Water, a water solution of a chemical agent, or a micron-sized foam can be applied to the surface of a partic-

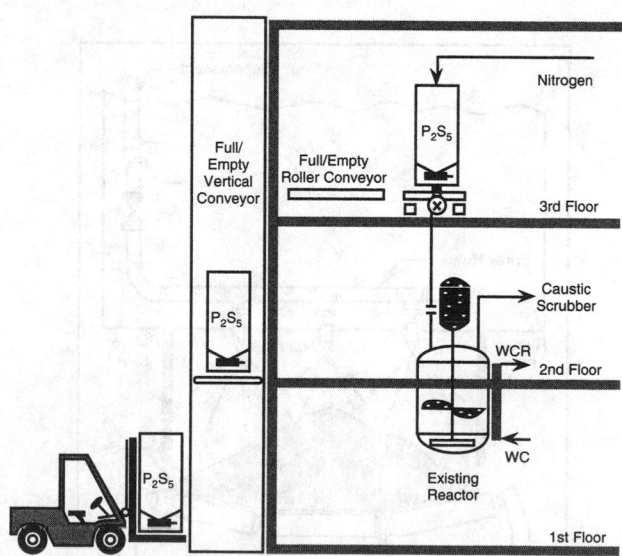

FIG. 5.23.1 Inverted IBCs elevated to the third floor for charging reactors, eliminating worker exposure to hazardous P_2S_5. (Reprinted, with permission, from B.O. Paul, 1994, Material handling system designed for 1996 regulations, *Chemical Processing* [July].)

ulate-generating material. This measure prevents (or suppresses) the fine particles contained in the material from leaving the surface and becoming airborne. The chemical agents used in wet suspension can be either surfactants or foaming agents. Figure 5.23.2 shows a suppression system at a crusher discharge point.

The use of foam injection to control dust from material handling and processing operations is a recently developed method to augment wet suppression techniques. The foam is generated by a proprietary surfactant compound added to a small quantity of water, which is then vigorously mixed to produce a small-bubble, high-energy foam in the 100 to 200 μm size range. The foam uses little liquid volume and when applied to the surface of a bulk material, wets the fines more effectively than untreated water does. Foam has been successfully used in controlling the emissions from belt transfer points, crushers, and storage-pile load-ins.

CAPTURE AND REMOVAL

Most industrial process, fugitive particulate emissions are controlled by capture and collection or industrial ventilation systems. These systems have three primary components:

A hood or enclosure to capture emissions that escape from the process

A dry dust collector that separates entrained particulates from the captured gas stream

A ducting or ventilation system to transport the gas stream from the hood or enclosure to the APCD

Before designing a dust collection system, the environmental engineer must thoroughly understand the process and its operations and the requirements of operating personnel. The engineer must evaluate the materials in the dust in addition to the particle size and define the characteristics of the material including the auto-ignition temperature, explosive limits, and the potential for electrostatic buildup in moving these materials. The engineer must locate the pickup and select the components to be used in the system.

Each operation generates various amounts of dust. The environmental engineer uses the amount of dust, the particle size, and the density of the dust in determining the capture velocity of the dust, which affects the pickup or hood design. Normal capture velocity is a function of particle size and ranges from 6 to 15 ft/sec (Opila 1993). In practice, the closer the hood is to the dust source, the better the collection efficiency is and the less air is needed for the collection process. The engineer should design enclosures and hoods at suitable control velocities to smooth the air flow. Examples of dust collection follow (Kashdan, et al. 1986):

A bag tube packer (see Figure 5.23.3) where displaced air is treated from the feed hopper supplying the packer, from the bag itself, and from the spill hopper below the bagger for any leakage during bag filling

Open-mouth bag filling (see Figure 5.23.4) where the dust pickup hood collects the air displaced from the bag

Barrel or drum filling (see Figure 5.23.5) that uses a dust pickup design with the same contour of the drum being filled. The pickup is designed for a single drum diameter; the design becomes complicated if 15, 30, and 55 gal drums are filled at the same filling station. The height and diameter of these drums vary affecting the capture velocity of the air.

The pickup of dust created when material flows down a chute onto a conveyor belt (see Figure 5.23.6)

The pickup points required for dust-free operation of a flatdeck screen (see Figure 5.23.7)

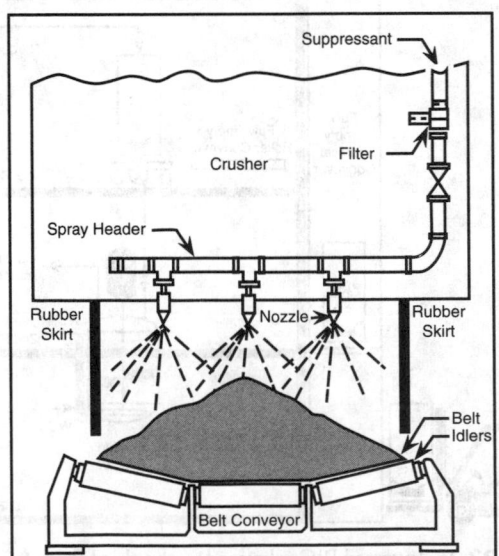

FIG. 5.23.2 Wet suppression system at a crusher discharge point. (Reprinted from C. Cowherd, Jr. and J.S. Kinsey, 1986, *Identification, assessment, and control of fugitive particulate emissions,* EPA-600/8-86-023, Research Triangle Park, N.C.: U.S. EPA.)

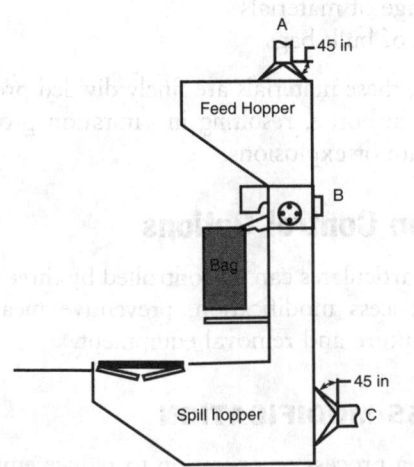

FIG. 5.23.3 A bag tube packer. (Reprinted, with permission, from Robert L. Opila, 1993, Carefully plan dust collection systems, *Chem. Eng. Prog.* [May].)

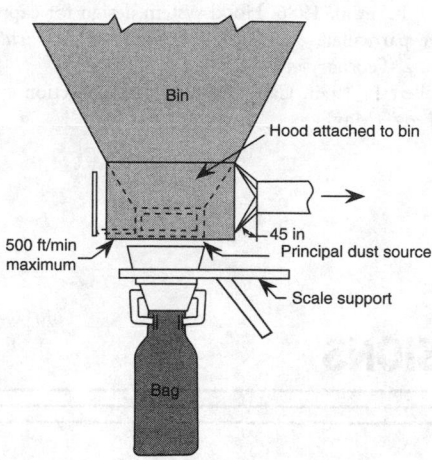

FIG. 5.23.4 Open-mouth bag packer. (Reprinted, with permission, from Opila, 1993.)

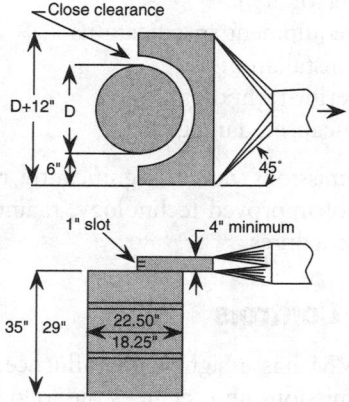

FIG. 5.23.5 Dust pickup for drum filling. (Reprinted, with permission, from Opila, 1993.)

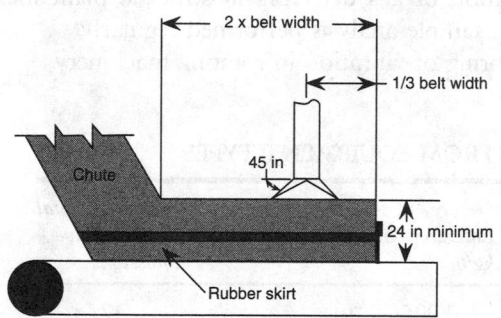

FIG. 5.23.6 Dust pickup for chute discharging onto a belt conveyor. (Reprinted, with permission, from Opila, 1993.)

A slot-type hood used with a ribbon mixer (see Figure 5.23.8)

Maintaining the conveying velocity throughout the collection system is important. If conveying velocities are too low, saltation (deposition of particles within a duct) occurs until the duct is reduced to a point where conveying velocity is obtained. This material buildup can be trou-

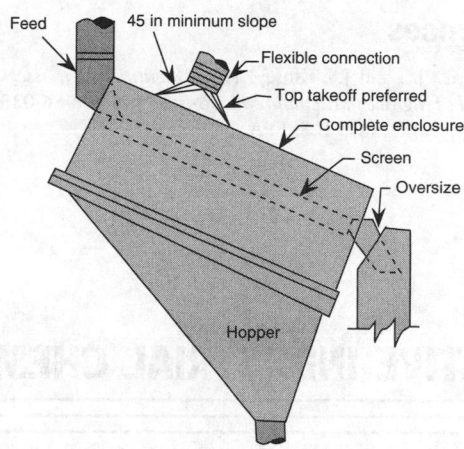

FIG. 5.23.7 Dust pickup on a flatdeck screen. (Reprinted, with permission, from Opila, 1993.)

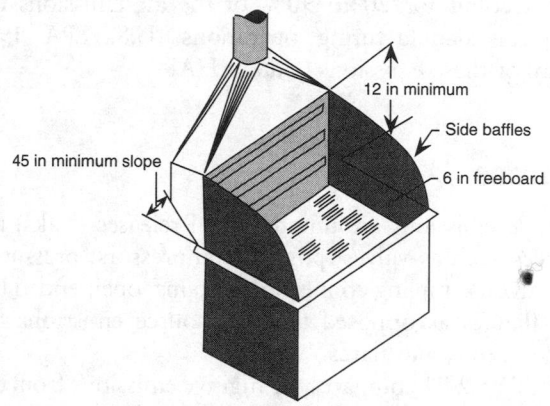

FIG. 5.23.8 Slot type hood mounted on a ribbon mixer. (Reprinted, with permission, from Opila, 1993.)

TABLE 5.23.1 MINIMUM AIR VELOCITIES IN DUCTS TO PREVENT DUST SETTLING

Type of Dust	Velocity, ft/min
Low density (gases, vapors, smoke, flour, lint)	2000
Medium-low density (grain, sawdust, plastic, rubber)	3000
Medium-high density (cement, sandblast, grinding)	4000
High-density (metal turnings, lead dust)	5000

Source: J.A. Danielson, ed., 1973, *Air pollution engineering manual,* 2d ed., AP–40. (Washington, D.C.: U.S. EPA).

blesome for a sanitary plant operation. Table 5.23.1 provides guidelines for choosing a velocity. In addition, dry and wet cleaning of the dust collection system must be provided.

—*David H.F. Liu*

References

Cowherd, C., Jr., and J.S. Kinsey. 1986. *Identification, assessment, and control of fugitive particulate emissions.* EPA-600/8-86-023, Research Triangle Park, N.C.: U.S. EPA.

Kashdan, E.R., et al. 1986. Hood system design for capture of process fugitive particulate emissions. *Heating/Piping/Particulate Emissions* 58, no. 2 (February):47–54.

Opila, Robert L. 1993. Carefully plan dust collection systems. *Chem. Eng. Prog.* (May).

5.24
FUGITIVE INDUSTRIAL CHEMICAL EMISSIONS

This section is concerned with unintentional equipment leaks of VOCs from industrial plants. The quantity of these fugitive emissions is hard to measure, but in some cases they account for 70 to 90% of the air emissions from chemical manufacturing operations (U.S. EPA 1984). Many of these emissions contain HAPs.

Sources

Fugitive emissions are unintentional releases (leaks) from sources such as valves, pumps, compressors, pressure-relief valves, sampling connection systems, open-ended lines, and flanges as opposed to point-source emissions from stacks, vents, and flares.

Table 5.24.1 compares the fugitive emissions from these sources in terms of source emissions factors and total source contributions of VOCs. A count of all equipment multiplied by the emission factors estimates the total emissions.

Fugitive emissions do not occur as part of normal plant operations but result from the following:

Malfunctions
Wear and tear
Lack of proper maintenance
Operator error
Improper equipment specifications
Improper installation
Use of inferior technology
Externally caused damage

Fugitive emissions can be significantly reduced with the adoption of improved technology, maintenance, and operating procedures.

Source Controls

Good O&M has a significant influence on lowering the fugitive emissions and includes the following:

Daily inspection for leaks by plant personnel
Immediate leak repair
Installation of gas detectors in strategic plant locations, with sample analysis performed regularly
Monitoring of vibration in rotating machinery

TABLE 5.24.1   COMPARISON OF FUGITIVE EMISSIONS FROM EQUIPMENT TYPES

Equipment Type	Process Fluid	Emission Factor, kg/hr	Percent of Total VOC Fugitive Emissions
Valve	Gas or vapor	0.0056 ⎤	47
	Light liquid	0.0071 ⎬	
	Heavy liquid	0.00023 ⎦	
Pump	Light liquid	0.0494 ⎤	16
	Heavy liquid	0.0214 ⎦	
Compressor	—	0.228	4
Pressure-relief valve	Gas or vapor	0.1040	9
Sampling connection	—	0.0150	3
Open-ended line	—	0.0017	6
Flange	—	0.00083	15
			100

Source: U.S. Environmental Protection Agency (EPA), 1982, *Fugitive emission sources of organic compounds,* EPA 450/3-80/010, Research Triangle Park, N.C.

Minimization of pipe and connector stresses caused by the vibration of pumps and compressors

Inspection and testing of relief-valves and rupture disks for leaks

Reduction in the number and volume of samples collected for control purposes

Inspection and periodic replacement of seals and gaskets

Because the emission control of equipment-related sources is unit-specific, this section discusses the control techniques by source category.

VALVES

Except for check and pressure-relief valves, industrial valves need a stem to operate. An inadequately sealed stem is a source of fugitive gas emissions. Valves with emissions greater than 500 ppmv are considered *leakers*. The 500-ppmv-limit threshold can place excessive demands on many low-end valves. Figure 5.24.1 shows the primary maintenance points for a packed, stemmed valve. The cost of keeping these valves in compliance can be expensive over time compared to another valve of greater initial cost.

The following special valves are designed to control fugitive emissions:

Bellows-type stems for both rising stem and quarter-turn valves show almost zero leakage and require no maintenance during their service life (Gumstrup 1992). However, because bellows seals are more costly than packed seals, they are typically used in lethal and hazardous service. Figure 5.24.2 shows a typical design of a bellows-sealed valve.

Diaphragm valves and magnetically actuated, hermetically sealed control valves are two other valves least prone

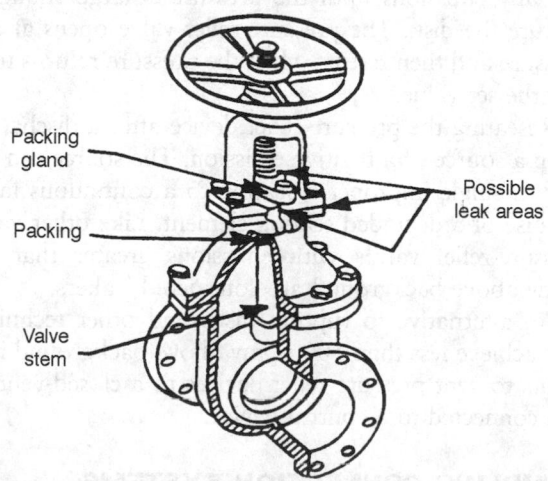

FIG. 5.24.1 Primary maintenance points for a valve stem. (Reprinted from U.S. Environmental Protection Agency, 1984, *Fugitive VOC emissions in the synthetic organic chemicals manufacturing industry,* EPA 625/10-84/004, Research Triangle Park, N.C.: U.S. EPA, Office of Air Quality and Standards.)

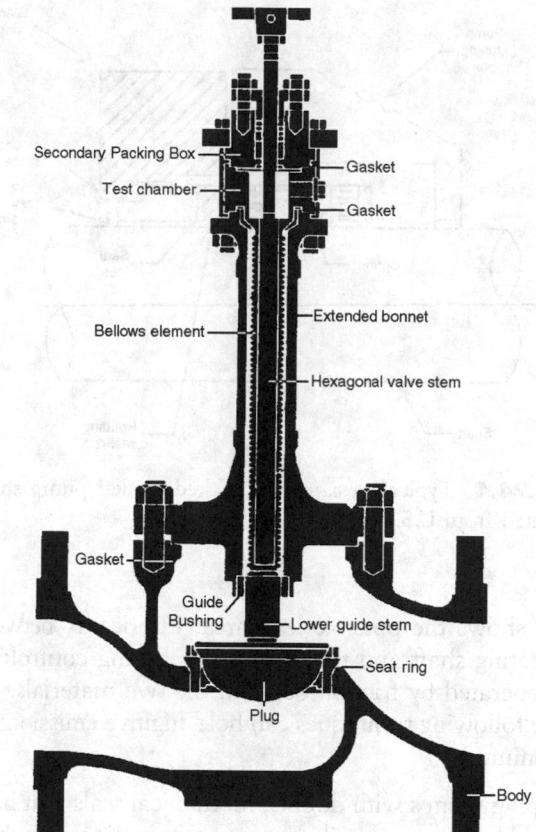

FIG. 5.24.2 Typical design of a bellows-sealed valve.

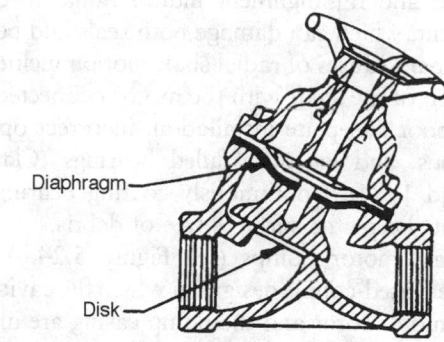

FIG. 5.24.3 Typical design of diaphragm valves. (Reprinted from U.S. EPA, 1984.)

to leaking. Figure 5.24.3 shows a typical diaphragm valve.

For services not requiring zero leakage on rising and rotary valve stem seals, a new group of improved packed seals is available. Most new seals claim to reduce fugitive emissions to a maximum leakage of less than 100 ppmv (Ritz 1993). Gardner (1991) discusses other types of valves for emission control.

PUMPS

The failure of a sealant where a moving shaft meets a stationary casting is a source of fugitive gas emissions. Figure

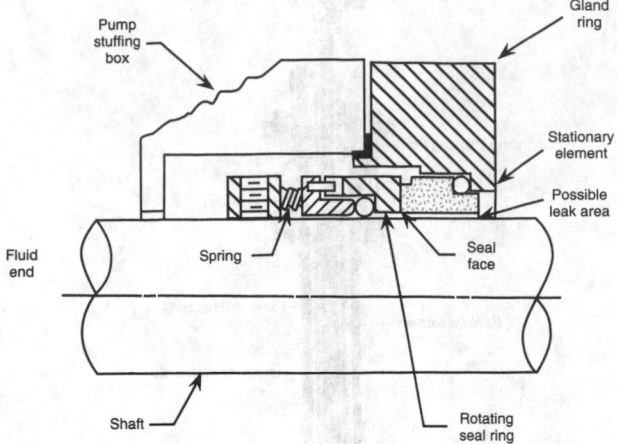

FIG. 5.24.4 Typical design of a packed, sealed pump shaft. (Reprinted from U.S. EPA, 1984.)

5.24.4 shows the possible leak area. Lubricants between the rotating shaft and the stationary packing control the heat generated by friction between the two materials.

The following techniques can hold fugitive emissions to a minimum:

Equipping pumps with double, mechanical seals that have liquid buffer zones and alarms or automatic pump shut-offs for seal failures (see Figure 5.24.5). The effective use of mechanical seals requires closed tolerances. Shaft vibration and misalignment induce radial forces and movement, which can damage both seals and bearings. The primary causes of radial shaft motion include poor alignment of the shaft with the motor connected to the pump, poor baseplate installation, incorrect operating conditions, and loose or failed bearings (Clark and Littlefield 1994). Continuously coating bearings with lubricants keeps the surfaces free of debris.

Using canned-motor pumps (see Figure 5.24.6). These units are closed-couple designs in which the cavity housing the motor rotor and the pump casing are interconnected. As a result, the pump bearings run in the process liquid, and all seals are eliminated.

Adopting diaphragm and magnetic-drive pumps. These two pumps do not require a sealant to control leakage.

Adams (1992) discusses the selection and operation of mechanical seals.

COMPRESSORS

The standard for compressors requires (1) using mechanical seals equipped with a barrier fluid system and control degassing vents or (2) enclosing the compressor seal area and venting emissions through a closed-vent system to a control device. These systems provide control efficiencies approaching 100% (Colyer and Mayer 1991).

Most concepts to control fugitive emissions from pumps

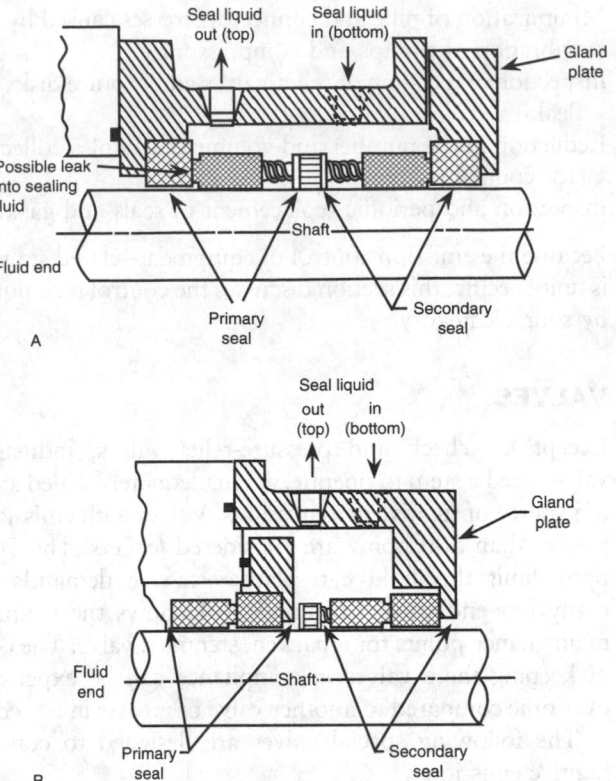

FIG. 5.24.5 Two arrangements of dual mechanical pump seals. (A) Back-to-back arrangement; (B) tandem arrangement (Reprinted from U.S. EPA, 1984).

apply to compressors. Likewise, analyzing the vibration characteristics anticipates pending problems.

PRESSURE-RELIEF DEVICES

Figure 5.24.7 shows a pressure-relief valve and rupture device to control fugitive emissions. This device does not allow any emissions until the pressure is large enough to rupture the disk. The pressure-relief valve opens at a set pressure and then reseats when the pressure returns to below the set value.

Reseating the pressure-relief device after a discharge is often a source of a fugitive emission. The source can vary from a single improper reseating to a continuous failure because of a degraded seating element. Like other valves, pressure-relief valves with emissions greater than 500 ppmv above background are considered leakers.

An alternative to rupture disks and other techniques that achieve less than 500 ppmv above background is for plants to vent pressure-relief devices to a closed-vent system connected to a control device.

SAMPLING CONNECTION SYSTEMS

MACT consists of closed-purge sampling, closed-loop sampling, and closed-vent vacuum systems in the rules for sampling connections. These systems are described as follows:

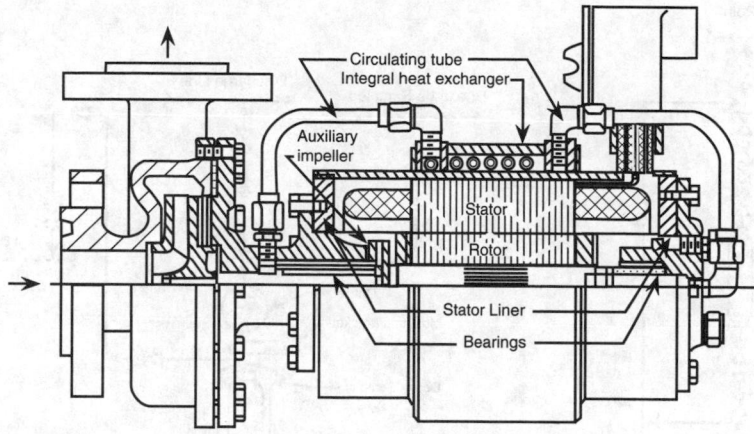

FIG. 5.24.6 Chempump canned-motor pump.

A closed-purge sampling system eliminates emissions due to purging either by returning the purge material directly to the process or by collecting the purge in a system that is not open to the atmosphere for recycling or disposal.

A closed-loop sampling system also eliminates emissions due to purging by returning process fluid to the process through an enclosed system that is not directly vented to the atmosphere.

A closed-vent system captures and transports the purged process fluid to a control device.

Figure 5.24.8 shows two such systems.

OPEN-ENDED LINES

Enclosing the open end of a valve or line with a cap, plug, or a second valve eliminates emissions except when the line is used for draining, venting, or sampling operations. The control efficiency associated with these techniques is approximately 100% (Colyer and Mayer 1991).

FLANGES AND CONNECTORS

Flanges are significant sources of fugitive emissions, even at well-controlled plants, due to the large number of flanges and connectors. In most cases, tightening the flange bolts on the flanged connectors, replacing a gasket, or correcting faulty alignment of a surface eliminates a leak. The use of all welded construction minimizes the number of flange joints and screwed connections.

Unsafe-to-monitor connectors can expose personnel to imminent hazards from temperature, pressure, or explosive conditions. During safe-to-monitor periods, plant personnel should monitor these connectors, especially the critical ones, as frequently as possible using leak detectors.

AGITATORS

Limited screening data indicate that agitators are a significant source of emissions. Agitators are technologically similar to pumps so emissions are controlled using seal technology. However, agitators have longer and larger diameter shafts than pumps and produce greater tangential loading. Therefore, the performance of pump seal systems cannot estimate agitator seal performance. A leak from an

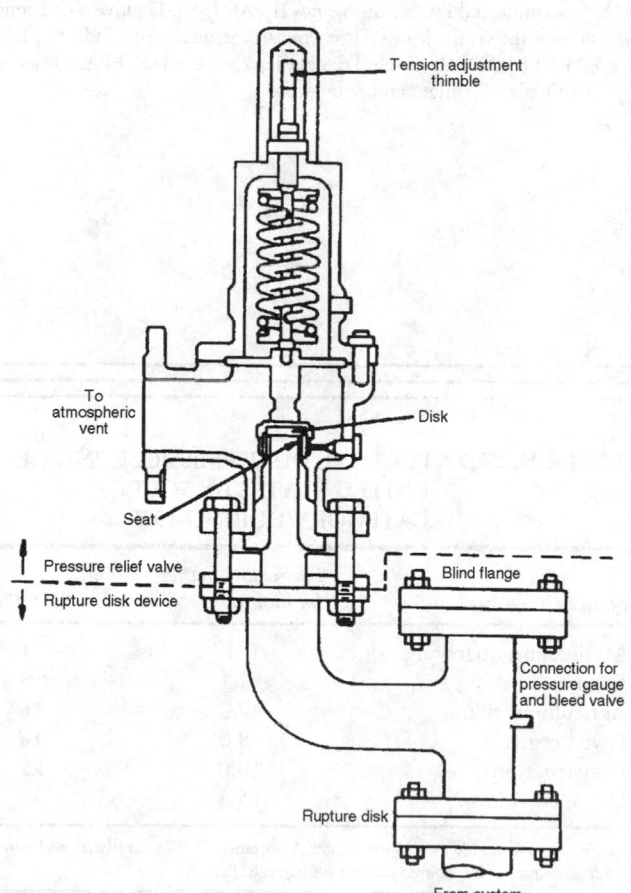

FIG. 5.24.7 Pressure-relief valve mounted on a rupture disk device.

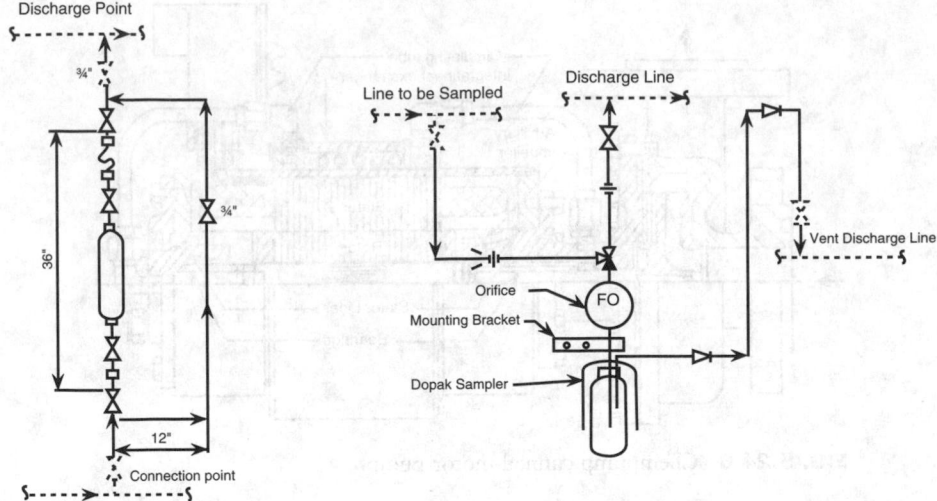

FIG. 5.24.8 Two arrangements for sampling systems.

agitator is defined as a concentration of 10,000 ppmv or greater.

—*David H.F. Liu*

References

Adams, W.V. 1992. Controlling fugitive emissions from mechanical seals. *Hydrocarbon Processing* (March):99–104.

Clark, E., and D. Littlefield. 1994. Maximize centrifugal pump reliability. *Chemical Engineering* (February).

Colyer, R.S., and J. Mayer. 1991. Understanding the regulations governing equipment leaks. *Chem. Eng Prog.* (August).

Gardner, J.F. 1991. Selecting valves for reduced emissions. *Hydrocarbon Processing* (August).

Gumstrup, B. 1992. Bellows seal valves. *Chemical Engineering* (April).

Ritz, G. 1993. Advances in control valve technology. *Control* (March).

U.S. Environmental Protection Agency (EPA). 1984. Fugitive VOC emissions in the synthetic organic chemicals manufacturing industry. EPA 625/10-84/004. Research Triangle Park, N.C.: U.S. EPA, Office of Air Quality Planning and Standards.

5.25
FUGITIVE DUST

Fugitive dust that deposits large particles on buildings, vehicles, and materials is a nuisance. Inhalation of dust can be debilitating, especially when the dust contains toxic elements and minerals. In addition, fugitive dust reduces visibility.

Fugitive dust consists of geological material suspended into the atmosphere by wind action and by human activities. Most of this dust soon deposits within a short distance of its origin, yet a portion of it can be carried many miles by atmospheric winds.

Sources

Table 5.25.1 shows the approximate emissions of fugitive dust from manmade sources in the United States for

TABLE 5.25.1 FUGITIVE DUST EMISSIONS IN THE UNITED STATES BY SOURCE CATEGORY DURING 1990

Source Category	Particulate matter (million tn/yr)	Total (%)
Mining and quarrying	0.37	1
Erosion	4.1	9
Agricultural tilling	7.0	16
Paved roads	8.0	18
Construction	10.0	22
Unpaved roads	15.5	34

Source: Air and Waste Management Association, 1992, *Standards and nontraditional particulate source control* (Pittsburgh, Pa.).

TABLE 5.25.2 CONTROL TECHNIQUES FOR VARIOUS SOURCES

Fugitive Emission Source	Chemical Stabilizers	Vegetative Cover	Watering	Windscreens	Wind Barriers/Berms	Plantings	Pile Shaping and Orientation	Paving and Gravel	Sweeping and Cleaning	Reduced Speed	Curbing and Stabilizing Shoulders	Operations Change	Reduced Drop Distance	Water Sprays and Foggers	Electrostatic Curtains	Partial or Complete Enclosure	Hooding and Ducting	Covers	Wheel Washes	Foams
Paved roads			X	X	X	X			X	X	X									
Unpaved roads	X		X	X	X	X		X		X	X									
Unpaved parking lots	X		X	X	X	X		X		X	X									
Active storage piles			X	X	X	X	X					X				X		X		
Inactive storage piles	X	X	X	X	X	X	X					X				X		X		
Exposed areas	X	X	X	X	X	X		X												
Construction sites			X	X	X			X				X				X				
Conveyor transfer				X		X						X	X	X	X	X	X			X
Drop points				X								X	X	X	X	X	X			X
Loading and unloading				X	X							X	X	X	X	X	X			X
Vehicle carryout								X	X										X	
Truck and rail spills								X	X	X								X		
Crushing and screening			X	X	X							X	X	X	X	X	X	X		
Waste sites	X	X		X		X		X				X								
Tilling operations			X									X								
Feed lots	X	X	X									X								

Source: Adapted from E.T. Brookman and D.J. Martin, 1981, A technical approach for the determination of fugitive emission source strength and control requirements, *74th Annual APCA Meeting, Philadelphia, June 1981.*

1990. These estimates are derived from standardized U.S. EPA emission factors that relate soil characteristics, meteorological conditions, and surface activities to emission rates. These estimates do not include all potential emitters (for example, natural dust emissions, such as wind devils, are omitted as are many from industrial activities). Mass emissions are restricted to particles that are less than 10 μm.

Prevention and Controls

Table 5.25.2 is a partial listing of control techniques for reducing dust emissions from various sources. Table 5.25.3 presents an approximate categorization of control system capabilities. When more than one control technique can be used in series, the combined efficiency is estimated by a series function.

Partially or completely closing off the source from the atmosphere to prevent wind from entraining the dust can effectively reduce emissions. Frequently, confinement plus suppression is an optimum combination for effective, economical dust control.

WIND CONTROL

Preventing the wind from entraining dust particles is accomplished by keeping the wind from blowing over materials. This prevention involves confinement or wind control. Tarps that cover piles are a wind control procedure for storage piles, truck, railcars, and other sources (see Figure 5.25.1).

A variation of the source enclosure method to control fugitive dust emissions involves applying wind fences (also referred to as windscreens). Porous wind fences significantly reduce emissions from active storage piles and exposed ground areas. The principle of a windscreen is to provide a sheltered region behind the fenceline where the mechanical turbulence generated by ambient winds is significantly reduced. The downwind extent of the protected region is many times the physical height of the fence. This sheltered region reduces the wind erosion potential of the exposed surface in addition to allowing gravitational settling of larger particles already airborne. Figure 5.25.2 is a diagram of a portable windscreen used at a coal-fired power plant.

For unpaved routes and ditches, special fabrics placed over exposed surfaces prevent dust from becoming air-

TABLE 5.25.3 APPROXIMATE CATEGORIZATION OF CONTROL SYSTEM CAPABILITIES

	Type of Control system							
	RACT			BACT			LAER	
Source	Control	Efficiency, %		Control	Efficiency, %		Control	Efficiency, %
Unpaved roads	Wetting agent (water)	50		Wetting agent (other than water)	60–80		Paving and sweeping	85–90
	Speed control	25–35		Drastic speed control	65–80			
				Soil stabilization	50			
				Apply gravel	50			
				Road carpet	80			
Active storage piles	Wetting agent (water)	50–75		Wetting agents (other than water)	70–90		Encrusting agents	90–100
	Pile orientation	50–70		Pile orientation	50–70		Tarp cover	100
	Leading slope angle	35		Wind screens	60–80			
Inactive storage piles	Vegetation	65		Chemical stabilization plus vegetation	80–90		Tarp cover	100
Transfer points	Water sprays	35		Wetting agent sprays	55		Enclosure with sprays	90–100
				Fogging sprays	80		Electrostatic-enhanced fogging sprays (EEFS)	80–95
Conveyors	Water sprays	35		Wetting agent sprays	55		Enclosure with sprays	90–100
				Fogging sprays	80		EEFS	80–95
Car dumpers	Water sprays	35		Wetting agent sprays	40		Enclosure with sprays	85–90
				Fogging sprays	75		EEFS	75–90
Construction activities	Watering	50		Chemical stabilization	80		Enclosure	90

Source: H.E. Hesketh and F.L. Cross, Jr., 1983, *Fugitive emission and controls* (Ann Arbor, Mich.: Ann Arbor Science).
RACT-Reasonable available control technology
BACT-Best available control technology
LAER-Lowest available emission rate

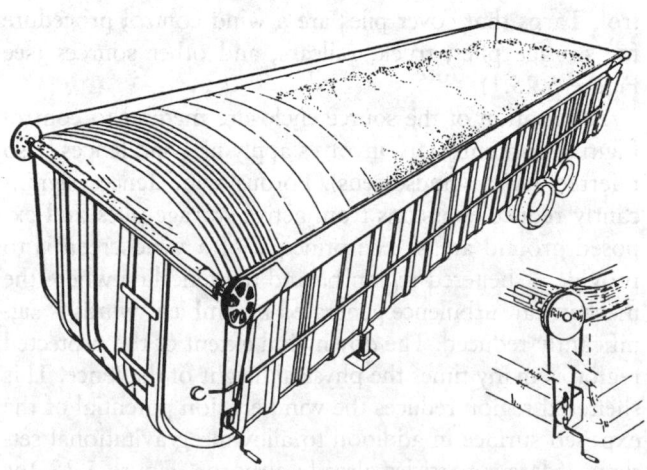

FIG. 5.25.1 Complete truck tarp; tarp movement and hold-down system. (Reprinted, with permission, from Aero Industries.)

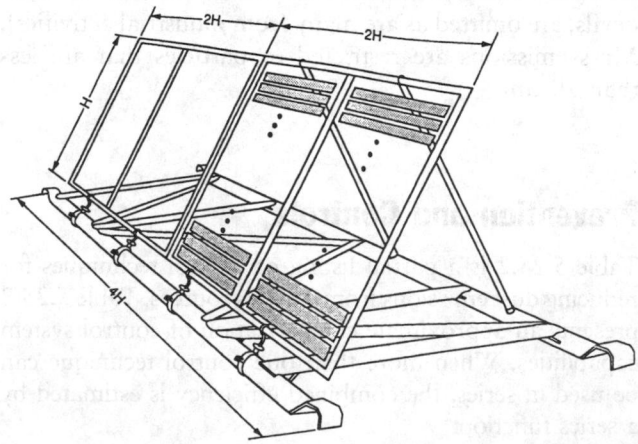

FIG. 5.25.2 Diagram of a portable windscreen. (Reprinted, with permission, from C. Cowherd, Jr. and J.S. Kinsey, 1986, *Identification, assessment, and control of fugitive particulate emissions*, EPA 600/8-86-023, Research Triangle Park, N.C.: U.S. EPA.)

TABLE 5.25.4 CLASSIFICATION OF TESTED CHEMICAL SUPPRESSANTS

Dust Suppressant Category	Trade Name
Salts	Peladow
	LiquiDow
	Dustgard
	Oil well brine
Lignosulfonates	Lignosite
	Trex
Surfactants	Biocat
Petroleum-based	Petro Tac
	Coherex
	Arco 2200
	Arco 2400
	Generic 2 (QS)
Mixtures	Arcote 220/Flambinder
	Soil Sement

Source: U.S. Environmental Protection Agency (EPA), 1987, *Emission control technologies and emission factors for unpaved road fugitive emission,* EPA 625/5-87/002 (Cincinnati: U.S. EPA, Center for Environmental Research Information).

borne from wind or machinery action. These fabrics are referred to as *road carpets,* and alternatively they prevent erosion by rain runoff.

WET SUPPRESSION

Wet suppression systems apply either water, a water solution of a chemical agent, or a micron-sized foam to the surface of a particulate-generating material. This measure prevents (or suppresses) the fine particles in the material from leaving the surface and becoming airborne.

The chemical agents used in wet suppression systems are either surfactants or foaming agents for materials handling and processing operations or various dust palliatives applied to unpaved roads. In either case, the chemical agent agglomerates and binds the fines to the aggregate surface, thus eliminating or reducing its emission potential.

In agricultural fields, electrostatic foggers provide electrostatically charged water droplets that agglomerate suspended particles, thereby increasing the particle size and deposition velocity.

VEGETATIVE COVER

Vegetation reduces wind velocity at the surface and binds soil particles to the surfaces. Of course, an area with vegetative cover should not be disturbed after planting (U.S. EPA 1977). Likewise, tilling implements, orientations, and frequencies should aim to limit the suspension of surface dust in agricultural fields.

TABLE 5.25.5 IMPLEMENTATION ALTERNATIVES FOR CHEMICAL STABILIZATION OF AN UNPAVED ROAD

Cost Elements	Implementation Alternatives
Purchase and ship chemical	Ship in railcar tanker (11,000–22,000 gal/tanker)
	Ship in truck tanker (4000–6000 gal/tanker)
	Ship in drums via truck (55 gal/drum)
Store chemical	Store on plant property
	In new storage tank
	In existing storage tank
	Needs refurbishing
	Needs no refurbishing
	In railcar tanker
	Own railcar
	Pay demurrage
	In truck tanker
	Own truck
	Pay demurrage
	In drums
	Store in contractor tanks
Prepare road	Use plant-owned grader to minimize ruts and low spots
	Rent contractor grader
	Perform no road preparation
Mix chemical and water in application truck	Put chemical in spray truck
	Pump chemical from storage tank or drums into application truck
	Pour chemical from drums into application truck, generally using forklift
	Put water in application truck
	Pump from river or lake
	Take from city water line
Apply chemical solution via surface spraying	Use plant-owned application truck
	Rent contractor application truck

Source: U.S. EPA, 1987.

CHEMICAL STABILIZATION

Dust from unpaved roads or uncovered areas can be reduced or prevented by chemical stabilization. Chemical suppressants can be categorized as salts, lignin sulfonates, wetting agents, petroleum derivatives, and special mixtures. Manufacturers generally provide information for typical applications, dilution and application rates, and costs. Table 5.25.4 is a partial list of chemical stabilizers.

Table 5.25.5 lists alternatives for implementing chemical stabilization of an unpaved road.

—*David H.F. Liu*

Reference

U.S. Environmental Protection Agency (EPA). 1977. *Guideline for development of control strategies in areas with fugitive dust problems.* EPA 450/2-77/029. Washington, D.C.: U.S. EPA.

Odor Control

5.26
PERCEPTION, EFFECT, AND CHARACTERIZATION

Odor Terminology

An odor is a sensation produced by chemical stimulation of the chemoreceptors in the olfactory epithelium in the nose. The chemicals that stimulate the olfactory sense are called *odorants* although people frequently refer to them as *odors*.

Several dimensions of human responses to the odor sensation can be scientifically characterized. These dimensions are threshold, intensity, character, and hedonic tone.

THRESHOLD

Threshold, or detectability, refers to the theoretical minimum concentration of odorant stimulus necessary for perception in a specified percentage of the population, usually the mean. A threshold value is not a fixed physiological fact or a physical constant but is a statistical point representing the best estimate from a group of individual scores. Two types of thresholds can be evaluated: detection and recognition.

The *detection threshold* is the lowest concentration of odorant that elicits a sensory response in the olfactory receptors of a specified percentage, usually 50%, of the population being tested (see Figure 5.26.1). The detection threshold is the awareness of an odor without necessarily recognizing it.

The *recognition threshold* is the minimum concentration recognized as having a characteristic odor quality by a specified percentage of the population, usually 50%. It differs from the detection threshold because it is the point at which people can describe a specific odor character to the sensory response.

In the measurement of environmental odors, which are generally complex mixtures of compounds, the threshold is not expressed as a concentration level. Instead, the threshold is expressed as dilution-to-threshold ratios (D/T); it is dimensionless. A D/T ratio of 100 means that a given volume (i.e., 1 cu ft) of odorous air requires 100 volumes (i.e., 100 cu ft) of odor-free air to dilute it to threshold, or a barely detectable odor. This concept is shown in Figure

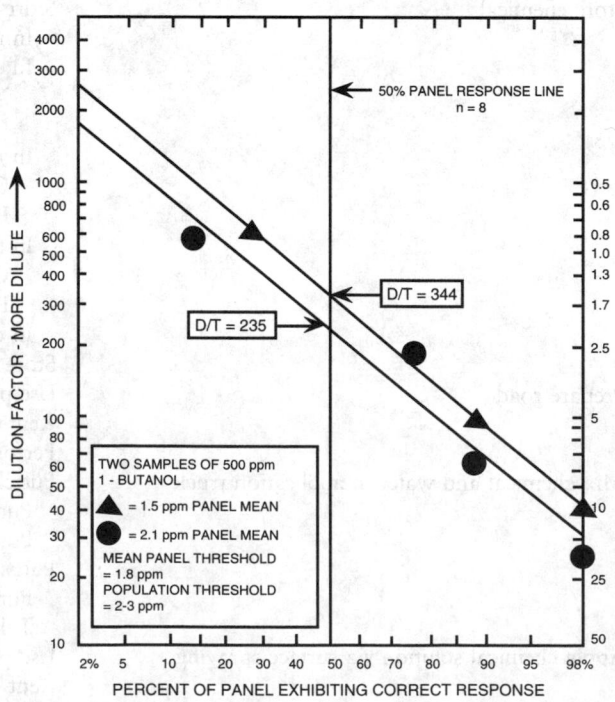

FIG. 5.26.1 Odor threshold ratio.

5.26.1 where a 50% panel response occurs at 235 D/T for one sample and 344 D/T for the other sample.

Over the years, many terms have been used to express the concentration of odor including the following:

ODOR UNIT—one volume of odorous air at the odor threshold; often the volume is defined in terms of cubic feet as follows:

$$\frac{\text{odor units}}{\text{cu ft}} = \frac{\text{volume of sample diluted to threshold (cu ft)}}{\text{original volume of sample (cu ft)}}$$

5.26(1)

ODORANT QUOTIENT—Expressed by the following equation. The Z is for Zwaardermaker who was the earliest investigator to use dilution ratios for odor measurement (ASTM 1993).

$$Z_t = \frac{C_o}{C_t} = \frac{\text{odorant concentration of a sample}}{\text{odorant concentration at threshold}}$$

5.26(2)

Both odor units/cu ft and Z_t are dimensionless and are synonyms of the D/T; the odor unit is a unit of volume and is not a synonym of D/T (ASTM 1993; Turk 1973).

Another term, the *odor emission rate,* is often used to describe the severeness of the downwind impact, i.e., the problem a typical odor source creates at downwind locations. The odor emission rate, expressed in unit volume per minute, is the product of the D/T value of the odorous air and the air flowrate (cfm or its equivalent).

Listings of threshold values for pure chemicals are common in many environmental resource references. However, the reported threshold values are limited and vary as much as several orders of magnitude, suggesting that thresholds are limited in usefulness. However, a review of threshold value methodology finds that when threshold values are subjected to basic methodological scrutiny, the range of experimentally acceptable values is considerably reduced. Two recent documents of odor threshold reference guides are available. One published by the American Industrial Hygiene Association (1989) is based on a review for 183 chemicals with occupational health standards. The other document is published by the U.S. EPA (1992). It compiles odor thresholds for the 189 chemicals listed in the CAAAs.

INTENSITY

Odor intensity refers to the perceived strength of the odor sensation. Intensity increases as a function of concentration. The relationship of the perceived strength (intensity) and concentration is expressed by Stevens (1961) as a psychophysical power function as follows:

$$S = K I^n$$

5.26(3)

where:

S = perceived intensity of the odor sensation
I = physical intensity of stimulus (odorant concentration)
n = constant
K = constant

This equation can be expressed in logarithm as follows:

$$\log S = \log K + n \log I$$

5.26(4)

where:

n = slope
K = y-intercept

Figure 5.26.2 shows an intensity function on logarithm coordinates of the standard odorant 1-butanol. The slope of the function, also called the *dose-response function,* varies with the type of odorant. Odor pollution control is concerned with the dose-response function, or the degree of dilution necessary to decrease the intensity. This function can be described by the slope. A low slope value indicates that the odor requires greater dilution for it to dissipate; a high slope value indicates that the intensity can be reduced by dilution more quickly. Compounds with low slope values are hydrogen sulfide, butyl acetate, and amines. Compounds with high slope values are ammonia and aldehydes. This function explains why hydrogen sulfide, butyl acetate, and amines can be detected far away from the odor origin. On the other hand, ammonia and aldehydes cause odor problems at locations near the origin.

Category Scales

Category scales were the first technique developed to measure odor intensity. One widely used scale was developed for a 1930 study of odor used as gas alarms. The scale has the following six simple categories:

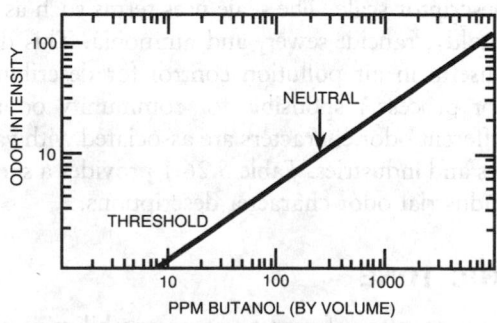

$$S = kI^n \text{ on log–log coordinates}$$

S = PERCEIVED INTENSITY
k = Y–INTERCEPT
I = ODORANT CONCENTRATION
n = SLOPE OF PSYCHOPHYSICAL FUNCTION

FIG. 5.26.2 Butanol reference scale.

0 No odor
1 Very faint odor
2 Faint odor
3 Easily noticeable odor
4 Strong odor
5 Very strong odor

The limitations of category scaling are that the number of categories to choose from is finite and they are open to bias through subjective number preferences or aversions.

Magnitude Estimation

In magnitude estimation, observers create their own scales based on a specific reference point. The data from several observers is then normalized to the reference point. Magnitude estimation is a form of ratio scaling and has many advantages over category scaling (i.e., an unlimited range and greater sensitivity); however, it does require a more sophisticated observer and statistical analysis.

Reference Scales

The reference scale for measuring odor intensity is standardized as ASTM Standard Practice E 544 (ASTM 1988). The method uses a standard reference odorant, 1-butanol, set in a series of known concentrations. The advantages of this method are that it 1) allows the comparison of subjective odor intensities between laboratories, 2) allows for odor control regulations to be expressed in terms of perceived intensity rather than odor thresholds, and 3) allows cross modality comparisons (i.e., sound and odor). One disadvantage is that some people find it difficult to compare odors that have a different odor character than the reference standard.

CHARACTER

Another dimension of odor is its *character,* or what the odor smells like. ASTM publication DS 61 (Dravnieks 1985) presents character profiles for 180 chemicals using a 146 descriptor scale. The scale uses terms such as fishy, nutty, moldy, rancid, sewer, and ammonia. This dimension is useful in air pollution control for describing the source or process responsible for community odors because different odor characters are associated with various processes and industries. Table 5.26.1 provides a short list of the industrial odor character descriptions.

HEDONIC TONE

Hedonic tone, also referred to as acceptability, is a category judgment of the relative pleasantness or unpleasantness of an odor. In the context of air pollution field work, the hedonic tone is often irrelevant. Perception of an odor is based on the combination of frequency of occurrence, odor character, and odor intensity. Even pleasant odors can become objectionable if they persist long enough.

TABLE 5.26.1 INDUSTRIAL ODOR DESCRIPTORS

Odor Character Descriptor	Potential Sources
Nail polish	Painting, varnishing, coating
Fishy	Fish operation, rendering, tanning
Asphalt	Asphalt plant
Plastic	Plastics plant
Damp earth	Sewerage
Garbage	Landfill, resource recovery facility
Weed killer	Pesticide, chemical manufacturer
Gasoline	Refinery
Airplane glue	Chemical manufacturer
Household gas	Gas leak
Rotten egg	Sewerage, refinery
Rotten cabbage	Pulp mill, sewage sludge
Cat urine	Vegetation

Human Response to Odors and Odor Perception

Olfactory acuity in the population follows a normal bell curve distribution of sensitivity, ranging from hypersensitive to insensitive and anosmic (unable to smell). Individual odor threshold scores can be distributed around the mean value to several orders of magnitude.

Olfactory studies have revealed interesting information regarding odor perception. Two sensory channels are responsible for human detection of inhaled chemical substances in the environment. These are *odor perception* and the *common chemical sense* (CCS). Odor perception is a function of cranial nerve I, whereas the CCS is a function of cranial nerve V. The CCS is described as pungency and associated sensations such as irritation, prickliness, burning, tingling, and stinging. Unlike the olfactory structure, the CCS lacks morphological receptor structures and is not restricted to the nose or oral cavity (Cometto-Muñiz and Cain 1991).

Human responses to odor perception follow patterns associated with both the olfactory and the CCS functions. These are discussed next to facilitate the understanding of what prompts odor complaints and the difficulties associated with odor identification and measurement.

SENSITIZATION, DESENSITIZATION, AND TOLERANCE OF ODORS

Repeated exposure to an odor can result in either an enhanced reaction described as sensitization or a diminishing reaction defined as tolerance. When people become sensitized to an odor, the complaints regarding the odor can increase. On the other hand, tolerance of an odor can attribute to a person's unawareness of the continuous exposure to a potentially harmful substance. As an illustration of the complexity of odor perception, pungency stimulation shows an increased response to odors from a continuous or quickly repetitive stimulus during a short term exposure. Also, a person can be more sensitive to one odorant than another. Such differences are often caused by repeated exposure to an odor. For example, tolerance is not uncommon for chemists or manufacturers who are exposed to an odorant daily over a period of years. Fatigue also affects odor perception; repeated exposure to an odorant can result in a desensitization to the odorant, where an observer can no longer detect an odor although it is strongly detectable by another.

ODOR MIXTURES

An additional problem in odor identification is that ambient odors are generally mixtures of compounds in different concentration levels. A comparison of the perceived intensity of two chemicals presented alone at a concentration and the two chemicals in a mixture at the same concentrations usually shows that the perceived intensity of either chemical in the mixture is lower than it is alone. This response is known as hypoadditivity. On the other hand, the CCS responses show that the perception of the odor is mainly additive, i.e., equal in the mixture to the perception of either chemical alone (Cometto-Muñiz and Cain 1991).

Questions of safety with chemical mixtures are more successfully answered by experimentally determining the odor threshold of the mixture and then relating it to the chemical composition of the mixture. Some mixtures can contain a highly odorous but relatively nontoxic chemical together with a nonodorous but highly toxic chemical. The chemicals in a mixture can also disassociate through aging or different chemical processes, and the judgment of whether a toxic component is present can be incorrect if it is based solely on the detection of the odorous component. Assumptions about the relationship between odor and risk can only be made for specific cases for chemical mixtures.

OTHER FACTORS AFFECTING ODOR PERCEPTION

Human response to odor is also based on the community, meteorology, and topography interactions. An odor with an acceptable hedonic tone and low intensity can generate complaints when the odor character is unfamiliar, causing concern with the toxicity. Conversely, some high intensities of hedonically unpleasant odors are tolerated because they are considered socially acceptable. Even pleasant odors, such as perfumes or roasting coffee, are considered objectionable if they persist for a long time or are frequently present.

Demographically, awareness of air pollution problems increases with education, income and occupational level, and age. Some of this awareness is also based on economic factors, such as property values.

ODOR AND HEALTH EFFECTS

Odors can warn of potentially dangerous materials. The warning property of odor has long been recognized. In the Middle Ages, odors were held responsible for disease, rather than being the result of disease and poor hygiene. Physicians protected themselves by theoretically purifying the air in a sick room with perfumed water. In this century, the property of odor as a warning agent has been used on a worldwide scale through the addition of pungent odorants such as ethyl mercaptan to odorless natural gas. The perceived connection between odors and disease persists. With national attention focused on waste and chemical spills, in the absence of specific information to the contrary, the average person concludes that a bad smell is unhealthy.

The relationship between odor and health effects should be clarified. Many odorants are perceptible at concentrations far below harmful concentration levels. For example, hydrogen sulfide (H_2S) has been detected at concentrations as low as 0.15 ppb, whereas the acceptable exposure limit or the threshold limit value–time-weighted average (TLV–TWA) recommended by the American Conference of Governmental Industrial Hygienists (ACGIH) for this compound is 10 ppm, 6 orders of magnitude greater. Thus, the detection of the "rotten egg" character of H_2S is not necessarily an indicator of a potential health effect. Indeed, at concentrations near the TLV, the odor intensity of H_2S is unbearable. Similarly, the TLVs for creosol and ethyl acrylate are 4 orders of magnitude greater than their threshold values.

The nose is not a suitable screening device to determine the presence or the absence of a health risk. Although detecting an odor indicates that a chemical exposure has occurred, a more detailed sampling investigation should be conducted.

Studies that have reviewed community odor and health problems reveal that a variety of common ailments are related to chemical exposure. However, in most cases, the identified chemicals were well below the thresholds for toxicity. This evidence suggests that detecting unpleasant odors can cause adverse physiological and neurogenic responses such as nausea, stress, and low concentration levels and that these effects are a result of chemical exposure. Therefore, further studies are necessary to define allowable exposures and how they relate to odor detection.

By definition, chemicals that are hazardous to health are considered toxic whether odorous or not. They are therefore controlled under existing laws such as the Toxic Substances Act and the CAA and their related regulations.

—*Amos Turk*
Samuel S. Cha

References

American Industrial Hygiene Association (AIHA). 1989. *Odor Thresholds for chemicals with established occupational health standards*. AIHA.

American Society for Testing and Materials (ASTM). 1988. *Standard practice for referencing suprathreshold odor intensity.* E 544-75 (Reapproved 1988). ASTM.

———. 1993. *Standard terminology relating to sensory evaluation of materials and products.* E 253-93a. ASTM.

Cometto-Muñiz, J.E. and W.S. Cain. 1991. Influence of airborne contaminants on olfaction and the common chemical sense. Chap. 49 in *Smell and taste in health and disease,* edited by T.V. Getchell et al. New York: Raven Press.

Dravnieks, A. 1985. Atlas of odor character profiles. *ASTM Data Series 61,* ASTM.

Stevens, S.S. 1961. The psychophysics of sensory function. In *Sensory communications,* edited by Rosenblith. Cambridge, Mass.: MIT Press.

Turk, A. 1973. Expressions of gaseous concentration and dilution ratios. *Atmospheric Environment,* vol. 7:967–972.

U.S. Environmental Protection Agency (EPA). 1992. *Reference guide to odor thresholds for hazardous and pollutants listed in the Clean Air Amendments of 1990.* EPA 600/R-92/047.

5.27
ODOR CONTROL STRATEGY

Odor control is similar to any other air pollution control problem. Choosing a new material with less odor potential, changing the process, using an add-on pollution control system such as a scrubber or an afterburner, or raising the stack height all achieve community odor control. However, since odor is a community perception problem, the control strategies must be more flexible. When the tolerance level of a community changes, the odor control requirements also change. In addition, unique local topography and weather mean that each odor control problem usually requires its own study. An adopted control strategy for a similar problem at another place may not be suitable under different circumstances.

As far as control strategy is concerned, eliminating the odor-causing source is always better than using add-on control equipment because the equipment can become an odor source itself. For example, when a fume incinerator controls solvent odor, the incinerator, if not properly operated, can produce incomplete combustion by-products with a more offensive odor and lower odor threshold than the original odor. When multiple sources are involved, the control effort should be targeted to sources with higher odor emission rates, lower slopes of the dose-response function, and unpleasant odor characters.

Odor control techniques fall into the following categories:

Activated carbon adsorption
Adsorption with chemical reaction
Biofiltration
Wet scrubbing
Combustion
Dispersion

This section discusses the advantages and disadvantages of each technique. The details of each approach to pollution abatement are covered in other sections of this handbook.

Activated Carbon Adsorption

Activated carbon adsorption is a viable method in many odor control problems. Due to the nonpolar nature of its surface, activated carbon is effective in adsorbing organic and some inorganic materials. In general, organics having molecular weights over 45 and boiling points over 0°C are readily adsorbed.

The service life of activated carbon is limited by its capacity and the contaminating load. Therefore, provisions must be made for periodic renewal of the activated carbon. The renewal frequency is determined on the basis of performance deterioration (breakthrough) or according to a time schedule based on previous history or calculations of expected saturation. The exhausted carbon bed can be discarded or reactivated. Reactivation involves passing superheated steam through the carbon bed until sufficient material is desorbed. Recondensation of the material can recover it for reuse. This process is most suitable when only single compounds are involved. (Turk 1977).

Adsorption with Chemical Reaction

Impregnating filter media such as granular activated carbon or activated alumina with a reactive chemical or a catalyst can convert the adsorbed contaminants to less odorous compounds. For example, an air filter medium consisting of activated alumina and potassium permanganate can remove odorous hydrogen sulfide, mercaptans, and other sulfur contaminants. Another approach, patented by Turk and Brassey, can remove hydrogen sulfide from oxygen containing gas. The process involves adding ammonia to the air stream containing hydrogen sulfide prior to its passage through activated carbon. The ammonia catalyzes the oxidation of hydrogen sulfide to elemental sulfur. Impregnating activated carbon with sodium or potassium hydroxide can also neutralize hydrogen sulfide and form elemental sulfur. Both the ammonia/granular activated carbon system and the caustic/granular activated carbon system perform better than unimpregnated carbon.

Biofiltration

Biofiltration is an odor control technology that uses a biologically active filter bed to treat odorous chemical compounds. Materials such as soil, leaf compost, peat, and wood chips can be used for the filter bed. The filter bed provides an environment for microorganisms to degrade and ultimately remove the odorous chemical compounds.

Ideal operating requirements for a filter bed include high adsorption capacity, low pressure drop, high void fraction, high nutrient content, neutral to slightly alkaline pH, moderate temperatures, and adequate moisture content. Hydrogen sulfide, ammonia, and most organic components can be broken down in a biological filter, while most inorganic components cannot.

Wet Scrubbing

Wet scrubbing is another widely used technology for odor control. Most wet scrubbers for odor control employ reactive chemicals. Reactive scrubbing involves the removal of odorous materials by neutralization, oxidation, or other chemical reactions. Odor removal by any liquid–gas process is a function of the solubility of the chemical compound in the liquid phase, the total effective gas–liquid contact area, the concentration of the odorous chemical compound in the gas stream, and the residence time of the gas stream in the scrubber.

Atomized mist scrubbers and packed scrubbers are both capable of providing large gas–liquid contact areas for gas absorption. In a mist scrubber, pneumatic nozzles use a high-velocity compressed air stream to atomize a chemical solution into 5- to 20-μm-diameter drops, providing a large contact area. Unlike packed scrubbers, mist scrubbers do not clog even if the gas stream contains large particles. In a packed scrubber system, the chemical solution is often recirculated. As a result, the water and chemical consumption is significantly reduced.

For more complex odor problems caused by a mixture of odorants, multiple stages of scrubbers utilizing different chemicals in each stage are often used. Figure 5.27.1 shows a multiple-stage scrubber installed to control odorous emissions from a wastewater sludge treatment facility. The system uses sulfuric acid in the first-stage, cocurrent, coarse-packing, packed-tower scrubber to remove ammonia and particulates; a caustic/hypochlorite solution in the second-stage, cocurrent, horizontal-mist scrubber to remove or reduce sulfur compounds and VOCs; and a weak caustic solution in the third-stage, cross-flow, packed-bed scrubber to remove the remaining odorants, especially those chlorine compounds generated in the second-stage scrubber. The control system also uses a 50-ft-tall stack to enhance the dilution of the remaining odorants during plume dispersion (Ponte and Aiello 1993).

Combustion

Combustion is an effective technique for odor control. The odor control efficiency of a combustion system depends largely on the level of complete combustion. Incomplete combustion can actually increase an odor problem by forming more odorous chemical compounds. Three major factors influence the design principles for odor emission control: temperature, residence time, and mixing. All three factors are interrelated to each other, and variation in one can cause changes in the others. Generally, temperatures of 1200 to 1400°F and residence times of 0.3 to 0.5 sec with good mixing conditions effectively destroy the odorous chemical compounds in a combustion chamber. If moisture and corrosion are not of concern, facilities often use the boiler onsite for odorant destruction.

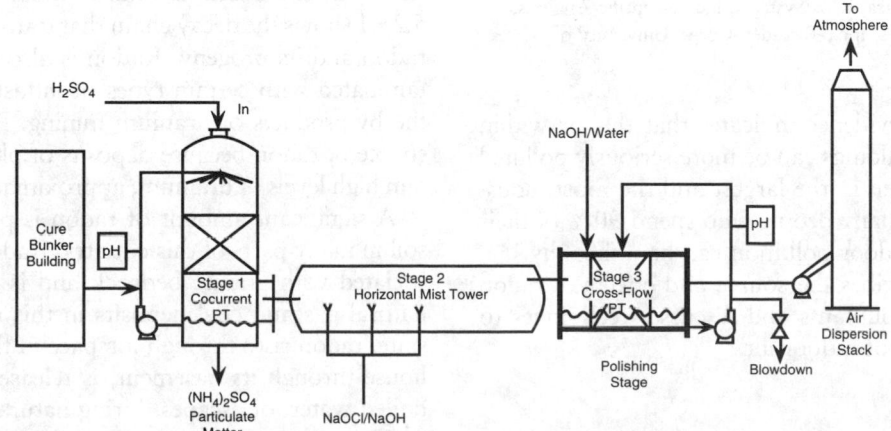

FIG. 5.27.1 Multiple-stage scrubber system schematic diagram.

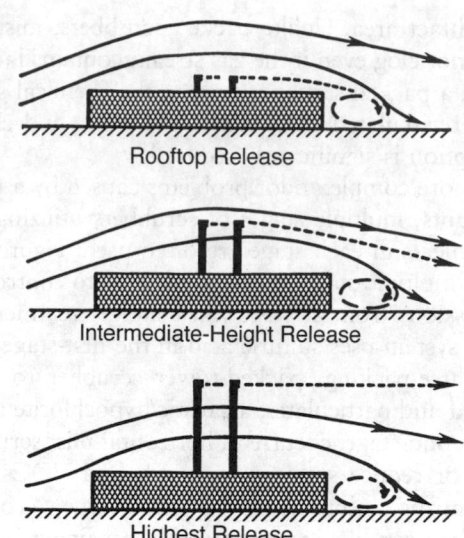

FIG. 5.27.2 Wake effect and stack height.

Dispersion

The use of elevated emission points can reduce the odorant concentration at downwind ground level because of dispersion or dilution. As shown in Figure 5.27.2, aerodynamic building wake effects often downwash emissions from rooftop sources. When the stack height is increased, emissions avoid the building wake, disperse at a higher level, and are diluted during transport before reaching the ground.

—*Amos Turk*
Samuel S. Cha

References

Ponte, M., and K. Aiello. 1993. Odor emission and control at the world's largest chemical fixation facility. Presented at the New York Water Pollution Control Association, Inc., 65th Annual Winter Meeting.

Turk, A. 1977. Adsorption. In *Air Pollution,* 3rd ed. Edited by A.C. Stern. New York: Academic Press.

Turk, A., and J.M. Brassey. 1986. Removal of hydrogen sulfide from air streams. U.S. Patent 4,615,714.

Indoor Air Pollution

5.28
RADON AND OTHER POLLUTANTS

PARTIAL LIST OF SUPPLIERS

Aircheck; Amersham Corporation; Electro-Mechanical Concepts, Inc.; Health Physics Associates, Ltd.; R.A.D. Service and Instruments Ltd.; Radon Inspection Service; Radon Testing Corporation of America; Ross Systems, Inc.; Scientific Analysis, Inc.; Teledyne Isotopes, Inc.; Terradex Corp.; University of Pittsburgh.

Growing scientific evidence indicates that the air within homes and other buildings can be more seriously polluted than outdoor air even in the largest and the most industrialized cities. For many people who spend 90% of their time indoors that indoor pollution can be unhealthy.

This section describes the source and effects of radon and other indoor pollutants and discusses techniques to improve the quality of indoor air.

Radon

This section is concerned with radon including its source and effects and control techniques.

SOURCE AND EFFECTS

Radon gas is produced by the decay of naturally occurring uranium found in almost all soils and rocks. Figure 5.28.1 shows the decay chain that transforms uranium into radon and its progeny. Radon is also found in soils contaminated with certain types of industrial waste, such as the by-products of uranium mining. Phosphate rock is a source of radon because deposits of phosphate often contain high levels of uranium, approximately 50 to 150 ppm.

A significant amount of radon is present in wells and soil in many parts of this country. Radon is commonly associated with granite bedrock and is also present in the natural gas and coal deposits in this rock. In its natural state, radon rises through airspace in the soil and enters a house through its basement, is released from agitated or boiled water, or escapes during natural gas use.

The most common pathways through which radon gas seeps in from the soil include cracks in concrete floors and walls, drain pipes, floor drains, sumps, and cracks or pores

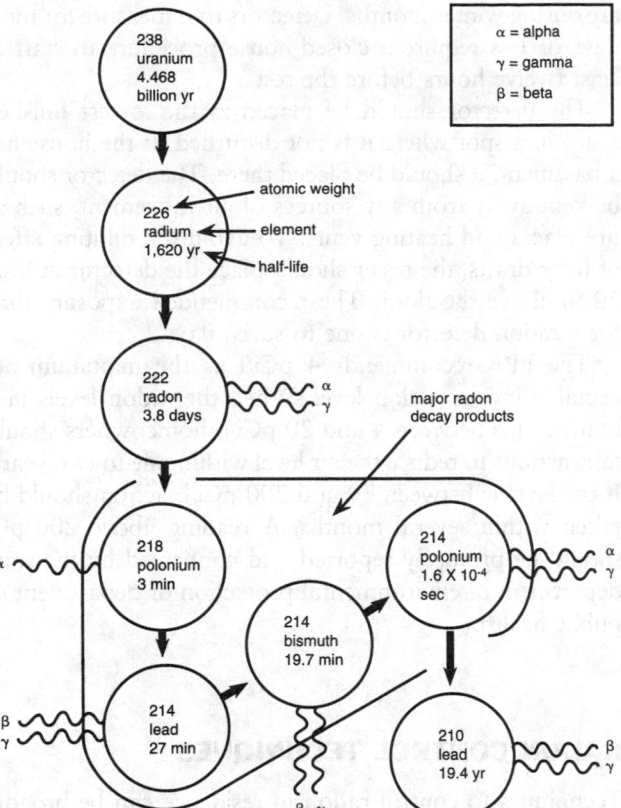

FIG. 5.28.1 Radium decay chart. About halfway through its decay sequence, uranium becomes a gas, radon, which as it disintegrates gives off radioactive particles of polonium, bismuth, and lead. Also noted is the half-life of each material—the time required for half of its radioactivity to dissipate.

in hollow block walls. Radon is drawn in by reduced air pressure, which results when the interior pressure drops below the pressure in the ground. This pressure drop is commonly caused by a warmer indoor climate; kitchen or attic exhaust fans; or consumption of interior air by furnaces, clothes dryers, or other appliances.

Radon is a colorless, odorless, almost chemically inert, radioactive gas. It is soluble in cold water, and its solubility decreases with increasing temperature. This characteristic of radon causes it to be released during water-related activities, such as taking showers, flushing toilets, and general cleaning.

Scientists and health officials express fears that the reduced infiltration of fresh air from the outside to increase energy efficiency is eliminating the escape route for radon and making a bad indoor pollution problem worse. Other factors to consider include the inflow rate of radon which depends on the strength of the radon source beneath the house and the permeability of the soil.

Since radon is naturally radioactive, it is unstable, giving off radiation as it decays. The radon decay products, radon progeny, or radon daughters, which are formed, cling to dust. If inhaled, the dust can become trapped in the lung's sensitive airways. As the decay products break down further, more radiation is released which can dam-

age lung tissue and lead to lung cancer after a period of ten to thirty years. Outside, radon dissipates quickly. However, in an enclosed space, such as a house, it can accumulate and cause lung cancer. Scientists believe that smoking increases any cancer risk from radon.

Figure 5.28.2 shows how the lung cancer risks of radon exposure compare to other causes of the disease. Scientists at the U.S. EPA estimate that if 100 individuals are exposed to a level of 4 picocuries per liter (pCi/l) over seventy years, between one and five of them will contract lung cancer. If these same individuals live in houses with levels of 200 pCi/l for only ten years, the number of anticipated lung cancer deaths would rise to between four and forty-two out of 100.

Working Levels (WL) and pCi/l

By definition, a curie is the decay rate of one gram of radium—37 billion decays per sec. Radioactivity in the environment is usually measured in trillionth of a curie or a picocurie (pCi). With a concentration of 1 pCi/l of air,

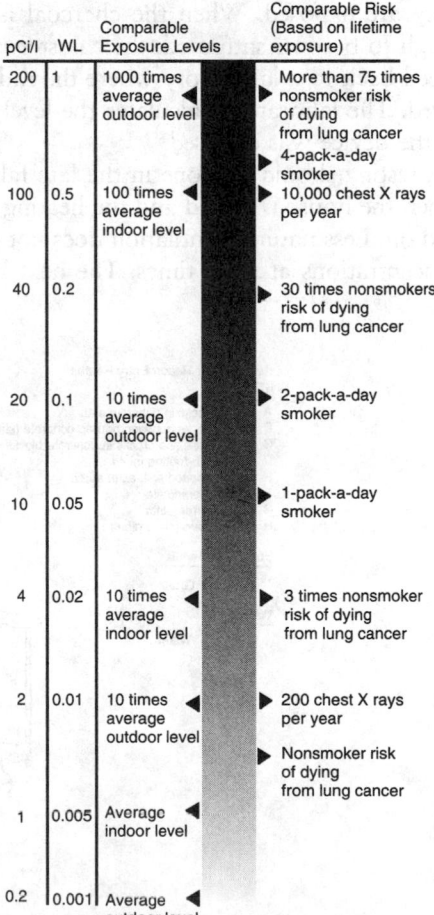

FIG. 5.28.2 Radon risk evaluation chart. This chart shows how the lung cancer risks of radon exposure compare to other causes of the disease. For example, breathing 20 pCi/l poses about the same lung cancer risk as smoking two packs of cigarettes a day. (Reprinted from U.S. Environmental Protection Agency.)

about 2 alpha particles are emitted per minute from radon atoms in each liter of air.

Another unit, the WL derived from mine regulations, is sometimes used. Actually, the WL is a measure of the concentration of radon daughters rather than of radon itself. Approximately, 1 pCi/l of radon gas is equivalent to 0.005 WL, or 1 WL is equivalent to about 200 pCi/l of radon gas.

RADON DETECTION

The two most common radon testing devices are the charcoal canister and the alpha-check detector. Alpha-check detectors are ideal for making long-term measurements but are not suited for quick results (Lafavore 1987). A charcoal-adsorbent detector is the most practical approach for most. This test method is low-cost; the price for a single unit ranges from $10 to 50 (Cohen 1987). However, a disadvantage of charcoal is its sensitivity to temperature and humidity.

Charcoal-adsorbent detectors consist of granules of activated charcoal that adsorb gases (including radon) to which they are exposed. When the charcoal is exposed long enough to become saturated, the canister is resealed and shipped back to a laboratory where the radioactivity is measured. The laboratory calculates the level of radon to which the device was exposed.

Ideally, testing should be done in the late fall or early spring when the house is closed and the heating system is not turned on. Less natural ventilation does not dilute the radon concentrations at these times. The next best times are during winter months. Detectors that measure for three days or less require a closed-house procedure to start at least twelve hours before the test.

The detector should be placed in the lowest finished space in a spot where it is not disturbed. If the house has a basement, it should be placed there. The detector should be kept away from any sources of air movement, such as fire places and heating vents. To avoid the diluting effect of floor drafts, the tester should place the detector at least 20 in above the floor. The recommended exposure time for a radon detector is one to seven days.

The EPA recommends 4 pCi/l as the maximum acceptable indoor radon level. When the radon levels in a home range between 4 and 20 pCi/l, homeowners should take actions to reduce the air level within one to two years. If the level is between 20 and 200 pCi/l, action should be taken within several months. A reading above 200 pCi should be promptly reported and confirmed by the state department of environmental protection or department of public health.

RADON CONTROL TECHNIQUES

Techniques to control radon in residence can be broadly classified as follows:

Source removal (new construction considerations)
Source control
Sealing major radon source
Sealing radon entry routes (see Figure 5.28.3)
Subslab ventilation (Figure 5.28.4)

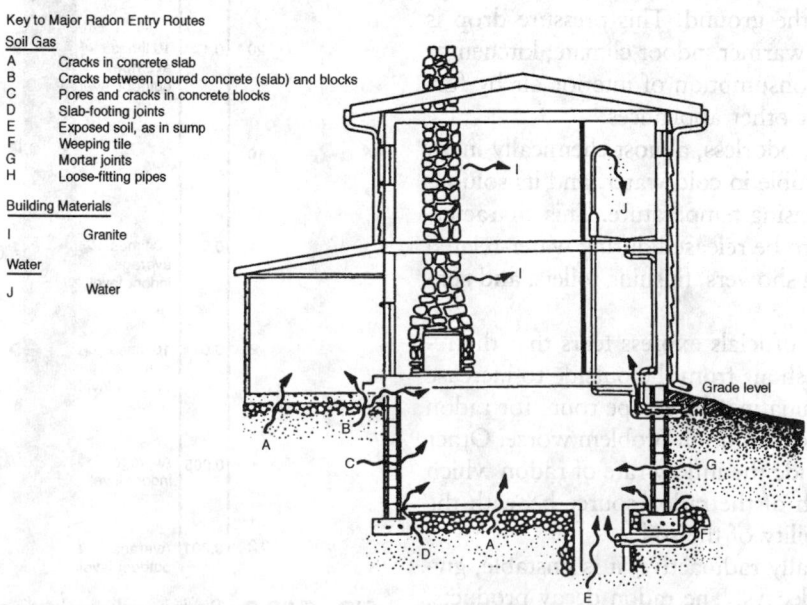

Key to Major Radon Entry Routes

Soil Gas

A   Cracks in concrete slab
B   Cracks between poured concrete (slab) and blocks
C   Pores and cracks in concrete blocks
D   Slab-footing joints
E   Exposed soil, as in sump
F   Weeping tile
G   Mortar joints
H   Loose-fitting pipes

Building Materials

I   Granite

Water

J   Water

FIG. 5.28.3 Major radon entry routes into detached houses (Reprinted from U.S. Environmental Protection Agency (EPA), 1986, *Reduction techniques for detached houses*, EPA 625/5-86/019, Research Triangle Park, N.C.: U.S. EPA.)

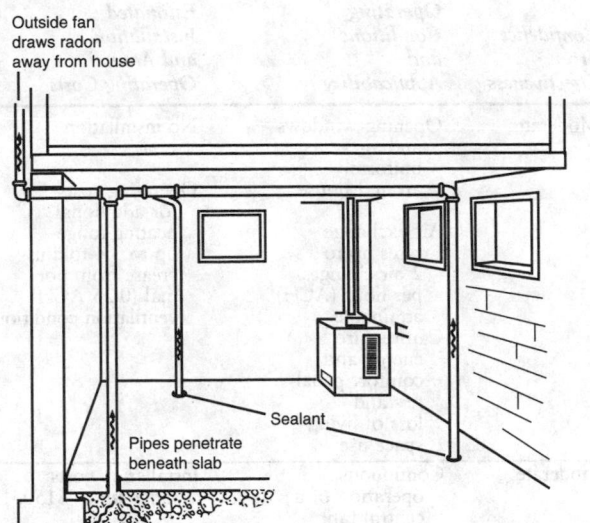

FIG. 5.28.4 Individual pipe variation of subslab ventilation. (Reprinted from U.S. EPA, 1986.)

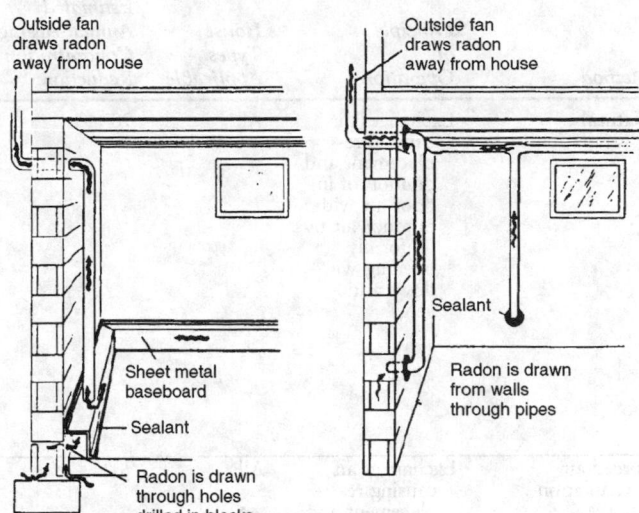

FIG. 5.28.6 Two variations of wall ventilation: the baseboard method and the single-point pipe method.

Drain-tile soil ventilation (Figure 5.28.5)
Active ventilation of hollow block basement walls (see Figure 5.28.6)
Avoidance of house depressurization
Ventilation of indoor radon concentration
Natural circulation
Forced-air ventilation
Heat recovery ventilation

However, the effectiveness of indoor air ventilation decreases with increased ventilation rates. Also, the radon source concentration and radon entry rate can reach a fi-

nite level where high ventilation rates are not productive (see Figure 5.28.7).

Table 5.28.1 summarizes the methods of reducing the radon level in a residence. No two houses have identical radon problems. Therefore, a routine method for reducing radon levels does not exist. Most homes usually require more than one of the nine methods listed to significantly reduce their radon levels. Before deciding on an approach, a homeowner should consider the unique characteristics of the house and consult a contractor specializing in the remediation of radon problems.

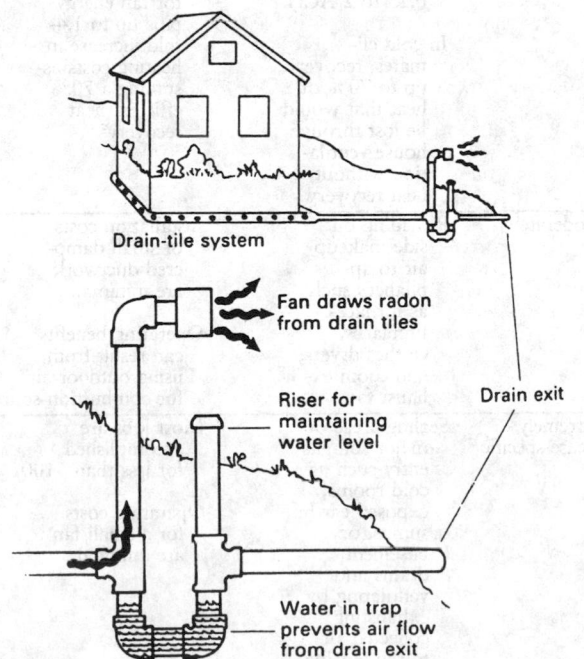

FIG. 5.28.5 Drain-tile soil ventilation system draining to remote discharge area.

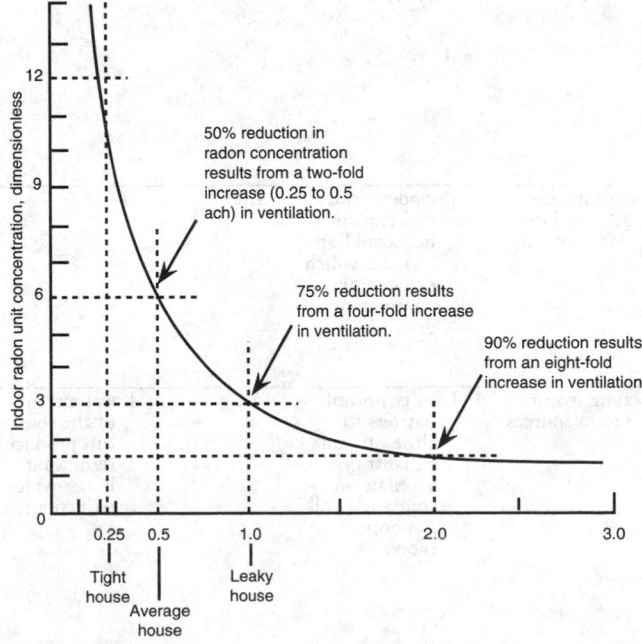

FIG. 5.28.7 Effect of ventilation on indoor radon concentrations.

TABLE 5.28.1 SUMMARY OF RADON REDUCTION TECHNIQUES

Method	Principle of Operation	House Types Applicable	Estimated Annual Average Concentration Reduction, %	Confidence in Effectiveness	Operating Conditions and Applicability	Estimated Installation and Annual Operating Costs
Natural ventilation	Exchanges air causing replacement and dilution of indoor air with outdoor air by uniformly opening windows and vents	All[a]	90[b]	Moderate	Opening windows and air vents uniformly around house Air exchange rates up to 2 air changes per hour (ACH) attainable Can require energy and comfort penalties and loss of living space use	No installation cost Operating costs for additional heating range up to 3.4-fold increase from normal (0.25 ACH) ventilation conditions.
Forced-air ventilation	Exchanges air causing replacement and dilution of indoor air with outdoor air by fans located in windows or vent openings	All	90[b]	Moderate	Continuous operation of a central fan with fresh air makeup, window fans, or local exhaust fans Forced air ventilation used to increase air exchange rates to 2 ACH Can require energy and comfort penalties and loss of living space use	Installation costs range up to $150. Operating costs range up to $100 for fan energy and up to 3.4-fold increase in normal (0.25 ACH) heating energy costs[c].
Forced-air ventilation with heat recovery	Exchanges air causing replacement and dilution of indoor air with outdoor air by a fan-powered ventilation system	All	90[b]	Moderate to high	Continuous operation of units rated at 25–240 cfm Air exchange increase from 0.25 to 2 ACH In cold climates, recovery up to 70% of heat that would be lost through house ventilation without heat recovery	Installation costs range from $400 to 1500 for 25–240-cfm units. Operating costs range up to $100 for fan energy plus up to 1.4-fold increase in heating costs assuming a 70% efficient heat recovery[c].
Active avoidance of house depressurization	Provides clean makeup air to household appliances which exhaust or consume indoor air	All	0–10[e]	Moderate[f]	Providing outside makeup air to appliances such as furnaces, fireplaces, clothes dryers, and room exhaust fans	Installation costs of small dampered ductwork are minimal. Operating benefits can result from using outdoor air for combustion sources.
Sealing major radon sources	Uses gas-proof barriers to close off and exhaust or ventilate sources of soil-gas-borne radon	All	Local exhaust of the source can produce significant house-wide reductions.	Extremely case-specific	Sealing areas of major soil-gas entry such as cold rooms, exposed earth, sumps, or basement drains and ventilating by exhausting collected air to the outside	Most jobs are accomplished for less than $100. Operating costs for a small fan are minimal.

Continued on next page

TABLE 5.28.1 *Continued*

Method	Principle of Operation	House Types Applicable	Estimated Annual Average Concentration Reduction, %	Confidence in Effectiveness	Operating Conditions and Applicability	Estimated Installation and Annual Operating Costs
Sealing radon entry routes	Uses gas-proof sealants to prevent soil-gas-borne radon-entry	All	30–90	Extremely case-specific	Sealing all noticeable interior cracks, cold joints, openings around services, and pores in basement walls and floors with appropriate materials	Installation costs range between $300 and 500.
Drain-tile soil ventilation	Continuously collects dilutes and exhausts soil-gas-borne radon from the footing perimeter of houses	BB[a] PCB[a] S[a]	Up to 98	Moderate[g]	Continuous collection of soil-gas-borne radon using a 160-cm fan to exhaust a perimeter drain tile. Applicable to houses with a complete perimeter, footing level, drain tile system and with no interior block walls resting on subslab footings	Installation cost is $1200 by contractor. Operating costs are $15 for fan energy and up to $125 for supplemental heating.
Active ventilation of hollow-block basement walls	Continuously collects, dilutes, and exhausts soil-gas-borne radon from hollow-block basement walls	BB[a]	Up to 99+	Moderate to high	Continuous collection of soil-gas-borne radon using one 250-cm fan to exhaust all hollow-block perimeter basement walls. Baseboard wall collection and exhaust system used in houses with French (channel) drains	Installation costs for a single, suction-and-exhaust-point system is $2500 (contractor-installed in unfinshed basement). Installation cost for a baseboard wall collection system is $5000 (contractor-installed in unfinished basement). Operating costs are $15 for fan energy and up to $125 for supplemental heating.
Subslab soil ventilation	Continuously collects and exhausts soil-gas-borne radon from the aggregate or soil under the concrete slab	BB[a] PCB[a] S[a]	80–90, as high as 99 in some cases	Moderate to high	Continuous collection of soil-gas-borne radon using one fan (~100 cfm, ≥0.4 in H_2O suction) to exhaust aggregate or soil under slab. For individual suction point approach, roughly one suction point per 500 sq ft of slab area. Alternate approach of piping network under slab, permitting adequate ventilation without a power-driven fan	Installation cost for individual suction point approach is about $2000 (contractor installed). Installation costs for retrofit subslab piping network are over $5000 (contractor installed). Operating costs are $15 for fan energy (if used) and up to $125 for supplemental heating.

Source: U.S. Environmental Protection Agency (EPA), 1986, *Reduction for detached houses*, EPA 625/5-86/019 (Research Triangle Park, N.C.: U.S. EPA).
Note: [a]BB (block basement) houses with hollow-block (concrete block or cinder block) basement or partial basement, finished or unfinished
PCB (poured concrete basement) houses with full or partial, finished or unfinished poured-concrete walls
C (crawl space) houses built on a crawl space
S (slab or slab-on-grade) houses built on concrete slabs

Other Indoor Pollutants

This section discusses the source and effects and control techniques for other indoor pollutants.

SOURCE AND EFFECTS

Other indoor pollutants include asbestos, bioaerosols, carbon dioxide and carbon monoxide, formaldehyde, nitrogen oxides, ozone, inhalable particulates, and VOCs.

Asbestos

Asbestos is a silicate mineral fiber that is flexible, durable, and incombustible and makes good electrical and thermal isolators. It has been used as insulation for heating, water, and sewage pipes; sound absorption and fireproofing materials; roof, siding, and floor tiles; corrugated paper; caulking; putty; and spackle. In short, asbestos was used extensively in all types of construction until about 1960. Once released from its binding material (by erosion, vibration, renovation, or cleaning) the fibers can remain airborne for long periods.

Bioaerosols

Biological contaminants include animal dander, cat saliva, human skin scales, insect excreta, food remnants, bacteria, viruses, mold, mildew, mites, and pollen. The sources of these contaminants include outdoor plants and trees, people, and animals. Pollens are seasonal; fungal spores and molds are prevalent at high temperatures. A central air-handling system can distribute these contaminants throughout a building.

Carbon Dioxide and Carbon Monoxide

Carbon monoxide, a colorless, odorless, toxic gas formed by the incomplete combustion of fossil fuels, is the most prevalent and dangerous indoor pollutant. It results from poorly ventilated kitchens, rooms over garages, and unvented combustion appliances (stoves, ovens, heaters, and the presence of tobacco smoke).

Formaldehyde

Formaldehyde, the simplest of aldehydes, is a colorless gas that is emitted from various building materials, household products, or combustion processes. Indoor sources include pressed-wood products, including particle board, paneling, fiberboard, and wallboard; textiles, such as carpet backings, drapes and upholstery fabrics, linens, and clothing; urea–formaldehyde foam insulation; adhesives; paints; coatings; and carpet shampoos.

Minimal outgassing by each product can significantly increase the formaldehyde level when the ventilation rate is low. Hot and humid conditions usually cause formaldehyde to outgas at a greater rate. Product aging diminishes its emission rate although in some cases, this process can take several years.

Nitrogen Oxides

Nitrogen oxides are combustion by-products produced by the burning of natural gas or oil in oxygen-rich environments such as kitchen stoves and ovens, furnaces, and unvented gas and kerosene heaters. When a fireplace or wood stove is used, some of these pollutants enter the room. Cracks in the stovepipe, downdrafts, or wood spillage from a fireplace can exacerbate the condition. Coal burning adds sulfur dioxide to the nitrogen oxides.

Ozone

Ozone is recognizable by its strong, pungent odor. Indoors, significant ozone can be produced by electrostatic copying machines, mercury-enhanced light bulbs, and electrostatic air cleaners. Poorly ventilated offices and rooms housing photocopying machines can accumulate significant levels of ozone.

Inhalable Particulates

Particulates are not a single type of pollutant; they describe the physical state of many pollutants—that is, all suspended solid or liquid particles less than a few hundred μm in size. Among the pollutants that appear as particulates are asbestos and other fibrous building materials, radon progeny, smoke, organic compounds, infectious agents, and heavy metals, such as cadmium in cigarette smoke.

Because of the diversity of particulates and their chemical nature, considering the adverse health effects of this category as a whole is not possible.

VOCs

VOCs are chemicals that vaporize readily at room temperature. High levels of organic chemicals in homes are attributed to aerosols, cleaners, polishes, varnishes, paints, pressed-wood products, pesticides, and others.

CONTROL TECHNIQUES

The basic control techniques to improve the quality of indoor air are source removal, ventilation, isolation, and air cleaners.

Source Removal

This technique involves removing or modifying the source of pollution and replacing it with a low-pollution substitute. Asbestos pipe insulation should be encased securely or replaced by nonasbestos pipe insulation if possible. Limiting the use of formaldehyde insulation, particle boards, carpets, fabric, or furniture containing formaldehyde can limit exposure to formaldehyde. Replacing kerosene and gas space heaters with electric space heaters can eliminate exposure to carbon dioxide, carbon monoxide, and nitrogen oxide. Biological contaminants and VOCs can be controlled by source removal.

Ventilation

Increasing ventilation can remove the offending pollutants, such as VOCs, and dilute the remaining pollution to a safer concentration. For example, gas stoves can be fitted with range hoods and exhaust fans that draw the air and effluent in over the cooking surface and blow carbon dioxide, carbon monoxide, and nitrogen oxides outdoors. However, range hoods that use charcoal filters to clean the air and then revent it to the room are ineffective in controlling carbon dioxide, carbon monoxide, and nitrogen oxides. One solution is to open a window near the stove and fit it with a fan to blow out the pollutants.

Forced ventilation with window fans works the same as natural ventilation but insures a steadier and more reliable ventilation rate. Figure 5.28.8 shows heat exchangers that are the basis of a heat-recovery ventilation system.

One way to measure the success of a ventilation system is to use a monitor that detects carbon dioxide (a gas that causes drowsiness in excess amounts). If carbon dioxide levels are high, other pollutants are also likely to be present in excessive amounts. Monitors are available that trigger a ventilation system to bring in more fresh air when needed (Soviero 1992).

Isolation

Isolating certain sources of pollution and preventing their emissions from entering the indoor environment is best. Formaldehyde outgassing from urea–formaldehyde foam insulation can also be partially controlled by vapor barriers; wallpaper or low-permeability paint applied to interior walls; and plywood coated with shellac, varnish, polymeric coatings, or other low-diffusion barriers. These barriers contain the formaldehyde outgasses, which are seemingly reabsorbed by their source rather than released into the home.

Air Cleaners

The air cleaners for residential purposes are based on filtration, adsorption, and electrostatic precipitation as follows:

Filters made of charcoal, glass fibers, and synthetic materials are used to remove particles. Pollen or lint, which are relatively large particles, are easily trapped by most

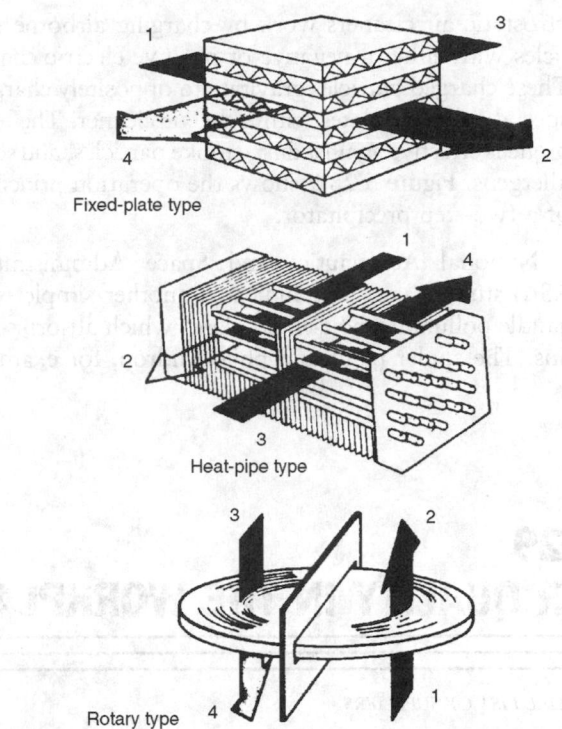

FIG. 5.28.8 Heat exchangers for heat-recovery ventilator. A heat-exchange element is the basis of a heat-recovery ventilator. Fresh outdoor air (1) is warmed as it passes through the exchanger and enters the house (2). Stale indoor air (3) leaving the house is cooled as it transfers heat to the exchanger and is vented outside (4). In the fixed-plate type, heat is transferred through plastic, metal, or paper partitions. The turning wheel of the rotary type picks up heat as it passes through the warm air path and surrenders heat to the cold air stream half a rotation later. Liquid refrigerant in the pipes of the heat-pipe type evaporates at the warm end and condenses at the cold end, transferring heat to the cold air.

filters. High-efficiency particulate air (HEPA) filters can remove particles larger than 0.3 μm, which include bacteria and spores, but not viruses.

Whereas filters trap larger particles, adsorbents react with the molecules. Three common adsorbents are activated charcoal, activated alumina, and silica gel. Adsorbents remove gases, such as formaldehyde, and ammonia.

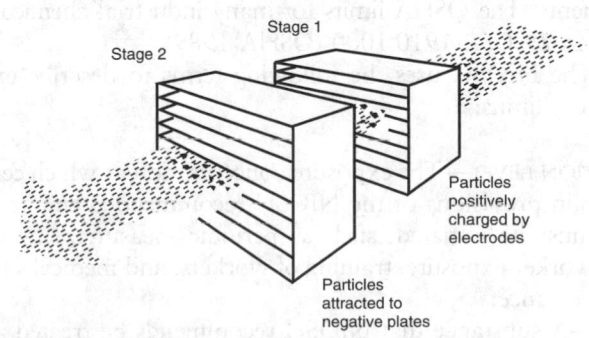

FIG. 5.28.9 Two-step precipitator.

Electrostatic air cleaners work by charging airborne particles with either a negative or positive electric charge. These charged particles gravitate to oppositely charged, special collector plates within the air cleaner. The technique is effective against dust, smoke particles, and some allergens. Figure 5.28.9 shows the operation principles of a two-step precipitator.

The National Aeronautics and Space Administration (NASA) studies have long indicated another simpler way to handle pollutants: household plants which absorb some toxins. The spider plant and philodendron, for example, remove formaldehyde and carbon dioxide from indoor air (Soviero 1992).

—*David H.F. Liu*

References

Cohen, Bernard. 1987. Radon: *A homeowner's guide to detection and control.* (August). New York: Consumers Union.
Lafavore, Michael. 1987. Radon: *The invisible threat.* Emmaus, Pa.: Rodale Press.
Soviero, Marcelle M. 1992. Can your house make you sick? *Popular Science* (July).

5.29
AIR QUALITY IN THE WORKPLACE

PARTIAL LIST OF SUPPLIERS

Altech; American Gas & Chemical Ltd; Bacharach Inc.; CEA Instruments Inc.; Coleman & Palmer Instrument Co.; Du Pont; Enmet Analytical Corp.; Gas Tech Inc.; Gilian; Lab Safety Supply; MDA Scientific Inc.; 3 M; National Draeger Inc.; Sensidyne Inc.; Sierra Monitor Inc.

OSHA requires employers to provide a safe and healthy working environment for employees. Of greatest concern are gases and vapors in the extremely toxic category, having TLV, permissible exposure limit (PEL), short-term exposure limit (STEL), or TLV-C values of 10 ppm or less. These standards are usually developed by the National Institute of Occupational Safety and Health (NIOSH).

This section focuses on exposure limits, continuous dosage sensors, and the material safety data sheet (MSDS), which contains standardized information about the properties and hazards of toxic substances.

Exposure Limits

Table 5.29.1 lists some commonly occurring toxic components. The OSHA limits for many industrial chemicals are in 29 *CFR* 1910.1000 (OSHA 1989).

The NIOSH uses the following terms to describe exposure limits:

ACTION LEVEL—The exposure concentration at which certain provisions of the NIOSH-recommended standard must be initiated, such as periodic measurements of worker exposure, training of workers, and medical surveillance.

CA—A substance that NIOSH recommends be treated as a potential human carcinogen.

CEILING—A description usually used with a published exposure limit that refers to a concentration that should not be exceeded, even for an instant.

IMMEDIATELY DANGEROUS TO LIFE OR HEALTH (IDLH)—A level defined for respiratory protection that represents the maximum concentration from which, in the event of respiratory failure, a person can escape within 30 min without experiencing any impairing or irreversible health effects.

PEL—An exposure limit published and enforced by OSHA as a legal standard.

RECOMMENDED EXPOSURE LIMIT (REL)—The exposure recommended by NIOSH not to be exceeded.

STEL—The maximum concentration to which workers can be exposed for 15 min four times throughout the day with at least 1 hr between exposures.

TWA—The average time, over a work period, of a person's exposure to a chemical or agent determined by sampling for the contaminant throughout the time period.

Occupational Exposure Monitoring

Table 5.29.2 lists selected workplace safety and health hazard standards. Monitoring can be accomplished through the use of color change badges, color detector tubes, and other monitoring techniques.

COLOR CHANGE BADGES

Breathing zone monitoring can be performed via a small media (monitor) fastened to the worker's collar or lapel for periods of time corresponding to the STEL and TWA

TABLE 5.29.1 TOXIC GASES AND VAPORS (10 PPM AND BELOW)

Name of Element or Compound	TLV-TWA, ppm	STEL, ppm	TLV-C, ppm	Human Carcinogenity Status
Acetic acid (vinegar)	10			
Acrolein	0.1	0.3		
Acrylonitrile	2			Suspected
Aniline	2			
Arsine	0.05			
Benzene	10			Suspected
Biphenyl	0.2			
Boron trifluoride			1	
Bromine	0.1	0.3		
Bromine pentafluoride	0.1			
Bromoform	0.5			
1,3 Butadiene	10			Suspected
Carbon disulfide	10			
Carbon tetrabromide	0.1			
Carbon tetrachloride	5			Suspected
Carbonyl difluoride	2	5		
Chlorine	0.5	1		
Chlorine dioxide	0.1	0.3		
Chlorine trifluoride			0.1	
Chloroform	10			Suspected
bis (Chloromethyl)ether	0.001			Confirmed
Diborane	0.1			
Dichloroacetylene			0.1	
Dimethyl hydrazine	0.5			Suspected
Dimethyl sulfate	0.1			Suspected
Ethylene oxide	1			Suspected
Fluorine	1	2		
Formaldehyde			0.3	Suspected
Germane	0.2			
Hydrazine	0.1			Suspected
Hydrogen bromide		3		
Hydrogen chloride		5		
Hydrogen cyanide		10		
Hydrogen fluoride		3		
Hydrogen selenide	0.05			
Hydrogen sulfide	10			
Methyl isocyanate	0.02			
Methyl mercaptan	0.5			
Naphthalene	10	15		
Nickel carbonyl	0.05			
Nitric acid	2	4		
Nitrobenzene	1			
Nitrogen dioxide	3	5		
2 Nitropropane	10			Suspected
Osmium tetroxide	0.0002	0.0006		
Oxygen difluoride			0.05	
Ozone			0.1	
Pentborane	0.005			
Phenylhydrazine	0.1			Suspected
Phosgene	0.1			
Phosphine	0.3	1		
Phosphorus oxychloride	0.1			
Phosphorus pentachloride	0.1			
Phosphorus trichloride	0.2	0.5		
Silane	5			
Sodium azide			0.11	
Stibine			0.1	
Sulfur dioxide	2	5		
Sulfur tetrafluoride			0.1	
Tellurium hexafluoride	0.02			
Thionly chloride			1	
Toluidine	2			Suspected
Vinyl bromide	5			Suspected
Vinyl chloride	5			Confirmed

TABLE 5.29.2 SELECT WORKPLACE SAFETY AND HEALTH
STANDARDS

Contaminant	Concentration[a] (ppm)	Concentration[a] (mg/m³)
Ammonia	50	35
Carbon dioxide	5000	9000
Carbon monoxide	50	55
Cresol	5	22
Formaldehyde	2	3
Furfuryl alcohol	50	200
Nitric oxide	25	30
Nitrogen dioxide	5	9
Octane	500	2350
Ozone	0.1	0.2
Propane	1000	1800
Sulfur dioxide	5	13
TCA	50	240
Inert or nuisance dust, respirable fraction	—	5
Asbestos	b	b
Coal dust	—	2.4

Source: Occupational Safety and Health Administration (OSHA), 1989, Air contaminants—
Permissible exposure limits standard, *Code of Federal Regulations,* Title 29, sec. 1910.1000, App. 2
(Washington, D.C.: U.S. Government Printing Office).
   Notes: [a]Values are 8-hr, time-weighted averages, except values for nitrogen dioxide, which are
ceiling values.
   [b]Fewer than two fibers longer than 5 μm in each cubic centimeter.

exposure limits. Table 5.29.3 lists the available badges and badge holders, including a smoke detector. When the exposed badge is treated with a developing agent, a color change results, which can be interpreted by electronic monitors. Suppliers analyze the exposed badges within 24 hr and return an analysis report.

COLOR DETECTOR (DOSIMETER) TUBES

Color detector tubes (CDT) determine contaminant concentrations in work areas without pumps, charts, or train-

TABLE 5.29.3 PERSONNEL-MONITORING TOXIC
GAS EXPOSURE BADGES

Description	OSHA PEL[b]
Ethylene oxide	1 ppm
Ethylene oxide, STEL[a]	5 ppm
Xylene	100 ppm
Formaldehyde	0.3 ppm
Carbon monoxide	50 ppm
Smoke-check	—

Source: Cole–Parmer Instrument Co.
   Notes: [a]Short-term exposure limits for 15-min exposure; all others require 8 hr.
   [b]OSHA permissible exposure levels.

ing. The testor simply opens the detector tube, inserts it into the tube holder, and pulls a predetermined amount of sample air through the CDT. The color progression on the packed adsorption bed indicates the concentration of a particular toxic gas. Disposable CDTs are inexpensive, and some 250 different tubes are available. Table 5.29.4 lists some of the available tubes.

OTHER MONITORING TECHNIQUES

Membrane filters with air sampling pumps monitor nuisance dust, lead silica, zinc, mineral oil, mist, and more.

The following activities are recommended for an employee exposure control program:

Assign workers the correct monitors to determine their exposure level.

Take enough samples to get an accurate, representative sampling.

Evaluate the results to determine compliance with current OSHA standards.

If the results show overexposure, determine the cause, and adjust the process and procedures of the task involved.

If no overexposure is found, a decision should be made on the frequency of routine monitoring.

Employees should be monitored for possible new exposures any time the process or procedure changes.

Document all results.

TABLE 5.29.4 FEATURES AND CAPABILITIES OF COLOR-CHANGING DOSIMETER TUBES

Description	Measuring Range (ppm)	Typical Applications	Shelf Life (yr)
Ammonia (NH_3)	50 to 900	Process control	3
Ammonia (NH_3)	10 to 260	Industrial hygiene, leak detection	3
Benzene (C_6H_6)	5 to 200	Industrial hygiene	1
Carbon dioxide (CO_2)	0.05 to 1.0%	Air contamination, concentration control	2
Carbon monoxide (CO)	10 to 250	Blast furnace, garage, combustion	1
Hydrogen sulfide (H_2S)	100 to 2000	Industrial raw gases, metallurgy	3
Hydrogen sulfide (H_2S)	3 to 150	Metal or oil refinery, chemical lab	3
Methyl bromide (CH_3Br)	5 to 80	Insert fumigation for mills and vaults	1
Nitrogen dioxide (NO_2)	0.5 to 30.0	Arc welding, acid dipping of metal products	1
Sulfur dioxide (SO_2)	20 to 300	Metal refining, waste gas analysis	2
Smoke tubes	—	Ventilation and air flow determination	—

Source: Cole-Parmer Instrument Co.

MSDSs

A major area that OSHA regulations address is the Hazard Communication Standard. The overall goal of the standard is to implement risk management and safety programs by regulated employers. According to the standard, employers must instruct employees on the nature and effects of the toxic substances with which they work, either in written form or in training programs. The instruction must include the following:

The chemical and common name of the substance
The location of the substance in the workplace
Proper and safe handling practices
First aid treatment and antidotes in case of overexposure
The adverse health effects of the substance
Appropriate emergency procedures
Proper procedures for cleanup of leaks or spills
Potential for flammability, explosion, and reactivity
The rights of employees under this rule

Most of this information is available from the MSDS, 29 CFR 1910.1200, U.S. Department of Labor.

Employers must keep copies of MSDSs for each hazardous chemical in the workplace readily accessible to their employees. If the nature of the job is such that employees travel between different workplaces during a work shift, the MSDSs can be kept at a central location at the employer's primary facility.

—*David H.F. Liu*

References

Occupational Safety and Health Administration (OSHA). 1989. Air contaminants—Permissible exposure limits standards. *Code of Federal Regulations,* Title 29, sec. 1910.1000. Washington, D.C.: U.S. Government Printing Office.

Bibliography

Adams, W.V., et al. 1990. *Guidelines for meeting emissions regulations for rotating machinery with mechanical seals.* STLE Publication SP-30 (October).

Air and Waste Management Association. 1992. *Air pollution engineering manual.* Edited by A.J. Buonicore and W.T. Davis. New York: Van Nostrand Reinhold.

American Conference of Governmental Industrial Hygienists (ACGIH). 1967. *Air sampling instruments.* Cincinnati, Ohio.

———. 1978. *Air sampling instruments for evaluation of atmospheric contaminants.* 5th ed.

———. 1993. *1993–1994 Threshold limit values.* Cincinnati, Ohio.

American Society for Testing and Materials (ASTM). 1968. *Basic principles of sensory evaluation.* STP 433. ASTM.

———. 1968. *Manual on sensory testing methods.* STP 434. ASTM.

———. 1971. *1971 Annual book of ASTM standards.* Part 23: Water: Atmospheric analysis. Philadelphia: ASTM.

———. 1981. *Guidelines for the selection and training of sensory panel members.* STP 758. ASTM.

Amoore, J.E., and E. Hautala. 1983. Odor as an aid to chemical safety: Odor thresholds compared with threshold limit values and volatiles for 214 industrial chemicals in air and water dilution. *J. Applied Toxicology* 3:272–290.

Benitez, J. 1993. *Process engineering and design for air pollution control.* PTR Prentice Hall.

Bohn, H., and R. Bohn. 1988. Soil beds weed out air pollutants. *Chem. Eng.* (25 April).

Bohn, H.I., et al. 1980. Hydrocarbon adsorption by soils as the stationary phase of gas-solid chromatography. *J. Environ. Quality* (October).

Buonicore, J. and W.T. Davis. 1992. *Air pollution engineering manual.* New York: Van Nostrand Reinhold.

Calvert, S., and H.M. Englund, eds. 1984. *Handbook of air pollution control.* John Wiley and Sons, Inc.

Cha, S.S., and K.E. Brown. 1992. Odor perception and its measurement—An approach to solving community odor problems. *Operations Forum,* vol. 9, no. 4:20–24.

Cheremisinoff, P.N., and A.G. Morressi. 1977. *Environmental assessment and impact statement handbook.* Ann Arbor Science Publishers Inc.

Cooper, C. David, and F.C. Alley. 1986. *Air pollution control—A design approach.* Prospect Heights, Ill.: Waveland Press, Inc.

Cothern, C.R., and J.E. Smith, Jr., eds. 1987. *Environmental radon.* New York: Plenum Press.

Crocker, B.B. 1974. Monitoring plant air pollution. *Chemical Engineering Progress* (January).

Downing, T.M. 1991. Current state of continuous emission monitoring technology. Presented to Municipal Utilities Conference of the American Public Power Association Meeting, Orlando, FL, February 1991.

Dravnieks, A. 1972. Odor perception and odorous air pollution. *Tappi* 55:737–742.

Eklund, B. 1992. Practical guidance for flux chamber measurements of fugitive volatile organic emission rates. *Journal of the Air and Waste Management Association.* (December).

Franz, J.J., and W.H. Prokop. 1980. Odor measurement by dynamic olfactometry. *J. Air Pollution Association* 30:1283–1297.

Gas detectors and analyzers. 1992. *Measurements and Control* (October).

Helmer, R. 1974. Desodorisierung von geruchsbeladend abluft in bodenfiltern. *Gesundheits-Ingenieur*, vol. 95, no. 1:21–26.

Hesketh, H. 1974. Fine particle collection efficiency related to pressure drop, scrubbant and particle properties, and contact mechanisms. *J. APCA* 24:939–942.

Hines, A.L., T.K. Ghosh, S.K. Loyalka, and R.C. Warder, Jr. 1993. *Indoor air: Quality and control.* Englewood Cliffs, N.J.: PRT Prentice Hall.

Hochheiser, S., F.J. Burmann, and G.B. Morgan. 1971. Atmospheric surveillance, the current state of air monitoring technology. *Environmental Science and Technology* (August).

Hodges, D.S., V.F. Medina, R.L. Islander, and J.S. Devinny. 1992. Biofiltration: Application for VOC emission control. 47th Annual Purdue Industrial Waste Conference, West Lafayette, Indiana, May 1992.

Jacobs, M.B. 1967. *The analytical toxicology of industrial inorganic poisons.* Vol. 22, *Chemical analysis.* New York: Interscience.

Laird, J.C. 1978. Unique extractive stack sampling. 1978 ISA Conference, Houston.

Laznow, J., and T. Ponder. 1992. Monitoring and data management of fugitive hazardous air pollutants. ISA Conference, Houston, October 1992.

Licht, W. 1980. *Air pollution control engineering.* New York: Marcel Dekker, Inc.

Liptak, B. 1994. *Analytical instrumentation.* Radnor, Pa.: Chilton Book Company.

Lord, H.C., and R.V. Brown. 1991. Open-path multi-component NDIR monitoring of toxic combustible or hazardous vapors. 1991 ISA Conference, Anaheim, California, Paper #91-0401.

Meyer, Beat. 1983. *Indoor air quality.* Reading, Mass.: Addison-Wesley Publishing Co.

National Air Pollution Control Administration. 1969. *Air quality criteria for sulfur oxides.* Pub. No. AP-50. Washington, D.C.: 161–162.

National Research Council, Committee on Odors from Stationary and Mobile Sources. 1979. *Odors from stationary and mobile sources.* Washington, D.C.: National Academy of Sciences.

Pevoto, L.F., and L.J. Hawkins. 1992. Sample preparation techniques for very wet gas analysis. ISA Conference, Houston, October 1992.

Reynolds, J.P., R.R. Dupont, and L. Theodore. 1991. *Hazardous waste incinerator calculations, problems and software.* John Wiley and Sons Inc.

Sax, N.J., and R.J. Lewis. 1986. *Rapid guide to hazardous chemicals in the workplace.* New York: Van Nostrand.

Seinfeld, J.H. 1975. *Air pollution—Physical and chemical fundamentals.* McGraw-Hill Book Company.

Stephan, D.G., G.W. Walsh, and R.A. Herrick. 1960. Concepts in fabric air filtration. *Am. Ind. Hyg. Assoc. J.* 21:1–14.

Stern, A.C., ed. 1968. *Air pollution.* 2d ed. Vol. II, *Analysis, monitoring, and surveying.* Chap. 16. New York: Academic Press.

Stern, A.C., H.C. Wohlers, R.W. Boubel, and W.P. Lowry. 1973. *Fundamentals of air pollution.* New York: Academic Press.

Theodore, L., and J. Reynolds. 1987. *Introduction to hazardous waste incineration.* John Wiley and Sons Inc.

Turk, A. 1977. Adsorption. In *Air pollution*, 3d ed. Edited by A.C. Stern. New York: Academic Press.

U.S. Consumer Products Safety Commission. *Status report on indoor pollution in 40 Tennessee homes.*

U.S. Environmental Protection Agency (EPA). 1977. *Technical guidance for the development of control strategies in areas with fugitive dust problems.* EPA 450/3-77/010. Washington, D.C.: U.S. EPA.

———. 1991. Burning of hazardous waste in boilers and industrial furnaces: Final rule corrections; Technical amendments. *Federal Register* 56, no. 137 (17 July):32688–32886.

———. 1992. *National air quality emissions trends report, 1991.* EPA-450-R-92-001. Research Triangle Park, N.C.: Office of Air Quality Planning and Standards (October).

Unterman, R. 1993. Biotreatment systems for air toxics. AIChE Central Jersey Section Seminar on Catalysis and Environmental Technology, May 20, 1993.

Vesilind, P.A. 1994. *Environmental engineering.* 3d ed. Butterworth Heinemann.

Weiss, M. 1994. Environmental res complicate CEM success. *Control* (March).

Wood, S.C. 1994. Select the right NO control technology. *Chem. Eng. Prog.* (January).

6

Noise Pollution

David H.F. Liu | Howard C. Roberts

6.1
THE PHYSICS OF SOUND AND HEARING

Sound can be defined as atmospheric or airborne vibration perceptible to the ear. *Noise* is usually unwanted or undesired sound. Consequently, a particular sound can be noise to one person and not to others, or noise at one time and not at other times. Sound loud enough to be harmful is called noise without regard to its other characteristics. Noise is a form of pollution because it can cause hearing impairment and psychological stress.

This section introduces the subject of sound in engineering terms and includes appended references which provide detailed back-up material. It includes the general principles of sound production and propagation, a description of the ear and its functions, a description of the effects of noise on the hearing apparatus and on the person, and an introduction to hearing measurement and hearing aids.

Sound Production and Propagation

Audible sound is any vibratory motion at frequencies between about 16 and 20,000 Hz; normally it reaches the ear through pressure waves in air. Sound is also readily transmissible through other gases, liquids, or solids; its velocity depends on the density and the elasticity of the medium, while attenuation depends largely on frictional damping. For most engineering work, adiabatic conditions are assumed.

Sound is initially produced by vibration of solid objects, by turbulent motion of fluids, by explosive expansion of gases, or by other means. The pressures, amplitudes, and velocities of the components of the sound wave within the range of hearing are quite small. Table 6.1.1 gives typical values; the sound pressures referenced are the dynamic excursions imposed on the relatively constant atmospheric pressure.

In a *free field* (defined as an isotropic homogeneous field with no boundary surfaces), a point source° of sound produces spherical (Beranek 1954) sound waves (see Figure 6.1.1). If these waves are at a single frequency, the instantaneous sound pressure $(P_{r,t})$ at a distance r and a time t is

$$P_{r,t} = [(\sqrt{2}P)/r] \cos[\omega(t - r/t)] \text{ dynes/cm}^2 \qquad 6.1(1)$$

where the term $\sqrt{2P}$ denotes the magnitude of peak pressure at a unit distance from the source, and the cosine term represents phase angle.

In general, instantaneous pressures are not used in noise control engineering (though peak pressures and some non-sinusoidal pulse pressures are, as is shown later), but most sound pressures are measured in root-mean-square (RMS)

values—the square root of the arithmetic mean of the squared instantaneous values taken over a suitable period. The following description refers to RMS values.

For spherical sound waves in air, in a free field, RMS pressure values are described by

$$P_r = P_o/r \text{ dynes/cm}^2 \qquad 6.1(2)$$

where P_r denotes RMS sound pressure at a distance r from the source, and P_o is RMS pressure at unit distance from the source. (Meters in metric units, feet in English units.) Acoustic terminology is based on metric units, in general, though the English units of feet and pounds are used in engineering descriptions.

A few other terms should be defined, and their mathematical relationships noted.

Sound intensity I is defined as the acoustic power W passing through a surface having unit area; and for spherical waves (see Figure 6.1.1), this unit area is a portion of a spherical surface. Sound intensity at a distance r from a source of power W is given by

$$I_r = W/4\pi r^2 \text{ watts/cm}^2 \qquad 6.1(3)$$

Sound intensity is also given by

$$I_r = P_o^2/r^2\rho c \text{ watts/cm}^2 \qquad 6.1(4)$$

where ρ is the adiabatic density of the medium, and c is the velocity of sound in that medium. Similarly, the following equation gives the sound pressure if the sound is radiated uniformly:

$$P_r = (1/r)\sqrt{W\rho c/4\pi} \qquad 6.1(5)$$

If the radiation is not uniform but has directivity, the term ρc is multiplied by a directivity factor Q. To the noise-control engineer, the concept of intensity is useful principally because it leads to methods of establishing the sound power of a source.

The term ρc is called the acoustic impedance of the medium; physically it represents the rate at which force can be applied per unit area or energy can be transferred per unit volume of material. Thus, acoustic impedance can be expressed as force per unit area per second (dynes/cm²/sec) or energy per unit volume per second (ergs/cm³/sec).

Table 6.1.1 shows the scale of mechanical magnitudes represented by sound waves. Amplitude of wave motion at normal speech levels, for example, is about 2×10^{-6} cm, or about 1 micro inch; while amplitudes in the lower part of the hearing range compare to the diameter of the hydrogen atom. Loud sounds can be emitted by a vibrat-

TABLE 6.1.1 MECHANICAL CHARACTERISTICS OF SOUND WAVES

	RMS Sound Pressure (dynes/cm²)	RMS Sound Particle Velocity (cm/sec)	RMS Sound Particle Motion at (1,000 Hz cm)	Sound Pressure Level (dB 0.0002 bar)
Threshold of hearing	0.0002	0.0000048	0.76×10^{-9}	0
	0.002	0.000048	7.6×10^{-9}	20
Quiet room	0.02	0.00048	76.0×10^{-9}	40
	0.2	0.0048	760×10^{-9}	60
Normal speech at 3'	2.0	0.048	7.6×10^{-6}	80
Possible hearing impairment	20.0	0.48	76.0×10^{-6}	100
	200	4.80	760×10^{-6}	120
Threshold of pain	2000	48.0	7.6×10^{-3}	140
Incipient mechanical damage	20×10^3	480	76.0×10^{-3}	160
	200×10^3	4800	760×10^{-3}	180
Atmospheric pressure	2000×10^3	48000	7.6	200

ing partition even though its amplitude is only a few micro inches.

REFLECTION, DISPERSION, ABSORPTION, AND REFRACTION

Sound traversing one medium is reflected when it strikes an interface with another medium in which its velocity is different; the greater the difference in sound velocity, the more efficient the reflection. The reflection of sound usually involves dispersion or scattering.

Sound is dispersed or scattered when it is reflected from a surface, when it passes through several media, and as it passes by and around obstacles. Thus, sound striking a building as plane waves usually is reflected with some dispersion, and plane waves passing an obstacle are usually somewhat distorted. This effect is suggested by Figure 6.1.1. The amount of dispersion by reflection depends on the relationship between the wavelength of the sound and the contour of the reflecting surface.

The absorption of sound involves the dissipation of its mechanical energy. Materials designed specifically for that purpose are porous so that as the sound waves penetrate, the area of frictional contact is large and the conversion of molecular motion to heat is facilitated.

Sound waves can be refracted at an interface between media having different characteristics; the phenomenon can be described by Snell's law as with light. Except for events taking place on a large scale, refraction is usually distorted by dispersion effects. In the tracking of seismic waves and undersea sound waves, refraction effects are important.

In engineering noise analysis and control, reflection, refraction, and dispersion have pronounced effects on directivity patterns.

WAVE CHARACTER

Since sound is a wave motion, it can be focussed by reflection (and less easily by refraction), and interference can

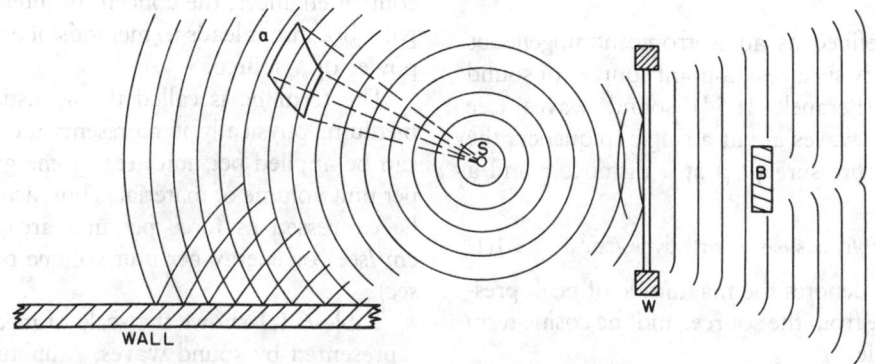

FIG. 6.1.1 Sound sources. A point source at S produces a calculable intensity at a. The sound waves can set an elastic membrane or partition (like a large window) at W into vibration. This large source can produce roughly planar sound waves, which are radiated outward with little change in form but are distorted and dispersed as they pass the solid barrier B.

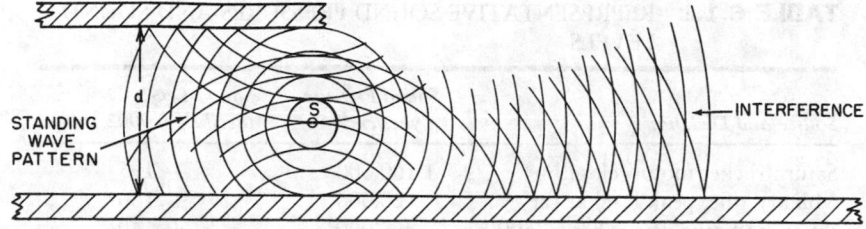

FIG. 6.1.2 Reflection of sound waves. If the distance d between two parallel walls is an integral number of wavelengths, standing waves can occur. Interaction between direct waves from a source S and the reflected waves can produce interference.

occur, as can standing wave patterns. These effects are important in noise control and in auditorium acoustics. Another wave–motion phenomenon, the coincidence effect, affects partition behavior.

When two wave forms of the same frequency are superimposed, if they are inphase, they add and reinforce each other; while if they are of opposing phase, the resultant signal is their difference. Thus, sound from a single source combined with its reflection from a plane surface can produce widely varying sound levels through such interference. If reflective surfaces are concave, they can focus the sound waves and produce high sound levels at certain points. Dispersion often partially obscures these patterns.

Sound from a single source can be reinforced by reflection between two walls if their separation is a multiple of the wavelength; this standing-wave pattern is described by Figure 6.1.2.

These phenomena are important in auditorium design, but they cannot be ignored in noise control work. Reinforcement by the addition of signals can produce localized high sound levels which can be annoying in themselves and are also likely to produce mechanical vibrations—and thus new, secondary noise sources.

Random noise between parallel walls is reinforced at a series of frequencies by the formation of standing waves; this reinforcement partially accounts for the high noise level in city streets.

ENERGY RELATIONSHIPS IN SOUND

The magnitudes most used to describe the energy involved in sound or noise are sound pressure and sound power. Pressure, either static (barometric) or dynamic (sound vibrations), is the magnitude most easily observed. Sound pressure is usually measured as an RMS value—whether this value is specified or not—but peak values are sometimes also used.

From the threshold of hearing to the threshold of pain, sound pressure values range from 0.0002 to 1000 or more dynes per square centimeter (Table 6.1.1). To permit this wide range to be described with equal resolution at all pressures, a logarithmic scale is used, with the decibel (dB) as its unit. Sound pressure level (SPL) is thus defined by

$$SPL = 20 \log_{10} (P/P_{ref}) \text{ dB} \qquad 6.1(6)$$

where P is measured pressure, and P_{ref} is a reference pressure. In acoustic work this reference pressure is 0.0002 dynes/cm². (Sometimes given as 0.0002 microbars, or 20 micronewtons/meter². A reference level of 1 microbar is sometimes used in transducer calibration; it should not be used for sound pressure level.) Table 6.1.2 lists a few representative sound pressures and the decibel values of sound pressure levels which describe them.

This logarithmic scale permits a range of pressures to be described without using large numbers; it also represents the nonlinear behavior of the ear more convincingly. A minor inconvenience is that logarithmic quantities cannot be added directly; they must be combined on an energy basis. While this combining can be done by a mathematical method, a table or chart is more convenient to use; the accuracy provided by these devices is usually adequate.

Table 6.1.3 is suitable for the purpose; the procedure is to subtract the smaller from the larger decibel value, find the amount to be added in the table, and add this amount to the larger decibel value. For example, if a 76 dB value is to be added to an 80 dB value, the result is 81.5 dB (80 plus 1.5 from the table). If more than two values are to be added, the process is simply continued. If the smaller of the two values is 10 dB less than the larger, it adds less than 0.5 dB; such a small amount is usually ignored, but if several small sources exist, their combined effect should be considered.

The sound power of a source is important; the magnitude of the noise problem depends on the sound power. Sound power at a point (sound intensity) cannot be measured directly; it must be done with a series of sound pressure measurements.

The acoustic power of a source is described in watts. The range of magnitudes covers nearly 20 decimal places; again a logarithmic scale is used. The reference power level normally used is 10^{-12} watt, and the sound power level (PWL) is defined by

$$PWL = 10 \log_{10} (W/10^{-12}) \text{ dB} \qquad 6.1(7)$$

or, since the power ration 10^{-12} means the same as -120 dB, the following equation is also correct:

$$PWL = 10 \log_{10} W + 120 \text{ dB} \qquad 6.1(8)$$

In either case, W is the acoustic power in watts.

TABLE 6.1.2 REPRESENTATIVE SOUND PRESSURES AND SOUND LEVELS

Source and Distance	Sound Pressure (dynes/cm²)	Sound Level (decibels 0.0002 μ bar)
Saturn rocket motor, close by	1,100,000	195
Military rifle, peak level at ear	20,000	160
Jet aircraft takeoff; artillery, 2500'	2000	140
Planing mill, interior	630	130
Textile mill	63	110
Diesel truck, 60'	6	90
Cooling tower, 60'	2	80
Private business office	.06	50

Source	Acoustic Power of Source	
Saturn rocket motor	30,000,000	watts
Turbojet engine	10,000	watts
Pipe organ, forte	10	watts
Conversational voice	10	microwatts
Soft whisper	1	millimicrowatt

Sound power levels are established through sound pressure measurements; in a free field, sound is radiated spherically from a point source, thus

$$PWL = SPL + 20 \log_{10} r + 0.5 \text{ dB} \qquad 6.1(9)$$

For precise work, barometric corrections are required. In practical situations, a directivity factor must often be introduced. For example, if a machine rests on a reflecting surface (instead of being suspended in free space), reflection confines the radiated sound to a hemisphere instead of a spherical pattern, with resulting SPL readings higher than for free-field conditions.

Actual sound power values of a source, in watts, can be computed from PWL values using Equation 6.1(8).

In all cases, the units should be stated when sound pressure or sound power values are listed (dynes/cm², watts), and the reference levels should be made known when

TABLE 6.1.3 ADDITION OF DECIBEL VALUES

Difference Between the Two Decibel Values	Amount to be Added to the Higher Level
0	3.0
1	2.5
2	2.0
3	2.0
4	1.5
5	1.0
6	1.0
7	1.0
8	0.5
9	0.5
10	0

sound pressure levels or sound power levels are listed (0.0002 dynes/cm², and 10^{-12} watt).

The Hearing Mechanism

Sound reaches the ear usually through pressure waves in air; a remarkable structure converts this energy to electrical signals which are transmitted to the brain through the auditory nerves. The human ear is capable of impressive performance. It can detect vibratory motion so small it approaches the magnitude of the molecular motion of the air. Coupled with the nerves and brain, the ear can detect frequency differences and combinations, magnitude, and direction of sound sources. It can also analyze and correlate such signals. A brief description of the ear and its functioning follows.

Figure 6.1.3 shows the anatomical division of the ear. The external human ear (called the auricle or the pinna) and the ear opening (the external auditory canal or meatus) are the only parts of the hearing system normally visible. They gather sound waves and conduct them to the eardrum and inner drum. They also keep debris and objects from reaching the inner ear.

The working parts of the ear include the eardrum and organs which lie behind it; they are almost completely surrounded by bone and are thus protected.

The sound transducer mechanism is housed in the middle ear (Figure 6.1.4). The eardrum or tympanic membrane is a thin, tough membrane, slightly oval in shape and a little less than 1 cm in mean diameter; it vibrates in response to sound waves striking it. The vibratory motion is transmitted through three tiny bones, the ossicles (the malleus, the incus, and the stapes; or the hammer, anvil, and stirrup), to the cochlea; it enters the cochlea at the oval window.

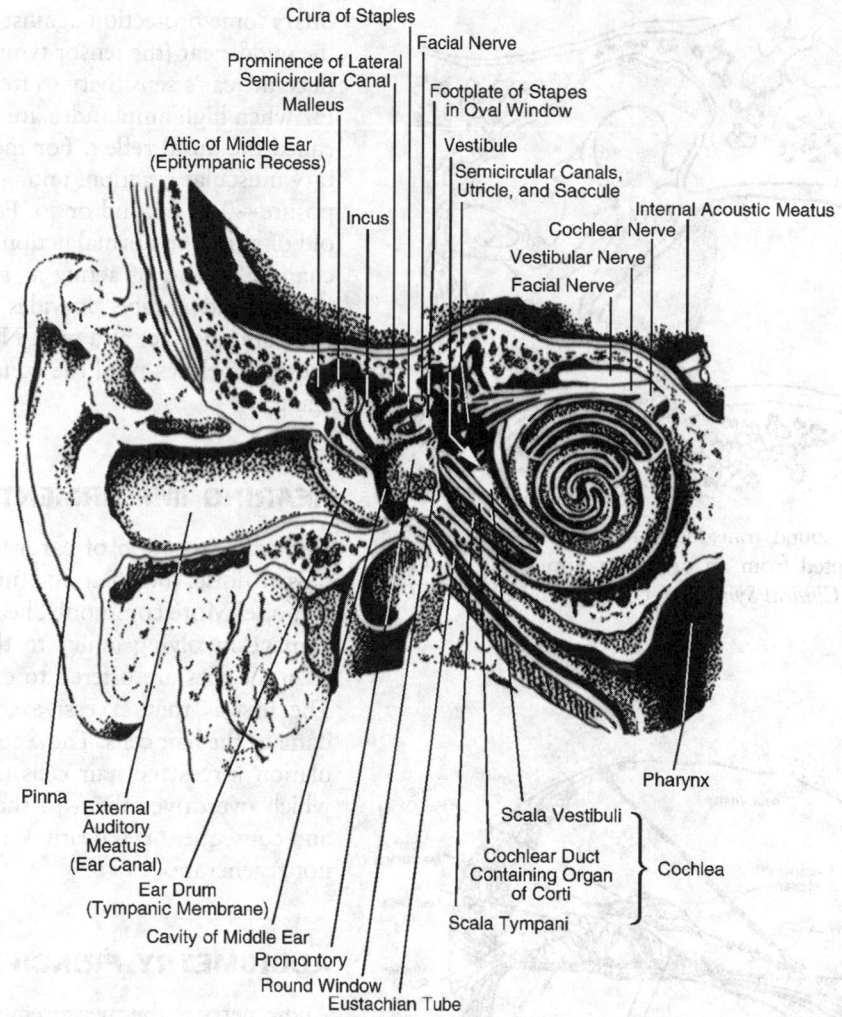

Crura of Staples
Facial Nerve
Prominence of Lateral
Semicircular Canal
Footplate of Stapes
in Oval Window
Malleus
Vestibule
Attic of Middle Ear
(Epitympanic Recess)
Semicircular Canals,
Utricle, and Saccule
Internal Acoustic Meatus
Incus
Cochlear Nerve
Vestibular Nerve
Facial Nerve

Pinna
Pharynx
External
Auditory
Meatus
(Ear Canal)
Scala Vestibuli
Ear Drum
(Tympanic Membrane)
Cochlear Duct
Containing Organ
of Corti
Cochlea
Cavity of Middle Ear
Scala Tympani
Promontory
Round Window
Eustachian Tube

FIG. 6.1.3 Anatomical divisions of the ear. (© Copyright 1972 CIBA Pharmaceutical Company, Division of CIBA-GEIGY Corporation. Reproduced, with permission, from *Clinical Symposia,* illustrated by Frank H. Netter, M.D. All rights reserved.)

The ossicles are in an air-filled space called the middle ear; close to the middle ear are small muscles which act on them and on the tympanum. The principal function of the ossicles seems to be to achieve an impedance match between the external auditory canal and the fluid-filled cochlea. The principal function of the middle-ear muscles seems to be to control the efficiency of the middle ear by controlling tension of the eardrum and the mechanical advantage of the ossicles as a lever system. The middle ear is connected through the Eustachian tube with the nasal passages so that it can accommodate to atmospheric pressures; without this connection, changing atmospheric pressure would apply a steady force to the eardrum and prevent its free vibration.

The cochlea or cochlear canal functions as a transducer; mechanical vibrations enter it; electrical impulses leave it through the auditory nerve. The cochlea is a bone shaped like a snail, coiled two and one-half times around its own axis (Figure 6.1.3). It is about 3 cm long and 3 mm in di-

ameter at its largest part. It is divided along most of its length by the cochlea partition, which is made up of the basilar membrane, Reissner's membrane, and the organ of Corti.

A cross section through the cochlea (Figure 6.1.5) reveals three compartments: the scala vestibuli, the scala media, and the scala tympani. The scala vestibuli and the scala tympani are connected at the apex of the cochlea. They are filled with a fluid called perilymph in which the scala media floats. The hearing organ (organ of Corti) is housed in the scala media. The scala media contains a different fluid, endolymph, which bathes the organ of Corti.

The scala media is triangular in shape and is about 34 mm in length (Figure 6.1.5). Cells grow up from the basilar membrane. They have a tuft of hair at the end and are attached to the hearing nerve at the other end. A gelatinous membrane (tectoral membrane) extends over the hair cells and is attached to the limbus spiralis. The hair cells are embedded in the tectoral membrane.

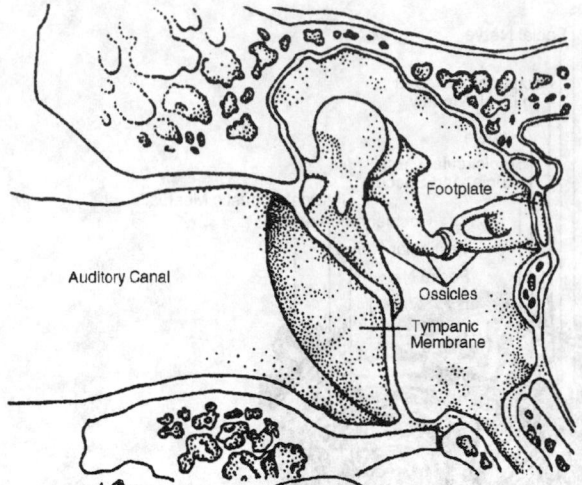

FIG. 6.1.4 The sound transducer mechanism housed in the middle ear. (Adapted from an original painting by Frank H. Netter, M.D., for *Clinical Symposia*, copyright by CIBA-GEIGY Corporation.)

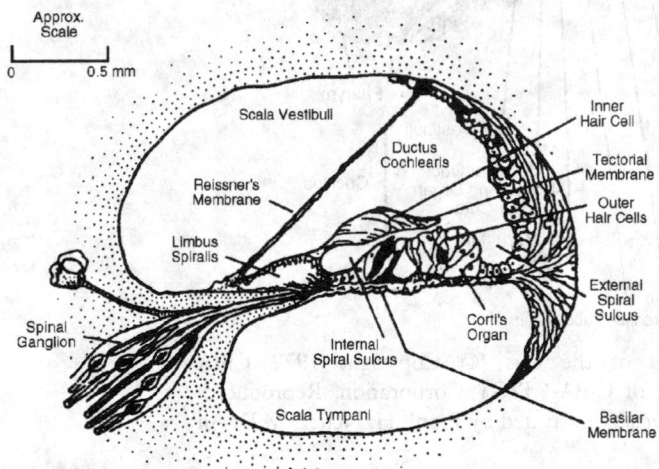

FIG. 6.1.5 Cross section through the cochlea.

Vibration of the oval window by the stapes causes the fluids of the three scala to develop a wave-like motion. The movement of the basilar membrane and the tectoral membrane in opposite directions causes a shearing motion on the hair cells. The dragging of the hair cells sets up electrical impulses which are transmitted to the brain in the auditory nerves.

The nerve endings near the oval and round windows are sensitive to high frequencies. Those near the apex of the cochlea are sensitive to low frequencies.

Another structure of the inner ear is the semicircular canals, which control equilibrium and balance. Extremely high noise levels can impair one's sense of balance.

The ear has some built-in protection; since it is almost entirely surrounded by bone, a considerable amount of mechanical protection is provided. The inner-ear mechanism offers some protection against loud noises. The muscles of the middle ear (the tensor tympanus and stapedius) can reduce the ear's sensitivity to frequencies below about 1000 Hz when high amplitudes are experienced; this reaction is called the aural reflex. For most people, it is an involuntary muscular reaction, taking place a short time after exposure—0.01 second or so. For sounds above the threshold of pain, the normal action of the ossicles is thought to change; instead of acting as a series of levers whose mechanical advantage provides increased pressure on the eardrum, they act as a unit. Neither of these protective reactions operates until the conditions are potentially damaging.

HEARING IMPAIRMENT

With the exception of eardrum rupture from intense explosive noise, the outer and middle ear are rarely damaged by noise. More commonly, hearing loss is a result of neural damage involving injury to the hair cells (Figure 6.1.6). Two theories are offered to explain noise-induced injury. The first is that excessive shearing forces mechanically damage the hair cells. The second is that intense noise stimulation forces the hair cells into high metabolic activity, which overdrives them to the point of metabolic failure and consequent cell death. Once destroyed, hair cells cannot regenerate.

AUDIOMETRY PRINCIPLES

Audiometry is the measurement of hearing; it is often the determination of the threshold of hearing at a series of frequencies and perhaps for the two ears separately, though more detailed methods are also used. Audiometric tests are made for various reasons; the most common to determine the extent of hearing loss and for diagnosis to permit hearing aids to be prescribed.

In modern society a gradual loss in hearing is normal and occurs with increasing age; Figure 6.1.7 shows this condition. These curves show the average loss in a number of randomly selected men and women (not selected solely from noisy occupations), and these data are accepted as representing typical presbycusis conditions. (*Presbycusis* refers to the normal hearing loss of the elderly.) For all persons tested, the effect increases with age and is more pronounced at high frequencies than at low.

Men normally show the effect to a greater degree than women. In the last decade or so, women have experienced more presbycusis than formerly. Experts disagree as to whether noise is the predominant factor; but evidence shows that presbycusis and other processes of aging take place faster when noise levels and other social stresses are high. Another term, *sociocusis*, is being used to describe the hearing loss from exposure to the noises of modern society.

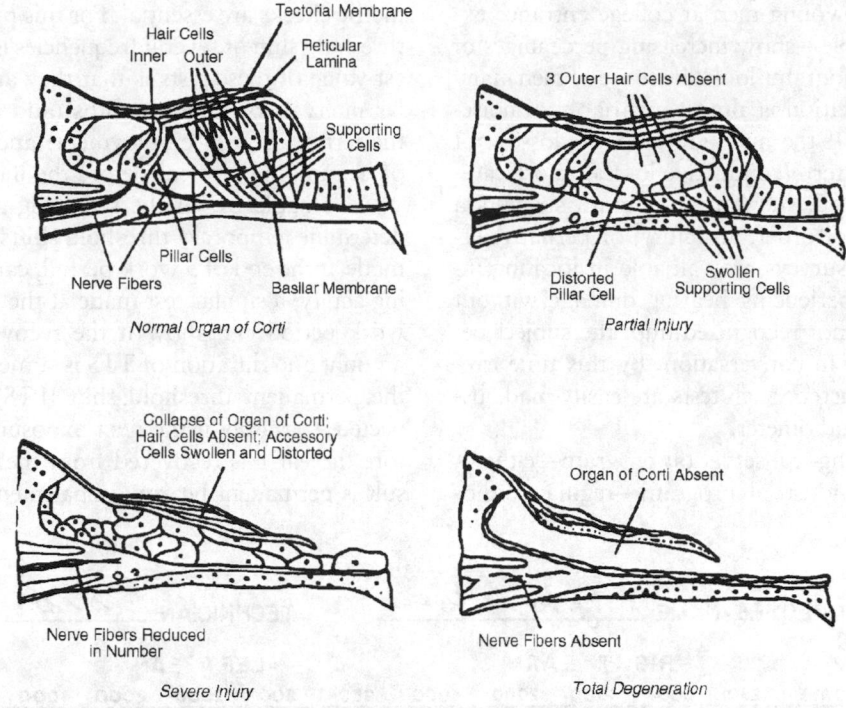

FIG. 6.1.6 Various degrees of injury to the hair cells.

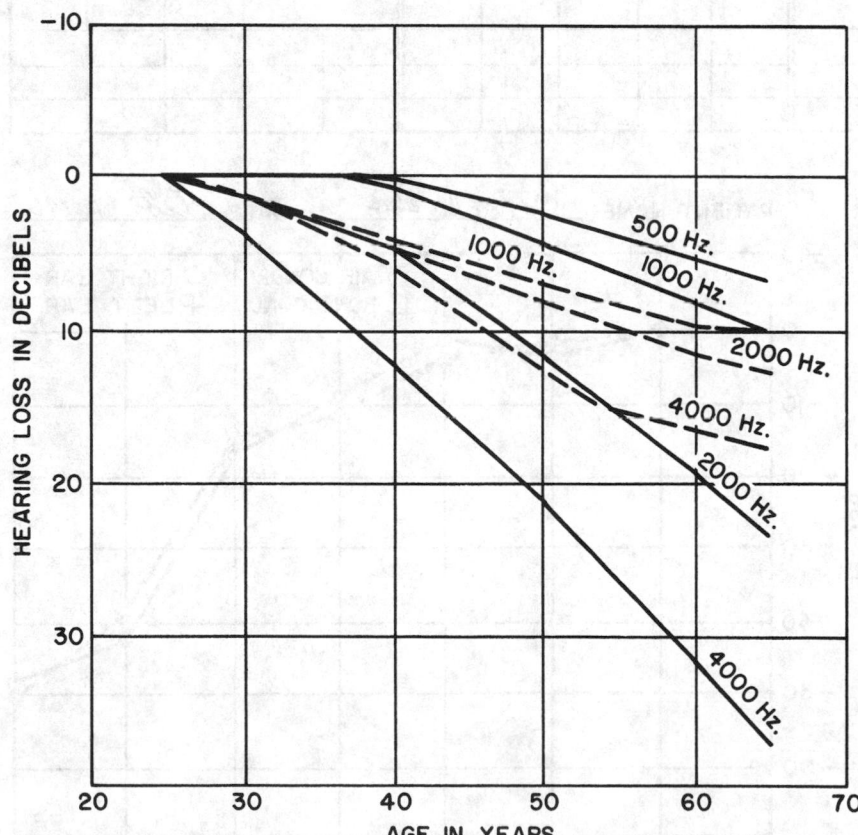

FIG. 6.1.7 Normal presbycusis curves. Statistical analysis of audiograms from many people show normal losses in hearing acuity with age. Data for men are represented by solid lines; those for women by dotted lines.

Group surveys—of young men at college entrance examinations, for example—show increasing percentages of individuals whose audiograms look like those of men many years older. This indication is almost invariably of noise-induced hearing loss. If the audiogram shows losses not conforming to this pattern (conductive losses), more careful checking is indicated; such an audiogram suggests a congenital or organic disorder, an injury, or perhaps nervous damage. Group surveys are valuable in locating individuals who are experiencing hearing damage without realizing it; it is often not recognized until the subject begins to have difficulty in conversation. By this time irremediable damage occurred. Such tests are easily made using a simple type of audiometer.

As a part of a hearing–conservation program—either a public health or an industrial program—regular audiometric checks are essential. For this purpose, checking only threshold shift at several frequencies is common. The greatest value of these tests is that they are conducted at regular intervals of a few months (and at the beginning and the termination of employment) and can show the onset of hearing impairment before the individual realizes it.

A valuable use of the screening audiometric test is to determine temporary threshold shifts (TTS). Such a check, made at the end of a work period, can show a loss of hearing acuity; a similar test made at the beginning of the next work period can show if the recovery is complete. The amount and duration of TTS is somewhat proportional to the permanent threshold shift (PTS) which must be expected. Certainly if the next exposure to noise occurs before the ear has recovered from the last, the eventual result is permanent hearing impairment.

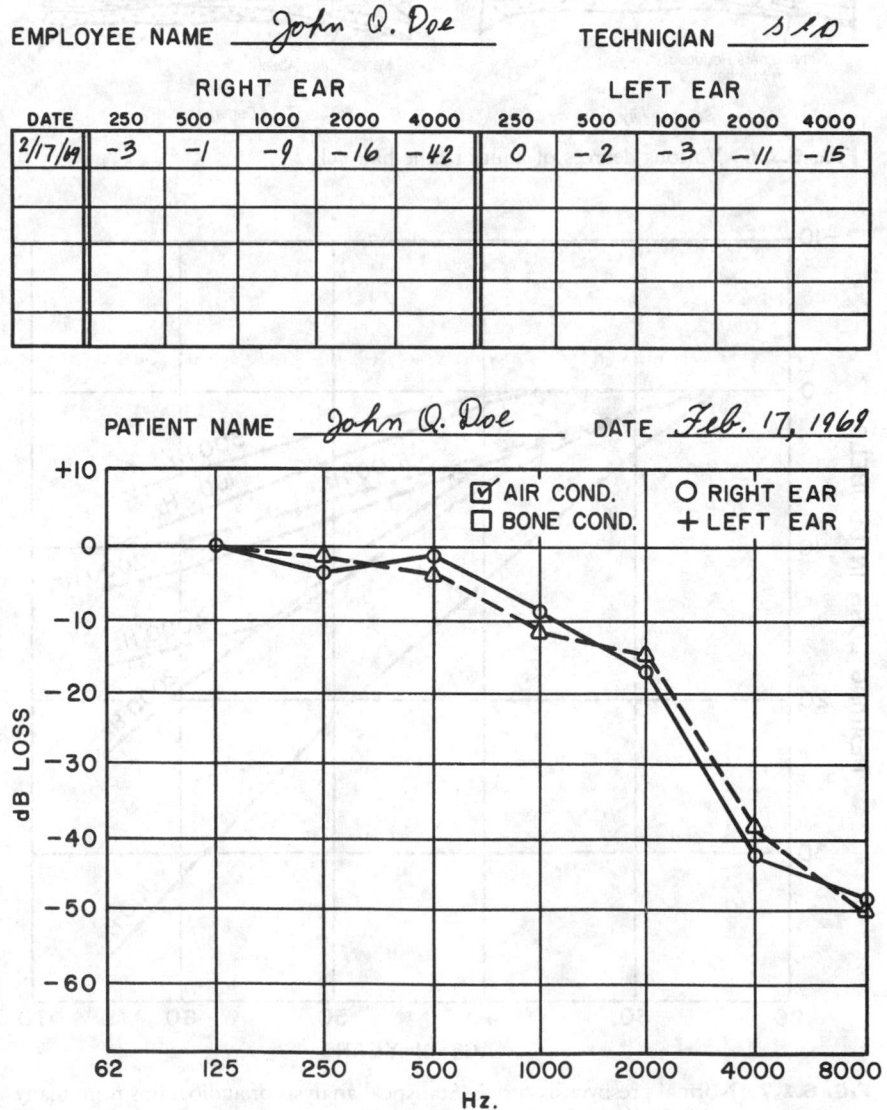

EMPLOYEE NAME __John Q. Doe__   TECHNICIAN __SLO__

| | RIGHT EAR | | | | | LEFT EAR | | | | |
DATE	250	500	1000	2000	4000	250	500	1000	2000	4000
2/17/69	-3	-1	-9	-16	-42	0	-2	-3	-11	-15

PATIENT NAME __John Q. Doe__   DATE __Feb. 17, 1969__

FIG. 6.1.8 Recording of audiometer data. Typical forms for recording audiometric data are either like the simplified table or like the audiometric curve. More data are normally included than are shown here.

AUDIOMETRIC PRACTICES

A typical audiometer for this use consists of an audio-frequency source with amplifier, attenuator, and headset (air-conduction earphone, perhaps also a bone-conduction unit). The following are the standard test frequencies: 62, 135, 250, 500, 1000, 2000, 3000, 4000, 6000 and 8000 Hz. Not all of them are available on all audiometers. The sound output is adjusted so that at each frequency the level at the ear represents the hearing norm. Suitable controls are provided; a graphic recording device may also be used. To speed up group testing, more than one set of earphones can be provided.

The basic procedure is simple; for each ear and test frequency, the sound level is slowly raised until the subject hears the tone; the level reached is the threshold and is so recorded. Typical forms are shown in Figure 6.1.8; the lower form is graphic and shows the response of both ears superimposed. The upper is an abbreviated tabular form convenient for keeping permanent records of employees.

In routine testing in industrial locations, regular tests are made only at frequencies of 500, 1000, 2000, 4000, and 6000 Hz, with occasional tests over the entire range.[1] This abbreviated test takes less time than the comprehensive one; in addition, testing at the upper and lower extremes of frequency is difficult and subject to error. At the highest frequencies, problems of coupling between earphone and eardrum often occur. Differences or scatter of 5 dB in audiograms is not uncommon; it occurs except under the best laboratory conditions.

While an audiometric testing laboratory for the best types of clinical work is an elaborate installation, a facility for routine tests can be set up in an industrial plant occupying less than 100 square feet. It can be located in a first-aid station or even in a personnel office using a commercially available isolating booth for the audiometer and the subject.

Systematic differences in threshold levels have been found when one audiometric technique is changed to another; these differences have been as large as 10 dB. Though some of these differences cannot be entirely explained, these points should be remembered: changes disclosed in a continuing series of audiograms are more likely to be reliable than any single audiogram, and uniformity and consistency in technique are essential.

HEARING AIDS

As long as the cochlea and the auditory nerve survive, hearing loss can usually be compensated with an electronic hearing aid. Many of these are available. In principle, all are alike; a microphone picks up sound, an amplifier provides more energy, and an earpiece directs it to the hearing mechanism. Even if the eardrum and middle ear are damaged, a bone-conduction unit can often carry energy to the cochlea.

If the loss in hearing acuity is considerable, speech communication may no longer be satisfactory. In such circumstances (if not sooner), the use of a hearing aid should be considered. Loss in intelligibility is the usual result of loss in high-frequency sensitivity—which is often the result of continued exposure to noise. The frequency response of the hearing aid should be tailored to compensate for the specific deficiencies of the ear; if everything through the audible spectrum is simply made louder, the ear may be so affected by the low frequencies that no gain is realized in intelligibility.

—*Howard C. Roberts*
David H.F. Liu

References

Beranek, L.L. 1954. *Acoustics*. New York: McGraw-Hill.

Jerger, James, ed. 1963. *Modern developments in audiology*. New York: Academic Press.

---

1. In audiometric screening, a rule of thumb is that if the threshold at 500, 1000 and 2000 Hz is no higher than 25 dB, no hearing impairment is assumed since many normal people show this condition. If the subject's thresholds are above 40 dB, he needs amplification to hear speech properly. Obviously, this test does not check rate of hearing loss.

6.2
NOISE SOURCES

Noise is found almost everywhere, not just in factories. Thunder is perhaps the loudest natural sound we hear; it sometimes reaches the threshold of discomfort. Jet aircraft takeoffs are often louder to the listener. Some industrial locations have even louder continuous noise. Community noise is largely produced by transportation sources—most often airplanes and highway vehicles. Noise sources are also in public buildings and residences.

Typical Range of Noise Levels

Variation in noise levels is wide. In rural areas, ambient noise can be as low as 30 dB; even in residential areas in or near cities, this low level is seldom achieved. In urban areas, the noise level can be 70 dB or higher for eighteen hours of each day. Near freeways, 90 to 100 dB levels are not unusual. Many industries have high noise levels. Heavy industries such as iron and steel production and fabricating and mining display high levels; so do refineries and chemical plants, though in the latter few people are exposed to the highest levels of noise. Automobile assembly plants, saw-mills and planing mills, furniture factories, textile mills, plastic factories, and the like often employ many people in buildings with high noise levels throughout. Hearing impairment of such employees is probable unless corrective measures are taken.

The construction industry often exposes its employees to hazardous noise levels and at the same time adds greatly to community noise. Community noise may not be high enough to damage hearing (within buildings) and yet have an unfavorable effect on general health.

Transportation contributes largely to community noise. The public may suffer more than the employees—the crew and passengers of a jetliner do not receive the high noise level found along the takeoff and approach paths. The drivers of passenger cars often are less bothered by their own noise than are their fellow drivers, and they are less annoyed than residents nearby for psychological reasons.

Noise levels high enough to be harmful in their immediate area are produced by many tools, toys, and other devices. The dentist's drill, the powder-powered stud-setting tool used in building, home workshop tools, and even hi-fi stereo headphones can damage the hearing of their users. They are often overlooked because their noise is localized.

Some typical noise sources are listed in Table 6.2.1 and are classified by origin.

TABLE 6.2.1 NOISE LEVELS FROM VARIOUS AREAS

Noise Sources[a]	Noise Levels dB or 0.0002 μ bar
Industrial	
Near large gas-regulator, as high as	150
Foundry shake-out floor, as high as	128
Automobile assembly line, as high as	125
Large cooling tower (600')	120–130
Construction and mining	
Bulldozer (10')	90–105
Oxygen jet drill in quarry (20')	128
Rock drill (jumbo)	122
Transportation	
Jet takeoff (100')	130–140
Diesel truck (200')	85–110
Passenger car (25')	70–80
Subway (in car or on platform), as high as	110
Community	
Heavy traffic, business area, as high as	110
Pneumatic pavement-breaker (25')	92–98
Power lawn mower (5'), as high as	95
Barking dog (250'), as high as	65
Household	
Hi-fi in living room, as high as	125
Kitchen blender	90–95
Electric shaver, in use	75–90

[a]Figures in parentheses indicate listening distance. Where a range is given, it describes the difference to be expected between makes or types.

Characteristics of Industrial Noise

Industrial noise varies in loudness, frequency components, and uniformity. It can be almost uniform in frequency response (white noise) and constant in level; large rotating machines and places such as textile mills with many machines in simultaneous operation are often like this. An automobile assembly line usually shows this steady noise with many momentary or impact noises superimposed on it. Other industries show continuous background noise at relatively low levels with intermittently occuring periods of higher noise levels.

Such nonuniform noises are likely to be more annoying and more fatiguing than steady noise, and they are more difficult to evaluate. The terms used to describe them are sometimes ambiguous. Usually the term *intermittent* refers to a noise which is *on* for several seconds or longer—perhaps for several hours—then *off* for a comparable time.

TABLE 6.2.2 PULSE LOUDNESS COMPARED TO CONTINUOUS NOISE

Sound Press Level (SPL) Increase in Continuous Noise to Give Equal Loudness (dB)	Pulse Width (ms)
25–30	0.02
19–22	1.0
13	5.0
10	10
4	50
2	100
0	500 or longer

The term *interrupted* usually has approximately the same meaning except that it implies that the *off* periods are shorter than the *on* periods. Intermittent or interrupted noises can be measured with a standard sound level meter and a clock or stopwatch.

Sounds whose duration is only a fraction of a second are called impulsive, explosive, or impact sounds. The terms are often used interchangeably for pulses of differing character, alike only in that they are short. They must be measured with instruments capable of following rapid changes or with instruments which sample and hold peak values.

The wave form of the noise can be modified appreciably by reflection before it reaches the ear, but it is usually described as either single-spike pulses or rapidly damped sinusoidal wave forms. Such wave forms can be evaluated fairly accurately by converting the time-pressure pattern into an energy spectrum and then performing a spectral analysis. A more accurate evaluation of the effect of intermittent but steady-level noise is possible through computation based on the ratio of on-to-off times.

The ear cannot judge the intensity of extremely short noise pulses or impact noises since it seems to respond more to the energy contained in the pulse than to its maximum amplitude. Pulses shorter than $\frac{1}{2}$ second, therefore, do not sound as loud as continuous noise having the same sound pressure level; the difference is as much as 20 dB for a pulse 20 ms long. (See Table 6.2.2.) Thus, the ear can be exposed to higher sound pressures than the subject realizes from sensation alone; a short pulse with an actual sound pressure level of 155 to 160 dB might seem only at the threshold of discomfort, 130 to 135 dB for continuous noise. Yet this momentary pressure is dangerously near that at which eardrum rupture or middle-ear damage can occur.

Interruptions in continuous noise provide brief rest periods which reduce fatigue and the danger of permanent hearing impairment. Conversely, intermittent periods of high noise during otherwise comfortable work sessions are annoying and tend to cause carelessness and accidents.

Industrial noises also vary in their frequency characteristics. Large, slow-moving machines generally produce low-frequency noises; high-speed machines usually produce noise of higher frequency. A machine such as a large motor-generator produces noise over the entire audible frequency range; the rotational frequency is the lowest (1800 RPM produces 30 Hz) but higher frequencies from bearing noise (perhaps brush noise too), slot or tooth noise, wind noise, and the like are also present.

A few noise spectra are shown in Figure 6.2.1, in octave-band form. Curve No. 1 of a motor-generator set shows a nearly flat frequency response; it is a mixture of many frequencies from different parts of the machine. Curve No. 2, for a large blower, shows a predominantly low-frequency noise pattern; its maximum is around 100 or 120 Hz and can be caused by the mechanical vibration of large surfaces excited by magnetic forces. Curve No. 3 is for a jet plane approaching land; it contains much high-frequency energy and sounds like a howl or scream, while the blower noise is a rumble. Curve No. 4 describes the high-pitched noise caused by turbulence in a gas-reducing valve; it is mechanically connected to pipes which readily radiate in the range of their natural frequencies of vibration. Octave-band analyses have only rather broad resolution and are suited to investigate the audible sound characteristics; the mechanical vibrations causing the noise are best analyzed by a continuously variable instrument.

The radiating area of a source affects the amount of sound emitted; not only does the total amount of acoustic energy radiated increase roughly in proportion to the area in vibration, but a pipe or duct passing through a wall emits sound on both sides of the wall. The vibration amplitude can be only a few microinches yet produce loud sounds. If the natural frequency of an elastic member is near the frequency of the vibration, its amplitude can become large unless the member is damped or the driving force isolated.

INDUSTRIAL NOISE SOURCES

In rotating and reciprocating machines, noise is produced through vibration caused by imperfectly balanced parts; bearing noise, wind noise, and other noises also exist. The amplitude of such noises varies with operating speed, usually increasing exponentially with speed. Noise frequencies cover a wide range since normally several harmonics of each fundamental are produced.

Electrical machines produce noise from magnetic as well as mechanical forces. Alternating current machines convert electrical to mechanical energy by cyclically changing magnetic forces which also cause vibration of the machine parts. These magnetic forces change in magnitude and direction as the machine rotates and air gaps and their magnetic reluctance change. The noise frequencies thus produced are related both to line frequency and its harmonics and to rotational speed. The entire pattern is quite com-

SPL (dB re 0.0002 DYNES/CM²)
100
90
80
70
60
50
40
⑥
②
⑤
③
①
④
20 75
75 150
150 300
300 600
600 1200
1200 2400
2400 4800
4800 10000
FREQUENCY BAND (Hz.)

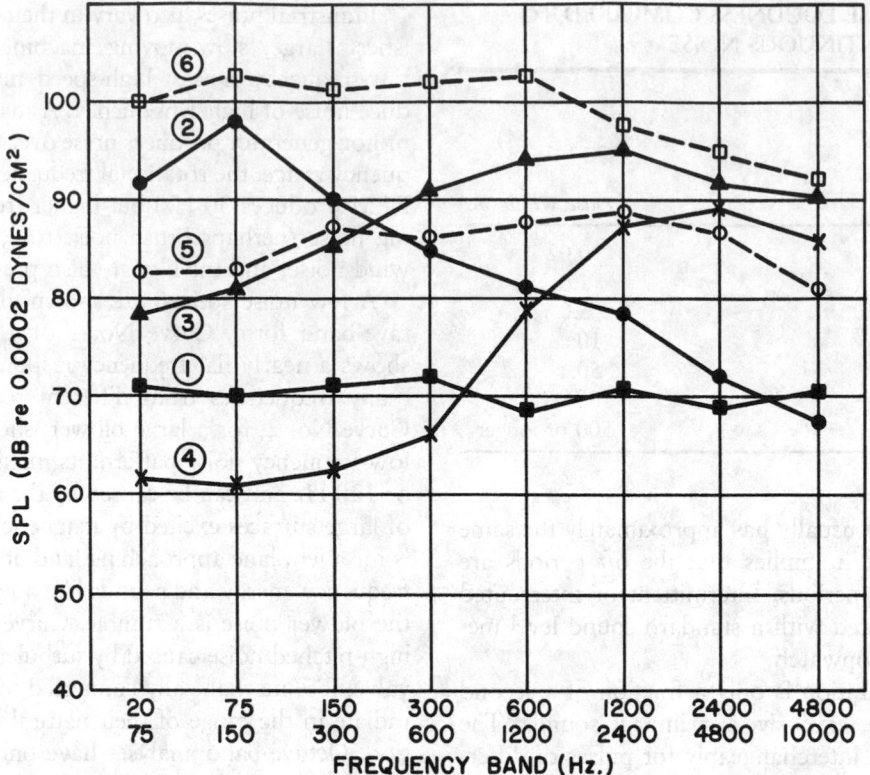

FIG. 6.2.1 Octave-band spectra of noises. *1.* Large motor-generator set (SPL 79 dBC); *2.* 150-HP blower, measured at inlet (SPL 102 dBC); *3.* Jet aircraft in process of landing, at 200 meters altitude (SPL 101 dBC); *4.* High-pressure reducing valve (SPL 91 dBC); *5.* 100-HP centrifugal pump (SPL 93 dBC); *6.* 600-HP diesel engine at 100 feet (SPL 112 dBC).

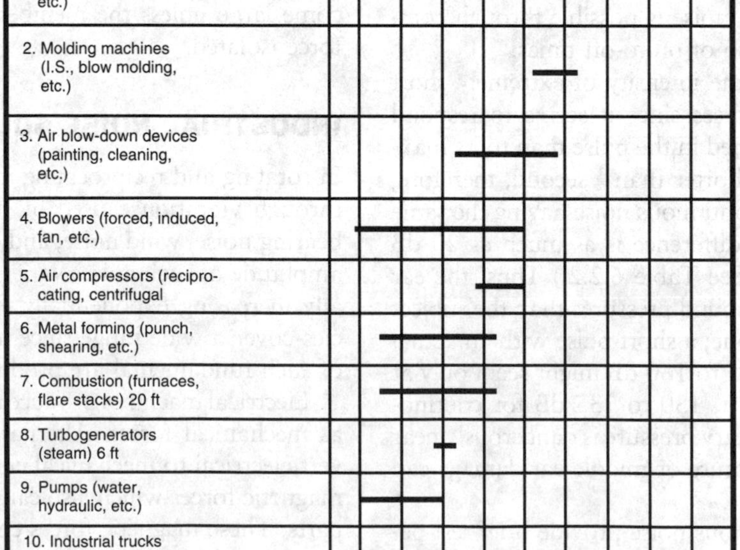

Noise levels, dB(A)
80 85 90 95 100 105 110 115 120
1. Pneumatic power tools (grinders, chippers, etc.)
2. Molding machines (I.S., blow molding, etc.)
3. Air blow-down devices (painting, cleaning, etc.)
4. Blowers (forced, induced, fan, etc.)
5. Air compressors (reciprocating, centrifugal
6. Metal forming (punch, shearing, etc.)
7. Combustion (furnaces, flare stacks) 20 ft
8. Turbogenerators (steam) 6 ft
9. Pumps (water, hydraulic, etc.)
10. Industrial trucks (LP gas)
11. Transformers

FIG. 6.2.2 Range of industrial plant noise levels at operator's position. (Reprinted from U.S. Environmental Protection Agency)

plicated. In nonrotating machines (transformers, magnetic relays, and switches), the noise frequencies are the line frequency and its harmonics and the frequencies of vibration of small parts which are driven into vibration when their resonant frequencies are near some driving frequency.

In many machines, more noise is produced by the material being handled than by the machine. In metal-cutting or grinding operations, much noise is produced by the cutting or abrading process and is radiated from both workpiece and machine.

Belt and screw conveyors are sometimes serious noise sources; they are large-area sources; their own parts vibrate and cause noise in operation, and the material they handle produces noise when it is stirred, dropped, or scraped along its path of motion. Vibration from conveyors is conducted into supports and building structure as well. Feeding devices, as for automatic screw machines, often rattle loudly.

Jiggers, shakers, screens, and other vibrating devices produce little audible noise in themselves (partly because their operating frequency is so low), but the material they handle produces much higher frequency noise. Ball mills, tumblers, and the like produce noise from the many impacts of shaken or lifted-and-dropped pieces; their noise frequencies are often low, and much mechanical vibration is around them.

Industry uses many pneumatic tools. Some air motors are quite noisy, others less so. Exhausting air is a major noisemaker, and the manner in which it is handled has much to do with the noise produced. Exhausting or venting any gas (in fact, any process which involves high velocity and pressure changes) usually produces turbulence and noise. In liquids, turbulent flow is noisy because of cavitation. Turbulence noise in gas is usually predominantly high frequency; cavitation noise in liquids is normally midrange to low frequency. Both types of noise can span several octaves in frequency range.

Gas and steam turbines produce high-frequency exhaust noise; steam turbines (for improved efficiency) usually exhaust their steam into a condenser; gas turbines sometimes feed their exhausts to mufflers. If such turbines are not enclosed, they can be extremely noisy; turbojet airplane engines are an example.

Impact noises in industry are produced by many processes; materials handling, metal piercing, metal forming, and metal fabrication are perhaps most important. Such noises vary widely because of machine design and location, energy involved in the operation, and particularly because of the rate of exchange of energy.

Not all industrial noises are within buildings; cooling towers, large fans or blowers, transformer substations, external ducts and conveyor housings, materials handling and loading, and the like are outside sources of noise. They often involve a large area and contribute to community noise. Bucket unloaders, discharge chutes, and carshakers, such as those used for unloading ore, coal, and gravel, pro-

duce noise which is more annoying because of its lack of uniformity.

Figure 6.2.2 summarizes a range of industrial plant noise levels at the operator's position. Table 6.2.3 gives some industrial equipment noises sources.

MINING AND CONSTRUCTION NOISE

Both mining and construction employ noisy machines, but construction noise is more troublesome to the general public because of its proximity to urban and residential areas.

Motor trucks, diesel engines, and excavating equipment are used in both kinds of work. Welders and rivetters are

TABLE 6.2.3   INDUSTRIAL EQUIPMENT NOISE
SOURCES

System	Source
Heaters	Combustion at burners
	Inspiration of premix air at burners
	Draft fans
	Ducts
Motors	Cooling air fan
	Cooling system
	Mechanical and electrical parts
Air Fan Coolers	Fan
	Speed alternator
	Fan shroud
Centrifugal Compressors	Discharge piping and expansion joints
	Antisurge bypass system
	Intake piping and suction drum
	Air intake and air discharge
Screw Compressors (axial)	Intake and discharge piping
	Compressor and gear casings
Speed Changers	Gear meshing
Engines	Exhaust
	Air intake
	Cooling fan
Condensing Tubing	Expansion joint on steam discharge
Atmospheric Vents, Exhaust and Intake	Discharge jet
	Upstream valves
	Compressors
Piping	Eductors
	Excess velocities
	Valves
Pumps	Cavitation of fluid
	Loose joints
	Piping vibration
	Sizing
Fans	Turbulent air-flow interaction with the blades and exchanger surfaces
	Vortex shredding of the blades

widely used in construction, especially in steel-framed buildings and shipbuilding. Pneumatic hammers, portable air compressors, loaders, and conveyors are used in both mining and construction. Crushing and pulverizing machines are widely used in mining and in mineral processing. A Portland cement plant has all of these plus ball or tube mills, rotary kilns, and other noisemakers as well. Highway and bridge construction use noisy earth-moving equipment; asphalt processing plants produce offensive fumes as well as burner noises; concrete mixing plants produce both dust and noise.

The actual noise-producing mechanisms include turbulence from air discharge; impact shock and vibration from drills, hammers, and crushers; continuous vibrations from shakers, screens, and conveyors; explosive noise; and exhaust turbulence from internal combustion engines.

Transportation Noise

Motor vehicles and aircraft are estimated to cause more urban and community noise than all other sources combined, and 60 to 70% of the U.S. population lives in locations where such transportation noise is a problem. The number of workers exposed to hazardous noise in their daily work is estimated at between 5 and 15% of the population; most of them are also exposed to the annoying, sleep-destroying general urban noise.

Table 6.2.4 lists some typical values for aircraft, motor vehicles, railways, and subways. These values are not the absolute maximum but high typical values; actual noise levels for motor vehicles, for example, are modified by the condition of the vehicle, condition of the pavement, manner of driving, tires, and surroundings. Both motor vehicles and aircraft are directional sources in that the location of the point of measurement affects the measured noise-level values.

Of all sources, aircraft noise probably causes the most annoyance to the greatest number of people. Airports are located near population centers, and approach and take-off paths lie above residential areas. Residential buildings are especially vulnerable to aircraft noise since it comes from above striking roofs and windows, which are usually vulnerable to noise penetration. The individual resident feels that he is pursued by tormenting noise against which he has no protection and no useful channel for protest.

Railway equipment has a high noise output but causes less annoyance than either highway and street traffic or air traffic. Railway noise is confined to areas adjacent to right-of-way, usually comes from extended sources, and is predictable. It is basically low frequency, thus less annoying than aircraft. Since railway equipment stays on its established routes, protection to residential areas is easily provided. Subway trains can be extremely annoying to their passengers; their noise levels are high; tunnel and station surfaces are highly reflective; and many passengers are present. Newer subway construction is less noisy than in the past (when 100 to 110 dBA was common). Subways, trolleys, and city buses all contribute considerably to urban noise and vibration.

TABLE 6.2.4 TRANSPORTATION NOISES

Source[a]	Levels (ref 0.0002 dynes/cm²)[b]	
4-engine turbojet, 400′ altitude, takeoff	117 dBC	115 PNdB
4-engine turbofan, 400′ altitude, approach	105 dBC	122 PNdB
4-engine turbojet, 4000′ laterally, takeoff	96 dBC	105 PNdB
4-engine turbofan, 4000′ laterally, approach	70 dBC	79 PNdB
Engine run-up, small business jet (1000′)	106 dBA	119 PNdB
Noise level inside airplane in flight, as high as	90 dBA	
Noise levels inside helicopter in flight, as high as	92 dBA	
Noise levels inside city bus, as high as	88 dBA	
Noise 100′ from interstate freeway	60–100 dBA	
Passenger cars, road speed (50′)	66–72 dBA	
Passenger cars, accelerating (50′)	75–91 dBA	
Motorcycles, road speed (50′)	65–87 dBA	
Motorcycles, accelerating (50′)	75–100 dBA	
Dump trucks, road speed (50′)	78–90 dBA	
Tractor-trailer, road speed (50′)	95 dBA	105 dBC
Chicago subway platform	100–110 dBA	
Chicago subway car	95–110 dBA	
New York subway platform	100–110 dBA	
Diesel freight train (500′) as high as	80 dBA	

[a]Figures in parentheses are distance to listener.
[b]Dual weightings indicate that the character of the noise does not conform to usual annoyance criteria.

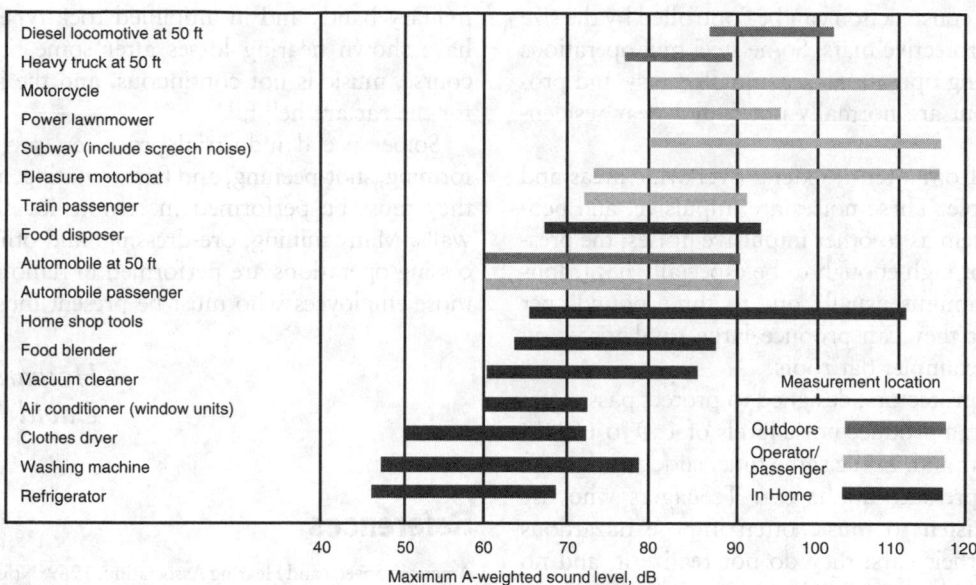

FIG. 6.2.3 Range of community noise levels. (Reprinted from U.S. Environmental Protection Agency, 1978, *Protective noise levels,* EPA 550/9-79/100 [Washington, D.C.])

Pumphouses and pipeline distributing terminals compare to other industrial locations, but the pipelines themselves present no noise problem.

Urban Noise

The distribution patterns for urban noise are quite complex and differ from city to city; yet, in general, common factors describe them.

A noise base exists twenty-four hours per day, consisting of household noises, heating and ventilating noises, ordinary atmospheric noises, and the like; this noise base is usually of low level, from 30 to 35 dB. Here and there are somewhat louder sources of noise: electrical substations, powerplants, shopping centers with roof-mounted equipment, hotels, and other buildings which do not change with the night hours.

During the day and evening hours this base level increases because of increased residential activity and also because of general widespread city traffic. A new pattern appears: in busy downtown areas traffic is heavy, on throughways and main streets extremely dense traffic occurs during rush periods with heavy traffic continually, some factories are at work, etc. Noise levels in the streets can rise to 85 to 95 dB locally. An intermittent pattern is added from emergency vehicles, aircraft, and the like. The general noise level for the entire city can increase by 10 to 20 dB. The highest noise levels remain local; after a few blocks, the noise is attenuated through scattering and reflections among buildings, and the many sources blend into the general noise pattern.

The intermittent noise pattern is usually more disturbing than the steady pattern, especially at night. Measurements near main highways and freeways show general traffic noise levels at a distance of 30 meters to be in the 65 to 80 dB range with frequent excursions up to 100 dB or even higher almost always caused by trucks but sometimes by motorcycles.

Only at the edges of urban areas does the noise level drop appreciably; and even there main highways, airports, and such can prevent a reduction. In most cities, no place is further than a few blocks from some part of the grid of principal streets carrying heavy traffic.

Important contributions are made by entertainment installations. These noise sources include music on streets and in shopping centers, amusement parks and racetracks, paging and public address systems, schools, athletic fields, and even discotheques where performances indoors are often audible several blocks away. Other offenders include sound trucks, advertising devices, and kennels or animal shelters.

Because noise sources are distributed over an urban area, the sound-power output of a source can be more informative than the noise level produced at a specific distance. Figure 6.2.3 lists some urban sources with their approximate sound-power ratings.

Specific Noise Sources

Some noise sources are so intense, so widespread, or so unavoidable that they must be characterized as specific cases.

Pile driving and building demolition involve violent impacts and large forces and are often done in congested urban areas. Some piles can be sunk with less noisy methods, but sometimes the noise and vibration must simply be tolerated; however, these effects can be minimized and the working hours adjusted to cause the least disturbance.

Blasting for such construction can be controlled by the size of charges and protective mats. Some steel mill operations and scrap-handling operations are equally noisy and produce vibration but are normally not found near residential areas.

Sonic booms from aircraft extend over wide areas and affect many people. These noises are impulsive, and people respond to them as to other impulsive noises; the pressure levels are not high enough to be especially hazardous to hearing (maximums usually one to three pounds per square foot), but they can produce large total forces on large areas, for example, flat roofs.

Some air-bag protectors, designed to protect passengers in automobiles, can produce noise levels of 140 to 160 dB inside a closed car and, at the same time, sudden increases in atmospheric pressure at the ear. Teenagers who use headphones to listen to music often impose hazardous sound levels on their ears; they do not realize it, and no one else hears it. Hearing aids have been known to produce levels so high that in an attempt to gain intelligibility, actual harm is done. In these instances, the frequency response of the unit should produce the high levels only in the frequency range where they are needed.

In construction work, explosive-actuated devices can produce high noise levels at the operator's ear. Chain saws and other portable gasoline-powered tools are used close to the operator and, thus, their noise readily reaches the ear.

The dentist's drill, used several hours per day close to the ear, is a hearing hazard. Even musicians (especially in military bands and in amplified rock-type music groups) have shown hearing losses after some time. Usually, of course, music is not continuous, and the intervals of rest for the ear are helpful.

Some special industrial processes, such as explosion-forming, shot-peening, and flame-coating, are so noisy that they must be performed in remote locations or behind walls. Many mining, ore-dressing, and other mineral-processing operations are performed in remote locations; but those employees who must be present must be protected.

—*Howard C. Roberts*
David H.F. Liu

References

American Speech and Hearing Association. 1969. Noise as a public health hazard; Conference proceedings. Report no. 4. Washington, D.C.

Sperry, W.C., J.O. Powers, and S.K. Oleson. 1968. Status of the aircraft noise abatement program. *Sound and Vibration* (August): 8–21.

U.S. Environmental Protection Agency. 1974. *Information on levels of environmental noise requisite to protect health and welfare with an adequate margin of safety.* EPA 550/9-74/004. Washington, D.C.

———. 1978. *Protective noise levels.* EPA 550/9-79/100. Washington, D.C.

———. 1981. *Noise in America: The extent of noise problems.* EPA 550/9-81/101. Washington, D.C.

———. Office of Noise Abatement and Control. 1971. *Noise from industrial plants.* Washington, D.C.

University of Washington Press. 1970. Transportation noises; a symposium on acceptability criteria. Seattle, Washington.

6.3
THE EFFECTS OF NOISE

Human response to noise displays a systematic qualitative pattern, but quantitative responses vary from one individual to another because of age, health, temperament, and the like. Even with the same individual, they vary from time to time because of change in health, fatigue, and other factors. Variation is greatest at low to moderately high sound levels; at high levels, almost everyone feels discomfort. A detailed investigation of the physiological damage to human ears is difficult, but controlled tests on animals indicate the probable type of physiological damage produced by excessively high noise levels.

Reactions to Noise

Specific physiological reactions begin at sound levels of 70 to 75 dB for a 1000 Hz pure tone. At the threshold of such response, the observable reaction is slow but definite after a few minutes. These reactions are produced by other types of stimulation, so they can be considered as reactions to general physiological stress. First the peripheral blood vessels constrict with a consequent increase in blood flow to the brain, a change in breathing rate, changes in muscle tension, and gastrointestinal motility and sometimes glandular reactions detectable in blood and urine. Increased stimulation causes an increase in the reaction, often with a change in form. These reactions are sometimes called *N-reactions*—nonauditory reactions. If the stimulus continues for long, adaptation usually occurs with the individual no longer conscious of the reaction, but with the effect continuing. Auditory responses occur as well as these nonauditory or vegetative ones. If exposure is continued long enough, TTS can occur, and a loss of some hearing acuity usually results with increasing age. Some workers refer to a "threshold of annoyance to intermittent noise" at 75 to 85 dB.

At a slightly higher level, and especially for intermittent or impulsive noise, another nonauditory response appears—the *startle effect*. Pulse rate and blood pressure change, stored glucose is released from the liver into the bloodstream (to meet emergency needs for energy), and the production of adrenalin increases. The body experiences a fear reaction. Usually psychological adaptation follows, but with changed physiological conditions.

At noise levels above 125 dB, electroencephalographic records show distorted brain waves and often interference with vision.

Most of these nonauditory reactions are involuntary; they are unknown to the subject and occur whether he is awake or sleeping. They affect metabolism; and since body chemistry is involved, an unborn baby experiences the same reactions as its mother. Sounds above 95 dB often cause direct reaction of the fetus without the brief delay required for the chemical transfer through the common bloodstream.

Most people find that under noisy conditions, more effort is required to maintain attention and that the onset of fatigue is quicker.

AUDITORY EFFECTS

Within 0.02 to 0.05 seconds after exposure to sound above the 80 dB level, the middle-ear muscles act to control the response of the ear. After about fifteen minutes of exposure, some relaxation of these muscles usually occurs. This involuntary response of the ear—the auditory reflex—provides limited protection against high noise levels. It cannot protect against unanticipated impulsive sounds; it is effective only against frequencies below about 2000 Hz. And in any case, it provides only limited control over the entrance of noise. These muscles relax a few seconds after the noise ceases.

Following exposure to high-level noise, customarily a person has some temporary loss in hearing acuity and often a singing in the ears (tinnitus). If it is not too great, this temporary loss disappears in a few hours. But if, for example, the TTS experienced in one work period has not been recovered at the start of the next work period, the effect accumulates; permanent hearing damage is almost certain if these conditions persist.

Important variables in the development of temporary and permanent hearing threshold changes include the following:

Sound level: Sound levels must exceed 60 to 80 dBA before the typical person experiences TTS.
Frequency distribution of sound: Sounds having most of their energy in the speech frequencies are more potent in causing a threshold shift than are sounds having most of their energy below the speech frequencies.

Duration of sound: The longer the sound lasts, the greater the amount of threshold shift.
Temporal distribution of sound exposure: The number and length of quiet periods between periods of sound influences the potentiality of threshold shift.
Individual differences in tolerance of sound may vary among individuals.
Type of sound—steady-state, intermittent, impulse, or impact: The tolerance to peak sound pressure is reduced by increasing the duration of the sound.

PTS

A direct relationship exists between TTS and PTS. Noise levels that do not produce TTS after two to eight hours of exposure do not produce PTS if continued beyond this time. The shape of the TTS audiogram resembles the shape of the PTS audiogram.

Noise-induced hearing loss is generally first characterized by a sharply localized dip in the hearing threshold limit (HTL) curve at frequencies between 3000 and 6000 Hz. This dip commonly occurs at 4000 Hz (Figure 6.3.1). This dip is the *high frequency notch*.

The progress from TTS to PTS with continued noise exposure follows a fairly regular pattern. First, the high-frequency notch broadens and spreads in both directions. While substantial losses can occur above 3000 Hz, the individual does not notice any change in hearing. In fact, the individual does not notice any hearing loss until the speech frequencies between 500 and 2000 Hz average more than a 25 dB increase in HTL on the ANSI–1969 scale. The onset and progress of noise-induced permanent hearing loss is slow and insidious. The exposed individual is unlikely to notice it. Total hearing loss from noise exposure has not been observed.

ACOUSTIC TRAUMA

The outer and middle ear are rarely damaged by intense noise. However, explosive sounds can rupture the tympanic membrane or dislocate the ossicular chain. The permanent hearing loss that results from brief exposure to a very loud noise is termed *acoustic trauma*. Damage to the outer and middle ear may or may not accompany acoustic trauma. Figure 6.3.2 is an example of an audiogram that illustrates acoustic trauma.

Damage-Risk Criteria

A damage-risk criterion specifies the maximum allowable exposure to which a person can be exposed if risk of hearing impairment is to be avoided. The American Academy of Ophthalmology and Otolaryngology defines hearing impairment as an average HTL in excess of 25 dB (ANSI–1969) at 500, 1000, and 2000 Hz. This limit is called the *low fence*. Total impairment occurs when the

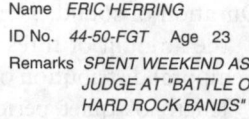

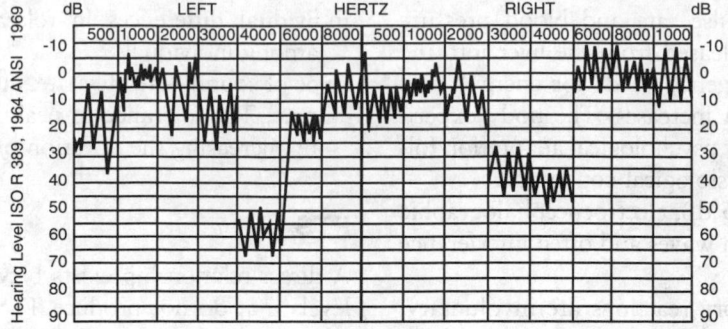

FIG. 6.3.1 An audiogram illustrating hearing loss at the high-frequency notch.

average HTL exceeds 92 dB. Presbycusis is included in setting the 25 dB ANSI low fence. Two criteria have been set to provide conditions under which nearly all workers can be repeatedly exposed without adverse effect on their ability to hear and understand normal speech.

Psychological Effects of Noise Pollution

SPEECH INTERFERENCE

Noise can interfere with our ability to communicate. Many noises that are not intense enough to cause hearing impairment can interfere with speech communication. The interference or *masking* effect is a complicated function of the distance between the speaker and listener and the frequency components of the spoken words. The speech interference level (SIL) is a measure of the difficulty in communication that is expected with different background noise levels. Now analysis talk in terms of A-weighted background noise levels and the quality of speech communication (Figure 6.3.3).

ANNOYANCE

Annoyance by noise is a response to auditory experience. Annoyance has its base in the unpleasant nature of some sounds, in the activities that are disturbed or disrupted by noise, in the physiological reactions to noise, and in the responses to the meaning of the messages carried by the noise. For example, a sound heard at night can be more annoying than one heard by day, just as one that fluctuates can be more annoying than one that does not. A sound that resembles another unpleasant sound and is perhaps threatening can be especially annoying. A sound that is mindlessly inflicted and will not be removed soon can be

more annoying than one that is temporarily and regretfully inflicted. A sound, the source of which is visible, can be more annoying than one with an invisible source. A sound that is new can be less annoying. A sound that is locally a political issue can have a particularly high or low annoyance.

The degree of annoyance and whether that annoyance leads to complaints, product rejection, or action against an existing or anticipated noise source depend upon many

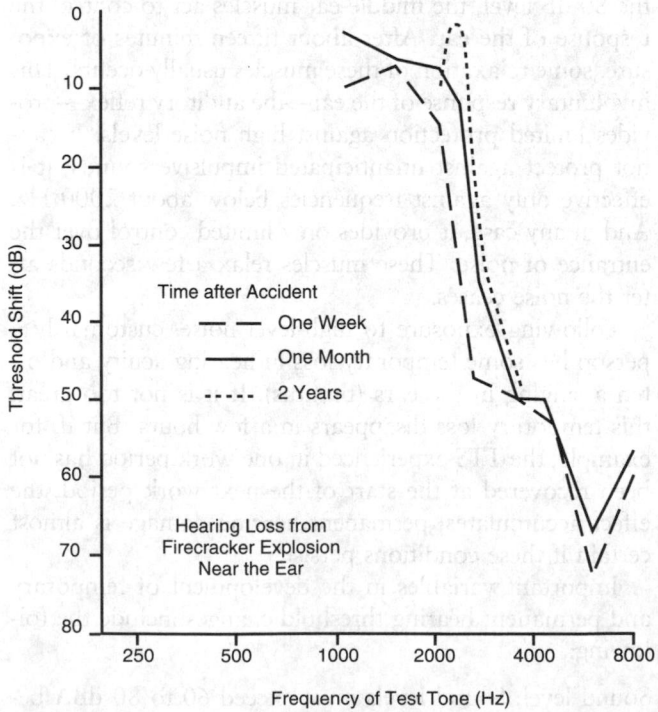

FIG. 6.3.2 An example audiogram illustrating acoustic trauma. (Reprinted, by permission, from W.D. Ward and Abram Glorig, 1961, A case of firecracker-induced hearing loss. *Laryngoscope* 71, Copyright by Laryngoscope.)

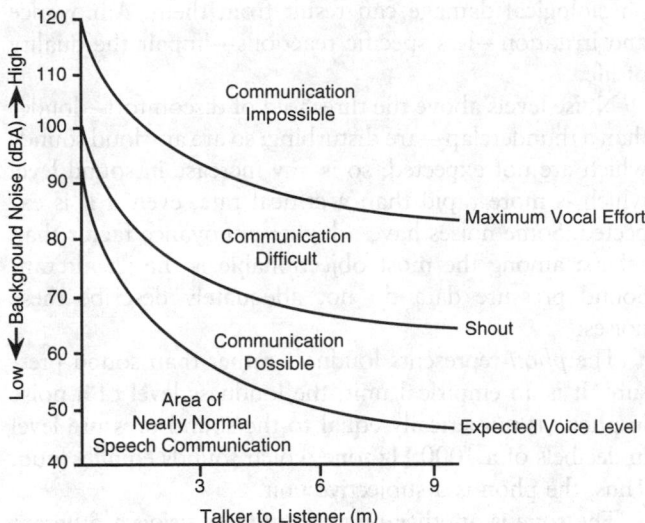

FIG. 6.3.3 Quality of speech communication as a function of sound level and distance. (Reprinted from James D. Miller, 1971, *Effects of noise on people,* U.S. Environmental Protection Agency Publication No. NTID 300.7 [Washington, DC: U.S. Government Printing Office.])

factors. Some of these factors have been identified, and their relative importance has been assessed. Responses to aircraft noise have received the greatest attention. Less information is available concerning responses to other noises, such as those of surface transportation and industry and those from recreational activities. Many of the noise rating or forecasting systems in existence were developed to predict annoyance reactions.

SLEEP INTERFERENCE

Sleep interference is a category of annoyance that has received much attention and study. Everyone has been wakened or kept from falling to sleep by loud, strange, frightening, or annoying sounds. Being wakened by an alarm clock or clock radio is common. However, one can get used to sounds and sleep through them. Possibly, environmental sounds only disturb sleep when they are unfamiliar. If so, sleep disturbance depends only on the frequency of unusual or novel sounds. Everyday experience also suggests that sound can induce sleep and, perhaps, maintain it. The soothing lullaby, the steady hum of a fan, or the rhythmic sound of the surf can induce relaxation. Certain steady sounds serve as an acoustical shade and mask disturbing transient sounds.

Common anecdotes about sleep disturbance suggest an even greater complexity. A rural person may have difficulty sleeping in a noisy urban area. An urban person may be disturbed by the quiet when sleeping in a rural area. And how is it that a parent wakes to a slight stirring of his or her child, yet sleeps through a thunderstorm? These observations all suggest that the relations between expo-

sure to sound and the quality of a night's sleep are complicated.

The effects of relatively brief noises (about three minutes or less) on a person sleeping in a quiet environment have been studied the most thoroughly. Typically, presentations of the sounds are widely spaced throughout a sleep period of five to seven hours. Figure 6.3.4 presents a summary of some of these observations. The dashed lines are hypothetical curves that represent the percent awakenings for a normally rested young adult male who adapted for several nights to the procedures of a quiet sleep laboratory. He has been instructed to press an easily reached button to indicate that he has awakened and has been moderately motivated to awake and respond to the noise.

While in light sleep, subjects can awake to sounds that are about 30–40 dBs above the level they can detect when conscious, alert, and attentive. While in deep sleep, subjects need the stimulus to be 50–80 dBs above the level they can detect when conscious, alert, and attentive to awaken them.

The solid lines in Figure 6.3.4 are data from questionnaire studies of persons who live near airports. The percentage of respondents who claim that flyovers wake them or keep them from falling asleep is plotted against the A-weighted sound level of a single flyover. These curves are for approximately thirty flyovers spaced over the normal sleep period of six to eight hours. The filled circles represent the percentage of sleepers that awake to a three-

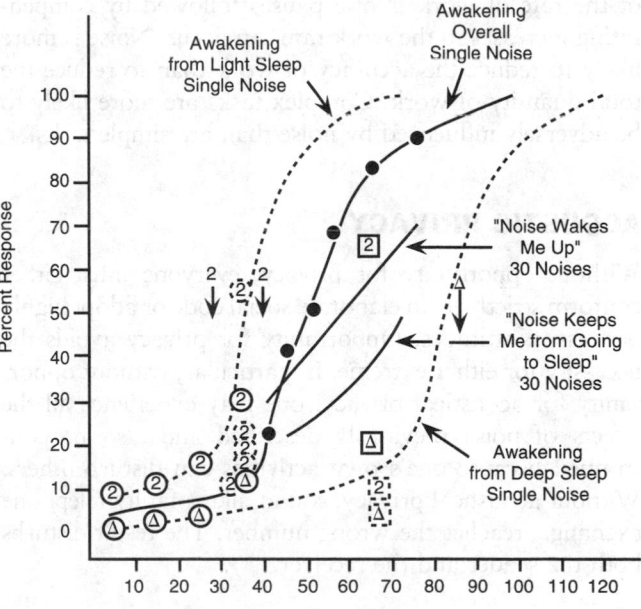

FIG. 6.3.4 Effects of brief noise on sleep. (Reprinted from J.D. Miller, 1971, U.S. Environmental Protection Agency Publication No. NTID 300.7 [Washington, DC: U.S. Government Printing Office].)

minute sound at each A-weighted sound level (dBA) or lower. This curve is based on data from 350 persons, tested in their own bedrooms. These measures were made between 2:00 and 7:00 A.M. Most of the subjects were probably roused from a light sleep.

EFFECTS ON PERFORMANCE

When a task requires the use of auditory signals, speech or nonspeech, noise at any level sufficient to mask or interfere with the perception of these signals interferes with the performance of the task.

Where mental or motor tasks do not involve auditory signals, the effects of noise on their performance are difficult to assess. Human behavior is complicated, and discovering how different kinds of noises influence different kinds of people doing different kinds of tasks is difficult. Nonetheless, the following general conclusions have emerged. Steady noises without special meaning do not seem to interfere with human performance unless the A-weighted noise level exceeds about 90 dBs. Irregular bursts of noise (intrusive noise) are more disruptive than steady noises. Even when the A-weighted sound levels of irregular bursts are below 90 dBs, they can interfere with the performance of a task. High-frequency components of noise, above about 1000–2000 Hz, produce more interference with performance than low-frequency components of noise.

Noise does not seem to influence the overall rate of work, but high levels of noise can increase the variability of the rate of work. Noise pauses followed by compensating increases in the work rate can occur. Noise is more likely to reduce the accuracy of work than to reduce the total quantity of work. Complex tasks are more likely to be adversely influenced by noise than are simple tasks.

ACOUSTIC PRIVACY

Without opportunity for privacy, everyone must either conform strictly to an elaborate social code or adopt highly permissive attitudes. Opportunity for privacy avoids the necessity for either extreme. In particular, without opportunity for acoustical privacy, one may experience all the effects of noise previously described and also be constrained because one's own activities can disturb others. Without acoustical privacy, sound, like a faulty telephone exchange, reaches the wrong number. The result disturbs both the sender and the receiver.

SUBJECTIVE RESPONSES

Except when it is a heeded warning of danger, a noise which excites a fear reflex is psychologically harmful; noises which prevent rest or sleep are a detriment to health and well-being. These reactions are psychological, yet physiological damage can result from them. Annoyance and irritation—less specific reactions—impair the quality of life.

Noise levels above the threshold of discomfort—louder than a thunderclap—are disturbing; so are any loud sounds which are not expected; so is any increase in sound level which is more rapid than a critical rate, even if it is expected. Some noises have a higher annoyance factor than others; among the most objectionable is the jet aircraft. Sound pressure data do not adequately describe these noises.

The *phon* represents loudness rather than sound pressure. It is an empirical unit; the loudness level of a noise in phons is numerically equal to the sound pressure level in decibels of a 1000 Hz tone which sounds equally loud. Thus, the phon is a subjective unit.

The *sone* is another unit of loudness using a different scale. A loudness level of 40 phons represents 1 sone, and each 10-phons increase doubles the number of sones. The change in sensation of loudness is better represented by

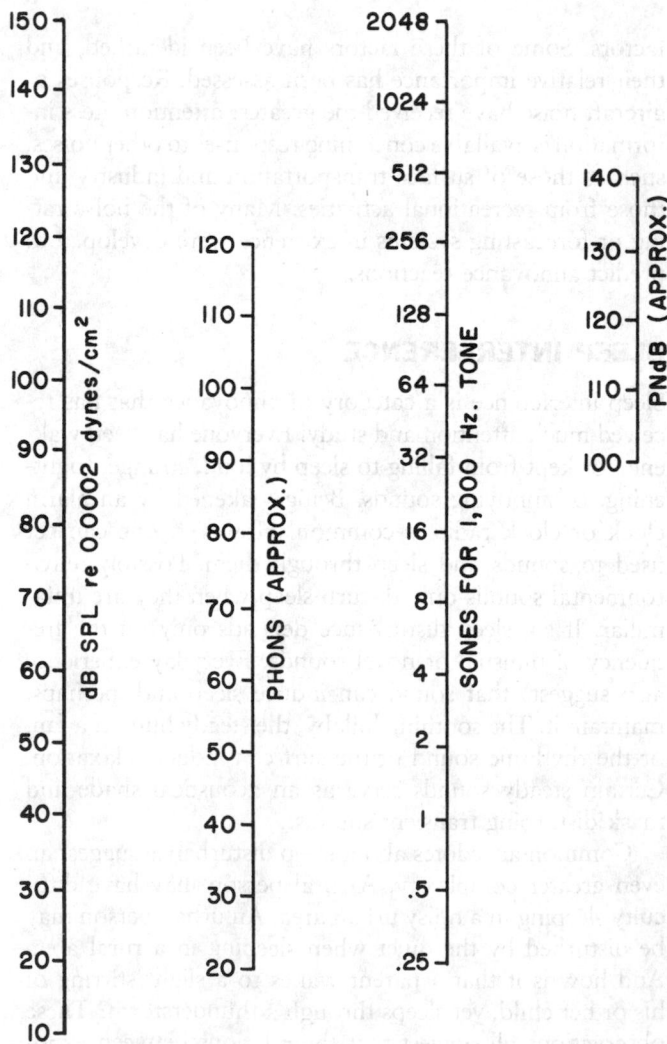

FIG. 6.3.5 Comparison of objective and subjective noise scales.

sones than by phons.

For the combined high frequencies and high noise levels produced by jet aircraft, other criteria have been developed; one is the unit of noisiness called the *noy*. This unit is used to express the perceived noisiness (PN) or annoyance in dBs, as PNdB. Such PNdB values can be conveniently approximated with a standard sound-level meter.

In Figure 6.3.5 the relative values for sound levels in objective and subjective units are compared; the chart is for comparison, not conversion.

These units deal with continuous noise, but fluctuating or intermittent noise is more annoying. To deal with these characteristics (for transportation noise), a procedure can calculate a "noise pollution level." A composite noise rating also describes the noise environment of a community over twenty-four hours of the day. Like the others, it accounts for loudness and frequency characteristics, as well as fluctuations and frequency and impulsive noises, in its calculations.

One of the most disturbing elements of noise within buildings is its impairment of privacy; voices or other noises penetrating a wall, door, or window are especially annoying. This reaction is psychological; the annoyance is in response to an intrusion, which seems impertinent.

—*Howard C. Roberts*
David H.F. Liu

References

American Standards Association Subcommittee Z24-X-2. 1954. *Relation of hearing loss to noise exposure.* (January) New York.

Botsford, J. and B. Lake. 1970. Noise hazard meter. *Journal of the Acoustical Society of America* 47:90.

Miller, J.D. 1971. *Effect of noise on people.* U.S. Environmental Protection Agency Publication no. NTID 300.7. Washington, D.C.: U.S. Government Printing Office.

Kovrigin, S.D. and A.D. Micheyev. 1965. *The effect of noise level on working efficiency.* Rept. N65-28927. Washington, D.C.: Joint Publications Research Service.

Kryter, Karl. 1970. *The effects of noise on man.* New York: Academic Press.

U.S. Department of Health, Education, and Welfare, National Institute for Occupational Safety and Health. 1972. *Criteria for a recommended standard: Occupational exposure to noise.* Washington, D.C.: U.S. Government Printing Office.

6.4
NOISE MEASUREMENTS

Available Instruments
Portable and precision sound-level meters, sound monitors, noise-exposure integrators, audiometers, octave-band analyzers, graphic and wide-band recorders, loudness, computers, and others.

Power Required
Portable instruments are usually battery powered. Laboratory instruments or computing-type loudness meters are supplied from normal power lines with 117 volts, 60 cycle ac. Their power demand is less than 500 volt-amperes.

Range of Operations
Usually from 30 to 140 dB. Usually from 20 to 20,000 Hz.

Cost
Sound-level meters, from $500 up. Frequency analyzers, from $2000 up.

Partial List of Suppliers
Ametek, Inc., Mansfield and Green Div.; B & K Instruments, Inc.; Larsen David Laboratory; Quest Technologies; H.H. Scott.

Noise measurements are usually conducted for one of three purposes:

To understand the mechanisms of noise generation so that engineering methods can be applied to control the noise
To rate the sound field at various locations on a scale related to the physiological or psychological effects of noise on human beings
To rate the sound power output of a source, usually for future engineering calculations, that can estimate the sound pressure it produces at a given location

This section describes a few frequently used terms and units proposed for the study of sound and noise; most are quite specialized. It also describes techniques and instruments to measure noise.

Basic Definitions and Terminology

Sound waves in air can be described in terms of the cyclic variation in pressure, in particle velocity, or in particle displacement; for a complete description, frequency and wave-form data are also required.

Sound pressure is the cyclic variation superimposed upon the steady or atmospheric pressure; usually it is the RMS value. An RMS value is determined by taking the square root of the arithmetic mean of the instantaneous values over one complete cycle for a sine wave, or for as many cycles of a nonsinusoidal wave form as are necessary for a reliable sample. The units of sound pressure are force per unit area—dynes per square centimeter or newtons per square meter. Particle displacement is in centimeters. Most sound pressures are given in RMS values, and most sound level meters display RMS values.

To describe the range of sound pressures in a logarithmic scale is convenient, the unit of SPL is the dB, described by

$$SPL = 20 \log_{10} (P/P_{ref}) \text{ dB} \qquad 6.4(1)$$

where P is measured sound pressure, and P_{ref} is the reference pressure ordinarily used. The customary reference pressure is 0.0002 dynes/cm², or 0.0002 μ bars. (One standard atmosphere is equal to 1,013,250 microbars, so 1 microbar is nearly 1 dyne/cm². The reference level should always be stated when sound pressure levels are given, as dB re 0.0002 dynes/cm².)

Sound power—the *acoustic power* produced by a source—is described in watts. Again a logarithmic scale is used to accomodate the wide range involved, without inconveniently clumsy figures. The unit again is the dB. The PWL is expressed by

$$PWL = 10 \log_{10} (W/W_{ref}), \text{ or } 10 \log_{10} (W/10^{-12}) \text{ dB} \qquad 6.4(2)$$

where W is the acoustic power in watts, and W_{ref} is the reference level which should always be stated; the reference level ordinarily used is 10^{-12} watt. Since the power ratio 10^{-12} can also be written as 120 dB, Equation 6.4(3) is convenient to write as:

$$PWL = 10 \log_{10} W + 120 \text{ dB re } 10^{-12} \text{ watt} \qquad 6.4(3)$$

Sound pressures and sound power values are physical magnitudes, expressed in physical terms. SPLs and PWLs are ratios (the ratio of a measured value to a reference value) expressed in logarithmic terms called dBs.

Other terms and quantities used in noise control work will be defined as they are used. Sound pressures and sound powers are basic. The ear responds to sound pressure waves, and nearly all sound magnitude measurements are in terms of sound pressure. Sound power determines the total noise produced by a machine and, thus, is important in machine design.

Frequency Sensitivity and Equal Loudness Characteristics

The ear is most sensitive in the range of frequencies between 500 and 4000 Hz, less sensitive at higher frequencies and much less sensitive at low frequencies. This range of greatest sensitivity coincides with the range for voice communication.

Through listening tests, this variation in sensitivity has been evaluated. The curves in Figure 6.4.1 present these data. These curves are commonly called *equal loudness curves*, but *equal sensation curves* describes them more accurately. They describe several of the ear's characteristics. The lowest curve represents the threshold of hearing for the healthy young ear. The dotted portions of the curves (not a part of the ISO recommendation from which these

data come) indicate the change in hearing which occurs with increasing age; they show not noise-induced presbycusis but sociocusis. These dotted curves describe a loss in hearing acuity within the intelligibility range of frequencies.

The curves also show that as the amount of energy increases, the difference in sensitivity almost disappears. The thresholds of discomfort and pain (not part of this figure) actually fall quite close to the 120 and 140 dB levels respectively.

OBJECTIVE AND SUBJECTIVE VALUES

Sound pressure and sound power are objective values; they show physical magnitudes as measured by instruments. However, while almost anyone subjected to noise exposure beyond recognized levels experiences some hearing impairment, some psychophysiological reactions (annoyance in particular) vary with the individual. They are subjective values; they must be determined in terms of human reactions.

Loudness is a subjective magnitude. Although it depends primarily on signal intensity, or sound pressure, frequency and wave form are also important. (See Figure 6.4.1) Through listening tests, a unit of loudness level has been established; it is the *phon*. The loudness level of a sound in phons is numerically equal to that SPL in dBs of a 1000 Hz continuous sine wave sound, which sounds equally loud. For most common sounds, values in phons do not differ much from SPLs in dBs.

To classify the loudness of noises on a numerical scale, the *sone* was devised as the unit of loudness. It is related to the loudness level in phons in this way: a noise of 40 phons loudness level has a loudness of 1 sone, and for each increase in level of 10 phons, the value in sones is doubled. The sone has the advantage that a loudness of 64 sones sounds about twice as loud as 32 sones; it provides a better impression of relative loudness than the dB.

Sone values are not measured directly, but they can be obtained by computation. Two general methods are accepted—one by Stevens and one by Zwicker. Their values differ slightly, but either seems satisfactory for most uses. (The Zwicker method is built into a commercial instrument; it involves separating the sound into a group of narrow-band components, then combining the magnitudes of these components mathematically.) Table 6.4.1 compares some values of sound pressure, loudness, and noisiness.

Jet airplane noise, with its broad band but predominantly high-frequency spectrum, is one of the most annoying. It is also one of the loudest continuous noises. Values in sones or phons are not adequate to describe noises of this character, and the concept of *perceived noise* has developed. Noisiness in this system is usually expressed in dBs, as PNdB. Perceived noise values, like values in sones, are not directly measurable, but are computed from measured data. For many uses, analysts can make acceptable approximations by taking measurements with a stan-

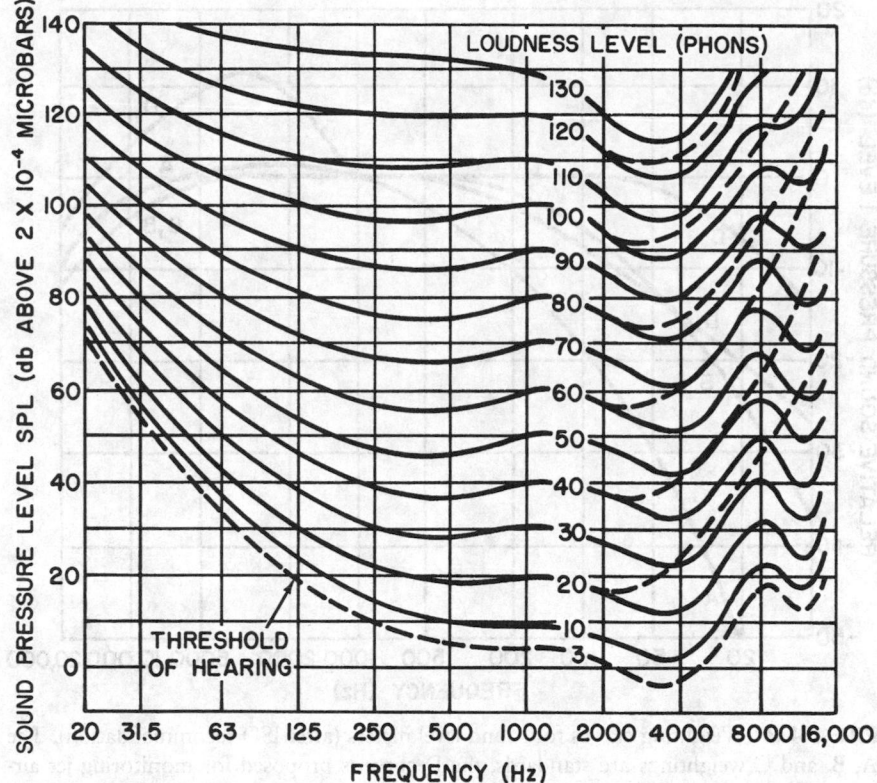

FIG. 6.4.1 Equal loudness curves. These curves display the varying sensitivity of the normal ear with both frequency and average level. The interrupted extensions at high frequencies show the typical loss in hearing acuity with age (Reprinted partially from ISO recommendation 226).

dard sound level meter, using D weighting, and adding 7 to the observed values. For jet plane noises above 90 dBA, analysts can secure useful approximate values by taking sound-level readings using A weighting, and adding 12 to the indicated value. Since not all sound-level meters have D weighting, this latter method is often used.

WEIGHTING NETWORKS

SPL measurements are made with instruments which respond to all frequencies in the audible range; but since the sensitivity of the ear varies with both frequency and level, the SPL does not accurately represent the ear's response. This condition is corrected by weighting characteristics in sound level meters.

Weighting networks modify the frequency response of the instrument so that its indications simulate the ear's sensitivity. One, for A weighting, gives readings representing the ear's response to sounds near the 40 dB level; another, B weighting, approximates the response of the ear at about 70 dB values, and C weighting is used for levels near 100 dB or higher. Readings taken with these weightings are

TABLE 6.4.1   COMPARISON OF NONEQUIVALENT NOISE UNITS

Loudness Level[a] (phons)	Description	Loudness (sones)	Sound Level (dBA)	Perceived Noise Level (PNdB)
140	Threshold of pain	1,024	140	153
125	Automobile assembly line	362	125	138
120	Jet aircraft	256	120	133
100	Diesel truck	64	100	112
80	Motor bus (50′)	16	80	
60	Low conversation	4	60	
40	Quiet room	1	40	
20	Leaves rustling	0.25	20	

[a]Only sones and phons are rigorously related mathematically; the other values are for comparison only.

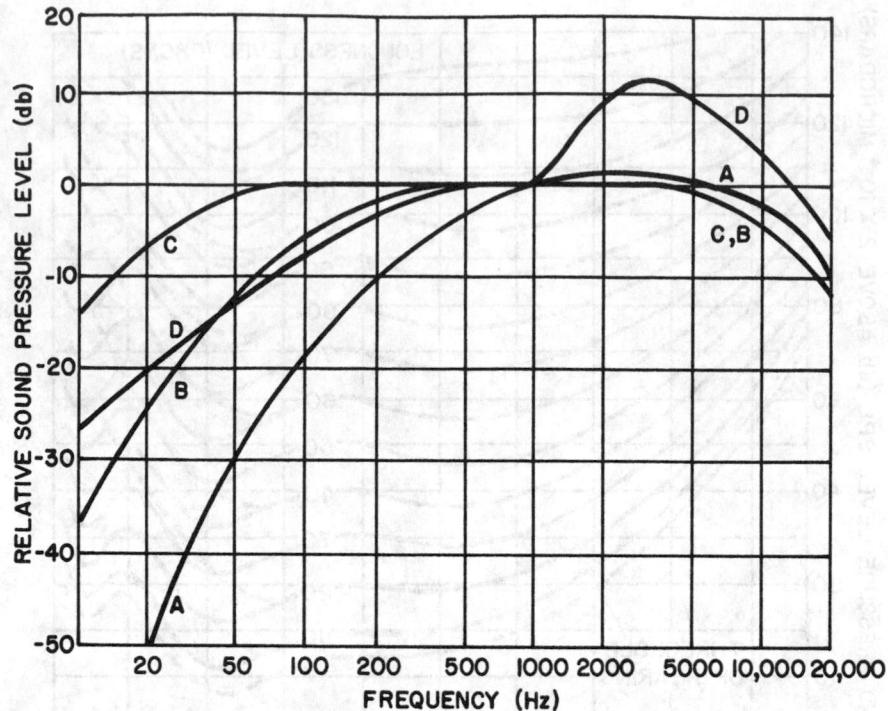

FIG. 6.4.2 Weighting curves for sound level meters (after ISO recommendation). The A, B, and C weightings are standards; the D curve is proposed for monitoring jet aircraft noise.

identified as dBA, dBB, dBC, etc; they are called sound level readings to distinguish them from SPL readings which are not modified by frequency.

Many sound level meters have a fourth or *flat* characteristic which is slightly more uniform than C weighting. It is used in frequency analysis. Still another characteristic, D weighting, is proposed for jet plane noise measurements. Figure 6.4.2 shows the response curves provided by these weighting networks.

FREQUENCY ANALYSIS OF NOISE

The frequency characteristics of sound are important; they describe its annoyance factor as well as its potential for hearing damage. They indicate to the noise control analyst probable sources of noise and suggest means for confining them. To the mechanical or design engineer, frequency analysis can show the source of machine vibrations which produce noise and contribute to damage. Noise is often a symptom of malfunctioning, and the frequency analysis can sometimes describe the malfunction. Spectral characteristics are also useful in describing the transmission of sound through a wall or its absorption by some material.

Octave-band analyses are not difficult to make, even in the field, and are adequate for noise control work, though not for machine design. They are usually made with sets of bandpass filters attached to a sound-level meter using the flat-frequency response. The filters are designed so that each passes all frequencies within one octave; but at the edge of the pass-band, the transmission falls off sharply. The transmission at one-half the lower band-edge frequency (or at twice the upper) is at least 30 dB below the pass-band transmission; and it is at least 50 dB below pass-band transmission at one-fourth the lower (or four times the upper) edge of the pass-band frequency.

In computing loudness from measured data, one-tenth, one-third, one-half, and full octave-band analyzers are used. In describing the noise-transmission characteristics of objects such as walls and doors and classifying noise environments for speech interference, analysts ordinarily use full octave-band analyzers. Table 6.4.2 lists the commonly used center frequencies and the frequency ranges for octave-band and one-third octave filters. Separate bandpass filter sets are often used for these analyzers; for one-tenth octave and narrower band analyzers, tunable electronic units are widely used. Narrow-band analyzers are used in studying the characteristics of machines.

SPEECH INTERFERENCE AND NOISE CRITERIA (NC) CURVES

Interference with the intelligibility of speech is a serious problem caused by noise; it impairs comfort, efficiency, and safety. The amount of such interference depends on both frequency and level of sound; and a family of curves has been developed to describe various noise environments. These are called *NC curves* and are shown in Figure 6.4.3.

TABLE 6.4.2 BAND PASS FILTER DATA; PREFERRED NUMBER SERIES[A]

Octave Band Frequencies (Hz)			One-third Octave Band Frequencies (Hz)		
Low Band Edge	Center Frequency	High Band Edge	Low Band Edge	Center Frequency	High Band Edge
22	31.5	44	14	16	18
44	63	88	18	20	22
88	125	176	22	25	28
176	250	353	28	31.5	35
352	500	706	35	40	45
706	1000	1414	45	50	56
1414	2000	2828	56	63	71
2828	4000	5656	71	80	90
5656	8000	11,312	90	100	112
11,312	16,000	22,614	112	125	140
			140	160	179
			179	200	224
			224	250	280
			280	315	353
			353	400	448
			448	500	560
			560	630	706
			706	800	897
			897	1000	1121
			1121	1250	1401
			1401	1600	1794
			1794	2000	2242
			2242	2500	2803
			2803	3150	3531
			3531	4000	4484
			4484	5000	5605
			5605	6300	7062
			7062	8000	8968
			8968	10,000	11,210
			11,210	12,500	14,012
			14,012	16,000	17,936
			17,936	20,000	22,421

[a]ISO Recommendation 266 and USAS SL6-1960 listed sets of preferred numbers which were recommended for use in acoustic design. Most filters are now designed on this basis. The tables give octave-band and one-third octave-band filter characteristics, using preferred numbers as the center frequencies from which the band-edge frequencies were computed. Older filter sets used slightly different frequencies; they are interchangeable in all ordinary work.

Each NC curve describes a set of noise conditions; the acoustic environment suitable for a need is specified with a single number. For example, a concert hall or auditorium should be NC-25 or better; a private office is NC-35. That is, octave-band SPLs measured in these areas should not exceed, at any frequency, the values specified by the appropriate NC curve.

These criteria are related to the psychological characteristics of the ear and, consequently, the shape of the curves is like the equal-sensation curves. For broad-band noise, if 7 to 9 units is subtracted from the sound level measured in dBA, the difference approximates the NC-curve rating. If the ear can detect some frequency which seems to dominate an otherwise uniform background noise, this rough criterion is not acceptable. An octave-band analysis is needed, especially if specifications are being checked.

Vibration and Vibration Measurement

Noise is usually accompanied by vibration; noise is caused by vibration; noise causes vibration. The physiological effects of vibration have not been studied as intensively as those of sound, but usually discomfort provides warning so that hazard is easily anticipated. However, exposure to vibration often contributes to fatigue and, thus, to loss in efficiency and to accidents. Vibration study is an important part of noise control, especially in analyzing problems.

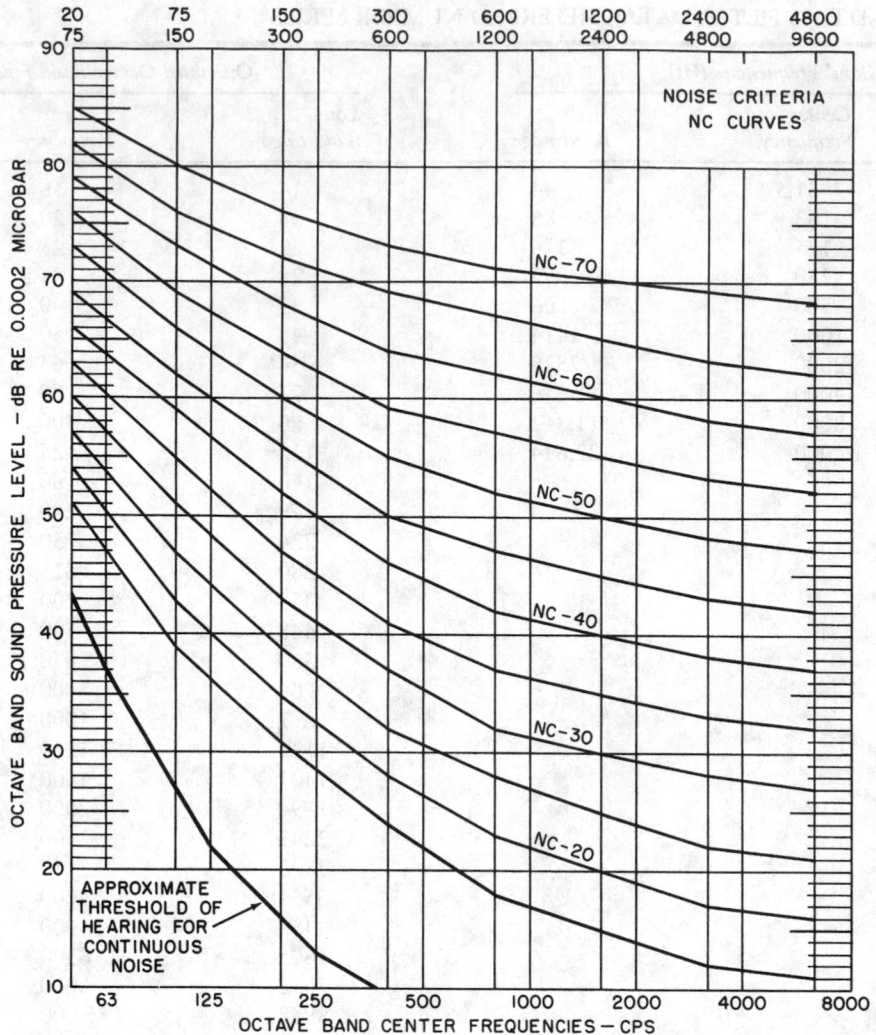

FIG. 6.4.3 Noise criteria curves. The abscissa shows both the older bandlimit frequencies (at top) and the band-center frequencies based on the preferred-number series (at bottom). Both are still in use.

Roughly speaking, vibration becomes perceptible when its amplitude reaches about 2×10^{-3} millimeters at 50 Hz and seems intolerable at about 8×10^{-2} millimeters at 50 Hz. These estimates assume exposure for rather long periods and close contact to the vibrating surface—as riding in a vehicle. Vibration of large surfaces at these amplitudes produces high sound pressures. Individuals vary in their tolerance of vibration, but probably few people are disturbed by vibrations with accelerations less than 0.001 times that of gravity, or 1 cm/sec², and many can tolerate ten times that much.

Vibration can be described in terms of its frequency and either its acceleration, velocity, or amplitude as

$$a = -2\pi f v = -(2\pi f)^2 x \qquad 6.4(4)$$

where a is acceleration, v velocity, x is amplitude of displacement, and f is frequency.

Vibration is measured by mechanical, optical, or electrical means (Lipták 1970) depending on conditions. The low-frequency, large-amplitude vibrations of an elastically suspended machine might be measured with a ruler, by eye. The high-frequency, small-amplitude vibrations of a high-speed motor are best measured with an electrical instrument which is sensitive to acceleration, velocity, or displacement.

General purpose vibration meters often use a measuring element sensitive to acceleration over a range of frequencies. The instrument also shows velocity by performing one integration or displacement by performing two integrations on the acceleration signal. The integration is done electronically, within the instrument.

Vibratory force applied to an elastic membrane produces vibratory motion, thus sound pressure waves cause a window or partition to vibrate. Vibratory force applied to an elastically supported mass produces vibratory motion, as in a machine with vibration-isolating supports. In both cases, the amplitude of vibration is affected by the ratio of the frequency of the applied force—the *forcing fre-*

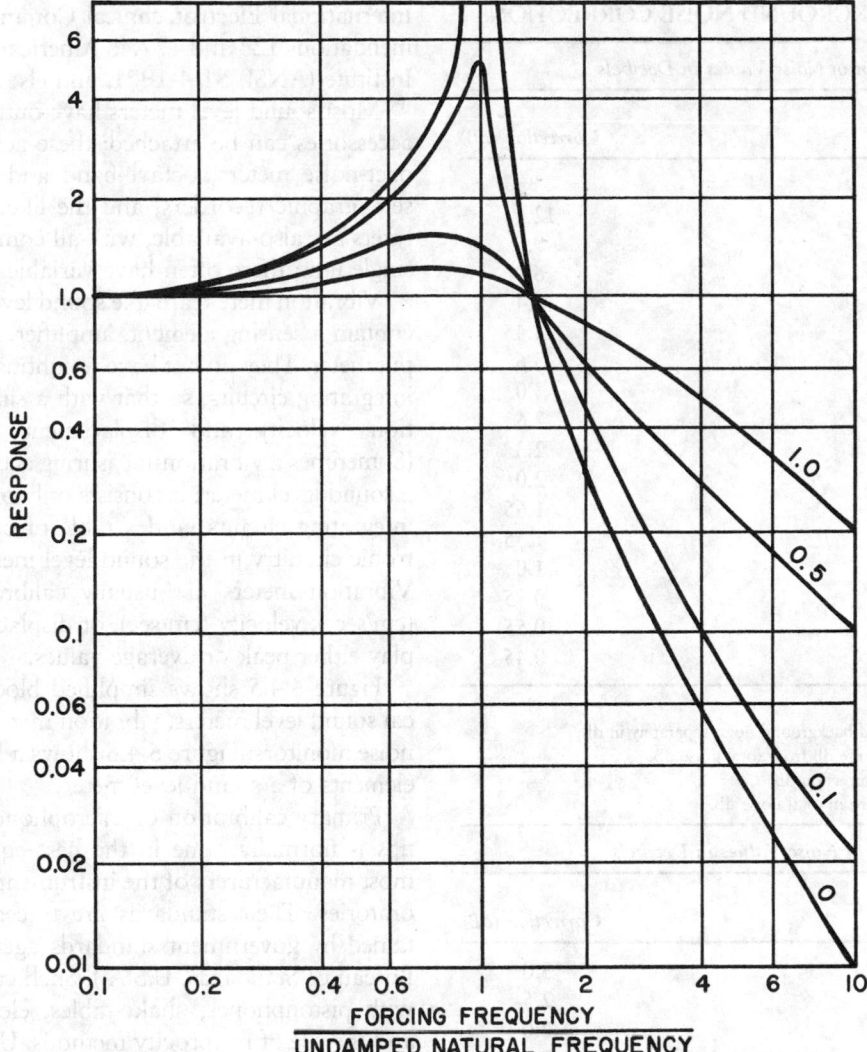

FIG. 6.4.4 Response of a resonant system. The response of an elastic system excited by frequencies near its undamped resonance frequency varies with both the forcing frequency and the amount of damping. Damping from 0 to 1 times critical is shown.

quency—to the natural frequency of the object. Other important parameters are the resisting force (proportional to effective mass) and the damping ratio (proportional to the amount of energy dissipated through friction). Figure 6.4.4 shows the relation between the frequency ratio and the response of the system (no units given for response in this figure).

Measuring Noise

BACKGROUND CORRECTIONS

Sound-measuring instruments do not distinguish between the noise of principal interest and any background noise present. Background noise corrections are needed to determine the contribution of a specific source or a group of sources.

The simplified procedure is as follows: the noise level is measured with the unknown source(s) in operation; then with the other conditions unchanged, the unknown source is stopped and the background level measured. (If several different sources are being measured, they can be turned off in different combinations.) Analysts must evaluate the difference between the readings on an energy basis, not simply by taking numerical differences.

Table 6.4.3 provides a convenient means for adding or subtracting dB values, accurate enough for most work. Background noise corrections subtract the background from the total noise. For example, if the total noise measures 93 dB and the background noise measures 89 dB, the level for the unknown source is 90.7 dB. Using these tables, analysts can also add dB values; the values for more than one source are measured separately, then combined in pairs. Three sources, of 85, 87, and 89 dB, add to give 92 dB. In both addition and subtraction, the dB values must be obtained in the same manner, including weighting.

TABLE 6.4.3 BACKGROUND NOISE CORRECTIONS

Subtraction of Noise Values in Decibels

Total Noise Minus Background Noise (dB)	Correction (dB)
0	∞
0.25	12.5
0.50	9.6
1.0	6.8
1.5	5.4
2.0	4.45
2.5	3.6
3.0	3.0
3.5	2.6
4.0	2.3
4.5	2.0
5.0	1.65
6.0	1.35
7.0	1.0
8.0	0.75
9.0	0.55
10.0	0.45

Procedure:
1. Measure total noise and background noise separately in dB.
2. Subtract background noise dB from total noise dB.
3. Consult table to find correction dB.
4. Subtract correction dB from total noise dB.

Addition of Noise Values in Decibels

Difference Between the Two Values (dB)	Correction (dB)
0	3.0
1	2.5
2	2.0
3	2.0
4	1.5
5	1.0
6	1.0
7	1.0
8	0.5
9	0.5
10	0.0

Procedure:
1. Measure the two sources separately in dB.
2. Subtract the smaller from the larger value.
3. Consult table to find correction dB.
4. Add correction to the higher of the two values.

INSTRUMENTS FOR MEASURING NOISE

The basic instrument for measuring noise levels is the sound level meter, sensitive to RMS sound pressures between about 20 and 20,000 Hz. It is equipped with weighting networks, fast and slow response, an attenuator with 10-dB steps, and an indicating meter which spans 16 dBs, from −6 to +10 dB. It operates over a total range of about 30 to 140 dB sound pressure level. [Minimum specifications for general-purpose sound level meters are in the International Electrotechnical Commission (IEC) Recommendations 123 and 177, in American National Standards Institute (ANSI) S1-4-1971, and elsewhere.]

Most sound level meters have output terminals so that accessories can be attached; these accessories include impact-noise meters, octave-band and ⅓ octave-band filter sets, graphic recorders, and the like. Self-contained analyzers are also available, with all components housed in a single unit; these often have variable width settings.

Vibration meters are like sound level meters in that they contain a sensing element, amplifier, attenuator, and output meter. They do not have weighting circuits but do have integrating circuits, so that with a single pickup acceleration, velocity and displacement can be measured. (Sometimes a vibration-measuring accessory operates with a sound level meter; it consists only of a vibration pickup, integrating circuits, and a table of conversions; the electronic circuitry in the sound level meter operates with it.) Vibration meters are usually calibrated in acceleration (cm/sec^2), velocity (cm/sec), or displacement (cm) and display either peak or average values.

Figure 6.4.5 shows simplified block diagrams of typical sound level meters, vibration meters, audiometers, and noise monitors; Figure 6.4.6 shows a block diagram of the elements of a sound level meter.

Primary calibration of microphones or vibration pickups is normally done in the best equipped laboratories; most manufacturers of the instruments maintain such laboratories. Their standards are traceable to those maintained by government standards agencies (e.g., National Bureau of Standards, U.S.A.). Such calibrations are made with pistonphones, shake-tables, electrostatic actuators, and the use of reciprocity methods. Until recent years, the Rayleigh disk was widely used. Reciprocity calibrators are now easily available, and small electronic generators are widely used for field checks.

International standards and recommendations are available from the International Organization for Standardization, Geneva, Switzerland; U.S. standards are available from the U.S. ANSI, New York City.

Audiometers are basically audio-frequency oscillators, adjustable in frequency and in output level, with headsets for the subject. The subject determines the existence of the threshold. Some audiometers are more complex, having recording devices and masking noise facilities.

Since loudness is a subjective magnitude, it cannot be measured with a simple instrument. Arbitrary systems have been developed for determining loudness by instrumental measurements, using rather complex empirical criteria and techniques. A self-contained instrument is available, which analyzes the sound as it is received, breaks it up into *critical bands*—actually ⅓ octave bands—performs a number of calculations on these data, and produces a single-figure output. This equipment has been incorporated into a system for continuously monitoring impulse and continuous noise and for printing out loudness levels and occurrence times.

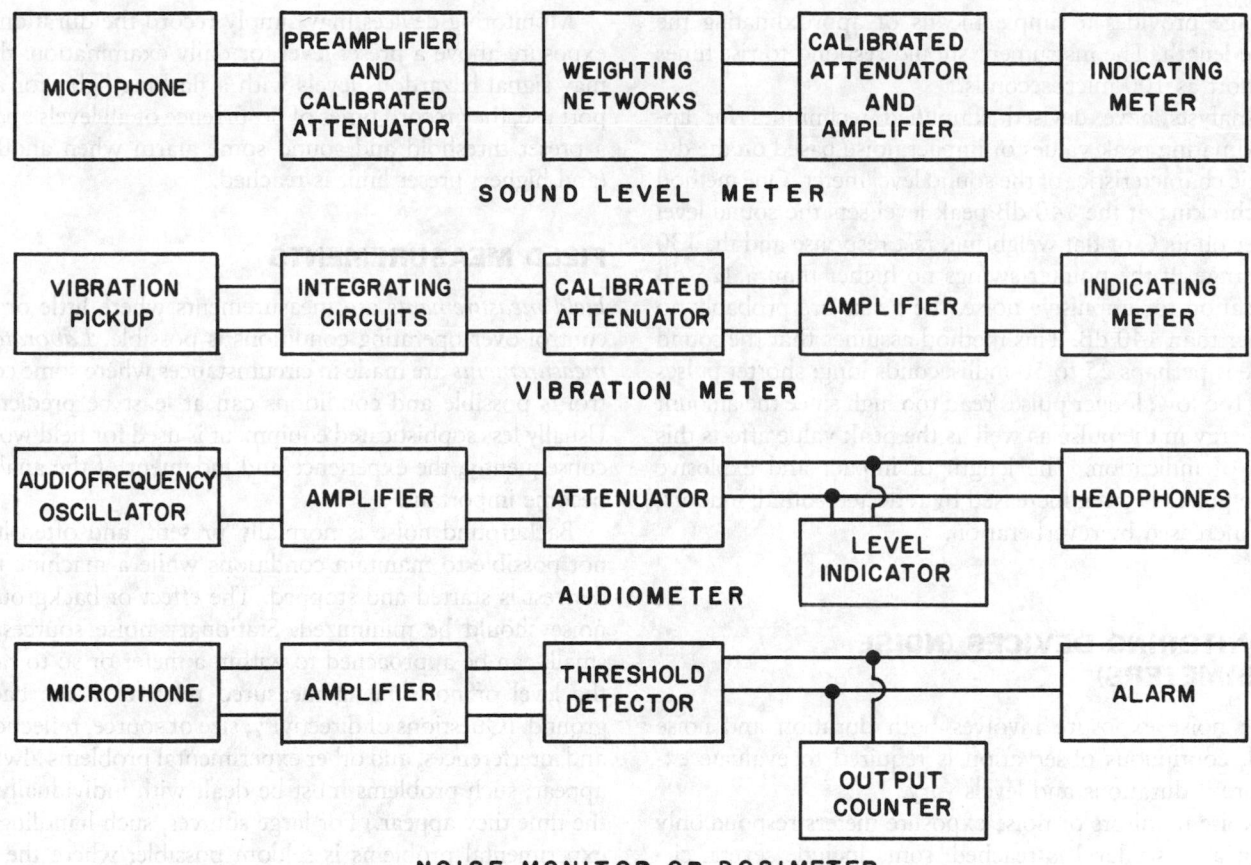

FIG. 6.4.5 Simplified block diagram of noise-measuring instruments.

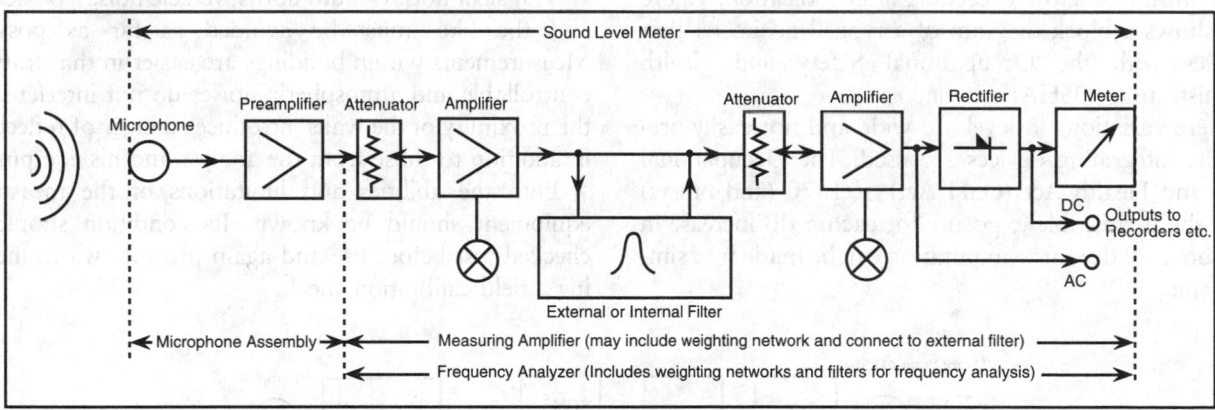

FIG. 6.4.6 Block diagram of the elements of a sound level meter.

IMPACT AND IMPULSE MAGNITUDES

The brief noise pulses (lasting only a fraction of a second) caused usually by impacts or explosions require special means of measurement. Even the fast setting of a standard sound level meter usually requires about 0.2 second to reach its deflection, and an overshoot also occurs.

Analysts can ideally measure true impulse or impact magnitudes by recording the wave form on an oscillograph for detailed analysis, and then determining frequencies and total energy as well as peak magnitudes. Most impulse noise meters are somewhat simpler.

A typical impact-noise measuring instrument can determine the peak pressure (by a sample-and-hold circuit or by a peak-reading voltmeter) and the time–average (the average voltage over a chosen period of time). The result is indicated on a pointer-type instrument, where the reading is retained long enough to permit accurate readings. The time–average mode of operation can average over any selected period of time from 2 milliseconds to 1 second; this average provides a simple means of approximating the pulse length. The instrument should respond to rise times as short as 100 microseconds.

Analysts have devised simplified techniques for approximating peak values of impact noise based on the dynamic characteristics of the sound level meter. One method for checking at the 140 dB peak level sets the sound level meter on its C or flat weighting, fast response and the 130 dB range. If the pointer swings no higher than a 125 dB indication for impulsive noise, the peaks are probably no higher than 140 dB. This method assumes that the sound pulse is perhaps 25 to 50 milliseconds long; shorter pulses read too low; longer pulses read too high since the amount of energy in the pulse as well as the peak value affects this type of indication. The length of impact and explosive noise pulses is often increased by reflected sound; indoors, it is increased by reverberation.

MONITORING DEVICES (NOISE DOSIMETERS)

Since noise exposure involves both duration and noise level, continuous observation is required to evaluate exposure if durations and levels vary.

Noise monitors or noise exposure meters respond only when a preset level is reached; some include several circuits with different preset values. They are available in small portable packages to be worn by a worker or in larger (and more accurate) models which are not restricted to the within-limits or exceeding-limit indication. Figure 6.4.7 shows a block diagram of a typical dosimeter that complies with the Occupational Safety and Health Administration (OSHA) criteria.

Where variations in level are wide and not easily predictable, integrating devices are used. The Occupational Safety and Health Act (OSH Act) of 1970 (and others) halves the permissible exposure for each 5 dB increase in level above 90 dB; this computation can be made by a simple circuit.

Monitoring devices may simply record the duration of exposure above a preset level for daily examination; they may signal hazardous levels with a flashing light; for airport use, they record times of occurrence of all levels above a preset threshold and sound some alarm when another (and higher) preset limit is reached.

FIELD MEASUREMENTS

Field measurements are measurements where little or no control over operating conditions is possible. *Laboratory measurements* are made in circumstances where some control is possible and conditions can at least be predicted. Usually less sophisticated equipment is used for field work; consequently, the experience and judgment of the analyst become important.

Background noise is normally present, and often it is not possible to maintain conditions while a machine under test is started and stopped. The effect of background noise should be minimized. Stationary noise sources, if small, can be approached to within a meter or so to raise the level of noise being measured relative to the background. (Questions of directivity, size of source, reflections and interferences, and other experimental problems always appear; such problems must be dealt with individually at the time they appear.) For large sources, such handling of experimental problems is seldom possible; where the effect of an entire building or a large cooling tower on a neighborhood, for example, is being studied, the ingenuity and experience of the analyst is taxed to the utmost.

Transient noises—auto horns, vehicle noises, passersby, and the like—must be avoided as far as possible. Measurements within buildings are easier in that traffic is controllable and atmospheric noises do not interfere, but the proximity of the walls introduces danger of reflections in addition to those from the analyst and his equipment.

Both the abilities and limitations of the measuring equipment should be known. Its condition should be checked just before use and again just afterward, including a field calibration check.

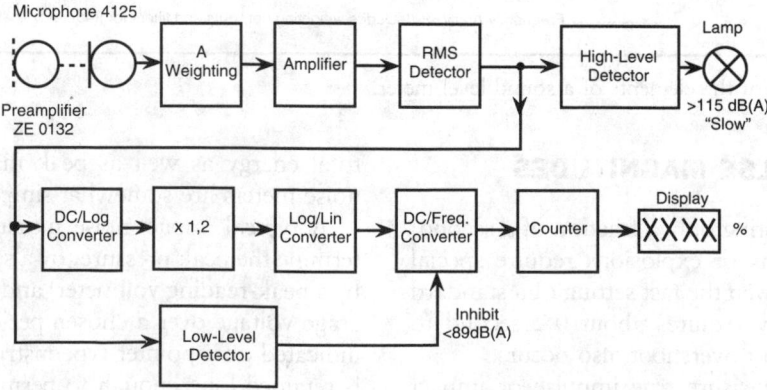

FIG. 6.4.7 Simplified block diagram of a personal noise dose meter, complying with OSHA criteria.

PRACTICAL PROBLEMS

Whether indoors or out, in good conditions or bad, some problems are common to almost all noise measurements.

The convenient measurable range of sound levels extends from about 10 dB above the ambient level (which can be as low as 20 dB in good test rooms) to about 140 dB or so. Whenever measurements are made with background noise within 10 dB of the test noise level, errors tend to increase. By whatever means is convenient, the difference between the test noise level and background should be increased. Work outside these limits is possible, but not easy.

The analyst can increase the sensitivity of the measuring instrument at low levels by using a preamplifier (but with danger of introducing electrical noise) or reduce it at high levels with attenuators (but the mechanical vibration in high-noise fields can affect the operation of the instrument).

Microphones vary in their characteristics. Condenser types have wide frequency response, good sensitivity, and withstand high temperatures and high-sound levels. They are often deranged by high humidity, they require some auxiliary equipment; and their cost is rather high.

Piezoelectric microphones (made with synthetic ceramic) are excellent for general purpose use. Their stability and sensitivity are good, but their frequency response is less wide than that of condenser microphones. Their cost is relatively low. They are not usable at extremely low or extremely high temperatures.

Dynamic microphones can be used at both lower and higher temperatures than the piezoelectric and at higher humidities than the condenser types. They sometimes lose calibration over long periods of time, however, and their frequency response can vary from temperature change. Their construction employs acoustic resonating chambers, which must remain clean and dust free. None of the three types is completely nondirectional.

Frequently the acoustic power of a source is needed but is not supplied by the maker. Ideally this kind of data should be determined in a quiet room or in a free-field situation with low background noise, but practical values can often be approximated by field measurements. For outside sources, if reflecting surfaces (except the ground or pavement) are not close by and the device is the principal noise source, the following equation gives the power level:

$$PWL = SPL + 20 \log_{10} r + 0.5 - Q \text{ dB} \qquad 6.4(5)$$

[See also Section 6.1 and Equation 6.1(9).] Here r is the distance in feet from the source to the sound pressure meter, and Q is a directivity factor.

The directivity factor varies from about 2 for a hard pavement to about 0.5 for absorbing grass; the sound radiated by the machine is usually somewhat directional in itself, too. Several SPL measurements should be taken to strike an average. Values obtained in this way are seldom highly accurate; usually field measurements give low values for PWL. A simpler procedure (for rough estimates) is to measure the SPL at a distance of 1 meter; the PWL value is about 15 units higher. (This procedure can be used on small sources only.) Neither of these procedures works well on extended sources such as cooling towers, large trucks, or the like, nor for distances greater than about 60 meters.

The term here called *directivity factor* is sometimes called *directional gain* or DG. It is the factor by which the power of the source should be multiplied, if its sound radiation were nondirectional, to give the measured level in the direction of measurement. The directivity factor must be determined for the location and direction in question for each measurement or machine.

Field measurements can be disturbed by wind noise. Analysts can reduce this problem by using a windscreen. Windscreens have various forms, but a common method is to surround the microphone with a skeletal spherical frame about 15 to 20 cm in diameter and stretch fine-meshed cloth over it.

At extreme distances, sound behavior becomes unpredictable; but noise control work seldom requires measurements beyond a mile or so. At high altitudes, low barometric pressure, or high humidity, the density change in the atmosphere affects calculations. For most work, these effects can be ignored.

Multiple sources are usually present, as well as multiple transmission paths; and as in other measurements, the presence of the measuring device can affect the local conditions. Interpretation in both raw and tabulated data can be a problem. Judgment, based on experience, is of the greatest importance in noise-control work.

—Howard C. Roberts
David H.F. Liu

References

Acoustics handbook. Application Note 100. California: Hewlett-Packard Co.

Cheremisinoff, Paul N. 1993. *Industrial noise control.* Englewood Cliffs, N.J.: Prentice Hall, Inc.

Lipták, B.G. 1970. *Instrument engineers' handbook.* Vol. I, sec. 9.1, Radnor, Penna.: Chilton Book Co.

Peterson, A.P.G. and E.E. Gross, Jr. *Handbook of noise measurements.* Concord, Mass.: General Radio Co.

Yerges, Lyle F. 1969. *Sound, noise, and vibration control.* New York: Van Nostrand Reinhold Co.

6.5
NOISE ASSESSMENT AND EVALUATION

During the years after passing the OSH Act of 1970, Congress worked to define the section concerning workplace noise. Their work resulted in the promulgation of the Hearing Conservation/Noise Amendment in 1983. While OSHA regulations receive the most coverage, mining and military organizations have their own safety and health regulations. In addition, many states have regulations similar to OSHA's and occasionally more stringent. Besides workplace noise exposure, noise remains a predominant issue in communities all over the country and the world, and complaints to local officials are rising.

The amendment provides coverage for all employees exposed to a time-weighted average of 85 dBA. It outlines the following components of a hearing conservation program: (1) noise assessment, (2) hearing testing, (3) hearing protection, (4) education and training, and (5) record keeping. Workers' compensation laws established in each state may also specify noise assessment similar to OSHA's.

This section focuses on noise assessment and analysis of noise-level data in the workplace, community noise, and noise-level criteria.

Workplace Noise

A hearing conservation program begins with determining a worker's noise exposure. It is the first step in identifying those employees who must be included in the total hearing conservation program. Several criteria suggest the need for noise assessment:

- If any area has a past record of excessive noise
- If employees complain of discomfort or temporary hearing loss
- If employees are unable to converse easily, without shouting, at a distance of 2 feet.

A dosimeter is the most important instrument in noise assessment. It determines the noise level to which employees are exposed by measuring sound over time and analyzing the information to produce a noise dose, expressed in a percentage. A noise dose, D, is defined as:

$$D = \frac{C}{T} \qquad 6.5(1)$$

where D is the noise dose; C, actual duration of exposure in hours; T, noise exposure limit in hours. The allowable exposure time (Table G-16 of the OSHA regulations) is listed in Table 6.5.1. The employee exposure exceeds the OSHA limits if the noise dose, D, exceeds unity, or 100%.

Where the daily exposure is due to more than one noise level, the ratios for each level are added to compute the total noise dose as follows:

$$D = \frac{C_1}{T_1} + \frac{C_2}{T_2} + \dots \frac{C_n}{T_n} \qquad 6.5(2)$$

NOISE DOSIMETERS

The OSHA amendment of 1983 pushed the development of noise instrumentation to higher capability levels. Noise dosimeters are programmable, data-logging, multithreshold instruments. They communicate directly with printers and personal computers. They provide information such as multithreshold doses, average sound level or LAvg, peak levels, histograms, statistical distributions, and projections.

A dosimeter is a small unit which can be attached to a belt or placed in a shirt pocket. The microphone is placed at the ear level, about 3 to 5 inches from the head, to avoid shadow error. Dosimeters must integrate all noise—continuous, variable, and impulse or impact—in the range between 80 to 130 dBs, using a 5-dB doubling or exchange rate; and they must be A weighted. (See Section 6.4)

The following definitions explain these dosimeter terms and keys:

THRESHOLDS—That sound level below which the instrument assumes no noise.

TABLE 6.5.1   OSHA HEARING CONSERVATION TABLE[a]

A-Weighted Sound Level	Duration (Hours)
80 dB*	32
85†	16
90‡	8
95	4
100	2
105	1
110	0.5
115	0.25
120	0.125
125§	0.063
130§	0.031

Important Levels:
*Measuring Threshold
†Hearing Conservation Begins-50% Dose
‡Eight-Hour Criteria Level
§Minimum Upper Range
a = Table G-16a (Abbreviated)

LTL—Low threshold level, when set at 80 dB, measures only 80 dB and above; assumes 0s below that level. Used for OSHA hearing conservation compliance.

HTL—High threshold level, when set at 90 dB, measures only 90 dB and above; assumes 0s below that level. Used for OSHA engineering control compliance.

CRITERION LEVEL—90 dB, when measured for eight hours, reads 100% dose.

5DB EXCHANGE RATE—Used by OSHA; every 5 dB increase or decrease either doubles or halves the dose.

A WEIGHTING—Used by OSHA (and others); reads the way the human ear hears sound. Basically the low frequencies and some high are not read as loud as they acoustically are.

LAVG—Average sound level for the actual time measured based on other than a 3 dB exchange rate, ie. 5 dB-OSHA, 4 dB-DOD, or a 3 dB exchange rate with a threshold. Useful for short-term samples or for making projections.

LEQ—Average sound level for the actual time measured based on a 3 dB exchange rate and no threshold.

TWA—An eight-hour dB average regardless of the sample time length. For example, in a four-hour sample measurement, TWA assumes the last four hours are 0s and averages them in the overall reading, making the average dB level lower than it should be. Appropriate to use if someone works other than an eight-hour day, ie. twelve-hour shifts. TWA takes twelve hours of exposure and condenses it into eight giving the appropriate TWA for eight hours. Not appropriate to use for short samples. (Note: LAvg is always greater than TWA in samples less than eight hours. Samples of exactly eight hours result in equal LAvg and TWA, and samples longer than eight hours produce TWAs that exceed LAvg.)

DOSE—The accumulated exposure obtained, expressed in percent allowable over eight hours.

SEL—A one-second average of a noise occurence that is any duration in length. (Used to compare several noise events with different time durations.)

The capability for multithreshold dose measurements is helpful. One threshold level is 90 dBA set by OSHA engineering regulations. A concern arises when an employee works in an environment which is always below 90 dBA but where the average sound level is in the high 80s. The dosimeter indicates this area as having no hazard (0% dosage) while the worker is within a few decibels of maximum allowable exposure (100% exposure). Having a second available threshold set at 80 dBA allows the analyst to detect marginal environments while assuring compliance with both the engineering regulation and the hearing conservation amendment. In addition, having these data available provides clues on how to attack noise problems through engineering efforts.

An analyst can determine the effect of using hearing protection devices in the workplace environment by subtracting the NRR value of a device (furnished by the manufacturer) from the LAvg. Histograms, or sound-level–time history, in the printed format are quite useful. The benefits of histograms include identification of noise patterns—and significant break in patterns—during the workday. Patterns help identify sound sources which require corrective action to lower total exposure to the worker. The histogram identifies the exact time when a significant break in these patterns occurs. This information is the basis of documentation presented at compensation hearings.

SOUND LEVEL METERS

This instrument is used to spot-check for excessive noise. When coupled with an octave-band filter, it can also be used to assess the frequency content of the noise. This information is useful for noise control.

Integrating Sound Level Meters

Integrating sound level meters measure average dB levels and are useful for evaluating rapidly changing noise environments. They present a direct A-weighted dB (dBA) reading.

Community Noise

A study conducted by the Department of Housing and Urban Development indicates that the number one concern among city dwellers in the United States is noise. The agencies responsible for enforcing community noise laws are the public health department, code enforcement agencies, and the police or county sheriff departments.

Before considering whether instrumentation is needed, the agency must review the local ordinance. Most ordinances include sections that address measurement requirements and techniques as well as permissible noise-level tables. These sections usually indicate indirectly the minimum instruments required to sample noise sources.

NOISE RATING SYSTEMS

An example of the wording in the measurement techniques section of a noise ordinance might read: "The measurement must be made for a period of at least 10 minutes and any noise source that exceeds those levels set forth in Table X at least 10% of the time (L10) shall be deemed in violation of the law." An L10 or L_{10} is known as an *exceedance level* whereby a given dB level is exceeded 10% of the measurement period.

Other terms found in noise ordinances and their definitions follow.

LDN OR L_{dn} (level day night)—A noise exposure level computed on a twenty-four-hour daily basis whereby 10 dB is added to all readings between the hours 10 P.M. (2200 hours) and 7 A.M. (0700 hours). This adjustment penalizes noise exposure levels that occur during what is considered normal sleeping hours for a community.

LEQ OR L_{eq} (level equivalent continuous equal energy)— That constant noise level that, over a given period of time, expends the same amount of energy as the fluctuating level over the same time period. It is expressed as follows:

$$L_{eq} = 10 \log \sum_{i=1}^{i=n} 10^{Li/10} t_i \qquad 6.5(3)$$

where n is the total number of samples taken, L is the noise level in dBA of the ith sample, t is the fraction of the total sample time.

Ldn and Leq are used in the U.S. Environmental Protection Agency (EPA) published noise criteria levels that it deemed necessary to health and welfare of U.S. citizens (Table 6.5.2). An Lnd of 45 provides a fair margin of safety. These terms appear in the Federal Housing Authority (FHA) noise standards for construction (Table 6.5.3).

LNP OR L_{np} (level noise pollution)—The sum of the A-weighted SPLs (dBA characteristics of each piece of equipment) added together using the Leq equation plus 2.56 times the standard deviation of the SPL to account for annoyance due to fluctuation. Figure 6.5.1 shows the use of the Lnp in the provisional criteria relating the noise pollution level (NPL) to community noise level acceptability at construction site boundaries.

TNI (or traffic noise index)—Derived from the equation

$$TNI = L(90) + 4d - 30 \text{ dB; and } d = L(10) - L(90) \qquad 6.5(4)$$

Some older ordinances call for octave-band analysis. These ordinances were derived from research which showed that some frequencies have a higher annoyance factor and therefore should have a lower allowable dB level. These laws are not as common since they demand a rather arduous measurement procedure and require a specialized instrument.

INSTRUMENTATION

Generally one of the following types of instruments is indicated for community noise:

NOISE LOGGING DOSIMETERS/ANALYZER—Used for longer measurements, possibly twenty-four hours or more. These instruments should have the capability to measure exceedance levels and Ldn levels and to store all data for later printout or downloading to a personal computer.

INTEGRATING L_{eq} SOUND LEVEL METER—Used for measuring noise for shorter periods of time where the noise

TABLE 6.5.2   YEARLY ENERGY AVERAGE L_{EQ} REQUISITE TO PROTECT PUBLIC HEALTH WITH ADEQUATE SAFETY MARGIN

	Measure	Indoor		To Protect Against Both Effects (b)	Outdoor		To Protect Against Both Effects (b)
		Activity Interference	Hearing Loss Consideration		Activity Interference	Hearing Loss Consideration	
Residential with outside space	L_{dn}	45		45	55		55
and farm residences	$L_{eq(24)}$		70			70	
Residential with no	L_{dn}	45		45			
outside space	$L_{eq(24)}$		70				
Commercial	$L_{eq(24)}$	(a)	70	70(c)	(a)	70	70(c)
Inside transportation	$L_{eq(24)}$	(a)	70	(a)			
Industrial	$L_{eq(24)(d)}$	(a)	70	70(c)	(a)	70	70(c)
Hospitals	L_{dn}	45		45	55		55
	$L_{eq(24)}$		70			70	
Educational	$L_{eq(24)}$	45		45	55		55
	$L_{eq(24)(d)}$		70			70	
Recreational areas	$L_{eq(24)}$	(a)	70	70(c)	(a)	70	70(c)
Farm land and general	$L_{eq(24)}$				(a)	70	70(c)
unpopulated land							

Note: Explanation of identified level for hearing loss: The exposure period that results in hearing loss at the identified level is a period of forty years.
Code:
(a) Since different types of activities appear to be associated with different levels, identification of a maximum level for activity interference may be difficult except in those circumstances where speech communication is a critical activity.
(b) Based on lowest level.
(c) Based only on hearing loss.
(d) An $L_{eq(8)}$ of 75 dB can be identified in these situations so long as the exposure over the remaining sixteen hours per day is low enough to result in negligible contribution to the twenty-four-hour average, that is, no greater than an L_{eq} of 60 dB.

TABLE 6.5.3 FHA NOISE STANDARDS FOR NEW CONSTRUCTION

Land Use Category	Exterior Design Noise Level dBA[a]		Description of Land Use Category
	L_{eq}	L_{10}	
A	57	60	Tracts of lands in which serenity and quiet are of extraordinary significance and serve an important public need, and where the preservation of those qualities is essential if the area is to continue to serve its intended purpose. For example, such areas could include amphitheaters, particular parks or portions of parks, or open spaces, which are dedicated or recognized by appropriate local officials for activities requiring special qualities of serenity and quiet.
B	67	70	Residences, motels, hotels, public meeting rooms, schools, churches, libraries, hospitals, picnic areas, recreation areas, playgrounds, active sports areas, and parks.
C	72	75	Developed lands, properties, or activities not included in categories A and B above.
D	Unlimited	Unlimited	Undeveloped lands
E	52 (Interior)	55 (Interior)	Public meeting rooms, schools, churches, libraries, hospitals, and other such public buildings.

[a]Either L_{eq} or L_{10} may be used, but not both. The levels are to be based on a one-hour sample.

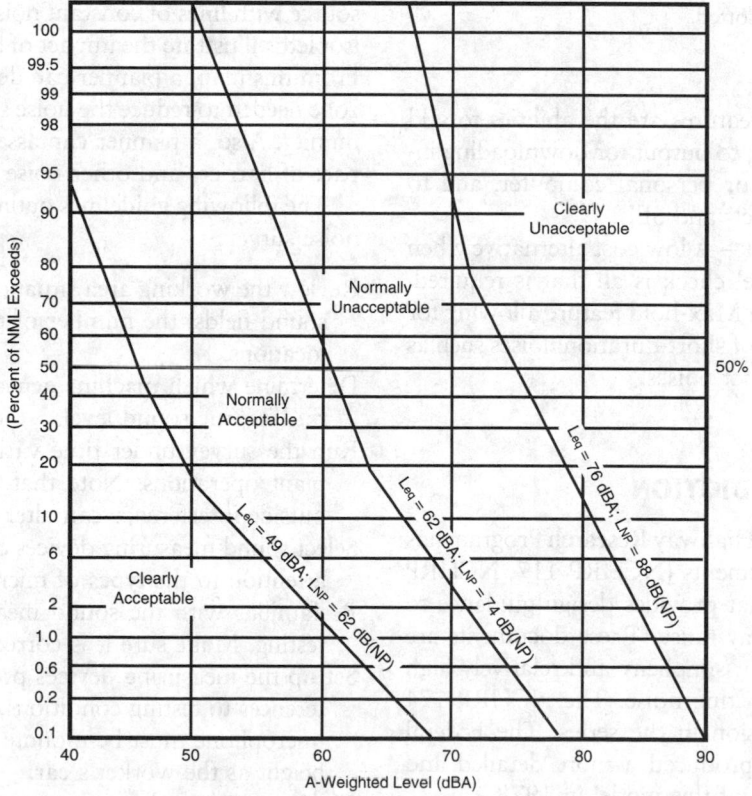

FIG. 6.5.1 Provisional criteria relating NPL to community noise level acceptability. (Reprinted from U.S. Environmental Protection Agency, 1971, *Noise from construction equipment*, [Washington, D.C.: EPA], 17.)

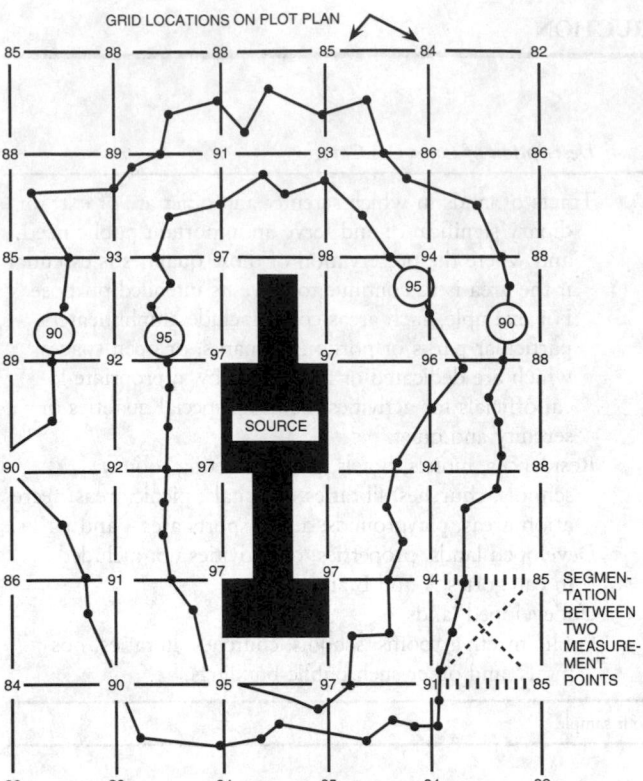

GRID LOCATIONS ON PLOT PLAN

FIG. 6.5.2 Grid plot of noise source and surrounding areas. Isopleths of noise levels are plotted.

is variable. Additional features are the abilities to add an octave-band filter set, to output for downloading information to a printer or personal computer, and to measure exceedance levels and SEL.

SIMPLE SOUND LEVEL METER—A low cost alternative when a quick on-the-spot level check is all that is required. The meter should have a Max-hold feature allowing for accurate measurements of short-duration noises such as moving vehicles or impact noises.

TRAFFIC NOISE PREDICTION

The National Cooperative Highway Research Program has developed a series of documents (NCHRP 117, NCHRP 144, and NCHRP 174) that provides design guidance to predict and control highway noise. These documents are widely used because of their simplicity and relatively high success in accurately predicting noise. The NCHRP 174 procedure is the last revision in the series. The Federal Highway Administration produced a more detailed and empirically correct version of this model in 1978.

Computer programs are available to predict noise levels generated by a proposed project. They can be used to determine the noise impact on surrounding areas. These programs (hand calculations are possible for simple cases) can determine the noise levels at distances away from the

project and for various atmospheric effects, barriers, and topographical features. Equations for these prediction methods are well documented.

Plant Noise Survey

In addition to the objectives previously mentioned, a noise survey can be conducted to define a baseline for noise-level analysis. Planning for a noise survey includes:

Determining the type of acoustical data and time of day measurements are to be taken

Determining the sampling locations and the significant sound sources

Using a map to identify significant topographical features and other structures and sound barriers

Considering the effect of meterological conditions, such as temperature, barometric pressure, relative humidity, and wind speed and direction on noise levels

Mapping significant information such as residential, commercial, or industrial zones; population densities; special areas (hospitals); and areas of unique noise characteristics

Figure 6.5.2 shows a map of a region surrounding a source with lines of constant noise levels (isopleths). These isopleths illustrate the impact of a noise source on an area. From this map, a planner can determine the size of buffer zone needed to reduce the noise source impact on the community. Also, a planner can assess the environmental impact of barriers and other noise reduction techniques.

The following guidelines outline how to conduct a plant noise survey:

Review the working area situation thoroughly; the type of sound fields, the number of people affected, and their locations.

Determine which machine generates the most sound and find its true sound level.

Run the survey under time variations as well as normal plant operations. Note that a change in humidity or outside interference can alter results.

Select sound measuring devices carefully, giving particular attention to the types of microphones necessary.

Be familiar with the sound measuring equipment before testing. Make sure it is correctly calibrated.

Set up the measuring devices properly and have no interferences to testing conditions, if possible. Note that the microphone must be mounted on a tripod at the same height as the worker's ear.

Make sure that all equipment aiding in measuring—the meter, recorder, and correcting apparatus—is outside the testing area.

—David H.F. Liu

References

Banach, Jim. 1992. Noise measurement: The age of technology and information. In *Best's Safety Directory*. 1993 ed. Oldwick, N.J.: A.M. Best Company, Inc.

Cheremisinoff, Paul N. 1993. *Industrial noise control*. Englewood Cliffs, N.J.: Prentice-Hall, Inc.

Cheremisinoff, Paul N. and Angelo C. Morresi. 1977. *Environmental assessment & impact statement handbook*. Ann Arbor, Mich.: Ann Arbor Science Publishers, Inc.

Greenberg, Michael R. 1979. *A primer on industrial environment*. New Brunswick, N.J.: Rutgers—The State University of New Jersey, 19–96.

Kugler, B. Andrew, Daniel E. Commins, and Williams J. Galloway. 1976. *Highway noise: A design guide for prediction and control*. National Highway Research Program Report 174.

Olishifski, J.B. and E.R. Harford. 1975. *Industrial noise and hearing conservation*. National Safety Council.

U.S. Department of Transportation, Federal Highway Administration. 1973–1979. Noise standard and procedure, *Code of Federal Regulations*. Title 23, part 772. Washington, D.C.: U.S. Government Printing Office.

U.S. Environmental Protection Agency. 1971. *Noise from construction equipment*. NTID 300.1. Washington, D.C.

———. 1974. *Information on levels of environmental noise requisite to protect health and welfare with an adequate margin of safety*. EPA 550/9-74/004. Washington, D.C.

6.6
NOISE CONTROL AT THE SOURCE

Source-Path-Receiver Concept

To solve a noise problem, one must find out something about what the noise is doing, where it comes from, how it travels, and what can be done about it. A straightforward approach is to examine the problem in terms of its three basic elements; that is, sound arises from a source, travels over a path, and affects a receiver or listener.

The source can be one or any number of mechanical devices that radiate noise or vibratory energy. Such a situation occurs when several machines are operating at the same time.

The most obvious transmission path by which noise travels is a direct line-of-sight air path between the source and the listener. For example, aircraft flyover noise reaches an observer on the ground by the direct line-of-sight air path.

Noise also travels along structural paths. Noise can travel from one point to another via any one path or a combination of several paths. Noise from a washing machine operating in one apartment can be transmitted to another apartment along air passages such as open windows, doorways, corridors, or duct work. Direct physical contact of the washing machine with the floor or walls sets these building components into vibration. This vibration is transmitted structurally throughout the building causing walls in other areas to vibrate and to radiate noise.

The receiver may be, for example, a single person, or a suburban community.

The solution of a noise problem requires alteration or modification of any or all of the following three basic elements:

- Modifying the source to reduce its noise output
- Altering or controlling the transmission path and the environment to reduce the noise level reaching the listener
- Providing the receiver with personal protective equipment

Modifying the source to reduce noise output involves noise-level specifications, process substitution, machines substitution, and systems design.

NOISE-LEVEL SPECIFICATIONS

The best way of controlling noise at its source is to buy quieter machines. Buying quieter machines is almost always more economical than trying to reduce noise by modifying the machine after purchase. Everyone profits from quieter machines: the employees' hearing is better protected, work is performed more efficiently, and the employer gains from increased production and product quality. The purchase order should specify the maximum permissible noise levels for equipment as listed in the Table 6.6.1.

PROCESS SUBSTITUTION

Substituting a quieter process, machine, or tool is another method of controlling noise. Operations such as riveting, punching, shearing, and metal-forming are often performed by impact when a slower energy application is equally effective. Welding is a quieter substitute for riveting, drilling for punching, pressing or rolling for forging, hot forming for cold forming, grinding of castings for chip-

TABLE 6.6.1   EQUIPMENT NOISE-LEVEL LIMITS

Description	dB (re. 0.0002 μ Bar) Octave Band Center Frequency									Measurement Location
	63	125	250	500	1000	2000	4000	8000	A	
General Work Area	104	97	90	84	81	79	80	82	90	Any location 3' from equipment
Speech and Work				Interference: Room Type						
Control Room	73	69	66	63	59	56	53	53	65	Inside
Office	70	66	62	59	55	51	48	48	61	Inside
Equipment										
Heaters	99	92	85	79	76	74	75	77	85	Any location 3' from heater
Steam Generators	99	92	85	79	76	74	75	77	85	Any location 3' from heater
Air Fins	99	92	85	79	76	74	75	77	85	3' under unit
Pumps	94	74	80	80	79	79	74	68	85	3' from unit
Motors	94	74	74	81	81	78	74	73	85	3' from unit
Turbines	94	78	71	74	77	80	77	74	85	3' from unit
Gear Units	87	78	69	70	75	83	75	61	85	3' from unit
Relief Valves	109	99	99	93	99	106	103	95	110	3' from exit
Control Valves	84	74	74	68	74	81	78	70	85	3' from valve
Compressors	84	74	70	71	81	80	73	65	85	3' from unit
Fans, Blowers	99	92	85	79	76	74	75	77	85	3' from unit
Flares	104	97	90	84	81	79	80	82	90	50' Horizontal and 5' off grade
Ejectors	84	74	74	68	74	81	78	70	85	3' from unit
Cooling Towers	104	97	90	84	81	79	80	82	90	Any location 3' from equipment
Silencers: To Atms.	104	97	90	84	81	79	80	82	90	10' from exit
In-Line	99	92	85	79	76	74	75	77	85	3' from exit
Machinery	94	74	74	81	81	78	74	73	85	3' from unit
Diesel Engine	77	78	79	80	80	79	77	72	85	3' from unit

ping, and hydraulic and pneumatic equipment for mechanical equipment.

MACHINE SUBSTITUTION

Noise reduction can be significant when belt drives are used instead of gears. If using gears is necessary, rotating gears should be substituted for square gears; nylon gears for metallic gears. Other recommendations for reducing noise are described in the subsection on control of noise sources by design and in Section 6.7.

SYSTEMS DESIGN

Besides engineering controls, noise reduction and isolation can be approached through machine mounting or by architectural means. If machines are laid out too closely, the operator may be exposed to a high dB level. However, if machines are spaced adequately apart, noise levels can be within acceptable limits. Noise can be confined within a restricted area by architectural means: building location

and arrangement, design, use of suitable building materials, and location of noise-producing and noise-sensitive areas. Sound control for ceilings in offices must also be planned at the architectural stage.

Holes should not be placed back to back immediately next to each other. Electrical boxes should be staggered, at least one stud space. A nonhardening, nonskinning, resilient caulking material should be used to seal all cutouts, such as around electrical and telephone outlets. Also, all intersections with the adjoining structure, such as under-floor and ceiling runner tracks, around the perimeter where the assembly meets the floor, ceiling, and partitions, should be sealed. Using center-of-gravity mounting whenever feasible prevents translational modes of vibration from coupling to rotational modes.

Control of Noise Source by Design

This section describes controlling the noise source by design including reducing impact forces, reducing speeds and pressures, reducing frictional resistance, reducing the ra-

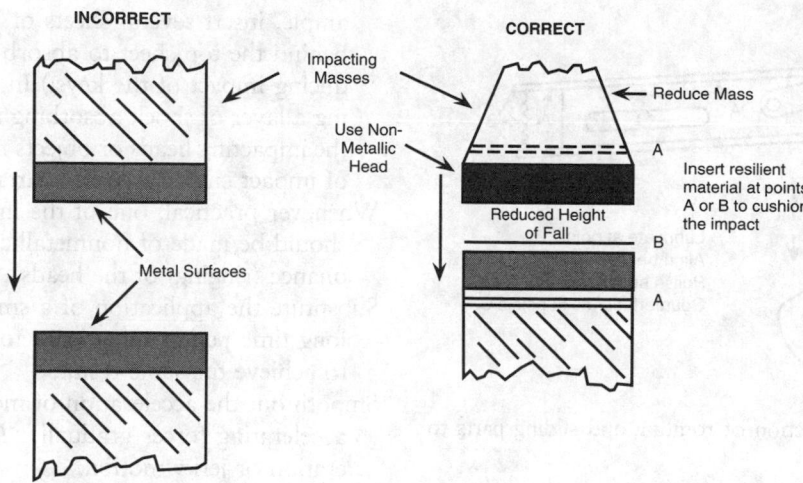

FIG. 6.6.1 Methods of reducing impact forces to lower noise radiation.

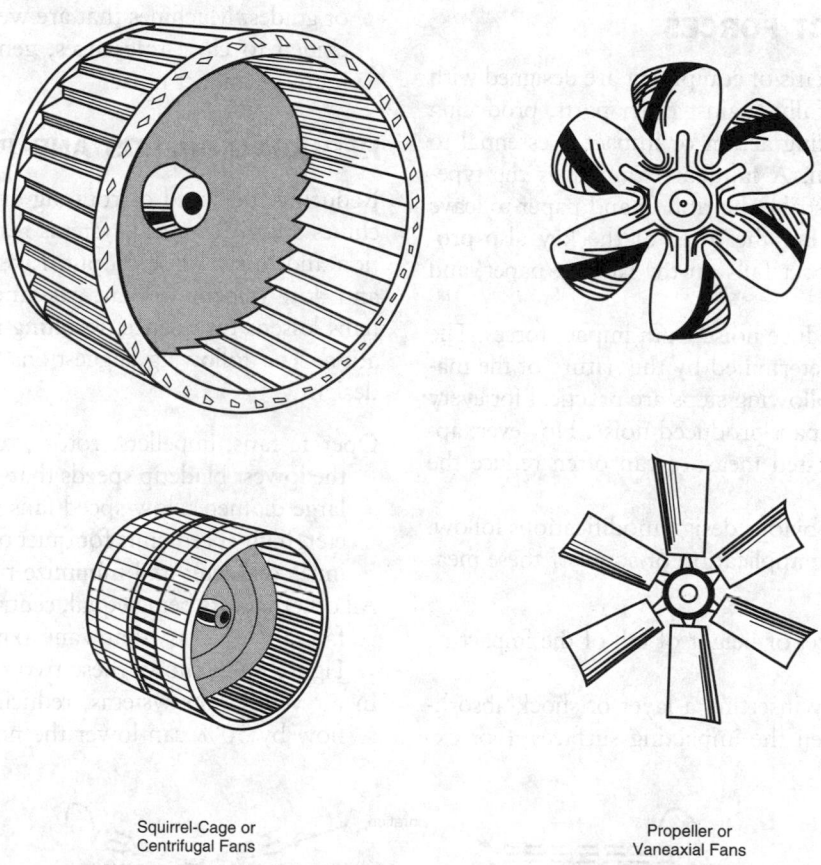

Squirrel-Cage or
Centrifugal Fans

Propeller or
Vaneaxial Fans

FIG. 6.6.2 For a given mass flow, squirrel-cage fans are generally less noisy than propeller-type fans.

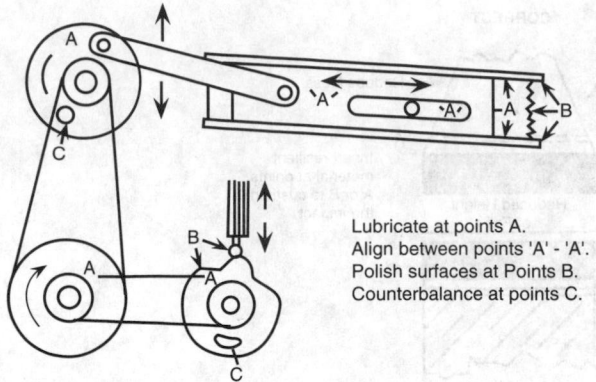

Lubricate at points A.
Align between points 'A' - 'A'.
Polish surfaces at Points B.
Counterbalance at points C.

FIG. 6.6.3 Reducing friction of rotating and sliding parts to decrease noise radiation.

diating area, reducing noise leakage, and isolating and damping vibrating elements.

REDUCING IMPACT FORCES

Many machines and items of equipment are designed with parts that strike forcefully against other parts, producing noise. Often, this striking action or impact is essential to the machine's function. A familiar example is the typewriter—its keys must strike the ribbon and paper to leave an inked impression. But the force of the key also produces noise as the impact falls on the ribbon, paper, and platen.

Several steps can reduce noise from impact forces. The particular remedy is determined by the nature of the machine. Not all of the following steps are practical for every machine and every impact-produced noise. However, applying even one suggested measure can often reduce the noise appreciably.

Some of the more obvious design modifications follow. Figure 6.6.1 shows the application of some of these measures.

Reduce the weight, size, or height of fall of the impacting mass.

Cushion the impact by inserting a layer of shock-absorbing material between the impacting surfaces. (For ex-

ample, insert several sheets of paper in the typewriter behind the top sheet to absorb some of the noise-producing impact of the keys.) In some situations, inserting a layer of shock-absorbing material behind each of the impacting heads or objects reduces the transmission of impact energy to other parts of the machine.

Whenever practical, one of the impact heads or surfaces should be made of nonmetallic material to reduce resonance (ringing) of the heads.

Substitute the application of a small impact force over a long time period for a large force over a short period to achieve the same result.

Smooth out the acceleration of moving parts by applying accelerating forces gradually. Avoid high, jerky acceleration or jerky motion.

Minimize overshoot, backlash, and loose play in cams, followers, gears, linkages, and other parts. To achieve this measure, reduce the operational speed of the machine, make better adjustments, or use spring-loaded restraints or guides. Machines that are well made, with parts machined to close tolerances, generally produce a minimum of impact noise.

REDUCING SPEEDS AND PRESSURES

Reducing the speed of rotating and moving parts in machines and mechanical systems results in smoother operation and lower noise output. Likewise, reducing pressure and flow velocities in air, gas, and liquid circulation systems lessens turbulence, resulting in decreased noise radiation. The following suggestions can be incorporated in design:

Operate fans, impellers, rotors, turbines, and blowers at the lowest bladetip speeds that still meet job needs. Use large-diameter, low-speed fans rather than small-diameter, high-speed units for quiet operation. In short, maximize diameter and minimize tip speed.

All other factors being equal, centrifugal squirrel-cage type fans are less noisy than vane axial or propeller type fans. Figure 6.6.2 shows these two types of fans.

In air ventilation systems, reducing the speed of the air flow by 50% can lower the noise output by 10 to 20

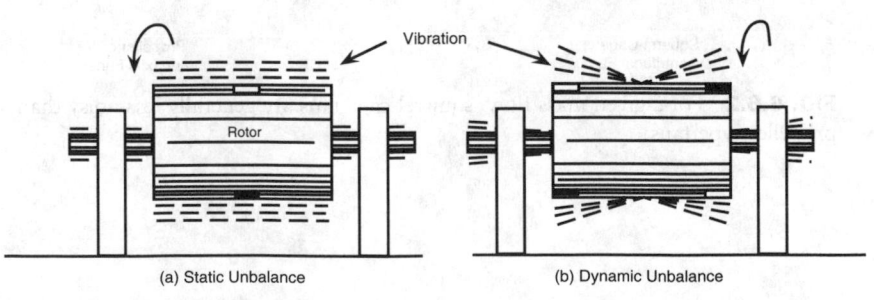

(a) Static Unbalance

(b) Dynamic Unbalance

■ Black blocks are heavy parts of rotor that cause vibration. ▭ White blocks are locations where counterweights must be placed to eliminate the vibration.

FIG. 6.6.4 Effects of static and dynamic unbalance on rotor operation.

dB, or roughly one-quarter to one-half of the original loudness. Air speeds less than 3 m/s measured at a supply or return grille produce a level of noise that usually is unnoticeable in residential or office areas. To reduce air speed in a given system operate lower motor or blower speeds, install more ventilating grilles, or increase the cross-sectional area of the existing grilles.

REDUCING FRICTIONAL RESISTANCE

Reducing friction between rotating, sliding, or moving parts in mechanical systems frequently results in smoother operation and lower noise output. Similarly, reducing flow resistance in fluid distribution systems results in less noise radiation.

Four of the more important factors that should be checked to reduce frictional resistance in moving parts are the following (see Figure 6.6.3):

Alignment: Proper alignment of all rotating, moving, or contacting parts results in less noise output. Good axial and directional alignment in pulley systems, gear trains, shaft coupling, power transmission systems, and bearing and axle alignment are fundamental requirements for low noise output.

Polish: Highly polished and smooth surfaces between sliding, meshing, or contacting parts are required for quiet operation, particularly where bearings, gears, cams, rails, and guides are concerned.

Balance: Static and dynamic balancing of rotating parts reduces frictional resistance and vibration, resulting in lower noise output (see Figure 6.6.4).

Eccentricity (out-of-roundness): Off-centering of rotating parts such as pulleys, gears, rotors, and shaft and bearing alignment causes vibration and noise. Likewise, out-of-roundness of wheels, rollers, and gears causes uneven wear resulting in flat spots that generate vibration and noise.

The key to effective noise control in fluid systems is *streamline flow*. This fact holds true for all their systems including air flow in ducts or vacuum cleaners and water flow in plumbing systems. Streamline flow is simply smooth, nonturbulent, low-friction flow.

Two factors that determine whether flow is streamline or turbulent are the speed of the fluid and the cross-sectional area of the flow path, that is, the pipe or duct diameter. The rule of thumb for quiet operation is to use a low-speed, large-diameter system to meet a specified flow capacity requirement. However, even such a system can inadvertently generate noise if certain aerodynamic design features are overlooked or ignored. A system designed for quiet operation employs the following features (see Figure 6.6.5):

Low fluid speed: Low fluid speeds avoid turbulence, one of the main causes of noise.

Smooth boundary surfaces: Duct or pipe systems with smooth interior walls, edges, and joints generate less turbulence and noise than systems with rough or jagged walls or joints.

Simple layout: A well-designed duct or pipe system with a minimum of branches, turns, fittings, and connectors is less noisy than a complicated layout.

Long-radius turns: Changes in flow direction should be gradual and smooth. A recommendation is for turns to have a curve radius equal to about five times the pipe diameter or major cross-sectional dimension of the duct.

Flared sections: Flaring of intake and exhaust openings, particularly in a duct system, tends to reduce flow speeds at these locations, often with substantial reductions in noise output.

Streamlined transition in flow path: Changes in flow path dimensions or cross-sectional areas should be gradual and smooth with tapered or flared transition sections to avoid turbulence. A good rule of thumb is to keep the cross-sectional area of the flow path as large and uniform as possible throughout the system.

Minimal obstacles: The greater the number of obstacles in the flow path, the more tortuous, turbulent, and noisy the flow. All other required and functional devices in the path, such as structural supports, deflectors, and control dampers, should be as small and streamlined as possible to smooth out the flow patterns.

REDUCING RADIATING AREA

Generally speaking, the larger the vibrating part or surface, the greater the noise output. The rule of thumb for quiet machine design is to minimize the effective radiating surface areas of the parts without impairing their operation or structural strength. This design includes making parts smaller, removing excess material, or cutting openings, slots, or perforations in the parts. For example, replacing a large, vibrating sheet-metal safety guard on a machine with a guard made of wire mesh or metal webbing might substantially reduce noise because of the reduced surface area of the part (see Figure 6.6.6).

REDUCING NOISE LEAKAGE

In many cases, machine cabinets can be made into effective soundproof enclosures through simple design changes and the application of some sound-absorbing treatment. Adopting some of the following recommendations may lead to substantial reductions in noise output:

Caulk all unnecessary holes or cracks, particularly at joints.

Seal electrical or plumbing penetrations of the housing or cabinet with rubber gaskets or a suitable nonsetting caulk.

If practical, cover all other functional or required openings or ports that radiate noise with lids or shields edged with soft rubber gaskets to effect an airtight seal.

NOISY DESIGN

Ragged Joints

Small Diam.

Fast Turbulent Flow

Rough Surfaces

QUIET DESIGN

Smooth Joints

Large Diam.

Slow Streamline Flow

Smooth Surfaces

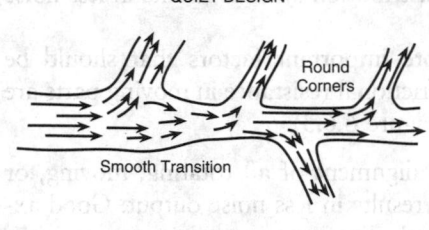

NOISY DESIGN

Sharp Corners

Sharp Transition

Complicated Layout

QUIET DESIGN

Round Corners

Smooth Transition

Simple Layout

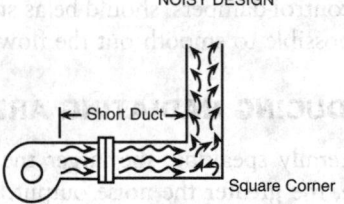
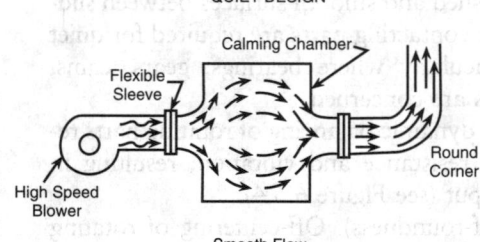

NOISY DESIGN

Short Duct

Square Corner

Turbulent Flow

QUIET DESIGN

Calming Chamber

Flexible Sleeve

High Speed Blower

Round Corner

Smooth Flow

NOISY DESIGN

Turbulence Caused by Rectangular Devices

QUIET DESIGN

Remove or Streamline Objects in Flow Path

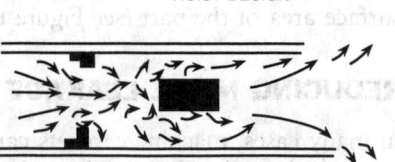

FIG. 6.6.5 Design of quiet flow systems.

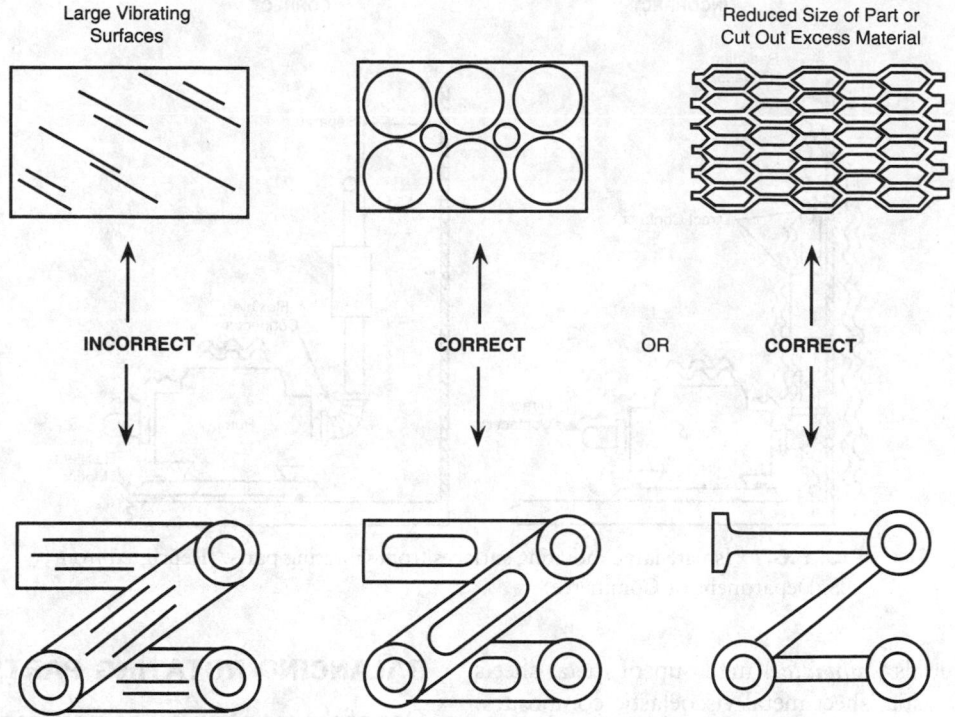

FIG. 6.6.6 Reduce the area of vibrating surfaces to lower noise radiation. (Reprinted from U.S. Department of Commerce)

Equip other openings required for exhaust, cooling, or ventilation purposes with mufflers or acoustically lined ducts.

Direct openings away from the operator and other people.

ISOLATING AND DAMPING VIBRATING ELEMENTS

In all but the simplest machines, the vibrational energy from a specific moving part is transmitted through the machine structure, forcing other component parts and surfaces to vibrate and radiate sound—often with greater intensity than that generated by the originating source itself.

Generally, vibration problems have two parts. First, energy transmission must be prevented between the source and surfaces that radiate the energy. Second, the energy must be dissipated or attenuated somewhere in the structure. The first part of the problem is solved by *isolation*. The second part is solved by *damping*.

The most effective method of vibration isolation involves the resilient mounting of the vibrating component on the most massive and structurally rigid part of the machine. All attachments or connections to the vibrating part, in the form of pipes, conduits, and shaft couplers, must have flexible or resilient connectors or couplers. For example, pipe connections to a pump that is resiliently mounted on the structural frame of a machine should be made of resilient tubing and mounted as close to the pump

as possible. Resilient pipe supports or hangers may also be required to avoid bypassing the isolated system (see Figures 6.6.7 and 6.6.8).

Damping material or structures are those that have some viscous properties. They tend to bend or distort slightly, thus consuming part of the noise energy in molecular motion. The use of spring mounts on motors and laminated galvanized steel and plastic in air-conditioning ducts are examples.

When the vibrating noise source is not amenable to isolation, as in ventilation ducts, cabinet panels, and covers, then damping materials can be used to reduce the noise.

The type of material best suited for a particular vibration problem depends on several factors such as size, mass, vibrational frequency, and operational function of the vibrating structure. Generally speaking, the following guidelines should be observed in the selection and use of such materials to maximize vibration damping efficiency (see Figure 6.6.9):

Damping materials should be applied to those sections of a vibrating surface where the most flexing, bending, or motion occurs. These areas are usually the thinnest sections.

For a single layer of damping material, the stiffness and mass of the material should be comparable to that of the vibrating surface to which it is applied. Therefore, single-layer damping materials should be about two or three times as thick as the vibrating surface to which they are applied.

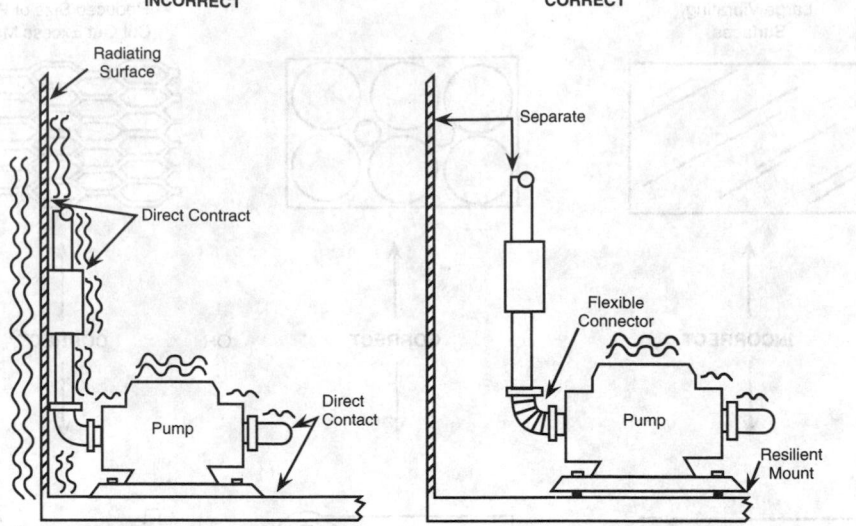

FIG. 6.6.7 Isolate large radiating surfaces from vibrating parts. (Reprinted from U.S. Department of Commerce)

Sandwich materials (*laminates*) made up of metal sheets bonded to mastic (sheet metal viscoelastic composites) are more effective vibration dampers than single-layer materials; the thickness of the sheet-metal constraining layer and the viscoelastic layer should each be about one-third the thickness of the vibrating surface to which they are applied. Ducts and panels can be purchased fabricated as laminates.

Control of Noise Source by Redress

The best way to solve noise problems is with good design of the source. However, frequently an existing source is a noise problem either because of age, abuse, or poor design. Then the problem must be redressed or corrected as it exists. The following sections identify measures for redressing or correcting the source.

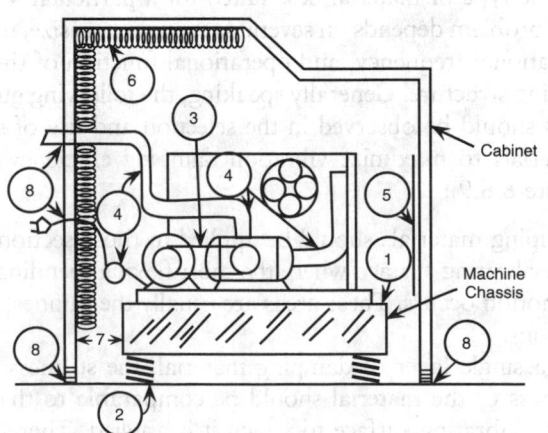

FIG. 6.6.8 Techniques to reduce the generation of airborne and structure-borne noise in machines. (Adapted from U.S. Department of Commerce)

BALANCING ROTATING PARTS

One of the main sources of machinery noise is structural vibration caused by the rotation of poorly balanced parts, such as fans, fly wheels, pulleys, cams, and shafts. Measures used to correct this condition involve adding counterweights to the rotating unit or removing some weight from the unit. A familiar noise caused by imbalance is in the high-speed spin cycle of washing machines. The imbalance results from clothes being unevenly distributed in the tub. By redistributing the clothes, balance is achieved, and the noise ceases. This same principle of balance can be applied to furnace fans and other common sources of such noise.

REDUCING FRICTIONAL RESISTANCE

A well-designed machine that has been poorly maintained can become a serious source of noise. General cleaning and lubrication of all rotating, sliding, or meshing parts at contact points go a long way toward fixing the problem.

APPLYING DAMPING MATERIALS

Since a vibrating body or surface radiates noise, applying any material that reduces or restrains the vibrational motion of that body decreases its noise output. Three basic types of redress vibration damping materials are available:

Liquid mastics, which are applied with a spray gun and harden into relatively solid materials, the most common being automobile undercoating

Pads of rubber, felt, plastic foam, leaded vinyls, adhesive tapes, or fibrous blankets, which are glued to the vibrating surface

Sheet metal viscoelastic laminates or composites, which are bonded to the vibrating surface

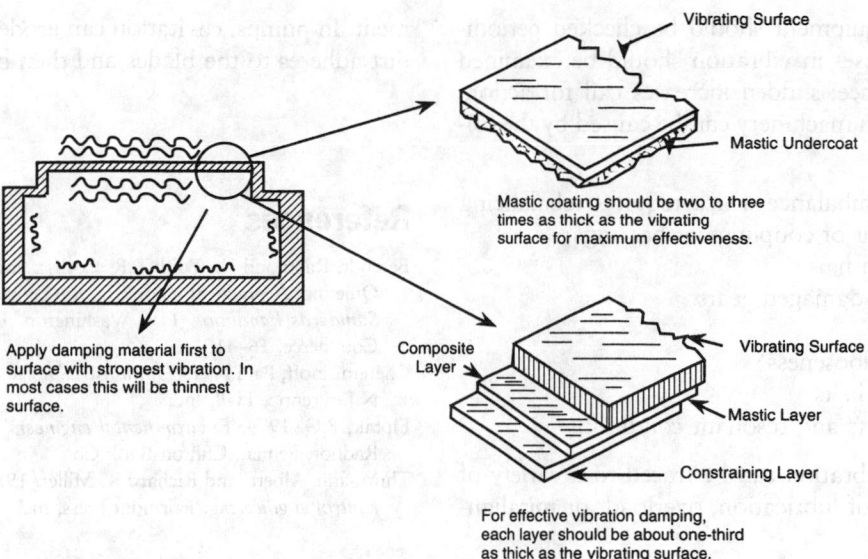

FIG. 6.6.9 Reducing vibration with damping materials.

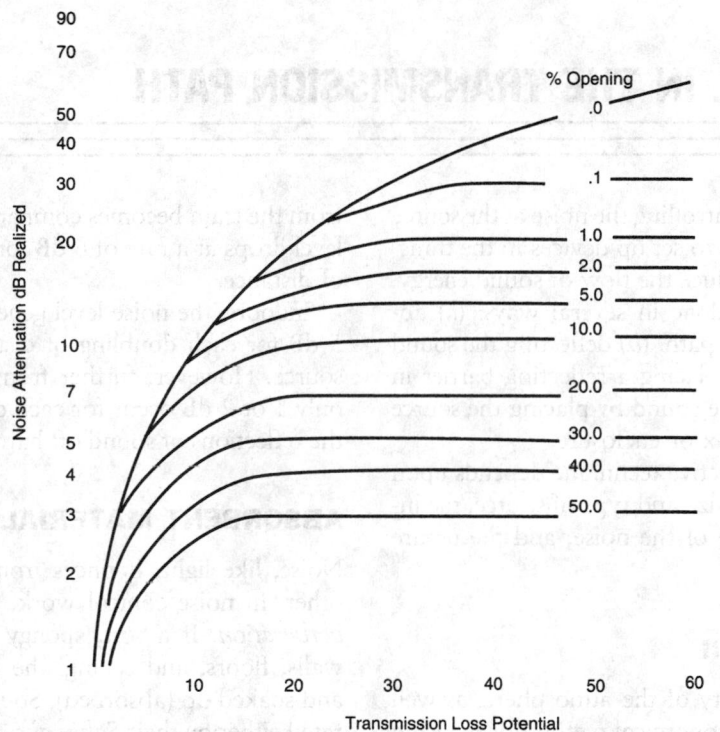

FIG. 6.6.10 Transmission loss potential versus transmission loss realized for various opening sizes as a percent total wall area. (Reprinted from Bell, 1973, *Fundamentals of industrial noise control,* [Trumbull, CN: Harme Publications].)

SEALING NOISE LEAKS

Small holes in an otherwise noise-tight structure can reduce the effectiveness of the noise control measures. As shown in Figure 6.6.10, if the designed transmission loss of an acoustical enclosure is 40 dB, an opening that comprises only 0.1% of the surface area reduces the effectiveness of the enclosure by 10 dB.

PERFORMING ROUTINE MAINTENANCE

The noise of a worn muffler is familiar. Likewise, studies of automobile tire noise in relation to pavement roughness show that maintenance of the pavement surface is essential to keep noise at minimum levels. Normal road wear can yield noise increases on the order of 6 dBA.

Faulty installation and maintenance can result in excessive vibration. Equipment should be checked periodically. Gradual increases in vibration should be examined in routine maintenance; sudden increases call for action. Increased vibration in machinery can be caused by the following:

- Rotational imbalance which requires rebalancing
- Misalignment of couplings or bearings
- Eccentric journals
- Defective or damaged gears
- Bent shafts
- Mechanical looseness
- Faulty drive belts
- Rubbing parts and resonant conditions

Rapid increases in vibration can be traced to a variety of causes, such as lack of lubrication, overload, or misalign-

ment. In pumps, cavitation can erode an impeller. In fans, dirt adheres to the blades and then breaks off unevenly.

—*David H.F. Liu*

References

Berendt, Raymond D., Edith L.R. Corliss, and Morris S. Ojalvo. 1976. Quieting: A practical guide to noise control. In *National Bureau of Standards handbook 119*. Washington, D.C.: U.S. Department of Commerce, 16–41.

Cheremisinoff, Paul N. 1993. *Industrial noise control*. Englewood Cliffs, N.J.: Prentice Hall, Inc.

Lipták, B.G. 1974. *Environmental engineers' handbook*. Vols. 3, 4, 5. Radnor, Penna.: Chilton Book Co.

Thurmann, Albert, and Richard K. Miller. 1986. *Fundamentals of noise control engineering*. Fairmont Press, Inc.

6.7
NOISE CONTROL IN THE TRANSMISSION PATH

After all possible ways of controlling the noise at the source are tried, the next defense is to set up devices in the transmission path to block or reduce the flow of sound energy. This noise control can be done in several ways: (*a*) absorbing the sound along the path, (*b*) deflecting the sound in some other direction by placing a reflecting barrier in its path, or (*c*) containing the sound by placing the source inside a sound-insulating box or enclosure.

Selection of the most effective technique depends upon various factors, such as the size and type of source, the intensity and frequency range of the noise, and the nature and type of environment.

Acoustical Separation

Using the absorptive capacity of the atmosphere, as well as divergence, is a simple, economical method of reducing the noise level. Air absorbs high-frequency sounds more effectively than it absorbs low-frequency sounds. However, if enough distance is available, even low-frequency sounds are absorbed appreciably.

When the distance from a point source is doubled, the sound pressure level is lowered by 6 dB. It takes about a 10 dB drop to halve the loudness. If the line source is a railroad train, the noise level drops by only 3 dB for each doubling of distance from the source. The main reason for this lower rate of attenuation is that line sources radiate sound waves that are cylindrical in shape. The surface area of such waves only increases two-fold for each doubling of distance from the source. However, when the distance

from the train becomes comparable to its length, the noise level drops at a rate of 6 dB for each subsequent doubling of distance.

Indoors, the noise level generally drops only from 3 to 5 dB for each doubling of distance in the vicinity of the source. However, further from the source, reductions of only 1 or 2 dB occur for each doubling of distance due to the reflections of sound off hard walls and ceiling surfaces.

ABSORBENT MATERIALS

Noise, like light, bounces from one hard surface to another. In noise control work, this bouncing is called *reverberation*. If a soft, spongy material is placed on the walls, floors, and ceiling, the reflected sound is diffused and soaked up (absorbed). Sound-absorbing materials are rated either by their *Sabin absorption coefficients* (α_{SAB}) at 125, 500, 1000, 2000, and 4000 Hz or by a single number rating called the *noise reduction coefficient* (NRC).

If a unit area of open window is assumed to transmit all and reflect none of the acoustical energy that reaches it, it is assumed to be 100% absorbent. This unit area of totally absorbent surface is called a *sabin*. The absorptive properties of acoustical materials are then compared with this standard. The performance is expressed as a fraction or percentage of the sabin (α_{SAB}). The NRC is the average of the α_{SAB}s at 250, 500, 1000, and 2000 Hz rounded to the nearest multiple of 0.05. The NRC has no physical meaning. It is a useful means of comparing similar materials.

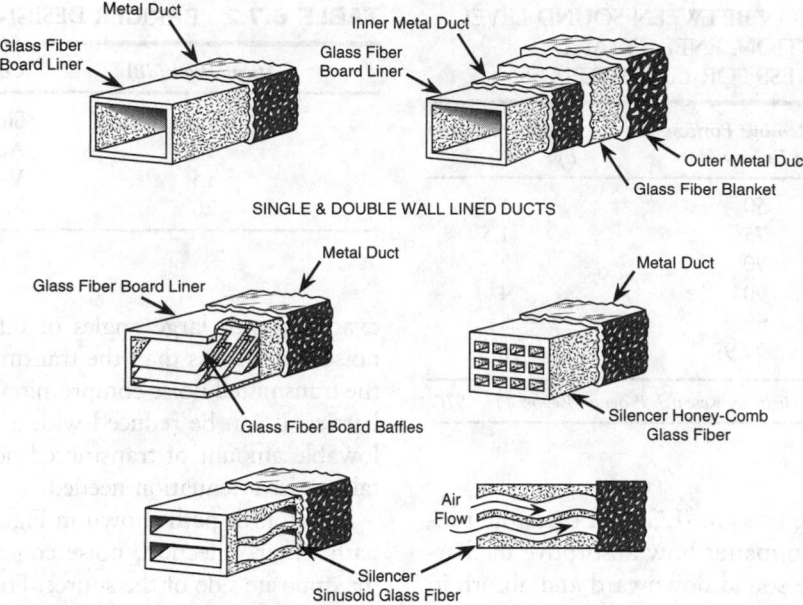

FIG. 6.7.1 Various types of acoustical duct lining, baffles, and silencers.

Sound-absorbing materials such as acoustical tile, carpets, and drapes placed on ceiling, floor, or wall surfaces can reduce the noise level in most rooms by about 5 to 10 dB for high-frequency sounds, but only by 2 or 3 dB for low-frequency sounds. Unfortunately, such treatment does not protect an operator of a noisy machine who is in the direct noise field. For greatest effectiveness, sound-absorbing materials should be installed as close to the noise source as possible.

If only a small or limited amount of sound-absorbing material is available, putting it in the upper trihedral corners of the room, formed by the ceiling and two walls, is the most effective use of it in a noisy room. Due to the process of reflection, the concentration of sound is greatest in the trihedral corners of a room. Additionally, the upper corner locations also protect lightweight fragile materials from damage.

Because of their light weight and porous nature, acoustical materials are ineffectual in preventing the transmission of either airborne or structure-borne sound from one room to another. In other words, if you can hear people walking or talking in the room or apartment above, installing acoustical tile on your ceiling will not reduce the noise transmission.

ACOUSTICAL LININGS

Lining the inside surfaces of ducts, pipe chases, or electrical channels with sound-absorbing materials can effectively reduce the noise transmitted through such passageways. In typical duct installations, noise reductions of 10 dB/m for an acoustical lining 2.5 cm thick are well within reason for high-frequency noise. A comparable degree of noise reduction for the lower frequency sounds is more difficult

to achieve because it usually requires at least a doubling of the thickness and length of the acoustical treatment. Figure 6.7.1 shows various types of duct lining baffles and silencers.

Physical Barriers

BARRIERS AND PANELS

Placing barriers, screens, or deflectors in the noise path is an effective way of reducing noise transmission, provided that the barriers are large enough in size and depending upon whether the noise is high-frequency or low-frequency. High-frequency noise is reduced more effectively than low-frequency noise.

The effectiveness of a barrier depends on its location, its height, and its length. Figure 6.7.2 shows that the noise can follow five different paths.

First, noise follows a direct path to receivers who can see the source well over the top of the barrier. The barrier

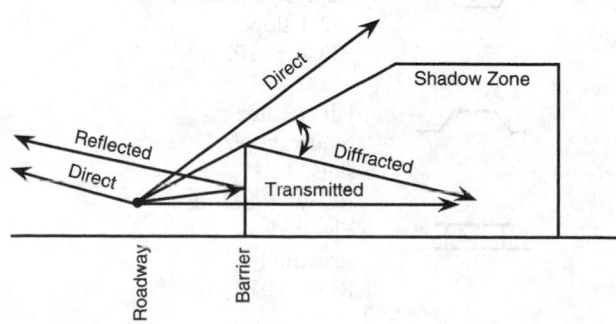

FIG. 6.7.2 Noise paths from a source to a receiver. (Reprinted from *National Cooperative Highway Research Program Report 174*, 1976.)

TABLE 6.7.1   RELATION BETWEEN SOUND LEVEL REDUCTION, ENERGY, AND LOUDNESS FOR LINE SOURCES

To Reduce A-level by dB	Remove Portion of Energy (%)	Divide Loudness by
3	50	1.2
6	75	1.5
10	90	2
20	90	4
30	99.9	8
40	99.99	16

Source: National Cooperative Highway Research Program Report 174, 1976.

TABLE 6.7.2   BARRIER DESIGN

Attenuation (dB)	Complexity
5	Simple
10	Attainable
15	Very difficult
20	Nearly impossible

does not block their line of sight (L/S) and therefore provides no attenuation. No matter how absorptive the barrier is, it cannot pull the sound downward and absorb it.

Second, noise follows a diffracted path to receivers in the shadow zone of the barrier. The noise that passes just over the top edge of the barrier is diffracted (bent) down into the apparent shadow shown in the figure. The larger the angle of diffraction, the more the barrier attenuates the noise in this shadow zone. In other words, less energy is diffracted through large angles than through smaller angles.

Third, in the shadow zone, the noise transmitted directly through the barrier is significant in some cases. For

example, with large angles of diffraction, the diffracted noise may be less than the transmitted noise. In this case, the transmitted noise compromises the performance of the barrier. It can be reduced with a heavier barrier. The allowable amount of transmitted noise depends on the total barrier attenuation needed.

The fourth path shown in Figure 6.7.2 is the reflected path. After reflection, noise concerns only a receiver on the opposite side of the source. For this reason, acoustical absorption on the face of the barrier can sometimes reduce this reflected noise; however, this treatment does not benefit receivers in the shadow zone.

In most practical cases, reflected noise is not important in barrier design. If the source of noise is represented by a line of noise, another short-circuit path is possible. Part of the source may be unshielded by the barrier. For example, the receiver might see the source beyond the ends of the barrier if the barrier is not long enough. This noise from around the ends may compromise, or short-circuit,

TABLE 6.7.3   NOISE REDUCTIONS FOR VARIOUS HIGHWAY CONFIGURATIONS

Highway Configuration[a]		Height or Depth (m)	Truck Mix (%)	Noise Reduction[b] at Distance from ROW (dBA)	
Sketch	Description			30 m	152 m
	Roadside barriers 7.6 m from edge of shoulders; ROW = 78 m wide	6.1	0	13.9	13.3
			5	13.0	12.1
			10	12.6	11.7
			20	12.3	11.3
	Depressed roadway w/2:1 slopes ROW = 102 m	6.1	0	9.9	11.4
			5	8.8	10.3
			10	8.4	9.8
			20	8.1	9.4
	Fill elevated roadway w/2:1 slopes; ROW = 102 m	6.1	0	9.0	6.3
			5	7.6	2.7
			10	7.1	1.8
			20	6.7	1.1
	Elevated structure; ROW = 78 m	7.3	0	9.8	6.0
			5	9.6	2.4
			10	9.3	1.5
			20	8.8	0.8

Source: B.A. Kugler, D.E. Commins, and W.J. Galloway, 1976, Highway noise: Generation and control, National Cooperative Highway Research Program Report 173.

[a]Assumes divided 8 lanes with 9.1 m median.
[b]Based on observed 1.5 m above grade.

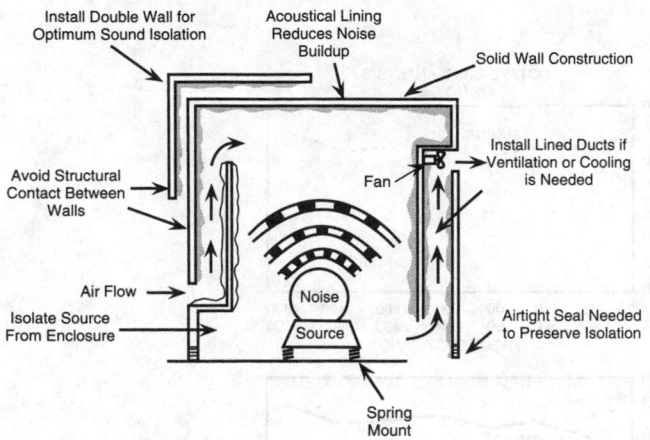

FIG. 6.7.3 Enclosures for controlling noise. (Reprinted from *National Bureau of Standards Handbook 119, 1976.*)

barrier attenuation. The required barrier length depends on the total net attenuation needed. When 10 to 15 dB attenuation is needed, barriers must be long. Therefore, to be effective, barriers must not only break the line of sight to the nearest section of the source, but also to the source far up and down the line.

Of these four paths, the noise diffracted over the barrier into the shadow zone represents the most important parameter from the barrier design point of view. Generally, determining barrier attenuation or barrier noise reduction involves only calculating the amount of energy diffracted into the shadow zone. The procedures presented in the barrier nomograph used to predict highway noise are based on this concept.

Another general principle of barrier noise reduction is the relation between noise attenuation expressed in (1) dBs,

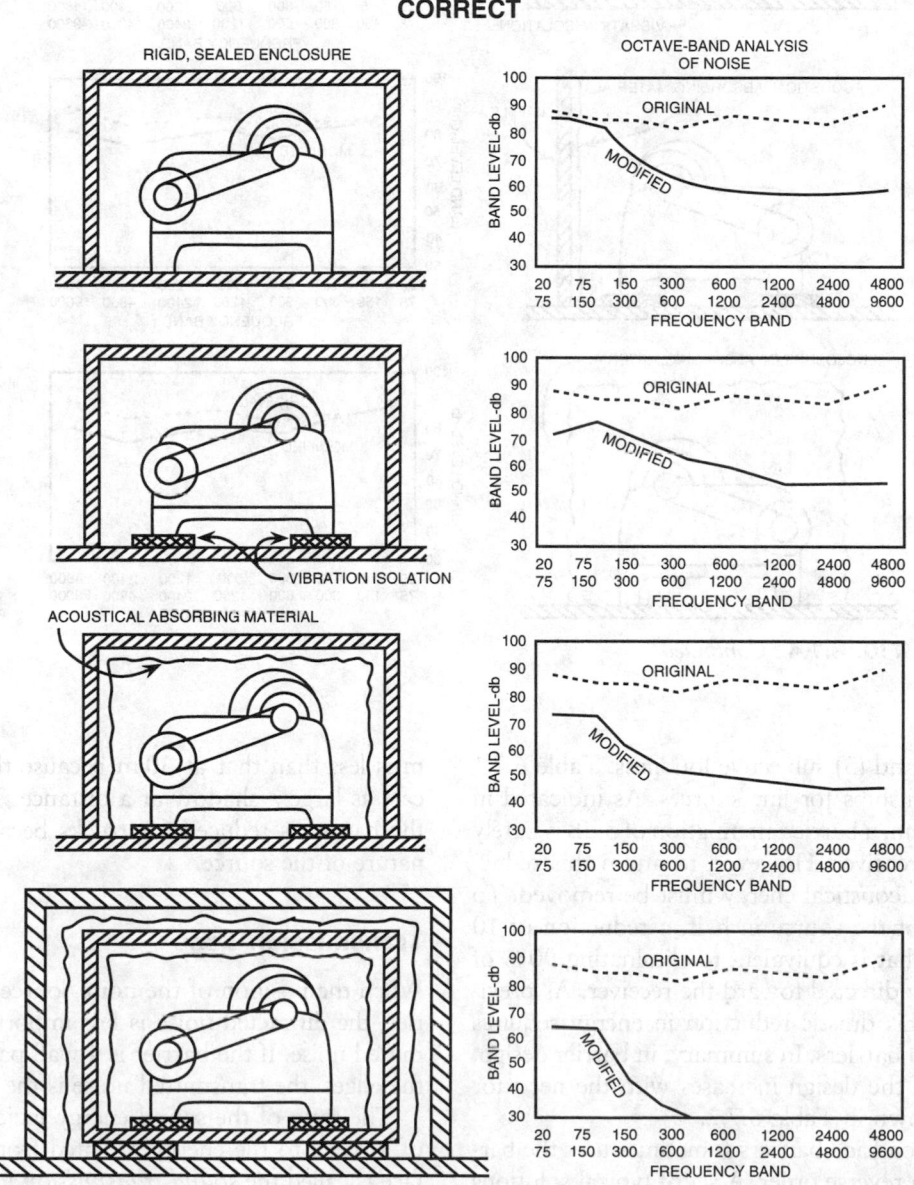

FIG. 6.7.4 Effectiveness of various noise reduction techniques. (Adapted by permission of General Radio Company.) *Continued on next page*

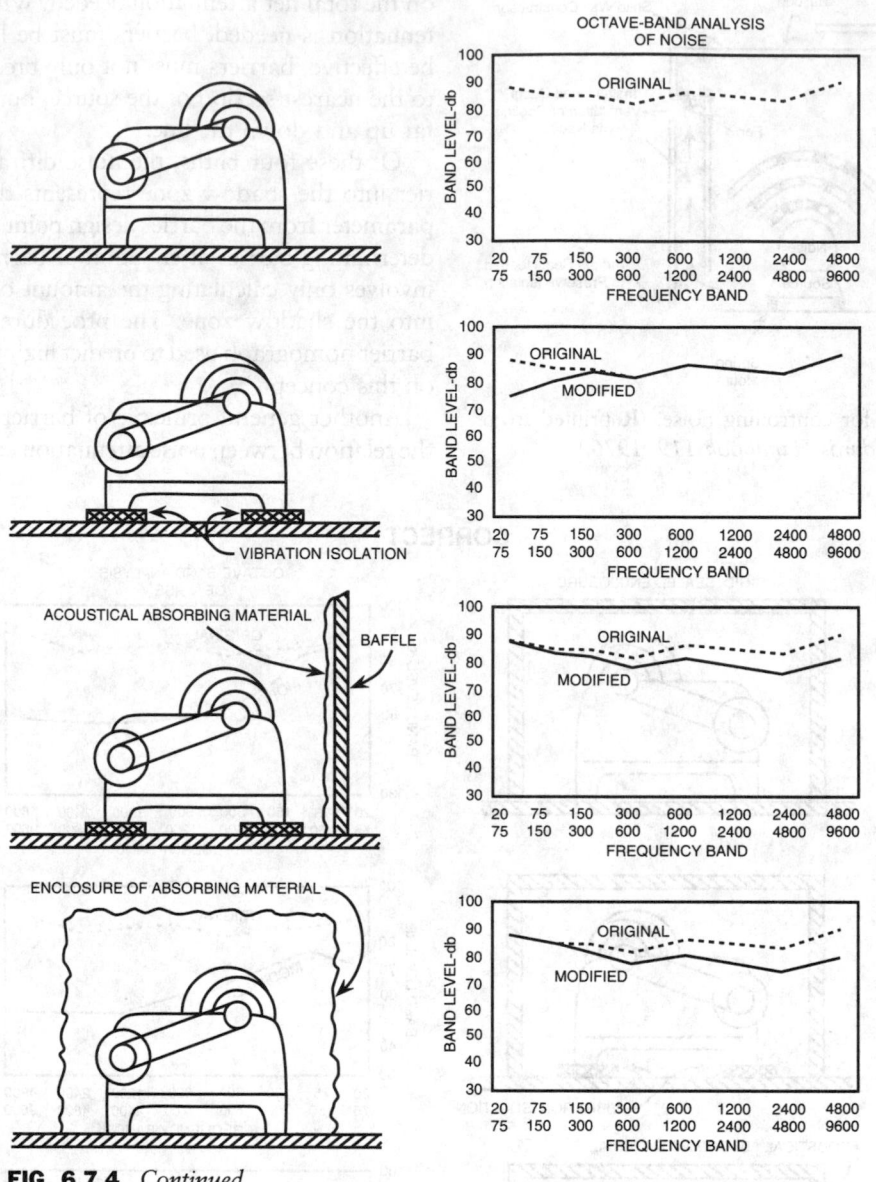

INCORRECT

FIG. 6.7.4 *Continued*

(2) energy terms, and (3) subjective loudness. Table 6.7.1 gives these relationships for line sources. As indicated in the loudness column, a barrier attenuation of 3 dB is barely discerned by the receiver. However, to attain this reduction, 50% of the acoustical energy must be removed. To cut the loudness of the source in half, a reduction of 10 dB is necessary. That is equivalent to eliminating 90% of the energy initially directed toward the receiver. As previously indicated, this drastic reduction in energy requires very long and high barriers. In summary, in barrier design, the complexity of the design increases with the need for attenuation as shown in Table 6.7.2.

The design of roadside barriers sometimes uses the barrier nomograph in reverse order. A set of typical solutions is summarized in Table 6.7.3. The noise reduction at 152

m is less than that at 30 m because the barrier does not cast as large a shadow at a distance. The effectiveness of the barrier is reduced for trucks because of the elevated nature of the source.

Transmission Loss

When the position of the noise source is close to the barrier, the diffracted noise is less important than the transmitted noise. If the barrier is a wall panel that is sealed at the edges, the transmitted noise is the only concern.

The ratio of the sound energy incident on one surface of a panel to the energy radiated from the opposite surface is called the *sound transmission loss* (TL). The actual energy loss is partially reflected and partially absorbed.

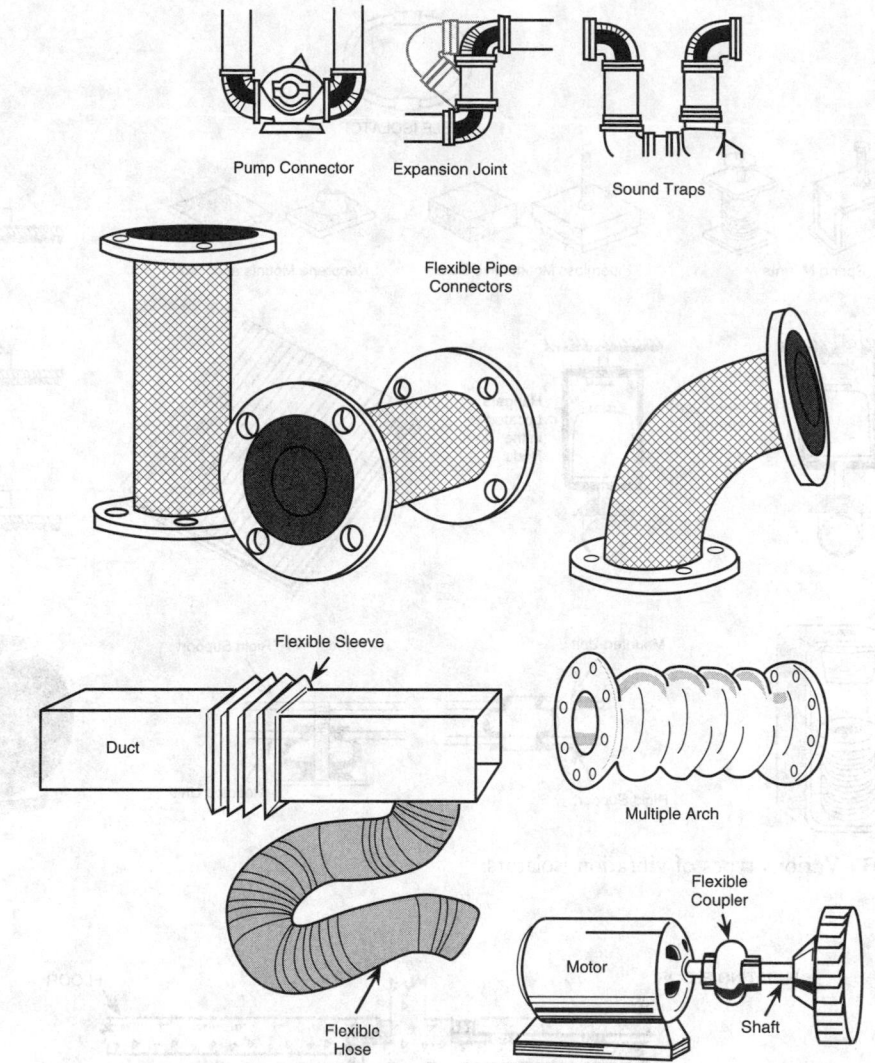

FIG. 6.7.5 Various types of flexible connectors.

Since TL is frequency-dependent, only a complete octave or one-third octave band curve fully describes the performance of the barrier.

ENCLOSURES

Sometimes enclosing a noisy machine in a separate room or box is more practical and economical than quieting it by altering its design, operation, or component parts. The walls of the enclosure should be massive and airtight to contain the sound. Absorbent lining on the interior surfaces of the enclosure can reduce the reverberant buildup of noise within it. Structural contact between the noise source and the enclosure must be avoided, or the source vibration can be transmitted to the enclosure walls and thus short-circuit the isolation. For maximum effective noise control, all of the techniques illustrated in Figure 6.7.3 must be employed. Figure 6.7.4 shows the design and effectiveness of various enclosure configurations.

Isolators and Silencers

VIBRATION ISOLATORS AND FLEXIBLE COUPLERS

If the noise transmission path is structure-borne in character, vibration isolators in the form of resilient mountings, flexible couplers, or structural breaks or discontinuities should be interposed between the noise source and receiver. For example, string mounts placed under a machine can prevent the floor from vibration, or an expansion joint cut along the outer edge of a floor in a mechanical equipment room can reduce the amount of vibration transmitted to the structural frame or walls of a building. These measures are shown in Figures 6.7.5 and 6.7.6; several often-used procedures, in Figure 6.7.7.

MUFFLERS AND SILENCERS

No distinction exists between mufflers and silencers. They are often used interchangeably. They are in effect acoustic

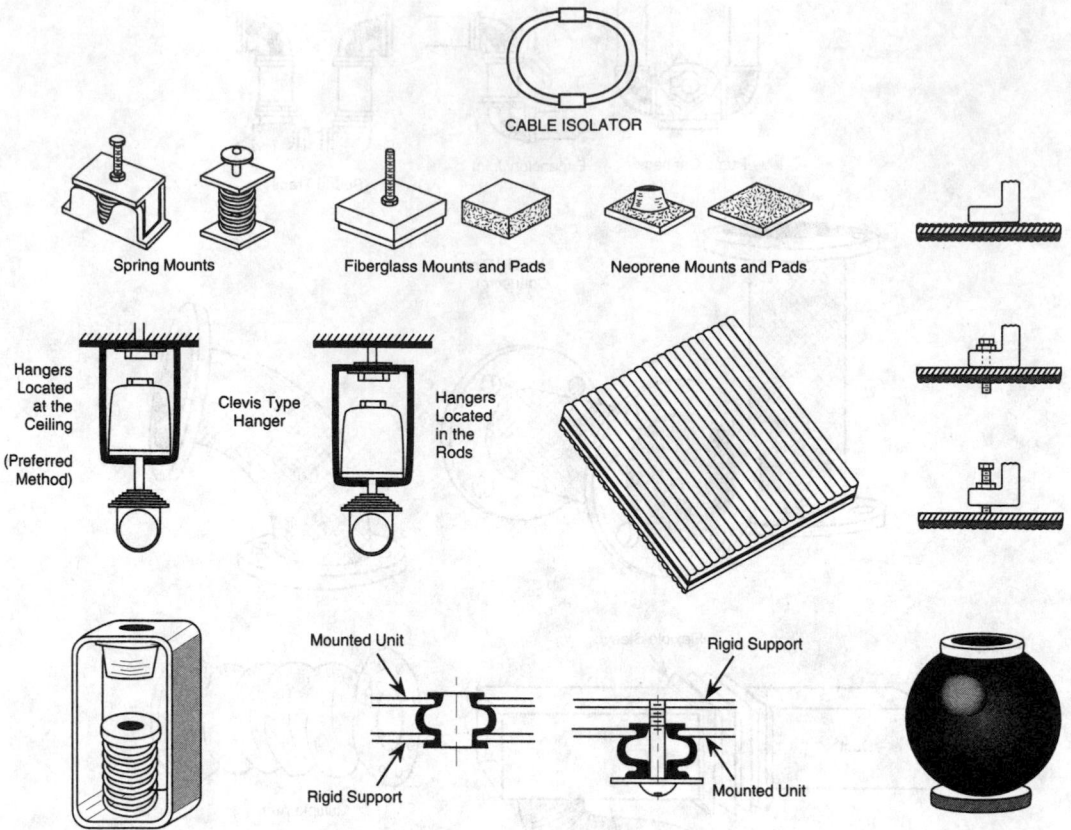

CABLE ISOLATOR

Spring Mounts

Fiberglass Mounts and Pads

Neoprene Mounts and Pads

Hangers Located at the Ceiling

(Preferred Method)

Clevis Type Hanger

Hangers Located in the Rods

Mounted Unit

Rigid Support

Rigid Support

Mounted Unit

FIG. 6.7.6 Various types of vibration isolators.

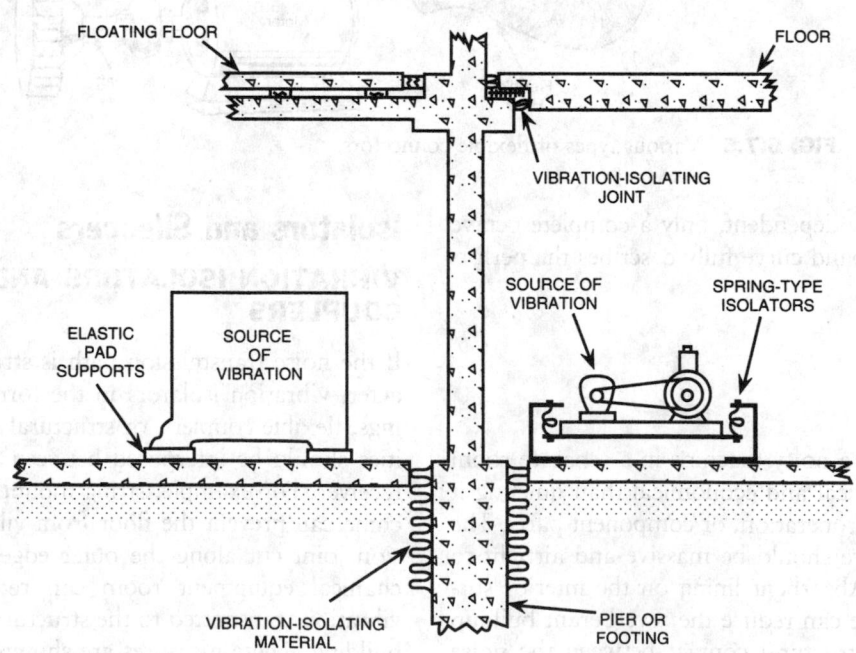

FLOATING FLOOR

FLOOR

VIBRATION-ISOLATING JOINT

ELASTIC PAD SUPPORTS

SOURCE OF VIBRATION

SOURCE OF VIBRATION

SPRING-TYPE ISOLATORS

VIBRATION-ISOLATING MATERIAL

PIER OR FOOTING

FIG. 6.7.7 Isolation of machinery vibration.

Absorptive silencer. This silencer is the most common type and takes the form of a duct lined on the interior with a sound-absorptive material.

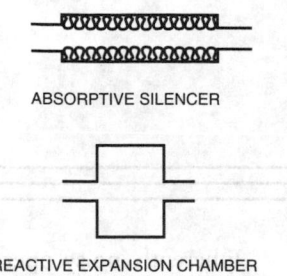

ABSORPTIVE SILENCER

Reactive expansion chamber. This type reflects sound energy back toward the source to cancel some of the oncoming sound energy.

REACTIVE EXPANSION CHAMBER

Reactive resonator. This type functions in approximately the same way as the reactive expansion chamber type.

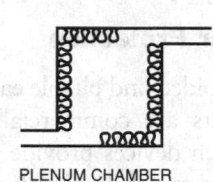

REACTIVE RESONATOR

Plenum chamber. This device allows the sound to enter a small opening in the chamber: that sound which has not been absorbed by the chamber's acoustical lining leaves by a second small opening, generally at the opposite end of the chamber.

PLENUM CHAMBER

Lined bend. Sound energy flowing down a passage is forced to turn a corner, the walls of which are lined with acoustical material. The sound energy is thus forced to impinge directly on a sound-absorbing surface as it reflects its way around the corner; each successive impingement takes sound energy from the traveling wave.

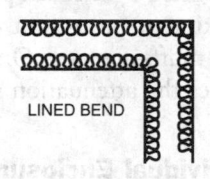

LINED BEND

Diffuser. This device does not actually reduce noise. In effect, it prevents the generation of noise by disrupting high-velocity gas streams.

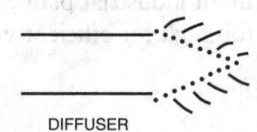

DIFFUSER

FIG. 6.7.8 Basic silencers.

filters and are used to reduce fluid flow noise. Figure 6.7.8 shows six basic types of silencers.

The devices can be classified into:

Dissipative or absorptive silencers whose noise reduction is determined by fibrous or porous sound absorbing liners. Where hot gases are handled, the absorbent may

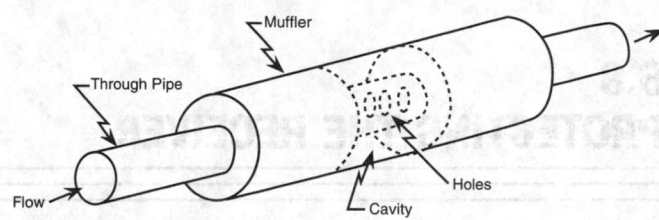

FIG. 6.7.9 Muffler.

be metal or even ceramic. These devices are good for high-frequency sounds.

Reactive silencers whose noise reduction is determined mainly by geometry. These devices are shaped to reflect or expand the sound waves with resultant self-destruction. Reactive silencers are used for low-frequency applications.

Figure 6.7.9 shows a typical muffler that is designed to attenuate sound waves with minimal back pressure. It includes a cylindrical-type unit. The outer portion of the through-pipe conduit contains a number of cavities where noise suppression occurs. A porous packing is sometimes used to increase efficiency. Airflow to the cavities is regulated by the size and number of holes from the center section. Mufflers are effective for high- and middle-frequency noise control.

— *David H.F. Liu*

References

Berendt, Raymond D., Edith L.R. Corliss, and Morris S. Ojalvo. 1976. Quieting: A practical guide to noise control. In *National Bureau of Standards handbook 119*. Washington, D.C.: U.S. Department of Commerce, 16–41.

Cheremisinoff, Paul N. 1993. *Industrial noise control*. Prentice Hall, Inc.

Davis, Mackenzie, and David A. Cornwell. 1991. *Introduction to environmental engineering*. 2d ed. McGraw-Hill, Inc.

Lipták, B.G. 1974. *Environmental engineers' handbook*. Vols. 3, 4, 5. Radnor, Penna.: Chilton Book Co.

Sabin, H.J. 1942. Notes on acoustic impedance measurement. *Journal of the Acoustical Society of America* 14:143.

Thurmann, Albert, and Richard K. Miller. 1986. *Fundamentals of noise control engineering*. Fairmont Press, Inc.

6.8
PROTECTING THE RECEIVER

When neither reducing noise at its source nor reducing its transmission from the source to the receiver is satisfactory, the receiver must be protected. Protection is likely to be needed when conditions change frequently or when several requirements must be met at the same time.

Work Schedules

The amount of continuous exposure to high noise levels must be limited. For hearing protection, scheduling noisy operation for short intervals of time each day over several days is preferable to a continuous eight-hour run for a day or two.

In industrial or construction operations, an intermittent work schedule benefits not only the operator of the noisy equipment but also other workers in the vicinity. If an intermittent schedule is not possible, workers should have relief time during the day. They should take their relief time at a low-noise-level location and should not trade this time for more pay, paid vacation, or an early out at the end of the day!

Inherently noisy operations, such as street repair, municipal trash collection, factory operation, and aircraft traffic, should be curtailed at night and early morning to avoid disturbing the sleep of the community.

Equipment and Shelters

Ear Protection

Molded and pliable earplugs, cup-type protectors, and helmets are commercially available as hearing protectors. Such devices provide noise reductions from 15 to 35 dB (Figure 6.8.1). Earplugs are effective only if they are properly fitted by medical personnel. As shown in Figure 6.8.1, maximum protection can be obtained when both plugs and muffs are used. Only muffs with certification that stipulates the attenuation should be used.

Individual Enclosures or Noise Shelters

In an industrial plant, large areas where the noise level is too high for efficient work often exist. Any kind of office

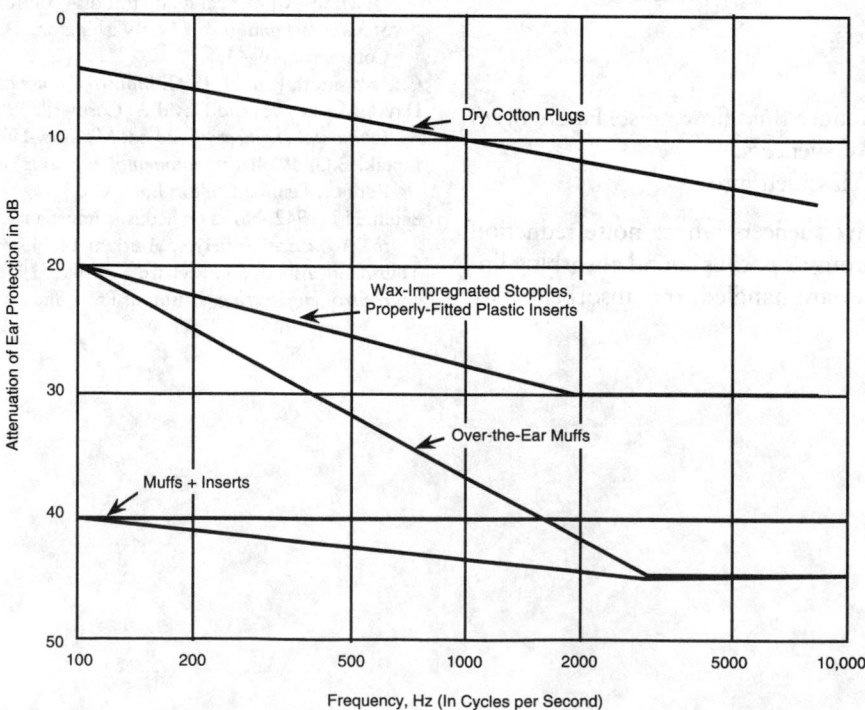

FIG. 6.8.1 Sound attenuation characteristics of various types of ear protectors.

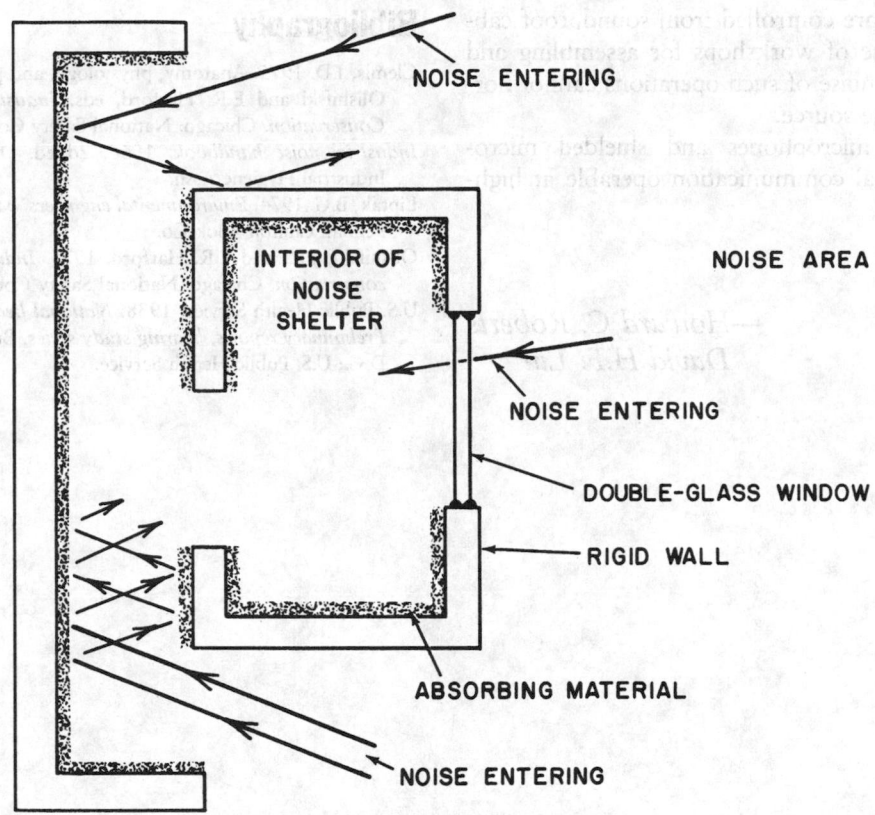

FIG. 6.8.2 Principle of the labyrinth noise shelter.

work is impractical at such noise levels, yet inplant offices are needed.

Noise shelters provide an effective solution for these problems. They may be fully enclosed rooms with separate heating and ventilating systems, which protect from dust and odors as well as noise. Plant offices are often designed and constructed this way based on the principles described in Section 6.7.

When easy access to a noise shelter is needed, the labyrinth principle can be applied. This principle is useful in isolating areas where specialized work is done (such as, inspection and final adjustment) and where both people and work move continually in and out. Figure 6.8.2 shows how this principle is constructed. Such a noise shelter can be used for the operator's station when the machine controls are within the shelter and vision is through a large window area. In such construction, the wall surrounding the sheltered area should have a high-sound-transmission loss, and double glass suitably spaced and mounted to reduce noise transmission in the window.

Noise does not enter the space because it is reflected—preferably two or more times—from surfaces covered with absorbing material. At each reflection, appreciable noise reduction occurs. The noise shelter is also lined with absorbing material to absorb what noise does enter. The performance quality of such enclosures depends on their design and construction.

The labyrinth principle effectively prevents the passage of noise from one work area to another. The labyrinth principle allows free passage between work areas without the interference of closed doors.

Other kinds of individual enclosures include the cabs used on agricultural and earth-moving machines. These cabs and motor trucks provide shelter from noise as well as weather conditions. Industrial crane cabs also provide this protection since safety in their areas depends on the alertness and ability of the operator.

Other Possibilities

When the noise within a confined area is too high to allow workers into it even with personal protection devices, the operation might be automated. An automated process is supervised from observation posts, that is, from remote-control stations where workers are adequately protected. A remote-control post receives information via a closed-circuit television, or it can be a highly insulated area within the department. Mechanical devices handle the production procedures under operator or computer control. For example, rolling mills are controlled from soundproof cabins. The same is true of workshops for assembling and testing engines. The noise of such operations cannot normally be stifled at the source.

Noise-cancelling microphones and shielded microphones keep electrical communication operable at high-noise levels.

—*Howard C. Roberts*
David H.F. Liu

Bibliography

Clemis, J.D. 1975. Anatomy, physiology, and pathology of the ear. J.B. Olishifski and E.R. Harford, eds. *Industrial Noise and Hearing Conservation.* Chicago: National Safety Council.

Industrial noise handbook. 1966. 2d ed. Detroit, Mich.: American Industrial Hygiene Assn.

Lipták, B.G. 1974. *Environmental engineers' handbook.* Vol. 3, Radnor, Penna.: Chilton Book Co.

Olishifski, J.B., and E.R. Harford. 1975. *Industrial noise and hearing conservation.* Chicago: National Safety Council.

U.S. Public Health Service. 1938. *National health survey (1935–1936): Preliminary reports, hearing study series.* Bulletins 1-7. Washington, D.C.: U.S. Public Health Service.

7

Wastewater Treatment

Carl E. Adams, Jr. | Donald B. Aulenbach | L. Joseph Bollyky | Jerry
L. Boyd | Robert D. Buchanan | Don E. Burns | Larry W. Canter |
George J. Crits | Donald Dahlstrom | Stacy L. Daniels | Frank W.
Dittman | Wayne F. Echelberger, Jr. | Ronald G. Gantz | Louis C.
Gilde, Jr. | Brian L. Goodman | Negib Harfouche | R. David
Holbrook | Sun-Nan Hong | Derk T.A. Huibers | Frederick W.
Keith, Jr. | Mark K. Lee | Béla G. Lipták | János Lipták | David
H.F. Liu | Francis X. McGarvey | Thomas J. Myron, Jr. | Van T.
Nguyen | Joseph G. Rabosky | LeRoy H. Reuter | Bernardo Rico-
Ortega | Chakra J. Santhanam | E. Stuart Savage | Frank P.
Sebastian | Gerry L. Shell | Wen K. Shieh | John R. Snell | Paul L.
Stavenger | Michael S. Switzenbaum

Organics, Salts, Metals, and Nutrient Removal   773

Sludge Disposal   893

Sources and Characteristics

7.1
NATURE OF WASTEWATER

The nature of wastewater is described by its flow and quality characteristics. In addition, wastewater discharges are classified based on whether they are from municipalities or industries. Flow rates and quality characteristics of industrial wastewater are more variable than those for municipal wastewater.

Flow Rates

Municipal wastewater is comprised of domestic (or sanitary) wastewater, industrial wastewater, infiltration and inflow into sewer lines, and stormwater runoff. Domestic wastewater refers to wastewater discharged from residences and from commercial and institutional facilities (Metcalf and Eddy, Inc. 1991). Domestic water usage, and the resultant wastewater, is affected by climate, community size, density of development, community affluence, dependability and quality of water supply, water conservation requirements or practices, and the extent of metered services. Metcalf and Eddy, Inc. (1991) provide details on the influence of these factors. Additional factors influencing water use include the degree of industrialization, cost of water, and supply pressure (Qasim 1985). One result of the combined influence of these factors is water use fluctuations. Table 7.1.1 summarizes such fluc-

tuations (Metcalf and Eddy, Inc. 1991). About 60 to 85% of water usage becomes wastewater, with the lower percentages applicable to the semiarid region of the southwestern United States (Metcalf and Eddy, Inc. 1991).

Environmental engineers can use unit flow rate data to develop estimates for wastewater flow rates from residential areas, commercial districts, and institutional facilities. Tables 7.1.2 through 7.1.4 depict data for these use categories, respectively. Industrial wastewater flow rates vary and are a function of the type and size of industry. For estimation purposes, typical design flows from industrial areas that have little or no wet-process-type industries are 1000 to 1500 gal/acre per day (9 to 14 m³/ha · d) for light industrial developments and 1500 to 3000 gal/acre per day (14 to 28 m³/ha · d) for medium industrial developments (Metcalf and Eddy, Inc. 1991). Better estimates for industries can be developed with industry-specific information.

Wastewater volume generated in a municipality depends on the population served, the per capita contribu-

TABLE 7.1.1   TYPICAL FLUCTUATIONS IN WATER USE IN COMMUNITY SYSTEMS

Water Use	Percentage of Average for Year	
	Range	Typical
Daily average in maximum month	110–140	120
Daily average in maximum week	120–170	140
Maximum day	160–220	180
Maximum hr	225–320	270[a]

Source: Metcalf and Eddy, Inc., 1991, *Wastewater engineering*, 3d ed. (New York: McGraw-Hill).
Note: [a]1.5 × maximum day value.

TABLE 7.1.2   TYPICAL WASTEWATER FLOW RATES FROM RESIDENTIAL SOURCES

Source	Unit	Flow, gal/unit · d	
		Range	Typical
Apartment:			
High-rise	Person	35–75	50
Low-rise	Person	50–80	65
Hotel	Guest	30–55	45
Individual residence:			
Typical home	Person	45–90	70
Better home	Person	60–100	80
Luxury home	Person	75–150	95
Older home	Person	30–60	45
Summer cottage	Person	25–50	40
Motel:			
With kitchen	Unit	90–180	100
Without kitchen	Unit	75–150	95
Trailer park	Person	30–50	40

Source: Metcalf and Eddy, Inc., 1991.
Note: l = gal × 3.7854.

TABLE 7.1.3   TYPICAL WASTEWATER FLOW RATES FROM COMMERCIAL SOURCES

Source	Unit	Flow, gal/unit·d Range	Flow, gal/unit·d Typical
Airport	Passenger	2–4	3
Automobile service station	Vehicle served	7–13	10
	Employee	9–15	12
Bar	Customer	1–5	3
	Employee	10–16	13
Department store	Toilet room	400–600	500
	Employee	8–12	10
Hotel	Guest	40–56	48
	Employee	7–13	10
Industrial building (sanitary waste only)	Employee	7–16	13
Laundry (self-service)	Machine	450–650	550
	Wash	45–55	50
Office	Employee	7–16	13
Restaurant	Meal	2–4	3
Shopping center	Employee	7–13	10
	Parking space	1–2	2

Source: Metcalf and Eddy, Inc., 1991.
Note: l = gal × 3.7854.

TABLE 7.1.4   TYPICAL WASTEWATER FLOW RATES FROM INSTITUTIONAL SOURCES

Source	Unit	Flow, gal/unit·d Range	Flow, gal/unit·d Typical
Hospital, medical	Bed	125–240	165
	Employee	5–15	10
Hospital, mental	Bed	75–140	100
	Employee	5–15	10
Prison	Inmate	75–150	115
	Employee	5–15	10
Rest home	Resident	50–120	85
School, day			
With cafeteria, gym, and showers	Student	15–30	25
With cafeteria only	Student	10–20	15
Without cafeteria and gym	Student	5–17	11
School, boarding	Student	50–100	75

Source: Metcalf and Eddy, Inc., 1991.
Note: l = gal × 3.7854.

tion, and other nondomestic sources such as industrial wastewater discharges. Environmental engineers may need to use population forecasting to project future rates of wastewater generation in the service area of a wastewater treatment plant. Some mathematical or graphical methods used to project population data to a design year include (Qasim 1985):

- arithmetic growth
- geometric growth
- decreasing rate of increase
- mathematical or logistic curve fitting
- graphical comparison with similar cities
- ratio method
- employment forecast
- birth cohort

Water usage exhibits daily, weekly, and seasonal patterns; and wastewater flow rates can also exhibit such patterns. Figures 7.1.1 and 7.1.2 show typical hourly, daily,

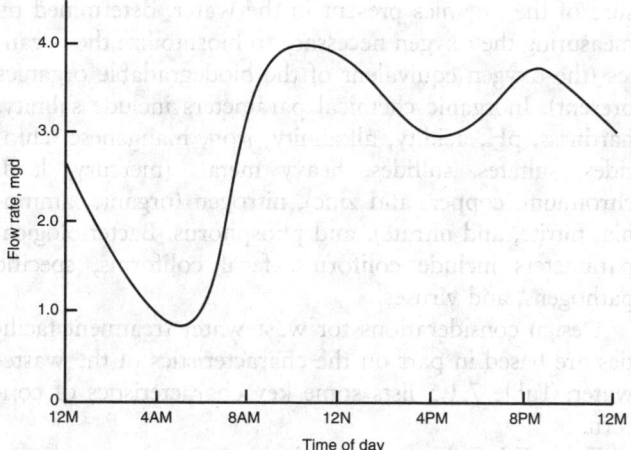

FIG. 7.1.1 Typical pattern of hourly variations in domestic wastewater flow rates. (Reprinted, with permission, from Metcalf and Eddy, Inc., 1991.)

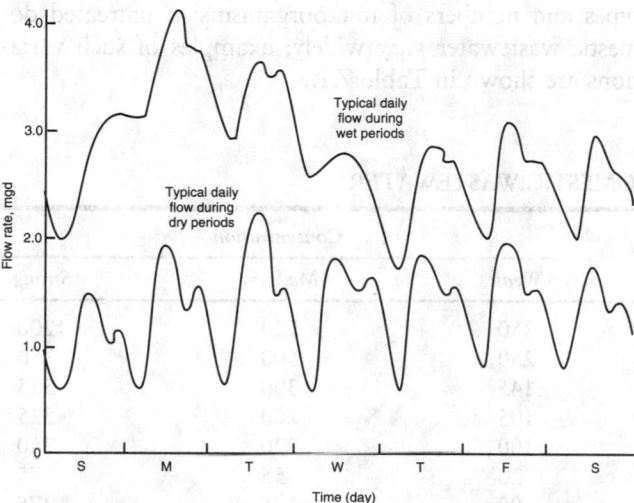

FIG. 7.1.2 Typical patterns of daily and weekly variations in domestic wastewater flow rates. (Reprinted, with permission, from Metcalf and Eddy, Inc., 1991.)

and weekly wastewater flow rates, respectively (Metcalf and Eddy, Inc. 1991).

Wide variations of wastewater flow rates can occur within a municipality. For example, minimum to maximum flow rates range from 20 to 400% of the average daily rate for small communities with less than 1000 people, from 50 to 300% for communities with populations between 1000 and 10,000, and up to 200% for communities up to 100,000 in population (Water Pollution Control Federation and American Society of Civil Engineers 1977). Large municipalities have variations from 1.25 to 1.5 average flow. When storm water runoff goes into municipal sewerage systems, the maximum flow rate is often two to four times the average dry-weather flow (Water Pollution Control Federation and American Society of Civil Engineers 1977).

Flow rate information needed in designing a wastewater treatment plant includes (Metcalf and Eddy, Inc. 1991):

AVERAGE DAILY FLOW—The average flow rate occurring over a 24-hr period based on total annual flow rate data. Environmental engineers use average flow rate in evaluating treatment plant capacity and in developing flow rate ratios.

MAXIMUM DAILY FLOW—The maximum flow rate occurring over a 24-hr period based on annual operating data. The maximum daily flowrate is important in the design of facilities involving retention time, such as equalization basins and chlorine-contact tanks.

PEAK HOURLY FLOW—The peak sustained hourly flow rate occurring during a 24-hr period based on annual operating data. Data on peak hourly flows are needed for the design of collection and interceptor sewers, wastewater pumping stations, wastewater flowmeters, grit chambers, sedimentation tanks, chlorine-contact tanks, and conduits or channels in the treatment plant.

MINIMUM DAILY FLOW—The minimum flow rate that occurs over a 24-hr period based on annual operating data. Minimum flow rates are important in sizing conduits where solids deposition might occur at low flow rates.

MINIMUM HOURLY FLOW—The minimum sustained hourly flow rate occurring over a 24-hr period based on annual operating data. Environmental engineers need data

TABLE 7.1.5 PHYSICAL, CHEMICAL, AND BIOLOGICAL WASTEWATER CHARACTERISTICS CONSIDERED FOR DESIGN

Physical	Chemical	Biological
Solids	Organics	Plants
Temperature	Proteins	Animals
Color	Carbohydrates	Viruses
Odor	Lipids	
	Surfactants	
	Phenols	
	Pesticides	
	Inorganics	
	pH	
	Chloride	
	Alkalinity	
	Nitrogen	
	Phosphorus	
	Heavy metals	
	Toxic materials	
	Gases	
	Oxygen	
	Hydrogen sulfide	
	Methane	

Source: Water Pollution Control Federation and American Society of Civil Engineers, 1977, *Wastewater treatment plant design* (Washington, D.C.), 10.

on the minimum hourly flow rate to determine possible process effects and size wastewater flow meters, particularly those that pace chemical-feed systems.

SUSTAINED FLOW—The flow rate value sustained or exceeded for a specified number of consecutive days based on annual operating data. Data on sustained flow rates may be used in sizing equalization basins and other plant hydraulic components.

Wastewater Characteristics

Wastewater quality can be defined by physical, chemical, and biological characteristics. Physical parameters include color, odor, temperature, solids (residues), turbidity, oil, and grease. Solids can be further classified into suspended and dissolved solids (size and settleability) as well as organic (volatile) and inorganic (fixed) fractions. Chemical parameters associated with the organic content of wastewater include the biochemical oxygen demand (BOD), chemical oxygen demand (COD), total organic carbon (TOC), and total oxygen demand (TOD). BOD is a measure of the organics present in the water, determined by measuring the oxygen necessary to biostabilize the organics (the oxygen equivalent of the biodegradable organics present). Inorganic chemical parameters include salinity, hardness, pH, acidity, alkalinity, iron, manganese, chlorides, sulfates, sulfides, heavy metals (mercury, lead, chromium, copper, and zinc), nitrogen (organic, ammonia, nitrite, and nitrate), and phosphorus. Bacteriological parameters include coliforms, fecal coliforms, specific pathogens, and viruses.

Design considerations for wastewater treatment facilities are based in part on the characteristics of the wastewater; Table 7.1.5 lists some key characteristics of concern.

Table 7.1.6 shows the typical concentration range of various constituents in untreated domestic wastewater. Depending on the concentrations, wastewater is classified as strong, medium, or weak. Table 7.1.7 shows typical mineral increases resulting from domestic water use. The types and numbers of microorganisms in untreated domestic wastewater vary widely; examples of such variations are shown in Table 7.1.8.

TABLE 7.1.6 TYPICAL COMPOSITION OF UNTREATED DOMESTIC WASTEWATER

Contaminants	Unit	Concentration		
		Weak	Medium	Strong
Total solids (TS)	mg/l	350	720	1200
total dissolved solids (TDS)	mg/l	250	500	850
fixed	mg/l	145	300	525
volatile	mg/l	105	200	325
suspended solids (SS)	mg/l	100	220	350
fixed	mg/l	20	55	75
volatile	mg/l	80	165	275
Settleable solids	mL/l	5	10	20
BOD, mg/l:				
5-day, 20°C (BOD$_5$, 20°C)	mg/l	110	220	400
TOC	mg/l	80	160	290
COD	mg/l	250	500	1000
Nitrogen (total as N)	mg/l	20	40	85
organic	mg/l	8	15	35
free ammonia	mg/l	12	25	50
nitrites	mg/l	0	0	0
nitrates	mg/l	0	0	0
Phosphorus (total as P)	mg/l	4	8	15
organic	mg/l	1	3	5
inorganic	mg/l	3	5	10
Chlorides[a]	mg/l	30	50	100
Sulfate[a]	mg/l	20	30	50
Alkalinity (as CaCO$_3$)	mg/l	50	100	200
Grease	mg/l	50	100	150
Total coliform	no/100 ml	10^6–10^7	10^7–10^8	10^7–10^9
Volatile organic compounds (VOCs)	μg/L	<100	100–400	>400

Source: Metcalf and Eddy, Inc., 1991, *Wastewater engineering*, 3d ed. (New York: McGraw-Hill).
Notes: °F = 1.8(°C) + 32.
[a]Values should be increased by the amount present in the domestic water supply; see Table 7.1.7.

TABLE 7.1.7 TYPICAL MINERAL INCREASE FROM DOMESTIC WATER

Constituent	Increment Range,[a] mg/l
Anions	
Bicarbonate (HCO$_3$)	50–100
Carbonate (CO$_3$)	0–10
Chloride (Cl)	20–50[b]
Nitrate (NO$_3$)	20–40
Phosphate (PO$_4$)	5–15
Sulfate (SO$_4$)	15–30
Cations	
Calcium (Ca)	6–16
Magnesium (Mg)	4–10
Potassium (K)	7–15
Sodium (Na)	40–70
Other constituents	
Aluminum (Al)	0.1–0.2
Boron (B)	0.1–0.4
Fluoride (F)	0.2–0.4
Manganese (Mn)	0.2–0.4
Silica (SiO$_2$)	2–10
Total alkalinity (as CaCO$_3$)	60–120
TDS	150–380

Source: Metcalf and Eddy, Inc., 1991.
Notes: [a]Reported values do not include commercial and industrial additions.
[b]Excluding the addition from domestic water softeners.

The remainder of this subsection focuses on selected quality characteristics of wastewater. Details related to the analytical procedures for quality characteristics are available in *Standard methods for the examination of water and wastewater* (Clesceri, Greenberg, and Trussell 1989).

TABLE 7.1.8 TYPES AND NUMBER OF MICROORGANISMS TYPICALLY FOUND IN UNTREATED DOMESTIC WASTEWATER

Organism	Concentration, number/ml
Total coliform	10^5–10^6
Fecal coliform	10^4–10^5
Fecal streptococci	10^3–10^4
Enterococci	10^2–10^3
Shigella	Present[a]
Salmonella	10^0–10^2
Pseudomonas aeroginosa	10^1–10^2
Clostridium perfringens	10^1–10^3
Mycobacterium tuberculosis	Present[a]
Protozoan cysts	10^1–10^3
Giardia cysts	10^{-1}–10^2
Cryptosporidium cysts	10^{-1}–10^1
Helminth ova	10^{-2}–10^1
Enteric virus	10^1–10^2

Source: Metcalf and Eddy, Inc., 1991.
Note: [a]Results for these tests are usually reported as positive or negative rather than quantified.

The TS content of wastewater can be defined as the matter that remains as residue upon evaporation at 103 to 105°C; Figure 7.1.3 shows the categories of solids in wastewater. Table 7.1.9 shows particle sizes related to different categories. Clesceri, Greenberg, and Trussell (1989) provide detailed information related to testing procedures. Figure 7.1.4 shows typical values for solids found in medium strength wastewater.

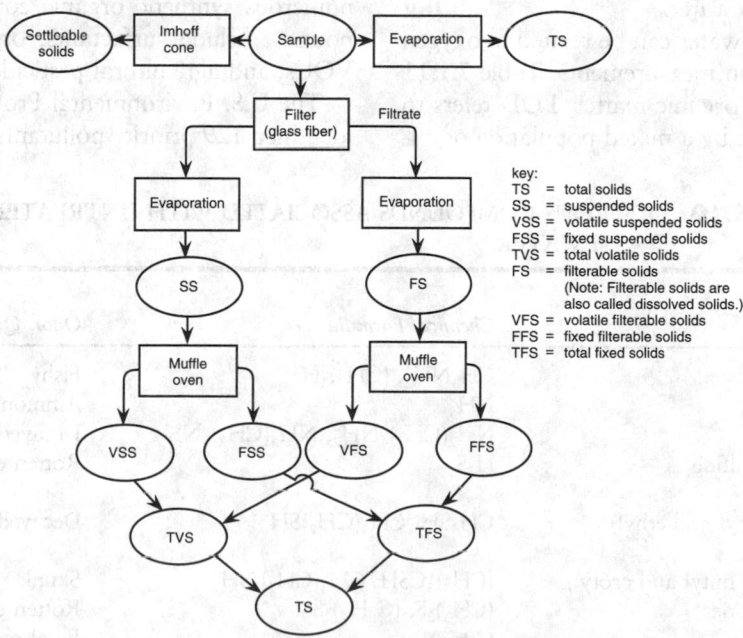

key:
TS = total solids
SS = suspended solids
VSS = volatile suspended solids
FSS = fixed suspended solids
TVS = total volatile solids
FS = filterable solids
(Note: Filterable solids are also called dissolved solids.)
VFS = volatile filterable solids
FFS = fixed filterable solids
TFS = total fixed solids

FIG. 7.1.3 Interrelationships of wastewater solids (also called dissolved solids). (Reprinted, with permission, from Metcalf and Eddy, Inc., 1991.)

TABLE 7.1.9 GENERAL CLASSIFICATION OF WASTEWATER SOLIDS

Particle Classification	Particle Size, mm
Dissolved	Less than 10^{-6}
Colloidal	10^{-6} to 10^{-3}
Suspended	Greater than 10^{-3}
Settleable	Greater than 10^{-2}

Source: R.A. Corbitt, 1990, Wastewater disposal, Chap. 6 in *Standard handbook of environmental engineering*, edited by R.A. Corbitt (New York: McGraw-Hill Publishing Company).

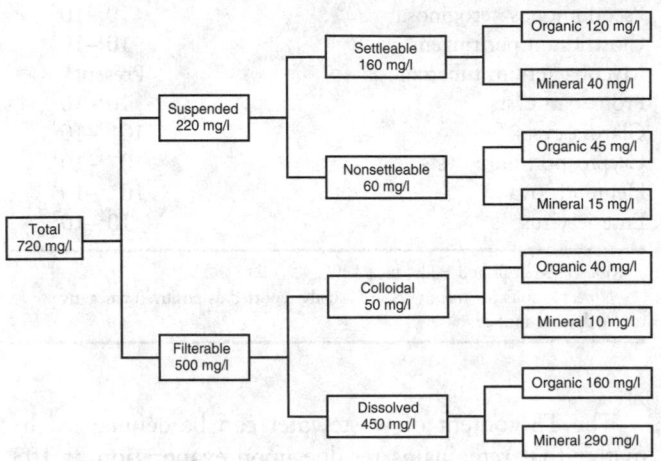

FIG. 7.1.4 Classification of solids found in medium-strength wastewater. (Metcalf and Eddy, Inc., 1991.)

Odors in wastewater are caused by gases from decomposition or by odorous substances within the wastewater. Table 7.1.10 lists examples of odorous compounds associated with untreated wastewater.

Organic matter in wastewater can be related to oxygen demand and organic carbon measurements. Table 7.1.11 shows several measures of organic matter. BOD refers to the amount of oxygen used by a mixed population of mi-

croorganisms under aerobic conditions that stabilize organic matter in the wastewater. The 5-day BOD is primarily composed of carbonaceous oxygen demand, while the 20-day BOD includes both carbonaceous and nitrogenous oxygen demands. Figure 7.1.5 is a typical BOD curve, developed via laboratory measurements, for domestic wastewater. The following relationships and definitions are associated with Figure 7.1.5 (Qasim 1985):

$$
\begin{aligned}
UOD &= L_o + L_n \\
y &= L_o(1 - e^{-Kt}) \\
L_{oT} &= L_o(0.02T + 0.60) \\
K_T &= K(1.047)^{T-20}
\end{aligned}
\qquad 7.1(1)
$$

where:

UOD = ultimate oxygen demand, mg/l
L_n = nitrogenous oxygen demand or second stage BOD, mg/l
L_o = ultimate carbonaceous BOD, or first stage BOD at 20°C, mg/l (for domestic wastewater BOD_5 is approximately equal to $\frac{2}{3} L_o$)
y = carbonaceous BOD at any time t, mg/l
L_{oT} = ultimate carbonaceous BOD at any temperature T°C, mg/l
K = reaction rate constants at 20°C, d^{-1} (for domestic wastewater K = 0.2 to 0.3 per d)
K_T = reaction rate constant to any temperature T°C, d^{-1}

For medium-strength domestic wastewater, about 75% of SS and 40% of FS are classified as organic. The main groups of organic substances in wastewater include proteins (40–60%), carbohydrates (25–50%), and fats and oils (10%) (Metcalf and Eddy, Inc. 1991). Urea, a constituent of urine, is also found in wastewater along with numerous synthetic organic compounds. Synthetic compounds include surfactants, organic priority pollutants, VOCs, and agricultural pesticides.

The U.S. Environmental Protection Agency (EPA) has identified 129 priority pollutants in wastewater; these pol-

TABLE 7.1.10 ODOROUS COMPOUNDS ASSOCIATED WITH UNTREATED WASTEWATER

Odorous Compound	Chemical Formula	Odor, Quality
Amines	CH_3NH_2, $(CH_3)_3H$	Fishy
Ammonia	NH_3	Ammoniacal
Diamines	$NH_2(CH_2)_4NH_2$, $NH_2(CH_2)_5NH_2$	Decayed flesh
Hydrogen sulfide	H_2S	Rotten eggs
Mercaptans (e.g., methyl and ethyl)	CH_3SH, $CH_3(CH_2)SH$	Decayed cabbage
Mercaptans (e.g., T = butyl and crotyl)	$(CH_3)_3CSH$, $CH_3(CH_2)_3SH$	Skunk
Organic sulfides	$(CH_3)_2S$, $(C_6H_5)_2S$	Rotten cabbage
Skatole	C_9H_9N	Fecal matter

Source: Metcalf and Eddy, Inc., 1991, *Wastewater engineering*, 3d ed. (New York: McGraw-Hill).

TABLE 7.1.11 OXYGEN DEMAND AND ORGANIC CARBON PARAMETERS

Parameter	Description
BOD_5	Biochemical or biological oxygen demand exerted in 5 days; the oxygen consumed by a waste through bacterial action; generally about 45–55% of THOD.
COD	Chemical oxygen demand. The amount of strong chemical oxidant (chromic acid) reduced by a waste; results are expressed in terms of an equivalent amount of oxygen; generally about 80% of THOD.
THOD	Theoretical oxygen demand. The amount of oxygen theoretically required to completely oxidize a compound to CO_2, H_2O, PO_4^{-3}, SO_4^{-2}, and NO_3.
TOC	Total organic carbon; generally about 30% of THOD.
BOD_L	The ultimate BOD exerted by a waste in an infinite time.
IOD	Immediate oxygen demand. The amount of oxygen consumed by a waste within 15 min (chemical oxidizers and bacteria not used).

Source: R.A. Corbitt, 1990, Wastewater disposal, Chap. 6 in *Standard handbook of environmental engineering,* edited by R.A. Corbitt (New York: McGraw-Hill Publishing Company).

lutants are subject to discharge standards. Examples of priority pollutants and their health-related concerns are listed in Table 7.1.12.

Pathogenic organisms in wastewater can be categorized as bacteria, viruses, protozoa, and helminths; Table 7.1.13 lists examples of such organisms present in raw domestic wastewater. Because of the many types of pathogenic organisms and the associated measurement difficulties, coliform organisms are frequently used as indicators of human pollution. On a daily basis, each person discharges from 100 to 400 billion coliform organisms, in addition to other kinds of bacteria (Metcalf and Eddy, Inc. 1991). In terms of the indicator concept, the presence of coliform organisms indicates that pathogenic organisms may also be present, and their absence indicates that the water is free from disease-producing organisms (Metcalf and Eddy, Inc. 1991). Total coliform and fecal coliform are often used as indicators of wastewater effluent disinfection.

The quantities of fecal coliform (FC) and fecal streptococci (FS) discharged by humans are significantly different from the quantities discharged by animals. As a result, the ratio of the FC count to the FS count can show whether the suspected contamination derives from human or animal waste. Table 7.1.14 gives example data on the ratio of FC to FS counts for humans and various animals. The FC/FS ratio for domestic animals is less than 1.0; whereas the ratio for humans is more than 4.0.

Some differences can be delineated between municipal and industrial wastewater discharges. For example, fluctuations in industrial wastewater flow rates typically exceed those for municipal wastewater. Industrial plants may not operate continuously; there may be daily, weekly, or seasonal variations in operations, reflected by flow rate variations. In addition, the number and types of contaminants in industrial wastewater can vary widely, and the concentrations can range from near zero to 100,000 mg/l

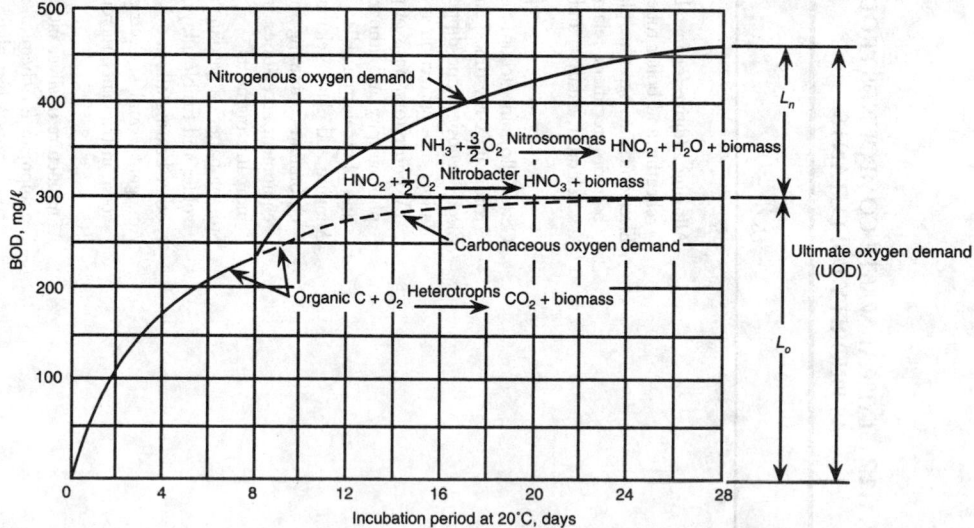

FIG. 7.1.5 Typical BOD curve for domestic wastewater showing carbonaceous and nitrogenous oxygen demands. (Reprinted, with permission, from S.R. Qasim, 1985, *Wastewater treatment plants—Planning, design, and operation,* New York: Holt, Rinehart and Winston.)

TABLE 7.1.12 TYPICAL WASTE COMPOUNDS PRODUCED BY COMMERCIAL, INDUSTRIAL, AND AGRICULTURAL ACTIVITIES AND CLASSIFIED AS PRIORITY POLLUTANTS

Name (Formula)	Use	Concern
Nonmetals		
Arsenic (As)	Alloying additive for metals, especially lead and copper as shot, battery grids, cable sheaths, and boiler tubes. High purity (semiconductor) grade	Carcinogen and mutagen. *Long term*—can cause fatigue, loss of energy and dermatitis.
Selenium (Se)	Electronics, xerographic plates, TV cameras, photocells, magnetic computer cores, solar batteries, rectifiers, relays, ceramics (colorant for glass) steel and copper, rubber accelerator, catalyst, and trace element in animal feeds	*Long term*—red staining of fingers, teeth, and hair; general weakness; depression; and irritation of the nose and mouth.
Metals		
Barium (Ba)	Getter alloys in vacuum tubes, deoxidizer for copper, Frary's metal, lubricant for anode rotors in X-ray tubes, and spark-plug alloys	Flammable at room temperature in powder form. *Long term*—increased blood pressure and nerve block.
Cadmium (Cd)	Electrodeposited and dipped coatings on metals, bearing and low-melting alloys, brazing alloys, fire protection systems, nickel-cadmium storage batteries, power transmission wire, TV phosphors, basis of pigments used in ceramic glazes, machinery enamels, fungicide photography and lithography, selenium rectifiers, electrodes for cadmium-vapor lamps, and photoelectric cells	Flammable in powder form. Toxic by inhalation of dust or fume. *Long term*. A carcinogen. Soluble compounds of cadmium highly toxic. *Long term*—concentrates in the liver, kidneys, pancreas, and thyroid; hypertension suspected effect.
Chromium (Cr)	Alloying and plating element on metal and plastic substrates for corrosion resistance, chromium-containing and stainless steels, protective coating for automotive and equipment accessories, nuclear and high-temperature research and constituent of inorganic pigments	Hexavalent chromium compounds are carcinogenic and corrosive to tissue. *Long term*—skin sensitization and kidney damage.
Lead (Pb)	Storage batteries, gasoline additive, cable covering, ammunition, piping, tank linings, solder and fusible alloys, vibration damping in heavy construction, foil, babbit, and other bearing alloys	Toxic by ingestion or inhalation of dust or fumes. *Long term*—brain and kidney damage and birth defects.
Mercury (Hg)	Amalgams, catalyst electrical apparatus, cathodes for production of chlorine and caustic soda, instruments, mercury vapor lamps, mirror coating, arc lamps, and boilers	Highly toxic by skin absorption and inhalation of fume or vapor. *Long term*—toxic to central nervous system, and can cause birth defects.
Silver (Ag)	Manufacture of silver nitrate, silver bromide, photo chemicals; lining vats and other equipment for chemical reaction vessels, water distillation, etc.; mirrors, electric conductors; silver plating electronic equipment; sterilant; water purification; surgical cements; hydration and oxidation catalyst special batteries, solar cells, reflectors for solar towers; low-temperature brazing alloys; table cutlery; jewelry; dental, medical, and scientific equipment; electrical contacts; bearing metal; magnet windings; and dental amalgams. Colloidal silver used as a nucleating agent in photography and medicine, often combined with protein	Toxic metal. *Long term*—permanent grey discoloration of skin, eyes, and mucus membranes.

Organic Compounds

Name	Uses	Hazard
Benzene (C_6H_6)	Manufacturing of ethylbenzene (for styrene monomer); dodecylbenzene (for detergents); cyclohexane (for nylon); phenol; nitrobenzene (for aniline); maleic anhydride; chlorobenzene hexachloride; benzene sulfonic acid; and as a solvent	A carcinogen. Highly toxic. Flammable, and a dangerous fire risk
Ethylbenzene ($C_6H_5C_2H_5$)	Intermediate in the production of styrene and as a solvent	Toxic by ingestion, inhalation, and skin absorption; irritant to skin and eyes. Flammable and a dangerous fire risk
Toluene ($C_6HC_5H_3$)	Aviation gasoline and high-octane blending stock; benzene, phenol, and caprolactam; solvent for paints and coatings, gums, resins, most oils, rubber, vinyl organosols; diluent and thinner in nitrocellulose lacquers; adhesive solvent in plastic toys and model airplanes; chemicals (benzoic acid, benzyl and bezoyl derivatives, saccharine, medicines, dyes, perfumes); source of toluenediisocyanates (polyurethane resins); explosives (TNT); toluene sulfonates (detergents); and scintillation counters	Flammable and a dangerous fire risk. Toxic by ingestion, inhalation, and skin absorption

Halogenated Compounds

Name	Uses	Hazard
Chlorobenzene (C_6H_5Cl)	Phenol, chloronitrobenzene, aniline, solvent carrier for methylene diisocyanate, solvent, pesticide intermediate, heat transfer	Moderate fire risk. Recommend avoiding inhalation and skin contact
Chloroethene (CH_2CHCl)	Polyvinyl chloride and copolymers, organic synthesis, and adhesives for plastics	An extremely toxic and hazardous material by all avenues of exposure. A carcinogen
Dichloromethane (CH_2Cl_2)	Paint removers, solvent degreasing, plastics processing, blowing agent in foams, solvent extraction, solvent for cellulose acetate, and aerosol propellant	Toxic. A carcinogen and a narcotic
Tetrachloroethene (CCl_2CCl_2)	Dry cleaning solvent, vapor-degreasing solvent, drying agent for metals and certain other solids, vermifuge, heat transfer medium, and manufacture of fluorocarbons	Irritant to eyes and skin

Pesticides, Herbicides, Insecticides[a]

Name	Uses	Hazard
Endrin ($C_{12}H_6OCl_6$)	Insecticide and fumigant	Toxic by inhalation and skin absorption and a carcinogen
Lindane ($C_6H_6Cl_6$)	Pesticide	Toxic by inhalation, ingestion, and skin absorption
Methoxychlor ($Cl_3CCH(C_6H_4OCH_3)_2$)	Insecticide	Toxic material
Toxaphene ($C_{10}H_{10}Cl_6$)	Insecticide and fumigant	Toxic by ingestion, inhalation, and skin absorption
Silvex ($Cl_3C_6H_2OCH(CH_3)COOH$)	Herbicides and plant growth regulator	Toxic material; use restricted

Source: Metcalf and Eddy, Inc., 1991, *Wastewater engineering*, 3d ed. (New York: McGraw-Hill).
Note: [a]Pesticides, herbicides, and insecticides are listed by trade name. The compounds listed are also halogenated organic compounds.

525

TABLE 7.1.13 INFECTIOUS AGENTS POTENTIALLY PRESENT IN RAW DOMESTIC WASTEWATER

Organism	Disease	Remarks
Bacteria		
Escherichia coli (enteropathogenic)	Gastroenteritis	Diarrhea
Legionella pneumophila	Legionellosis	Acute respiratory illness
Leptospira (150 spp.)	Leptospirosis	Jaundice, and fever (Weil's disease)
Salmonella typhi	Typhoid fever	High fever, diarrhea, and ulceration of small intestine
Salmonella (~1700 spp.)	Salmonellosis	Food poisoning
Shigella (4 spp.)	Shigellosis	Bacillary dysentery
Vibrio cholerae	Cholera	Extremely heavy diarrhea and dehydration
Yersinia enterolitica	Yersinosis	Diarrhea
Viruses		
Adenovirus (31 types)	Respiratory disease	
Enteroviruses (67 types, e.g., polio, echo, and Coxsackie viruses)	Gastroenteritis, heart anomalies, and meningitis	
Hepatitis A	Infectious hepatitis	Jaundice and fever
Norwalk agent	Gastroenteritis	Vomiting
Reovirus	Gastroenteritis	
Rotavirus	Gastroenteritis	
Protozoa		
Balantidium coli	Balantidiasis	Diarrhea and dysentery
Cryptosporidium	Cryptosporidiosis	Diarrhea
Entamoeba histolytica	Amebiasis (amoebic dysentery)	Prolonged diarrhea with bleeding and abscesses of the liver and small intestine
Giardia lamblia	Giardiasis	Mild to severe diarrhea, nausea, and indigestion
Helminths[a]		
Ascaris lumbricoides	Ascariasis	Roundworm infestation
Enterobius vericularis	Enterobiasis	Pinworm
Fasciola hepatica	Fascioliasis	Sheep liver fluke
Hymenolepis nana	Hymenolepiasis	Dwarf tapeworm
Taenia saginata	Taeniasis	Beef tapeworm
T. solium	Taeniasis	Pork tapeworm
Trichuris trichiura	Trichuriasis	Whipworm

Source: Metcalf and Eddy, Inc., 1991.
Note: [a]The helminths listed are those with a worldwide distribution.

(Nemerow and Dasgupta 1991). Some industrial wastewater contaminants are toxic; while others exhibit deoxygenation rates about five times greater than that for municipal wastewater (Nemerow and Dasgupta 1991).

Permits and Effluent Limitations

The *Federal Water Pollution Control Act Amendments of 1972* (PL 92-500) established basic water quality goals and policies for the United States (U.S. Congress 1972). The objective of the *Clean Water Act of 1987* (also known as the *Water Quality Act of 1987*) was to restore and main-

tain the chemical, physical, and biological integrity of the nation's waters; this objective was also in the precursor laws, including PL 92-500. Three key components of the *Clean Water Act of 1987* relevant to wastewater discharges include water quality standards and planning, discharge permits, and effluent limitations.

New point sources of wastewater discharge must apply for National Pollutant Discharge Elimination System (NPDES) permits under the auspices of Section 402 of the *Clean Water Act of 1987* and its precursors back to 1972. Permits typically address pertinent effluent limitations (discharge standards) for conventional and toxic pollutants, monitoring and reporting requirements, and schedules of

TABLE 7.1.14 ESTIMATED PER CAPITA CONTRIBUTION OF INDICATOR MICROORGANISMS FROM HUMANS AND SOME ANIMALS

Animal	Average Indicator Density/g of Feces		Average Contribution/capita·24h		
	Fecal Coliform, 10^6	Fecal Streptococci, 10^6	Fecal Coliform, 10^6	Fecal Streptococci, 10^6	Ratio FC/FS
Chicken	1.3	3.4	240	620	0.4
Cow	0.23	1.3	5400	31,000	0.2
Duck	33.0	54.0	11,000	18,000	0.6
Human	13.0	3.0	2000	450	4.4
Pig	3.3	84.0	8900	230,000	0.04
Sheep	16.0	38.0	18,000	43,000	0.4
Turkey	0.29	2.8	130	1300	0.1

Source: Metcalf and Eddy, Inc., 1991.
Note: lb = g × 0.0022.

compliance (Miller, Taylor, and Monk 1991). Effluent limitations can be based on control technologies or required removal (treatment) efficiencies, or they can be based on achieving water quality standards. Water quality-based effluent limitations can require greater treatment levels as determined via a waste load allocation study for the relevant stream segment.

Section 402(p) of the *Clean Water Act of 1987* requires NPDES permits for storm water (runoff water) discharges associated with industrial activity, discharges from large municipal separate storm water systems (systems serving a population of 250,000 or more), and discharges from medium municipal separate storm water systems (systems serving a population of 100,000 or more, but less than 250,000). NPDES permits for storm water from industrial areas require the development of a pollution prevention plan to reduce pollution at the source (U.S. EPA 1992b). Detailed information on developing pollution prevention plans for construction activities in urban or industrial areas is also available (U.S. EPA 1992a).

—*Larry W. Canter*

References

Clesceri, L.S., A.E. Greenberg, and R.R. Trussell, eds. 1989. *Standard methods for the examination of water and wastewater.* 17th ed. Washington, D.C.: American Public Health Association, American Water Works Association, and Water Pollution Control Federation.

Metcalf and Eddy, Inc. 1991. *Wastewater engineering.* 3d ed. 36, 50–57, 65–70, 93–96, 100–101, and 108–112. New York: McGraw-Hill, Inc.

Miller, L.A., R.S. Taylor, and L.A. Monk. 1991. *NPDES permit handbook.* 3d Printing. Rockville, Md.: Government Institutes, Inc.

Nemerow, N.L. and A. Dasgupta. 1991. *Industrial and hazardous waste treatment.* 10. New York: Van Nostrand Reinhold.

Qasim, S.R. 1985. *Wastewater treatment plants—Planning, design, and operation.* 9, 40–41. New York: Holt, Rinehart and Winston.

U.S. Congress. 1972. *Federal Water Pollution Control Act Amendments of 1972.* PL 92-500, 92nd Congress, S. 2770, 18 October 1972.

U.S. Environmental Protection Agency (EPA). 1992a. *Storm water management for construction activities—Developing pollution prevention plans and best management practices.* EPA 832-R-92-005. Washington, D.C.: Office of Water. (September).

———. 1992b. *Storm water management for industrial activities—Developing pollution prevention plans and best management practices.* EPA 832-R-92-006. Washington, D.C.: Office of Water. (September).

Water Pollution Control Federation and American Society of Civil Engineers. 1977. *Wastewater treatment plant design.* 10. Washington, D.C.

7.2
SOURCES AND EFFECTS OF CONTAMINANTS

Sources of Contaminants

Contaminants in municipal wastewater are introduced as a result of water usage for domestic, commercial, or institutional purposes; water usage for product processing or cooling purposes within industries discharging liquid effluents into municipal sewerage systems; and infiltration/inflow and/or stormwater runoff. Table 7.2.1 shows examples of typical physical, chemical, and biological characteristics of municipal wastewater, along with potential sources. Table 7.2.2 summarizes selected sources and effects of industrial wastewater constituents.

POINT AND NONPOINT SOURCES

Two main sources of water pollutants are point and nonpoint. Nonpoint sources are also referred to as area or diffuse sources. Nonpoint pollutants are substances introduced into receiving waters as a result of urban area, industrial area, or rural runoff; e.g., sediment and pesticides or nitrates entering surface water due to surface runoff from agricultural farms. Point sources are specific discharges from municipalities or industrial complexes; e.g., organics or metals entering surface water due to wastewater discharge from a manufacturing plant. In a surface water body, nonpoint pollution can contribute significantly to total pollutant loading, particularly with regard to nutrients and pesticides. To illustrate the relative contributions, Figure 7.2.1 shows the estimated nationwide loadings of four key water pollutants.

Municipal and industrial wastewater discharges are primary contributors to point source discharges in the United States. Table 7.2.3 provides quantitative information on BOD, total suspended sediments, and phosphorus discharges from municipal treatment facilities and several industrial categories.

Over the past two decades, more than $75 billion in federal, state, and local funds were used to construct municipal sewage treatment facilities, and the private sector has spent additional billions to limit discharges of conventional pollutants (Council on Environmental Quality 1992). Between 1972 and 1988, the number of people served by municipal treatment plants with secondary treatment or better increased from 85 million to 144 million; Figure 7.2.2 depicts these changes in population served. These trends suggest that further reduc-

tions in pollutant loadings from point sources will occur.

VIOLATIONS OF WATER QUALITY STANDARDS

From a holistic perspective, a major source of wastewater discharges is associated with increases in the violations of receiving water quality standards. The 1990 National Water Quality Inventory in the United States assessed, in relation to applicable water quality standards and designated beneficial uses of the water, about one-third of the total river miles, half of the acreages of lakes, and three-quarters of the estuarine square miles. Table 7.2.4 summarizes these results relative to supporting designated uses. About one-third of the three assessed water resources did not fully meet their respective designated uses (Council on Environmental Quality 1993). Table 7.2.5 indicates the causes and sources of pollution for the three types of water resources.

Figure 7.2.3 shows pollution sources for impaired river miles; Figure 7.2.4 shows pollution in estuarine waters. While they are not the only sources of pollution, municipal and industrial wastewater discharges contribute significantly to impaired water resources.

Table 7.2.6 summarizes violation rates of water quality criteria in U.S. rivers and streams from 1975–1989. FC bacteria violations are well above the rates for other constituents for each year. In addition, the percent of violations has declined for all listed parameters since 1975.

Effects of Contaminants

The effects of pollution sources on receiving water quality are manifold and depend on the type and concentration of pollutants (Nemerow and Dasgupta, 1991). Soluble organics, as represented by high BOD waste, deplete oxygen in surface water. This results in fish kills, the growth of undesirable aquatic life, and undesirable odors. Trace quantities of certain organics cause undesirable tastes and odors, and certain organics can be biomagnified in the aquatic food chain.

SSs decrease water clarity and hinder photosynthetic processes; if solids settle and form sludge deposits, changes in benthic ecosystems result. Color, turbidity, oils, and floating materials influence water clarity and photosynthetic processes and are aesthetically undesirable.

TABLE 7.2.1 PHYSICAL, CHEMICAL, AND BIOLOGICAL CHARACTERISTICS OF WASTEWATER AND THEIR SOURCES

Characteristic	Sources
Physical Properties	
Color	Domestic and industrial wastes and natural decay of organic materials
Odor	Decomposing wastewater and industrial wastes
Solids	Domestic water supply, domestic and industrial wastes, soil erosion, and inflow/infiltration
Temperature	Domestic and industrial wastes
Chemical Constituents	
ORGANIC	
Carbohydrates	Domestic, commercial, and industrial wastes
Fats, oils, and grease	Domestic, commercial, and industrial wastes
Pesticides	Agricultural wastes
Phenols	Industrial wastes
Proteins	Domestic, commercial, and industrial wastes
Priority pollutants	Domestic, commercial, and industrial wastes
Surfactants	Domestic, commercial, and industrial wastes
VOCs	Domestic, commercial, and industrial wastes
Other	Natural decay of organic materials
INORGANIC	
Alkalinity	Domestic wastes, domestic water supply, and groundwater infiltration
Chlorides	Domestic wastes, domestic water supply, and groundwater infiltration
Heavy metals	Industrial wastes
Nitrogen	Domestic and agricultural wastes
pH	Domestic, commercial, and industrial wastes
Phosphorus	Domestic, commercial, and industrial wastes and natural runoff
Priority pollutants	Domestic, commercial, and industrial wastes
Sulfur	Domestic water supply and domestic, commercial, and industrial wastes
GASES	
Hydrogen sulfide	Decomposition of domestic wastes
Methane	Decomposition of domestic wastes
Oxygen	Domestic water supply and surface-water infiltration
Biological Constituents	
Animals	Open watercourses and treatment plants
Plants	Open watercourses and treatment plants
Protists:	
Eubacteria	Domestic wastes, surface-water infiltration, and treatment plants
Archaebacteria	Domestic wastes, surface-water infiltration, and treatment plants
Viruses	Domestic wastes

Source: Metcalf and Eddy, Inc., 1991, *Wastewater engineering*, 3d ed. (New York: McGraw-Hill, Inc.).

Excessive nitrogen and phosphorus lead to algal overgrowth with concomitant water treatment processes. Chlorides cause a salty taste in water; and in sufficient concentration, water usage must be limited. Acids, alkalies, and toxic substances can cause fish kills and create other imbalances in stream ecosystems.

Thermal discharges can also cause imbalances and reduce the stream waste assimilative capacity. Stratified flows from thermal discharges minimize normal mixing patterns in receiving streams and reservoirs. Table 7.2.7

provides an overview of certain contaminants and their impact in surface waters. Table 7.2.8 summarizes the impact of certain pollutants with regard to use impairment of the water.

ECOLOGICAL EFFECTS

The effects of pollutant discharges in municipal or industrial wastewaters can be considered from an ecological perspective. For example, Welch (1980) discusses the effects

TABLE 7.2.2 EXAMPLES OF SOURCES AND EFFECTS OF WASTEWATER CONSTITUENTS

Component Group	Effects	Typical Sources
Biooxidizables expressed as BOD_5	Deoxygenation, anaerobic conditions, fish kills, odors	Large amounts of soluble carbohydrates: sugar refining, canning, distilleries, breweries, milk processing, pulping, and paper making
Primary toxicants: As, CN, Cr, Cd, Cu, F, Hg, Pb, and Zn	Fish kills, cattle poisoning, plankton kills, and accumulations in flesh of fish and mollusks	Metal cleaning, plating, and pickling; phosphate and bauxite refining; chlorine generation; battery making; and tanning
Acids and alkalines	Disruption of pH buffer systems and disordering previous ecological system	Coal-mine drainage, steel pickling, textiles, chemical manufacture, wool scouring, and laundries
Disinfectants: Cl_2, H_2O_2, formalin, and phenol	Selective kills of microorganisms, taste, and odors	Bleaching of paper and textiles; rocketry; resin synthesis; penicillin preparation; gas, coke, and coal-tar making; and dye and chemical manufacture
Ionic forms: Fe, Ca, Mg, Mn, Cl, and SO_4	Changed water characteristics: staining, hardness, salinity, and encrustations	Metallurgy, cement making, ceramics, and oil-well pumpage
Oxidizing and reducing agents: NH_3, NO_2^-, NO_3^-, S^-, and SO_3^{-2}	Altered chemical balances ranging from rapid oxygen depletion to overnutrition, odors, and selective microbial growths	Gas and coke making, fertilizer plants, explosive manufacture, dyeing and synthetic fiber making, wood pulping, and bleaching
Evident to sight and smell	Foaming, floating, and settleable solids; odors; anaerobic bottom deposits; oils, fats, and grease; and waterfowl and fish injuries	Detergent wastes, tanning, food and meat processing, beet sugar mills, woolen mills, poultry dressing, and petroleum refining
Pathogenic organisms: *B. anthracis*, *Leptospira*, fungi, and viruses	Infections in humans, reinfection of livestock, plant diseases from fungi-contaminated irrigation water and risks to humans slight	Abattoir wastes, wool processing, fungi growths in waste treatment works, and poultry-processing waste waters

Source: R.A. Corbitt, 1990, Wastewater disposal, Chap. 6 in *Standard handbook of environmental engineering,* edited by R.A. Corbitt (New York: McGraw-Hill Publishing Company).

of wastewater discharges on the ecological characteristics of the receiving water environment; his topics include the specific effects on phytoplankton, zooplankton, periphyton, macrophytic rooted plants, benthic macroinvertebrates, and fish.

Material and energy flow diagrams demonstrate biogeochemical cycles and system interrelationships. For example, Figure 7.2.5 shows the material and energy flow in an aquatic ecosystem. Food web (or food chain) relationships and energy flow considerations indicate the dynamic aspects of the biological environment. They are also used to develop qualitative and quantitative models of aquatic or terrestrial systems, useful in predicting aquatic impacts of wastewater discharges. Wetland loss or degradation from municipal or industrial wastewater discharges illustrates an ecosystem effect. Such loss or degradation also occurs as a consequence of other human activities or natural occurrences (Mannion and Bowlby 1992).

To analyze the potential effects of wastewater discharges, the engineer may consider environmental cycling

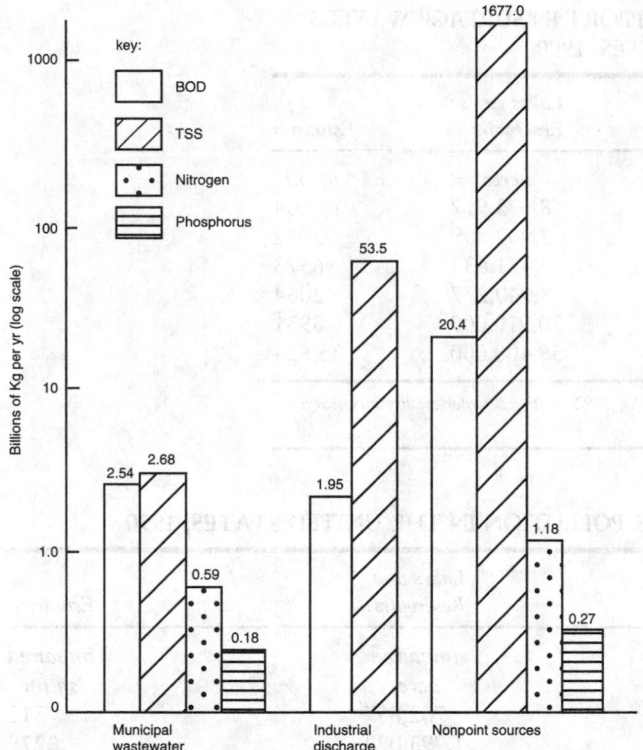

FIG. 7.2.1 Estimated nationwide loadings of selected water pollutants. (Reprinted, with permission, from Corbitt, 1990, *Wastewater disposal*, Chap. 6 in Standard handbook of environmental engineering, edited by R.A. Corbitt, New York: McGraw-Hill Publishing Company.)

	1960	1978	1988
Not served	70.0*	66.0	69.9
Raw discharge	na†	na	1.5
Secondary treatment	na	56.0	78.0
No discharge	na	na	6.1
Less than secondary treatment	36.0	na	26.5
Greater than secondary treatment	4.0	49.0	65.7

*Numbers represent millions of people
†na = not available.

FIG. 7.2.2 Population served by municipal wastewater treatment systems. (Reprinted from Council on Environmental Quality 1992, *Environmental trends,* Washington, D.C.).

of specific pollutants. Figure 7.2.6 shows various transfer routes and processes related to pollutant movement within the hydrosphere and biosphere. Examples of specific biogeochemical cycles for one nutrient (nitrogen) and one metal (mercury) are shown in Figures 7.2.7 and 7.2.8, respectively. The information in these examples indicates that the biological environment is a dynamic system that can be stressed as a result of various wastewater discharges.

Environmental engineers can use chromium to illustrate changes within aqueous systems since it is a transition metal that exhibits various oxidation states and behavior patterns (Canter and Gloyna 1968). Trivalent chromium is generally present as a cation, $Cr(OH)^+$, and is chemically reactive, tending to sorb on suspended materials and subsequently settle from the liquid phase. Hexavalent chromium is anionic (CrO_4^-) and chemically unreactive,

TABLE 7.2.3 QUANTITATIVE POLLUTANT LOADINGS INTO SURFACE WATERS

Pollutant	Point Sources[a] Total (10^6 tn/yr)	Source Contribution[b] (% of total)	Nonpoint Sources[a] Total (10^6 tn/yr)
BOD	1.87	MTF = 72.3 AFI = 21.6 CMI = 5.8 MMI = 0.3	14
TSS	2.13	MTF = 61.5 AFI = 13.0 CMI = 8.9 MMI = 16.7	3130
Phosphorus	2.66	MTF = 81.8 AFI = 18.0 CMI = 0.05 MMI = 0.05	515

Source: Council on Environmental Quality, 1989, *Environmental trends* (Washington, D.C.).
Notes: [a]Point source information is from mid-1980s; nonpoint source information is from 1980.
[b]MTF = municipal treatment facilities, AFI = agriculture and fisheries industry, CMI = chemical and manufacturing industry, and MMI = minerals and metals industry.

TABLE 7.2.4 DESIGNATED USE SUPPORT IN SURFACE WATERS
OF THE UNITED STATES, 1990

Designated Use Support	Rivers	Lakes & Reservoirs	Estuaries
	mi	acres	sq mi
Fully supporting	407,162	8,173,917	15,004
Threatened	43,214	2,902,809	3052
Partially supporting	134,472	3,471,633	6573
Not supporting	62,218	3,940,277	2064
Not assessed	1,153,000	20,910,000	8931
Total	1,800,000	39,400,000	35,624

Source: U.S. Environmental Protection Agency (EPA), 1992, National water quality inventory:
1990, Report to Congress (Washington, D.C.).

TABLE 7.2.5 CAUSES AND SOURCES OF SURFACE WATER POLLUTION IN THE UNITED STATES, 1990

Causes of Pollution	Rivers	Lakes and Reservoirs	Estuaries
	impaired mi	impaired acres	impaired sq mi
Siltation	67,059	702,857	312
Nutrients	51,747	1,793,022	3279
Organic enrichment	47,545	1,072,184	1876
Pathogens	35,151	129,286	1781
Metals	28,287	2,672,427	431
Salinity	21,914	243,482	nr
Habitat modification	20,258	nr	nr
Pesticides	20,701	nr	105
Priority organics	nr	247,317	917
SS	20,819	731,993	467
Flow alteration	15,565	677,413	nr
pH	9368	192,153	36

Sources of Pollution	Rivers	Lakes and Reservoirs	Estuaries
	impaired mi	impaired acres	impaired sq mi
Agriculture	103,439	1,996,772	1074
Municipal	27,994	593,518	2038
Habitat modification	24,884	1,407,827	307
Resource extraction	24,015	301,398	91
Storm sewers and runoff	18,129	973,077	1790
Industrial	15,568	318,446	615
Silviculture	15,459	106,502	89
Construction	9810	106,398	640
Land disposal	7188	846,892	1137
Combined sewers	2836	3015	359
Unknown	9266	403,080	189

Source: U.S. EPA, 1992.

Notes: nr = not reported. In addition to the causes and sources listed, thermal modifications impair 9970 river mi. Taste and odor impairments affect 105,288 acres of lakes and reservoirs, and noxious aquatic plants impair 711,323 acres. Additional causes of pollution in estuaries are ammonia (50 sq mi), oil and grease (36 sq mi), and unknown (109 sq mi). Estimates of impairment are based on the sums of partially and not supporting designated uses in Table 7.2.4 which represent 9.5% of total U.S. river mi, 8.9% of total U.S. lake acres, and 16.7% of total U.S. estuary square mi.

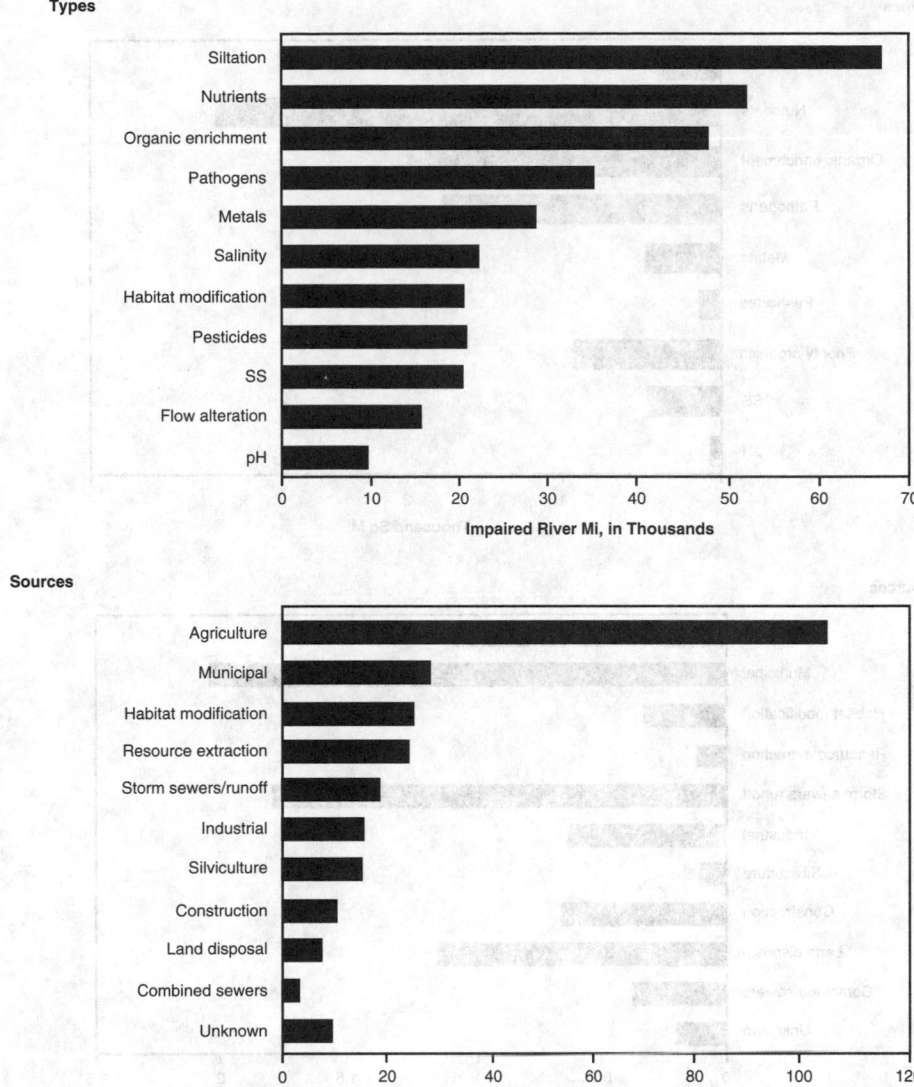

FIG. 7.2.3 Types and sources of pollution in rivers of the United States. Figure based on river mi monitored in 1990, which represent 9.5% of total U.S. river mi. (Reprinted from U.S. Environmental Protection Agency (EPA), 1992, *National water quality inventory: 1990*, Report to Congress, Washington, D.C.)

thus tending to remain in solution. Changes can occur in the chromium oxidation state due to stream water quality. For example, hexavalent chromium can be chemically reduced to trivalent chromium under anaerobic conditions, whereas trivalent chromium can be oxidized to hexavalent chromium under aerobic conditions. This information qualitatively predicts the impact of chromium discharges into river systems.

An additional concern is the potential reconcentration of pollutant materials (e.g., heavy metals and pesticides) into aquatic organisms and their subsequent harvesting and consumption by man. One example is the study from 1974–1990 of pesticide contaminant levels in herring gull eggs from the five Great Lakes (Council on Environmental Quality 1992). Another example is the closure of shellfish

beds due to bacterial contamination of the water and shellfish (Council on Environmental Quality 1993).

TOXICITY EFFECTS

Depending on their constituents, some municipal or industrial wastewater discharges exhibit toxicity effects on aquatic organisms. As a result, biomonitoring can be required for effluent discharges. In biomonitoring, indicator organisms are chosen to represent all segments of the aquatic community of the water body under study (U.S. Department of Energy 1985). Five groups of organisms have traditionally been studied as indicators of water quality (U.S. Department of Energy 1985): bacteria, fish, plankton, periphyton, and macroinvertebrates. The choice of an

Types

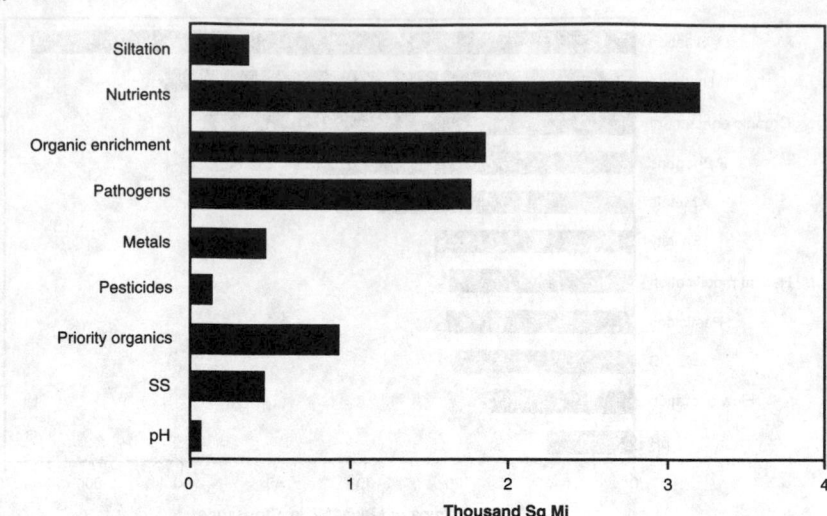

Sources

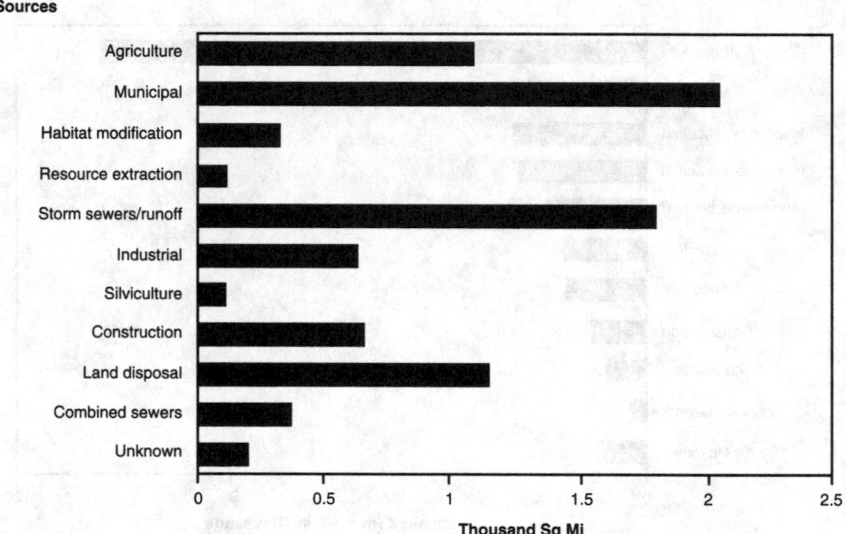

FIG. 7.2.4 Types and sources of pollution in estuaries of the United States. Figure based on sq mi of estuaries monitored in 1990, which represent 16.7% of total U.S. estuaries. An estuary is a tidal body of water, formed where a river meets the sea. Estuaries, such as bays, have a measurable quantity of salt. (Reprinted from U.S. EPA 1992.)

indicator species is important in a biomonitoring study because it can affect the results of the study phase and toxicity characterization procedures.

Bacteria are indicators of fecal contamination and the presence of pathogenic organisms (U.S. Department of Energy 1985). Fish are useful in assessing water quality because in aquatic communities, they often represent the highest trophic level. If fish are lacking or exist only in small numbers, decreased water quality may be affecting the fish directly or indirectly by affecting those organisms on which the fish feed.

Plankton are good indicators of the water quality because they float passively with the currents (U.S. Department of Energy 1985). Estimating the plankton

population indicates, in part, the nutrient-supplying capability of that water body (U.S. Department of Energy 1985). Periphyton, because of their stationary or sessile existence, also indicate nutrient availability in an aquatic system; furthermore, they represent the integration of the physical and chemical conditions of water passing through their locations (U.S. Department of Energy 1985).

Benthic species are excellent indicators of water quality because they are generally restricted to the area where they are found (U.S. Department of Energy 1985). Benthic macroinvertebrates, because of their varying environmental requirements, form communities characteristic or associated with physical and chemical conditions. For example, the presence of immature insects, certain mollusks,

TABLE 7.2.6 NATIONAL AMBIENT WATER QUALITY IN RIVERS AND STREAMS: VIOLATION RATES, 1975–1989

Year	FC Bacteria	Dissolved Oxygen	Total Phosphorus	Total Cadmium, Dissolved	Total Lead, Dissolved
		percent of all measurements exceeding water quality criteria			
1975	36	5	5	*	*
1976	32	6	5	*	*
1977	34	11	5	*	*
1978	35	5	5	*	*
1979	34	4	3	4	13
1980	31	5	4	1	5
1981	30	4	4	1	3
1982	33	5	3	1	2
1983	34	4	3	1	5
1984	30	3	3	<1	<1
1985	28	3	3	<1	<1
1986	24	3	3	<1	<1
1987	23	2	2	<1	<1
1988	22	2	2	<1	<1
1989	20	3	3	<1	<1

Source: Council on Environmental Quality, 1993, *Environmental quality, 23rd annual report,* (Washington, D.C. [January]).

Note: Violation levels are based on U.S. EPA water quality criteria: FC bacteria—above 200 cells per 100 mi; dissolved oxygen—below 5 mg/l; total phosphorus—above 1.0 mg/l; cadmium, dissolved—above 10 μg/l; and lead, dissolved—above 50 μg/l. * = base figure too small to meet statistical standards for reliability of derived figures.

TABLE 7.2.7 IMPORTANT WASTEWATER CONTAMINANTS BASED ON POTENTIAL EFFECTS AND CONCERNS IN TREATMENT

Contaminants	Reason for Importance
SS	SS can lead to the development of sludge deposits and anaerobic conditions when untreated wastewater is discharged to the aquatic environment.
Biodegradable organics	Composed principally of proteins, carbohydrates, and fats, biodegradable organics are commonly measured in terms of BOD and COD. If discharged to the environment untreated, their biological stabilization can deplete natural oxygen resources and cause septic conditions.
Pathogens	Communicable diseases can be transmitted by pathogenic organisms in wastewater.
Nutrients	Both nitrogen and phosphorus, along with carbon, are essential nutrients for growth. When discharged to the aquatic environment, these nutrients can lead to the growth of undesirable aquatic life. When discharged in excessive amounts on land, they can also lead to groundwater pollution.
Priority pollutants	These pollutants include organic and inorganic compounds selected on the basis of their known or suspected carcinogenicity, mutagenicity, teratogenicity, or high acute toxicity. Many of these compounds are found in wastewater.
Refractory organics	These organics tend to resist conventional wastewater treatment. Typical examples include surfactants, phenols, and agricultural pesticides.
Heavy metals	Heavy metals are usually added to wastewater from commercial and industrial activities and may have to be removed if the wastewater is to be reused.
Dissolved inorganics	Inorganic constituents such as calcium, sodium, and sulfate are added to the original domestic water supply as a result of water use and may have to be removed if the wastewater is to be reused.

Source: Metcalf and Eddy, Inc., 1991, *Wastewater engineering,* 3d ed. (New York: McGraw-Hill, Inc.).

and crayfish usually indicates relatively clean water; while communities of sludge worms, air breathing snails, midges, and aquatic earthworms indicate the presence of oxygen-consuming materials and deteriorating conditions.

Toxicity tests, the second category of biomonitoring, are less expensive and labor intensive than ecological surveys. Toxicity tests use indicator organisms in a controlled situation to examine water and effluent toxicity. The toxic effects of concern are death, immobilization, serious incapacitation, reduced fecundity, or reduced growth. Toxicity tests can be used to assess acute or chronic toxicity. Acute toxicity is a severe toxic effect resulting from a brief exposure, while chronic toxicity results from prolonged exposure. Tests for acute toxicity can be conducted within

TABLE 7.2.8   WATER USE LIMITS DUE TO WATER QUALITY DEGRADATION

	Use						
Pollutant	Drinking Water	Aquatic Wildlife, Fisheries	Recreation	Irrigation	Industrial Uses	Power and Cooling	Transport
Pathogens	xx	0	xx	x	xx[1]	na	na
SS	xx	xx	xx	x	x	x[2]	xx[3]
Organic matter	xx	x	xx	+	xx[4]	x[5]	na
Algae	x[5,6]	x[7]	xx	+	xx[4]	x[5]	x[8]
Nitrate	xx	x	na	+	xx[1]	na	na
Salts[9]	xx	xx	na	xx	xx[10]	na	na
Trace elements	xx	xx	x	x	x	na	na
Organic micropollutants	xx	xx	x	x	?	na	na
Acidification	x	xx	x	?	x	x	na

Source: D. Chapman, ed., 1992, *Water quality assessments—A guide to the use of biota, sediments, and water in environmental monitoring,* 9 (London: Chapman and Hall, Ltd.).

Notes:

xx Marked impairment causing major treatment or excluding desired use
x Minor impairment
0 No impairment
na Not applicable
+ Degraded water quality can be beneficial for this specific use.
? Effects not fully realized
[1] Food industries
[2] Abrasion

[3] Sediment settling in channels
[4] Electronic industries
[5] Filter clogging
[6] Odor and taste
[7] In fish ponds, higher algal biomass can be accepted.
[8] Development of water hyacinth (*Eichhomia crassiodes*)
[9] Also includes boron and fluoride
[10] Ca, Fe, and Mn in textile industries

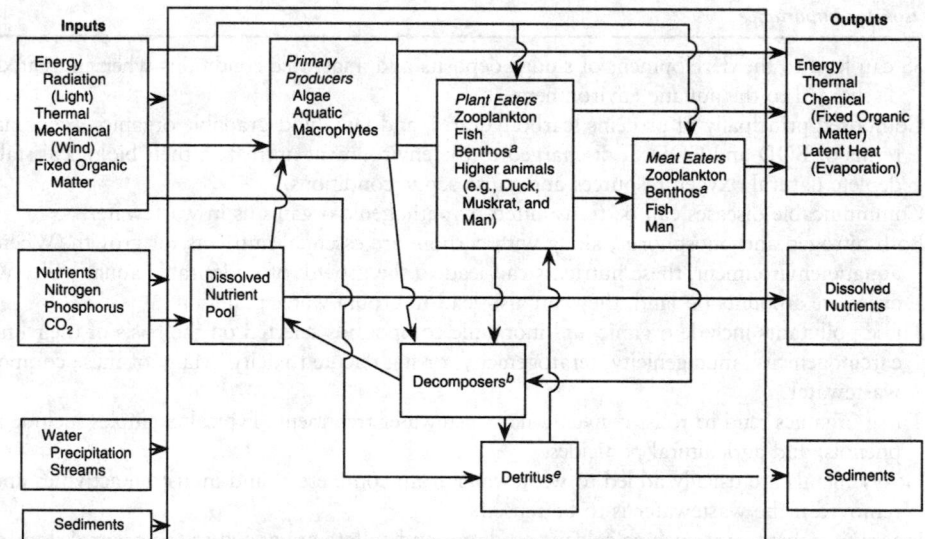

FIG. 7.2.5   Material and energy flow in an aquatic ecosystem: [a]Organisms living at or on the bottom of bodies of water, [b]Fungi and bacteria, [c]Small particles of organic matter. (Reprinted, with permission, from D.C. Watts and D.L. Loucks, 1969, *Models for describing exchange within ecosystems,* Madison, Wis.: Institute for Environmental Studies, University of Wisconsin.)

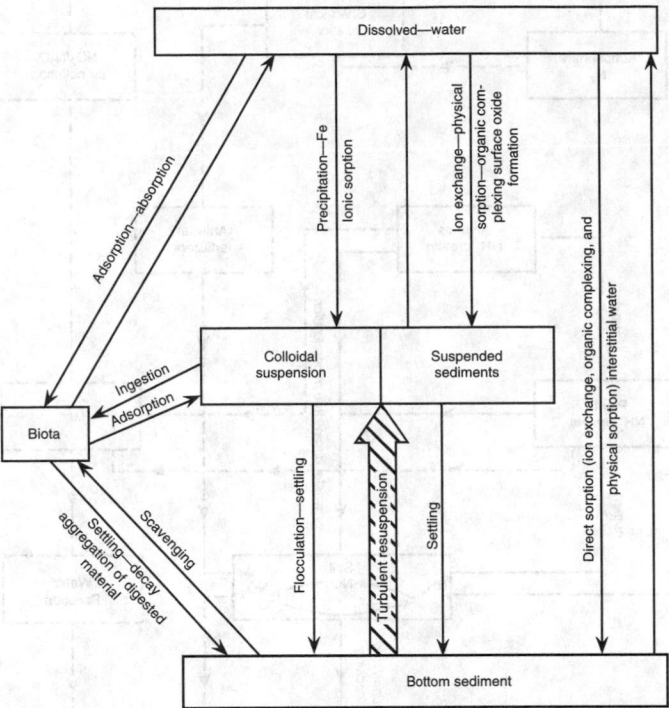

FIG. 7.2.6 Pollutant transfer routes and processes within the hydrosphere and biosphere. (Reprinted, with permission, from R.A. Corbitt, 1990, Wastewater disposal, Chap. 6 in *Standard handbook of environmental engineering*, edited by R.A. Corbitt, New York: McGraw-Hill Publishing Company.)

24–96 hr, and chronic tests for toxicity are conducted usually in four or more days (Clesceri, Greenberg, and Trussell 1989).

Toxicity testing can be further divided into ambient and laboratory tests. Ambient toxicity tests are conducted in situ. In such tests, indicator organisms are kept in cages within the water under study, or they are exposed to the test water in chambers at the site. The water in the latter test is renewed daily with fresh water from the study sites.

Laboratory tests are dissimilar because they are conducted offsite in a laboratory. These tests are conducted within a set of conditions such as those described in *Short-term methods for estimating the chronic toxicity of effluents and receiving waters to freshwater organisms* (U.S. EPA 1989). In such tests, the test water is fractionally diluted with synthetic water. Replicate groups of indicator species are exposed to different concentrations of a dilution series, and toxic effects are recorded for each dilution. The recorded data are then analyzed using a series of statistical tests to determine the effective or lethal concentrations. Effective and lethal concentrations are those point estimates of the toxicant concentration that cause an observable adverse effect (such as death, immobilization, serious incapacitation, reduced fecundity, or reduced growth) in a percentage of the test organisms (U.S. EPA 1989).

EFFECTS OF CONTAMINANTS ON WASTEWATER TREATMENT PLANTS

As a final example of wastewater discharge effects, industrial wastewater can affect wastewater treatment plants if such wastewater is introduced into municipal sewerage systems. The pollution characteristics of industrial wastewater with definable effects on sewers and treatment plants include

1. BOD
2. SS
3. floating and colored material
4. volume
5. other harmful constituents

Excessive BOD and volume can cause organic and hydraulic overloads, respectively. SS can create operational problems due to excess solids and sludge production. Floating and colored material are related to visible pollution. Examples of other harmful industrial wastewater constituents include (Nemerow and Dasgupta 1991):

Toxic metal ions (Cu^{++}, Cr^{+6}, Zn^{++}, and Cn^{-}), which interfere with biological oxidation by tying up the enzymes required to oxidize organic matter

Acids and alkalis, which can corrode pipes, pumps, and treatment units, interfere with settling, upset the bio-

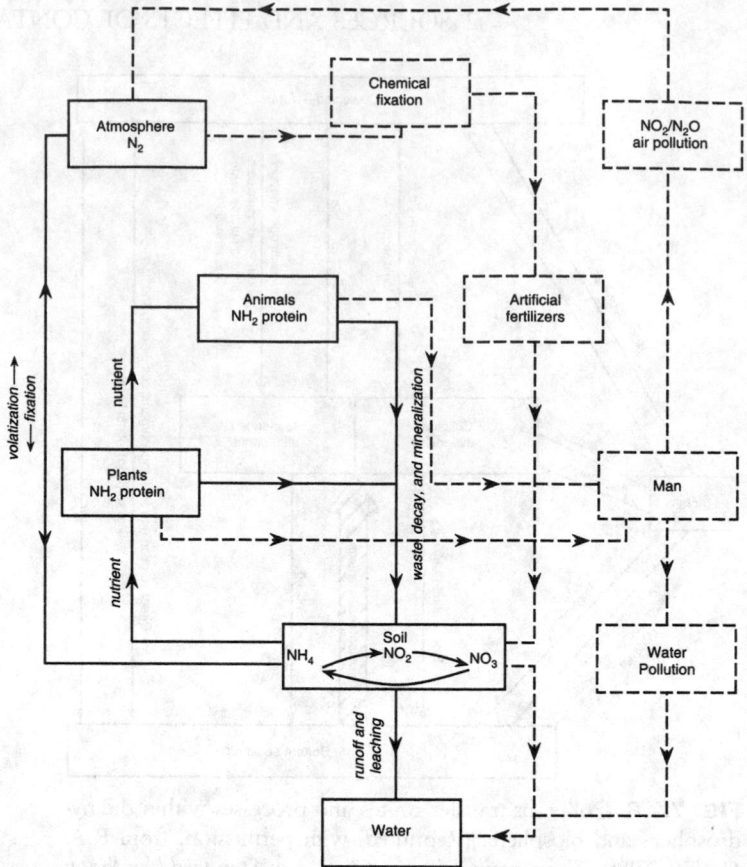

FIG. 7.2.7 The biogeochemical cycle for nitrogen. Dotted lines denote human intervention. (Reprinted, with permission, from D. Drew, 1983, *Man–environment processes,* London: George Allen and Unwin, Ltd.)

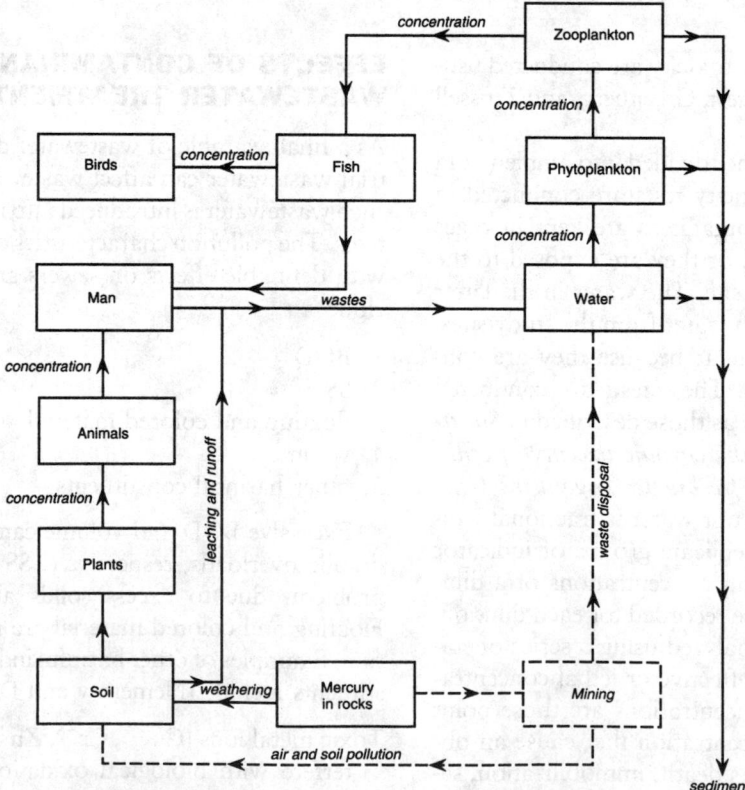

FIG. 7.2.8 The biogeochemical cycle for mercury. Special reference is given to the biological concentration in the aquatic environment. Broken lines show areas of human intervention. (Reprinted, with permission, from Drew, 1983.)

logical purification of sewage, release odors, and intensify color

Detergents, which cause foaming of aeration units

Phenols and other toxic organic material

—Larry W. Canter

References

Canter, L.W. and E.F. Gloyna. 1968. Transport of chromium-51 in an organically polluted environment. *Proc. 23rd Purdue Industrial Waste Conference.* 374–387. Lafayette, Ind.: Purdue University.

Clesceri, L.S., A.E. Greenberg, and R.R. Trussell, eds. 1989. *Standard methods for the examination of water and wastewater.* 17th ed. 10-1 to 10-201. Washington, D.C.: American Public Health Association, American Water Works Association, and Water Pollution Control Federation.

Council on Environmental Quality. 1992. *Environmental quality, 22nd annual report.* 187 and 263–266. Washington, D.C. (March).

———. 1993. *Environmental quality, 23rd annual report.* 225–227 and 320–324. Washington, D.C. (January).

Mannion, A.M. and S.R. Bowlby. 1992. *Environmental issues in the 1990s.* 219–221. West Sussex, England: John Wiley and Sons, Ltd.

Nemerow, N.L. and A. Dasgupta. 1991. *Industrial and hazardous waste treatment.* 7–10. New York: Van Nostrand Reinhold.

U.S. Department of Energy. 1985. *Analysis of biomonitoring techniques to supplement effluent guidelines: Final report.* DOE/PE/16036—TI. 1–35. Washington, D.C. (November).

U.S. Environmental Protection Agency (EPA). 1989. *Short-term methods for estimating the chronic toxicity of effluents and receiving waters to freshwater organisms.* 2d ed. EPA 600–4–89–001. Cincinnati, Ohio. (March).

Welch, E.B. 1980. *Ecological effects of waste water.* 1–2. Cambridge: Cambridge University Press.

7.3
CHARACTERIZATION OF INDUSTRIAL WASTEWATER

Origins of Industrial Wastewater

Industrial wastewater is discharged from industries and associated processes utilizing water. Industrial water users in the United States discharge over 285 billion gal of wastewater daily (Corbitt 1990). Water is used in industrial cooling, product washing and transport, product generation, and other purposes. Although variable between industries and plants within the same industry, about two-thirds of the total wastewater generated from U.S. industries results from cooling operations (Corbitt 1990).

WASTEWATER FROM PROCESS OPERATIONS

Water used for process operations (noncooling purposes) can become degraded as a result of introducting nutrients, suspended sediments, bacteria, oxygen-demanding matter, and toxic chemicals. The degree of pollutant (contaminant) loading is a function of the type of industry, specific unit processes, and the extent of wastewater minimization practices employed. Table 7.3.1 summarizes the origin of contaminant introductions in various industries and lists some key contaminant characteristics.

Several books summarize the characteristics of wastewater from various industries; examples include Nemerow (1978) and Nemerow and Dasgupta (1991). Nemerow and Dasgupta (1991) provide information on the types and quantities of water pollutants from major private sector categories such as the apparel, food, materials, chemical, and energy industries.

WASTEWATER FROM PETROLEUM REFINING

As an example of industrial wastewater discharges, this section presents brief information on the pollutant discharges from petroleum refineries. The total amount of water used in a petroleum refinery is estimated to be 770 gal per barrel of crude oil (Nemerow 1978). Approximately 80 to 90% of the water is used for cooling purposes only and is not contaminated except by leaks in the lines. Process wastewaters, comprising 10 to 20% of the total, can include free and emulsified oil from leaks, spills, tank draw-off, and other sources; waste caustic, caustic sludges, and alkaline waters; acid sludges and acid waters; emulsions incident to chemical treatment; condensate waters from distillate separators and tank draw-off; tank-bottom sludges; coke from equipment tubes, towers, and other locations; acid gases; waste catalyst and filtering clays; and special chemicals from by-product chemical manufacturing.

Both conventional and toxic pollutants are found in petroleum refinery wastewater. Conventional pollutants are those that have received historical attention, while toxic pollutants relate to parameters receiving increasing atten-

Industries Producing Wastes	Origin of Major Wastes	Major Characteristics
Apparel		
Textiles	Cooking of fibers and desizing of fabric	Highly alkaline, colored, high BOD and temperature, and high SS
Leather goods	Unhairing, soaking, deliming, and bating of hides	High total solids, hardness, salt, sulfides, chromium, pH, precipitated lime, and BOD
Laundry trades	Washing of fabrics	High turbidity, alkalinity, and organic solids
Dry cleaning	Solvent cleaning of clothes	Condensed, toxic, and organic vapors
Food and Drugs		
Canned goods	Trimming, culling, juicing, and blanching of fruits and vegetables	High in SS, colloidal, and dissolved organic matter
Dairy products	Dilutions of whole milk, separated milk, buttermilk, and whey	High in dissolved organic matter, mainly protein, fat, and lactose
Brewed and distilled beverages	Steeping and pressing of grain; residue from distillation of alcohol; and condensate from stillage evaporation	High in dissolved organic solids containing nitrogen and fermented starches or their products
Meat and poultry products	Stockyards; slaughtering of animals; rendering of bones and fats; residues in condensates; grease and wash water; and picking of chickens	High in dissolved and suspended organic matter, blood, other proteins, and fats
Beet sugar	Transfer, screening, and juicing waters; drainings from lime sludge; condensates after evaporator; and juice and extracted sugar	High in dissolved and suspended organic matter containing sugar and protein
Pharmaceutical products	Mycelium, spent filtrate, and wash waters	High in suspended and dissolved organic matter, including vitamins
Yeast	Residue from yeast filtration	High in solids (mainly organic) and BOD
Pickles	Lime water; brine, alum and turmeric, syrup, seeds, and pieces of cucumber	Variable pH, high SS, color, and organic matter
Coffee	Pulping and fermenting of coffee beans	High BOD and SS
Fish	Rejects from centrifuge; pressed fish, and evaporator and other wash water wastes	Very high BOD, total organic solids, and odor
Rice	Soaking, cooking, and washing of rice	High BOD, total and suspended solids (mainly starch)
Soft drinks	Bottle washing; floor and equipment cleaning; and syrup storage tank drains	High pH, suspended solids, and BOD
Bakeries	Washing and greasing of pans and floor washing	High BOD, grease, floor washing, sugars, flour, and detergents
Materials		
Pulp and paper	Cooking, refining, washing of fibers, and screening of paper pulp	High or low pH, color, high suspended, colloidal, and dissolved solids, and inorganic fillers
Photographic products	Spent solutions of developer and fixer	Alkaline containing various organic and inorganic reducing agents
Steel	Coking of coal, washing of blast-furnace flue gases, and pickling of steel	Low pH, acids, cyanogen, phenol, ore, coke, limestone, alkali, oils, mill scale, and fine SS
Metal-plated products	Stripping of oxides and cleaning and plating of metals	Acid, metals, toxic, low volume, and mainly mineral matter
Iron-foundry products	Wasting of used sand by hydraulic discharge	High SS, mainly sand; and some clay and coal
Oil fields and refineries	Drilling muds, salt, oil, and some natural gas; acid sludges; and miscellaneous oils from refining	High dissolved salts from field and high BOD, odor, phenol, and sulfur compounds from refinery

Continued on next page

TABLE 7.3.1 *Continued*

Industries Producing Wastes	*Origin of Major Wastes*	*Major Characteristics*
Petrochemicals	Contaminated water from chemical production and transportation of second generation oil compounds	High COD, TDS, metals, COD/BOD ratio, and cpds. inhibitory to biologic action
Cement	Fine and finish grinding of cement, dust leaching collection, and dust control	Heated cooling water, SS, and some inorganic salts
Chemicals		
Acids	Dilute wash waters and many varied dilute acids	Low pH and low organic content
Detergents	Washing and purifying soaps and detergents	High in BOD and saponified soaps
Cornstarch	Evaporator condensate or bottoms when not reused or recovered, syrup from final washes, and wastes from bottling-up process	High BOD and dissolved organic matter; mainly starch and related material
Explosives	Washing trinitrotoluene (TNT) and guncotton for purification and washing and pickling of cartridges	TNT, colored, acid, odorous, and containing organic acids and alcohol from powder and cotton, metals, acid, oils, and soaps
Pesticides	Washing and purification products such as 2,4-D and dichlorodiphenyl trichloroethane (DDT)	High organic matter, benzenering structure, toxic to bacteria and fish, and acid
Phosphate and phosphorus	Washing, screening, floating rock, and condenser bleedoff from phosphate reduction plant	Clays, slimes and tall oils, low pH, high SS, phosphorus, silica, and fluoride
Plastic and resins	Unit operations from polymer preparation and use and spills and equipment washdowns	Acids, caustic, and dissolved organic matter such as phenols and formaldehyde
Fertilizer	Chemical reactions of basic elements and spills, cooling waters, washing of products, and boiler blowdowns	Sulfuric, phosphorous, and nitric acids; mineral elements, P, S, N, K, Al, NH_3, and NO_3; Fl; and some SS
Toxic chemicals	Leaks, accidental spills, and refining of chemicals	Various toxic dissolved elements and compounds such as Hg and polychlorinated biphenyls (PCBs)
Mortuary	Body fluids, washwaters, and spills	Blood salt, formaldehydes, high BOD, and infectious diseases
Hospital and Research Laboratories	Washing, sterilizing of facilities, used solutions, and spills	Bacteria and various chemicals and radioactive materials
Chloralkali wastes	Electrolytic cells and making chlorine and caustic soda	Mercury and dissolved metals
Organic chemicals	Various chemical productive processes	Varied types of organic chemicals
Energy		
Steam power	Cooling water, boiler blowdown, and coal drainage	Hot, high volume, and high inorganic and dissolved solids
Scrubber power plant wastes	Scrubbing of gaseous combustion products by liquid water	Particulates, SO_2, impure absorbents or NH_3 and NaOH
Coal processing	Cleaning and classification of coal and leaching of sulfur strata with water	High SS (mainly coal), low pH, high H_2SO, and $FeSO_4$

Source: N.L. Nemerow and A. Dasgupta, 1991, *Industrial and hazardous waste treatment* (New York: Van Nostrand Reinhold).

TABLE 7.3.2 SUBCATEGORIES OF THE PETROLEUM REFINING INDUSTRY REFLECTING SIGNIFICANT DIFFERENCES IN WASTEWATER CHARACTERISTICS

Topping: Topping and catalytic reforming whether the facility includes any other process in addition to topping and catalytic process. This subcategory is not applicable to facilities that include thermal processes (e.g., coking and visbreaking) or catalytic cracking.

Cracking: Topping and cracking, whether the facility includes any processes in addition to topping and cracking, unless specified in one of the following subcategories listed.

Petrochemical: Topping, cracking, and petrochemical operations, whether the facility includes any process in addition to topping, cracking, and petrochemical operations[a], except lube oil manufacturing operations.

Lube: Topping, cracking, and lube oil manufacturing processes, whether the facility includes any process in addition to topping, cracking, and lube oil manufacturing processes, except petrochemical operations.[a]

Integrated: Topping, cracking, lube oil manufacturing processes, and petrochemical operations, whether the facility includes any processes in addition to topping, cracking, lube oil manufacturing processes, and petrochemical operations.[a]

Source: U.S. Environmental Protection Agency (EPA), 1980, *Treatability manual,* Vol. II—Industrial descriptions, Sec. II.14—Petroleum refining, EPA-600/8-80-042b (Washington, D.C. [July]).

Notes: [a]The term petrochemical operations means the production of second generation petrochemicals (i.e., alcohols, ketones, cumene, and styrene) or first generation petrochemical and isomerization products (i.e., benzene/toluene/xylene [BTX], olefins, and cyclohexane) when 15% or more of refinery production is as first generation petrochemicals and isomerization products.

tion due to their potential environmental toxicity (U.S. EPA 1980). Five refinery subcategories, based on throughputs and process capacities, delineate information on wastewater characteristics as defined in Table 7.3.2. Table 7.3.3 presents ranges and median loadings in raw wastewater of conventional pollutants for the petroleum refining industry subcategories. Raw wastewater is the effluent from the oil separator, which is an integral part of refinery process operations for product and raw material recovery prior to wastewater treatment. Table 7.3.4 lists the toxic pollutants that have been measured in wastewater generated at petroleum refineries; no concentration data are included.

Wastewater Discharge Standards

The flow rates and quality characteristics of wastewater within and between types of industries vary widely. Therefore, wastewater discharge standards are related to industry type. For example, Table 7.3.5 summarizes the effluent limitations based on best practicable treatment (BPT) for point sources associated with the cracking subcategory of petroleum refining.

Effluent limitations can be technology-based or water quality-based, with the former considering BPT, best available technology economically achievable (BAT), best conventional pollutant control technology (BCT), or new source performance standards (NSPSs). BPT emphasizes end-of-pipe controls and reflects the average of the best for the industry category; it deals primarily with conventional pollutants such as BOD, oil and grease, solids, pH, and some metals. BAT can include pollution prevention through process control and end-of-pipe technology; it deals primarily with toxics such as organics and heavy metals. BCT is used with BAT. NSPSs are based on the best available demonstrated control technology (BADCT); it is typically similar to BAT with BCT.

Water quality-based effluent limitations can require greater levels of treatment as dictated by a waste load allocation scheme based on the total maximum daily load for the stream segment. Adjustment factors for the effluent limitations, which account for facility size and specific processes, are in the regulations (40 *CFR* Chap. 1, part 419). Effluent limitation information is also available for BAT, BCT, and NSPS for the cracking subcategory and for BPT, BAT, BCT, and NSPS for the topping subcategory, petrochemical subcategory, lube subcategory, and integrated subcategory (40 *CFR* Chap. 1, part 419)

Effluent discharge standards have been developed for wastewater from many industry types. This section does not address multiple types of industries; however, as an illustration of variability, Table 7.3.6 lists the water quality parameters for fourteen types of industries (Corbitt 1990). As the table shows the parameters are industry-specific.

Wastewater Characterization Surveys

Environmental engineers use wastewater characterization surveys, also referred to as industrial waste surveys, to establish flows, quality characteristics, and pollutant loadings at an individual industrial plant. They use the results of such surveys in (1) determining the treatment level necessary to meet effluent discharge standards, (2) selecting treatment processes, (3) making the discharge permit application for the facility, (4) establishing pretreatment requirements for the facility prior to discharge into municipal sewerage systems, and (5) developing a wastewater flow and loading minimization program. They also use survey information in establishing industrial user charges for discharges into municipal systems (Water Pollution Control Federation and American Society of Civil Engineers 1977).

TABLE 7.3.3 RAW WASTEWATER[a] LOADINGS IN NET KILOGRAMS/1000 M³ OF FEEDSTOCK THROUGHPUT BY SUBCATEGORY IN PETROLEUM REFINING

	Topping Subcategory		Cracking Subcategory		Petrochemical Subcategory		Lube Subcategory		Integrated Subcategory	
Characteristics	Range[b]	Median	Range[b]	Median	Range[b]	Median	Range[b]	Median	Range[b]	Median
Flow[c]	8.00–558	66.6	3.29–2,750	93.0	26.6–443	109	68.6–772	117	40.0–1,370	235
BOD$_5$	1.29–217	3.43	14.3–466	72.9	40.9–715	172	62.9–758	217	63.5–615	197
COD	3.43–486	37.2	27.7–2,520	217	200–1,090	463	166–2,290	543	72.9–1,490	329
TOC	1.09–65.8	8.01	5.43–320	41.5	48.6–458	149	31.5–306	109	28.6–678	139
TSS	0.74–286	11.7	0.94–360	18.2	6.29–372	48.6	17.2–312	71.5	15.2–226	59.1
Sulfides	0.002–1.52	0.054	0.01–39.5[d]	0.94[d]	0.009–91.5	0.86	0.00001–20.0	0.014	0.52–7.87[d]	2.00[d]
Oil and grease	1.03–88.7	8.29	2.86–365	31.2	12.0–235	52.9	23.7–601	120	20.9–269	74.9
Phenols	0.001–1.06	0.034	0.19–80.1	4.00	2.55–23.7	7.72	4.58–52.9	8.29	0.61–22.6	3.78
Ammonia	0.077–19.5	1.20	2.35–174	28.3	5.43–206	34.3	6.5–96.2	24.1	0.12–1.92	0.49
Chromium	0.0002–0.29	0.007	0.0008–4.15	0.25	0.014–3.86	0.234	0.002–1.23	0.046		

Source: U.S. EPA, 1980.

Notes: [a]After refinery oil separator.

[b]Probability of occurrence less than or equal to 10 or 90% respectively.

[c]1000 m³/1000 m³ of feedstock throughput.

[d]Sulfur.

543

TABLE 7.3.4 QUALITATIVE LISTING OF TOXIC WATER POLLUTANTS POTENTIALLY IN PETROLEUM REFINERY WASTEWATERS

Metals and inorganics	Polycrylic aromatic
Antimony	hydrocarbons
Arsenic	Acenaphthene
Asbestos	Acenaphthylene
Beryllium	Anthracene
Cadmium	Benzo(a)pyrene
Chromium	Chrysene
Copper	Fluoranthene
Cyanide	Flourene
Lead	Naphthalene
Mercury	Phenanthrene
Nickel	Pyrene
Selenium	Polychlorinate biphenyls and
Silver	related compounds
Thallium	Aroclor 1016
Zinc	Aroclor 1221
Phthalates	Aroclor 1232
Bis(2-ethylhexyl) phthalate	Aroclor 1242
Di-n-butyl phthalate	Aroclor 1248
Diethyl phthalate	Aroclor 1254
Dimethyl phthalate	Aroclor 1260
Phenols	Halogenated aliphatics
2-Chlorophenol	Carbon tetrachloride
2,4-Dichlorophenol	Chloroform
2,4-Dinitrophenol	Dichlorobromomethane
2,4-Dimethylphenol	1,2-Dichloroethane
2-Nitrophenol	1,2-Trans-dichloroethylene
4-Nitrophenol	Methylene chloride
Pentachlorophenol	1,1,2,2-Tetrachloroethane
Phenol	Tetrachloroethylene
4,6-Dinitro-o-cresol	1,1,1-Trichloroethane
Parachlorometa cresol	Trichloroethylene
Aromatics	Pesticides and metabolites
Benzene	Aldrin
1,2-Dichlorobenzene	α-BHC
1,4-Dichlorobenzene	β-BHC
Ethylbenzene	δ-BHC
Toluene	γ-BHC
	Chlordane
	4,4'-DDE
	4,4'-DDD
	α-Endosulfan
	β-Endosulfan
	Endosulfan sulfate
	Heptachlor
	Isophorone

Source: U.S. EPA, 1980.

TABLE 7.3.5 BPT EFFLUENT LIMITATION GUIDELINES FOR CRACKING SUBCATEGORY POINT SOURCES FROM PETROLEUM REFINING

Pollutant or pollutant property	BPT Effluent Limitations	
	Maximum for any 1 day	Average of daily values for 30 consecutive days shall not exceed
	Metric Units (kg per 1000 m^3 of feedstock)[3]	
BOD5	28.2	15.6
TSS	19.5	12.6
COD[1]	210	109
Oil and grease	8.4	4.5
Phenolic compounds	0.21	0.10
Ammonia as N	18.8	8.5
Sulfide	0.18	0.082
Total chromium	0.43	0.25
Hexavalent chromium	0.035	0.016
pH	(2)	(2)
	English Units (lb per 1000 bbl feedstock)	
BOD5	9.9	5.5
TSS	6.9	4.4
COD[1]	74.0	38.4
Oil and grease	3.0	1.6
Phenolic compounds	0.074	0.036
Ammonia as N	6.6	3.0
Sulfide	0.065	0.029
Total chromium	0.15	0.088
Hexavalent chromium	0.012	0.0056
pH	(2)	(2)

Source: Code of Federal Regulations, Title 40, Chap. 1, part 419—Petroleum refining point source category, 419–457 (1 July 1991).

Notes: [1]In any case where the applicant can demonstrate that chloride ion concentration in the effluent exceeds 1000 mg/l (1000 ppm), the Regional Administrator can substitute TOC as a parameter in lieu of COD. Effluent limitations for TOC shall be based on effluent data from the plant correlating TOC to BOD5. If the Regional Administrator judges that adequate correlation data are not available, the effluent limitations for TOC shall be established at a ratio of 2.2 to 1 to the applicable effluent limitations on BOD5.
[2]Within the range of 6.0 to 9.0.
[3]Feedstock denotes the crude oil and natural gas liquids fed to the topping units.

Figure 7.3.1 shows the components of a comprehensive industrial wastewater survey. The two main elements in a survey are (1) definition of the physical characteristics of the plant's sewer systems and (2) development of individual wastewater stream profiles. Defining the physical systems is required before information on flow characteristics and composition of individual wastewater streams can be obtained.

A survey of aircraft paint stripping wastewater generated at U.S. Navy facilities is an example of industrial wastewater characterization. The survey was conducted at six Naval Air Rework Facilities (NARFs). The field survey at each NARF was conducted for periods ranging from 5 to 7 days (Law, Olah, and Torres 1985). The waste-

TABLE 7.3.6 PARAMETERS ADDRESSED IN DISCHARGE STANDARDS FOR SELECTED INDUSTRIAL WASTEWATERS

	Industry													
Parameter	Auto-mobile	Bever-age	Can-ning	Fertil-izer	Inor-ganic Chemi-cals	Or-ganic Chemi-cals	Meat Prod-ucts	Metal Finish-ing	Plas-tics & Syn-thetics	Pulp & Paper	Petro-leum Refin-ing	Steel	Tex-tiles	Dairy
BOD₅	x	x	x	x	x	x	x		x	x	x	x	x	x
COD	x		x	x	x	x		x	x	x	x		x	x
TOC			x			x				x	x			x
TOD					x									
pH	x	x	x	x	x	x	x		x	x	x	x	x	x
Total solids	x				x									
SS	x	x	x	x	x	x	x	x	x	x	x	x	x	x
Settleable solids		x			x		x							
TDS		x	x	x	x	x	x		x	x	x		x	
Volatile SS														
Oil and grease	x	x		x		x	x	x	x	x	x	x	x	
Heavy metals, general				x		x		x		x	x		x	
Chromium	x			x	x						x	x	x	
Copper	x										x	x		
Nickel	x											x		
Iron	x			x	x						x	x		
Zinc	x			x					x		x	x		
Arsenic				x	x									
Mercury					x									
Lead	x				x						x			
Tin	x				x							x		
Cadmium				x										
Calcium				x										
Fluoride				x	x									

Continued on next page

TABLE 7.3.6 Continued

Parameter						Industry								
	Automobile	Beverage	Canning	Fertilizer	Inorganic Chemicals	Organic Chemicals	Meat Products	Metal Finishing	Plastics & Synthetics	Pulp & Paper	Petroleum Refining	Steel	Textiles	Dairy
Cyanide	x				x	x		x	x		x	x		
Chloride	x			x	x	x					x	x		x
Sulfate	x			x	x				x		x	x		
Ammonia	x			x	x	x	x		x	x	x	x		
Sodium				x	x									
Silicates					x									
Sulfite										x				
Nitrate	x			x	x	x			x	x	x			x
Phosphorus			x	x	x	x	x		x	x	x			x
Urea or organic nitrogen				x										
Color		x	x				x			x	x		x	x
Total coliform		x	x				x			x				
FC		x								x				
Toxic materials		x			x		x		x	x	x		x	
Temperature		x		x	x				x	x	x	x	x	x
Turbidity		x					x			x	x		x	x
Foam		x												x
Odor											x			
Phenols	x				x	x			x	x	x	x	x	
Chlorinated benezoids & polynuclear aromatics					x				x					
Mercaptan sulfide									x		x		x	

Source: R.A. Corbitt, 1990, Wastewater disposal, Chap. 6 in *Standard handbook of environmental engineering*, edited by R.A. Corbitt (New York: McGraw-Hill, Inc.).

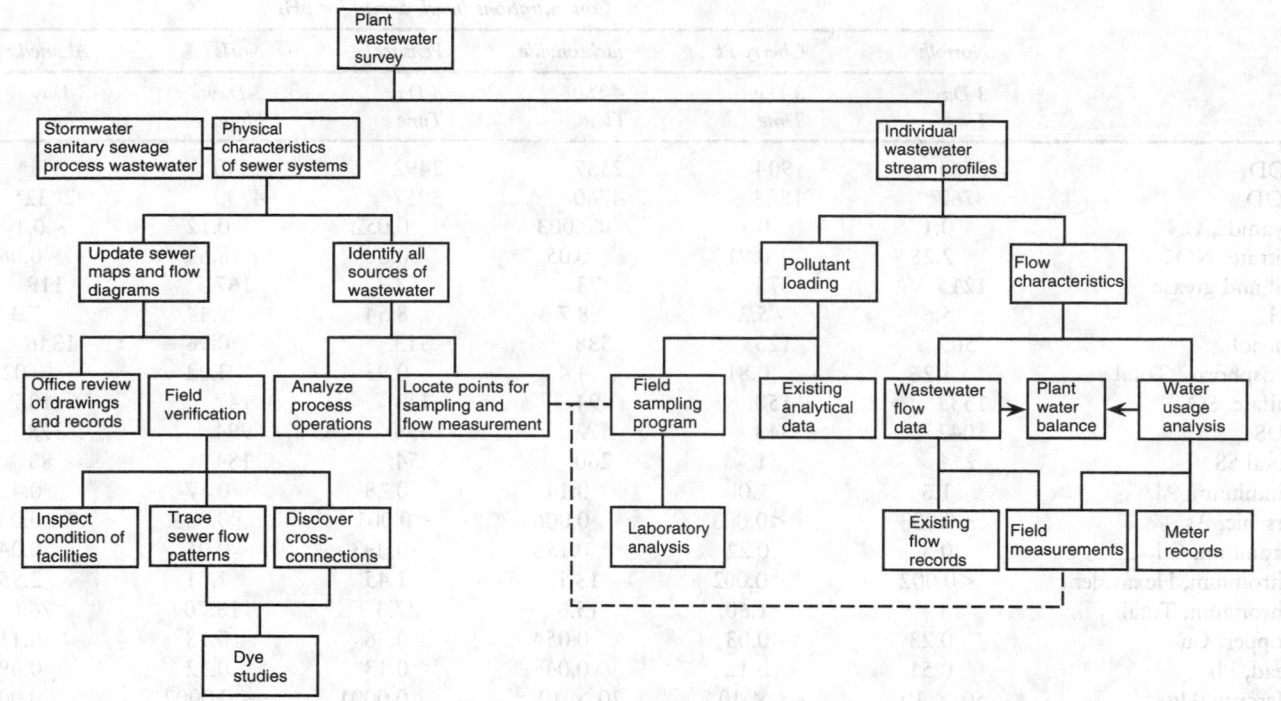

FIG. 7.3.1 Components of an industrial wastewater survey. (Reprinted, with permission, from R.A. Corbitt, 1990, Wastewater disposal, Chap. 6 in *Standard handbook of environmental engineering*, edited by R.A. Corbitt (New York: McGraw-Hill, Inc.)

water characteristics at these NARFs varied due to differences in missions and operations. Thus, the survey required characterization of wastewaters from all six NARFs to adequately represent aircraft paint stripping wastewater generated by the U.S. Navy.

The survey team selected 24 test parameters to characterize the quality of the aircraft paint stripping wastewater. Some parameters were important in monitoring chemical or biological treatment, while others were related to monitoring requirements for discharge regulations. The four parameters identified as most important, either because of high concentrations or limitations imposed by regulatory agencies, were oil and grease, phenol, chromium, and total toxic organics (TTO). The TTO parameter includes up to 110 toxic organics identified by the U.S. EPA. It is being incorporated into discharge permits by various regulatory agencies. The EPA set a value of 2.13 mg/l for TTO as the wastewater pretreatment standard for metal finishing industries, and aircraft paint stripping wastewater is in this category. The 2.13 mg/l value is based on the summation of all quantifiable concentrations greater than 0.01 mg/l for the 110 listed toxic organics.

To reduce sample analysis costs, the TTO analyses conducted during the survey excluded organics not present in paint stripping wastewater, including PCBs and pesticides. Thus, the analyses for TTO included only volatile organics (EPA Methods 601 and 602), acid extractables, and base neutral extractable organics (EPA Method 625). A total of eighty-four toxic organic compounds were analyzed from the list of 110 compounds; the twenty-six compounds not analyzed were PCBs and pesticides.

In nearly two months of site visits to the six NARFs, eighty-three aircraft paint stripping wastewater samples were collected, including nineteen composite samples (collected by automatic samplers) and sixty-four grab samples. For each sample, twenty-three analytical tests were performed. In addition, the TTO tests were performed on seventeen grab samples.

Table 7.3.7 shows the results of the analyses. Naval aircraft paint stripping wastewater is characterized by high organic pollutant contents and, except for chromium, low concentrations of heavy metals. The concentration level of oil and grease is generally lower than 500 mg/l; however, for one NARF, it averaged as high as 1215 mg/l. The average pH values varied from 5.2 to 9.4. Averages for phenol concentrations were typically in the hundreds range, but it was as high as 800 to 1300 mg/l for two NARFs. Due to the use of a nonphenolic paint stripper, one NARF (North Island) was able to keep its wastewater phenol concentration to around 1 mg/l; however, at this NARF, both stripper and labor usage increased due to the reduced effectiveness of the phenol-free stripper. The average TDS and SS values are typical for many wastewaters. Total chromium levels varied from 1.6 to 76 mg/l, while hexavalent chromium levels ranged from below 0.002 to about 13 mg/l. For the TTO parameter, values ranged from 124 to 2765 mg/l.

TABLE 7.3.7 COMPARISON OF COMPOSITE SAMPLE AVERAGES FROM THE NARFs

	Concentrations (mg/l, except for pH)					
	Norfolk	Cherry Pt	Jacksonville	Pensacola	NorIs	Alameda
	3-Day Time	3-Day Time	4-Day Flow	3-Day Time	3-Day Flow	3-Day Time
BOD$_5$	3564	904	2537	2492	1540	311[a]
COD	8767	1823	4760	3957	4730	32032[a]
Cyanide, CN$^-$	0.1	0.1	<0.003	0.052	0.127	<0.1
Nitrate, NO$_3^-$	2.28	0.91	0.05	0.60	8.6	0.067
Oil and grease	1215	375	73	53.3	167.5	119
pH	5.6	5.2	8.7	8.54	9.44	7.3
Phenol	509	123	838	513	0.96	1346
Phosphorus, Total	1.26	0.81	4.4	0.95	3.02	<0.025[a]
Sulfate, SO$_4$	1533	150	94	113	147	10.7
TDS	1042	343	572	696	994	378
Total SS	214	51	260	54	184	88
Aluminum, Al	1.6	1.0	0.14	0.28	0.87	0.456
Arsenic, As	<0.003	<0.003	0.006	<0.001	0.003	0.0059
Cadmium, Cd	0.5	0.22	0.156	0.093	0.05	0.0437
Chromium, Hexavalent	<0.002	<0.002	13.1	1.45	7.71	2.95
Chromium, Total	13	1.60	15.6	27.3	18.70	76.0
Copper, Cu	0.23	0.03	0.054	0.36	0.18	0.119
Lead, Pb	0.51	0.12	0.04	0.13	0.12	0.099
Mercury, Hg	50×10^{-5}	48×10^{-5}	20×10^{-5}	0.0001	<0.0002	0.0006
Nickel, Ni	0.77	0.04	<0.03	0.022	<0.05	0.039
Selenium, Se	<0.003	<0.003	<0.005	<0.001	<0.003	0.0053
Silver, Ag	<0.01	0.01	0.005	0.001	0.01	0.009
Zinc, Zn	1.22	0.26	0.28	0.88	0.277	0.365
TTO[b]						
1. 601/602						
2. 625	124	312	1328	490	2765	986

Source: A. Law, N.J. Olah, and T. Torres, 1985, *Navy aircraft paint stripping waste characterization,* Technical memorandum 71-85-30 (Port Hueneme, Calif.: Naval Civil Engineering Laboratory).

Notes: [a]Alameda's data are suspected.
[b]Grab sample analysis averages.

TABLE 7.3.8 U.S. AIR FORCE AND U.S. NAVY PAINT STRIPPING WASTEWATERS

	Concentrations (mg/l)	
Parameter	Air Force[a]	Navy[b]
1. COD	9200–36,400	1800–8800
2. Oil and grease	8.4–66.3	50–1300
3. pH	8–8.6	5–9.5
4. Phenol	1040–4060	<1–1300
5. Total phosphorus	10–28	0.8–4.5
6. TSS	107–303	50–300
7. Total chromium	17.5–59.5	1.5–80
8. TTO	NM	124–2765
9. Methylene chloride	75–2000	70–2760[c]
10. TOC	1710–14,400	NM

Source: Law, Olah, and Torres, 1985.
Notes: NM = Not measured.
[a]From Perrotti (1975).
[b]Approximate range from Table 7.3.7.
[c]Approximate range from TTO tests.

Table 7.3.8 compares the data on paint stripping wastewater characteristics reported by Perrotti (1975) for the U.S. Air Force and the data procured by Law, Olah, and Torres (1985). The significant differences between the two are related to the concentration ranges of COD, phenol, oil and grease, total chromium, and pH. The Navy and Air Force paint stripping wastewaters are comparable with regard to the concentration ranges of total SS and methylene chloride. The TTO concentration was not measured in the Air Force study. These data show that the Navy's aircraft paint stripping wastewater is different from that generated at the corresponding Air Force facilities. These differences have implications in terms of potential treatment technologies.

As noted throughout this section, quality characteristics of industrial wastes vary considerably depending on the type of industry. A useful parameter in describing industrial wastes is an organic matter-based population equivalent defined as:

$$PE = \frac{A \times B \times 8.34}{0.17} \qquad 7.3(1)$$

where:

PE = population equivalent based on organic constituents in the industrial waste
A = industrial waste flow, mg/day
B = industrial waste BOD, mg/l
8.34 = number of lb/gal
0.17 = number of lb BOD/person–day

Similar population equivalent calculations can be made for SS, nutrients, and other constituents. Expressing all waste loadings on a similar basis requires population equivalent calculations to be made for various pollutants from both point and nonpoint sources in a geographical area.

—*Larry W. Canter*

References

Code of Federal Regulations, Title 40, Chap. 1, Part 419—Petroleum refining point source category. 419–457. (1 July 1991).

Corbitt, R.A. 1990. Wastewater disposal. Chap. 6 in *Standard handbook of environmental engineering*, edited by R.A. Corbitt. 6.26–6.33. New York: McGraw-Hill, Inc.

Law, A., N.J. Olah, and T. Torres. 1985. *Navy aircraft paint stripping waste characterization*. Technical Memorandum 71-85-30. Port Hueneme, Calif.: Naval Civil Engineering Laboratory.

Nemerow, N.L. 1978. *Industrial water pollution: Origin, characteristics and treatment*. 529–549. Reading, Mass.: Addison-Wesley Publishing Company.

Nemerow, N.L. and A. Dasgupta. 1991. *Industrial and hazardous waste treatment*. 387–391. New York: Van Nostrand Reinhold.

Perrotti, A.E. 1975. *Activated carbon treatment of phenolic paint stripping wastewater*. AFCEC-TR-75-14. Tyndall AFB, Fla.: U.S. Air Force Civil Engineering Center. (May).

U.S. Environmental Protection Agency (EPA). 1980. *Treatability manual*. Vol. II, Industrial descriptions. Section II.14, Petroleum refining. EPA-600/8-80-042b. Washington, D.C. (July).

Water Pollution Control Federation and American Society of Civil Engineers. 1977. *Wastewater treatment plant design*. 9–10. Washington, D.C.

7.4
WASTEWATER MINIMIZATION

Wastewater minimization can be considered in terms of volume (or flow) reduction, strength (or pollutant concentration) reduction, or combinations of these reductions. Emphases on flow rate reductions are generally associated with both municipal and industrial wastewater, while pollutant concentration reductions are the focus of attention within industries. Municipalities also emphasize decreases in pollutant concentration via pretreatment ordinances for industries discharging wastewater into municipal sewerage systems.

Benefits resulting from wastewater minimization include the following:

- Reductions of flow and wastewater loadings on existing treatment plants with finite available capacities
- Minimization of hydraulic or organic overloads and toxicity effects on existing treatment facilities
- Reductions in unit process sizes, and possibly the types of processes, for new treatment facilities
- Reductions in initial and operation and maintenance costs for wastewater treatment
- Reductions in water costs for domestic and industrial users and in chemical or manufacturing costs for industries
- Reductions in energy requirements within municipalities and industries
- Reductions in water usage demands on limited water supplies in specific areas
- Facilitation of compliance with treatment plant effluent discharge standards
- Reductions in undesirable impacts on receiving water quality and aquatic ecology

Municipal Wastewater Flow Reduction

Flow reductions related to municipal wastewater include: (1) water conservation; (2) reuse of water in homes; (3) reduction of infiltration and inflow; and (4) reduction in stormwater runoff via best management practices. Water savings of up to 20 to 30% can be accomplished in homes and businesses using flow reduction devices and practicing simple water conservation measures (Qasim 1985). Table 7.4.1 gives examples of home water savings devices and their potential for water use reduction. Table 7.4.2 provides brief descriptions of flow reduction devices and water-efficient appliances. Municipal ordinances can require water saving devices for new homes and businesses.

Qasim (1985) suggests that home water reuse can lead to a 30 to 40% reduction in water consumption and a 40 to 50% reduction in wastewater volume. Wastewater from sinks, bathtubs, showers, and laundry can be treated onsite and reused for toilet flushing and lawn sprinkling.

Reducing infiltration and inflow into sewer lines also reduces wastewater flow rates arriving at a treatment facility. Infiltration refers to the volume of groundwater entering sewers and building sewer connections from the soil and through defective joints, broken or cracked pipes, improper connections, and manhole walls. Inflow denotes the volume of water discharged into sewer lines from sources such as roof leaders, cellar and yard area drains, foundation drains, commercial and industrial clean water discharges, and drains from springs and swampy areas (Sullivan et al. 1977). Wet weather flows containing large volumes of infiltration and inflow can create hydraulic overload difficulties at treatment plants (Qasim 1985). For existing sewers, a program involving cleaning, inspection, testing, and rehabilitation measures may be necessary; rehabilitation includes root control, grouting, and pipe lin-

ing (Sullivan et al. 1977). New, smaller diameter sewers can also be placed inside existing sewers that have major infiltration and inflow problems. New sewers should be designed, constructed, and inspected to remain within an infiltration limit of 20 gal/in diameter/mi/day (185.2 l/cm diameter/km/day) (Sullivan et al. 1977).

The 1987 *Clean Water Act* requires the use of best management practice (BMP) for minimizing the flow and pollutional characteristics of storm water runoff from industrial areas and municipalities with populations of 100,000 or more. The act emphasizes minimizing nonpoint pollution from the introduction of storm water into sanitary sewer systems or from direct discharge into receiving bodies of water.

BMP refers to a combination of practices that are most effective in preventing or reducing pollution generated by nonpoint sources to a level compatible with water quality goals (Novotny and Chesters 1981). After problem assessment, examination of alternative practices, and public participation, the state (or designated area-wide planning agency) determines the BMP based on technological, economic, and institutional considerations. Examples of BMPs include spill prevention and response, sediment and erosion control measures, and runoff management measures (U.S. EPA 1992a,b). Applying these measures can reduce infiltration in areas with separate (storm water and sanitary) sewer systems and reduce water flows and pollutional characteristics in areas with combined sewer systems.

Industrial Wastewater Flow Reduction

Industrial plants can achieve wastewater volume reductions by using

1. classification (segregation) of wastewater according to quality characteristics

TABLE 7.4.1   COMPARISON OF HOME WATER USE WITH AND WITHOUT CONSERVATION DEVICES

| | *Flow, gal/capita · day* | | |
| | Without Conservation | With Conservation Devices | |
Use	Devices	Level 1[a]	Level 2[a]
Baths	7	7	7
Dishwashers	2	1	1
Faucets	9	9	8
Showers	16	12	8
Toilets	22	19	14
Toilet leakage	4	4	8
Washing machines	16	14	13
Total	76	66	59

Source: Metcalf and Eddy, Inc., 1991, *Wastewater engineering*, 3d ed. (New York: McGraw-Hill, Inc.).

Note: [a]Level 1 uses retrofit devices such as flow restrictors and toilet dams. Level 2 uses water-conserving devices and appliances such as low-flush toilets and low-water-use washing machines.

TABLE 7.4.2 SUMMARY OF FLOW REDUCTION DEVICES AND APPLIANCES

Device/Appliance	Description and Application
Faucet aerators	Increases rinsing power of water by adding air and concentrating flow, reducing the amount of wash water used
Limiting-flow shower heads	Restricts and concentrates water passage by means of orifices that limit and divert shower flow for optimum use by the bather
Low-flush toilets	Reduces the discharge of water per flush
Pressure-reducing valve	Maintains home water pressure at a lower level than the water distribution system; decreases the probability of leaks and dripping faucets
Retrofit kits for bathroom fixtures	Consists of shower-flow restrictors, toilet dams, or displacement bags, and toilet leak detector tablets
Toilet dam	A partition in the water closet that reduces the amount of water per flush
Toilet leak detectors	Tablets that dissolve in the water closet and release dye to indicate leakage of the flush valve
Water-efficient dishwasher	Reduces water used
Water-efficient clothes washer	Reduces water used

Source: Metcalf and Eddy, Inc., 1991.

2. conservation of water use within industrial processes
3. production decreases
4. reuse of municipal and industrial treatment plant effluents as a water supply, and
5. elimination of batch or slug discharges of process wastewater (Nemerow and Dasgupta 1991).

Plants can achieve wastewater classification by considering the flow rates, quality characteristics, and treatment needs of wastewater used for manufacturing processes, cooling purposes, and sanitary uses.

Water conservation within industrial processes occurs when the industry changes from open to closed systems (Nemerow and Dasgupta 1991). For example, the sequential reuse of water within canning plants can achieve savings up to 20 to 25% and the use of shutoff valves on hoses and water lines can lead to another 20 to 25% reduction in water use. Nemerow and Dasgupta (1991) give additional examples of production decreases, effluent reuse, and elimination of slug discharges.

Pollutant Concentration Reduction

Clean technology, pollution prevention, and waste minimization are technical and managerial activities that can reduce the pollution emissions from industrial operations (Freeman et al. 1992; Hirschhorn and Oldenburg 1991; and Office of Technology Assessment 1986). Clean technology refers to applying technical processes to minimize waste material from the processes themselves (Johansson 1992). Pollution prevention relates to approaches that prevent pollution from occurring, including the incorporation of clean technology. Other housekeeping and conservation practices can be included in pollution prevention. Waste minimization tries to minimize negative impacts on the environment by reducing the amount of waste material from

operations. Such waste reductions include applying pollution control technologies, chemical substitutions, clean technologies, and other activities that minimize the waste generated.

Chapter 3 provides additional information on pollution prevention.

—Larry W. Canter

References

Freeman, H.M., T. Harten, J. Springer, P. Randall, M.A. Curran, and K. Stone. 1992. Industrial pollution prevention: A critical review. *Journal of Air and Waste Management Association* 42, no. 5 (May): 618–656.

Hirschhorn, J.S. and K.U. Oldenburg. 1991. *Prosperity without pollution: The prevention strategy for industry and consumers.* New York: Van Nostrand Reinhold.

Johansson, A. 1992. *Clean technology.* Boca Raton, Fla.: Lewis Publishers, Inc.

Nemerow, N.L. and A. Dasgupta. 1991. *Industrial and hazardous waste treatment.* 101–105. New York: Van Nostrand Reinhold.

Novotny, V. and G. Chesters. 1981. *Handbook of nonpoint pollution.* New York: Van Nostrand Reinhold.

Office of Technology Assessment. 1986. *Serious reduction of hazardous waste: For pollution prevention and industrial efficiency.* OTA-ITE-317. 3–20. Washington, D.C. (September).

Qasim, S.R. 1983. *Wastewater treatment plants—Planning, design, and operation.* 32–37. New York: Holt, Rinehart and Winston.

Sullivan, R.H., M.M. Cohn, T.J. Clark, W. Thompson, and J. Zaffle. 1977. *Sewer system evaluation, rehabilitation, and new construction.* EPA 600–2-77–017d. iv-vii and 1. Cincinnati, Ohio: U.S. EPA. (December).

U.S. Environmental Protection Agency (EPA). 1992a. *Storm water management for construction activities—Developing pollution prevention plans and best management practices.* EPA 832-R-92-005. Washington, D.C.: Office of Water. (September).

———. 1992b. *Storm water management for industrial activities—Developing pollution prevention plans and best management practices.* EPA 832-R-92-006. Washington, D.C.: Office of Water. (September).

7.5
DEVELOPING A TREATMENT STRATEGY

In-plant treatments are a set of cost-effective onsite unit operations and processes installed in an industrial facility to remedy the production or discharge of hazardous or conventional waste to the environment. These remedial techniques consist of preliminary and primary process equipment, instrumentation, and control units related to the industry type and wastewater characteristics.

Factors affecting unit process selection are influent water characteristics, effluent quality required, reliability, sludge handling, and costs. Figure 7.5.1 shows the proposed strategy for wastewater management at a manufacturing complex.

Evaluating Compliance

The ultimate goal of any wastewater treatment system is to comply with regulations in a cost-effective manner. Common outlets for wastewater discharges are as follows:

Discharge to Surface Water. Effluent from wastewater treatment operations is piped directly to a surface water body and is subject to NPDES regulations. Effluent limitations depend on the ambient water quality criteria, the condition of the receiving stream, and the amount of mixing available. Discharge to surface water is usually a viable outlet for effluents containing benign contaminants or being treated to a level guaranteeing that the receiving stream is not impacted.

Discharge to the Sewer. Effluent from wastewater treatment operations is sent to the sewer, which is connected to a POTW. The wastewater is subject to municipal pretreatment regulations. Typically this outlet is good for effluents containing constituents that the POTW can effectively degrade, principally biodegradable organics of moderate strength. The capacity of the POTW to accept the waste must be considered.

Offsite Disposal. Effluents and other residues (sludge) from wastewater treatment operations are transported to an offsite treatment facility. The handler determines the level of pretreatment required for off-site disposal. This method is appropriate for low-volume, high-toxicity effluents and residuals. Effluents and residuals in this category are usually prohibited from discharge through other outlets (NPDES outfalls or municipal sewers).

Compliance evaluations can have either of the following forms:

Assessment of whether a plant's wastewater treatment operations are meeting effluent discharge limitations. If a facility is consistently not in compliance on a critical NPDES permit parameter, such as a primary pollutant concentration, this noncompliance is more urgent than an occasional minor deviation from a composite parameter such as BOD. The former situation requires immediate revision of a facility's wastewater management strategy, involving significant modifications or even complete replacement of existing treatment units.

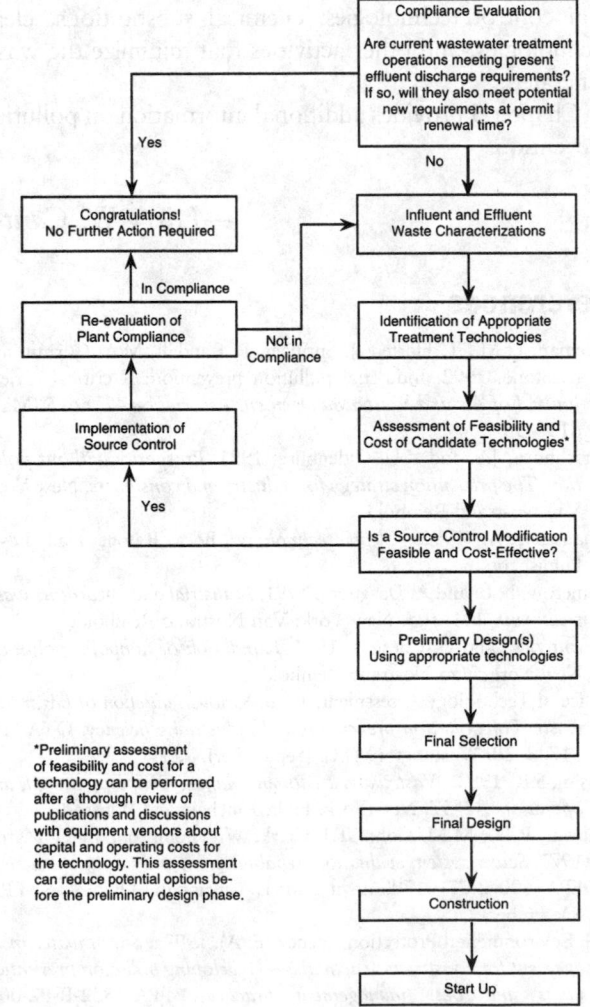

FIG. 7.5.1 Developing a wastewater treatment strategy. (Reprinted, with permission, from L.A. McLaughlin, H.S. McLaughlin, and K.A. Groff, 1992, Develop an effective wastewater treatment strategy, Chem. Eng. Prog. [September].)

Assessment of whether the facility can meet newer, more restrictive discharge limitations to be imposed when NPDES permits are renewed. These facilities should examine their current permits and allow adequate time before renewal to determine whether they can meet the anticipated discharge limitations.

Characterizing Wastewater

Figure 7.5.1 is a guide to the decisions involved in developing an appropriate waste management strategy.

The most important step in developing a wastewater management strategy is to completely characterize the wastewater. Although compliance monitoring indicates current compliance status, it is not an adequate starting point for the cost-effective design of a wastewater treatment system. Environmental engineers must characterize both the source to and effluent from current wastewater treatment operations. Understanding how the wastewater is produced is as important as knowing what contaminants are present (McLaughlin, McLaughlin, and Groff 1992). A review of manufacturing processes provides the knowledge base needed to evaluate the best place to reduce, recover, or treat individual waste streams.

The data should include the following information:

- All production activities within the facility, i.e., raw materials used and production records
- Detailed drawings of the plant showing the locations of processing units, their water distribution, and wastewater production and collection systems
- The quantity, analysis, frequency, and flow rate of the waste stream discharge from each unit process
- The frequency, extent, and type of monitoring and sampling used in accordance with the nature and variability of each waste stream
- The flow measurement and location of sample collection points within the facility indicating the type of monitoring stations (permanent or temporary) used

The constituents to be assayed and quantified in the influent and effluent depend on manufacturing process characteristics and should be determined on a case-by-case basis (McLaughlin, McLaughlin, and Groff 1992). In general, environmental engineers should analyze the constituents to assess compliance with current and future regulatory requirements and consider:

- Options for treating individual wastewater resources
- Potentials for modifying the manufacturing process to reduce, eliminate, or modify contaminants.

Table 7.5.1 lists the constituents and parameters that should be analyzed.

To assess whether current treatment systems require modification or replacement, environmental engineers must be aware of the target compounds for treatment and the additional constraints of individual treatment processes. For example, surfactants in metal waste streams must be analyzed because these compounds can chelate with metals and deteriorate conventional metal-removal technologies. Another example is evaluating the presence of salts when recycling is considered since equipment restrictions (such as preventing scaling) can limit the level of salt during recycling. Thus, environmental engineers should develop the list of constituents to analyze with both current and potential treatment technologies in mind.

Environmental engineers should also characterize wastewater in terms of flow rate. A comprehensive understanding of flow rates and patterns of flow to wastewater treatment operations is critical in the design of either a new system or system modifications. A good waste characterization accounts for all components of the final discharge including all resources and losses of water and the constituents present.

The best way to completely characterize wastewater is to develop a mass balance augmented by an understanding of the manufacturing process that generates the wastewater streams.

Selecting Treatment Technologies

Before implementing any in-plant controls or pretreatment alternatives, the industry should first explore ways to reduce production of specific pollutants and then examine the feasibility of recycling or reusing the wastewater generated during production. For example, the concentrated solution obtained from cleanup operations can be recycled as part of the starting materials for the next production run. Additional steps for reducing wastewater requiring treatment include good housekeeping practices; spill control measures, such as spill containment enclosures and drip trays around tanks; and eliminating wet floor areas.

The principal pollutants affected by modifying industrial manufacturing processes and in-plant treatment methods are as follows:

- Insoluble substances that can be separated physically with or without flocculation
- Organic substances separable by adsorption
- Substances separable by precipitation
- Substances that can be precipitated as insoluble iron salts or that can be chelated
- Substances separable by degassing or stripping
- Substances requiring a redox reaction
- Acids and bases
- Substances that can be concentrated by ion exchange or reverse osmosis
- Substances treatable by biological methods

TABLE 7.5.1 EXAMPLES OF COMMON POLLUTANTS AND OTHER PARAMETERS FOR WHICH EFFLUENTS SHOULD BE CHARACTERIZED

Constituent/Parameter	Description
VOCs Acid-extractable organics Base- and Neutral-extractable organics Metals, total and metals, soluble	Priority pollutants. Concentrations of these compounds are typically regulated on both sewer and NPDES permits.
BOD COD TOC TSS Temperature pH	Conventional pollutants. Permissible levels and values are also typically regulated on both sewer and NPDES permits.
Whole-effluent toxicity (LC_{50})	A relatively new parameter, it is usually only evaluated for NPDES permits.
Surfactants	Potential interfering agents
Ammonia Nitrate, nitrite Phosphorus	Nutrients. Determination is needed to adequately evaluate the potential for biological treatment.
Sulfate Chloride Sodium	Inorganic salts. Potential interfering agents
Flow rates	Necessary to perform a mass balance on the facility

Source: McLaughlin, McLaughlin, and Groff, 1992.

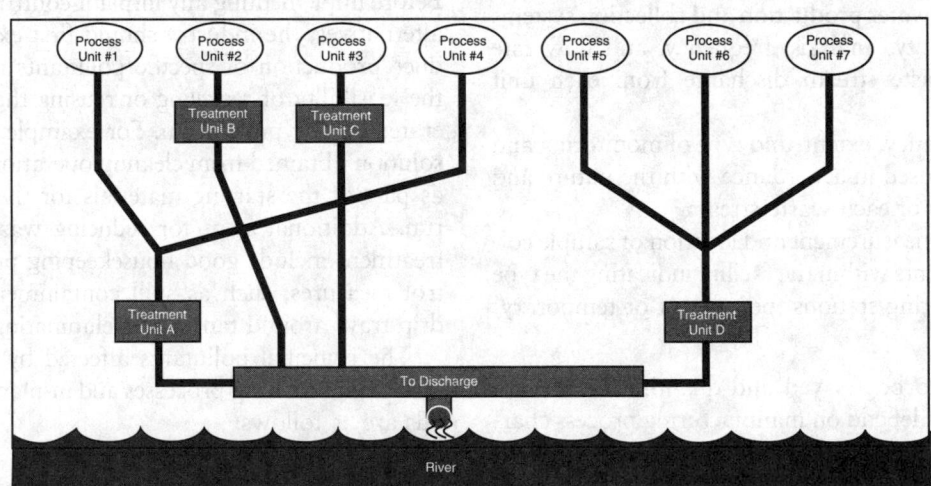

FIG. 7.5.2 Combining streams that use the same treatment technology and treating other streams at the source. (Reprinted, with permission, from McLaughlin, McLaughlin, and Groff, 1992.)

Environmental engineers should identify viable technologies for individual wastewater streams. Then, they can combine, on paper, the streams using the same technologies to create composite waste treatment trains. They then compare the resulting wastewater treatment trains to the current manufacturing and waste treatment practices to identify possible candidates for waste segregation and independent treatment (see Figure 7.5.2).

Treatability testing may be needed, especially when a plant that already has a physical-chemical or biological treatment facility is confronted with new wastes. For example, at a large chemicals complex, wastewater is screened for treatability as follows. The stream is pre-treated to remove heavy metals and SS, and pH is adjusted. It is then fed to a batch-activated sludge reactor, and primed with biomass from the plant treatment facility. If the wastewater degrades quickly, as it should, it can be fed into the plant's main flow. If it does not, the choices are in-plant pretreatments, PAC addition to the bioreactor, or granular actuated carbon (GAC) treatment of the effluent.

The problem associated with combining two streams that require different technologies is that the cost of treating the combined stream is almost always more than individual treatment of the separate streams. This is because the capital cost of most treatment operations is proportional to the total flow of the wastewater, and the operating cost for treatment increases with a decreasing concentration for a given mass of contaminant.

Thus, if two waste streams use the same treatment, combining them improves the economics of scale for capital investment and similar operating costs. In contrast, if two treatment operations are required, combining the two streams increases capital costs for both treatment operations. In addition, if the streams are combined before the treatment, both treatments have lower contaminant concentrations for the same net contaminant mass, resulting in higher operating costs per lb of contaminant removed (McLaughlin et al. 1992).

—*Negib Harfouche*

Reference

McLaughlin, L.A., H.S. McLaughlin, and K.A. Groff. 1992. Develop an effective wastewater treatment strategy. *Chem. Eng. Prog.* (September).

Monitoring and Analysis

7.6
FLOW AND LEVEL MONITORING

The control and monitoring of flows and levels in the wastewater treatment industry involve the measurement of water, biological sludge, solid and liquid additives, and reagent flows. This section discusses methods of flow detection followed by a summary of wastewater-related level detection techniques.

Flow Sensors for the Wastewater Industry

Flow detection applications in the wastewater treatment industry include the measurement of large flows in partially filled pipes using weirs, flumes, or ultrasonic sensors. When water is flowing in regular pipelines, magnetic flowmeters, venturi tubes, flow nozzles, and pitot tubes are the usual sensors. In smaller pipelines, orifice plates, vortex flowmeters, or variable area flowmeters are used. For sludge services, doppler-type ultrasonic and magnetic flowmeter (provided with electrode cleaners), V-cone detector, and segmental wedge-type detector can be used. Gas, liquid, or solid additives can be charged by Coriolis

mass flowmeters (gas or liquid), metering pumps, turbine or positive displacement meters (liquids), variable-area flowmeters (gas or liquid), or gravimetric feeders (solids). Table 7.6.1 summarizes flowmeter features and capabilities. The following sections provide a brief summary of the features and capabilities of the flowmeters used in the wastewater treatment industry.

Magnetic Flowmeters

DESIGN PRESSURE

Varies with pipe size. For a 4 in (100 mm) unit, the maximum pressure is 285 psig (20 bars); special units are available with pressure ratings up to 2500 psig (172 bars).

DESIGN TEMPERATURE

Up to 250°F (120°C) with Teflon liners and up to 360°F (180°C) with ceramic liners.

MATERIALS OF CONSTRUCTION

Liners: ceramics, fiberglass, neoprene, polyurethene, rubber, Teflon, vitreous enamel, and Kynar; Electrodes: platinum, Alloy 20, Hastelloy C, stainless steel, tantalum, titanium, tungsten carbide, Monel, nickel, and platinum-alumina cermet.

TABLE 7.6.1 ORIENTATION TABLE FOR FLOW SENSORS

Type of Design	Clean Liquids	Viscous Liquids	Slurry	Gas	Solids	Direct Mass—Flow Sensor	Volumetric Flow Detector	Flow Rate Sensor	Inherent Totalizer	Direct Indicator	Transmitter Available	Linear Output	Rangeability	Pressure Loss Thru Sensor	Approximate Straight Pipe-Run Requirement (Upstream Diameter/Downstream Diameter)	Accuracy * ±% Full Scale, ** ±% Rate, *** ±% Registration	Flow Range Units
Elbow taps	√	L	L	√			√	√			√	SR	3:1⑦	N	25/10②	5–10*	gpm—m³/hr; scfm—Sm³/hr
Jet deflection	√	√		√			√	√			√	√	25:1	M	20/5②	2*	scfm—Sm³/hr
Laminar flowmeters	√②	√②	√②	√			√	√			√	√	10:1	H	15/5	½–5*⑧	gpm—m³/hr; scfm—Sm³/hr
Magnetic flowmeters	√②	√②	√				√	√			√	√	10:1⑧	N	5/3	½**–2*	gpm—m³/hr
Mass flowmeters and miscellaneous coriolis	√	√	L	√	SD		√	√	SD / SD	SD / SD			100:1 / 20:1	A / H	N / N	½**–½** / 0.15–½**	lbm/hr—kgm/hr; scfm—Sm³/hr
Metering pumps	√	√						√	√		SD	√	20:1	—	N	1/10–1*	gpm—m³/hr
Orifice (plate or integral cell)	√	L	L	√			√	√			√	SR	3:1②	H	20/5②	1**–2*	gpm—m³/hr; scfm—Sm³/hr
Pitot tubes	√	√	L	√			√	√		√	√	SR	3:1⑧	M	30/5②	0.5–5*	gpm—m³/hr; scfm—Sm³/hr
Positive displacement gas meters				√			√	√	√	√	SD	√	10:1 to 200:1	M	N	½–1***	scfm—Sm³/hr

Flowmeter type	Units	Rangeability①			
Positive displacement liquid meters	gpm—m³/hr	10:1①	H	N	0.1-2**
Segmental wedge	gpm—m³/hr	3:1	M	15/5	3**
Solids flowmeters	lbm/hr-kgm/hr	20:1	—	5/3	½**-4*
Target meters	gpm—m³/hr / SCFM—Sm³/hr	4:1	H	20/5	0.5*-5*
Thermal meters (mass flow)	gpm—m³/hr / SCFM—Sm³/hr	20:1①	A	5/3	1-2*
Turbine flowmeters	gpm—m³/hr / SCFM—Sm³/hr	10:1⑪	H	15/5¹	1** / ¼
V-cone flowmeter	gpm—m³/hr / ACFM—Sm³/hr	3:1②	M	2/5	½-2**
Ultrasonic flowmeters — Transit	gpm—m³/hr³	20:1	N	15/5⑦	1**-2*
Ultrasonic flowmeters — Doppler	SCFM—Sm³/hr	10:1	N	15/5⑦	2-3*
Variable-area flowmeters	gpm—m³/hr / SCFM—Sm³/hr	5:1	A	N	½**-10**
Venturi tubes	gpm—m³/hr³	3:1②	M	15/5②	½**-1*
Flow nozzles	SCFM—Sm³/hr	3:1②	H	20/5②	1**-2*
Vortex shedding	gpm—m³/hr³	10:1⑥	H	20/5	0.5-1.5**
Fluidic	ACFM—Sm³/hr	20:1⑥	H	20/5	1-2**
Oscillating	gpm—m³/hr³	10:1①	H	20/5	0.5*
Weirs and flumes	gpm—m³/hr³	100:1	M	See Text	2-5*

L = Limited
SD = Some Designs
H = High
A = Average
M = Minimal
N = None
SR = Square Root

- - - - = Nonstandard Range

① = The data in this column is for general guidance only.
② = The inherent rangeability of the primary device is substantially greater than shown. The value used reflects the limitation of the differential pressure sensing device when 1% of the actual flow accuracy is needed. With multiple-range intelligent transmitters, the rangeability can reach 10:1.
③ = The pipe size establishes the upper limit.
④ = Practically unlimited with the probe-type design.
⑤ = Must be conductive.
⑥ = Can be reranged over 100:1.
⑦ = Varies with upstream disturbance.
⑧ = Can be more at high Re. No. services.
⑨ = Up to 100:1 with high precision design.
⑩ = Commercially available gas flow elements can be ±1% of rate.
⑪ = More for gas turbine meters.

TYPE OF FLOW DETECTED

Volumetric flow of conductive liquids, including slurries and corrosive or abrasive materials.

MINIMUM CONDUCTIVITY REQUIRED

The majority of designs require 1 to 5 μS/cm. Some probe types require more. Special designs can operate at 0.05 or 0.1 μS/cm.

FLOW RANGES

From 0.01 to 100,000 gpm (0.04 to 378,000 liters per minute (lpm)).

SIZE RANGES

From 0.1 to 96 in (2.5 mm to 2.4 m) in diameter.

VELOCITY RANGES

0–0.3 to 0–30 ft/sec (0–0.1 to 0–10 m/sec).

ERROR (INACCURACY)

±1% of actual flow with pulsed direct current (dc) units within a range of up to 10:1 if flow velocity exceeds 0.5 ft/sec (0.15 m/sec). ±1% to ±2% full-scale with alternating current (ac) excitation.

COST

The probe designs are least expensive, at a cost of about $1500. A 1-in (25-mm) ceramic tube unit can be obtained for under $2000. A 1-in (25-mm) metallic wafer unit can be obtained for under $3000. An 8-in (200-mm) flanged meter that has a Teflon liner and stainless electrodes and is provided with 4 to 20 mA dc output, grounding ring, and calibrator costs about $8000. The scanning magmeter probe used in open-channel flow scanning costs about $10,000.

PARTIAL LIST OF SUPPLIERS

ABB Kent-Taylor Inc.; AccuDyne Systems Inc.; Accurate Metering Systems Inc.; ADE-Applied Digital Electronics; Badger Meter Inc.; Baily Controls Co.; Brooks Instrument Div. of Rosemount; Colorado Engineering Experimental Station; Dantec Electronics; H.R. Dulin Co.; Dynasonics Inc. (probe-type); Edinboro Computer Instruments Corp.; Electromagnetic Controls Corp.; Endress + Hauser Instruments; Engineering Measurements Co.; Fischer & Porter Co.; Foxboro Co.; Harwil Corp.; Honeywell, Industrial Controls Div.; Instrumark International Inc.; Johnson Yokogawa Corp.; K & L Research Co. (probe-type); Krone-America Inc.; Marsh-McBirney Inc. (probe-type); Meter Equipment Mfg.; Mine Safety Appliances Co.; Monitek Tech. Inc.; Montedoro Whitney; MSR Magmeter Manufacturing Ltd. (probe-type); Omega Engineering; Rosemount Inc.; Sarasota Measurements & Controls; Schlumberger Industries Inc.; Signet Industrial (probe-type); Sparling Instruments Co.; Toshiba International; Turbo Instruments Inc.; Vortab Corp.; Wallace & Tiernan Inc.; Wilkerson Instrument Co.; XO Technologies Inc.; Yokogawa Electric Corp.

Magnetic flowmeters use Faraday's Law of electromagnetic induction for measuring flow. Faraday's Law states that when a conductor moves through a magnetic field of given strength, a voltage level is produced in the conductor that depends on the relative velocity between the conductor and the field. This concept is used in electric generators. Faraday foresaw the practical application of the principle to flow measurement because many liquids are adequate electrical conductors. In fact, he attempted to measure the flow velocity of the Thames River using this principle. He failed because his instrumentation was not adequate, but 150 years later, the principle is successfully applied in magnetic flowmeters.

THEORY

Figure 7.6.1 shows how Faraday's Law is applied in the electromagnetic flowmeter. The liquid is the conductor that has a length equivalent to the inside diameter of the flowmeter D. The liquid conductor moves with an average velocity V through the magnetic field of strength B. The induced voltage is E. The mathematical relationship is:

$$E = BDV/C \qquad 7.6(1)$$

where:

C is a constant to take care of the proper units

When the pair of magnetic coils is energized, a magnetic field is generated in a plane mutually perpendicular to the axis of the liquid conductor and the plane of the electrodes. The velocity of the liquid is along the longitudinal axis of the flowmeter body; therefore, the voltage induced within the liquid is mutually perpendicular to the velocity of the liquid and the magnetic field.

The liquid should be considered as an infinite number of conductors moving through the magnetic field with each element contributing to the voltage that is generated. An increase in the flow rate of the liquid conductors moving through the field increases the instantaneous value of the voltage generated. Also, each of the individual generators contributes to the instantaneously generated voltage.

Whether the profile is essentially square (characteristic of a turbulent velocity profile), parabolic (characteristic of a laminar velocity profile), or distorted (characteristic of poor upstream piping), the magnetic flowmeter is excellent at averaging the voltage contribution across the metering cross section. The sum of the instantaneous voltages generated represents the average liquid velocity because each increment of liquid velocity within the plane of the electrode develops a voltage proportional to its local velocity. The signal voltage generated is equal to the average velocity almost regardless of the flow profile. The mag-

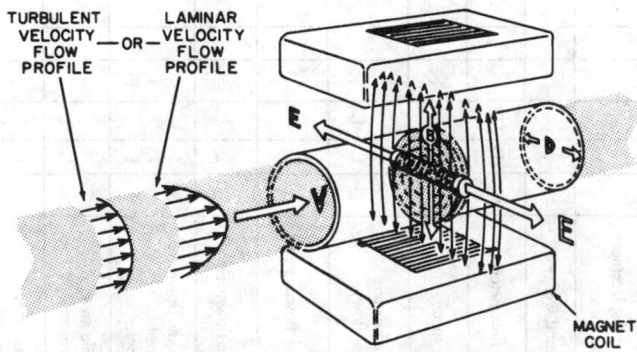

FIG. 7.6.1 Schematic representation of the magnetic flowmeter.

netic flowmeter detects the volumetric flow rate by sensing the linear velocity of the liquid.

The equation of continuity (Q = VA) is the relationship that converts the velocity measurement to volumetric flow rate if the area is constant. The area must be known and constant and the pipe must be full for a correct measurement.

DESIGNS AND APPLICATIONS

Magnetic flowmeters are available in conventional (see Figure 7.6.2), ceramic (see Figure 7.6.3), and probe (see Figure 7.6.4) constructions.

Most liquids or slurries are adequate electrical conductors to be measured by electromagnetic flowmeters. If the liquid conductivity is equal to 20 μS per cm or greater, most conventional magnetic flowmeters can be used. Special designs are available to measure the flow of liquids with threshold conductivities as low as 0.1 μS.

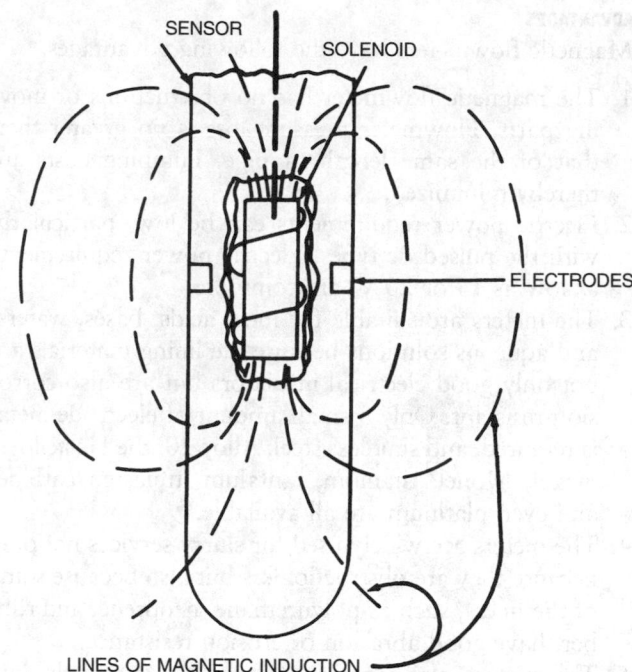

FIG. 7.6.4 The probe-type magnetic flowmeter.

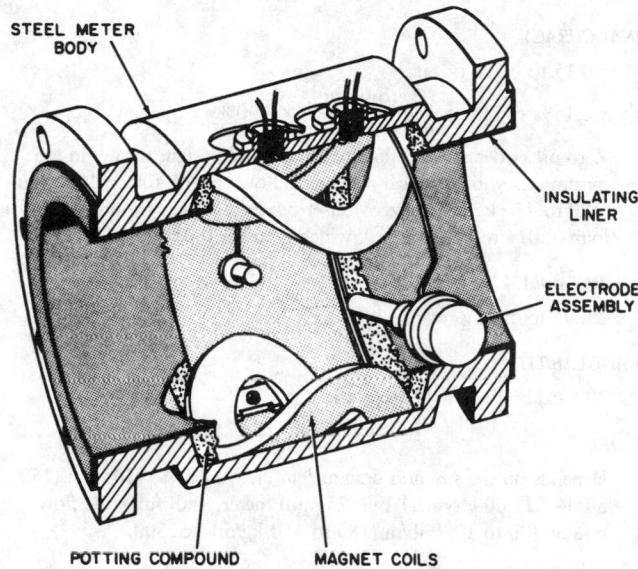

FIG. 7.6.2 The short-form magnetic flowmeter.

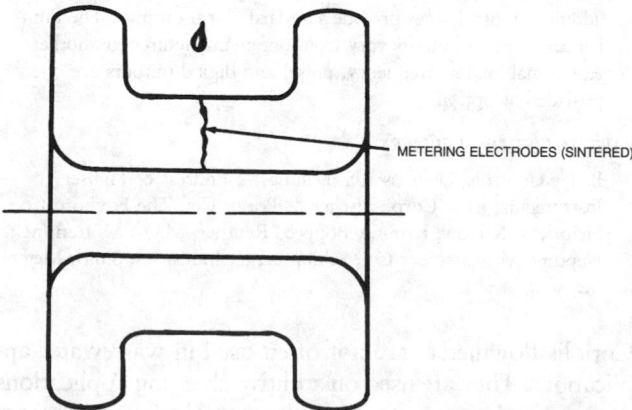

FIG. 7.6.3 The ceramic insert-type magnetic flowmeter.

Magnetic flowmeters are not affected by viscosity or consistency (referring to Newtonian and nonNewtonian fluids, respectively). Changes in the flow profile due to changes in Reynolds numbers or upstream piping do not greatly affect the performance of magnetic flowmeters. The voltage generated is the sum of the incremental voltages across the entire area between the electrodes, resulting in a measure of the average fluid velocity. Nevertheless, the meter should be installed with five diameters of straight pipe before and three diameters of straight pipe following the meter.

Magnetic flowmeters are bidirectional. Manufacturers offer converters with output signals for both direct and reverse flows.

The magnetic flowmeter must be full to assume accurate measurement. If the pipe is only partially full, the electrode voltage, which is proportional to the fluid velocity, is still multiplied with the full cross section, and the reading will be high. Similarly, if the liquid contains entrained gases, the meter measures them as liquid, the reading will be high.

The meter's electrodes must remain in electrical contact with the fluid being measured and should be installed in the horizontal plane. In applications where a buildup or coating occurs on the inside wall of the flowmeter, periodic flushing or cleaning is recommended.

Special meters for measuring sewage sludge flow are designed to prevent the buildup and carbonizing of sludge on the meter electrodes. They use self-heating to elevate the metering body temperature to prevent sludge and grease accumulation.

ADVANTAGES

Magnetic flowmeters have the following advantages:

1. The magnetic flowmeter has no obstructions or moving parts. Flowmeter pressure loss is no greater than that of the same length of pipe. Pumping costs are thereby minimized.
2. Electric power requirements can be low, particularly with the pulsed dc types. Electric power requirements as low as 15 or 20 W are common.
3. The meters are suitable for most acids, bases, waters, and aqueous solutions because the lining materials are not only good electrical insulators but are also corrosion-resistant. Only a small amount of electrode metal is required, and stainless steel, Alloy 20, the Hastelloys, nickel, Monel, titanium, tantalum, tungsten carbide, and even platinum are all available.
4. The meters are widely used for slurry services not only because they are obstructionless but also because some of the liners, such as polyurethane, neoprene, and rubber, have good abrasion or erosion resistance.
5. The meters are capable of handling extremely low flows. Their minimum size is less than $\frac{1}{8}$ in (3.175 mm) inside diameter. The meters are also suitable for high volume flow rates with sizes as large as 10 ft (3.04 m).
6. The meters can be used as bidirectional meters.

LIMITATIONS

Magnetic flowmeters do have some specific application limitations:

1. The meters work only with conductive fluids. Pure substances, hydrocarbons, and gases cannot be measured. Most acids, bases, water, and aqueous solutions can be measured.
2. The conventional meters are relatively heavy, especially in larger sizes. Ceramic and probe-type units are lighter.
3. Electrical installation care is essential.
4. The price of magnetic flowmeters ranges from moderate to expensive. Their corrosion resistance, abrasion resistance, and accurate performance over wide turndown ratios can justify the cost. Ceramic and probe-type units are less expensive.
5. Periodically checking the zero on ac-type magnetic flowmeters requires block valves on either side to bring the flow to zero and keep the meter full. Cycled dc units do not have this requirement.

Coriolis Mass Flowmeters

SIZES

$\frac{1}{16}$ to 6 in (1.5 to 150 mm).

FLOW RANGE

0 to 25,000 lb/m (0 to 11,340 kg/m).

FLUIDS

Liquids, slurries, compressed gases, and liquified gases; not gas-liquid mixtures or gases at below 150 psig (10.3 bars).

OUTPUT SIGNAL

Linear frequency, analog, digital, scaled pulse, and display.

DETECTOR TYPES

Electromagnetic, optical, and capacitive.

OPERATING PRESSURE

Depends upon tube size and flange rating: 1800 psig (124 bars) typical standard; 5000 psig (345 bars) typical high-pressure.

PRESSURE DROP REQUIRED

From under 10 psig (0.7 bars) to over 100 psig (6.9 bars) as a function of viscosity and design.

OPERATING TEMPERATURE

Depends on the design: −100 to 400°F (−73 to 204°C) typical standard; 32 to 800°F (0 to 426°C) high-temperature.

MATERIALS OF CONSTRUCTION

Stainless steel, Hastelloy, titanium, and NiSpan C as standard; tantalum and Tefzel-lined as special.

INACCURACY

±0.15 to 0.5% of rate

$$\pm 0.15\% \text{ of rate} \pm \frac{\text{zero offset}}{\text{mass flow rate}} \times 100\%$$

Zero offset depends on flowmeter size and design; for a 1 in (25 mm) meter with a typical maximum flow rate of 400 to 1000 lb/m (180 to 450 kg/m), the zero offset typically ranges from 0.03 to 0.1 lb/m (0.014 to 0.045 kg/m), which is under 0.01%.

REPEATABILITY

±0.05 to ±0.2% of rate.

RANGEABILITY

20:1 calibration range (typical).

COST

Depends on the size and design: $\frac{1}{16}$ in (1.5 mm)—$3950; 6 in (150 mm)—$21,000; typical 1 in (25 mm) meter, with full-scale flow rate of 400 to 1000 lb/m (180 to 450 kg/m)—$5300.

A typical flowmeter comes standard with one pulse or frequency output that represents flow rate; one analog output configurable for flow rate, density, or temperature; and a display or digital output that provides flow rate, density, temperature, and flow total. In addition, most devices provide standard alarm outputs. The number and type of outputs vary from one manufacturer to another. Additional analog, frequency, pulse, and digital outputs are often provided as options.

PARTIAL LIST OF SUPPLIERS

Bailey Controls; Danfoss A/S (Denmark); Endress & Hauser Instruments; Exac Corp.; Fischer & Porter Co.; The Foxboro Co.; Heinrichs, K-Flow; Krohne, Bopp & Reuther; Micro Motion Inc.; Neptune Measurement Co.; Schlumberger Industries; Smith Meter Inc.

Coriolis flowmeters are not often used in wastewater applications. They are used on additive charging applications where the chemical is added on a weight basis or where

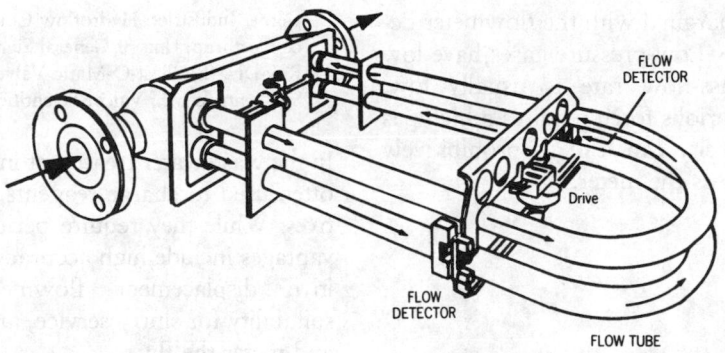

FIG. 7.6.5 Coriolis mass flowmeter.

their capability to detect both the mass flow and density of slurry streams is an advantage.

Since the appearance of the first commercial meters in the late 1970s, Coriolis flowmeters (see Figure 7.6.5) have become widely used. Their ability to measure mass flow directly with high accuracy and rangeability and to measure a variety of fluids makes Coriolis flowmeters the preferred flow measurement instrument for many applications. Coriolis flowmeters are also capable of measuring process fluid density and temperature. Since Coriolis flow measurement is a relatively new technology, many of the subtleties of its operation are still being investigated.

ADVANTAGES

Coriolis flowmeters have the following advantages:

1. They are capable of measuring a range of fluids that are often incompatible with other flow measurement devices. The operation of the flowmeter is independent of the Reynolds number; therefore, extremely viscous fluids can also be measured. A Coriolis flowmeter can measure the flow rate of Newtonian fluids, all types of nonNewtonian fluids, and slurries. Compressed gases and cryogenic liquids can also be measured by some designs.

2. Coriolis flowmeters provide a direct mass-flow measurement without the addition of external measurement instruments. While the volumetric flow rate of the fluid varies with changes in density, the mass-flow rate of the fluid is independent of density changes.

3. Coriolis flowmeters have outstanding accuracy. The base inaccuracy is commonly 0.2%. In addition, the flowmeters are extremely linear over their entire flow range.

4. The rangeability of the flowmeters is usually 20:1 or greater. Coriolis flowmeters have been successfully applied at flow rates 100 times lower than their rated full-scale flow rate.

5. A Coriolis flowmeter is capable of measuring mass-flow rate, volumetric flow rate, fluid density, and temperature—all from one instrument.

6. The operation of the flowmeter is independent of flow characteristics such as turbulence and flow profile. Therefore, upstream and downstream straight run requirements and flow conditioning are not necessary. They can also be used in installations that have pulsating flow.

7. Coriolis flowmeters do not have internal obstructions that can be damaged or plugged by slurries or other types of particulate matter in the flow stream. Entrained gas or slugs of gas in the liquid do not damage the flowmeter. The flowmeter has no moving parts that wear out and require replacement. These design features reduce the need for routine maintenance.

8. The flowmeter can be configured to measure flow in either the forward or reverse direction. In reverse flow, a time or phase difference occurs between the flow detector signals, but the relative difference between the two detector signals is reversed.

9. Coriolis flowmeter designs are available for use in sanitary applications and for the measurement of shear sensitive fluids. Materials are available that permit the measurement of corrosive fluids.

LIMITATIONS

Coriolis flowmeters have the following limitations:

1. They are not available for large pipelines. The largest Coriolis flowmeter has a maximum flow rating of 25,000 lb/min (11,340 kg/min) and is equipped with 6-in. (15-cm) flanges. Larger flow rates require more than one flowmeter mounted in parallel.

2. Some flowmeter designs require high fluid velocities to achieve significant time or phase difference between the flow detector signals. These high velocities can result in high pressure drops across the flowmeter.

3. Coriolis flowmeters are expensive. However, the cost of a Coriolis meter is often comparable to (or below) the cost of a volumetric meter plus a densitometer used together to determine the mass-flow rate.

4. Coriolis flowmeters have difficulty measuring the flow rate of low-pressure gas. Applications with pressures

less than 150 psig are marginal with the flowmeter designs currently available. Low-pressure gases have low density, and their mass-flow rate is usually low. Generating sufficient Coriolis force requires a high gas velocity. This high velocity can lead to prohibitively high pressure drops across the meter.

Metering Pumps

TYPES

A. Peristaltic
B. Piston or plunger types (provided with packing glands)
C. Diaphragm or glandless types (mechanical, hydraulic, double-diaphragm, and pulsator designs)

CAPACITY

A. 0.0005 cc/min to 20 gpm (90 lpm)
B. 0.001 gph to 280 gpm (0.005 lph to 1250 lpm)
C. Mechanical diaphragms: from 0.01 to 50 gallons per hour (gph) (0.05 to 3.7 lpm); mechanical bellows: from 0.01 to 250 gph (0.05 to 18 lpm); and others: from 0.01 to 800 gph (0.05 liters per hour (lph) to 60 lpm); pulsator pumps: from 30 to 1800 gph (2 to 130 lpm)

ERROR (INACCURACY)

A. ±0.1 to ±0.5% of full scale over a 10:1 range
B & C. ±0.25 to ±1% of full scale over a 10:1 range; can be as good as ±0.1% full scale at 100% stroke and tends to drop as stroke is reduced

MAXIMUM DISCHARGE PRESSURE

A. 50 psig (3.5 bars)
B. 50,000 psig (3450 bars)
C. Mechanical bellows: up to 75 psig (5 bars); mechanical diaphragm: up to 125 psig (8.5 bars); hydraulic Teflon diaphragm: 1500 psig (104 bars); pulsator pumps: up to 5000 psig (345 bars); and hydraulic metallic diaphragms: up to 40,000 psig (2750 bars)

MAXIMUM OPERATING TEMPERATURE

A. 70 to 600°F (−57 to 315°C)
B. Jacketed designs: up to about 500°F (260°C)
C. Units containing hydraulic fluids can handle from −95 to 360°F (−71 to 182°C), Teflon and Viton diaphrams are limited to 300°F (150°C), and neoprene and Buna N are limited to 200°F (92°C). The metal bellows and the remote head designs can operate from cryogenic to 1600°F (870°C).

MATERIALS OF CONSTRUCTION

A. Neoprene, Tygon, Viton, and silicone
B. Cast iron, steel, stainless steel, Hastelloy C, Alloy 20, Carpenter 20, Monel, nickel, titanium, glass, ceramics, Teflon, polyvinyl chloride (PVC), Kel-F, Penton, polyethylene, and other plastics
C. Polyethylene, Teflon, PVC, Kel-F, Penton, steel, stainless steel, Carpenter 20, Monel, Hastelloy B & C

COST

A. $200 to $800
B. $1000 to $6000
C. $1000 to $12,000

PARTIAL LIST OF SUPPLIERS

American LEWA Inc. (A,B,C); Barnant Co. (A); Blue White Industries; Bran & Luebbe Inc.; Clark-Cooper Corp. (B,C); Cole-Parmer Instrument Co.; Flo-Tron Inc. (B); Fluorocarbon Co.; Gerber Industries; Hydroflow Corporation; LDC Analytical; Leeds & Northrup, Unit of General Signal; Liquid Metronics Inc. Milton Roy Div. (B); Plast-O-Matic Valves Inc.; Ruska Instrument Corp.; S J Controls Inc.; Valcor Scientific; Wallace & Tiernan Inc. (B,C)

In the wastewater treatment industry, metering pumps are often used to charge reagents, coagulants, or other additives. While they require periodic recalibration, their advantages include high accuracy (similar to turbine or positive displacement flowmeters), high rangeability, suitability for slurry service, and the ability to both pump and meter the fluid.

Orifices

DESIGN PRESSURE

For plates, limited by the readout device only; integral orifice transmitter to 1500 psig (10.3 MPa)

DESIGN TEMPERATURE

Function of the associated readout system when the differential pressure unit must operate at the elevated temperature. For the integral orifice transmitter, the standard range is −20 to 250°F (−29 to 121°C).

SIZES

Maximum size is the pipe size.

FLUIDS

Liquids, vapors, and gases

FLOW RANGE

From a few cc/min using integral orifice transmitters to any maximum flow; limited only by pipe size

MATERIALS OF CONSTRUCTION

No limitation on plate materials. Integral orifice transmitter wetted parts can be obtained in steel, stainless steel, Monel, nickel, and Hastelloy.

INACCURACY

The orifice plate, if the bore diameter is correctly calculated and prepared, can be accurate to ±0.25 to ±0.5% of the actual flow. When a conventional d/p cell is used to detect the orifice differential, that adds a ±0.1 to ±0.3% of the full-scale error. The error contribution of smart d/p cells is only 0.1% of the actual span.

INTELLIGENT D/P CELLS

Inaccuracy of ±0.1%, rangeability of 40:1, the built-in proportional integral and derivative (PID) algorithm

RANGEABILITY

If rangeability is defined as the flow range within which the combined flow measurement error does not exceed ±1% of the actual flow, then the rangeability of conventional orifice installations is 3:1. When intelligent transmitters with automatic switching capability between the high and low spans are used, the rangeability can approach 10:1.

COST

A plate only is $50 to $300, depending on size and materials. For steel orifice flanges from 2 to 12 in (50 to 300 mm), the cost ranges from $200 to $1000. For flanged meter runs in the same size range, the cost ranges from $400 to $3000. The cost of electronic or pneumatic integral orifice transmitters is between $1500 and $2000. The cost of d/p transmitters ranges from $900 to $2000, depending on type and intelligence.

PARTIAL LIST OF SUPPLIERS

ABB Kent-Taylor Inc. (includes integral orifices); Crane Manufacturing Inc.; Daniel Flow Products Inc. (orifice plates and plate changers); Fischer & Porter Co. (includes integral orifices); Fluidic Techniques, a Div. of FTI Industries; Foxboro Co. (includes integral orifices); Honeywell Industrial Div.; Lambda Square Inc.; Meriam Instrument, Div. Scott & Fetzer (orifice plates); Rosemount Inc.; Vickery-Simms, a Div. of FTI Industries. In addition, orifice plates, flanges, and accessories can be obtained from most major instrument manufacturers.

The orifice plate, when installed in a pipeline, causes an increase in flow velocity and a corresponding decrease in pressure. The flow pattern shows an effective decrease in the cross-section beyond the orifice plate, with a maximum velocity and minimum pressure at the vena contracta (see Figure 7.6.6). This location can be from .35 to .85 pipe diameters downstream from the orifice plate depending on the β ratio and the Reynolds number.

This flow pattern and the sharp leading edge of the orifice plate (see Figure 7.6.6) that produces it are important. The sharp edge results in an almost pure line contact between the plate and the effective flow, with negligible fluid-to-metal friction drag at this boundary. Any nicks, burrs, or rounding of the sharp edge can result in large measurement errors.

When differential pressure is measured at a location close to the orifice plate, friction effects between the fluid and the pipe wall upstream and downstream from the orifice are minimized so that pipe roughness has a minimum effect. Fluid viscosity, as reflected in the Reynolds number, has a considerable influence, particularly at low Reynolds numbers. Since the formation of the vena contracta is an inertial effect, a decrease in the ratio of inertial to frictional forces (decrease in Reynolds number), and the corresponding change in flow profile, results in less constriction of flow at the vena contracta and an increase of the flow coefficient. In general, the sharp edge orifice plate should not be used at pipe Reynolds numbers under 10,000. The minimum recommended Reynolds number varies from 10,000 to 15,000 for 2-in (50-mm) through 4-in (102-mm) pipe sizes for β ratios up to 0.5 and from 20,000 to 45,000 for higher β ratios. The Reynolds number requirement increases with pipe size and β ratio and can range up to 200,000 for pipes 14 in (355 mm) and larger. Maximum Reynolds numbers can be 10^6 for 4-in (102-mm) pipe and 10^7 for larger sizes.

WASTEWATER APPLICATIONS

If the water is dirty, containing solids or sludge, the pressure taps must be protected by clean water purging or by use of chemical seals and the orifice plates should be the segmental or eccentric orifice type (see Figure 7.6.7). Annular orifices and V-cone meters are also applicable to dirty services. Because the pressure recovery of orifices is low, they are not recommended to measure larger flows

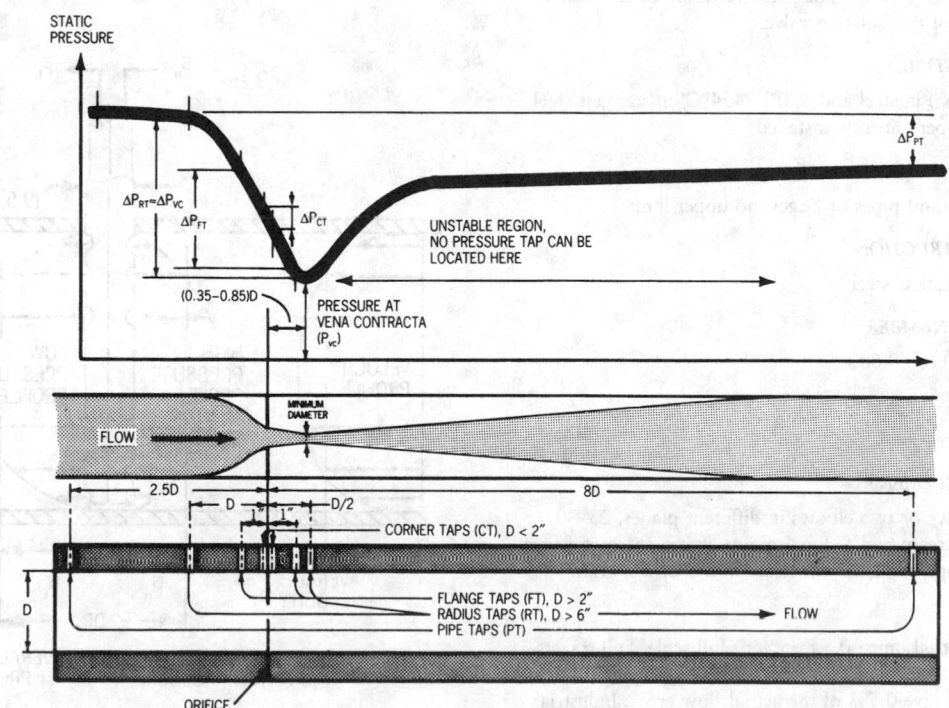

FIG. 7.6.6 Pressure profile through an orifice plate and the different methods of detecting the pressure drop.

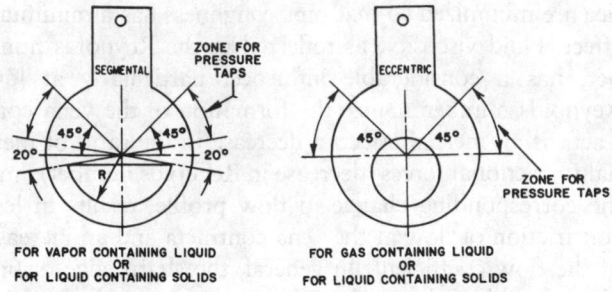

FIG. 7.6.7 Segmental and eccentric orifice plates.

due to the excessive pumping costs. In these applications, venturi-type, high-recovery flow elements should be used.

The main advantages of orifices are their familiarity, simplicity, and the fact that they do not need calibration. The disadvantages include their low rangeability, low accuracy, high pressure drop, and potential plugging.

Pitot Tubes

TYPES

A. Standard, single-port
B. Multiple-opening, averaging
C. Area averaging for ducts

APPLICATIONS

Liquids, gases, and steam

OPERATING PRESSURE

Permanently installed carbon or stainless steel units can operate at up to 1400 psig (97 bars) at 100°F (38°C) or 800 psig (55 bars) at approximately 700°F (371°C). The pressure rating of retractable units is a function of the isolating valve.

OPERATING TEMPERATURE

Up to 750°F (399°C) in steel and 850°F (454°C) in stainless steel construction when permanently installed

FLOW RANGES

Generally 2-in (50-mm) pipes or larger; no upper limit

MATERIALS OF CONSTRUCTION

Brass, steel, and stainless steel

MINIMUM REYNOLDS NUMBER

Range from 20,000 to 50,000

RANGEABILITY

Same as orifice plates

STRAIGHT-RUN REQUIREMENTS

Downstream of valve or two elbows in different planes, 25–30 pipe diameters upstream and 5 downstream; if straightening vanes are provided, 10 pipe diameters upstream and 5 downstream

INACCURACY

For standard industrial units: 0.5 to 5% of full scale. Full-traversing Pitot Venturis under National-Bureau-of-Standards-type laboratory conditions can give 0.5% of the actual flow error. Industrial Pitot Venturis must be individually calibrated to obtain 1% of range performance. Inaccuracy of individually calibrated multiple-opening averaging pitot tubes is claimed to be 2% of the range

when the Reynolds numbers exceed 50,000. Area-averaging duct units are claimed to be between 0.5 and 2% of the span. The error of the d/p cell is additional to the errors listed.

COSTS

A 1-in-diameter averaging pitot tube in stainless steel costs $750 if fixed and $1400 if retractable for hot-tap installation. The cost usually doubles if the pitot tube is calibrated. Hastelloy units for smokestack applications can cost $2000 or more. A local pitot indicator costs $400; a d/p transmitter suited for pitot applications with 4 to 20 mA dc output costs about $1000.

PARTIAL LIST OF SUPPLIERS

ABB Kent-Taylor Inc. (A); Air Monitor Corp. (C); Alnor Instrument Co. (A); Andersen Instruments Inc. (A); Blue White Industries (A); Brandt Instruments (C); Davis Instrument Mfg. Co. (A); Dietrich Standard, a Dover Industries Company (Annubar—B); Dwyer Instruments Inc. (B); Fischer & Porter Inc. (A); Foxboro Co. (Pitot Venturi—A); Land Combustion Inc. (A); Meriam Instrument, a Scott Fetzer Company (B); Mid-West Instrument (Delta Tube—B); Preso Industries (Elliptical—B); Sirco Industries Ltd. (A); Ultratech Industries Inc. (A); United Electric Controls Co. (A)

While pitot sensors are low-accuracy and low-rangeability detectors, they do have a place in wastewater treatment-related flow measurement. Pitot tubes should be used when the measurement is not critical, the water is reasonably clean, and a low cost measurement is needed. These sensors can be inserted in the pipe without shutdown and can also be removed for periodic cleaning while the pipe is in use. Multiple-opening pitot tubes (see Figure 7.6.8) are less sensitive to flow velocity profile variations than single-opening (see Figure 7.6.9) tubes. In some dirtier applications, purged pitot tubes are also used.

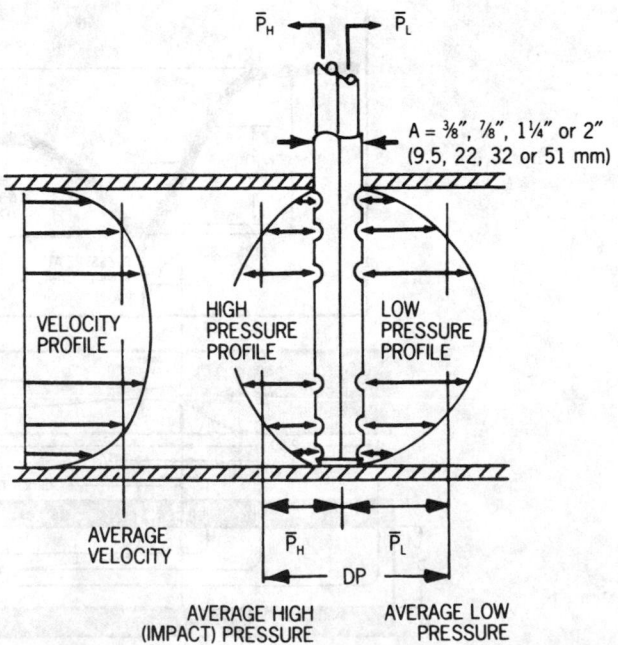

FIG. 7.6.8 The design of an averaging pitot tube. (Reprinted, with permission, from Dietrich Standard, a Dover Industries Company.)

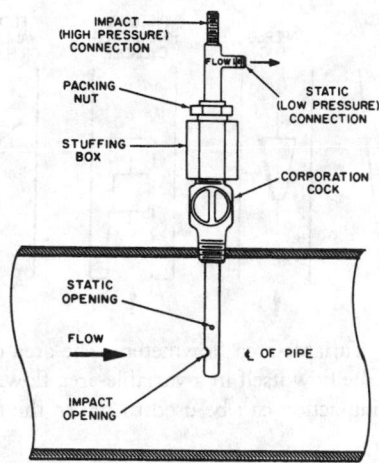

FIG. 7.6.9 Schematic diagram of an industrial device for sensing static and dynamic pressures in a flowing fluid.

Segmental Wedge Flowmeters

APPLICATIONS

Clean, viscous liquids or slurries and fluids with solids

SIZES

1- to 12-in (25.4- to 305-mm) diameter pipes

DESIGNS

For smaller sizes (1 and 1.5 in), the wedge can be integral; for larger pipes, remote seal wedges are used with calibrated elements.

WEDGE OPENING HEIGHT

From 0.2 to 0.5 of the inside pipe diameter

PRESSURE DROPS

25 to 200 in H_2O (6.2 to 49.8 kPa)

MATERIALS OF CONSTRUCTION

Carbon or stainless steel element; stainless or Hastelloy C seal; special wedge materials like tungsten carbide are available.

DESIGN PRESSURE

300 to 1500 psig (20.7 to 103 bars) with remote seals

DESIGN TEMPERATURE

−40 to 700°F (−40 to 370°C) but also used in high-temperature processes up to 850°F (454°C)

INACCURACY

The elements are individually calibrated; the d/p cell error contribution to the total measurement inaccuracy is 0.25% of full scale. The error over a 3:1 flow range is usually not more than 3% of the actual flow.

COST

A 3-in (75-mm) calibrated stainless steel element with two stainless steel chemical tees and an electronic d/p transmitter provided with remote seals is about $3500.

PARTIAL LIST OF SUPPLIERS

ABB Kent-Taylor Inc.

The segmental wedge flow element provides a flow opening similar to that of a segmental orifice, but flow ob-

struction is less abrupt (more gradual), and its sloping entrance makes the design similar to the flow tube family. It is primarily used on slurries. Its main advantage is its ability to operate at low Reynolds numbers. While the square root relationship between the flow and pressure drop in sharp-edged orifices, venturis, or flow nozzles requires a Reynolds number above 10,000, segmental wedge flowmeters require a Reynolds number of only 500 or 1000. For this reason the segmental wedge flowmeter can measure flows at low flow velocities and when process fluids are viscous. In that respect, it is similar to conical or quadrant edge orifices.

For pipe sizes under 2 in (50 mm), the segmental wedge flow element is made by a V-notch cut into the pipe and a solid wedge welded accurately in place (see Figure 7.6.10). In sizes over 2 in, the wedge is fabricated from two flat plates that are welded together before insertion into the spool piece. On clean services, regular pressure taps are located equidistant from the wedge (see Figure 7.6.10), while on applications where the process fluid contains solids in suspension, chemical tees are added upstream and downstream of the wedge flow element. The chemical seal element is flush with the pipe, eliminating pockets and making the assembly self-cleaning. The seals are made of corrosion-resistant materials and are also suited for high-temperature services. Some users have reported applications on processes at 3000 psig (210 bars) and 850°F (454°C).

Variable-Area Flowmeters

Variable-area flowmeters are used to regulate purge flow and as flow indicators or transmitters.

PURGE FLOWMETER

One variety of variable-area flowmeters is the purge flowmeter (see Figure 7.6.11). The features and characteristics of these instruments are summarized next.

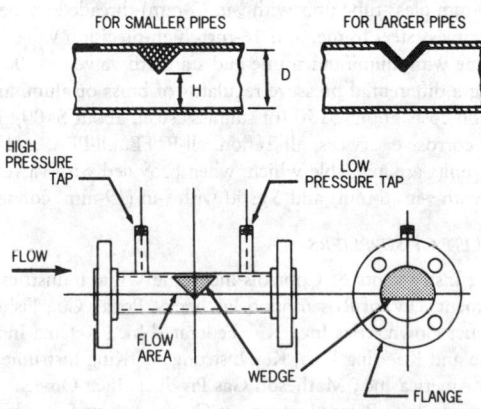

FIG. 7.6.10 The segmental wedge flowmeter designed for clean fluid service.

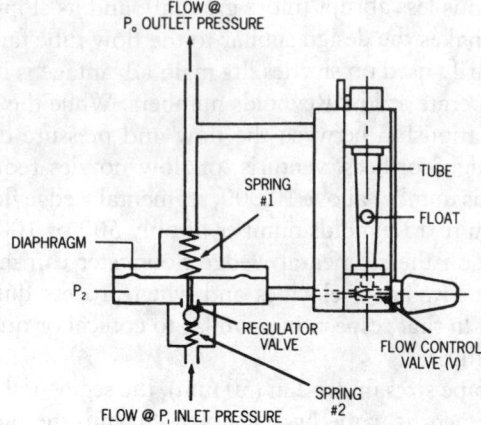

FIG. 7.6.11 A purge flow regulator consisting of a glass tube rotameter, an inlet needle valve, and a differential pressure regulator. (Reprinted, with permission, from Krone America Inc.)

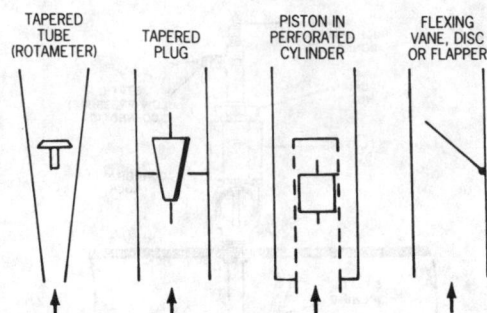

FIG. 7.6.12 Variable-area flowmeters. The area open to flow is changed by the flow itself in a variable-area flowmeter. Either gravity or spring action can be used to return the float or vane as flow drops.

VARIABLE-AREA FLOWMETERS

In the wastewater treatment industry, variable-area flowmeters are also used as flow indicators or transmitters if the process fluid is clean. Figure 7.6.12 shows their operating principles, and their features and capabilities are listed next.

TYPES

A. Rotameter (float in tapered tube)
B. Orifice/rotameter combination
C. Open-channel variable gate
D. Spring and vane or piston

STANDARD DESIGN PRESSURE

A. 350 psig (2.4 MPa) average maximum for glass metering tubes, dependent on size. Up to 720 psig (5 MPa) for metal tubes and special designs to 6000 psig (41 MPa)

STANDARD DESIGN TEMPERATURE

A. Up to 400°F (204°C) for glass tubes and up to 1000°F (538°C) for some models of metal tube meters

END CONNECTIONS

Female pipe thread or flanged

FLUIDS

Liquids, gases, and vapors

FLOW RANGE

A. 0.01 cc/min to 4000 gpm (920 m³/hr) of liquid
0.3 cc/min to 1300 scfm (2210 m³/hr) of gas

INACCURACY

A. Laboratory rotameters can be accurate to ±½% of actual flow; most industrial rotameters perform within ±1 to 2% of full scale over a 10:1 range, and purge or bypass meters perform within ±5 to 10% of full range.
B and D. ±2 to ±10% of full range
C. ±7.5% of actual flow

MATERIALS OF CONSTRUCTION

A. TUBE: Borosilicate glass, stainless steel, Hastelloy, Monel, and Alloy 20. FLOAT: *Conventional type*—brass, stainless steel, Hastelloy, Monel, Alloy 20, nickel, titanium, or tantalum, and special plastic floats. *Ball type*—glass, stainless steel, tungsten carbide,

APPLICATIONS

Low flow regulation for air bubblers, for purge protection of instruments, for purging electrical housings in explosion-proof areas, and for purging the optical windows of smokestack analyzers

PURGE FLUIDS

Air, nitrogen, and liquids

OPERATING PRESSURE

Up to 450 psig (3 MPa)

OPERATING TEMPERATURE

For glass tubes up to 200°F (93°C)

RANGES

From 0.01 cc/min for liquids and from 0.5 cc/min and higher for gases. A ¼-in (6-mm) glass tube rotameter can handle 0.05 to 0.5 gpm (0.2 to 2 lpm) of water or 0.2 to 2 scfm (0.3 to 3 cmph) of air

INACCURACY

Generally 2 to 5% of the range (laboratory units are more accurate)

COSTS

A 150-mm glass-tube unit with ⅛-in (3-mm) threaded connection, 316 stainless steel frame, and 16-turn high-precision valve is $260; the same with aluminum frame and standard valve is $100. Adding a differential pressure regulator of brass or aluminum construction costs about $150 (of stainless steel, about $500). For highly corrosive services, all-Teflon, all-PTFE, all-PFA, and all-CTFA units are available which, when provided with valves, cost $550 with ¼-in (6-mm) and $1300 with ¾-in (19-mm) connections.

PARTIAL LIST OF SUPPLIERS

Aaborg Instruments & Controls Inc.; Blue White Industries; Brooks Instrument, Div. of Rosemount; Fischer & Porter Co.; Fisher Scientific; Flowmetrics Inc.; ICC Federated Inc.; Ketema Inc. Schutte and Koerting Div.; Key Instruments; King Instrument Co.; Krone America Inc.; Matheson Gas Products Inc.; Omega Engineering Inc.; Porter Instrument Co. Inc.; Scott Specialty; Wallace & Tiernan Inc.

sapphire, or tantalum. END FITTINGS: Brass, stainless steel, or alloys for corrosive fluids. PACKING: The generally available elastomers are used and O-rings of commercially available materials; Teflon is also available.

COST

A $\frac{1}{4}$-in (6-mm) glass tube purge meter starts at $100. A $\frac{1}{4}$-in stainless steel meter is about $300. Transmitting rotameters start at about $1000; with 0.5% of rate accuracy, their costs are over $2000. A 3-in (75-mm) standard bypass rotameter is about $500; a 3-in stainless steel tube standard rotameter is about $2000. A 3-in tapered-plug variable-area meter in aluminum construction is about $1000; the same unit in spring and vane design is around $750.

PARTIAL LIST OF SUPPLIERS

Aaborg Instruments & Controls Inc. (A); Aquamatic Inc. (B); Blue White Industries (A); Brooks Instrument Div. of Rosemount (A); Dwyer Instruments Inc. (A); ERDCO Engineering Corp. (D); ESKO Industries Ltd. (A); Fischer & Porter Co. (A); Flowmetrics Inc. (A); Gilflo Metering & Instrumentation Inc. (D); Gilmont Instruments Div. of Barnant Co. (B); Headland Div. of Racine Federated Inc. (D); ICC Federated Inc. (A); ISCO Environmental Div. (C); Ketema Inc. Schutte and Koerting Div. (A); Key Instruments (A); King Instrument Co. (A); Kobold Instruments Inc.; Krone America Inc. (A); Lake Monitors Inc.; Matheson Gas Products Inc. (A); McMillan Co.; Meter Equipment Mfg. Inc. (D); Metron Technology (A); Omega Engineering Inc. (A); G. A. Planton Ltd. (D); Porter Instrument Co. Inc. (A); Turbo Instruments Inc. (D); Universal Flow Monitors Inc. (D); Wallace & Tiernan Inc. (A); Webster Instruments (D)

Venturi and Flow Tubes

DESIGN TYPES

A. Venturi tubes; B. Flow tubes; C. Flow nozzles

DESIGN PRESSURE

Usually limited only by the readout device or pipe pressure ratings

DESIGN TEMPERATURE

Limited only by the readout device if the operation is at very low or high temperature

SIZES

A. 1 in (25 mm) up to 120 in (3000 mm)
B. 4 in (100 mm) up to 48 in (1200 mm)
C. 1 in (25 mm) up to 60 in (1500 mm)

FLUIDS

Liquids, gases, and steam

FLOW RANGE

Limited only by minimum and maximum beta (β) ratio and available pipe size range

INACCURACY

Values given are for flow elements only; d/p cell and readout errors are additional.
A. ±0.25% of rate if calibrated in a flow laboratory and ±0.75% of rate if uncalibrated
B. Can range from ±0.5 to ±3% of rate depending upon the design and variations in fluid operating conditions
C. ±1% of rate when uncalibrated to ±0.25% when calibrated

MATERIALS OF CONSTRUCTION

Virtually unlimited. Cast venturi tubes are usually cast iron, but fabricated venturi tubes can be made from carbon steel, stainless steel, most available alloys, and fiberglass plastic composites. Flow nozzles are commonly made from alloy steel and stainless steel.

PRESSURE RECOVERY

90% of the pressure loss is recovered by a low-loss venturi when the beta (β) ratio is 0.3, while an orifice plate recovers only 12%. (The corresponding energy savings in a 24-in (600-mm) waterline is about 20 hp.)

REYNOLDS NUMBERS

Venturi and flow tube discharge coefficients are constant at Re > 100,000. Flow nozzles are used at high pipeline velocities (100 ft/sec or 30.5 m/sec), usually corresponding to Re > 5 million. Critical-flow venturi nozzles operate under choked conditions at sonic velocity.

COSTS

Flow nozzles are less expensive than venturi or flow tubes but cost more than orifices. American Society of Mechanical Engineers (ASME) gas flow nozzles in aluminum for 3- to 8-in (75- to 200-mm) lines cost from $200 to $750. Epoxy–fiberglass nozzles for 12- to 32-in (300- to 812-mm) lines cost from $750 to $2500. The relative costs of Herschel venturis and flow tubes in different sizes and materials are as follows:

	6-in Stainless Steel	8-in Cast Iron	12-in Steel
Herschel venturi	$8000	$5500	$6000
Flow tube	$3600	$2100	$2900

PARTIAL LIST OF SUPPLIERS

ABB Kent Taylor (B); Badger Meter Inc. (A,B); Bethlehem Corp. (B); BIF Products of Leeds & Northrup (A,B,C); Daniel Flow Products Inc. (A,C); Delta-T Co. (C); Digital Valve Co. (critical-flow venturi nozzles); Fielding Crossman Div. of Lisle Metrix Ltd. (A,C); Fischer & Porter Co. (B); Flow Systems Inc. (B); Fluidic Techniques Inc. (A); Fox Valve Development Corp. (A); F.B. Leopold Co. (A,B); Permutit Co. Inc. (A,C); Perry Equipment Corp. (B); Henry Pratt Co. (A,B); Preso Industries (A,B); Primary Flow Signal Inc. (A,C); STI Manufacturing Inc.; Tri-Flow Inc. (A); Vickery-Simms Div. of FTI Industries (A); West Coast Research Corp.

In applications where the flows of large volumes of water are measured, considerations of the measurement pumping costs often outweigh the initial cost of the sensor. Because the venturi flowmeters (see Figure 7.6.13) require less pressure drop than any other d/p-type flow sensor, their designs (see Figure 7.6.14) are frequently used in the wastewater treatment industry.

LIMITATIONS

The main limitation of venturi tubes is cost, both for the tube itself and often for the long piping required for the larger sizes. However, the energy cost savings attributable to their higher pressure recovery and reduced pressure loss usually justify the use of venturi tubes in larger pipes.

Another limitation is the high minimum Reynolds number required to maintain accuracy. For venturis and flow tubes, this minimum is around 100,000; while for flow

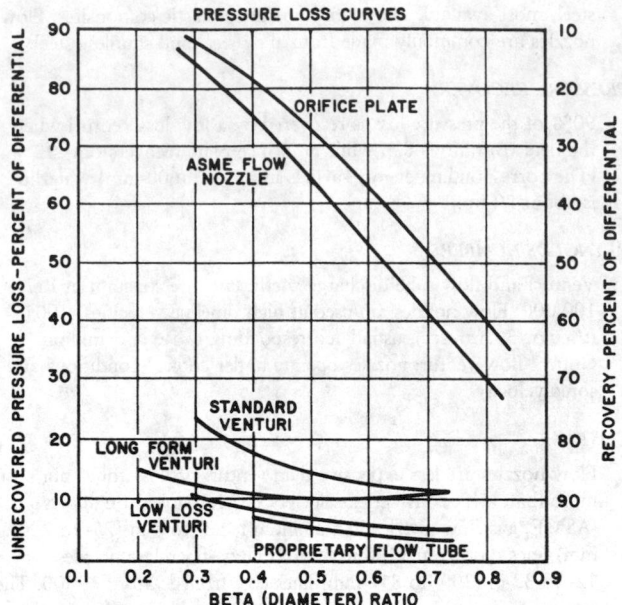

FIG. 7.6.13 Pressure loss curves.

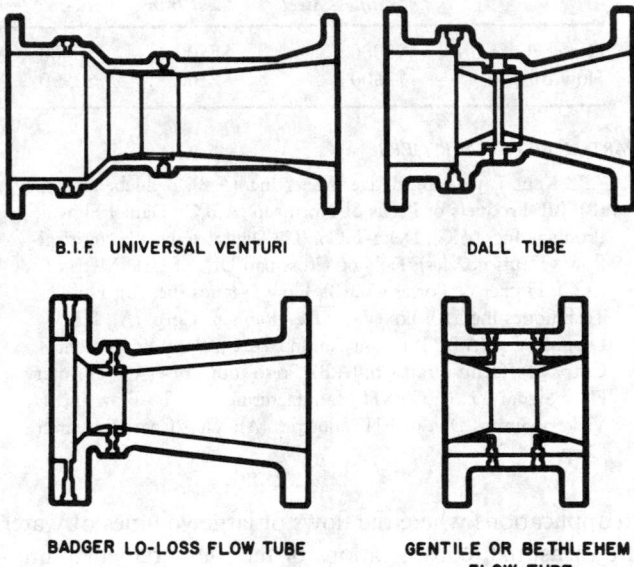

FIG. 7.6.14 Proprietary flow tubes.

nozzles, it is over 1 million. Correction factors are available for Reynolds numbers below these limits, and measurement performance also suffers.

Cavitation can also be a problem. At high flow velocities (corresponding to the required high Reynolds numbers) at the vena-contracta, static pressure is low, and when it drops below the vapor pressure of the flowing fluid, cavitation occurs. Cavitation destroys the throat section of the tube since no material can stand up to cavitation. Possible ways of eliminating cavitation include relocating the meter to a point in the process where the pressure is higher and the temperature is lower, reducing the pressure drop

across the sensor, or replacing the sensor with one that has less pressure recovery.

Due to their construction, venturis, flow tubes, and flow nozzles are difficult to inspect. Providing an inspection port on the outlet cone near the throat section can solve this problem. An inspection port is important when dirty (erosive) gases, slurries, or corrosive fluids are metered. On dirty services where the pressure ports are likely to plug, the pressure taps on the flow tube can be filled with chemical seals that have stainless steel diaphragms installed flush with the tube interior.

ADVANTAGES

The main advantages of these sensors include their high accuracy, good rangeability (on high Reynolds number applications), and energy-conserving high-pressure recovery. For these reasons, in higher velocity flows and larger pipelines (and ducts), many users still favor venturis in spite of their high costs. Their hydraulic shape also contributes to greater dimensional reliability and therefore to better flow-coefficient stability than that of orifice-type sensors, which depend on the sharp edge of the orifice for their flow coefficient.

The accuracy of a flow sensor is defined as the uncertainty tolerance of the flow coefficient. Calibration can improve accuracy. Table 7.6.2 gives accuracy data in percentage of actual flow, as reported by various manufacturers. These values are likely to hold true only for the stated ranges of beta ratios and Reynolds numbers, and they do not include the added error of the readout device or d/p transmitter.

Vortex Flowmeters

TYPES

   A. Vortex
   B. Fluidic shedding coanda effect
   C. Oscillating vane in orifice bypass

SERVICES

   A. Gas, steam, and clean liquids
   B and C. Clean liquids

SIZE RANGES AVAILABLE

   A. 0.5 to 12 in (13 to 300 mm), also probes
   B. 1 to 4 in (25 to 100 mm)
   C. 1 to 4 in (25 to 100 mm)

DETECTABLE FLOWS

   A. Water—2 to 10,000 gpm (8 lpm to 40 m³/hr)
   Air—3 to 12,000 scfm (0.3 to 1100 scmm)
   Steam (D&S at 150 psig [10.4 bars])—25 to 250,000 lbm/hr (11 to 113,600 kg/hr)
   B. Water—1 to 1000 gpm (4 to 4000 lpm)
   C. Water—5 to 800 gpm (20 to 3024 lpm)

FLOW VELOCITY RANGE

   A. Liquids—1 to 33 ft/sec (0.3 to 10 m/sec)
   Gas and steam—20 to 262 ft/sec (6 to 80 m/sec)

TABLE 7.6.2   VENTURI, FLOW TUBE, AND FLOW NOZZLE INACCURACIES (ERRORS) IN PERCENT OF ACTUAL
FLOW FOR VARIOUS RANGES OF BETA RATIOS AND REYNOLDS NUMBERS

Flow Sensor		Line Size in Inches (1 in = 25.4 mm)	Beta Ratio	Pipe Reynolds Number Range for Stated Accuracy	Inaccuracy in % of Actual Flow
Herschel standard	Cast[1]	4–32	.30–.75	2×10^5 to 1×10^6	±0.75%
	Welded	8–48	.40–.70	2×10^5 to 2×10^6	±1.5%
Proprietary true venturi	Cast[2]	2–96	.30–.75	8×10^4 to 8×10^6	±0.5%
	Welded	1–120	.25–.80	8×10^4 to 8×10^6	±1.0%
Proprietary flow tube	Cast[3]	3–48	.35–.85	8×10^4 to 1×10^6	±1.0%
ASME flow nozzles[4]		1–48	.20–.80	7×10^6 to 4×10^7	±1.0%

[1]No longer manufactured because of long laying length and high cost.
[2]Badger Meter Inc.; BIF Products of Leeds & Northrup; Fluidic Techniques, Inc.; F.B. Leopold Co.; Permutit Co., Inc.; Henry Pratt Co.; Primary Flow Signal, Inc.; Tri-Flow Inc.
[3]Badger Meter Inc.; Bethlehem Corp.; BIF Products of Leeds & Northrup; Fischer & Porter Co.; F.B. Leopold Co.; Henry Pratt Co.; Preso Industries.
[4]BIF Products of Leeds & Northrup; Daniel Flow Products, Inc.; Permutit Co., Inc.; Primary Flow Signal, Inc.

MINIMUM REYNOLDS NUMBERS

A. Under Reynolds number of 8000 to 10,000, meters do not function at all; for best performance, Reynolds number should exceed 20,000 in sizes under 4 in (100 mm) and 40,000 in sizes above 4 in.

B. Reynolds number = 3000

OUTPUT SIGNALS

A, B, and C. Linear pulses or analog

DESIGN PRESSURE

A. 2000 psig (138 bars)

B. 600 psig (41 bars) below 2 in (50 mm); 150 psig (10.3 bars) above 2 in

C. 300 psig (30.6 bars)

DESIGN TEMPERATURE

A. −330 to 750°F (−201 to 400°C)

B. 0 to 250°F (−18 to 120°C)

C. −14 to 212°F (−25 to 100°C)

MATERIALS OF CONSTRUCTION

A. Mostly stainless steel; some in plastic

B. 316 stainless steel with Viton A O-rings

C. Wetted body is Kynar, sensor is Hastelloy C

RANGEABILITY

A. Reynolds number at maximum flow divided by minimum Reynolds number of 20,000 or more

B. Reynolds number at maximum flow divided by minimum Reynolds number of 3000

C. 10:1 for Reynolds number at maximum flow divided by minimum Reynolds numbers of 14,000 for 1 in, 28,000 for 2 in, 33,000 for 3 in, and 56,000 for 4 in

INACCURACY

A. 0.5 to 1% of rate for liquids and 1 to 1.5% of rate for gases and steam with pulse outputs; for analog outputs, add 0.1% of full scale.

B. 1 to 2% of actual flow

C. 0.5% of full scale over 10:1 range

COST

A. Plastic and probe units cost about $1500; stainless steel units in small sizes cost about $2500; insertion-types cost about $3000.

C. The sensor with only unscaled pulse output in 1-, 2-, 3-, and 4-in sizes costs $535, $625, $875, and $1295, respectively. The additional cost of a scaler is $250 and of a 4–20 mA transmitter is $350.

PARTIAL LIST OF SUPPLIERS

ABB Kent (A); Alphasonics Inc. (A); Badger Meter Inc. (C—proximity switch sensor); Brooks Div. of Rosemount (A—ultrasonic); EMC Co. (A—dual piezoelectric sensor); Endress + Hauser Instruments (A—capacitance sensor); Fischer & Porter Co. (A—internal strain gauge sensor); Fisher Controls (A—dual piezoelectric sensor); Flowtec AG of Switzerland (A); Foxboro Co. (A—piezoelectric sensor); Johnson Yokogawa Corp. (A—dual piezoelectric sensor); J-Tec Associates Inc. (A—retractable design available, ultrasonic sensor); MCO/Eastech (A—including insertion-type, mechanical, thermal, or piezoelectric sensors); Moore Products Co. (B); Nice Instrumentation Inc. (A—dual piezoelectric sensor); Oilgear/Ball Products (A—vortex velocity); Sarasota Automation Inc. (A); Schlumberger Industries Inc. (A—dual piezometric sensor); Turbo Instruments Inc. (A); Universal Flow Monitors Inc. (A—plastic body, piezoelectric sensor); Universal Vortex (A—piezoelectric sensor)

Weirs and Flumes

TYPES

These devices measure open-channel flow by causing level variations in front of primaries. Bubblers, capacitance, float and hydrostatic and ultrasonic devices are used as level sensors. These devices can also measure open-channel flows without primaries by calculating the flow from depth and velocity data obtained from ultrasonic and magnetic sensors.

OPERATING CONDITIONS

Atmospheric

APPLICATIONS

Waste or irrigation water flows in open channels

FLOW RANGE

From 1 gpm (3.78 lpm)—no upper limit

RANGEABILITY

Most devices provide 75:1, V-notch weirs can reach 500:1.

INACCURACY

2 to 5%

COSTS

Primaries used as pipe inserts cost under $1000. A 6-in (150-mm) Parshall flume costs about $1500, and a 48-in (1.22-m) Parshall flume costs about $5000. Primaries for irrigation applications are usually field-fabricated. Manual depth sensors can be obtained for $200; local bubbler or float indicators for $750 to $1500; and programmable transmitting capacitance, ultrasonic, or bubbler units from $1800 to $3000. Open-channel flowmeters, when calculating flow based on depth and velocity, range from $5000 to over $10,000.

PARTIAL LIST OF SUPPLIERS

ABB Kent Taylor Inc. (primaries); American Sigma Inc. (bubbler); Badger Meter Inc. (Parshall or manhole flume, ultrasonic and open-channel computing); Bernhar Inc. (ultrasonic for partially filled pipes); Bestobell/Mobrey (ultrasonic); BIF Unit of Leeds & Northrup (primary and detector); Drexelbrook Engineering Co. (capacitance for flumes); Endress + Hauser Inc. (ultrasonic and capacitance); Fischer & Porter Co. (ultrasonic); Free Flow Inc. (primaries); Greyline Instruments Inc. (ultrasonic); Inventron Inc. (ultrasonic); ISCO Inc. (bubbler, hydrostatic, and ultrasonic); Key-Ray/Sensall Inc. (ultrasonic); Leeds & Northrup BIF (flow nozzles); Leupold & Stevens Inc. (float); Manning Environmental Corp. (primaries); Marsh-Mcbirney Inc. (electromagnetic); Mead Instruments Corp. (velocity probe); Milltronics Inc. (ultrasonic); Minitek Technologies Inc. (open-channel magmeter and ultrasonic); Montedoro-Whitney Corp. (open-channel flow by ultrasonics); MSR Magmeter Mfg. Ltd. (robotic magmeter probe for open channel); N.B. Instruments Inc. (computer monitoring of sewers); Plasti-Fab Inc. (primaries); Princo Instruments Inc. (capacitance); J.L. Rochester Co. (manual depth sensor); Sparling Instruments Co. (primaries); TN Technologies Inc. (ultrasonic)

In the wastewater treatment industry, the flow in large, open pipes or channels must be measured. The weir and flume designs, particularly the Parshall flume (see Figure 7.6.15), make such measurements. The common feature of all these flow sensors is that they detect the level rise in front of a restriction in the flow channel.

DETECTORS FOR OPEN-CHANNEL SENSORS

The level rise generated by flumes or weirs can be measured by any level detector including simple devices such as air bubblers.

The flow in open channels can also be detected without using flumes, weirs, or any other primary devices. One such design computes flow in round pipes or open channels by ultrasonically measuring the depth, calculating the flowing cross-sectional area on that basis, and multiplying the area by the velocity to obtain volumetric flow (see Figure 7.6.16).

Another open-channel flowmeter that does not need a primary element uses a robotically operated magnetic flowmeter probe to scan the velocity profile in the open channel (see Figure 7.6.17). In this design, the computer

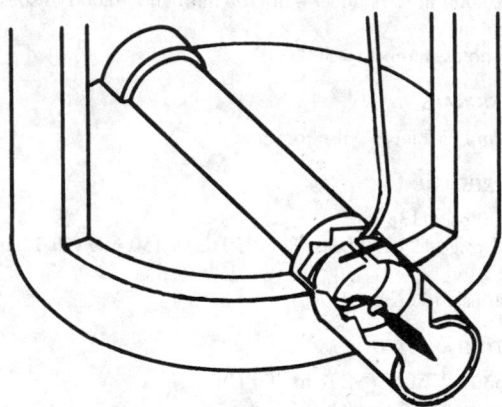

FIG. 7.6.16 Volumetric flow computer measuring depth and velocity in an open channel without a primary device. (Reprinted, with permission, from Montedoro-Whitney Corp.)

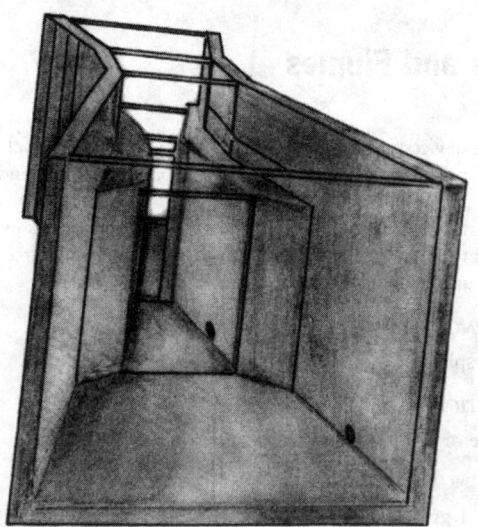

FIG. 7.6.15 Dual-range Parshall flume. (Reprinted, with permission, from Fischer & Porter Co.)

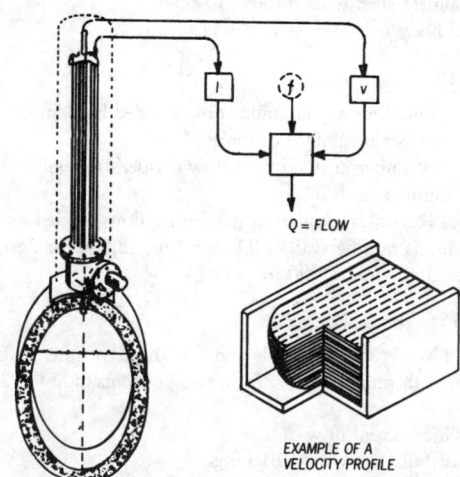

FIG. 7.6.17 Robotically operated magmeter probe sensor used to compute channel flow. (Reprinted, with permission, from MSR Magmeter Mfg. Ltd.)

algorithm separately calculates and adds the flow segments through each slice of the velocity profile as the velocity sensor moves down to the bottom of the channel.

Level Sensors

Most level sensors used in the wastewater industry do not need to be very accurate; reliable operation, rugged design, and low maintenance are more important. For these reasons, the newer level detector designs (laser, microwave, radar, gamma radiation, and time-domain reflectometry types) are seldom used. Similarly, the designs that use mechanical motion (float, displacer, or tape designs) are used infrequently since the solid-state or force-balance designs are more maintenance free.

On clean water level applications for local level indication, reflex-type level gauges, resistance tapes, and bubbler gauges are used most often. For high- and low-level switches, conductivity, capacitance, vibrational, ultrasonic and thermal level switches are used. For level transmitter applications, d/p and ultrasonic designs are often used.

For dirty or sludge-type level measurement, extended-diaphragm-type or purged d/p sensors, capacitance probes, and ultrasonic detectors are usually used. Lately, electronic load cells have also been used to detect the level on the basis of weight measurement in some larger tanks. For sludge or oil interface detection, ultrasonic, optical, vibrational, thermal, and microwave level switches work well. Table 7.6.3 provides an overall summary of the features and capabilities of all level measuring devices.

INTERFACE MEASUREMENT

When detecting the interface between two liquids, the measurement can be based on the difference of densities, dielectric constants, electric or thermal conductivities, opacity, or the sonic and ultrasonic transmittance of the two fluids. Environmental engineers should base their measurement on the process property with the largest step change between the upper and lower fluids. If, instead of a clean interface, a rag layer (a mix of the two fluids) exists between the two fluids, the interface detector cannot change that fact (it cannot eliminate the rag layer); but if properly selected, the interface detector can signal its beginning and end and thereby measure its thickness.

Interface level switches are usually ultrasonic, optical (Figure 7.6.18), capacitance, float, conductivity, thermal, microwave, or radiation designs. The ultrasonic switch described in Figure 7.6.19 uses a gap-type probe installed at a 10-degree angle from the horizontal. At one end of the gap is the ultrasonic source, at the other end is the receiver. As long as the probe is in the upper or lower liquid, the detector receives the ultrasonic pulse.

When the interface enters the gap, the pulse is deflected, and the switch is actuated. This switch can detect the interface between water and oil or other hydrocarbons, such as vinyl-acetate. If the thickness of the light layer rather than the location of the interface in the tank is of interest, the ultrasonic gap sensor can be attached to a float as shown in Figure 7.6.20.

Continuous measurement of the interface between two liquids can be detected by d/p transmitters if P_1 is detected in the heavy liquid and P_2 in the light liquid. In atmospheric vessels, three bubbler tubes can achieve the same interface measurement. The configuration shown in Figure 7.6.21 is appropriate for applications where the density of the light layer is constant and the density of the heavy liquid is variable. In these differential pressure-type systems, the movement of the interface level must be large enough to cause a change that satisfies the minimum span of the d/p transmitters. If the difference between the dielectric constants is substantial, such as in crude oil desalting, capacitance probes can also serve as continuous interface detectors.

On clean services, float- and displacer-type sensors can also be used as interface level detectors. For float-type units a float density heavier than the light layer but lighter than the heavy layer must be selected. In displacer-type sensors, the displacer must always be flooded, the upper connection of the chamber must be in the light liquid layer, and the lower connection must be in the heavy liquid layer. In this arrangement, the displacer becomes a density sensor. Therefore, the smaller the difference between the densities of the fluids and the smaller the range within which the interface can move, the larger displacer diameter will be required. Displacer density can be the same or more than that of the heavy layer.

Bubblers

APPLICATIONS

Usually local indicator on open tanks containing corrosive, slurry, or viscous process liquids. Can also be used on pressurized tanks but only up to the pressure of the air supply.

OPERATING PRESSURE

Usually atmospheric.

OPERATING TEMPERATURE

Limited only by pipe material; purging has also been used on high-temperature, fluidized-bed combustion processes to detect levels.

MATERIALS

Any pipe material available.

COSTS

$100 to $500 depending on accessories.

INACCURACY

Depends on readibility of pressure indicator, usually ±0.5% to 2% of full scale.

RANGE

Unlimited.

TABLE 7.6.3 ORIENTATION TABLE FOR LEVEL DETECTORS

Type	Level Range	Maximum Temperature (°F) °C = (°F − 32)/1.8	Available as Noncontact	Inaccuracy (1 in = 25.4 mm)	Cost Under $1000	Cost $1000–$5000	Cost Over $5000	Switch	Local Indicator	Transmitter	Clean	Viscous	Slurry/Sludge	Interface	Foam	Powder	Chunky	Sticky	Limitations
Air bubblers	—	UL		1–2% FS	√				√	√	G	F	P	F					Introduces foreign substance to process; high maintenance
Capacitance	—	2000	√	1–2% FS	√				√	√	G	F-G	F	G-L	P	P	F	P	Problem with interface between conductive layers and detection of foam
Conductivity switch (Point sensor)		1800		⅛ in	√			√			F	P	F	L	L	L	L	L	Can detect interface only between conductive and nonconductive liquids; field effect design for solids
Diaphragm	—	350		0.5% FS	√			√			G	F	F			F	F	P	Switches only for solids service
Differential pressure	—	1200		0.1% AS	√				√	√	E	G-E	G	P					Plugging eliminated by only extended diaphragm seals or repeaters. Purging and sealing legs also used
Displacer	—	850		0.5% FS	√			√	√	√	E	P	P	F-G					Not recommended for sludge or slurry service
Float	—	500		1% FS	√			√	√	√	G	P	P	F					Most designs limited by moving parts to clean service. Only preset density floats following interfaces
Laser	—	UL	√	0.5 in			√	√		√	L	G	G		F	F	F	F	Limited to cloudy liquids or bright solids in tanks with transparent vapor spaces
Level gauges (Point sensor)		700		0.25 in	√				√		G	F	P	F		G	G	G	Glass not allowed in some processes
Microwave switch (Point sensor)		400	√	0.5 in	√			√			G	G	G	G		G	G	F	Thick coating
Optical switches (Point sensor)		260	√	0.25 in	√			√			G	F	F	F-G	F	F	P	F	Refraction-type for clean liquids only; reflection-type requires clean vapor space
Radar	—	450	√	0.12 in			√		√	√	E	P	F	P	F	P	P	P	Interference from coating, agitator blades, spray, or excessive turbulence
Radiation	—	UL	√	0.25 in			√	√	√	√	G	E	E	G	E	F	G	E	Requires a Nuclear Regulatory Commission (NRC) license

Level Range scale (in feet): 1 3 6 12 24 48 96 100 150 200
Level Range scale (in meters): 0.3 1 2 4 8 16 32 34 50 67

Technology		Number	Accuracy									Limitations
Resistance tape		225	0.5 in		√	√	G	G	G	G		Limited to liquids under near-atmospheric pressure and temperature conditions
Rotating paddle switch	Point sensor	500	1 in	√	√	√				G	F P	Limited to detection of dry, noncorrosive, low-pressure solids
Slip tubes		200	0.5 in	√	√	√	F	P	P			An unsafe manual device
Tape-type level sensors		300	0.1 in		√	√	E	F	P	G	G F F	Only the inductively coupled float suited for interface measurement. Float hangup a potential problem with most designs
Thermal		850	0.5 in	√	√	√	G	F	F	P	F	Foam and interface detection limited by the thermal conductivities involved
TDR/PDS		221	3 in	√	√	√	F	F	F		G G	Limited performance on sticky process materials
Ultrasonic		300	1% FS	√	√	√	F–G	G	G	F–G	F F G	Presence of dust, foam, dew in vapor space; performance limited by sloping or fluffy process material
Vibrating switches	Point sensor	300	0.2 in	√	√	√	F	G	G	F	F G G	Operation limited by excessive material buildup can prevent

TDR = Time Domain Reflectometry
PDS = Phase Difference Sensors
AS = in % of actual span
E = Excellent
F = Fair
FS = in % of full scale
G = Good
L = Limited
P = Poor
UL = Unlimited

573

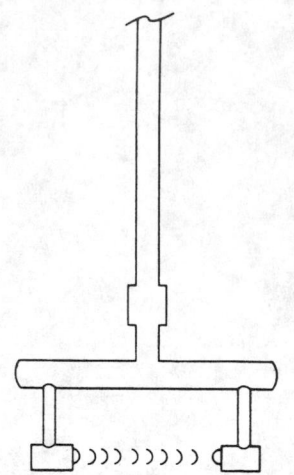

FIG. 7.6.18 Optical or ultrasonic sludge level or interface switch. (Courtesy of Sensall Inc.)

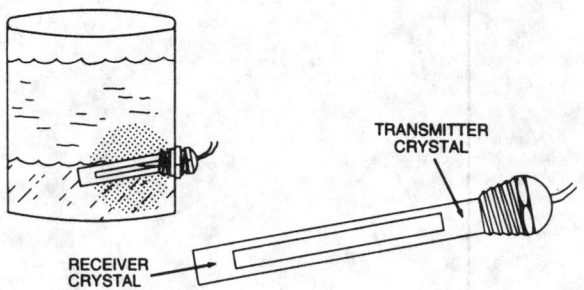

FIG. 7.6.19 Ultrasonic interface level switch. (Courtesy of Sensall Inc.)

PARTIAL LIST OF SUPPLIERS

Automatic Switch Co.; Computer Instruments Corp.; Davis Instrument Mfg.; Dwyer Instruments Inc.; Fischer & Porter Co.; King Engineering Corp.; Meriam Instrument Div. of Scott & Fetzer; Petrometer Corp.; Scannivalve Corp.; Time Mark Corp.; Trimount Div. of Custom Instrument Components; Uehling Instrument Co.; Wallace & Tiernan Inc.

Capacitance Probes

SERVICE

Point and continuous level measurement of solids and liquids (both conductive and nonconductive) using both the wetted probe and the noncontacting proximity designs.

DESIGN PRESSURE

Up to 4000 psig (28 MPa)

DESIGN TEMPERATURE

PTFE insulation can be used from −300 to 500°F (−185 to 296°C). Uncoated bare probes can be used up to 1800°F (982°C). Alumina insulation can be used up to 2000°F (1128°C). Proximity designs can also be used to measure the level of molten metals.

EXCITATION

A few MHz

MATERIALS OF CONSTRUCTION

Generally stainless steel for nonconductive and Teflon-coated stainless steel for both conductive and nonconductive services, but

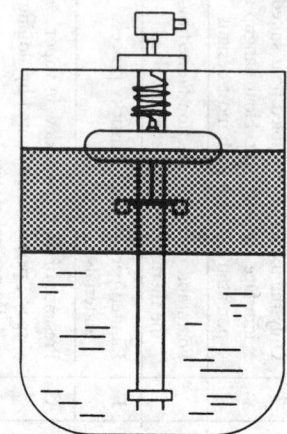

FIG. 7.6.20 Detecting the thickness of the top layer.

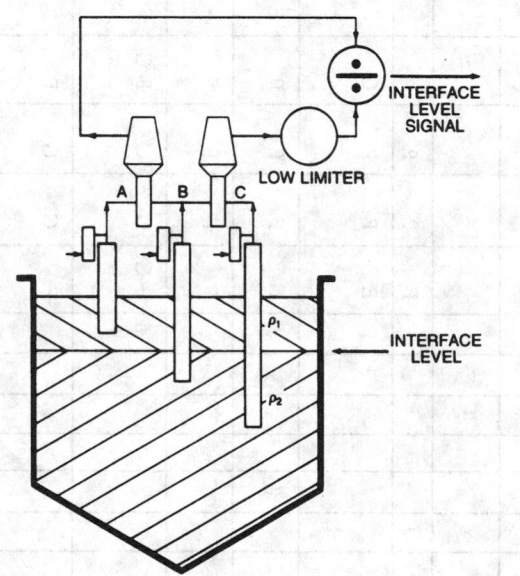

FIG. 7.6.21 Interface detection with bubbler tubes. (Courtesy of Fischer & Porter Co.)

higher alloys, ceramics, PVC, Kynar, and other plastic coatings are also available.

SPANS

From 0.25 to 4000 picofarad (pf). Because of sensitivity limitations, a minimum span of 10 pf is preferred.

INACCURACY

On–off point sensors usually actuate within $\frac{1}{4}$ in (6 mm) of their setpoints. For continuous level detection, dividing the sensitivity by the span calculates the minimum percentage error of 1 to 2% of full scale.

SENSITIVITY AND DRIFT

Depending on design, sensitivities vary from 0.1 to 0.5 pf, while the drift per 100°F (56°C) temperature change can vary from 0.2 to 5 pf.

RANGE

Proximity devices can be used from a fraction of an inch to a few feet; probes can be used up to 20 ft (6 m) and cables up to 200 ft (61 m).

DEADBAND AND TIME DELAY

Capacitance-type level switches are usually provided with deadband settings adjustable over the full span of the unit and time delays adjustable over a 0- to 25-sec range.

COST

From $600 for a simple level switch with power supply and output relay, plus $600 for a continuous indicator. Microprocessor-based intelligent units with special probe configurations start at $2000.

PARTIAL LIST OF SUPPLIERS

(* indicates that the supplier also markets proximity probes.)
*ADE Corp.; Aeroquip Corp.; Agar Corp. Inc.; Amprodux Corp. Inc.; *Arjay Engineering Ltd.; ASC Computer Systems; ASI Instruments Inc.; Babbitt International Inc.; Bailey Controls Co.; Bedford Control Systems; Bernhard Inc.; Bindicator; Controlotron Corp.; *Custom Control Sensors Inc.; *Delavan Inc.; Delta Controls Corp.; *Drexelbrook Engineering Co.; *Electromatic Controls Corp.; Endress + Hauser Instruments; Enraf-Nonius; ETA Control Instruments; Fischer & Porter; Fowler Co.; Free Flow Inc.; *FSI/Fork Standards Inc.; Great Lakes Instruments Inc.; HITech Technologies Inc.; Hyde Park Electronics; Hydril P.T.D.; Invalco; KDG Mobrey Ltd.; Lumenite Electronic Co.; Magne-Sonics; Magnetrol International; Monitor Manufacturing Co.; *MTI Instruments Div.; Omega Engineering; Penberthy Inc.; Princo Instruments Inc.; *Robertshaw Controls Co.; Rosemount Inc.; Systematic Controls; Transducer Technologies Inc.; TVC Instruments Co.; Vega B.V.; Zi-Tech Instrument Corp.

Conductivity Probes

APPLICATIONS

Point or differential level detection of conductive liquids or slurries with dielectric constants of 20 or above. For electric types, the maximum fluid resistivity is 20,000 ohm/cm; electronic types can work on even more resistive fluids. Field effect probes are used on both conductive and nonconductive solids and liquids.

DESIGN PRESSURE

Up to 3000 psig (21 MPa) for conductivity probes and 100 psig (6.9 bars, or 0.69 MPa) for field conductivity probes.

DESIGN TEMPERATURES

From $-15°F$ ($-26°C$) to 140°F (60°C) for units with integral electronics and from $-15°F$ ($-26°C$) to 1800°F (982°C) for units with remote electronics when detecting conductivity. Field effect probes can operate up to 212°F (100°C).

MATERIALS OF CONSTRUCTION

Conductivity probes are made of 316 stainless steel, Hastelloy, titanium, or Carpenter 20 rods with Teflon, Kynar, or PVC sleeves. The housing is usually corrosion-resistant plastic or aluminum for NEMA 4 and 12 service. The field effect probe has a Ryton probe and aluminum housing.

PROBE LENGTHS

$\frac{1}{4}$-in (6-mm) solid rods are available in lengths up to 6 ft (1.8 m); $\frac{1}{16}$-in (2-mm) stainless steel cables can be obtained in lengths up to 100 ft (30 m) for conductivity applications. Field effect probes are 8 in (200 mm) long.

SENSITIVITY

Adjustable from 0 to 50,000 ohms for conductivity probes

INACCURACY

$\frac{1}{8}$ in (3 mm)

COST

From $50 to $400. The typical price of an industrial conductivity switch is about $300.

PARTIAL LIST OF SUPPLIERS

BL Tec.; Burt Process Equipment; B/W Controls—Magatek Controls; Conax Buffalo Corp.; Control Engineering Inc.; Delavan Inc. Division Colt Industries; Delta Controls Corp.; Electromatic Controls Corp.; Endress + Hauser Instruments; Great Lakes Instruments Inc.; Invalco Inc.; Lumenite Electronic Co.; Monitor Mfg.; National Controls Corp.; Revere Corp. of America; Vega B.V.; Warrick Controls Inc.; Zi-Tech Instrument Corp.

D/P Cells

DESIGN PRESSURE

To 10,000 psig (69 MPa)

DESIGN TEMPERATURE

To 350°F (175°C) for d/p cells and to 1200°F (650°C) for filled systems; others to 200°F (93°C). Standard electronics are generally limited to 140°F (60°C).

RANGE

d/p cells and indicators are available with full-scale ranges as low as 0 to 5 in (0 to 12 cm) H_2O. The higher ranges are limited only by physical tank size since d/p cells are available with ranges over 433 ft H_2O (7 MPa or 134 m H_2O).

INACCURACY

±0.5 to 2% of full scale for indicators and switches. For d/p transmitters, the basic error is from ±0.1 to 0.5% of the actual span. Added to this error are the temperature and pressure effects on the span and zero. In intelligent transmitters, pressure and temperature correction is automatic, and the overall error is ±0.1 to 0.2% of the span with analog outputs and even less with digital outputs.

MATERIALS OF CONSTRUCTION

Plastics, brass, steel, stainless steel, Monel, and special alloys for the wetted parts. Enclosures and housings are available in aluminum, steel, stainless steel, and fiberglass composites, with aluminum and fiberglass the most readily available.

COST

$200 to $1500 for transmitters in standard construction and $100 to $500 for local indicators. Add $400 to $800 for extended diaphragms and $300 to $600 for smart features such as communications and digital calibration. Expert tank systems cost approximately $1500 for the basic transmitter plus $3500 to $4500 for the interface unit and $1500 to $4000 for software plus a handheld communicator.

PARTIAL LIST OF D/P CELL SUPPLIERS

ABB Kent-Taylor; Dresser Industries; Enraf Nonius; Fischer & Porter Co.; Foxboro Co.; Honeywell, Inc.; ITT Barton; Johnson Yokogawa; King Engineering Corp.; L&J Engineering Inc.; Major Controls, Inc.; Rosemount Inc., Measurement Div., Varec Div.; Schlumberger Industries, Statham Div.; Smar International; Texas Instruments; Uehling Instrument

PARTIAL LIST OF TANKFARM PACKAGE SUPPLIERS

The Foxboro Co.; King Engineering Corp.; L&J Engineering Inc.; Sarasota M&C Inc.; Texas Instruments Inc.; Varec, a Rosemount Div.

The level measurement device used most often on slurry and sludge services is the extended-diaphragm-type differential pressure transmitter (see Figure 7.6.22). The diaphragm extension eliminates the dead-ended cavity in the nozzle, where materials accumulate, and brings the sensing diaphragm flush with the inside surface of the tank. The sensing diaphragm is sometimes coated with Teflon to further minimize material buildup. One of the best methods of keeping the low-pressure side of the d/p cell clean is to insert another extended-diaphragm device in the upper nozzle. This device can be a pressure repeater, which can repeat both vacuums and pressures within the range of the available vacuum and plant or instrument air supply pressures. When air or vacuum is unavailable at the process pressures, extended-diaphragm-type chemical seals can be used (see Figure 7.6.23) if properly compensated for ambient temperature variations and sun exposure.

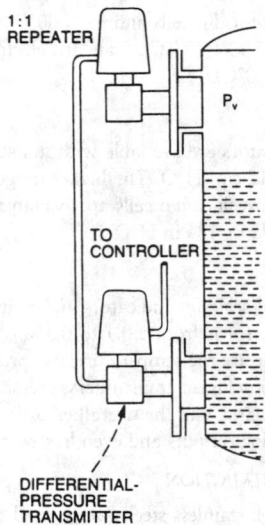

FIG. 7.6.22 Schematic diagram showing the clean and cold air output of the repeater repeating the vapor pressure (P_v) in the tank.

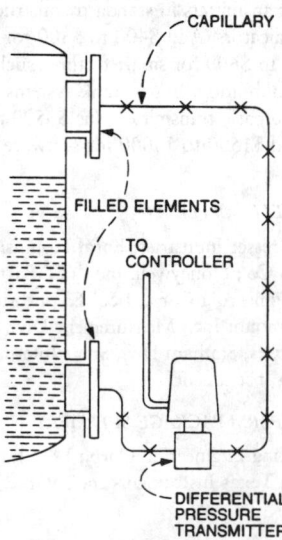

FIG. 7.6.23 Schematic diagram that shows how the temperature compensated, extended-diaphragm-type, chemical seals protect the d/p cell from plugging.

Level Gauges

TYPES

Tubular glass, armored reflex, or transparent and magnetic gauges

DESIGN PRESSURE

Tubular gauge glasses are usually limited to 15 psig (1 bar). At 100°F (38°C), armored-reflex gauges can be rated to 4000 psig (270 bars = 27 MPa); transparent gauges to 3000 psig (200 bars = 20 MPa), and bullseye units up to 10,000 psig (690 bars = 69 MPa). Magnetic level gauges are available up to 3500 psig (230 bars = 23 MPa).

DESIGN TEMPERATURE

Tubular gauge glasses are usually limited to 200°F (93°C). Armored gauges can be used up to 700°F (371°C), and magnetic gauges are available from −320 to 750°F (−196 to 400°C).

MATERIALS OF CONSTRUCTION

The wetted parts of armored gauges are available in steel, stainless steel, and tempered borosilicate glass. Magnetic level gauges are available with steel flanges and stainless steel, K-monel, Hastelloy-B, and solid PVC chambers. Available chamber and float liner materials include Kynar, Teflon, and Kel-F.

RANGE

For armored gauges, the visible length of a section is 10 to 20 in (250 to 500 mm). A maximum of four sections per column is recommended with a maximum total distance between gauge connections of 5 ft (1.5 m).

INACCURACY

Level gauges can be provided with scales. The reading accuracy is limited by visibility (foaming and boiling), and the height of the liquid column in the gauge can also differ from the process level. If the liquid in the gauge is warmer, it is also lighter, and therefore the error is on the high side; if the liquid in the gauge is colder (heavier), the indication is low. Readout wafer size limits magnetic gauge display accuracy to $\frac{1}{4}$ in (6 mm).

COSTS

Excluding the cost of shutoff valves or pipe stands, the per-ft (300 mm) unit cost of tubular glass gauges is about $25; armored-reflex and transparent gauges cost about $150/ft and $200/ft, respectively, while magnetic level gauges in stainless steel construction cost about $500/ft.

PARTIAL LIST OF SUPPLIERS

Daniel Industries Inc.; Essex Brass Co.; Imo Industries Inc. (magnetic); Jerguson Gauge and Valves, Div. of the Clark Reliance Corp. (regular and magnetic); Jogler Inc.; Kenco Engineering Co. (magnetic); Krohne America Inc.; K-Tek Corp. (magnetic); MagTech Div. ISE of Texas Inc. (magnetic); Metron Technology (magnetic); Oil-Rite Corp.; Penberthy Inc. (regular and magnetic)

Optical Sensors

TYPES

Visible or infrared (IR) light reflection (noncontacting type usually for solids and laser type for molten glass applications), light transmission (usually for sludge level), and light refraction in clean liquid level services

APPLICATIONS

Point sensor probes for liquid, sludge, or solids (some continuous detectors also available)

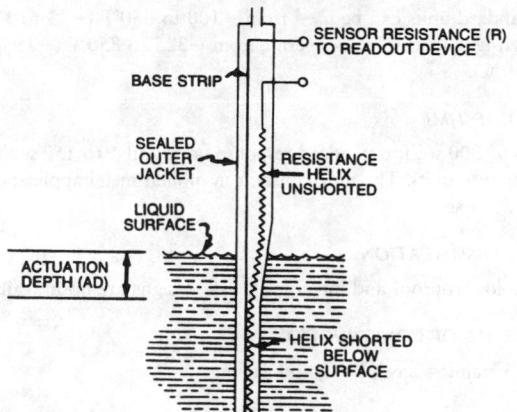

FIG. 7.6.24 Schematic diagram of resistance tape sensor operation.

DESIGN PRESSURE

Up to 150 psig (10.3 bars) with polypropylene, polysulfone, PVDF, or Teflon and up to 500 psig (35 bars) with stainless steel probes

DESIGN TEMPERATURE

Between 150 and 200°F (66 to 93°C) with plastic probes and up to 260°F (126°C) with stainless steel probes

MATERIALS OF CONSTRUCTION

Quartz reflectors with Viton-A or Rulon seals, mounted in polypropylene, polysulfone, Teflon, polyvinyl fluoride, phenolic, aluminum, or stainless steel probes

HOUSINGS

Can be integral with the probe or remote. Explosion-proof enclosures and intrinsically safe probes are both available. With remote electronics, the fiber-optic cable can be from 50 to 250 ft (15 to 76 m) long.

DIMENSIONS

Refraction probe lengths vary from 1 to 24 in (25 to 600 mm), and the probe diameter is usually 0.5 to 1 in (12 to 25 mm).

COSTS

Fiberoptic level switches cost from $100 and $300. Portable sludge level detectors cost $900. Continuous transmitters to measure sludge depth or sludge interface cost $4000 and up.

PARTIAL LIST OF SUPPLIERS

Automata Inc. (noncontacting IR); BTG Inc. (IR); Conax Buffalo Corp. (fiber optic); Enraf Nonius Tank Inventory Systems Inc. (IR); Gems Sensors Div. IMO Industries Inc. (fiber optics); Genelco Div. of Bindicator Inc. (IR switch); Kinematics & Controls Corp. (switch); Markland Specialty Engineering Ltd. (IR for sludge); OPW Division of Dover Corp.; 3M Specialty Optical Fibers; Zi-Tech Instrument Corp. (switch)

Resistance Tapes

APPLICATIONS

Liquids including slurries but not solids. Can also measure temperature

RESOLUTION

$\frac{1}{8}$ in, which is the distance between helix turns

ACTUATION DEPTH (AD)

The depth required to short out the tape varies with the specific gravity (SG) as follows: AD (in inches) = 4/(SG). Therefore, AD at the minimum SG of 0.5 is 8 in (200 mm).

TEMPERATURE EFFECT

A 100°F (55°C) change in temperature changes the resistance of the unshorted tape by 0.1%. Temperature compensation is available.

INACCURACY

0.5 in if the AD is zeroed out and both AD and temperature are constant. If SG varies, a zero shift based on AD − 4/(SG) occurs. Cold temperature also raises the AD.

WETTED MATERIAL

Fluorocarbon polymer film

ALLOWABLE OPERATING PRESSURE

From 10 to 30 psia (0.7 to 2.1 bars absolute)

OPERATING TEMPERATURE RANGE

−20 to 225°F (−29 to 107°C)

COSTS

Resistance tape unit cost varies with service and with tape length. A 10 ft (3 m) tape with breather and transmitter for water service costs from $600 to $1000. The added cost for longer tapes is $25 or more per foot, depending on the service.

SUPPLIERS

Metritape Inc.; R-Tape Corp.; Sankyo Pio-Tech

The resistance tape (see Figure 7.6.24) for continuous liquid level measurement was invented in the early 1960s, initially for water well gauging and subsequently for marine and industrial usage. The sensor is a flat, coilable strip (or tape), ranging from 3 to 100 ft (1 to 30 m) in length, suspended from the top of the tank. It is small enough in cross-section to be held within a perforated pipe (diameter of 2 to 3 in), which also supports the transducer and acts as a stilling pipe when the process is turbulent.

While resistance tapes are not widely used in the wastewater treatment industry today, their low cost, low maintenance, and adaptability for multipoint scanning makes them a candidate for use in new plants.

Thermal Switches

TYPES

Switches operate on either thermal difference or thermal dispersion. Transmitters utilize the thermal conductivity difference between liquids and vapors. Metal mold level controllers use direct temperature detection.

APPLICATIONS

Liquid, interface, and foam level detection. Special units are available for molten metal level measurement.

DESIGN PRESSURE

Up to 3000 psig (207 bars = 20.7 MPa)

DESIGN TEMPERATURE

Standard units can be used from −100 to 350°F (−73 to 177°C); high-temperature units operate from −325 to 850°F (−198 to 490°C).

RESPONSE TIME

10 to 300 sec for standard response units and 1 to 150 sec for fast response units. The time constant in molten metal applications is under 1 sec.

AREA CLASSIFICATION

Explosion-proof and intrinsically safe designs are both available.

MATERIALS OF CONSTRUCTION

316 stainless steel, PVC, and Teflon

INACCURACY

The repeatability is 0.25 in (6 mm) for side-mounted and 0.5 in (13 mm) for top-mounted level switches. Transmitters are less accurate. Molten metal level error depends on thermocouple spacing.

COST

The cost of a thermal level switch is about $250. Transmitters cost about $1000. Mold level systems are field-installed.

PARTIAL LIST OF SUPPLIERS

Chromalox Instruments and Control; Delta M Corp. (transmitter); Fluid Components Inc. (switch and monitor); Intek Inc. Rheotherm Div. (switch); Scientific Instruments Inc.; Scully Electronic Systems, Inc. (switch)

Ultrasonic Detectors

APPLICATIONS

Wetted and noncontacting switch and transmitter applications for liquid level or interface and solids level measurement. Also used as open-channel flow monitors.

DESIGN PRESSURE

Probe switches are used up to 3000 psig (207 bars = 20.7 MPa); transmitters are usually used for atmospheric service up to 7 psig (0.5 bar), but some special units are available for use up to 150 psig (10.3 bars).

DESIGN TEMPERATURE

Switches from −100 to 300°F (−73 to 149°C); transmitters from −30 to 150°F (−34 to 66°C)

MATERIALS OF CONSTRUCTION

Aluminum, stainless steel, titan, Monel, Hastelloy B & C, Kynar, PVC, Teflon, polypropylene, PVDF, and epoxy

RANGES

For tanks and silos (pulse usually travels in vapor space), up to 200 ft (60 m) for some special designs and up to 25 ft (7.6 m) for most standard systems. For wells (usually submerged), up to 2000 ft (600 m)

INACCURACY

$\frac{1}{8}$ in (3 mm) for a horizontal probe switch. For transmitters, the error varies from 0.25 to 2% of full scale depending on the dust and dew in the vapor space and the quality of the surface that reflects the ultrasonic pulse.

COSTS

Level switches cost from $200 to $500; transmitters cost from under $1000 to $2500, with the average cost around $1800.

PARTIAL LIST OF SUPPLIERS

Bindicator Co.; Contaq Technologies Corp.; Controltron Corp.; Crane/Pro-Tech Environmental Instruments; Delavan Inc. Process Instrumentation Operations; Delta Controls Corp.; Electro Corp.; Electronic Sensors Inc.; Endress + Hauser Inc.; Enterra; Fischer and Porter Co.; Genelco Div. Bindicator; Gordon Products Inc.; Greyline Instruments Inc.; HiTech Technologies Inc. (fly ash application); Hyde Park Electronics Inc.; Introkek, Subsidiary of Magnetrol International; Inventron Inc.; Kay Ray/Sensall Inc.; KDG Mobrey Ltd.; Kistler-Morse Corp.; Krone America Inc. (sludge interface); Magnetrol International; Markland Specialty Engineering Ltd. (sludge level); Marsh-McBirney Inc.; Massa Products Corp.; Microswitch/Honeywell, Milltronics Inc.; Monitek Technologies Inc.; Monitor Mfg.; Monitrol Mfg. Co.; Panametrics Inc.; Penberthy; Sirco Industrial Ltd.; SOR Precision Sensors; TN Technologies Inc.; Ultrasonic Arrays Inc. (thickness, texture, surface reflectivity); United Sensors Inc.; Vega B.V.; Zevex Inc.

As is shown in Figures 7.6.18 and 7.6.19, ultrasonic level sensors are used widely on sludge level and sludge interface detection services. Ultrasonic sludge blanket detectors can also be lowered periodically into the tank for transmittance measurements, or they can be permanently positioned for echo detection. In the newer designs, targets or sounding pipe ridges are used for automatic calibration. Even more recently, flexural sensors are installed to measure the transit time or echo in the tank wall instead of through the process liquid.

Vibrating Switches

TYPES

A. Tuning fork
B. Vibrating probe
C. Vibrating reed

APPLICATIONS

Liquid, slurry, and solids level switches

DESIGN PRESSURE

A and B. To 150 psig (10.3 bars = 1 MPa)
C. Up to 3000 psig (207 bars = 20.7 MPa)

DESIGN TEMPERATURE

A. −45 to 200°F (−43 to 93°C)
B. 8 to 176°F (−10 to 80°C)
C. From −150 to 300°F (−100 to 149°C)

MATERIALS OF CONSTRUCTION

Aluminum, steel, and stainless steel

MINIMUM BULK DENSITY

A and C. Down to 1.0 lbm/ft³ (16 kg/m³)
B. Requires an apparent specific gravity of 0.2

INACCURACY

The repeatability of type C is $\frac{1}{8}$ in (3 mm)

COST

   Standard type A, $300; other designs up to $500

PARTIAL LIST OF SUPPLIERS

   Automation Products Inc.; Bindicator Co.; Endress + Hauser Inc.;
   KDG Mobrey Ltd.; Monitor Mfg.; Monitrol Bin Level

Manufacturing Co.; Nohken Co. Ltd.; Vega B.V.; Zi Tech
Instrument Corp.

—Béla G. Lipták

7.7
pH, OXIDATION-REDUCTION PROBES (ORP) AND ION-SELECTIVE SENSORS

Because the goal of the wastewater treatment industry is to purify and neutralize industrial and municipal waste streams, sensors are needed to detect the activity and concentration of various ionic substances. An important water parameter is the pH, which indicates the activity of the hydrogen ion and describes the acidity or alkalinity of the stream. Ion selective electrodes detect the activity of other ions, while ORPs describe the chemical or biological processes in progress. This section describes the features and capabilities of these three sensor types.

Probes and Probe Cleaners

In wastewater applications, environmental engineers use analytical probes to detect concentrations in the sludge layers situated in the lower parts of scraped bottom tanks. These probes are installed on pivoted hinges so that the mechanical scraper assembly can pass (see Figure 7.7.1).

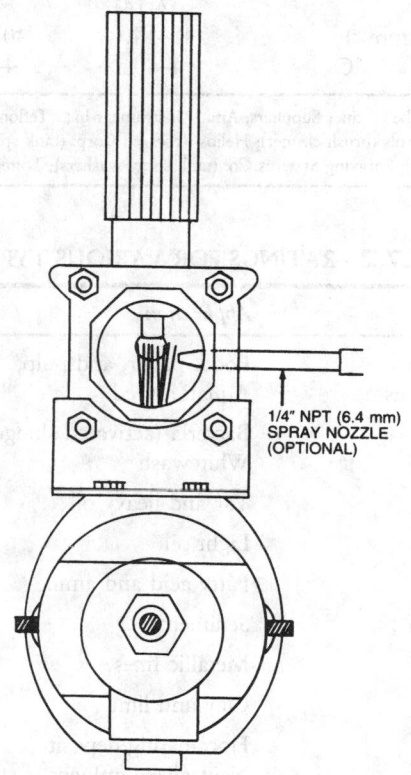

FIG. 7.7.2 Probe cleaner mounted in sight-flow glasses for good visibility. (Courtesy of Aimco Instruments Inc.)

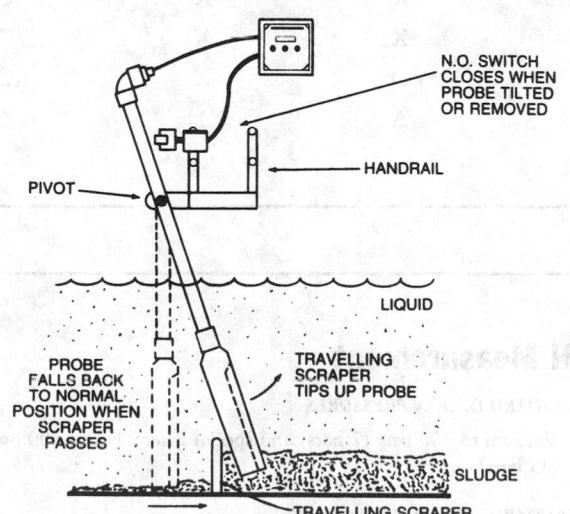

FIG. 7.7.1 Probe-type sensors used to detect the composition of sludge and slurry layers in clarifiers. (Courtesy of Markland Specialty Engineering Ltd.)

While probe-type in-line analyzers eliminate the transportation lag and sample deterioration problems associated with offline analysis, they illustrate the need for efficient probe cleaners. A probe cleaner should be placed inside a sight glass so that clearer performance can be continuously observed by the operator (see Figure 7.7.2). A variety of probe-cleaning devices are available. Table 7.7.1 lists features and capabilities for the removal of various coatings and Table 7.7.2 lists suppliers.

If no sampling system is used, sample integrity is automatically guaranteed, and sensors that penetrate the

TABLE 7.7.1 SELECTION OF AUTOMATIC PROBE CLEANERS

| | Applicable Choice of Probe Cleaner | | | | | | |
| | Mechanical | | Chemical | | | Hydro-Dynamic (self-cleaning) | Acoustical or Ultrasonic |
	Brush	Rotary Scraper	Acid	Base	Emulsifier		
Service							
Oils and fats		√			√		√
Resins (wood and pulp)				√			√
Latex emulsions		√					
Fibers (paper and textile)						√	
SS						√	
Crystalline precipitations (carbonates)	√	√	√				
Amorphous precipitations (hydroxides)	√	√	√				√
Material of construction	Stainless steel (brush pH 7–14)	Stainless steel	PVC	PVC	PVC	Stainless steel	Polypropylene, stainless steel
Temperature °F	40–140	40–140	40–140	40–140	40–140	40–250	40–195
°C	4–60	4–60	4–60	4–60	4–60	4–120	4–90

Note: Probe Cleaner Suppliers; Amico Instrument Inc. (Teflon brush); Branson Cleaning Equipment Co. (ultrasonic cleaners); Fetterolf Corp. (spray rinse valve); Graphic Controls (brush cleaner); Helios Research Corp. (tank spray washers); Polymetron, Div. of Uster Corp. (probe cleaners); Sybron/Gamlen, Gamajet Div. (tank spray washers); Spraying Systems Co. (tank spray washers); Toftejorg Inc. (tank spray washers).

TABLE 7.7.2 RATINGS FOR VARIOUS TYPES OF CLEANERS

	Application	Ultrasonic	Water-jet	Brushing	Chemical
Slime Microorganism	Food, paper, and pulp, Aquatic weed	X	O	O	△
	Bacteria (activated sludge) Whitewash	△	O	O	△
Oil	Tar and heavy oil	X	X	X	△
	Light oil	O	△	△	O
	Fatty acid and amine	X	O	X	O
Suspension	Sediment	O	X	X	O
	Metallic fines	△	X	X	O
	Clay and lime	△	O	X	O
Scale	Flocculating deposit Neutralized effluent CaCO₃	△	△	△	△

Source: Horiba Instruments.
Notes: O: Recommend, △: Applicable, X: Not applicable.

process pipe with a retractable, cleanable probe are preferred. Probe sensors, either solid or membrane, require periodic cleaning. This can be done manually, by withdrawing the probe through an isolating valve so that the process is not opened when the electrode is cleaned; or automatically, using automatic probe-cleaning devices such as pressurized liquid or gas jets, and thermal, mechanical, or ultrasonic cleaning and scraping devices.

pH Measurement

STANDARD DESIGN PRESSURES

Vacuum to 100 psig (7 bars) and special assemblies to 500 psig (35 bars)

STANDARD DESIGN TEMPERATURES

Generally −5 to 100°C (23 to 212°F); sterilizable, −30 to 130°C (−22 to 266°F); Glasteel, −5 to 140°C (<5 pH) (23 to 284°F)

MATERIALS OF CONSTRUCTION

Electrode hardware: stainless steel, Monel, Hastelloy, titanium, epoxy, Kynar, halar, PVC, chlorinated polyvinyl chloride, polyethylene, polypropylene, polyphenylene sulfide, ryton, Teflon, and various elastomer materials

ASSEMBLIES

Flow-through, submersion, insertion, and retractable

CLEANERS

Ultrasonic, jet washer (chemical and water), and brush

INACCURACY

Electrodes 0.02 pH; lab meters and displays 0.01 pH; transmitters 0.02 mA; and installation effects 0.2 pH

RANGE

0 to 14 pH

COSTS

Electrodes cost $100 to $500 (Glasteel is $2000); lab meters, $200 to $800; transmitters, $500 to $2000; assemblies, $200 to $1000; and cleaners, $400 to $2000. A brush cleaner in 316 SS is $1900, and a retractable cleaner in 316 SS is $1750. A fiber-optic pH assembly is $15,000, with associated electronics costing an additional $10,000.

PARTIAL LIST OF SUPPLIERS

Amico-Instruments; Bailey-TBI; Beckman Instruments (Process Instruments and Control Group); Broadley-James; Custom Sensors & Technology; Electro-Chemical Devices; Foxboro Analytical; George Fischer Signet; Great Lakes Instruments; Horiba Instruments; Ingold; Innovative Sensors; Johnson Yokogawa Electrofac; Lakewood Instruments; Leeds & Northrup Instruments; McNab; Monitek; Orion Industrial Division of Orion Research; Pfaudler; Phoenix; Uniloc Division of Rosemount; SensoreX; Van London

An important step in wastewater treatment is neutralization. The neutralization process includes the reagent delivery system, the mixing equipment, the reaction and equalization tanks, and the associated controls. In general, a single stirred-reaction vessel can neutralize influents between 4 and 10 pH. If the influent pH varies from 2 to 12 pH, one stirred and one attenuation tanks are needed. If the influent pH drops below 2 or rises above 12 pH, two stirred and one attenuation tanks are needed. Section 7.41 describes these aspects of the overall pH control system. This section discusses only pH measurement probes.

The measurement of pH covers a wide range of dilute acid and base concentrations (see Figure 7.7.3). For strong acids and bases, these measurements can track changes from one to one millionth percent. Thus, pH is a sensitive indicator of deficient and excess acid and base reactant concentrations for chemical reactors and scrubbers. For example, a few millionths of a percent of excess of sodium hydroxide (a strong base) is needed for chlorine destruction with sodium bisulfite. pH measurement can reduce the addition of sodium hydroxide to a minimum and still ensure complete use of sodium bisulfite.

Biological reactors use acids and bases to supply food or neutralize the waste products of organisms. Cells are

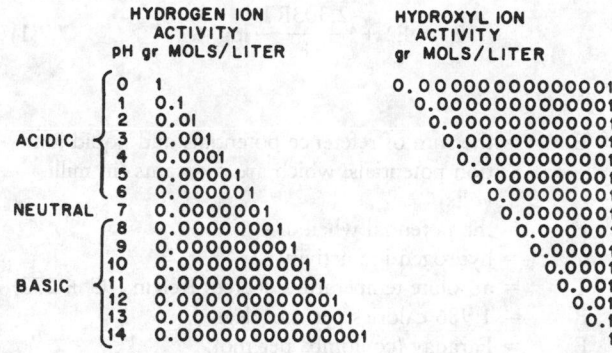

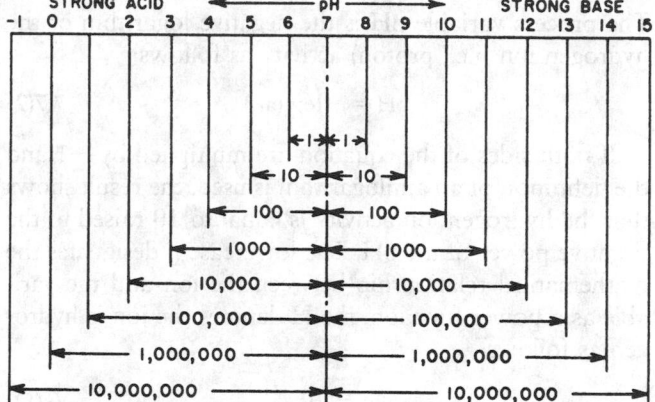

GRAPH OF REAGENT DEMAND. REAGENT ADDITION UNITS ARE 10^{-6} MOLS/LITER.

FIG. 7.7.3 The logarithmic nature of pH.

extremely sensitive to pH fluctuations. Genetically engineered bacteria tend to be weak and need tight pH control. Thus, pH is critical to the cell growth rate, enzyme reactions, and the extraction of intercellular products. The sensitivity of cells to pH has even wider significance in that any food, drink, or drug ingested or injected and any waste discharged to the environment must have pH specifications to prevent damage to living matter and ecological systems. Stricter environmental regulations have increased the number and importance of pH measurements.

Some environmental regulations have instantaneous limits on pH. An excursion outside the acceptable range for a fraction of a second can be a recordable violation. The pH measurement system must be designed to prevent violations from spurious readings due to installation effects. For most discharges, the acceptable range lies between 6 and 9 pH.

THEORETICAL REVIEW

In pH measurement systems, a pH responsive glass takes up hydrogen ions and establishes a potential at the glass surface with respect to the solution. This potential is related to the hydrogen ion activity of the solution by the Nernst relationship as follows:

$$E_g = E_g^o + \frac{2.303RT}{F} \log_{10}a \qquad 7.7(1)$$

where:

E_g	=	the sum of reference potentials and liquid junction potentials, which are constants (in millivolts)
E_g^o	=	the potential when a = 1
a	=	hydrogen ion activity
T	=	absolute temperature degrees Kelvin (°C + 273)
R	=	1.986 calories per mol degree
F	=	Faraday (coulombs per mol)
2.303	=	logarithm conversion factor

The process variable pH is the negative logarithm of the hydrogen ion (i.e., proton) activity as follows:

$$pH = -\log(a_H) \qquad 7.7(2)$$

If both sides of the equation are multiplied by −1 and the definition of an antilogarithm is used, the result shows that the hydrogen ion activity is equal to 10 raised to the negative power of the pH. The lowercase p designates the mathematical relationship between the ion and the variable as a power function; the H denotes the ion is hydrogen as follows:

$$a_H = 10^{-pH} \qquad 7.7(3)$$

For dilute aqueous (water) solutions, the activity coefficient is approximately unity, and the hydrogen ion concentration is essentially equal to the hydrogen ion activity. As the concentrations of acids, bases, and salts increase, the crowding effect of the ions reduces hydrogen ion activity. Thus, an increase in salt concentration can increase the pH reading even though the hydrogen ion concentration is constant.

An acid is a proton donor, and a base is a proton acceptor. When an acid dissociates (breaks apart into its component ions), it yields a hydrogen ion and a negative acid ion. When a base dissociates, it gives a positive base ion and a hydroxyl ion that is a proton acceptor. When water dissociates, the result is both a hydrogen ion (proton) and hydroxyl ion (proton acceptor). Thus, water acts as both an acid and a base. Neutralization is the association of hydrogen ions from acids and hydroxyl ions from bases to form water.

pH measurement can track fourteen decades of hydrogen ion concentration and detect changes as small as 10^{-14} (at 14 pH). The concentration changes of strong acids and bases also follow the decade change per pH unit within this range. No other concentration measurement has such rangeability and sensitivity. These characteristics have profound implications for pH control.

Concentrated strong acids and bases have a pH that lies outside this range. For example, concentrated sulfuric acid has a pH of −10, and concentrated sodium hydroxide has a pH of 19 as measured by a hydrogen electrode. However, the set point of a pH loop is usually well within the 0 to 14 range. Some feedforward pH loops can require measurements outside this range, but the shortened life expectancy and increased error from the electrode at range extremes make such measurements impractical.

The neutral point is where the hydrogen ion concentration equals the hydroxyl ion concentration. At 25°C, this point occurs at 7 pH (see Table 7.7.3).

MEASUREMENT ELECTRODE

The hydrogen ion does not exist alone in aqueous solutions. It associates with a water molecule to form a hy-

TABLE 7.7.3 CONCENTRATIONS OF ACTIVE HYDROGEN AND HYDROXYL IONS AT 25°C AT DIFFERENT pH VALUES (IN GRAM-MOLES/LITER) AND SOME FLUIDS WITH CORRESPONDING pH VALUES

pH	Fluid Example	Hydrogen Ions	Hydroxyl Ions
0	4% Sulfuric acid	1.0	0.000000000000
1		0.1	0.000000000000
2	Lemon juice	0.01	0.000000000001
3		0.001	0.00000000001
4	Orange juice	0.0001	0.0000000001
5	Cottage cheese	0.00001	0.000000001
6	Milk	0.000001	0.00000001
7	Pure water	0.0000001	0.0000001
8	Egg white	0.00000001	0.000001
9	Borax	0.000000001	0.00001
10	Milk of magnesia	0.0000000001	0.0001
11		0.00000000001	0.001
12	Photo developer	0.000000000001	0.01
13	Lime	0.0000000000001	0.1
14	4% sodium hydroxide	0.00000000000001	1.0

dronium ion (H_3O^+). The glass measurement electrode (see Figure 7.7.4) develops a potential when hydronium ions get close enough to the glass surface for hydrogen ions to associate with hydronium ions in an outer layer of the glass surface. This thin hydrated gel layer is essential for electrode response. The input to the pH measurement circuit is a potential difference between the external glass surface exposed to the process (E_1) and the internal glass surface wetted by a 7 pH solution (E_2). If the external glass surface is in exactly the same condition as the internal glass surface, the Nernst equation states that the potential difference in millivolts is proportional to the deviation of the process pH from 7 pH at 25°C.

Flat glass electrodes minimize glass damage and maximize a sweeping action to prevent fouling. A small button flat glass electrode has a range of 0 to 10 pH, and a large flush flat glass electrode has a range of 2 to 12 pH. High sodium ion concentrations and low hydrogen ion activity have a larger effect on flat glasses.

The photometer-type pH sensor shown in Figure 7.7.5 uses a fiber-optic sensor. It should be free of problems due to sodium ions, temperature, coating, and abrasion in a properly designed sample system. The time delay caused by the temperature lag in the sampling system and the higher cost and maintenance are disadvantages. Also, the color dyes are sensitive to oxidants.

CLEANERS

Significant natural self-cleaning by turbulent eddies requires a velocity of 5 or more feet per second (fps) past the electrode. A velocity of greater than 10 fps can cause excessive measurement noise and sensor wear. The area obstructed by the electrode must be subtracted from the total cross-sectional area when estimating the total area open to flow around the electrode. The pressure drop at the restricted cross section should be calculated to ensure

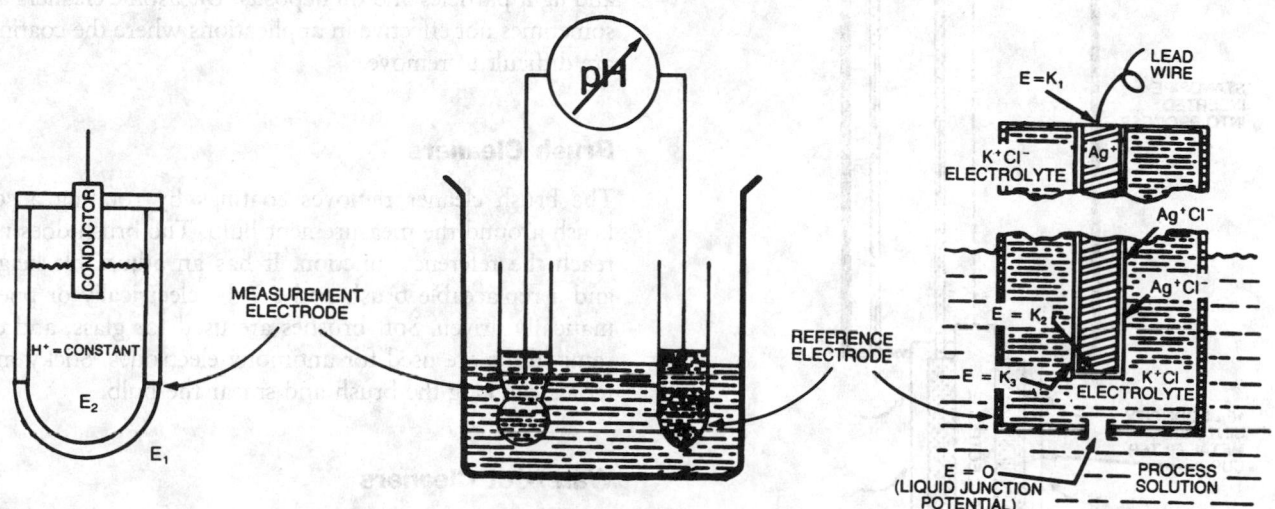

FIG. 7.7.4 The traditional configuration of a glass measurement electrode and a flowing junction reference electrode.

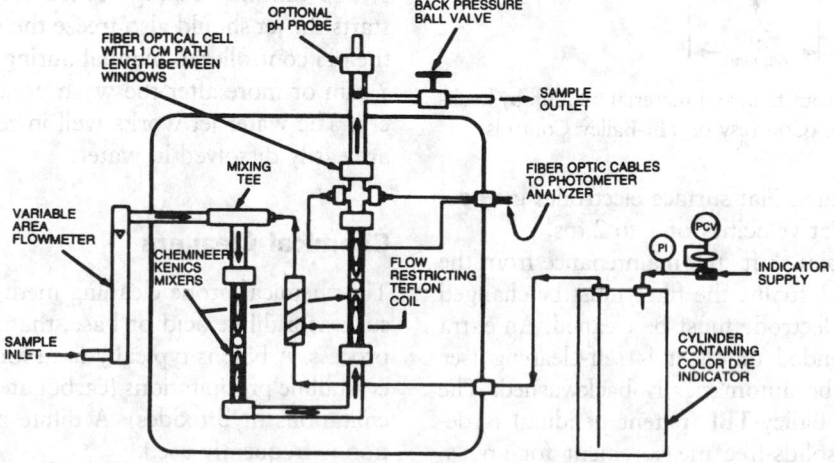

FIG. 7.7.5 Fiber-optic photometer-type pH detector. (Courtesy of Custom Sensors & Technology)

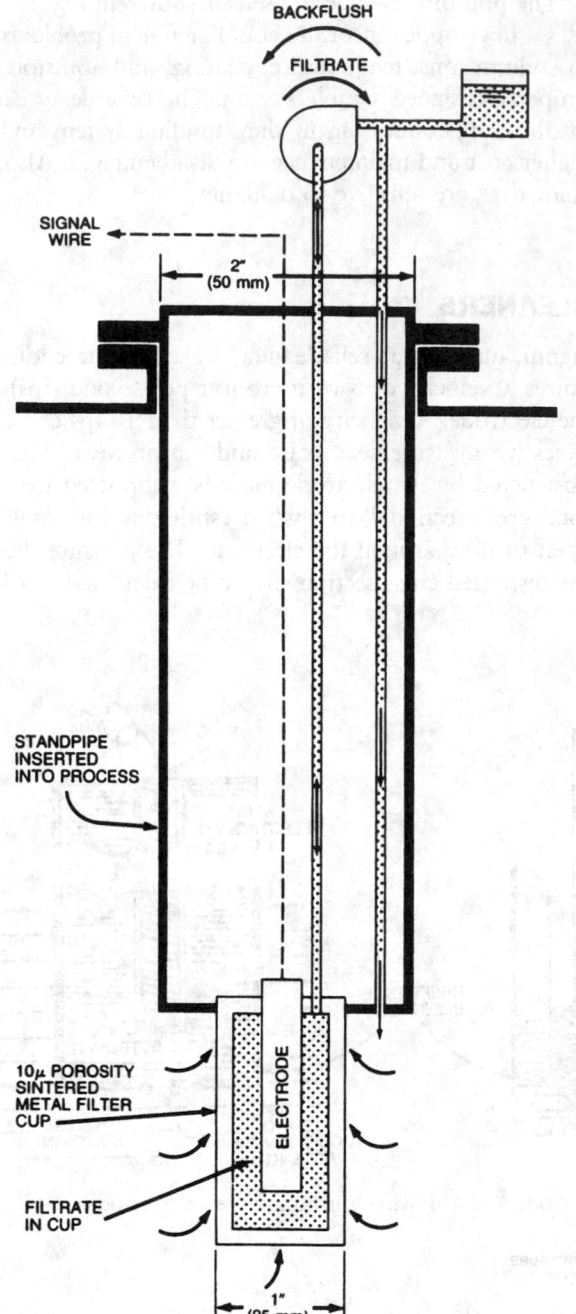

FIG. 7.7.6 Electrode protected from material buildup by backflushed porous filter cup. (Courtesy of TBI-Bailey Controls)

that no cavitation occurs. Flat surface electrodes get adequate cleaning action at velocities of 1 to 2 fps.

The addition of filters shifts the maintenance from the electrode to the filter. Usually, the filter must be changed more often than the electrode must be cleaned. An extra filter is not recommended unless it is self-cleaning (see Figure 7.7.6) or can be automatically backwashed. The Filtrate Master from Bailey-TBI (patent pending) is designed to provide a solids-free measurement for an assembly submerged in a slurry. It reverses flow and pulses loose particles caught in the 10-μm metal filter.

Four types of automatic cleaners are ultrasonic, brush, water-jet, and chemical. Table 7.7.4 shows the performance ratings for various applications, and Figure 7.7.7 shows the components of these assemblies. These methods concentrate on the removal of coatings from the measurement bulb. Particles and material clogged in the porous reference junction are generally difficult to dislodge. The impedance of plugged reference junctions can get so high that it approaches an open circuit, and the pH reading exceeds the scale.

Ultrasonic Cleaners

Ultrasonic cleaners use ultrasonic waves to vibrate the liquid near electrode surfaces. Effectiveness depends on the vibration energy and fluid velocity past the electrodes. Heavy-duty electrodes are needed to withstand the ultrasonic energy. The ultrasonic cleaner works well in processes where fine particles and easily supersaturated sediments are formed or in suspension. It can move loose and light particles and oil deposits. Ultrasonic cleaners are sometimes not effective in applications where the coatings are difficult to remove.

Brush Cleaners

The brush cleaner removes coatings by rotating a soft brush around the measurement bulb. The brush does not reach the reference junction. It has an adjustable height and a replaceable brush and can be electrically or pneumatically driven. Soft brushes are used for glass, and ceramic disks are used for antimony electrodes. Sticky materials can clog the brush and smear the bulb.

Water-Jet Cleaners

The water-jet cleaner directs a high-velocity water jet to the measurement bulb. The reading of the loop becomes erratic during washing. Therefore, the cycle timer that starts the jet should also freeze the pH reading and switch the pH controller to manual during the wash cycle and for 2 min or more after the wash period for electrode recovery. The water jet works well in removing materials that are easily dissolved in water.

Chemical Cleaners

The chemical probe cleaning method uses a chemical jet, such as a dilute acid or base, that is compatible with the process. A base is typically used for resins and an acid for crystalline precipitations (carbonates) and amorphous precipitations (hydroxides). A dilute hydrochloric acid solution is frequently used.

Chemical cleaning tends to be the most effective method, but acid and base cleaners chemically attack the

TABLE 7.7.4 RATINGS FOR VARIOUS CLEANERS

	Application	Ultrasonic	Water-jet	Brushing	Chemical
Slime Microorganism	Food, paper, and pulp, Aquatic weed	X	O	O	△
	Bacteria (activated sludge) Whitewash	△	O	O	△
Oil	Tar and heavy oil	X	X	X	△
	Light oil	O	△	△	O
	Fatty acid and amine	X	O	X	O
Suspension	Sediment	O	X	X	O
	Metallic fines	△	X	X	O
	Clay and lime	△	O	X	O
Scale	Flocculating deposit Neutralized effluent CaCO₃	△	△	△	△

Source: Horiba Instruments.
Notes: O: Recommend, △: Applicable, X: Not applicable.

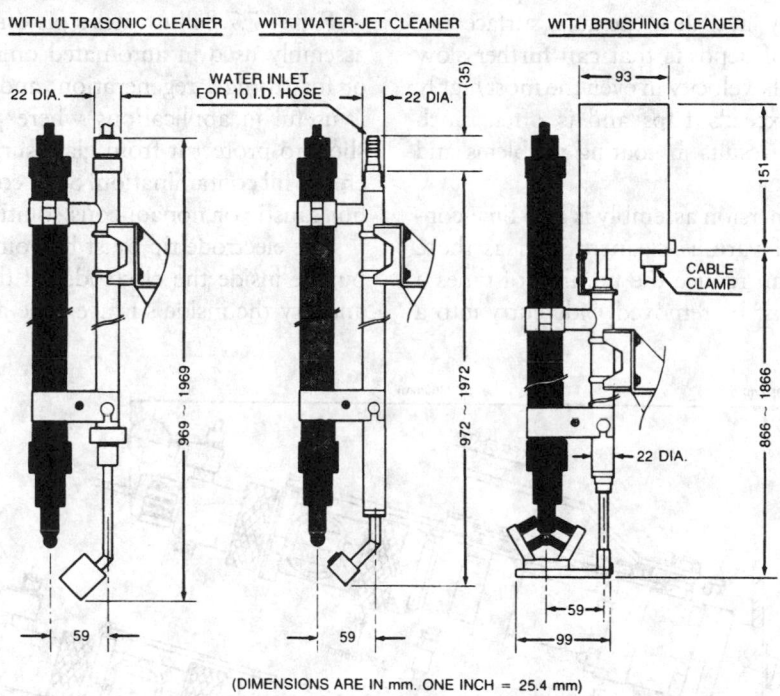

FIG. 7.7.7 Submersion assemblies with various cleaners. (Courtesy of Horiba Instruments)

glass. In addition, cleaning cycles that are too frequent or too long can cause premature failure of the glass electrode. As with the water jet, the cycle timer must hold the last pH reading and suspend control action during the wash cycle. Redundant pH sensors can be installed in parallel so that while one electrode assembly is being reconditioned, the other is in control. In such installations, the reconditioned assembly automatically returns to control after the wash cycle.

Wastewater treatment facilities often manually clean electrodes by soaking them in a dilute hydrochloric acid solution for several hours. Soaking electrodes for 1 min in a dilute solution of hydrofluoric acid in a nonglass container can reactivate electrodes that are sluggish or have too small a span or efficiency. The reactivation occurs by the hydrofluoric acid dissolving part of the aged gel layer. The electrode should then be soaked overnight in its normal storage solution (typically a 4-pH buffer).

INSTALLATION METHODS

Installation design has two main principles. First, for pH control, the sensor location and assembly must minimize transportation delays and sensor time constant. The additional dead time from a delayed and slow measurement increases the loop's period, control error, and sensitivity to nonlinearity. Second, the installation must minimize the number of times the electrodes must be removed for maintenance (e.g., calibration and cleaning). Removal and manual handling increase error and reduce electrode life. The fragile gel layer is altered by handling, and the equilibrium achieved by the reference junction is upset.

Submersion Assemblies

Sample systems are undesirable because they add transportation delay and increase cost, and problems arise with winterization and plugging. Therefore, a submersion assembly is best for control. However, velocities below 1 fps dramatically slow the electrode measurement response due to the increased boundary layer near the glass surface and promote the formation of deposits that can further slow the measurement. The bulk velocity in even the most highly agitated vessels rarely exceeds 1 fps and is often much lower. This low velocity results in coating problems and a slow response.

The removal of a submersion assembly is also time-consuming. The addition of various cleaners such as those shown in Figure 7.7.7 can reduce the number of times a submersion assembly must be removed. Side entry into a vessel with a retractable probe is the standard installation for fermentors, as shown in Figure 7.7.8.

Retractable and Brush-Cleaned Units

The best location for most probes or assemblies, except in the most abrasive services, is in a recirculation line close to the vessel outlet. Installing the probe downstream of the pump is preferred because the strainer blocks and the pump breaks up clumps of material that could damage the electrodes. The retractable electrode is the most straightforward and economical solution. However, accidents caused by removing the restraining strap or omitting tubing ferrules have caused these assemblies to be banned from many plants.

Many wastewater treatment facilities use flow-through assemblies or direct-probe insertions with block, drain, and bypass valves. The flow is returned to the suction of the pump or vessel. If the flow chamber has a cross-sectional area much larger than the process connections, the velocity drops too low, response time slows, and coating problems increase.

Figure 7.7.8 illustrates a piston-actuated, retractable pH assembly used in automated online cleaning applications or for storage, regeneration, and calibration. This design is useful in applications where probe exposure must be short to protect it from glass surface deterioration or reference fill contamination. Such contamination is caused by hot caustic or nonaqueous solutions.

The electrode tip must be pointed down so that the air bubble inside the electrode fill does not reside in the tip and dry the inside surface. The air bubble provides some

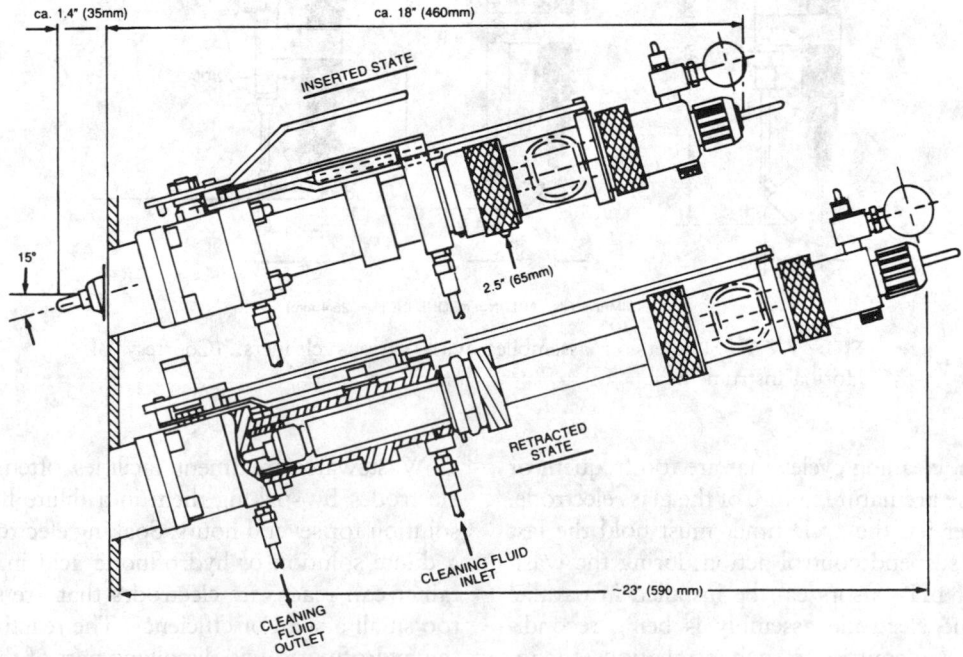

FIG. 7.7.8 Retractable side entry probes often used on fermentors. (Courtesy of Ingold)

compressibility to accommodate thermal expansion. An installation angle of 15 degrees or more from the horizontal is sufficient to keep the bubble out of the tip. Some electrode designs eliminate the bubble and provide a flexible diaphragm for fill contraction and expansion.

Section 7.41 discusses pH control systems.

ORP Probes

DESIGN PRESSURES

Vacuum to 150 psig (10.6 bars) is standard; special assemblies are available for up to 500 psig (35 bars).

DESIGN TEMPERATURES

Generally from 23 to 212°F (−5 to 100°C)

MATERIALS OF CONSTRUCTION

Mounting hardware is available in stainless steel, Hastelloy, titanium, PVC, CPVC, polyethylene, polypropylene, epoxy, polyphenylene sulfide, Teflon, and various elastomer materials; electrodes are available in platinum or gold.

ASSEMBLIES

Submersion, insertion, flow-through, and retractable

CLEANERS

Ultrasonic, water- or chemical-jet washer, and brushes

RANGE

Any span between −2000 mV and +2000 mV

INACCURACY

Typically, inaccuracy is ±10 mV and is a function of the noble metal electrode condition and reference electrode drift; repeatability is about ±3 mV.

COSTS

Electrodes cost from $100 to $500; portable or bench-top laboratory display and control units range from $300 to $1000; transmitters range from $500 to $2000; and cleaners are available from $500 to $2000.

PARTIAL LIST OF SUPPLIERS

Broadley James Corp.; Capital Controls Inc.; Electro-Chemical Devices Inc.; Foxboro Analytical Co.; George Fischer Signet Inc.; Great Lakes Instruments Inc.; Johnson Yokogawa Corp.; Lakewood Instruments; Leeds & Northup, Unit of General Signal; Polymetron; Rosemount Analytical Inc. Uniloc Div.; Sensorex; TBI-Bailey Controls Co.

ORP measurement is important in wastewater treatment applications. Examples of these applications are the removal of heavy metals, such as chromium, from metal finishing wastewater and cooling tower blowdown streams. ORP sensors are also used in cyanide removal, which is often required when heavy metals are removed. In sanitary wastewater treatment, ORP measurement controls the addition of an oxidant for odor control. Also, in most aerobic or anaerobic biological digestion processes, bacterial health can be judged on the basis of ORP measurements.

Section 7.43 discusses various ORP control systems; this section describes the ORP detector only.

PRINCIPLES OF ORP MEASUREMENT

In oxidation-reduction reactions, the substances involved gain or lose electrons and show different electron configurations before and after the reaction. Oxidation is the overall process by which a specie in a chemical reaction loses one or more electrons and increases its state of oxidation. An oxidant is a substance capable of oxidizing a chemical specie; it acquires the electron or electrons lost by the specie and is itself reduced in the overall process. Reduction is the overall process in which a specie in a chemical reaction gains one or more electrons and decreases its state of oxidation. A reductant is a substance capable of reducing a chemical specie; it loses the electrons gained by the specie and is itself oxidized in the overall process.

An ORP reaction, therefore, involves an electron exchange capable of doing work. This capability is expressed in terms of potential for a half-cell, or electron, reaction. Table 7.7.5 lists the potentials for standard conditions, that is, where reactants and products are at unit activity. The voltages in this table are referenced to the standard hydrogen electrode (SHE), which is assigned the value of 0.000 V.

Note that the reactions in Table 7.7.5 are written as reductions, which is the almost universally used convention. In this section, the term ox/red indicates the oxidized form on the left side of the equation and the reduced form on the right. For example, the standard potential for ferric iron, Fe^{3+}, reduced to ferrous iron, Fe^{2+}, is written as $E°Ox/Red = +0.770$ V.

$E_{Red/Ox} = -E_{Ox/Red}$ simply means that the polarity is reversed when the reaction is written as an oxidation reaction. For example, $Fe^{2+} = Fe^{3+} + e^-$. $E°_{Red/Ox} = -0.770$ V.

For example, in a common industrial process, hexavalent chromium is reduced with a ferrous sulfate solution. The half-reactions are:

TABLE 7.7.5 REDUCTION POTENTIALS OF SOLUTION IN ORP MEASUREMENT

Reduction	E°, Volts
$O_3 + 2H_3O^+ + 2e^- = O_2 + 3H_2O$	+2.070
$Cr_2O_7^{2-} + 14H_3O^+ + 6e^- = 2Cr^{3+} + 21H_2O$	+1.330
$ClO^- + H_2O + 2e^- = Cl^- + 2OH^-$	+0.890
$Fe^{3+} + e^- = Fe^{2+}$	+0.770
Ag/AgCl electrode 4 M KCl	+0.199
$2H_3O^+ + 2e^- = H_2 + 2H_2O$	0.000
$Zn^{2+} + 2e^- = Zn$	−0.763
$CNO^- + H_2O + 2e^- = CN^- + 2OH^-$	−0.970
$Na^+ + e^- = Na$	−2.711

$$Cr_2O_7^{2-} + 14\,H_3O^+ + 6e^- = 2\,Cr^{3+} + 21\,H_2O$$

$$E^\circ_{Ox_2/Red_2} = 1.330$$

$$Fe^{2+} = Fe^{3+} + 1\,e^-$$

$$\overline{Cr_2O_7^{2-} + 6\,Fe^{2+} + 14\,H_3O^+ = 2Cr^{3+} + 6Fe^{3+} + 21\,H_2O}$$

$$E_{Red_1/Ox_1} = -0.770 \qquad\qquad 7.7(4)$$

The reaction is not complete at pH 4.0 but is substantially complete at pH 2.0, assuming that a concentration of 10^{-6} M represents completion.

In industrial and laboratory work, ORP cell potentials are measured, not against a SHE but against an Ag/AgCl, 4 M KCl reference, $E^\circ_{Ox/Red} = +0.199$ V, or a saturated calomel electrode (SCE) whose $E^\circ = 0.244$ V. The following equation converts to the potential measured, designated as $E_{AgCl\ ref}$:

$$E_{meas} = E_{cell} - E_{AgCl} \qquad\qquad 7.7(5)$$

For example, the E° for the cell is:

$$Fe^{3+} + e^- = Fe^{2+}\ \ E_{Ox/Red} = 0.770\ \text{vs SHE} \qquad 7.7(6)$$

and the potential measured versus the silver–silver chloride reference is:

$$\begin{aligned}E_{meas/AgCl\ ref} &= +.770\ \text{V} - .199\ \text{V}\\ &= 0.571\ \text{V}\end{aligned} \qquad\qquad 7.7(7)$$

Absolute potentials in ORP measurement are not always used. Most equipment manufacturers use slightly modified pH analyzers for ORP measurement. These instruments normally have the standard or zero adjustment of the parent pH meter. Furthermore, in general, the reverse of polarity is achieved merely by reversing the inputs.

Microprocessor-based ORP analyzers generally have wide rangeability with high-resolution digital displays, alarm–control setpoints, and output signal scaling limits.

Most chemical reaction systems involving electron exchange are controlled near the equivalence point with a controlled excess of added reagent so that the reactions are driven to completion. Thus, most ORP reactions are controlled just beyond the steep portion of the titration curve.

EQUIPMENT FOR ORP MEASUREMENT

The basic instrumentation for ORP measurement closely parallels that for pH measurement. In fact, many instrument suppliers use slightly modified pH analyzers, with a changed sensitivity and an mV scale in place of a pH scale. The hardware used to install the electrodes in the process stream is generally the same as that used in pH systems (see Figure 7.7.9).

Two major differences exist between an ORP system and a pH system. One difference is the sensing electrode, which is normally a noble metal, typically platinum or gold, although other metals and carbon have been used.

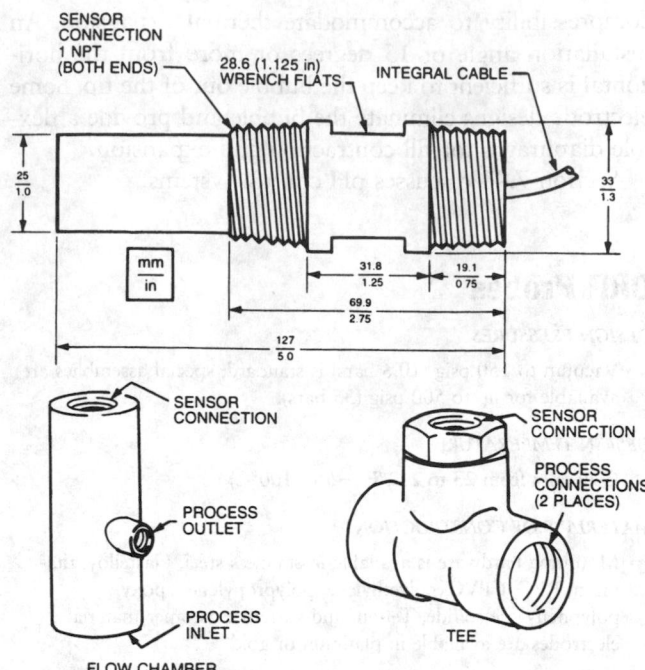

FIG. 7.7.9 ORP and pH probes packaged and mounted in the same way. (Courtesy of The Foxboro Co.)

The second major difference is in temperature compensation. Process pH systems are typically temperature-compensated, whereas ORP systems are almost never temperature-compensated.

Basic thermodynamics apply to pH and ORP as expressed by the classic Nernst equation. For oxidation-reduction half-cell reactions, this equation can be represented as:

$$E_{cell} = E^\circ_{Ox/Red} + \frac{2.303\,RT}{nF}\log\left[\frac{Ox}{Red}\right] \qquad 7.7(8)$$

where:

E° = the potential under standard conditions of unit activity referred to the SHE
R = the gas constant, 1.986 cal per mol degree
F = Faraday's constant
T = temperature in °K
n = number of electrons exchanged in the reaction

Table 7.7.5 shows how the n values change from reaction to reaction. These changes, plus the fact that a given ORP reaction can encompass side reactions, reveal why it is difficult, if not impossible, to temperature-compensate an ORP reaction. In the Nernstian representation of pH, n always equals 1.

In Equation 7.7(8), the standard potential $E^0_{Ox/Red}$ is found in tables in handbooks and in the value relative to the standard hydrogen electrode. Therefore, the E_{cell} for the prevailing concentration is also relative to the SHE.

Often, ORP and pH electrodes can be mounted in the same tees or flow-through chambers and can have identical appearances. The electrode shown in Figure 7.7.9 is a ruggedized flat-glass electrode with identical dimensions for both pH and ORP services; the only difference is the gold or platinum wire tip in the ORP version.

PRACTICAL APPLICATION OF ORP

The chromium reaction typically takes place at a pH of 2.0 to 2.5. At this pH, the smell of sulfur dioxide is present when the sulfite ion is in slight excess. An experienced operator can adjust his control point potential to attain a slight odor of sulfur dioxide and then make further adjustments based on laboratory analysis for hexavalent chromium. Setting up a system based on calculation is possible when all reactants and products are known. However, this case is rare in industrial processes, and most applications require analytical verification of results.

The ORP responses in two common applications are illustrated by the titration curves in Figures 7.7.10 and 7.7.11. These curves are only examples. The responses can vary considerably from one installation and process composition to another. The actual control points must be finely adjusted after system startup.

CARE OF AN ORP SYSTEM

Maintaining an ORP measuring and control system is comparable to maintaining a similar pH system. However, because standards analogous to pH buffers are less common, maintenance is sometimes cut short. The most pressing maintenance problem with ORP is the noble metal sensing electrode. It is subject not only to coating but also to poisoning, both of which can result in sluggish or inaccurate measurement of the potential. This inaccuracy can result in improper demand for the reagent because a control point not representative of the reagent excess is held. Jones reported on the use of standards using quinhydrone-saturated pH buffers to establish known potentials as a check

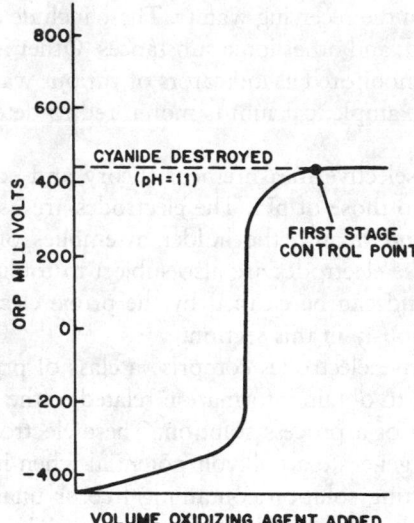

FIG. 7.7.11 Cyanide oxidation titration curve.

on the condition and response of the electrode system. The recommended treatment for either a change of span or a shift in potential is cleaning with aqua regia.

Ion-Selective Electrodes

TYPE OF ELECTRODE

  Glass, solid-state, solid matrix, liquid-ion exchanger, and gas sensing

STANDARD DESIGN PRESSURE

  Generally dictated by the electrode holder; 0 psig for the solid-state and liquid-ion exchanger, 0 to 100 psig (0 to 7 bars) for most electrode types, and over 100 psig (over 7 bars) for solid-state designs

STANDARD DESIGN TEMPERATURE

  32 to 122°F (0 to 50°C) for the solid matrix and liquid-ion exchanger; 23 to 176°F (−5 to 80°C) for most others, with 212°F (100°C) intermittent exposure permissible

RANGE

  From fractional ppm to concentrated solutions

RELATIVE ERROR

  For direct measurements, an absolute error of ±1.0 mV is equivalent to a relative error of ±4% for monovalent ions and ±8% for divalent ions; for end-point detection or batch control, ±0.25% or better is possible; and for expanded scale commercial amplifiers, the error is better than ±1% of full scale.

COST

  Similar to those of pH installations; electrodes, $300 to $700; systems, $2000 to $10,000

PARTIAL LIST OF SUPPLIERS

  Corning; Fisher Scientific; The Foxboro Co.; Great Lakes Instruments, Inc.; HNU Systems, Inc.; Horiba Instruments, Inc.; Ingold Electrodes, Inc.; Leeds and Northrup; Orion Research, Inc., Radiometer; Rosemount Analytical, Inc.

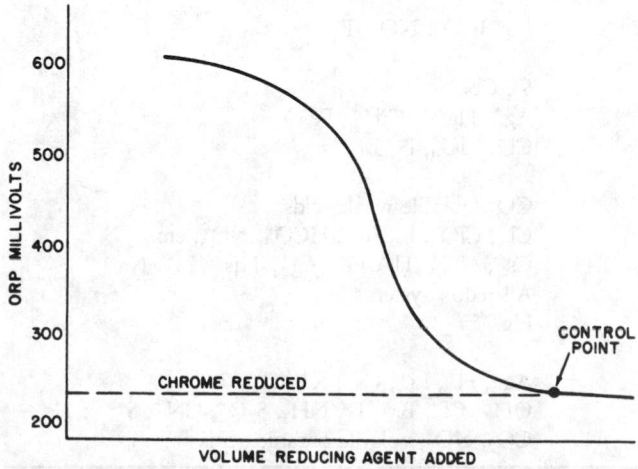

FIG. 7.7.10 Chrome reduction titration curve.

Wastewater treatment plant effluents must be monitored for all ionic substances that are deleterious to humans or animal life in the receiving waters. These include cyanides, sulfides, lead, and other ionic substances. Other ionic substances are monitored as indicators of various water properties. For example, calcium is monitored to detect water hardness.

The ion-selective measurement theory and equipment are similar to those of pH. The electrodes are usually the same size and fit into the holder assemblies of the pH probes. These electrodes are also subject to fouling by oil or slimes, and can be cleaned by the probe cleaners discussed previously in this section.

Ion-selective electrodes comprise a class of primary elements used to obtain information related to the chemical composition of a process solution. These electrochemical transducers generate a millivolt potential when immersed in a conducting solution containing free or unassociated ions to which the electrodes are responsive. The potential magnitude is a function of the logarithm of measured ion activity (*not* the total concentration of that ion) as expressed by the Nernst equation (see Equation 7.7[9]). The familiar pH electrode for measuring hydrogen ion activity is the best known ion selective electrode and was the first to be commercially available. The potential developed across an ion-selective membrane is related to the ionic activity as shown by the Nernst equation as follows:

$$E = \frac{2.3\,RT}{nF} \log \frac{a_1}{a_{int}} \qquad 7.7(9)$$

where:

E	= the potential developed across the membrane
a_1	= the activity of the measured ion in the sample or process
a_{int}	= the activity of the same ion in the internal solution
2.3 RT/nF	= the Nernst slope, or the slope of the calibration curve, and is a function of the absolute temperature T and the charge on the ion being measured n
R	= the gas law constant

With few exceptions—notably the silver-billet electrode for halide measurements and the sodium-glass electrode—the pH electrode was the only satisfactory electrode available to the process industry prior to 1966. Currently, more than two dozen electrodes are suitable for industrial use. Table 7.7.6 lists several electrodes commercially available for

TABLE 7.7.6 ION-SELECTIVE ELECTRODES

Ion/Specie	Type of Electrode	Lower Detectable Limit, ppm	Principal Interferences
Ammonia	Gas-sensing	0.009	Volatile amines
Bromide	Solid-state	0.04	CN^-, I^-, $S^=$
Cadmium	Solid-state	0.01	Ag^+, Hg^{++}, Cu^{++}, Fe^{++}, Pb^{++}
Calcium	Solid matrix/ liquid membrane	0.2	Zn^{++}, Fe^{++}, Pb^{++}, Cu^{++}, Ni^{++}, Sr^{++}, Mg^{++}, Ba^{++}
Carbon dioxide	Gas-sensing	0.4	Volatile weak acids
Chloride	Solid-state	0.2	Br^-, CN^-, $S^=$, SCN^-, I^-
Chloride	Liquid membrane	0.2	ClO_4^-, Br^-, I^-, NO_3^-, OH^-, F^-, OAc^-, $SO_4^=$, HCO_3^-
Copper(II)	Solid-state	0.006	Ag^+, Hg^{++}, Fe^{+++}
Cyanide	Solid-state	0.01	$S^=$, I^-
Divalent cation*	Solid matrix/ liquid membrane	0.001	—
Fluoroborate (BF_4^-), (boron)	Liquid membrane	0.11	I^-, HCO_3^-, NO_3^-, F^-
Iodide	Solid-state	0.006	$S^=$, CN^-
Lead	Solid-state	0.2	Ag^+, Hg^{++}, Cd^{++}, Fe^{++}
Nitrate	Solid matrix/ liquid membrane	0.3	Cl^-, ClO_4^-, I^-, Br^-
Nitrite	Gas-sensing	0.002	CO_2, volatile weak acids
Perchlorate	Liquid membrane	0.7	Cl^-, ClO_3^-, I^-, Br^-, HCO_3^-, NO_3^-, etc.
Potassium	Liquid membrane	0.04	Cs^+, NH_4^+, Tl^+, H^+, Ag^+, $Tris^+$, Li^+, Na^+
Redox (platinum)	Solid-state	Varies	All redox systems
Silver/sulfide	Solid-state	0.01 Ag 0.003 S	Hg^{++}
Sodium	Glass	0.02	Ag^+, H^+, Li^+, Cs^+, K^+, Tl^+
Thiocyanate	Solid-state	0.3	OH^-, Cl^-, Br^-, I^-, NH_3, $S_2O_3^=$, CN^-, $S^=$
Sulfur dioxide	Gas-sensing	0.06	CO_2, NO_2, volatile organic acids

*The water hardness electrode is also known as the divalent cation electrode.

process applications. Other research sensors are also available.

Electrode Types

Electrodes are classified by the sensing membrane used. Environmental engineers use glass electrodes to detect sodium, ammonium, and potassium. They can use pH electrodes for carbon dioxide or ammonia detection by covering the glass membrane with a permeable membrane sac filled with a pH buffer. In this case, the gas diffuses in and out of the permeable membrane, and the resulting pH change is related to gas activity.

Solid-state electrodes are made from crystalline membranes. The fluoride electrode has a single crystal, which is dropped in lanthanum fluoride to form a sensing membrane. Silver and sulfide membranes are silver sulfide pellets. These membranes are sealed in epoxy (see Figure 7.7.12). Table 7.7.7 lists some solid-state electrodes and the composition of their membranes. Metal can be deposited on and an electrical lead can be connected to the surface of some pressed pellets and single crystalline silver-salt membranes (see Figure 7.7.13).

Liquid Ion Exchange Electrodes

A membrane can be saturated with an organic ion exchange material dissolved in an organic solvent. The electrode in Figure 7.7.14 has an internal aqueous filling so-

TABLE 7.7.7 SOLID-STATE ELECTRODES AND THEIR MEMBRANE COMPOSITION

Electrode	Membrane	Form
Fluoride	LaF_3	Single crystal
Silver/sulfide	Ag_2S	Pressed pellet
Chloride,	AgX^*	Single crystal
bromide	AgX-Ag_2S	Pressed pellet
or iodide		
Cyanide	AgI-Ag_2S	Pressed pellet

*X = Cl, Br or I.

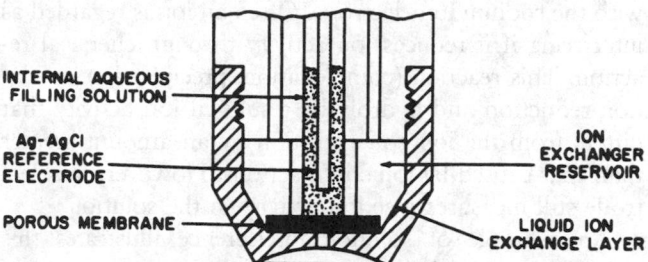

FIG. 7.7.14 Cross section of a divalent cation electrode tip.

lution, in which the reference electrode is immersed, and an ion exchange reservoir of nonaqueous water-immiscible solution that wicks into the porous membrane. The bottom layer of the material is the unknown process solution, the center is the nonaqueous liquid ion exchange solution, and the top layer is the internal aqueous solution. The electrodes cannot be used in nonaqueous solutions because the liquid ion exchanger would dissolve. In the solid-state version of this probe, the ion exchanger is permanently embedded in a plastic matrix with a nonporous membrane.

INTERFERENCES

All ion selective electrodes are similar in operation and use. They differ only in the process by which the ion to be measured moves across the membrane and by which other ions are kept away. Therefore electrode interferences must be discussed in terms of membrane materials.

Glass electrodes and solid-matrix/liquid-ion exchange electrodes both function by an exchange of mobile ions within the membrane, and ion exchange processes are not specific. Reactions occur among many ions with similar chemical properties, such as alkali metals, alkaline earths, or transition elements. Thus, a number of ions can produce a potential when an ion selective electrode is immersed in a solution. Even the pH glass electrode responds to sodium ions at a high pH (low hydrogen ion activity). Fortunately, an empirical relationship can predict electrode interferences, and a list of selectivity ratios for the interfering ions is available from the manufacturers' specifications or other chemical publications.

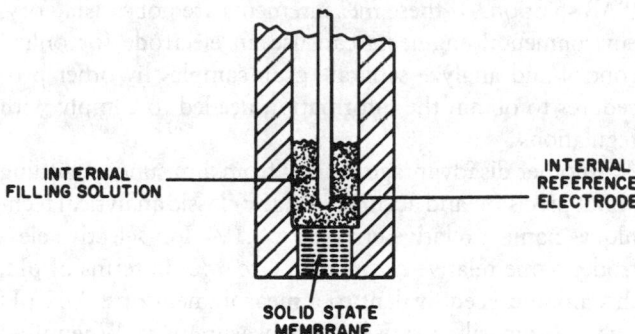

FIG. 7.7.12 Solid-state membrane electrode.

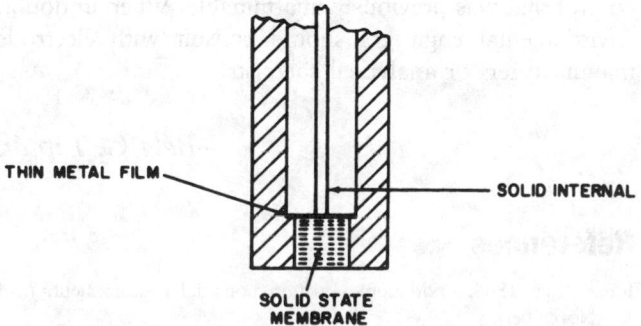

FIG. 7.7.13 Solid-state membrane electrode with solid internals.

Solid-state matrix electrodes are made of crystalline materials. Interferences resulting from ions moving into the solid membrane are not expected. Interference usually occurs from a chemical reaction with the membrane. An interference with the silver-halide membranes (for chloride, bromide, iodide, and cyanide activity measurements) involves a reaction with an ion in the sample solution, such as sulfide, to form a more insoluble silver salt.

A true interference produces an electrode response that can be interpreted as a measure of the ion of interest. For example, the hydroxyl ion, OH^-, causes a response with the fluoride electrode at fluoride levels below 10 ppm. Also, the hydrogen ion, H^+, creates a positive interference with the sodium ion electrode. Often an ion is regarded as interfering if it reduces ion activity through chemical reaction. This reaction (complexation, precipitation, oxidation-reduction and hydrolysis) results in ion activity that differs from the ion concentration by an amount greater than that caused by ionic interactions. However, the electrode still measures true ion activity in the solution.

An example of solution interference illustrates this point. A silver ion in the presence of ammonia forms a stable silver–ammonia complex that is not measured by the silver electrode. Only the free, uncombined silver ion is measured. Environmental engineers can obtain the total silver ion from calculations involving the formation constant of the silver–ammonia complex and the fact that the total silver equals the free silver plus the combined silver. Alternately, they can draw a calibration curve relating the total silver (from analysis or sample preparation) to the measured activity. The ammonia is *not* an electrode interference.

Most confusion stems from the fact that analytical measurements are in terms of concentration without regard to the actual form of the material in solution, and electrode measurements often disagree with the laboratory analyst's results. However, the electrode reflects what is actually taking place in the solution at the time of measurement. This information can be more important in process applications than the classic information. With suggested techniques, environmental engineers can reconcile the two measurements.

ADVANTAGES AND DISADVANTAGES

Compared to other composition-measuring techniques, such as photometric, titrimetric, chromotographic, or automated-classic analysis, ion-selective electrode measurement has several advantages. Electrode measurement is simple, rapid, nondestructive, direct, and continuous. Therefore, it is easily applied to closed-loop process control. In this respect, it is similar to using a thermocouple for temperature control. Electrodes can also be used in opaque solutions and viscous slurries. In addition, the electrodes measure the free or active-ionic species under process conditions and the status of a process reaction.

However, several disadvantages exist. The specificity of ion-selective electrodes is not as good as that of the glass pH electrode. Interferences vary from minor to major; environmental engineers must consult publications and manufacturers' data on limitations for each electrode. Also, the electrodes do not measure total ion concentration, although this parameter is often requested. Prior to the introduction of electrodes, concentration information was the only information available from the chemists' classic measurement techniques. Control laboratory chemists and process engineers do not think in terms of activity, even when making pH measurements. This habit may disappear as the ion-selective technique becomes more popular.

Sometimes concentration is a beneficial measurement, for example, in material-balance calculations or pollution control. Knowledge of material balance allows engineers to predict where a process reaction will occur. This information is necessary if a process is to be controlled by introducing changes that nullify those predicted.

In pollution control, environmental engineers believe that many ions, even in the combined state, are detrimental to life forms. For example, fluorides, cyanides, and sulfides are deleterious to fish and humans in many combined forms. However, they are not detected by ion-selective electrodes in the combined state. Consequently, pollution control agencies usually require concentration information. Electrodes can be used for concentration measurement if they are calibrated with solutions matching the process or ISAB solutions. If these measurements are not satisfactory, environmental engineers can use an electrode for online control and analyze separate grab samples by other procedures to obtain the information needed to comply with regulations.

Another disadvantage derives from a misunderstanding about precision and accuracy. Many classic analytical techniques name a relative error of $\pm 0.1\%$. Ion-selective electrodes name relative errors of ± 4 to 8%. In terms of pH, this amount is equivalent to a measurement of ± 0.02 pH units—ordinarily a satisfactory measurement. When used with understanding, ion-selective electrodes can supply satisfactory composition information and afford closed-loop control that was previously unattainable. When in doubt, environmental engineers should consult with electrode manufacturers or analytical chemists.

—*Béla G. Lipták*

References

Jones, R.H. 1966. Oxidation reduction potential measurement. *JISA* (November).

Lipták, Béla G. 1995. *Analytical Instrumentation.* Radnor, Pa.: Chilton.

7.8
OXYGEN ANALYZERS

In the aquatic life cycle, oxygen plays a critical role. If the water is transparent, algae and other plant life generate oxygen as they build their body cells through photosynthesis. The other half of this cycle is respiration, in which bacteria and other animal life forms consume algae and other larger organic molecules while using the dissolved oxygen (DO) in the water and exhaling carbon dioxide. Therefore, two kinds of oxygen measurements are required in the operation of wastewater treatment plants:

The DO concentration of receiving waters must be monitored because the DO amount signals the life-supporting capacity of water. When the DO content drops to zero, the water body can no longer support aerobic life, bacteria and other animals suffocate, and the water becomes an open sewer.

The second oxygen measurement is made on the wastewater effluent discharged into receiving water. Here, the measurement determines the amount of damage that the discharged effluent will do to the receiving water. This damage is measured in the milligrams of DO that a liter of effluent will take from the receiving water, as bacteria decomposes the organic material in the effluent. This measurement is called the BOD of the wastewater effluent.

In addition to measuring the oxygen demand biologically (by bacteria), environmental engineers can measure it chemically; this demand is called COD. Environmental engineers also measure the carbon content of the effluent using total carbon analyzers or total organic carbon analyzers.

This section briefly discusses the use of in-place probes versus sampling and sample filtering versus homogenization. Then it describes DO detection probes and lists BOD and other oxygen demand sensors.

Bypass Filters and Homogenizers

When detecting DO, an environmental engineer can use a probe and insert it directly into the process or take a sample and deliver it to a DO analyzer. The probe eliminates the dead time of sample transportation, which degrades closed-loop control. However, the probe requires effective and unattended cleaning attachments. The advantage of sampling is that loop components are more accessible. Therefore, DO probes are used for online control, and sampling systems are used only for monitoring.

In DO detection, the measured parameter is in the liquid phase and the solids can be filtered out. However, in BOD measurement, oxygen demand occurs mostly in the solids. Therefore, they must be included in the evaluation.

Consequently, most BOD and COD analyzers require a sampling system in which homogenization (liquification of solids), is used instead of filters, to prevent plugging and retain sample integrity.

SLIPSTREAM AND BYPASS FILTERS

To minimize transportation lag, the flow rate system takes a large slipstream from the process and tubes it to the analyzer. Because the sample flow to the analyzer is small, the analyzer uses only a small portion of this stream and returns the bulk to the process (see Figure 7.8.1). This arrangement permits the high-flow rate system to continuously sweep the main volume of the filter, minimizing lag time; at the same time, only the low-flow stream to the analyzer is filtered, maximizing filter life.

A slipstream filter requires inlet-to-outlet ports at opposite ends of the filter element to allow the high flow rate of the bypassed material to sweep the surface of the filter element and reservoir. It also requires a third port connected to the low-flow rate line to the analyzer so that filtered samples can be withdrawn from the filter reservoir.

If bubble removal from a liquid is required, this function can be combined with slipstream filtration since the recommended flow direction for bubble removal is outside-to-inside and the separated bubbles are swept out of the housing by the bypass stream. In this case, the liquid feed should enter at the bottom of the housing, and the bypass liquid should exit at the top of the housing.

Some samples can be separated using cyclone separators. In this device (see Figure 7.8.2), the process stream enters tangentially to provide a swirling action, and the cleaned sample is taken near the center. The transportation lag can be kept to less than 1 min, and the unit is applicable to both gas and liquid samples. This type of cen-

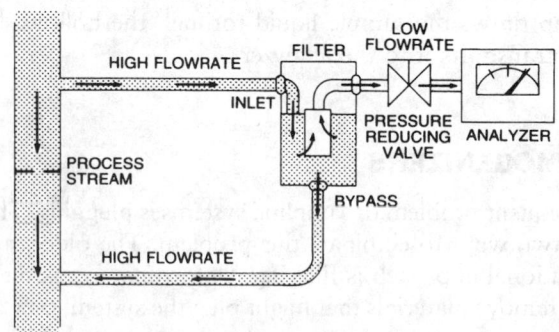

FIG. 7.8.1 Slipstream or bypass filtration.

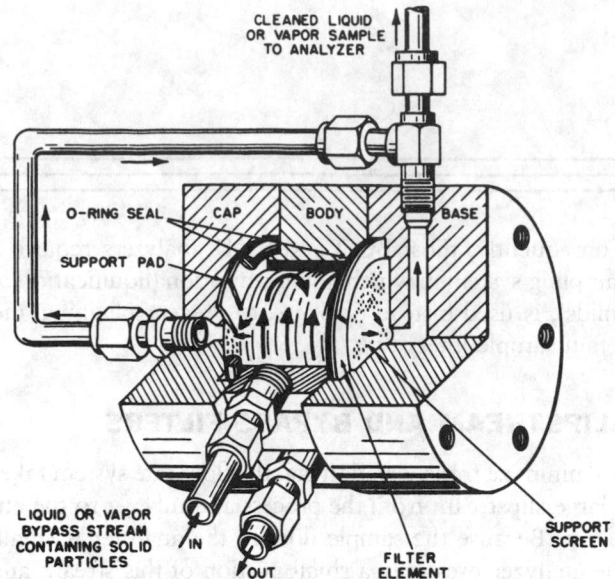

FIG. 7.8.2 Bypass filter with cleaning action amplified by the swirling tangentially entering sample.

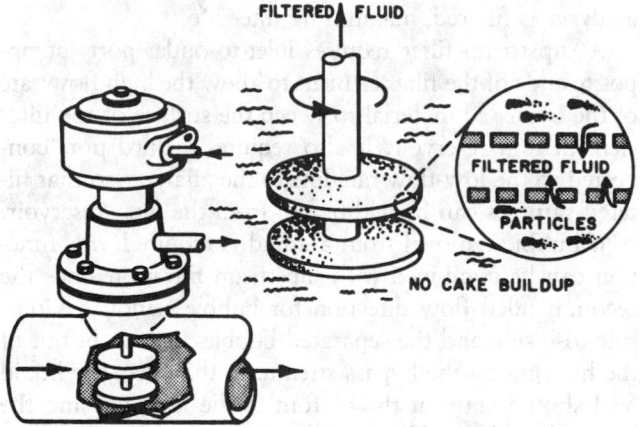

FIG. 7.8.3 Rotary disc filter.

trifuge can also separate sample streams by gravity into their aqueous and organic constituents.

Another good filter design is the rotary disc filter (see Figure 7.8.3). Here, the filtered liquid enters through the small pores in the self-cleaning disc surfaces. The sample pump draws the sample liquid through the hollow shaft and transports it to the analyzer.

HOMOGENIZERS

A frequent problem of sampling systems is plugging. There are two ways to eliminate this problem. The older, more traditional approach is filtering. Unfortunately, as the filters remove materials that might plug the system, they also remove process constituents and make the sample less representative.

The newer approach is to eliminate plugging potential by reducing solid particle size (homogenization) while maintaining sample integrity. Thus, when pulverizers replace filters, analyzer samples become more representative.

Homogenizers disperse, disintegrate, and reduce the solid particle size, reducing agglomerates and liquifying the sample. Homogenizers can be mechanical, using rotor-stator-type disintegrator heads. In this design, the rotor acts as a centrifugal pump to recirculate the slurry, while the shear, impact collision, and cavitation at the disintegrator head provide homogenization.

In ultrasonic homogenizers, high-frequency mechanical vibration is introduced into a probe (horn), which creates pressure waves as it vibrates in front of an orifice (see Figure 7.8.4). As the horn moves away, it creates large numbers of microscopic bubbles (cavities). When it moves forward, these bubbles implode, producing powerful shearing action and agitation due to cavitation. Such homogenizers are available with continuous flow-cells for flow rates up to 4 gph (16 lph) and can homogenize liquids to less than 0.1-μm particle sizes. The flow cell is made of stainless steel and can operate at sample pressures of up to 100 psig (7 bars).

AUTOMATIC LIQUID SAMPLERS

Automatic liquid samplers collect intermittent samples from pressurized pipelines and deposit them in sample containers. Samples are collected on a time-proportional or flow-proportional basis. Figure 7.8.5 shows a sampler that withdraws a predetermined volume of sample every time the actuator piston is stroked. In time-proportional mode,

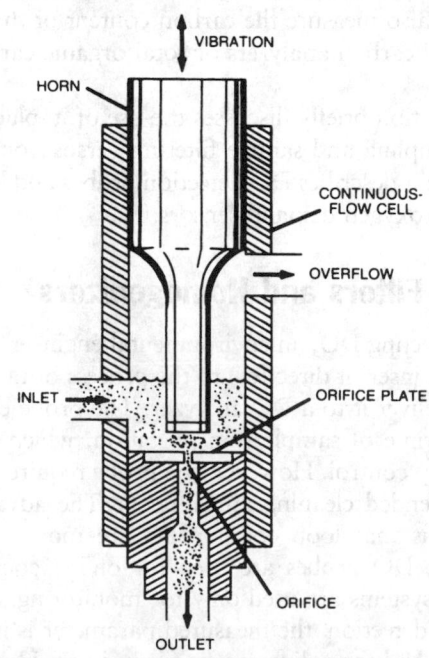

FIG. 7.8.4 Ultrasonic homogenizer. (Reprinted, with permission, from Cole-Parmer Instrument Co.)

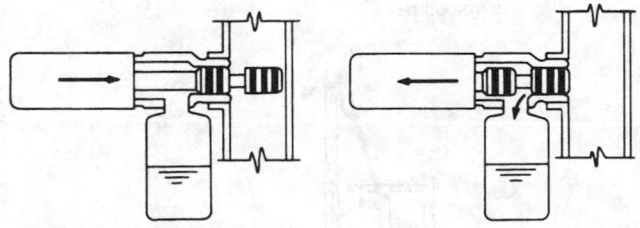

FIG. 7.8.5 Intermittent collection of samples. (Reprinted, with permission, from Bristol Equipment Co.)

this sampling frequency is constant, while in flow-proportional mode, this unit varies sampling frequency as a function of flow.

In some automatic liquid samplers, sampling frequency is adjusted by pneumatic pulse relays or electronic controls. Pulse duration is usually adjustable from 0.25 sec to 1 min, while pulse frequency is adjusted from a few sec up to an hr.

DO Analyzers

TYPES

A. Polarographic; B. Galvanic; C. Coulometric; D. Multiple anode; E. Thallium

OPERATING PRESSURE

Up to 50 psig (3.5 bars) or submersion depths of up to 25 ft (8.3 m)

OPERATING TEMPERATURE RANGE

32 to 122°F (0 to 50°C); special designs up to 175°F (80°C)

FLOW VELOCITY AT SENSING MEMBRANE

Preferably in excess of 1 fps (0.3 m/sec); some can operate down to 0.2 fps (0.06 m/sec).

MATERIALS OF CONSTRUCTION

Typical material for sensor housing is PVC; for electrodes, gold and silver or copper; and for membrane assembly, ABS plastic or stainless-steel, mesh-reinforced Teflon membrane.

SPEED OF RESPONSE

90% in 30 sec; 98% in 60 sec

RANGES

Common ranges are 0 to 5, 0 to 10, 0 to 15, and 0 to 20 ppm; special units are available with ranges up to 0 to 150 ppm or down to the 0 to 20 ppb range used on boiler feedwater applications. Systems can also be calibrated in partial pressure units.

INACCURACY

Generally ±1 to ±2% of the span; industrial transmitter errors are generally within 0.02 ppm over a 0 to 20 ppm range. Thallium cells are available with a 0 to 10 ppb range and can read the DO within an error of 0.5 ppb.

COSTS

Portable, battery-operated, 1.5 to 2% FS units that also read temperature cost from $300 to $700; replacement probes cost about $250; 1% FS, microprocessor-based, portable benchtop units for laboratory or plant service cost from $1000 to $2000; industrial-quality (0.02 ppm error limit) DO probe and a 4- to 20-mA transmitter cost about $3500; and cleaning assemblies cost from $500 to $2000.

PARTIAL LIST OF SUPPLIERS

ABB Kent Inc.; Ametek Inc.; Cole-Parmer Co.; Delta F Corp.; Electro-Nite Co.; Enterra Instrument Technologies Inc.; Fischer & Porter Co.; Foxboro Co.; Great Lakes Instruments Inc.; Hays Republic Corp.; Honeywell Industrial Controls; Horiba Ltd.; Ingold Electrodes Inc.; Leeds & Northrup Co.; Milton Roy Co.; MTL (B); Ohmart Corp.; Orbisphere Laboratories (Switzerland); Robertshaw Controls Co.; Rosemount Analytical, Uniloc Div.; Royce Instrument Corp.; Teledyne Analytical Instruments; Waltron Ltd.; Yokogawa Corp. of America; YSI Inc.

In wastewater treatment applications, the DO concentration is often measured within the slimy sludge layer. Therefore, the membrane surface must be kept clean. Keeping the membrane surface clean is achieved partly by coating the membrane with a growth-inhibiting chemical, toxic to bacteria, and partly by attaching a vibratory paddle-cleaner, which cleans the membrane by back-and-forth motion close to the surface (see Figure 7.8.6).

POLAROGRAPHIC CELL

The basic polarographic cell shown in Figure 7.8.7 has two noble-metal electrodes and requires a polarizing voltage to reduce oxygen. Sample DO diffuses through the membrane into the electrolyte, which is usually an aqueous KCl solution. If the polarizing voltage is constant (usually 0.8 V supplied by a mercury battery) across the electrodes, the oxygen is reduced at the cathode, and the resulting current flow is directly proportional to the electrolyte oxygen content.

Polarographic cells, like galvanic cells, are affected by temperature. Therefore, they require controlled sample temperature or temperature compensation to attain high-precision measurements of ±1 to 2% in accuracy. If sample temperature varies between 32 and 110°F (0 and 43°C), measurement error rises to approximately ±6% in some designs.

Both galvanic and polarographic cells require a minimum sample flow velocity. This velocity eliminates stagnant layers of sample over the membrane, which would otherwise interfere with continuous oxygen transfer into the cell. Higher sample velocities are also beneficial because of their scrubbing action. Some suppliers provide a combination cell and pump unit where the flow velocity of 5 fps (1.5 m/sec) is directed against the membrane for maximum cleaning effect.

GALVANIC CELL

The ranges of the galvanic-cell DO analyzer are as low as 0 to 20 ppb for applications such as measuring DO content in boiler feedwater.

All galvanic cells consist of an electrolyte and two electrodes (see Figure 7.8.8). Electrolyte oxygen content is

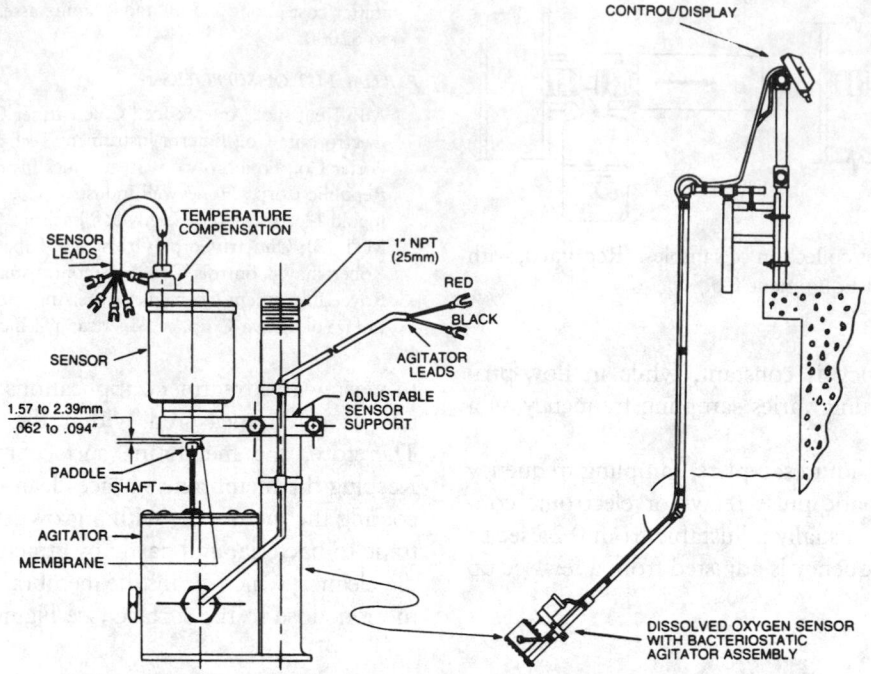

FIG. 7.8.6 Cleaner assembly for freeing the membrane surface of a DO probe of buildup or biological growths. (Reprinted, with permission, from Robertshaw Controls Co.)

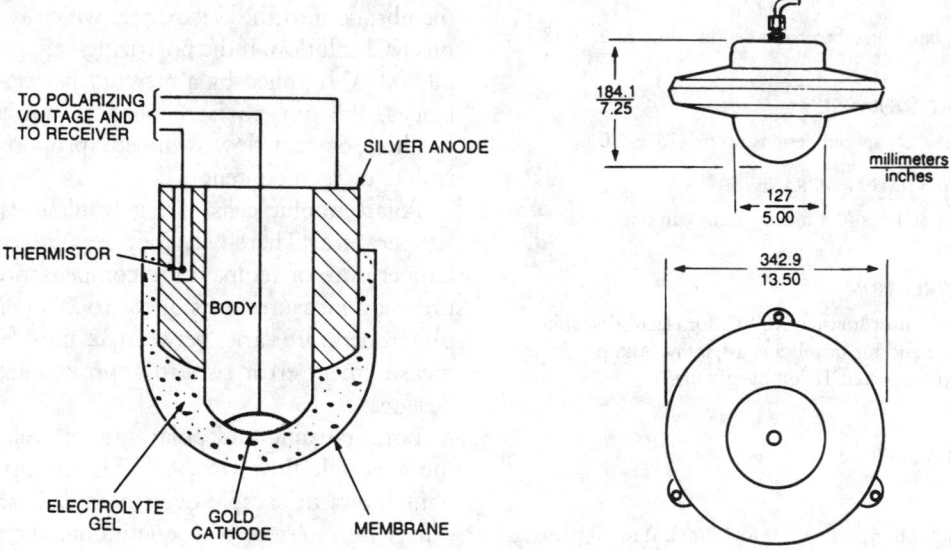

FIG. 7.8.7 Probe-type polarographic cell oxygen detector and flotation collar mount.

brought into equilibrium with that of the sample. The electrodes are polarized by an applied voltage that causes electrochemical reactions when oxygen contacts the electrodes. In this reaction, the cathode reduces oxygen into hydroxide, thus releasing four electrons for each molecule of oxygen. These electrons cause a current flow through the electrolyte, with magnitude in proportion to the electrolyte oxygen concentration.

The following gases are likely to contaminate the cell: chlorine and other halogens, high concentrations of carbon dioxide, hydrogen sulfide, and sulfur dioxide.

Special cells have been developed to minimize the effect of background gases. When an acid gas (such as CO_2) that would neutralize a potassium hydroxide electrolyte solution is present in the background, a potassium bicarbonate electrolyte can be used. Special cells are also available

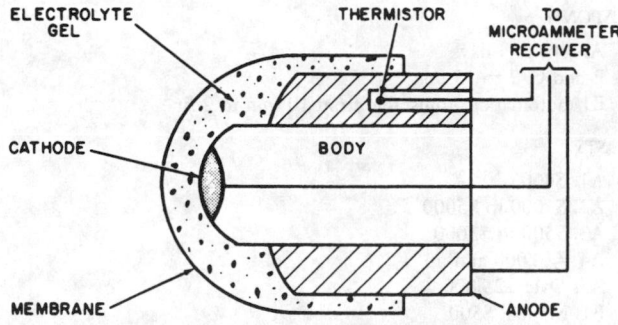

FIG. 7.8.8 Probe-type galvanic cell oxygen detector.

for measuring oxygen in acetylene and fuel gases.

In flow-through cell designs, sampling systems bring the process stream to the analyzer and filter it, scrub it with caustic, or otherwise prepare it for measurement. The probe-type membrane design does not require a sampling system if it can be located in a representative process area where process stream pressure, temperature, and velocity

are compatible with the cell's mechanical and chemical design.

Probe Design

In this design (see Figure 7.8.8), the electrodes are wetted by an electrolytic solution retained by a membrane (usually Teflon). This membrane acts as a selective diffusion layer, allowing oxygen to diffuse into the sensor while keeping foreign matter out. The sensor is usually mounted in a thermostatically controlled housing; therefore, the thermistor compensates for minor temperature variations.

Membrane characteristics are critical to performance. The ideal membrane is inert, stable, strong, permeable to oxygen, and impermeable to other ions and water molecules. In most cases, a compromise solution is accepted.

Figure 7.8.9 shows the design of a gold–copper electrode, galvanic (amperometric) cell and its rail mounting installation. The maintenance requirements of this design

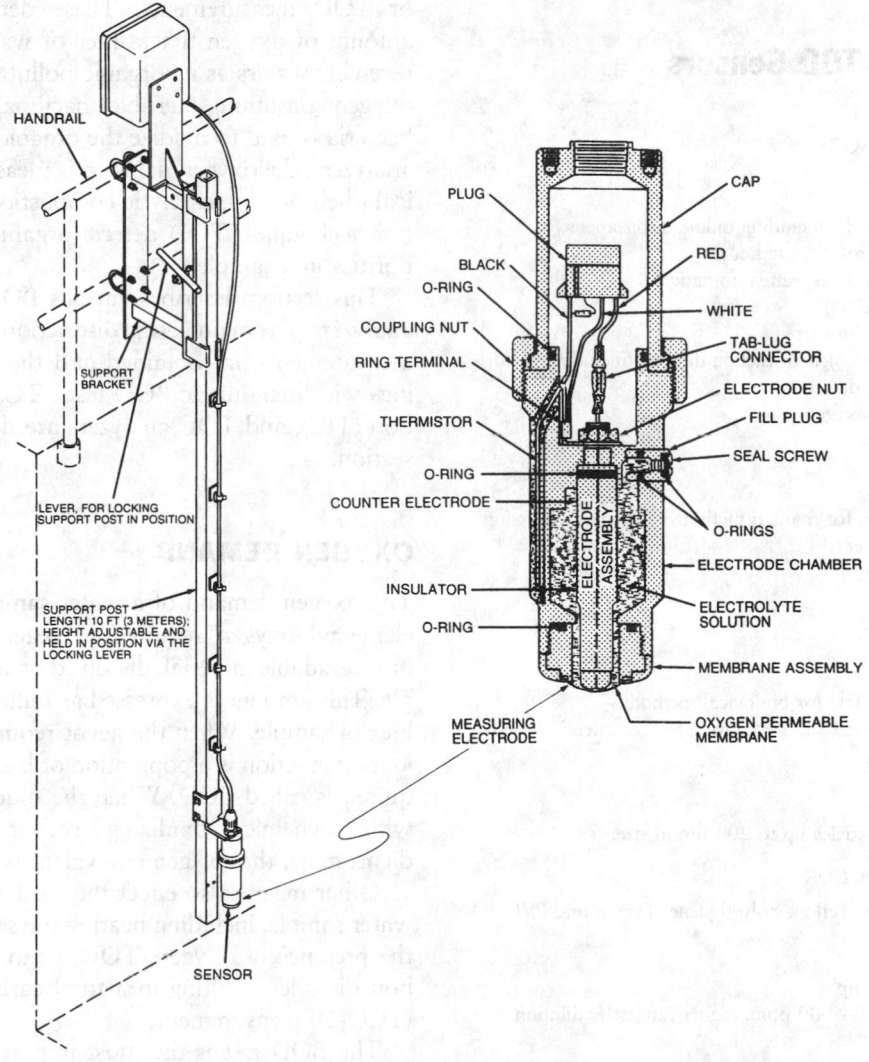

FIG. 7.8.9 Galvanic DO cell and rail mounting installation design. (Reprinted, with permission, from Fischer & Porter Co.)

are reduced with an electrolyte supply that lasts for 2 to 3 years and an easily replaceable membrane assembly. These analyzer systems are available in weatherproof housing, 1% of span inaccuracy, and with 4 to 20 mA of transmitter output.

Flow-Through Design

In these cells, the process sample stream bubbles through the electrolyte. Therefore, the oxygen concentration of the electrolyte is in equilibrium with the sample oxygen content, and the resulting ion current between electrodes represents this concentration.

In some trace analyzer designs, the cathode is made of a porous metal, and the sample gas passes through this electrode, immersed in the electrolyte. Oxygen reduction is usually complete within the pores of this electrode.

Sampling systems are usually provided with these cells, consisting of filtering and scrubbing components and flow, pressure, and temperature regulators.

BOD, COD, and TOD Sensors

TYPES OF MEASUREMENTS

A. Biological agency (BOD)
A1. Winkler titration
A2. DO sensor
A3. Manometric methods (including online respirometer)
A4. Coulometric (electrolysis) methods
A5. BOD for eleven samples, semiautomatic
B. Chemical agency (COD)
B1. Oxidation with dichromate
B2. Combustion (catalytic) with carbon dioxide (including nondispersive IR [NDIR] detector)
B3. Combustion with oxygen
C. TOD

SAMPLING TECHNIQUE

A and B. Grab samples for manual methods; automatic sampling for continuous instruments

SAMPLE PRESSURE

Essentially atmospheric

SAMPLE TEMPERATURE

A. 20°C (68°F) during test for biological methods
B. 150 to 1000°C (302 to 1832°F) during test for chemical methods

SUSPENDED SOLIDS

Sample can contain particles up to 200 μm in size.

MATERIALS OF CONSTRUCTION

A and B. Glass, quartz, Teflon, polyethylene, Tygon, and PVC

RANGES

A and B. 0.1 mg/l and up
C. Standard 0–100 to 0–5000 ppm; higher ranges by dilution

INACCURACY

A and B. 3 to 20% depending on method
C. ±2% of range at the 95% confidence level

RESPONSE

A. 2 hr to 5 days
B and C. 2 to 15 min
B1 (automatic). Adjustable from 10 min to 2 hr

COSTS

A1. $400
A2. $1000 to $5000
A3. $500 to $3000
A4. $10,000 and up
A5. Over $25,000
B1. Manual $500
B2. $18,000
C. Over $15,000

PARTIAL LIST OF SUPPLIERS

Anatel Corp.; Badger Meter Inc. (A); Bran and Luebbe Analyzing, Technicon Industrial Systems (B); Horiba Instruments Inc. (A); Ionics Inc. (B,C); Robertshaw Controls Co. (A); Tech-Line Instruments, Div. of Artech International Inc. (A); Xertex Corp. Delta Analytical (A); YSI Inc. (A)

The total damage caused by discharging wastewater into lakes or rivers is expressed and quantified by BOD, COD, or TOD measurements. These detectors measure the amount of oxygen that a liter of wastewater takes from receiving waters as its organic pollutants are degraded by oxygen-consuming (aerobic) bacteria. In BOD analyzers, bacteria is used to oxidize the organic pollutants; in COD analyzers, the oxygen demand is measured through chemical (dichromate), catalytic combustion, or direct combustion techniques. TOD detects organic and inorganic impurities in a sample.

This section describes various BOD, COD, and TOD analyzers. The main design distinction is the speed at which measurements are obtained and the correlation of readings with manomeric BOD tests. TC, total inorganic carbon (TIC), and TOC analyzers are described later in this section.

OXYGEN DEMAND

The oxygen demand of a water sample is the amount of elemental oxygen required to react with oxidizable or biodegradable material, dissolved or suspended in the sample. This amount is expressed as milligrams of oxygen per liter of sample. When the agent required to effect the oxidation reaction is a population of bacteria, the oxygen required is called BOD. When the oxidation is carried out with a chemical oxidizing reagent such as potassium dichromate, the oxygen equivalent is called the COD.

Other means also effect the oxidation of material in a water sample, including heating the sample in a furnace in the presence of oxygen, TOD, or in the presence of carbon dioxide, resulting in a total carbon dioxide demand (TCO$_2$D) measurement.

The BOD test is the most important oxygen demand measurement for analyzing effluents and receiving waters (streams, lakes, and rivers). The BOD test measures the amount of oxygen used by microorganisms feeding on organic water pollutants under aerobic conditions.

In this test, a bacterial culture is added to the sample under well-defined conditions, and oxygen utilization is measured. Although test procedures are carefully defined, obtaining reproducible results is difficult and the procedure is subject to the influence of many variables, particularly when the wastewater contains a variety of complex materials.

Factors contributing to variations in BOD results are:

- Biological seed used
- pH if not near neutrality
- Temperature if other than 20°C (68°F) (as shown in Figure 7.8.10)
- Sample toxicity
- Incubation time

When the incubation time and temperature are not stated, the general assumption is that the test was run at 20°C for a period of 5 days. For example, in Figure 7.8.10, the BOD test result can be stated as $BOD_5 = 100$ mg/l. Figure 7.8.10 also shows that bacteria first consume carbonaceous material, and only when the carbonaceous materials are all oxidized (around the 15th day) are nitrogenous material consumed.

Five-Day BOD Procedure

If a water sample BOD at 20°C (68°F) is measured as a function of time, a curve such as the one in Figure 7.8.10 is obtained. For the first 10 to 15 days, the curve is approximately exponential, but at about the fifteenth day, a sharp increase occurs that then falls to a steady BOD rate. Because of the length of time and because the curve does not flatten, environmental engineers have universally adopted a standard test period of 5 days for the BOD pro-

cedure. This laboratory procedure requires some skill and training to obtain concordant results.

In the procedure, the environmental engineer mixes a measured sample portion to be analyzed with seeded dilution water so that after 5 days of incubation, the DO in the mixture is still sufficient for biological oxidation of the material in the sample. Of course, this portion cannot be known beforehand; consequently, the environmental engineer must run several dilutions simultaneously for an unknown sample, or use experience as a guide for well-defined samples. Seeded dilution water contains a phosphate buffer (including ammonium chloride), magnesium sulfate, calcium chloride, and ferric chloride, as well as a portion of seeding material. The former group of inorganic materials is frequently referred to as nutrients. The latter group is a suspension of bacteria in water, usually supernatant liquor from a domestic sewage plant.

Seeds can also be prepared from soil, developed from cultures in the laboratory, or obtained from receiving water 2 to 5 mi downstream of the discharge. The environmental engineer determines the DO content of the mixture at the start of the test and again after 5 days of incubation at 20°C in a special BOD bottle. The DO can be determined by the Winkler titration method or instrumentally with a DO membrane electrode. The difference in DO after 5 days is used to calculate the BOD of the original sample. Corrections must be applied for the immediate oxygen demand (due to inorganic reducing materials) and for the oxygen required by the bacteria for sustaining life (endogenous metabolism).

No standard exists for measuring the accuracy of the BOD test. The precision of the method is also difficult to ascertain because of the many variables. However, environmental engineers have tested the single-operator precision of the method using a standard glucose–glutamic acid solution. Using eight types of seed materials, they found the single-operator precision to be 11 mg/l at a level of 223 mg/l, or about 5%. These results were obtained with highly skilled personnel under well-controlled laboratory conditions.

A semiautomatic instrument can measure the BOD of as many as eleven samples. The samples must be manually placed on the instrument turntable, and the controls must be manually set. The instrument provides for the automatic reaeration of samples in which the DO has fallen to low values. The polarographic DO sensor measures the DO on a preset time schedule. The automatic reaeration capability for low DO eliminates the need for dilution, leading to improved precision in BOD results.

The instrument consists of a measuring unit (DO probe, aerator, water-sealing mechanism, unplugging mechanism, sample bottle, and turntable) and a control unit by which all operations are programmed. The measuring unit is housed in a chamber maintained at 20°C. The instrument can store DO data on each sample and calculate the BOD from the DO values as just described.

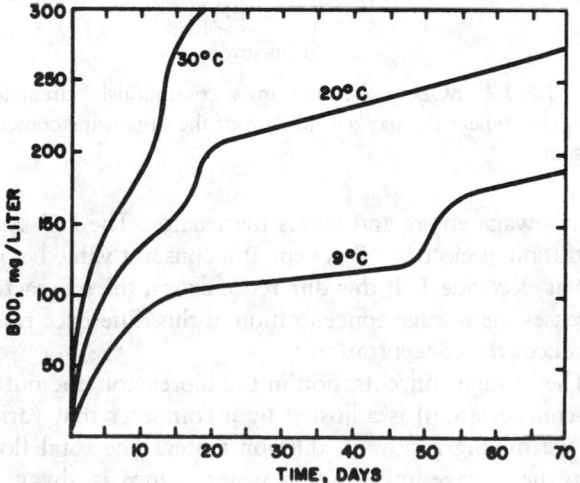

FIG. 7.8.10 Progress of BOD at 9°, 20°, and 30°C (48, 68, and 86°F). The break in each curve corresponds to the onset of nitrification.

Manometric BOD Test

In the standard dilution method, all oxygen required must be inside the BOD bottle since it is sealed in a gas-tight manner at the initiation of the incubation period and air can not enter the sample. In the manometric procedure, the seeded sample is confined in a closed system that includes an appreciable amount of air. As the oxygen in the water is depleted, it is replenished by the gas phase. A potassium hydroxide absorber within the system removes any gaseous carbon dioxide generated by bacterial action. The oxygen removed from the air phase causes a drop in pressure that is measured by a manometer. This drop is then related to sample BOD.

The manometric method lends itself to automatic recording of oxygen utilization since the pressure can be monitored continuously. This monitoring is accomplished in a commercially available automatic respirometer (see Figure 7.8.11). The manometric method introduces the sample, from 1 to 4 liters, into a closed system containing air. The countercurrent circulation of both air and water in the system insures equilibrium between the dissolved and gaseous oxygen. The system provides a carbon dioxide scrubber in the gas-circulation line. A recording manometer detects the utilization of oxygen, and the test is run for several hours. Published data indicate a correlation between 4-hr respirometer BOD and standard BOD. Laboratory and automatic online versions of this instrument are also available.

BOD measurement is inherently a time-consuming process and ill-suited to the requirements of process monitoring or control. The shortest period for the automatic respirometer is 2 hr, much too long for an effective control instrument. However, it is an excellent device for laboratory studies since it can simulate the activated sludge process.

BOD Assessment in Minutes

When BOD concentration is determined in groundwater, waiting 5 days for the results may be acceptable. However, in the control and operation of sewage treatment plants, it is not. The hold-up capacity of industrial and municipal wastewater treatment facilities and the need for closed-loop control necessitates faster sensors. Figure 7.8.12 shows such an analyzer.

In this design, a bioreactor is filled with several plastic rings, the interior of which are protected against mechanical abrasion to provide a growth surface for organisms. A circulation pump quickly distributes the sewage in the bioreactor and keeps the plastic rings in continuous motion. The sewage concentration (nutrient level) in the reactor is at a constant low value, resulting in an oxygen demand of about 3 mg/l. The bioreactor measures and maintains a constant oxygen demand by detecting a decrease in the oxygen concentration at points where the di-

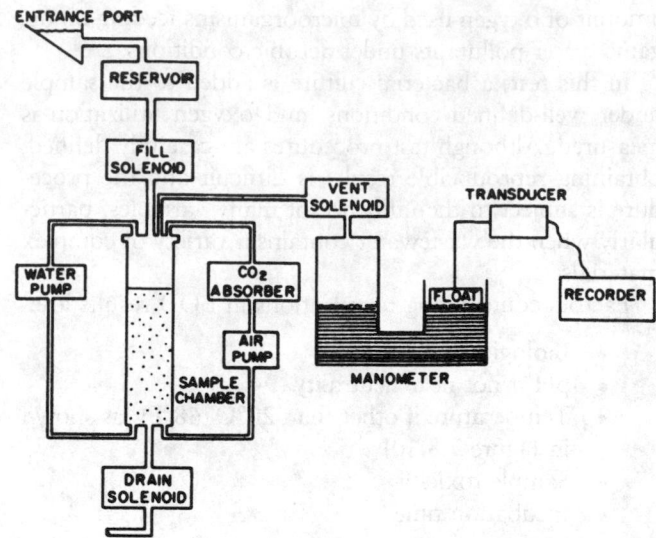

FIG. 7.8.11 BOD determination by an automatic respirometer.

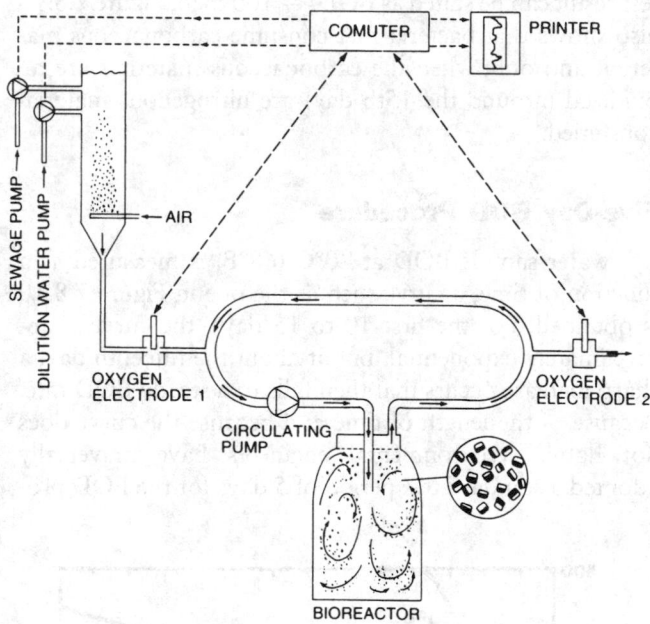

FIG. 7.8.12 BOD assessment in a continuously circulated bioreactor where the oxygen take-up of the organisms controls dilution.

luted sewage enters and leaves the reactor. The DO concentration at electrode 2 is kept at a constant value below that at electrode 1. If this difference drops, the bioreactor increases the sewage concentration; if this difference rises, it reduces the concentration.

The sewage concentration in the bioreactor (the nutrient concentration) is adjusted by a computer that varies sewage mixing ratio and dilution water. The total flow from the sewage and dilution water pumps is always 1 l/min, and the ratio of the two streams is modulated. Therefore, this pumping ratio indicates sewage sample BOD concentration. Environmental engineers have found the correlation between this fast BOD measurement and the 5-day BOD obtained through conventional methods acceptable. Pipe fouling was minimal, and weekly recalibration of oxygen electrodes was satisfactory.

COD

This laboratory method requires skill and training similar to that required for the BOD test. The environmental engineer heats a sample to its boiling point with known amounts of sulfuric acid and potassium dichromate. Using a reflux condenser minimizes the loss of water. After 2 hr, the environmental engineer cools the solution and determines the dichromate amount that reacted with the oxidizable material in the water sample by titrating the excess potassium dichromate with ferrous sulfate using ferrous 1,10-phenanthraline (ferroin) as the indicator. The environmental engineer calculates the dichromate consumed as to the oxygen equivalent for the sample and reports it as oxygen mg/l of the sample.

Interpretations of COD values are difficult since this method of oxidation is markedly different from the BOD method. Although ultimate BOD values can agree with COD values, a number of factors can prevent this concordance including:

1. Many organic materials are oxidizable by dichromate but not biochemically oxidizable and vice versa. For example, pyridine, benzene, and ammonia are not attacked by the dichromate procedure.
2. A number of inorganic substances such as sulfide, sulfites, thiosulfates, nitrites, and ferrous iron are oxidized by dichromate creating an inorganic COD that is misleading when the organic content of wastewater is estimated. Although the seed acclimation factor gives erroneously low results on BOD tests, COD results do not depend on acclimation.
3. Chlorides interfere with the COD analysis and their effect must be minimized for consistent results. The standard procedure provides for a limited amount of chlorides in the sample. Despite these limitations, the dichromate COD is useful in the control of wastewater effluents containing caustic and chlorine, dyeing and textile effluents, organic and inorganic chemicals, paper, paints, plating, plastics, steel, aluminum, and ammonia.

COD Detector

The COD method usually refers to the laboratory dichromate oxidation procedure. It is also applied to other procedures that differ from the dichromate method but involve chemical reaction. These methods are embodied in instruments both for manual operation in the laboratory and for automatic operation online. They have the advantage of reducing the analysis time from days (5-day BOD) and hours (dichromate and respirometer) to minutes.

Automatic Online Designs

Figure 7.8.13 shows an online analyzer with COD ranges from 0–100 ppm to 0–5000 ppm and adjustable measurement cycle times from 10 min to 5 hr. The sample flow can be continuous at rates to 0.25 gpm (1.0 lpm) and can contain solid particles to 100 μ.

The automatic COD analyzer periodically injects a 5-cc sample from the flowing process stream into the reflux chamber, after mixing it with dilution water (if any) and two reagents: dichromate solution and sulfuric acid. The reagents also contain an oxidation catalyst (silver sulfate) and a chemical that complexes chlorides in the solution (mercuric sulfate). The heater boils the mixture at 302°F (150°C), and the cooling water in the reflux condenser recondenses the vapors. The solution is refluxed for a preset time during which the dichromate ions are reduced to trivalent chromic ions as the oxygen-demanding organics are oxidized in the sample.

The chromic ions give the solution a green color. The environmental engineer measures the COD concentration from the amount of dichromate converted to chromic ions by measuring green color intensity through a fiber-optic detector. The microprocessor-controlled package is available with automatic zeroing, calibration, and flushing features.

In one instrument, a 20-μl water sample is manually injected into a carbon dioxide carrier stream and swept

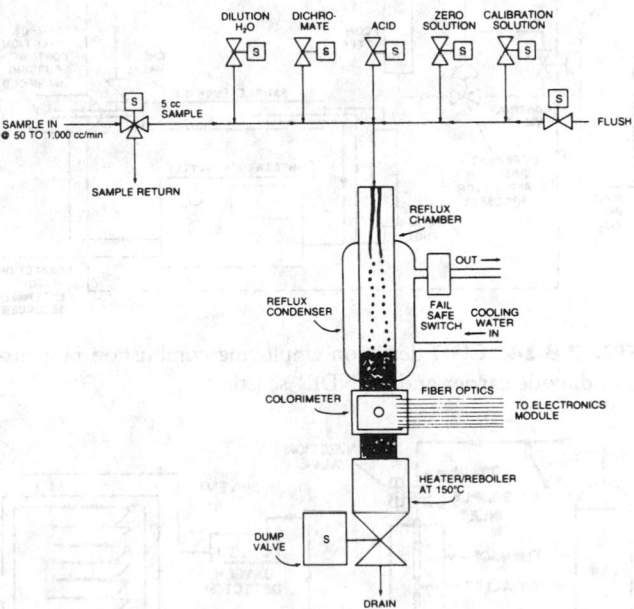

FIG. 7.8.13 Automatic COD analyzer. This analyzer uses a dichromate reagent and fiber-optic colorimeter detector and provides features of adjustable reflux time and autocalibration. (Courtesy of Ionics Inc.)

through a platinum catalyst combustion furnace. In this furnace, pollutants are oxidized to carbon monoxide and water, and the water is removed from the stream by a drying tube. Then, the reaction products receive a second platinum catalytic treatment. An NDIR detector then measures carbon monoxide concentration. The environmental engineer can convert the readings to COD using a calibration chart. An analysis can be completed in 2 min. This instrument is commercially available for manual operation (see Figure 7.8.14). Data obtained on domestic sewage indicate excellent correlation between this method (frequently called CO_2D) and the standard COD method.

TOD

The TOD method is based on quantitative measurement of the oxygen used to burn the impurities in a liquid sample. Thus, it is a direct measure of the oxygen demand of the sample. Measurement is by continuous analysis of the oxygen concentration in the combustion gas effluent (see Figure 7.8.15).

The TOD analyzer converts oxidizable components in a liquid sample in its combustion tube into stable oxides by a reaction that disturbs oxygen equilibrium in the carrier gas stream. An oxygen detector detects momentary depletion in the oxygen concentration in the carrier gas and records it as a negative oxygen peak on a potentiometric recorder. The analyzer obtains sample TOD by comparing recorded peak heights to peak heights of standard TOD calibration solutions, e.g., potassium acid phthalate (KHP).

Prepurified nitrogen from a cylinder passes through a fixed length of oxygen permeable tube into the combustion chamber, gas scrubber, and oxygen detector. The environmental engineer can vary the baseline oxygen concentration, determined as the nitrogen passes through the temperature-controlled permeation tube, to accommodate different TOD ranges by changing the nitrogen flow rate.

The combustion chamber is a length of Vicor tubing or quartz tube containing a platinum catalyst mounted in an electric furnace and held at a temperature of 900°C (1652°F). The aqueous sample is injected into this chamber and the combustible components are oxidized.

CORRELATION AMONG BOD, COD, AND TOD

Many regulatory agencies recognize only the BOD or COD measurements of the pollution load as the basis for pollution control. They are concerned with the pollution load on the receiving water, which is related to lowering the DO due to bacterial activity. Thus, if environmental engineers use other methods to satisfy the legal requirements of pollution load in effluents or measure BOD removal, they need an established correlation between the other methods and BOD or COD (preferably BOD).

A correlation of the various methods begins with the assumption that the BOD is the standard reference method. The salient features of this method are (1) a property measurement of the sample, i.e., the amount of oxygen required for bacterial oxidation of the bacterial food in the water, the BOD; (2) the dependence of oxygen demand on the nature of the food as well as on its quantity; and (3) the dependence of oxygen demand on the nature and amount of the bacteria.

The variation in OD due to variation in the amount (lb/gal) of food in the wastewater is expected; variation in OD when the amount of food is constant but changes occur in BOD requirements is difficult to predict. The same observations apply to the bacterial seed. Thus, variation in OD due to variation in the number or activity of bacteria, or changes in the nature of bacterial food leads to systematic or bias errors in BOD measurement that cannot be predicted or corrected for.

Therefore, the standard reference method is inherently variable and subject to analytical error. Researchers in an interlaboratory comparative study employing a synthetic waste found standard deviations around the mean of ±20% for BOD and ±10% for COD.

Another extensive study (Ford, Eller, and Gloyna 1971) made the following conclusions:

1. A reliable statistical correlation between wastewater BOD and COD and the corresponding TOC or TOD

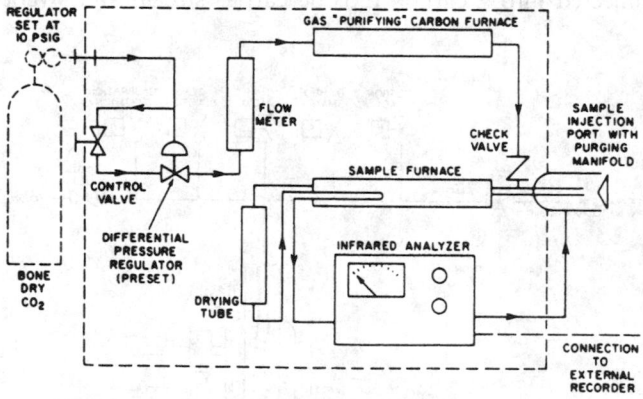

FIG. 7.8.14 COD detection employing combustion in a carbon dioxide carrier and an NDIR sensor.

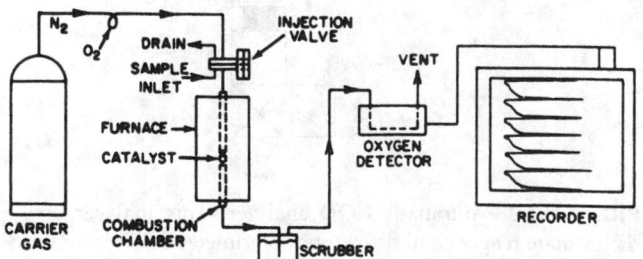

FIG. 7.8.15 Basic components of a TOD analyzer.

can frequently be achieved, particularly when organic strength is high and diversity in dissolved organic constituents is low.

2. The relationship is best described by a least squares regression with the degree of fit expressed by the correlation coefficient—this relationship applies to the characterization of individual chemical-processing and oil-refining wastewaters, not to all types of samples across the board.

3. The observed correspondence COD–TOD was better than COD–BOD for the wastewater mentioned (generally, correlating BOD with TOD was difficult, particularly when the wastewater contained low concentrations of complex organic materials).

4. The BOD–COD or BOD–TOC ratios of untreated wastewater indicate the biological treatment possible with wastewater. As these ratios increase, higher organic removal treatment efficiencies occur by biological methods.

Several papers indicate high correlation between BOD and other methods. This correlation is achieved when the nature of the pollutant is constant and only its amount changes. For complex and varying mixtures, obtaining good correlations is difficult.

An interesting example is given in the work of Nelson, Lysyj, and Nagano (1970), who discuss a pyrolytic method combined with flame ionization detection (FID). Values from the new method agreed with BOD values within ±15% for BOD values greater than 100 ppm on raw sewage and primary effluent. However, they found discrepancies of several hundred percent when the BOD was 20 ppm or less. These poor results can be attributed to a marked variation in biodegradability of carbonaceous products in the secondary effluent compared with the products before treatment as well as to the small amount of total material left.

Total Carbon Analyzers

METHODS OF DETECTION

A. NDIR; B. FID; C. Aqueous conductivity with ultraviolet (UV) irradiation; D. Colorimetry.

SAMPLES

Laboratory samples range from 0.01 to 10 cc. For in-line applications, sample flow rates range from 0.25 to 30 cc/min on continuous units. When a continuous bypass sample is used on cyclically analyzed injection samples, the bypass flow is from 50 to 1000 cc/min.

FLOWING SAMPLE SOLIDS CONTENT

Up to 1000 mg/l; size of particles up to 200 μ in diameter

MATERIALS OF CONSTRUCTION

Glass, quartz, Teflon, stainless steel, Hastelloy, polyethylene, and PVC

MEASUREMENT CYCLE TIME

A and B. 2 to 15 min (for type A, TC requires 2.5 min, and TOC requires 6 min)
C. Continuous with speed of response of 90 sec

UTILITIES OR REAGENTS REQUIRED

Air (10 atmospheric cubic feet per hour [acfh], or 4.6 alpm), oxygen, nitrogen (carrier gas flow is 100 cc/min at 50 psig, or 3.5 bars), hydrogen, mineral acid (1.0 gal per month of sulfuric or phosphoric), oxydizing reagent, buffer, and cooling water

RANGES

A and B. 0–2 ppm to 0–30,000 ppm (0 to 3%)
C. 0–100 ppb to 0–1 ppm

SENSITIVITY

A and B. 0.1 ppm or 0.5% of full-scale, whichever is greater

INACCURACY

2 to 5% full-scale as a function of design, sample size, and range

COSTS

$10,000 to $20,000; NDIR TC is $20,000, and NDIR TOC with differential detectors is $25,000.

PARTIAL LIST OF SUPPLIERS

Anatel Corp., Astro International (A); Bran and Luebbe Analyzing, Technicon Industrial Systems (D); Ionics (A,C); Rosemount Analytical Inc. Dohrmann Div. (A,B); Xertex Corp. Delta Analytical (C)

The damage done by discharged wastewater treatment effluent to the environment is a function of its organic loading because organic molecule decomposition consumes the DO of the receiving waters. BOD analysis measures all molecules that exert an oxygen demand, however it is slow, and readings vary depending on the bioassay used. COD analysis is faster but is affected by variations in chemical oxidation efficiencies. Carbon analyzers (TC and TOC) further increase speed but do not detect the load represented by nitrogen-based molecules. A direct correlation between BOD, COD, and TOC is not possible. Yet, TOC provides a rapid and reasonably accurate indication of pollution levels.

TOC, TC, AND TIC

To determine the TOC content of a sample, an environmental engineer must use one of two techniques to eliminate the TIC usually present but of little or no interest to the TOC analysis. The TIC in a water sample is usually in the form of inorganic bicarbonates and carbonates. One technique analyzes these components independently and then subtracts them from the TC. The TOC is then determined by the difference between TC and TIC (TC − TIC = TOC).

The other technique acidifies the sample to a pH of 2 to 3 followed by a brief gas sparging to drive off the carbon dioxide formed by the acidification. Any carbon remaining after the sparging should be TOC. Thus, the TC in the sparged sample is equal to the TOC content of the sample. A weakness in this technique is the possible loss of some VOCs that may be in the sample. Further techniques can account for these VOCs.

Development of TOC Analyzers

The TOC method was introduced in 1964 as a single-channel TC analyzer using a catalytic oxidation combustion technique followed by analysis of the resulting carbon dioxide. This method removed inorganic carbon (IC) by acid sparging or determined its concentration by titration. A few years later a second channel was added to the method that permitted parallel determination of IC in a second heated-reaction chamber. Several other techniques have since appeared with various changes in methodology and detection.

Because of the rapid acceptance and usefulness of TOC analysis as a laboratory method, online TOC analyzers became available in the late 1960s. Their success was limited by the relative complexity of these continuous analyzers. Today, there are a half dozen distinctly different methodologies and means of detection for TOC analysis.

Automatic Online Design

Figure 7.8.16 shows the catalytic-oxidation-type TOC analyzer for continuous online operation. This design incorporates an acid injection system that converts IC into carbon dioxide and removes it in a sparging chamber before it reaches the analyzer. If this sparging portion of the sam-

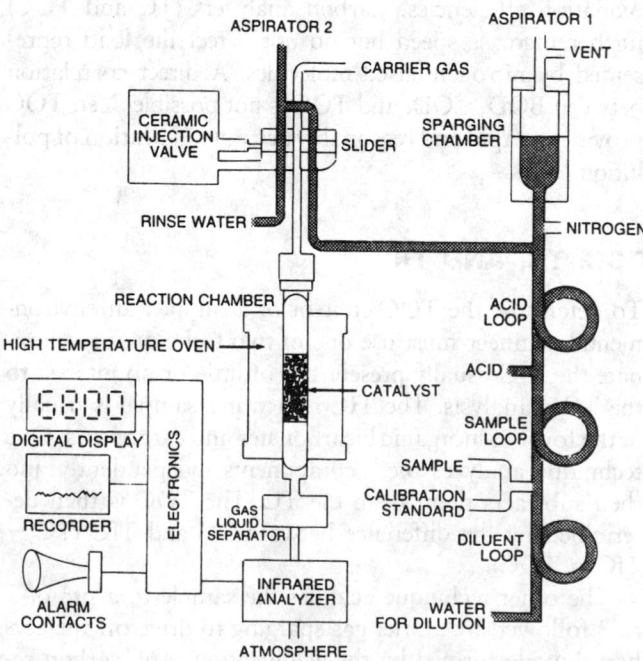

FIG. 7.8.16 Online, NOIR-type TOC analyzer. (Courtesy of Ionics Inc.)

pling system is left off, the analyzer becomes a TC analyzer. If a second, low-temperature reaction chamber is added in parallel with the one shown, the analyzer becomes a TOC differential analyzer.

The cycle time of this instrument is 2.5 min in the TC mode and 6 min in the TOC mode. It can operate unattended indoors or outdoors in an analyzer house. It can handle samples with solids content to 1000 ppm and particle sizes to 200 μ because an automatic water rinse is applied after each measurement cycle.

The design of the ceramic injection valve guarantees the accuracy of the sample size. The calibration of the instrument is automatically checked each time a known standard is introduced. During autocalibration, the analyzer runs three consecutive calibration standards, averages the results, and adjusts instrument calibration within preset limits or activates an alarm. The analyzer also has a dilution and automatic range change capability for concentrations that exceed the operating range. Carrier gas (nitrogen) consumption is 100 cc/min at 50 psig (3.5 bars), while acid consumption in the TOC mode is about 1.0 gal (3.8 l) per month.

FID

In the FID analyzer, a small acidified sample is transported in the presence of an oxidizer through a heated vaporization zone. Here, the IC, in the form of CO_2 plus any VOC, is driven off. The residual sample is sent through a pyrolysis zone to convert the remaining TOC to CO_2. The CO_2 is subsequently converted to methane in a nickel-reduction step. An FID detector measures the resulting methane.

The VOC is separated from the CO_2 in a bypass column, reduced to methane, and routed to the same FID for an additional VOC analysis to be added to the dissolved organic carbon value.

Figure 7.8.17 shows another method that uses the FID to analyze the VOC directly after the TIC (CO_2) is removed. In this method, catalytic oxidation combustion is replaced by a wet oxidation method. This method adds persulfate to the sample and exposes the solution to UV radiation to enhance oxidation efficiency. The resulting CO_2 is sparged and converted in the nickel-reduction methanator, and its concentration is measured in the FID analyzer. This wet oxidation technique is available as an online analyzer.

AQUEOUS CONDUCTIVITY

Another method employs wet oxidation and UV irradiation of the sample, which is contained in a recirculating stream of demineralized water. A conductivity cell located in this stream measures the increase in conductivity due to the CO_2 resulting from the TOC. A larger sample of water is acidified and sparged with the carrier air. As CO_2 is driven from the sample due to the TIC, it is dissolved in

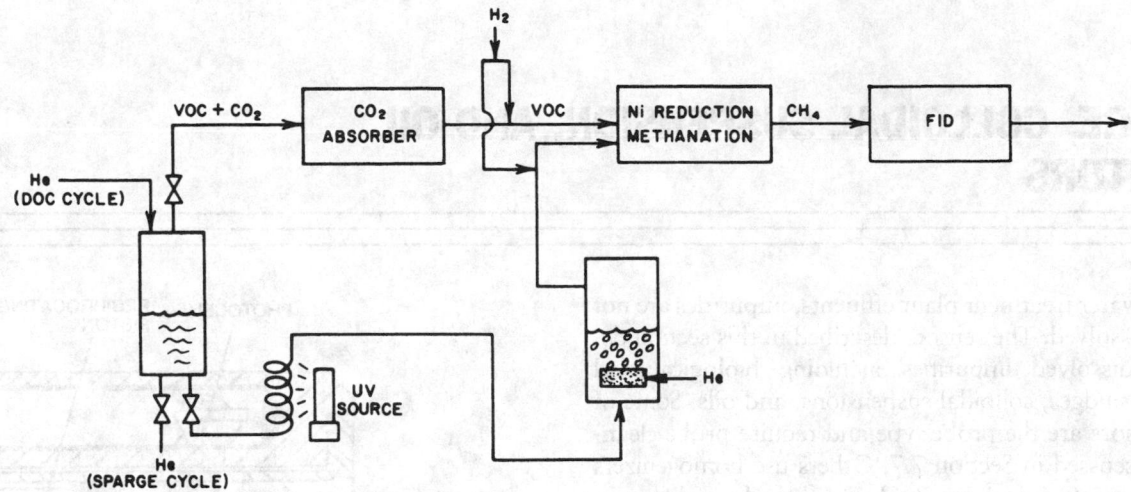

FIG. 7.8.17 FID analyzer with wet oxidation.

18.3 mΩ-cm demineralized water. The resulting increase in conductance becomes the new baseline for the next step, which entails oxidation of the TOC remaining in the water.

This oxidation is accomplished by UV radiation which oxidizes the TOC to CO_2. The added CO_2 raises conduc-

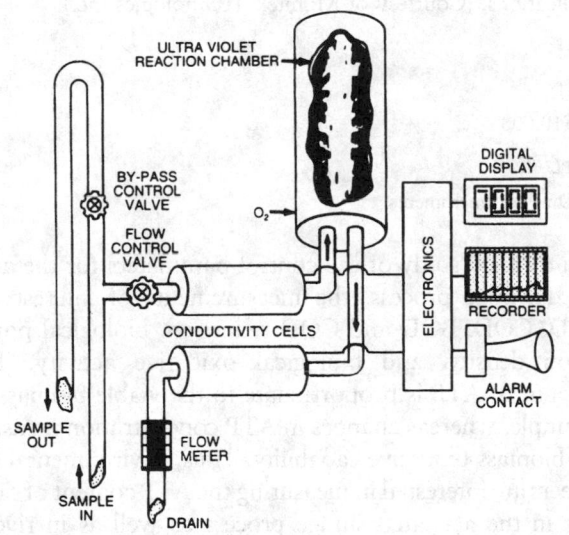

FIG. 7.8.18 Differential-conductivity-type online organic carbon analyzer capable of detecting ppb levels of TOC in high-purity water. (Reprinted, with permission, from Ionics Inc.)

tivity to a logarithmically higher level in proportion to the TOC present. The water is then automatically demineralized as it passes through an ion exchange resin bed to prepare it for the next analysis.

This method is best suited for the measurement of low-TOC levels in solids-free samples, such as in drinking water. This method claims sensitivities in the ppb range.

Figure 7.8.18 shows a high-sensitivity, online TOC analyzer used in boiler feedwater; condensate return; or semiconductor, nuclear, or pharmaceutical plant water supply applications where ultrapure water is required. This analyzer takes a 25 cc/min continuous sample from the process water, mixes it with oxygen, and irradiates it with UV light in the reaction chamber. In the presence of oxygen and catalyzed by the UV radiation, the carbon molecules are oxidized into carbon dioxide. As the carbon dioxide is dissolved in the water, its conductivity increases. The analyzer interprets the difference between the conductivities of inlet and outlet water as a measure of the TOC.

This analyzer is continuous, fast (90 sec response time), and sensitive. Its span can be as narrow as 0 to 100 ppb. The sample can contain solids, but their particle size must be under 200 μ.

—*Béla G. Lipták*

7.9
SLUDGE, COLLOIDAL SUSPENSION, AND OIL MONITORS

In wastewater treatment plant effluents, impurities are not always dissolved. The sensors described in this section detect nondissolved impurities including biological and chemical sludges, colloidal suspensions, and oils. Some of these sensors are the probe-type and require probe cleaners, as discussed in Section 7.7. Others use homogenizers (see Section 7.8), which liquify the solids and simplify sample handling in sample-based analyzers.

SS and Sludge Density Sensors

Figure 7.9.1 shows a probe-type SS detector widely used in biological sludge applications. The probe contains a reciprocating piston. This piston expels a sample (every 15 to 40 sec) during its forward stroke while wiping clean the optical glass of its internal measurement chamber, and pulls in a fresh sample during its return stroke. This device measures the total attenuation of light, which in biological sludge applications is mostly due to SS. Due to its self-cleaning capabilities, cleaning and maintenance are minimal.

This unit is available with a 4- to 20-mA transmitter output that is updated every 15 to 40 sec and has a full-scale inaccuracy of 5% over SS ranges between 0 to 0.1% and 0 to 10%.

Activated Sludge Monitors

METHOD OF DETECTION

Photometric measurement of light emitted by chemical reaction

SAMPLE PRESSURE

Atmospheric

SAMPLE TEMPERATURE

Ambient

SAMPLE TYPE

Grab sample

MATERIALS OF CONSTRUCTION

Glass

RANGE

10^{-7} to 10^{-2} μg adenosine triphosphate (ATP) per 10 ml sample of bacterial extract. Sensitivity to 10^{-7} μg per 10 μl sample. Calibratable for number of bacteria per μg ATP

RESPONSE

Laboratory method: minutes after starting reaction

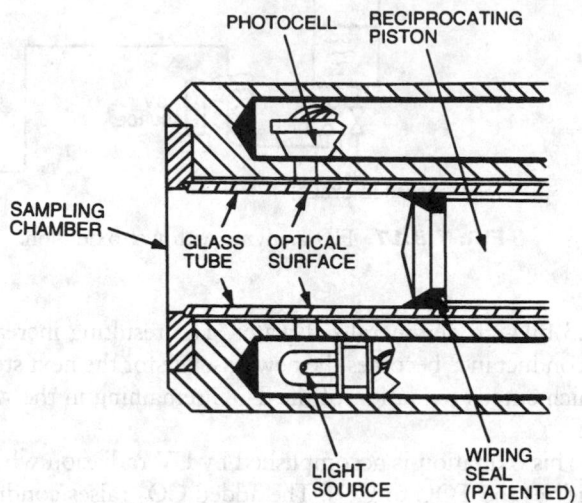

FIG. 7.9.1 Probe-type SS detector used on biological sludge applications. (Courtesy of Monitek Technologies Inc.)

COST

$10,000

SUPPLIER

Du Pont Instruments

In a detailed study of the control parameters for the activated sludge process, the measurements of interest are BOD, COD, BOD and COD reduction, biological population density, and biological oxidative activity. The amount of ATP is proportionate to the viable biomass in a sample, whereas changes in ATP concentration measure the biomass oxidative capability. Thus, environmental engineers are interested in measuring the ATP content of samples in the activated sludge process as well as in rivers, lakes, and other receiving waters.

An ATP assay procedure has been developed based on the reactions just described. Briefly, the procedure involves the rapid killing of live bacterial cells and the immediate extraction of ATP into an aqueous solution. The latter is then treated with firefly lantern extract, and a photometer measures the light emission of the resultant solution. The firefly lantern extract and the ATP required for calibration are commercially available. Du Pont is the only supplier who has designed a manually operated instrument specifically for this measurement.

The instrument is supplied with the required reagents. The lab technician dissolves a tablet containing buffer and magnesium sulfate in water and adds a homogeneous powder of luciferin and luciferase. Then, the technician filters the sample through a coarse filter to remove solid matter, which is discarded. The filtrate passes through a bacterial filter to catch the living bacteria. The technician treats the bacteria on the filter with butanol, which ruptures the cell walls and releases the ATP. To take the filtrate up to volume, the technician adds water and a microliter aliquot to the prepared reagent already in a cuvette. He then places the cuvette in the instrument to read its light emission. The instrument automatically converts the light flash to ATP or microorganism concentration per milliliter, depending on how it is calibrated.

Slurry Consistency Monitors

TYPES

Blade, rotary, polarized light, probe, level detector, and flow bridge

ELEMENT MATERIALS

316 stainless steel

NORMAL DESIGN TEMPERATURE

Up to 250°F (120°C)

NORMAL DESIGN PRESSURE

Up to 125 psig (8.6 bars)

RANGE

1.75 to 8% consistency

SENSITIVITY

0.01 to 0.03% consistency

REPEATABILITY

0.5% of reading

INACCURACY

Function of empirical calibration, usually 1% of reading

COST

Laboratory units cost $1000 to $2000; continuous industrial units cost $3000 to $5000. SS transmitters cost $4000 to $5000.

PARTIAL LIST OF SUPPLIERS

Automation Products, Dynatrol Div.; Berthold Systems Inc.; C.W. Brabender; BTG Inc.; DeZurik, a Unit of General Signal; EG&G Chandler Engineering; The Foxboro Co.; Gam Rad West Inc.; IRD Mechanalysis Inc.; Kajaani Electronics Ltd. (Finland); LT Industries Inc.; Markland Specialty Engineering Ltd.; Measurex Corp.; Monitek Technologies Inc.; Ronan Engineering Co.; Schlumberger Industries, Solarotron; TECO, Thompson Equipment Co.; Testing Machines Inc.; Valmet Automation Inc.

While *density* is mass per unit volume, *consistency* is mass per unit mass. It is a percentage obtained by dividing the weight of the solids by the unit weight of the wet sample. The most direct method of measuring consistency is to dry a sample unit weight and measure the weight of the dried solids. Consistency should not be confused with *basis weight,* which is the weight of a unit area of a sheet product. *Freeness* expresses how readily a slurry releases water. Consistency should be measured at a constant velocity because it affects the reading. Consistency increases with freeness or alkalinity (pH) and decreases with temperature and inorganic material content.

INLINE CONSISTENCY MEASUREMENT

Ideally, the complete process stream should be exposed to the sensor, but in large flows this exposure is not practical. Therefore, samples are taken. The sample should be taken from the center of the pipe, preferably from the discharge of a centrifugal pump, so that separation or settling of solids is minimized (see Figure 7.9.2).

Consistency-measuring instruments detect the consistency of the process fluid as shear forces acting on the sensing element. Two basic types of consistency detectors are the fixed and rotary. In the latter, the shear force is reflected as the torque required to maintain a rotary sensor at constant speed, as the imbalance of a strain-gauge resistance bridge, or as a turning moment. The instruments are calibrated inline; thus the output is not in terms of dry consistency but rather some arbitrary, reproducible value.

Fixed sensors depend on the process flow for measurement, and for such instruments, the output is affected by the velocity of the flow. The sensor contour minimizes the flow effects on the output over the operating flow range. On the other hand, rotating sensors do not depend on process flow for measurement. While these units are also sensitive to flow velocity variations, they can be used over wider flow ranges. In addition, the rotary motion of the sensor produces some self-cleaning action while fixed sensors depend solely on a properly designed contour to prevent material obstructions.

The sensing element of this instrument is a blade, specially shaped to minimize the effects of velocity. The instrument can be mounted on any line 4 in (100 mm) or larger. The mounting is through a 2-in (50-mm) flange supplied with the instrument.

A variation of this design uses a shaped float inserted through a pipeline tee. The shear forces acting on the float are transmitted to the force bar of a pneumatic transmitter mounted on top of the tee. The unit can only be installed in a vertical pipe, with 5 pipe diameters of straight

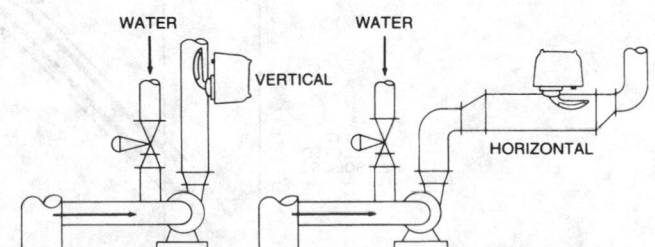

FIG. 7.9.2 Installation of a blade-type consistency transmitter in vertical and horizontal pipelines. (Courtesy of DeZurik, a Unit of General Signal)

run required on the upstream side. The minimum line size is 6 in (150 mm).

PROBE TYPE

This sensor transmitter functions as a resistance-bridge strain gauge. The bridge elements are bonded to the inner wall of a hollow cylinder that is inserted into the process. The shear force acting in the cylinder, due to the consistency of the process fluid, causes an imbalance of the resistance bridge. The amount of imbalance is proportional to the shear force and the consistency of the process fluid. The resistance bridge is powered from a recorder that also contains ac potentiometer electronics.

The sensor is mounted through a threaded bushing furnished with the unit. The flowing velocity must be between 0.5 and 5.0 ft/sec (0.15 and 1.5 m/sec) for repeatability of around 0.1% of bone dry consistency.

OPTICAL SENSOR

Optical consistency meters use a sample cell through which polarized light is passed. Because only solids scatter polarized light, the amount of depolarization is a measure of SS concentration or consistency. Another optical consistency detector operates in the IR region and uses a self-cleaning, multiple fiber-optic probe as its sensor (see Figure 7.9.3). The probes can be inserted into the pipeline to adjustable depths through an isolating ball valve (2 in or 50 mm) and can monitor the consistency in the range of 2 to 6%.

Summary

While convenient from an installation standpoint, inline instruments are sensitive to flow variations. Fixed sensors are often plagued by material buildup, particularly if the sample contains fibers. Rotating sensors are self-cleaning

because the sensor motion spins off any material; however, variations in shaft seal friction can be troublesome. The flow-bridge method of consistency measurement is applicable to a range of materials with better accuracy than the other instruments. Since this instrument is not installed inline, the process flow need not be shut down for instrument maintenance.

Sludge and Turbidity Monitors

TYPES

Laboratory units can be manual or flow-through; turbidity transmitters marketed for the process industry are available in probe and flow-through designs.

DESIGN PRESSURES

Up to 250 psig (17 bars)

DESIGN TEMPERATURES

250°F (120°C); 450°F (232°C) special

CONSTRUCTION MATERIALS

Stainless steel, glass, and plastics

RANGES

In ppm silica units, from 0–0.5 to 0–1000; backscattering designs available from 10–5000 ppm to 5–15%. Ranges in Jackson turbidity units (JTU) units from 0–0.1 to 0–10,000; in nephelometric turbidity units (NTU) units from 0–1 to 0–200; in Formazin turbidity units (FTU) units from 0–3 to 0–1100. A sludge density probe with reciprocating piston has a range from 0–0.1% to 0–10% of SS.

INACCURACY

0.5 to 2% of full scale for most and 5% of full scale for reciprocating-piston, probe-type, sludge SS sensor

COSTS

Standard solutions for calibration cost $100 per bottle; laboratory turbidity meters cost from $600 to $1000; laboratory nephelometers with a continuous-flow attachment cost $1500; and process industry transmitters cost from $2000 to $4000 depending on the features and materials of construction. A sludge-density-detecting, self-cleaning probe with an internal reciprocating piston and indicating transmitter costs $6700.

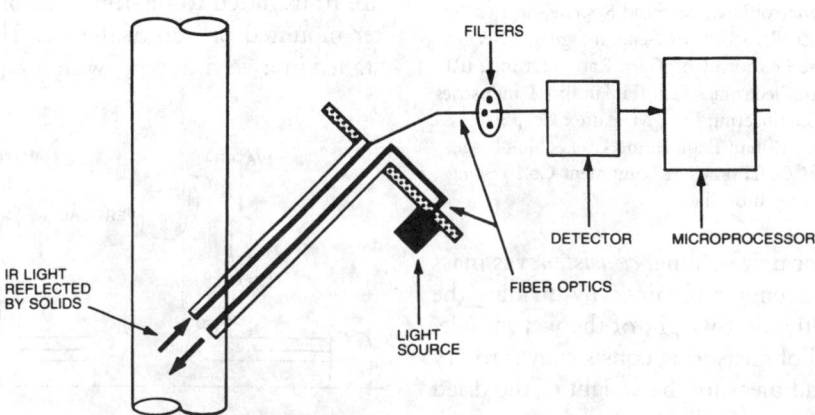

FIG. 7.9.3 Self-cleaning, fiber-optic probe used in consistency measurement. (Courtesy of Kajaani Electronics Ltd., Finland)

PARTIAL LIST OF SUPPLIERS

Bailey Controls Co. Div. Babcock & Wilcox; BTG Inc.; Cole-Parmer Instrument Co.; Custom Sensor & Technology; Du Pont Co. Instrument Systems; Fischer & Porter Co.; Foxboro Co.; Gam Rad West Inc.; Great Lakes Instruments Inc.; Honeywell Industrial Controls; HF Scientific Inc.; Interocean Systems Inc.; Kajaani Electronics Ltd. (Finland); Kernco Instrument Co.; Lisle-Matrix Ltd.; Markland Specialty Engineering Ltd.; Maselli Measurements Inc.; McNab Inc.; Merlab (Hungary); Monitek Technologies Inc.; Ohmart Corp.; Photronic Inc.; Rosemount Analytical Inc.; Sigrist-Photometer AG; Turner Design; Wedgewood Technology Inc.

Turbidity is a measure of water cloudiness caused by finely dispersed SS that scatter visible or IR light. The higher the turbidity (cloudier the fluid), the more scattering occurs. Therefore, transmitted light intensity is reduced while the scattered light intensity (detected at a 90° angle to the light path) increases. Turbidity can also be detected indirectly by colorimeters, activated sludge monitors, or consistency meters.

TURBIDITY UNITS

Different turbidity instruments detect light intensity differently. The three main techniques are *perpendicular* scattering (nephelometry), *backscattering*, and *forward* scattering. Different turbidity units have evolved in connection with different designs. The JTU is a purely optical scale and correlates with forward scattering measurements. The value of one JTU corresponds to the turbidity of a liter of distilled water with 1 mg (1 ppm) of suspended diatomaceous fullers earth (an inert material).

NTUs are based on a U.S. EPA–approved stable polymeric suspension standard and correlate with perpendicular scattering designs.

FTUs use a Formazin polymer standard and also correlate with perpendicular scattering designs. Two Formazin scales are used, and according to some sources, the NTU reference standards are more stable and last longer than the FTU standards. Turbidity measurement error cannot be less than the accuracy at which the standard calibrating solution is available. In Formazin standards, this variation can approach 1%.

All turbidity units measure the amount of solid particles in suspension. Parts per million (ppm) units refer to the weight of the solids in suspension. However, because this measurement requires individual calibration, they usually refer to ppm of silica (silicon dioxide). Therefore, if the cloudiness (turbidity) of the process sample is the same as the turbidity resulting when 1 mg of silica is mixed in a liter of distilled water, the turbidity reading is 1 ppm on the silica scale.

FORWARD SCATTERING OR TRANSMISSION TYPES

In the forward scattering or transmission design (see Figure 7.9.4), the turbidity meter light source is on one side of the process sample, and the detector is on the other. This design determines the total attenuation. When attenuation is due to color absorption, the unit is a colorimeter; when attenuation is caused by light scattered by solid particles, the unit is a turbidity meter.

Dual-Beam Design

When both color and solids are present, the total attenuation is the sum of absorption and scattering effects. Therefore, the single-beam turbidity analyzer can only be used if no color is present or the color is constant and its effect can be zeroed out. When background absorbance or color varies, a dual-beam or split-beam analyzer is needed. Such a unit is described in Figure 7.9.5.

This unit uses two light paths, one passing through the unfiltered process sample cell and the other through a reference sample cell containing filtered process fluid. Analyzer output is proportional to the difference of optical absorbance between the two cells, which corresponds to the solid particles present in the sample but not in the reference cell.

The design shown in Figure 7.9.5 has an oscillating mirror that alternately directs the light beam (alternating 600 times per sec) to the measuring and reference cells. The photocell detector converts the intensity differential of the two beams into a photocurrent that modulates the opening of a mechanical shutter so that the differential is zero.

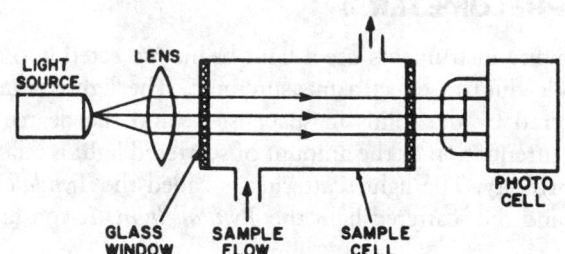

FIG. 7.9.4 Schematic diagram of a transmission-type turbidity meter.

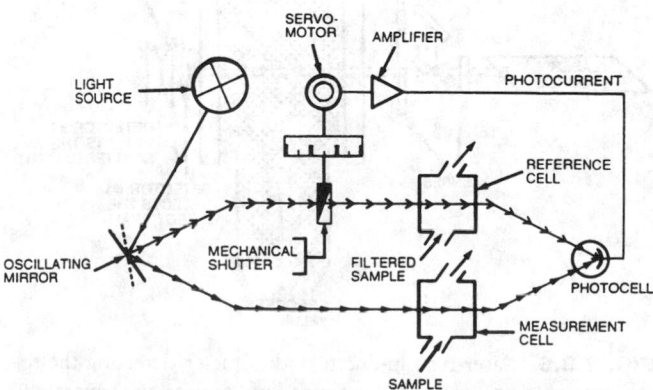

FIG. 7.9.5 Oscillating dual-beam, forward-scattering turbidity analyzer. (Courtesy of Sigrist Photometer AG)

Therefore, the more solid particles in the sample, the more the shutter needs to be closed, and the position of the shutter can be read as turbidity. If the reference cell is filled by other reference materials, the same instrument can measure other properties such as color and fluorescence.

Laser Type

In the laser-type, in-line turbidity meter shown in Figure 7.9.6, a thin ribbon of light is transmitted across the process stream. This light is attenuated by the process fluid and then falls on detector 1. If there are solids in the process fluid, some of the light is scattered. This scattered light is collected and falls on detector 2. The ratio of the two detector signals relates to the amount of solids in the process stream (turbidity); being a ratio signal, it is unaffected by light source aging, line voltage variations, or background light intensity variations. The laser-type detector is less sensitive to interference by gas bubbles than other turbidity meters because the laser-based light ribbon is so thin (about 2 mm). Therefore, when a bubble passes through it, it causes a pulse, which can be filtered out.

In general, in-line turbidity meters are less subject to bubble interference than turbidity analyzers that require sampling. Because in-line units do not lower the operating pressure of the stream, dissolved gases are not encouraged to come out of solution.

SCATTERED LIGHT DETECTORS (NEPHELOMETERS)

Turbidity instruments use a light beam projected into the sample fluid to effect a measurement. The light beam is scattered by the solids in suspension, and the degree of light attenuation or the amount of scattered light is related to turbidity. The light scattering is called the *Tyndall effect* and the scattered light the *Tyndall light*. A constant-candlepower lamp provides a light beam for measurement, and one or more photosensors convert the measured light intensity to an electrical signal for readout.

Usually the photosensor comes with a heater and thermostat to maintain a constant temperature because the device output is temperature sensitive. The supply voltage to the lamp must be regulated to at least $\frac{1}{2}\%$. This regulation eliminates errors due to source intensity variations because the measured light is referenced to the source. Because deposits formed on the flow chamber windows by the sample interfere with measurement, the windows require frequent cleaning or automatic compensation.

Transmission-Type Design

Instruments measuring scattered light vary in design. One type uses a flow chamber similar to the one in Figure 7.9.4, except that the window for the measured light is located at 90° to the window for the incident light (see Figure 7.9.7). One window transmits light beams into the measuring chamber and the other, at right angles to the first, transmits scattered light to the photosensor. A light trap is located opposite the incident light window to eliminate reflection.

With this arrangement, dissolved colors do not affect the measurement; however, instrument sensitivity decreases with the presence of color because some light is absorbed. Variations of the basic unit include the use of two source beams and two photosensors in conjunction with two pairs of opposed windows.

Some designs use a separate photosensor to monitor lamp output and adjust the lamp supply voltage through a feedback circuit to maintain constant light intensity.

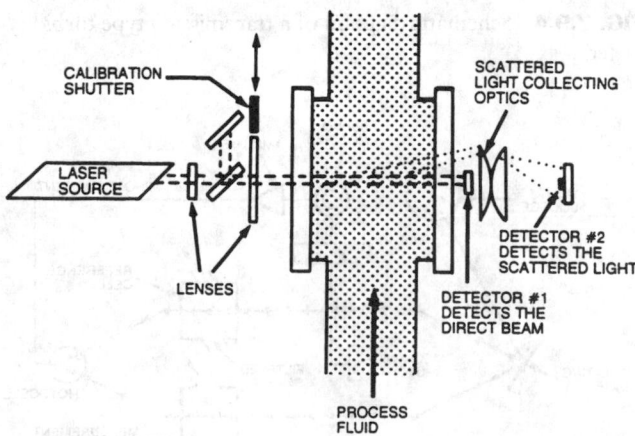

FIG. 7.9.6 Laser-type in-line turbidity meter detecting the total attenuation and the amount of scattering separately. (Reprinted, with permission, from ACSI)

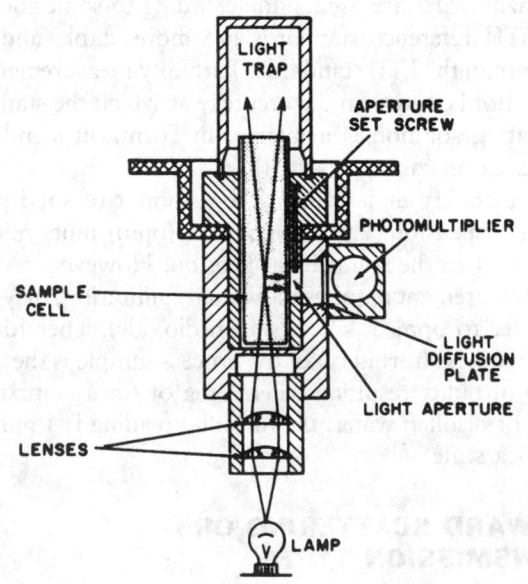

FIG. 7.9.7 Light-scattering turbidity meter.

Probe Design

For wastewater and biological sludge applications, probe-type turbidity transmitters are preferred. One design (see Figure 7.9.8) uses an IR light source and measures the resulting 90° scattered light intensity. These microprocessor-based units are provided with built-in compensators for ambient light variations, and wipers for automatic cleaning of the dip or insertion probe. Cleaning frequency is adjustable between 1 and 6 hr. During the wiper action of the cleaner, the transmitter output signal is held at its last value.

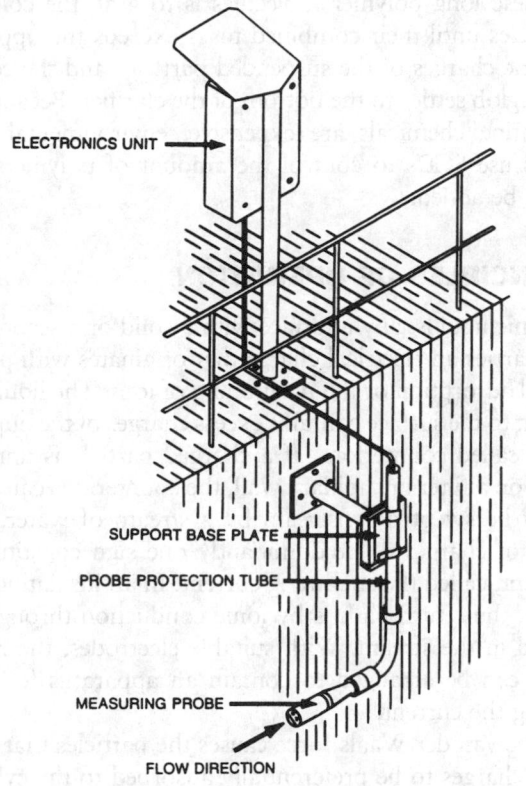

FIG. 7.9.8 Automatically cleaned, 90° scatterer light-detecting turbidity transmitter. (Courtesy of BTG Inc.)

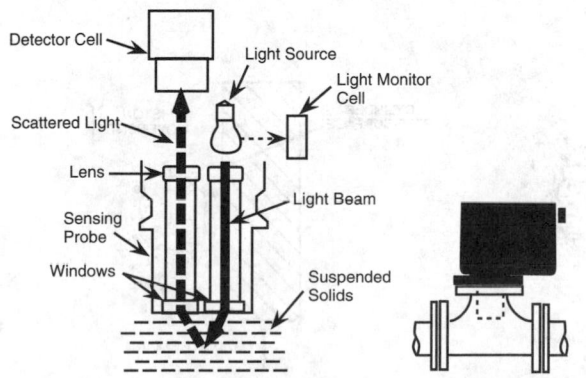

FIG. 7.9.9 In-line, backscatter-type turbidity analyzer. (Courtesy of Gam-Rad Inc.)

BACKSCATTER TURBIDITY ANALYZERS

Figure 7.9.9 shows the in-line version of the backscatter-type turbidity analyzer, which can be installed in either pipes or vessels. Here, the 180° backscatter effect is measured. The units are suited for high-temperature applications (up to 450°F or 232°C) and for high concentrations of solids. Ranges from 10 to 5000 ppm to 5 to 15% on the silica scale are available. A backscattering design using fiber-optic light cables is also available (see Figure 7.9.3).

Summary

Turbidity measurement is fairly simple in theory; the most serious practical problems are posed by light source intensity changes, deposits on optical windows, and the presence of dissolved colors in the sample. Units are available that automatically correct for these effects and variations in the ambient light intensity as well as for gas bubbles. Self-cleaning probe design units are also available. Selection should be made on the basis of information needed (transmission or 90° or 180° scatter), and on the nature and concentration of the solids to be detected and the material of construction for each type.

Installing the Sludge Monitor

Before designing the sludge monitoring installation, environmental engineers must answer two questions:

1. Is the information in the liquid or solid phase of the sludge?
2. Should the measurement be made online (in-place) or should a sample be delivered to the analyzer?

The answers to these questions will direct the environmental engineer to the correct installation.

ONLINE MONITORING

The main advantage of online monitoring is eliminating the sampling system. Without a sampling system, transportation lag time is eliminated, allowing good closed-loop control. Maintenance is also reduced, since many delicate sampling system components are eliminated. Finally, online monitoring detects the unaltered real process; with sampling, sample integrity can be impeded by filtering, condensation, and leakage.

Online monitoring requires an analyzer (usually a probe) that is clean and in good working order. Therefore, probe cleaners (see Tables 7.7.1 and 7.7.2) and placing the probe inside a sight-glass for convenient visual inspection (see Figure 7.7.2) are important. The capability of removing the probe from the pipe or tank without a process shutdown is also beneficial. This is accomplished using retractable probes (see Figure 7.7.8), which are periodically (and automatically) withdrawn from the process for un-attended cleaning and recalibration.

If the probe measures only the composition of the liquid phase, a periodically backflushed, porous filter cup (see Figure 7.7.6) can protect it from being coated with solids.

SAMPLING-BASED SLUDGE MONITORING

Before a sludge sample is transported to the analyzer, the environmental engineer must determine if the information of interest is in the liquid or solids phase of the sludge. If only the liquid phase must be monitored, solids can be removed from the sample by self-cleaning filters (see Figures 7.8.2 and 7.8.3). When a sample is collected over a long time period, intermittent sample collectors should be used for monitoring (see Figure 7.8.5).

If the information is in the solids phase and a sampling system is used, the size of the solid particles must be reduced through homogenization (see Figure 7.8.4). Once the slurry is liquified, it can flow through the sampling system without plugging it.

Colloidal Suspension Monitors

APPLICATIONS

Batch operations, titrations, or continuous monitoring; can control the clarification of beverages, dewatering, thickening of suspensions, addition of coagulant chemicals, or treatment demand by measuring the surface charge on particles

MATERIALS OF CONSTRUCTION

Stainless steel, silver, and Teflon

SAMPLE SIZE REQUIRED

About 10 cc

APPROXIMATE COST

$12,000

PARTIAL LIST OF SUPPLIERS

Komline-Sanderson Engineering Corp.; Leeds and Northrup; Mütek GmbH; Panametrics, Inc.

In wastewater treatment, clarification is a major step. Certain materials such as clay do not settle out because the static electric charges of the individual particles keep the clay particles uniformly dispersed. In such *colloidal suspensions,* gravitational forces alone cannot cause settling because the opposing forces caused by the like electrical charges of the particles are stronger than the gravitational forces acting on them.

Clarification of colloidal suspensions involves measuring particle surface charge and then adding coagulating chemicals in proportion to that charge. The surface charge is detected by streaming current detectors (SCDs), while the coagulating chemicals are usually polymers. The role of these long polymer molecules is to grab the colloidal particles until their combined mass exceeds the opposing electric charges of the suspended particles and the coagulated glob settles to the bottom of the clarifier. Because coagulating chemicals are expensive, environmental engineers use SCDs to control the amount of polymers that must be added.

PRINCIPLES OF OPERATION

In ionic liquids, any interface with a solid or a second liquid carries an electrical charge that originates with preferential adsorption or the positioning of ions. The liquid adjacent to the surface contains excess charges of the opposite sign, called counterions. If a charged particle is immobilized on a filter or capillary wall, the counterions can physically be swept downstream by a stream of water. This flow of charges of predominantly one sign constitutes a current, called the streaming current. In an insulating capillary, the return path is by ionic conduction through the liquid in the stream. With suitable electrodes, the return path can be arranged to contain an apparatus for measuring the current.

The van der Waals force causes the particles that carry high charges to be preferentially adsorbed to the cylinder and piston surfaces as shown in Figure 7.9.10. When such a particle moves upward by piston movement, its counter-

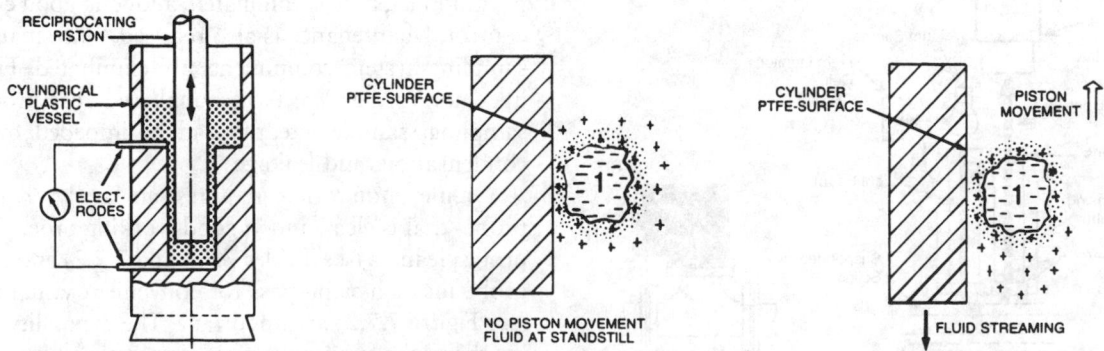

FIG. 7.9.10 A cloud of counterions sheared off by the streaming fluid in the boundary of the diffuse layer at the cylinder surface. (Courtesy of Mütek GmbH)

ions are sheared off as the fluid moves in the opposite direction relative to the piston. The totality of these ions is the streaming potential detected by this analyzer.

APPLICATIONS

Most applications involve either titration of the sample or prior treatment in the plant since a single reading on untreated material provides little information. SCD readings are almost independent of the concentration of SS. Titrations can be made with as little sample as will submerge the active part of the instrument.

To estimate the unit treatment demand of a liquid, the environmental engineer must first titrate a volumetric sample with the contemplated treating chemical at a known concentration until he obtains a zero signal. To compare alternative treating chemicals, the environmental engineer titrates identical samples of material with various chemicals. When the effect of a change in pH on treating requirements is studied, the environmental engineer titrates identical samples at various pH levels. This effect can be significant, with chemicals differing considerably in their tolerance of low or high pH.

In the usual treatment plant, the SCD can continuously control the feed of cationic chemicals. The need for changing the rate of adding these chemicals arises from variations in the stream flow rate, changes in the SS loading, the unit demand of the solids, or any combination of these factors. The main advantage of SCD control is its early response to changes. Charge neutralization occurs almost as soon as the treating chemical is dispersed in the stream; therefore, samples can be taken 1 or 2 min after addition of the chemical.

Batch samples taken to the SCD should be adequate to permit rinsing the apparatus several times. Skimming or decanting removes sand, larger solid particles, or oil globules. Since charge is a surface phenomenon and the fines have most of the surface, removing larger particles has little effect. For a continuous sample, a self-cleaning bypass filter should be used (Figures 7.8.2 and 7.8.3). Periodic backflushing or cleaning can also be required.

For continuous control, the SCD measurement signal is fed to a two-mode controller that modulates the chemical feed pump or control valve (see Figure 7.9.11). The maximum and minimum valve opening is limited as a defense against sample loss, which can cause an open loop. Pressure regulators serve as adjustable settings to limit the controller output to a range between the minimum and maximum expected demand.

The SCD controller set point is based on the downstream turbidity measurement. If an existing flow proportioning controller throttles the chemical feed, the SCD controller can influence its ratio set point in a cascade arrangement. Often, more than one chemical is involved, and the environmental engineer must consider a sequence of additions and attendant interactions.

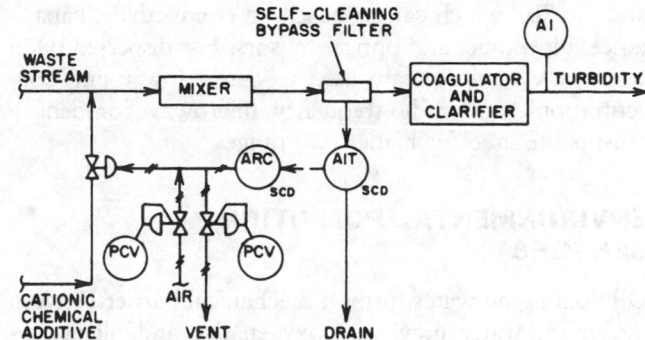

FIG. 7.9.11 Chemical addition control using a streaming current detector.

Oil Monitors

TYPES OF DESIGNS

A. Reflected light oil slick detector (on–off)
B. Capacitance—available in probe form for interface detection, in a flow-through design, or in a floating plate configuration for measuring oil thickness
C. UV
D. Microwave (radio frequency)—available as an interface probe, as a tape-operated tank profiler, or as an oil in the water content detector
E. Conductivity probes for interface detection

RANGE

A. Generally from 0–50 ppm to 0–100%
B. The flow-through, dual-concentric detector from 5 to 15% water in oil
C. 0–10 ppm to 0–150 ppm of oil in water
D. The oil content is detectable from 0 to 100%

INACCURACY

Generally from 1 to 5% of full scale
B. The flow-through, dual-concentric detector has a sensitivity of about 0.05 to 0.1% water.
C. 0.1 ppm for a 0- to 10-ppm range
D. Interface detected to 5%, tank profile to 1%, and water concentration to 0.1%

COSTS

C. $12,000 to $20,000 for dual-wavelength unit with auto-zero and 0–10 ppm to 0–150 ppm range

PARTIAL LIST OF SUPPLIERS

(For capacitance, conductivity, and ultrasonic probe suppliers see *The Instrument Engineers' Handbook: Process Measurement,* Third Ed.; Agar Corp. (D); Amprodux Inc.; Bailey Controls Div.; Bernhard & Scholtissek, Veba Oel AG; Delta C Technologies (B); Du Pont Instrument Systems (C); Endress + Hauser Instruments (B); Foxboro Co. (B); Invalco (C); Spatial Dynamics Applications Inc.; Teledyne Analytical Instruments (C)

Oil and grease must be removed from wastewater plant effluents before they are discharged into the environment. Oil can either float on top of the aqueous effluent or be dispersed in it. With an oil layer, the monitoring task is an interface level measurement (see Figures 7.6.18, 7.6.19, and 7.6.20), which can be based on conductivity, capacitance, ultrasonic, and optical sensors. For dispersed oil in water, UV analyzers are used to detect low (ppm) concentrations, and radio-frequency microwave or density sensors are used for higher (%) ranges.

ENVIRONMENTAL POLLUTION SENSORS

Oil floating on water forms a mechanical barrier between the air and water, preventing oxygenation and killing oxygen-producing vegetation on the banks of streams. By coating the gills of fish, these materials prevent breathing and cause fish to suffocate. Therefore, ships and municipal and industrial waste treatment plants must monitor outfalls and control oil removal to prevent oil-bearing wastes from entering natural waves. Continuous monitors are available to detect any hydrocarbon floating on the surface of water.

Oil in the water is equally undesirable. It contributes to the BOD and can also be toxic to aquatic biota, fish food in water, and fish themselves. Optical detection methods for both types of contamination require regular, conscientious maintenance for continuous, reliable performance. The capacitance approach for monitoring oil film thickness on water appears to require less maintenance but is limited to detecting floating oil. Environmental engineers must evaluate each application separately considering the limited capabilities of available instrumentation.

On–Off Oil-on-Water Detector

This device (an application of nephelometry) detects a visible oil (hydrocarbon) slick on fresh or salt water. It consists of two parts: a sensing head and a controller. The sensing head, in an explosion-proof housing supported on pontoons, floats on the body of water. An S-shaped baffle directs flowing water past the sensing head. A beam of light is focused through a lens onto the water's surface. Reflected light is refocused by a second lens onto a photocell. In the absence of oil on the water, minimum light reflection occurs. In the presence of floating oil, the reflected light intensity increases.

Measurement is based on the difference between the reflected light photocell output and a reference photocell measuring light source output. Alarm functions and an output signal proportional to reflected light intensity are available from the controller.

Oil-Thickness-on-Water Detector

The device just described measures the presence or absence of oil floating on water. The oil-thickness-on-water detector measures the oil layer thickness. It consists of a floating sensing head connected by shielded cable to a remote controller. The sensor measures the thickness of an oil layer on water by capacitance measurement (see Figure 7.9.12).

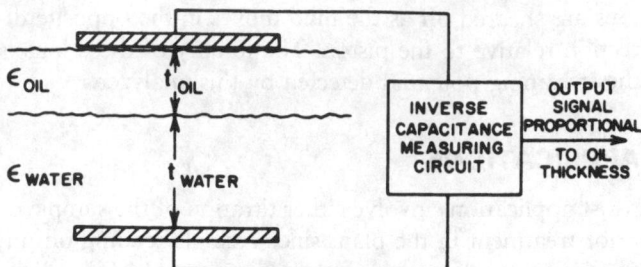

FIG. 7.9.12 Parallel-plate capacitor detecting the thickness of an oil layer on water.

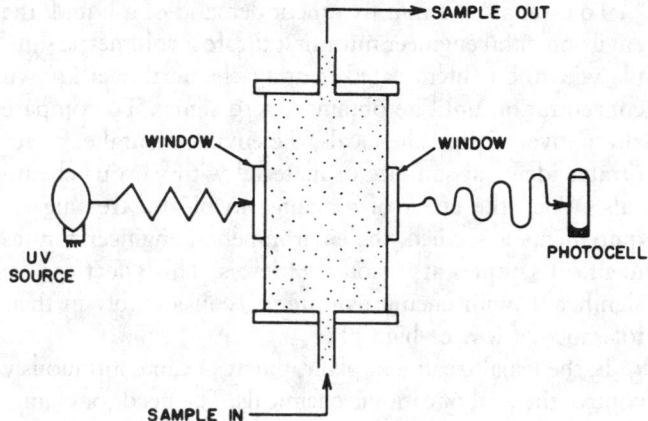

FIG. 7.9.13 Oil-in-water detector.

The inverse capacitance is proportionate to the oil thickness. The circuit generates a dc voltage in proportion to the inverse capacitance, which is in direct proportion to the oil thickness and is available for remote transmission. The sensor depends on the large differential in dielectric constants between oil and water for its operation. Manufacturers claim that the sensor is not confused by emulsified sludge, which has a large dielectric constant, or by oily froth, which cannot pass under the float.

Oil-in-Water Detector

When a contaminated water sample stream is irradiated with UV waves at a peak intensity of 365 nm, the oil contaminant emits visible radiation. This radiation can be measured by a photocell. Visible radiation increases with increasing concentrations of the fluorescent substance. The relationship between the concentration and the visible radiation emitted is substantially linear in low concentrations (below 15×10^{-6}). In higher concentrations, some nonlinearity occurs as a result of a saturation effect.

The most common measurement method is to pass a sample through the sensing head in an upflow direction (see Figure 7.9.13). The head is equipped with two windows set at right angles that minimize the intensity of direct radiation from the source striking the photocell and also reduce the multiple scattering of visible radiation effect. Optical filters at the incident and emergent windows (not shown) reduce this effect to a negligible level.

To detect the oil concentration in water, a falling-stream-type detector is also available. With this device, the sample stream is shaped into a rectangle and falls through the viewing field of the UV beam and the photocell. Efficient optical filtration is important to overcome the unavoidable effects of the direct reflection of incident radiation from the surface of the shaped stream.

UV Oil-in-Water Analyzer

Figure 7.9.14 shows the sampling system of a continuous oil-in-water analyzer used to monitor steam condensate, recycled cooling water, and refinery or offshore drilling effluents. This system uses a single-beam, dual-wavelength UV analyzer, superior to the single-wavelength designs because it compensates for variations in sample's sediment content, turbidity, algae concentration, or window coatings. The cell operates according to Beer's law, which relates oil concentrations to UV energy absorption by the fixed-length cell. The UV measuring band is centered at 254 nm, and the readings are sensitive to 0.1 ppm with a range of 0 to 10 ppm and provide a 90% response in 1 sec.

The automatic-zero feature of the instrument is provided by sending sample water to both the measurement and zeroing sides of the conditioning system. When the sample is in the measurement mode, it is sent through a high-speed, high-shear homogenizer, which disperses all suspended oil droplets and oil adsorbed onto foreign matter so that the sample sent to the analyzer becomes a uniform and true solution.

Once an hour, the analyzer is automatically rezeroed. In this mode, the sample water is sent through a filter that removes all oil and after sparging, the sample water is sent to the analyzer. This oil-free, zero-reference sample still contains the other compounds found in the measurement sample and therefore can be used for zeroing out this background.

Radio-Frequency (Microwave) Sensors

When a cup of water and oil is placed in a microwave oven, the water heats up, while the oil does not. This occurs because shortwave radio-frequency energy is absorbed more efficiently by water than oil. In the radio-wave detector, the transmitter produces fixed-frequency and con-

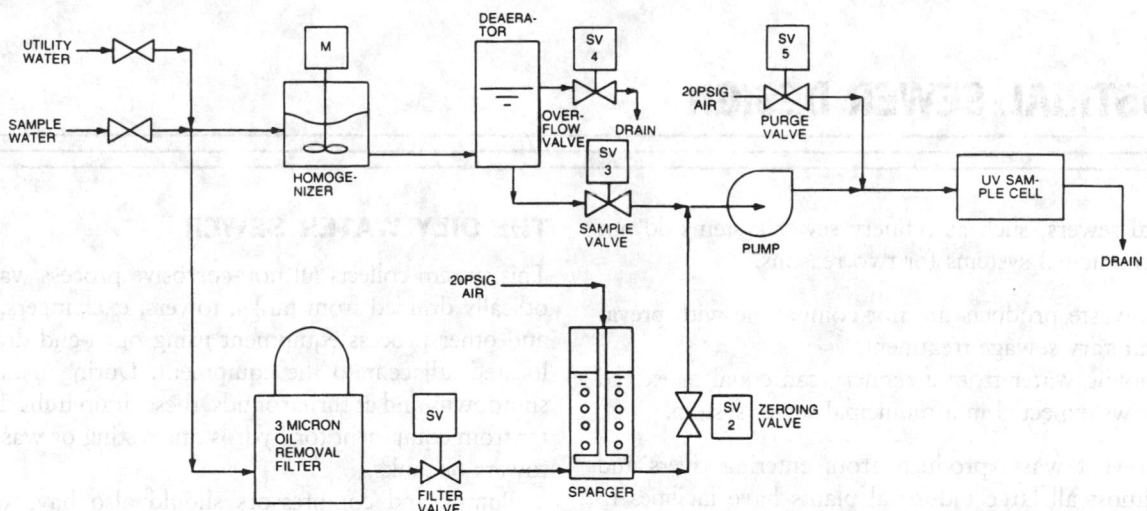

FIG. 7.9.14 UV oil-in-water analyzer with automatic-zero feature. (Courtesy of Teledyne Analytical Instruments).

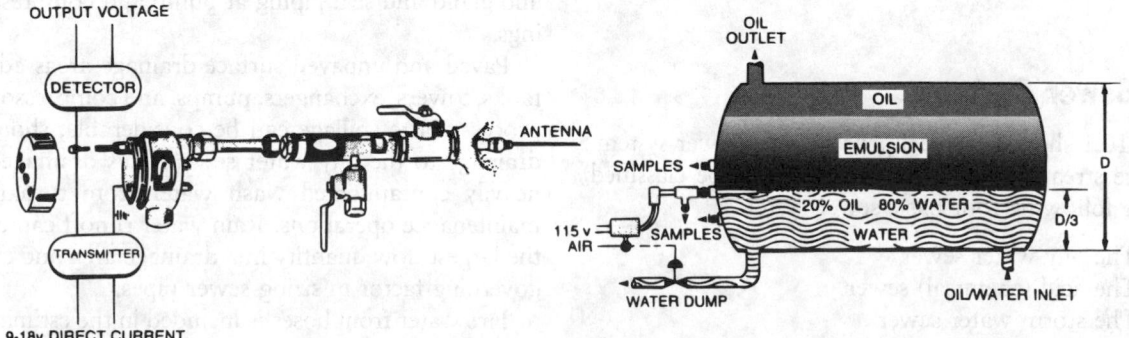

FIG. 7.9.15 Radio-wave oil–water interface detector probe. (Courtesy of Agar Corp.)

stant-energy waves. The more energy is absorbed by the process fluid (the more water in the mixture), the lower the voltage at the detector. The advantages of this design, compared to capacitance systems, include a wider range (0 to 100%), lower sensitivity to buildup, insensitivity to temperature and salinity variations, and suitability for higher temperature operations (up to 450°F or 232°C).

Radio-wave, oil-in-water sensors are available as probe-type sensors for water–oil interface control. A typical application is the free-water knockout (see Figure 7.9.15), where the probe is installed horizontally at one-third of the diameter from the bottom and is set to open the water dump valve when the emulsion concentration drops below 20% oil (80% water). In this way, the emulsion (rug layer) builds up above the probe, while only clean water is dumped. The probe can also provide a 4- to 20-mA

transmitted output signal that signals the water concentration within an error of 5%.

An available portable tank profiler also uses the same principle of operation. Here, the radio-wave element supported by a tape, is lowered into the tank, which can be 100 ft (30 m). As the sensor is lowered, it measures both the interface location (within an error of 0.12 in or 3 mm) and the emulsion concentration throughout the tank from 0 to 100% within an error of 1%.

A water-in-oil monitoring probe is also available, which can detect water concentration over a 0 to 100% range within an error of 0.1% in tanks or pipelines. All these devices are available in explosion-proof construction and with digital displays.

—*Béla G. Lipták*

Sewers and Pumping Stations

7.10
INDUSTRIAL SEWER DESIGN

Industrial sewers, such as refinery sewer systems do not tie into municipal systems for two reasons:

Refinery waste products are not compatible with prevalent sanitary sewage treatment.
Spent cooling water from a refinery can equal or exceed the flows expected in a municipal sewer system.

To prevent waste products from entering rivers and lakes, almost all large industrial plants have facilities to separate and collect waste products. Refinery and chemical plant sewers flow by gravity and are usually partially filled.

Basic Sewer Systems

Figure 7.10.1 shows a typical process plant sewer system. The waste streams in most large plants can be classified under the following four basic sewer systems:

- The oily water sewer
- The acid (chemical) sewer
- The storm water sewer
- The sanitary sewer

THE OILY WATER SEWER

This system collects all non-corrosive process waste periodically drained from tanks, towers, exchangers, pumps, and other process equipment using open-end drain hubs located adjacent to the equipment. During maintenance shutdowns and at turnarounds, these drain hubs drain water from equipment for hydrostatic testing or washing out towers or tanks.

Pumps and compressors should also have open-end drain hubs located at the ends of foundation blocks. These open-end drain hubs collect drainage from pump bedplates and gland and seal piping at pump and compressor bearings.

Paved and unpaved surface drainage areas adjacent to tanks, towers, exchangers, pumps, and compressors, where process waste spillage can be considerable, should divert drainage to the oily water sewer. This drainage includes heavily contaminated wash water from turnaround or maintenance operations. Rain water runoff can constitute the largest flow quantity in a drainage area and can be the governing factor in sizing sewer pipes.

Fire water from hoses is included in the estimated maximum flow quantities within unit areas containing haz-

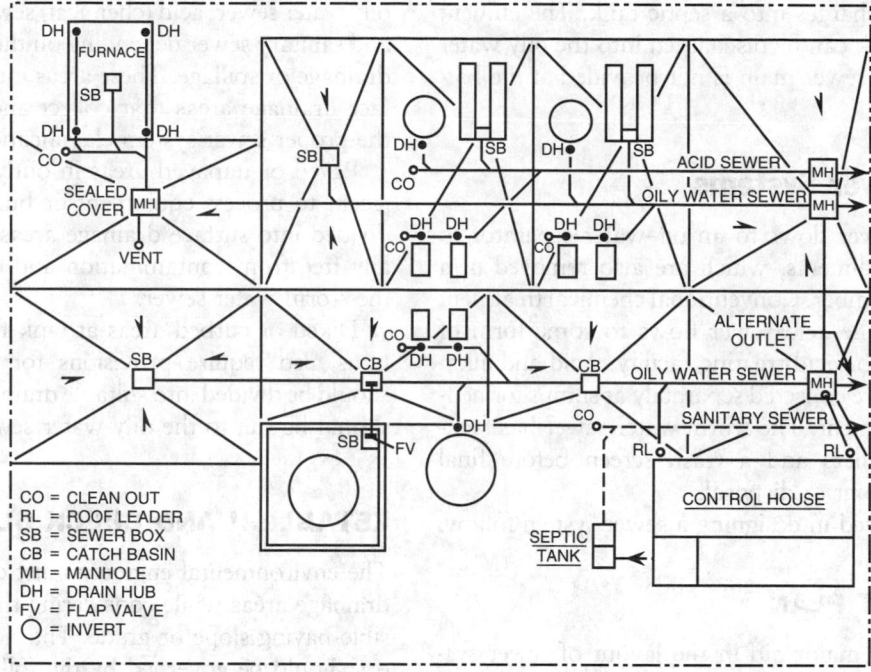

FIG. 7.10.1 Typical process plant sewer system.

CO = CLEAN OUT
RL = ROOF LEADER
SB = SEWER BOX
CB = CATCH BASIN
MH = MANHOLE
DH = DRAIN HUB
FV = FLAP VALVE
○ = INVERT

ardous hydrocarbon or chemical equipment. Where fire water is included, it can vary from 500 to 1000 gpm. The amount depends on the size and number of equipment pieces as well as the number of sewer boxes or drains within the area.

The oily water sewer main should be run to the battery limit as a separate system. There it should be connected to the oily water trunk sewer that runs to an oil–water separator.

ACID (CHEMICAL) SEWER

This sewer collects heavily contaminated, corrosive, process chemical waste that occurs as spillage, leakage, and valved drains at process equipment and pumps.

Open-end drain hubs located at all tanks, towers, exchangers, and associated equipment facilitate draining. Large drains at towers or tanks can be handled more conveniently by an acid-proof concrete or acid-brick-line sewer box rather than an open-end drain hub.

Pump blocks should have an open-end drain hub to collect pump casing drains, drainage from gland and seal piping at pumps, and drains in pump suction and discharge piping.

Acid areas that collect corrosive process waste usually have acid-resistant curbed paving to confine and collect any acid drainage or spillage within these areas. Curbed and paved areas should be provided in locations where pump groups, storage, and handling areas are subject to spillage and wash-down water.

Wash water collected in these surface drainage areas should be collected in the acid sewer. However, where pos-

sible, storm water surface drainage should not be run into the acid sewers.

Acid waste should be run in a separate sewer from alkaline waste. Acid and alkaline waste should be run as two separate sewer systems to the battery limit and the acid treating facility or a neutralizing sump.

STORM WATER SEWER

The storm water sewer collects maximum surface drainage including rainfall, wash water that is not contaminated, and cooling water that is not returned in return headers to cooling water facilities.

Storm water runoff is calculated on the basis of 100% runoff for all paved areas and 50% runoff for unpaved areas. The remaining 50% in unpaved areas is assumed to be absorbed into the ground.

Rainfall data for various geographic locations are readily available from the government, state, and city weather bureau records, and other published data. Storm water accumulation for each in of rainfall/hr/sq ft is equal to 0.0104 gpm.

Fire water from hoses should be included in the estimates of storm water runoff if flooding would cause damage to installations and present a hazard during fire fighting operations. The storm water main should be run to the battery limit and connected to the trunk sewer.

SANITARY SEWER

The sanitary sewer constitutes a separate sewer system into which only wastes of sanitary facilities are permitted. The sanitary sewer discharges into a septic tank. The effluent from the septic tank can be discharged into the oily water sewer if a sanitary sewer main is not provided at the battery limit.

Designing Sewer Systems

The oily water sewer flows to an oil–water separator to remove oil and sediments, which are also removed in a sludge disposal chamber. Conventional chemical treatment is also required. The acid sewer flows to some form of neutralizing sump or acid-treating facility. Acid and alkaline sewer wastes are collected separately at sumps for neutralization or treatment. The storm water sewer has facilities for oil skimmers and a trash screen before final discharge at the point of disposal.

The steps involved in designing a sewer system follow.

DEVELOP PLOT PLAN

The plot plan is a major aid in the layout of sewer systems. The plot plan indicates the locations of all pumps, exchangers, tanks, and towers. It also indicates the extent of paved areas, roadways, and underground utilities (water and electric) and the locations and inverts of sewer tie-ins to the sewer mains at the battery limits.

LAY OUT SYSTEM

An environmental engineer can begin the layout of a sewer system by indicating all major equipment foundations, taking their locations from the plot plan. The layout should indicate all pipe rack columns, lighting poles, and minor footings that can interfere with the sewers. The environmental engineer should integrate underground cooling water systems into the sewer system layout as an additional system, as well as any underground electrical utilities to avoid any interferences.

The task now is to design sewer systems into an intricate layout that has as many as four separate sewer systems underground. The layout must provide gravity flow in each of the systems, maintain the given inverts of the sewer systems at the battery limits, and be free of interferences at the cross-overs of sewer systems and underground water and electrical systems. Existing underground piping, electrical trenches, and other encumbrances further complicate the sewer layout.

CLASSIFY AND TYPE SURFACE DRAINAGE

The plot plan, having furnished the locations of all pumps, exchangers, compressors, tanks, and towers, also shows the extent of paved and curbed areas. The paved and curbed areas adjacent to process equipment should be segregated into proper sewer system classifications, namely, oily water sewer, acid (chemical) sewer, storm water sewer, and sanitary sewer depending on the type of process waste drainage or spillage. These areas must be divided into surface drainage areas that collect and channel drainage to the proper sewer system classification.

Paved or unpaved areas in outlying locations, not adjacent to process equipment or buildings, should also be divided into surface drainage areas. These areas are usually free from contamination and can be channeled into the storm water sewer.

Diked or curbed areas at tank farm and storage locations also require provisions for surface drainage and should be divided into suitable drainage areas. These drains should be run to the oily water sewer main.

ESTABLISH AND CHECK SLOPE

The environmental engineer must design and size surface drainage areas while considering the limits of the permissible paving slope or grade. The slope of all drainage areas should be governed by the following limits so that a hazardous or tripping condition is avoided:

The paving inside a building should have a minimum slope of 1 in over 10 ft and a maximum elevation drop of 3 in for each drainage area.

Paved and unpaved areas outside buildings should have a sewer box or catch basin for each surface drainage area. The maximum difference in elevation between the high point of grade and the grade at the catch basin should not be more than 6 in with a slope of 1 in over 10 ft.

The total number of surface area divisions depends on the drop in elevation of surface drainage areas.

DESIGN SEWER

After the surface drainage areas are divided into areas based on slope and drop in elevation, the environmental engineer must segregate and run them to the proper sewer classification.

The divided areas are provided with a sewer box or catch basin. The separate sewer systems must now be connected to the proper classified sewer. The outlet connections at the sewer box or catch basin can be at the bottom or side depending on the limits of sewer inverts (see Figure 7.10.2). The *invert* is the elevation of the bottom inside surface of the sewer pipe.

The sewer design should provide for ample future expansion of the plant and unit areas. Sewer mains, in particular, should be sized to include the estimated flow of any expansion.

Storm water runoff is calculated in rainfall in based on 100% runoff for all paved areas and 50% runoff for unpaved areas. Obtain design rainfall in amount of in per hr for the specific area where the plant is to be located; oth-

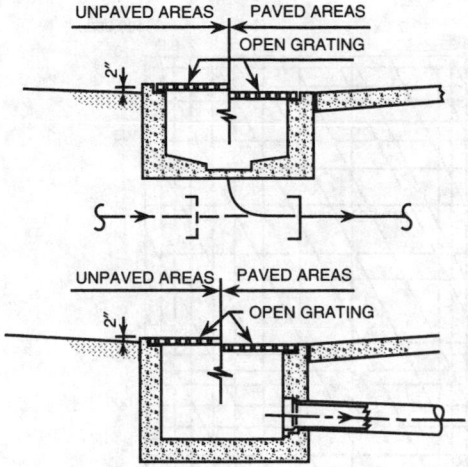

FIG. 7.10.2 Sewer box outlet detail.

erwise, the general chart shown in Figure 7.10.3 can be used.

Storm water for each in of rainfall/hr/sq ft equals 0.0104 gpm.

DETERMINE SEWER PIPE SIZE

The minimum size of underground sewer pipe, branch or drain hub, should be 4 in. The minimum size of the sewer main (collecting two or more 4-in sewer branches or drain hubs) should be 6 in. The minimum size of the sewer pipe in any curbed or diked area should be 8 in.

Sewer size depends on plot size, amount of rainfall, and quantity of process waste, fire water, and any other liquids requiring disposal. As previously stated, the sewer mains should be sized to include the estimated flow of future expansion.

Velocities used in sewer system design should have a minimum of 3 ft per sec and a maximum of 7 ft per sec. Flow capacities, velocities, and slopes (for sewers running

$\frac{3}{4}$ full) can be coordinated so that the curves shown in Figure 7.10.4 give the sewer line size required for the slope and velocity for the flow capacity in gallons per minute.

The flow capacities, velocities, and slopes of sewer lines are contingent on plant site grades available. The flat grades necessary at most plant sites are a determining factor in the slope and velocity of the sewer lines. Where possible, and if gradients permit, the maximum velocity should be used.

The following design example determines the sublateral size for a sewer system:

Given: Rainfall = 3 in/hour
    Process waste = 100 gpm
    Fire water = 250 gpm per catch basin
    Runoff in unpaved areas = 40%
Computations:
Storm water
    Area involved: Paved, 2 areas, 3280 sq ft total.
            Unpaved, 3 areas, 6920 sq ft total.

$$\text{gpm} = \frac{(0.31 * 3280) + (0.31 * 6920 * 0.40)}{0.75}$$

gpm = 251
Process Waste + Storm Water = 100 + 251 = 351 gpm
Process Waste + Fire Water = 100 + (4*250) = 1100 gpm

Using 1100 gpm and the sewer sizing chart (see Figure 7.10.4) shows that a 10-in line sloped .017 ft/ft with a flow velocity of 4.4 ft/sec can handle this quantity.

DETERMINE INVERT

The invert of the sewer inlet and outlet at catch basins and manholes should be at the same elevation when both in-

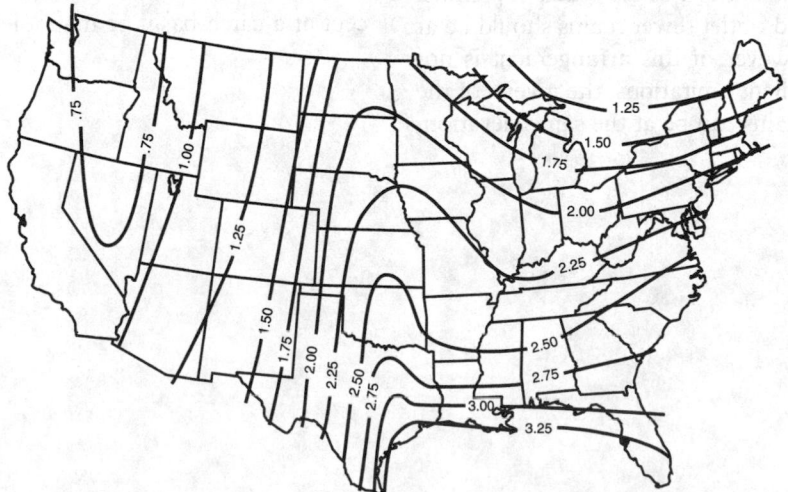

FIG. 7.10.3 General rainfall chart. This chart should be used only when the design rainfall of the job area is not specified.

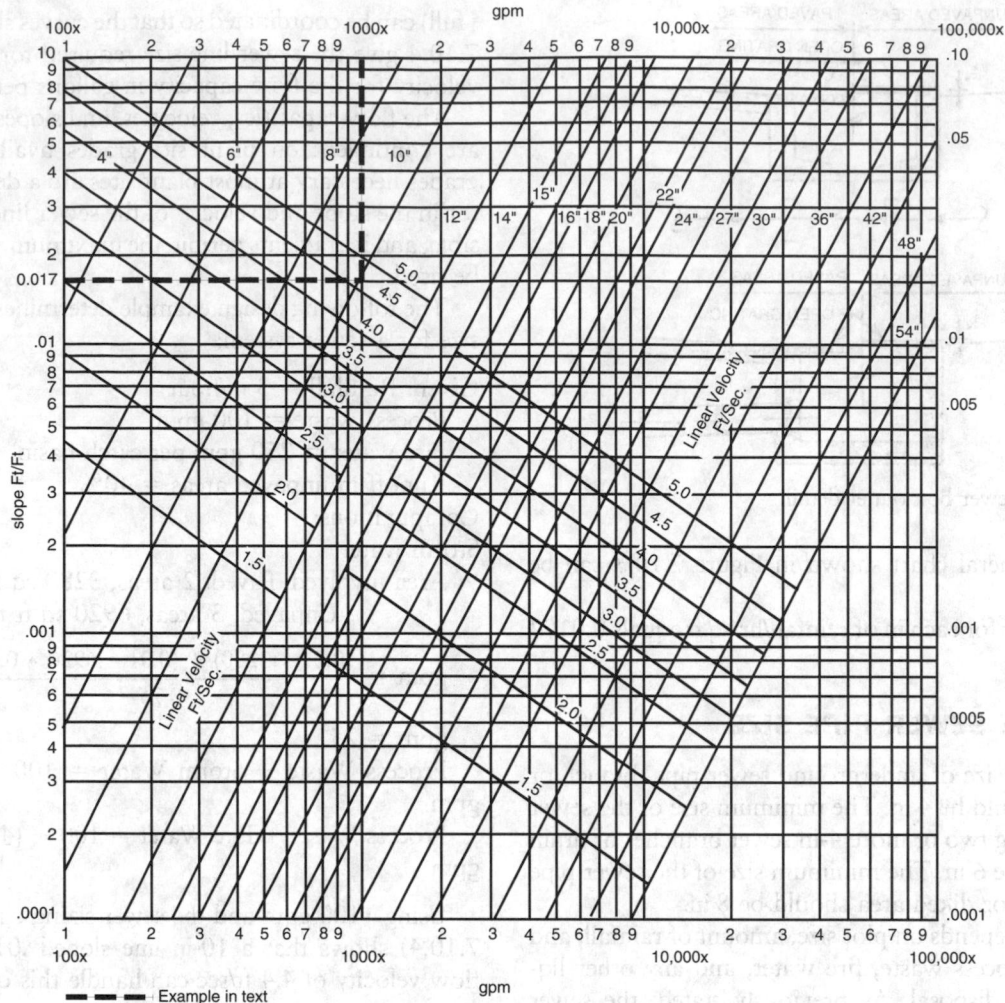

FIG. 7.10.4 Sewer sizing chart. The chart shows the linear velocity and slope for tile, concrete, and cast iron sewer pipe during gravity flow. The chart is based on Manning's formula for circular pipes flowing full for values of N = .015 for 10-in pipe and smaller and N = .013 for 12-in pipe and larger.

let and outlet sewer mains are the same size. Where the sewer size increases at the outlet and the gradient permits, the tops of both inlet and outlet sewer mains should be at the same elevation. However, if this arrangement is not practical because of gradient limitations, the inverts of the inlet and outlet sewer mains can be at the same elevation, preferably with the outlet slightly lower. Straight runs of sewer mains should not change in pipe diameter (size) except at a catch basin or manhole.

—*Mark K. Lee*

7.11
MANHOLES, CATCH BASINS, AND DRAIN HUBS

This section discusses appurtenances in sanitary and industrial sewers. Sewer appurtenancese include manholes, building connections, junction boxes, drain hubs, catch basins, and inverted siphons. Additional information on sewer appurtenances is in publications by Metcalf and Eddy, Inc. (1991), Steel and McGhee (1979), National Clay Pipe Institute (1978) and WPCF (1970).

Sanitary Sewer Appurtenances

Figure 7.11.1 shows a plan and profile of a sanitary sewer and its laterals with enlarged sections of sewer trenches and manholes.

MANHOLES

Manholes should be of durable structure, provide easy access to sewers for maintenance, and cause minimum interference to sewage flow. Manholes for small sewers are usually about 1.2 m (4 ft) in diameter. Sewers larger than 60 mm (24 in) in diameter should have larger manhole bases although a 1.2-m barrel can still be used.

Manholes should be located at the end of the line (called terminal cleanout), at sewer intersections, and at changes in grade and alignment except in curved sewers as shown in Figure 7.11.1. The maximum spacing of manholes is 90–180 m (300–600 ft) depending on the size of the sewer

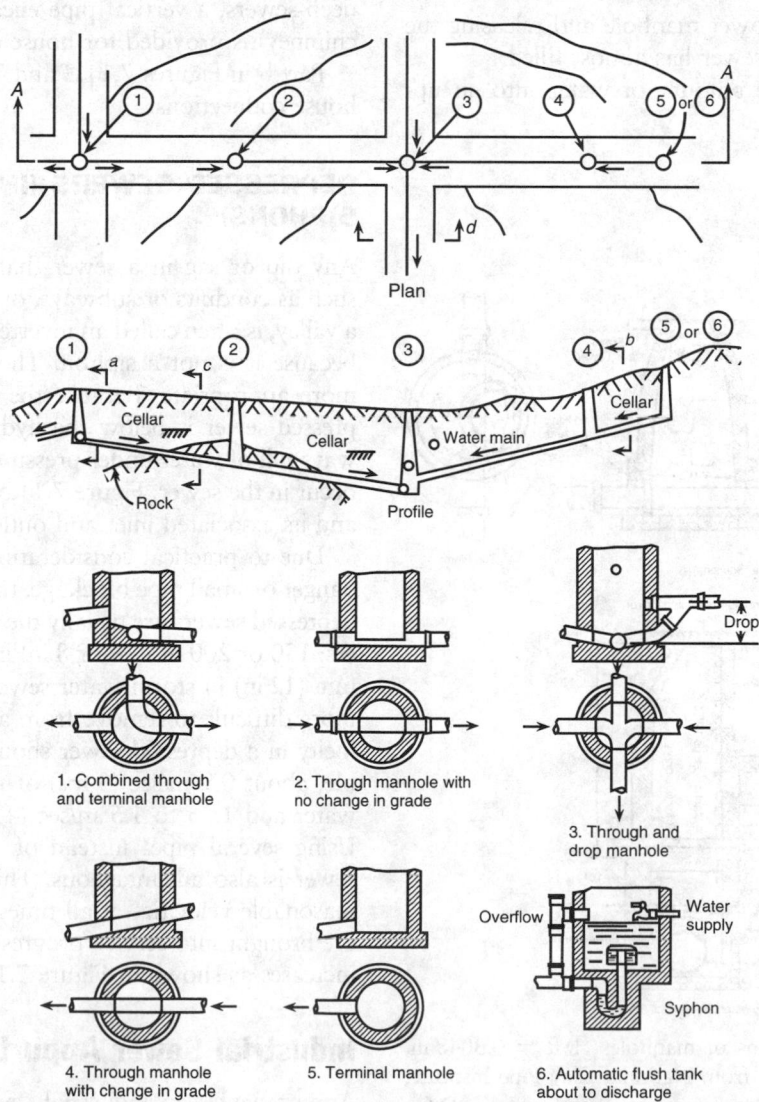

FIG. 7.11.1 Plan, profile, and construction details of sanitary sewers.

and the sewer cleaning equipment. Manholes, however, should not be located in low places where surface water can enter. If such locations are unavoidable, special water-tight manhole covers should be provided.

Part a in Figure 7.11.2 shows a typical manhole; part a in Figure 7.11.3 shows the details of a terminal cleanout.

Drop Manholes

A drop manhole reduces the turbulence in the manhole when the elevation difference between incoming and outflow sewers is greater than 0.5 m (1.5 ft). Turbulence due to a sudden drop of wastewater can cause splashing, release of odorous gases, and damage to the manhole. Part b in Figure 7.11.2 shows the details of a drop manhole.

Flushing Manholes

In their upper reaches, most sewers receive so little flow that they are not self-cleaning and must be flushed from time to time. This flushing is done by the following means:

Damming the flow at a lower manhole and releasing the stored water after the sewer has almost filled.
Suddenly pouring a large amount of water into an upstream manhole.

Providing a flushing manhole at the uppermost end of the line. A flushing manhole is filled with water through a fire hose attached to a nearby hydrant before a flap valve, shear gate, or similar quick-opening device leading to the sewer is opened.
Installing an automatic flush tank that fills slowly and discharges suddenly. Apart from the cost and maintenance difficulties, the danger of backflow from the sewer into the water supply is a negative feature of automatic flush tanks.

BUILDING CONNECTIONS

Building sewers are generally 10–15 cm (4–6 in) in diameter and constructed on a slope of 0.2 m/m. Building connections are also called house connections, service connections, or service laterals. Service connections are generally provided in municipal sewers during construction. While the sewer line is under construction, connections are conveniently located in the form of wyes or tees and plugged tightly until service connections are made. In deep sewers, a vertical pipe encased in concrete (called a chimney) is provided for house connections.

Part b in Figures 7.11.3 and 7.11.4 show the details of house connections.

DEPRESSED SEWERS (INVERTED SIPHONS)

Any dip or sag in a sewer that passes under structures, such as conduits or subways, or under a stream or across a valley, is often called an inverted siphon. It is a misnomer because it is not a siphon. The term *depressed sewer* is more appropriate. Because the pipe constituting the depressed sewer is below the hydraulic grade line, it is always full of water under pressure although little flow may occur in the sewer. Figure 7.11.5 shows a depressed sewer and its associated inlet and outlet chambers.

Due to practical considerations, such as the increased danger of small pipe blockage, the minimum diameters for depressed sewers are usually the same as for ordinary sewers: 150 or 200 mm (6 or 8 in) in sanitary sewers and 300 mm (12 in) in storm water sewers. Since obstructions are more difficult to remove from a depressed sewer, the velocity in a depressed sewer should be as high as practicable, about 0.9 m/sec (3 fps) or more for domestic wastewater and 1.25 to 1.5 m/sec (4 to 5 fps) for stormwater. Using several pipes instead of one pipe for a depressed sewer is also advantageous. This arrangement maintains reasonable velocities at all times because additional pipes are brought into service progressively as wastewater flow increases as shown in Figure 7.11.5.

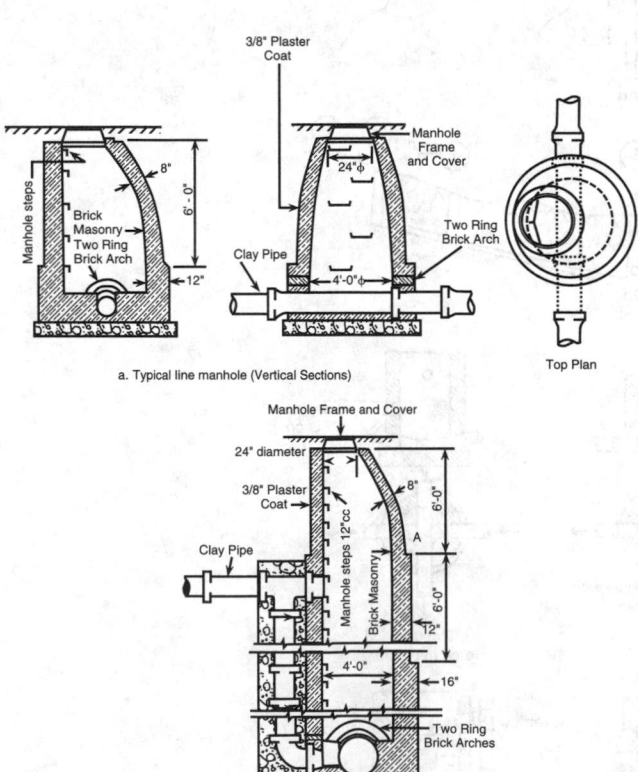

a. Typical line manhole (Vertical Sections)

b. Drop manhole (Vertical Section)

FIG. 7.11.2 Typical designs of manholes. 1 ft = 0.3048 m. (Reprinted, with permission, from National Clay Pipe Institute, 1978, *Clay pipe engineering manual*, Washington, D.C.: National Clay Pipe Institute.)

Industrial Sewer Appurtenances

Appurtenances in industrial sewers are shown in Figure 7.10.1 in the preceding section.

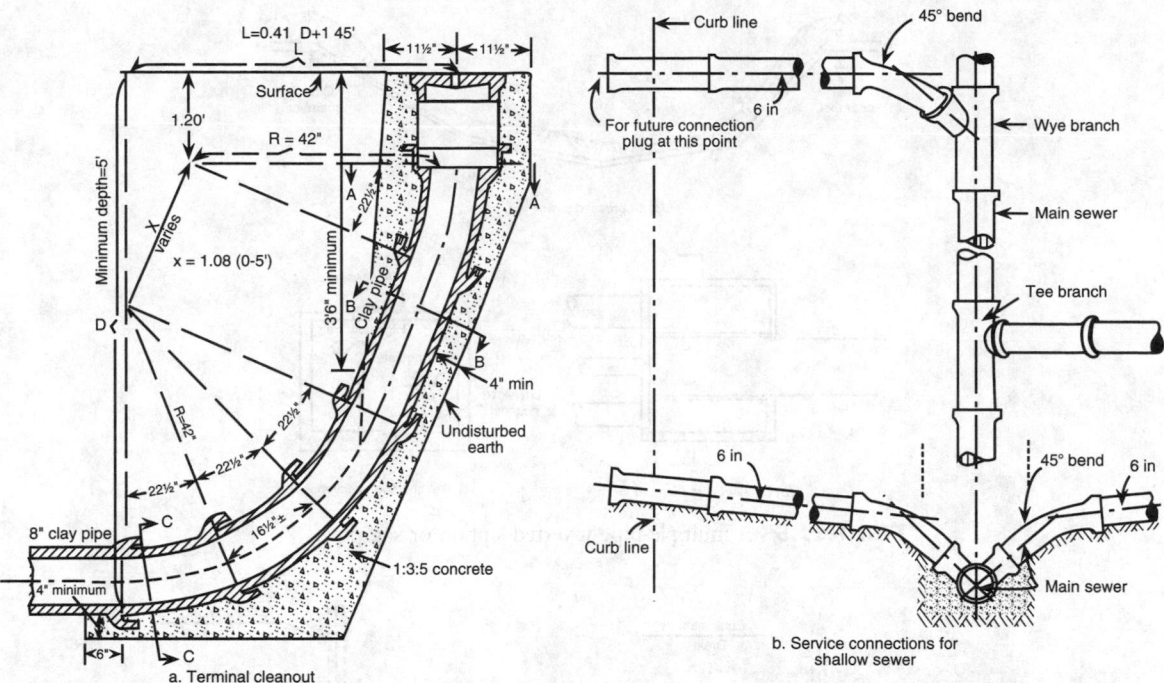

FIG. 7.11.3 Typical terminal cleanouts and house connections. 1 ft = 0.3048 m. (Reprinted, with permission, from National Clay Pipe Institute, 1978; Water Pollution Control Federation and American Society of Civil Engineers).

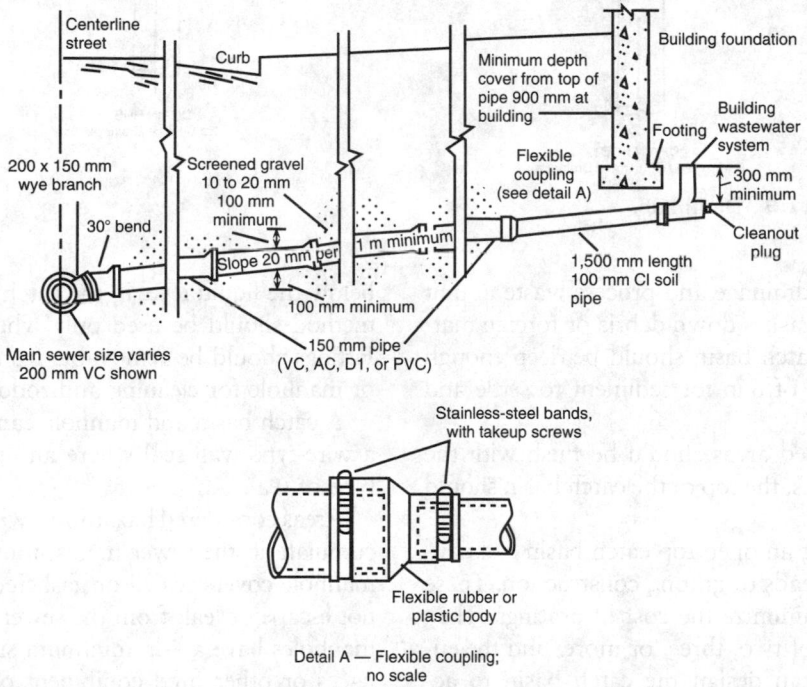

FIG. 7.11.4 Typical house connection. Note: mm × 0.03937 = in.

DRAIN HUBS

Drain hubs (see Figure 7.11.6) collect drainage from equipment above the grade or paving and run it to the proper classified sewer. Drain hubs need not consist of an actual hub attached to the pipe. A piece of pipe projecting 2 in above grade or paving is sufficient. Drain hubs extending 2 in above the grade or paving prevent surface drainage from entering the sewer systems.

CATCH BASINS

Catch basins (see Figure 7.11.7) are used as a junction for changes of direction of sewer branch lines. The location of sewer branch junctions may coincide so that a catch basin can be substituted for a sewer box, which is provided for surface drainage. A catch basin is also used as a junction for a change in diameter (size) of all sewer mains.

Catch basins with open tops covered with grating are

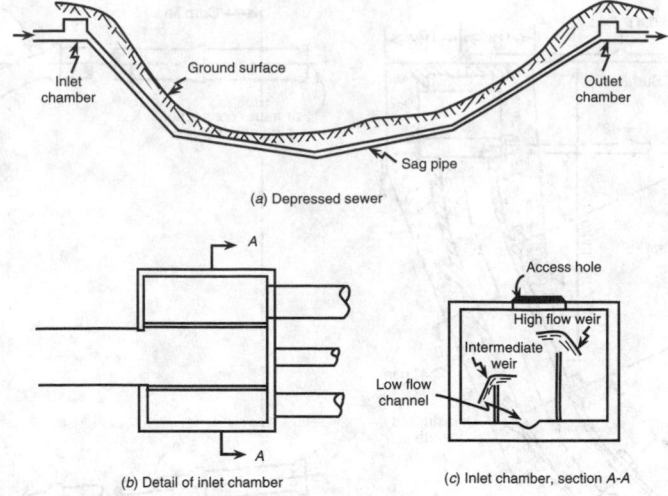

FIG. 7.11.5 A multiple-pipe inverted siphon or sag pipe.

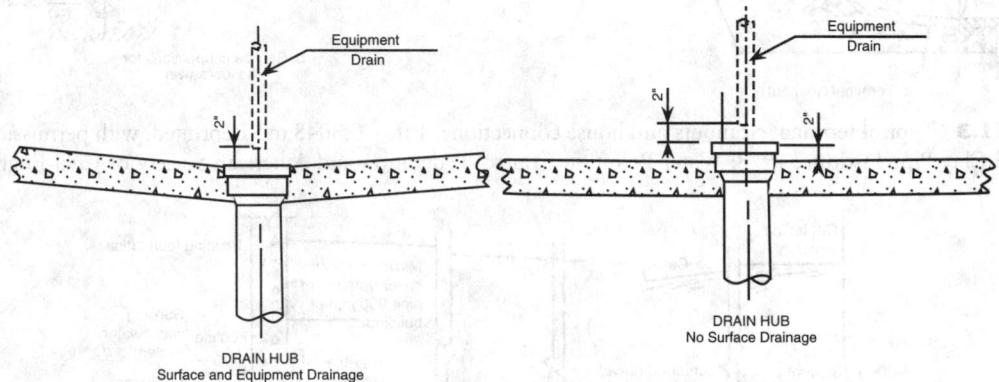

FIG. 7.11.6 Drain hub.

used to collect surface drainage and process waste in unit areas where drainage washes down debris or foreign matter. The bottom of a catch basin should be deep enough to provide a minimum of 6 in for sediment to settle and separate.

Catch basins in paved areas should be flush with the paving; in unpaved areas, the top of the catch basin should be 2 in above the grade.

The grating covering an open-top catch basin can consist of standard stair treads of grating construction. These standard stair treads minimize the cost of gratings. They can be used in groups of two, three, or more; and the environmental engineer can design the catch basin to accommodate the number of gratings used.

Seals must be provided on all sewer branch inlets and mains connecting to a catch basin or manhole (see Figures 7.11.7 and 7.11.8). A seal should consist of an elbow or a tee with an outlet extending downward to provide a minimum of a 6-in seal. Some refineries use a special combination seal and clean-out fitting inserted into the the catch basin wall or manhole when the concrete is poured.

A simple method for providing a seal is to run the inlet pipe into the catch basin or manhole at a sharp downward angle so that the upper edge of the inlet pipe is 6 in

below the liquid level in the catch basin or manhole. This method should be used only when the pipeline is short. Fittings should be removable from inside the catch basin or manhole for cleaning and rodding.

A catch basin and manhole can be constructed to have a wire-type wall seal where an application warrants this type of seal.

Areas considered hazardous, where flammable gases accumulate in the sewer mains, must have catch basin and manhole covers sealed or gasketed so that these gases do not escape or leak from the sewer lines. Catch basins and manholes have a 4-in minimum size vent. Areas with furnaces or other fired equipment or ignition sources must run underground vents at least 100 ft from the source of ignition and 10 ft above grade in a safe location.

FLOOR DRAINS

Floor drains inside buildings can be used in floors that only handle water. A sewer box or catch basin must be used where process waste spillage can occur inside a building.

Floor drains should not be used in control rooms, switch rooms, or lavatories, because sewer gas accumulations may

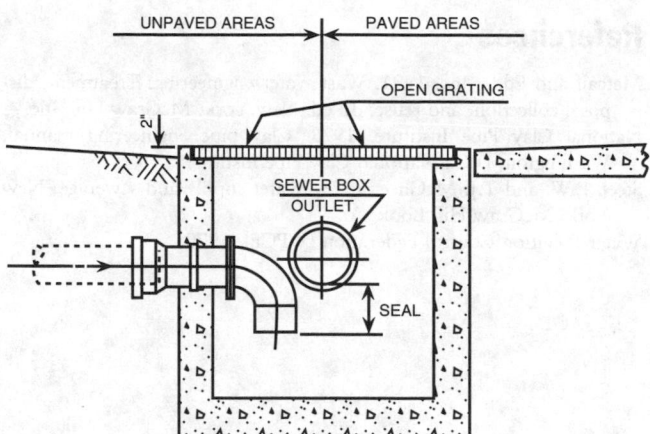

FIG. 7.11.7 Catch basin.

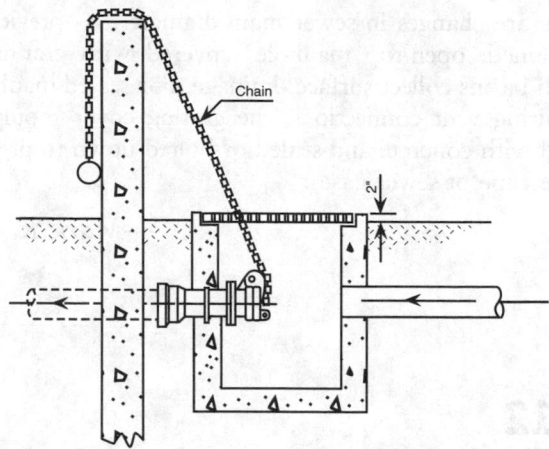

FIG. 7.11.9 Flap valve.

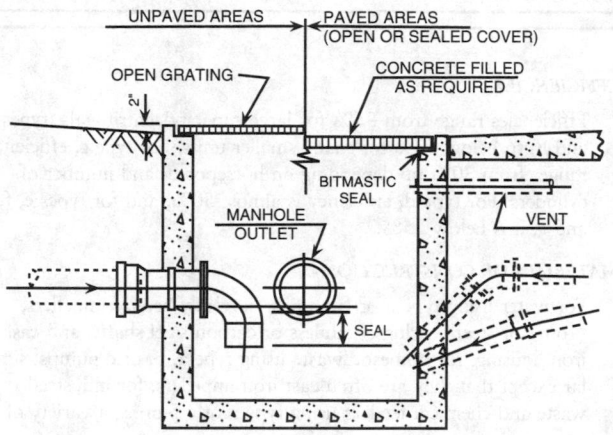

FIG. 7.11.8 Manhole.

back up through the floor drains causing explosions and flash fires. A water source for the seals in running traps, normally provided in sewer branches running from buildings, is not available because little, if any, water is available from washing floors or any other source. Because the floors usually receive only an occasional mopping, the sewer line and trap seal can dry up allowing sewer gases to enter the buildings.

Sewer branch mains from a pump house, compressor house, or any enclosed building should have connections to sewer systems at manholes or catch basins with an inlet provided with a seal. Where these connections are not feasible, building drains and separately defined areas should have running clean-out traps (P trap) located outside building walls with accessible clean-out plugs to avoid disturbing the paving.

Running a sewer branch or subbranch in one plane from a drain hub to a catch basin and from a catch basin to a manhole, prevents the sewer line from clogging and facilitates rodding and cleaning. Whenever possible, all sewer system branch connections should be at a catch basin or manhole.

Curbed or diked storage areas should have sufficient catch basins to accommodate surface drainage. The sewer pipe runs from the catch basin through the curb or dike

wall to the nearest manhole. The sewer outlet connection inside the catch basin should use a flap valve (see Figure 7.11.9).

A flap valve operates, by chain, at the curb or dike wall. It should be closed at all times except to drain surface water from the enclosure. A gate valve should be installed in the sewer line outside the enclosure and have an extension stem for operation at grade level.

When a leak or break occurs in the storage tank or piping, the flap valve prevents loss of tank contents and allows for the recovery of tank contents retained within the diked area. It also prevents large amounts of hazardous and flammable liquids from entering the sewer system, which can create a hazardous fire condition if tank contents are volatile.

Drain lines that collect hot, noncorrosive waste drainage above 210°F, such as boiler blow-off, steam trap discharges, and hot waste drainage, without receiving cooling quench from other streams should be constructed of steel pipe and fittings. The steel sewer pipe should be run to the nearest catch basin or manhole. Hot acid waste drainage should be run in acid-resistant, alloy material sewer pipe and fittings.

A valve box for underground pipelines should have side walls with drainage only through a gravel bed at the open bottom of the valve box.

Sewer pipes running under roadways and trucking areas must be protected from damage by heavy concrete slabs or protective pipe sleeves. Sewer pipes embedded in concrete foundations should be constructed of steel pipe and fittings.

MANHOLES

Manholes should be installed in sewer mains at intervals of 300 ft maximum for sewer sizes to 24 in and at 500 ft maximum intervals for sizes above 24 in (see Figure 7.11.8). Manholes installed at dead ends of sewer mains can act as junctions to connect sewer branches. Manholes are also installed in sewer main runs as junctions, where there are changes in sewer main diameter. As previously explained, open-top manholes covered with grating or catch basins collect surface drainage. For sealed manholes requiring vent connections, the grating cover should be filled with concrete and sealed or bolted down to prevent the escape of sewer gases.

—*Mark K. Lee*

References

Metcalf and Eddy, Inc. 1991. Wastewater engineering: Treatment, disposal collection, and reuse. 3d ed. New York: McGraw-Hill, Inc.

National Clay Pipe Institute. 1978. Clay pipe engineering manual. Washington, D.C.: National Clay Pipe Institute.

Steel, E.W. and T.J. McGhee. 1979. Water supply and sewerage. New York: McGraw-Hill Book Co.

Water Pollution Control Federation (WPCF). 1970.

7.12
PUMPS AND PUMPING STATIONS

TYPES OF PUMPS

  a. Radial-flow centrifugals
  b. Axial-flow and mixed-flow centrifugals
  c. Reciprocating pistons or plungers
  d. Diaphragm pumps
  e. Rotary screws
  f. Pneumatic ejectors
  g. Air-lifts

Other pumps and pumping devices are available, but their use in environmental engineering is infrequent.

SELECTION OF PUMPS

See Table 7.12.1.

EFFICIENCIES

Efficiencies range from 85% for large capacity centrifugals (types a and b) to below 50% for many smaller units. For type c, efficiency ranges from 30% up depending on horsepower and number of cylinders. For type d, efficiency is almost 30%, and for types e, f, and g, it is below 25%.

MATERIALS OF CONSTRUCTION

For water using type a or b pumps, normally bronze impellers, bronze or steel bearings, stainless or carbon steel shafts, and cast iron housing; for domestic waste using type a, b, or c pumps, similar except that they are often cast iron impellers; for industrial waste and chemical feeders using type a or c pumps, a variety of

TABLE 7.12.1 PUMP SELECTION

Feature Summary Type Designation	Type of Pump	Clear Liquids—Low Viscosity	Clear Liquids—High Viscosity	Thin Slurries or Suspensions	Raw or Partially Treated Sewage and Heavy Suspensions	Viscous or Thick Slurries and Sludges	For Capacity (GPM)	For Ft of Head
a	Radial-flow centrifugals	✔		✔	✔[1]	✔[1]		
b	Axial-flow and mixed-flow centrifugal	✔		✔	✔			
c	Reciprocating pistons and plungers	✔[2]	✔	✔	✔[2]	✔		
d	Diaphragm pumps	✔[2]	✔[2]	✔	✔	✔		
e	Rotary screws		✔[2]		✔[2]	✔		
f	Pneumatic ejectors				✔			
g	Airlift pumps			✔	✔			

Notes: ✔ Suitable for normal use. [1] See text for limitations. [2] Not used for this purpose in environmental engineering (with some exceptions). If not checked, either not suitable or not normally used for this purpose.

materials depending on corrosiveness; type d similar except that the diaphragm is usually rubber; types e, f, and g normally steel components

PUMPING STATIONS

Designed with pump in liquid chamber (wet well design) or pump in dry pit with wet well before it (dry well design). Most are designed for a specific application, but for lower capacity pumping facilities (up to 60 hp), prefabricated stations are common.

PARTIAL LIST OF SUPPLIERS

Allweiler Pump Inc. (e); Aurora Pump Unit, General Signal Corp. (a,b); Barnes Pumps Inc. (a,d); Crane Co. (a,d); Dresser Pump Div. (a,c); Duriron Co. (a,c,d); Fairbank Morse Pump Corp. (a,b); Flygt Corp. (a); Gorman-Rupp Industries Div. (a,b,d); Goulds Pumps (a,c); Ingersoll-Rand Co. (a,c,d); Komline-Sanderson Engineering Corp. (c,d); Lakeside Equipment Corp. (e); Marlow Div., ITT (a,c,d); Moyno Pump, Robbins & Myers Inc. (c,e); Smith & Loveless (a,d,f); Vanton Pump & Equipment Corp. (a,c,d,e); Wallace & Tiernan (c,d); Weil Pump Co. (a,f); Wemco Div., Envirotech (a); Zimpro-Passavant Inc. (c,d,e)

Pumping applications at wastewater treatment facilities include pumping (1) raw or treated wastewater, (2) grit, (3) grease and floating solids, (4) dilute or well-thickened raw sludge or digested sludge, (5) sludge or supernatant return, and (6) chemical solutions. Pumps and lift stations are also used extensively in the collection system. Each pumping application requires specific design and pump selection considerations.

Pumping stations are often required for pumping (1) untreated domestic wastewater, (2) storm water runoff, (3) industrial wastewater, (4) combined domestic wastewater and storm water runoff, (5) sludge at a wastewater treatment plant, (6) treated domestic wastewater, and (7) circulating water systems at treatment plants.

This section briefly discusses various pumps and their applications. Because centrifugal pumps are commonly used for raw wastewater pumping, centrifugal pumps and pump selection, the design procedure for a raw wastewater pumping station, specifications of pumping, and control equipment are described.

Pump Types and Applications

Figure 7.12.1 shows the basic configurations of some pump installations.

CENTRIFUGAL PUMPS

This classification is the most common type of pump, including radial-flow centrifugals and axial- and mixed-flow centrifugals. In the form of tall, slender, deep-well submersibles, they pump clear water from depths greater than 2000 ft. Horizontal centrifugals with volutes almost the size of a man can pump 9000 gpm of raw sewage through municipal treatment plants. Few applications are beyond their range, including flow rates of 1 to 100,000 gpm and process fluids from clear water to the densest sludge.

Radial-Flow Centrifugals

Radial-flow pumps throw the liquid entering the center of the impeller or diffuser out into a spiral volute or bowl. The impellers can be closed, semiopen, or open depending on the application (see Figure 7.12.2). Closed impellers have higher efficiencies and are more popular than the other two types. They can be readily designed with non-

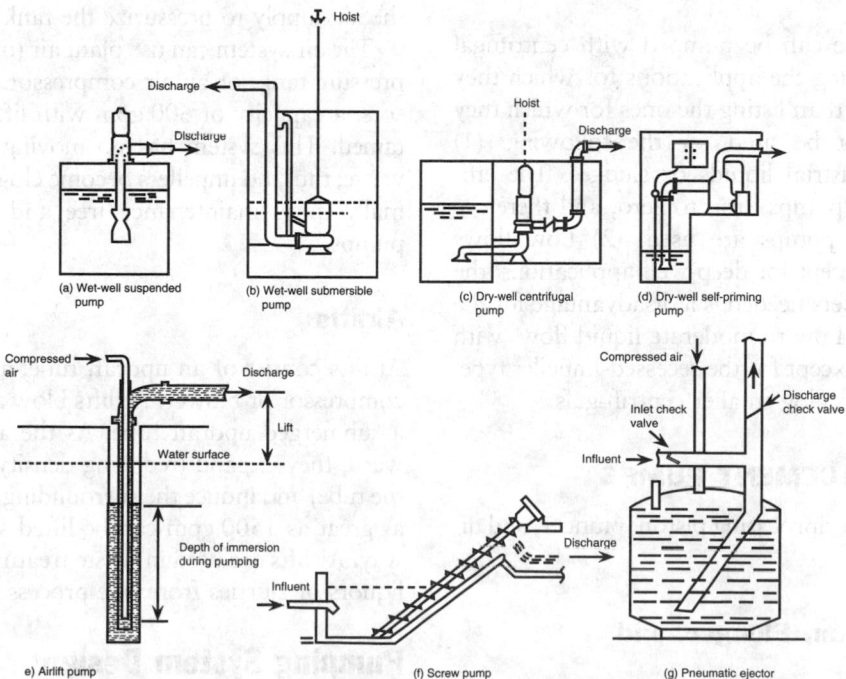

FIG. 7.12.1 Types of pumps and pumping stations.

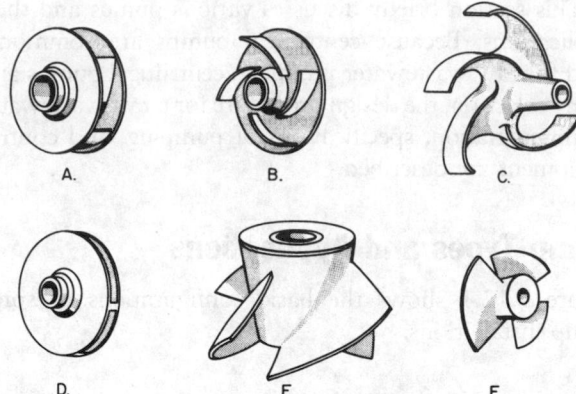

FIG. 7.12.2 Types of centrifugal pump impellers. A. Closed impeller; B. Semiopen impeller; C. Open impeller; D. Diffuser; E. Mixed flow impeller; F. Axial flow impeller.

clogging features. In addition using more than one impeller can increase the lift characteristics. These pumps can have a horizontal or vertical design.

Axial- and Mixed-flow Centrifugals

Axial-flow propeller pumps, although classed as centrifugals, do not truly belong in this category since the propeller thrusts rather than throws the liquid upward. Impeller vanes for mixed-flow centrifugals are shaped to provide partial throw and partial push of the liquid outward and upward. Axial- and mixed-flow designs can handle large capacities but only with reduced discharge heads. They are constructed vertically.

Applications

Most water and waste can be pumped with centrifugal pumps. Therefore, listing the applications for which they are not suited is easier than listing the ones for which they are. They should not be used for the following: (1) Pumping viscous industrial liquids or sludges. The efficiencies of centrifugal pumps drop to zero, and therefore positive displacement pumps are used. (2) Low flows against high heads. Except for deep-well applications, the large number of impellers needed is a disadvantage for the centrifugal design. (3) Low to moderate liquid flows with high-solids contents. Except for the recessed-impeller type, rags and large particles clog smaller centrifugals.

POSITIVE DISPLACEMENT PUMPS

These pumps include reciprocating piston, plunger, and diaphragm pumps.

Reciprocating Piston, Plunger, and Diaphragm Pumps

Almost all reciprocating pumps used in environmental engineering are metering or power pumps. The steam-driven pump is rarely used in water or wastewater processing. Frequently, a piston or plunger is used in a cylinder, which is driven forward and backward by a crankshaft connected to an outside drive. Adjusting metering pump flows involves merely changing the length and number of piston strokes. A diaphragm pump is similar to a reciprocating piston or plunger, but instead of a piston, it contains a flexible diaphragm that oscillates as the crankshaft rotates.

Applications

Plunger and diaphragm pumps feed metered amounts of chemicals (acids or caustics for pH adjustment) to a water or waste stream. They also pump sludge and slurries in waste treatment plants.

ROTARY SCREW PUMPS

In this type, a motor rotates a vaned screw or rubber stator on a shaft to lift or feed sludge or solid waste material to a higher level or the inlet of another pump.

AIR PUMPS

These pumps include pneumatic ejectors and airlifts.

Pneumatic Ejectors

In this pumping method waste flows into a receiver pot, and an air pressure system then blows the liquid to a treatment process at a higher elevation. A controller is usually included, which keeps the tank vented while it is being filled. When the tank is full, the level controller energizes a three-way solenoid valve to close the vent port and open the air supply to pressurize the tank.

The air system can use plant air (or steam), a pneumatic pressure tank, or an air compressor. With large compressors, a capacity of 600 gpm with lifts of 50 ft can be obtained. This system has no moving parts in contact the waste; thus, no impellers become clogged. Ejectors are normally more maintenance free and operate longer than pumps.

Airlifts

Airlifts consist of an updraft tube, an air line, and an air compressor or blower. Airlifts blow air into the bottom of a submerged updraft tube. As the air bubbles travel upward, they expand (reducing density and pressure within the tube) and induce the surrounding liquid to enter. Flows as great as 1500 gpm can be lifted short distances in this way. Airlifts are used in waste treatment to transfer mixed liquors or slurries from one process to another.

Pumping System Design

To choose the proper pump, the environmental engineer must know the capacity, head requirements, and liquid characteristics. This section addresses the capacity and head requirements.

CAPACITY

To compute capacity, the environmental engineer should first determine average system flow rate, then decide if adjustments are necessary. For example, when pumping wastes from a community sewage system, the pump must handle peak flows roughly two to five times the average flow, depending on community size. Summer and winter flows and future needs also dictate capacity, and population trends and past flow rates should be considered in this evaluation.

HEAD REQUIREMENTS

Head describes pressure in terms of feet of lift. It is calculated by the expression:

$$\text{Head in feet} = \frac{\text{Pressure (psi)} \times 2.31}{\text{Specific gravity}} \qquad 7.12(1)$$

The discharge head on a pump is a sum of the following contributing factors:

STATIC HEAD (h_d)—The vertical distance through which the liquid must be lifted (see Figure 7.12.3).

FRICTION HEAD (h_f)—The resistance to flow caused by friction in the pipes. Entrance and transition losses can also be included. Because the nature of the fluid (density, viscosity, and temperature) and the nature of the pipe (roughness or straightness) affect friction losses, a careful analysis is needed for most pumping systems although tables can be used for smaller systems.

VELOCITY HEAD (h_v)—The head required to impart energy into a fluid to induce velocity. Normally this head is quite small and can be ignored unless the total head is low.

SUCTION HEAD (h_s)—Reduces the pressure differential that the pump must develop when a positive head is on the suction side (a submerged impeller). If the water level is below the pump, the suction lift plus friction in the suction pipe must be added to the total pressure differential required.

TOTAL HEAD (H)—Expressed by the following equation:

$$H = h_d + h_f + h_v \pm h_s \qquad 7.12(2)$$

SUCTION LIFT

The amount of suction lift that can be handled must be carefully computed. As shown in Figure 7.12.4, it is limited by the barometric pressure (which depends on elevation and temperature), the vapor pressure (which also depends on temperature), friction and entrance losses on the suction side, and the net positive suction head (NPSH)—a factor that depends on the shape of the impeller and is obtained from the pump manufacturer.

SPECIFIC SPEED

The impeller's rotational speed affects the capacity, efficiency, and extent of cavitation. Even if the suction lift is within permissible limits, cavitation can be a problem and should be checked. The specific speed of the pump is determined with the following equation:

$$\text{Specific speed, } N_s = \frac{\text{RPM} \times \sqrt{\text{Capacity (gpm)}}}{H^{3/4}} \qquad 7.12(3)$$

Charts are available showing the upper limits of specific speed for various suction lifts.

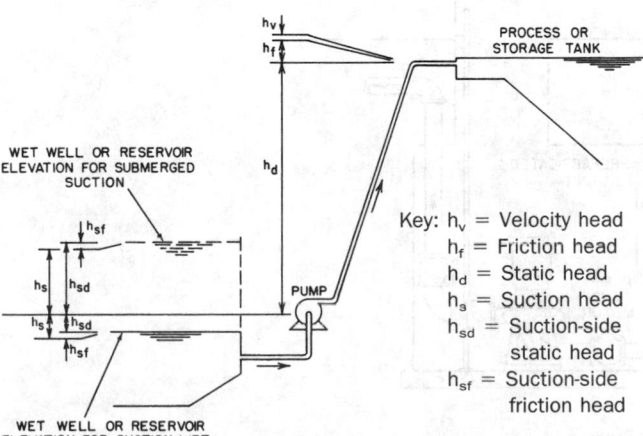

Key: h_v = Velocity head
h_f = Friction head
h_d = Static head
h_s = Suction head
h_{sd} = Suction-side static head
h_{sf} = Suction-side friction head

FIG. 7.12.3 Determination of pump discharge head requirements.

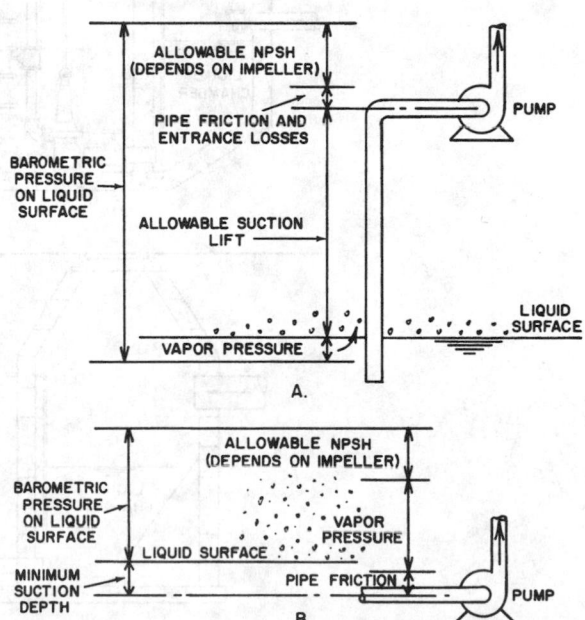

FIG. 7.12.4 Role played by NPSH in determining allowable suction lift. **A.** Pump with suction lift. **B.** Pump with submerged suction but high vapor pressure (possibly hot water).

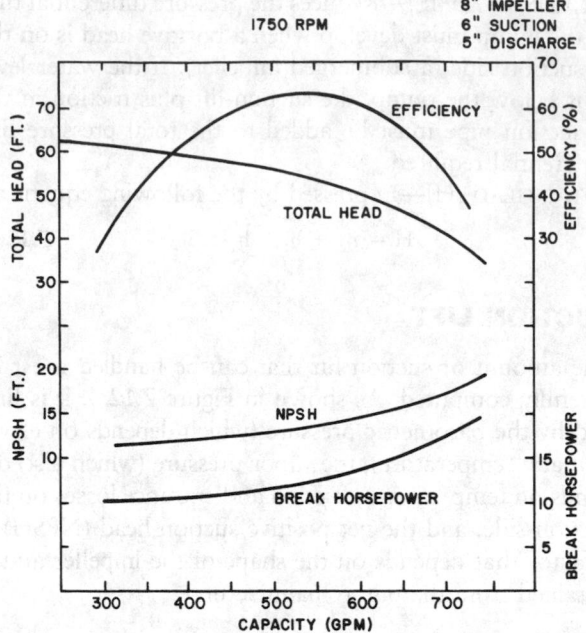

FIG. 7.12.5 Typical pump curve for a single impeller.

HORSEPOWER

The horsepower required to drive the pump is called *brake horsepower* (bhp). The following equation determines the brake horsepower:

$$bhp = \frac{Capacity\ (gpm) \times H\ (ft) \times Sp.\ Gr.}{3960 \times Pump\ efficiency} \qquad 7.12(4)$$

PUMP CURVES

Essential pump features are described by performance curves. Charts or tables that summarize pump curve data are also available. Figure 7.12.5 shows a typical centrifugal pump curve.

Pumping Station Design

Figure 7.12.6 shows typical pump station designs. Figure 7.12.7 illustrates a pneumatic ejector package. In selecting the best design for an application, environmental engineers should consider the following factors:

Many gases are formed by domestic waste, including some that are flammable. When pumps or other equipment are located in rooms below grade, the possibility of ex-

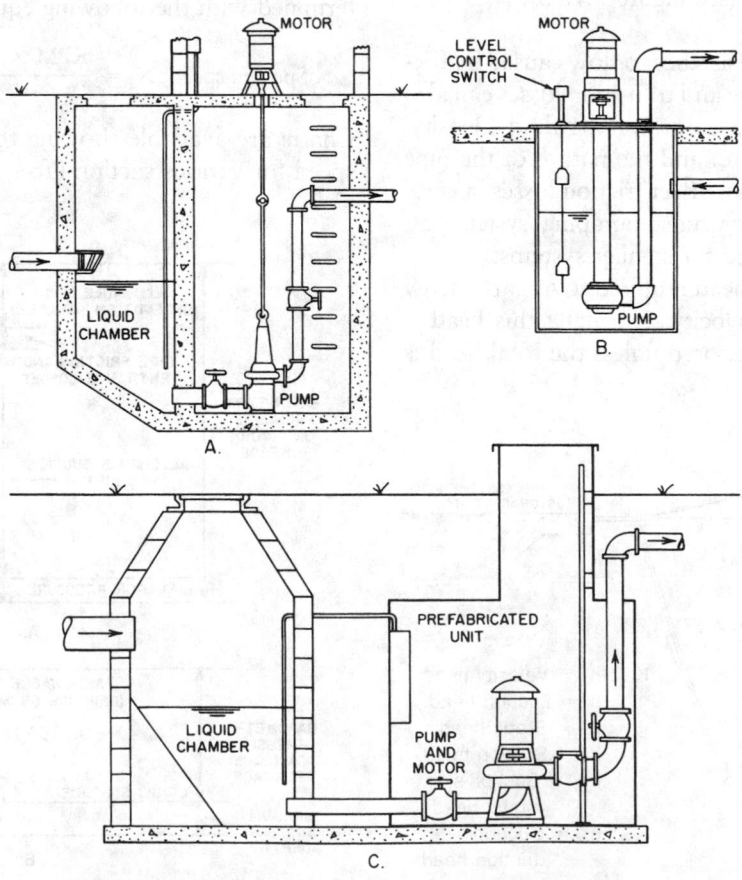

FIG. 7.12.6 Pumping stations. **A.** Dry-well design; **B.** Wet-well design; **C.** Prefabricated pumping station.

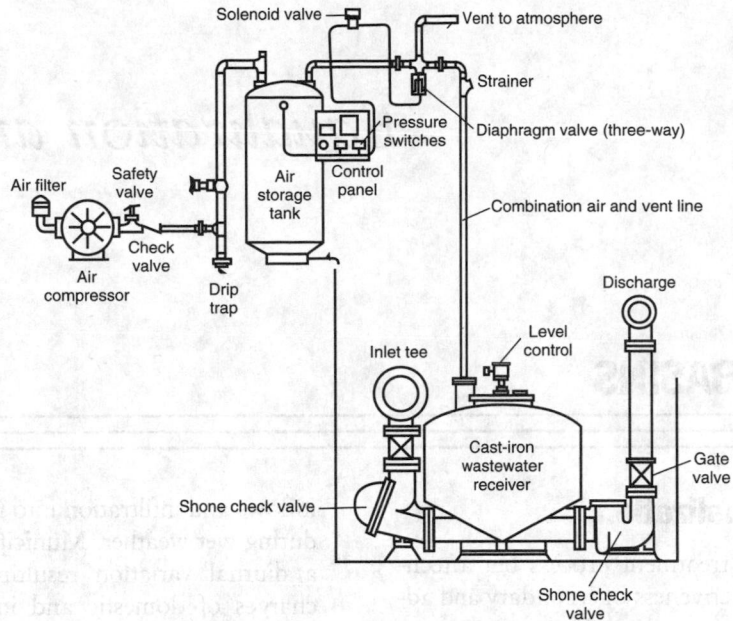

FIG. 7.12.7 Pneumatic ejector and associated piping.

plosion or gas buildup exists, and ventilation is extremely important.

When wastewater is pumped at high velocities or through long lines, the hammering caused by water can be a problem. Valves and piping should be designed to withstand these pressure waves. Even pumps that discharge to the atmosphere should use check valves to cushion the surge.

Bar screens and comminutors are not recommended, but for small centrifugal pump stations, they can be necessary.

Pump level controls are not fully reliable because rags can short electrodes and hang on floats. Purged-air systems (air bubblers) require less maintenance but need an air compressor that operates continuously. Therefore, maintenance-free instrumentation must be provided.

Charts and formulas are available for sizing wet wells, but infiltration and runoff must also be considered.

Sump pumps, humidity control, a second pump with an alternator, and a pump hoisting mechanism are recommended.

Most states prefer dry-well designs.

—R.D. Buchanan
David H.F. Liu

Equalization and Primary Treatment

7.13 EQUALIZATION BASINS

Purpose of Flow Equalization

Flow equalization is not a treatment process but a technique that improves the effectiveness of secondary and advanced wastewater treatment processes. Flow equalization levels out operation parameters such as flow, pollutant levels, and temperature over a time frame (normally 24 hr), minimizing the downstream effects of these parameters. Environmental engineers determine the need for flow equalization primarily based on the potential effects of the waste stream on the receiving waters or treatment facility.

This effect is determined by the following key components:

- The variability of operating parameters to be equalized (including toxicity)
- The volume of the flow being discharged

In defining the need for flow equalization, environmental engineers need sufficient background information on these factors as well as information on the relative cost of constructing and implementing effective flow equalization facilities and the cost savings by reducing the effects on downstream equipment.

This section provides information on flow equalization processes used to pretreat industrial waste streams, considers the effects of each process, and provides basic design criteria for each.

The following locations are suitable for flow equalization:

Near the head end of treatment work. Flow equalization usually involves constructing large basins to collect and store wastewater flow, from which wastewater is pumped to the treatment plant at a constant rate. These basins are normally located near the head end of the treatment work, preferably downstream of pretreatment facilities such as bar screens, comminutors, and grit chambers.

Prior to discharge. Wastewater flows have a diurnal variation from less than $\frac{1}{2}$ to more than 200% of the average flowrate. In addition, daily volumes increase from

inflows and infiltration into the sewer collection system during wet weather. Municipal waste strength also has a diurnal variation resulting from nonuniform discharges of domestic and industrial waste. Industrial waste entering a municipal system can cause excessive flows and peak organic loads. Therefore, facilities should be installed at industrial sites for flow smoothing prior to discharge.

Prior to advanced waste treatment operations. Many advanced operations, such as filtration and chemical clarification, are adversely affected by flow variation and sudden changes in solid loading. Maintaining a uniform influent improves chemical feed control and process reliability. The costs saved by installing smaller units for chemical precipitation and filtration, together with reduced operating expenses, can compensate for the added costs of flow equalization facilities.

Offline in a collection system. Figure 7.13.1 shows the treatment scheme using side-line flow equalization. This facility uses biological–chemical processing followed by multimedia filters. The flow equalization basin is a circular concrete tank with a volume of 315,000 gal, which is equivalent to 15% of the 2.1 million-gallons-per-day

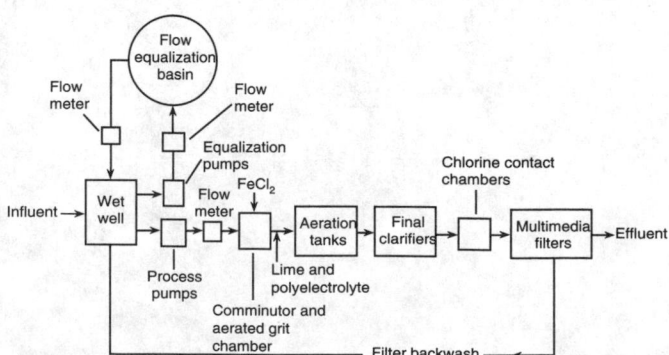

FIG. 7.13.1   Process diagram for biological–chemical treatment followed by filtration using side-line flow equalization, Walled Lake–Novi Waste Water Treatment Plant. (Reprinted from U.S. Environmental Protection Agency (EPA), 1974, *Flow equalization*, Technology Transfer, 19, U.S. EPA [May].)

(mgd) design flow. The process pumps transfer a constant preset flow from the wet well for treatment, and variable-speed pumps deliver excess flow to the equalization basin. During periods of low influent flow, wastewater is released from the basin to the wet well to maintain the established flow through the plant.

As in-line units. Equalization chambers can also be in-line units that pass all wastewater through the basins. Although the normal placement is between grit removal and primary settling, holding tanks can be placed at other points in the treatment. For example, a basin serving as a pump suction pit can be located just ahead of the filters to dampen hydraulic surges without providing complete flow equalization.

Flow Equalization Processes

Four basic flow equalization processes are as follows:

Alternating Flow Diversion.   The alternating-flow diversion system (see Figure 7.13.2) collects the total flow of an effluent for a time period (normally 24 hrs) while a second basin is discharging. The basins alternate between filling and discharging for successive time periods. Thorough mixing is maintained so that the discharge maintains constant pollutant levels with a constant flow. This system provides a high degree of equalization for a basin size by leveling all discharge parameters. A disadvantage of this system is the high construction cost associated with storing the waste stream volume for the time period used.

Intermittent Flow Diversion.   The intermittent-flow diversion system (see Figure 7.13.3) diverts significant variance in stream parameters to an equalization basin for short durations. The diverted flow is then bled into the stream at a controlled rate. The rate at which the diverted flow is fed back to the main stream depends on the volume and variance of the diverted water, reducing downstream effects.

Completely Mixed, Combined Flow.   The completely mixed, combined-flow system (see Figure 7.13.4) completely mixes multiple flows combined at the front end of the facility. This system reduces the variance in each stream by thoroughly mixing with the other flows. This system assumes that the flows are compatible and can be combined without creating additional problems.

Completely Mixed, Fixed Flow.   The completely mixed, fixed-flow system (see Figure 7.13.5) is a large, completely mixed, holding basin located before the wastewater facility that levels variations of the influent stream parameters and provides a constant discharge.

Each of these systems requires different design criteria. Therefore, the first step in the selection process is to de-

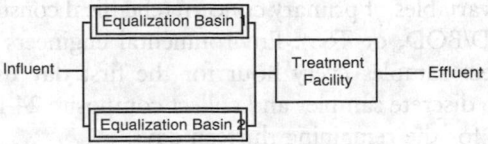

FIG. 7.13.2 Alternating-flow diversion equalization system.

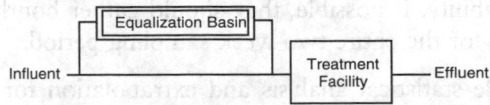

FIG. 7.13.3 Intermittent-flow diversion system.

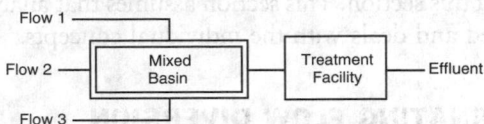

FIG. 7.13.4 Completely mixed, combined-flow system.

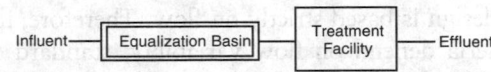

FIG. 7.13.5 Completely mixed, fixed-flow system.

fine the type of variability the system must equalize. Then, the facility can be designed with appropriate criteria.

Design of Facilities

The design of equalization facilities begins with a detailed study to characterize the nature of the wastewater and its variability. This study should also include gathering data on flow and pollutants of consequence.

A primary consideration is the effect of the effluent on downstream facilities. The most significant quantity is the mass flow rate; therefore, data on both flow and concentration (in terms of BOD_5, TSS, or other variables) must be measured on a time-series basis. Previous studies indicate that this type of data is normally distributed; therefore, the average mass flow from the sampled values is an estimate of the true average mass flow.

Because the collected data is time-series, obtaining random samples is difficult. Time-series data are by nature not random. Therefore, the study must contain sufficient samples for proper characterization of the statistical parameters as follows (McKeown and Gellman 1976):

For cyclical data, a minimum of two cycles must be collected. The spacing of data should be small enough to have a reasonable probability of measuring peak or minimum values.

Where seasonable considerations are important, at least one sampling program should be conducted during each season.

For flow equalization design, a minimum recommendation for industry waste sampling is two weeks of data for variables of primary concern (chemical constituents, COD/BOD, or TSS). Environmental engineers should collect samples every hour for the first day using an auto discrete sampler and collect composite 24-hr samples for the remaining thirteen days.

Environmental engineers can also use strip flow chart flow recordings and real-time TOC analyzers to determine variability. If possible, they should gather hourly flow data for the entire two-week sampling period.

While statistical analysis and extrapolation for confidence levels are important in determining the effects of variance (on reducing potential effects), they are not the focus of this section. This section assumes that analysis has occurred and deals with the individual concepts.

ALTERNATING FLOW DIVERSION

Because the alternating flow diversion system is intended to hold the total flow for a fixed time period (normally 24 hr), its design is based strictly on flow. Therefore, the design criteria depend on flow variability, standard deviation, and maximum flow for the time frame.

For example, an industrial facility has a total daily flow and pollution profile as shown in Table 7.13.1. If a thirty-day period is assumed to represent one month, the equalization basin can be designed using Table 7.13.1 and a management design criterion of 110% of maximum flow. Each of the equalization basins is designed to hold 686.98 m^3. Therefore, the management design criterion (that is, 110% of maximum flow) becomes the dominant variable in this equation.

This variable is given to the design engineers by plant management or assumed to be based on prior experience. The following equation applies:

$$V_t = D_c F_c Tk \qquad 7.13(1)$$

where:

V_t = Volume of the equalization basin, m^3
T = Time period of equalization
D_c = Management design criteria, %
F_c = Management flow criteria (f_a, average flow for a time period, or f_m, maximum flow for a time period), m^3/hr
k = Units conversion constant

INTERMITTENT FLOW DIVERSIONS

Intermittent flow diversion systems are more complex as design criteria include the variance of pollutants being diverted, the average length of the variance, and the rate of discharge back to the system. Environmental engineers must evaluate each of these factors with respect to the effect on downstream processes, especially if biological systems follow. This equalization system is best used when variances are easily detectable and infrequent and can have

TABLE 7.13.1   INDUSTRIAL FACILITY DAILY FLOW PROFILE

Day of Month	Total Flow, m^3/d	Phenol Levels, $\mu g/l$ (ppb)
1	377.44	45
2	471.46	393
3	411.96	421
4	254.35	433
5	350.64	683
6	464.19	822
7	339.29	123
8	624.52	467
9	569.57	682
10	47.15	732
11	420.13	398
12	238.46	541
13	553.42	868
14	487.81	558
15	241.18	656
16	562.75	329
17	272.52	822
18	229.83	771
19	237.55	613
20	348.47	400
21	134.44	821
22	0.00	0
23	143.98	160
24	610.90	214
25	97.65	670
26	398.78	362
27	560.94	303
28	253.44	245
29	574.11	120
30	525.06	251
Average daily	360.59 (0.25 m^3/min)	463
Minimum	0.00	0
Maximum	624.52	868

a dramatic effect on downstream processes. An example of this variance is the phenol levels in an effluent stream.

The following steps should be applied to the design of an intermittent flow diversion system.

Step 1: Determine the frequency and duration of the variance to be diverted (to design of the equalization basin)
Step 2: Calculate the controlled release rate of the diverted flow to maintain normal operation.
Step 3: Use the diverted volume to calculate the surge basin volume to maintain continuous flow to the treatment facility.
Step 4: Verify that the equalized flow meets discharge limits.

As stated earlier in this section, data collection and system profiling are keys to an effective design for this type of equalization system. An effective system is automated based on electronic monitoring of the stream, with diversion occurring as necessary. Three examples of this technology are pH sensors to monitor pH for excursions, on-line gas chromatographs to monitor phenol excursions, and conductivity sensors to monitor TDS. Variations of these parameters can cause substantial damage to biological systems or receiving waters (especially when only primary treatment is used).

For example, in Table 7.13.1, the phenol levels vary substantially from day to day because of variance in plant operation. With the variability of plant operation, diverting the flow at this facility to prevent violation, and bleeding the diverted flow back as concentrations allow is necessary.

The phenol level in Table 7.13.1 shows 24 hr composite samples with a discharge limit of 500 parts per billion (ppb). Further analysis of individual samples indicates that the problem was generated during two periods over the course of the day, lasting about 3 hr each. Also, during this time frame, flow rate increased to 0.473 m/min.

Therefore, the total volume to be diverted is calculated as follows:

$$V_D = F_D T_D f_D k \qquad 7.13(2)$$

where:

V_D = Volume of flow to be diverted per time period, m

F_D = Flow rate diverted, m/min

T_D = Time of diversion, hr

f_D = Frequency of diversion, number/day

k = Conversion constant for unit, min/hr

Therefore, V_D = (0.473 m^3/min)(3 hr)(2/day) (60 min/hr) = 170.28 m^3/day.

The control discharge rate can be established as follows:

$$f_C = V_D/Tk \qquad 7.13(3)$$

where:

f_C = Controlled discharge rate, m^3/min

V_D = Volume diverted, m^3

T = Time period for return, hr

k = Conversion factor for unit

Therefore, f_C = (170.28 m^3/24 hr)(1 hr/60 min) = 0.118 m^3/min.

The volume of the surge basin can now be calculated. As calculated in Equation 7.13(2), 170.28 m^3 of the total flow will be diverted and fed back to the stream at a constant rate. Therefore, the average flow for the remainder of the time is (360.09 − 170.28) = 189.81 m^3 for the 18-hr period. This amount equates to 0.1318 m^3/min on a 24-hr basis. Correspondingly, maintaining this flow for the 6-hr diversion period requires a surge basin equal to the volume for the diversion time frame (6 hr in this case) at an average flow rate for the remaining period. This volume can be calculated as follows:

$$V_S = F_A T_D k \qquad 7.13(4)$$

where:

V_S = Volume of the surge basin, m^3

F_A = Average flow rate without diversion flow, m^3/min

T_D = Diversion time period, hr

k = Conversion factor

Therefore, V_S = (0.175 m^3/min)(6 hr)(60 min/hr) = 63.22 m^3.

Any excesses in design capacity determined by management as part of the design criteria are not represented in calculation.

Combining the return of the diverted flow with the mainstream can be accomplished with in-line mixing or flash mixing just before downstream processes. The total combined flow (f_T) is calculated as follows:

$$f_T = f_A + f_C \qquad 7.13(5)$$

where:

f_A = Average flow rate without diversion, m^3/min

f_C = Controlled discharge rate, m^3/min

Therefore, f_T = (0.118 + 0.176) m^3/min = 0.294 m^3/min.

COMPLETELY MIXED COMBINED FLOW

The completely mixed, combined flow, equalization system addresses the variability resulting from multiple flows from different sections of a plant. This variability often generates impulse or step input changes to the wastewater treatment facility. The primary purpose of this system is to trim impulse variance or provide a more gradual change in operating parameters.

Again, the volume of the equalization basin is determined based on the effects the change in operating parameters has on downstream systems. Because this situation is more complex, this discussion approaches the design details from the simplest perspective: time and combined flows.

Therefore, the volume of the equalization basins V_e is calculated as follows:

$$V_e = (\Sigma f_i) T_e k \qquad 7.13(6)$$

where:

f_i = Individual flow rates, m^3/min

T_e = Time for equalization, hr

k = Conversion factor for units

For example, if three flows come into the equalization basin with flow rates of 1.98, 0.567, and 0.189 m^3/min, respectively, and the required equalization time is 1 hr, the following equation applies:

$$\begin{aligned} V_e &= (f_1 + f_2 + f_3)T_e k \\ &= (1.98 + 0.567 + 0.189)(1 \text{ hr})(60 \text{ min/hr}) \\ &= 164.16 \text{ m}^3 \qquad 7.13(7) \end{aligned}$$

From here, the environmental engineer can calculate the relative change in each operating parameter using the following formulas and converting the variability of the individual stream to the variability in the total flow:

$$\text{Var}_T = (\text{Var}_{pi}) \frac{f_i}{f_t} \qquad \textbf{7.13(8)}$$

where:

Var$_T$ = Variance in the concentration of the total stream, ppm or ppb (mg/l or μg/l)
Var$_{pi}$ = Variance in the concentration of the individual stream, ppm or ppb (mg/l or μg/l)
f_i = Flow of the individual stream, m^3/min
f_t = Flow of the total stream, m^3/min

For example, if the concentration of a pollutant in an individual stream changes by 50 mg/l, the total stream changes as follows:

$$\text{Var}_T = 50 \,(150/700) = 10.7 \text{ mg/l} \qquad \textbf{7.13(9)}$$

This variance can be used in the calculation as a change in the concentration of the combined stream and a potential effect on the downstream system.

A typical industrial waste problem involves constant flow with wastewater concentration as the only variable. Environmental engineers can use the following method in designing facilities to reduce this kind of concentration variability. The method assumes the data are normally distributed.

For example, if a completely mixed, constant flow tank has a variable concentration input and discrete samples are collected at a uniform interval time interval Δt, the influent variance (s^2) can be estimated as follows:

$$s^2 = [(C_i - C)^2]/(n - 1) \qquad \textbf{7.13(10)}$$

where:

C_i = Influent concentration at the i time interval
C = Mean concentration
n = Number of samples

The influent coefficient of variation (v_o) is as follows:

$$v_o = s/C \qquad \textbf{7.13(11)}$$

An estimate of required equalization time based on the variation of concentration and sampling interval is calculated as follows:

$$\Theta = \Delta t/2 [(v_o/v_t)/v_e]^2 \qquad \textbf{7.13(12)}$$

where:

Θ = Required equalization time, hr
Δt = Sampling interval, hr
v_o = Influent coefficient of variation of concentration, mg/l
v_t = Average influent concentration, mg/l
v_e = Effluent variability coefficient, mg/l

Both the influent and effluent coefficients of variability are based on discrete samples collected at uniform time intervals Δt.

Normally, raw wastewater characteristics, providing v_o and Δt, are the only information available. The effluent variability v_e must be related to the downstream requirements and therefore is the primary design variable. The environmental engineer must select it based on subsequent treatment units and effluent standards. Where specific limits on acceptable variability do not exist, engineering judgment must be exercised.

The effluent variability V_e can be estimated as follows:

$$V_e = \{[(C_e \text{ max}/C) - 1]/C\}/N \qquad \textbf{7.13(13)}$$

where:

(C) max = The equalization tank effluent concentration not to be exceeded
C = Mean value of concentration
N = Cumulative standard normal for the required confidence level (confidence level is the probability that a specified concentration will not be exceeded.)

Cumulative standard N can be selected from the abbreviated Table 7.13.2. Application of this method is illustrated in the example based on data in Table 7.13.3: v_t = 698 mg/l and v_o = 158.6 mg/l.

If downstream conditions (for example, the next treatment unit in line) restrict effluent variability to 10% (v_e = 0.1), using Equation 7.13(12) gives the required equalization time as follows:

$$\Theta = 1/2 [(158.6/698)/0.1]^2 \qquad \textbf{7.13(14)}$$

Specific restrictions on variability are uncommon; a more realistic problem is to design an equalization tank so that the effluent does not exceed a specified value.

These analyses can ultimately produce the type of curve shown in Figure 7.13.6. This graph allows the subjective analysis of a particular tank size to determine how well it suits the requirements.

If a detention time of 3 hr has tentatively been selected based on the foregoing analysis and physical considerations at a plant, the effluent from this size tank is not expected to exceed the value of 800 mg/l approximately 5% of the time, or about eight samples per week. Reducing this expectation to fewer than two samples per week (that is, a confidence level of 99%) exceeding 800 mg/l means increasing the detention time to approximately 7 hr.

TABLE 7.13.2   SELECTION OF CUMULATIVE STANDARD NORMAL FOR A DESIRED CONFIDENCE LEVEL

Confidence Level	Cumulative Standard Normal N
90.0	1.282
95.0	1.645
99.0	2.327
99.9	3.091
99.99	3.719

TABLE 7.13.3 HOURLY INFLUENT COD (MG/L) DATA FOR FOUR-DAY PERIOD

Hour of Day	First Day	Second Day	Third Day	Fourth Day
7	413	565	485	723
8	468	612	409	765
9	510	536	466	864
10	568	637	482	844
11	487	536	507	669
Noon	600	684	631	711
1	674	644	695	879
2	638	615	545	847
3	638	662	660	876
4	648	468	545	890
5	584	752	736	890
6	697	738	666	1030
7	629	752	704	1090
8	606	655	625	920
9	626	695	730	823
10	684	800	679	1030
11	742	738	853	1050
Midnight	729	380	612	1010
1	884	708	504	736
2	638	678	606	882
3	677	648	599	812
4	1210	608	5651	832
5	995	738	590	867
6	780	662	631	775

Source: A.T. Wallace and D.M. Zellman, 1971, Characterization of time varying organic loads, *J. Sanit. Eng. Div.*, *Proc. ASCE* 97:257.

Notes: Average = 698 mg/l
Maximum = 1210 mg/l
Standard deviation = 158.6 mg/l

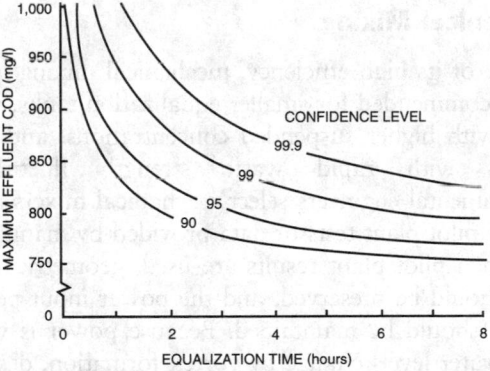

FIG. 7.13.6 Maximum effluent values as a function of equalization time and confidence level.

Similarly, if the confidence level of 90% (one sample in ten or roughly two samples per day exceeding 800 mg/l) is acceptable, the size of the tank can be reduced to yield a detention time of about 2 hr.

CUMULATIVE FLOW CURVE

Equalization basins for individual facilities can be sized based on a cumulative flow or mass diagram. This well-known method has long been used to determine the storage required for water reservoirs. The graphic technique consists of plotting cumulative flow versus time for one complete cycle (24 hr for municipal facilities). Two parallel lines, with slopes representing the rate of pumping or flow of the equalization tank, are drawn tangent to the high and low points of the cumulative flow curve. The required tank size is the vertical distance between the two tangent lines. The method is shown in Table 7.13.4 and Figure 7.13.7.

The preceding procedure provides the tank size for the flow-time trace of one day. The variability and thus the

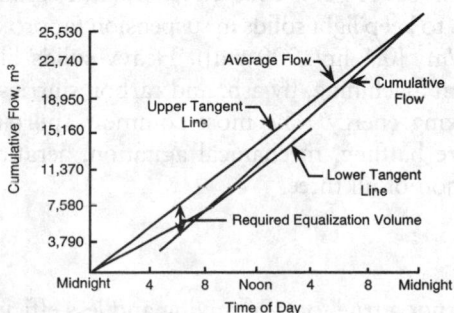

FIG. 7.13.7 Cumulative flow curve.

TABLE 7.13.4 FLOW DATA FOR EXAMPLE PROBLEM

Time	Flow Rate,* m³/hr	Cumulative Flow, m³	Time	Flow Rate,* m³/hr	Cumulative Flow, m³
Midnight	946	0	1	1110	12,830
1	901	901	2	1400	14,290
2	799	1700	3	1310	15,600
3	753	2453	4	1490	17,090
4	738	3191	5	1350	18,440
5	719	3910	6	1100	19,540
6	749	4659	7	1370	20,910
7	780	5439	8	1420	22,330
8	1000	6439	9	1370	23,700
9	1370	7809	10	1100	24,800
10	1280	9089	11	1270	26,070
11	1230	10,320	Midnight	1230	27,300
Noon	1400	11,720			

Note: Flow from Ewing Township, New Jersey, WWTP.

amount of equalization required changes from day to day. Therefore, environmental engineers must select a day or flow rate that represents the flow conditions to be equalized.

Operational Considerations

This section discusses the operational considerations of equalization systems including mixing and draining and cleaning requirements.

MIXING REQUIREMENTS

The contents of an equalizing vessel must be mixed. Typically, continuous mechanical mixing is best although inlet arrangements sometimes provide the necessary homogeneity of soluble waste constituents. If settleable and floatable solids are present, the wastewater must be mixed to maintain a constant effluent concentration and prevent accumulations. For biodegradable waste, the equalization tank will develop odor problems unless aeration is provided. Aeration and mixing systems can be combined (for example, floating surface aerators).

Although mixing power levels vary with basin geometry, 0.3 l/m³ sec (18 cfm/1000 cu ft) of basin volume is the minimum to keep light solids in suspension (approximately 0.02 kW/m³ [0.1 hp/1000 gal]). Heavy solids like grits, swarf from machining, fly ash, and carbon slurries require more mixing energy. The most common approaches to mixing are baffling, mechanical agitation, aeration, or a combination of all three.

Baffling

Although not a true form of mixing and less efficient than other methods, baffling prevents short-circuiting and is the most economical. Over-and-under or around-the-end baf-

fles can be used. Over-and-under baffles are preferable in wide equalization tanks because they provide more efficient horizontal and vertical distribution.

The influent should be introduced at the tank bottom so that the entrance velocity prevents SS in the wastewater from sinking and remaining on the bottom. Additionally, a drainage valve should be located on the influent side of the tank to allow drainage of the tank when necessary. Normally, baffling is not recommended for wastewater that has a high concentration of settleable solids.

Mechanical Mixing

Because of its high efficiency, mechanical mixing is typically recommended for smaller equalization tanks, wastewater with higher suspended concentrations, and waste streams with rapid waste strength fluctuations. Environmental engineers select mechanical mixers on the basis of pilot plant tests or data provided by manufacturers. When pilot plant results are used, geometrical similarity should be preserved, and the power input per unit volume should be maintained. Because power is wasted when water levels change by vortex formation, designers should avoid creating a vortex by mounting the mixer off center or at a vertical angle or by extending baffles out from the wall.

Because both mechanical and diffused-aeration systems must have a minimum depth to maintain mixing, extra volume should be provided below the low water level.

Aeration

Mixing by aeration is the most energy-intensive of the equalization methods. In addition to mixing, aeration provides chemical oxidation of reducing compounds as well as physical stripping of volatile chemical compounds. Some states require an air discharge permit for discharging volatile organic emissions to the atmosphere or classifying an equalization tank as a process tank.

Waste gases can be used for mixing if no harmful substance is added to the wastewater. Flue gas containing large quantities of carbon dioxide can be used to mix and neutralize high-pH wastewater.

DRAINING AND CLEANING

Equalization systems should be sloped to drain, and a water supply should be provided for flushing without hoses;

otherwise, the remains after draining can cause odor and health problems.

—David H.F. Liu

Reference

McKeown, J.J. and I. Gellman. 1976. Characterizing effluent variability from paper industry wastewater treatment processes employing biological oxidation. *Prog. Water Technol.* 8:147.

7.14
SCREENS AND COMMINUTORS

PARTIAL LIST OF SUPPLIERS

Screens
   American Well Works; BIF Sanitrol, a unit of General Signal; Chain Belt Co.; Envirex Inc., a Rexnord Co.; FMC Corp. Materials Handling Div.; Hycor Corp.; Keene Corp., Water Pollution Control; Lakeside Equipment Corp.; Link Belt Co.; LYCO; Walker Process Corp.; Welker Equipment Co.; Wemco

Comminutors
   Chicago Pump Co.; Clow Corporation; Infilco; Worthington Pump Corp.; Yeomans

Screens are usually installed at the entrance of the wastewater treatment plant to protect mechanical equipment, avoid interference with plant operations, and prevent objectionable floating materials such as rags or rubber from entering the primary settling tanks. Screening devices intercept floating or suspended larger material. The retained material is then removed and disposed of by burial or in-

cineration or is returned into the waste flow after grinding.

Trash racks or coarse racks are screening devices constructed of parallel rectangular or round steel bars with clear openings, usually 2 to 6 in (5.1 to 15.2 cm). They protect combined sewer systems from large objects and are usually followed by regular bar screens or comminutors.

Bar screens are racks of inclined or vertical flat bars installed in a channel and are basically protective devices. They are used ahead of mechanical equipment such as raw sewage pumps, grit chambers, and primary sedimentation tanks.

Where mechanical screening, comminuting devices, or both are used, manually cleaned, auxiliary bypass bar screens should also be provided. These bypass screens provide automatic diversion of the entire sewage flow should the mechanical equipment fail (see Figure 7.14.1). Velocity

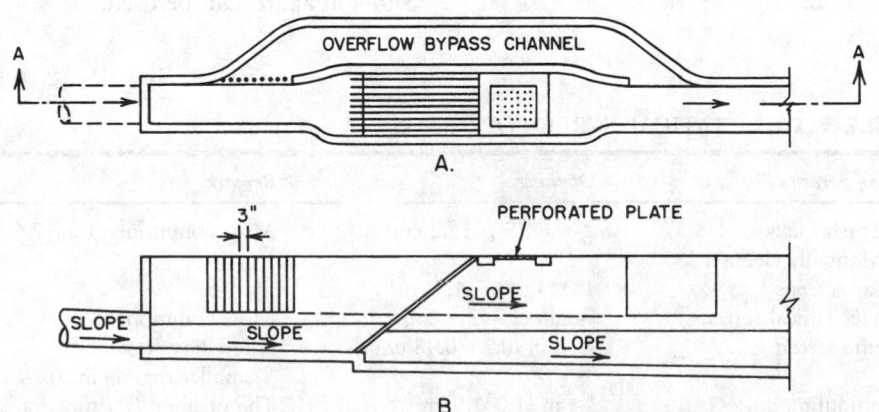

FIG. 7.14.1 Hand-cleaned bar screen with overflow bypass. **A.** Plan view; **B.** A–A section.

distribution in approach and discharge channels is important to screen operation, and a straight arrangement for approach channels is recommended.

The wastewater treatment plant should have gates installed to divert the flow from mechanical screens and comminutors or distribute the flow. Provisions should also be made for dewatering each unit. To compensate for head loss through the racks and prevent jetting action behind the screen, the rack chamber floor should be 3 to 6 in below the approach channel invert.

Environmental engineers often determine the width of the bar screen on the basis that the net submerged area of openings below the crown of the incoming sewer line be not less than 150 to 250% of the cross-sectional area of the influent sewer.

Screen Openings and Hydraulics

Screen openings should be narrow enough to retain sticks, rags, and other trash but wide enough to allow excreta and toilet paper to pass. Table 7.14.1 lists common screen openings. The lower the velocity through the screen, the greater the amount of material removed from the waste. Deposition of solids in the channel, however, prohibits reducing the velocity beyond certain limits.

The *Ten states' standards* (Great Lakes-Upper Mississippi River Board of State Sanitary Engineers 1968) require that the average rate of flow velocity through manually raked bar screens should be approximately 1 fps and the maximum velocity during wet weather periods through mechanically cleaned bar screens should not exceed 2.5 fps. The velocity should be calculated from a vertical projection of the screen openings on the cross-sectional area between the channel invert and the flowline.

Head loss for screens varies with the quantity and nature of the screenings that accumulate between cleanings. Environmental engineers can calculate the head loss created by a clean screen by considering the flow and the effective area of the screen openings as follows:

$$H = \frac{V^2 - v^2}{2g} \times \frac{1}{0.7} \text{ or } H = 0.0222(V^2 v^2) \quad \textbf{7.14(1)}$$

where:

H = head loss, ft
V = velocity through screen, fps
v = velocity of incoming waste, fps
g = acceleration due to gravity

The minimum allowance for loss through a hand-cleaned screen is 6 in, assuming frequent attention to the screens by operating personnel. The maximum head loss through clogged racks should be kept below 2.5 ft.

Material collected on screens impedes the flow. Excessive back ups in the incoming line can cause pounding and deposition of putrefying solids. When screens are cleaned, high flow surges occur that can cause hydraulic and treatment problems at the plant. The use of steeper grades in the influent pipe preceding the screen can reduce these problems.

Mechanically cleaned screens are usually protected by enclosures. Efficient ventilation is important and prolongs the life of both the equipment and the enclosure. The enclosure should have separate outside entrances. Convenient access and ample working space are important. Convenient unloading and handling of rackings can be provided by screw conveyors, belt conveyors, containers, or buckets.

Screen Types

HAND-CLEANED BAR SCREENS

Many bar screens have no mechanical cleaning devices and are cleaned periodically by hand rakes (see Figure 7.14.2). Manually cleaned screens, except those used for emergency, should be placed on a slope of 30° to 45° with the horizontal. This positioning increases screening surface by 40 to 100%, facilitates cleaning, and prevents excessive head loss by clogging. The clear opening between bars should be 1 to $1\frac{3}{4}$ in, and operators should remove screening as often as necessary (from two to five times a day) to secure free flow in the sewer.

When excessive head loss is expected, an overflow by-pass channel equipped with a trash rack (vertical bars set 3 to 4 in apart) can be used.

TABLE 7.14.1 TYPICAL SCREEN OPENINGS

Type of Screen	Opening	Remarks
a) Trash racks	2–6 in (5.1–15.2 cm)	Most commonly 3 in (7.6 cm)
b1) Manually cleaned bar screens	1–$1\frac{3}{4}$ in (2.5–4.4 cm)	
b2) Mechanical screens	$\frac{5}{8}$–1 in (1.4–2.5 cm)	Most commonly $\frac{3}{4}$ in (1.9 cm)
e) Fine screen	$\frac{3}{32}$–$\frac{3}{16}$ in (0.24–0.48 cm)	A few have openings smaller than $\frac{3}{32}$ in (0.24 cm).
Comminuting devices	$\frac{1}{4}$–$\frac{3}{4}$ in (1.0–1.9 cm)	The opening is a function of the hydraulic capacity.

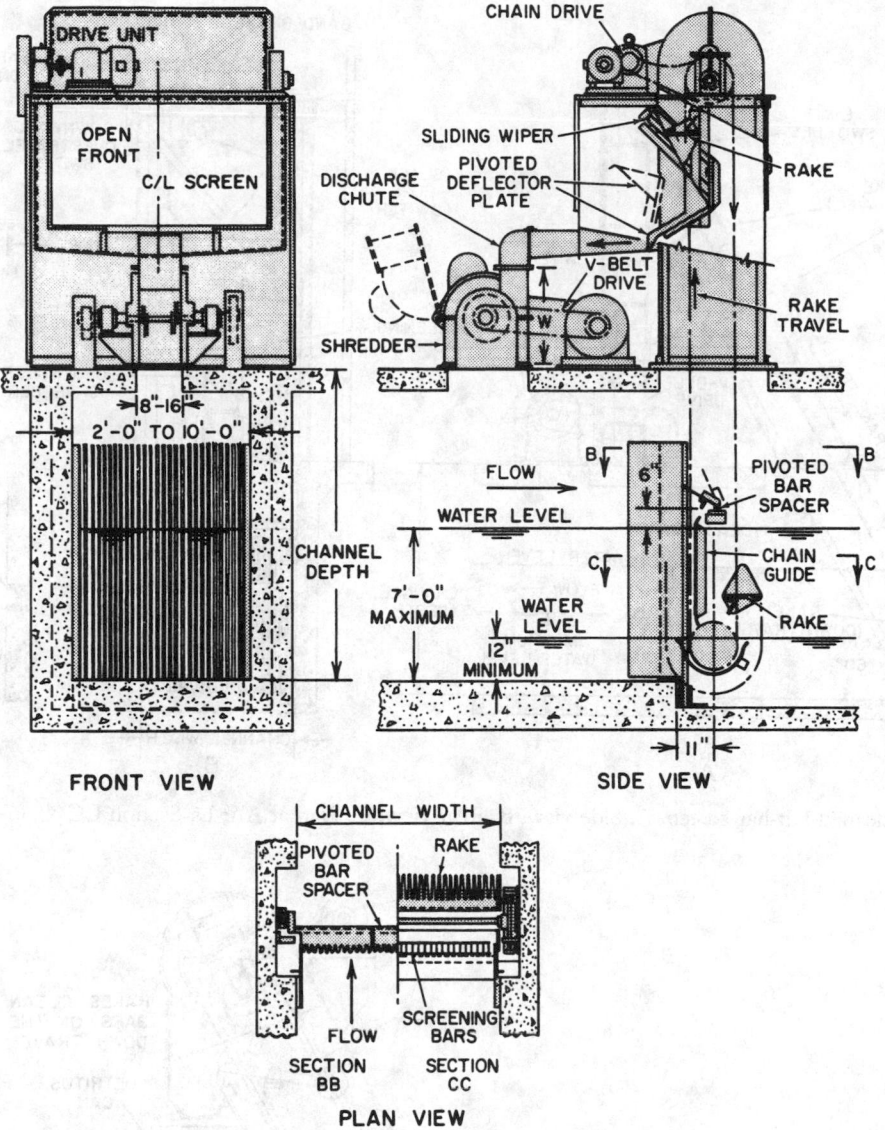

FIG. 7.14.2 Backcleaned flat-bar screen.

MECHANICALLY CLEANED BAR SCREENS

These screens are also called mechanical screens or mechanical rakes. Almost all large- and medium-size treatment plants use mechanically cleaned bar screens. The clear opening for mechanical screens is between 1 and $\frac{5}{8}$ in. This size is why mechanical screens are sometimes called fine racks.

The controls that operate mechanically cleaned screens include a manual start and stop switch, a clock-operated automatic start and stop switch, a high-water-level switch with or without an audible alarm; a head-loss- (screen pressure drop) actuated start and stop switch, and an overload switch with or without an audible alarm.

All motors and controls should be explosion proof. The racks are cleaned with long-tined rakes that fit into the openings. Cross bars or bolts should be located so that they do not interfere with raking.

Mechanical flat-bar screens can be back cleaned or front cleaned. The rakes travel at a rate of 7 to 20 fpm and can be adjusted to rest at the top for a period from 3 sec to 60 min. Backcleaning mechanisms are not subject to jamming at the bottom by trash deposits because they are also protected by the screen which is cleaned.

Discharge of screenings can be at the front or back. The front discharge of screenings is preferable because any raking dropped upstream of the screen can be recovered.

In *backcleaned flat-bar screens,* the raking mechanism consists of a rake or series of rakes with the ends attached to a pair of endless chains. These chains continuously move the rake slowly upward over the back face, or effluent side of the screen, carrying the rakings to the top of the screen. There they are dropped into a conveyor or bucket or onto a screening platform (see Figure 7.14.2).

In *frontcleaned flat-bar screens,* the cleaning mechanism is located in front of the bar screens. As the rake moves

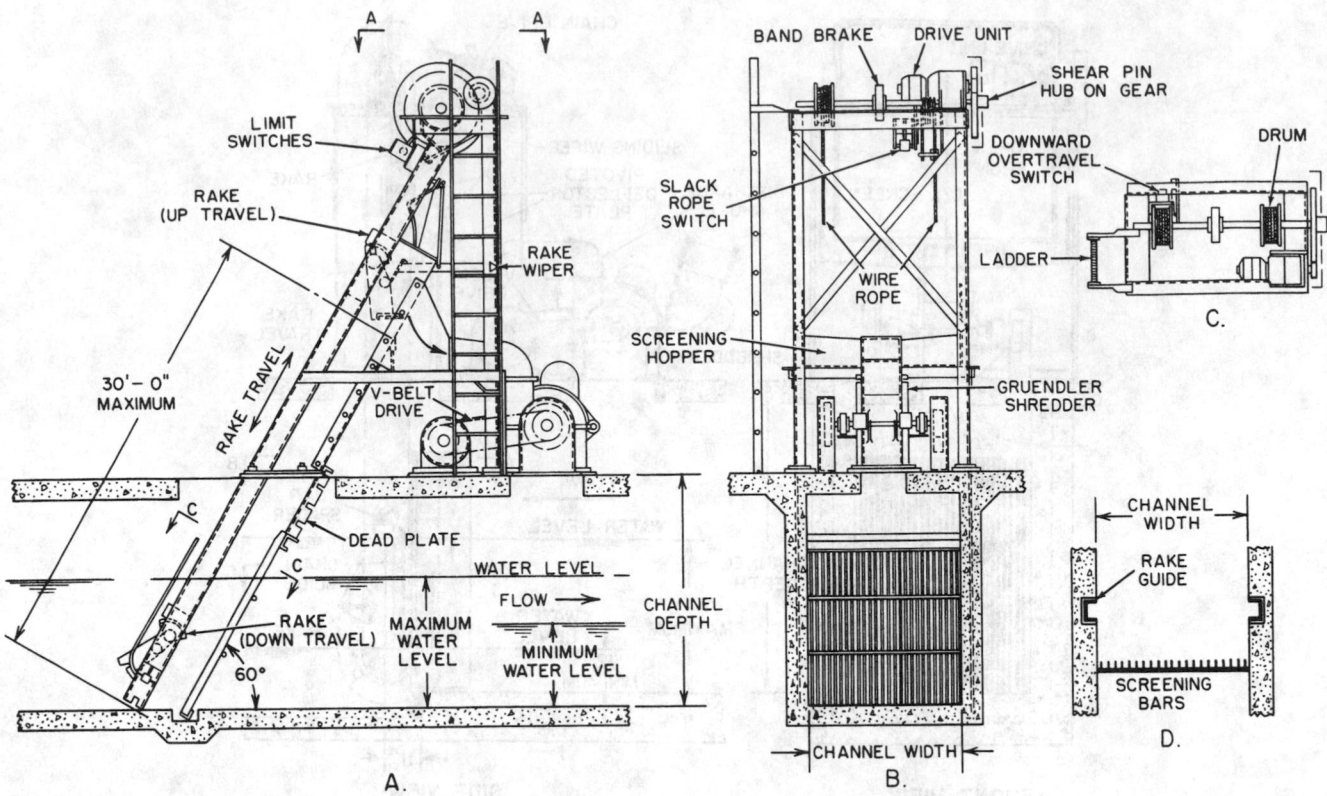

FIG. 7.14.3 Frontcleaned flat-bar screen. **A.** Side view; **B.** Rear view; **C.** Plan at AA; **D.** Section CC

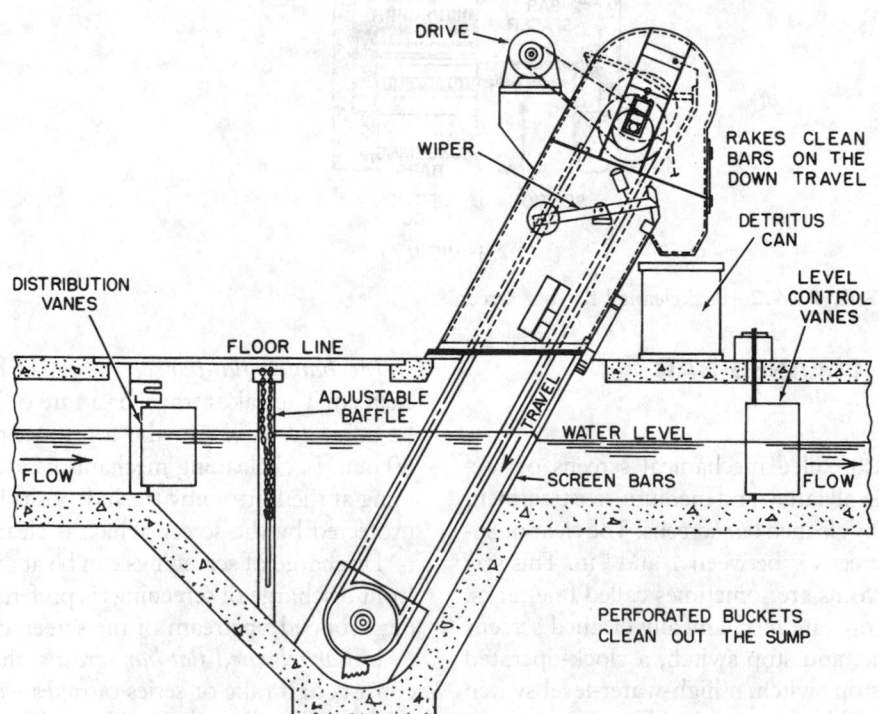

FIG. 7.14.4 Mechanical bar screen and grit collector.

on the face or influent side of the screen, the raking is carried upward (see Figure 7.14.3). At the discharge point, a wiper cleans the rakes, which are operated by wire cables.

For shallow channels, *curved-bar screens* with bars formed as segments of a circle and installed concave to the flow direction are available. The rakes revolve slowly around a horizontal shaft located in the center axis of the screen curvature. After passing the top of the bars, the raking plate contacts a metal rocking apron that retains the screenings on the plate until it clears the apron. The screenings are then removed by a scraper bar.

MECHANICAL BAR SCREEN AND GRIT COLLECTOR

A combined mechanical screen and grit collector is available for small- and medium-sized plants (see Figure 7.14.4). The unit is similar to a frontcleaned mechanical screen except that the rakes are attached to one or more perforated buckets and a steep hopper to collect the grit is ahead of the screen. The buckets travel downward, and the grit is dewatered on upward travel by perforations in the buckets. The disadvantage of this combined solution, however, is that the screenings and grit are mixed.

COARSE-MESH SCREENS

Instead of bar screens or comminutors, a few plants use coarse-mesh screens. These screens have a basket of wires or rods into which the waste is discharged. The water flows through, leaving the coarse suspended matter in the basket. Mesh size is 1 in (2.54 cm) or more. The basket is raised at intervals by hand or crane, emptied, then replaced.

FINE SCREENS

Screens, and particularly fine screens, were among the first wastewater treatment devices. At the turn of this century fine screens were often installed where wastewater was discharged into large rivers or wide tidal estuaries to remove unsightly floating matter and increase the efficiency and economy of chlorination.

Since the SS removal efficiency of fine screens is only 10 to 20% compared to the 60% or more achieved by sedimentation, fine screening is no longer an acceptable alternative to sedimentation. However, some industries use fine screens successfully to remove solids from waste of processes for meat packing, canning (causes excessive scum or foaming in digesters), wool, and textiles. When possible, fine screens should be installed at the source of the industrial waste.

Instead of sand filters, some wastewater treatment plants use fine screens to remove fine suspended matter from treatment plant effluent when it is discharged into streams that are likely to reach recreational areas.

Fine screens are also used in front of biological treatment units. The required net area of submerged openings is commonly 2 sq ft per mgd (0.186 m^2 per 3785 m^3 per day or 20,300 m^3 per day per m^2) for domestic waste and 3 sq ft per mgd (0.280 m^2 per 3785 m^3 per day or 13,500 m^3 per day per m^2) for combined waste. Fine screens are also used preceding trickling filters to reduce clogging of the distributor nozzles.

Design and Cleaning of Fine Screens

Because of the small size of the openings, fine screen cleaning must be continuous. Cleaning can be accomplished by brushes or scrapers, water, steam, or air jets forced through the openings from the back side. The efficiency of a fine screen depends on the fineness of the openings and the velocity of sewage flow through the openings. The most commonly used screening media are as follows:

Slotted perforated plates with $\frac{1}{32}$- to $\frac{3}{32}$-in- (0.8- to 2.4-mm-) wide slots. Where brushing or scraping is used, plates are more practicable than wire screens.

Wire mesh with approximately $\frac{1}{8}$-in openings

Woven wire cloth with openings usually less than $\frac{1}{8}$ in. One type is made of bars that are wedge shaped in cross section, with the large end of the bar in the face of the screen. The wedge-shaped bars are set in slots of a series of U-bars to maintain spacing and hold the bars in place. The standard openings range from $\frac{1}{100}$ to $\frac{1}{4}$ in, and the slots are continuous.

Wedge-shaped wire. The wire is pressed into a wedge-shaped section built into flat panels that have openings varying from 0.005 to $\frac{3}{16}$ in. The screens are made of corrosion-resistant material, preferably stainless steel.

Revolving-Drum Screen

In this screen, a stainless-steel no. 12 to no. 20 woven-wire mesh is applied to a cylindrical frame. The cylinder rotates around its axes while it is between one-third and two-thirds submerged. Revolving drum screens require a fairly constant water level; therefore, a weir plate usually maintains water elevation.

Revolving-Drum Screen with Outward Flow

With this screen, the waste flow can approach the drum from a direction parallel to the revolving axis. The liquid flows into the interior of the drum at one end, passes through the filter media, and flows out at a right angle to the axis (see Figure 7.14.5). The solids, which are retained on the inside surface of the screen, are raised above the liquid level as the drum slowly rotates and are usually removed by a water spray.

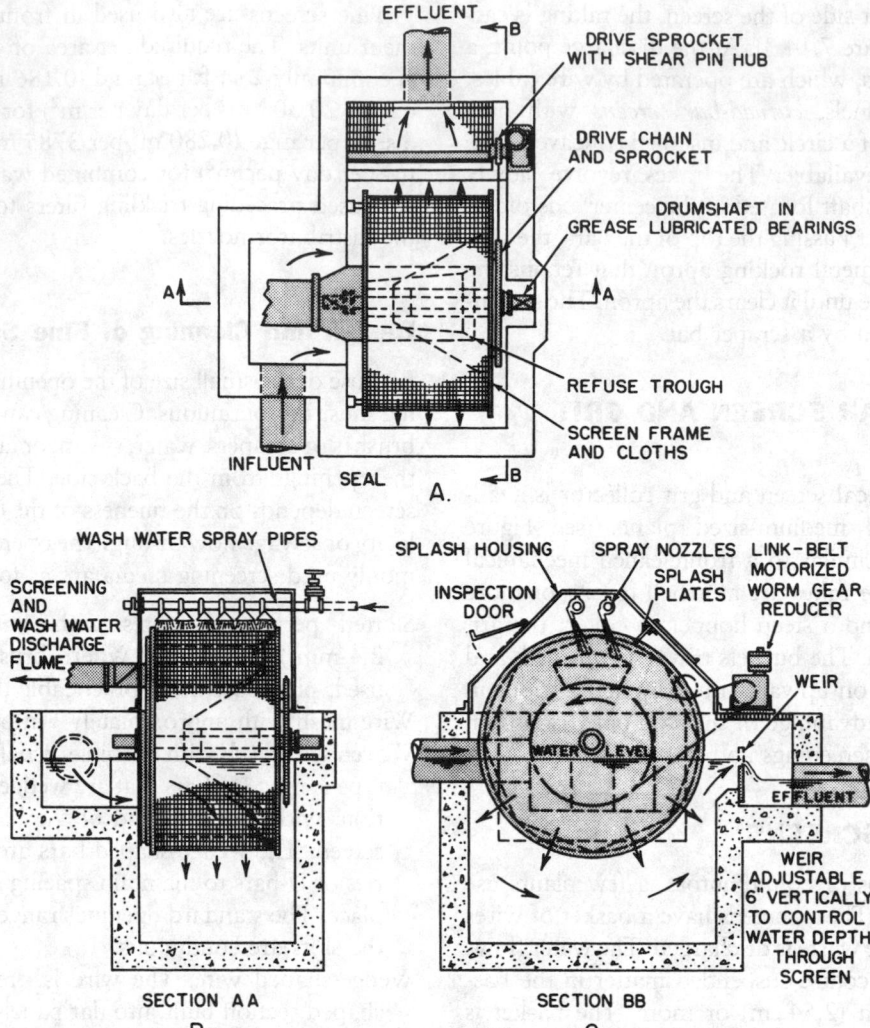

FIG. 7.14.5 Revolving-drum screen with outward flow. A. Plan view; B. Side view; C. End view.

Revolving-Drum Screen with Inward Flow

With this screen, the waste flow approaches the drum perpendicular to its axis. The liquid passes through the screen and flows out at one end. The solids, which are retained on the outside surface of the drum, are raised above the liquid level as the drum rotates and are removed by brushes; scrapers; or backwashing with water, air, or steam.

A disadvantage of removing screenings by water spray is that the removed solids are mixed with large amounts of spray water.

Revolving-Vertical-Disk Screens

In operating principle, this screen is similar to the revolving drum screen except that instead of a drum, a slowly revolving disk screen is placed in the approach channel completely blocking the flow so that it must pass through

the screen (see Figure 7.14.6). As the liquid passes through the screen, solids are retained, elevated above the water level, and flushed by a water spray to a trough.

This screen is not suitable to remove larger objects or excessive amounts of suspended matter; neither is it suitable to handle greasy, gummy, or sticky solids. It requires a constant water level, which is usually secured by using a weir. The screening medium can be 2- to 60-mesh stainless steel wire cloth. Screenings are mixed with large amounts of spray water.

Inclined Revolving-Disk Screens

This screen consists of a round, flat plate revolving on an axle inclined 10° to 25° from the vertical (see Figure 7.14.7). The disk consists of several bronze plates containing slots, generally $\frac{1}{16}$ to $\frac{1}{32}$ in (1.6 to 0.8 mm) wide. The waste flows through the lower two-thirds of the plates. As

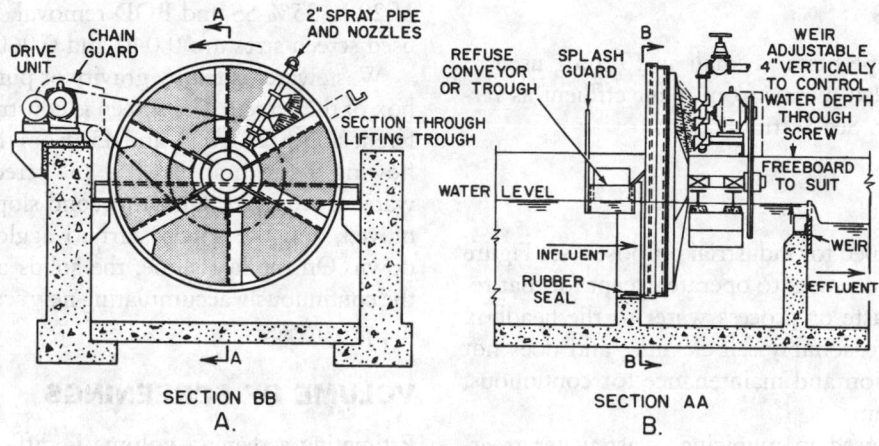

FIG. 7.14.6 Revolving-vertical-disk screen. **A.** Front view; **B.** Side view.

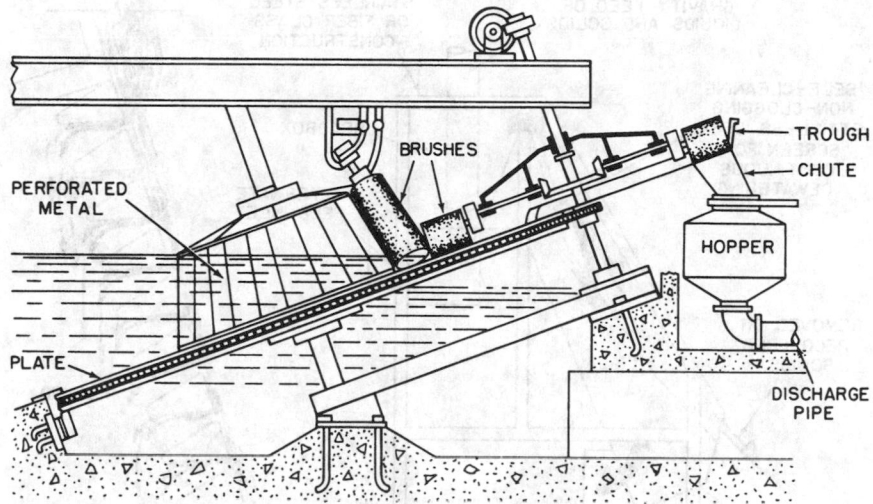

FIG. 7.14.7 Inclined revolving-disk screen.

the plate rotates, retained solids are brought above the liquid level where brushes remove them for disposal.

Traveling-Water Screen

This screen has had limited use in sewage treatment to remove solids from plant effluent. This screen uses a series of tilted or inclined overlapping screen trays mounted on two strands of steel chain. The head wheel is motor-driven, moving screen trays out of the sewage for solids removal by jets of water, then returning the trays into the wastewater flow.

Endless Band Screen

This screen operates in basically the same way as the revolving-vertical-disk screen. Instead of a vertical revolving disk, a screen in the form of an endless band is installed

in the approach channel and the incoming flow is forced to pass through it. This endless screen moves slowly around top and bottom drums while half or more of it is submerged.

Vibrating Screens

Vibrating screens are used in the food packing industry to remove grease and meat particles, eliminate manure, recover animal hair, remove feathers from poultry processing, and remove vegetable and fruit particles from canning waste.

Vibrating screens are flat and covered by fine stainless steel cloth of 20- to 100- or even 200-mesh supported by rubber-covered bars or stronger stainless steel, coarse-wire mesh. Vibration reduces the blinding and clogging of the fine screens. A manual or automatic spray washer with steam or detergents can reduce blinding and clogging and facilitate handling of greasy or sticky material.

MICROSCREENS

Microscreens with apertures as small as 20 μ are used to remove fine suspended matter from plant effluent as *tertiary treatment units*. See Section 7.33.

HYDRASIEVES

This unit was developed for industrial purposes (see Figure 7.14.8). It requires no power to operate except for that required to lift the waste or process water to the headbox of the screen. It is reasonably self-cleaning and does not require much attention and maintenance for continuous, trouble-free operation.

Hydrasieves are used in municipal wastewater treatment plants where the manufacturer claims an efficiency of 20 to 35% SS and BOD removal. The most frequently used screen sizes are 0.040 and 0.060 in (1 and 1.5 mm).

Wastewater is fed by gravity or pumping into the headbox of the screen. The screen is constructed with three different slopes, 25°, 35°, and 45° from the vertical. The overflowing waste first hits the 25° screen, where most free water is removed. On the second slope, more water is removed, and the solids start to agglomerate as they roll down. On the last slope, the solids are pushed down by the continuously accumulating new screenings while draining progresses.

VOLUME OF SCREENINGS

Estimating screening volume is difficult because the volume depends on the size of the screen and on infiltration,

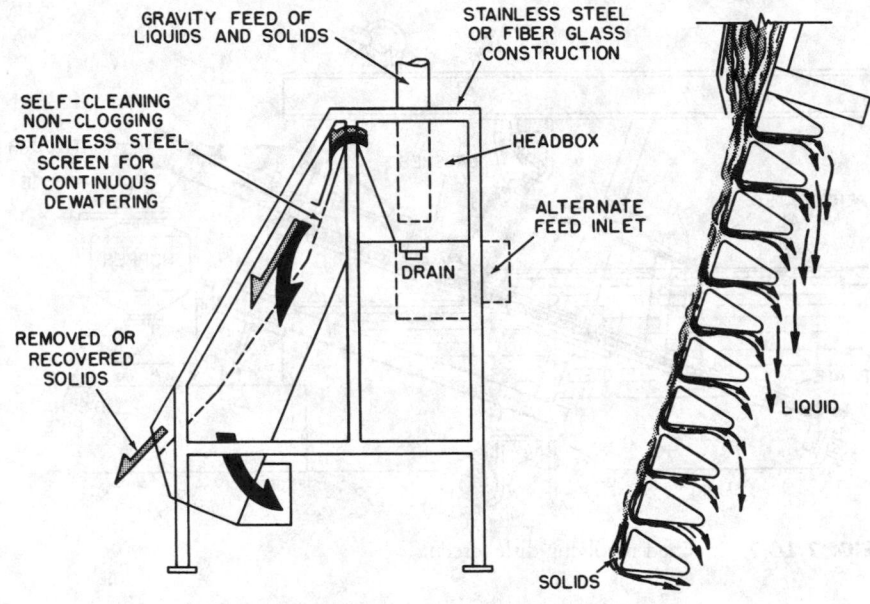

ENLARGED SECTION OF THE SCREEN

FIG. 7.14.8 Hydrasieve.

storm water quantity in combined sewers, the habits of the contributing population (e.g., garbage grinders), and the character of industrial waste received.

The solids removed from municipal waste by fine screens ranges from 10 to 35 ft³ per mgd (0.28 to 0.99 m³ per 3785 m³ per day) or from 5 to 20% of the SS in raw waste depending on screen size and the nature of the waste.

DISPOSAL OF SCREENINGS

Disposal methods include burial, incineration, digestion, and grinding. Open-area disposal is prohibited. Most smaller plants dispose their screenings by burial. Each day they add a cover of approximately 6 in over buried screenings to prevent fly and odor problems.

Large plants often use incinerators to dispose screenings alone or mixed with dewatered sludge. Dewatering the screenings with presses is usually recommended. The heating value of screenings is between 5000 and 8000 Btu per pound of dry solids.

Several smaller treatment plants have made unsuccessful attempts to burn screenings with sewage sludge. The digestion of screenings, separately or combined with sewage sludge, is also a satisfactory disposal method. Screening grinders are beneficial for medium-size plants. Reduced-size solids are returned to raw sewage or mixed with sewage sludge depending on grinder location related to the treatment units.

Hammer-Mill-Type Screening Grinder

These screening grinders use swing hammers. Organic solids are ground to a pulp and returned by a water spray into the waste stream. Power requirements are higher than with other designs.

Comminutors

To eliminate the problems associated with collection, removal, storage, and handling of screenings, wastewater treatment plants install devices that continuously intercept, shred, and grind large floating material in the waste flow into small pieces. These cutting and shredding devices are called *comminutors*. *Comminution* reduces solid particles to smaller sizes.

The use of comminutors reduces odors, flies, and unsightliness. They normally operate continuously and are generally located between the grit chamber and the primary settling tanks. When only one comminutor exists, a bypass channel with a manually-cleaned bar screen and flow-diverting gates should be constructed.

Comminutors with rotating cutters can be divided into two groups. In one group, the screen rotates and has cutters; in the other group, the screen is stationary and only the cutters rotate.

ROTATING CUTTERS

The rotating-screen-type comminutor consists of a motor-driven, revolving, vertical drum almost completely submerged in the wastewater flow. Water passes into the drum through slots and out the bottom (see Figure 7.14.9).

Material that is too large to pass through the slots is cut into pieces by the cutting members acting like shears. The angles between the cutting members are designed to eject iron or other hard materials. This comminuting de-

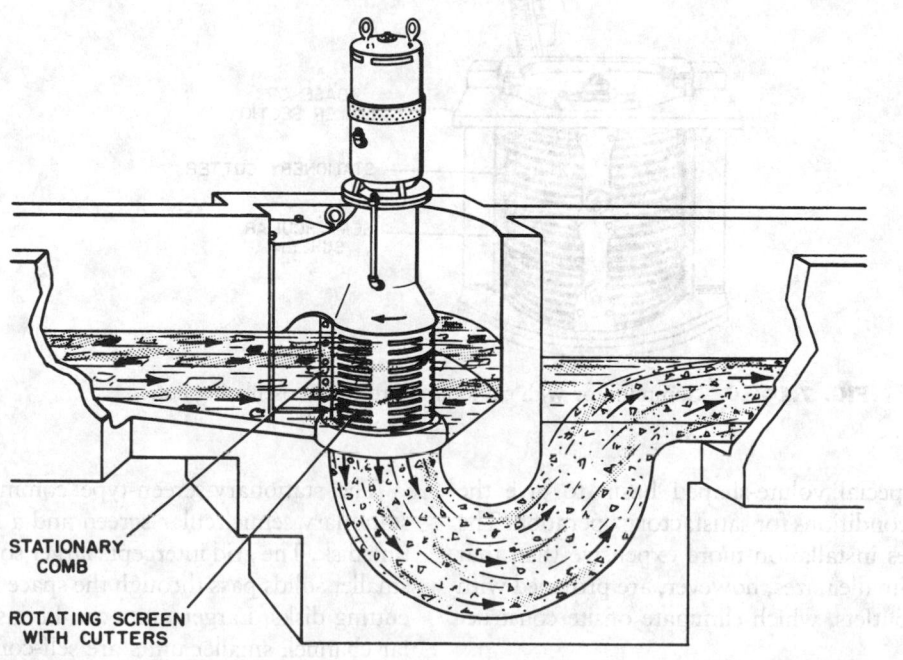

STATIONARY COMB

ROTATING SCREEN WITH CUTTERS

FIG. 7.14.9 Comminutor with rotating-screen cutter.

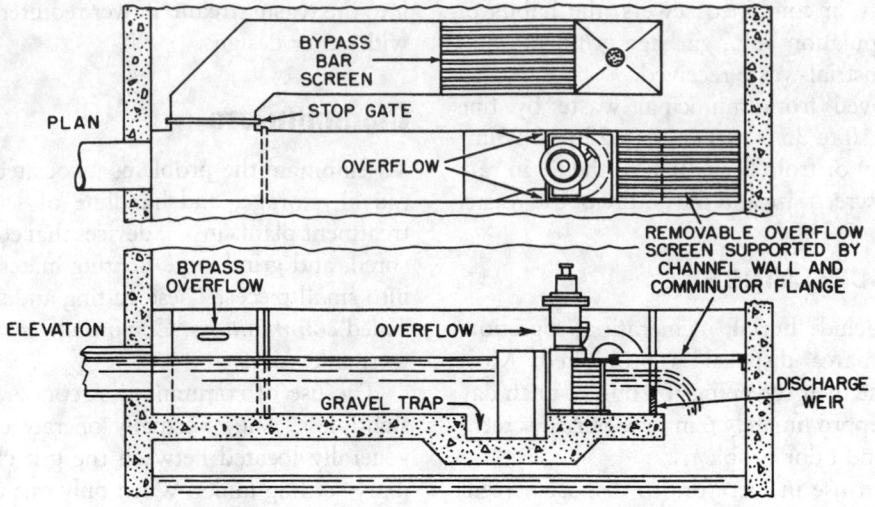

TYPICAL WET WELL INSTALLATION

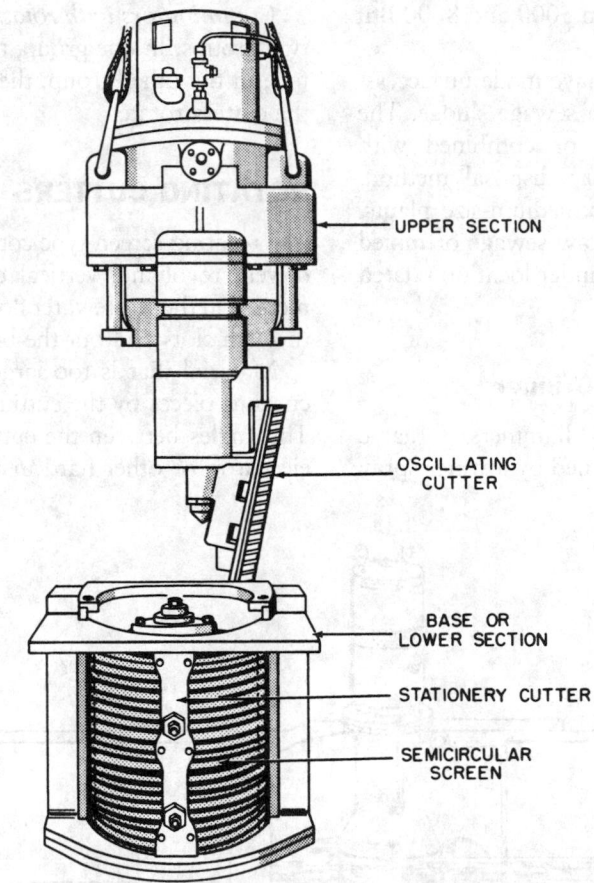

FIG. 7.14.10 Comminutor with stationary screen and oscillating cutter.

vice requires a special volute-shaped basin to give the proper hydraulic conditions for satisfactory operation. The basin shape makes installation more expensive than that of other devices. Smaller sizes, however, are provided with special cast-iron outlets, which eliminate onsite construction work.

The stationary-screen-type comminutor consists of a stationary semicircular screen and a rotating circular cutting disk. The grid intercepts larger solid particles, whereas smaller solids pass through the space between the grid and cutting disks. Larger units can be installed in a rectangular channel; smaller units are self-contained.

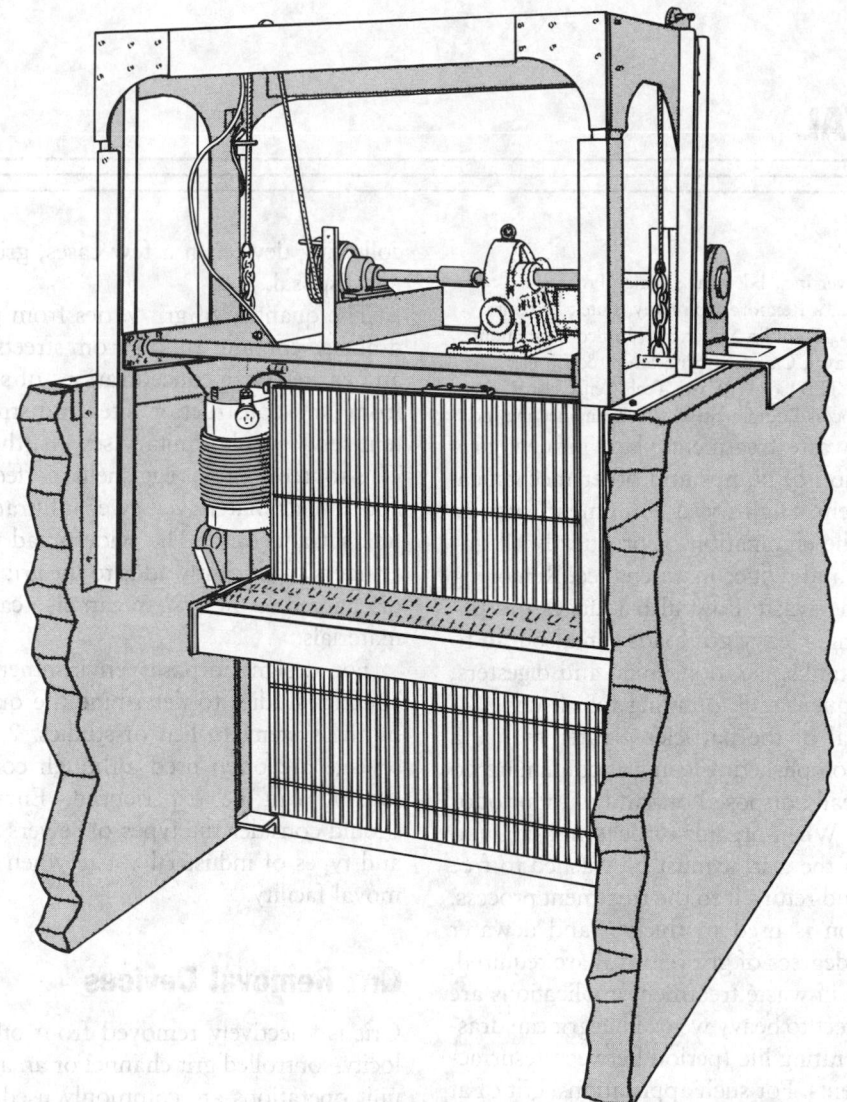

FIG. 7.14.11 Barminutor.

OSCILLATING CUTTERS

This comminutor consists of a semicircular screen with horizontal slots set in a sewage channel concave to the approaching flow and a stationary vertical cutter in its center. A motor-driven, vertical arm with a rack of cutting teeth (which mesh with those of the stationary cutter) oscillates 180° around the screen's center and carries the screenings to the stationary cutter bar where they are shredded (see Figure 7.14.10).

BARMINUTORS

This comminuting unit is used for flows exceeding 10 mgd. This combined screening and cutting machine consists of

a specially designed bar screen with small openings (see Figure 7.14.11). The machine has rotating cutters comprising a comminuting unit that travels up and down the screen, cutting the retained solids. The screen is designed for use in a rectangular channel.

—*János Lipták*
David H.F. Liu

Reference

Great Lakes–Upper Mississippi River Board of State Sanitary Engineers. 1968. *Recommended standards for sewage works (Ten states' standards)*. Health Education Service.

7.15
GRIT REMOVAL

PARTIAL LIST OF SUPPLIERS

Aerators, Inc.; Dorr-Oliver Inc.; Edex Inc.; Eimco Process Equipment; Envirex, Inc., a Rexnord Company; Eutek Systems; Fairfield Engineering Co.; Franklin Miller Inc.; Infilco Degremont Inc.; Krebs Engineers; Laval, Claude, Corp.; Smith & Loveless, Inc.; Techniflo Systems; TLB Corp.; Waste Tech/Centrisys; Wemco Operation of Eimco Process Equipment; Westech Engineering Inc.

Removing grit from waste treatment plant influent prevents wear and abrasion of pumps and other mechanical machinery in the system. High-speed equipment, such as centrifuges, requires the elimination of practically all grit to prevent rapid wear and reduce maintenance. Removing grit from the incoming waste flow also reduces the potential for pipe plugging. Heavy grit loads can also lead to deposition in settling tanks, aeration units, and digesters, which can require frequent tank draining for cleaning.

Usually the removal of the particles greater than 0.2 mm or 65 mesh is accomplished. Clean grit containing no organic material is ideally disposed on land without odor or nuisance problems. When organic or decomposing material is removed with the grit, it must be washed to free the organic material and return it to the treatment process.

When centrifugation is used to thicken and dewater waste sludges, higher degrees of grit removal are required. Most centrifuges used in waste treatment applications are hard-finished and subject to heavy wear. Fine grit can drastically shorten the operating life (period between resurfacings) of these instruments. For such applications, grit of at least 150-mesh (0.10 mm) size should be removed from the waste stream, and occasionally detritus as fine as 325 mesh (0.04 mm) must also be removed to protect centrifugal machines.

Characteristics of Grit

Grit is the heavy mineral material in raw sewage, and may contain sand, gravel, silt, cinders, broken glass, seeds, small fragments of metal, and other small inorganic solids. It is generally nonputrescible. Grit settles more rapidly than organic or putrescible material in sewage, allowing a reasonably clean separation from the waste stream under normal conditions.

Grit is an inert material. Once drained of most of its water, it can be spread on the ground and used on roadways and sand drying beds. Ideally, clean grit contains less than 3% putrescible material. If it is not washed and cleaned, it presents a nuisance problem by causing foul odors, which attract rodents. Grit with high-putrescible levels must be buried after being removed from the grit collecting device. In a few cases, grit is burned before final disposal.

The quantity of grit varies from plant to plant. Storm drainage contains runoff from streets and open land areas and carries large concentrations of soil, sand, and cinders from land construction sites and street sanding. In a system serving only sanitary sewers, the grit load is smaller. This source contains egg shells, coffee grinds, broken glass, and similar materials. Sewer infiltration can also bring in fine silt and sand. The widespread use of home garbage grinders significantly adds to the grit load. Industrial waste discharged to the system can also carry a variety of gritty materials.

For design purposes environmental engineers should conduct studies to determine the quantity of grit carried by the system. In lieu of studies, 2 to 12 ft^3 per mgd of sewage are often used although considerably larger grit loading can be experienced. Environmental engineers should consider the types of sewers used and the amount and types of industrial waste when designing the grit removal facility.

Grit Removal Devices

Grit is selectively removed from other organics in a velocity-controlled grit channel or an aerated chamber. Both unit operations are commonly used. A newer, more efficient approach to grit removal is the use of hydrocylones.

GRAVITY SETTLING

The grit in wastewater has a specific gravity in the range of 1.5–2.7. The organic matter in wastewater has a specific gravity around 1.02. Therefore, differential sedimentation is a successful mechanism for separating grit from organic matter. Also, grit exhibits discrete settling, whereas organic matters settle as flocculant solids (see Section 7.21).

The velocity-controlled grit channel is a long, narrow, sedimentation basin with better flow control through velocity. Some wastewater treatment plants control the velocity by using multiple channels. A more economical arrangement and better velocity control is achieved by the use of control sections on the downstream of the channel. These control sections maintain constant velocity in the channel for a range of flows by using proportional weirs, Parshall flumes, and parabolic flumes.

To design effective grit removal facilities, environmental engineers must know the volume of the sewage flow and quantity of grit. The quantity of grit can be variable; therefore, a safety factor must be allowed. Multiple channels are usually provided when manual grit cleaning is used.

The typical values of detention time, horizontal velocity, and settling velocity for a 65-mesh (0.21-mm diameter) material are 60 sec 0.3 m/sec, and 1.15 m/min, respectively. The theoretical length is 18 m (60 ft). The depth of flow is governed by the volume of sewage flow. The width is not critical but is normally small so that the channels are long and narrow.

The head loss through a velocity-controlled channel is 30–40% of the maximum water depth in the channel. The effect of scouring the settled grit surface at this velocity washes away much of the putrescible material that settles out with the grit. The grit removed by this design usually requires burial to prevent odor.

Grit chambers can be cleaned manually or mechanically. Manual cleaning is usually used only in smaller and older plants where manual methods are used to rake, shovel, or bucket the grit from the chamber. During this operation, flow to the chamber is shut off or diverted to another channel, and the grit tank is drained for cleaning. Designing the grit chamber with the necessary depth provides grit storage.

Mechanical grit removal equipment consists of moving bucket scrapers, horizontal and circular moving rake scrapers, or screw conveyors. Once the grit is removed from the grit collection tank the facility dewaters it using screw or rake classifiers, screens, or similar devices. Hydraulic ejectors, jets, and air lift pumps are also used. Mechanically cleaned tanks require a smaller grit storage volume. Aeration is occasionally used in the grit chamber to wash out organic material from the grit. Figure 7.15.1 shows a mechanical grit removal design.

AERATED GRIT CHAMBER

Aerated grit chambers are widely used for selective removal of grit. The spiral roll of the aerated grit chamber liquid drives the grit into a hopper located under the air diffuser assembly (see Figure 7.15.2). The shearing force of the air bubbles strips the inert grit of much of the organic material that adheres to its surface.

Aerated grit chamber performance is a function of roll velocity and detention time. Adjusting the air feed rate controls the roll velocity. Nominal air flow values are 0.15 to 0.45 m^3/min of air per meter of tank length (m^3/min · m). The liquid detention time is usually about 3 min at the maximum flow. Length-to-width ratios range from 2:5 to 5:1 with depths of 2 to 5 m.

The grit that accumulates in the chamber varies depending on the type of sewer system (combined type or separate type) and the efficiency of the chamber. For combined systems, 90 m^3 of grit per million cubic meters of sewage (30 $m^3/10^6 \ m^3$) is not uncommon; in separate systems, the amount is something less than 30 $m^3/10^6 \ m^3$. Deposited grit is normally recovered by air lift or screw conveyor. The grit is buried in a sanitary landfill.

Aerated grit chambers are extensively used at medium- and large-size treatment plants. They offer the following advantages over velocity-controlled grit channels:

An aerated chamber can also be used for chemical addition, mixing, and flocculation before primary treatment.

Wastewater is freshened by air, reducing odors and removing additional BOD_5.

Minimal head loss occurs through the chamber.

Grease removal can be achieved if skimming is provided.

By controlling the air supply, the chambers can remove grit of low-putrescible organic content.

By controlling the air supply, the chambers can remove grit of any specified size. However, due to the variable specific gravity and the size and shape of the particles, some limitations on removal may exist.

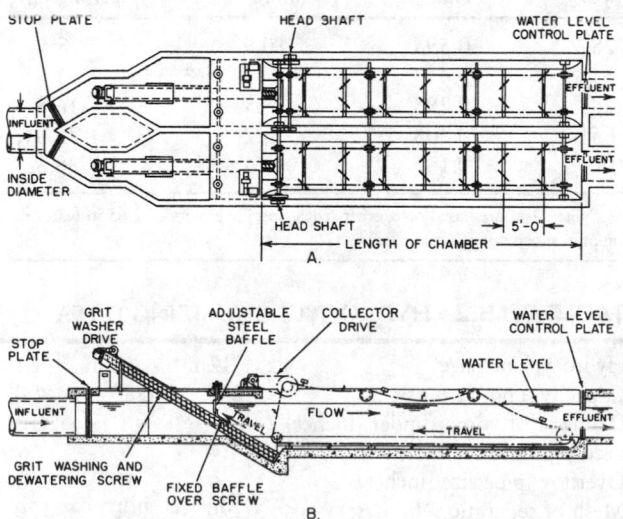

FIG. 7.15.1 Mechanical grit removal facility. **A.** Plan view; **B.** Longitudinal section.

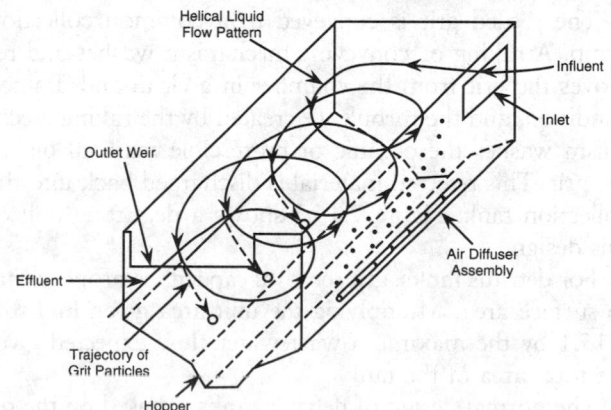

FIG. 7.15.2 Aerated grit chamber. (Reprinted, with permission, from Metcalf & Eddy, Inc. and G. Tchobanoglous. 1979, *Wastewater engineering: Treatment, disposal, reuse* (New York: McGraw-Hill).

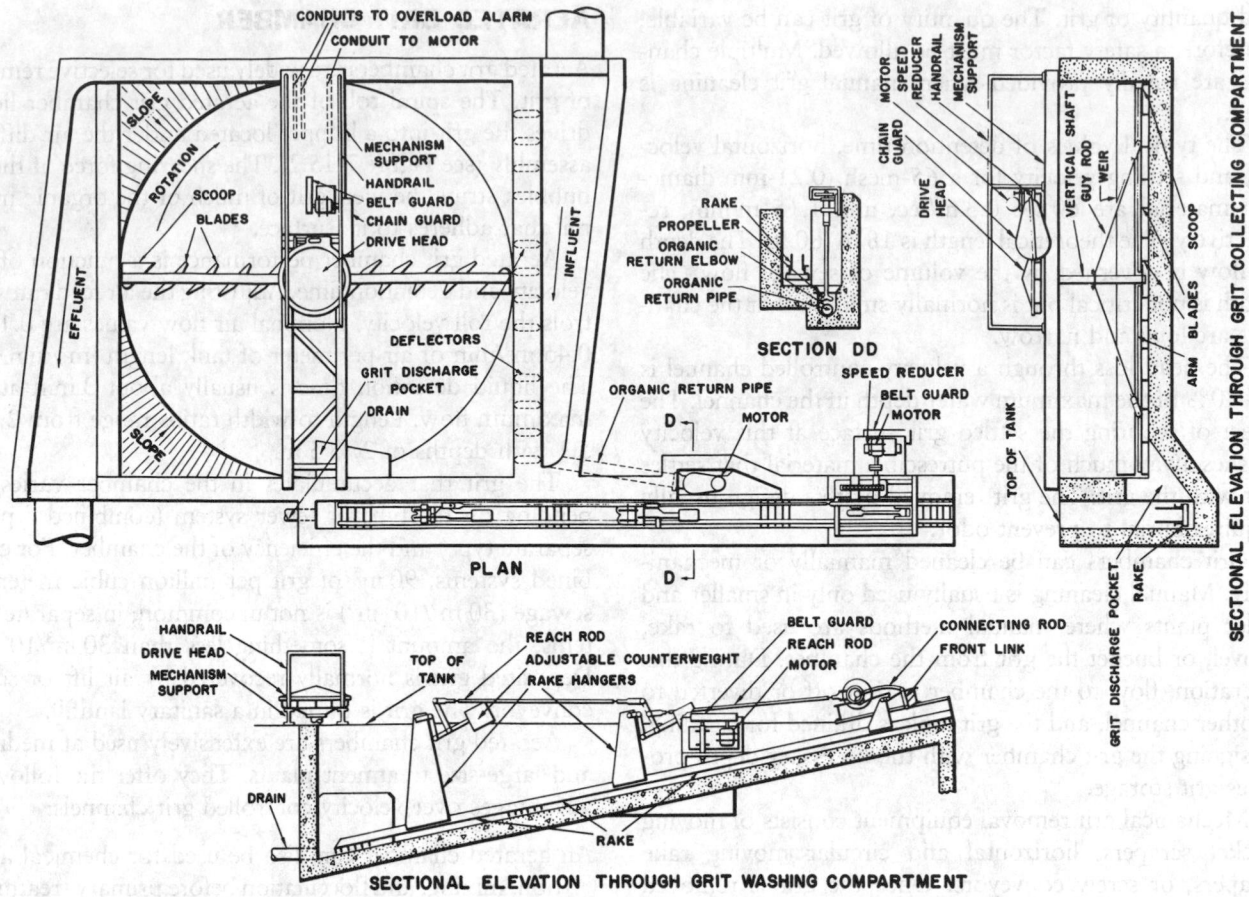

FIG. 7.15.3 Detritus tank and grit washer.

DETRITUS TANKS

Detritus tanks remove a mixture of grit and organic matter. The width and shape of the grit chamber are not critical with this design, but the surface area (which relates to settling velocity) is. The area requirements are proportional to the settling velocity of the grit particle and to the wastewater flow rate.

The settled grit is conveyed to a common collection sump. A raking or conveying mechanism washes and removes the grit from the chamber in a clean and drained condition, and the turbulence created by the raking mechanism washes the organic or putrescible material out of the grit. This rejected material is discharged back into the collection tank. Figure 7.15.3 shows a degritting unit of this design.

For detritus tanks, grit removal capacity is proportional to surface area. Multiplying the unit area given in Table 7.15.1 by the maximum wastewater flow expected gives the total area of the tank.

The normal design of detritus tanks is based on the removal of at least 65-mesh-size particles with a unit area requirement of 38.6 ft² per mgd. At this design rate, 95% of all grit coarser than 65-mesh is removed. Minimum velocities are not critical. Settled organic matter is washed from the grit in the classifier and returned to the process.

TABLE 7.15.1   AREA REQUIREMENTS FOR GRIT REMOVAL BY DETRITUS TANKS

Particle Size		Settling Rate (ft per min)	Unit Area Required per mgd (ft)
Mesh	mm		
28	0.595	10.9	8.5
35	0.417	7.8	11.9
48	0.295	5.5	16.8
65	0.208	3.7	25.0
100	0.147	2.4	38.6

Note: Data are based on a grit particle specific gravity of 2.65 in water flowing at a velocity of 1.2 fps or less.

TABLE 7.15.2   HYDROCYCLONE SIZING DATA

	12 in	18 in	24 in
Hydrocyclone (size)			
Capacity, gpm at 6 psig	205	580	800
Diameter of vortex finder (inches)°	5	8	10
Feed pipe size (inches)	4	6	6
Overflow pipe size (inches)	6	10	10
Mesh of separation°°	270	200	170

Notes: ° Other sizes and capacity ranges are available.

°° Smallest particle size removed; design is based on 95% removal of 150-mesh or larger grit.

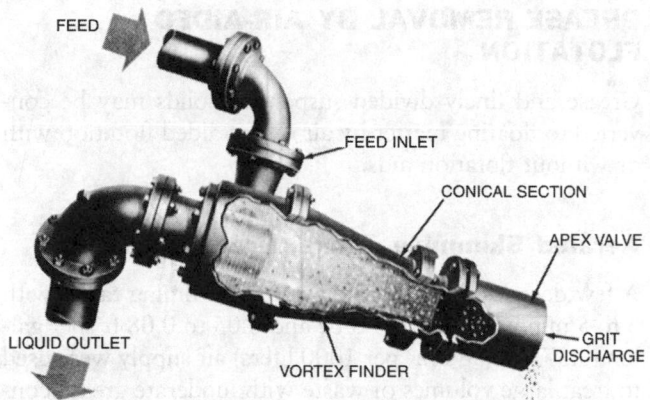

FIG. 7.15.4 Hydrocyclone grit separator.

HYDROCYCLONES

Hydrocyclones are used for sewage sludge degritting in applications requiring high efficiency and a high degree of grit removal. These requirements are particularly prevalent where high-speed centrifuges or close-tolerance equipment such as positive displacement pumps are used.

The hydrocyclone is similar to a conventional dust cyclone in that the feed is introduced tangentially to a cylindrical feed section and the liquid slurry develops a rotational movement and passes into a conical section. The centrifugal force created by cyclonic liquid movement forces heavier solid particles to the outer wall. Solids move along this wall and out the apex of the cone.

A vented overflow opening in the top of the cylindrical section insures that atmospheric pressure exists at the axis of the cyclone. The liquid and lighter solid materials pass up the center of the vortex and out the overflow. Shearing forces are high due to the change in tangential velocity across the diameter of the cyclone, and scouring of the lighter organic material from the grit occurs.

The minimum head requirement for feeding the cyclone is approximately 14 ft (6 psig) for developing sufficient pressure to create centrifugal forces. The normal design of the units is based on 95% removal of 150-mesh and coarser grit at maximum flow.

Hydrocyclones can degrit raw sewage, the primary clarifier underflow prior to thickening, the underflow of pretreatment units such as grit chambers or detritus tanks, and other flows where degritting is required. The grit underflow from the unit can be drained and dewatered in a grit bin or in a screw or rake classifier. Figure 7.15.4 shows the operation of a hydrocyclone.

Hydrocyclone advantages include smaller space requirements, lower cost, finer mesh separation, minimum number of moving parts, and low maintenance. Major disadvantages of this unit are the requirements for a high inlet pressure and a constant feed flow rate.

The size and number of hydrocyclone units is based on the maximum wastewater flow to be handled. The available wastewater pressure determines the flow through each unit and the size of the grit particles that can be removed. If supply pressures are increased, higher flow capacities and improved removal efficiency of smaller grit particles result. Table 7.15.2 gives the typical size and capacity data for a hydrocyclone.

Environmental engineers must know the quantity of grit removed in sizing the grit collection system for the underflow of the hydrocyclone. The hydrocyclone requires a constant feed flow rate for efficient operation. The overflow of the hydrocyclone must be vented for proper unit operation. For variable flow conditions, multiple units (operating as many units as the incoming flow requires) or an on–off timer control should be used.

—János Lipták
David H.F. Liu

7.16
GREASE REMOVAL AND SKIMMING

TYPES OF DESIGNS

a) Grease Interceptors, b) Flotation Units, b1) Aeration Type Units, b2) Pressure Type Units, b3) Vacuum Type Units, b4) Combined Treatment Units, c) Slotted Pipe Skimmers for Square Tanks, d) Rotating Arm Skimmer for Circular Tanks.

The terms grease and oil as used in wastewater treatment denote a variety of materials, including fats, waxes, free fatty acids, calcium and magnesium soaps, mineral oils and other nonvolatile materials that are soluble in and can be extracted by hexane from an acidified sample. In average domestic waste, grease constitutes 10 percent of the total organic matter, and the per capita contribution is estimated to be 0.033 lb. (15 gm) per day. Meat packing, dairy, laundry, garage-machine shop and oil refinery wastes also have high grease-oil content.

It is usually required to reduce the grease content of the industrial wastes below 100 mg per liter before it is discharged to a municipal system. Under quiescent conditions, some portion of the grease settles with the sludge and some floats to the surface, where it may be removed by skimming.

The term scum, as used in wastewater treatment, denotes *all floating* material collected or collectable by skimming, including floating grease, septic sludge raised to the surface, wood pieces, rubber and plastic bottles. Usually, however, floating grease constitutes the bulk and is the most putrescible part of the scum. The terms grease, grease and oil, or scum are often used interchangeably.

GREASE TRAPS AND GREASE INTERCEPTORS

Grease traps or grease interceptors collect grease and floating material from households, garages, restaurants, small hotels, hospitals and the like. Frequent cleaning is essential if operation is to be efficient. Owing to the danger of explosion or fire, motor oil, gasoline and similar light mineral oils should not be allowed to enter sewer systems.

Where the wastewater contains large amounts of greasy kitchen waste, grease traps should be used. Their minimum capacity is 3.0 gal (11.5 l) per capita and should be no less than 30 gal (0.115 m³) per unit (see Figure 7.16.1). The influent line should terminate at least 6 in (15 cm) below the water line, and the effluent pipe should take off near the bottom of the tank.

Commercial grease interceptors are also available for restaurants and smaller industrial establishments, which are connected to public sewer systems.

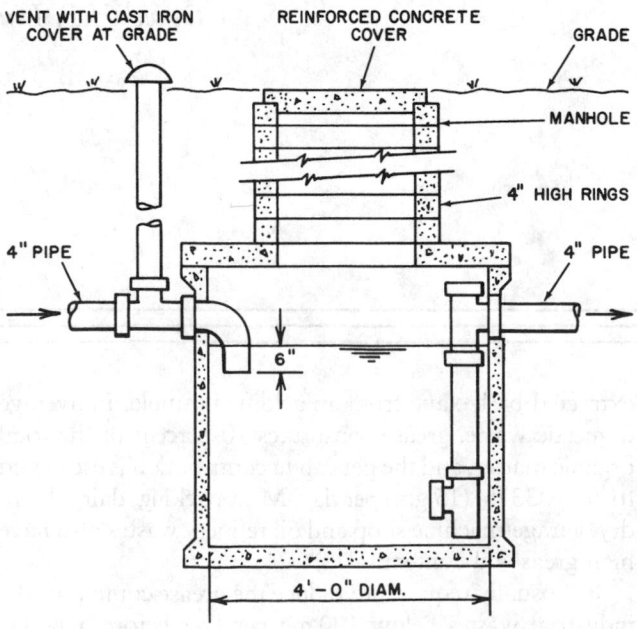

FIG. 7.16.1 Concrete grease trap, 300-gallon capacity.

GREASE REMOVAL BY AIR-AIDED FLOTATION

Grease and finely divided suspended solids may be converted to floating matter by air or gas-aided flotation with or without flotation aids.

Aerated Skimming Tanks, Preaeration

A few decades ago separate aerated skimming tanks with 3 to 5 minutes detention time and 0.05 to 0.08 ft³ per gallon (0.375 to 0.60 m³ per 1000 liters) air supply were used to treat large volumes of waste with moderate grease content. Compressed air containing 1 to 1.5 mg per liter chlorine gas was often used to increase efficiency.

The current version of this treatment (preaeration, that is, aeration before primary treatment) accomplishes more than grease removal. It also facilitates sedimentation and helps to refresh septic waste, which combined with increased floating and suspended solid removal improves the BOD reduction.

For grease removal 5 to 15 minutes aeration, using 0.01 to 0.1 ft³ per gallon (0.075 to 0.75 m³ air per 100 liters) of air is usually sufficient.

Manufactured Skimming Tanks

Although aeration facilitates grease removal, alone it is not very effective. Therefore, for manufactured skimming units, more efficient, mechanized flotation processes are used.

Pressure and vacuum flotation techniques generate air bubbles by reducing the pressure of a supersaturated air-waste mixture or by applying vacuum to the mixture, which is saturated under atmospheric pressure. The liberated minute bubbles tend to form around and attach themself to suspended particles in the waste. With this type of equipment, capital and operating costs are both relatively high. Therefore this type of flotation unit is most popular for pretreatment of industrial wastes at the source, where the flowrate is less and the concentration is high.

PRESSURE TYPE UNITS

This process consists of pressurizing the wastewater with air, usually in a separate air-saturation tank, to 1 to 3 atmospheres and venting the tank to the atmosphere (see Figure 7.16.2). When the pressure on the liquid is reduced, the dissolved air, in excess of saturation at atmospheric pressure, is released in extremely fine bubbles. The rising air bubbles attach themselves to the solid particles in the waste and carry them to the surface. Overflow rates range from 2000 to 6000 gpd per ft². This type of flotation involves high operating power costs.

VACUUM TYPE UNITS

This process consists of saturating the wastewater with air by aerating in a tank, or by permitting air to enter on the

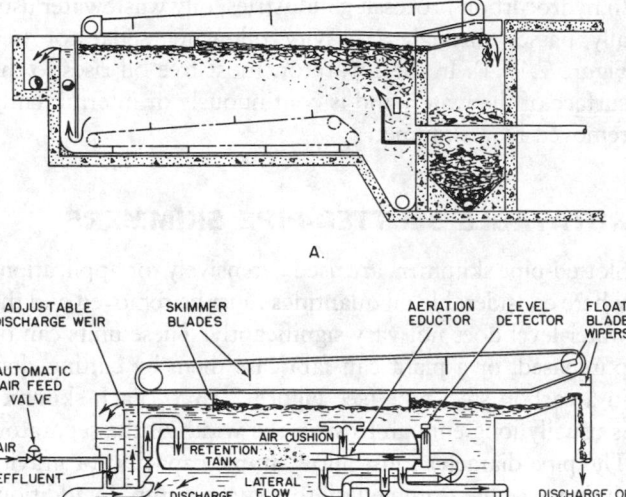

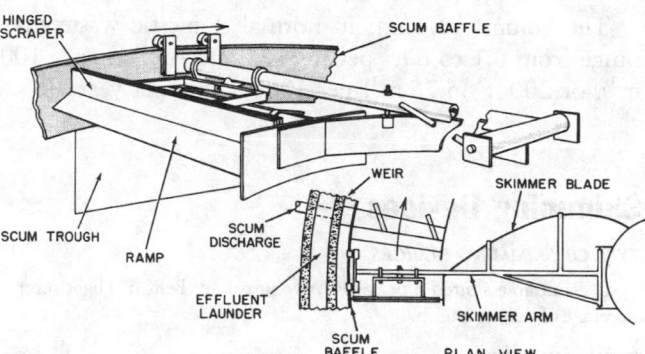

FIG. 7.16.3 Manual scum removal with rotating radial arm skimmer.

FIG. 7.16.2 Flotation unit with skimmers. A. Flotation unit; B. Pressure flotation unit.

suction side of the waste transfer feed pump. Under vacuum the solubility of air in the waste is decreased and air is released in minute bubbles. The rising air bubbles attach themselves to the solid particles in the waste, carrying them to the surface.

Because of the fairly high vacuum levels involved (9 in Hg vacuum), this type of flotation unit requires an expensive, airtight construction. Grease removal of up to 50%; suspended solid removal of 35 to 55%; BOD removal of 17 to 35% may be achieved at surface loadings of 4000 to 6000 gpd per ft². The air requirement is 2.5 to 5 ft³ per 100 gallons.

SCUM COLLECTION

Scum collecting and removal facilities, including baffling are desirable ahead of the outlet weirs on all settling tanks. Horizontal spraying with water under pressure may be employed to collect the scum for hand removal, if no mechanical skimmers are installed.

Square Settling Tanks

The straight scum baffle is installed ahead of the effluent weir-troughs and is submerged to a minimum depth of 18 in. Scum collectors usually move the scum toward the effluent side of the tank. Treatment plants are often equipped with hand-operated revolving slotted pipe skimmers installed horizontally across the tank ahead of the scum baffle (see Figure 7.16.2). When scum is to be removed, the skimmer pipe is rotated until one edge of the slot is sub-

merged slightly below the waste surface. The scum, mixed with waste, flows into it and is discharged through the pipe to a scum pit located outside the tank.

If the scum is to be removed *mechanically,* cross collectors consisting of endless chains above the surface are used. These carry flights which move the scum into the scum-trough, or in case of the helicoid spiral skimmer, the collector slowly turns and sweeps (with its rubber blades) the scum over the curving back edge of the tank into the scum-trough.

Circular Settling Tanks

The circular scum baffle is installed ahead of the circular effluent weir-trough and is submerged to a depth of at least 18 in.

Scum removal in circular tanks is usually performed by a *radial arm skimmer,* which is attached to and rotates with the sludge-removal equipment (see Figure 7.16.3). A skimmer blade moves the scum to the periphery, and hinged scraper blades or neoprene wipers sweep the scum up on a ramp and into a scum-trough. Some tanks are also equipped with automatic flushing devices.

Air Skimmers

Air skimmers may be used to serve small treatment plants. They work on the same principle as airlifts. The only difference is that the suction end of the airlift pipe is turned up by 180°, terminating slightly below the tank's water level. It is provided with a funnel-type extension to collect the scum.

Scum Disposal

The scum is generally collected in a separate scum sump, which can be provided with dewatering facilities. The scum is usually combined with the primary sludge and is disposed of by 1) digestion, yielding gas with high fuel value; 2) by vacuum filtration or incineration; and 3) by burial.

The volume of scum in normal domestic waste may range from 0.1 to 6 ft³ per mg (0.75 to 45 liters per 100 m³), or 200 ft³ (5.75 m³) per 1000 capita per year.

Skimming Devices

TYPE OF SKIMMING DEVICES

a) Rotatable slotted-pipe, b) Revolving roll, c) Belt, d) Flight scrapers, e) Floating pump.

SKIMMING RATE

Types a and d are essentially unlimited. Types b and c handle up to 1 gpm per foot of roll length or belt width; therefore, they typically handle up to 10 and 3 gpm, respectively. Type e handles up to 500 gpm.

WATER CONTENT OF RECOVERED OIL

Oils recovered with types a and e typically contain 80 to 90% water. Oils from b and c contain only 5 to 10% water.

MATERIALS OF CONSTRUCTION

Type a is carbon steel. Types b and e are usually stainless steel, aluminum, fiberglass, or plastic. Type c has a synthetic rubber or nickel belt. Type d has wooden flights.

In hydrocarbon processing industries, oily wastewater usually passes through a gravity oil–water separator (see Figure 7.16.4). In the separator, most free oil rises to the surface of the water, and is continuously or intermittently removed by a skimmer.

ROTATABLE SLOTTED-PIPE SKIMMERS

Slotted-pipe skimmers are used extensively for applications where considerable oil quantities must be removed and the water level does not vary significantly. These units can be purchased, or a plant can fabricate them by cutting slots in a carbon steel pipe (see Figure 7.16.5). Each skimmer is usually long enough to span the width of the separator. The pipe diameter must allow ample capacity for gravity drainage of the skimmed oil to a sump pump. In addition, the diameter must be large enough to allow each edge of the open slots to be rotated from well above and below the liquid level. This diameter allows adjustment or termination of the skimming rate. Due to their simplicity, slotted pipes are inexpensive and essentially maintenance-free.

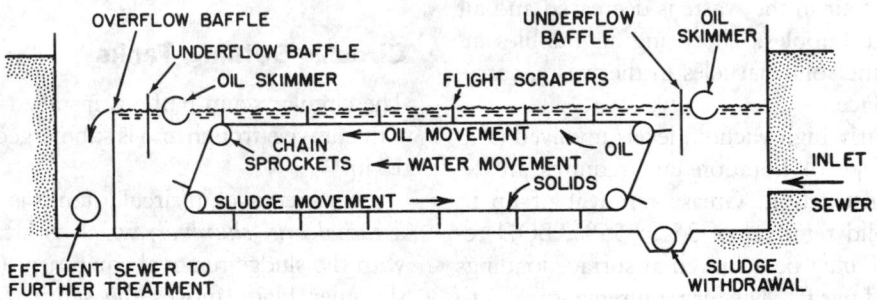

NOTE: THE FLIGHT SCRAPER MECHANISM IS OPTIONAL AND NOT USED IN MANY CASES.

FIG. 7.16.4 Oil–water separator.

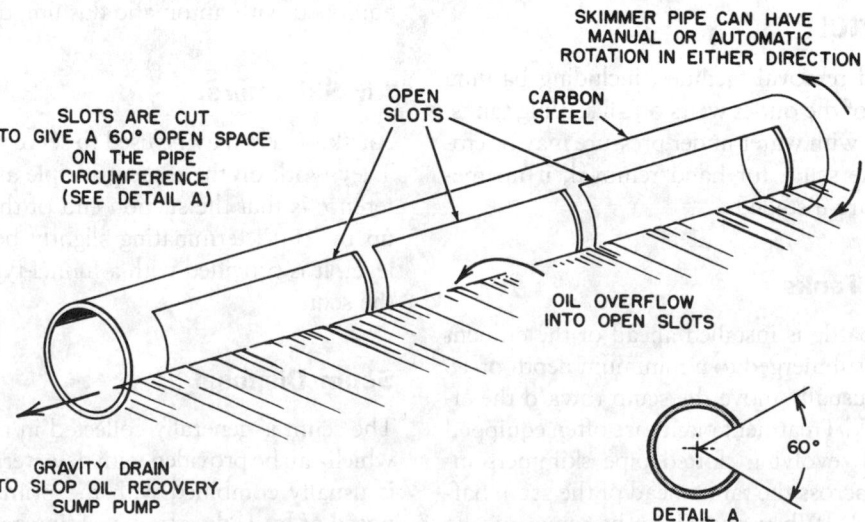

FIG. 7.16.5 Typical rotatable, slotted-pipe oil skimmer.

The major disadvantage of this skimmer is the high percentage of water collected with the skimmed oil. When a thick layer of oil accumulates prior to skimming, the initial oil recovered contains only small quantities of water. However, unless the skimmer is constantly adjusted, the water content averages 80 to 90%. The best solution to this problem is to pump the recovered mixture to a large tank for phase separation. The water phase can then be drained back to the separator inlet, and the oil can be reclaimed for further processing.

REVOLVING ROLL SKIMMERS

Revolving roll skimmers are used for applications where the quantity of oil for recovery is less than 10 gpm and the water collected does not exceed 5 to 10%. The rolls generally have a smooth surface and a shape similar to slotted pipe skimmers except that the diameter is slightly larger. Each cylinder is positioned horizontal to the liquid surface and is immersed at least 0.5 in. As the skimmer revolves, a thin oil film is adsorbed onto the roll surface and is rotated to a scraper blade. The capacity of a roll skimmer depends on oil viscosity, rotation speed, and cylinder length. If the liquid level in the separator varies, roll skimmers can be mounted on floating pontoons.

BELT SKIMMERS

Belt skimmers operate on the same surface adsorption principle as roll skimmers; however, the hardware is different in shape and design as shown in Figure 7.16.6. These units are ideally suited for separators or lagoons with variable liquid levels and less than a 3-gpm-oil-removal rate. The belt length is sized so that the lower end is always immersed. Widths are available in sizes of 12, 18, 24, and 36 in, and the amount of water picked up is only 5 to 10%.

The biggest problem with belt skimmers is that the floating oil does not always migrate to the skimmer for pickup. Skimmers are usually located to take advantage of the prevailing winds and water currents. A minor problem is that heavy greases and some highly oxidized hydrocarbons do not cling to the belt.

FLIGHT SCRAPERS

Flight scrapers are generally used to move oil and sludges to pickup points (see Figure 7.16.4). However, they can also be operated as skimmers when designed to push float-

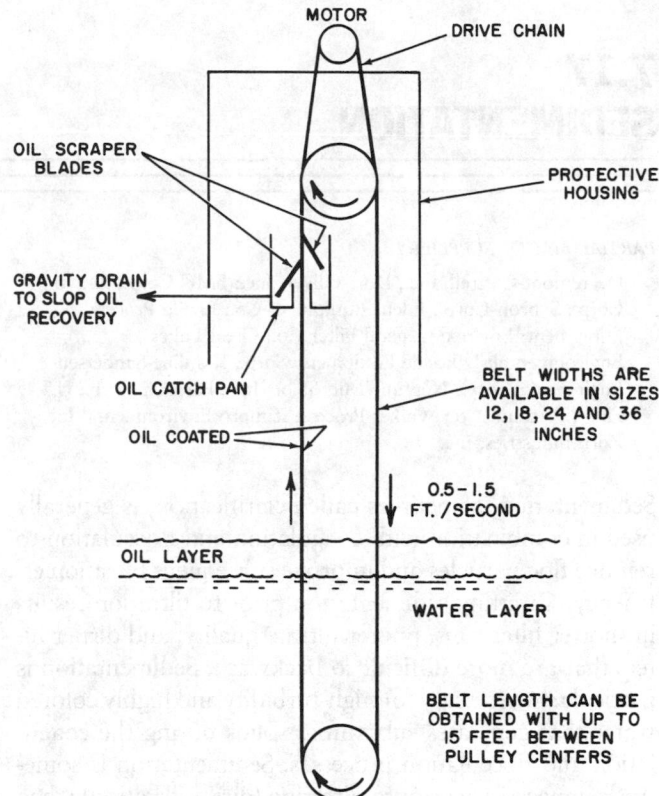

FIG. 7.16.6 Belt oil skimmer.

ing oil up an inclined ramp to a collecting pan. Most units have a chain and sprocket drive mechanism with wooden flights. Their cost is usually higher than other skimmers.

FLOATING-PUMP SKIMMERS

Several types of floating-pump oil skimmers are on the market. Each model usually has a corrosion-resistant float filled with polyurethane, a peripheral overflow weir for oil, a small oil collection sump, and a low-head centrifugal pump. The discharge from the pump passes through a flexible hose to a tank located nearby. The drive is usually a gasoline-powered engine or an electrical motor. These lightweight skimmers allow easy movement from one location to another. Their biggest drawback is the 80 to 90% water concentration in the recovered oil. They can be purchased in sizes up to 500 gpm.

—*R.G. Gantz*
János Lipták

7.17
SEDIMENTATION

PARTIAL LIST OF SUPPLIERS

   Degremont-Cottrell, Inc.; Dorr-Oliver, Inc.; FMC Corp.; Dravo
   Corp.; Sybron Corp.; Edens Equipment Co.; Eimco Process
   Equipment; Envirex; General Filter Co.; Great Lakes
   Environmental; Lakeside Equipment Corp.; Komline-Sandersen
   Engineering Corp.; Neptune MicroFloc, Inc.; Parkson Corp.; U.S.
   Filter, Permutit Co.; Walker Process; Zimpro Environmental Inc.;
   Zurn Industries, Inc.

Sedimentation, sometimes called clarification, is generally used in combination with coagulation and flocculation to remove floc particles and improve subsequent filtration efficiency. Omitting sedimentation prior to filtration results in shorter filter runs, poorer filtrate quality, and dirtier filters that are more difficult to backwash. Sedimentation is particularly necessary for high-turbidity and highly colored water that generates substantial solids during the coagulation and flocculation processes. Sedimentation is sometimes unnecessary prior to filtration (direct filtration) when the production of flocculation solids is low and filtration can effectively handle solids loading.

Sedimentation is sometimes used at the head of a water treatment plant in a presedimentation basin, which allows gravity settling of denser solids that do not require coagulation and flocculation to promote solid separation. The application of a presedimentation basin is most common where surface water has a high silt or turbidity content. Some wastewater treatment plants use coagulation before presedimentation basins.

Types of Clarifiers

The design of most clarifiers falls into one of the following categories: horizontal flow, solids contact, or inclined surface.

HORIZONTAL-FLOW CLARIFIERS

In horizontal-flow clarifiers, sedimentation occurs in specially designed basins. These basins are known as settling tanks, settling basins, sedimentation tanks, sedimentation basins, or clarifiers. They can be rectangular, square, or circular. The most common basins are rectangular tanks and circular basins with a center feed.

In rectangular basins (see part A in Figure 7.17.1), the flow is in one direction and is parallel to the basin's length. This is called *rectilinear flow*. In center-feed circular basins (see part B in Figure 7.17.1), the water flows radially from the center to the outside edges. This is called *radial flow*. Both basins are designed to keep the velocity and flow distribution as uniform as possible so that currents and eddies do not form and keep the suspended material from

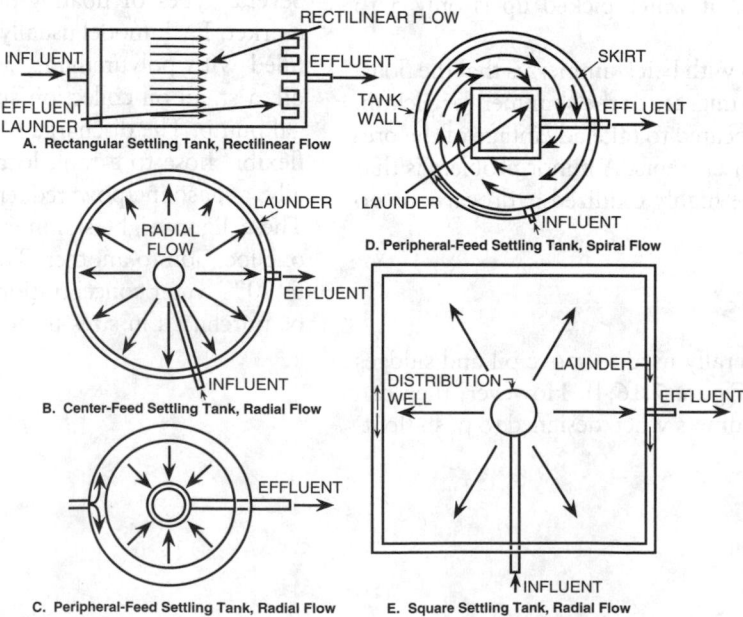

FIG. 7.17.1 Flow patterns in sedimentation basins.

settling. Other flow patterns are shown in parts C, D, and E in Figure 7.17.1.

Basins are usually made of steel or reinforced concrete. The bottom slopes slightly to make sludge removal easier. In rectangular tanks, the bottom slopes toward the inlet end, whereas in circular or square tanks, the bottoms are conical and slope toward the center of the basin.

The selection of any shape depends on the following factors:

- Size of installation
- Regulation preference of regulatory authorities
- Local site conditions
- Preference, experience, and engineering judgement of the designer and plant personnel

The advantages and disadvantages of rectangular clarifiers over circular clarifiers follow.

ADVANTAGES:

- Less area occupied when multiple units are used
- Economic use of common walls with multiple units
- Easy covering of units for odor control
- Less short circuiting
- Lower inlet–outlet losses
- Less power consumption for sludge collection and removal mechanisms

DISADVANTAGES:

- Possible dead spaces
- Sensitivity to flow surges
- Collection equipment restricted in width
- Multiple weirs required to maintain low-weir loading rates
- High upkeep and maintenance costs of sprockets, chains, and fliers used for sludge removal

Square clarifiers combine the common-wall construction of rectangular basins with the simplicity of circular sludge collectors. These clarifiers have generally not been successful (Montgomery 1985). Because effluent launderers are constructed along the perimeter of basins, the corners have more weir length per degree of radial arc. Thus, the flow is not distributed equally, resulting in large sludge

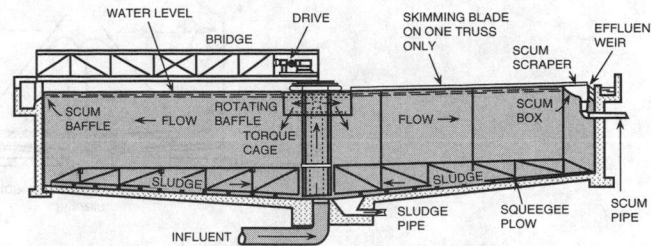

FIG. 7.17.3 Parts of a circular basin. (Courtesy of the FMC Corp., Material Handling Systems Division)

depositions in basin corners. Corner sweeps added to circular sludge collection mechanisms to remove sludge settling in the corners have been a source of mechanical difficulty. Because of these problems, few square basins are constructed for water treatment.

Circular settling tanks are often chosen because they use a trouble-free, circular sludge removal mechanism and, for small plants, can be constructed at a lower capital cost per unit surface area.

Figures 7.17.2 and 7.17.3 show the details of rectangular and circular horizontal flow clarifiers.

SOLID-CONTACT CLARIFIERS

Part A in Figure 7.17.4 shows the operational principles of solid-contact clarifiers. Incoming solids are brought in contact with a suspended sludge layer near the bottom. This layer acts as a blanket, and the incoming solids agglomerate and remain enmeshed within this blanket. The liquid rises upward while a distinct interface retains the solids below. These clarifiers have hydraulic performance and a reduced retention time for equivalent solids removal in horizontal flow clarifiers.

INCLINED-SURFACE CLARIFIERS

Inclined-surface basins, also known as a high-rate settler, use inclined trays to divide the depth into shallower sections. Thus, the depth of all particles (and therefore the settling time) is significantly reduced. Wastewater treatment plants frequently use this concept to upgrade the existing overloaded primary and secondary clarifiers. Part B

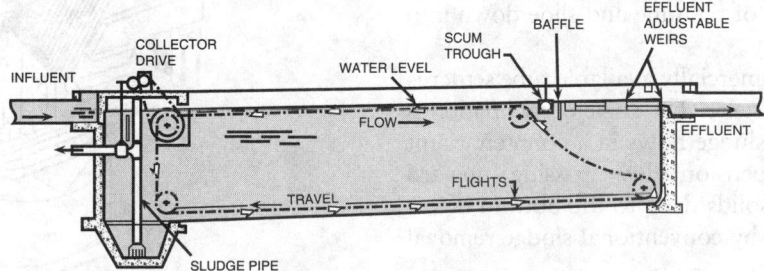

FIG. 7.17.2 Parts of a rectangular basin. (Courtesy of the FMC Corp., Material Handling Systems Division)

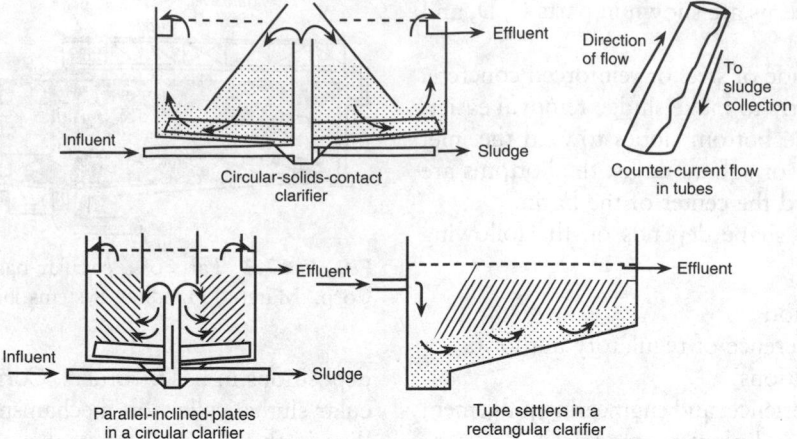

FIG. 7.17.4 Types of clarifiers. A. Circular-solids-contact clarifier. B. Parallel inclined plates in a circular clarifier. C. Tube settlers in a rectangular clarifier. D. Counter-current flow in tubes.

in Figure 7.17.4 shows the operating principles of inclined surface clarifiers.

Inclined-surface clarifiers provide a large surface area, reducing clarifier size. No wind effect exists, and the flow is laminar. Many overloaded, horizontal-flow clarifiers are upgraded with this concept. The major drawbacks of the inclined-surface clarifiers include:

- Long periods of sludge deposits on the inner walls can cause septic conditions.
- The effluent quality can deteriorate when sludge deposits slough off.
- Clogging of the inner tubes and channels can occur.
- Serious short-circuiting can occur when the influent is warmer than the basin temperature.

Two design variations to the inclined-surface clarifiers are tube settlers and parallel-plate separators.

Tube Settlers

In these clarifiers, the inclined trays are constructed with thin-wall tubes. These tubes are circular, square, hexagonal, or any other geometric shape and are installed in an inclined position within the basin. The tubes are about 2 ft long and are produced in modules of about 750 tubes. The incoming flow enters these tubes and flows upward. Solids settle on the inside of the tube and slide down into a hopper.

The most popular commercially available tube settler is the steeply inclined tube settler. The angle of inclination is steep enough so that the sludge flows in a countercurrent direction from the suspension flow passing upward through the tube. Thus, solids drop to the bottom of the clarifier and are removed by conventional sludge removal mechanisms.

Test results for alum-coagulated sludge indicate that solids remain deposited in the tubes until the angle of inclination increases to 60° or more from the horizontal.

Parallel Plate Separators

Parallel-plate separators have parallel trays covering the entire tank. The operational principles for these separators are the same as those for the tube settlers.

Other Inclined-Surface Separators

Another design of shallow-depth sedimentation uses *lamella plates* (see Figure 7.17.5), which are installed parallel at a 45° angle. In this design, water and sludge flow in the same direction. The clarified water is returned to the top of the unit by small tubes.

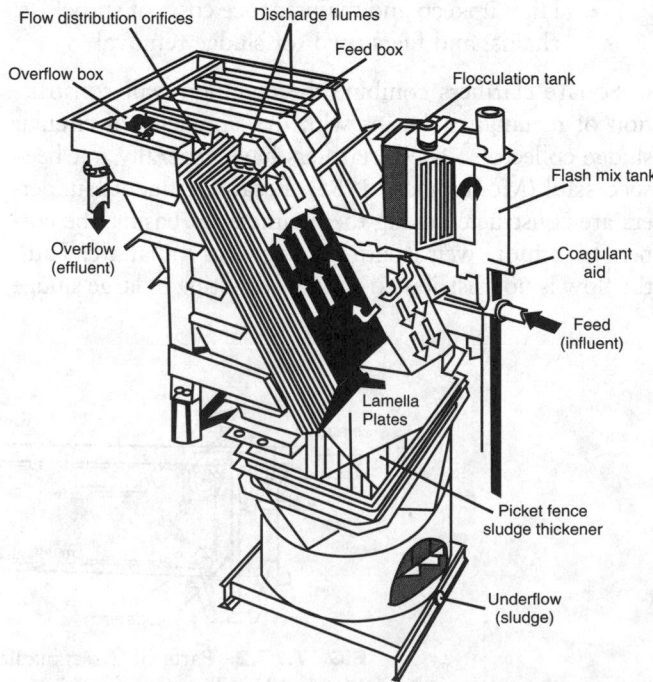

FIG. 7.17.5 Lamella plates. (Courtesy of Parkson Corp.)

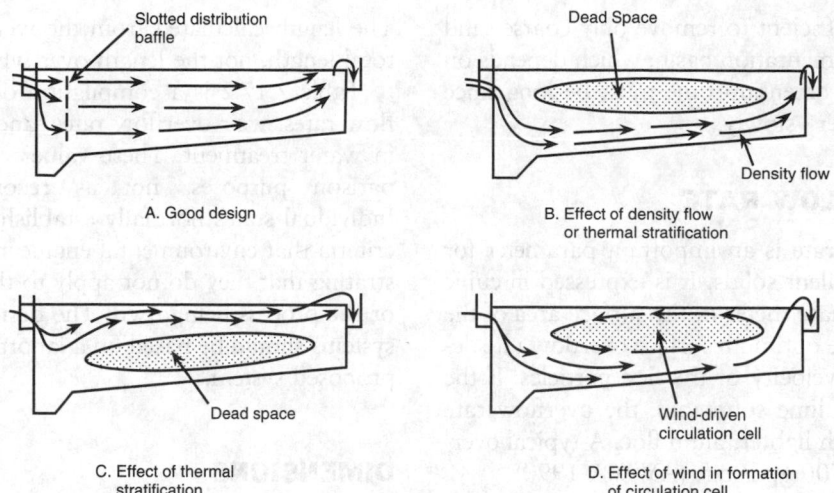

FIG. 7.17.6 Flow patterns in rectangular sedimentation tanks.

Design Factors

The design objective of primary sedimentation is to produce settled water with the lowest possible turbidity. For effective filtration, the turbidity of settled water should not exceed 10 NTU. Since effective sedimentation is closely linked with coagulation and flocculation, the wastewater treatment plant must ensure that the best possible floc is formed.

The flow should be distributed uniformly across the inlet of the basin (see Figure 7.17.6). The solids removal efficiency of a clarifier is reduced by the following conditions:

- Eddy currents induced by the inertia of the incoming fluid
- Surface current produced by wind action (see part D in Figure 7.17.6). The resulting circulating current can short-circuit the influent to the effluent weir and scour settled particles from the bottom.

- Vertical currents induced by the outlet structure
- Vertical convection currents induced by the temperature difference between the influent and the tank contents (see parts B and C in Figure 7.17.6).
- Density currents causing cold or heavy water to underrun a basin, and warm or light water to flow across its surface (see part B in Figure 7.17.6).
- Currents induced by the sludge scraper and sludge removal system

Therefore, factors such as the overflow rate, detention period, weir-loading rate, shape and dimensions of the basin, inlet and outlet structures, and sludge removal system affect the design of a sedimentation basin (Table 7.17.1).

DETENTION TIME

The detention time depends on the purpose of the basin. In a mechanically cleaned presedimentation basin, the de-

TABLE 7.17.1 TYPICAL WATER TREATMENT CLARIFIER DESIGN DETAILS

Type of Basin	Detention Time, hr	Weir Overflow Rate		Surface Overflow Rate	
		$m^3/(m \cdot day)$	$gal/(ft \cdot day)$	m/day	$gal/(ft^2 \cdot day)$
Presedimentation	3–8				
Standard basin following:					
Coagulation and flocculation	2–8	250	20,000	20–33	500–800
Softening	4–8	250	20,000	20–40	500–1000
Upflow clarifier following:					
Coagulation and flocculation	2	175	14,000	55	1400
Softening	1	350	28,000	100	2500
Tube settler following:					
Coagulation and flocculation	0.2				
Softening	0.2				

tention time can be sufficient to remove only coarse sand and silt. In a plain sedimentation basin, which depends on removing fine SS, the detention time must be long since small particles settle very slowly.

SURFACE OVERFLOW RATE

The surface overflow rate is an important parameter for basins clarifying flocculent solids. It is expressed in cubic meters per day per square meter of the surface area of the tank or gal/ft²-day. The optimum surface overflow rate depends on the settling velocity of the floc particles. If the floc is heavy (as with lime softening), the overflow rate can be higher than with lighter, alum floc. A typical overflow for alum floc is 500 gpd/sq ft (AWWA 1990).

The surface overflow rate can be determined by jar test studies in which the best coagulant, optimum dosage, and best flocculation are used. However, the environmental engineer must usually rely on past empirical experience and estimate a safe basin overflow rate based on representative water analyses and estimated coagulant use. Changing seasons and changing water quality pose additional problems.

WEIR-OVERFLOW RATE

Weir-loading rates have some effect on the removal efficiency of sedimentation basins. These rates are expressed in cu m/m or gal/ft length of the weir. The higher the weir overflow rate, the more influence the outlet zone can have on the settling zone. To minimize this impact, environmental engineers should not use a rate exceeding 20,000 gpd/ft. For light alum floc, the rate may have to be decreased to 14,000 gpd/ft or 10 gpm/ft. Typical weirs consist of 90°V notches approximately 50 mm (2 in) deep placed from 100 to 300 mm (4 to 12 in) on the center.

TABLE 7.17.2   DIMENSIONS OF RECTANGULAR AND CIRCULAR BASINS

Clarifier	Range	Typical
Rectangular		
Length, m	10–100	25–60
Length-to-width ratio	1.0–7.5	4
Length-to-depth ratio	4.2–25.0	7–18
Sidewater depth, m	2.5–5.0	3.5
Width, m[a]	3–24	6–10
Bottom slope, %	1	1
Circular		
Diameter, m[b]	3–60	10–40
Side depth, m	3–6	4
Bottom slope, %	8	8

[a]Most manufacturers build equipment in width increments of 61 cm (2 ft). If the width is greater than 6 m (20 ft), multiple bays may be necessary.
[b]Most manufacturers build equipment in 1.5-m (5-ft) increments of diameter.

The length calculated from the weir overflow rate is the total length, not the length over which flow occurs.

Table 7.17.2 is a compilation of typical surface overflow rates, weir overflow rates, and detention times used in water treatment. These values are provided for comparison purposes, not as recommended standards. Individual states normally establish recommended design criteria that environmental engineers can alter by demonstrating that they do not apply to the water being treated or the process being used. The design of water treatment systems should be based on a laboratory evaluation of the proposed system.

DIMENSIONS

The dimensions of a sedimentation basin must accommodate standard equipment supplied by the manufacturer. Also, environmental engineers must consider the size of the installation, local site conditions, regulations of local water pollution control agencies, the experience and judgment of the designer, and the economics of the system. Table 7.17.2 summarizes the basic dimensions of rectangular and circular clarifiers.

For any wastewater supply that requires coagulation and filtration to produce safe water, a minimum of two basins should be provided.

INLET STRUCTURE

Water that by-passes the normal flow path through the basin and reaches the outlet in less than normal detention time occurs to some extent in every basin. It is a serious problem, causing floc to be carried out of the basin due to the shortened sedimentation time.

The major cause of short-circuiting is poor inlet baffling. If the influent enters the basin and hits a solid baffle, a strong current and short-circuit result. The ideal inlet reduces entrance velocity to prevent development of currents toward the outlet, distribute water uniformly across the basin, and mixes it with water already in the tank to prevent density current. A near-perfect inlet consists of several small openings (100–200-mm diameter, circular [4–8-in or equivalent]) distributed through the width and depth of the basin. In these openings, the head loss is large compared to the variation in head between the dif-

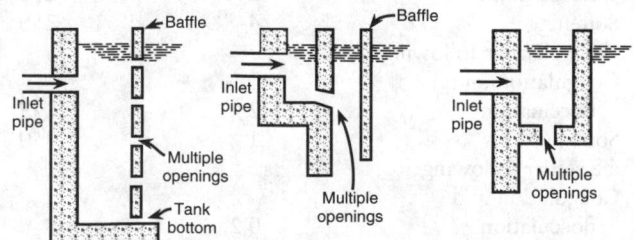

FIG. 7.17.7   Typical sedimentation tank inlets.

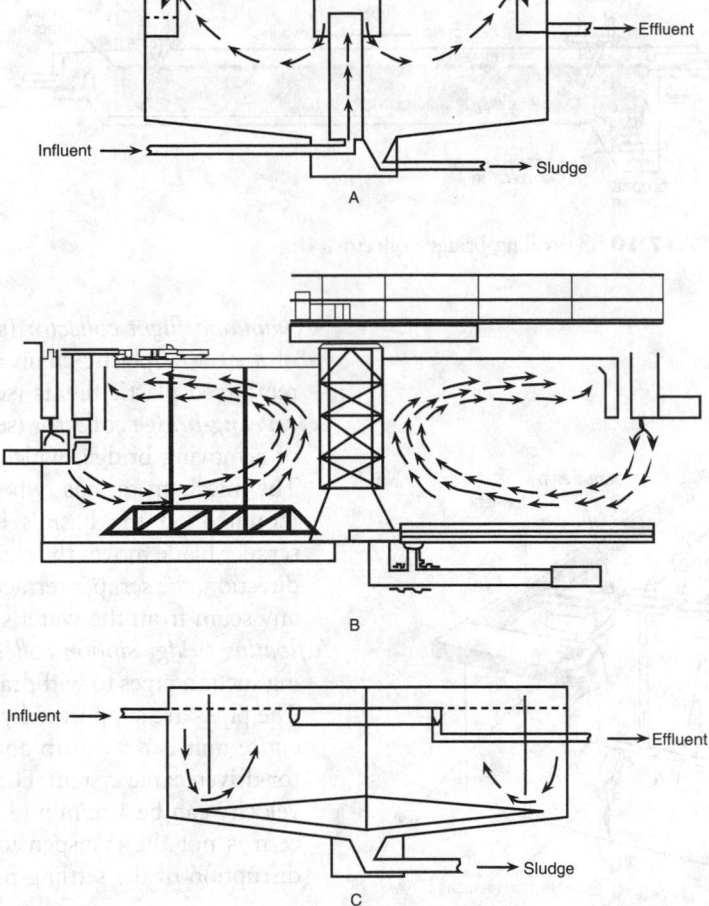

FIG. 7.17.8 Influent and effluent structures for circular clarifiers. (Reprinted, with permission, from Envirex Inc., a Rexnord Company)

ferent openings. Figure 7.17.7 shows some typical designs that compromise between simplicity and function.

Based on inlet structures, circular clarifiers are classified as center- and peripheral-feed. In center-feed, circular clarifiers, the inlet is at the center, and the outlet is along the periphery. A concentric baffle distributes the flow equally in radial directions. The advantages of center-feed clarifiers are low upkeep cost and ease of design and construction. The disadvantages include short-circuiting, low detention efficiency, lack of scum control, and loss of sludge into the effluent. Part A in Figure 7.17.8 shows the flow scheme of a center-feed, circular clarifier.

In peripheral-feed clarifiers, the flow enters along the periphery. These clarifiers are considerably more efficient and have less short-circuiting than center-feed clarifiers. Peripheral-feed clarifiers have two major variations. These variations are shown in parts B and C in Figure 7.17.8.

OUTLET STRUCTURES

Effluent structures are designed to do the following:

- Provide uniform distribution of flow over a large area

- Minimize lifting of the particles and their escape into the effluent
- Reduce floating matter from escaping into the effluent

The most common effluent structures for rectangular and circular tanks are weirs that are adjustable for leveling. These weir plates are long enough to avoid the high heads that can result in updraft currents and particle lifting.

Both straight-edge and V notches on either one or both sides of the trough have been used in rectangular and circular tanks. V notches provide uniform distribution at low flows. A baffle in front of the weir stops floating matter from escaping into the effluent. Normally, weirs in rec-

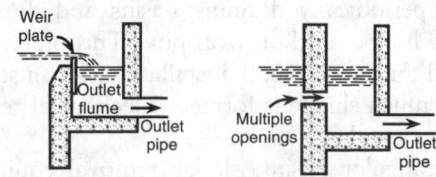

FIG. 7.17.9 Outlet details of sedimentation tanks.

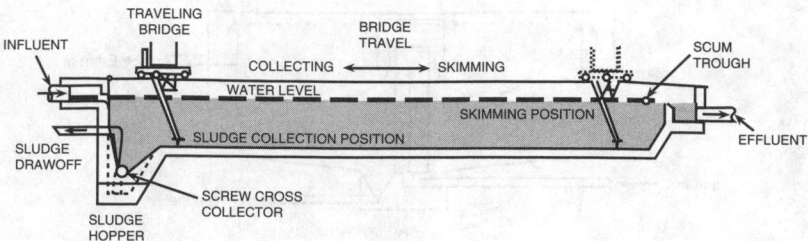

FIG. 7.17.10 Traveling-bridge collector.

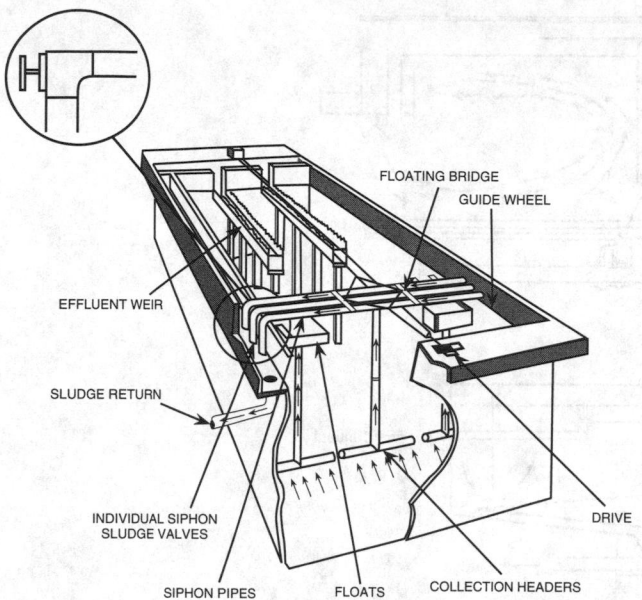

FIG. 7.17.11 Floating-bridge siphon collector. (Courtesy of Leopold Co., Division of Sybron Corp.)

tangular tanks are on the opposite end of the inlet structure. Environmental engineers can use different weir configurations in rectangular basins to obtain a beneficial weir length. Figure 7.17.9 shows typical sedimentation tank outlets.

In circular clarifiers, the outlet weir can be near the center of the clarifier or along the periphery as shown in Figure 7.17.8. The center weir generally provides a high-velocity gradient that can result in solids carryover.

SLUDGE REMOVAL

As solids settle to the bottom of a basin, a sludge layer develops. This layer must be removed because the solids can become resuspended or tastes or odors can develop. Wastewater treatment plants can manually remove the sludge by periodically draining basins and flushing the sludge to a hopper and drawoff pipe. This practice is recommended only for small installations or installations where not much sludge is formed. Mechanical removal is usually warranted.

For rectangular basins, sludge removal equipment is usually one of the following mechanisms (AWWA 1990):

A *chain and flight collector* (see Figure 7.17.1) consisting of a steel or plastic chain and redwood- or fiberglass-reinforced plastic flights (scrapers).

A *traveling-bridge collector* (see Figure 7.17.10) consisting of a moving bridge, which spans one or more basins. The mechanism has wheels that travel along rails mounted on the basin's edge. In one direction, the scraper blade moves the sludge to a hopper. In the other direction, the scraper retracts, and the mechanism skims any scum from the water's surface.

A *floating-bridge siphon collector* (see Figure 7.17.11) using suction pipes to withdraw the sludge from the basin. The pipes are supported by foam plastic floats, and the entire unit is drawn up and down the basin by a motor-driven cable system. For suction sludge removal, the velocity can be 1 m/min (3 fpm) because the main concern is not the resuspension of settled sludge but the disruption of the settling process.

To keep solids from returning to the cleaned liquid, scrapers should operate at velocities below 1 fpm. The power requirements are about 1 hp per 10,000 sq ft of tank area, but straight-line collectors must have motors about ten times that strong to master the starting load (Fair, Geyer, and Okun 1968).

Circular basins are usually equipped with scrapers or plows, as shown in Figure 7.17.3. These slant toward the center of the basin and sweep sludge toward the center of the basin, then to the effluent hopper or pipe. The bridge can be fixed as illustrated, or it can move with the truss.

Regardless of the collection method, the sludge is washed or scraped into a hopper. It is then pumped to sludge discharge treatment facilities.

—*David H.F. Liu*

References

American Water Works Association (AWWA). 1990. *Water quality and treatment: A handbook of community water supplies.* American Water Works Association, McGraw Hill, Inc.

Fair, G.M., J.C. Geyer, and D.A. Okun. 1968. *Water and water engineering.* New York: John Wiley & Sons.

Montgomery, J. McKee. 1985. *Water treatment principles and design.* John Wiley & Sons, Inc.

7.18
FLOTATION AND FOAMING

PROCESS PERFORMANCE

Foaming partially removes oil, SS, and 80 to 90% of surface active compounds.

Flotation is 65 to 95% effective in removing SS; 65 to 98% effective for fats, oils, and grease; and 25 to 98% effective in reducing BOD_5.

FLOTATION SYSTEM DATA

Vessel area: 1 to 5000 ft^2
Capacity: 1 to 3600 gpm
Operating pressure: 25 to 90 psig
Most flotation system components are made of surface-coated mild steel or concrete. Stainless steel and other corrosive-resistant metals are available.

Flotation is a unit operation which removes solid or liquid particles from a liquid (such as oil droplets removed from water). Adding a gas (usually air) to the system facilitates separation. Rising gas bubbles either adhere to or are trapped in the particle structure of the SS, thereby decreasing its specific gravity relative to the liquid phase and affecting separation of the suspended particles.

Methods of floation include dispersed- and the dissolved-gas flotation. Dispersed-gas flotation, commonly referred to as froth flotation, is not widely used in wastewater treatment. Although many design criteria and removal mechanisms apply to both dispersed- and dissolved-gas flotation, this section emphasizes dissolved-gas flotation.

Gas–Particle Contact

Before flotation of the SS can be accomplished, the particle must be in contact with the gas. Figure 7.18.1 shows gas–particle contact possibilities. The first type of contact is by precipitation of the gas bubble on the suspended particle or by collision of the rising gas bubble with the suspended particle. The angle of contact between the gas bubble and the suspended particle determines whether the gas bubble attaches or remains attached to the suspended particle. The second mechanism of attachment is trapping the rising gas bubble in a floc structure. The third mechanism is entrapment of a gas bubble within a floc structure as it forms.

Gas Solubility and Release

According to Henry's law, gas solubility in water is directly proportional to partial gas pressure and inversely proportional to water temperature. Table 7.18.1 lists the solubility of air, nitrogen, and oxygen in water and their respective densities with respect to temperature.

The following equation determines the quantity of gas that can be dissolved in water at an elevated pressure:

$$S = Sg\left(\frac{P}{Pg}\right) \qquad 7.18(1)$$

where:

S = soluble gas concentration at an elevated pressure, mg/l
Sg = soluble gas concentration at one atmosphere, mg/l
P = elevated system pressure, mm Hg
Pg = atmospheric pressure, 760 mm Hg

The amount of gas that can be released from solution when the system is reduced to atmospheric pressure is calculated as follows:

$$S_r = Sg\left(\frac{P}{Pg - 1}\right) \qquad 7.18(2)$$

where:

S_r = gas quantity released, mg/l

Modifying Equation 7.18(2) to account for the gas solution and release efficiency of the pressurization process yields the following:

$$S_r = Sg\left(\frac{Pf}{Pg} - 1\right) \qquad 7.18(3)$$

where:

f = pressurization system gas dissolving and release efficiency, fraction

Conventional pressurization systems usually attain 50% gas dissolving efficiency, whereas packed- or mixed-retention tank designs produce up to 90% efficiency.

An additional correction in gas solubility may be required when gases are dissolved in wastewater with a high-dissolved-solids content. Gas solubility reductions of as much as 20% from that listed in Table 7.18.1 have been observed.

Pressurization Systems

Flotation system performance depends not only on supplying sufficient gas for flotation but also on the manner in which gas is delivered to the flotation vessel. The pressurization system generally consists of a pressurization pump, a retention tank, and a gas supply. The pump in-

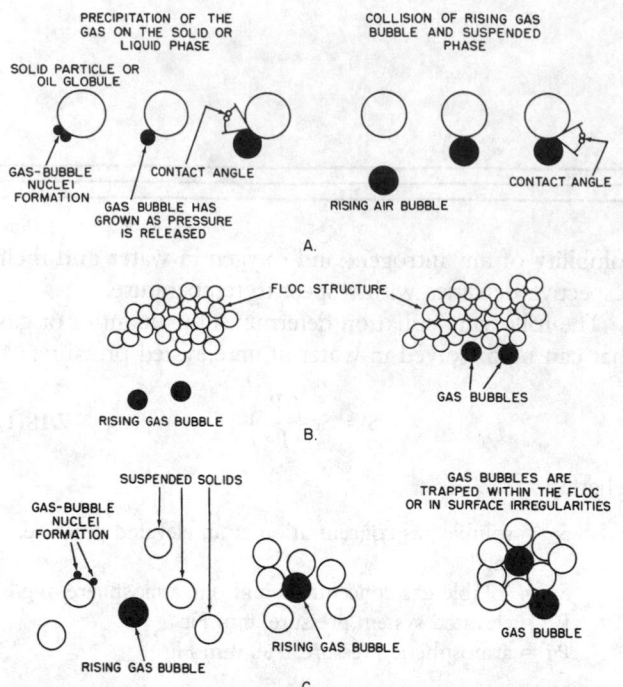

FIG. 7.18.1 Three methods of dissolved-air flotation. A. Adhesion of a gas bubble to a suspended liquid or solid phase; B. Trapping of gas bubbles in a floc structure as the gas bubbles rise; C. Absorption and adsorption of gas bubbles in a floc structure as the floc structure is formed.

creases wastewater pressure while the retention tank provides adequate time for gas to transfer into liquid and also for excess gas applied to the system to be released.

Three general types of pressurization systems can dissolve gases. Figure 7.18.2 shows the position of the pressurization system in each flotation method. Full-flow pressurization transfers gas to the total-feed flow. The pressure

of this system is generally 30 to 40 psig. This technique is applied when enough gas can be dissolved for flotation above pressure and when the wastewater flow passing through a centrifugal pump does not impair separation efficiency of the flotation process.

Partial-flow pressurization is used when a portion of the wastewater flow passing through the pressurization system does not impair the separation efficiency of the flotation system and when enough gas can be dissolved to affect flotation with the bypass stream pressurized to 60 to 75 psig. Using a partial rather than a full-flow pressurization system can frequently yield savings in the cost of the pressurization system.

Recycle-flow pressurization is favored when a natural or chemically formed floc is separated from the wastewater. In this system, a portion of the clarified flotation effluent is recycled to the pressurization system. This recycled flow then becomes the carrier of the dissolved gas later released for flotation. Recycle-flow pressurization is being applied increasingly in flotation applications. Recycle-flow pressurization systems are favored when dissolved air flotation is used for thickening biological sludges.

Process Design Variables

In addition to the process variables that govern the design of gravity clarification and thickening equipment, the performance of pressurization flotation equipment is affected by the gas-to-solids weight ratio and the point of chemical addition. Figure 7.18.3 shows a flotation unit. The circular tank shown includes a feed entrance baffling area, bottom sludge scraping, and an effluent discharge overflow weir. The float skimming equipment and float re-

TABLE 7.18.1 GAS SOLUBILITIES AND DENSITIES

Temperature C°	Solubility* mg/l			Density† g/l		
	Air	N₂	O₂	Air	N₂	O₂
0	37.2	29.4	69.6	1.293	1.251	1.429
10	29.3	23.3	54.3	1.249	1.209	1.375
20	24.3	19.3	44.3	1.206	1.168	1.330
30	20.9	16.8	37.4	1.166	1.129	1.287
40	18.5	14.9	33.2	1.130	1.091	1.247
50	17.0	13.8	30.2	1.093	1.060	1.208
60	15.9	12.9	28.3	1.061	1.027	1.170
70	15.3	12.4	26.8	1.030	0.997	1.137
80	15.0	12.3	25.8	1.000	0.970	1.105
90	14.9	12.3	25.4	0.974	0.944	1.073
100	15.0	12.3	25.3	0.949	0.918	1.047

*Weight solubilities in water at one atm (760 mm Hg) in the absence of water vapor.
†Gas densities at one atm (760 mm Hg) in the absence of water vapor.

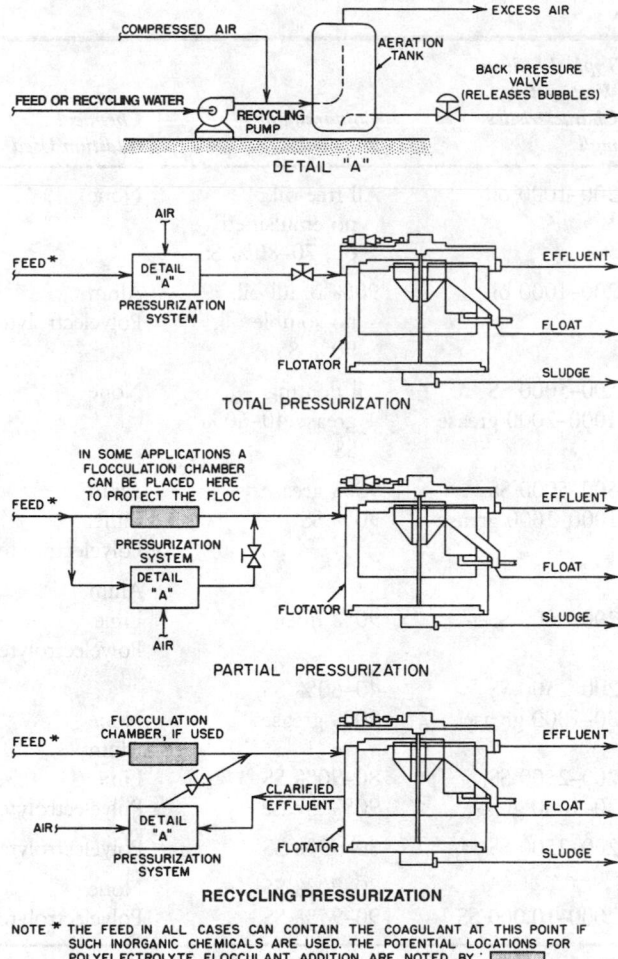

FIG. 7.18.2 Pressurization methods applied in flotation.

NOTE * THE FEED IN ALL CASES CAN CONTAIN THE COAGULANT AT THIS POINT IF SUCH INORGANIC CHEMICALS ARE USED. THE POTENTIAL LOCATIONS FOR POLYELECTROLYTE FLOCCULANT ADDITION ARE NOTED BY :

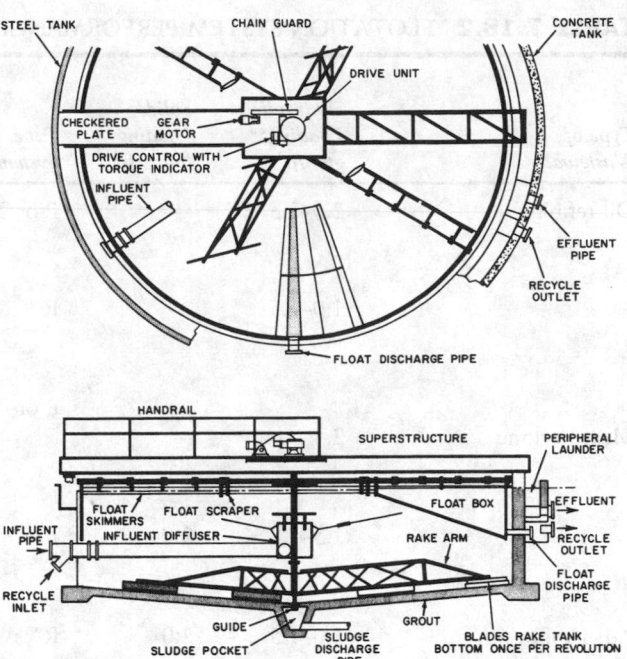

FIG. 7.18.3 Typical flotation unit.

tention baffle are the only additions that are different from a conventional clarification unit.

The gas-to-solids weight ratio is a major design consideration. This ratio determines the operating pressure of the pressurization system. This weight ratio varies from 0.01 to 0.06, but 0.02 is sufficient in most applications. Environmental engineers should use bench-scale tests or, if possible, pilot-plant tests to determine the gas-to-solids weight ratio for each application. An increase in the gas-to-solids weight ratio increases (to a limiting value) the separation rate of the gas–solids mass. This relationship cannot be adequately predicted and must be established by testing.

Full- and partial-flow pressurization systems do not affect the size of the flotation vessel since the flow to the flotation vessel does not increase. However, recycle-flow pressurization systems increase the feed flow rate to the flotation vessel by the recycled amount. Thus, flotation vessel size must be larger for recycle-flow pressurization

systems than for full- or partial-flow pressurization systems.

Hydraulic and Solids Loadings

Hydraulic and solids loadings determine flotation vessel surface area as they do in gravity settling. In most cases, hydraulic loading is the limiting design criterion. Table 7.18.2 lists the removal efficiencies and the approximate hydraulic and solids loadings involved in treating industrial and municipal wastewater. The feed flow rate to a flotation unit using recycle-flow pressurization includes the recycled flow in addition to the raw feed flow. Therefore, flotation vessel surface area should be sized for total flow when hydraulic loading is the limiting design criterion.

Solids loading is directly related to raw feed flow rate, solids concentration, and flotation vessel area. Only in cases of sludge thickening by dissolved-gas flotation does solids loading determine flotation vessel surface area. In these cases, only recycle-flow pressurization systems are capable of supplying the gas quantities required at conventional operating pressures to achieve the required gas-to-solids ratio for thickening. A recycle flow-rate of twice the feed flow rate is common.

Chemical Additions

Wastewater treatment facilities are using chemicals in an increasing number of gravity and flotation clarification applications to reduce SS content of the effluent. They add

TABLE 7.18.2 FLOTATION SYSTEM PERFORMANCE DATA

Type of Wastewater	Hydraulic Loading** gpm/ft²	Solids Loading lb/hr·ft²	Type System*	Typical Influent Wastewater Characteristics mg/l	Contaminant Removal	Chemical Addition Used
Oil refining	2.0–2.5	—	P or T	200–1000 oil	All free oil, no emulsified oil, 70–80% SS	None
	1.0–1.5	—	R	200–1000 oil	90% of all oil, no soluble oil, 90% SS	Alum Polyelectrolyte
			P or T	500–5000 SS 1000–2000 grease	All floating grease 40–60% SS	None
Meat packing	2.5	—				
	1.5–1.8	2.0	R	500–5000 SS 1000–2000 grease	90% grease, 90% SS	Alum Lime Polyelectrolyte Alum
Paper mill	1.0–1.5	2.0	R	200–3000 SS	90% fiber	Alum Lime Polyelectrolyte
Poultry processing	1.5–2.0	—	P, T, or R	200–2500 SS 30–1000 grease	40–60% SS, 90% grease	None Alum
	1.0–1.5	—	R	200–2500 SS 30–1000 grease	80–90% SS 90% grease	Lime Polyelectrolyte
Fruit cannery	0.5–1.5	—	R	200–2500 SS	80–90% SS	Polyelectrolyte
Waste-activated sludge thickening municipal	1.25–1.5	2.0	R	2000–10,000 SS	80–90% SS 90–95% SS	None Polyelectrolyte

Notes: *Pressurization System Type
P = Partial flow
T = Total flow
R = Recycle flow
**In nonrecycling flotation units, it is the maximum influent rate per unit of tank surface area.

alum, ferric chloride, lime, and various polymeric compounds to form stable floc particles or break oil emulsions. These chemicals are usually added to pretreatment flash mixers and flocculators ahead of the flotation unit. The chemically treated wastewater then flows by gravity to the flotation system where gas is added by recycle-flow pressurization. Full- and partial-flow pressurization systems are infrequently used because the floc structure in the chemically treated wastewater degrades in the pressurization system and does not reform in the flotation unit.

Foaming Process

Foam fractionation separates a solution containing a surface-active solute into two fractions: the foam fraction containing a high concentration of surface-active solute and a drain fraction depleted of the same solute. Foam formation also collects SS and oils and separates them from the wastewater being treated.

Figure 7.18.4 shows the basic concept of foam fractionation. Gas, usually air, is diffused into the bottom section of the fractionation unit while the feed flow enters above the gas inlet but below the liquid surface. Foam is generated and lifted upward by the gas in the unit. This process discharges the foam from the fractionation unit and collapses it by heating it or spraying it with previously collapsed foam. Heating to facilitate foam collapsing is preferred when SS are separated by this process. The effluent is discharged near the bottom of the fractionation unit.

Process Variables

Variables affecting foam fractionation process efficiency include gas-to-liquid volume ratio, gas-to-solute concentration weight ratio, gas bubble size, type of solute, and foam characteristics. Increasing the gas-to-liquid flow rate ratios results in increased solute removal due to an in-

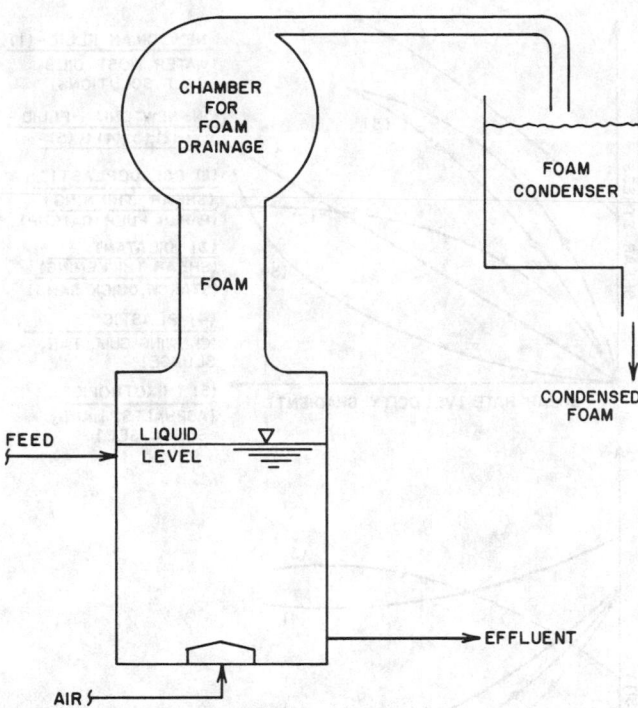

FIG. 7.18.4 Foam fractionation unit.

creased gas surface area but yields a wetter foam. Decreasing the gas-to-liquid flow rate ratio results in a lower foam volume. Reducing the gas bubble size results in a greater bubble surface area for greater solute removal at lower gas flow rates. Effective removal of surface-active agents with high volume reduction requires a stable foam from which the excess liquid can be drained rapidly. In some cases, the wastewater treatment facility must add a surfactant to the process feed to attain this stable foam. Empirical equations relating process variables and removal efficiencies have been developed.

Equipment geometry affects fractionation unit efficiency. The method of gas dispersion, feed point, foam-discharge point, liquid depth, and vessel shape all affect process efficiency.

—David H.F. Liu

7.19
SLUDGE PUMPING AND TRANSPORTATION

Sludges from wastewater treatment vary in composition and ease of dewatering. Because of the rheological characteristics of sludge, transport through closed conduits is a difficult engineering task, and the design is usually based on empirical knowledge.

Solids are removed from clarifier sludge wells using gravity or pumps. In most cases, raw sludge and digested primary sludge are treated by positive displacement pumps, whereas activated sludge is easily moved by less-expensive centrifugal units.

Types of Sludge

Sludge describes a number of slurries in wastewater and water treatment.

RAW SLUDGE

Raw sludge is obtained from the primary clarifier. It is usually a vile, putrescible material, containing from 1 to 12% solids. It cannot be disposed of without further treatment, which can consist of aerobic or anaerobic digestion,

incineration, or wet combustion. It dewaters reasonably well in a centrifuge because it consists mostly of large solid particles (smaller solids do not settle in the primary clarifier). Therefore, centrifuges are frequently used ahead of incinerators. Vacuum filtration of raw sludge is not recommended because of odor.

ACTIVATED SLUDGE

Waste-activated sludge results from the overproduction of microbial organisms in the activated sludge process. Wastewater treatment facilities periodically discharge this material to maintain the recommended SS concentration of mixed liquor in the aeration tank.

Waste-activated sludge creates a handling problem in many treatment plants. It is a light, fluffy material, composed of bacteria, rotifers, protozoa, and enough filamentous organisms to make concentration difficult. The underflow from a final clarifier may only contain 1% solids. Pumping this much water to digesters is inefficient and can lead to digester failure. Accordingly, wastewater treatment facilities use thickening devices to concentrate the sludge to 5 to 7% solids. In addition to gravitational thickeners used in the past, flotation and centrifugal thickeners are also used.

FILTER HUMUS

Trickling filter humus is obtained from the final clarifiers following trickling filters. The quantity of filter humus for a facility is significantly lower than the quantity of waste-activated sludge. Usually, wastewater treatment facilities do not thicken filter humus prior to pumping to the digester.

CHEMICAL SLUDGE

Chemical sludge is a new problem for wastewater treatment plant operators. Chemicals like alum or lime are now used for nutrient removal as well as for increasing plant efficiency without adding hardware. Centrifugation effectively removes calcium carbonate formed by the addition of lime. Metal hydroxides (e.g., magnesium hydroxide) are light precipitates that resist removal and compaction. Waste alum sludge from water treatment is difficult to thicken or dewater; only recently has it been recognized as a serious problem. Future water treatment plants must provide the means to treat and dispose of waste alum sludge. Currently, no effective method exists.

Physical Characteristics of Sludge

Sludges are almost always characterized by their solids concentration. Often, environmental engineers use the sludge volume index (SVI) to describe sludge settling characteristics. Unfortunately, the SVI is an unsatisfactory parameter and should only be used for plant control and other applications in which comparing two or more sludges is not necessary. A better means of describing sludge physical characteristics is sludge rheology.

Almost all slurries, especially wastewater treatment sludges, are thixotropic and pseudoplastic fluids (see Figure 7.19.1). As shown in Figure 7.19.2, they exhibit an apparent yield strength, a parameter that environmental engineers can use to describe sludge behavior. Rheograms (see Figure 7.19.2) can be constructed for any sludge provided that the proper tools are used. Unfortunately, rheological data are difficult to obtain because viscometers (Liptak 1995) must be modified before the sludge can be analyzed. Nevertheless, environmental engineers should measure the rheological properties and relate them to other physical or biological characteristics under study.

Transport of Sludge in Closed Conduits

Because of the nonNewtonian nature of sludge (Dick and Ewing 1967), environmental engineers cannot usually ap-

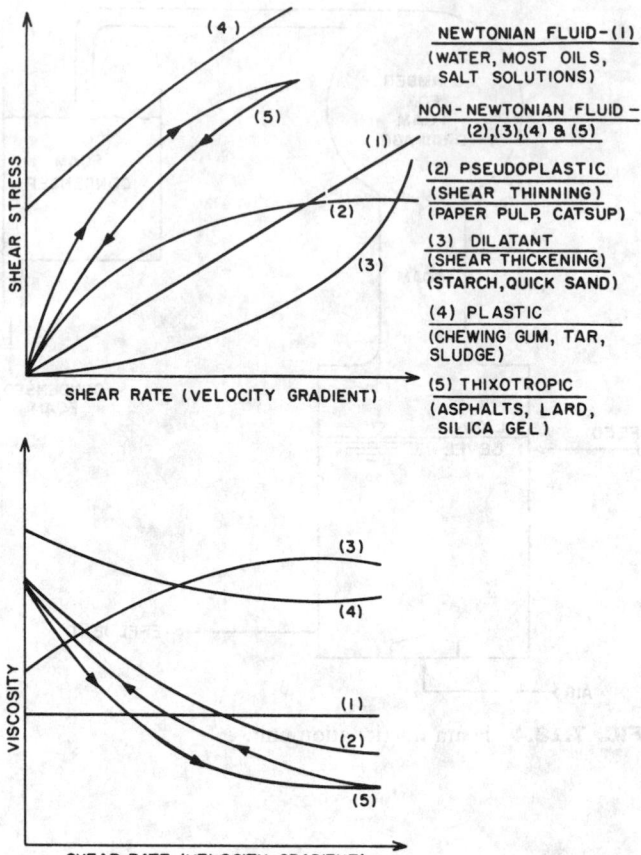

FIG. 7.19.1 Shear diagram of Newtonian and nonNewtonian fluids.

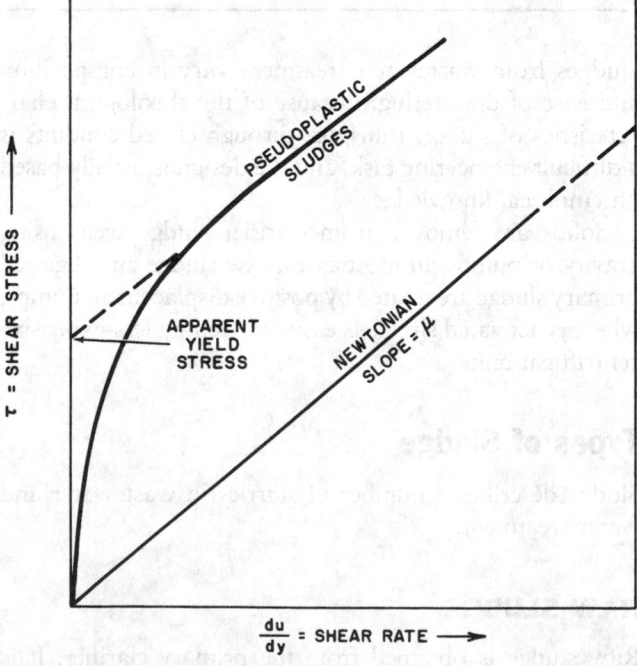

FIG. 7.19.2 Typical rheogram for a sludge. Projecting the straight line portion of the pseudoplastic curve to the zero shear rate gives the apparent yield stress.

ply standard hydraulic formulas for fluid friction without correction. The viscosity term is meaningful for pseudoplastic materials only at a known and fixed shear rate.

For light sludges, such as activated sludge, errors occur when the viscosity is assumed to be that of water and the common formulas are used. For thickened sludge or raw sludge, however, considerable errors can result (Zandi 1971). The simplest method is for environmental engineers to use the Hazen-Williams formula and adjust coefficient C as required.

In one study (Brisbin 1957), raw sludge was pumped through a 4-in and 6-in line, and the head losses were measured. Table 7.19.1 shows the resulting data.

For valves and fittings (minor losses), environmental engineers should calculate equivalent lengths of pipe for pure water and apply the correction factors to the equivalent pipe length. The following equation gives the Hazen-Williams formula:

$$V = 1.318 \, C \, R^{0.63} \, S^{0.54} \qquad 7.19(1)$$

where:

V = velocity in feet per second
R = hydraulic radius in feet
S = slope of energy gradient in feet per feet

The friction coefficient C is a function of the pipe material and age. The formula was originally developed for water, and its application for sludge, as shown on Table 7.19.1, requires modifications of C.

Velocity in pipes is also important because at low velocities laminar flow can be attained. For pseudoplastic materials, this velocity forces the flow into the central core of the pipe only, with stagnant sludge near the wall. Such plug flow can cause troublesome operation. Velocities between 5 and 8 fps should be maintained.

Clogged sludge lines are one of the most annoying problems for wastewater treatment plant operators. All lines should be as large as possible (8 or 6 in is usually a minimum size), and cleanouts should be provided wherever possible. Long radius elbows and sweep tees also should be used. High and low points should be avoided; the high points result in gas pockets, and the low points clog easily with large or heavy objects.

TABLE 7.19.1 HYDRAULIC FLOW COEFFICIENT OF RAW SLUDGE

Total Solids in Raw Sludge (%)	Apparent Hazen-Williams Coefficient C, Based on C = 100 for Water
0	100
2	81
4	61
6	45
8.5	32
10	25

Sludge Pumps

Sludge pumping is an important part of wastewater treatment plant operations. Raw sludge must be moved to digesters, activated sludge must be returned and periodically discharged, scum must be pumped from scum pits, and pumping sludge is often used for mixing digesters. The type and consistence of sludges vary, and environmental engineers must individually analyze each pumping requirement.

PISTON PUMP

Plunger (positive displacement) pumps are used mostly for pumping raw sludge (see Figure 7.19.3). They can pump heavy solids without clogging and are easy to clean if clogging does occur. Pumping at low rates is possible, and suction lifts can be accommodated. The disadvantages of plunger pumps are that they can be messy and noisy, but their trouble-free service more than compensates for these disadvantages.

DIAPHRAGM PUMP

Diaphragm positive displacement pumps are often used for small installations (see Figure 7.19.4). Changing the length of the strokes varies the capacities. Counting the strokes and knowing the pump volume determines the sludge pumping rate (Zandi 1971). Both diaphragm and plunger pumps can be used as scum pumps and pumps for digested sludge.

SCREW PUMP

Screw-feed and rotary pumps are positive displacement units that produce a steady flow, as opposed to the pulsating flow from diaphragm and plunger pumps (see Figure 7.19.5). Screw and rotary units are always used to pump sludge to the dewatering equipment. The centrifuge requires a steady, uninterrupted flow. A pulsating or im-

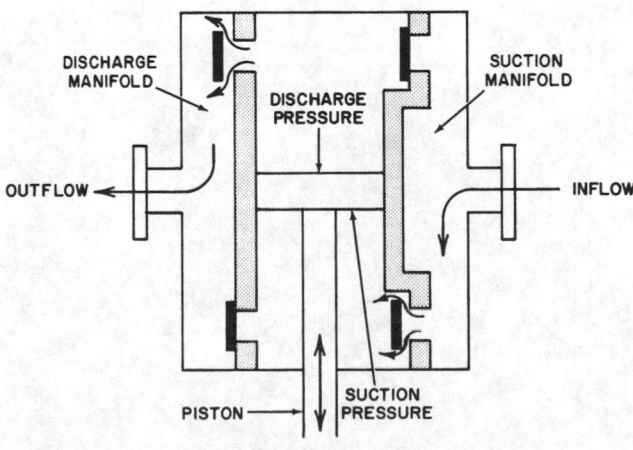

FIG. 7.19.3 Plunger-type sludge pump.

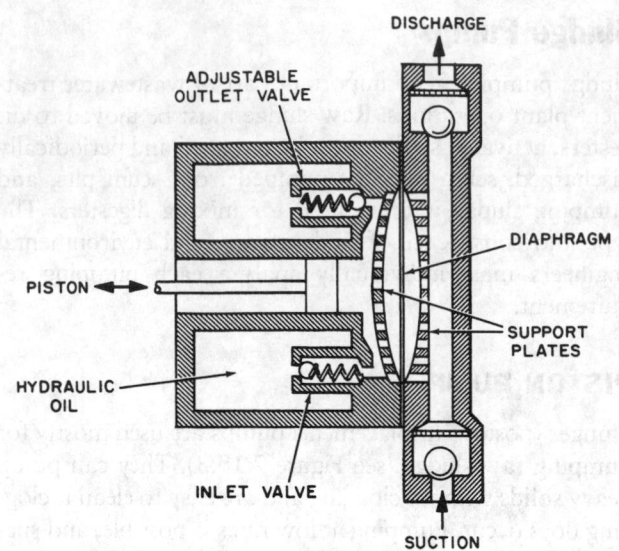

FIG. 7.19.4 Diaphragm-type sludge pump.

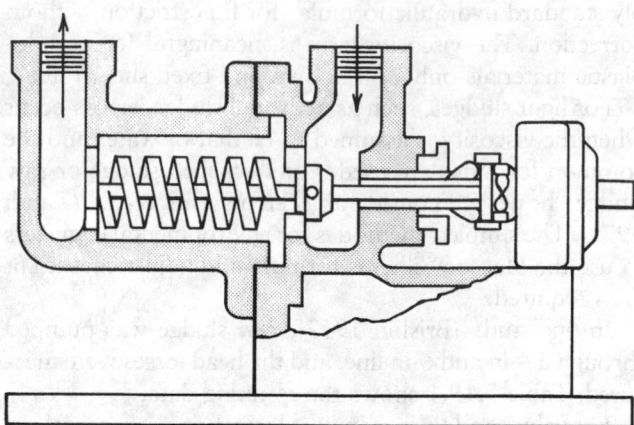

FIG. 7.19.5 Screw-type sludge pump.

peller-type centrifical pump should never be used in front of a centrifuge.

AIR-LIFT PUMP

The air-lift pump is an excellent means of returning activated sludge in small, activated-sludge plants. It is almost trouble-free and requires no extra power.

CENTRIFUGAL PUMP

Centrifugal pumps, although less expensive than positive displacement pumps, suffer from clogging problems, both in the pump and in the line. Centrifugal pumps are used most often with activated sludge since the possibility of fouling is minimized on that service. The use of open, bladeless impellers prevents clogging problems.

References

Brisbin, S.G. 1957. Flow of concentrated raw sewage sludges in pipes. *Journal, Sanitary Engineering Division, ASCE* 83, SA3 (June).
Lipták, B.G. (ed.). 1995. *Instrument engineers' handbook. Process Measurement and Analysis.* Radnor, Pa.: Chilton.
Dick, R.I. and B.B. Ewing. 1967. The rheology of activated sludge. *Journal Water Pollution Control Federation* 39:10.4 (April).
Zandi, I. (ed.). 1971. *Advances in solid-liquid flow in pipes and its application.* New York: Pergamon.

Conventional Biological Treatment

7.20
SEPTIC AND IMHOFF TANKS

SEPTIC TANK DESCRIPTION

A settling tank provides treatment for raw sewage. Flow-through settling and sludge digestion take place in the same chamber. Treatment is anaerobic.

IMHOFF TANK DESCRIPTION

A settling tank provides treatment for raw sewage. The flow-through settling chamber is separate from the quiescent sludge digestion compartment. Sludge digestion is anaerobic.

CAPACITIES (GALLONS PER DAY [gpd])

Septic tanks: 500 to 10,000 gpd
Imhoff tanks: 3000 to 300,000 gpd
(These tanks are used for larger installations.)

LOADING RATES TO DISTRIBUTION ON SAND FILTERS

0.1 to 0.15 mgd per acre of Imhoff effluent

EFFICIENCIES

Septic tanks: 15 to 25% removal of BOD and 40 to 60% SS removal.
Imhoff tanks: 25 to 35% BOD removal and 40 to 60% SS removal.

MATERIALS OF CONSTRUCTION

Septic tanks: usually concrete (often prefabricated), occasionally prefabricated steel
Imhoff tanks: concrete

ADDITIONAL TREATMENT REQUIRED

Septic tanks: usually subsurface drainage fields, occasionally intermittent sand filters or lagoons
Imhoff tanks: usually trickling filters, occasionally intermittent sand filters or lagoons

DISPOSAL OR TREATMENT OF SLUDGE

Septic tanks: normally to municipal treatment plant or burial by private contractor
Imhoff tanks: usually sludge drying beds

PRIMARY USE

Septic tanks: dwellings, camps, and small institutions
Imhoff tanks: communities and small cities

Treatment Characteristics

Raw sewage can be treated in either septic tanks or Imhoff tanks.

SEPTIC TANKS

Septic tanks receive raw sewage, allow it to settle, and pass the relatively clear liquid to the adsorption field, which is the next stage of treatment. The remaining solids digest slowly in the bottom of the tank. Septic tanks are inexpensive, but because of their incomplete treatment, they are suitable only for small flows.

Anaerobic decomposition, which takes place in the absence of free oxygen in a septic tank, is a slow process. To maintain practical detention times, the reactions cannot be carried far. Therefore, the effluent is often malodorous, containing a multitude of microorganisms and organic materials that require further decomposition.

IMHOFF TANKS

The process that takes place in an Imhoff tank is similar except that the tank is designed so that the flow-through upper chamber is separate from the lower digestion chamber, resulting in a two-story tank. The upper compartment acts only as a settling zone where little or no decomposition takes place. This chamber often remains aerobic, and its effluent has a lower BOD than the effluent from a septic tank. Anaerobic digestion takes place in the lower chamber. Because the effluent is a higher quality, the process is suitable for communities and small cities. Additional treatment for further decomposition of organic matter in the effluent is required.

Septic Tank Design

In septic tank design, environmental engineers must consider the treatment following the septic tank as a part of the septic tank system (U.S. Department of Health, Education, and Welfare 1967). A two-compartment design arranged in a series is preferred (see Figure 7.20.1). The first chamber should contain two-thirds and the second chamber should contain one-third of the total volume. The liquid depth should be between 4 and 7 ft.

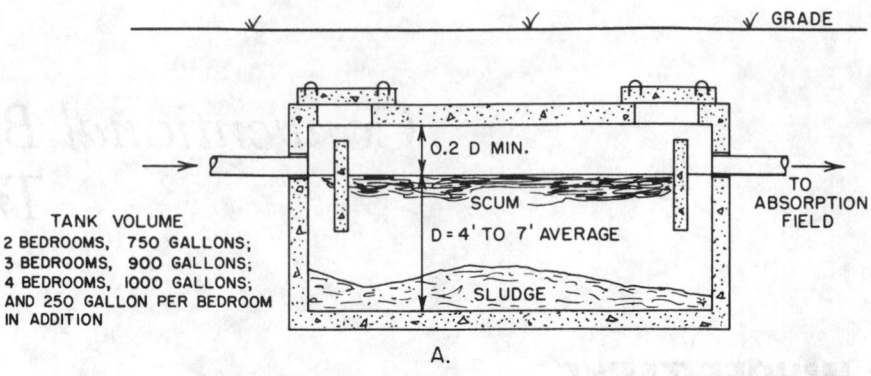

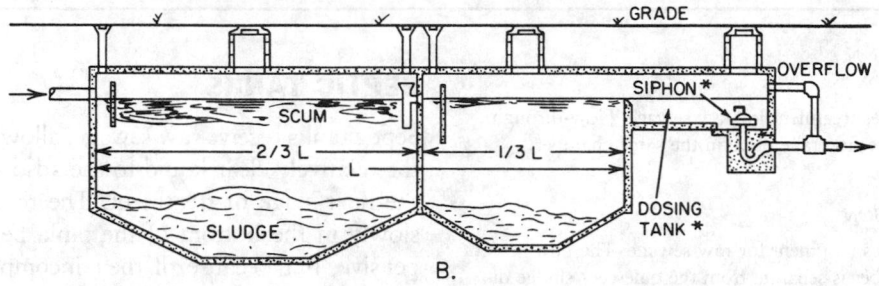

FIG. 7.20.1 Septic tank configurations. **A.** Typical household septic tank; **B.** Typical large institutional septic tank with dosing siphon. For large fields, uniform distribution is obtained by periodic flooding of the field followed by periodic drying. Dosing tanks are used to flood these fields; they collect the sewage, and automatic bell siphons or pumps transport the waste to the field.

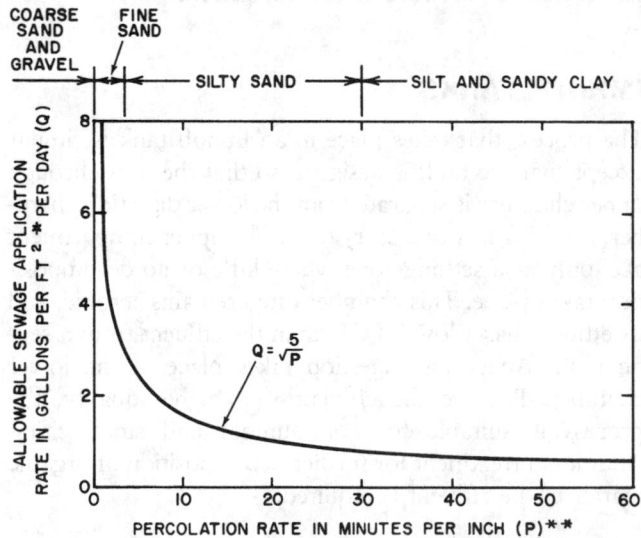

* SURFACE AREA OF ABSORPTION TRENCHES (TRENCH BOTTOM AREA PLUS ALLOWANCE FOR AREA OF SIDEWALLS.)

** MINUTES REQUIRED FOR THE WATER IN THE TEST HOLE TO RECEDE ONE INCH.

FIG. 7.20.2 Relationship between allowable sewage application rate and soil percolation rates for soil absorption trenches or seepage pits. (Reprinted from U.S. Department of Health, Education, and Welfare, 1967, *Manual of septic tank practice*, Public Health Service Publication no. 526, Washington, D.C.)

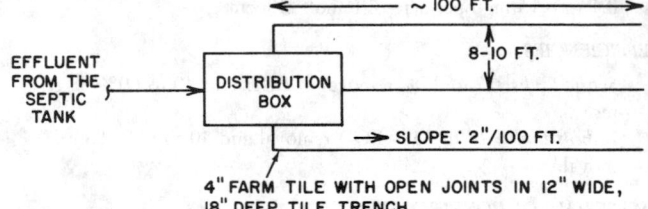

FIG. 7.20.3 Septic tank absorption field. Trench surface area required: If the water in the test hole takes 1 min to recede 1 in, 70 ft² per bedroom is needed. If the water in the test hole takes 30 or 60 min, 250 or 330 ft² per bedroom is required.

The minimum effective tank capacity should be as follows: for flows up to 1500 gpd, $1\frac{1}{2}$ times the daily sewage flow; for flows in excess of 1500 gpd, the volume V in gallons can be calculated from the following equation:

$$V = 1125 + 0.75 \, Q \qquad 7.20(1)$$

where:

Q = The daily sewage flow

Environmental engineers must check soil porosity and base the design of the soil absorption field on the rate of percolation. This rate is the number of minutes required for the effluent to recede 1 in in a test hole that has been

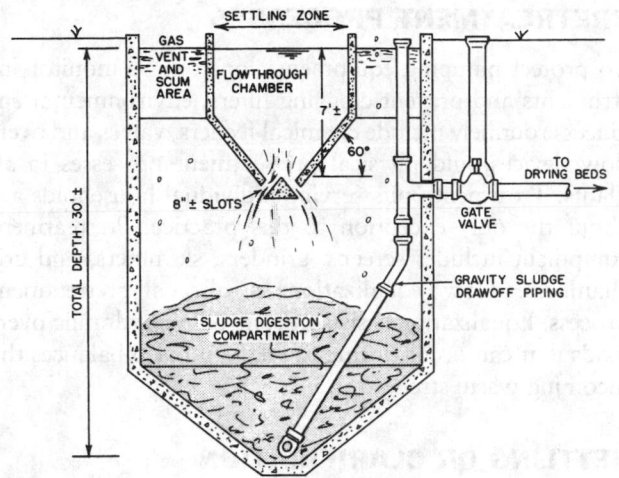

FIG. 7.20.4 Imhoff tank configuration.

bored, filled with water, and allowed to swell the day previous to the test (see Figure 7.20.2). Figure 7.20.3 provides information on absorption area requirements.

If the septic tank must handle wastewater from garbage grinders and automatic washing machines, the soil absorption area (see Figure 7.20.3) should be increased by 60%, and the tank volume should be increased (see Part A in Figure 7.20.1) by 25%. If the soil absorption trench required exceeds 500 ft^2 or the septic tank is larger than 1500 gal, a dosing tank is needed.

IMHOFF TANK DESIGN

Surface loading of the settling zone should be 600 gpd per ft^2 with detention times of 1½ to 2 hr and velocities below 0.75 in per sec. The effective settling zone depth should be about 7 ft and its length can be from 25 to 50 ft. The gas-vent and scum area should be 20% of the total surface area. Total depths average around 30 ft (see Figure 7.20.4).

Septic tanks are suitable only for isolated facilities with low waste flows where the soil can be used as an absorption field. Their use should be avoided except when an alternative is not available and the site conditions are favorable.

The operation of Imhoff tanks is not complex. They are less efficient than settling basins and heated-sludge digestion tanks. The newer treatment methods offer more efficient alternatives to Imhoff tanks, but in small treatment units, they do provide efficient solids separation without mechanical or electrical equipment.

—*R.D. Buchanan*

Reference

U.S. Dept. Health, Education, and Welfare. 1967. *Manual of septic tank practice*. Public Health Service Publication. No. 526. Washington, D.C.

7.21
CONVENTIONAL SEWAGE TREATMENT PLANTS

This section describes a conventional sewage treatment plant, emphasizing the total plant concept.

Conventional plants are best identified by what they do not achieve, namely nutrient removal, demineralization, and the removal of trace organics. Therefore, the conventional plant's performance is usually measured by reductions in suspended matter, BOD, and bacteria.

The processes in conventional plants include 1) pretreatment, 2) settling, 3) chemical treatment, 4) biological oxidation, 5) disinfection, and 6) sludge conditioning and disposal processes. Figure 7.21.1 shows the interrelationships among these processes. This section does not cover plants that treat special industrial wastes by processes other than those used for domestic waste. In addition, an individual municipal treatment plant may not use all processes discussed in this section.

Selection of Specific Processes

In the selection of specific processes and plant design, environmental engineers must consider the existing and potential regulatory standards; plant operators' requirements and availability; existing and projected sewage flow; flow pattern and waste characteristics; climate, topography, and availability of land; plant location within the community; and all aspects of cost. They must also assess the life-cycle cost of the plant. Lower construction costs can frequently be offset by high maintenance and increased operation costs.

The plant design cannot be optimized for all of these factors, consequently the environmental engineer must make difficult decisions in the face of uncertainty. The problem is complicated because the design life of the fa-

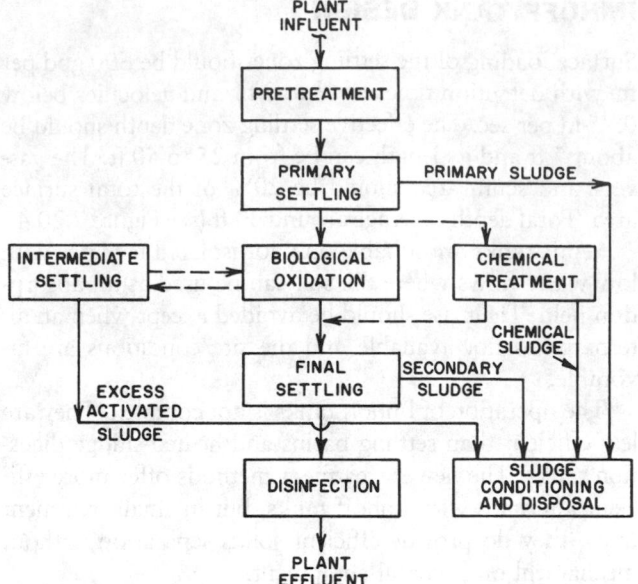

FIG. 7.21.1 Conventional domestic wastewater treatment process flow sheet.

cility is normally twenty-five years or more. A typical approach is for the design to minimize cost while achieving a treatment level for a set of constraints represented by the legal and physical aspects of the project.

Although many treatment plants, particularly large facilities, to be constructed include advanced forms of wastewater treatment, most plants in operation use conventional processes. For several more decades, these plants will serve a large percentage of the population. Upgrading existing facilities offers a promising alternative for improving treatment compared to completely new construction and the total loss of an existing plant's capabilities.

A principal consideration in overall plant design is to provide flexibility for expansion and upgrading of the initial plant. Flexibility for efficient plant operation allows the operator to overcome problems and provide maintenance and repair with minimum effect on plant performance. The design should also include the capability for implementing new ideas that may improve plant performance. Even minor modifications are sometimes difficult and more costly to make than the same capability built into the initial design and construction.

Anticipating plant expansion can also save money and avoid disruption. Land availability and wise plant layout are two basic considerations. Space allowances in buildings for future needs—pumps, instrumentation, and pipelines in the initial construction—are usually only a minor cost factor. For a small additional cost, environmental engineers can select chemical feeders and chlorinators for the initial project with sufficient capacity to handle anticipated increased rates. Often, the tanks designed for one purpose can be effectively used for a different process and accommodate plant expansion or upgrading.

PRETREATMENT PROCESSES

To protect pumping equipment, control and monitor instruments and prevent clogging filters, environmental engineers routinely include chemical feeders, valves and overflow devices, and physical pretreatment processes in all plants. Treatment units serving individual households are about the only exception to this practice. Pretreatment equipment includes screens, grinders, skimmers, and grit chambers. Flow equalization is also a pretreatment process. Equalization assists in controlling hydraulic overloads that can occur during the day and also balances the incoming waste strength.

SETTLING OR CLARIFICATION

Settling processes remove settleable solids by gravity settling either prior to or after biological or chemical treatment and between multiple-stage biological or chemical treatment steps. In larger tanks, mechanical scrapers accumulate the solids at an underflow withdrawal point, whereas in smaller and some older systems, a hopper bottom is used for solids collection. Solids move down the sloped tank bottom by gravity in hopper-bottom tanks. Both circular and rectangular tank shapes are used. A rectangular or square tank uses the land area more efficiently and environmental engineers can save construction costs by nesting units and using common walls. With circular tanks, this cannot be done.

Settling tanks are commonly designed based on the overflow rate, the unit volume of flow per unit of time divided by the unit of tank area (gallons per day per square foot). Typical overflow rates are 600 gpd per ft^2 for primary settling, 1000 gpd per ft^2 for intermediate settling, 800 to 1000 gpd per ft^2 for final clarifiers after activated-sludge units, and 700 to 1000 gpd per ft^2 for final clarifiers after trickling filters. The detention times for settling range from 1 to 2.5 hr for average flows depending on the processes before or after the settling step.

CHEMICAL TREATMENT

Traditional chemical precipitation uses either iron or aluminum salts to form a floc, which is then settled. Lime also clarifies. This process step can reduce the SS up to 85%. The accumulated chemical sludge is removed by gravity flow or pumping to conditioning or disposal or both. The chemicals and sewage are flash-mixed in a mixing tank that has only a few minutes detention time followed by 30 to 90 min detention in a flocculation tank that is slowly agitated to aid floc growth. As shown in Figure 7.21.1, settling follows the flocculation tank.

BIOLOGICAL OXIDATION

Two basic techniques, fixed-bed and fluid-bed, are used in conventional biological treatment. The trickling filter has a fixed-bed of stone or plastic packing material that provides a growth surface for zoogleal bacteria and other organisms. The intermittent-sand filter and the spray irrigation system are other examples of the fixed-bed technique.

The activated-sludge processes and sewage lagoons are fluid-bed systems. The activated-sludge process uses mechanical aeration and returns a percentage of the active sludge to the process influent. Lagoons or stabilization ponds and oxidation ditches do not routinely waste sludge, but multipond systems can have recirculation. Septic tanks and Imhoff tanks combine the settling and biological oxidation processes in a single tank.

Activated Sludge

Activated-sludge processes use continuous agitation and artificially supplied aeration of settled sewage together with recirculation of a portion of the active sludge that settles in a separate clarifier back to the aeration tanks. These processes vary in detention time, the method of mixing and aeration, and the technique of introducing the waste and recirculated sludge into the aeration tank.

Figure 7.21.2 is a conventional activated-sludge plant flow diagram. A return of activated sludge at a rate equal to about 25% of the incoming wastewater flow is normal; however, plants operate with recirculation rates from 15 to 100%. The mixture of primary clarifier overflow and activated sludge is called *mixed liquor*. The detention time is normally 6 to 8 hr in the aeration tank.

In a conventional plant, the oxygen demand is greatest near the influent end of the tank and decreases along the flow path. Plants built before the process was well understood provided uniform aeration throughout the tank. A conventional plant cannot accommodate variations in hydraulic and organic loadings effectively, and the final clarifier must be sized to handle a heavy solids load. Usually aeration units are in parallel so that a shutdown of one unit does not totally disrupt plant operation. Modifications have evolved as the activated-sludge plant has become more widely used and are described in the following paragraphs.

One technique that furnishes more uniform oxygen demand throughout the aeration tank is introducing the primary settled waste at several points in the aeration tank instead of at a single point as in the conventional process. This modification is *step aeration*, and Figure 7.21.3 is a typical flow diagram. The percentage of settled, activated sludge returned to the aeration tank is usually greater than in the conventional process (about 50% typically), and the detention time is reduced to 3 or 4 hr since the loading is more evenly distributed in the tank. Additional piping and pumps are required to distribute the waste to several locations; however, the improved performance is considered to be worth the expense.

A less popular alternative to distributing the load to the aeration tank is to provide different quantities of oxygen along the tank length, related to the oxygen demand that gradually decreases along the tank length. The flow sheet for *tapered aeration* is the same as that in Figure 7.21.2. The disadvantage of tapered aeration is that although it is more economical due to reduced air quantities, it can only be designed for one loading.

Figure 7.21.4 shows the *extended aeration* treatment process. In extended aeration, as the name implies, the activated-sludge detention time is increased by a factor of 4 or 5 compared to conventional activated sludge. The pri-

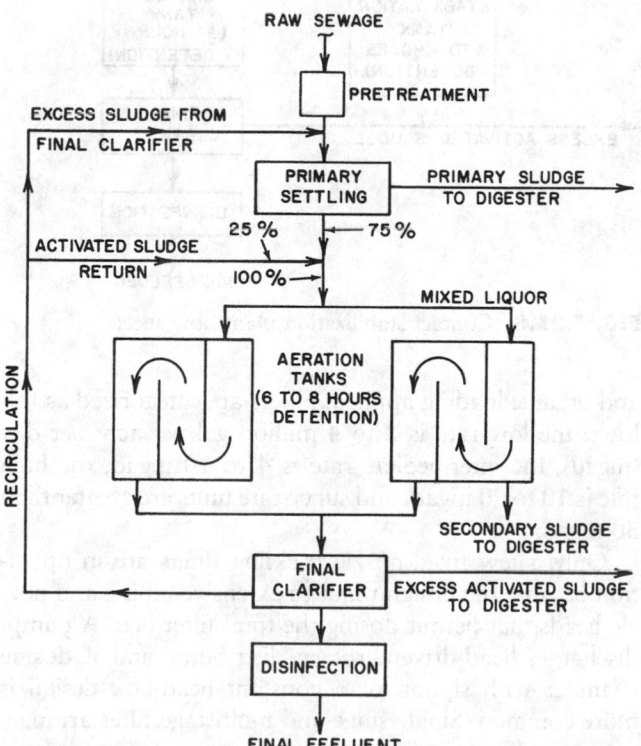

FIG. 7.21.2 Conventional activated-sludge plant flow sheet.

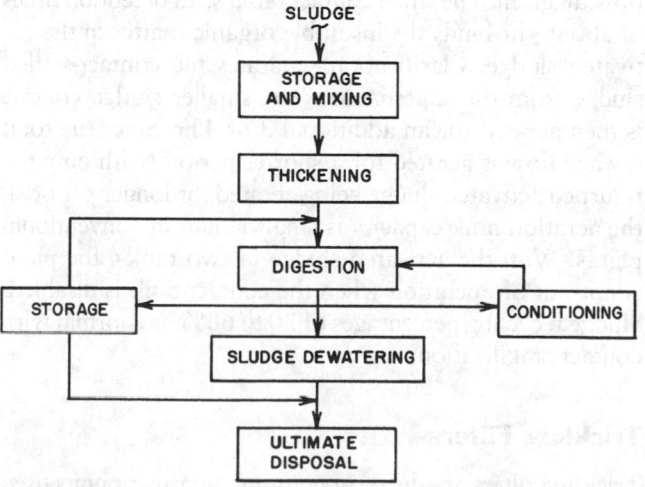

FIG. 7.21.3 Step-aeration type activated-sludge plant flow sheet.

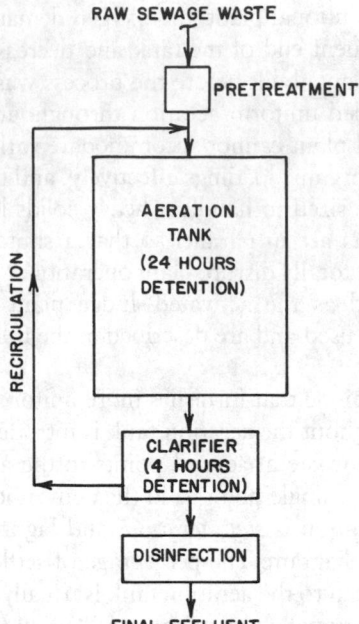

FIG. 7.21.4 Extended-aeration plant flow sheet.

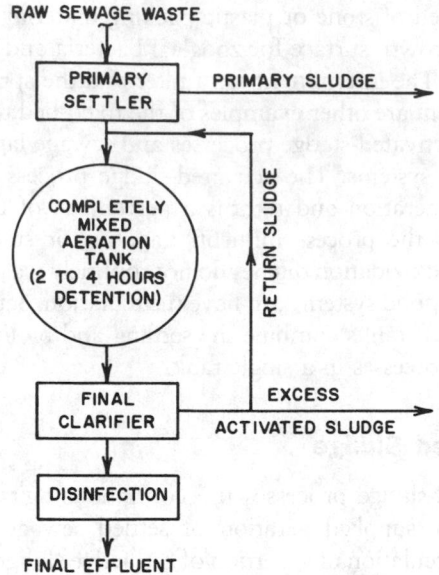

FIG. 7.21.5 Completely mixed, activated-sludge treatment flow sheet.

mary clarifier is eliminated. At a surface settling rate of 350 to 700 gpd per ft², 4 hr final settling is typical. The extended-aeration period reduces or eliminates the requirement for disposing excess sludge and is therefore a popular system for small plants.

Figure 7.21.5 is a flow diagram for the *completely mixed, activated-sludge* system. This process is an extension of step aeration and provides a uniform oxygen demand throughout the aeration tank. Mechanical aerators also provide mixing for this unit. The SS concentration in the mixed liquor is two to three times the concentration in most conventional plants. Aeration detention times are reduced to 2 to 4 hr. The sludge recycling ratio is generally high because the greater flow improves mixing.

Figure 7.21.6 is a typical *contact stabilization* process flow diagram. The small contact tank, with detention times of about ½ hr binds the insoluble organic matter in the activated sludge. Clarification separates the contact-settled sludge from the supernatant. The smaller sludge volume is then aerated for an additional 3 or 4 hr. Since the total sewage flow is aerated for a shorter period (with only the returned activated sludge being aerated for longer periods), the aeration tank capacity is smaller than in conventional plants. With the activated sludge in two tanks, the plant is not out of operation when the contact tank is disabled. Sludge recycling percentages of 30 to 60% are normal with contact stabilization.

Trickling Filters

Trickling filters are the most common treatment units used by municipalities to provide aerobic biological treatment. Trickling filters are classified according to the hydraulic

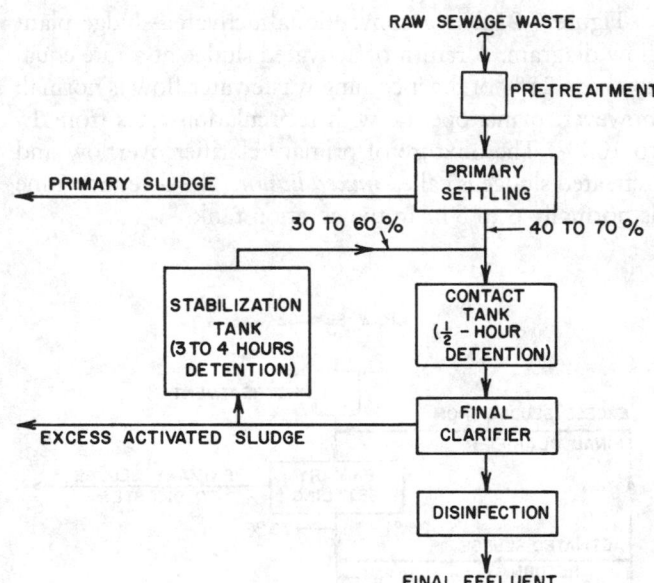

FIG. 7.21.6 Contact stabilization plant flow sheet.

and organic loading applied. Filters are categorized as follows: the low rate is 2 to 4 million gal per acre per day (mgad), the intermediate rate is 4 to 10 mgad, the high rate is 10 to 30 mgad, and super-rate units are greater than 30 mgad.

Only a few fixed-nozzle trickling filters are in operation because the design requires extensive piping and nozzle heads that permit dosing the total filter bed. A pump-discharge, head-driven, rotary distributor and a dosing chamber with siphon or a constant head-box design is more common. Single-stage and multistage filter arrangements are both used because many recirculation schemes and options of intermediate settling between multistage filters are available. The recirculation method is much less a factor in plant performance than the recirculation ratio.

Figure 7.21.7 shows the more common single-stage filter recirculation flow diagrams. Sludge from the final clarifier is usually recirculated to a point before the primary settling tank. The recirculation flow is also taken from in front of and behind the final clarifier to a point either before or after the primary settling. All units must be designed for total hydraulic flow and organic loading.

Figure 7.21.8 shows several flow routings used for multistage filters. All sludge returned from the intermediate and final clarifiers that is not wasted (excess) is returned to a point before the primary settling tanks.

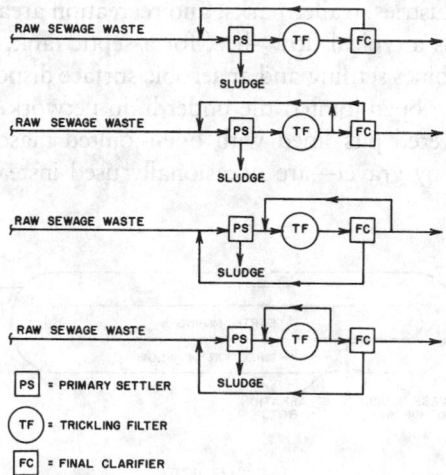

FIG. 7.21.7 Typical single-stage, trickling filter, recirculation flow sheets.

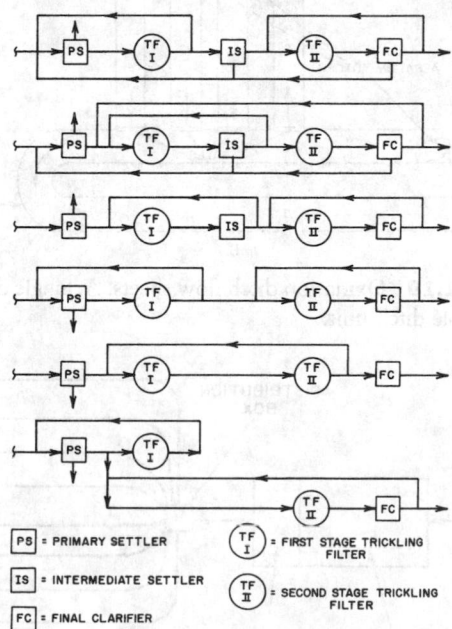

FIG. 7.21.8 Typical multistage, trickling-filter, recirculation flow sheets.

Lagoon and Oxidation Ponds

Lagoons and ponds have many applications ranging from complete raw waste treatment to polishing a secondary plant's effluents. The applications have certain characteristics in common: they are each engineer-designed and uncovered and do not use metal or concrete tanks. A *lagoon* is a pond of engineering design that receives waste that has not been settled or exposed to biological oxidation prior to entering it.

Figure 7.21.9 is a simple flow-sheet representing the raw sewage lagoon. Simplicity is the main feature of the raw sewage lagoon. Since it is constructed by excavation and diking, it is a low-cost system that can be constructed rapidly. Operator attention is minimal, and the flow through the system is usually by gravity unless recirculation is provided. The raw sewage lagoon usually has a bar screen placed in the influent and can have a Parshall flume with a drum recorder to determine the inflow to the lagoon.

Recirculation can reduce the buildup of bottom solids near the inflow entrance point into the pond. The raw sewage pond is usually a facultative aerobic system, which means that anaerobic conditions exist at and near the bottom and aerobic conditions prevail in the upper layers of the pond most of the time. Facultative organisms can function under either aerobic or anaerobic conditions.

A series of ponds is frequently used when it comprises the sole treatment. The number and size of the ponds are functions of the effluent quality, incoming waste load, temperature, and climate. Part B in Figure 7.21.9 shows a mul-

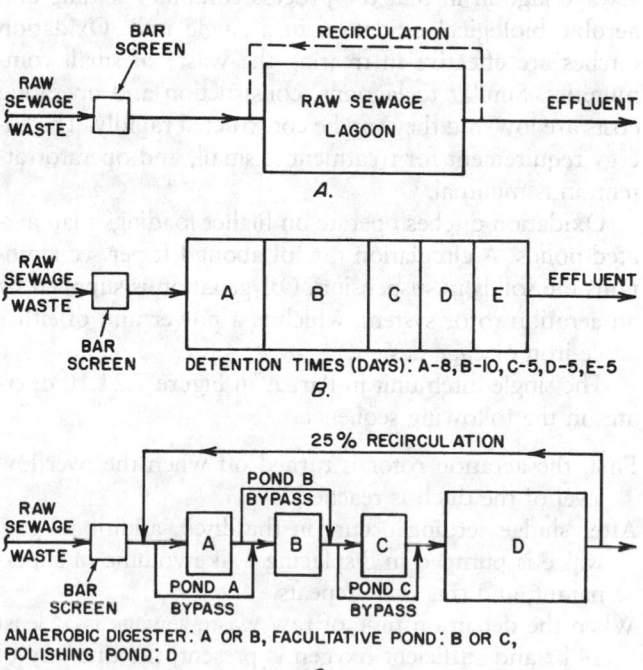

FIG. 7.21.9 Raw sewage lagoon flow sheets. **A,** Single pond system; **B,** Multipond facultative aerobic lagoon system; **C,** Anaerobic-aerobic pond system.

tipond facultative system flow sheet with the corresponding detention times.

The primary pond designed as an anaerobic pond is becoming more popular. Part C in Figure 7.21.9 is a typical flow sheet for an anaerobic–aerobic pond system. Ponds A, B, and C are each anaerobic ponds, and the flow arrangement provides flow through any two of the anaerobic ponds in the series (AB, AC, and BC). This arrangement permits one pond to serve as an anaerobic digester.

The second anaerobic pond produces a higher quality effluent than does a single pond, thus reducing the load and size of the facultative pond. An anaerobic pond is normally used for six months to a year as the anaerobic digester. A pumped recirculation of about 25% of the total flow is common. The raw sewage lagoon can have additional ponds (D in Figure 7.21.9) in the series after the facultative pond for additional polishing treatment.

The oxidation pond, as opposed to the raw sewage lagoon, receives influent that has undergone primary treatment. A maturation pond provides a final, polishing treatment step that follows some form of secondary treatment. Therefore, the maturation pond is a form of tertiary treatment. Mechanically aerating the oxidation pond improves treatment and reduces the pond size. When mechanical aeration is provided, floating surface aerators are almost universally used. The series flow in ponds buffers against shock loadings.

Oxidation Ditch

The oxidation ditch is a variation of the aerated raw sewage lagoon in that the process combines settling and aerobic biological oxidation in a single unit. Oxidation ditches are effective in treating the waste of small communities. Similar to lagoons, construction and operating costs are low and they can be constructed rapidly. The energy requirement for treatment is small, and operator attention is minimal.

Oxidation ditches operate on higher loadings than aerated ponds. A circulation rate of about 1 ft per sec maintains the solids in suspension. Oxygenation is supplied by an aeration rotor system, which is a power unit of either angle-iron or cage design.

The single-ditch unit in Part A in Figure 7.21.10 operates in the following sequence:

First, the aeration rotor is turned off when the overflow level of the ditch is reached.
After sludge settling occurs in the ditch, additional raw waste is pumped in displacing a like volume of supernatant, and this cycle repeats.
When the detention time of raw waste sewage is at least 24 hr and sufficient oxygen is present, the quantity of excess sludge is small.

Part B in Figure 7.21.10 shows the multiple-ditch configuration. Ditches B and C are alternately used for set-

tling, while ditch A operates continuously. When ditch B is used for settling, the gates connecting pond A with B are closed, and the aeration rotor in ditch B is shut off. When the ditch is not used for settling, the aeration rotor is turned on, and the ditch functions in an auxiliary treatment capacity. After settling occurs in either ditch B or C, the gates to the ditch are opened, and the supernatant is discharged as in the single-ditch unit. After the supernatant is discharged, the settled sludge in the ditch is resuspended and distributed by the aeration rotor in that ditch.

Septic Tank

The septic tank continues to serve as the wastewater disposal system for millions of households and numerous small industries, trailer parks, and recreation areas. Figure 7.21.11 is a typical flow sheet for a septic tank. The system combines settling and anaerobic surface disposal, usually by an open-jointed tile underdrain network. Seepage pits—covered pits lined with open-jointed masonry surrounded by gravel—are occasionally used instead of the

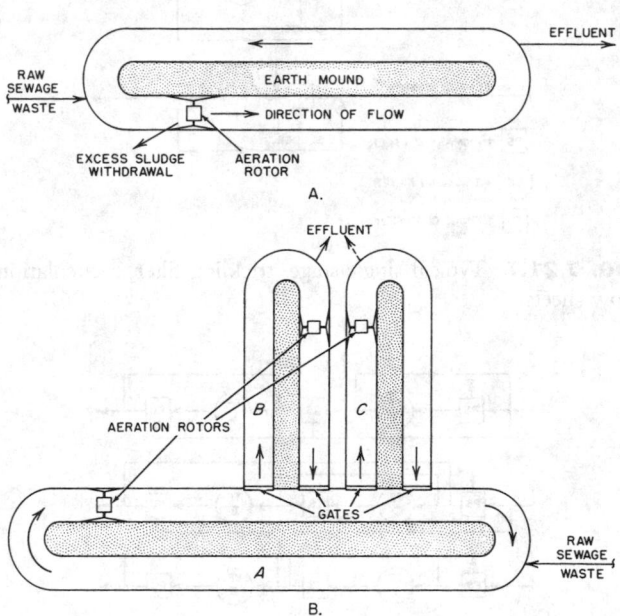

FIG. 7.21.10 Oxidation ditch flow sheets. A, Single ditch unit, B, Multiple ditch unit.

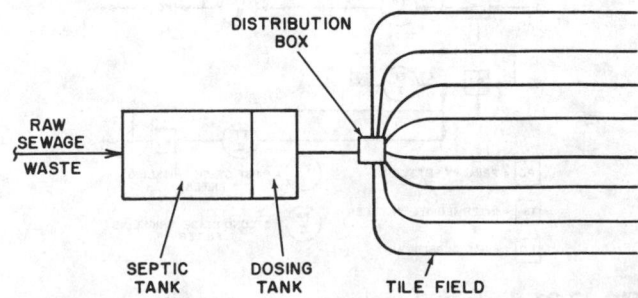

FIG. 7.21.11 Septic tank and disposal field flow sheet.

tile field for disposal. At least two seepage pits are provided, and they are alternately dosed.

The function of the dosing tank is to furnish a sufficient flow rate to use the full tile field or seepage pit. When a dosing chamber is absent, the head reach of the field tends to become overloaded. Since the septic tank almost always operates without power, an automatic siphon is used for dosing by discharging the chamber contents each time the level reaches a fixed point. The distribution box proportions sewage flow between the individual tile lines.

Imhoff Tank

The Imhoff tank, a two-story tank, uses the upper chamber for settling and the lower chamber for sludge digestion. It can be followed by additional process steps to improve plant effluent quality. The Imhoff-tank–intermittent-sand-filter combination (see Figure 7.21.12) were popular for small municipalities prior to the wide use of the trickling filter and activated-sludge processes. Some units serve homes and recreation areas.

The tank has the advantage of being a nonmechanical device; however, it does require deep excavation unless it is built above ground, which requires more expensive construction and pumping of the raw sewage. Since heating of the digester is uneconomical due to the heat losses in the settling section of the tank, the digester is sized for unheated operation. For uniform sludge distribution to the digestion section of the tank, the tank periodically reverses

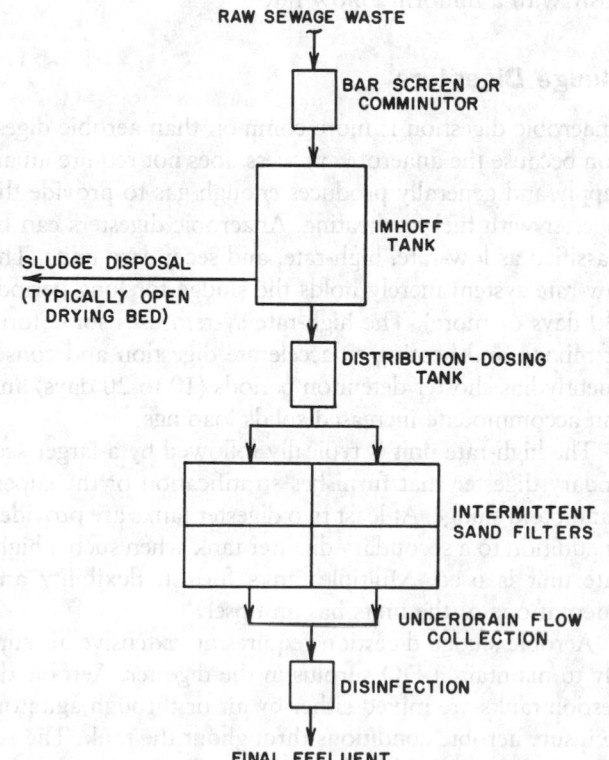

FIG. 7.21.12 Imhoff-tank–sand-filter flow sheet.

the flow pattern. Both the tank bottom and the division between the settling and digestion sections have steep slopes so that the sludge can slide down the walls.

Sludge is normally withdrawn through a pipeline to open sludge drying beds by hydrostatic pressure. The gas produced by digestion is inhibited from entering the settling chamber by an overlap on the chamber dividing walls. It escapes through ventilation compartments that make up at least 20% of the tank surface area. Gases are usually not collected because the digester is not heated and no fuel is needed. A rectangular tank is the most common, although circular tanks are also used.

Intermittent Sand Filters

Although the intermittent sand filter was popular in small plants early in this century, it is being phased out. Figure 7.21.12 shows a flow diagram of an intermittent sand filter that furnishes additional treatment to an Imhoff tank effluent. Because both sand filters and Imhoff tanks were popular during the same period, their combined use was also common. Requirements for better quality plant effluents and the availability of other processes that do not require as much operator attention or produce the objectional odors associated with sand filters have influenced their being phased out.

Even small plants must use several sand filter units to permit time for drying, cleaning, and replacing the media. They frequently use a dosing tank with an automatic siphon to dose the filter in use, usually two to four times a day. A loading rate of about 100,000 to 150,000 gal per acre per day is practical for Imhoff tank effluent. Where more effective treatment precedes the sand filters, the application rate can be increased. Selecting the sand filter to be dosed is usually a manual operation controlled by valves.

A sand filter usually consists of 3 to 4 in of sand laid over 6 to 12 in of gravel. Tile underdrains collect the effluent, which is discharged to the receiving waters. A uniform flow distribution to sand filters is important and is usually achieved by trough distribution or a rotary-arm distribution device.

DISINFECTION

Conventional treatment plants use chlorination as the final treatment process to reduce bacteria concentration. Prechlorination, performed on the plant influent, is used if the incoming sewage is septic or the flows are low and the holdup time in the plant is long enough that the waste can become septic.

Prechlorination is usually at a fixed dosage, and a residual chlorine level is not maintained. Postchlorination provides 15 to 30 min detention time in a baffled, closed tank to prevent short-circuiting and dissipation of the chlorine. Wastewater treatment facilities do not frequently use chlorination to reduce the BOD in the effluent. Ordinarily, a combined chlorine residual between 0.2 and 1 mg/l is the target for the final effluent.

SLUDGE THICKENING, CONDITIONING, AND DISPOSAL

Settled solid material from various treatment processes is called *sludge*. For excess quantities of sludge to be disposed of economically and with minimal objections, the raw sludge is digested and dewatered. Controlled digestion reduces the quantity of complex organic material, increases the number of ultimate sludge disposal alternatives, and decreases the undesirability of the sludge.

Septic tanks and lagoons are not supplied with separate digesters or sludge conditioning equipment. Imhoff tanks can have a separate digester depending on unit size, waste and effluent character, and plant location. Disposing excess sludge from these units involves either periodic manual cleaning (septic tanks and lagoons) or, more commonly, pumping or the gravity-flow transfer of the sludge to open drying beds for Imhoff tanks. Clarifiers, activated-sludge-type units, and trickling filters all require separate sludge digesters and/or disposal systems.

Sludge Storage

Wastewater treatment facilities route excess quantities of activated sludge and sediments from primary, intermediate, and final clarifiers through some or all of the steps outlined in Figure 7.21.13. Plants can have facilities for disposal only; dewatering and disposal; digestion, dewatering, and disposal; or thickening, digestion, dewatering, and disposal. Wet sludge quantities, depending on the waste and treatment processes, can constitute as much as 1.5% of influent flow.

Some form of short-term storage before the digestion step is essential to prevent overloading, regulate sludge flow to the digesters, and allow collection and mixing of the sludge. Some plants mix some final effluent with the sludge (in the mixing tank) to improve its thickening characteristics. Storage tanks should be open so that gas does not buildup in the tank.

Sludge Thickening

Gravity thickening in a deep, circular, open tank is frequently used before anaerobic digestion. Some wastewater treatment facilities also use polyelectrolytes to improve gravity thickening, but their use for this purpose is not widespread.

Air flotation thickening of activated sludge is an alternative. This process uses dissolved or diffused air and sometimes a coagulant to float the sludge to the surface where it is removed by a mechanically driven skimmer. The effluent drawn near the bottom of the flotation tank

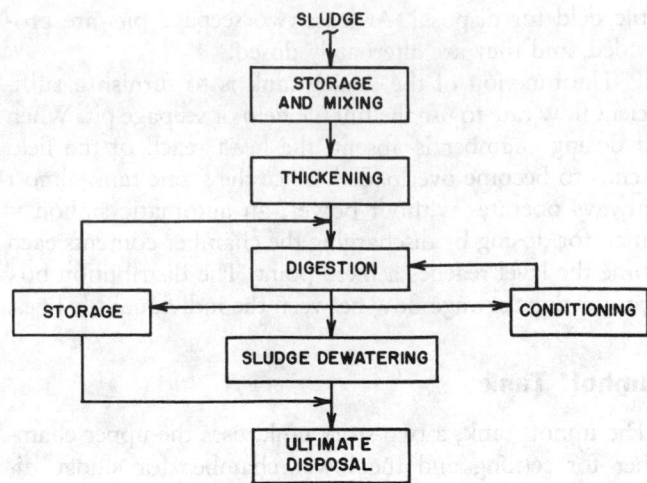

FIG. 7.21.13 Process flow sheet for sludge digestion and disposal.

is combined with incoming plant influent and passes through the treatment cycle. Most flotation units recycle a portion of the effluent through the flocculation-air chamber because the recycled liquid enhances flocculation.

Although flotation is effective on activated sludge because of its density, it is not used on clarifier sludges because they are more difficult to float.

The primary purpose of mixing tanks is to provide a satisfactory mixture of clarifier sludge and activated sludge when a single thickening process is used. A mixing–storage tank also permits the thickener to operate continuously with a uniform inflow rate.

Sludge Digesters

Anaerobic digestion is more common than aerobic digestion because the anaerobic process does not require an air supply and generally produces enough gas to provide the digester with fuel for heating. Anaerobic digesters can be classified as low-rate, high-rate, and secondary units. The low-rate system merely holds the sludge for long periods (30 days or more). The high-rate system uses some form of mixing and heating to accelerate digestion and consequently has shorter detention periods (10 to 20 days) and can accommodate increased solids loadings.

The high-rate unit is typically followed by a larger secondary digester that furnishes stratification of the supernatant and sludge. At least two digester tanks are provided in addition to a secondary digester tank when such a high-rate unit is used. Multiple tanks furnish flexibility and safety if one of the units has an upset.

Aerobic sludge digestion requires an extensive air supply to maintain a DO surplus in the digester. Aerobic digestion tanks are mixed either by air or through agitators to insure aerobic conditions throughout the tank. The supernatant from aerobic digestion, similar to an anaerobic system, is mixed with the raw plant influent. The advan-

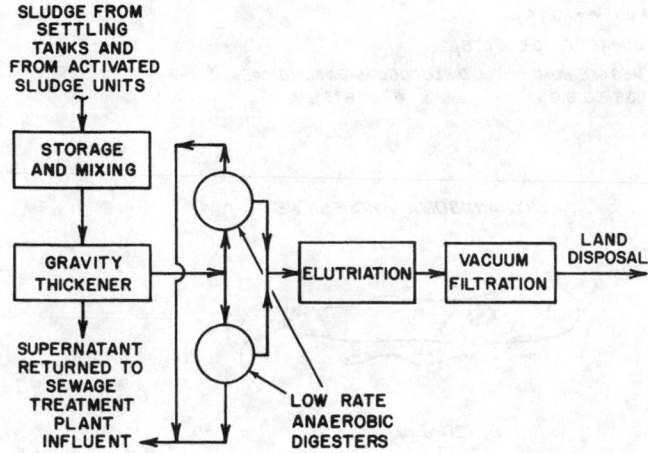

FIG. 7.21.14 Conventional sludge disposal with gravity thickener, elutriation, and vacuum filtration.

tages of aerobic tanks is that they are not subject to upsets and produce a more treatable supernatant. The cost of air makes the operating cost of aerobic digestion higher than that of anaerobic digestion.

Sludge Conditioning

Elutriation and chemical addition are the two sludge conditioning alternatives. Conditioning is an intermediate process between primary and secondary anaerobic digesters. It also improves dewatering of digested sludge.

Iron, aluminum salts, and lime are the most common chemicals for conditioning.

Elutriation is the washing of suspended sludge held in suspension by air or stirring. It reduces alkalinity and makes the sludge more filterable. Single and multitank (countercurrent) elutriation are both used. The single-tank system uses repeated sludge washing, whereas in the countercurrent system, fresh wash water is added to the last sludge tank. The wash water overflow from the last-stage tank furnishes the wash for the previous tank. Although the countercurrent system requires additional tanks and piping, it uses less makeup wash water.

Sludge Dewatering

Rotating-drum vacuum filters are the conventional sludge dewatering equipment. The filtrate is returned to the plant influent, and the sludge cake is disposed. Sludge drying beds are common in smaller plants for disposal. Weather conditions are an important factor when open drying beds are used. Some wastewater treatment facilities use glass-covered drying beds to exclude precipitation. Drying beds have underdrain networks of tile fields laid in sand and gravel. The discharge from this network should be disinfected when it is not returned to the treatment plant. Sludge treatment and disposal are usually performed in separate treatment plants (or plant sections). Figure 7.21.14 shows a typical sludge treatment plant.

—L.H. Reuter

Secondary Treatment

7.22
WASTEWATER MICROBIOLOGY

The objectives of biological wastewater treatment are to coagulate and remove nonsettlable colloidal solids and to stabilize organic matter. For example, in municipal wastewater treatment, the objective is usually to reduce organic content and, if necessary, to remove nutrients such as nitrogen and phosphorus. In some cases, trace concentrations of toxic organic compounds also require removal.

In industrial wastewater treatment, reduction or removal of organic and inorganic compound concentrations is essential. Microorganisms (see Figure 7.22.1) play a ma-

jor role in decomposing waste organic matter, removing carbonaceous BOD, coagulating nonsettlable colloidal solids, and stabilizing organic matter. These microorganisms convert colloidal and dissolved carbonaceous organic matter into various gases and cell tissue. The cell tissue, having a specific gravity greater than water, can then be removed from treated water through gravity settling. Thus, wastewater treatment facilities use these microorganisms in biological wastewater treatment processes to dispose of wastes in a nontoxic and sanitary manner.

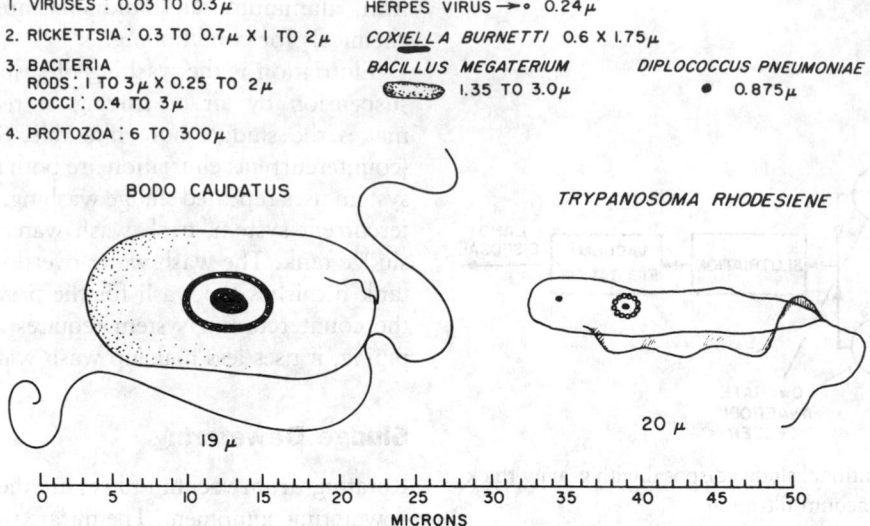

FIG. 7.22.1 Species from classes of organisms.

Nutritional Requirements

This section discusses the fundamentals of wastewater microbiology by examining microorganisms' nutritional requirements, enzymic reactions involved in their activities, environmental parameters affecting their growth and activities, and microbial groups associated with various biological wastewater treatment processes.

Nutritional Requirements

For microorganisms, nutrients (1) serve as an energy source for cell growth and biosynthetic reactions, (2) provide the material required for synthesis of cytoplasmic materials, and (3) serve as acceptors for the electrons released in energy-yielding reactions. To sustain reproduction and proper function, microorganisms require an energy source, a carbon source for synthesis of new cellular material, and inorganic nutrients such as nitrogen, phosphorus, sulfur, potassium, calcium, and magnesium. In addition, organic nutrients (growth factors) may also be required for cell synthesis. Table 7.22.1 lists the primary nutritional requirements.

TABLE 7.22.1 CLASSIFICATION OF NUTRIENT REQUIREMENTS

Function	Sources
Energy Source	Organic compounds, inorganic compounds, and sunlight
Carbon Source	Carbon dioxide, bicarbonate, and organic compounds
Electron Acceptor	Oxygen, organic compounds, and combined inorganic oxygen (nitrate, nitrite, sulfate)

Source: Adapted from L.D. Benefield and C.W. Randall, 1980, *Biological process design for wastewater treatment* (Englewood Cliffs, N.J.: Prentice-Hall).

The nutritional requirements of microorganisms provide a basis for classification. Microorganisms are classified on the basis of the form of carbon they require:

Autotrophic: These microorganisms use carbon dioxide or bicarbonate as their sole source of carbon, from which they construct all their carbon-containing biomolecules.

Heterotrophic: These microorganisms require carbon in the form of complex, reduced organic compounds, such as glucose.

Microorganisms are also classified on the basis of their required energy source:

Phototrophs: These microorganisms use light as their energy source.

Chemotrophs: These microorganisms use oxidation-reduction reactions to provide their energy.

Chemotrophic microorganisms can be further classified on the basis of the type of chemical compounds that they oxidize, i.e., on the basis of their electron donor. For example, chemoorganotrophs use complex organic molecules as their electron donors, while chemoautotrophs use simple inorganic molecules such as hydrogen sulfide (H_2S) or ammonia (NH_3^+). Table 7.22.2 summarizes microorganism classification by sources of energy and cell carbon.

In addition to energy and carbon sources, microorganisms require principal inorganic nutrients such as nitrogen, sulfur, phosphorus, potassium, magnesium, calcium, iron, sodium, and chloride. Minor nutrients of importance include zinc, manganese, molybdenum, selenium, cobalt, copper, nickel, and tungsten (Metcalf and Eddy, Inc. 1991). Microorganisms also require organic nutrients, known as growth factors, as precursors or constituents of organic cell material that cannot be synthesized from other carbon sources. These growth factors differ from one or-

TABLE 7.22.2   GENERAL CLASSIFICATION OF MICROORGANISMS BY SOURCES
OF ENERGY AND CARBON

Classification	Energy Source	Carbon Source
Autotrophic:		
Photoautotrophic	Light	Carbon dioxide
Chemoautotrophic	Inorganic oxidation-reduction reactions	Carbon dioxide
Heterotrophic:		
Photoheterotrophic	Light	Organic carbon
Chemoheterotrophic	Organic oxidation-reduction reactions	Organic carbon

Source: Adapted from Metcalf and Eddy, Inc., 1991, *Wastewater engineering: Treatment, disposal, and reuse,* 3d ed., (New York: McGraw-Hill).

ganism to the next, but they fall within one of the following three categories: (1) amino acids, (2) purines and pyrimidines, and (3) vitamins (Metcalf and Eddy, Inc. 1991).

Microbial Enzymes

All microbial cell activities depend upon food utilization, and all chemical reactions involved are controlled by enzymes. Enzymes are proteins produced by a living cell that act as a catalyst to accelerate specific reactions in accordance with rate equations. Enzymes are specific in that they catalyze only certain kinds of reactions, and they act on only one kind of substance. Few hundredths of a second elapse while enzymes combine with chemicals undergoing change; chemical reactions occur, and new compounds are formed. Enzymes have little affinity to new compounds; thus, they are free to combine with other molecules of the substance for which they have specificity.

Microbial enzymes catalyze three types of reactions: hydrolytic, oxidative, and synthetic. Hydrolytic reactions involve enzymes hydrolyzing insoluble substrates into simple soluble components that pass through cell membranes into a cell by diffusion. These enzymes are extracellular, that is, they are released into the medium, while intracellular enzymes are released after cell disintegration.

Reactions that yield energy for growth and cell maintenance are catalyzed by intracellular enzymes. These reactions involve oxidation and reductions, that is, the addition or removal of oxygen or hydrogen. Most microorganisms oxidize by the enzymatic removal of hydrogen from molecules. Hydrogen is removed from compounds one atom at a time by dehydrogenases. Then, it is passed from one enzyme system to another until it is used to reduce the final hydrogen acceptor, otherwise known as the electron acceptor.

The electron acceptor is determined by the nature of the surrounding environment and the character of the relevant cells. Thus, in aerobic reactions, oxygen is the electron acceptor, while an oxidized compound is the electron acceptor in an anaerobic reaction resulting in a reduced compound. The oxidation process releases energy, and the

reduction process consumes energy. Thus, a positive net energy output in a reaction is used in growth and cell maintenance.

Intracellular enzymes also catalyze the synthesis of cellular material for cell maintenance and new cell production. These enzymes are synthetic enzymes, and are required to produce the types of complex compounds found in a microbial cell. Synthetic reactions obtain the required large amount of energy from oxidation reactions occurring during the microorganism's energy metabolism.

Environmental Factors Affecting Microbial Growth

Enzyme activity is affected by environmental conditions, which also affects the activity of the corresponding microorganisms. Environmental parameters influencing the growth and performance of microorganisms include temperature, pH, and oxygen concentration.

TEMPERATURE EFFECTS

Since microbial growth is controlled mostly by chemical reactions, and the nature and rate of chemical reactions are affected by temperature, the rate of microbial growth and total biomass growth are affected by temperature. The microbial growth rate increases with temperature to a certain maximum where the corresponding temperature is the optimum temperature (see Figure 7.22.2). Then, growth does not occur after a small increase in temperature above the optimum value, followed by a decline in the growth rate with an increase in temperature beyond the optimum.

For example, bacteria can be divided into three different classes on the basis of their temperature tolerance: psychrophilic, mesophilic, and thermophilic. Psychrophilic bacteria tolerate temperatures in the range of -10 to $30°C$, with the temperature for optimum growth in the range of 12 to 18°C. The mesophilic group tolerates temperatures in the range of 20 to 50°C, with an optimum temperature between 25 and 40°C, while thermophilic bacteria survive in a temperature range of 35 to 75°C and have optimum growth at temperatures in the range of 55 to 65°C (Metcalf

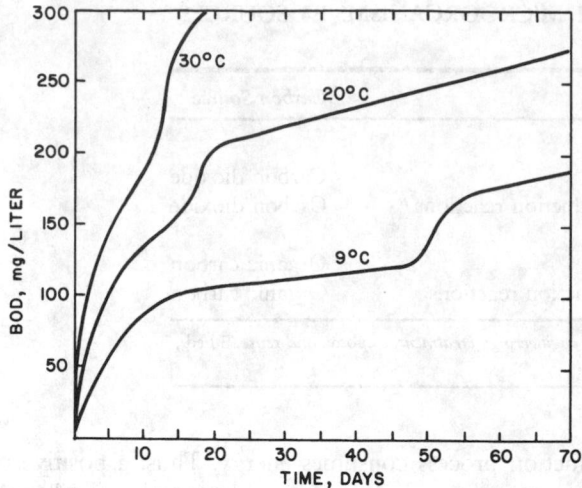

FIG. 7.22.2 Progress of BOD at 9°, 20°, and 30°C. The break in each curve corresponds to the onset of nitrification.

and Eddy, Inc. 1991). In their respective classes, facultative thermophiles and facultative psychrophiles are bacteria that have optimum temperatures that extend into the mesophilic range. Optimum temperatures for obligate thermophiles and obligate psychrophiles are outside the mesophilic range.

The van't Hoff rule provides a generalization of the effect of temperature on enzyme reaction rates stating that the reaction rate doubles for a 10°C temperature increase. Also, according to Arrhenius, the following equation describes the relationship between reaction-rate constants and temperature:

$$d(\ln K)/dt = (E_a/R)(1/T^2) \qquad 7.22(1)$$

where:

K = the reaction-rate constant
E_a = the activation energy, cal/mole
R = the ideal gas constant (1.98 cal/mole-°K)
T = the reaction temperature, °K

Integrating Equation 7.22(1) yields the following equation:

$$\ln K = -(K_a/R)(1/T) + \ln B \qquad 7.22(2)$$

where B is an integration constant.

Equation 7.22(2) can be integrated between two temperature boundaries T_2 and T_1 to yield the following relationship that estimates the effect of temperature over a limited range:

$$\ln (K_2/K_1) = (E_a/R)[(T_2 - T_1)/(T_2T_1)] \qquad 7.22(3)$$

In biological treatment, the activation energy E_a can range from 2000 to 20,000 cal/mole. For most biological treatment cases, the term $(E_a/R)/(T_2T_1)$ is constant; therefore, the following equation applies:

$$K_2/K_1 = \Theta^{(T2-T1)} \qquad 7.22(4)$$

where Θ is the temperature coefficient.

pH EFFECTS

Since enzymes are responsible for microorganism activity, pH effects on enzymes translate to effects on the corresponding microorganism. Some enzymes like acidic environments, while some like a medium environment, and others prefer an alkaline environment. When the pH increases or decreases beyond the optimum, enzyme activity decreases until it disappears. For most bacteria, the extremes of the pH range for growth are 4 and 9, while the optimum pH for growth is within the range of 6.5 to 7.5. Bacteria, in general, prefer a slightly alkaline environment; in contrast, algae and fungi prefer a slightly acidic environment. Biological treatment processes, however, rarely operate at optimum growth. Full-scale, extended-aeration, activated sludge and aerated lagoons can successfully operate at pH levels between 9 and 10.5; however, both systems are vulnerable to a pH less than 6 (Benefield and Randall 1980).

OXYGEN REQUIREMENTS AND MICROBIAL METABOLISM

Microorganisms can also be classified on the basis of whether they use an electron acceptor in the generation of energy. Organisms that generate energy by the enzyme-mediated electron transport from an electron donor to an external electron acceptor carry out respiratory metabolism. Fermentative metabolism, on the other hand, does not involve an external electron acceptor. Fermentation is less efficient in yielding energy than respiration. Hence, heterotrophic microorganisms that are strictly fermentative are characterized by smaller growth rates and cell yields than respiratory heterotrophs.

Microorganisms using molecular oxygen as electron acceptors are called *aerobes*, while those using molecules other than oxygen for electron acceptors are called *anaerobes*. Facultative microorganisms can use oxygen or another chemical compound as electron acceptors. Facultative microorganisms can be divided into two subgroups based on metabolic abilities. True facultative anaerobes can switch from fermentative to aerobic respiratory metabolism depending on the presence of molecular oxygen. Aerotolerant anaerobes, however, have a strictly fermentative metabolism but are insensitive to the presence of molecular oxygen. Obligate aerobes cannot grow in the absence of molecular oxygen, and obligate anaerobes are poisoned by an oxygen presence.

Oxidized inorganic compounds such as nitrate and nitrite can function as electron acceptors for some respiratory organisms in the absence of molecular oxygen. The biological treatment processes that exploit these microorganisms are often referred to as *anoxic*. In addition, those microorganisms that grow best at low molecular oxygen concentrations are termed *microaerophiles*.

The principal significance of the electron acceptors used

TABLE 7.22.3 TYPICAL ELECTRON ACCEPTORS IN BIOLOGICAL WASTEWATER TREATMENT BACTERIAL REACTIONS

Environment	Electron Acceptor	Process
Aerobic	Oxygen	Aerobic metabolism
Anaerobic	Nitrate	Denitrification*
	Sulfate	Sulfate reduction
	Carbon dioxide	Methanogenesis

Source: Adapted from Metcalf and Eddy, Inc., 1991, *Wastewater engineering: Treatment, disposal, and reuse,* 3d ed., (New York: McGraw-Hill).
Note: *Also known as anoxic denitrification.

by microorganisms involves the completeness of the resulting reaction and therefore the amount of energy available for cell growth and maintenance. Aerobes and facultative microorganisms completely oxidize the electron donors, while anaerobes, sometimes referred to as fermenters, do not. Table 7.22.3 gives some typical electron acceptors.

Microbial Populations

Microorganisms are commonly classified on the basis of cell structure and function as eukaryotes, eubacteria, and archaebacteria. Eubacteria and archaebacteria are prokaryotes—cells whose genomes are not contained within a nucleus. Eukaryotes have a membrane-bound nucleus that stores the genome of the cell as chromosomes composed of deoxyribonucleic acid (DNA). Prokaryotes are generally referred to as bacteria. Eukaryotic organisms involved in biological treatment include fungi, protozoa and rotifers, algae, and invertebrates.

BACTERIA

Bacteria are members of a diverse and ubiquitous group of prokaryotic, single-celled organisms. They are the only living organisms that use all possible metabolic pathways. Bacteria can be classified based on their shapes: spherical, cylindrical (rods), and helical (spiral). Most bacteria reproduce by binary fission although some reproduce sexually or by budding. Bacteria range in size from 0.5 to 15 μ depending on their shape: 0.5–1.0 μ for spherical-shaped species, 1.5–3.0 μ for rod-shaped species, and 6–15 μ for spiral-shaped species. The interior of a typical bacteria cell—known as the cytoplasm—contains a colloidal suspension of proteins, carbohydrates, and other complex organic compounds. The cytoplasm also houses ribonucleic acid—responsible for protein synthesis—and the nuclear area that contains the DNA—carrying the information necessary for cell reproduction. Bacteria are approximately 80% water and 20% dry material, of which 90% is organic and 10% is inorganic.

Bacteria can be generally classified into two groups, aerobic bacteria and anaerobic bacteria, which is defined later in this section. In the aerobic bacteria class (see Figure 7.22.3), two ecological groups are of concern: the floc-forming microorganisms, which can propagate in an activated-sludge system, and the biofilm-forming microorganisms, which grow attached to surfaces—a feature that is exploited in wastewater treatment processes such as the trickling filter. Apparently, the ability to form bacterial floc is associated with the ability to attach to surfaces, and these two ecological groups overlap to a large extent. Among the well-known names of genera of bacteria that belong to these groups are *Pseudomonas, Zooglea, Bacillus, Flavobacterium,* and *Nocardia.*

The anaerobic group (see Figure 7.22.4) includes the fermentative bacteria such as *Clostridium, Propionibacterium, Streptobacterium, Streptococcus, Lactobacillus,* and *Enterobacter.* Other common genera in the anaerobic group include the sulfate-reducing bacterium, *Desulfovibrio,* and methanogens such as *Methanosarcinia* and *Methanothrix.* Anaerobic degradation of organic matter usually requires a complex, interactive community with many different species.

FUNGI

Another group of decay organisms is fungi. Fungi are diverse, widespread, unicellular (e.g., yeasts) and multicellular (e.g., molds possessing a filamentous mass termed mycelium) eukaryotic organisms, lacking in chlorophyll and usually bearing spores and often filaments. Hence, they are nonphotosynthetic, heterotrophic protists. They are classified on the basis of their mode of reproduction: sexually or asexually, fission, budding, or spore formation. Fungi are strict aerobes that are tolerant of low pH levels and nitrogen-limiting conditions. Because of their ability

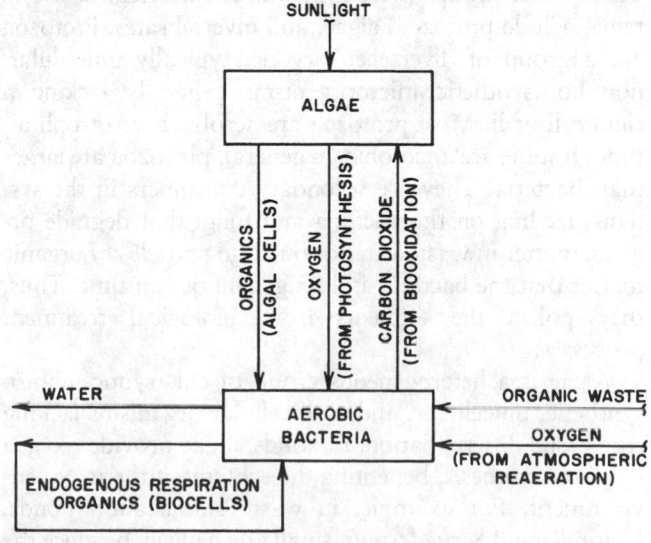

FIG. 7.22.3 Algal–bacterial interplay in an aerobic lagoon.

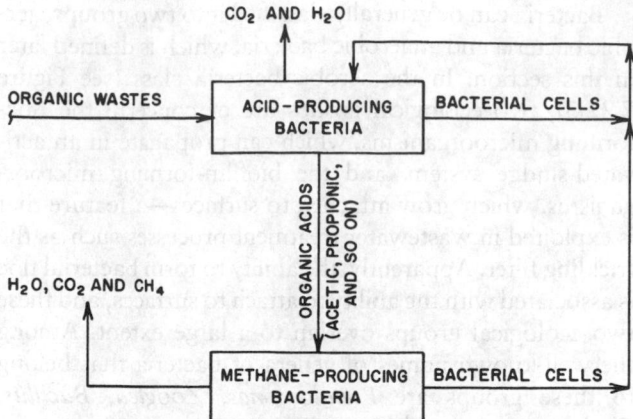

FIG. 7.22.4 Anaerobic degradation process.

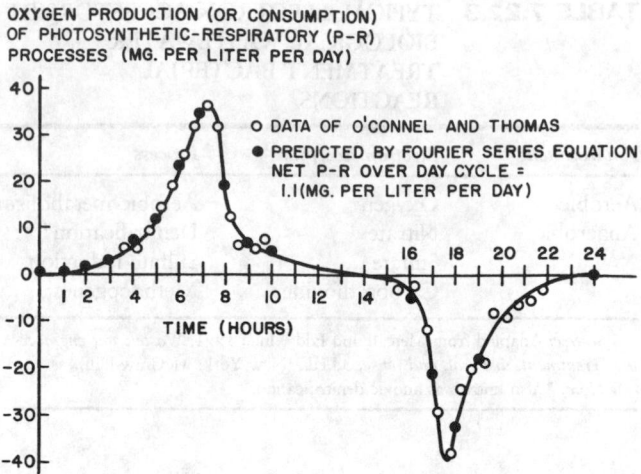

FIG. 7.22.5 Daily cycle of algal activity related to net oxygen production. (Data from R.L. O'Connell and N.A. Thomas, 1965, Effect of benthic algae on stream dissolved oxygen. *Journal of the American Society of Civil Engineers* 91, no. SA3:1).

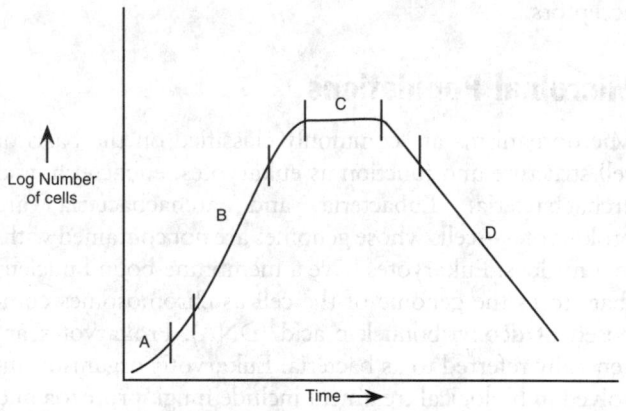

Key:
A: Lag Phase
B: Log-Growth Phase
C: Stationary Phase
D: Log-Death Phase

FIG. 7.22.6 Batch bacterial growth curve.

to degrade cellulose, fungi are important in the biological treatment of some industrial wastes and in composting of organic solid waste.

Compared to the research on waste degradation by bacteria, much less exists concerning the active role of fungi in waste degradation. Fungi are present in suspended-growth systems, but their role is not well known. In attached-growth systems, they are a major component of the biota and may be responsible for forming the base film to which other microorganisms attach. Most of the fungi that have been recovered from wastewater treatment systems are the imperfect stages of Ascomycetes. Microorganisms that can grow as either single cells; yeast; or as filaments; *Candida, Rhodotorula, Oedidendron, Geotrichum,* and *Tricosporon*; are common in waste systems along with many common molds.

PROTOZOA, ALGAE, AND INVERTEBRATES

Three other groups present in wastewater treatment systems include protozoa, algae, and invertebrates. Protozoa are a group of diverse eukaryotic, typically unicellular, nonphotosynthetic microorganisms generally lacking a rigid cell wall. Most protozoa are aerobic heterotroph although some are anaerobic. In general, protozoa are larger than bacteria. They are secondary consumers in the systems, feeding on the bacteria and fungi that degrade organic matter in wastewater or on large particles of organic matter that the bacteria and fungi cannot consume. Thus, they polish the effluents from biological treatment processes.

Algae is a heterogeneous group of eukaryotic, photosynthetic, unicellular, and multicellular organisms lacking true tissue differentiation. In ponds, algae provide oxygen by photosynthesis, benefiting the ecology of the water environment. For example, in waste stabilization ponds, *Chlorella* and *Scenedesmus*, small green algae, produce the oxygen (see Figure 7.22.5) that is required by aerobic, het-

erotrophic bacteria. However, algae can be a problem in blooms where excessive algal growth in the receiving water can deplete the oxygen supply to the animal population below the water's surface.

Invertebrates are secondary or tertiary consumers. Invertebrates in wastewater treatment systems include rotifers, crustacea, insect larvae, nematodes, and worms. Rotifers are aerobic, heterotrophic, and multicellular animals. A rotifer possesses two sets of rotating cilia on its head, providing mobility and the ability to feed. It is effective in consuming dispersed and flocculated bacteria and small particles of organic matter. The presence of rotifers in an effluent indicates a highly efficient biological purification process.

GENERAL GROWTH PATTERN OF BACTERIA

Figure 7.22.6 shows the general bacterial growth pattern. Bacterial growth is comprised of four phases: lag phase, log-growth phase, stationary phase, and log-death phase. During the lag-phase, microorganisms acclimate to their new environment and begin to reproduce. In the log-growth phase, bacterial cells multiply at a rate determined by their generation time and ability to process the substrate. When the microorganisms enter the stationary phase, they have exhausted the substrate necessary for growth, and their population is at a standstill. If no new substrate is added, the microorganisms begin to die; hence, in the log-death phase, the death rate exceeds the production of new cells.

The death rate is usually a function of the viable population and environmental characteristics. In some cases, the log-death phase is the inverse of the log-growth phase (Metcalf and Eddy, Inc. 1991). Moreover, a phenomenon occurs when the concentration of available substrate is at a minimum: the microorganisms are forced to metabolize their own protoplasm without replacement. This process,

known as lysis, occurs when dead cells rupture and the remaining nutrients diffuse out to furnish the remaining cells with food. This type of cell growth is sometimes referred to as cryptic growth and occurs in the endogenous phase.

While bacteria play a primary role in waste degradation and stabilization, other groups of microorganisms described previously also take part in waste stabilization. The position and shape of the growth curve, with respect to time, of a microorganism in a mixed-culture system depend on the available substrate and nutrients and environmental factors such as temperature, pH, and oxygen concentrations.

—*Wen K. Shieh*
Van T. Nguyen

References

Benefield, L.D., and C.W. Randall. 1980. *Biological process design for wastewater treatment.* Englewood Cliffs, N.J.: Prentice-Hall.

Metcalf and Eddy, Inc. 1991. *Wastewater engineering: Treatment, disposal, and reuse.* 3d ed. New York: McGraw-Hill.

7.23
TRICKLING FILTERS

FILTER TYPE
  a. Low rate
  b. Intermediate rate
  c. High rate
  d. Super high rate
  e. Roughing
  f. Two stage

FILTER MEDIUM
  a and b: Rock and slag; c: rock; d: plastic; e: plastic and redwood; and f: rock and plastic

HYDRAULIC LOADING (gal/ft²-min)
  a: 0.02–0.06; b: 0.06–0.16; c: 0.16–0.64; d: 0.2–1.2; e: 0.8–3.2; and f: 0.16–0.64

BOD₅ LOADING (lb/ft³-day)
  a: 0.005–0.025; b: 0.015–0.03; c: 0.03–0.06; d: 0.03–0.1; e: 0.1–0.5; and f: 0.06–0.12

BOD₅ REMOVAL (%)
  a: 80–90; b: 50–70; c: 65–85; d: 65–80; e: 40–65; and f: 85–95

DEPTH (ft)
  a: 6–8; b: 6–8; c: 3–6; d: 10–40; e: 15–40; and f: 6–8

RECIRCULATION RATIO
  a: 0; b: 0–1; c: 1–2; d: 1–2; e: 1–4; and f: 0.5–2

FILM SLOUGHING
  a and b: intermittent and c–f: continuous

NITRIFICATION
  a and f: well; b: partial; c and d: little; and e: none

Process Description

Trickling filters have been used for wastewater treatment for nearly 100 years. A trickling filter (see Figure 7.23.1) is an attached-growth, biological process that uses an inert medium to attract microorganisms, which form a film on the medium surface. Table 7.23.1 lists the physical properties of trickling filter media.

A rotary or stationary distribution mechanism distributes wastewater from the top of the filter percolating it through the interstices of the film-covered medium. As the wastewater moves through the filter, the organic matter is adsorbed onto the film and degraded by a mixed population of aerobic microorganisms (see Figure 7.23.2). The oxygen required for organic degradation is supplied by air circulating through the filter induced by natural draft or ventilation.

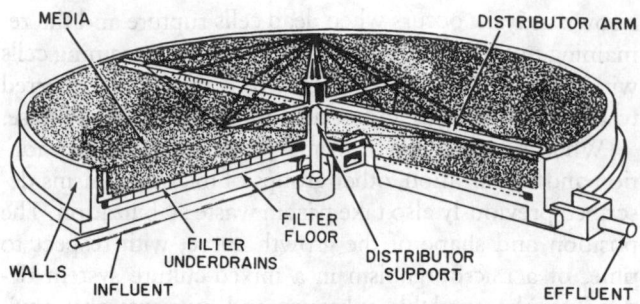

FIG. 7.23.1 Cross section of a stone media trickling filter.

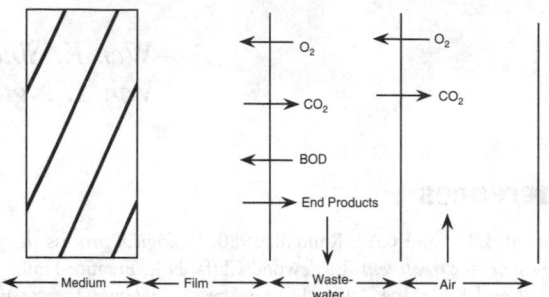

FIG. 7.23.2 A schematic representation of the biological film in a trickling filter.

A light-weight, highly-permeable medium with a large specific surface area (e.g., plastic modules) is conducive to microorganism buildup and ensures unhindered movement of wastewater and air. A porous underdrain system at the bottom of the filter collects treated effluent and circulates air. The filter recirculates and mixes a portion of the effluent with the incoming wastewater to reduce its strength and provide uniform hydraulic loading (Metcalf and Eddy, Inc. 1991).

As the film thickness increases, the region of the film near the medium surface can be deprived of organic matter, reducing the adhesive ability of the microorganisms. Therefore, a thick film is more susceptible to the sloughing effects caused by wastewater flow. Furthermore, the inner portion of a thick film can become anaerobic because oxygen may be unavailable. As a result, the release of gases can weaken the film and increase the sloughing effects. Once the thick film is removed, a new film starts to grow on the medium surface, signaling the beginning of a new growth cycle (Characklis and Marshall 1990).

Process Microbiology

The microorganism population in a trickling filter consists of aerobic, anaerobic, and facultative bacteria, fungi, algae, and protozoans. Also present are higher forms such as worms, insect larvae, and snails. The predominating microorganisms in the trickling filter are the facultative bacteria. *Achromobacter, Flavobacterium, Pseudomonas,* and *Alcaligenes* are among the bacterial species commonly associated with the trickling filter. Filamentous forms such as *Sphaerolitus natans* and *Beggiatoa* are found in the slime layer, while *Nitrosomonas* and *Nitrobacter* are present in the lower reaches of the filter.

Fungi in the filter are responsible for waste stabilization. Their presence becomes important in industrial wastewater treatment where pH levels are low. Various fungal species identified include *Fusazium, Muco, Pencillium, Geotrichum, Sporatichum,* and various yeasts. Fungi, however, are often responsible for clogging filters and preventing ventilation due to their rapid growth.

Algae are also found in trickling filters, albeit only in the upper reaches of the filter where sunlight is available. Their main role is not in degrading the organic matter but in providing oxygen during the daytime to the percolating wastewater. Some of the algae species commonly found in trickling filters include *Phormidium, Chlorella,* and *Ulothrix.* Algae can also clog the filter surface, resulting in undesirable odors.

Protozoans in trickling filters, as in activated-sludge processes, are responsible for keeping the bacterial population in check rather than for waste stabilization. The ciliates are the predominating species among protozoa; they

TABLE 7.23.1 COMPARATIVE PHYSICAL PROPERTIES OF TRICKLING FILTER MEDIA

Types of Media	Nominal Size	Units per ft³	Unit Weight lb/ft³	Specific Surface Area ft²/ft³	Void Space %
Granite	1–3 in	—	90	19	46
	4 in	—	—	13	60
Blast Furnace Slag	2–3 in	51	68	20	49
Aeroblock (vitrified tile)	6 in × 11 in × 12 in	2	70	20–22	53
Raschig Rings (ceramic)	1½ in × 1½ in	340	40.8	35	68.2
Dowpac 10	21 in × 37½ in	2	3.6–3.8	25	94
Dowpac 20	21½ in × 38½ in	2	6	25	94

include the *Vorticella, Opercularia,* and *Epistylis* species. Along with the protozoans, the higher animal forms in the filter—such as snails, worms, and insects—feed on the biological film, keeping the bacterial population in a state of high growth and rapid substrate utilization. Thus, these higher forms are not commonly found in high-rate trickling filter towers.

Changes in organic loading, hydraulic retention time, pH, temperature, air availability, influent wastewater composition, and other factors vary the populations of each type of microbial community throughout the filter depth.

Process Flow Diagrams

Figures 7.21.7 and 7.21.8 are examples of common process flow diagrams for single- and multistage trickling filters. Wastewater treatment facilities often recirculate the treated effluent from the clarifier to (Atkinson and Ali 1976; Metcalf and Eddy, Inc. 1991):

- Reduce the possibility of organic shock loadings by diluting the incoming wastewater
- Maintain uniform hydraulic loadings especially under low and intermittent flow conditions
- Achieve an extensive film coverage and a relatively uniform film thickness through the filter
- Reduce the nuisances of odor and flies

However, such benefits are achieved at the expense of higher hydraulic loadings.

Trickling filters are classified by their hydraulic loadings. Typical hydraulic loadings for low-rate (without recirculation) and high-rate (with recirculation) trickling filters are 1.17–3.52 and 9.39–37.55 m^3/m^2-day, respectively. The corresponding loadings for super high-rate trickling filters are as high as 70.41 m^3/m^2-day. The effluent from a low-rate trickling filter is usually low in BOD and well nitrified. Wastewater treatment facilities commonly use two-stage trickling filters for treating high-strength wastewater and achieving nitrification at hydraulic loadings comparable to those for high-rate trickling filters (Tebbutt 1992).

Process Design

Despite recent advances in attached-growth biological wastewater treatment processes, the design and analysis of trickling filters are still largely based empirical models. Some of these empirical models are presented next (McGhee 1991; Metcalf and Eddy, Inc. 1991).

VELZ EQUATIONS

The following equation is used for a single-stage system and the first stage of a two-stage system:

$$S_{e1} = [(S_i + r_1 S_{e1})/(1 + r_1)] \exp[(-kDA^n/Q^n)(1.035^{T-20})]$$

$$7.23(1)$$

The following equation is used for the second stage of a two-stage system:

$$S_{e2} = [(S_e + r_2 S_{e2})/(1 + r_2)] \exp[-kDA^n S_{e1}/Q^n S_i)(1.035^{T-20})]$$

$$7.23(2)$$

where:

S_e	= the effluent BOD from the filter, mg/l
S_i	= the influent BOD, mg/l
r	= the ratio of recirculated flow to wastewater flow
D	= the filter depth, m
A	= the filter plan area, m^2
Q	= the wastewater flow, m^3/min
T	= the wastewater temperature, °C
k, n	= empirical coefficients (for municipal wastewaters, k = 0.02 and n = 0.5)
subscript i (i = 1,2)	= the stage number

NRC EQUATIONS

The following equations apply to a single-stage system and the first stage of a two-stage system:

$$1 - (S_{e1}/S_i) = 1/[1 + 0.532(QS_i/V_1 F_1)^{0.5}] \quad 7.23(3)$$

$$F_1 = [(1 + r_1)/(1 + 0.1r_1)^2] \quad 7.23(4)$$

The following equations apply to the second stage of a two-stage system:

$$1 - (S_{e2}/S_{e1}) = 1/[1 + 0.532(Q/S_{e1} V_2 F_2)^{0.5}] \quad 7.23(5)$$

$$F_2 = [(1 + r_2)/(1 + 0.1r_2)^2] \quad 7.23(6)$$

where:

V = the filter volume, m^3
F = the recirculation factor

ECKENFELDER EQUATION (PLASTIC MEDIA)

The Eckenfelder equation used for plastic media is as follows:

$$S_e/S_i = \exp[-KS_a^m D(Q/A)^{-n}] \quad 7.23(7)$$

where:

K	= the observed rate constant for a given filter depth, ft/day
S_a	= the specific surface area of filter, ft^2/ft^3
D	= the filter depth, ft
Q	= the wastewater flow, ft^3/day
A	= the filter plan area, ft^2
m and n	= empirical coefficients

GERMAIN/SCHULTZ EQUATIONS (PLASTIC MEDIA)

The Germain/Schultz equations used for plastic media are as follows:

$$S_e/S_i = \exp\left[-k_{20,i}D_i(Q/A)^{-n}\right] \qquad 7.23(8)$$

$$k_{20,2} = k_{20,1}(D_1/D_2)^x \qquad 7.23(9)$$

where:

$k_{20,i}$ = the treatability constant corresponding to a specific filter depth D_i at 20°C, $(gal/min)^n ft$

Q = the wastewater flow, gal/min

n and x = empirical constants (n is usually 0.5; x is 0.5 for rock and 0.3 for cross-flow plastic media)

UNDERDRAINS

The underdrains used in trickling filters support the filter medium, collect the treated effluent and the sloughed biological solids, and circulate the air through the filter. Precast blocks of vitrified clay or fiberglass grating arranged on a reinforced concrete floor can be used as the underdrain system for a rock-media trickling filter. Precast concrete beams supported by columns or posts can be used as the underdrain and support system for a plastic-media trickling filter (McGhee 1991). The floor should be sloped towards either central or peripheral collection channels at a 1 to 5% grade for improved liquid flow. The minimum flow velocity in the collection channel should be 0.6 m/sec (2 ft/sec) at the average daily flowrate (Metcalf and Eddy, Inc. 1991). The liquid flow in underdrains and collection channels should not be more than half full for adequate air flows.

FILTER MEDIA

The ideal medium used in a trickling filter should have the following properties: high specific surface area, high void space, light weight, biological inertness, chemical resistance, mechanical durability, and low cost. Table 7.23.2 summarizes the properties of some commercially available media suitable for trickling filter applications. Plastic media are reported to be highly effective for BOD and SS removal over a range of loadings (Harrison and Daigger 1987). Furthermore, lighter and taller filter structures can be constructed to house plastic media, reducing land requirements.

CLARIFIERS

Clarifiers used in trickling filters remove large and heavily sloughed biological solids or humus without providing thickening functions. Therefore, the design of these clari-

TABLE 7.23.2 PROPERTIES OF FILTER MEDIA

Packing Type	Nominal Size (in)	Density (lb/ft³)	Specific Surface Area (ft²/ft³)	Porosity (%)
Redwood	48 × 48 × 20	9–11	12–15	70–80
Blast Furnace	2.0–3.2 (small)	56–75	17–21	40–50
Slag	3.0–5.0 (large)	50–62	14–18	50–60
River Rock	1.0–2.6 (small)	78–90	17–21	40–50
	4.0–5.0 (large)	50–62	12–15	50–60
Plastic				
Surpac*	—	3.6	25	94
Koroseal*	—	2.7–3.5	40	94
Flocor*	—	4.1	29	95
PVC Tubes				
Cloisonyle*	—	—	69	—
Raschig Rings				
Ceramic	0.25–4.00	36–60	14–217	62–80
Carbon	0.25–3.00	23–46	122–212	85–95
Steel	0.50–3.00	25–75	20–122	85–95
Pall Rings				
Ceramic	2.00–3.00	38–40	20–29	74
Steel	1.00–2.00	24–30	31–63	94–96
Polypropylene	1.00–3.50	4.25–5.50	26–63	90–92
Ceramic Berl Saddles	0.25–2.00	39–56	32–274	60–72
Ceramic Intalox Saddles	0.25–3.00	37–54	28–300	75–80

Source: Adapted from A.Y. Li, 1984, *Anaerobic processes for industrial wastewater treatment,* Short Course Series, no. 5 (Taichung, Taiwan: Department of Environmental Engineering, National Chung Hsing University).

Note: *Trade name.

fiers is similar to the design of primary settling tanks. The overflow rates are 400–600 gal/ft²-day at average flow and 1000–1200 gal/ft²-day at peak flow, respectively. The overflow rate is based on the influent flow plus the recirculation flow.

Clarifier depth ranges from 10 to 15 ft. Details on the design of secondary clarifiers are presented in Section 7.29.

DESIGN PROCEDURES

Trickling filters are used extensively in the treatment of municipal wastewater and to a lesser extent in industrial wastewater. Using synthetic media has increased the capability of the trickling filter, and using multistage, synthetic-media filters has achieved a high degree of treatment in industrial wastewater.

Table 7.23.3 lists the treatability factor obtained on settled sewage by trickling filters with various media. The treatability factor (K) is a characteristic of the wastewater, and the n value is a characteristic of the trickling filter media. However, Figure 7.23.3 indicates that the treatability factor is influenced by the n value. If filter media with higher n values are used, the treatability constant is reduced for settled sewage.

The treatability factor varies with wastewater type. Selected organic compounds such as phenol and compounds containing the cyanide complex show high treatability factors, but typical industrial wastewater with or-

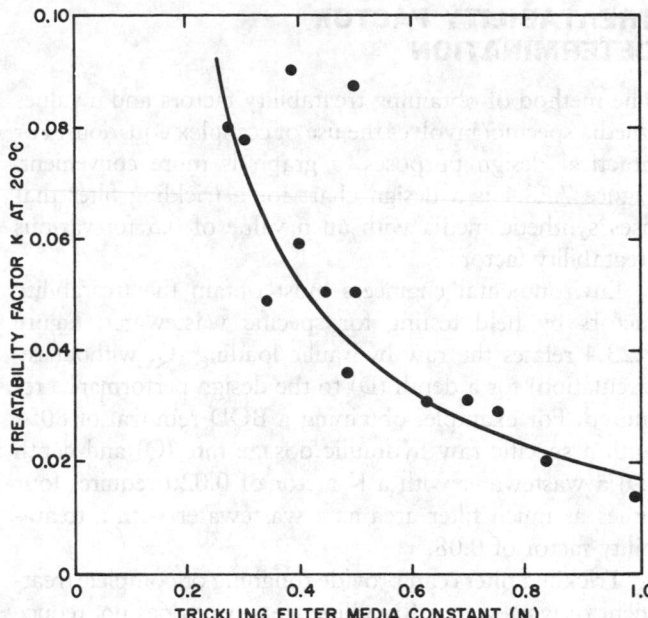

FIG. 7.23.3 Effect of the n value on the treatability factor of settled sewage.

ganics in solution has a treatability factor considerably below that of domestic sewage. This variability indicates the need for confirming the K factor with a pilot-plant evaluation prior to the final design of the trickling filter. Suitable pilot facilities are usually available from manufacturers of synthetic media.

TABLE 7.23.3 BOD TREATABILITY FACTORS OF SETTLED SEWAGE IN TRICKLING FILTERS WITH VARIOUS MEDIA

Media Used	Depth (ft)	Range of Influent BOD Concentration (mg/l)	Applied Hydraulic Loading (gal/min/ft²)	n Factor	Treatability Factor (K)* at 20°C (min⁻¹)
1½ in Flexirings	8	65–90	0.196–0.42	0.39	0.09
2½ in Clinker	6	220–320	0.015–0.019	0.84	0.021
1½–2½ in Slag	6	112–196	0.08–0.19	1.00	0.014
2½ in Slag	6	220–320	0.015–0.019	0.75	0.029
2½–4 in Rock	12	200	0.48–1.47	0.49	0.036
1–3 in Granite	6	186–226	0.031–0.248	0.4	0.059
¾ in Raschig Rings	6	186–226	0.031–0.248	0.7	0.031
1 in Raschig Rings	6	186–226	0.031–0.248	0.63	0.031
1½ in Raschig Rings	6	186–226	0.031–0.248	0.306	0.078
2¼ in Raschig Rings	6	186–226	0.031–0.248	0.274	0.08
Straight Block	6	186–226	0.031–0.248	0.345	0.048
Surfpac	21.6	200	0.49–3.9	0.5	0.05
Surfpac	12.0	200	0.97–3.9	0.45	0.05
Surfpac	21.5	—	—	0.50	0.045†
Surfpac	21.5	—	—	0.50	0.088‡

Notes: *The treatability factor is calculated with formula $\dfrac{L_e}{L} = e - \dfrac{KD}{Q_i^n}$. This formula gives the K and n values. The treatability factor relates to the degree of ease of treating wastewater. The treatability factor K has the unit min⁻¹ when the flow rate is expressed as gal per min per ft². The n factor is related to the type of media and is function of the specific surface and configuration of media. The n factor is a dimensionless constant.

†Dissolved BOD only
‡Total BOD

TREATABILITY FACTOR DETERMINATION

The method of obtaining treatability factors and n values (media specific) involves the use of complex equations. For practical design purposes, a graph is more convenient. Figure 7.23.4 is a design chart for a trickling filter that uses synthetic media with an n value of 0.5 for various treatability factors.

Environmental engineers must obtain the treatability factors by field testing for specific wastewater. Figure 7.23.4 relates the raw hydraulic loading (Q, without recirculation) for a depth (D) to the design performance required. For example, obtaining a BOD removal of 80% with a specific raw hydraulic dosage rate (Q) and depth (D) a wastewater with a K factor of 0.020 requires four times as much filter area as a wastewater with a treatability factor of 0.08.

Trickling filters can provide roughing or complete treatment of wastewater. Roughing treatment does not reduce the treatability of the filter effluent in a subsequent, activated-sludge, process step. A synthetic-media trickling filter is suited for roughing treatment when high temperature is involved because the cooling effect of the filter makes the effluent more amenable to activated-sludge treatment.

Synthetic-media trickling filters are not economically suitable to obtain high treatment levels for soluble organic matter with low treatability factors (below 0.05) because of the large filter volume required.

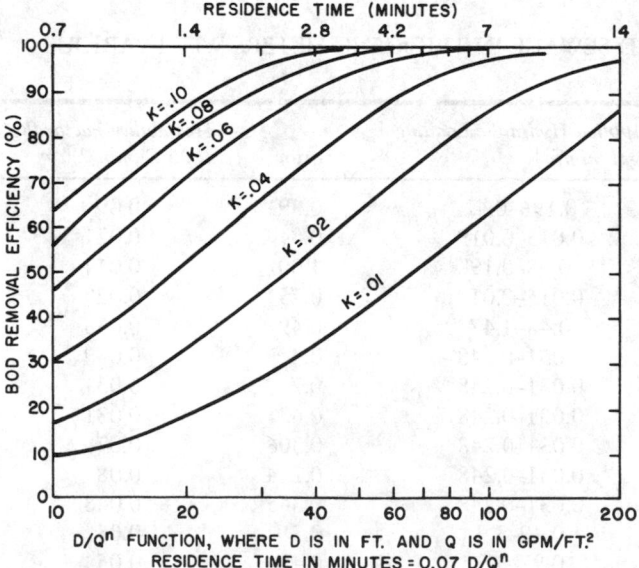

FIG. 7.23.4 Effect of residence time on BOD removal efficiency. The n is a dimensionless exponent function of the trickling filter media. Its value is 0.67 for conventional rock media, 0.50 for most synthetic media, and intermediate values for other types of trickling filter media.

LIST OF ABBREVIATIONS

The following abbreviations apply to the design of trickling filters:

E_1 Percent BOD removal efficiency through first-stage filter and clarifier

E_2 Percent BOD removal efficiency through second-stage filter and clarifier

W   BOD loading, in lb per day, to first-stage filter, not including recycling

W_1 BOD loading, in lb per day, to second-stage filter, not including recycling

L_D Removable BOD at depth D, in mg/l

L   Total removable BOD, in mg/l

D   Depth of filter, in ft

K_1 Constant

L_e Unsettled filter effluent BOD, in mg/l

L_i Filter influent BOD, in mg/l

i   Influent flow, in mgd

r   Recirculation flow, in mgd

a   Filter radius, in ft

T   Wastewater temperature, in °C

Q_1 Hydraulic load, in gal per min per sq ft (not including recirculation)

K_3 or K   Treatability constant

Q   Flow, in mgd

L_o Influent BOD (including recirculation), in mg/l

A   Area of filter, in acres

m,n   Constants for media

K_{20} Treatability constant at 20°C

V   Volume of filter, in acre-feet

F   Recirculation factor

$$F = \frac{1 + R}{(1 + 0.1R)^2} \qquad 7.23(10)$$

where R is the recirculation ratio

—*Wen K. Shieh*
Van T. Nguyen

References

Atkinson, B., and M.E. Ali. 1976. Wetted area, slime thickness and liquid phase mass transfer in packed bed biological film reactors (trickling filters). *Trans. Instr. Chem. Engrs.* 54, no. 239.

Characklis, W.G., and K.C. Marshall, eds. 1990. *Biofilms.* New York: John Wiley & Sons.

Harrison, J.R., and G.T. Daigger. 1987. A comparison of trickling filter media. *J. Water Poll. Control Fed.* 59, no. 679.

McGhee, T.J. 1991. *Water supply and sewerage.* 6th ed., New York: McGraw-Hill.

Metcalf and Eddy, Inc. 1991. *Wastewater engineering: Treatment, disposal, and reuse.* 3d ed. New York: McGraw-Hill.

7.24
ROTATING BIOLOGICAL CONTACTORS

TREATMENT LEVEL

  a. Secondary
  b. Combined carbon oxidation/nitrification
  c. Nitrification

HYDRAULIC LOADING (gal/ft²-day)

  a: 2–4; b: 0.75–2; and c: 1–2.5

SOLUBLE BOD₅ LOADING (lb/ft²-day)

  a: 0.75–2; b: 0.5–1.5; and c: 0.1–0.3

LIQUID RETENTION TIME (LRT) (hrs)

  a: 0.7–1.5; b: 1.5–4; and c: 1.2–2.9

BOD₅ REMOVAL (%)

  a–c: 85–95

EFFLUENT NH₃ (mg/l)

  b: <2 and c: 1–2

Process Description

A rotating biological contactor (RBC) is an attached-growth, biological process that consists of a basin(s) in which large, closely spaced, circular disks mounted on horizontal shafts rotate slowly through wastewater (see Figure 7.24.1). The disks are made of high-density polystyrene or PVC for durability and resistance. Corrugation patterns increase surface area and structural integrity (Metcalf and Eddy, Inc. 1991).

Bacterial growth on the surface of the disks leads to the formation of a film layer that eventually covers the entire wetted surface of the disks. The rotating disks are partially submerged in the wastewater. In this way, the film layer is alternatively exposed to the wastewater from which the organic matter is adsorbed and the air from which the oxygen is absorbed.

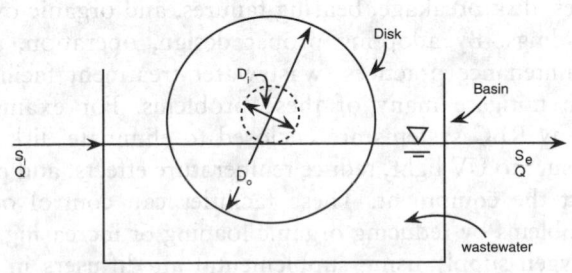

FIG. 7.24.1 A schematic of an RBC system.

The mechanisms of organic degradation in an RBC film layer are similar to those shown in Figure 7.23.2. Rotation also provides a means for removing excess bacterial growth on the disks' surfaces and maintaining suspension of sloughed biological solids in wastewater. A final clarifier removes sloughed solids.

Partially submerged RBCs are used for carbonaceous BOD removal, combined carbon oxidation and nitrification, and nitrification of secondary effluent (Grady and Lim 1980; Metcalf and Eddy, Inc. 1991). Completely submerged RBCs are used for denitrification (Grady and Lim 1980).

Process Flow Diagrams

Figure 7.24.2 shows typical arrangements of RBCs. In general, an RBC system is divided into a series of independent stages or compartments by baffles in a single basin (see Part A in Figure 7.24.2) or separate basins arranged in series (see Part B in Figure 7.24.2).

Compartmentalization creates a flow pattern with little longitudinal mixing in the flow direction (i.e., a plug-flow pattern), increasing overall removal efficiency of an RBC system (Tchobanoglous and Schroeder 1985). It can also promote separation of bacterial species at different stages, achieving optimal performance. For example, autotrophic bacteria responsible for nitrification can concentrate at later stages in an RBC system designed for combined carbon removal and nitrification where the mixed liquor BOD is low. Consequently, nitrification performance is more reliable and stable.

Process Design

RBC design is often based on empirical design curves supplied by RBC manufacturers. Once the environmental engineer estimates the surface loading L (gal/ft²-day) required to achieve a BOD removal efficiency, the required disk surface area A (ft²) for a total flow Q (gal/day) is calculated as follows:

$$A = Q/L \qquad \text{7.24(1)}$$

A more rational design approach, which considers the mass balances for both the substrate and biomass at a specific stage, has been proposed by Ramalho (1983). The re-

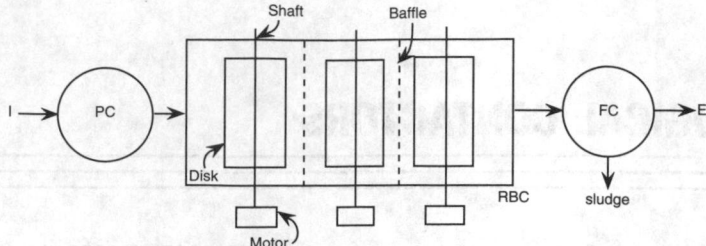

A. Compartmentalization in a single basin using baffles

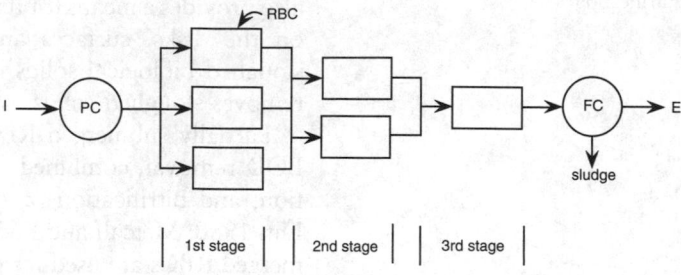

B. Basins arranged in series

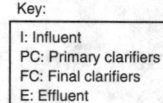

Key:
I: Influent
PC: Primary clarifiers
FC: Final clarifiers
E: Effluent

FIG. 7.24.2 Typical arrangements of RBCs. **A,** Compartmentalization in a single basin using baffles; **B,** Basins arranged in series.

sulting design equations for a single-stage RBC are as follows:

$$Q(S_i - S_e) = PAS_e/(K_s + S_e) \qquad 7.24(2)$$

$$P = kX_A\delta \qquad 7.24(3)$$

$$A = \pi(D_o^2 - D_i^2)/2N \qquad 7.24(4)$$

where:

Q = wastewater flowrate, m^3/day
S_i = influent BOD, mg/l
S_e = effluent BOD, mg/l
A = required wetted surface area to achieve the required BOD reduction from S_i to S_e, m^2
K_s = half-velocity constant, mg/l (see Section 7.25)
k = maximum BOD removal rate, l/day (see Table 7.25.4)
X_A = dry density of the film layer, mg/l
δ = film layer thickness, m
D_o = diameter of the disk, m
D_i = diameter of the circle that is never submerged (see Figure 7.24.1), m
N = number of disks per stage (compartment)

Environmental engineers can determine both P and K_s by performing a laboratory- or pilot-scale treatability study using the same wastewater. With an equal wetted surface area per stage for an n-stage RBC system, the following equation applies:

$$S_{n-1} - S_n = PAS_n/(K_s + S_n) \qquad 7.24(5)$$

The solution of Equation 7.24.5 requires a trial-and-error approach (McGhee 1991). The total wetted surface area required to achieve a given BOD reduction in a multistage RBC system is less than that in a single-stage RBC system.

OPERATING PROBLEMS

Many RBC operating problems are caused by shaft failures, disk breakage, bearing failures, and organic overloadings. By adopting proper design, operation, and maintenance practices, wastewater treatment facilities can mitigate many of these problems. For example, many RBC systems are enclosed to eliminate disk exposure to UV light, reduce temperature effects, and protect the equipment. These facilities can control odor problems by reducing organic loading or increasing the oxygen supply using supplemental air diffusers in the basin.

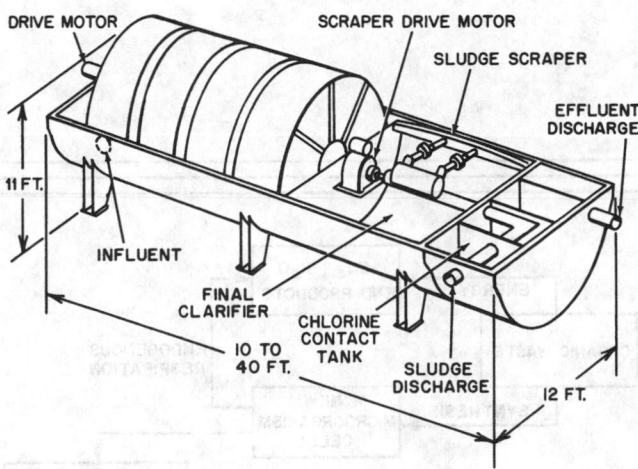

FIG. 7.24.3 Packaged biological disc unit.

BIOLOGICAL DISCS

A biological disc unit consists of a series of closely spaced, large-diameter, expanded polystyrene discs mounted on a horizontal shaft. The discs are partially immersed in wastewater and rotated.

From the microorganisms present in the wastewater, a biological growth develops on the surface of the discs. As the discs rotate, the bacteria alternately passes through the wastewater and the air. Operating in this manner, the discs provide support for microbial growth and alternately contact this growth with organic wastewater pollutants and air.

The rotational speed, which controls the contact intensity between the biomass and the wastewater, and the rate of aeration can be adjusted according to the organic load in the wastewater. Biological disc units are available in sizes up to 12 ft in diameter.

The residence time of the wastewater in the disc sections and the rotational speed of the discs determine unit BOD removal efficiency. Installing a number of discs in a series of stages improves the residence time distribution and yields a greater BOD removal efficiency.

Staged operation is advantageous when the wastewater contains several types of biodegradable materials because staging enhances the natural development of different biological cultures in each stage. For example, the discs in the later stages are dominated by nitrifying bacteria that oxidize ammonia after most of the carbonaceous BOD has been removed. Staged operation also permits the use of intermediate solids separation units at strategic points. The process is claimed to be stable under hydraulic surges and intermittent flows.

The approximate cost of a bio-disc unit is 25¢ per gal per day of domestic sewage, excluding site work. Because of the high buoyancy of the disc materials and the low rotational speeds, the power consumption is low. For domestic wastewater, power consumption can be 10 hp per mgd. This rate is approximately equivalent to removing 4.2 lb of BOD per hp-hr invested.

For sewage flow capacities of 5000 to 120,000 gpd, a package unit is available (see Figure 7.24.3). It includes a feed mechanism, a section of bio-disk surfaces, an integral clarifier tank with sludge removal mechanism, and a chlorine contact section. Depending on the nature of the influent sewage and on the total flow capacity, the length of the package unit varies from 10 to 40 ft.

—*Wen K. Shieh*
Van T. Nguyen

References

Grady, C.P.L., Jr., and H.C. Lim. 1980. *Biological wastewater treatment: Theory and applications.* New York: Marcel Dekker.

McGhee, T.J. 1991. *Water supply and sewerage.* 6th ed. New York: McGraw-Hill.

Metcalf and Eddy, Inc. 1991. *Wastewater engineering: Treatment, disposal, and reuse.* 3d ed. New York: McGraw-Hill.

Tchobanoglous, G., and E.D. Schroeder. 1985. *Water quality.* Reading, Mass.: Addison-Wesley.

7.25
ACTIVATED-SLUDGE PROCESSES

PROCESS TYPE

    a. Conventional
    b. Completely mixed
    c. Step feed
    d. Contact stabilization
    e. High-purity oxygen
    f. Oxidation ditch
    g. Sequencing batch reactor
    h. Deep shaft

LRT (hrs)

a: 4–8; b: 3–5; c: 3–5; d: 0.5–1 (contact tank), 3–6 (stabilization tank); e: 1–3; f: 8–36; g: 12–50; and h: 0.5–5

SOLIDS RETENTION TIME (SRT) (days)

a–d: 5–15; e: 3–10; f: 10–30; and g: not applicable

F/M (lb BOD$_5$/lb MLVSS-day)

a & c: 0.2–0.4; b & d: 0.2–0.6; e: 0.25–1; f & g: 0.05–0.3; and h: 0.5–5

VOLUMETRIC LOADING (lb BOD$_5$/ft^3-day)

a: 0.02–0.04; b: 0.05–0.12; c: 0.04–0.06; d: 0.06–0.075; e: 0.1–0.2; f: 0.005–0.03; and g: 0.005–0.015

BOD$_5$ REMOVAL (%)

85–95

MLSS (mg/l)

a: 1500–3000; b: 2500–4000; c: 2000–3500; d: 1000–3000 (contact tank), 4000–10,000 (stabilization tank); e: 2000–5000; f: 3000–6000; and g: 1500–5000

RECYCLING RATIO

a & c: 0.25–0.75; b: 0.25–1; d: 0.5–1.5; e: 0.25–0.5; f: 0.75–1.5; and g: not applicable

AERATION TYPE

Diffused aeration and mechanical aeration.

Process Description

The activated-sludge process, first developed in England in 1914, has been used widely in municipal and industrial wastewater treatment. Although many process variations have been developed for specific applications, biodegradation of organic matter in the activated-sludge process can be illustrated using a typical flow diagram as shown in Figure 7.21.2.

Clarified wastewater discharged from the primary clarifier is delivered into the aeration basin where it is mixed with an active mass of microorganisms (referred to as activated sludge) capable of aerobically degrading organic matter into carbon dioxide, water, new cells, and other end products (see Figure 7.25.1). Diffused or mechanical

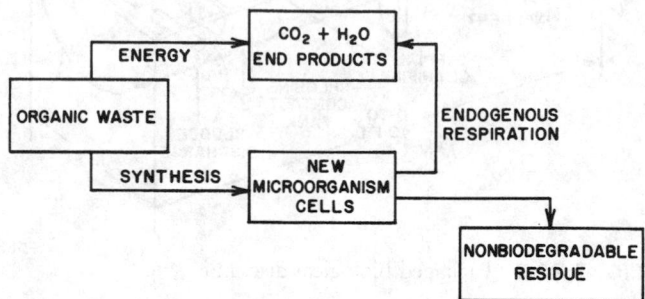

FIG. 7.25.1 Aerobic biological oxidation of organic wastes.

aeration maintains the aerobic environment in the basin and keeps reactor contents (referred to as mixed liquor) completely mixed.

After a specific treatment time, the mixed liquor passes into the secondary clarifier, where the sludge settles under quiescent conditions and a clarified effluent is produced for discharge. The process recycles a portion of settled sludge back to the aeration basin to maintain the required activated-sludge concentration (expressed in terms of mixed-liquor, volatile SS [MLVSS] concentration). The process also intentionally wastes a portion of the settled sludge to maintain the required SRT for effective organic (BOD) removal.

Process Microbiology

The activated-sludge process is an aerobic, continuous-flow, secondary treatment system that uses sludge-containing, active, complex populations of aerobic microorganisms to break down organic matter in wastewater. Activated sludge is a flocculated mass of microbes comprised mainly of bacteria and protozoa.

In the activated-sludge process, bacteria are the most important microorganisms in decomposing the organic material in the influent. During treatment, aerobic and facultative bacteria use a portion of the organic matter to obtain energy to synthesize the remaining organic material into new cells. Only a portion of the original waste is actually oxidized to low-energy compounds such as nitrate, sulfate, and carbon dioxide; the remainder of the waste is synthesized into cellular material. In addition, many intermediate products are formed before the end products.

The group of bacteria involved in activated-sludge systems belongs primarily to the Gram negative species, including carbon oxidizers and nitrogen oxidizers, floc-formers and nonfloc-formers, and aerobes and facultative anaerobes. In general, the bacteria in the activated-sludge process include those in the genera *Pseudomonas, Zoogloea, Achromobacter, Flavobacterium, Nocardia, Bdellovibrio, Mycobacterium, Nitrosomonas,* and *Nitrobacter.* An adequate population of the nitrifying bacteria, *Nitrosomonas* and *Nitrobacter,* must be maintained. These are slow-growing species; therefore, maintaining the sludge wasting rate ensures that they do not wash out.

Although floc-formers are mainly selected by the settling and recycling process, activated sludge can become dominated by filamentous bacteria. This situation is frequently associated with poor settlement characteristics. Researchers have shown that increasing the mean residence time of the cells enhances settling characteristics of biological floc (Forster 1985).

Another bacteria group found in activated sludge is the actinomycetes group, in particular *Nocardia* and *Rhodococcus.* These species are blamed for the formation of stable foams on activated-sludge tanks. The reason for the proliferation of these species is not known, and control methods have yet to be established (Forster 1985).

The protozoan population in activated sludge includes flagellates, amoebae, and ciliates. Over 200 different species of protozoa have been found in activated sludges (Forster 1985). Ciliates are the most prevalent type in activated sludge, with species such as *Vorticella* and *Opercularia* comprising up to one-third of the ciliate population. These species attach themselves to the sludge flocs. Another significant type of ciliates includes *Aspidisca* and *Trachelophylum*—species that creep over the sludge surface.

The balance of bacteria and protozoans in activated sludge depends on the nature of the wastewater and the plant operation. Protozoans are more susceptible to toxins and heavy metals than bacteria, and disruption of the protozoan population has been attributed to poor plant operation (Forster 1985). The role of the protozoan is not to stabilize the waste but to control the bacterial population, feeding on free-swimming bacteria that would otherwise produce a turbid effluent. However, carnivorous ciliates maintain a check on the bacteria-feeding population. Hence, protozoans are important in determining effluent quality.

Other microorganisms in activated sludge include fungi, nematodes, and rotifers. Fungi appear to have two roles: consumers of organic matter and predators for nematodes and rotifers. The role of fungi as a consumer of organic matter is the more common, especially in systems with low pH where bacterial growth is inhibited.

A proliferation of fungi usually imparts poor settleability to the sludge. Nematodes, like protozoans, also consume bacteria, while rotifers ingest sludge flocs, removing small particles that would otherwise cause turbidity. Rotifers also break up large flocs, providing available adsorption sites. Nevertheless, the effluent from an activated-sludge system can be high in biological solids as

a result of poor design of the secondary settling unit, poor operation of the aeration units, or the presence of filamentous microorganisms such as *Sphaerotilus, E. coli,* and fungi (Metcalf and Eddy, Inc. 1991).

Process Flow Diagrams

Figure 7.25.2 shows a conventional activated-sludge process flow diagram. This process is primarily used in the treatment of municipal wastewater. The process uses long, rectangular aeration tanks with minimal longitudinal mixing that creates plug-flow patterns (see Figure 7.25.3). The wastewater is mixed with the recycled sludge at the head end of the aeration tank and then flows through the tank where organic matter is progressively removed. As a result, a BOD concentration profile is established through the tank that can diminish when the recycled sludge flow is significant. Air application is generally uniform through the tank.

The conventional activated-sludge process is susceptible to shock and toxic loading conditions since longitudinal mixing is absent in aeration tanks. The tapered-aeration, activated-sludge process (see Figure 7.25.4) and the step-feed-aeration, activated-sludge process (see Figure 7.21.3) are two process variations of the conventional activated-sludge process. The aeration rate decreases along the tank length in the tapered-aeration, activated-sludge process and matches the BOD concentration profile to improve process economy. The aeration equipment is spaced unevenly through the tank.

Settled wastewater enters at several points in the aeration tank in the step-feed-aeration, activated-sludge process, equalizing loading and oxygen demand. This operation mode increases the flexibility of the process to handle shock and toxic loading conditions.

On the other hand, mixing intensity in the aeration tank of the completely-mixed, activated-sludge process (see Figure 7.21.5) is sufficiently high to yield a uniform mixed

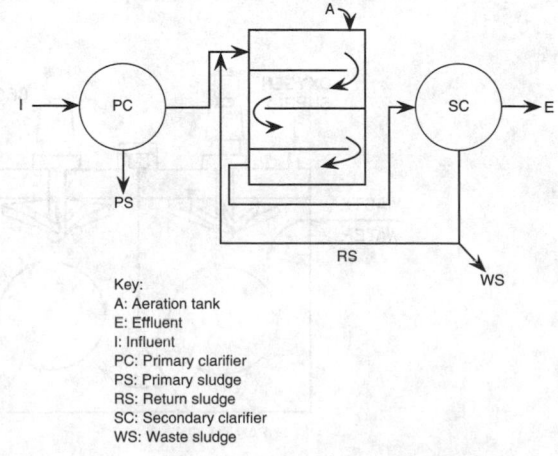

Key:
A: Aeration tank
E: Effluent
I: Influent
PC: Primary clarifier
PS: Primary sludge
RS: Return sludge
SC: Secondary clarifier
WS: Waste sludge

FIG. 7.25.2 Conventional activated-sludge process.

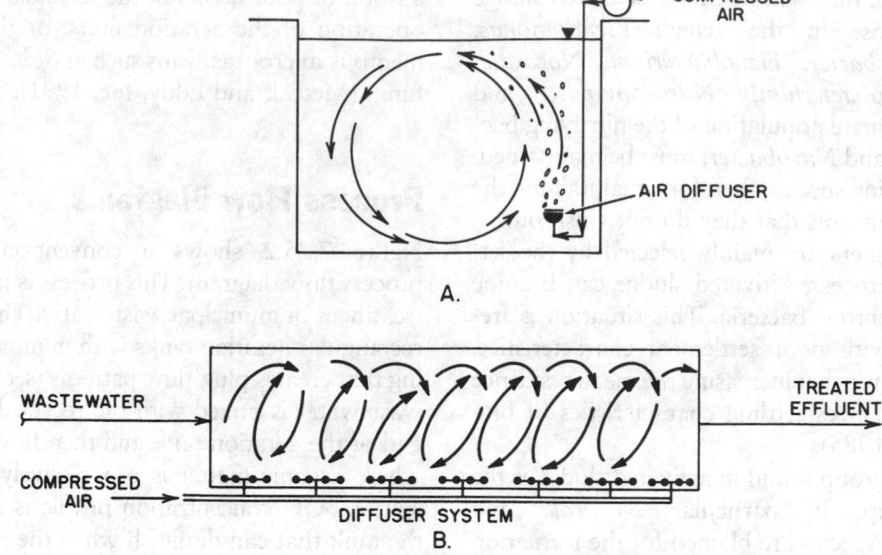

FIG. 7.25.3 Conventional activated-sludge process showing plug-flow and spiral-flow diffused aeration. **A,** End view; **B,** Top view.

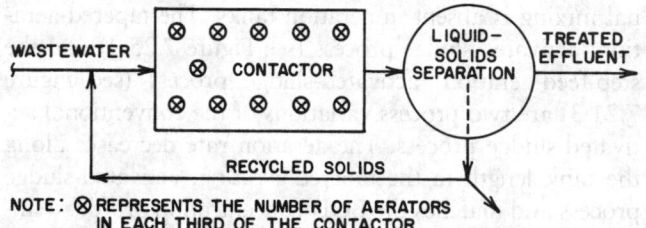

FIG. 7.25.4 Tapered-aeration, activated-sludge process.

liquor that can smooth out and dilute load variations. As a result, the completely-mixed, activated-sludge process is resistant to shock and toxic loadings and is used widely for treating industrial wastewater. The aeration equipment is equally spaced for good mixing.

The contact-stabilization, activated-sludge process (see Figure 7.21.6) uses two separate tanks or compartments (contact and reaeration) to treat wastewater. This process first delivers the wastewater (usually without primary set-

tling) into the aerated contact tank where it mixes with the stabilized sludge that rapidly removes suspended, colloidal, and a portion of the dissolved BOD (entrapment of suspended BOD in sludge flocs and adsorption of colloidal and dissolved BOD by sludge flocs). These reactions yield approximately 90% removal of BOD within 15 min of contact time (Eckenfelder 1980).

The mixed liquor then passes into the secondary clarifier where sludge is separated from clarified effluent. The settled sludge is recycled back to the reaeration tank where organic matter stabilization occurs. The resulting total aeration basin volume is typically 50% less than that of the conventional activated-sludge process (Metcalf and Eddy, Inc. 1991).

The oxygen-activated-sludge process uses high-purity oxygen instead of air (see Figure 7.25.5). The aeration tanks are usually covered, and the oxygen is recirculated, reducing the oxygenation requirements. This process must vent a portion of the gas accumulated inside the aeration

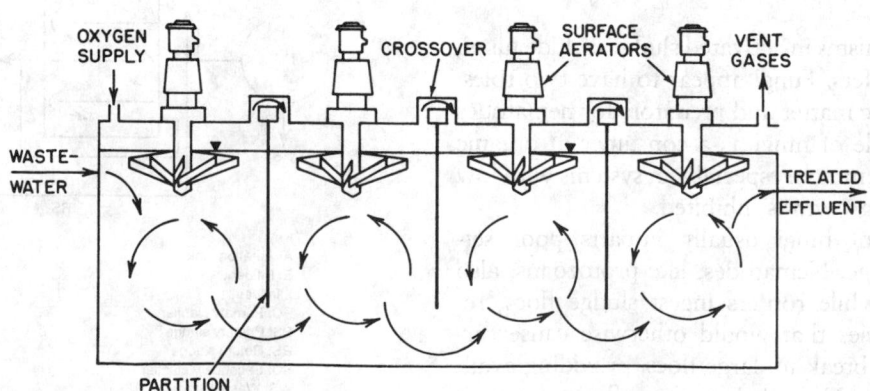

FIG. 7.25.5 Activated-sludge contactor using pure oxygen.

tank to remove carbon dioxide. However, adjusting the pH of the mixed liquor may still be needed.

Since the amount of oxygen added in the oxygen-activated-sludge process is approximately four times greater than that available in the conventional activated-sludge process, the BOD loading applied is higher, yielding a small aeration basin volume. Experimental evidence also indicates that oxygen-activated sludge settles better than air-activated sludge. A facility for generating and supplying high-purity oxygen is needed at a treatment site.

The extended-aeration process (see Figure 7.21.4) is similar to the conventional activated-sludge process except it operates in the endogenous respiration phase to reduce excess process sludge. As a result, the aeration basin is generally much larger. Only preliminary wastewater treatment to remove coarse materials is needed to protect treatment equipment. The extended-aeration process is designed for the treatment of wastewater generated from small installations and communities. Section 7.26 presents a detailed discussion on the extended-aeration process.

The oxidation ditch (see Part A in Figure 7.25.6) is a process variation of the extended-aeration process that uses a ring- or oval-shaped channel as the aeration basin. Mechanical aeration devices, such as aeration rotors, aerate and mix the mixed liquor. An alternating anoxic and oxic environment is established in the channel depending on the distance from the aeration device. Consequently, the oxidation ditch can achieve good nitrogen removal via nitrification and denitrification. Some oxidation ditches use intrachannel clarifiers to separate the sludge from the mixed liquor.

The deep-shaft, activated-sludge process (see Part B in Figure 7.25.6) uses a deep annular shaft (400 to 500 ft deep) as the reactor that provides the dual function of primary settling and aeration. The process forces mixed liquor and air down the center of the shaft and allows it to rise through the annulus. Oversaturation of oxygen occurring in the deep-shaft, activated-sludge process significantly increases oxygen transfer efficiency. Since gas bubbles are formed as the mixed liquor rises through the annulus, this process uses air flotation instead of gravity settling to separate sludge from the clarified effluent.

The sequencing batch reactor (SBR) is a single, fill-and-draw, completely-mixed reactor that operates under batch conditions. Recently, SBRs have emerged as an innovative wastewater treatment technology (Irvine and Ketchum 1989; U.S. EPA 1986). SBRs can accomplish the tasks of primary clarification, biooxidation, and secondary clarification within the confines of a single reactor. A typical treatment cycle consists of the following five steps: fill, react, settle, draw, and idle (U.S. EPA 1986). Depending on the mode of operation, SBRs can achieve good BOD and nitrogen removal. SBRs are uniquely suited for wastewater treatment applications characterized by low or intermittent flow conditions.

Design Concepts

In designing activated-sludge processes, environmental engineers must consider the organic loading, microorganism concentration, contactor retention time, artificial aeration, liquids–solids separation, effluent quality, and process costs.

ORGANIC LOADING

The basic criterion of design is the organic loading. The organic loading or food to microorganism (F/M) ratio is the amount of biodegradable organic material available to an amount of microorganisms per unit of time. This ratio can be expressed more concisely as follows:

$$F/M = \frac{(\text{Organic concentration})(\text{Wastewater flow})}{(\text{Microorganism concentration})(\text{Contactor volume})}$$

7.25(1)

or

$$F/M = \frac{(\text{BOD}_5)Q}{(\text{MLSS})V}$$

7.25(2)

where:

F/M = Organic loading, lb BOD$_5$ per lb mixed-liquor SS (MLSS) day
BOD$_5$ = Biological oxygen demand, mg/l
MLSS = Mixed-liquor SS, mg/l
V = Contactor volume, million gal
Q = Wastewater flow, mgd

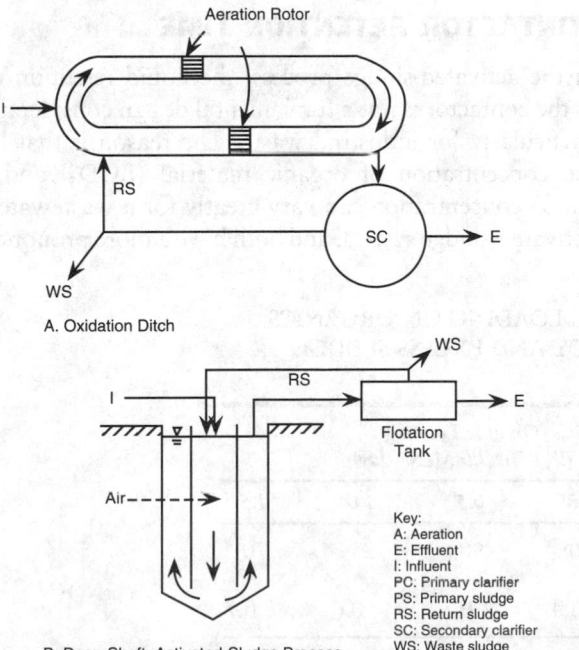

A. Oxidation Ditch

B. Deep-Shaft, Activated-Sludge Process

Key:
A: Aeration
E: Effluent
I: Influent
PC: Primary clarifier
PS: Primary sludge
RS: Return sludge
SC: Secondary clarifier
WS: Waste sludge

FIG. 7.25.6 Activated-sludge process flow diagrams. **A,** Oxidation ditch; **B,** Deep-shaft activated sludge process.

To use this analytical expression of organic loading, environmental engineers must collect or assume data on the wastewater to be treated. The concentration of biodegradable organic material is expressed as BOD_5. For municipal wastewater, the BOD_5 ranges from 100 to 300 mg/l. The volume of wastewater to be treated is based on historical flow measurements plus an estimation of any increase or decrease anticipated during the life of the treatment plant.

The viable microorganisms in the activated-sludge process are expressed in terms of MLSS. MLSS is not the concentration of viable microorganisms but an indication of the microorganisms present in the system. Environmental engineers use the MLSS concentration because measuring the actual number of viable organisms in the system is difficult. The organic loading equation represents the ratio of the weight of organic material fed to the total weight of microorganisms available for oxidation.

Environmental engineers choose the organic loading on the basis of the desired effluent quality. If the organic loading is maintained at a high level, the effluent quality is poor, and solids (excess microorganisms) production is high. As the organic loading is reduced, however, the quality of the effluent improves, and the sludge production decreases. Table 7.25.1 shows the effect of organic loading.

MICROORGANISM CONCENTRATION

Since the concentration of microorganisms (MLSS) maintained in the contactor has a direct effect on the oxidation of organic pollutants, the liquid–solids separation characteristics of these solids are important. The SVI value indicates the ability of microorganisms separate from the wastewater after contact.

The SVI is defined analytically as the volume in milliliters occupied by 1 g of MLSS after a 1-l sample has settled in a graduated cylinder for 30 min. The SVI value for an activated-sludge system varies with the concentration of microorganisms maintained in the contactor. Table 7.25.2 reflects this point by listing identical settling characteristics as indicated by SVI values for various MLSS concentrations.

The table shows that the same SVI value of 100 can be observed for MLSS concentration from 500 to 8000 mg/l, yet the volume occupied by the MLSS after 30 min of settling is in the same proportion as the MLSS concentration. Therefore, the SVI value is meaningful only in indicating separation characteristics of solids at a particular concentration. If the same 30-min MLSS volume were required for a concentration of 8000 mg/l compared to 500 mg/l, the SVI value would have to be 6 compared to 100 at 500 mg/l MLSS.

The SVI value is of operational importance since it reflects changes in the treatment system. Any increase of SVI with no increase of MLSS concentration indicates that the solids settling characteristics are changing and a plant upset can occur.

Figure 7.25.7 shows the relationship between the MLSS concentration, SVI, and the recycling ratio (R/Q). The amount of recycled flow depends largely on the settling characteristics of the MLSS. For example, if the SVI value is 400 and the required MLSS concentration is 2000 mg/l, a recycling ratio of about 3.5 is required. On the other hand, if the SVI is 50, the recycling ratio required is about 0.2. This relationship demonstrates that the settling characteristics of the formed biological solids are important to the successful operation of the activated-sludge process.

For municipal wastewater, environmental engineers use an SVI value of approximately 150 and a MLSS concentration of 2000 mg/l for design. To achieve the required MLSS concentration in the contactor they use a recycling ratio of about 0.5.

CONTACTOR RETENTION TIME

In the activated-sludge process, the liquid retention time in the contactor is not a fundamental design consideration, particularly for industrial waste. The reason is that both the concentration of organic material (BOD_5) and the MLSS concentration can vary greatly for a wastewater or activated-sludge system, and both have a more pronounced

TABLE 7.25.1 EFFECT OF ORGANIC LOADING ON ORGANICS REMOVAL EFFICIENCY AND EXCESS SLUDGE PRODUCTION

Design Parameter	Organic Loading (lb BOD_5/lb MLSS-day)				
	0.1[a]	0.3[b]	0.5[b]	1.0[c]	1.5[c]
BOD_5 removal efficiency	95	90	90	75	70
Excess sludge produced (lbs/lb BOD_5 removed)	0.2	0.4	0.5	0.6	0.7

Notes: [a]Extended-aeration, activated sludge
[b]Conventional activated sludge
[c]High-rate, activated sludge

TABLE 7.25.2 EFFECT OF MLSS CONCENTRATION ON SETTLED VOLUME FOR A CONSTANT SVI VALUE

MLSS mg/l	SVI (ml/g of settled MLSS)	Volume* (ml)
500	100	50
1000	100	100
2000	100	200
4000	100	400
8000	100	800

Note: *MLSS volume after 30 min settling.

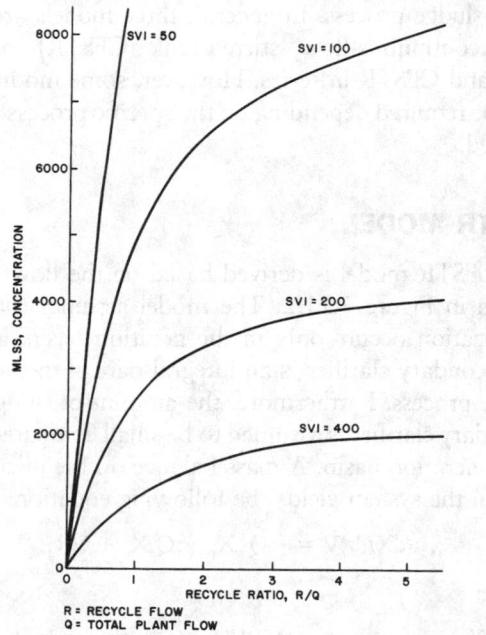

FIG. 7.25.7 Relationship between SVI, recycling ratio, and MLSS concentration.

effect on the process results than does the liquid retention time. The basic design parameters are the organic strength (BOD_5) of the wastewater and the MLSS concentration. Of these two parameters, only the MLSS can be varied by operation. Environmental engineers base their designs on the organic loading or F/M ratio, incorporating both the BOD_5 and MLSS concentration.

ARTIFICIAL AERATION

The activated-sludge process design must provide oxygenation and mixing to achieve efficient results. Current methods of accomplishing both oxygenation and mixing include 1) compressed-air diffusion, 2) sparge-turbine aeration, 3) low-speed surface aerators, and 4) motor-speed surface aerators.

Air diffusers were the earliest aeration devices used (see Figure 7.25.8). These devices compress air to the hydrostatic pressure on the diffuser (3 to 10 psig) and release it as small air bubbles. The larger the number and the smaller the size of the air bubbles produced, the better the oxygen transfer. Releasing air bubbles beneath the surface also results in airlift mixing of the contactor contents.

Combining compressed-air and turbine mixing eliminates the problems of clogging experienced with diffusers and adds versatility to the mixing and oxygen transfer. With the sparge-turbine aerator, the mixing and oxygenation can be varied independently within an operating range.

The additional development of aeration devices resulted in the elimination of compressors. The low-speed surface aerator uses atmospheric oxygen by causing extreme liquid turbulence at the surface. It is nearly twice as efficient in oxygen transfer as diffusers or sparge turbines.

The motor-speed surface aerator is the latest aeration device. This device operates at the liquid surface but does not have a gear reducer between the motor and impeller.

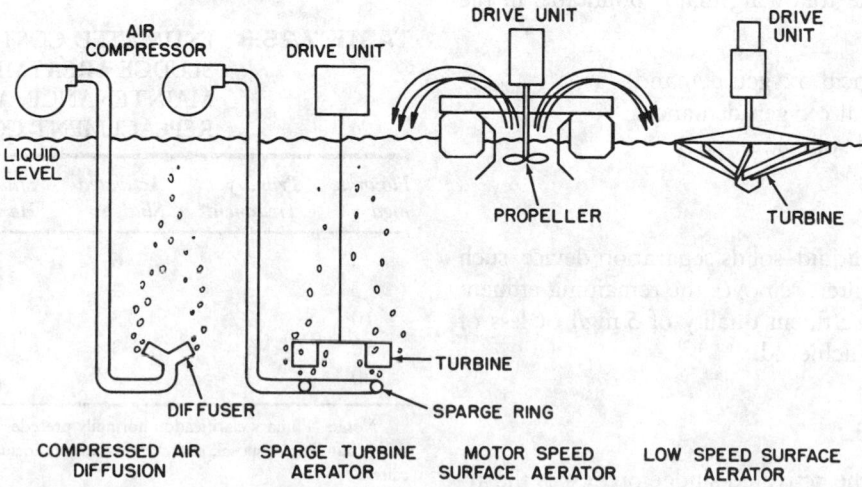

FIG. 7.25.8 Artificial oxygenation and mixing devices.

Because no gear reducer is used, the cost is significantly less than the low-speed surface aerator. Unfortunately, the oxygen transfer efficiency and liquid pumpage rate are also significantly reduced. The device has been used extensively to supplement oxygen requirements for oxidation ponds.

LIQUID–SOLIDS SEPARATION

Since the key to the activated-sludge process is maintaining a high concentration of microorganisms in the contactor, an efficient liquid–solids separation device must be employed. A major operational problem associated with the activated-sludge process is sludge bulking, a condition in which the settling characteristics of the solids make the liquid–solids separation inordinately difficult.

Among the environmental conditions causing sludge bulking are a high concentration of carbohydrates in the wastewater or a nutrient or oxygen deficiency in the system. These conditions can be rectified if they are quickly identified. Unfortunately, sludge bulking and contributing factors are not always easily identifiable, and difficult liquid–solids separation develops periodically. During these critical periods, the liquid–solids separation device must effectively separate bulking material from the wastewater and allow the solids to recycle back to the contactor. For gravity settling, a hydraulic separation rate of 250 to 500 gal per day per sq ft should be used. At a low hydraulic overflow rate, critical periods of sludge bulking can usually be handled without loss of gross solids into the effluent.

EFFLUENT QUALITY

Activated-sludge process design is based on the desired effluent quality. Successful operation of the activated-sludge process with an F/M ratio of 0.35 lb BOD_5 per lb MLSS per day and efficient liquid–solids separation should yield effluent containing an average of 20 mg/l of SS and 20 mg/l of BOD_5. For municipal wastewater, activated-sludge treatment removes the following major pollutants in the percentages listed:

90+% BOD_5 (biological oxygen demand)
70+% COD (chemical oxygen demand)
90+% SS (suspended solids)
30+% P (phosphorus)
35% N (nitrogen)

If a more efficient liquid–solids separation device, such as a granular-media filter, removes the remaining effluent solids (~20 mg/l), an effluent quality of 5 mg/l or less of BOD_5 and SS can be achieved.

PROCESS COSTS

The costliest item in the activated-sludge process is the artificial aeration device in the contactor. This equipment represents a large initial capital outlay, with high operation and maintenance costs. Treatment associated with most activated-sludge treatment plants includes primary clarification, sludge handling, and chlorination. Table 7.25.3 shows an estimated cost breakdown for the total activated-sludge treatment facility for plants varying in size from 1 to 100 mgd. The table shows the approximate cost distribution in such a facility together with the approximate cost for various plant sizes and each phase of the total treatment facility.

Process Kinetic Models

Environmental engineers have applied the principle of reactor engineering in the analysis and design of the activated-sludge process. In general, three models are widely used: continuous-flow, stirred-tank (CFSTR), plug-flow (PF), and CFSTR-in-series. However, some modifications may be required depending on the specific process diagram selected.

CFSTR MODEL

The CFSTR model is derived based on the flow diagram shown in Figure 7.21.2. The model assumes that waste stabilization occurs only in the aeration basin, although the secondary clarifier is an integral part of the activated-sludge process. Furthermore, the amount of sludge in the secondary clarifier is assumed to be small compared to that in the aeration basin. A mass balance on the microorganisms in the system yields the following equation:

$$(dX/dt)V = -Q_wX_r - Q_eX_e + VR_g \qquad 7.25(3)$$

where:

X	= MLVSS concentration in the aeration basin, mg/l
V	= aeration basin volume, l
Q_w	= sludge waste rate, l/day

TABLE 7.25.3 ESTIMATED COST OF ACTIVATED-SLUDGE TREATMENT (OPERATION, MAINTENANCE, AND REPLACEMENT COSTS)†

Flowrate mgd	Primary Treatment*	Activated-Sludge	Sludge Handling	Chlorination	Total
1	4	6	11	1	22
3	3	4	8	1	16
10	2	3	6	1	12
30	2	3	5	1	11
100	2	3	4	1	10

Notes: *Primary clarification normally precedes the activated-sludge process; sludge handling, disposal, and chlorination are included in the total treatment facility.

†Treatment costs are in cents per 400 gal of wastewater.

Q_e or $(Q - Q_w)$ = effluent rate, l/day
X_r = MLVSS concentration in the recycled sludge flow, mg/l
R_g = net growth rate of microorganisms, mg MLVSS/l-day

Under steady-state conditions, Equation 7.25(3) can be simplified to the following equations:

$$(Q_wX_r + Q_eX_e)/VX = 1/\theta_c = YU - k_d \qquad 7.25(4)$$

$$U = Q(S_i - S_e)/XV = (S_i - S_e)/X\theta \qquad 7.25(5)$$

where:

Y = bacterial yield
U = specific utilization rate, l/day
k_d = bacterial decay rate, l/day
S_i = influent BOD (after primary settling), mg/l
S_e = effluent BOD (soluble), mg/l
θ_c = SRT, days
θ = LRT in the aeration basin, days

The parameter θ_c defines the average residence time of microorganisms in the system.

A mass balance of system BOD under steady-state conditions yields the following equation:

$$(S_i - S_e)/\theta = kXS_e/(K_s + S_e) = \mu_mXS_e/Y(K_s + S_e) \qquad 7.25(6)$$

where:

k = maximum utilization rate, l/day
K_s = half-velocity constant, mg/l
μ_m = maximum growth rate, l/day

Equation 7.25(6) assumes that a hyperbolic (Monod or Michaelis-Menten) relationship exists between k (or μ_m) and the mixed-liquor BOD concentration (S) (see Figure 7.25.9). From Equations 7.25(4) to 7.25(6) the following equations can be written:

$$X = \theta_cY(S_i - S_e)/\theta(1 + k_d\theta_c) \qquad 7.25(7)$$

$$S = K_s(1 + k_d\theta_c)/[\theta_c(Yk - k_d) - 1] \qquad 7.25(8)$$

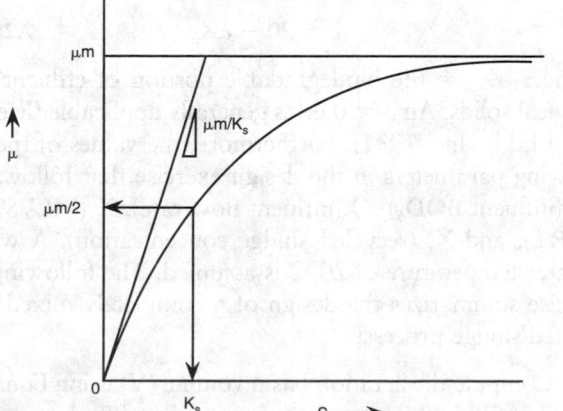

FIG. 7.25.9 Plot showing the effects of the substrate (BOD) concentration (S) on growth rate (μ). Subscript m indicates the maximum value.

$$Y_o = Y/(1 + k_d\theta_c) \qquad 7.25(9)$$

where Y_o = observed bacterial yield. The F/M ratio is a widely used term in the analysis and design of the activated-sludge process and is calculated as follows:

$$F/M = S_i/X\theta \qquad 7.25(10)$$

$$U = (F/M)E/100 \qquad 7.25(11)$$

where E = BOD removal efficiency, %.

PF MODEL

The following equations describe the PF model proposed by Lawrence and McCarty (1970):

$$1/\theta_c = Yk(S_i - S_e)/[S_i - S_e)$$
$$+ (1 + r)K_s\ln(S^*/S_e)] - k_d \qquad 7.25(12)$$
$$S^* = (S_i + rS_e)/(1 + r) \qquad 7.25(13)$$

where r = recycling ratio (Q_r/Q). Equation 7.25(12) is valid for $\theta_c/\theta > 5$.

CFSTR-IN-SERIES MODEL

The CFSTR-in-series model is used as an approximation to the PF model. A PF reactor is divided into n compartments arranged in series, with each compartment modelled as a CFSTR. Since S_e is usually small compared to K_s, the hyperbolic relationship depicted in Figure 7.25.9 can be simplified to a linear relationship with $K = k/K_s$. Therefore, Equation 7.25(6) becomes the following equation:

$$S_e/S_i = 1/(1 + KX\theta) \qquad 7.25(14)$$

Applying Equation 7.25(14) to each compartment of a PF reactor yields the following equations:

$$S_{e1}/S_i = 1/(1 + K_1X_1\theta_1)$$
$$S_{e2}/S_{e1} = 1/(1 + K_2X_2\theta_2) \qquad 7.25(15)$$

or

$$S_{e2}/S_i = 1/(1 + K_1X_1\theta_1)(1 + K_2X_2\theta_2) \qquad 7.25(16)$$

$$S_{en}/S_i = S_e/S_i = 1/(1 + K_1X_1\theta_1)(1 + K_2X_2\theta_2)..(1 + K_nX_n\theta_n)$$
$$7.25(17)$$

If all n compartments are sized for equal volume and mean K and X values are adopted for all compartments, then Equation 7.25(17) can be simplified to the following equation:

$$S_e/S_i = 1/[1 + KX(\theta/n)]^n \qquad 7.25(18)$$

Table 7.25.4 summarizes typical kinetic coefficients for municipal wastewater applications. Environmental engineers must correct all kinetic coefficients used in design equations to account for temperature effects (see Equation 7.22[4]).

TABLE 7.25.4 TYPICAL ACTIVATED-SLUDGE
KINETIC COEFFICIENTS (MUNICIPAL
WASTEWATER)

Coefficient	Range*	Typical*
k (l/day)	2–10	5
K_s (mg BOD_5/l)	25–100	60
(mg COD/l)	15–70	40
Y (mg MLVSS/mg BOD_5)	0.4–0.8	0.6
k_d (l/day)	0.025–0.075	0.06

Source: Adapted from Metcalf and Eddy, Inc., 1991, *Wastewater engineering: Treatment, disposal, and reuse,* 3d ed. New York: McGraw-Hill.

*20°C values. Equation 7.22(4) accounts for temperature effects. A typical θ value for municipal wastewater is 1.04.

The kinetic models previously derived can be applied directly in designing a variety of activated-sludge processes. However, some modifications may be necessary since the flow diagram involved can be different. The next subsections use the design of contact-stabilization and step-aeration activated-sludge processes as examples to show the modifications involved (Ramalho 1983).

CONTACT-STABILIZATION ACTIVATED-SLUDGE PROCESS

A steady-state mass balance on the microorganisms around the stabilization tank (see Figure 7.21.6) yields the following equations:

$$rQX_r + YQ(1 - \alpha)S_{iT} + YrQS_e - k_dV_sX_s - rQX_s = 0$$

or

$$V_s = [rQ(X_r - X_s + YS_e) + YQ(1 - \alpha)S_{iT}]/k_dX_s \quad \text{7.25(19)}$$

$$r = X_c/(X_s - X_c) \quad \text{7.25(20)}$$

where:

  S_{iT} = total influent BOD, mg/l
  V_s = stabilization tank volume, l
  X_s = MLVSS concentration in the stabilization tank, mg/l
  X_c = MLVSS concentration in the contact tank, mg/l
  α = soluble fraction of influent BOD

All other terms are defined previously. The example assumes that all insoluble BOD (suspended and colloidal) entrapped and adsorbed by microorganisms is stabilized in the stabilization tank.

STEP-AERATION ACTIVATED-SLUDGE PROCESS

A two-stage, step-aeration activated-sludge process (see Figure 7.21.3) shows the design procedures involved (Ramalho 1983). This example assumes that the feed

stream is equally divided between the two stages and the MLVSS concentration is constant throughout the tank. The steady-state mass balances on the microorganisms yield the following equations:

$$rQX_r - Q(r + 0.5)X = YQ(S_i - S_{e1}) - k_dXV_1 \quad \text{7.25(21)}$$

$$Q(r + 1)X - Q(r + 0.5)X = 0.5QX = YQ(S_{e1} - S_{e2}) - k_dXV_2$$
$$\text{7.25(22)}$$

A steady-state mass balance on the microorganisms around the entire system yields the following equations:

$$(Q - Q_w)X_e + Q_wX_r = YQ[(S_i - S_{e1}) + (S_{e1} - S_{e2})] - 2k_dXV$$
$$\text{7.25(23)}$$

or

$$(Q - Q_w)X_e + Q_wX_r = YQ(S_i - S_e) - 2k_dXV \quad \text{7.25(24)}$$

where $V_1 = V_2 = V$ (equal volumes for each stage) and $S_{e2} = S_e$.

The SRT and recycling ratio (r) of the process are as follows:

$$\theta_c = 2XV/[Q_wX_r + (Q - Q_w)X_e] \quad \text{7.25(25)}$$

$$r = [Y_o(S_i - S_e) - X]/(X - X_r) \quad \text{7.25(26)}$$

Equation 7.25(25) can be generalized for an n-stage, step-aeration, activated-sludge process as follows:

$$1/\theta_c = [YQ(S_i - S_e)/nXV] - k_d \quad \text{7.25(27)}$$

whereas Equation 7.25(26) is valid for any numbers of stages.

Process Design

The design of a completely mixed, activated-sludge process illustrates the general procedures involved (Metcalf and Eddy, Inc. 1991). When a BOD_5 (5-day BOD) is used and the effluent produced is to have ≤20 mg/l BOD_5, the following equation applies:

$$S_e = 20 - \alpha X_e \quad \text{7.25(28)}$$

where αX_e = the biodegradable portion of effluent biological solids. An $\alpha = 0.63$ is generally applicable (Metcalf and Eddy, Inc. 1991). Furthermore, the values of the following parameters in the design exercise that follows are S_i (influent BOD_5), Q (influent flow rate), X (MLVSS), θ_c (SRT), and X_r (recycled sludge concentration). A wastewater temperature of 20°C is assumed. The following exercise summarizes the design of a completely mixed activated-sludge process:

1. Compute the aeration basin volume (V) using Equation 7.25(7). In this case, the values of X and θ_c are assumed, and the values of kinetic coefficients are taken from Table 7.25.4. Note that θ (LRT) is defined as V/Q, where Q is the influent flow rate.

2. Compute the amount of sludge to be wasted per day. The amount of MLVSS produced due to the removal of BOD (P_x) is as follows:

$$P_x = Y_oQ(S_i - S_e)10^{-6} \text{ (kg/day)} \qquad 7.25(29)$$

Y_o is defined in Equation 7.25(9). The amount of sludge lost in the effluent is QX_e10^{-6} (kg/day). Therefore, the amount of sludge to be wasted per day is $P_x - QX_e10^{-6}$ (kg/day).

3. Compute the sludge waste rate (Q_w) using Equation 7.25(4).
4. Compute the LRT (θ) of the aeration basin.
5. Check the F/M ratio and BOD removal efficiency (E) using Equations 7.25(10) and 7.25(11), respectively.
6. Compute the oxygen requirements. The amount of BOD_L (ultimate BOD) removal is $Q(S_i - S_e)10^{-6}/f$, where f is the conversion factor for converting BOD_5 to BOD_L. (Note that f = 0.65–0.68 for municipal wastewater.) Therefore, the amount of oxygen required (kg/day) is $(Q/f)(S_i - S_e)10^{-6} - 1.42P_x$.
7. Compute actual air requirements (diffused aeration). The design air requirements (m³/day) are as follows:

$$[(SF)(Q/f)(S_i - S_e)10^{-6} - 1.42P_x]/(\gamma_a)(0.232)(\beta_d) \qquad 7.25(30)$$

where:

SF = safety factor (usually 2)
γ_a = specific weight of the air, kg/m³
β_d = oxygen transfer efficiency of the diffused aeration equipment
γ_a = p/RT
p = sum of atmospheric pressure and air-diffuser discharge pressure
R = universal gas constant
T = absolute temperature

For mechanical aeration equipment, the power requirement (kW) is calculated as follows:

$$[(SF)(Q/f)(S_i - S_e)10^{-6} - 1.42P_x]/24N \qquad 7.25(31)$$

where N = field oxygen transfer capacity of the mechanical aeration equipment (kg O₂/kW-hr). Table 7.25.5 summarizes the typical ranges of field oxygen transfer capacities of various mechanical aerators.

8. Calculate the required recycling ratio (r) as $r = X/(X_r - X)$.

The secondary clarifier design is an integral part of the overall activated sludge process design. Details on the secondary clarifier design are presented in Section 7.32.

Operational Problems

Major operational problems of the activated-sludge process are caused by bulking sludge, rising sludge, and Nocardia foam (Metcalf and Eddy, Inc. 1991). A bulking sludge has poor settleability and compactability and is usually caused by excessive growth of filamentous microor-

TABLE 7.25.5 OXYGEN TRANSFER EFFICIENCIES OF MECHANICAL AERATORS UNDER FIELD CONDITIONS

Aerator	Oxygen Transfer Rate (kg O₂/kW-hr)
Surface low-speed	0.73–1.46
Surface low-speed with draft tube	0.73–1.28
Surface high-speed	0.73–1.22
Surface downdraft turbine	0.61–1.22
Submerged turbine with sparger	0.73–1.09
Submerged impeller	0.73–1.09
Surface brush and blade (aeration rotor)	0.49–1.09

Source: Adapted from Metcalf and Eddy, Inc., 1991, *Wastewater engineering: Treatment, disposal, and reuse,* 3d ed., (New York: McGraw-Hill).
Note: Field conditions are wastewater temperature, 15°C; altitude, 152 m (500 ft); oxygen-transfer correction factor, 0.85; salinity-surface-tension correction factor, 0.9; and operating DO concentration, 2 mg/l.

ganisms. Factors such as waste characteristics and composition, nutrient contents, pH, temperature, and oxygen availability can cause sludge bulking. The absence of certain components in the wastewater such as nitrogen, phosphorus, and trace elements can lead to the development of a bulking sludge. This absence is critical when industrial wastes are mixed with municipal wastewater for combined treatment.

Wide fluctuations in pH and DO are also known to cause sludge bulking. At least 2 mg/l of DO should be maintained in the aeration basin under normal operating conditions. Wastewater treatment facilities should check the F/M ratio to insure that it is within the recommended range. They should also check the additional organic loads received from internal sources such as sludge digesters and sludge dewatering operations to avoid internal overloading conditions, especially under peak flow conditions.

Chlorination of the return sludge effectively controls filamentous sludge bulking. Chlorine doses in the range of 2–3 mg/l of Cl₂ per 1000 mg/l of MLVSS are suggested. However, high doses can be necessary under severe conditions (8 to 10 mg/l of Cl₂ per 1000 m/l of MLVSS).

Rising sludge is usually caused by the release of gas bubbles entrapped within sludge flocs in the secondary clarifier. Nitrogen gas bubbles formed by denitrification of nitrite and nitrate under anoxic secondary clarifier conditions are known to cause sludge rising. Oversaturation of gases in the aeration tank can also cause sludge rising in the secondary clarifier, especially when aeration tank depth is significantly deeper than that of the secondary clarifier (Li 1993). Reducing the sludge retention time in the secondary clarifier is effective in controlling rising sludge. Close monitoring and control of aeration in the aeration tank can also reduce rising sludge in the secondary clarifier.

Nocardia foam is associated with a slow-growing, filamentous microorganisms of the *Nocardia* genus. Some factors causing Nocardia foaming problems are low F/M ratio, long SRT, and operating in the sludge reaeration mode (Metcalf and Eddy, Inc. 1991). Reducing SRT is the most common means of controlling Nocardia foaming problems.

—*Wen K. Shieh*
Van T. Nguyen

References

Eckenfelder, W.W., Jr. 1980. *Principles of water quality management.* Boston: CBI Publishing Co. Inc.

Forster, C.F. 1985. *Biotechnology and wastewater treatment.* Cambridge: Cambridge University Press.

Irvine, R.L., and L.H. Ketchum, Jr. 1989. Sequencing batch reactor for biological wastewater treatment. *CRC Crit. Rev. Environ. Control* 18, no. 255.

Lawrence, A.W., and P.L. McCarty. 1970. Unified basis for biological treatment design and operation. *J. San. Eng. Div.,* ASCE 96, no. 757.

Li, A.L. 1993. Dynamic modeling of the deep tank aeration process. Ph.D. diss., Department of Systems, University of Pennsylvania, Philadelphia.

Metcalf and Eddy, Inc. 1991. *Wastewater engineering: Treatment, disposal, and reuse.* 3d ed. New York: McGraw-Hill.

Ramalho, R.S. 1983. *Introduction to wastewater treatment processes.* 2d ed. New York: Academic Press, Inc.

U.S. Environmental Protection Agency (EPA). 1986. *Sequencing batch reactors.* EPA/625/8-86/011. Cincinnati: U.S. EPA, Center for Environmental Research Information.

7.26
EXTENDED AERATION

LRT (hr)
　18–36

SRT (days)
　20–30

F/M (lb BOD₅/lb MLVSS-day)
　0.05–0.15

VOLUMETRIC LOADING (lb BOD₅/ft³-day)
　0.01–0.25

BOD₅ REMOVAL (%)
　85–95

MLSS (mg/l)
　3000–6000

RECYCLING RATIO
　0.5–1.5

AERATION TYPE
　Diffused aeration and mechanical aeration

Process Description

The extended-aeration process (see Figure 7.21.4) is a modification of the conventional activated-sludge process. It is commonly used to treat the wastewater generated from small installations (e.g., schools, resorts, and trailer parks) as well as small and rural communities.

In extended aeration activated-sludge detention time is increased by a factor of four or five compared to conventional activated sludge. A final settling of 4 hr is typical at a surface settling rate of 350 to 700 gpd per sq ft. The main advantage of the extended-aeration process is that the amount of excess biological solids (sludge) produced is eliminated or minimized. Wastewater treatment facilities minimize this amount by operating the process in the endogenous respiration phase with the SRT maintained in the range of 20–60 days. As a result, the cost incurred with sludge disposal is reduced.

The extended-aeration process is further simplified since only preliminary influent wastewater treatment is required to remove coarse materials; the primary clarifier is eliminated. However, the size of the aeration basin is much larger than that of the conventional activated-sludge process. This larger basin accommodates a longer LRT in the aeration basin (i.e., 16–36 hr). The effluent produced is generally low in BOD and well nitrified (Ramalho 1983). An operational problem related to nitrification is a drop in pH which treatment facilities can correct by adding lime slurry to the aeration basin.

Although the amount of excess sludge in the extended-aeration process is significantly reduced, secondary clarification is needed to remove the accumulated nonbiodegradable portion of sludge and the influent solids that are not degraded or removed. The design of secondary clarifiers is discussed in Section 7.29.

Process Flow Diagrams

In addition to the conventional extended-aeration process shown in Figure 7.21.4, a variation known as the oxidation ditch (see Figure 7.21.10) is also widely used. The aeration rotor provides the dual function of aeration and flow velocity. An alternating anoxic or oxic environment occurs in the oxidation ditch depending on the distance from the aeration rotor. As a result, an oxidation ditch achieves good nitrogen removal via nitrification and denitrification.

The process separates sludge from the treated effluent by using either an external secondary clarifier or an intra-channel clarifier. Primary clarification is usually not provided.

Several manufactured extended-aeration units are available (Viessman and Hammer 1993).

Process Design

The following equations show the design of an extended-aeration process without nitrification and with zero-sludge yield (Ramalho 1983):

$$0.77YQ(S_i - S_e) = k_dXV \qquad 7.26(1)$$

$$\theta = 0.77Y(S_i - S_e)/k_dX \qquad 7.26(2)$$

where:

Y = sludge yield
Q = influent flow rate, l/day
S_i = influent BOD, mg/l
S_e = effluent BOD, mg/l

k_d = sludge decay (endogenous respiration) rate, l/day
X = MLVSS concentration in the aeration basin, mg/l
V = basin volume, l
θ = LRT, day

These equations incorporate 0.77 since approximately 77% of the sludge produced is biodegradable.

The recycling ratio (r) is calculated as follows:

$$r = [X - 0.23Y(S_i - S_e)]/(X_r - X) \qquad 7.26(3)$$

where X_r = MLVSS concentration in the recycled sludge flow, mg/l.

—Wen K. Shieh
Van T. Nguyen

References

Ramalho, R.S. 1983. *Introduction to wastewater treatment processes.* 2d ed. New York: Academic Press, Inc.

Viessman, W., Jr., and M.J. Hammer. 1993. *Water supply and pollution control.* 5th ed. New York: Harper Collins.

7.27
PONDS AND LAGOONS

TYPE

a. Aerobic (low-rate)
b. Aerobic (high-rate)
c. Aerobic (maturation)
d. Facultative
e. Anaerobic
f. Aerated lagoon

FLOW REGIME

a–c: intermittent mixing; d: mixed (surface layer); e: no mixing; and f: completely mixed

SURFACE AREA (acres)

a: <10; b: 0.5–2; c: 2–10; d: 2–10; e: 0.5–2; and f: 2–10

DEPTH (ft)

a: 3–4; b: 1–1.5; c: 3–5; d: 4–8; e: 8–16; and f: 6–20

LRT (days)

a: 10–40; b: 4–6; c: 5–20; d: 5–30; e: 20–50; and f: 3–10

BOD₅ LOADING (lb/acre-day)

a: 60–120; b: 80–160; c: ≤15; d: 50–180; and e: 200–500

BOD₅ REMOVAL (%)

a, b, d, & f: 80–95; c: 60–80; and e: 50–85

ALGAL CONCENTRATION (mg/l)

a: 40–100; b: 100–260; c: 5–10; d: 5–20; and e: 0–5

EFFLUENT SS (mg/l)

a: 80–140; b: 150–300; c: 10–30; d: 40–60; e: 80–160; and f: 80–250

Process Microbiology

This section briefly discusses two low-cost, suspended-growth wastewater treatment systems—stabilization ponds and aerated lagoons.

Stabilization ponds possess a similar biological community as activated-sludge with the addition of an algal population. Oxygen is supplied in an aerobic photosynthetic pond by natural reaeration from the atmosphere and algal photosynthesis. The oxygen released by photosynthetic algae is used by bacteria to degrade organic matter (see Figure 7.22.3). Degradation by bacteria releases carbon dioxide and nutrients used by algae. Higher life forms such as rotifers and protozoa are also present in the pond and function primarily as polishers of the effluent. Temperature has a significant effect on aerobic pond operation. Organic loading, pH, nutrients, sunlight, and degree of mixing also affect each microbial group's population throughout the pond.

In facultative ponds, two different biological communities exist. The microbial community of the pond's upper layer is similar to that of an aerobic pond, whereas microorganisms in the lower and bottom layers are facultative and anaerobic (see Figure 7.27.1). Respiration occurs in the presence of sunlight; however, the net effect is oxygen production, i.e., photosynthesis. During photosynthesis, algae uses carbon dioxide, resulting in high pH levels in low alkalinity wastewater. In facultative ponds, algae can use bicarbonate as a carbon source for cell growth; when this occurs, there is a diurnal fluctuation in pH. However, at high-pH levels, carbonate and hydroxide species predominate. In wastewater containing a high concentration of calcium, calcium carbonate precipitates preventing the pH from rising any higher.

The microbiology involved in an aerated-lagoon process is similar to that of an activated-sludge process. However, differences arise because the large surface area of lagoons can cause more temperature effects than normally encountered in conventional activated-sludge processes. Aerobic digestion—a process that treats organic sludges produced from various treatment operations—is similar to the activated-sludge process.

When the available substrate supply is depleted, microorganisms consume their own protoplasm to obtain energy for cell maintenance and enter the endogenous phase. Cell tissue is aerobically oxidized to carbon dioxide, water, and ammonia. The cell tissue actually oxidized is 70 to 80%, with inert components and nonbiodegradable organic matter remaining. The ammonia produced in cell tissue oxidation is eventually oxidized to nitrate digestion proceeds.

Stabilization Ponds

A stabilization pond is a low-cost treatment process widely used in small communities and industrial facilities. It is a shallow body of wastewater contained in an earthen basin, using a completely mixed biological process without solids return. Mixing is usually provided by natural processes such as wind, heat, or fermentation; however, mixing can be induced by mechanical or diffused aeration.

One of three types of environmental conditions—fostering three corresponding types of biological activity—

can prevail in a stabilization pond process: aerobic, aerobic–anaerobic, and anaerobic. Aerobic ponds are used primarily for treating soluble organic wastes and effluents from wastewater treatment plants. Facultative ponds (see Figure 7.27.1), in which aerobic–anaerobic conditions exist, are the most common type and are used to treat domestic waste and a variety of industrial waste. Anaerobic ponds are applied where rapid stabilization of strong organic waste is required (see Figure 7.22.4). Wastewater treatment facilities commonly use anaerobic ponds in series with facultative ponds to provide complete treatment.

Aerobic and facultative ponds are biologically complex. Figure 7.27.2 shows the general reactions that occur. Part of the organic matter in the influent is oxidized by bacteria, producing ammonia, carbon dioxide, sulfate, water, and other end products of aerobic metabolism. These products are subsequently used by algae during daylight to produce oxygen. Bacteria use this supplemental oxygen and the oxygen provided by wind action to decompose the other part of the organic matter.

In states where stabilization-pond-treatment processes are commonly used, regulations govern pond design, installation, and operation. A minimum retention time of 60 days is often required for flow-through facultative ponds receiving untreated wastewater (Metcalf and Eddy, Inc. 1991). Frequently, retention times as high as 120 days are specified. However, even with a low retention time of 30 days, a high degree of coliform removal is ensured (Metcalf and Eddy, Inc. 1991). Other typical standards (see Figure 7.27.3) include embankment slopes (1:3 to 1:4), organic loading rate (2.2 to 5.5 g BOD/m^2-day, depending on climate), and permissible seepage through the bottom (0 to 6 mm/day). In some climates, treatment facilities can operate ponds without discharge to surface waters (McGhee 1991).

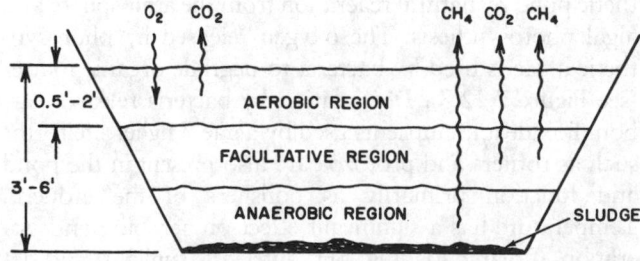

FIG. 7.27.1 Elevation diagram of facultative lagoon strata and operation.

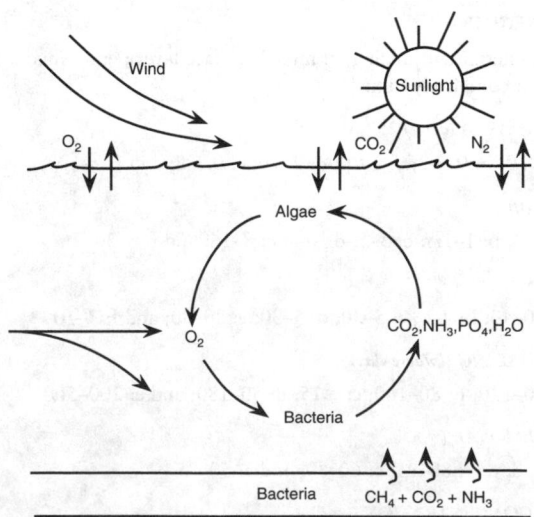

FIG. 7.27.2 Schematic of a stabilization pond.

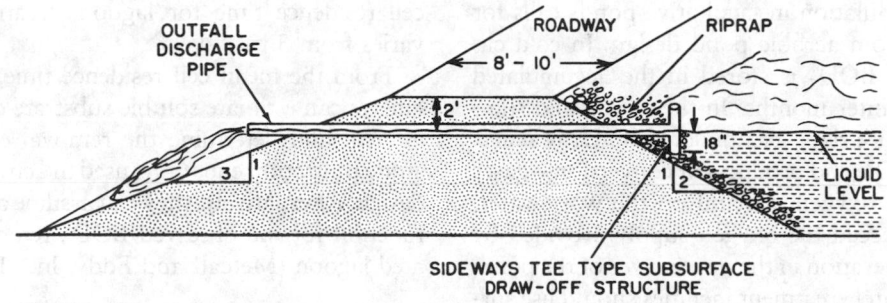

FIG. 7.27.3 Levee and outfall structure.

AEROBIC PONDS

Of all the biological treatment process designs, the stabilization pond design is the least defined. Aerobic stabilization ponds contain bacteria and algae in suspension under aerobic conditions. Aerobic ponds are of two basic types. In one type, the objective is to maximize algae production. These aerobic ponds generally operate at depths of 0.15 m to 0.45 m. In the other type of aerobic ponds, the amount of oxygen produced is maximized, and depths range to 1.5 m. Shallower depths encourage rooted aquatic plant growth, interfering with the treatment process. However, greater depths can interfere with mixing and oxygen transport from the surface. To achieve the best results with aerobic ponds, wastewater treatment facilities should provide mixing with pumps or surface aerators.

Environmental engineers usually base the aerobic pond process design on the organic loading rates and hydraulic retention times derived from pilot-plant studies and observations of operating systems. They adjust the pond loading rate to reflect the oxygen available from photosynthesis and atmospheric reaeration. Frequently, environmental engineers design large aerobic pond systems as completely mixed reactors, with two or three reactors in series.

Another design approach involves the use of a first-order, removal-rate equation developed by Wehner and Wilhelm (Metcalf and Eddy, Inc. 1991). This equation describes the substrate removal for an arbitrary flow-through pattern that lies somewhere between completely mixed and plug-flow as follows:

$$S/S_o = 4ae^{-(1/2d)}/[(1 + a)^2 e^{-(a/2d)} - (1 - a)^2 e^{-(a/2d)}] \quad \text{7.27(1)}$$

where:

S = effluent substrate concentration, mg/l
S_o = influent substrate concentration, mg/l
a = $(1 + 4ktd)^{1/2}$
d = dispersion factor (D/uL)
D = axial dispersion coefficient, (m^2/h)
u = fluid velocity (m/h)
L = characteristic length (m)
k = first-order reaction constant (h^{-1})
t = retention time (h)

The term kt in Equation 7.27(1) can be plotted as a function of S/S_o for various dispersion factors—varying from zero for PF reactors to infinity for completely mixed reactors—to yield a graph that facilitates the use of the equation in designing ponds (Metcalf and Eddy, Inc. 1991). The dispersion factor ranges from 0.1 to 2.0 for most stabilization ponds. For aerobic ponds, the dispersion factor is approximately 1.0 since completely mixed conditions usually prevail in these ponds for high performance. Depending on the operational and hydraulic characteristics of the pond, typical values for the overall first-order BOD_5 removal-rate constant k range from 0.05 to 1.0 per day (Metcalf and Eddy, Inc. 1991).

Although aerobic pond efficiency is high—up to 95%—and most soluble BOD_5 is removed from influent wastewater, bacteria and algae in the effluent can exert a BOD_5 higher than that of the original waste. Hence, wastewater treatment facilities must apply methods of removing biomass from the effluent.

FACULTATIVE PONDS

In facultative ponds, waste conversion is performed by a combination of aerobic, anaerobic, and facultative bacteria. As shown in Figures 7.27.1 and 7.27.2, the facultative pond is comprised of three zones: (1) a surface zone where algae and bacteria thrive symbiotically, (2) an anaerobic zone at the bottom sludge layer where accumulated organics are decomposed by anaerobic bacteria, and (3) an aerobic–anaerobic zone in the middle where facultative bacteria are responsible for waste conversion. Using the oxygen produced by algae growing near the surface, aerobic and facultative bacteria oxidize soluble and colloidal organics, producing carbon dioxide. This carbon dioxide is used by the algae as a carbon source. Anaerobic waste conversion in the bottom zone produces dissolved organics and gases such as CH_4, CO_2, and H_2S that are either oxidized by aerobic bacteria or released to the atmosphere.

Facultative stabilization pond designs (see Figure 7.21.9) are similar to those of aerobic ponds; i.e., they are usually based on loading factors developed from field experience. Unlike aerobic ponds, facultative ponds promote settling of organics to the anaerobic zone. Therefore, quiescent conditions are required, and dispersion factors in facultative ponds vary from 0.3 to 1.0 (Metcalf and Eddy, Inc. 1991).

The sludge accumulation in facultative ponds calls for another deviation from aerobic pond design. In cold climates, a portion of BOD_5 is stored in the accumulated sludge during the winter months. In the spring and summer as the temperature rises, accumulated BOD_5 is anaerobically converted. The end products of conversion—gases and acids—exert an oxygen demand on the wastewater. This demand can exceed the oxygen supply provided by algae and surface reaeration in the upper layer of the pond. In this case, wastewater treatment facilities should use surface aerators capable of satisfying 175 to 225% of the incoming BOD_5. The accumulation of sludge in the facultative pond can also lead to a higher SS concentration in the effluent, reducing overall pond performance.

ANAEROBIC PONDS

Anaerobic ponds (see Figure 7.21.9) treat high-strength wastewater with a high solids concentration. These are deep earthen ponds with depths to 9 m to conserve heat energy and maintain anaerobic conditions. Influent waste settles to the bottom, and partially clarified effluent is discharged to another treatment process for further treatment. Anaerobic conditions are maintained throughout the depth of the pond except for the shallow surface zone. Waste conversion is performed by a combination of precipitation and anaerobic metabolism of organic wastes to carbon dioxide, methane and other gases, acids, and cells. On the average, anaerobic ponds achieve BOD_5 conversion efficiencies to 70%, and under optimum conditions, 85% efficiencies are possible (Metcalf and Eddy, Inc. 1991).

Aerated Lagoons

Aerated lagoons are basins where wastewater can be treated in a flow-through only manner or without solids recycling. Lagoon depths vary from 1 to 4 m (Ramalho 1983). Oxygenation of the wastewater in lagoons is usually accomplished by surface, turbine, or diffused aeration. The turbulence created by aeration keeps lagoon contents suspended. Depending on the retention time, the aerated lagoon effluent contains approximately one-third to one-half the value of the incoming BOD in the form of cellular mass. Wastewater treatment facilities can use a settling basin (see Figure 8.1.1 in Chapter 8) or tank for solids removal, by settling, from the effluent prior to discharge.

In designing an aerated lagoon, environmental engineers must incorporate the following parameters: (1) BOD removal, (2) effluent characteristics, (3) temperature effects, and (4) oxygen requirements. The design basis for a lagoon can be the mean cell residence time since the aerated lagoon is a completely mixed reactor without recycling. Selected mean cell residence time should ensure that the suspended biomass has good settlement properties, and be high enough to prevent cell wash-out. Typical design mean cell residence time for lagoons treating domestic waste varies from 3 to 6 days.

From the mean cell residence time, environmental engineers can estimate soluble substrate concentration in the effluent and determine the removal efficiency from substrate utilization equations used in activated-sludge process design. Alternatively, they can assume a first-order removal function for the observed BOD_5 removal in a single aerated lagoon (Metcalf and Eddy, Inc. 1991) as follows:

$$S/S_o = 1/[1 + k(V/Q)] \qquad 7.27(2)$$

where:

S = effluent BOD_5 concentration, mg/l
S_o = influent BOD_5 concentration, mg/l
k = overall, first-order, BOD_5, removal-rate constant, day^{-1}
V = volume, l
Q = flow rate, l/day

The values for the removal-rate constant k vary from 0.25 to 1.0. Effluent characteristics of significance are the BOD_5 and the SS concentration. Environmental engineers can estimate both of these characteristics using the equations presented in Section 7.25 for calculating similar parameters in an activated-sludge effluent.

The effect of temperature on biological activity is described in Section 7.22. When influent wastewater temperature, ambient temperature, lagoon surface area, and wastewater flow rate are known, environmental engineers can estimate the resulting temperature in the lagoon using the following equation (Metcalf and Eddy, Inc. 1991):

$$(T_i - T_w) = [(T_w - T_a)fA]/Q \qquad 7.27(3)$$

where:

T_i = influent wastewater temperature
T_w = lagoon wastewater temperature
T_a = ambient temperature
f = a proportionality factor that incorporates heat transfer coefficients and the effects of surface area increase due to aeration, wind, and humidity (typical value for the eastern United States is 0.5 in SI units)
A = lagoon surface area
Q = wastewater flow rate

Oxygen requirements are computed as outlined in the design calculations for aeration in the activated-sludge process (see Section 7.25).

Anaerobic Lagoons

Anaerobic lagoons produce noxious odors that result when the acid-producing bacteria reduce sulfate compounds to hydrogen sulfide (H_2S). As the acid producers deplete the nonsulfate sources of combined oxygen, they start to reduce sulfate for oxygen with the liberation of H_2S, which has the odor of rotten eggs. At low concentrations, H_2S is merely obnoxious and therefore a nuisance, but at higher concentrations, it attacks painted surfaces and is also deleterious if inhaled for an extended period. To operate an anaerobic lagoon, wastewater treatment facilities must minimize the liberation of H_2S to eliminate these effects. They can accomplish this task by controlling the concentration of sulfate compounds in the waste contents.

If a sulfate concentration of less than 100 mg/l is maintained in the influent to the lagoon, no significant odor problems occur. If odor is a problem, the facility can add nitrate to the lagoon to alleviate it temporarily. When nitrate is applied, the acid producers switch to nitrate for oxygen. Sulfate reduction is thus stopped. This measure is only temporary; the only real long-term solution is to limit sulfate concentration in the influent. Due to the odor problem, anaerobic lagoons must be located in remote areas if sulfate starvation is not practiced.

DIMENSIONS

An anaerobic lagoon is similar in construction to an aerobic lagoon in levee dimensions and construction materials. The anaerobic lagoon, however, usually requires less surface area than the aerobic facility. Since oxygen transfer from the atmosphere is not important, the anaerobic lagoon can be as deep as is practical. A depth of at least 15 ft is recommended whenever groundwater considerations and area geology permit. The relative depth of an anaerobic lagoon provides improved heat retention. The lagoon should be as long as practical (an efficient length to width ratio is 2:1).

The BOD loading rate in anaerobic lagoon design is 500 to 1000 lb BOD per acre per day with an expected BOD removal efficiency of 50 to 80%. The required detention time is between 30 and 50 days. The ideal pH range for the anaerobic process is 6.6 to 7.6, but lagoon efficiency is not significantly hampered if pH is gradually increases to 9.0. Above pH 9.0, the efficiency drops off rapidly. Sudden bursts of high and low pH also hinder lagoon performance.

Because of the buffering effect provided by liberating carbon dioxide in the anaerobic process, the lagoon can also act as an effective neutralization system. It is capable of neutralizing approximately 0.5 lb of caustic per lb of BOD removed while the lagoon is buffered at a pH of roughly 8.0.

The anaerobic process functions optimally over two temperature ranges: the mesophylic range of 85° to 100°F and the thermophilic range of 120° to 135°F. Only the mesophylic range, however, applies to an unheated lagoon. The lagoon is optimally effective when the temperature range for mesophylic operation is not violated. However, as temperatures decrease below 85°F, the lagoon efficiency decreases only slightly until a temperature of about 60°F is reached, at which point the efficiency drops off rapidly.

This temperature requirement is why the lagoon should be as deep as possible, i.e., to maximize heat retention.

APPLICATIONS

The aerobic lagoon is applicable to lower strength wastes (usually with a BOD of less than 200 mg/l), which are not toxic to an algal system. The anaerobic lagoon is applicable to high-strength wastes (usually greater than 500 mg/l of BOD) and applications in which a highly purified effluent is not required. In anaerobic lagoons, either the sulfate concentration must be low (less than 100 mg/l), or the lagoon must be in a remote location. The facultative lagoon is applicable to wastes of approximately 200 to 500 mg/l BOD concentration. The waste cannot be toxic to algae or contain a large sulfate concentration.

In all three lagoons, a large amount of land must be available since each lagoon requires many acres for construction. Lagoons are much less susceptible to upsets from accidental discharges or large loading variations than other methods of biological treatment. Therefore, they are applicable in these situations.

Frequently, more than one type of lagoon is used. For example, additional effluent treatment from an anaerobic lagoon can be provided in a facultative or aerobic lagoon. The initial treatment in an anaerobic lagoon often renders the waste more amenable to aerobic treatment. The use of two lagoons in series further purifies the effluent and requires less land area.

COSTS

The primary investment associated with constructing a lagoon is the cost of the land and the excavation and earthmoving costs in constructing the basin. If the soil where the lagoon is constructed is permeable, an additional cost for lining is incurred.

In the midwest region of the United States, except for major population centers, the price of land is about $1500 per acre. Excavation costs vary and depend on whether dirt must be introduced or hauled away. If the dirt removed from the lagoon floor can be used for levee construction, excavation costs are roughly $2.00 per cu yd of dirt excavated. Levees are frequently compacted sufficiently by earthmoving equipment, but compacting equipment, if required, costs from $3 to $5 per cu yd. Synthetic lining material is expensive, and its use should be avoided wherever possible. The price for most plastic liners is about $1 per sq yd.

Operating costs are almost zero. In most cases, neither pumps nor any other electrically operated device is required. Therefore, power costs are usually nonexistent. Although some analytical work is required to assure proper operation, the extent of such a program is minimal compared to other methods of biological and chemical treatment. An extensive sampling system is usually not required to obtain samples for analysis. Due to the equalization effect of a large facility, a daily grab sample usually produces the necessary operational information.

Operational personnel are not required except for sampling, analysis, and general upkeep; the system is virtually maintenance-free.

—Wen K. Shieh
Van T. Nguyen

7.28
ANAEROBIC TREATMENT

REACTOR TYPE

    a. Anaerobic contact process
    b. Anaerobic upflow sludge blanket (USB) reactor
    c. Anaerobic filter (downflow or upflow)
    d. Anaerobic fluidized-bed reactor (AFBR)

LRT (hrs)

    a: 2–10; b: 4–12; c: 24–48; and d: 5–10

ORGANIC LOADING (lb COD/ft³-day)

    a: 0.03–0.15; b: 0.25–0.75; c: 0.06–0.3; and d: 0.3–0.6

COD REMOVAL FOR INDUSTRIAL WASTEWATER (%)

    a: 75–90; b: 75–85; c: 75–85; and d: 80–85

OPTIMAL TEMPERATURE (°C)

    30–35 (mesophilic) and 49–57 (thermophilic)

OPTIMAL pH

    6.8–7.4

OPTIMAL TOTAL ALKALINITY (mg/l as CaCO₃)

    2000–3000

OPTIMAL VOLATILE ACIDS (mg/l as acetic acid)

    50–500

Process Description

Anaerobic treatment applies to both wastewater treatment and sludge digestion. This section discusses only anaerobic wastewater treatment. Anaerobic wastewater treatment is an effective biological method for treating many organic wastes. The microbiology involved in the process includes facultative and anaerobic microorganisms, which, in the absence of oxygen, convert organic materials into gaseous end products such as carbon dioxide and methane.

Anaerobic wastewater treatment was discovered in the middle of the last century; however, environmental engineers have only seriously considered it in the last twenty years (Forster 1985). Despite intense research in this field

References

McGhee, T.J. 1991. *Water supply and sewerage.* 6th ed. New York: McGraw-Hill.
Metcalf and Eddy, Inc. 1991. *Wastewater engineering: Treatment, disposal, and reuse.* 3d ed. New York: McGraw-Hill.
Ramalho, R.S. 1983. *Introduction to wastewater treatment processes.* 2d ed. New York: Academic Press, Inc.

in the past few decades, much research is still needed in several areas. These areas include (Forster 1985):

Microbiology—Further research on the biochemistry and genetics related to the anaerobic microbial species is required.

Start up procedures—Optimal procedures to minimize the lag time between the commissioning of a reactor and its placement into full operation must be investigated.

Optimization of process engineering—Further optimization of the anaerobic treatment process is required, especially involving ancillary equipment, small-scale reactors, and support media (where applicable).

The major advantages of anaerobic treatment over aerobic treatment are as follows:

The biomass yield for anaerobic processes is much lower than that for aerobic systems; thus, less biomass is produced per unit of organic material used. This reduced biomass means savings in excess sludge handling and disposal and lower nitrogen and phosphorus requirements.

Since aeration is not required, capital costs and power consumption are lower.

Methane gas produced in anaerobic processes provides an economically valuable end product.

The savings from lower sludge production, electricity conservation, and methane production range from $0.20 to $0.50 per 1000 gal of domestic sewage treatment (Jewell 1987). The reduction of sludge and aeration energy consumption each result in savings that are greater than the cost of the energy required by the anaerobic process (Jewell 1987). In addition, a substantial part of the energy requirements for anaerobic processes can be obtained from exhaust gas.

Higher influent organic loading is possible for anaerobic systems than for aerobic systems because the anaerobic process is not limited by the oxygen transfer capability at high-oxygen utilization rates in aerobic processes.

However, some disadvantages are associated with the anaerobic process as follows:

Energy is required by elevated reactor temperatures to maintain microbial activity at a practical rate. (Generally, the optimum temperature for anaerobic processes is 35°C.) This disadvantage is not serious if the methane gas produced by the process can supply the heat energy.

Higher detention times are required for anaerobic processes than aerobic treatment. Thus, an economical treatment time can result in incomplete organic stabilization.

Undesirable odors are produced in anaerobic processes due to the production of H_2S gas and mercaptans. This limitation can be a problem in urban areas.

Anaerobic biomass settling in the secondary clarifier is more difficult to treat than biomass sedimentation in the activated-sludge process. Therefore, the capital costs associated with clarification are higher.

Operating anaerobic reactors is not as easy as aerobic units. Moreover, the anaerobic process is more sensitive to shock loads (Benefield and Randall 1980).

Process Microbiology

The end products of anaerobic degradation are gases, mostly methane (CH_4), carbon dioxide (CO_2), and small quantities of hydrogen sulfide (H_2S) and hydrogen (H_2). The process involves two distinct stages: acid fermentation and methane fermentation.

In acid fermentation, the extracellular enzymes of a group of heterogenous and anaerobic bacteria hydrolyze complex organic waste components (proteins, lipids, and carbohydrates) to yield small soluble products. These simple, soluble compounds (e.g., triglycerides, fatty acids, amino acids, and sugars) are further subjected, by the bacteria, to fermentation, β-oxidations, and other metabolic processes that lead to the formation of simple organic compounds, mainly short-chain (volatile) acids (e.g., acetic [CH_3COOH], propionic [CH_3CH_2COOH], butyric [CH_3-CH_2-CH_2-$COOH$]) and alcohols. In the acid fermentation stage, no COD or BOD reduction is realized since this stage merely converts complex organic molecules to short-chain fatty acids, alcohols, and new bacterial cells, which exert an oxygen demand.

In the second stage, short-chain fatty acids (other than acetate) are converted to acetate, hydrogen gas, and carbon dioxide—a process referred to as *acetogenesis*. Subsequently, several species of strictly anaerobic bacteria bring about *methanogenesis*—a process in which hydrogen produces methane from acetate and carbon dioxide reduction. In this stage, the stabilization of the organic material truly occurs. Figure 7.28.1 shows the two stages of

anaerobic treatment as sequential processes; however, both stages occur simultaneously and synchronously in an active, well-buffered system.

The main concern of a wastewater treatment facility in operating an anaerobic system is that the various bacterial species function in a balanced and sequential way (Forster 1985). Hence, although other types of microorganisms may be present in the reactors, attention is focussed mostly on the bacteria.

The major groupings of bacteria, as numbered in Figure 7.28.1, and the reactions they mediate are as follows (Pavlostathis and Giraldo-Gomez 1991): (1) fermentative bacteria, (2) hydrogen-producing acetogenic bacteria, (3) hydrogen-consuming acetogenic bacteria, (4) carbon-dioxide-reducing methanogens, and (5) aceticlastic methanogens. Two common genera of aceticlastic methanogens are *Methanothrix* and *Methanosarcina*; and species from the *Methanobacterium* group are commonly known to produce methane by hydrogen reduction of carbon dioxide.

Environmental Factors

Facultative and anaerobic bacteria associated with the acid fermentation process are tolerant to changes in pH and temperature and have a higher growth rate than the methanogenic bacteria from the second stage. Hence, the methane fermentation stage is the rate-limiting step in anaerobic processes. Since methane fermentation controls the process rate, maintaining optimal operating conditions in this stage is important.

Within the pH range of 6.0–8.5, the rate of methane fermentation is somewhat constant; outside this range, the rate drops dramatically (Benefield and Randall 1980). Other research has shown that the optimum pH range is 6.8 to 7.4 (Ramalho 1983). The alkalinity produced from the degradation of organic compounds in the anaerobic process helps control the pH by buffering the anaerobic system. The alkalinity, at typical fermentation pH levels of approximately 7, is primarily in the form of bicarbonates. Carbon dioxide comprises 30–40% by volume of the

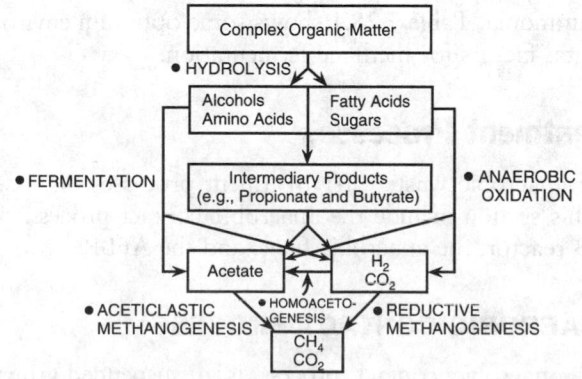

FIG. 7.28.1 Reaction pathways of anaerobic treatment of complex organic matter.

off gas from anaerobic treatment. Thus, within the operating pH range of 6.6–7.4, the alkalinity concentration can vary from 1000 to 5000 mg/l as calcium carbonate.

Another parameter requiring control is the reactor retention time for methanogens; it must be adequate to prevent cell wash-out. Research shows that the required retention time varies from 2 to 20 days (Ramalho 1983).

In the methanogenesis stage, approximately 70% of methane produced is formed from the methyl group of the acetate by acetophilic methanogens, while the remainder of the methane is formed from the oxidation of hydrogen by hydrogenophilic methanogens. The partial pressure of hydrogen is thought to regulate both intermediate fatty acid catabolism and methane formation (Forster 1985). Thus, the methanogens must maintain a low hydrogen concentration.

Hydrogen is also an inhibitory substance in methane production when a high concentration of sulfate ions is present. Sulfate-reducing bacteria, such as *Desulfovibrio*, compete for acetate and hydrogen and use them more effectively than methanogens to convert sulfate to sulfide. Therefore, the methane production is diminished. A secondary inhibition of methanogenesis occurs if the soluble sulfide ion concentration becomes greater than 200 mg/l (Forster 1985).

Cation concentration has been shown to affect the rate of methane formation (Benefield and Randall 1980). At low concentrations, cations stimulate the fermentation rate. However, the rate decreases when the optimum concentration is exceeded. The intensity of rate reduction depends on the extent that the optimum concentration is exceeded. For example, concentrations of calcium within the range of 100–200 mg/l have a stimulatory effect, while concentrations between 2500–4500 mg/l are moderately inhibitory, and concentrations of 8000 mg/l or higher are strongly inhibitory to methane fermentation (Benefield and Randall 1980).

Ammonia concentrations have a similar effect on the rate of methane fermentation as cation concentrations. However, one distinction is that the process pH determines the distribution between free ammonia and the ammonium ion. High pH levels favor free ammonia—the toxic form of ammonia. Table 7.28.1 shows some optimum environmental factors for methane fermentation.

Treatment Processes

The anaerobic wastewater treatment processes discussed in this section include the anaerobic contact process, the USB reactor, the anaerobic filter, and the AFBR.

ANAEROBIC CONTACT PROCESS

The anaerobic contact process is a suspended-growth process, similar in design to the activated-sludge process except that anaerobic conditions prevail in the former

TABLE 7.28.1 ENVIRONMENTAL FACTORS FOR METHANE FERMENTATION

Parameter	Optimum	Extreme
Temperature (°C)	30–35	25–40
pH	6.8–7.4	6.2–7.8
Volatile acids (mg/l as acetic acid)	50–500	2000
Alkalinity (mg/l as $CaCO_3$)	2000–3000	1000–5000

Source: Adapted from L.D. Benefield and C.W. Randall, 1980, *Biological process design for wastewater treatment* (Englewood Cliffs, N.J.: Prentice-Hall).

process. Part A in Figure 7.28.2 shows the process schematic.

The anaerobic contact process is comprised of two parts. The contact part involves thorough mixing of the wastewater influent with a well-developed anaerobic sludge culture. The separation part involves the settling out of anaerobic sludge from the treated wastewater and recycling back to the contact reactor. The process usually has a vacuum degasifier placed following the aerobic reactor to eliminate gas bubbles that cause SS in the clarifier to float.

BIOENERGY and *ANAMET* are two commercially available, proprietary anaerobic contact processes. *BIOENERGY* is a conventional anaerobic contact process that uses a thermal shock procedure to facilitate sludge separation. As the mixed liquor, at 35°C, flows from the contact reactor to the settling unit, a series of heat exchangers rapidly decreases its temperature to 25°C. This temporarily interrupts gasification allowing effective sludge–solids separation by gravity. The temperature of the recycled sludge is increased before it is returned to the contact unit.

In the *ANAMET* process, an aerobic biological treatment polishing step follows the anaerobic contact process to provide near-complete organics removal. The process recycles the sludge produced in the aerobic treatment process back to the anaerobic reactor to reduce excess sludge production across the entire system and increase biogas yield. Also, recirculation of the nutrient-containing sludge from the aerobic reactor reduces external nutrient requirements in the anaerobic reactor (Shieh and Li 1987).

Design considerations for the anaerobic contact unit are similar to those described for the activated-sludge process in Section 7.25. In addition, separation unit design can follow the prescriptions for secondary clarifiers described in Section 7.29.

USB REACTOR

USB reactor is essentially a suspended-growth reactor, but it is also a fixed-biomass process. Part B in Figure 7.28.2 shows the process schematic. This USB system is based on the development of a sludge blanket. In this sludge blanket, the component particles are aggregated to withstand the hydraulic shear of the upwardly flowing wastewater without being carried upwards and out of the reactor. The sludge flocs must be structurally stable so that hydraulic shear forces do not break them into smaller portions that can be washed out, and they should also have good settlement properties.

The wastewater is fed at the bottom of the reactor, and active anaerobic sludge solids convert the organics into methane and carbon dioxide. The anaerobic biomass is distributed over the sludge blanket and a granular sludge bed. The sludge solids concentration in the sludge bed is high—100,000 mg/l SS—and does not vary over a range of process conditions (Shieh and Li 1987). The sludge solids concentration in the sludge blanket is lower and depends on process conditions (Li 1984). The reactor can include an internal baffle system, usually referred to as a gas–liquid separator, above the sludge blanket to separate the biogas, sludge, and liquid. A patented USB reactor called the *BIOTHANE* process was developed by the Biothane Corporation in the United States.

In general, the USB process can achieve high COD removal efficiency at volumetric COD loadings up to 2.0 lb/ft^3/day and hydraulic retention times of 4 to 24 hr for a variety of wastewater (Shieh and Li 1987). The formation of granular sludge particles is essential to adequate reactor performance (Shieh and Li 1987). Granular sludges have a good settlement rate and can form a compact sludge bed with a solids concentration of 40–150 g/l (Forster 1985). Researchers report that the presence of calcium ions, adequate mixing in the sludge zone, and a low concentration of poorly flocculating suspended matter in the wastewater contribute to the formation of granular sludge particles with favorable qualities (Shieh and Li 1987).

The design of an USB reactor must provide an adequate sludge zone since most of the biomass is retained there. The sludge zone is completely mixed because the wastewater is fed into the reactor through a number of regularly spaced inlet ports (Shieh and Li 1987). Hence, the volume of the sludge zone can be determined with Equation 7.25(6).

The sludge blanket zone is another completely mixed contact unit in sequence with the sludge zone. If no waste conversion occurs in the gas–liquid separator and because the sludge blanket zone receives the effluent COD concentration from the USB reactor, the volume of the sludge blanket zone can also be determined with Equation

Key:
AR: Anaerobic reactor
B/MS: Biofilm/media separator
CZ: Clarification zone
E: Effluent
G: biogas
G/LS: Gas–liquid separator
I: Influent
RS: Return sludge
SC: Secondary clarifier
SZ: Sludge zone
WS: Waste sludge

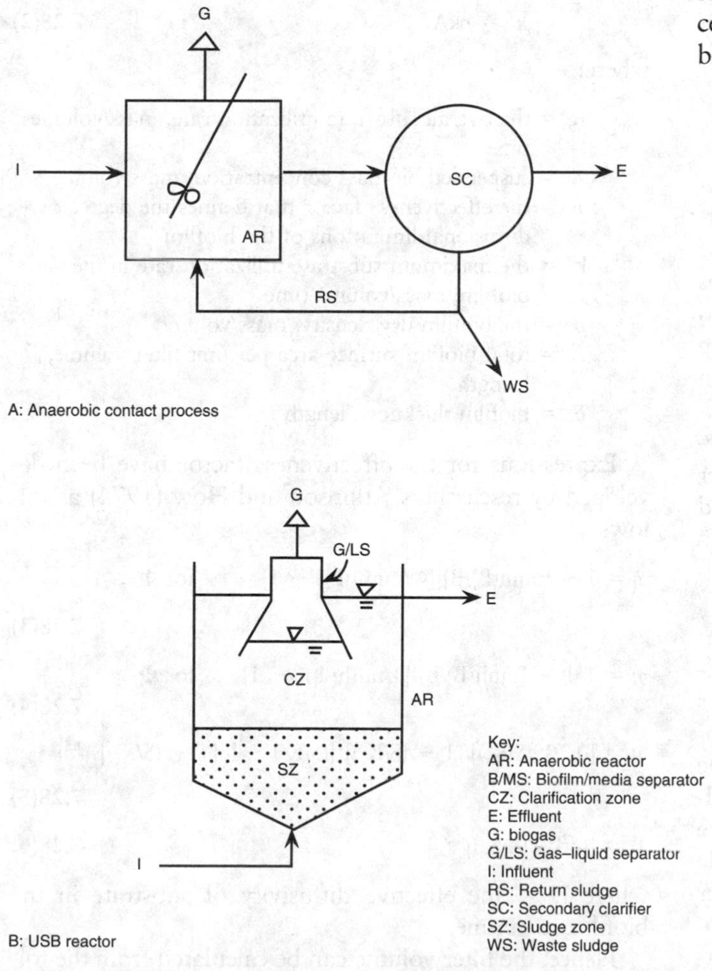

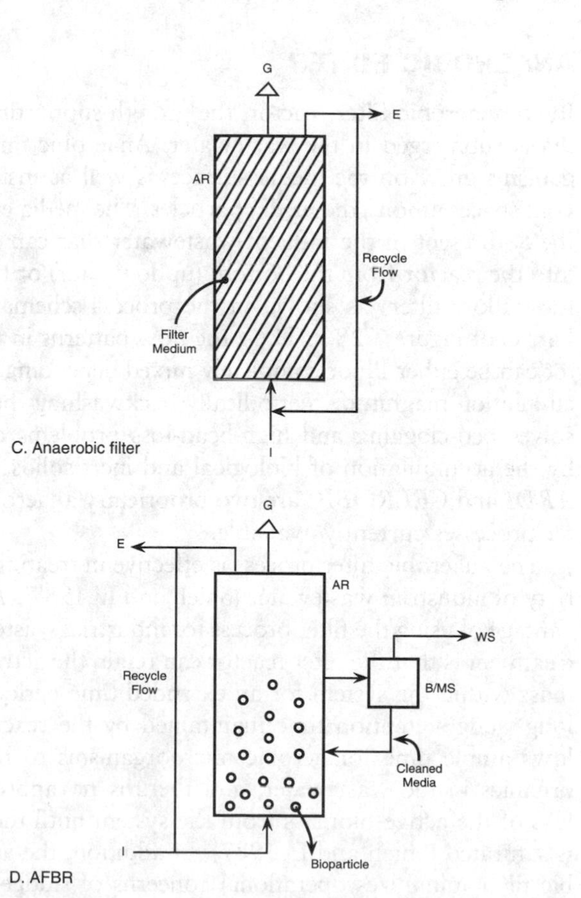

FIG. 7.28.2 Process schemes of anaerobic treatment processes. **A,** Anaerobic contact process; **B,** Upflow sludge blanket reactor; **C,** Anaerobic filter; **D,** Anaerobic fluidized bed reactor.

7.25(6). Thus, the total volume of the USB reactor, not including the gas–liquid separator, is the sum of the sludge zone volume and the sludge blanket zone volume.

The use of a gas–solids separation system in the upper portion of a USB reactor is claimed to be an essential feature, regardless of the settlement characteristics of the sludge (Forster 1985). An investigation of the effects of hydraulic and organic loading rates on the solids and design values of a particular plant is necessary to prove this claim. An increase in organic loading results in increased gas production, reduced floc density, and a greater tendency for floc flotation. The net result is a greater probability of solids wash-out (Forster 1985). This condition is exaggerated at high hydraulic loading rates. Hence, an evaluation of whether solids wash-out significantly depletes sludge solids and whether the solids concentration is tolerated in the final effluent is necessary to determine if a gas–liquid separator is required.

Researchers have reported that USB reactor performance is limited by the ability of the gas–liquid separator to retain sludge solids in the system (Hamoda and Van der Berg 1984). They maintain that the amount of sludge solids retained in the reactor increases with an increased gas–liquid volume in the separator. Lettinga et al. (1980) detail design information for a gas–liquid separator and the start up guidelines for the USB reactor.

ANAEROBIC FILTER

In an anaerobic filter reactor, the growth-supporting media is submerged in the wastewater. Anaerobic microorganisms grow on the media surface as well as inside the void spaces among the media particles. The media entraps the SS present in the influent wastewater that can be fed into the reactor from the bottom (upflow filter) or the top (downflow filter) as shown in the process schematics in Part C of Figure 7.28.2. Thus, the flow patterns in the filter can be either PF or completely mixed depending on recirculation magnitude. Periodically backwashing the filter solves bed-clogging and high-head-loss problems caused by the accumulation of biological and inert solids. *BACARDI* and *CELROBIC* are two proprietary anaerobic filter processes currently available.

The anaerobic filter process is effective in treating a variety of industrial wastewater (Shieh and Li 1987). An advantage of using the filter process for industrial wastewater treatment is that the filter reactor can retain the active biomass within the system for an extended time period. The long sludge-retention time maintained by the reactor allows ample time for aerobic microorganisms to remove organics in the wastewater, and there is no appreciable loss of the active biomass from the system until the filter is saturated (Shieh and Li 1987). In addition, the anaerobic filter minimizes operational concerns of sludge wasting and disposal because the synthesis rate of excess biomass under anaerobic conditions is low (Young and McCarty 1968).

Because it can retain a high concentration of active biomass within the system for an extended time period, the anaerobic filter can easily adapt to varied operating conditions (e.g., without significant changes in effluent quality and gas production due to fluctuations in parameters such as pH, temperature, loading rate, and influent composition). Also, intermittent shutdowns and complications in industrial treatment will not damage the filter since it can be fully recovered when it is restarted at a full load (Shieh and Li 1987).

A problem associated with the filter's ability to retain the biomass for a long time period is the close control of biomass holdup. Although periodic backwashing of the filter is a feasible method for maintaining the biomass holdup at the required level, more efficient techniques are needed.

Environmental engineers can determine the size of an anaerobic filter from the volumetric loading approach or from the biofilm kinetic theory approach described next. From the design approach commonly used in heterogeneous catalytic processes, the following expressions describe the overall substrate utilization rate for a completely mixed anaerobic filter:

$$R_o = (kSX_s)/(K_s + S) + (\eta k'S)/(K_s + S) \qquad 7.28(1)$$

$$k' = \rho kA\delta \qquad 7.28(2)$$

where:

R_o = the overall substrate utilization rate, mass/volume-time
X_s = suspended biomass concentration, mass/volume
η = the effectiveness factor that defines the degree of diffusional limitations of the biofilm
k = the maximum substrate utilization rate in the biofilm, mass/volume-time
ρ = the biofilm dry density, mass/volume
A = total biofilm surface area per unit filter volume, l/length
δ = biofilm thickness, length

Expressions for the effectiveness factor have been developed by researchers Atkinson and How (1974) as follows:

$$\eta = 1 - [\tanh(B)/B]\{[\Phi/\tanh(\Phi)] - 1\} \qquad \text{for } \Phi \leq 1$$

$$7.28(3)$$

$$\eta = 1/\Phi - [\tanh(B)/B]\{[1/\tanh(\Phi)] - 1\} \qquad \text{for } \Phi \geq 1$$

$$7.28(4)$$

$$\Phi = [0.709(S/K_s)]/[1 + (S/K_s)] \{(S/K_s) - \ln[1 + (S/K_s)]^{-0.5}\}$$

$$7.28(5)$$

$$B = [(\delta^2 k')/(DK_s)]^{0.5} \qquad 7.28(6)$$

where D = the effective diffusivity of substrate in the biofilm, area/time.

Hence, the filter volume can be calculated from the following equation:

$$V = (QS_o)/R_o \qquad 7.28(7)$$

For a PF anaerobic filter, environmental engineers can calculate the overall substrate utilization rate using the CFSTR-in-series model (see Section 7.25) as follows:

$$V_i = (QS_{i-1})/(R_{oi}) \quad \text{for } i = 1, 2, 3, \ldots, n \quad 7.28(8)$$

$$V = \Sigma V_i \quad 7.28(9)$$

Applying Equations 7.28(1) to 7.28(6) to each reactor i calculates R_{oi}.

AFBR

The AFBR is an expanded-bed reactor that retains media in suspension from drag forces exerted by upflowing wastewater. Part D in Figure 7.28.2 shows the process schematic. Fluidization of the media particles provides a large surface area where biofilm formation and growth can occur.

The media particles have a high density resulting in a settling velocity that is high enough so that high-liquid-velocity conditions can be maintained in the reactor. However, the media particles' overall density decreases as biomass growth accumulates on the surface area. The decrease in density can cause the bioparticles to rise and be washed out of the reactor. To prevent this situation, the reactor controls fluidized-bed height at a required level by wasting a corresponding amount of overgrown bioparticles. The wasted bioparticles can then be received by a mechanical device that separates the biomass from the wasted media particles. The cleaned particles can then be returned to the reactor, while the separated biomass is wasted as sludge.

The AFBR combines a suspended-growth system and an attached-growth system since biomass growth attaches to the media particles which are suspended in the wastewater. The reactor recycles a portion of the effluent flow ensuring uniform bed fluidization and sufficient substrate loading.

An advantage of the AFBR is that it employs small fluidized media that provide a high biomass holdup in the reactor, reducing hydraulic retention time. The AFBR also prevents bed-clogging and high-pressure drops—complications associated with anaerobic filters. Due to the flexibility provided by bed-height control in an AFBR, a constant biomass concentration can be maintained in the reactor independent of substrate loadings (Shieh and Keenan 1986). Another advantage of the AFBR is that it is insensitive to variations in influent pH, temperature, and waste loading because it maintains a high biomass holdup and completely mixed conditions inside the reactor.

Some commercially available AFBR processes include the *ANITRON* system developed by Dorr-Oliver, Inc.; the *BIOJET* process, which employs an AFBR with an enlarged top section; and the *ENSO-FENOX* process, which combines an AFBR with a trickling filter. The AFBR has been applied to a variety of industrial treatment processes with substrates such as molasses, synthetic sucrose, sweet whey, whey permeate, glucose, and acid whey.

Since high circulation rates are used in AFBR operation (especially for treating high-strength industrial wastewater) the reactor can be designed as a completely mixed, heterogenous process. Because most of the active biomass is retained in the fluidized-bed, the contribution of suspended biomass growth to overall reactor performance is insignificant. Thus, the first term in Equation 7.28(1) can be removed, and the remaining expression calculates the overall substrate utilization rate in an AFBR as follows:

$$R_o = (\eta k'S)/(K_s + S) = (\eta kXS)/(K_s + S) \quad 7.28(10)$$

$$V = (QS_o)/R_o \quad 7.28(11)$$

Environmental engineers can use Equations 7.28(3) through 7.28(6) to determine the effectiveness factor associated with an AFBR in a similar manner as for the anaerobic filter. To determine the reactor biomass concentration (X), they can use solid–liquid fluidization correlations. These correlations link the particle concentration in the fluidized state to the measurable physical characteristics of a fluidized-bed process. The following Richardson–Zaki correlation is widely used:

$$U/U_t = \varepsilon^n \quad 7.28(12)$$

where:

U = superficial upflow velocity of the wastewater through an AFBR, distance/time
U_t = bioparticle terminal settling velocity, distance/time
ε = bed porosity
n = the expansion index

Shieh and Chen (1984) propose two empirical correlations relating U_t and n to the Galileo number (N_{Ga}) that defines the physical characteristics of an AFBR as follows:

$$U_t = 5753.71(N_{Ga})^{-0.8222} \quad 7.28(13)$$

$$n = 47.36(N_{Ga})^{-0.2576} \quad 7.28(14)$$

$$N_{Ga} = [8(r_p)^3(\rho_p - \rho)\rho g]/\mu^2 \quad 7.28(15)$$

$$\rho_p = \rho_m(r_m/r_p)^3 + [\rho/(1 + P)][1 - (r_m/r_p)^3] \quad 7.28(16)$$

where:

ρ_p = bioparticle density, mass/volume
ρ = wastewater density, mass/volume
g = gravitational acceleration, distance/time^2
μ = wastewater dynamic viscosity, mass/time–distance
ρ_m = media density, mass/volume
P = biofilm moisture content
r_m = support media radius, length
r_p = bioparticle radius, length

The following equation calculates the AFBR biomass concentration:

$$X = \rho(1 - \varepsilon)[1 - (r_m/r_p)^3] \quad 7.28(17)$$

The choice for media types should be based on the following media characteristics:

- A large surface area for microbial growth

- A large void space to accommodate the accumulation of biological and inert solids and minimize short-circuiting
- Inertness to biological and chemical reactions
- Resistance to abrasion and erosion
- A light weight

Small media should be used since they provide large surface-to-volume ratios, and thus, a greater surface area for biofilm growth without increasing reactor volume. Small media are also easier to fluidize, reducing the circulation requirements which decreases the shearing effects and allows a more quiescent environment for optimal biofilm growth. Silica sand, anthracite coal, activated carbon, stainless-steel wire spheres, and reticulated polyester foams are some of the media that can be considered for AFBR applications. Yee (1990) reports on the effects of various media types on AFBR performance and kinetics.

ANAEROBIC LAGOONS

Anaerobic microorganisms do not require DO in the water to function. They obtain their oxygen requirement from the oxygen chemically contained in organic materials.

Anaerobic decomposition involves two separate but interrelated steps. First, the acid-producing bacteria decompose the dissolved organic waste to organic acids, such as acetic, propionic, and butyric acid (see Figure 7.22.4). The organic acids are then further decomposed by methane-producing bacteria to the end products of methane, carbon dioxide, and water. Effective operation requires a balance between acid production and breakdown because methane producers are sensitive to the concentration of volatile acids.

As a general rule, inhibition occurs at volatile acid concentrations in excess of 2000 mg/l. This tolerance level also depends on the concentration of ammonia and other cations. The maximum alkalinity concentration is approximately 2000 mg/l as $CaCO_3$. As an operational guide, the alkalinity concentration should be greater than 1.67 times the volatile acids concentration.

The decomposition of organic acids to methane and carbon dioxide can be generalized as follows:

$$C_nH_aO_b + \left(n - \frac{a}{4} - \frac{b}{2}\right) H_2O \longrightarrow \left(\frac{n}{2} - \frac{a}{8} + \frac{b}{4}\right) CO_2$$

$$+ \left(\frac{n}{2} + \frac{a}{8} - \frac{b}{4}\right) CH_4 \quad 7.28(18)$$

At a standard temperature and pressure, 1 lb of BOD removed yields 5.62 ft^3 methane. Anaerobic decomposition processes are summarized in Figure 7.22.4.

A portion of the waste material is used by the anaerobic biosystem as a source of energy and in the synthesis of new bacterial cells. Cell synthesis is affected by the type of waste being treated, but generally, for every pound of BOD destroyed by the anaerobic process, approximately 0.1 lb of new cells is produced compared to 0.5 lb in the aerobic process. Therefore, the sludge or solids buildup is less in the anaerobic system. When anaerobic lagoon contents are black in color, this indicates that the lagoon is functioning properly.

—*Wen K. Shieh*
Van T. Nguyen

References

Atkinson, B., and S.Y. How. 1974. The overall rate of substrate uptake reaction by microbial films. II. Effect of concentration and thickness with mixed microbial films. *Trans. Instr. Chem. Engrs.* 52, no. 260.

Benefield, L.D., and C.W. Randall. 1980. *Biological process design for wastewater treatment.* Englewood Cliffs, N.J.: Prentice-Hall.

Forster, C.F. 1985. *Biotechnology and wastewater treatment.* Cambridge: Cambridge University Press.

Jewell, W.J. 1987. Anaerobic sewage treatment. *Environ. Sci. Technol.* 21, no. 14.

Lettinga, G., A.F.M. Van Velsen, S.W. Hobma, W. de Zeeuw, and A. Klapwijk. 1980. Use of the upflow blanket (USB) reactor concept for biological wastewater treatment, especially for anaerobic treatment. *Biotechnol. Bioeng.* 22, no. 669.

Li, A.Y. 1984. *Anaerobic process for industrial wastewater treatment.* Short Course Series, no. 5. Taichung, Taiwan: Department of Environmental Engineering, National Chung Hsing University.

Pavlostathis, S.G., and E. Giraldo-Gomez. 1991. Kinetics of anaerobic treatment: A critical review. *CRC Crit. Rev. Environ. Control* 21, no. 411.

Ramalho, R.S. 1983. *Introduction to wastewater treatment processes.* 2d ed. New York: Academic Press.

Shieh, W.K., and J.D. Keenan. 1986. Fluidized bed biofilm reactor for wastewater treatment. In *Advances in biochemical engineering/biotechnology 33,* edited by A. Flechter. Berlin: Springer-Verlag.

Shieh, W.K., and A.Y. Li. 1987. High-rate anaerobic treatment of industrial wastewaters. In *Global bioconversions 3,* edited by D.L. Wise, 41–79. Boca Raton, Fla.: CRC Press, Inc.

Yee, C.J. 1990. Effects of microcarriers on performance and kinetics of the anaerobic fluidized bed biofilm reactor. Ph.D. diss., Department of Systems, University of Pennsylvania, Philadelphia.

Young, J.C., and P.L. McCarty. 1968. The anaerobic filter for wastewater treatment. *J. Water Poll. Control Fed.* 41, no. R160.

7.29
SECONDARY CLARIFICATION

TREATMENT TYPE

    a. Following air-activated-sludge
    b. Following oxygen-activated-sludge
    c. Following extended-aeration
    d. Following trickling filter
    e. Following secondary RBC
    f. Following nitrification RBC

TANK GEOMETRY

    Rectangular, circular, or square

OVERFLOW RATE (gal/ft²-day)

AVERAGE

    a, b, & e: 400–800; c: 200–400; d & f: 400–600

PEAK

    a, b, d, & e: 1000–1200; c: 600–800; f: 800–1000

SOLIDS LOADING (lb/ft²-day)

AVERAGE

    a & e: 0.8–1.2; b: 1–1.4; c: 0.2–1; d & f: 0.6–1

PEAK

    a, b, & e: 2; c: 1.4; d & f: 1.6

DEPTH (ft)

    a–c: 12–20; d–f: 10–15

An important aspect of biological wastewater treatment is secondary clarification. First, biological solids (sludge) produced during biological wastewater treatment must be separated from treated effluent prior to final discharge. Therefore, a secondary clarifier must have an adequate clarification capacity to insure that SS discharge requirements are met. Second, since maintaining proper SRTs is important in the operation of activated-sludge processes, secondary clarifiers in the activated-sludge process must have an adequate thickening capacity to produce the required underflow density for sludge recirculation.

Process Description

Biological solids removal is essentially accomplished by gravity settling (see Figure 7.16.5). However, biological solids can settle differently depending on their origins and characteristics. The sloughed solids produced from trickling filters and RBCs are generally large and heavy. Therefore, their settling motion is discrete (i.e., not influenced by the motion of neighboring particles, as shown in Figure 7.16.6) and can be described by Stokes' Law (Metcalf and Eddy, Inc. 1991).

On the other hand, biological flocs produced in activated-sludge processes undergo some flocculation with neighboring particles during the settling process. As flocculation occurs, the mass of particles increases and settles faster (see Figure 7.16.7). As a result, the settling process is classified as flocculant settling (Metcalf and Eddy, Inc. 1991). Since many complex mechanisms are involved during flocculation and their interactions are difficult (if not impossible) to define, the analysis of flocculant settling requires experimental data obtained from settling tests (see Figure 7.16.8).

Design

TRICKLING FILTERS AND RBCs

Secondary clarifiers used with trickling filters and RBCs provide effluent clarification by removing large, sloughed solids. Therefore, the design criteria are based on particle size and density.

Environmental engineers can formulate the gravitational force (F_G) and the frictional drag force (F_D) acting on a spherical particle settling through a liquid using the classic laws of Newton and Stokes, respectively, as follows:

$$F_G = (\rho_s - \rho)gV_s \qquad 7.29(1)$$

$$F_D = C_D A_s \rho v^2/2 \qquad 7.29(2)$$

where:

  ρ_s = particle density
  ρ = liquid density
  g = gravitational acceleration
  V_s = particle volume
  C_D = drag coefficient
  A_s = projected area of the particle perpendicular to the direction of settling
  v = particle settling velocity

The particle settling velocity v becomes the terminal settling velocity v_c when $F_G = F_D$. Therefore, the following equation applies

$$v_c = [4gd(\rho_s - \rho)/3\rho C_D]^{0.5} \qquad 7.29(3)$$

where d = particle diameter ($1.5V_s/A_s$).

Under laminar flow conditions, Equation 7.29(3) can be modified as follows:

$$v_c = gd^2(\rho_s - \rho)/18\mu \qquad 7.29(4)$$

where μ = liquid viscosity.

For the design of continuous-flow secondary clarifiers, v_c is designated as the surface overflow rate as follows:

$$v_c = (Q + Q_r)/A = D/\theta_{sc} \qquad 7.29(5)$$

where:

Q = influent wastewater flow rate
Q_r = recirculation flow rate
A = surface area of the secondary clarifier
D = depth of the secondary clarifier
θ_{sc} = LRT in the secondary clarifier

The total fraction of biological solids removed F is calculated as follows:

$$F = (1 - \chi_s) + \int_0^{x_s} v_s dx \qquad 7.29(6)$$

where $(1 - \chi_s)$ = fraction of particles with settling velocities $\geq v_c$, and the second term at the right side of Equation 7.29(6) indicates the fraction of particles removed with settling velocities $v_s < v_c$.

ACTIVATED-SLUDGE PROCESSES

The minimum surface area required for clarification in the secondary clarifier A_c can be calculated as follows:

$$A_c = (Q + Q_r)/v_z \qquad 7.29(7)$$

where v_z = the zone settling velocity, which can be determined from the following procedure. The mixed liquor with a MLVSS concentration of X is placed in a settling column. Activated-sludge begins to settle under quiescent conditions, and an interface forms between the surface of the blanket of settling sludge and the clarified liquid above. Plotting the height of this interface as a function of settling time generates a settling curve corresponding to the MLVSS concentration X. The slope of a tangent drawn to the initial portion of the settling curve yields v_z.

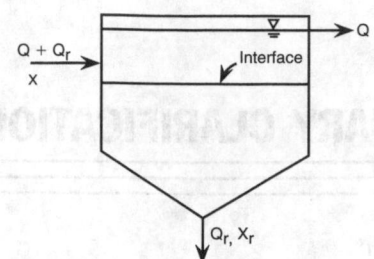

FIG. 7.29.1 Schematic representation of a continuous-flow, secondary clarifier under steady-state conditions.

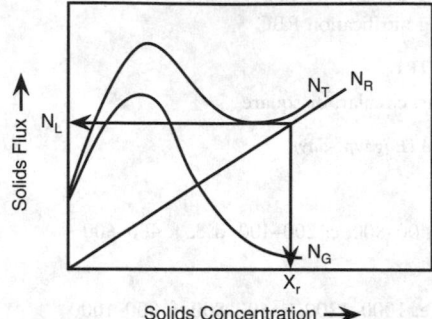

FIG. 7.29.2 Definition sketch of the solids flux analysis technique.

Applying the technique of solid flux analysis (Dick and Ewing 1967; Dick and Young 1972; Dick 1976) determines the minimum surface area required for thickening in the secondary clarifier A_T. Figure 7.29.1 is a schematic representation of a secondary clarifier operated under steady-state conditions. The mass flux of biological solids due to gravity settling N_G is calculated as follows:

$$N_G = X_i v_i \qquad 7.29(8)$$

TABLE 7.29.1 DESIGN CRITERIA FOR SECONDARY CLARIFIERS

Biological Treatment	Overflow Rate (gal/ft²-day)	Solids Loading (lb/ft²-day)	Depth (ft)
Air-Activated Sludge	400–800 (avg)	19.2–28.8 (avg)	12–20
	1000–1200 (peak)	48.0 (peak)	
Oxygen-Activated Sludge	400–800 (avg)	24.0–33.6 (avg)	12–20
	1000–1200 (peak)	48.0 (peak)	
Extended Aeration	200–400 (avg)	12.0–24.0 (avg)	12–20
	600–800 (peak)	33.6 (peak)	
Trickling Filters	400–600 (avg)	14.4–24.0 (avg)	10–15
	1000–1200 (peak)	38.4 (peak)	
RBCs			
Secondary Effluent	400–800 (avg)	19.2–28.8 (avg)	10–15
	1000–1200 (peak)	48.0 (peak)	
Nitrified Effluent	400–600 (avg)	14.4–24.0 (avg)	10–15
	800–1000 (peak)	38.4 (peak)	

Source: Adapted from Metcalf and Eddy, Inc., 1991, *Wastewater engineering: Treatment, disposal, and reuse,* 3d ed. (New York: McGraw-Hill).

where:

X$_i$ = biological solids concentration at location i
v$_i$ = settling velocity of biological solids at concentration X$_i$

The mass flux of biological solids due to sludge recirculation N$_R$ is calculated as follows:

$$N_R = X_i Q_r / A \qquad 7.29(9)$$

Therefore, the following equation calculates the total mass flux of biological solids N$_T$:

$$N_T = N_G + N_R \qquad 7.29(10)$$

Figure 7.29.2 plots Equations 7.29(8) to 7.29(10) as a function of biological solid concentration X. If a horizontal line is drawn tangent to the low point on the N$_T$ curve, its intersection on the y-axis yields the limiting mass flux of biological solids N$_L$. Therefore, A$_T$ is calculated as follows:

$$A_T = (Q + Q_r) X / N_L \qquad 7.29(11)$$

The larger value between A$_C$ and A$_T$ is the design value. The required sludge concentration in the recycled flow is X$_r$. Table 7.29.1 summarizes typical design information for secondary clarifiers.

7.30
DISINFECTION

GEOMETRIC CONFIGURATIONS

Secondary clarifiers have various geometric configurations; the most common ones are either circular (see Figure 7.16.9) or rectangular (see Figure 7.16.10). Other types include tray clarifiers, tube settlers (see Figure 7.16.11), lamella (parallel-plate) settlers, and intrachannel clarifiers (in oxidation ditches). Wastewater treatment facilities can increase existing clarifier capacity by installing inclined tubes or parallel plates.

—*Wen K. Shieh*
Van T. Nguyen

References

Dick, R.I. 1976. Folklore in the design of final settling tanks. *J. Water Poll. Control Fed.* 48, no. 633.

Dick, R.I., and B.B. Ewing. 1967. Evaluation of activated sludge thickening theories. *J. San. Eng. Div.*, ASCE 93, no. 9.

Dick, R.I., and K.W. Young. 1972. Analysis of thickening performance of final settling tanks. Presented at the 27th Annual Purdue Industrial Conference, West Lafayette, IN.

Metcalf and Eddy, Inc. 1991. *Wastewater engineering: Treatment, disposal, and reuse.* 3d ed. New York: McGraw-Hill.

DISINFECTION MEANS

a. Chemical (chlorination, ozonation, and acid and alkaline treatments)
b. Physical (heating, UV irradiation, filtration, and settling)
c. Radiation (electromagentic and acoustic)

CHLORINE DOSAGE (mg/l)

8–15 mg/l (secondary effluent)

CONTACT TIME (min)

>30 minutes (peak hourly flow)

CHLORINATION TANK FLOW CONFIGURATION

PF

MAXIMUM CHLORINE RESIDUALS

0.1–0.5 mg/l (undiluted effluent)

Disinfection is the selective destruction of disease-causing organisms. Disinfection of effluents prior to discharge insures that bacteria, viruses, and amoebic cysts are reduced to acceptable levels. Many means can accomplish the disinfection of effluents: chemical agents, physical agents, mechanical means, and radiation.

The most common chemical agents used in disinfection are chlorine and its compounds. Ozone is highly effective but it does not leave residual. Acids and alkalies are sometimes used since pH >11 or pH <3 are toxic to most bacteria. Bromine, iodine, phenols, alcohols, and hydrogen peroxide are other common chemical disinfection agents. Heat and light (especially UV light) are effective physical disinfection agents. However, using heat and UV light to disinfect large quantities of effluents is cost prohibitive. The presence of suspended matter in effluents can also reduce the efficacy of UV radiation.

Preliminary and primary treatment processes used in wastewater treatment (e.g., coarse and fine screens, grit chambers, and primary clarifiers) are capable of removing or destroying a large number of bacteria (Metcalf and Eddy, Inc. 1991). Up to 75% of the bacteria in incoming wastewater can be removed or destroyed by the settling mechanism alone. However, removal and destruction of bacteria are the by-products instead of the primary functions of these treatment processes. Wastewater treatment facilities can use electromagnetic, acoustic, or particle radiation to disinfect water, wastewater, and sludge. However, these applications are limited due to the high costs involved.

Bacteria and viruses are removed or killed by disinfection and sterilization. Numerous disinfection and sterilization techniques are available. Tables 8.2.1 and 8.2.2 in Chapter 8 compare their effectiveness, advantages, and disadvantages. This section discusses disinfection with chlorine since it is the most common disinfectant used.

Chlorine Chemistry

Chlorine is an active element that reacts with many chemical compounds in water and wastewater to form new and often less offensive components. Hydrolysis and ionization occur when chlorine gas is added to water which forms $HOCl$ and OCl^-, the free available chlorine. Hypochlorite salts such as $Ca(OCl)_2$ and $Na(OCl)$ can be added to water to form free chlorine. $HOCl$ is predominant at a pH <7.0 which is beneficial since its disinfection power is approximately 40–80 times of that of OCl^-.

$HOCl$ reacts with ammonia in wastewater to form various chloramines (NH_2Cl, $NHCl_2$, and NCl_3), the combined available chlorine (Sawyer and McCarty 1978). Since chloramines are less effective disinfectants, additional chlorine is needed to insure the presence of a free chlorine residual. Figure 7.30.1 is a schematic illustration of the stepwise reaction phenomena that result when chlorine is added to wastewater containing ammonia (breakpoint chlorination curve).

An even more important characteristic of chlorine is that it is toxic to most pathogenic microorganisms (see Figure 7.30.2). Chlorine acts as an oxidizing agent to change the character of an offending chemical. Chlorine

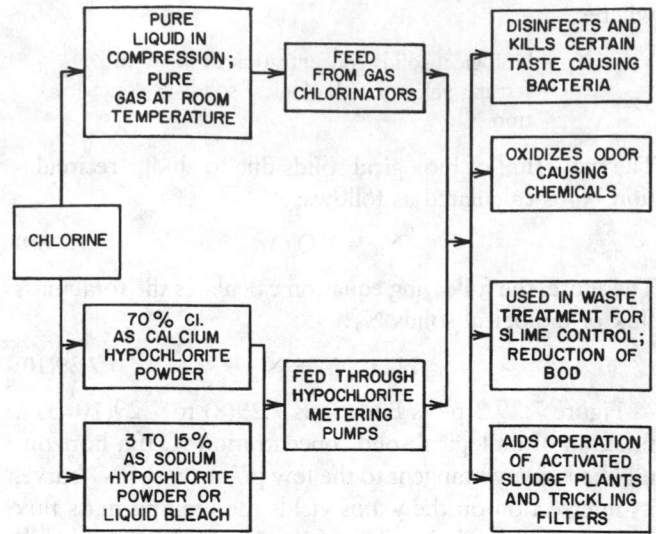

FIG. 7.30.2 Utilization of chlorine.

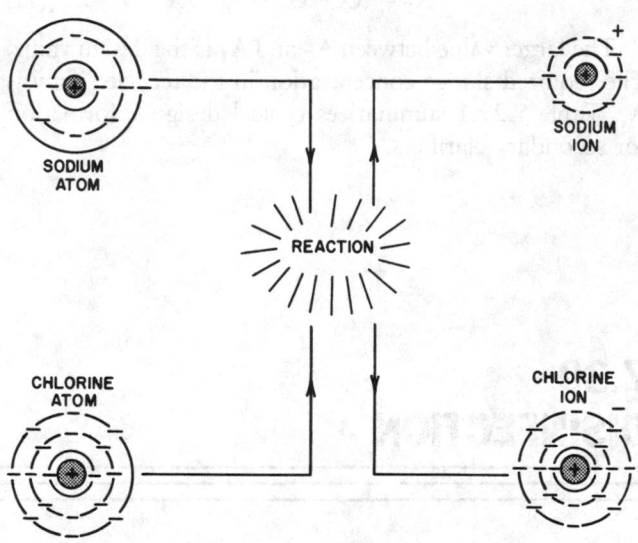

FIG. 7.30.3 A typical oxidation reaction.

is a strong oxidizing agent because its atoms are constructed with three shells of electrons and the outer shell (with seven electrons) has a strong tendency to acquire an eighth electron for stability. Oxidation is a process in which an atom loses electrons (see Figure 7.30.3).

PRECHLORINATION/POSTCHLORINATION

Conventional treatment plants use chlorination as the final treatment process to reduce bacterial concentrations. Prechlorination is performed on plant influent if the incoming sewage is septic or the flows are low and plant holdup time allows waste to become septic. Prechlorination is usually at a fixed dosage, and a residual chlorine level is not maintained. Postchlorination provides 15 to 30 min detention time in a baffled, closed tank to prevent short-circuiting and dissipation of the chlorine. Chlori-

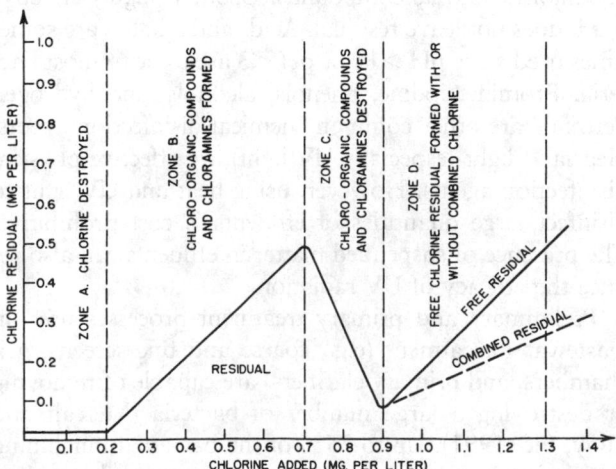

FIG. 7.30.1 Breakpoint chlorination.

nation is not frequently used to reduce the BOD in the plant effluent. Ordinarily, a combined chlorine residual between 0.2 and 1 mg/l is the target for the final effluent.

Disinfection should kill or inactivate all disease-producing (pathogenic) organisms, bacteria, and viruses of intestinal origin (enteric). These microscopic entities can survive in water for weeks. The amount of chlorine required depends on the chlorine demand of the water being disinfected, the amount of disinfectant required as residual, and the detention time during which the disinfectant acts on the organisms. A free-chlorine residual of at least 2 ppm (mg/l) should be attained. Chlorinator capacities should be based on at least 30 min contact time, and water flow rates should coincide with anticipated maximum chlorine demands.

Chlorinators should be accurate over the entire feed range, and standby machines should be available. The wastewater treatment facility should determine the chlorine demand of the raw water frequently enough to adjust the chlorine feed. This procedure is best accomplished by automatic equipment that continuously determines the residual chlorine content of treated water and adjusts the chlorine dosage accordingly.

A free-chlorine residual of not less than 0.3 ppm should be maintained in the active parts of the water distribution system. Where large, open (uncovered) reservoirs of finished water exist, auxiliary disinfection must be provided. The disinfection of newly laid or extensively repaired water mains is also required, and procedures for this operation are prescribed in standards.

Disinfection by chlorine, in addition to eliminating pathogenic organisms, also controls taste and odor through breakpoint chlorination. This process is the addition of sufficient chlorine to destroy or oxidize all substances that create a chlorine demand with excess chlorine remaining in the free-residual state. Figure 7.30.1 shows this process and its effects. A free-chlorine residual is that part of the total residual remaining in the water (after a specified contact period) that reacts chemically and biologically as hypochlorous acid or a hypochlorite ion.

The effectiveness of chlorine disinfection is a function of contact time, pH, and temperature. Figure 7.30.4 shows these relationships. Providing adequate contact time and dosage is especially important with respect to viruses. Virus particles have been isolated that have escaped treatment processes, but clarification and disinfection of water afford a high measure of protection against viruses.

FACTORS AFFECTING CHLORINE DISINFECTION

The factor Ct determines chlorine disinfection efficacy, where C is the disinfectant concentration in mg/l and t is the contact time in minutes. The degree of disinfection remains unchanged at a constant Ct value. This relationship is true when either of the two variables is increased while the other one is simultaneously decreased. Consequently, increasing the contact time increases the efficiency of a less effective disinfectant. Table 7.30.1 summarizes the ranges of Ct values of disinfectants for 99% inactivation of various microorganisms at 5°C.

Effective mixing of chlorine with wastewater is essential for effective bacterial kill. Diffusers inject chlorine directly into the path of the wastewater flow for effective

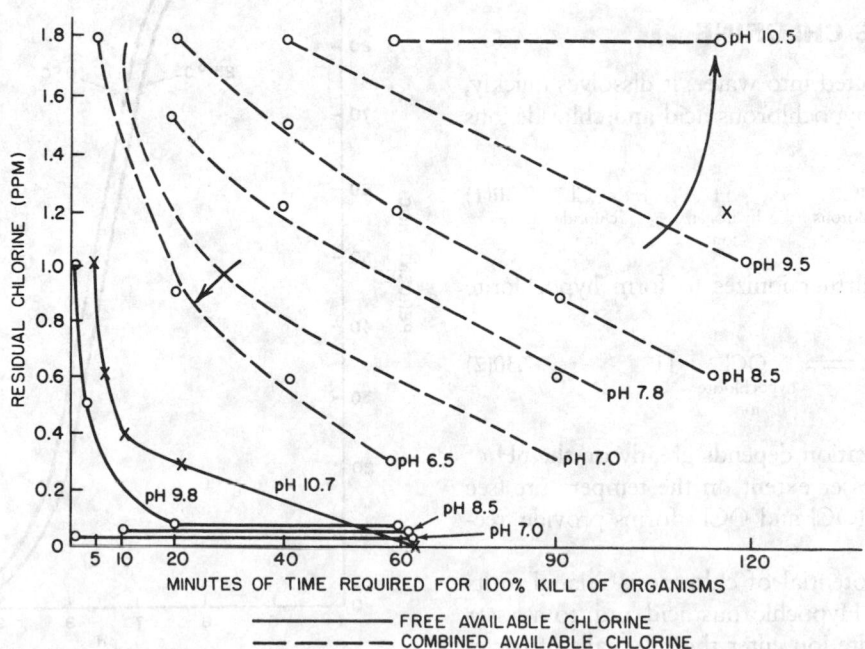

FIG. 7.30.4 Relationship between time, amount of chlorine, and bactericidal action. (Reprinted from *USPHS report.*)

TABLE 7.30.1   RANGES OF CT VALUES OF DISINFECTANTS FOR 99% INACTIVATION OF MICROORGANISMS AT 5°C

Microorganism	Free Chlorine (pH 6–7)	Chloramines (pH 8–9)	Chlorine Dioxide (pH 6–7)	Ozone (pH 6–7)
E. coli	0.034–0.05	95–180	0.4–0.75	0.02
Polio 1	1.1–2.5	770–3700	0.2–6.7	0.1–0.2
Rotavirus	0.01–0.05	3800–6500	0.2–2.1	0.006–0.06
G. lamblia cysts	47–>150	—	—	0.5–0.6
G. muris cysts	30–630	—	7.2–19	1.8–2.0

Source: Adapted from W. Viessman, Jr. and M.J. Hammer, 1993, *Water supply and pollution control*, 5th ed. (New York: Harper Collins).

mixing. Mechanical mixing devices insure rapid and complete chlorine mixing with effluent.

Because of the importance of contact time, wastewater treatment facilities usually use a PF chlorination tank with a back-and-forth-flow pattern similar to that shown in Figure 7.25.2 to insure that at least 80 to 90% of the effluent is retained in the tank for the specified contact time. A minimum contact time of 30 min at the peak hourly flow is recommended. A chlorine dosage of 8–15 mg/l is adequate for a well-designed tank for chlorinating secondary effluents.

Wastewater treatment facilities should keep the maximum chlorine residuals in undiluted effluents at 0.1–0.5 mg/l to protect receiving surface water systems. Consequently, dechlorination can be required to reduce chlorine residual toxicity. Sulfur dioxide added at the end of the chlorination tank oxidizes both free chlorine and chloramines to chloride. Activated-carbon adsorption of free- and combined-chlorine residuals is effective but expensive.

FREE-AVAILABLE CHLORINE

When chlorine is injected into water, it dissolves quickly, hydrolyzing to form hypochlorous acid and chloride ions as follows:

$$Cl_2 + H_2O \rightleftharpoons HOCl + H^+ + Cl^- \qquad 7.30(1)$$

chlorine gas → hypochlorous acid + hydrogen ion + chloride ion

Hypochlorous acid further ionizes to form hypochlorite ions as follows:

$$HOCl \rightleftharpoons OCl^- + H^+ \qquad 7.30(2)$$

hypochlorite ion

The extent of ionization depends greatly on the pH of the water and to a lesser extent on the temperature (see Figure 7.30.5). The HOCl and OCl⁻ forms provide free-available chlorine.

The disinfection potential of chlorine is related to its oxidation properties. Hypochlorous acid and, to a lesser extent, the hypochlorite ion enter the cell walls of bacteria, oxidizing certain enzymes and other organic cellular material essential to the bacteria's life processes.

Chlorine is ordinarily purchased as a liquid compressed in pressurized tanks or cylinders. Small installations sometimes use solutions of sodium or calcium hypochlorite. These chemicals also ionize to produce the hypochlorite ion. For calcium hypochlorite, the following reaction occurs:

$$Ca(OCl)_2 \longrightarrow Ca^{2+} + 2\,OCl^- \qquad 7.30(3)$$

calcium hypochlorite

CHLORAMINES AND COMBINED AVAILABLE CHLORINE

When ammonia or organic amines are present (as they usually are in wastewater effluents), chlorination produces a class of compound called chloramines. They have the

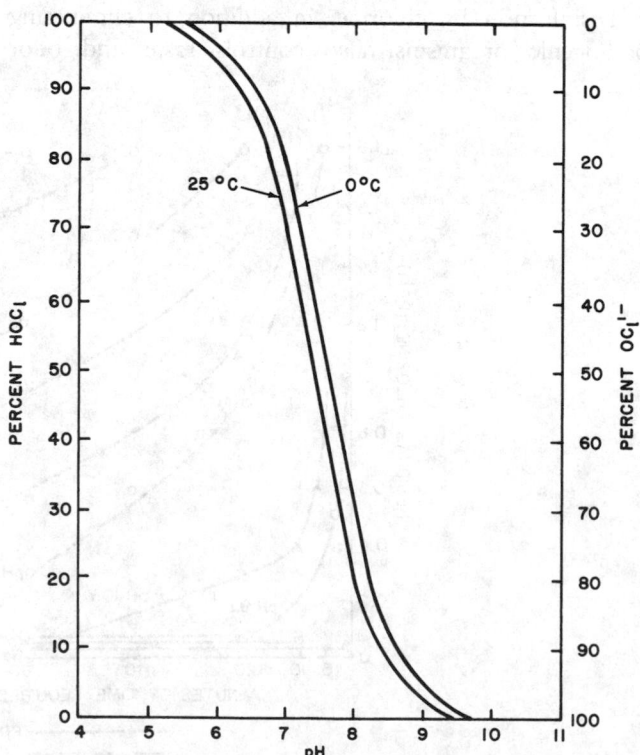

FIG. 7.30.5 Extent of hypochlorous acid (HOCl) ionization into ClO¹⁻ as a function of pH and temperature.

prefix *mono*, *di*, or *tri* depending on their final form as follows:

$$NH_3 + HOCl \longrightarrow \underset{\text{monochloramine}}{NH_2Cl} + H_2O \qquad 7.30(4)$$

$$NH_3 + 2\,HOCl \longrightarrow \underset{\text{dichloramine}}{NHCl_2} + 2\,H_2O \qquad 7.30(5)$$

$$NH_3 + 3\,HOCl \longrightarrow \underset{\text{trichloramine}}{NCl_3} + 3\,H_2O \qquad 7.30(6)$$

Chloramines are often present with hypochlorites, and together they comprise *combined available chlorine*. Chloramines are sometimes beneficial because a combined-available-chlorine residual lasts longer than a free-chlorine residual. However, the killing power of the free chlorine is much greater.

RATE OF KILL

Figure 7.30.6 shows the rate of kill (microorganisms) for combined- and free-chlorine residuals. This curve is based on the rate of kill expressed in the following equation (unique for chlorine):

$$\frac{dN}{dt} = -kNt \qquad 7.30(7)$$

where:

N = population of living microorganisms
t = time
k = rate constant

Integrating and converting to base 10 yields the following equation:

$$t_1^2 = \frac{4.6}{k} \log \frac{N_0}{N_1} \qquad 7.30(8)$$

where:

N_0 = initial population
N_1 = population at t_1

Kill rates are usually measured in terms of coliform bacteria present. Figure 7.30.7 shows the relationship between coliform bacteria and chlorine residuals for sewage; Figure 7.30.8 shows the relationship between chlorine dosages and detention times.

CHLORINE DOSAGE

In determining the amount of chlorine required for disinfection, environmental engineers must consider the pH. The state of New York recommends that for drinking water at a pH of 7.0, the concentration of free-available-chlorine residual after a 10-min detention time should be 0.2 mg/l. For a pH of 8.0 with the same detention time, the concentration of free-chlorine residual should be 0.4 mg/l. At pH values of 7 and 8, the recommended combined-available-chlorine residuals are 1.5 and 1.8 mg/l, respectively, after a contact period of 60 min.

For waste treatment plant effluents, a combined-chlorine residual of 0.5 mg/l after 15 min is generally acceptable (although some states require as much as 2 mg/l). Providing a residual of 0.5 mg/l requires a much larger dosage, depending on the efficiency of prior treatment, be-

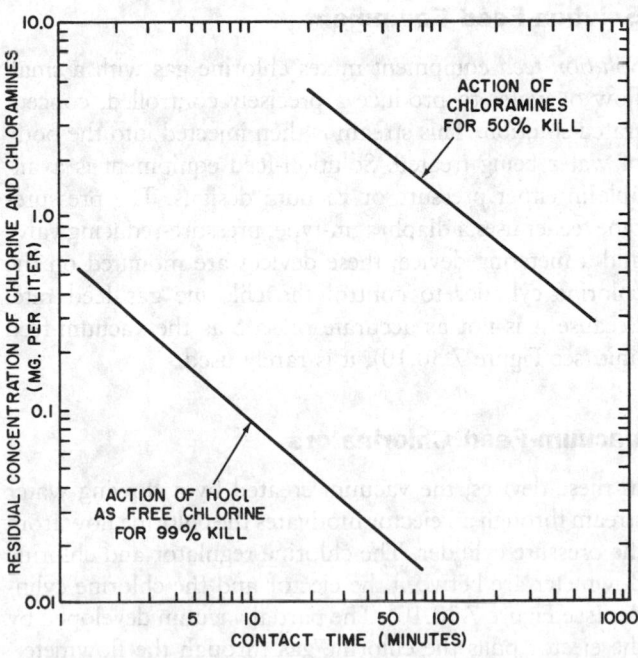

FIG. 7.30.6 Toxicity to microorganisms as a function of hypochlorous acid and chloramine contact time and concentration. Actual curves depend on pH and temperature.

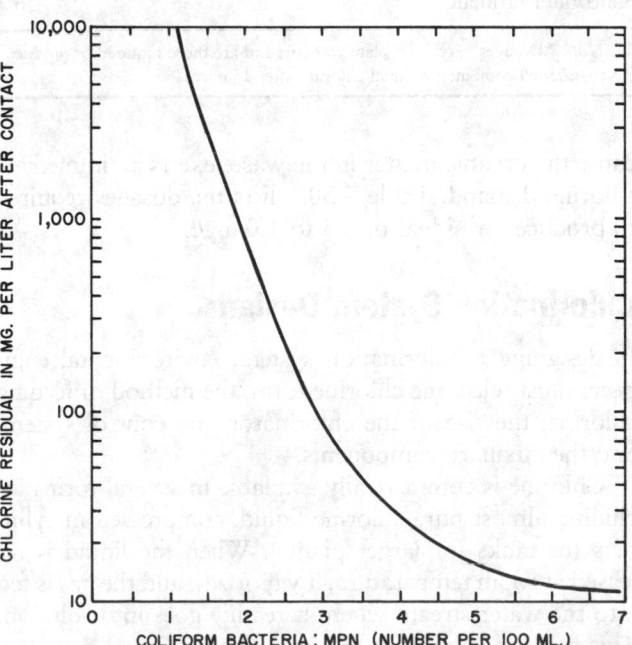

FIG. 7.30.7 Coliform kill versus chlorine residual concentration for biological waste treatment plant effluent.

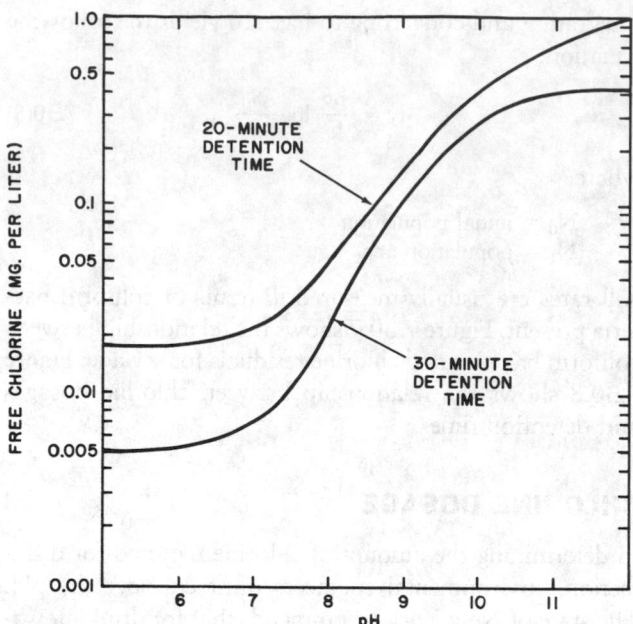

FIG. 7.30.8 Chlorine required to kill 99% of coliform bacteria at two detention times and various pH levels.

TABLE 7.30.2 CHLORINE DOSAGES FOR WASTE TREATMENT PLANT EFFLUENTS

Type of Treatment	Approximate Dosage* (mg/l)
Primary plant effluent	20 to 25
Trickling filter plant effluent	15
Chemical precipitation effluent	15
Activated-sludge plant effluent	8
Sand filter effluent	6

Note: *Dosages vary from plant to plant and are those required to produce 0.5 to 1.0 mg/l combined residual chlorine after 15 min.

cause the organic matter in the waste exerts an immediate chlorine demand. Table 7.30.2 lists the dosages required to produce a residual of 0.5 to 1.0 mg/l.

Chlorination System Design

In designing a chlorination system, environmental engineers must select the chlorine form, the method of feeding chlorine, the size of the chlorinator, the control system, and the auxiliary components.

Chlorine is commercially available in several forms including almost pure chlorine liquid, compressed in cylinders (or tanks for larger plants). When the liquid is released at room temperature, it vaporizes, and the gas is fed into the water stream where it readily goes into solution. This form of chlorine is the least expensive.

Another form is a powder, usually calcium hypochlorite, which is mixed with water in a plastic drum. The third form is sodium hypochlorite. Commercial liquid bleaches can also be used but are expensive since they contain only 15% or less of available chlorine.

METHODS OF FEEDING

Chlorine can be fed into the wastewater by several methods, including direct feed, solution-feed equipment, hypochlorite feeders, and vacuum-feed chlorinators.

Direct Feed

In *direct feed*, chlorine gas is fed directly to the stream or body of water being treated, and the need for electric or hydraulic power is eliminated. However, because of the risk of leakage, it is a dangerous technique and is used only in remote locations or under special circumstances.

Hypochlorite Feeders

Hypochlorite feeders are positive displacement diaphragm or metering pumps. Because of the cost of the chlorine compounds (five to eight times the equivalent cost of liquid chlorine), their use is limited to isolated areas where chlorine cylinders are difficult to obtain and the water flows are low, usually less than 100 gpm at peaks. The hypochlorite powder is mixed in plastic drums to the necessary solution strength (see Figure 7.30.9). The wastewater treatment facility can precisely adjust the feed rate, and the controls can be either intermittent or ratioed to the variable flow through the water line.

Solution-Feed Equipment

Solution-feed equipment mixes chlorine gas with a small flow of water to produce a precisely controlled, concentrated solution. This stream is then injected into the body of water being treated. Solution-feed equipment is available in either pressure or vacuum designs. The pressure-type feeder uses a diaphragm-type, pressure-reducing valve and a metering device; these devices are mounted on the chlorine cylinder to control the chlorine gas feed rate. Because it is not as accurate or safe as the vacuum-feed unit (see Figure 7.30.10), it is rarely used.

Vacuum-Feed Chlorinators

In these devices, the vacuum created by a flowing water stream through an ejector motivates the chlorine flow from the pressure cylinder. The chlorine regulator and chlorine flowmeter are between the ejector and the chlorine cylinder (see Figure 7.30.10). The partial vacuum developed by the ejector pulls the chlorine gas through the flowmeter. The regulator prevents the flowmeter inlet pressure from reaching atmospheric, eliminating the hazard of chlorine leakage. The rotameter and the rate setting valve are nor-

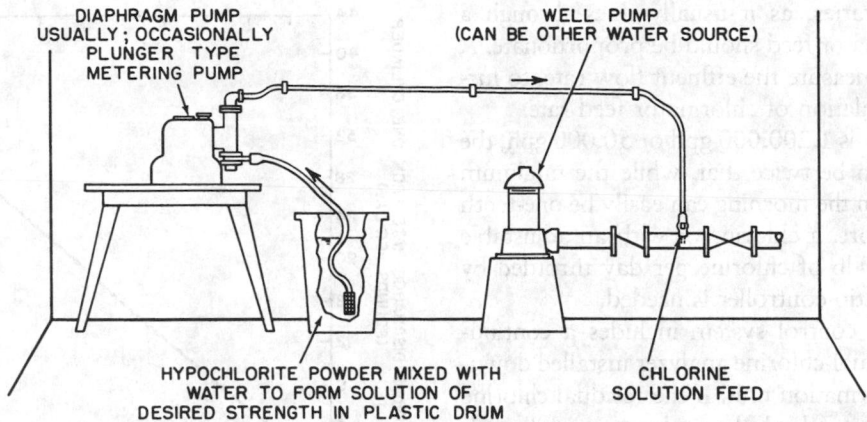

DIAPHRAGM PUMP
USUALLY; OCCASIONALLY
PLUNGER TYPE
METERING PUMP

WELL PUMP
(CAN BE OTHER WATER SOURCE)

HYPOCHLORITE POWDER MIXED WITH
WATER TO FORM SOLUTION OF
DESIRED STRENGTH IN PLASTIC DRUM

CHLORINE
SOLUTION FEED

FIG. 7.30.9 Calcium hypochlorite chlorination system.

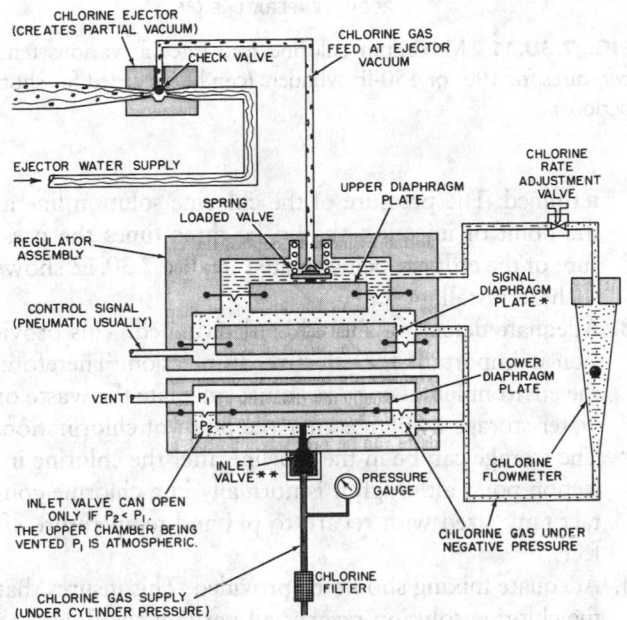

CHLORINE EJECTOR
(CREATES PARTIAL VACUUM)

CHECK VALVE

CHLORINE GAS
FEED AT EJECTOR
VACUUM

EJECTOR WATER SUPPLY

SPRING
LOADED VALVE

UPPER DIAPHRAGM
PLATE

CHLORINE
RATE
ADJUSTMENT
VALVE

REGULATOR
ASSEMBLY

CONTROL SIGNAL
(PNEUMATIC USUALLY)

SIGNAL
DIAPHRAGM
PLATE *

VENT

P_1

LOWER
DIAPHRAGM
PLATE

P_1
P_2

INLET
VALVE **

PRESSURE
GAUGE

INLET VALVE CAN OPEN
ONLY IF $P_2 < P_1$.
THE UPPER CHAMBER BEING
VENTED P_1 IS ATMOSPHERIC.

CHLORINE
FLOWMETER

CHLORINE GAS UNDER
NEGATIVE PRESSURE

CHLORINE GAS SUPPLY
(UNDER CYLINDER PRESSURE)

CHLORINE
FILTER

FIG. 7.30.10 Vacuum chlorinator regulator and meter. *Required when remote control signal is used; **this valve closes the chlorine supply if the pressure in the lower chamber rises to atmospheric.

mally located on the face of the chlorine control cabinet, allowing for ready adjustment.

The chlorine enters the throat of the ejector venturi, where it dissolves in the flowing water. This water solution is fed to the stream or tank being treated. Most states allow only this type of gas chlorinator to be used because the regulator and meter in this design have few moving parts and are trouble free for long periods. Also, because the chlorine (except at the cylinder) is under vacuum, the chance of it escaping to the atmosphere is reduced.

This type of gas chlorinator can be obtained with feed-rates of less than 1 to 8000 lb per 24 hr. The meters can be manifolded for any combination of capacities.

In addition to manual control with adjustment of the rotameter outlet valve, automatic remote control from a residual chlorine analyzer or other controller is also feasible. The control signal (pneumatic) is introduced over the middle (signal) plate of the regulator and causes a variation of the chlorine vacuum as it enters the rotameter.

CHLORINATOR SIZING

Chlorination system sizing involves determining the dosage required. This amount depends on the chlorine demand of the water or waste and on its flow rate.

The following example shows the chlorine dosage required for the effluent from an activated-sludge waste treatment plant. The average flow through the plant is 1.2 million gpd. Table 7.30.2 shows that the average dosage required to produce 0.5 mg/l of combined residual chlorine after 15 min contact time is about 8 mg/l or 8 ppm. The weight of daily waste flow through the plant is as follows:

1,200,000 gal per day × 8.34 lb per gal
$$= 9,998,000 \text{ lb of liquid effluent per day}$$

The weight of chlorine gas needed per day is as follows:

$$\frac{x \text{ lb of chlorine}}{9,998,000 \text{ lb of effluent}} = \frac{8 \text{ parts}}{1,000,000 \text{ parts}}$$

$$x = 8 \times 9.99 \cong 80 \text{ lb of chlorine per day}$$

A typical chlorinator model feeds chlorine gas at rates adjustable from 10 to 200 lb per day.

CONTROL METHOD SELECTION

In the design of chlorinators, environmental engineers must also select the control system. The simplest control system is a manual rate of feed control. Almost as simple is a control system with an intermittent start and stop feature. This method is unsatisfactory unless the stream being treated has a constant flow rate when started (such as from a pump).

When the flow varies, as it usually does through a sewage plant, chlorinator feed should be proportionate. A ratio controller can measure the effluent flow rate, to furnish automatic modulation of chlorinator feed rate.

If the average flow is 1,200,000 gpd or 50,000 gph, the hourly peak flow can be twice that, while the minimum flow at four o'clock in the morning can easily be one-tenth that amount. Therefore, a chlorinator with an adjustable feed rate of 8 to 160 lb of chlorine per day throttled by an automatic flow ratio controller is needed.

A more complete control system includes a continuously recording, residual-chlorine analyzer installed downstream from the chlorination unit. If the residual chlorine level drops below the standard, the analyzer controller adjusts the chlorine flow upward. The wastewater treatment facility can vary the rate of feed (see the rate valve opening on Figure 7.30.10) to compensate for changes in the flow through the plant and adjust the dosage (see the control signal port on Figure 7.30.10) to compensate for changes in effluent chlorine demand.

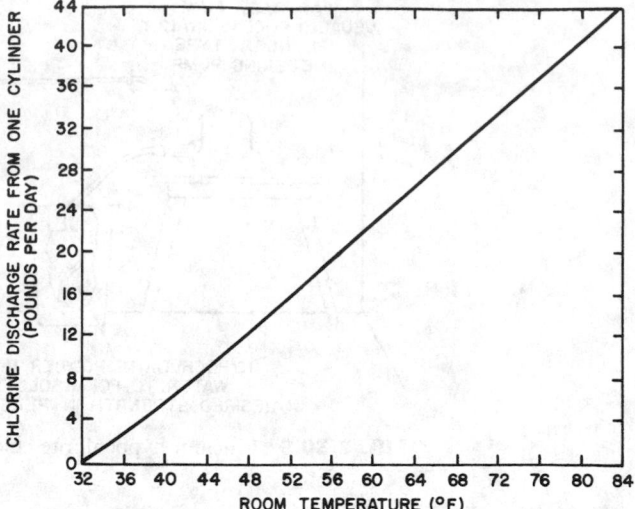

FIG. 7.30.11 Maximum chlorine feed rates at various temperatures for 100- or 150-lb cylinders (can be exceeded for short periods).

AUXILIARY COMPONENTS

Environmental engineers should also consider the auxiliary components of the system, including the following:

1. A separate room or building should be designed for chlorination. Gas chlorinators should be isolated from other equipment units or areas where personnel work. (Hypochlorite feeders need not be isolated.) The chlorination room should be heated to 60°F or higher to assure that chlorine remains in vapor form when released from the cylinder. Figure 7.30.11 shows the relationship between the cylinder withdrawal rate and room temperature. In the earlier example, several cylinders connected by a manifold were needed to provide the capacity.

2. The necessity of a booster pump is to inject the chlorine solution into the stream to be treated should be determined. The pressure of the chlorine solution line at the point of injection should be three times the pressure of the effluent being treated. Figure 7.30.12 shows such an installation.

3. Adequate detention should be determined. This provision is important for effective disinfection. Therefore, the environmental engineer must calculate the waste or water storage available after the point of chlorination. The storage can be in the pipeline after the chlorine injection point although it is normally in a chlorine contact tank sized with regard to pH and temperature effects.

4. Adequate mixing should be provided. This insures that the chlorine solution reaches all parts of the flow.

5. Chlorination in two stages should be provided, particularly for wastewater streams. This type provides better results than a single-stage system.

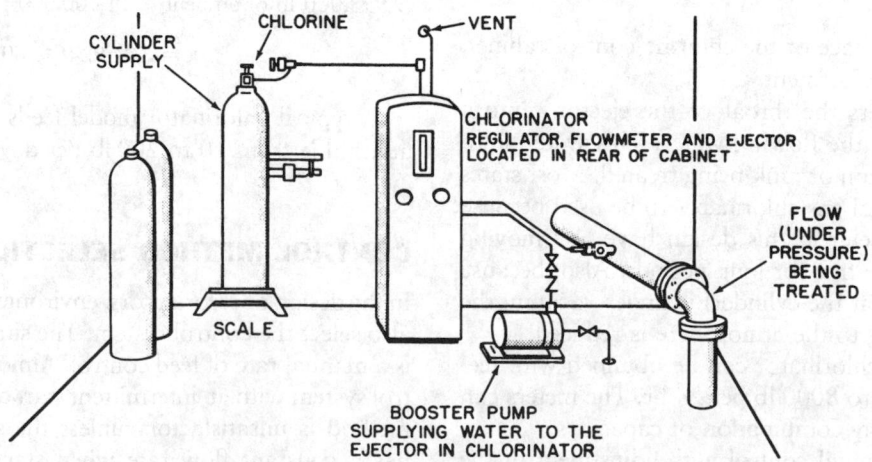

FIG. 7.30.12 Chlorinator with booster pump.

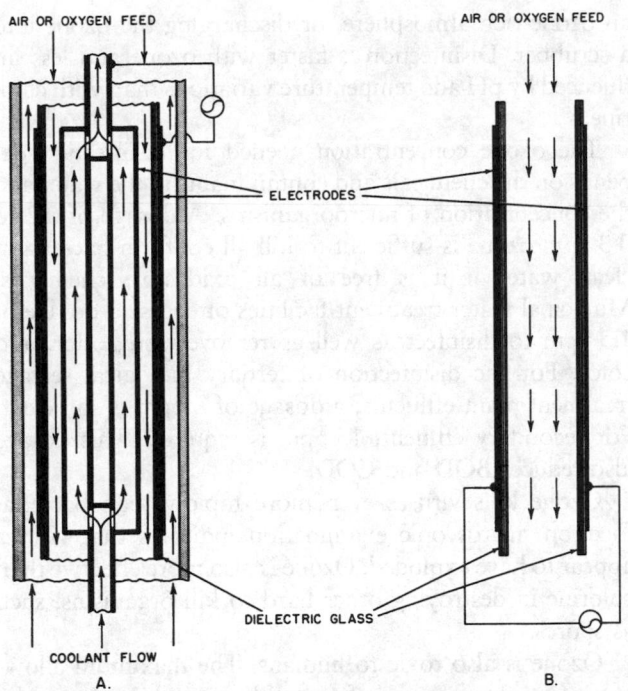

FIG. 7.30.13 Ozone generators. A, Tube type; B, Plate type. Ozone is generated in an electric discharge field by passing air or oxygen between two electrodes charged with high-voltage alternating current. A dielectric material, usually glass, is placed between the two electrodes to prevent direct discharges.

6. Adequate ventilation must be provided. Chlorine gas is extremely poisonous. Adequate mechanical ventilation of the chlorine room and venting of the chlorinator is mandatory. A gas mask in a case by the door and leak detection chemicals are also beneficial.
7. The remainder of the chlorine should be weighed. For smaller plants, a simple way to keep track of the chlorine liquid in the cylinder is to place it on a platform scale.

Ozonation

Ozone, a triatomic allotrope of oxygen, is produced industrially in an electric discharge field generator from dry air or oxygen at the site of use (see Figure 7.30.13). The ozone generator produces an ozone–air or ozone–oxygen mixture containing 1 and 2% ozone by weight. Wastewater treatment facilities introduce this gas mixture into the water by injecting or diffusing it into a well-baffled mixing chamber or scrubber or by spraying the water into an ozone atmosphere.

OZONE CHARACTERISTICS

Ozone is a powerful oxidizing agent (see Table 7.30.3). The mechanism of its bactericidal action is believed to be diffusion through the cell membrane followed by the irreversible oxidation of cell enzymes. The disinfection is unusually rapid and requires low ozone concentrations.

The viricidal action of ozone is even faster than its bactericidal effect. The mechanism by which the virus is destroyed is not yet understood. Ozone is also more effective than chlorine against spores and cysts such as *Endamoeba histolytica.*

Ozonation can accomplish disinfection and color, taste, and odor control in a single treatment step. Ozone reacts rapidly with all oxidizable organic and inorganic materials in water.

The ozone dosage for disinfection depends on the pollutant concentration in raw water. An ozone dose of 0.2 to 0.3 ppm is usually sufficient for bactericidal action only. The ozone dosage for secondary activated-wastewater-treatment-effluent disinfection is 6 or more ppm.

Ozonation leaves no disinfection residue; therefore, ozonation should be followed by chlorination in drinking water supply treatment applications. For optimum drinking water, the raw water should first be ozonated to re-

TABLE 7.30.3 OXIDATION POTENTIALS OF CHEMICAL DISINFECTANTS

Disinfectant		Oxidation Potential (volts)
Ozone	$O_3 + 2H^+ + 2e^- \longrightarrow O_2 + H_2O$	2.07
Permanganate	$MnO_4^- + 4H^+ + 3e^- \longrightarrow MnO_2 + 2H_2O$	1.67
Hypobromous acid	$HOBr + H^+ + e^- \longrightarrow \frac{1}{2}Br_2 + H_2O$	1.59
Chlorine dioxide	$ClO_2 + e^- \longrightarrow ClO_2^-$	1.50
Hypochlorous acid	$HClO + H^+ + 2e^- \longrightarrow Cl^- + H_2O$	1.49
Hypoiodous acid	$HIO + H^+ + e^- \longrightarrow \frac{1}{2}I_2 + H_2O$	1.45
Chlorine gas	$Cl_2 + 2e^- \longrightarrow 2Cl^-$	1.36
Oxygen	$O_2 + 4H^+ + 4e^- \longrightarrow 2H_2O$	1.23
Bromine	$Br_2 + 2e^- \longrightarrow 2Br^-$	1.09
Hypochlorite	$ClO^- + H_2O + 2e^- \longrightarrow Cl^- + 2OH^-$	0.94
Chlorite	$ClO_2^- + 2H_2O + 4e^- \longrightarrow Cl^- + 4OH^-$	0.76
Iodine	$I_2 + 2e^- \longrightarrow 2I^-$	0.54

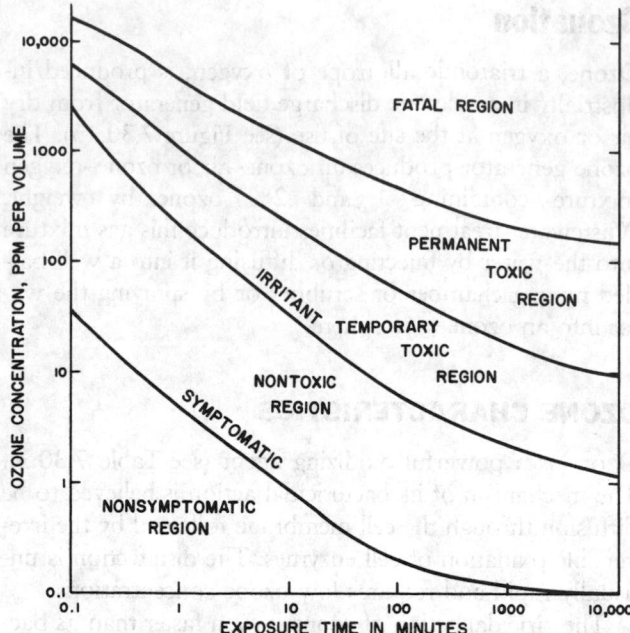

FIG. 7.30.14 Human tolerance for ozone.

move color, odor, and taste and to destroy bacteria, viruses, and other organisms. Then, the water should be lightly chlorinated to prevent recontamination.

DOSAGE AND PERFORMANCE

Wastewater treatment facilities can introduce the ozone-containing air or oxygen mixture produced by the ozone generator (1 or 2% ozone) into the water by injecting or diffusing it into a mixing chamber, spraying the water into

an ozone-rich atmosphere, or discharging the ozone into a scrubber. Disinfection is faster with ozone and less influenced by pH and temperature variations than with chlorine.

The ozone concentration needed for disinfection depends on the chemicals and contaminants in the water and the concentration of microorganisms. A dosage of 0.2 to 0.3 ppm ozone is sufficient to kill all coliform bacteria in clean water if it is free of all oxidizable chemicals. Municipal water treatment facilities often use a dosage of 1.5 ppm to disinfect as well as remove taste, odor, and color. For the disinfection of tertiary biological sewage treatment plant effluents, a dosage of 6 ppm is sufficient. For secondary effluent, 15 ppm is required. This dosage also reduces BOD and COD.

Ozone kills viruses even more rapidly than bacteria. Electron microscopic examination indicates that viruses appear to have exploded. Ozone is also more effective than chlorine in destroying other hard to kill organisms, such as spores.

Ozone is also toxic to humans. The maximum allowable concentration in air for an 8-hr period is 0.1 ppm by volume (see Figure 7.30.14).

—Wen K. Shieh
Van T. Nguyen

References

Metcalf and Eddy, Inc. 1991. *Wastewater engineering: Treatment, disposal, and reuse.* 3d ed. New York: McGraw-Hill.
Sawyer, C.N., and P.L. McCarthy. 1978. *Chemistry for environmental engineering.* 3d ed. New York: McGraw-Hill.

Advanced or Tertiary Treatment

7.31
TREATMENT PLANT ADVANCES

ADVANCED OR SPECIAL PROCESSES

  a) *Biological*—1) Completely mixed, activated sludge, 2) Ultrafiltration membrane in activated sludge, 3) Use of pure oxygen, 4) Multistage activated sludge
  b) *Physical*—1) Electrodialysis, 2) Reverse osmosis, 3) Carbon adsorption, 4) Centrifugation
  c) *Chemical*—1) Hydrolysis, 2) Ozone, 3) Precipitation, 4) Coagulation, 5) Flocculation

  d) *Heat Treatment*—1) Heat treatment alone, 2) Wet-air oxidation
  e) *Recovery Methods*—1) Recalcination, 2) Activated carbon recovery

Advanced treatment removes nutrients, trace organics, and dissolved minerals. Physicochemical treatment techniques used in conjunction with biological processes are an effective means of guaranteeing high-quality, reusable effluents. Because physicochemical methods are not overly effective in reducing soluble organic BOD on such wastewater biological processes are universally applied.

Advanced wastewater treatment plants usually consist of several unit processes operating in series. Each unit process represents an additional treatment step on the way to a treated, reusable effluent. By combining differing unit processes, these plants can attain almost any purity of effluent to meet the use of the treated wastewater.

A convenient method of describing advanced treatment plants is to use engineering flow sheets to graphically present the combination of unit processes and the direction of waste flow through them. The flow sheets in this section show treatment processes that are practical methods of treating wastewater. Some of these processes may only have been used in pilot plants; others are operating in full-scale applications.

Treatment processes can be classified according to the degree of treatment required. Advanced treatment plants have high treatment efficiencies, i.e., they produce water suitable for reuse or nonpolluting disposal.

One of the most widely used indicators of treatment efficiency is BOD. Because organic material usually exerts the largest component of BOD, BOD removal is a direct measure of the organic material removed from wastewater. Another important parameter is phosphate removal. Phosphates act as fertilizers in natural water and promote algae and plant growth. Such growth ultimately leads to anaerobic and polluted conditions.

Common categories of treatment levels are primary, secondary, and tertiary and the physicochemical treatment, which combines primary and tertiary processes. Conventional treatment usually consists of primary (SS removal) and secondary (BOD reduction) steps. Advanced treatment refers to specialized or tertiary processes. The process categories discussed in this section are subdivided into secondary equivalent (80 to 90% BOD reduction), secondary equivalent (80 to 90% BOD reduction) with phosphate removal, and tertiary equivalent (95% BOD reduction with phosphate removal).

Secondary Equivalent (80 to 90% BOD Reduction)

Most of these treatment processes were discussed in Section 7.21 because while the effluent produced is of high quality, it is not to be directly reused. Conventional primary and secondary treatment implies a primary stage of settling followed by biological treatment of a trickling filter or the activated-sludge-process type. The biological stage is usually followed by another settling step, and sludge disposal is by incineration, biological digestion, or landfill.

The San Mateo, California sewage treatment plant is an example of a primary system updated with sludge incineration. Because of odor problems from digesters and increased loads, the city converted a digester to a building to house a multiple-hearth incinerator and installed a 54-in concrete pipeline to carry the effluent into the deep channel of San Francisco Bay. While this treatment plant is not advanced, the city plans to include secondary and tertiary treatment.

Figure 7.31.1 shows this upgraded, 13-mgd primary plant. The sewage is pumped into clarifiers. The clarifier scrapes floating grease and scum from the surface and feeds it by pumping to a multiple-hearth incinerator. Sludge removed from the bottom of the clarifiers is thickened (250 gpm) to about 6% and dewatered by centrifuges to 30% solids. A combination of three parts raw sludge and one part digested sludge is pumped to the centrifuges. A conveyor belt moves sludge to the top of the multiple-hearth furnace—housed in an obsolete 65-ft diameter tank. Natural gas burners supplement the heating value of the sludge plus grease and skimmings. The sterile, odor-free ash is cooled and trucked to a landfill.

PURE OXYGEN

Conventional secondary treatment usually involves an extensive aeration step, mixing wastewater with a bacterial seed. The use of pure oxygen makes the aeration steps more efficient by providing an oxygen-rich environment for the entire process.

The principle used in this process is that gas solubility in water increases linearly with the partial pressure of that gas (Henry's law). Where normal aeration can only maintain a DO level of 1 to 2 mg/l, pure oxygen provides 8 to 10 mg/l of oxygen accompanied by an increase in the dri-

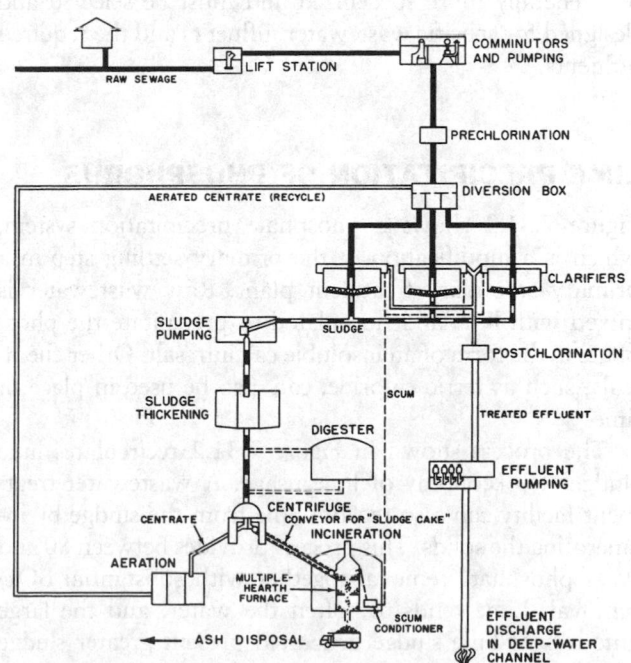

FIG. 7.31.1 Sewage treatment plant at San Mateo, California.

ving force to replenish any oxygen used. Thus, less aeration time is needed without large increases in sludge production as occurs with other high-rate systems. The system is also less susceptible to overloading. In addition, less mixing energy is needed, and a better settling flow is formed due to the reduced turbulence. Equipment manufacturers recommend a normal scaleup factor of 0.5 when equipment for full-scale applications based on laboratory data is sized.

MOVING CARBON BED ADSORPTION

In this process, the wastewater after SS removal flows through activated carbon beds, which adsorb a large portion of soluble organic molecules.

The carbon is continuously reactivated in a furnace and returned to the top of the vertical beds. This physicochemical treatment process does not rely on bacteria to degrade organic matter. For industrial wastewater in which no SS are present, wastewater treatment facilities can omit the coagulation–flocculation step and treat the water only with activated carbon.

Secondary Equivalent plus Phosphate Removal

These processes produce an effluent with 80 to 90% BOD reduction and phosphate removal by inserting chemical and physical treatment steps either in place of or in addition to secondary treatment. The quality of effluent produced is often suitable for reuse in industry. The processes are generally more specialized and must be selected and designed for specific wastewater influents and the required effluents.

LIME PRECIPITATION OF PHOSPHORUS

Figure 7.31.2 shows a phosphate precipitation system, which is a modification of the primary settling step in a primary–secondary treatment plant. Raw wastewater is mixed with lime and flocculated to precipitate the phosphate in the form of an insoluble calcium salt. Other chemicals, such as ferric chloride, can also be used in place of lime.

The process shown in Figure 7.31.2 recirculates lime sludges for economy of lime usage. A wastewater treatment facility can also reclaim lime from the sludge by incinerating the solids. This process provides between 80 and 95% phosphate removal together with substantial BOD removal. Lime tends to soften the water, and the large amounts of lime sludge generated present greater sludge handling problems.

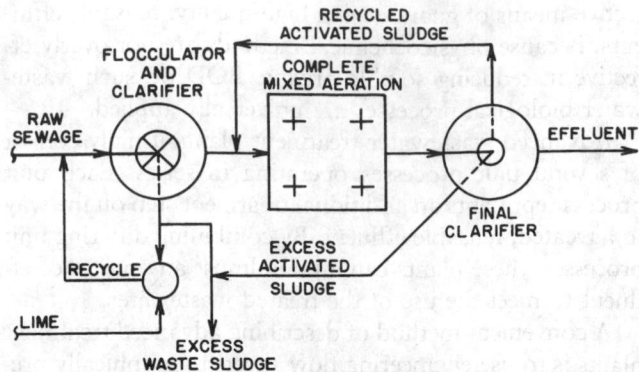

FIG. 7.31.2 Lime precipitation of phosphorus.

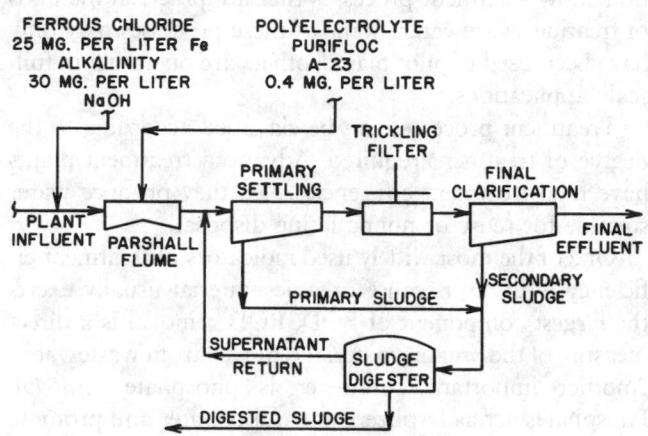

FIG. 7.31.3 Ferrous precipitation of phosphate.

FERROUS PRECIPITATION OF PHOSPHATE

Figure 7.31.3 shows the use of a metal ion, such as in ferrous chloride (pickling waste), to convert soluble phosphate in wastewater into an insoluble form. A polyelectrolyte flocculant is used to help settle out phosphate complexes in the primary settling basin. Either ferric chloride or sodium aluminate can also be used in place of ferrous chloride.

The concentration of metal ion added is between 15 and 20 mg/l, and the polymer addition is about 0.4 mg/l. The process removes from 70 to 80% of the influent phosphate. When combined with biological treatment, this process achieves removals as high as 90% in full-scale trials.

The economy of the process depends largely on the cost of the metal salts. Ferrous chloride, a waste from steel plants, is economical if it does not have to be transported far from its source. This process also reduces the BOD load on secondary treatment units. Corrosion-resistant sludge handling equipment is required due to ferric or ferrous sludges. The chemical sludge produced is not compatible with all methods of dewatering, and the chloride ion content of the effluent increases as a result of this treatment process.

ALUM AND FERRIC PRECIPITATION OF PHOSPHATE

Figure 7.31.4 shows a process in which metal salts (alum or ferric) added to the secondary effluent produce phosphate precipitates that are trapped in the granular media filter. Periodic backwashing of the filter yields a waste phosphorus sludge that is uniformly mixed in an equalization tank. This sludge is then recycled back to the primary settling basin in proportion to the plant influent flow rate. The waste sludge adsorbs more phosphorus and thereby reduces the phosphate load on the filter.

Because of the reduced phosphorus concentration in the filter influent, chemical costs can be reduced by as much as 30%. Flow- and phosphorus-monitoring instruments can be used to automate this process. Laboratory and pilot-plant operations have produced effluent phosphorus concentrations of less than 1 mg/l.

Figure 7.31.5 and Table 7.31.1 describe the components and performance of two additional wastewater treatment systems. In system A for the 10-mgd flow shown, two 34-ft diameter by 10-ft high granular media filters are required primarily for SS removal, but they can also remove phosphorus if precipitating chemicals are added. Additional BOD and COD removal is provided by five 20-ft diameter by 20-ft high carbon columns plus one standby column of the same size. Granular carbon can be regenerated in a furnace. Furnaces can also recalcine the lime sludge (burn the $CaCO_3$ into CaO).

In System B, chemical coagulation takes place in two 90-ft diameter by 17-ft sidewall depth columns, which provide high-rate solids-contact treatment and inorganic phosphorus removal.

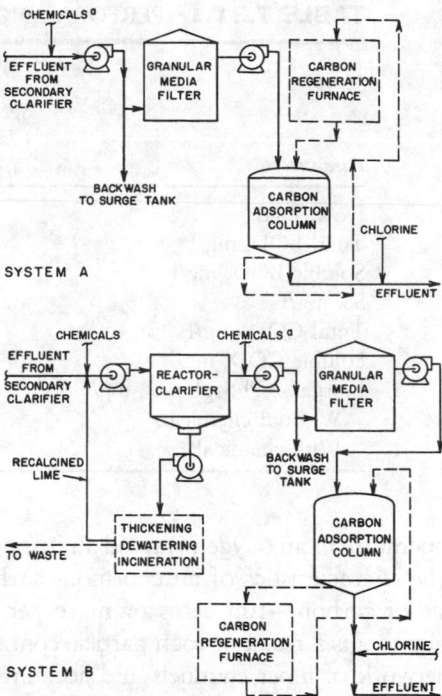

FIG. 7.31.5 Granular filtration, carbon adsorption, and chemical coagulation units. Chemicals can be added to achieve greater BOD, SS, and phosphorus removal.

SLUDGE HEAT TREATMENT PROCESS

Heat treatment of sludges to facilitate dewatering is discussed in Section 7.51 and shown in Figure 7.51.1. After the sludge has been conditioned with heat and pressure, it has a lower COD and can be readily dewatered without chemicals in a decanting and vacuum filter unit.

This process grinds the sludge and pumps through a heat exchanger into a reactor where it is held for 30 to 60 min while steam heats it to 380°F and maintains a high pressure of 160 to 250 psig. The system is a continuous process, and the high-temperature sludge leaving the reactor is used to heat the incoming sludge. The process was developed in Europe by Porteous in the 1930s and is now being applied in the United States. No chemical additives are required to make the treated sludge compatible with vacuum filtering.

PHYSICOCHEMICAL TREATMENT (PCT)

In PCT, the two most important unit processes are chemical coagulation and adsorption on activated carbon. Coagulation is similar to phosphate removal methods in that metal ions are added to the wastewater flow. Flocculation uses polyelectrolytes to form the heavy organic floc that easily settles out.

Activated carbon has been used for years in industrial processing and is now being applied in wastewater treatment. The carbon is a product of the combustion of carbonaceous material (bituminous coal or coconut shells) at

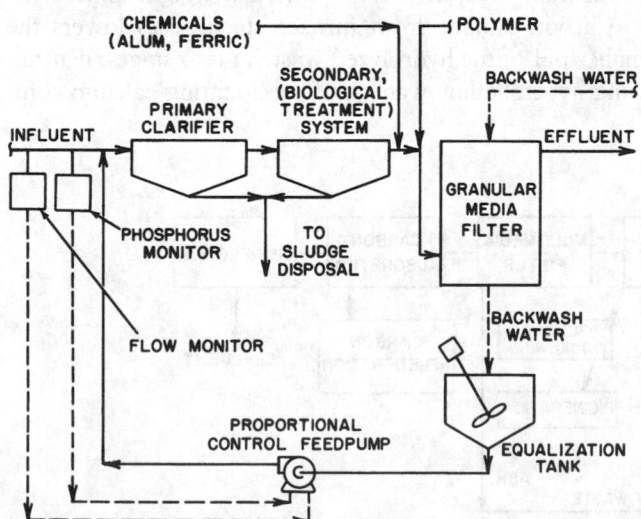

FIG. 7.31.4 Alum or ferric precipitation of phosphate.

TABLE 7.31.1 PERFORMANCE OF SYSTEM A AND B IN FIGURE 7.31.5

Parameters	System A		System B		
	Secondary Clarifier Effluent	Carbon-Adsorber Effluent	Reactor-Clarifier Effluent	SVG Filter Effluent	Carbon-Adsorber Effluent
Flow, mgd	10	10	10	10	10
Total BOD, mg/l	25	5	12	8	5
Soluble BOD, mg/l	10	5	10	8	5
SS, mg/l	30	2 0	5	2	
Total COD, mg/l	50	10	25	25	10
Soluble COD, mg/l	20	10	20	16	10
Inorganic Phosphorus, mg/l					
Without chemical	7	7	—	—	—
With chemical	7	0.5	0.5	0.4	0.4

a high temperature in an oxygen-starved atmosphere. This char has the characteristics of an enormous surface area per granule of carbon—100 acres or more per lb. This large surface area is a result of each particle containing an intricate network of inner channels and accounts for the carbon's great adsorptive capacity.

The treatment uses granular carbon in the form of a bed. Wastewater flows through this bed much like a sand filter. Organic molecules are attracted by the carbon and adsorbed on its surface. When the carbon becomes loaded with the adsorbed molecules and loses its adsorptive ability, it can be removed from the system and regenerated.

Figure 7.31.6 shows a PCT flow sheet. This figure shows all waste flow routes together with the backwash lines and the regeneration equipment shown in Figure 7.31.7. This treatment has great flexibility with interconnected carbon columns because any number of columns can be used in series while others are being refilled with fresh carbon.

In evaluating some processes, the EPA reports that the products of these processes (including carbon treatment) equal or exceed that of a well-operated conventional biological plant. The flow sheets in Figures 7.31.6 and 7.31.7 show major unit processes in PCT systems. Following pretreatment, which includes screening and grit removal, these

systems add and mix a coagulant such as lime. Recalcifying the sludge in a furnace can supply part of the lime coagulant dosage needed. The supernatant from the clarifier passes through a multimedia filter of sand and anthracite coal to remove the remaining SS, which is an inexpensive way to protect the carbon from solids carryover. The activated-carbon beds remove refractory colloidal and some soluble organic molecules.

After disinfection with chlorine, the treated effluent can be discharged. The removal efficiencies of this type of plant are 95 to 99% of organic material and 85 to 99% of phosphorus. The costs are significantly higher if a coagulant other than lime is used since lime can be recovered on site.

HYDROLYSIS–ADSORPTION PROCESS

This process uses lime to hydrolyze large organic molecules into molecules small enough to be adsorbed by activated carbon. The dosage of lime (200 to 600 mg/l) is governed by the molecular nature of the wastewater contaminants and the pH level (11.1 to 12.2 pH), which produces hydrolysis of high-molecular-weight molecules.

Carbon dioxide from furnace stack gases lowers the higher pH of the hydrolyzed waste in two stages, neutralizing hyperalkaline water and precipitating calcium com-

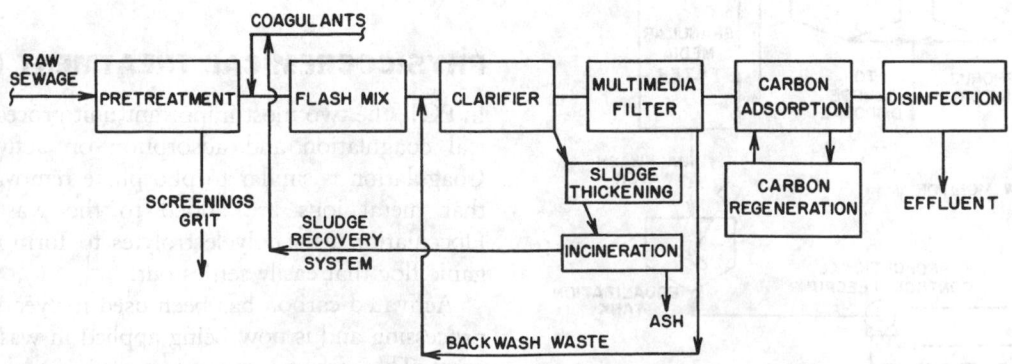

FIG. 7.31.6 PCT of wastewater.

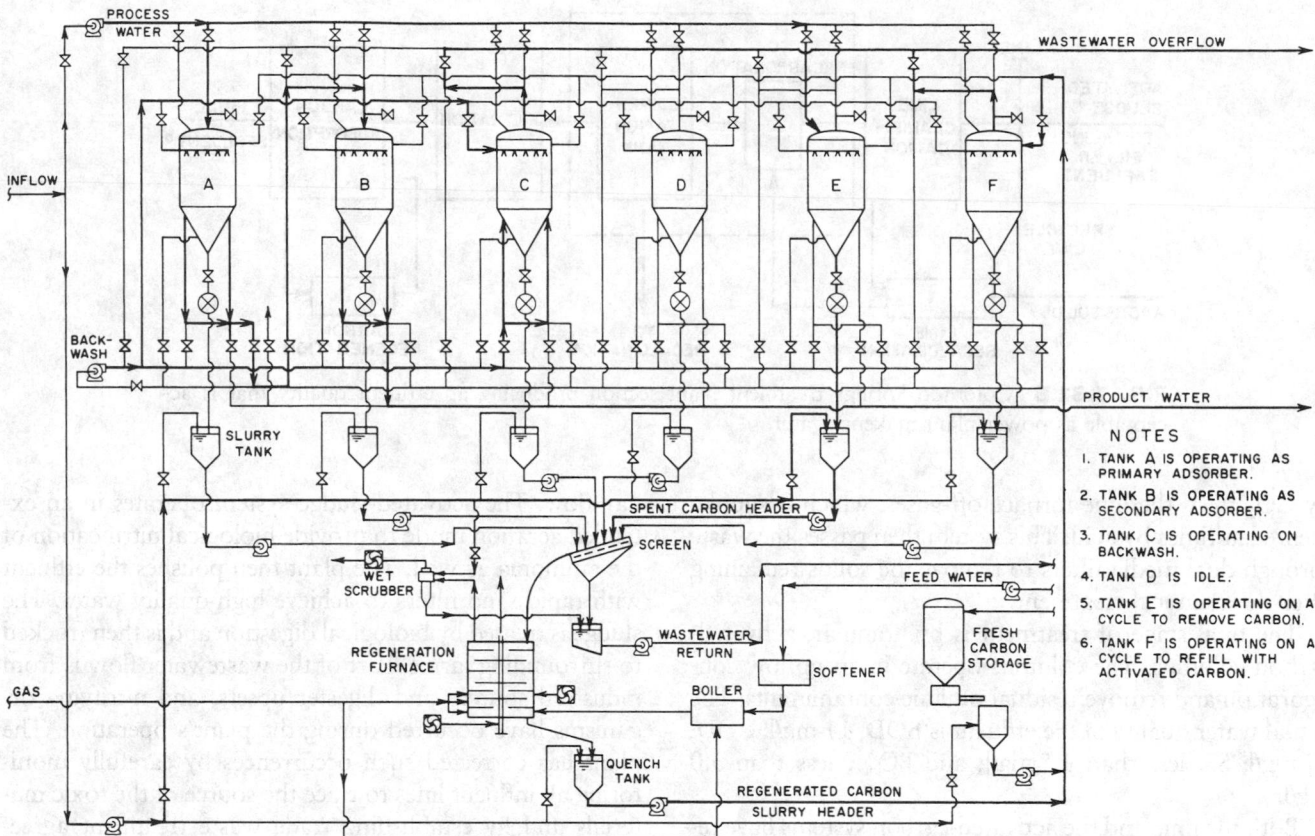

FIG. 7.31.7 Multiple-carbon columns in the PCT of wastewater.

NOTES

1. TANK A IS OPERATING AS PRIMARY ADSORBER.

2. TANK B IS OPERATING AS SECONDARY ADSORBER.

3. TANK C IS OPERATING ON BACKWASH.

4. TANK D IS IDLE.

5. TANK E IS OPERATING ON A CYCLE TO REMOVE CARBON.

6. TANK F IS OPERATING ON A CYCLE TO REFILL WITH ACTIVATED CARBON.

pounds. The hydrolysis of large molecules into smaller ones allows the carbon to adsorb more organic materials, which lowers its detention time to between 7.5 to 15 min. The hydrolysis–adsorption process can operate at 90% BOD and COD removal efficiencies and at 97% phosphorus removal efficiencies.

PCT plants are not subject to upsets and efficiency losses from toxic wastes, and they use many recycling methods to recover some of the chemical treatment agents. Their land requirements are also less, and phosphorus removal is part of their total treatment performance.

Recycling calcium in the sludge can reduce the sludge disposal problems associated with the large quantities of chemicals. Heavy metals are also precipitated into the lime sludge. Wastewater treatment facilities can expand their plant size by adding modular units such as carbon columns or reactor-clarifiers.

Tertiary Equivalent (95% BOD Reduction plus Phosphate Removal)

A tertiary treatment plant removes practically all solid and organic contaminants from wastewater, thereby producing drinkable water direct from sewage. Many tertiary treatment plants include nitrogen removal systems using either bacterial nitrification, air stripping, or ion exchange.

In times of scarce water resources, the treatment systems described may represent the future source of water for semiarid or highly populated regions as well as the means to eliminate water pollution.

COLORADO SPRINGS TREATMENT PLANT

A full, tertiary treatment plant is in operation in Colorado Springs, Colorado. The plant was designed with the objective of producing a high-quality effluent that is acceptable both as irrigation water and makeup water for power station cooling towers. The plant has a dual design. Both systems use the effluent from an existing trickling filter treatment plant.

The plant produces irrigation water by filtering this effluent in four coarse media filters that are hydraulically loaded at a rate of 10 to 20 gpm per sq ft. These filters are effective at removing the gross particulate matter at a rate of 8 mgd before the waste is used as irrigation water.

The system that produces cooling tower makeup water is more extensive, reflecting the more stringent effluent quality requirements (see Figure 7.31.8). The trickling filter effluent first enters a solids-contact clarifier where a slurry of mostly recycled lime coagulates the SS and precipitates the phosphates. The effluent is then neutralized

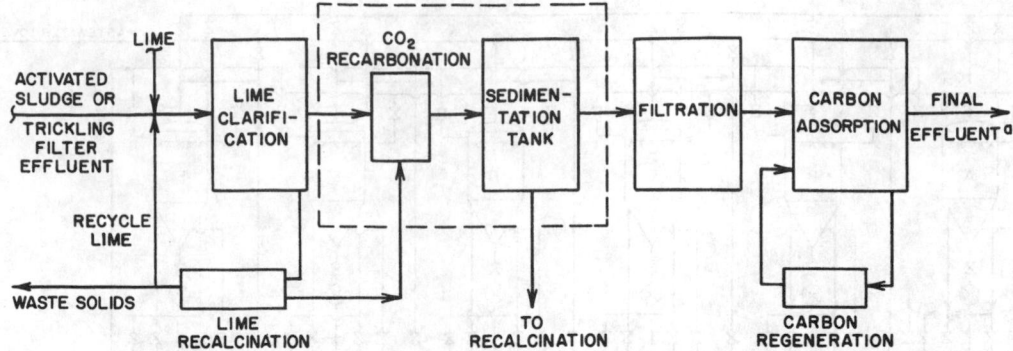

FIG. 7.31.8 Colorado Springs treatment plant section producing an effluent quality that is acceptable as power plant makeup water.

by CO_2 from the lime furnace off-gases, which is supplemented at times by acid. This system then passes the waste through dual media filters to remove the solids remaining after the chemical treatment.

The final stage of treatment is by granular, activated-carbon columns. The columns operate in an upflow configuration and remove residual organic contaminants. The actual water quality of the effluent is BOD, 11 mg/l; COD, 17 mg/l; SS, less than 1.5 mg/l; and PO_4^{3-}, less than 3.0 mg/l.

Both the lime and the activated-carbon systems have recycling loops that use multiple-hearth furnaces to regenerate some of the chemicals used. The recycled lime was found to be more effective in raising the pH than fresh lime. The recycled lime dosage is 280 mg/l, and the dosage for new lime is 325 mg/l.

The retention time in the contact clarifier is 1.25 hr, and the anthracite coal and sand filters are hydraulically loaded at a rate of 20 gpm per sq ft. Plant operating expenses are reduced from the sale of the effluent as power plant makeup water at a production rate of 2 mgd. The biological sludge in this plant is the first application of the Porteous heat treatment process in the United States.

RYE MEADS TREATMENT PLANT

A wastewater treatment plant operating since 1956 at Rye Meads, England, produces the highest quality effluent in the United Kingdom. The purpose of this regional plant is to reduce the sewage flow into the River Lee, which downstream supplies 19% of the water for the city of London. Thus, this example has water reuse aided by a short stretch of natural river course.

The design criteria were a 99% removal of SS and a 75% reduction in ammoniacal nitrogen (as N) to lower the oxygen demand in the river. Both criteria have been met almost continuously since plant startup. No chlorination (which is a widely used disinfection method in the United States) or any other form of disinfection is used.

The plant uses a diffused-air, activated-sludge system to reduce organic matter in the 10 million imperial gal per

day flow. The activated-sludge system operates in an extended aeration mode to provide biological nitrification of the ammonia as well. The plant then polishes the effluent with rapid sand filters to achieve high-quality water. The sludge is treated by biological digestion and is then trucked to surrounding farms. Part of the wastewater flow is from industrial sources and digester upsets, and nitrifying organisms have occurred during the plant's operation. The plant has corrected such occurrences by carefully monitoring all influent lines to trace the source of the toxic materials and by establishing trade waste treatment agreements. Goldfish in tanks in the maintenance shop are used as monitors for toxic materials.

SOUTH LAKE TAHOE RECLAMATION PLANT

The water reclamation plant at South Lake Tahoe, California, is a large and advanced treatment plant. Lake Tahoe in northern California and Nevada is in a natural basin that has been largely undeveloped until recent years. The lake, one of the three clearest lakes in the world, was destined to become another polluted body of water unless the nutrient inflow from sewage disposal was stopped and other sources of pollution were greatly retarded.

To meet this challenge, the South Lake Tahoe Public Utility District, with the cooperation of the U.S. government and industry, built this plant. The treated effluent water, which meets U.S. and World Health Organization (WHO) drinking water standards, is pumped out of the basin and creates the Indian Creek Reservoir, which has been approved by the California Department of Public Health for water sports and has been stocked with rainbow and rainbow hybrid trout by the California Department of Fish and Game.

The flow sheet in Figure 7.31.9 shows the unit processes in the 7.5 mgd South Lake Tahoe plant. The first treatment step is conventional primary settling followed by activated-sludge treatment. The sludges from these steps are centrifuged, dewatered, and incinerated to an ash. The

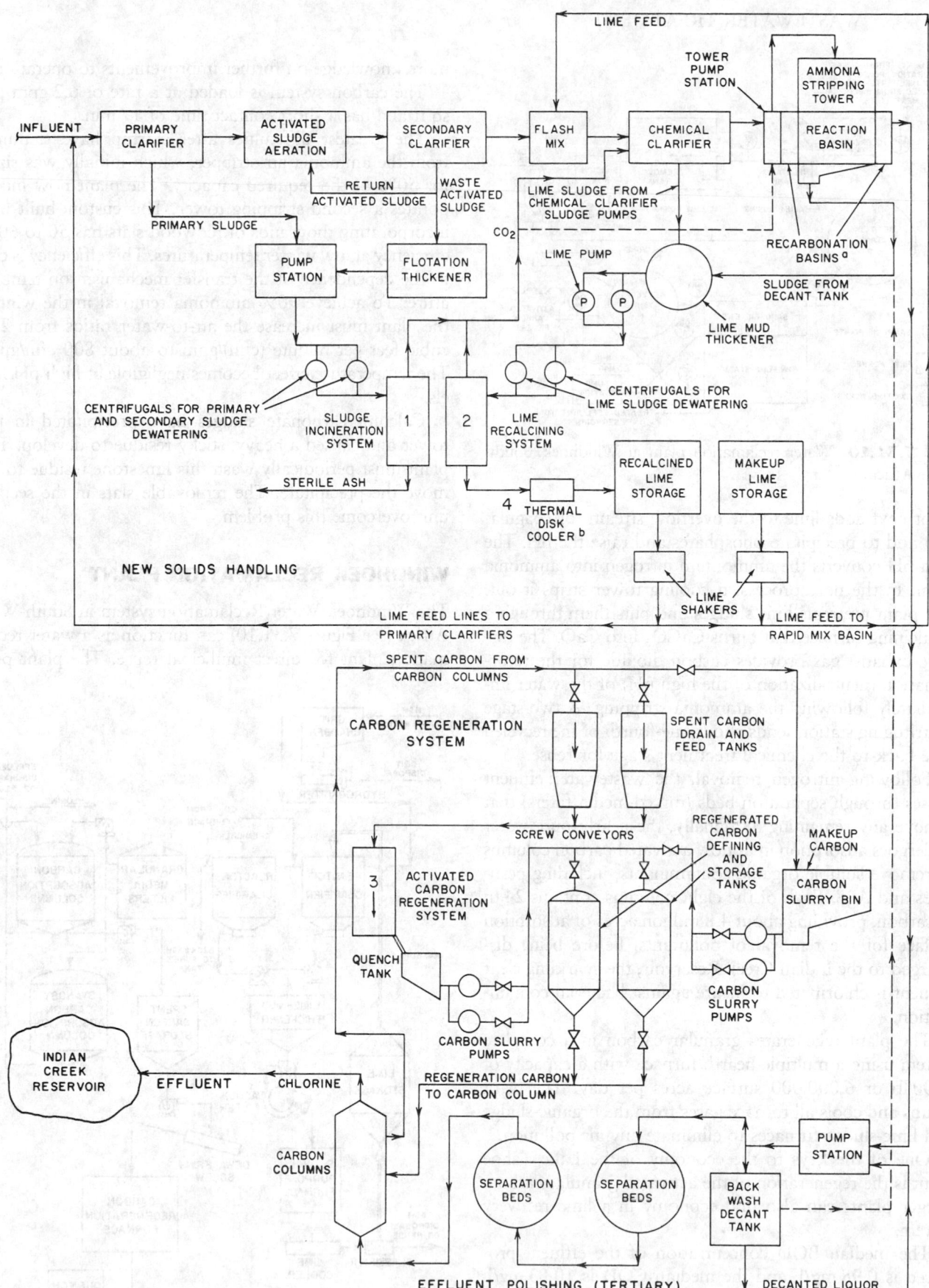

FIG. 7.31.9 Public utility district water reclamation plant at South Lake Tahoe, California. [a]Carbon dioxide is added to water in the reaction basin after it has passed through the ammonia-stripping column. [b]Thermal disk is a processing unit with a series of hollow disks filled with a heat-transfer medium. The solid, reclaimed lime is cooled by the disks as it passes between them.

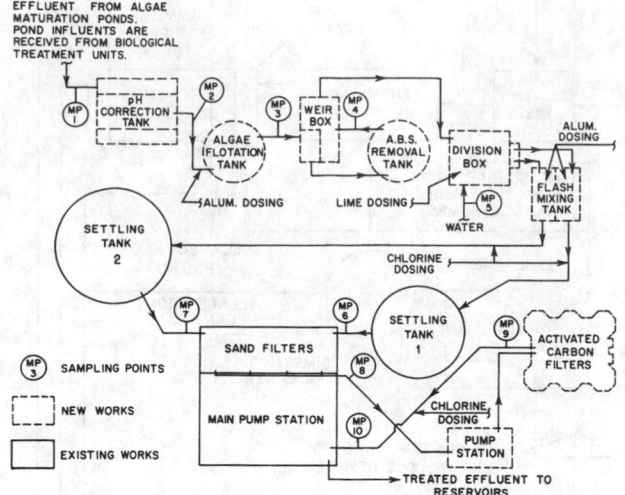

FIG. 7.31.10 Water reclamation plant at Windhoek, South-West Africa.

plant next adds lime to the overflow stream for coagulation and to precipitate phosphates and raise the pH. The high pH converts the ammonium nitrogen into ammonia form. In the next process, a cooling tower strips it out. The plant pretreats lime sludges and puts them through a recalcining furnace that burns $CaCO_3$ into CaO. The furnace exhaust gas provides carbon dioxide for the recarbonation (neutralization of the high pH) of the water immediately following the ammonia stripping. A two-stage centrifuging station sends about one-fourth of the recycled lime back to the chemical treatment stage for reuse.

Following nitrogen removal, the wastewater effluent passes through separation beds (mixed-media filters) that remove any remaining SS. Finally, the nearly pure water undergoes adsorption in upflow activated-carbon columns to remove soluble organic contaminants, including pesticides and ABS. Each of the eight columns contains 24 tn of carbon, providing about 4.8 million acres of adsorption surface for the removal of pollutants. Before being discharged to the Indian Creek Reservoir, the sparkling clear effluent is chlorinated to insure against bacterial contamination.

The plant regenerates granular carbon in a complete system using a multiple-hearth furnace with a capacity of 6000 lb or 6,000,000 surface acres per day. The plant scrubs and cools all furnace gases from the organic-sludge and lime-sludge furnaces to eliminate any air pollution.

One of the keys to the economy of the Lake Tahoe plant is the regeneration of the activated granular carbon. Larger plants can also find economy in a lime recovery system.

The median BOD concentration of the effluent produced is 0.98 mg/l, and the median COD is 10.83 mg/l. SS consists mostly of carbon fines in the effluent, amounting to 0.53 mg/l in concentration. The lowest phosphorus concentration (0.09 mg/l) was achieved when waste streams were recycled to the lime basin for 6 months. All systems are efficient, and testing continues to generate

more knowledge on further improvements to operation.

The carbon system is loaded at a rate of 6.2 gpm per sq ft and has a short contact time of 17 min.

The greatest difficulties after startup have stemmed from the ammonia air stripper, which initially was sized for 50% of the required capacity. The plant now incorporates a second stripping tower. This custom-built unit incorporating thousands of redwood slats has 50 to 60% efficiency at low winter temperatures. This efficiency is due to the dependence of the transfer mechanism on temperature. To achieve 90% ammonia removal in the winter, the plant must increase the air-to-water ratios from 250 cubic feet per minute (cfm)/gpm to about 800 cfm/gpm. The temperature effect becomes negligible at high pH levels.

Calcium carbonate sludges have precipitated in the tower and caused a heavy, sticky residue to develop. The plant must periodically wash this limestone residue to remove the precipitate. The removable slats in the second unit overcome this problem.

WINDHOEK RECLAMATION PLANT

The Windhoek Water Reclamation system in South-West Africa (see Figure 7.31.10) can function as a water reclamation plant for direct municipal reuse. The plant pro-

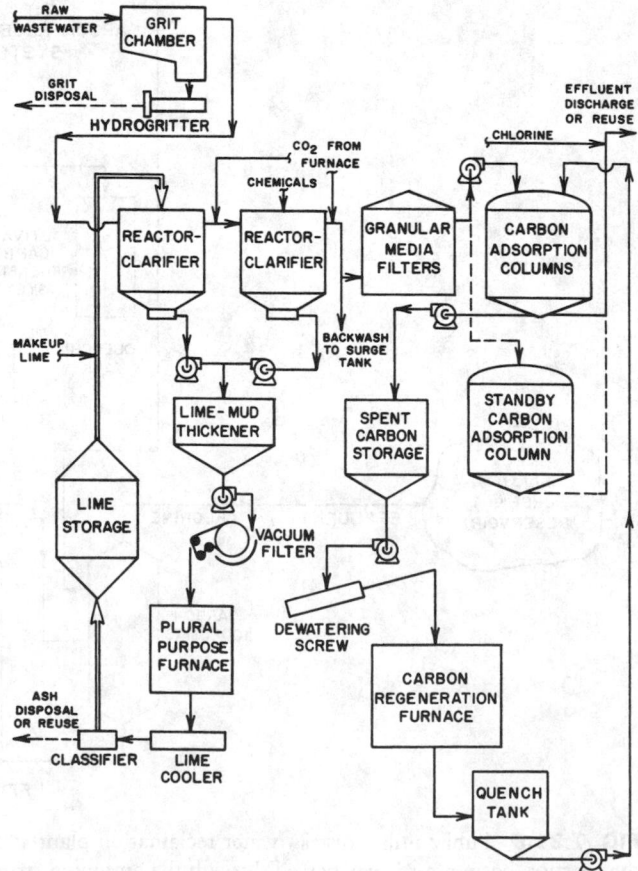

FIG. 7.31.11 Hydrolysis adsorption process for water reclamation.

TABLE 7.31.2 WASTEWATER CHARACTERISTICS FOR A 10-MGD PLANT

Parameters	Influent Quality	Effluent Quality
Flow, mgd	10	9.9
BOD, mg/l	250	5.0
SS, mg/l	250	5.0
COD, mg/l	500	10.0
Total Inorganic Phosphorus, mg/l	10	0.3

duces drinking water for a city. The primary and secondary treatment system uses biofilters and digesters. Effluent from the biofilters is held for about 14 days in algae maturation ponds where nitrogen and phosphorus are used by the algae.

Reusing the treated sewage in the ponds was determined to be the most economical source of drinking water. The algae-laden pond water is pumped to a purification plant at a rate of 1.2 mgd. Recarbonation (lowering of the pH) to 7.2 is accomplished by the submerged combustion of propane gas. Then, a unique flotation system removes 90% of the algae by adding alum and mixing it rapidly. Detergents are removed by a foam fractionator.

The plant adds alum, lime, and chlorine to the treated water which is settled and then filtered by sand and adsorption on granular activated carbon to remove trace amounts of organic molecules. Currently, the carbon is not regenerated. Breakpoint chlorination provides additional nitrogen removal. Salt buildup in the drinking water has been kept to a maximum of 180 mg/l. This specialized system for Windhoek is particularly suitable for hot climates because the algae beds do not function well at low temperatures.

HYDROLYSIS ADSORPTION PROCESS

The hydrolysis adsorption process produces drinking quality water. Figure 7.31.11 and Table 7.31.2 provide a more detailed version of system B in Figure 7.31.5. Achieving the higher quality water (drinkable) only requires a longer contact time in the carbon adsorbers. Nitrogen removal can be easily added since the high pH established by lime treatment is also a requirement for the ammonia-stripping unit process.

—*F.P. Sebastian*

7.32
CHEMICAL PRECIPITATION

TYPES OF CHEMICAL COAGULANTS

a) Alum, b) Lime, c) Hydrated lime, d) Sulfuric acid, e) Anhydrous ferric chloride

SS REMOVAL EFFICIENCY

Up to 85%

FLOCCULATION AIDS

f) Polyelectrolytes

CHEMICAL DOSE REQUIRED (mg/l)

80 to 120 (e), 100 to 150 (a), 350 to 500 (c)

NEUTRALIZING REAGENTS

a) *Acidic*—a1) sulfuric acid, a2) hydrochloric acid, a3) carbon dioxide, a4) sulfur dioxide, a5) nitric acid; b) *Basic*—b1) caustic soda, b2) ammonia, b3) soda ash, b4) hydrated lime, b41) dolomitic hydrated quicklime, b42) high calcium hydrated quicklime, b5) limestone.

SOLUBILITY AND pH OF LIME SOLUTIONS

For a concentration range of 0.1 to 1.0 gm. of CaO per liter the corresponding pH rises from 11.5 to 12.5 at 25°C. Saturation occurs at about 1.2 gm. per liter.

LIME AVAILABILITY

Dry forms with several particle size distributions, or liquid forms usually as a slurry with +20 wt percent solids.

CAUSTIC AVAILABILITY

In dry form as 75 percent Na_2O or as a 50 percent NaOH solution; precipitates formed are usually soluble.

HYDRAULIC LOADING (GPM PER FT² OF CLARIFIER)

0.2 to 0.4 (a), 0.3 to 0.4 (e), 0.5 to 0.8 (c)

QUANTITY OF CHEMICAL SLUDGE PRODUCED (LB PER MG)

250 to 500 (a), 350 to 700 (e), 4000 to 7000 (c)

DETENTION TIMES (MIN)

2 to 5 in mixing tank, 30 to 90 in flocculation tank

Process Description

Historically, SS have been removed from wastewater by gravity sedimentation. The removal efficiency of this unit operation is a function of the presence of readily settleable solids. Typical municipal wastewater contains about 50 to

100 mg/l of difficult-to-settle SS. These very small particles have densities approaching that of the suspending medium (water). Typically, these solids are bacteria, viruses, colloidal organic substances, and fine mineral solids. The precipitation of chemical agents causes these difficult-to-settle solids to flocculate (particle growth) and become settleable.

Chemical treatment can also precipitate certain dissolved pollutants, forming a settleable suspension. For example, phosphate is precipitated by aluminum (Al^{3+}), and heavy metals are precipitated as hydroxides when the pH is raised.

During the middle and late 1800s, chemical treatment was implemented in about 200 sewage treatment plants in England. During the 1930s, the United States became interested in chemical treatment. However, these users recognized that chemical treatment alone does not remove putrescible dissolved organics and is therefore not a complete treatment process. Interest in chemical treatment of municipal wastewater increased in the 1960s due to new and stringent requirements concerning phosphorus removal.

Types of Chemicals Used

Table 7.32.1 shows the properties of several common chemicals used in the removal of SS from wastewater.

Hydrolyzable trivalent metallic ions of aluminum and iron salts are established coagulation chemicals. The hydrolysis species of Al^{3+} or Fe^{3+} destabilize colloidal pollutants and render them amenable to flocculation (particle growth), which enhances settleability. If excessive amounts of these metallic ions are used, the gelatinous floc precipitated can enmesh some small pollutant particles, allowing them to be removed by settling. Both Al^{3+} and Fe^{3+} are effective phosphorus precipitating chemicals.

Alum can also break oil emulsions. Thus, alum treatment converts the colloidal emulsion to a suspension amenable to clarification by dissolved-air flotation.

pH ADJUSTING CHEMICALS

The solubility of several chemical pollutants is pH dependent. Thus, adjusting the pH to minimum solubility often allows efficient precipitation of these pollutants. Figure 7.32.1 shows the following three fundamental types of pH responses:

Case I shows the occurrence of minimum solubility or maximum solids precipitation at optimum pH value. Typical wastes producing this type of response include iron contained in acid mine drainage and fluoride and arsenic-bearing wastes.

Case II shows a neutral or acidic waste containing certain heavy metals. The optimum pH is at the knee of the precipitated solids generation curve. Lime treatment of metal finishing waste is typical of a Case II response.

Case III is the inverse of Case II; the chemical pollutant is precipitated due to pH depression, usually with sulfuric acid. A latex-bearing waste is typical of a Case III response.

TABLE 7.32.1 PROPERTIES OF CHEMICAL COAGULANTS

Common or Trade Name	Chemical Name	Chemical Symbol	Shipping Containers	Weight lb/ft³	Suitable Handling Materials	Commercial Strength
Alum	Aluminum sulfate	$Al_2(SO_4)$ · 14 H_2O	300–400 lb bbl, carload bulk; carload, barrels	60–63 / 62–67	Dry: Iron and steel Solution: lead, rubber, silicon, iron, and asphaltum	15–22% Al_2O_3
Quicklime	Calcium oxide	CaO	50 lb bags, 100 lb barrels, bulk carload	40–70 / 26–48	Rubber, iron, steel, cement, and asphaltum	63–73% CaO
Hydrated lime	Calcium hydroxide	$Ca(OH)_2$	50 lb bags, 100 lb barrels, bulk carload	40–70 / 26–48	Rubber, iron, steel, cement, and asphaltum	85–99% $Ca(OH)_2$ / 63–73% CaO
Oil of vitriol	Sulfuric acid	H_2SO_4	Drums, bulk	——	Concentrated: Steel and iron	93% H_2SO_4
					Dilute: lead, porcelain, glass and rubber	78% H_2SO_4
Anhydrous ferric chloride	Ferric chloride	$FeCl_3$	300 lb bbl (crystals)	——	Glass, stoneware, and rubber	59–61% $FeCl_3$
			500 lb casks, 300–400 lb kegs		Glass, stoneware, and rubber	98% $FeCl_3$

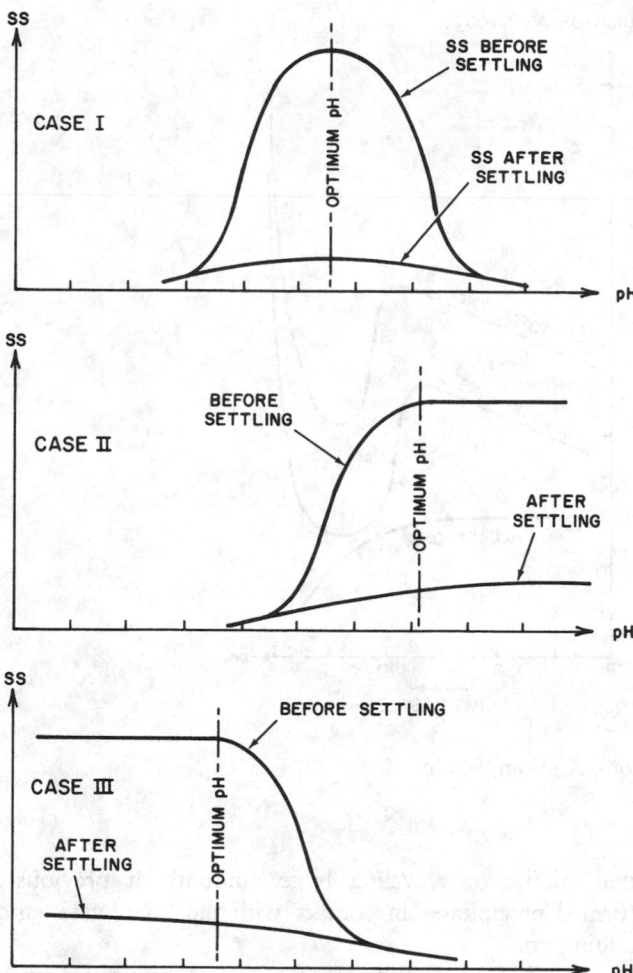

FIG. 7.32.1 Precipitation of chemical pollutants by pH adjustment.

Adding lime to municipal wastewater elevates the pH and causes insolubilization of calcium phosphate, calcium carbonate, and magnesium hydroxide. The hydroxide precipitated at an elevated pH acts as a coagulant, destabilizing and enmeshing colloidal pollutants. The chemistry of lime treatment is generally described by water softening reactions. However, the presence of dissolved organic matter and condensed phosphates in municipal wastewater causes some interference with water softening reactions, such as the precipitation of calcium carbonate. Figure 7.32.2 shows the solubility of the total and calcium hardness ions in a typical municipal wastewater.

FLOCCULATION AIDS

Organic polyelectrolyte flocculation aids are effective in promoting SS removal. The addition of polyelectrolytes does not produce a chemical precipitation per se but can drastically promote particle growth. Improvement in the removal of finely divided solids by gravity settling is the result. Polyelectrolytes are effective both for wastewater SS and for precipitates formed by chemical treatment.

Polyelectrolytes in suspensions become attached to two or more particles and provide bridging between them.

Chemical Sludge Production

An inherent burden with the improved SS removal by chemical treatment is the production of chemical sludge. The thickening and dewatering properties of chemical–sewage sludge are worse than those of the sewage sludge alone because of the presence of hydroxide sludges and the increased amounts of colloidal pollutants.

The addition of alum to a basic wastewater containing alkalinity produces a chemical floc. Generally, 1 lb of SS (chemical floc) is produced for each 0.25 to 0.40 lb of aluminum added. Figure 7.32.2 is a typical example of the sludge production and SS reduction by alum treatment.

The chemical sludge produced by adding lime to municipal wastewater depends on the chemical characteristics of the water, the pH level, and the method of operation. The net chemical sludge produced is a result of an interaction of these three parameters.

Lime treatment of municipal wastewater with low hardness ($\leq$200 mg/l as $CaCO_3$) and low alkalinity ($\leq$150 mg/l as $CaCO_3$) should be accomplished in two stages, with the corresponding pH levels at 11 to 11.5 and 9.5 to 10, respectively. The chemical sludge produced under these conditions is about 4000 to 5500 lb per mg. For municipal wastewater, with high hardness ($\geq$350 mg/l as $CaCO_3$), and high alkalinity ($\geq$250 mg/l as $CaCO_3$), single-stage treatment at a pH of 10.5 to 11.0 is recommended. The chemical sludge produced under these conditions is about 5500 to 6500 lb per mg.

The use of organic polyelectrolytes does not produce significant amounts of chemical sludge.

The nature and quantity of the chemical sludge produced by adjusting the pH of industrial waste depends mainly on the initial concentration of the chemical pollutants and the efficiency of the liquid–solids separation step. Generating the data shown in Figure 7.32.2 provides data on sludge production.

Unit Operations

SS removal through chemical treatment is accomplished by three series unit operations: rapid-mixing, flocculation, and settling.

RAPID MIXING

The chemical reagent must first be completely dispersed throughout the wastewater. This requirement is especially important when an inorganic coagulant such as alum is used because the precipitation reactions occur immediately. In lime treatment, the lime slurry should be dispersed throughout the wastewater in the presence of the previously formed precipitate (recycled sludge). The sludge pro-

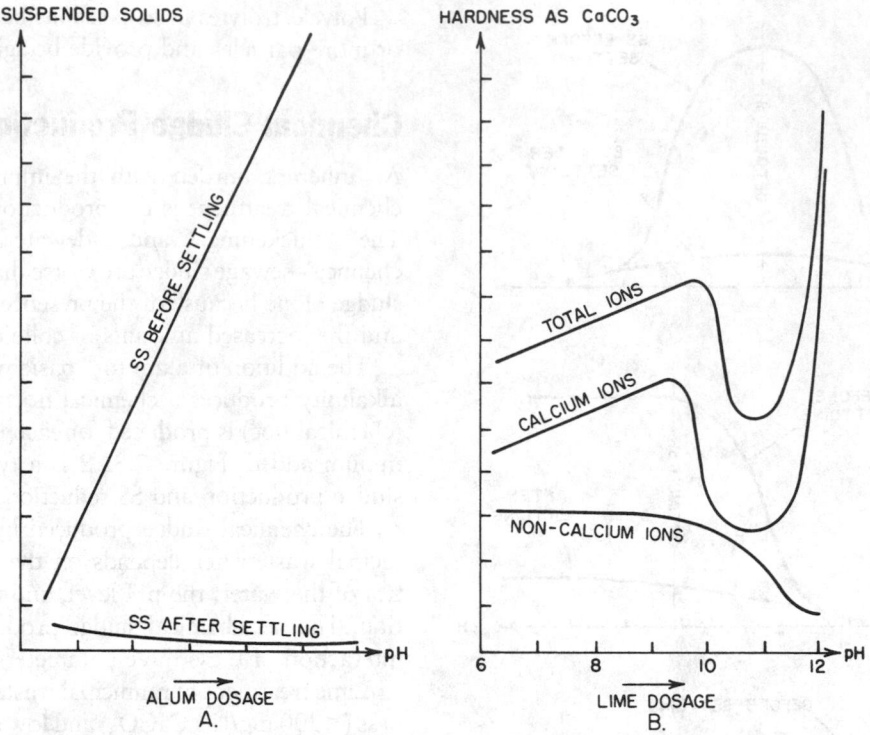

FIG. 7.32.2   Effects of alum and lime additions. A, Alum; B, Lime.

vides an abundant surface area on which large amounts of chemical precipitates can form. When the sludge is not recycled, gross deposits (scaling) of calcium carbonate develop on the tank walls and other surfaces.

Rapid mixing occurs in 10 to 30 sec in a basin with a turbine mixer. About 0.25 to 1 hp per mgd is used for rapid mixing. A mean temporal velocity gradient in excess of 300 ft per sec per ft is recommended.

FLOCCULATION

After effective coagulation-precipitation reactions (rapid mix) occur, promoting particle size growth through flocculation is the next step. The purpose of flocculation is to bring coagulated particles together by mechanically inducing velocity gradients within the liquid.

Flocculation takes 15 to 30 min in a basin containing turbine or paddle-type mixers. Mean temporal velocity gradients of 40 to 80 ft per sec per ft are recommended. The lower value is for fragile floc (aluminum or iron floc), and the higher value is for lime-treatment floc.

SOLIDS CONTACT

Solids contacting is especially beneficial for lime treatment because it reduces the deposition problems inherent in once-through, rapid-mix, flocculation systems. Wastewater treatment facilities provide solids-contacting by

maintaining or recycling large amounts of previously formed precipitates in contact with the wastewater and adding lime.

Several types of solids-contact treatment units are available. These units were originally developed for lime-softening water treatment and are effective for lime treatment of wastewater.

LIQUID–SOLIDS SEPARATION

After flocculation, the final step is clarification by gravity settling. The conventional clarifier design is suitable for this purpose. However, wastewater treatment facilities should provide positive-sludge withdrawal to prevent problems associated with the formation of septic sewage sludge. Figure 7.32.3 shows a once-through and a solids-contacting system for enhancing SS removal with chemical treatment.

If a highly clarified effluent of less than 10 mg/l SS is needed, an additional liquid–solids separation step is required. The unit operation recommended for effluent polishing is granular media filtration, e.g., dual-media filtration. Here, the clarifier effluent containing 20 to 40 mg/l of finely divided floc is passed through the filter. For peak hydraulic loadings of less than 5 gpm per ft^2, the filtered effluent contains not more than 5 to 10 mg/l of SS. The filter bed in this case consists of 2 ft of 1.0- to 1.5-mm anthracite coal over 1 ft to 0.5- to 0.8-mm sand.

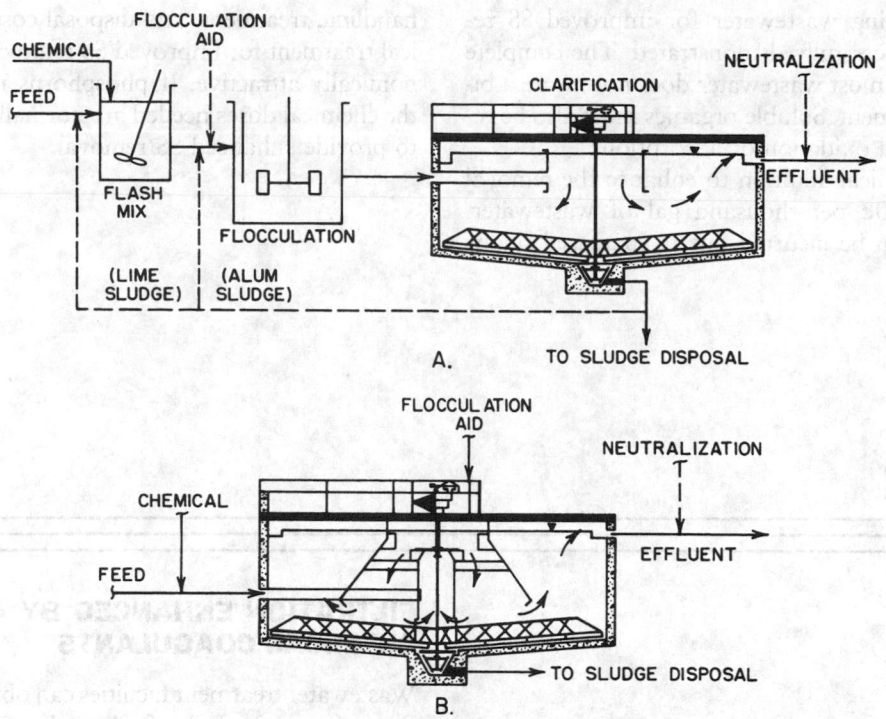

FIG. 7.32.3 Chemical treatment systems. **A**, Once-through system; **B**, Solids-contact system.

Design Considerations

The principal cost in enhancing SS removal by chemical treatment is the chemicals. Environmental engineers can estimate the chemical dose requirements from laboratory jar tests and/or pilot-plant studies. Daily, weekly, and seasonal variations in wastewater characteristics require that wastewater treatment facilities must adjust the chemical dose during plant operation to minimize additive use while providing the required solids removal. Table 7.32.2 lists the typical chemical doses.

Generally, the hydraulic loading of clarification equipment determines the SS removal efficiency. Because high removal efficiencies are sought, conservative hydraulic loadings should be used. Table 7.32.2 also lists the recommended average hydraulic loadings for 80 to 90% average SS removal from raw municipal wastewater.

Although polyelectrolyte flocculation aids at nominal doses of 0.25 mg/l can at least double the chemical-sewage floc settling rates, providing twice the clarifier capacity is normally less expensive than using the polyelectrolyte. For new plant construction, environmental engineers should make a thorough analysis of polyelectrolyte use versus reduced clarifier hydraulic loading. For upgrading existing facilities, wastewater treatment facilities can justify the use of polyelectrolytes to meet effluent quality requirements.

TABLE 7.32.2 DESIGN CRITERIA FOR 80 TO 90% SS REMOVAL FROM RAW MUNICIPAL WASTEWATER BY CHEMICAL TREATMENT

Criteria	Ferric Chloride $FeCl_3$	Alum	Hydrated Lime $Ca(OH)_2$
Dose, mg/l	80–120	100–150	350–500
Hydraulic loading,° gpm/ft^2 of clarifier	0.3–0.4	0.2–0.4	0.5–0.8
Chemical sludge production, lb/mg	350–700	250–500	4000–7000
Chemical cost, ¢/lb of chemical	4–5	3–4	1
Treatment chemical cost, ¢/1000 gal of wastewater	2½–5	2½–5	3–4

Note: °Without use of polyelectrolytes

Chemically treating wastewater for improved SS removal has been successfully demonstrated. The complete removal of SS from most wastewater does not insure a biologically stable effluent. Soluble organics must also be removed by biological oxidation or adsorption.

The cost of chemical addition to enhance the removal of SS is around 10¢ per thousand gal of wastewater. Additional costs can be incurred due to increased sludge

handling, treatment, and disposal costs. Therefore, chemical treatment for improved SS removal is not usually economically attractive. If phosphorus removal is required, the chemical doses needed are normally more than enough to provide enhanced SS removal.

—D.E. Burns

7.33
FILTRATION

Special Filters

FILTER TYPES

   a) Diatomite
   b) Microstrainers
   c) Gravity
   d) Pressure
   e) Deep-bed
   f) Multilayers
   g) Cartridge
   h) Continuous Moving Bed
   i) Membrane

HYDRAULIC LOADINGS (GPM/FT²)

0.5 to 3 (a,g); 2 to 6 (b,c,cf,d,df,ef); 6 to 10 (b,d,df,e,ef,f); and 10 to 30 (e)

SOLIDS REMOVAL CAPACITY (LB PER SQ FT OF FILTER SURFACE)

0.1 to 0.5 (a,g); 0.3 to 0.5 (c,d); 0.5 to 1 (c,cf,e,ef); 1 to 2 (e,ef,f); and 2 to 5 (e,ef,f)

FILTER BED DEPTH (FT)

1 to 2 (cf,df); 2 to 3 (c,d,df); 2 to 6 (df,ef); and 3 to 12 (e)

FILTER MEDIA USED

Acetate (g); Barium Sulfate (e); Cellulose (a,g); Coal (c,cf,d,df,f); Coke (e); Diatomite Fibers (a); Garnet (b); Gravel (ef); Nylon (g); Polypropylene (g); Sand (c,cf,d,df,f); and Stone (g)

FILTER OPENINGS OR FILTER MEDIA SIZES (MM)

0.5 to 100 μ (g); 0.05 to 0.2 (a); 0.4 to 0.6 (c,cf,d,df,f); 0.6 to 0.8 (c,d); 0.8 to 1.4 (cf,df,f); 0.8 to 4.0 (e); and 6 to 12 (e-coarse layer)

FILTER BACKWASH RATE

15 gpm per ft² with multilayer design having siliceous media

AIR PURGE RATE RECOMMENDED FOR CLEANING

3 to 7.5 scfm per sq ft

This section begins with a discussion of special filters, concentrating on water filters operating on influent SS concentrations of 100 mg/l or less.

FILTRATION ENHANCED BY APPLIED CHEMICAL COAGULANTS

Wastewater treatment facilities can obtain completely clear filter effluents only by feeding chemical coagulants, such as alum, iron, or polyelectrolytes, to the wastewater prior to filtration. Adding chemicals to clarifiers reduces the solids load on the clarifier overflow effluent filters.

When a facility is considering adding chemicals into the filter influent, they should perform coagulation tests or pilot-filtration tests. Chemical coagulation can result in 100% solids removal including colloidal particle sizes as small as 0.05μ. When stain-free filter effluent is required, chemical feeds are absolutely necessary. Without chemical feeds, the SS removal efficiency of filters generally ranges from 50 to 80%.

However, adding chemical coagulants increases the solids load; therefore, filter runs are shortened. Surface-type and ultra-high-rate filters are drastically affected in this regard. Adding coagulants can be costly not only because of shortened runs, but also because of the cost of chemicals, pumping, and labor.

In multilayered or in-depth filter applications, the influent must contain flocculant particles ranging from $\frac{1}{32}$ to $\frac{1}{4}$ in, whether these particles are natural particles, biological flocs, or freshly coagulated flocs. The larger flocs are removed by the coarse filter layers, and the smaller particles are removed by the fine sand layers below. If only smaller flocs are present, they are deposited primarily in the fine lower layers, resulting in a solids removal capacity no better than that of single-layer filters.

SELECTION AND OPERATION OF FILTERS

Waste flows have cyclic variations; for example, the peak flow of most domestic wastewater sources is about twice the average flow. Equalizing or surge basins are needed as part of the filter system to accommodate these flow variations.

When domestic waste is mixed with industrial waste containing metals or inorganic salts such as iron, copper, or aluminium, the filtration characteristics of the secondary effluents (after biological processes) can be enhanced. In such cases, fewer colloidal particles and larger or stronger flocs can be expected. These effluents can be treated with high-rate filtration or surface-type microscreening with removal efficiencies of over 80%, without the use of additional chemical coagulants. When domestic waste is aerated for more than 10 hr, an easily filtered floc generally forms without the need for chemicals.

FILTRATION ENHANCED BY EQUIPMENT DESIGN

When the SS concentration in filter influents approaches 120 mg/l, single-layer or surface filtration equipment is usually quickly overloaded, and the run lengths between cleanings become short, ranging from 1 to 3 hr with solids removal capacities of about $\frac{1}{4}$ to $\frac{1}{2}$ lb per sq ft.

Multilayer or deep-bed filters were developed for such applications. They enhance the filter capacity by providing more or deeper coarse layers where more settling and storage of solids can take place. Roughly, an extra $\frac{1}{2}$ lb of solids capacity is obtained with each additional layer or foot of depth in the filter.

GENERAL DESIGN PARAMETERS

Horizontal or vertical pressure filters are recommended in wastewater applications because they handle higher solids loads and pressure heads and are more compact and less costly. The freeboard (open space above the filter bed) for anthracite- (carbon and plastic) containing filters should be a minimum of 50%. For sand filters, a 30% minimum is required.

Environmental engineers can best determine the filter backwash rate based on the operating temperature and available bed expansion. For dual or multilayered filters containing siliceous media, a minimum of 15 gpm per sq ft backwash rate should be used. An air scour (purging), applied from the bottom, gives superior cleaning to surface washers or subsurface washers. Air purging also saves 30 to 50% of the wash water requirement. The recommended air rate is 3 to 7.5 scfm per sq ft. Underdrain graded gravel or siliceous layers should be a minimum of 16 in deep, with sizes ranging from $1\frac{1}{2}$ in to 6 × 10 mesh.

Plastic strainer underdrain nozzles screwed into flat steel decks, cemented into glazed blocks, or screwed into header laterals offer underdrains for either gravity or pressure filters without graded gravel layers. The application of these nozzles for wastewater must be carefully considered because of the possibility of clogging the fine strainer openings.

Filter strainers can also be fitted with long stems or air-metering tubes for uniform air distribution during the scouring (backwash) cycle. The air is introduced under the filter deck and forms a cushion, as shown in Figure 7.33.1. After the air pocket forms it allows air seepage to flow through the stem slots in proportion to the back pressure that develops.

The total bed depth of single or dual layers should be at least 24 in, with the sand or anthracite layers preferably 16 in each (12 in minimum). Deeper beds offer more storage space and thus longer runs. Table 7.33.1 gives the media specification ranges.

Backwash Water Source

A clear backwash water storage tank should be provided for at least a 7-min-backwash period at 15 gpm per sq ft or more. For a battery of three or more filters, the clear backwash water supply can be the filtered effluent from other filters.

Filter runs are terminated when (1) the head pressure loss across the filter reaches a predetermined value (10 ft H_2O), (2) the effluent turbidity exceeds the acceptable

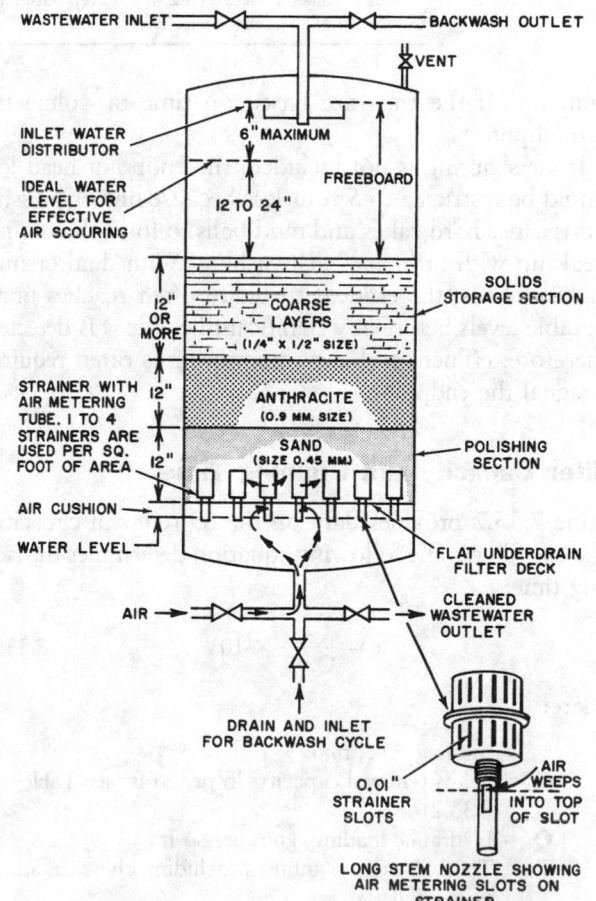

FIG. 7.33.1 Multilayered filter showing a flat underdrain deck with long-stem nozzles for washing with backwash and air.

TABLE 7.33.1 FILTER MEDIA SELECTION

| Type of Filter Design | Filter Media Size (mm unless otherwise noted) | | | | |
	Sand	Anthracite Coal	Garnet	Coarse, Activated Carbon	Plastic Granules
Single-Layer Beds	0.4–0.6	0.6–0.8	—	—	—
Multilayer Beds	0.4–0.6	0.8–1.4	—	10×4 mesh[a]	$\frac{1}{4} \times \frac{1}{2}$ in[b]
Ultra-high Rates	0.8–4.0	0.8–4.0	—	—	—
Garnet for Tightening Sand Beds	—	—	0.15–0.3	—	—

Notes: [a]Bulk weight of 16 lb per cu. ft or less.
[b]Specific gravity of 1.02 to 1.07.

TABLE 7.33.2 CAPACITY OF FILTER TYPES

Filter Type[a]	SS Removal Capacity (lb per sq ft)
Surface filtration, septum or leaf	$\frac{1}{8}$ to $\frac{1}{4}$
Single-layer granular, sand or coal	$\frac{1}{4}$ to $\frac{1}{2}$
Mixed or dual-layer, sand and coal	$\frac{1}{2}$ to 1
Multilayered with coarse top layer (and upflow type)	1 to 2
Deep, high-rate, coarse media without chemical flocs	$\frac{1}{2}$ to 2

Note: [a]Values are based on 2 to 3 ft deep beds. For deeper layers or beds, the solids removal capacity increases by 0.5 lb per ft of additional coarse layer.

limit, or (3) the runs are based on time or volumetric throughput.

If air scouring is not included, the endpoint head loss should be restricted to 5 ft or less because higher pressure drops cause hard cakes and mud-balls to form that do not break up with ordinary backwashing. With dual or multilayered beds, the effluent turbidity often reaches unacceptable levels before any significant head loss is detected. Therefore, effluent turbidity monitoring is often required to signal the endpoint.

Filter Capacity and Running Time

Table 7.33.2 provides data on the SS removal capacities of some filters. The following equation determines the running time:

$$t = \frac{1.2C}{Q\,Ts} \times 10^5 \qquad 7.33(1)$$

where:

- t = running time, min
- C = Solids removal capacity, lb per sq ft (see Table 7.33.2)
- Q = Hydraulic loading, gpm per sq ft
- Ts = Total SS in filter influent including chemical additives, mg/l

Table 7.33.3 provides data on hydraulic loading based on a 4-hr filtering period.

DIATOMACEOUS EARTH FILTERS

The diatomaceous earth filter is available both as a vacuum and as a pressure filter. Both are designed with either septums or leaves covered with screening or fine slots that have openings of 0.003 to 0.005 in. These septums are precoated with a matt of 0.10 to 0.15 lb per sq ft of diatomaceous earth.

The head loss at the end of the run can be 35 to 100 psi with the pressure units. The solids removal capacity is rather low at about $\frac{1}{8}$ lb per sq ft. For a feed with an SS concentration of 10 mg/l, the running time at 1 gpm per sq ft hydraulic loading is about 24 hr.

Diatomaceous earth filtration produces clear effluents with removal efficiencies of over 90%. However, the costs are also high. Colloidal substances are usually not removed (as with ordinary granular filters) unless coagulants are added. Coagulants are seldom added to diatomaceous earth filter influents because they shorten the runs.

MICROSCREENING

Microscreening has been applied in domestic water treatment, sewage waste water filtration, and filtering industrial effluents. Microscreening uses a special woven metallic or plastic filter fabric mounted on the periphery of a revolving drum provided with continuous backwashing. The drum operates submerged in the flowing wastewater to approximately two-thirds of its depth.

TABLE 7.33.3 MAXIMUM RECOMMENDED HYDRAULIC LOADING OF FILTERS WITH 4 HR OF RUNNING TIME IN GPM-PER-SQ-FT UNITS

Total (including alum) SS Concentration of Filter Influent (mg/l)	Type of Filter Design			
	Standard Single Layer	Mixed or Dual Layers	Multilayers	Deep Beds[a]
10	5	10	10	15
25	5	8	10	15
50	4	6	8	10
100	2.5	4.5	8	8
150	2.0	3	5	6
200	—	2.5	4	5
250	—	2	3	4
300	—	—	3	4

Note: [a]Beds at least 48 in deep with noncoagulated feeds and a SS removal efficiency of 50 to 80%.

TABLE 7.33.4 MICROSTRAINER COST AND PERFORMANCE ON THE OVERFLOW FROM SECONDARY SEWAGE TREATMENT UNITS

Microscreen Size Diameter × Width (ft)	Range of Total Capacity (mgd)	Required bhp	
		Drive	Backwash Pump
5 × 1	0.1–0.5	0.5	1
5 × 3	0.3–1.5	0.75	3
7.5 × 5	0.8–4.0	2	5
10 × 10	3–10	5	7.5

Screen Size (ft)	Screen Opening (μ)	Hydraulic Loading of Submerged Area (gpm/ft)	Total Maximum Capacity (mgd)	Removal Efficiency (%)	
				SS	BOD
7.5 × 5	23	10	3	70–80	60–70
7.5 × 5	35	13	3	50–60	40–50
10 × 10	23	10	50	70–80	60–70
10 × 10	35	13	50	50–60	40–50

Wastewater enters through the open upstream end of the drum and flows radially outward through the microfabric leaving behind the SS. The deposited solids are carried upward on the inside of the fabric beneath a row of wash water jets. From there, they are flushed into a waste hopper mounted on a hollow axle of the drum.

Water for backflushing is drawn from the filtered water effluent and pumped through the jets spanning the full width of the screen fabric. Depending on the rotation speed and the size of the screen openings, only about one-half of the applied wash water actually penetrates the screen.

The drum rotation and backwash are continuous and adjustable. Either manual or automatic control based on the differential pressure can be provided. The pressure head develops due to the intercepted solids, which build up on the inside of the microfabric and create a filtration mat capable of removing particles smaller than the mesh aperture size.

Microscreen openings vary between 23 and 60 μ, which corresponds to 165,000 to 60,000 openings per sq in of surface area. The stainless steel wire cloth used in microstrainers is generally more successful than the plastic type.

The flow capacity of a size of microscreen depends on the rate of fabric clogging, drum speed, area of submergence, and head loss. The rate of screen blockage under standard head and flow conditions is called the *filterability index*, which can be determined experimentally. The amount of backwash water used ranges from 2 to 5% of the total hydraulic loading, which is in the range of 5 to 30 gpm per sq ft. Table 7.33.4 provides performance data for such units.

MOVING-BED FILTERS

Moving-bed filters are also applied to wastewater filtration. The filter involves the intermittent removal of the most heavily clogged portion of the sand filter media from the filtration zone without interrupting the filtration process.

As the influent wastewater passes through the face of the sand bed, the entire filter bed is periodically pushed in the opposite direction. The face or clogged portion of the sand bed is then periodically washed into a sludge hopper by a stream of cutter water. From there, the sludge plus the dirty sand is moved by eductors to a washing section and storage tower. The clean sand is gravity fed into the filter after each stroke.

The largest filter unit available is rated at a maximum of 250,000 gpd and has a 350-sq-ft total area. The wash water requirements represent about $7\frac{1}{2}\%$ of the influent flow rate. A 9 hp motor is required for this unit.

The moving-bed filter operates at a maximum hydraulic loading of 7 gpm per sq ft and can remove 70 to 90% of SS from secondary, primary, and rural wastewater.

Because the intermittent movement of sand creates periodic upsets in the effluent quality and SS content, these units are classed as rough filtration or straining devices.

MEMBRANE FILTRATION

Surface filtration at high pressures (50 to 1000 psig) and low flow rates through the films or dynamically formed membranes is termed *membrane filtration*.

In membrane filtration, porous membranes with flux rates (hydraulic loadings) over 500 gpd per sq ft at 50 psig are used for polishing effluents from other filters. Membranes with accurately controlled porosities of 0.01, 0.1, 0.22, 0.45 and higher μ openings are available. Environmental engineers use the 0.45-μ membrane in evaluating filter effluents for trace concentrations of colloids, color, metallic oxides, and bacteria.

In *ultrafiltration*, tighter or less porous ultramembranes, with flux rates (hydraulic loadings) initially ranging from 50 to 300 gpd per sq ft at 50 psig, are capable of rejecting high-molecular-weight (2000 and above), soluble, organic substances, but not salt.

In *hyperfiltration* (reverse osmosis), specially prepared membranes or hollow fibers with flux rates at 5 to 50 gpd per sq ft at 400 to 800 psig affect salt, soluble organic matter, colloidal or soluble silica, and phosphate removal at 80 to 95% efficiency.

All membrane processes are considered to be final polishing filters, with common particulate removals in excess of 99%. In so doing, they foul easily, and their flux flow rate declines logarithmically with running time. Therefore, wastewater treatment facilities must protect membrane filters from fouling by pretreating the feeds using coagulation and rough filtration.

Ultrafiltration Membranes in Activated-Sludge Processes

BOD REMOVAL EFFICIENCY

+95%

FILTER EFFLUENT QUALITY

BOD below 1 mg/l, COD between 20 and 30 mg/l, no SS or coliform bacteria, essentially colorless and odorless, and removal of molecules with molecular weights of 8000 to 45,000 together with some viruses

MEMBRANE FILTER PROTECTION AGAINST FOULING

By screening of the influent and maintaining high-flow velocities at the membrane surface

AVAILABLE TREATMENT CAPACITY RANGE

3000 to 30,000 gpd

MEMBRANE SURFACE VELOCITY REQUIRED

3 to 8 fps provided by recirculation.

PORE SIZES

3 to 100 Å

MEMBRANE PRESSURE DROP

2 to 40 psid

HYDRAULIC LOADING (FLUX) RANGE

5 to 30 gpd per sq ft

Continuous biological oxidation processes combined with ultrafiltration membranes are used as a means of effluent separation. The membranes filter out the biological cells while allowing passage of the treated effluent.

These membranes can achieve BOD removals in excess of 99% on a commercial scale with an influent containing 200 to 600 mg/l BOD. The membrane filter produces an essentially colorless and odorless effluent having both zero SS and coliform bacterial counts. These systems consist of an activated-sludge reactor and a membrane filter and are available in treatment capacities ranging from 3000 to 30,000 gpd.

ULTRAFILTRATION MEMBRANES

Ultrafiltration membranes are thin films cast from organic polymer solutions. The film thickness is 5 to 10 mils (0.005 to 0.01 in). The film is anisotropic, i.e., it has thin separation layer on a porous substructure. The thin working or separation layer has a thickness of 0.1 to 10 μ. Figure 7.33.2 shows a film cross section. The separation layer contains pores of closely controlled sizes ranging from 3 to 100 Å.

ULTRAFILTRATION PROCESS

In ultrafiltration devices, the separation layer is adjacent to a pressurized chamber containing the filter influent. When pressure is applied, small molecules pass through the membrane and exit on the other side; larger molecules

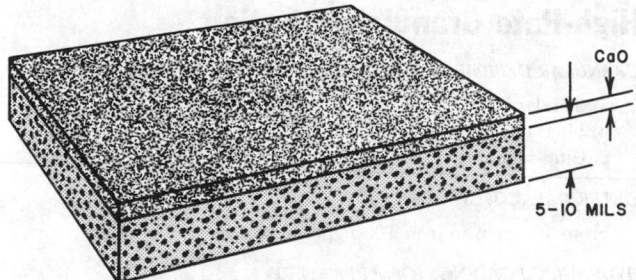

FIG. 7.33.2 Anisotropic, diffusive ultrafilter.

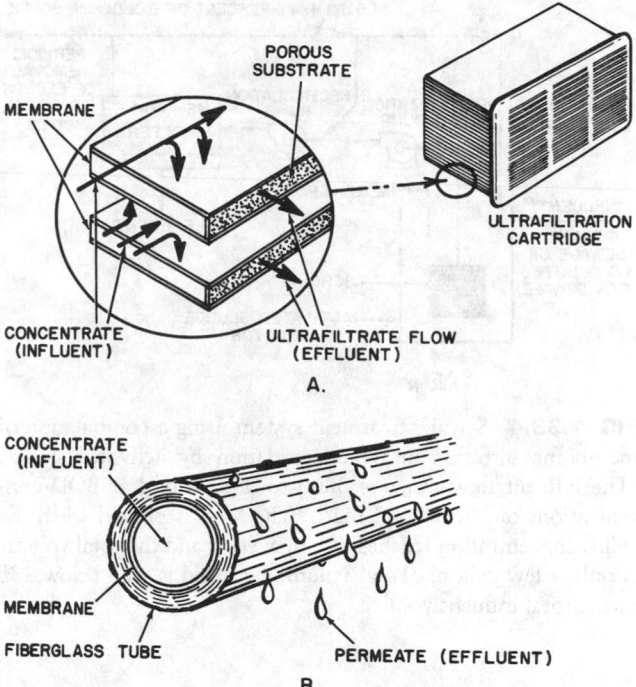

FIG. 7.33.3 Ultrafilter membrane packaging configurations. A. Plate device; B. Tubular device.

are retained within the pressurized chamber. The pressure drop across the membrane ranges from 2 to 40 psid. If the chamber is continuously fed with new influent, the concentration on the feed side gradually increases. This concentrate is continuously bled from the pressurized side of the membrane.

Typical ultrafiltration membranes used in these systems exhibit useful operating fluxes (hydraulic loadings) from 5 to 30 gpd per sq ft at pressure drops of 2 to 30 psid. The membranes filter out protein molecules with molecular weights from 8000 to 45,000; consequently, most viruses are also retained.

ULTRAFILTRATION DEVICES

The objective in the mechanical design of an ultrafiltration device is to provide the largest working area of membrane surface per unit of filter volume. Provisions must be made for a pressurized channel on the feed side of the membrane, support of the membrane film, draining and collecting the filtered effluent that permeates the membrane, and mechanically supporting the whole structure.

Environmental engineers determine the dimensions of the feed channel based on the size of the particles contained in the feed streams and hydrodynamic considerations to provide sufficient flow past the membrane surface to minimize concentration polarization (a concentrated layer developing at the membrane surface). Flow velocities in the range of 3 to 8 ft per sec are used. Because of the high solids content in the reaction systems and the presence of large particles in the feed, large feed channels are required. The membranes are packaged in either plate configurations having channel dimensions of approximately 0.090 in or tubular configurations having inside diameters of $\frac{1}{4}$ to 1 in. Figure 7.33.3 is a schematic representation of both types.

The plate device is comprised of sheets of porous support material on which the membrane is cast. The sheets are in a parallel array and terminate in a collection header or manifold. Feed material (influent) passes between the sheets, and the effluent permeates the membrane and passes into and up the porous support member to the exit header.

With the tubular configuration, a support tube manufactured from sintered, porous, polymeric materials or fabricated as a composite from fiberglass and polyester or epoxy materials forms the pressure vessel. The membrane is cast or placed on the inside of the tube. Feed material (influent) flows through the inside of the tube, and due to operating pressure, the effluent permeates the membrane, passes into the porous supporting substrate, and is collected in a manifold. Series and parallel arrays of tubes are available (as in a shell and tube-heat exchanger) guaranteeing adequate flow past the membrane to minimize concentration polarization. Groupings of membrane modules in series and parallel can also provide the feed recirculation rates required to minimize concentration polarization.

SEWAGE TREATMENT APPLICATIONS

As shown in Figure 7.33.4, the effluent from the activated-sludge reactor is continuously withdrawn through a screening device to the recirculation pump of a membrane loop. The membranes separate the large molecules, and the retained solids are returned to the activated-sludge reactor. Because only small molecules can pass through the membrane separation device, a high concentration of biological solids develops in the activated-sludge reactor. Wastewater treatment facilities characteristically run this type of activated-sludge process with biological solids concentrations from 15,000 to 40,000 mg/l. Operating with such high solids content has two advantages: (1) the large biomass quickly degrades organics that can enter with the

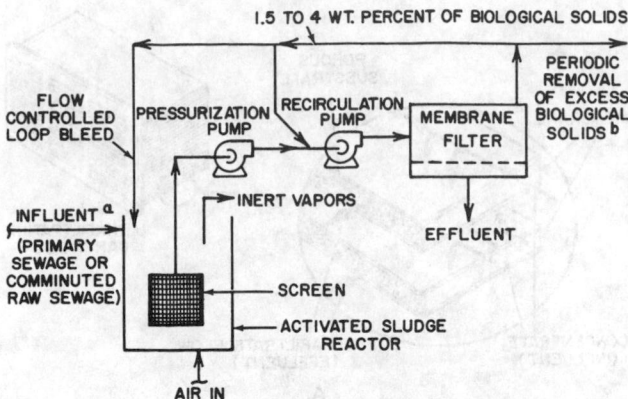

FIG. 7.33.4 Sewage treatment system using a combination of membrane filtration and biodegradation by activated sludge. [a]The influent flow range is 3000 to 30,000 gpd at BOD concentrations of 200 to 600 mg/l. [b]The loop is purged when the solids concentration reaches 4% by weight and the total volume is only a few gallons. Total volume removed is well below 1% of the total influent volume.

feed and prevents fouling of the membrane surfaces, and (2) the activated-sludge reactor size is much reduced.

The membrane filter in the system guarantees a practically infinite detention time for the slow biodegradable components in the sewage because they cannot exist from the system. The biological solids are also totally contained by the membranes. Such a sewage treatment system operates with almost no discharge of excess activated-sludge solids. Practically all feed materials are converted to carbon dioxide, water, and inorganic salts. Some inert materials do accumulate in the reactor, and purging (see Figure 7.33.4) of the contents of the reaction system is periodically necessary.

Performance of Sewage Treatment Applications

Commercial-scale systems with capacities of 3000 to 30,000 gpd have operated for thousands of hours. These systems handle sanitary waste containing 200 to 600 mg/l BOD. Hydraulic loading (flux) of the membranes is sustained at approximately 10 gpd per sq ft.

These systems achieve essentially 95% BOD removal. The BOD that does pass through is in the form of dissolved solids. The BOD content of the effluent is below 1 mg/l, with COD ranging from 20 to 30 mg/l. Coliform bacteria counts in the effluent have been zero, and additional sterilization of the effluent is not practiced.

The processes are stable when occasionally exposed to toxic materials in the feed that would upset other, less-concentrated, biological systems. This stability can be explained partially by the large biomass in the process and partially by the fact that the membrane is a positive barrier preventing the exit of the dead biomass.

High-Rate Granular Filtration

GRANULAR DEEP-BED FILTER TYPES

   a) Standard
   b) High rate
   c) Ultra-high rate

FILTER CAPACITIES

   From 100 gpm to over 1000 gpm of water

HYDRAULIC LOADINGS (GPM PER SQ FT)

   2 to 5 (a), 5 to 15 (b), and over 15 (c)

BED DEPTHS

   4 to 8 ft

SS REMOVAL EFFICIENCY

   50 to 75% (c), 80 to 90% (b), and 90 to 98% (a).

SOLIDS LOADINGS (LB PER SQ FT)

   1 to 10

MAXIMUM ACCEPTABLE SOLIDS CONCENTRATIONS IN INFLUENT

   100 mg/l

MAXIMUM ACCEPTABLE FIBER CONCENTRATION IN INFLUENT

   10 to 25 mg/l

MAXIMUM ACCEPTABLE SOLIDS PARTICLE SIZE IN INFLUENT

   200 μ

MAXIMUM ACCEPTABLE OIL CONCENTRATION IN INFLUENT

   25 to 75 mg/l

High-rate, granular filtration systems are used for SS removal from a variety of water and wastewater streams. Applications include industrial process water; municipal potable water; final polishing of sewage treatment effluents; removal of mill scale and oil from hot rolling-mill cooling water; removal of residual oil from American Petroleum Institute (API) separator and dissolved-air flotation effluents; and pretreatment of water and wastewater for advanced forms of treatment, such as carbon adsorption, reverse osmosis, ion exchange, electrodialysis, and ozonation.

Recent applications include the simultaneous removal of SS and suspended or dissolved phosphorus and nitrogen nutrients using combination biological–physical–chemical processes in granular-filtration-type systems. The basic purpose of high-rate granular filtration processes is to remove low concentrations of SS from large volumes of water (from 100 gpm to usually more than 1000 gpm).

DEFINITIONS

The following terms apply to high-rate granular filtration:

HIGH-RATE FILTRATION—A filtration process designed to operate at unit flow rates (hydraulic loadings) between 5 and 15 gpm per sq ft.

Granular (deep-bed) filtration—A filtration process that uses one or more layers of granular filter material coarse enough for the SS to penetrate into the filter bed to a depth of 12 in or more. Total depth of the single or composite beds is defined arbitrarily as a minimum of 4 ft.

Mono-media—A deep-bed filtration system that uses only one type of granular media for the filtration process, exclusive of gravel support layers.

Dual-media—A deep-bed filtration system that uses two separate and discrete layers of dissimilar media, e.g., anthracite and sand, placed on top of each other for filtration.

Mixed-media—A deep-bed filtration system that uses two or more dissimilar granular materials, e.g., anthracite, sand, and garnet, blended by size and density to produce a composite filter media graded hydraulically after backwash from coarse to fine in the direction of the flow.

Specific solids loading—The weight of the SS that can be removed by a filter before backwashing is required; usually expressed as lb per sq ft per cycle. Specific loading is a function of filter application, media size, media depth, unit flow rate (hydraulic loading), and influent solids concentration.

Optimization—Designing and operating a filtration system to produce the maximum quantity of acceptable filtrate at the minimum capital and operating cost. In an operationally optimized system, SS breakthrough occurs at the maximum available pressure head loss.

REMOVAL MECHANISMS

The predominant mechanisms responsible for removing SS in granular filtration systems have been previously discussed. As shown in Figure 7.33.5, surface screening is not a predominant removal mechanism in this type of system, and solids are retained deep within the voids of the media. This system requires intensive backwashing (see Figure 7.33.6) to dislodge the entrapped solids that accumulate during the filtration process.

Surface wash systems are acceptable only if screening is the predominant removal mechanism. Otherwise, high-energy backwashing is required throughout the depth of deep-bed granular filters to completely clean the media after each filtration cycle.

REMOVAL EFFICIENCY

The removal efficiency of a granular filtration system is a function of the hydraulic loading (unit filtration rate), the grain size and depth of the media used, and the filterability of the solids to be removed. Concentration, particle size, density, and shape of the SS and water temperature also affect filtration efficiency.

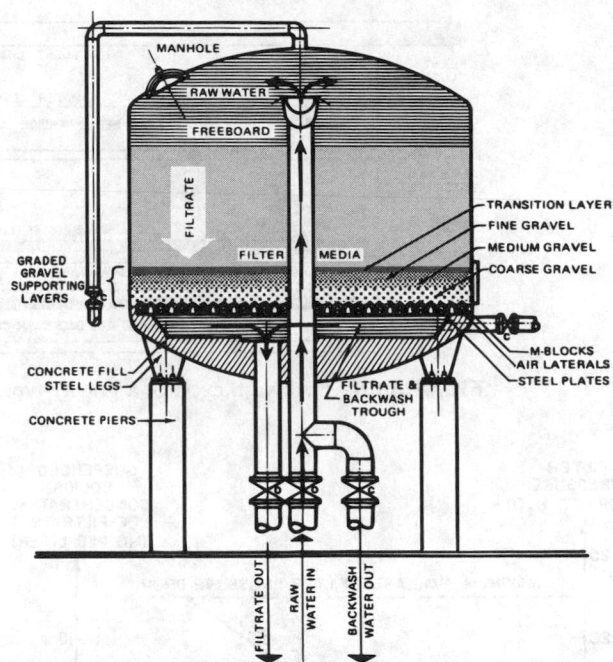

FIG. 7.33.5 Pressure-type, deep-bed, granular filter in forward flow operation. (Forward flow corresponds to the filtering, and back flow corresponds to the wash cycle.)

Ultra-high-rate filters remove only 50 to 75% of applied solids. Filters with less than 15-gpm-per-sq-ft-hydraulic loading provide at least 90% solids removal, and 98% or more is also possible.

Coarse media are used in the diameter range of 0.5 to 5 mm, with most of the media in the 1- to 3-mm range. The specific solids loading ranges are between 1 and 10 lb per sq ft. Coagulants and polyelectrolytes can be fed directly into the filter.

For example, if a specific loading of 5 lb per sq ft can be obtained, a 12.5-ft-diameter filter can remove 600 lb of SS before backwashing is required. This specific loading permits the addition of chemical coagulants into the filter influent to improve the removal efficiency without excessively reducing the filtration cycle period.

SYSTEM COMPONENTS AND DESIGN

The use of air scouring (purging to assist the cleaning action) during backwashing is almost universal practice (see Figure 7.33.7). With mono-media filters, wastewater treatment facilities can use high air and backwash water rates without disrupting the bed layers or carrying the lighter components of the bed out of the filter.

Downflow filters eliminate the tendency of upflow units to fluidize or break through, especially near the end of the filtration cycle when pressure drop across the filter is substantial. Both gravity and pressure-motivated filters are used in waste treatment, but pressure units have deeper beds, higher capacities, and the ability to be completely

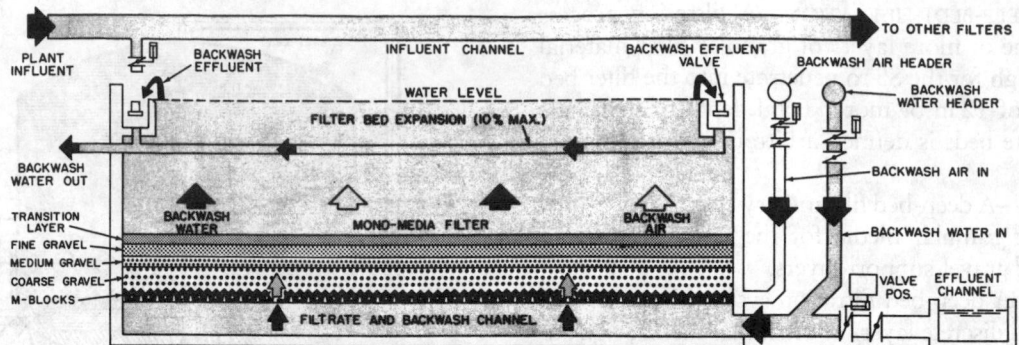

FIG. 7.33.6 Backwash cycle of a gravity-type, deep-bed, granular filter.

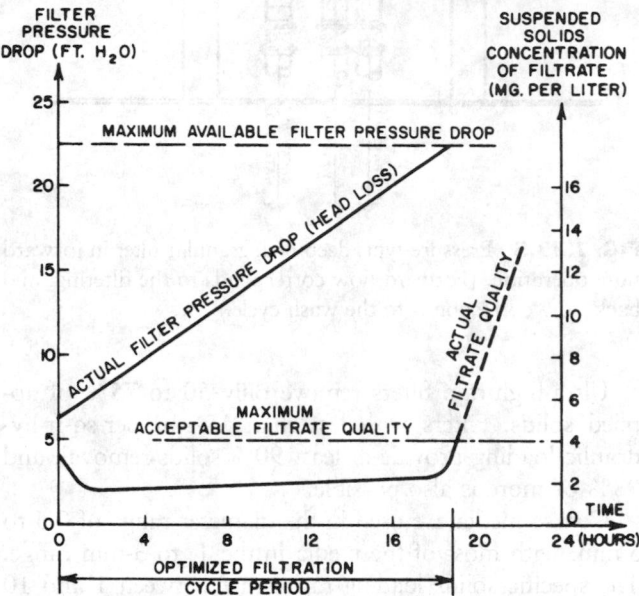

FIG. 7.33.7 Optimization (economic) of filter operation.

automated. Some regulatory agencies, however, prefer gravity systems because the filters can be observed during filtration and backwashing.

OPTIMIZATION

For many years, filters were used as insurance against contamination of potable water. Historically, in wastewater treatment, the effluent quality obtained from clarifiers was sufficient to meet discharge requirements, and effluent filtration was unnecessary. However, with the advent of effluent standards requiring more than 90% removal of the SS, wastewater filtration has become common practice.

Once the physical system is designed and installed, environmental engineers can try to optimize the operational cost. The criterion is that the maximum amount of acceptable quality filtrate is produced per unit of operating cost.

During filtration, the pressure required to maintain flow across the filter increases gradually due to solids accumu-

lating in the filter media. The effluent quality is slightly better at the start of each filtration and remains relatively constant until media voids are full. Then, a sudden breakthrough of solids into the filtrate occurs. A filter is optimized when the maximum available filter pressure drop coincides in time with the breakthrough of the SS (see Figure 7.33.7).

APPLICATIONS

High-rate, granular filtration is used for effluent polishing, pretreatment, phosphate removal, and nitrogen removal.

Effluent Polishing

Deep-bed, granular filters are applicable to polishing effluents from physical, chemical, or biological wastewater treatment systems. These filters use coarse-bed media and permit high hydraulic and solids loadings (see Table 7.33.5). The effluent from a typical municipal or industrial activated-sludge plant contains approximately 20 mg/l of SS. Effluent polishing by granular filtration can remove another 80 to 90% of this contaminant. Similarly, these filters can remove 50 to 90% of the free oil (and its associated BOD) from API or rolling-mill system effluents.

Pretreatment

Many advanced wastewater treatment processes require pretreating the waste to reduce its SS content. This requirement is especially true for the adsorption processes (activated carbon and ion exchange) and of some of the membrane processes (reverse osmosis and electrodialysis).

Prefiltration prior to the carbon or resin columns can eliminate the need for backwashing and reduce the loss by attrition. Occasionally, wastewater treatment facilities can reduce the number of columns because taking a column out of service for backwashing is unnecessary. Reverse osmosis membrane systems reduce membrane fouling by eliminating SS from the feed.

In pretreatment, essentially complete removal of SS is necessary; consequently, deep beds of fine media, e.g., 4 ft

TABLE 7.33.5 FILTRATION APPLICATIONS FOR GRANULAR, MONO-MEDIA, DEEP-BED UNITS

Applications	Coarse Bed Media Size (mm diameter)	Bed Media Depth (ft)	Hydraulic Loading or Filter Rate (gpm/ft)
Activated-sludge effluents, gravity	2–3	4–6	3–5
Activated-sludge effluents, pressure	2–3	6	5–10
API effluents (oil removal)	1–2	6–8	3–5
Chemical treatment effluents	1–2	4–6	5–8
Denitrification with effluent polishing	3–6	6–10	2–5
Hot-strip-mill effluents, carbon steel	2–3	6	8–14
Hot-strip-mill effluents, alloy steels	1–2	6–8	6–10
Impounded supply, process water	2–3	4–6	10–20
Phosphate removal	2–3	4–6	5–8
Pretreatment	1–2	4–8	5–12
Primary sewage effluent, storm water	2–3	4–6	5–15
River water filtration, process water	1–2	6–8	4–8
Side arm filtration, cooling towers	2–3	4–6	10–20
Trickling filter effluents	1–2	4–6	4–10

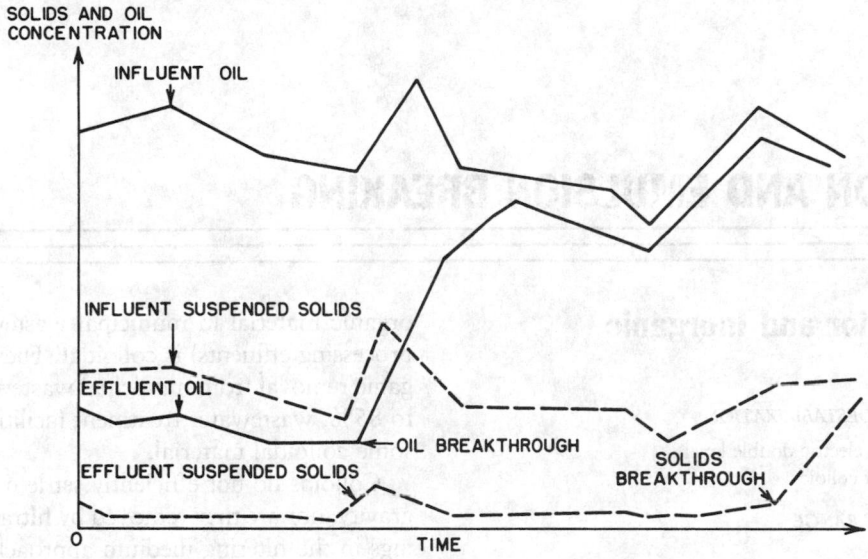

FIG. 7.33.8 The relationship between oil and SS breakthrough in granular filters.

of mixed media or 6 ft of 1.0-mm, mono-media, and low filter rates (2 to 4 gpm per sq ft) are used.

Phosphate Removal

Phosphate removal by chemical precipitation, either as a separate process or in combination with biological processes, does not require filtration. However, when either low residual levels of phosphorus or maximum utilization of chemicals is desired, filtration of the chemically treated effluent is desirable. As much as 2 mg/l phosphorus can be removed by granular filtration of the effluent from the phosphate precipitation processes.

Nitrate Removal

Nitrogen removal with biological denitrification is enhanced by granular effluent filtration. Slow growing denitrifying organisms are easily washed out of a biological system by hydraulic surges or clarifier upsets. Granular filtration prevents the loss of nitrifying or denitrifying organisms and returns them to the system with the backwash water. This form of postfiltration is especially important when wastewater temperatures are below 65°F because biological reaction rates slow down at such temperatures.

Wastewater treatment facilities can combine granular filters for SS removal with columnar denitrification units in a single system. In such combined systems, larger media are used, and special backwashing procedures are required to obtain 80% removal of the nitrate without substantially reducing the SS removal efficiency.

LIMITATIONS

Oil, fiber, and SS can be tolerated in granular filtration systems, within limits. The scale-forming tendencies of the wastewater must also be eliminated or controlled. These systems generate backwash water as a natural consequence of the process and wastewater treatment facilities must provide for its proper handling.

Fiber in concentrations greater than about 10 to 25 mg/l can cause operating problems in granular filters. Long fibers mat and blind off the filter surface. Short fibers accumulate in the underdrains and plug the bottom of the filter unless special designs are used to prevent it.

Oil can plug the underdrains or prevent complete backwashing, especially when fine-bed media (less than 1.0 mm diameter) are used. Water-wash filters can have difficulty operating with 25 mg/l of free oil, while heavy-duty, air-scouring backwash systems can operate satisfactorily with up to 50 to 75 mg/l of oil. Oil reduces the specific solids loading and breaks through the filter sooner than SS. Figure 7.33.8 shows this effect.

An SS concentration in the influent of about 100 mg/l can be tolerated in a properly designed granular filter provided that the size of the solids does not prevent penetration into the media. The maximum feasible particle size is about 200 μ although minor concentrations (5 mg/l) of oversized solids, such as leaf fragments and bits of paper, can be tolerated. High SS concentrations reduce the filter cycle time, which becomes a problem when the operating cycle is so short that the filters cannot be washed in time to keep the system in operation. For example, if six filters are in a continuous system and a total of $\frac{1}{2}$ hr is required to wash a filter, the minimum usable cycle time is 3 hr.

—*G.S. Crits*
P.L. Stavenger
E.S. Savage

7.34
COAGULATION AND EMULSION BREAKING

Colloidal Behavior and Inorganic Coagulants

METHODS OF COLLOID DESTABILIZATION

   a) Modification of the electric double layer
   b) Polymer bridging of colloids

COLLOID MATERIAL SIZE RANGE

   0.01 to 10 μ

COLLOID CONCENTRATION IN SEWAGE

   15 to 25% of all organic material

COAGULANTS

   Aluminum salt, ferric iron salts, and magnesium

OPTIMUM pH LEVELS

   5 to 7 pH for aluminum

CHEMICAL SLUDGE PRODUCTION BY LIME TREATMENT

   4500 to 6500 lb per mg of water

COLLOIDAL POLLUTANTS

Colloids are generally defined as particulate matter in the 0.01 to 10 μ size range. Approximately 15 to 25% of the organic material in municipal wastewater (more in food processing effluents) is colloidal. Therefore, to achieve organic removal from municipal wastewater greater than 75 to 85%, wastewater treatment facilities must also remove some colloidal material.

Colloids do not efficiently settle under the influence of gravity, nor are they removed by filtration unless the openings in the filtering medium approach the size of the colloids themselves (see Section 7.33). For effective removal of colloidal material from wastewater by gravity settling or in-depth filtration, wastewater treatment facilities must use coagulation and flocculation. Understanding the significance of these unit processes requires an understanding of the physical, chemical, and electrical properties of colloidal suspensions.

COLLOIDAL PROPERTIES

Two broad classes of colloidal material are inorganic and organic. Inorganic colloids generally consist of inert mineral particles such as silt, clay, and dust. These pollutants do not generally pose a hazard to human health, but they do have an undesirable esthetic quality when they are discharged to receiving waters in high concentration.

TABLE 7.34.1   RELATIVE SIZE OF POLLUTANTS IN WASTEWATER

Particle Diameter (meters)	Designation	Typical Substance	Relative Size
$<10^{-8}$	Ions and molecules	Glucose and chloride	1
$>10^{-8} \, <10^{-5}$	Colloids	Bacteria, phages, clay, and macro-molecules	1 to 1000
$>10^{-5} \, <10^{-3}$	Fine particulates	Silt, fine sands, and clays	1000 to 100,000
$>10^{-3}$	Coarse particulates	Coarse sand	Greater than 100,000

Organic colloids generally consist of bacteria, viruses, phages, fragments of cellular material from living organisms, waste food, and a variety of materials from the chemical and food processing industries. These organic colloidal pollutants can be deleterious. In addition, they also consume DO and create a potential for putrefaction during their biological degradation.

The most important property of colloids is their very small size. Table 7.34.1 shows the size range in which colloidal materials are normally classed. Because of their size, colloids have a large surface area per unit volume, or per unit mass of material. The properties of this surface, or more specifically the interfacial region between the solid colloid and the bulk liquid (water) phases, governs the action of a colloidal suspension.

The electrical properties at the solid–liquid interface depend on the origin of the solid surface and the physicochemical properties of the solid and liquid phases. Solid particles encountered in wastewater treatment originate from three general sources including degradation of larger particles, biological agents, and condensation of small particles forming larger ones.

When suspended in water, the surfaces of these solids exhibit a surface charge that can arise from the specific adsorption of potential determining ions, dissociation of ionic species at the surface, internal atomic defects in the solids phase, or other causes. This surface charge is counterbalanced by oppositely charged ions in the liquid adjacent to the solid–liquid interface. The distribution of ions in the region adjacent to the interface is different from that in the bulk of the solution and is described by the electrical double-layer theory.

Figure 7.34.1 is a diagram of an electrical double layer based on Stern's modification of the Gouy–Chapman model. A layer of fixed, nearly immobile ions adjacent to the negatively charged surface reduces the surface potential to the Stern potential. Outside this fixed layer (called the Stern layer) is a diffuse layer of ions, whose concentration is described by a Boltzman distribution. In the diffuse layer, an excess of positive counterions exists so that the total charge in the diffuse layer equals the total surface charge minus the total charge in the Stern layer.

When two charged surfaces are brought together so that their diffuse layers overlap, an electrical force exists be-

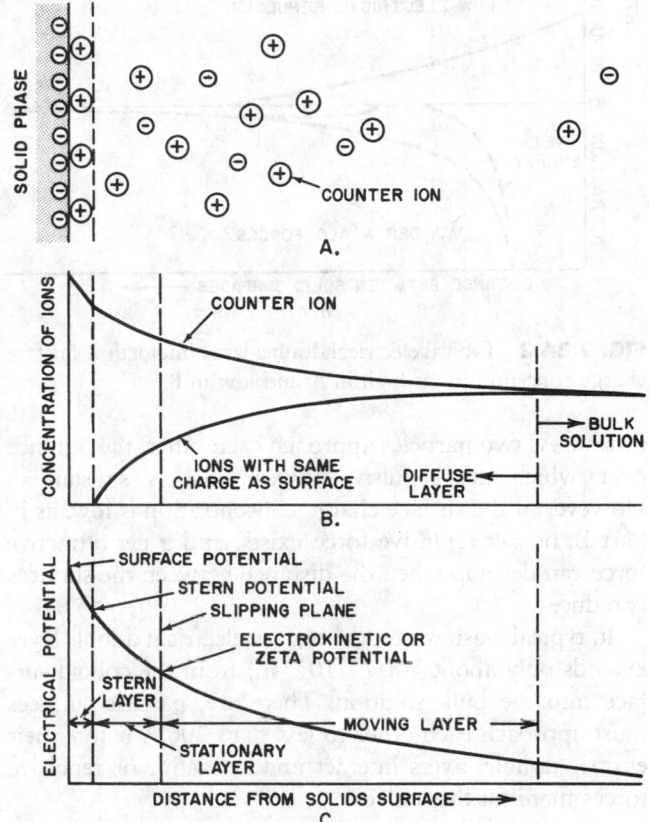

FIG. 7.34.1 Model of electrical double layer for an electronegatively charged surface.

tween the two surfaces. For surfaces with similar potentials, a repulsive force exists; for surfaces with opposite potentials, an attractive force exists. The magnitude of the repulsive or attractive force is related to the magnitude of their respective potentials and the distance of separation between the surfaces.

In addition to electrical forces between surfaces, van der Waals universal attractive forces between atoms are also significant. These attractive forces arise from interacting, fixed and induced atomic dipoles and fundamental dispersion forces.

Figure 7.34.2 shows the results of an interaction of charged surfaces according to the electrical double-layer model. Part A shows surfaces with a high charge concen-

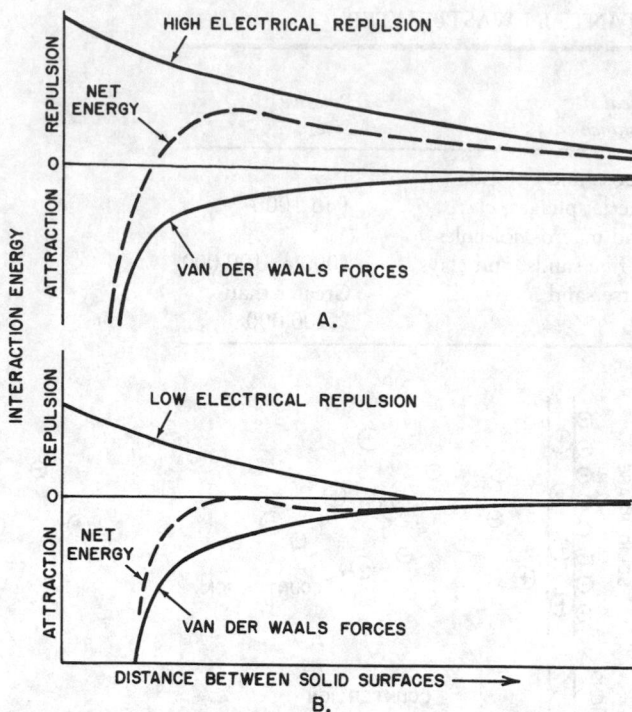

FIG. 7.34.2 Typical electrical double layer interaction. Surface charge concentration is high in A and low in B.

tration. As two particles approach each other, the distance over which net repulsive forces exist is substantial. However, if the surface charge concentration is low, as in Part B, no net repulsive force exists, and a net attractive force can develop when the distance between the surfaces is reduced.

In typical wastewater, the diffuse electrical double layer extends only about 100 Å (10^{-8} m) from the colloid surface into the bulk solution. Therefore, particle surfaces must approach each other to less than 200 Å before their electric double layers interact and attractive or repulsive forces manifest themselves.

DESTABILIZATION OF COLLOIDAL SUSPENSIONS

For definition purposes, a stable colloidal suspension exists when particles subjected to flocculation, i.e., bringing particles together, do not adhere to one another to form agglomerates of primary colloidal particles. The most important property causing colloidal stability is the existence of net repulsive forces when electrical double layers interact. Conversely, a destabilized colloidal suspension exists when particles subjected to flocculation adhere to one another, forming agglomerates of primary colloidal particles.

Modification of the Electric Double Layer

This method of destabilizing colloidal suspensions modifies the electrical double layer. Reducing the surface or Stern potential or compressing the diffuse electrical double layer diminishes the repulsive interaction forces, which can result in destabilization.

In practice, many colloids have hydrogen- and hydroxyl-potential-determining ions. Adjusting the solution pH alone reduces their surface potential, thus diminishing the repulsive electrical forces and resulting in destabilization. This condition is shown by both curves in Figure 7.34.3.

An indifferent electrolyte does not contain surface-charge-potential-determining ions. Adding an indifferent electrolyte to a stable colloidal suspension causes destabilization by compressing the electric double layer, which reduces or eliminates a net repulsive interaction barrier. Figure 7.34.3 shows the effect of indifferent electrolyte concentration changes where moving from point A to B requires the addition of an electrolyte and results in destabilization.

According to the Stern modification of the Gouy-Chapman electric, double-layer model, destabilizing a col-

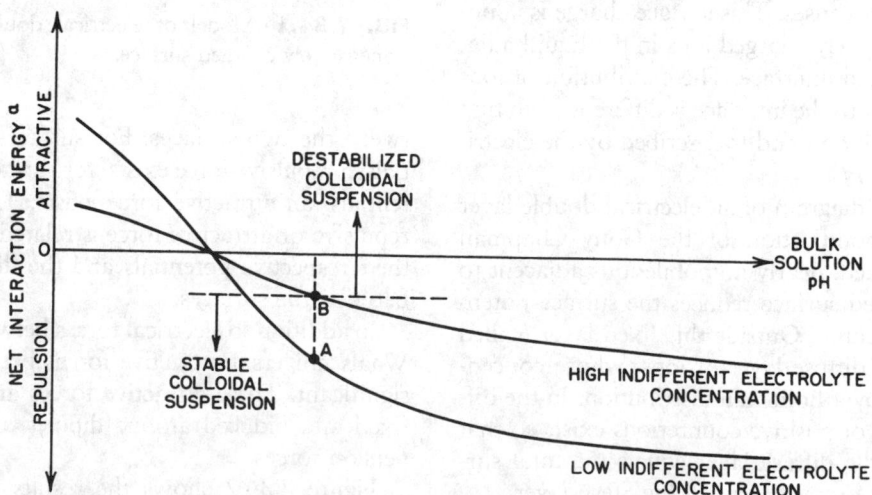

FIG. 7.34.3 Effect of pH and an indifferent electrolyte on colloidal suspension stability (for surfaces with hydrogen- and hydroxide-potential-determining ions). [a]At a given separation distance where electric double layers interact.

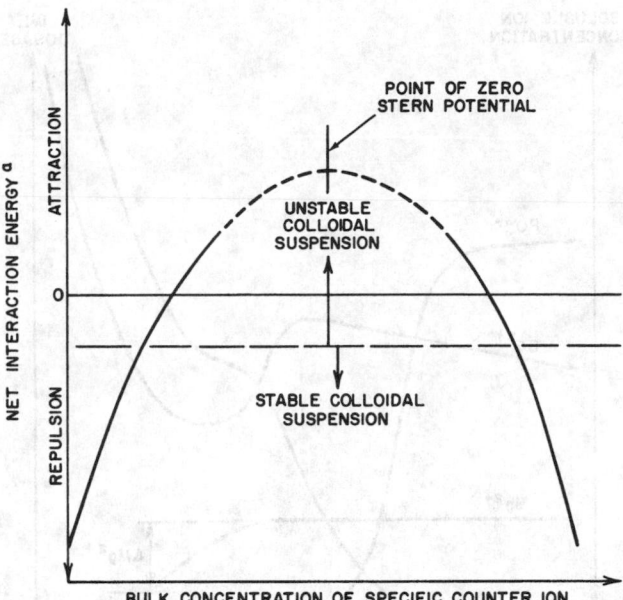

FIG. 7.34.4 Effect of specific counter ion concentration on colloidal suspension stability. [a]At some given separation distance where electric double layers usually interact.

loidal suspension by specific adsorption of specific counter ions at the colloid surface is possible. Here, the surface potential remains unchanged, but the Stern and zeta potential can be reduced or even reversed in charge. In this process, adsorptive rather than electrostatic, forces must be operative. Figure 7.34.4 shows the effect of Stern potential reduction and reversal and gives a range of specific counter ion bulk solution concentrations where destabilization occurs.

Chemical Bridging of Colloids

Since the early 1950s, environmental engineers have used natural and synthetic polymers to agglomerate or flocculate finely divided, suspended material in water and wastewater treatment. From the beginning, they observed the anomalous behavior of the electric double-layer model. Therefore, they developed the chemical bridging model to explain the action of polymers on colloidal suspension stability.

In its simplest form, the chemical bridging model suggests that a polymer can attach itself to the surface of a colloid at one or more sites, with a significant length of the polymer extending into the bulk solution. Reaction A in Figure 7.34.5 shows this condition. That such action causes destabilization is suggested in reaction B, where two colloids with attached and extended polymers agglomerate when flocculated.

The key aspect of the bridging model is that adsorption of polymers on colloid surfaces involves more than coulombic forces. Postulated interactions include hydrogen bonding, coordinate covalent bonding and linkages, van der Waals forces, and polymer–solvent solubility considerations.

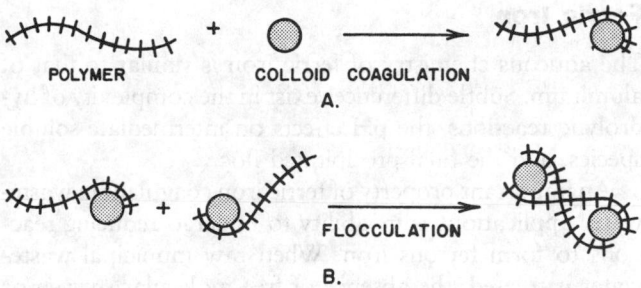

FIG. 7.34.5 Colloidal suspension destabilization with high-molecular-weight organic polymers. A. Destabilization; B. Agglomeration.

INORGANIC COAGULANTS

Significant coagulants in wastewater treatment are aluminum, ferric iron salts, and magnesium, which is active in lime treatment.

Aluminum

When aluminum is added to wastewater, usually in the form of dissolved aluminum sulfate, it undergoes several reactions with the naturally present alkalinity and various anionic ligands. A series of hydrolytic reactions occurs, proceeding from the formation of a simple hydroxo complex, $Al(OH)^{2+}$, through the formation of soluble poly-hydroxy polynuclear inorganic polymers, to the formation of a colloidal precipitate. These reactions are rapid, occurring within a fraction of a second.

The soluble, polymeric, kinetic intermediates are generally believed to be the causative agent, i.e., coagulant species, in destabilizing colloidal suspensions. Because the adsorption of these coagulating species on the colloid surface involves more than coulombic effects, the electrical double-layer model cannot be explicitly invoked.

Since OH^- is the most important ligand in the soluble polymeric aluminum hydroxo complex, the effectiveness of aluminum coagulation is strongly pH dependent. Generally, an optimum pH is in the range of 5 to 7. The presence of significant amounts of the competing ligand PO_4^{3-} reduces optimum coagulation pH.

Since aluminum reacts with wastewater alkalinity (usually HCO_3^- results), wastewater treatment facilities must consider the possible excessive depression of pH in a weakly buffered water. They may have to add alkalinity, usually in the form of hydrated lime ($Ca[OH]_2$), to prevent excessive pH depression.

Using aluminum as a coagulant in wastewater treatment involves dosages in excess of the coagulation requirements, which results in precipitation of excess alum floc. This excess floc results in substantial removal of finely divided SS by entrapment in the floc structure during flocculation and gravity settling.

Ferric Iron

The aqueous chemistry of ferric iron is similar to that of aluminum. Subtle differences exist in the complexity of hydrolytic reactions, the pH effects on intermediate soluble species, and the final precipitated floc.

An important property of ferric iron coagulant in wastewater applications is its ability to undergo reducing reactions to form ferrous iron. When raw municipal wastewater is treated, the absence of free molecular oxygen or any other oxidation agent results in a reducing environment. Research shows that precipitated ferric-iron hydrolysis species release iron as a reduced ferrous compound when they are maintained in an anaerobic environment. This property is significant in applications where removal of phosphorus compounds and the destabilization of colloidal pollutants is required. However, its significance for destabilization only is unknown.

Magnesium

Lime treatment for phosphorus precipitation and SS removal from raw and biologically treated municipal wastewater has become an accepted sanitary engineering unit process. The chemistry of lime treatment related to colloidal suspension destabilization is not a well-developed technology. However, certain observations based on laboratory and full-scale plant operating results can be made.

Adding hydrated lime [$Ca(OH)_2$] to wastewater simply increases the solution pH. At increasing pH levels, calcium phosphate, calcium carbonate, and magnesium hydroxide are insoluble as shown in Figure 7.34.6. The phosphate and carbonate precipitates undergo some agglomerative and adsorptive interactions with soluble and colloidal pollutants.

However, the coagulant species of significance is the magnesium hydroxide precipitate. Thus, wastewater treatment facilities must adjust the pH upward to a value that causes magnesium precipitation. This pH level varies with different wastewaters. For example, the lower the initial magnesium concentrations, the higher the pH required to precipitate the amount of magnesium necessary for destabilization. The incremental precipitation of the magnesium coagulant is shown as Mg^{2+} in Figure 7.34.6.

Two factors are significant in the lime treatment of wastewater. The first is the excessive amount of chemical sludge (precipitates) produced. Depending on the amount of Ca^{2+}, HCO_3^-, and PO_4^{3-} present, chemical sludge production ranges from 4500 to 6500 lb per MG. Second, a variety of precipitates are formed. These precipitates range in character from colloidal hydroxylapatite [$Ca_5OH-(PO_4)_3$] to dense $CaCO_3$ to voluminous $Mg(OH)_2$ floc. The proportions of these precipitates and the originally present SS determine the net floc settling rate, concentration of settled sludge, and sludge dewatering properties.

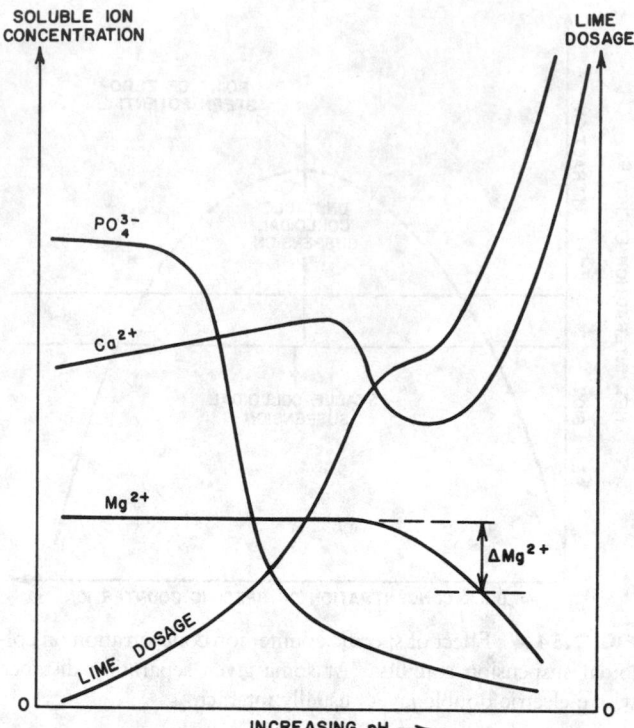

FIG. 7.34.6 Soluble ion distribution at increasing pH values.

PRACTICAL APPLICATIONS

The inorganic coagulant doses required to achieve colloid destabilization depend on the mixing and the liquid–solids separation method.

Coagulant Mixing

The hydrodynamics of inorganic coagulant–wastewater mixing are important. The rapid reaction of aluminum and ferric iron and the fact that soluble kinetic intermediates, the effective coagulating species, are adsorbed on colloid surfaces necessitate rapid, intense dispersion of coagulants into the wastewater. Inadequate mixing causes localized pH and metal ion concentrations, which require increased coagulant dosages to achieve colloid destabilization. Wastewater treatment facilities must add pH adjusting chemicals, e.g., $Ca(OH)_2$, and have them completely reacted before coagulant dispersion to assure the proper pH coagulation level. In addition, they must consider seasonal variations in the wastewater temperature. As the temperature decreases, they must increase the mixing energy to maintain the level of mixing intensity.

Liquid–Solids Separation

Definitive inorganic coagulant dosage is required to effect colloidal suspension destabilization. However, in the practical application of coagulation–flocculation to remove colloidal material from wastewater, the inorganic coagulant dose required depends on the method of flocculation and the liquid–solids separation employed.

Figure 7.34.7 shows the condition that exists when the unit process serves to coagulate and clarify biologically treated wastewater. The inorganic coagulant dose required to coagulate the colloidal material prior to removal by granular media filtration, e.g., coal-sand filters, is one-half to one-sixth that required for removal by mechanical flocculation followed by gravity settling. The reason for this phenomenon is that the deep-bed filter is an efficient flocculation device (bringing destabilized colloids together). On the other hand, gravity settling requires an excess of metal hydroxide precipitate (above that required for destabilization) to produce a settleable floc.

LABORATORY DETERMINATION OF COAGULANT DOSES

No universally accepted laboratory procedure exists for determining the inorganic coagulant doses required for colloidal suspension destabilization in a plant-scale operation. The main problem is that a laboratory model of prototype unit operations of continuous flow through flash mixing, mechanical flocculation, and gravity settling is not available.

Jar Test

A qualitative method used extensively in the water treatment industry is the jar test. In this test, environmental en-

gineers add different coagulant doses to several rapidly mixing samples of wastewater (usually ≤1.5 l) and continue mixing for about 1 min. The sample is then slowly mixed (to simulate flocculation) for 10 to 30 min and allowed to settle quiescently for an additional 10 to 30 min. Environmental engineers then make qualitative observations such as time for visible floc formation, floc size, and floc settling rates.

They also make a direct or indirect measurement of the supernatant SS concentration. They approximate the coagulant dose requirement based on their judgement of these observations. The effective use of jar test information is an art and does not represent a true model of prototype operations.

Zeta Potential Test

A more theoretically based method of determining the required inorganic coagulant dose involves the use of a microelectrophoretic device. Laboratory technicians intensely mix varying doses of an inorganic coagulant into samples of wastewater. They then place aliquots (fractions of the sample) in a small glass chamber and apply an appropriate voltage gradient across the solution.

Laboratory technicians measure the rate of particle migration by visual observation using a calibrated eyepiece in a compound microscope. They then compute the rate of migration divided by the voltage gradient, microns per second per volts per centimeter. This resulting value is the electrophoretic mobility and is proportional to the electrokinetic or zeta potential (see Figure 7.34.1).

Based on the electrical double-layer model, a certain coagulant dosage results in a zero electrophoretic mobility corresponding to a zero electrokinetic potential. At zero electrokinetic potential, the electrical repulsive interaction of the electric double layers is minimized. In practice, the optimum destabilization can exist at a slightly positive or negative electrophoretic mobility.

Since hydrogen and hydroxyl ions are potential determining ions for most wastewater colloids as well as the inorganic coagulant species, an optimum coagulation pH exists. Figure 7.34.8 shows the coagulant-dose–pH-interaction on electrophoretic mobility. From these data, environmental engineers can estimate an optimum economic combination of pH adjustment chemical and coagulant.

Flocculation with Organic Polyelectrolytes

POLYELECTROLYTE FLOCCULANT TYPES

  a) Anionic
  b) Cationic
  c) Nonionic
  d) Variable charge

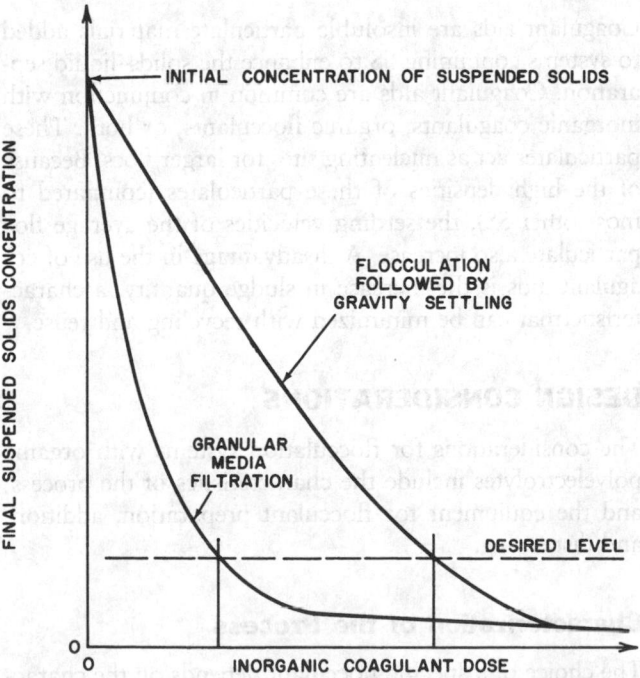

FIG. 7.34.7 The effect of the liquid–solids separation technique on the coagulant dose required for efficient suspended colloidal material removal.

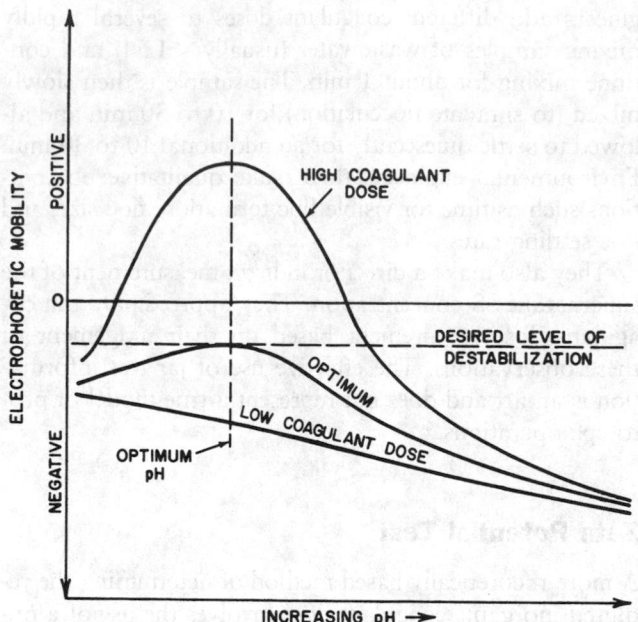

FIG. 7.34.8 Determining the optimum coagulant dose and pH by electrokinetic measurements.

MINIMUM SOLIDS CONCENTRATION FOR EFFECTIVE FLOCCULATION

50 mg/l

FLOCCULANT FEED SOLUTION CONCENTRATION BY WEIGHT

0.25 to 0.5%

TIME REQUIRED FOR FLOCCULATION PRIOR TO SETTLING

Under 5 min

POLYELECTROLYTE ADDITION RATES FOR FLOCCULATION

0.1 to 1 mg/l of raw sewage or 1 to 10 lb per tn of dry solids in sludge treatment applications

FLOCCULANTS

Flocculants are water-soluble, organic polyelectrolytes that are used alone or in conjunction with inorganic coagulants or coagulant aids to agglomerate solids suspended in aqueous systems. The large dense flocs resulting from this process permit more rapid and efficient solids–liquid separation.

Separating SS from raw water and wastewater for purification generally involves gravity settling in large clarifiers operating at low velocity gradients prior to a secondary biological process, a tertiary physicochemical process, or both. This primary physical process is enhanced by coagulation and flocculation of the initially fine colloidal particles into larger and more dense aggregates that settle more rapidly and completely. Coagulation and flocculation are sequential processes distinguished primarily by the types of chemicals used for initiation and the size of the particles developed.

Coagulation is the conversion of finely dispersed colloids into small floc with the addition of electrolytes like inorganic acids, bases, and salts. The salts of iron, aluminum, calcium, and magnesium are inorganic electrolytes. Partial coagulation can also result from naturally occurring processes, such as biological growth, chemical precipitation, and physical mixing. Flocculation is the agglomeration by organic polyelectrolytes (or by mechanical means) of the small, slowly settling floc formed during coagulation into large floc that settles rapidly.

Polyelectrolyte flocculants are linear or branched organic polymers. They have high molecular weights and are water soluble. Compounds similar to polyelectrolyte flocculants include surface-active agents and ion-exchange resins. The former are low-molecular-weight, water-soluble compounds used to disperse solids in aqueous systems. The latter are high-molecular-weight, water-soluble compounds that selectively replace certain ions in water with more favorable or less noxious ones.

Polyelectrolytes can be natural or synthetic in origin. Naturally occurring polyelectrolytes include various starches, polysaccharides, gums, and other plant derivatives. Table 7.34.2 lists various types of synthetic polyelectrolyte flocculants. A variety of products are available among the individual types, which are shipped either as dry granular powders in bags or in bulk or as concentrated viscous liquids in drums or tank cars.

COAGULANT AIDS

Coagulant aids are insoluble particulate materials added to systems containing SS to enhance the solids–liquid separation. Coagulant aids are common in conjunction with inorganic coagulants, organic flocculants, or both. These particulates act as nucleating sites for larger flocs. Because of the high densities of these particulates (compared to most other SS), the settling velocities of the average floc particulate also increase. A disadvantage in the use of coagulant aids is the increase in sludge quantity, a characteristic that can be minimized with recycling and reuse.

DESIGN CONSIDERATIONS

The considerations for flocculation systems with organic polyelectrolytes include the characteristics of the process, and the equipment for flocculant preparation, addition, and dispensing.

Characterization of the Process

The choice of a specific flocculant depends on the characteristics of the process system to be flocculated. The density of the suspending liquid (usually water) and the effective density of the suspended particles must be

TABLE 7.34.2 CLASSIFICATION OF POLYELECTROLYTE FLOCCULANTS

Type	Ionic Charge	Examples
Anionic	Negative	Hydrolyzed polyacrylamides, polyacrylic acid, poly-acrylates, and polystyrene sulfonate
Cationic	Positive	Polyalkylene polyamines, polyethylenimine, polydi-methylaminomethyl polyacrylamide, polyvinyl-benzyl trimethyl ammonium chloride, and polydimethyl diallyl ammonium chloride
Nonionic	Neutral	Polyacrylamides and polyethylene oxide
Miscellaneous	Variable	Alginic acid, dextran, guar gum, and starch derivatives

sufficiently different to permit separation. Sand and grit particles are heavy and compact. They can have effective densities more than twice that of water. Biological solids and hydrated inorganic precipitates are hydrophilic, i.e., associated with surface-bound and internally contained water. Their densities can be only slightly greater than that of water. The density of water is affected slightly by temperature and more significantly by salt content.

The salt content and pH of suspending water affect the surface charge of the SS. The sign, magnitude, and distribution of this surface charge strongly influence the type and quantity of the flocculant to be used. Negatively charged solids can be flocculated by cationic flocculants; positively charged solids can be flocculated by anionic flocculants. Negatively charged solids can also be coagulated by inorganic cations and then flocculated by an anionic flocculant.

The size, shape, and concentration of solid particles also affect flocculation and settling. Large particles settle faster than small particles. Irregularly shaped particles settle slower than smooth, spherical particles. Flocculation effectiveness is reduced if the solids concentration is low (<50 mg/l) since the probability for contact among particles is reduced. Settling is hindered at high solids concentrations (>2000 mg/l) because of excessive interparticle contact. Most suspensions subjected to settling in water and wastewater treatment settle freely, with the exception of concentrated biological sludges.

Equipment for Flocculant Preparation and Addition

Polyelectrolyte flocculants are soluble in water and compatible with most dissolved materials at the concentrations normally found in tapwater. To prepare flocculant solutions, wastewater treatment facilities should use water that has a low solids content to avoid the formation of insoluble sludges. The water's pH should be nearly neutral.

Concentrated solutions of metallic salts and anionic polyelectrolytes should never be prepared in the same tank, since sufficient concentrations of trivalent cations or certain divalent cations can cause partial precipitation of anionic flocculants. Flocculant solutions are essentially non-corrosive, and standard materials of construction, such as PVC, black iron, and mild steel, can be used for all equipment.

High-molecular-weight polymers require time to completely solubilize since water must hydrate the long entwined molecules before they uncoil in solution. The preparation time decreases when dry particles are evenly distributed throughout the solvent water. Wastewater treatment facilities can distribute the dry product with a mechanical mixer and more easily with an eductor. Figure 7.34.9 shows a typical disperser for preparing solutions of dry flocculant. This disperser operates on an aspiration principle and distributes individual flakes of flocculant throughout the water where they are readily dissolved.

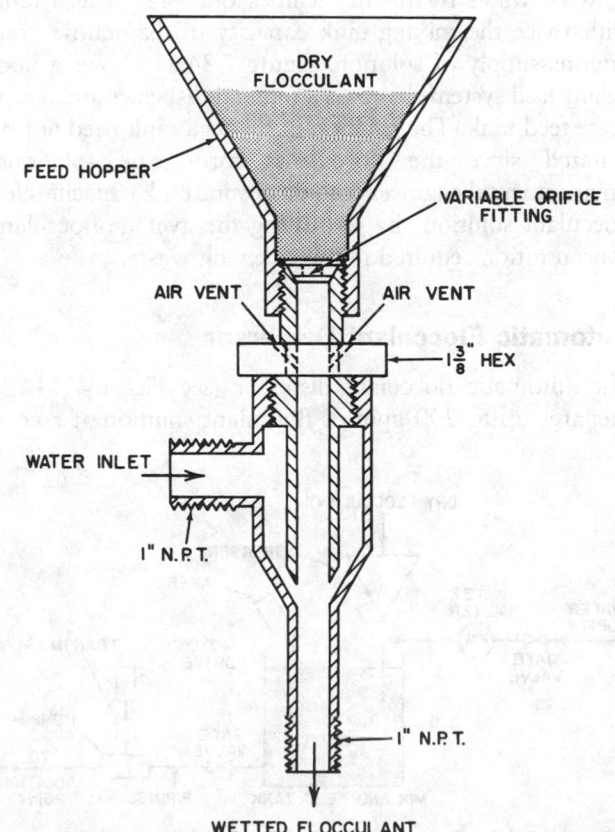

FIG. 7.34.9 Manual flocculant disperser.

After dispersing the solid flocculant, wastewater treatment facilities need only use minimum agitation to assure a uniform solution concentration. A low flow of compressed air or a mechanical mixer can be used. Mechanical mixers or air spargers can be used for agitation during solution preparation. Dissolving dry flocculants does not require violent agitation by high-speed mechanical mixers; the mixing equipment need only generate a moderate rolling action throughout the makeup tank. Figure 7.34.10 shows a typical flocculant feed system using a manual disperser.

Flocculant solutions are highly viscous, and ordinary flow regulating valves and meters are usually not adequate to control the small volumes of these solutions. Wastewater treatment facilities should feed flocculant solutions with accurate chemical metering pumps. Positive displacement pumps such as progressive cavity, rotary gear, or piston pumps are all suitable.

The chemical feed pump should have a variable flow rate control mechanism. For treating less than 700 gpm of water or waste, wastewater treatment facilities can use pumps equipped with dial-controlled, variable-speed drive that can be adjusted while in operation. They should select the pump size so that normal operations use 30 to 50% of the pump capacity. This size provides freedom to decrease or increase the pumping rates as required. Wastewater treatment facilities can use automatic flow ratio control systems for treating larger volumes.

Wastewater treatment facilities often use a feed tank with twice the mixing tank capacity to maintain a continuous supply of solution. Figure 7.34.11 shows a flocculant feed system that uses a manual disperser and a separate feed tank. The solution in the feed tank need not be agitated since the flocculants form true solutions. Environmental engineers can determine tank capacities for flocculant solutions by estimating the average flocculant concentration required in the receiving waste.

Automatic Flocculant Disperser

The automatic flocculant disperser (see Figure 7.34.12) prepares up to 200 gpm of flocculant solution at a con-

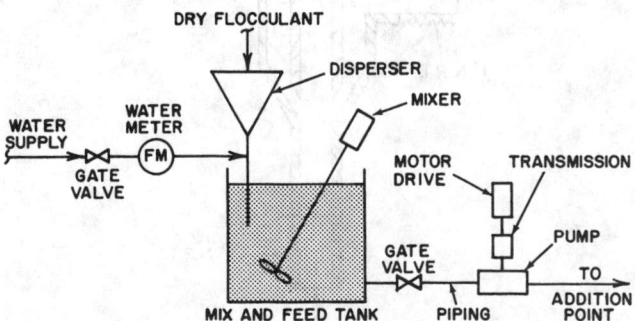

FIG. 7.34.10 Flocculant feed system with manual dispersing equipment.

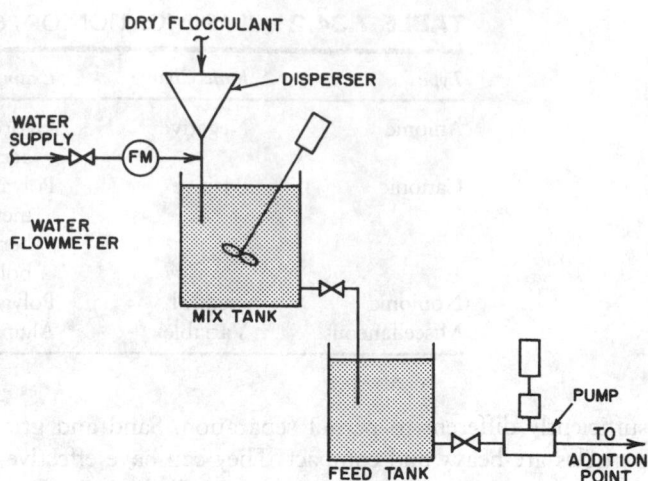

FIG. 7.34.11 Flocculant feed system with manual dispersing equipment and separate feed tank.

centration of 0.25% by weight. A screw feeder adds dry flocculant at a variable rate (≤10 lb per min) into the vortex formed within a mixing bowl. Water at 40 psig and 25 to 100 gpm first passes through a strainer and a water meter. Then it reaches a solenoid valve activated by a float control that monitors the liquid level in the feedtank receiving the prepared flocculant solution.

The water flow is then split, with most of it passing through a valve and rotameter controlling the flow through the mixing bowl. The remaining water passes through a small internal reservoir equipped with a float valve. This maintains the required water level in the bottom of the mixing bowl for optimum dry flocculant dispersion. The combined flow from the mixing bowl and the reservoir is then pumped through a pump at 25 to 100 gpm directly to the feed tank or combined with another stream of dilution water (≤100 gpm) before delivery to the feed tank.

The flocculant solution is further mixed in the feed tank before it is displaced by the incoming flow and delivered to the point of addition in the treatment plant. The flocculant disperser can be operated either manually or mechanically. Other simplified units are commercially available for preparing flocculant solutions in the intermediate range of 0.5 to 25 gpm. The manual disperser is limited to the preparation of small quantities of flocculant solution.

Wastewater treatment facilities should consider an automatic flocculant disperser (see Figure 7.34.12) when treating larger volumes of water flow. The hopper is filled manually with dry flocculant on a daily to weekly basis, depending on plant flow rate. An automatic dry flocculant addition system should be incorporated when large volumes of water (>100 mgd) are treated. Figure 7.34.13 shows a flocculant feed system with an automatic disperser that incorporates a level control on the feed tank.

Flocculants are ordinarily prepared as dilute solutions that are then dispersed into the water or waste to be

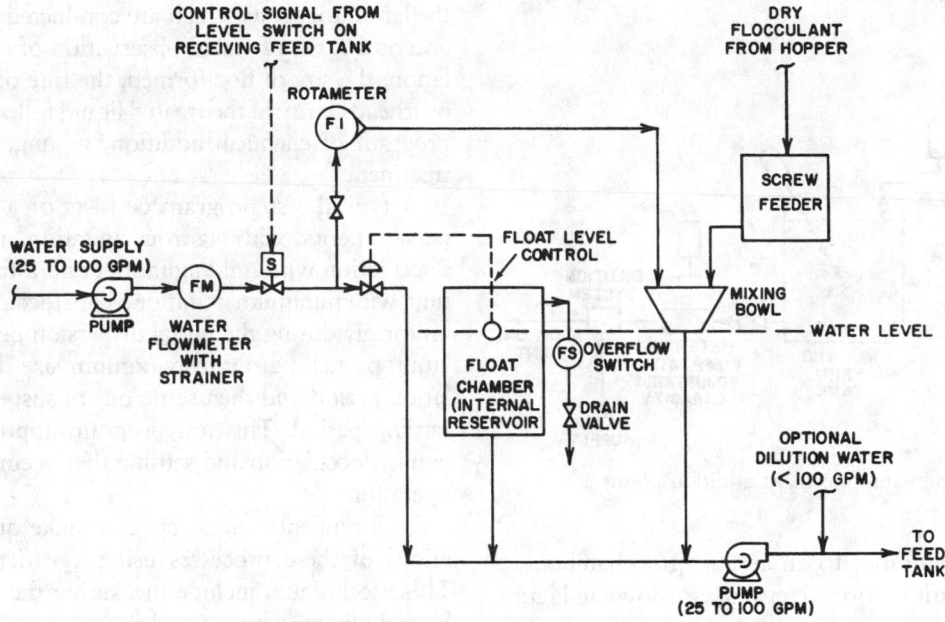

FIG. 7.34.12 Automatic flocculant disperser.

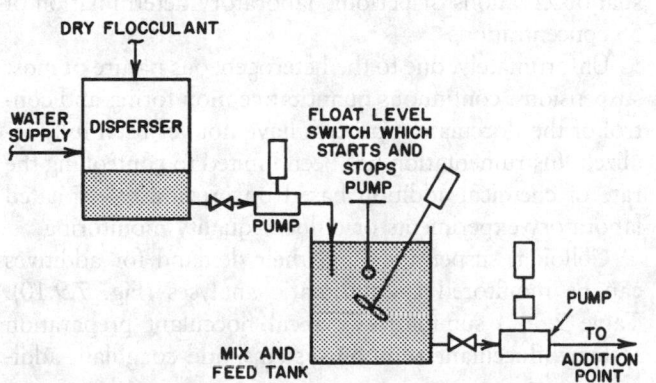

FIG. 7.34.13 Flocculant feed system with automatic dispersing equipment.

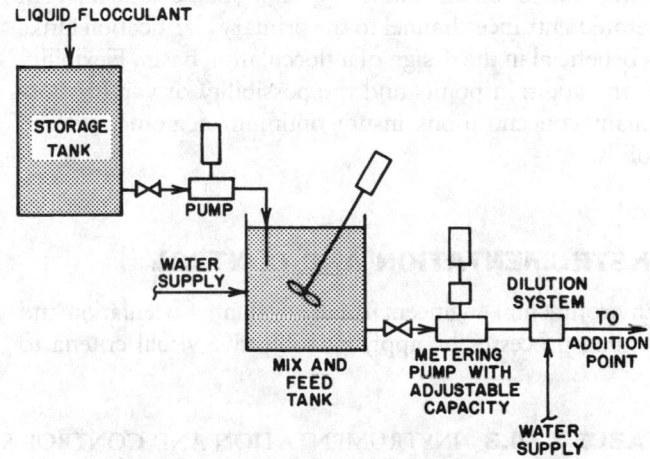

FIG. 7.34.14 Liquid flocculant feed system.

treated. The average feed solution concentration for most flocculants is 0.25 to 0.5% by weight. In applications where liquid flocculants are added at high concentrations, the wastewater treatment facility prepares the feed solution by diluting the concentrated bulk solution as shipped. Figure 7.34.14 shows a liquid flocculant feed system.

Wastewater treatment facilities can apply flocculants by diluting a concentrated solution and then feeding the resulting dilute solution to the stream to be treated. Preparing large volumes of dilute solutions is not necessary. A small volume of the concentrated feed solution can be metered and diluted to a lesser concentration immediately before it is added to the stream to be treated. Environmental engineers should consider the chemical time requirement for completing the dilution step, the storage tank size, and the chemical pump capacities when determining the feed solution concentration.

The flocculation of SS, precipitated inorganic salts, and organic complexes adsorbed or absorbed to particulates is not subject to the same design criteria as conventional water treatment that uses inorganic coagulants alone. The time for flocculation after the flocculant addition and prior to settling can be considerably less (under 5 min) than the time recommended for water treatment plants, which frequently exceeds 30 min.

Experience shows that prolonging gentle mixing after the initial rapid dispersion of the flocculant does not increase the removal and can partially destroy the floc particles. Environmental engineers should also consider the charging period of adding a metal coagulant and adding the flocculant if both agents are used.

The flocculant is usually added at some turbulent point in the plant flow, such as at the throat of a flume, ahead

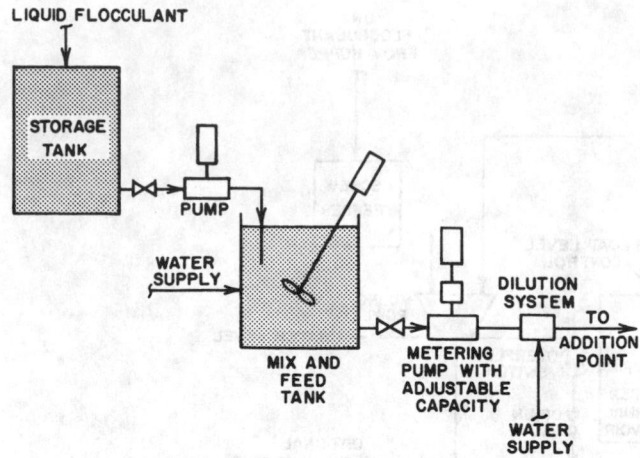

FIG. 7.34.15 Coagulant–flocculant addition points.

of a weir, at the entrance to an aerated-grit chamber, or at some other location prior to settling, as shown in Figure 7.34.15.

Adding the flocculant before low-lift pumps can cause floc destruction. However, the suction side of these pumps is acceptable for the addition of inorganic coagulants. An aerated-entrance channel to the primary clarification tanks is beneficial in the design of a flocculation basin. Flexibility in the addition points and the possibility of varying flocculant concentrations insure optimum agglomeration of solids.

INSTRUMENTATION AND CONTROL

Environmental engineers usually evaluate flocculation and settling processes by applying subjective visual criteria to the laboratory tests which are conducted in parallel. These criteria consist of visual observation of the rate of flocculation, the size of floc formed, the rate of settling, and the overhead clarity of the treated liquid following a prescribed program of chemical addition, mixing, and duration of treatment.

A typical test program consists of a short initial dispersion period with vigorous agitation, a longer period of flocculation with mild agitation, and a final period of settling with minimum agitation. The flocculant is distributed uniformly during the initial dispersion period. The flocculating particles grow to maximum size during the flocculation period and then settle out of suspension during the settling period. This test program approximates the dynamic flocculation and settling that occur during full-scale operation.

Environmental engineers can make quantitative evaluations of these processes using a variety of techniques. These techniques include measuring the SS concentration by turbidimetric means and the surface charges of the particles by streaming current or zeta potential. For control purposes, these techniques are preferred to subjective visual observations or periodic laboratory determination of SS concentrations.

Unfortunately, due to the heterogeneous nature of most suspensions, continuous quantitative monitoring and control of the flocculation process have not yet been fully realized. Instrumentation has been limited to controlling the rate of chemical addition based on previously estimated laboratory experiments or effluent quality monitoring.

Colloidal suspensions and their demand for additives can be monitored by automatic analyses (Fig. 7.9.10). Table 7.34.3 summarizes typical flocculant preparation systems, flocculant addition systems, and coagulant addi-

TABLE 7.34.3 INSTRUMENTATION AND CONTROL SYSTEMS FOR THE FLOCCULATION PROCESS

Average Plant Flow (mgd)	Flocculant Preparation System	Flocculant Addition System	Coagulant Addition System
<1	Manual operation; batch preparation in tank	Manual operation; variable-speed device with pump calibration or rotameter, or both	Similar to the flocculant addition system
1–10	Automatic operation; flocculant disperser holding tank	Automatic operation; variable-ratio control of feed to sewage flow	Similar to the flocculant addition system
10–25	Same as above	Same as above with feedback correction and flow totalization	Similar to the flocculant addition system
25–100	Same as above	Same as above plus tank-level indicators and malfunction alarms	Similar to the flocculant addition system plus automatic influent analyzers and transmitters
>100	Same as above; alternate bulk handling	Same as above	Similar to the flocculant addition system plus automatic analyzer and transmitter for both influent and effluent: density transmitter and use of empirical design equations

tion systems for five ranges of total plant flow. The overall accuracy of instrumentation should increase with increased plant size.

Average Flow: <1 mgd

For this size flow, the wastewater treatment facility manually prepares the flocculant solution in a small storage tank. The addition rates are manually set by variable-speed devices such as positive displacement pumps. For flow rate indication, this system uses the pump calibration curve or a glass tube rotameter. The coagulant solution is fed by manually set, variable-speed pumps similar to the flocculant controls.

Average Flow: 1 to 10 mgd

For this flow, the flocculant solution is automatically prepared and added to the influent flow by a variable-ratio control system that varies the speed of the pump proportional to the influent wastewater flow rate. The coagulant solution addition system is similar. The influent flow

should be sensed by a magnetic flowmeter or a Venturi tube.

Average Flow: 10 to 25 mgd

For this flow, the flocculant solution is automatically prepared as in the 1 to 10 mgd system and automatically added to the influent flow by a variable-flow-ratio controller that varies the speed of the pump proportional to the influent flow rate. The actual amount of flocculant added is detected and used as a feedback signal. The controller determines the difference between what should be added and what is being added, and adjusts the pump speed to eliminate the deviation. The coagulant addition system is similar. Flow totalization of each stream is recommended to provide data for material balances.

Average Flow: 25 to 100 mgd

This size plant automatically prepares the flocculant solution as in the 10 to 25 mgd system. It automatically adds the flocculant solution to the influent based on the detected influent flow rate multiplied by the required variable ratio as described for 10 to 25 mgd plants.

This system can automatically control the coagulant addition on the basis of influent composition analyses and influent flow rate. Automatic analyzers and transmitters are usually required. A variable-ratio controller controls the coagulant addition rate. The system can include continuous level transmitters on coagulant storage tanks with high- or low-level switches, or both, to simplify loading and unloading. Flow recorders and totalizers can be included to provide data for material balances.

Average Flow: >100 mgd

Flocculant solutions for systems of this size are automatically prepared using an automatic flocculant disperser and suitable holding tanks. A bulk handling system is an alternative to batch flocculant transfer. This system automatically proportions the flocculant solution to the raw waste flow by multiplying influent flow rate by a variable ratio.

The coagulant feed is automatically proportioned to influent waste flow or it can also be based on empirical design equations for a specific waste stream. A density transmitter on the coagulant feed determines the weight concentration of the coagulant in solution which is multiplied by the coagulant flow rate, resulting in a coagulant mass-flow-rate signal. The performance of the system is detected usually by a turbidity analyzer (ARC in Fig. 7.34.16). The required coagulant addition rate is adjusted by this ARC, which adjusts the set-point of a variable-ratio controller (FRC) that receives the coagulant mass-flow-rate as its measurement and operates a variable-speed pump to deliver the calculated amount of coagulant.

Flow recorders and totalizers provide a material balance (see Figure 7.34.16). Electronic DCS instrumentation

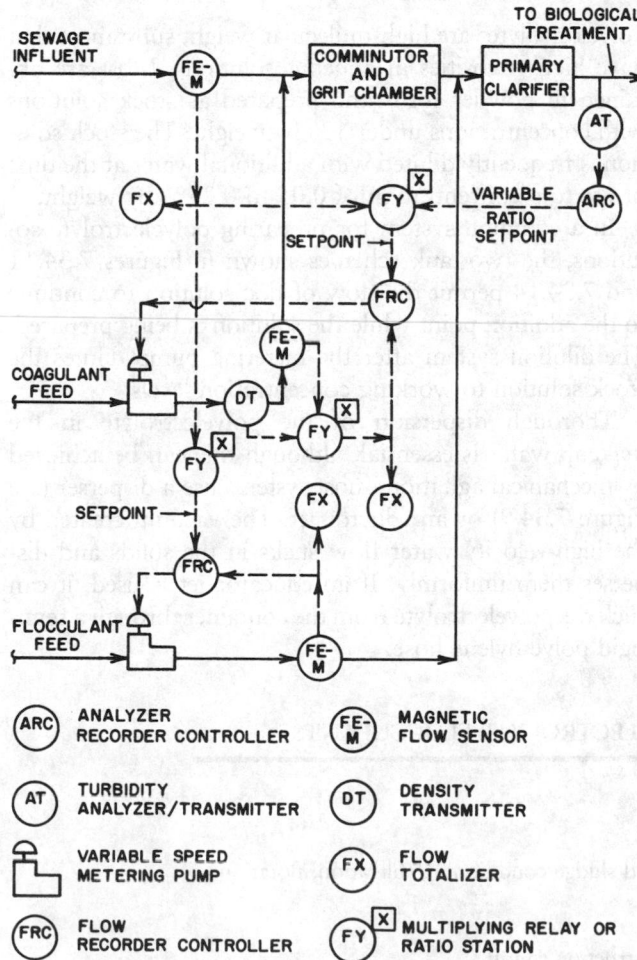

FIG. 7.34.16 Control system for automatic flocculant and coagulant additions.

is preferable because of the long transmission distances in many waste treatment plants.

APPLICATIONS

Environmental engineers have used organic flocculants to improve solids–liquid separations in the applications listed in Table 7.34.4. Municipal water treatment performance improves when flocculants supplement inorganic coagulants in removing solids from raw water and in improving the solids removal capabilities of vacuum filters and centrifuges. Municipal wastewater treatment uses flocculants to remove SS, flocculate precipitated nutrients such as phosphorus, and condition sludges for dewatering processes such as flotation, elutriation, filtration, centrifugation and sand-bed dewatering. Environmental engineers treat surface waters with flocculants to remove solids that are periodically received from storm runoff, dredging, or construction. Treating industrial waste with flocculants removes a variety of suspended pollutants and enhances the separation and removal of some emulsified and dissolved pollutants.

Coagulation Systems

TYPES OF COAGULATION–FLOCCULATION SYSTEMS

  a) Gravity Settling
  b) Flotation

COAGULATION APPLICATIONS

Neutralize and agglomerate oil droplets or particles in the $50\,\mu$ or smaller size range

MATERIALS OF CONSTRUCTION FOR FLOCCULATION EQUIPMENT

Stainless steel or epoxy-reinforced fiberglass

SYSTEM DESCRIPTION

A coagulation system has the following functions:

1. Preparation of solutions of coagulants, flocculants, and coagulant aids. Usually employed are alum, ferric chloride, or other inorganic coagulants; cationic polyelectrolyte flocculants; or anionic or nonionic polyelectrolyte flocculants.

2. Rapid mixing of coagulants with wastewater.
3. Growth of floc to produce large agglomerates under gentle mixing conditions.
4. Separation of floc from water, usually by settling or flotation.

Figure 7.34.16 shows a complete, fully automated coagulation–flocculation system. The example shown involves coagulation aided by an inorganic coagulant followed by the addition of a polyelectrolyte flocculant. This system disperses the polyelectrolyte in the main wastewater line by injecting the solution into the pipeline. At high velocities in the line, rapid mixing is produced. The polyelectrolyte flocculant is added just before the feed well of the clarifier, which has a slow-moving rake that keeps the settled sludge distributed and induces gentle mixing to enhance floc formation.

The design and nature of coagulation systems are based on the coagulation characteristics of the wastewater and the nature of the coagulant used.

PREPARATION OF COAGULANT SOLUTIONS

Polyelectrolytes are high-molecular weight substances that have high viscosities in aqueous solutions. Most are obtained in powder form and prepared as stock solutions with concentrations under 1% by weight. The stock solution is frequently diluted with additional water at the time of use to a concentration of 0.05 to 0.25% by weight.

In a common system for preparing polyelectrolyte solutions, the two-tank schemes shown in Figures 7.34.11 and 7.34.14 permit the flow of floc solution to continue to the addition point while the solution is being prepared. The dilution system after the metering pump dilutes the stock solution to working concentration levels.

Thorough dispersion of the polyelectrolyte in the makeup water is essential, although this can be achieved by mechanical agitation, most systems use a disperser (see Figure 7.34.9) or an eductor jet. The vacuum created by the high-velocity water flow sucks in the solids and disperses them uniformly. If an educator jet is used, it can suck the polyelectrolyte from the container through a semi-rigid polyethylene hose.

TABLE 7.34.4 APPLICATIONS FOR POLYELECTROLYTE FLOCCULANTS

Municipal Water Treatment
  Primary flocculation and sludge conditioning
Municipal Waste Treatment
  Raw waste flocculation, phosphorus removal, and sludge conditioning (filtration, flotation, and centrifugation)
Surface Water Treatment
  River clarification, dredging operations, and construction runoff
Industrial Water Treatment
  Raw water (influent), process water (internal), and wastewater (effluent)

After the polyelectrolyte is dispersed in water, it must be mechanically agitated to dissolve the polyelectrolyte. A typical agitator for a 400-gal tank requires a 1-hp motor and a marine propeller with a 200- to 400-rpm agitation velocity. The agitation time varies with the concentration of the stock solution and can range from 30 to 120 mins. Stock solution can remain in a storage tank for only several days because stock solutions are not stable for more than 3 to 4 weeks.

Because most polyelectrolytes are acidic, flocculant solutions should be prepared in lined steel, epoxy-reinforced fiberglass, 316SS, or other corrosion-resistant tanks. However, construction materials in the wastewater system downstream of the flocculant solution injection point are not affected by the corrosive effect of the polyelectrolytes since their dilution is in the ppm range or less.

ENGINEERING DESIGN CONSIDERATIONS

The coagulation process starts with the rapid, high-intensity mixing of the coagulant with wastewater. Since complex reactions are involved, environmental engineers must use laboratory tests such as the zeta potential and jar tests to establish proper pH and dosage ranges.

Wastewater treatment facilities can rapidly mix inorganic coagulants with the wastewater either inline or in a separate flash-mixing basin. The detention time requirement ranges from 10 to 60 sec. Polyelectrolyte flocculants are added after the rapid-mix phase or downstream from the coagulant addition point because rapid mixing can break up the floc. Mixing during the flocculation stage should be gentle, with a mean velocity gradient of 50 to 175 determined as follows:

$$\sqrt{\frac{W}{\mu}} = \text{Mean velocity gradient (sec}^{-1}\text{)} \qquad 7.34(1)$$

where:

    W = power per unit volume of fluid (lb/sec-sq ft)
    μ = absolute viscosity of wastewater (lb/sq ft)

SETTLING OR FLOTATION

The separation mechanism is the same for settling and flotation. Stokes' law governs both methods of separation. When the settling velocity (Vs) is positive, the particle settles; when it is negative, the solid particle rises.

When a coagulated waste can be treated by either settling or flotation, wastewater treatment facilities can achieve higher separation rates and higher solid concentrations through dissolved air flotation. This process requires smaller basins and results in smaller sludge volumes with greater water recovery. On the other hand, flotation systems require more operator attention and in some cases extra maintenance.

In addition to their use with settling and flotation, wastewater treatment facilities often use flocculation or coagulation to enhance other separation operations, including filtration and centrifugation.

COAGULATION FOLLOWED BY SETTLING

Coagulation systems used in combination with settling are available in two basic design configurations: conventional systems and sludge-blanket systems. In conventional system, the rapid-mix step is completed before the water enters the large settler, where the flocculation and clarification steps are completed. In a sludge-blanket-type unit, the coagulant mixing, flocculation, and settling steps all take place in a single unit.

Figure 7.34.15 shows a conventional system where rapid mixing of the cationic inorganic coagulant is done in the comminutor (a mix tank can also be used), and the water is then fed to the primary clarifier. The anionic polyelectrolyte flocculant is injected into the wastewater line before the clarifier unit. A moving scraper collects the settled floc and removes it to the sludge outlet. Recycling part of the floc usually improves clarification efficiency.

The flocculation step can also occur in a separate flocculation tank with slow-moving, horizontal paddles located upstream to the conventional settler.

Sludge-blanket units combine coagulation, settling, and upward filtration in one compact unit (see Figure 7.34.17). In this design, influent water is mixed with coagulation chemicals as it is fed to the inner chamber. Floc formation begins in this chamber. The formed floc slurry then moves to the outer chamber where the water rises through a bed of previously formed floc which functions as a filter bed, effectively retaining the floc.

The velocity of the water decreases as it moves up the outer chamber because the cross-sectional area increases. This decreased velocity assists in the separation. The clear effluent water flows out through a weir at the top while draining after blowoff removes the sludge collected at the bottom. This type of unit has a small space requirement.

COAGULATION FOLLOWED BY FLOTATION

Total pressurization is used when the material to be separated is not broken down by shearing forces in the pressurization system or the floc reforms quickly after the pressure is released. This system (see Figure 7.34.18) produces a maximum amount of air bubbles and requires a small flotation compartment. Wastes with heavy solid loads or those that rapidly form strong flocs are compatible with this method.

Chemical additives can be added upstream or in the pressurization system itself. The system operates by pressurizing the wastewater stream to 30 to 60 psig and mix-

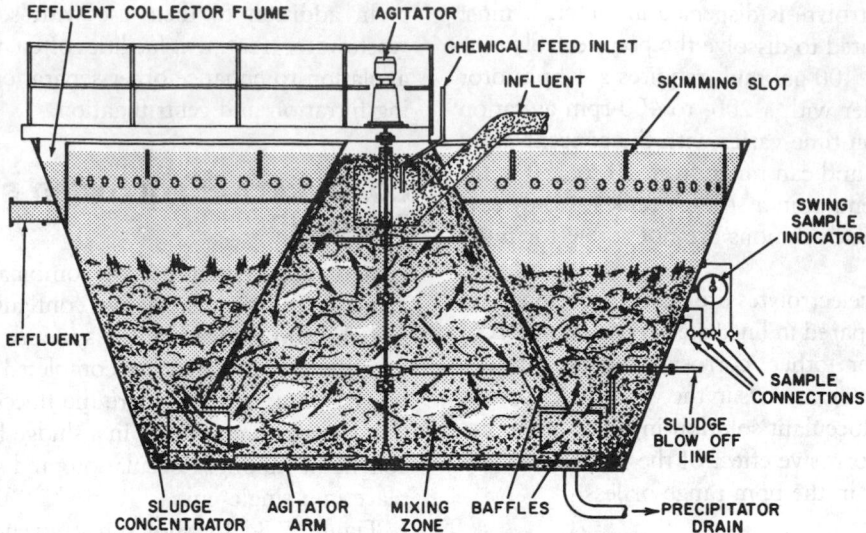

FIG. 7.34.17 Combined coagulation and settling apparatus. (Reprinted, with permission, from Permutit Co.)

ing it with compressed air. A DO level approaching saturation is achieved in the retention tank. The pressure of the air-saturated stream is reduced in a backpressure valve on the tank discharge. This reduced pressure initiates the formation of microscopic air bubbles which assist in the flotation process.

Partial pressurization is used when the solid load is light. As shown in Figure 7.34.18, only a fraction of the waste stream is pressurized by air. This scheme frequently requires a separate flocculation chamber in the bypassed stream. In this system, the pressurized stream solids must be absorbed on the preformed floc when mixed in the inlet compartment. Otherwise, a secondary addition of flocculant solutions is required in the inlet compartment of the flotator.

Recycling pressurization is used when the floc formed cannot be pressurized and large quantities of air must be dissolved. In this scheme (see Figure 7.34.18), the addition of coagulants and flocculants precedes the flotation step. A side stream of the clarified effluent is air-pressurized. When extended floc formation time is required, this method is particularly applicable. The pressurization components are less prone to solids build-up in this system.

COAGULATION IN SLUDGE HANDLING

Coagulation is also used to condition slurries or sludges. Coagulation is a highly effective sludge conditioner and is often used ahead of vacuum or gravity-type filters. It also aids the separation in solid-bowl centrifuges. In these applications, the coagulant solution is added to the pool either within the centrifuge or externally in the feed line.

Coagulant solutions can be used to increase the drying rate on sludge drying beds. The coagulant solution can be added to the sludge as it is applied to sand beds.

Emulsion Breaking

TYPES OF EMULSIONS

    a) Oil-in-water
    b) Water-in-oil

VOLUME CONCENTRATION OF EMULSIONS

For oil–water systems, between 26 and 74% volume concentrations

TREATMENT TECHNIQUES USED

Emulsion breaking or oily waste treatment is applied at over 1% oil concentration. Below 1%, clarification and coalescing techniques apply

SPECIFIC UNIT OPERATIONS USED

Gravity separation, air flotation, centrifugation, filtration, electrical dehydration, and chemical treatment

Emulsions are stable, heterogeneous systems consisting of at least one immiscible liquid dispersed in another in the form of microscopically visible droplets. Oil–water emulsions are formed in sewage systems as a result of contact between oil, water, and emulsifying agents, or formed directly as industrial by-products.

In refineries, the direct formation of emulsions can result from chemical treatment of lubricating oils, waxes, and burning oils; barometric condensers; tank drawoffs; desalting operations; acid sludge recovery processes; and wax deoiling. Oily wastes from general petrochemical plants are usually treated in oil–water separators. The effluent from separators that still contains small amounts of emulsion is treated before final discharge to the sewer.

Steel manufacturing and finishing operations produce mixtures of waste-soluble oils containing cleaning solutions, which are stable emulsions. The automobile industry and steel mills discharge spent machining and cutting oils, coolants, and drawing compounds.

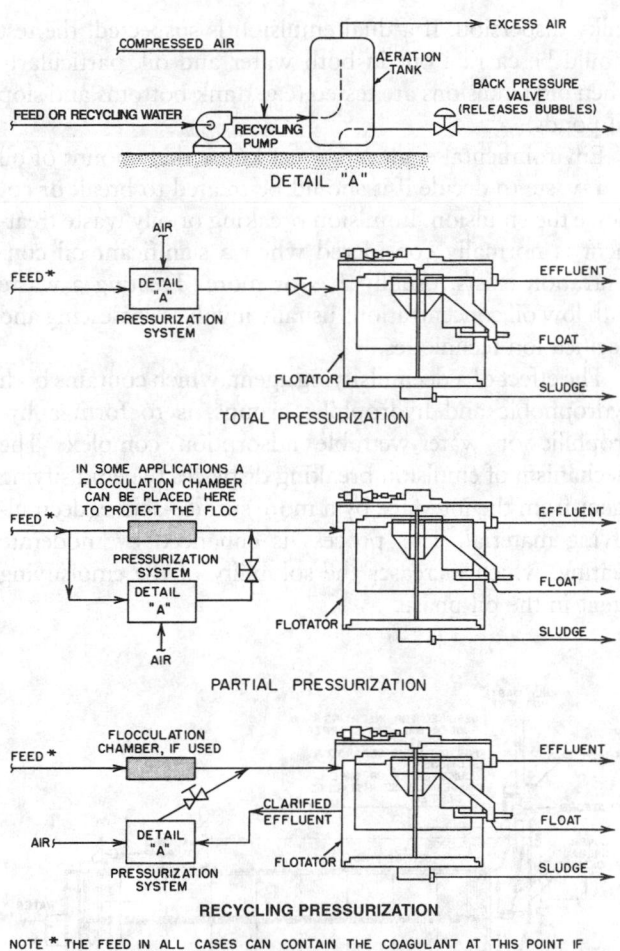

FIG. 7.34.18 Pressurization methods applied in flotation.

TYPES OF EMULSIONS

Emulsions consist of two liquid phases: the disperse and continuous phases. The phase in the form of finely divided droplets is the disperse or internal phase; the phase forming the matrix in which these droplets are suspended is the continuous or external phase. Based on the liquid phases, a typical water–oil emulsion can exist either as an oil-in-water (oil is the disperse phase) or water-in-oil emulsion. These types are abbreviated as o/w and w/o, respectively.

When an assembly of spheres of equal radii are in a position of densest packing, they occupy 74% of the total volume; the remaining 26% is empty space. Theoretically, this relationship means that for a given system, o/w and w/o emulsions are both possible between the phase volume concentrations 0.26 and 0.74. Below 0.26 and above 0.74, only one form can exist. In the intermediate range of volume concentrations, no one form of emulsion is favored. These systems are described as *multiple* or *dual* emulsions with the disperse phase containing globules of the other phase.

STABILITY OF EMULSIONS

The stability of an emulsion is based on its physical and electrical properties.

Physical Properties

The properties of an emulsion depend largely on its *composition* and its mode of *formation*. Physical properties also control the stability of these systems. The formation of an emulsion is a function of the boundary tensions between the two liquid phases determining the type of emulsion as either w/o or o/w. *Interfacial tensions* also exist between the liquid phases and between the liquid and solid phases. The latter is usually a lower magnitude than that of a liquid–liquid interface.

The flow resistance of an emulsion is one of its most important properties. Viscosity measurements provide considerable information about the structure of emulsions and their stability. The viscosity of the continuous or external phase is of prime importance in overall emulsion viscosity. The viscosity of the oil components in a w/o emulsion is usually indicative of emulsion stability but has little significance in an o/w system.

Miscibility determines the emulsion type. An emulsion is readily dilutable by the liquid that constitutes the continuous phase. Therefore, o/w emulsions are readily miscible with water; conversely, w/o systems are readily miscible with oil. An emulsion remains stable as long as the interfacial film and emulsifying agents are not materially affected.

Electrical Properties

Since oil is an insulator, o/w emulsions conduct an electrical current, whereas w/o emulsions ordinarily do not. Researchers have studied the conductivity of emulsions by measuring the current flowing between two fixed platinum electrodes immersed in the emulsion. The dielectric properties of emulsion systems are different from the average of the individual phases. The dielectric constant is important because of its intimate relationship to emulsion stability. The dielectric properties of an emulsion can be measured in a single parameter defined as the zeta potential (see Figure 7.34.1).

Degree of Stabilization

The concentration of droplets of the disperse phase toward the top or bottom of the emulsion results in a difference in density between the liquid phases. This phenomenon is known as *creaming*. Creaming does not necessarily represent a breaking of the emulsion, but with large droplet sizes, it can eventually lead to emulsion breaking.

Another type of emulsion instability results from flocculation or clumping of individual droplets to form larger aggregates. In this case, the emulsion has inverted due to a sudden change from o/w to w/o and vice versa. The progressive coarsening of the dispersion leads to a complete separation or breaking of the emulsion. Emulsion breaking is an irreversible process, often preceded by creaming or inversion.

EMULSION BREAKING METHODS

Several physical methods can separate oils and SS from wastewater, including gravity separation, dissolved air flotation, centrifugation, filtration, and electrical dehydration. Selecting the method depends on the nature of the wastewater and the degree of treatment required. Chemical methods of breaking water–oil emulsions are based on the addition of chemicals that destroy the protective action of hydrophobic or hydrophilic emulsifying agents and allow the water globules and oil to coalesce.

Figure 7.34.19 shows a typical API separator. This two-stage separator can be expanded to include additional sections. Figure 7.34.20 shows a common treatment arrangement that uses a combination of mechanical process and a chemical process. System operation is as follows:

1. Piping design and feeding facilities allow the operator to pump into either or both tanks and add an emulsion-breaking chemical to either tank or to the waste enroute from the API separator.
2. Steam coils are in both tanks since heat improves the speed and efficiency of phase separation after emulsion resolution.
3. This system fills one tank with waste from a bottom inlet until an upper-level limit is reached. The flow then switches to the other tank. The filled tank is allowed to settle for about 60 min and is then inspected.
4. The system pumps any oil that has risen to the top to reclaimed oil storage. It draws water from the bottom and routes it to the separator inlet. After this step, normally about three-fourths of a tank of w/o emulsion are left.
5. The system heats the emulsion with the steam coils. After the proper temperature is reached, it mixes it by rolling the tank with gas. The emulsion-breaking chemical is added during the mixing.
6. After thorough mixing, this system shuts off the gas and steam so that the treated emulsion can cool and settle until phase separation is complete.
7. The water phase, containing some SS but virtually oil-free, is routed to a settling pond. The oil goes to reclaimed-oil storage and is recycled to the refinery crude unit at a steady rate.

TREATMENT OF WASTE EMULSIONS

Environmental engineers can determine the emulsion type by using the phase dilution method. The procedure is to place one drop of the emulsion in about 20 cc. of water with gentle stirring. A w/o emulsion shows no dispersion, and the water remains clear. An o/w emulsion forms a milky dispersion. If a dual emulsion is suspected, the test should be carried out in both water and oil, particularly when old emulsions are tested (e.g., tank bottoms and slop oil ponds).

Environmental engineers must know the amount of oil in a waste to decide if it should be treated to break or coalesce the emulsion. Emulsion breaking or oily waste treatment is normally considered when a significant oil concentration exists, usually 1% or more. Treating a waste with low oil concentrations usually involves coalescing and clarification techniques.

The effect of a deemulsifying agent, which contains both hydrophobic and hydrophilic groups, is to form a hydrophilic or water-wettable adsorption complex. The mechanism of emulsion breaking displaces the emulsifying agent from the interface by a more surface-active, deemulsifying material. This process is enhanced by moderate heating, which increases the solubility of the emulsifying agent in the oil phase.

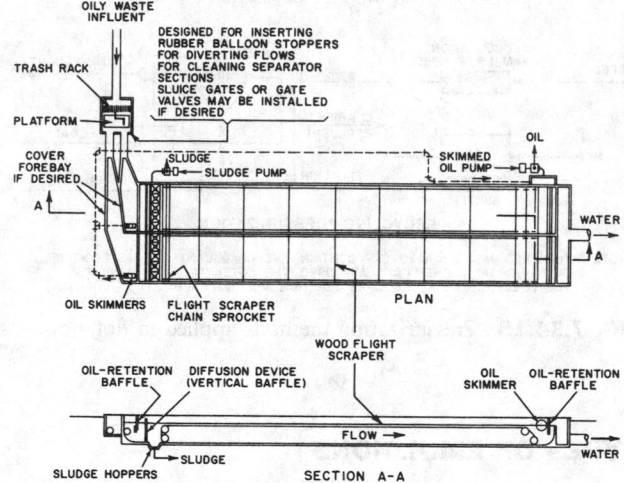

FIG. 7.34.19 API oil-water separator.

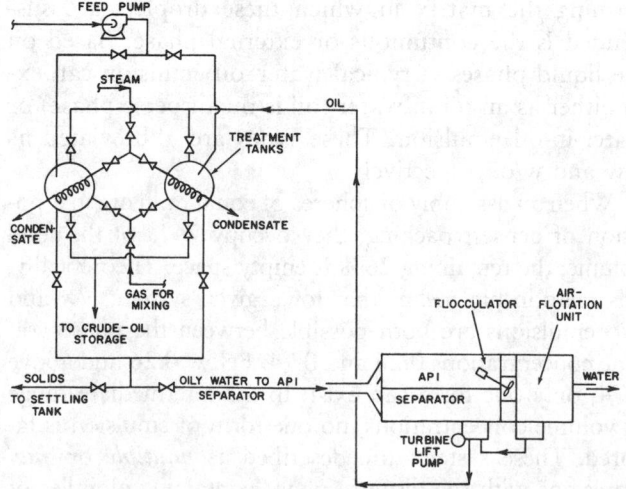

FIG. 7.34.20 Emulsion-breaking installation.

FUTURE OF EMULSION BREAKING

The removal of suspended and floating oil—generally known as primary treatment—can be accomplished by physical means. More advanced treatment is necessary to remove either dispersed or chemically emulsified oil. Chemical treatment can be an economical means of waste emulsion handling. A properly designed, emulsion-breaking process produces recoverable oil and treated water of adequate quality. The small volume of chemical sludge produced during the emulsion-breaking process is amenable to ultimate disposal.

—D.E. Burns
Donald Dahlstrom
Gerry L. Shell
S.L. Daniels
C.J. Santhanam
B. Rico-Ortega

Organics, Salts, Metals, and Nutrient Removal

7.35
SOLUBLE ORGANICS REMOVAL

The activated-sludge process, developed in England in the early 1900s, is remarkably successful at removing soluble organics from wastewater (Junkins, Deeny, and Eckhoff 1983; Tchobanoglous and Burton 1991; Vesilind and Pierce 1982). In an air-sparged tank (see Section 7.25), live microorganisms rapidly adsorb, then slowly oxidize these organics to carbon dioxide and water. At the same time, these organisms reproduce. The process removes the microorganism sludge by settling, while digestion of adsorbed organics continues, which activates the sludge for recycling.

Excess activated sludge is disposed after a dewatering attempt. Its disposal is the most difficult and costly aspect of this wastewater treatment. Even the best sludge dewatering equipment produces a sludge cake with not more than 14–18% dry solids.

Improvements are aimed at making the activated-sludge process more efficient. Advances in biotechnology and fluid dynamics allow stricter environmental standards to be attained. Three major advances are as follows:

1. The use of oxygen instead of air facilitates maintaining the required DO levels. It also makes sludge separation easier (see Figure 7.35.1). Therefore, an oxygen-sparged plant can operate at higher sludge levels with a smaller aeration tank, and sludge disposal is reduced.
2. Replacing the air or oxygen-sparged tank with a fluidized-bed reactor allows the biomass to be adsorbed on small particles kept in suspension by the circulating effluent. This biofilm promotes high solids retention. The biomass concentration in the fluidized-bed reactor (15,000 ppm) is ten times greater than in the standard activated-sludge process. This increased concentration allows the fluid bed reactor to operate at a high, contaminant-removal efficiency.
3. Eliminating oxygen and using anaerobic methane fermentation of municipal wastewater in a fluidized-bed reactor is an even more efficient process. It is capable of operating with high biomass levels (30,000–100,000 ppm) at COD removal rates of 5–50 kg/m^3/day.

Using Oxygen Instead of Air

EFFLUENT BOD CONCENTRATION

11 to 23 mg/l

RETENTION TIME

1 to 3 hr

BOD REDUCTION

88 to 92%

ORGANIC LOADING PER 10,000 GAL OF AERATOR TANK

150 to 250 lb of BOD per day

COD REDUCTION

71 to 84%

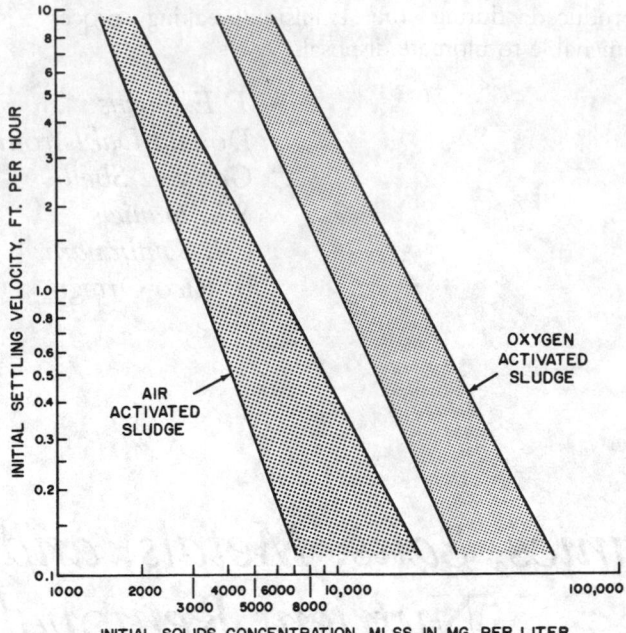

FIG. 7.35.1 Settling velocities of aerated and oxygenated activated-sludge particles.

RECYCLED SLUDGE SOLIDS CONCENTRATION

20,000 to 40,000 mg/l

EXCESS ACTIVATED-SLUDGE PRODUCTION

0.3 to 0.45 lb of VSS per pound of BOD removed

OXYGEN UTILIZATION

+90%

POWER REQUIREMENTS OF OXYGEN PRODUCTION FROM AIR

3.6 to 4.5 lb of O_2 per hp-hr

POWER REQUIREMENTS FOR MIXING AND RECIRCULATION

3 to 5 hp-hr per mgd of feed or about 2 lb O_2 transferred per hp-hr

TANK SIZES IN OXYGENATION UNITS

The aeration tank is smaller (by up to a factor of three) but needs to be gas tight; the clarifier is the same size.

GAS VOLUME USED IN OXYGENATION PROCESSES

1% of gas volume in aeration systems

OPERATING DO LEVEL IN MIXED LIQUOR

6 to 10 mg/l

Figure 7.35.2 compares the use of air and oxygen in activated-sludge wastewater treatment. An air-sparged plant treating 1 mgd of wastewater with a BOD level 300 ppm and a BOD removal efficiency of 90% removes about 2250 lb/day of BOD. It produces about 1350 lb/day of MLVSS in 10,000 lb/day of sludge cake.

In this system a 35-hp compressor injects 1200-scfm air in the 48,000-cu-ft aeration tank (Nogaj 1972). This air contains 1250 lb/hr of oxygen or 13.33 lb of oxygen per

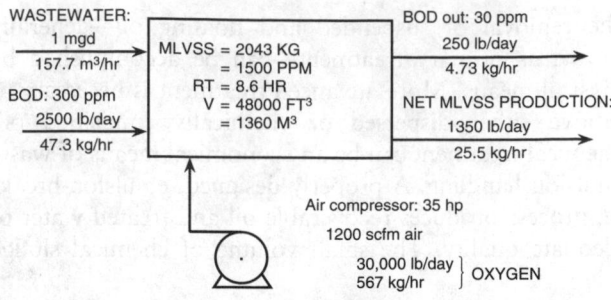

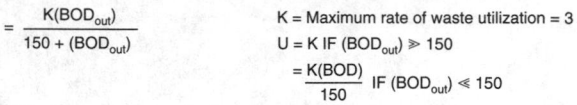

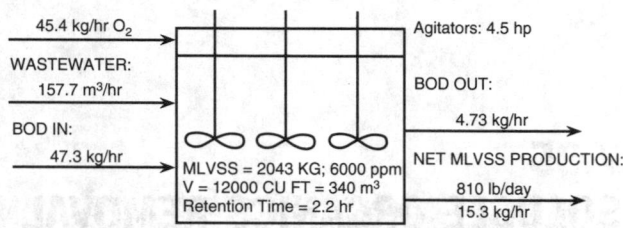

FIG. 7.35.2 Comparison of air and oxygen in activated-sludge wastewater treatment.

lb of BOD removed. The low oxygen efficiency of 7.5% shows that mixing rather than dissolving oxygen is the determining factor of air sparging. Good mixing is essential for keeping bacteria in suspension, breaking up bacteria flocs, and promoting maximum contact of bacteria with the organics they use as a food source. A well-mixed aeration basin prevents raw wastewater from flowing directly from inlet to outlet. Sludge activation works better with pure commercial oxygen.

Pure oxygen accelerates the oxidative processes by at least a factor of four. Rapid oxygen transfer allows the plant to operate at high sludge concentrations. With oxygen, the sludge is less slimy and settles more readily, as shown in Figure 7.35.1. The SS can be increased threefold to 6000 ppm for the same settling velocity of 7 ft/hr. Figure 7.35.2 shows that at 6000 ppm MLVSS, a 1-mgd plant needs an aeration tank of only 12,000 cu ft. The clarifier can be operated at mass loadings of 45–65 lb/day/sq ft and hydraulic overflow rates of 600–1200 gpd/sq ft. In effect, the plant settler using air and the one using oxygen are the same size in spite of different MLVSS concentrations. For 1-mgd plants, these settlers have a surface of 1100 sq ft, a volume of 16,000 cut ft, and a mean hydraulic residence time of 2.9 hr. Better sludge settling also reduces sedimentation chemical use. Often these savings can pay for the cost of oxygen.

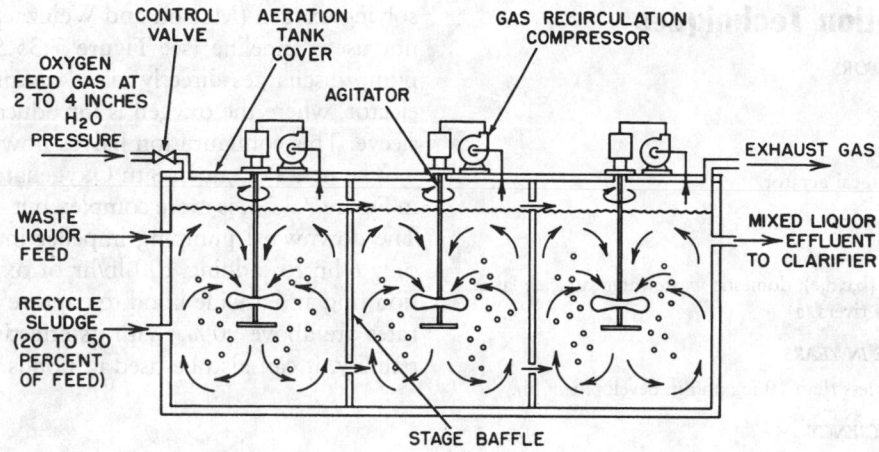

FIG. 7.35.3 Three-stage oxygenation system that uses rotating spargers and recirculation compressors.

The Linde Division of Union Carbide (now independently operated as Praxair Inc.) demonstrated 95% oxygen utilization in the 1968 study of the Batavia, New York sewage works (Albertson et al., 1971; Gross 1976). (In 1980, the process named UNOX was acquired by Lotepro Corp., the U.S. subsidiary of Linde A.G. of Munich, Germany.) The Batavia plant covered its conventional aeration basins to prevent costly oxygen loss (see Figure 7.35.3). Efficient, upward-pumping, slow-speed, low-shear, agitator impellers kept the activated-sludge microbes in suspension. These agitators reduced energy needs for the 1-mgd plant to 4–5 hp, down from the previously estimated 35 hp for air sparging.

Table 7.35.1 shows test results of the UNOX process applied to industrial wastewater streams (Gross 1976). Oxygen is used in more than 250 wastewater treatment plants with an average capacity of 32 mgd.

The four following factors accelerate oxygen use:

1. Tighter legal water and air discharge limits. Tightening the legal discharge limits requires increased bacterial activity, better clarification, and lower excess sludge production. By maintaining optimum metabolic rates, pure oxygen cuts sludge production in half. With pure oxygen, odor problems are greatly reduced or eliminated.
2. New air separation technologies. Noncryogenic air-separation technologies, particularly membrane and pressure swing adsorption (PSA), make lower-cost oxygen available to customers that do not require the highest purity.
3. Better oxygen application techniques. These techniques are discussed later.
4. Wastewater treatment plant capacity shortages. Many plants are looking for effective ways to expand capacity or handle periodic overloads.

TABLE 7.35.1 TEST RESULTS OF UNOX PROCESS APPLIED TO INDUSTRIAL WASTEWATER

Type of Production Facility and Character of Wastewater	Influent BOD (mg/l)	Retention Time (hr)	MLVSS (mg/l)	Food-to-Biomass Ratio (U), lb BOD/day-lb MLVSS	BOD$_R$, %	lb Oxygen Required/lb. BOD
Pesticides—Organic solvents and alcohols plus other biologically resistant organics	530	5.0	6000	0.41	90	2.45
Petrochemical—Low-molecular-weight organic acids and alcohols	330	3.3	4300	0.58	92	1.02
Petrochemical—Low-molecular-weight organic acids	1750	16	8900	0.31	96	0.98
Petrochemical—Glycols, glycol ethers, alcohols, alkanol amines, and acetone	570	3.5	4800	0.82	84	1.12
Food Processing—citric acid	4824	32.0	6200	0.67	98	1.04
Brewery Distillery	1010	6.1	6100	0.7	98	1.0
Pulp & Paper—Kraft pulp mill	291	2.0	5000	0.69	89	1.23

Source: R.W. Gross Jr., 1976, The UNOX process, *Chem. Eng. Progr.* 72, no. 10.

Oxygen Application Techniques

TYPES OF SPECIAL AERATORS

  a) U-tube
  b) Jet
  c) Draft tube air lifts
  d) Self-priming mechanical aerator
  e) Biological discs

APPLICATIONS

Industrial wastewater (b,c,d,e); domestic wastewater (a,b,c,e); lakes and reservoirs (c), and rivers (a)

OPERATING EXPERIENCE IN YEARS

More than 10 (a,b,e); less than 10 (c); and in development (d).

OXYGEN TRANSFER EFFICIENCY

In units of theoretical lb of oxygen per hp-hr of power invested: 0.3 to 3 (a), 2 to 6 (c), 3 to 5 (b), 15 to 17 (d). In units of lb of BOD removed per hp-hr of power invested: 2 to 8 (e), 15 (d).

In 1971, Air Reduction (now part of the British Oxygen Company) introduced a pipeline reactor as an efficient means of contacting wastewater with activated sludge and oxygen (Hover, Huibers, and Serkanic 1971). Turbulent flow causes effective mixing of wastewater and sludge with oxygen, especially in the froth flow regime (Baker 1958).

Praxair's affiliate Societa Italiana Acetilene and Derivati developed the MIXFLO system (see Figure 7.35.4). This 2- to 4-atm, side stream pipeline pumping system dissolves up to 90% of the injected oxygen (Storms 1993); 60% in the pipeline and 30% in the bulk liquid after exiting the dispersion ejector. This system exchanges ease of operation for a high-power input (38 hp at an oxygen-use rate of 100 lb/hr). The system is used in over 150 activated-sludge plants.

Liquid Air Corporation (L'Air Liquide) supplies the VENTOXAL system and the AIROXAL process for dis-solving oxygen (Matson and Weinzaepfel 1992) that does not use a pipeline (see Figure 7.35.5). The submersible pump discharges directly via a venturi mixer or primary ejector, where the oxygen is introduced into a distributor sleeve. This configuration halves power input, to 20 hp.

The new Praxair In-Situ Oxygenator system, as shown in Figure 7.35.6, is more complex but uses even less power. The downward pumping impeller in the draft tube uses only 6 hp for adding 100 lb/hr of oxygen. It has a flotation ring and a wide hood to capture off gas. Oxygen use rates are above 90%. With its effective solids mixing capability, it can also be used in ponds.

Use of Oxygen in Bioremediation

The MIXFLO system was successfully used for the $54-million bioremediation of the petroleum and petrochemical waste lagoon at the French Limited Superfund Site in Crosby, Texas (Bergman, Greene, and Davis 1992; *French Ltd.* 1992; Sloan 1987). In about five months, the chlorinated organics and mono- and polyaromatics were reduced from 1000+ to 100− ppm. The average rate of this aerobic treatment was 0.36 kg BOD/m^3/day, which is about half the rate of air or aerobic municipal sewage digestion. This rate is remarkable, considering contaminant character. Benzopyrene was reduced from 2000 ppm at day 100 to 20 ppm at day 200, benzene from 800 to 20, vinylchloride from 300 to 25, and total PCBs from 80 to 9. In November 1992 at day 300, all levels were below 8 ppm.

Project costs included $1.3 million for 18,000 tn of liquid oxygen delivered to the site at $72/tn; $0.7 million for pumping power at $0.05/kWh, and $0.8 million for other power costs. Onsite incineration would have cost $125 million.

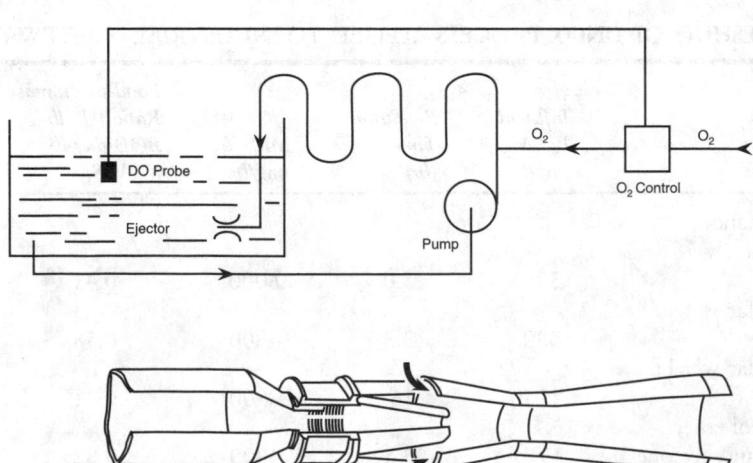

FIG. 7.35.4 Schematic diagram of the MIXFLO oxygenation system and ejector. (Reprinted, with permission, from G.E. Storms, 1993, *Oxygen dissolution technologies for biotreatment applications*, Tarrytown, N.Y.: Praxair Inc.)

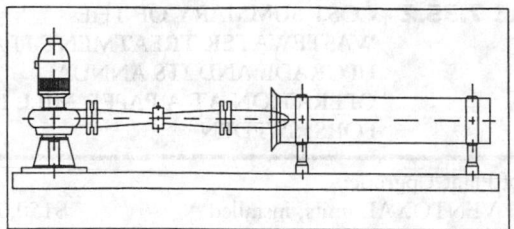

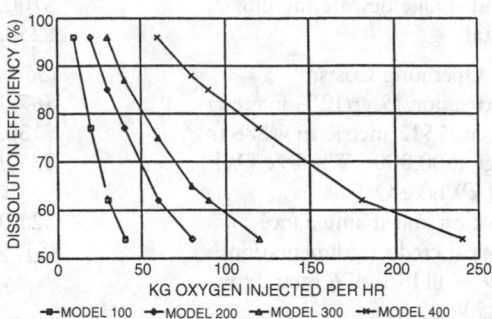

FIG. 7.35.5 Schematic diagram of the VENTOXAL system and its oxygenation efficiency. (Reprinted, with permission, from M.D. Matson and B. Weinzaepfel, 1992, Use of pure oxygen for overloaded wastewater treatment plants: The Airoxal Process, *Water Environment Federation, 65th Annual Conference and Expo, New Orleans, LA, Sept. 25, 1992*, paper AC92-028-007.)

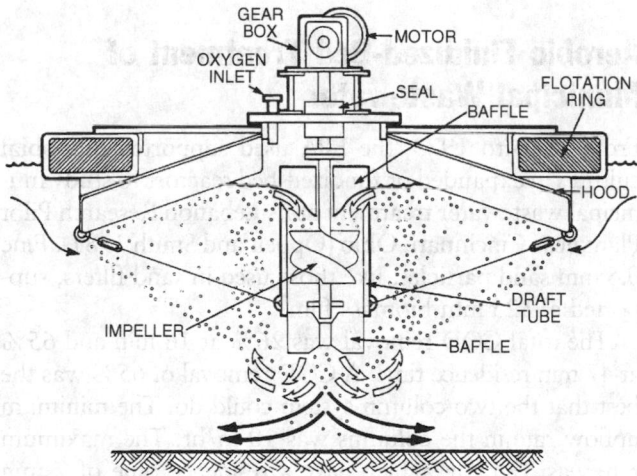

FIG. 7.35.6 The Praxiar Inc. In-Situ Oxygenator (Reprinted, with permission, from G.E. Storms, 1993, *Oxygen dissolution technologies for biotreatment applications*, Tarrytown, N.Y.: Praxair Inc.)

Use of Oxygen in Handling Industrial Wastewater

The VENTOXAL system effectively handles overload conditions, as illustrated by a case study of a paper mill wastewater treatment plant in Fors, Sweden. This paper mill

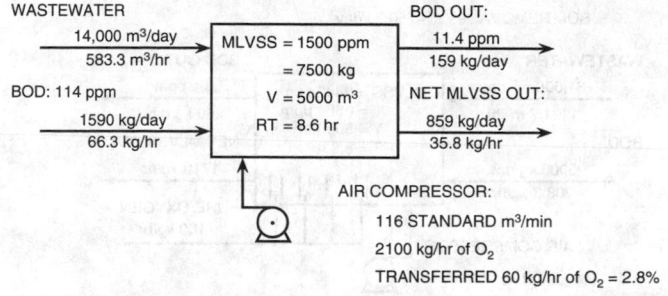

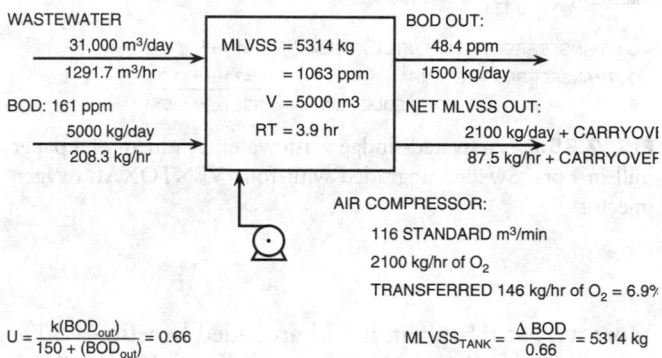

FIG. 7.35.7 Activated-sludge waste wastewater treatment at a paper mill in Fors, Sweden.

produces newsprint and linerboard from thermo-mechanical pulp (TMP) and chemico-thermo-mechanical paper (CTMP). The periodic production of CTMP caused an effluent overload.

The activated-sludge plant had been designed for 14,000 m^3/day of clarified TMP effluent with 1590 kg/day BOD. CTMP production raised this BOD load to 5000 kg/day. The schematic of the treatment plant in Figure 7.35.7 shows that the increased BOD load reduced its removal efficiency from 90 to 70%. More than twice the BOD was removed at the overload conditions than at the design conditions; 3500 versus 1431 kg/day. The mean hydraulic residence time fell from 8.6 to 3.9 hr, and the MLVSS level fell from 1500 to 1063 ppm. The effluent settler was less effective in handling the increased sludge load.

Four 300 m^3/hr VENTOXAL systems were capable of adding 180 kg/hr of oxygen. This addition doubled the aeration capacity, raising the BOD removal efficiency to 95%. Figure 7.35.8 shows that the MLVSS concentration in the aeration tank now reached almost 7000 ppm. As previously discussed and shown in Figure 7.35.1, the use of pure oxygen enhances sludge settling velocities.

PLANT UPGRADE:

PLANT EQUIPPED WITH FOUR VENTOXAL 300 m³/hr OXYGEN INJECTORS

BOD REMOVAL 95%: 4750 kg/day

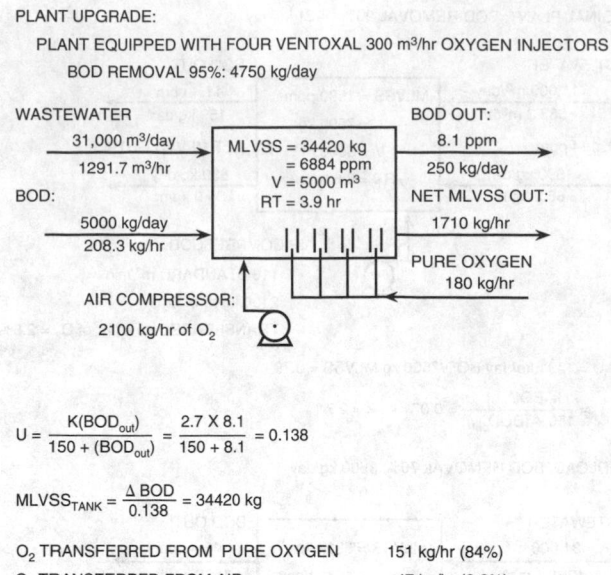

$$U = \frac{K(BOD_{out})}{150 + (BOD_{out})} = \frac{2.7 \times 8.1}{150 + 8.1} = 0.138$$

$$MLVSS_{TANK} = \frac{\Delta BOD}{0.138} = 34420 \text{ kg}$$

O₂ TRANSFERRED FROM PURE OXYGEN	151 kg/hr (84%)
O₂ TRANSFERRED FROM AIR	47 kg/hr (2.2%)
BOD REDUCTION	198 kg/hr (95%)

FIG. 7.35.8 Activated-sludge wastewater treatment of a paper mill in Fors, Sweden upgraded with four VENTOXAL oxygen injectors.

TABLE 7.35.2 COST SUMMARY OF THE WASTEWATER TREATMENT PLANT UPGRADE AND ITS ANNUAL OPERATION AT A PAPER MILL IN FORS, SWEDEN

Cost of Plant Upgrade:	
Four VENTOXAL units, installed (including control panels)	$150,000
Second sludge dewatering unit	$100,000
Total	$250,000
Annual Operating Costs:	(350 days/yr)
Amortization (5 yr/12% interest)	$67,000
Oxygen (1512 metric tn @$86/tn)	$130,000
Energy (500,000 kWh @7¢/kWh)	$35,000
(0.33 kWh/kg O₂)	
Operation and maintenance	$25,000
Chemical credit (sedimentation)	($119,000)
(1589 − 1113 = 476 metric ton	
@ $250/tn)	
Total	$138,000

Extra BOD Removal: 3319 kg/day;
  1162 metric tn/yr @ $120/metric tn BOD removed

Source: Matson and Weinzaepfel, 1992.

Moreover, at this plant, it had an added benefit—a 30% reduction of sedimentation chemicals from 4540 to 3180 kg/day. These savings covered the oxygen costs, as shown in Table 7.35.2. This table summarizes the capital and operating costs of the plant upgrade.

Biological Fluidized-Bed Wastewater Treatment

The biological, fluidized-bed, wastewater treatment process is a new adaptation of the fixed-film, biological reactor—the trickling filter. In this process, the adsorbent particles are small and kept in suspension by the circulating effluent (see Figure 7.35.9). This improved process is considered the most significant development in wastewater treatment in the last fifty years.

The mixed microbial cultures associated with wastewater treatment have excellent adhesion characteristics. They form continuous layers of immobilized biomass on any support material, especially when food (BOD) is limiting. This biofilm is a dense matrix of bacteria and polysaccharides, similar to activated-sludge but less sensitive to perturbations in substrate conditions, toxic compounds, and the food-to-biomass ratio (U).

The biofilm promotes high solids retention. Therefore, the biomass concentration in the fluidized-bed reactor is typically 15,000 ppm, ten times greater than in the standard activated-sludge process. This concentration allows the fluidized bed reactor to operate at a high contaminant

removal efficiency. In the standard activated-sludge process, a low U leads to poor floc formation.

Aerobic Fluidized-Bed Treatment of Municipal Wastewater

From 1970 to 1973, the EPA used supported microbial cultures in expanded or fluidized-bed reactors to study municipal wastewater treatment at its Lebanon Research Pilot Plant near Cincinnati, Ohio (Oppelt and Smith 1981). Fine 0.5 mm sand particles, like those used in sand filters, supported a 0.25 mm biomass film.

The total COD removal was 26% at 16 min and 65% at 47 min residence time. A COD removal of 65% was the best that the two-column system could do. The minimum upflow rate in the columns was 10 m/hr. The maximum rate was 39 m/hr, which gave a residence time of 7 min per two-column pass. Sand loss was a major problem. It occurred mainly from the transport of oxygen bubbles.

The EPA obtained better results with eight reactors in series. They installed bubble trap devices at the top of each column to collapse the bubbles and allow the sand to drop back. Oxygen was added in fine bubble diffusers at the base of the downleg between the reactor stages.

In a once-though operation with an empty-bed retention time of 44 min, COD removal increased to 75%, and BOD removal increased to 89% (effluent COD 48.8 ppm from 196.3 and effluent BOD 13 from about 118). The high biomass concentration of MLVSS = 15,000 ppm increased the COD treatment efficiency from 3.0 kg/m³/day

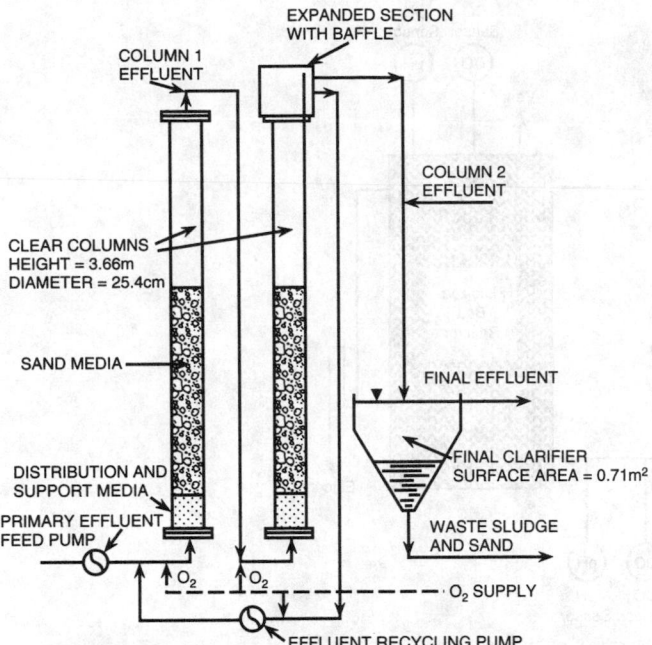

FIG. 7.35.9 Schematic diagram of two fluidized-bed reactors in series for biological wastewater treatment. (Reprinted, with permission, from E.T. Oppelt and J.M. Smith, 1981, U.S. Environmental Protection Agency research and current thinking on fluidised-bed biological treatment, in *Biological fluidised bed treatment of water and waste water,* edited by P.F. Cooper and B. Atkinson, Chichester: Ellis Horwood Ltd., Publishers.)

to 4.8 kg/m³/day for a standard oxygen-activated sludge process. The low food-to-biomass mass ratio of U = 0.3 kg COD/kg VSS/day achieved the low effluent BOD of 13.

The effluent BOD was a function of U. At U = 1.1, the effluent BOD was 30. Ecolotrol Inc. of Westbury, New York used this value to design a large 10-mgd plant (1577 m³/hr) with five, parallel, fluidized-bed reactors at the Bay Park Wastewater Treatment Plant in Nassau County, New York (Jeris, Owens, and Flood 1981). These five reactors with a combined volume of 1000 m³ operated at an up-flow rate of 37 m/hr and a recycling ratio of 4.6. A PSA oxygen generator provided 6 tn/day of oxygen (5443 kg/day). The plant had various operating difficulties; controlling microbial growth was the most problematic. The plant was shut down in 1991.

Dorr–Oliver, a Connecticut engineering company, developed the Oxitron system for fluidized-bed wastewater treatment on the basis of the Ecolotrol design (Sutton, Shieh, and Kos 1981). Dorr–Oliver added an influent distributor and a proprietary influent oxygenator with a 20-sec hydraulic retention time to allow the use of at least 50 mg/l of oxygen per pass. They controlled the biofilm thickness on the fluidized sand by keeping the bed expansion at a specific level. The particles with more biofilm are lighter and raise the bed level. When that situation happens, this system pumps them with a rubber-lined pump to a hydroclone and sand washer and returns sand to the reactor.

In the pilot plant shown in Figure 7.35.10, tests were conducted with wastewater having a median BOD of 67 ppm and a median COD of 175 ppm. The best results, 78% BOD removal to 15 ppm, were obtained at an up-flow rate of 20.3 m/hr, a hydraulic retention time of 37 min, and a food-to-biomass ratio of U = 0.16 kg BOD/kg VSS/day.

Dorr–Oliver designed full-scale plants for municipal wastewater treatment at Haywood, California (70,000 m³/day) and Sherville, Indiana (35,000 m³/day). Based on operating costs, the OXITRON system is favored for

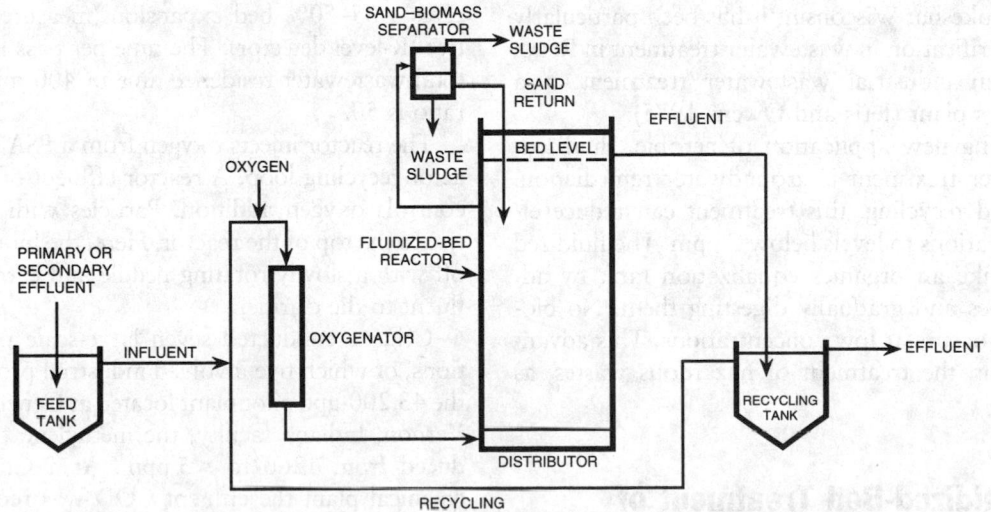

FIG. 7.35.10 Schematic diagram of the Dorr–Oliver Oxitron System pilot plant in Orillia, Ontario. (Reprinted, with permission, from P.M. Sutton, W.K. Shieh, and P. Kos, 1981, Dorr–Oliver's Oxitron system fluidised-bed water and wastewater treatment process, in *Biological fluidised-bed treatment of water and wastewater,* edited by P.F. Cooper and B. Atkinson, Chichester: Ellis Horwood Ltd., Publishers.)

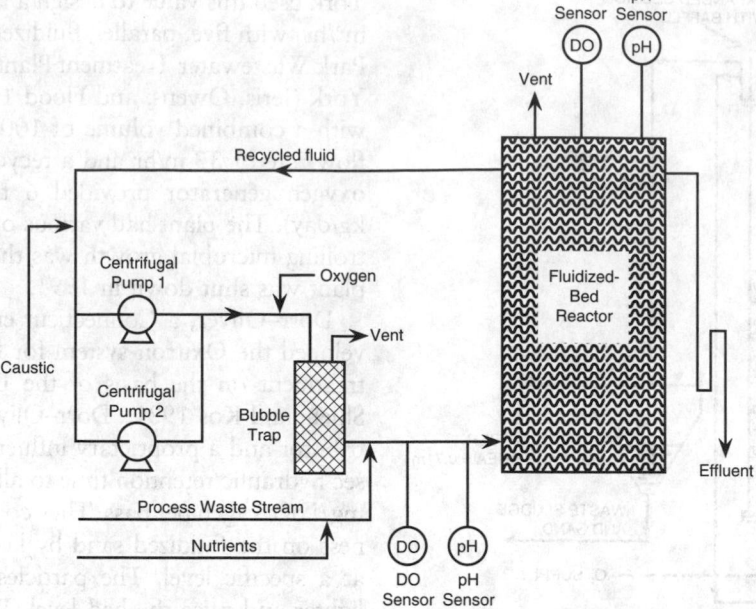

FIG. 7.35.11 Schematic diagram of the Celgene biotreatment process. (Reprinted, with permission, from W. Gruber, 1993, Celgene's biotreatment technology: Destroying organic compounds with an in-line, on-site treatment system, *EI Digest* [November].)

plants with flows above 20,000 m³/day. At this size, the operating costs, including the present value of the facility discounted at 7% for 20 yr, is $4.15 million. These costs do not include sludge disposal; they are based on a BOD reduction of 90% from 200 to 20 ppm and a SS reduction from 140 to 30 ppm. The Haywood, California plant experienced operating problems similar to the Nassau County, New York plant. This plant was also shut down.

The OXITRON system design is now available from Envirex in Waukesha, Wisconsin. It has been particularly useful for denitrification in wastewater treatment in Reno, Nevada and in industrial wastewater treatment at a General Motors plant (Jeris and Owens 1975).

An interesting new application of aerobic, fluidized-bed, wastewater treatment is groundwater remediation. With continued recycling, this treatment can reduce effluent concentrations to levels below 5 ppm. The fluidized biomass acts like an organics equalization tank by adsorbing organics and gradually digesting them. No biomass washout occurs at low concentrations. This advantage is useful in the treatment of hazardous wastes, as described next.

Aerobic Fluidized-Bed Treatment of Industrial Wastewater

Recently, Celgene Corporation introduced the fluidized-bed reactor with its *in-process, biotreatment system*

(Gruber 1993; Sommerfield and Locheed 1992). Celgene's expertise was in selecting and applying the most suitable microbes for metabolizing specific organics, e.g., trichloroethylene, methylene chloride, other organic chlorides, ketones, and aromatics on EPA's list of seventeen hazardous materials targeted for 50% industrial emission reduction by 1995.

Figure 7.35.11 shows the Celgene fluidized-bed reactor. This reactor suspends carbon particles with the immobilized biomass in an upflow of 29 m/hr. This flow gives a 35–50% bed expansion (measured with a reflective IR-level detector). The time per pass is 8 min. With a total wastewater residence time of 400 min, the recycling ratio is 50.

The reactor injects oxygen from a PSA unit into the reactor recycling loop. A reactor effluent of DO = 35 ppm controls oxygen addition. Particles with excess biomass rise to the top of the reactor. Here, the biomass is knocked off with a slowly rotating peddle and carried with the effluent to the clarifier.

Celgene conducted seven large-scale pilot demonstrations, of which five involved industrial process streams. In the 43,200-gpd pilot plant located at General Electric's Mt. Vernon, Indiana facility, the methylene chloride was reduced from 1260 to <5 ppm. At a Gulf Coast petrochemical plant the effluent COD was reduced from 210 to 40 ppm with 0.3 ppm phenols, <5 ppm aromatics, and <10 ppm SS. Treatment costs are about $10 per 1000 gal. The commercialization of the process has been taken over by Sybron Chemicals.

Manville and Louisiana State University used a fluidized-bed system at Ciba–Geigy's St. Gabriel plant site to lower the sodium chloroacetate level in a 3–4% saline waste stream from 6000 to 10 ppm (Attaway et al. 1988). They used a 0.25-inch, diatomaceous-earth carrier with a pore structure optimized for microbe immobilization in two bioreactors in series with a volume of 141.3 gal each. At a throughput of 0.25 gpm, they observed biological activity in both reactors. The effluent of the first reactor had a sodium chloroacetate level of 2400 ppm.

Anaerobic Wastewater Treatment with Attached Microbial Films

Twenty years ago, no evidence existed to suggest that dilute wastewater could be treated anaerobically at an ambient temperature. Prodded by the 1973 energy crisis, Jewell (1981) and co-workers converted a fluidized-bed reactor for anaerobic methane fermentation and studied municipal sewage treatment (Jewell, Switzenbaum, and Morris 1979).

This study led to the discovery of an efficient process capable of operating at high biomass concentrations of 30,000 ppm. At retention times of <30 min, this anaerobic process removed most biodegradable organics from municipal wastewater, reducing the COD to <40 ppm and the SS to <5 ppm. The volumetric density of the anaerobic film exceeded 100 kg VSS/m³. The net yield of the biomass produced was Y = 0.1 kg VSS/kg COD. This density gave a solids residence time (SRT) that was three to eight times that of the aerobic process. Ultimately, the biomass concentration of the anaerobic attached-film-expanded bed (AFEB) could approach 100,000 ppm. This concentration raises the COD removal rates to >50 kg/m³/day and still maintains an SRT >30 days.

An aerobic AFEB cannot operate at these high-rate conditions without washout. However, an aerobic AFEB can ultimately produce a lower effluent COD and nitrify ammonia. A series treatment with an anaerobic AFEB followed by an aerobic AFEB, with each unit having a 15-min retention time, resulted in a superb effluent quality with a COD = 10, an SS = 1, and a turbidity of 2.

Figure 7.35.12 contrasts the aerobic and anaerobic AFEB processes. Up to an organic loading of 2 kg/m³/day, the two processes are about equal. At higher loading rates, the anaerobic AFEB process produces a better effluent COD.

—*Derk T.A. Huibers*

References

Albertson, J.G., J.R. McWirter, E.K. Robinson, and N.P. Valdieck. 1971. *Investigation of the use of high purity oxygen aeration in the con-*

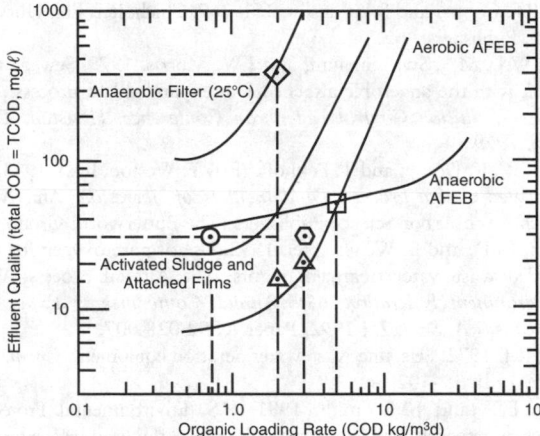

FIG. 7.35.12 Relationship of COD removal efficiency and organic loading capacity for standard and AFEB aerobic and anaerobic treatment processes.

ventional activated sludge process. FWQA Department of the Interior program no. 17050 DNW, Contract no. 14-12-465 (May) and Contract no. 14-12-867 (September).

Attaway, H., D.L. Eaton, T. Dickenson, and R.J. Portier. 1988. Waste stream detoxified with immobilized microbe system. *Pollution Engineering* (September): 106–108.

Baker, O. 1958. Multiphase flow in pipelines. *The Oil and Gas Journal* (10 November): 156–167.

Bergman Jr., T.J., J.M. Greene, and T.R. Davis. 1992. An in-situ slurry-phase bioremediation case with emphasis on selection and design of a pure oxygen dissolution system. *Air & Waste Management Association and EPA Risk Reduction Laboratory, In-Situ Treatment of Contaminated Soil and Water Symposium, Cincinnati, Ohio Feb. 4–6, 1992.*

Fouhy, K. 1992. Biotowers treat highly contaminated streams. *Chem. Eng.* (December): 101–102.

French Ltd.: A successful approach to bioremediation. 1992. *Biotreatment News* (October, November, and December).

Gross Jr., R.W. 1976. The Unox Process. *Chem. Eng. Progr.* 72, no. 10: 51–56.

Gruber, W. 1993. Celgene's biotreatment technology: Destroying organic compounds with an in-line, on-site treatment system. *EI Digest* (November): 17–20.

Hover, H.K., D.T.A. Huibers, and L.J. Serkanic Jr. 1971. *Treatment of secondary sewage.* U.S. Patent 3,607,735 (21 September).

Jeris, J.S., R.W. Owens, and F. Flood. 1981. Secondary treatment of municipal wastewater with fluidized-bed technology. In *Biological fluidised bed treatment of water and wastewater,* edited by P.F. Cooper and B. Atkinson, 112–120. Chichester: Ellis Horwood Ltd., Publishers.

Jeris, J.S., and R.W. Owens. 1975. Pilot-scale high-rate biological denitrification. *J. Water Pollution Control Fed.* 47:2043.

Jewell, W.J. 1981. Development of the attached film expanded bed (AFEB) process for aerobic and anaerobic waste treatment. In *Biological fluidised bed treatment of water and wastewater,* edited by by P.F. Cooper and B. Atkinson, 251–267. Chichester: Ellis Horwood Ltd., Publishers.

Jewell, W.J., M.S. Switzenbaum, and J.W. Morris. 1979. Sewage treatment with the anaerobic attached film expanded bed process. *52nd Water Pollution Control Federation Conference, Houston, Texas, Oct. 1979.*

Junkins, R., K. Deeny, and T. Eckhoff. (Roy F. Weston, Inc.). 1983. *The activated sludge process: Fundamentals of operation.* Ann Arbor, Mich.: Ann Arbor Science Publishers (The Butterworth Group).

Matson, M.D. and B. Weinzaepfel. 1992. Use of pure oxygen for overloaded wastewater treatment plants: The Airoxal process. *Water Environment Federation, 65th Annual Conference & Expo. New Orleans, LA, Sept. 25, 1992.* Paper AC92-028-007.

Nogaj, R.J. 1972. Selecting wastewater aeration equipment. *Chem. Eng.* (17 April): 95–102.

Oppelt, E.T. and J.M. Smith. 1981. U.S. Environmental Protection Agency research and current thinking on fluidised-bed biological treatment. In *Biological fluidised bed treatment of water and waste water,* edited by P.F. Cooper and B. Atkinson, 165–189. Chichester: Ellis Horwood Ltd., Publishers.

Sloan, R. 1987. Bioremediation demonstrated at hazardous waste site. *Oil & Gas Journal* (14 September): 61–66.

Sommerfield, T., and T. Locheed. 1992. Hungry microorganisms devour methylene chloride: Fluidized bed bioreactor reduces emissions to less than 100 ppb, keeps operating costs down. *Chem. Proc.* 55, no. 3: 41–42.

Storms, G.E. 1993. Oxygen dissolution technologies for biotreatment applications. Tarrytown, N.Y.: Praxair Inc.

Sutton, P.M., W.K. Shieh, and P. Kos. 1981. Dorr–Oliver's Oxitron system fluidised-bed water and wastewater treatment process. In *Biological fluidised bed treatment of water and wastewater,* edited by P.F. Cooper and B. Atkinson, 285–300. Chichester: Ellis Horwood Ltd., Publishers.

Tchobanoglous, G., and F.L. Burton (Metcalf & Eddy Inc.). 1991. *Wastewater engineering: Treatment, disposal and reuse.* 3d ed. New York: McGraw-Hill.

Vesilind, P.A., and J.J. Peirce. 1982. *Environmental engineering.* Boston: Butterworth.

7.36
INORGANIC SALT REMOVAL BY ION EXCHANGE

PARTIAL LIST OF SUPPLIERS

Advanced Separation Technologies; Aquatech International Company; Arrowhead Industrial Water; Cochrane Division of Crane Corp.; Culligan International; Degramont Infilco; Dow Chemical; Gregg Water Conditioning Inc.; Craver Water Conditioning; HOH Systems; Hungerford and Terry; Illinois Water Treatment; Ionics Inc.; Ion Pure; Permutit; Purolite; Rohm and Haas; Sybron Chemicals Inc.; Vaponics; U.S. Filter Company.

The ion-exchange reaction is the interchange of ions between a solid phase and a liquid surrounding the solid. Initially, ion exchange was confined to surface reactions, but these reactions were gradually replaced by gel-type structures where exchange sites were available throughout the particle. Figure 7.36.1 shows this process graphically.

Ion-Exchange Reaction

Exchange sites exhibit an affinity for certain ions over others. This phenomenon is helpful in removing objectionable ionic materials from process streams. Environmental engineers have studied the affinity relationships and have identified certain simple rules. First, ions with multiple charges are held more strongly than those of lower charges. Ions with the same charge are held according to their atomic weight with heavier elements held more strongly. The affinity relationships for cation and anion exchangers are as follows:

Cation Exchangers:
- Monovalent—$Cs > Rb > K > Na > Li$
- Divalent—$Ra > Ba > Sr > Ca > Mg \gg Na$

Anion Exchangers:
- Monovalent—$I > Br > NO_3 > Cl > HCO_3 > F > OH$
- Divalent—$CrO_4 > SO_4 > CO_3 > HPO_4$

The affinity relationship can also be expressed with the following equilibrium (selectivity) equations based on the reversibility of ion-exchange reactions and the law of mass action:

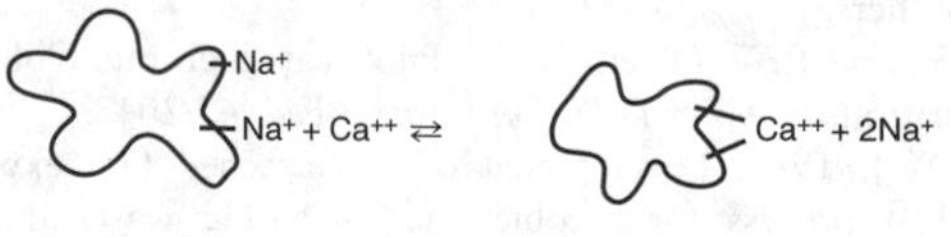

$$RNa^+ + H^+ \rightleftharpoons RH + Na^+ \qquad 7.36(1)$$

$$K_H^{Na} = \frac{[RH][Na]}{[RNa][H]} = \frac{y(1-x)}{x(1-y)} \qquad 7.36(2)$$

FIG. 7.36.1 The ion-exchange reaction.

The following equations are for the divalent–monovalent reactions:

$$2RNa + Ca^{++} \rightleftharpoons RCa + 2Na^+ \qquad 7.36(3)$$

$$K_{Na}^{Ca} = \frac{[RCa]\,[Na^+]^2}{[RNa]^2\,[Ca^{++}]} \qquad 7.36(4)$$

$$\frac{KQ}{C_o} = \frac{y(1-x)^2}{x(1-y)^2} \qquad 7.36(5)$$

In these equations, the brackets represent the ion concentration in the resin and the liquid phase. The y, x notation expresses the reactions as equivalent ratios. For the divalent–monovalent reaction, Q is the resin capacity, and C_o is the total concentration of the electrolyte in solution. Plotting these equations, usually as y versus x plots, shows the exchange processes that occur in the exchange zone or a batch contactor.

Structural Characteristics

Modern ion-exchange materials are prepared from synthetic polymers such as styrene-divinyl-benzene copolymers that have been sulfonated to form strongly acidic cation exchangers or aminated to form strongly basic anion exchangers. Weakly basic anion exchangers are similar to the strong base except for the choice of amines.

Weakly acidic cation exchangers are usually prepared from cross-linked acrylic copolymers.

Figure 7.36.2 shows the reactions involved in the preparation of anion exchange resins. Figure 7.36.3 shows the reactions for cation exchange resins. Figure 7.36.4 shows the structure of a typical chelation resin.

Ion-Exchange Characteristics

The resins prepared with the proper synthetic procedures have characteristics related to the percentage cross-linkage in the copolymers. They also have particle size ranges consistent with the 16–50 U.S. mesh size with a uniformity coefficient not exceeding 1.5. (The uniformity coefficient is the 40% size divided by the 90% size.) The effective size (90% size) should be not less than 0.35 mm as measured from a probability plot of size as a function of the accumulative percentage.

The values summarized in Table 7.36.1 are averages. The total number of functional groups per unit weight or volume of resins determines the exchange capacity, whereas the types of functional groups influence ion selectivity and equilibrium. Many specifications deal with these resins, including the whole-bead content, bead strength, attrition values, special sizes, effluent quality, and

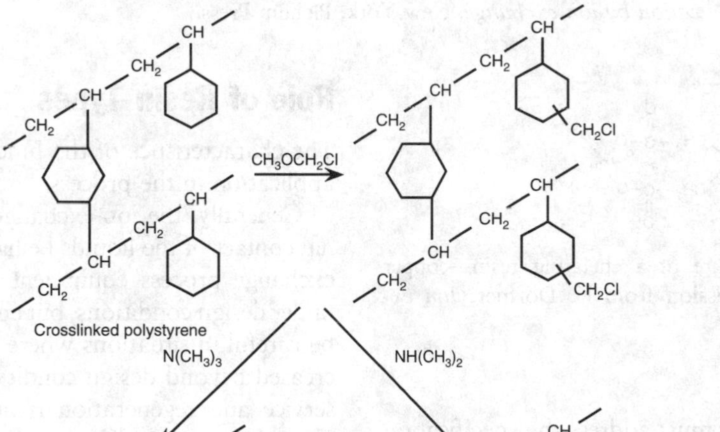

FIG. 7.36.2 Preparation of anion exchange resins. (Reprinted, with permission, from T.V. Arden, 1968, *Water purification by ion exchange,* 22, New York: Plenum Press.

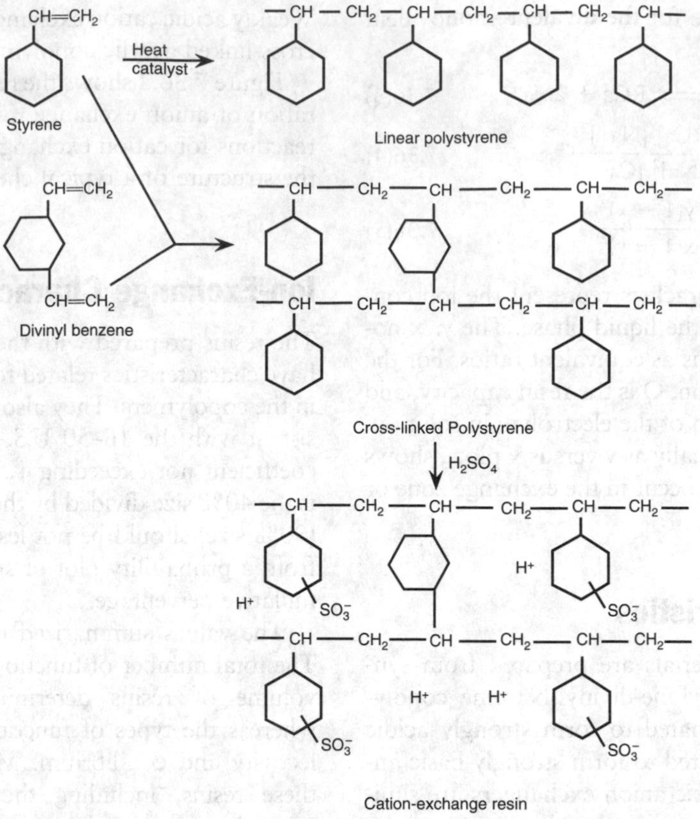

FIG. 7.36.3 Preparation of sulphonic-acid, cation-exchange resin. (Reprinted, with permission, from T.V. Arden, 1968, *Water purification by ion exchange,* New York: Plenum Press.)

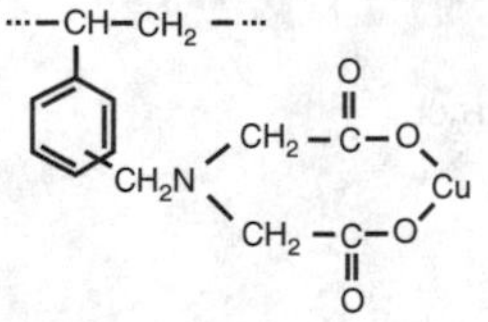

FIG. 7.36.4 Typical structure of a chelating resin—copper form. (Reprinted, with permission from K. Dorfner, *Ion exchange,* 38.)

porosity. The resin supplier must address the specification requirements for various conditions.

Process Applications

Environmental engineers base their selection of the proper process for a water treatment on criteria mentioned previously. Their flow sheet is based on resin properties. Resins are used in the following processes:

- Softening
- Dealkalization
- Desilicizing
- Organic scavenging
- Deionization
- Metal waste treatment

Role of Resin Types

The characteristics of the functional group determine the application in the process.

Generally, the ion-exchange process involves a columnal contact of the liquids being treated. The design of ion-exchange process equipment provides good distribution under design conditions, but environmental engineers must be careful in situations where the flow is decreased or increased beyond design conditions. The flow direction for service and regeneration is an important consideration. Two designs are offered: cocurrent and countercurrent. Generally, cocurrent is practiced when service and regenerant flows are in the same direction, usually downward.

Countercurrent normally services downflow and regeneration upflow. For example, the performance of a two-bed system is closely related to the way the cation exchange resin is regenerated. Alkalinity and sodium control are also closely associated with leakage. The advantages and disadvantages of these process steps can be summarized as follows:

STRONG-ACID-CATION-EXCHANGE RESIN—Removes all cations regardless of with which anion they are associated. This resin has moderate capacity and requires a strong acid regenerant such as hydrochloric or sulfuric acid.

TABLE 7.36.1 ION-EXCHANGE RESIN CHARACTERISTICS

Property	Strong Acid	Strong Base Type 1	Weak Base	Weak Acid
Water Retention, %	44–48	48–53	45–50	46–52
Capacity, meq/g dry	4.25	4.0	3.8	11
meq/ml	2.2	1.5	1.6	3.5–4.0
True Density, gm/cc	1.26–1.30	1.06–1.10	1.15	1.22–1.26
Apparent Density, lb/ft^3	50–52	44–46	44–47	48–50
Particle Range, U.S. Mesh	16–50	16–50	16–40	16–40
Effective Size, mm	0.45–0.60	0.4–0.48	0.36–0.48	0.35–0.55
Swelling, %	5	10	25	50
Shipping Weight, lb	52	45	45	48

WEAK-ACID-CATION-EXCHANGE RESIN—Removes only cations associated with alkalinity, such as bicarbonate, carbonate, or hydroxide. Any cations associated with chloride and sulfate are not removed.

STRONG-BASE-ANION-EXCHANGE RESIN—Removes all dissociated anions such as bicarbonate, sulfate, chloride, and silica. This resin exhibits low to moderate capacity and must be regenerated with a strong alkali such as sodium hydroxide. This resin can remove CO_2 and silica from water.

WEAK-BASE-ANION-EXCHANGE RESIN—Removes only anions associated with the hydrogen ion that are strong acid formers such as sulfate, chloride, or nitrate. Any anions associated with cations other than hydrogen pass through unaffected.

COCURRENT

Figure 7.36.5 shows a typical cocurrent operation for a strong-acid-cation-exchange bed operated in the acid cycle. In this operation, there is a residual band of unregenerated resin at the bottom of the bed. This band is stripped by the acidity, and a small amount of monovalent cations in the bottom band strips as shown in Figure 7.36.6. The ratio in this figure refers to the influent concentration of the ion of least affinity divided by the influent concentration.

Cocurrent operation has a fixed bed, so that particle motion is unlikely. Once the bed is classified, it stays in place, and the facility can backwash during each cycle without disturbing flow patterns.

COUNTERCURRENT

Figure 7.36.7 shows the location of the bands where regenerant passes upward, and influent water passes downward. In this operation, the residual band is at the top, and stripping of the band does not occur. Leakage depends on the end point selected. However, the bed must be held in place without mixing so that a clean band exists in the bottom of the bed. This clean band is usually achieved ei-

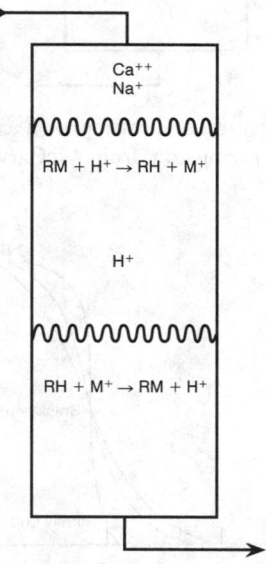

FIG. 7.36.5 Cation exchange with cocurrent regeneration. (Reprinted, with permission, from Frank McGarvey. *Introduction to industrial ion exchange,* Sybron Chemicals Inc.)

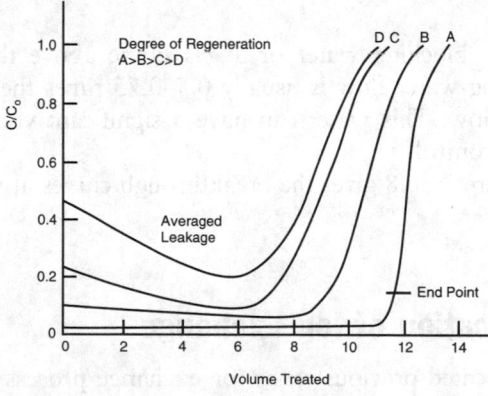

FIG. 7.36.6 Exhaustion curves for cation exchange with cocurrent operation—sodium chloride influent. (Reprinted, with permission, from McGarvey.)

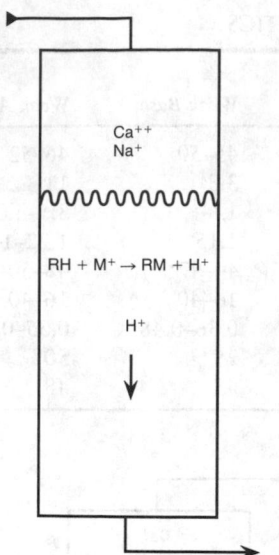

FIG. 7.36.7 Countercurrent operation of a cation-exchange resin. (Reprinted, with permission, from McGarvey.)

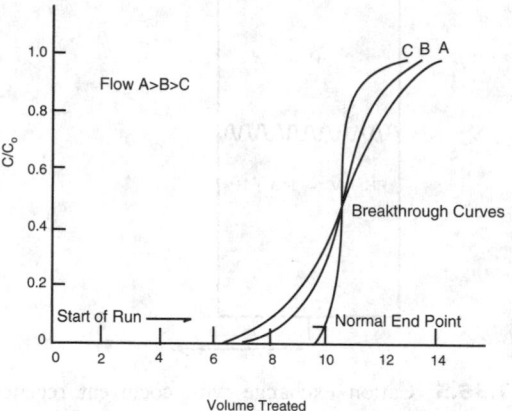

FIG. 7.36.8 Typical breakthrough curves for the countercurrent operation. (Reprinted, with permission, from McGarvey.)

ther by blocking water or a gas dome above the bed. Blocking water flow is usually 0.5–0.75 times the regenerant flow. This water can have a significant volume in waste control.

Figure 7.36.8 gives the breakthrough curves at various flows.

Application of Ion Exchange

As indicated previously, the ion-exchange process is normally practiced in columns. The actual arrangement is influenced by the following factors:

- Influent ion content
- Treated liquor specification
- Process economy based on size
- Waste restrictions
- Concentrations

When the total concentration of impurities exceeds 1000 ppm, other processes including electrodialysis, reverse osmosis, and evaporation are likely to be considered.

Since ion exchange is used to treat wastewater, the technology related to the process is based on the performance of ion exchange with the components in water. Table 7.36.2 shows the substances generally involved. These compounds usually occur in ppm (mg/l) concentration. The common compounds in water can also be a background in a metal waste application.

Understanding the technical units used in the United States is useful. For water treatment, the capacity of the resin is expressed as kilograins of $CaCO_3/ft^3$. A capacity of one chemical equivalent per liter is expressed as 21.8 kgr of $CaCO_3/ft^3$. Thus, this factor is used in the conversion from chemical units to American units.

The exchange capacity of individual ions in feed solutions is expressed as meq/l of resin or in the English system as parts per million of calcium carbonate. When the capacity of a resin is expressed as meq/unit volume, the required resin volume can be calculated from the daily equivalent of ions to be removed and the length of cycle per regeneration. The regenerant consumption is available from the resin manufacturer. The rinse water and spent regenerant volumes are also available from the manufacturer.

The selection of resins used in wastewater treatment depends on the ions in solution. A common arrangement is the two-bed system. In this system, strong acid is followed by a weak base or strong base as shown in Figure 7.36.9. Sodium leakage can be controlled by the regeneration level and countercurrent operation. Wastewater treatment facilities can handle an increased concentration of a metal by using a merry-go-round where three or more units operate in series as shown in Figure 7.36.10.

The water quality is also influenced by the selection of the anion-exchange resin. A weakly basic resin does not pick up weakly acidic substances such as carbon dioxide and silica, and the water is acidic (pH 5–6). On the other hand, a strong base removes weakly acidic materials and splits salts. This effect is shown in the effluent by a pH increase.

With a two-bed system, wastewater treatment facilities can use several combinations of beds to correct problems and make the process more economical including:

TABLE 7.36.2 COMMON SUBSTANCES IN WATER

Cation	Exchange Valence	Anion	Exchange Valence
Calcium	2	Bicarbonate	1
Magnesium	2	Chloride	1
Sodium	1	Sulfate	2
Potassium	1	Nitrate	1
		Silica	1

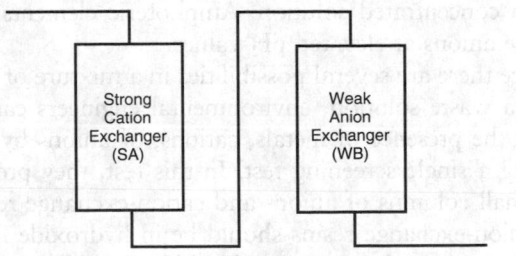

FIG. 7.36.9 Typical two-bed system. (Reprinted, with permission, from McGarvey.)

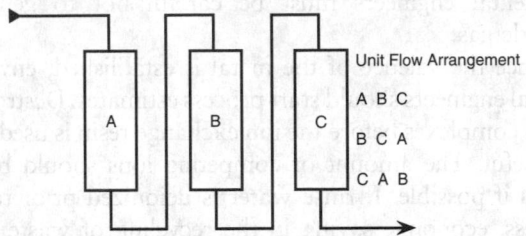

FIG. 7.36.10 Merry-go-round processes. (Reprinted, with permission, from McGarvey.)

- WA-SA-WB-SB
- SA-DEG→SB
- SA-SB-MB
- SA-SB-WA

where:

    WA = Weak-acid-cation exchanger
    SA = Strong-acid-cation exchanger
    WB = Weak-base-anion exchanger
    SB = Strong-base-anion exchanger
    MB = Mixture SA-SB
    DEG = Degasification

Weakly acidic and weakly basic ion exchangers are used since they regenerate efficiently with an acid and base, respectively. Information on ion exchange resin capacity and other performance is available from ion-exchange resin manufacturers. Capacity and leakage values are based on water treatment components. These values are generally available on computer programs. Since waste control is important, most ratings give neutralization estimates so that wastewater treatment facilities can balance acid and alkaline usage to give neutral waste. This consideration is important in plant design.

The engineering parameters listed in Table 7.36.3 show hydraulic and rate values for a typical ion-exchange design.

Waste Metal Processing

The ion-exchange process using ion-exchange resins is well known, and complete process designs have been developed for many processes including softening brackish waters,

TABLE 7.36.3 OPERATIONAL AND DESIGN PARAMETERS (VALUES RELATE TO DEIONIZATION AND SOFTENING DESIGN)

Parameter	Value
Hydraulic loading, gal/min/sq ft	6–15
Resin bed depth, ft	2–7
Expansion allowance, % of bed depth	75–100
Resin life expectancy, Cation exchanger	2–5 yr
Anion exchanger	1–3 yr
Softener	5–10 yr
Regeneration frequency, hr	8–12
Resin capacity, eq/l, Cation exchanger	0.70–1.2
Anion exchanger WB	0.8–1.2
Anion exchanger SB	0.5–1.0
Cation exchanger WA	1–2
Regeneration level,	
lb./ft³ H_2SO_4, Cation exchanger SA	3.6–7.6
lb./ft³ HCl, Cation exchanger SA	2–6
lb./ft³ NaOH, Anion exchanger SB	3–10
lb./ft³ NaOH, Anion exchanger WB	2.5–4
lb./ft³ HCl, Cation exchanger WA	3.0–6.5
Wastewater (rinse and backwash), gal/ft³	30–80
Chemical regenerant consumption, % theory	
Cation and anion SA/SB	200%
Cation and anion WA/WB	100%
Resin Cost, $/ft³, Strong-acid-cation exchanger	$75–100
Strong-base-anion exchanger	$150–200
Special resin	$300–500

condensate polishing, and processing ultrapure water used in the semiconductor industry. Processes for metal recovery have also been developed, but the information is still far from complete.

A complete analysis of a waste stream must be available before any process considerations can be made. An understanding of the solution chemistry for a metal is also required before a test program can be started. Copper is a common waste metal and is an example of a divalent metal. Copper is usually found as a divalent cation having good affinity for cation exchange resins. It also forms complexes with amines. It also occurs as an anion when complexed with EDTA or other chelating agents.

A waste liquor with copper, calcium, and sodium shows that the capacity is reduced by the presence of divalent cations other than copper. A selective resin based on EDTA removes copper from high concentrations of sodium chloride and to a lesser extent from calcium chloride. Cyanide complexes form stable complexes with strong base resins, and the metals can be recovered after the complex is destroyed with acid. In process engineering, environmental engineers should avoid such reactions unless they make provision for handling free hydrogen cyanide.

An experimental program to develop process information is assisted by reference-to-affinity orders that are available for most resins. The affinity series helps to establish

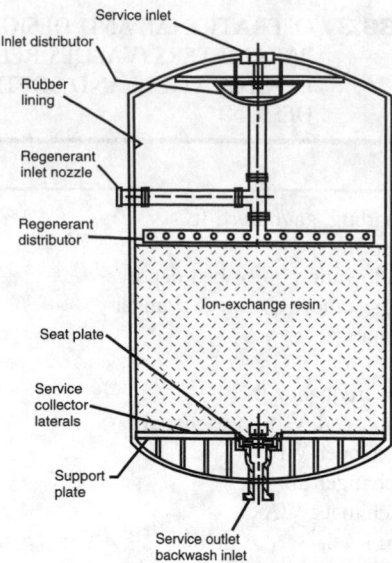

FIG. 7.36.11 Internal assembly of an ion-exchange column. (Reprinted, with permission, from Bolto and Pawkowski, 1982, *Waste water treatment by ion exchange.*)

from a concentrated solution. Amphoteric elements also become anions at elevated pH values.

Since there are several possibilities in a mixture of metals in a waste solution, environmental engineers can establish the presence of metals, cations, or anions by performing a single screening test. In this test, they prepare two small columns of anion- and cation-exchange resins. The anion-exchange resins should be in hydroxide form, and the cation exchange resins should be in acid form. The waste and the effluent sample for the metals should be fed into the columns at about 10 bed volumes per hr. Environmental engineers must be careful not to generate cyanide gas.

Once the valence of the metal is established, environmental engineers should start process estimates. Destroying metal complexes before the ion exchange resin is used may be useful. The amount of competing ions should be reduced if possible. If rinse water is deionized prior to the process, economic saving in the recycling of waste rinse waters and the recovery of valuable metals can be achieved.

When precipitation of toxic metals is the preferred process, environmental engineers usually use ion-exchange resins to polish the effluent to meet regulatory requirements. This process is a good place to use chelation resins that can pick up metals with a large salts background. However, precipitation can cause colloidal metal formation which is not easily removed by resins. Adjusting the pH and retention time can prevent this problem.

Metal recovery can be accomplished by electrolysis of the spent regenerant. Since ion exchange only concentrates

the likely displacement of metals when they are present as the metals, but does not give the potential for complex formations particularly when chlorides are involved. For example, Fe^{3+} forms a complex with chlorides in high concentration so that iron can be removed from concentrated hydrochloric acid as an anion, $FeCl_4^-$. By dilution, the complex is destroyed, and the iron comes off the resin bed. Other cations including aluminum can also be removed

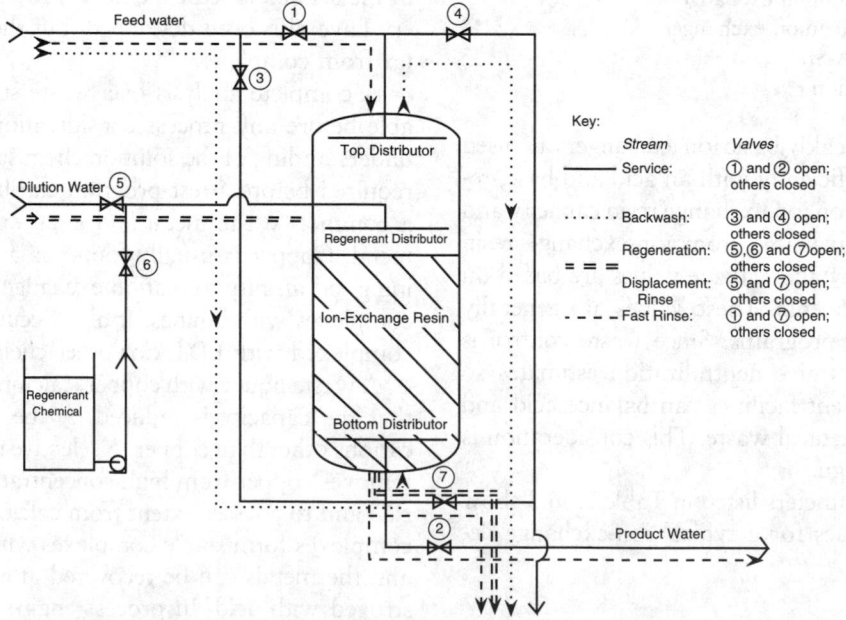

FIG. 7.36.12 Simplified ion-exchange operations cycle. The water used for backwash, dilution water, or displacement rinse can be feed water, softened water, decationized water, or DI water depending on the ion-exchange resin used and the quality of water produced in the service cycle. (Reprinted, with permission, from Dean Owens, 1985, *Principles of ion exchange and ion exchange water treatment*, 89.)

trace metals, regeneration conditions are important for economical recovery. For example, using sulfuric acid strips the cation exchange of copper so that electrolysis can be achieved without generating chlorine if chlorides are present.

Ion-Exchange Equipment

Ion-exchange resins are installed in tanks that have support plates and distribution piping. This piping distributes fluids across a packed bed of resin that rests on a support plate. A collector is installed in the support plate. All components should be constructed of corrosion-resistant materials. Figure 7.36.11 shows a cross section of a typical unit.

Many detail differences exist among manufactured equipment but they all have common flow streams that require control. Figure 7.36.12 details the operation of an ion-exchange plant. The end point can be determined with resistivity measurement and chemical metals analysis such as calcium, magnesium, and copper. Plants sometimes operate with two beds in series with the second bed operating as a polisher.

Process Conclusion

Once environment engineers have collected the information outlined previously, they can design a successful waste treatment plant. However, they should also consider the possible contamination of resin beds with oil, special organic chemicals, and SS, e.g., sludge. Generalizing from individual plants is difficult since many plating and anodizing systems operate differently.

Environmental engineers must recognize that ion exchange is a concentrating process. The elution and recovery of metals becomes part of other procedures such as electrodialysis, precipitation, and filtration. They must also consider varying local discharge regulations.

—*Francis X. McGarvey*

7.37
DEMINERALIZATION

Demineralization by Distillation

TYPES OF DISTILLATION SYSTEMS

a) Small units for high-purity water production
a1) Single stills
a2) Multistage stills
b) Large desalination systems for potable water production
b1) Multistage flash
b2) Vertical tube evaporator
b3) Vapor compression

DEMINERALIZED WATER QUALITY (ppm TDS)

Below 1 (a2), down to 1 (a1), and 500 (b)

APPLICATIONS

Small sizes (a1), up to 1 mgd (b3), and over 1 mgd (b1,b2)

MATERIALS OF CONSTRUCTION

Tin (a) and cupronickel (b)

THEORETICAL MINIMUM ENERGY REQUIREMENT

3 kWh per 1000 gal with 3.5% NaCl feedwater

ACTUAL PROCESS OPERATION EFFICIENCY

Less than 10% of thermodynamic optimum

RATIO OF WATER PRODUCED PER POUND OF STEAM INVESTED

Up to 20 (b1)

The Office of Saline Water (OSW), U.S. Department of Interior, funded much development work on sea water desalination. This office is a clearinghouse for technical information on these efforts and maintains a list of companies that have engineered desalting facilities under its program.

Small stills used for producing high-purity water are usually under 100 gph in capacity. They are single-stage stills, usually made of tin. Such units are expensive and unlikely to have significant application in wastewater treatment. Distillation processes used for large-scale desalination offer savings in energy compared to a simple still.

Distillation is an expensive method of demineralization and is not recommended except when one of the following conditions exist: (1) Potable water is required, and the only source is sea water; (2) a high degree of treatment is required; (3) contaminants cannot be removed by any other method; or (4) inexpensive waste heat is available.

Large-scale systems have been tested and are being commercially used in brackish and sea water desalination. They have not been applied to wastewater treatment, but demonstration plants have been evaluated in treating mine drainage waters.

DISTILLATION PROCESSES

Distillation operations are as varied as evaporator types and methods of using and transferring heat energy. The following types have been studied or used: (1) Boiling with submerged tube heating surface; (2) boiling with long-tube, vertical evaporator; (3) flash evaporation; (4) forced circulation with vapor compression; (5) solar evaporation; (6) rotating-surface evaporation; (7) wiped-surface evaporation; (8) vapor reheating process; (9) direct heat transfer using an immiscible liquid; and (10) condensing-vapor-heat transfer by vapor other than steam.

Of these types only (2), (3), and (4) are commercially important in desalination. The theoretical minimum energy required for a completely reversible process to obtain pure water from 3.50% NaCl salt water is about 3 kWh per 1000 gal at 25°C. Unfortunately, the thermodynamic minimum energy requirement has little practical relevance due to the many irreversibilities in an actual distillation process. These include pressure and force differences to overcome friction, temperature differences in heat exchangers and between system and the surroundings, and concentration differences for mass transfer. Actual processes operate at less than 10% of the optimum thermodynamic efficiency.

CORROSION AND SCALING

Due to temperature increases, inorganic salts come out of solution and precipitate on the inside walls of pipes and equipment. Scales due to calcium carbonate, calcium sulfate, and magnesium hydroxide are the most important in desalination processes. Controlling the pH minimizes carbonate and hydroxide scales. Most inorganic solutions are corrosive. Cupronickel alloys are most commonly used in sea water desalination. Other metals used are aluminum, titanium, and monel.

A significant development in distillation processes is the use of flash enhancers. These devices permit a closer approach to equilibrium flashing and can substantially improve the overall efficiency by increasing the liquid–vapor contact area through thermosiphon techniques.

All distillation processes reject part of the influent water as waste. Hence, all of these processes have concentrated waste disposal problems. The permissible maximum concentration in the waste depends on the solubility, corrosion, and vapor pressure characteristics of the wastewater. Therefore, the waste concentration is an important process optimization criterion.

FLASH DISTILLATION SYSTEMS

Multistage flash evaporation systems have been used commercially in desalination for many years. Conceptual designs for 1000-mgd plants are based on the flash principle. In the multistage flash process (see Figure 7.37.1), after the influent water has the SS removed and is deaerated, it is pumped through heat transfer units in several stages of

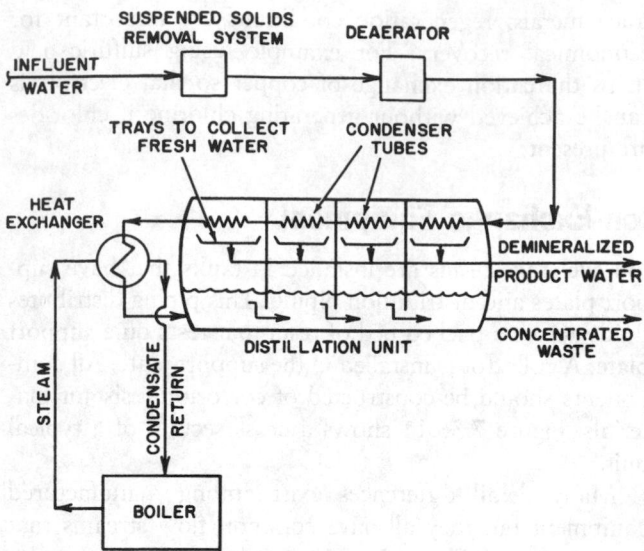

FIG. 7.37.1 Multistage flash process.

the distillation system. Evaporating influent water condenses on the outside of the tubes. The concentrated waste water cascades from one stage to the next as a result of the pressure differential maintained. In each stage, the flashed water condenses on the tubes and is collected in trays (see Figure 7.37.1). When the concentrated wastewater reaches the lowest pressure stage, it is pumped out.

From a thermodynamic point of view, multistage flash is less efficient than ordinary evaporation. On the other hand, it has the advantage of many stages combined into a single unit, resulting in less expensive construction and the elimination of external piping. The largest flash units have a performance ratio (lb water produced per lb steam used) of 20.

VERTICAL TUBE EVAPORATORS

Vertical-tube evaporators give 15 to 20% higher performance ratios and have fewer scale problems. They are used in large desalination plants.

In a three-stage, vertical-tube evaporator (see Figure 7.37.2), after the influent water is pretreated, it enters the heat exchanger in the last stage (No. 2) and progressively warms as it goes through the heat exchangers in the other effects (other stages). As the water moves through the heat exchangers, it condenses the water vapor emanating from the various effects. When the progressively warmed influent water reaches the first stage, it flows down the internal periphery of vertical tubes in a thin film, which is heated by steam. The wastewater feed to the second effect comes from the bottom of the first effect. Up to 15 effects are used in desalination plants.

VAPOR COMPRESSION EVAPORATORS

In these evaporators, the influent water is heated under nonboiling conditions, and the evaporated vapor is com-

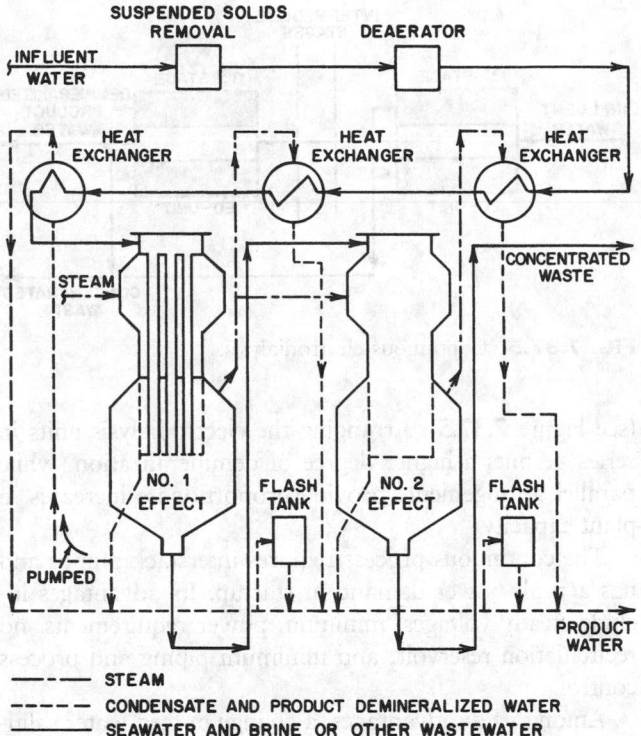

FIG. 7.37.2 Vertical-tube evaporator.

------ STEAM
----- CONDENSATE AND PRODUCT DEMINERALIZED WATER
——— SEAWATER AND BRINE OR OTHER WASTEWATER

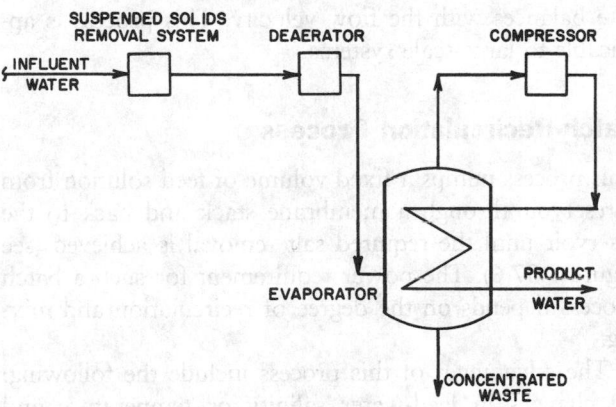

FIG. 7.37.3 Vapor-compression evaporator.

pressed and returned to serve as the heating medium (see Figure 7.37.3).

The vapor compression method has been tested and evaluated on brackish water, and the commercial applicability of such designs is limited by the capital costs of large vapor compressors. For the removal of inorganic salts with distillation, solar energy can also be used in regions where this source is continuously available.

Demineralization by Electrodialysis

TYPES OF SYSTEM DESIGNS
  a) Continuous
  b) Batch
  c) Feed and bleed
  d) Internally staged

MATERIALS OF CONSTRUCTION
  Cationic membranes made of sulfonic acid derivatives and anionic membranes made of quaternary amine compounds

APPLICATIONS
  Demineralization of brackish waters with up to 10,000 ppm salt concentration

This technique separates only ionized materials from water. Electrodialysis involves the use of electromotive forces to transport ionized material through semipermeable membranes that separate two or more solutions. The development of membranes made of ion exchange materials has led to the commercialization of electrodialysis systems, particularly for applications in the desalination of brackish water. Ion-exchange membranes transport ions of one charge only and have low electrical resistances.

PRINCIPLES OF ELECTRODIALYSIS

Electrodialysis units consist of several chambers made up of alternating anionic and cationic membranes arranged between two electrodes (see Figure 7.37.4). The solution containing cations and anions is fed through the chambers, and electromotive forces move the cations toward the cathode and the anions toward the anode.

Alternating anode and cathode membranes permit the passage of only one type of ion. Hence, after passing from one feed chamber to the next, the ions are blocked by an impermeable membrane. With this process, concentrated waste accumulates in every second chamber, and feed streams are purified in the others.

Electrodialysis requires that the membranes have sufficient ion-exchange capacity in addition to small-size (30Å) pores so that they repel electrostatically oppositely charged ions. The system controls the rate of water and current flow for optimum salt removal. Excessive current density results in acidic solutions being collected on the cathode side and basic solutions on the anode side of the membranes.

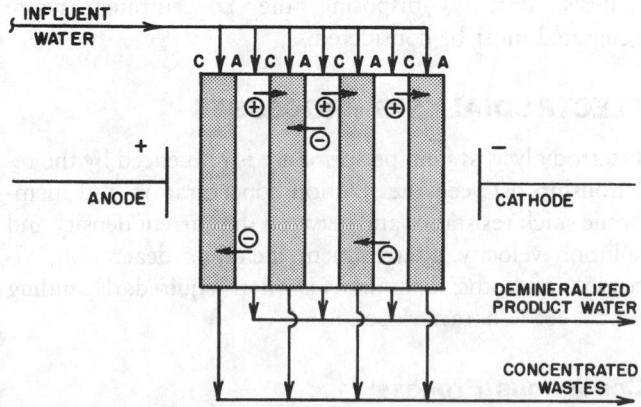

FIG. 7.37.4 Electrodialysis system. A = anion-permeable membrane; C = cation-permeable membrane.

Properties of Membranes

The membrane properties include electrical conductivity and selectivity for ion transport. Ion-exchange membranes are usually prepared from strong, hydrated electrolytes. Some common resins are polystyrene cross-linked with divinyl benzene and polyethylene- or fluorocarbon-base resins.

The electrical area resistance of membranes is up to 40 ohm-cm². Selectivity is characterized by permselectivity, which is defined with reference to transport numbers of ions as follows:

$$Sp = \frac{tm - ts}{1 - ts} \qquad 7.37(1)$$

where:

    tm = Transport number of ions within the membrane
    ts = Transport number of ions in free solution
    Sp = Permselectivity

Thus, the permselectivity indicates the permeability of the ion to which the membrane is impervious. This property of selectivity is caused by equilibrium between fixed ionic groups in the membranes and the solution wetting the membranes.

System Performance

The electrodialysis system in Figure 7.37.4 is inefficient for several reasons. As the ions migrate outward from the product chamber, the chamber progressively becomes depleted of ions, and its electrical resistance rises rapidly. This resistance determines the lower salinity limit in the product.

Increasing the operating temperature decreases resistance and thus improves the system performance. The build-up of solutes at the membrane–solid interface (concentration polarization) also increases electrical resistance. Ions in the feed water that precipitate due to a change in pH or are irreversibly adsorbed by the membranes can cause fouling. The back diffusion of salts due to an increase in salt concentration in one chamber can also reduce the separation efficiency. In addition, in all electrodialysis systems, disposing the concentrated waste generated must be considered.

ELECTRODIALYSIS PROCESSES

Electrodialysis system performance is influenced by the relationship between the solution concentration and membrane stack resistance and between the current density and solution velocity. Pretreatment including deaeration, filtration, and other operations is often required, depending on the feed characteristics.

Continuous Process

Fully continuous systems give demineralization performances and capacities beyond the range of modular units

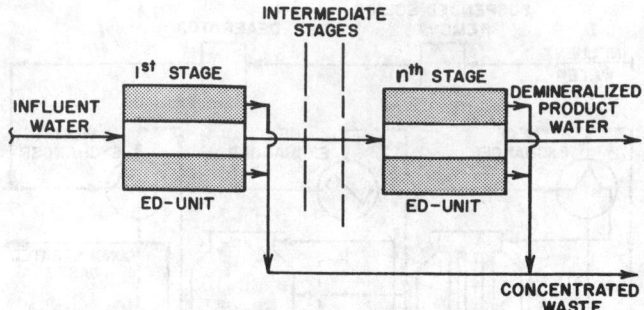

FIG. 7.37.5 Continuous electrodialysis.

(see Figure 7.37.5). Arranging the electrodialysis units in series permits a higher degree of demineralization, while parallel arrangements provide proportionate increases in plant capacity.

The continuous process requires interstack pumps and has a peak power demand at startup. Its advantages include steady voltages, minimum power requirements, no recirculation reservoir, and minimum piping and process control.

Among its disadvantages, a change in feed water salinity or temperature requires system adjustment, and the process is sensitive to increases in membrane resistance to flow. The continuous process requires that the production rate balances with the flow velocity. This process is applicable to large-scale systems.

Batch-Recirculation Process

This process pumps a fixed volume of feed solution from a reservoir through a membrane stack and back to the reservoir until the required salt removal is achieved (see Figure 7.37.6). The power requirement for such a batch process depends on the degree of recirculation and mixing.

The advantages of this process include the following: (1) changes in feed-water salinity or temperature and changes in membrane properties only modify the production rate but not the effluent quality; and (2) optimum velocity is independent of the production rate.

This process has the following disadvantages: (1) Higher power is needed; (2) recirculation reservoirs are required; (3) membranes do not operate at one equilibrium point; (4) current density through membrane also varies; and (5) more piping and control hardware are needed.

Feed-and-Bleed Process

The feed-and-bleed process (see Figure 7.37.7) can be used when continuous product flow is required, the degree of demineralization is small, and changes in the feed concentration are substantial.

This process continuously blends and recycles a portion of the product stream with the fresh feed. The power demand of this electrodialysis system is constant under equi-

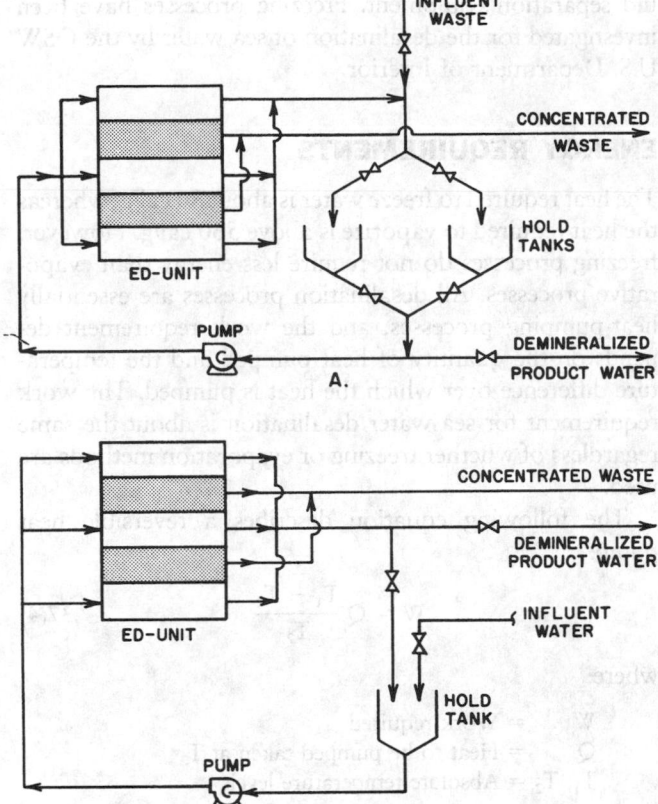

FIG. 7.37.6 Batch-recirculation electrodialysis. **A,** No mixing; **B,** Complete mixing.

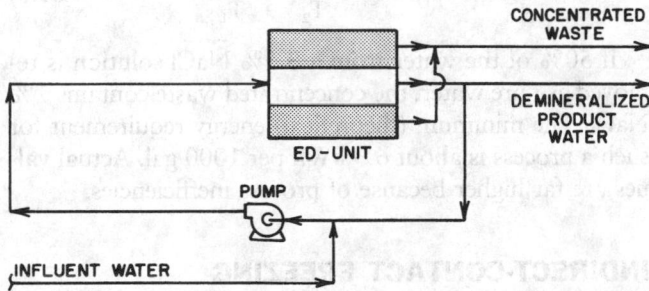

FIG. 7.37.7 Feed-and-bleed electrodialysis.

librium conditions. The advantages of the process include the following: (1) it provides a continuous product and can accept feed water with any TDS concentration; (2) the membranes are at equilibrium condition; and (3) a minimum current density is required.

The disadvantages of this system include high power consumption; and the requirement for sophisticated process control. The recirculation rate can be high with this process, and the process can have several feed-and-bleed loops arranged in series to give the effect of multiple-staging with lower recycling rates.

Internally Staged Process

For small units, internal staging within a stack provides all the advantages of a multistage series process without

requiring recirculating pumps or interconnecting piping (see Figure 7.37.8). In this design, the product stream makes several passes in series between a single set of electrodes, and therefore the degree of demineralization is higher than with a single pass. The process is continuous with a constant power demand.

Its advantages include a high degree of demineralization, no need for repressurizing pumps, and the need for only a single set of electrodes. This design has the following disadvantages:

A large membrane area is required per unit of product capacity.
The performance is sensitive to changes in flow rate.
The pump must overcome a high pressure drop.

SIZING OF ELECTRODIALYSIS UNITS

The following equations approximate the inorganic salt removal capacity and the power consumption of electrodialysis units:

$$T_\rho = 8.22 \times 10^{-5} \times \varepsilon \left(\frac{I}{A}\right) nA \qquad 7.37(2)$$

$$E = \frac{12.2 I(R_p + R_s)}{\varepsilon A} \qquad 7.37(3)$$

where:

E = Energy consumption, kWh lb-equivalent weight
A = Membrane area, ft^2
I = Current density, amps per ft^2
n = Total number of cell pairs
R$_p$ = Membrane resistance of one cell pair, ohm-ft^2
R$_s$ = Solution resistance, ohm-ft^2
ε = Current efficiency, a fraction
T$_\rho$ = Plant capacity, lb equivalent per hr

MEMBRANES

Ion-exchange resins comprise 60 to 70% or more of the membranes. These membranes are solid, hydrated, strong electrolytes. Many membranes are produced by graft polymerization of ionic monomers on a film base. Homogeneous membranes do not require a separate support and can be made of polyethylenes, aminated copolymers, or modified styrene-divinyl benzene compounds.

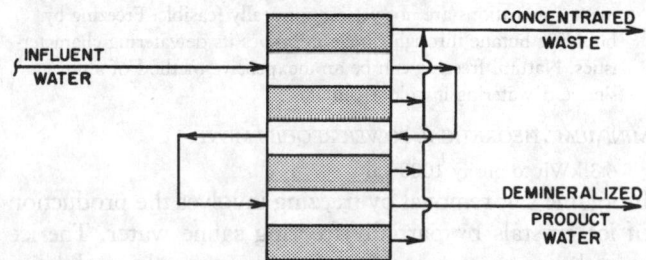

FIG. 7.37.8 Internally staged electrodialysis.

Heterogeneous membranes require a separate support. Dynel is the most common support material, while glass and other materials have also been used. These membranes are either fluorinated polymers or vinyl-divinyl benzene copolymers.

Extensive bibliographies and discussions on the methods of preparing membranes are available. Inorganic ion-exchange membranes are also being developed.

APPLICATIONS

The desalination of brackish waters is the main commercial use of electrodialysis systems. Units serving a range of users from individual homes to whole communities have been commercialized. The OSW has conducted extensive tests on the electrodialysis of brackish waters at salt concentrations of 10,000 ppm or below.

The EPA evaluated the use of electrodialysis in removing inorganic compounds from sewage effluents and found it potentially attractive. The treatment involves the use of diatomaceous earth filtration, granular carbon adsorption, and electrodialysis.

Other areas in which electrodialysis is used or appears to be promising include the following:

- Demineralizing whey by electrodialysis to produce the desalted whey used in infant formulas
- Desalting wastewater from fishmeal plants to recover protein
- Treating sulfonic acid pickle liquors to permit water recycling
- Recovering pulping chemicals and lignin products in the pulp and paper industry

Electrodialysis systems do not become much less expensive as the size becomes larger. Therefore, they are more likely to be used in small- and moderate-size applications. Advances in polymer science are likely to improve membrane selectivity and performance.

Demineralization by Freezing

TYPES OF FREEZING PROCESSES
a) Indirect contact
b) Direct contact
c) Hydrate process

APPLICATIONS
Food concentration, oil dewaxing; salination; and wastewater treatment applications are not yet economically feasible. Freezing by bubbling butane through sludge improves its dewatering characteristics. Natural freezing can be an inexpensive method of alum sludge dewatering in colder climates.

MINIMUM THEORETICAL POWER REQUIREMENT
6.3 kWh to purify 1000 gal
Inorganic salt removal by freezing involves the production of ice crystals by partially freezing saline water. The ice crystals are pure water and can be separated in solid–liquid separation equipment. Freezing processes have been investigated for the desalination of sea water by the OSW U.S. Department of Interior.

ENERGY REQUIREMENTS

The heat required to freeze water is about 80 cal/g, whereas the heat required to vaporize is above 560 cal/g. However, freezing processes do not require less energy than evaporative processes. All desalination processes are essentially heat-pumping processes, and the work requirement depends on the quantity of heat pumped and the temperature difference over which the heat is pumped. The work requirement for sea water desalination is about the same regardless of whether freezing or evaporation methods are used.

The following equation describes a reversible heat pump:

$$W = Q \frac{T_1 - T_2}{T_2} \qquad 7.37(4)$$

where:

W = Work required
Q = Heat to be pumped taken at T_2
T_1, T_2 = Absolute temperature levels

If T_0 is the ambient temperature, the following equation describes the ideal freezing process:

$$W = Q \frac{(T_1 - T_2)}{T_2} \frac{(T_0)}{T_1} \qquad 7.37(5)$$

If 50% of the water from a 3.5% NaCl solution is removed as pure water, the concentrated waste contains 7% NaCl. The minimum (theoretical) energy requirement for such a process is about 6.3 kWh per 1000 gal. Actual values are far higher because of process inefficiencies.

INDIRECT-CONTACT FREEZING PROCESS

In this process, heat transfer takes place through a metal wall. The process uses scraped-surface heat exchangers for both heat transfer and crystallization (see Figure 7.37.9). The feed solution is cooled in the surface crystallizers to below 32°F, and ice crystals are formed. Refrigeration is supplied by either ethylene glycol or an evaporative refrigerant, such as butane, on the shell side of the crystallizer.

The slurry containing ice crystals is sent to a continuous centrifuge where the ice crystals are separated from the mother liquor and sent to the melter tank after being washed. This process can obtain the heat for melting by precooling the incoming feed stream, thereby reducing the load on the refrigeration unit.

This process is used to concentrate coffee and fruit juices, but because the scraped-surface exchangers and

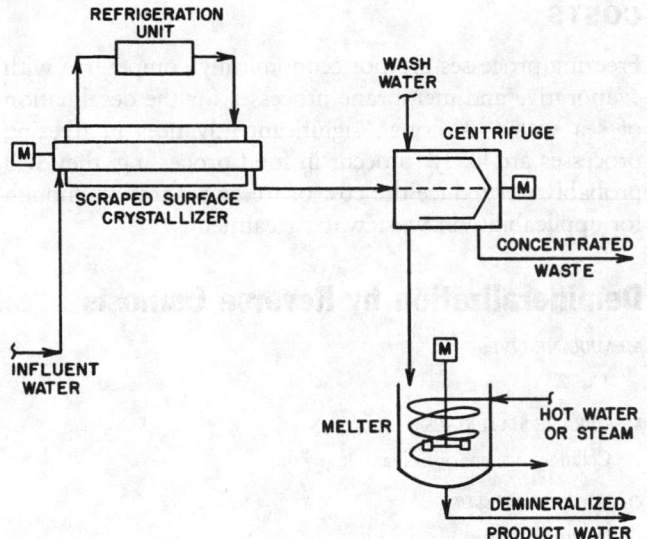

FIG. 7.37.9 Indirect-contact freezing process.

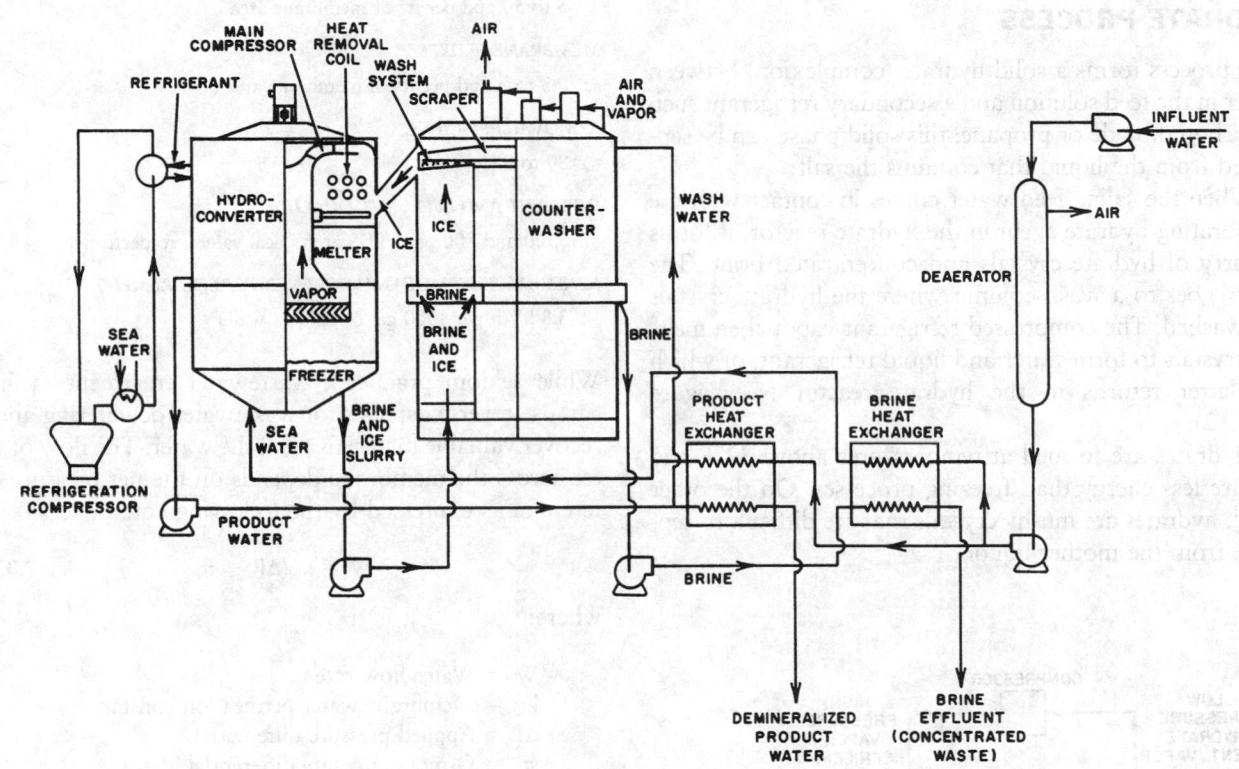

FIG. 7.37.10 Direct-contact freezing process with wash column.

FIG. 7.37.11 Desalination process using vacuum freezing and vapor compression.

other components are expensive, it is not used for wastewater treatment.

DIRECT-CONTACT FREEZING PROCESSES

Direct-contact processes have been investigated for large-scale seawater desalination applications (see Figure 7.37.10). Direct-contact processes use either a wash column or a vacuum-freezing, vapor-compression process.

The direct-contact freeze process uses direct contact with a refrigerant, such as butane, to effect crystallization. Therefore, the expensive scraped-surface units are unnecessary, and the process can produce larger crystals by using lower ΔT for crystallization.

The slurry containing the crystals moves to the wash column where the ice crystals rise due to their buoyancy, and countercurrent washing of the ice bed by fresh water takes place. The pure ice crystals are harvested off the top of the column and melted to produce demineralized water.

The vacuum-freezing, vapor-recompression process uses a hydroconverter and a washer (see Figure 7.37.11), and the feed solution enters the hydroconverter freezing chamber at the bottom. The refrigerant compressor causes vaporization in the vacuum chamber (3 to 5 mm Hg absolute), and the removal of the vaporization heat causes the formation of ice crystals in the water.

The slurry from the hydroconverter goes to the washer where the brine drains through the bottom while the crystals are returned to the hydroconverter at the top. They are melted by compressed vapors into pure demineralized water. The compressed vapors condense during this step and are drained with the melted ice as product effluent.

All direct-contact freezing processes require influent water pretreatment. The unavailability of large, axial, vapor compressors to handle the vapor volume at low pressures is the main limiting factor in the vacuum-freezing process.

HYDRATE PROCESS

This process forms a solid hydrate (complexion) between water in the feed solution and a secondary refrigerant such as carbon dioxide or propane; this solid phase can be separated from the liquid that contains the salt.

When the saline feed water comes in contact with the evaporating hydrate agent in the hydrate reactor, it forms a slurry of hydrate crystals and concentrated brine. The slurry goes to a wash column where the hydrate crystals are washed. The compressed refrigerant vapor then melts the crystals to form water and liquid refrigerant, of which the latter returns to the hydrate reactor (see Figure 7.37.12).

Hydrates are formed at temperatures above 32°F and require less energy than freezing processes. On the other hand, hydrates are mushy crystals that are difficult to separate from the mother liquor.

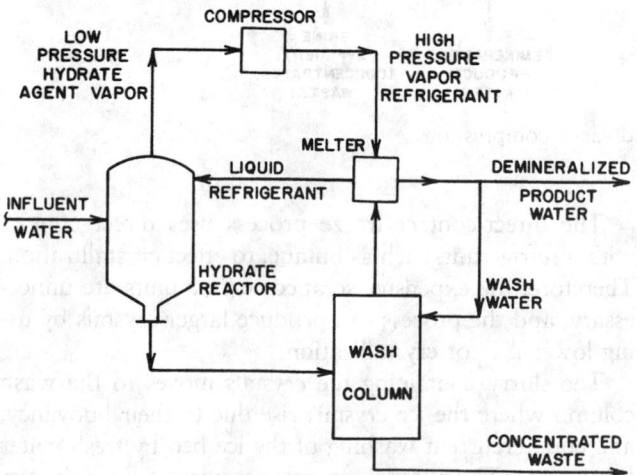

FIG. 7.37.12 Hydrate process.

COSTS

Freezing processes are not economically competitive with evaporative and membrane processes for the desalination of sea water. Although significant advances in freezing processes are likely to occur in food processing, they will probably not reduce the cost of freeze separation enough for applicability in wastewater treatment.

Demineralization by Reverse Osmosis

MEMBRANE LIVES

1 to 2 yr

MEMBRANE MATERIALS

Cellulose acetate or polyamide (nylon)

OPERATING PRESSURE

Up to 1500 psig

HYDRAULIC LOADINGS

5 to 50 gpd per ft² of membrane area

MEMBRANE FLUXES

0.3 to 5 gpd per ft² of membrane area

REMOVAL EFFICIENCY

80 to 95%

INFLUENT AND EFFLUENT QUALITY

3500 and 500 ppm TDS are typical values, respectively.

MINIMUM (THERMODYNAMIC) ENERGY REQUIREMENT

3.8 kWh per 1000 gal of effluent water

While seldom practiced, wastewater treatment facilities can use reverse osmosis for wastewater dewatering and to recover valuable materials from the water. The flow of water across the membrane depends on the net pressure differential as expressed by the following equation:

$$W = k_w(\Delta P - \pi) \qquad 7.37(6)$$

where:

W = Water flow rate
k_w = Membrane water permeation constant
ΔP = Applied pressure differential
π = Osmotic pressure differential

A small amount of salt permeates across the membranes. This salt flux is determined by the salt concentration on the feed and product water sides and is independent of the applied pressure.

The total energy requirement for reverse osmosis includes the minimum thermodynamic energy, which is about 3.8 kWh per 1000 gal for sea water at 25°C. In addition, power is required to overcome the irreversible losses, including pressure losses due to friction and concentration polarization. Concentration polarization is caused by a build up of solute ions at the surface of the membrane. Concentration polarization occurs because the water preferentially permeates the membrane, whereas the solute can only move from the membrane surface by back diffusion. This process depletes some pressure energy and is a major cause of inefficiency.

The following equation gives the salt permeation rate:

$$S = K_s(C_f - C_p) \qquad 7.37(7)$$

where:

S = Salt flow rate
K_s = Salt permeation constant
C_f = Concentration of salt in the influent solution
C_p = Concentration of salt in the effluent solution

Equations 7.37(6) and 7.37(7) show that the water effluent produced per unit of membrane surface increases with the pressure differential across the membrane, whereas the salt flux remains constant. Thus, an increase in operating pressure increases the ratio of water to salt and improves the efficiency of demineralization.

PRETREATMENT

Reverse osmosis usually requires pretreatment to remove contaminants including SS in sizes above 3 to 5μ; removing these contaminants prevents fouling the membrane surfaces. Because many saline waters contain salts such as calcium carbonate in concentrations near saturation, lowering the pH of the water is necessary to prevent scaling. Wastewater treatment facilities usually adjust the pH by adding a mineral acid in the range of 5 to 6.

Wastewater treatment facilities often add biocides, such as chlorine, to prevent microbial growth that could foul membrane surfaces. They also add threshold inhibitors to prevent salt precipitation from fouling membrane surfaces.

After pretreatment, the feed solution is pumped by a high-pressure pump to the reverse osmosis unit at about 500 to 800 psig (see Figure 7.37.13). Wastewater treatment facilities can use a number of individual reverse osmosis modules in series and parallel-flow configurations. The back-pressure valves on each assembly control the pressure within the membrane chamber. The pressure energy from the pressurized reject brine can be recovered in a turbine generator.

Wastewater treatment facilities can recover 60 to 75% of the water as pure effluent from the dilute wastes, leading to a three- to four-fold increase in the solute concentration in waste streams.

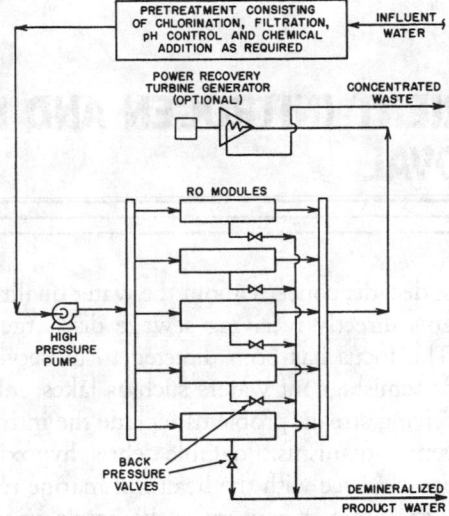

FIG. 7.37.13 Reverse osmosis.

DESALINATION EXPERIENCE

Most membranes are made of cellulose acetate or polyamides. Cellulose acetate is the most common material because of its high salt rejection rate and high water flux. Except for hollow-fiber types, all design configurations require separate support for membranes. Hollow-fiber-polyamide membranes are compact but require factory fabrication of the modules and a higher level of pretreatment to remove SS.

WASTEWATER APPLICATIONS

Reverse osmosis has been considered for the production of fresh water from acid mine wastes containing 2000 to 5000 ppm dissolved acids and salts. In addition to recovering pure water, the reverse osmosis unit produces a concentrated waste that is easier to neutralize. Other applications include the concentration of fruit juices and sulfite pulp liquor. The use of reverse osmosis to remove organic matter from sewage effluents is being tested. Problems yet to be resolved include surface fouling and the internal breakdown of the membranes.

—*C.J. Santhanam*

7.38
NUTRIENT (NITROGEN AND PHOSPHORUS) REMOVAL

In the past decade, concern about the water quality of natural systems directly receiving sewage discharge has increased. The focus has been directed to preserving nonflowing or semistagnant waters such as lakes, inlets, and bays. Receiving stream problems include the introduction of pathogenic organisms, floatable debris, hypoxic conditions, or interference with the health of marine resources. However, the greatest concern is the acceleration of eutrophic conditions in these surface waters.

Eutrophication is the natural aging process of a body of water as biological activity increases (Water Pollution Control Federation 1983). Eutrophic waters are characterized by high concentrations of aquatic weeds and algae. These organisms eventually die, sink to the bottom, and decay. Consequently, this cycle increases the sediment oxygen demand which decreases the DO in the lower water levels. Additionally, eutrophication is enhanced by the large day–night cycling of DO that accompanies increased photosynthesis and respiration (Metcalf and Eddy, Inc. 1991). The acceleration of eutrophication is directly linked to increased nutrient loadings from sewage treatment plant discharges.

Phosphorus and nitrogen are the two major nutrients contributing to eutrophication. In most cases, these nutrients are growth-limiting; algae can no longer grow if these nutrient pools are depleted. Therefore, environmental engineers consider the removal of phosphorus and nitrogen from point sources, such as sewage treatment plants, a cost-effective and appropriate method for controlling the level and extent to which eutrophication occurs.

However, eutrophication is not the only problem caused by these nutrients. Ammonia is toxic in small concentrations to some aquatic life. The oxidation of ammonia to nitrite/nitrate can severely deplete the DO concentration in a body of water. Nitrite, which has a greater affinity for hemoglobin than oxygen and thus replaces it in the bloodstream, has been found to cause methemoglobinemia, or "blue baby" disease, in infants (U.S. EPA 1975; Peavy, Rowe, and Tchobanoglous 1985). Phosphates, in concentrations as low as 0.2 mg/l, interfere with the chemical removal of turbidity in drinking water (Walker 1978).

Due to increased nutrient loadings as well as elevated public awareness and consequent demand for protection of the world's water resources, the research and development of processes that remove phosphorus and nitrogen from wastewater have advanced considerably. Most of the interest has been in the manipulation of ambient conditions to enhance biological mechanisms responsible for nutrient removal. Consequently, both municipal and industrial facilities use many wastewater treatment processes to comply with federal and state regulations.

Nutrient removal processes can be grouped into two main categories: biological and physiochemical systems. Biological processes can be further divided into fixed-film and suspended-growth systems. Recently, many treatment facilities have been required to incorporate some degree of nutrient removal since the majority of plants built in the United States during the 1970s were for organic and SS (BOD and TSS, respectively) removal only. Although applications exist for fixed-film and physiochemical nutrient removal processes, suspended-growth systems have received the most attention. Therefore, this section focuses on suspended-growth activated-sludge processes. It reviews the biological mechanisms responsible for nutrient removal as well as different treatment processes.

Nutrient Removal Mechanisms

Biological nutrient removal in wastewater involves manipulation of the process environment. Bacteria responsible for waste treatment either proliferate or decay as ambient conditions change. The degree of treatment depends on the number of specific genera of bacteria. Therefore, environmental engineers must account for certain influencing parameters during process design and operation. This section reviews the theories of phosphorus, ammonia, and nitrogen removal. Design parameters are reviewed with specific attention to actual operating experience.

PHOSPHOROUS REMOVAL

The phenomenon of enhanced biological phosphorus removal (EBPR) by activated sludge in excess of normal metabolic requirements was documented as early as the 1960s and was referred to as luxury uptake (Bargman et al. 1971; Levin and Shapiro 1965; Connell and Vacker 1967; Wells 1969; Borchardt and Azad 1968). Since the early 1970s, numerous studies have been conducted on various aspects of the mechanisms that control this biological process (McLaren et al. 1976; Hong et al. 1982; Spector 1977; Tracy and Flammino 1987; Fuchs and Chen 1975; Davelaar et al. 1978; Nicholls and Osborn 1979; Berber and Winter 1984; Aruw et al. 1988; Claete and

Steyn 1988). Even though environmental engineers have not obtained a confirmed, quantitative knowledge based on fundamental behavioral patterns and kinetics, they have designed activated-sludge systems to achieve the required degree of biological phosphorus removal (Hong 1982; Wentzel et al. 1985; 1988; Hong et al. 1989; Kang et al. 1985; DeFoe et al. 1993). Many full-scale plants, especially in the United States, have been installed in the past ten years.

To induce enhanced phosphorous uptake by activated sludge, the system must subject the process biology to a period of anaerobiosis prior to aeration; a prerequisite for EBPR is the proper selection of organisms. Environmental engineers can accomplish anaerobiosis by subjecting the sludge or mixed liquor to anaerobic/aerobic cycling (Davelaar et al. 1978; Nicholls and Osborn 1979; Berber and Winter 1984). Consequently, microorganisms capable of storing phosphorus as polyphosphate in their cell mass proliferate in the system.

Figure 7.38.1 shows typical concentration profiles for BOD and phosphorus. The release of orthophosphate mirrors the rapid uptake of organic substrate (BOD) in the anaerobic environment. In the aerobic zone, orthophosphate uptake to very low concentrations and continued removal of BOD occur.

Several explanations for these concentration profiles have been offered. Among them, the hypothesis offered by Hong et al. (1982) gives a plausible account of the biological mechanisms. The breakdown of polyphosphate in the anaerobic zone to generate energy for active transport of substrate into the cells of phosphate-accumulating organisms explains the phosphorus and BOD concentration changes in the anaerobic environment. Consequently, this environment serves as a selection zone; phosphate-accumulating bacteria, identified as the genus *Acinetobacter*, have a competitive advantage and proliferate in the system (Fuchs and Chen 1975; Claete and Steyn 1988).

As phosphate-accumulating and other organisms reach the aerobic environment, most readily available BOD is contained inside the high-phosphate bacteria. As the trans-

ported BOD is oxidized, the resulting energy resynthesizes polyphosphate and forms new cells. The new cells are responsible for net phosphorous removal in the system. The result of the alternating anaerobic/aerobic operation is wastewater with a low-effluent concentration of BOD and soluble phosphorus, and a waste sludge high in bound-polyphosphate content.

Tracy and Flammino (1987) postulated a more detailed biochemical pathway to further explain the apparent relationship between substrate metabolism and polyphosphate storage. Figure 7.38.2 shows a simplified version of this postulated pathway. The stored organics are reported to be polyhydroxybutyrate (PHB) (Bordacs and Chiesa 1988; Hong et al. 1983).

Environmental engineers have conducted microscopic examinations on mixed liquor samples from EBPR and conventional activated-sludge processes (Hong et al. 1982). They treated the samples with Neisser stain, which detects polyphosphate-containing volutin granules. Figure 7.38.3 shows photomicrographs of the stained samples (U.S. EPA 1987). With the Neisser treatment, darkly stained matter indicates the presence of polyphosphate. As the figure shows, the mixed liquor sample from the EBPR process contains a denser volume of stored polyphosphate.

The rate and extent of phosphate removal is related to the type and quantity of soluble substrate in the influent wastewater. Studies show that low-molecular-weight fatty acids (VFAs), such as acetate, are the preferred carbon source (Wentzel et al. 1985; 1988). Additionally, cationic species such as potassium and magnesium, are required for polyphosphate synthesis (Hong et al. 1983). Typical municipal wastewater contains sufficient soluble substrates and cations for biological phosphorus removal.

With given influent wastewater characteristics, an EBPR system designed for a higher F/M ratio generally achieves a higher degree of phosphorus removal. A system operating with a higher F/M ratio produces a larger amount of cell mass (sludge) to incorporate phosphorus. Since sludge production and wasting are responsible for net phosphorus removal, this parameter is key to successful biological phosphorus removal.

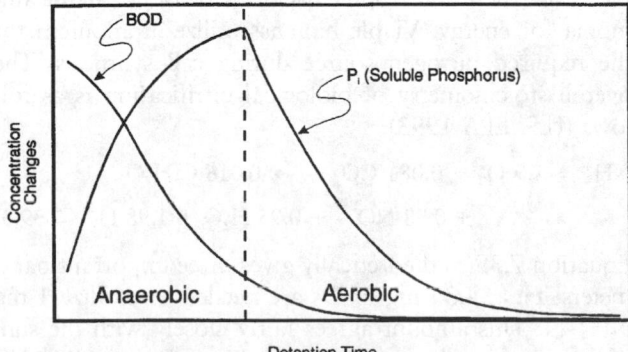

FIG. 7.38.1 Biological explanation for changes in BOD and phosphorus in an EBPR process.

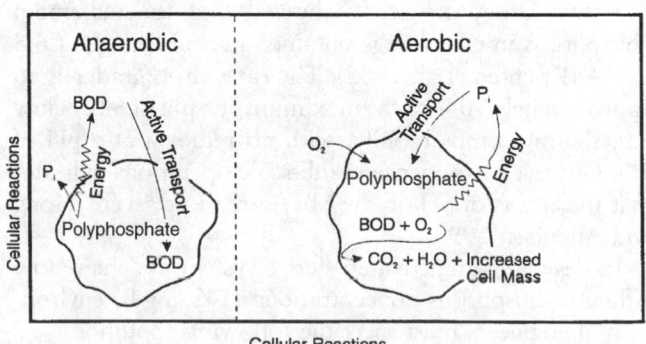

FIG. 7.38.2 A simplified biological pathway of a nutrient removal process.

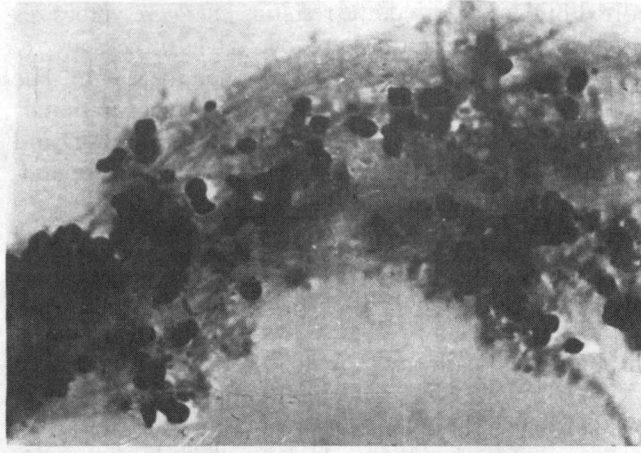

FIG. 7.38.3 Photomicrographs of stained, mixed liquors. *top*, EBPR sludge showing high concentration of polyphosphate deposits (1000×); *below*, control sludge showing absence of polyphosphate deposits (1000×).

The temperature and pH effects on EBPR processes have seldom been studied. Sell et al. (1981) postulated that the organisms responsible for phosphorus removal are facultative psycrophiles. Laboratory results indicate that phosphorus removal is enhanced as the operating temperature decreases.

Like most biological reactions, the organisms in the EBPR process favor a near-neutral pH. The results of an extensive laboratory study showed that the maximum phosphorus uptake rate is obtained in a pH range of 6.8 to 7.4 (Krichten et al. 1985). The rates drop gradually to approximately 70% of the maximum at a pH of 6.0. Below 6.0, the rate drops rapidly, with no removal at a pH of 5.0. However, experience in full-scale operations indicates that the effects of pH on the EBPR are more severe (Hong and Andersen 1993).

In designing an activated-sludge system to achieve low effluent phosphorus concentrations (1–2 mg/l), environmental engineers must meet the following conditions:

1. An anaerobic zone with a detention time of 1 to 2 hr
2. An influent soluble BOD/soluble P ratio of 12 or greater

3. Operation at the highest F/M and shortest SRT that still permits nitrification, if nitrification is required

NITROGEN REMOVAL

The nitrogen in municipal wastewater is predominately in the organic and ammonium forms. Based on the amount of degradation prior to reaching the treatment plant, the respective fractions vary. Typically, municipal wastewater contains 60% ammonium nitrogen, 40% organic nitrogen, with negligible concentrations of nitrate or nitrite (Joint Task Force of the Water Environment Federation and ASCE 1991). Untreated domestic wastewater typically contains 20–50 mg/l of total nitrogen (Metcalf and Eddy, Inc. 1991).

Ultimately, all activated-sludge treatment processes realize some degree of net nitrogen removal since nitrogen is required for the synthesis of a new, viable cell mass. Both ammonium and nitrate can be used as the nitrogen source. The degree of net nitrogen removal due to cell growth is a function of the organic loading, sludge age (SRT), and endogenous respiration (Marais and Ekema 1976). Typically, assimilation removes 20–30% of the total influent nitrogen (Van Haandel, Ekama, and Marais 1981). However, all biological nitrogen removal systems use two processes to achieve the required effluent quality: nitrification and denitrification. Nitrogen can also be removed from wastewater by ion exchange.

Nitrification

Biological nitrification oxidizes nitrogen from ammonia (NH_4^+) to nitrate (NO_3^-). The overall reaction is a two-step process. In the first step, ammonia is oxidized to nitrite (NO_2^-) and mediated by the genus *Nitrosomonas*. The second step converts nitrite to nitrate, controlled by the genus *Nitrobacter*. The conversional process is as follows:

$$NH_4^+ + 1.5\ O_2 \xrightarrow{\quad Nitrosomonas \quad} NO_2^- + 2H^+ + H_2O$$

$$7.39(1)$$

$$NO_2^- + 0.5\ O_2 \xrightarrow{\quad Nitrobacter \quad} NO_3^- \qquad 7.39(2)$$

However, these autotrophic bacteria do not use all the ammonia for energy. Viable biomass utilize ammonium for the required nitrogen source during cell synthesis. The overall stoichiometry of biological nitrification is as follows (U.S. EPA 1993):

$$NH_4^+ + 1.9\ O_2 + 0.081\ CO_2 \longrightarrow 0.016\ C_5H_7O_2N$$
$$+ 0.98\ NO_3{-} + 0.95\ H_2O + 1.98\ H^+ \quad 7.39(3)$$

Equation 7.39(3) theoretically gives three important parameters. First, 4.34 mg of O_2 are needed to oxidize 1 mg NH_4^+-N. This amount agrees fairly closely with the sum of the two preceding equations (4.57 mg O_2/mg NH_4^+-N). Secondly, 7.07 mg of alkalinity (as $CaCO_3$) are consumed per mg of NH_4^+-N nitrified. Finally, 0.13 mg of viable biomass are produced per mg NH_4^+-N converted. These parameters vary according to process operation and conditions, i.e., sludge age, organic loading, pH, and temperature.

Nitrification warrants treatment system adjustments and can occasionally be a difficult process to achieve reliably. The aerobic autotrophs responsible for nitrification are more sensitive to ambient conditions, toxics, and inhibitors than the competing carbonaceous heterotrophs (Sharma and Ahlert 1977; Sutton, Murphy, and Jank 1974). Additionally, some confusion exists over the biological kinetics during typical wastewater treatment conditions (Charley, Hooper, and McLee 1980).

As shown in the stoichiometry, nitrification requires more oxygen than just the amount required by carboneous heterotrophs for carbon oxidation (BOD removal). A longer sludge age must be maintained due to the slower growth rate of autotrophic nitrifiers. Also, if sufficient alkalinity is not present in wastewater, the pH can drop substantially. This low pH limits the extent of nitrification.

Table 7.38.1 shows typical process parameters for single-stage nitrifying systems.

Denitrification

Biological denitrification reduces nitrate (NO_3^-) to nitrogen gas (N_2), nitrous oxide (N_2O) or nitric oxide (NO). This nitrogen removal process is the one most widely used in municipal wastewater treatment (Water Pollution Control Federation 1983). Denitrifying organisms are primarily facultative aerobic heterotrophs that can use nitrate in the absence of DO. Many genera of bacteria are capable of denitrification: *Achromobacter, Bacilus, Brevibacterium, Enterobacter, Micrococcus, Pseudomonas,* and *Spirlilum* (Davies 1971; Prescott, Harley, and Klein 1990). Several conditions enhance the amount of biological denitrification: nitrate, a readily available carbon source, and a low DO concentration. Low DO is the most critical condition since denitrification is simply several modifications of the aerobic pathway used for BOD oxidation (U.S. EPA 1975).

The stoichiometric reaction describing this biological reaction depends on the carbon source involved as follows (Randall, Barnard, and Stensel 1992):

TABLE 7.38.1 TYPICAL PROCESS PARAMETERS FOR SINGLE-STAGE NITRIFYING SYSTEMS

Parameter	Range
MLSS, mg/l	2500–5000
Hydraulic retention time, hr	6–15
Sludge age, days	5–15
Residue DO, mg/l	1.0–2.0
pH	6.5–8.0
Temperature, C	10–30

Carbon Source	Theoretical Stoichiometry

methanol
$$NO_3^- + 0.83\ CH_3OH \longrightarrow 0.5\ N_2 + 0.83\ CO_2 + H_2O + OH^- \quad \text{7.39(4)}$$

acetic acid
$$NO_3^- + 0.63\ CH_3COOH \longrightarrow 0.5\ N_2 + 1.3\ CO_2 + 0.75\ H_2O + OH^- \quad \text{7.39(5)}$$

sewage
$$NO_3^- + 0.1\ C_{10}H_{19}O_3N \longrightarrow 0.5\ N_2 + CO_2 + 0.3\ H_2O + 0.1\ NH_3 + OH^- \quad \text{7.39(6)}$$

methane
$$NO_3^- + 0.63\ CH_4 \longrightarrow 0.5\ N_2 + 0.63\ CO_2 + 0.75\ H_2O + OH^- \quad \text{7.39(7)}$$

The denitrification rate is a function of, among others, the temperature of the wastewater and the type of carbon source used for electron transfer. However, determining which carbon substrate is most suitable for wastewater treatment has been difficult (Tam, Wong, and Leung, 1992).

The following equation describes overall synthesis reaction using methanol as the carbon source and nitrate as the nitrogen source (U.S. EPA 1975):

$$NO_3^- + 1.08\ CH_3OH + 0.24\ H_2CO_3 \longrightarrow$$
$$0.056\ C_5H_7O_2N + 0.47\ N_2 + 1.68\ H_2O + HCO_3^- \quad \text{7.39(8)}$$

Theoretically, 2.47 mg of methanol are required to reduce 1 mg of nitrate. Also, the equation predicts that 3.57 mg of alkalinity (as $CaCO_3$)/mg nitrate-nitrogen are produced. However, studies show that this estimate may be aggressive and that 3.0 mg of alkalinity are more likely (U.S. EPA 1993; Wanielista and Eckenfelder 1978). The alkalinity production capability of denitrification enables a combined nitrification–denitrification system to maintain a more stable pH. Additionally, a system that uses influent BOD as the carbon source and has the anoxic zone located before the aerated portion of the process will require less energy for aeration. This energy savings is because a portion of the waste is consumed by denitrification (anoxic stabilization).

Ion Exchange

FERTILIZER PLANT WASTEWATER

500 mg/l nitrogen concentration, 50% in ammonium and 50% in nitrate form at about 1 mgd total wastewater flow rate

TREATED EFFLUENT QUALITY

2 to 3 ppm NH_3 and 7 to 11 ppm NO_3

CONCENTRATED BY-PRODUCT STREAM

20% solids

CHEMICAL REGENERANT CONSUMPTION (PERCENT OF STOICHIOMETRIC)

135 to 150% acid and 110% base

NITROGEN REMOVAL EFFICIENCY

99%

Industrial wastewater discharges, especially from the ammonium nitrate fertilizer industry, typically contain concentrations of 500 mg/l of nitrogen, equally divided between ammonium and nitrate forms. Environmental engineers use the ion-exchange process to successfully remove such high concentrations of nitrogenous compounds. Ion exchange can purify the wastewater to a quality that complies with zero-pollutant-discharge criteria or that permits complete recycling of the wastewater. The ion-exchange process can also completely recover plant products lost into the waste stream and can efficiently recycle the recovered products into the plant processes.

Ion-exchange has been used successfully to treat the wastewater from a large nitrogen fertilizer plant producing 140,000 tn of prilled ammonium nitrate and 190,000 tn of nitrogen solutions annually. A captive ammonia plant producing 175,000 tn annually and a captive nitric acid plant producing 195,000 tn annually provide the feed stock for primary products. The wastewater from this typical plant contained a concentration of about 500 mg/l nitrogen, equally divided between ammonium and nitrate forms, in about 900,000 gal of effluent per day.

The fertilizer plant chose ion exchange as the treatment process at the conclusion of a two-year study which showed that no other process had the capability of approaching the zero-pollutant-discharge goal or permitted total water reuse and recycling. As a treatment process, ion exchange can provide effluent water of adequate quality for reuse or discharge, provided that the collected contaminant ions can be properly disposed, preferably as a recovered product.

Because of the nature of the manufacturing processes, the principal pollutant in nitrogen fertilizer plant effluent is ammonium nitrate. This salt is satisfactorily recovered from solution by ion exchange. Also, by regenerating cationic resin with nitric acid, and anionic resin with aqua ammonia, the ion-exchange process can provide for complete product recovery since the excess regenerants combine to form more principal product. Because the regenerants are inplant products, they are obtained at minimum cost.

The other ions resulting from resin backwash are principally those from water hardness. The ion-exchange process readily recycles recovered principal product, ammonium nitrate, and other recovered water impurities into the nitrogen solution production of typical nitrogen fertilizer plant. The resulting deionized water can also be reused.

The fertilizer plant selected continuous, countercurrent, moving-bed-type of ion exchange for this application since the system provides the low leakage of exchanged ions and high concentration of regenerant backwash favoring product recovery.

Table 7.38.2 shows an analysis of a typical nitrogen fertilizer plant wastewater as influent to the ion exchange treatment system together with analysis of the deionized effluent stream. Table 7.38.3 shows an analysis of the contaminants recovered from this wastewater.

In addition to the principal material, ammonium nitrate, the recovered ions constitute micronutrients for

TABLE 7.38.2 NITROGEN FERTILIZER PLANT WASTE AND TREATED WATER ANALYSIS

Component	Influent		Effluent
		ppm	ppm
NH_3		340	2–3
Mg		5	—
Ca		60	—
Na		0	—
NO_3		1240	7–11
Cl		53	—
SO_4		72	—
OH		—	—
pH		5–9	5.9–6.4
SiO_2		15	15

Note: The ammonium nitrate removal is 99.4%.

TABLE 7.38.3 AMMONIUM NITRATE PRODUCT STREAM ANALYSIS

Component	%
NH_4NO_3	17.83
Ca	0.260
Mg	0.020
NH_3	4.070
NO_3	14.880
Cl	0.220
SO_4	0.3120
Total Solids	19.762
Water	80.238

TABLE 7.38.4 DAILY QUANTITY OF MATERIAL RECOVERED (FROM 900,000 GAL WASTEWATER)

	Cation lb/d	Anion lb/d	Blended lb/d
$Ca(NO_3)_2$	1848	—	1848
$Mg(NO_3)_2$	222	—	222
NH_4Cl	—	600	600
$(NH_4)_2SO_4$	—	699	699
NH_4NO_3	12,021	12,021	31,211
Water	82,255	56,780	139,035
HNO_3	5585	—	—
NH_3	1290[a]	294	—

Total Product Ammonium Nitrate Solution 173,615
Note: [a]Adding NH_3 neutralizes excess HNO_3.

plants. However, a high concentration of chlorides makes extensive evaporation of the solution hazardous because of the probability of explosion. Consequently, the most feasible disposal of the recovered material is recycling into the production of nitrogen solutions.

Table 7.38.4 shows the materials recovered daily from 900,000 gal of nitrogen fertilizer plant wastewater. Table

7.38.2 shows that 15 ppm of silica, SiO_2, is in the wastewater. The ion-exchange treatment system does not remove silica since the anion resin used is a weak-base type. Consequently, the deionized water produced by the ion-exchanger is not entirely satisfactory for use as cooling-tower makeup water or boiler feed water because of the silica content.

The plant can add silica polisher (a standard fixed-bed exchanger using a strong-base anion resin) to the system to remove silica and produce deionized water satisfactory for all inplant use. This addition permits complete wastewater recycling and reuse in a closed-loop system.

Phosphorus Removal Processes

Phosphorus removal processes were used in full-scale applications in the 1970s. Since phosphorus is typically the limiting nutrient for algae growth, removal from point sources can potentially halt eutrophication. The phosphorus content of typical wastewater process bacteria ranges between 1.5–3% (on a dry-weight basis). However, the bacteria responsible for EBPR contain a larger amount (4–6% on a dry-weight basis). This section reviews the major biological phosphorus removal processes and discusses the advantages, disadvantages, and flow schematics.

ANAEROBIC/OXIC (A/O) PROCESS

Part a in Figure 7.38.4 is a schematic representation of the A/O process. The unique feature of this aerobic (oxic) activated-sludge process is an anaerobic (both oxygen- and nitrate-deficient) zone at the influent end of this process. The anaerobic and aerobic zones are each divided into several equally sized compartments. Influent wastewater and return activated sludge (RAS) are fed to the first compartment of the anaerobic zone. Typically, either centrally mounted or submersible mixers provide gentle mixing in the anaerobic zone. Wastewater treatment facilities can use various aeration methods, such as fine-bubble diffusers, surface mechanical aerators, and oxygen aeration to meet the oxygen demands in the oxic zone.

The A/O process is a high-rate process characterized by low hydraulic detention times (2.5–3.5 hr) and high F/M ratios (0.5–0.9 1/day). Wastewater treatment facilities can adapt this process for simultaneous phosphorus removal and nitrification by simply adjusting for the aerobic sludge age. Depending on climatic conditions and influent characteristics, especially influent TSS, the hydraulic detention time in the oxic zone can range from 4.0 to 8.0 hr, while the anaerobic detention time remains approximately 1 hr. More than sixty plants in the United States have used this process configuration. Figure 7.38.5 shows the operating and performance data of the North Plant at Titusville, Florida (no nitrification required), while Figure 7.38.6 shows the operating data at a plant in Lancaster, Pennsylvania (nitrification using pure oxygen aeration).

Even though simultaneous phosphorus and nitrogen removal has been successful, excessive nitrate concentrations in the RAS stream are a concern in the anaerobic zone. Therefore, environmental engineers recommend a modified basin configuration (see Part b in Figure 7.38.4). This process removes the nitrate in the return sludge before it is mixed in the anaerobic zone.

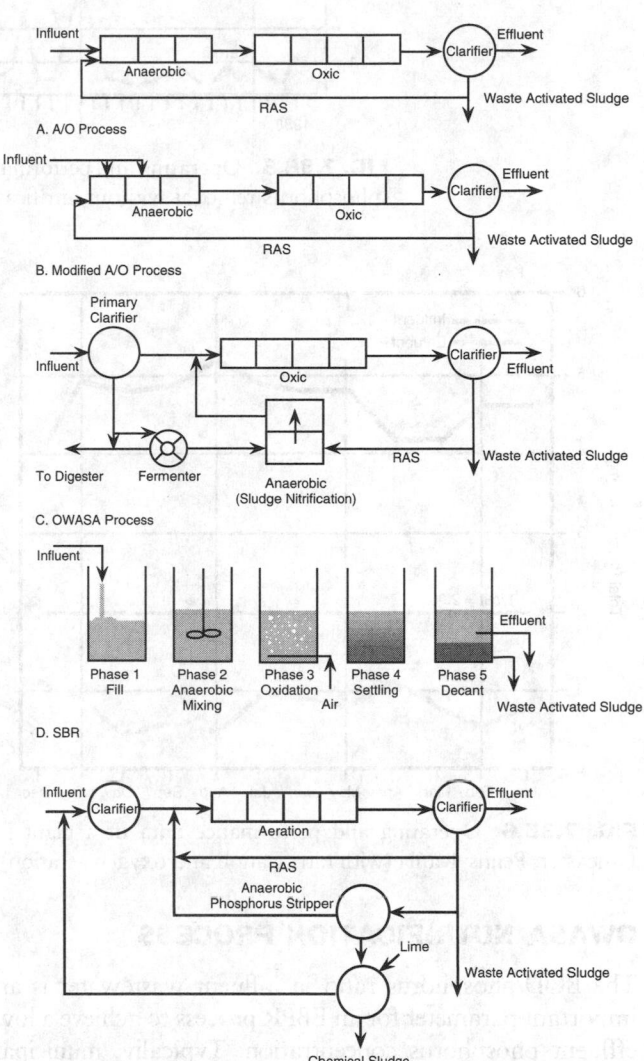

FIG. 7.38.4 Schematic diagrams of the phosphorous removal processes. (Reprinted, with permission, from S.N. Hong, 1982, A biological wastewater treatment system for nutrient removal, presented at the EPA Workshop on Biological Phosphorus Removal, Annapolis, Md., 1982); C.S. Block and S.N. Hong, 1984, *Treatment of wastewater containing phosphorus compounds*, U.S. patent 4,488,967; K. Kalb et al., 1990, Nutrified sludge—An innovative process for removing nutrients from wastewater, presented at the 63rd WPCF Conference, Washington, D.C. 1990; J.F. Manning and R.L. Irvine, 1985, The biological removal of phosphorus in sequencing batch reactors, *J. WPCF* 57:87; and R.T. Irvine et al., 1982, Summary report—workshop on Biological Phosphorus Removal in Municipal Wastewater Treatment, Annapolis, Md.)

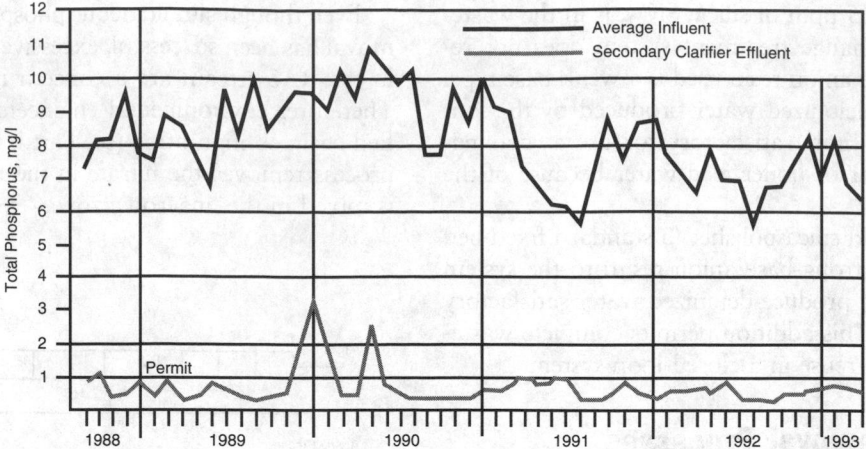

FIG. 7.38.5 Operating and performance data of the North Plant in Titusville, Florida (phosphorus removal without nitrification).

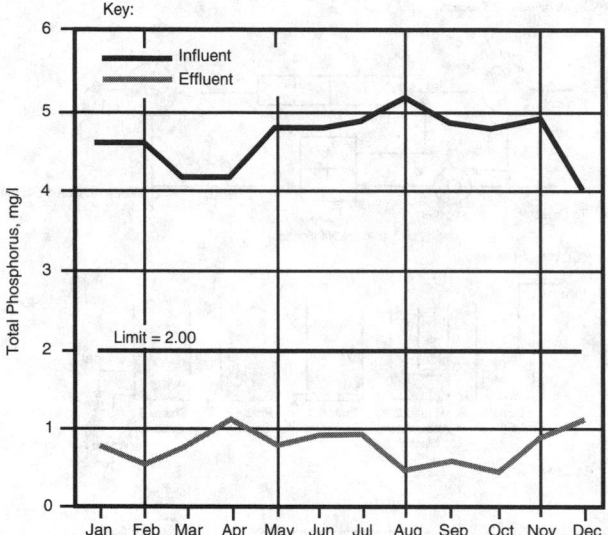

FIG. 7.38.6 Operating and performance data of a plant in Lancaster, Pennsylvania (with nitrification and oxygen aeration).

OWASA NUTRIFICATION PROCESS

The BOD/phosphorus ratio in influent wastewater is an important parameter for an EBPR process to achieve a low effluent phosphorus concentration. Typically, municipal wastewater contains a sufficient BOD/phosphorus ratio for biological phosphorus removal. However, in cases where excessive infiltration by rainwater into sewer lines or substrate consumption in pretreatment processes exist, the influent BOD concentration for biological phosphorus removal can be insufficient. Therefore, separate unit processes for side-stream production of BOD, mainly VFAs, supplement BOD requirements. Different methods for producing VFAs have been reported (Rabinowitz et al. 1987).

The OWASA process (see Part c in Figure 7.38.4) uses anaerobic fermentation of primary sludge to produce VFAs. This process mixes the VFA-enriched supernatant stream with the RAS in an anaerobic zone. The trickling-filter-treated influent wastewater flows directly into the oxic zone together with anaerobically conditioned RAS. Since the required BOD is supplemented by fermentation, a sufficient BOD/phosphorus ratio is maintained. Phosphorus removal performance is reliable and consistent with an effluent phosphorus concentration of 0.4 mg/l or less. Wastewater treatment facilities can also use this process for simultaneous biological phosphorus and nitrogen removal by increasing detention time in the oxic zone. Currently, one U.S. plant is in operation, located in Carrboro, North Carolina.

SEQUENCING BATCH REACTORS

All laboratory studies of the A/O process were performed with automatic-batch-fill-drawing (ABFD) apparatus with sequential operations of anaerobic mixing, aeration, and clarification in the same reactor. This concept has been used for full-scale operation and is an SBR. Part d in Figure 7.38.4 shows the sequence of operations in an SBR. An SBR can accomplish biological phosphorus removal alone or with nitrification (Manning and Irvine 1985). However, the application of SBRs is limited to low flow rates, typically 5 mgd or less.

PHOSTRIP PROCESS

Part e in Figure 7.38.4 shows a schematic configuration of the PhoStrip process. This process combines both biological and chemical phosphorus removal. Similar to other EBPR processes, this process subjects the biomass to alternating anaerobic/oxic environments to cultivate phosphorus-accumulating species. However, the process diverts a side-stream of phosphorus-enriched, activated sludge into a stripper where anaerobic conditions promote the release of intracellular phosphate. Then, the process treats the phosphorus-enriched, supernatant stream with lime to chemically precipitate phosphorus. The activated sludge, stripped of phosphorus, returns to the aeration basin for further uptake of phosphorus into the biomass under oxic conditions.

This process was used in several installations during the late 1970s and early 1980s (Peirano 1977; Northrop and Smith 1983). These installations reported good operating data but also had long periods of operation and maintenance problems. In some cases, the traditional chemical precipitation of phosphorus had lower operating costs than the PhoStrip process (Peirano 1977). These higher costs resulted in many installations terminating the PhoStrip process.

CHEMICAL PRECIPITATION

TYPES OF PRECIPITATING CHEMICALS

  a) Alum
  b) iron: (b1) ferrous, (b2) ferric
  c) lime

PHOSPHORUS DISCHARGES PER CAPITA PER YR

  In the form of human waste, 1.2 lb; in the form of detergents, 2.3 lb

RANGE OF ALUM CONCENTRATION USED

  50 to 300 mg/l

IDEAL pH RANGE FOR PRECIPITATION

  4 to 6 (a), 4.5 to 5 (b2), 7 to 8 (b1), and 10 to 11 (c)

WEIGHT RATIO OF ALUM REQUIRED TO PHOSPHORUS REMOVED

  22:1 for 95% removal; 16:1 for 85% removal

ALUM REQUIREMENT TO REMOVE 85% OF A 10 MG/L PHOSPHORUS CONTENT

  160 mg/l or 1335 lb per million gal

Wastewater treatment facilities can effectively use metal salts such as alum, ferrous sulfate, and ferric chloride to precipitate phosphorus in wastewater in conventional activated-sludge processes. Alum and ferric chloride are typically added into mixed liquor at the end of the aeration basin. Ferrous sulfate is added toward the front, allowing oxidation of ferrous into ferric ions. The molar ratio of metal ions to phosphorus depends on the required effluent phosphorus concentration and wastewater pH. Typically, the lower the effluent phosphorus concentration, the higher the molar ratio.

Environmental engineers often incorporate chemical precipitation into the design of EBPR processes as a standby or emergency measure to supplement biological processes in overcoming upset situations or augment phosphorus removal beyond the capacity of the biological process in meeting low effluent limits.

Many ionic forms effectively precipitate phosphorus from solution. The most notable are aluminium, calcium, and iron due to their low cost and general availability. Table 7.38.5 shows that all three of these ionic materials

(multivalent metallic cations) form insoluble precipitates with phosphorus. In general, the degree of phosphorus removal by chemical precipitation is a function of the following factors:

- initial phosphorus concentration
- precipitating cation concentration
- concentration of other anions competing with phosphorus for precipitating cations
- wastewater pH

The tendency of aluminum and iron to hydrolyze in aqueous solution creates competition between the hydroxide and phosphate ions for the precipitating metal ions. Thus, the efficiency of phosphorus removal depends on the relative concentrations of these two anions in solution and is consequently pH dependent. A decrease in pH or hydroxide favors phosphate precipitation with metallic cations. When calcium is the precipitant, competition for calcium is predominantly between the phosphate and carbonate anions. As Table 7.38.5 shows, hydroxylapatite, $Ca_{10}(OH)_2(PO_4)_6$, is the most stable calcium phosphate solid phase.

Nitrifying/Nitrogen Removal Systems

The demand and applications for nitrogen removal from wastewater have steadily increased. For example, wastewater treatment plants discharging directly to aquifers are required to remove nitrates to limit drinking water contamination. A large amount of full-scale experience is available in nitrifying/denitrifying systems. This section reviews nitrogen removal systems in two categories: systems that do and do not use internal recycling streams. It also provides flow schemes.

A common problem for plants that cannot completely nitrify is their inability to maintain a sufficient aerobic (oxic) sludge retention time (SRT). Aerobic SRT is the average amount of time that a single microorganism spends

TABLE 7.38.5 EQUILIBRIUM CONSTANTS RELATED TO PHOSPHATE PRECIPITATION WITH METAL IONS

Reaction	Log equilibrium constant,[a] 25°C
$Fe^{3+} + PO_4^{3-} = FePO_4(s)$	23
$3Fe^{2+} + 2PO_4^{3-} = Fe_3(PO_4)_2(s)$	30
$Al^{3+} + PO_4^{3-} = AlPO_4(s)$	21
$Ca^{2+} + 2H_2PO^-_4 = Ca(H_2PO_4)_2(s)$	1
$Ca^{2+} + HPO^-_4 = CaHPO_4(s)$	6
$10Ca^{2+} + 6PO_4^{3-} + 2OH^-$ $= Ca_{10}(OH)_2(PO_4)_6(s)$	90

Note: [a]The higher the equilibrium constant, the more stable the precipitate formed.

in the aerated portion of the process. To increase this parameter, wastewater treatment facilities can make several operational adjustments including increasing the operating mixed liquor suspended solids concentration (MLSS); or increasing the influent detention time (IDT) in the aerobic section of the treatment process.

Most activated-sludge treatment processes, can realize nitrification when a sufficient sludge age is maintained, DO is available, and no inhibitors are present in the influent wastewater (such as specific compounds or pH extremities). Additionally, nitrification and phosphorus removal can be realized in the same process; the A/O, OWASA, SBRs, and PhoStrip processes can all achieve nitrification.

SYSTEMS WITHOUT INTERNAL RECYCLING

The Wuhrman process, Ludzack–Ettinger process, and oxidation ditches are nitrogen removal systems without internal recycling.

Wuhrmann Process

The Wuhrmann process, or post-denitrification, achieves nitrification and carbonaceous oxidation before the wastewater enters the anoxic zone for denitrification (see Part a in Figure 7.38.7). Endogeneous respiration provides the required carbon source since all available extracellular carbon has been removed.

Although the efforts of Wuhrmann helped to develop other single sludge nitrification/denitrification systems, this process has never been used in full scale. Operational problems include high turbidity levels of the clarified effluent, ammonia release from cell lysis in the anoxic zone, and high nitrate levels due to low denitrification rates (U.S. EPA 1993).

Ludzack–Ettinger Process

Part b in Figure 7.38.7 shows the Ludzack–Ettinger process. Because influent wastewater is directed first into an anoxic zone followed by an aerobic zone, this process is called pre-denitrification. Since nitrification occurs after the anoxic zone, the RAS stream recycles the nitrates. As such, this process typically operates with a high RAS return rate (75–150% Q). The influent wastewater serves as the carbon source for denitrification and thus has a higher denitrification rate than the Wuhrmann process.

Oxidation Ditches

Oxidation ditches successfully control total nitrogen effluent concentrations either by encouraging simultaneous biological nitrification/denitrification (SBND) in the same reactor or alternating biological processes between reactors.

Wastewater treatment facilities can achieve SBND by strategically locating the aeration equipment in the process reactors. By doing so, they can create alternating aerobic and anoxic zones. Phased isolation ditch technology, used extensively in Europe, alternates biological nitrification and denitrification in separate reactors. However, this technology differs from SBRs since wastewater is continuously discharged from the system. The BioDenitro process, developed by Krüger A/S of Denmark, has demonstrated applicability and potential in the United States (Tetreault et al. 1987). Part c in Figure 7.38.7 is a schematic diagram of the BioDenitro process.

SYSTEMS USING INTERNAL RECYCLING STREAMS

The modified Ludzack–Ettinger (MLE) process and the four-stage Bardenpho process remove nitrogen using internal recycling streams.

MLE Process

The MLE process is based on the premise that an insufficient amount of nitrate is returned in the RAS stream. By adding an internal recycle (typically 100–400% Q) from the end of the aerobic zone, this process increases the amount of nitrate returned to the anoxic zone for subsequent denitrification. As such, the MLE process realizes a greater degree of total nitrogen (TN) reduction. Part d in Figure 7.38.7 is a flow schematic diagram of the MLE process.

Four-Stage Bardenpho Process

The Bardenpho process provides a TN removal capability that cannot be obtained in the MLE process. The four-stage Bardenpho process subjects the nitrate that was not recycled from the primary aerobic zone to anoxic conditions in a secondary anoxic zone (see Part e in Figure 7.38.7).

Although biological denitrification is occurring in both anoxic reactors, the carbon source is different; the carbon source of the primary anoxic zones is supplied by influent wastewater, whereas endogenous respiration is responsible for any denitrification in the secondary anoxic zone. A small secondary aerobic zone prior to secondary clarification strips away any nitrogen gas entrained in the solids and nitrifies any ammonia released from cell lysis.

Simultaneous Phosphorus and Nitrogen Removal Processes

Several versions of biological processes for simultaneous phosphorus removal and nitrification/denitrification are installed and operated throughout the world. Table 7.38.6 summarizes the capability of each process, and Figure 7.38.8 summarizes their configurations. The following paragraphs describe each process.

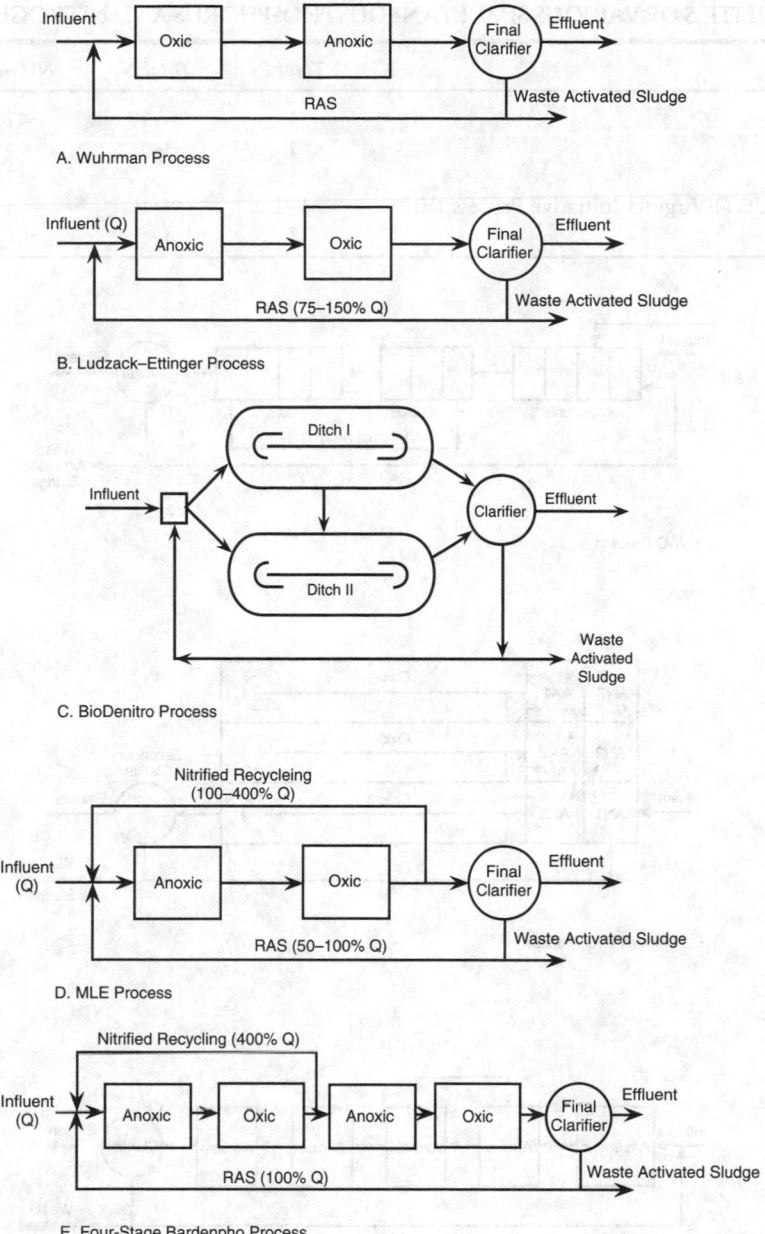

FIG. 7.38.7 Schematic diagrams of the nitrogen removal processes without internal recycling.

A²/O PROCESS

Part a in Figure 7.38.8 is a typical schematic representation of the A²/O process. The process consists of three zones in series: anaerobic, anoxic, and oxic. Each zone is further divided into several compartments. The mechanisms of phosphorus removal and nitrification are described in the preceding sections. This process recycles the nitrified mixed liquor containing nitrate and the nitrite at the end of the oxic zone. This internal recycled stream goes to the anoxic zone for denitrification at a rate of 100 to 200% of the influent wastewater flow rate. The process uses the organics in the influent wastewater as the carbon source for denitrification.

Influent detention times for the anaerobic and anoxic zones are generally 1 to 2 hr, while the oxic zone is 4 to 8 hr, depending on the influent wastewater characteristics. This process typically uses submersible mixers for solids mixing in the anaerobic and anoxic zones. Different types of aeration devices are used to satisfy oxygen demands in the oxic zone.

The A²/O process can achieve an effluent quality with 1 mg/l or less in total phosphorus and ammonia. However, effluent NO_x–N concentration is typically limited to 6 to 10 mg/l and depends on the NH_3–N concentration in the influent wastewater as well as the internal recycling rate. This process is used in more than twenty installations in

TABLE 7.38.6 CAPABILITIES OF VARIOUS SIMULTANEOUS PHOSPHORUS AND NITROGEN REMOVAL PROCESSES

Process	Total P	Total N	NH$_3$–N	Years in Operation, yr
A^2/O	1	6–10	<1	>15
Improved A^2/O	1	3–5	<1	1
Modified Bardenpho	1	3	<1	>15
University of Cape Town (UCT)/Virginia Initiative Process (VIP)	1–2	8–10	<1	3
BioDenipho	1	3–5	<1	>15

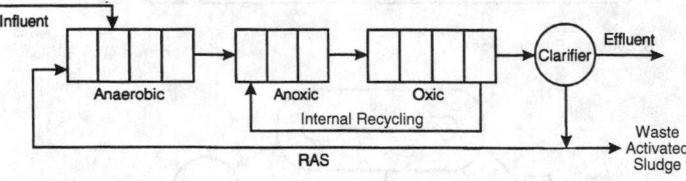

A. A^2/O Process

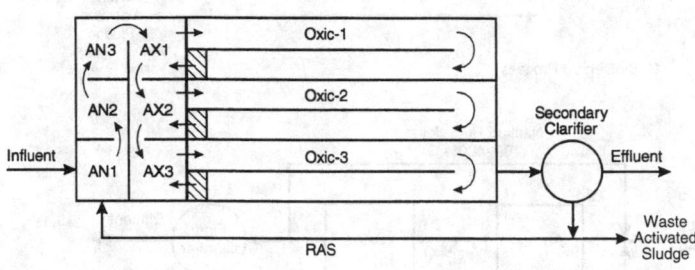

B. Improved A^2/O Process

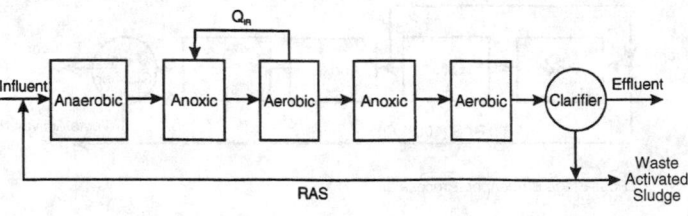

C. Modified Bardenpho Process

FIG. 7.38.8 Schematic diagrams of simultaneous phosphorus and nitrogen removal processes.

the United States. Table 7.38.7 shows the design parameters of some of these A^2/O systems.

IMPROVED A2/O PROCESS

This process further optimizes the NO$_x$–N removal in the A^2/O process. It recycles nitrified mixed liquor from each oxic stage to an anoxic stage for denitrification. Compared to the A^2/O process, this process maintains higher BOD concentrations in the first two anoxic stages and realizes higher denitrification rates. Furthermore, this process has an exhauster with 15 to 30 min detention time installed

at the end of each oxic stage to deplete the DO in the mixed liquor prior to recycling to the anoxic stage. This scheme further improves the denitrification rate.

This process can achieve an effluent NO$_x$–N of 3 to 6 mg/l. Part b in Figure 7.38.8 is a schematic diagram of this process.

MODIFIED BARDENPHO PROCESS

This process, also known as the Phoredox process, is composed of an anaerobic zone preceding the Bardenpho process to achieve biological phosphorus removal. Part c

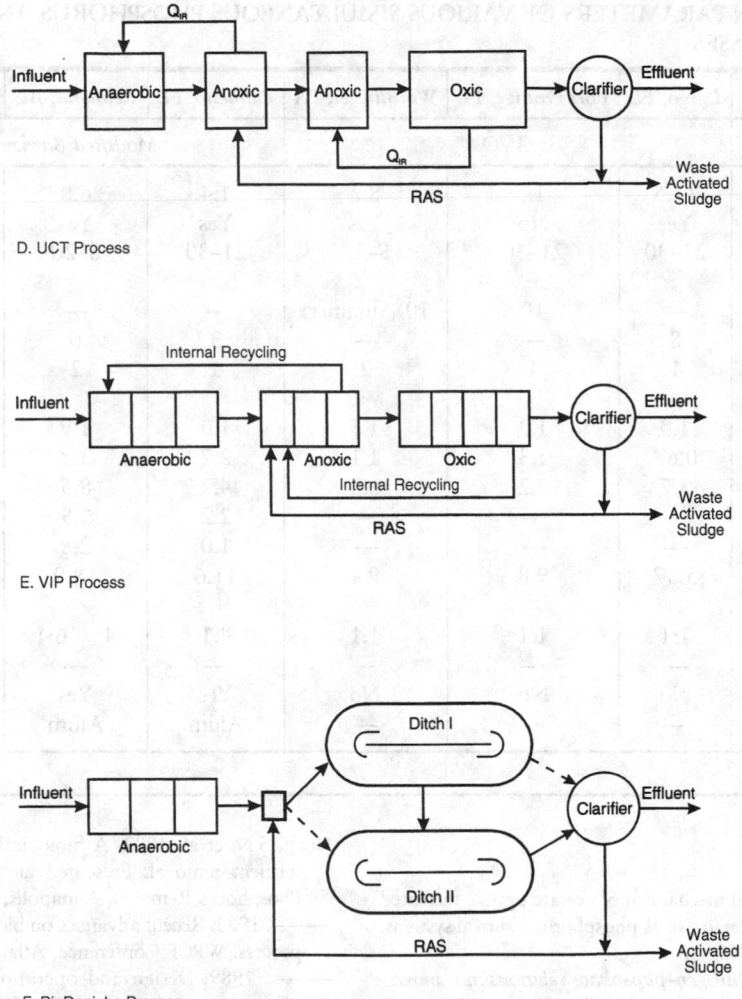

FIG. 7.38.8 *Continued*

in Figure 7.38.8 is a schematic diagram of the process. It is similar to the Bardenpho process in that the design includes an influent detention time of 18 to 24 hr, an internal recycling rate of 200 to 400% of the influent flow rates, and no partitioning of each zone.

This process can achieve an effluent NO_x–N of 2 mg/l or less with proper detention time in the primary and secondary anoxic zones. However, the inherently low F/M of the process requires chemical polishing to achieve an effluent total phosphorus concentration of 1 mg/l. This process is used by more than 10 installations in the United States. Table 7.38.7 shows the design parameters.

UCT/VIP PROCESSES

Parts d and e in Figure 7.38.8 are schematic diagrams of the UCT and VIP processes. The flow arrangement of the two processes is identical except that the VIP process is partitioned in the anaerobic, anoxic, and oxic zones. Both processes discharge RAS into the anoxic zone instead of the anaerobic zone to avoid the detrimental effects of re-

cycling NO_x–N on biological phosphorus removal. An additional return stream recycles MLSS from the end of the anoxic zone to the anaerobic zone. Currently, only one installation in the United States uses these processes.

BIODENIPHO PROCESS

Phased isolation-ditch technologies are also used for simultaneous phosphorus and nitrogen removal. The BioDenipho process includes an anaerobic zone that precedes the BioDenitro process to achieve enhanced biological phosphorus removal (EBPR). Part f in Figure 7.39.8 is a schematic diagram of the BioDenipho process. This process typically operates at a low F/M with a long hydraulic detention time and can achieve low effluent TN requirements.

—*Sun-Nan Hong*
R. David Holbrook

TABLE 7.38.7 DESIGN PARAMETERS OF VARIOUS SIMULTANEOUS PHOSPHORUS AND NITROGEN REMOVAL PROCESSES

	Largo, FL	*Port Orange, FL*	*Warminster, PA*	*Palmetto, FL*	*Kelowna, BC*	*Orange Co., FL*	*Louisburg, NC*
Process		*A²/O*			*Modified Bardenpho*		*BioDenipho*
Plant Capacity, mgd	12.5	12	8.2	1.4	6.0	7.5	1.5
Primary Clarification	Yes	No	Yes	Yes	Yes	No	No
Wastewater Temp, °C	21–30	21–30	9–22	21–30	8–20	21–30	10–20
Discharge Limitation							
NO_x–N	—	10	10 (Summer)	—	—	—	—
Total N	8	—	—	3	6	3	Future limit
Total P	1	1	2	1	2	1	2
Influent Detention Time, hr							
Anaerobic	1.0	1.3	1.5	1.0	1.9	1.9	1.5
Anoxic 1	0.67	1.3	1.1	2.7	3.8	3.4	—
Oxic 1	3.7	7.2	6.8	4.7	8.5	10.7	22.5
Anoxic 2	—	—	—	2.2	1.9	1.9	—
Oxic 2	—	—	—	1.0	2.8	0.3	—
Total	5.37	9.8	9.4	11.6	18.9	18.2	24[1]
Internal Recycle							
First Anoxic	1:1	1:1	1:1	4:1	4 ~ 6:1	4 ~ 6:1	—
Anaerobic	—	—	—	—	—	—	—
Chemical Addition	No	No	No	Yes	Yes	Yes	No
Type	—	—	—	Alum	Alum	Alum	—

Note:[1]State mandated.

References

Aruw, V., et al. 1988. Biological mechanism of acetate uptake mediated by carbohydrate consumption in excess phosphorus removal systems. *Water Research* 22, no. 5:565.

Bargman, R.D., et al. 1971. *Nitrogen–phosphate relationships and removals obtained by treatment processes at the Hyperion Treatment Plant.* Pergamon Press Ltd.

Berber, A., and C.T. Winter. 1984. The influence of extended anaerobic retention time on the performance of phoredox nutrient removal plant. *Water Science Technology* 17, no. 81.

Borchardt, J.A., and H.A. Azad. 1968. Biological extraction of nutrients. *J. WPCF* 40, no. 10:1739.

Bordacs, K., and S.C. Chiesa. 1988. Carbon flow patterns in enhanced biological phosphorus accumulating activated sludge cultures. IAW-PRC Conference, Brighton, England.

Charley, R.C., D.G. Hooper, and A.G. McLee. 1980. Nitrification kinetics in activated sludge at various temperatures and dissolved oxygen concentrations. *Water Research,* 14:1387–1396.

Claete, T.E., and P.L. Steyn. 1988. The role of acinetobactor as a phosphorus removing agent in activated sludge. *Water Research* 22, no. 8.

Connell, C.H., and D. Vacker. 1967. Parameters of phosphate removal by activated sludge. *Proceedings 7th Industrial Water and Waste Conference, University of Texas, Austin, Texas.* II-28–37.

Davelaar, D., et al. 1978. The significance of an anaerobic zone for the biological removal of phosphate from wastewaters. *Water SA* 4, no. 2:54.

Davies, T. 1971. Population description of a denitrifying microbial system. *Water Research,* 5:553.

DeFoe, R.W., et al. 1993. Large scale demonstration of enhanced biological phosphorus removal with nitrification at a major metropolitan wastewater treatment plant. WEF Conference, Anaheim, CA.

Fuchs, G.W., and M. Chen. 1975. Microbial basis of phosphate removal in the activated sludge process for the treatment of wastewater. *Microbial Ecology,* no. 119.

Hong, S.N. et al. 1982. A biological wastewater treatment system for nutrient removal. Presented at the EPA workshop on Biological Phosphorus Removal, Annapolis, Maryland.

———. 1983. Recent advances on biological nutrient control by the A/O process. WPCF Conference, Atlanta, GA.

———. 1989. Design and operation of a full-scale biological phosphorus removal system. The 52nd WPCF Conference, Houston, TX.

Hong, S.N., and K.L. Andersen. 1993. Converting a single sludge oxygen activated sludge system for nutrient removal. The 66th WEF Conference, Anaheim, CA.

Joint Task Force of the Water Environment Federation and the American Society of Civil Engineers (ASCE). 1991.

Kang, S.J., et al. 1985. Full-scale biological phosphorus removal using A/O process in a cold climate. In *Management strategies for phosphorus in the environment.* London: Selper Ltd.

Krichten, D.J., et al. 1985. Applied biological phosphorus removal technology for municipal wastewater treatment by the A/O Process. In *Management strategies for phosphorus in the environment.* London: Selper Ltd.

Levin, G.V., and J. Shapiro. 1965. Metabolic uptake of phosphorus by wastewater organisms. *J. WPCF* 37, no. 6:800.

Manning, J.F., and R.L. Irvine. 1985. The biological removal of phosphorus in sequencing batch reactors. *J. WPCF* 57, no. 87.

Marais, G.V.R., and G.A. Ekema. 1976. The activated sludge process: Steady state behavior. *Water S.A.* 2:162–300.

McLaren, et al. 1976. Effective phosphorus removal from sewage by biological means. *Water SA,* no. 1.

Metcalf and Eddy, Inc. 1991. *Wastewater engineering: Treatment, disposal and reuse.* 3d ed. McGraw-Hill.

Nicholls, H.A., and P.W. Osborn. 1979. Bacterial stress: Prerequisite for biological removal of phosphorus. *J. WPCF* 51, no. 4:557.

Northrop, J., and D.A. Smith. 1983. Cost and process evaluation of phostrip at Amherst, NY. WPCF Conference, Atlanta, GA.

Peavy, M.S., D.R. Rowe, and G. Tchobanoglous. 1985. *Environmental Engineering.* McGraw-Hill.

Peirano, L.E. 1977. Low-cost phosphorus removal at Reno-Sparks, Nevada. *J. WPCF* 49:568.

Prescott, L.M., J.P. Harley, and D.P. Klein. 1990. *Microbiology*. Wm. C. Brown Publishers.

Rabinowtiz, B., et al. 1987. A novel operational model for a primary sludge fermenter for use with the enhanced biological phosphorus removal process. IAWPRC Specialized Conference, Rome, Italy.

Randall, C.W., J.L. Barnard, and M.D. Stensel. 1992. *Design and retrofit of wastewater treatment plants for biological nutrient removal*. Vol. 5. Technomic Publishing Co.

Sell, R.L., et al. 1981. Low temperature biological phosphorus removal. The 54th WPCF Conference, Detroit, MI.

Sharma, B., and A.C. Ahlert. 1977. Nitrification and nitrogen removal. *Water Research* 11:897–925.

Spector, M.L. 1977. Production of non-bulking activated sludge. U.S. Patent 4,056,465.

Sutton, P.M., K.L. Murphy, and B.C. Jank. 1974. Biological nitrogen removal—The efficiency of the nitrification step. 47th Annual WPCF Conference, Denver, Colorado, October, 1974.

Tam, N.F.Y., Y.S. Wong, and G. Leung. 1992. Effects of exogenous carbon sources on removal of inorganic nutrient by the nitrification–denitrification process. *Water Research* 26:9, no. 9:1229–1236.

Tetreault, M.S., B. Rusten, A.M. Benedict, and J.F. Kreissel. 1987.

Assessment of phased isolation ditch technologies. *J. WPCF* 59, no. 9:833–840.

Tracy, K.D., and A. Flammino. 1987. *Biochemistry and energetics of biological phosphorus removal*. Pergamon Press.

U.S. Environmental Protection Agency (EPA). 1975. *Process design manual for nitrogen control*.

———. 1987. *Summary report—The causes and control of activated sludge bulking and foaming*. EPA 625–8-87–012.

———. 1993. *Nitrogen control manual*. EPA 625–12–93–010.

Van Haandel, A.C., Ekema, G.A. and G.V.R. Marais. 1981. The activated sludge process—3. *Water Res.* 15:1135–1152.

Wanielista, M.P., and W.W. Eckenfelder, Jr. 1978. *Advances in water and wastewater treatment—Biological nutrient removal*. Ann Arbor Science.

Water Pollution Control Federation. 1983. *Nutrient control—Manual of practice No. FD-7, Facilities Design*.

Wells, N.W. 1969. Differences in phosphate uptake rates exhibited by activated sludges. *J. WPCF* 41, no. 5:765.

Wentzel, W.C., et al. 1985. Kinetics of biological phosphorus release. *Water Sci. Tech.* 17, no. 57.

———. 1988. Enhanced polyphosphate organism cultures in activated sludge systems—Part I: Enhanced culture development. *Water SA*, 14 no. 81.

Chemical Treatment

7.39
NEUTRALIZATION AGENTS AND PROCESSES

NEUTRALIZING REAGENTS

a) Acidic: a1) Sulfuric acid, a2) Hydrochloric acid, a3) Carbon dioxide, a4) Sulfur dioxide, a5) Nitric acid
b) Basic: b1) Caustic soda, b2) Ammonia, b3) Soda ash, b4) Hydrated lime, b41) Dolomitic hydrated quicklime, b42) High-calcium, hydrated quicklime, b5) Limestone

SOLUBILITY AND pH OF LIME SOLUTIONS

For a concentration range of 0.1 to 1.0 gm of CaO per liter, the corresponding pH rises from 11.5 to 12.5 at 25°C. Saturation occurs at about 1.2 gm/l

LIME AVAILABILITY

Dry forms with several particle-size distributions or liquid forms usually as a slurry with +20 wt % solids

CAUSTIC AVAILABILITY

In dry form as 75% Na_2O or as a 50% NaOH solution, precipitates formed are usually soluble.

TOXICITY OF REAGENTS

See Tables 7.40.7 and 7.40.8.

REAGENT REQUIREMENTS FOR NEUTRALIZATION

See Figure 7.40.5.

Process Description

Neutralization is the restoration of the hydrogen (H^+) or hydroxyl (OH^-) ion balance in solution so that the ionic concentrations of each are equal. Conventionally, the notation pH (puissance d'hydrogen) describes the hydrogen ion concentration or activity present in a solution as follows:

$$pH = -\log_{10} [H^+] \qquad 7.39(1)$$

$$pH = -\log_{10} a_{H^+} \qquad 7.39(2)$$

where:

$[H^+]$ = the hydrogen ion concentration, gmol per liter
a_{H^+} = hydrogen ion activity

For a dilute solution of strong acids, i.e., acids considered completely dissociate (ionized in solution), the following equation applies:

$$a_{H^+} = [H^+] \qquad 7.39(3)$$

At neutrality, the concentration of hydrogen and hydroxyl ions is equal. The product of their ion concentration (K_w) at 25°C is as follows:

$$(-\log_{10} [H^+])(-\log_{10} [OH^-]) = -\log_{10}K_w$$

$$K_w = 1.008 \times 10^{-14} \qquad 7.39(4)$$

At neutrality, the following equations apply:

$$pH = \frac{-\log_{10}K_w}{2} = 7.0 = pOH \qquad 7.39(5)$$

$$pH + pOH = 14 \qquad 7.39(6)$$

Thus, if a solution has a pH = 2.0 at 25°C, hydrogen ion concentration is 1×10^{-2} moles H^+ per liter, pOH = 12, and hydroxyl ion concentration is 1×10^{-12} moles OH^- per l. The ion product of water depends highly on temperature, changing approximately two orders of magnitude over a 60°C span (see Figure 7.39.1).

The pH notation as a means of expressing the hydrogen ion concentration is logarithmic. A pH change from 2.0 to 1.0 does not mean that the ion concentration has doubled; a change of one pH unit is an order of magnitude change. Thus, if an acid influent changes by three pH units, the $[H^+]$ is changing by a factor of one thousand. This logarithmic nature becomes an important consideration when reagent delivery systems are sized because if the ion load to be neutralized changes by a factor of 1000, the reagent delivery system must have the same turndown (rangeability).

The need to neutralize, or at least place limits on, the pH variation of environmental waters has resulted in the promulgation of water quality standards legislation in virtually every state. The physical well-being of all life forms depends not only on the absolute value of the pH but also on the frequency of pH variation. Thus, for example, the lacrimal fluid of the human eye has a nominal pH of 7.4 and a high buffering capacity, i.e., it resists changes in pH. Variations of the lacrimal fluid as low as 0.1 pH unit can result in eye irritation. The Federal Water Pollution Control Administration has published a report that details not only the pH requirements for water of a designated end use but also the requirements for twenty other ions, as well as organic chemical limitations and physical and microbiological properties. Table 7.39.1 summarizes the preferred or acceptable pH ranges for various water quality categories.

Common Neutralization Reagents

Wastewater treatment facilities must counter the hydrogen or hydroxyl ion imbalance in a waste effluent by adding a material that restores the ion balance. Thus, if the waste effluent is acidic, i.e., pH < 7.0, they must blend a reagent having basic characteristics with the waste to achieve neutrality. Conversely, if the waste effluent is basic, i.e., pH > 7.0, they must use a reagent having acid characteristics. Table 7.39.2 lists common neutralization reagents.

In addition to the reagents listed in Table 7.39.2, waste acids and bases can also serve as neutralizing reagents. In some cases, particularly in ion-exchange resin regeneration, in which the resin bed is treated first with a caustic solution and then with an acid solution, wastewater treatment facilities can store these solutions and then blend them to achieve a neutral solution rather than discharge them to the sewer immediately after use.

Four widely used reagents are sulfuric acid, caustic soda, hydrated chemical lime, and (to a limited degree) limestone. The main reasons for their popularity are economy and ease of handling.

LIME

Chemical lime is produced by the calcination of high-quality limestone which produces either high-calcium quicklime or dolomitic quicklime. Further treatment of high-calcium quicklime and dolomitic quicklime produces their hydrated counterparts. Figure 7.39.2 shows the chemical lime production process.

Table 7.39.3 shows typical analyses of the two types of quicklimes. The difference in chemical composition between high-calcium and dolomitic quicklimes results in varying reaction rates and reactivities when these materials treat the same wastewater. Figure 7.39.3 demonstrates the reaction rate characteristics of high-calcium and dolomitic lime (2% excess) at 25°C when reacted with 0.1N H_2SO_4.

The reaction rate curve for high-calcium lime in the figure was obtained with 2% excess lime. Lesser amounts of

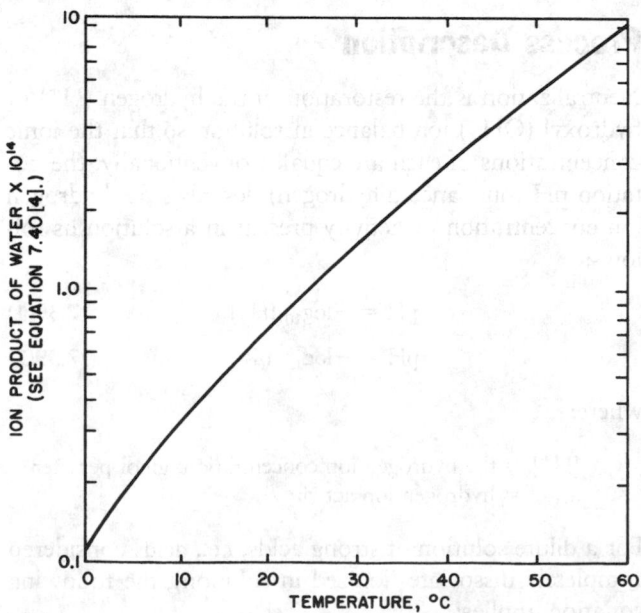

FIG. 7.39.1 Ion product of water as a function of temperature.

TABLE 7.39.1   pH RANGES FOR VARIOUS WATER QUALITY CATEGORIES

Water Quality Category	pH Range Preferred	pH Range Acceptable	Restrictions and Comments
I Recreation and Aesthetics	6.5–8.3	5.0–9.0	8.3 < pH < 6.5 discharges to have limited buffering capacity
II Public Water Supplies	Unspecified	6.0–8.5	
III Fish, Aquatic, and Wildlife	—	6.0–9.0	
A. Marine and estuarine organisms	—	6.7–8.5	Bulk location[a] change ≤ 0.1 pH
B. Wildlife	—	7.0–9.2	
C. Fresh water organisms	—	6.0–9.0	Alkalinity ≥ 20 mg $CaCO_3$ per liter
IV Agricultural Uses	6.0–8.5	5.5–9.0	
A. Farmstead	6.8–8.5	—	
B. Livestock	None Specified	Fish indicator ponds encouraged at terminal watershed locations	
C. Irrigation Water Supplies	—	4.5–9.0	
V Industrial	Highly dependent on the type of industry		

Source: Federal Water Pollution Control Administration, U.S. Department of Interior, 1968, *Report of the Committee on Water Quality Criteria* (Washington, D.C.: U.S. Government Printing Office [1 April]).

Note:[a] The entire water body receiving the water should not change its pH by more than 0.1 as a result of wastewater pH variations.

TABLE 7.39.2   COMMON NEUTRALIZATION REAGENTS

Acid Reagents	Base Reagents
Concentrated (66°Bé) sulfuric acid	Caustic soda (NaOH)
Concentrated (20 or 22°Be) hydrochloric acid	Ammonia
Carbon dioxide	Soda ash (Na_2CO_3)
Sulfur dioxide	Hydrated chemical lime ($Ca[OH]_2$)
Nitric acid	Limestone ($CaCO_3$)

excess lime result in longer reaction times. This consideration becomes important when environmental engineers size neutralization vessels. Due to the continuing reaction of lime, pH measured at the neutralization tank discharge may not be the final effluent pH downstream of the discharge, if the nominal time constant of the neutralization vessel is too short.

The solubility of lime (CaO) in aqueous solutions decreases with increasing temperature, e.g., 0.131 gm per 100 cc at 10°C and 0.07 gm per 100 cc at 80°C. Figure 7.39.4 plots the pH range of lime solutions up to saturation at 25°C. Lime as a neutralizing reagent is normally supplied in slurries or as a dry feed along with water to a small, agitated, holding container that overflows to the neutralization process. Normally, 20% or less lime slurries are used. The lower the lime concentration, the easier

the handling in terms of abrasion and clogging. Quicklime is available in bulk or bag form in reasonably standard sizes (see Table 7.39.4).

In addition to the dry form, a hydrated lime produced as a by-product in the manufacture of acetylene is available in slurry form having a nominal 35 wt % solids content. This lime slurry is delivered in tankwagons having a nominal capacity of 4500 gal. Table 7.39.5 gives a typical chemical analysis—particle size distribution—of this lime slurry. The cost of this lime form depends on the transportation cost.

Chemical reactions associated with hydrated limes depend on the type of acid(s) being neutralized. The reaction between sulfuric acid and hydrated, high-calcium and normal, dolomitic quicklimes shown in the following equations is typical:

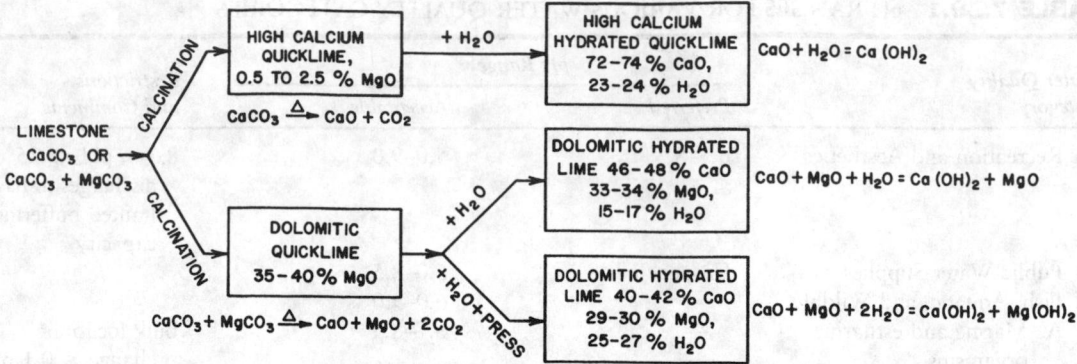

FIG. 7.39.2 Chemical lime production process. (Reprinted from National Lime Association, *Chemical lime facts,* Bulletin 214, Washington, D.C.)

TABLE 7.39.3   TYPICAL ANALYSES OF COMMERCIAL QUICKLIMES

Component	High-Calcium Quicklime, %	Dolomitic Quicklime, %
CaO	93.25–98.00	55.50–57.50
MgO	0.30–2.50	37.60–40.80
SiO_2	0.20–1.50	0.10–1.50
Fe_2O_3	0.10–0.40	0.05–0.40
Al_2O_3	0.10–0.50	0.05–0.50
H_2O	0.10–0.90	0.10–0.90
CO_2	0.40–1.50	0.40–1.50

Source: National Lime Association, *Chemical lime facts,* Bulletin 214 (Washington, D.C.).
Note: The range of values are not necessarily minimum and maximum percentages.

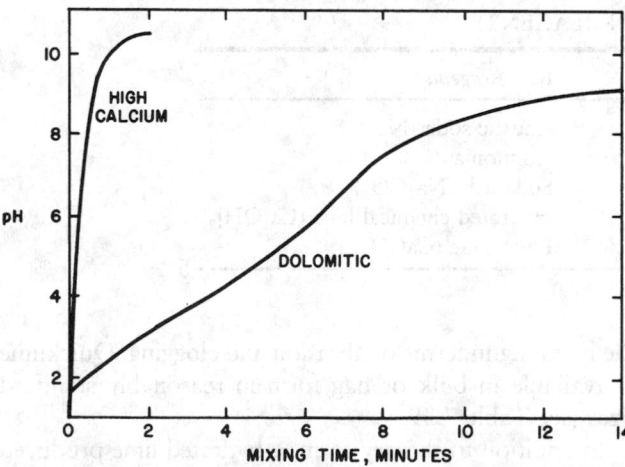

FIG. 7.39.3 Reaction rate characteristics of high-calcium and dolomitic quicklimes at 25°C for the neutralization of 0.1N sulfuric acid with a lime dosage of 2% in excess of the theoretical stoichiometric requirements. (Reprinted, with permission, from R.D. Hoak et al., 1948, *Ind. and Eng. Chem.* 40:2062; W.A. Parson, 1965, *Chemical treatment of sewage and industrial wastes,* 58, National Lime Association, Washington, D.C.)

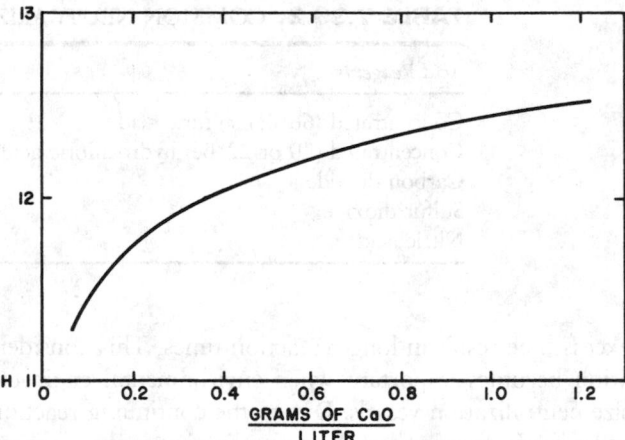

FIG. 7.39.4 pH of various lime concentrations in an aqueous solution. (Reprinted, with permission, from F.M. Lea and G.E. Bessey, 1937, *J. Chem. Soc.,* 1612.)

TABLE 7.39.4 AVAILABLE QUICKLIME SUPPLIES

| | Percentage Passing Through the Noted Mesh-Size Screen | | | | | |
| | Mesh Size and Opening | | | | | |
Name	No. 8 (0.093 in)	No. 20 (0.0328 in)	No. 100 (0.0058 in)	No. 200 (0.0029 in)	No. 325 (0.0017 in)	Remarks
Lump Lime	—	—	—	—	—	Up to 8 in diameter
Crushed or pebble lime	—	—	—	—	—	$\frac{1}{4}$ to $2\frac{1}{2}$ in diameter
Ground lime	100	60–100	40–60	—	—	—
Pulverized lime	100	100	85–90	—	—	—
Dried, hydrated lime	100	100	100	95	—	—
Air-classified lime	100	100	100	100	99.5	—

TABLE 7.39.5 CHEMICAL ANALYSIS AND PARTICLE-SIZE DISTRIBUTION OF HYDRATED CHEMICAL LIME FROM ACETYLENE PRODUCTION

| | Particle-Size Distribution | | | |
| | Percentage of Particles That Pass the Noted Mesh-Size Screens | | | |
Chemical Analysis	No. 20	No. 48	No. 100	No. 325
95% $Ca(OH)_2$ 1.5% $CaCO_3$ 0.25% MgO 1.6% Fe_2O_3 and Al_2O_3 1.1% Insolubles	99.9	99.2	97	85

Source: Chemline Corporation.

$$H_2SO_4 + Ca(OH)_2 \longrightarrow CaSO_4\downarrow + 2H_2O \qquad 7.39(7)$$

$$2H_2SO_4 + Ca(OH)_2 + MgO \longrightarrow$$
$$CaSO_4\downarrow + MgSO_4 + 3H_2O \quad 7.39(8)$$

These reactions produce an insoluble product ($CaSO_4$); however, for influent pH values of 2.0 or higher, the quantity of $CaSO_4$ produced is insufficient to cause precipitation. Hydrofluoric acid reacted with lime produces a reaction product that is about two orders of magnitude less soluble than $CaSO_4$.

LIMESTONE

Limestone ($CaCO_3$) is an effective means of neutralizing waste acids. The process involves flowing the waste acids (mixtures of HCl and H_2SO_4) through a bed of limestone granules. The bed can be 3 ft deep at waste flows of from 1.3 to 2.2 gpm/sq ft of bed area. Recycling the treated effluent (from one to three times) and aeration can also improve performance.

This approach to waste neutralization appears to be financially attractive but has limited reliability. The main reason for decreased reliability is the contamination of limestone by materials such as oil and grease with the subsequent loss of bed activity.

SODIUM HYDROXIDE (CAUSTIC)

The use of sodium hydroxide (caustic) is about equal to lime as a neutralizing agent. Although caustic is more expensive than lime, its reaction characteristics (virtually instantaneous) and handling convenience are factors behind its widespread use. Researchers evaluated the reactivities of various basic reagents on the same chemical system by mixing them with pickle liquor (60 gm iron and 20 gm sulfate per liter). They measured the effectiveness of reagent reactions with pickle liquor in terms of iron gm remaining in solution after 6 hr. These experiments were conducted at room temperature and at 60°C with and without aeration (agitation). Table 7.39.6 shows the results of these experiments.

Concerning the speed of reaction of NaOH (see Table 7.39.6) and the reaction rate curve for quick and dolomitic limes (see Figure 7.39.3), the data show that when reaction time is important, the efficacy of NaOH, particularly in solution form, is virtually instantaneous. The ease of delivery to the neutralization process is another advantage of NaOH solutions. However, NaOH solutions are corrosive, and safety showers located in the process area are suggested. Personnel working with this material should use eye and skin protective devices.

Sodium hydroxide is available in solid (75% Na_2O) or solution (50% NaOH) form. In concentrated solution form, heating containers and lines may be required for transfer operations during cold weather. Most large suppliers of inorganic chemicals can furnish solutions of any strength in tank trucks and pump them directly to the user's reagent storage tank. Users should equip reagent storage tanks located outdoors with heaters and appro-

TABLE 7.39.6 IRON CONCENTRATION (gr/L) REMAINING AFTER SIX HOURS OR LESS

Neutralizing Agent	Test Conditions / Room Temperature and No Aeration	60°C and No Aeration	Room Temperature with Aeration	60°C with Aeration
NaOH	Reaction Is Practically Instantaneous			
Na$_2$CO$_3$	0 in 0.75 hr	0 in 0.75 hr	0 in 0.75 hr	0 in 0.5 hr
CaO	0 in 0.25 hr	0 in 0.25 hr	0 in 5 min	0 in 5 min
CaO · MgO	1.88	3.14	1.04	0.30
Ca(OH)$_2$	0 in 0.5 hr	0 in 0.5 hr	0 in 0.5 hr	0 in 0.5 hr
Ca(OH)$_2$ · MgO	1.23	1.53	0.55	0 in 3.5 hr
Acetylene Sludge[a]	1.66	1.04	0 in 3.5 hr	0 in 3.5 hr
CaCO$_3$[b]	20.40	18.80	2.95	0.03

Source: R.D. Hoak, 1950, *Sewage and Ind. Wastes* 22:212.
Notes: [a]Ca(OH)$_2$ waste product. From $CaC_2 + 2H_2O \longrightarrow C_2H_2 \uparrow + Ca(OH)_2$.
[b]Pulverized limestone.

priate control equipment to protect against freezing during winter. As with lime, neutralization reactions depend on the acid or acids being neutralized as follows:

$$2NaOH + H_2SO_4 \longrightarrow Na_2SO_4 + 2H_2O \qquad 7.39(9)$$

$$NaOH + HCl \longrightarrow NaCl + H_2O \qquad 7.39(10)$$

$$NaOH + HNO_3 \longrightarrow NaNO_3 + H_2O \qquad 7.39(11)$$

For the reactions described by Equations 7.39(9), (10), and (11), all products are highly soluble, and unlike lime, at least for high H$_2$SO$_4$ concentrations, precipitate accumulation is not a problem.

SODA ASH

Soda ash (Na$_2$CO$_3$) is not widely used for neutralization. It does, however, have widespread use in applications requiring minor pH adjustment (one unit or less) and in water softening. Ordinarily, it is supplied in solid form, which means solid feeders or solution makeup equipment.

In addition to being about five times as expensive as hydrated chemical lime, soda ash produces carbon dioxide in reaction for applications where strong acids are treated. Carbon dioxide can cause frothing, particularly in agitated neutralization vessels. Like lime, minimal safety hazards are connected with handling this material.

AMMONIA

Ammonia, like soda ash, is not a prime neutralizing reagent. It is about twice as expensive as lime, and is the most toxic of all the alkaline neutralization reagents discussed, having a toxic hazard rating of 3 in all three acute local categories, i.e., irritant, ingestion, and inhalation, and U in the acute systemic category. Table 7.39.7 outlines these toxic hazard rating codes.

Ammonia is widely used in the petroleum industry where it is added to crude oil to neutralize acid constituents. The most serious deterrent for ammonia use is that it produces ammonium salts that supply nutrients (nitrogen) for algal growth.

The most important alkaline reagents are lime and sodium hydroxide. The choice between them depends largely on the economics of the total neutralization or treatment facility. When large volumes of waste are treated, requiring large amounts of reagents, wastewater treatment facilities chose lime because the cost is a significant portion of the total treatment cost. Since lime usually requires a substantial investment in slakers, tanks, pumps, and additional agitators, the small wastewater treatment facility can not afford such equipment, and a single tank filled (on a scheduled basis) with caustic solution by a local chemical supplier is the logical selection.

ACIDIC REAGENTS

Table 7.39.8 outlines the toxic hazard associated with acid reagents. Although five acids are listed as reagents, sulfuric acid, especially the 66°Bé (93.2% H$_2$SO$_4$), is by far the most widely used. It is the least expensive in highly concentrated solutions (66°Bé), it is noncorrosive (eliminating the need for special construction materials), and efficient in highly concentrated solutions. This last reason can cause problems if waste flows are small because the availability of small reagent delivery equipment is limited and small valves, pumps, or feeders can be plugged by small amounts of scale accumulating in storage equipment.

Hydrochloric and nitric acids are more costly than sulfuric acid and are highly corrosive requiring special construction materials. Carbon dioxide and sulfur dioxide, besides being expensive, must be dissolved in wastewater to produce carbonic acid and sulfurous acid.

TABLE 7.39.7 TOXIC RATING CODE

Toxic Rating	Description of Rating
0(a)	*None:* No harm under any conditions
0(b)	*None:* Harmful only under unusual conditions or in overwhelming dosages
1	*Slight:* Causes readily reversible changes that disappear after end of exposure
2	*Moderate:* Can involve both irreversible and reversible changes; not severe enough to cause death or permanent injury
3	*High:* Can cause death or permanent injury after very short exposure to small quantities
U	*Unknown:* Available information with respect to man is considered invalid.

Note: Calcium compounds (chemical quicklime, etc.) have a toxic rating of 1 in all local categories and U in the systemic category; sodium hydroxide has a rating of 3, 3, 2 in all local categories and U in the systemic category; sodium carbonate has a rating of 2, 2, 2 in all local categories and U in the systemic category.

TABLE 7.39.8 TOXIC RATING OF ACID REAGENTS

Acids	Acute Local			Acute Systemic
	Irritant	Ingestion	Inhalation	
H_2SO_4	3	3	3	U
HCl	3	3	3	U
HNO_3	3	3	3	3
SO_2	3	3	3	U
CO_2	0	0	0	1, inhalation

Source: N.I. Sax, *Dangerous properties of industrial materials,* 2d ed. (New York: Rhinhold.)

To obtain efficient gas dispersion in large vessels requires concentric draft tubes and turbine-type blades, while a single draft tube (if any) and axial flow blades are used to efficiently blend waste and reagent for neutralization.

Figure 7.39.5 shows the alkali requirements for neutralizing various acids and the acid requirements for neutralizing various alkalis.

—*T.J. Myron*

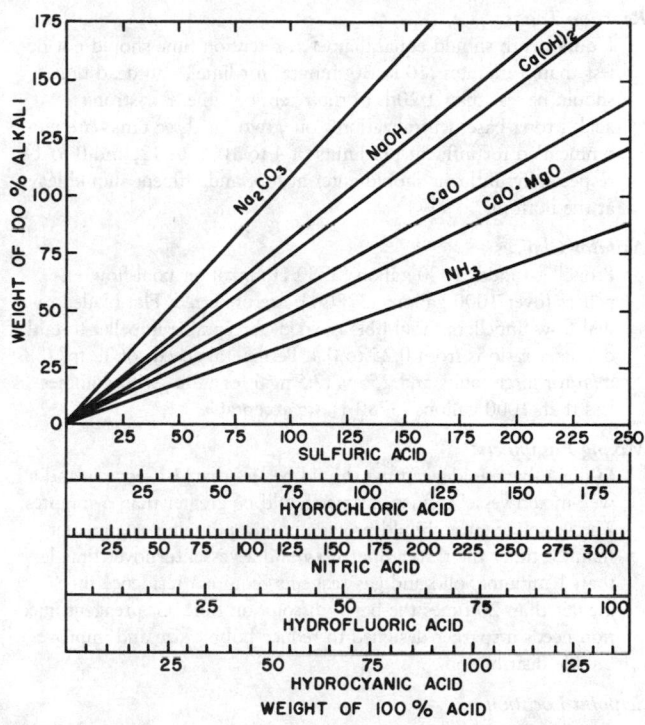

FIG. 7.39.5 Alkali neutralization graph. (Reprinted from W.A. Parson, 1965, *Chemical treatment of sewage and industrial wastes,* 58, Washington, D.C.: National Lime Association.)

7.40
pH CONTROL SYSTEMS

Final Element Rangeability
Requirement is extraordinary and depends upon titration curve and influent flow variability. Metering pumps are capable of 20:1 to 200:1, valves with positioners are capable of 50:1 or more, and a pair of split-ranged valves are capable of 1000:1 or more. pH swing that valves can handle can be determined as the base 10 logarithm of the rangeability for unbuffered titrations. If range-ability is 1000:1, the controllable pH swing is 3; for 100,000:1, it is 5

Final Element Precision and Characteristics
Requirement is exceptional and depends upon titration curve and desired control band. Electronically set metering pump and valves with positioners have repeatability from 0.1 to 2.0%. Linear valve characteristics are generally preferred.

Control Loop Dynamics

Effect of extreme nonlinearity and sensitivity of a pH process is diminished by a reduction in loop dead time. Ratios of loop dead time to time constant less than 0.02 are needed for setpoints on the steep portion of the titration curve to dampen oscillations. Reagent delivery delay is often the largest source of loop dead time.

Design Considerations

For proper mixing, process requires either an in-line mixer with 0.2 minute measurement filter and upstream and downstream volumes or a vertical, well-mixed vessel for attenuation of oscillations. Close-coupled control valve with ram valve or check valve to injection point and reagent dilution is needed to reduce reagent delivery delay. Additional stage of neutralization is needed for an inlet pH more than 2 pH units away from a small control band or whenever the final element rangeability or sensitivity requirement is excessive.

Reaction Tanks

Liquid depth should equal diameter, retention time should not be less than 5 minutes (10 to 30 minutes for lime), and dead time should be less than 1/20th of the retention time. For strong acid–strong base neutralizations, one, two, or three tanks are recommended for influent pH limits of 4 to 10, 2 to 12, and 0 to 14, respectively. Influent should enter at top, and effluent should leave at the bottom.

Agitator Choices

Propeller (under 1000 gallon [3780 1] tanks) or axial-flow impellers (over 1000 gallons [3780 1]) are preferred. Flat-bladed radial flow impellers should be avoided. Acceptable impeller-to-tank diameter ratio is from 0.25 to 0.4. Peripheral speeds of 12 fps (3.6 m/s) for large tanks and 25 fps (7.5 m/s) for tanks with volumes less than 1000 gallons (3780 1) are acceptable

Mixing Equipment

In-line mixer residence time should be less than 10 seconds and a well-mixed vessel residence time should be greater than 5 minutes. The vessel agitator should provide both a pumping rate greater than 20 times the throughput flow and a vessel turnover time less than 1 minute. Solid and gas reagents require a residence time greater than 20 times the batch dissolution time. Gas reagent injection needs a sparger designed to reduce bubble size and improve bubble distribution.

Setpoint Location

A setpoint on the flat portion of a titration curve reduces pH process oscillation and sensitivity and the control valve precision requirement.

pH Sensor Location

Insertion assemblies in pumped recirculation lines are preferred for increased speed of response, decreased coating, improved accessibility, and auto on-line washing and calibration. Insertion in recirculated lines is preferred by some.

Continuous Control Techniques

Flow feedforward for high loop dead time or rapid flow upsets. pH feedforward is only effective for influent pH on a relatively steep portion of the titration curve. A head start of reagent flow is needed for first stage when flow feedforward is not used. Signal linearization of measurement is beneficial for a constant titration curve. Self-tuning is helpful

Section 7.7 treats the subject of pH measurement. This section begins with an explanation of the difficult nature of the pH process; next, the process equipment used in pH control systems is described, including such topics as the selection of reagent delivery systems, mixing equipment, tank sizing, and other considerations. A discussion of the pH controller and its tuning rounds out the first half of the section, and the second half describes the various pH control applications.

Nature of the pH Process

The difficulty of pH control stems from the exceptionally wide range of the pH measurement, which for a 0 to 14 pH range covers 14 orders of magnitude of hydrogen ion concentration (Figure 7.40.1). It is commonly relied upon to detect changes as small as 10^{-7} in hydrogen ion concentration at mid-range. This incredible rangeability and sensitivity is the result of the nonlinear logarithmic relationship of pH to hydrogen ion activity as defined in Section 7.7. The process control implications are most severe for a process with only strong acids and based because the hydrogen ion concentration is proportional to the manipulated acid or base flow. The titration curve for such a system at 25°C is illustrated in Figure 7.40.2. The ordinate is the controlled variable (pH) and the abscissa is the ratio of the manipulated variable (reagent flow) to influent flow. Since the acids and bases are strong (completely ionized), the abscissa is also the hydrogen ion concentration.

As shown in Figure 7.40.2, change in pH for a change in reagent flow is 10^7 times larger at 7 pH than at 0 pH. The slope and hence the process gain changes by a factor of 10 for each pH unit deviation from the equivalence point at 7 pH. An expanded view of the apparently straight steep portion of the titration curve reveals another S-shaped curve (see Figs. 7.40.5 and 7.40.6). The controller gain for stability must be set inversely proportional to this process gain [as is shown in Equations 7.40(17) through 7.40(21)]. Therefore, changes in the operating pH require drastic changes in controller tuning. Even if a controller has a low enough gain to provide stability on the steepest portion of the titration curve, its response to upsets elsewhere will be so sluggish that the controller will only be able to handle disturbances that last over days. Such a controller response approaches the behavior of an integral-only mode and can be viewed more as an optimizer rather than a regulator, which should be the first line of defense against disturbances.

Seemingly insignificant disturbances are magnified by the steep portion of the titration curve, as shown by a small oscillation in the abscissa resulting in a large oscillation in the ordinate of Figure 7.40.2. The oscillations in the abscissa could be caused by an upset in the influent flow, influent concentration, reagent pressure, reagent concentration, or control valve dead band, or by the controller's reaction to noise. Even if the influent conditions were truly at steady state, just the commissioning of a pH loop can cause unacceptable fluctuations in pH if the setpoint is on the steep portion of the titration curve.

For an influent wastewater which is received with a pH

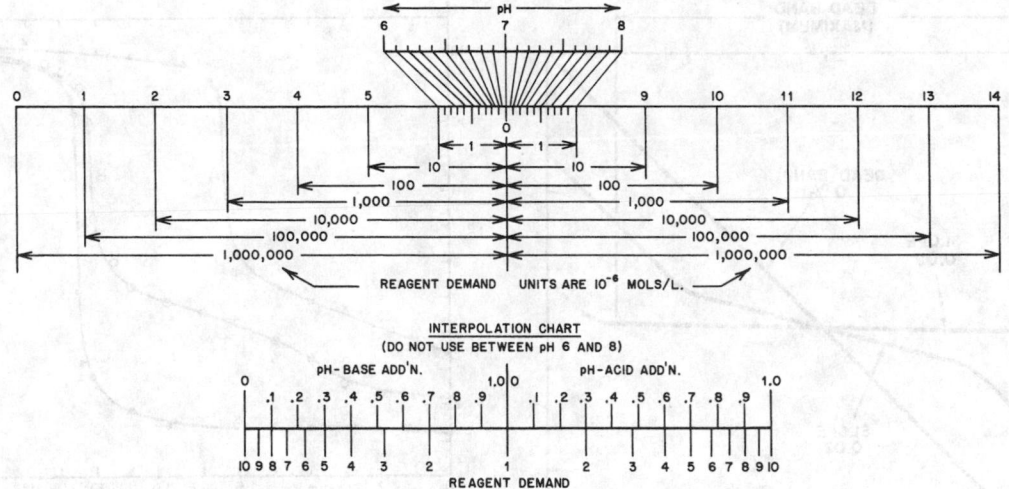

FIG. 7.40.1 pH versus reagent demand: strong acid–strong base

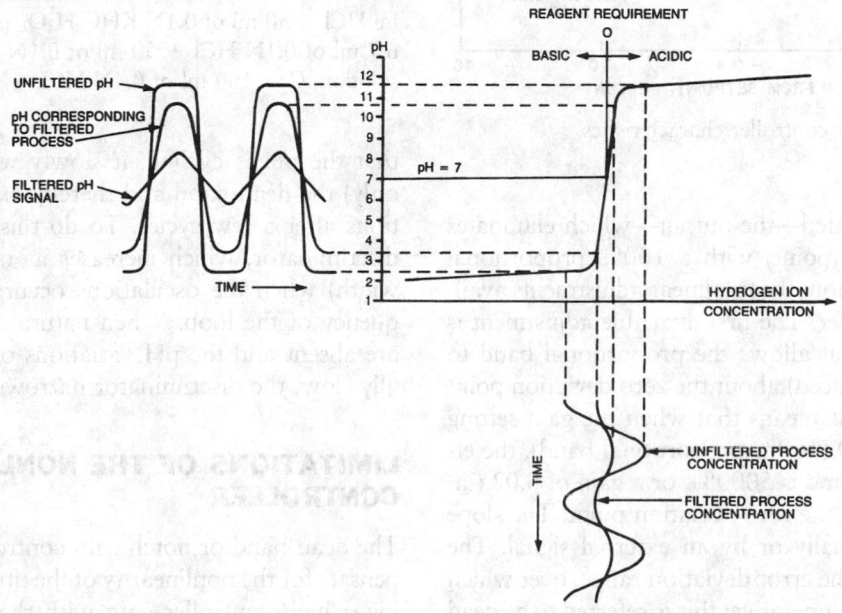

FIG. 7.40.2 The process gain of the neutralization process drops by a factor of 10 for each unit of pH from neutrality.

between 0 and 6, a valve or other final element with a rangeability of 10,000,000:1 and with a precision of better than 0.00005% is required to control the neutralization process within 1 pH of setpoint for the titration curve of Figure 7.40.2. Because a single valve can not provide this control, it is necessary to have three stages of neutralization with split-ranged valves.

If the total loop dead time was zero, which also implies zero valve dead band, and if the control valve trim characteristics and positioning were perfect, and if the measurement error and noise was zero, control using a single valve would be possible. Perfect control in general is possible only in a loop having no dead time and no instrument error. Such a loop could immediately see and correct for any disturbance and would never stray from setpoint. While such perfect control is not possible, it does

demonstrate that the goal for extremely tough loops such as the pH loop shown in Figure 7.40.2 should be to reduce dead time and instrument error as much as possible. As the dead time approaches zero, the detrimental effects of high process sensitivity and nonlinearity are also greatly reduced.

Nonlinear Controllers

Special controllers have been developed to compensate for the nonlinearity of most pH neutralization processes. These nonlinear controllers change their gain characteristics proportionally to the ion load (pH) of the process. The characteristics of the controller are as shown in Figure 7.40.3. The diagonal line represents the error-output relationship for the controller (in response to an error, a cor-

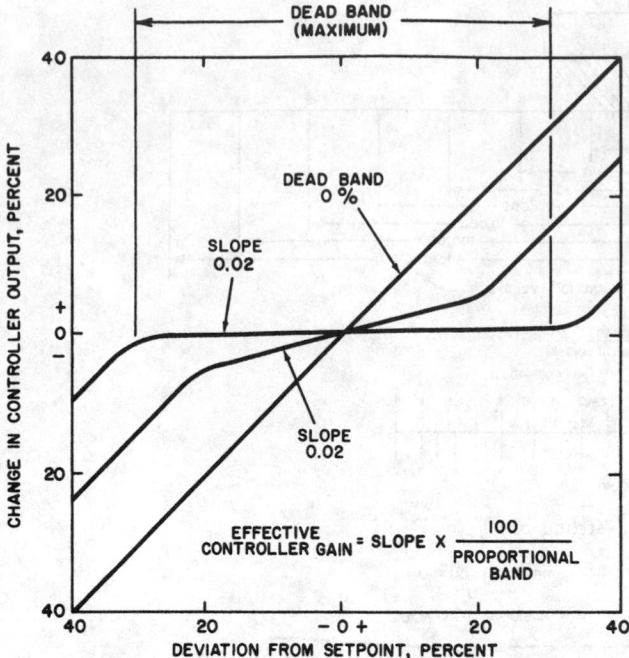

FIG. 7.40.3 Nonlinear controller characteristics.

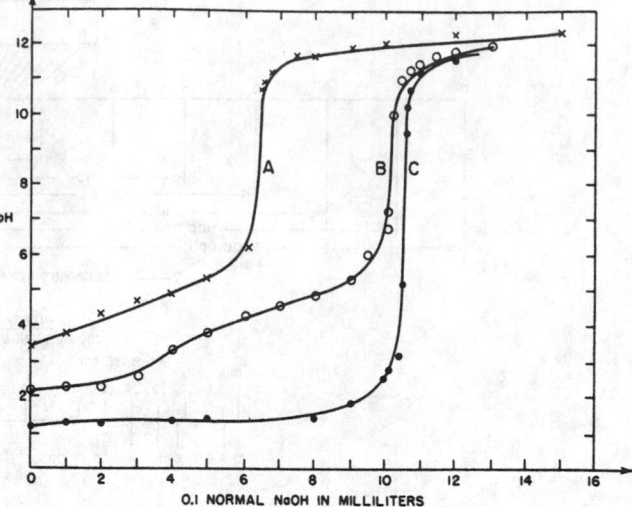

FIG. 7.40.4 Typical acid–base titration curves: *Key:* A = 9.9 ml HCl + 50 ml of 0.1N KHC$_8$H$_4$O$_4$ per 100 ml solution; B = 6.7 ml of 0.1N HCl + 50 ml of 0.1N KHC$_8$H$_4$O$_4$ per 100 ml solution; C = 100 ml of 0.1N HCl.

rective signal is generated—the output—which eliminates the deviation from setpoint) with a 100% proportional band (gain = 1.0) without the nonlinear adjustments available with this controller. The first available adjustment is a slope adjustment that allows the proportional band to be increased (gain reduced) about the zero deviation point by a factor of 50. This means that when the gain setting of the controller is 1.0 (100% proportional band), the effective proportional band is 5000%, or a gain of 0.02 (insensitive controller) at the zero deviation point. The slope can be adjusted manually or by an external signal. The second adjustment is the error deviation range, over which the slope adjustment is operative; this is referred to as dead band.

The dead band is adjustable from 0 to ±30% error (deviation from setpoint). This latter feature allows the gain of the control loop to be adapted proportionally to the ion load. If the process to be controlled resembles Figure 7.40.4, a reagent flow rate or valve position signal can automatically adjust the dead band. At high ion loadings (curve A) the controller gain will be low, a desirable condition when the process gains are high. At lower ion loadings (curve B) the dead band can be reduced, thereby increasing the gain of the controller, a condition that is desirable when the process and valve gains are low. The effectiveness of this type of controller and the benefits achieved by adapting the control loop characteristics to those of the process have been demonstrated on operating installations.

Adaptive controllers are also available for the automatic adjustment of the dead-band width, based on the condition of the pH loop. When the adaptive controller notices that the pH is cycling, it slowly widens (integral action only) the dead band and thereby extinguishes the oscillations after a few cycles. To do this the controller uses a discriminator, which increases its output (the dead-band width) when the oscillations occur near the natural frequency of the loop. When natural frequency oscillations are absent and the pH variations of the loop are unusually slow, the discriminator narrows the dead band.

LIMITATIONS OF THE NONLINEAR CONTROLLER

The dead-band or notch-gain controller attempts to compensate for the nonlinearity of the titration curve by matching a high controller gain with the leading and existing tails of the titration curve and a low controller gain with the steep center section of a titration curve. While this may appear to be a good fit when one looks at Figure 7.40.5, an expanded view of the region around the setpoint (neutrality) in Figure 7.40.6 reveals the inadequacy of the compensation. The nonlinear controllers also often assume the titration curve has a symmetrical S shape, which is not necessarily the case.

A more effective compensation technique uses line segments to approximate the curve, as shown in Figures 7.40.5 and 7.40.6. More line segments are used in the expanded view near the setpoint. Line segments are preferred to polynomials because smooth transitions from one polynomial to another are difficult to obtain and high order polynomials often have small humps and gain reversals which prevent the use of high controller gains or the use of rate action. The line segments are used to compute the abscissa (the ratio of reagent flow to influent flow) from the ordinate (pH). The abscissa is used as the controlled

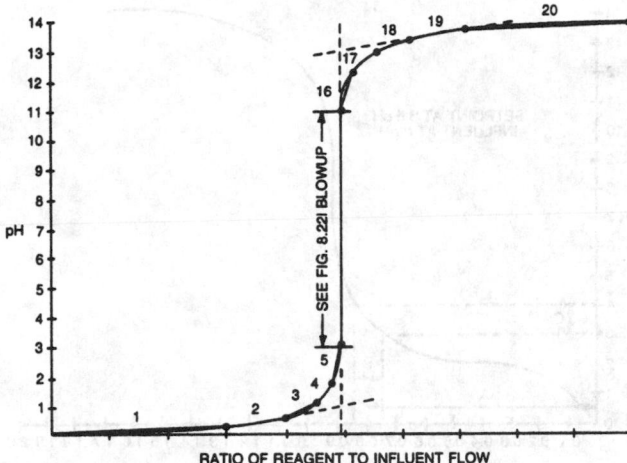

FIG. 7.40.5 Overall titration curve of a strong acid strong base combination.

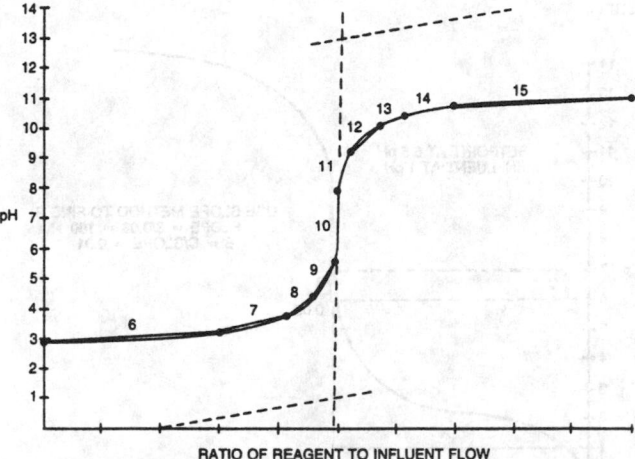

FIG. 7.40.6 Expanded view of the titration curve in Figure 7.40.5.

variable. If the curve is constant and accurate, the process gain becomes 1. Combined with a linear final element and a constant feed, the result is a nearly linear loop. The curve can be adjusted for temperature effects. The pH setpoint also goes through the same line-segment calculation. The display for the operator can be kept in units of pH, by the use of an output tracking strategy which provides a dummy controller for the operator interface.

Process Equipment and Reagent Delivery

Composition processes, whether pH or "pIon" (such as pCl and pAg), should be recognized as having two distinct aspects, one chemical and the other physical.

Several physical or process design considerations are associated with composition (pH or pIon) control applications. The most important is the primary device used to mix the reagent with the process stream. This can be as simple as reagent addition upstream of a pump or in-line

mixer with a downstream measurement point or as complex as two mixed reaction vessels followed by an attenuation vessel.

Where reaction vessels are required, design decisions must be made to determine (1) size and number, (2) baffling, (3) agitation (how much and what type), (4) measurement probe location(s), and (5) reagent addition point location.

The other and equally important physical aspect of pH control is the design of the reagent delivery system. Both of these aspects will be discussed here.

The ease or difficulty of most industrial control applications is closely related to a property of the process referred to as dead time. Analogous terms, such as "transport time," "pure delay," and "distance velocity lag," describe the same effect. Dead time is defined as the interval between the introduction of an input disturbance to a process and when a measuring device *first* corrects for the effect of that disturbance. Qualitatively, the relationship between dead time and controllability is simple: The more dead time, the more difficult the problem of control. The presence of dead time in pH or pIon processes is extremely detrimental to controllability. The major reason is the severe sensitivity of the measurement of interest at the control point. One of the major goals of system design is to eliminate the dead time or to reduce it to an absolute minimum.

REAGENT DELIVERY SYSTEMS

Reagent addition requirements can be handled in diverse ways, depending on the process loads (flow of material to be neutralized) into the neutralization facility and the variation of the hydrogen or hydroxyl ion concentration, or both, in that flow. It should be recognized at the outset that because of the logarithmic nature of the pH measurement, a pH change of one unit can cause a tenfold change in load, whereas a 100 to 300 GPM (378 to 1134 l/m) change in flow (assuming no change in pH) is only a threefold change. Thus, the consequences of flow variations in waste streams can be relatively minor in comparison with ion concentration variations.

The equipment used to deliver reagents to the process under automatic control includes a metering device or a control valve. Metering pumps as a choice for reagent delivery are very accurate; however, delivery rangeability capability is limited to approximately 20:1 if speed is manipulated. Both speed and stroke can be manipulated to yield 200:1 rangeability, but the resulting relationship is squared and may require characterization. This means that where speed alone is manipulated, pH variations for a strong acid–strong base reaction greater than ±0.65 will result in cyclic or inadequate control. (A pH change of 1.3 means a 20-fold change in reagent requirement.) When pH load variations are minor (0.4 or less) and flow variations are less than 4:1, the choice of a metering pump with speed control is sufficient.

Control valves, like the metering pump, have limited rangeability. In this category two types of internal plug forms are usually considered for throttling service. They are the linear and the equal-percentage throttling characteristics. The term "equal percentage" means that the valve will produce a change in flow rate corresponding to a unit change in lift (valve plug movement), which is a fixed percentage of the flow rate at that point.

Both valves are available with minimum turndown of 50:1, and some have recently been developed (mainly in the smaller sizes) with claimed rangeabilities as high as 500 to 1.

Digital valves that can furnish rangeabilities of 2000:1 are also available. Cost, size, complexity, and materials of construction limit the application of these devices.

It is desired that the installed characteristic of the final element be linear so as not to introduce another nonlinearity into the pH loop. For control valves, the pressure drop available typically doesn't change much and is large compared to the system drop due to the low reagent flow rates normally associated with pH control. The result is an installed characteristic close to the inherent characteristic. Consequently, linear trim is preferred over equal-percentage trim for most pH applications to provide a more constant gain. The actual gain deviates from the theoretical constant gain, particularly at valve openings exceeding 80%. Low reagent flow (C_v less than 2.0) can cause valve sizing problems since the flow may not be completely turbulent. When the flow is viscous, the fully turbulent C_v for a control valve should be multiplied by a dimensionless coefficient Fr, which is a function of Reynolds number for the valve and which reduces valve capacity when viscosity is high. Since the hydraulic friction losses in the valve external to the trim can be assumed to be negligible, the flow is fixed by conditions within the trim and by the geometry of the plug and seat. A Mikroseal packless valve (available from H.D. Baumann Assoc., Inc.) having maximum C_vs between 0.0006 and 0.7, forces laminar flow. Such valve designs provide high rangeability but also nonlinearity, since flow is proportional to the third power of valve position. An accurate current-to-air (I/P) converter and output signal characterization within a microprocessor-based controller is needed to linearize the valve characteristic.

Valve Specification

The maximum valve capacity, rangeability, and precision requirements can be found by determining the parameters A and B from the titration curve, shown in Figure 7.40.7 for a setpoint on the flat portion and in Figure 7.40.8 for a setpoint on the steep portion of the titration curve and by inserting them into Equations 7.40.1, 7.40.2, and 7.40.3. Parameter A is the distance along the x axis from the influent pH to the setpoint. It is multiplied by the maximum influent flow for a given operating condition. The

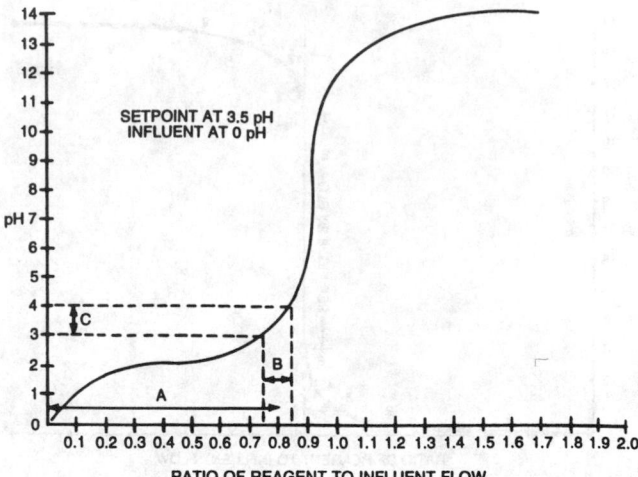

FIG. 7.40.7 Data required for reagent control valve specification when the pH setpoint is in a flat portion of the titration curve.

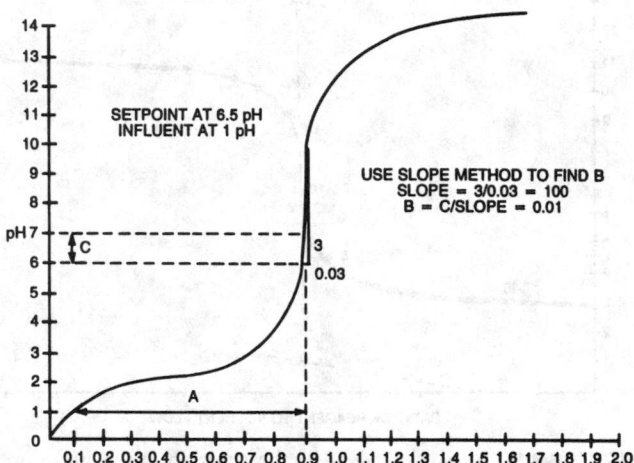

FIG. 7.40.8 Control valve specification data for application with pH setpoint in the steep portion of the titration curve.

$(A_{max})(F_{imax})$ combination that yields the largest product (increased by 25% to improve the valve gain uniformity) is used to set the maximum capacity for sizing the control valve. The maximum rangeability of reagent flow (R_{rmax}) is obtained by dividing the maximum product pair by the minimum product pair. The precision of the control valve in terms of its repeatability and dead band requirements is proportional to the ratio of the minimum B value to the maximum A value. The 80% value in Equation 7.40.3 corresponds to the 25% extra capacity used for sizing. For steep curves, B may be too small to estimate unless the titration curve is expanded in the control region (Figure 7.40.8).

$$F_{rmax} = (1.25)(A_{max})(F_{imax}) \qquad 7.40(1)$$

$$R_{rmax} = (1.25) \frac{(A_{max})(F_{imax})}{(A_{min})(F_{imin})} \qquad 7.40(2)$$

$$E_{rmin} = (80\%) \left(\frac{B_{min}}{A_{max}} \right) \qquad 7.40(3)$$

where:

A_{max} = maximum abscissa distance influent pH → setpoint (ratio)

A_{min} = minimum abscissa distance influent pH → setpoint (ratio)

B_{min} = minimum control band translated to abscissa (ratio)

E_{rmin} = minimum control repeatability and dead band (%)

F_{imax} = maximum influent flow (gpm)

F_{imin} = minimum influent flow (gpm)

F_{rmax} = maximum reagent valve capacity for sizing (gpm)

Figure 7.40.9 illustrates the determination of the tolerable reagent flow variations. A strong acid has been added to water to achieve a pH of 6 and a pH of 2 (curves A and B). The reagent flow (10% NaOH) requirement for each solution is plotted on separate scales (lower and upper abscissas). Assuming a control specification of pH 7.0 ± 0.5, a reagent flow of ±28% variation can be tolerated when the pH of the inlet material is 6.0 (Curve B). When the pH of the inlet material is 2.0, the tolerable reagent flow variation is ±0.0028% (Curve A) and the problem is 10,000 times more difficult.

Valve Linearization

The drawback of using equal-percentage valves is a high gain contribution to the overall loop gain, especially at high reagent flows (sensitivity if a function of valve opening). One approach to countering this variable gain is by a characterizer with an input-output characteristic which

is opposite that of the equal-percentage valve. This approach is illustrated in Figure 7.40.10. The resultant valve characteristic is approximately linear, which is highly desirable from an automatic control point of view, since the variable gain nature of the process makes the control problem difficult enough.

Using Multiple Valves

When the rangeability exceeds what can be obtained from a single control valve, a small and a large control valve can be split-ranged and sequenced so the transition point is in the usable throttling range of each valve. The split-range or switchover point should be chosen to keep the final element gain relatively constant. Equation 7.40(4) calculates the switchover point for the more general case when the split-ranged valves control different types or concentrations of reagents. The split-range computation should be done in the microprocessor-based controller for greater accuracy, flexibility, and standardization of positioner calibrations. This way, separate outputs are created for each valve and displayed on the operator interface.

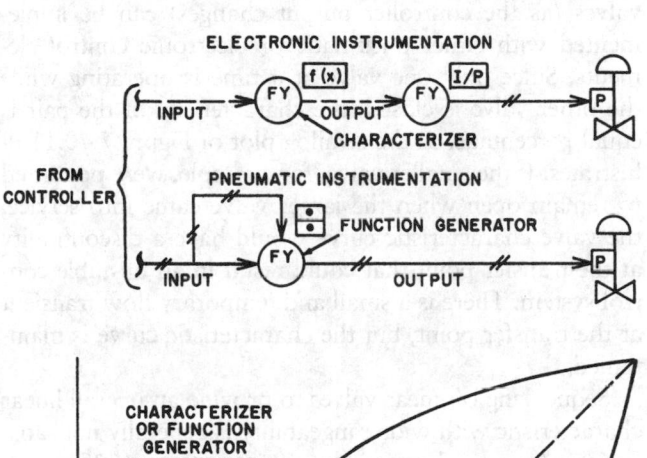

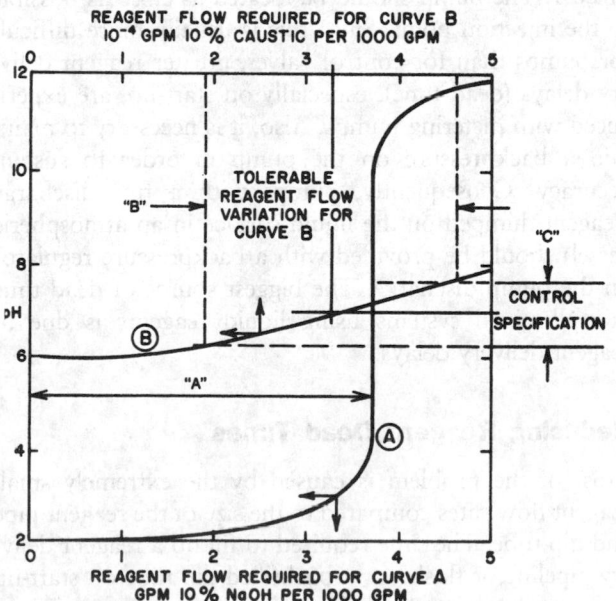

FIG. 7.40.9 Relationship of accuracy and rangeability to ion loading.

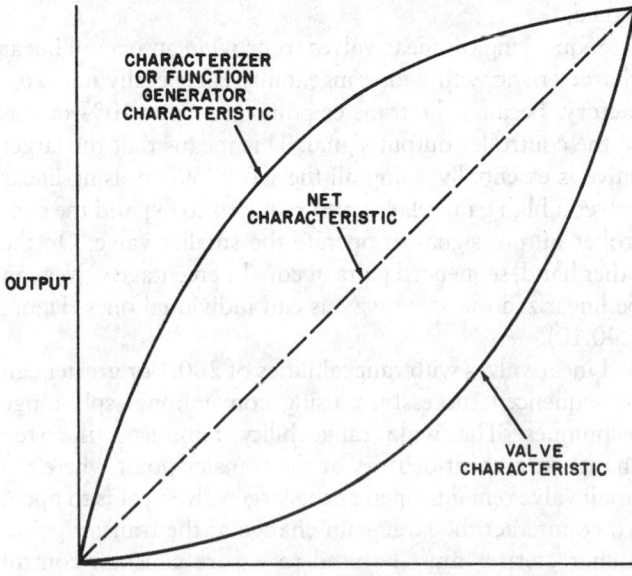

FIG. 7.40.10 Linearization of equal-percentage valves.

$$S = 100 \Big/ \left(1 + \frac{(F_{2max})(C_2)}{(F_{1max})(C_1)} \right) \qquad 7.40(4)$$

where:

F_{1max} = maximum flow of reagent valve 1
F_{2max} = maximum flow of reagent valve 2
C_1 = concentration of reagent 1 (normality)
C_2 = concentration of reagent 2 (normality)
S = best split-range point (% of controller output signal)

Conventional Valve Sequencing

Conventional analog hardware can also be used for control valve sequencing. Sequencing of a pair of equal-percentage valves can achieve an overall rangeability approaching the product of the individual valves rangeabilities, e.g., $50 \times 50 = 2500$. The loss of rangeability is mainly caused by the amount of overlap between valves. A plot of the performance characteristics of this pair of valves is shown in Figure 7.40.11. The valve positioner of the smaller sequenced valve is calibrated for full stroke over 0 to 52% controller output signal (closed at 0%; fully open at 52%). The positioner of the larger sequenced valve is calibrated for full stroke over the range of 48 to 100% of controller output. Transfer between the valves (as the controller output changes) can be implemented with either pneumatic or electronic control elements. Since only one valve at a time is operating while the other valve is closed, the characteristic of the pair is equal percentage, as the semilog plot of Figure 7.40.11 illustrates. If the smaller valve, for example, were permitted to remain open when the larger valve came into service, the valve characteristic curve would have a discontinuity at the transfer point that could result in an unstable control system. There is a small and temporary flow transient at the transfer point, but the characteristic curve is maintained.

Sequencing of linear valves to provide an overall linear characteristic with wide rangeability is generally not satisfactory, because the transfer point occurs at 10% or less of the controller output signal. This means that the larger valve is essentially doing all the work. When using linear valves a high gain relay is also required to expand the controller output signal to operate the smaller valve. On the other hand, sequenced pairs of equal-percentage valves can be linearized the same way as can individual ones (Figure 7.40.10).

Linear valves with rangeabilities of 200:1 or greater can be sequenced successfully using conventional split-range techniques. The wide rangeability minimizes the step change in reagent delivery at the transfer point where the small valve remains open as the large valve begins to open. To counteract the large gain change at the transfer point, a characterizer must be used to ensure constant control loop gain.

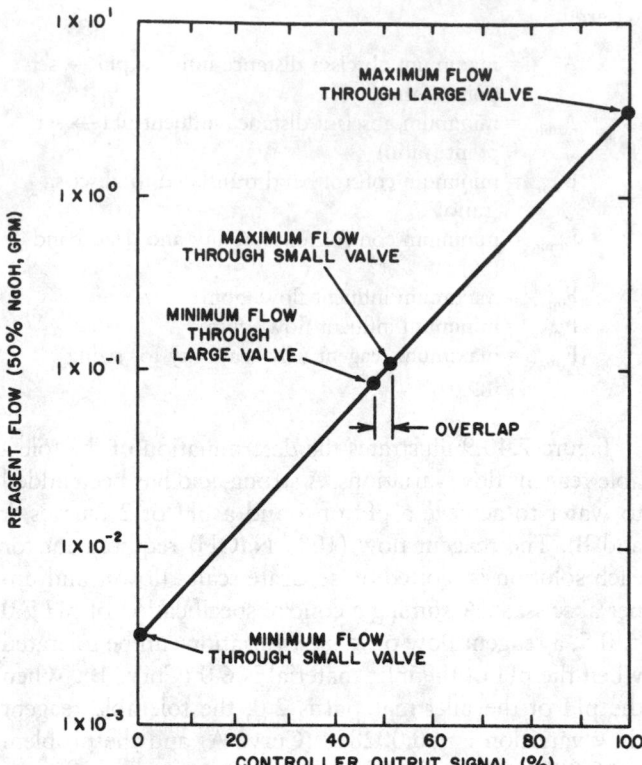

FIG. 7.40.11 Reagent flow using sequenced valves.

Metering Pumps

The specifications for metering pumps (Section 7.6) should be carefully checked for rangeability and precision. The pump speed and stroke should be electronically set rather than pneumatically. The mechanism should be chosen that requires the least maintenance and meets the precision and rangeability requirements. Many flexible diaphragm pumps have a repeatability of only 2% and a turndown of 20:1. The pump should be located as close as possible to the injection point. Since this is usually more difficult for pumps than for control valves, greater reagent delivery delays (dead time), especially on start-up, are experienced with metering pumps. Also, it is necessary to maintain a backpressure on the pump in order to sustain accuracy. Consequently, pumps with a free discharge (reagent dumped on the liquid surface in an atmospheric vessel) should be provided with a backpressure regulator on the pump discharge. The biggest source of dead time in well-mixed systems using liquid reagents is due to reagent delivery delays.

Reducing Reagent Dead Times

Most of the problem is caused by the extremely small reagent flow rates compared to the size of the reagent pipe and dip tube. The time required to fill up a reagent delivery pipeline or flush out a backfilled dip tube on start-up is the total volume divided by the reagent flow rate. Thus, a few gallons of volume and a one gallon per hour reagent flow can result in a several hours delay before reagent is injected into the process. In gravity flow of reagent, a similar delay occurs whenever the reagent valve changes its opening. Such systems are also subject to control valve capacity variations caused by head changes that may aggravate the rangeability problems. Here are some possible methods to reduce the reagent dead time:

1. Locate throttle valve at injection point.
2. Mount on-off valve (preferably ram type) at injection point.
3. Reduce diameter and length of injector or dip tube.
4. Add a reliable check valve to injector or dip-tube tip.
5. Dilute the reagent upstream.
6. Inject the reagent into vessel side just past baffles.
7. Inject reagent into recirculation line at vessel entry point.
8. Inject reagent into influent line at vessel entry point.

Reagents such as ammonium hydroxide, calcium hydroxide, and magnesium hydroxide are weak bases and have pKa values close enough to many control bands to provide some flattening of the titration curve and some increased controllability from the standpoint of reduced sensitivity to disturbances. However, ammonium hydroxide can flash and create gas bubbles that escape from vessels or choke in-line systems or travel downstream undissolved. Calcium hydroxide (lime) and magnesium hydroxide in solid form may take 15 to 30 minutes to dissolve and in slurry form 5 to 10 minutes to dissolve. This slowness of reagent response adds tremendous dead time to the system. Also, the residence time must be 20 times greater than the dissolution time to insure that less than 1% remains undissolved in the effluent. Consequently, large volumes must be used for pH control, which further increase the loop dead time.

Reagent Delivery Hysteresis

Consider a reagent delivery device such as a control valve, a metering pump, or a dry feeder. The smallest incremental change that these devices can make is approximately 1%. Converted to the logarithmic pH scale and using an influent pH of 14 and a setpoint of 7, 1% excess acid produces a pH of 2, and 1% too little yields pH 12. These values were derived from Figure 7.40.1, where 1×10^6 reagent units are required to neutralize pH 14 to 7. One percent of this total is 10,000 reagent units, which correspond to pH 2 and 12.

In a similar fashion, the effect of the same error can be estimated for any other setpoint. Using a setpoint of 12, for example, $1 \times 10^6 - 10,000$, or 990,000, reagent units are required for neutralization from a 14 pH influent. One percent excess acid (9900 reagent units) corresponds to 10,000 − 9900, or 100 reagent units, which is pH 10. One percent too little acid corresponds to 19,900 reagent

units, or approximately 12.3 pH using the interpolation chart on Figure 7.40.1.

The same procedure is used to illustrate the sensitivity of the process to hysteresis. Increasing the error to 1.5% for a setpoint of 12 gives a low pH of 3.13 and a high pH of 12.39 (10,000 − 14,850 = 4850 on the acid side and 10,000 + 14,850 on the caustic side).

Methods of reducing valve hysteresis, such as pulse interval control and the uses of digital valves, have been proposed. Although these techniques add cost and complexity to the control system, they should be investigated as alternatives to the installation of stirred tanks.

If hysteresis cannot be eliminated, it can profoundly influence pH loop performance unless some other element can be introduced to smooth out this incremental response or, in effect, to reduce the gain of the control loop.

MIXING AND AGITATION

The two types of mixing that are important to the control system are intermixing and backmixing. The reagent must be intermixed with the process stream to furnish complete elimination of the areas of unreacted reagent or untreated influent. Adequate intermixing between influent and reagent can be readily achieved by adding the reagent at a point of small cross-sectional area where there is some turbulence. Figure 7.40.12 illustrates the reagent being added in the pipeline before the influent enters the treatment facility. This is a desirable practice because it eliminates poor intermixing, which can cause a noisy signal to be observed in the effluent pH. A loop seal arrangement, particularly when long reagent transfer lines are required, allows the reagent line to remain full up to the point of introduction to the process and thus eliminates a potential source of process dead time.

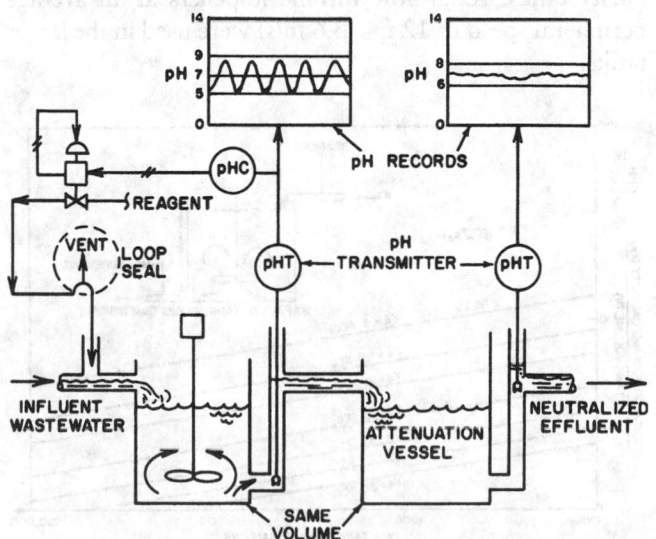

FIG. 7.40.12 Effect of attenuation vessel.

Backmixing is more important than intermixing for close pH control. The treated stream must be held in a vessel sufficiently long for the reagent to react and be backmixed. In general, the degree of backmixing can be defined in terms of the pumping capacity of an agitator with respect to the flow and volume of the neutralization vessel. In practice, however, this definition has limited usefulness because of variables such as agitator construction and blade pitch, baffling of the neutralization vessel, and placement of inlet and outlet measuring electrodes. Experience shows that the best way to define backmixing for control purposes is by the ratio of the system dead time to retention time of the neutralization vessel. The retention time is the volume of the vessel divided by the flow through the vessel. A ratio of dead time to retention time equal to 0.05 is adequate for good control.

Suitable baffles or agitator positioning should be used in mixed neutralization vessels to avoid a whirlpool effect. The power supplied by the impeller must be used to turn the contents of the vessel over, not to whirl them about. With these effects in mind, a propeller or an axial-flow impeller should be selected to direct the flow of the vessel contents toward the bottom of the tank. The flat bladed radial-flow impeller should be avoided, since it generally tends to divide the vessel into two sections and increases system dead time.

Figure 7.40.13 is a plot of tank size against agitator pumping capacity per unit tank volume on logarithmic coordinates. The family of curves shown for various dead times was developed from empirical data in tanks with capacities of 200, 1000, 10,000, and 18,000 gallons (756, 3780, 37,800, and 68,040 l). They apply to baffled tanks of cubic dimensions with the inlet at the surface and the outlet at the bottom on the opposite side of the tank. The ratio of impeller diameter to tank diameter varies from 0.25 to 0.4. Square pitch propellers at an average peripheral speed of 25 fps were used in up to 1000-gallon capacity tanks. Axial-flow turbine impellers at an average peripheral speed of 12 fps (3.6 m/s) were used in the larger tanks.

To be classified as a well-mixed vertical tank by the standards of pH control applications, the liquid height should be between 100% and 150% of the vessel width or diameter. The vessel walls should have baffles to prevent liquid rotation, the agitation pattern should be axial, and the agitator pumping rate calculated by Equation 7.40(5) should be at least 20 times the influent flow rate. The agitation should be great enough to break the surface and pull down the reagent (injected near the surface to minimize dip-tube length and reagent delivery delay) but not enough to cause air entrainment.

Agitator Dead Time and Time Constants

The dead times and time constants from mixing in this type of vessel can be estimated by Equations 7.40(6) and 7.40(7). For horizontal vessels and sumps, these equations do not hold, because in horizontal tanks plug flow regions exist, short circuiting occurs, and a significant amount of the residence time shows up as dead time. Holdup and averaging volumes (with inappropriate geometries) are beneficial upstream and downstream of all pH control loops, but their inappropriate geometry can be disastrous when used for tanks where difficult pH control is to take place.

$$F_a = (7.48) \frac{0.4}{\left(\frac{D_i}{D_t}\right)^{0.55}} (N_i)(D_i^3) \qquad 7.40(5)$$

$$\tau_d = \frac{V}{F_i + F_a} \qquad 7.40(6)$$

$$\tau_1 = \frac{V}{F_i} - \tau_d \qquad 7.40(7)$$

for $\tau_d \gg \tau_i$

$$\tau_1 = \left(\frac{F_a}{F_i}\right)(\tau_d) \qquad 7.40(8)$$

where:

D_i = impeller diameter (ft)
D_t = tank internal diameter (ft)
F_i = influent flow (gpm)
F_a = agitator pumping rate (gpm)
N_i = impeller speed (rpm)
τ_1 = mixing time constant (minutes)
τ_2 = mixing dead time (minutes)
V = vessel liquid volume (gallons)

Unfortunately, for steep titration curves most of the mixing time constant is lost due to acceleration of the pH measurement. Figure 7.40.14 shows how a 19-minute time constant is reduced to 0.04 minutes for a strong acid and strong base system by translating the points of 63% and 100% of the pH change for an upset to the abscissa. By the translation of the controlled variable from pH to the abscissa of the titration curve, linearization of the pH signal can restore the time constant to its original value.

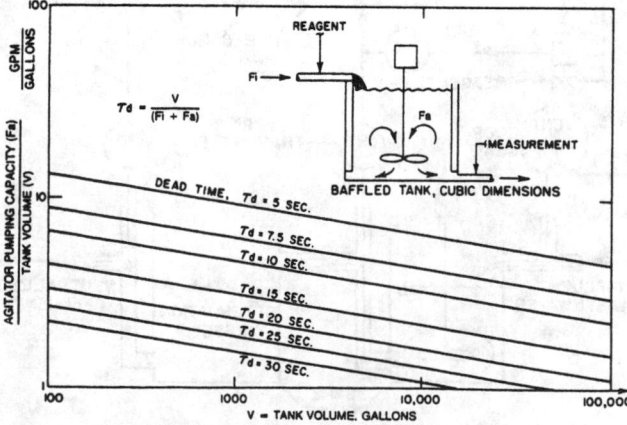

FIG. 7.40.13 Dead time (τ_d) as a function of mixing intensity.

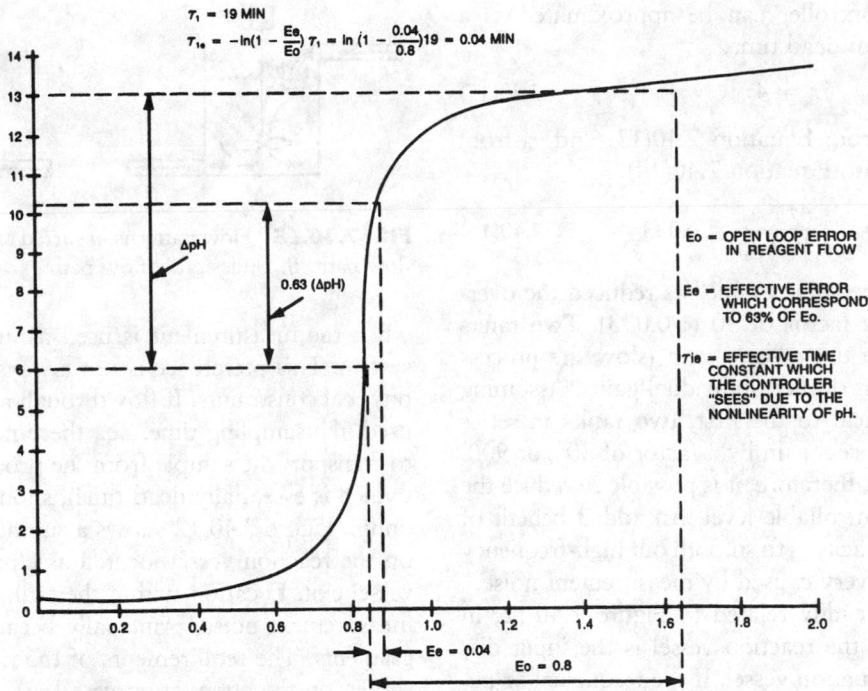

FIG. 7.40.14 When the titration curve is steep, the mixing time constant is much reduced (from $\tau_1 = 19$ min to $\tau_{1e} = 0.04$ min) due to the acceleration of the pH measurement. Linearization (Figures 7.40.5 and 7.40.6) restores the time constant.

CONTROL DYNAMICS

The performance of a stirred tank to periodic disturbances can be evaluated by consideration of the dead time and time constant properties of the tank.

For example, if the total system dead time is τ_{dt}, it can be defined as:

$$\tau_{dt} = \tau_{d1} + \tau_{d2} \qquad 7.40(9)$$

where:

τ_{d1} = tank dead time, inlet to outlet
τ_{d2} = remaining loop dead time (sampling system and control valve motor)

Given

$$\tau_{dt} = 0.05 \ V/F \qquad 7.40(10)$$

where:

V = vessel volume
F = flow through vessel

The time constant (τ_1) for an agitated vessel with dead time (τ_{d1}) can be expressed as:

$$\tau_1 = V/F - \tau_{d1} \qquad 7.40(11)$$

Assuming that the stirred tank has the minimum 3.0-minute time constant previously mentioned and that the total dead time is divided 80% to (τ_{d1}) and 20% to τ_{d2}, Equation 7.40(11) can be restated:

$$\tau_1 = 0.96 \ V/F \qquad 7.40(12)$$

Expressing τ_1 in terms of dead time by combining Equations 7.40.10 and 7.40.12:

$$\tau_1 = 19.2 \ \tau_{dt} \qquad 7.40(13)$$

The dynamic gain of a stirred tank to periodic disturbances is given by Equation 7.40(14):

$$G_d = \frac{\tau_0}{2\pi\tau_1} \qquad 7.40(14)$$

where:

G_d = dynamic gain of the stirred tank 5

$= \dfrac{\text{percent change in output}}{\text{percent change in input}}$

τ_0 = period of oscillation of the disturbance
τ_1 = first-order time constant of the tank; approximately equal to (tank volume/flow through the tank system dead time)

To visualize the effect of dynamic gain, consider a flowing stream whose pH falls from 7 to 4 and returns to 7 in one minute. If the stream flowed through a tank with one minute retention time (volume/flow), the spike in pH would pass through virtually unchanged, and the effluent pH would closely track the influent pH. If, however, the stream flowed through a tank with 60 minutes retention time, practically no upset would be observed in the effluent pH because of the capacity effect of the large volume.

The period of oscillation, τ_0, of a typical composition process under closed-loop control with an optimally tuned (controller settings adjusted to match the process it controls) three-mode controller can be approximated as a function of the system dead time.

$$\tau_0 \simeq 4\tau_{dt} \qquad 7.40(15)$$

Substituting for τ_1 from Equation 7.40(13) and τ_0 from Equation 7.40(15) into Equation 7.40(14):

$$G_d = \frac{4\tau_{dt}}{2\pi(19.2\tau_{dt})} = 0.033 \qquad 7.40(16)$$

In this example the stirred tank has reduced the overall process gain by a factor of 30 (1/0.033). Two tanks used in series reduce the process gain (slow the process down) by the product of their individual gains. Assuming a second tank identical to the first, two tanks in series would reduce the process gain by a factor of 30^2, or 900. With the stirred tank, therefore, it is possible to reduce the process gain to a controllable level. An added benefit of an increased tank capacity is to smooth out high-frequency errors in reagent delivery caused by measurement noise.

This example is readily related to Figure 7.40.12, in which the output of the reaction vessel is the input disturbance in the attenuation vessel. If the frequency or period (τ_0) of the input disturbance can be kept short (on the order of seconds) by virtue of a *tight* control loop around the reaction vessel, then the dynamic gain number of the attenuation vessel will be very low (0.033 for the example), thereby increasing its attenuation capability. This results in a stable effluent pH that averages the input disturbance.

TANK CONNECTION LOCATIONS

The inlet and outlet in the treatment vessel should be located at opposite sides—one high and one low—with respect to the bottom of the tank. Generally, it is most convenient to introduce the influent stream on the surface of the tank and to locate the outlet at the bottom of the vessel.

Variations in the location of the inlet and outlet can considerably change the dead time. Reversing the flow through the tank so that the inlet is on the bottom and the outlet is at the surface, for example, causes the dead time to increase by a factor of 2 or 3. Examination of the flow patterns in the tanks (Figure 7.40.15) shows that the path from inlet to outlet can be doubled by this change. The additional dead time is attributable to the swirl effect of the agitator, which is minimized, but not eliminated, by baffling.

SENSOR LOCATIONS

The location of the measuring electrodes also deserves serious consideration. The general guidelines are that the locations should be responsive and the information supplied by them should be timely. Submersible or recirculation pipeline insertion type electrode assemblies are preferred

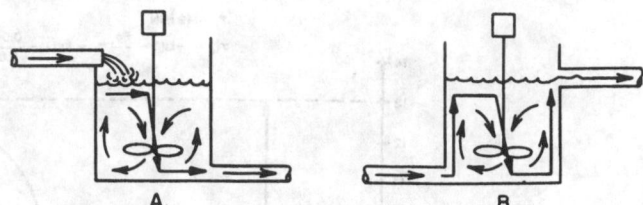

FIG. 7.40.15 Flow patterns in stirred tanks. *A.* Recommended flow path. *B.* Undesirable flow path.

when the measurement is used as an input to a control system. This preference is not always possible because of physical constraints. If flowthrough assemblies have to be used, the sampling time, i.e., the time required physically to transport the sample from the process to the electrodes (which is essentially dead time), should be kept to a minimum. Figure 7.40.12 shows a submersible-type assembly on the reaction vessel located as close as possible to the vessel exit. Location within the tank proper increases the measurement noise, principally because of concentration gradients. The requirements of the monitoring electrodes shown on the attenuation vessel are not as severe. Either flowthrough or submersible detectors can be used. The information supplied by these electrodes provides a clean record for any regulatory agencies involved.

EQUALIZATION TANKS

Upstream of a stirred neutralization vessel, a lagoon or a holding tank can be very useful because it serves to smooth out upsets in influent pH and flow, thus allowing the use of a simple feedback system rather than a more costly feedforward control system. A lagoon can also be used to store the material that is bypassed around the neutralization process in case of failure, a very important consideration if off-specification effluent causes a plant shutdown. The one thing that a lagoon cannot do is replace a mixed vessel as part of a control system. Any attempt to control the pH of a lagoon by closed-loop feedback control can only result in an effluent pH value on the opposite side of neutrality. The period of oscillation of such pH swings will depend on the dead time of the lagoon, but typically it will be on the order of hours.

Controller Tuning

The tuning of the pH controller can be approximated from three key parameters: the open-loop gain (K_o), the largest time constant of the loop (τ_1), and the total dead time (τ_d) in the loop. The total loop time delay is the most important of these terms. It is the sum of the dead times from valve dead band, reagent dissolution time, reagent piping transportation delay, the mixing equipment turnover time, mixing equipment transportation delay, sample transportation delay, electrode lag, transmitter damping (normally negligible), and digital filters and digital system scan update time. It is the time required for a disturbance to be recognized by the controller and the corrective reaction by the controller arrive at the entry point of that disturbance. Regardless of where the disturbance enters, the total loop time delay is the time it takes the disturbance effect to traverse the loop in Figure 7.40.16. The controller integral time and derivative time settings depend upon the loop dead time as shown in Equations 7.40(18), (19), and (21). The largest time constant slows down the excursion and gives the controller time to compensate for the upset. The largest time constant in a well-designed installation which is in excellent working condition and is provided with substantial back-mixed volumes is the process time constant (τ_1). The controller gain is proportional to the ratio of this time constant to the loop dead time (τ_d) multiplied by the open-loop gain (Ko) per Equation 7.40(17), if the dead time (τ_d) is less than the time constant (τ_1). Kc is proportional to the open-loop gain (Ko) per Equation 7.40(20) for systems where the dead time is greater than the time constant.

$$Kc = (\tau_1/\tau_d)/Ko \qquad 7.40(17)$$

$$Ti = 2\ \tau_d \qquad 7.40(18)$$

$$Td = 0.5\ \tau_d \qquad 7.40(19)$$

For $\tau_d > \tau_1$ and when using a PI controller:

$$Kc = 0.3\ Ko \qquad 7.40(20)$$

$$Ti = 1.0\ \tau_d \qquad 7.40(21)$$

where:

Kc = controller gain
Ko = the open-loop steady-state gain (dimensionless)
τ_1 = largest time constant with titration curve effect (minutes)
τ_d = total loop dead time (minutes)
Td = derivative time setting (minutes)
Ti = integral time setting (minutes/repeat)

DEAD TIME EXCEEDING TIME CONSTANT

The loop dead time can become larger than the largest time constant for in-line mixer installations, because these units provide mostly axial mixing instead of back-mixing. τ_d can also exceed τ_1 in several situations: in poorly mixed tanks; when the reagent dip tubes are poorly designed; in systems where transportation time exceeds turnover time; and when electrodes are improperly located or severely fouled. τ_d can exceed τ_1 in seemingly well-designed and well-mixed vessels also if the setpoint falls on a particularly steep section of the titration curve because (as was illustrated in Figure 7.40.14), most of the time constant is lost due to rapid movement of pH. Under such conditions, the measurement can actually accelerate and the process can appear to be non-self-regulating to the controller. For this case, dead-time dominance causes the window of allowable gains to close, and loop instability occurs regardless of tuning. Also, for non-self-regulating processes, it is important to maximize derivative or rate action and minimize the use of integral or reset action. Neither of these steps is possible when dead time exceeds the time constant in the loop.

PID CONTROLLER TUNING ($\tau_d < \tau_1$)

Tuning of the controller must be done at the normal setpoint and under the normal operating conditions due to the potentially severe nonlinearity of the process, as illustrated by the titration curve. Time constants predicted from equipment volume and agitator pumping rate must be corrected for the effect of the titration curve as shown in Figure 7.40.14 per the discussion in the section on mixing equipment. Open-loop tests will show the effect. Closed-loop methods of tuning, such as the Ziegler-Nichols ultimate oscillation method, must be done carefully to distinguish the gain where oscillations first start, because the loop will show nearly equal amplitude oscillations for a wide range of gains as it bounces back and forth between the flat por-

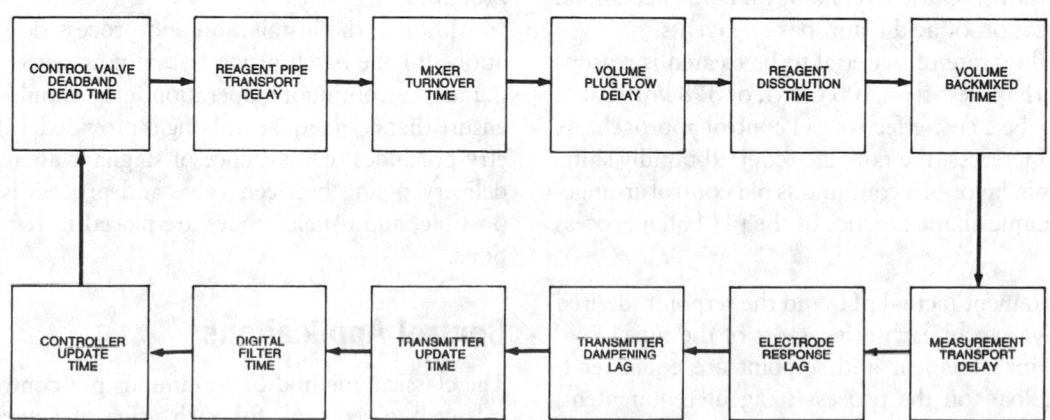

FIG. 7.40.16 The total loop time delay (τ_1) is the sum of all dead times and time delays (lags) in the loop.

tions of the titration curve. Increases in amplitude are difficult to detect once the oscillation moves outside the steep slope region of the titration curve.

The peak error for continuous pH control is proportional to the ratio of dead time to time constant multiplied by the open-loop error (the error with the loop in manual), if the dead time is less than the time constant. If τ_d exceeds τ_1, the peak error is proportional to the full open-loop error. The integrated error is proportional to this peak error multiplied by the dead time. It is critical to mark these errors along the abscissa of the titration curve and translate them to the ordinate (pH axis). Peak errors will be much larger than expected for excursions along the steep slope of the curve.

BATCH CONTROLLER TUNING

For batch control, the offset from setpoint can be made smaller than the peak error for continuous pH control if the sequential requirements of batch pH control are recognized and addressed.

Batch pH control is analogous to the titration done in chemistry lab. If the student has enough patience to use sequentially smaller doses and to wait longer as the pH approaches the endpoint, the final pH can end up within the measurement error of the endpoint. The increased difficulty of continuous pH control could be simulated by cutting a hole in the side of the beaker and adding a sample of variable flow and concentration.

If three vessels are provided, which are sequenced to fill, treat, and drain influent, plus if sufficient processing time and a strategy for variable dosing which duplicates the lab titration procedure is provided, the results will give good batch control of pH. The processing time must be long enough and the reagent dose sizes must be small enough to provide several doses even when the target setpoint is on a steep section of the titration curve and the wait time between charging reagent doses exceeds the loop dead time as the pH approaches the setpoint. The use of integral action (PI or PID control) for reagent addition to a batch volume will cause overshoot and will necessitate cross-neutralization of acidic and basic reagents.

When the flow rate of material to be treated is reasonably small (perhaps less than 100 GPM, or 378 l/m), batch treatment may be a cost-effective pH control approach. As the flow rate increases, the tankage required rapidly shifts the economics in favor of a continuous pH control arrangement. Two unique characteristics of the pH batch process are:

1. The measurement (actual pH) and the setpoint (desired pH) are away from each other most of the time.
2. When the measurement and setpoint are equal (endpoint), the load on the process (reagent requirement) and, hence, the controller output are zero.

The controller characteristics for the batch pH control application should be proportional plus derivative. Reset must not be used, since reset windup will result in overshoot of the controlled variable. In a proportional controller, the corrective action generated is proportional to the size of the error; in a reset controller, to the area under the error curve; and in a rate controller, to the rate at which the error is changing. Once the measurement goes past the setpoint there is no way for the control system to bring it back to the setpoint, unless, of course, two controllers and two reagent supplies are used. In the absence of the reset control mode (proportional-only), a controller is usually supplied with a 50% bias so that the controller output is 50% when the measurement and the setpoint are equal. For the batch application with a proportional plus derivative controller, the bias must be 0% so that when measurement and setpoints are equal, the controller output is 0%.

The effect of secondary lags in the valve, process vessel, and measurement are compensated for by the derivative action of the controller. If, for example, reagent is added but its effect has not yet been seen by the pH electrode, when measurement and set points are equal, then too much reagent will have been added. With the derivative-time setting properly adjusted, the controller will shut off the reagent valve while the measurement is still away from setpoint, thereby allowing the process to come gradually to equilibrium.

Too much derivative time in the controller is preferable to too little. When there is too much, the valve will close prematurely but will open again when the measurement does not reach setpoint. Too little derivative allows the valve to remain open too long, resulting in overshooting the desired pH target.

The variable gain characteristic of the equal-percentage valve is an asset to this type of control system. When the measurement is far away from setpoint, the valve will be wide open, permitting essentially unrestricted reagent flow to the process. As the measurement approaches setpoint and the valve closes, the decreasing gain of the valve counters the increasing gain of the process. Figure 7.40.17 illustrates the measurement-valve behavior of the batch process.

Although the installation and process design considerations for the batch process are not as severe or demanding as the continuous operation, care should be taken to ensure that (1) adequate mixing is provided, (2) tank geometry precludes the existence of stagnant areas, (3) reagent delivery piping between valve and process is as short as possible, and (4) electrodes are placed in responsive locations.

Control Applications

The classical method of continuous pH control is a vertical, well-mixed tank for each stage of neutralization, as depicted in Figure 7.40.18. Each control loop should have a time-constant-to-dead-time ratio of at least 20:1 and a total loop dead time of less than a minute. This 20:1 ra-

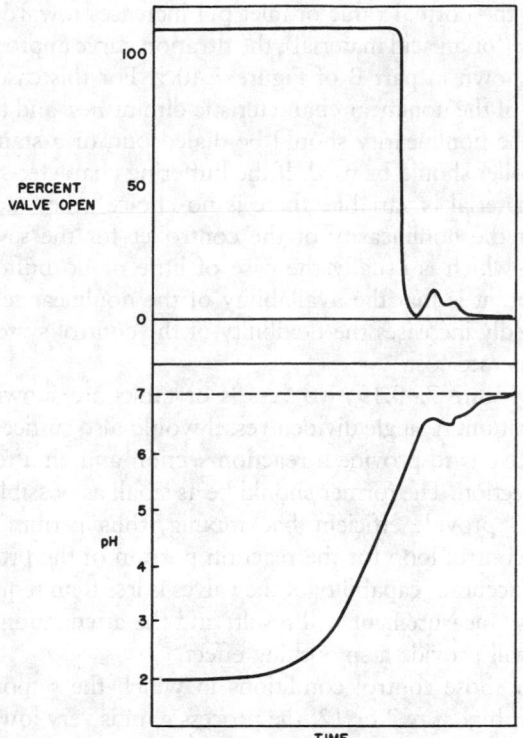

FIG. 7.40.17 Measurement and valve opening behavior for a batch process.

tio can be achieved by a ratio of agitator pumping rate to throughput flow of 20:1 [per Equation 7.40(8), developed earlier]. In other words, the sole source of dead time is assumed to be mixing. However, practical experience indicates that there are many other sources of dead time, as depicted in Figure 7.40.16, and that even vessels that are well-designed often have excessive dead time from a mixing standpoint. Thus, the 20:1 or better ratio can only be achieved by proper attention to final element selection, reagent piping design, vessel geometry, agitation patterns, electrode location, electrode cleaning, and scanning or update times. The vessels used in each stage typically are of

different sizes to protect against having equal periods of oscillation of the loops. For a control loop the toughest disturbance to handle is one with the same frequency, because oscillations which are in phase can get magnified. If the ratio of agitator pumping rate to throughput flow is kept constant, the dead time is proportional to vessel volume, per Equation 7.40(6). This would suggest a requirement that the vessels not be the same size, assuming that all the loop dead time comes from mixing turnover time. It is a good practice to use different tank volumes; however, special efforts must be made to ensure that the other sources of dead time are kept smaller for the smaller volumes.

In Figure 7.40.18 the first stage should have the least dead time and be the fastest loop so that it can react quickly and compensate for disturbances before they affect the downstream stages. This requirement is similar to the principle that the inner loop should be the fastest for cascade control. This way the integrated error and volumes of off-spec material from the first stage for a given upset are significantly reduced as they are proportional to the dead time squared. However, rapid fluctuations of the influent above and below the setpoint are more effectively averaged out in a large volume with considerable savings in reagent usage. For these situations it is particularly advantageous to introduce an attenuation volume upstream of the pH control systems. If this is not possible, it may be best to use the largest vessel for the first stage and the smallest vessel for the second stage.

The setpoint of the first stage in Figure 7.40.18 is selected to be sufficiently far from the incoming pH along the abscissa to ensure that the bulk of the reagent is added in the first stage, but this setpoint should still be on the flat portion of the titration curve as the first stage must bear the full brunt of the disturbances. The last stage should have the smallest final element under normal operation and a larger one to deal with failures or bypasses of the preceding stages. A conservative method of estimation of the number of stages needed is to require one stage

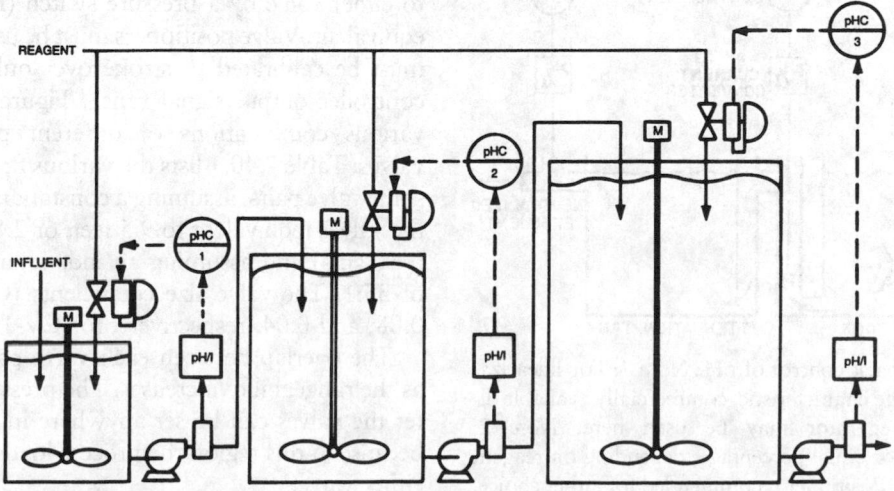

FIG. 7.40.18 Classical three-stage pH control system.

for every two pH units outside the control band. Influent at 1 pH and a control band from 7 to 8 pH places the influent 6 pH units away; therefore, the estimated requirement is three stages of neutralization. Feedforward control or signal linearization combined with valves of sufficient rangeability and precision can sometimes eliminate one stage.

FEEDBACK CONTROL SYSTEMS

Feedback control can be used very effectively in wastewater neutralization, provided the process is not subjected to dramatic or frequent load variations, or both. Maintained step changes in either load or setpoint can be handled effectively. Figure 7.40.19 illustrates a feedback control system in which the reagent flow rangeability requirements are not severe and can be handled by a single valve having linear characteristics. This system can accommodate (for a strong acid–strong base) inlet pH variations of approximately ±0.9 units around some normal value. If a linear valve is unavailable because of material or size limitations, an equal-percentage valve can be used but should be characterized to provide linear reagent delivery as shown in Figures 7.40.10 and 7.40.19. A valve positioner is required to eliminate valve hysteresis (difference in opening and closing characteristics) and to provide responsive valve movement.

The feedback controller in Figure 7.40.19 is a nonlinear controller with the characteristics shown in Figure 7.40.3. The overall loop stability depends on the characteristics of the treatable process material. For a process with a titration curve like that shown in part A of Figure 7.40.9, there is no question that the nonlinear characteristic will be helpful in achieving loop stability.

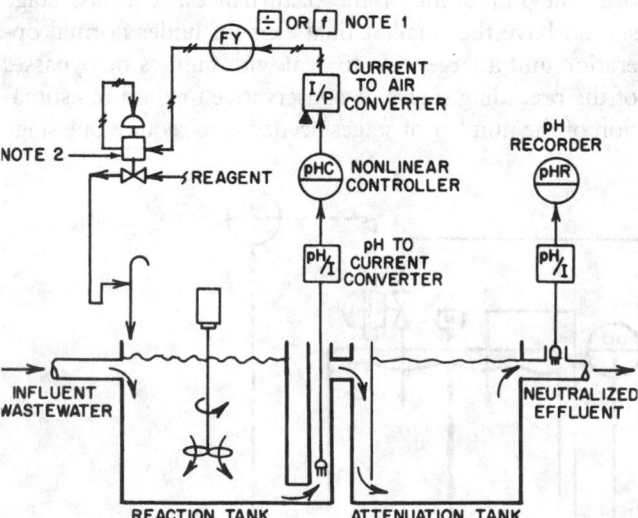

FIG. 7.40.19 Feedback control of pH. *Note 1:* For linearization of equal-percentage characteristic, commercially available divider or function generator may be used here. *Note 2:* Characteristics linear or equal percentage, depending on reagent delivery requirements. Positioner recommended for either choice.

As the normal value of inlet pH increases toward neutrality (for an acid material), the titration curve approaches that shown in part B of Figure 7.40.9. For this case, the value of the nonlinear characteristic diminishes, and therefore the nonlinearity should be dialed out, or a standard controller should be used. If the buffering characteristic of the material is variable, there is no choice other than to adjust the nonlinearity of the controller for the severest case—which is usually the case of little or no buffering. The point is that the availability of the nonlinear feature markedly increases the flexibility of the control system at a moderate cost.

In Figure 7.40.19 two vessels or tanks are shown for illustration. A single divided vessel would also suffice. The objective is to provide a reaction section and an attenuation section. The former should be as small as possible but should provide efficient backmixing, thus permitting a tight control loop for the reaction portion of the process. If the accuracy capability of the valves is less than required, a noisy measurement will result, and the attenuation portion will provide a smoothing effect.

For those control conditions in which the setpoint is low or high, say 2 or 12, the process gain is very low, i.e., it takes a large change in reagent flow to cause a small change in measured pH, and a linear controller with a high gain (high sensitivity, narrow proportional band) suffices. In fact, on-off control (reagent valve is either fully open or closed) may be adequate. Low values of pH setpoint are used for the destruction of hexavalent chromium (Figures 7.42.8 and 9). The destruction proceeds rapidly when the pH is controlled at about a value of 2.5. Higher values of pH lengthen the process.

Sequenced Valves

A wider reagent delivery capability can be obtained by using the sequenced valve approach (Figure 7.40.20). The arrangement is virtually the same as that shown in Figure 7.40.19, except that the controller output can be switched to either valve by a pressure switch (PS) or its electronic equivalent. Valve positioners must be used, since each valve must be calibrated to stroke over only a portion of the controller output signal range. Figure 7.40.21 illustrates various combinations of different pairs of sequenced valves. Table 7.40.1 lists the various flow rangeabilities for some valve pairs, assuming a constant pressure drop across the valves (equivalent to 9.5 feet, or 2.85 m, of 66° Bé sulfuric acid) and assuming an individual valve rangeability of 35:1. The valve size coefficients (C_vs) are 1.13, 0.14, 0.08, and 0.04, respectively, for CV-1, 2, 3, and 4.

The overlap between each valve pair becomes smaller as the rangeability increases. The pressure switch to transfer the valves can be set anywhere in the overlap region, because in this region the process loads can be satisfied by either valve.

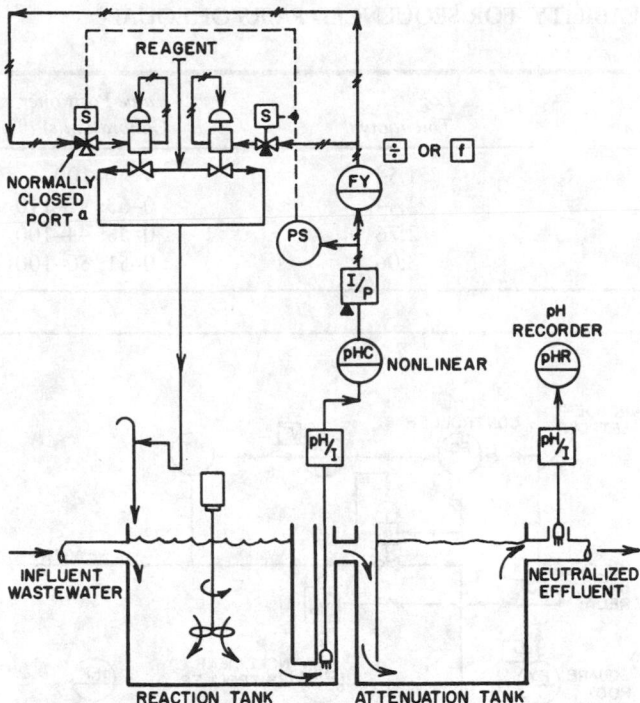

FIG. 7.40.20 Wide-range feedback control of pH. ªThis port is closed when the coil is de-energized because the pressure switch (PS) did not close the electric circuit to supply current to the solenoid coil.

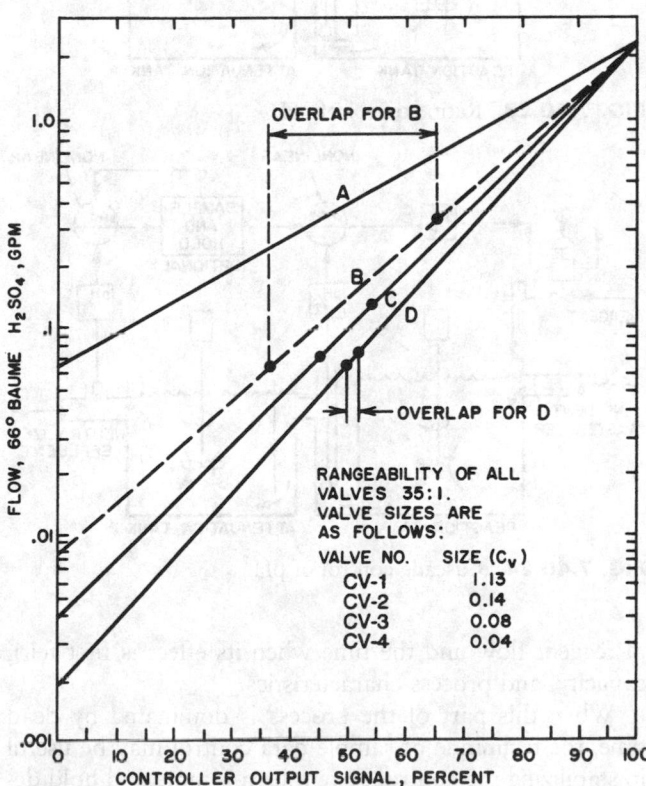

FIG. 7.40.21 Delivery capability for various valve pairs. *Key:* A = CV-1 alone; B = CV-1 + CV-2; C = CV-1 + CV-3; D = CV-1 + CV-4.

Two Reagent Systems

Situations may arise wherein the influent may enter the system on either side of neutrality. Figure 7.40.22 illustrates the two-sided feedback control system. Although only one valve for each side is shown, it would be possible to have a sequenced pair for one side of neutrality and a single valve for the other, or a sequenced pair for both sides. Since this is a feedback control system, load changes cannot be frequent or severe in order for this system to give acceptable performance. For those applications in which load changes are frequent and severe, a combination feedforward-feedback should be considered. If sequencing is used, the reagent delivery system will have a high gain characteristic, since the stroking of the pair (moving from closed to open) is accomplished with only half the controller output signal, thereby doubling the gain (making it twice as sensitive). The valve gain will vary with the turndown, and a characterizer will be required for each set of sequenced valves to provide uniform loop gain.

Ratio Control Systems

Ratio control of pH can be extremely effective when the process flow rate is the major load variable, and the objective is to meet increased flow with a corresponding increase in reagent. Since flow measurements may be in error and reagent concentration may vary, a means for on-line ratio adjustment must be provided. Figure 7.40.23 illustrates a ratio control system in which the reagent set point is changed proportionally to changes in process flow. A feedback signal supplied by the feedback controller (pHC) also adjusts the reagent flow setpoint proportionally to a nonlinear function of the deviation between desired and actual effluent pH.

Note that the rangeability of the ratio system is limited by that of the flowmeters, typically 4:1 for orifice meters to 30:1 for some turbine meters.

Cascade Control Systems

Cascade control (the output of one controller—the master or primary—is the setpoint of another) as applied to pH control systems can take two forms. In addition to the usual condition in which the output of one controller serves as the setpoint to another controller, it is also possible to have two vessels arranged in series, each with its own control system. The latter arrangement is referred to as cascaded residences.

The conventional cascade control system is shown in Figure 7.40.24, wherein the output of controller pHC-1 is the setpoint of the slave, or secondary controller, pHC-2. This arrangement is particularly useful when lime is the reagent. In this instance, because of the finite reaction time between the acid and reagent, the setpoint of pHC-2 may have to be lower than the desired pH of the final effluent

TABLE 7.40.1   REAGENT DELIVERY TURNDOWN (RANGEABILITY) FOR SEQUENCED PAIRS OF EQUAL-PERCENTAGE VALVES

Valve Pair	Line on Figure 7.40.21	Turndown	Log Towndown*	Valve Positioner Calibration(s) (%)
CV-1 (alone)	A	35:1	1.54	0–100
CV-1 + CV-2	B	275:1	2.44	0–63; 37–110
CV-1 + CV-3	C	570:1	2.76	0–58; 44–100
CV-1 + CV-4	D	1150:1	3.06	0–51; 50–100

*Signifies the approximate pH swing that valves will accommodate.

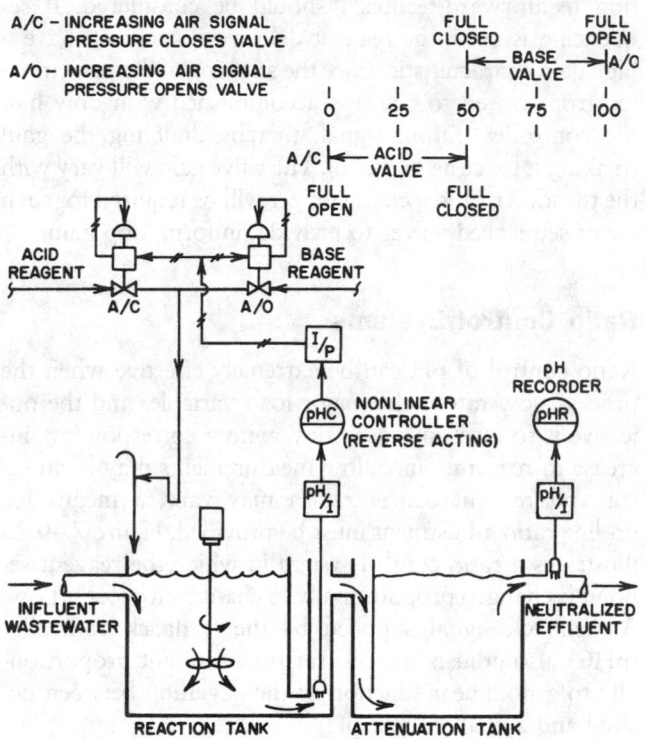

FIG. 7.40.22 Two-sided feedback control of pH.

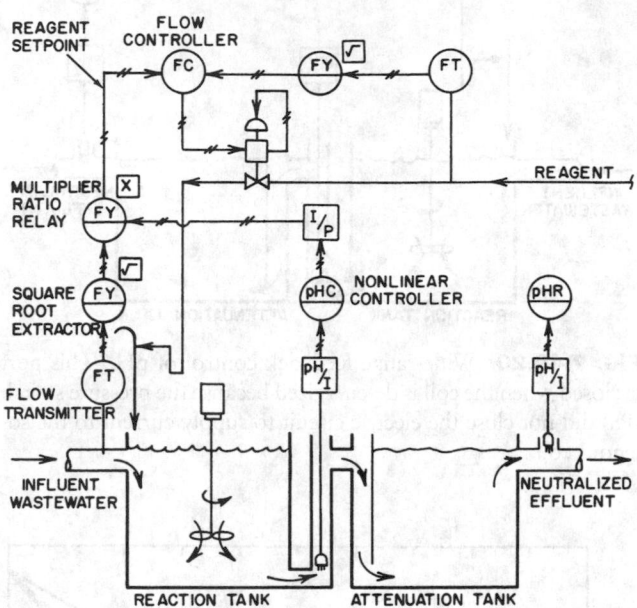

FIG. 7.40.23 Ratio control of pH.

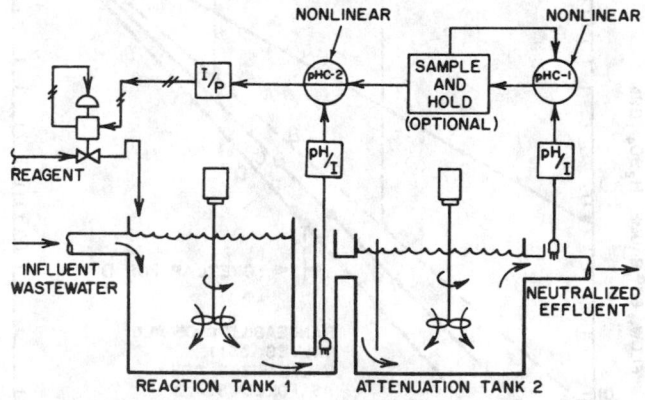

FIG. 7.40.24 Cascade control of pH.

because the materials are still reacting with each other after they have left the first tank. If the setpoint pHC-2 is too high, the pH of the final stream will be greater than desired. When flocculation is to be carried out downstream of the pH treatment facility, stable pH values can be extremely important.

A delicate balance must be struck in this type of system with respect to the size of the first vessel. A long residence time in the first tank ensures long contact time between reagents, thereby producing an effluent pH which is close to the desired value, but at the same time it may result in a sluggish control loop around this vessel. For efficient cascade control, response of the inner loop (control loop around the first tank) must be fast. The other control loop (pHC-2), sometimes referred to as the master, or primary, control loop, is usually tuned (control mode adjustments such as proportional band are set) so as to be less responsive than the inner loop. The tuning of pHC-1 will be a result of the dead time (a delay between a change

in reagent flow and the time when its effect is first felt), capacity, and process characteristics.

When this part of the process is dominated by dead time, the technique of sample data control may be useful in stabilizing the control system by a sample and hold device (Figure 7.40.24). This device may be a timer that automatically switches the controller between automatic and manual modes of operation. This can allow the controller to be in automatic for a fraction (x) of the cycle time (t) and then can switch it to a fixed-output, manual condition for the rest of the cycle $(1 - x)t$.

In the second form of cascade (cascaded residences) (Figure 7.40.25), each vessel has its own feedback loop. This approach is recommended when the incoming material is very strongly acidic or basic (pH values less than 1 or greater than 13). The first stage controls the effluent at a pH of approximately 4 or 10, and the second stage brings the effluent to its final value, near 7. The choice of pH setpoint for the first stage depends on the characteristics of the material. For example, material having a titration characteristic like that of A in Figure 7.40.9 may have a setpoint of approximately 3.5. The purpose is to make the control problem as simple as possible by staying on as linear a portion of the titration curve as possible.

This approach is logical when one considers the process gain characteristic as well as the accuracy limitation of a reagent delivery system. The remainder of the neutralization control problem is then similar to that illustrated in part B of Figure 7.40.9. In this manner, the control system does not have to cope with the entire nonlinear characteristic of the process all at once. A sequenced pair of valves is shown in conjunction with the first stage in order to handle the pH load variations. A single valve would probably suffice for the second stage, because its influent is pH-controlled. Depending on the valve sizes and on the individual valve rangeabilities, this three-valve arrangement has a maximum possible reagent flow turndown of approximately 125,000:1 ($50 \times 50 \times 50$).

FEEDFORWARD CONTROL SYSTEMS

A feedforward control (the consequences of upsets are anticipated and counteracted before they can influence the process) system is dedicated to initiating corrective action as soon as changes occur in process load. The corrective action is implemented using a control system that is essentially a mathematical model of the process. Ordinarily, the inclusion in the model of each and every load to which the process is subjected is neither possible nor economically justifiable. This means that a feedback control loop (usually containing the nonlinear controller for pH applications) is required in conjunction with the feedforward system. The function of the feedback controller is to trim and correct for minor inaccuracies in the feedforward model.

FEEDBACK-FEEDFORWARD COMBINATION CONTROL

The fractional control signal (X_B) to the valves is a function of the characterized flow signal f(F′). Therefore, the total output signal to the split-ranged valves (X_B) becomes:

$$X_B = f(F') + \text{feedforward output} - 2(\text{feedback output})$$

$$7.40(22)$$

Figure 7.40.26 illustrates a feedforward control system arrangement in which the flow characterization of the influent wastewater flow signal and the use of the dead-band adjustment feature previously discussed (Figure 7.40.3) are shown. A system as outlined in Figure 7.40.26 has demonstrated the need for the combination of feedforward-feedback control, because each type of control was tested individually and was found to be unsatisfactory.

A combination of feedforward-feedback on one side of neutrality and conventional feedback on the other is also possible. The nature and characteristics of the problem to be solved will indicate the nature of the solution.

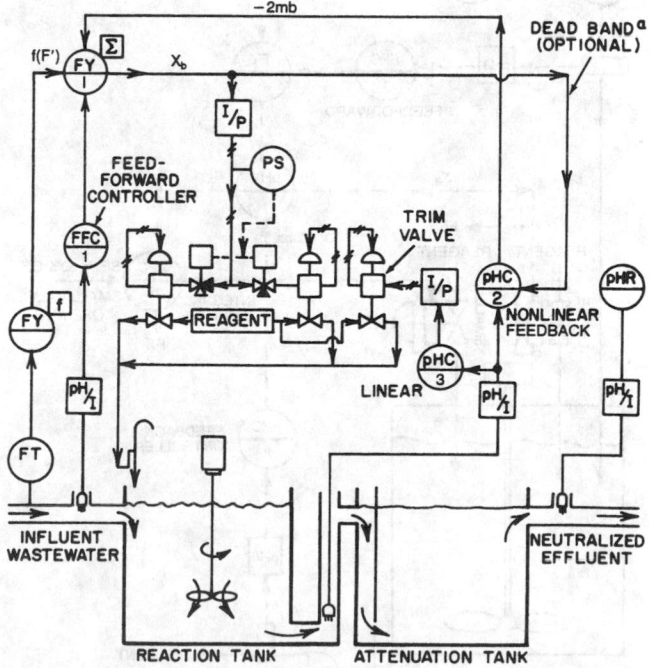

FIG. 7.40.26 Three-valve feedforward pH control system. [a]If pHC-2 is provided with a dead band, it will be inactive when the trimming controller (pHC-1) alone is operating.

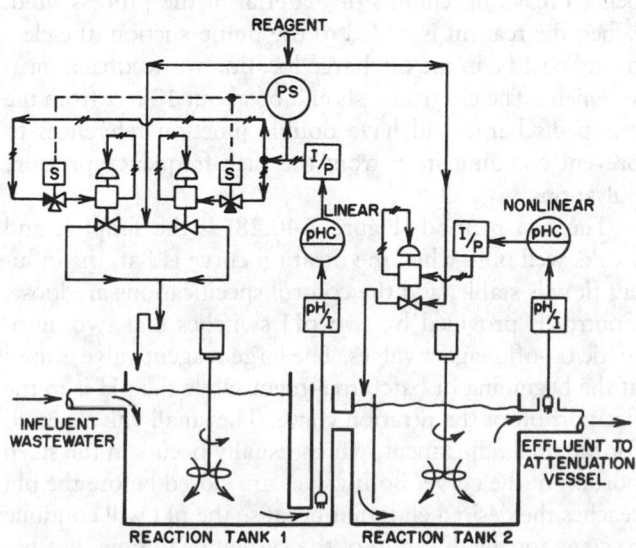

FIG. 7.40.25 Cascade residence control of pH.

Figure 7.40.27 shows how pH feedforward is used to rapidly position a large valve for major upsets. In this configuration an integral-only valve position controller (VPC) slowly optimizes the large valve position to keep the trim valve in the middle of its throttling range. Since the feedback manipulation of the big valve must be slow to prevent interaction between the two valves, the pH feedforward action provides a performance edge of rapid action for major upsets. Other methods to coordinate the movement of a large valve and a trim valve involve the use of output tracking strategies between two pH controllers, dedicated and tuned for each valve, so that both controllers are not in service at the same time.

OPTIMIZATION AND AUTO-START-UP

The existence of many volumes and the presence of numerous reagent addition points provides some potentials for optimization. The goal can be the minimization of reagent usage and of salt production (from cross-neutralization of reagents) while keeping the pH within acceptable limits for the process materials of construction. Fuzzy logic can be useful in such optimization schemes as follows: the magnitude and direction of changes in the pH of various tank volumes without pH control loops and the magnitude and direction of reagent consumption in volumes with pH control loops can be used to turn on or off the addition of upstream reagent. For example, if a down-

stream volume has either a high or increasing rate of base addition, the existence of an intervening volume of low or decreasing pH would result in shutting off the upstream reagent flow.

The automated start-up and shutdown of pH loops is feasible if redundant sensors configured into voting or median selector systems protect against measurement failure. These operations can be smooth and reliable enough to eliminate the need for operator attention. Most pH loops are too sensitive and nonlinear for manual manipulation, and the elimination of operator actions in most cases greatly improves the performance of the system.

BATCH CONTROL OF pH

Figures 7.40.28 through 31 show four major methods of batch control of pH. Each of the figures shows four possible locations for the electrodes. The submersion assembly which enters from the top is difficult to remove, decontaminate, and maneuver. Since the bulk velocity even in well-mixed tanks rarely exceeds 1 fps, this electrode response is likely to be slow and prone to problems due to coating or fouling. The side-entry electrode tip should be close to the agitator impeller to take advantage of the local increase in fluid velocity. Retractable insertion assemblies with ball valves for isolation and with optional flush connections (see Section 7.7) are used to allow withdrawal while the vessel is full. Locating the electrodes in a recirculation pipeline is the best from the standpoint of probe response, self-cleaning action, and ease of access, but this approach is generally more expensive due to the need for block, drain, and bypass valves (not shown) for retractable installations or because of the cost of the flow assemblies when the pipeline can be emptied for removal of the electrodes. Electrodes installed on the discharge side of the pump have a slightly larger transportation delay than do those on the pump suction but are less likely to be damaged, because the pump strainer catches and the pump impeller breaks up clumps of material in the process fluid. When the reagent is added to the pump suction, the electrode must be in the discharge location for feedback measurement. The electrodes should be about 10 feet from the pump discharge and have double junction references to prevent contamination from the high-frequency pressure pulsations.

The first method (Figure 7.40.28) is the simplest and works well only when the titration curve is flat, the influent flow is stable, and the control specifications are loose. Control is provided by two pH switches and two automatic on-off reagent valves. The large reagent valve is used at the beginning of batch treatment while the pH is in the flat portion of the titration curve. The small valve is used for the final adjustment, which usually occurs in the steep portion of the curve. Both valves are closed before the pH reaches the desired endpoint because the pH will continue to coast for the duration of the system dead time. If a basic batch is being neutralized the large valve might close

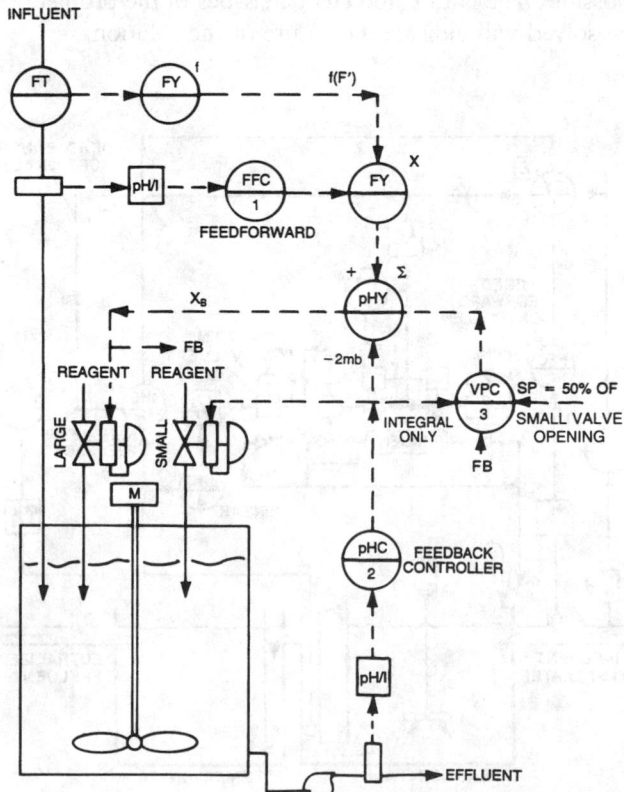

FIG. 7.40.27 Feedback-feedforward control system which keeps the small reagent valve near 50% open.

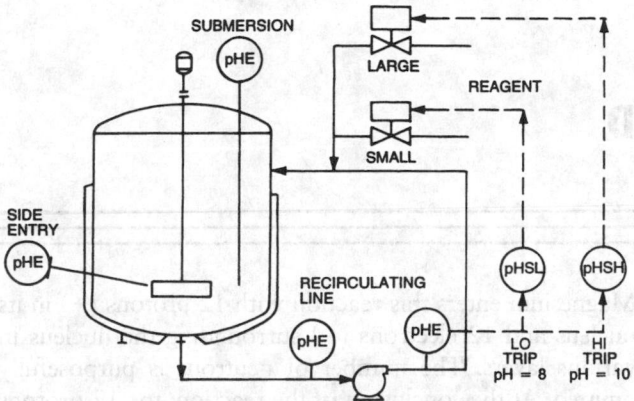

FIG. 7.40.28 Batch neutralization can be controlled by two on-off valves. If the batch is basic the large valve might close at pH = 10 while the small one closes at pH = 8.

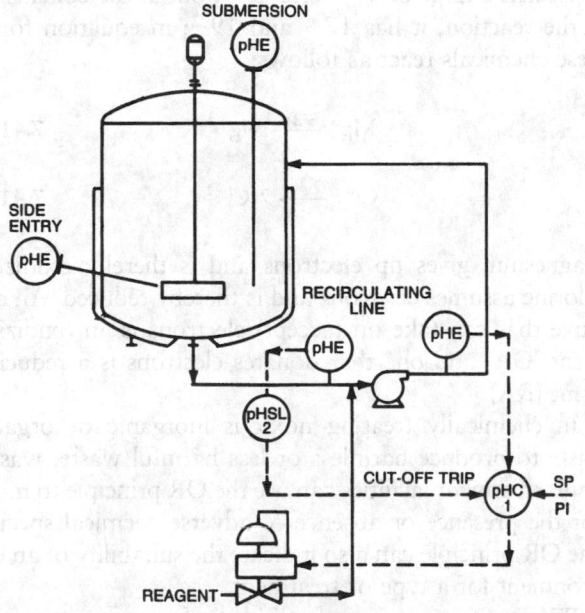

FIG. 7.40.29 Batch pH control can be configured with a proportional and integral in-line controller (pHC-1) and a safety trip cut-off (pHSI 2) guaranteeing that the batch pH does not drop too low.

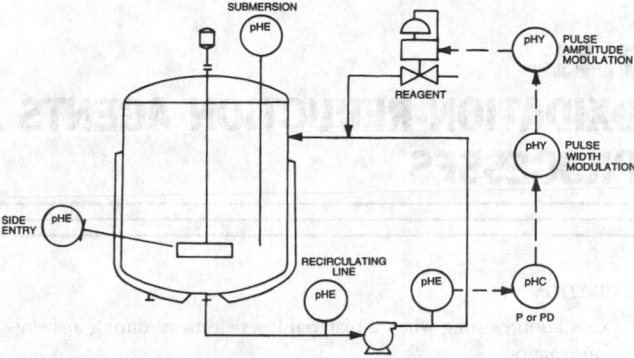

FIG. 7.40.30 Batch pH control with pulse width and amplitude modulation of the controller output.

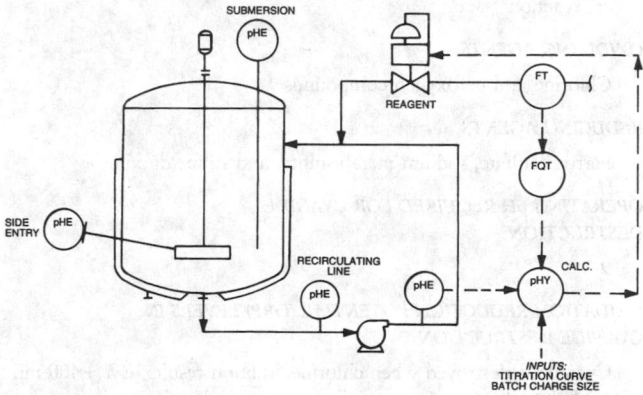

FIG. 7.40.31 Batch pH control can be based on the size of the batch and the required cut-off of reagent based on the titration curve.

at a pH of 10 while the small valve closes at pH of 8. The second method (Figure 7.40.29) uses in-line sensors. The in-line controller (PHC-1) setpoint is biased to kept the reagent valve open beyond its low output limit and thereby to minimize batch treatment time. A suction or vessel pH switch (pHSL-2) terminates the addition of reagent as the pH approaches the endpoint. While the reagent valve is temporarily shut, a vessel pH reading is obtained from the electrodes at the pump discharge. When needed, the in-line system can be momentarily restarted if the pH coasts to a value which is short of the endpoint. Variations of this strategy are used to operate continuous pH systems in a semi-batch mode. This can be done when sufficient ves-

sel capacity is available so that the vessel discharge valve can be closed while the pH of the vessel contents is being adjusted during start-up or after a big upset.

The third method (Figure 7.40.30) uses a proportional-only or proportional-plus-derivative controller (Fig. 7.40.17) with pulse width and amplitude modulation of the controller output to mimic the laboratory titration process. The further away the pH is from the endpoint, the bigger and longer are the reagent flow pulses. Manual mode outputs bypass the pulsation algorithms to facilitate manual stroking of the valve. While a variety of customized strategies can be developed to achieve the same result, this scheme has the advantage of a typical controller interface and tuning adjustments for operations and maintenance.

The last method (Figure 7.40.31) uses the titration curve and the vessel volume to predict required charge of reagent and the setpoint for a totalizer. It depends heavily upon the accuracy of the curve and may best be implemented by partitioning the curve into segments and by using multiple charges. The titration curve should be corrected for temperature and for composition variations. Titration of samples taken just prior to the batch being charged could be used to verify the curve.

7.41
OXIDATION-REDUCTION AGENTS AND PROCESSES

OXIDATION

Condition existing when a material loses electrons during a chemical reaction

REDUCTION

Condition existing when a material gains electrons during a chemical reaction

OXIDIZING AGENTS

Chlorine and peroxygen compounds

REDUCING AGENTS

Ferrous sulfate, sodium metabisulfite, and sulfur dioxide

OPERATING pH REQUIRED FOR CYANIDE DESTRUCTION

9

OXIDATION-REDUCTION POTENTIAL (ORP) LEVELS IN CYANIDE DESTRUCTION

Cyanide is destroyed when chlorine addition results in a +400 millivolts (mV) solution potential, and cyanate destruction is accomplished at +600 mV.

REAGENT REQUIREMENTS OF CYANIDE DESTRUCTION

Each part of cyanide requires 9.56 parts of chlorine to oxidize it to carbon dioxide, and each part of chlorine requires 1.125 parts of sodium hydroxide to neutralize it. The actual requirements can exceed the stated stoichiometric values by a factor of 2 or 3.

OPERATING pH REQUIRED FOR HEXAVALENT CHROME REDUCTION

Between 2 and 3 pH

ORP LEVELS IN CHROME REDUCTION

See Table 7.41.5.

Process Description

Oxidation-reduction (OR) refers to a class of chemical reactions in which one of the reacting species gives up electrons (oxidation), while another species in the reaction accepts electrons (reduction). At one time, the term oxidation was restricted to reactions involving oxygen; similarly, the term reduction was restricted to reactions involving hydrogen. Current chemical technology has broadened the scope of these terms to include all reactions in which electrons are given up and assumed by reacting species; in fact, electron donating and accepting must take place simultaneously. Thus, magnesium can burn in chlorine as well as in oxygen as shown in the following equation:

$$Mg + Cl_2 \xrightarrow{\Delta} MgCl_2 + \text{heat and light} \qquad 7.41(1)$$

Magnesium enters this reaction with 12 protons (+) in its nucleus and 12 electrons (−) surrounding the nucleus in various layers. The number of neutrons is purposefully omitted. At the conclusion of the reaction, the 12 protons remain, but now only 10 electrons surround the nucleus. The magnesium is no longer electrically neutral because it has an excess of protons. Similarly, the chlorine enters with an electric charge of 17+ and 17−, but at the conclusion of the reaction, it has 17+ and 19−; in equation form, these chemicals react as follows:

$$Mg^\circ \xrightarrow{-2e} Mg^{+2} \qquad 7.41(2)$$

$$Cl^\circ_2 \xrightarrow{+2e} 2\ Cl^{-1} \qquad 7.41(3)$$

Magnesium gives up electrons and is thereby oxidized; chlorine assumes electrons and is thereby reduced. An additive that can take on (accept) electrons is an oxidizing agent (OA) and one that donates electrons is a reducing agent (RA).

In chemically treating noxious inorganic or organic waste to produce harmless or less harmful waste, wastewater treatment facilities can use the OR principle to monitor the presence or absence of adverse chemical species. The OR principle can also indicate the suitability of an environment for a type of treatment.

Wastewater treatment facilities perform monitoring by measuring the electrical potential of the chemical system with respect to a known reference before and after treatment and holding the electrical potential constant by adding a suitable reagent. They keep the electrical potential at a value that indicates that no adverse species is present or that it has been destroyed. The measurement is a voltage (emf) usually referred to as the ORP.

The emf measurement for a system has no specificity, i.e., it indicates neither the presence nor absence of a particular ion. The emf measurement indicates the activity ratio of the oxidizing species present to that of the reducing species present. The pH electrode is an example of a measurement that specifies the activity of a particular ion, i.e., ionized hydrogen in solution. The ORP electrode in conjunction with a reference electrode, develops an emf value as given by the Nernst equation as follows:

$$E_{meter} = E_o + \frac{0.0591}{n} \log \frac{[\text{Ox.}]}{[\text{ReD.}]} \qquad 7.41(4)$$

where:

E$_o$ = a constant dependent on the choice of the reference electrode and half-cell potential of the reaction

n = number of electrons in the OR reaction

[Ox.] and [ReD.] = activities of the oxidized and reduced species, respectively

Application of the Nernst Equation

The following equations show the application of the Nernst equation with a simple reaction system, such as the conversion of ferrous to ferric ions by the addition of a solution containing ceric ions:

$$Fe^{2+} \longrightarrow Fe^{3+} + e \qquad 7.41(5)$$

$$Ce^{4+} + e \longrightarrow Ce^{3+} \qquad 7.41(6)$$

$$Fe^{2+} + Ce^{4+} \longrightarrow Fe^{3+} + Ce^{3+} \qquad 7.41(7)$$

Equations 7.41(5) and (6) are half-cell reactions. Equation 7.41(5) is written in the oxidized form, i.e., the charge(s) appear to the right of the arrow, and Equation 7.41(6) is written in the reduced form, i.e., the charge(s) appear to the left of the arrow. Each half-cell reaction has a standard or half-cell electrode potential.

The terms *standard electrode potential* and *half-cell potential* are closely related; the major distinction between them is conventional. If the half-cell reaction is written in the reduced form, as in Equation 7.41(6), the standard electrode potential and the half-cell are the same. If the reaction is written in the oxidized form (Equation 7.41[5]), the sign of the reported potential must be changed for the half-cell potential. Butler provides more information on the sign convention of OR reactions. Table 7.41.1 gives the standard electrode potentials for Equations 7.41(5) and (6).

The Nernst equation for the two reactions is as follows:

$$E_{meter\ Fe} = 0.771 + \frac{0.0591}{n} \log \frac{[Fe^{3+}]}{[Fe^{2+}]} \qquad 7.41(8)$$

$$E_{meter\ Ce} = 1.61 + \frac{0.0591}{n} \log \frac{[Ce^{4+}]}{[Ce^{3+}]} \qquad 7.41(9)$$

TABLE 7.41.1 ELECTRODE POTENTIALS FOR EQUATIONS 7.41(5) AND (6) WITH RESPECT TO THE STANDARD HYDROGEN ELECTRODE

Half-Cell Reaction	Standard Potential (V), $E_n°$
$Fe^{3+} + e \longrightarrow Fe^{2+}$	+0.771
$Ce^{4+} + e \longrightarrow Ce^{3+}$	+1.61

Source: The Chemical Rubber Company, Handbook of Chemistry and Physics, 45th ed. (Cleveland, Ohio).

TABLE 7.41.2 POTENTIALS OF STANDARD REFERENCE ELECTRODES RELATIVE TO THE STANDARD HYDROGEN ELECTRODE AT 25°C

Half-Cell Reaction	Standard Potential (V), $3°_{Ref}$
$AgCl + e \longrightarrow Ag° + Cl^-$ (1M KCl)	+0.235
$AgCl + e \longrightarrow Ag° + Cl^-$ (4M KCl)	+0.199

where n = 1 and the standard hydrogen electrode is the reference electrode. For process applications, the standard hydrogen electrode is not used; instead the silver–silver chloride electrode in 1 M or 4 M KCl solution is the reference electrode. Table 7.41.2 lists the half-cell potentials of these standard reference electrodes.

The following equation gives the value of the Nernst E$_o$ term:

$$E_o = E_n° - E_{ref}° \qquad 7.41(10)$$

where:

$E_n°$ = the standard potential of the reaction written in reduced form. Subscript *n* is the chemical symbol for the atom, molecule, or radical being considered.

Rewriting Equations 7.41(8) and (9) for use with the 1 M KCl electrode gives the following equations:

$$(E_{meter})_{Fe} = (0.771 - 0.235) + 0.0591 \log \frac{[Fe^{3+}]}{[Fe^{2+}]}$$

$$(E_{meter})_{Fe} = 0.536 + 0.0591 \log \frac{[Fe^{3+}]}{[Fe^{2+}]} \qquad 7.41(11)$$

$$(E_{meter})_{Ce} = (1.61 - 0.235) + 0.0591 \log \frac{[Ce^{4+}]}{[Ce^{3+}]}$$

$$(E_{meter})_{Ce} = 0.375 + 0.0591 \log \frac{[Ce^{4+}]}{[Ce^{3+}]} \qquad 7.41(12)$$

Note that $(E_{meter})_{Fe}$ and $(E_{meter})_{Ce}$ equal their respective Nernst E$_0$ values (0.536 and 0.375 V) when the ratio of [Ox.] to [ReD.] = 1. This ratio occurs at the half-titration point.

The following equation gives the ORP value at the equivalence or endpoint of the reaction:

$$E = \frac{n_1 E_{Fe}° + n_2 E°}{n_1 + n_2} \qquad 7.41(13)$$

where:

E = ORP, V

n_1 and n_2 = number of electrons in the oxidizing and reducing reactions, respectively

For the preceding example where $n_1 = n_2 = 1$, E is as follows:

$$E = \tfrac{1}{2}(E^\circ_{Fe} + E^\circ_{Ce}) = \tfrac{1}{2}(0.536 + 0.375)$$
$$E = +0.456 \text{ V}$$

Titration Curve

Figure 7.41.1 shows the approximate titration curve for the Fe^{2+}/Ce^{4+} system in two sections. Section A proceeds from the top left to the equivalence point, and section B proceeds from the equivalence point down to the right.

In the following example, Fe^{2+} is a noxious chemical effluent of a plant, and Ce^{4+} is a reagent added to produce the less noxious or inert Fe^{3+} form. A set of ORP electrodes monitors the reaction. What value of ORP indicates that the reaction is complete? If the process operates to the end or equivalence point, the least amount of titrating reagent is used; however, a slight deficiency in the amount of Ce^{4+} reagent signifies that Fe^{2+} is still present. If the process operates to the E°_{Ce} point, a large excess of Ce^{4+} reagent is required, which is costly and can require adding a substance to the treated wastewater that is just as adverse as the original waste.

The best operating potential for this process is about +0.400 V. Only a slight excess of reagent is required, but all of the Fe^{2+} is oxidized. For ORP applications, operation away from the equivalence point is advisable, insuring that the adverse species is completely eliminated.

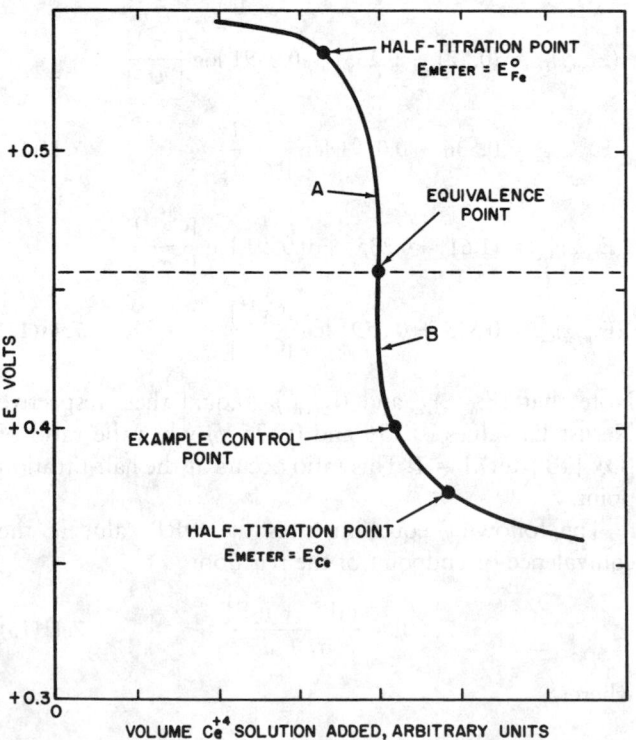

FIG. 7.41.1 Approximate titration curve for the Fe^{2+}/Ce^{4+} system.

The chemistry of this Fe^{2+}/Ce^{4+} reaction has been simplified to demonstrate the handling of the Nernst equation and the applicable sign convention. In an actual Fe^{2+}/Ce^{4+} system, the Fe^{2+} is usually in a highly acid solution.

Oxidation Processes—Cyanide Destruction

The destruction of cyanide waste by alkaline chlorination is an example of an oxidation process. The process is a two-stage operation, i.e., cyanide to cyanate and cyanate to carbon dioxide (usually as sodium bicarbonate) and nitrogen.

FIRST-STAGE REACTION

The first-stage reaction equation, assuming a sodium cyanide waste and liquid chlorine oxidizing reagent, is as follows:

$$NaCN + 2NaOH + Cl_2 \longrightarrow NaCNO + 2NaCl + H_2O$$

$$7.41(14)$$

The pH level of the reaction must be around 9 to preclude the formation of toxic gases such as cyanogen chloride (COCl) and hydrogen cyanide (HCN). When cyanide is in excess, which occurs when the untreated wastewater leaves the plant, the following reduction potential applies:

$$CNO^- + H_2O + 2e \longrightarrow CN^- + 2OH^- \quad 7.41(15)$$
$$\text{oxidized} \longrightarrow \text{reduced}$$

This reaction has a standard potential of -0.97 V with respect to the standard hydrogen electrode. Assuming a Ag–AgCl (4M KCl) reference electrode (see Table 7.41.2) gives the following:

$$E_0 = -0.970 - 0.199 = -1.169 \text{ V or } -1169 \text{ mV}$$

The ratio of oxidants to reductants can be written using Equation 7.41(15) as follows:

$$\frac{[Ox.]}{[ReD.]} = \frac{[CNO^-]}{[CN^-][OH^-]^2} \quad\quad 7.41(16)$$

Substituting the value of E_0 already calculated and Equation 7.41(16) into the Nernst equation gives the following:

$$(E_{meter})_{CN} = -1.169 + \frac{0.0591}{n} \log \frac{[CNO^-]}{[CN^-][OH^-]^2} \quad 7.41(17)$$

where $n = 2$. As stated in Equation 7.39(4) that $[OH^-][H^+] = K_w = 1 \times 10^{-14}$ at 25°C and $-\log [H^+] = pH$, Equation 7.41(17) can be written as follows:

$$(E_{meter})_{CN} = -1.169 + \frac{0.0591}{2} \log \frac{[CNO^-]}{[CN^-]} + 0.0591(14 - pH)$$

$$7.41(18)$$

Equation 7.41(18) shows the importance of the pH in the ORP process. If the pH is unknown and controlled, the meter reading is useless. When the process is controlled at pH = 9; $(E_{meter})_{CN}$ at the half-titration point can then be calculated as follows:

$$(E_{meter})_{CN} = -1.169 + 0 + 0.0591 (14-9)$$

$$(E_{meter})_{CN} = -0.873 \text{ V} \qquad 7.41(19)$$

SECOND-STAGE REACTION

The second-stage treatment is a continuation of the alkali–chlorination process in which the cyanate is destroyed as follows:

$$2NaCNO + 4NaOH + 3Cl_2 \longrightarrow$$
$$6NaCl + 2CO_2 + N_2 + 2H_2O \quad 7.41(20)$$

The reaction occurs at a controlled pH level, usually between 8 and 9.5, for a reaction time of 30 min or less. Table 7.41.3 shows the retention time relationship to pH requirements for the second-stage reaction.

The retention time requirement should not be confused with the vessel hold time constant; the latter is the ratio of volume divided by flow (V/F). The V/F ratio gives a nominal time constant. To determine the variance about the nominal value, environmental engineers should use a statistical technique to insure that all of the material treated has spent the required amount of time in the reaction vessel.

This second-stage reaction can be treated in a manner similar to the first-stage treatment. The reduction potential is estimated as +0.4 V as follows:

$$2CO_2 + N_2 + 2H_2O + 6e \longrightarrow 2CNO^- + 4OH^- \quad 7.41(21)$$

$$E_m = E_o + \frac{0.0591}{6} \log \frac{[Ox.]}{[ReD.]}$$

$$E_o = 0.4 - 0.199 = +0.201 \text{ V}$$

$$\log \frac{[Ox.]}{[ReD.]} = \log \frac{[Ox.]}{[CNO^-]^2[OH^-]^4}$$

$$= \log \frac{[Ox.]}{[CNO^-]^2} - 4 \log [OH^-]$$

$$= \log \frac{[Ox.]}{[CNO^-]^2} + 4(14 - pH)$$

$$(E_{meter})_{CNO} = 0.201 + \frac{0.0591}{6} \log \frac{[Ox.]}{[CNO^-]^2}$$

$$+ 4(14 - pH)\frac{0.0591}{6} \qquad 7.41(22)$$

When the pH is controlled at 9.0, the half-titration voltage $(E_m)_{CNO}$ can be estimated as follows:

$$(E_{meter})_{CNO} = 0.201 + 0 + 4(14 - 9)\frac{0.0591}{6} = 0.201 + 0.197$$

$$(E_{meter})_{CNO} = +0.398 \text{ V}$$

TABLE 7.41.3 RETENTION TIME REQUIREMENTS AS A FUNCTION OF pH LEVEL IN CYANATE DESTRUCTION

pH	Retention Time (min)
8.0	8
8.5	11
9.0	13
9.5	27
9.9	80
10	Infinite

The voltage at the equivalence point resulting from the half-cell reactions described by Equations 7.41(15) and (21) can now be computed. Since this system is unsymmetrical, i.e., n_1 electrons are transferred in one half-cell reaction and n_2 electrons are transferred by the other half-cell reaction, the following equation gives the voltage at the equivalence or endpoint:

$$E = \frac{n_1(E_{meter})_{CN} + n_2(E_{meter})_{CNO}}{n_1 + n_2}$$

$$E = \frac{2(-0.873) + 6(+0.398)}{6 + 2} = +0.0805 \text{ V} \qquad 7.41(23)$$

A typical value of the ORP control setpoint is +0.400 V, which is beyond the equivalence point value indicating the elimination of CN^-. Once the cyanate is undergoing destruction, the following equation gives the operative half-cell reaction:

$$ClO^- + H_2O + 2e \longrightarrow Cl^- + 2OH^- \qquad 7.41(24)$$

which has a standard potential of +0.89 V. When the reaction is controlled at pH = 9, the half-titration voltage of reaction in Equation 7.41(24) can be estimated as follows:

$$(E_{meter})_{Cl} = (0.89 - 0.199) + \frac{0.0591}{2} \log \frac{[ClO^-]}{[Cl^-]}$$
$$+ 0.0591(14 - pH)$$

$$= 0.691 + 0 + 0.296$$

$$(E_{meter})_{Cl} = +0.987 \text{ V}$$

The equivalence voltage of the half-cell reaction pair, Equations 7.41(21) and (24), is estimated from the half-titration voltages as follows:

$$E = \frac{6(E_{meter})_{CNO} + 2(E_{meter})_{Cl}}{2 + 6} = \frac{6 \times 0.398 + 2 \times 0.987}{8}$$

$$= +0.545 \text{ V}$$

For cyanate destruction, the normal control point is about +0.600 V.

Figure 7.41.2 shows the reactions from cyanide to cyanate to cyanate destruction in titration form. The re-

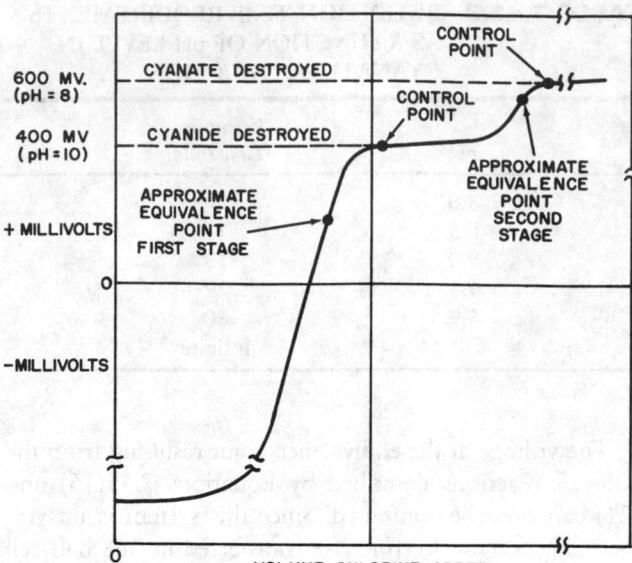

FIG. 7.41.2 Approximate path for cyanide destruction.

actions for the alkali chlorination of sodium cyanide are as follows:

1st Stage

$Cl_2 + 2NaOH \longrightarrow NaOCl + NaCl + H_2O$
$NaOCl + NaCN + H_2O \longrightarrow CNCl + 2NaOH$
$CNCl + 2NaOH \longrightarrow NaCNO + NaCl + H_2O$
$\overline{Cl_2 + NaCN + 2NaOH \longrightarrow NaCNO + 2NaCl + H_2O}$

2nd Stage

$3Cl_2 + 6NaOH \longrightarrow 3NaOCl + 3NaCl + 3H_2O$
$3NaOCl + 2NaCNO + H_2O \longrightarrow 2NaHCO_3 + N_2 + 3NaCl$
$\overline{3Cl_2 + 2NaCNO + 6NaOH \longrightarrow 2NaHCO_3 + N_2 + 2H_2O}$
$\hspace{8cm} + 6NaCl$

OXIDIZING REAGENTS

Other chlorine compounds are also oxidizing reagents. They are HOCl, NaOCL, and NH_2Cl_2. Liquid chlorine is used most often because of its convenient form (pure liquid shipped in cylinders) and the availability of chlorination equipment. As an economic measure, wastewater treatment facilities should consider the use of high-calcium lime as an alternate alkali for pH control when treating cyanide waste with greater than 200 ppm concentration.

Another process for cyanide destruction in zinc and cadmium electroplating operations uses a DuPont proprietary peroxygen compound in the presence of formalin to oxidize cyanide to cyanate. The endpoint for this process is determined by the cyanide ion electrode rather than by ORP measurement.

The quantity of alkaline material required to initially adjust the pH depends on the chemical and physical characteristics of each waste. No practical way is available for calculating this quantity, but it can be easily established by laboratory tests. Each part of cyanide requires 2.73

parts of chlorine to convert it to cyanate and 6.83 parts to oxidize it to carbon dioxide (as sodium bicarbonate) and nitrogen. In addition, each part of chlorine requires 1.125 parts of sodium hydroxide to neutralize the chlorine produced.

The actual chlorine quantities can be two or three times the theoretical requirements due to the chlorine demand on organic compounds, wetting agents, and so forth. However, either sodium or calcium hypochlorite is used instead of chlorine, the acids formed by hydrolysis are already neutralized, and little or no additional caustic is required.

Chlorine is also effective in oxidizing slaughterhouse waste, in which the treatment endpoint indication is visual rather than by instrument and is followed by coagulation. Phenolic wastes have also been successfully oxidized with chlorine. Chlorine dioxide, ozone, and potassium permanganate are alternate choices, but chlorine is the reagent of choice mainly because its cost per unit of oxidizing equivalent is lower than that of the alternates. Ammonia can increase chlorine consumption since the chlorine preferentially reacts with the ammonia before reacting with the phenol.

Reduction Processes—Hexavalent Chrome Removal

Reduction can be illustrated by the reaction in which toxic hexavalent chrome (Cr^{6+}) is reduced to the trivalent form (Cr^{3+}). In the latter, toxicity is reduced by a factor of about 100. The Cr^{3+} is then precipitated as the hydroxide and removed as a sludge. The following equation gives the half-cell reaction in reduced form:

$$Cr_2O_7^- + 14H^+ + 6e \longrightarrow 2Cr^{3+} + 7H_2O \qquad \textbf{7.41(25)}$$

This reaction has a reduction potential of roughly 1.33 V with respect to the standard hydrogen electrode. The conversion time of $Cr^{6+} + Cr^{3+}$ is pH dependent, as shown in Figure 7.41.3. To keep the reaction time under 30 min

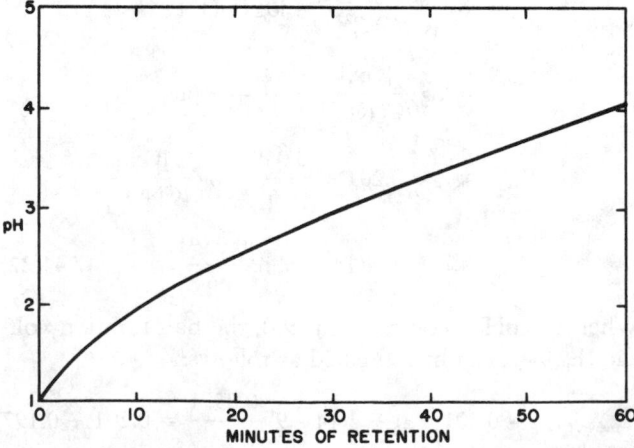

FIG. 7.41.3 Time–pH dependence of Cr^{6+} to Cr^{3+} conversion.

TABLE 7.41.4 CHROME REDUCTION AND PRECIPITATION REACTIONS

Ferrous sulfate ($FeSO_4$)	$2H_2CrO_4 + 6FeSO_4 + 6H_2SO_4 \longrightarrow$ $\qquad Cr_2(SO_4)_3 + 3Fe_2(SO_4)_3 + 8H_2O$ $Cr_2(SO_4)_3 + 3Ca(OH)_2 \longrightarrow 2Cr(OH)_3 + 3CaSO_4$
Sodium metabisulfite ($Na_2S_2O_5$)	$Na_2S_2O_5 + H_2O \longrightarrow 2NaHSO_3$ $2H_2CrO_4 + 3NaHSO_3 + 3H_2SO_4 \longrightarrow$ $\qquad Cr_2(SO_4)_3 + 3NaHSO_4 + 5H_2O$ $Cr_2(SO_4)_3 + 3Ca(OH)_2 \longrightarrow 2Cr(OH)_3 + 3CaSO_4$
Sulfur dioxide (SO_2)	$SO_2 + H_2O \longrightarrow H_2SO_3$ $2H_2CrO_4 + 3H_2SO_3 \longrightarrow Cr_2(SO_4)_3 + 5H_2O$ $Cr_2(SO_4)_3 + 3Ca(OH)_2 \longrightarrow 2Cr(OH)_3 + 3CaSO_4$

most wastewater treatment facilities conduct Cr^{6+} reduction at pH levels less than 3.0; however, if the tank capacity is available, higher operating pH levels can reduce the consumption of acid reagents.

The chromium in the treated waste is generally in the form of chromic acid, chromate, or dichromate. Table 7.41.4 shows the reduction from the Cr^{6+} to the Cr^{3+} form with ferrous sulfate, sodium metabisulfite, and sulfur dioxide. The table also shows precipitation of the Cr^{3+} form is with $Ca(OH)_2$.

Of the three reagents in Table 7.41.4, sulfur dioxide has economic and handling advantages. The first two reagents require auxiliary acid to hold the pH at the proper level for a low reduction time. If auxiliary acid is not used,

an excess of reagent chemicals is usually required to complete the reaction. For ferrous sulfate, about 250% excess is required; for sodium metabisulfate about 75% excess is required. The sulfurous acid formed when sulfur dioxide is hydrolized is normally sufficient to maintain the process at a low pH value, thereby precluding the need for excess acid.

Additionally, sulfur dioxide is easier to handle and feed due to its availability in bulk or large cylinders and standard feed-regulating equipment. Ferrous sulfate and sodium metabisulfite are dry powders that require feeders and mix tanks. Wastewater treatment facilities can require from 10 to 200% excess reducing reagent to overcome the quasi-buffering effect of oxidants in the process stream, especially if the stream is well aerated.

Figures 7.41.4 and 7.41.5 show titration curves for the Cr^{6+} reduction process in which the reducing reagents are sodium sulfite and sodium bisulfite, respectively. The reaction associated with the sodium sulfite reagent is as follows:

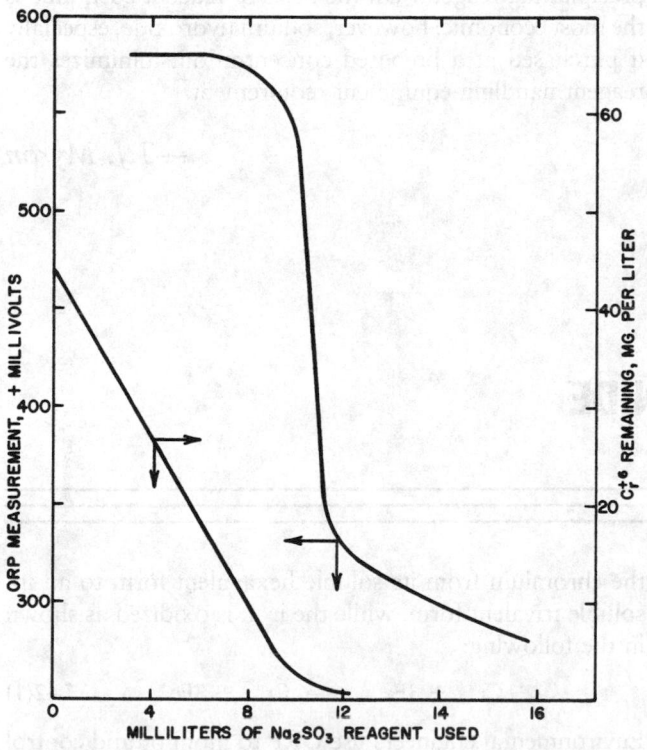

FIG. 7.41.4 Approximate path for Cr^{6+} to Cr^{3+} reduction. (Reprinted, with permission, from G.B. Hill, 1969, Complete removal of chronic acid waste with the aid of instrumentation, *Plating* 172 [February].)

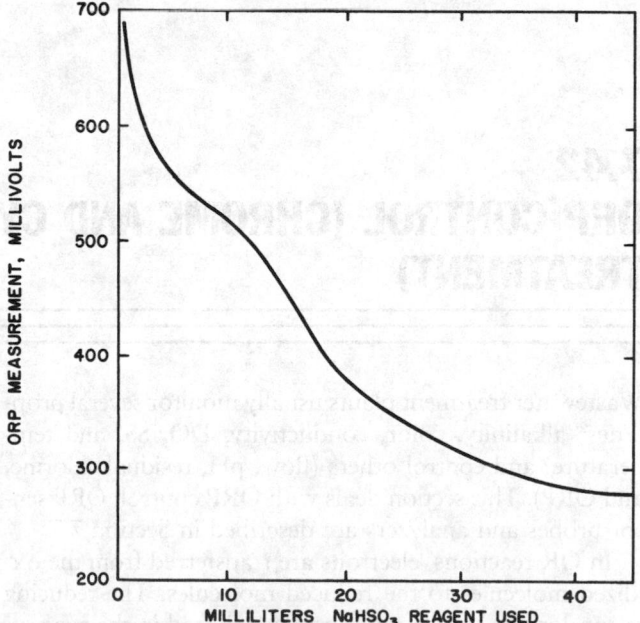

FIG. 7.41.5 Typical titration curve for Cr^{6+} to Cr^{3+} reduction at 2.5 pH. (Reprinted, with permission, from Hill, 1969.)

TABLE 7.41.5 CHROME REDUCTION ORP ENDPOINT VALUES FOR THREE REDUCING AGENTS

Reagent	pH	ORP Value (mV) Standard Hydrogen Electrode	ORP Value (mV) AgAgCl (1M KCl) Reference Electrode
Fe_2SO_4	2.0	500	265
$Na_2S_2O_5$	2.5	380	145
SO_2	2.9	165	-70

$$3Na_2SO_3 + 3H_2SO_4 + 2H_2CrO_4 \longrightarrow$$

$$Cr_2(SO_4)_3 + 3Na_2SO_4 + 5H_2O \quad \text{7.41(26)}$$

Like the cyanide oxidation process, the control point of the chrome reduction reaction is also monitored by an OR electrode combined with a reference electrode. Environmental engineers should carefully review published control point values for this process to ascertain the reference electrode used. If a standard hydrogen electrode is the basis rather than an industrial-type electrode, such as the AgAgCl (1M KCl) electrode, the reported values must be corrected by about −235 mV. Table 7.41.5 lists the approximate endpoint ORP values at the pH levels shown. The endpoint (equivalence point) and control point are not usually the same as shown by the Fe^{2+}/Ce^{4+} and cyanide examples (see Figure 7.41.1).

Wastewater treatment facilities should titrate each waste stream individually with the reagent to establish the correct control point value and verify the results by chemical analysis to make sure that the hexavalent chrome has been reduced to the required level.

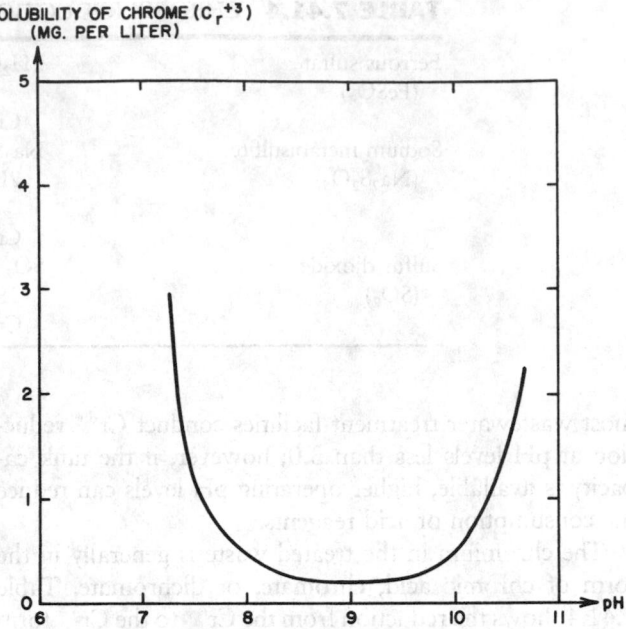

SOLUBILITY OF CHROME (C_r^{+3}) (MG. PER LITER)

FIG. 7.41.6 pH dependence of Cr^{3+} solubility.

The precipitation reactions in Table 7.41.4 show lime, $Ca(OH)_2$, as the hydroxide source. Figure 7.41.6 shows the pH dependence of this reaction. The trivalent chromium concentration in solution is at a minimum at a pH of 9. Sodium hydroxide is equally acceptable as the precipitating reagent. On the basis of reagent cost, lime is the most economic; however, sodium hydroxide, especially if purchased at a prepared concentration, minimizes the reagent handling equipment requirement.

—*T.J. Myron*

7.42
ORP CONTROL (CHROME AND CYANIDE TREATMENT)

Wastewater treatment plants usually monitor several properties (alkalinity, color, conductivity, DO, SS, and temperature) and control others (flow, pH, residual chlorine, and ORP). This section deals with ORP control. ORP sensor probes and analyzers are described in Section 7.7.

In OR reactions, electrons are transferred from the oxidized molecules to the reduced molecules. The reducing agent donates the electrons and is oxidized in the process. The donated electrons are accepted by the reduced chemical. An example of such a process is the reduction of chromium by ferrous ions. Here, the ferrous ions reduce

the chromium from its soluble hexavalent form to an insoluble trivalent form, while the iron is oxidized as shown in the following:

$$Cr^{+6} + 3Fe^{+2} \longrightarrow Cr^{+3} + 3Fe^{+3} \quad \text{7.42(1)}$$

Environmental engineers use ORP to monitor and control biological or chemical reactions by quantitatively determining the oxidizing or reducing properties of solutions and the amount of ions present. The most widely used applications for ORP control in wastewater treatment are chrome reduction and cyanide destruction. This section discusses both continuous and batch versions of these processes. Other ORP applications include digestion and other biological processes, spa and swimming pool water treatment, and paper bleaching.

ORP Measurement

During the chemical reaction shown in Equation 7.42(1), an inert metal electrode placed in contact with the solution detects the solution's ability to accept or donate electrons. The resulting ORP redox potential is directly related to reaction progress. A reducing ion (ferrous) provides electrons and makes the electrode reading more negative. An oxidizing ion (Cr^{+6}) accepts electrons and makes the electrode reading more positive. The resulting net electrode potential is related to the ratio of concentrations of oxidizing and reducing ions in solution.

ORP is sensitive in measuring the degree of treatment provided by the reaction. However, it cannot be related to a definite concentration (only the ratio) and, therefore, it cannot be used as a monitor of final effluent concentration.

Wastewater treatment facilities can theoretically calculate the exact potential to ensure complete treatment, but in practice, this potential is subject to variations in reference electrode potential, pH, the presence of other waste stream contaminants, temperature, the purity of reagents, and so forth. Therefore, they usually determine it empirically by testing the treated wastewater for trace levels of the material to be eliminated. The optimum control point (ORP reading) occurs when just enough reagent is added to complete the reaction. Suggested control points are given later in this section; however, those control points are approximate and should be verified online by sample testing.

ORP instruments are calibrated like voltmeters, measuring absolute mV, although a standardized (zero) adjustment is often available on instruments designed for pH measurement also.

To verify the operation of electrodes, wastewater treatment facilities should have a known ORP solution composition using quinhydrone and pH buffer solutions. These solutions must be made up fresh to prevent air oxidation and deterioration. A more stable ORP reference solution consists of 0.1 M ferrous ammonium sulfate, 0.1 M ammonium sulfate, and 1.0 M sulfuric acid. Its ORP is +476 mV when measured with a silver–silver chloride, saturated potassium chloride reference electrode.

ORP electrodes must have a very clean metal surface. Routine cleaning of electrodes with a soft cloth, dilute acids, or cleaning agents promotes a fast response.

ORP Control

In designing ORP control (like pH control), the environmental engineer must recognize the chemistry of the process. In addition to vessel size, vessel geometry, agitation requirements (needed to guarantee uniform composi-

tion), and reagent delivery systems, they must also consider solid removal problems.

In some cases where one or both of the half-reactions (redox reaction) involve hydrogen ions, ORP measurements also become pH-dependent. The potential changes measured by the ORP electrode continue to vary with the redox ratio, but the absolute potential also varies with pH. Therefore, the wastewater treatment facility must determine the control point experimentally and use both pH and ORP measurements to control the process.

As with pH, reliable ORP control requires vigorous mixing to ensure uniform composition throughout the reaction tank. For continuous control, the tank should provide adequate retention time (process flow rate divided by filled tank volume), typically 10 min or more.

Complete treatment requires a slight excess of reagent and a control point slightly beyond the steep portion of the titration curve. Control in this plateau area, where process gain is low, is often provided by simple on–off control. Reagent feeders are typically metering pumps or solenoid valves. Wastewater treatment facilities can use a needle valve in series with a solenoid valve to set reagent flow more accurately and improve on–off control.

Chemical Oxidation

Water and wastewater are treated by chemical oxidation in specific cases when the contaminant can be destroyed, its chemical properties altered, or its physical form changed. Examples of chemicals that can be destroyed are cyanides and phenol. Sulfides can be oxidized to sulfates, thus changing their characteristics completely.

Iron and manganese can be oxidized from the soluble ferrous or manganous state to the insoluble ferric or manganic state, respectively, permitting their removal by sedimentation. Strong oxidants, such as chlorine, chlorine dioxide, ozone, and potassium permanganate are used. Chlorine is preferred because it is the least expensive and is readily available. Ozone is a strong second choice and is favored by some facilities because its excess converts to oxygen, while excess chlorine can react with industrial waste to produce cancer-causing substances.

All these chemical reactions are pH-dependent in relation to the reaction time required to proceed to completion for the required end products. Wastewater treatment facilities use residual oxidant or ORP measurement to control the process.

Cyanide Waste Treatment

Metal-plating and metal-treating industries produce the largest amounts of cyanide waste; however, other industries also use cyanide compounds as intermediates. Cyanide solutions are used in plating baths for brass, copper, silver, gold, and zinc. The toxic rinse waters and dumps from these operations require cyanide destruction before discharge.

The most frequently used technique for cyanide destruction is a one- or two-stage chemical treatment process. The first stage raises the pH and oxidizes cyanide to less toxic cyanate. When required, the second stage neutralizes and further oxidizes cyanate to harmless bicarbonate and nitrogen. Neutralization also allows the metals to be precipitated and separated from the effluent.

OR implies a reversible reaction. Since these reactions are carried to completion and are not reversible, the term is misleading. In practice, control is by electrode potential readings. An illustration is the oxidation of cyanide into cyanate with chlorine, according to the following reaction:

$$
\begin{array}{ll}
2Cl_2 & \text{Chlorine} \\
+ & \\
4NaOH & \text{Sodium hydroxide} \\
+ & \\
2NaCN & \text{Sodium cyanide} \\
\downarrow & \\
2NaCNO & \text{Sodium cyanate} \\
+ & \\
4NaCl & \text{Sodium chloride} \\
+ & \\
2H_2O & \text{Water} \qquad \qquad 7.42(2)
\end{array}
$$

The electrode potential of the cyanide waste solution is −200 to −400 mV. After sufficient chlorine is applied to complete the reaction according to Equation 7.42(2), the electrode potential is +200 to +450 mV. The potential value does not increase until all cyanide is oxidized. Control of the pH is essential, with the minimum being 8.5. The reaction rate is faster at higher values.

Complete oxidation (destruction of cyanide) is a two-step reaction. The first step is oxidation to the cyanate level as described in Equation 7.42(2). The endpoint of the second-stage reaction is again detected by electrode potential readings and is +600 to +800 mV. The overall reaction is as follows:

$$
\begin{array}{ll}
5Cl_2 & \text{Chlorine} \\
+ & \\
10NaOH & \text{Sodium hydroxide} \\
+ & \\
2NaCN & \text{Sodium cyanide} \\
\downarrow & \\
2NaHCO_3 & \text{Sodium bicarbonate} \\
+ & \\
N_2 & \text{Nitrogen} \\
+ & \\
4H_2O & \text{Water} \qquad \qquad 7.42(3)
\end{array}
$$

BATCH TREATMENT OF CYANIDE

Wastewater treatment facilities can perform the two-step cyanide destruction process in either batch or continuous processes. Because of the toxicity of cyanide and the rigid regulations on cyanide effluent discharges, some facilities prefer batch treatment, which guarantees completion of treatment before discharge.

Figure 7.42.1 describes the batch oxidation of cyanide. This process charges chlorine at a constant rate under flow control, while a pH control loop maintains the batch at a pH of around 9.5 by adding caustic. As the cyanide is oxidized into cyanate, the ORP probe senses a rise in millivolts from about −400 to over +400. At that point, all cyanide is destroyed, and the second stage of the reaction, cyanate destruction, begins.

At a millivolt reading of between +600 and +750, all cyanate is also destroyed, and the batch is done. At this point, the ORP switch actuates a 30-min timer. If the ORP has dropped at the end of that period, indicating that further reaction has taken place, the cycle is repeated. Otherwise, the batch is ready to be discharged. While batch treatment is in progress, a separate collection tank stores the untreated cyanide waste.

Figure 7.42.2 shows another arrangement for batch treatment with one pH and one ORP controller. This arrangement sequences the steps, changing the pH and ORP setpoints to obtain the required treatment while ensuring that treatment is complete before the next step begins.

First caustic is added to raise the pH to 11. Then hypochlorite to raise the ORP to approximately +450 mV, while simultaneously adding more caustic, as required, to maintain a pH of 11. An interlock prevents the addition of acid before the oxidation of all cyanide to cyanate is complete. Then, adding acid neutralizes the batch, and further hypochlorite oxidation completes cyanate-to-bicarbonate conversion.

This system can include a settling period to remove solids, or the batch can be pumped to another tank or pond for settling.

CHLORINATOR CONTROLS

Figure 7.42.3 is a schematic typical of systems with variable quality and flow rate. The chlorinator has two oper-

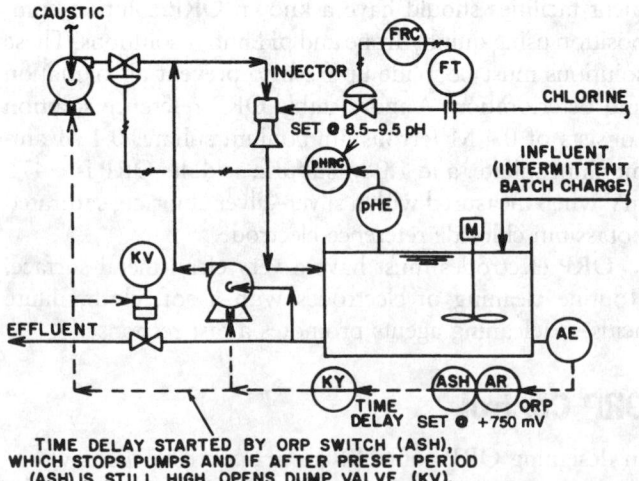

TIME DELAY STARTED BY ORP SWITCH (ASH), WHICH STOPS PUMPS AND IF AFTER PRESET PERIOD (ASH) IS STILL HIGH, OPENS DUMP VALVE (KV).

FIG. 7.42.1 Batch oxidation of cyanide waste with chlorine.

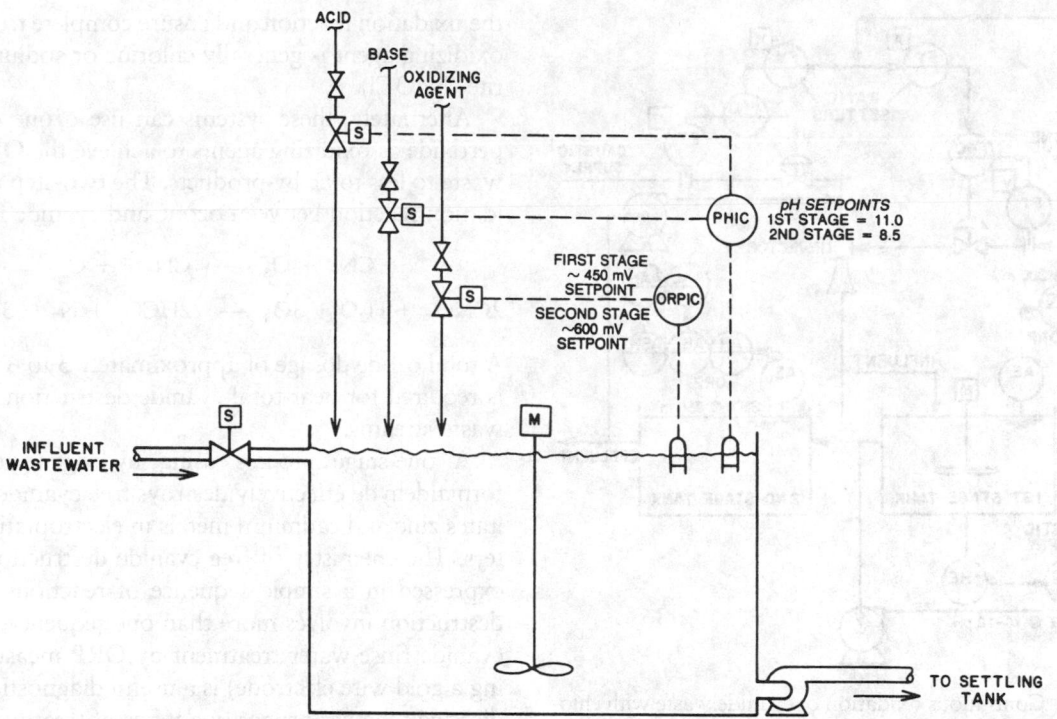

FIG. 7.42.2 Batch cyanide treatment.

ators: one controlled by feedforward, the other by a feed-back loop.

Most reactions are completed within 5 min. Except for cyanide treatment, most other chemical oxidation operations occur simultaneously with other unit operations, such as coagulation and precipitation, which govern the pH value. Although pH value affects the rate of reaction, it is seldom controlled solely for the oxidation process.

CONTINUOUS CYANIDE DESTRUCTION

Continuous flow-through systems have the advantage of reduced space requirements but require additional process equipment. In the system shown in Figure 7.42.4, the two reaction steps are separated.

In the first step, the ORP controller setpoint is approximately +300 mV. It controls the addition of chlorine to oxidize the cyanide into cyanate. The pH is maintained at approximately 10. The reaction time is approximately 5 min.

Since the second step (oxidation of the cyanates) requires an additional amount of chlorine charged at nearly the same rate as in the first step, the wastewater treatment facility obtains the chlorine flow rate signal by measuring chlorine feed rate to the first step and multiplying it by a constant to control the chlorine feed rate in the second step. The caustic requirement depends solely on the chlorine rate (pH control is not necessary); therefore, the facility can use the same signal to adjust caustic feed. The ORP instrument sampling the final effluent can signal

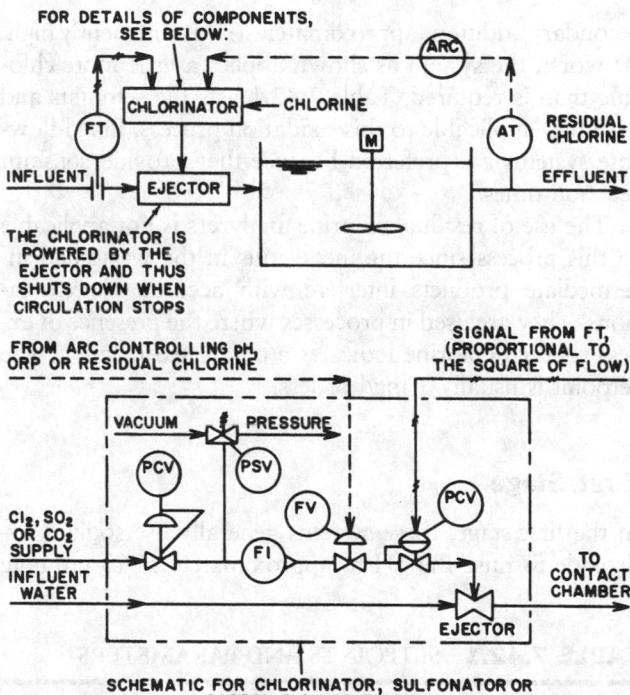

FIG. 7.42.3 Variable quality and flow rate of waste oxidized by chlorine.

process failure if the potential level drops below approximately +750 mV.

Although a feedback loop from the second ORP analyzer to the second chlorinator seems beneficial, practice has not shown this loop to be necessary because the ratioing accuracy between the first-stage chlorine rate and

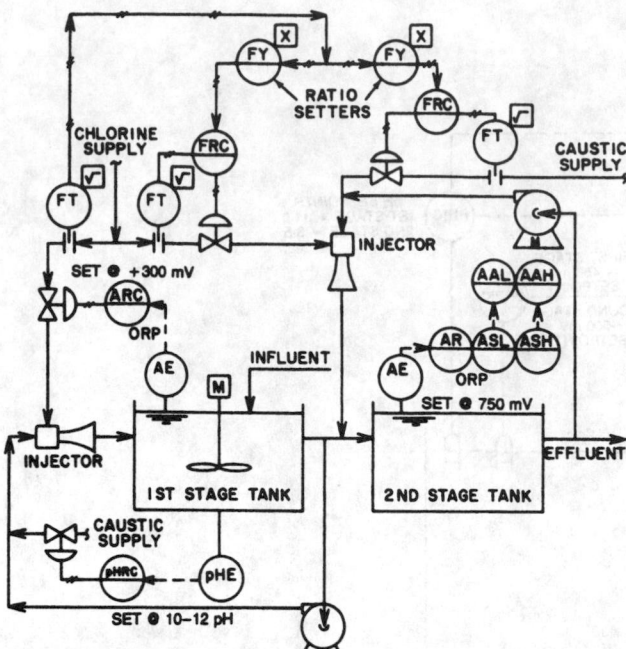

FIG. 7.42.4 Continuous oxidation of cyanide waste with chlorine. The influent has a continuous constant flow rate and variable quality.

secondary addition (approximately 1:1) is sufficiently high. At worst, the system as shown applies a little more chlorine than is required. Table 7.42.1 lists the setpoints and variables applicable to this oxidation process. Fixed-flow-rate systems are preferred because they provide constant reaction times.

The use of residual chlorine analyzers is not applicable to this process since the metal ions in the waste and intermediate products interfere with accurate determinations. They are used in processes where the presence of excess residual chlorine indicates a completed reaction. The setpoint is usually 1 mg/l or less.

First Stage

In the first stage, these systems generally use sodium hydroxide to raise the pH to approximately 11 to promote

TABLE 7.42.1 SETPOINTS AND PARAMETERS

Parameters	Cyanide to Cyanate	Destruction of Cyanate
pH	10–12	8.5–9.5
Reaction time (min)	5	45
ORP setpoint (mV)	+300	+750
Maximum concentration of cyanide (cyanate) that can be treated (mg/l)	1000	1000

the oxidation reaction and ensure complete treatment. The oxidizing agent is generally chlorine or sodium hypochlorite ($NaOCl$).

Alternately, these systems can use ozone or hydrogen peroxide as oxidizing agents to achieve the OR of cyanide waste to less toxic by-products. The two-step chemical oxidation reaction between ozone and cyanide is as follows:

$$CN^- + O_3 \longrightarrow CNO^- + O_2 \qquad 7.42(4)$$

$$2CNO^- + H_2O + 3O_3 \longrightarrow 2HCO_3 + N_2 + 3O_2 \qquad 7.42(5)$$

A total ozone dosage of approximately 3 to 6 O_3/ppm CN is required for near-total cyanide destruction in industrial waste streams.

A one-stage process using hydrogen peroxide and formaldehyde effectively destroys free cyanide and precipitates zinc and cadmium metals in electroplating rinse waters. The chemistry of free cyanide destruction cannot be expressed in a simple sequence of reactions because the destruction involves more than one sequence. Monitoring cyanide rinse-water treatment by ORP measurement (using a gold wire electrode) is a useful diagnostic tool for indicating whether proper quantities of treatment chemicals have been added.

The following equation gives the overall reaction for the first stage using sodium hypochlorite ($NaOCl$), with cyanide expressed in ionic form (CN^-) and the result expressed as sodium cyanate ($NaCNO$) and chloride ion (Cl^-):

$$NaOCl + CN^- \longrightarrow NaCNO + Cl^- \qquad 7.42(6)$$

For cases when the oxidizing agent is chlorine, refer to Equations 7.42(2) and (3).

As shown in Figure 7.42.5, the first-stage reduction is monitored and controlled by independent control loops: base addition by pH control and oxidizing agent addition by ORP control. The pH controller adds base whenever the pH falls below 11. The ORP controller adds oxidizing agent whenever the ORP falls below approximately +450 mV.

The ORP titration curve (see Figure 7.42.6), shows the mV range covered when cyanide is treated in batches. Continuous treatment maintains operation in the oxidized, positive region of the curve near the +450 mV setpoint. Wastewater treatment facilities can determine the exact setpoint empirically by measuring the potential when all cyanide is oxidized but no excess reagent is present. They can verify this point with a sensitive colorimetric test.

In this reaction, the pH has a strong inverse effect on the ORP. Thus, wastewater treatment facilities must closely control the pH to achieve consistent ORP control, especially if they use hypochlorite as the oxidizing agent. Adding hypochlorite raises the pH, which, if unchecked, lowers the ORP, calling for additional hypochlorite. Controlling the pH at a setting above the pH level where hypochlorite has an influence and separating the ORP elec-

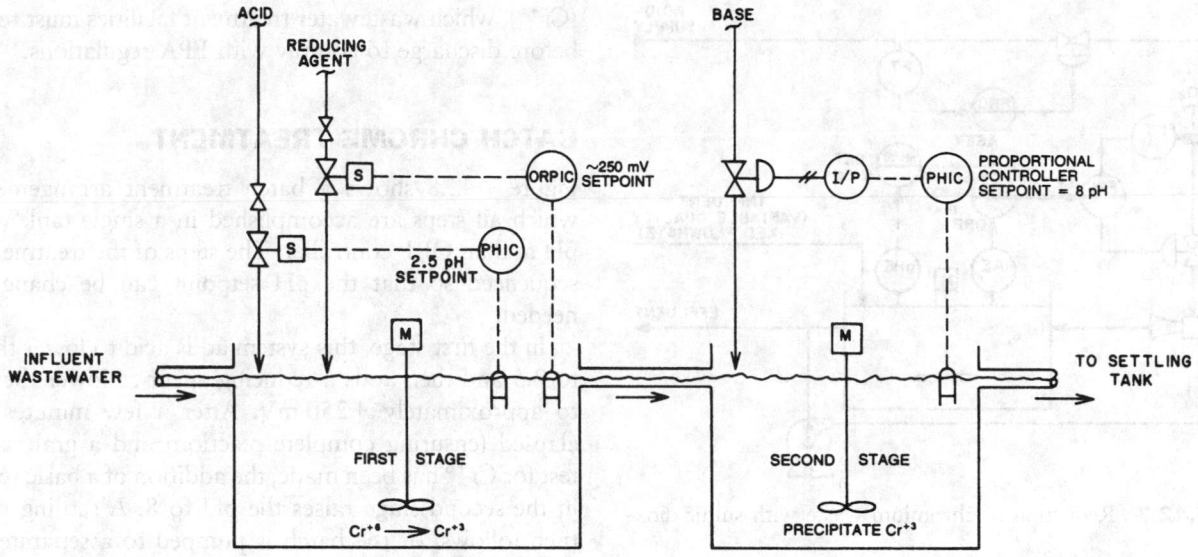

FIG. 7.42.5 Continuous cyanide treatment.

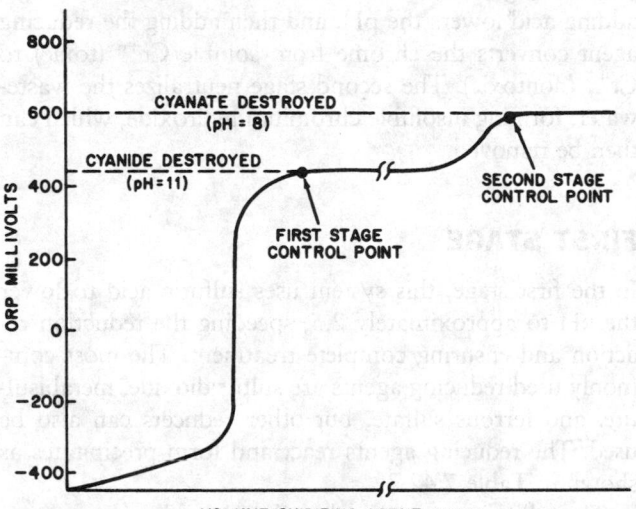

FIG. 7.42.6 Cyanide oxidation titration curve.

trodes from the hypochlorite addition point can prevent this situation.

Gold ORP electrodes give more reliable measurement than platinum for this application. Platinum can catalyze some additional reactions at its surface and is more subject to coating than gold. The solubility of gold in cyanide solutions is not a problem since it contacts primarily with cyanate. Any loss of gold acutally keeps the electrode clean.

Second Stage

In this stage, the system neutralizes wastewater to promote additional oxidation as well as to meet the discharge pH limits. Sulfuric acid is typically used to lower pH to ap-

proximately 8.5, where the second oxidation occurs more rapidly. Acid addition must have a fail-safe design because below neutrality (pH = 7), highly toxic hydrogen cyanide can be generated if first-stage oxidation has not been completed.

The system adds hypochlorite either in proportion to that added in the first stage or by separate ORP control to complete the oxidation to sodium bicarbonate ($NaHCO_3$) as follows:

$$2NaCNO + 3NaOCl + H_2O \longrightarrow 2NaHCO_3 + N_2 + 3NaCl$$

7.42(7)

ORP control in the second stage is similar to that in the first except that the control point is near +600 mV. In the second stage, pH control is more difficult than in the first because the control point is closer to the sensitive region of neutrality. The pH controller can be proportional only.

A subsequent settling tank or filter can remove suspended metal hydroxides although further treatment may be required.

Chemical Reduction

Chemical reduction is similar to chemical oxidation except that reducing reactions are involved. Commonly used reductants are sulfur dioxide and its sodium salts, such as sulfite, bisulfite, and metabisulfite. Ferrous iron salts are infrequently used. Typical examples are the reduction of hexavalent chromium, dechlorination, and deoxygenation.

Figure 7.42.7 is a schematic diagram of a typical system for the reduction of highly toxic hexavalent chromium to the innocuous trivalent form according to the following reaction:

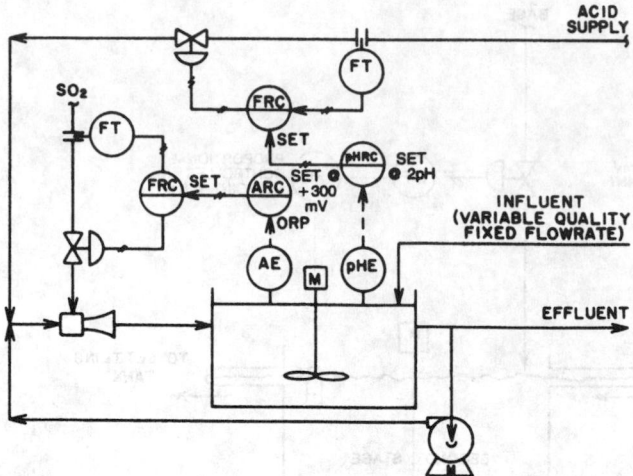

FIG. 7.42.7 Reduction of chromium waste with sulfur dioxide.

$3SO_2$	Sulfur dioxide	
+		
$2H_2CrO_4$	Chromic acid	
↓		
$Cr_2(SO_4)_3$	Chromic sulfate	
+		
$2H_2O$	Water	7.42(8)

Most hexavalent chrome wastes are acid, but the reaction rate is much faster at low pH values. For this reason, pH control is essential. Sulfuric acid is preferred because it is less expensive than other mineral acids. The setpoint of the pH controller is approximately 2.

As in the treatment of cyanide, the chemical reaction is not reversible, and the control of sulfur dioxide addition is by electrode potential level, using ORP instrumentation. The potential level of hexavalent chromium is +700 to +1000 mV, whereas that of the reduced trivalent chrome is +200 to +400 mV. The setpoint on the ORP controller is approximately +300 mV.

The control system consists of feedback loops for both pH and ORP. The common and preferred design is for fixed-flow systems.

The system removes the trivalent chromic sulfate from solution by subsequently raising the pH to 8.5, at which point it precipitates as chromic hydroxide. The control system for this step can be identical with the one used in Figure 7.42.11. Table 7.42.2 summarizes the critical process control factors.

Chrome Waste Treatment

Chromates are used as corrosion inhibitors in cooling towers and various metal finishing operations, including bright dip, conversion coating, and chrome plating. The resulting wastewater from rinse tanks, dumps, or cooling tower blowdown contains the toxic and soluble chromium ion

(Cr^{+6}), which wastewater treatment facilities must remove before discharge to comply with EPA regulations.

BATCH CHROME TREATMENT

Figure 7.42.8 shows a batch treatment arrangement in which all steps are accomplished in a single tank with a pH and an ORP controller. The steps of the treatment are sequenced so that the pH setpoint can be changed as needed.

In the first stage, this system adds acid to lower the pH to 2.5 and then adds a reducing agent to lower the ORP to approximately +250 mV. After a few minutes have elapsed (ensuring complete reaction) and a grab sample test for Cr^{+6} has been made, the addition of a basic reagent in the second stage raises the pH to 8. A settling period then follows, or the batch is pumped to a separate tank or pond for settling.

The most frequently used technique for chrome removal is a two-stage chemical treatment process. In the first stage, adding acid lowers the pH, and then adding the reducing agent converts the chrome from soluble Cr^{+6} (toxic) to Cr^{+3} (nontoxic). The second stage neutralizes the wastewater, forming insoluble chromium hydroxide, which can then be removed.

FIRST STAGE

In the first stage, this system uses sulfuric acid to lower the pH to approximately 2.5, speeding the reduction reaction and ensuring complete treatment. The most commonly used reducing agents are sulfur dioxide, metabisulfite, and ferrous sulfate, but other reducers can also be used. The reducing agents react and form precipitates as shown in Table 7.42.3.

The following equation describes the reduction reaction with chrome expressed as chromic acid CrO_3, which has a +6 charge on the chromium. The reducing agent is expressed as sulfurous acid (H_2SO_3), generated by sulfites at a low pH. The result is chromium sulfate, $Cr_2(SO_4)_3$, which has a +3 charge on the chromium, as follows:

$$2CrO_3 + 3H_2SO_3 \longrightarrow Cr_2(SO_4)_3 + 3H_2O \qquad 7.42(9)$$

Equation 7.42(9) describes the reaction when sulfur dioxide is the reducing agent.

TABLE 7.42.2 CRITICAL PROCESS CONTROL FACTORS

Variable	Setpoints and Parameters Value
pH	2.0
ORP setpoint (mV)	+300
Reaction time	10 at pH 2.0
(min)	5 at pH 1.5

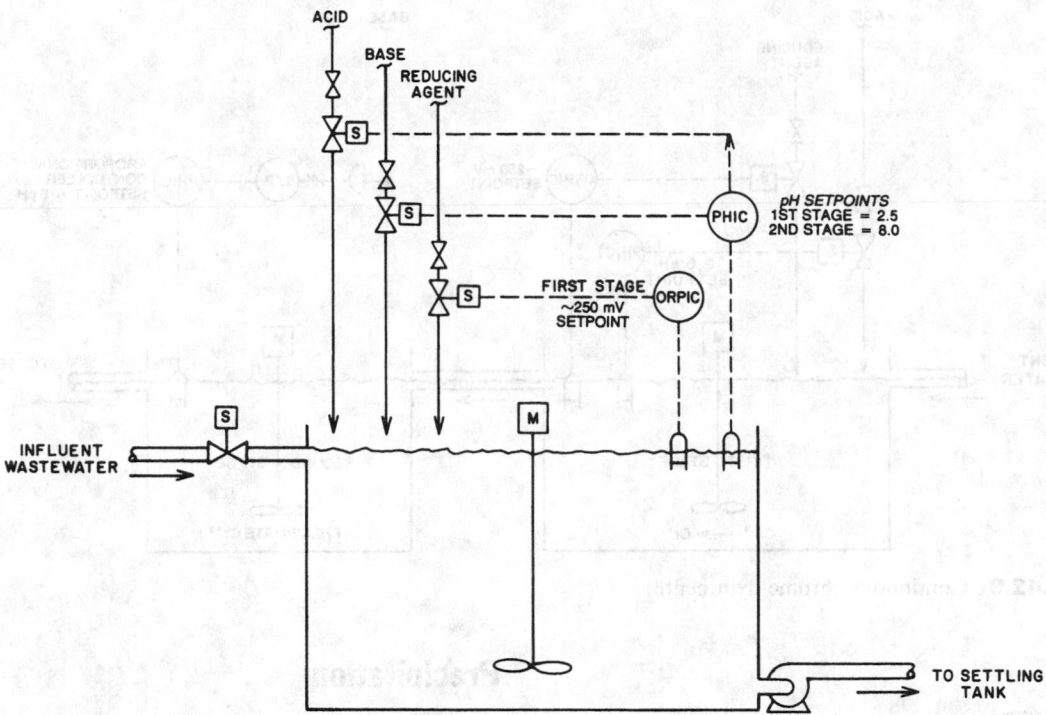

FIG. 7.42.8 Batch chrome treatment.

TABLE 7.42.3 CHROME REDUCTION AND PRECIPITATION REACTIONS

Reducing Agent	Reaction
Ferrous sulfate ($FeSO_4$)	$2H_2CrO_4 + 6FeSO_4 + 6H_2SO_4 \longrightarrow$ $Cr_2(SO_4)_3 + 3Fe_2(SO_4)_3 + 8H_2O$ $Cr_2(SO_4)_3 + 3Ca(OH)_2 \longrightarrow$ $2Cr(OH)_3 + 3CaSO_4$
Sodium metabisulfite ($Na_2S_2O_5$)	$Na_2S_2O_5 + H_2O \longrightarrow 2NaHSO_3$ $2H_2CrO_4 + 3NaHSO_3 + 3H_2SO_4 \longrightarrow$ $Cr_2(SO_4)_3 + 3NaHSO_4 + 5H_2O$ $Cr_2(SO_4)_3 + 3Ca(OH)_2 \longrightarrow$ $2Cr(OH)_3 + 3CaSO_4$
Sulfur dioxide (SO_2)	$SO_2 + H_2O \longrightarrow H_2SO_3$ $2H_2CrO_4 + 3H_2SO_3 \longrightarrow$ $Cr_2(SO_4)_3 + 5H_2O$ $Cr_2(SO_4)_3 + 3Ca(OH)_2 \longrightarrow$ $2Cr(OH)_3 + 3CaSO_4$

As shown in Figure 7.42.9, the system monitors and controls the first-stage reaction with independent control loops: acid addition by pH control and reducing agent addition by ORP control. The system adds acid under pH control whenever the pH rises above 2.5; it adds the reducing agent under ORP control whenever the ORP rises above approximately +250 mV.

The ORP titration curve (see Figure 7.42.10) shows the mV range covered when Cr^{+6} chrome is treated in batches. With continuous treatment, however, operation is main-tained in the completely reduced portion of the curve near the nominal +250 mV control point. The exact setpoint for an installation should be at a potential where all the Cr^{+6} is reduced but without excess sulfite consumption, which is accompanied by sulfur dioxide odor.

Chrome reduction is slow enough that 10 to 15 min can be required to complete the reaction. The reaction time increases if the pH is controlled at higher levels. Variations in pH also affect measured ORP readings; thus, the pH must be stable for consistent ORP control.

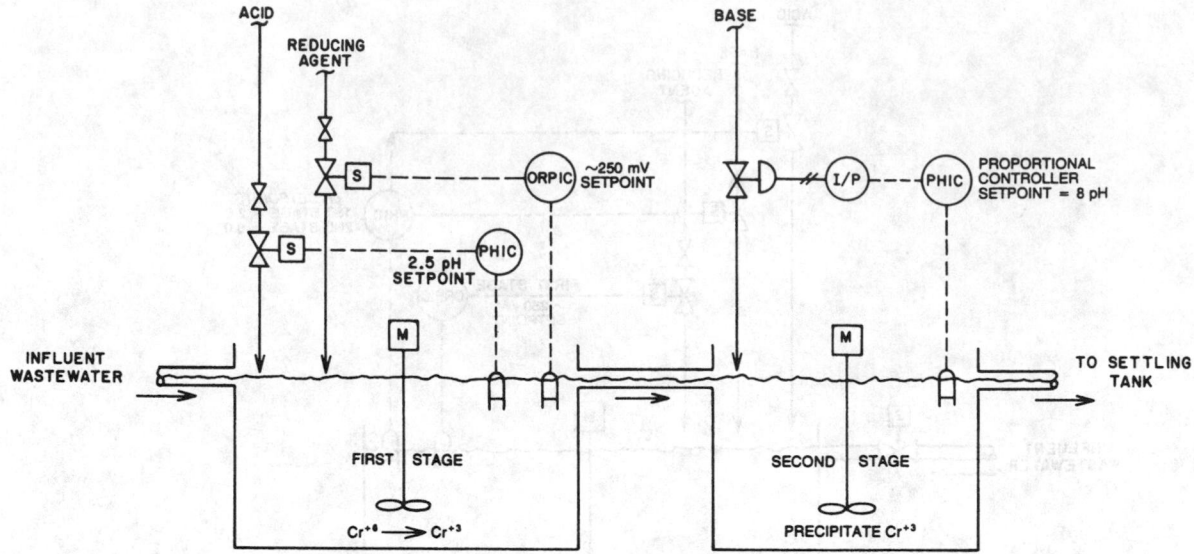

FIG. 7.42.9 Continuous chrome treatment.

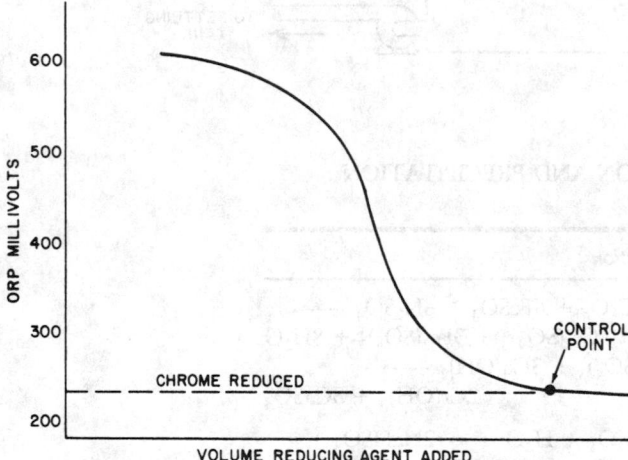

FIG. 7.42.10 Chrome reduction titration curve.

SECOND STAGE

In the second stage, the system neutralizes the wastewater to precipitate Cr^{+3} as insoluble chromium hydroxide, $Cr(OH)_3$, and to meet the discharge pH limits. The system uses sodium hydroxide or lime $[Ca(OH)_2]$ to raise the pH to 7.5 to 8.5, as shown in the following reaction:

$$Cr_2(SO_4)_3 + 6NaOH \longrightarrow 3Na_2SO_4 + 2Cr(OH)_3 \quad 7.42(10)$$

In the second stage, providing pH control is more difficult than in the first because the control point is closer to the sensitive region near neutrality. Although the second-stage reaction is fast, a retention time of at least 10 min is usually needed for continuous treatment processes to achieve stable operation. The pH controller is proportional only in this stage.

A subsequent settling tank or filter removes the suspended chromium hydroxide. Flocculating agents can assist in this separation.

Precipitation

Precipitation is the creation of insoluble materials by chemical reactions that provides treatment through subsequent liquid–solids separation. The removal of sulfates, removal of trivalent chromium, and softening of water with lime are typical precipitation operations. A variation of this process following the treatment discussed previously, in connection with the chemical oxidation process, removes iron and manganese. Lime softening is a common process and is used as an example.

The following reaction is involved:

$Ca(HCO_3)_2$	Calcium bicarbonate
+	
$Ca(OH)_2$	Calcium hydroxide
↓	
$2CaCO_3$	Calcium carbonate
+	
$2H_2O$	Water 7.42(11)

Calcium carbonate formed by this reaction is insoluble and can be removed by gravity separation (settling). The typical settling time is 30 min or less, but most systems are designed for continuous operation with typical detention times of 1 hr.

Water treatment with this process is called *excess-lime softening* because the application of lime is in excess of that required for the reaction described in Equation 7.42(11). Control consists of adding sufficient calcium hydroxide to maintain an excess hydroxide alkalinity of 10 to 50 mg/l, as shown in the following equation:

$$2P = MO + 10 \text{ to } 50 \qquad 7.42(12)$$

where P is the phenolphthalein alkalinity and MO is the methyl orange alkalinity. This addition results in a pH value of 10 to 11, but pH control is not satisfactory for economical operations. In this example, suitable analyti-

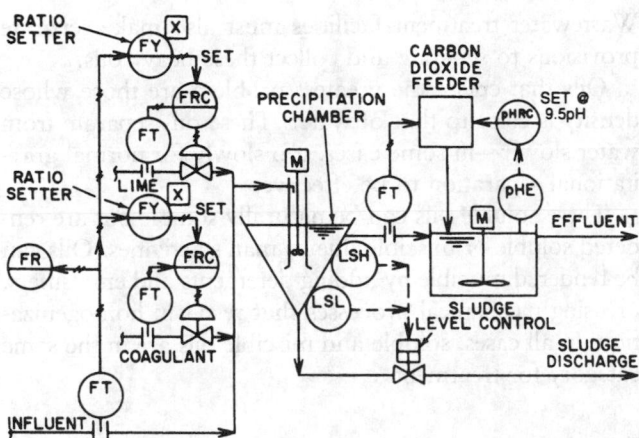

FIG. 7.42.11 Calcium carbonate precipitation control system. Details of the carbon dioxide feeder are shown in Figure 7.42.3.

cal instrumentation is not available for continuous system control. If the quality of the untreated water varies, operator control of the lime dosage is essential. Wastewater treatment facilities usually pace the manual dosage by feedforward control from the flow rate.

A factor in calcium carbonate precipitation is a chemical phenomenon known as *crystal seeding*. This factor involves the acceleration of carbonate crystal formation by the presence of previously precipitated crystals. Wastewater treatment facilities accomplish crystal seeding in practice by passing the water being treated through a sludge blanket in an upflow treatment unit, shown schematically in Figure 7.42.11. The resulting crystals of calcium carbonate are hard, dense, and discrete, and they separate readily.

When colloidal suspended material is also to be removed, which occurs when surface waters are softened, water treatment facilities also add a coagulant of aluminum, iron salts, or polymers to precipitate the colloids.

The dosage varies depending on the quantity of suspended material. The application of both coagulant and calcium hydroxide is controlled by flow–ratio modulation.

The resulting sludge, consisting of calcium carbonate, aluminum, or iron hydroxides, and the precipitated colloidal material are discharged to waste continuously. As previously noted, the presence of some precipitated carbonate is beneficial to remove all sludge. The use of an automatic level control system (see Figure 7.42.11) controls the sludge level at an optimum.

Water softened by excess-lime treatment is saturated with calcium carbonate and is therefore unstable. Water treatment facilities can achieve stability by adding carbon dioxide to convert a portion of the carbonates into bicarbonate, according to the following equation:

$$
\begin{array}{ll}
CO_2 & \text{Carbon dioxide} \\
+ & \\
CaCO_3 & \text{Calcium carbonate} \\
+ & \\
H_2O & \text{Water} \\
\downarrow & \\
Ca(HCO_3)_2 & \text{Calcium bicarbonate} \qquad 7.42(13)
\end{array}
$$

In contrast with the softening reaction, this process is suited to automatic pH control. Figure 7.42.11 shows this control system. The carbon dioxide feeder has two operators: one controlled by feedforward on the influent flow and the other by feedback on the effluent pH. The setpoint is about 9.5 pH.

Electrode fouling occurs as a result of the precipitation of crystallized calcium carbonate. Daily maintenance is required unless automated cleaners are used. The farther downstream (from the point of carbon dioxide application) the electrodes are placed, consistent with acceptable loop time delays, the less maintenance required.

—*Béla G. Lipták*

7.43
OIL SEPARATION AND REMOVAL

OIL SOLUBILITY

Decreased by either high base or high acid conditions

CHEMICALS USED TO ADJUST THE pH OF OIL SOLUTIONS

Sulfuric acid, lime, and soda ash

CHEMICALS USED TO COAGULATE AND MAKE OILS INSOLUBLE

Alum at a pH of 8 to 9 and ferrous sulfate at a pH of 8 to 10

PHYSICAL OIL SEPARATION TECHNIQUES

Settling, filtration, centrifuging, and flotation

Oil Sources

The two principal sites of oil waste are at the oil field and oil refinery. At the former, numerous problems are associated with well drilling. Drilling techniques frequently use water under pressure to bring material from the drill hole to the surface. In some cases, the drilling process passes through a significant amount of low-yield, oil-bearing strata prior to reaching the high oil-bearing region, resulting in oil being carried to the surface with the drilling mud. Also, on reaching the high oil-bearing stratum, the drilling process can regurgitate a small amount of highly concentrated oil.

Where there is naturally occurring or artificially induced high pressure on the oil-bearing areas, the oil can be forced up and out of the well. This situation represents a high concentration of oil that must be properly disposed. Brines are also associated with oil in the ground, and large amounts of brine contaminated with oil can reach the surface. The oil must be separated before the brine can be disposed in a satisfactory manner.

At the refinery, nearly all sources of oil waste result from spills. Although accidents can occur anywhere in the plant, the prime sources of spills are at loading and unloading sites. These areas should be curbed and separately sewered with provisions for separating oil from surface water runoff. Spills within the refinery present specific problems as a function of the oil type being processed at that point in the refinery. Treatment must be based on the type of oil involved.

Many other sources of oil reach the environment due to man's activities. Many of these are related to oil transport. Another widely dispersed source of oil spills is the automotive transportation system, with the local gasoline service station the focus of the problem. Due to federal taxation, reprocessing used oil for reuse is no longer economical.

Scavengers are reluctant to collect oil, and frequently oil reaches the local sewer system. The machine tool industry uses considerable volumes of oil for lubrication and cutting. All mechanized industries require oil for lubrication. In the steel industry, fabricated metals are frequently dipped in oil to prevent rust during storage. Runoff from highways and parking lots contains measurable amounts of oil leaked from motor vehicles. Another source of oil is the common two-cycle engine used frequently for lawn mowers and outboard motors. Tests show that at low speeds, two-cycle outboard motors bypass as much as one-third of the total fuel–oil mixture.

Numerous natural oil sources also exist. These sources occur frequently in marshy areas, primarily near natural oil-bearing strata. Natural oil seeps also exist in areas along the Pacific Coast. Generally, these natural sources are insignificant as pollution problems. However, if man or earthquakes disrupt the fissures, they can become enlarged to the point where a problem is created.

Oil Properties

Most oils are insoluble in water, aiding in their separation. Furthermore, most insoluble oils are lighter than water, and therefore they float on its surface. A few oils are heavier than water, and these settle to the bottom of water.

Wastewater treatment facilities must also make separate provisions to separate and collect these heavy oils.

Oils that create the greatest problem are those whose density is close to that of water. These oils separate from water slowly—in some cases, too slowly for normal gravitational separation to be effective.

These soluble oils can be naturally soluble but are rendered soluble or miscible due to man's activities. Oils can be rendered miscible by adding detergents and emulsifiers, or using mechanical processes that result in homogenization. In all cases, soluble and miscible oils are in the same category for treatment.

Treatment

The removal of oil from water involves the separation by chemical or physical means and the ultimate disposal of oil.

CHEMICAL TREATMENT

Adding chemicals can improve the separation of oils whose density is close to that of water or oils that are slightly soluble. Generally, oils are less soluble under extreme acid and extreme base conditions. Commonly, sulfuric acid is used; however, lime at a pH of 8.4 is also effective.

To a degree, chemical dispersants are broken down by adding an acid or a base, but specific dispersants require specific chemicals. Truly soluble oils are difficult to render insoluble with acids or bases alone, and chemical coagulation at an appropriate pH is usually required.

Studies show that alum in the pH range of 8.0 to 9.0 and ferrous sulfate in the pH range of 8.0 to 10.0 are effective in coagulating soluble oil. Because controlling the pH in this range with lime is more difficult, wastewater treatment facilities generally use soda ash (Na_2CO_3) to raise the pH to the required level. If the facility uses chemical coagulation at an elevated pH, prior treatment with acid to improve separation is not recommended since this treatment requires larger amounts of alkali for final pH adjustment. Polyelectrolytes with the other coagulants can improve separation.

A recycling system has been devised to recover both added chemicals and separated oil. This system adds ferric chloride, lime, ferric sulfate, and a polymer to the oily liquid waste. It treats the precipitated sludge containing oils with sulfuric acid at 90°F (32.2°C) which releases the oil from the sludge. This oil is then separated and reused, and the ferric sulfate is also reclaimed from the sludge and reused.

PHYSICAL TREATMENT

Insoluble oils lighter than water can be easily separated in a settling tank with an adjustable skimming weir. These oils readily float to the surface, and the depth of the weir is adjusted according to the amount of oil in the water.

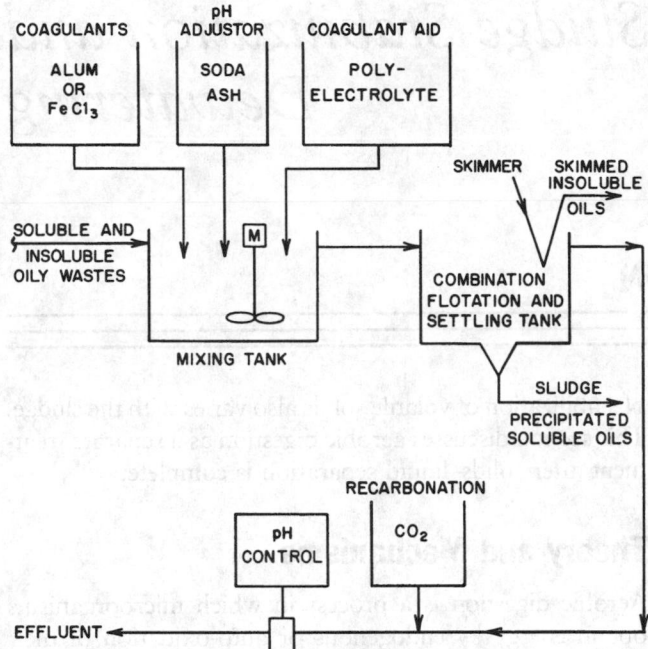

FIG. 7.43.1 Combined soluble and insoluble oil removal.

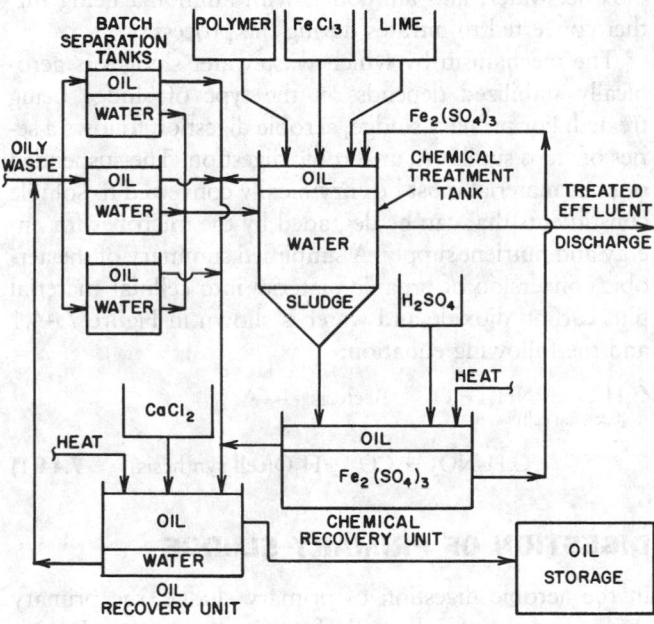

FIG. 7.43.2 Oily waste treatment system that uses chemical reuse and oil recovery.

Insoluble oils heavier than water can be recovered in a settling basin with a bottom sludge separator. The withdrawal rate of the sludge from the bottom must be proportional to the amount of heavy oil separated from these liquids. All coagulation techniques use the settling princi-

ple; the oil is rendered insoluble and heavier than the water so that it can be removed by settling with sludge removal.

For coagulated solids, wastewater treatment facilities use filtration through various media. Sand provides a satisfactory filter medium, but diatomaceous earth is more commonly used due to the disposability of the diatomaceous earth filter cake. Continuous centrifuging also separates oils from water, but this technique is expensive. Also, for varying densities and volumes of oil in waste, wastewater treatment facilities must continuously adjust the centrifuge system.

Dispersed air flotation is also effective in separating oil from water. This technique has been practiced with and without prior chemical coagulation. Sometimes breaking down the foam after it has carried the oil from the water is difficult. Dispersants can break down the foam so that the oil can be properly disposed of.

ULTIMATE DISPOSAL

The ultimate disposal of oil frequently creates a problem. Occasionally, reclaimed oil can be reprocessed in a refinery. In other cases, oils are mixtures that are not favorable for reclamation. In other cases, reclamation is not economical.

Some wastewater treatment facilities dispose oil on land, but they must take proper precautions to prevent the oil from contaminating the groundwater. To a limit, soil microorganisms can break down oil and render it inoffensive. Facilities for burning waste oil are also available. Problems arise from the differing volatilities of the oils collected. Reclamation of the heat produced is beneficial but not always practical due to variations in the amount of oil combusted.

TYPICAL TREATMENT SYSTEMS

In a typical combination treatment system for soluble and insoluble oil removal (see Figure 7.43.1), the soluble oils are precipitated by coagulants, elevated pH, and coagulant aids. Both soluble and insoluble oils are separated in the combination flotation and settling tank. The effluent may require recarbonation with appropriate pH control before being discharged to a receiving stream.

Figure 7.43.2 shows a system for chemical treatment with separation and recovery of oil and ferric sulfate. The proper application of chemical treatment methods can result in satisfactory separation and removal of oil from liquid waste.

—D.B. Aulenbach

Sludge Stabilization and Dewatering

7.44
STABILIZATION: AEROBIC DIGESTION

TYPES OF SLUDGES: a) Excess activated; b) Primary with activated or trickling-filter sludge

TOTAL QUANTITY OF WET SLUDGE: From an average treatment plant, between 1 and 2% of the sewage volume (influent) treated

VOLATILE SOLIDS REDUCTION EFFICIENCY: 40 to 60%

REDUCTION EFFICIENCY OF DEGRADABLE VSS: 85%

EXCESS SLUDGE PRODUCED: 0.2 lb per lb of BOD_5 removed

TYPICAL INFLUENT CONCENTRATION: Over 10,000 mg/l SS

DETENTION TIMES (DAYS): 10 to 15 (a); over 20 (b)

ORGANIC SOLIDS LOADINGS RATE (LB/1000 FT³): Below 100 (b); generally in the range of 40 to 120

AIR REQUIREMENTS (CFM/1000 FT³): 15 to 20 (a); 25 to 30 (b)

DIGESTOR TANK DESIGN: Similar to activated-sludge tanks; they are usually not covered or insulated

AERATION METHODS: Same as in conventional aeration tanks; mechanical surface aerators are superior to diffused aeration systems.

TYPICAL APPLICATIONS: Industrial and small municipal waste treatment plants in which mostly excess activated sludge is treated; cannot be used in cold climates

The aerobic stabilization of biological sludges, generated from wastewater treatment, is the basis for modifications in the activated-sludge process known as total oxidation and extended aeration. In many treatment plants, separate aerobic digestors stabilize mixtures of excess activated and primary sludge. The major objective of aerobic digestion is to produce a biologically stable end product suitable for disposal or subsequent treatment in a variety of processes. Aerobic digestion is generally more suited to the treatment of excess biological sludge than to primary sludge.

Primary sludge settles from raw waste prior to biological treatment. Secondary, biological sludge consists primarily of flocculated microorganisms and suspended organic material trapped or biosorbed to the floc. Secondary sludge is either excess activated sludge or trickling-filter humus. The mechanism of microbial degradation is different for various discrete mixtures of sludge. The degree of stabilization of volatile solids also varies with the sludge. This section discusses aerobic digestion as a separate treatment after solids–liquid separation is complete.

Theory and Mechanisms

Aerobic digestion is a process in which microorganisms obtain energy by endogenous or auto-oxidation of their cellular protoplasm. The biologically degradable constituents of cellular material are slowly oxidized to carbon dioxide, water, and ammonia, with ammonia being further converted to nitrates during the process.

The mechanism by which wastewater sludges is aerobically stabilized depends on the type of sludge being treated. For primary sludge, aerobic digestion follows a series of steps similar to anaerobic digestion. The suspended organic material must be enzymically converted to soluble constituents that can be degraded by the microbes for energy and nutrient supply. A simplified summary of the aerobic conversion of organic material into cellular material plus carbon dioxide and water is shown in Figure 7.44.1 and the following equation:

$$C_xH_yO_z + NH_3 + O_2 + \text{Bacteria} \longrightarrow$$
(bacterial cell)

$$C_5H_7NO_2 + CO_2 + H_2O(\text{cell synthesis}) \qquad \textbf{7.44(1)}$$

DIGESTION OF PRIMARY SLUDGE

In the aerobic digestion of primary sludge, the primary sludge acts as a food supply for microorganisms. During the initial stages of aerobic stabilization, assuming an unlimited food supply with sufficient nutrients, the bacterial growth is limited only by the microbial rate of reproduction (log growth phase). Figure 7.44.2 shows this phenomenon. The oxygen uptake rate continually increases due to the increasing population of new bacteria.

As organic matter oxidation continues, the microorganisms enter a declining growth stage due to the limited food supply. The oxygen uptake rate also declines during this period. As the food supply becomes depleted, the organisms are forced to depend on internal storage products as a source of energy, and endogenous metabolism or respiration becomes prevalent. If endogenous bacterial cells

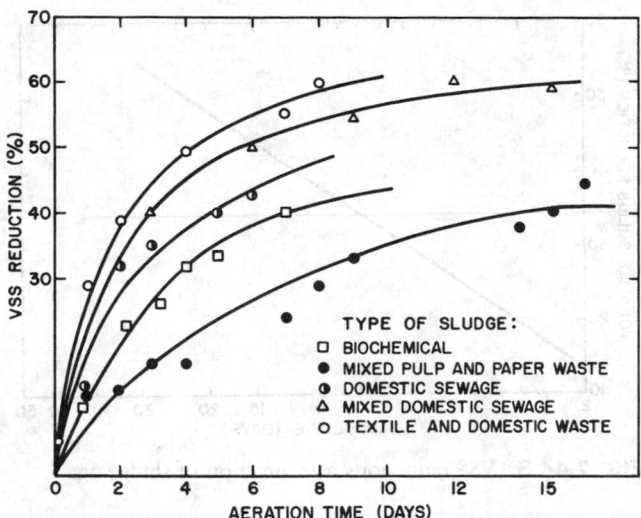

FIG. 7.44.1 Typical VSS reduction by aerobic digestion.

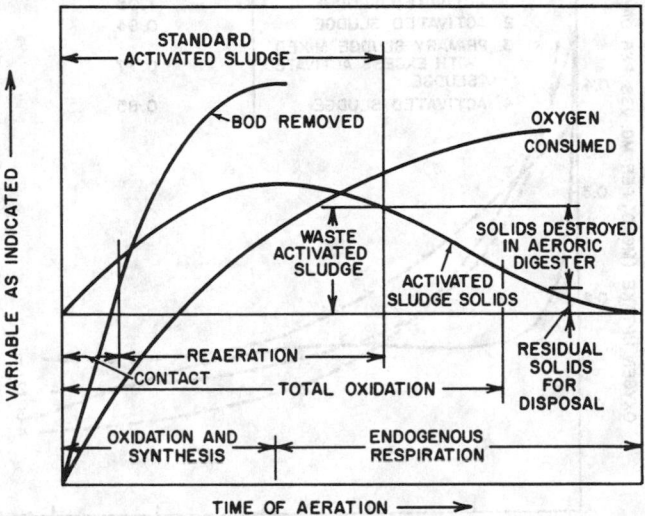

FIG. 7.44.2 Aerobic biological treatment.

are represented by the chemical formula $C_5H_7NO_2$, then the following equation gives the cellular destruction through aerobic digestion:

$$C_5H_7NO_2 + 5O_2 \longrightarrow 2H_2O + NH_3 + 5CO_2 \qquad 7.44(2)$$

As the surrounding food supply in the aerobic system becomes depleted and cannot supply sufficient organic matter for synthesis and energy, the rate of cellular destruction exceeds the organism growth rate. Although this phase of the metabolic cycle is generally referred to as the endogenous phase, it cannot be regarded simply as a period for individual cells using internal cellular carbon sources. The reason for this phenomenon is because a heterogeneous population of microorganisms is present, representing a complex ecosystem in which various microbial species serve as food for other members of the population. Eventually some organisms undergo cellular lysis, releasing protoplasm into the environment; the protoplasm is then used by other bacteria.

Aerobic stabilization of primary sludge results in a high F/M ratio in the aeration basin. Therefore, organic material in the primary sludge converts to bacterial cell material by synthesis, and the resulting change in the total VSS concentration is minimal. Hence, destruction of the bacterial cellular material requires long detention times.

DIGESTION OF SECONDARY SLUDGE

The aerobic digestion of excess secondary sludge is a continuation of the activated-sludge process. The F/M ratio is low, and little cell synthesis occurs. The major reaction is the oxidation and destruction of cellular constituents by lysis and auto-oxidation.

The cellular wall is composed of a polysaccharide-like material that is resistant to decomposition, resulting in a residual VSS concentration from the aerobic digestion process. However, this residual volatile portion is stable and does not produce problems in subsequent sludge handling operations or land disposal.

GENERAL BACKGROUND CONSIDERATIONS

The following parameters affect the aerobic digestion process:

1. Nature and characteristics of the sludge
2. Rate of sludge oxidation
3. Sludge age
4. Sludge loading rate
5. Temperature
6. Oxygen requirements

Generally, the majority of VSS digestion in the aerobic process occurs during the first 10 to 15 days of aeration. Figure 7.44.1 shows both the VSS reductions for a variety of sludges and the major portion of the reduction that occurs during the initial 10 days. The mixed pulp and paper waste sludge is most resistant to digestion (typical of paper mill sludge due to their high content of lignins and cellulose material).

Figures 7.44.3 and 7.44.4 show that maximum stabilization occurs within the 15 days of digestion. Figure 7.44.4 also reflects the increased reduction of volatile matter with increasing temperature.

The sludge age is defined as the ratio of the weight of VSS in the digestor to the weight of VSS added daily. The maximum sludge age is when no significant reduction occurs in the concentration of VSS. Laboratory studies indicate that VSS removal efficiency correlates with sludge age. One such study, summarized in Figure 7.44.5, was conducted at organic loadings between 42 and 112 lb per 1000 ft^3 and detention times of 15 to 30 days. The waste was a mixture of activated sludge and primary sludge. The following equation describes the relationship between the sludge age and VSS removal efficiency:

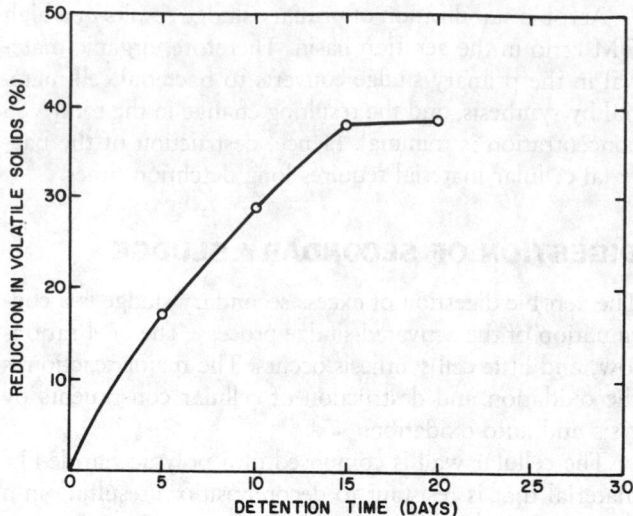

FIG. 7.44.3 Effect of detention time on aerobic digestion of activated sludge.

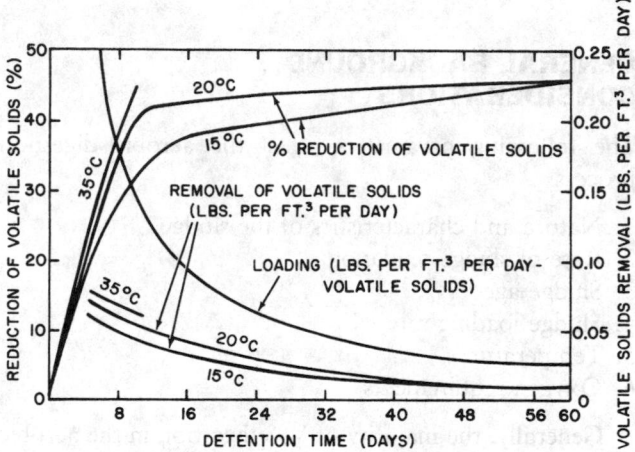

FIG. 7.44.4 Temperature effects on aerobic digestion.

$$\%\text{VSS Reduction} = 2.84 + 35.07 \log_{10}(\text{sludge age}) \quad 7.44(3)$$

Specific oxygen uptake rates (gm O_2 used per gram VSS per day) vary with detention time. Figure 7.44.6 shows the data reported for excess activated sludge.

DESIGN CONSIDERATIONS

Aerobic digestion is usually applied to extended-aeration or contact-stabilization, activated-sludge plants. However, the process is also suitable for many industrial and municipal biological sludges, including trickling-filter humus and excess activated sludge. Information pertaining to design criteria is not abundant. However, the principal design considerations are as follows:

1. Estimated daily quantity of sludge entering the digester
2. Specific oxygen requirements supplied by diffused or mechanical surface aerators
3. Digester detention time

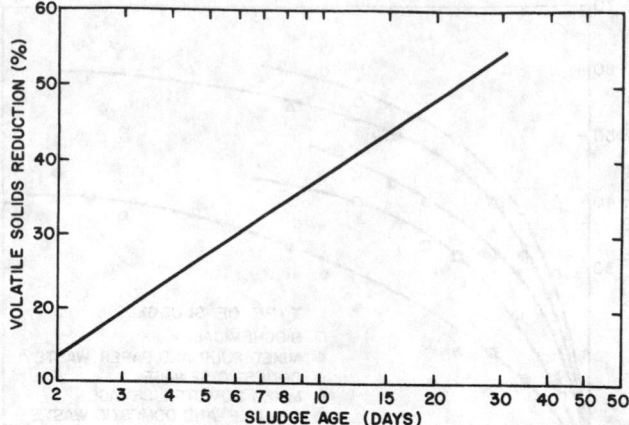

FIG. 7.44.5 VSS reductions as a function of sludge age.

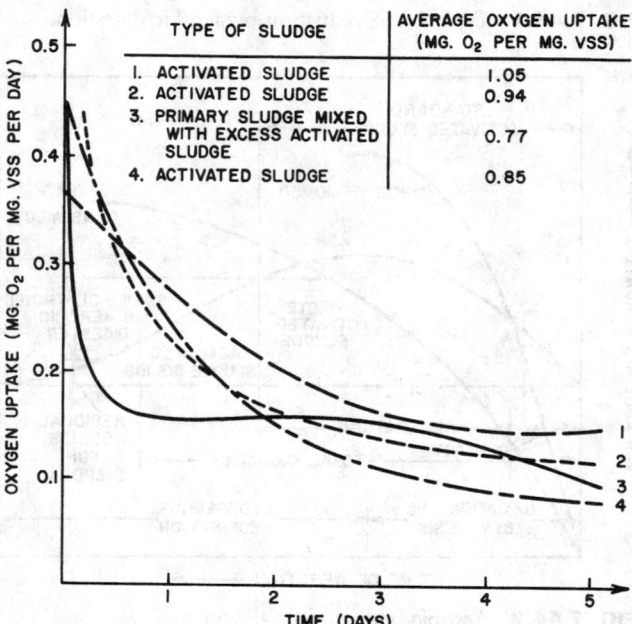

FIG. 7.44.6 Typical oxygen requirements for the aerobic digestion of sludge.

4. Efficiency of VSS reduction required
5. Solids loading rate

Recommended loadings for aerobically treated mixtures of primary and activated sludge or primary and trickling-filter sludge are less than 100 lb per 1000 ft³ with a minimum recommended detention time of 20 days. The suggested minimum detention time for excess activated sludge is 10 days and preferably 15 days. If the temperature in the digestion basin is less than 60°F, additional capacity should be provided.

Recommended oxygen (air) requirements are 15 to 20 cfm per 1000 ft³ of tank capacity; however, if only primary sludge is treated or waste-activated sludge is withdrawn directly from the final clarifier, the air supply should be 25 to 30 cfm per 1000 ft³ of tank capacity.

Aerobic digesters are similar to conventional activated-sludge tanks in that they are not covered or insulated. Thus, they are generally more economic to construct than covered, insulated, and heated anaerobic digesters. Similar to conventional aeration tanks, if diffused aeration is used, the aerobic digesters can be designed for spiral-roll or cross-roll aeration. Environmental engineers frequently use surface mechanical aeration in the design of aerobic digesters. The mixing qualities and oxygen transfer capability of surface aerators are superior to diffused-aeration systems per unit horsepower input.

DEVELOPMENT OF DESIGN CRITERIA

Because of the varying nature of sludge and the lack of sufficient design information, environmental engineers should initiate laboratory or pilot studies to develop the design information. Specifically, they should obtain the rate of VSS destruction, the maximum percent of expected VSS reduction, and the oxygen requirements for various degrees of VSS destruction.

To simplify laboratory procedures, environmental engineers can estimate the destruction rate coefficient and oxygen requirements from a batch study by using the following procedure:

1. Obtain three or four batch units with approximately 2 to 4 l of excess activated sludge in each. Vary the concentration in the batch units to cover the spectrum of anticipated concentrations in the proposed digester.
2. Aerate the units and after they are completely mixed, perform the following analysis on each:
   a. SS, mg/l
   b. VSS, mg/l
   c. Oxygen uptake, mg/l per day
3. Continue to aerate the systems for 25 to 30 days and perform the preceding analyses every 3 days on each unit (see Table 7.44.1).
4. Plot the VSS and SS concentration remaining versus the sludge age (aeration time). From this plot, approximate the oxidizable or degradable fraction of the solids, i.e., the residual VSS remaining after 25 or 30 days of digestion can be taken as the nondegradable portion of the volatile matter (1775 mg/l in Table 7.44.1). The volatile solids destroyed during this aeration are the maximum degradable portion of the volatile matter.
5. For each sampling period, recalculate the remaining degradable VSS. Plot the degradable VSS remaining as a function of detention time (see Figure 7.44.7), and calculate the reaction rate coefficient (k). Note the effect of initial VSS on the destruction rate, and indicate this relationship, if noticeable.
6. Record the oxygen uptake rate as a function of sludge age or aeration time (see Figure 7.44.8). When sizing aeration equipment, estimate the oxygen utilization value as the average value exerted during aeration for

the required VSS destruction level. This average oxygen requirement may be slightly greater or lower than the actual demand at equilibrium, but this average number is adequate for design purposes.

Experimental batch units should be operated at the lower anticipated temperature in the field under winter conditions. A design procedure using laboratory data uses a mass balance of degradable SS through an aerobic digester as follows:

Degradables in − Degradables out = Degradables destroyed

$$Q(X_1 - X_n) - Q(X_2 - X_n) = (dx/dt)V \qquad 7.44(4)$$

TABLE 7.44.1 SUMMARY OF AEROBIC DIGESTION LABORATORY TEST RESULTS

Time of Aeration (days)	VSS (mg/l)	O_2 Uptake (mg/l/hr)	Oxidizable VSS Remaining (mg/l)
0	6115	—	4355
1	4220	33.5	2460
3	2770	22.6	1010
5	2510	—	750
7	2280	15.0	520
9	1975	—	215
11	2105	12.0	345
13	1925	—	165
15	1760	8.2	0
17	1775	—	15

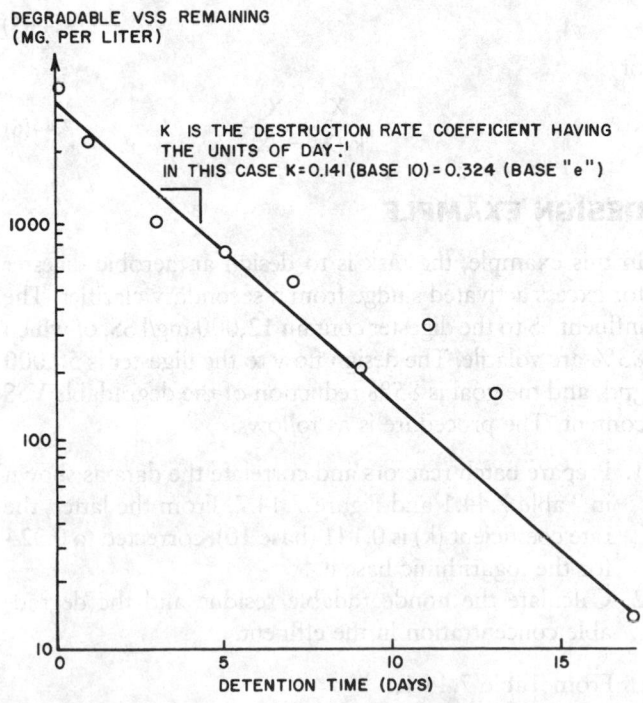

DEGRADABLE VSS REMAINING (MG. PER LITER)

K IS THE DESTRUCTION RATE COEFFICIENT HAVING THE UNITS OF DAY⁻¹
IN THIS CASE K=0.141 (BASE 10) = 0.324 (BASE "e")

FIG. 7.44.7 Determination of batch rate coefficients in an aerobic digester.

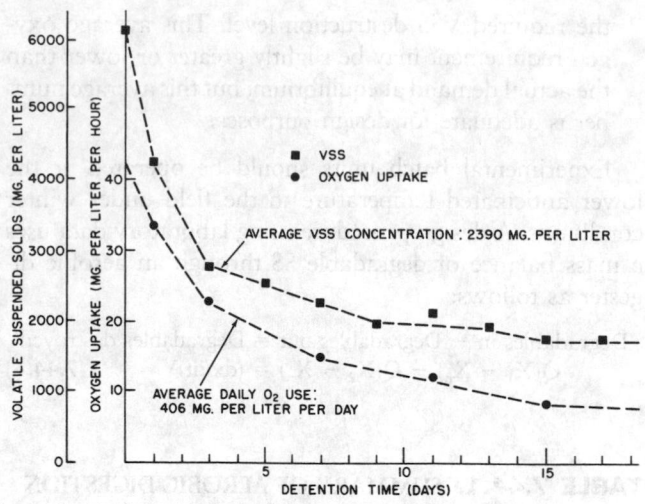

FIG. 7.44.8 Chronological data from a batch aerobic digester.

where:

X_1 = total VSS in influent (mg/l)
X_2 = total VSS in effluent (mg/l)
X_n = total nondegradable VSS, assumed identical for influent and effluent (mg/l)
Q = hydraulic flow (mgd)
V = reactor volume (mg)
$\dfrac{dx}{dt}$ = rate of destruction of degradable influent VSS ($dx/dt = k_b(X_2 - X_n)$
k_b = destruction rate coefficient of degradables, base e (day^{-1})

Thus:

$$(X_1 - X_n) - (X_2 - X_n) = k_b(X_2 - X_n)V/Q = k_bt(X_2 - X_n)$$

7.44(5)

or

$$t = \frac{X_1 - X_2}{k_b(X_2 - X_n)}$$

7.44(6)

DESIGN EXAMPLE

In this example, the task is to design an aerobic digester for excess activated sludge from a secondary clarifier. The influent SS to the digester contain 12,000 mg/l SS, of which 83% are volatile. The design flow to the digester is 50,000 gpd, and the goal is 85% reduction of the degradable VSS content. The procedure is as follows:

1. Prepare batch reactors and correlate the data as shown in Table 7.44.1 and Figure 7.44.7. From the latter, the rate coefficient (k) is 0.141 (base 10), corrected to 0.324 for the logarithmic base e.

2. Calculate the nondegradable residue and the degradable concentration in the effluent.

From Table 7.44.1,
nondegradables: $(X_n) = 1775/6115$
                    = 29% of X_1

Influent VSS (X_1):   = 0.83 (12,000)
                          = 10,000 mg/l
thus: $X_n = 0.29 (10,000)$
                          = 2900 mg/l
Influent degradables:
                $(X_1 - X_n) = 10,000 - 2900$
                            = 7100 mg/l
Effluent degradables:
                $(X_2 - X_n) = 0.15 (7100)$
                            = 1070 mg/l
Total effluent VSS:
                     $X_2 = 1070 + 2900$
                         = 3970 mg/l

3. Calculate the required detention time:

$$t = \frac{X_1 - X_2}{k(X_2 - X_n)}$$

$$= \frac{10,000 - 3970}{0.324(1070)}$$

= 17.4 days. Use 18 days.

Required volume = 0.05 mgd (18)
                = 0.9 mg
                = 120,000 ft^3

4. Calculate the oxygen requirements. Compute the average oxygen requirement by calculating the area under the curve in Figure 7.44.8 and dividing it by the design detention time.

Area under the oxygen curve   = 7310 mg/l O$_2$
Average daily use   = 7310/18
                              = 406 mg/l day

Similarly, estimate the average VSS during a test period to determine the specific oxygen requirements.

Area under curve   = 43,040 day-mg/l
Average VSS concentration = 2390 mg/l
Specific oxygen utilization   = 406/2390
                              = 0.17 mg/mg/day

The total oxygen requirements based on an average VSS concentration X_2 in the digester effluent of 3970 mg/l are then:

$$\left(0.17 \frac{\text{mg O}_2}{\text{mg VSS-day}} \times 3970 \frac{\text{mg VSS}}{1}\right)\left(\frac{8.34 \text{ lb}}{\text{MG-mg/l}}\right)(0.9 \text{ mg})$$

                = 5050 lb O$_2$/day
                = 210 lb O$_2$/hour

Check the power requirements for mixing to see if mixing or oxygen requirements control the design.

Summary

Aerobic digestion is practiced in many industrial and a few municipal treatment facilities. However, because specific technical data are lacking, the inference is that aerobic digestion is a new, unproven technique for solids handling. However, the aerobic digestion process has many potential benefits and warrants attention by design engineers.

Advantages

Aerobic digestion has the following advantages:

1. A biologically stable end product is produced.
2. The stable end product is relatively noxious; hence, land disposal by holding lagoons or spray irrigation is recommended.
3. Due to simplicity of construction, capital costs for the aerobic system are low compared with anaerobic digestion and other solid handling schemes.
4. Aerobically digested sludge generally possesses efficient dewatering characteristics. It drains well when placed on a sand bed and resists rewetting during rainfall.
5. Volatile solids reduction equivalent to the anaerobic digestion results is possible with aerobic systems in treating secondary sludges.
6. Supernatant (floating) liquors from aerobic digestion possess a lower BOD content than those from anaerobic digestion. The aerobic supernatant commonly possesses a BOD of less than 100 mg/l. This advantage is significant because many conventional biological treatment plants are overloaded due to recycling of high-BOD supernatant liquors from anaerobic digestors.
7. Fewer operation problems occur in aerobic digestion processes than in the more complex, anaerobic process due to the higher stability of the aerobic system. Therefore, lower maintenance costs and less skillful labor are needed with an aerobic facility.

8. Aerobically digested sludge has a higher fertilizer value than that resulting from anaerobic digestion.

Disadvantages

The process has the following disadvantages:

1. High power costs generate higher operating costs compared to anaerobic digestion. The difference in operating cost is not significant with smaller treatment plants but is important with large facilities.
2. Gravity thickening processes following aerobic digestion generate high solids concentrations in the supernatant.
3. Some aerobically digested sludges do not dewater easily in vacuum filtration equipment.
4. No methane gas is produced as a by-product because the process is aerobic.
5. The solids reduction efficiency of the aerobic digester can vary with extreme changes in ambient temperature, which subsequently affects the aeration basin temperature.

The aerobic digestion process is well suited for industrial sludge treatment and small, municipal, activated-sludge plants. The industrial community favors aerobic digestion because of the low capital investment and simple operation. Industry often uses mechanical aerators in inexpensive open tanks followed by holding or disposal lagoons. Although a difference in emphasis exists at municipal waste treatment plants with regard to economics, logically, aerobic digestion should be evaluated, particularly for activated-sludge facilities.

—*C.E. Adams*

7.45
STABILIZATION: ANAEROBIC DIGESTION

TYPES OF DESIGNS

    a) Conventional (unmixed and unheated) or low rate
    a1) Conventional primary (digester for most VSS reduction)
    a2) Conventional secondary (digester for solids–liquid separation)
    b) High-rate (completely mixed and heated)
    b1) High-rate primary (digester for most VSS reduction)
    b2) High-rate secondary (digester for solids–liquid separation)

CONDITIONING OPERATIONS APPLIED BETWEEN PRIMARY AND SECONDARY DIGESTERS

    Elutriation and chemical additive treatment

VOLATILE SOLIDS REDUCTION EFFICIENCY

    50 to 60% average

TYPICAL INFLUENT CONCENTRATION

    65 to 70% VSS

DETENTION TIME (DAYS)

    10 to 20 (b); 30 to 60 (a); under 15 (b1); and up to 30 (a2)

ORGANIC SOLIDS (VSS) LOADING RATES (LBS/1000 FT3/DAY)

    30 to 70 (a); 100 to 400 (b)

OPERATING TEMPERATURE

    Heated to 85 to 95°F (a1, b1)

OPERATION

    Intermittent (a,b); continuous (b)

DIGESTER DESIGNS

Primary digester with fixed; secondary with floating covers

GAS PRODUCTION

15 ft³ per lb VSS destroyed or 1 ft³ per capita per day

DIGESTER GAS HEATING VALUE

640 to 700 Btu/ft³

DIGESTER HEAT LOSS

1.3 to 2.6 Btu/hr/ft³ as a function of geographic location

Solids removed from wastewater by treatment plants and those accumulating from the conversion of degradable organic compounds to microbial mass contain substantial amounts of biologically degradable matter. Wastewater treatment facilities frequently use anaerobic digestion to reduce this organic material to carbon dioxide, methane, and other inert end products. In addition to the stabilization of degradable organic matter, anaerobic digestion also substantially decreases the weight of solids that require subsequent processing or disposal. In addition, anaerobic digestion produces a combustible gas that can be used for heating and incineration.

Conventional and High-Rate Digesters

Wastewater treatment facilities use both conventional and high-rate anaerobic digesters to reduce volatile solids in wastewater treatment sludge (see Figure 7.45.1).

Conventional digesters are loaded at a rate of 30 to 100 lb VSS per 1000 ft³ per day (commonly 30 to 70 lb per 1000 ft³ per day). Both feeding and sludge withdrawal are intermittent. Detention times of 30 to 60 days are common. Conventional primary digesters are heated to 85° to 95°F. Primary digesters are often followed by one or more secondary, unheated digesters, where primary digester detention time can be shorter. Although primary digesters are frequently equipped with fixed covers, secondary digester covers are often the floating type. Mixing is not used in primary digesters unless the digesters are followed by secondary, unmixed digester units.

High-rate digesters are loaded at a rate of 100 to 500 lb VSS per 1000 ft³ per day (commonly 100 to 400 lb per

1000 ft³ per day). Feeding and sludge withdrawal can be intermittent or continuous. High-rate primary digesters are heated to 85° to 95°F. High-rate units can operate in series, or they can be followed by conventional, unheated digesters, sludge holding and separation tanks, or other means of solids–liquid separation. Mixing by either gas recirculation or mechanical means is used with high rate units.

Physical Design

The high-rate design is best for primary digesters with fixed covers, water heaters, external heat exchangers, and gas recirculation or mechanical mixing means. Secondary digesters should have floating covers. Interconnecting primary and secondary gas piping is recommended. Accessories, such as gas scrubbers for hydrogen sulfide removal, waste-gas burners, gas meters, gas compressors, gas holders, condensate–sediment traps, and regulators, should also be provided.

The detention time for a primary digester (high-rate) is usually 15 days or less. The maximum detention time for a secondary digester is 30 days.

DIGESTER DIMENSIONS AND ACCESSORIES

Many wastewater treatment facilities use a 3 in 12 (3 ft vertical, 12 ft horizontal) digester bottom slope (14°). The digester design can be based on a 15-ft minimum side wall height with 2 ft of freeboard. Medium and large wastewater treatment facilities should consider duplicate digestion units. A means for supernatant liquid drawoff should be provided for secondary digesters only if the primary digesters are the high-rate type. The digesters should be equipped with flame arresters and vacuum relief devices.

The gas holder (usually a sphere) is designed on the basis of the following relationship:

$$C = V \left(\frac{P - P'}{14.7} \right) \left(\frac{520}{T + 460} \right) \qquad 7.45(1)$$

where:

C = Capacity of the gas holding sphere (ft³ of gas at 14.7 psia and 60°F)
V = Volume of sphere, ft³
P = Maximum operating pressure, psig
P′ = Minimum operating pressure, psig
T = Temperature of stored gas, °F

GAS PRODUCTION AND HEATING REQUIREMENTS

The storage capacity of the gas holder should provide for a detention time of 0.25 to 0.50 days. The normal gas production is 0.9 to 1.0 ft³ per capita of population per day

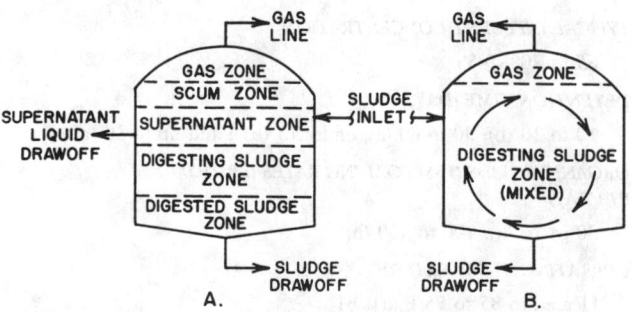

FIG. 7.45.1 Types of anaerobic digesters. A, Conventional; B, High rate.

when mixed primary and activated sludges are digested. Gas production is also estimated as 15 ft^3 per lb VSS destroyed. The Btu value of digester gases is generally 640 to 703 Btu/ft^3.

The following equation gives the digester heat loss:

$$Q = U(A)(T_2 - T_1) \qquad 7.45(2)$$

where:

Q = heat loss, Btu/hr
U = Overall coefficient of heat transfer, Btu/hr/ft^2/°F (see Table 7.45.1).
A = Area (ft^2) normal to direction of heat flow
T$_2$ = Digestion temperature, °F
T$_1$ = Average external temperature for coldest two-week period, °F

Alternatively, the heat loss is 1°F per day for the entire digester contents, or 2600 Btu/hr/1000 ft^3 of tank contents in the northern United States. For the southern United States, half of this rate can be assumed.

The following equation gives the sludge heat requirement:

$$Q = WC(T_2 - T_1) \qquad 7.45(3)$$

where:

Q = Heat required to raise sludge to the digestion temperature, Btu per day
W = Sludge added to the digester daily, lb
C = Mean specific heat of sludge, ≈1.0, Btu/lbm-°F
T$_2$ = Digestion temperature, °F
T$_1$ = Sludge temperature, coldest two-week period, °F

In the absence of other information, the sludge temperature can be assumed to be 50°F in the southern United States, 45°F in the central parts of the United States, and 40°F in the northern United States.

The following equation gives the length of the external, jacketed, sludge pipe, heat exchanger:

$$L = \frac{A}{\pi \dfrac{D}{12}} = \frac{H}{\pi U\, \Delta T_{LM} D/12} \qquad 7.45(4)$$

TABLE 7.45.1 OVERALL HEAT TRANSFER COEFFICIENTS FOR ANAEROBIC DIGESTERS

Digester Type	Overall Coefficient of Heat Transfer (Btu/hr/ft^2/°F)
Concrete roof	0.5
Floating cover	0.24
Concrete wall air space	0.35
Concrete wall in wet earth	0.25
Concrete wall in dry earth	0.18
Floor	0.12

where:

H = Total heat requirement, Btu/hr
U = Overall heat transfer coefficient (a value of 140 is used in the absence of other information) Btu/hr/°F/ft^2
A = Heat exchange surface area, ft^2
D = Diameter of sludge pipe, in
ΔT$_{LM}$ = Logarithmic mean of the temperature difference $\left(\dfrac{\Delta T_1 - \Delta T_2}{\log_e \Delta T_1/\Delta T_2}\right)$

Digester Sizing

The following equations predict the VSS destruction:
Conventional digesters:

$$V_d = 30 + \frac{t}{2} \qquad 7.45(5)$$

High-rate digesters:

$$V_d = 13.7 \, \text{Log}_e (t) + 18.94 \qquad 7.45(6)$$

where:

V$_d$ = VSS destroyed, %
t = Time of digestion, days

These equations specifically apply to sludge with an initial VSS content of 65 to 70%. The quantity of solids remaining in the primary digester at the end of the digestion period can be calculated as follows:

$$M_{T_t} = M_{T_w}(t_1)\left(1 - \frac{\%V(V_d)}{2}\right) \qquad 7.45(7)$$

where:

M$_{T_t}$ = Solids remaining in digester, lb
M$_{T_w}$ = Daily solid waste input into digester, lb
t$_1$ = Digestion period, days
%V = VSS in the sludge received by the digester, %

The following equations calculate the required primary digester volume:

$$M_{T_{m1}} = \frac{[M_{T_w}(t) + M_{T_t}]/2}{t(V_w)(8.34)} \qquad 7.45(8)$$

$$V_1 = \frac{M_{T_t}}{M_{T_{m1}}(62.5)} \qquad 7.45(9)$$

where:

M$_{T_{m1}}$ = Mean solids concentration in primary digester, %
V$_w$ = Daily sludge volume sent to the digester, gal
V$_1$ = Volume of primary digester(s), ft^3
M$_{T_w}$, M$_{T_t}$ t = See Equation 7.45(7).

The secondary digester volume can be calculated as follows:

$$V_2 = \frac{t_2\left(\dfrac{M_{T_t}}{t_1}\right)}{M_{T_{m2}}(7.5)(8.34)} \qquad 7.45(10)$$

where:

V_2 = Volume of secondary digester(s), ft³

t_1, t_2 = Detention time in primary and secondary digesters, days

$M_{T_{m2}}$ = Mean solids concentration in secondary digesters, %

The following equation calculates digester diameters:

$$r = \sqrt{\frac{V/n}{h\pi}} \qquad 7.45(11)$$

where:

r = Radius, ft
V = Total volume required, ft³
n = Number of digesters
π = 3.1416
h = Mean liquid depth, ft

ANAEROBIC DIGESTER DESIGN EXAMPLE

In this example, high-rate, primary digesters (heated and completely mixed) followed by secondary digesters (unheated and unmixed) provide anaerobic digestion of primary and excess activated sludge from a 10 mgd wastewater treatment plant. Secondary digesters provide a maximum 30-day detention time and are intended mainly for solids–liquid separation, sludge thickening, and sludge storage. The daily loading of solids to the primary digesters is as follows:

1. Primary (raw) sludge: 12,010 lb per day, (65% VSS)
2. Excess activated sludge: 8162 lb per day, (77% VSS)

    Total solids = 20,172 lb per day, (70% VSS)
    Volatile solids = 14,091 lb per day

The daily loading in terms of sludge volume is as follows:

1. Primary (raw sludge):   20,000 gpd (7.2% solids)
2. Excess activated sludge: 15,495 gpd (6.3% solids)

    Total sludge volume = 35,495 gpd (6.8% solids)

VSS reduction in the primary digesters with a 15-day digestion period can be calculated in accordance with Equation 7.45(6) as follows:

$$V_d = 13.7 \, Log_e \,(15) + 18.94 = 56\%$$

The quantity of solids remaining in the primary digesters at the end of the 15-day digestion period can be determined with Equation 7.45(7) as follows:

$$M_{T_t} = 20,172(15)\left(1 - \frac{[.70][0.56]}{2}\right)$$
$$= 243,274 \text{ lb}$$

Equations 7.45(8) and (9) determine the required volume V_1 of the primary digesters without the specific gravity correction as follows:

Mean solids concentration =

$$\frac{([15][20,172] + 243,274)/2}{(15)(35,495)(8.34)} = 6.15\%$$

$$V_1 = \frac{243,274}{0.0615(62.5)} = 63.291 \text{ ft}^3$$

The digester loading is therefore as follows:

$$\frac{14,091 \text{ lb. VSS per day}}{63,291 \text{ ft.}^3} = 220 \text{ lb VSS per 1000 ft}^3 \text{ per day}$$

The amount of solids withdrawn to the secondary digesters on a daily basis is as follows:

$$\frac{243,274 \text{ lb.}}{15 \text{ days}} = 16,218 \text{ lb per day}$$

If the average sludge solids concentration in the secondary digesters (achieved through supernatant decantation) is 8%, Equation 7.45(10) calculates the required digester volume V_2 as follows:

$$V_2 = \frac{(30)(243,274)/15}{0.08(7.5)(8.34)} = 97,230 \text{ ft}^3$$

This calculation assumes no digestion in the secondary digesters.

DIGESTER DIMENSIONS

With two primary and two secondary digesters, a mean water depth of 17.5 ft, a 15-ft sidewall depth, and 2 ft of freeboard, Equation 7.45(11) determines the required individual digester sizes as follows:

Primary digesters each:

$$r = \sqrt{\frac{63,291/2}{(3.1416)(17.5)}} \cong 24 \text{ ft}$$

Secondary digesters each:

$$r = \sqrt{\frac{97,230/2}{(3.1416)(17.5)}} \cong 30 \text{ ft}$$

GAS HOLDER SIZING

The gas holder can be designed as a sphere with sufficient volume to provide 0.5-day storage capacity. A maximum gas compression of 30 psig and a minimum pressure of 10 psig at a stored gas temperature of 80°F can be the design basis. The gas production rate is assumed to be 15 ft³ per lb of VSS destroyed as follows:

$$14,091(0.56)(15) = 118,364 \text{ scfd}$$

Solving Equation 7.45(1) gives the required gas sphere capacity as follows:

$$0.5(118,364) = V\left(\frac{30-10}{14.7}\right)\left(\frac{520}{80+460}\right)$$

$$V = 45,190 \text{ ft}^3$$

The sphere diameter, therefore, is as follows:

$$D = 2r = 2\sqrt[3]{\frac{V}{(4\pi)/3}} = 2\sqrt[3]{\frac{45,190}{4.189}} \cong 45 \text{ ft}$$

The normal range of gas production is $\approx 0.9 - 1.0 \text{ ft}^3$ per capita per day when mixed primary and activated sludges are digested. In this example, the production is as follows:

$$\frac{118,364 \text{ ft}^3 \text{ per day}}{100,000 \text{ population}} = 1.2 \text{ ft}^3 \text{ per capita per day}$$

PRIMARY DIGESTER HEATING REQUIREMENTS

For the purposes of this example, a digestion temperature of 95°F and a plant location in the northern U.S. where the lowest (two-week mean) daily temperature is 10°F are used. All digester sidewall areas are earth-covered, with 75% of the areas covered by dry earth and the remaining 25%, as well as the floor area, in groundwater. The mean winter ground temperature is 25°F and the groundwater temperature is 40°F. Equation 7.45(2) calculates the heat losses for both primary digesters as follows:

1. Fixed cover (concrete): $Q = (\pi)(r^2)(2)(U)(T_2 - T_1)$

$$Q = 3.1416(24)^2(2)(0.5)(95 - 10)$$
$$= 153,813 \text{ Btu/hr}$$

2. Dry-carth-covered sidewall area:

$$Q = 0.75(3.1416[48][15][2][0.18][95 - 25])$$
$$= 42,751 \text{ Btu/hr}$$

3. Wet sidewall area:

$$Q = 0.25(3.1416[48][15][2][0.25][95 - 40])$$
$$= 15,551 \text{ Btu/hr}$$

4. Floor area:

$$Q = 3.1416(24)^2(2)(0.12)(95 - 40) = 23,886 \text{ Btu/hr}$$

5. Total primary digester area heat loss:

$$
\begin{array}{r}
153,813 \\
42,751 \\
15,551 \\
+23,886 \\
\hline
236,001 \text{ Btu/hr}
\end{array}
$$

Alternatively, the approximate digester heat loss in the northern United States can be computed as 2600 Btu/hr/1000 ft³ of digester volume as follows:

$$2600(63.3) = 164,580 \text{ Btu/hr}$$

This estimate, however, must be modified for the prevailing conditions, actual winter temperatures, or the presence of groundwater, e.g., at the plant site.

6. Sludge heating requirement using Equation 7.45(3): (The sludge temperature as it enters the primary digester is 50°F for this example.)

$$Q = \frac{20,172}{0.068}(1)(95 - 50)$$
$$= 13,349,118 \text{ Btu/day}$$
$$= 556,213 \text{ Btu/hr}$$

7. Total primary digester heating requirement:

Tank	236,001
Sludge	+556,213
Total	792,214 Btu/hr

For digester heating, a gas-fired heat exchanger can use the digester gas as fuel. The digester gas is assumed to have a heating value of 650 Btu per ft³. Therefore, the excess heat can be calculated as:

1. 118,364 ft³/day (650 Btu/ft³) = 76,936,600 Btu/day
   = 3,205,692 Btu/hour

3,205,692 − 792,214 = 2,413,478 Btu/hr excess heat is available for sludge incineration, building heat, and so forth.

2. Heat exchanger length: In this example, an external heat exchanger with 4-in sludge tubes jacketed by 6-in water tubes is used. Water enters the exchanger at 145°F and exits at 100°F. The sludge enters at 80°F and exits at 95°F. An overall heat transfer coefficient of 140 can be assumed. Equation 7.45(4) calculates the required heat exchanger length is calculated as follows:

$$\Delta T_1 = 145 - 80 = 65°F$$
$$\Delta T_2 = 100 - 95 = 5°F$$

$$\Delta T_{LM} = \frac{65 - 5}{\text{Log}_e\ 65/5} = 23.4°F$$

$$A = \frac{H}{U\ \Delta T_{LM}} = \frac{792,214}{(140)(23.4)} = 242 \text{ ft}^2$$

$$L = \frac{242}{(3.1416)(4/12)} = 231 \text{ ft}$$

—B.L. Goodman

7.46
SLUDGE THICKENING

TYPE OF THICKENER
  a) Gravity settling
  b) Pressure flotation

TYPICAL DEGREE OF SLUDGE THICKENING ACHIEVED
  By a factor of 3 (a); or 4 to 6% solids in thickened stream (a,b)

TYPICAL SOLIDS FLUX RATE (LB/FT²/HR)
  0.5 (a)

TYPICAL OVERFLOW RATE (GPD/FT²)
  200 (a); and 2500 to 5000 (b)

HYDRAULIC DETENTION TIME (HRS)
  0.3 (b); and from 6 and up (a)

Excess activated sludge and trickling-filter humus are almost always withdrawn at a solids concentration of less than 1% by weight. The underflow solids concentrations of primary clarifiers are frequently less than 5% by weight. To decrease the hydraulic loading on sludge digestion and dewatering units and thus increase their efficiency in terms of the weight of solids processed per unit area or volume per unit time, wastewater treatment facilities commonly use thickeners.

Gravity Thickening

Gravity-type thickeners (see Figure 7.46.1), in which waste sludge is continuously fed and thickened sludge is continuously withdrawn, are commonly used in wastewater treatment systems. The required thickener area de-

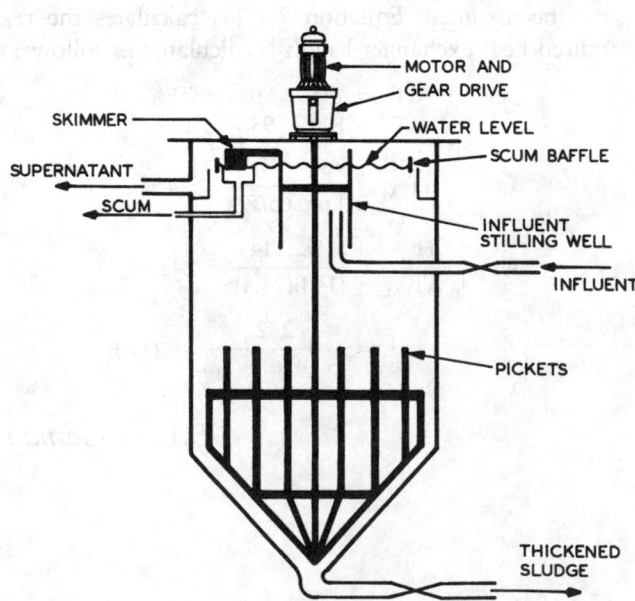

FIG. 7.46.1 Gravity thickener.

pends on the rate of solids arrival at the thickener bottom, which depends on the settling velocity of the sludge solids and the concentration of the thickened sludge removed from the unit. Solids move downward through the thickener at the combined flux rate given in the following equation:

$$G = C_{in}(V_s + V_u) \qquad 7.46(1)$$

where:

  G = Solids flux rate, lb/ft²/hr
  C_{in} = Solids concentration in thickener influent, lb/ft³
  V_s = Solids settling velocity due to gravity, ft/hr
  V_u = Solids downward transport velocity due to underflow solids removal, ft/hr

The following equation expresses the downward transport velocity (V_u) due to concentrated solids withdrawal:

$$V_u = \frac{G}{C_{out}} \qquad 7.46(2)$$

where:

  C_{out} = Concentration of thickened solids removed, lb/ft³

Thus, a limiting solids flux rate G_L exists and can be determined for any combination of values for the pertinent variables. The following equation gives the required thickener area A_T (in ft²):

$$A_T = \frac{M_{T_w}}{G_L} \qquad 7.46(3)$$

where:

  M_{T_w} = hourly solid waste input into thickener, lb/hr

Environmental engineers can derive the required design data from settling tests conducted in cylinders with diameters of at least 3 to 6 in, and preferably greater, having a depth of from 3 ft to full-scale depth. These cylinders should be equipped with a stirring device having a tip speed of about 40 ft/hr. Figure 7.46.2 plots the data derived from these tests.

GRAVITY THICKENER DESIGN EXAMPLE

For the purposes of this example, the solids to be thickened are excess activated sludge with an initial solids concentration of 7 gm/l (0.434 lb/ft³). This sludge continuously enters the thickener, and the thickened sludge is continuously withdrawn at a concentration of 21 gm/l (1.302 lb/ft³). The G_L is determined from data plots (see Figure 7.46.2) 0.52 lb/ft²/hr. The volume of sludge to be concentrated is 5200 gph (693.3 ft³/hr). The required

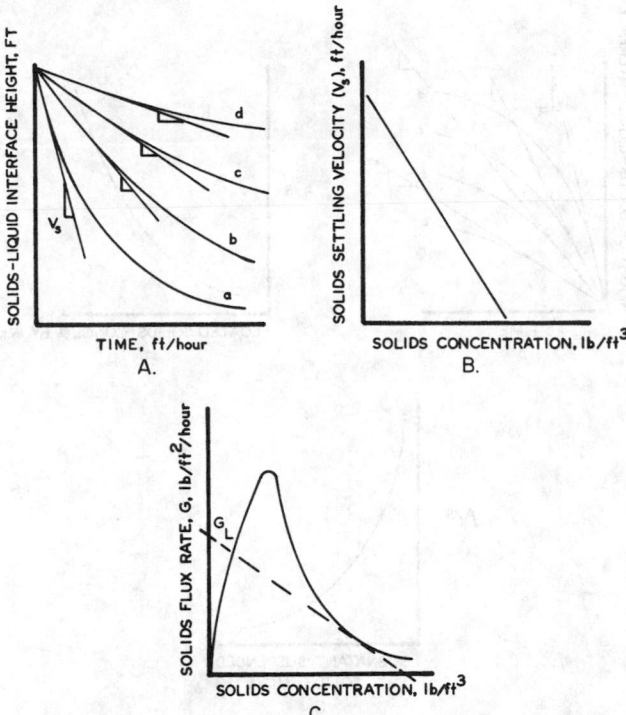

FIG. 7.46.2 Gravity thickener design data plots. Data plot procedure: 1. Conduct settling tests at various initial solids concentrations (a through d in panel A), covering the range of expected thickener influent solids concentrations; 2. Determine the settling velocity V_s for each concentration as the slope of a line tangent to the initial lineal settling rate portion of each curve; 3. Plot V_s values versus solids concentration, and construct a line of best fit; 4. Compute values of the flux rate G for a variety of solids concentrations (G = solids concentration, lb/ft^3 × settling velocity, ft/hr), and plot as in panel C; and 5. Determine the value of the limiting flux rate G_L as the intercept on the flux rate axis of a line tangent to the flux curve drawn from the desired underflow solids concentration.

thickener area can be determined by using Equation 7.46(3) as follows:

$$A_T = \frac{0.434(693.3)}{0.52} = 578.6 \text{ ft}^2 = \frac{D^2\pi}{4}$$

$$\text{Diameter (D)} = \sqrt{\frac{578.6}{0.7854}} = 27.14 \text{ ft}$$

A good selection for this application is to provide two 27- to 30-ft diameter units equipped with vertical pickets. With one unit in service, the unit overflow rate (rate of effluent discharge) is as follows:

$$\frac{124,800 \text{ gal/day}}{578.6 \text{ ft}^2} = 215.7 \text{ gal/ft}^2/\text{day}$$

Flotation Thickening

A variety of pressurized, air flotation units are commercially available as complete packages. The basic units are either rectangular or round, with rectangular units more common in wastewater treatment systems (see Figure 7.46.3). Both direct pressurization and pressurized recycling systems are available. Wastewater treatment facilities use the later when the liquid contains flocculant or other fragile particles that disperse or suffer attrition when subjected to the high shear forces in the pressurization process.

The performance of pressure flotation units is related to the ratio of the weight of the air used to the weight of the solids supplied to the unit, A/S. However, this approach neglects both the number of bubbles into which the air is divided after the pressure release and the number of particles into which the weight of incoming solids is divided.

No ready means exists to assess the potential impact of these variables. The degree of flocculation of an activated sludge can vary in response to changes in influent waste characteristics or process management. Therefore, wastewater treatment facilities should use conservative designs. Alternatively, they can add flocculating chemicals to the unit to insure a constant number of particles per unit weight of applied solids. Polymers are usually selected as the flocculating chemicals. Facilities can substantially increase the particle rise rate with low polymer dose rates.

Wastewater treatment facilities can use the following equation to select pressurized recycling flotation units:

$$\frac{A}{S} = \frac{C(s_a)(f')(p-1)R}{C_{in}(Q_w)} \qquad 7.46(4)$$

where:

A/S = ratio of lb of air to lb of incoming solids
C   = A conversion factor, 0.834
s_a = Solubility of air in water, 18.68 cc/l at 20°C
f'   = Percent saturation achieved (decimal fraction)

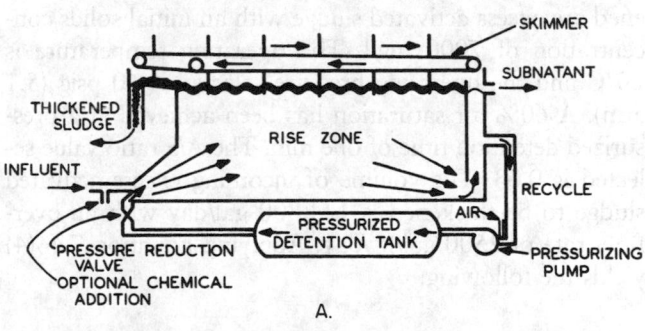

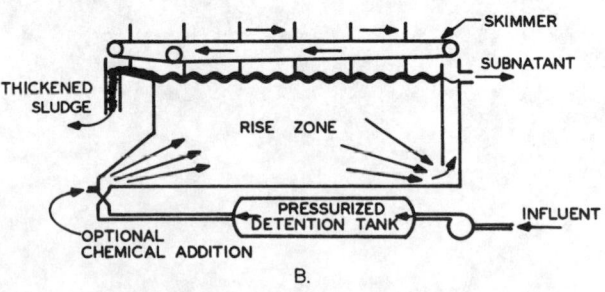

FIG. 7.46.3 Flotation thickeners. **A,** Pressurized recycle system; **B,** Direct pressurization system.

p = Operating pressure, atmospheres
R = Pressurized recycling rate, mgd
C_{in} = Solids concentration in thickener influent, mg/l
Q_w = Waste sludge influent flow rate, mgd

Wastewater treatment facilities can use the following equation to select direct pressurization flotation units:

$$\frac{A}{S} = \frac{C(s_a)(f')(p-1)}{C_{in}} \qquad 7.46(5)$$

Environmental engineers usually select flotation units that provide a surface overflow rate of 2500 to 5000 gal/ft²/day and a hydraulic detention time of about 20 min. They should also make provisions for dual units, bypassing, dewatering, and flocculant addition.

Environmental engineers develop design data by pressurizing either the waste influent (direct pressurization) or a simulated recycled sample, such as settled or filtered waste or sludge supernatant, and releasing it into an open graduated cylinder. For the pressurized recycling system, they release various amounts of recycled liquid into the required amount of waste or sludge to achieve the required recycling ratio (see Figure 7.46.4). Both the time of pressurization and the pressure are based on the latitude of design permitted by the commercial equipment. For custom designs, environmental engineers should study a range of values. The effect of temperature should also be studied, especially if the flotation units are not housed or influent temperatures vary.

Pressurized Recycling Design Example

For the purposes of this example, the solids to be thickened are excess activated sludge with an initial solids concentration of 7000 mg/l. The operating temperature is 20°C and the operating pressure selected is 60 psig (5.1 atm). A 60% air saturation has been achieved at a pressurized detention time of one min. The A/S ratio value selected is 0.03. The volume of incoming excess activated sludge to be thickened is 124,800 gal/day with an overflow rate of 2500 gal/ft²/day. Applying Equation 7.46(4) yields the following:

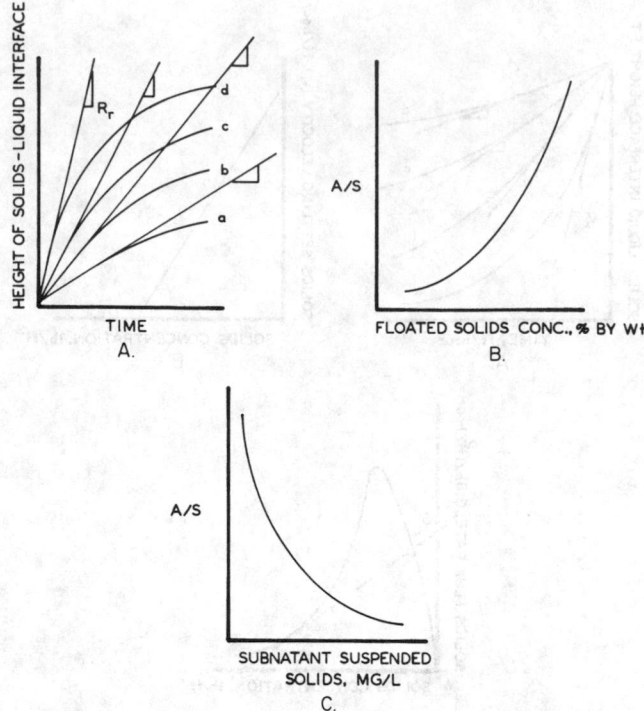

FIG. 7.46.4 Pressure flotation design data plots. Data plot procedure: 1. Conduct rise rate tests at various initial solids concentrations (a through d in panel A) covering the range of expected unit influent solids concentrations; 2. Determine the initial rise rate R_r for each concentration as the slope of a line tangent to the initial lineal rise rate portion of each curve; 3. Plot the concentration of floated solids (% by weight) versus the A/S for each influent solids level; 4. Plot the subnatant SS concentration (unit effluent SS) versus the A/S for each influent solids level; and 5. Select the design A/S ratio to achieve the solids capture or concentration objective, or alternatively, optimize for maximum concentration and capture.

$$0.03 = \frac{(0.834)(18.68)(0.60)(5.1-1)R}{7000(0.1248)}$$

$$R = 0.684 \text{ mgd}$$
$$= 475 \text{ gpm}$$

The required unit surface area is as follows:

$$A = \frac{124{,}800 \text{ gal/day}}{2500 \text{ gal/ft}^2/\text{day}} \cong 50 \text{ ft.}^2$$

—*B.L. Goodman*

7.47
DEWATERING FILTERS

TYPES OF FILTERS

    (a) Rotary drum vacuum filters (RDVF)
    (b) Filter press
    (c) Continuous belt filter press
    (d) Sand drying beds

MATERIALS OF CONSTRUCTION

    Outdoor sand-bed filters: Reinforced concrete, cast-iron piping, and cast steel valves; Pressure vacuum, or deep-bed filters: Wetted parts under tensile loading: stainless steels; rubber-coated or polymer-coated steel; and copper-bearing alloys for a pH near 7; Wetted parts for compressive loading only, such as filter press plates: aluminum alloys, bronzes, cast steel, lead-coated or polymer-coated steel, wood, and stainless steel; Wetted parts, very low load, such as filter cloth support grid: polymers like polypropylene; Nonwetted parts: carbon steel filter media for waste treatment filters: Cellulose (e.g., cotton), polyamide fibers (nylon and nomex), polyesters, acrylics, polyolefins, fluorocarbons, fiberglass, mineral fibers, and woven metals

TOTAL COSTS

    About $20 per ton of dry solids

OPERATING COSTS

    $10 or more per ton of dry solids

FILTER SIZES AVAILABLE

    Up to 550 ft^2 (a,d); up to 3500 ft^2 (d)

NUMBER OF DRIVE MECHANISMS

    2(c,d); 6(a,b)

SLUDGE CONCENTRATIONS

    Primary sludge: 6 to 11% solids; excess activated sludge: 2000 to 4000 mg/l VSS; anaerobic digester effluent: 4.5 to 7% VSS

FILTER CAKE SOLIDS CONTENT

    20 to 30%

VACUUM FILTER FILTRATION RATES (LB/HR/FT^2)

    On primary sludge, 5 to 7; on excess activated sludge after heat or chemical treatment, 1 to 4; on lime precipitated chemical sludge, 2 to 6; on aerobically digested sludge, 1 to 2

PARTIAL LIST OF SUPPLIERS

    Andritz-Ruther (b,c); Arlat Inc. (b); Ashbrook-Simon-Hartley (b,c); Eimco Process Equipment (a,c); Dorr-Oliver Inc. (a,c); Humbold Deacanter (b); Komline-Sanderson (a,b,c); Westech Engineering Inc. (a)

Table 7.47.1 provides an orientation on filter selection and application. Screens are discussed in Section 7.14. This section discusses those filters frequently used in waste treatment plants.

Domestic Wastewater

The distinguishing characteristic of nearly all domestic liquid waste entering municipal treatment plants is the low concentration of dissolved and suspended contaminants and the large volume of wastewater that must be treated. This characteristic means that domestic and industrial water is readily available and inexpensive. In the U.S., the per capita domestic water consumption is about 100 gallons per day; the solids concentration of generated wastes is 0.1 wt% or less.

Environmental engineers have proposed processes to reduce daily water consumption without jeopardizing public health. Such measures may eventually be required by law and would increase the solids concentration in waste streams. Meanwhile, wastewater treatment facilities must treat these dilute waste streams to protect public health, the water supply, and the environment.

Incoming domestic waste always contains large solid objects that are difficult or impossible to treat. Such tramp solids include bottle caps, cigarette filter tips, clumps or cloth or lint, and rubber goods, and they must be removed initially by coarse screens (see Section 7.14). Table 7.47.2 lists the typical composition of raw domestic waste after

TABLE 7.47.1 FILTRATION METHODS FOR WASTE TREATMENT APPLICATIONS

Filtration Application	Filter Design Selected
Separation of bacterial slurry, activated sludge, or submicron inorganic slurry, 0.5 to 5% solids, into unwashed solids plus cloudy filtrate	Continuous rotary vacuum belt filter with special cake discharge method; continuous rotary pressure filter
Separation of digested primary sludge into wet solids plus cloudy filtrate	Coil filter, belt filter
Separation of valuable chemical liquid from waste solids, with extensive cake washing	Batch pressure filters with or without automatic cake discharge
Separation of slurry into clean liquid plus solids mixed with filter aid	Rotary drum or batch-type pre-coat filter
Separation of slurry into clear liquid plus heavy slurry of separated solids	Deep-bed filter with backwash

TABLE 7.47.2   COMPOSITION OF RAW DOMESTIC SEWAGE AND MUNICIPAL SEWAGE

Constituent	Domestic Sewage[2] (mg/l)	Municipal Sewage (mg/l)			Combined Industrial and Municipal Sewage (mg/l)	
		Maximum	Mean	Minimum	Plant A	Plant B
TDS	200	396	305	237	—	—
Total SS	300	264	148	85	320	92
TS	500	640	453	322	—	—
TOC	120	—	—	—	—	—
BOD (dissolved and suspended)	200	276	147	75	429	208
COD (dissolved and suspended)	350	436	288	159	—	—
BOD/SS ratio	0.67	1.04	1.95	1.87	1.34	2.26

initial screening. This table also gives the composition of municipal sewage based on long term measurements.

FILTRATION VERSUS CLARIFICATION OF PRIMARY DOMESTIC WASTE

Filtration of dilute waste (sewage) is not feasible because of the high investment. In addition, the filter cake is impossible to form. Primary clarifiers usually separate solids from most of the liquid and produce more concentrated solids for filtration and additional processing (see Figure 7.47.1). Other low-power devices, such as sand-bed filters, are also used for this purpose. Solids-concentrating or solids-separating devices requiring high velocities, pressures, and power consumption, such as centrifuges, hydrocyclones, filter presses, or leaf filters, are ordinarily uneconomical for primary municipal waste treatment but are sometimes used for industrial pretreatment on a smaller scale.

The mechanical separation (filtration or clarification) of primary sludge is only partially effective as a treatment because 30 to 40% of BOD and COD are water soluble and cannot be so removed.

PRIMARY SLUDGE

The SS content of the primary clarifier or thickener underflow (primary sludge) varies depending on the detention time. Researchers found volatile solids of 8.2 to 10.7% weight in undigested primary sludge going from thickeners to high-rate anaerobic digesters (Jeris 1968). They measured material with a solids content of 6 to 10% weight in the same stream when vacuum filtration equipment was fed from thickeners. This sludge is usually hydrophilic and nondewaterable. One of the purposes of the anaerobic digester is to convert it into a readily dewaterable form.

Researchers reported a filtration rate of 6.9 lb/hr/ft² for settled primary sludge on an RDVF (Eckenfelder 1970). Other results include a filtration rate of 5 lb/hr/ft² for settled primary sludge on the vacuum filter (dry basis), with 32% solids in the filter cake. Undigested raw sludge should

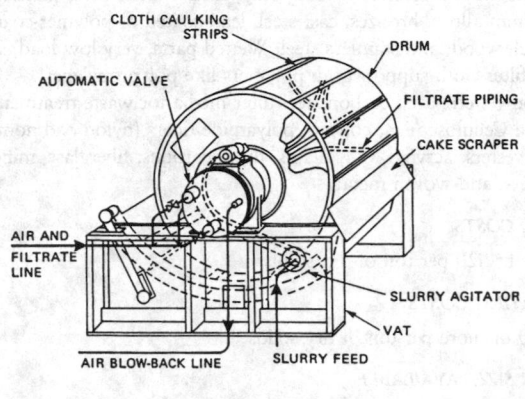

A. CUTAWAY VIEW OF A DRUM OR SCRAPER-TYPE RDVF

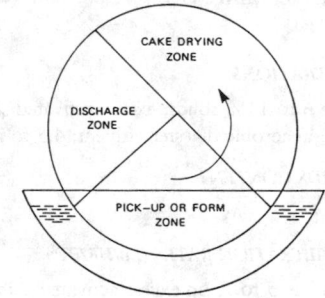

B. OPERATING ZONES OF A RDVF

FIG. 7.47.1   Schematic diagram of vacuum filter dewatering equipment.

not be filtered because of odors, cleaning problems, and potential health hazards.

ACTIVATED AND DIGESTED SLUDGE

Derived solid sludge from anaerobic digestion of settled raw sludge is readily dewaterable by filtration. The SS concentration in the digester product is lower (4.6 to 6.7% VSS) than that in the feed because of the anaerobic conversion of some organic material to methane. This sludge can be successfully filtered on coil-type RDVFs (see Figure 7.47.2).

Sludge from secondary treatment (activated-sludge process) consists of bacteria, dead cells, and associated material formed under oxidizing conditions. Suspended bacterial solids in the aeration tank are usually in the range of 2500 to 4000 mg/l. These solids can be removed on a deep-bed filter (see Figure 7.47.3) which can take the full flow of the plant.

Ordinarily, however, the aeration tank suspension is continuously circulated to a clarifier, whose underflow (bottom discharge sludge) contains 1 to 3% dry solids concentration or higher by weight depending on the settling time and properties. When excess activated sludge is conditioned with ferric chloride (10 lb per 100 lb dry solids), filtration usually proceeds well on vacuum belt filters at solids rates of 1 to 4 lb/hr/ft^2, or filtrate rates of 15 to 40 gph/ft^2 (see Table 7.47.3). Figure 7.47.4 shows the special cake discharge methods.

Activated sludge does not filter as well as digested primary sludge as on the coil filter. The latter contains enough fibrous material to bridge the openings in the coils, whereas activated sludge does not.

Industrial Wastewater

A comparison of five industrial wastes shows that the SS content varies from 0.28 to 400 lb per unit of production (Eckenfelder 1966). The majority (65 to 75%) of nonpolymeric chemical substances are soluble in water at 25°C

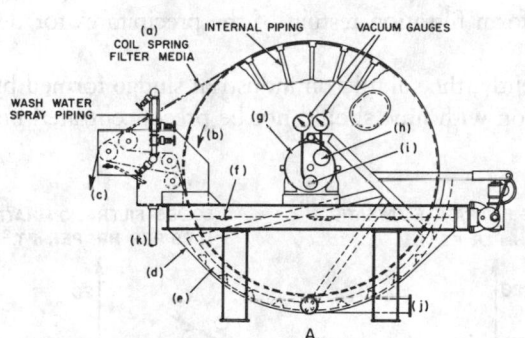

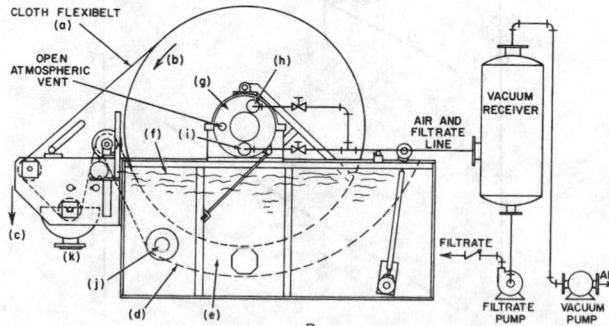

FIG. 7.47.2 Continuous RDVF. A, Coil filter; B, Belt filter. Sizes range to 550 ft^2 per unit. *Legend:* a, filter medium (coil or cloth); b, filter drum; c, cake discharge; d, slurry vat; e, agitator; f, slurry level; g, filter valve; h, drying air outlet; i, filtrate outlet; j, vat drain; k, belt wash drain.

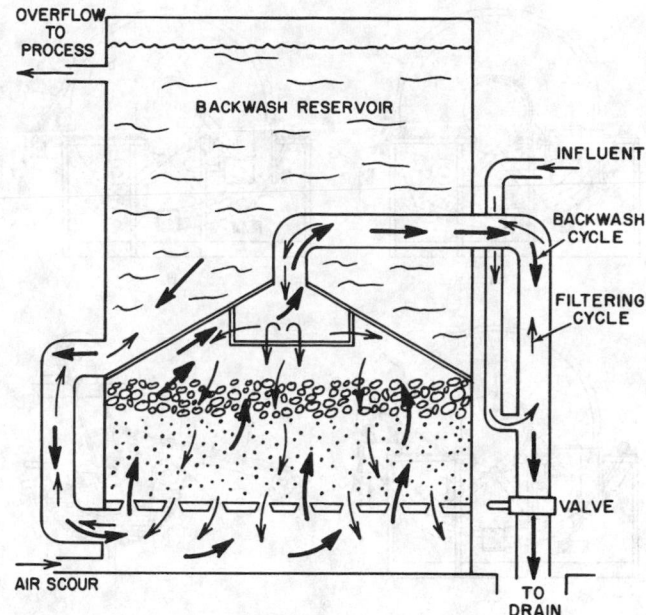

FIG. 7.47.3 Deep-bed filter with intermittent backwash. Water supply for backwashing occupies the upper part of the filter tank. When the drop across the filter reaches 4 ft of water, an air scour blows through the bed, carrying collected matter from the media. Then, the stored water flows through the bed in reverse, leaving the sand and anthracite clean.

TABLE 7.47.3　CONTINUOUS RDVF DATA FOR ACTIVATED SLUDGE

Parameter	Range of Values
Solids concentrations in slurry, % dry basis weight	0.64 to 3.43
Filtration operating vacuum level, ″ Hg	17.7 to 25.0
Slurry temperature, °C	11.5 to 15.5
Filtrate rates, gph/ft^2	18.2 to 27.4
Solids rates, dry basis; lb/hr/ft^2	0.47 to 3.47°
Cake thickness, in	$\frac{1}{64}$ to $\frac{3}{32}$
Drum diameter, ft	3.0
Drum width, ft	1.0
Drum speed, rpm	0.40 to 0.909

Note: °This rate is lower than the solids rate obtained from slurry flow because of solids passing through cloth.

and a pH of 7. Therefore, as the proportion of industrial waste in a system increases, the ratio of total BOD (dissolved and suspended) to SS increases. High polymers in industrial waste forms colloidal suspensions or solutions.

FILTER AIDS

In addition to inert chemical substances, raw industrial waste can contain microorganisms—bacteria, molds, and yeasts. For this group, the settling rates are often low, and the filtration resistance is high; yet frequently, the concentrations (1 to 10% by weight) make them unaccept-

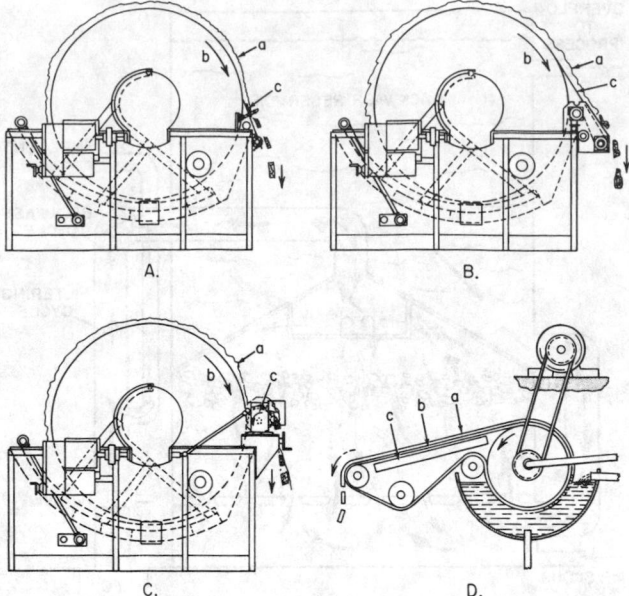

FIG. 7.47.4 RDVFs—alternate types of cake discharging mechanisms. **A,** Scraper discharge: filter cake (a) is removed from cloth on drum (b) by doctor blade (c), and air blow (2 to 4 psig) aids in dislodging cake; **B,** String discharge: filter cake (a) is lifted from cloth on drum (b) by parallel strings (c) tied completely around drum, about $\frac{1}{2}$ in apart; **C,** Roll discharge: filter cake (a), consisting of cohesive materials like clay or pigment, is lifted from cloth on drum (b) by adhesion to discharge roll (c); **D,** Heated belt discharge: filter cake (a), which adheres to cloth belt (b) because of adhesive effect of filtrate, is separated from cloth by heater (c), which dries the cloth; cake does not need drying.

able in public collection systems. Many industries solve this problem by filtration with generous amounts ($\frac{1}{2}$ to 1 lb per pound dry solids) of filter aid, usually diatomaceous earth. However, this solution creates a solids disposal problem.

Industries can minimize their use of filter aids by using the heated-belt discharge method on the rotary vacuum belt filter. Activated sludge (a derived solid consisting entirely of bacteria) can be successfully filtered this way without any filter aid when it is coagulated with ferric chloride. Consequently, attempts should also be made to coagulate other bacterial slurries.

Any coagulant used in the filtration of disposable waste requires Food and Drug Administration (FDA) approval, and ferric chloride is not acceptable. Nevertheless, an acceptable coagulant should be sought from the approved list to lower pretreatment costs.

Unless an industrial plant's effluent contains less than the maximum allowable solids content established by the regulatory agency, filtration or other pretreatment can be required. When an industry pretreats an effluent to meet other requirements, such as pH, the pretreatment can result in the precipitation of additional solids, a step that adds to the filtration or separation load.

SOLIDS CONCENTRATED BY PHYSICAL METHODS

Processes for physically concentrating solids from industrial waste include coagulation and aeration, which achieve either faster settling or flotation rates. Pretreatment by filtration applies only to particulate solids that settle in water. Because settleable solids vary widely in particle size, properties, and composition, filtration tests are needed. In selecting and designing filtration equipment, environmental engineers should also consider the toxicity, health, and fire hazards of the materials to be filtered.

SOLIDS DERIVED FROM CHEMICAL REACTIONS

Chemical reactions during pretreatment produce additional SS in industrial wastewater. The most common precipitating agent is lime slurry (10% calcium hydroxide plus H_2O). It precipitates a variety of calcium compounds, both organic and inorganic in water. Dolomitic lime ($Ca[OH]_2 \cdot Mg[OH]_2$) is a superior precipitating agent for some waste material. These two forms of lime represent the best combination of availability, economy, and versatility.

Compounds precipitated by lime usually settle well and can be filtered at a rate of 2 to 6 lb/hr/ft^2 on an RDVF (see Figure 7.47.5). Lime adds little to the dissolved solids content, and most calcium compounds have low solubility. Regardless of the alkali used, environmental engineers must perform filtration testing of the precipitates for design.

Frequently, the submicron industrial sludge formed by the reaction with lime should not be preconcentrated be-

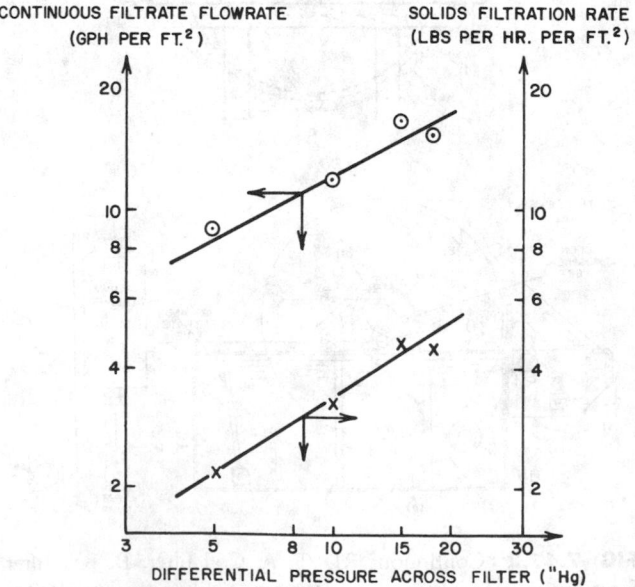

FIG. 7.47.5 Continuous RDVF of acid neutralization slurry ($CaSO_4 + Fe[OH]_3$ in H_2O).

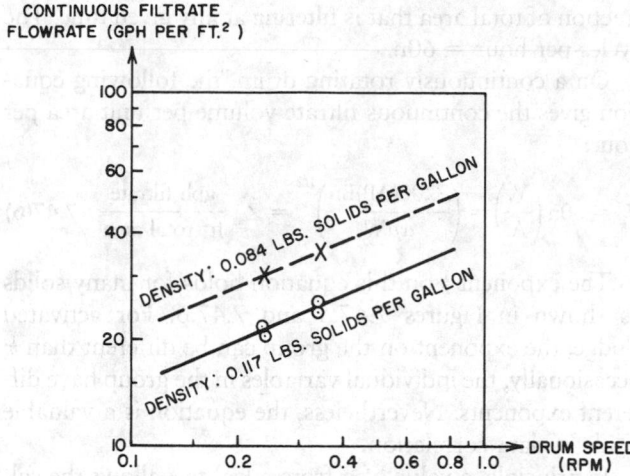

CONTINUOUS FILTRATE
FLOWRATE (GPH PER FT.²)

FIG. 7.47.6 Effect of drum speed and slurry density on filtrate flow rate.

cause filtration proceeds better on a dilute slurry (see Figure 7.47.6).

Combined Domestic and Industrial Wastewaters

In combined waste from many sources, the SS content is still too low (0.1%) to be filtered. The most common problem in solids removal from municipal plus industrial waste is an increased proportion of dissolved impurities. Table 7.47.2 shows typical figures (yearly averages) for two currently operating plants (A and B) handling high proportions of industrial waste. Little experience is available in filtration of this type of settled primary sludge.

Anaerobic digested sludge derived only from domestic waste is easily dewatered on the coil or belt filter (see Figure 7.47.2), but the same sludge derived from mixed domestic and industrial waste often has filtration problems. When the primary digesters receive excessive quantities of toxic chemicals, hair, fibers, bristles, grease, polymers, or gelatinous proteins, the digestion process can be retarded or interrupted, creating filtration problems. When digesters operate continuously on primary settled sludge containing a high percentage of industrial waste and the anaerobic bacteria cannot tolerate the industrial chemicals, only partial digestion is achieved. In this case, little gas is produced, and the sludge dewaters poorly.

Secondary aerobic treatment (activated sludge) is successful and versatile, producing easily filtered bacterial solids from all sorts of dissolved domestic and industrial chemicals (see Table 7.47.3). The current trend in domestic–industrial treatment plant design is to alleviate the problem of primary sludge by mixing it with activated sludge. The mixture is then concentrated by continuous filters or centrifuges and incinerated or disposed at a solids content of 20% weight or more. Figure 7.47.7 shows a flow diagram of this type.

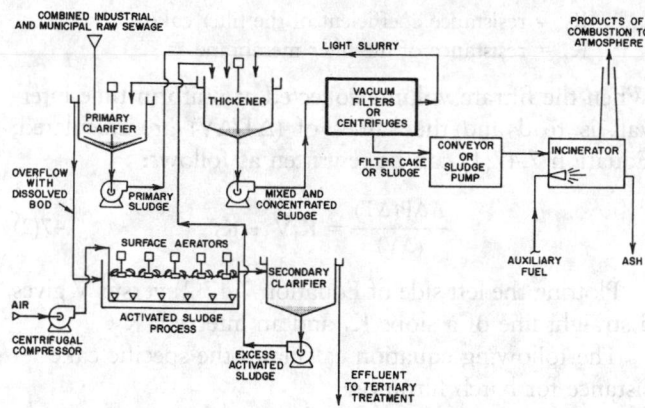

FIG. 7.47.7 Filtration and incineration of mixed waste sludge.

Filtration Tests and Sizing Calculations

While Buchner funnel tests often determine SS filtration characteristics, numerous improvements have been made in the method (see Figure 7.47.8). The filter cloth and its support grid should be the same as those on the full-scale or pilot-scale filter. To simulate an RDVF, this test arrangement holds the leaf downward in the slurry and gently moves it about during filtration. To produce a thicker cake, as on a filter press, wastewater treatment facilities can convert the leaf into a cylindrical funnel by adding a section of PVC pipe 4 in long.

These tests are batch types performed at a constant pressure. The following equation models these tests:

$$\frac{dV}{AdT} = \frac{\Delta P}{K_1 V + K_2} \qquad 7.47(1)$$

where:

V = filtrate volume
T = time (measured from the beginning of test)
A = filter leaf area
ΔP = pressure differential, maintained constant across the leaf

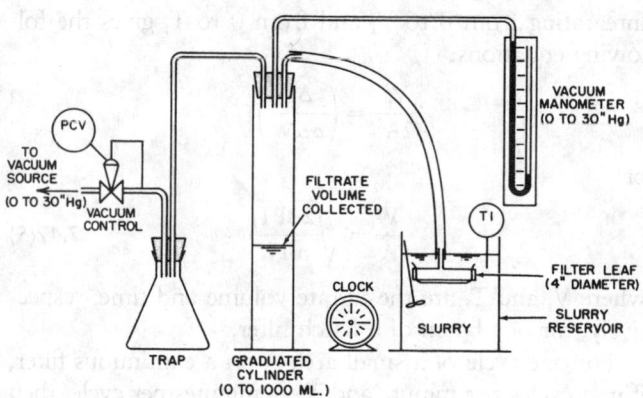

FIG. 7.47.8 Preferred leaf test arrangement.

K_1 = resistance coefficient of the filter cake
K_2 = resistance of the filter membrane

When the filtrate volume collected at uniform time intervals is read and the values of $(\Delta T/\Delta V)$ are calculated, Equation 7.47(1) can be rewritten as follows:

$$\frac{A\Delta P(\Delta T)}{(\Delta V)} = K_1 V + K_2 \qquad \text{7.47(2)}$$

Plotting the left side of Equation 7.47(2) versus V gives a straight line of a slope K_1 and an intercept K_2.

The following equation calculates the specific cake resistance for batch filtration:

$$\text{Let } K_1 = \alpha\mu W/A \qquad \text{7.47(3)}$$

where:

α = specific resistance of the cake (to be calculated)
μ = viscosity of the filtrate (from liquid properties)
W = mass of dry solids/volume of slurry (from drying procedure)

Equation 7.47(3) neglects the filtrate content of the filter cake and calculates the value of α for the cake from A, μ, W, and K_1.

CONTINUOUS ROTARY FILTERS

For continuous rotary filters, environmental engineers must calculate the volumetric filtrate rate, test with mini-filters, and determine the solids retention and cake thickness.

Volumetric Filtrate Rate

If $K_2 \cong 0$, the following equations apply:

$$\frac{dV}{AdT} = \frac{\Delta P}{K_1 V} = \frac{\Delta P}{\left(\dfrac{\alpha\mu W}{A}\right)V} \qquad \text{7.47(4)}$$

and

$$\frac{VdV}{A^2} = \left(\frac{\Delta P}{\alpha\mu W}\right)dT$$

Integrating from 0 to V_f and from 0 to T_f gives the following equations:

$$\frac{V_f^2}{2A^2} = \left(\frac{\Delta P}{\alpha\mu W}\right)T_f$$

or

$$\frac{V_f}{A} = \sqrt{\frac{2\Delta P T_f}{\alpha\mu W}} \qquad \text{7.47(5)}$$

where V_f and T_f are the filtrate volume and time, respectively, for one batch of a batch filter.

For one cycle of a small area A on a continuous filter, if n = cycles per minute and T_c = minutes per cycle, then n = $1/T_c$. Also, $T_f = BT_c$ when $0 < B < 1.00$. B is the fraction of total area that is filtering at any given time. The cycles per hour = 60n.

On a continuously rotating drum, the following equation gives the continuous filtrate volume per unit area per hour:

$$Z_c = 60n\left(\frac{V_f}{A}\right) = \left(\frac{7200(\Delta P)Bn}{\alpha\mu W}\right)^{1/2} = Z_c, \frac{\text{gph filtrate}}{\text{ft}^2 \text{ total area}} \qquad \text{7.47(6)}$$

The exponent $\frac{1}{2}$ on this equation holds for many solids as shown in Figures 7.47.5 and 7.47.6. For activated sludge, the exponent on the group can be different than $\frac{1}{2}$; occasionally, the individual variables in the group have different exponents. Nevertheless, the equation is a valuable guide in data correlation.

In principle, a value of α from a leaf test allows the calculation of total full-scale filter area A required to achieve a total filtrate rate V' in gph, i.e., $A = V'/Z_c$. In practice, leaf tests give only approximate values of α, and the accuracy is usually not enough for precise design. This inaccuracy is partially because the filter cake does not pack and compress the same way on a test leaf as on a continuous filter. The wall effects in the filter funnel are also important.

Mini-filters

For reliable filter design, the best guides are previous experience with the same slurry and tests with miniature, continuous rotary filters (mini-filters) of the belt or coil type. These filters are available on a rental basis.

Solids Retention and Cake Thickness

On a continuous rotary filter, if W = the solids content of feed slurry (lb dry solids per gallon filtrate), the following equation gives the continuous solids rate in the feed slurry:

$$Z_c W = \sqrt{\frac{7200B\Delta PnW}{\alpha\mu}}, \frac{\text{lb}}{\text{hr} \times \text{ft}^2} \qquad \text{7.47(7)}$$

If the solids content of the filtrate = W_f (lb per gallon), then $Z_c W_f$ represents the solids that pass through the cloth. If ρ_c is the cake density, the following equation calculates the cake thickness as discharged (L_c, ft):

$$L_c = \frac{\text{cake volume, ft}^3}{\text{hr} \times \text{drum area} \times \text{drum speed, revolutions per hr}}$$

$$= \frac{\text{cake mass rate per cake density}}{\text{hr} \times \text{drum area} \times \text{drum speed, revolutions per hr}}$$

$$= \frac{Z_c(W - W_f)}{\rho_c(60n)} = \frac{1}{60\rho_c n}\left(\sqrt{\frac{7200B(\Delta P)nW}{\alpha\mu}} - Z_c W_f\right) \qquad \text{7.47(8)}$$

A comparison of Equations 7.47(7) and (8) indicates that as the drum rotation speed increases, the solids rate also increases, but the filter cake becomes thinner. The passage of fine solids through the filter cloth is often unavoidable in continuous filtration and can require recycling the filtrate to the clarifier. Cake washing is of minor importance in waste treatment filters, but when it is required, batch filters are usually superior to continuous ones.

FILTER PRESSES AND OTHER BATCH PRESSURE FILTERS

With a set of operating data on a continuous rotary filter, environmental engineers can calculate the specific resistance α using Equation 7.47(6), in which ΔP is the differential pressure on a continuous filter.

Based on time in minutes, the following equation applies:

$$Z_c/60 = nV_f/A = \sqrt{\frac{2(\Delta P)Bn}{\alpha \mu W}}$$
$$= K_3 B^{1/2}(\Delta P_c)^{1/2} n^{1/2}, \text{ gpm per sq foot} \quad 7.47(9)$$

where:

$$K_3 = \sqrt{\frac{2}{\alpha \mu W}} = \sqrt{\frac{2}{K_1 A}} \quad 7.47(10)$$

Equation 7.47(7) gives the continuous dry solids rate (assuming complete retention of solids by the filter cloth).

For nearly incompressible filter cakes, the following equation calculates the approximate required area of a batch filter to replace a continuous filter for the same slurry at the same average rate. The filtrate liquid per unit area per batch is as follows:

$$\frac{V_{fb}}{A} = \sqrt{\frac{2\Delta P_b T_{fb}}{\alpha \mu W}} = K_3 \Delta P_b^{1/2}(T_{fb})^{1/2} \quad 7.47(11)$$

For constant-pressure, batch filtration, V_{fb} is related to $\sqrt{T_{fb}}$. If the following conditions exist,

T_{cb} = total batch cycle time, min
   = filter cake form time + discharge time
   = $T_{fb} + T_{db}$

then the average batch filtrate rate is as follows:

$$Z_b = \frac{V_f}{T_{cb}A} = \frac{K_3(\Delta P_b)^{1/2}(T_{fb})^{1/2}}{(T_{fb} + T_{db})}, \frac{\text{gpm}}{\text{ft}^2} \quad 7.47(12)$$

The following equation calculates the relative rates on batch and continuous filters for the same slurry at the same temperature:

$$\frac{Z_c}{Z_b} = \frac{\text{average rate, continuous filter}}{\text{average rate, batch filter}}$$

$$= \frac{\Delta P_c^{1/2}(Bn)_c^{1/2}}{\left[\dfrac{\Delta P_b^{1/2}(T_{fb})^{1/2}}{T_{fb} + T_{db}}\right]} \quad 7.47(13)$$

For quick-cleaning batch filters $T_{db} \ll T_{fb}$, then $T_{cb} \simeq T_{fb}$,

and Equation 7.47(13) simplifies to the following equations:

$$\frac{Z_c}{Z_b} = \left(\frac{\Delta P_c}{\Delta P_b}\right)^{1/2} (Bn)_c^{1/2}(T_c)_b^{1/2}$$

$$= \left(\frac{\Delta P_c}{\Delta P_b}\right)^{1/2} \left(\frac{B}{T_c}\right)_c^{1/2} (T_{cb})^{1/2}$$

or

$$\frac{Z_c}{Z_b} = \left(\frac{\Delta P_c}{\Delta P_b}\right)^{1/2} \left[\frac{(T_c)_b}{(T_c)_c}\right]^{1/2} B^{1/2} \quad 7.47(14)$$

FILTER SIZING EXAMPLE

In this example, the following operating data are for a continuous rotary vacuum filter that separates a solid from suspension at 22 in Hg vacuum (10.8 psi) with speed = $\frac{1}{2}$ rpm, drum area = 100 ft², B = 0.33 (B = T_f/T_c), and continuous cycle time = $(T_c)_c$ = 2 min = $\frac{1}{30}$ hr.

Environmental engineers can calculate the size of filter press or other batch pressure filter required to filter the same slurry at the same average rate when the filter cake is assumed to be incompressible. This example assumes that a multistage centrifugal pump supplies the filter press with slurry at an average differential pressure of 150 psid.

To fit the working schedule, this example uses cycle times of 2 hr, 4 hr, and 8 hr for the filter press. Due to automatic cleaning equipment (see Figures 7.47.9 and 7.47.10), the cleaning time is negligible compared to the cycle time. The equipment and piping resistances on the filter press are also negligible.

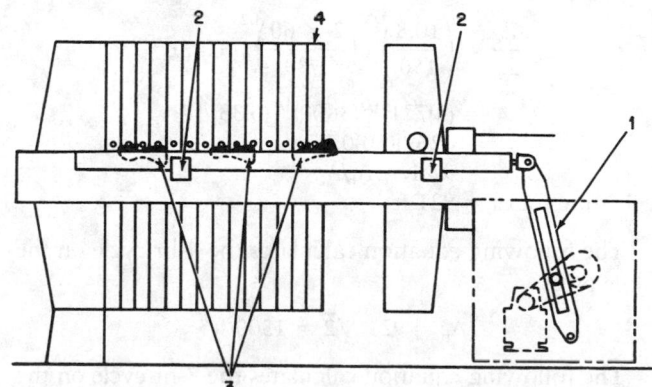

FIG. 7.47.9 Filter press with automatic cleaning. Sizes are available up to 3500 ft² per unit. At the end of the cycle, the opening and closing gear moves the end of the press to the right and starts the plate moving cycle. The reciprocating mechanism (1) drives the slide bars mounted on bearing blocks (2) to the right. One pair of pawls (3) mounted in the slide bars picks up only the first plate (4) and moves it to the right. The filter cake falls out. The mechanism reverses and moves the next plate. No separate plates or frames exist. Each plate is recessed on each side to provide cake space. During filtration, the slurry enters and filtrate exits through an internal passage (not shown).

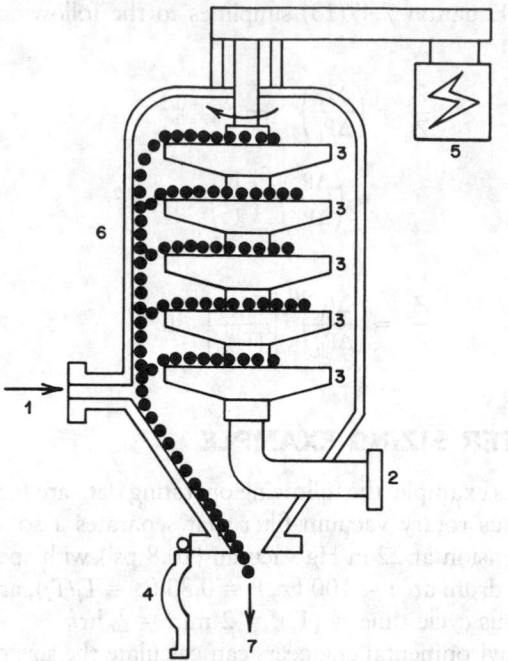

FIG. 7.47.10 Rotating-leaf filter with automatic cleaning. During filtration, the slurry enters at (1), the filtrate exists at (2), and solids are retained on leaves (3) and covered with a filter cloth. Upon completion of filtration, the washing and drying bottom closure (4) opens. The drive motor (5) starts and rotates the stack of filter leaves. Centrifugal force causes the solids to move off the filter leaves, strike the inside wall of the tank (6), and flow down to the solids exit (7). Sizes are available up to 540 ft² per unit.

The following equation calculates the 2-hr cycle on the filter press:

$$\frac{Z_c}{Z_b} = \left(\frac{10.8}{150}\right)^{1/2} \left(\frac{2 \times 60}{1/30}\right)^{1/2} (0.33)^{1/2}$$

$$= (.0721)^{1/2}(3600)^{1/2}(0.33)^{1/2}$$
$$= (.269)(60)0.57$$
$$= 9.21 = A_b/A_c$$

Filter press area = 921 ft²

The following equation calculates the 4-hr cycle on the filter press:

$$A_b = 921 \sqrt{2} = 1300 \text{ ft}^2$$

The following equation calculates the 8-hr cycle on the filter press:

$$A_b = 921 \sqrt{4} = 1842 \text{ ft}^2$$

This procedure is only approximate if the pressure varies during the batch, and graphic integration using the pump characteristic curve provides better results.

The results are not valid for batch pressure filters if the cake is appreciably compressible. For either batch or continuous filtration of compressible solids, environmental engineers should conduct tests at various average pressures.

For batch filters, in addition to adequate filtration area the design must provide sufficient cake space to contain the total volume of the wet filter cake deposited during the proposed cycle time; otherwise the cycle can come to a premature end. This specification requires knowledge of the wet cake density and solids content, which should be measured in the laboratory and not assumed.

If T_{db} is not negligible compared to T_{fb}, environmental engineers can include its value (the batch cake discharge time) in the batch equations, and the calculation is more precise. However, high precision should not be expected in filtration calculations. As in the rotary filter case, environmental engineers obtain the best results for a batch filter in laboratory measurements using a mini-filter of the same general design as the full-scale filter.

This example shows that a batch filter requires a larger area than a continuous filter for the same slurry and average filtrate rate. The batch filter, however, has only two electrical drives—the feed pump and the discharge mechanism. The continuous rotary drum filter usually has six drives, including the feed pump, pan mixer, drum drive, belt discharge drive, vacuum pump, and filtrate pump.

Batch versus Continuous Filters

In municipal waste treatment plants, batch filters have not been used. The large scale of the operation, the compressibility of the sludge, and the lack of automatic cleaning devices are probably the reasons for this general lack of use.

For pretreatment units in an individual plant where the waste volume is small and the sludge is incompressible, plant management should compare the use of batch filters with automatic cleaning (see Figures 7.47.9 and 7.47.10) with continuous rotary filters.

CAKE WASHING

Cake washing, which is unimportant in large municipal–industrial treatment plants, can be beneficial in industrial pretreatment when a valuable material dissolved in the filtrate is separated from waste solids. Better washing can be obtained on batch filters than on continuous filters.

A thorough theoretical and experimental study of cake washing concludes that removing 80% of a solute dissolved in the filtrate left in the void spaces of the filter cake requires at least six void volumes of wash liquid and sometimes more. This statement explains why batch filters are superior to continuous rotary filters when extensive washing is required.

BATCH TESTS IN PRESSURE FILTERS WITH VARIABLE PRESSURE

The previous calculation of K_1 and K_2 requires a constant controlled pressure. An actual pressure filter, pilot-scale or full-scale, is usually supplied with slurry by a centrifugal pump at a varying pressure. If the pump suction and filter discharge are both at atmospheric pressure and the pipe and valve losses are negligible, the pump discharge pressure always equals the batch filter differential pressure (ΔP). The filtrate volume (V), pressure (ΔP_b), and time (T) data can then be plotted on companion plots (see Figure 7.47.11).

The true mean ΔP_b and true mean flow rate can be obtained by graphic integration. The required filtration area for any total mean flow rate can then be calculated.

DEEP-BED FILTER DESIGNS

Filters with deep beds of sand, diatomaceous earth, coke, charcoal, and other inexpensive packing materials have been successfully used in the filtration of potable water and can also be used in the small-scale treatment of dilute wastewater (see Figure 7.47.3). Their best applications are in polishing the filtrate from a continuous filter or the overflow from a primary or secondary settling tank. Without preseparation, the bed becomes loaded quickly. When the particles and bacteria in sizes smaller than the interstices of the bed, plus suspended BOD, are removed from the

liquid, exceptional clarity is obtained. The dissolved substances, including dissolved BOD, are not removed.

Environmental engineers can predict the variation of pressure loss and flow rate with time from small-scale tests if they determine K_1 and K_2 experimentally. The sand bed and waste material tested must be the same as that used in the process. Predicting the initial pressure-loss for various flow rates is possible.

The Ergun equation for the initial pressure loss in a deep-bed filter is particularly useful for granular materials because it covers both laminar and turbulent regimes as follows:

$$\left(\frac{\Delta P}{L}\right)g_c = \frac{150\mu u_0}{D_p^2} \cdot \frac{1-\varepsilon^2}{\varepsilon^3} + 1.75\frac{\rho u_0^2}{D_p} \cdot \frac{1-\varepsilon}{\varepsilon^3} \qquad 7.47(15)$$

where:

ΔP = pressure loss (frictional only) in direction of flow, lb/ft^2
L = length of granular bed in direction of flow, ft
μ = fluid viscosity, lb/sec, ft
D_p = mean particle diameter, ft
ε = void fraction of bed, dimensionless
ρ = fluid density, lb/ft^3
u_0 = fluid velocity based on empty cross section, ft/sec
g_c = gravitational acceleration

The general shapes of curves constructed from Equation 7.47(15) agree with experimental values; however, the coefficients vary appreciably for different fluid–solid systems. Therefore, environmental engineers should supplement the equation with tests when possible. Impurities deposited in the bed during batch filtration must be removed periodically, whenever the pressure loss rises and the flow rate decreases to the limiting acceptable values. Backwashing the bed to remove impurities creates a by-product of contaminated liquid, which must be disposed.

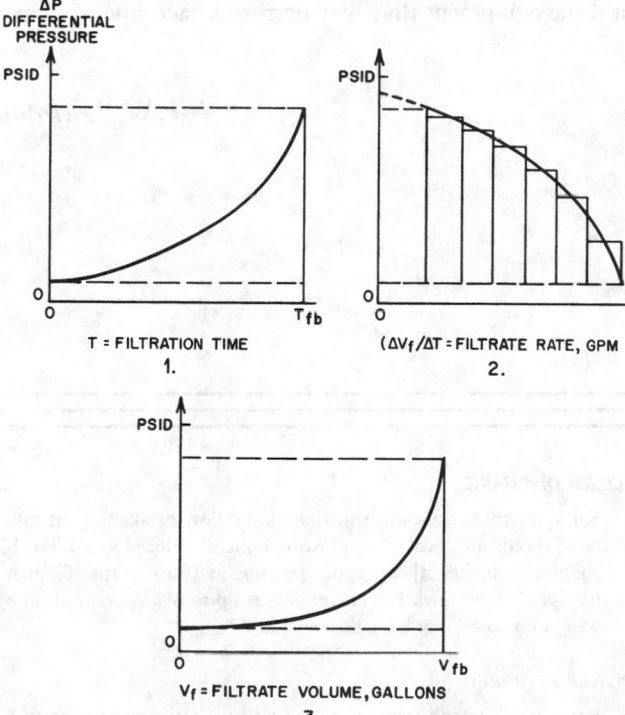

FIG. 7.47.11 Batch pressure filtration with variable pressure. Graphs 1 and 3 are generalized graphs of the original data taken during a run; graph 2 can be calculated from 1 and 3 or obtained from a flowmeter. The curved line on 2 is on or close to the characteristic curve of the pump. At any time, the following equation calculates the specific resistance α:

$$\alpha = \frac{(area)^2(psid)}{\mu(W)(V)(\Delta V/\Delta T)}$$

The variability or constancy of α indicates the compressibility or lack of it in the filter cake.

PRECOAT FILTERS

Environmental engineers must use variations of these deep-bed design methods wherever batch leaf precoat filters and rotary vacuum precoat filters are used. These precoats usually consist of diatomaceous earth or asbestos fibers preapplied to the filter medium. Since these materials do not have measurable particle diameters, D_p is unknown and the Ergun equation is not applicable. Therefore, test data are required even for initial pressure loss.

On a batch leaf precoat filter, the differential pressure increases continuously to the maximum allowable value. The precoat and impurities are then washed off, and a new precoat is applied. These filters can be operated without a vacuum pump when the filtrate is continuously removed with a centrifugal pump capable of operating at a low NPSH.

On a continuous rotary precoat filter (similar to the one in Figure 7.47.4), a slowly advancing knife continuously shaves off a few thousandths of an inch of the precoat plus the accumulated impurities as the drum rotates. As a result, the pressure loss is kept fairly constant throughout a run, and the run can be extended to one week when a 2- to 4-in precoat of processed diatomaceous earth is used. Because replacing the precoat only requires a few hours, the operation is virtually continuous. These filters are successful in removing SS and oil from slightly polluted water (0 to 200 ppm impurities). They are not used on sewage.

OUTDOOR SAND-BED FILTERS

Large, inexpensive, deep-bed filters for sewage treatment, constructed at ground level from naturally occurring sand, can be used on raw sewage, but presettling of solids is advisable. Isolated locations are recommended for these filters because some odors are inevitable and large land areas are required.

The filtered sludge leaves a mat of solids on the sand bed, which must be removed periodically. Operation and maintenance are simple, and operating costs are low. These units are suitable for small communities and isolated institutions, if approval from the authorities can be obtained. Table 7.47.4 shows data on the deep-bed sand filter.

TANK-TYPE, DEEP-BED FILTERS

Deep-bed filters within closed tanks are commercially available. They can be filled with sand or any other inert granular material appropriate for the filtration process.

TABLE 7.47.4 MAXIMUM ALLOWABLE LOADING OF INTERMITTENT SAND FILTER

Type of Sewage	Gallons per day per acre
Raw	20,000–80,000
Settled (overflow)	50,000–125,000
Biologically treated	Up to 500,000

Backwashing is carried out in batches and continuously. The operator selects a backwashing velocity that fluidizes the granular bed without transporting it out of the equipment while the backwash removes the waste solids. The bed packing particles must have a higher settling velocity and usually a higher density than the particles or flocs of solid waste being removed.

Because of the variability of finely divided SS, buying filters cannot be done in the same way as pumps, motors, and instruments, i.e., on the basis of design specifications only. The buyer should participate in filtration tests long enough to obtain a statistical probability that the filter will work. Table 7.47.1 summarizes the types of filtration tasks and the equipment that best performs each one.

—*F.W. Dittman*

7.48
DEWATERING: CENTRIFUGATION

CENTRIFUGE DESIGN TYPE
  a) Imperforate bowl knife
  b) Solid bowl conveyor
  c) Disc centrifuge with nozzle discharge

CENTRIFUGE SELECTION AND APPLICATION
  See Table 7.48.1

HYDRAULIC CAPACITY
  For a, up to 60 gpm for feed solids ≤1% decreasing to 40 gpm at 3 to 5% solids; for b, to 400 gpm on lime sludges, as low as 75 gpm on some sewage sludges for the largest unit; for c, from 20 to 300 gpm normal and 400 gpm maximum

SOLIDS HANDLING CAPACITY
  For a, to 12 tn/hr wet cake on paper mill waste; seldom limited by solids loading in waste treatment applications

CAKE HOLDING CAPACITY
  For a, practical maximum about 14 ft³ for centrifugal acceleration > 1200 × gravity

SOLIDS DISCHARGE
  For a, intermittent; solids showing plastic flow by skimmer at full-bowl speed; stiffer cakes by plowing knife at reduced speed. For b, continuous by helical conveyor operating at 10 to 30 rpm differential speed from bowl. For c, continuous up to 6% concentration by weight on secondary biological sludge.

POWER REQUIRED
  For a, windage and friction plus about ⅛ hp per gpm, totaling to 45 hp at maximum rates. For b, to 250 hp, but varies directly with liquid and solids loading. For c, from 3 gpm per hp at high rates to 1½ gpm per hp at low rates.

MATERIALS OF CONSTRUCTION
  For a, normally stainless steel on waste treatment; carbon steel, special alloys, and various coatings available. For b, commonly stainless steel in waste treatment; carbon steel available; monel or titanium for special corrosion problems. For c, bowls always stainless steel, usually type 316; covers usually stainless steel, but carbon or cast steel sometimes used.

TABLE 7.48.1 ORIENTATION TABLE FOR CENTRIFUGATION

Type of Sludge or Waste°°	Centrifuge Characteristics									
	Suitability of Centrifuge by Type°				Capacity Range			Discharged Cake Concentration		
	Good	Fair	Poor	No	High	Medium	Low	High	Medium	Low
Sewage°°, primary raw	B		A	C	B		A	B	A	
Sewage, primary anaerobic digested	B	A		C	B		A		B	A
Sewage, primary aerobic digested	A	B		C		A,B			A,B	
Sewage, primary raw limed	B		A	C	B		A	A,B		
Sewage, secondary biological (activated and humus)	A,C		B		C	A	B	A	B, C	
Sewage, secondary biological (activated) with alum	A, C	B			C	A,B		A	B, C	
Sewage, primary + secondary biological, cosettled	A	B		C		A, B			B	A
Sewage, primary + secondary biological, anaerobic digested	B	A		C		B	A	B		A
Sewage, whole, modified biological (activated)	A	B	C		C	A	B		A, B	C
Sewage, whole, aerobic digested	A	B	C		C		A, B		A, B	C
Sewage, primary, heat-treated raw	A, B			C	B	A		B	A	
Sewage, secondary biological, heat-treated raw	A	B		C		A, B			A, B	
Sewage, primary + secondary biological, heat-treated raw	A, B			C	B	A		B	A	
Sewage, tertiary with lime (for phosphate)	B		A	C		B	A	A, B		
Sewage, teritary with alum (for phosphate)	A	B, C			C	A, B		A	B	C
Industrial, coarse solids	B	A		C	B		A	B	A	
Industrial, clean biological	A, C		B		A,C	B			A, B	C
Industrial, hydrous or flocculant solids	A	B, C			A, C	B			A, B	C
Industrial, oil–water emulsion	C	A		B		C	A			
Industrial, fine solids	A, C	B			A	C	B		A, B	C
Water treatment, lime softening	B		A	C	B		A	A, B		
Water treatment, alum	A, C	B			C	A, B			A, B	C
Classification of lime and hydrous solids	A, B			C	A, B			B	A	
Classification of digested sewage (aerobic or anaerobic)	A, B			C	B	A		B	A	

Notes: °Type of Centrifuge: A = Imperforate bowl with knife and skimmer discharge; B = Solid bowl with conveyor discharge; C = Disk bowl with nozzle discharge.

°°In each of these waste streams, the feed to the centrifuge is the sludge produced by the operation. For example, the first waste listed in the table is a raw sludge taken from the underflow of a primary clarifier.

DESIGN PRESSURE

Normally atmospheric for a, b, and c. Design b is also available at 150 psig rating in vertical construction.

DESIGN TEMPERATURE

Up to 200° to 225°F without special design for a, b, and c.

The three types of centrifuges applied in waste treatment processes rely on settling to separate solids and differ only in discharging the settled solids. Environmental engineers base their centrifuge selection on the particle size, concentration, feed rate, performance required, disposal methods, and costs. Table 7.48.1 lists three criteria (1) suitability based on the anticipated type of solids and minimal coagulant requirements, (2) capacity range based on the largest commercial units at the same recovery of SS, and (3) discharged cake concentration relative to other dewatering equipment at the same recovery level.

The conveyor centrifuge is used for coarse and heavy loadings of solids and normally requires a coagulant for reasonable recoveries. The disc centrifuge with nozzle discharge is restricted to sludges containing no coarse particles and usually having low concentrations and large volumes. The imperforate bowl with knife and skimmer can handle fairly low feed rates at high recoveries, usually without coagulant (see Figure 7.48.1).

Imperforate Bowl (Basket) Knife Centrifuge

This centrifuge, the simplest type in general use, is suited for soft or fine solids that are difficult to filter and waste that varies in concentration and solids characteristics.

That design can be either a top-driven suspended bowl or an underdriven bowl with three-point casing suspension; the drive can be a hydraulic or an electric motor. A 12- or 14-in diameter bowl is often used for test work, whereas 30 to 48 in is the range of commercial units. Older designs operate below 1000 G (G is the ratio of centrifugal acceleration to the acceleration of gravity), whereas current units use 1300 G. Cake concentration and clarification both improve with increased acceleration.

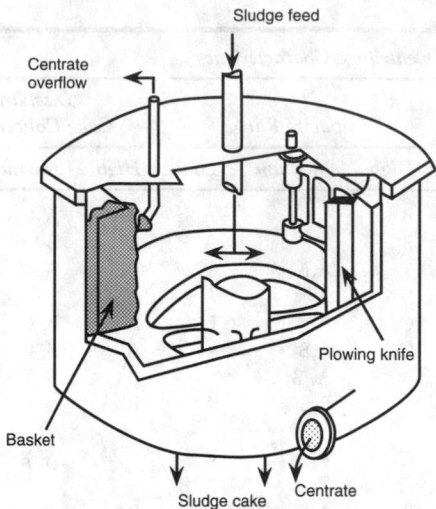

FIG. 7.48.1 Basket centrifuge. (Reprinted from U.S. Environmental Protection Agency [EPA], 1987, *Design manual: Dewatering municipal wastewater sludges,* EPA/625/1-87/014 [September].)

A fully automated knife centrifuge for waste treatment (see Figure 7.48.2) includes a rotating bowl, covers to collect clarified effluent, a feed distributor, an overflow lip to maintain an annular layer of liquid in the bowl, a cake-level detector, a skimming tube with drive, a plowing knife with axial or rotary drive, and an open-bottom bowl through which the plowed cake can drop. Feed entering the top and introduced near the bottom of the bowl flows axially, with the clarified effluent discharged at the top. The SS settle on the bowl wall and eventually impede clarification. This centrifuge then interrupts the feed for $1\frac{1}{2}$ to $2\frac{1}{2}$ min to discharge the cake. The total centrifugation cycle usually takes 6 to 30 min.

Instrumentation automatically initiates the discharge cycle. During the cycle, the skimmer moves outward to remove supernatant liquid and may continue to remove soft solids. Coarser or fibrous solids are then plowed out with the knife while the bowl is held at a reduced speed. A concentration gradient exists across the cake from the softest material at the inside to the heaviest and coarsest at the bowl wall. The cake can be removed as a whole, or it can be classified into concentration fractions for separate handling when the actions of skimmer and knife are preset.

APPLICATIONS

Since clarification occurs in a quiescent zone undisturbed by moving elements such as a conveyor, settling is effective, and recovery is good. Because the bowl speed is low, shearing of flocculant material is minimal. These centrifuges can efficiently capture (90%) of even difficult biological or alum sludge solids without coagulants.

Applications for these centrifuges include metal hydroxide waste, aerobic sewage sludge, and water treatment alum sludge. Figure 7.48.3 gives the typical performance

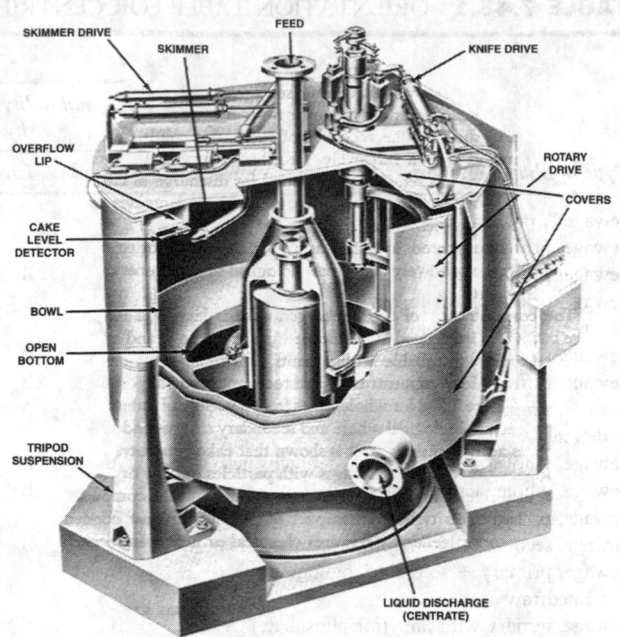

FIG. 7.48.2 Imperforate bowl knife centrifuge with skimming tube.

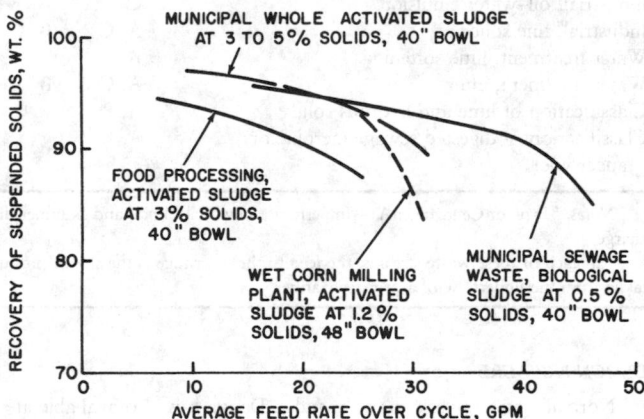

FIG. 7.48.3 Clarification of biological sludges in imperforate bowl knife centrifuge.

curves for four biological sludges showing the recovery of SS as a function of the feed rate averaged over a full cycle. Recovery refers to the proportion of the entering solids that are retained in the bowl for discharge as cake.

As with all centrifuge designs, increasing the feed rate decreases the residence time and recovery. These centrifuges produce recoveries of 90% or better over a range of feed rates. Industrial activated sludge is often more difficult to clarify than the consistent biological sludge generated in municipal sewage plants.

Concentration of discharged cake is a function of the sludge but is also influenced by the residence time in the bowl and the G level. Residence time, controllable within limits, is a function of centrifuge size, throughput rate, feed concentration, and recovery. Figure 7.48.4 gives typical performance data for whole cakes from several industrial activated sludges, municipal biological whole and sec-

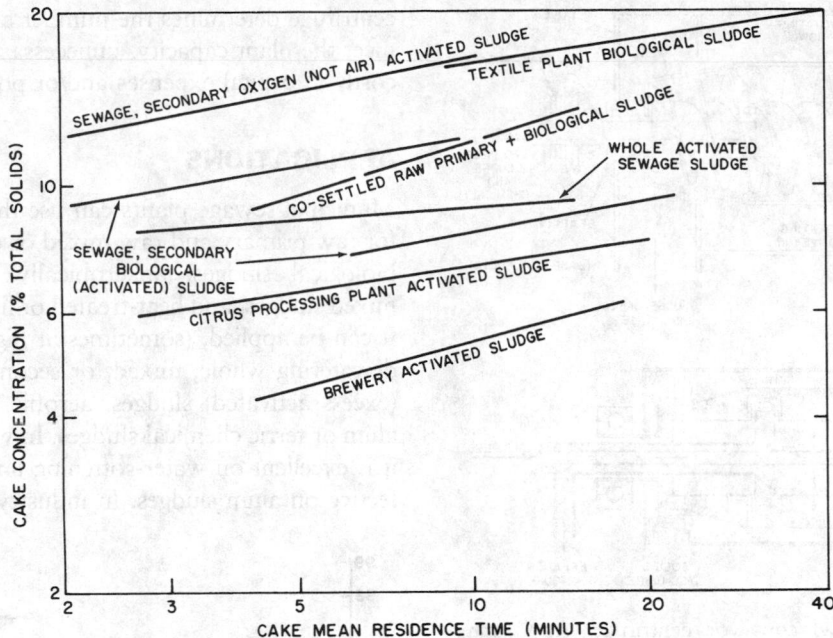

FIG. 7.48.4 Whole cake concentration from imperforate bowl knife centrifuge applied to various sludges.

ondary sludges, and raw co-settled primary plus activated sludges.

The figure shows that cake concentration increases with residence time. For sludges with particles too fine or too gelatinous for liquid drainage, concentration depends on the compaction characteristics; hydrogels typically compact to 5 to 15% concentration. Further dewatering generally requires chemical or heat treatment.

LIMITATIONS AND MAINTENANCE

Although imperforate bowl knife centrifuges have strong capabilities on difficult sludges, the hydraulic parameters or the frequency of cake discharge limits these units to low concentration feeds at about 50 gpm maximum.

These centrifuges can handle wastes that are incapable of gravity-thickening to more than a few percent and that are subject to natural flocculation. These centrifuges adapt readily to changes in the feed rate, concentration, and solids characteristics. The complementary action of the skimmer and knife during cake discharge classifies the cake.

Because the operation is at low speeds compared to other centrifuges, the maintenance is low, with the bearing life in the range of 1 to 3 years. Lubrication is simple, power consumption is low, and little replacement of parts is needed.

Solid Bowl Conveyor Centrifuge

A conveyor centrifuge is used when the SS are coarse, high concentration cakes are required, solids loading is high, or classified solids recovery is required. The unit is flexible, easy to operate over a range of feed rates and concentrations, and requires little attention. The combination cylindrical–conical bowl ranges from a 6-in diameter for testing to 14 to 36 in for commercial units. Long bowls at more than a 4-to-1 length-to-diameter ratio have higher capacities and are commonly applied on waste sludge at 1000 to 2700 G acceleration.

The rotating assembly of a conveyor centrifuge consists of a bowl shell and a conveyor supported between two sets of bearings. This bowl and conveyor are linked through a planetary gear system designed to rotate the bowl and conveyor at slightly different speeds. The bowls are belt-driven. The rotating assembly is covered by a stationary casing for safety, odor, and noise control.

In the usual countercurrent design (see Part A in Figure 7.48.5), the feed enters through a pipe in the hollow shaft of the conveyor, accelerates through ports, and enters the annular pond of liquid at the wall of the bowl close to the conical section. Liquid flows to the far end of the bowl and discharges as clarified effluent (centrate) over adjustable weirs that regulate the depth of the pond. The conveyor moves the settled solids to the conical section where they drain on the unsubmerged beach before discharge.

In the cocurrent design (see Part B in Figure 7.48.5), the feed is introduced at the end of the bowl. The liquid and settled solids move together toward the conical beach. An adjustable skimmer removes the supernatant liquid, while the solids drain on the beach before discharge.

The abrasive nature of waste sludges ranges from extremely abrasive pulp mill waste to clean biological sludge. Paper mill and primary sewage sludges have necessitated

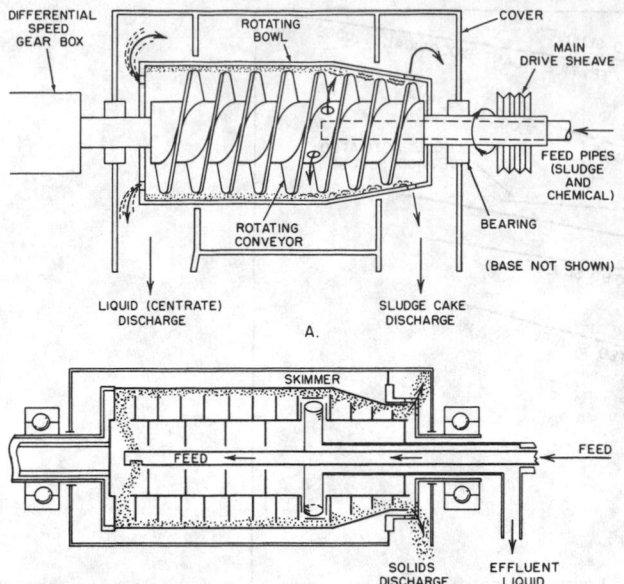

FIG. 7.48.5 Solid bowl conveyor centrifuge. **A,** Counter-current; **B,** Cocurrent.

the use of abrasion-resistant, hard-surfacing materials for direct application or as replaceable inserts.

OPERATING VARIABLES

Design variables affect clarification and cake dryness. The design and application determine the bowl diameter, length, beach angle, and conveyor configuration. The bowl capacity is proportional to the length at a given diameter; the angle of the conical beach, normally 5 to 10°, can affect the discharge of solids.

Wastewater treatment facilities can adjust the bowl speed, relative conveyor speed, and pond depth to change the performance after installation. Relative conveyor speed is basically set by gearbox design but is also controlled by an external drive or braking assembly attached to the pinion shaft of the box. The lowest differential speed compatible with adequate solids removal is the optimum choice because it decreases turbulence, increases drainage time, and decreases the linear velocity between the conveyor, bowl, and solids and, consequently, the associated abrasion. The pond depth adjustment is a compromise between better clarification in a deeper bond and a dryer cake from a shallower pond.

The process variables are primarily the feed rate and the addition of chemicals. Increasing the feed rate characteristically reduces recovery (see Figure 7.48.6) and increases the cake dryness due to the selective loss of finer material.

Adding polymeric or other coagulants in 0.1 to 0.2% solution internally to the pond or before the centrifuge improves recovery but at a cost (see Figure 7.48.7). Cationic polymers are usually best for sewage unless alum or lime addition requires a shift to anionic. Since the feed rate per

centrifuge determines the number and size of the units to meet the plant capacity, unnecessarily high recoveries are costly in capital expenses and/or polymer usage.

APPLICATIONS

Municipal sewage plants can use the conveyor centrifuge for raw primary and raw mixed or cosettled primary plus biological sludges, anaerobically digested primary or mixed sludges, and heat-treated or limed chemical sludges. It can be applied, (sometimes at high coagulant costs) to dewatering whole, mixed, or secondary waste biological (excess activated) sludges, aerobic digested sludges, and alum or ferric chemical sludges. In water treatment plants, it is excellent on water-softening lime sludges but less effective on alum sludges. In industry, applications include

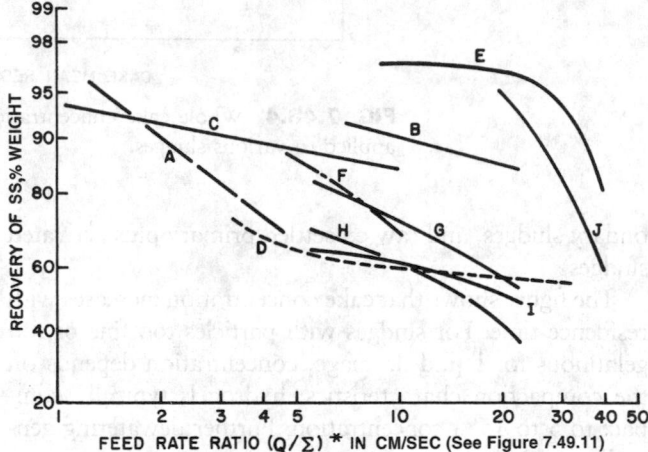

FIG. 7.48.6 Recovery–capacity performance curves for conveyor centrifuge. *Sludge Description Legend: A,* converter paper mill, coated paper; *B,* kraft mill, boxboard and coated paper; *C,* cold rolling mill, clarifier sludge; *D,* sewage, raw limed primary clarifier, classified without polymer; *E,* sewage, raw limed primary clarifier, treated with 1.5 lbs. anionic polymers added per ton dry feed solids; *F,* raw tannery waste; *G,* sewage, raw primary clarifier; *H,* sewage, raw primary plus secondary biological sludge; *I,* sewage, anaerobic digested primary plus secondary biological; *J,* basic oxygen furnace scrubber liquor.

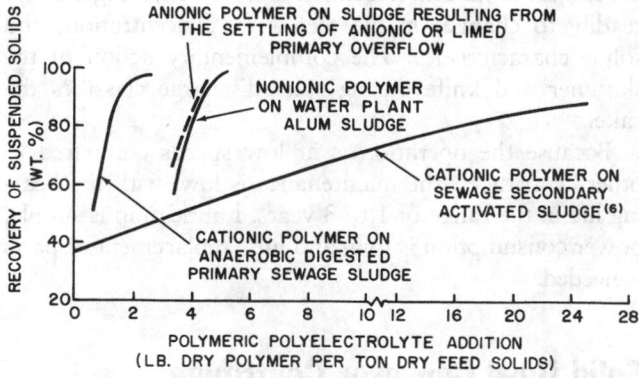

FIG. 7.48.7 Recovery response to polyelectrolyte addition in a conveyor centrifuge.

clarifying scrubber liquors, recovering solids from packing-house and metals-treating waste, roughing out solids upstream of disc centrifuges on refinery sludges, and dewatering potash and mining tailings. Controlled settling allows the recovery of calcium carbonate from limed sludge.

Figure 7.48.6 shows the expected solids recovery against feed rate ratios for a number of wastes. On municipal sewage sludges, comparing the G and H curves shows the effect of adding secondary biological sludge to primary sludge. Anaerobic digestion of the latter does not change the clarification characteristics markedly (I). The cake concentration depends largely on the sludge and can vary (see Table 7.48.4).

AUXILIARY EQUIPMENT

Degritting, even down to 150 mesh, prior to centrifugation is strongly recommended. Treatment facilities obtain more consistent results with less feed-zone plugging by comminuting the feed solids. The chemical coagulant facilities range from a batch-mixing tank and variable-speed pump to a fully automated system for dosages up to 12 lb of dry polymer per ton of dry feed solids.

MAINTENANCE

The gearboxes and bearings have self-contained or circulating lubrication. Fault detection systems can locate problems causing automatic shutdown, e.g., torque overload, hot bearings, inadequate oil circulation, and vibration. The maintenance frequency is a function of the abrasiveness of the solids; an initial inspection is suggested after 2000 hr of operation with subsequent inspections as required.

Disc Centrifuge with Nozzle Discharge

This centrifuge is specifically suited to the thickening of excess activated sludge or alum sludge that are free of coarse solids. It can handle high feed rates with effective clarification for very fine particles at low concentrations. Its high Gs make it effective for liquid–liquid and liquid–liquid–solid emulsion separations.

A disc centrifuge bowl can be suspended from or bottom-supported on a flexible spindle that permits the bowl to seek its natural axis of rotation under small unbalanced loads. Commercial units, generally belt-driven at speeds of 3000 to 9000 rpm, develop Gs of 2500 to 8000 with bowl diameters from 9 in for test units to 32 in for industrial units.

Because corrosion is seldom tolerated in the highly stressed bowls, stainless steel is the standard material of construction. The bowl is suspended inside a set of covers (see Figure 7.48.8) containing two decks that separately receive the clarified effluent and thickened sludge. For an emulsion, a third deck collects the second liquid phase.

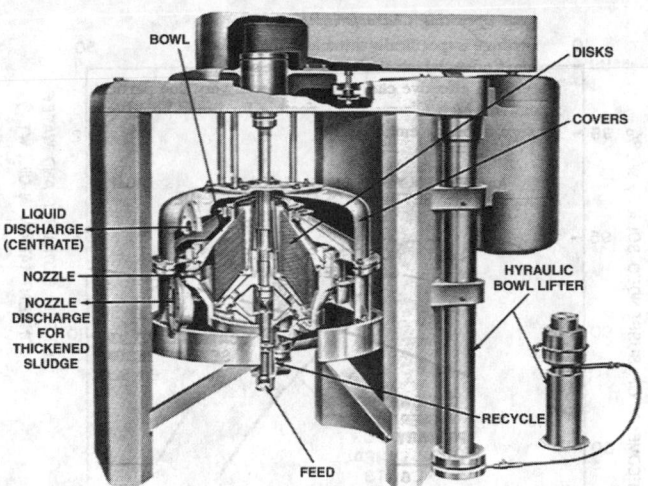

FIG. 7.48.8 Cutaway diagram of a disk bowl centrifuge with nozzle discharge and recycling.

The distinguishing feature of the bowl is a stack of cones or discs arranged so that the feed stream is divided among them, reducing the particle settling distance. The angle of the cone causes aggregated solids to slide and settle on the inner walls of the bowl that slope toward a peripheral zone containing orifices, called nozzles. The number of nozzles varies with the bowl size, but the size of the nozzle orifice is usually 0.030 to 0.080 in.

For efficient performance, the solids should not be fibrous or pack into a hard cake, and the incoming feed must be screened. The solids concentration in the nozzle discharge can reach ten to twenty times that in the feed. Recycling some of the nozzle discharge stream back into the bowl can control underflow concentrations (thickened sludge).

APPLICATIONS

The clarification of biological sludge (excess activated) from feed concentrations of 0.3 to 1.0% SS is the most important application of these centrifuges. Figure 7.48.9 shows the capacity range of a large disc centrifuge used at six municipal plants for excess activated-sludge thickening to $5\frac{1}{2}$% solids concentration without the use of coagulants.

Contrary to some reports for midrange feed rates of disc centrifuges, varying the SVI from 40 to 137 or altering the centrifuge underflow concentration below $6\frac{1}{2}$% has little effect on the solids recovery (see Figure 7.48.10). Figure 7.48.9 also shows the direct effect of the feed rate on the quality of clarification of oil separated from an aged petroleum refinery waste emulsion.

LIMITATIONS AND MAINTENANCE

The disc centrifuge should only be used for continuous extended running because the bowl must be cleared of solids

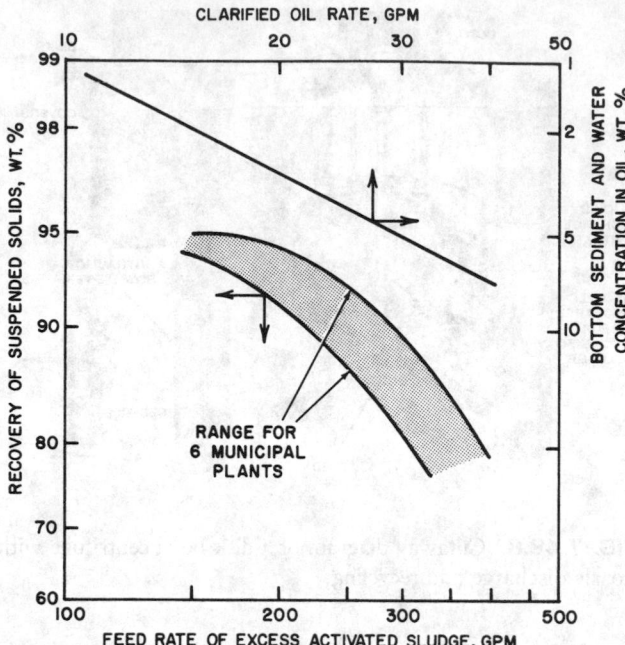

FIG. 7.48.9 Recovery–capacity performance curves for large disk centrifuges.

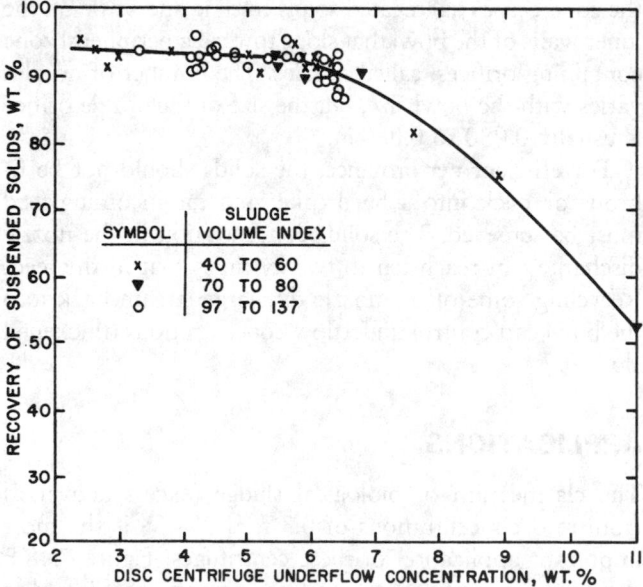

FIG. 7.48.10 Effect of the SVI and underflow concentration on recovery for mid-capacity flows of municipal excess activated sludge in disk centrifuges.

before the centrifuge is restarted after a shutdown. Maintenance is nominal since bearings in smaller units are grease-packed, whereas larger units use circulating oil or spray-mist systems. On abrasive applications, some areas require hard-surfacing; nozzle bushings, on the other hand, are replaceable.

AUXILIARY EQUIPMENT

Adequate feed screening is vital to keep these units online as long as 8 weeks without cleaning. Inline self-cleaning

strainers are generally satisfactory. Due to changes in the feed, control is recommended for consistency of thickened sludge. This control can be manual or fully automated, which uses a viscosity sensing loop to detect the sludge discharge consistency and modulate the recycling rate.

PERFORMANCE CORRELATION

Theoretically, the Σ concept allows performance comparison between similar centrifuges operating on the same feed material. Since Σ reflects a fixed maximum bowl capacity, for any flow rate (Q), the Q/Σ ratio represents the hydraulic loading.

In theory, any two similar centrifuges treating the same sludge should show the same settling performance at the same value of Q/Σ, and differences reflect nontheoretical factors. Since these efficiency factors are nearly identical in hydrodynamically similar centrifuges, i.e., in two sizes of disc centrifuge bowls of identical dimensional ratios, the clarification performance in one bowl can be directly related by the Σ ratio to the anticipated performance of a different size bowl.

When the centrifuges are not similar, as in a conveyor bowl compared to a disc bowl, the Q/Σ plots indicate relative performance, but environmental engineers cannot use the performance curve of one type for scaleup to the other type. Nevertheless, a Q/Σ plot showing flow ratios against solids recovery represents a generalized correlation and is a standard method of evaluating centrifuge performance.

Figure 7.48.11 is a generalized plot for a municipal excess activated sludge in disc, conveyor, and knifing centrifuges. The low Q/Σ value for the disc centrifuge results from its high shear with consequent feed degradation. The conveyor and basket centrifuges cause less floc degradation; the former has a scrolling inefficiency on a soft cake. Even considerable polymer addition cannot entirely overcome this effect.

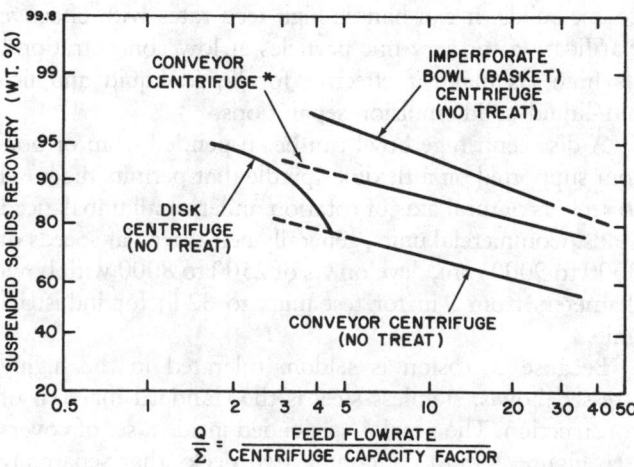

FIG. 7.48.11 Relative performance of centrifuges on municipal excess activated sewage sludge. °12 pounds of polymer added per ton of dry solids in feed.

LABORATORY TESTING

The only valid method for testing a limited amount of sludge for centrifuge performance is to run the material in a pilot unit. These tests generally require feed rates up to 10 gpm but also permit the evaluation of chemical acids. The clarification, capacity, and cake dryness scaleup predictions are good. For many municipal sewage sludges, sufficient information is available for prediction. For industrial wastes, testing is recommended.

SELECTION OF CENTRIFUGE

The operating curves in Figure 7.48.11 show one aspect of centrifuge selection. For example, if a wastewater treatment facility thickens municipal excess activated sludge at 90% recovery to limit the return of solids with the recirculated centrifuge liquid discharge (centrate), the feed flow rate must be large enough to require one or more of the largest units of any centrifuge type.

The conveyor centrifuge without polymer addition cannot reach 90% recovery and cannot be considered in this example. For the other three curves in Figure 7.48.11, the ratios of Q/Σ at 90% recovery are listed in Table 7.48.2 together with the ratios of Σ for the respective large units. The relative capacities of these units are determined by the product $\Sigma \times Q/\Sigma$.

The disc centrifuge shows a 4-to-1 advantage over the other units on this basis despite its lower Q/S value. The ratios of the unit capital costs do not differ markedly even when auxiliaries are included. The installation costs are not listed but are similar for the three types. The relationship of capital cost per unit capacity shows the disc centrifuge has more than a 3-to-1 advantage. The table also lists the solids contents of the thickened sludges.

Environmental engineers must consider other selection factors. The method and cost of disposal determine the required sludge concentration level. For smaller flows, the selection may favor discontinuous operation using the knifing (basket) centrifuge. Some industrial biological sludges with a high inert material content do well in the conveyor centrifuge. The maintenance, polymer, and operating costs must also be considered in the selection.

TWO-STAGE CENTRIFUGATION

Studies have shown mathematically and experimentally that multiple-centrifuge units operate better in parallel than in series in optimizing recovery and cake dryness when only one type of centrifuge or a single step of clarification is needed (Murkes 1969). However, two-stage centrifugation in series is beneficial under other circumstances. For example, an aerobic digested sewage sludge at $2\frac{1}{2}$% concentration consisting of one part primary solids to three parts secondary biological solids is difficult to clarify and dewater in a conveyor centrifuge without appreciable polymer addition. Table 7.48.3 shows the operating conditions for 90% recovery.

If a conveyor centrifuge operates as a first stage without polymer addition, the recovery is approximately 50% because the coarser particles and some flocculent material are removed. If the effluent is then directed to an imperforate bowl knifing centrifuge (basket), which readily recovers over 90% of the remaining fine solids without coagulant, this two-stage system operates with no polymer costs at an overall recovery of 97%. The combined cakes are almost as dry as those from a single-stage operation at a lower recovery. With two-stage centrifuging, the second-stage centrifuge represents a capital cost investment that is often covered by the savings on polyelectrolyte in 1 to 2 years.

HEAT TREATMENT PROCESSES

Modifying the sludge dewatering characteristics by wet oxidation or by heat treatment at temperatures up to 400°F and commensurate pressures of about 250 psig can reduce the sludge and yield a drier final cake. The corrosive tendencies of the liquor usually necessitate stainless steel equipment. Therefore, reducing the equipment size with a

TABLE 7.48.2 COMPARISON OF CENTRIFUGE CAPITAL COSTS ON THICKENING OF EXCESS ACTIVATED SLUDGE

	Type of Centrifuge		
Centrifuge Performance and Cost Factors	*Disc*	*Conveyor°*	*Basket (skimmer and knife)*
Ratio of Q/Σ at 90% recovery	1.0	2.0	5.5
Ratio of largest Σ factors	22.8	2.3	1.0
Relative capacities as $\Sigma \times Q/\Sigma = Q$	22.8	4.6	5.5
Ratios of capital costs	1.3†	1.2‡	1.0
Ratios of capital cost per unit capacity	1.0	4.6	3.2
Sludge concentration range, % by weight	5–6	4–12	9–11

Notes: °With polymer addition; 90% recovery is impossible without this additive.
†Includes cost of prescreening and concentration control equipment.
‡Does not include feed comminution, polymer preparation equipment, or polymer costs.

TABLE 7.48.3   COMPARISON OF ONE-STAGE AND TWO-STAGE CENTRIFUGING OF AEROBIC DIGESTED SEWAGE SLUDGE

| | Centrifuge Operation | |
Performance Factors	Single Stage	Two Stage
Single or First Stage		
Usage of polymer, lb/tn of dry solids	8–12	0
Solids recovery, %	90	50
Solids in cake, %	12–13	13–14
Second Stage		
Usage of polymer, lb/tn of dry solids		0
Overall solids recovery, %		97
Solids in combined (first and second stage) cakes, %		10–11

preconcentration of the feed at 6 or 7% solids is economically advantageous.

Wastewater treatment facilities can preconcentrate the feed by combining gravity-settled primary sludge with disc-centrifuged secondary sludge concentrate or by partially dewatering cosettled primary and secondary sludges in a conveyor or knife (basket) centrifuge depending on the flow rates. Sludge concentrations higher than 7% are difficult to pump and require an increase in the heat transfer surface. Conveyor centrifuges and RDVFs can dewater the treated sludge to cakes containing 35 to 45% solids without chemical addition.

RECOVERY OF CHEMICAL ADDITIVES

Figure 7.48.6 shows the performance of centrifuging for the selective recovery of calcium carbonate solids from a limed sewage sludge. When the economics of plant size and location warrant calcining and recycling, such recovery is readily made from limed sewage sludges, tertiary limed sludges for phosphate removal, water-softening lime sludges, and limed industrial sludges, frequently at efficiencies of 70 to 85% of the concentration of the calcium carbonate in the feed.

The calcium carbonate content is often 40 to 70% of the SS. The SS content of flocculated phosphate complexes and magnesium hydroxide varies from almost zero to 30%. Figure 7.48.12 shows the relative recoveries of magnesium hydroxide and calcium carbonate for two limed sludges. Calcium hydroxyapatite recovery approximates that of the hydroxide. Selective recovery of calcium carbonate also assures a reduced content of unfavorable solids in the recycled material.

The cake is deposited in an imperforate bowl knifing centrifuge (basket) with a high concentration gradient due to both differences in residence time and the relative densities of the collected solids. Whole sewage from a contact stabilization plant can produce a cake with a 5% solids content at the inner edge and 30% at the bowl wall, with a nonlinear concentration gradient. Because the lighter and lower concentration material is largely organic, returning to stabilization by selective removal with the skimming

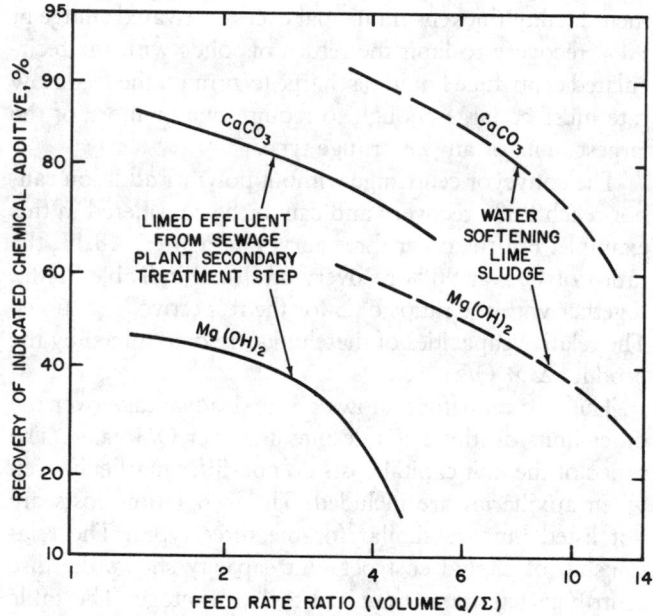

FIG. 7.48.12   Relative recoveries of magnesium hydroxide and calcium carbonate from limed sludges.

tube is sometimes feasible. The knifed-out residue then has a higher concentration and is more suitable for incineration. Adjusting the cut point between the skimmed and knifed material can accommodate changes in the feed and required dryness of the heavier discharge.

Selected Applications and Trends

Trends to improve performance in municipal and industrial waste treatment plants lead to new processes and more complex combinations of sludge. Primary and secondary sludges can be dewatered separately or mixed. The former often results in a higher initial cost but also gives a dryer combined cake and lower operating cost. Table 7.48.4 shows the effect on dewatering of adding secondary to primary sludge. Many smaller plants using only aerobic contact stabilization have low sludge rates that are handled at somewhat higher capital costs in knifing centrifuges without coagulant use.

TABLE 7.48.4   TYPICAL CENTRIFUGE PERFORMANCES ON WASTE SLUDGE

Type of Centrifuge	Type of Waste Sludge	Solids Recovery, %	Solids in Cake, %	Chemical Addition, lb/tn°
Conveyor	Sewage, raw primary	65–80	30–35	0
Conveyor	Sewage, raw primary	80–95	25–35	2–8
Conveyor	Sewage, digested primary	75–85	30–35	0
Conveyor	Sewage, digested primary	80–95	22–30	2–8
Conveyor	Sewage, raw primary and biological	50–70	18–22	0
Conveyor	Sewage, raw primary and biological	80–95	15–20	4–12
Knifing	Sewage, raw primary and biological	90–97	11–14	0
Conveyor	Sewage, digested primary and biological	55–70	23–30	0
Conveyor	Sewage, digested primary and biological	80–95	20–25	4–10
Conveyor	Sewage, raw secondary biological	85–95	5–15	8–15
Knifing	Sewage, raw secondary biological	90–95	9–11	0
Disk	Sewage, raw secondary biological	85–90	5–7	0
Conveyor	Sewage, aerobic digested	80–90	10–18	6–20
Knifing	Sewage, aerobic digested	90–95	9–12	0
Conveyor	Sewage, limed primary	55–70	40–60	0
Conveyor	Sewage, limed primary	80–95	15–30	1–5
Conveyor	Sewage, tertiary lime (phosphate)	35–70	55–70	0
Conveyor	Water-softening, lime	60–90	40–60	0
Conveyor	Water-softening, lime	90–100	35–50	2–8
Knifing	Water treatment, alum	90–95	8–20	0
Conveyor	Pulp and paper, Kraft	80–85	25–40	0
Conveyor	Pulp and paper, groundwood	85–90	15–20	0
Conveyor	Pulp and paper, deinking	85–90	15–25	0
Conveyor	Pulp and paper, box board	85–95	25–35	0
Conveyor	Pulp and paper, secondary biological	60–70	12–14	0
Conveyor	Pulp and paper, secondary biological	90–95	10–13	2–10
Conveyor	Rolling mill waste, $Fe(OH)_3$	95	12–18	1

Note: °Pounds of polymer added per ton of dry solids in the feed.

European practice currently recovers 90 to 95% of solids during dewatering. Because the need for such high recoveries has not been fully proven, the United States practice cannot justify the increased equipment and polymer costs and the resulting wetter cakes. In the United States, the usual requirements are in the range of 80 to 90% recovery, with a trend to lower recoveries, particularly in connection with wholly aerobic plants in which the return sludge quantities are large.

Many waste treatment plants are providing for chemical treatment to remove phosphate. Lime is frequently favored in many western states, and alum or iron salts is favored in the eastern states. The addition of lime at any point in the process produces sludge that is best treated in conveyor centrifuges either for high recovery with anionic polymers or for classification. Alum is commonly added before the secondary biological sludge clarifier but has little effect on centrifugal dewatering except for the shift to an anionic polymer. Alum (tertiary) treatment of secondary biological effluent produces sludge similar to water-treatment alum sludge.

Sludge containing appreciable quantities of iron salts often is amenable to clarification and dewatering in a con-veyor centrifuge. Lime sludge from water softening is readily dewatered in conveyor centrifuges producing dry cakes (see Table 7.48.4). In simple water treatment by alum, resultant sludge quality and cake concentration obtained in an imperforate bowl knife centrifuge seem to be a function of the turbidity of the raw water.

ALTERNATE DEWATERING METHODS

Other methods of sludge dewatering in addition to centrifugation are available. Because the power requirements for centrifuges are a direct function of volumetric throughput, treatment facilities preconcentrate sludge, usually by gravity to the point of hindered settling (see Figure 7.48.4) and compaction, to reduce operating costs.

Flotation can preconcentrate large waste streams. Coagulant and labor costs are appreciable for concentrations of 4% or more. Centrifugation produces higher concentrations. RDVFs, when used to dewater biological sludge, have higher labor, maintenance, and coagulant costs. Their power requirements per unit throughput are lower, but their space and building requirements are higher. A vacuum filter produces a greater clarity filtrate but a slightly wetter cake.

Filter presses, recently automated, are used widely in Europe, but their capital costs frequently exceed those for centrifugation. No coagulant is used, but flyash or an equivalent filter aid is normally needed at ratios from 1:1 to 5:1 relative to the sludge solids. Filter presses reduce the moisture load on the incinerator but require additional

cake-handling facilities. Continuous-gravity or pressure-screening devices are often more economical than centrifuges for a population of 10,000 or less due to their low operating speed and low maintenance and power costs. Their high coagulant use does not alter their overall economy.

—*F. W. Keith*

7.49
HEAT TREATMENT AND THERMAL DRYERS

Porteous Heat Treatment

OPERATING PRESSURE: 180 psig

OPERATING TEMPERATURE: 360°F

UNIT CAPACITIES: Up to 100 gpm

COOKING DETENTION TIME: 30 min

IMPROVEMENT IN DEWATERING BY HEAT TREATMENT: 200%

SOLIDS CONTENT OF THICKENED, HEAT-TREATED SLUDGE: 7 to 11%

CLARIFIED EFFLUENT BOD: 4500 mg/l or more

COD REDUCTION BY HEAT TREATMENT: 22%

VOLATILE SOLIDS REDUCTION BY HEAT TREATMENT: 28%

MATERIALS OF CONSTRUCTION FOR EQUIPMENT: Carbon steel

Heat treatment followed by filtration is economical for dewatering sludge without using chemicals. Heating sludge at elevated temperatures and pressures causes the cells to rupture, releasing bound water from the sludge particle and the water of hydration. The gel system structure is irreversibly destroyed, and the solids lose affinity for water and are readily separated from the liquid by decantation. The technical term for this process is *heat syneresis.*

The exact nature of mechanisms that release bound water and water of hydration from the sludge is not fully understood. It *is* known that coalescence of particles occurs, and that the high-velocity particle collisions (as a function of the energy level at high pressure [180 psig] and temperature [360°F]) disturb the molecules and assist in particle structure breakdown.

PROCESS

The continuous, automated process is sized on the basis of the flow rate. Figure 7.49.1 shows a basic flow sheet and the components of the heat treatment process. This process draws the sludge (any type) from the storage tank

through a grinder, reducing particle size and fiber lengths. Then, a feed pump pressurizes the sludge to approximately 250 psig and passes it through a water-to-sludge heat exchanger that elevates its temperature to approximately 300°F before it enters the reactor.

The process simultaneously injects steam from a boiler into the reactor with the incoming preheated sludge, raising sludge temperature to approximately 360°F. The sludge cooks in the reactor for about 30 min at the corresponding pressure of 180 to 200 psig and is then discharged through the cooling exchanger discharge valve to the decant thickener tank. The cooled sludge temperature is approximately 110°F.

For high pressure pumps, the triplex ram (a three-cylinder, single-acting displacement pump without rings) or progressive cavity-type (a precision screw conveyor pump,

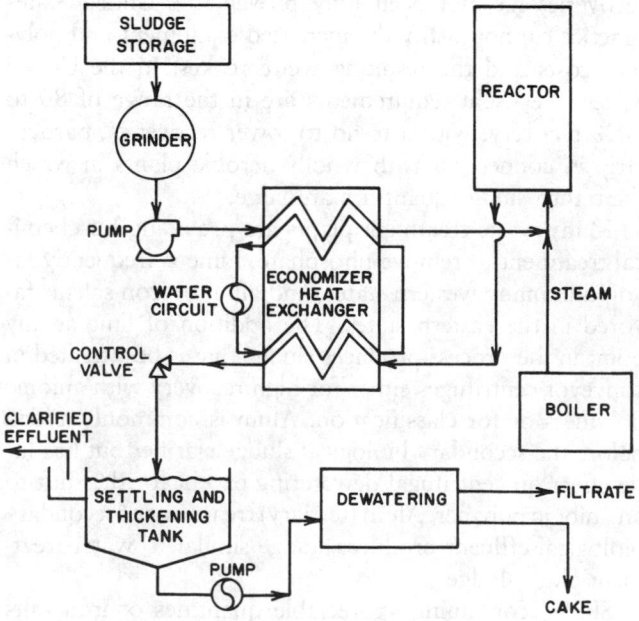

FIG. 7.49.1 Diagram of the Porteous process.

such as Moyno) pumps are commonly used. The heat exchanger has removable end caps and water-to-sludge circuitry to eliminate tube fouling and maintenance. The process removes grit from the influent sludge to minimize tube plugging and protect pumps, valves, and pipes from erosion.

Sludge flows both to and from the reactor through the heat exchanger inner tube. The exchanger is a concentric tube-in-tube design. A closed water circuit system with pressure tank and pump transfers heat from the reactor effluent to the incoming feed sludge. Water flows through the tube jacket, and the heat exchanger efficiency in this design is above 80%.

The reactor vessel is designed in accordance with the ASME unfired pressure vessel code requirements. Radiation or probe-type level sensors control the reactor sludge level. The process cools and liquifies gases released from the reactor to avoid odor formation. Also, the enclosed settling tank prevents odors from leaving the tank, and a small afterburner destroys odoriferous vapors.

The settling tank is a conventional, picket-type thickener where solids and liquids separate and readily settle to produce a sludge with 8 to 14% solids content. The thickened sludge from the thickener is pumped to the dewatering device.

Separate sludge mixing tanks should be provided when treated sludge is stored for extended periods because supersettling to a 40% solids content can occur, creating pumping problems. Vacuum filters, centrifuges, filter presses, and horizontal vacuum extractors have all been successfully applied to dewater and decanted sludge.

Dewatering Results

Three field-proven techniques for dewatering heat-treated sludge are vacuum filtration, centrifugation, and pressure filtration. Table 7.49.1 gives the process characteristics of each technique. This table shows vacuum filter performance data obtained from the 4500 gph Porteous installation in Colorado Springs, Colorado. The filter yield on heat-treated sludge averages 12 lb/ft²/hr compared to the 3- to 5-lb/ft²/hr rate achieved by chemically conditioned sludge. The moisture content of the heat-treated sludge filter influent is 50 to 60% compared to 80% for the chemical sludge, and the volume of filter cake also decreased by 50%.

The filter feed solids concentration from the thickener varies from 7 to 11%. Figure 7.49.2 shows the effects on cake yield. The filter rate also varies as a function of the ratio of primary sludge to secondary sludge. Filtration rates of 4 to 10 lb/ft²/hr can be obtained on activated sludge and 12 to 26 lb/ft²/hr on primary sludge. The filtration performance on mixed raw primary and activated sludge is proportional to the quantity, fiber content, and individual rate of each sludge.

Wastewater treatment facilities have achieved centrifugation results of 50 to 65% moisture in cake concentrations and an efficiency of 70 to 80% in solids capture. With polymer additions in test work, the Aire Plant in Geneva, Switzerland, reported a 95% capture of solids. Filter pressing is used extensively in England and continental Europe. The pressing time is directly related to the cooking temperature (see Figure 7.49.3).

Performance Factors

Variables that determine the dewatering properties of heat-treated sludge include process variables (temperature, pressure, steam flow, and detention time) and sludge variables (feed solids concentration, sludge type, volatiles-in-feed sludge, chemical content of sludge, and fiber content of sludge). Heat treatment improves the dewatering characteristics of all sludges, and environmental engineers should use laboratory studies to determine the optimal operating conditions.

The relationship of time, pressure, and temperature controls the degree of sludge solubilization and the BOD and COD of the effluent liquor and filtrate.

Liquor Treatment

The clarified effluent liquor and dewatered filtrate are returned to the treatment plant (see Figure 7.47.7) and produce a biological load on secondary oxidation treatment

TABLE 7.49.1   DEWATERING OF HEAT-TREATED SLUDGE

Equipment Used	Cake Concentration Produced (% TS)	Efficiency of Solids Removal° (%)	Dewatering Capacity Increase Relative to Nonheat-treated Sludge (%)
Vacuum filtration	35–45	95	200–300
Centrifuge	40–55	80–90	200–300
Pressure filter	50–65	100	200–300

Note: °Assuming that sludge does not include wash water solids.

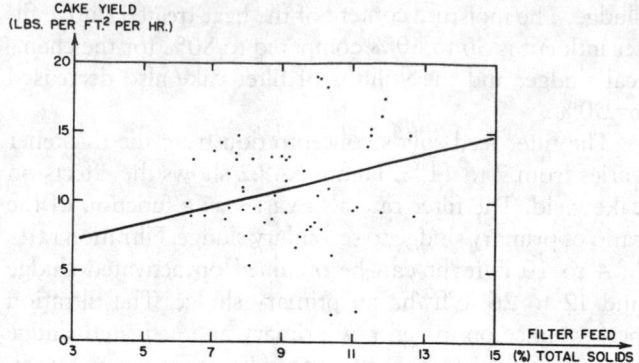

Fig. 7.49.2 Relationship between filter feed concentration and cake yield.

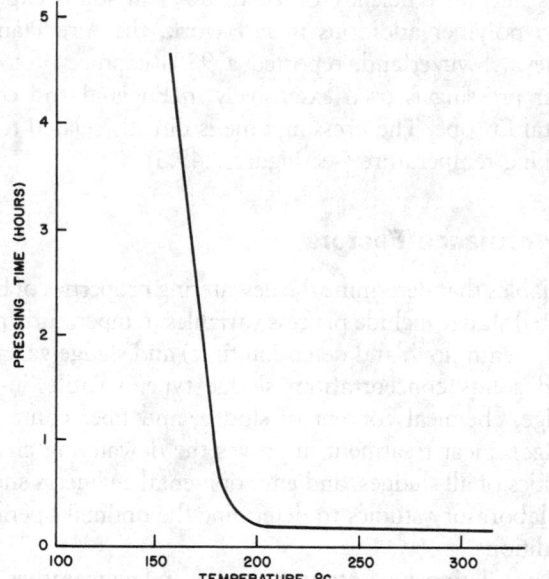

FIG. 7.49.3 Relationship of filter pressing time to cooking temperature.

processes. Table 7.49.2 shows typical liquor by-product compositions utilizing data of European plants. Figure 7.49.4 shows the BOD and COD values of the liquors produced at the first installation at Colorado Springs as a function of feed (to the grinder) solids concentration.

From both sources, the typical clarified effluent BOD concentration is approximately 4500 mg/l. These plants usually return this liquor to the sewage influent daily over several hours to minimize the effects of high BOD concentrations. They achieved BOD reductions of 78 to 97% in activated-sludge, pilot-plant tests.

The heat-treatment process accomplishes a COD reduction of 22%, which closely agrees with the volatile solids reduction (about 28%). The COD reduction is almost entirely due to the oxidation of the solids rather than the liquor.

The sewage plant's secondary treatment must handle the total increase in BOD load because most of the BOD load in the liquor is soluble and is not removed in the clarification step. Each pound of BOD recycled to the treatment plant produces roughly 0.5 to 0.6 lb of sludge. Plastic and rock trickling filters used in English plants as pretreatment units prior to the heat-treatment step reduce the BOD level by 52%. The pH of the clarified effluent liquor is about 0.5 pH units less than the feed pH. Wet-combustion systems generally reduce the feed sludge pH by 3 to 5 units.

Materials of Construction

All components of the heat-treatment system, except for the stainless steel balls in the valves and the reactor pressure control line, are constructed of carbon steel. Carbon steel materials have been used at the Halifax, England installation since 1935 without corrosion problems. Tube

TABLE 7.49.2   ANALYSES OF SOME LIQUOR BY-PRODUCTS FROM SLUDGE DEWATERING PROCESSES IN EUROPEAN PLANTS (RESULTS IN MG/L)

	Type of Liquor				
Constituent	Heat-Treatment Liquor, Halifax (Secondary sludge)	Heat Treatment Liquor, Luton (All sludge)	Press Liquor, Bradford Sewage Department	Press Liquor, Halifax (Primary sludge)	Supernatant Liquor, Halifax Experimental Digestion Plant
TS	10,820	7500	10,430	10,440	8610
Total ash	1460	1700	4240	6300	2750
Oxygen absorbed in 4 hr	2610	1500	698	288	1600
5-day BOD	4620	4500	4170	3780	1480
Organic nitrogen	1110	410	504	150	—
Ammoniacal nitrogen (NH₃ ion)	418	830	190	220	—
Albuminoid nitrogen (protein)	506	—	135	17.3	—

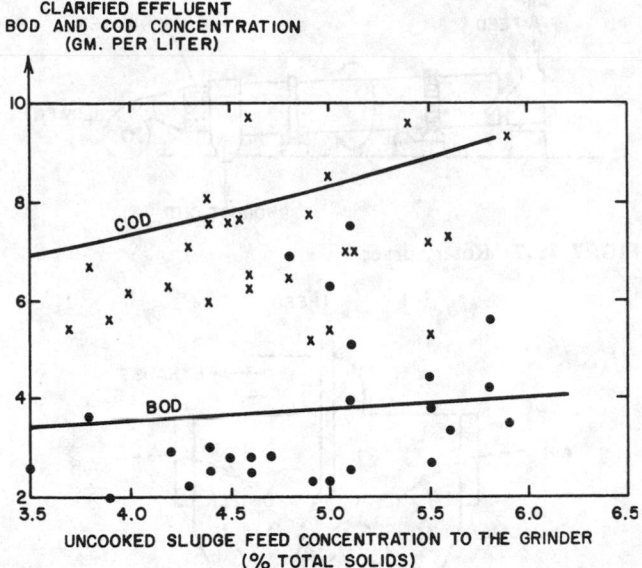

FIG. 7.49.4 Clarified effluent BOD and COD concentration as a function of solids content of heat treated sludge into the thickener.

turns and sludge piping are fabricated of heavy thickness due to the erosive forces of grit and sand in the feed sludge.

Wet-combustion systems are fabricated from stainless steel.

Table 7.49.3 shows operating data for the Porteous process.

As previously mentioned, this process is more economic than the chemical conditioning of sludges to improve their dewatering characteristics. Heat-treatment fuel costs are further reduced if the dewatered cake is incinerated and the waste heat is recovered. The final sludge cake is ultimately disposed as a low-grade soil conditioner, land fill material, or fuel. The high temperature used in the process sterilizes the sludge from pathogenic germs and weed seeds.

Thermal Dryers

TYPES OF THERMAL DRYERS

a) Flash dryers
b) Screw conveyor dryers
c) Multiple-hearth dryers
d) Rotary dryers
e) Atomized or spray dryers

Thermal drying of sludge is economical only if a market for the product is available. Although dried sludge makes a good soil conditioner and fertilizer and is convenient to use, especially for the home gardener, the cost of preparing and packaging it is seldom recouped from the profits. Accordingly, sludge drying is seldom used in the United States and does not represent an economical alternative to incineration or other disposal processes. However, if a municipality looks beyond economics and considers sludge as a natural resource, they have a strong argument for drying sludge and using it as a soil conditioner and fertilizer.

DRYER TYPES

Several types of thermal dryers used by the chemical process industry can be applied to sludge drying. The sludge is always dewatered prior to drying, regardless of the type of dryer selected.

Flash dryers operate by promoting contact between the wet sludge and a hot gas stream (see Figure 7.49.5). Drying takes place in less than 10 sec of violent action, either in a vertical tube or in a cage mill. A cyclone, with a bag filter or wet scrubber, if necessary, separates the solids from the gas phase. The vapors are returned through preheaters to the furnace, minimizing odor problems. A portion of the solid product is often returned to precondition the wet sludge.

The *screw conveyor dryer* uses a hollow shaft and blades through which hot gas or water is pumped (see

TABLE 7.49.3 PORTEOUS PROCESS DATA

Porteous System: gph	1000	2000	3000	4500	6000	7500	9000
Total dry solids feed, lb per day	10,000	20,000	30,000	45,000	60,000	75,000	90,000
Percent dry solids	5	5	5	5	5	5	5
Total wet sludge feed, gpd	24,000	48,000	72,000	108,000	144,000	180,000	216,000
Operation period, hr per week	168	168	168	168	168	168	168
Porteous system rate, gph	1000	2000	3000	4500	6000	7500	9000
System horsepower	12	20	35	48	64	70	80
Power demand, Kwh per day (hp)(0.60)(24)	173	280	505	692	922	1008	1152
Fuel required, million Btu per day (gal/day)(800 Btu)(10^{-6})	19.2	38.4	57.6	86.4	115.2	144.0	176.0
Boiler feed water, gpd (0.08 gal)(gpd)	1920	3840	5760	8640	11,520	14,400	17,600
Flushing water, gpd (3 hr/day)(gph)	3000	6000	9000	13,500	18,000	22,500	27,000

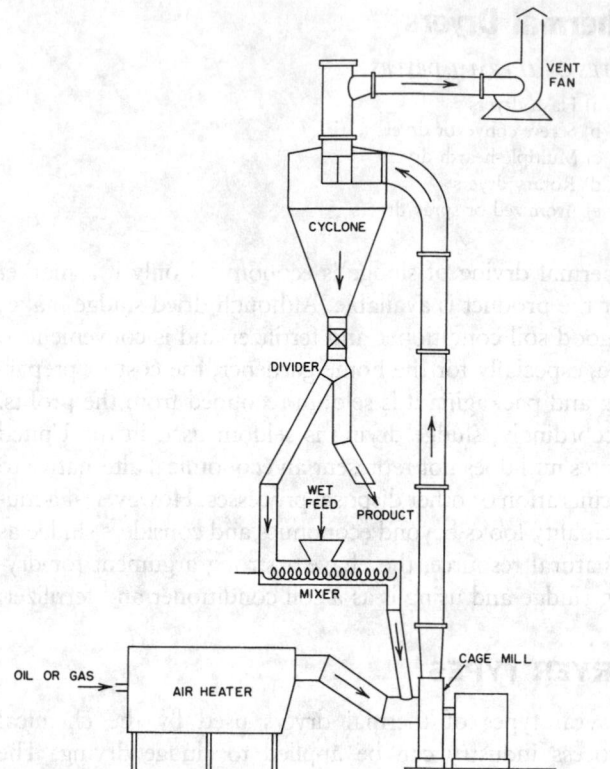

FIG. 7.49.5 Flash dryer.

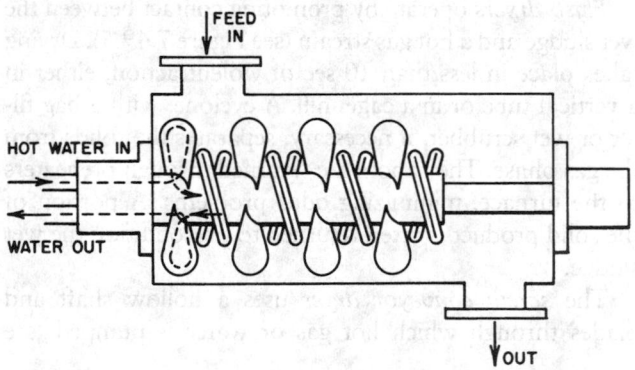

FIG. 7.49.6 Screw dryer.

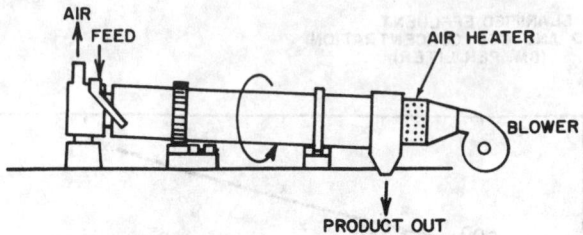

FIG. 7.49.7 Rotary dryer.

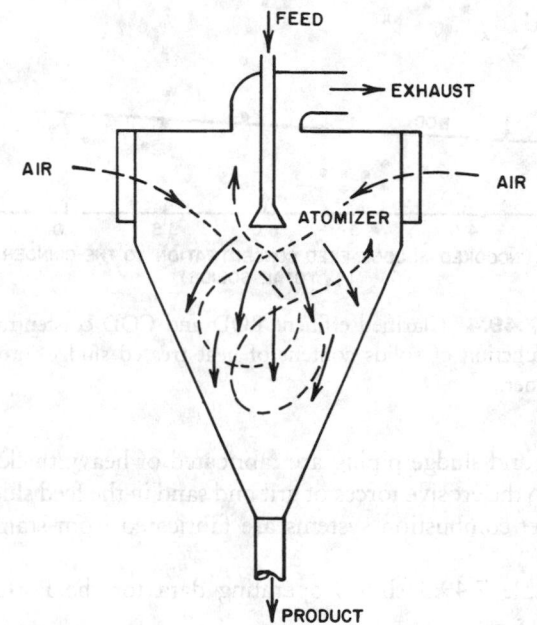

FIG. 7.49.8 Atomized or spray dryer.

Figure 7.49.6). The heat is transferred to the sludge as it is conveyed through the dryer.

Multiple-hearth dryers are converted multiple-hearth furnaces. The wet sludge can be mixed with dry product as it descends through the furnace. Fuel burners are located both on top and bottom.

Rotary dryers consist of a rotating cylinder through which the sludge moves (see Figure 7.49.7). Various types of blades or flights are installed in the dryers depending on the type of material being dried. Drying takes place by direct contact with heated air.

The chemical process industry has used *atomized or spray dryers* for many years (see Figure 7.49.8). For example, they make detergents by spraying the wet slurry into a heated tower. Spraying solids countercurrently into a downward draft of hot gas dries them although concurrent spray dryers are also used in the chemical industry.

OPERATION AND ECONOMICS

Experience shows that purchasers of dried sludge reduce their offers once a plant is in operation. Lack of competition for the product is usually the reason. Waste heat from refuse incineration can be used for drying sludge. Having an incinerator allows a plant to burn the dry sludge if no market exists for it. This burning eliminates the need for stockpiling.

Operational problems with heat drying (flash and rotary types used on sludge) include explosions due to grease accumulation, storage problems due to the absorption of moisture, and the fine-powder form of the product which is difficult to handle and apply. Pelletizing is a method of alleviating the last problem.

Air pollution, in the form of dust and odor, is another problem. Expensive control equipment is often required to meet local air quality standards.

Sludge Disposal

7.50
SLUDGE INCINERATION

Multiple-Hearth Radiant Incineration

PER CAPITA DRY SLUDGE PRODUCTION

   0.2 lb per day

SLUDGE FEED CONCENTRATION

   Up to 75% moisture

CAPACITY IN UNITS OF WET FEEDRATE

   13 in operating diameter (OD)–16 lb/hr; 30 in OD–300 lb/hr; 22 ft
   3 in OD–30,000 lb/hr; 28 ft 3 in OD–over 50,000 lb/hr

INCINERATOR ASH PRODUCTION

   10% of feed sludge volume

LOADING RATE

   7 to 12 lb/ft^2/hr

EXHAUST GAS TEMPERATURE

   700° to 800°F (without afterburner)

Since its initial use in 1934, the multiple-hearth incineration system has become the most widely used means of ultimate sewage sludge disposal, with more than 200 installations in the United States accounting for a capacity of over 50,000 wet tons of sludge per day. Multiple-hearth process development has kept pace with the increasing concern over air pollution, odor control, and resource reuse. New developments include the use of multiple-purpose furnaces for sludge incineration and chemical reclamation.

Following the various dewatering processes, waste solids still laden with water remain for disposal. More stringent laws and codes have reduced the choices for disposing such waste solids or sludge.

PROCESS DESCRIPTION

Multiple-hearth furnaces were originally developed in 1889 to roast pyrites for the manufacture of sulfuric acid. Modern multiple-hearth systems have some 120 proven uses, including:

- Burning (or drying) raw sludge, digested sludge, and sewage greases
- Recalcining lime sludge (the burning of $CaCO_3$ into CaO) and waste pond lime from sugar manufacturing

- Pyrolysis: the manufacture and regeneration of activated-granular carbon and the regeneration of diatomaceous earth; pyrolyzing fruit, nut, and lumber waste (peach pits, walnut shells, almond shells, sawdust, and bark) for charcoal briquettes
- Reclaiming oily chips from boring machines to metal briquettes and reclaiming cryolite in aluminum smelting operations
- Other roasting uses like mercury, molybdenum sulfide, carbon, magnesium oxide, uranium yellow cake, and nickel

The multiple-hearth furnace is a simple piece of equipment, consisting primarily of a steel shell lined with refractory on the inside. The refractory can be either castable or brick, depending on the size of the furnace.

The interior is divided by horizontal brick arches into separate compartments called hearths. Alternate hearths have holes at the periphery to allow the feed solids to drop onto the hearth below. The center shaft, driven by a variable-speed motor, rotates the rabble arms situated on each hearth (see Figure 7.50.1). The rabble teeth on these arms are at an angle so that the material is moved inward and then outward on alternate hearths. The shaft and rabble arms are cooled by air introduced at the bottom; this air is recycled as required by the thermal process.

The sludge is fed through the furnace roof by a screw feeder or a belt and flapgate. The rotating rabble arms and rabble teeth push the sludge across the hearth to drop holes where it falls to the next hearth and continues downward until the sterile phosphate-laden ash is discharged at the bottom.

The multiple-hearth system has the following three distinct operating zones:

1. The top hearths where the feed is partially dried
2. The incineration and deodorization zone where temperatures of 1400° to 1800°F (760° to 982°C) are maintained
3. The cooling zone where the hot ash gives up heat to the incoming combustion air.

The warmed air rises to the combustion zone in counterflow to the sludge flow, and the hot combustion gases

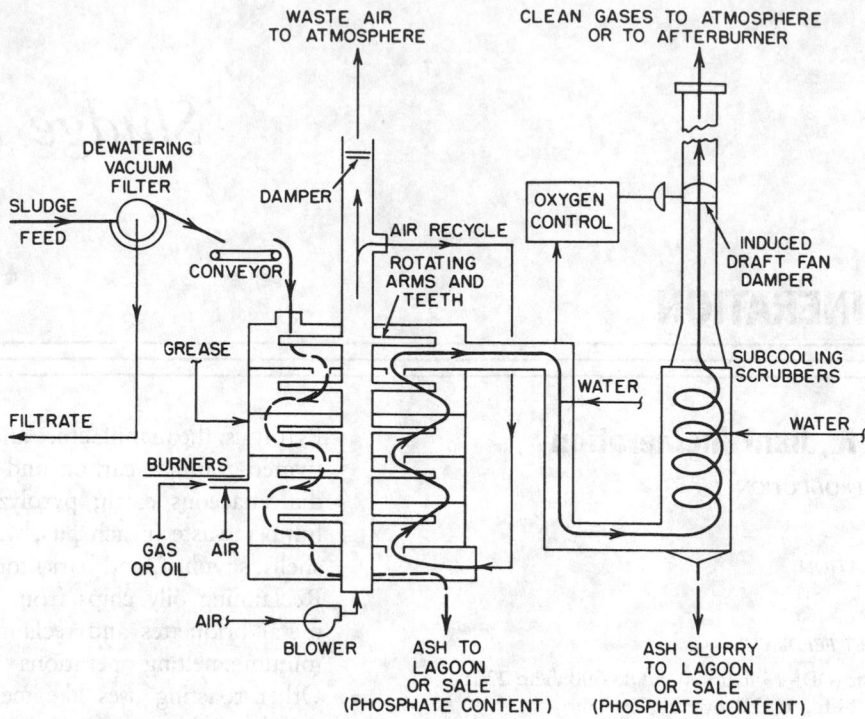

FIG. 7.50.1 Multiple-hearth sludge incineration.

sweep over the cold incoming sludge, evaporating the sludge moisture to about 48% At this moisture content, a phenomenon called *thermal jump* can occur in the combustion zone. This beneficial energy exchange allows the generation of odor-free exhaust gas at temperatures of 500° to 1100°F (260° to 593°C). Table 7.50.1 gives the typical temperature profile across the sludge furnace.

CONSTRUCTION

The steel shell of the incinerator is either of welded or bolted construction. Furnaces with diameters less than 10 ft are shop-welded and shipped in one piece. The larger sizes require field assembly and are usually bolted.

AUXILIARY EQUIPMENT

Belt conveyors, ribbon screws, and bulk-flow conveyors are commonly used. They discharge to a hopper with a

TABLE 7.50.1 SLUDGE FURNACE TEMPERATURE PROFILE

Hearth No.	At Approximately Half Capacity (°F)	Nominal Design Capacity (°F)
1	670	800
2	1380	1200
3	1560	1650
4	1450	1450
5	1200	1200
6	325	300

counterweighted flapgate that intermittently drops the material into the furnace. Screw feeders and bulk-flow conveyors often discharge directly into the furnace.

The burners can handle all common liquid and gaseous fuels, including distillate and residual oils, natural gas, sewage digestion gas, propane, or combinations of these fuels. The design of burner tiles and boxes is important because slagging can occur within the tiles.

EQUIPMENT SELECTION AND PROCESS DESIGN

At one time, multiple-hearth furnaces were sized by a series of slide-rule and chart computations. Now a computer routine normally performs sizing. The size of the multiple-heart furnace is determined by the processing rate of the wet feed per square foot of furnace area, i.e., the loading rate equals 7 to 12 lb/ft²/hr. Table 7.50.2 lists some standard sizes.

EXHAUST GASES—AIR POLLUTION

Exhaust gases at most installations pass from the incinerator furnaces through refractory-lined flues and enter three-stage, impingement-type scrubbers. Frequently, making the stack discharges essentially invisible is favorable. Accordingly, following the scrubber, some systems subcool the gases (below saturation temperature) to 110°F to condense the water vapor and reduce the acid formation in wet gases.

In advanced waste treatment systems that use lime, the use of CO_2 gases to recarbonate the effluent also reduces

TABLE 7.50.2 STANDARD MULTIPLE-HEARTH FURNACE SIZES

Unit Size	OD for Wall Thickness of: 6"	9"	13½"	Column Height	Square Feet of Effective Hearth Area and Normal Shell Height — Hearths 1	2	3	4	5	6	7	8	9	10	11	12
13"	° 18" °@2½"			1'	1	2¼ 1'8"	2¼	3 3'4"								
30"	° 3' °@3"	44" °@7"		1½'	4	8 2'2"	12	16 3'10"	20	24 6'	28	32 7'8"	36	40 9'4"		
39"	4'3"	4'9"	5'6"	2½'	7	14 3'6"	19	28 6'	32	37 8'6"	42	48 11'0"	54	61 13'6"		
54"	5'6"	6'	6'9"	4'	15	31 4'2"	42	63 7'4"	74	85 10'6"	98	112 13'8"	126	140 16'10"		
5½'	6'6"	7'	7'9"	4'	24	47 4'8"	63	94 8'3"	110	125 11'10"	145	166 15'5"	187	208 19'0"		
7'	8'0"	8'6"	9'3"	5'	32	65 6'3"	96	130 10'10"	161	193 15'5"	225	256 20'0"	288	319 24'7"	351	383
8½'	9'6"	10'	10'9"	6½'	47	94 6'3"	138	188 10'8"	235	276 15'1"	323	364 19'5"	411	452 23'10"	510	560
12'	13'0"	13'6"	14'3"	6½'	97	195 6'9"	287	390 11'8"	487	575 16'7"	672	760 21'5"	857	944 26'4"	1041	1128 31'2"
14½'	15'6"	16'	16'9"	7'	143	286 8'0"	422	573 13'2"	716	845 18'7"	988	1117 24'1"	1260	1400 29'6"	1540	1675 35'0"
16½'	17'6"	18'	18'9"	7'	181	363 8'4"	534	727 14'3"	908	1068 20'2"	1249	1410 26'0"	1591	1752 31'11"	1933	2090 37'9"
18'	19'0"	19'6"	21'3"	8'	215	431 8'4"	634	863 14'4"	1078	1268 20'2"	1483	1660 26'1"	1875	2060 31'11"	2275	2464 37'10"
20'	21'0"	21'6"	23'3"	8'	269	538 9'6"	790	1077 16'1"	1346	1580 22'9"	1849	2084 29'6"	2350	2600 36'2"	2860	3120 42'11"
23½'	24'6"	25	25'9"	8'	382	764 11'4"	1145	1528 18'9"	1909	2292 26'2"	2674	3056 33'7"	3438	3818 41'	4200	4584 48'5"
26'	27'0"	27'6"	28'3"	8'	463	926 12'7"	1389	1852 20'9"	2315	2778 28'10"	3241	3704 36'11"	4167	4630 45'	5093	5556 53'1"

Note: °OD corresponding to noted wall thickness.

the amount of air pollution caused by the incinerator. In these systems, the higher degrees of wastewater treatment also reduce air pollution. Thus, cleaner water and air go with the lime PCT process and the hydrolysis adsorption process.

ODOR CONTROL

The operating and exhaust temperature of fluidized-bed incinerators is usually high enough to destroy odorous substances.

The thermal jump helps the sewage sludge to bypass the temperature zone where offensive odors are distilled and sometimes enables the multiple-hearth furnace to also effectively burn sewage solids without producing odors even though the gas outlet temperatures are as low as 500°F.

The malodorous volatile organic acids, butyric and caproic, exist in low concentrations in ordinary sludge; they exist as nonvolatile salts in raw sludge conditioned with lime and ferric chloride because of the high pH. Consequently, they are not released during ordinary drying operations but are carried into the burning zone where they are destroyed.

While the possibility of odor production always exists, wastewater treatment facilities can usually accomplish odor-free incineration. As a precautionary measure, they can add standby afterburners on the top hearth.

ASH RESIDUE

The final output of the multiple-hearth sludge incineration system is sterile, inert, and free of putrescible material and obnoxious odors. The ash volume amounts to approximately 10% of the furnace feed, based on a sludge cake containing 75% moisture with 70% of the sludge solids volatile. The ash is dry and contains less than 1% combustible matter, which is normally fixed carbon. This ash can be transferred hydraulically, mechanically, or pneumatically and can be used for landfill or roadfill. It is also being used experimentally to make bricks and concrete blocks.

NEW DEVELOPMENTS

For advanced waste treatment processes using lime, wastewater treatment facilities can enhance the economics of onsite lime recalcination by using the organic sludge as a partial fuel source for lime and recalcining in a single, multipurpose furnace.

CHARACTERISTICS OF MULTIPLE-HEARTH SYSTEMS

The following statements summarize the characteristics of multiple-hearth systems.

1. They generally handle sludges containing up to 75% moisture without auxiliary fuel.
2. They can incinerate or pyrolyze a variety of sludge materials, individually or in combination, including difficult-to-handle materials such as scum, grease, and ground refuse.
3. They can be used (or retrofitted) as a reclamation furnace for chemicals such as lime in combination with sludge or separately.
4. They can operate continuously or intermittently over a range of sludge feed capacities because the excess air for combustion and exit-gas temperatures can vary to suit conditions.
5. They have a long life and low maintenance costs over decades of operation.
6. They have a single feed point without a pressure feed requirement.

Reuse of Incinerator Ash

Various forms of slag tiles, bricks, and concrete blocks are made from ash residues. The city of Tokyo sells a large tonnage of its multiple-hearth furnace ash to C. Itoh Fertilizer Sales Company, Ltd.

The source of the ash is Tokyo's Odai treatment plant. This plant services a drainage area of 11,248 acres, with a planned treatment capacity of 111 mgd. The Odai plant extends over an area of 23.4 acres and uses a 100-tn/day, multiple-hearth furnace. An increasing portion of the city of Tokyo's total sewage sludge is processed by incineration.

The ash from the Odai plant is marketed under the trade name Vitalin (the Japanese word "lin" means phosphorus). A bag of Vitalin has the following percentage composition:

- Silica oxide   30.00
- Magnesium oxide   3.30
- Calcium oxide   30.00
- Phosphoric oxide   6.20
- Ferric oxide   18.20
- Potassium   1.00
- Nitrogen   0.20
- Manganese   0.06
- Copper   0.61
- Boron   200.00 ppm

Vitalin is sold under the special fertilizer category because material containing less than 12% phosphate cannot be classed as fertilizer. (Some states also require that the material have a nitrogen content of 6% or a total NPK range of 20 to 25%.)

In addition to Odai sewage sludge ash, the city of Nagoya has sold sludge ash from a multiple-hearth furnace under the name of Hormolin.

The phosphate in incinerator ash can be used in plant metabolism even though the P_2O_5 is insoluble above a pH of 3. In acidic soils, silicate and lime increase the pH of the soil. For such purposes, Japanese farmers use mixtures of organic SiO_2 and CaO. Thus, the components in the sludge ash are valuable to the soil for this purpose even if the phosphate content has limitations.

Table 7.50.3 contains an analysis of the ash from the South Lake Tahoe Water Reclamation plant. The Tahoe ash has more phosphate (7 to 10%) than the Japanese products. The lime used in the tertiary (phosphate removal) phase is removed with the sludge stream and is present in the ash in concentrations of 30 to 35%.

The city of Osaka has also used ash as a base material for roads around the Nakahama Sewage Treatment plant, but the ash is not used commercially in that manner.

With increased interest in resource recycling, the need for alternatives to ocean dumping of sludge, and a projected U.S. phosphate supply of only 80 years, the prospect of using the phosphate contained in sewage merits further investigation. Sludge or sludge ash containing appreciable quantities of metals such as zinc and chromium can damage crops (not grass or cereals) by heavy and repeated applications. However, the toxic effects are manifested only on acid soils, and sludge containing lime can probably offset some of the harmful effects.

The cumulative effect of boron (200 ppm in the Odai ash) requires further investigation, and monitoring and controlling the toxic material content in all sludge streams is advisable. Phosphate-rich sludge ash provides opportunities for recycling materials and recapturing value to defray the cost of advanced sewage treatment plant operations.

Fluidized-Bed Incineration

TYPES OF PROCESSES

a) Combustion
b) Pyrolysis

TABLE 7.50.3 TYPICAL ANALYSIS OF ASH FROM TERTIARY-QUALITY, ADVANCED WASTE TREATMENT SYSTEMS

| | Percent of Total | | | |
Content	Sample 1 Lake Tahoe	Sample 2 Lake Tahoe	Minneapolis–St. Paul	Cleveland
Silica (SiO$_2$)	23.85	23.72	24.87	28.85
Alumina (Al$_2$O$_3$)	16.34	22.10	13.48	10.20
Iron oxide (Fe$_2$O$_3$)	3.44	2.65	10.81	14.37
Magnesium oxide (MgO)	2.12	2.17	2.61	2.13
Total calcium oxide (CaO)	29.76	24.47	33.35	27.37
Available (free) calcium oxide (CaO)	1.16	1.37	1.06	0.29
Sodium (Na)	0.73	0.35	0.26	0.18
Potassium (K)	0.14	0.11	0.12	0.25
Boron (B)	0.02	0.02	0.006	0.01
Phosphorus pentoxide (P$_2$O$_5$)	6.87	15.35	9.88	9.22
Sulfate ion (SO$_4$)	2.79	2.84	2.71	5.04
Loss on ignition	2.59	2.24	1.62	1.94

FLUIDIZED-BED TEMPERATURE

1300° to 1500°F is normal; 2000°F is maximum.

TEMPERATURE VARIATIONS THROUGHOUT THE BED

5° to 10°F

HEAT BALANCE

Combustion releases; pyrolysis consumes heat energy.

MAXIMUM SLUDGE MOISTURE CONTENT WITHOUT NEED OF AUXILIARY FUEL

65% without air preheater, 73% with preheater

Although incineration using fluidized beds is relatively new, the fluidization of solid particle media has been used for some time in the chemical and petrochemical industries. The petroleum industry has used fluidized-bed reactors for catalytic cracking of complex hydrocarbons to produce simple, molecular hydrocarbon structures. If the fluidized medium consists of catalyst particles (a substance that enhances a chemical reaction without being consumed), chemical reactions are promoted within the medium.

Incineration in a fluidized bed consists of injecting substances, namely solid waste or liquid sludge and air, into a catalytic bed that causes a chemical reaction to occur. In this reaction, as in the cracking process, simpler products are formed. Frequently, the catalyst for this reaction is a bed of heated, fluidized sand.

PROCESS DESCRIPTION

The fluidized-bed incinerator is a vertical cylinder with an air distribution plate containing many small openings near the bottom (see Figure 7.50.2). This plate, which allows air to pass into the media, also supports the sand or other bed media. An external air source forces the air into the bottom of the cylinder. Once the air is distributed, the bed expands, i.e., it becomes fluidized (see Figure 7.50.3).

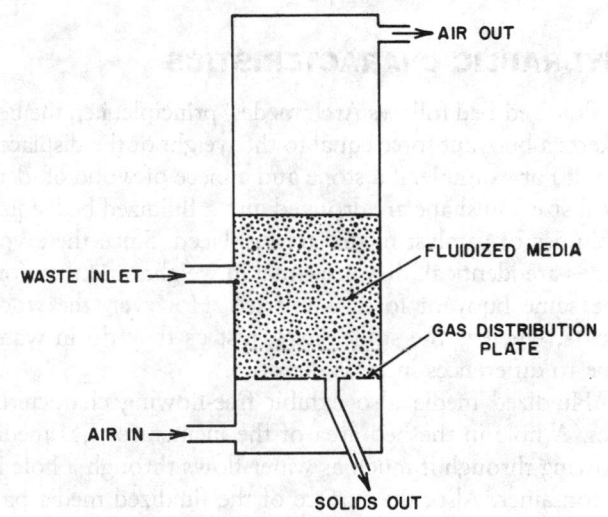

FIG. 7.50.2 Fluidized-bed incinerator.

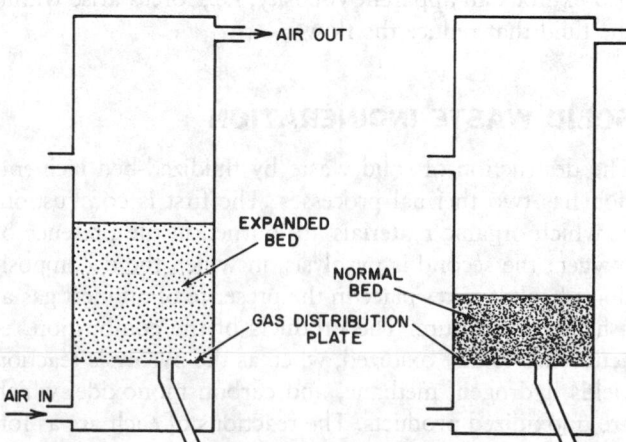

FIG. 7.50.3 Comparison of expanded and normal bed.

The volume of air that can pass through the bed is limited. If air is admitted at a low rate, it travels through the tortuous channels among the particles and escapes from the top of the bed without causing the individual particles of the bed to move, and the particles remain in their original packed configuration. This condition exists until the force exerted by the air overcomes the weight of the particles.

As the airflow increases, the bed expands. Initially, the expansion is such that the particles remain in contact with each other, however, an additional increase in airflow results in sufficient bed expansion so that the catalyst particles no longer remain in contact. When this expansion occurs, each particle is surrounded by air, and true fluidization begins. If the airflow or velocity continues to increase, the bed expands further. Additional velocity increases result in entrainment of the media particles in the discharging air stream until eventually all are carried out of the bed.

HYDRAULIC CHARACTERISTICS

A fluidized-bed follows Archimedes' principle, i.e., the bed exerts a buoyant force equal to the weight of the displaced fluid. For example, if a stone and a piece of wood of identical size and shape are dropped into a fluidized bed, equal volumes of catalyst media are displaced. Since these volumes are identical, they are equal in weight, and therefore the same buoyant force is exerted. However, the wood floats, whereas the stone sinks, just as they do in water due to differences in mass densities.

Fluidized media also exhibit free-flowing characteristics. A hole in the bed area of the incinerator has media flowing through it much as water flows through a hole in a container. Also, the surface of the fluidized media parallels that of the earth; if its container is tilted, the bed surface remains parallel with the ground. Lastly, the fluidized bed exhibits an apparent viscosity, i.e., forces arise within the fluid that reduce the flow.

SOLID WASTE INCINERATION

The destruction of solid waste by fluidized-bed incineration has two thermal processes. The first is combustion, in which organic materials are burned in the presence of oxygen; the second is pyrolysis, in which the decomposition of solids takes place in the presence of an inert gas at a high temperature. The products of the combustion reaction are totally oxidized, whereas the pyrolysis reaction yields hydrogen, methane, and carbon monoxide, which are unoxidized products. The reactions of each are as follows:

$$\text{Solid waste} + \text{Oxygen} \xrightarrow{\text{Combustion}} CO_2 + H_2O + \text{Ash} + \text{Heat}$$

$$7.50(1)$$

$$\text{Solid waste} + \text{Inert gas} \xrightarrow{\text{Pyrolysis}}$$

$$H_2 + CO + CO_2 + \underset{\text{(methane)}}{CH_4} + \text{Charcoal heat} \quad 7.50(2)$$

Since these reactions take place in fluidized beds at about 1400°F, the heat produced from the combustion reaction helps maintain the bed temperature, while pyrolysis requires heat.

The advantages of using fluidized beds for the destruction of solid waste materials include the following:

1. The heated particles store large quantities of readily available heat.
2. Particle movement throughout the bed prevents the formation of hot spots or temperature zones.
3. High heat-transfer rates result in rapid combustion.
4. Bed agitation prevents solids stratification.
5. Unfavorable gases and products can undergo total combustion, eliminating the need for expensive air pollution control equipment.
6. Few moving parts are located within the bed, reducing maintenance.
7. Temperature variations are minimal (less than 5° to 10°F) throughout the bed.

The disadvantages of using fluidized beds for solid waste incineration are as follows:

1. The maximum temperatures cannot exceed 2000°F when sand is used as a bed medium because sand softens at this temperature.
2. The power costs are high.
3. Equipment is necessary to recover fine solids because the catalyst media become entrained.
4. Auxiliary fuel is usually necessary because the composition and heating value of solid waste vary.

SLUDGE INCINERATION

Fluidized-bed incinerators can be used for the combustion of both organic and inorganic sludge. Sewage sludge with a high concentration of organics has been incinerated, and sludge with high inorganic content, e.g., salts of sodium and calcium, can be disposed in this manner. The pulp and paper industry destroys waste liquors using the fluidized-bed technique.

Two important factors that control sludge destruction in the bed are airflow and operating temperature. Gas passes upward through a bed of sand that contains solid sludge particles less than $\frac{1}{4}$ in in size. Vigorous mixing within the bed optimizes the reaction. Since air velocity is the source of bed agitation, treatment facilities can increase or decrease the reaction rates by controlling the air supply. Temperatures in these beds normally range from 1300° to 1500°F, and the amount of excess air required is 25%. These two factors insure complete combustion so that odorous materials are completely destroyed.

Previously dewatered sludge, which sometimes contains as much as 35% total solids, is normally introduced into the freeboard area of the furnace above the fluidized-bed material (see Figure 7.50.4). Hot gases from the bed evaporate the water in the sludge while the sludge solids enter the fluidized medium.

Once in the bed, the organic material oxidizes to carbon dioxide and water vapor, which exit from the reactor as exhaust gas. Inorganic material either deposits on the bed particles (which is called the agglomerative operation) or leaves the bed in the exhaust gas, which is the nonagglomerative system. Then, dust collection equipment removes the particles in the exhaust gas. For the agglomerative system, continuously or intermittently withdrawing excess bed material maintains a constant bed volume.

The advantages and disadvantages of liquid incineration using fluidized beds are similar to those listed for solid waste disposal. Waste incineration is a new application for fluidized beds. The total combustion of the waste does not produce obnoxious gaseous products. The fluidized-bed method is applicable for both solid and liquid waste disposal.

Wet Oxidation

ACCEPTABLE FEED COD RANGE

25 to 150 gm/l

MINIMUM WASTE HEATING VALUE

To keep the reactor thermally self-sustaining requires 1500 Btu/gal, which corresponds to about 7% solids or 80 gm/l COD. With lower heating values, external heat is needed.

COD REDUCTION

Increases with temperature, reaction time, and feed concentration; ranges from 5 to 80%

VSS REDUCTION

Ranges from 30 to 98%

MATERIALS OF CONSTRUCTION

Stainless steel

OPERATING TEMPERATURE

Cannot exceed 705°F; usually between 300° and 600°F

OPERATING PRESSURE

Usually between 300 and 2000 psig

The patented Zimmermann process involves flameless or wet combustion in aqueous solutions or dispersions. In aqueous dispersions, this process can oxidize a range of organic and industrial wastes to carbon dioxide and water by adding air or oxygen. The wet-air oxidation process (this term preferred by the patentees) does not require dewatering prior to combustion and oxidizes the combustible matter in the liquid phase by applying heat and pressure.

The advantages of this process are that it creates no air pollution and generates sterile, easily filtered, and biodegradable end products. The inherent cleanliness of the fully enclosed system and the potential to generate or recover steam, power, and chemicals are other advantages.

The disadvantages include the need to use stainless steel construction materials and the complex equipment that requires high capital investment and well-trained operators.

PROCESS DESCRIPTION

This process mixes the waste liquor with air and preheats it by steam during startup and by the reactor effluent during operation to about 300° to 400°F (see Figure 7.50.5). At this reactor inlet temperature, oxidation starts with the heat release which further increases the temperature as the liquor–air mixture moves through the reactor. The higher the operating temperature, the greater the COD reduction for the same contact time period (see Figure 7.50.6). The operating temperature cannot exceed the critical temperature of water (705°F) because the continuous presence of a liquid water phase is essential.

The operating temperature in the reactor is the temperature of the saturated steam at the partial pressure of steam in the air–steam mixture. The air–steam mixture is

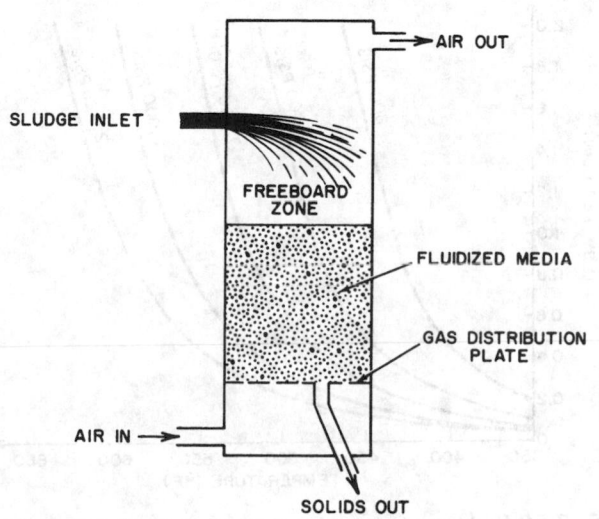

FIG. 7.50.4 Fluidized-bed sludge incinerator.

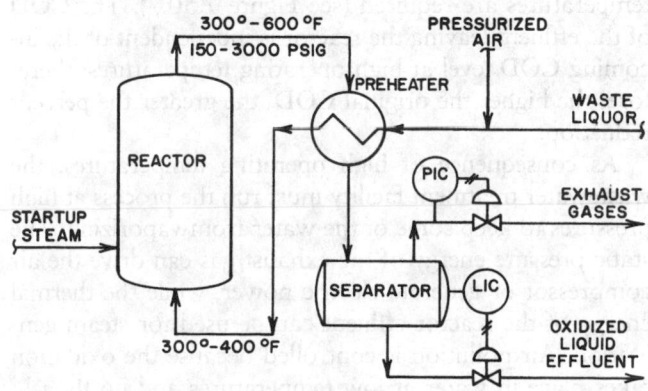

FIG. 7.50.5 Wet oxidation process.

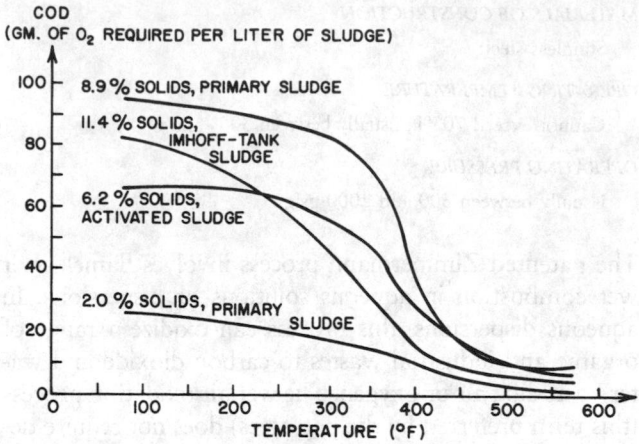

FIG. 7.50.6 COD reduction from sludge exposed to excess air for 1 hr at various temperatures.

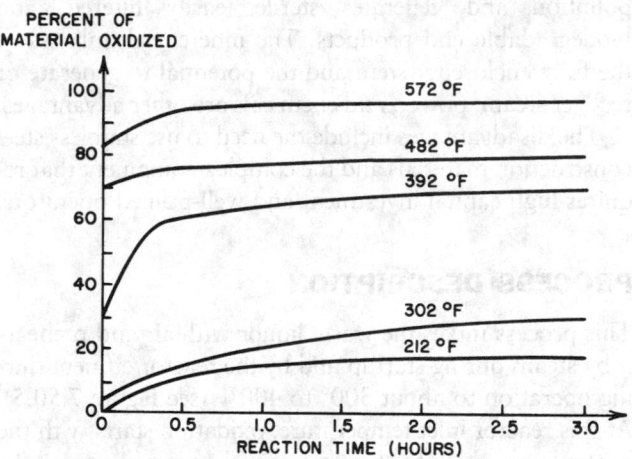

FIG. 7.50.7 High COD reduction and low reaction time with high operating temperatures.

a gas with an increasing CO_2 and a decreasing O_2 content as the material moves through the reactor.

The wastewater treatment facility selects the operating temperature according to the required COD reduction and reaction time. When the reaction time must be short and the reduction in COD must be substantial, high operating temperatures are required (see Figure 7.50.7). The COD of the effluent leaving the reactor is independent of the incoming COD level at high operating temperatures; therefore, the higher the original COD, the greater the percent reduction.

As consequence of high operating temperatures, the wastewater treatment facility must run the process at high pressures to keep some of the water from vaporizing. The static pressure energy of the exhaust gas can drive the air compressor or generate electric power, while the thermal energy of the reactor effluent can be used for steam generation. Air pollution is controlled because the oxidation takes place in water at low temperatures and no fly ash, dust, sulfur dioxide, or nitrogen oxide is formed.

LIMIT ON AIR USAGE

Figure 7.50.8 shows the weight ratio of steam to air at saturation in the reactor vapor space for various operating pressures and temperatures. These curves are based on the following equation:

$$S/A = (53.3\ T)/144\ (Pt-Ps)Vs \qquad 7.50(3)$$

where:

S = steam (lbm)
A = air (lbm)
T = temperature ($°R = °F + 460$)
Pt = total pressure (psia)
P_s = saturated steam pressure (psia)
V_s = specific volume of saturated steam (ft^3/lbm)

The curves in Figure 7.50.8 show the maximum amount of air that can be added per gallon of waste liquor without elimination of the liquid phase in the reactor. The steam-to-air ratio is 2 lbm per lbm at 2000 psig and 595°F, 1500 psig and 553°F, or 1000 psig and 510°F. At this ratio, if each gallon of waste liquor contains 8 lb of water, the addition of 4 lb air per gallon of waste vaporizes all water. Therefore, the wastewater treatment facility must select a lower ratio, such as 3.5 lbm of air per gallon of waste, to maintain some water in the liquid phase.

SEWAGE SLUDGE APPLICATIONS

The wet combustion process has been used in sewage sludge treatment since the early 1960s, and in this period, both continuous and batch plants were installed for either high-pressure or low-pressure operation. Table 7.50.4 describes the capabilities of this process based on data provided by the patentee. The tabulation assumes that the waste has 3.5% solids and a COD of 43 gm O_2/l before oxidation. The table lists three levels of oxidation corre-

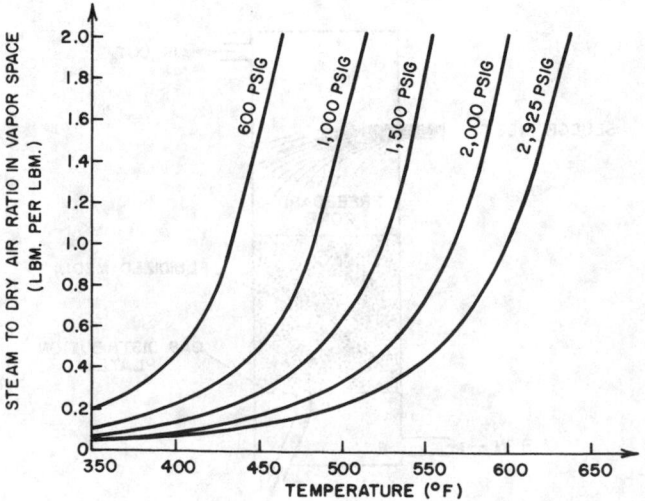

FIG. 7.50.8 Steam-air ratio at saturation in the reactor vapor space for various operating temperatures and pressures.

TABLE 7.50.4 SEWAGE SLUDGE CHARACTERISTICS FOLLOWING DIVERSE LEVELS OF WET-AIR OXIDATION

	Before Oxidation	After Low Oxidation			After Intermediate Oxidation			After High Oxidation		
	Raw Sludge	Oxidized Slurry	Filtrate	Filter Cake	Oxidized Slurry	Filtrate	Filter Cake	Oxidized Slurry	Filtrate	Filter Cake
% COD Reduction	0	5–15	—	—	40–50	—	—	70–80	—	—
% Insoluble Volatile Solids Reduction	0	30–50	—	—	70–80	—	—	92–98	—	—
Filtration Resistance (sec²/gm × 10⁷)	3500	6	—	—	6	—	—	10	—	—
Volume, l	1.00	°1.05	1.00	0.05	1.00	0.97	0.03	1.00	0.97	0.03
TS, gm/l	35.1	29.8	8.2	21.6	21.6	7.1	14.5	16.0	5.5	10.5
Volatile Solids, gm/l	23.4	18.1	5.2	12.9	9.9	4.1	5.8	4.3	3.1	1.2
Ash, gm/l	11.7	11.7	3.0	8.7	11.7	3.0	8.7	11.7	2.4	9.3
pH	6.0	5.2	—	—	4.9	—	—	6.2	—	—
Moisture in Filter Cake, %	—	—	—	64	—	—	58	—	—	50
Drained Cake Dry Weight (lb/ft³) Packed	—	—	—	20	—	—	35	—	—	55
Settled Volume after 4 Hr (%)	100	37	—	—	15	—	—	7.5	—	—
Phosphorus as P°°, gm/l	0.61	0.61	0.18	0.43	0.61	0.06	0.55	0.61	0.02	0.59
Total Nitrogen, gm/l	1.49	1.49	1.15	0.34	1.49	1.42	0.07	1.49	1.44	0.05
Ammonia Nitrogen, gm/l	0.57	0.75	0.72	0.03	1.00	0.98	0.02	1.20	1.19	0.01
Total Sulfur, gm/l	0.30	0.30	0.20	0.10	0.30	0.25	0.05	0.30	0.28	0.02
SO₄ as S, gm/l	—	0.17	0.14	0.03	0.30	0.25	0.05	0.30	0.28	0.02

Source: Zimpro Division of the Sterling Drug Incorporated of New York.
Notes: °Increased volume due to steam injection.
°°× 3.065 = PO_4 content.

sponding to ranges of 5 to 15%, 40 to 50% and 70 to 80% reduction in COD.

The wet oxidation process can accommodate sludge concentrations between COD values of 25 and 150. It reduces the filtration resistance and improves the sludge draining characteristics so that it becomes compatible with vacuum filtration. Sulfur is oxidized to sulfate with no sulfur dioxide leaving with the gas exhaust. The process removes most phosphorus with the filter cake and can accomplish complete precipitation without lime addition. The filtrate liquid contains short-chain, water-soluble organic compounds that arc biodegradable. Nitrogen is not oxidized to nitrite or nitrate, and no nitrogen dioxide (NO_2) leaves with the exhaust gas because the organic nitrogen is degraded to ammonia.

OTHER APPLICATIONS

Powdered, activated-carbon regeneration is practical using the Zimmermann process. The spent carbon slurry at 6 to 8% solids enters the wet oxidation process. The reaction temperature allows the adsorbed organic compounds to be oxidized without destruction of the activated carbon. The application of powdered carbon in wastewater treatment can benefit from the economic carbon regeneration.

Wet oxidation can render plastics, detergents, insecticides, and other nonbiodegradable materials compatible with conventional sewage treatment processes. When the waste contains both paper and plastic material, these need not be separated because both are decomposed at the same reaction temperature.

Wastes that are deficient in nitrogen require the addition of this element for satisfactory biological treatment. Wet oxidation of the sludge can reduce the cost of adding nitrogen because this process returns nitrogen to the biological treatment step as ammonia. Other potential applications of the Zimmermann process include the treatment of tannery, glue factory, plating, sulfide, phenol, paper, cyanide, textile mill, brewery, or photochemical waste and the recovery of chrome, magnesium, titanium, and silver.

Flash Drying or Incineration

FUEL REQUIREMENTS FOR SLUDGE DRYING

6500 Btus/lb sludge with 18% solids

The flash drying or incineration of sewage sludge was a new process in the mid 1930s. It allowed the plant operator to either incinerate the sludge or flash dry it using conventional fuels, such as coal, fuel oil, or natural gas. With this flexibility, the operator produced only the amount of dried sludge that the market could bear and incinerated the remainder.

The odor and dust problems with this process resulted in less use, and in some cases, plant expansions use alternate processes, such as the multihearth incinerator or the fluidized-bed incinerator. Both these methods, however, lack the flexibility of drying or incinerating or both.

The flash-drying system is less costly if no dried finished sludge is produced or the pollution controls are minimized. With new improvements and scrubbers and an increased demand for dried sludge, this process should have an increased demand.

PROCESS DESCRIPTION

The flash-drying method has become common for drying or incinerating sewage solids because of its low capital costs and flexibility. The pretreatment includes sludge thickening in some conventional manner such as by vacuum filters. Then, this method mixes this dewatered sludge with previously dried sludge to reduce its moisture content and its effective particle size. The preconditioned mixture is fed into the drying system where it moves at a velocity of several thousand ft/min in a stream of gas having a temperature of 1000° to 1200°F. The sludge passes through this high-temperature, turbulent zone in a few seconds during which time its moisture content reduces to 10% or less. Next, a cyclone separator separates the hot gases from the fine, fluffy, heat-dried sludge.

If incineration is used, the system introduces the dried sludge produced in the flash dryer into the furnace through special sludge burners and burns it at about 1400°F. The heat from this burning process is recycled into the drying operation.

If the sludge is conserved and sold or sold in bulk or in bags, this method must burn significant quantities of auxiliary fuel in the furnace either separately or with a small quantity of dried sewage sludge. Environmental engineers estimate that this method required 0.4 lb coal plus 0.94 ft³ of natural gas (6500 Btu) to produce a dry pound of sewage sludge when starting with a thick, liquid sludge of about 18% solids.

FLASH DRYING VERSUS OTHER PROCESSES

The multiple hearth process cannot dry sludge without incinerating it. This process is the main competitor to the flash-dryer system and dominates it in use. The fluidized-bed unit is essentially a high-temperature furnace with hot sand fluidized by air jets. Sludge enters the top of the chamber and falls into the hot bed and combusts during the violent mixing of sludge and hot sand. The use of this process is increasing over the flash-dried system although it does

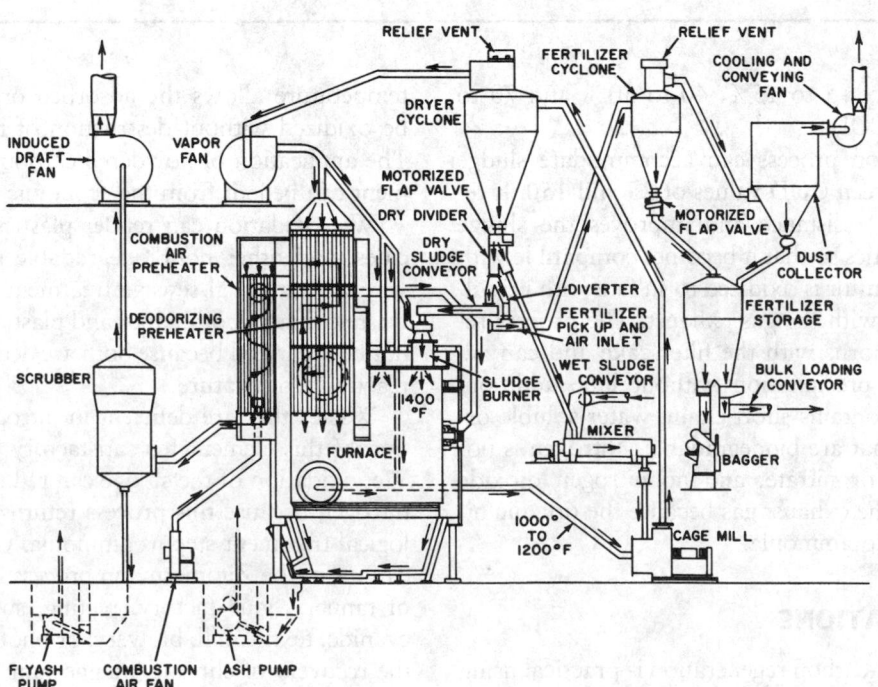

FIG. 7.50.9 Sludge drying and incineration using a deodorized flash-drying process.

not have the flexibility of either drying or incinerating the sludge.

Composting sewage sludge is an innovation because no commercial installations exist in which sewage solids are composted alone. The process has potential both ecologically and economically but requires using shredded municipal refuse. When sludge is composted alone, without shredded refuse, successful treatment requires recycling the drier, already composted, sewage solids with the raw, wetter, dewatered solids, such as occurs in the second step of the flash-drying method. This recycling produces a drier, more porous sludge, which is necessary for good composting.

Although flash drying of sewage sludge is not common due to its previous odor and air pollution problems (which have since been corrected), it has the advantage of complete flexibility in incinerating or drying the sludge at a reasonable price. Thus, a sewage plant using this process can be flexible with one piece of equipment and dry only the amount of sludge that the market demands and incinerate the rest.

Figure 7.50.9 shows the operation of a flash-dryer incinerator system. Flash dryers are also used in the paper industry and to dry sewage sludge.

VALUE OF HEAT-DRIED SLUDGE

Heat-dried sludge compared to sludge dried on sand beds is free of pathogens and weed seeds and is therefore safer to use. However, under normal conditions, heat-dried sludge is more powdery and more difficult to spread and mix with soil than conventional sludge. It is initially repellent to water although once it becomes partially moist, it readily absorbs more water. Therefore, heat-dried sludge must be further treated before being sold as a fertilizer.

It sells in 50- to 65-lb sacks. Although Milorganite, due to its guaranteed 6% nitrogen content, (coming from Milwaukee's beer waste) is still in demand, a similar product made in Chicago (only $3\frac{1}{2}$ to 4% nitrogen) is not as popular, and only a small percentage of their current supply is being sold.

The destruction of pathogens during heat drying is such that only 2 coliform bacteria/gr remain in over 100 samples. Bacterial, parasitic, and viral enteric pathogens commonly found in sewage have the same order of heat sensitivity as coliforms.

—*F.P. Sebastian*
J.G. Rabosky
Béla G. Lipták

7.51
LAGOONS AND LANDFILLS

EFFECT OF WET SLUDGE BEING APPLIED TO LAND

Nitrogen Loss to Leachate
60 mg/l for 1-in layer per year, 1070 mg/l for 12-in layer per year

Other Pollutants
Bacteria can travel up to 100 ft through granular soil; heavy metals migrate only after the soil is saturated.

SLUDGE LAGOON LOADING RATES
400 to 1000 tn dry solids/acre/yr

The disposal of waste sludge through lagoons and landfills is an economical means of ultimate sludge disposal. The lagoons can receive undigested primary sludge, excess activated sludge, or digested sludge as either an interim process in the total sludge handling scheme or as a method of ultimate sludge disposal. Normally, landfills are the ultimate disposal locations for dried (dewatered) sludge, and this disposal method can be economical depending on the haul distance from the wastewater treatment plant to the landfill.

In considering the location, design, operation, and maintenance of sludge disposal lagoons and landfills, environmental engineers must consider the sludge loading criteria, possible health effects through groundwater pollution, the potential for heavy metal accumulation in the soil and groundwater, the possibility of fertilizer nutrients like nitrogen and phosphorus reaching the surface water, and general nuisance developments.

The land availability and climate are important considerations when lagooning is considered as a dewatering technique. Large land areas are generally required. Poor dewatering occurs in cold and rainy climates. Operation and maintenance costs are associated mainly with the removal of dried sludge. Lagoons should not be used as a final treatment for coagulant sludge if the ultimate disposal requires solids concentrations greater than 9%.

Design and Operation

Lagoons are typically earthen constructions equipped with an inlet control device and overflow structures (Montgomery 1985). Building lagoons involves enclosing a land surface with dikes or excavation (Masschelein 1992). Impermeable liners placed in the bottoms of lagoons minimize drainage (Borchardt et al. 1981). Drying occurs by removal of the supernatant and evaporation. Wastewater treatment facilities often use lagoons as a storage and sludge-thickening step prior to further dewatering and ultimate disposal (Westerhoff et al. 1978).

The volume requirements in lagoon design are a function of both the volume of water being treated as well as the degree of dewatering achieved within the lagoon. Figure 7.51.1 shows information on the volume requirements on the basis of these two parameters. For softening sludges, practical experience in several midwestern cities indicates that 0.45 to 0.65 acre–ft of lagoon volume are required per 1 mgd of water treated per 100 mg hardness (as $CaCO_3$) removed (Faber et al. 1969). This design estimate assumes an average sludge concentration within the lagoon of 50% dry solids.

Lagoons can operate either as continuous fill (permanent lagoons) or fill and dry (dewatering lagoons). However, operating lagoons as a thickening process as opposed to an ultimate disposal method is best (Montgomery 1985).

Continuous-Fill Lagoons

For a continuous-fill lagoon, a side water depth of 8 to 13 ft with a 3- to 5-yr capacity is recommended (Montgomery 1985; Faber et al. 1969). Multiple cells equipped with decanting devices are also preferred (Faber et al. 1969). In these lagoons, the sludge is applied in layers, and the supernatant is removed periodically for air drying. This process is repeated until the lagoon is filled with solids. The lagoon is then covered, and the land is reclaimed. The type of disposed sludge dictates any future use of the land.

Lime sludge can reach 40% solids when settled through ponded water. Solids concentrations as low as 20 to 30% are also reported. If the supernatant is allowed to flow off the site or is removed with decanting equipment, up to 50% solids can be achieved (Faber et al. 1969).

Alum and iron coagulant sludge can reach 10 to 15% solids, with iron sludge dewatering faster than alum sludge. Neither is suitable for landfill at this dry-solids content. After 2 to 3 years, 30 to 40% solids can be reached (Masschelein 1992). The percent solids values are overall averages. The percent dry solids concentration varies with depth, being greater near the bottom of the lagoon.

Fill and Dry Lagoons

When wastewater treatment facilities use fill and dry lagoons, they require multiple lagoons to alternate filling, draining and drying the supernatant, and removing the dried sludge. The lagoons are sized based on variables such as sludge production and characteristics and average air temperature. Side water depths of approximately 3 to 6 ft contain sludge discharges from 1 to 3 ft.

The sludge is then allowed to settle. The wastewater treatment facility decants off the supernatant periodically to increase the sludge-to-air contact until the sludge is sufficiently dry. They repeat this process until the lagoon is filled with approximately 4 to 12% solids for coagulant sludge and 40 to 50% for lime-softening sludge.

Several months to more than a year can be required to achieve these solids concentrations (Montgomery 1985). The dried sludge is then removed with a dragline, clamshell, or front-end loader, and the lagoon is used again. Because of the low solids concentrations achieved with coagulant sludge, lagooning typically requires further dewatering for landfill disposal.

Freeze and Thaw

In climates with long periods of temperatures below freezing, freezing and thawing offers a method of further concentrating coagulant sludge (Krasauskas 1969). Freezing releases the hydration water from the aluminum hydroxide complex producing a volume as small as $\frac{1}{6}$ the original volume. Thawing produces small granular particles like coffee grounds that dry to a brown powder which is easily dewatered and disposed (Krasauskas and Streicher 1969).

At least two lagoons are necessary. One lagoon is left to dewater prior to winter and then allowed to freeze. Another lagoon receives the sludge. Wastewater treatment facilities can achieve a final solids concentration of 17.5% for alum sludge with the freeze–thaw method.

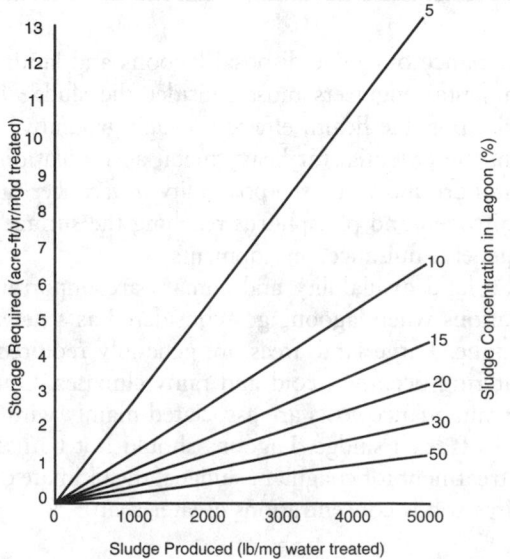

FIG. 7.51.1   Relationship between water volume treated, solids concentration, and required lagoon storage volume. (Reprinted, with permission, from L. Streicher et al., 1972, Disposal of water treatment plant wastes, *Jour. AWWA* [December].)

Health Considerations

Concerning the possible health effects associated with these methods of sludge disposal, the information to date indicates that the potential hazard of disease transmission by pathogenic organisms originating from sludge disposal on land is not significant. Bacteria normally do not travel distances greater than 100 ft through granular soils, and viruses do not pass through 2 ft of clean sand at moderate liquid application rates during 7 months.

The fate of heavy metal pollutants (such as iron, cobalt, nickel, copper, and zinc) in soils is not known, but their chemistry indicates that they generally form insoluble precipitates. The major mechanism for the retention of heavy metals in soil may be sorption on hydrous oxides of iron and manganese, thus significantly retarding the migration of these pollutants in the soil. As the capacity of soil to retain heavy metal elements is exceeded, a breakthrough to the groundwater occurs, and environmental engineers must consider this possibility when planning a permanent lagoon or landfill disposal facility.

The possibility of fertilizer nutrient buildup in groundwater and surface water from sludge lagoon and landfill leachates does exist. Therefore, environmental engineers should be concerned with this possibility because high concentrations of nitrates in drinking water can have toxic effects on humans and nitrogen and phosphorus contribute to eutrophication in surface water.

The nitrogen loss to leachate can be as high as 60 mg/l as N for 1 in of wet sludge applied to land per year and can increase to 620 and 1070 mg/l for 6- and 12-in application rates, respectively. Because sludge lagoons are ordinarily flooded, biological denitrification processes may have a minimizing effect on the nitrogen loss to lagoon leachate.

The potential phosphorus pollution of ground and surface water by leachate is not too serious since soils normally have a high capacity to retain phosphates. Only when the phosphate retention capacity of the soil is exceeded can this type of pollution create a problem; this capacity can eventually be exceeded in conjunction with permanent sludge lagooning operations. Soil erosion in lagoon and landfill areas can also be a source of nutrient pollution by supplying soil-adsorbed phosphorus to surface water.

Odors and troublesome insects are common complaints of the general public concerning the location and operation of sludge lagoons and landfills. Proper operation and maintenance of the disposal sites through restricted access, weed control, and effective landfill covering minimizing these adverse concerns.

Engineering Considerations

Sludge lagoons have been classified as 1) thickening, storage, and digesting lagoons; 2) drying lagoons; and 3) permanent lagoons. Class 1 is specified when the capacity of conventional sludge handling facilities is exceeded, but digestion in a lagoon is a long process and can become a nuisance. With this type of facility, the sludge eventually has to be removed and properly dried.

Ordinarily, drying lagoons are a substitute for sludge drying beds, and the wastewater treatment facility must remove the dried sludge prior to refilling the lagoon. This treatment can require multiple-lagoon units.

The permanent lagoon, from which the sludge is not removed, is the most inexpensive method of sludge disposal provided that adequate land is available close to the waste treatment plant. Facilities for the removal of the supernatant liquid are suggested for this type of lagoon operation.

The engineering layout and design of sludge lagoons should include provisions for uniform distribution of the applied sludge and a convenient method for removing the dry sludge, if necessary. A discharge system that restricts sludge travel to 200 ft has been suggested along with embankments at an exterior slope of 1:2, an interior slope of 1:3, and a top width adequate for maintenance vehicle travel. For the different sludge lagoon operations, the solids loading rates vary from 400 to 1000 tn of dry solids per acre per year.

The higher loading rate is for dewatering lagoons. Sludge lagoons are commonly located on the wastewater treatment plant grounds and can also be constructed in permeable soils when ground and surface water pollution by lagoon leachate is not a potential difficulty. If adequate land is not available on the treatment site, pumping the sludge to locations within 5 or 10 mi of the plant can be economic. Meteorological parameters such as temperature, precipitation, and evaporation influence the operation of sludge lagoons and should be considered in the location and design of these facilities.

Operation

The operation of sludge lagoons and landfills can be based on a 3-yr cycle. In such an operation, the lagoon is filled for a year and then allowed to dry for 18 months followed by cleaning. The supporting soil lies fallow for 6 months before the lagoon returns to operation.

For a dewatering lagoon, filling to a depth of 2.5 to 4 ft is suggested. Wastewater treatment facilities can do this filling by first adding a layer of 1 ft of sludge, then switching temporarily to a second lagoon to allow drying, and then adding the remaining layers to the first lagoon. With one wet year, a 4-ft depth can provide 2 to 3 years of capacity with this procedure.

Sludge lagooning for land reclamation does not require removal of the sludge. Landfilling with dewatered wastewater sludge mixed with municipal solid waste can improve the operation of sanitary landfills by accelerating degradation in the landfill and thereby shortening the time until the landfill can be used.

The principal advantages of lagooning and landfilling for ultimate waste sludge disposal are the low operating and maintenance costs. Among the disadvantages, large areas are required, nuisance difficulties may exist, and ground and surface water can be polluted by leachate. Landfill operations also require an adequate earth supply for covering the fill area.

—W.F. Echelberger

7.52
SPRAY IRRIGATION

TYPES OF SPRAY IRRIGATION SYSTEMS: a) Infiltration type; b) Overland type

NITROGEN AND PHOSPHORUS REDUCTION: Up to 90%

BOD REDUCTION: About 99%

MAXIMUM ACCEPTABLE SODIUM CONTENT IN WASTEWATER: 1000 ppm

REQUIRED WASTEWATER PUMPING PRESSURE: 60 to 100 psig

PRETREATMENT OF WASTEWATER: Screening and grease removal only

HYDRAULIC LOADING BY SPRAY IRRIGATION: 1 in/day (a), 3 in/week (b)

ORGANIC LOADING IN FOOD INDUSTRY: 100 to 250 lb BOD/acre/day is normal; 500 to 1000 is maximum.

TYPICAL INFLUENT AND EFFLUENT CONCENTRATIONS (MG/L): Wastewater BOD: 600 to 700; Underdrain effluent BOD: less than 10

Spray irrigation is a modification of the system used in agriculture for irrigating crops. However, the objective is the disposal of liquid waste rather than providing moisture and nutrients to harvestable crops.

The first operative spray irrigation system in the United States was located in Pennsylvania in 1947. Since 1947, spray irrigation systems have been used for the disposal of waste from paper mills, kraft and neutral sulfite semi-chemical pulp mills, vegetable and fruit canneries, straw-board mills, dairies, fine chemical fermentations, and milk bottling plants. The acceptance of this method of waste disposal is verified by its use in Indiana, Wisconsin, Michigan, Ohio, Illinois, Oregon, New Jersey, Texas, Ontario, Kentucky, Tennessee, Pennsylvania, Minnesota, and Iowa. The system is attractive because of its flexibility and total treatment of applied fluids.

Flexibility in expanding or contracting the capacity of the treatment facilities is especially beneficial with fermentation waste because of the changing quantity and quality of this waste. Total treatment is a solid asset when highly concentrated waste, such as fermentation residue, is handled. Typical fermentation waste with 65,000 mg/l BOD after 98% treatment in a conventional, complete-treatment, biological facility still contains 1300 mg/l BOD, which is usually unacceptable for discharge.

After pretreatment and grease removal, this system sprays wastewater through a sprinkler system onto land that is planted with special grasses. The wastewater infiltrates the ground where soil microorganisms convert the organic waste into inorganic nutrients. The purified water is collected by an underground perforated pipe and sent to a final polishing pond.

The removal of soluble organic wastes by spray irrigation is a highly efficient process for treating industrial waste. Until recently, most of these systems required good soil infiltration characteristics. The system required a site where large volumes of water—as much as 1 in per day—could be applied and where the hydrological characteristics of the soil permitted water transfer underground and laterally out of the area of application. In many cases where the infiltration was adequate, the lack of sufficient lateral movement either limited the rate of application or created flooding. Usually, overcoming these problems involved installing artificial drainage similar to farm tile drainage except for extra precautions in the spacing and the means of avoiding siltation of the collection system because of the high application rates.

Techniques were also developed that use impervious soils for purification by overland flow. These systems are used with impervious clay-type soils in which significant infiltration is not possible.

Physical and Biological Nature

Spray irrigation of wastewater should not be confused with farm irrigation. Water can be purified either as it flows overland or as it percolates through the soil. Purification, in either case, occurs biologically and depends on the biota and organic litter on and in the soil. Pure sand, without organic debris, provides only mechanical filtration without reducing the soluble organic matter.

Most spray irrigation systems rely on water percolating through the soil and flowing away from the irrigation area by an underground route. Ordinarily, this underground flow is natural drainage, but it can also be enhanced by artificial drainage. When hydrological characteristics limit the lateral movement of underground water, wastewater treatment facilities can substitute an overland flow technique.

The grasses planted on the treatment field are multifunctional; they protect the soil surface from erosion and compaction and retard the flow of water across the slope in overland flow systems. They also provide a protective habitat for microorganisms and a vast surface area for adsorption, mass biological activity, and treatment of the impurities in the water. When the grass is cut for hay, it is a valuable crop that can effectively reclaim the plant nutrients released to the soil during decomposition of the organic waste material. Almost any species of grass is satisfactory provided that it produces abundantly, is water tolerant, and forms a turf.

Of all the interacting phenomena in the natural filtration system, microbiological activities are the most important and are carried on by all molds, fungi, bacteria, earthworms, snails, and insects that feed directly or indirectly on the organic waste (see Figure 7.52.1). The mi-

crobial populations in the disposal field, although specific for the plant effluent, comprise a highly complex community.

Organisms use the organic waste products from both carbohydrates and proteinaceous matter as nutrients. The carbon dioxide and water released by the degradation of carbohydrates escape into the air. The ammonia released upon decomposition of protein can be 1) released to the atmosphere, 2) used directly by microorganisms, and 3) converted into nitrite and nitrate. Microbial use of the organic effluent constituents converts a portion of this material into new forms of organic material that, if not removed, are used by different microbial populations. The process is repetitive, and a portion of the organic matter is converted into carbon dioxide, water, and ammonia at each cycle.

Tests show an evolutionary or seeding process whereby microorganisms specific for an effluent develop on the disposal site. The time required for the evolutionary process may be one reason for the greater capacity of older disposal systems. Also, the maturing of the system may be hastened when it is seeded with specific organisms.

Temperature and Shock Load Effects

A spray irrigation system continues to purify water when temperatures are near freezing. Since the respiration of microorganisms slows down as temperature decreases, researchers believed that the impurities were being adsorbed on the surface of the vegetation and held there until the weather grew warm again. However, biological studies show that as the temperature decreases, the number of organisms increases, thus maintaining a constant level of mass activity. Figure 7.52.2 shows this phenomenon.

Spray irrigation systems have the outstanding capability of handling shock loads as well as periods of long shutdowns and immediate startups, producing excellent results in either case. In addition, variations in effluent composition, such as the results of night cleanup, produce no adverse effects. The effluent pH of a spray irrigation system stays between 6.8 and 7.0 although the waste applied at night reaches a pH of 12 for approximately 1 hour and sometimes for as long as 3 hours.

Figure 7.52.3 provides evidence of this dampening effect, showing the diurnal variations of electrical conductivity of both the wastewater and the field effluent for each season. The higher conductivity in runoff during the summer months is due to the increase in evapotranspiration and the decrease in the runoff volume.

For an overland flow system, if a single terraced slope is accidentally overloaded due to mechanical failure, the effluent treatment continues in the other terraces and waterways before the effluent reaches the receiving stream. This capability makes the spray irrigation method a safe technique.

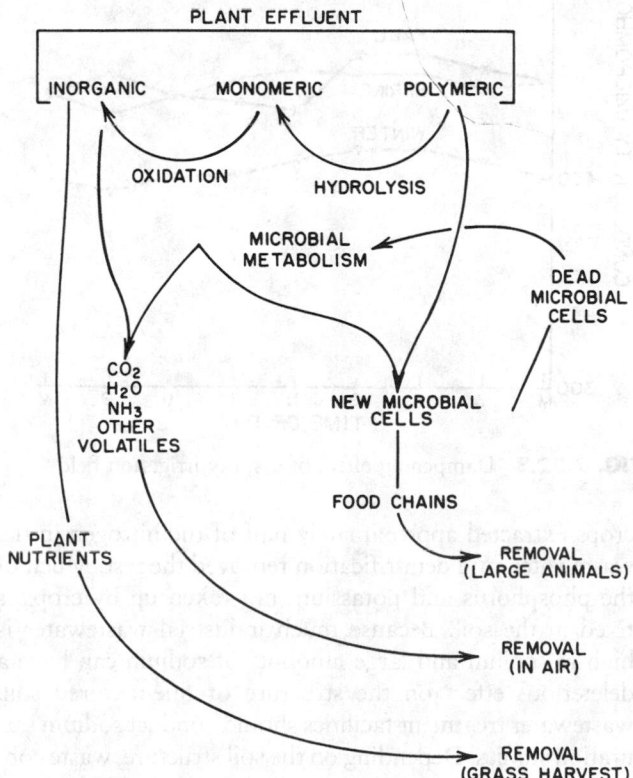

FIG. 7.52.1 Population succession cycle.

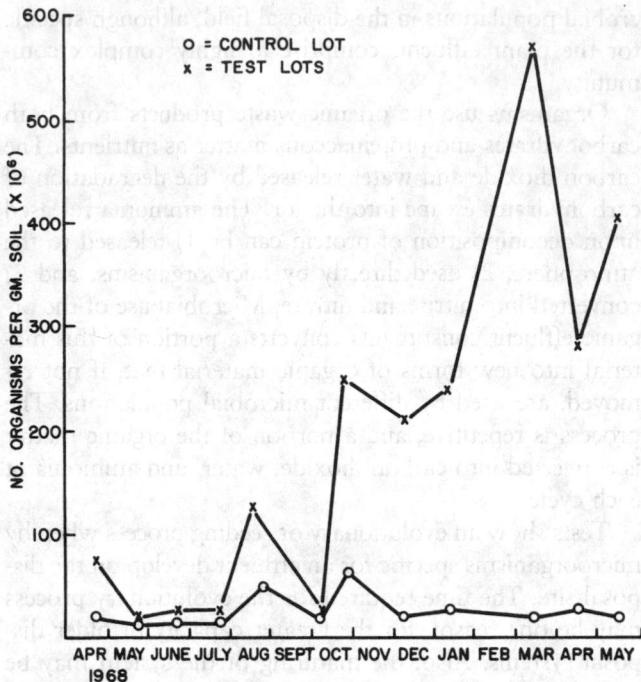

FIG. 7.52.2 Total microbial population on control and test lots.

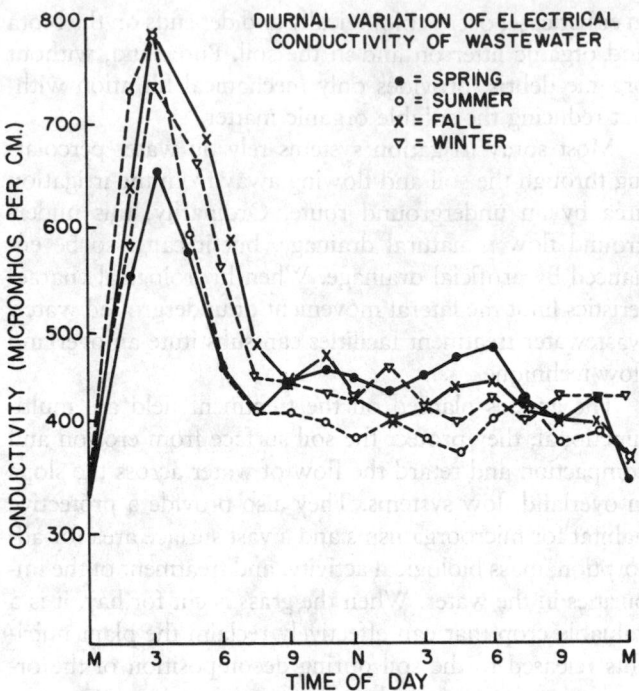

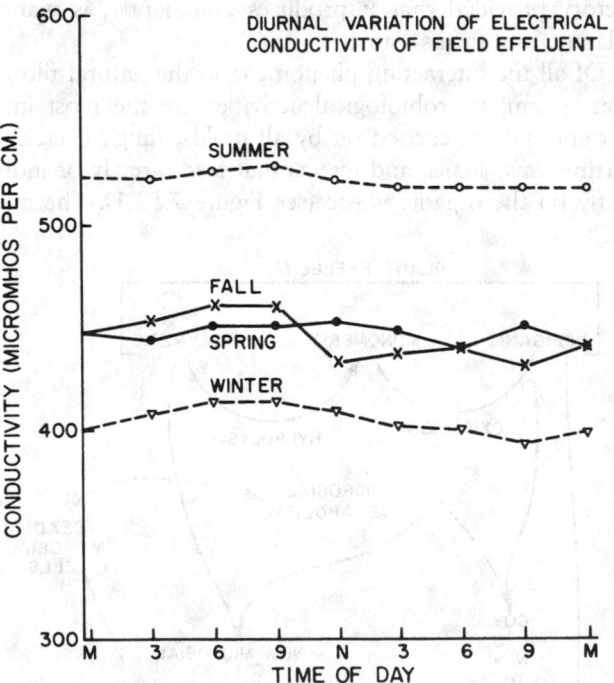

FIG. 7.52.3 Dampening effect of a spray irrigation field.

Wastewater Pretreatment

An advantage of spray irrigation systems is the small degree of pretreatment needed. For most industrial waste, this procedure consists of coarse screening and possible grease recovery. Screening can be achieved on screens with openings of 1 to 2 mm. Fine-mesh screening is not normally warranted since the sprinklers usually have openings of $\frac{3}{16}$-in or larger diameter. Grease separation by gravity is normally adequate for removing most floatable solids and reducing the problems caused by floating material in the pump reservoir. When the wastewater contains sand or fine grit, the use of a detritus separator is also advisable.

For most soluble organic waste, rapid handling and minimum detention time avoid the anaerobic decomposition of the organic matter that creates odors. Pumping is accomplished with high-head pumps, with most designs operating in the range of 60 to 100 psig. The minimum pressure at the sprinklers is 35 psig for efficient spray distribution and breakup.

Sprinkler systems are the most convenient method of uniformly applying waste to the ground without damaging the vegetative cover or soil structure. The organic loading varies depending on the type of wastewater applied. Most food processing waste applications operate at 100 to 250 lb BOD/acre/day although several systems operate in the 500- to 1000-lb/acre/day loading range.

Environmental engineers have conducted studies to determine the fate of nitrates, phosphates, potassium, and sodium added through effluent irrigation. In several cases, crops extracted approximately half of the nitrogen in the wastewater, and denitrification removed the rest. Much of the phosphorus and potassium not taken up by crops is fixed in the soil. Because much industrial wastewater is high in sodium and large amounts of sodium can have a deleterious effect on the structure of fine-textured soil, wastewater treatment facilities should conduct sodium saturation studies. Depending on the soil structure, waste containing more than 1000 ppm sodium is unsuitable for long-term irrigation of fine-textured soils.

System Description

In most cases, waste is pumped through lateral piping and sprayed through sprinklers or spray nozzles located at intervals (see Figures 7.52.4 and 7.52.5). The waste percolates through the soil, and the organic compounds undergo biological degradation. The liquid is either stored in the soil layer or discharged to the groundwater. Maintaining a cover crop, such as grass, vines, trees, or other vegetation, maintains porosity in the upper soil layers. As much as 10% of the waste flow evaporates or is absorbed by the roots and leaves of plants. Trees develop a high-porosity soil cover and yield high transpiration rates. A small elm tree can take up as much as 3000 gpd under arid conditions.

Normally, the waste is first pumped to a holding reservoir, which can be merely an earthen pit but serves to equalize fluctuations in daily operations. The waste is then pumped from the reservoir to the spray nozzles by a header and lateral piping system. The piping system can be permanent, using valves to direct the waste to various locations within the spray field (see Figure 7.52.6), or the piping can be temporary and can be physically moved for spraying different areas. The permanent piping method requires a greater initial investment; the temporary system entails a greater operating cost.

A combination of the two systems is a sound compromise. A permanent headerline from the waste reservoirs across the spray field combined with lightweight, movable laterals with quick-disconnect connections (see Figure 7.52.7) works well if the spray field is passable. A tough and durable steel pipe can be used for the headerline. Lightweight pipe, such as aluminum or plastic, is required for the laterals since they are usually moved daily or weekly.

If the pipe is laid above ground, it must be arched occasionally. Because spraying is done only during daylight, the header and laterals must be drained at night in subfreezing weather. Quick-disconnect connections make system draining easier. The spray nozzles should be durable

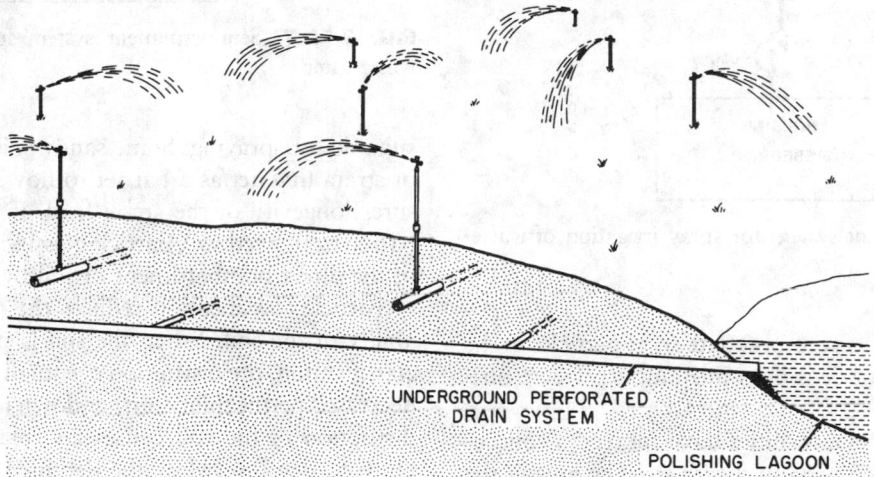

FIG. 7.52.4 Artificial underground drainage system design.

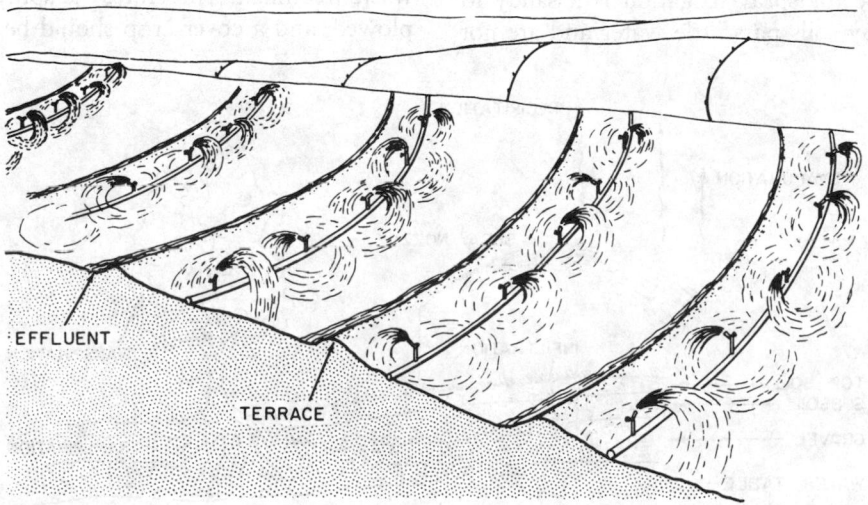

FIG. 7.52.5 Overland technique of spray irrigation.

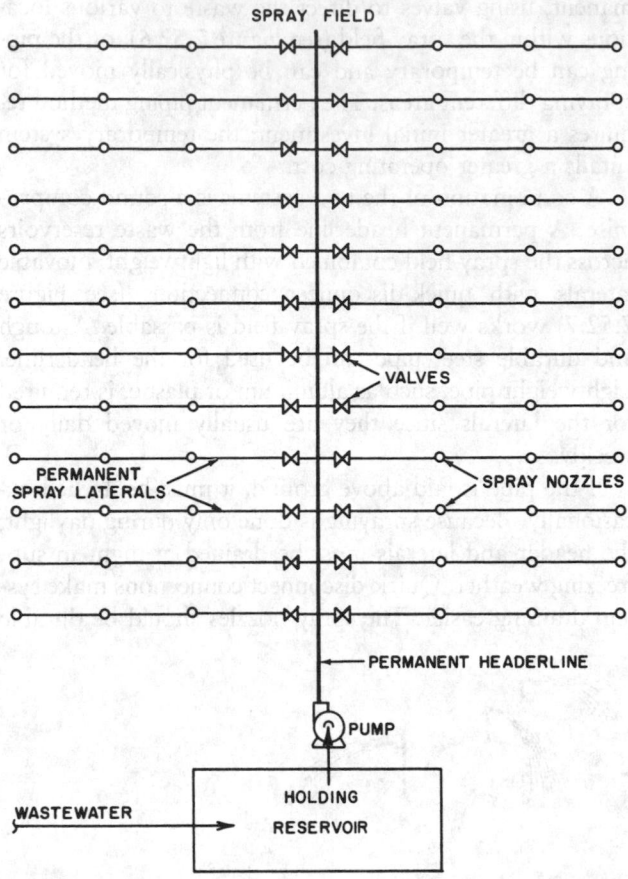

FIG. 7.52.6 Permanent system for spray irrigation of wastewater.

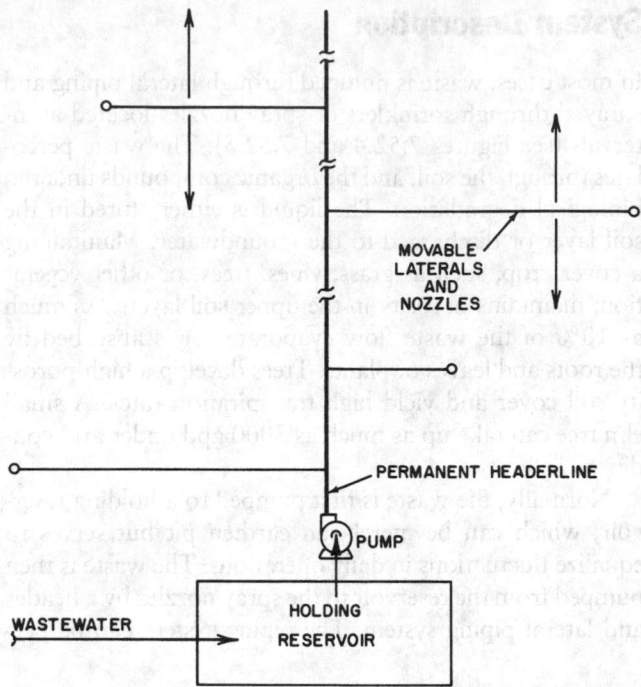

FIG. 7.52.7 Semipermanent system for spray irrigation of wastewater.

and sufficiently large to accommodate solid particles in the waste. An efficient nozzle is the $\frac{5}{8}$-in Rainburg 1-acre nozzle.

Wastewater treatment facilities have sprayed a concentration of as much as 5% SS in the waste. The greater the volume a nozzle can spray, the less chance the nozzle will freeze up.

The ideal soil type for spray irrigation is a sandy to sandy-loam type. Clay soils pass little water and are not suitable for spraying. Some sandy soils exhibit clay lenses or strata that act as a barrier to flow. A ground cover assures longevity of the spray field. If wastes is applied to barren or plowed soil, particle classification occurs because of the direct bombardment of soil by the liquid.

Particle classification reduces the porosity and permeability of the soil. When the soil is plowed after particle classification, the disruption eventually extends the area of decreased permeability farther down in the soil. Continued disruption of the soil causes an area of fluid-flow resistance to develop below the depth of plowing or subsoiling.

Once this condition has occurred, the area is useless for waste treatment. Therefore, the spray field should not be plowed, and a cover crop should be maintained. Experi-

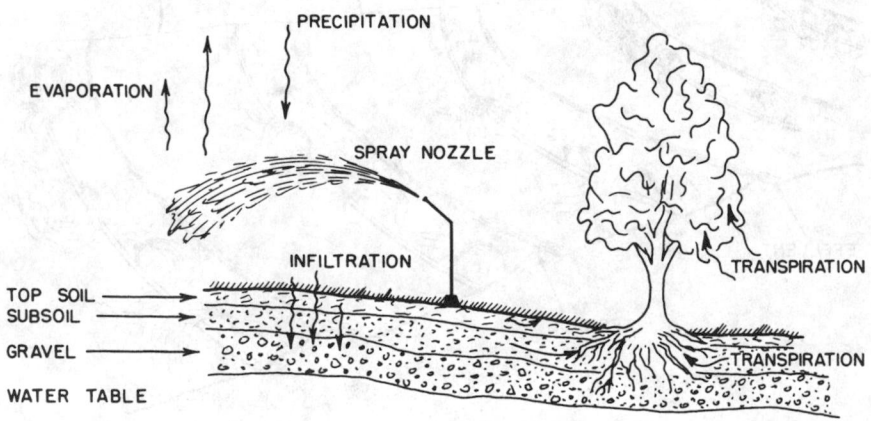

FIG. 7.52.8 Spray irrigation of wastewater.

ments using waste woodchips and bark from a papermill operation as a cover for the spray field are proving successful. The bark chips prevent direct bombardment of the soil and thus particle classification. Figure 7.52.8 shows the phenomenon used in a spray irrigation system.

Wastewater treatment facilities should also install a monitoring system to safeguard the groundwater in the area. Test wells located at various depths and locations can measure the increase and the spread of contamination. Usually, measuring one or two of the major contaminants in the waste (sulfate or nitrate) suffices for groundwater analysis. If those parameters do not increase, the wastewater treatment facility can assume that the groundwater is not being contaminated.

An excessively alkaline or acid waste is harmful to the cover crop and hampers operation. High salinity impairs the growth of a cover crop and causes sodium to replace calcium and magnesium by ion exchange in clay soils. This alteration causes soil dispersion and results in poor drainage and aeration in the soil. A maximum salinity of 0.15% can eliminate these problems.

Design Criteria and Parameters

The capacity of the soil to absorb liquid is proportional to the overall coefficient of permeability for the soil between the ground surface and the groundwater table as follows:

$$Q = 0.328 \ KS \qquad 7.52(1)$$

where:

Q = gallons per minute (gpm) per acre
K = overall coefficient of permeability, feet per minute (fpm)
S = degree of saturation of soil (near 1.0 for a steady-rate application)

The coefficient of permeability K depends on the soil characteristics (see Table 7.52.1).

The application rate depends on the soil structure, the land contour, the waste characteristics, the local evaporation–precipitation rate, and the supervision afforded the spray irrigation system. An efficient initial and design guide is the following schedule for the use of 1-acre spray plots:

1. Spray for 10 hr at 90 gpm; rest the plot for 2 weeks.
2. Spray for a second 10 hr; rest the plot for 2 weeks.
3. Spray for a third 10 hr, and discontinue use of plot after 30 hr dosage (162,000 gal) for approximately 90 days before restarting schedule.

The schedule is a guide; actual practice dictates which spray plots can receive more or less. Applying waste at too great a rate causes ponding and surface runoff. Ponding can result in anaerobic decomposition and thus cause odor problems. In addition, it renders the spray fields impass-

TABLE 7.52.1 COEFFICIENTS OF SOIL PERMEABILITY

Soil Type	Permeability Coefficient K (fpm)
Trace fine sand	1.0–0.2
Trace silt	0.8–0.04
Coarse and fine silt	0.012–0.002
Fissured clay soils and organic soils	0.0008–0.0004
Dominating clay soils	<0.0002

able for pipe moving; therefore, ponding should be avoided.

Environmental engineers should size the holding reservoir to contain at least 2 days of waste production. They should design pumping, piping, and nozzles in both number and size to apply at least one day's waste production during daylight.

Infiltration Techniques

To determine the soluble organic removal capability of a soil, environmental engineers must study the infiltration capacity and permeability of the soil. This capability is a function of the soil texture and depends on the nature of the vegetation, moisture content of the soil, and temperature. Infiltration rates vary from $\frac{1}{10}$ in/hr for low-organic-content clay soils to $\frac{1}{2}$ in/hr for sandy silt loam to more than 1 in/hr for deep sand.

The movement of treated water in soil is a function of the soil pore size, root structure, and evapotranspiration by plants. When the limitations of infiltration or lateral moisture movement are exceeded, ponding and (frequently) failure of the irrigation system occur. Therefore, environmental engineers must tailor the application rates and the hours during which the waste is applied daily to the soil drainage capacity at the site.

When the lateral transmissibility in the soil is not high enough to provide rapid drainage from the application area, an artificial underdrainage system can be installed. Figure 7.52.4 shows such a system. It consists of a bitumastic, impregnated-paper-fiber, underdrain pipe with $\frac{3}{8}$-in perforations at 9-in intervals. The pipe is 5 to 7 ft below the ground surface and is wrapped in a $\frac{1}{2}$-in fiberglass mat used as a filter guard to prevent siltation of the drain pipe by soil particles.

The field in Figure 7.52.4 has a high infiltration rate for the first 5 to 7 ft but is then underlain by dense substrata. Without an underdrainage system, the field can become a morass or marsh.

The spray irrigation system automatically applies approximately a 1-in layer of wastewater to the field per day. The sprinklers are spaced for complete overlap. The underground perforated pipe collects the purified water and

TABLE 7.52.2   TREATMENT EFFICIENCY OF OVERLOAD SPRAY IRRIGATION SYSTEM

| Parameter | Mean Concentration mg/l | | Percent Removal | |
	Wastewater	Section Effluent	Concentration Basis	Mass Basis
TSS	263	16	93.5	98.2
TOC	264	23	90.8	—
BOD	616	9	98.5	99.1
Total phosphorus	7.6	4.3	42.5	61.5
Total nitrogen	17.4	2.8	83.9	91.5

sends it to the final polishing pond. Such a system that begins with raw waste sprayed on the field at an average 635 ppm BOD concentration results in an underdrain effluent concentration of less than 10 ppm BOD. The percent BOD reduction on a concentration basis is almost 99%.

Overland Techniques

The best-documented system of overland filtration is a food industry installation in Paris, Texas. The soil structure is gray clay underlaid by red clay at varying depths that are highly erodable. Little infiltration occurs in this system which is laid on a land so contoured that the waste flows in a thin sheet across the surface. The treatment depends on the microbiological activity of soil organisms to purify the water as it travels across the field. The treated water is collected in terraces, and four or five sprinkler lines are laid out on a hillside as shown in Figure 7.52.5. This installation applies the wastewater at approximately 0.6 in/day or 3 in/5-day week.

Studies indicate that 175 ft of downhill slope provide effective purification. Sprinklers normally blanket an area 100 ft in diameter, and the downslope requirement is 50 ft beyond the perimeter of the sprinklers. For maximum efficiency, the degree of slope should be between 2 and 6%. Flatter slopes encourage puddling and subsequent anaerobic conditions, whereas the retention time on a steep slope is insufficient for complete treatment at normal application rates.

The primary grass in this system is Reed Canary, which yields a large quantity of high-quality hay, containing up to 23% crude protein with twice the mineral content of other good-quality hay. In feeding tests, cattle preferred the hay grown on the disposal site to other types of hay.

By relating potential evapotranspiration to the quality and quantity of the hay crop environmental engineers can predict the time of year or stage of growth when the highest-value hay can be harvested. They can use the relationship between potential evapotranspiration and soil tractionability to plan a hay harvest that least disrupts the disposal system's normal operation. Poor soil traction-

ability can interfere with a planned harvest, whereas in other areas with lighter soils, an optimum harvest can result in highest-crop value and equipment utilization. Two or more harvests of Reed Canary grass per growing season may be feasible.

This installation has achieved removal rates of up to 90% phosphorus and nitrogen in the wastewater. The subsequent reclaiming of most of these nutrients by the hay crop extends the finite capacity of the soil to store nutrients.

While the soil concentration of TDS and sodium is increasing at this site (Paris, Texas), it has not reached the point that is injurious to plants. Some signs exist that the rate of increase is lessening and a state of equilibrium is being approached. Nitrates percolating through the groundwater reserve are not expected to build to a harmful level.

Table 7.52.2 summarizes the treatment performance of the Paris overland flow system. Early observations indicated that while BOD and nitrogen removal were high—99 and 90%, respectively—the phosphorus removal was low, about 45% (see Table 7.52.2). A change in the operating procedure that provided a longer rest period between applications, with no change in the total volume, increased the phosphorus removal to nearly 90% without affecting the BOD or nitrogen removal efficiency.

An analysis of groundwater samples showed that while mineral salts had increased over a 5-year period, the total accumulation was not critical and the rate of increase appeared to be dropping off. Based on the data accumulated to date, no significant disturbance to the soil structure is anticipated for 35 to 50 years, and an equilibrium stage (due to rainfall) will probably be reached sometime in the interim.

Conclusions

Industrial wastewater disposal by spray irrigation provides a low-cost method of waste treatment in many areas. Soils have an enormous capacity to absorb pollutants and convert them into plant nutrients or inorganic substances through microbiological activity. Placing soluble organic matter on soil oxidizes it to carbon dioxide or converts it to humus. Phosphates are held by the soil, and nitrates are taken up by the plants. Nitrates are also denitrified by soil microorganisms. Thus, a high-quality water effluent is produced.

The least-expensive systems are those on soil with high infiltration capacity and efficient hydrological characteristics. However, successful installations can be made on flat, poorly drained areas or gently sloping land.

—*T.F. Brown, Jr.*
L.C. Gilde, Jr.

7.53
OCEAN DUMPING

Dumping waste in the ocean, seas, estuaries, or inland lakes is regulated by emission standards for abating pollution of the oceans. The concept of an infinite ocean (a mile and a half deep on the average around the world) has given way to the reality that the ocean is a limited and valuable resource. This resource must be protected, otherwise it can become like Lake Erie and the Baltic Sea, and irreversible oceanic life systems may create an uninhabitable environment for people as well as for marine life. To quote the famous French underwater explorer–scientist Jacques Cousteau, "In 30 years of diving, I have seen this slow death everywhere underwater. In the past 20 years, life in our oceans has diminished 40%".

Some calculations suggest that the cycle time of oceans (the time required for an ocean's waters to evaporate, form clouds, return to the land in the form of rain, percolate down into the groundwater, and eventually return to the rivers and back to the ocean) is about 2000 years. Cousteau was reporting only the first visible consequences of ocean pollution. It will take 2000 years to learn the total impact of the pollution to date.

Most scientists say that more research is needed, but in the meantime, caution is essential. A research facility (New York Ocean Science Laboratory in Montauk, Long Island) is emphasizing the study of ocean pollution, including the effects of organic matter and heavy metals, particularly mercury.

Worldwide, losses from pollution are due to reduced eatable seafood despite the addition of organic matter and nutrients.

Effects

Some authorities argue that adding organic nutrients benefits the sea, citing statistics on the increased yield of fish. Increased fish yields in controlled environments such as nutrient-rich fish ponds support this view. Others argue that ocean dumping of sludge is safe because sludge contains only treated and stabilized biodegradable substances without any floatables.

Some also argue that as long as enough DO is in the water to support animal life and decompose organic waste, sludge dumping does not upset the ecological balance of the receiving water. With this logic, the ocean can be viewed as a great sink, capable of absorbing almost anything that is thrown into it. If this view were correct, San Francisco Bay could handle the waste of 200 million people since the tidal action in the bay replaces the water twice a day.

Some frequently argue that sludge, the end product of the sewage treatment process, is a benign substance. This statement is not completely true. Unfortunately, the sludge from a city like New York also contains toxic industrial waste because industrial plants frequently dump their waste in the municipal sewage system. The synergistic effects of the many synthetic chemicals, toxic substances, pesticides, PCBs, heavy metals, and medical wastes containing viruses and bacteria are not fully understood and are likely to be harmful to the ecology of the receiving water. Many experts feel that neither the DO level nor the organic-waste-assimilating capacity of oceans is a safe criterion for waste disposal. The effects and interrelationships are more complex, and the consequences are not understood well enough to accept such simplistic arguments.

Since dumping began at the 106 Mile Site, which receives New York's sludge, Rhode Island fishermen and lobstermen have reported diseased lobsters and crabs and a general drop in their catch of bottom-dwelling fish. Some controversy exists concerning the fishermen's reports that sludge is driving the fish away or that it causes shell disease (burn-spot disease) in lobsters and crabs. The National Oceanic and Atmospheric Administration (NOAA) indicated that the ocean dilution is sufficient to eliminate the harmful effects; however, it made that statement without studying the fish in the area.

Scientists have reported a proliferation of certain forms of sea life at the 106 Mile Site. Part of the reason why damage to the receiving water is reduced, if not eliminated, is due to the great depth and large area of this site. The heavy fraction of the sludge takes 3 to 4 days to sink to the bottom, while the lighter particles take up to a year. This delay allows more time for microorganisms to decompose the sludge. While scientists have reported that the ocean bottom at the dump site teems with sea life, they have not yet determined if this sea life is contaminated with bacteria, viruses, or heavy metals that might enter the food chain.

Biological studies on the Chesapeake Bay relate primarily to the disposal of bottom deposits dredged from the area and show the degree of pollution and its drift and harmful effect on the environment. For example, the total phosphates and nitrogen levels increased by a factor of 50 to 100 over normal levels as spoil material deposited over an area five times that designated for the disposal. The studies indicate that life was adversely affected, particularly that of the bottom organisms. A mathematical approach is useful to the understanding of benthic sludge decomposition and the degree and rate of purification of waste deposited on the sea bottom surrounding outfalls.

Regulation

Ocean dumping of sludge and other solid waste is widely practiced in Europe and Japan. For many decades in the United States, sewage sludge was barged to approved areas on the Gulf and Atlantic coasts. Toxic waste to be dumped in the ocean was usually put in containers and shipped to more remote locations. While many cities, such as San Diego and San Francisco, have banned ocean dumping, others continue to barge their sludge into the sea.

The marine disposal of radioactive waste was terminated in the United States in 1967. Yet in 1968, the yearly quantity of other waste dumped in the sea was still close to 50 million tn (see Table 7.53.1). Even though Congress banned ocean dumping of sewage sludge in 1992, this form of waste disposal will probably not stop completely until the turn of the century.

Unregulated disposal beyond the boundaries of the territorial sea imperils the waters, resources, and beaches of the maritime nations. Specific legislation is needed to give national and international authorities the responsibility for preventing ocean pollution and protecting ocean resources. Creating such authorities and enforcement methods is a slow, difficult process, and no effective policy for ocean management has evolved on either the national or international level.

ILLEGAL DUMPING

The incident of the *Khian Sea* shows the state of international controls on ocean dumping. In September 1986, the *Khian Sea*, owned by a Bahamian company, the Amalgamated Shipping Corporation, loaded in Philadelphia with 28 million lb of toxic incinerator ash. The ship attempted, unsuccessfully, to discharge its load in the Bahamas, the Dominican Republic, Honduras, Costa Rica, Guinea-Bissau, and the Cape Verde Islands. In February 1987, it discharged 4 million of its 28 million lb of ash in Haiti but then was ordered out of that country.

In September 1988, the *Khian Sea* was sighted in the Suez Canal with a new name, *Felicia,* and a new owner, Romo Shipping, Inc. In October 1988, Captain Abdel Hakim, vice-president of Romo Shipping, sent a message to Amalgamated Shipping (which has since gone out of business) indicating that the ash had been discharged; Hakim did not say where it had been discharged. The resolution of this case and who will answer for what in front of which legal authority is not clear. But this incident illustrates the chaotic state of international control over dumping in international water.

Illegal dumping is not limited to international water. In December 1988, the State Attorney General of New York accused the General Marine Transport Corporation of illegal dumping in the Raritan River, the Hudson Bay, and the coastal waters of New York City. The lawsuit also names four officers of the corporation and recommends placing environmental police on barges.

Some sludge haulers do not take their loads to designated areas but dump them closer inshore. To control this situation, the EPA now requires that each load of sludge be accompanied by a black box, which is dumped with the load. This requirement allows the EPA to protect against cheating. Another EPA requirement is that the barge must dump the load slowly to maximize dispersal.

Illegal dumping also includes medical waste, which has caused New York area beaches to close. Until this dumping occurred, no government agency was charged with tracking the safe disposal of hospital waste from the point of generation to the point of disposal. After the beach closings, Congress introduced bills requiring the EPA to create a paper trail to control the disposal of medical waste. New regulations in New York State require hazardous and infectious medical debris, including needles, to be placed in strong, moisture-resistant, red bags conspicuously labeled as infectious. Medical practitioners must either carry their infectious waste to an approved hospital incinerator or deliver it to a certified trucker. Regulations have also established an elaborate record-keeping system to track waste from the source to disposal.

Sludge Pipelines and Marine Fills

Some coastal cities have found that piping sewage sludge into the sea is less expensive than barging it. In the late 1960s, Los Angeles reduced its sludge-handling costs from $14 to $2/dry ton by constructing a 7-mi-long, 22-in-diameter pipe and discharging the sludge through the pipe

TABLE 7.53.1 ESTIMATED MARINE DISPOSAL FOR 1968

	Pacific Coast		Atlantic Coast		Gulf Coast		Total		Percent of Total	
Type of Waste	Annual Tonnage	Estimated Cost, $	Annual Tonnage	Estimated Cost, $	Annual Tonnage	Estimated Cost, $	Annual Tonnage	Estimated Cost, $	Tonnage	Cost
Dredging spoils	7,320,000	3,175,000	15,808,000[a]	8,608,000	15,300,000	3,800,000	38,428,000	15,583,000	80%	53%
Industrial										
waste	981,000	991,000	3,011,000	5,406,000	690,000	1,592,000	4,682,000	7,989,000	10%	27%
containerized	300	16,000	2200	17,000	6000	171,000	8500	204,000	<1%	1%
Refuse[b]	26,000	392,000					26,000	392,000	<1%	<1%
Sludge[c]			4,477,000	4,433,000			4,477,000	4,433,000	9%	15%
Miscellaneous	200	3000					200	3000	<1%	<1%
Construction and demolition debris			574,000	430,000			574,000	430,000	1%	2%
Explosives			15,200	235,000			15,200	235,000	<1%	<1%
Total waste[d]	8,327,500	4,577,000	23,887,400	19,129,000	15,986,000	5,563,000	48,210,090	29,269,000	100%	100%

Notes: [a]Includes 200,000 tn of flyash

[b]At San Diego, 4700 tn of vessel garbage at $280,000 dumped in 1968 (discontinued in November 1968).

[c]Tonnage on wet basis. Assuming average 4.5% dry solids, this amount is about 200,000 tn/yr. of dry solids being barged to sea.

[d]Radioactive waste omitted because sea-disposal operations were terminated in 1967.

at a depth of 320 ft on the edge of a submarine canyon. Later, Los Angeles mixed some of its sludge with sawdust and sold it as compost.

In designing outfalls in the ocean, environmental engineers try to achieve good mixing between the heavier saltwater and the lighter sewage effluent. Good mixing is essential to ensure that the waste does not surface like an oil slick and pollute the shoreline and beaches. Environmental engineers can use the following equation in designing an outfall:

$$N = (K)(Q^2)/(Y)(X^2) \qquad 7.53(1)$$

where:

N = The maximum tolerable shore pollution expressed as the arithmetic mean or 80 percentile
Q = The average sludge flow
Y = The depth at which the sludge is discharged into the ocean
X = The distance of the discharge point from the shore
K = A constant that varies from 5 to 10 million, when the units are in feet and gallons

Another method of disposing sludge and municipal solid waste is ocean landfills. Hong Kong has built marine fills for disposing its refuse in the estuaries of the bay. The marine fill is surrounded by a solid rock dyke, and refuse is loaded and compacted into it. Tidal action causes some leaching of the marine fill. After about 3 years, the refuse converts into compost and is used on nearby farms.

Disposal Methods

The most common ocean disposal method is to thicken the waste to a sludge or solid and barge it to the point of disposal. When the waste is toxic, it is usually put in containers and dropped in more remote places. So-called approved areas exist on the East, Gulf, and Atlantic coasts for waste disposal.

Detailed oceanographic studies indicate that inversion areas, in the water above the outfalls, of piped sludge limit the spread of solids and coliform bacteria to the surface although evidence exists that some digested sludge travels as much as 6 miles.

Wastewater treatment facilities can also dispose sludge at sea through a pipeline either by diluting digested sludge with the treated effluent from the plant or reducing the solids content and allowing the solids to be diffused into the ocean with the sewage. The advantage of removing the solids and digesting them prior to disposal is that this treatment results in an 80% reduction of volatile solids and more than a 99% removal of coliform and pathogenic bacteria.

The disposal of sludge and other solid waste in the ocean is more prevalent in other countries than in the United States (particularly in the industrialized areas of Japan and Europe).

Abstracts appearing regularly in the *Journal for Water Pollution Control Federation* stress the need for stricter regulations for ocean dumping as well as pollution abatement on a world scale. A low-cost alternative to ocean disposal is the disposal of digested sludge on croplands.

—*J.R. Snell*
Béla G. Lipták

7.54
AIR DRYING

LAND AREA REQUIRED

   1 to 2 ft^2 per capita

LAYER THICKNESS ON SAND BEDS

   7 to 8 in of digested primary sludge with 6 to 8% solids

MOISTURE CONTENT OF DRIED SLUDGE

   60 to 70%

SALE PRICE OF DRIED SLUDGE

   Usually free

For a small community, air drying digested sludge is the accepted, most common, and most economic process for sludge treatment and disposal. The advantages of simplicity and economy overshadow the disadvantages of potential nuisance, susceptibility to adverse weather, residual pathogens, weed seeds, and insect populations. Design criteria are well established for various parts of the United States. Wastewater treatment facilities have replaced most of the sand with wide strips of pavement to facilitate the mechanical removal of the sludge. Except in dry regions of the country, this pavement has reduced draining and greatly increased the drying time.

Various additives used to reduce the drying period (alum, lime, and polyelectrolytes) are not practical. Wastewater treatment facilities can improve drying in open sand beds by the following means:

1. Making the sand bed uncompacted and smooth prior to flooding
2. Providing 7 to 8 in of wet-sludge depth for optimum results in an uncovered dry bed
3. Providing 12 to 14 in of wet sludge in covered beds
4. Providing a prethickened sludge, which is better than a thin sludge

Except for odor control in developed areas or areas with an extremely cold climate, beds need not be covered. Final preparation of the dried sludge for public use, such as shredding, windrow composting, or heat drying, increases its value.

Sand-bed drying has many advantages for smaller communities. Over 70% of communities with a population less than 5000 use sludge drying beds. This use drops to 25% for municipalities with a population between 5000 and 25,000, and to 5% for cities with more than 25,000 inhabitants. The only economic alternative to sludge drying beds for a small community is a sludge lagoon or land disposal.

When communities use composting in digesting the organic part of municipal refuse, introducing a gravity-thickened, raw sludge is economically competitive. For larger coastal communities, ocean disposal has been used but is being scrutinized as polluting. Perhaps the strongest advantage of using the drying beds is their simplicity; no special skills are needed to operate them.

The greatest disadvantages of this technique are the large areas required, the potential nuisance from odors and insects, and the cost of labor to remove the sludge after drying. Open drying beds are susceptible to adverse weather, while covering them (except under unusual circumstances) is impractical. The weathering process of drying on an open sand bed causes some nitrogen loss. Pretreatment through anaerobic or aerobic digestion prior to dewatering is necessary to stabilize the sludge. Weed seeds and pathogens are not destroyed, and open sand drying beds attract insect populations.

Area Needed

The area required for open sand beds depends on the weather conditions (humidity, temperature, and rainfall). Tables 7.54.1 and 7.54.2 give the area needed for sludge drying beds under different circumstances.

Mechanical Sludge Removal

The minimum design for sludge removal should include two concrete strips leading into each sludge drying bed so that a pickup truck can back up close enough to all parts of the bed for an operator to cast the dried sludge directly onto the truck with a pitchfork. A depth of 9 to 18 in of well-washed sand should be used in the bed. A uniformity coefficient not over 4 and preferably under 3½ should be used, and the effective size of the sand should be between 0.3 and 0.75 mm.

Underdrains should have 4-in diameter pipes spaced 8 to 10 ft apart so that they do not fill with sand. Backfilling with gravel around underdrains enhances their effectiveness. Partially covering the bed with asphalt surfaces sloping 1 to 2 in/ft allows front-end loaders to remove sludge.

Paved drying areas allow the use of mechanical equipment but decrease the drying rate compared to drainable, sand drying beds. An open-type, asphalt bed with a small layer of sand over it is a good compromise for maximizing drainage while supporting equipment.

CHEMICAL AND PHYSICAL CONDITIONING

Properly administered, chemical pretreatment of sludge before drying can reduce the drainage and drying time by

TABLE 7.54.1 SLUDGE DRYING BED AREA NEEDED FOR DEWATERING DIGESTED SLUDGE

	Area (ft²/capita)[a]	
Type of Sludge	Open Beds	Covered Beds
Primary	1.00	0.75
Trickling filters	1.50	1.25
Activated sludge	1.75	1.35
Chemical coagulation	2.00	1.50

Note: [a]South of 40° north latitude, these figures can be reduced by 25%; and north of 45° north latitude, they should be increased by 25%.

TABLE 7.54.2 SLUDGE DRYING BED AREA NEEDED FOR DEWATERING DIGESTED SLUDGE—RANGE OF VALUES

	Area (ft²/capita)[a]	
Type of Sludge	Open Beds	Covered Beds
Primary digested	1.0 to 1.5	0.75 to 1.0
Primary and humus digested	1.25 to 1.75	1.0 to 1.25
Primary and activated digested	1.75 to 2.5	1.25 to 1.5
Primary and chemically precipated digested	2.0 to 2.25	1.25 to 1.5

[a]With glass-covered beds, more sludge drawings per year are obtained because of protection against rain and snow; a combination of open and enclosed beds provides maximum use. Open beds can evaporate cake moisture faster than covered beds under favorable weather conditions.

50%. Most operators do not use chemicals under normal circumstances; they use them only when extra quantities of sludge must be disposed on overcrowded facilities. The usual chemical choices are ferric chloride, chlorinated copperas, and alum. The dosage for alum is about 1 lb of commercial grade in 100 gal of digested sludge. It reacts not only with the hydroxide ion to produce a floc of $Al(OH)_3$ but also with carbonated salts in digested sludge to release carbon dioxide, causing the sludge particles to float and the liquor moiety to drain away more readily.

The best coagulant and optimum dose for a sludge are best determined in the laboratory and then tried on a pilot scale in the field. The sludge should be freshly drawn from the tank and properly mixed with the liquid coagulant just prior to the sludge running onto the sand drying bed.

Both inorganic coagulants and organic polyelectrolyte flocculants can be used. Polyelectrolytes save not only money but also floor space and improve safety and reduce operating time. The findings for polymer use in sludge preflocculating before vacuum filters also apply to preflocculating prior to sand bed drying.

The drying rate of well-digested sludge can be classified into two phases: (1) a constant-rate drying period and (2) the falling-rate drying period, which on the average is 5% greater than the evaporation rate of free water. Formulas for determining the drying time of well-digested sludge are believed to be applicable to both sludge drying beds and lagoons.

Fly ash can also be used in wastewater treatment and sludge conditioning. Since fly ash is generally a waste product from most power plants, it can be obtained for the cost of hauling. Fly ash alone or combined with lime can successfully be used a coagulant for either wastewater or sludge. Sawdust and coal have also been used with some success as an aid in sludge dewatering. Introducing these solid additives is more difficult than adding liquid ones.

FREEZING

The phenomenon of a rapid water release after sludge has been slowly frozen and thawed is well known. The problem is that freezing and thawing sludge on sand beds is difficult. It requires the installation of freezing tubes and a ready way of adding and removing insulation to and from the surface. Plant operators can frequently take advantage of freezing and thawing by placing sludge on the bed late in the fall and removing it early in the spring or during a winter thaw.

VACUUM UNDERDRAINS

Theoretically, evacuating the underdrains of sand drying beds speeds up the initial dewatering of sludge. As soon as water drains to the point at which air can enter, the vacuum does little or no good unless the surface of the sludge bed is covered with a plastic or membrane seal. The practicality of speeding dewatering by this method is arguable but should be further explored.

OPTIMUM DEPTH FOR VARIOUS SLUDGE

A depth of 7 to 8 in is optimum on open sand beds dosed with well-digested primary sludge of 6 to 8% solids. Greater depths can be used under certain conditions, and depths as high as 12 to 14 in are possible with covered sand beds. A thinner sludge can be placed in a deeper layer than a thickened sludge. The same may be true of sludge pretreated with chemicals or polymers. Trickling-filter or activated sludge alone or combined with primary sludge should not be placed in as thick a layer as primary sludge alone because of the slime present.

DEGREE OF DRYNESS

The purpose of a sludge drying bed is to reduce the moisture content of digested sludge from 4 to 8% solids to 20 to 40% solids. Figure 7.54.1 shows the relationship between the percent solids of sludge applied to the bed and the expected pounds of solids that can be obtained from each square foot of bed annually. Generally, mechanical equipment can remove sludge with as much as 70 to 80% moisture. Hand removal with a pitchfork with tines about 1 in apart is the most common way of removing dried sludge in smaller plants. The moisture content should be between 60 and 70% for best results. At this moisture content, the sludge begins to peal away from the sand but does not yet crumble.

The more moisture the sludge contains, the more weight must be removed. If sludge is shredded and stored for use by the citizens, the moisture content at the time of removal is critical. If time permits and ample sand bed capacity exists, the operator can sometimes reduce the moisture content to as low as 50%. If heat drying is used prior to bagging, the sludge should be dried to at least 50% moisture. The same is true if composting is used as a pretreatment to bagging or use by citizens. However, if the sludge is taken to a landfill or farm, it can be removed with a moisture concentration as high as 70 to 80%.

COMPOSTING PARTIALLY DRIED SLUDGE

Equipment for composting animal manures should be applicable for composting sewage sludge.

MECHANIZED REMOVAL

In large cities, mechanical equipment removes the sewage sludge after drying on sand beds. Most larger modern installations use vacuum filters or centrifuges to dewater the sludge prior to ultimate disposal. Intermediate-size plants, where labor costs have increased the cost of hand sludge

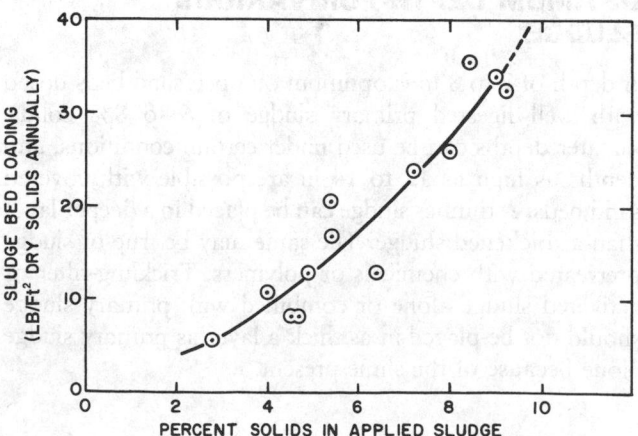

FIG. 7.54.1 Sludge volume reduction at a variable rate during drying, with the greatest volume reduction occurring at a 70 to 60% moisture content.

removal, have installed lightweight, front-end loaders and scrapers.

To prevent crushing underdrains or disturbing the sand, installations use pavement in strips or fractional areas. In other cases, flotation-type equipment is used directly over the sand. Occasionally, specially designed equipment spans the sludge drying bed and is supported on rails on either side.

MAINTENANCE OF SAND BEDS

Installations should consider the following recommendations for good care of drying beds:

1. A coarse, 2.0-mm-effective-size sand should be used on the surface.
2. The sludge should be drawn slowly so that a hole is not pulled in the digester sludge blanket.
3. The sludge bed should be thoroughly cleaned, removing all small bits of dried sludge before the bed is reflooded.
4. The bed (preferably 8 in in depth) should be rotated if possible.

A report from Birmingham, England, showed a 9 in depth to be optimum after depths from 6 to 18 in were tried. These researchers also found that only 11% of the volume of sludge left the beds through the underdrains while the rest evaporated (Bowers 1957).

For tropical and semitropical areas, sludge drying beds should be substantially smaller than those in moderate or cooler climates. Because sand beds give only four loadings per year in Winnipeg, Canada and these loadings must be during the summer season, they are impractical. Drained lagoons in the same location generate no obnoxious odors and produced efficient drainage and drying in an 18-month period when poured to a depth of about 3 ft (Bubbis 1953). Without underdrains, 24 months are required. Drainage rates increase with coarser types of drainage media although at a cost of less solids in the filtrate.

COSTS AND SALES

No appreciable revenue can be expected from the sale of digested primary sludge dried on open sand beds if the quantities are small. The labor cost of removing and preparing this sludge for use is substantial. The purpose of encouraging the public to use it is more for sound public relations than to subsidize the treatment and disposal costs.

People generally like to have some humus on their gardens. They look to the city sewage plant as a place where it is readily available and are usually willing to go there for it but do not expect to pay for it. Because of the low quality of this type of sludge, it is generally not bagged and sold, as is heat-dried sludge.

—*J.R. Snell*

7.55
COMPOSTING

Composting is the aerobic, theromophilic, biological decomposition of organic material under controlled conditions. It is essentially the same process that is responsible for the decay of organic matter in nature except that it occurs under controlled conditions.

Over the past 20 years, legislative actions have imposed strict limits on the disposal of organic waste such as sewage sludge, municipal solid waste, and agricultural waste due to the potentially severe environmental problems associated with the management of such residuals. For example, 10 years ago, most sludge produced in the United States was disposed in oceans or landfills. Ocean disposal is now illegal, and landfills are rapidly closing. Also, increasing, stringent, air pollution regulations make incineration less attractive.

At the same time, increased amounts of organic wastes are being generated. This increase is especially true of sewage sludge because of the upgrading of wastewater treatment plants and the expansion of services (U.S. EPA 1993).

As a consequence, new practices are being encouraged that include the treatment of organic waste with resource recovery. Composting is one of these practices (Kuchenrither et al. 1987). Composting is a method of solid waste treatment in which the organic component of the solid waste stream is biologically decomposed under controlled aerobic conditions to a state in which it can be safely handled, stored, and applied to the land without adversely affecting the environment. It is a controlled, or engineered, biological system.

Composting can provide pathogen kill, volume reduction and stabilization, and resource recovery. Properly composed waste is aesthetically acceptable, essentially free of human pathogens, and easy to handle. Compost can improve a soil's structure, increase its water retention, and provide nutrients for plant growth. As Hoitink and Keener (1993) note "It is not surprising therefore that composting of wastes has resulted in a variety of beneficial effects in agriculture as the Western World progressed from a 'throw-away' mentality to a more environmentally friendly society."

Golueke (1986) points out that composting relies more on scientific principles as time passes. At the same time, advances have been made in the technology used for the process, such as the static pile (Epstein et al. 1976), and the development of in-vessel systems (U.S. EPA 1989). Coupling the need for waste diversion practices (from landfills) with the advances in the fundamental science and process technology has increased the use of composting

for wastewater treatment. Approximately 159 sludge composting facilities operate in the United States (Goldstein, Riggle, and Steuteville 1992).

This section concentrates on the fundamentals of sludge composting. Section 10.14 focuses on municipal solid waste composting. Several excellent sources provide more detail on composting principles (Hoitink and Keener 1993; Rynk 1992; Haug 1993).

Process Description

Numerous types of composting systems exist, but for the most part, composting systems can be divided into three categories: windrow, static pile, and in-vessel. Windrow systems are composed of long, narrow rows of sludge mixed with a bulking agent. The rows are typically trapezoidal in shape, 1 to 2 m high and 2 to 4.5 m wide at the base. The rows are usually uncovered but can be protected by simple roofs. The sludge mixture is aerated by convective air movement and diffusion. Wastewater treatment facilities periodically turn the rows using mechanical means to expose the sludge to ambient oxygen, dissipate heat, and refluff the rows to maintain good free air space. Windrows can also be aerated by induced aeration (Hay and Kuchenrither 1990). Windrows are space-intensive but mechanically simple.

Static pile systems are also composed of a sludge–amendment mixture but are aerated by forced-aeration systems installed under the piles (Epstein et al. 1976). The aeration can be either positive or negative. Finstein, Miller, and Strom (1986) stress the need for positive aeration for process control. Others note the advantage of negative (suction) aeration for better possibilities of capturing the process air for odor control. Currently, most facilities in the United States use the static pile method (Goldstein, Riggle, and Steuteville 1992).

In-vessel composting takes place in either partially or completely enclosed containers. A variety of schematics use various forced aeration and mechanical turning technologies (Tchobanoglous and Burton 1991). In-vessel composting is space efficient but more mechanically complex than the other two system categories. They offer excellent possibilities for process and odor control. Among the facilities commissioned within the past few years, a greater percentage are using in-vessel methods (Goldstein, Riggle, and Steuteville 1992).

Each of the system categories is capable of producing a good compost in a reliable and efficient manner. The choice of any given system depends on the site location, available space, and other local conditions.

Each system is composed of common basic steps. As shown in Figure 7.55.1, the basic steps of the composting process include the following:

1. Mixing dewatered sludge with a bulking agent
2. Aerating the composting pile by either the addition of air or mechanical turning
3. Further curing
4. Recovering the bulking agent
5. Final distribution

The first and second steps are critical to the process success. Recovering the bulking agent is an optional step that relates to system economy (reuse of the bulking agent) and product quality (the product compost with or without wood chips). The curing stage also relates to product quality because it influences compost stability. During this period, which can last as long as 30 to 60 days, further product stabilization with pathogen die-off and degassing occurs (Rynk 1992). Final disposal depends on the market for the product compost. The intended market for the compost dictates the need for bulking agent recovery as well as the length of the curing stage, and any other final operations (such as bagging).

While composting is a simple process, facilities must operate in a careful manner to ensure the production of a good-quality, stable compost while minimizing adverse environmental aspects, such as odor production. To ensure the production of a stable compost in a reliable and efficient manner while minimizing odor production, wastewater treatment facilities must operate any system to promote the growth of the microbial population and maintain these organisms under proper environmental conditions for a sufficient amount of time for the reactions (of stabilization) to occur.

The diagram proposed by Rynk (1992), as shown in Figure 7.55.2, shows the composting process. As described by Rynk (1992), the following conditions must be established and maintained:

- Organic materials appropriately mixed to provide the nutrients needed for microbial activity and growth, including a balanced supply of carbon and nitrogen (C:N ratio)

- Oxygen at levels that support aerobic organisms
- Enough moisture to permit biological activity without hindering aeration
- Temperatures that encourage vigorous microbial activity from thermophilic microorganisms

Process Fundamentals

The factors affecting the composting process include oxygen and aeration, nutrients (C:N ratio), moisture, porosity, structure, texture and particle size, pH, temperature, and time. These conditions are developed and maintained by process management. The following considerations are important for process management:

- Raw material selection and mixture
- Moisture management
- Aeration
- Time

These considerations are explained next.

RAW MATERIAL SELECTION AND MIXTURE

Wastewater treatment facilities add amendments to sludge to adjust the moisture and other characteristics (such as nutrient level) to improve composting. Bulking agents are added for structural support. The goal is to create a mixture of sludge and bulking agent and amendment with the proper characteristics to support aerobic digestion. The choice of material depends on the characteristics of the sludge, in particular the moisture content (which depends on the degree and type of dewatering process) and the nitrogen content of the sludge. The proper mixture has an appropriate C:N balance, proper porosity (to ensure aerobic conditions), and proper moisture content. Haug (1993) points out that the amount of free air space in the mixture is more crucial than the porosity, which is the amount of space not occupied by solids or water.

In terms of porosity and moisture content, the latter is usually the determining factor. A typical value for a mixed pile is about 60% moisture and 40% solids. High moisture levels lower the free air space of the pores and thus

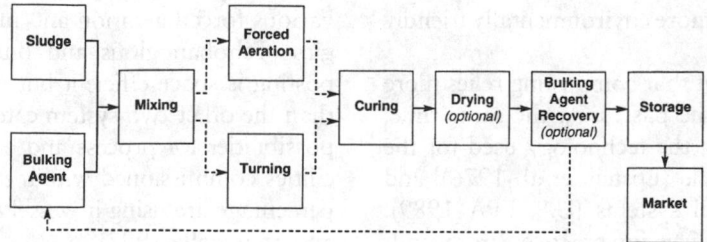

FIG. 7.55.1 Composting process flow diagram.

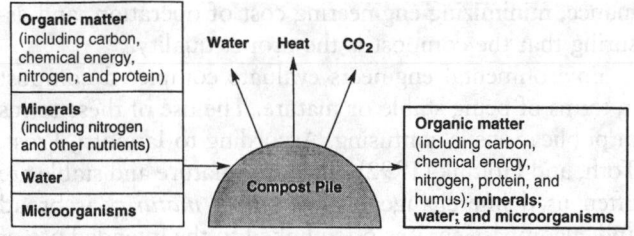

FIG. 7.55.2 Composting process.

inhibit aerobic activity, while low moisture levels do not support sufficient biological activity.

MOISTURE MANAGEMENT

Moisture management is an important part of composting. As stated previously, typically the initial moisture content of the sludge mixture is adjusted to about 60%. During the composting process, water is lost via evaporation. Water loss is driven by diffusion, air exchange, and heat generation. Some water can leach out of the mixture. Water is gained by precipitation (for uncovered systems) and as a product of respiration. In general, a net loss of water occurs. The final mixture has a moisture content of about 40%. As noted, both too high and too low levels are problems.

AERATION

Aeration serves three interdependent functions of composting. Aeration adds stoichiometric oxygen for respiration, removes water vapor, and dissipates heat. Finstein and Hogan (1993) note that heat removal determines the rate of aeration and stress that this removal is important for process control. For proper pathogen removal, the temperature must reach at least 55°C. However, allowing a composting system to reach temperatures of 70 to 80°C is self-limiting, results in poor operation, and leads to the production of unstable compost.

TIME

The length of time the composting process runs depends on the degree of stability of the compost being produced. In other words, the end point is variable. The degree of stability depends on the use of the compost end product. The time varies depending on the type of reactor and sludge mixture, but in general, the active composting time is 3 to 4 weeks. This time does not include additional curing time.

Design

Many factors must be considered in the design of a composting system. These factors are summarized by Tchobanoglous and Burton (1991) and are listed in Table 7.55.1.

TABLE 7.55.1 DESIGN CONSIDERATIONS FOR AEROBIC SLUDGE COMPOSTING PROCESSES

Item	Comment
Type of sludge	Both untreated and digested sludge can be composted successfully. Untreated sludge has a greater potential for odors, particularly for windrow systems. Untreated sludge has more energy available, degrades more readily, and has a higher oxygen demand.
Amendments and bulking agents	Amendment and bulking agent characteristics, such as moisture content, particle size, and available carbon, affect the process and quality of the product. Bulking agents should be readily available. Wood chips, sawdust, recycled compost, and straw can be used.
C:N ratio	The initial C:N ratio should be in the range of 25:1 to 35:1 by weight. Checking the carbon ensures that it is easily biodegradable.
Volatile solids	The volatile solids of the composting mix should be greater than 50%.
Air requirements	Air with at least 50% oxygen remaining should reach all parts of the composting material for optimum results, especially in mechanical systems.
Moisture content	The moisture content of the composting mixture should be not greater than 60% for static pile and windrow composting and not greater than 65% for in-vessel composting.
pH	The pH of the composting mixture should generally be in the range of 6 to 9.
Temperature	The optimum temperature for biological stabilization is between 45 and 55°C. For best results, the temperature should be maintained between 50 and 55°C for the first few days and between 55 and 60°C for the remainder of the composting period. If the temperature increases beyond 60°C for a significant period of time, the biological activity is reduced.
Mixing and turning	Mixing or turning the material being composted on a regular schedule or as required prevents drying, caking, and air channeling. The frequency of mixing or turning depends on the type of composting operation.
Heavy metals and trace organics	Heavy metals and trace organics in the sludge and finished compost should be monitored so that the concentrations do not exceed the applicable regulations for end use of the product.
Site constraints	Factors in selecting a site include the available area, access, proximity to the treatment plant and other land uses, climatic conditions, and availability of a buffer zone.

Numerous references provide additional information on the design of composting systems. In particular, Rynk (1992), Haug (1993), and U.S. EPA (1985) provide information for general design, and U.S. EPA (1989) provides information for the in-vessel system.

Special Considerations

With restrictions on the disposal of sludge, beneficial use of biosolids has become a significant trend. Composting is the leading beneficial reuse technology in terms of manufacturing a product for application to the land. Of course, the success of composting depends on marketing the final compost product. In other words, a use must exist for the compost generated from wastewater sludge.

In addition, the public must accept the composting process. Donovan (1992) notes that the most difficult challenge municipalities face in implementing sludge plans is facility siting. The general public is apprehensive concerning any waste handling facility, and specific concerns about odor, health, traffic, and land values have slowed or stopped many projects.

Composting is basically a simple process; it is quite robust and therefore a forgiving process. It can be managed in many cases (such as the backyard compost pile) with little or no technical knowledge. However, as composting applications increase and broaden in scope, the need exists for more sophistication in the design and operation of composting facilities. This need is underscored by the emphasis on aspects such as odor control and compost product quality. Such concerns demand a higher level of technology and management.

STABILITY AND PRODUCT QUALITY

In composting operations, the objective is decomposition rather than complete stabilization. The degree of decomposition, however, is not an absolute state since it depends on the final product use. In some cases, the degree is one where the material does not cause nuisances when stored even if it is wetted. If the final product is used on a plant system, the compost should not be phytotoxic.

Currently, many parameters can be used for composting process control and final product quality including the final drop in temperature; degree of self-heating capacity; amount of decomposable and resistant organic material; rise in redox potential; oxygen uptake and carbon dioxide evolution; starch test; color, odor, appearance, and texture; pathogen and indicator organisms; and inhibition of germination of cress seeds (Finstein, Miller, and Strom 1986; Inbar et al. 1990). This list covers many possibilities, but which are best for measuring the completion of composting is unclear. The optimal parameter or group of parameters is important for maximizing process performance, minimizing engineering cost of operation, and assuring that the compost is the proper quality.

Environmental engineers evaluate completed compost in terms of being stable or mature. The use of these terms in publications is confusing. According to Iannotti, Frost, Toth, and Hoitink (1992), the terms mature and stable are often used interchangeably. *Compost maturity* is broad and encompassing; it is often linked to the intended use of the compost and is therefore subjective. *Compost stability* is readily definable by its biological property of microbial activity. As such, Iannotti, Frost, Toth, and Hoitink (1992) propose a stability assay based on DO respirometry. Nonetheless, a simple, yet reliable, and universally acceptable analytical tool for evaluating compost stability does not exist.

In addition to stability, pathogen destruction is an important characteristic defining product quality. Other characteristics used for compost product specification include the concentration of specific constituents (e.g., metals and nutrients), particle size, texture, pH, moisture content, odor, weed seed inactivation, phytotoxicity, reduction of volatile solids, and product consistency (U.S. EPA 1989). The choice of characteristics depends on the compost use.

The major compost uses include large-scale landscaping (golf courses, public works projects, highway median strips), local nurseries, industries (as potting material), greenhouses, urban gardeners, land reclamation projects (strip mines), and landfill (daily and final cover).

Often the criteria used are legal regulations such as those for heavy metals and pathogens. Recently, federal regulations have been issued for the use and disposal of sewage sludge, including compost (U.S. EPA 1993). States are now in the process of adopting these regulations or formulating more stringent regulations.

PATHOGENS

Wastewater sludge is known to contain pathogens including bacteria, viruses, parasites, and helminths. Epstein and Donovan (1992) note that pathogens can be grouped under three major headings: primary pathogens, secondary or opportunistic pathogens, and endotoxins. They further note that the major concerns with pathogens related to composting wastewater sludge are product disinfection, worker health, and public health as impacted by facility location.

The U.S. EPA (1979; 1993) in the previous 40 *CFR* Part 257 regulations and the new 40 *CFR* Part 503 regulations is primarily concerned with product quality and safety of the compost. The possible presence of pathogens is a major concern. The previous regulation for pathogen control was technology based. Under 40 *CFR* Part 257, minimum standards were issued for processes to significantly reduce pathogens (PSRP). Compost that had been subject to PSRP could be used but was limited to certain restrictions. The previous regulations also defined processes to further reduce pathogens (PFRP). Fewer restrictions were placed on the use of PFRP compost.

Both PSRP and PFRP are based on a time–temperature requirement. For example, if a composting process reached at least 40°C for at least 5 consecutive days, and 55°C for at least 4 hr during that time, it met PSRP. If aerated, static piles and in-vessel systems maintained a temperature of at least 55°C for 3 consecutive days (in the coolest part of the pile), then that compost met PSRP. Such sludge was subjected to less restriction for distribution and marketing.

The new regulations regulate the product compost as well as the process. To obtain a compost that can be widely distributed or marketed (now called Class A), processors must use a PFRP time–temperature standard (or equivalent processing) and produce a product with less than or equal to 1000 fecal coliforms/g dry solids or less than or equal to 3 salmonella/g dry solids.

Numerous studies have been conducted for pathogen levels at composting facilities and in the final compost product (Epstein and Donovan 1992). Most of these studies focused on indicator organisms (fecal coliforms) and salmonellae. Yanko (1988) notes that composting is an effective method for the disinfection of sludge although considerable variability exists among the data related to the method of composting and system design and operation. In other words, proper disinfection requires careful system design and operation. In addition, the possibility exists of a repopulation of organisms in disinfected compost (Farrell 1993).

Several authors have reported the potential for the repopulation of salmonellae in composted wastewater sludge (Epstein and Wilson 1975; Brandon, Burge, and Enkiri 1977; Brandon and Neuhauser 1978; Burge et al. 1987). Brandon et al. (1977; 1978) relate the repopulation to the moisture content of the compost. Russ and Yanko (1981) evaluate the factors affecting salmonellae repopulation in sludge compost. They found that the moisture level, temperature, and nutrient content of the composted solids affect repopulation. They further report that repopulation is transient with population peaks occurring around 5 days followed by a subsequent die-off. Other authors note the importance of microbial competition for minimizing repopulation (Hussong, Burge, and Enkiri 1985; Yeager and Ward 1981).

The most important parameter for pathogen destruction is temperature. Adequate temperature can be reached, but it depends on proper facility design and operation. The most important considerations are preparing a good initial mix and maintaining aerobic conditions. These conditions are a function of the mix properties, moisture control, aeration, and the C:N ratio. As previously noted, these factors impact the composting process and affect pathogen destruction.

ODORS AND ODOR CONTROL

Odor control has become a major concern in the successful operation of any composting facility. Indeed, some operating facilities have been closed due to odor problems (Libby 1991). Numerous papers have been published identifying the causes of odors and management strategies to control odors (Hentz et al. 1992; Miller 1993; Goldstein 1993; Van Durme, McNamara, and McGinley 1992).

Many potential sources of odors exist at composting facilities. While the process air coming off a compost pile is most odorous, environmental engineers must evaluate all potential sources of odors. Therefore, a proper inventory of the potential sources of odors is necessary including liquid sludge and dewatered sludge facilities.

Haug (1990) states that odors are part of the composting process and cannot be eliminated, but they can be managed. Finstein et al. (1986; 1993) point out that controlling the composting process is crucial in minimizing odor production. This process control includes good aeration and maintaining the proper temperature, moisture, and structure of the piles.

A variety of compounds cause odors including fatty acids, amines, aromatics, inorganic and organic sulfur compounds, and terpenes (Miller 1993). They vary in their chemical properties. Some are acidic; some are basic. Some can be oxidized; others cannot. Differences exist in their solubility and adsorbability. Also, the compounds change over the course of time of operation. Therefore, any treatment system must have a broad spectrum of removal mechanisms.

Typical odor management (in addition to good process operation) involves containment of process gases, collection of gases, gas treatment, and proper dispersion. Other management possibilities include the dilution of odors with large volumes of air and the use of masking agents. Gas treatment options include oxidation processes, chemical scrubbers, and biofilters.

Summary and Conclusions

Composting is a cost effective and environmentally sound alternative for the stabilization and ultimate disposal of wastewater sludge. It produces compost—a stable, humus-like material which is a soil conditioner. Thus, the process can achieve waste treatment with resource recovery and represents a beneficial use of sludge.

Recent advances have been made in the basic fundamental science associated with composting along with the technology used for the process. These advances have increased the use of the process for wastewater sludge management.

While the composting process is simple in concept, it must be regarded as an engineered unit process. As such, it must be based on sound scientific principles, designed with good engineering, and operated with care by well-trained and motivated operators. With these practices, wastewater treatment facilities can produce a safe compost of consistently good quality in an environmentally sound manner.

—*Michael S. Switzenbaum*

References

Brandon, J.R., W.D. Burge, and N.K. Enkiri. 1977. Inactivation by ionizing radiation of *Salmonella enteritides* serotype montevideo grown in composted sewage sludge. *Applied and Environmental Microbiology* 33: 1011–1012.

Brandon, J.R., and K.S. Neuhauser. 1978. *Moisture effects on inactivation and growth of bacteria and fungi in sludges.* Pub. no SAND 78-1304. Albuquerque, N.M.: Sandia Lab.

Burge, W.D., P.D. Miller, N.K. Enkiri, and D. Hussong. 1987. *Regrowth of Salmonellae in composted sewage sludge.* EPA/600/S2-86/016.

Donovan, J.F. 1992. Developments in wastewater sludge management practices in the United States. Paper presented at the New Developments in Wastewater Policy, Management and Technology Conference, Sydney, Australia, May 18, 1992.

Epstein, E., and J.F. Donovan. 1992. Pathogens in compost and their fate. Paper presented at the WEF Seminar on Pathogens in Sludge, New Orleans, LA September 1992.

Epstein, E., and G.B. Wilson. 1975. Composting raw sludge. In *Municipal sludge management and disposal.* Rockville, Md.: Information Transfer Inc.

Epstein, E., G.B. Wilson, W.D. Burge, D.C. Mullen, and N.K. Enkiri. 1976. A forced aeration system for composting of wastewater sludge. *Journal Water Pollution Control Federation* 48: 688–694.

Farrell, J.B. 1993. Fecal pathogen control during composting. In *Science and engineering of composting,* edited by H.A.J. Hoitink and H.M. Keener. Hinesburg, Vt.: Upper Access Books.

Finstein, M.S., and J.A. Hogan. 1993. Integration of composting process microbiology, facility structure, and decision making. In *Science and engineering of composting,* edited by H.A.J. Hoitink and H.M. Keener. Hinesburg, Vt.: Upper Access Books.

Finstein, M.S., F.C. Miller, and P.F. Strom. 1986. Monitoring and evaluating composting process performance. *Journal Water Pollution Federation* 58: 272–278.

Goldstein, N. 1993. Odor control progress for sludge composting. *BioCycle* 34, no. 3: 56–59.

Goldstein, N., D. Riggle, and R. Steuteville. 1992. Sludge composting maintains growth. *BioCycle* 33, no. 12: 49–54.

Golueke, C.G. 1986. Compost research accomplishments and needs. *BioCycle* 27, no. 4: 40–43.

Haug, R.T. 1990. An essay on the elements of odor management. *BioCycle* 31, no. 10: 60–67.

———. 1993. *The practical handbook of compost engineering.* Boca Raton, Fla.: Lewis Publishers.

Hay, J.C., and R.D. Kuchenrither. 1990. Fundamentals and application of windrow composting. *Journal Environmental Engineering (ASCE)* 116: 746–763.

Hentz, L.H., Jr., C.M. Murray, J.L. Thompson, L.L. Gasner, and J.B. Dunson Jr. 1992. Odor control research at the Montgomery County Regional Composting Facility. *Water Environment Federation* 64: 13–18.

Hoitink, H.A.J., and H.M. Keener (eds). 1993. *Science and engineering of composting.* Hinesburg, Vt.: Upper Access Books.

Hussong, D., W.D. Burge, and N.K. Enkiri. 1985. Occurrence, growth, and suppression of Salmonellae in composted sewage sludge. *Applied and Environmental Microbiology* 50: 887–893.

Iannotti, D., Frost, B.L. Toth, and H.A.J. Hoitink. 1992. Compost stability. *BioCycle* 33: no. 11: 62–66.

Inbar, Y., Y. Chen, Y. Hadar, and H.A.J. Hoitink. 1990. New approaches to compost maturity. *BioCycle* 31, no. 12: 64–69.

Kuchenrither, R.D., D.M. Diemer, W.J. Martin, and F.J. Senske. 1987. Composting's role in sludge management: A national perspective. *Journal Water Pollution Control Federation* 59: 125–131.

Libby K. 1991. Lessons from a closed MSW composting plant. *BioCycle* 32, no. 12: 48–52.

Miller, F.C. 1993. Minimizing odor production. In *Science and engineering of composting,* edited by H.A.J. Hoitink and H.M. Keener. Hinesburg, Vt.: Upper Access Books.

Russ, C.F., and W.D. Yanko. 1981. Factors affecting Salmonellae repopulation in composted sludges. *Applied and Environmental Microbiology* 41: 597–602.

Rynk, R. ed. 1992. *On-farm composting handbook.* NRAES-54. Ithaca, N.Y.: Northeast Agricultural Engineering Service.

Tchobanoglous, G., and F.L. Burton. 1991. *Wastewater engineering,* 3d ed. New York, N.Y.: McGraw-Hill Inc.

U.S. Environmental Protection Agency (EPA). 1979. Criteria for classification of solid waste disposal facilities and practices. *Code of Federal Regulations.* Title 40, Part 257. *Federal Register* 179, (13 September 1979): 53438.

———. 1985. *Seminar publication on composting of municipal wastewater sludges.* EPA/625/4-85/014.

———. 1989. *In-vessel composting of municipal wastewater sludge.* EPA/625/8-89/016. Cincinnati, Ohio: CERI.

———. 1993. Standards for the use or disposal of sewage sludge. *Code of Federal Regulations.* Title 40, *Federal Register 58,* (19 February 1993); 9248.

Van Durme, G.P., B.F. McNamara, and C.M. McGinley. 1992. Bench-scale removal of odor and volatile organic compounds at a composting facility. *Water Environmental Research* 64: 19–27.

Yanko, W.A. 1988. *Occurence of pathogens in distribution and marketing municipal sludges.* EPA/600/1-87/014. Research Triangle Park, N.C.: HERL.

Yeager, J.G., and R.L. Ward. 1981. Effects of moisture content on long-term survival and regrowth of bacteria in wastewater sludge. *Applied and Environmental Microbiology* 41: 1117–1122.

Bibliography

Alkhatib, E.A., and L.T. Thiem. 1991. Wastewater oil removal evaluated. *Hydrocarbon Processing* (August).

American Society of Civil Engineers (ASCE). 1977. Wastewater treatment plant design. In *Water Pollution Control Federation Manual of Practice No. 8.* ASCE Manuals and Reports on Engineering Practice No. 36.

Bauman, E.R., and H.E. Babbit. 1953. *An investigation of the performance of six small septic tanks.* University of Illinois Engineer Experimentation Station, Bulletin Series No. 409.

Brown, J.D. and G.T. Shannon. 1963. *Design guide to refinery sewers.* Presented to the API Div. of Refining, Philadelphia, 14 May, 1963.

Buchanan, R.D. 1974. Pumps and pumping stations. In *Environmental engineering handbook,* edited by B.G. Liptak. Radnor, Pa.: Chilton Book Company.

Cotteral, J.A., and D.P. Norris. 1969. Septic tank systems. *J. Sanit. Engineer. Div. Amer. Soc. Civ. Engineer.* 95:715, August, 1969.

Eckenfelder, W.W. Jr. 1966. *Industrial water pollution control.* New York: McGraw-Hill.

Eckenfelder, W.W., J. Patoczka, and A.T. Watkin. 1985. Wastewater treatment. *Chem. Eng. Prog.* (September).

Fair, G.M., J.C. Geyer, and D.A. Okun. 1970. *Elements of water supply and wastewater disposal.* New York: John Wiley and Sons, Inc. (November).

La Grega, M.D. and J.D. Keenan. 1974. Effects of equalizing wastewater flows. Journal of Water Pollution Control Fed. 46, no. 1: 123.

Linsley, R.K., J.B. Franzini, D.L. Freyberg, and G. Tchobanoglous. 1992. *Water-resources engineering*. 4th ed. New York: McGraw-Hill, Inc.

Lipták, B.G., ed. 1974. *Environmental engineers' handbook. vol. 1, Water pollution*. Radnor, Pa.: Chilton.

Lipták, B.G. 1995. Pumps as control elements. In *Instrument engineers' handbook*. 3d ed. Radnor, Pa.: Chilton Book Company.

Lipták, J. 1974. Grit removal. Vol. I of *Environmental engineers' handbook*, edited by B.G. Lipták. Radnor, Pa.: Chilton Book Co.

Lipták, J. 1974. Screening devices and comminutors. In *Environmental engineers' handbook*, edited by B.G. Lipták. Radnor, Pa.: Chilton Book Co.

Metcalf & Eddy, Inc. 1981. *Wastewater engineering: Collection and pumping of wastewater*. New York: McGraw-Hill Book Company.

Metcalf & Eddy, Inc. 1991. *Wastewater engineering, treatment, disposal and reuse*. 3d ed. New York: McGraw-Hill Book Company.

Metcalf, L. and H.P. Eddy. 1935. *American sewerage practice (Vol. III). Disposal of sewage*. New York: McGraw-Hill.

Nemerow, N.L. 1962. *Theories and practices of industrial water treatment*. Reading, Mass.: Addison-Wesley.

Novotny, V., K.R. Imhoff, M. Olthof, and P.A. Krenkel 1989. *Karl Imhoff's handbook of urban drainage and wastewater disposal*. New York: John Wiley.

Phelps, E.B. 1944. *Stream sanitation*. New York: Wiley.

Qasim, S. 1985. *Wastewater treatment plant: Planning, design, and operation*. New York: Holt, Rinehart, Winston.

Rich, L.G. 1961. *Unit operations of sanitary engineering*. New York: Wiley.

Sawyer, C.N., and P.L. McCarty. 1978. *Chemistry for environmental engineering*. 3d ed. New York: McGraw-Hill.

Shieh, W.K., and C.Y. Chen. 1984. Biomass hold-up correlations for a fluidized bed biofilm reactor. *Chem. Eng. Des. Res.* 62, no. 133.

U.S. Environmental Protection Agency (EPA). 1975. Process design manual for suspended solids removal. EPA 625-I-75-003a. Washington, D.C.: U.S. EPA.

Yee, C.J. 1990. Effects of microcarriers on performance and kinetics of the anaerobic fluidized bed biofilm reactor. Ph.D. Dis., Department of Systems, University of Pennsylvania, Philadelphia.

8

Removing Specific Water Contaminants

I.M. Abrams | D.B. Aulenbach | E.C. Bingham | L.J. Bollyky | T.F. Brown, Jr. | B. Bruch | R.D. Buchanan | L.W. Canter | C.A. Caswell | R.A. Conway | G.J. Crits | E.W.J. Diaper | J.W.T. Ferretti | R.G. Gantz | W.C. Gardiner | L.C. Gilde, Jr. | E.G. Kominek | D.H.F. Liu | A.F. McClure, Jr. | F.L. Parker | R.S. Robertson | D.M. Rock | C.J. Santhanam | L.S. Savage | S.E. Smith | F.B. Taylor | C.C. Walden | R.H. Zanitsch

8.1
REMOVING SUSPENDED SOLID CONTAMINANTS

Algae Control

The types of algae and the concentration in wastewater depend on residence time, climate and weather, amount of pollutants entering the pond, and dimensions of the pond. Normally, small unicellular types of algae develop first, e.g., *Chlorella*. Because of their physical dimensions they are difficult to remove by the processes listed in Table 8.1.1. Longer residence times lead to the development of larger algae and other plankton, which is more readily removed. The algae concentration affects the choice of removal process and the rate of treatment. Because of their light density, the dried weight of suspended solids is not an efficient measure of concentration. Algae are normally measured in volumetric or areal standard units (Anon. 1971). In surface water supplies, concentrations may be as high as 30,000 cells per milliliter (ml), this can be much higher in nutrient-rich waste treatment effluents. A combination of processes may be the best treatment, e.g., copper sulfate addition and microstraining, as used on surface water supplies in London, England.

Carbon Particles

Carbon particulate matter suspended in waste effluent must be either controlled or removed prior to discharge. Wastes associated with the carbon black and acetylene industries are of concern. These wastes may contain up to 1000 milligrams per liter (mg/l) carbon particles in suspension; in most cases this carbon concentration must be reduced to less than 50 mg/l suspended solids. Usually, these solids settle readily and are removed by gravity settling and/or flotation.

Individual particle sizes range from a submicron to larger than 100 micron (μ). Larger particles settle, whereas smaller particles float. Transition size particles remain suspended almost indefinitely unless forced out of suspension by mechanical or chemical means. Unless a highly clarified effluent is required, suspended matter may not have to be removed as it amounts to a small proportion of total solids concentration.

GRAVITY SETTLING

Two types of gravity systems are available: (1) settling Lagoons, which provide retention time for solid particles to settle as sludge. These must be cleaned periodically; and (2) mechanical Clarifiers, which remove suspended solids and also rid bottom sludges mechanically.

The settling lagoon requires a minimum capital investment. Cleanout costs are high compared with the mechanical clarifier operating costs.

Settling devices are usually designed on the basis of overflow rate, gal per day (gpd) per sq ft of surface area. According to the Ten State Standards (Great Lakes-Upper Mississippi River Board of State Sanitary Engineers 1968), this rate should be in the range of 600 to 1000 gpd/sq ft. In designing the carbon settling lagoon, frequency of lagoon cleaning must be considered, and the lagoon must be sized accordingly. Carbon sludge will settle to a density of 5–20% solids.

TABLE 8.1.1 ALGAE REMOVAL PROCESSES: MERITS AND FLAWS

Algae Removal Process	Advantages	Limitations
Copper sulfate	Simple and inexpensive	Creates toxicity; only some algal forms attacked
Chlorine	Simple and inexpensive	High doses needed; not all algae attacked
Coagulation and settling	Positive removal of all types of algae	High chemical doses needed; difficult sludges produced
Sand filters	Positive removal of all types of algae	Rapid filter clogging may occur
Microstraining	Simple and inexpensive	Not all algal forms removed
Air flotation	Positive removal of all types of algae	Not all algal forms removed; sludges may be difficult to handle

TABLE 8.1.2 EXAMPLE: SETTLING LAGOON FILL TIME
CALCULATION

Settling Lagoon Data:

Area = 5 acres
Depth = 5 ft
Flow = 10 million gal/day (mgd)
Influent concentration = 1000 mg/l
Effluent concentration = 50 mg/l
Sludge density = 5%
Carbon deposited per day:

$$(1000 - 50) \times 10 \times 8.34 = 80{,}000 \text{ lb/day}$$

Lagoon volume:

$$V = 5 \text{ acre} \times 5 \text{ ft} = 25 \text{ acre-ft} = 8.3 \times 10^6 \text{ gal}$$

Solids capacity of lagoon at 5% sludge density:

$$5\% = 50{,}000 \text{ mg/l} = 0.42 \frac{\text{lb}}{\text{gal}}$$

$$\text{Capacity} = 0.42 \frac{\text{lb}}{\text{gal}} \times 8.3 \times 10^6 \text{ gal} = 3.5 \times 10^6 \text{ lb solids}$$

Time required to fill lagoon with sludge:

$$T = \frac{3.5 \times 10^6 \text{ lb}}{8 \times 10^4 \text{ lb/day}} = 44 \text{ days}$$

As an example, a 5-acre lagoon, 5 ft deep, with an influent suspended solids concentration of 1000 mg/l and an effluent concentration of 50 mg/l at a flowrate of 10 mgd will retain almost 80,000 lb of solids per day. If the solids settle to a 5% sludge density, the lagoon will be filled with sludge in less than two months, as indicated by the calculations in Table 8.1.2. A settling lagoon design for this application would probably be based on cleaning frequency rather than on overflow rates.

The outfall structure of a settling system should retain floating material and maintain laminar flow to prevent solids from resuspending at discharge due to turbulence. An underflow-overflow weir (Figure 8.1.1) efficiently provides such an outfall. According to the Ten State Standards (Great Lakes–Upper Mississippi 1968), weir loading rates should not exceed 10,000 gpd per linear ft of weir to assume minimum resuspension of settled matter from turbulent flow. For the example in Table 8.1.2, a weir 1000 ft long would be required.

SOLIDS DISPOSAL

Whether a mechanical clarifier, a settling lagoon or other means of solids removal is utilized, concentrated carbon slurry or sludge must be disposed of. Disposal methods include incineration, landfill disposal, reuse, and dewatering. Removal and disposal of concentrated solids slurry is the most difficult part of the carbon clarification system.

Eliminating waste at the source is ideal. Tightening production controls and modifying the process can drastically reduce waste losses and should be investigated before any removal system is developed. No treatment system is justifiable without assurance that waste production is minimized at the source. Frequently, waste carbon is a product loss, and recovery is valuable. Keeping carbon out of wastewater prevents problems in waste treatment.

Foundry Sand

Foundry melting emissions contain solid particles ranging from coarse dust to fines of submicron size. Cupola emissions are much coarser than electric furnace emissions, which are generally less than 5 μ.

Foundry melting dusts include combustibles containing 20–30% carbonaceous material. Iron oxides account for nearly 60% of collected dusts; silica and miscellaneous metallic oxides account for smaller quantities.

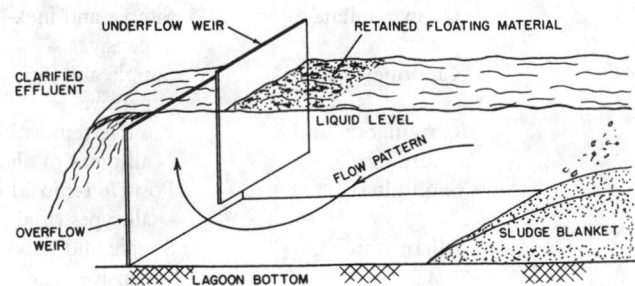

FIG. 8.1.1 Settling lagoon outfall structure.

Water curtains and scrubbers are used to remove solids from foundry stack gases. Wet scrubbers also remove acidic compounds. Scrubber water is treated to neutralize acids and to remove solids prior to recirculation. Settled solids are vacuum filtered prior to disposal. Most foundries have a number of scrubbers working on different operations, and all effluents are combined and treated together. In grinding and shakeout areas, the scrubber may be either cyclonic or water curtain, which tolerates dirty feedwater. However, abrasive materials of +200 mesh should be removed to avoid abrasion of circulating pumps.

For complete solids removal—down to smoke particles from cupola emission gas—high-energy scrubbers such as Venturis are required, which need clean water. Cupola cooling water should also be clean to prevent heat exchange surface fouling. If water is used for slag quenching, a mass of porous particles up to $\frac{1}{4}$ in is produced. These usually float. Casting washing produces a slurry with +150 mesh sand. Most of these materials can be separated on a vibrating screen of approximately 50 mesh.

Depending on the recirculation system, grit separators, settling basins, or clarifiers are used. A hydroseparator removes fine sand down to approximately 50 μ. Removal of finer solids requires chemical treatment with lime, alum, and possibly a polyelectrolyte to produce clarified effluent containing 10–20 mg/l of suspended solids. Disc, drum, or belt filters are used for dewatering foundry waste solids. Filter rates range from 25–40 lb of dry solids/hr/sq ft.

Some foundries have sand scrubber wastes. This differs from dust collection water as it settles more slowly. Overflow rates of no more than 0.3–0.5 gpm/sq ft can be used. Filtration rates for sand scrubber wastes vary from 3–10 lb of solids/hr/sq ft.

Laundry Wastes

THE PROBLEM OF COMMERCIAL WASTE

Commercial coin-operated laundry installations pose problems when sewers are not available, and septic tank or leach field systems are utilized. Because of the small amount of land available for liquid waste discharge, additional treatment is necessary. Treated effluent reuse should also be considered.

Table 8.1.3 indicates typical waste flow (Flynn and Andres 1963) from laundry installations on Long Island, N.Y. A typical installation of 20 machines produces 4,000 gpd. Depending on soil conditions, this volume might require a much larger disposal area than is available. Table 8.1.4 describes typical laundry waste properties and composition as resembling weak sewage with the exception of high alkyl benzyl sulfonate (ABS) and phosphate contents.

Large quantities of water are required for washing, therefore alleviating both water supply and waste disposal

TABLE 8.1.3 TYPICAL WASTE FLOW FROM A COIN-OPERATED WASHING MACHINE

Average wastewater flow	89–240 gal/day
Maximum average flow	587 gal/day
Minimum design basis for treatment based on a 12–hr day	550 gal/machine

problems via partial or complete recycling of treated wastewater effluents should be considered.

TREATMENT SYSTEMS

Septic Tanks

Septic tanks followed by leach field systems are often inadequate to process the quantity and quality of water to be disposed.

Physical Methods

All laundry waste should be strained in a removable basket so that lint does not clog pumps and other equipment in the treatment system.

Plan settling of laundry waste removes the heavier grit particles washed out of clothes. Most biological oxygen demand (BOD) is soluble, therefore settling has little effect on the BOD and chemical oxygen demand (COD) of the waste.

Several types of filtration units are used to treat laundromat wastes. A sand filter efficiently removes particulate matter. Pressures and filters usually require less space than gravity sand filters. The latter is used following other treatment methods and is little different from filtration through soil. Filtration through diatomaceous earth filter cake is highly recommended, since it removes bacteria and some viruses, and is particularly effective in separating chemical sludges. In diatomaceous earth filtration, prior settling or sand filtration lengthens filter runs but will not result in a better quality effluent.

TABLE 8.1.4 TYPICAL QUALITY OF LAUNDRY WASTES

Parameter	Concentration, mg, per liter	
	Average	Range
pH	7.13	5.0–7.6
BOD	120	50–185
COD	315	136–455
ABS (methylene blue active substance)	33	15–144
Total Dissolved Solids	700	390–1450
Phosphate (PO_4^{3-})	146	84–199
Acidity as $CaCO_3$	91	73–124
Alkalinity as $CaCO_3$	368	340–420

Chemical Methods

Coagulation or precipitation followed by settling and/or filtration has proven effective in treating laundromat wastes. Alum alone at a pH of 4–5 may result in a 75% reduction in ABS and an 85% reduction in phosphate content of the waste. Iron salts effect a similar reduction, whereas calcium chloride can reduce ABS by 85%, but this results in only a 50% reduction in phosphate content at high doses.

In addition, ABS may be completely neutralized, using a cationic detergent. Tests must be performed to provide exact equalization with no excess of either detergent. Substances to perform this are commercially available. Phosphates are effectively removed by precipitation techniques. Alum, iron salts, and calcium salts at high pH offer a high degree of phosphate removal. Better than 90% phosphate removal can be obtained by calcium chloride combined with adjusting the pH to 10, or by lime, both followed by filtration in a diatomaceous earth filter.

Physicochemical Methods

Considered a physicochemical process, *ion exchange* has not been successful in producing high quality water for reuse from laundry waste.

Residual organic matter may be effectively removed by contact with *activated carbon*. Granular carbon in an upflow pressure tank seems to be most efficient, although adding powdered activated carbon to other chemicals prior to filtration can also be effective. Activated carbon is also effective in removing anionic detergents. However, high ABS concentration exhausts the capacity of activated carbon to remove other organic matter, therefore prior treatment to reduce ABS should be applied.

Biological Methods

When soluble organic material is present, it is difficult to reduce BOD by more than 60% through chemical precipitation and filtration. To achieve high degrees of BOD removal, biological treatment may be required. Although there is an adequate bacteria food supply of carbon and phosphorus in the waste, total nitrogen content may be deficient for biological treatment.

Solids Disposal

Chemical precipitation solids and diatomaceous earth solids are amenable to landfill disposal. Biological sludges are treated similarly to septic tank sludges. The sludge holding tank should be conveniently located for periodic pumping by a local scavenging firm.

Suggested Treatment System

A schematic flow diagram for a suggested laundromat waste treatment system is shown in Figure 8.1.2. After screening lint, waste is stored in a holding tank to equalize flow and provide sufficient volume for operating the treatment system during normal daytime hours. A pump can deliver waste to the chemical mixing tank where the appropriate chemicals are added. A settling tank removes the bulk of precipitated solids prior to diatomaceous earth filtration. A pump is required to provide pressure for filtration in the diatomaceous earth filter. Recycling to the chemical mixing tank would be required during the filter precoat operation.

Following filtration, activated carbon adsorption may be practiced as needed. A final storage tank is provided for adding chlorine if needed or for holding effluent for future use. Settling tank sludges and diatomaceous earth filter discharges should be collected in a sludge holding

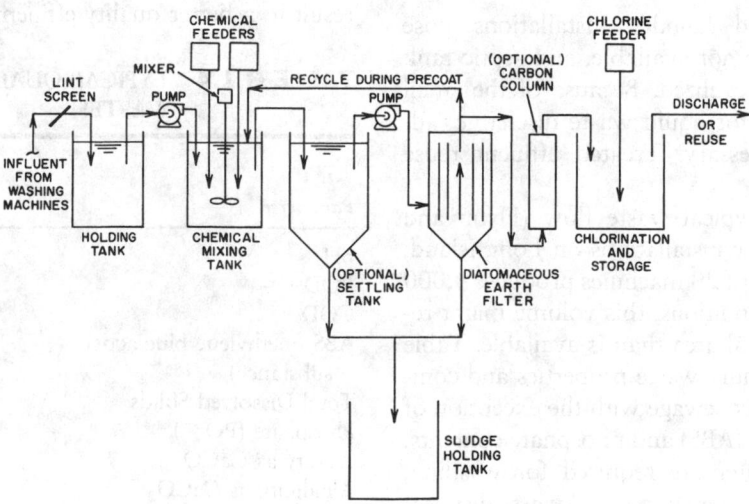

FIG. 8.1.2 Laundry waste treatment

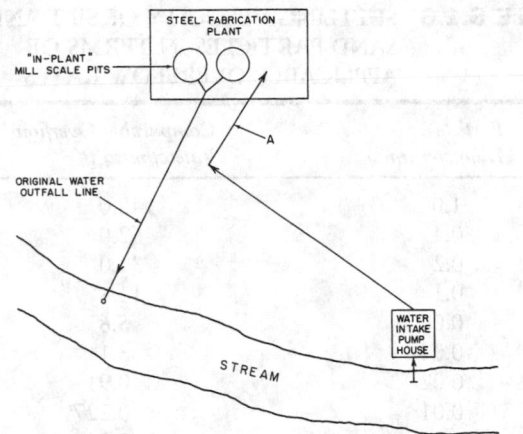

FIG. 8.1.3 Original water supply layout. A. Original plant water supply line. (Raw river water was used without pretreatment for mill scale removal process.)

tank and pumped out periodically by a scavenger system. This system should provide effluent satisfactory for discharge or partial reuse.

QUALITY OF EFFLUENT

Chemically precipitated and filtered wastes can be disposed in a subsurface system, provided that there is adequate land to accommodate the hydraulic load. Biological treatment may be necessary to improve water quality before discharge into a small stream.

Water reuse should be considered because of the large volume. Since chemical coagulants increase total dissolved solids in water, complete reuse and recycle would continuously increase total dissolved solids. Thus, chemicals should be limited to prevent excess. Because the water is still warm, heat energy can be saved by recycling treated effluent. To control total solids buildup, an ion exchange system is theoretically applicable. However, experience shows that this system is not effective in treating laundry waste effluents. Other uses for the treated water may be found, depending on the water requirements of nearby industries. Recharging water into the soil uses the soil's natural treatment ability and maintains a high water level in the aquifer, providing water for the laundromat.

Mill Scale

This is a case history of the design, construction, and operation of a wastewater treatment system established to remove mill scale from water contaminated by steel mill scale removal operation and to provide a closed system enabling reuse of water for the mill scale removal operation.

The installed cost of the total system was approximately $600,000, including two parallel treatment sys-

tems assuring continuous 24-hr operation via available alternate flow patterns for necessary equipment repair or maintenance.

DESIGN PARAMETERS

To define the problem, existing system elements were reviewed (Figure 8.1.3). The original design specified a once-through system capable of processing an existing flow of 3500 gpm with the capability to handle 7000 gpm in the future. Effluent quality was to meet stringent state requirements for discharge to the waterway. Applying knowledge of stream quality to the original design requirements raised question about the once-through concept. It was noted that if process utilization of this water did not require a higher quality supply than the polluted raw river water presently used, the need for a once-through system was questionable.

A system to treat this wastewater to meet stage discharge standards would be very expensive. However, it cost much less to treat this wastewater only to the extent required by the process. Historically, this requirement was met by the quality of a badly polluted stream. The cost difference between a reuse system and a once-through discharge system is substantial. Water quality design standards were key factors in system cost.

Table 8.1.5 lists the design parameters. Provisions were also made for sludge and recovered oil handling with minimal expense and minimal personnel time required. The original process flowsheet is shown in Figure 8.1.4. A closed system of this type is susceptible to three primary problems: algal accumulation, dissolved solids buildup, and heat buildup.

Solving these problems requires bactericide and/or algicide additives, blowdown and addition of makeup water, and a system cooling tower. The original design included a cooling tower hookup, if required, together with a chemical feed system. However, makeup water from the

TABLE 8.1.5 DESIGN PARAMETERS FOR MILL SCALE WATER TREATMENT PLANT[a]

Wastewater Flow	3500 gpm existing
	7000 gpm design capability
Primary Pollutants	Iron solids (fines)
	Oil
	Heat
Treated Effluent Quality Required	Continuous 24-hr reuse capability
Acceptable Pollutant Content in Effluent	Iron (suspended solids)
	600 ppm
	Oil 150 ppm (plus freefloating oil)

[a]System to be as fully automatic as possible.

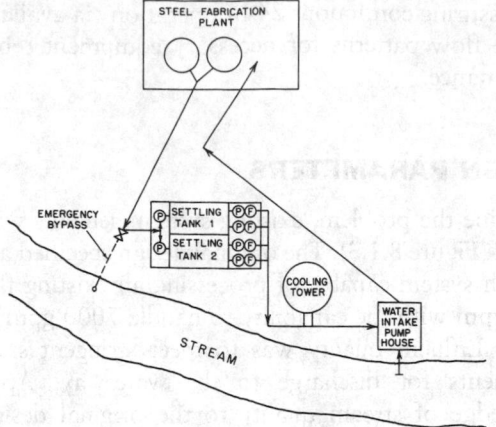

FIG. 8.1.4 Reuse system on steel plant water. (P = pump; F = filter)

river was thought sufficient to compensate for evaporative losses and to control dissolved solids buildup. Dissolved solids presented no serious problem.

OPERATIONAL HISTORY

In operation, the system is entirely satisfactory. The cooling tower was not installed originally because heat loss through the system—due to the length of the lines and the surface area of the tanks—was considered sufficient. During most of the operating time, this was true. However, during summer when ambient surface air temperatures occasionally reach 110° to 115°F in this region, Joliet, Ill., heat loss was not enough to maintain comfort for personnel manning the spray nozzles in the plant. During such periods, return water temperature rose to 114°F for a few days. Therefore, a cooling tower was installed.

The sludge averages 50 to 60% solids, about the minimum water content for the sludge to slide easily from the discharge chutes into catch buckets.

TABLE 8.1.6 SETTLING VELOCITY OF SILT AND SAND PARTICLES IN TERMS OF APPLICABLE OVERFLOW RATES

Particle Diameter (mm)	Comparable Overflow Rate cpm/sq ft
1.0	148.0
0.4	62.0
0.2	31.0
0.1	11.8
0.06	5.6
0.04	3.1
0.02	0.91
0.01	0.227
0.004	0.036

Oil-skimming devices are rotary cylinder units mounted at the water surface level in the tanks. These units require heat protection to prevent freezing in the winter. The sludge is recovered; since it consists primarily of mill scale, it can be sold as blast furnace charging material.

Strainers are 0.005 in units with 5,000 gpm capacity each. These are in the system for insurance in the event of heavy overloading of the settling tanks. This might occur if one of the two parallel systems was shut down for pump or ejection mechanism repairs when the mill is operating at peak capacity.

Until now, the system has performed well, except for minor startup and training problems. Mill operating personnel are pleased, because return water quality is far better than the raw river water they were using.

Mineral Tailings

Wastewater from mining or ore beneficiation contains suspended particles of fine sand, silt, clay, and possible limestone. A large percentage of solids may be colloidal due to their nature or as a result of milling and flotation pro-

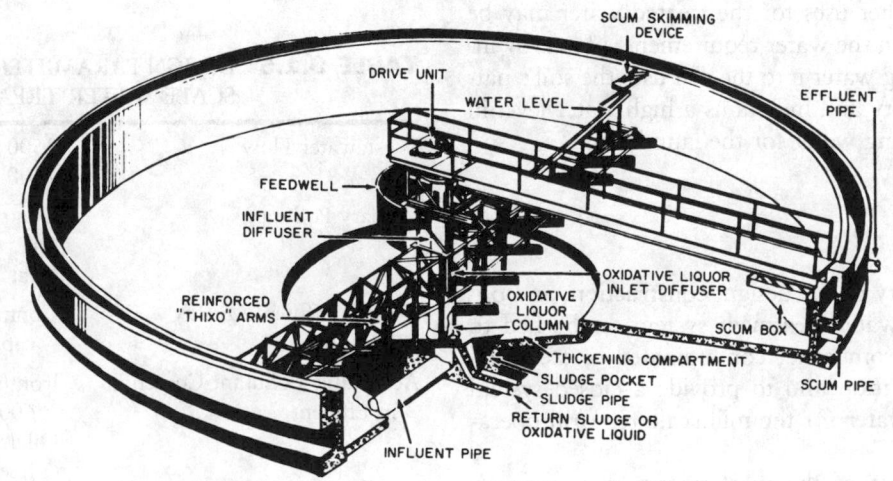

FIG. 8.1.5 Thickener for mineral tailings

cessing with reagents added to disperse the solids. Table 8.1.6 shows the velocities at which particles of sand and silt subside in still water (American Water Works Association 1969) at 50°F.

Collodial particles cannot be removed by settling without chemical treatment. Because of the chemicals added in milling and during flotation, it is virtually impossible to economically clarify mineral tailings, and mineral tailing overflows from thickener clarifiers are usually retained indefinitely. Figure 8.1.5 illustrates thickener design used in alumina, steel, coal, copper, and potash processing.

*—E.W.J. Diaper, T.F. Brown, Jr.,
E.G. Kominek, D.B. Aulenbach,
C.A. Caswell*

References

American Water Works Association, Inc. 1969. *Water treatment plant design.* New York, N.Y.

Anon. 1971. *Standard Methods for the Examination of Water and Wastewater.* 13th ed.

Aulenbach, D.B., P.C. Town, and M. Chilson. 1970. *Treatment of laundromat wastes, Part I.* Proceedings, 25th Industrial Waste Conference. Purdue University, Lafayette, Ind. (May 5–7).

Aulenbach, D.B., M. Chilson, and P.C. Town. 1971. *Treatment of Laundromat Wastes, Part II.* Proceedings, 26th Industrial Waste Conference. Purdue University, Lafayette, Inc. (May 4–6).

Burns and Roe, Inc. 1971. *Process design manual for suspensed solids removal.* Environmental Protection Agency Technology Transfer.

Flynn, J.M. and B. Andres. 1963. Launderette waste treatment processes. *J.W.P.C.F.,* 35:783.

Great Lakes–Upper Mississippi River Board of State Sanitary Engineers. 1968. *Recommended standards for sewage works.*

8.2
REMOVING ORGANIC CONTAMINANTS

Aldehydes

Aldehydes have several properties important to water pollution control. Saturated aldehydes are readily biodegraded and represent a rapid oxygen demand on the ecosystem, whereas unsaturated aldehydes can inhibit biological treatment systems at low concentrations. Aldehyde volatility makes losses through air stripping an important consideration.

BIOLOGICAL OXIDATION

Aldehyde amenability to biodegradation is indicated by high biochemical oxygen demand (BOD) levels reported by several investigators. At a low test concentration, formaldehyde, acetaldehyde, butyraldehyde, crotonaldehyde, furfural, and benzaldehyde all exhibited substantial biooxidation (Heukelekian and Rand 1955; Lamb and Jenkins 1952). An olefinic linkage in the α,β position usually renders the material inhibitory (Stack 1957). The levels inhibitory to unacclimated microorganisms for acrolein, methacrolein and crotonaldehyde were 1.5, 3.5, and 14 mg. per liter (mg/l), respectively, whereas levels for acetaldehyde, propionaldehyde and butyraldehyde were 500 mg/l or above. Formaldehyde was inhibitory at 85 mg/l.

Bacteria can develop adaptive enzymes to allow biological oxidation of many potentially inhibitory aldehydes to proceed at high influent levels. Stabilization by acclimated organisms of several organic compounds typical of petrochemical wastes has been investigated (Hatfield 1957). For organisms acclimated to 500 mg/l formaldehyde, approximately 3 hr aeration time was required to bring the effluent concentration to zero. However, effluent organic concentration after this interval was still high, indicating oxidation to formic acid or Cannizzaro dismutation to methanol and formic acid. Eight to ten hr of aeration were required for the effluent BOD to approach zero. Removals of acetaldehyde (measured as BOD) were from an initial concentration of 430 to 35 mg/l after a 5 hr aeration time. Propionaldehyde removals were from 410–25 mg/l after five hr. The oxidation pattern of paraformaldehyde, the polymer of formaldehyde, resembled its precursor.

Data collected through Warburg respirometer studies using seed sludges from three waste treatment plants (Gerhold and Malaney 1966) showed that aldehydes were oxidized to an extent second only to corresponding primary alcohols. Only formaldehyde exhibited toxicity to all three sludges. Branching in the carbon chain increased resistance to biooxidation.

AIR STRIPPING

Kinetic data for air stripping of propionaldehyde, butyraldehyde, and valeraldehyde have been presented (Gaudy, Engelbrecht and Turner 1961). Removal of propionaldehyde in model units at 25°C followed first-order reaction kinetics; removals calculated from residual aldehyde and residual chemical oxygen demand (COD) analyses were parallel, indicating that no oxidation of the acid

TABLE 8.2.1 CARBON ADSORPTION OF ALDEHYDES

	Aldehyde Removal from 1000 mg/l Solution at 5 gm/l Carbon Dose	
	Equilibrium Loading mg/g Carbon	Removal Level, %
Formaldehyde	19	9
Acetaldehyde	22	12
Propionaldehyde	57	28
Butyraldehyde	106	53
Acrolein	61	31
Crotonaldehyde	92	46
Benzaldehyde	188	94
Paraldehyde	148	74

CARBON ADSORPTION

Aldehydes, due to their low molecular weight and hydrophilic nature, are not readily adsorbed onto activated carbon. Typical data from Freudlich isotherm tests of adsorbability at various carbon dosage levels are presented in Table 8.2.1. On a relative basis, aldehydes were less amenable to adsorption than comparable undissociated organic acids but were more amenable than alcohols (Giusti 1971). However, none of the low molecular weight, polar, highly volatile materials were readily adsorbed.

Cellulose Pulp

All pulp mill effluents contain wood extractives, a highly diverse, ill-defined chemical group that varies widely according to wood species and origin. Chemical pulping wastes also contain hydrolyzed hemicelluloses and lignin, solubilized during cooking. Since various pulp processes vary considerably in mill design and operation, effluents are extremely diverse.

WASTEWATER VOLUME

Problems arise due to the tremendous volumes discharged (Table 8.2.2). Newer installations recycle process waters. Much market pulp is bleached, with bleach plant discharges as large as those from pulping. Since mills with 500–1000 ton/day capacity are not uncommon, volumes discharged at a single point may be abnormally high.

EFFLUENT CHARACTERISTICS

Pulp effluents usually have an abnormal pH, a variable loading of suspended fibrous solids, and an appreciable oxygen demand (Table 8.2.2). Older mills may have even heavier loadings. Kraft pulping produces alkaline wastes,

occurred. However, at 40°C stripping was not described by first-order kinetics, and propionaldehyde oxidation to less volatile propionic acid was apparent when removals measured as COD were less than those measured as aldehyde.

Stripping of butyraldehyde and valeraldehyde at 25°C did not follow first-order kinetics, indicating oxidation of aldehyde to acid may also be occurring. Removals after an 8 hr aeration time at 25°C and an air flow of 900 ml/min/l, were 85% for propionaldehyde and butyraldehyde, and 98% for valeraldehyde. In a biological system all three removal mechanisms would exist: biological oxidation and synthesis, air stripping, and air oxidation. The magnitude of each means would depend primarily on the activity of the bacterial culture and the degree of gas-liquid contact.

TABLE 8.2.2 EFFLUENT CHARACTERISTICS OF CELLULOSE PULPING WASTES[a]

Unit Process	Water Volume U.S. gal/ton	pH	BOD₅[a] lb/ton	Suspended Solids lb/ton
Hydraulic debarking	500–10,000	4.6–8.0	5–20	30–50
Groundwood	6,500–10,000	6.0–6.5	10–40	15–80
Neutral sulfite semichemical pulping (with recovery)	3,000–20,000	6.5–8.5	30–60	<10
Kraft pulping	6,000–20,000	7.5–10.0	10–50	<20
Sulfite pulping (no recovery)	20,000–30,000	2.5–3.5	550–750	150–200
Sulfite pulping (with recovery)	20,000–30,000	2.5–4.0	50–100	40–60
Bleaching	20,000–40,000	2.0–5.0	10–25	14–25

[a]Oxygen consumed at 20°C during a 5-day incubation with acclimated microorganisms.

whereas sulfite pulping and bleaching plant wastes are acidic. Chemical recovery is essential in keeping oxygen-depleting materials low. Large calcium bisulfite mill effluents may have oxygen demands equivalent to 2,000,000 or 3,000,000 people. Effluents display some toxicity to aquatic fauna, albeit of a low order. Neutral and higher pH value effluents are darkly colored, which is aesthetically undesirable and inhibits photosynthesis. In smaller streams, fish downstream from pulp mill outfalls can have tainted flesh. Odor and taste imparted to receiving waters can also interfere with the subsequent use of the stream for drinking water. Wind and wave action can create foam on receiving waters, and inorganic salt content may prevent use in irrigation.

METHODS OF TREATMENT

No process can alleviate all pulping effluent problems. Abnormal pH is neutralized with slaked lime, calcium carbonate or sodium hydroxide, since integrated pulping effluents are usually acidic (Laws and Burns 1960; Charles and Decker 1970). Settling removes suspended solids except for some mechanically ground "fines."

All microbiological oxidation systems reduce pulp effluent oxygen demand, but concurrent removal of acute toxicity is not related to operating parameters for these systems. Microbiological treatment may not completely remove substances responsible for tainting fish flesh or causing odor, foam, and taste in drinking water. Microbiological treatment does not remove color, however color bodies can be precipitated by massive lime treatment (EPA 1970).

RESEARCH PROBLEMS

Originally, pulping waste treatments were the same as those used in domestic sewage treatment. Problems arise with pulping effluents because of their variable nature. In short-term microbiological oxidation systems, sludge recycling difficulties may occur. Biologists emphasize the need to remove sublethal toxicity, however the responsible chemical entities are largely unknown, and means of measurement are lacking. Massive lime treatment has technical and economic limitations, and specific information concerning unresolved problems is lacking. Thus, a considerable impetus exists for in-process changes or new processes to minimize current wastewater problems.

Food Processing Wastes

Water is absolutely necessary in food processing. Through conservation and reuse, liquid waste is reduced, cutting the pollution load. The National Canners Association has set four conditions governing the use of reclaimed waters in contact with food products:

1. the water must be free of microorganisms of public health significance
2. the water must contain no chemicals in concentrations toxic or otherwise harmful to man
3. the water must be free of any materials or compounds that could impart discoloration, off-flavor or odors to the product or otherwise adversely affect quality
4. the water appearance and content must be aesthically acceptable

WATER REUSE

Historically, water reuse was given little consideration. Water is relatively abundant in nature and reuse was considered hazardous due to bacterial contamination. Contamination potential (Figure 8.2.1) shows that, in washing fruit, unless 40% of the water is exchanged each hour, the growth rate of bacteriological organisms becomes extremely high. To overcome this, other means of control such as chlorination must be used. The importance of chlorination in maintaining satisfactory sanitary conditions is graphically shown in Figure 8.2.2. When chlorination was discontinued, the bacterial count more than doubled. As soon as chlorination resumed, bacterial counts were again brought under control.

Water conservation can be achieved through counterflow reuse systems. Figure 8.2.3 outlines a counterflow system for reuse of water in a pea cannery. At the upper right, fresh water is used for the final product wash before the peas are canned. From this point, the water is reused and carried back in successive stages for each preceding washing and fluming (the transport of the fruits by flowing water in an open channel) operation. As the water flows coun-

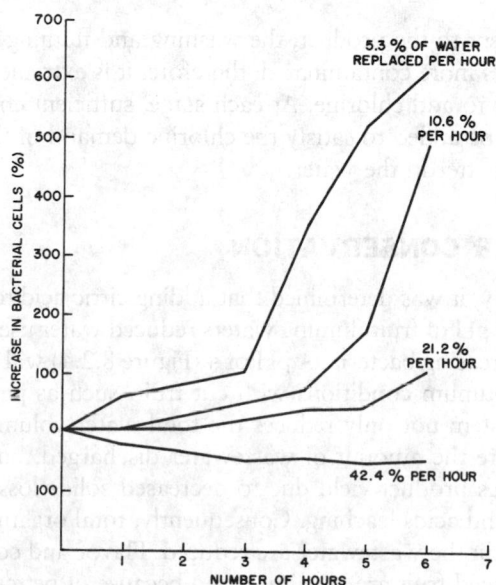

FIG. 8.2.1 Effect of rate of water replacement on growth of mesophilic bacteria at 90°F.

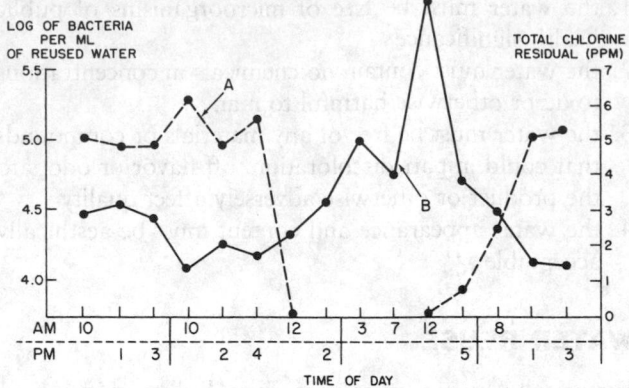

FIG. 8.2.2 Effect of chlorine concentration on bacterial counts in reused water. A. Chlorine concentration; B. Bacterial counts.

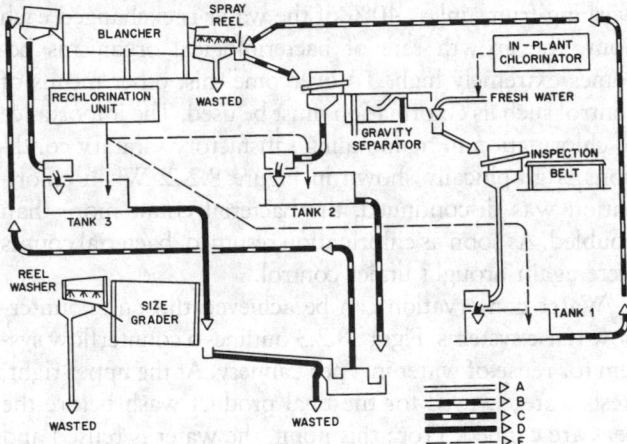

FIG. 8.2.3 Four-stage counterflow system in a pea cannery. A. First use of water; B. Second use of water; C. Third use of water; D. Fourth use of water; E. Concentrated chlorine water.

tercurrent to the product, the washing and fluming water becomes more contaminated; therefore, it is extremely important to add chlorine. At each stage, sufficient chlorine should be added to satisfy the chlorine demand of the organic matter in the water.

WATER CONSERVATION

Recently, it was determined that adding citric acid to control the pH of fruit fluming waters reduced water use without increasing bacteria. A pH of 4 (Figure 8.2.4) will maintain optimum conditions with cut fruit, such as peaches. The system not only reduces the total water volume and therefore the amount of wastewater discharged, but also increases product yield due to decreased solids loss from sugar and acids leaching. Consequently, total organic pollutants in the wastewater are reduced. Flavor and color of the canned fruit are also improved because of better soluble solid retention.

Closed loop systems, such as the hydrostatic cooker-cooler for canned product, are another conservation

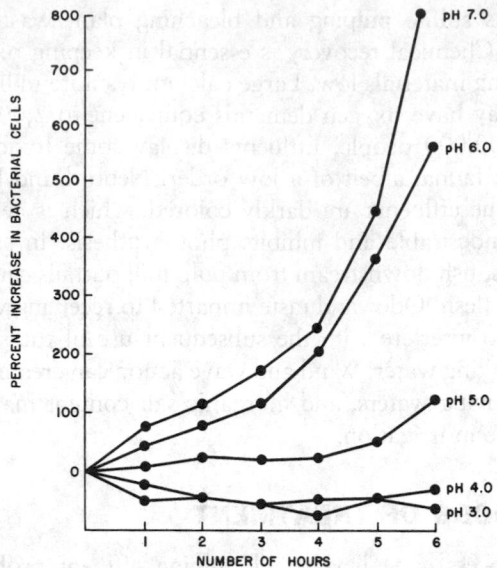

FIG. 8.2.4 Effect of pH control on bacterial cell growth.

method. The water is reused continuously, with fresh makeup water added only to offset minor losses from evaporation. Closed loop systems not only conserve water but also reclaim much heat and can result in significant economic savings.

It is not the intent of this section to describe the enormous array of concepts and ramifications used in the food processing industry to reduce water and waste loads while maintaining product quality. Many factors determine the final effectiveness of proper water use. For example, tomatoes spray-washed on a roller belt where they are turned are almost twice as clean as the same tomatoes washed on a belt of wire mesh construction. In another example, warm water is approximately 40% more effective in removing contaminants than the same volume of cold water.

There is a delicate balance between water conservation and sanitation, with no straightforward or simple formula for the least water use. Each process must be evaluated with the equipment used to arrive at a satisfactory procedure for water use, chlorination, and other factors, such as detergents.

ELIMINATION OF WATER USE

Eliminating water in certain operations eliminates attendant wastewater treatment problems. Wherever possible, food should be handled by either a mechanical belt or pneumatic dry conveying system. If possible, the food should be cooled in an air system. Recent studies by the National Canners Association in comparing hot air blanching of vegetables with conventional hot water blanching show that both product and environmental quality were improved by using air. Blanching, used to deactivate enzymes, produces a very strong liquid waste. For

TABLE 8.2.3 HOT AIR VS HOT WATER BLANCHING

Product	Blanching System	Wastewater gal/ton	COD Produced lb/ton	SS Produced lb/ton
Green peas	Hot water	1,000.0	32.70	1.42
Green peas	Hot air	0.018	Not measured	Not measured
Green beans	Hot water	1,710.0	4.70	0.11
Green beans	Hot air	0.25	0.002	0.0002
Corn on the cob	Hot water	1,223.0	4.70	0.041
Corn on the cob	Hot air	0.013	5.6×10^{-5}	1×10^{-6}
Red beets	Hot water	1,333.0	4.11	0.16
Red beets	Hot air	0.089	0.001	7.4×10^{-6}
Spinach	Hot water	1,430.0	2.6	3×10^{-1}
Spinach	Hot air	3.6×10^{-2}	3.0×10^{-4}	3×10^{-7}

pea processing, this small volume of wastewater is estimated to be responsible for 50% of the entire wasteload BOD; for corn, 60%; and for beets with peelings, 80%. Preliminary results show a reduced pollution load (Table 8.2.3), while improving product nutrients, vitamins, and mineral content.

WASTEWATER TREATMENT

Preprocessing

Proper management of food processing wastes requires consideration of individual operations from harvest through waste disposal as integrated subunits of the total process. Every effort should be made to eliminate wastes and to avoid bringing wastes from the farm into the processing plant. Where possible, preprocessing should occur in the field, returning the organic materials to the land. In the processing plant, wastewater volume and strength should be reduced at each step. This principle applies to all food processing wastes, including fruit, vegetables, meat and poultry, and dairy.

Waste segregation within a plant is important in optimizing the least-cost approach to treatment. In a typical brewery (Figure 8.2.5), where 3% of the flow contains

59% of the BOD, it is less expensive to treat this small flow separately than to mix it with the entire plant waste flow. This is effective when a plant treats its own wastes or releases waste to a municipality with surcharges for high-strength waste.

Food processing wastes are amenable to biological treatment, and they frequently provide nutrients essential to efficient biological treatment. Although various waste treatment methods are available to the food processor (Figure 8.2.6) there is no simple guide for the most practical and economical method.

Lagoons and Land Disposal Systems

Since food wastes contain suspended and soluble organic contaminants, they are readily treated in lagoons and land disposal systems. The lagoons may be complete storage ponds, frequently used by seasonal processors for waste containment. In four to six months, the waste is stabilized, with up to 90% BOD reduction. If large lagoon acreage is available, aerobic conditions are maintained by limiting organic loadings to less than 100 lb of BOD per acre per day. When extremely strong wastes are encountered, a combination of anaerobic and aerobic lagoons provides an excellent means of treatment on less land, since the anaerobic system may reduce BOD from 60% to 90%, reducing the aerobic lagoon acreage required to achieve desired effluent quality.

Anaerobic lagoons are odorous and require an artificial or natural cover. In meat products, the high grease content forms a natural cover. Aerobic lagoons can also cause odors if overloaded and lacking sufficient dissolved oxygen. Various mechanical aeration methods have reduced required lagoon acreage, but these increase power costs.

Land disposal can be achieved by flooding; however, the most efficient means is conventional farm spray irrigation equipment. Sandy soil with a high infiltration rate offers no surface runoff, and no discharge to a receiving stream. Recently, an overland flow technique has been developed as an equivalent of tertiary treatment.

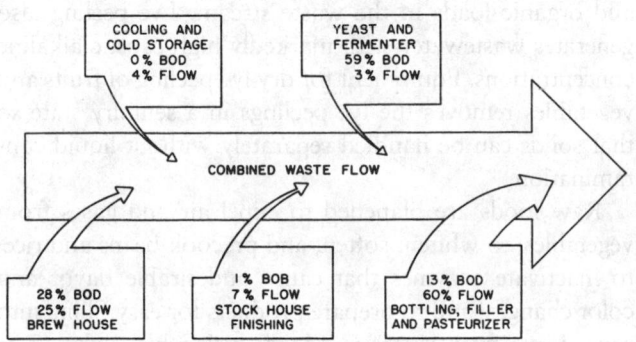

FIG. 8.2.5 Source and relative strength of brewery wastes

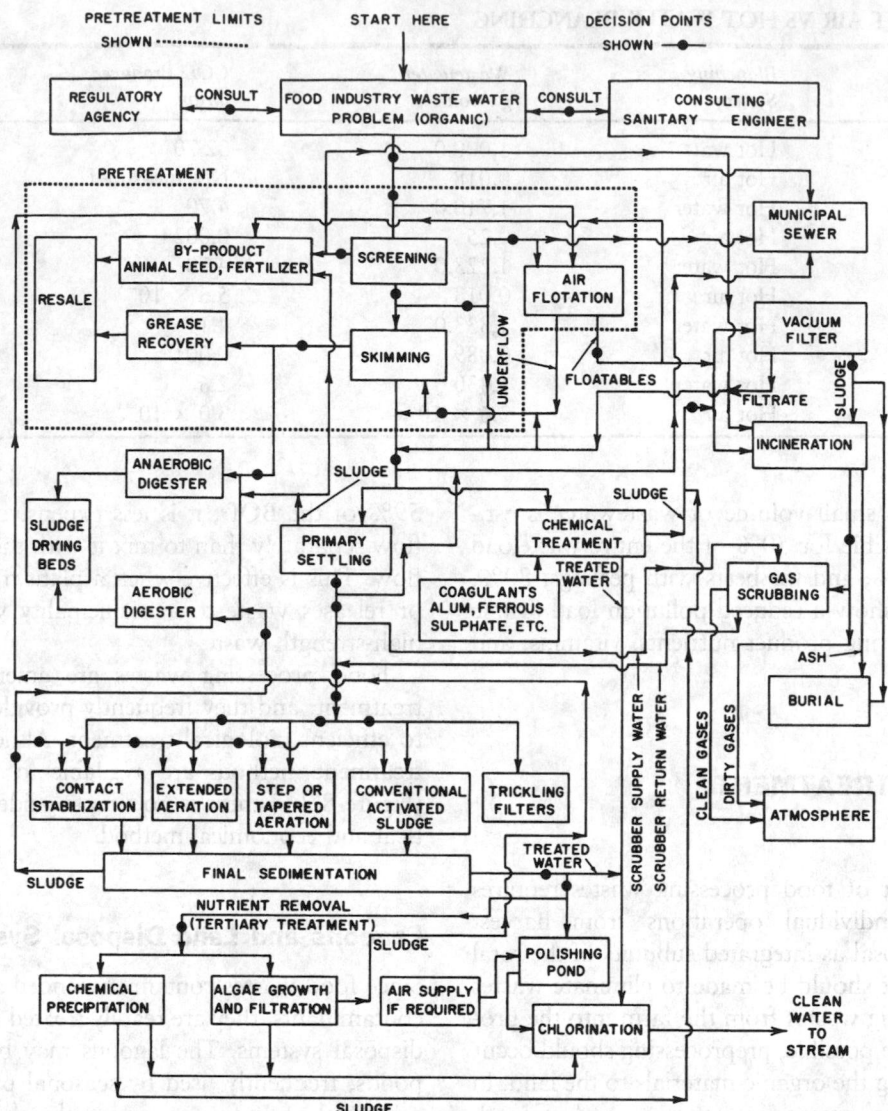

FIG. 8.2.6 Wastewater treatment maze (for organic waste from food processing industries). The diagram illustrates the many options open to solving waste treatment problems. The best route through the maze is suggested by an engineering study and report. Such a report discloses possible treatment methods, anticipated influent properties, effluent requirements and costs. Most important, the report serves as a mutually agreed-upon criterion with regulatory agencies. Designing a waste treatment system should not be considered without such a study and report.

Canning Wastes

The canning industry uses an estimated 50 billion gal of water per year to process one billion cases of food. Liquid waste is normally screened as a first step in any treatment process. Solids from these screens can be trucked away as garbage or collected in a by-products recovery program.

Food product washing is the greatest source of liquid waste. The water used is normally reclaimed in a counterflow system, with a final discharge high in soluble organic matter and containing suspended solids—much of it inorganic—from the soil. Other wastes come from peeling operations. The amount of suspended matter varies with the type of peeling. The type of peeler—steam, lye, or abrasive—has an effect on the nature of the waste generated.

Normal practices utilize large volumes of water to wash away loosened peelings, creating tremendous suspended and organic loads in the waste stream. Lye peeling also generates wastewater with markedly high caustic alkaline concentrations. Equipment for dry lye peeling of fruits and vegetables removes the lye peelings in a semidry state so that solids can be handled separately without liquid contamination.

Raw foods are blanched to expel air and gases from vegetables; to whiten, soften, and precook beans and rice; to inactivate enzymes that cause undesirable flavor and color changes; and to prepare products for easy filling into cans. Little fresh water is added during blanching (8-hr shift), therefore the organic material concentration becomes high due to leaching of sugars, starches, and other soluble materials. Although low in volume, blanch water is highly concentrated and frequently represents the largest load of soluble wastes in the entire food processing operation. The amount of dissolved and colloidal organic matter varies, depending on the equipment used.

The last major source of liquid wastes is the washing of equipment, utensils, and cookers, as well as washing of floors and food preparation areas. This wastewater may contain a large concentration of caustic, increasing the pH above the level experienced during food processing.

After cooking, the cans are cooled, which requires a large volume of water. The cooling water is clean and warm and should be reused for washing.

Meat and Poultry Wastes

Feed lot, stockyard, and poultry receiving area wastes consist primarily of manure, unconsumed feed, feathers, and straw, together with common dirt and drain water. Pollution can be reduced if solid wastes are not diluted by water.

In killing operations blood must be collected separately and prevented from entering sewer or waste treatment systems, since blood has an extremely high waste strength of about 100,000 ppm BOD. In poultry plants, various processes must be isolated to avoid cross-contamination from live birds or wastes of previous operations. As the bird goes through the plant on shackles, feathers are removed and flumed away. A major incision is made, entrails and major organs are pulled out, and inedible viscera are discarded in a flowaway flume system. The lungs and other material remaining in the carcass are removed by vacuum suction.

Flowaway systems (for feathers, entrails and offal) create an increased organic load, and it is desirable to use a dry conveying system. Most plants use the flowaway system as a more convenient and nuisance-free operation. After the offal flowaway leaves the area, it must be screened in order to remove solids. These solids and wastes from other operations are then sent to a rendering plant where they are utilized in making chicken feed.

Meat packing houses generate a strong waste. These wastes are amenable to treatment, as are poultry wastes. Before releasing processing wastewaters into city sewers or private waste treatment systems, screening and grease removal should be provided to recover solids for by-product use. Removal of large solids and free floating grease is also important to avoid clogging sewer lines and fouling biological treatment systems.

Dairy Wastes

Among waste generating operations in the dairy industry are receiving stations, bottling plants, creameries, ice cream plants, cheese plants, and condensed and dried milk prod-

uct plants. Wastes include separated milk, buttermilk, or whey, as well as occasional batches of sour milk. Diverse methods are being explored for reclamation and concentration of materials, such as reverse osmosis for whey. Unfortunately, there is no simple economical method to reclaim and utilize these materials as byproducts. Indiscriminate dumping of these materials into sewers should be avoided, and where possible these extremely strong wastes should be treated separately or eliminated by hauling.

Milk wastes are normally treated in municipal plants, since most dairies are located in communities. The wastes are amenable to biological treatment, and screening is commonly provided; grit removal is sometimes necessary, as well.

Solid Waste Disposal

Most solid wastes from food processing are generated in processing raw materials. Some materials, such as packaging, faulty or damaged containers, office or warehouse papers, and refuse from laboratories, should be kept separate from the food solids. Solid food waste is produced in growing and harvesting raw crops, in food processing, and by the retailer and consumer.

Many food processing operations are seasonal and generate large quantities of organic solid wastes in a short time. The putrescible nature of the wastes requires quick handling in utilization or disposal. Land disposal operations—by far the most common method of disposal—must be rigidly controlled to prevent odor production and fly breeding. It is apparent that the food processing industry must recycle and recover more of its by-products.

Utilization of food processing waste as animal feed is a widely used method of disposal. In some areas, seafood canning waste is pressed into fish meal for animal feed or into fertilizer material. Tomatoes are pressed and dehydrated for use as dog food and cattle food. Pea vines, corncobs, and corn husks are also used as feed. Citrus peel waste may be pressed for molasses, which may then be processed, dried, and sold as cattle feed. Certain types of pits and nutshells have been converted to charcoal.

Other possibilities exist, such as producing alcohol from fruit wastes and composting fruit waste solids, but usually it is much cheaper to dump, landfill, spread on the land, or discharge at sea than to attempt reclamation. There does not appear to be much chance of a change in this area unless prevailing economic conditions can be altered through new legal restrictions or some form of subsidy program.

Hydrocarbons

A bulk oil handling terminal stores and tranships petroleum products, petrochemicals, animal fats, greases and food grade vegetable oils. In addition they often accept and dispose of ballast wastewaters from marine tankers that deliver to the terminal or pick up cargo for transhipment. A biological treatment system is appropriate because of the wide range of physical and chemical characteristics of the various types of oils and petrochemicals; mechanical and/or chemical means of separation and neutralization are too expensive to install and operate.

The equipment used in the system includes (1) a collection system for the wastewater flow; (2) an API separator; (3) a high-rate oxidation pond (or "aerated lagoon") with a 150,000 gal capacity; (4) a secondary settling or "polishing pond" with a capacity of 450,000 gal; (5) a recirculation system; and (6) an 800,000 gal storage tank for ship ballast holding and for surge flow equalization.

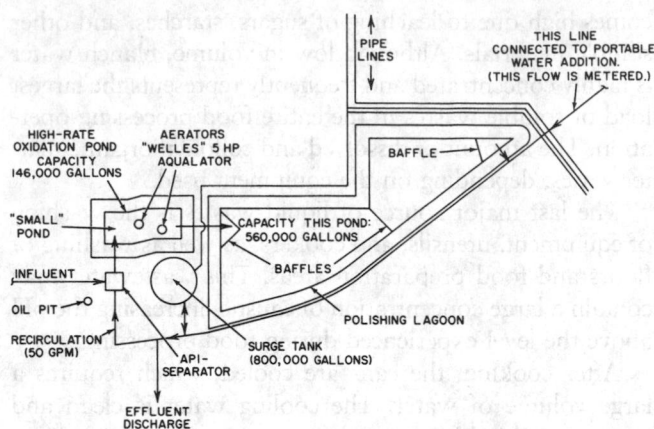

FIG. 8.2.7 Bulk oil-handling terminal waste treatment system.

DESIGN BASIS

Biological treatment was chosen because some oils float, some sink, some are "soluble," and some saponifiable. Thus, a broad-spectrum treatment was required. No municipal sewerage system was available, therefore the effluent had to meet waterway discharge requirements. This specified effluent concentration limits (mg/l): including biological oxygen demand (BOD) of 20 or less; hexane solubles of 15 or less; suspended solids of not over 25; and a pH range of 6 to 10. In addition, effluent had to be substantially color free. Influent characteristics were as follows:

Average daily flow	20 gpm
Average BOD	400 ppm
Average hexane solubles	300 ppm
Average suspended solids	100 ppm
Average pH range	5 to 12

Maximum aeration requirements were calculated to provide (1) sufficient flexibility to vary input air in response to extreme pollutant load variations; and (2) excess hydraulic mixing capacity to increase suspended solids oxidation and reduce the volume of sludge accumulating in the system.

The use of 3–5 hp floating aerators provides a total available oxygen transfer rate of 7.5 lb oxygen per lb of

BOD, according to the manufacturer. Under most terminal operating conditions, only two aerators were required to provide 95% BOD removal. Sludge accumulation was below 350 lb wet sludge (7 lb dry) per day. The system has never had an odor problem.

A recirculating system was established for peak waste loads in oil handling terminal operations (Figure 8.2.7). The 800,000 gal ballast tank gives an additional ten days of holding time for recirculation when pollutant loadings far exceed design capacity.

OPERATIONAL HISTORY

The BOD of the high-rate oxidation pond ("small pond") at startup was 2420 ppm (mg/l), and the hexane soluble content was 2040 mg/l. Both ponds were covered with about 6 in of floating oil and grease (see Figure 8.2.8 for the rate of stabilization).

The system was set on a recirculation rate of 50 gpm. Three days later, when the pH showed no further erratic swings, dried bacterial cultures (special species of saprophytic and facultative bacteria that consume oil) were added to create a biomass specifically for oil and grease reduction. The initial dosage was 5 lb, followed by 1 lb/day addition for 14 days. After this initiation, the system was

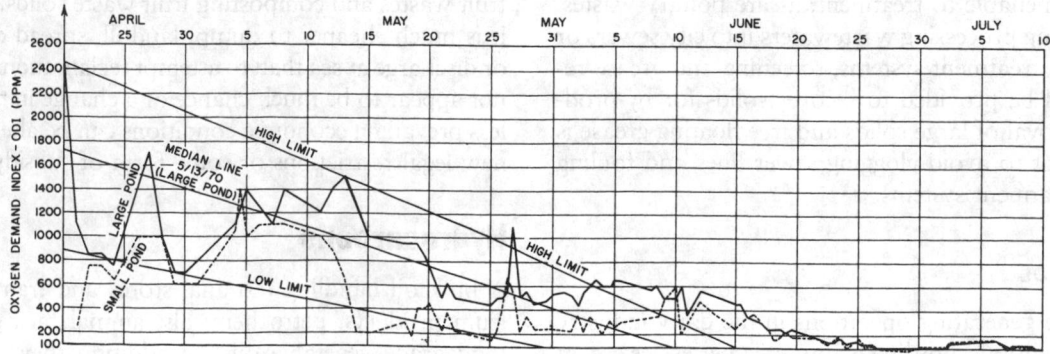

FIG. 8.2.8 BOD reduction in ponds as a function of time after startup. (BOD is usually 50% of ODI.)

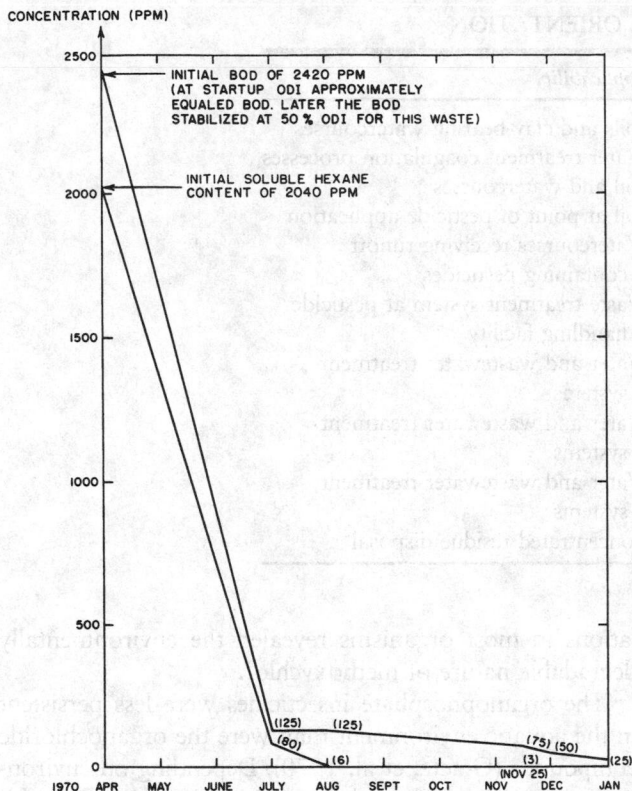

CONCENTRATION (PPM)

INITIAL BOD OF 2420 PPM
(AT STARTUP ODI APPROXIMATELY
EQUALED BOD. LATER THE BOD
STABILIZED AT 50 % ODI FOR THIS WASTE)

INITIAL SOLUBLE HEXANE
CONTENT OF 2040 PPM

FIG. 8.2.9 Polishing pond performance from startup. A = initial BOD of 2420 ppm (at startup ODI roughly equals BOD; later BOD is stabilized at 50 percent ODI for this waste); B = initial soluble hexane content of 2090 ppm.

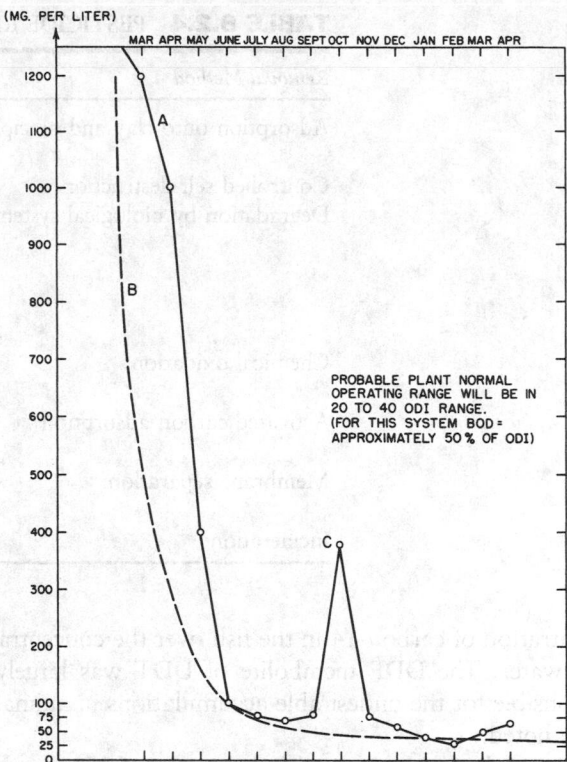

ODI (MG. PER LITER)

PROBABLE PLANT NORMAL
OPERATING RANGE WILL BE IN
2O TO 40 ODI RANGE.
(FOR THIS SYSTEM BOD=
APPROXIMATELY 50 % OF ODI)

FIG. 8.2.10 Theoretical vs actual performance. A. Rate of pollutant addition reducers; B. Standard theoretical curve for rate of pollutant reduction by biological treatment systems; C. Curve distortion due to exceptional load condition. System gave 97% reduction in 30 days.

maintained by the addition of $\frac{1}{2}$ lb of the dried culture three times a week. Figure 8.2.9 illustrates initial reduction of the hexane soluble content and continuing control since the beginning of plant operation.

The effectiveness of a biological treatment to control oily wastewater is also shown in Figure 8.2.10 where theoretical and actual performances are compared.

Pesticides

Since pesticides enter the aquatic environment in runoff from agricultural areas as well as from point sources, control must be based on a multiphased approach:

1. Controlled application in minimum quantities over areas where specifically needed
2. Degradation in soil and watercourses
3. Removal at plants producing potable water
4. Treatment of wastes from pesticide handling facilities and sewered areas

The various mechanisms for removing pesticides entering the environment are discussed in this section as outlined in Table 8.2.4, and the chemical structures of the pesticides are shown in Figure 8.2.11.

PESTICIDE REMOVAL IN NATURAL AQUATIC SYSTEMS

Pesticide occurrence in surface waters can be traced to several sources: agricultural runoff, industrial discharge, purposeful application, cleaning of contaminated equipment, and accidental spillage. Chlorinated hydrocarbons in aqueous solutions are readily adsorbed by clay materials. After adsorption, small fractions of some pesticides are gradually desorbed into the overlying water where the pesticide concentration is maintained at a dynamic equilibrium level. Drainage of clay-bearing waters from agricultural areas represents a continuous supply of pesticides to the aqueous solution. Desorption rates are not significantly affected by pH, temperature, salt and organic levels (Huang 1971).

The introduction of many new pesticides in recent years has created the need for reliable evaluation of the effects on the aquatic biota. The model ecosystem for these evaluations consists of glass aquaria arranged in a sloping soil-air-water interface (Metcalf, Sangha and Kapoor 1971). A food chain of plant and animal organisms, compatible with the environmental conditions simulated in the aquarium, is chosen for following radiolabeled DDT (labeled in the aryl rings with C^{14}) and methoxychlor. Average data presented in Table 8.2.5 show a 13,000-fold increase in con-

TABLE 8.2.4 PESTICIDE REMOVAL ORIENTATION

Removal Method	Applicability
Adsorption onto clay and precipitates	Soils and clay-bearing watercourses
	Water treatment coagulation processes
Controlled self-destruction	Soil and watercourses
Degradation by biological systems	Soil at point of pesticide application
	Watercourses receiving runoff containing pesticides
	Waste treatment system at pesticide handling facility
Chemical oxidation	Water and wastewater treatment systems
Activated carbon adsorption	Water and wastewater treatment systems
Membrane separation	Water and wastewater treatment systems
Incineration	Concentrated residue disposal

centration of carbon-14 in the fish over the concentration in water. The DDE metabolite of DDT was largely responsible for the undesirable accumulations in animal tissue noted.

In studies with tritium-labeled methoxychlor, accumulations of the pure compound and its degradation products in fish were of the order of 0.01 those for DDT (Metcalf, Sangha and Kapoor 1971). The presence of several degradation products and the relatively low accumu-lations in most organisms revealed the environmentally degradable nature of methoxychlor.

The organophosphate insecticides were less persistent in the aquatic environment than were the organochloride compounds (Graetz, et al. 1970). Depending on environmental conditions, degradation is by chemical or microbiological means, or both. Chemical degradation involves hydrolysis of the ester linkages. Hydrolysis can be either acid-catalyzed, e.g., ciodrin, or base-catalyzed, e.g., malathion. Microbial degradation can be by hydrolysis or oxidation. Partial degradation is often the case, although for diazinon, chemical hydrolysis of the thiophosphate linkage attached to the heterocyclic ring results in 2-iso-propyl-4-methyl-6-hydroxypyrimidine, which is degraded rapidly by soil microorganisms. Among the orthophosphates, parathion is one of the most resistant to chemical hydrolysis, but microbial degradation to aminoparathion can proceed.

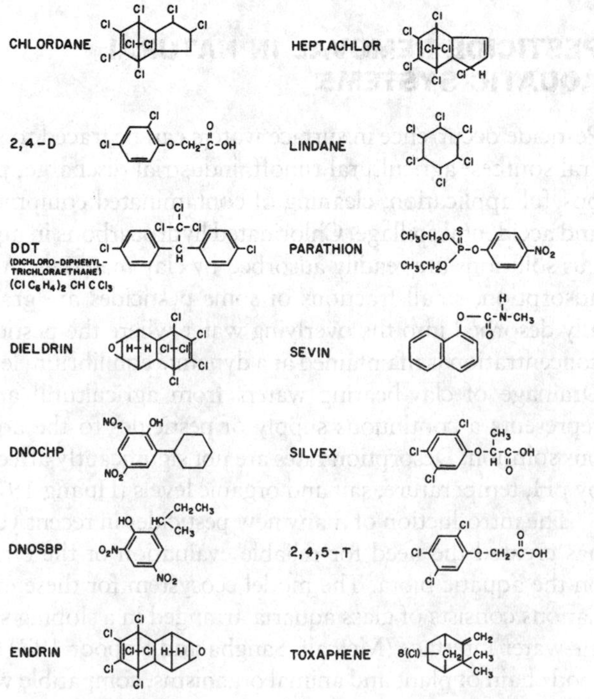

FIG. 8.2.11 Chemical structures of key pesticides. A. Chlordane; B. 2,4-D; C. DDT; D. Dieldrin; E. DNOCHP; F. DNOSBP; G. Endrin; H. Heptachlor; I. Lindane; J. Parathion; K. Sevin; L. Silvex; M. 2,4,5-T; N. Toxaphene.

TABLE 8.2.5 DISTRIBUTION OF DDT IN MODEL ECOSYSTEM

	Distribution		
	Water	Snail	Fish
Total Carbon-14 Content, mg. per liter	0.003	20	38
Distribution, %			
as DDT[a]	5	31	31
as DDE[b]	7	47	56
as DDD[c]	8	11	12
as polar metabolites	74	7	1
Unclassified	6	4	0

[a]DDT, Dichlorodiphenyltrichloroethane
[b]DDE, Dichlorodiphenyldichloroethylene
[c]DDD, Dichlorodiphenyldichloroethane

Standard biochemical oxygen demand tests involving glucose incubation with a carbaryl insecticide, Sevin, indicate no inhibition of bacterial oxidation of glucose up to a Sevin concentration of 100 mg/l. In fact, Sevin was biooxidized to a considerable extent at this level; oxidation was enhanced after a period of acclimatization.

BIODEGRADABLE REPLACEMENT AND CONTROLLED SELF-DESTRUCTION

Biodegradable substitutes have been developed for some hard pesticides. One approach is to substitute aromatic chlorine atoms in the DDT molecule (Anon., Chemical Week 109:36 1971). The new compounds reportedly do not build up in animal tissue and concentrate at higher levels in the food chain.

A mildly acid reduction by zinc will speed degradation of DDT and other pesticides in natural systems (EPA 1970). A copper catalyst speeds up the reduction. Effective degradation of DDT to bis(p-chlorophenyl) ethane appears possible in soil by using micron-sized particles of the reductant in close proximity to the DDT. Thin, slowly soluble wax or silyl coatings on the reductant can delay the reaction. A second technique for delayed reaction involves controlled air oxidation to sulfur to produce the required acidity. Effective degradation of DDT in aqueous systems was also achieved using reduction techniques. The procedure was reported effective in substantially degrading dieldrin, endrin, aldrin, chlordane, toxaphene, Kelthane, methoxychlor, Perthane and lindane.

BIOLOGICAL TREATMENT PROCESSES

The waste flow from a parathion production unit undergoes activated sludge treatment (Coley and Stutz 1966) with a residence time of 7–10 days, providing nearly complete breakdown of parathion and paranitrophenol as well as over 95% reduction in organic matter as measured by chemical oxygen demand (COD).

Studies were also conducted in designing a wastewater treatment facility for production of organic phosphorus pesticides (Lue-Hing and Brady 1968). Although treatability studies showed the waste to be biodegradable, shock loads caused stresses at up to 6000 mg/l solids. Consequently, a two-stage activated sludge system was chosen in which the first stage is a dispensable, low-solids, detoxification unit. Removal of dissolved organic matter measured as biochemical oxygen demand was 90–98% in the pilot plant.

The oxidation of Sevin carbaryl insecticide by an activated sludge culture is depicted in Figure 8.2.12. No adverse effects on bacteria, protozoa and rotifers were noted. Biological degradation studies (Leigh 1969) of lindane indicated no significant removal of this pesticide from microbial activity following 28 days of acclimatization in sta-

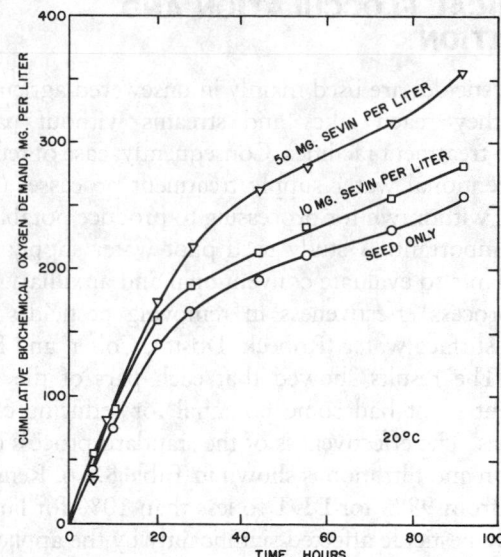

FIG. 8.2.12 Oxidation of Sevin carbaryl insecticide by acclimated bacteria.

tically aerated cultures. Removals in unseeded controls (reference samples) were approximately 46% while biological removals averaged only 41%. The biodegradability of heptachlor could not be deduced from similar studies because analyses of aqueous solutions of this pesticide indicated partial degradation to 1-hydroxyl chlordene and an undetermined compound. Removals of as high as 99.4% were attained within four days for heptachlor, but volatilization losses were considered significant.

The degradation of chlorinated hydrocarbon pesticides was studied under anaerobic conditions (Hill and McCarty 1966) such as lake and stream bottoms, lagoon treatment systems, and digestion systems. Lindane and DDT were rapidly decomposed, the latter to DDE which degraded more slowly. Heptachlor and endrin also formed intermediate degradation products within short periods. The rate of decomposition of aldrin was similar to that for DDD; only slight degradation of heptachlor epoxide occurred, and dieldrin remained unchanged. Anaerobic conditions were more favorable than aerobic conditions for pesticide degradation. Sorption of chlorinated hydrocarbon pesticides was found to be greater on algae than on bentonite or fine sand; the process was partially reversible and the degree of sorption was inversely related to the solubility of the pesticide.

Lindane was degraded anaerobically in pure culture; only 0.5% of the lindane present after 1 hr incubation was found in the reaction mixture after 27 hr incubation (MacRae, Raghu and Bautista 1969). The covalently linked chlorine of the lindane molecule was released. A detected intermediate product reached a maximum level after about 4 hr incubation and diminished to undetectable levels after 27 hr incubation.

CHEMICAL FLOCCULATION AND OXIDATION

Since pesticides are used mainly in unsewered agricultural areas, they reach lakes and streams without passing through treatment facilities. Consequently, ease of removal in conventional water supply treatment processes (when water is withdrawn for processing to produce potable water) is important. A study used pilot water supply treatment plants to evaluate conventional and auxilliary treatment process effectiveness in removing pesticides from natural surface water (Robeck, Dostal, Cohen and Kreiss 1965). The results showed that each part of the water treatment plant had some potential for reducing certain pesticides. The effectiveness of the standard process of coagulation and filtration is shown in Table 8.2.6. Removals ranged from 98% for DDT to less than 10% for lindane. The only pesticide affected significantly by the application of chlorine or potassium permanganate (1–5 mg/l) was parathion, 75% of which was oxidized to paroxon, a more toxic material. At high dosages, ozone (10–38 mg/l) reduced chlorinated hydrocarbons; by-products of unknown toxicity were formed.

In full-scale evaluations (Nicholson, Grzenda and Teasley 1968), the standard processing steps of coagulation, settling, rapid sand filtration, and chlorination were successful in reducing DDT and DDE levels but not toxaphene and lindane levels. Side tests with a 25-μ filter removed DDT and DDE more effectively than toxaphene and lindane, indicating that the latter materials were transported in solution.

Chemical degradability of frequently used chlorinated hydrocarbon insecticides has also been investigated (Leigh 1969). Lindane and endrin were not removed by either chlorine or potassium permanganate at oxidant dosages ranging from 48 to 61 mg/l, contact times of 48 hr and a wide range of pH values. Heptachlor was removed by $KMnO_4$ to the extent of 88% with only slight variation due to pH adjustment. Heptachlor and DDT were both partially removed by chlorine, and DDT was partially removed by $KMnO_4$ with slightly higher removals at lower pH levels. Maximum removals by potassium persulfate, attained only for lindane and DDT, were 9.4% and 18.5%, respectively, at higher pH values.

Several physical and chemical treatments for removing the herbicide 2,4-D and its ester derivatives from natural waters have also been investigated (Aly and Faust 1965). Chemical coagulation of 1 mg/l solutions by 100 mg/l aluminum sulfate showed no promise with the herbicides and derivatives studied. Activated carbon studies indicated carbon requirements for reducing 2,4-D concentrations from 1 to 0.1 mg/l were 31 mg/l for sodium salt, 14 mg/l for isopropyl ester, 15 mg/l for butyl ester and 16 mg/l for isooctyl ester. Potassium permanganate dosed at 3 mg/l did not oxidize 1 mg/l of these same compounds. However, 0.98 mg/l of 2,4-DCP was completely oxidized by 1.25 mg/l $KMnO_4$ in 15 min. Ion exchange studies indicated that strongly basic anion-exchange resins more effectively removed the compounds studied than cation exchange resins.

Strong oxidants to degrade chlorinated hydrocarbon pesticides (Buescher, Dougherty and Skrinde 1964) have also been studied. Preliminary studies with lindane and aldrin showed negligible removals with hydrogen peroxide and sodium peroxide at 40 mg/l dosages and four-hr contact times. Chlorination had negligible effects on lindane, but completely oxidized aldrin, while potassium permanganate ($KMnO_4$) oxidized lindane to approximately 12% and aldrin, fully. Further studies of potassium permanganate added in varying doses from 6 to 40 mg/l to lindane solution indicated that the excessive time and oxidant dosages required for removals greater than 40% made this treatment unfeasible. Complete removal for aldrin could be attained in 15 min at 1 mg/l dosage of $KMnO_4$.

Due to the relatively small fraction of ozone in the air stream used for ozonation, pesticide removals from air stripping were measured, as well as removals from oxidation. Up to 75% of lindane was removed by ozonation, whereas aeration alone had no measurable effect. Dieldrin and aldrin were completely removed almost at once, but aeration studies also showed fairly rapid removals.

ACTIVATED CARBON ADSORPTION

Considerable data on the adsorption of several pesticides and related nitrophenols on activated carbon have been reported (Weber and Gould 1966). Carbon loadings of 40–53% indicate economic feasibility for removal of trace quantities of these persistent compounds. Rate and Langmuir equilibrium constants for the pesticides are shown in Table 8.2.7. The quantity of pesticide adsorbed per gm of carbon at complete monolayer coverage of the carbon surface (X_m values) indicates high ultimate carbon loadings. B^{-1} values, which relate to energies of adsorption, indicate that relatively high residual concentrations

TABLE 8.2.6 REMOVAL OF PESTICIDES IN WATER TREATMENT PLANT OPERATIONS

Pesticide (10 ppb dosage)	Coagulation-Filtration	Removal, percent Carbon Slurry	
		5 ppm	20 ppm
Lindane	<10	30	80
Endrin	35	80	94
Dieldrin	55	75	92
2,4,5-T Ester	65	80	95
Parathion	80	>99	>99
DDT	98	Not Tested	Not Tested

TABLE 8.2.7 CARBON ADSORPTION CONSTANTS FOR ORGANIC PESTICIDES

	Relative Rate Constant (μmoles/g)^2 per hr $\times 10^{-4}$	Limiting Monolayer Carbon Loading (X_m), mg per g	b^{-1} (relates to energy of adsorbtion) mg/l
2,4-D	1.44	387	2.32
2,4,5-T	1.00	448	1.71
Silvex	0.71	464	1.86
DNOSBP	1.35	444	1.39
DNOCHP	1.12	500	1.81
Sevin	1.64	—	—
Parathion	1.49	530	0.24

Note 1: Experimental Conditions: $C_0 = 10$ μmoles per liter, 0.273 mm. Columbia carbon, 25°C

Note 2: Symbols relate to Langmuir isotherm: $x = \dfrac{X_m bC}{1 + bC}$

(Reprinted with permission, from I.C. MacRae, K. Raghu, and E.M. Bautista, 1969, *Nature* 221:859.

are required for all but parathion to attain saturation capacity.

Additional studies (Dedrick and Beckman 1967) indicate that adsorption of 2,4-dichlorophenoxyacetic acid (2,4-D) can be correlated by both the Freundlich and the Langmuir isotherms; however, two sets of correlating constants are required for each of the low and high concentration ranges. No significant differences in carbon capacities were noted between granular and powdered carbon. Carbon loadings of approximately 60% by weight of the herbicide were attained at liquid concentrations 95% of saturation, or about 740 mg/l.

Carbon adsorption studies using a slurry approach showed parathion to be most amenable and lindane least amenable (Table 8.2.6) to removal by activated carbon. Use of a granular bed at 0.5 gpm/cu ft resulted in almost complete removal of all pesticides.

REVERSE OSMOSIS

Specific chemical permeation through a cellulose acetate membrane has also been reported (Hindin, Bennett and Narayanan 1969). The membranes were immersed in water at 82°C for 30 min prior to use. At a pressure differential of 100 atm, a temperature of 25°C, flux rates on the order of 15 gal/sq ft/day, and feed concentrations of about 500 mg/l, reduction of lindane was 73% while DDT and TDE (DDD) were rejected above 99%. High reductions were obtained for those chemical species existing primarily in the colloidal, aggregate, micelle, or macromolecular form. If the chemical species existed both as an aggregate in dispersion and as a discrete molecule in true solution where vapor pressure of the discrete molecule in true solution was appreciably greater than that of water, the range of reduction was 50–80%. Where discrete molecules more volatile than water were tested, range of reductions was 14–40%.

INCINERATION

Along with deep-well injection, incineration of concentrated pesticide waste is an alternative to treatment and disposal in surface waters. Solid wastes are burned in a rotary kiln or other incinerator at 1600°–2200°F (Anon. Chemical Week 108:37 1971). Afterburners can be used to reach temperatures of 2800°F. A scrubber is used to clean exhaust gases.

RESEARCH TRENDS

Since outlawing DDT and other pesticides that build up in the foodchain seems imminent in many developed areas, replacements must be found, or there will be a recrudescence of health problems. For example, malaria and Venezuelan equine encephalomyelitis resurge in areas where mosquito control is lax or mosquitos become resistant to the pesticides used. In the case of mosquito control, malathion and propoxur are recommended as replacements for DDT as resistance grows (Anon. Chemical Week 109:36 1971). Although fenitrothion, iodofenphos, phenothoate and Landrin show promise, all are more expensive and less effective than DDT.

Until suitable replacements are developed, much remains to be done in the realm of pesticide removal from waters—both prior to discharge of wastewater and in treating water for human use. Although the literature on the effects and measurement of pesticides is voluminous, articles on removal techniques for pesticides are relatively few.

Phenol

Although phenol (C_6H_5OH) has been detected in decaying organic matter and animal urine, its presence in a surface stream is attributed to industrial pollution. Petroleum refineries, coke plants, and resin plants are major industrial phenolic waste sources. Phenolic compounds and their derivatives are used in coatings, solvents, plastics, explosives, fertilizer, textiles, pharmaceuticals, soap, and dyes.

Treatment methods for phenol removal include biological (activated sludge, trickling filter, oxidation pond, and lagoon); chemical oxidation (air, chlorine, chlorine dioxide, ozone, and hydrogen peroxide); physical (activated carbon adsorption, solvent extraction, and ion exchange); and physicochemical (incineration and electrolytic oxidation).

SOLVENT EXTRACTION

For wastewaters containing high phenol concentrations, solvent extraction reduces the phenol to acceptable levels. Occasionally, recovered phenol is reused in the manufacturing process or solid as a by-product. In solvent extraction, two immiscible or partially soluble liquids are brought into contact for transfer of one or more components. Using a solvent such as benzene, phenol can be extracted from the wastewater. The extracted phenol is then washed out with caustic to form the sodium salt, and the benzene is reused. In the petroleum industry, light catalytic cracking oils are used as extractors, and in the coking industry, coke oven light oils are used as extractors. Process efficiency depends on solvent choice and system design.

BIOLOGICAL TREATMENT

The microorganisms capable of degrading phenol are highly specialized and require a controlled, stable environment. Under ideal conditions several weeks are required to develop the proper biological sludge. The efficiency of an acclimated biological system treating phenolic wastes depends strongly on temperature, pH, nutrients (nitrogen, phosphorus, minerals), oxygen concentration, phenol concentration, and other organics concentrations in the wastewater.

To degrade phenol, the microorganism population must be stable. Fluctuation in any of the preceding variables shifts the balance of this population, reducing system efficiency and possibly killing the biological organisms. Optimum phenol removal occurs at neutral pH (7.0), 70°F and constant phenol concentration.

Biological methods of phenol removal include activated sludge, trickling filters, oxidation ponds, and lagoons. Efficiency ranges from 65–90% removal, depending on the ability of the particular wastewater treatment system to control the process variables listed. Activated sludge, trickling filters, and oxidation ponds are all capable of high phenol removal if properly designed and operated; however, the trickling filter process is regarded as being more capable of withstanding slug loads without loss of performance. Lagoons for treating phenolic wastes are designed to avoid overflow, with evaporation and seepage used to balance the influent flow. This method is less desirable, due to the possibility of ground water pollution, odor, and overflows from rainfall.

Frequently, phenolic wastes are diluted with sanitary wastes and treated at the local municipal plant (Muller and Covertry 1968). Combined municipal-industrial treatment buffers the dilution and provides an ample supply of nutrients and microorganisms should the system be upset. Phenolic wastewaters should be neutralized prior to discharge to the municipal sewer system.

CARBON ADSORPTION

Activated carbon in the powdered and granular forms is used to remove phenolic tastes and odors from drinking water supplies. In wastewater treatment applications, where phenol content is considerably greater than in potable water applications and the flow is continuous, granular carbon systems are more economical.

Depending on the concentration of phenol and other organic compounds in the wastewater, activated carbon will adsorb from 10 to 25 lb of phenol per 100 lb of carbon. This capacity can be determined from isotherm and column test data. In general, phenol adsorption improves as the pH decreases.

Adsorption at high pH is poor, since phenolate salt forms and is difficult to adsorb. This is an advantage in applications where phenol recovery is worthwhile. The phenol is adsorbed at the low pH and reclaimed as sodium salt by chemical regeneration, using hot caustic. If the phenolate cannot be reused, regenerant disposal is a problem. Also, if quantities of other organic substances are present in the waste stream, they too will be adsorbed. These organic compounds may not be desorbed during caustic regeneration, which will decrease the phenol capacity of the carbon upon subsequent regeneration. If chemical regeneration does not sufficiently recover the phenol capacity of the carbon, thermal reactivation will be required.

Figure 8.2.13 is a flow diagram of a granular carbon system for phenol removal employing chemical regeneration and phenol recovery. Pretreatment consists of acidification to pH 4.2 to precipitate the suspended solids and clarify the overflow. The phenol content of the feedstream ranges from 400 to 2500 mg/l, and the effluent objective is less than 1 mg/l phenol (Gould and Taylor 1969).

CHEMICAL OXIDATION

Air, chlorine, ozone, and other chemical oxidizing agents are used to destroy phenol, which is first converted to hydroquinone and then to quinone. Additional oxidation destroys the aromatic ring, forming organic acids and eventually carbon dioxide and water (Eisenhauer 1968).

Air is an inexpensive oxidizing agent but reactions are slow. Phenol can be completely decomposed by chlorina-

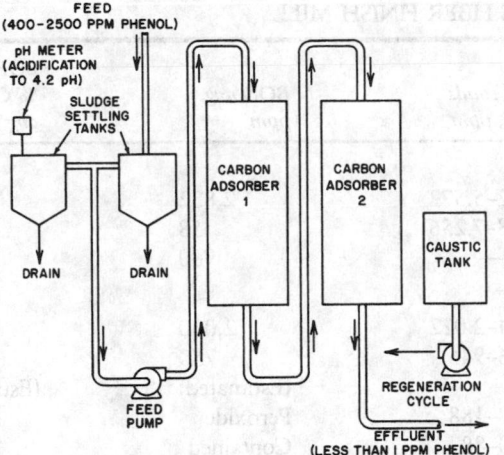

FIG. 8.2.13 Granular carbon systems for phenol removal

tion at pH 7.7, provided that the stoichiometric amount of chlorine is added. This is accomplished in water treatment plants by superchlorination. The major portion of the chlorine applied consumes other organic compounds and destroys ammonia. Approximately 42 parts of chlorine per part of phenol are required (Ohio River Valley Sanitation Commission 1951).

Ozonation effectively oxidizes phenol. However, the initial cost of producing ozone is high. Ammonia does not interfere in ozonation, and approximately 5.8 parts of ozone are required per part of phenol (Ohio River Valley Sanitation Commission 1951).

Starch

Starch wastes are produced by food processing operations, including starch manufacturing from corn, potatoes, and wheat. The wastes are essentially carbohydrates with a high oxygen demand.

BIOLOGICAL TREATMENT

Starch wastes respond to biological treatment using trickling filters, aerated lagoons, or activated sludge processes. Waste pH should be adjusted to between 6.0 and 9.0, suspended solids should be removed and, if necessary, nutrients should be added to maintain a BOD-nitrogen-phosphorous ratio of 100 to 5 to 1.

Starch is almost completely oxidized biologically, provided that the loading is maintained within the limits of the biological activity. If an activated sludge process is used, it is important to maintain an F to M (BOD to mixed liquor suspended solids) ratio of less than 0.3 (per day) to minimize propagation of filamentous organisms that interfere with solids separation.

Oxygen Requirements

In activated sludge operations it is necessary to supply oxygen to sustain the process and to provide intimate mixing and contact of activated sludge with the organic matter and nutrients. (A low-speed turbine-type surface aerator is shown in Figure 8.2.14.) Oxygen requirements depend on BOD removal and on process loading. The oxygen requirement is expressed by equation 8.2(1):

lb of oxygen required per lb BOD removed

$$= A + \left(B \times \frac{\text{lb mixed liquor volatile suspended solids}}{\text{lb BOD applied per day}} \right)$$

8.2(1)

In equation 8.2(1), "A" is related to the oxygen requirement for synthesis of new cells, and "B" is related to the oxygen requirement for respiration. The value of "A" ranges from 0.35 to 0.55, and "B" ranges from 0.05 to

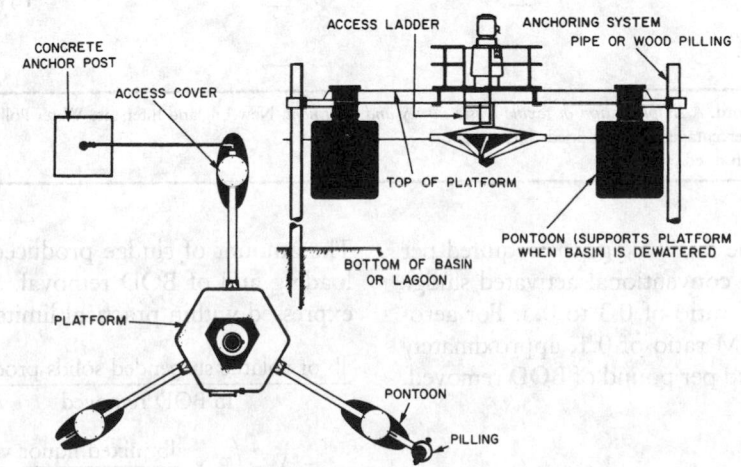

FIG. 8.2.14 Low-speed surface aerator installation

TABLE 8.2.8 COMPOSITION OF WASTES FROM A SYNTHETIC FIBER FINISH MILL

	pH range	Total solids range, ppm	BOD avg. ppm	BOD % OWF[b] avg[a]
Rayon processing				
Scour and dye	8.2–9.0	1.012–5.572	2,832	5.7
Salt take-off	6.8–6.9	3.388–7.256	58	0.1
Waterproof	—	—	960	1.9
Acetate processing				
Scour and dye	8.3–8.5	1.534–2.022	2,000	5.0
Scour and bleach	8.9–9.6	766–946	750 (Estimated)	1.8 (Estimated)
First rinse	7.0–9.1	108–188	Peroxide	
Second rinse	6.8–7.3	80–88	Contained peroxide	0.0
Nylon processing				
Scour	9.3–12.6	1.492–2.278	1,360	3.4
First rinse	8.2–10.7	150–954	90	0.2
Second rinse	6.5–8.2	106–932	25	0.1
Dye	7.8–9.0	318–1,016	368	0.9
Last rinse	7.3–7.6	106–134	11	0.0
Waterproof	—	—	450	1.1
Orlon processing				
First scour	9.5–10.0	1.350–2.470	2,190	6.6
First rinse	6.4–8.7	102–294	109	0.4
First dye	2.2–6.5	170–1.950	175	0.5
Second rinse	4.1–6.5	116–300	42	0.1
Second dye	1.3–1.7	130–3.002	995	3.0
Second scour	5.9–7.7	612–1.824	688	2.0
Third rinse	6.3–7.4	82–152	50	0.2
Waterproof	3.7–4.3	896–2.318	2,110	6.3
Dacron processing		(Estimated from OWF concentrations as listed)		
Scour	—	—	650	
Dyes				
o-phenylphenol (10% OWF)	—	—	6,000	18.0
benzoic acid (40% OWF)	—	—	27,000	81.0
salicylic acid (40% OWF)	—	—	24,000	72.0
phenylmethylcarbinol (30% OWF)	—	—	19,000	57.0
monochlorobenzene (40% OWF)	—	—	480	1.4

From Masselli, Masselli, and Burford. *A simplification of textile waste survey and treatment.* New England Interstate Water Pollution Control Commission.
[a]% on weight of fiber, a weight percentage based on dried cloth weight.
[b]OWF, weight percentage based on dried cloth.

0.10. As a general rule, one lb of oxygen is required per lb of BOD removed under conventional activated sludge operations with an F to M ratio of 0.3 to 0.5. For aerobic digestion with an F to M ratio of 0.1, approximately 1.5 lb of oxygen are required per pound of BOD removed.

Sludge Production

In the activated sludge process, soluble organic matter is converted to suspended solids in the form of bacterial cells.

The amount of sludge produced is a function of process loading and of BOD removal. Sludge production can be expressed within practical limits by equation 8.2(2):

$$\frac{\text{lb of volatile suspended solids produced}}{\text{lb BOD removed}}$$

$$= A - \left(B \times \frac{\text{lb mixed liquor volatile suspended solids}}{\text{lb BOD applied per day}} \right)$$

8.2(2)

The value of "A" varies from 0.4 to 0.9, and the value of "B" from 0.01 to 0.1, depending on the waste being treated. An approximate expression for sludge production in many treatment applications is given in equation 8.2(3):

$$\frac{\text{lb volatile suspended solids produced}}{\text{lb BOD removed}}$$

$$= 0.75 - \left(0.05 \times \frac{\text{lb mixed liquor volatile suspended solids}}{\text{lb BOD applied per day}}\right)$$

$$8.2(3)$$

Based on conventional activated sludge operations, between 0.5 and 0.6 lb of excess sludge are produced per lb of BOD removed. With aerobic digestion, approximately 0.2 lb of excess sludge are produced per lb of BOD removed.

Aerobically *digested* sludge can be dewatered on vacuum filters with loadings of approximately 1 lb/sq ft/hr. Dewatering excess sludge from conventional activated sludge operations requires a heat treatment for sludge conditioning or a heavy dosage of conditioning chemicals to form a filter cake that will dewater and separate from a filter cloth.

Textile Industry Wastes

Textile industry wastes are categorized by their source. Man-made fibers constitute approximately 80% of the fibers used. Table 8.2.8 lists wastewater compositions from synthetic fiber finish mills, and Table 8.2.9 reflects performance data of the various treatment methods in reducing BOD, SS, color, grease, and alkalinity.

In textile wastes the suspended solids concentration is minute, the BOD range can attain 3000 ppm, and color can sometimes reach as high as 3000 APHA color units. Electroflocculation removes most color by electrolytically inducing flotation and collection of foam. Thereafter, biological or chemical oxidation can be utilized to polish the effluent and reduce the BOD to 25—virtually eliminating color. Such textile mill effluent is of sufficient quality to be recycled and reused.

Viruses and Bacteria

Bacteria and viruses are removed or killed by disinfection and sterilization. Disinfection destroys all harmful microorganisms, while sterilization kills all living organisms. Disinfection of drinking water protects public health by preventing microorganism growth in the pipelines. Disinfection of wastewater treatment effluents protects marine life. Sterilization provides water suitable for medical and pharmaceutical use. Numerous disinfection and sterilization techniques are available, and Tables 8.2.10 and 8.2.11 compare the effectiveness, advantages, and disadvantages.

TABLE 8.2.9   TREATMENT PROCESS REMOVAL EFFICIENCIES

Treatment method	Normal reduction %				
	BOD	Grease	Color	Alkalinity	Suspended Solids
Grease recovery					
Acid cracking	20–30	40–50	0	0	0–50
Centrifuge	20–30	24–45	0	0	40–50
Evaporation	95	95	0	0	
Screening	0–10	0	0	0	20
Sedimentation	30–50	80–90	10–50	10–20	50–65
Flotation	30–50	95–98	10–20	10–20	50–65
Chemical coagulation					
$CaCl_2$	40–70	—	—	—	80–95
Lime + $CaCl_2$	60	97	—	—	80–95
CO_2 + $CaCl_2$	15–25	—	—	—	80–95
Alum	20–56	—	75		
Copperas	20				
H_2SO_4 + alum	21–83				
Urea + alum	32–65				
H_2SO_4 + $FeCl_2$	59–84				
$FeSO_4$	50–80				
Activated sludge	85–90	0–15	10–30	10–30	90–95
Trickling filtration	80–85	0–10	10–30	10–30	90–95
Lagoons	0–85	0–10	10–30	10–20	30–70

Reprinted, from FWPCA. 1967. The cost of clean water, vol. III. Industrial Waste Profile, No. 4. *Textile Mill Products*. September.

TABLE 8.2.10 DISINFECTION TREATMENT METHODS

Features	Chlorination (using liquid Cl_2)	Ozonation	Ultraviolet	Heating	Halogens; (bromine; iodine)	Metal Ions (silver; mercury; copper)
Required dosage[a] (ppm)	1–3 (A); 2–5 (B)	1.5–4.0 (A); 2.5–5.0 (B)	—	—	—	—
Contact time required (minutes)	10–30	5–10	Minimum	15–20	10–30	120
Effectiveness against Bacteria	Yes	Yes	Yes	Yes	Yes	Yes
Virus	Some	Yes	Some	Yes	Some	No
Spores	No	Yes	No	No	No	No
Advantages	Inexpensive and well-developed technology, which provides lasting protective residual.	Rapid method of removing color, taste and odor while destroying viruses and spores; generated on site; oxidation products are non-toxic.	Fast method which requires no chemicals.	Requires no special equipment.	Similar to chlorine except less irritating to the eye.	Has long-lasting bactericidal effect.
Disadvantages	Not effective against some spores and viruses; can, in high concentrations, produce products that are toxic to marine life and can cause undesirable taste and odor.	More expensive and less developed than chlorine and it does not leave a protective residue.	Leaves no protective residue, expensive, not applicable on large scale and requires pretreatment for turbidity removal.	Slow and expensive.	Slower and more expensive than chlorine.	Slow and expensive. Amines and other pollutants interfere with its effectiveness.
Other Remarks	Most frequently utilized method in the United States.	Frequently used in Europe; combined with chlorination, it can produce high-quality drinking water.	Mostly used on special laboratory and small industrial applications	Excellent household emergency method.	Sometimes used as swimming pool disinfectant.	—

Note: A = requirements for drinking water disinfection.
B = requirements for the disinfection of secondary (activated) wastewaters treatment effluent.

Disinfection should kill or inactivate all disease-producing (pathogenic) organisms, bacteria, and viruses of intestinal origin (enteric).

Pathogenic organisms include (1) bacteria of the coliform group, both fecal and nonfecal, such as *Escherichia coli, Aerobacter aerogenes,* and *Escherichia freundii;* (2) bacteria of the fecal streptococcus group; (3) other microorganisms such as Salmonella, Shigella, and the cyst *Endamoeba histolytica;* and (4) enteric viruses such as the etiologic agents of polio and infectious hepatitis. Test procedures, developed for their identification, are usually involved and time consuming. Therefore, the identifications (Metcalf, Wallis and Melmick 1972) of one group of bacteria (coliform) is usually taken as an indication of water quality and a measure of effectiveness of bacteria disinfection. It is assumed that the absence of coliform bacteria indicates the absence of all pathogenic bacteria.

Enteric viruses in the drinking water are reported to be responsible for hepatitis, poliomyelitis, and other epidemic diseases. Viruses are substantially more resistant to chlorine than bacteria, and the absence of coliform bacteria does not necessarily indicate the absence of viruses. Virology is not developed to the point that routine identification and assay tests are possible. The development of a portable virus concentrator, making routine identification and assay of viruses in water and wastewater more practical, has been reported. The concentrator first removes suspended solids through filtration and absorbs viruses on a cellulose adsorption column. The viruses are then eluted from the adsorption column and subjected to standard laboratory assay. (1972).

The probability of disease (D) when a pathogenic organism is brought into contact with a human water consumer (host) is proportional to the number of organisms (N) and their virulence (V) and inversely proportional to the resistance (R) of the host. The purpose of disinfection is to minimize N and V in equation 8.2(4).

$$D = \frac{NV}{R} \qquad 8.2(4)$$

Disinfection treatments utilize oxidation, surface active chemicals, acids and bases, metal ions, ultraviolet radiation, and physical treatment.

CHLORINATION

Chlorination is by far the most frequently used disinfection method in United States municipal drinking water treatment plants. The acting disinfectant may be chlorine or a chlorine derivative, such as hypochlorous acid (most commonly), chloramines, or chlorine dioxide. Several treatment methods have been developed. *Simple chlorination* involves adding chlorine after filtration or as the only treatment. *Chlorine-ammonia treatment* utilizes the addition of both ammonia and chlorine and the germicidal action of chloramines. *Residual chlorination* is applied to provide residual chlorine in the water. *Breakpoint chlorination* adds sufficient chlorine to react with ammonia and all other chemicals present as well as to assure a free chlorine residue.

Liquid chlorine is the least expensive form of chlorine. It was used in most large municipal water works until several large cities restricted or prohibited transportation and storage of large volumes of liquid chlorine to prevent accidental release into the atmosphere. Chlorine can be used and stored more safely in its solid form as $Ca(OCl)_2$. However, the cost is substantially higher.

Bactericidal and Viricidal Action

The bactericidal action of chlorine is the result of its strong oxidizing power. The formation of hypochlorous acid, the strongest disinfecting agent among the chlorine derivatives, is shown by equation 8.2(5). The bacteria-killing mechanism is believed to involve diffusion of hypochlorous acid through the cell membrane and oxidation of the cell enzymes.

$$Cl_2 + H_2O \rightleftharpoons HOCl + HCl \qquad 8.2(5)$$

The viricidal action of hypochlorous acid is substan-

TABLE 8.2.11 STERILIZATION TREATMENT METHODS

Treatment	Operating Conditions	Advantages	Disadvantages
Heating in autoclaves	121°C for 15 min	Reliable	Slow heatup
Ozonation	4–5 ppm for 15 min	Effective against all microorganisms	Fast
Superchlorination-Dechlorination	5–6 ppm for 2–3 hr	Effective against most microorganisms; inexpensive	Long contact time, need to dechlorinate after treatment
Ultraviolet		Fast; no chemical added	Not effective against spores

tially slower and less effective than its bactericidal action. The killing mechanism is believed to involve attacking many protein sites rather than one critical site of the virus. The chlorine treatment, designed to kill bacteria, does not necessarily kill viruses. Chlorine is not effective in normal concentrations to kill the cyst *Endamoeba histolytica,* the cause of amoebic dysentery, a protozoan disease that invades the body by a parasitic organism through the intestinal tract. Fortunately, it is a relatively rare disease.

Chlorine is also ineffective against nematodes, a free-living microorganism present in surface water supplies. Nematodes, although nonpathogenic, are capable of ingesting and harboring potentially dangerous organisms.

Minimum bactericidal chlorine residual was determined by the Public Health Service in terms of free available chlorine, using a 10-min contact time, and in terms of combined available chlorine (free chlorine and chloramines), using a 60-min contact time. The free available chlorine necessary for disinfection is 0.2 ppm at pH 6–8 and 0.4 ppm at pH 8–9. The corresponding concentrations with combined available chlorine are 1.5 and 1.8 ppm.

OZONATION

Ozone, a triatomic allotrope of oxygen, is produced industrially in an electric discharge field generator from dry air or oxygen at the site of use. The ozone generator produces an ozone-air or ozone-oxygen mixture containing 1 and 2% ozone by weight. This gas mixture is introduced into the water by injection or diffusion into a well-baffled mixing chamber or scrubber, or by spraying the water into an ozone atmosphere.

Ozone is a powerful oxidizing agent. The mechanism of its bactericidal action is believed to be diffusion through the cell membrane followed by the irreversible oxidation of cell enzymes. Disinfection is unusually rapid and requires only low ozone concentrations.

The viricidal action of ozone is even faster than its bactericidal effect. The mechanism by which the virus is destroyed is not yet understood. Ozone is also more effective than chlorine against spores and cysts such as *Endamoeba histolytica.*

Disinfection, color, taste, and odor control can be accomplished in a single treatment step by ozonation. Ozone reacts rapidly with all oxidizable organic and inorganic materials present in the water.

The ozone dosage necessary for disinfection depends on pollutant concentration in the raw water. An ozone dose of 0.2 to 0.3 ppm is usually sufficient for bactericidal action only. The ozone dosage necessary for secondary activated wastewater treatment effluent disinfection is 6 or more ppm. Ozonation leaves no disinfection residue, and therefore ozonation should be followed by chlorination in drinking water supply treatment applications. To obtain optimum drinking water, raw water should first be ozonated to remove color, odor, and taste and to destroy bacteria, viruses and other organisms. Then the water should be chlorinated lightly to prevent recontamination.

Aquarium and Fish Farm Water Disinfection

Ozonation should be selected as a disinfection treatment for marine applications where residual disinfecting agents or toxic oxidation products (chlorinated amines) cannot be tolerated. Ozone is unstable in water and decomposes slowly, with a half-life of approximately 30 min at 25°C. The decomposition rate is dependent on water quality. The half-life of the ozone at 25°C is 50 min in distilled water and 20 min in tap water. Decomposition is substantially accelerated by hydroxyl ions, transition metals and free radicals. The oxidation products of ozonation are usually nontoxic and biodegradable. Furthermore, ozonation leaves the water saturated with dissolved oxygen, important in fish hatcheries or fish farms.

Ozonation disinfects water and saturates it with dissolved oxygen. Ozonation can reduce organic contaminants and waste in fish farm water, allowing water recycling. Ozone concentrations higher than 0.1 ppm should be avoided because they can harm the fish. Research from the National Marine Fisheries Service demonstrates that ozonation destroys undesirable microorganisms with no harmful effects to the fish.

Other Disinfectants. For a discussion of the merits and drawbacks of ultraviolet irradiation, heating, chemical oxidants and metal ions, see Table 8.2.10.

—*R.A. Conway, C.C. Walden, L.C. Gilde, Jr., C.A. Caswell, R.H. Zanitsch, E.G. Kominek, J.W.T. Ferretti, L.J. Bollyky*

References

Aly, O.M., and S.D. Faust. 1965. Removal of 2,4-dichlorophenoxyacetic acid derivatives from natural waters. *J. Am. Water Works Assoc.* 65(2):221.

Anon. 1971. Technology newsletter. *Chemical Week* 33. (4 August).

Anon. 1971. Mosquitos repel insecticides. *Chemical Week* 109:36 (4 August).

Anon. 1971. Treating chemical wastes. *Chemical Week* 108:37 (17 April).

Buescher, C.A., J.H. Dougherty, and R.T. Skrinde. 1964. Chemical oxidation of selected organic pesticides. *J. Water Pollution Control Fed.* 36(8):1005.

Charles, G.E., and G. Decker. 1970. Biological treatment of bleach plant wastes. *J. Water Poll. Contr. Fed.* 42:1725.

Coley, G., and C.N. Stutz. 1966. Treatment of parathion wastes and other organics. *J. Water Pollution Control Fed.* 38(8):1345.

Dedrick, R.L., and R.B. Beckman. 1967. Kinetics of adsorption by activated carbon from dilute aqueous solution. *Chem. Engr. Prog. Symp. Ser.* 63(74):68.

Eisenhauer, H.R. 1968. Dephenolization of water and wastewater. *Water and Pollution Control* 106(9). (September) p. 34.

Gaudy, A.F., R.S. Engelbrecht, and B.G. Turner. 1961. Stripping kinetics of volatile components of petrochemical wastes. *J. Water Pollution Control Fed.* 33(4):383.

Gerhold, R.M., and G.W. Malaney. 1966. Structural determinations in the oxidation of aliphatic compounds by activated sludge. *J. Water Pollution Control Fed.* 38(4):562.

Giusti, D.M. 1971. *Amenability of petrochemical waste constituents to activated carbon adsorption.* Master's Thesis, West Virginia University. Morgantown, W.Va.

Gould, M., and J. Taylor. 1969. Temporary water clarification system. *Chemical Engineering Progress* 65(12). (December) p. 47.

Graetz, D.A., G. Chesters, T.C. Daniel, L.W. Newland, and G.B. Lee. 1970. Parathion degradation in lake sediments. *J. Water Pollution Control Fed.* 42(2):R 76.

Hatfield, R. 1957. Biological oxidation of some organic compounds. *Industrial and Engineering Chemistry* 49(2):192.

Heukelekian, H., and M.C. Rand. 1955. Biochemical oxygen demand of pure organic compounds. *J. Water Pollution Control Federation* 27(9):1040.

Hill, D.W., and P.C. McCarty. 1966. *The anaerobic degradation of selected chlorinated hydrocarbon pesticides.* Paper presented at Annual Meeting of Water Pollution Control Federation. Kansas City, Kansas. (September).

Hindin, E., P.J. Bennett, and S.S. Narayanan. 1969. Organic compounds removed by reverse osmosis. *Water & Sewage Works* 116(12):466.

Huang, J. 1971. Effect of selected factors on pesticide sorption and desorption in the aquatic environment. *J. Water Pollution Control Fed.* 43(8):1739.

Lamb, C.R., and G.F. Jenkins. 1952. BOD of Synthetic Organic Chemicals. *Proc. 7th Ind. Waste Conference.* Purdue University. Lafayette, Ind.

Laws, R.L., and O.B. Burns, Jr. 1960. Recent developments in the application of the activated sludge process for the treatment of pulp and paper mill wastes. *Pulp and Paper Magazine of Canada* 61:T507–T513.

Leigh, G.M. 1969. Degradation of selected chlorinated hydrocarbon insecticides. *J. Water Pollution Control Fed.* 41(11):R 450.

Lue-Hing, C., and S.D. Brady. 1968. Biological Treatment of Organic Phosphorus Pesticide Wastewaters. *Proc. 23rd Purdue Industrial Waste Conference,* Purdue Univ. Eng. Extension Series, No. 132, 1166.

MacRae, I.C., K. Raghu, and E.M. Bautista. 1969. *Nature* 221:859.

Metcalf, R.L., G.K. Sangha, and I.P. Kapoor. 1971. Model ecosystem for the evaluation of pesticide biodegradability and ecological magnification. *Environmental Sci. Tech.* 5(8):709.

Metcalf, C.J., C. Wallis, and J.L. Melmick. 1972. *Concentrations of viruses from sea water.* Proceedings from the 6th International Conference on Water Pollution Research. Jerusalem, Israel (June).

Muller, J.M., and F.L. Covertry. 1968. Disposal of coke plant waste in the sanitary water system. *Blast Furnace and Steel Plant.* 56(5) (May). p. 400.

Nicholson, H.P., A.R. Grzenda, and J.I. Teasley. *J. South-East Sect., Am. Water Works Assoc.* 32(1):21.

Ohio River Valley Sanitation Commission. 1951. *Phenol wastes—treatment by chemical oxidation.* (June).

Robeck, G.C., K.A. Dostal, J.M. Cohen, and J.F. Kreissl. 1965. Effectiveness of water treatment processes in pesticide removal. *J. Am. Water Works Assoc.* 57(2):181.

Stack, V.T., Jr. 1957. Toxicity of alpha, beta-unsaturated carbonyl compounds to microorganisms. *Industrial and Engineering Chemistry* 49(5):913.

U.S. Environmental Protection Agency (EPA). 1970. *Investigation of means for controlled self-destruction of pesticides.* Aerojet-General Corporation Report for the Water Quality Office. 16040 ELO (June).

U.S. Environmental Protection Agency (EPA). 1971. *Research, development and demonstration projects: 1970 grant and contract awards.* Environmental Protection Agency, Water Quality Office, Office of Research and Development. Washington, D.C.

Wallis, C., and J.L. Melmick. 1972. *A portable virus concentrator for use in the field.* Proceedings of the 6th International Conference on Water Pollution Research. Jerusalem, Israel (June).

Wallis, C., A. Homma, and J.L. Melmick. Development of an apparatus for concentration of viruses from large volumes of water. *J. Amer. Water Works Assoc.*

Weber, W.J., Jr., and J.P. Gould. 1966. Sorption of Organic Pesticides from Aqueous Solution. In Gould, R.F. (ed.). Organic Pesticides in the Environment. *Advances in Chemistry Series,* 60. Washington, D.C.: ACS Publication. 280.

8.3
REMOVING INORGANIC CONTAMINANTS

Aluminum

Aluminum may be present in acid wastes as the trivalent aluminum ion or in alkaline wastes as an aluminate ion. Aluminum is precipitated as the hydroxide or hydrolysis species of polymeric aluminum. The precipitation and conditioning of precipitated solids has an important effect on the separation rate and on the settling and dewatering characteristics of the precipitate. Precipitation in the presence of previously formed solids produces denser and more rapidly settling floc particles. The addition of a polyelectrolyte also improves settling characteristics. Low velocity gradients, with agitator peripheral speeds as low as 5 ft/sec, are desirable to avoid shearing the floc into small particles that settle slowly.

Depending on the method of precipitation and solids separation, aluminum hydroxide can be concentrated to 1.0 to 2.0% by weight. Adding suitable polyelectrolytes, it can be further dewatered by centrifugation or vacuum filtration. However, without preconditioning of the aluminum hydroxide sludge, either precoat vacuum filtration or filter presses are required for sludge dewatering.

Bicarbonate

Bicarbonate is the principal alkaline form in natural (untreated) water, although carbonate and hydroxide are found in lime or lime-soda treated waters, and phosphates and silicates may also contribute to alkalinity in wastewaters. Adverse effects of high alkalinity in boiler feedwater include corrosion from liberated carbon dioxide and foaming—with resultant carry-over of contaminants. Scaling can also occur in cooling water systems due to the formation of insoluble calcium carbonate. For drinking water, the U.S. Public Health Service Standards limit alkalinity to 35 ppm over the hardness level.

REMOVING BICARBONATE ALKALINITY

Methods for reducing bicarbonate alkalinity (Table 8.3.1) are divided into chemical addition and ion exchange techniques. Chemical methods include cold lime process and acidification. Ion exchange methods involve strong-acid cation exchange, weak-acid cation exchange, or strong-base anion exchange in the chloride cycle.

Cold Lime Process

Bicarbonate is removed by lime addition in accordance with the following reactions:

$$Ca(HCO_3)_2 + Ca(OH)_2 \longrightarrow 2\,CaCO_3\downarrow\ +\ 2\,H_2O \qquad 8.3(1)$$

$$Mg(HCO_3)_2 + 2\,Ca(OH)_2 \longrightarrow$$
$$Mg(OH)_2\downarrow + 2\,CaCO_3\downarrow + 2\,H_2O \quad 8.3(2)$$

$$2\,NaHCO_3 + Ca(OH)_2 \longrightarrow$$
$$CaCO_3\downarrow + Na_2CO_3 + 2\,H_2O \quad 8.3(3)$$

As indicated by reactions 8.3(1) and 8.3(2), alkalinity associated with hardness is removed by precipitation. However, sodium bicarbonate is converted to carbonate, which remains soluble and contributes to phenolphthalein alkalinity, as shown in 8.3(3).

Acidification

By adding sulfuric acid to water, calcium bicarbonate is converted to calcium sulfate, minimizing scaling by the less soluble calcium carbonate. The reaction may be represented as follows:

$$Ca(HCO_3)_2 + H_2SO_4 \longrightarrow CaSO_4 + 2\,CO_2\uparrow\ +\ 2\,H_2O$$
$$8.3(4)$$

Carbon dioxide formed is removed by aeration. Care must be exercised in adding acid to avoid corrosive conditions. With waters already high in sulfate, the solubility product of calcium sulfate may be exceeded, and unwanted precipitation is likely to occur.

Strong Acid Cation Exchange

By passing water through a sulfonic acid cation exchanger in the hydrogen form, the following reaction occurs:

$$RSO_3H + NaHCO_3 \longrightarrow RSO_3Na + CO_2\uparrow\ +\ H_2O \quad 8.3(5)$$

(where R represents the resin matrix). Neutral salts are converted to free mineral acids, and the cation exchanger is regenerated with an excess of acid. The process may be

TABLE 8.3.1 PROCESSES FOR REMOVING BICARBONATE ALKALINITY

Process	Results, Comments	Relative Capital Cost	Relative Operating Cost	Indicated Application
Cold Lime Process	Reduces bicarbonate, calcium and magnesium; increases pH (>10); residual hardness 35–90 ppm; supersaturated $CaCO_3$ may form unless recarbonated.	High	Low	Municipal-large scale, where bicarbonate hardness exceeds 100–150 ppm, 250 gpm and up.
Acidification and Aeration	Partial reduction of alkalinity; avoid excess to prevent corrosion; easily automated; use HCl or HNO_3 on high-sulfate waters.	Low	Low	Cooling water systems for scale prevention.
Split Stream Process (Strong-Acid Cation Exchange and Aeration)	Partial or complete hardness removal; alkalinity controlled by proportioning flow through dealkalizer; excess acid disposal; excess salt if softener used.	High	Moderate	Industrial; low-pressure boiler-feed water, especially where ion-exchange softener already exists; 25–300 gpm.
Weak-Acid Cation Exchange and Aeration	Effluent alkalinity from 0–20% of influent; "temporary" hardness completely removed; high acid efficiency, minimum disposal; may be combined with salt-regenerated softener.	Moderate	Low	Industrial; process and boiler-feed water; 25–300 gpm.
Chloride-Cycle Anion Exchange	Exchange of bicarbonate, sulfate, phosphate and nitrate for chloride; no reduction in dissolved solids; regenerated with NaCl + NaOH.	Low	Moderate	Small industrial for low-pressure boiler-feed water; 50 gpm; also municipal—where removal of nitrate, phosphate and color is required.

used in tandem with a conventional sodium cycle (salt-regenerated) cation exchange softener. Either the raw or softened water is blended with the acidified effluent to give a *split-stream* which controls alkalinity.

Weak Acid Cation Exchange

Weak acid resins also remove both alkalinity and hardness, as illustrated by the following reaction:

$$2 \ RCOOH + Ca(HCO_3)_2 \longrightarrow$$

$$(RCOO)_2Ca + 2 \ CO_2\uparrow + 2 \ H_2O \quad 8.3(6)$$

In contrast to the strong acid resin, the weak acid exchanger converts little, if any, of the neutral salts (chlorides and sulfates) to free mineral acidity. Neutralization is not required, although degasification or aeration is practiced. Regeneration of the weak acid resin is more efficient than with strong acid cation exchangers, minimizing acid waste disposal. Combining a salt-regenerated softener with an acid-regenerated weak acid resin provides complete softening and dealkalization with a single column.

Chloride Cycle Anion Exchange

Strong base anion exchangers remove bicarbonate, as illustrated by the following reaction:

$$RCl + NaHCO_3 \rightleftharpoons RHCO_3 + NaCl \quad 8.3(7)$$

The principal regenerant is sodium chloride. Capacities can be increased by adding a small proportion of caustic soda to the salt. This process eliminates the need for acid-proof equipment and offers convenience, low cost, and a relatively small space requirement. In addition to bicarbonate removal, reductions in sulfate, nitrate, phosphate, anionic surfactants and color are also achieved. However, the effluent chloride level increases in the exchange.

Cadmium

Cadmium, a relatively rare element, is extensively used not only in protecting other metals, but also in manufacturing primary batteries and standard electrochemical cells; in producing pigments with outstanding properties; and in production of phosphors, semiconductors, electrical contactors, and special purpose low-temperature alloys.

SOURCES OF CADMIUM-BEARING WASTEWATERS

Because the largest consumption of cadmium (60%) is for plating, performed in aqueous baths, there is a drag-out of plating chemicals from the plating bath to the following rinse bath. The amount of drag-out is a function of the size of the article being plated, its intricacy, the presence of blind holes, and the duration of pause to drip over the plating tank.

Cadmium is a by-product of zinc production and is a valuable source of revenue for the zinc smelter. During zinc smelting, evolved cadmium fumes are collected. Consequently, if the gases from electric furnaces, autobody incineration, and certain domestic products are water scrubbed, cadmium is found in the scrubbing water. Whenever zinc or brass is electroplated, the drag-out also contains cadmium, as these plating tanks serve as cadmium concentrators.

In the manufacture, incineration, and careless disposal of primary cells, there is cadmium loss.

The 1962 USPHS Drinking Water Standards set a cadmium limit of 0.01 mg/l. The toxicity of cadmium and certain disease manifestations necessitate treatment of wastewaters containing cadmium to reduce treated effluent concentration to the level of 0.01 mg/l.

TREATMENT METHODS

The solubility product of cadmium sulfide is 3.6×10^{-29}. Its solubility is 8.6×10^{-10} mg/l. As cadmium electroplating is performed in cyanide baths, the drag-out is alkaline. Therefore, alkaline carbonates and sulfides can remove cadmium as an insoluble salt. The hydroxide is too soluble, resulting in cadmium concentrations of 5 to 10 mg/l. If carbon dioxide is subsequently absorbed before neutralization, additional cadmium will be removed. Cadmium, even when present in trace concentrations, is strongly coprecipitated with calcium carbonate.

Removal to concentration levels around 0.01 mg/l requires the removal of particulate carbonates or sulfides, since residual soluble cadmium can be expected to be within limits. The particulates are very small and settle very slowly, requiring digestion to increase particle size followed by settling or filtration to remove the fines.

A treatment solution containing $NaOH$, Na_2CO_3, Na_2S and CaO will effect satisfactory treatment, but sulfide release may result in disagreeable odors when final effluent pH is reduced to low values, if the sulfide is not destroyed (e.g., by sodium hypochlorite, which is used for cyanide destruction).

Ion exchange, reverse osmosis, electrodialysis, distillation, and flotation processes can all remove cadmium from wastewaters.

Calcium

Calcium may be present in water solution as bicarbonate, sulfate, chloride, or nitrate. It may also be produced in water solution when lime is used to neutralize waste acid.

With the exception of calcium carbonate and calcium sulfate, most calcium compounds are very soluble. The solubility of calcium sulfate compounds varies with temperature. Gypsum ($CaSO_4 \cdot 2H_2O$) has a solublity, in mg/l, of about 1800 at 32°F, 2100 at 100°F, and about 1700 at 212°F. The solublity of calcium carbonate in pure water is small—about 15 mg/l. However, when precipitated, it produces supersaturated solutions that are relatively stable at water temperatures below 200°F.

Precipitation in the presence of a common ion—either calcium or carbonate—reduces solubility. Precipitation in the presence of about 5% by weight of calcium carbonate virtually eliminates supersaturation. This same phenomenon is noted when calcium fluoride is precipitated. Precipitated calcium compounds are crystalline and relatively easy to dewater by vacuum filtration at rates from 20 to 50 lb/sq ft/hr.

Chromium

Hexavalent chromium salts occur as pollutants in industrial effluents from leather, aluminum anodizing, and metal plating. Chemical plant effluents contain these from extensive use of chromium salts as corrosion inhibitors in cooling systems.

In industrial effluents, chromium wastes are treated by reduction and precipitation, removing the pollutant, or by ion exchange in which chromate salt is recovered and the deionized water is reused. The latter treatment recovers the pollutant chromium for economical reuse.

REDUCTION AND PRECIPITATION

Hexavalent chromium is first reduced to the trivalent state by adding a reducing agent, with proper adjustment for acidity. This is followed by precipitation of the reduced chromium as the hydroxide, which is then physically removed from the system by settling. The reactions are:

$$\underset{\substack{SO_2 \text{ or} \\ Na_2S_2O_5}}{\text{Reducing Agent:}} \quad Cr^{6+} + Fe^{2+} + H^+ \longrightarrow Cr^{3+} + \underset{SO_4^{2-}}{Fe^{3+}} \qquad 8.3(8)$$

$$Cr^{3+} + OH^- \longrightarrow Cr(OH)_3 \qquad 8.3(9)$$

Reaction 8.3(8) proceeds almost instantaneously at a pH of 2.0 or less. Each reducing agent shown in the reaction is effective; Fe^{2+}, however, requires an excess of about $2\frac{1}{2}$ times the stoichiometric quantity, resulting in an excess of $Fe(OH)_3$ sludge from neutralization. In small treatment systems, sodium metabisulfite ($Na_2S_2O_5$) is usually the preferred reagent. In water it hydrolyzes to sodium bisulfite, and sulfuric acid must be added to lower the pH for the reducing reaction. Excess reagent must be added if dissolved oxygen is present in the wastewater. Larger systems, on a batch or continuous basis, use sulfur dioxide, which hydrolyzes to sulfurous acid. Additional acid for pH adjustment is not always required.

ION EXCHANGE

Hexavalent chromium is recovered by ion exchange for reuse as a chromate-rich solution. This solution can be re-

cycled into the cooling tower water treatment system, and the resulting chromate-free water may be disposed of or further demineralized and reused.

A successful process contacts the chromate-laden wastewater after proper pH adjustment, with a weak-base anion exchange resin in the sulfate form. The chromate (CrO_4^{2-}) ion exchanges with the sulfate (SO_4^{2-}) ion and is incorporated in the resin. The chromate is recovered as a mixture of sodium chromate (Na_2CrO_4) and sodium dichromate ($Na_2Cr_2O_7$) upon regeneration of the resin. The regenerant is a 5% (by weight) solution of caustic soda (NaOH) added in an overall quantity equivalent to 10% in excess of the stoichiometric amount.

Sodium hydroxide restores the chromate as sodium salts and temporarily places the resin in the hydroxyl form. The sodium hydroxide on the resin is neutralized by adding the stoichiometric quantity of 0.1N sulfuric acid. This neutralization step also restores the resin to the sulfate form. The reactions are as follows:

$$\underset{R^*H}{\overset{R^*H}{\diagdown}}SO_4 + Na_2Cr_2O_7 \rightleftharpoons \underset{R^*H}{\overset{R^*H}{\diagdown}}Cr_2O_7 + Na_2SO_4$$

$$8.3(10)$$

$$\underset{R^*H}{\overset{R^*H}{\diagdown}}Cr_2O_7 + NaOH \rightleftharpoons (Na_2Cr_2O_4 \rightleftharpoons$$

$$Na_2CrO_4) + 2\ R^\circ H - OH \quad 8.3(11)$$

$$2\ R^*H{-}OH + H_2SO_4 \rightleftharpoons \underset{R^*H}{\overset{R^*H}{\diagdown}}SO_4 + 2H_2O \qquad 8.3(12)$$

where:

$R^\circ = R_3N$, a weakly basic macroporous resin

A typical flow diagram using the Higgins-type, continuous, countercurrent ion exchange system is shown in Figure 8.3.1.

Cyanides

The major portion of cyanide-containing wastewater comes from metal finishing and metal plating plants. Photo-processing plants also contribute significantly. Cyanides are extremely poisonous, especially at acidic pH levels, where they are present as hydrocyanic acid, a powerful poison. Cyanide-containing wastewater should be treated prior to discharge into sewer lines, streams, or rivers. Treatment processes usually involve either partial oxidation of the cyanide to the substantially less toxic cyanate or complete oxidation to carbon dioxide and nitrogen. Frequently used oxidizing agents include chlorine, ozone, and electrolytic oxidation. The cyanide concentra-

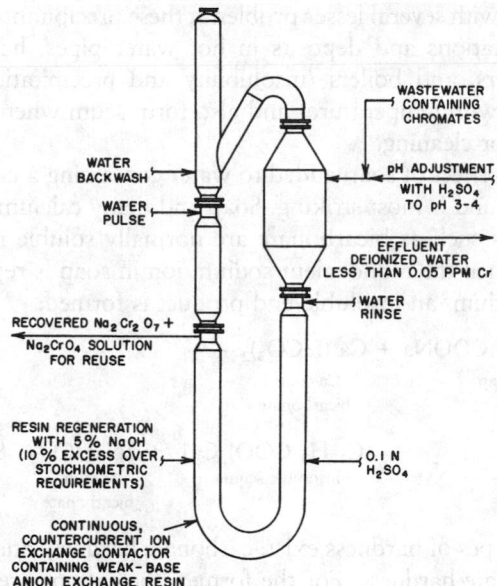

FIG. 8.3.1 Chromate recovery by ion exchange.

tion in the effluent should be less than 0.2–1.0 ppm when the receiving body is a sewer line, stream, or river.

CHLORINATION

Oxidation of cyanide to cyanate occurs in the pH range of 8–9 and requires only minutes (equation 8.3[13]). Further oxidation to carbon dioxide and nitrogen is much slower, requiring hours (equation 8.3[14]).

$$NaCN + Cl_2 + 2NaOH \longrightarrow NaCNO + 2NaCl + H_2O$$

$$8.3(13)$$

$$2NaCNO + 3Cl_2 + 8NaOH \longrightarrow$$
$$N_2 + 2Na_2CO_3 + 6NaCl + 4H_2O \quad 8.3(14)$$

Chlorine is added as gaseous chlorine or as a hypochlorite solution. Special equipment is required for safe and efficient addition of chlorine gas. For smaller plants, hypochlorite solution is recommended since metering and handling is simpler and less hazardous. Sludge formation usually accompanies chlorination. The sludge consists of hydroxides of metal ions, always present in plating solution.

OZONATION

Ozone oxidation of cyanides is best carried out in the pH range of 9–10, and the oxidation of cyanide to cyanate is extremely rapid (equation 8.3[15]). Further reaction of cyanates is much slower. The addition of copper (2+) salt catalysts accelerates the reaction. A typical ozone control system is shown in Figure 8.3.2.

$$CN^- + O_3 \longrightarrow CNO^- + O_2 \qquad 8.3(15)$$

Cyanide oxidation can also be carried out electrolytically. The more toxic sodium and potassium cyanides can also

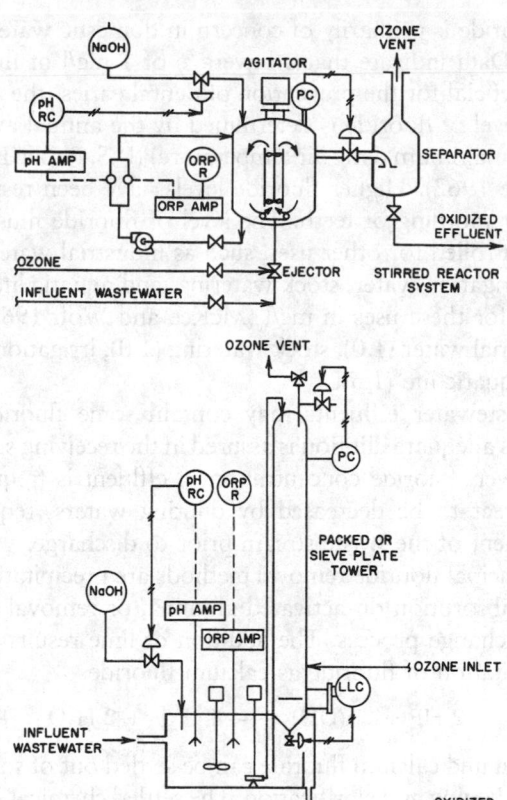

FIG. 8.3.2 Cyanide waste oxidation control systems utilizing ozone as the oxidant. *Key:* pHRC = pH recording controller; ORPR = ORP recorder; PC = pressure controller; LLC = low-level control; AMP = amplifier

be converted to substantially less toxic ferrocyanide complexes by adding ferrous sulfate. This process is not recommended, however, because ferrocyanide releases cyanide when exposed to sunlight.

Chlorination is the most frequently used and best developed process. The addition of chlorine gas is hazardous and requires storage of large quantities of chlorine. Ozone is a faster, more powerful oxidizing agent, requiring smaller holding and reaction tanks. The relative amounts required for each process are shown in Table 8.3.2.

Fluoride

Fluoride occurs naturally in some U.S. waters. Discharges from some industrial plants also contain fluoride. The level

TABLE 8.3.2 CHEMICAL ADDITIVE REQUIREMENTS OF CYANIDE REMOVAL

Chemical	Pound Chemical Required per lb of Cyanide Removed
Chlorine gas	2.7–6.8
Sodium hypochlorite	2.9–7.2
Calcium hypochlorite	2.8–6.9
Ozone	1.8–4.6

of fluoride is primarily of concern in domestic water supplies. Data indicate that an average of 1 mg/l of fluoride is beneficial for the prevention of dental caries (the allowable level of fluoride is determined by the annual average of the maximum daily air temperature) (U.S. Public Health Service 1962). Higher fluoride levels have been responsible for mottling of teeth. The level of fluoride must also be controlled for other uses, such as industrial water supply, irrigation water, stock watering, and aquatic life. The limits for these uses in mg/l (McKee and Wolf 1963) are industrial water (1.0), stock watering (1.0), irrigation (10), and aquatic life (1.5).

Wastewater effluents may contain some fluoride, as long as adequate dilution is assured in the receiving stream. However, fluoride concentration in effluent is frequently too great to be decreased by diluting waters, requiring treatment of the waste stream prior to discharge.

Principal flouride removal methods are precipitation by lime, absorption on activated alumina, or removal by an ion exchange process. The addition of lime results in the precipitation of fluoride as calcium fluoride:

$$2\,HF + Ca(OH)_2 \longrightarrow CaF_2\downarrow + 2\,H_2O \qquad 8.3(16)$$

Precipitated calcium fluoride can be settled out of solution by thickening and clarification. The settled chemical sludge can then be treated as other sludges and dewatered utilizing vacuum filtration or centrifugation. The limiting factor for this process is the solubility of calcium fluoride, which is 7.8 mg/l (as F). There are indications that lime high in magnesium can further reduce the fluoride solubility concentration (Rohrer 1971).

Another method of fluoride removal is the use of an aluminum compound to bind the aluminum and fluoride as a complex. While filter alum (aluminum sulfate) has been investigated, it has not been effective, as other anions in the water tend to reduce effectiveness. Activated alumina can be used to reduce fluoride concentration to the 1–2 mg/l range. The capacity of activated alumina for storing fluoride is about 0.1 lb/cu ft. Flowrates on the order of 3–5 gpm/sq ft are possible. The activated alumina can be regenerated with caustic soda, aluminum sulfate, or sulfuric acid with little apparent loss of capacity or activated alumina volume.

Ion exchange materials have also been investigated for defluoridation of water. Anion exchange materials regenerated with a caustic soda solution have been utilized, but this is an expensive process if fluoride removal is the only requirement.

Hardness

Hardness is caused by divalent cations (ions with a positive charge of 2+). Usually the offending cations are calcium (Ca^{2+}) and magnesium (Mg^{2+}). These and similar cations react with compounds containing monovalent cations (usually sodium, Na^+) to form insoluble products.

Along with several lesser problems, these precipitants form encrustations and deposits in hot water pipes, heat exchangers, and boilers (insolubility and precipitation increase with temperature) and also form scum when using soap for cleaning.

The effect of soap added to water containing a calcium compound is most striking. Soap and many calcium compounds such as bicarbonate are normally soluble in water. When the monovalent sodium ion in soap is replaced by calcium, an insoluble end product is formed:

$$2C_{17}H_{35}COONa + Ca(HCO_3)_2 \longrightarrow$$

Soap $\qquad$ Calcium bicarbonate

$$(C_{17}H_{35}COO)_2Ca\downarrow + 2NaHCO_3 \qquad 8.3(16)$$

Insoluble scum $\qquad$ Sodium bicarbonate

Two types of hardness exist: carbonate hardness and noncarbonate hardness. For the former, the cations are combined with either bicarbonate or carbonate. For noncarbonate hardness, the cations are combined with chlorides, sulfates, and other anions.

ION EXCHANGE

To eliminate hardness, various resins known as *zeolites* (Ze) are used. These resins usually contain monovalent cations (usually Na^+), but since they prefer divalent cations for stability, they exchange the Na^+ for calcium or magnesium.

$$Ca^{2+} + Na_2Ze \longrightarrow CaZe + 2Na^+ \qquad 8.3(17)$$

Sodium zeolite

The reaction may be reversed by adding a large quantity of monovalent ions (Na^+).

LIME AND LIME–SODA ASH SOFTENING

To reduce hardness to 80 or 100 mg/l, the lime or lime–soda ash softening processes may be used. These processes are used when some hardness can be tolerated, as in domestic water supplies. The operational cost of these processes is much less than for the ion exchange process. In lime softening, calcium is removed as follows:

$$Ca(HCO_3)_2 + Ca(OH)_2 \longrightarrow 2CaCO_3\downarrow + 2H_2O \qquad 8.3(18)$$

Calcium bicarbonate $\qquad$ Lime $\qquad$ Calcium carbonate

The calcium carbonate is insoluble and precipitates out. If noncarbonate hardness such as calcium sulfate is also present, soda ash must be added:

$$Ca^{2+} + SO_4^{2-} + Na_2CO_3 \longrightarrow CaCO_3\downarrow + 2Na^+ + SO_4^{2-}$$

Sulfate $\qquad$ Soda ash

$$8.3(19)$$

Similar reactions are involved in magnesium precipitation. Since lime is used in excess, the softened water still con-

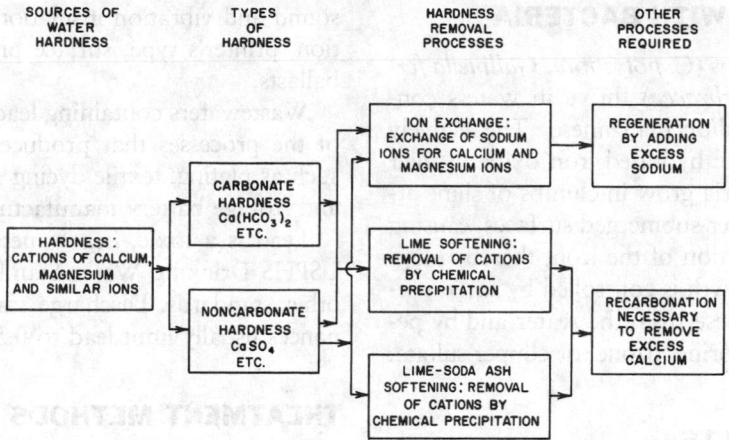

FIG. 8.3.3 Summary of hardness removal processes.

tains Ca^{2+} and OH^- ions that must be stabilized (Figure 8.3.3). This can be done by bubbling carbon dioxide through the water (recarbonation). Water softening operations are usually followed by flocculation, settling, and filtration.

Iron

Iron usually occurs with manganese in groundwater. The presence of these metals in excess of 0.1 ppm and 0.05 ppm, respectively, is unacceptable for public water supplies and for most industrial uses. Above these concentrations, precipitates are formed on contact with air; residues stain fixtures and interfere with clothes washing and most manufacturing processes. The iron may be a water soluble ferrous salt or iron bacteria, i.e., hydrated iron oxide enclosed in the cell structure of filamentous microorganisms, such as *Crenothrix polyspora*. Dissolved inorganic iron is usually removed by aeration, chemical precipitation, or ion exchange. Iron bacteria removal requires destruction of cell membranes by strong oxidizing agents such as ozone or chlorine (Table 8.3.3).

The oxygen-poor, carbon dioxide–rich lower layers of water reservoirs reduce and dissolve iron salts in the soil as ferrous salts. Similarly, the relatively oxygen-free acidic groundwater dissolves iron deposits.

TABLE 8.3.3 PROCESSES FOR IRON AND MANGANESE REMOVAL

Nature of Contaminants in the Influent Wastewater	Process[a]	Treatment Operating pH	Oxidation	Remarks
Iron—no organic matter	A-ST-SF	6.5	Yes	Easy to operate
Iron and manganese; little organic matter	A_{cat}-ST-SF	6.5	Yes	Easy to operate; requires double pumping
Iron and manganese bound to organic matter; no excessive organic acids	A-F_{cat}	6.5	Yes	Easy to control; requires double pumping "sniffler" valve
Iron and manganese bound to organic matter; no excessive organic acid or carbon dioxide	F_{cat}	6.5	Yes	No aeration but filter reactivated by chlorination or by permanganate
Iron and manganese loosely bound to organic matter	A-Cl-ST-SF	7.0–8.0	Yes	Aeration reduces chlorine requirement
Iron and manganese in combination with organic matter and organic acids	A-L-ST-SF	8.5–9.6	Yes	pH control required
Iron and manganese in colored turbid surface water containing organic matter	A-Co-L-ST-SF	8.5–9.6	Yes	Laboratory control required
Iron and manganese in oxygen-free well water containing about 1.5 to 2.0 ppm iron and manganese	Cation exchange	6.5	No	Periodic regeneration with salt solution
Iron in soft well water; iron is present as ferrous bicarbonate	L-ST-SF	8.0–8.5	No	Iron is precipitated as ferrous carbonate in the absence of oxygen

[a]A = aeration; ST = settling; SF = sand filtration; A_{cat} = catalytic aeration; F_{cat} = filtration over oxidation catalysts; Cl = chlorination; L = lime treatment; Co = coagulation.

CONTROLLING IRON WITH BACTERIA

Filamentous microorganisms (*C. polyspora, Gallinella ferruginea* and *Leptothrix ochracea*) thrive in waters containing traces of iron and/or manganese. The actively growing bacteria precipitate hydrated iron oxide in their cell structures. These bacteria grow in clumps of slime attached to pipe walls or other submerged surfaces, causing slow corrosion and dissolution of the iron, thereby plugging the pipe. Bacterial growth is controlled by careful removal of iron and manganese from the water and by periodic disinfection with chlorine, ozone, or copper sulfate.

REMOVING IRON SALTS

The treatment for removing dissolved iron salts usually involves (1) oxidation by air, chlorine, or ozone followed by filtration; (2) chemical precipitation followed by filtration; or (3) ion exchange. The capacity of the treatment plant, the pH of the water, and the presence of other contaminants determine which process is the most economical. Iron is usually removed more readily than iron and manganese together. The removal of dissolved iron chelated to organic compounds is usually accomplished by coagulation followed by settling and filtration.

Oxidation is accomplished most economically by aeration. Aeration also purges carbon dioxide from water, which keeps iron dissolved as ferrous carbonate. Iron oxides may be removed by settling, filtration often is necessary. If the iron is loosely bound to organic matter, the aeration process is slow and must be accelerated by iron oxide or manganese dioxide catalysts deposited on sand, crushed stone, or coke. Chlorine and ozone effectively oxidize iron at low pH in the presence of a high organic content.

Chemical precipitation by lime is usually effective if the iron is present as ferric humates. Above a pH of 9.6, most iron is removed as ferric hydroxide. Treatment is followed by coagulation, settling, and filtration. Ion exchange effectively removes ferrous and manganous salts using sodium zeolite. Air (oxygen) must be excluded in this operation to prevent oxidation to iron and manganese oxides, which can form precipitates and plug the ion exchange column. This process also removes other salts in the water and decreases hardness.

Aeration is the most economical iron removal method in large-capacity municipal treatment plants. Chemical precipitation is frequently used in beverage and food processing plants. Ozonation can selectively remove iron and manganese and preserve the mineral taste of water.

Lead

Lead is used mainly in various solid forms, as pure metal and in several compounds. Major uses are storage batteries, bearings, solder, waste pipelines, radiation shielding, sound and vibration insulation, cable covering, ammunition, printer's type, surface protection, and weights and ballasts.

Wastewaters containing lead originate from only a few of the processes that produce lead-containing products, such as plating, textile dyeing and printing, photography, and storage battery manufacturing and recycling.

Lead is a toxic, heavy metal limited to 0.05 mg/l by USPHS Drinking Water Standards, and to 0.10 mg/l by other standards. Discharge standards in sewer use ordinances usually limit lead to 0.5 mg/l.

TREATMENT METHODS

Chemical methods of treatment include batch, continuous flow, or integrated with the production process. Table 8.3.4 lists several insoluble lead compounds and their corresponding solubilities at room temperature. The anions in these compounds and their sources are listed in Table 8.3.5.

Aluminum hydroxide from alum use aids in settling the lead sulfate formed. A combination of hydroxide, carbonate, and sulfide results in a buffered treatment solution, allowing a check on the effectiveness of clarification due to the formation of black lead sulfide. Hypochlorite can also be used to prepare the insoluble quadrivalent oxide:

$$Pb^{2+} + 2OH^- \longrightarrow Pb(OH)_2 \qquad 8.3(20)$$

$$Pb(OH)_2 + ClO^- \longrightarrow PbO_2 + H_2O + Cl^- \quad 8.3(21)$$

At high pH, lead exists as the plumbate ion, PbO_2^{2-} which can also be oxidized by hypochlorite.

$$PbO_2^{2-} + ClO^- + H_2O \longrightarrow PbO_2 + Cl^- + 2OH^- \quad 8.3(22)$$

In reality, wastewaters contain other substances that also require removal. Therefore, a given treatment method may also remove other substances, and a treatability study is needed before selecting a treatment method.

Expected effluent quality in terms of lead concentration for batch or continuous flowthrough treatment is reported

TABLE 8.3.4   ROOM TEMPERATURE SOLUBILITIES
OF LEAD COMPOUNDS

Compound	Solubility in mg. per liter
Pb Cl$_2$	11.0
Pb SO$_4$	4.2×10^{-2}
Pb (OH)$_2$	1.9×10^{-2}
Ph CO$_3$	6.0×10^{-4}
PbCr O$_4$	4.3×10^{-5}
PbS	4.9×10^{-12}
Pb SO$_3$	Insoluble
2Pb CO$_3$ · Pb(OH)$_2$	Insoluble

TABLE 8.3.5 ANION SOURCES FOR LEAD COMPOUNDS

Anion	Source of Anion
Sulfate (SO_4^{2-})	Sulfuric acid or alum
Hydroxide (OH^-)	Caustic soda or lime
Carbonate (CO_3^{2-})	Soda ash
Chromate (CrO_4^{2-})	Spent chrome plating bath or chrome plating rinse
Sulfide (S^{2-})	Sodium sulfide
Sulfite (SO_3^{2-})	Sulfur dioxide

to be 0.5 mg/l, whereas that for integrated treatment is 0.01 mg/l.

The amphoteric nature of lead compounds requires careful control of pH for both precipitation and handling (dewatering) of sludges resulting from treatment. Each stage requires a different operating pH control range. As with all precipitation reactions, nucleation and crystal growth are important, although the high molecular weight of lead aids in particulate settling.

Physical methods such as electrodialysis, ion exchange, and reverse osmosis can also remove lead from wastewaters. Lead may also be removed by deliberately introducing the wastewater to acclimated biological treatment plants for complexing with biologically formed organic substances. A combination of chemical and biological methods, the *in process* treatment, can be used, with the lead chemically complexed and removed from the biological process.

Magnesium

Magnesium is usually present in water or brine as bicarbonate, sulfate, or chloride. It may also be produced in water solutions, when dolomitic lime is used to neutralize waste acid. With the exception of magnesium hydroxide, magnesium compounds are very soluble. The solubility of magnesium hydroxide is about 8 mg/l at ambient water temperatures. However, when precipitated without an excess of hydrogen ion, solubility, including supersaturated mangesium hydroxide, rises to about 20 mg/l.

If precipitation is carried out in the presence of a high concentration—up to 5% by weight of previously precipitated hydroxide—supersaturation is reduced. Magnesium hydroxide usually precipitates as a flocculant material, which settles slowly and will only concentrate to about 1% by weight. However, when precipitated in the presence of previously precipitated solids, the settling rate and density of the settled sludge increase considerably.

Magnesium is not considered a contaminant in wastewater unless it is present in a brine (saltwater). However, concentrations in excess of 125 mg/l can exert a cathartic

and diuretic effect. In addition, magnesium salts break down on heating to form boiler scale.

Manganese

The limit for manganese in drinking water is 0.05 ppm. Above that concentration it stains fixtures and interferes with laundering and chemical processing. Manganese usually occurs with iron in ground and surface waters, and many iron-removing processes (Table 8.3.3) will also remove manganese. The treatment process usually oxidizes the water-soluble manganous salts to insoluble manganese dioxide by catalytic air oxidation in the pH range of 8.5–10, by chlorination at a pH of 9–10, or by ozonation at neutral pH. Ion exchange is also effective, but it removes other salts that may or may not be desirable.

Manganese, like iron, is extracted from the bottom of deep reservoirs or from the ground by carbon dioxide–rich, oxygen-poor water as manganous bicarbonate.

Catalytic air oxidation at an alkaline pH is the most economical method for large treatment plants. Aeration occurs on contact beds of coke or stone coated with manganese dioxide or on beds of pyrolusite. Oxidation is more rapid when the pH is adjusted between 8.5 and 10.0 by lime or caustic (Table 8.3.6). One ppm dissolved oxygen oxidizes approximately 7 ppm manganese. The insoluble manganese dioxide is usually removed by settling and filtration.

Oxidation of manganese can also be carried out at a neutral pH using ozone as oxidant, and excessive ozone dosages can oxidize the manganese to pink permanganate. Chlorine oxidation of manganese requires no catalyst.

Ion exchange processes using sodium or hydrogen exchange resin remove manganous salts effectively, together with other salts. The exchange resin has to be regenerated periodically with sodium chloride or sulfuric acid solution. Aeration at the pH range of 8.5–10 is usually the most economical process for large water treatment plants. Ozonation removes manganese and iron selectively, preserving the mineral taste of the water. Ion exchange is the

TABLE 8.3.6 PRECIPITATION OF MANGANESE AND IRON AS A FUNCTION OF PH

	Manganese			Iron		
	Residue	Precipitated		Residue	Precipitated	
pH	(ppm)	(ppm)	(%)	(ppm)	(ppm)	(%)
7.0	5.49	0.0	0.0	5.59	0.0	0.0
8.0	5.49	0.0	0.0	1.53	4.06	72.6
8.8	5.49	0.0	0.0	0.47	5.12	91.6
9.0	1.90	3.59	65.4			
9.2	1.00	4.49	81.7			
9.4	0.11	5.38	98.0	0.06	5.53	98.9
9.6	0.03	5.46	99.4	0.00	5.59	100.0

treatment of choice when both manganese removal and softening are necessary.

Mercury

Mercury exists in water in metallic, mercurous (mercury I), and mercuric (mercury II) forms, both in solution or suspension. Removal methods include total and partial recycling, impounding, reduction and filtration, sulfide treatment, ferrous chloride treatment, and the use of activated carbon resins. No treatment to prevent methylation of sediments is known. Mercury can be present in solution, suspension, and floating, as the metal, soluble, and insoluble compounds, and complexes.

PROPERTIES

Mercury is a silvery liquid; its density is 13.546 at 20°C, its molecular weight is 200.61, its boiling point is 356.9°C, and its freezing point is −38.87°C. It has an appreciable vapor pressure (Othmer and Steinden 1967), and reacts with halogens, sulfur, and oxygen, to give corresponding halides, sulfides, and oxides. It does not react with water, alkali, or weak acids. It is oxidized by concentrated nitric acid, releasing nitric oxide, and by hot concentrated sulfuric acid, releasing sulfur dioxide. It forms alloys called amalgams with metals and with ammonium ions. It solubility in air-free water is 20–30 ppb at 30°C, but this increases in the presence of air, chlorides, and alkali (Linke).

The mercurous ion is the univalent form of mercury and it is diatomic. It disproportionates in the presence of sulfide, hydroxy, and cyanide ions (Linke), according to equation 8.3(23).

$$Hg_2^{2+} + S^{2-} = Hg° + HgS \qquad 8.3(23)$$

Mercury(I) salts have low solubility except for the nitrate, chlorate, and perchlorate. They behave as strong electrolytes. Mercury(I) is the mercurous or univalent form of mercury and shows no tendency to form covalent bonds.

Mercury(II), the bivalent form of mercury, forms predominantly covalent bonds and is readily complexed by inorganic and organic ligands such as $HgCl_4^{2-}$, HgS_2^{2-} and $Hg(CN)_4^{2-}$. Mercury(II) forms oxides, sulfides and all the common salts.

Mercury(II) forms organomercury compounds in which there is at least one C—Hg bond (as distinguished from mercury salts of organic acids and organic complexes). The organomercury bond is relatively stable. These compounds are moderately insoluble in water and may be decomposed by hydrolysis or oxidation. Metallic mercury and its inorganic mercury compounds are converted to organomercury compounds as well as the reverse, by geochemical and biochemical processes.

SOURCES OF CONTAMINATION

The sources of mercury contamination (USGS 1970) are summarized in Table 8.3.7. Mercury vaporizes from ore deposits and is washed back to earth by rain. The concentration of mercury in rainwater averages about 0.2 ppb. Mercury is held tightly by the upper 2 in or so of soil. Surface waters generally contain less than 0.1 ppb of mercury. Underground waters are higher due to longer contact with minerals. Hot springs and exposed ore bodies contribute higher concentrations in solution and by erosion. Man-created sources add to the natural sources. Suspended matter and sediments may contain 5–25 times as much mercury as the surrounding water bodies.

METHYLATION OF INORGANIC MERCURY

Certain microorganisms present in sediments convert inorganic mercury to methyl mercury, which is soluble (5 gpl), readily assimilated by aquatic life, and also more toxic than inorganic forms of mercury (Royal Society of Canada 1971). It also becomes more concentrated as it passes up the food chain.

This biological conversion process was regarded as an anaerobic process but was later found to occur even more efficiently under aerobic conditions. The process was first identified with microorganisms, but higher organisms such as chickens also seem to have this conversion capacity.

METHODS OF REMOVAL FROM WATER

In nature, soil particles such as clays, oxides, peat moss, and humus adsorb mercury from rainfall and remove it

TABLE 8.3.7   SOURCES OF MERCURY CONTAMINATION

Natural

Mercury ores, volatilization, solution, and erosion; rocks, volcanos, hot springs; rainfall; sediments and methylation of inorganic mercury

Man-Created

Mining and metallurgy; recovery and purification of mercury; burning fossil fuels; sewage and garbage

Industrial

Electrical apparatus (batteries, lamps, and power tubes); electrolytic chloralkali; paint (mildew proofing and antifouling); instruments (switches, relays, gauges, meters, barometers, thermometers, manometers, and pump seals); catalysts; agriculture (seed dressing, fungicides); dental preparations; general laboratory use; pharmaceuticals (diuretics, antiseptics, preservatives, and skin preparations); pulp and paper-slimicide; and miscellaneous

from the cycle. The tendency of mercury to sink rapidly and combine with sulfide in anaerobic bottom sediments to form cinnabar (HgS) appears to be a major scavenging mechanism. Reaction of mercury with organic matter is another such mechanism.

For industrial and metallurgical effluents, the removal treatment must fit the concentration of mercury and the quantity and composition of effluent. The following steps and principles apply for preliminary handling:

1. Isolate the mercury-bearing water from mercury-free effluent.
2. Minimize the use and quantity of water.
3. Recycle contaminated water. In order to accomplish this some simple treatment such as cooling, settling, or filtering may be necessary.
4. Use V-shaped rather than rectangular trenches or floor drains.
5. Use a series of traps to collect solids.
6. Impound wastewater in a tank or leak-proof pond for temporary storage and to even out the variations in flowrates and composition.
7. Observe that pumping underground may be possible in some cases.
8. Settle, filter, or both.

Total Recycle

Recycling mercury-contaminated water is recommended; however, accumulated sludges and/or concentrated solutions may be more difficult to treat for mercury removal than the original wastewater. Evaporation was tested and abandoned as impractical because of the volatility of mercury and the buildup of contaminated bittern or brines.

Reduction

Reduction of mercury to the metallic state followed by filtration may be suitable for small volumes of concentrated effluent, and can be accomplished by electrolysis, reduction with a less noble metal, or a reducing agent. Mercury can be recovered in a relatively pure state at the cathode of a special electrolytic cell. When metallic reducing agents such as copper, iron, zinc (New Jersey Zinc Co. 1971), aluminum, and sodium amalgam are used, the mercury is recovered as an amalgam or as droplets coalescing on the surface of the metal. Mercury is recovered in the pure state by distillation. Mercury ions are replaced by the reducing metal ions according to equations 8.3(24) and 8.3(25).

$$Zn + Hg_2^{2+} \longrightarrow 2Hg + Zn^{2+} \qquad 8.3(24)$$

$$2NaHg_x + Hg_2^{2+} \longrightarrow (2x + 2)Hg + 2Na^+ \quad 8.3(25)$$

In the latter case, mercury is recovered in the amalgam form.

Reducing agents such as hydrazine, hydroxylamine, hypophosphorous, formaldehyde, and sodium borohydride (Anon. Chem Eng. News 48 1970) are suggested. The mercury is recovered by coalescence and/or filtration. Mercury concentrations remaining in the filter effluent (filtrate) are reported in the 100 ppb range after treatment through a 5-micron filter.

Sulfide Treatment

Waste treatment with sodium hydrosulfide (NaHS) or sodium sulfide (Na₂S) and a flocculant has been suggested and widely used as a primary treatment in the chlor-alkali industry (Bouveng and Ullman 1969) in the United States (Figure 8.3.4).

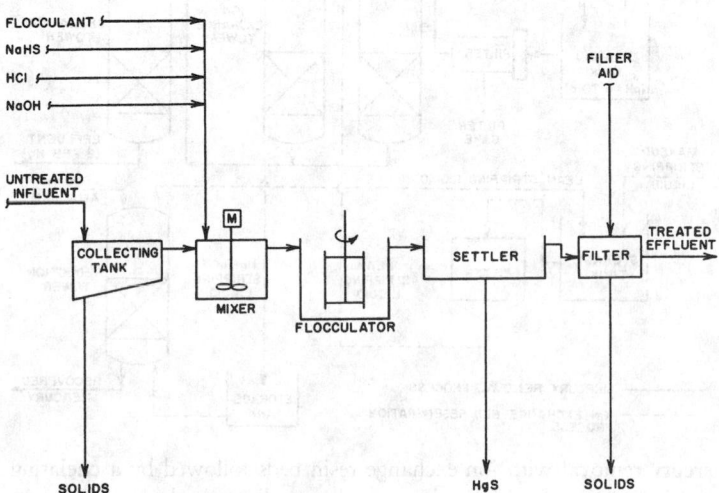

FIG. 8.3.4 Sulfide treatment for mercury removal—a continuous process.

Wastewater containing mercury reacts with sulfide according to equation 8.3(26).

$$Hg + Hg_2^{2+} + Hg^{2+} + 2S^{2-} \longrightarrow 2Hg + 2HgS \qquad 8.3(26)$$

Mercury and mercuric sulfide can be recovered by settling and/or filtration. However white mercuric sulfide is very insoluble and readily forms a soluble complex with excess sulfide.

$$HgS + S^{2-} \longrightarrow HgS_2^{2-} \qquad 8.3(27)$$

This effect is most severe at high pH—adjustment to a pH of 7 or 7.5 is required. Buffering with sodium carbonate (2 gpl) helps to correct flowrate variations. Automatic control of pH and sulfide concentration is possible using a glass pH electrode and a silver sulfide specific ion electrode. Sulfide is readily oxidized particularly by chlorine.

Efficient operation gives 50–60 ppb mercury concentration in the effluent but under usually occurring upset conditions, mercury concentration in the effluent increases to 200–500 ppb.

Ferrous Chloride Treatment

Ferrous chloride reduces mercury salts in wastewaters to insoluble compounds as shown in equation 8.3(28).

$$2\,Fe^{2+} + 2\,Hg^{2+} + 8OH^- \longrightarrow 2\,Fe(OH)_3 + Hg_2O + H_2O$$

$$8.3(28)$$

Wastewater is adjusted to a pH of 9–9.5 and a 50 ppm excess (over stoichiometric) and FeCl$_2$ is added. After settling in a pond for several days, the effluent contains 5–6 ppb of mercury.

Adsorption on Activated Carbon

Activated carbon is used as filter-aid and also in granular beds. Removal performances from 100–200 ppb (influent) to 10–20 ppb (effluent) are reported. For the filtration of 50% caustic soda from mercury cells, a precoat type filter with activated carbon precoat, such as Nuchar 722, is used. At the start of a cycle, mercury concentration is 10 ppb in the filter effluent, but increases to the Codex (National Academy of Sciences 1971) limit of 500 ppb when the filter is backwashed. Mercury is recovered from the activated carbon by distillation.

Ion Exchange and Chelating Resins

Anion exchange and chelating resins are commercially used in chlor-alkali plants in Japan and Sweden (Gardiner 1971; Aktiebolaget Billingsfors-Långed; Ajinomoto Co., Inc.; Terraneers, Ltd.). Untreated influent is first adjusted for pH and oxidation reduction potential and is filtered. Anion resins remove the mercury down to an effluent concentration of 100–200 ppb. Chelating resins further reduce mercury concentration to 2 ppb. Anion resins and some chelating resins can be regenerated with sodium sulfite, sodium hydrosulfite, or hydrochloric acid. Mercury can be recovered from backwash stripping solutions by electrolysis or with sodium amalgam, or the mercury-rich solution can be used as is. Mercury is recovered from some chelating resins and filter solids by distillation.

The untreated influent wastewater is adjusted to a pH between 6 and 8, and chemicals are added to destroy oxidizing substances (Figure 8.3.5). Then the effluent is filtered. The use of an activated carbon filter removes some mercury and destroys oxidants that might exist because of

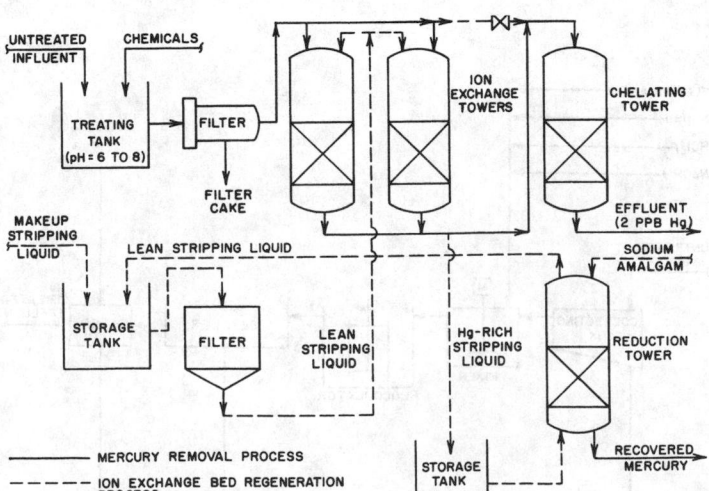

FIG. 8.3.5 Treatment for mercury removal with ion exchange resin beds followed by a chelating resin bed. Mercury is stripped from the ion exchange resin by stripping liquid and then is reduced by sodium amalgam and recovered. A chelating resin is a heterocyclic compound that will combine with a metallic ion to form a chelate, i.e., a compound with the metallic ion attached by covalent bonds to two or more nonmetallic atoms in the same compound.

faulty pretreatment. Precipitates may accumulate in the resin beds; consequently, occasional backwashing with water and air is required.

Counteracting Methylation of Mercury

The discovery that microorganisms in sediments convert inorganic mercury to the soluble and more poisonous methyl mercury makes it important to inactivate the mercury already present in waterways. Suggestions include dredging; covering with fine material (which is effective if not disturbed); changing environmental pH and salinity; finding a catalyst to change methyl mercury chloride to dimethyl mercury, which is volatile and would escape from the water; and inactivating the sediment with mercaptans.

Nickel

Nickel may be present in plating wastes or in the waste from nonferrous metallurgical plants.

Nickel plating is carried out at a pH range of 1.5 to 6.0, with the majority of solutions operative between 2.0 and 4.5. The concentration of nickel in rinse waters following nickel plating varies widely, depending on the method to minimize drag-out and whether flowthrough or countercurrent rinsing is used. As a general rule, a three-stage counterflow rinsing operation reduces rinse water consumption by 90–95%, making it more economical to recover nickel for reuse.

Nickel may be reclaimed from the rinse tank by evaporation, and the concentrated solution returned to plating. The condensate is recovered and reused as makeup to the rinse system. Ion exchangers are also used for the recovery of nickel and water. The rinse water is passed through cation and anion exchangers in series, with deionized water recycled into the rinse tanks. The cation exchanger, which removes the nickel ions, is periodically regenerated with acid. The regenerating solution containing the concentrated nickel salts can then be treated and reused in plating operations.

If it is uneconomical to segregate the nickel rinse water for recovery, or if waste contains other metals that would interfere with recovery operations, nickel can be completely removed by precipitation with lime at a pH of 8.0 or higher. Settling characteristics depend on the technique used for precipitation, flocculation, and settling. Designs should achieve a separation rate of less than 0.5 gpm/sq ft, with the clarifier effluent filtered to remove nickel hydroxide present as suspended solids in the clarifier overflow.

Precipitated hydroxide can only be concentrated to 1.0–2.0% by plain gravity settling. Sludge dewatering requires precoat filtration or plate and frame filtration.

Silica

INSOLUBLE SILICA

Wastes with fine sand, silica gels, activated silica sols, and colloidal silica, as well as silica-containing substances such as silt, clay, fly ash, diatomaceous earth, diatoms, and minerals can be removed by clarification or by inline pressure coagulation-filtration. Without chemical flocculation, filtration can only remove 50–80% of silica, or perhaps none if particle size is sub-μ. Macroreticular colloidal removal ion exchange resin has been effective in removing virtually 100% of colloidal silica (Kunin and Hetherington 1969).

SOLUBLE SILICA

Soluble silica is not an environmental contaminant. It originates from well or surface water supplies and has concentrations of 1–120 mg/l as SiO_2. Above 120 mg/l, colloidal or crystalline flakes of insoluble silica may be visible in neutral water (pH < 8.0). Higher concentrations of silica may be solubilized by having a carbonate or hydroxide alkalinity of at least 1.5 times the SiO_2 concentration. Soluble silica in neutral water (pH 5–8) is present as a mixture of bisilicate, $HSiO_3^-$; as weak silicic acid, H_2SiO_3; and perhaps as SiO_2—compounds analogous to HCO_3^-, H_2CO_3, and CO_2 in water (Camp 1963).

Removal of soluble silica from wastewater may be required for a few processes (State Water Pollution Control Board 1952) or in case of high pressure boiler feed, when a turbine is present (Crits 1968). Soluble silica can be removed

1. By warm or hot lime precipitation, with the addition of MgO (or naturally occurring magnesium, if enough is present) to reduce silica concentration by 80–95%—down to 1.0 mg/l (Applebaum 1968) in the effluent
2. From soft water using a desilicizer, a strongly basic anion resin in the hydroxide form (Applebaum 1968; Zunino 1962). Silica concentration in the effluent may be as low as 0.2 mg/l.
3. By strongly basic anion resins used in demineralization. Silica concentration in the effluent can be reduced to 0.002 mg/l if required.

Strontium

Strontium is an alkaline-earth element with a chemistry similar to that of calcium and other divalent cations in Group IIa of the periodic table of the elements. The metal is used in some alloys of tin and lead; various strontium salts are used in pyrotechnics, refining beet sugar, glass, paints, ceramics, and some medicines; and strontium radioisotopes are generated from fission reactions at nuclear installations (McKee and Wolf 1963). The major liquid industrial wastes containing strontium are nuclear wastes including Sr[89] and Sr[90].

Strontium removals can be achieved through lime-soda ash softening. When stoichiometric chemical dosages are used, a 65–75% removal efficiency of dissolved strontium can be obtained. Increased removals result from chemical dosages greater than stoichiometric amounts.

Phosphate coagulation can also be utilized since strontium cations form relatively insoluble phosphate compounds at high pH. At a pH of 11.3 or greater, with a phosphate to calcium ratio greater than 2.2 to 1, more than 97% of strontium can be removed (Landerdale 1951). Other methods of chemical precipitation include coprecipitation with aluminum and cesium from acid aluminum wastes, and scavenging by tannic acid or calcium oxalate (Straub 1964).

Inorganic strontium can also be removed from water through cationic ion exchange materials. Processes such as electrodialysis and reverse osmosis also show promise for strontium removal application.

Sulfate

Sulfate removal may be required to meet recommended limits for drinking water, which suggests a maximum concentration of 250 mg/l for sulfates and 500 mg/l for total solids (U.S. Public Health Service 1963). Sulfate may be removed at the source for reuse or to prevent downstream biological reduction that can produce odors.

The choice of a sulfate removal system depends on the initial sulfate ion content and the final quality of water desired. Alternative means of removal are ion exchange, evaporation and crystallization, reverse osmosis and bacterial reduction.

ION EXCHANGE

Anion exchange resins can be used for removal of sulfate ions. For example, a new deionization method (Kunin and Downing 1971) has shown over 9% efficiency in removing waste ions from acid mine waters. Sulfate ions are removed by an anion exchange resin functioning in the bicarbonate cycle according to the following reaction:

$$2 \, (R\text{—}NH)HCO_3 + FeSO_4 \longrightarrow (R\text{—}NH)_2SO_4 + Fe(HCO_3)_2$$

$$8.3(29)$$

Resin regeneration is accomplished with dilute ammonium hydroxide followed by carbonation with carbon dioxide.

EVAPORATION AND CRYSTALLIZATION

When the sulfate ion content is sufficiently high to warrant recovery for reuse or sale, a crystallization approach should be evaluated. Water can be removed by evaporation and the sodium sulfate crystallized out (Cosgrove) at about 5°C as $Na_2SO_4 \cdot 10 \, H_2O$ (Glauber's salt). The crystals can then be removed by filtration. Metal impurities can be removed by neutralizing with sodium hydroxide solution and refiltering, evaporating, crystallizing, and drying. Depending on the relative concentrations of metal ions and sulfates, an alternative approach first removes heavy metals by ion exchange.

REVERSE OSMOSIS

Rejection of sulfate ions across a modified, semipermeable cellulose acetate membrane is reported at above 90% (Hindin and Bennett 1969). This process is justified by the need for water reuse in water-deprived areas and by the available means of disposal or recovery of the high sulfate-containing concentrate.

BIOLOGICAL REDUCTION

Sulfate-reducing bacteria form hydrogen sulfide under reducing conditions, i.e., in the absence of molecular oxygen and other proton acceptors like nitrate ions. Organic matter is oxidized to acetic acid by the most common sulfate-reducing bacteria; however, complete oxidation to carbon dioxide and water also can occur. If sulfate levels are high, the objectionable odors resulting from the release of hydrogen sulfide gas can be minimized by (1) precipitation of metallic sulfide salts; (2) oxidation of the sulfides to sulfates either by anaerobic photosynthetic bacteria or by microaerophilic sulfur-oxidizing bacteria; and (3) maintaining an alkaline pH level.

Sulfide

Water containing sulfides in excess of a 0.5 ppm concentration has an offensive (rotten eggs) odor and is also very corrosive. The sulfide is present at acidic pH levels as hydrogen sulfide gas and at alkaline pH values as sulfide salt. Occasionally, when the water also contains iron, the sul-

TABLE 8.3.8 OXIDIZING WASTEWATER CONTAINING 5 PPM OF H_2S

Oxidant	Dosage	
	Dose[a] Required (ppm)	Total Weight of Oxidant Required (lbs/MG)
Ozone	5	41.8
50% Hydrogen Peroxide	10	83.6
Chlorine	40	334.4

[a]The dosage requirements differ because ozone is introduced as 1% O_3 in air; the oxygen present in air helps to keep the system aerobic, preventing H_2S formation; and the dosage indicated for chlorine is sufficient to prevent further H_2S formation, while the ozone dosage is not.

fide may be present as a finely divided black (FeS) precipitate. Hydrogen sulfide is usually removed by aeration at an acidic pH, followed by a final oxidation treatment with chlorine, ozone, or hydrogen peroxide.

Aeration at a low pH removes hydrogen sulfide by purging rather than by oxidation. The aerators are usually constructed of wood rather than metal to avoid corrosion. Bacterial growth aids in sulfide removal. Chemical oxidation of 1–2 ppm of hydrogen sulfide in water can be carried out most economically with ozone. Hydrogen peroxide and chlorine are also effective. The oxidant dosages necessary to remove 1 ppm hydrogen sulfide are ozone 1 ppm, hydrogen peroxide (50%) 2 ppm and chlorine 8 ppm. Table 8.3.8 lists treatment for a 1 MGD wastewater flow containing 5 ppm of hydrogen sulfide.

Zinc

Spectrographic analysis of 969 river samples by the FWPCA from 1963 to 1965 showed zinc in 80% of all the samples. Concentrations (Pickering 1968) varied from 0.003 to 1.080 mg/l, with 0.136 mg/l reported as the mean concentration in drinking water from 37 U.S. locations (Kehoe, Cholak and Largent 1944). In greater than trace concentrations, zinc is harmful to aquatic organisms. The toxic concentration varies with pH and hardness. The 96-hr median tolerance limit (TL_m) for fathead minnows is 4.7 mg/l zinc concentration with 8 pH and 50 mg/l hardness, and it is 35.5 mg/l with 6 pH and 200 mg/l hardness (Mount 1966). Long-term tests on minnows in water with 200 mg/l hardness show egg production reduced 50% with

the zinc concentration at 0.009 times the 96-hr TL_m value (Brungs 1969), an amount lower than that found in some natural streams. Soluble zinc concentrations can be reduced by ion exchange and precipitation processes.

ION EXCHANGE

Using a recycling regeneration process, it is possible to obtain an average recovery of 92% (Aston 1968). Hydrogen and sodium ions effect loss of exchange capacity (Blake and Randle 1967). The loss amounts to 50% when these ions total 15 g/l. On cooling tower water with 6 mg/l zinc concentration, a residual of below 1 mg/l is reported in the treated effluent. If zinc is being removed for recovery, the ability of ion exchangers to adsorb impurities may necessitate subsequent purification.

PRECIPITATION

Precipitation of the hydroxide neutralizes acid zinc solutions and reduces zinc in the same process. If only lime is added in a single step after extended settling, there is still less than 1% hydroxide in the sludge. Often such sludges must be stored indefinitely in lagoons because they dry very slowly. A novel system (Figure 8.3.6) precipitates zinc by repeated adsorption on hydroxide particles (Rock 1971). These become dense spheroids concentratable to 5–10% hydroxide. Residual zinc in the effluent is less than 1 mg/l, independent of the zinc concentration in the feed (Chamberlain and Anderson 1971). The zinc is reused in a rayon plant. Other processes obtain improved precipi-

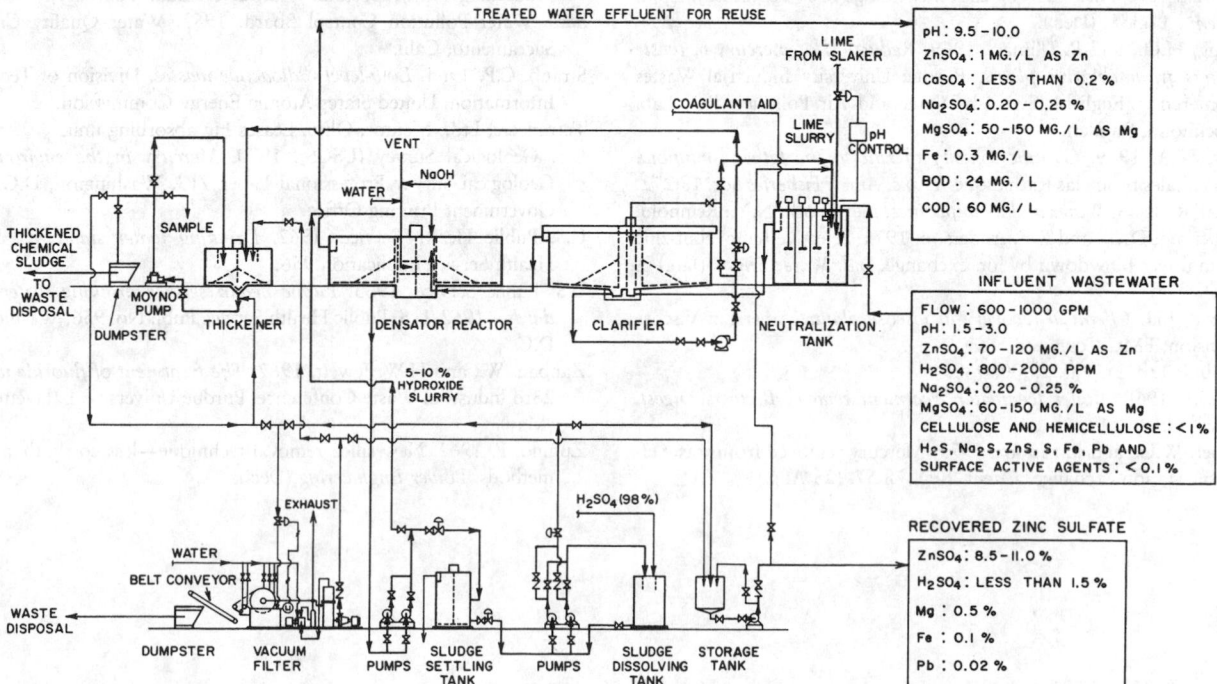

FIG. 8.3.6 Zinc recovery flowsheet (Reprinted from the U.S. Environmental Protection Agency (EPA). 1971. *Zinc precipitation and recovery from viscose rayon wastewater.* [American Enka Co. Project No. 12090 ESG. January.])

tates of zinc carbonate (Courtaulds Ltd.) and sometimes use inert nuclei. Others reduce the zinc by adsorption and precipitation in activated sludge systems (Offhaus 1968) if the waste sludge is removed without additional digestion.

For the process in Figure 8.3.6 the operating and maintenance costs for recovery of the soluble waste zinc depend on the sulfuric acid-zinc sulfate ratio in the waste and on the amount of zinc recovered daily. When recovering 2000 lb Zn daily from a waste with a ratio of 5 to 6, the operating and maintenance costs are 12.5–14.0 cents/lb of Zn. The cost of purchased zinc oxide is 15.6 cents/lb equivalent zinc.

—E.G. Kominek, I.M. Abrams, S.E. Smith,
E.C. Bingham, L.J. Bollyky, A.F. McClure, Jr.,
R.D. Buchanan, W.C. Gardiner, G.J. Crits,
L.W. Canter, R.A. Conway, D.M. Rock

References

Ajinomoto Co., Inc. Resinous mercury adsorbent. Tokyo, Japan.

Aktiebolaget Billingsfors-Långed, Billingsfors, Sweden, Q-13 anion resin and Q-Sorb chelating resin.

Anon. 1970. Process removes mercury in plant wastes. *Chem. Eng. News* 48 (14 Dec.).

Applebaum, S.B. 1968. *Demineralization by Ion Exchange*. New York, N.Y. Academic.

Aston, R.S. 1968. *Recovery of zinc from viscose rayon effluent*. Proc. 23rd Ind. Waste Conf., May, Purdue University, Lafayette, Ind. (132) 63–74.

Blake, W., and J. Randle. 1967. *Removal of Zn^{2+} from the ternary system $Zn^{2+}N—Na^+—H^+$ by cation-exchange resin columns. *J. Appl. Chem.*, 17:358. (Dec.).

Bouveng, H.O., and P. Ullman. 1969. *Reduction of mercury in waste waters from chlorine plants*. Purdue University Industrial Wastes Conference (English). Swedish Water and Air Pollution Res. Lab. Stockholm, Sweden.

Brungs, W.A. 1969. *Chronic toxicity of zinc to the fathead minnow*. Pimephales promelas Rafinesque. *Trans. Amer. Fisheries Soc.* 18:272.

Camp, T.R. 1963. *Water and its impurities*. New York, N.Y.: Reinhold.

Chamberlain, D.G., and R.E. Anderson. 1971. Selective removal of zinc from tower blowdown by ion exchange. *Ind. Water Engin.* (Jan.) p. 33.

Cosgrove, J.H. *Chemical recovery in viscose plants*. American Viscose Division, FMC Corp.

Courtaulds Ltd. Fr. Pat. 2,000, 930.

Crits, G.J. 1968. *Boiler feedwater treatment review*. Brewers Digest. April.

Gardiner, W.C., and F. Munoz. 1971. Mercury removed from waste effluent via ion exchange. *Chem. Eng.* 78:57. (23 Aug.).

Hindin, E., and P.J. Bennett. 1969. Water reclamation by reverse osmosis. *Water and Sewage Works* 116(2):67.

Kehoe, R.A., J. Cholak, and E.J. Largent. 1944. The concentrations of certain trace metals in drinking water. *J. Amer. Water Wks. Assoc.*, 36:637.

Kunin, R., and R. Hetherington. 1969. *A progress report on the removal of colloids from water by macroreticular ion exchange resins*. 30th International Water Conference. Pittsburgh, Penna. (Oct.).

Kunin, R., and D.G. Downing. 1971. Ion exchange system boasts more pulling power. *Chemical Engineering*, 78(15):67.

Landerdale, R.A., Jr. 1951. Treatment of radioactive wastes by phosphate precipitation. *Ind. Engr. Chem.* 43:1538.

Linke, W.F., *Solubilities of inorganic and metal organic compounds* (4th ed.). Princeton, N.J.: VanNostrand.

McKee, J.E., and H.W. Wolf. 1963. *Water quality criteria*, 2d ed. Resources Agency of California. State Water Quality Control Board Publication 3-A.

Mount, D.I. 1966. The effect of total hardness and pH on acute toxicity of zinc to fish. *Air and Water Pollut. Int. J.* 10:49.

National Academy of Sciences—National Research Council. 1971. New specifications (limits of impurities) for mercury in Potassium Hydroxide Solution and Sodium Hydroxide. *Food chemicals codex*, 1st ed. Washington, D.C. (1 February). p. 13 and p. 16.

New Jersey Zinc Co. 1971. *Chem. Eng.* 78:63. (22 Feb.).

O'Connor, J.T., and C.E. Renn. 1964. Soluble-adsorbed zinc equilibrium in natural waters. *J. Amer. Water Wks. Assoc.*, 56:1055.

Offhaus, K. 1968. Zinkgehalt und Toxizität in den Abwässern der Chemiefaserindustrie. *Wass. Abwass. Forschung.* (1):7.

Othmer, D.F., and A. Steinden. 1967. *Encyclopedia of chemical technology* (2d ed.). Vol. 13. Mercury and Mercury Compounds. New York, N.Y.: Interscience Publishers.

Pickering, Q.H. 1968. *Water research*. London, England: Pergamon.

Rock, D.M. 1971. Hydroxide precipitation and recovery of certain metallic ions from wastewaters. Water-1970. *Chem. Eng. Prog. Symp. Series.* 67:107. pp. 442–444.

Rohrer, K.L. 1971. An integrated facility for treatment of lead and fluoride wastes. *Industrial Wastes* 118:36. (September/October).

Royal Society of Canada. 1971. *Mercury in man's environment*. Proceedings of Symposium. Ottawa, Canada. (February).

State Water Pollution Control Board. 1952. Water Quality Criteria. Sacramento, Calif.

Straub, C.P. 1964. *Low-level radioactive wastes*. Division of Technical Information. United States Atomic Energy Commission.

Terraneers, Ltd., Mentor, Ohio. Leases Hg absorbing unit.

U.S. Geological Survey (USGS). 1970. *Mercury in the environment*. Geological Survey Professional Paper 713. Washington, D.C.: U.S. Government Printing Office.

U.S. Public Health Service. 1962. *Drinking water standards*. Public Health Service Publication 956.

U.S. Public Service. 1963. *Public health service drinking water standards—1962*. U.S. Public Health Service Publ. No. 956, Washington, D.C.

Zabban, W., and H.W. Jewett. 1967. *The treatment of fluoride wastes*. 23rd Industrial Waste Conference. Purdue University. Lafayette, Ind. (May).

Zunino, F. 1962. New silica removal technique—less costly than other methods. *Power Engineering* (Dec.).

8.4
INORGANIC NEUTRALIZATION AND RECOVERY

Boiler Blowdown Water

Blowdown is water discharged from boiler systems containing a relatively high concentration of suspended and dissolved solids. The discharged blowdown is replaced by fresh (usually demineralized) low-solids feedwater. Excessive solids buildup in boiler water can cause carryover into the steam drum in the absence of suitable antifoam agents, or scaling when salt solubility is exceeded. Boiler blowdown water is generally alkaline and often contains suspended matter from sludges of insoluble sulfates and carbonates. High temperature, dissolved solids, and alkalinity present disposal problems for untreated blowdown water, unless a larger wastewater stream is available for dilution before discharge into receiving water (IUPAC 1963). For petrochemical plants, the blowdown stream is a minor contributor to the overall plant effluent disposal problem (Beychok 1967).

The rate of blowdown at equilibrium must ensure that solids introduced into the boiler by the feedwater are totally removed. Since chlorides are soluble and none are intentionally added to the boiler feedwater, they provide a means for measuring total salts in boiler water. If chloride and total soluble salt concentrations in the feedwater are known, the ratio of feedwater chloride to boiler water chloride indicates the ratio of feedwater total salts to boiler water total salts. The equilibrium total dissolved solids allowable in the boiler water are a direct function of the operating pressure of the steam-generating system (De Lorenzi 1951). A simple equation for determining the required rate of blowdown is given by Equation 8.4(1)

$$X = \frac{100 \, C_f}{C_b - C_f} \qquad 8.4(1)$$

where:

> X = % blowdown (based on steam produced)
> or
> $\frac{100 \, X}{100 + X}$ = % blowdown (based on water fed)
> C_f = total chloride concentration in the feedwater
> C_b = total chloride concentration in the boiler water

Boiler blowdown requirements are dictated by the limits on total allowable dissolved solids concentration in boiler water; by the economics of heat transfer for the boiler system; and by the methods of disposal or treatment for recovery of blowdown stream (Hamer, Jackson, and Thurston 1961). Blowdown may be intermittent or continuous. For intermittent blowdown, volumes are small, dissolved solids are controlled by frequency and duration of blowdown, and settled sludge is properly removed. Continuous blowdown is preferable because it provides steady control of dissolved and suspended solids concentration in boiler water. It can be used if the continuous blowdown rate is above the minimum of 200 l/hr (1 gpm). Heat recovery is also easier with continuous blowdown. Intermittent blowdown, however, is still required at longer intervals to remove sludge accumulation. Antifoam agents permit higher dissolved solids concentrations in the boiler, decreasing the rate of blowdown.

Spent Caustics from Refineries

Most refineries use caustic treatment for various product streams to remove hydrogen sulfide, mercaptans, phenolic compounds, thiophenols (sulfur-bearing aromatics), and naphthenic acid impurities. Caustic treatment of catalytically and thermally cracked gasoline produces a spent caustic that is rich in phenolates and thiophenolates, with lesser amounts of sulfides and mercaptides, commonly referred to as *phenolic caustic*. Treatment of fuel gas, LPG, and straight run gasoline produces *sulfidic caustic,* rich in sulfides and mercaptides, with only small amounts of phenolates. Both types of spent caustics emit a foul odor and pose serious disposal problems. If dumped into a receiving stream, they will discolor water, impart an objectionable taste, exert a high chemical oxygen demand, and poison fish.

PHENOLIC

Several methods are utilized for recovery and disposal of spent phenolic caustics. Table 8.4.1 briefly describes and compares the major methods.

The best approach to phenolic caustic disposal is to recover the valuable cresylic acids (phenol, cresols, xylenols) and thiophenols for use as base materials in making plastics, wire insulation, lubricating oil additives, rubber reclaiming agents, and other products. Recovery can be achieved by neutralizing (springing) spent caustic with flue gas or mineral acids, forming an aqueous salt solution from which the cresylic acids and thiophenols can be decanted. A typical flue gas neutralization system is shown in Figure 8.4.1.

Both neutralization methods present obstacles. The sodium salt solution is too concentrated for release to a freshwater stream; in addition, it normally contains 5,000

TABLE 8.4.1 DISPOSAL OR RECOVERY OF PHENOLIC SPENT CAUSTICS

Method[a]	Description	Operating Information or Conditions	Beneficial Characteristics	Possible Adverse Characteristics
Sale of Phenolic Caustics[b]	Phenolic caustics are sold to a commercial plant specializing in the recovery of valuable cresylic acids[c], thiophenols, aromatic disulfides, and sodium salts.	Concentrated caustic (25–30°Be') should be used for gasoline treatment to minimize spent caustic shipping charges. Phenolic caustics should be segregated from sulfidic caustics.	Phenolic caustics can be sold at a profit for the refiner. A commercial recovery plant is more capable of preventing pollution problems from developing during processing of spent caustics.	Shipping charges may be excessive for some locations. Gasoline treatment facilities may need to be revised in some refineries before switching to concentrated virgin caustic.
Flue Gas Neutralization[d]	After contacting with flue gas in a batch or continuous system, cresylic acids and thiophenols phase separate from the Na_2CO_3 which is formed. Sulfides and mercaptans are carried overhead with the flue gas.	Neutralization tower operates at about 180°F and 2–5 psig. Flue gas and steam strip most of the H_2S and mercaptans from solution. Final pH is above 8.5.	Flue gas is more economical than mineral acids. Cresylic acids and thiophenols are recovered for sale. The final pH is above 8.5 and corrosion problems are not as severe as when using mineral acids.	The Na_2CO_3 solution is usually contaminated with phenols, thiophenols and H_2S. If this solution cannot be sold, a serious disposal problem still exists. The process is subject to foul odors. Burning of H_2S and mercaptan off gases produces SO_2 emissions.
Mineral Acid Neutralization[e]	After neutralization, cresylic acids and thiophenols phase separate from the sodium salt solution. Considerable H_2S and mercaptans stay in solution, unless stripping steam is used. Process can be batch or continuous.	The pH is usually reduced to 3–4. Stripping steam is used to ensure maximum liberation of H_2S and mercaptans. Operating pressure is essentially atmospheric.	Valuable cresylic acids and thiophenols are recovered for sale. Resulting sodium salt solution may be contaminated than with flue gas neutralization.	Low pH requires corrosion resistant construction material. Reagent costs are higher. If the contaminated salt solution cannot be sold, a serious disposal problem still exists. This process has the same odor and SO_2 problems as flue gas neutralization.
Burning	Phenolic caustics can be burned simultaneously with other oily wastes in a fluid bed incinerator.	Incinerator operates at 1200°–1500°F. Na_2CO_3, Na_2SO_4, and ash are withdrawn from the incinerator and hauled to a landfill.	Sulfur, in the thiophenols, H_2S and mercaptans, reacts to form Na_2SO_4. Odors are not a problem since complete combustion occurs.	Installation of a fluid bed incinerator is very expensive. Flue gas must be cleaned to prevent particulate emissions. Na_2CO_3 and Na_2SO_4 may leach into the soil during rainfall.
Deep Well Disposal	Spent caustics are pumped into a deep underground formation with impervious rock above and below.	Caustics must initially be tested to determine if they are compatible with the formation.	Daily operating attention is minimized. There are no by-products to be sold.	Groundwater contamination is possible if formation is not tightly sealed. Plugging may occur in the formation if chemical reactions occur.
Biological Oxidation	Small quantities of dilute caustics can be mixed with contaminated wastewaters and be biologically degraded.	Aerated lagoons, activated sludge, and trickling filter units can be utilized.	This method is limited to small quantities of dilute caustics.	Spent caustics are toxic to biological organisms at low concentrations. This method increases dissolved solids in the plant effluent and creates severe odor problems.
Ocean Dumping	Spent caustics are dumped about 100 miles offshore in federally approved areas.	Wastes are diluted with sea water or discharged so they sink.	Low cost	Spent caustic is toxic to some sea creatures if concentrated sufficiently.

[a]Methods are listed in descending order of overall ability to recover or dispose of phenolic caustics.
[b]The Merichem Company purchases phenolic caustics.
[c]Cresylic acids = phenols, cresol isomers, and xylenol isomers.
[d]Some $NaHCO_3$ is formed simultaneously with Na_2CO_3.
[e]H_2SO_4 or HCl can be utilized.

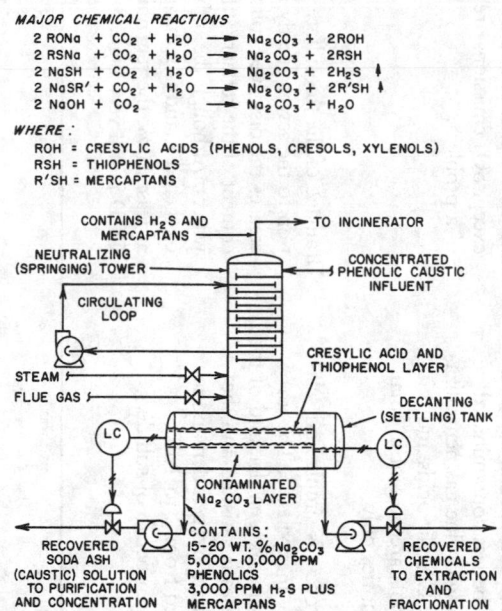

MAJOR CHEMICAL REACTIONS

$$2\,RONa + CO_2 + H_2O \longrightarrow Na_2CO_3 + 2ROH$$
$$2\,RSNa + CO_2 + H_2O \longrightarrow Na_2CO_3 + 2RSH$$
$$2\,NaSH + CO_2 + H_2O \longrightarrow Na_2CO_3 + 2H_2S\downarrow$$
$$2\,NaSR' + CO_2 + H_2O \longrightarrow Na_2CO_3 + 2R'SH\downarrow$$
$$2\,NaOH + CO_2 \longrightarrow Na_2CO_3 + H_2O$$

WHERE:

ROH = CRESYLIC ACIDS (PHENOLS, CRESOLS, XYLENOLS)
RSH = THIOPHENOLS
R'SH = MERCAPTANS

FIG. 8.4.1 Typical flue gas neutralization system for phenolic caustic.

to 10,000 (Beychok 1967) ppm phenolics and 3,000 ppm H_2S. Therefore, it must be purified and concentrated into a salable product. H_2S and mercaptan off gases may create air pollution in the form of sulfur dioxide if burned. These gases, plus thiophenols, create severe odor problems if leaks or spills occur.

These problems can be solved by large commercial plants specializing in the recovery and marketing of phenolic caustic products. Such firms purchase concentrated spent caustics from numerous refineries. In most cases the refiner can actually profit from the sale of the waste materials (Price 1967).

In caustic treatment, a fluidized bed incinerator converts caustic to sodium carbonate and sodium sulfate without creating air pollution (EPA).

SULFIDIC

Sulfidic spent caustics present a more difficult disposal problem. Recovered products have little or no market value, especially if contaminated with undesirable materials. A summary of the available treatment and disposal processes is given in Table 8.4.2. One popular disposal method is to oxidize the sulfides to thiosulfate and the mercaptans to disulfides (API 1969), reducing toxicity and oxygen demand significantly. The disulfides are decanted and the thiosulfate solution is released to the plant sewer system. However, this increases dissolved solids concentrations in refinery effluent waters.

Steel Mill Pickle Liquor

For a metallic surface to receive a high quality protective coating, it must be cleaned. Steel surface cleaning usually involves a detergent wash, a rinse and/or acid wash, followed by a rinse. In this process, called pickling, the acid wash cuts through surface oxide layers to expose bright base metal. Pickling, continuous or batch prepares a surface suitable for plating, galvanizing, and other surface treatments. Since about 50% of integrated steel mill products may be acid pickled, and because most plating lines utilize acid pickling, the steel industry accounts for most acid consumption in the United States.

THE PICKLING PROCESS

Sulfuric acid was historically the acid of choice; however, hydrochloric acid has displaced it. Hydrochloric acid, although more expensive, pickles much faster than sulfuric acid, with less base metal loss. New automatic high-speed steel mills require the integration of rapid systems, including pickling. Operating speeds of 600 ft/min are reported.

Sulfuric acid is fed to the pickling bath at a concentration of about 20% and it is considered spent when half its acid value is replaced by ferrous sulfate, $FeSO_4$. Hydrochloric acid fed at about 20% concentration is almost completely consumed before it is "spent." The spent liquor is estimated as 35–45 lb of pickling acid per ton of steel, resulting in 8–15 gal of spent pickle liquor per ton of steel. For an annual steel production of 50 million tons, an estimated 500 million gal of spent liquor are produced annually.

The pickling area is followed by the rinsing-neutralization area. This may consist of several baths in series, with pickle liquor flowing cocurrent or countercurrent to the steel. Pickle bath heating may be performed by steam injection, which causes some dilution of the liquor, or the steel ware may be preheated. Heated tanks, maintained at temperatures as high as 200°F, tend to concentrate the liquor due to evaporation. Balancing the acid makeup and drag-out, and returning some drag-out to the pickle tank can control concentration.

During sulfuric acid pickling, hydrogen is released, which attacks the fresh surface and causes embrittlement. Chemicals to pacify the surface or inhibit the rate of hydrogen attack can be used. In addition, chemicals called *accelerators*—which promote the removal of scale without affecting the rate of base metal attack—are utilized. The spent sulfuric acid pickle liquor, half iron and half free acid, may also contain a variety of additional chemicals.

Hydrochloric acid does not cause hydrogen embrittlement and has a high intrinsic pickling rate with little base metal attack; additives are not usually needed. Additional advantages of hydrochloric acid include:

- no need for scale breaking
- better surface finish
- faster reaction rate than sulfuric acid
- prevention or reduction of overpickling due to slower attack rate on base metal

TABLE 8.4.2 DISPOSAL OR RECOVERY OF SULFIDIC SPENT CAUSTICS

Method[a]	Description	Operating Information or Conditions	Beneficial Characteristics	Possible Adverse Characteristics
Sale of Sulfidic Caustics[b]	Sulfidic caustics are sold to a commercial plant which recovers sodium sulfide, sodium hydrosulfide and disulfides.	Concentrated caustic (25°–30°Bé') should be used in the refinery to minimize shipping costs for spent caustics. Sulfidic caustics should be segregated from phenolic caustics.	Sulfidic caustics may not return a profit for some refineries; however, sales should pay for the transportation charges and eliminate the need for in-plant disposal facilities.	Shipping costs may be prohibitive for some locations. Caustic treating facilities within the refinery may need to be revised before switching to concentrated virgin caustic.
Burning	c	c	c	c
Deep Well Disposal	c	c	c	c
Continuous Regeneration	A light hydrocarbon stream is in continuous contact with caustic in an absorption tower. The spent caustic is regenerated by contact with steam in another tower.	Typical operating conditions in the regenerator are 215°–240°F and 1–10 psig. Mercaptan off gases released during regeneration are burned.	Process is simple and relatively inexpensive.	Process is limited to absorption of mercaptans. Any sulfides absorbed will not be removed during regeneration. Consequently, caustic eventually becomes spent and poses a disposal problem.
Flue Gas Neutralization[d]	Contact with flue gas liberates H_2S and mercaptans with simultaneous formation of a sodium salt solution. Only small amounts of cresylic acids and thiophenols are present.	Operating conditions are the same as for phenolic caustics. However, the quantity of H_2S and mercaptans in the overhead gas is much greater. Sulfur recovery is necessary in most cases.	Same as for phenolic caustics, except cresylic acids and thiophenols are present in small quantities only.	Same as for phenolic caustics, except SO_2 emissions are more of a problem.
Mineral Acid Neutralization[e]	Neutralization and steam stripping liberate H_2S and mercaptans with simultaneous formation of a sodium salt solution. Only small amounts of cresylic acids and thiophenols are present.	Operating conditions are the same as for phenolic caustics. However, the quantity of H_2S and mercaptans in the overhead gas is much greater. Sulfur recovery is necessary in most cases.	Resulting sodium salt solution *may* be less contaminated than with flue gas neutralization. Cresylic acids and thiophenols are present in small quantities only.	Same as for phenolic caustics except SO_2 emissions are more of a problem.
Oxidation	Sulfidic caustics are continuously in contact with air and steam in a packed column. Sulfides are oxidized to thiosulfate and mercaptans are oxidized to disulfides.	Typical operating conditions are 165°–225°F and 60–85 psig. Disulfides phase separate and are decanted. The thiosulfate solution is then released to the sewer system.	The process is simple and relatively inexpensive. The intermediate oxygen demand of the sulfides is satisfied and the ultimate oxygen demand is reduced from 2 to 1 lb of oxygen per lb of sulfide.	Drainage of the thiosulfate solution to the sewer greatly increases the dissolved solids content of the refinery effluent waters. Although the ultimate oxygen demand is reduced at least 50%, the remaining COD is still very high.
Ocean Dumping	c	c	c	c

[a]Methods are listed in descending order of overall ability to recover or dispose of sulfidic caustics without creating additional pollution problems.
[b]The Merichem Company purchases sulfidic caustics.
[c]Same as for phenolic caustics. Refer to Table 8.4.1.
[d]Some $NaHCO_3$ is formed simultaneously with Na_2CO_3.
[e]H_2SO_4 and HCl can be utilized.

- lower acid content of spent acid
- improvement of subsequent processes due to solubility of compounds formed during pickling
- cleaner rinsing
- better drying
- improved overall quality of steel produced

Hydrochloric acid pickling is not without problems. The acid is more expensive than sulfuric acid, and more corrosive. When converting from sulfuric acid to hydrochloric acid operation, storage, pickle tanks, and fume hoods must be replaced.

DISPOSITION OF SPENT LIQUOR

Spent pickle liquor disposal methods include discharge to a waterway, hauling by a contractor, deep well disposal, recovery of acid values, neutralization, and regeneration of both acid and iron values.

Recovery of Acid Values

Recovery of acid values refers to the removal of iron salts from the spent liquor by concentrating then cooling the liquor, which crystallizes the ferrous salts. The resulting mother liquor contains less iron, and its acid strength is returned to operating levels by the addition of concentrated acid.

Ferrous sulfate crystallizes as the heptahydrate. If the crystal is collected, dehydrated to the monohydrate, and added in excess to the sulfuric acid spent liquor, the liquor dehydrates and the heptahydrate forms and can be crystallized out of the cooled liquor.

Gaseous chlorine can be fed into the spent hydrochloric acid liquor to oxidize ferrous iron to the ferric state and crystallize the ferric chloride.

Recovered iron salts can be used as soil stabilizers and in water and wastewater treatment plants.

Neutralization

Neutralization is performed in an aqueous solution, forming an iron hydroxide that may be recovered for return to steel melting furnaces. With neutralization, iron values are recovered as iron oxide.

Lime, as ground limestone, calcium carbonate, or cement clinker flue dust; lime slurry from acetylene manufacture; and powdered slag from electric furnaces have been used to treat sulfuric acid liquors. If the liquor is neutralized, a mixture of gypsum ($CaSO_4 \cdot 2\,H_2O$) and iron oxide is formed, which can be separated if the pH is controlled. Neutralization to a pH of 0.6 to 2.0 results in the production of gypsum (hydrated calcium sulfate) which precipitates out or is centrifuged from the liquor containing most of the iron. When the pH is increased to a range of 6 to 10, the iron is precipitated as ferrous hydroxide

and can be separated from the additional gypsum produced by a hydrocyclone or differential settling. This finer crystal can then be used to seed first stage crystallization. Titration curves (Figure 8.4.2) illustrate the neutralization characteristics of iron in its two oxidation states.

A substantially more compact iron precipitate is produced from ferric hydroxide. Figure 8.4.3 shows the effect of iron oxidation state on chemical sludge settling rate. The solubility product of $Fe(OH)_2$ is 1.64×10^{-14}, whereas that of $Fe(OH)_3$ is 1.1×10^{-36}.

Ferrous iron is oxidized to ferric iron by blowing air, oxygen, and nitrous oxide (prepared from the catalytic oxidation of ammonia) through it. This oxidation can be carried out on the hot spent liquor or the cooled colloidal suspension of ferrous hydroxide, resulting in black magnetic iron oxide (Fe_3O_4), ferric oxide (Fe_2O_3), or ferriferrous oxide.

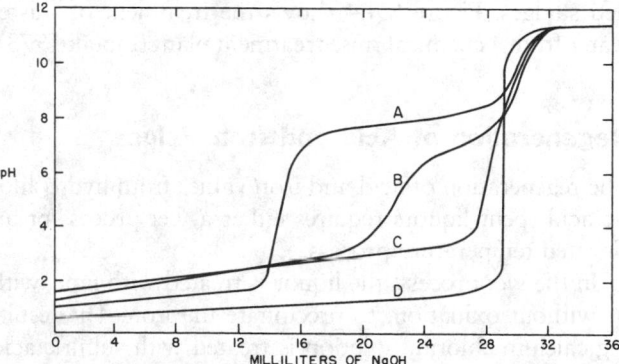

FIG. 8.4.2 Effect of oxidation state of iron on neutralization characteristics of waste pickle liquor as exhibited by titration with sodium hydroxide. *Key:* A = curve for mixture of ferrous sulfate and sulfuric acid; B = curve for mixture of ferrous sulfate plus ferric sulfate plus sulfuric acid; C = curve for mixture of ferric sulfate and sulfuric acid; D = curve for pure sulfuric acid

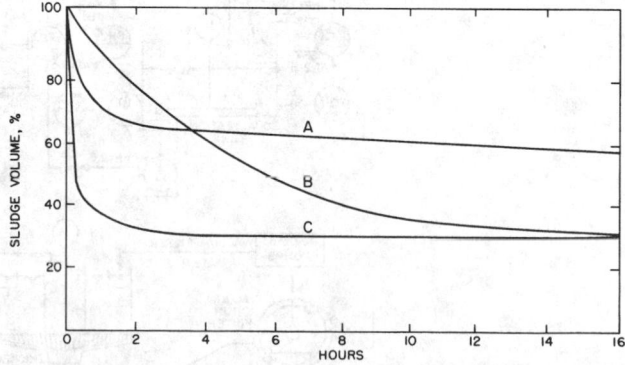

FIG. 8.4.3 Effect of oxidation state of iron on sludge volume produced upon neutralization of waste pickle liquor by sodium hydroxide. *Key:* A = curve for 100 percent ferric iron, 0 percent ferrous iron; B = curve for 0 percent ferric iron, 100 percent ferrous iron; C = curve for 65 percent ferric iron, 35 percent ferrous iron

Ammonia neutralization involves a single step process wherein the sulfate is recovered as ammonium sulfate crystal, which can be used as fertilizer.

Spent hydrochloric acid liquors can be neutralized with calcium compounds or ammonia. Sodium hypochloride is also used as an oxidizing agent and to raise the pH to precipitate a ferric oxide.

The small pickler (using a batch, not a continuous, pickling process) will first concentrate spent liquor by a factor of 10; after adding lime, the heat of neutralization vaporizes the water, drying the mass into a pourable solid. The steam from the initial concentration should be condensed and reused for pickling, because it can pollute if released to the atmosphere.

Problems associated with neutralization include oxidation rates from ferrous to ferric iron, settling kinetics of the two hydroxides and their mixtures, separation of neutralization products from liquors, and final disposition of residues obtained. Synthetic flocculants may be of value in settling the hydroxide and in vacuum filtration of the settled sludges. Figure 8.4.4 shows the treatment of wastewater from a chemical rinse treatment plant (Lipták 1973).

Regeneration of Acid and Iron Values

The regeneration of acid and iron values from hydrochloric acid spent liquors requires either a wet process or an elevated temperature process.

In the wet process, the liquor is treated with lime, with or without oxidation, to precipitate the iron. The resulting calcium chloride solution is treated with sulfuric acid to precipitate gypsum and to produce hydrochloric acid for recycle to the pickle line.

The thermal processes are usually two-step. The first step involves drying the ferrous chloride by evaporation of free water and hydrogen chloride; it is then oxidatively decomposed to Fe_2O_3 or Fe_3O_4 with further release of the chloride (as hydrogen chloride) by the following endothermic reaction:

$$2\ FeCl_2 + 2\ H_2O + \tfrac{1}{2}\ O_2 \longrightarrow Fe_2O_3 + 4\ HCl \qquad 8.4(2)$$

In some thermal units, the two processes occur simultaneously, while in others the processes occur in two different areas, which are heated to different temperatures. Evaporation can be performed at 500° to 600°C. Roasters, fluidized beds, and multiple-hearth incinerators are used, producing an iron oxide that is removed as an ash from the vapor stream by cyclones. The water and hydrogen chloride in the vapor are condensed to produce 20% hydrochloric acid for recycling to the pickling line. Various economizers are used to cool the leaving gases and to preheat the air and spent liquor feed.

Miscellaneous Treatment Methods

Ion exchange processes exchange ferrous iron for hydrogen ions. Regeneration results in a concentrated iron salt that can be processed by one of the several methods described earlier. Electrodialysis segregates the returnable liquor from a concentrated iron salt. Ion exchange membranes separate the various compartments in this process.

Spent nitric, phosphoric, and hydrofluoric acid liquors and their various mixtures can be treated by one of the several processes already described. Lime will remove fluorides, but hydrogen fluoride can be distilled from an acidic solution at about 200°C. If silicon enters the liquor, fluorosilicic acid will also distill off with the hydrogen fluoride.

—*R.A. Conway, R.G. Gantz, S.E. Smith*

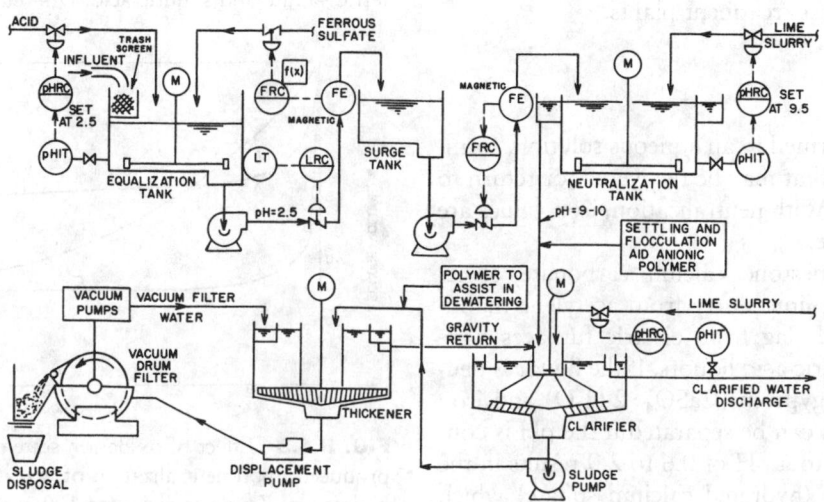

FIG. 8.4.4 Treatment of wastewater from a chemical rinse treatment plant. *Key:* pHRC = pH recording controller; pHIT = pH indicating transmitter; FRC = flow recording controller; FE = flow element; LRC = level recorder controller; LT = level transmitter

References

American Petroleum Institute (API). 1969. *Manual on disposal of refinery wastes*. Volume on liquid wastes. Chapter 11, p. 6.

Beychok, M.R. 1967. *Aqueous wastes from petroleum and petrochemical plants*. London: Wiley.

De Lorenzi, O. (ed.). 1951. *Combustion engineering* (1st ed). New York: Superheater, Inc.

Hamer, P., J. Jackson, and E.F. Thurston. 1961. *Industrial waste treatment practice*. London: Butterworths.

International Union of Pure and Applied Chemistry (IUPAC). 1963. *Reuse of water in industry*. A Contribution to the Solution.

Lipták, B.G. (ed.). 1973. *Instrumentation in the Processing Industries*. Chapter VI. Philadelphia, Pa.: Chilton.

Parsons, W.A. 1965. *Chemical treatment of sewage and industrial wastes*. National Lime Association.

Price, A.R. 1967. Turn treating costs into profit. *Hydrocarbon Processing* 46(9): 149.

U.S. Environment Protection Agency (EPA). Fluid Bed Incineration of Petroleum Refinery Wastes. Project No. 12050EKT. Washington, D.C.: U.S. Government Printing Office.

8.5
OIL POLLUTION

Effects on Plant and Animal Life

About one million metric tons of oil are dumped each year into the sea (Audubon 1971) from shipping accidents in port alone. Such large spills affect the health of the marine ecosystem. Ninety percent of atmospheric oxygen is manufactured in the sea by phytoplankton. The sea also furnishes food and minerals and controls the weather.

Oil pollution also concerns those who take their living from the sea. When sea animals are killed or made dangerous to man, fishing is curtailed. In areas of major spills, fishing and hotel industries are affected.

TOXICITY

Effects on the marine community may be immediate or long range. Immediate effects include the killing of up to 95% of organisms in the area of a spill or leak. Oil coats surface organisms and may also coat benthic or bottom-dwelling organisms. Wave action and currents spread oil horizontally, mixing hydrocarbons with water. Even after external manifestations are gone, oil may still be found in the organisms. Bacteria detoxify poisons in the oil, but they influence the least toxic products first, rendering the remaining oil more poisonous than before.

Oil absorbed by marine organisms is stored in the organism's lipid pool and is not available for further degradation. It will only be destroyed when the animal dies. This oil may be passed on through the food chain as one animal preys upon another, ultimately reaching man (Figure 8.5.1). Oil pollution also destroys delicate sea habitat. As plants die, currents cause erosion. An animal can never recover if the habitat is gone. Oil causes especially severe damage to salt marshes and estuaries.

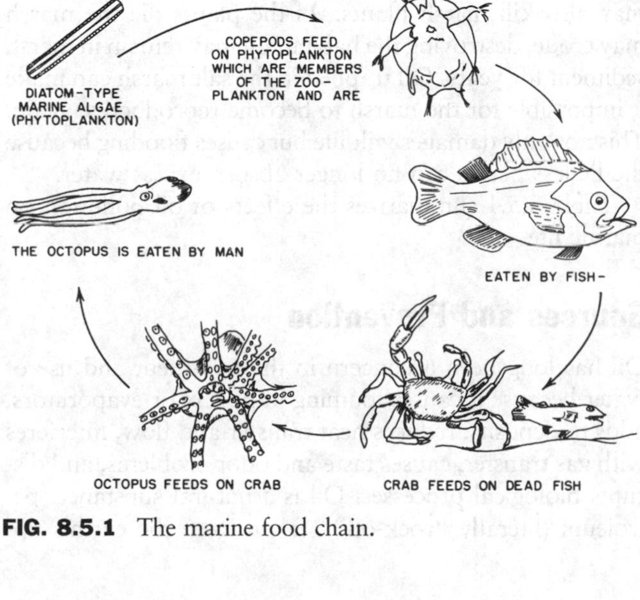

MARINE ORGANISMS

Mammals

Few mammals actually live in the sea—whales, seals, and sea otters are the only important mammalian members of the marine community. Like all mammals they possess body hair for insulation. Oil pollution causes the hair to become matted, reducing its effectiveness as a thermal insulator. The animal either freezes or succumbs to diseases because of lowered resistance. Mammals are also affected by oil through the food chain. This is especially important in whales since they are plankton feeders. Reduction in the plankton population may cause whales to relocate or starve.

Birds

Oil mats the feathers so that the birds cannot fly or stay afloat. This matting also hampers insulation, causing many

COPEPODS FEED ON PHYTOPLANKTON WHICH ARE MEMBERS OF THE ZOO-PLANKTON AND ARE

DIATOM-TYPE MARINE ALGAE (PHYTOPLANKTON)

THE OCTOPUS IS EATEN BY MAN

EATEN BY FISH—

OCTOPUS FEEDS ON CRAB   CRAB FEEDS ON DEAD FISH

FIG. 8.5.1 The marine food chain.

birds to freeze. The birds try to preen and ingest the oil, which can poison them. The oil can also cause blindness if it penetrates the eyes. Birds that manage to return to their nests in breeding season carry oil to the eggs, preventing them from hatching. If the bird survives the initial onslaught of the oil, it may die later because of lowered resistance or habitat damage. Only 3–5% of birds survive oil pollution even if they receive treatment (White and Blair 1971).

Fish

Much oil pollution occurs in offshore waters vital to fish production. In one experiment, fish larvae that came into contact with oil hatched prematurely and died. Nutrition in fish is affected by oil in two ways, including blocking of taste receptors, and imitating the natural chemical messages that alert fish to their prey. Respiration is interfered with either by blocking the gills with oil or by effects of trace metals and hydrocarbons on brachial cells. Emulsified oil is more harmful to fish than surface oil.

Mollusks

As members of the benthic (bottom) community, mollusks are more harmed than other marine organisms. They live in large beds on the ocean floor and have no way to escape the oil. In the immediate area of oil pollution, mollusk beds are smothered as oil is spread vertically. In one area, long-range pollution caused mussels to become sterile. Mollusks tend to assimilate hydrocarbons into their lipid pool and store them. Oysters removed to unpolluted water do not show any reduction in unnatural hydrocarbon levels. These stored hydrocarbons are passed through the food chain and can ultimately be passed to man.

Zooplankton

Zooplankton are microorganisms living in the upper surface of the ocean. They include the larvae of fish, lobsters, and other marine invertebrates and are the food source for the majority of larger ocean organisms. In the immediate area of oil pollution, plankton are killed. This has effects beyond the polluted area, since many fish larvae in plankton are migratory species. Exposure to diesel oil increases the death rate of exposed organisms.

Benthic Organisms

Ocean bottom dwellers cover a wide range of organisms, including crustaceans, corals, sponges, sea anemones, worms, amphipods, and mollusks. They are smothered by oil sinking to the ocean floor. Bottom dwellers are also adversely affected when the bottom plants can no longer hold back erosion, causing the bottom habitat to be destroyed. The amphipods of the family Ampeliscidae are highly sus-

ceptible to oil pollution. On the other hand, the annelid worm *Capitella capitata* is not. It was noted that after the amphipods were dead and unable to repopulate because of long-term oil pollution, the *C capitata* population was increasing. In one typical oil spill area, a trawl in 10 ft of water disclosed that 95% of bottom life was dead or dying. Organisms included lobsters, crabs, snails, and clams.

PLANTS AND OIL

Algae

Algae, the counterpart of grass, plants, and trees in the water, provides food and shelter and helps to prevent erosion of the sea bottom. As oil spreads, it covers the algae and prevents photosynthesis. As the algae deteriorates, so does the environment. Erosion not only destroys the bottom habitat; it spreads the oil further. A small concentration of oil causes the destruction of native algae and permits foreign species to take hold, lowering the quality of the environment.

Phytoplankton

Phytoplankton consists of microscopic algae living on the ocean surface. These tiny plants manufacture 90% of the atmospheric oxygen and also form the basis of the marine food chain. High oil concentrations will kill phytoplankton. Lower concentrations stop cell division and cause slower death. A 19% extract of crude oil retarded cell division in *Phaedactylum tricornatum* after a 4-day exposure. It was observed that cell membranes were damaged by hydrocarbons so that the cytoplasm leaked out.

Marsh Plants

Oil washed into tidal marshes can destroy rushes, sedges, and grasses if the area is covered. Oil clogs the stomata and makes photosynthesis impossible. Poisons in the oil may also kill marsh plants. As the plants die the marsh may erode, destroying the habitat; oil may remain in marsh sediment for years. Oil trapped in the salt marsh can make it impossible for the marsh to become reproductive again. This not only damages wildlife but causes flooding because the lifeless plants can no longer absorb excess water.

Table 8.5.1 summarizes the effects of oil pollution on marine life.

Sources and Prevention

Oil has long been a concern in the treatment and use of water because it causes foaming in boilers or evaporators, adds to deposits, reduces heat transfer and flow, interferes with gas transfer, causes taste and odor problems, and disrupts biological processes. Oil is a natural substance, petroleum (literally "rock-oil"), taken from the earth's up-

TABLE 8.5.1 OIL POLLUTION EFFECTS ON MARINE ORGANISMS

Organism	Immediate Area of Spill	Periphery of Spill	Possible Long-Range Damage
Zooplankton	Killed	Increased death rate	Unknown
Benthic Organisms	Smothered	Disorder in marine communities	Unknown; possibility of stored hydrocarbons building up in the food chain
Molluscs	Smothered	Shellfish are sterilized; hydrocarbons are stored in the muscles	Unknown; possibility of stored hydrocarbons building up in the food chain
Fish	Hatch prematurely; if in stage may also be killed	Nutrition is upset	Unknown; possibility of stored hydrocarbons building up in the food chain
Birds	Blindness; birds may also freeze	Resistance is lowered	Unknown; possibility of stored hydrocarbons building up in the food chain
Mammals	May freeze	Resistance is lowered	Unknown; possibility of stored hydrocarbons building up in the food chain
Phytoplankton	Killed	Retardation of cell division; damage to cell membranes	Unknown; possibility of stored hydrocarbons building up in the food chain
Algae	Killed	Replacement of native species with resistant foreign species	Unknown; possibility of stored hydrocarbons building up in the food chain
Higher Plants	Killed	Gradual deterioration of the environment due to storage of oil in the sediment	Unknown

per strata, and its by-products have many beneficial uses. Some are called oils, such as fuel oil. However, oily materials are also derived from coal, animal, and vegetable matter and are made synthetically. Oil is also a paraffin or hydrocarbon, i.e., composed mainly of the elements carbon and hydrogen.

Although no definition is complete, oily materials have several common properties. They are generally liquid at room temperature and are less dense than and usually not miscible with water; they spot brown paper and are flammable; they tend to spread on water, producing slicks; and they are persistent and can produce troublesome emulsions.

The analytical method for measuring the amount of oily material in a substance also defines the substance. The most useful methods involve extracting the materials into a solvent, such as hexane, from a quantity of water. The separated solvent layer is evaporated, leaving residues of oily material related to its concentration in the sample. These residues are *solvent extractables*. Examples of other solvents are carbon tetrachloride, chloroform, benzene, and dichlorodifluoroethylene. Each solvent has somewhat different results owing to differences in properties and in the nature of the materials extracted. Non-oily matter such as organic acids and esters may also be extracted. Phenols, resins, sulfur, and some dyes can be included. Lighter hydrocarbons, if present, may volatilize when the solvent is evaporated.

Other methods of analysis employed are volumetric, refractive index, thin-layer and gas chromatography, infrared and ultraviolet spectrophotometry (Lipták 1972). Samples for analysis must be representative of the water under study. The tendency of oily materials to float, stick to surfaces, and separate from water makes taking good samples difficult, and it is in this area that many automatic samplers fail. Adding mixing chambers at sampling points in lines aids in proper sampling. Water standards that set quantitative limits on oil or oily matter should also specify the method of measurement. If a standard is based on an oily appearance, i.e., evidence of a floating layer, slick, or iridescence, then the term extractables has little meaning.

OILY MATERIALS

Oily materials may be classified according to their properties. Physically, light hydrocarbons or solvents, such as kerosene or gasoline, are less viscous and more volatile than heavy hydrocarbons like tars and residual fuel oils. Chemically, some materials are classed as aliphatic, i.e., straight or branched chain, saturated or unsaturated. Still others are classed as aromatic (unsaturated and with ring structure). They may have chemical functionality. Their structure may contain acid, carbonyl or other functional groups and have elements other than carbon, hydrogen and oxygen, such as nitrogen or sulfur.

Oily materials can be classified as to use, such as fuels, lubricants, coatings, cleaners, solvents, cutting and rolling fluids, hydraulic fluids, carriers, and cooking fats and oils. Many uses are both domestic and industrial. Light hydrocarbons and solvents are found in industrial waste streams, as a result of degreasing or extraction, cleaning, painting, and coating operations. Their vapors represent potential fire and explosion hazards, and their presence makes the removal of heavier oily materials more difficult. The pollution potential of all hydrocarbons is moderately high; however light hydrocarbons are more readily oxidized biologically than heavier fuels, tars, and residues. They are also a potential source of air pollution.

Industrial Sources

One of the principal industrial sources of oily wastes is the petroleum industry. Oily wastes result from producing, refining, storing, or transporting operations, or in the use of this industry's products. Another major oil source is the metals industry. Most oily wastes result from metalworking or forming operations. Oily materials lubricate and cool instruments or the metals being worked. Emulsified oily materials and finely divided suspended solids make these wastes difficult to handle.

Coke plants generate much oily wastewater, most of which is derived from cooling, quenching coke, or scrubbing gases. Much of the contaminated water is reused or consumed elsewhere in the mill. The wastewater contains phenols, cresol, and related extractable materials. Phenolic-type extractable materials may also be found in water used to cool or wash cupola stack gases at foundries. In processing meat, fish and poultry, oily wastes are produced from cleaning, slaughtering, and processing by-products. A major source of oily wastes is the rendering process. Cooking plant tissues, seeds, grains, and nuts aids in extracting their oils for commercial purposes; cooking and extraction processes result in oily wastewaters.

Most oily wastes in the textile industry result from scouring fibers in an early process step, especially scouring wool. The waste liquor yields valuable lanolin, but the wastewater is also high in extractables and difficult to process.

Oily wastes in the transportation industry result from leaks, spills, or cleaning operations. Tankers, barges, and tank trucks transporting oily materials must be cleaned to prevent possible product contamination. The cleaning solutions contain oily materials and create pollution if discharged without treatment. Latex in wastewaters may be extractable. Latex in the rubber industry is generally removed from waste streams with little difficulty. However, in the paint industry the presence of solvents, resins, and emulsifiers can make removal very difficult.

Some oily materials are introduced into water systems during heating or cooling steps. Oily materials may be derived from leaks in seals, condensers, or heat exchangers from the process side of the equipment. When steam used for direct heating of fatty or oily products, the recovered condensate will likely be contaminated. Run-off from industrial areas following storms may be contaminated with oily materials. The rain washes processing units, walkways, buildings, and surrounding grounds, carrying away oily materials deposited there.

Municipal Sources

The major sources of oily wastes are from food preparation, garbage disposal, and cleaning. Cleaning includes laundry, car washing, and general cleaning jobs where water is the main solvent and carrier. Grease and oily materials are removed at sewage treatment plants. Road oil and degraded asphalt are washed from roads into storm sewers and streams. Rainwater also contains soot and various hydrocarbons washed from the faces of buildings and other structures in communities.

Natural Sources

Coniferous trees and shrubs contribute oily materials to run-off water, particularly in areas wooded with pine trees.

DETECTION, IDENTIFICATION, AND SURVEYS

A survey of potential sources of oily materials is initiated with an inventory of known oily products or by-products used or produced in the area. Processes, machinery, and storage areas are checked for leaks, drips, and potential for spills or accidental contamination. Both cooling and condensate-return water are examined. Stack gases are tested for hydrocarbons and soot. Cleaning and wash-down procedures are observed; run-off water is tested and methods of handling the water noted; and potential problems from startup and shutdown of processes or equipment are probed. Field personnel should be familiar with each process, system, unit, and individual piece of equipment. In certain cases material balances are helpful.

Oil-soluble dyes help to check for leaks in complex systems or to locate oil-water interfaces. Ultraviolet light detects as little as 0.02 ppm of some oils by fluorescence in the dark. Dip sticks and sonic probes are useful to check for multiple layers in tanks and sumps. When an oily material is isolated, its source can often be identified by infrared analysis or gas chromatography. Analysis of trace metal content may also help indicate its source. Many oily materials are identified by their characteristic odor. More and more oily materials will be *tagged*[1] for easier identification.

PREVENTION

Early process control can reduce the quantity of treatable waste. Segregation as a part of early control can simplify treatment processes. The presence of emulsifiers, wetting agents, soaps, deflocculants, and dispersants, as well as finely divided suspended solids, makes separation of oily materials and the treatment of wastes more difficult. Advantages can be realized from high temperatures and low pH levels, and the presence of substances that make necessary pH adjustments impractical may be avoided.

Control at the source may reduce raw material and product losses. Oily materials recovered downstream in a treatment plant are usually more contaminated and require more costly refining for reuse. The ability of one process over another to produce less oily or more easily handled wastes must be considered in initial planning, along with plant site selection, availability of suitable raw materials, labor, product markets, utilities, transportation, and wa-

ter supply. Questions concerning waste treatment methods and facilities must be thoroughly examined. Plant layout must be the best compromise between efficient production, storage of materials, and segregation, collection, and treatment of wastes.

In selecting plant equipment, pollution potential should be considered in addition to cost, performance, and service life. Raw materials, processing aids, and cleaners should be selected to simplify oily waste treatment problems. Cleaners containing oils or solvents that might end up in waste streams should be avoided. Materials that form difficult-to-break emulsions should be avoided. Impurities in raw materials can add to oily waste problems, yielding bottoms, tailings, or unusable by-products from production. Poor quality raw materials may lead to off-spec finished products, which must be wasted, blended off, or sold at lower profit. If an oily waste or used solvent cannot be reused, it may be taken by a jobber for re-refining or by a scavenger. Finally, it may be burned, perhaps recovering its heating value. In general, wastes containing high concentrations of oily materials should not be discharged to lagoons or through deep well disposal.

Often oily materials recovered early in a process can be reused following a simple cleanup step such as filtration. At other times it may be necessary to reconstitute the oily product by reblending certain components depleted in use or during re-refining. This service is also offered by some suppliers or jobbers.

One of the most important factors in preventing or reducing oily wastes is housekeeping and maintenance. Good practices are largely a matter of adequate planning, training, and followup. Employees need to know the accepted practices in handling each oily waste. Regular inspections of operations must be made. Procedures should be reviewed on a regular basis for possible revision. A good preventative maintenance program will do much to prevent accidental losses. Safe procedures for special emergencies should be developed, particularly for handling sudden releases of large quantities of oily materials. Proper plant design and layout helps, but specialized equipment may be required and employees must know how to use it.

Continuing Needs

Improved recovery and re-refining methods for oily materials should increase their reuse, and attention should be given to developing useful by-products from oily wastes. The food industry already uses some waste fats and oils in animal feeds, and the pulping industry recovers useful tall oil which was once wasted with black liquor.

Improved methods of cleaning tanks and vessels are greatly needed. More efficient use of dry cleaning techniques should be employed. Cleaning solutions should be kept segregated and renovated or fortified for reuse wherever possible. The dry-cleaning industry has learned this in the reuse of its cleaning fluids. Disposable tank liners

---

1. A *tag* substance is any material not normally found in oils, which will stay with the oily phase and be detectable in low concentrations. Such substances can be dyes, radioactive materials or, if there are no other sources of lead in the process, lead.

or inserts should find use in some instances; ultrasonic cleaning techniques might help reduce the volume of cleaning wastes.

Finally, there is a continuing need for simple, rapid methods of detection and identification of oily materials. Some progress is being made with continuous ultraviolet and infrared detectors and by the addition of tag materials.

Methods of Control

METHODS OF OIL SPILL CONTROL

a) Mechanical Containment
   a1) floating booms, a2) bubble barriers and a3) current barriers

b) Mechanical Recovery
   b1) weirs and suction devices, b11) floating weirs, b12) suction heads, b13) free vortex, b2) lifting surfaces, b21) rotating discs, b22) rotating drums and b23) moving belts

c) Application Agents
   c1) dispersants, c2) sinking agents, c3) collecting agents, c4) herding agents, c5) burning agents and c6) biodegradation

Listing of Oil Sinking Agents
   See Table 8.5.2

CHARACTERISTICS AND COMPOSITION

Spreading Rate of Oil Spills

Several factors need to be vectorially combined to define the oil spill spreading pattern. The current will drift the unrestrained oil at about the same velocity as the water. Wind adds a component of about 3–4% of the wind velocity, and natural spreading acts concentrically to disperse the slick. This is initially caused by the oil's hydrostatic head balanced by the oil's inertia. Typically, for an impulsive 500–2,500,000 gal spill of 0.9 specific gravity (SG) oil, this acts for about $\frac{1}{4}$–1 hr until the oil reaches about $\frac{1}{4}$ in thickness and a 500–3,000 ft diameter. At this point, pressure spreading is primarily balanced by viscous drag in the underlying water, and the slick diameter grows at about 300 ft/hr. As the gravity head decreases, the net surface tension spreading pressure (water to air-oil to air-oil to water), usually about 20 dynes per centimeter, continues to disperse the oil until typically a 0.01–0.001 in thickness is reached.

Variations in spreading rate depend on the oil's specific gravity, surface tension, characteristic evaporation, solubility in water, emulsification of water into the oil, and pour point. In a confined area, natural surface active agents in the oil can spread into a monomolecular film holding up to $\frac{1}{4}$ in of even low specific gravity oil.

Specific Gravity

Oil spill specific gravities range from 0.75 to 1.03. The lower values represent highly refined products such as gas-

oline, kerosene, and diesel fuels. The upper values represent residual oils. Crude oils have specific gravities between 0.8 and 1.0; however, this increases rapidly when the light ends evaporate. Also, with a low sea state, crudes and oils containing asphaltines readily form water-in-oil emulsions, raising pollutant specific gravity from 0.85 to 0.95 in several days.

An oil spill's buoyant hydrostatic head is inversely proportional to the difference between the water and oil specific gravities. The lower specific gravity oils spread rapidly, but once captured by a containment boom their buoyancy resists entrainment into a current stream passing under the oil and boom. Also, specific gravity can limit removal of the oil from a contained pool. In this case, recovery is proportional to the recovery device frontal length and the gravity-inertial feedrate per unit length (Q).

$$Q = \frac{2Ho}{3} \sqrt{\frac{2g \, \Delta \, Ho}{3}} \qquad 8.5(1)$$

where:

   Ho = Oil Depth,
   g = gravitational constant,
   Δ = water SG − oil SG

Thus, between the specific gravities of 0.75 and 1.0 the recovery feedrate will vary thirty-fold for a given thickness. Conversely almost a tenfold increase in thickness is required to have the same effect on recovery feedrate.

Viscosity

Spill viscosities range from 0.7 to over 20,000 centistokes (cst). Residual oils, weathered emulsions and high pour point crudes can even reach a semisolid state. The emulsions strongly deviate from Newtonian characteristics, frequently exhibiting very high viscosities at low shear rates and much lower viscosity at higher shear rates, such as those generated in transfer pumps. Viscosities for crudes weathered for up to a day and emulsified by moderate seas are between 300 and 1000 cst. There is no direct relationship between viscosity and specific gravity. However they tend to increase together (Figure 8.5.2).

Emulsification

Water-in-oil emulsions are unstable and difficult to form in highly refined oils. However, most crudes and all residual oils contain asphaltines, resins, cresols, phenols, organic acids, metallic salts, and other surface-active agents that concentrate at the interface between entrained water droplets and the oil. A crude spill can become a 40% water emulsion in a single day due to open sea action. In 5 days, this can increase to 80%. Increased shearing rates and action decreases water droplet size and increases emulsion stability. Pumping emulsions with free water may re-

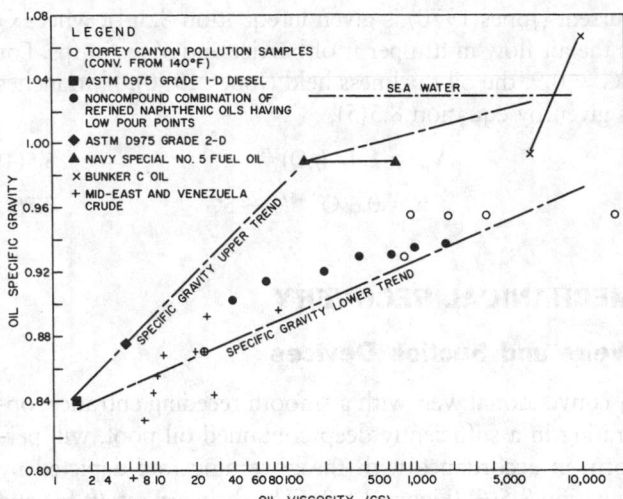

FIG. 8.5.2 Oil specific gravity vs viscosity at 60°F.

sult in up to 98% water in the oil emulsions which are so formed.

Flammability

Crudes frequently contain 30–70% gasoline and benzine, presenting a significant fire and explosion hazard. These conditions are rapidly mitigated by evaporation and mass transport when the wind is blowing, due to their large surface area. Residual oils and weathered emulsions present only a minimal fire hazard.

MECHANICAL CONTAINMENT

Floating Containment Barriers (Booms)

Desirable characteristics for containment booms include low cost; compact storage; easy and rapid deployment; durability; easy cleanup; and performance compatible with the environment. Boom construction materials must be protected from extended immersion in water or oil. Relatively compatible plastic materials include polyethelene, polyurethane, polyvinylchloride, polypropylene, epoxies, polyesters, nylon, and neoprene.

Even in calm, still water, the boom's draft and freeboard must be adequate and balanced to account for the oil's buoyant head. Likewise, although there is a pressure balance across the boom at the bottom of the contained slick, the boom must be strong enough to hold the hydrostatic head differential above that point.

$$D = 0.5 \; \rho v^2 A C_D \qquad 8.5(2)$$

$$M = 0.5 \; \rho v^2 d C_D X \qquad 8.5(3)$$

The flat plate drag, D, of a containment boom's immersed area, A, normal to a current of density, ρ, and velocity, v, is given by equation 8.5(2), where C_D is the drag coefficient and D equals the drag tension loads that result. Test

measurements indicate that C_D can exceed 3 for a 2–ft draft boom filled with 0.9 SG oil in a 1–kt (knot) current. The boom's rolling moment about its lower edge is given by equation 8.5(3), where X is the distance to the center of pressure, and X/d ≤ 0.5 without oil and ≤ 0.7 when filled with 0.9 SG oil. This moment must be counteracted by ballast and/or roll flotation. A large number of commercially available booms have the requisite strength and stability to operate in up to a 2–kt current in calm water.

CURRENT ACTION

If the boom is held at its ends, upstream from the oil spill, it will assume a catenary shape and capture the oil that drifts against it. The oil will form a stagnant pool with its maximum depth near the boom. If critical depth is exceeded at the boom, there may be a rapid drainage failure, during which most of the captured oil is lost. In a 1–kt current, with oils of 0.8, 0.9 and 0.98 specific gravity, tests indicate that boom depth must exceed 3, 10, and 55 in, respectively, to prevent drainage. In a 2–kt current the minimum depth almost triples. From near the boom to near the upstream edge, oil thickness decreases parabolically. At the leading edge, there is a rapid thickening due to a gravity or head wave.

When the current exceeds $\frac{1}{3}$ kt, oil can be torn off the bottom of the head wave and entrained in the current flowing under the captured pool. If the distance from the barrier, the barrier's draft, and the oil's buoyancy are insufficient, the entrained oil will sweep under the boom and will not resurface to coalesce with the captured pool. For a 2–ft draft boom and for oil specific gravities of 0.9 and 0.98, the respective critical current speeds are 2 and 0.9 kts. If the current is higher and there is sufficient sea room, the boom must drift with the current to keep the relative speed below critical. In a tidal channel or river, the boom can be streamed at an angle such that the current velocity component normal to the boom is below the critical speed. In this way, oil can be diverted without major loss from midchannel to a lower current area or to the river bank.

WAVE ACTION

Waves cause cross-coupled surge, heave, and roll of the boom. If wave surge and current tension loads are taken through an independent bridle, they will have a minimum tendency to constrain the boom's heave response. Minimizing boom mass tends to increase the natural heave frequency and minimize the lag in heave response. Thus, ballasting for stability or even adding weight for flotation can be critical to heave. Flotation placement away from the faces of the boom yields stability and decreased lag in heave motion. If natural frequency exceeds wave frequency, the boom will follow the wave and minimize surge forces.

If the boom does not readily heave or surge, the local force will be very high. Also, as the free water line meanders from the boom's still waterline, there are changes in roll movement. This is accentuated by current and wave surge. Thus, such a heavy-duty, slow sea response boom must be exceedingly strong, rugged and have a generous freeboard and draft. An example is the Merrit-Navy boom, consisting of 10–ft sections, each with 3–ft draft and 3–ft freeboard. It is constructed of 4–ft by 8–ft by $\frac{3}{4}$–in marine plywood with a 100–lb ballasted fabric skirt. Four oil drums strapped on for flotation add to the section weight. All of this requires two $\frac{1}{2}$–in wire ropes to be used for the tension bridle.

An independent bridle takes the tension loads out of a number of independent boom segments. Each segment can be connected to the next by a slack skirt. Lines trailing from the bridle to the boom segment permit the best heave response but provide no roll resistance. Multiple lines from the bridle to the top and bottom of each segment can limit roll at the sacrifice of heave. Such a design was developed by Johns-Manville for the U.S. Coast Guard. This boom is compliant and responsive to sea action. It has survived 15–ft waves with swells breaking on the barrier.

A number of booms have been employed on open ocean spills but there are no quantitative reports of their performance as a function of sea state and current. The U.S. Coast Guard boom is of minimum weight and is highly transportable (Hoult et al. 1970). Their 1972 tests in open seas of over State 3[2] were the first full-scale instrumented evaluations conducted. At up to 1–kt current, the oil leakage rate was very low. In these tests it held over 25,000 gal of oil in a thickness of 2$\frac{1}{2}$ in.

Bubble and Current Barriers

A surface current generated by spray nozzles or by the upwelling from a bubble stream can oppose oil spreading pressures. The simple bubble barrier consists of a submerged pipe with numerous air discharge holes along its length. Its primary use is to encircle fuel or oil loading terminals, or to protect ship berths. However, it cannot be used if the natural current is over the $\frac{1}{4}$–kt normal to the barrier. In this case, the turbulent countercurrent caused by the bubbles will form a head wave and carry the oil down into the main current, which then passes the oil through the bubble stream. When the bubbles emerge from the pipe, they expand to a maximum size of $\frac{1}{4}$–in diameter as they rise. The bubbles will break up before becoming much larger. Thus, the air pressure required needs only to exceed the pipe's hydrostatic pressure. The number of holes in the pipe and their size is not critical. The rising bubble column creates an upward flow of water that upwells and then generates a surface current. Up to 4-in thick layers of oil with SG under 0.92 have been held under static conditions by this technique. The maximum surface

2. From *Sea State Table* by Wilbur Marks, David Taylor Model Basin: small waves, becoming larger; fairly frequent white horses; moderate breeze (14 to 16 kts); 4–ft significant wave height (5–ft average of $\frac{1}{10}$th highest); 65–ft average wave length; 28-naut. mi minimum fetch; 5.2 hr minimum duration of wind.

current (Jones 1970) is given in equation 8.5(4), where Q is the air flow in ftm per ft of barrier and V_m is in fps. For SG $\leq$ 0.9, the oil thickness held (Jones 1970), h, in inches is given by equation 8.5(5),

$$V_m = 1.47 \, (gQ)^{1/3} \qquad \qquad 8.5(4)$$

$$h = 0.6 \, Q^{0.45}/1 - SG \qquad \qquad 8.5(5)$$

MECHANICAL RECOVERY

Weirs and Suction Devices

A conventional weir with a smooth receding entrance, operating in a sufficiently deep contained oil pool, will perform in accordance with the governing rate depicted by equation 8.5(6) (Figure 8.5.3). The recovery rate/ft for one edge of a weir in a static pool of oil is plotted as a function of the weir immersion depth. The plot of the spreading rate/ft for typical oils shows how deep the oil must be to support a given removal rate. In this case the oil must be more than twice as deep as the weir immersion depth. If it is not, oil flow will be broken and the weir will flood with underlying water. This explains the importance of setting these devices to avoid water flooding. It also shows that thin layers of oil cannot be recovered without large amounts of excess water. The same principle applies to suction devices.

In a current, weir flowrate over the lip increases in accordance with the rate equation 8.5(6). However, the flowpipe below the weir lip can act as a barrier (Figure 8.5.4).

$$Q = \frac{2}{3} \, CL_w \, \sqrt{2g} \left[\left(H_w + \frac{V_0^2}{2g} \right)^{3/2} \left(\frac{V_0^2}{2g} \right)^{3/2} \right] \qquad 8.5(6)$$

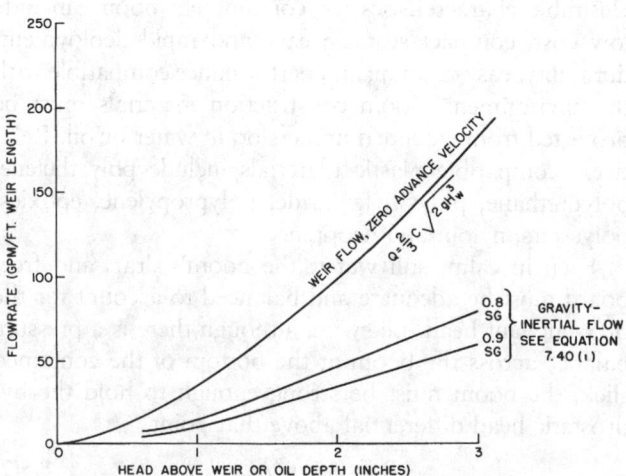

FIG. 8.5.3 Weir flow and gravity inertial flow vs oil depth. *Key:* Q = volume rate of flow over weir, in ft^3/sec; L_w = length of weir, in ft; C = discharge coefficient $\approx$ 0.6 (light oil with beveled, inward shaped weir); g = acceleration of gravity, in ft/sq sec; H_w = head, undisturbed level about weir, in ft; V_0 = velocity of advance of weir, in ft/sec.

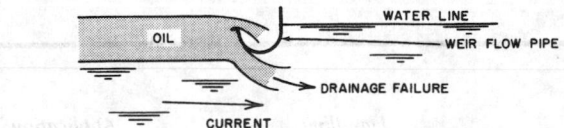

FIG. 8.5.4 Weir drainage failure in current.

If care is not taken, oil accumulates in front and drainage failure occurs. This becomes relevant when wave surge is present. Failure can be caused by a vortex at the weir ends through which the whole pool of oil drains. If current is not present and surge is reasonably low, small waves increase the average rate of flow due to $\frac{3}{2}$ power dependence on head. However, care must be taken to avoid flooding. The usual practice is to provide a large flotation response area to match heave to wave motion and to provide for surge motion to avoid drainage conditions.

To overcome weir difficulty of operating in thin slicks, various techniques of increasing oil depth may be used. The inverted weir, consisting of a shallow shaped barrier backed up by a second barrier, is one method used. In operation, thin slicks are diverted under the first barrier, and the current entraining the oil is recirculated to the surface as part of a vertical and countercurrent flow caused by the second barrier. In a more refined and improved design, secondary currents are avoided and oil resurfacing for concentration in the protected basin occurs due to the oil's buoyancy. This still requires a conventional weir or suction device to provide final oil removal from the thickened pool between the barriers.

Many weirs are custom designed. The U.S. Coast Guard developed a prototype for open sea use consisting of a basin formed by two parallel booms joined by a flexible bottom. Oil and water flow over the first boom or through slots in it. The water gravity separates from the oil and escapes through slots in the flexible bottom. A sonic sensor detects the oil level in the basin, activating weir removal of the oil by the second boom.

Lifting Surfaces

This class of recovery devices depends on moving an oleophilic, hydrophobic surface through the floating oil layer. The motion creates a viscous shear and attaches an oil boundary layer to the surface. The surface is then withdrawn above the waterline, where the oil is extracted and pumped to storage. The advantage of this device over the weir is that it will not flood with water when the oil is thin or the immersion depth varies.

One lifting surface device, the vertical disc, can be efficiently packaged into a small volume with disc spacing of 2 in or less. The device was originally available from Centri-Spray Corp. for calm water applications. An improved version called "Clean Sweep" using discs and vanes was commercially introduced by Lockheed (Bruch and Maxwell 1971). Model test data indicated that the mod-

ification will function at over 2 kts current or forward way in as high as Sea State 4[3] and will recover all oil encountered to a thickness of at least 0.1 in. These tests were conducted with disc diameters from 10 in to 8 ft.

In a 4-ft diameter it will recover 100 GPM of 1000 cst oil per ft of device in a 1-kt current. It can recover oils with over 10,000 cst viscosity. The U.S. Coast Guard awarded Lockheed a contract to develop a prototype air-transportable system for oil recovery use in up to Sea State 4.

Moving belt units were successfully used to clean up weed and peat moss contaminated residual oil from a spill in cold Canadian waters. A small pond type unit is also available. Similar to the moving belt is the rotating drum. The drum is usually employed only in calm water applications. One type, developed by Amoco and the subject of several improvement versions, uses an absorbent polyurethane foam-covered drum. Different density foam covers are used for various viscosity oils to the maximum recommended 2500 cst.

APPLICATION OF AGENTS

Dispersants

Many agents will disperse oil in water; generally, their use is prohibited by the EPA unless they protect human life, property from fire, or an endangered species from more direct damage. Both biological toxicity information and oxygen requirements for biodegradation are necessary to obtain an intelligent appraisal of their safety.

Sinking Agents

Sinking agents are used only with approval in waters more than 100 meters deep, where there are no fisheries or onshore currents. Approval is only given when no other feasible means of control is available. They present a significant logistical problem, since a pound of agent will only sink $\frac{1}{2}$–5 pounds of oil. An efficient agent should be permanently hydrophobic, permanently oleophilic; high in density; inexpensive; readily available; and easy to spread. Table 8.5.2 compares the characteristics of a number of commercially available materials, of which only stearic acid-treated chalk has been used in quantity.

Collecting Agents

The two major types of collecting agents are gels and absorbents. Although several types of gels exist, they are moderately expensive and difficult to apply, and there is

3. Moderate waves that take a pronounced long form, many white horses; Fresh Breeze (18 to 19 kts); 6.5 ft. significant wave height (8.3 ft. average of $\frac{1}{10}$th highest); 95 ft. average wave length; 55 nautical mile minimum fetch and an 8.3 hour minimum duration of wind.

TABLE 8.5.2 TYPICAL OIL-SINKING AGENTS

Materials with Relatively Permanent Oil Retention Characteristics	lb Oil per Pound Agent	SG	Availability	Handling Hazards	Application Method
Barite treated with latex	1.3	3.0	Limited, need treatment facility	None known	Dry sprinkle only
Chalk treated with stearic acid	1.5	2.7	5 ton; NYC; Europe	None known	Dry sprinkle & agitate
Asbestos (treated 100% hydrophobic)	4; must agitate to sink	2.4	Unlimited; with time for treatment	Asbestos is minimum if distributed in solution to avoid accidental dust inhalation	Spray in solution of water
Carbonized treated sand	0.4	2.6	Limited; need treatment facility	Possible silicosis	
Fly ash, chlorosilane or silane treated and neutralized	0.5	2.7			Dry sprinkle only
Sand, 120 g silane treated and neutralized		2.7			

no existing equipment suitable for recovery. The most frequently applied absorbents are hay and straw, because they are inexpensive, widely available, and easy to apply with mulching equipment. However, their application to floating oil requires slow trash bucket and rake recovery since all of the mechanical, weir, and suction-type recovery devices are clogged by the straw or their recovery rate is drastically reduced. Their best use is for cleaning up oil on beaches or in tidal pools. Straw will absorb about five times its weight in oil.

Polyurethane foams have the largest oil sorption capacity. They can hold 30–80 times their own weight in oil. Urea formaldehyde foam is equally effective. Polyethylene fibers and shredded polystyrene foam are only about half as effective. However, a patented grafted expanded polystyrene is reportedly as effective as polyurethane. Of the natural products, wood cellulose fiber and shredded redwood fiber are one-third to one-fifth as efficient as polyurethane. Ground corncobs are about as effective as straw. Foams take more time to absorb higher viscosity oils. On the open sea, wind prevents efficient distribution of the expensive low density absorbents. The more expensive materials also require that squeezing out the oil and reprocessing be economical.

Herding Agents

A novel method of oil recovery involves magnetism to separate the oil from the water and to lift it. A ferrofluid, consisting of a stable colloidal dispersion of super paramagnetic particles, is mixed with the oil. However, there is no apparent advantage over mechanical methods where surface tension wetting and fluid shear attachment are sufficient for both separation of oil from the water and recovery. The latter approach avoids all expense for both treatment agent and its application.

Another method is the use of a nontoxic, biodegradable surface active agent to surround an oil spill and to change significantly the water-air surface tension. The U.S. Naval Research Laboratory reported on the use of sorbitan esters of fatty acids and polyoxyethylene alkyl ethers as monomolecular surface films having 40 dyne per centimeter spreading pressure. Also, Shell Oil Co. commercially introduced a similar acting proprietary chemical called "Oil Herder." The recommended application rate is 20 gal/mi of spill perimeter.

Burning Agents

Because of the rapid loss of volatile substances from an oil spill, ignition and the support of combustion requires assistance. Despite the cost implication and other problems, burning is attractive because it provides rapid disposal of a large quantity of oil with a minimum of material handling. Most burning agents are nontoxic or inert. The major problem is distributing the agent and the need for a better ignition technique. Among several agents evaluated by the EPA are straw, cellular glass beads and silane-treated fumed silica. All of the agents appeared capable of supporting combustion, even in extremely cold waters, by acting as a wicking agent. Based on very limited testing, the beads gave the best performance, followed by straw, and finally the fumed silica. Wind creates a severe problem in applying the agents, including the silica, which was mixed with water. Wind and waves also break up uncontained pools and cut off continued combustion. Both the beads and the fumed silica, when not used in combustion, appear recoverable by normal mechanical recovery systems. The straw presents a special problem.

Biodegradation

There are many toxic products in oil, most of which are associated with the lighter volatile compounds. Many microorganisms can utilize hydrocarbons as an energy source to convert them into cell mass. On a pilot plant basis, an oil company has produced protein from residual paraffin base oil by bacterial fermentation. One source estimates that consumption rates will be limited eventually by the 2–lb oxygen demand to convert 1 lb of oil into cell mass. They further estimate that under moderate conditions there are about 25,000 lb of oxygen absorbed per sq mi of sea per day. Hence, normal oxygenation may limit oil consumption to 12,500 lb/sq mi/day.

There may have to be an addition of nitrogen and phosphorus in the form of ammonium nitrate and potassium phosphate. Industrial waste systems normally consume 20 lb of nitrogen and 40 lb phosphorus with this much oxygen, whereas the sea normally has less than half this amount available. Experiments by the Department of Oceanography, Florida State University, showed that selected microbial cultures can accelerate the removal of paraffinic crudes at a rate twice that of evaporation. The rate almost doubled with a 10°C rise in temperature; surfactants were produced, hastening emulsification.

—L.S. Savage, R.S. Robertson, B. Bruch

References

Audubon. 1971. *Oil pollution.* Vol. 73:3 (May). p. 101.

Bruch, B., and K.R. Maxwell. 1971. *Lockheed oil spill recovery device.* Joint Conference on Prevention and Control of Oil Spills. Washington, D.C. (June 15–17).

Hoult, D.P., R.H. Cross, J.H. Milgram, E.G. Pollak, and H.J. Reynolds. 1970. *Concept development of a prototype lightweight oil containment system.* U.S. Coast Guard Rpt. No. 714102/A/003. (June).

Jones, W.T. 1970. *Air barriers as oil spill containment devices.* SPE 3050, 45th Annual Fall Meeting of the Society of Petroleum Engineers. Houston, Tex. (Oct. 4–7).

White, P.T., and J.S. Blair. 1971. Bare-handed battle to cleanse the bay. *National Geographic* 139. (June). p. 877.

8.6
PURIFICATION OF SALT WATER

Conversion Processes

Three principal methods of augmenting the world's supply of potable water are:

1. Cloud seeding to furnish artificial production of rain
2. Advanced waste treatment to render wastewaters directly reusable
3. Conversion of salt water by a variety of processes

The first has been used with some success in arid areas, but the results are not always predictable. The second has passed from a laboratory procedure to the construction of full-size plants which are treating water in certain South African communities and returning it directly to the distribution systems to deliver drinking water to the population.

Conversion of salt water to fresh water is not a new idea. It is, in fact, the oldest and most extensive process known. Each day the power of the sun evaporates millions of tons of water from the oceans, which returns to the earth as rain. Desalted water has chemical and physical properties similar to rainwater. It is low in total solids, corrosive, clear, generally odorless, and somewhat insipid. It may contain dissolved gases, and water desalted by the membrane process may still contain a major part of trace elements (such as boron) originally in salt or brackish water.

The great attraction of desalting water is that over 97% of the world's supply of water is saline; furthermore, much of it is contiguous to arid regions. An example of the importance of desalting is the experience at Kuwait. During the early 1950s, the Kuwait Oil Company built an oil refinery dependent on desalted water from the Gulf. In 1953, the Kuwait government put into service a then-large capacity (1.2 mgd) desalting plant of 10 triple-effect, submerged tube evaporators. This installation is also an example of economics favorable to desalting, as the heat source for the evaporative process was waste or natural gas.

Oceans. The oceans and seas contain 97% of the world's 326 million cu mi of water. One cu mi is equal to one trillion gal. Sea water contains about 35,000 ppm of salts.

Salt Lakes. In a few places there are lakes or seas with no outlet. The resultant removal of water by evaporation has left behind a liquid in which the salt content is near saturation. Examples are the Dead Sea in Israel and the Great Salt Lake of Utah. It is reported that when water is low in the Great Salt Lake, the water contains 250,000 ppm (25%) of common salt.

Ground High-Salinity. Some groundwater in the western United States and in the vicinity of oil fields may be classified as brine for it contains upward of 10,000 ppm salt.

Surface-Brackish Water. Some surface streams and estuaries contain salt of 2000 to 5000 ppm and are therefore unsuitable to drink. They may be ideal as sources of desalted water because of the low initial salt content.

Three basic types of desalting processes include evaporative, membrane, and freezing, and each of these has subtypes or alternative methods.

Evaporative (Distillation) Processes

The four subtypes in this group are multistage flash (MSF), submerged tube (ST), long tube vertical (VTE), and vapor compression (VC). The MSF process contains three flow-streams, namely influent seawater, recycled brine and product effluent water. Cold seawater is pumped through a heat exchanger where the heat gained by the seawater is furnished in part as the heat loss of the condensing product water. Partly heated seawater then goes through an atmospheric degassing tank and joins the brine in the first evaporative stage. The brine is pumped to the second stage, passing through condensers around which more product water is formed by condensation of fresh water evaporated from concentrated brine in the second stage. Condensed product water from both stages accumulates in trays and flows to a product water sump, from which it is pumped to points of use.

The vertical-tube evaporator (VTE) system is somewhat simpler. It consists of vertical tubes installed in heat exchangers arranged in series. Seawater falls through the tubes in the first section where it absorbs heat and condenses fresh water.

The submerged tube (ST) process operates somewhat like a conventional boiler. Typified by small capacity distillation equipment on ocean-going vessels, the VC process operates as a vertical tube evaporator either with boiling brine inside the tubes or flashing above the tubes. It owes its name to the fact that the fresh-water steam is pressurized.

Electrodialysis and Reverse Osmosis

Electrodialysis and reverse osmosis both use membranes. In the electrodialysis process salt water passes between layers of membranes that are selectively permeable or impermeable to ions in the salt water, depending on the membrane charge. Brine is produced in one part of the device and fresh water in the other, and the energy of separation comes from an electric current.

The heart of the reverse osmosis process is a semipermeable membrane separating salt and fresh water. The normal phenomenon of the flow of fresh water through the membrane to the salt water side is reversed by applying pressure to the salt water side.

Freezing

If saline waters are cooled sufficiently, they freeze, and the resulting ice is fresh water. The ice crystals are separated from the brine, cleansed and melted.

Desalination Plants

Data on desalting plants by type of process are given in Table 8.6.1. The selection of any one or combination of processes can be justified by economic and other factors as listed below:

1. The salt content of the source
2. The salt concentration acceptable in the desalted plant effluent
3. Logistics, i.e., the location of the proposed plant, its size, the labor market, and alternative sources of water
4. Available sources and costs of energy
5. The costs of alternative methods of supplying water
6. Provisions for the disposal of brine: is there a local salt market?
7. How much fresh water is required

Salt Content of Plant Influent and Effluent

Where there is a choice of influent concentration, the selection of the water with the lowest salt content reduces energy requirements and costs. The most abundant source of salt water is the ocean, which on average contains about 35,000 ppm salt. Utilizing the sea and converting it to fresh water provides water supplies to islands in the Aegean Sea and in the South Pacific, where solar stills are used.[1] These are of the greenhouse type, requiring a rather extensive area.

1. Transparent plastic is supported on a framework arranged so that as the vapors rise from the pool of salt water under the canopy, they condense on the underside of the plastic and run into drains or troughs to a storage point.

TABLE 8.6.1 DESALTING PLANTS BY TYPE OF PROCESS[a]

Type of Process Used	Number of Plants	Total Plant Capacity (MGD)
Evaporation		
multistage flash (MSF)	229	146.3
long tube vertical (VTE)	96	54.4
submerged tube (ST)	302	38.7
vapor compression (VC)	19[b]	2.2[b]
	646	241.6
Membrane		
electrodialysis (ED)	34	5.1
reverse osmosis (RO)	3	0.2
	37	5.3
Freezing		
vacuum freezing	3	0.3
vapor compression (VC)	3	0.3
Totals	686	247.2

[a]Adapted from Office of Saline Water Special Report on Status of Desalting, November 1970.

[b]Does not include 3000 vapor compression units on vessels sailing the oceans; each of these can produce several thousand gal per day.

The U.S. Public Health Service Drinking Water Standards recommend a limit of 250 ppm for chloride and 500 ppm for total dissolved solids. When water is produced by desalination in areas governed by these standards, this level of salt and dissolved solids is specified as an upper limit. The International Standards for drinking water set the maximum allowable total dissolved solids limit at 1500 ppm.

Plant Location and Energy Source

The location and construction of a desalting plant often involves transporting construction materials for great distances. It is most economical to build a plant on the edge of the sea, providing large amounts of energy to convert the salt water to fresh water and pump the fresh water inland. The cost of fuel is a critical factor in the choice of any distillation process, and the cost of electric power is equally important in electrodialysis processes.

An inevitable by-product of a desalting plant is brine (concentrated salt water). When a plant is located at the edge of the sea, the brine is emptied into the ocean itself. The point of discharge must be chosen to minimize adverse effects upon the ecology through temperature elevation and salinity at the point of outfall. In some areas of the world, there is a market for the salt produced by additional evaporation of the brine.

The requirements for fresh water determine both the type of plant and the choice of fuel. For example, on an ocean-going vessel the demand for fresh water may be only 20,000 gal/day. There are places in Australia where solar

distillation is used to produce as little as 600 to 1200 gal/day. At the other end of the spectrum are the water needs of large cities, which exceed millions of gal/day.

Desalting Processes

Optimization of energy utilization is one of the principal factors in the design of a desalting process. Energy—either thermal, mechanical, nuclear or electrical—is required. The cost of energy may represent as much as 50% of the total water production cost. The percent of recovery of fresh water from salt water, a basic design criterion, rests on considerations of energy utilization. Starting with seawater having a concentration of 35,000 ppm of sodium chloride, the minimum isothermal work required at 70°F to separate 1000 gal of seawater is 2.6 kw-hr. However, this assumes removing water from an infinitely large volume of seawater, or pumping an infinite amount of raw seawater through the plant with little recovery of fresh water. Theoretical work for separation increases with increasing recovery (recovery is the ratio of product to feed expressed as a percentage), while pumping cost decreases with increasing recovery. The total energy required as a function of percent recovery is expected to follow the curve (Gilliland 1955) shown in Figure 8.6.1.

The optimum level of recovery for seawater is about 30–50% or about 2–3 volumes of feed to 1 volume of product water. For brackish water the optimum is about 3–8 volumes of feed to 1 volume of product. The higher percent recovery for brackish water feed is desirable because it is available in limited quantities compared to seawater, and also because disposal of concentrated brine is usually a problem.

The principal factors in the design of an evaporation system (Othmer 1970; Gilliland 1955; Rubin 1963) are

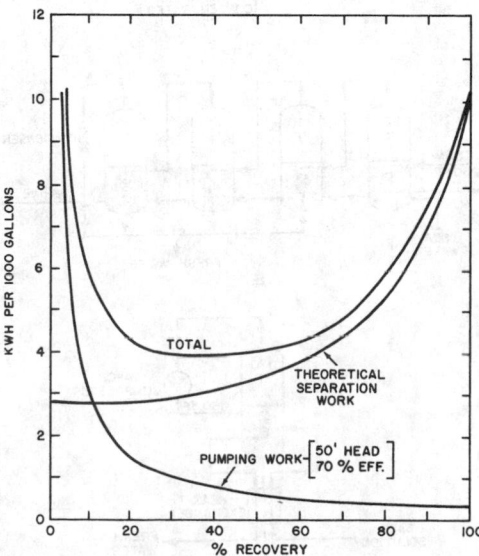

FIG. 8.6.1 Work requirements for the separation of seawater as a function of recovery. Product per feed in percent.

conservation of energy, separation of evolved vapors from brine, heat transfer, and prevention of fouling of the heating surfaces. Of these, conservation of energy is paramount. The cost of water would be prohibitive if steam was used on a once-through basis, as in a single-effect evaporator (Figure 8.6.2).

Figures 8.6.2 and 8.6.3 show evaporation systems that permit reuse of latent heat of vaporization of the prime steam supplied by the boiler. In multieffect evaporation (Figure 8.6.2), water evaporates at the highest pressure in the first effect. This vapor is condensed in the second effect to evaporate an approximately equal amount of vapor. Proceeding in this manner, the amount of steam and cooling water used per lb of fresh water produced is reduced.

In vapor compression evaporation (Figure 8.6.2), the water vapor evolved is superheated. This vapor is compressed to a pressure or saturation temperature sufficient to provide the necessary heat transfer. The condensed water vapor is the product water, and the residue brine is used to preheat the incoming salt water.

In multistage flash evaporation (MSF) (Figure 8.6.3), the operating principle is described. Multiple-effect and MSF operations not only reduce the amount of steam required and the cost of the boiler, they also save cooling water; hence the capital cost of the condenser and cooling water supply system is also reduced. Vapor compression operation eliminates condenser cooling water and almost eliminates the need for a steam boiler.

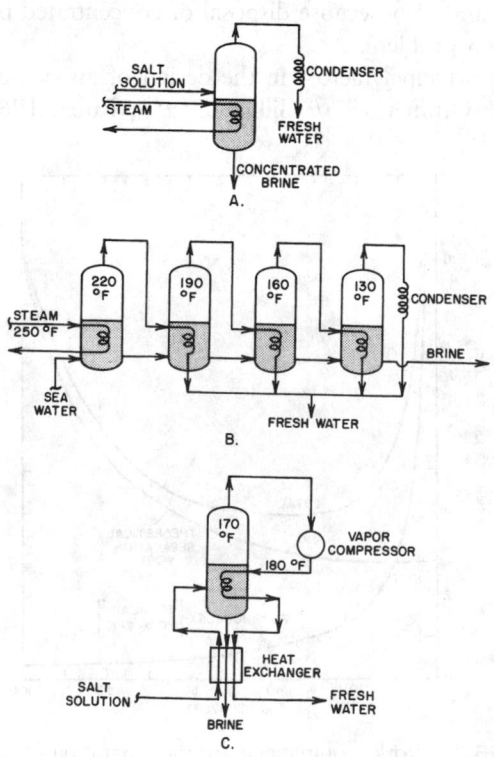

FIG. 8.6.2 Evaporators. A. Single-effect; B. Four-effect; C. Vapor compressor

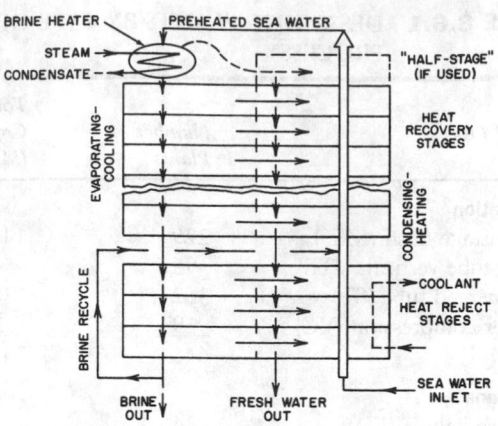

FIG. 8.6.3 Multistage flash evaporation. After being heated to the highest temperature in the brine heater at the top of the diagram, the seawater passes while flash evaporating and cooling at successively lower pressures (vertical arrows down on left side). The vapors leaving each flash evaporation (horizontal arrows) pass to preheat the seawater in the condensing-heating tubes (wide vertical arrow on right side). Freshwater condensate also passes from higher to lower stages for evaporation and cooling, to discharge at the bottom. Vapors may also be withdrawn from the brine heater to be condensed in the half-stage (lines upper right) to increase production of fresh water.

MULTIEFFECT EVAPORATION

Figure 8.6.4 is the flowsheet of the 1 million gpd 12-effect evaporator OSW Demonstration Unit (Guccione 1962) installed at Freeport, Texas. In this plant, seawater feed is heated to about 130–135°F. To prevent deposits on evaporator heat transfer surfaces, sulfuric acid is added to lower the feed pH and to react with the scale-former:

$$CaCO_3 + 2H^+ \longrightarrow CO_2 + H_2O + Ca^{2+} \qquad 8.6(1)$$

$$Mg(OH)_2 + 2H^+ \longrightarrow Mg^{2+} + 2H_2O \qquad 8.6(2)$$

Gases in the seawater and carbon dioxide formed are stripped in the deaerator. To prevent corrosion, pH is then adjusted to about 8 by adding caustic. This treated water is further preheated, charged to the top of the first vertical tube evaporator (VTE), passing through a distributor plate into long tubes (extreme right of Figure 8.6.4). Steam injected midway on the shell side of the tube bundle is condensed, heating the seawater. The seawater feed flows to the lower section of the vaporizer where the vapor is disengaged. The generated vapors enter the shell side of the next effect (to the left) and condense to give fresh water. Brine from the first effect becomes the feed to the top of the second effect. The remaining effects operate in the same manner at progressively lower temperatures and pressures. Finally, fresh water from each effect is piped through a heat exchange system for preheating incoming seawater.

With 12 effects and some 20 heat exchangers, about 10 lb of fresh water is obtained per lb of prime steam used. By operating at temperatures no higher than 261°F and at a concentration factor below 3, the scale problem is con-

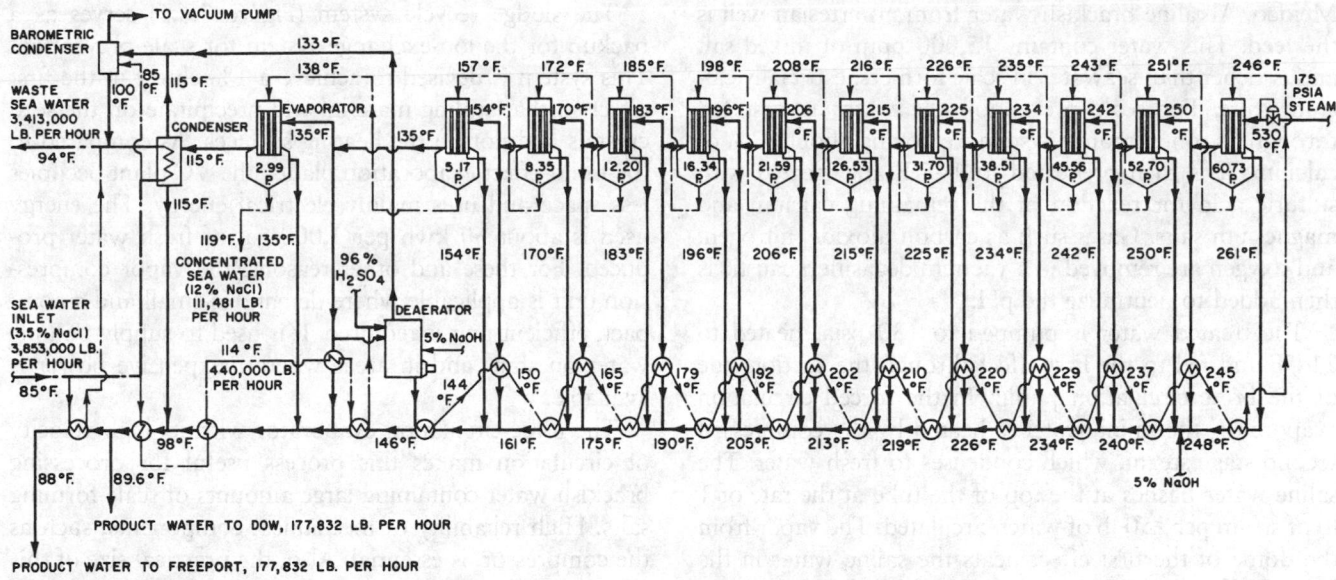

FIG. 8.6.4 Multieffect evaporation. Flowsheet of 1 million GPD OSW demonstration plant at Freeport, Texas.

trolled. The production rate of this plant exceeded design capacity.

Among the main drawbacks of a vertical tube evaporation system such as the one described are: a large number of metallic tubular surfaces are required for evaporation and condensation; and many heat exchanger surfaces are needed for recovery of heat from the streams of salt water and fresh water. These are all expensive.

Two developments may remove these drawbacks. One is the development of a high-performance fluted tube by the Oak Ridge National Laboratory for film evaporation

in VTE and for condensation. Increasing effectiveness of the tube's heat-transfer surface may reduce the cost of the VTE system. The other development is an MSF system to recover the heat. This VTE-MSF process may reduce the cost of heat-recovery surface and obtain additional water in the MSF section (Browning 1970).

VAPOR COMPRESSION EVAPORATION

Figure 8.6.5 is the flowsheet (Browning 1970) of the 1 million gpd OSW Demonstration Plant at Roswell, New

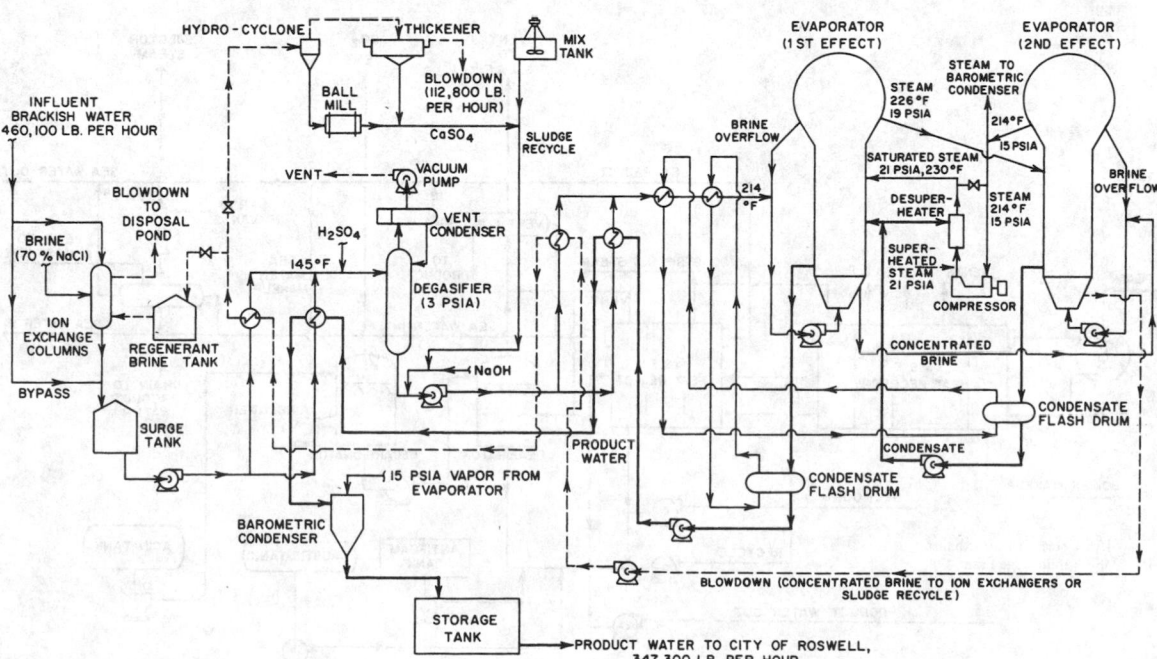

FIG. 8.6.5 Forced circulation vapor compression plant. Flowsheet of 1 million GPD OSW demonstration plant at Roswell, New Mexico.

Mexico. Alkaline brackish water from an artesian well is the feed. This water contains 15,000 ppm of mixed salt and is richer than seawater in scale formers, especially calcium salts. For scale prevention the water is first fed through an ion-exchange system, removing about 87% of calcium. It is then preheated to 145°F and treated with sulfuric acid for reaction of the remaining calcium and magnesium salts. Gases such as carbon dioxide, nitrogen, and oxygen are removed in a vacuum degasifier; caustic is then added to neutralize the pH.

The treated water is pumped to 130 psia, heated to 214°F and fed to the first-effect through the suction side of the brine circulation pump. In the forced circulation evaporator, the saline water is heated by the compressed second-stage steam, which condenses to fresh water. The saline water flashes at the top of the tube at the rate of 1 lb of steam per 250 lb of water circulated. The vapor from the dome of the first effect heats the saline water in the second effect and condenses to give fresh water. The vapor from the dome of the second effect is compressed to heat the first effect. The compressor operating across the two effects is a five-stage axial-flow machine driven by a 2,000-hp electric motor.

The high circulation rate of water in the forced circulation evaporator is intended to obtain a heat transfer coefficient as high as possible. Thus, the necessary heat transfer is satisfied without excessively large surface areas, while the required temperature difference is minimized. The relationship between this temperature difference and the power consumption by the compressor and the circulation pump has an impact on plant and water costs (Othmer 1969; Gilliland 1955).

The sludge recycle system (Figure 8.6.5) serves as a backup for the ion-exchange system for scale prevention. This system is devised to achieve a 1% slurry in the first effect so that scaling material will precipitate on the seed crystals and not on the heating surfaces. As compared to the multi-effect evaporation plants, the VC plant occupies less space and uses mainly electrical energy. The energy used is about 60 kwh per 1,000 gal of fresh water produced. For these and other reasons, the vapor compression unit is applicable where demand is small and a compact, efficient unit is required. It is used to supply potable water on ships and in areas where inexpensive power is available.

The force-circulation evaporator with its high velocity of circulation makes this process useful for processing brackish water containing large amounts of scale-forming salts. High reliability of mechanical components, such as the compressor, is essential. Also, the practical size of this compressor limits the size of the vapor compressor evaporator. The high performance heat-transfer surfaces developed for the VTE system may also be helpful. Likewise, the MSF process may recover heat from brine blowdown in large-scale operations.

MULTIFLASH EVAPORATORS

In a 2.5 million gpd plant designed for OSW as a standard or Universal Desalination Plant (Othmer 1970) (Figure 8.6.6), seawater at 85°F is first sent to the heat rejection stages (Figure 8.6.3) where it is heated to 97.7°F. Part of this seawater is discharged, while the other part is treated for scale, corrosion, and foam prevention. The

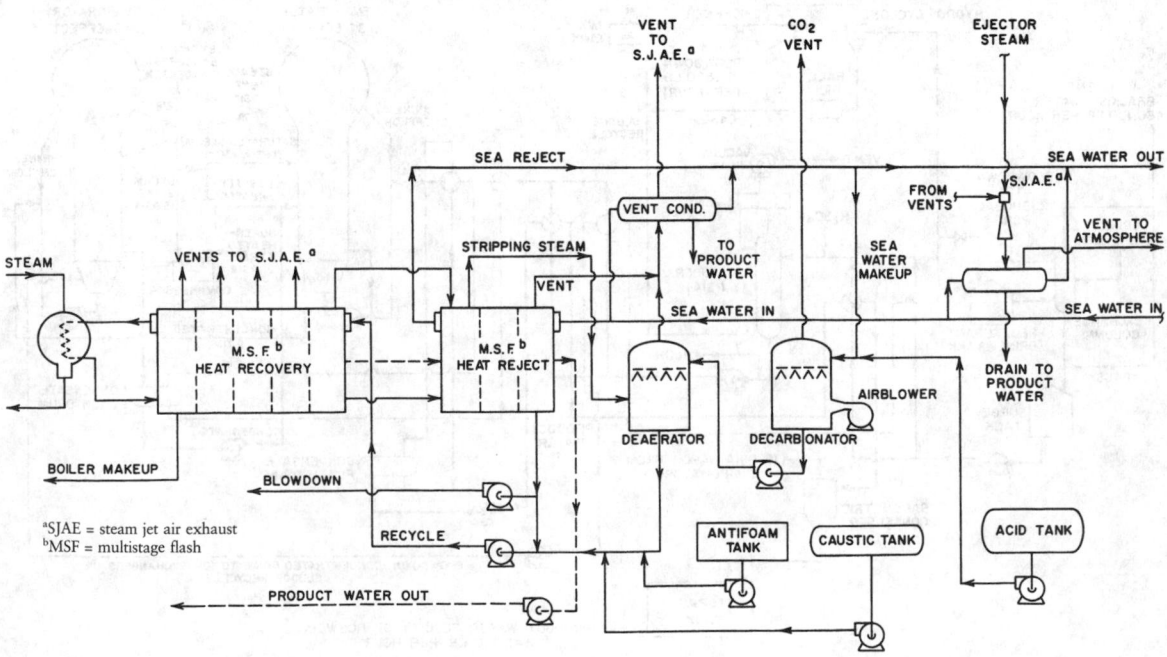

FIG. 8.6.6 Universal desalination plant (2.5 million GPD).

treated seawater joins the discharge from the heat reject stages and goes to the lowest temperature stage of the heat recovery system. It passes through all the stages and is heated to 250°F in the brine heater with prime steam from the boilers. The hot seawater returns to the MSF where it evaporates and cools at successively lower pressures, first in the recovery stages, then in the heat reject stages. The exit brine stream is split, part for recycle and part for blowdown. The vapor released in flashing in each stage condenses to heat the incoming seawater and gives fresh water.

The MSF plants now supply 97% of fresh water derived from seawater. In a single-effect MSF (Figure 8.6.6) with 40–50 stages, as high as 8–11 lb of fresh water are produced per lb of prime steam used. With a multirecycle, three-effect MSF at Chula Vista, California, the gain ratio increased (Othmer 1970) from 8 to 11 to 20.

Scale formation on heating surfaces, a troublesome problem in the two preceding evaporation processes, is minimized here. Evaporation takes place by flashing at successively lower pressure, therefore no heat transfer surface is needed for evaporation. However, precipitation of salts, such as calcium sulfate scale in the tubes in high temperature parts of the system, is still a problem. The solubility of these salts in water decreases with increasing temperature.

For heat conservation, equilibrium between the cooler seawater from each stage and the vapor formed without salt entrainment is desired for this process. In practice it is not possible to obtain this in the present MST (Othmer 1970). Another drawback is the use of large numbers of flash stages, usually forty or more. The large number of stages is required to reduce individual temperature drop or flash temperature and to reduce violence of ebullition (boiling), which causes entrainment.

Variations in MSF design for increasing the gain ratio or lb of fresh water/lb of steam used, without addition in plant cost, are possible (Othmer 1969). Controlled Flash Evaporation (Roe and Othmer 1971) (CFE) would allow equilibrium between vapor and liquid throughout the stages and higher flashing temperature ranges without transport losses of pressure and temperature and salt entrainment.

Freezing Processes

Freezing processes involve cooling incoming seawater, freezing it to obtain freshwater ice, separating the ice and brine liquor, melting the ice to give fresh water, and using the purified water and concentrated brine to chill the incoming seawater.

Freezing has several advantages over evaporation, most important that latent heat of water fusion is only about one-seventh the latent heat of evaporation—thus freezing processes hold the promise of low-energy power requirement; and low temperature operation minimizes the main-

tenance associated with scale formation and corrosion. However, freezing processes have some inherent disadvantages, including the fact that freezing time is longer than that for vaporization of water; and cleaning the salt from the ice and handling the ice crystals is quite difficult.

VACUUM-FREEZE VAPOR COMPRESSION

Figure 8.6.7 shows the Zarchin-Colt process in which water is the refrigerant. Seawater, after prechilling in a heat exchanger to a temperature approaching the freezing point of brine, enters the freezer. Evaporation is induced by the suction of the compressor, which absorbs heat from the remaining brine. Ice crystals are formed, growing to about 0.5 mm. The ice-brine slurry is pumped to the ice decanter where ice crystals float to the top. There they are washed with a portion of the product water made. This washwater moves downward and carries away salt. A rotating scraper trims the ice layer at the top and sends the washed crystals to the freezer, where they come into contact with the compressed water vapors. The water vapor condenses on the ice crystals suspended on a rotating perforated tray and melts them. (The ammonia refrigeration system shown in Figure 8.6.7 removes heat gain through insulation.) This melted ice is the desired fresh-water product. The cold fresh water and cold brine from the ice decanter are routed through a heat exchange unit to chill the incoming seawater. This process was successfully operated in a 100,000 gpd pilot plant. The total power cost is estimated, for an improved operation, to be as low as 27.3 kwh per 1,000 gal in a combination of $2\frac{1}{2}$ million gpd units.

One drawback of this process is the large vapor volume prevailing under vacuum (about $\frac{1}{200}$th atm) in the freezer. This requires a large diameter compressor, which

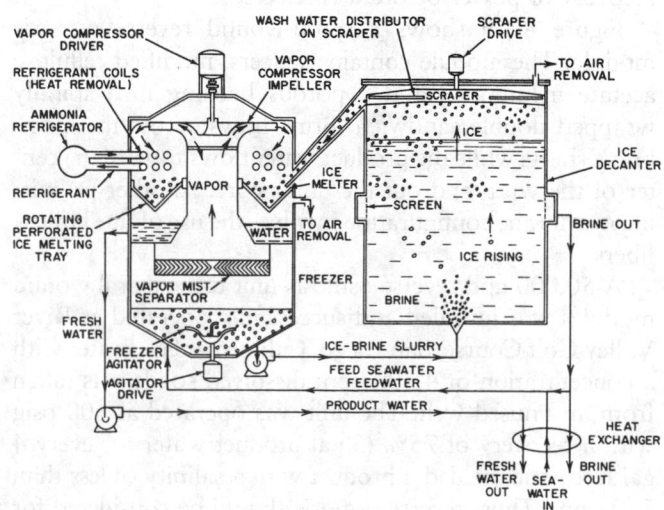

FIG. 8.6.7 Zarchin-Colt freezing process for seawater desalting.

limits the size of a single unit to about 200,000 gpd. Larger capacity plants must consist of duplicate units. To circumvent this, a process was developed involving a secondary refrigerant. Instead of water vapor, a refrigerant such as butane, which has a vapor pressure above atmospheric at the freezing temperature of brine, is used. Liquid butane is mixed with the brine in the freezer and, after absorbing the heat of ice fusion, it vaporizes. After compression, the butane vapor is used to melt the ice crystals. However, this development work was terminated because projected costs for commercial plants were no lower than costs for evaporation.

Reverse Osmosis

In this process, water from a salt solution is forced across a selectively permeable membrane by a pressure difference. The membranes allow water to pass through but not salt ions. The pressure applied must be greater than osmotic pressure. The osmotic pressure in a freshwater-salt water system separated by a selectively permeable membrane is a direct function of salt concentration: about 25 atm for sea water and 1.4 atm for brackish water with 2,000 ppm solids.

Figure 8.6.8 shows a typical reverse osmosis system. Salt water is first pumped through a filter to remove gross particles and iron, and then it is subjected to additional pretreatment as required to prevent fouling of the membrane surface. Some examples of brackish water pretreatment include lime-soda treatment or the addition of sequestering agents to prevent calcium precipitation. The salt water is then pressurized to a level high enough to reverse normal osmotic pressure and to provide driving force across the system, including the membrane. The salt water is fed into reverse osmosis cells (modules). Part of the water permeates the membrane and is collected as fresh water. The remainder (brine) passes through a turbine for recovery of power before it is rejected.

Figure 8.6.9 shows a spiral-wound reverse osmosis module. The module contains spacers, modified cellulose acetate membrane, and a porous backing in a spirally wrapped double sandwich. Brine flows across the membrane sheets while the product water flows toward the center of the wrap and out the unit's core. Another promising membrane configuration involves the use of fine hollow fibers.

A 50,000 gpd reverse osmosis unit using spiral wound modules was installed and successfully operated at River Valley Golf Course, San Diego, California. Feedwater with a concentration of 4,500 ppm dissolved solids was taken from an unused well. The unit was operated at 600 psig with a recovery of 75% (3 gal product water for every 4 gal feed) and yielded a product with a salinity of less than 350 ppm. Thus, reverse osmosis should be considered for converting brackish water to potable water. The future of reverse osmosis processes for large-scale installations

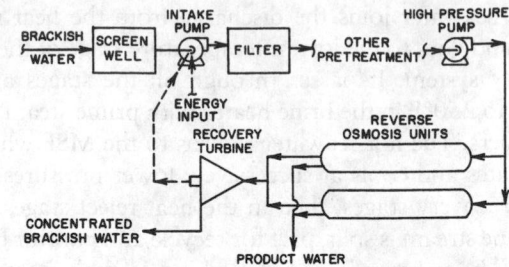

FIG. 8.6.8 Reverse osmosis for seawater desalting.

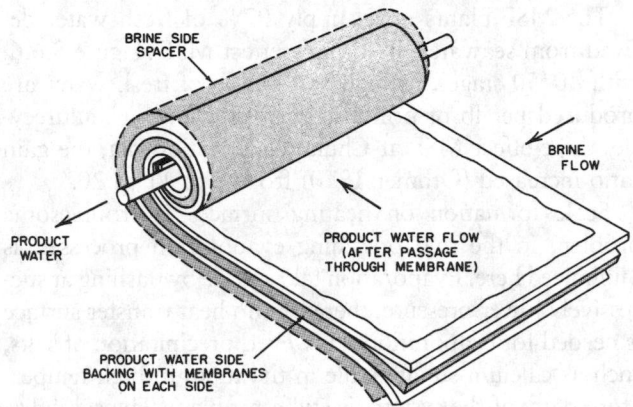

FIG. 8.6.9 Spiral membrane module for a reverse osmosis unit.

hinges on the reduction of membrane replacement costs, a major item in total water cost. The objective is either reducing basic membrane cost or extending service life. A large part of the present work on reverse osmosis is directed toward increasing product-water flux while maintaining salt properties (Browning 1970).

Electrodialysis

Electrodialysis is based on the development of membranes that are selective for the passage of ions of a given charge. Two different membranes are used: one is more selective to anions, and the other is more selective to cations. Electric current aids the diffusion of these ions, and the electric energy required is proportional to the concentration of salts in the saline water. Therefore, the process is more attractive for desalting of brackish (low salt concentration) waters.

Figure 8.6.10 shows a multicellular arrangement for desalting brackish water. The unit consists of alternative anion-permeable and cation-permeable membranes, and salt solution is passed through all compartments. The adjacent cells have the anion-permeable and cation-permeable membranes on the opposite side. The imposition of an electric potential causes the cations to migrate to the cathode, and the anions to the anode. However, neither cations nor anions, after passing from the feedcells into the adjacent brine cells, can pass through more than one cell toward the electrodes because they are blocked by impermeable

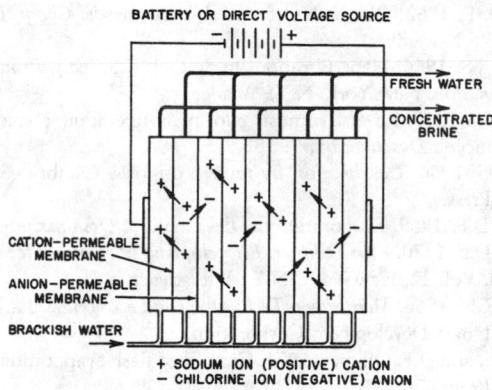

FIG. 8.6.10 Electrodialysis used to remove salt from brackish water.

membranes. Thus, the feedstream is depleted while the adjacent stream is enriched in ions. The plant produces a potable water stream and a salt-rich stream which is rejected.

Ion-exchange resins, which comprise 60–70% of the membrane, are solidly hydrated, strong electrolytes and might be regarded as solid sulfuric acid or as caustic solid. The resin most commonly used is polystyrene cross-linked with divinylbenzene. The ion-exchange resin permeable to cations is made by sulfonating the polystyrene resins; the resin permeable to anions contains a quaternary ammonium group attached to the polystyrene resin.

Electrodialysis is an established process for desalting brackish water. Units with capacities from 10,000 to 650,000 gpd have been installed. The process has major advantages for brackish water but is considered too costly in electric power requirements for desalting seawater.

Keeping the membrane surface clean is a major problem. Prefiltering and chemical treatment of feed has kept plants operating, but a thorough study of brine composition should be made and special pretreatment methods should be developed as required. Membrane replacement costs are a major part of producing fresh water by this method. Current research includes studies to improve selectivity of the anionic-permeable and cationic-permeable

membranes to increase the maximum allowable current per unit membrane area and to develop feedwater pretreatment processes for the removal of various membrane-blocking contaminants.

Table 8.6.2 summarizes the most favorable feed properties and capacities for each of the six processes described in this section.

Processes based on evaporation are operated in large-scale plants for desalting seawater. Out of the several evaporative processes, MSF has been most widely used in very large-scale installations, while VC is commonly used in small-scale plants. Vapor compression with forced circulation evaporators offers the possibility for desalting brackish water.

Processes based on semi-permeable membranes are used mainly for desalting brackish water. Electrodialysis is already a well-established process, although reverse osmosis is also feasible. Controlled flash evaporation appears to be an attractive alternative to the MSF process.

The Future of Desalination

The most successful evaporative desalting method is the MSF process (Figures 8.6.3 and 8.6.6). These units are built with as many as 69 stages and produce up to 20 lb of water/lb of steam. Research continues to improve the efficiency of this process by flash enhancers and by combining the MSF system with multiple-effect evaporation process steps.

Problems common to all desalting processes include the decreasing but still high combined cost of equipment and operation (see Conversion Processes) and the low load factors associated with these plants. The load factor refers to the percentage of time that the plant is in operation, and this seldom exceeds 60% for desalting plants. Low load factors are partially caused by corrosion problems and partially by scaling problems. For seawater, scaling becomes a problem at 160°F, although methods have been developed to control scaling at temperatures as high as 350°F. In addition to maintenance problems, scaling also degrades the heat transfer efficiency of heat exchange surfaces.

TABLE 8.6.2 SELECTION AND APPLICABILITY OF DESALTING PROCESS

Type of Desalting Process	Favorable Feedwater	Feed Concentration[a] Range (ppm)	Plant Capacity Range (MGD)
Multistage flash	Cold-soft	5,000–35,000	>1
Verticle tube evaporator	Cold-soft	5,000–35,000	>5
Vapor compression	Warm-soft	5,000–35,000	1–20
Vacuum-freeze vapor-compression	Cold-hard	5,000–35,000	0.25–5
Reverse osmosis	Warm-soft	3,000–10,000	0.1–10
Electrodialysis	Warm-soft	1,000–4,000	0.1–10

[a]Dissolved salts

In membrane-type desalting processes, the research has been directed toward both reducing membrane cost and increasing membrane life span.

Today there are over 1000 desalination plants in operation, converting over a billion gallons of saltwater per day into potable water.

—*F.B. Taylor, D.H.F. Liu, C.J. Santhanam*

References

Anon. 1971. Why hollow-fiber reverse osmosis won the top CE price for DuPont. *Chem. Eng.* 78: 54. (29 November).

Brennan, P.J. 1963. Fresh water from vapor-compression evaporation. *Chem. Eng.* 70: 170. (14 October).

Browning, J.E. 1970. Zero in on desalting. *Chem. Eng.* 77: 64. (23 March).

Ellwood, P. 1970. *Chem. Eng.* (2 Nov.) pp. 46–48.

Gilliland, E.R. 1955. Fresh water for the future. *Ind. Eng. Chem.* 47(12): 2410.

Guccione, E. 1962. Old method for fresh-water needs. *Chem. Eng.* 69: 102 (26 November).

Johnson, J.S. 1966. Hyperfiltration. In Spiegel, K.S. (ed.), *Principles of desalination.* New York, N.Y.: Academic.

Larson, T.J. 1970. Reverse osmosis pilot plant operation: a spiral module concept. *Desalination,* 7: 187.

Merten, U. 1966. *Desalination by reverse osmosis.* Cambridge, Mass.: MIT Press.

Othmer, D.F. 1969. Evaporation for desalination. *Desalination,* 6: 13.

Othmer, D.F. 1970. *Kirk-Othmer: Encyclopedia of chemical technology.* 2d ed. Vol. 22, New York, N.Y.: Interscience.

Rickles, R.N. 1966. *Membrane: Technology and economics.* Park Ridge, N.J.: Noyes Development Corporation.

Roe, R.C., and D.F. Othmer. 1971. Controlled flash evaporation. *Chem. Eng. Prog.* 67(7): 77.

Rubin, F.L. 1963. Perry's Chemical Engineers' Handbook. 4th ed. New York, N.Y.: McGraw-Hill.

Weismantel, G.E. 1968. Recycle boost desalting efficiency. *Chem. Eng.* 75: 86 (15 July).

Wilson, J.R. (ed.). 1960. *Dimineralization by electrodialysis.* London: Butterworth.

8.7
RADIOACTIVE LIQUID WASTE TREATMENT

Radioactive liquid wastes usually contain high, medium, or low amounts of radioactivity. The means of treating high-volume–low-activity waste is very different from the means of treating small-volume–high-activity waste. Medium-level waste is treated to convert it into high- and low-activity waste. Low-activity wastes are treated to remove radioactivity, then discharged to the environment in amounts well below permissible limits. This is the *dilute and disperse* philosophy. High-activity wastes, because of their hazard, must be concentrated, contained, and removed from man's environment. Definitions of waste categories (ASI 1967; IAEA 1970a) are given in Table 8.7.1.

Low-Activity Wastes
PRECIPITATION

Low-activity radioactive wastes are collected and mixed for a more uniform effluent or segregated for specific treatment of individual components. In the first approach,

TABLE 8.7.1 RADIOACTIVE WASTE DEFINITIONS

	IAEA[a]		ASA[a]		Common	
Category	Activity Level $A(\mu Cl/ml)$	Category	Activity Level	Category	Activity Level (Cl/l)
1	$A < 10^{-6}$	A	$A < MPC_p$ [b]	Low	$\sim 10^{-6}$
2	$10^{-6} < A < 10^{-3}$	B	$MPC_p < A < MPC_0$ [b]		
				Intermediate	$\sim 10^{-3}$
3	$10^{-3} < A < 10^{-1}$	C	$MPC_0 < A < 10^4 \, MPC_0$		
4	$10^{-1} < A < 10^4$	D	$10^4 \, MPC_0 < A < 10^8 \, MPC_0$		
				High	~ 1
5	$10^4 < A$	E	$10^8 \, MPC_0 < A$		

[a]IAEA, International Atomic Energy Agency; ASA, United States of America Standards Institute.
[b]MPC_p, maximum permissible concentration for members of the population at large; MPC_0, maximum permissible concentration for 40-hr work week occupational exposure.

wastewater flocculation, precipitation, sorption, filtration, and ion exchange can be adapted to radioactive wastes. Typical removals (Straub 1964) of mixed fission products and individual nuclides are shown in Table 8.7.2. Common methods for mixed fission product precipitation are aluminum salts, iron salts, tannic acid with lime, phosphate with lime, ferrocyanides, and excess lime-soda ash.

When wastes are segregated, specific treatments for radioactive isotopes include strontium, combined calcium, nonradioactive strontium-iron phosphate, or hydroxide at pH 11.5, tannic acid, and nickel ferrocyanide; cesium, nickel ferrocyanides, nickel ferrocyanides, and copper and iron ferrocyanides; and ruthenium, nickel, copper, or iron ferrocyanide. Processes are chosen based on local considerations to obtain a high degree of radioactivity removal at a high floc settling rate, with a minimum sludge volume and a maximum degree of economy.

Decontamination factors of 10 may be obtained for mixed fission products and decontamination factors of 200 for specific isotopes with a specific treatment. Provision must be made for discharging decanted water and for drying, packaging, and storing sludge from coagulation and precipitation. Residues are presently sent to an approved commercial burial site.

ION EXCHANGE

If the total solids content of the wastes is low (less than 1000 ppm), the volume of waste is small, or a final polishing of effluents is necessary, ion exchange is a suitable treatment method. At commercial power plants, ion exchange, filtration, and evaporation are the major processes used for liquid radioactive waste treatment. Both sulphonic and phenolic-carboxylic resins are used. Typical ion exchange units for waste treatment at boiling water reactor power plants include one 200 gpm mixed bed with no regeneration; one 75 gpm mixed bed, with no regeneration; and one 50 gpm mixed bed with no regeneration. Some boiling water reactors use very fine, 90% less than 325 mesh, ion exchangers as filters and ion exchange beds with no regeneration (Goldman 1968).

At pressurized water reactor power stations, typical ion exchange treatments include one 12 gpm mixed bed unit, no ion exchange in waste disposal system; 4 mixed bed units; and 45 ft.^{3} cation exchangers (Goldman 1968). All of the reactors use ion exchangers in coolant purification operations. On June 7, 1971, the Atomic Energy Commission published a schematic diagram of the general concept of radioactive waste handling systems for light water-cooled nuclear power reactors (Figure 8.7.1).

TABLE 8.7.2 TREATMENT PROCESSES FOR REMOVAL OF RADIOACTIVE WASTES

Process	Decontamination Factors[a]	
	Individual Radionuclides	Mixed Fission Products[b]
Conventional		
Coagulation and settling	0–100+	2–9.1
Clay addition, coagulation and settling	0–100	1.1–6.2
Sand filtration	1–100	
Coagulation, settling and filtration	1–50	1.4–13.3
Lime-soda ash softening	2–100	
Ion exchange, cation	1.1–500	2.0–6.1
Ion exchange, anion	0–125	
Ion exchange, mixed bed	11–3300	50–100
Solids-contact clarifier	1.9–15	2.0–6.1
Evaporation	1.00–10,000	
Nonconventional		
Phosphate	1.2–1000	125–250
Metallic dusts	1.1–1000	1.1–8.6
Clay treatment	0–100+	
Diatomaceous earth	1.1–∞	
Sedimentation	<1.05	
Activated sludge	1.03–8.2	4.8–9.8
Trickling filter	1.05–37	3.5–6.1
Sand filter	8.3–100	1.9–50
Oxidation ponds	<1.1–20	

[a]Decontamination factor = $\dfrac{\text{initial concentration}}{\text{final concentration}}$

[b]No data listed implies lack of information, not unsuitability of the process.

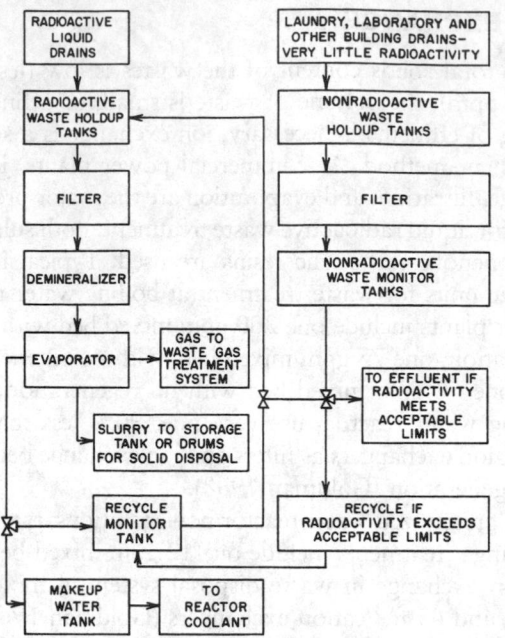

FIG. 8.7.1 Nuclear power reactor for liquid waste handling system.

EVAPORATORS

Evaporators can obtain high decontamination factors, 10^4 to 10^6, if carryover is eliminated or if the evaporator is followed by ion exchange of condensate.

The evaporator at the National Reactor Testing Station is typical of evaporators used at research and production sites. The continuous evaporator is constructed of stainless steel (Type 347) and is a thermosyphon type with an external heat exchanger of 48 sq m area, rated at 800,000 kcal/hr heat duty. Vapor passes to an entrainment chamber with 4 bubble cap trays, where it is scrubbed with clean water. The evaporator is capable of processing 1800 l/hr. The tritium in the wastes is not concentrated at all, but the other radionuclides are concentrated by a factor of 50, and the condensate is decontaminated (Lohse, Rhodes and Wheeler 1970) by a factor of 2000.

DILUTION AND RELEASE

All processes require some decontaminated waste to be released into the ground, local waterways, or the atmosphere, or that waste is recycled. No plants completely recycle wastes, although some newer reactors will do so. Studies at the Oak Ridge National Laboratory show that it is possible to treat low-activity waste by zeta-potential controlled additions of alum and activated silica to remove colloids. This is followed by demineralization and decontamination by cation and anion exchange resins, the passage through a column of activated carbon to remove cobalt and organic materials (Blanco 1966). The residues, exhausted resins, and activated carbon, must be packaged and sent to a storage facility.

Ground disposal of decontaminated liquid waste took place at many sites, but the practice lost favor because of the uncontrolled nature of the release and the irreversibility of the process (IAEA 1967). The process takes advantage of the slow movement of groundwater (enabling shorter half-lived radioisotopes to decay), the ion exchange properties of soil, and if above the water table, the capacity of the unsaturated soil to store moisture. In addition, there is slow dispersion of wastes in the groundwater system.

HYDROFRACTURE

One method of low and intermediate concentrate disposal to the ground that maintains control of wastes is hydrofracture. In this method, a well is drilled to the desired geological formation and cased. Then, the casing is perforated at the specific depth desired, pressure is applied, and the formation is fractured. After the formation is fractured, a radioactive waste mixture containing portland cement, fly ash, attapulgite, illite, delta gluconolactone, and tributyl phosphate is injected into the space and spread as a thin sheet parallel to the bedding (Figure 8.7.2). The radioactive waste forms a dense solid with improved cesium retention on the illite and improved strontium retention on the fly ash. The delta gluconolactone retards set times, and the attapulgite is a suspender, reducing the quantity of cement required.

This technique was first used in December 1966, when the first ultimate disposal of radioactive waste took place with the injection of 72,000 gal of intermediate waste containing 20,000 curies of cesium-137 in a shale formation 870 ft below the ground surface. The method was also routinely used at Oak Ridge National Laboratory and was demonstrated at the Nuclear Fuel Services site.

Hydrofracture techniques are used where there are thick formations of shale in flat-lying, well-bedded, sedimentary rock for intermediate activity wastes and possibly for low and high activity wastes with suitable modifications.

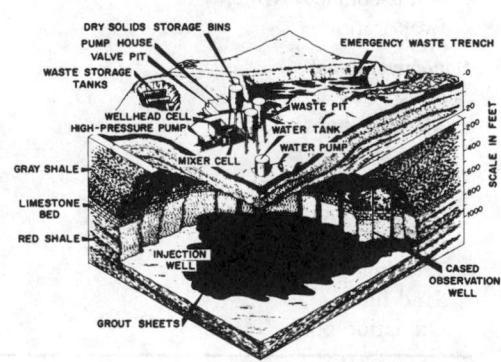

FIG. 8.7.2 Disposal of intermediate activity wastes by hydrofracture.

BITUMINIZATION

Although not used in the United States because of the adequacy of present techniques and the high costs of conversion to new methodologies, bituminization is favored for installations in other countries for solidification and immobilization of low and intermediate activity sludges and residues (IAEA 1970b). The process is easy, cheap and not dependent on waste type or storage location. Wastes are introduced into asphalt or emulsified asphalt and the water is removed. Radiation levels of 10^8 rads (see glossary) do not cause soft asphalts to swell significantly, nor do they increase the leach rate.

High-Activity Wastes

GENERATION

High-activity wastes are generated when irradiated fuel elements are reprocessed to recover unfissioned uranium and to remove the fission products with large neutron absorption cross-sections.

Spent fuel elements are cooled for 150 days or more, allowing shorter-lived fission products, particularly I^{131}, to decay. The end pieces of the fuel element are cut off and the main element sheared into small pieces and leached with hot nitric acid to dissolve the UO_2. The leached hulls are rinsed and sealed in 30-gal drums and buried as solid waste. The radiation level is 10,000 R/hr. The nitric acid-uranium-fission product impurities solution then goes through the Purex solvent extraction process to recover and decontaminate uranium and plutonium from the heterogeneous solution. The solvent is tributyl phosphate dissolved in n-dodecane, which complexes preferentially with uranium and plutonium, which are in solution in the organic phase. The plutonium is separated from the uranium by a nitric acid solution containing ferrous sulfonate and is removed in the aqueous phase. Further purification of the plutonium and uranium streams is required. The nitric acid-fission product stream is evaporated for recycling concentration and storage in tanks.

STORAGE IN TANKS

All high-activity liquid radioactive wastes are now stored in tanks below the surface. At older U.S. sites, such as Hanford, Savannah River, and the NFS fuel reprocessing plant, wastes were stored as alkaline solutions in carbon steel tanks with a capacity of more than 100,000,000 gal in over 200 tanks. Cooling is provided by coils or by condensation of boiling waste vapors. A schematic of a newer tank for high-activity storage is shown in Figure 8.7.3.

Storage in the acid form is now the preferred method because of smaller volumes, no history of leaks, and less difficulty with precipitated solids. Current U.S. regulations allow a 5-yr storage of liquid high-activity wastes in tanks at reprocessing sites before solidifying and shipping to a government repository.

Some of the older tanks at the Hanford site were used to store salt cake remaining after cesium and strontium are removed from high-activity wastes. In these units, a 3000 cfm of 1200°F airflow to an airlift circulator, a 4000 kw electric immersion heater, and a conventional steam heated tube bundle evaporator of 6 million btu/hr capacity evaporates the water and causes the remaining salts to crystallize (IAEA 1967). The process is shown schematically for the electric immersion heater in Figure 8.7.4.

CONVERSION TO SOLIDS

At the Idaho Chemical Processing plant, high-activity wastes were converted to calcined solids in a fluidized bed process (Figure 8.7.5). About 250,000 cu ft of alumina and zirconium wastes containing about 5×10^7 curies have been solidified between 1963 and 1970. The calcine powder is blown to stainless steel storage bins. Because the temperature previously had been limited to 400°C, the resultant solid was moderately leachable, and it was feared that some fission products, nitrates, and mercuric oxide might volatilize. Subsequent studies showed that they were entrapped in the powder. The solids were incorporated

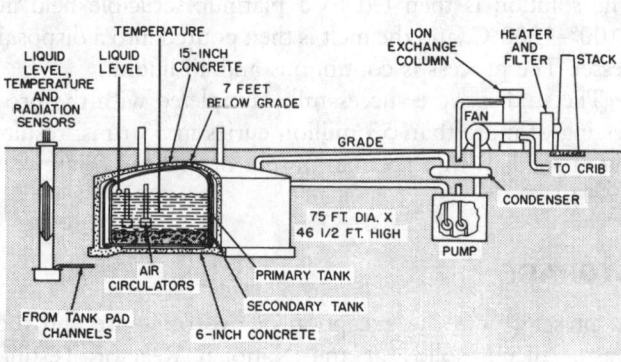

FIG. 8.7.3 Storage of high-activity nuclear wastes in tanks.

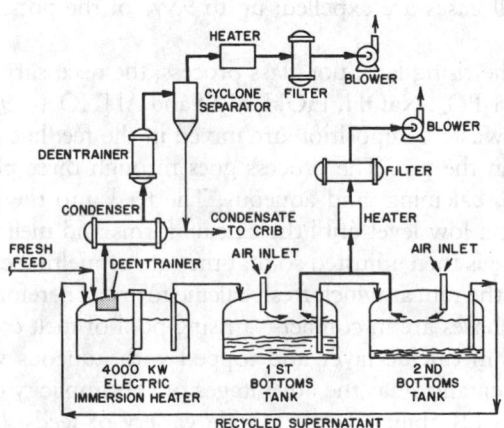

FIG. 8.7.4 In-tank conversion of radioactive wastes to salt cake.

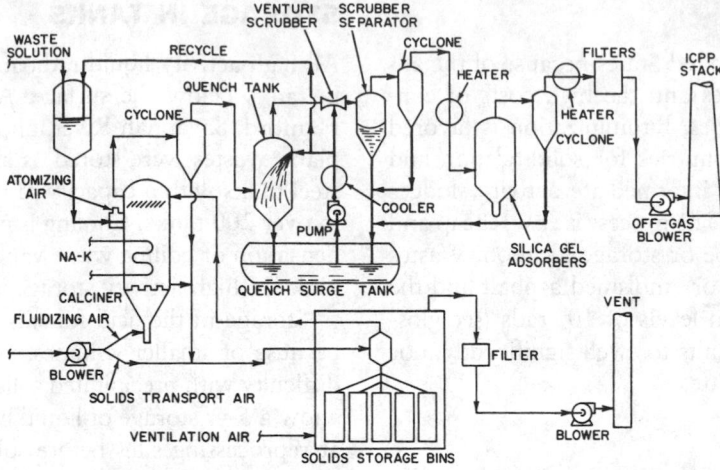

FIG. 8.7.5 Conversion of high-activity waste to calcine solid.

into a glass matrix or a pot glass system to reduce leachability and mobility.

Work on the solidification of high-activity wastes at high temperatures to obtain a less leachable solid culminated in a demonstration at the Waste Solidification Engineering Prototype facility at Hanford (Parker 1969). At this plant the batch pot solidification scheme of the Oak Ridge National Laboratory, the phosphate glass solidification process of Brookhaven National Laboratory and the radiant heat spray solidification process of Battelle Northwest Laboratory were demonstrated at full-scale (light water reactor wastes from fuels irradiated at 45,000 mw days/ton at a power level of 30 mw/ton and liquid metal fast breeder reactor wastes for fuels irradiated at 100,000 mw days/ton at 200 mw/ton).

In the pot calcination process, waste feed is batchfed to a heated process vehicle which also serves as the final disposal vessel. After the pot is filled, heating continues until the waste is converted into a calcine at about 900°C. The pot operates at a constant liquid level and the calcine is deposited radially in the pot. The rate of feed decreases as the calcined material thickens. The pot is kept at 900°C until all gases are expelled; up to 95% of the pot can be filled.

In the rising-level pot glass process, the necessary additives (H_3PO_4, NaOH, LiOH, H_2O and Al $[NO_3]_3 \cdot 9H_2O$) to the waste composition are mixed in the feedline or directly in the pot. The process goes through three phases: molten, calcining, and aqueous. The feed into the pot is kept at a low level until the calcine forms and melts. The feedrate is then adjusted so that the rate of melting calcine equals the rate at which fresh calcine forms. Therefore, the three phases are in contact—a rising pool of melt covered by a thin calcine layer and topped with aqueous waste. Pot calcination has the advantages of (1) simplicity of operation; (2) ability to use a wide variety of feeds; (3) reduction of nitrate content to low levels; (4) minimum of gas but not constant production; and (5) use of process

vessel as disposal vessel. Its disadvantages include (1) batch process; (2) poor heat conduction as calcine builds up on the walls; and (3) hazard of high organic matter concentration.

In the spray calcination process, liquid waste is atomized by spraying with steam or air through nozzles at the top of a stainless steel column, with column walls kept between 600° and 800°C by three-zone heating. The suspension of droplets falling down the column dry and are calcined into a powder. The powder falls into the melter, and process gases and some finer waste powders are blown into the filter chamber. The powder collects on a porous metal filter and is occasionally blown off by high pressure steam, falling into platinum melters where borosilicate glass balls are added to make the melt. The melter operates between 700° and 1300°C, and the molten waste flows over a weir into the disposal vessel. The advantages of the spray calciner are: (1) short residence time in the calciner (safer with thermodynamic, unstable feeds); (2) minimum volume at constant flow of off gases; and (3) utility for wide range of feed composition.

In the continuous phosphate glass solidification process, liquid waste is mixed with phosphoric acid and water, and nitric acid is volatilized at 130°–160°C for a volume reduction of about 10, and a nitrate removal of about 90%. The solution is then fed to a platinum crucible held at 1100°–1200°C, and the melt is then poured into a disposal vessel. The process is continuous and all liquid.

The studies were successfully completed with the processing of more than 53 million curies in 33 runs, resulting in solids with a thermal output of 193 kw (Blasewitz 1971).

STORAGE

At present, with the exception of hydrofracture and the waste calcine solids at the National Reactor Testing Station, all high-activity wastes is stored in tanks. Owing to federal regulations, all high-activity wastes must be converted to solids and stored in a federal repository (order issued on November 14, 1970). The most likely sites for storage appear to be geologic formations, particularly bedded salt.

Geological formations are favored because it is believed that if their integrity can be maintained, wastes will remain in place for geologic periods of time. Field scale demonstrations of 5,000,000 curies of stored fission products in irradiated fuel elements at a depth of 1000 ft in a salt mine at Lyons, Kansas, indicate that salt storage, if the integrity of the geologic formation can be maintained, will be successful (Bradshaw and McClain 1971). Salt was favored as the geologic material because even though highly soluble, it is self-sealing at moderate increases in temperature and pressure, and salt formations are widely available in the United States. Other geologic materials such as basalt, gneiss, and schists are also being considered for storage of solidified wastes. The possibility of high-activity liquid waste storage in caverns excavated in basement rock below the Savannah River plant is being vigorously pursued.

However, because of uncertainties about long-term (>1000 years) geologic behavior and the effects of stored wastes, more serious consideration is being given to short-term (<100 years) storage of solidified wastes in man-made structures for easier control, maintenance, and retrieval, if necessary.

—*F.L. Parker*

References

Blanco, R.E., et al. 1966. Recent developments in treating low and intermediate level radioactive wastes in the United States of America. In *Practices in the treatment of low and intermediate level radioactive wastes.* International Atomic Energy Agency. Vienna.

Blasewitz, A.G. (ed.). 1971. *Research and development activities fixation of radioactive residues.* (BNWL-1557). Battelle Northwest Laboratory. Richland, Wash. (February).

Bradshaw, R.L., and W.C. McClain, (eds.). 1971. *Project salt vault: A demonstration of the disposal of high activity solidified wastes in underground salt mines* (ORNL-4555). Oak Ridge National Laboratory. Oak Ridge, Tenn. (April).

De Laguna, W. 1968. *Engineering development of hydraulic fracturing as a method for permanent disposal of radioactive wastes* (ORNL-4259). Oak Ridge National Laboratory. Oak Ridge, Tenn. (August).

Goldman, M.I. 1968. United States practice in management of radioactive wastes at nuclear power plants. *Management of radioactive wastes at nuclear power plants.* International Atomic Energy Agency. Vienna.

Harvey, R.W., and W.C. Schmidt. 1971. *Radioactive waste management at Hanford.* Atlantic Richfield Hanford Company. Richland, Wash. (March).

International Atomic Energy Agency (IAEA). 1967. *Disposal of radioactive wastes into the ground.* International Atomic Energy Agency. Vienna.

International Atomic Energy Agency (IAEA). 1970b. *Management of low and intermediate level radioactive wastes.* International Atomic Energy Agency. Vienna.

International Atomic Energy Agency (IAEA). 1970a. *Standardization of radioactive waste categories.* International Atomic Energy Agency. Vienna.

Lohse, G.E., D.W. Rhodes, and B.R. Wheeler. 1970. Preventing activity release at the Idaho Chemical Processing Plant. In *Management of low and intermediate level radioactive wastes.* International Atomic Energy Agency. Vienna.

Parker, F.L. 1969. Status of radioactive waste disposal in U.S.A. *J. Sanit. Engineer. Div., American Society of Civil Engineers* 95: SA3. (June).

Straub, C.P. 1964. *Low level radioactive wastes.* U.S. Atomic Energy Commission.

United States of America Standards Institute (ASI). 1967. *Proposed definitions of radioactive waste categories.* American Institute of Chemical Engineers. New York, N.Y.

References

Groundwater and Surface Water Pollution

GROUNDWATER POLLUTION CONTROL

Yong S. Chae | Ahmed Hamidi

Groundwater Cleanup and Remediation   1086

STORM WATER POLLUTANT MANAGEMENT

David H.F. Liu | Kent K. Mao

Principles of Groundwater Flow

9.1
GROUNDWATER AND AQUIFERS

This section defines groundwater and aquifers and discusses the physical properties of soils, liquids, vadose zones, and aquifers.

Definition of Groundwater

Water exists in various forms in various places. Water can exist in vapor, liquid, or solid forms and exists in the atmosphere (atmospheric water), above the ground surface (surface water), and below the ground surface (subsurface water). Both surface and subsurface waters originate from precipitation, which includes all forms of moisture from clouds, including rain and snow. A portion of the precipitated liquid water runs off over the land (surface runoff), infiltrates and flows through the subsurface (subsurface flow), and eventually finds its way back to the atmosphere through evaporation from lakes, rivers, and the ocean; transpiration from trees and plants; or evapotranspiration from vegetation. This chain process is known as the hydrologic cycle. Figure 9.1.1 shows a schematic diagram of the hydrologic cycle.

Not all subsurface (underground) water is groundwater. Groundwater is that portion of subsurface water which occupies the part of the ground that is fully saturated and flows into a hole under pressure greater than atmospheric pressure. If water does not flow into a hole, where the pressure is that of the atmosphere, then the pressure in water is less than atmospheric pressure. Depths of groundwater vary greatly. Places exist where groundwater has not been reached at all (Bouwer 1978).

The zone between the ground surface and the top of groundwater is called the *vadose zone* or *zone of aeration*. This zone contains water which is held to the soil particles by capillary force and forces of cohesion and adhesion. The pressure of water in the vadose zone is negative due to the surface tension of the water, which produces a negative pressure head. Subsurface water can therefore be classified according to Table 9.1.1.

Groundwater accounts for a small portion of the world's total water, but it accounts for a major portion of the world's freshwater resources as shown in Table 9.1.2.

Table 9.1.2 illustrates that groundwater represents about 0.6% of the world's total water. However, except for glaciers and ice caps, it represents the largest source of freshwater supply in the world's hydrologic cycle. Since much of the groundwater below a depth of 0.8 km is saline or costs too much to develop, the total volume of readily usable groundwater is about 4.2 million cubic km (Bouwer 1978).

Groundwater has been a major source of water supply throughout the ages. Today, in the United States, groundwater supplies water for about half the population and supplies about one-third of all irrigation water. Some three-fourths of the public water supply system uses groundwater, and groundwater is essentially the only water source for the roughly 35 million people with private systems (Bouwer 1978).

Aquifers

Groundwater is contained in geological formations, called *aquifers,* which are sufficiently permeable to transmit and yield water. Sands and gravels, which are found in alluvial deposits, dunes, coastal plains, and glacial deposits, are the most common aquifer materials. The more porous the material, the higher yielding it is as an aquifer material. Sandstone, limestone with solution channels, and other Karst formations are also good aquifer materials. In general, igneous and metamorphic rocks do not make good aquifers unless they are sufficiently fractured and porous.

Figure 9.1.2 schematically shows the types of aquifers. The two main types are *confined aquifers* and *unconfined aquifers.* A confined aquifer is a layer of water-bearing material overlayed by a relatively impervious material. If the confining layer is essentially impermeable, it is called an *aquiclude.* If it is permeable enough to transmit water vertically from or to the confined aquifer, but not in a horizontal direction, it is called an *aquitard.* An aquifer bound by one or two aquitards is called a *leaky* or *semiconfined aquifer.*

Confined aquifers are completely filled with groundwater under greater-than-atmospheric pressure and therefore do not have a *free water table.* The pressure condition in a confined aquifer is characterized by a *piezometric surface,* which is the surface obtained by connecting equilibrium water levels in tubes or piezometers penetrating the confined layer.

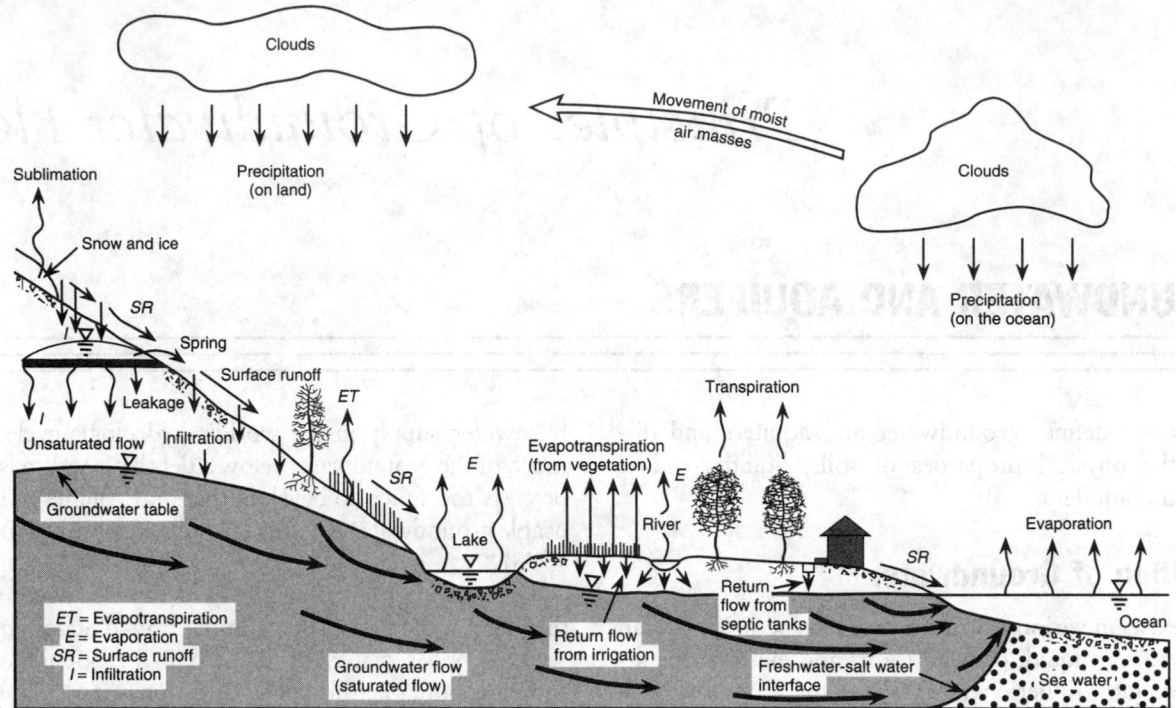

FIG. 9.1.1 Schematic diagram of the hydrologic cycle. (Reprinted from J. Bear, 1979, *Hydraulics of groundwater*, McGraw-Hill, Inc.)

An unconfined aquifer is a layer of water-bearing material without a confining layer at the top of the groundwater, called the *groundwater table*, where the pressure is equal to atmospheric pressure. The groundwater table, sometimes called the *free* or *phreatic surface*, is free to rise or fall. The groundwater table height corresponds to the equilibrium water level in a well penetrating the aquifer. Above the water table is the vadoze zone, where water

pressures are less than atmospheric pressure. The soil in the vadoze zone is partially saturated, and the air is usually continuous down to the unconfined aquifer.

Physical Properties of Soils and Liquids

The following discussion describes the physical properties of soils and liquids. It also defines the terms used to describe these properties.

PHYSICAL PROPERTIES OF SOILS

Natural soils consist of solid particles, water, and air. Water and air fill the pore space between the solid grains. Soil can be classified according to the size of the particles as shown in Table 9.1.3.

Soil classification divides soils into groups and subgroups based on common engineering properties such as *texture, grain size distribution,* and Atterberg limits. The most widely accepted classification system is the unified classification system which uses group symbols for identification, e.g., SW for well-graded sand and CH for inorganic clay of high plasticity. For details, refer to any standard textbook on soil mechanics.

Figure 9.1.3 shows an element of soil, separated in three phases. The following terms describe some of the engineering and physical properties of soils used in groundwater analysis and design:

TABLE 9.1.1 CLASSIFICATION OF SUBSURFACE WATER

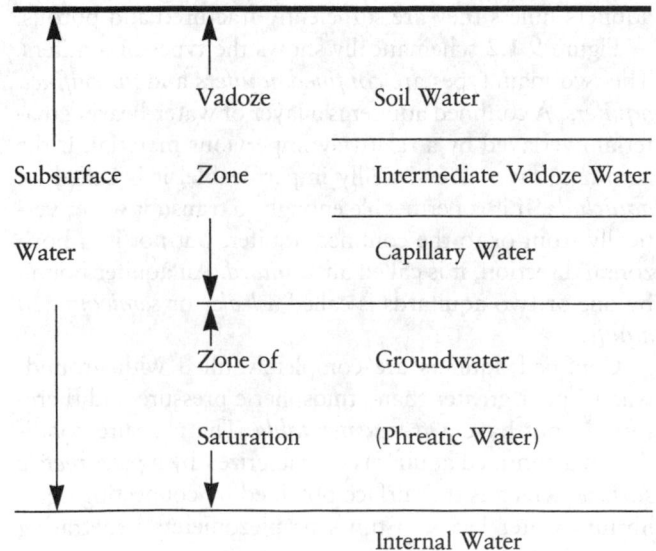

	Vadoze	Soil Water
Subsurface	Zone	Intermediate Vadoze Water
Water		Capillary Water
	Zone of	Groundwater
	Saturation	(Phreatic Water)
		Internal Water

TABLE 9.1.2 ESTIMATED DISTRIBUTION OF WORLD'S WATER

	Volume 1000 km³	Percentage of Total Water
Atmospheric water	13	0.001
Surface water		
Salt water in oceans	1,320,000	97.2
Salt water in lakes and inland seas	104	0.008
Fresh water in lakes	125	0.009
Fresh water in stream channels (average)	1.25	0.0001
Fresh water in glaciers and icecaps	29,000	2.15
Water in the biomass	50	0.004
Subsurface water		
Vadose water	67	0.005
Groundwater within depth of 0.8 km	4200	0.31
Groundwater between 0.8 and 4 km depth	4200	0.31
Total (rounded)	1,360,000	100

Source: H. Bouwer, 1978, *Groundwater hydrology* (McGraw-Hill, Inc.).

POROSITY (n)—A measure of the amount of pores in the material expressed as the ratio of the volume of voids (V_v) to the total volume (V), $n = V_v/V$. For sandy soils n = 0.3 to 0.5; for clay n > 0.5.

VOID RATIO (e)—The ratio between V_v and the volume of solids V_S, $e = V_v/V_S$; where e is related to n as $e = n/(1 - n)$.

WATER CONTENT (ω)—The ratio of the amount of water in weight (W_W) to the weight of solids (W_S), $\omega = W_W/W_S$.

DEGREE OF SATURATION (S)—The ratio of the volume of water in the void space (V_W) to V_v, $S = V_W/V_v$. S varies between 0 for dry soil and 1 (100%) for saturated soil.

COEFFICIENT OF COMPRESSIBILITY (α)—The ratio of the change in soil sample height (h) or volume (V) to the change in applied pressure (σ_v)

$$\alpha = -\frac{1}{h}\frac{dh}{d\sigma_v} = -\frac{1}{V}\frac{dV}{d\sigma_v} \qquad 9.1(1)$$

The α can be expressed as

$$\alpha = \frac{(1 + \mu)(1 - 2\mu)}{E(1 - \mu)} = \frac{1}{B + \frac{4}{3}G} \qquad 9.1(2)$$

where:

E = Young's modulus
μ = Poisson's ratio
B = bulk modulus
G = shear modulus

Clay exists in either a dispersed or flocculated structure depending on the arrangement of the clay particles with

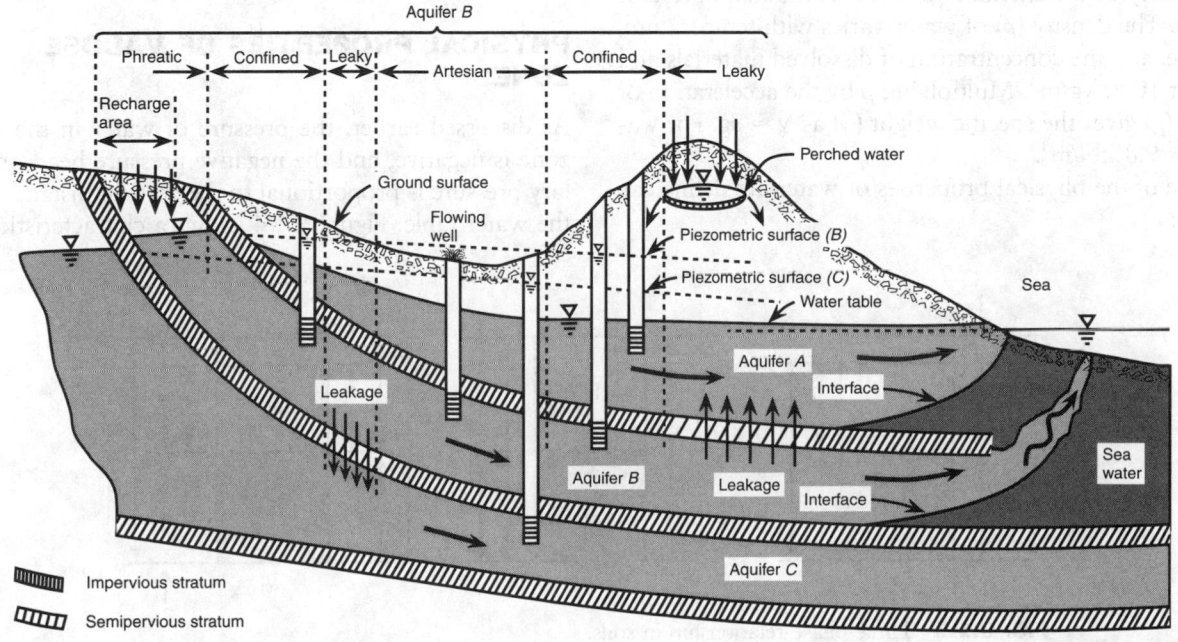

FIG. 9.1.2 Types of aquifers. (Reprinted from J. Bear, 1979, *Hydraulics of groundwater*, McGraw-Hill, Inc.)

TABLE 9.1.3 USUAL SIZE RANGE FOR GENERAL SOIL
CLASSIFICATION TERMINOLOGY

Material	Upper, mm	Lower, mm	Comments
Boulders, cobbles	1000^+	75^-	
Gravel, pebbles	75	2–5	No. 4 or larger sieve
Sand	2–5	0.074	No. 4 to No. 200 sieve
Silt	0.074–0.05	0.006	Inert
Rock flour	0.006		Inert
Clay	0.002	0.001	Particle attraction, water absorption
Colloids	0.001		

Source: J.E. Bowles, 1988, *Foundation analysis and design,* 4th ed. (McGraw-Hill).

the type of cations that are adsorbed to the clay. If the layer of adsorbed cation (such as C_a^{++}) is thin and the clay particles can be close together, making the attractive van der Waals forces dominant between the particles, then the clay is flocculated. If the clay particles are kept some distance apart by adsorbed cations (such as N_a^+), the repulsive electrostatic forces are dominant, and the clay is dispersed. Since clay particles are negatively charged, which can adsorb cations from the soil solution, clay can be converted from a dispersed state to a flocculant condition through the process of cation exchange (e.g. $N_a^+ \rightarrow C_a^{++}$) which changes the adsorbed ions. The reverse, changing from a flocculated to a dispersed clay, can also occur. Clay structure change is used to handle some groundwater problems in clay because the hydraulic properties of soil are dependent upon the clay structure.

PHYSICAL PROPERTIES OF WATER

The density of a material is defined as the mass per unit volume. The density (ρ) of water varies with temperature, pressure, and the concentration of dissolved materials and is about 1000 kg/m^3. Multiplying ρ by the acceleration of gravity (g) gives the specific weight (γ) as $\gamma \approx \rho g$. For water, $\gamma \approx 9.8$ kN/m^3.

Some of the physical properties of water are defined as follows:

DYNAMIC VISCOSITY (μ)—The ratio of shear stress (τ_{yx}) in x direction, acting on an x–y plane to velocity gradient (dv$_x$/dy); $\tau_{yx} = \mu\, dv_x/dy$. For water, $\mu = 10^{-3}$ kg/m·s.

KINEMATIC VISCOSITY (v)—Related to μ by $v = \mu/\rho$. Its value is about 10^{-6} m^2/s for water.

COMPRESSIBILITY (β)—The ratio of change in density caused by change in pressure to the original density

$$\beta = \frac{1}{\rho}\frac{d\rho}{dp} = -\frac{1}{V}\frac{dV}{dp}$$

$$\beta \approx 0.5 \times 10^{-9}\ \text{m}^2/\text{N} \qquad 9.1(3)$$

The variation of density and viscosity of water with temperature can be obtained from Table 9.1.4.

Physical Properties of Vadose Zones and Aquifers

A description of the physical properties of vadose zones and aquifers follows.

PHYSICAL PROPERTIES OF VADOSE ZONES

As discussed earlier, the pressure of water in the vadose zone is negative, and the negative pressure head or capillary pressure is proportional to the vertical distance above the water table. Figure 9.1.4 shows a characteristic curve

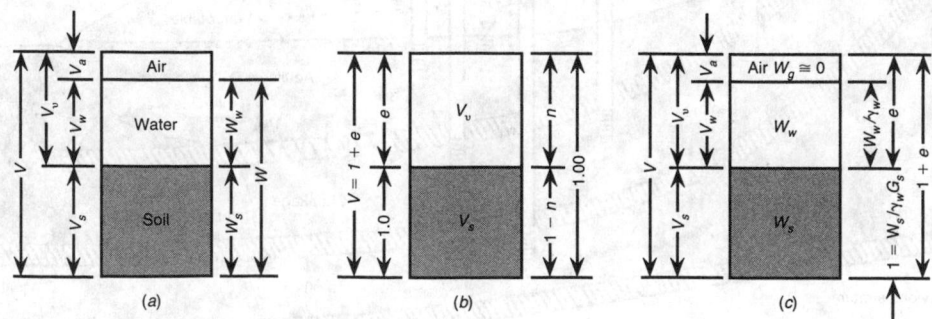

FIG. 9.1.3 Three-phase relationship in soils.

TABLE 9.1.4 VARIATION OF DENSITY AND VISCOSITY OF WATER WITH TEMPERATURE

Temperature (°C)	Density (kg/m³)	Dynamic Viscosity (kg/m s)
0	999.868	1.79×10^{-3}
5	999.992	1.52×10^{-3}
10	999.727	1.31×10^{-3}
15	999.126	1.14×10^{-3}
20	998.230	1.01×10^{-3}

Source: A. Verrjuitt, 1982, *Theory of groundwater flow,* 2d ed. (Macmillan Publishing Co.).

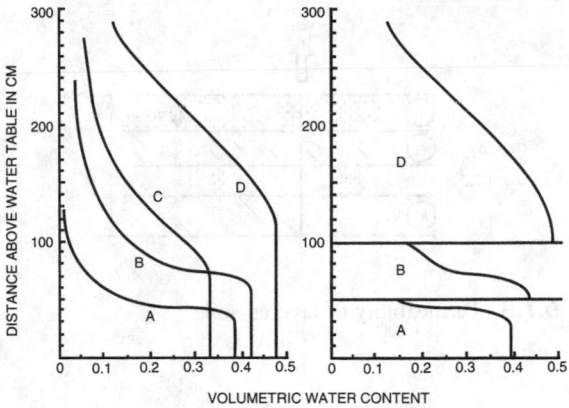

FIG. 9.1.4 Schematic equilibrium water-content distribution above a water table (left) for a coarse uniform sand (A), a fine uniform sand (B), a well-graded fine sand (C), and a clay soil (D). The right plot shows the corresponding equilibrium water-content distribution in a soil profile consisting of layers of materials A, B, and D. (Reprinted from H. Bouwer, 1978, *Groundwater hydrology,* McGraw-Hill, Inc.)

of the relationship between volumetric water content and the negative pressure head (height above the water table or capillary pressure).

For materials with relatively uniform particle size and large pores, the water content decreases abruptly once the air-entry value is reached. These materials have a well-defined capillary fringe. For well-graded materials and materials with fine pores, the water content decreases more gradually and has a less well-defined capillary fringe.

At a large capillary pressure, the volumetric water content tends towards a constant value because the forces of adhesion and cohesion approach zero. The volumetric water content at this state is equal to the *specific retention.* The specific retention is then the amount of water retained against the force of gravity compared to the total volume of the soil when the water from the pore spaces of an unconfined aquifer is drained and the groundwater table is lowered.

PHYSICAL PROPERTIES OF AQUIFERS

As stated before, an aquifer serves as an underground storage reservoir for water. It also acts as a conduit through which water is transmitted and flows from a higher level to a lower level of energy. An aquifer is characterized by the three physical properties: *hydraulic conductivity, transmissivity, and storativity.*

Hydraulic Conductivity

Hydraulic conductivity, analogous to electric or thermal conductivity, is a physical measure of how readily an aquifer material (soil) transmits water through it. Mathematically, it is the proportionality between the rate of flow and the energy gradient causing that flow as expressed in the following equation. Therefore, it depends on the properties of the aquifer material (porous medium) and the fluid flowing through it.

$$K = k \frac{\gamma}{\mu} \qquad \text{9.1(4)}$$

where:

K = hydraulic conductivity (called the coefficient of permeability in soil mechanics)
k = intrinsic permeability
γ = specific weight of fluid
μ = dynamic viscosity of fluid

For a given fluid under a constant temperature and pressure, the hydraulic conductivity is a function of the properties of the aquifer material, that is, how permeable the soil is. The subject of hydraulic conductivity is discussed in more detail in Section 9.2.

Transmissivity

Transmissivity is the physical measure of the ability of an aquifer of a known dimension to transmit water through it. In an aquifer of uniform thickness d, the transmissivity T is expressed as

$$T = \bar{K}d \qquad \text{9.1(5)}$$

where $\bar{K}$ represents an average hydraulic conductivity. When the hydraulic conductivity is a continuous function of depth

$$\bar{K} = \frac{1}{d} \int_o^d Kz \, dz \qquad \text{9.1(6)}$$

When a medium is stratified, either in horizontal (x) or vertical (y) direction with respect to hydraulic conductivity as shown in Figure 9.1.5, the average value $\bar{K}$ can be obtained by

$$\bar{K}_x = \sum_{m=1}^{n} \frac{K_m d_m}{d} \qquad \text{9.1(7)}$$

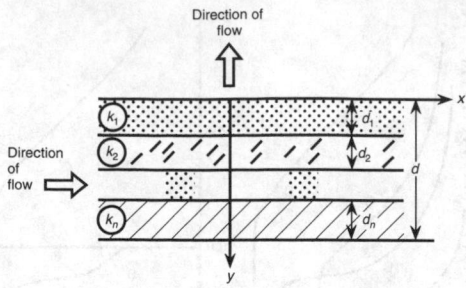

FIG. 9.1.5 Permeability of layered soils.

$$\bar{K}_y = \frac{d}{\sum\limits_{m=1}^{n} \frac{d_m}{K_m}} \qquad 9.1(8)$$

Storativity

Storativity, also known as the *coefficient of storage* or *specific yield*, is the volume of water yielded or released per unit horizontal area per unit drop of the water table in an unconfined aquifer or per unit drop of the piezometric surface in a confined aquifer. Storativity S is expressed as

$$S = \frac{1}{A} \frac{dQ}{d\phi} \qquad 9.1(9)$$

where:

dQ = volume of water released or restored
dϕ = change of water table or piezometric surface

Thus, if an unconfined aquifer releases 2 m³ water as a result of dropping the water table by 2m over a horizontal area of 10 m², the storativity is 0.1 or 10%.

—*Y.S. Chae*

Reference

Bouwer, H. 1978. *Groundwater hydrology.* McGraw-Hill, Inc.

9.2
FUNDAMENTAL EQUATIONS OF GROUNDWATER FLOW

The flow of water through a body of soil is a complex phenomenon. A body of soil constitutes, as described in Section 9.1, a solid matrix and pores. For simplicity, assume that all pores are interconnected and the soil body has a uniform distribution of phases throughout. To find the law governing groundwater flow, the phenomenon is described in terms of average velocities, average flow paths, average flow discharge, and pressure distribution across a given area of soil.

The theory of groundwater flow originates with Henry Darcy who published the results of his experimental work in 1856. He performed a series of experiments of the type shown in Figure 9.2.1. He found that the total discharge Q was proportional to cross-sectional area A, inversely proportional to the length Δs, and proportional to the head difference $\phi_1 - \phi_2$ as expressed mathematically in the form

$$Q = KA \frac{\phi_1 - \phi_2}{\Delta s} \qquad 9.2(1)$$

where K is the proportionality constant representing hydraulic conductivity. This equation is known as Darcy's equation. The quantity Q/A is called *specific discharge* q. If $\phi_1 - \phi_2 = \Delta\phi$ and $\Delta s \rightarrow 0$, Equation 9.2(1) becomes

$$q = -K \frac{d\phi}{ds} \qquad 9.2(2)$$

This equation states that the specific discharge is directly proportional to the derivative of the head in the direction of flow (hydraulic gradient). The specific discharge is also known as Darcy's velocity. Note that q is not the actual flow velocity (seepage velocity) because the flow is limited to pore space only. The seepage velocity v is then

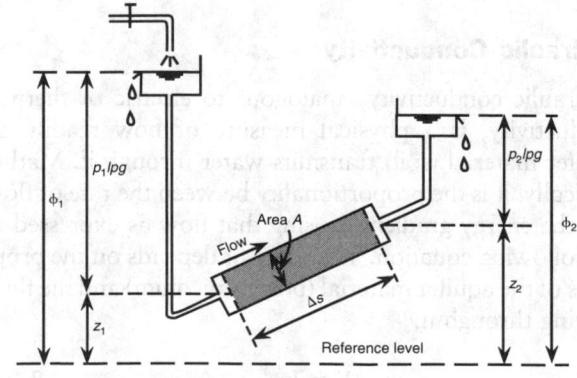

FIG. 9.2.1 Darcy's experiment.

$$v = \frac{Q}{n \cdot A} = \frac{q}{n} \qquad 9.2(3)$$

where n is the porosity of the soil. Note that v is always larger than q.

Intrinsic Permeability

The hydraulic conductivity K is a material constant, and it depends not only on the type of soil but also on the type of fluid (dynamic viscosity μ) percolating through it. The hydraulic conductivity K is expressed as

$$K = k\frac{\gamma}{\mu} \qquad 9.2(4)$$

where k is called the intrinsic permeability and is now a property of the soil only. Many attempts have been made to express k by such parameters as average pore diameter, porosity, and effective soil grain size. The most familiar equation is that of Kozeny-Carmen

$$k = Cd^2 \frac{n^3}{(1-n)^2} \qquad 9.2(5)$$

where:

   n = porosity
   d = the effective pore diameter
   C = a constant to account for irregularities in the geometry of pore space

Another equation by Hazen states

$$k = CD^2 = C_1D_{10}^2 \qquad 9.2(6)$$

where:

   D = the average grain diameter
   D_{10} = the effective diameter of the grains retained

Values of hydraulic conductivity can be obtained from empirical formulas, laboratory experiments, or field tests. Table 9.2.1 gives the typical values for various aquifer materials.

Validity of Darcy's Law

Darcy's law is restricted to a specific discharge less than a certain critical value and is valid only within a laminar

TABLE 9.2.1 THE ORDER OF MAGNITUDE OF THE PERMEABILITY OF NATURAL SOILS

	$k\ (m^2)$	$K\ (m/s)$
Clay	10^{-17} to 10^{-15}	10^{-10} to 10^{-8}
Silt	10^{-15} to 10^{-13}	10^{-8} to 10^{-6}
Sand	10^{-12} to 10^{-10}	10^{-5} to 10^{-3}
Gravel	10^{-9} to 10^{-8}	10^{-2} to 10^{-1}

Source: A. Verrjuit, 1982, *Theory of groundwater flow*, 2d ed. (Macmillan Publishing Co.).

flow condition, which is expressed by Reynolds number R_e defined as

$$R_e = \frac{qD\rho}{\mu} = \frac{qD}{\nu} \qquad 9.2(7)$$

Experiments have shown the range of validity of Darcy's law to be

$$R_e \leq 1 \sim 10 \qquad 9.2(8)$$

In practice, the specific discharge is always small enough for Darcy's law to be applicable. Only cases of flow through coarse materials, such as gravel, deviate from Darcy's law. Darcy's law is not valid for flow through extremely fine-grained soils, such as colloidal clays.

Generalization of Darcy's Law

In practice, flow is seldom one dimensional, and the magnitude of the hydraulic gradient is usually unknown. The simple form, Equation 9.2(2), of Darcy's law is not suitable for solving problems. A generalized form must be used, assuming the hydraulic conductivity K to be the same in all directions, as

$$q_x = -K\frac{\partial \phi}{\partial x}$$

$$q_y = -K\frac{\partial \phi}{\partial y}$$

$$q_z = -K\frac{\partial \phi}{\partial z} \qquad 9.2(9)$$

For an anisotropic material, these equations can be written as

$$q_x = -K_{xx}\frac{\partial \phi}{\partial x} - K_{xy}\frac{\partial \phi}{\partial y} - K_{xz}\frac{\partial \phi}{\partial z}$$

$$q_y = -K_{yx}\frac{\partial \phi}{\partial x} - K_{yy}\frac{\partial \phi}{\partial y} - K_{yz}\frac{\partial \phi}{\partial z}$$

$$q_z = -K_{zx}\frac{\partial \phi}{\partial x} - K_{zy}\frac{\partial \phi}{\partial y} - K_{zz}\frac{\partial \phi}{\partial z} \qquad 9.2(10)$$

In the special case that $K_{xy} = K_{xz} = K_{yx} = K_{yz} = K_{zx} = K_{zy} = 0$, the x, y, and z directions are the principal directions of permeability, and Equations 9.2(10) reduce to

$$q_x = -K_{xx}\frac{\partial \phi}{\partial x} = -K_x\frac{\partial \phi}{\partial x}$$

$$q_y = -K_{yy}\frac{\partial \phi}{\partial y} = -K_y\frac{\partial \phi}{\partial y}$$

$$q_z = -K_{zz}\frac{\partial \phi}{\partial z} = -K_z\frac{\partial \phi}{\partial z} \qquad 9.2(11)$$

This chapter considers isotropic soils since problems for anisotropic soils can be easily transformed into problems for isotropic soils.

Equation of Continuity

Darcy's law furnishes three equations of motion for four unknowns (q_x, q_y, q_z, and ϕ). A fourth equation notes that the flow phenomenon must satisfy the fundamental physical principle of conservation of mass. When an elementary block of soil is filled with water, as shown in Figure 9.2.2, no mass can be gained or lost regardless of the pattern of flow.

The conservation principal requires that the sum of the three quantities (the mass flow) is zero, hence when divided by $\Delta x \cdot \Delta y \cdot \Delta z$

$$\frac{\partial(\rho q_x)}{\partial x} + \frac{\partial(\rho q_y)}{\partial y} + \frac{\partial(\rho q_z)}{\partial z} = 0 \qquad 9.2(12)$$

When the density is a constant, then Equation 9.2(12) is reduced to

$$\frac{\partial q_x}{\partial x} + \frac{\partial q_y}{\partial y} + \frac{\partial q_z}{\partial z} = 0 \qquad 9.2(13)$$

This equation is called the equation of continuity.

Fundamental Equations

Darcy's law and the continuity equation provide four equations for the four unknowns. Substituting Darcy's law Equation 9.2(9) into the equation of continuity Equation 9.2(13) yields

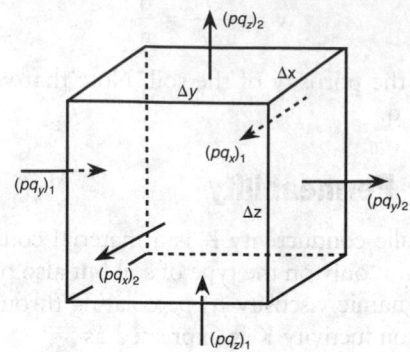

FIG. 9.2.2   Conservation of mass.

$$\frac{\partial^2\phi}{\partial x^2} + \frac{\partial^2\phi}{\partial y^2} + \frac{\partial^2\phi}{\partial z^2} = 0 \qquad 9.2(14)$$

or

$$\nabla^2\phi = 0 \qquad 9.2(15)$$

which is Laplace's equation in three dimensions.

Solving groundwater flow problems amounts to solving Laplace's equation with the appropriate boundary conditions. It is essentially a mathematical problem. Sometimes a problem must be simplified before it can be solved, and these simplifications involve considering the physical condition of groundwater flow.

—*Y.S. Chae*

9.3
CONFINED AQUIFERS

This section discusses groundwater flow in confined aquifers including one-dimensional horizontal flow, semi-confined flow, and radial flow. It also discusses radial flow in a semiconfined aquifer.

One-Dimensional Horizontal Flow

One-dimensional horizontal confined flow means that water is flowing through a confined aquifer in one direction only. Figure 9.3.1 shows an example of such a flow. Since $q_y = q_z = 0$, the governing Equation 9.2(14) reduces to

$$\frac{d^2\phi}{dx^2} = 0 \qquad 9.3(1)$$

and the general solution of this equation is $\phi = Ax + B$. Using the boundary conditions from Figure 9.3.1 of

$$x = 0 \qquad \phi = \phi_1$$
$$x = L \qquad \phi = \phi_2$$

gives

$$\phi = \phi_1 - \frac{\phi_1 - \phi_2}{L}x \qquad 9.3(2)$$

Equation 9.3(2) indicates that the piezometric head ϕ decreases linearly with distance. The specific discharge q_x is

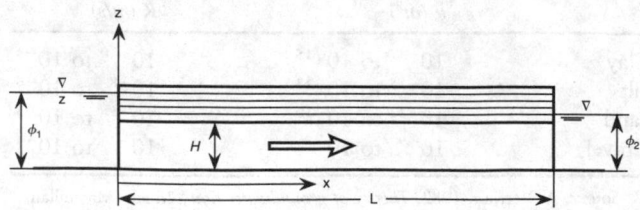

FIG. 9.3.1   One-dimensional flow in a confined aquifer.

then found using Darcy's law

$$q_x = -K \frac{\partial \phi}{\partial x} = K \frac{\phi_1 - \phi_2}{L} \qquad 9.3(3)$$

which follows that the specific discharge does not vary with position. The discharge flowing through the aquifer Q_x per unit length of the river bank is then

$$Q_x = q_x \cdot H = KH \frac{\phi_1 - \phi_2}{L} \qquad 9.3(4)$$

Semiconfined Flow

If an aquifer is bound by one or two aquitards which allow water to be transmitted vertically from or to the confined aquifer as shown in Figure 9.3.2, then a semiconfined or leaky aquifer exists, and the flow through this aquifer is called *semiconfined flow*. Small amounts of water can enter (or leave) the aquifer through the aquitards of low permeability, which cannot be ignored. Yet in the aquifer proper, the horizontal flow dominates ($q_z = o$ is assumed).

The fundamental equation of semiconfined flow is derived from the principle of continuity and Darcy's law as follows:

Consider an element of the aquifer shown in Figure 9.3.2. The net outward flux due to the flow in x and y directions is

$$-K \left(\frac{\partial^2 \phi}{\partial x^2} + \frac{\partial^2 \phi}{\partial y^2} \right) \Delta x \cdot \Delta y \cdot H \qquad 9.3(5)$$

The amount of water percolating through the layers per unit time is

$$K_1 \frac{\phi - \phi_1}{d_1} \Delta x \cdot \Delta y$$

$$K_2 \frac{\phi - \phi_2}{d_2} \Delta x \cdot \Delta y \qquad 9.3(6)$$

Continuity now requires that the sum of these quantities be zero, hence

$$KH \left(\frac{\partial^2 \phi}{\partial x^2} + \frac{\partial^2 \phi}{\partial y^2} \right) - \frac{\phi - \phi_1}{c_1} - \frac{\phi - \phi_2}{c_2} = 0 \qquad 9.3(7)$$

where $c_1 = d_1/K_1$ and $c_2 = d_2/K_2$, which are called *hydraulic resistances* of the confining layers. The terms $(\phi - \phi_1)/c_1$ and $(\phi - \phi_2)/c_2$ represent the vertical leakage through the confining layers.

Defining *leakage factor* $\lambda = \sqrt{Tc}$ where $T = KH$, the transmissivity of the aquifer, Equation 9.3(7), can be written as

$$\frac{\partial^2 \phi}{\partial x^2} + \frac{\partial^2 \phi}{\partial y^2} - \frac{\phi - \phi_1}{\lambda^2_1} - \frac{\phi - \phi_2}{\lambda^2_2} = 0 \qquad 9.3(8)$$

This equation is the fundamental equation of semiconfined flow. When the confining layers are completely impermeable ($K_1 = K_2 = 0$), Equation 9.3(8) reduces to Equation 9.2(14).

Radial Flow

Radial flow in a confined aquifer occurs when the flow is symmetrical about a vertical axis. An example of radial flow is that of water pumped through a well in an open field or a well located at the center of an island as shown in Figure 9.3.3. The distance R, called the *radius of influence zone*, is the distance to the source of water where the piezometric head ϕ_0 does not vary regardless of the amount of pumping. The radius R is well defined in the case of pumping in a circular island. In an open field, however, the distance R is theoretically infinite, and a steady-state solution cannot be obtained. In practice, this case does not occur, and R can be obtained by empirical formula or measurements.

The differential equation governing radial flow is obtained when the cartesian coordinates used for rectilinear flow are transformed into polar coordinates as

$$\frac{\partial^2 \phi}{\partial x^2} + \frac{\partial^2 \phi}{\partial y^2} = \frac{\partial^2 \phi}{\partial r^2} + \frac{1}{r} \frac{\partial \phi}{\partial r} + \frac{1}{r^2} \frac{\partial^2 \phi}{\partial r^2} + \frac{1}{r^2} \frac{\partial^2 \phi}{\partial \theta^2} \qquad 9.3(9)$$

Since ϕ is independent of angle θ, the last term of this equation can be dropped. The fundamental equation of

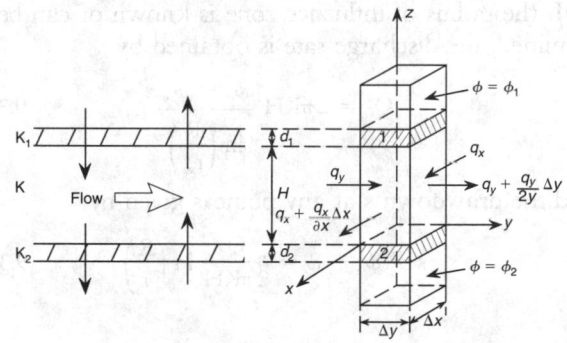

FIG. 9.3.2 Semiconfined flow.

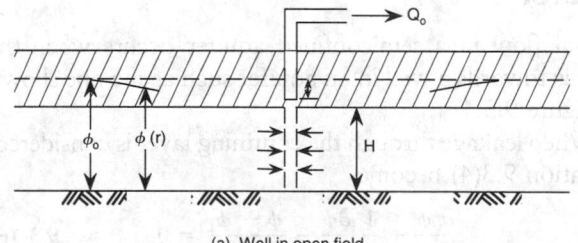

(a) Well in open field

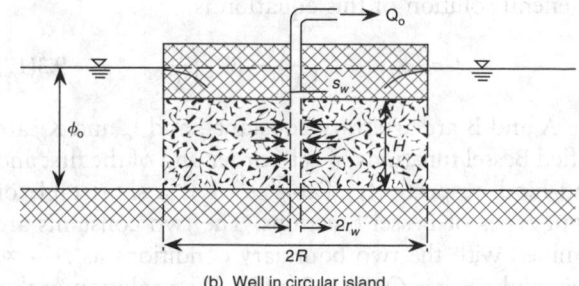

(b) Well in circular island

FIG. 9.3.3 Radial flow in a confined aquifer.

radial flow is then

$$\frac{\partial^2 \phi}{\partial r^2} + \frac{1}{r} \frac{\partial \phi}{\partial r} = 0 \qquad \textbf{9.3(10)}$$

or

$$\frac{1}{r} \frac{d}{dr} \left(r \frac{d\phi}{dr} \right) = 0 \qquad \textbf{9.3(11)}$$

The solution of this differential equation with boundary conditions (Gupta 1989) yields

$$\phi = \frac{Q}{2\pi KH} \ln \frac{r}{R} + \phi_o \qquad \textbf{9.3(12)}$$

This equation is known as the Thiem equation.

To calculate the head at the well ϕ_w using Equation 9.3(12), substitute the radius of the well r_w for r, which gives

$$\phi_w = \frac{Q}{2\pi KH} \ln \left(\frac{r_w}{R} \right) + \phi_o \qquad \textbf{9.3(13)}$$

Since the flow is confined, the head at the well must be above the upper impervious boundary (ϕ must be greater than H). Otherwise, the flow in that situation becomes unconfined flow, and Equation 9.3(13) is not applicable.

If the radius of influence zone is known or can be determined, the discharge rate is obtained by

$$Q_o = 2\pi KH \frac{\phi_o - \phi_w}{\ln \left(\dfrac{R}{r_w} \right)} \qquad \textbf{9.3(14)}$$

and the drawdown s at any point is given by

$$s = \phi_o - \phi = \frac{Q}{2\pi KH} \ln \left(\frac{R}{r} \right) \qquad \textbf{9.3(15)}$$

Radial Flow in a Semiconfined Aquifer

Radial flow in a semiconfined aquifer occurs when the flow is towards a well in an aquifer such as the one shown in Figure 9.3.4.

When leakage through the confining layer is considered, Equation 9.3(4) becomes

$$\frac{\partial^2 \phi}{\partial r^2} + \frac{1}{r} \frac{\partial \phi}{\partial r} - \frac{\phi - \phi_1}{\lambda_1^2} = 0 \qquad \textbf{9.3(16)}$$

The general solution of this equation is

$$\phi = \phi_o + AI_o \left(\frac{r}{\lambda} \right) + BK_o \left(\frac{r}{\lambda} \right) \qquad \textbf{9.3(17)}$$

where A and B are arbitrary constants, and I_o and K_o are modified Bessel functions of zero order and of the first and second kind, respectively. Table 9.3.1 is a short table of the four types of Bessel functions. The two constants are determined with the two boundary conditions as r → ∞, $\phi = \phi_o$ and $r - r_w$, $Q_o = -2\pi r H q_r$. The solution of this equation is then

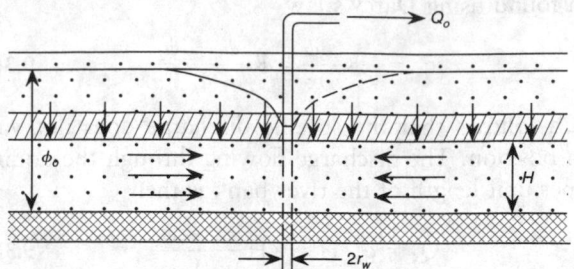

FIG. 9.3.4 Radial flow in an infinite semiconfined aquifer. (Reprinted from A. Verrjuit, 1982, *Theory of groundwater flow*, 2d ed., Macmillan Pub. Co.)

TABLE 9.3.1 BESSEL FUNCTIONS

x	$I_0(x)$	$I_1(x)$	$K_0(x)$	$K_1(x)$
0.0	1.0000	0.0000	∞	∞
0.1	1.0025	0.0501	2.4271	9.8538
0.2	1.0100	0.1005	1.7527	4.7760
0.3	1.0226	0.1517	1.3725	3.0560
0.4	1.0404	0.2040	1.1145	2.1844
0.5	1.0635	0.2579	0.9244	1.6564
0.6	1.0920	0.3137	0.7775	1.3028
0.7	1.1263	0.3719	0.6605	1.0503
0.8	1.1665	0.4329	0.5653	0.8618
0.9	1.2130	0.4971	0.4867	0.7165
1.0	1.2661	0.5652	0.4210	0.6019
1.1	1.3262	0.6375	0.3656	0.5098
1.2	1.3937	0.7147	0.3185	0.4346
1.3	1.4693	0.7973	0.2782	0.3726
1.4	1.5534	0.8861	0.2436	0.3208
1.5	1.6467	0.9817	0.2138	0.2774
1.6	1.7500	1.0848	0.1880	0.2406
1.7	1.8640	1.1963	0.1655	0.2094
1.8	1.9896	1.3172	0.1459	0.1826
1.9	2.1277	1.4482	0.1288	0.1597
2.0	2.2796	1.5906	0.1139	0.1399
2.1	2.4463	1.7455	0.1008	0.1228
2.2	2.6291	1.8280	0.0893	0.1079
2.3	2.8296	2.0978	0.0791	0.0950
2.4	3.0493	2.2981	0.0702	0.0837
2.5	3.2898	2.5167	0.0624	0.0739
2.6	3.5533	2.7554	0.0554	0.0653
2.7	3.8416	3.0161	0.0493	0.0577
2.8	4.1573	3.3011	0.0438	0.0511
2.9	4.5028	3.6126	0.0390	0.0453
3.0	4.8808	3.9534	0.0347	0.0402
3.1	5.2945	4.3262	0.0310	0.0356
3.2	5.7472	4.7342	0.0276	0.0316
3.3	6.2426	5.1810	0.0246	0.0281
3.4	6.7848	5.6701	0.0220	0.0250
3.5	7.3782	6.2058	0.0196	0.0222
3.6	8.0277	6.7927	0.0175	0.0198
3.7	8.7386	7.4358	0.0156	0.0176
3.8	9.5169	8.1404	0.0140	0.0157
3.9	10.3690	8.9128	0.0125	0.0140
4.0	11.3019	9.7595	0.0112	0.0125

$$\phi = \phi_o - \frac{Q_o}{2\pi T} K_o\left(\frac{r}{\lambda}\right) \qquad 9.3(18)$$

When r approaches 4λ, K_o (4) approaches zero which means that at $r > 4\lambda$, drawdown is practically negligible. Note that when $r/\lambda << 1$, $K_o(r/\lambda) \approx -\ln(r/1.123\lambda)$, ϕ becomes

$$\phi = \phi_o + \frac{Q_o}{2\pi T} \ln\left(\frac{r}{1.123\lambda}\right) \qquad 9.3(19)$$

This equation is similar to the governing equation for a confined aquifer, Equation 9.3(13), with the equivalent radius R_{eq} equal to 1.123λ. Therefore, the equation can be rewritten as

$$\phi = \phi_o + \frac{Q_o}{2\pi T} \ln\left(\frac{r}{R_{eq}}\right) \qquad 9.3(20)$$

Equation 9.3(20) indicates that the drawdown near the well s_w can be expressed as

$$s_w = \phi_o - \phi_w = -\frac{Q_o}{2\pi T} \ln\left(\frac{r}{1.123\lambda}\right) \qquad 9.3(21)$$

Basic Equations

The fundamental equations of groundwater flow can be derived in terms of the discharge vector Q_i rather than the specific discharge q_i. For two-dimensional flow, the discharge vector has two components Q_x and Q_y and is defined as

$$Q_x = Hq_x$$
$$Q_y = Hq_y \qquad 9.3(22)$$

With the use of Darcy's law

$$Q_x = Hq_x = H\left(-K\frac{\partial \phi}{\partial x}\right)$$

$$Q_y = Hq_y = H\left(-K\frac{\partial \phi}{\partial y}\right) \qquad 9.3(23)$$

These equations can be rewritten as

$$Q_x = -\frac{\partial(KH\phi)}{\partial x}$$

$$Q_y = -\frac{\partial(KH\phi)}{\partial y} \qquad 9.3(24)$$

With the substitution of a new variable Φ, defined as

$$\Phi = KH\phi + C_c \qquad 9.3(25)$$

where C_c is an arbitrary constant, Equations 9.3(24) can be simplified since the derivatives of C_c with respect x and y are zero as

$$Q_x = -\frac{\partial \Phi}{\partial x}$$

$$Q_x = -\frac{\partial \Phi}{\partial y} \qquad 9.3(26)$$

The function Φ is referred to as the *discharge potential* for horizontal flow or simply as the *potential*.

Now the governing equation for horizontal confined flow, Equation 9.2(13), expressed in terms of the head ϕ is

$$\frac{\partial^2 \phi}{\partial x^2} + \frac{\partial^2 \phi}{\partial y^2} = 0 \qquad 9.3(27)$$

and can be written in terms of the potential Φ as

$$\frac{\partial^2 \Phi}{\partial x^2} + \frac{\partial^2 \Phi}{\partial y^2} = 0 \qquad 9.3(28)$$

or

$$\nabla^2 \Phi = 0 \qquad 9.3(29)$$

Solutions to horizontal confined flow can be obtained when Φ is determined from this Laplace's equation with proper boundary conditions satisfied.

The following equations give solutions for horizontal confined flow in terms of Φ.

(1) One-dimensional flow

$$\Phi = KH\phi = \Phi_1 - \frac{\Phi_1 - \Phi_2}{L} x \qquad 9.3(30)$$

(2) Radial flow

$$\Phi = KH\phi = \frac{Q}{2\pi} \ln\frac{r}{R} + \Phi_o \qquad 9.3(31)$$

Two-dimensional flow problems expressed by the differential Equation 9.3(29) are discussed in more detail in Section 9.6.

—*Y.S. Chae*

Reference

Gupta, R.S. 1989. *Hydrology and hydraulic systems.* Prentice-Hall, Inc.

9.4
UNCONFINED AQUIFERS

As defined in Section 9.1, an unconfined aquifer is a water-bearing layer whose upper boundary is exposed to the open air (atmospheric pressure), as shown in Figure 9.4.1, known as the phreatic surface. Problems with such a boundary condition are difficult to solve, and the vertical component of flow is often neglected. The Dupuit-Forchheimer assumption to neglect the variation of the piezometric head with depth ($\partial\phi/\partial z = 0$) means that the head along any vertical line is constant ($\phi = h$). Physically, this assumption is not true, of course, but the slope of the phreatic surface is usually small so that the variation of the head horizontally ($\partial\phi/\partial x$, $\partial\phi/\partial y$) is much greater than the vertical value of $\partial\phi/\partial z$. The basic differential equation for the flow of groundwater in an unconfined aquifer can be derived from Darcy's law and the continuity equation.

Discharge Potential and Continuity Equation

The discharge vector, as defined in Section 9.3, is the product of the specific discharge q and the thickness of the aquifer H. For an unconfined aquifer, the aquifer thickness h varies, and thus

$$Q_x = q_x h = -Kh\frac{\partial\phi}{\partial x}$$

$$Q_y = q_y h = -Kh\frac{\partial\phi}{\partial y} \qquad 9.4(1)$$

Since $h = \phi$ and K is a constant, Equation 9.4(1) becomes

$$Q_x = -\frac{\partial}{\partial x}\left(\frac{1}{2}K\phi^2\right)$$

$$Q_y = -\frac{\partial}{\partial y}\left(\frac{1}{2}K\phi^2\right) \qquad 9.4(2)$$

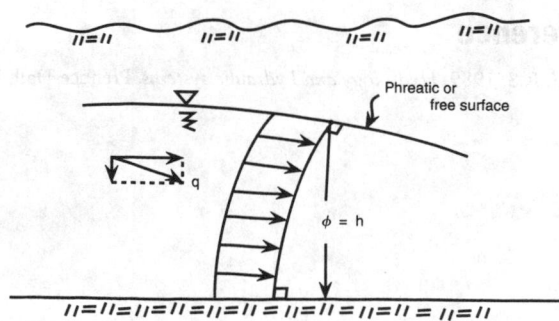

FIG. 9.4.1 Unconfined aquifer.

the discharge potential for unconfined flow introducing as

$$\Phi = \frac{1}{2}K\phi^2 + C_u \qquad 9.4(3)$$

where C_u is an arbitrary constant. Now Equations 9.4(2) can be rewritten as

$$Q_x = -\frac{\partial\Phi}{\partial x}$$

$$Q_y = -\frac{\partial\Phi}{\partial y} \qquad 9.4(4)$$

These equations are the same as those derived for confined flow, Equation 9.3(26).

The continuity equation for unconfined flow, without regard for inflow or outflow along the upper boundary due to precipitation or evaporation, is the same as that for confined flow as

$$\frac{\partial Q_x}{\partial x} + \frac{\partial Q_y}{\partial y} = 0 \qquad 9.4(5)$$

Basic Differential Equation

The governing equation for unconfined flow is obtained when Equation 9.4(4) is substituted into Equation 9.4(5) as

$$\frac{\partial^2\Phi}{\partial x^2} + \frac{\partial^2\Phi}{\partial y^2} = 0 \qquad 9.4(6)$$

The governing equation for both confined and unconfined flows is the same, in terms of the discharge potential, and problems can be solved in the same manner mathematically. The only difference between confined and unconfined flows lies in the expression for Φ as

$$\Phi = KH\phi + C_c \quad \text{for confined flow} \qquad 9.4(7)$$

and

$$\Phi = \frac{1}{2}K\phi^2 + C_u \quad \text{for unconfined flow} \qquad 9.4(8)$$

One-Dimensional Flow

The simplest example of unconfined flow is that of an unconfined aquifer between two long parallel bodies of water, such as rivers or canals, as shown in Figure 9.4.2. In this case, ϕ is a function of x only, and the differential Equation 9.4(6) reduces to

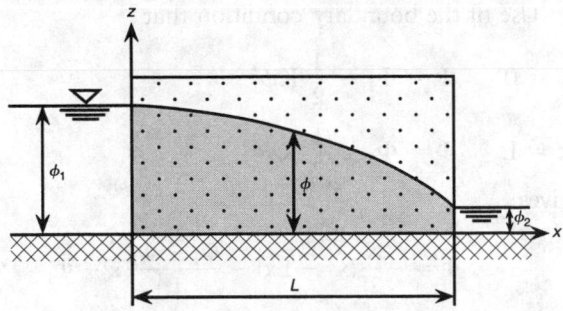

FIG. 9.4.2 One-dimensional flow in an unconfined aquifer.

$$\frac{d^2\Phi}{dx^2} = 0 \qquad 9.4(9)$$

with the general solution

$$\Phi = Ax + B \qquad 9.4(10)$$

Constants A and B can be found from the boundary conditions

$$x = 0, \qquad \Phi = \Phi_1 \qquad B = \Phi_1$$

$$x = L, \qquad \Phi = \Phi_2 \qquad A = \frac{\Phi_2 - \Phi_1}{L}$$

Substitution of A and B into Equation 9.4(10) yields

$$\Phi = \frac{\Phi_2 - \Phi_1}{L} x + \Phi_1 \qquad 9.4(11)$$

An expression for the head ϕ can be found by

$$\frac{1}{2} K\phi^2 = \frac{\frac{1}{2} K(\phi_2^2 - \phi_1^2)}{L} x + \frac{1}{2} K\phi_1^2 \qquad C_u = o$$

$$\phi^2 = \frac{\phi_2^2 - \phi_1^2}{L} x + \phi_1^2 \qquad 9.4(12)$$

This equation shows that the phreatic surface varies parabolically with distance (Dupuit's parabola).

The discharge Q_x is now

$$Q_x = -\frac{\partial\Phi}{\partial x} = \frac{\Phi_1 - \Phi_2}{L} \qquad 9.4(13)$$

or

$$Q_x = \frac{K(\phi_1^2 - \phi_2^2)}{2L} \qquad 9.4(14)$$

Radial Flow

In the case of radial flow in an unconfined aquifer as shown in Figure 9.4.3, the results obtained for confined flow can be directly applied to unconfined flow because the governing equations are the same in terms of the discharge potential. From Equation 9.3(31), the governing equation for radial unconfined flow is

$$\Phi = \frac{Q}{2\pi} \ln\left(\frac{r}{R}\right) + \Phi_o \qquad 9.4(15)$$

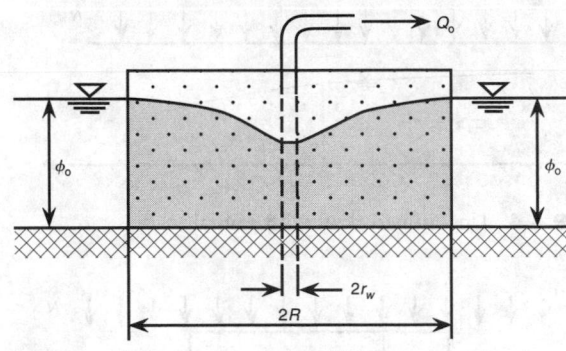

FIG. 9.4.3 Radial flow in an unconfined aquifer.

The governing equation in terms of the head ϕ is

$$\frac{1}{2} K\phi^2 = \frac{Q}{2\pi} \ln\left(\frac{r}{R}\right) + \frac{1}{2} K\phi_o^2$$

$$\phi^2 = \frac{Q}{\pi K} \ln\left(\frac{r}{R}\right) + \phi_o^2 \qquad 9.4(16)$$

or

$$\phi = \sqrt{\frac{Q}{\pi K} \ln\left(\frac{r}{R}\right) + \phi_o^2} \qquad 9.4(17)$$

Note that the expression for the head ϕ for radial unconfined flow is different from that for radial confined flow even though the discharge potential for both types of flow is the same. Also, the principle of superposition applies to Φ but not to ϕ. Superposition of two solutions in Equation 9.4(15), therefore, is allowed, but not in Equation 9.4(17).

The introduction of the drawdown s as $s = \phi_o - \phi$ means $\phi^2 = (\phi_o - s)^2 = \phi_o^2 - 2\phi_o s + s^2 = \phi_o^2 - 2\phi_o s (1 - s/2\phi_o)$. Hence, Equation 9.4(16) can be written as

$$s\left(1 - \frac{s}{2\phi_o}\right) = -\frac{Q}{2\pi K\phi_o} \ln\left(\frac{r}{R}\right) \qquad 9.4(18)$$

If drawdown s is small compared to ϕ_o, then $s/2\phi_o \approx 0$, and Equation 9.4(18) can be written as

$$s = \frac{Q}{2\pi K\phi_o} \ln\left(\frac{R}{r}\right) \qquad s \ll \phi_o \qquad 9.4(19)$$

This equation is identical to the drawdown equation for confined flow, Equation 9.3(15). This fact is true only if the drawdown is small compared to the head ϕ_o. However, Equation 9.4(19) can be accurate enough as a first approximation.

Unconfined Flow with Infiltration

Water can infiltrate into an unconfined aquifer through the soil above the phreatic surface as the result of rainfall or artificial infiltration. As shown in Figure 9.4.4, water percolates downward into the acquifer at a constant infiltration rate of N per unit area and per unit time.

The continuity equation for unconfined flow, Equation 9.4(5), can be modified to read

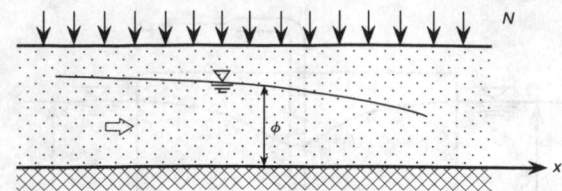

FIG. 9.4.4 Unconfined flow with rainfall.

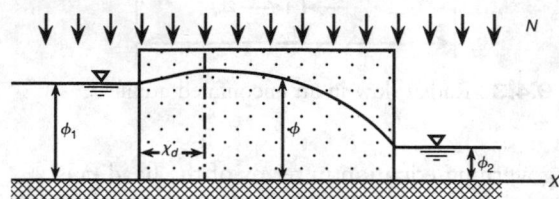

FIG. 9.4.5 One-dimensional unconfined flow with rainfall. (Reprinted from A. Verrjuit, 1982, *Theory of groundwater flow*, Macmillan Pub. Co.)

$$\frac{\partial Q_x}{\partial x} + \frac{\partial Q_y}{\partial y} - N = 0 \qquad 9.4(20)$$

Hence, the differential equation for the potential becomes

$$\frac{\partial^2 \Phi}{\partial x^2} + \frac{\partial^2 \Phi}{\partial y^2} + N = 0 \qquad 9.4(21)$$

In terms of ϕ, this equation reads

$$\frac{\partial^2 \phi}{\partial x^2} + \frac{\partial^2 \phi}{\partial y^2} + \frac{2N}{K} = 0 \qquad 9.4(22)$$

One-Dimensional Flow with Infiltration

For one-dimensional flow shown in Figure 9.4.5, Equation 9.4(21) becomes

$$\frac{d^2 \Phi}{dx^2} + N = 0 \qquad 9.4(23)$$

The general solution of this equation is

$$\Phi = -\frac{N}{2} x^2 + Ax + B \qquad 9.4(24)$$

Use of the boundary condition that

$$x = 0 \qquad \Phi = \Phi_1 = \frac{1}{2} K\phi_1^2$$

$$x = L \qquad \Phi = \Phi_2 = \frac{1}{2} K\phi_2^2$$

gives

$$\Phi = -\frac{N}{2}(x^2 - Lx) - \frac{\Phi_1 - \Phi_2}{L} x + \Phi_1 \qquad 9.4(25)$$

and

$$Q_x = -\frac{d\Phi}{dx} = Nx - \frac{NL}{2} + \frac{\Phi_1 - \Phi_2}{L} \qquad 9.4(26)$$

The location of the divide x_d, where ϕ is maximum, is obtained from

$$\frac{d\Phi}{dx} = 0 = Nx_d - \frac{NL}{2} + \frac{\Phi_1 - \Phi_2}{L} = 0$$

$$\therefore x_d = \frac{\Phi_1 - \Phi_2}{NL} + \frac{L}{2} \qquad (0 \leq x_d \leq L) \qquad 9.4(27)$$

Note that x_d could be larger than L or could be negative. In those cases, the divide does not exist, and the flow occurs in one direction throughout the aquifer.

Radial Flow with Infiltration

Figure 9.4.6 shows radial flow in an unconfined aquifer with infiltration. If a cylinder has a radius r, the amount of water infiltrating into the cylinder is equal to $Q_{in} = N\pi r^2$, and the amount of water flowing out of the cylinder is equal to $2\pi r \cdot hq_r = 2\pi r Q_r$. The continuity of flow requires that $2\pi r Q_r = N\pi r^2$, giving

$$Q_r = \frac{N}{2} r \qquad 9.4(28)$$

which can be written as

$$Q_r = -\frac{\partial \Phi}{\partial r} = \frac{N}{2} r \qquad 9.4(29)$$

yielding

$$\Phi = -\frac{N}{4} r^2 + C \qquad 9.4(30)$$

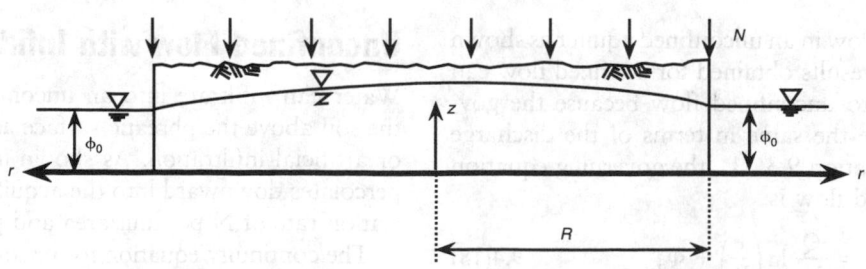

FIG. 9.4.6 Radial unconfined flow with infiltration. (Reprinted from O.D.L. Strack, 1989, *Groundwater mechanics*, Vol. 3, Pt. 3, Prentice-Hall, Inc.)

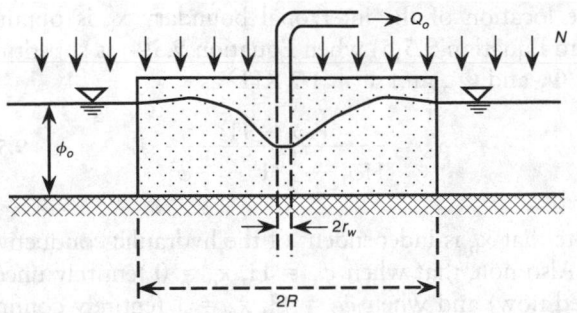

FIG. 9.4.7 Radial flow from pumping with infiltration. (Reprinted from A. Verrjuit, 1982, *Theory of groundwater flow*, 2d ed., Macmillan Pub. Co.)

The constant C in this equation can be determined from the boundary condition that $r = R$, $\Phi = \Phi_o$. The expression for Φ then becomes

$$\Phi = -\frac{N}{4}(r^2 - R^2) + \Phi_o \qquad 9.4(31)$$

The location of the divide is obviously at the center of the island where $d\Phi/dr = 0$ and $r_d = 0$.

Radial Flow from Pumping Infiltration

Figure 9.4.7 shows radial flow in an unconfined aquifer with infiltration in which water is pumped out of a well located at the center of a circular island.

The principle of superposition can be used to solve this problem. In the first case, the radial flow is from pumping alone; in the second, the flow is from infiltration. Since the differential equations for both cases are linear (Laplace's equation and Poisson's equation), the solution for each can be superimposed to obtain a solution for the whole with the sum of both solutions meeting the boundary conditions.

The addition of the two solutions, Equations 9.4(15) and 9.4(31), with a new constant C gives

$$\Phi = -\frac{N}{4}(r^2 - R^2) + \frac{Q}{2\pi}\ln\left(\frac{r}{R}\right) + C \qquad 9.4(32)$$

The constant C can be obtained from the boundary condition $r = R$, $\Phi = \Phi_o$. Hence,

$$\Phi = -\frac{N}{4}(r^2 - R^2) + \frac{Q}{2\pi}\ln\left(\frac{r}{R}\right) + \Phi_o \qquad 9.4(33)$$

The discharge Q_r is now obtained as

$$Q_r = -\frac{\partial\Phi}{\partial r} = \frac{N}{2}r - \frac{Q}{2\pi r} \qquad 9.4(34)$$

The divide r_d is a circle and occurs when $Q_r = \partial\Phi/\partial r = 0$ as

$$\frac{N}{2}r_d - \frac{Q}{2\pi r_d} = 0$$

$$\therefore \quad r_d = \sqrt{\frac{Q}{\pi N}} \quad (r_d \le R) \qquad 9.4(35)$$

—Y.S. Chae

9.5
COMBINED CONFINED AND UNCONFINED FLOW

As water flows through a confined aquifer, the flow changes from confined to unconfined when the piezometric head ϕ becomes less than the aquifer thickness H. This case is shown in Figure 9.5.1. At the interzonal boundary, the head ϕ becomes equal to the thickness H. The continuity of flow requires no change in discharge at the interzonal boundary. Hence, the following equation governing the discharge potential is the same throughout the flow region:

$$\frac{\partial^2\Phi}{\partial x^2} + \frac{\partial^2\Phi}{\partial y^2} = 0 \qquad 9.5(1)$$

where

$$\Phi = KH\phi + C_c \qquad \text{for} \qquad \phi \ge H$$

$$\Phi = \frac{1}{2}K\phi^2 + C_u \qquad \text{for} \qquad \phi < H$$

At the interzonal boundary, Φ yields the same value, giving

$$KH^2 + C_c = \frac{1}{2}KH^2 + C_u$$

$$C_c = C_u - \frac{1}{2}KH^2 \qquad 9.5(2)$$

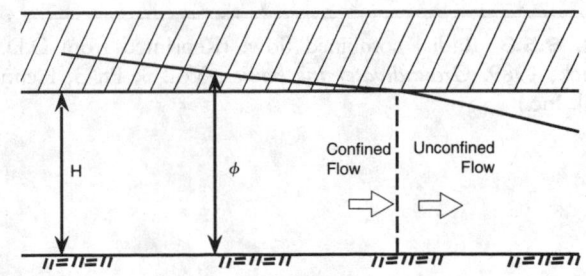

FIG. 9.5.1 Combined confined and unconfined flow.

If one of the two constants C_u is set to zero, then

$$C_c = -\frac{1}{2} KH^2, \qquad C_u = 0 \qquad 9.5(3)$$

The potential Φ can be expressed as

$$\Phi = KH\phi - \frac{1}{2} KH^2 \qquad (\phi \geq H)$$

$$\Phi = \frac{1}{2} K\phi^2 \qquad (\phi < H) \qquad 9.5(4)$$

One-Dimensional Flow

Figure 9.5.2 shows combined confined and unconfined flow in an aquifer of thickness H and length L. The aquifer is confined at $x = 0$ and unconfined at $x = L$.

The expression for the potential Φ is the same throughout the flow region as

$$\Phi = -(\Phi_1 - \Phi_2)\frac{x}{L} + \Phi_1 \qquad 9.5(5)$$

However, the expression for Φ in terms of ϕ is different for each zone as given in Equation 9.5(4). The expression for the discharge Q is

$$Q_x = \frac{\Phi_1 - \Phi_2}{L} = \frac{KH\phi_1 - \frac{1}{2} KH^2 - \frac{1}{2} K\phi_2^2}{L} \qquad 9.5(6)$$

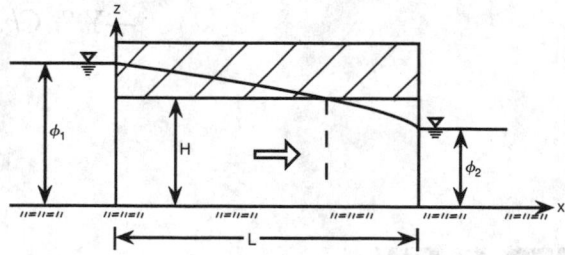

FIG. 9.5.2 One-dimensional combined flow.

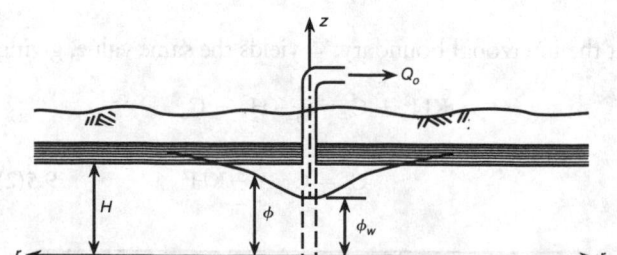

FIG. 9.5.3 Radial combined flow. (Reprinted from O.D.L. Strack, 1989, *Groundwater mechanics*, Vol. 3, Pt. 3, Prentice Hall, Inc.)

The location of the interzonal boundary x_b is obtained from Equation 9.5(5) when Equation 9.5(4) is substituted for Φ_1 and Φ_2, and $\Phi = 1/2\ KH^2$ as

$$x_b = \frac{H\phi_1 - H^2}{H\phi_1 - \frac{1}{2} H^2 - \frac{1}{2} \phi_2^2} \cdot L \qquad 9.5(7)$$

Note that x_b is independent of the hydraulic conductivity K. Also note that when $\phi_1 = H$, $x_b = 0$ (entirely unconfined flow) and when $\phi_2 = H$, $x_b = 1$ (entirely confined flow).

Radial Flow

If the drawdown near the well caused by pumping dips below the aquifer thickness H, then unconfined flow occurs in that region as shown in Figure 9.5.3. The expression for the potential Φ is the same for the entire flow region as

$$\Phi = \frac{Q}{2\pi} \ln\left(\frac{r}{R}\right) + \Phi_o \qquad 9.5(8)$$

In this equation, Φ_o is the potential at $r = R$ when the flow is confined. Hence

$$\Phi_o = KH\phi_o - \frac{1}{2} KH^2 \qquad (\phi_o > H) \qquad 9.5(9)$$

The potential at well Φ_w for unconfined flow is

$$\Phi_w = \frac{1}{2} K\phi_w^2 \qquad (\phi_w < H) \qquad 9.5(10)$$

Equation 9.5(8) can now be rewritten as

$$\frac{1}{2} K\phi_w^2 = \frac{Q}{2\pi} \ln\left(\frac{r_w}{R}\right) + KH\phi_o - \frac{1}{2} KH^2 \qquad 9.5(11)$$

and solving for Q gives

$$Q = \frac{2\pi\left(\frac{1}{2} K\phi_w^2 - KH\phi_o + \frac{1}{2} KH^2\right)}{\ln\left(\frac{r_w}{R}\right)} \qquad 9.5(12)$$

The distance r_b to the interzonal boundary, which is a circle, can be obtained from Equation 9.5(8) with $\Phi = 1/2\ KH^2$ as

$$\frac{1}{2} KH^2 = \frac{Q}{2\pi} \ln\left(\frac{r_b}{R}\right) + \Phi_o$$

$$\therefore \qquad r_b = R \cdot e^{2\pi KH(H-\phi_o)/Q} \qquad 9.5(13)$$

—Y.S. Chae

Hydraulics of Wells

9.6
TWO-DIMENSIONAL PROBLEMS

This section describes methods for handling two-dimensional groundwater flow problems including superposition, the method of images, and the potential and flow function.

Superposition

The differential equation for two-dimensional steady flow in a homogeneous aquifer is

$$\frac{\partial^2 \Phi}{\partial x^2} + \frac{\partial^2 \Phi}{\partial y^2} = 0 \qquad \text{9.6(1)}$$

Because this equation is a linear and homogeneous differential equation, the principle of superposition applies. The principle states that if two different functions Φ_1 and Φ_2 are solutions of Laplace's equation, then the function

$$\Phi(x,y) = c_1\Phi_1(x,y) + c_2\Phi(x,y) \qquad \text{9.6(2)}$$

is also a solution.

Superposition of solutions is valuable in several groundwater problems. For example, the case of groundwater flow due to simultaneous pumping from several wells can be solved by the superposition of the elementary solution for a single well.

A TWO-WELL SYSTEM

Consider the case of two wells in an infinite aquifer as shown in Figure 9.6.1, in which water is discharged (positive Q) from well 1 and is recharged (negative Q) into well 2. This case is referred to as a sink-and-source problem.

The potential Φ at a point which is located at a distance r_1 from well 1 and r_2 from well 2 can be expressed when the potential Φ_1 is superimposed with respect to well 1 and Φ_2 is superimposed with respect to well 2 as

$$\Phi = \Phi_1 + \Phi_2 = \frac{Q_1}{2\pi} \ell n \, r_1 - \frac{Q_2}{2\pi} \ell n \, r_2 + C \qquad \text{9.6(3)}$$

The constant $C = \Phi_0$ @ $r_1 = r_2 = R$. If $Q_1 = Q_2 = Q$ in a special case, then

$$\Phi = \frac{Q}{2\pi} \ell n \left(\frac{r_1}{r_2}\right) + \Phi_0 \qquad \text{9.6(4)}$$

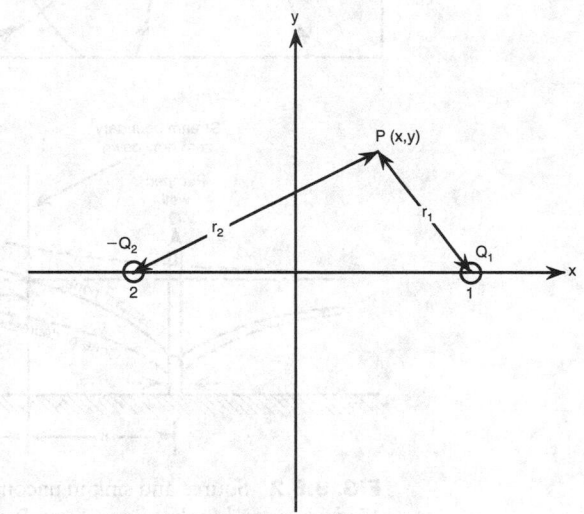

FIG. 9.6.1 Discharge and recharge wells.

or

$$\phi = \frac{Q}{2\pi T} \ell n \left(\frac{r_1}{r_2}\right) + \phi_0 \qquad \text{for a confined aquifer} \qquad \text{9.6(5)}$$

$$\phi^2 = \frac{Q}{\pi K} \ell n \left(\frac{r_1}{r_2}\right) + \phi_0^2 \qquad \text{for an unconfined aquifer} \qquad \text{9.6(6)}$$

Figure 9.6.2 shows the flow net for a two-well sink-and-source system. Equation 9.6(4) shows that along the y axis where $r_1 = r_2 = r_0$, Φ = constant. This statement means that the y axis is an equipotential line along which no flow occurs, and the drawdown is zero ($\phi = \phi_0$). This result occurs because the system is in symmetry about the y axis and the problem is linear. Note that the distance R does not appear in Equation 9.6(4). This omission is because the discharge from the sink is equal to the recharge into the source, indicating that the system is in hydraulic equilibrium requiring no external supply of water.

Another example of using the principle of superposition is the case of two sinks of equal discharge Q. Equation 9.6(3) now reads

$$\Phi = \Phi_1 + \Phi_2 = \frac{Q}{2\pi} \ell n \, (r_1 r_2) + C \qquad \text{9.6(7)}$$

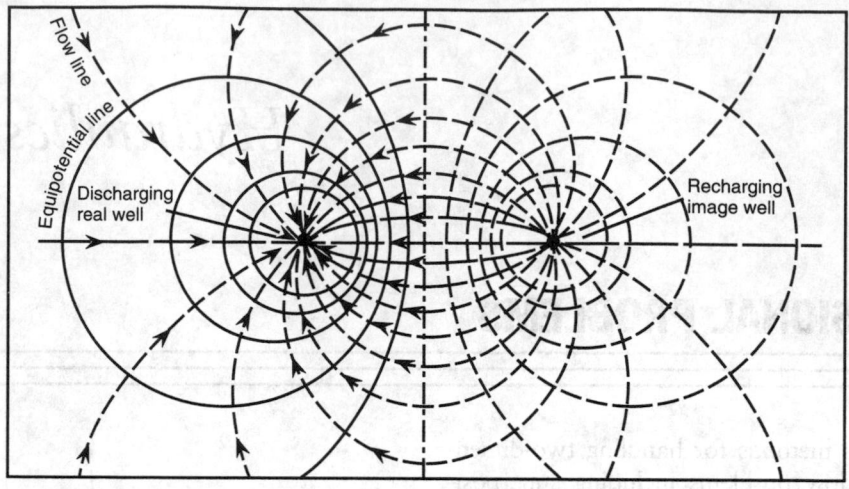

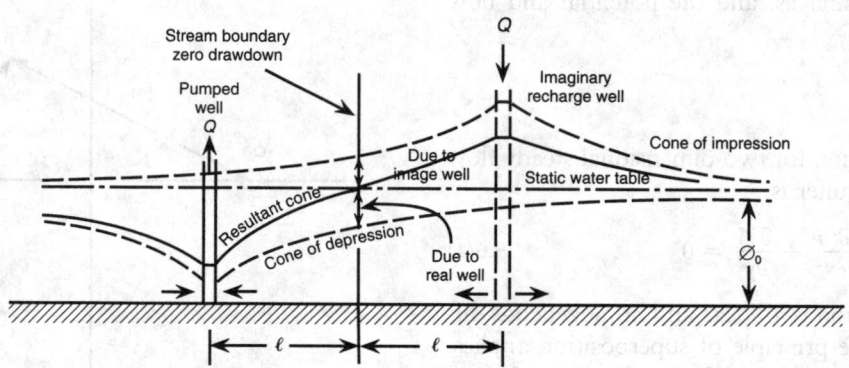

FIG. 9.6.2 Source and sink in unconfined flow. (Reprinted from R.S. Gupta, 1989, *Hydrology and hydraulic systems*, Prentice-Hall, Inc.)

Use of the boundary condition $r = R$, $\Phi = \Phi_o$ yields

$$\Phi = \frac{Q}{2\pi} \ell n\left(\frac{r_1 r_2}{R^2}\right) + \Phi_0 \qquad 9.6(8)$$

Figure 9.6.3 shows the flow net for a two-well sink-and-sink system. The y axis plays the role of an impervious boundary along which no water flows across. This result occurs because the flow at points on the y axis is directed along the axis due to the equal pull of flow from the two wells located equidistance from the points.

A MULTIPLE-WELL SYSTEM

The principle of superposition previously discussed for two wells can be applied to a system of multiple wells, n wells in number from i = 1 to n. The solution for such a system can be written with the use of superposition as

$$\Phi = \frac{1}{2\pi}\left[\sum_{i=1}^{n} Q_i \, \ell n\left(\frac{r_i}{R}\right)\right] + \Phi_o \qquad 9.6(9)$$

or

$$\phi = \frac{1}{2\pi T}\left[\sum_{i=1}^{n} Q_i \, \ell n\left(\frac{r_i}{R}\right)\right] + \phi_o \quad \text{for a confined aquifer}$$

$$9.6(10)$$

and

$$\phi^2 = \frac{1}{\pi K}\left[\sum_{i=1}^{n} Q_i \, \ell n\left(\frac{r_i}{R}\right)\right] + \phi_o^2 \quad \text{for an unconfined aquifer}$$

$$9.6(11)$$

The drawdown at the jth well is then

$$\phi_w = \frac{1}{2\pi T}\left[Q_j \, \ell n\left(\frac{r_w}{R}\right) + \sum_{i=1}^{n-1} Q_i \, \ell n\left(\frac{r_{i,j}}{R}\right)\right] + \phi_o$$

$$\text{for a confined aquifer} \quad 9.6(12)$$

and

$$\phi_w^2 = \frac{1}{\pi K}\left[Q_j \, \ell n\left(\frac{r_w}{R}\right) + \sum_{i=1}^{n-1} Q_i \, \ell n\left(\frac{r_{i,j}}{R}\right)\right] + \phi_o^2$$

$$\text{for an unconfined aquifer} \quad 9.6(13)$$

where $r_{i,j}$ is the distance between the jth well and ith wells. The quantities inside the brackets [ ] in these equations are called the *drawdown factors*, F_p at a point and F_w at a well, respectively. These equations can be rewritten as

$$\Phi = \Phi_o + \frac{1}{2\pi} F_p \quad \text{at a point} \qquad 9.6(14)$$

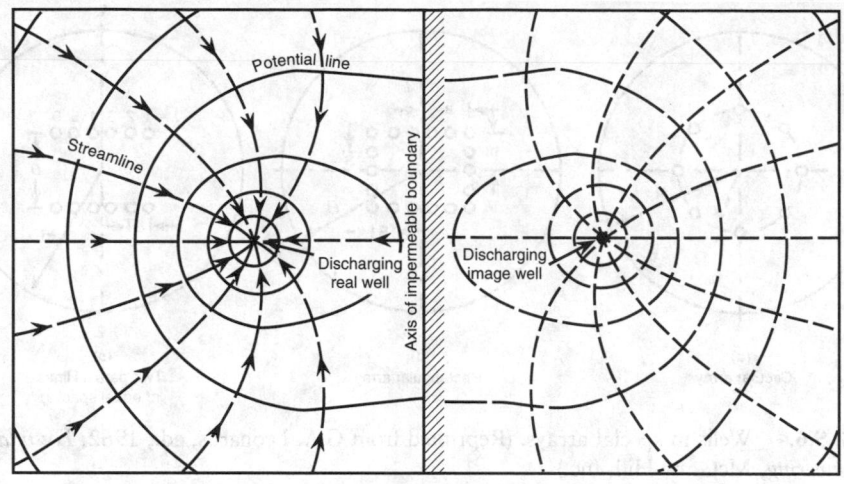

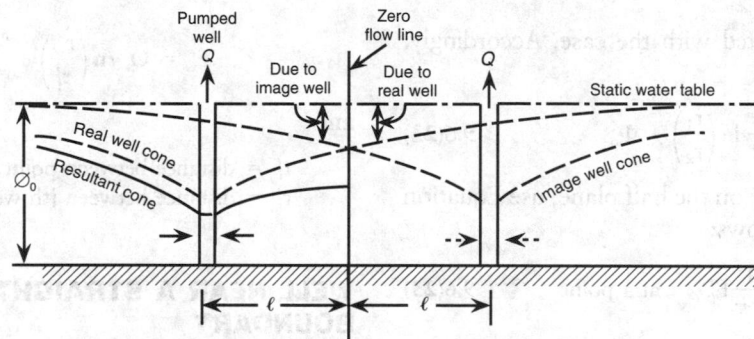

FIG. 9.6.3 Sink and sink in unconfined flow. (Reprinted from R.S. Gupta, 1989, *Hydrology and hydraulic systems,* Prentice-Hall, Inc.)

$$\Phi_w = \Phi_o + \frac{1}{2\pi} F_w \quad \text{at a well} \qquad 9.6(15)$$

where

$$F_p = \sum_{i=1}^{n} Q_i \, \ell n \left(\frac{r_w}{R}\right) \qquad 9.6(16)$$

$$F_w = Q_j \, \ell n \left(\frac{r_w}{R}\right) + \sum_{i=1}^{n-1} Q_i \, \ell n \left(\frac{r_{i,j}}{R}\right) \qquad 9.6(17)$$

The following examples give the drawdown factors of wells in special arrays:

a. Circular array, n wells in equal spacing (Figure 9.6.4a)

$$F_p = nQ \, \ell n \frac{\rho}{R} \qquad 9.6(18)$$

$$F_w = Q \, \ell n \frac{nr_w \rho^{n-1}}{R^n} \qquad 9.6(19)$$

b. Rectangular array (Figure 9.6.4b)
    • Approximate method:
        Equivalent radius $\rho_e = 4\sqrt{ab}/\pi$
        Then use Equation 9.6(11)
    • Exact method:
        Use Equation 9.6(9), 9.6(12) or 9.6(13)

c. Two parallel lines of equally spaced wells (Figure 9.6.4c)

$$F_c = 4Q \sum_{i=1}^{i=n/4} \ell n \frac{R}{\frac{1}{2} \cdot \sqrt{S^2(2i - 1)^2 + B^2}} \qquad 9.6(20)$$

$$F_w = 2Q \sum_{i=1}^{i=n/2} \ell n \frac{R}{\frac{1}{2} \cdot \sqrt{S^2(2i - 3)^2 + B^2}} \qquad 9.6(21)$$

Method of Images

A special application of superposition is the method of images. This method can be used to solve problems involving the flow in aquifers of relatively simple geometrical form such as an infinite strip, a half plane, or a quarter plane. The following problems are specific examples.

WELL NEAR A STRAIGHT RIVER

To solve the problem of a well near a long body of water (river, canal, or lake) shown in Figure 9.6.5, replace the half-plane aquifer by an imaginary infinite aquifer with an imaginary well placed at the mirror image position from the real well. This case now represents the sink and source problem discussed previously, and Equation 9.6(4) satis-

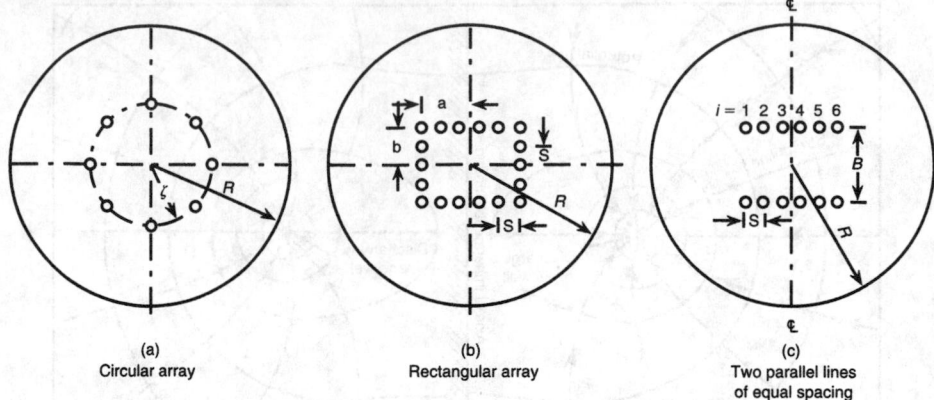

FIG. 9.6.4 Wells in special arrays. (Reprinted from G.A. Leonards, ed., 1962, *Foundation engineering*, McGraw-Hill, Inc.)

fies all conditions associated with the case. Accordingly, the solution is given by

$$\Phi = \frac{Q}{2\pi} \ln\left(\frac{r_1}{r_2}\right) + \Phi_o \qquad 9.6(22)$$

If n number of wells are on the half plane, use Equation 9.6(7) for solution as follows:

$$\Phi = \Phi_o + \frac{1}{2\pi} F'_p \quad \text{at a point} \qquad 9.6(23)$$

$$\Phi_w = \Phi_o + \frac{1}{2\pi} F'_w \quad \text{at a well} \qquad 9.6(24)$$

where

$$F'_p = \sum_{i=1}^{n} Q_i \ln\left(\frac{r_i}{r'_i}\right) \qquad 9.6(25)$$

$$F'_w = Q_i \ln\left(\frac{r_w}{r'_j}\right) + \sum_{i=1}^{n-1} Q_i \ln\left(\frac{r_{i,j}}{r'_{i,j}}\right) \qquad 9.6(26)$$

and

r'_i = distance between point and imaginary ith well.
$r'_{i,j}$ = distance between jth well and imaginary ith well.

WELL NEAR A STRAIGHT IMPERVIOUS BOUNDARY

The problem of a well near a long straight impervious boundary (e.g. a mountain ridge or fault) is solved in a similar manner as that of a well near a straight river. In this case, the type of image well is a sink rather than a source as shown in Figure 9.6.6.

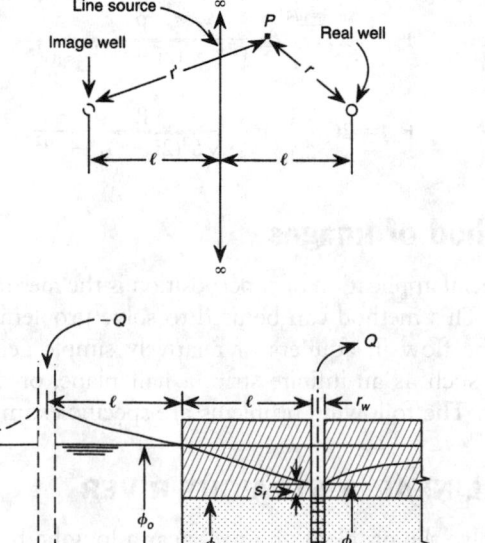

FIG. 9.6.5 Well near a straight river. (Reprinted from G.A. Leonards, ed., 1962, *Foundation engineering*, McGraw-Hill, Inc.)

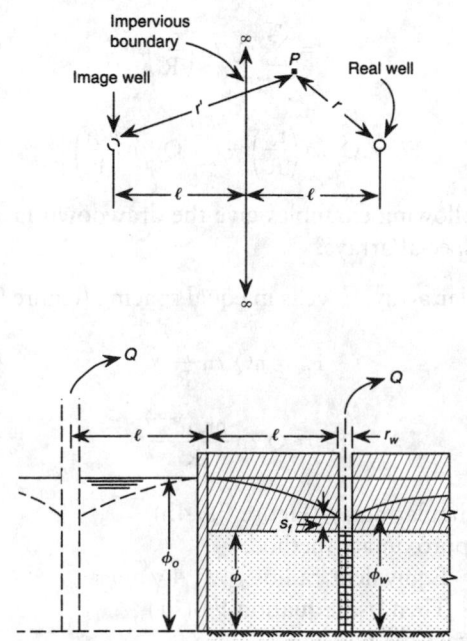

FIG. 9.6.6 Well near a straight impervious boundary.

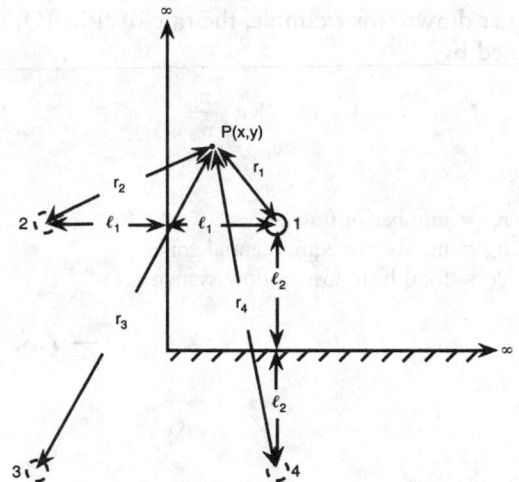

FIG. 9.6.7 Well in a quarter plane.

The solution for this case, therefore, is the same as the case of a sink-and-sink problem given by Equation 9.6(8), which is

$$\Phi = \frac{Q}{2\pi} \ell n \left(\frac{r_1 r_2}{R} \right) + \Phi_o \qquad 9.6(27)$$

WELL IN A QUARTER PLANE

Figure 9.6.7 shows the case of a well operating in an aquifer bounded by a straight river and an impervious boundary. To solve this problem, place a series of imaginary wells (wells numbered 2, 3, and 4), and use superposition. Figure 9.6.7 indicates that wells 2 and 3 are sources, and well 4 is a sink. Hence,

$$\Phi = \frac{Q}{2\pi} \left[\ell n \left(\frac{r_1}{R} \right) - \ell n \left(\frac{r_2}{R} \right) - \ell n \left(\frac{r_3}{R} \right) + \ell n \left(\frac{r_4}{R} \right) \right] + \Phi_o$$

$$9.6(28)$$

Potential and Flow Functions

In the Basic Equations section, the fundamental equation of groundwater flow expressed in terms of discharge potential Φ is:

$$\frac{\partial^2 \Phi}{\partial x_2} + \frac{\partial^2 \Phi}{\partial y^2} = 0 \qquad 9.6(29)$$

The potential $\Phi(x,y)$ is a single-value function everywhere in the x, y plane. Therefore, lines of constant Φ_1, Φ_2, . . ., called *equipotential lines,* can be drawn in the x, y plane as shown in Figure 9.6.8. When the lines are drawn with a constant interval between the values of the two successive lines ($\Delta\Phi = \Phi_1 - \Phi_2 = \Phi_2 - \Phi_3 = \ldots$), then an equal and constant amount of potential drop is between any two of the equipotential lines.

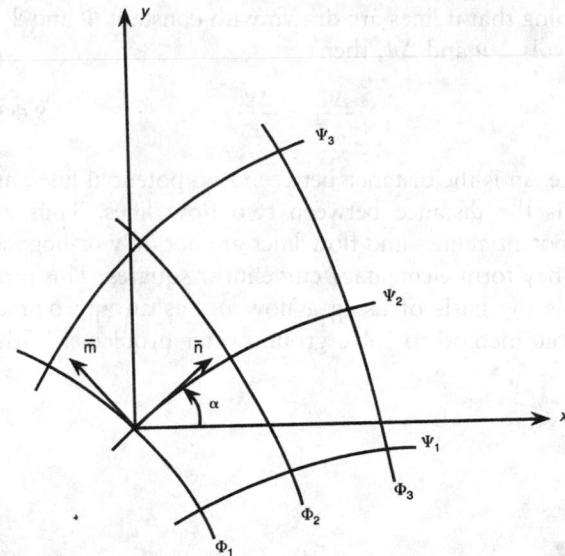

FIG. 9.6.8 Potential and flow lines.

At any arbitrary point on the equipotential line, flow occurs only in the direction perpendicular to the line (n direction), and no flow occurs in the tangential direction (m direction) as

$$q_n = \frac{\partial \Phi}{\partial \bar{n}} = -q; \qquad q_m = \frac{\partial \Phi}{\partial \bar{m}} = 0 \qquad 9.6(30)$$

Accordingly, lines can be drawn perpendicular to the equipotential lines as shown in Figure 9.6.8. These lines are called flow or stream lines.

At this point a second function, called flow or stream function Ψ, is introduced. Since the specific discharge vector must satisfy the equation of continuity, the function Ψ is defined by

$$q_x = -\frac{\partial \Psi}{\partial y}, \qquad q_y = \frac{\partial \psi}{\partial x} \qquad 9.6(31)$$

It now follows that

$$\frac{\partial^2 \Psi}{\partial x^2} + \frac{\partial^2 \Psi}{\partial y^2} = 0 \qquad 9.6(32)$$

or

$$\nabla^2 \psi = 0 \qquad 9.6(33)$$

which shows that Ψ is, like the potential Φ, a harmonic function and should satisfy Laplace's equation.

The directional function Ψ with respect to m and n can now be easily written as follows because no flow component is in m direction:

$$\frac{\partial \psi}{\partial \bar{m}} = -q, \qquad \frac{\partial \psi}{\partial \bar{n}} = 0 \qquad 9.6(34)$$

The lines of constant Φ and Ψ form a set of orthogonal curves called a flow net. Also,

$$q = -\frac{\partial \Phi}{\partial \bar{n}} = -\frac{\partial \psi}{\partial \bar{m}} \qquad 9.6(35)$$

meaning that if lines are drawn with constant Φ and Ψ at intervals $\Delta\Phi$ and $\Delta\psi$, then

$$\frac{\Delta\Phi}{\Delta\bar{n}} = \frac{\Delta\psi}{\Delta\bar{m}} \qquad 9.6(36)$$

where Δn is the distance between two potential lines, and Δm is the distance between two flow lines. Thus, the equipotential lines and flow lines are not only orthogonal, but they form elementary curvelinear squares. This property is the basis of using a flow net as an approximate graphic method to solve groundwater problems. With a flow net drawn, for example, the rate of flow (Q) can be obtained by

$$Q = K\phi_o \frac{n_f}{n_\phi} \qquad 9.6(37)$$

where:

n_f = number of flow zones
n_φ = number of equipotential zones
φ_o = total head loss in flow system

—*Y.S. Chae*

9.7
NONSTEADY (TRANSIENT) FLOW

Nonsteady or transient flow in aquifers occurs when the pressure and head in the aquifer change gradually until steady-state conditions are reached. During the course of transient flow, water can be either stored in or released from the soil. Storage has two possibilities. First, water can simply fill the pore space in soil without changing the soil volume. This storage is called *phreatic storage*, and usually occurs in unconfined aquifers as the groundwater table moves up or down. In the other storage, water is stored in the pore space increased by deformation of the soil and involves a volume change. This storage is called *elastic storage* and occurs in all types of aquifers. However, in confined aquifers, it is the only form of storage.

Transient Confined Flow (Elastic Storage)

In a completely saturated confined aquifer, water can be stored or released if the change in aquifer pressure results in volumetric deformation of the soil. The problem is complex because the constitutive equations for soil are highly nonlinear even for dry soil, and coupling them with groundwater flow increases the complexity.

The basic equation for the phenomenon is the storage equation (Strack 1989), as

$$\nabla^2\Phi = H\left[\frac{\partial\varepsilon_o}{\partial t} + n\beta\frac{\partial p}{\partial t}\right] \qquad 9.7(1)$$

where ε_o = volume strain, and β = compressibility of water. From soil mechanics

$$\frac{\partial\varepsilon_o}{\partial t} = m_v\frac{\partial p}{\partial t} \qquad 9.7(2)$$

where m_v = modulus of volume change. Equation 9.7(1) can then be written as

$$\nabla^2\Phi = H(m_v + n\beta)\frac{\partial p}{\partial t} \qquad 9.7(3)$$

When the variation of K, H, and ρ with time are neglected, then

$$\frac{\partial\Phi}{\partial t} = KH\frac{\partial}{\partial t}\left[\frac{p}{\rho g} + z\right] = \frac{KH}{\rho g}\frac{\partial p}{\partial t} \qquad 9.7(4)$$

so that Equation 9.7(3) can be written as

$$\nabla^2\Phi = \frac{S_s}{K}\frac{\partial\Phi}{\partial t} \qquad 9.7(5)$$

or

$$\frac{\partial\Phi}{\partial t} = \frac{K}{S_s}\nabla^2\Phi \qquad 9.7(6)$$

where S_s [(1/m)] is the coefficient of specific storage

$$S_s = \rho g(m_v + n\beta) \qquad 9.7(7)$$

If the compressibility of water β is ignored, then $S_s = \rho gm_v$.

Some typical values of m_v are given in Table 9.7.1. Equation 9.7(5) can also be written in terms of ϕ as

$$\nabla^2\phi = \frac{S_e}{T}\frac{\partial\phi}{\partial t} \qquad 9.7(8)$$

where S_e = coefficient of elastic storage = $S_s \cdot H$.

Transient Unconfined Flow (Phreatic Storage)

The vertical movement of a phreatic surface results in water being stored in soil pores without causing the soil to

TABLE 9.7.1 TYPICAL VALUES OF
COMPRESSIBILITY (m_v)

	Compressibility, $(m^2/N$ or $Pa^{-1})$
Clay	10^{-6}–10^{-8}
Sand	10^{-7}–10^{-9}
Gravel	10^{-8}–10^{-10}
Jointed rock	10^{-8}–10^{-10}
Sound rock	10^{-9}–10^{-11}
Water (β)	4.4×10^{-10}

Source: R.A. Freeze and J.A. Cherry, 1979, *Groundwater* (Prentice-Hall, Inc.).

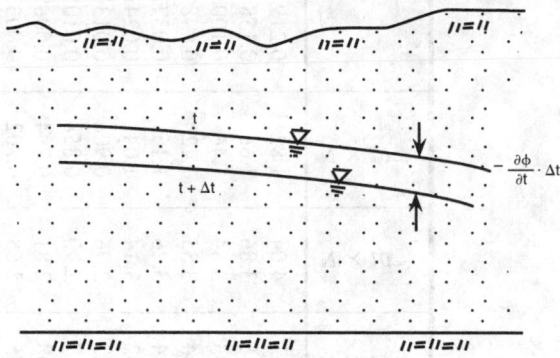

FIG. 9.7.1 Storage change due to unconfined flow.

deform. Phreatic storage is, therefore, several orders of magnitude greater than elastic storage, which can be ignored.

The basic differential equation for the transient unconfined flow (Strack 1989), such as shown in Figure 9.7.1, can be given as

$$\nabla^2 \Phi = S_p \frac{\partial \phi}{\partial t} \qquad 9.7(9)$$

where S_p = coefficient of phreatic storage.

Equation 9.7(9) can be linearized in terms of the potential Φ as

$$\nabla^2 \Phi = \frac{S_s}{K} \frac{\partial \Phi}{\partial t} \qquad 9.7(10)$$

or

$$\frac{\partial \Phi}{\partial t} = \frac{K}{S_s} \nabla^2 \Phi \qquad 9.7(11)$$

This equation is the same as that for transient confined flow. However, S_s is related to S_p as $S_s = S_p/\overline{\phi}$, where $\overline{\phi}$ is the average piezometric head in the aquifer.

Transient Radial Flow (Theis Solution)

The governing equation for the transient radial flow (flow toward a well in an aquifer of infinite extent) is obtained when Equation 9.7(10) is rewritten in terms of radial coordinate r as

$$\frac{\partial^2 \Phi}{\partial r^2} + \frac{1}{r} \frac{\partial \Phi}{\partial r} = \frac{S_s}{K} \frac{\partial \Phi}{\partial t} \qquad 9.7(12)$$

The solution to this equation is commonly given as

$$\Phi = -\frac{Q}{4\pi} Ei(u) + \Phi_o \qquad 9.7(13)$$

known as the Theis solution. Ei is the exponential integral, and u is a dimensionless variable defined by

$$u = \frac{S_s r^2}{4Kt} \qquad 9.7(14)$$

or

$$u = \frac{S_e r^2}{4Tt} \qquad \text{for confined flow} \qquad (T = KH) \qquad 9.7(15)$$

or

$$u = \frac{S_p r^2}{4Tt} \qquad \text{for unconfined flow} \qquad (T = K\overline{\phi}) \qquad 9.7(16)$$

The exponential integral $Ei(u)$ is referred to as the well function $W(u)$. $Ei(u)$ can be approximated by

$$Ei(u) = \left[-0.577216 - \ell n\, u + u - \frac{u^2}{2.2!} + \frac{u^3}{3.3!} - \cdots \right] \qquad 9.7(17)$$

Using the well function $W(u)$, the Theis solution can be written as

$$\Phi = -\frac{Q}{4\pi} W(u) + \Phi_o \qquad 9.7(18)$$

or in terms of the head ϕ as

$$\phi = -\frac{Q}{4\pi T} W(u) + \phi_o \qquad 9.7(19)$$

The drawdown s is obtained by

$$s = \frac{Q}{4\pi T} W(u) \qquad 9.7(20)$$

Values of $W(u)$ for different values of u are shown in Table 9.7.2. The drawdown s at a given distance r from the well at given time t can be calculated from Equation 9.7(20) and Table 9.7.2.

Figure 9.7.2, accompanied by Table 9.7.3, shows an example of drawdown versus a time curve for a transient radial flow in a confined aquifer with T = 1000 m^2/d and S = 0.0001 for a pumping rate of Q = 1000 m^3/d. The figure shows that even in a transient flow, the rate of drawdown (Δs) achieves a steady state after a short period of pumping, two days in this example.

If u is small (e.g., less than 0.01), only the first two terms of the brackets in Equation 9.7(17) are significant. Equation 9.7(19) can be simplified to

TABLE 9.7.2 VALUES OF W(U) FOR DIFFERENT VALUES OF U

u / N	$N \times 10^{-15}$	$N \times 10^{-14}$	$N \times 10^{-13}$	$N \times 10^{-12}$	$N \times 10^{-11}$	$N \times 10^{-10}$	$N \times 10^{-9}$	$N \times 10^{-8}$	$N \times 10^{-7}$	$N \times 10^{-6}$	$N \times 10^{-5}$	$N \times 10^{-4}$	$N \times 10^{-3}$	$N \times 10^{-2}$	$N \times 10^{-1}$	N
1	34.0	31.7	29.4	27.1	24.8	22.4	20.1	17.8	15.5	13.2	10.9	8.63	6.33	4.04	1.82	0.219
1.2	33.8	31.5	29.2	26.9	24.6	22.3	20.0	17.7	15.4	13.1	10.8	8.45	6.15	3.86	1.66	0.158
1.5	33.6	31.3	29.0	26.6	24.3	22.0	19.7	17.4	15.1	12.8	10.5	8.23	5.93	3.64	1.46	0.100
2	33.3	31.0	28.7	26.4	24.1	21.8	19.5	17.2	14.8	12.5	10.2	7.94	5.64	3.35	1.22	0.0489
2.2	33.2	30.9	28.6	26.3	24.0	21.7	19.4	17.1	14.8	12.4	10.1	7.84	5.54	3.26	1.15	0.0372
2.5	33.0	30.7	28.4	26.1	23.8	21.5	19.2	16.9	14.6	12.3	10.0	7.72	5.42	3.14	1.04	0.0249
3	32.9	30.6	28.3	26.0	23.7	21.3	19.0	16.7	14.4	12.1	9.84	7.53	5.23	2.96	0.906	0.0130
3.2	32.8	30.5	28.2	25.9	23.6	21.3	19.0	16.7	14.4	12.1	9.77	7.47	5.17	2.90	0.858	0.0101
3.5	32.7	30.4	28.1	25.8	23.5	21.2	18.9	16.6	14.3	12.0	9.68	7.38	5.08	2.81	0.794	0.00697
4	32.6	30.3	28.0	25.7	23.4	21.1	18.8	16.5	14.2	11.9	9.55	7.25	4.95	2.68	0.702	0.00378
4.2	32.5	30.2	27.9	25.6	23.3	21.0	18.7	16.4	14.1	11.8	9.50	7.20	4.90	2.63	0.670	0.00300
4.5	32.5	30.2	27.9	25.5	23.2	20.9	18.6	16.3	14.0	11.7	9.43	7.13	4.83	2.57	0.625	0.00207
5	32.4	30.0	27.7	25.4	23.1	20.8	18.5	16.2	13.9	11.6	9.33	7.02	4.73	2.47	0.560	0.00115
5.2	32.3	30.0	27.7	25.4	23.1	20.8	18.5	16.2	13.9	11.6	9.29	6.98	4.69	2.43	0.536	0.000909
5.5	32.3	30.0	27.7	25.3	23.0	20.7	18.4	16.1	13.8	11.5	9.23	6.93	4.63	2.38	0.503	0.000641
6	32.2	29.9	27.6	25.3	23.0	20.7	18.4	16.1	13.7	11.4	9.14	6.84	4.54	2.30	0.454	0.000360
6.2	32.1	29.8	27.5	25.2	22.9	20.6	18.3	16.0	13.7	11.4	9.11	6.81	4.51	2.26	0.437	0.000286
6.5	32.1	29.8	27.5	25.2	22.9	20.6	18.3	16.0	13.7	11.4	9.06	6.76	4.47	2.22	0.411	0.000203
7	32.0	29.7	27.4	25.1	22.8	20.5	18.2	15.9	13.6	11.3	8.99	6.69	4.39	2.15	0.374	0.000115
7.2	32.0	29.7	27.4	25.1	22.8	20.5	18.2	15.9	13.6	11.3	8.96	6.66	4.36	2.12	0.360	0.0000922
7.5	32.0	29.6	27.3	25.0	22.7	20.4	18.1	15.8	13.5	11.2	8.92	6.62	4.32	2.09	0.340	0.0000658
8	31.9	29.6	27.3	25.0	22.7	20.4	18.1	15.8	13.5	11.2	8.86	6.55	4.26	2.03	0.311	0.0000377
8.2	31.9	29.6	27.2	24.9	22.6	20.3	18.0	15.7	13.4	11.1	8.83	6.53	4.23	2.00	0.300	0.0000301
8.5	31.8	29.5	27.2	24.9	22.6	20.3	18.0	15.7	13.4	11.1	8.80	6.49	4.20	1.97	0.284	0.0000216
9	31.8	29.5	27.1	24.9	22.6	20.3	17.9	15.6	13.3	11.0	8.74	6.44	4.14	1.92	0.260	0.0000124
9.2	31.7	29.5	27.1	24.8	22.5	20.2	17.9	15.6	13.3	11.0	8.72	6.41	4.12	1.90	0.251	0.00000999
9.5	31.7	29.4	27.1	24.8	22.5	20.2	17.9	15.6	13.3	11.0	8.68	6.38	4.09	1.87	0.239	0.00000718
10	31.7	29.4		24.8	22.4	20.1	17.8	15.5	13.2	10.9	8.63	6.33	4.04	1.82	0.219	

Source: H. Bouwer, 1978, *Groundwater hydrology* (McGraw-Hill, Inc.).

TABLE 9.7.3 CALCULATION OF S IN RELATION TO T

	r = 100 m			r = 200 m		
t, Days	u	W(u)	s, m	u	W(u)	s, m
0.001	0.25	1.044	0.083	1	0.219	0.017
0.005	0.05	2.468	0.196	0.2	1.223	0.097
0.01	0.025	3.136	0.249	0.1	1.823	0.145
0.05	0.005	4.726	0.376	0.02	3.355	0.267
0.1	0.002 5	5.417	0.431	0.01	4.038	0.322
0.5	0.000 5	7.024	0.559	0.002	5.639	0.449
1	0.000 25	7.717	0.614	0.001	6.331	0.504
5	0.000 05	9.326	0.742	0.000 2	7.940	0.632
10	0.000 025	10.019	0.797	0.000 1	8.633	0.687

Source: H. Bouwer, 1978, *Groundwater hydrology* (McGraw-Hill, Inc.).

$$\phi = -\frac{Q}{4\pi T} \ell n \left(\frac{2.25Tt}{Sr^2}\right) + \phi_o \qquad 9.7(21)$$

then

$$s = \frac{Q}{4\pi T} \ell n \left(\frac{2.25Tt}{Sr^2}\right) \qquad 9.7(22)$$

Equation 9.7(21) can be rewritten as

$$\phi = \frac{Q}{2\pi T} \ell n \left[\frac{r}{\left(2.25\frac{Tt}{S}\right)^{1/2}}\right] + \phi_o \qquad 9.7(23)$$

$$\phi = \frac{Q}{2\pi T} \ell n \left(\frac{r}{R_{eq.}}\right) + \phi_o \qquad 9.7(24)$$

where

$$R_{eq.} = \left(2.25\frac{Tt}{S}\right)^{1/2}$$

Note that Equation 9.7(24) is similar in expression to the steady-state flow. Equation 9.7(22) allows direct calculation of drawdown in terms of distance r and time t for given aquifer characteristics T and S at a known pumping rate Q.

The exact solution of Equation 9.7(13) is difficult for unconfined aquifers because $\overline{T} = K\overline{\phi}$ is not constant but varies with distance r and time t. The average head $\overline{\phi}$ can be estimated and used in the Theis solution for small drawdowns. For large drawdowns, however, the use of $\overline{\phi}$ for the Theis solution is not valid.

For large drawdowns, Boulton (1954) presents a solution which is valid if the water depth in the well exceeds $0.5\ \phi_o$. Boulton's equation is:

$$s = \frac{Q}{2\pi K\phi_o}(1 + C_k)V(t', r') \qquad 9.7(25)$$

where V is Boulton's well function, and C_k is a correction factor. The t' and r' are defined as

$$r' = \frac{1}{\phi_o} \cdot r \qquad 9.7(26)$$

$$t' = \frac{K}{S_p\phi_o}t \qquad 9.7(27)$$

The values of V(r',t') and C_k are given in Table 9.7.4 and Table 9.7.5 respectively.

The head at the well ϕ_w can be calculated from the equation (Bouwer 1988) as

$$Q_w^2 = \phi_o^2 - \frac{Q}{\pi k} \ell n \left(1.5 \sqrt{\frac{Kt}{S_pr_w}}\right) \qquad 9.7(28)$$

which is valid if $t' = (Kt/\phi_oS) > 5$. If t' is smaller than 5, ϕ_w is calculated as

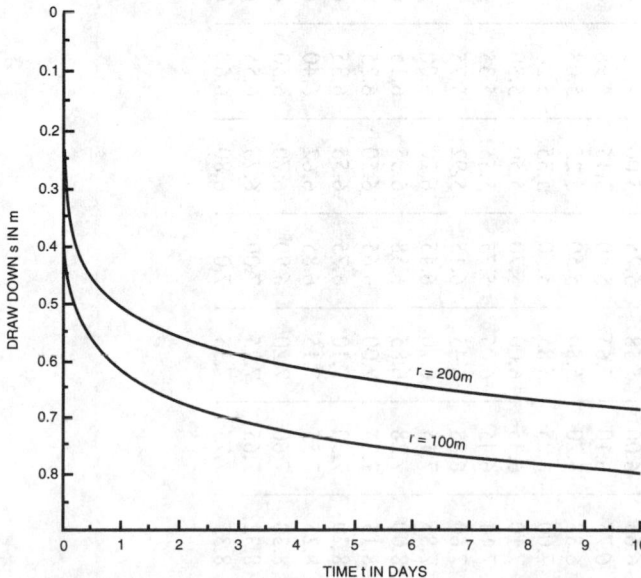

FIG. 9.7.2 Drawdown versus time due to pumping from a well. (Reprinted from H. Bouwer, 1978, *Groundwater hydrology*, McGraw-Hill, Inc.)

TABLE 9.7.4 VALUES OF THE FUNCTION V(T',R') FOR DIFFERENT VALUES OF T' AND R'

t' \ r'	0.001	0.002	0.003	0.004	0.005	0.006	0.007	0.008	0.009	0.01	0.02	0.03	0.04	0.05	0.06	0.07	0.08	0.09
0.01	2.99	2.30	1.90	1.64	1.42	1.28	1.15	1.04	0.95	0.875	0.474	0.322	0.240	0.192	0.158	0.135	0.118	0.104
0.02	3.68	2.97	2.58	2.30	2.09	1.92	1.76	1.64	1.52	1.42	0.860	0.610	0.468	0.378	0.316	0.270	0.236	0.210
0.03	4.08	3.40	3.00	2.70	2.46	2.28	2.13	2.00	1.88	1.79	1.18	0.860	0.675	0.555	0.465	0.400	0.350	0.310
0.04	4.35	3.68	3.26	2.98	2.75	2.58	2.42	2.29	2.17	2.06	1.42	1.07	0.850	0.710	0.600	0.525	0.460	0.410
0.05	4.58	3.90	3.49	3.20	2.96	2.79	2.64	2.50	2.38	2.28	1.60	1.24	1.010	0.850	0.725	0.630	0.560	0.500
0.06	4.76	4.06	3.65	3.36	3.15	2.96	2.80	2.68	2.56	2.45	1.78	1.40	1.15	0.970	0.840	0.735	0.650	0.585
0.07	4.92	4.20	3.80	3.51	3.30	3.12	2.96	2.82	2.70	2.60	1.91	1.54	1.28	1.09	0.950	0.835	0.740	0.670
0.08	5.08	4.34	3.94	3.65	3.42	3.24	3.09	2.95	2.84	2.72	2.04	1.65	1.39	1.20	1.04	0.925	0.825	0.750
0.09	5.18	4.47	4.05	3.75	3.54	3.35	3.20	3.05	2.95	2.84	2.14	1.75	1.50	1.29	1.14	1.02	0.910	0.825
0.1	5.24	4.54	4.14	3.85	3.63	3.45	3.30	3.15	3.04	2.94	2.25	1.85	1.58	1.38	1.22	1.09	0.985	0.890
0.2	5.85	5.15	4.78	4.50	4.28	4.10	3.93	3.80	3.66	3.56	2.87	2.46	2.20	1.98	1.80	1.65	1.52	1.42
0.3	6.24	5.50	5.12	4.85	4.61	4.43	4.28	4.14	4.01	3.90	3.24	2.84	2.54	2.32	2.14	1.98	1.85	1.74
0.4	6.45	5.75	5.35	5.08	4.85	4.67	4.50	4.38	4.26	4.15	3.46	3.05	2.76	2.54	2.36	2.20	2.07	1.96
0.5	6.65	6.00	5.58	5.25	5.00	4.85	4.70	4.55	4.45	4.30	3.65	3.24	2.95	2.72	2.52	2.38	2.24	2.14
0.6	6.75	6.10	5.65	5.40	5.15	4.98	4.82	4.68	4.56	4.45	3.76	3.37	3.09	2.85	2.67	2.50	2.38	2.26
0.7	6.88	6.20	5.80	5.50	5.25	5.08	4.92	4.80	4.68	4.55	3.90	3.50	3.20	2.99	2.80	2.64	2.50	2.38
0.8	7.00	6.25	5.85	5.60	5.35	5.20	5.00	4.90	4.80	4.65	3.96	3.55	3.26	3.05	2.86	2.71	2.58	2.46
0.9	7.10	6.35	6.00	5.70	5.50	5.30	5.12	5.00	4.90	4.75	4.05	3.65	3.36	3.15	2.96	2.80	2.66	2.55
1	7.14	6.45	6.05	5.75	5.55	5.35	5.20	5.05	4.95	4.83	4.10	3.74	3.45	3.22	3.04	2.90	2.75	2.64
2	7.60	6.88	6.45	6.15	5.92	5.75	5.60	5.50	5.35	5.25	4.59	4.18	3.90	3.68	3.50	3.34	3.20	3.09
3	7.85	7.15	6.70	6.45	6.20	6.00	5.85	5.75	5.60	5.50	4.82	4.42	4.12	3.90	3.72	3.57	3.45	3.31
4	8.00	7.28	6.85	6.58	6.35	6.15	6.00	5.90	5.75	5.70	4.95	4.55	4.26	4.04	3.86	3.70	3.59	3.46
5	8.15	7.35	7.00	6.65	6.50	6.25	6.10	6.00	5.85	5.80	5.05	4.68	4.40	4.19	4.00	3.85	3.71	3.60
6	8.20	7.50	7.10	6.75	6.55	6.35	6.20	6.10	5.95	5.85	5.20	4.78	4.50	4.26	4.09	3.92	3.80	3.69
7	8.25	7.55	7.15	6.85	6.62	6.40	6.30	6.20	6.05	5.95	5.25	4.85	4.58	4.35	4.18	4.00	3.90	3.78
8	8.30	7.60	7.20	6.90	6.70	6.50	6.35	6.25	6.10	6.05	5.30	4.92	4.65	4.40	4.25	4.10	3.95	3.82
9	8.32	7.65	7.25	7.00	6.75	6.55	6.40	6.30	6.15	6.10	5.35	5.00	4.70	4.49	4.30	4.15	4.00	3.90
10	8.35	7.75	7.35	7.05	6.80	6.60	6.45	6.35	6.20	6.14	5.40	5.02	4.80	4.52	4.35	4.19	4.05	3.92

Continued on next page

TABLE 9.7.4 *Continued*

t'\r'	0.1	0.2	0.3	0.4	0.5	0.6	0.7	0.8	0.9	1	2	3	4	5
0.01	0.093	0.0430	0.0264	0.0180	0.0132	0.0100	0.0078	0.0062	0.0049	0.0040	0.00057	0.00015		
0.02	0.187	0.0865	0.0530	0.0365	0.0268	0.0205	0.0160	0.0125	0.0100	0.0081	0.00118	0.00020		
0.03	0.278	0.130	0.0800	0.0550	0.0405	0.0310	0.0240	0.0190	0.0150	0.0122	0.00184	0.00032		
0.04	0.368	0.174	0.107	0.0735	0.0540	0.0415	0.0322	0.0255	0.0202	0.0165	0.00244	0.00043		
0.05	0.450	0.215	0.133	0.0920	0.0675	0.0520	0.0400	0.0320	0.0255	0.0206	0.00305	0.00055		
0.06	0.530	0.257	0.160	0.110	0.0810	0.0610	0.0478	0.0380	0.0305	0.0250	0.00365	0.00065		
0.07	0.610	0.298	0.186	0.130	0.0950	0.0725	0.0565	0.0450	0.0360	0.0292	0.00430	0.00078		
0.08	0.680	0.340	0.214	0.148	0.108	0.0825	0.0645	0.0510	0.0412	0.0336	0.00500	0.00090		
0.09	0.750	0.378	0.236	0.164	0.122	0.0930	0.0730	0.0585	0.0470	0.0380	0.00570	0.00105		
0.1	0.815	0.415	0.260	0.180	0.134	0.103	0.0805	0.0640	0.0515	0.0420	0.00635	0.00118		
0.2	1.32	0.750	0.500	0.359	0.268	0.208	0.165	0.132	0.107	0.0880	0.0145	0.00278		
0.3	1.64	1.02	0.700	0.515	0.392	0.308	0.246	0.200	0.164	0.135	0.0238	0.00490		
0.4	1.86	1.22	0.870	0.650	0.510	0.405	0.328	0.268	0.220	0.182	0.0350	0.00750	0.00160	0.00038
0.5	2.03	1.37	1.00	0.770	0.610	0.490	0.400	0.330	0.275	0.230	0.0450	0.0104	0.00240	0.00056
0.6	2.16	1.49	1.12	0.875	0.700	0.570	0.468	0.390	0.325	0.276	0.0580	0.0138	0.00320	0.00080
0.7	2.28	1.60	1.22	0.965	0.775	0.640	0.525	0.445	0.375	0.320	0.0715	0.0175	0.00425	0.00108
0.8	2.36	1.69	1.30	1.04	0.850	0.715	0.600	0.500	0.425	0.364	0.0840	0.0212	0.00525	0.00140
0.9	2.45	1.75	1.38	1.11	0.920	0.775	0.650	0.550	0.475	0.404	0.0980	0.0260	0.00630	0.00165
1	2.54	1.85	1.45	1.18	0.975	0.825	0.700	0.595	0.510	0.444	0.113	0.0310	0.00840	0.00235
2	2.97	2.29	1.88	1.60	1.38	1.22	1.07	0.950	0.840	0.750	0.259	0.0950	0.0330	0.0115
3	3.20	2.50	2.10	1.82	1.60	1.42	1.28	1.15	1.05	0.960	0.388	0.165	0.0700	0.0275
4	3.36	2.66	2.25	1.97	1.75	1.58	1.42	1.30	1.20	1.10	0.495	0.235	0.112	0.0535
5	3.49	2.78	2.38	2.09	1.87	1.69	1.54	1.42	1.30	1.21	0.580	0.300	0.150	0.0715
6	3.59	2.90	2.47	2.18	1.95	1.78	1.65	1.52	1.40	1.30	0.660	0.360	0.195	0.0990
7	3.66	2.96	2.55	2.25	2.04	1.85	1.70	1.58	1.48	1.38	0.730	0.415	0.230	0.125
8	3.74	3.00	2.60	2.32	2.11	1.94	1.79	1.66	1.55	1.44	0.790	0.465	0.272	0.155
9	3.80	3.09	2.67	2.39	2.17	2.00	1.85	1.72	1.60	1.50	0.850	0.515	0.307	0.182
10	3.84	3.12	2.74	2.45	2.24	2.05	1.90	1.77	1.65	1.55	0.890	0.550	0.340	0.210

Note: For $t' > 5$, $V(t', r')$ is about equal to $0.5W[(r')^2/4t]$, which is the well function in Table 9.7.2.

Source: From N.S. Boulton, 1954, The drawdown of water table under non-steady conditions near a pumped well in an unconfined formation, *Proc. Inst. Civ. Eng. (London)* 3, Pt. 2:564–579.

TABLE 9.7.5 CORRECTION FACTOR Ck

r'	0.03	0.04	0.06	0.08	0.1	0.2	0.4	0.6	0.8	1	2	4
C_k	−0.27	−0.24	−0.19	−0.16	−0.13	−0.05	0.02	0.05	0.05	0.05	0.03	0

$$\phi_w = \phi_o - \frac{Q}{2\pi K \phi_o}\left(m + \ell n \frac{\phi_o}{r_w}\right) \qquad 9.7(29)$$

where m is a function of t′ and can be obtained from a curve plotted through the following points:

t′	0.05	0.2	1	5
m	−0.043	0.087	0.512	1.288

—*Y.S. Chae*

References

Boulton, N.S. 1954. The drawdown of water table under non-steady conditions near a pumped well in an unconfined formation. Proc. Inst. Civ. Eng. (London) 3, pt 2:564–579.

Bouwer, H. 1978. *Groundwater hydrology*. McGraw-Hill, Inc.

Strack, O.D.L. 1989. *Groundwater mechanics*. Vol. 3, pt. 3:564–579. Prentice-Hall, Inc.

9.8
DETERMINING AQUIFER CHARACTERISTICS

Hydraulic conductivity K, transmissivity T, and storativity S are the hydraulic properties which characterize an aquifer. Before the quantities required to solve groundwater engineering problems, such as drawdown and rate of flow, can be calculated, the hydraulic properties of the aquifer K, S, and T must be determined.

Determining the hydraulic properties of an aquifer generally involves applying field data obtained from a pumping test. Other techniques such as auger-hole and piezometer methods can be used to determine K where the groundwater table or aquifers are shallow.

Pumping test technology is prominent in the evaluation of hydraulic properties. It involves observing the drawdown of the piezometric surface or water table in observation wells which are located some distance from the pumping well and have water pumped through them at a constant rate. Pumping test analysis applies the field data to some form of the Theis equation in general, such as

$$s = \frac{Q}{4\pi T} W(u, \alpha, \beta, \ldots) \qquad 9.8(1)$$

where $u = Sr^2/4Tt$ and α, β = dimensionless factors defining particular aquifer system conditions. In general, matching the field data curve (usually a plot of s versus r^2/t) with the standard curve (known as the *type curve*) drawn between W and u for various control values of $\alpha, \beta, \ldots$, calculates the values of S and T. This process is explained in the next section. Techniques requiring no matching have since been developed.

Various site conditions are associated with a pumping test in a well–aquifer system. The following list summarizes different site conditions (Gupta 1989):

I. Type of pumping
   A. Drawdown
   B. Recovery
   C. Interference
II. State of flow
   A. Steady-state
   B. Nonsteady (transient) state
III. Area extent of aquifer
   A. Aquifer of infinite extent
   B. Aquifer bound by an impermeable boundary
   C. Aquifer bound by a recharge boundary
IV. Depth of well
   A. Fully penetrating well
   B. Partially penetrating well
V. Confined aquifer
   A. Nonleaky aquifer
   B. Leaky confining bed releasing water from storage
   C. Leaky confining bed not yielding water from storage but transmitting water from overlying layer
   D. Leaky aquifer in which the head in the overlying aquifer changes
VI. Unconfined aquifer
   A. Aquifer in which significant dewatering occurs
   B. Aquifer in which vertical flow occurs near the well
   C. Aquifer with delayed yield

Selecting a proper type curve is essential for the data analysis. During the last decades, several contributors have developed type curves for various site conditions or combinations of categories. Starting with Theis, who made the original type curve concept, other contributors to this field include Cooper and Jacob (1946) and Chow (1952) for confined aquifers, and Hantush and Jacob (1955), Neuman and Witherspoon (1969), Walton (1962), Boulton (1963) and Neuman (1972) for unconfined aquifers.

Confined Aquifers

This section discusses the methods used in determining aquifer characteristics for confined aquifers.

STEADY-STATE

The Thiem equation, Equation 9.3(12), gives the drawdown between two points (s_1 and s_2) measured at distances r_1 and r_2, respectively, as

$$s_1 - s_2 = s = \frac{Q}{2\pi T} \ell n \left(\frac{r_2}{r_1}\right) \qquad 9.8(2)$$

Hence, T can be calculated by

$$T = \frac{Q}{2\pi(s_1 - s_2)} \ell n \left(\frac{r_2}{r_1}\right) \qquad 9.8(3)$$

or from Figure 9.8.1, T can be obtained by

$$T = \frac{2.3Q}{2\pi} \frac{\Delta \log r}{\Delta s} \qquad 9.8(4)$$

Figure 9.7.2 shows that the drawdown between two points $s_1 - s_2$ reaches a constant value after a day or two. Therefore, Equation 9.8(3) can be used to determine T before the flow achieves a steady state.

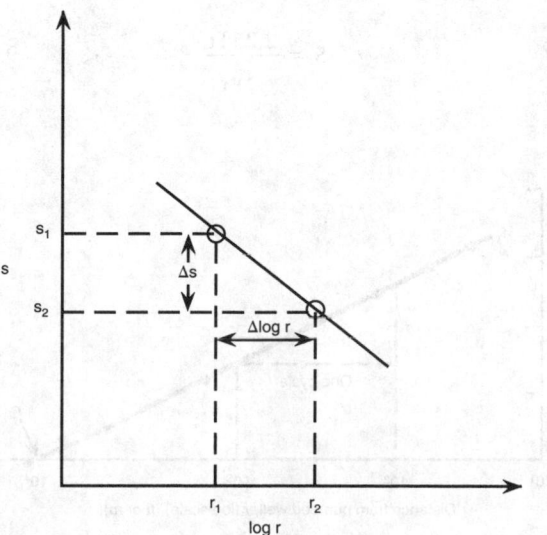

FIG. 9.8.1 Plot of drawdown s versus distance r.

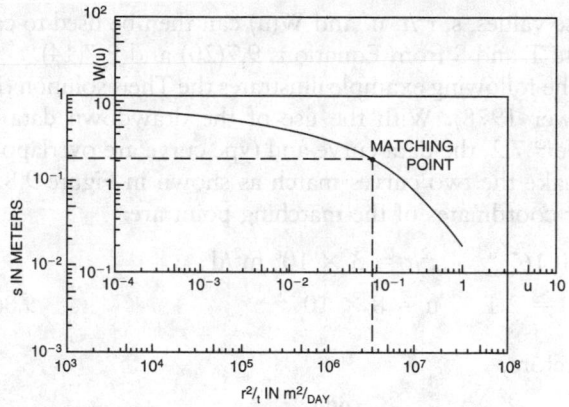

FIG. 9.8.2 Relations s versus r^2/t and W(u) versus u. (Reprinted from H. Bouwer, 1978, *Groundwater hydrology*, McGraw-Hill, Inc.)

Once T has been calculated, S can be determined with the transient-flow equations, Equations 9.7(14) and 9.7(20), as

$$W(u) = \frac{4\pi Ts}{Q} \longrightarrow T = \frac{Q}{4\pi s} W(u) \qquad 9.8(5)$$

$$u = \frac{r^2 S}{4Tt} \longrightarrow S = \frac{4Ttu}{r^2} \qquad 9.8(6)$$

Since T, Q, and s are known for a given r and t, W(u) can be obtained. With the use of Table 9.7.2, the corresponding value of u can be found. S can be calculated from Equation 9.8(6).

TRANSIENT-STATE

Three methods of analysis are the type-curve method (Theis), the Cooper–Jacob method, and the Chow method. These methods are briefly described.

Type Curve Method (Theis)

The Theis equation, Equations 9.7(20) and 9.7(14), can be written respectively as

$$\log s = \log \frac{Q}{4\pi T} + \log W(u) \qquad 9.8(7)$$

$$\log \frac{r^2}{t} = \log \frac{4T}{S} + \log u \qquad 9.8(8)$$

If these two equations are plotted on the same log–log paper, the resulting curves are the same shape but horizontally and vertically offset by the constants $Q/4\pi T$ and $4T/S$. If each curve is plotted on a separate sheet, the curves can be made to match when the sheets are overlapped as shown in Figure 9.8.2. An arbitrary point on the matching curve is selected, and the coordinates of this matching point are read horizontally and vertically on both graphs.

These values, s, r^2/t, u, and W(u) can then be used to calculate T and S from Equations 9.7(20) and 9.7(14).

The following example illustrates the Theis solution (H. Bouwer 1978). With the use of the drawdown data in Table 9.7.2, the data curve and type curve are overlapped to make the two curves match as shown in Figure 9.8.2. Four coordinates of the matching point are:

$$s = 0.167^m \qquad r^2/t = 3 \times 10^6 \text{ m}^2/\text{d}$$
$$W(u) = 2.1 \qquad u = 8 \times 10^{-2} \qquad \text{9.8(9)}$$

Therefore,

$$T = \frac{Q}{4\pi s} W(u) = \frac{1000}{4\pi(0.167)}(2.1) = 1001 \text{ m}^2/\text{d} \quad \text{9.8(10)}$$

$$S = \frac{4Tu}{r^2/t} = \frac{4(1001)(8 \times 10^{-2})}{3 \times 10^6} = 0.0001 \quad \text{9.8(11)}$$

Cooper-Jacob Method

Cooper and Jacob (1946) showed that when u becomes small (u << 1), the drawdown equation can be represented by Equation 9.7(22) as

$$s = \frac{2.3Q}{4\pi T} \log\left(\frac{2.25Tt}{Sr^2}\right) \qquad \text{9.8(12)}$$

On semilog paper, this equation represents a straight line with a slope of $2.3Q/4\pi T$. This equation can be plotted in three different ways: (1) s versus log t, (2) s versus log r, or (3) s versus log t/r^2 or log r^2/t.

DRAWDOWN–TIME ANALYSIS (s VERSUS log t)

The drawdown measurements s at a constant distance r are plotted against time as shown in Figure 9.8.3. The slope of the line is $2.3Q/4\pi T$ and is equal to

$$\frac{\Delta s}{\log \frac{t_2}{t_1}} = \frac{2.3Q}{4\pi T} \qquad \text{9.8(13)}$$

If a change in drawdown Δs is considered for one log cycle, then log $(t_2/t_1) = 1$, and this equation reduces to

$$\Delta s = \frac{2.3Q}{4\pi T} \qquad \text{9.8(14)}$$

or

$$T = \frac{2.3Q}{4\pi(\Delta s)} \qquad \text{9.8(15)}$$

When the straight line intersects the x axis, s = 0 and the time is t_o. Substituting these values in Equation 9.8(12) gives

$$0 = \frac{2.3Q}{4\pi T} \log \frac{2.25Tt_o}{r^2 S} \qquad \text{9.8(16)}$$

so

$$1 = \frac{2.25Tt_o}{r^2 S} \qquad \text{9.8(17)}$$

and

$$S = \frac{2.25Tt_o}{r^2} \qquad \text{9.8(18)}$$

Example: Figure 9.8.3 shows that $t_o = 1.6 \times 10^{-3}$ days and slope $\Delta s = 0.181$. These values yield T = 1011 m²/d and S = 0.00009, which agree with the values for T and S obtained by the Theis solution.

DRAWDOWN–DISTANCE ANALYSIS (s VERSUS log r)

The drawdown measurements s are plotted against distance r at a given time t as shown in Figure 9.8.4. From similar considerations as in drawdown–time analysis

$$T = \frac{2.3Q}{2\pi(\Delta s)} \qquad \text{9.8(19)}$$

$$S = \frac{2.25Tt}{r_o^2} \qquad \text{9.8(20)}$$

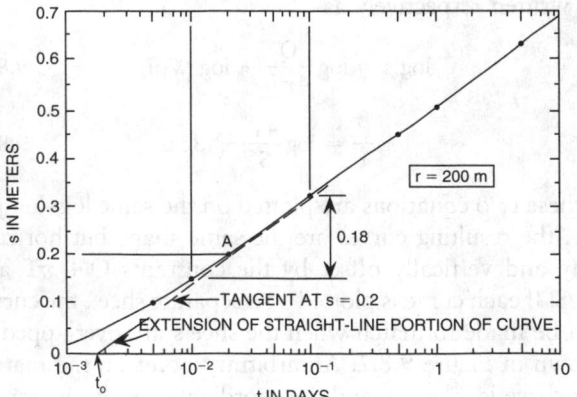

FIG. 9.8.3 Drawdown versus time plot. (Reprinted from H. Bouwer, 1978, *Groundwater hydrology,* McGraw-Hill, Inc.)

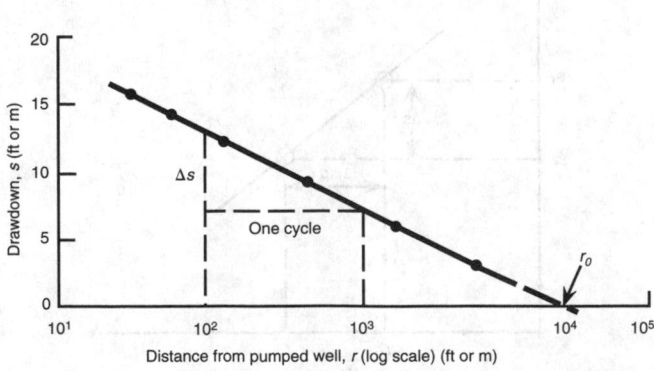

FIG. 9.8.4 Drawdown versus distance plot. (Reprinted from R.S. Gupta, 1989, *Hydrology and hydraulic systems,* Prentice-Hall, Inc.)

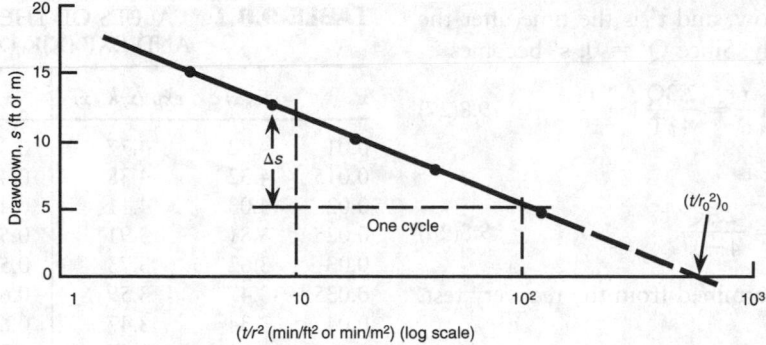

FIG. 9.8.5 Plot of drawdown versus combined time–distance. (Reprinted from R.S. Gupta, 1989, *Hydrology and hydraulic systems*, Prentice-Hall, Inc.)

DRAWDOWN–COMBINED-TIME–DISTANCE ANALYSIS (s VERSUS log r²/t)

The drawdown measurements in many wells at various times are plotted as shown in Figure 9.8.5. Similarly as before

$$T = \frac{2.3Q}{4\pi(\Delta s)} \qquad 9.8(21)$$

$$S = 2.25T\left(\frac{t}{r^2}\right)_0 \qquad 9.8(22)$$

Chow Method

Chow's procedure (1952) combines the approach of Theis and Cooper–Jacob and introduces the function

$$F(u) = \frac{W(u)e^u}{2.3} = \frac{s}{\Delta s/\log(t_2/t_1)} \qquad 9.8(23)$$

where s is the drawdown at a point. The relation between F(u), W(u), and u is shown in Figure 9.8.6. For one log cycle on a time scale

$$\log(t_2/t_1) = 1 \qquad 9.8(24)$$

and

$$F(u) = \frac{s}{\Delta s} \qquad 9.8(25)$$

From the drawdown–time curve, obtain s at an arbitrary point and Δs over one log cycle. The ratio s/Δs is equal to F(u) in the Equation 9.8(25). F(u), W(u), and u can be obtained from Figure 9.8.6. With W(u), u, s, and t known, T and S can be calculated with Equations 9.7(20) and 9.7(14).

> Example: Point A in Figure 9.8.3 gives s = 0.2^m and Δs = 0.18^m at r = 200^m and F(u) = 0.2/0.18 = 1.11. From Figure 9.8.6, W(u) = 2.2 and u = 0.065. Substituting into Equations 9.7(20) and 9.7(14) yields T = 875 m²/d and S = 0.00011, which reasonably agree with the values obtained by the two methods just described.

Recovery Test

Figure 9.8.7 schematically shows a recovery test in which the water level in the observation wells rises when pumping stops after the pumping test is complete. Since the principle of superposition applies, the drawdown s′ after the pumping test is complete can be expressed as

$$s' = \frac{Q}{4\pi T}\ell n\left(\frac{2.25Tt}{r^2S}\right) - \frac{Q - Q'}{4\pi T}\ell n\left(\frac{2.25Tt'}{r^2S}\right) \qquad 9.8(26)$$

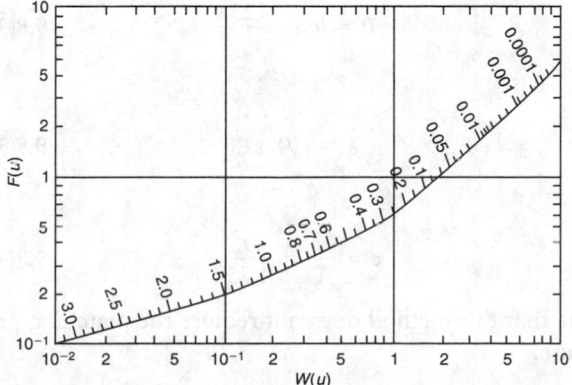

FIG. 9.8.6 Relations between F(u), W(u), and u. (Reprinted from V.T. Chow, 1952, On the determination of transmissivity and storage coefficients from pumping test data, *Trans. Am. Geoph. Union* 33:397–404.)

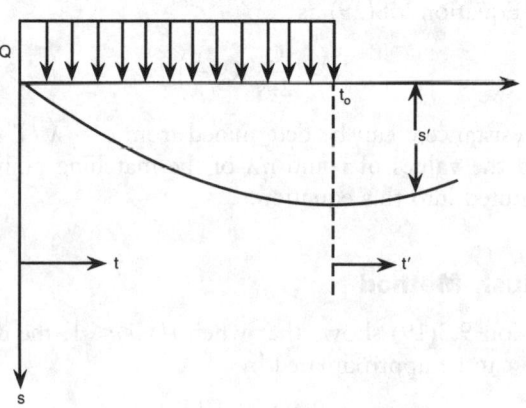

FIG. 9.8.7 Recovery test.

where Q' is the rate of flow, and t' is the time after the pumping stops, respectively. Since $Q' = 0$, s' becomes

$$s' = \frac{Q}{4\pi T} \ell n \frac{t}{t'} = \frac{2.3Q}{4\pi T} \log \frac{t}{t'} \qquad 9.8(27)$$

Thus, T can be calculated as

$$T = \frac{2.3Q}{4\pi \Delta s'} \qquad 9.8(28)$$

However, S cannot be determined from the recovery test.

Semiconfined (Leaky) Aquifers

This section discusses the methods used in determining aquifer characteristics for semiconfined (leaky) aquifers (Bouwer 1978).

STEADY-STATE

The DeGlee–Hantush–Jacob method (DeGlee 1930, 1951; Hantush and Jacob 1955) and the Hantush method (1956, 1964) are used to determine the aquifer characteristics in semiconfined aquifers under steady-state conditions.

De Glee–Hantush–Jacob Method

The drawdown in a semiconfined aquifer is given by Equation 9.3(18) as

$$s = \frac{Q}{2\pi T} K_o \left(\frac{r}{\lambda} \right) \qquad 9.8(29)$$

where $K_o(r/\lambda)$ = modified Bessel function of zero order and second kind, and $\lambda = \sqrt{Tc}$ as defined before. The values of $K_o(r/\lambda)$ versus r/λ are shown in Table 9.8.1. The value T can be determined as in a confined aquifer with the use of the matching procedure. The data curve is obtained from a plot of s versus r on log–log paper, and the type curve is obtained from a plot of $K_o(r/\lambda)$ versus r/λ. Overlapping these two plots matches the two curves, and four coordinates of an arbitrary selected print on the matching curve are noted. The value T is then calculated from Equation 9.8(29) as

$$T = \frac{Q}{2\pi s} K_o \left(\frac{r}{\lambda} \right) \qquad 9.8(30)$$

The resistance c can be determined from $c = \lambda^2/T$ when T and the values of r and r/λ of the matching point are substituted into this equation.

Hantush Method

Equation 9.3(19) shows that when $r/\lambda \ll 1$, the drawdown can be approximated by

$$s = \frac{2.3Q}{2\pi T} \log \frac{1.123\lambda}{r} \qquad 9.8(31)$$

TABLE 9.8.1 VALUES OF THE FUNCTIONS $K_0(X)$ AND EXP $(X)K_0(X)$

x	$K_0(x)$	exp $(x)K_0(x)$	x	$K_0(x)$	exp $(x)K_0(x)$
0.01	4.72	4.77	0.35	1.23	1.75
0.015	4.32	4.38	0.40	1.11	1.6
0.02	4.03	4.11	0.45	1.01	1.59
0.025	3.81	3.91	0.50	0.92	1.52
0.03	3.62	3.73	0.55	0.85	1.47
0.035	3.47	3.59	0.60	0.78	1.42
0.04	3.34	3.47	0.65	0.72	1.37
0.045	3.22	3.37	0.70	0.66	1.33
0.05	3.11	3.27	0.75	0.61	1.29
0.055	3.02	3.19	0.80	0.57	1.26
0.06	2.93	3.11	0.85	0.52	1.23
0.065	2.85	3.05	0.90	0.49	1.20
0.07	2.78	2.98	0.95	0.45	1.17
0.075	2.71	2.92	1.0	0.42	1.14
0.08	2.65	2.87	1.5	0.21	0.96
0.085	2.59	2.82	2.0	0.11	0.84
0.09	2.53	2.77	2.5	0.062	0.760
0.095	2.48	2.72	3.0	0.035	0.698
0.10	2.43	2.68	3.5	0.020	0.649
0.15	2.03	2.36	4.0	0.011	0.609
0.20	1.75	2.14	4.5	0.006	0.576
0.25	1.54	1.98	5.0	0.004	0.548
0.30	1.37	1.85			

Source: Adapted from M.S. Hantush, 1956, Analysis of data from pumping tests in leaky aquifers, *Transactions American Geophysical Union* 37:702–14 and C.W. Fetter, 1988, *Applied hydrogeology,* 2d ed., Macmillan.

A plot of s versus log r forms a straight line, the slope of which is $2.3Q/2\pi T$. If Δs is taken over one log cycle, then T can be calculated as

$$T = \frac{2.3Q}{2\pi(\Delta s)} \qquad 9.8(32)$$

Extending the straight line into the abscissa yields the intercept r_o where $s = 0$. Then, from Equation 9.8(31)

$$0 = \log \frac{1.123\lambda}{r_o} \qquad 9.8(33)$$

so

$$\lambda = r_o/1.123 \qquad 9.8(34)$$

and

$$c = \frac{\lambda^2}{T} = \frac{r_o^2}{1.25T} \qquad 9.8(35)$$

Note that this method does not require the matching procedure.

TRANSIENT-STATE

Hantush and Jacob (1955) showed that the drawdown in a semiconfined aquifer is described by

$$s = \frac{Q}{4\pi T} W\left(u, \frac{r}{\lambda}\right) \qquad 9.8(36)$$

where

$$u = \frac{r^2 S}{4Tt} \qquad 9.8(37)$$

Equation 9.8(36) is similar to Equation 9.7(20) for a confined aquifer except that the well function contains the additional term r/λ. The values of $W(u, r/\lambda)$ are given in Table 9.8.2.

Walton Method

Walton's solution (1962) of Equation 9.8(36) is similar to the Theis method for a confined aquifer. Plotting s versus t/r^2 gives the data curve. Plotting $W(u, r/\lambda)$ versus u for various values of r/λ gives several type curves. Figure 9.8.8 shows the type curves. The data curve is superimposed on the type curves to get the best fitting curve. Again, four coordinates of a match point are read on both graphs. The resulting values of $W(u, r/\lambda)$ and s are substituted into Equation 9.8(36) to calculate T. The value of S is obtained from Equation 9.8(37) when u, t/r^2, and T are substituted. The value c is calculated from $c = \lambda^2/T$ where λ is obtained from the r/λ value of the best fitting curve.

Hantush's Inflection Point Method

Hantush's procedure (1956) for calculating T, S, and c from pumping test data utilizes the halfway point or inflection point on a curve relating s to log t. The inflection point is the point where the drawdown s is one-half the final or equilibrium drawdown as

$$s = \frac{Q}{4\pi T} K_o\left(\frac{r}{\lambda}\right) \qquad 9.8(38)$$

The value u at the inflection point is

$$\frac{r}{2\lambda} = u = \frac{r^2 S}{4Tt_i} \qquad 9.8(39)$$

where t_i is t at the inflection point. The ratio between the drawdown and the slope of the curve at the inflection point Δs expressed as the drawdown per unit log cycle of t is derived as

$$2.3 \frac{s}{\Delta s} = e^{r/\lambda} \cdot K_o\left(\frac{r}{\lambda}\right) \qquad 9.8(40)$$

The values of function $e^{r/\lambda} \cdot K_o(r/\lambda)$ versus r/λ are in Table 9.8.1.

To determine T, S, and c from pumping test data, follow the following procedure:

1. Plot drawdown–time on semilog paper (s–log t).
2. Locate the inflection point P where s = 1/2 × final drawdown.
3. Draw a line tangent to the curve at point P, and determine the corresponding value of t_i and the slope Δs.

4. Substitute s and Δs values into Equation 9.8(40) to obtain $e^{r/\lambda} \cdot K_o(r/\lambda)$, and determine the corresponding value of r/λ and $K_o r/\lambda$ from Table 9.8.1.
5. Determine T from Equation 9.8(38).
6. Determine S from Equation 9.8(39).
7. Determine c from $c = \lambda^2/T$.

Unconfined Aquifers

This section discusses the methods used in determining aquifer characteristics for unconfined aquifers.

STEADY-STATE

As previously explained, the equation of groundwater flow for unconfined aquifers reduces to the same form as that for confined aquifers except that the thickness of the aquifer is not constant but varies as the aquifer is dewatered. Therefore, the flow must be expressed through an average thickness of the aquifer ϕ_{av}. The Thiem equation is then

$$Q = \frac{\pi K(\phi_2^2 - \phi_1^2)}{\ell n\left(\frac{r_2}{r_1}\right)} = \frac{\pi K 2\phi_{av}(\phi_2 - \phi_1)}{\ell n\left(\frac{r_2}{r_1}\right)}$$

$$= \frac{2\pi T_{av}(\phi_2 - \phi_1)}{\ell n\left(\frac{r_2}{r_1}\right)} \qquad 9.8(41)$$

where $\phi_2 = \phi_o - s_2$ and $\phi_1 = \phi_o - s_1$.

From Equation 9.8(41),

$$T_{av} = \frac{Q \ell n\left(\frac{r_2}{r_1}\right)}{2\pi \cdot (s_1 - s_2)} \qquad 9.8(42)$$

which is the same form as that for a confined aquifer. The transmissibility of the aquifer T is then

$$T = \frac{2\phi_o}{2\phi_o - s_1 - s_2} \cdot T_{av} \qquad 9.8(43)$$

Once T has been determined, S can be obtained in the same manner as a confined aquifer. Note that when the steady-state method is applied, pumping does not have to continue until true steady-state conditions are reached since $\Delta s = s_1 - s_2$ reaches an essentially constant value after a few days of pumping.

TRANSIENT-STATE

As explained previously, the transient flow of groundwater in an unconfined aquifer occurs from two types of storage: phreatic and elastic. As water is pumped out of the aquifer, the decline in pressure in the aquifer yields water due to the elastic storage of the aquifer storativity S_e, and the declining water table also yields water as it drains under gravity. Unlike the confined aquifer, the release of wa-

TABLE 9.8.2 VALUES OF W(U,R/A) FOR DIFFERENT VALUES OF U AND R/A

u \ r/λ	0.002	0.004	0.006	0.008	0.01	0.02	0.04	0.06	0.08	0.1	0.2	0.4	0.6	0.8	1	2	4	6	8
0	12.7	11.3	10.5	9.89	9.44	8.06	6.67	5.87	5.29	4.85	3.51	2.23	1.55	1.13	0.842	0.228	0.0223	0.0025	0.0003
0.000002	12.1	11.2	10.5	9.89	9.44														
4	11.6	11.1	10.4	9.88	9.44														
6	11.3	10.9	10.4	9.87	9.44														
8	11.0	10.7	10.3	9.84	9.43														
0.00001	10.8	10.6	10.2	9.80	9.42	8.06													
2	10.2	10.1	9.84	9.58	9.30	8.06													
4	9.52	9.45	9.34	9.19	9.01	8.03	6.67												
6	9.13	9.08	9.00	8.89	8.77	7.98	6.67												
8	8.84	8.81	8.75	8.67	8.57	7.91	6.67												
0.0001	8.62	8.59	8.55	8.48	8.40	7.84	6.67	5.87											
2	7.94	7.92	7.90	7.86	7.82	7.50	6.62	5.86	5.29										
4	7.24	7.24	7.22	7.21	7.19	7.01	6.45	5.83	5.29	4.85									
6	6.84	6.84	6.83	6.82	6.80	6.68	6.27	5.77	5.27	4.85									
8	6.55	6.55	6.54	6.53	6.52	6.43	6.11	5.69	5.25	4.84									
0.001	6.33	6.33	6.32	6.32	6.31	6.23	5.97	5.61	5.21	4.83	3.51								
2	5.64	5.64	5.63	5.63	5.63	5.59	5.45	5.24	4.98	4.71	3.50								
4	4.95	4.95	4.95	4.94	4.94	4.92	4.85	4.74	4.59	4.42	3.48								
6	4.54				4.54	4.53	4.48	4.41	4.30	4.18	3.43	2.23							
8	4.26				4.26	4.25	4.21	4.15	4.08	3.98	3.36	2.23							
0.01	4.04				4.04	4.03	4.00	3.95	3.89	3.81	3.29	2.23	1.55	1.13					
2	3.35				3.35	3.35	3.34	3.31	3.28	3.24	2.95	2.18	1.55	1.13					
4	2.68				2.68	2.68	2.67	2.66	2.65	2.63	2.48	2.02	1.52	1.13	0.842				
6	2.30				2.30	2.29	2.29	2.28	2.27	2.26	2.17	1.85	1.46	1.11	0.839				
8	2.03					2.03	2.02	2.02	2.01	2.00	1.94	1.69	1.39	1.08	0.832				
0.1	1.82						1.82	1.82	1.81	1.80	1.75	1.56	1.31	1.05	0.819	0.228			
2	1.22						1.22	1.22	1.22	1.22	1.19	1.11	0.996	0.857	0.715	0.227			
4	0.702						0.702	0.702	0.701	0.700	0.693	0.665	0.621	0.565	0.502	0.210			
6	0.454						0.454	0.454	0.454	0.453	0.450	0.436	0.415	0.387	0.354	0.177	0.0222		
8	0.311						0.311	0.310	0.310	0.310	0.308	0.301	0.289	0.273	0.254	0.144	0.0218		
1	0.219									0.219	0.218	0.213	0.206	0.197	0.185	0.114	0.0207	0.0025	0.0003
2	0.049										0.049	0.048	0.047	0.046	0.044	0.034	0.011	0.0021	0.0002
4	0.0038											0.0038	0.0037	0.0037	0.0036	0.0031	0.0016	0.0006	0
6	0.0004														0.0004	0.0003	0.0002	0.0001	0
8	0																		0

Source: From M.S. Hantush, 1956, Analysis of data from pumping tests in leaky aquifers, *Transactions American Geophysical Union* 37:702–14. Reference to the original article is made for more extensive tables and expression of W(u,r/λ) in more significant figures (See also M.S. Hantush, 1964, Hydraulics of wells, In Vol. 1 of *Advances in hydroscience*, edited by V.T. Chow [New York and London: Academic Press]:281–432) and H. Bouwer, 1978, *Groundwater hydrology*, McGraw-Hill, Inc.

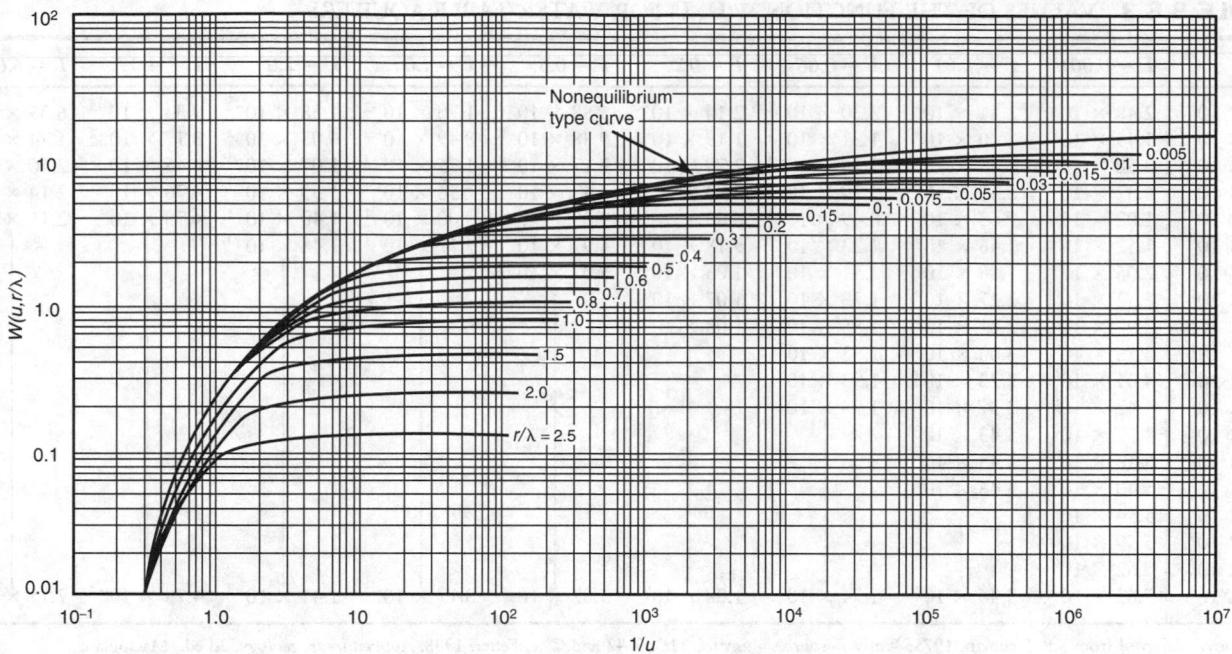

FIG. 9.8.8 Type curves for a leaky aquifer. (Reprinted from C.W. Fetter, 1988, *Applied hydrogeology,* 2d ed., Macmillan Pub. Co.)

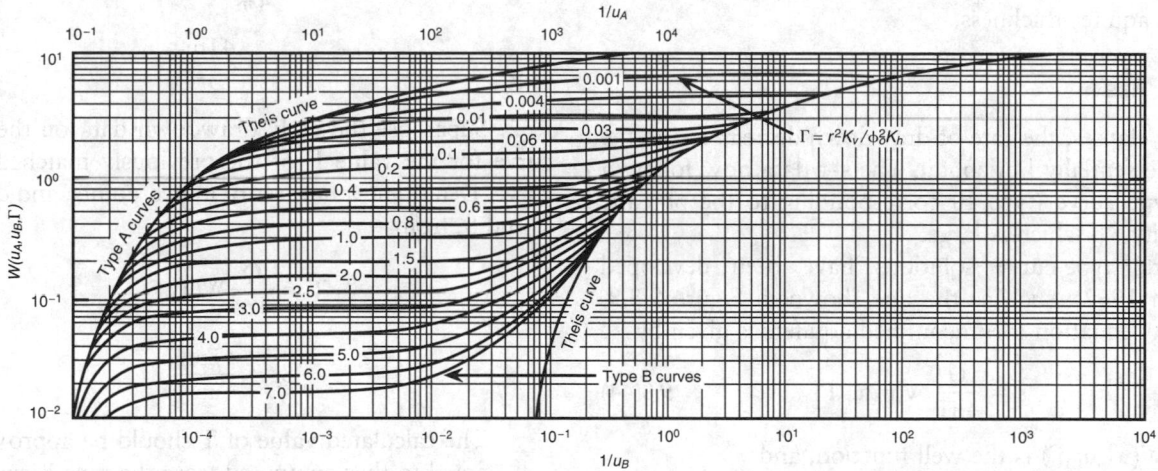

FIG. 9.8.9 Type curves and curves for a delayed yield. (Reprinted from C.W. Fetter, 1988, *Applied hydrogeology,* 2d ed., Macmillan Pub. Co.)

ter from storage is not immediate in response to the drop of the water table. The yield is delayed depending on the elastic and phreatic storativity of the aquifer. Accordingly, the delayed yield produces a sigmoid drawdown curve as shown in Figure 9.8.9.

Essentially, three distinct phases of drawdown–time (s–t) relations occur as shown in Figure 9.8.9: initial phase, intermediate phase, and final phase.

Initial Phase

As the pumping begins, a small amount of water is released from the aquifer under the pressure drop due to the compression of the aquifer. During this stage, the aquifer behaves as a confined aquifer, and the flow is essentially horizontal. The drawdown–time data follow a Theis-type curve (type A) for elastic storativity S_e, which is small.

Intermediate Phase

Following the initial phase, as the water table begins to decline, water is drawn primarily from the gravity drainage of the aquifer. The flow at this stage has both horizontal and vertical components, and the s–t relationship is a function of the ratio of the horizontal to vertical hydraulic con-

TABLE 9.8.3   VALUES OF THE FUNCTION $W(u_A, \Gamma)$ FOR WATER TABLE AQUIFERS

$1/u_A$	$\Gamma = 0.001$	$\Gamma = 0.01$	$\Gamma = 0.06$	$\Gamma = 0.2$	$\Gamma = 0.6$	$\Gamma = 1.0$	$\Gamma = 2.0$	$\Gamma = 4.0$	$\Gamma = 6.0$
4.0×10^{-1}	2.48×10^{-2}	2.41×10^{-2}	2.30×10^{-2}	2.14×10^{-2}	1.88×10^{-2}	1.70×10^{-2}	1.38×10^{-2}	9.33×10^{-3}	6.39×10^{-3}
8.0×10^{-1}	1.45×10^{-1}	1.40×10^{-1}	1.31×10^{-1}	1.19×10^{-1}	9.88×10^{-2}	8.49×10^{-2}	6.03×10^{-2}	3.17×10^{-2}	1.74×10^{-2}
1.4×10^{0}	3.58×10^{-1}	3.45×10^{-1}	3.18×10^{-1}	2.79×10^{-1}	2.17×10^{-1}	1.75×10^{-1}	1.07×10^{-1}	4.45×10^{-2}	2.10×10^{-2}
2.4×10^{0}	6.62×10^{-1}	6.33×10^{-1}	5.70×10^{-1}	4.83×10^{-1}	3.43×10^{-1}	2.56×10^{-1}	1.33×10^{-1}	4.76×10^{-2}	2.14×10^{-2}
4.0×10^{0}	1.02×10^{0}	9.63×10^{-1}	8.49×10^{-1}	6.88×10^{-1}	4.38×10^{-1}	3.00×10^{-1}	1.40×10^{-1}	4.78×10^{-2}	2.15×10^{-2}
8.0×10^{0}	1.57×10^{0}	1.46×10^{0}	1.23×10^{0}	9.18×10^{-1}	4.97×10^{-1}	3.17×10^{-1}	1.41×10^{-1}		
1.4×10^{1}	2.05×10^{0}	1.88×10^{0}	1.51×10^{0}	1.03×10^{0}	5.07×10^{-1}				
2.4×10^{1}	2.52×10^{0}	2.27×10^{0}	1.73×10^{0}	1.07×10^{0}					
4.0×10^{1}	2.97×10^{0}	2.61×10^{0}	1.85×10^{0}	1.08×10^{0}					
8.0×10^{1}	3.56×10^{0}	3.00×10^{0}	1.92×10^{0}						
1.4×10^{2}	4.01×10^{0}	3.23×10^{0}	1.93×10^{0}						
2.4×10^{2}	4.42×10^{0}	3.37×10^{0}	1.94×10^{0}						
4.0×10^{2}	4.77×10^{0}	3.43×10^{0}							
8.0×10^{2}	5.16×10^{0}	3.45×10^{0}							
1.4×10^{3}	5.40×10^{0}	3.46×10^{0}							
2.4×10^{3}	5.54×10^{0}								
4.0×10^{3}	5.59×10^{0}								
8.0×10^{3}	5.62×10^{0}								
1.4×10^{4}	5.62×10^{0}	3.46×10^{0}	1.94×10^{0}	1.08×10^{0}	5.07×10^{-1}	3.17×10^{-1}	1.41×10^{-1}	4.78×10^{-2}	2.15×10^{-2}

Source: Adapted from S.P. Neuman, 1975, *Water Resources Research* 11:329–42 and C.W. Fetter, 1988, *Applied hydrogeology*, 2d ed., Macmillan.

ductivity of the aquifer, the distance to the pumping well, and the aquifer thickness.

Final Phase

As time elapses, the rate of drawdown decreases, and the flow is essentially horizontal. The s–t data now follow a Theis-type curve (type B) corresponding to the phreatic storativity S_p, which is large.

Several type-curve solutions have been developed (Walton 1962), such as the one shown in Figure 9.8.9. The flow equation for unconfined aquifers is given by

$$s = \frac{Q}{4\pi T} W(u_A, u_B, \Gamma) \qquad 9.8(44)$$

where $W(u_A, u_B, \Gamma)$ is the well function, and

$$u_A = \frac{S_e r^2}{4Tt} \quad \text{(for early drawdown data)} \qquad 9.8(45)$$

$$u_B = \frac{S_p r^2}{4Tt} \quad \text{(for later drawdown data)} \qquad 9.8(46)$$

and

$$\Gamma = \frac{r^2 K_v}{\phi_o^2 K_h} \qquad 9.8(47)$$

The values of $W(u_A, \Gamma)$ and $W(u_B, \Gamma)$ are given in Tables 9.8.3 and 9.8.4. The type curves are used to evaluate the field data for drawdown and time with the use of the following procedure (Fetter 1988):

1. Superpose the late drawdown–time data on the type-B curves for the best fit. At any match point, determine the values of $W(u_B, \Gamma)$, u_B, t, and s. Obtain the value Γ from the type curve. Calculate T and S_p from

$$T = \frac{Q}{4\pi s} W(u_B, \Gamma) \qquad 9.8(48)$$

$$S_p = \frac{4Ttu_B}{r^2} \qquad 9.8(49)$$

2. Superpose the early drawdown data on the type-A curve for the value Γ of the previously matched type-B curve. Determine a new set of match points, and calculate T and S_e from

$$T = \frac{Q}{4\pi s} W(u_A, \Gamma) \qquad 9.8(50)$$

$$S_e = \frac{4Ttu_A}{r^2} \qquad 9.8(51)$$

The calculated value of T should be approximately equal to that computed from the type-B curve.

3. Determine K_h and K_v from

$$K_h = \frac{T}{\phi_o} \qquad 9.8(52)$$

$$K_v = \frac{\Gamma \phi_o^2 K_h}{r^2} \qquad 9.8(53)$$

Slug Tests

A slug test is a simple and inexpensive way of determining local values of aquifer properties. Instead of the well being pumped for a period of time, a volume of water is suddenly removed or added to the well casing, and recovery or drawdown are observed over time. Through careful evaluation of the drawdown curve and knowledge of the well screen geometry, the hydraulic conductivity of an aquifer can be derived (Bedient 1994).

TABLE 9.8.4 VALUES OF THE FUNCTION W(U_B,Γ) FOR WATER TABLE AQUIFERS

$1/u_B$	$\Gamma = 0.001$	$\Gamma = 0.01$	$\Gamma = 0.06$	$\Gamma = 0.2$	$\Gamma = 0.6$	$\Gamma = 1.0$	$\Gamma = 2.0$	$\Gamma = 4.0$	$\Gamma = 6.0$
4.0×10^{-4}	5.62×10^{0}	3.46×10^{0}	1.94×10^{0}	1.09×10^{0}	5.08×10^{-1}	3.18×10^{-1}	1.42×10^{-1}	4.79×10^{-2}	2.15×10^{-2}
8.0×10^{-4}								4.80×10^{-2}	2.16×10^{-2}
1.4×10^{-3}								4.81×10^{-2}	2.17×10^{-2}
2.4×10^{-3}								4.84×10^{-2}	2.19×10^{-2}
4.0×10^{-3}					5.08×10^{-1}	3.18×10^{-1}	1.42×10^{-1}	4.88×10^{-2}	2.21×10^{-2}
8.0×10^{-3}					5.09×10^{-1}	3.19×10^{-1}	1.43×10^{-1}	4.96×10^{-2}	2.28×10^{-2}
1.4×10^{-2}					5.10×10^{-1}	3.21×10^{-1}	1.45×10^{-1}	5.09×10^{-2}	2.39×10^{-2}
2.4×10^{-2}					5.12×10^{-1}	3.23×10^{-1}	1.47×10^{-1}	5.32×10^{-2}	2.57×10^{-2}
4.0×10^{-2}					5.16×10^{-1}	3.27×10^{-1}	1.52×10^{-1}	5.68×10^{-2}	2.86×10^{-2}
8.0×10^{-2}				1.09×10^{0}	5.24×10^{-1}	3.37×10^{-1}	1.62×10^{-1}	6.61×10^{-2}	3.62×10^{-2}
1.4×10^{-1}			1.94×10^{0}	1.10×10^{0}	5.37×10^{-1}	3.50×10^{-1}	1.78×10^{-1}	8.06×10^{-2}	4.86×10^{-2}
2.4×10^{-1}			1.95×10^{0}	1.11×10^{0}	5.57×10^{-1}	3.74×10^{-1}	2.05×10^{-1}	1.06×10^{-1}	7.14×10^{-2}
4.0×10^{-1}			1.96×10^{0}	1.13×10^{0}	5.89×10^{-1}	4.12×10^{-1}	2.48×10^{-1}	1.49×10^{-1}	1.13×10^{-1}
8.0×10^{-1}	5.62×10^{0}	3.46×10^{0}	1.98×10^{0}	1.18×10^{0}	6.67×10^{-1}	5.06×10^{-1}	3.57×10^{-1}	2.66×10^{-1}	2.31×10^{-1}
1.4×10^{0}	5.63×10^{0}	3.47×10^{0}	2.01×10^{0}	1.24×10^{0}	7.80×10^{-1}	6.42×10^{-1}	5.17×10^{-1}	4.45×10^{-1}	4.19×10^{-1}
2.4×10^{0}	5.63×10^{0}	3.49×10^{0}	2.06×10^{0}	1.35×10^{0}	9.54×10^{-1}	8.50×10^{-1}	7.63×10^{-1}	7.18×10^{-1}	7.03×10^{-1}
4.0×10^{0}	5.63×10^{0}	3.51×10^{0}	2.13×10^{0}	1.50×10^{0}	1.20×10^{0}	1.13×10^{0}	1.08×10^{0}	1.06×10^{0}	1.05×10^{0}
8.0×10^{0}	5.64×10^{0}	3.56×10^{0}	2.31×10^{0}	1.85×10^{0}	1.68×10^{0}	1.65×10^{0}	1.63×10^{0}	1.63×10^{0}	1.63×10^{0}
1.4×10^{1}	5.65×10^{0}	3.63×10^{0}	2.55×10^{0}	2.23×10^{0}	2.15×10^{0}	2.14×10^{0}	2.14×10^{0}	2.14×10^{0}	2.14×10^{0}
2.4×10^{1}	5.67×10^{0}	3.74×10^{0}	2.86×10^{0}	2.68×10^{0}	2.65×10^{0}	2.65×10^{0}	2.64×10^{0}	2.64×10^{0}	2.64×10^{0}
4.0×10^{1}	5.70×10^{0}	3.90×10^{0}	3.24×10^{0}	3.15×10^{0}	3.14×10^{0}	3.14×10^{0}	3.14×10^{0}	3.14×10^{0}	3.14×10^{0}
8.0×10^{1}	5.76×10^{0}	4.22×10^{0}	3.85×10^{0}	3.82×10^{0}	3.82×10^{0}	3.82×10^{0}	3.82×10^{0}	3.82×10^{0}	3.82×10^{0}
1.4×10^{2}	5.85×10^{0}	4.58×10^{0}	4.38×10^{0}	4.37×10^{0}	4.37×10^{0}	4.37×10^{0}	4.37×10^{0}	4.37×10^{0}	4.37×10^{0}
2.4×10^{2}	5.99×10^{0}	5.00×10^{0}	4.91×10^{0}	4.91×10^{0}	4.91×10^{0}	4.91×10^{0}	4.91×10^{0}	4.91×10^{0}	4.91×10^{0}
4.0×10^{2}	6.16×10^{0}	5.46×10^{0}	5.42×10^{0}	5.42×10^{0}	5.42×10^{0}	5.42×10^{0}	5.42×10^{0}	5.42×10^{0}	5.42×10^{0}
8.0×10^{2}	6.47×10^{0}	6.11×10^{0}	6.11×10^{0}	6.11×10^{0}	6.11×10^{0}	6.11×10^{0}	6.11×10^{0}	6.11×10^{0}	6.11×10^{0}
1.4×10^{3}	6.67×10^{0}	6.67×10^{0}	6.67×10^{0}	6.67×10^{0}	6.67×10^{0}	6.67×10^{0}	6.67×10^{0}	6.67×10^{0}	6.67×10^{0}
2.4×10^{3}	7.21×10^{0}	7.21×10^{0}	7.21×10^{0}	7.21×10^{0}	7.21×10^{0}	7.21×10^{0}	7.21×10^{0}	7.21×10^{0}	7.21×10^{0}
4.0×10^{3}	7.72×10^{0}	7.72×10^{0}	7.72×10^{0}	7.72×10^{0}	7.72×10^{0}	7.72×10^{0}	7.72×10^{0}	7.72×10^{0}	7.72×10^{0}
8.0×10^{3}	8.41×10^{0}	8.41×10^{0}	8.41×10^{0}	8.41×10^{0}	8.41×10^{0}	8.41×10^{0}	8.41×10^{0}	8.41×10^{0}	8.41×10^{0}
1.4×10^{4}	8.97×10^{0}	8.97×10^{0}	8.97×10^{0}	8.97×10^{0}	8.97×10^{0}	8.97×10^{0}	8.97×10^{0}	8.97×10^{0}	8.97×10^{0}
2.4×10^{4}	9.51×10^{0}	9.51×10^{0}	9.51×10^{0}	9.51×10^{0}	9.51×10^{0}	9.51×10^{0}	9.51×10^{0}	9.51×10^{0}	9.51×10^{0}
4.0×10^{4}	1.94×10^{1}	1.94×10^{1}	1.94×10^{1}	1.94×10^{1}	1.94×10^{1}	1.94×10^{1}	1.94×10^{1}	1.94×10^{1}	1.94×10^{1}

Source: Adapted from S.P. Neuman, 1975, *Water Resources Research* 11:329–42 and C.W. Fetter, 1988, *Applied hydrogeology*, 2d ed., Macmillan.

Hvorslev (1951) developed the simplest slug test method in a piezometer, which relates the flow rate $Q(t)$ at the piezometer at any time to the hydraulic conductivity and unrecovered head distance $H_o - h$ in Figure 9.8.10 by

$$Q(t) = \pi r^2 \frac{dh}{dt} = FK(H_o - h) \qquad 9.8(54)$$

where F is a factor that depends on the shape and the dimensions of the piezometer intake. If $Q = Q_o$ at $t = 0$, then $Q(t)$ decreases toward zero as time increases. Hvorslev defined the basic time lag $T_o = \pi r^2/FK$ and solved Equation 9.8(54) with initial conditions $h = H_o$ at $t = 0$. Thus

$$\frac{H - h}{H - H_o} = e^{-t/T_o} \qquad 9.8(55)$$

When recovery $H - h/H - H_o$ versus time is plotted on semilog paper, T_o is noted at t where recovery equals 37% of the initial change. For the piezometer intake length divided by radius, L/R greater than 8, Hvorslev has evaluated the shape factor F and obtained an equation for K as

$$K = \frac{r^2 \ell n(L/R)}{2LT_o} \qquad 9.8(56)$$

Several other slug test methods have been developed for confined aquifers by Cooper et al. (1967) and Papadopoulos et al. (1973). These methods are similar to Theis's in which a curve-matching procedure is used to obtain S and T for a given aquifer. Figure 9.8.11 shows the slug test curves developed by Papadopoulos for various values of variable α, defined as

$$\alpha = \frac{r_s^2}{r_c^2} S \qquad 9.8(57)$$

The obtained data are plotted and matched to the plotted type curves for a best match, from which α is selected for a particular curve. The vertical time axis t which overlays the vertical axis for $Tt/r_c^2 = 1.0$ is selected, and a value of T can then be found from $T = 1.0 r_s^2/t_1$. Then, the value of S can be found from the definition of α. The method is representative of the formation only in the immediate vicinity of the test hole and should be used with caution (Bedient 1994).

The most commonly used method for determining hydraulic conductivity in groundwater investigation is the Bouwer and Rice (1976) slug test shown in Figure 9.8.12. Although it was originally designed for unconfined

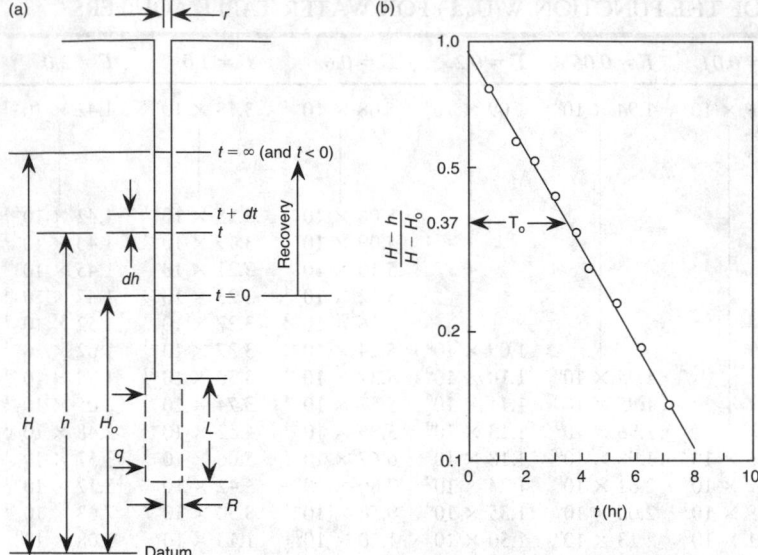

FIG. 9.8.10 Hvorslev piezometer test. (Reprinted from P.B. Bedient et al., 1994, *Groundwater contamination*, PTR Prentice-Hall, Inc.)

aquifers, it can be used for confined aquifers if the top of the screen is some distance below the upper confining layer. The method is based on the following equation:

$$K = \frac{r_c^2 \, \ell n(R_e/r_w)}{2L_e} \frac{1}{t} \, \ell n \frac{y_o}{y_t} \qquad 9.8(58)$$

where:

R_e = effective radial distance over which the head difference is dissipated

r_w = radial distance between the well center and the undisturbed aquifer (including gravel pack)

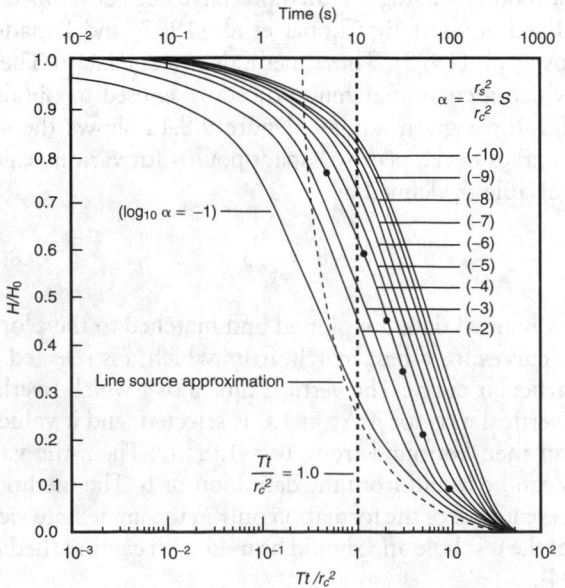

FIG. 9.8.11 Slug test type curves. (Reprinted from I.S. Papadopoulos, J.D. Bredehoeft, and H.H. Cooper, Jr., 1973, On the analysis of slug test data, *Water Resources Res.* 9, no. 4: 1087–1089.)

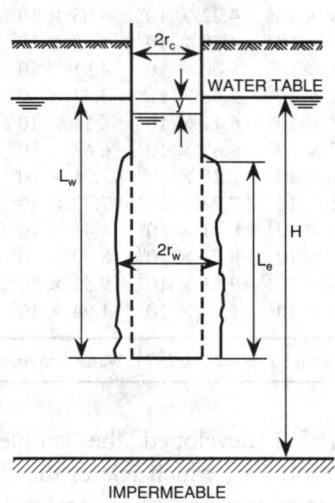

FIG. 9.8.12 Slug test setup. (Reprinted from H. Bouwer, 1978, *Groundwater hydrology*, McGraw-Hill, Inc.)

L_e = height of the perforated, screened, uncased, or otherwise open section of the well through which groundwater enters

y_o = y at time zero

y_t = y at time t

t = time sine y_o

In Equation 9.8(58), y and t are the only variables. Thus, if a number of y and t measurements are taken, they can be plotted on semilog paper to give a straight line. The slope of the best-fitting straight line provides a value for $\ell n(y_o/y_t)/t$. All other parameters in Equation 9.8(58) are known from well geometry, and K can be calculated.

—*Y.S. Chae*

References

Bedient, P.B. et al. 1994. *Groundwater contamination*, PTR Prentice-Hall, Inc.

Boulton, N.S. 1963. Analysis of data from non-equilibrium pumping tests allowing for delayed yield from storage. *Proc. Inst. Civ. Eng.* 16:469–482.

Bouwer, H. 1978. *Groundwater hydrology.* McGraw-Hill, Inc.

Bouwer, H., and R.C. Rice. 1976. A slug test for determining hydraulic conductivity of unconfined aquifers with completely or partially penetrating wells. *Water Resour. Res.* 12:423–428.

Chow, V.T. 1952. On the determination of transmissivity and storage coefficients from pumping test data. *Trans. Am. Geoph. Union* 33: 397–404.

Cooper, H.H., Jr., J.D. Bredcheoft, and I.S. Papadopoulos. 1967. Response of a finite-diameter well to an instantaneous charge of water. *Water Resour. Res.* 3:263–269.

Cooper, H.H., Jr., and C.E. Jacob. 1946. A generalized graphical method for evaluating formation constants and summarizing well-field history, *Trans, Am, Geoph. Union* 27:526–534.

DeGlee, G.J. 1930. Over grondwaterstromingen bij wateronttrekking door middel van putten. Doctoral dissertation, Techn. Univ., Delft. The Netherlands. Printed by J. Waltman.

DeGlee, G.J. 1951. Berekeningsmethoden voor de winning van grondwater. In *Drinkwaterroorzlening 3e Vacantie cursus,* 38–80. Moorman's periodieke pers. The Hague, Netherlands.

Fetter, C.W. 1988. *Applied hydrogeology.* 2d ed. Macmillan Pub. Co.

Gupta, R.S. 1989. *Hydrology and hydraulic systems.* Prentice-Hall, Inc.

Hantush, M.S. 1956. Analysis of data from pumping tests in leaky aquifers. *Trans. Am. Geophys. Un.* 37:702–714.

Hantush, M.S. 1964. Hydraulics of wells. In *Advances in Hydroscience.* Vol. 1: edited by V.T. Chow, 281–432, New York and London: Academic Press.

Hantush, M.S., and C.E. Jacob. 1955. Non-steady radial flow in an infinite leaky aquifer. *Am. Geophys. Un. Trans.* 36:95–100.

Hvorslev, M.J. 1951. *Time lag and soil permeability in groundwater observations.* U.S. Army Waterways Experiment Station Bull. 36.

Neuman, S.P. 1972. Theory of flow in unconfined aquifers considering delayed response of the water table. *Water Resources Res.* 8:1031–1045.

Neuman, S.P., and P.A. Witherspoon. 1969. Theory of flow in a confine two-aquifer system. *Water Resource Research* 5:803–816.

Papadopoulos, I.S., J.D. Bredehoeft, and H.H. Cooper, Jr. 1973. On the analysis of slug test data. *Water Resources Res.* 9, no. 4:1087–1089.

Walton, W.C. 1962. Selected analytical methods for well and aquifer evaluation. *Illinois State Water Surrey Bull.* 49.

9.9
DESIGN CONSIDERATIONS

In well design, well losses, specific capacity, and partially penetrating wells must be considered.

Well Losses

In the previous sections, the drawdown in a pumping well was assumed to be due only to head losses in the aquifer. In reality, however, additional drawdown is caused by head losses in the well system itself (screen or perforated casing) as water flows through it. The former is known as the drawdown due to the *formation loss* (s_w) and the latter as the drawdown due to the *well loss* (s_f) as shown in Figure 9.9.1. The total drawdown s_t is then $s_t = s_w + s_f$.

Since the flow in the aquifer is laminar, s_w varies linearly with $Q(s_w = C_wQ)$. However, the flow through the well system (screen and perforated casing) is turbulent, and thus s_f can vary with some power of $Q(s_f = C_fQ^n)$. The total drawdown can be expressed as

$$s_t = s_w + s_f = C_wQ + C_fQ^n \qquad 9.9(1)$$

where

$$C_w = \frac{1}{2\pi T}\ell n\left(\frac{R}{r_w}\right) \quad \text{for steady-state} \qquad 9.9(2)$$

$$C_w = \frac{1}{4\pi T}W\left(\frac{r_w^2S}{4Tt}\right) \quad \text{for unsteady-state} \qquad 9.9(3)$$

$$n = 2 \text{ to } 3.5$$

The best way to determine the values of C_w, C_f, and n is by experiment utilizing a step-drawdown test (Bear 1979;

Bouwer 1978). Well efficiency is related to the well loss and is defined as

$$E_w = \frac{s_w}{s_t} = 1 - \frac{s_f}{s_t} \qquad 9.9(4)$$

Specific Capacity

The *specific capacity* of a well is defined as the well flow per unit drop of water level in the well.

$$\text{Specific capacity} = \frac{Q}{s_t} = \frac{1}{C_w + C_fQ^{n-1}} \qquad 9.9(5)$$

Specific capacity decreases with the pumping rate and time as shown in Figure 9.9.2. It is a useful concept because it describes the productivity of both the aquifer and the well in a single parameter. A reduction of up to 40% in the specific capacity has been observed in one year in wells deriving water entirely from storage (Gupta 1989). The specific capacities of wells in an aquifer system can be used to calculate the transmissivity distribution of the aquifer based on pumping a well of a known diameter for a given period of time.

Partially Penetrating Wells (Imperfect Wells)

In practice, the underlying impermeable soil layer is often absent or is encountered at a great depth. Wells which do

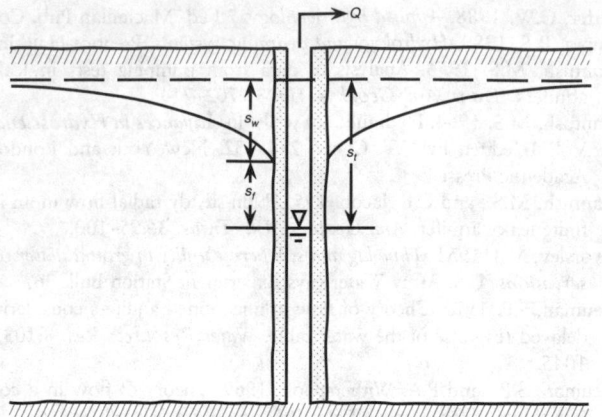

FIG. 9.9.1 Formation and well losses in a pumping well. (Reprinted from R.S. Gupta, 1989, *Hydrology and hydraulic systems,* Prentice-Hall, Inc.)

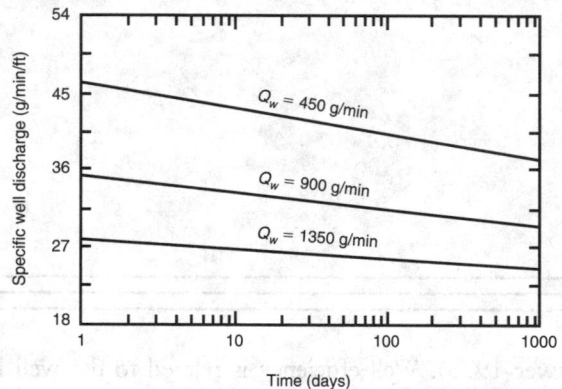

FIG. 9.9.2 Examples of variation of specific discharge with pumping rate and time. (Reprinted from J. Bear, 1979, *Hydraulics of groundwater,* McGraw-Hill, Inc.)

not penetrate through the entire thickness of the aquifer are called *partially penetrating wells* or *imperfect wells.* Imperfect wells are encountered more often in practice than perfect, fully penetrating, wells. Imperfect wells can also have open or closed ends. Figure 9.9.3 shows various configurations of wells in an unconfined aquifer (ordinary

wells), and Figure 9.9.4 shows configurations for wells in a confined aquifer (artesian wells).

Compared to a fully penetrating well, a partially penetrating well has an additional head loss, as shown in Figure 9.9.5, due to the convergence of flow lines and their extended length. Hence, for a given pumping rate Q, the drawdown in an imperfect well is larger than in a perfect well.

Starting with Forchheimer in 1898, numerous analytical and empirical equations have been developed to solve imperfect wells: Kozney (1933), Muskat (1937), Li et al, Hantush (1962), and Kirkham (1959) among others.

CONFINED AQUIFERS

As previously stated, an imperfect well requires more drawdown for a given Q than does a perfect well. The additional drawdown can be represented by

$$s_{wp} = \frac{Q}{4\pi T}\left(\ell n \frac{2.25Tt}{r^2S} + 2s_p\right) \qquad 9.9(6)$$

where s_{wp} is the drawdown at the well. The values of s_p as a function of H/r_w for various values of L_e/H can be obtained from Figure 9.9.6, which was developed by Sternberg (1967).

The performance of an imperfect well related to a perfect well is expressed as an efficiency, defined as the ratio of Q_p to Q at a given drawdown as

$$\frac{Q_p}{Q} = \frac{1}{1 + \dfrac{s_p}{s_w}\dfrac{Q}{2\pi T}} \qquad 9.9(7)$$

or

$$\frac{Q_p}{Q} = \frac{1}{1 + \dfrac{s_p}{\ell n\left(\dfrac{r}{r_w}\right)}} \qquad 9.9(8)$$

Equation 9.9(8) applies to wells with their perforated or open section at the top (Figure 9.9.5-c) or at the bot-

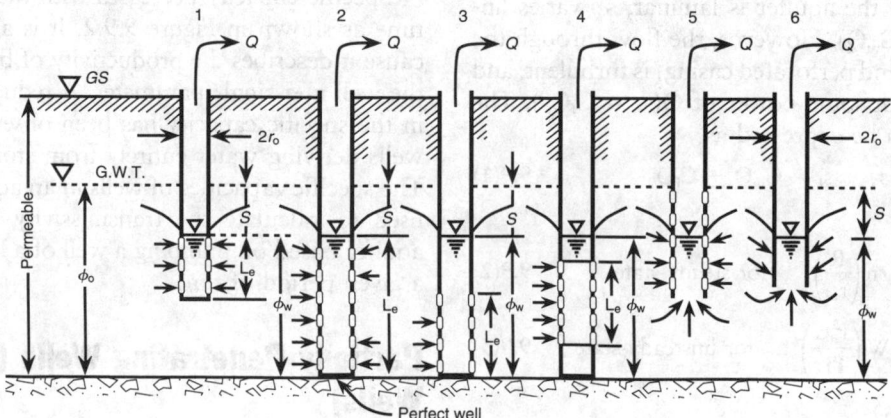

FIG. 9.9.3 Types of ordinary imperfect wells. (Reprinted from A.R. Jumikis, 1964, *Mechanics of soils,* Van Nostrand.)

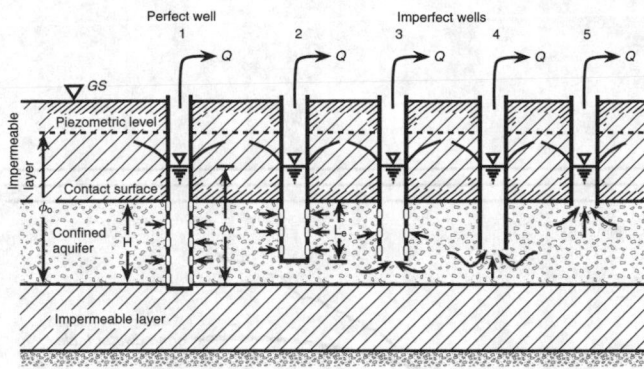

FIG. 9.9.4 Types of artesian imperfect wells. (Reprinted from A.R. Jumikis, 1964, *Mechanics of soils*, Van Nostrand.)

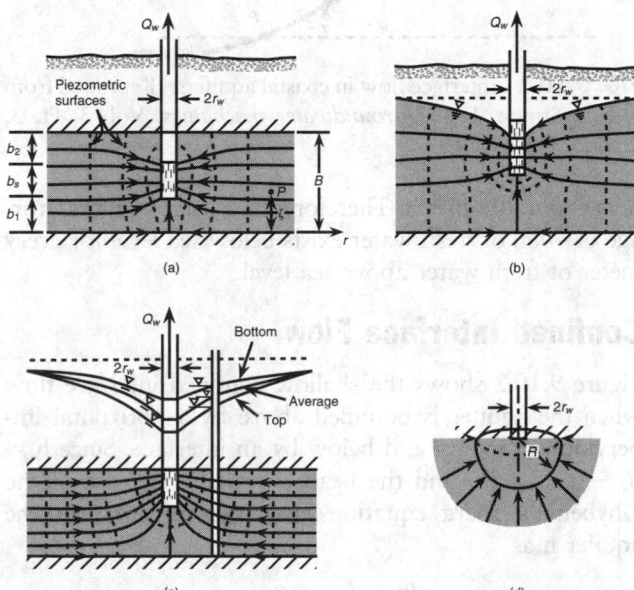

FIG. 9.9.5 Partially penetrating wells. (a) In a confined aquifer. (b) In a phreatic aquifer. (c) Drawdown curves along streamlines. (d) Zero penetration in a thick aquifer. (Reprinted from J. Bear, 1979, *Hydraulics of groundwater*, McGraw-Hill, Inc.)

tom. If the open section of the well is in the center of the aquifer (Figure 9.9.5-a), vertical flow components occur at both the top and bottom of the section. For this case, Q_p/Q can be obtained for the half-section when the section is split symmetrically along the midway.

UNCONFINED AQUIFERS

The s_p values in Figure 9.9.6 also give reasonable estimates of Q_p/Q for wells in unconfined aquifers, particularly if the drawdown is small compared to the aquifer thickness and the well has been pumped for some time (Bouwer 1978).

Forchheimer (1898) observed that the discharge Q_p of a well with a perforated casing and closed end becomes larger, at equal drawdowns, as the immersed depth L_e of the perforated well increases. He assumed that the increase with depth varies with the geometric mean between the parabolic and elliptic ordinates and showed an efficiency to be

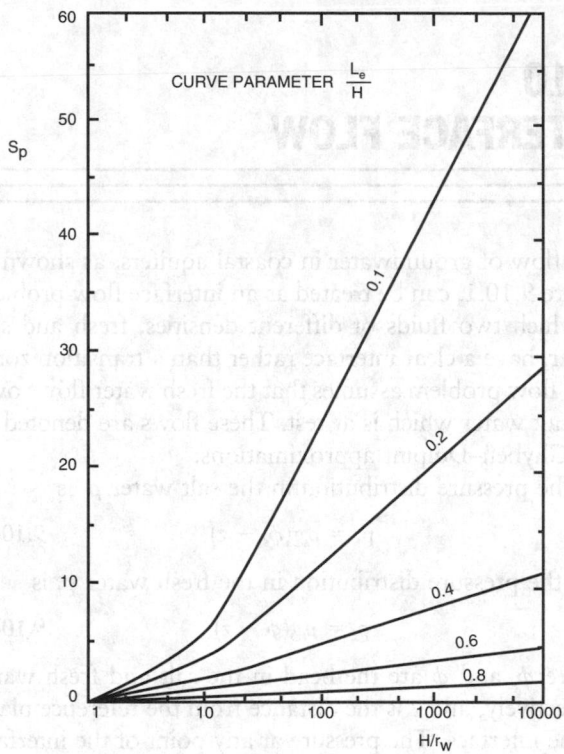

FIG. 9.9.6 Graph of s_p versus H/r_w. (Reprinted from Y.M. Sternberg, 1967, Efficiency of partially penetrating wells, *Ground Water* II, no. 3:5–8.)

$$\frac{Q_p}{Q} = \sqrt{\frac{L_e}{z}} \, \sqrt[4]{\frac{2z - L_e}{z}} \quad \text{for closed end} \qquad 9.9(9)$$

and

$$\frac{Q_p}{Q} = \sqrt{\frac{L_e + 0.5 r_w}{z}} \cdot \sqrt[4]{\frac{2z - L_e}{z}} \quad \text{for open end} \qquad 9.9(10)$$

where z = distance from the water level in the well to the impervious stratum in meters.

—*Y.S. Chae*

References

Bear, J. 1979. *Hydraulics of groundwater*. McGraw-Hill, Inc.

Bouwer, H. 1978. *Groundwater hydrology*. McGraw-Hill, Inc.

Forchheimer, Ph., 1898. Grundwasserspiegel bei Brunnenanlagen. *Zeitschrift des Osterreichischen Ingenieur-und Architekten-Vereins* 50, no. 45:645.

Gupta, R.S. 1989. *Hydrology and hydraulic systems*. Prentice-Hall, Inc.

Hantush, M.S. 1962. Aquifer tests on partially penetrating wells. Trans. Am. Soc. Civ. Eng. Vol. 127, pt. 1:284–308.

Kirkham, D. 1959. Exact theory of flow into a partially penetrating well. *J. Geophys. Research* 64:1317–1327.

Kozeny, J. 1933. Theorie und Berechnung der Brunnen. *Wasserkraft und Wasserwirtschaft* 28:104.

Li, W.H. et al. A new formula for flow into partially penetrating wells in aquifers. Trans. Am. Geophys. Union 35:806–811.

Muskat, M. 1937. *The flow of homogeneous fluids through porous media*. McGraw-Hill.

Sternberg, Y.M. 1967. Efficiency of partially penetrating wells. *Ground Water* 11, no. 3:5–8.

9.10
INTERFACE FLOW

The flow of groundwater in coastal aquifers, as shown in Figure 9.10.1, can be treated as an interface flow problem in which two fluids of different densities, fresh and salt water, have a clear interface rather than a transition zone. This flow problem assumes that the fresh water flows over the salt water which is at rest. These flows are denoted as the Ghyben–Duipint approximations.

The pressure distribution in the salt water ρ_s is

$$p_s = \rho_s g(\phi_s - z) \qquad 9.10(1)$$

and the pressure distribution in the fresh water p_f is

$$p_f = \rho_f g(\phi - z) \qquad 9.10(2)$$

where ϕ_s and ϕ are the head in the salt and fresh water respectively, and z is the distance from the reference plane to the interface. The pressure at any point of the interface must be a single value, that is $p_f = p_s$. Therefore, with Equations 9.10(1) and 9.10(2) and with $z = H_s - h_s$, then

$$\rho_s g(\phi_s - H_s + h_s) = \rho_f g(\phi - H_s + h_s) \qquad 9.10(3)$$

If $\phi_s = H_s$ and $\phi = H_s + h_f$, Equation 9.10(3) yields

$$h_s = \frac{\rho_f}{\rho_s - \rho_f} h_f = \alpha h_f \qquad 9.10(4)$$

This equation is known as the Ghyben–Herzberg equation. This equation is also valid for confined aquifers, in which the upper boundary of the aquifer is a horizontal impermeable boundary rather than a phreatic surface and h_f represents the piezometric head with respect to sea level. The ratio between the densities of salt and fresh water is of the order of 1.025. Then, Equation 9.10(4) shows that

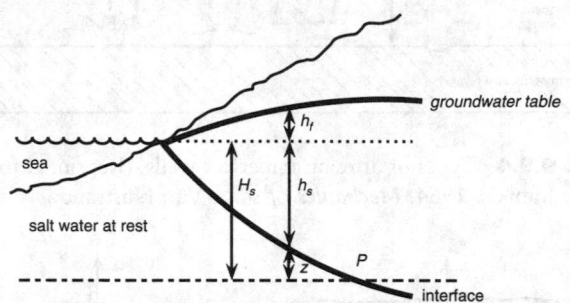

FIG. 9.10.1 Interface flow in coastal aquifers. (Reprinted from O.D.L. Strack, 1989, *Groundwater mechanics*, Vol. 3, Pt. 3, Prentice-Hall, Inc.)

h_s is about 40 times h. Therefore, in coastal aquifers, storage of 40 m of fresh water exists below sea level for every meter of fresh water above sea level.

Confined Interface Flow

Figure 9.10.2 shows the shallow confined interface flow when the aquifer is bounded above by a horizontal impervious boundary and below by an interface. Since $h = h_s - (H_s - H)$ and the head $\phi = h_f + H_s$, use of the Ghyben–Herzberg equation gives the thickness of the aquifer h as

$$h = \frac{\rho_f}{\rho_s - \rho_f} \phi - \frac{\rho_s}{\rho_s - \rho_f} H_s + H \qquad 9.10(5)$$

The elevation z of the interface above the reference level equals $z = H_s - h_s$. Use of the Ghyben–Herzberg equation then yields

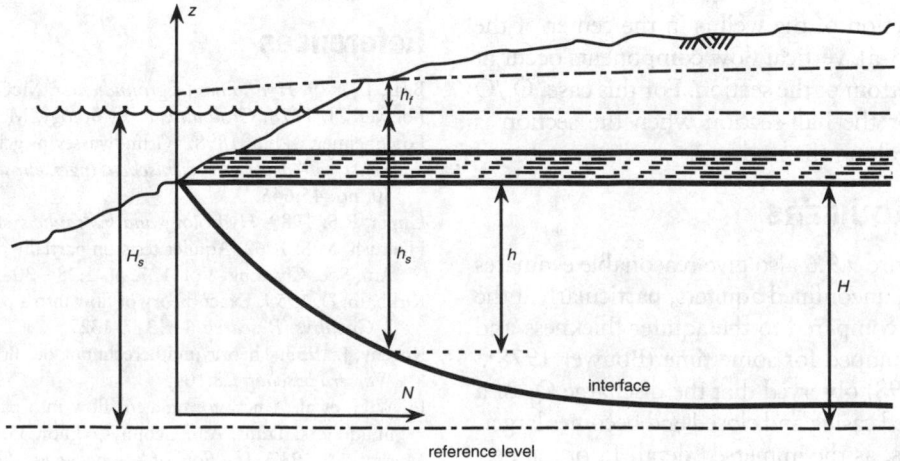

FIG. 9.10.2 Shallow confined interface flow. (Reprinted from O.D.L. Strack, *Groundwater mechanics*, Vol. 3, Pt. 3, Prentice-Hall, Inc.)

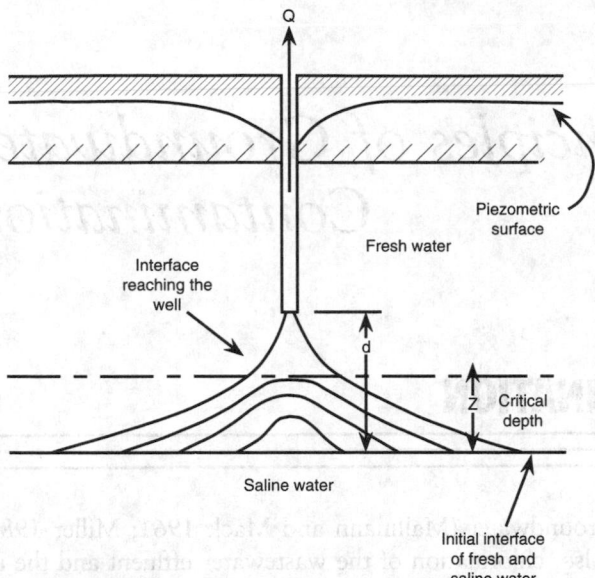

FIG. 9.10.3 Upconing of salt water under a pumping well. (Reprinted from R.S. Gupta, 1989, *Hydrology and hydraulic systems*, Prentice-Hall, Inc.)

TABLE 9.10.1 CONTROLLING SALTWATER INTRUSION OF VARIOUS CATEGORIES

Source or Cause of Intrusion	Control Methods
Seawater in coastal aquifer	Modification of pumping pattern
	Artificial recharge
	Extraction barrier
	Injection barrier
	Subsurface barrier
Upconing	Modification of pumping pattern
	Saline scavenger wells
Oil field brine	Elimination of surface disposal
	Injection wells
	Plugging of abandoned wells
Defective well casings	Plugging of faulty wells
Surface infiltration	Elimination of source
Saline water zones in freshwater aquifers	Relocation and redesign of wells

Source: D.K. Todd, 1980, *Groundwater hydrology* (John Wiley and Sons).

$$z = \frac{\rho_s}{\rho_s - \rho_f} H_s - \frac{\rho_f}{\rho_s - \rho_f} \phi \qquad 9.10(6)$$

Unconfined Interface Flow

A shallow unconfined aquifer is shown in Figure 9.10.2. The aquifer thickness h can now be expressed with the Ghyben–Herzberg equation as

$$h = \frac{\rho_s}{\rho_s - \rho_f} \phi - \frac{\rho_s}{\rho_s - \rho_f} H_s \qquad 9.10(7)$$

and the elevation of the interface z is obtained in the same way as the confined interface flow as

$$z = \frac{\rho_s}{\rho_s - \rho_f} H_s - \frac{\rho_f}{\rho_s - \rho_f} \phi \qquad 9.10(8)$$

Upconing of Saline Water

Figure 9.10.3 depicts a situation in which water is pumped from a freshwater zone underlaid by a saline water layer. The interface between fresh water and saline water rises toward the well in a cone shape as shown in the figure. This phenomenon is known as *upconing*.

The height of upconing under the steady-state condition (Gupta 1989) is given by

$$z = \frac{Q}{2\pi Kd} \frac{\rho_f}{\rho_s - \rho_f} \qquad 9.10(9)$$

where d = depth to the initial interface below the bottom of the well. Salt water reaches the well, contaminating the supply, when the rise becomes critical at z = 0.3 to 0.5 d. Thus, the maximum discharge that keeps the rise below the critical limit is obtained when z = 0.5 d is substituted in Equation 9.10(9) as

$$Q_{max} = \pi Kd^2 \frac{\rho_s - \rho_f}{\rho_f} \qquad 9.10(10)$$

In reality, brackish water occurs between fresh and salt water. Even with a low rate of pumping, some saline water inevitably reaches the pump. Increasing the distance d and decreasing the rate of pumping Q minimizes the upconing effect.

Protection Against Intrusion

Controlling the intrusion of saline water before it contaminates an aquifer system is desirable because removing it once it has developed is difficult. Years may be required to restore normal conditions. Table 9.10.1 summarizes many control methods suggested for various categories of problems.

—*Y.S Chae*

Reference

Gupta, R.S. 1989. *Hydrology and hydraulic systems*. Prentice-Hall, Inc.

Principles of Groundwater Contamination

9.11
CAUSES AND SOURCES OF CONTAMINATION

A groundwater contaminant is defined by most regulatory agencies as any physical, chemical, biological, or radiological substance or matter in groundwater. The contaminants can be introduced in the groundwater by naturally occurring activities, such as natural leaching of the soil and mixing with other groundwater sources having different chemistry. They are also introduced by planned human activities, such as waste disposal, mining, and agricultural operations. Because the contamination from naturally occurring activities is usually small, it is not the focus of this chapter. However, human activities are the leading cause of groundwater contamination and the focus of most regulatory agencies.

The most prevalent human activities that cause groundwater contamination are (1) waste disposal, (2) storage and transportation of commercial materials, (3) mining operations, (4) agricultural operations, and (5) other activities as shown in Figure 9.11.1.

This section discusses the principal sources and causes of groundwater contamination from these activities with regard to their occurrence and effects on groundwater quality.

Waste Disposal

Waste disposal includes the disposal of liquid waste and solid waste.

LIQUID WASTE

Underground or aboveground disposal practices of domestic, municipal, or industrial liquid waste can cause groundwater contamination. Among all disposal practices of domestic liquid waste, septic tanks and cesspools contribute the most wastewater to the ground and are the most frequently reported sources of groundwater contamination (U.S. EPA 1977). Septic tanks and cesspools contribute filtered sewage effluent directly to the ground which can introduce high concentrations of BOD, COD, nitrate, organic chemicals, and possibly bacteria and viruses into

groundwater (Mallmann and Mack 1961; Miller 1980). Also, chlorination of the wastewater effluent and the use of chemicals to clean septic systems can produce additional potential pollutants (Council on Environmental Quality 1980).

With regard to municipal liquid waste, land application of sewage effluent and sludge is perhaps the largest contributor to groundwater contamination. Treated wastewater and sludge have been applied to land for many years to recharge groundwater and provide nutrients that fertilize the land and stimulate plant growth (Bauer 1974; U.S. EPA 1983). However, land application of sewage effluent can introduce bacteria, viruses, and organic and inorganic chemicals into groundwater (U.S. EPA 1974).

Another major municipal source of groundwater contamination is urban runoff from roadway deicing. In many urban areas, large quantities of salts and deicing additives are applied to roads during the winter months. These salts and additives facilitate the melting of ice and snow; however, they can percolate with the water into the ground and cause groundwater contamination of shallow aquifers (Field et al. 1973). In addition, the high solubility of these salts in water and the relatively high mobility of the resulting contaminants such as chloride ions in groundwater can cause the zone of contamination to expand (Terry 1974).

With regard to industrial liquid waste, surface impoundments and injection wells are probably the largest contributors to groundwater contamination. As legislation to protect surface water resources has become more stringent, the use of surface impoundments and injection wells has become an attractive wastewater effluent disposal option for many industries. However, leakage of contaminants through the bottom of a surface impoundment or migration of fluids from an injection well into a hydrologically connected usable aquifer can cause groundwater contamination (Council on Environmental Quality 1981). The extent and severity of groundwater contamination from these sources is further complicated by the fact that, in addition to being hazardous, many of the organic and

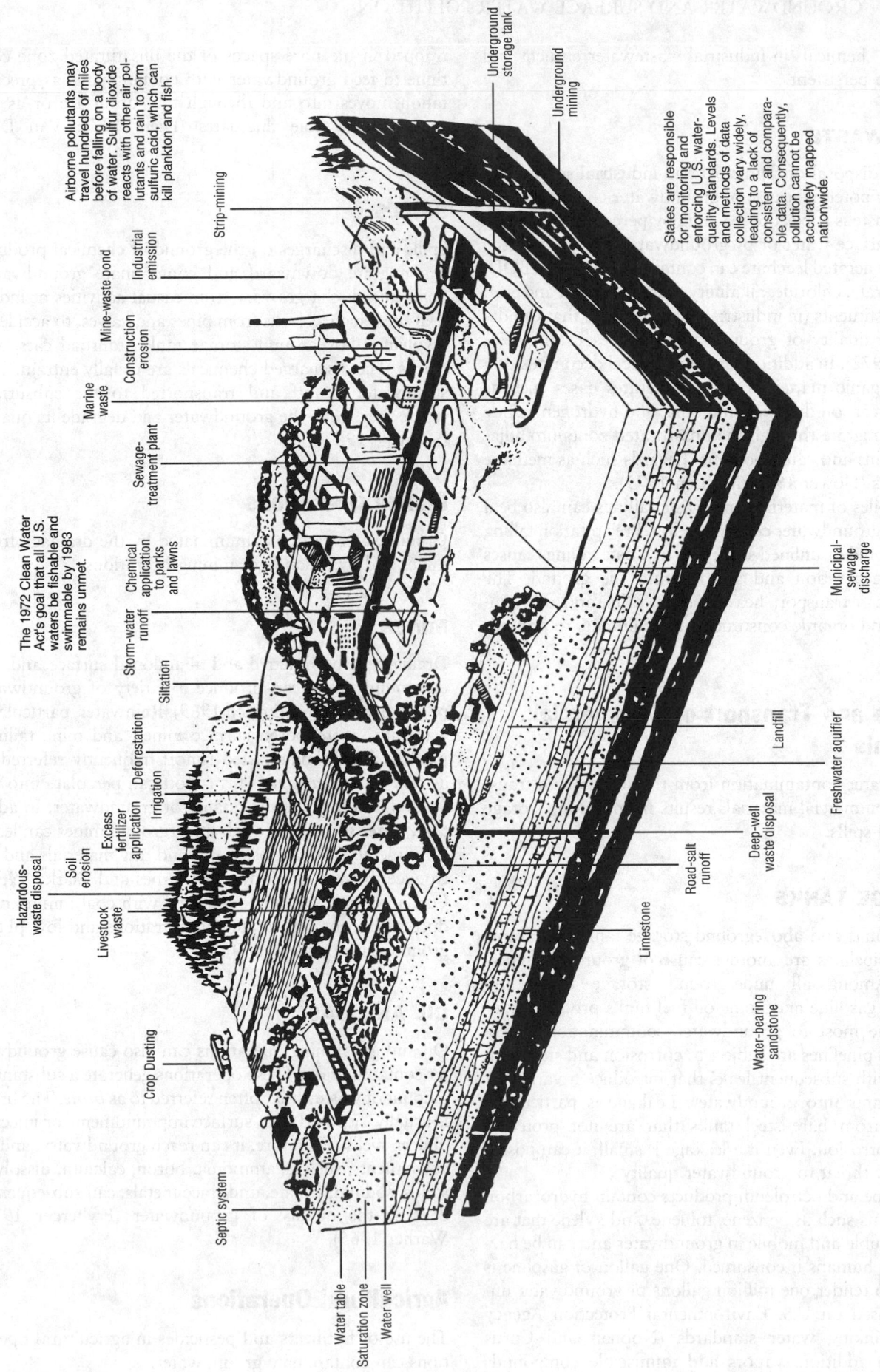

FIG. 9.11.1 Sources of groundwater contamination. (Reprinted from National Geographic, 1993, Water, *National Geographic Special Editions* [November], Washington, D.C.: National Geographic Society.)

The following labels appear in the figure:

Underground storage tank

Underground mining

States are responsible for monitoring and enforcing U.S. water-quality standards. Levels and methods of data collection vary widely, leading to a lack of consistent and comparable data. Consequently, pollution cannot be accurately mapped nationwide.

Airborne pollutants may travel hundreds of miles before falling on a body of water. Sulfur dioxide reacts with other air pollutants and rain to form sulfuric acid, which can kill plankton and fish.

Strip-mining

Industrial emission

Mine-waste pond

Marine waste

Construction erosion

Sewage-treatment plant

The 1972 Clean Water Act's goal that all U.S. waters be fishable and swimmable by 1983 remains unmet.

Chemical application to parks and lawns

Storm-water runoff

Siltation

Deforestation

Irrigation

Excess fertilizer application

Soil erosion

Hazardous-waste disposal

Livestock waste

Municipal-sewage discharge

Landfill

Freshwater aquifer

Deep-well waste disposal

Road-salt runoff

Limestone

Water-bearing sandstone

Crop Dusting

Septic system

Water table

Saturation zone

Water well

1051

inorganic chemicals in industrial wastewater effluent and sludge are persistent.

SOLID WASTE

The land disposal of municipal and industrial solid waste is another potential cause of groundwater contamination. Buried waste is subject to leaching by percolating rain water and surface water or by groundwater contact with the fill. The generated leachate can contain high levels of BOD, COD, nitrate, chloride, alkalinity, trace elements, and even toxic constituents (in industrial waste landfill) that can degrade the quality of groundwater (Hughes et al. 1971; Zanoni 1972). In addition, the biochemical decomposition of the organic matter in waste generates gases such as methane, carbon dioxide, ammonia, and hydrogen sulfide that can migrate through the unsaturated zone into adjacent terrains and cause potential hazards such as methane explosions (Flower 1976; Mohsen 1975).

Stockpiles of materials and waste tailings can also be a source of groundwater contamination. Precipitation falling on uncovered or unlined stockpiles or waste tailings causes leachate generation and seepage into the ground. The leachate can transport heavy metals, salts, and other inorganic and organic constituents as pollutants to groundwater.

Storage and Transport of Commercial Materials

Groundwater contamination from the storage and transport of commercial materials results from leaking storage tanks and spills.

STORAGE TANKS

Underground and aboveground storage tanks and transmission pipelines are another cause of groundwater pollution. Among all underground storage tanks and pipelines, gasoline and home oil fuel tanks probably contribute the most to groundwater contamination. These tanks and pipelines are subject to corrosion and structural failures with subsequent leaks that introduce a variety of contaminants into groundwater. Leakage is particularly frequent from bare steel tanks that are not protected against corrosion. Even if a leakage is small, it can pose a significant threat to groundwater quality.

Gasoline and petroleum products contain hydrocarbon components such as benzene, toluene, and xylene that are highly soluble and mobile in groundwater and can be hazardous to humans if consumed. One gallon of gasoline is enough to render one million gallons of groundwater unusable based on U.S. Environmental Protection Agency (EPA) drinking water standards (Noonan and Curtis 1990). In addition, vapors and immiscible compounds

trapped in the pore spaces of the unsaturated zone continue to feed groundwater with contaminants as precipitation moves into and through the subsurface or as the groundwater table fluctuates (Dietz 1971; Van Dam 1967).

SPILLS

Spills and discharges on the ground of chemical products can migrate downward and contaminate groundwater. Spills and discharges vary from casual activities at industrial sites, such as leaks from pipes and valves, to accidents involving aboveground storage tanks, railroad cars, and trucks. The discharged chemicals are usually entrained by stormwater runoff and transported to the subsurface where they reach the groundwater and degrade its quality (Scheville 1967).

Mining Operations

Groundwater can be contaminated by the drainage from mines and by oil and gas mining operations.

MINES

Drainage of both active and abandoned surface and underground mines can produce a variety of groundwater pollution problems (Emrich 1969). Rainwater, particularly acid rain, overexposed surface mines, and mine tailings produce highly mineralized runoff frequently referred to as *acid mine drainage*. This runoff can percolate into the ground and degrade the quality of groundwater. In addition, water seepage through underground mines can leach toxic metals from exposed ores and raw materials and introduce them to groundwater (Barnes and Clarke 1964). Oxidation and leaching connected with coal mining produce high iron and sulfate concentrations and low pH in groundwater (Miller 1980).

OIL AND GAS

Oil and gas mining operations can also cause groundwater contamination. These operations generate a substantial amount of wastewater, often referred to as *brine*. The brine is usually disposed of in surface impoundments or injected in deep wells. Therefore, it can reach groundwater, and its constituents, such as ammonia, boron, calcium, dissolved solids, sodium, sulfate, and trace metals, can subsequently degrade the quality of groundwater (Fryberger 1975; Warner 1965).

Agricultural Operations

The use of fertilizers and pesticides in agricultural operations can contaminate groundwater.

FERTILIZERS

Fertilizers are the primary cause of groundwater contamination beneath agricultural lands. Both inorganic (chemically manufactured) and organic (from animal or human waste) fertilizers applied to agricultural lands provide nutrients such as nitrogen, phosphorous, and potassium that fertilize the land and stimulate plant growth. A portion of these nutrients usually leaches through the soil and reaches the groundwater table. Phosphate and potassium fertilizers are readily adsorbed on soil particles and seldom constitute a pollution problem. However, only a portion of nitrogen is adsorbed by soil or used by plants, and the rest is dissolved in water to form nitrates in a process called *nitrification*. Nitrates are mobile in groundwater and have potential to harm infant human beings and livestock if consumed on a regular basis (Hassan 1974).

PESTICIDES

Pesticides, herbicides, and fungicides used for destroying unwanted animal pests, plants, and fungal growth can also cause groundwater contamination. When applied to land or disposed of in landfills, these chemicals degrade in the environment by a variety of mechanisms. However, their parent compounds and their byproducts persist long enough to adversely impact the soil and groundwater (California Department of Water Resources 1968).

Other Activities

Interaquifer exchange and saltwater intrusion are two other human activities that cause groundwater contamination.

INTERAQUIFER EXCHANGE

In interaquifer exchange, two aquifers are hydraulically connected. Contamination occurs when contaminants are transferred from a contaminated aquifer to a clean aquifer. Interaquifer exchange is common when a deep well penetrates more than one aquifer to provide increased yield or when an improperly cased or abandoned well serves as a direct connection between two aquifers of different potential heads and different water quality. The hydraulic connection (well or fractures) can allow contaminants from aquifers with the greatest hydraulic head to move to aquifers of less hydraulic head (Deutsch 1961).

SALTWATER INTRUSION

Saltwater intrusion, in which saline water displaces or mixes with fresh groundwater, is another source of groundwater contamination. Saltwater intrusion is usually caused when the hydrodynamic balance between the fresh water and the saline water is disturbed, such as when fresh groundwater is overpumped in coastal aquifers (Task Committee on Salt Water Intrusion 1969). Saltwater intrusion can also occur when the natural barriers that separate fresh and saline water are destroyed, such as in the construction of coastal drainage canals that enable tidal water to advance inland and percolate into a freshwater aquifer (Todd 1974).

—Ahmed Hamidi

References

Barnes, I., and F.E. Clarke. 1964. *Geochemistry of groundwater in mine drainage problems*. U.S. Geotechnical Survey Prof. Paper 473-A.

Bauer, W.J. 1974. Land treatment designs, present and future. Proceedings of the International Conference on Land for Waste Management, edited by J. Thomlinson. 343–346. Ottawa, Canada: National Research Council.

California Department Water Resources. 1968. *The fate of pesticides applied to irrigated agricultural land* Bv11.174-1. Sacramento, Calif.

Council on Environmental Quality. 1980. *The eleventh annual report of the Council on Environmental Quality*. December.

———. 1981. Contamination of groundwater by toxic organic chemicals. (January). Washington, D.C.: U.S. Government Printing Office.

Deutsch, M. 1961. Incidents of chromium contamination of groundwater in Michigan. Proceedings of Symposium on Groundwater Contamination, April. Cincinnati, Ohio: U.S. Dept. of Health, Education and Welfare.

Dietz, D.N. 1971. Pollution of permeable strata by oil components. In *Water pollution by oil*, edited by Peter Hepple, 128–142. Elsevier, Amsterdam.

Emrich, G.H., and G.L. Merritt. 1969. Effects of mine drainage on groundwater. *Groundwater* 7, no. 3:27–32.

Field, R. et al. 1973. *Water pollution and associated effects from street salting*. EPA-R2-73-257. Cincinnati, Ohio: U.S. EPA.

Flower, F.B. 1976. *Case history of landfill gas movement through soils*, edited by E.J. Genetilli and J. Cirello, 177–184. Cincinnati, Ohio: U.S. EPA.

Fryberger, J.S. 1975. Investigation and rehabilitation of a brine-contaminated aquifer. *Groundwater* 13, no. 2:155–160.

Hassan, A.A. 1974. Water quality cycle—reflection of activities of nature and man. *Groundwater* 12, no. 1:16–21.

Hughes, G. et al. 1971. *Pollution of groundwater due to municipal dumps*. Tech. Bull. no. 42. Ottawa, Ont.: Canada Dept. of Energy, Mines and Resources, Inland Waters Branch.

Mallmann, W.L., and W.N. Mack. 1961. Biological contamination of groundwater. Proceedings of Symposium on Groundwater Contamination. April. U.S. Department of Health, Education and Welfare.

Miller, D.W. 1980. *Waste disposal effects on groundwater*. Berkeley, Calif.: Premier Press.

Mohsen, M.F.N. 1975. Gas migration from sanitary landfills and associated problems. Ph.D. thesis, University of Waterloo.

Noonan, D.C., and J.T. Curtis. 1990. *Groundwater remediation and petroleum: A guide for underground storage tanks*. Chelsea, Mich.: Lewis Publishers.

Scheville, F. 1967. Petroleum contamination of the subsoil, a hydrological problem. In *The joint problems of the oil and water industries*, edited by Peter Hepple. 23–53. Elsevier, Amsterdam.

Task Committee on Salt Water Intrusion. 1969. Saltwater intrusion in the United States. *Journal of Hydraulics Division*, ASCE 95, no. Hy5:1651–1669.

Terry, R.C., Jr. 1974. *Road salt, drinking water, and safety.* Cambridge, Mass.: Ballinger.

Todd, D.K. 1974. Salt water intrusion and its controls. *Journal of AWWA* 66:180–187.

U.S. Environmental Protection Agency. 1974. Land application of sewage effluents and sludge, selected abstracts. Washington, D.C.: Government Printing Office.

———. 1977. *Waste disposal practices and their effects on groundwater.* Report to Congress, 81–107.

———. 1983. Process design manual: Land application of municipal

sludge. EPA-625/1-83-016. Cincinnati, Ohio: U.S. EPA, Municipal Environmental Lab.

Van Dam, J. 1967. The migration of hydrocarbons in water bearing stratum. In *The joint problems of the oil and water industries,* edited by Peter Hepple. London: Institute of Petroleum.

Warner, D.L. 1965. *Deep-well injection of liquid waste.* Publ. no. 999-WP-21. U.S. Public Health Service.

Zanoni, A.E. 1972. Groundwater pollution and sanitary landfills—a critical review. *Groundwater* 10, no. 1:3–16.

9.12
FATE OF CONTAMINANTS IN GROUNDWATER

When a contaminant is introduced in the subsurface environment, its fate and concentration are controlled by a variety of physical, chemical, and biochemical processes that occur between the contaminant and the constituents of the subsurface environment. A complete discussion and assessment of all these processes for all contaminants are beyond the scope of this chapter. However, this section illustrates some of the most important processes for several groups of contaminants and the impact of these processes on the concentration and mobility of contaminants.

Organic Contaminants

The physicochemical reactions that can alter the concentration of an organic contaminant in groundwater can be grouped into five categories as suggested by Arthur D. Little (1976) and Rao and Jessup (1982). These categories include (1) hydrolysis of the contaminant in water, (2) oxidation–reduction, (3) biodegradation of the contaminant by microorganisms, (4) adsorption of the contaminant by the soil, and (5) volatilization of the contaminant to the air present in the unsaturated zone. The relative importance of each of these reactions depends on the physical and chemical characteristics of the contaminant and on the specific conditions of the subsurface environment.

HYDROLYSIS

Hydrolysis is a chemical reaction in which an organic chemical (RX) reacts with water or a hydroxide ion (OH) as follows:

$$R - X + H_2O \longrightarrow R - OH + H^+ + X^- \quad \textbf{9.12(1)}$$

$$R - X + OH^- \longrightarrow R - OH + X^- \quad \textbf{9.12(2)}$$

During these reactions, a leaving group (X) is replaced by a hydroxyl ion (OH), and a new carbon–oxygen bond is

formed. The R represents the carbonium ion and the X the leaving group. Common leaving groups include halides (Cl^-, Br^-), alcohols ($R—O^-$), and amines ($R_1R_2N^-$). The acquisition of a new polar functional group increases the water solubility of the organic chemical.

Examples of hydrolysis include the following (Valentine 1986):

$$RCl + H_2O \longrightarrow ROH + H^+ + Cl^- \quad \textbf{9.12(3)}$$
an alkyl halide $\qquad$ an alcohol

$$R_1COOR_2 + H_2O \longrightarrow R_2OH + R_1COOH \quad \textbf{9.12(4)}$$
an ester $\qquad$ an alcohol + a carboxylic acid

$$RC(ON)R_1R_2 + H_2O \longrightarrow RCOOH + R_1R_2NH \quad \textbf{9.12(5)}$$
an amide $\qquad$ a carboxylic acid + an amine

$$RCH_2CN + H_2O \longrightarrow RCH_2COOH + NH_3 \quad \textbf{9.12(6)}$$
a nitrile $\qquad$ a carboxylic acid + ammonia

The hydrolysis of organic chemicals in water is generally considered first-order with respect to the organic chemical's concentration; thus, the rate of hydrolysis can be calculated with the following equation (Dragun 1988b):

$$k \cdot C = -\frac{dC}{dt} \quad \textbf{9.12(7)}$$

or

$$k = \frac{2.303}{t} \log\left(\frac{C_0}{C}\right) \quad \textbf{9.12(8)}$$

where:

$\quad k$ = rate constant, 1/time
$\quad t$ = time
$\quad C_0$ = initial concentration, ppm
$\quad C$ = concentration at time t, ppm

The time needed for half of the concentration to react, half-life, can be calculated if k is known with use of the following equation:

$$t_{1/2} = \frac{0.693}{k} \qquad \textbf{9.12(9)}$$

where $t_{1/2}$ is equal to the half-life.

Table 9.12.1 lists the hydrolysis half-lives for several organic chemicals. Half-lives vary from seconds to tens of thousands of years. Certain compounds such as alkyl halides, chlorinated amides, amines, carbamates, esters, epoxides, phosphonic acid esters, phosphoric acid esters, and sulfones are potentially susceptible to hydrolysis (Dragun 1988b). Other compounds such as aldehydes, alkanes, alkenes, alkynes, aliphatic amides, aromatic hydrocarbons and amines, carboxylic acids, and nitro fragments are generally resistant to hydrolysis (Dragun 1988b; Harris 1982).

For organic chemicals undergoing an acid- and base-catalyzed hydrolysis (in the case of acid or alkaline solutions), the total hydrolysis rate constant k_T can be expressed (Harris 1982; Mabey and Mill 1978) as

$$k_T = k_H [H^+] + k_N + k_{OH} [OH^-] \qquad \textbf{9.12(10)}$$

where:

k_T	= total hydrolysis rate constant
k_H	= rate constant for acid-catalyzed hydrolysis
$[H^+]$	= hydrogen ion concentration
$[OH^-]$	= hydroxyl ion concentration
k_N	= rate constant for neutral hydrolysis
k_{OH}	= rate constant for base-catalyzed hydrolysis

Several other parameters can affect the rate of hydrolysis including temperature, the pH of the soil particle surfaces, the presence of metals in soils, the adsorption of the organic chemical, and the soil water content (Burkhard and Guth 1981; Konrad and Chesters 1969).

After the hydrolysis rate constant k is estimated, the behavior of a compound can be modeled with a form of the advection–dispersion equation. Equation 9.13(1), that includes a first-order degradation term.

OXIDATION–REDUCTION

In organic chemistry, oxidation–reduction (redox) refers to the transfer of atoms rather than direct electrons as is the case of inorganic chemistry. Oxidation of an organic compound frequently involves a gain in oxygen and a loss in hydrogen atoms, and the reduction involves a gain in hydrogen and a loss in oxygen content. Oxidation–reduction reactions greatly affect contaminant transport and are usually closely related to the microbial activity and the type of substrates available to the organisms. Organic contaminants provide the reducing equivalents for the microbes. After the oxygen in the subsurface environment is depleted, the most easily reduced materials begin to react and, along with the reduced product, dictate the dominant potential.

The occurrence of oxidation in the subsurface is a function of the electrical potential in the reacting system (Dragun and Helling 1985). For oxidation to occur, the potential of the soil system must be greater than that of the organic chemical. Soil reduction potentials can be generally classified as follows:

Aerated soils:	+0.8 to +0.4 volts
Moderately reduced soils:	+0.4 to +0.1 volts
Reduced soils:	+0.1 to −0.1 volts
Highly reduced soils:	−0.1 to −0.5 volts

A number of organic chemicals can hydrolyze, oxidize, and reduce quickly and sometimes violently upon contact with groundwater. Table 9.12.2 lists several classes of organic chemicals that react rapidly and violently with groundwater.

The hydrolysis, oxidation, or reduction of one organic chemical usually results in the synthesis of one or more new organic chemicals. Organic chemistry textbooks identify the basic reaction products. Soil minerals can significantly influence the chemical structure of reaction products. In addition, certain organic chemicals can form significant amounts of residues that bind to the soil. Examples of such chemicals include anilines, phenols, triazines, urea herbicides, carbamates, organophosphates, and cyclodiene insecticides (Sax 1984).

BIODEGRADATION

Biodegradation of an organic chemical in soil is the modification or decomposition of the chemical by soil microorganisms to produce ultimately microbial cells, carbon dioxide, oxygen, and water.

Soil serves as the home for numerous microorganisms capable of degrading organic chemicals. The most predominant microorganisms in soil include bacteria, actinomycetes, and fungi. One gram of surface soil can contain from 0.1 to 1 billion cells of bacteria, 10 to 100 million cells of actinomycetes, and 0.1 to 1 million cells of fungi (Dragun 1988b). The microorganism population in soils is generally greatest in the surface horizons where temperature, moisture, and energy supply is favorable for their growth. As the depth increases, the number of aerobic microorganisms decreases; however, anaerobic microorganisms can exist depending on availability of nutrients and organic material.

The biodegradation of an organic chemical by a microorganism is catalyzed by enzymes which are produced as part of the metabolic activity of the living organism. The biotransformation occurs either inside the microorganism via intracellular enzymes or outside the microorganism by the action of extracellular enzymes. After an organic chemical and an enzyme collide, an enzyme–chemical complex forms. Then, depending on the alignment between the functional groups of the chemical and the enzyme, a reaction product (modified or decomposed organic chemical) is formed by the removal of one or more functional groups by oxidation or reduction reactions (Dragun

TABLE 9.12.1   HYDROLYSIS HALF-LIVES FOR VARIOUS ORGANIC CHEMICALS

Chemical	Half-life	Chemical	Half-life
acetamide	3,950 y	ethion	9.9 d
atrazine	2.5 h	N-ethylacetamide	70,000 y
azirdine	154 d	ethyl acetate	136 d
benzoyl chloride	16 s	ethyl butanoate	5.8 y
benzyl bromide	1.32 h	ethyl trans-buteonate	17 y
benzyl chloride	15 h	ethyl difluoroethanoate	23 m
benzylidene chloride	0.1 h	ethyl dimethylethanoate	9.6 y
bromoacetamide	21,200 y	ethyl methylthioethanoate	87 d
bromochloromethane	44 y	ethyl phenylmethanoate	7.3 y
bromodichloromethane	137 y	ethyl propanoate	2.5 y
bromoethane	30 d	ethyl propenoate	3.5 y
1-bromohexane	40 d	ethyl propynoate	17 d
3-bromohexane	12 d	ethyl pyridylmethanoate	0.41 y
bromomethane	20 d	fluoromethane	30 y
bromomethylepoxyethane	16 d	2-fluor-2-methylpropane	50 d
1-bromo-3-phenylpropane	290 d	hydroxymethylpropane	28 d
1-bromopropane	26 d	iodoethane	49 d
3-bromopropene	12 h	iodomethane	110 d
chloroacetamide	1.46 y	2-iodopropane	2.9 d
chlorodibromoethane	274 y	3-iodopropene	2.0 d
chloroethane	38 d	isobutyramide	7,700 y
chlorofluoriodomethane	1.0 y	isopropyl bromide	2.0 d
chloromethane	339 d	isopropyl ethanoate	8.4 y
chloromethylepoxyethane	8.2 d	malathion	8.1 d
2-chloro-2methylpropane	23 s	methoxyacetanide	500 y
2-chloropropene	2.9 d	N-methylacetamide	38,000 y
3-chloropropene	69 d	methyl chloroethanoate	14 h
cyclopentanecarboxamide	5,500 y	methyl dichloroethanoate	38 m
dibromoethane	183 y	methylepoxyethane	14.6 d
1,3-dibromopropane	48 d	methyl parathion	10.9 d
dichloroacetamide	0.73 y	methyl trichloroethanoate	<3.6 m
dichloroiodomethane	275 y	monomethyl phosphate	1.0 d
dichloromethane	704 y	parathion	17 d
dichloromethyl ether	25 s	phenyl dichloroethanoate	3.7 m
diethyl methylphosphonate	990 y	phenyl ethanoate	38 d
dimethoxysulfone	1.2 m	phosphonitrilic hexamide	46 d
1,2-dimethylepoxyethane	15.7 d	propadienyl ethanoate	110 d
1,1-dimethylepoxyethane	4.4 d	ronnel	1.6 d
diphenyl phosphate	20.6 d	tetrachloromethane	7,000 y (1 ppm)
epoxyethane	12 d	tribromomethane	686 y
3,4 epoxycyclohexene	6 m	trichloroacetamide	0.23 y
3,4 epoxycyclooctane	52 m	trichloromethane	3,500 y
1,3-epoxy-1-oxopropane	3.5	trichlorethylbenzene	19 s
		triethyphosphate	5.5 y
		tri(ethylthio)phosphate	8.5 y

Source: J. Dragun, 1988, *The soil chemistry of hazardous materials* (Silver Springs, MD: Hazardous Materials Control Research Institute).
d = days, h = hours, s = seconds, m = minutes, y = years.

1988a). Some typical biochemical reactions are as follows (Valentine and Schnoor 1986):

Decarboxylation:

$$R—OCH_3 \longrightarrow ROH + CO_2 \qquad \textbf{9.12(11)}$$

Oxidation of an amino group:

$$R—NH_2 \longrightarrow RNO_2 \qquad \textbf{9.12(12)}$$

Reductive dehalogenation:

$$R—CCl_2—R \longrightarrow RCHClR + Cl^- \qquad \textbf{9.12(13)}$$

Hydrolysis:

$$R—CH_2CN \longrightarrow RCHONH_3 \qquad \textbf{9.12(14)}$$

An organic chemical has two levels of the biodegradation. Primary degradation refers to any biologically in-

TABLE 9.12.2 ORGANIC CHEMICALS THAT MAY RAPIDLY REACT WITH SOIL WATER AND GROUNDWATER

acetic anhydride	methacrylic acid
acetyl bromide	2-methylaziridne
acetyl chloride	methyl isocyanate
acrolein	methyl isocyanoacetate
acrylonitrile	oxopropanedinitrile
3-aminopropiononitrile	perfluorosilanes
bis(difluoroboryl)methane	peroxyacetic acid
butyldichloroborane	peroxyformic acid
calcium cyanamide	peroxyfuroic acid
2-chloroethylamine	peroxpivalic acid
chlorosulfonyl isocyanate	peroxytrifluoriacetic acid
chlorotrimethylsilane	phenylphosphonyl dichloride
cyanamide	phosphorus tricyanide
2-cyanoethanol	pivaloyloxydiethylborane
cyanoformyl chloride	potassium bis(propynyl)palladate
cyanogen chloride	potassium bis(propynyl)platinate
dichlorodimethylsilane	potassium diethynylplatinate
dichlorophenylborane	potassium hexaethynylcobaltate
dicyanoacetylene	potassium methanediazoate
diethylmagnesium	potassium tert-butoxide
diethylzinc	potassium tetracyanotitanate
diketene	potassium tetraethynylnickelate
dimethylaluminum chloride	propenoic acid
dimethylmagnesium	sulfur trixoide-dimethylformamide
dimethylzinc	sulfinylcyanamide
diphenylmagnesiuim	2,4,6-trichloro-1,3,5-triazine
2,3-epoxypropionaldehyde oxime	trichlorovinylsilane
	triethoxydialuminum tribromide
N-ethyl-N-propylcarbamolyl chloride	vinyl acetate
glyoxal	
isopropylisocyanide dichloride	
maleic anhydride	

Source: J. Dragun, 1988, *The soil chemistry of hazardous materials* (Silver Springs, MD: Hazardous Materials Control Research Institute).

$$k \cdot C = -\frac{dC}{dt} \qquad 9.12(15)$$

or

$$k = \frac{2.303}{t} \log\left(\frac{C_0}{C}\right) \qquad 9.12(16)$$

where:

k = rate constant, 1/time
t = time
C_0 = initial concentration, ppm
C = concentration at time t, ppm

The time needed for half of the concentration to react, half-life, can be calculated if k is known with use of the following equation:

$$t_{1/2} = \frac{0.693}{k} \qquad 9.12(17)$$

where $t_{1/2}$ is equal to half-life.

Table 9.12.3 lists the biodegradation rates for many pesticides; the biodegradation rates for other organic chemicals are in Dragun (1988b). However, note that the estimate of biodegradation rates of organic chemicals may not be accurate. Biodegradation rates can be affected by many factors such as pH, temperature, water content, carbon content, clay content, oxygen, nutrients, microbial population, acclimation, and concentration. Most of these factors are interrelated. For example, the pH can affect both the availability of a substrate as well as the composition of the microbial community.

After the degradation rate constant k is estimated, the behavior of a specific compound can be modeled with use of a form of the advection–dispersion equation, Equation 9.13(1), that includes a first-order degradation term.

ADSORPTION

Adsorption is the bonding of an organic chemical to the soil mineral surfaces (clay) or to the organic matter surfaces. The bonding is usually temporary and is accomplished by ionic, ligand, dipole, hydrogen, or Van der Waal's bonds. Adsorption is important in the movement of organic chemicals in groundwater because it decreases the mobility and retards the migration of an organic chemical in groundwater. Furthermore, the adsorbed portion of an organic chemical may not be available in solution for other chemical reactions such as hydrolysis and biodegradation.

The degree and extent of adsorption of an organic chemical to soil is determined by the chemical's structure and the soil's physical and chemical characteristics. Organic chemicals with large molecular structures, such as PCBs, PAHs, toluene, and dichlorodiphenyl trichloroethane (DDT), tend to be extensively adsorbed onto soil (Landrum et al. 1984). Organic chemicals with positive charges, such as the herbicides paraquat and diquat, are readily adsorbed onto the cation exchange sites (clay min-

duced structural alteration in the organic chemical. Ultimate biodegradation refers to the degradation of the organic chemical into carbon dioxide, oxygen, water, and other inorganic products.

Primary biodegradation of an organic chemical can generate a variety of degradation products that can contaminate groundwater. For example, the degradation of trichloroethylene (TCE) can lead to dichloroethylenes (DCEs), dichloroethanes (DCAs), vinyl chloride, and chloroethane (Dragun 1988b; Alexander 1981; Goring and Hamaker 1972). The degradation of cyclic hydrocarbons can lead to aliphatic hydrocarbons, and aliphatic hydrocarbons can be converted in successive reactions into alcohols, aldehydes, and then aliphatic acids (Tabak et al. 1981).

The biodegradation of many organic chemicals is generally first-order with respect to the organic chemical's concentration (Scow 1982). As a result, the biodegradation rate constant can be calculated with use of the following first-order equation as in hydrolysis:

TABLE 9.12.3 BIODEGRADATION RATE CONSTANTS FOR ORGANIC COMPOUNDS IN SOIL

Compound	k (Day^{-1})
Aldrin, Dieldrin	0.013
Atrazine	0.019
Bromacil	0.0077
Carbaryl	0.037
Carbofuran	0.047
DDT	0.00013
Diazinon	0.023
Dicamba	0.022
Diphenamid	0.123[b]
Fonofos	0.012
Glyphosate	0.10
Heptachlor	0.011
Lindane	0.0026
Linuron	0.0096
Malathion	1.4
Methyl parathion	0.16
Paraquat	0.0016
Phorate	0.0084
Picloram	0.0073
Simazine	0.014
TCA	0.059
Terbacil	0.015
Trifluralin	0.008
2,4-D	0.066
2,4,5-T	0.035

Source: W. Mabey and T. Mill, 1978, Critical review of hydrolysis of organic compounds in water under environmental conditions, *Jour. Phys. Chem. Ref. Data* 17, no. 2:383–415.

eral surfaces). In addition, the adsorption of organic chemicals depends on the organic matter content of the soil. The relationship between the organic content of soil and the adsorption coefficient of organic chemicals is generally linear for soils with an organic carbon content greater than 0.1 (Hamaker and Thompson 1972).

The adsorption process is usually reversible. At equilibrium, the adsorption coefficient, which is the rate at which the dissolved organic chemical in water transfers into the soil, can be described with the linear Freundlich isotherm equation as

$$K_d = \frac{C_s}{C_w} \qquad 9.12(18)$$

where:

K_d = distribution coefficient
C_s = concentration adsorbed on soil surfaces, ug/g
C_w = concentration in water, ug/ml

Other nonlinear isotherm equations are also used (Lyman 1982), such as:

$$K_d = \frac{C_s}{C_w^{1/n}} \qquad 9.12(19)$$

where n is a constant usually between 0.7 and 1.1.

As in Equations 9.12(18) and 9.12(19), the distribution or adsorption coefficient K_d is directly proportional to the organic carbon content of the soil; thus, K_d can be written as

$$K_d = \frac{K_{oc}}{f_{oc}} \qquad 9.12(20)$$

where:

K_{oc} = normalized adsorption coefficient
f_{oc} = soil organic carbon content

The normalized adsorption coefficient can be estimated from the organic chemical's water solubility or octanol water partition coefficient with use of regression equations (Dragun 1988b), such as

$$\log(K_{oc}) = a \cdot \log(S) + b \qquad 9.12(21)$$

$$\log(K_{oc}) = c \cdot \log(K_{ow}) + d \qquad 9.12(22)$$

where:

S = water solubility
K_{ow} = octanol water partition coefficient
a,b,c,d = coefficients that depend on the organic chemical

Table 9.12.4 lists the adsorption coefficient K_{oc} for several organic chemicals. The regression coefficients a, b, c, and d for several chemicals are in Brown and Flagg (1981), Briggs (1973), and Keneya and Goring (1980). Therefore, after the distribution coefficient K_d is estimated, the effect of adsorption on the mobility of a specific compound can be calculated with use of a form of the advection–dispersion equation, Equations 9.13(1) and 9.13(3), that includes the retardation factor R.

VOLATILIZATION

Volatilization is the loss of chemicals in vapor form from the soil water (liquid phase) or the soil surfaces (solid phase) to the soil air (gas phase) of the unsaturated zone. Only the first type of volatilization, from the liquid phase to the gas phase, is discussed in this section. The volatilization from the solid phase to the gas phase is relatively small and usually neglected. However, information on this type of volatilization is presented by Mayer, Letey, and Farmer (1974); Baker and Mackay (1985); and Jury, Farmer, and Spencer (1984).

The extent of volatilization of an organic chemical from water to the soil air can be determined by Henry's law which states that when a solution becomes dilute, the vapor pressure of a chemical is proportional to its concentration (Thomas 1982) as

$$C_a = H \cdot C_w \qquad 9.12(23)$$

TABLE 9.12.4 MEASURED K_{oc} VALUES FOR VARIOUS ORGANIC CHEMICALS

Chemical	K_{oc}	Chemical	K_{oc}
acetophenone	35	ipazine	1,660
alachlor	190	isocil	130
aldrin	410	isopropalin	75,250
ametryn	392	leptophos	9,300
6-aminochrysene	162,900	linuron	813
anthracene	26,000	malathion	1,778
asulam	300	methazole	2,620
atrazine	148	methomyl	160
benefin	10,700	methoxychlor	80,000
alpha-BHC	1995	methylparathion	5,129
beta-BHC	1995	metobromuron	60
2,2'-biquinoline	10,471	metribuzin	95
bromacil	72	monolinuron	200
butraline	8,200	monuron	100
carbaryl	229	napthalene	1,300
carbofuran	105	napropamide	680
carbophenothion	45,400	neburon	2,300
chloramben	21	nitralin	960
chlorobromuron	460	nitrapyrin	458
chloroneb	1159	norflurazon	1,914
chloroxuron	3200	oxadiazon	3,241
chloropropham	589	parathion	4,786
chlopyrifos	13,490	pebulate	630
crotoxyphos	170	phenanthrene	23,000
cyanazine	200	phenol	27
cycloate	345	phorate	3,200
2,4-D	57	picloram	17
DBCP	129	profluralin	8,600
p,p'-DDT	129	prometon	350
diallate	1,900	prometryn	48
diamidaphos	32	pronamide	200
dicamba	0.4	propachlor	265
dichlobenil	235	propazine	158
dinitramine	4,000	propham	51
dinoseb	124	pyrazon	120
dipropetryn	1,170	pyrene	62,700
disulfoton	1,780	pyroxychlor	3,000
diruon	398	silvex	2,600
DMSA	770	simazine	135
EPTC	240	2,4,5-T	53
ethion	15,400	tebuthiuron	620
fenuron	27	terbacil	51

Source: J. Dragun, 1988, *The soil chemistry of hazardous materials* (Silver Springs, MD: Hazardous Materials Control Research Institute).

where:

C_a = concentration of the chemical in air
H = Henry's law constant
C_w = concentration of the chemical in water

Henry's law constant for a chemical can be calculated with the following equation:

$$H = \frac{P_v \cdot M_w}{760 \cdot S} \qquad 9.12(24)$$

where:

P_v = vapor pressure of the chemical in mmHg
M_w = molecular weight of the chemical
S = solubility in mg/l

Published texts report Henry's law constant in various units such as atm-m³/mole, atm-cm³/g, or dimensionless depending on the units used for C_a and C_w. Table 9.12.5 lists Henry's constants for several organic chemicals. According to Lyman and others (1982), if H is less than 10^{-7} atm-m³/mole, the substance has a low volatility. If H is less than 10^{-5} but greater than 10^{-7} atm-m³/mole, the substance volatilizes slowly. However, the volatilization becomes an important transfer mechanism when H is greater than 10^{-5} atm-m³/mole.

Several soil characteristics affect the volatilization of organic chemicals in groundwater. Volatilization decreases as the soil porosity decreases or as the soil water content increases. Soils with high clay content tend to have a high water content and hence low volatilization (Jury 1986).

Inorganic Contaminants

Comprehensive information on the behavior of most inorganic chemicals in groundwater is limited. Agriculturally important compounds have been studied for many years; however, inorganic compounds such as metals have only recently begun to attract widespread interest as groundwater and soil contamination become a concern. This section illustrates some of the most important processes for several groups of inorganic contaminants and the impact of these processes on the concentration and mobility of contaminants.

Inorganic constituents in the subsurface environment can be classified into the following four categories: nutrients, acids and bases, halides, and metals. The origin and sources of these inorganics are discussed in Section 9.11.

NUTRIENTS

Nutrients such as nitrogen, phosphorous, and sulfur are essential for plant and microorganism growth. They are either applied to the land surface to increase its fertility or discarded with waste streams that contain appreciable amounts of these nutrients. These nutrients, however, can have appreciable concentrations that can leach into the ground and adversely affect the quality of groundwater.

Nitrogen (N) is found in waste, soil, and the atmosphere in various forms such as ammonia, ammonium, nitrite, nitrate, and molecular nitrogen. Nitrogen is converted to ammonium (NH_4^+) by a process called *ammonification*. Because of its positive charge, ammonium can be held in the soil on cation exchange sites. Ammonium can also be converted temporarily to nitrite (NO_2^-) and then to nitrate (NO_3^-) by aerobic nitrifying organisms through a process called *nitrification*. Ammonification and nitrification normally occur in the unsaturated zone where microorgan-

TABLE 9.12.5 VALUES OF HENRY'S LAW CONSTANT FOR SELECTED CHEMICALS

Low Volatility ($H < 3 \times 10^{-7}$)	H attm-m^3	H' (non-dim.)	High Volatility ($H < 10^{-3}$)	H attm-m^3	H' (non-dim.)
3-Bromo-1-propanol	1.1×10^{-7}	4.6×10^{-6}	Ethylene dichloride	1.1×10^{-3}	4×10^{-2}
Diedrin	2×10^{-7}	8.9×10^{-6}	Naphthalene	1.15×10^{-3}	4.9×10^{-2}
			Biphenyl	1.5×10^{-3}	6.8×10^{-2}
Middle Range ($3 \times 10^{-7} < H < 10^{-3}$)			Aroctor 1254	2.7×10^{-3}	1.6×10^{-1}
			Methylene chloride	3×10^{-3}	1.3×10^{-1}
			Aroctor 1248	3.5×10^{-3}	1.6×10^{-1}
Lindane	4.8×10^{-7}	2.2×10^{-5}	Chlorobenzene	3.7×10^{-3}	1.65×10^{1}
m-Bromonitrobenzene	1.6×10^{-6}	7.4×10^{-5}	Chloroform	4.7×10^{-3}	2.0×10^{-1}
Pentachlorophenol	3.4×10^{-6}	1.5×10^{-4}	o-Xylene	5.1×10^{-3}	2.2×10^{-1}
4-tert-Butylphenol	9.1×10^{-6}	3.8×10^{-4}	Benzene	5.5×10^{-3}	2.4×10^{-1}
Triethylamine	1.3×10^{-5}	5.4×10^{-4}	Toluene	6.6×10^{-3}	2.8×10^{-1}
Aldrin	1.4×10^{-5}	6.1×10^{-4}	Aroclor 1260	7.1×10^{-3}	3.0×10^{-1}
Nitrobenzene	2.2×10^{-5}	9.3×10^{-4}	Perchloroethylene	8.3×10^{-3}	3.4×10^{-1}
Epichlorohydrin	3.2×10^{-5}	1.3×10^{-3}	Ethyl benzene	8.7×10^{-3}	3.7×10^{-1}
DDT	3.8×10^{-5}	1.7×10^{-3}	Trichloroethylene	1×10^{-2}	4.2×10^{-1}
Phenanthrene	3.9×10^{-5}	1.7×10^{-3}	Mercury	1.1×10^{-2}	4.8×10^{-1}
Acenaphthene	1.5×10^{-4}	6.2×10^{-3}	Methyl bromide	1.3×10^{-2}	5.6×10^{-1}
Acetylene tetrabromide	2.1×10^{-4}	8.9×10^{-3}	Cumene (isopropyl)	1.5×10^{-2}	6.2×10^{-1}
Aroclor 1242	5.6×10^{-4}	2.4×10^{-2}	1,1,1-Trichloroethane	1.8×10^{-2}	7.7×10^{-1}
Ethylene dibromide	6.6×10^{-4}	2.8×10^{-2}	Carbon tetrachloride	2.3×10^{-2}	9.7×10^{-1}
			Methyl chloride	2.4×10^{-2}	9.7×10^{-1}
			Ethyl bromide	7.3×10^{-2}	3.1
			Vinyl chloride	2.4	99
			2,2,4-Trimethyl pentane	3.1	129
			n-Octane	3.2	136
			Fluorotrichloromethane	5.0	—
			Ethylene	>8.6	~360

Source: R.G. Thomas, 1982, Volatilization from water, In *Handbook of chemical property estimation methods* (New York: McGraw-Hill, Inc.).

isms and oxygen are abundant, but nitrate can be readily leached from the soil into groundwater where it may present a health hazard; nitrate is highly mobile in groundwater because of its negative charge. *Denitrification* is a process whereby NO_3^- is reduced to nitrous oxide (N_2O) and elemental nitrogen (N_2) by facultative anaerobic bacteria (Downing, Painter, and Knowles 1964; Freeze and Cherry 1979; Bemner and Shaw 1958).

Phosphorous (P) is found in organic waste, rock phosphate quarries, fertilizers, and pesticides in concentrations high enough to potentially leach into groundwater. The decomposition of organic waste and dissolution of inorganic fertilizers provide soluble phosphorous, soluble orthophosphate, and a variety of condensed phosphates, tripolyphosphates, adsorbed phosphates, and crystallized phosphates (U.S. EPA 1983). The hydrolysis and mineralization of these products provide soluble phosphate which can be used by plants and microorganisms, adsorbed to soil particles, or leached to groundwater. Although phosphorous is not a harmful constituent in drinking water, its presence in groundwater is environmentally significant if the groundwater discharges to a surface water body where phosphorous can produce algae growth and cause eu-

trophication of the aquatic system (Freeze and Cherry 1979).

Sulfur (S) is found in appreciable amounts in waste streams from kraft mills, sugar refining, petroleum refining, and copper and iron extraction facilities (Overcash and Pal 1979). Aerobic bacteria can oxidize the reduced forms of sulfur to form sulfate which can be highly adsorbed to soil when the cation adsorbed on the clay is aluminum; moderately adsorbed when the cation is calcium; and weakly adsorbed when the cation is potassium (Tisdale and Nelson 1975). Leaching losses of sulfur to groundwater can be large because of the anionic structure of sulfur and the solubility of most of its salt. Leaching is greatest when monovalent cations such as potassium and sodium predominate; moderate when calcium and manganese predominate; and minimal when the soil is acidic and appreciable levels of exchangeable iron and aluminum are present (Tisdale and Nelson 1975).

ACIDS AND BASES

Industrial liquid wastes are comprised of large volumes of inorganic acids and bases that can alter the soil's properties. Acids can increase the amount of aluminum (Al), iron (Fe), and other cations in the water phase of the soil system as the hydrogen ion (H^+) cation competes for cation exchange sites. If significant amounts of H^+ are present, they can dissolve the more acid-solid minerals, releasing cations which are previously fixed to the mineral structure into the water phase (Dragun 1988b). In addition, acids can cause the dissolution of some of the clay minerals and generally increase soil permeability. Bases can increase the amount of cations in the water phase by dissolving the more base-soluble soil minerals. Bases can also cause the dissolution of some of the soil's predominant clay minerals and generally decrease soil permeability.

HALIDES

Halides are the stable anions of the highly reactive halogens: fluoride (F), chloride (Cl), bromine (Br), and iodine (I). Halides occur naturally in soils and are also present in many industrial waste streams.

Fluoride is present in phosphatic fertilizers, hydrogen fluoride, fluorinated hydrocarbons, and certain petroleum refinery waste. The leaching losses and mobility of fluoride can be large because of the anionic structure of fluoride and the solubility of some of its salt (Bemner and Shaw 1958). Sodium salts of fluoride (NaF) are soluble and result in high soluble fluoride levels in soils low in calcium. Calcium salts of fluoride (CaF_2), however, are relatively insoluble and limit the amount of fluoride leached to groundwater. Fluoride solubility depends on the kind and relative quantity of cations present in soil that has formed salts with the fluoride ion (F^-). Fluorosis disease can occur in animals who consume water containing 15 ppm of fluoride (Lee 1975).

Chloride (Cl) is present in chlorinated hydrocarbon production and chlorine gas production wastes as well as other wastes. Chloride is soluble and mobile in groundwater because of its anionic structure.

Bromide (Br) is present in synthetic organic dyes, mixed petrochemical wastes, photographic supplies, and pharmaceutical and inorganic wastes. Other forms of bromide such as bromate and bromic acid occur naturally in soils at smaller concentrations. Most bromide salts (CaBr, MgBr, NaBr, and Kbr) are soluble and readily leachable into water percolating through the soil and down to groundwater (U.S. EPA 1983).

Iodine (I) is present in pharmaceutical and chemical industrial wastes. Iodine is only slightly water soluble and tends to be retained in soils by forming complexes with organic matter and being fixed to phosphates and sulfates.

METALS

Metals are found in industrial wastes in a variety of forms. When these metals are introduced into the subsurface environment, they can react with water and soil in several physicochemical processes to produce appreciable concentrations that affect the quality of groundwater. The most important processes that affect the concentration and mobility of metals in groundwater include filtration, precipitation, complexation, and ion exchange.

Filtration occurs when dissolved and solid matter are trapped in the pore spaces clogging the pore spaces and decreasing the permeability of the soil system (Dragun 1988b).

Precipitation occurs when metal ions react with water to form reaction products which precipitate in soil as oxide and oxyhydroxide minerals or form oxyde and oxyhydroxide coatings on soil minerals. Precipitation of metals as hydroxides, sulfides, and carbonates is common (Dragun 1988b).

Complexation involves the formation of soluble, charged or neutral complexes between metal ions and inorganic or organic anions called *ligands*. The complexes formed influence the mobility and concentration of the metal in groundwater. For example, the mobility of zinc in groundwater is affected by the formation of complex species between the zinc ion and inorganic anions present in the water, such as HCO_3^-, CO_3^{2-}, SO_4^{2-}, Cl^-, F^-, and NO_3^- (Freeze and Cherry 1979). The complexation of cobalt-60 ions by synthetic organic compounds enhances its mobility in groundwater (Killey et al. 1984). Other metal species are reported to be highly mobile in groundwater after soluble complexes are formed with humic substances or organic solvents (Bradbent and Ott 1957; Griffin and Chou 1980).

The predominant complex species in an aqueous solution are influenced by the redox and pH of the soil. The relationship between the redox, pH, and the complex species is commonly expressed in Eh–pH diagrams for each metal; Eh is the electronic potential. Figure 9.12.1 shows an example of an Eh–pH diagram for mercury. Methods for calculating Eh–pH diagrams are discussed by several authors (Brookings 1980; Garrells and Christ 1965; Verink 1979).

Using Eh–pH diagrams, environmental engineers can qualitatively determine the most important complexes formed by the metal in water and estimate the concentration and mobility of the metal in groundwater. The concentration of cations reported in chemical analyses of groundwater normally represents the total concentration of each element in water. However, most cations exist in more than one molecular or ionic form. These forms can have different valences and, therefore, different mobilities due to different affinities to sorption and different solubility controls.

Adsorption is another process affecting the concentration and mobility of metals in groundwater. Positive adsorption involves the attraction of metal cations in water by negatively charged soil particles. Therefore, adsorption can decrease the concentration of dissolved metals in water and retard their movement. The cation exchange ca-

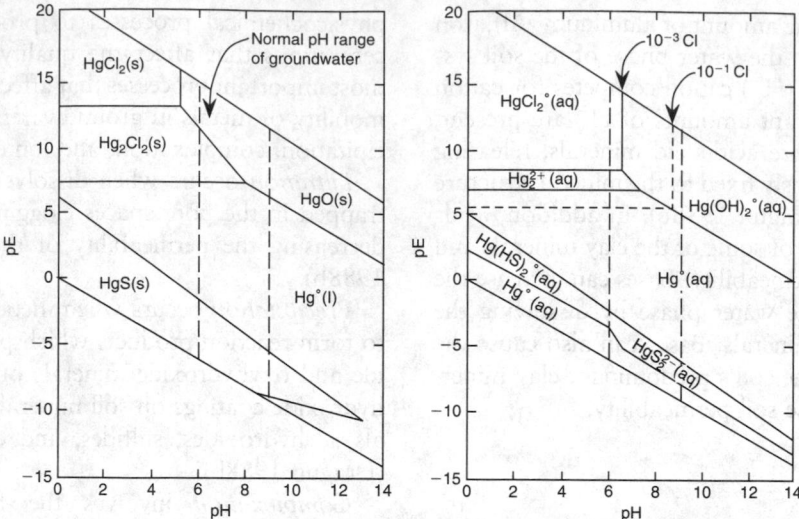

FIG. 9.12.1 Stability fields of solid phases and aqueous species of mercury as a function of pH and Eh at 1 bar total pressure. (Reprinted from J.D. Hem, 1967, *Equilibrium chemistry of iron in groundwater,* In *Principles and applications of water chemistry,* edited by S.D. Faust and J.V. Hunter, New York: John Wiley and Sons.)

pacity (CEC) of a soil, defined as the amount of cations adsorbed by the soil's negative charges, is usually expressed as milliequivalents (meq) per 100 grams of soil. In general, clay soils and humus have a higher CEC than other soils.

Some cations are more attracted to a soil surface than others based on the size and charge of their molecule. For example the Cu^{2+} cation in water can displace and replace a Ca^{2+} cation present at the soil surface through a process known as ion exchange. Also, trivalent cations are preferentially adsorbed over divalent cations which are preferentially adsorbed over monovalent cations. The release of

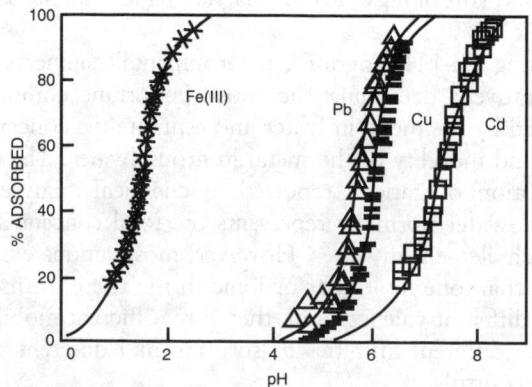

FIG. 9.12.2 Adsorption of metal ions on amorphous silica as a function of pH. (Reprinted from U.S. Environmental Protection Agency, 1989, *Transport and fate of contaminants in the subsurface,* Seminar Publication EPA/625/4-89/019, Cincinnati: U.S. EPA.)

ions by exchange processes can aggravate a contamination problem. For example, increases in water hardness resulting from the displacement of calcium and magnesium ions from geological materials by sodium or potassium in landfill leachate have been documented (Hughes, Candon, and Farvolden 1971).

The cation exchange is reversible, and its extent can be described by the adsorption or distribution coefficient (Dragun 1988b) as

$$K_d = \frac{C_s}{C_w} \qquad 9.12(25)$$

where:

K_d = adsorption or distribution coefficient
C_s = concentration adsorbed on soil surfaces (ug/g of soil)
C_w = concentration in water (ug/ml)

Table 9.12.6 lists the adsorption coefficients for several metals. The greater the coefficient K_d, the greater the extent of adsorption. Furthermore, changes in metal concentration, as well as pH, can have a significant effect on the extent of adsorption as shown in Figure 9.12.2.

A negative adsorption occurs when anions (negatively charged metal ions) are repulsed by negative soil particle charges. This repulsion causes high mobility and migration of anions in water. This process is also known as *anion exclusion.*

—*Ahmed Hamidi*

TABLE 9.12.6   RANGES FOR K_D FOR VARIOUS
ELEMENTS IN SOILS AND CLAYS

Element	Observed Range (ml/g)	Mean[a]	Standard Deviation[b]
Ag	10–1,000	4.7	1.3
Am	1.0–47,230	6.7	3.0
As(III)	1.0–8.3	1.2	0.6
As(V)	1.9–18	1.9	0.5
Ca	1.2–9.8	1.4	0.8
Cd	1.3–27	1.9	0.9
Ce	58–6,000	7.0	1.3
Cm	93–51,900	8.1	1.9
Co	0.2–3,800	4.0	2.3
Cr(III)	470–150,000	7.7	1.2
Cr(VI)	1.2–1,800	3.6	2.2
Cs	10–52,000	7.0	1.9
Cu	1.4–333	3.1	1.1
Fe	1.4–1,000	4.0	1.7
K	2.0–9.0	1.7	0.5
Mg	1.6–13.5	1.7	0.5
Mn	0.2–10,000	5.0	2.7
Mo	0.4–400	3.0	2.1
Np	0.2–929	2.4	2.3
Pb	4.5–7,640	4.6	1.7
Po	196–1,063	6.3	0.7
Pu	11–300,000	7.5	2.3
Ru	48–1,000	6.4	1.0
Se(IV)	1.2–8.6	1.0	0.7
Sr	0.2–3,300	3.3	2.0
Te	0.003–0.28	3.4	1.1
Th	2,000–510,000	11.0	1.5
U	11–4,400	3.8	1.3
Zn	0.1–8,000	2.8	1.9

Source: C.F. Baes III and R.D. Sharp, 1983, A proposal for estimation of soil leaching and leaching constants for use in assessment models, *Journal of Environmental Quality* 12, no. 1:17–28.

[a]Mean of the logarithms of the observed values.
[b]Standard deviation of the logarithms of the observed values.

References

Alexander, M. 1981. Biodegradation of chemicals of environmental concern. *Science* 211:132–138.

Arthur D. Little, Inc. 1976. *Physical, chemical, and biological treatment techniques for industrial wastes.* Report to U.S. EPA, Office of Solid Waste Management Programs, PB-275-054/56A Vol. 1 and PB-275-278/1GA Vol. 2.

Baker, L.W., and K.P. Mackay. 1985. Screening models for estimating toxic air pollution near a hazardous waste landfill. *Journal of the Air Pollution Control Association* 35, no. 11:1190–1195.

Bemner, J.M., and K. Shaw. 1958. Denitrification in soils-factors affecting denitrification. *Journal of Agricultural Science* 51, no. 1:22–52.

Bradbent, F.E., and J.B. Ott. 1957. Soil organic matter: Metal complexes—factors affecting various cations. *Soil Science* 83:419–427.

Briggs, G.G. 1973. A simple relationship between soil adsorption of organic chemicals and their octanaol/water partition coefficients. Proceedings of the 7th British Insecticide and Fungicide Conference. Vol. 1. Nottingham, Great Britain: The Boots Company, Ltd.

Brookings, D.G. 1980. *Eh–pH diagrams for elements of interest at the Oklo Natural Reactor at 25°C, 1 bar pressure and 200°C, 1 bar pressure.* Report to Los Alamos National Laboratory, CNC-11.

Brown, D.S., and E.W. Flagg. 1981. Empirical prediction of organic pollutant adsorption in natural sediments. *Journal of Environmental Quality* 10:382–386.

Burkhard, N., and J.A. Guth. 1981. Chemical hydrolysis of 2-chloro-4,6-bis (alkylamino)-1,3,5-triazine herbicides and their breakdown in soils under the influence of adsorption. *Pestic. Sci* 12:45–52.

Downing, A.L., H.A. Painter, and C. Knowles. 1964. Nitrification in the activated sludge process. *Journal and Proceedings of the Institute of Sewage Purification,* Part 2.

Dragun, J. 1988a. Microbial degradation of petroleum products in soil. Proceedings of a Conference on Environmental and Public Health, Effects of Soils Contaminated with Petroleum Products, October, New York: John Wiley and Sons.

———. 1988b. *The soil chemistry of hazardous materials.* Silver Springs, Md.: Hazardous Materials Control Research Institute.

Dragun, J., and C.S. Helling. 1985. Physicochemical and structural relationships of organic chemicals undergoing soil and clay catalyzed free-radical oxidation. *Soil Science* 139:100–111.

Freeze, R.A., and J.A. Cherry. 1979. *Groundwater.* Englewood Cliffs, N.J.: Prentice-Hall.

Garrells, R.M., and C.L. Christ. 1965. *Minerals, solutions, and equilibria.* New York: Harper and Row.

Goring, C.A.J., and J.W. Hamaker. 1972. *Organic chemicals in the soil environment.* Vols. 1 and 2. New York: Marcel Dekker.

Griffin, R.A., and S.F.J. Chou. 1980. *Attenuation of polybrominated biphenyls and hexachlorobenzene by earth materials.* Environmental Geology Notes 87, Illinois State Geological Survey. Urbana, Ill.

Hamaker, J.W., and J.M. Thompson. 1972. Adsorption. In *Organic chemicals in the soil environment,* edited by C.A.I. Goring and J.W. Hamaker. New York: Marcel Dekker.

Harris, J.C. 1982. Rate of hydrolysis. In *Handbook of chemical property estimation methods.* New York: McGraw-Hill.

Hughes, G.M., R.A. Candon, and R.N. Farvolden. 1971. *Hydrogeology of solid waste disposal sites in northern Illinois.* Solid Waste Management Series, report SW-124. U.S. EPA.

Jury, W.A. 1986. Volatilization from soil. In *Vadoze modeling of organic contaminants,* edited by Stephen Hern and Susan Melancon, 159–176. Chelsea, Mich.: Lewis Publishers.

Jury, W.A., W.J. Farmer, and W.F. Spencer. 1984. Behavior assessment model for trace organics in soils. *Journal of Environmental Quality* 13, no. 4.

Keneya, E.E., and C.A.I. Goring. 1980. Relationship between water solubility, soil-sorption, octanol-water partitioning, and bioconcentration of chemicals in biota. In *Aquatic toxicology,* ASTM STP 707. Philadelphia, Pa.: ASTM.

Killey, R.W. et al. 1984. Subsurface cobalt-60 migration from a low-level waste disposal site. *Environmental Science Technology* 18, no. 3:148–156.

Konrad, J.G., and G. Chesters. 1969. Degradation in soils of ciodrin and organophosphate insecticide. *J. Agr. Food Chem.* 17:226–230.

Landrum, P.F. et al. 1984. Reverse-phase separation method for determining pollutant binding to aldrich humic acid and dissolved organic carbon of natural waters. *Environmental Science and Technology* 18:187–192.

Lee, H.L. 1975. Trace elements in animal production. In *Trace elements in soil-plant-animal systems,* edited by D. Nicholas and R. Egan. New York: Academic Press.

Lyman, W. 1982. Adsorption coefficient for soils and sediments. In *Handbook of chemical property estimation.* New York: McGraw-Hill.

Lyman, W.J., W.F. Reehl, and D.H. Rosenblatt. 1982. *Handbook of chemical property estimation methods: Environmental behavior of organic compounds.* New York: McGraw-Hill.

Mayer, R., J. Letey, and W.J. Farmer. 1974. Models for predicting volatilization of soil-incorporated pesticides. Soil Science Society of America Proceedings. Vol. 38:563–568.

Mabey, W., and T. Mill. 1978. Critical review of hydrolysis of organic compounds in water under environmental conditions. *Jour. Phys. Chem. Ref. Data* 17, no. 2:383–415.

Overcash, M.R., and D. Pal. 1979. *Design of land treatment systems for industrial wastes; Theory and practice.* Ann Arbor, Mich.: Ann Arbor Science.

Rao, P.S.C., and R.E. Jessup. 1982. Development and verification of simulation models for describing pesticide dynamics in soils. *Ecol. Modeling* 16:67–75.

Sax, N.I. 1984. *Dangerous properties of industrial materials.* 6th ed. New York: Van Nostrand Reinhold.

Scow, K.M. 1982. Rate of biodegradation. In *Handbook of chemical property estimation methods.* New York: McGraw-Hill.

Tabak, H.N. et al. 1981. Biodegradability studies with organic priority pollutants compounds. *Journal Water Pollution Control Federation* 53:1503–1518.

Thomas, R.G. 1982. Volatilization from water. In *Handbook of chemical property estimation methods.* New York: McGraw-Hill.

Tisdale, S.L., and W.L. Nelson. 1975. *Soil fertility and fertilizers.* 3d ed. New York: Macmillan.

U.S. Environmental Protection Agency 1983. *Hazardous waste land treatment.* SW-874. Washington, D.C.: U.S. EPA, Office of Solid Waste and Emergency Response.

Valentine, R.L. 1986. *Vadoze zone modeling of organic pollutants,* edited by Stephen Hern and Susan Melancon, 233–243. Chelsea, Mich.: Lewis Publishers.

Valentine, R.L., and J.L. Schnoor. 1986. Biotransformation. In *Vadoze zone modeling of organic pollutants,* edited by Stephen Hern and Susan Melancon. Chelsea, Mich.: Lewis Publishers.

Verink, E.D. 1979. Simplified procedure for constructing pourbaix diagrams. *Journal of Education Modules Math. Sci. Eng.* 1:535–560.

9.13
TRANSPORT OF CONTAMINANTS IN GROUNDWATER

This section discusses the transport of contaminants in groundwater and describes the transport process and the behavior of the contaminant plume.

Transport Process

When a contaminant is introduced in groundwater, it spreads and moves with the groundwater as a result of (1) advection which is caused by the flow of groundwater, (2) dispersion which is caused by mechanical mixing and molecular diffusion, and (3) retardation which is caused by adsorption. The mathematical relationship between these processes can be written as follows (Javandel, Doughtly, and Tsang 1984):

$$\frac{\partial}{\partial x_i}\left[D_{ij}\frac{\partial C}{\partial x_j}\right] - \frac{\partial}{\partial x_i}(Cv_i) - \frac{C'W'}{n} = R\frac{\partial C}{\partial t} \qquad 9.13(1)$$

$$v_i = \frac{-K_{ij}}{n}\frac{\partial h}{\partial x_j} \qquad 9.13(2)$$

$$R = \left[1 + \frac{\rho_b K_d}{n}\right] \qquad 9.13(3)$$

where:

- C = contaminant concentration
- v_i = seepage or average pore water velocity in the direction x_i
- D_{ij} = dispersion coefficient
- K_{ij} = hydraulic conductivity
- C′ = solute concentration in the source or sink fluid
- W′ = volume flow rate per unit volume of the source or sink
- n = effective porosity
- h = hydraulic head
- R = retardation factor
- x_i = cartesian coordinate

The following discussion uses a simplified two-dimensional representation of Equation 9.13(1) to describe the transport of contaminants in groundwater. In a homogeneous, isotropic medium having a unidirectional steady-state flow with seepage velocity V, Equation 9.13(1) can be rewritten as

$$D_L\frac{\partial^2 C}{\partial x^2} + D_T\frac{\partial^2 C}{\partial y^2} - V\frac{\partial C}{\partial x} = R\frac{\partial C}{\partial t} \qquad 9.13(4)$$

where:

- C = contaminant concentration
- V = seepage or average pore water velocity
- D_L = longitudinal dispersion coefficient
- D_T = transversal dispersion coefficient
- R = retardation factor

ADVECTION

A contaminant moves with the flow of groundwater according to Darcy's law. Darcy's law states that the flow rate of water through soil from point 1 to point 2 is proportional to the head loss and inversely proportional to the length of the flow path as

$$Q = -K \cdot A \frac{h_2 - h_1}{L} \qquad 9.13(5)$$

where:

Q = groundwater flow rate
A = cross-sectional area of flow
$h_2 - h_1$ = head loss between point 1 and point 2
L = distance between point 1 and point 2
K = hydraulic conductivity

The actual seepage or average pore water velocity can be calculated as

$$V = \frac{Q}{n \cdot A} = -\frac{K}{n} \frac{h_2 - h_1}{L} \qquad 9.13(6)$$

where n is the effective porosity or percent of interconnected pore spaces that actually contributes to the flow.

The average pore water velocity calculated in Equation 9.13(6) is a conservative estimate of the migration velocity of the contaminant in groundwater. Therefore, when only advection is considered, a contaminant moves with the groundwater flow at the same rate as water, and no diminution of concentration is observed. In reality, however, the movement of the contaminant is also influenced by dispersion and retardation.

DISPERSION

Dispersion is the result of two processes, molecular diffusion and mechanical mixing.

Molecular diffusion is the process whereby ionic or molecular constituents move under the influence of their kinetic activity in the direction of their concentration gradients. Under this process, constituents move from regions of higher concentration to regions of lower concentration; the greater the difference, the greater the diffusion rate. Molecular diffusion can be expressed by Fick's law as

$$F = -D_f \frac{dC}{dx} \qquad 9.13(7)$$

where:

F = mass flux per unit area per unit time
D_f = diffusion coefficient
C = contaminant concentration
dC/dx = concentration gradient

Fick's law was derived for chemicals in unobstructed water solutions. When this law is applied to porous media, the diffusion coefficient should be smaller because the ions follow longer paths between solid particles and because of adsorption. This application yields an apparent diffusion coefficient D^* represented by

$$D^* = w \cdot D_f \qquad 9.13(8)$$

where w is an empirical coefficient less than 1. Perkins and Johnston (1963) suggest an approximate value of 0.707 for w. Bear (1979) suggests that w is equivalent to the tortuosity of the porous medium with a value close to 0.67. Values of D^* for major ions can be obtained from Robinson and Stokes (1965).

Mechanical mixing is the result of velocity variations within the porous medium. The velocity is greater in the center of the pore space between particles than at the edges. As a result, the contaminant spreads gradually to occupy an ever-increasing portion of the flow field. Mechanical mixing dispersion can occur both in the longitudinal direction of the flow as well as in the transverse direction. According to Bachmat and Bear (1964), the mechanical mixing component of dispersion can be assumed proportional to the seepage velocity as

$$D_{11} = a_L \cdot V \qquad 9.13(9)$$
$$D_{22} = a_T \cdot V \qquad 9.13(10)$$

where:

D_{11} = longitudinal mechanical mixing component of dispersion
D_{22} = transversal mechanical mixing component of dispersion
a_L = longitudinal dispersivity
a_T = transversal dispersivity
V = average linear pore water velocity

Finally the hydrodynamic dispersion coefficients can be written as

$$D_L = D_{11} + D_f = a_L \cdot V + D^* \qquad 9.13(11)$$
$$D_T = D_{22} + D_f = a_T \cdot V + D^* \qquad 9.13(12)$$

The dispersivity coefficients a_L and a_T are characteristic of the porous medium. Representative values of dispersivity coefficients can be determined from breakthrough column tests in the laboratory or tracer tests in the field (Anderson 1979).

Figure 9.13.1 shows how dispersion can cause some of the contaminant to move faster than the average groundwater velocity and some of the contaminant to move slower than the average groundwater velocity. The front of the contaminant plume is no longer sharp but rather smeared. Therefore, when dispersion is also considered,

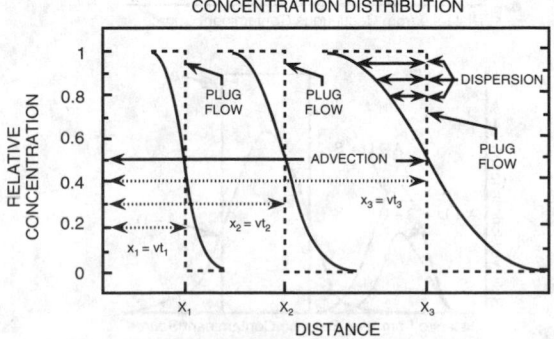

FIG. 9.13.1 Effect of dispersion–advection on concentration distribution. (Reprinted from U.S. Environmental Protection Agency, 1989, *Transport and fate of contaminants in the subsurface*, Seminar Publication EPA/625/4-89/019 (Cincinnati: U.S. EPA.)

the contaminant actually moves ahead of what would have been predicted by advection only.

RETARDATION

Retardation in the migration of contaminants in groundwater is due to the adsorption mechanism, which was described in Section 9.12 for both organic and inorganic constituents. The retardation coefficient can be calculated based on the distribution or adsorption coefficients of the contaminant and the characteristics of the porous medium as

$$R = \left[1 + K_d \frac{\rho_d}{n} \right] \qquad 9.13(13)$$

where K_d is the distribution or adsorption coefficient described previously. The values p_d and n are the bulk density and porosity of the soil. The velocity of the contaminant in groundwater can be calculated as follows:

$$V_c = \frac{V}{R} \qquad 9.13(14)$$

where V_c is the velocity of the contaminant movement in groundwater, V is the groundwater velocity, and R is the retardation factor. A high retardation factor, i.e., high adsorption coefficient, significantly retards the movement of the contaminant in groundwater. Figure 9.13.2 illustrates the effect of advection, dispersion, and retardation on the mobility of a contaminant in groundwater.

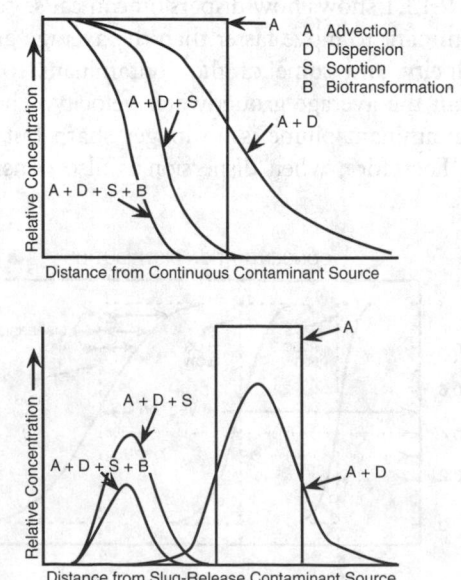

A   Advection
D   Dispersion
S   Sorption
B   Biotransformation

FIG. 9.13.2 Effect of advection, dispersion, and retardation on the mobility of a contaminant in groundwater. (Reprinted from M. Barcelona, 1990, *Contamination of groundwater: Prevention, assessment, restoration,* Pollution Technology Review No. 184, Park Ridge, N.J.: Nayes Data Corporation.)

Contaminant Plume Behavior

The behavior and movement of contaminants in groundwater depend on the solubility and density of the contaminant, groundwater flow regime, and the local geology. This section qualitatively discusses the effect of each of these factors on the contaminant plume.

CONTAMINANT DENSITY

Immiscible fluids such as oils do not readily mix with water; therefore, they either float on top of the water table or sink into the groundwater depending on their density. Immiscible fluids with densities less than water, also called light nonaqueous phase liquids (LNAPLs) or *floaters*, form a separate phase that can float on the groundwater table. For example, if a light-bulk hydrocarbon is released from a surface spill as shown in Figure 9.13.3, it migrates downward in the unsaturated zone due to gravity and capillary forces. If the volume of the released hydrocarbon is large, the hydrocarbon reaches the groundwater and forms a pancake on top of the water table. The pancake tends to spread laterally and in the downgradient direction until it reaches residual saturation. A portion of the pancake dis-

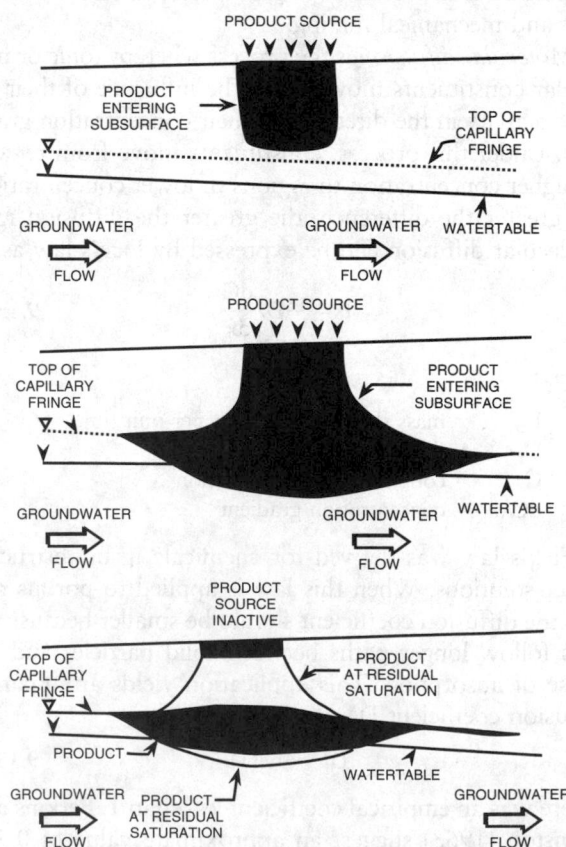

FIG. 9.13.3 Movement of LNAPLs into the subsurface. (Reprinted from U.S. Environmental Protection Agency, 1989, *Transport and fate of contaminants in the subsurface,* Seminar Publication EPA/625/4-89/019 (Cincinnati: U.S. EPA.)

solves in groundwater and eventually migrates with the water. The maximum spread of the pancake over the groundwater table can be estimated (CONCAWE Secretariat 1974) by

$$S = \frac{1000}{F}\left[V - \frac{A \cdot D}{K}\right] \qquad 9.13(15)$$

where:

S = maximum spread of the pancake, m^2
F = thickness of the pancake, mm
V = volume of infiltrating bulk hydrocarbon, m^3
A = area of infiltration, m^2
d = depth to groundwater, m
K = constant dependent on the soil's retention capacity for oil

Table 9.13.1 lists K values for different types of hydrocarbons and soil textures.

Immiscible fluids with densities greater than water, also called dense nonaqueous phase liquids (DNAPL) or *sinkers,* sink through the saturated zone and show a concentration gradient through the aquifer, becoming more concentrated near the aquifer base as shown in Figure 9.13.4. Fingering of the dense fluid into the water can also occur depending on the characteristics of the aquifer and the viscosity of the fluid (Dragun 1988). The downward migration of the sinker can continue until a zone of lower permeability, such as a clay confining layer or a bedrock surface, is encountered. Halogenated hydrocarbons and coal tars are the principal solvents possessing densities greater than that of water. Examples of DNAPLs include methylene chloride, chloroform, trichloroethylene (TCE), tetrachloroethylene or perchloroethylene (PCE), and various Freons.

Another important factor of both LNAPL and DNAPL plume behavior is residual contamination. As the plume migrates downward through the unsaturated or saturated zone, a small amount of fluid remains attached to soil particles and within the soil pore spaces via capillarity forces. This residual contamination can reside in the soil for many

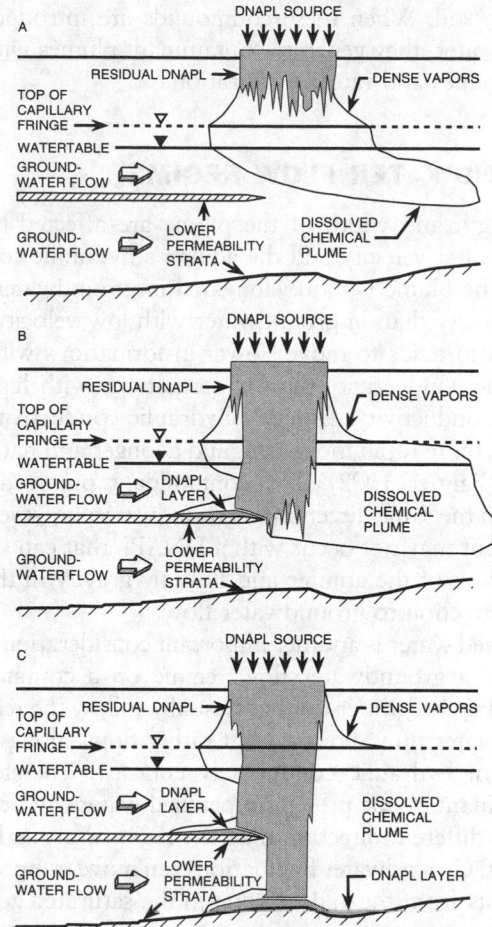

FIG. 9.13.4 Movement of DNAPLs into the subsurface. (Reprinted from S. Fenestra and J.A. Cherry, Dense organic solvents in groundwater: An introduction, In *Dense chlorinated solvents in groundwater,* Progress Report 0863985 (Ontario, Canada: Institute of Groundwater Research, University of Waterloo.)

years and serve as a continuous source of contamination. For more information on density flow, see Schwille (1988) and Fenestra and Cherry (1988).

CONTAMINANT SOLUBILITY

The solubility of a substance in water is defined as the saturated concentration of the substance in water at a given temperature and pressure. This parameter is important in the prediction of a contaminant plume in groundwater and in planning for its recovery. Substances with high water solubility have a tendency to remain dissolved in the water column, not adsorbed onto soil particles, and are more susceptible to biodegradation. Conversely, substances with low water solubility tend to adsorb onto soil particles and volatilize more readily from water. The water solubility of several substances is listed in Montgomery (1989). Several compounds, such as bulk hydrocarbons, are comprised of numerous individual chemicals and substances with different solubilities in water and different adsorption coeffi-

TABLE 9.13.1   TYPICAL VALUES FOR K FOR VARIOUS SOIL TEXTURES

Soil Texture	K		
	Gasoline	Kerosene	Light Fuel Oil
Stone & Coarse Gravel	400	200	100
Gravel & Coarse Sand	250	125	62
Coarse & Medium Sand	130	66	33
Medium & Fine Sand	80	40	20
Fine Sand & Silt	50	25	12

Sources: CONCAWE Secretariat, 1974, *Inland oil spill clean-up manual,* Report no. 4/74 (The Hague, The Netherlands: CONCAWE). D.N. Dietz, 1970, Pollution of permeable strata by oil components, In *Water pollution by oil,* edited by P. Hepple (New York: Elsevier Publishing and Institute of Petroleum).

cients in soil. When these compounds are introduced in groundwater, they generate contaminant plumes with different shapes and rates of migration.

GROUNDWATER FLOW REGIME

The length and width of the plume are affected by the groundwater velocity and the aquifer's hydraulic conductivity. The plume is more elongated in groundwater with high velocity than in groundwater with low velocity. The plume also tends to move slower in formations with low hydraulic conductivity than in formations with high hydraulic conductivity. A higher hydraulic conductivity can result in more rapid movement and a longer and narrower plume (Palmer 1992). The contaminant plume usually moves in the same direction as groundwater; however, this movement may not occur with a DNAPL that can sink to the bottom of the aquifer and flow by gravity in the opposite direction to groundwater flow.

Perched water is another important consideration in the effect of a groundwater flow regime on a contaminant plume. Perched water does not usually follow the regional groundwater flow direction but rather flows along an interface of hydraulic conductivity contrast. Therefore, a contaminant plume present in perched water can be moving in a different direction than the regional groundwater gradient. Groundwater fluctuations can move trapped contaminants from the vadose zone to the saturated zone.

GEOLOGY

The behavior of a contaminant plume depends largely on the type of geological profile through which it is moving. Geological structures such as dipping beds, faults, cross-bedding, and facies can affect the rate and direction of a migrating plume. Dipping beds can change the direction of a migrating plume. Faults can act as a barrier or a conduit to the contaminant plume depending on the material in the fault. Interbedded clay lenses in a permeable sand formation can split or retard a sinking contaminant plume and change its shape and course. Fractures and cracks in fractured bedrock formations can act as a conduit to the contaminant plume depending on their size and interconnections. Interaquifer exchange can move a plume of contamination from formations with the greatest hydraulic head to formations of a lesser hydraulic head (Deutsche 1961).

—Ahmed Hamidi

References

Anderson, M.P. 1979. Using models to simulate the movement of contaminants through groundwater systems. *CRC Crit. Rev. Env. Control* 9, no. 2:97–156.

Bachmat, Y., and J. Bear. 1964. The general equations of hydrodynamic dispersion in the homogeneous isotropic porous mediums. *J. Geophys. Res.* 69, no. 12:2561–2567.

Bear, J. 1979. *Hydraulics of groundwater.* New York: McGraw-Hill.

CONCAWE Secretariat. 1974. *Inland oil spill clean-up manual.* Report no. 4/74. The Hague, Netherlands: CONCAWE.

Deutsche, M. 1961. Incidents of chromium contamination of groundwater in Michigan. Proceedings of Symposium on Groundwater Contamination, April. Cincinnati, Ohio: U.S. Dept. of Health, Education and Welfare.

Dragun, J. 1988. *The soil chemistry of hazardous materials.* Silver Spring, Md.: Hazardous Materials Control Research Institute.

Fenestra, S., and J.A. Cherry. 1988. Subsurface contamination by dense-non aqueous phase liquid (DNAPL) chemicals. International Groundwater Symposium, May. Halifax, Nova Scotia: International Association of Hydrogeologists.

Javandel, I., C. Doughtly, and C.F. Tsang. 1984. *Groundwater transport: Handbook of mathematical models.* Water Resources Monograph 10. Washington, D.C.: American Geophysical Union.

Montgomery, J.H., and L.M. Wekom. 1989. *Groundwater chemicals desk reference.* Chelsea, Mich.: Lewis Publishers.

Palmer, C.M. 1992. Principles of contaminant hydrogeology. Chelsea, Mich.: Lewis Publishers.

Perkins, T.K., and O.C. Johnston. 1963. A review of diffusion and dispersion in porous media. *Soc. Pet. Eng. J.* 3:70–84.

Robinson, R.A., and R.H. Stokes. 1965. *Electrolytes solutions.* 2d ed. London: Butterworth.

Schwille, F. 1988. *Dense chlorinated solvents in porous and fractured media.* Chelsea, Mich.: Lewis Publishers.

Groundwater Investigation and Monitoring

9.14
INITIAL SITE ASSESSMENT

The purpose of a groundwater remedial investigation is to determine the nature and extent of contamination, identify current or potential problems caused by the contamination, and assist in the evaluation and selection of the remedial action. The remedial investigation generally has two phases. The first phase, called initial assessment, involves the use of existing site information and initial field screening techniques to identify potential sources of contamination; develop a conceptual understanding of the site and contamination process; and optimize subsequent, more intrusive, field investigation. The second phase involves a detailed subsurface investigation to assess the magnitude and extent of contamination and evaluate remedial actions.

Interpretation of Existing Information

Because the potential costs involved in groundwater remedial investigations are large, the best use of existing data and information must be made. Existing information and data can be site-specific, such as records of operations and records of previous investigations, or regional including surveys of geology, hydrology, surface soils, and meteorology. Existing data, however, can vary in quality; therefore, a thorough review and interpretation of these data prior to the investigation is necessary.

SITE-SPECIFIC INFORMATION

Existing data on site history can provide useful information on potential causes and sources of groundwater contamination. Data that should be collected include old maps and aerial photographs, interviews with present and former employees at the plant site, records of operations, records of product losses and spills, waste disposal practices, and the list of contaminants generated over the operating history of the site. The inventory must also include a history of the raw materials used and wastes disposed of over the years as industrial processes changed. Particular attention should be paid to potential sources of groundwater contamination such as locations of abandoned and active landfills and wastewater impoundments, buried product pipelines, old sewers, tanks, cesspools, dry wells, product storage areas, product loading areas, storm water collection areas, and previous spill areas.

In addition, foundation borings or construction details of supply wells can provide firsthand information on the types and characteristics of subsurface soils and groundwater at the site. Chemical data may be available from the results of previous monitoring activities at the site or at adjacent properties. These data should be analyzed and plotted on base maps and used to estimate background groundwater and soil quality.

REGIONAL INFORMATION

Regional information can be used to identify potential off-site sources of contamination and to provide background information on regional geology, hydrology, surface soils, and meteorology. This information can provide insight into the complexities of the groundwater contamination and help guide future site investigations.

A regional inventory of potential offsite sources of contamination can be developed through aerial photographs, land-use maps, and field inspections. Old aerial photographs are especially useful because they may be the only means of identifying abandoned facilities such as old landfills, lagoons, and industrial facilities. Land-use maps can identify unsewered residential areas that can be a potential source of contamination, especially where organic chemical septic tank cleaners have been used. Topographic maps can identify surface drainage patterns that can carry contaminants to the plant site and recharge the underlying groundwater system.

Regional geologic reports, maps, and cross sections can provide details on the regional subsurface geology including areal extent, thickness, composition, and structure of the geological units present in the region. Regional hydrogeologic reports can provide information on the regional groundwater flow direction and quality as well as the groundwater usage in the region. A survey of state files can reveal long-term groundwater quality problems in the general area of the plant site.

Soil maps can be used to evaluate the migration potential of contaminants through the unsaturated zone. Climatological data can be used to determine precipitation rates and patterns as well as surface runoff and groundwater recharge rates. In addition, climatological data can be used to determine evapotranspiration rates from shallow groundwater tables and their effect on the gradient and direction of groundwater.

An inventory of regional information is available from state and federal agencies such as the U.S. Geological Survey (USGS), the U.S. EPA, the U.S. Department of Agriculture (USDA), and the Soil Conservation Service (SCS). Other sources of information include computerized databases on environmental regulations and technical information on a variety of chemical compounds (Lynne et al. 1991). Examples of these databases include the Computer-Aided Environmental Legislative Data System (CELDS), which provides a collection of abstracted federal and state environmental regulations and standards; HAZARDLINE, which provides information on over 500 hazardous workplace substances as defined by the Occupational Safety and Health Administration (OSHA); and the Chemical Information System (CIS), which provides a variety of subjects related to chemistry.

Initial Field Screening

Data collection in a groundwater remedial investigation can begin with minimally intrusive techniques, called initial field screening techniques. These techniques are less expensive than the more intrusive techniques such as soil borings, test pits, and well monitoring. In addition, field screening techniques provide information which streamlines data collection and optimizes the use of intrusive techniques. The principal categories of initial field screening techniques include surface and downhole geophysical surveys and onsite chemical screening, such as a soil-gas survey.

SURFACE GEOPHYSICAL SURVEYS

Surface geophysical surveys are applied at the surface to provide a rapid reconnaissance of the hydrogeologic conditions at the site, such as depth to bedrock, degree of weathering, and the presence of clay lenses, fracture zones, or buried waste. In addition, surface geophysical surveys can be used to detect and map inorganic contaminant plumes, obtain the flow direction, and estimate the concentration gradients (Benson et al. 1985).

Surface geophysical surveys include electromagnetic conductivity, electrical resistivity, seismic refraction, and ground-penetrating radar as described in Pitchford, Mazzella, and Scarbrough (1988); Benson, Glaccum, and Noel (1984); and the U.S. EPA desk reference guide on subsurface characterization and monitoring techniques (1993a). A description of the most commonly used surface geophysical surveys follows.

Electromagnetic (EM) Methods

The EM methods use a transmitter coil to generate an electromagnetic field that induces eddy currents in the ground below the instrument. A receiver coil measures secondary electromagnetic fields created by the eddy currents and produces an output voltage that can be related to variations in subsurface conductivity as shown in Figure 9.14.1. Variations in subsurface conductivity may be caused by changes in the basic soil or rock types, thickness of the soil and rock layers, moisture content, fluid conductivity, and depth to the water table.

Environmental engineers can use EM surveys to obtain data by profiling or sounding. In profiling, the engineer makes measurements at a number of stations along a survey line to map lateral changes in the subsurface electrical conductivity to a given depth. In sounding mode, the engineer places the instrument at one location and takes measurements at increasing depths, by changing coil orientation or coil spacing, to map vertical changes in electrical conductivity and, therefore, the soil and rock type at that location.

An advantage of the EM methods is that the surveys can be done quickly because direct contact of the instrument with the ground is not required. The disadvantage, however, is that the EM surveys are susceptible to the presence of metals and powerlines on the surface of the ground.

Electrical Resistivity (ER) Methods

In ER methods, environmental engineers measure the resistivity of subsurface materials by injecting an electrical current into the ground through a pair of surface electrodes (current electrodes) and measuring the resulting potential field (voltage) from a second pair of electrodes (potential electrodes) as shown in Figure 9.14.2. Several types of electrode geometries can be used for resistivity measurements including the Wenner, Schlumberger, dipole, and others. The Wenner array is the simplest in terms of geometry and consists of four electrodes spaced equally in a line.

The ER measurements are a function of the soil or rock types, thickness of the soil and rock layers, moisture content, fluid conductivity, and depth to the water table. The ER of a geological formation is calculated based on the electrode separation, the geometry of the electrode array, the applied current, and the measured voltage.

As with the EM surveys, environmental engineers can use the ER surveys to obtain data by profiling or sounding. In profiling, engineers take measurements at a number of stations along a survey line to map lateral changes

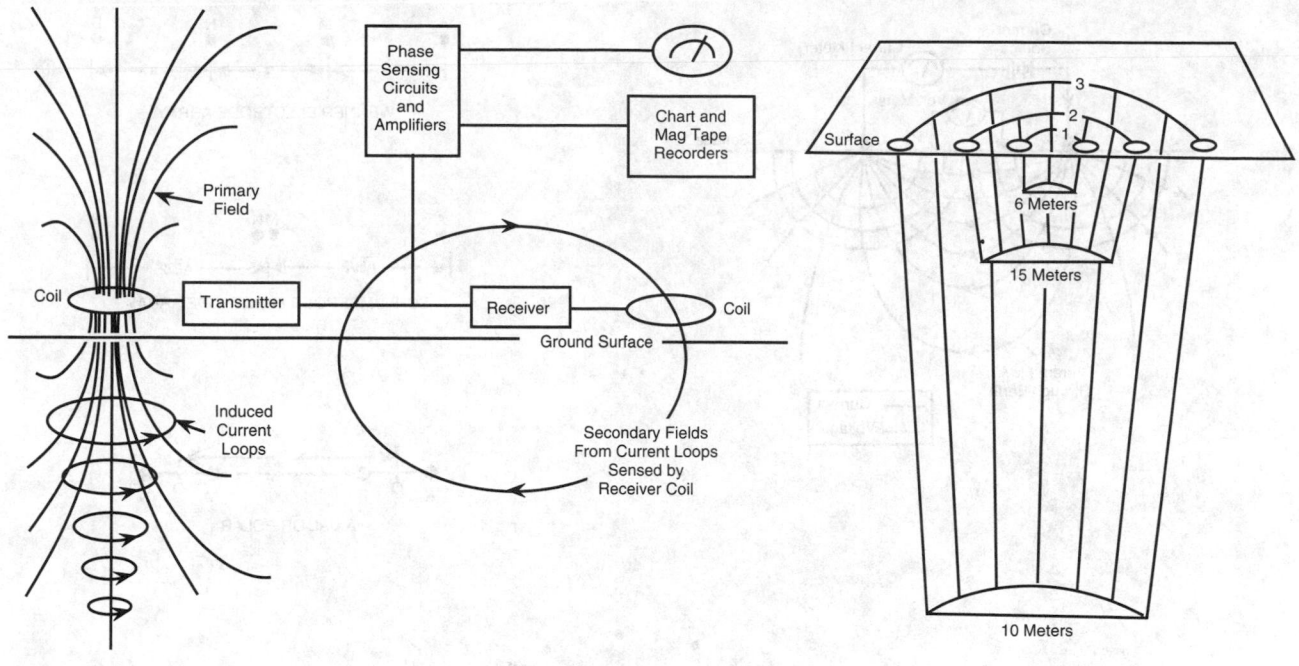

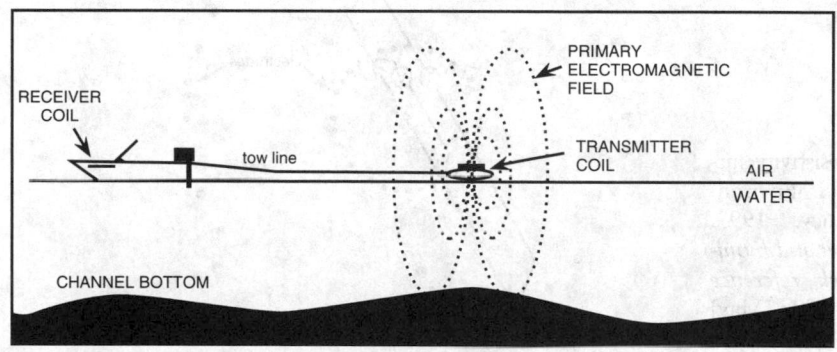

FIG. 9.14.1 Electromagnetic survey. (Reprinted from U.S. Environmental Protection Agency, 1993, *Subsurface characterization and monitoring techniques, a desk reference guide,* U.S. EPA/625/R-93/003a [May] U.S. EPA.)

in the subsurface electrical properties to a given depth. Then, they can use the data to delineate hydrogeological anomalies or map inorganic plumes. Sounding measurements, on the other hand, are made at increasing depths so that engineers can map vertical changes in electrical properties. Engineers use data from sounding measurements to determine the depth, thickness, and type of soil or rock layer at the site. The data from ER surveys can be interpreted with the use of computer models or master curves to create geoelectric sections (Orellana and Mooney 1966). These sections illustrate changes in the vertical and lateral resistivity conditions at the site.

The ER surveys are useful for identifying shallow contaminated groundwater bodies where (1) a significant contrast exists in water quality; (2) the water table is less than 40 feet deep; (3) the geology of the water table aquifer is relatively homogeneous; and (4) local interferences, such as buried pipelines, power lines, or metal fences, are not present.

The advantages of the ER methods are that they are well established and their equipment is inexpensive, mobile, and easy to operate and provides relatively rapid areal coverage. In addition, the ER methods are superior to the EM methods for detecting thin resistive layers. The disadvantage, however, is that continuous profiling is not possible, and the requirement for ground contact can cause problems in resistive material and generally makes the ER surveys slower to use than the EM surveys. Furthermore, use of the ER methods is limited in wet weather and on paved areas, and the methods are less sensitive to conductive pollutants than the EM methods.

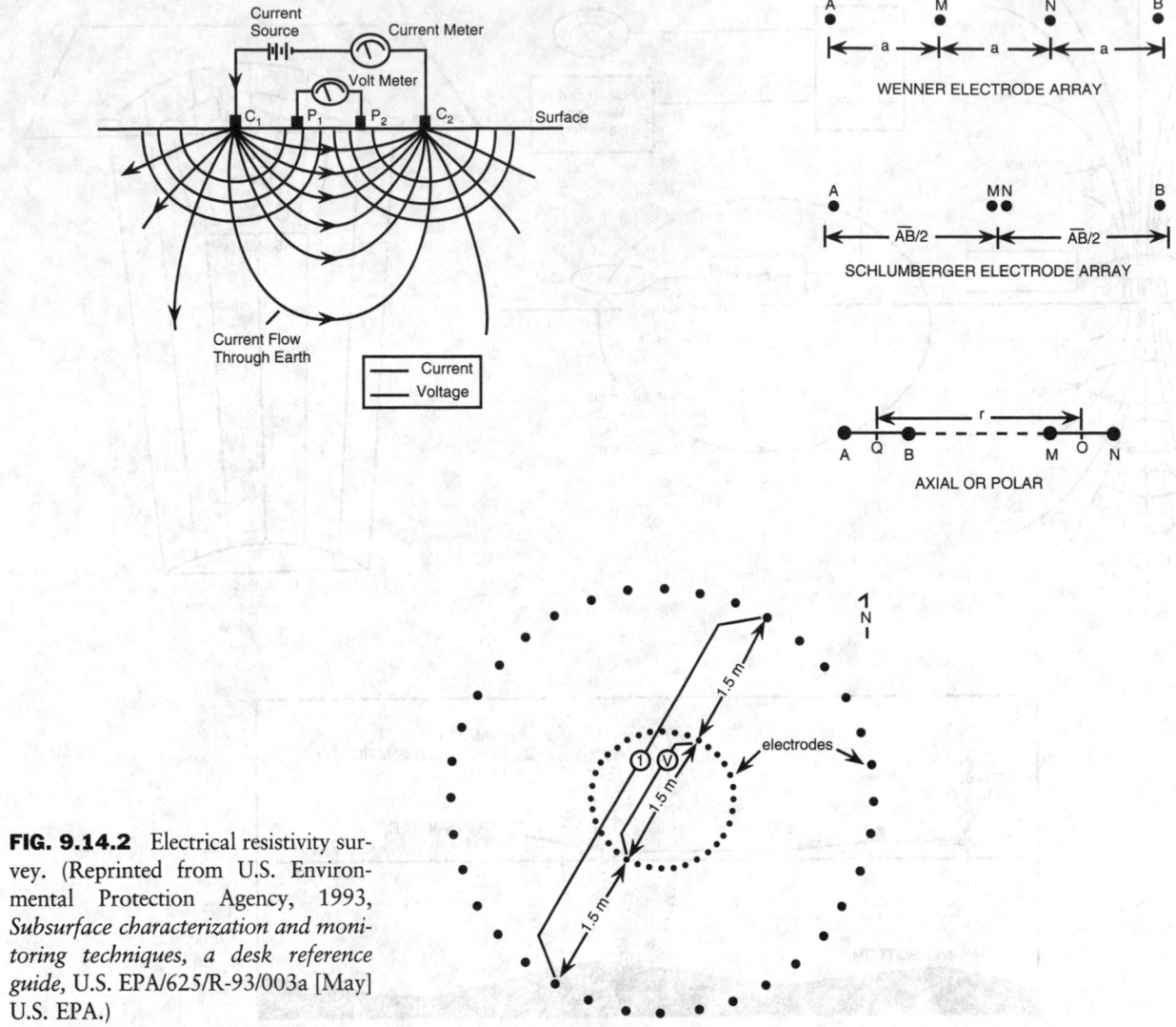

FIG. 9.14.2 Electrical resistivity survey. (Reprinted from U.S. Environmental Protection Agency, 1993, *Subsurface characterization and monitoring techniques, a desk reference guide,* U.S. EPA/625/R-93/003a [May] U.S. EPA.)

Seismic Refraction (SR) Methods

Environmental engineers often use the SR methods to determine the top of bedrock or depth of the water table, locate fractures or faults, and characterize the type of rock or degree of weathering. The SR methods are based on the fact that elastic waves travel through different earth materials at different velocities; the denser the material, the higher the wave velocity.

The elastic waves are initiated by an energy source (hammer or controlled explosive charge) at the ground surface. A set of receivers, called geophones, is set up in a line radiating outward from the energy source as shown in Figure 9.14.3. Waves initiated at the surface and refracted at the critical angle by a high-velocity layer at a depth reach the more distant geophones quicker than the waves that travel directly through the low-velocity surface layer. The time between the shock and the arrival of the elastic wave at a geophone is recorded on a seismograph. Using a set of seismograph records, engineers can derive a graph of arrival time versus distance from the shot point to the geophone. They can then analyze the line segments, slope, and break points in the graph to identify the number of layers and the depth of each layer. In addition, they can use typical seismic velocity ranges to determine the type of soil of each layer (U.S. EPA 1993a).

The advantages of SR methods are that the equipment is readily available, portable, and relatively inexpensive. In addition, the methods are accurate and provide rapid areal coverage with depths of penetration up to 30 meters. The disadvantage, however, is that the resolution might be obscured by layer sequences where the velocity of the layers decreases with depth, and thin layers, called blind zones, might not be detected. Furthermore, the methods are susceptible to noise from adjacent areas (such as construction activities) and do not detect contaminants in groundwater.

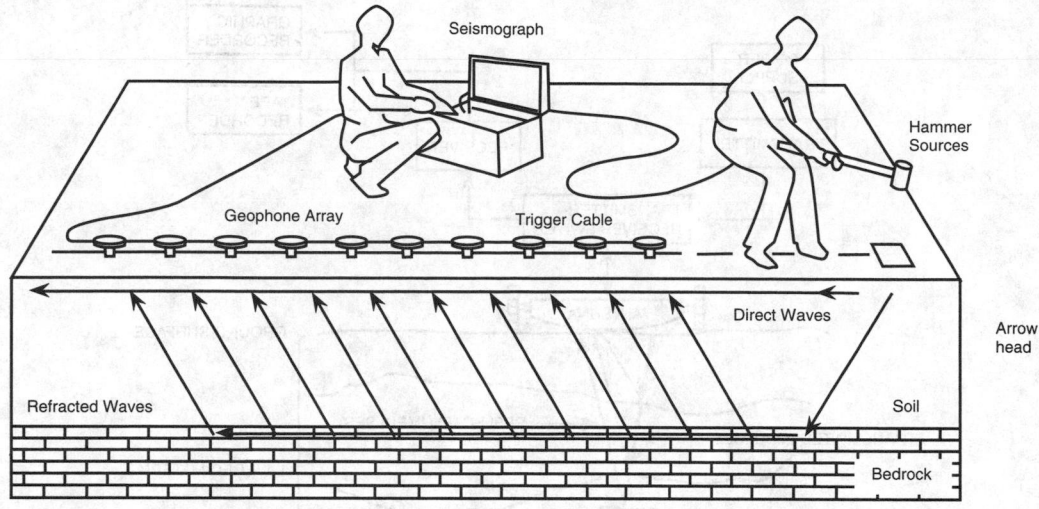

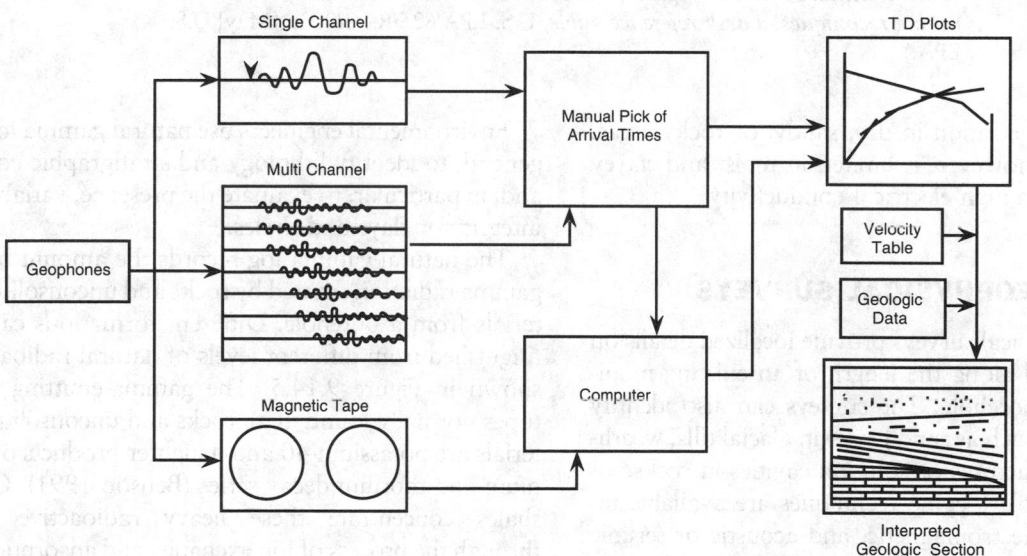

FIG. 9.14.3 Seismic refraction survey. (Reprinted from U.S. Environmental Protection Agency, 1993, *Subsurface characterization and monitoring techniques, a desk reference guide*, U.S. EPA/625/R-93/003a [May] U.S. EPA.)

Ground Penetrating Radar (GPR) Methods

Environmental engineers often use the GPR methods to locate buried objects, map the depth to shallow water tables, and delineate soil horizons. The principles involved in GPR technology are similar to those in seismic refraction, except that in GPR, electromagnetic energy is used instead of acoustic energy, and the resulting image is relatively easy to interpret.

In a GPR survey, a transmitting and a receiving antenna are dragged along the ground surface as shown in Figure 9.14.4. The small transmitting antenna radiates short pulses of high-frequency radio waves into the ground, and the receiving antenna records variations in the reflected return signal. The attenuation loss of the signal in the ground increases with ground conductivity and with frequency for a given material. Changes in ground electric conductivity are associated with natural hydrogeological conditions such as bedding cementation, moisture, clay content, voids, and fractures. Therefore, an interface between two soil or rock layers with sufficient contrast in electric conductivity shows up in the radar profile (Benson and Glaccum 1979).

The advantages of the GPR methods include rapid areal coverage, where site conditions are favorable, and great

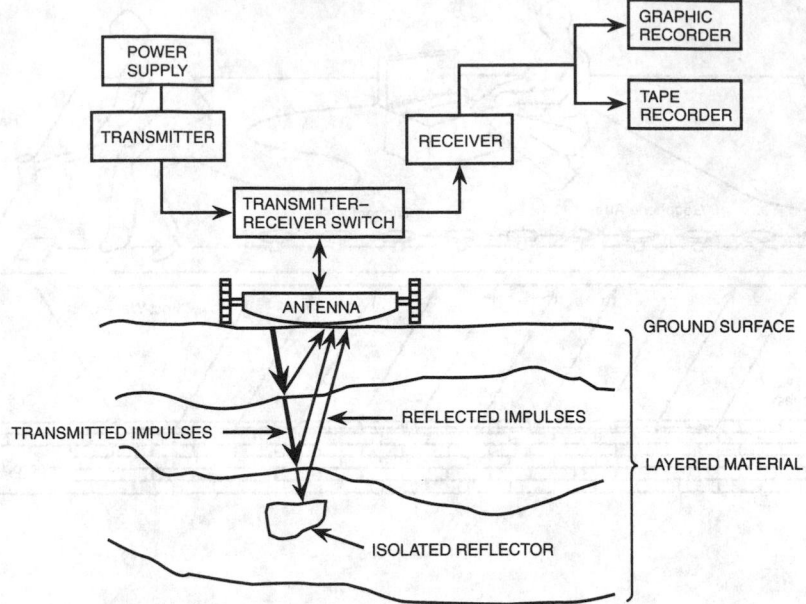

FIG. 9.14.4 Ground-penetrating radar survey. (Reprinted from U.S. Environmental Protection Agency, 1993, *Subsurface characterization and monitoring techniques, a desk reference guide,* U.S. EPA/625/R-93/003a [May] U.S. EPA.)

resolution and penetration in dry, sandy, or rocky areas. The use of GPR, however, is limited in moist and clayey soils and soils with high electrical conductivity.

DOWNHOLE GEOPHYSICAL SURVEYS

Downhole geophysical surveys provide localized details on soil, rock, or fluid along the length of an existing monitoring well or a borehole. The surveys can also identify permeable zones, such as sand lenses in glacial tills, weathered zones, and fractures or solution cavities in rocks.

Several downhole logging techniques are available including nuclear, electromagnetic, and acoustic or seismic as described in Keys and MacCary (1976) and the U.S. EPA desk reference guide on subsurface characterization and monitoring techniques (1993a). Some of these techniques provide measurements from inside plastic or steel casing, and some allow measurements in the unsaturated zone as well as the saturated zone. A description of the most commonly used logs follows.

Nuclear Logging Methods

Nuclear logging includes methods that detect the presence of unstable isotopes or create such isotopes in the vicinity of a borehole. Several nuclear logging techniques are available including natural gamma logs, gamma–gamma logs, and neutron–neutron logs. Natural gamma logs are probably the most common nuclear methods used in groundwater studies.

Environmental engineers use natural gamma logging, in general, to identify lithology and stratigraphic correlation and, in particular, to evaluate the presence, variability, and integrity of clays and shales.

The natural gamma log records the amount of natural gamma radiation emitted by rocks and unconsolidated materials from a borehole. Different formations can be distinguished from different levels of natural radioactivity as shown in Figure 9.14.5. The gamma-emitting radioisotopes normally found in all rocks and unconsolidated materials are potassium-40 and daughter products of the uranium and thorium decay series (Benson 1991). Clays and shales concentrate these heavy radioactive elements through the process of ion exchange and adsorption; therefore, their natural gamma activity is much higher than that of other materials.

The natural gamma log instrumentation is relatively simple and inexpensive and involves radiation detection only. However, only qualitative analysis is possible with this method, and the sensitivity of the probe is reduced by large diameter holes, drilling fluid, and casing (U.S. EPA 1993a).

Electromagnetic Logging Methods

As with the EM method, the electromagnetic logging method measures the electrical conductivity of soil or rock in open or polyvinyl-chloride- (PVC) cased boreholes above or below the water table. Environmental engineers use this method to perform lithological characterization, locate the zones of saturation, and perform chemical char-

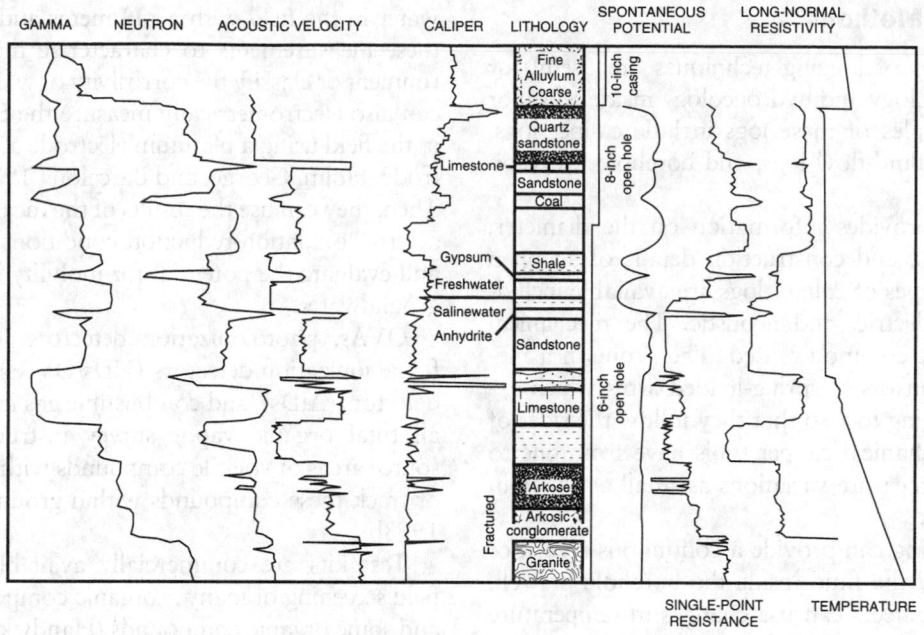

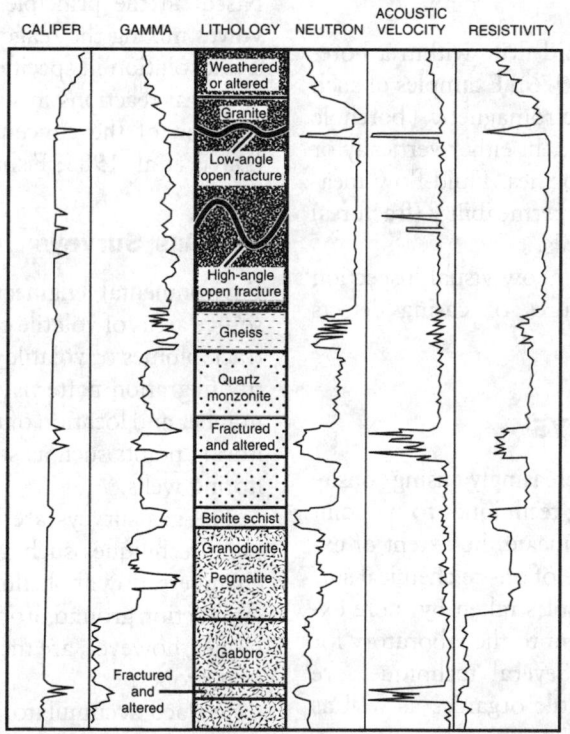

FIG. 9.14.5 Well log suites in sedimentary and fractured rocks. (Reprinted from U.S. Environmental Protection Agency, 1993, *Subsurface characterization and monitoring techniques, a desk reference guide*, U.S. EPA/625/R-93/003a [May] U.S. EPA.)

acterization of groundwater. Several electromagnetic logging techniques are available including induction logs, microwave-sensing logs, nuclear magnetic resonance logs, and surface-borehole logs. Induction logs are probably the most common electromagnetic methods used in groundwater studies.

The probe in an induction log contains a transmitter coil on the upper part, which induces eddy current in the formation around the borehole, and a receiver on the lower part. Engineers measure conductivity using the same principles as the EM methods. Because the response of the log is a function of the specific conductance of the pore fluids, it is an excellent indicator of the presence of inorganic contaminants (Benson 1991). Variations in conductivity with depth also indicate changes in clay content, permeability of a formation, or fractures.

Other Logging Methods

Several other types of logging techniques are useful for characterizing lithology and hydrogeology inside a well or a borehole. Examples of these logs include caliper logs, temperature logs, fluid-flow logs, and borehole television logs.

A caliper log provides information on the diameter, lithology, fractures, and construction details of an open borehole. Many types of caliper logs are available including mechanical, electric, and acoustic. The mechanical caliper is the most commonly used. The probe in a mechanical caliper consists of spring-loaded arms which extend from the logging tool so that they follow the sides of the borehole. Mechanical caliper tools have from one to six arms and can measure variations as small as $\frac{1}{4}$ inch in borehole diameter.

A temperature log can provide a continuous record of the temperature of the fluid inside the borehole or well. Environmental engineers can use changes in temperature to identify leaks in the casing where damage or corrosion has occurred.

Fluid-flow logs measure the fluid flow within a borehole or well (Keys and MacCary 1976). Examples of such logs include thermal and electromagnetic borehole flowmeters that sense water movement either vertically or horizontally (or both) at low velocities. Fluid-flow measurements can locate zones of high permeability (fractures) and areas of leakage in artisan wells.

Borehole television camera logs allow visual inspection of a borehole or well for fractures or casing defects (Morahan and Dorrier 1984).

ONSITE CHEMICAL SURVEYS

Environmental engineers are increasingly using onsite chemical surveys as field screening techniques to pinpoint source areas or approximately delineate the extent of existing contaminant plumes. The use of onsite chemical surveys optimizes the number of samples taken by more expensive intrusive techniques and sent to the laboratory for confirmatory chemical analysis. Several techniques are available for volatile and nonvolatile organics as well as for inorganic compounds.

Onsite chemical screening techniques vary from qualitative chemical analyses using indicators such as organic vapor analyzers (OVAs) or HNU meters to more quantitative soil-gas surveys using gas chromatography and mass spectrometry (GC/MS).

Qualitative Onsite Chemical Surveys

Generally, environmental engineers use these field screening techniques to collect preliminary site information and guide future and more intrusive field investigations. Engineers can measure the pH of the soil, waste, or ground-water in the field with a pH meter and use the results of these measurements to characterize the subsurface environment or classify the corrosivity of waste materials. They can also electrometrically measure the Eh of groundwater in the field using a platinum electrode and a reference electrode (Holm, George, and Barcelona 1986; Ritchey 1986). Then, they can use the results of the measurements to characterize oxidation-reduction conditions in the subsurface and evaluate the potential for mobility of heavy metals in groundwater.

OVAs, photoionization detectors (PID/HNU meter), flame ionization detectors (FIDs/OVAs), argon ionization detectors (AIDs), and combustible gas indicators (EDs) are all total organic vapor survey instruments that locate source areas of volatile compounds within the vadose zone or track these compounds within groundwater (U.S. EPA 1993b).

Test kits are commercially available for preliminary field screening of many inorganic compounds (Hatch kits) and some organic compounds (Handy kits). These kits are based on the principles of colorimetry. Colorimetry involves mixing the reagents of known concentrations with a test solution in specified amounts. This mixing results in chemical reactions in which the color of the solution is a function of the concentration of the analyte of interest (Davis et al. 1985; Fishman and Friedman 1989).

Soil-Gas Surveys

Environmental engineers use soil-gas surveys to locate source areas of volatile compounds within the vadose zone, track plumes of volatile compounds in groundwater, identify migration patterns of landfill gases, and optimize the number and location of more expensive and intrusive monitoring points such as soil borings and groundwater monitoring wells.

Soil-gas surveys are based on several in situ soil sampling techniques such as headspace analysis, surface flux chambers, downhole flux chambers, surface accumulators, and suction ground probes. The most commonly used techniques, however, are the surface accumulators and the suction probes.

Surface accumulators involve the passive sampling of soil gas by trapping volatile organic compounds (VOCs) onto an adsorbent contained within an inverted glass tube (Zdeb 1987). The inverted glass tube is buried in the soil for a few days to weeks. The adsorbent consists of a ferromagnetic wire coated with activated charcoal and is contained in an inverted test tube. The adsorbent passively collects diffusing VOCs which adsorb onto the activated charcoal. After a few days or weeks, the glass tube is sealed and taken to the laboratory for VOC analysis.

Ground probe sampling techniques for soil gas involve inserting a tube into the ground and pumping the soil gas with a vacuum pump. Engineers then analyze the extracted gas in the field for VOCs using portable analytical instruments. The probes can be manually or pneumatically driven or installed in boreholes. Grab samples can be taken at the same depth (or at different depths) at several locations for areal (or vertical) characterization of soil-gas concentrations.

The vertical and horizontal spacing of the probes can be affected by many factors such as soil moisture and organic matter content, presence of perched water, depth to groundwater, permeability of the subsurface materials, and the Henry's Law constant of the VOC in question (Silka 1986). The upward diffusion of vapors is usually blocked by soil strata containing a finer grained soil with a higher moisture content or higher organic carbon content.

—Ahmed Hamidi

References

Benson, R.C. 1991. Remote sensing and geophysical methods for evaluation of subsurface conditions. In *Practical handbook of groundwater monitoring,* edited by David Nielson. Chelsea, Mich.: Lewis Publishers.

Benson, R.C., and R.A. Glaccum. 1979. Radar surveys for geotechnical site assessment. Proceedings of the Geophysical Methods in Geotechnical Engineering, Specialty Session, 161–178. Atlanta, Ga.: American Society of Civil Engineers.

Benson, R.C., R.A. Glaccum, and M.R. Noel. 1984. *Geophysical techniques for sensing buried wastes and waste migration.* Worthington, Ohio: National Water Well Association.

Benson, R.C. et al. 1985. Correlation between geophysical measurements and laboratory water sample analysis. Proceedings of the National Water Well Association, Environment Protection Agency Conference on Surface and Borehole Geophysical Methods in Groundwater Investigation. National Water Well Association.

Davis, S.N. et al. 1985. *Introduction to groundwater tracers.* EPA/600/2-85/022, NTIS PB86-100591.

Fishman, M.J., and L.C. Friedman, eds. 1989. *Methods for determination of inorganic substances in water and fluvial sediments.* 3d ed. U.S. Geological Survey Techniques of Water Resources Investigations, TWRI 5-A1.

Holm, T.R., G.K. George, and M.J. Barcelona. 1986. *Dissolved oxygen and oxidation-reduction potentials in groundwater.* EPA/600/2-86/042, NTIS PB86-179678.

Keys, W.S., and L.M. MacCary. 1976. *Application of borehole geophysics to water resources investigations.* Techniques of Water Resources Investigations of the United States Geophysical Survey.

Lynne, M. et al. 1991. The overall philosophy and purpose of site investigation. In *Practical handbook of groundwater monitoring,* edited by David Nielson. Chelsea, Mich.: Lewis Publishers.

Morahan, T., and R.C. Dorrier. 1984. The application of television borehole logging to groundwater monitoring programs. *Groundwater Monitoring Review* 4, no. 4:172–175.

Orellana, E., and H.M. Mooney. 1966. *Master tables and curves for vertical electrical sounding over layered structures.* Madrid, Spain: Interciencia.

Pitchford, A.M., A.T. Mazzella, and K.R. Scarbrough. 1988. *Soil and geophysical techniques for detection of subsurface organic contamination.* USEPA/600/4-88-019, NTIS. U.S. EPA.

Ritchey, J.D. 1986. Electronic sensing devices used for in situ groundwater monitoring. *Groundwater Monitoring Review* 6, no. 2:108–113.

Silka, L.R. 1986. Simulation of the movement of volatile organic vapor through the unsaturated zone as it pertains to soil–gas surveys. Proceedings of the NWWA/API Conference on Petroleum Hydrocarbons and Organic Chemicals in Groundwater: Prevention, Detection, and Restoration. Dublin, Ohio: National Water Well Association.

U.S. Environmental Protection Agency. 1993a. *Subsurface characterization and monitoring techniques, a desk reference guide.* USEPA/625/R-93/003a (May). U.S. EPA.

———. 1993b. *Subsurface characterization and monitoring techniques, a desk reference guide,* Vol. 2. USEPA/625/R-93/003b (May). U.S. EPA.

Zdeb, T.F. 1987. Multi-depth soil–gas analysis using passive and dynamic sampling techniques. Proceedings of Petroleum Hydrocarbons and Organic Chemicals in Groundwater: Prevention, Detection, and Restoration. Dublin, Ohio: National Water Well Association.

9.15
SUBSURFACE SITE INVESTIGATION

The purpose of a subsurface investigation is to collect samples and obtain actual quantitative measurements of chemical concentrations, hydraulic parameters, and lithological data within a particular hydrogeologic strata or group of strata. Environmental engineers can use these samples and measurements to assess the magnitude and extent of groundwater or soil contamination and support the selection and design of engineering options for remediation.

Engineers can conduct subsurface investigations using temporary groundwater and soil sampling techniques such as HydroPunch, soil probes, and cone penetrometers (Edge and Cordy 1989) or more permanent techniques such as the installation of monitoring wells and soil borings. Temporary techniques are less expensive but less reliable; therefore, they are usually used for screening purposes and the optimization of the location and number of permanent systems. Permanent techniques, on the other hand, are more expensive and more reliable; therefore, their use is usually limited to confirm actual concentrations and subsurface conditions.

Subsurface investigations involve several field activities such as drilling, installation, development, and sampling of monitoring wells. These activities are intrusive to the subsurface environment; therefore, engineers should conduct them with care to prevent cross-contamination and obtain representative groundwater and soil samples that retain both the physical and chemical properties of the subsurface environment. A description of these field activities follows.

Subsurface Drilling

Subsurface drilling for groundwater remedial investigations uses much of the same technology as conventional geotechnical exploration but with some significant differences. Geotechnical exploration requires the collection of an intact physical specimen which can be tested for geotechnical properties. In comparison, groundwater remedial investigations require that the specimen also be representative of existing conditions and valid for chemical analysis. Therefore, the selection of drilling methods and sampling protocols in a groundwater remedial investigation is more restrictive and should be based upon site-specific conditions and the type of testing to be done.

The criteria used in the selection of a drilling method include the type of geological formation, depth of drilling, depth of screen setting, types of pollutants expected, accessibility to the site, and availability of drilling equipment. The following section briefly describes the drilling methods used in groundwater remedial investigations.

DRILLING METHODS

Several drilling methods are used in groundwater remedial investigations including air rotary, direct mud-rotary, reverse mud-rotary, hollow-stem augers, solid-stem augers, and cable tools among others (Davis, Jehn, and Smith 1991). The following discussion focuses on the two methods most commonly used for monitoring well installations: hollow-stem auger and direct mud-rotary.

Hollow-Stem Auger

The hollow-stem auger is a form of continuous flight auger usually used for drilling monitoring wells in unconsolidated materials. The auger consists of a tubular steel center shaft or axle around which is welded a continuous steel strip in the form of a helix, also known as flight, as shown in Figure 9.15.1. As the auger column rotates and axially advances in the ground, the dug material is simultaneously conveyed to the surface by the helix.

The main advantage of hollow-stem auger drilling is that no drilling fluids or lubricants are used; therefore, no contaminants are introduced into the aquifer. In addition, the hollow stem of the auger allows sampling of soil material as the borehole is advanced and installation of casings and screens for monitoring wells when the required depth has been reached. The drill head, or cutting bit, lo-

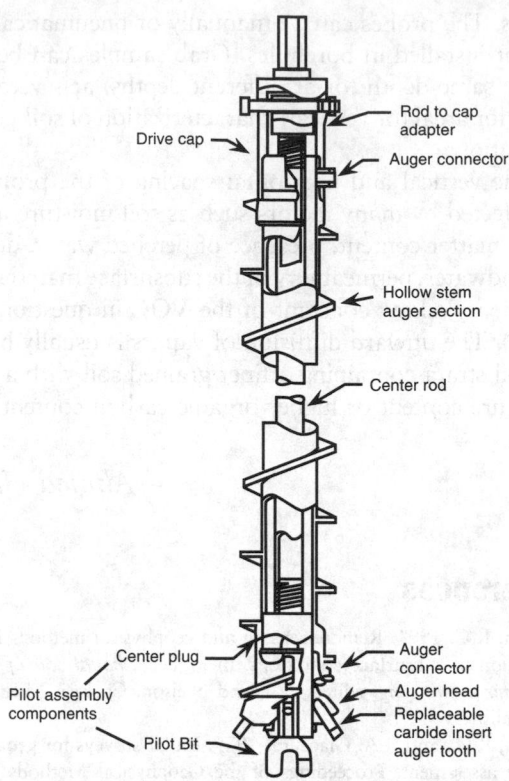

FIG. 9.15.1 Hollow-stem auger system. (Reprinted from U.S. Environmental Protection Agency, 1993, *Subsurface characterization and monitoring techniques, a desk reference guide,* Vol. 1, USEPA/625/R-93/003a [May] U.S. EPA.)

cated at the bottom of the auger can be removed (tripped) through the center of the auger to the surface. This feature allows the auger to stay in place providing an open, cased hole into which samplers, downhole drive hammers, casings, screens, and other instruments can be inserted.

The hollow-stem auger cannot be used, however, in consolidated, rock, or well-cemented formations. In addition, depths are usually limited to no more than 150 feet, and vertical leakage of water through the borehole during drilling is likely to occur.

Direct Mud-Rotary

Direct mud-rotary drilling is a drilling method in which a fluid is forced down the drill stem, out through the bit, and back up the borehole to remove the cuttings as shown in Figure 9.15.2. The cuttings are removed by settling in a sedimentation tank or pond, and the fluid is circulated back down the drill stem. The drilling fluid can be a liquid, such water or mud (water with special additives, e.g., bentonite and polymers), or it can be gas, such as air or foam (air with additives, e.g., detergents) (Davis, Jehn, and Smith 1991).

Mud-rotary drilling is a flexible and rapid drilling method in all types of geologic materials and depth ranges. The circulating fluid serves to cool and lubricate the bit,

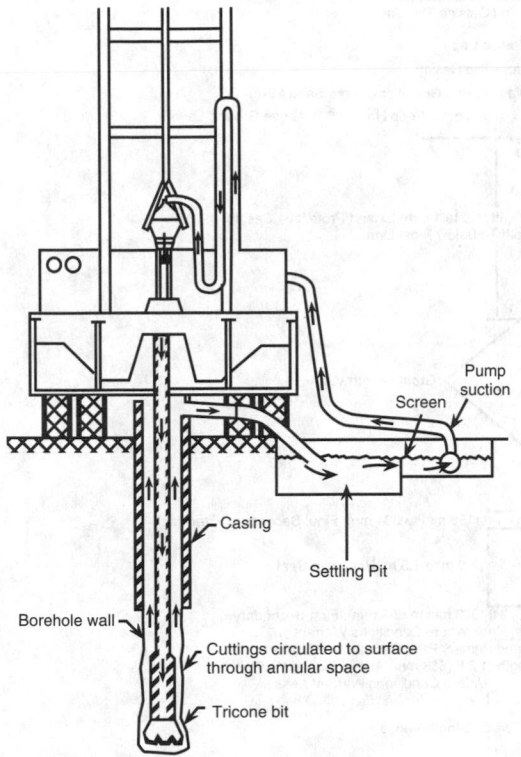

FIG. 9.15.2 Direct mud-rotary circulation system. (Reprinted from U.S. Environmental Protection Agency, 1993, *Subsurface characterization and monitoring techniques, a desk reference guide,* Vol. 1, USEPA/625/R-93/003a [May] U.S. EPA.)

stabilize the borehole, remove the cuttings, and prevent the inflow of formation fluids, thus minimizing cross-contamination of aquifers. In addition, samples can be obtained directly from the circulated fluid when a sample-collecting device is placed in the discharge pipe before the settling tank.

Mud-rotary drilling, however, requires the introduction of some foreign liquids into the aquifer, which can compromise the validity of subsequent monitoring well samples. In addition, contaminants might be circulated with the fluid, and the collection of representative samples is difficult due to the mixing of drill cuttings. Other limitations of mud-rotary drilling include the inability to provide information on the position of the water table and the loss of drilling fluids in fractured materials.

SOIL SAMPLING

Soil samples are usually taken at regular intervals during drilling to be analyzed for chemical composition and tested for physical properties such as particle size distribution, textural classification, and hydraulic conductivity. The samples are generally taken from the bottom of the borehole and at the necessary depth when the sampling device is driven with the aid of a 140-pound hammer. The hammer is connected to the sampling device by drill rods. The

number of hammer blows, usually counted for each 6-inch increment of the total drive, indicates the compaction and density of the formation being penetrated.

The most commonly used soil sampling devices are the split-spoon sampler and the Shelby tube. The split-spoon sampler is a 12- or 18-inch long hollow cylinder consisting of two equal semicylindrical halves held together at each end with threaded couplings. The sampler is lowered to the bottom of the borehole and driven with a hammer to the necessary depth. When the sampler is brought to the surface, it is disassembled, or split, to remove the soil sample. Split-spoon sampling provides representative soil samples for physical or chemical testing. The samples, however, are disturbed; therefore, the results of the analysis should be used with caution. When the same sampler is used to collect different samples, the engineer should decontaminate it after each sampling event to prevent cross-contamination.

The Shelby tube sampler is a thin-walled tube made of steel, aluminum, brass, or stainless steel. The cutting edge of the tube is sharpened, and the upper end is attached to a coupling head by cap screws. The sampler must meet certain criteria, such as a clearance ratio of 0.5 to 1.5 and end area ratio of 10, to ensure the least disturbance to the sample (U.S. EPA 1993). The sample collection procedure is similar to split-spoon sampling except that the tube is pushed into the soil by the weight of the drill rig rather than driven. When the sampler is brought to the surface, the sample is sealed and preserved for laboratory analysis. Shelby tube sampling is used for soil analyses that require undisturbed soil samples, such as hydraulic conductivity testing.

Monitoring Well Installation

The installation of groundwater monitoring wells involves selecting the location and number of wells, drilling the boreholes, installing the casings and screens, placing the filter pack materials and annular seals, and finally developing the wells. Each of these steps should be designed to meet the objectives of the monitoring program and suit the conditions of the site. A typical monitoring well design is shown in Figure 9.15.3.

WELL LOCATION AND NUMBER

The selection of the proper number and locations of monitoring wells is obviously one of the most important decisions in any groundwater monitoring program or groundwater remedial investigation. For monitoring purposes, most guidances suggest a minimum of four monitoring wells per potential source of groundwater contamination (U.S. EPA 1986). Three of these wells are placed downgradient of the potential source, and one is placed upgradient. The purpose of these wells is to provide information on the background quality of the groundwater

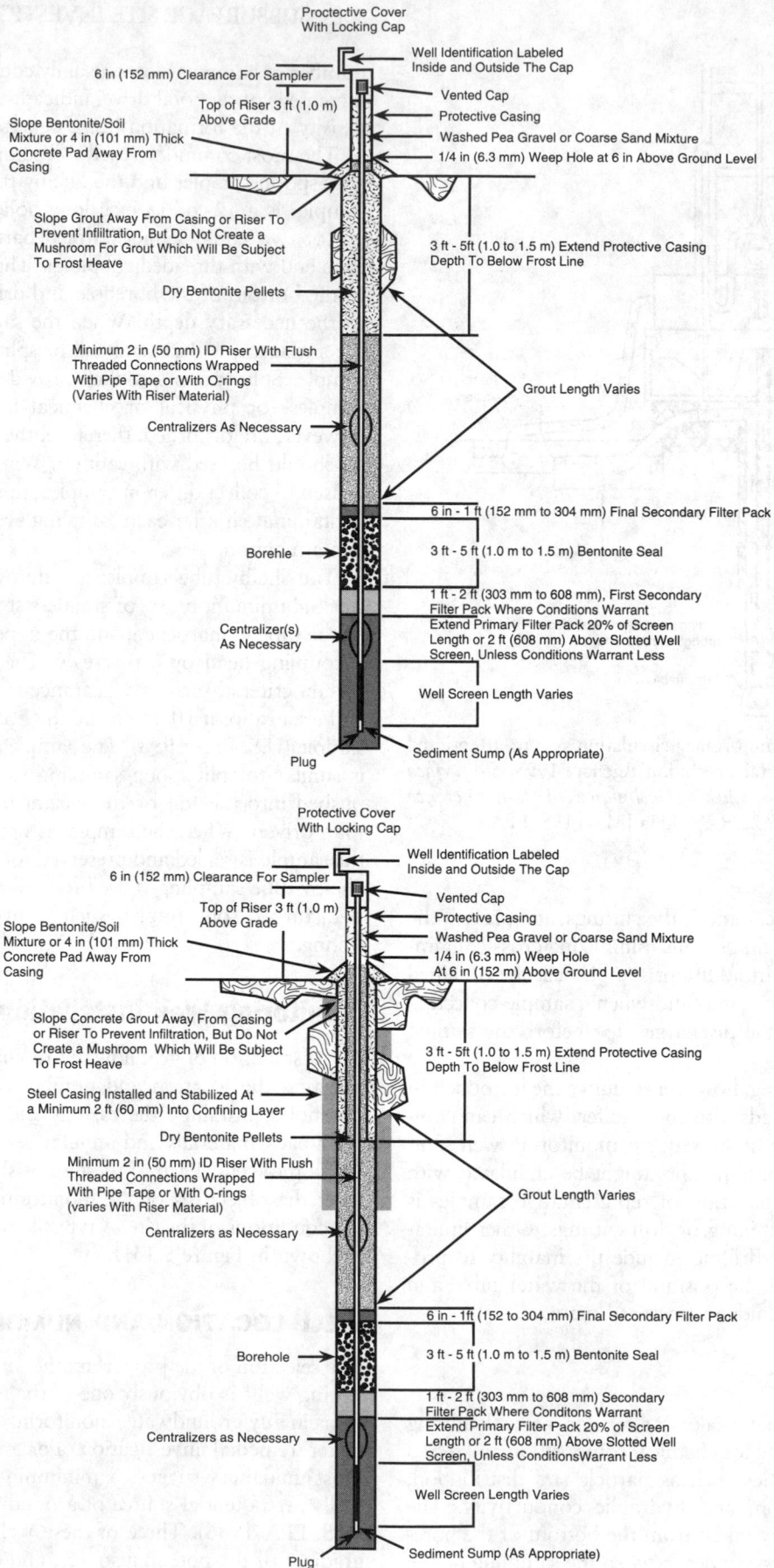

FIG. 9.15.3 Typical monitoring well design and construction detail. (Reprinted from U.S. Environmental Protection Agency, 1993, *Subsurface characterization and monitoring techniques, a desk reference guide*, Vol. 1, USEPA/625/R-93/003a [May] U.S. EPA.)

upgradient of the source and detect or monitor any contaminant plumes emanating from the source.

In a remedial investigation, however, the preliminary selection of the location and number of wells needed to delineate and monitor the plume is usually based on the results of initial field screening techniques such as gas surveys, HydroPunch, geophysical surveys, and borings. Environmental engineers use the data from these investigations to estimate the extent of the contaminant plume and establish the basic hydrogeologic parameters of the site. Once the hydrogeology of the site is understood and the migration path of the suspected contaminant plume is established, the location and number of wells can be finalized. In general, the more complicated the hydrogeology, the more complex the migration path of the contaminant plume and the greater the number of required monitoring wells (Barcelona et al. 1990).

CASINGS AND SCREENS

The purpose of the casing and screen in a groundwater monitoring well is to provide access from the surface to the groundwater in order to collect groundwater samples or measure groundwater elevations. The casing prevents geologic materials from collapsing into the borehole, while the screen allows groundwater to enter the monitoring well. The screen is generally attached to the subsurface end of the casing.

Several types of casing and screen materials are available including stainless steel, galvanized steel, carbon steel, PVC, Teflon, and aluminum. Selecting the monitoring well casing and screen material depends on the physical strength and chemical reactivity of the material under subsurface conditions. With regard to physical strength, the casing and screen should be able to withstand the forces exerted on them by the surrounding geologic materials. These forces can be significant for deep monitoring wells (greater than 30 meters). Nielson and Schalla (1991) provide data on the physical strength of different types of casing and screen materials.

With regard to chemical reactivity, the material of the casing and screen should neither adsorb nor leach chemical constituents which would bias the representativeness of the samples collected. In addition, the material must be durable enough to endure chemical attacks (corrosion or chemical degradation) from the natural chemical constituents or the contaminants in the groundwater. Teflon is probably the most chemically resistant material used in monitoring well installation, but the cost of Teflon is high (Barcelona et al. 1990). Stainless steel offers good strength and chemical resistance in most environments (except in highly acidic conditions), but it too is expensive. Galvanized steel is less expensive; however, it can impart iron, manganese, zinc, and cadmium to many waters. PVC has good chemical resistance except to low molecular weight

ketones, aldehydes, and chlorinated solvents (Miller 1982).

Two types of screens are commonly used in monitoring wells: machine-slotted pipes and continuous-clot wire-wound screens. Machine-slotted pipes are readily available and inexpensive, but the low amount of open area in these screens makes development of the well difficult. Continuous-slot screens, in contrast, are more efficient, but their cost is relatively high. The design of the slot size of the screen must be based on the characteristics of the filter pack material and the grain size of the stratum. The optimum slot size should provide maximum open area for water to flow through and minimum entry of fine particles into the well during piping (Nielson and Schalla 1991; Aller et al. 1991).

The depth of placement of the screen as well as its length are usually determined based on the depth and thickness of the water-bearing zone to be monitored. When the objective of the well is to monitor a potable water supply aquifer, then a longer screen, perhaps over the entire thickness of the aquifer, might be selected. On the other hand, when the objective of the well is to vertically delineate a plume, such as with cluster wells, then shorter screens at specific intervals might be selected.

The screen should be fully submerged to prevent contact between the contaminated groundwater and the atmosphere, particularly for volatile compounds. The screen is, however, extended above the water table for wells constructed to monitor floating products. In this case, the screen length and position must accommodate variations in water table elevation.

The casings are produced in various diameters (2, 4, 6, and 8 inches) and various lengths (5, 10, and 20 feet) that are joined by various coupling methods during installation. The casing diameter depends on the future use of the well, the type of pumping equipment, and the method of drilling. Small diameters (2 and 4 inches) are used for monitoring wells, while large diameters (6 and 8 inches) are used for recovery wells.

The casing must extend from the top of the screen to the ground surface level. The casing is protected at the surface by a metal protective casing or a manhole. Multiple casings are installed for wells penetrating more than one water-bearing formation. The purpose of multiple casings is to prevent a hydraulic connection and potential cross contamination between the water-bearing formations along the annular space produced by the installation of well casings. Figure 9.15.3 shows an example of a double-cased well installation where the outer casing is anchored into the confining layer before the borehole is advanced and the well is installed through the inner casing.

FILTER PACKS AND ANNULAR SEALS

Filter packs placed around the well screen allow groundwater to flow freely into the well while keeping fine particles from entering the well. Two types of filter packs are used in monitoring wells: naturally developed filter packs and artificially introduced filter packs. Naturally developed filter packs are produced in situ when the fine-grained materials around the screen are removed during the well development process. Environmental engineers construct artificial filter packs by backfilling the annular space surrounding the screen with a granular, relatively inert material such as clean silica sand.

In an artificially filter-packed well, the filter material can be selected for optimum efficiency of well operation, but the procedure of introducing the filter pack is time consuming and expensive. Furthermore, bridging can prevent complete filling around the well screen, and the filter pack material can introduce contaminants into the aquifer; a leaching test can determine whether this contamination is a problem. Naturally developed filter packs are, on the other hand, simpler, less expensive, and do not introduce new contaminants into the aquifer. However, well development for these filter packs is more difficult, and success is less assured.

Engineers can use a tremie pipe or a reverse circulation method to place the artificial filter pack. The tremie pipe method allows funneling of the material directly into the interval around the well screen. In a reverse circulation method, a mixture of sand and water is fed into the annulus around the screen, and the water entering the screen is pumped up to the surface through the casing. The engineer progressively pulls back the temporary casing (for hollow-stem augers) to expose the screen as the filter pack material builds up around the well screen.

Artificially introduced filter packs usually extend from the bottom of the screen to at least 3 to 5 feet above the bottom of the screen. This extension accounts for settlement of the filter pack material and allows a sufficient buffer zone between the well screen and the annular seal above.

After the filter pack is placed around the well screen, the engineer seals the annular space between the well casing and the formation to prevent upward or downward movement of water and contaminants along this pathway. In addition, the engineer places a surface seal of concrete around the protective casing to prevent surface drainage into the borehole. The annular seal is usually composed of bentonite or neat cement (Williams and Evans 1987). Bentonite is readily available and inexpensive but can cause constituent interference due to ion exchange. Neat cement is also readily available and inexpensive, but channeling between the casing and seal can develop due to temperature changes during the curing process (U.S. EPA 1993).

The engineer places the sealing mixture in the annular space using a side-discharge tremie pipe through which the grout is pumped from the surface. Complete sealing of the annular space is necessary to avoid potential bridging of the grout with formation material (Campbell and Lehr 1975).

WELL DEVELOPMENT

The purpose of well development is to remove the residues of drilling fluids and fine particles of filter packs so that subsequent sampling is representative of the groundwater. The development should be performed as soon as possible after the well is installed and the annular seal is cured.

Development methods include bailing, overpumping, air surging, and high-velocity jetting. In bailing, a bailer is dropped and retrieved in and out of the well causing an outward surge of water through the well screen and filter pack. Such surging forces the loosely bound fine particles through the screen and into the well where they can be removed by the bailer. Bailing has the advantage of being a simple technique which does not introduce new fluids into the aquifer. However, bailing is time consuming and ineffective in unproductive wells.

In overpumping, a submersible pump is lowered into the well and alternatively turned on and off, usually at a slightly higher rate than what the formation can deliver. This action, along with the repeated raising and lowering of the pump into the well, causes the water to move back and forth through the well screen, moving fine particles and drilling fluids into the well where they can be removed. Overpumping is convenient for small wells or poor aquifers; however, excessive pumping rates can cause well collapse, especially in deep wells.

Air surging consists of injecting compressed air in the well, causing the water column to lift almost to the surface, and shutting off the air supply to allow the column to fall back into the well. Repeated use of this technique causes an outward surging action in the well intake which forces the loosely bound fine particles through the screen and into the well where they can be removed. Environmental engineers must filter the injected air so that contaminants, such as lubricants of the compressor, are not introduced into the well.

High-velocity jetting uses nozzle devices to force a horizontal stream of water against the well screen opening. Engineers can remove the material that enters the screen in the backwash of the jet stream by pumping or bailing. High-velocity jetting is effective in removing the mud cake and breaking the bridges in the filter pack. However, this technique can introduce potential air and water contaminants to the aquifer.

Groundwater Sampling

The objective of any groundwater sampling program is to collect and analyze samples that are representative of existing groundwater conditions at the site. This goal is achieved with a sampling plan that incorporates sampling procedures designed to minimize sources of error or misrepresentation in each stage of the sampling process. The key stages of sampling involve well purging, sample collection and pretreatment, sample handling and preservation, and analysis and reporting of analytical data. A brief description of these sampling stages follows. Figure 9.15.4 presents a generalized flow diagram of groundwater sampling protocol.

PURGING

Before environmental engineers can sample a monitoring well, they must remove the water standing in the well to allow fresh water from the aquifer to enter the well. Purging is necessary because the stagnant water in the well is subject to chemical reactions from contact with well construction materials and the atmosphere for extended periods of time (Seanor and Brannaka 1983; Wilson and Dworkin 1984). The volume of water which should be removed from the well is based on the hydraulic characteristics of individual wells and geological settings (Gibb, Schuller, and Griffin 1981). A general rule is to remove three to five well volumes or to remove water until the water quality indicators, such as pH, conductance, and temperature are stable.

When purging a well, engineers should not allow the water level to drop below the level of the well screen to avoid aeration and loss of volatile or redox-sensitive compounds. In addition, the pumping rate should not exceed levels that might cause turbulent flow in the well and subsequent pressure changes and loss of dissolved gases (Meridith and Brice 1992). Overpumping can also dilute the sample or increase its turbidity because of the fine particles that may be drawn into the well.

Engineers should use the same equipment for purging and sampling to minimize the number of items that enter the well and therefore, the possibility of cross contamination. Furthermore, placing the purging device at the top of the well screen or at the top of the column of water ensures that all stagnant water is removed (Unwin and Huis 1983).

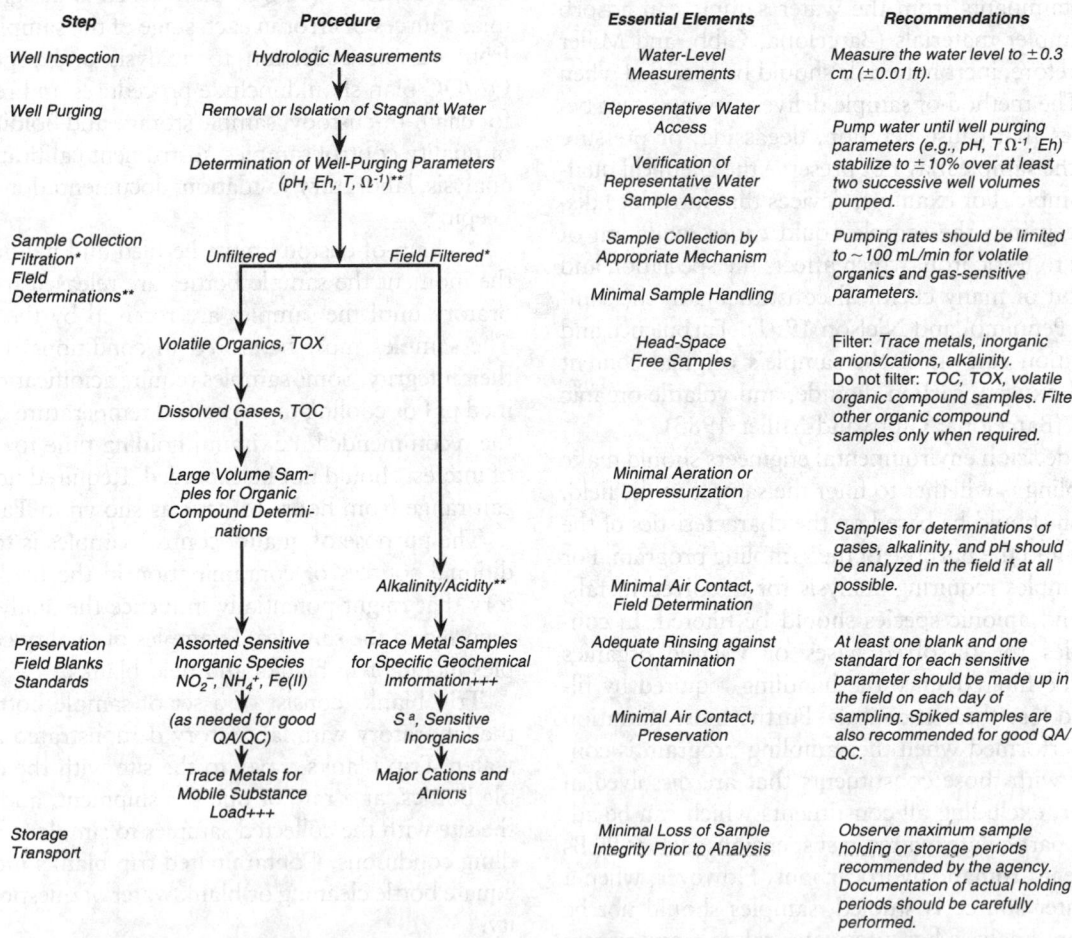

Step	Procedure		Essential Elements	Recommendations
Well Inspection	Hydrologic Measurements		Water-Level Measurements	Measure the water level to ±0.3 cm (±0.01 ft).
Well Purging	Removal or Isolation of Stagnant Water		Representative Water Access	Pump water until well purging parameters (e.g., pH, T Ω⁻¹, Eh) stabilize to ±10% over at least two successive well volumes pumped.
	Determination of Well-Purging Parameters (pH, Eh, T, Ω⁻¹)**		Verification of Representative Water Sample Access	
Sample Collection Filtration* Field Determinations**	Unfiltered Field Filtered*		Sample Collection by Appropriate Mechanism	Pumping rates should be limited to ~100 mL/min for volatile organics and gas-sensitive parameters.
			Minimal Sample Handling	
	Volatile Organics, TOX		Head-Space Free Samples	Filter: Trace metals, inorganic anions/cations, alkalinity. Do not filter: TOC, TOX, volatile organic compound samples. Filter other organic compound samples only when required.
	Dissolved Gases, TOC			
	Large Volume Samples for Organic Compound Determinations		Minimal Aeration or Depressurization	
				Samples for determinations of gases, alkalinity, and pH should be analyzed in the field if at all possible.
	Alkalinity/Acidity**		Minimal Air Contact, Field Determination	
Preservation Field Blanks Standards	Assorted Sensitive Inorganic Species NO₂⁻, NH₄⁺, Fe(II) Trace Metal Samples for Specific Geochemical Information+++		Adequate Rinsing against Contamination	At least one blank and one standard for each sensitive parameter should be made up in the field on each day of sampling. Spiked samples are also recommended for good QA/QC.
	(as needed for good QA/QC) Sᵃ, Sensitive Inorganics		Minimal Air Contact, Preservation	
	Trace Metals for Mobile Substance Load+++ Major Cations and Anions			
Storage Transport			Minimal Loss of Sample Integrity Prior to Analysis	Observe maximum sample holding or storage periods recommended by the agency. Documentation of actual holding periods should be carefully performed.

*   Denotes samples that should be filtered to determine dissolved constituents. Filtration should be accomplished preferably with inline filters and pump pressure or by N₂ pressure methods. Samples for dissolved gases or volatile organics should not be filtered. In instances where well development procedures do not allow for turbidity-free samples and may bias analytical results, split samples should be spiked with standards before filtration. Both spiked samples and regular samples should be analyzed to determine recoveries from both types of handling.

**   Denotes analytical determinations that should be made in the field.

+++   See Puls and Barcelona (1989).

FIG. 9.15.4 Generalized flow diagram of groundwater sampling protocol. (Reprinted from U.S. Environmental Protection Agency, 1993, *Subsurface characterization and monitoring techniques, a desk reference guide*, Vol. 1, USEPA/625/R-93/003a [May] U.S. EPA.)

COLLECTION AND PRETREATMENT

Groundwater samples can be collected with portable or dedicated in situ sampling equipment. Portable equipment includes bailers, syringes, suction-lift pumps, submersible pumps, and gas-driven devices. In situ sampling equipment includes cone penetrometer samplers (e.g., Hydropunch, BAT, CPT, or DMLS), chemical-sensitive probes, ion-selective electrodes, fiber-optic chemical sensors, multilevel capsule samplers, and multiport casings. A description of these types of equipment as well as their advantages and disadvantages is in the U.S. EPA desk reference guide on subsurface characterization and monitoring techniques (1993).

Selecting sampling equipment should be based on the purpose of the sampling as well as the construction materials of the sampling equipment and the method of sample delivery. The construction materials of the sampling device could affect the integrity of the sample because constituents can leach from the materials into the water samples or contaminants from the water sample can adsorb onto the sampler materials (Barcelona, Gibb, and Miller 1983). Therefore, inert materials should be specified when necessary. The method of sample delivery is important because devices that cause aeration, degassing, or pressure changes of the sample may not preserve the chemical quality of the sample. For example, devices that introduce dissolved oxygen into the sample could cause oxidation of ferrous iron to ferric iron, which affects the speciation and concentration of many chemical constituents in the sample (Hrzog, Pennimo, and Nielson 1991). Turbulence and depressurization can affect the sample's original content of dissolved oxygen, carbon dioxide, and volatile organic compounds (Barcelona, Gibb, and Miller 1983).

Another decision environmental engineers should make before sampling is whether to filter the sample in the field. This decision should be based on the characteristics of the constituents and the purpose of the sampling program. For example, samples requiring analysis for dissolved metals, alkalinity, and anionic species should be filtered. In contrast, samples for dissolved gases or volatile organics should not be filtered since the handling required by filtration could lose these chemicals. Furthermore, filtration should be performed when the sampling program is concerned only with those constituents that are dissolved in groundwater, excluding all constituents which can be adsorbed onto particulate matter in suspension, such as PCBs or polynuclear aromatic hydrocarbons. However, when a drinking water source is studied, samples should not be filtered before analysis because water taken from private wells is generally not filtered before use. In some instances, engineers must run parallel sets of filtered and unfiltered samples to determine the dissolved and adsorbed portions of the constituent of interest.

Filtration is accomplished by vacuum, pressure, or inline filtration devices. Stolzenburg and Nichols (1986) describe a variety of filtration equipment and their effects on sampling. The preferred device is the inline filter because it reduces the aeration and degassing of the sample as well as the potential of sample cross contamination caused by improper equipment decontamination.

To prevent cross contamination, engineers should decontaminate the equipment used for sample collection or pretreatment prior to and after each use. The decontamination should involve a minimum of scraping or brushing to remove any soil or residue from the device, washing with potable or deionized water, washing with detergents or cleaning fluids such as acetone, and pressure cleaning with a high-pressure steam cleaner.

QUALITY ASSURANCE AND QUALITY CONTROL

Groundwater sampling requires a quality assurance and quality control (QA/QC) plan which is designed to minimize sources of error in each stage of the sampling process, from sample collection to analysis and reporting. The QA/QC plan should include procedures and requirements for chain-of-custody, sample storage and holding time, use of quality control samples, instrument calibration, sample analysis, laboratory validation, documentation, and record keeping.

A chain-of-custody must be filed and maintained from the moment the sample bottles are released from the laboratory until the samples are received by the laboratory. The samples must be stored in conditions that preserve their integrity. Some samples require acidification to a specified pH or cooling to a specified temperature. In addition, the recommended maximum holding time for the analyte of interest should not be exceeded. Required holding times can range from hours to days as shown in Table 9.15.1.

The purpose of quality control samples is to detect additional sources of contamination in the field or laboratory that might potentially influence the analytical values reported in the samples. Examples of quality control samples include trip blanks and field blanks.

Trip blanks consist of a set of sample bottles filled at the laboratory with laboratory demonstrated analyte-free water. Trip blanks travel to the site with the empty sample bottles, at a rate of one per shipment, and back from the site with the collected samples to simulate sample handling conditions. Contaminated trip blanks indicate inadequate bottle cleaning or blank water of questionable quality.

Field blanks serve the same purpose as trip blanks but are also used to indicate potential contamination from ambient air or sampling instruments. At the field location, analyte-free water is passed through clean sample equipment and placed in an empty sample container for analysis. Therefore, by being opened in the field and transferred over a cleaned sampling device, the field blank can indi-

TABLE 9.15.1 REQUIRED HOLDING TIMES FOR SEVERAL ANALYTES

Parameters (Type)	Volume Required (mL) 1 Sample[a]	Containers (Material)	Preservation Method	Maximum Holding Period
Well purging				
pH (grab)	50	T,S,P,G	None; field det.	<1 hr[b]
Ω^4 (grab)	100	T,S,P,G	None; field det.	<1 hr[b]
T (grab)	1000	T,S,P,G	None; field det.	None
Eh (grab)	1000	T,S,P,G	None; field det.	None
Contamination Indicators				
pH, Ω^4 (grab)	As above	As above	As above	As above
TOC	40	G,T	Dark, 4°C	24 hr
TOX	500	G,T	Dark, 4°C	5 days
Water quality				
Dissolved gases ($O_2CH_2CO_2$)	10 mL minimum	G,S	Dark, 4°C	<24 hr
Alkalinity/acidity	100	T,G,P	4°C/None	<6 hr[b] <24 hr
	Filtered under pressure with appropriate media			
(Fe, Mn, Na$^+$, K$^+$, Ca^{++}, Mg^{++})	All filtered 1000 mL[f]	T,P	Field acidified to pH <2 with HNO_2	6 months[c]
(PO_4^-, Cl_3^- Silicate)	@50	(T,P,G glass only)	4°C	24 hr[f] 7 days[c], 7 days
NO_3^-	100	T,P,G	4°C	24 hr[d]
SO_4^-	50	T,P,G	4°C	7 days[c]
OH_4^+	400	T,P,G	4°C/H_2SO_4 to pH <2	24 hr[f] 7 days
Phenols	500	T,G	4°C/H_2PO_4 to pH <4	24 hours
Drinking Water suitability As, Ba, Cd, Cr, Pb, Hg, Se, Ag	Same as above for water quality cations (Fe, Mn, etc.)[f]	Same as above	Same as above	6 months
F	Same as chloride above	Same as above	Same as above	7 days
Remaining organic	As for TOX/TOC, except where analytical parameters method calls for acidification of sample			24 hours

Source: U.S. Environmental Protection Agency, 1993, *Subsurface characterization and monitoring techniques, a desk reference guide,* Vol. 1, USEPA 625/R-93/003a, May (U.S. EPA).

T = Teflon; S = stainless steel; P = PVC, polypropylene, polyethylene; G = borosilicate glass.

[a]It is assumed that at each site, for each sampling date, replicates, a field blank, and standards must be taken at equal volume to those of the samples.

[b]Temperature correction must be made for reliable reporting. Variations greater than ±10% can result from a longer holding period.

[c]In the event that NHO_2 cannot be used because of shipping restrictions, the sample should be refrigerated to 4°C, shipped immediately, and acidified on receipt at the laboratory. Container should be rinsed with 1:1 HNO_3 and included with sample.

[d]28-day holding time if samples are preserved (acidified).

[e]Longer holding times in U.S. EPA (1986b).

[f]Filtration is *not* recommended for samples intended to indicate the mobile substance lead. See Puis and Barcelona (1989a) for more specific recommendations for filtration procedures involving samples for dissolved species.

cate ambient and equipment conditions that can potentially affect the quality of the associated samples.

—*Ahmed Hamidi*

References

Aller, L. et al. 1991. *Handbook of suggested practices for the design and installation of groundwater monitoring wells.* EPA/600/4-89/034.

Barcelona, M. et al. 1990. Contamination of groundwater: Prevention, assessment, restoration. *Pollution Technology Review* 184. Park Ridge, N.J.: Noyes Data Corporation.

Barcelona, M.J., J.P. Gibb, and R.A. Miller. 1983. *A guide to the selection of materials of monitoring well construction and groundwater sampling.* Illinois State Water Survey Report 327.

Campbell, M.D., and J.H. Lehr. 1975. Well cementing. *Water Well Journal* 29, no. 7:39–42.

Davis, H.E., J. Jehn, and S. Smith. 1991. Monitoring well drilling, soil sampling, rock coring, and borehole logging. In *Practical handbook of groundwater monitoring,* edited by David Nielsen. Chelsea, Mich.: Lewis Publishers.

Edge, R.W., and K. Cordy. 1989. The HydroPunch: An in situ sampling tool for collecting groundwater from unconsolidated sediments. *Groundwater Monitoring Review* (summer):177–183.

Gibb, J.P., R.M. Schuller, and R.A. Griffin. 1981. *Procedures for the collection of representative water quality data from monitoring wells.* Cooperative Groundwater Report 7, Illinois State Water and Geological Surveys.

Hrzog, B., J. Pennimo, and G. Nielson. 1991. Groundwater sampling. In *Practical handbook of groundwater monitoring,* edited by David Nielsen. Chelsea, Mich.: Lewis Publishers.

Meridith, D.V., and D.A. Brice. 1992. Limitations on the collection of representative samples from small diameter monitoring wells. Groundwater Management II (6th NOAC): 429–439.

Miller, G.D. 1982. Uptake and release of lead, chromium, and trace level volatile organics exposed to synthetic well castings. Proceedings of Second National Symposium on Aquifer Restoration and Groundwater Monitoring, 26–28 May, Columbus, Ohio: NWWA.

Nielson, D.M., and R. Schalla. 1991. Design and installation of groundwater monitoring wells. In *Practical handbook of groundwater monitoring,* edited by David Nielsen. Chelsea, Mich.: Lewis Publishers.

Seanor, A.M., and L.K. Brannaka. 1983. Efficient sampling techniques. *Groundwater Age* 17, no. 8:41–46.

Stolzenburg, T.R., and D.G. Nichols. 1986. Effects of filtration methods and sampling on inorganic chemistry of sampled well water. Proceedings of the Sixth National Symposium and Exposition on Aquifer Restoration and Groundwater Monitoring, 216–234. Dublin, Ohio: National Water Well Association.

U.S. Environmental Protection Agency. 1986. *RCRA groundwater monitoring technical enforcement guidance document.* OSWER-9950.1. Washington, D.C.: U.S. Government Printing Office.

———. 1993. *Subsurface characterization and monitoring techniques, a desk reference guide,* Vol. 1. USEPA/625/R-93/003a (May). U.S. EPA.

Unwin, J.P., and D. Huis. 1983. A laboratory investigation of the purging behavior of small diameter monitor well. Proceedings of the Third Annual Symposium on Groundwater Monitoring and Aquifer Restoration, 257–262. Dublin, Ohio: National Water Well Association.

Williams, C., and L.G. Evans. 1987. Guide to the selection of cement, bentonite and other additives for use in monitor well construction. Proceedings of First National Outdoor Action Conference, 325–343. Dublin, Ohio: National Water Well Association.

Wilson, L.G., and J.M. Dworkin. 1984. Development of a primer on well water sampling for volatile organic substances. Bk. 1, Chap. D-2 of *U.S. Geological Survey techniques of water resources investigations.*

Groundwater Cleanup and Remediation

9.16
SOIL TREATMENT TECHNOLOGIES

Restoration or cleanup of a contaminated aquifer usually involves also addressing the contaminated soils at the vadose zone. The residual contaminants in the vadose need to be treated or removed to prevent the continuous feed of contaminants to groundwater due to leaching, rainfall percolation, or groundwater table fluctuations. Several techniques are available to treat contaminants in the vadose zone. These techniques include excavation and removal, physical treatments, biological treatments, thermal treatments, and stabilization treatments. Selecting the appropriate method depends on the volume of soils to be handled, the type of soils and contaminants, the regulatory requirements, and costs.

Excavation and Removal

Excavation and soil removal is one of the most common activities in groundwater remediation and cleanup. Excavation involves removing contaminated soil from the unsaturated zone to prevent further groundwater contamination by the residuals present in that zone. Excavation is often used at sites where site conditions preclude onsite treatment, stabilization, or capping of the contaminated unsaturated zone. Factors affecting excavation include the volume of soils to be handled, the location of the area to be excavated, the type of soils and contaminants, and the regulatory requirements. The excavated material is often disposed of at a permitted landfill or treated and reused.

Physical Treatment

The physical treatments for treating contaminants in the vadose zone include soil–vapor extraction, soil washing, and soil flushing.

SOIL–VAPOR EXTRACTION

This treatment technology removes volatile compounds from the vadose zone. Airflow is injected through extraction wells creating a vacuum and a pressure gradient that induces volatiles to diffuse through the extraction wells. The volatiles are collected as gases and treated aboveground. The technology is effective for halogenated volatiles and fuel hydrocarbons (U.S. EPA 1991a). The technology is also cost effective when large volumes of soil are involved; since treatment takes place onsite, the risks and costs associated with transporting large volumes of contaminated soils are eliminated.

The technology, however, is less effective in soils with low air permeability, low temperatures, or high carbon content. In addition, although this technology reduces the volume of the contaminants, the toxicity of the contaminants is not reduced.

SOIL WASHING

Soil washing removes adsorbed contaminants from soil particles. The process involves excavating the contaminated soil and washing it with a leaching agent, a surfactant, or chelating agency or adjusting the pH (U.S. EPA 1990c). Sometimes extraction agents are added to enhance the process. The process reduces the volume of contaminant; however, residual suspended solids and sludges from the process may need further treatment since they contain a higher concentration of contaminant than the original. The technology is effective for halogenated semivolatiles, fuel hydrocarbons, and inorganics (U.S. EPA 1993a).

The technology, however, is less effective when the soil contains a high percentage of silt and clay particles or high organic content. In addition, this technology reduces the volume of the contaminants, but the toxicity of the contaminants is unchanged.

SOIL FLUSHING

Soil flushing is an in situ process whereby environmental engineers apply a water-based solution to the soil to enhance the solubility of the contaminant (U.S. EPA 1991b). The water-based solution is applied through injection wells or shallow infiltration galleries. The contaminants are mobilized by solubilization or through chemical reactions with the added fluid. The generated leachate must be intercepted by extraction wells or subsurface drains and pumped to the surface for aboveground treatment. The technology is effective for nonhalogenated volatile organics and for soils with high permeability.

The technology, however, is less effective for soils with low permeability or with particles that strongly adsorb contaminants such as clays. In addition, special precautions are necessary to prevent groundwater contamination.

Biological Treatment

Slurry biodegradation, ex situ bioremediation and land farming, and in situ biological treatment are biological treatments for treating contaminants in the vadose zone. These treatment are discussed next.

SLURRY BIODEGRADATION

The slurry biodegradation process involves excavating the contaminated soil and mixing it in an aerobic reactor with water and nutrients. This process maximizes the contact between the contaminants and the microorganisms capable of degrading those contaminants. The temperature in the reactor is usually maintained at an appropriate level, and neutralizing agents are often added to adjust the pH to an acceptable range (U.S. EPA 1990b). After the treatment is complete, the slurry is dewatered, and the soil can be redeposited on site. This technology is effective for soils contaminated with fuel hydrocarbons (U.S. EPA 1993a). In addition, the contaminants can be completely destroyed and the soil reused.

The technology, however, is less effective for contaminants with low biodegradability. In addition, the presence of chlorides or heavy metals as well as some pesticides and herbicides in the soil can reduce the effectiveness of the process by inhibiting the microbial action.

EX SITU BIOREMEDIATION AND LANDFARMING

This process involves excavating the contaminated soil, piling it in biotreatment cells, and periodically turning it over to aerate the water (U.S. EPA 1993a). The moisture, heat, nutrients, oxygen, and pH are usually controlled in the process. In addition, volatile emissions as well as leachate from the biotreatment cells should be controlled. The technology is effective for soils contaminated with fuel hydrocarbons. Also, the contaminants can be completely destroyed and the soil reused.

IN SITU BIOLOGICAL TREATMENT

This process enhances the naturally occurring biological activities in the contaminated subsurface soil. Circulating either a nutrient and oxygen-enriched water-based solution or a forced air movement which provides oxygen in the soil enhances the naturally occuring microbes (U.S. EPA 1991a). In the latter process, also called *bioventing*, the air flow rate is lower than in vapor extraction since the objective is to deliver oxygen while minimizing volatilization and the release of contaminants to the atmosphere. The technology is effective for nonhalogenated volatiles and fuel hydrocarbons. In addition, the contaminant toxicity is reduced or even eliminated. The technology, however, is less effective for nonbiodegradable compounds and for soils with low permeability.

Thermal Treatment

Thermal treatment is used to treat contaminants in the vadose zone and includes incineration and thermal desorption. A brief description of these processes follows.

INCINERATION

Incineration is a process whereby organic compounds in contaminated soil are destroyed in the presence of oxygen at high temperatures (U.S. EPA 1990a). Three common types of incinerators are rotary kilns, circulating fluidized beds, and infrared incinerators. The excavated contaminated soil is fed into the incinerator and incinerated at temperatures ranging from 1600 to 2200°F. Because the residual ash may contain residual metals, it must be disposed of in accordance with appropriate regulations. In addition, the generated flue gases must be handled with appropriate air pollution control equipment.

Incineration is potentially effective for halogenated and nonhalogenated volatiles as well as fuel hydrocarbons and pesticides. Most organic contaminants are destroyed by this technology; however, metals are not destroyed and end up in the flue gases or the ash. In addition, certain types of soils such as clay soils or soil containing rocks may need screening prior to incineration.

THERMAL DESORPTION

Thermal desorption is a physical separation process in which the excavated contaminated soil is heated to a temperature at which the water and organic contaminants are volatilized (U.S. EPA 1991d). The volatilized contaminants are then sent to a gas treatment system. Low-temperature thermal desorption is potentially effective for halogenated semivolatiles, nonhalogenated volatiles, and pesticides (U.S. EPA 1993a). High-temperature thermal desorption

is effective for halogenated volatiles and semivolatiles as well as fuel hydrocarbons.

The contaminants, however, are not destroyed by this technology and require further gas treatment. In addition, the technology is less effective for tightly aggregated soils or those containing large rock fragments.

Stabilization and Solidification Treatment

Stabilization and vitrification treatments are also used to treat contaminants in the vadose zone. These treatments are described next.

STABILIZATION

The soil stabilization process can be used in either in situ or ex situ treatment. The process involves mixing the contaminated soil with binding materials such as cement, lime, or thermoplastic binders to bind the contaminants to the soil and reduce their mobility (U.S. EPA 1993b). Depending on the process and binding material, the final product ranges from a loose, soil-like material to concrete-like molded solids. Pretreatment is usually required for soils with high contents of oil and grease, surfactants, or chelating agents. The process is effective for soils, sludges, or slurries contaminated with inorganics.

The technology, however, is not effective for soils contaminated with organics or soils with high water or clay content. Organics, sulfates, or chlorides can interfere with the curing of the solidified product. Clay can interfere with the mixing process, adsorbing the key reactants and interrupting the polymerization chemistry of the solidification agents. Furthermore, the stabilization process increases the volume of treated soil since reagents are added.

VITRIFICATION

Soil vitrification is used in both in situ and ex situ treatment. The process involves inserting large graphite electrodes into the soil and applying a high current of electricity to the electrodes (U.S. EPA 1992). The electrodes are typically arranged in 30-foot squares and connected by graphite on the soil surface. The heat causes a melt that gradually works downward through the soil incorporating inorganic contaminants into the melt and paralyzing organic components. After the process is complete and the ground has cooled, the fused waste material is dispersed in a chemically inert, stable, glass-like product with low leaching characteristics.

The technology is potentially effective for halogenated and nonhalogenated volatiles and semivolatiles as well as fuel hydrocarbons, pesticides, and inorganics. The process reduces the mobility of the contaminants, and the vitrified mass resists leaching for geological time periods. The technology, however, is energy-intensive, and the off-gases must be collected and treated before release.

—*Ahmed Hamidi*

References

U.S. Environmental Protection Agency. 1990a. *Mobile/transportable incineration treatment.* EPA/540/2-90/014. Washington, D.C.: U.S. EPA.

———. 1990b. *Slurry biodegradation.* EPA/540/2-90/016 (September). Washington, D.C.: U.S. EPA.

———. 1990c. Soil washing treatment. *Engineering Bulletin* EPA/540/2-90/017 (September). Washington, D.C.: U.S. EPA.

———. 1991a. *Bioremediation in the field.* EPA/540/2-91/018. Washington, D.C.: U.S. EPA.

———. 1991b. *In situ soil flushing.* EPA/540/2-91/021. Washington, D.C.: U.S. EPA.

———. 1991c. *In-situ soil vapor extraction treatment.* EPA/540/2-91/006 (May). Washington, D.C.: U.S. EPA.

———. 1991d. *Thermal desorption treatment.* EPA/540/2-91/008 (May). Washington, D.C.: U.S. EPA.

———. 1992. *Vitrification technologies for treatment of hazardous and radioactive waste.* EPA/625/R-92/002 (May). Washington, D.C.: U.S. EPA.

———. 1993a. *Remediation technologies screening matrix and reference guide.* EPA/542/B-93/005 (July). Washington, D.C.: U.S. EPA.

———. 1993b. *Solidification/stabilization of organics and inorganics.* EPA/540/S-92/015 (May). Washington, D.C.: U.S. EPA.

9.17
PUMP-AND-TREAT TECHNOLOGIES

Pump-and-treat systems consist of a groundwater withdrawal system and an aboveground treatment system. The groundwater withdrawal system, also called the *containment system*, includes pumping wells or subsurface drains designed to remove the contaminants from the groundwater system and control the plume from further migration. In some cases, injection wells are used to inject treated water back into the aquifer. Aboveground treatment systems include chemical, physical, and biological treatment technologies.

Withdrawal and Containment Systems

As previously stated, the withdrawal and containment systems include well systems and subsurface drains. A description of these systems follows.

WELL SYSTEMS

Well systems remove contaminants from groundwater and stop the plume from further migration by manipulating the subsurface hydraulic gradients. Three general classes of well systems are well points, deep wells, and injection wells. Well points use suction lifting as the standard technique for pumping water; therefore, they can be used only for shallow aquifers where the suction lifting is less than 25 feet. Figure 9.17.1 shows several closely spaced point wells connected to a centrally located suction lift pump through a single main header pipe. Deep-well systems are used for greater depths and are usually pumped individually by submersible pumps. Dual pumps are used for floating product recovery as shown in Figure 9.17.2. In injection wells, the injection of clean or treated water into the aquifer flashes the aquifer or forms a barrier to groundwater flow.

Design Considerations

The design of a well system involves determining the number of wells needed, placing and spacing the wells, and determining the pumping cycles and rates of the wells. The number and spacing of the wells should completely capture the plume of contamination and produce as little uncontaminated water as possible to reduce treatment costs. In addition, the well's capture zones should intersect each other to prevent dead spots where contaminants stay stagnant or routes where the contaminant can escape the zone of capture. Environmental engineers determine the zone of capture by plotting the drawdowns within the radius of influence of each well on the potentiometric surface map of the site and calculating the cumulative drawdowns. The radius of influence of each well is determined by pumping test analysis as discussed in Section 9.8 or estimated from the following formulas when pumping test data are lacking (Kuffs et al. 1983):

Equilibrium:

$$R_0 = 3(H - h_w)(0.47K)^{1/2} \qquad \text{9.17(1)}$$

Nonequilibrium:

$$R_0 = r_w + (Tt/4790\ S)^{1/2} \qquad \text{9.17(2)}$$

where:

R_0 = radius of influence, ft
K = permeability, gpd/ft^2
H = total head, ft
h_w = head in well, ft

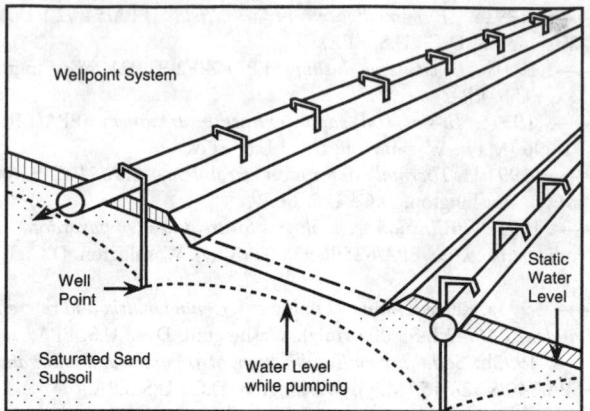

FIG. 9.17.1 Suction lift and a series of point wells. (Reprinted from S. Sommer and J.F. Kichens, 1980, *Engineering and development support of general decon technology for the DARCOM installation and restoration program, Task 1: Literature review on groundwater containment and diversion barriers*, Draft report by Atlantic Research Corp. to U.S. Army Hazardous Materials Agency, Contract No. DAK 11-80-C-0026, [October], Aberdeen Proving Ground.)

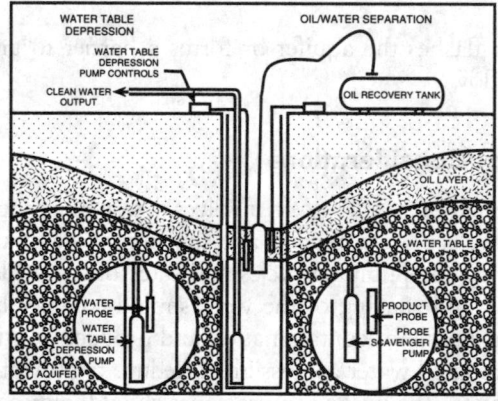

FIG. 9.17.2 Dual pumping wells. (Reprinted from E.K. Nyer, 1992, *Groundwater treatment technology*, 2d ed., New York: Van Nostrand Reinhold.)

Q = pumping rate, gpm
r_w = well radius, ft
T = transmissivity, gpd/ft
t = time, min
S = storage coefficient, dimensionless

Methods of Construction

The construction of a well system involves setting up the drilling equipment, drilling the well hole, installing casings and liners, grouting and sealing annular spaces, installing well screens and fittings, packing gravel and placing material, and developing the well. Detailed discussions on these aspects can be found in Johnson Division, UOP, Inc. (1975). In addition, Figure 9.17.3 shows typical well construction detail. In recent years, several innovative well installation techniques have been developed including in-

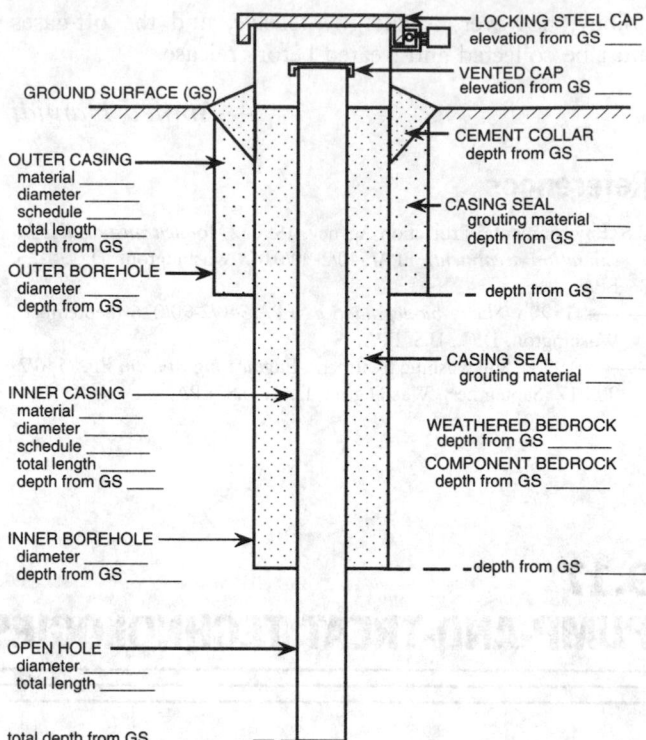

FIG. 9.17.3 Typical well installation.

stalling horizontal wells which act as subsurface drains but require less soil excavation and disturbances (Oakley et al. 1994).

Operation and Maintenance

Equilibrium pumping is often used for plume management; however, nonequilibrium pumping has advantages in cases of floating and sinking plumes and can be used to flush sorbed contaminants associated with the residual phase. Pulsed pumping of recovery wells can be used to washout residuals from unsaturated zones, allow contaminants to diffuse out of low permeability zones, and flush and bring stagnant zones into active flow paths. Pulsed pumping, however, incurs additional costs and concerns that must be evaluated for site specific conditions (U.S. EPA 1989).

Cost

The costs of well systems vary from site to site depending on the geology, the depth of the aquifer, the extent and type of contamination, the periods and durations of pumping, and the electrical power requirements. A cost analysis study for a variety of well systems can be found in Cambel and Lehr (1977) and in Powers (1981).

Advantages and Limitations

Well system technology is an efficient and effective means of assuring groundwater pollution control. Wells can be readily installed, or previously installed monitoring wells can sometimes be used as part of a well system. In addition, the technology provides high-design flexibility, and the construction costs can be lower than artificial barriers. However, wells require continued maintenance and monitoring after installation; therefore, operation and maintenance costs can be high. In addition, the application of this technology to fine soils is limited due to the low yield and small radius of influence in these soils.

SUBSURFACE DRAINS

Subsurface drains involve excavating a trench and placing a perforated pipe and coarse material such as gravel in the trench. The drain usually drains by gravity to a sump where the water is pumped to the surface for treatment. Subsurface drains essentially function like an infinite line of extraction wells, creating a continuous zone of depression in which groundwater flows towards the drain. Two types of subsurface systems are relief drains and interceptor drains. The major difference between these drains is that the drawdown created by an interceptor drain is proportional to the hydraulic gradient, whereas the drawdown created by a relief drain is a function of the hydraulic conductivity and depth to the impermeable layer below the drain.

Environmental engineers use *relief drains* primarily to lower the water table and prevent its contact with waste material or to contain a plume in place and prevent contamination from reaching a deeper aquifer. Relief drains can be installed in parallel on either side of a waste site or completely around the perimeter of the waste site as shown in Figure 9.17.4. The areas of influence of relief drains should overlap to prevent the contaminated groundwater from escaping between the drain lines.

Engineers use *interceptor drains* to intercept a plume hydraulically downgradient from its source and prevent the contamination from reaching wells and surface water located downgradient from the site. Interceptor drains are installed perpendicular to groundwater flow and downgradient of the plume of contamination as shown in Figure 9.17.5. In some cases, engineers use interceptor drains in conjunction with a barrier wall to prevent infiltration of clean water from downgradient of the drain thereby reducing treatment costs (see Figure 9.17.6). A series of interceptor drains or collector pipes (laterals) can be connected to a main pipe (header) as shown in Figure 9.17.7.

Design Considerations

The primary design components of a subsurface drain system are (1) the location of the drains, (2) the spacing of

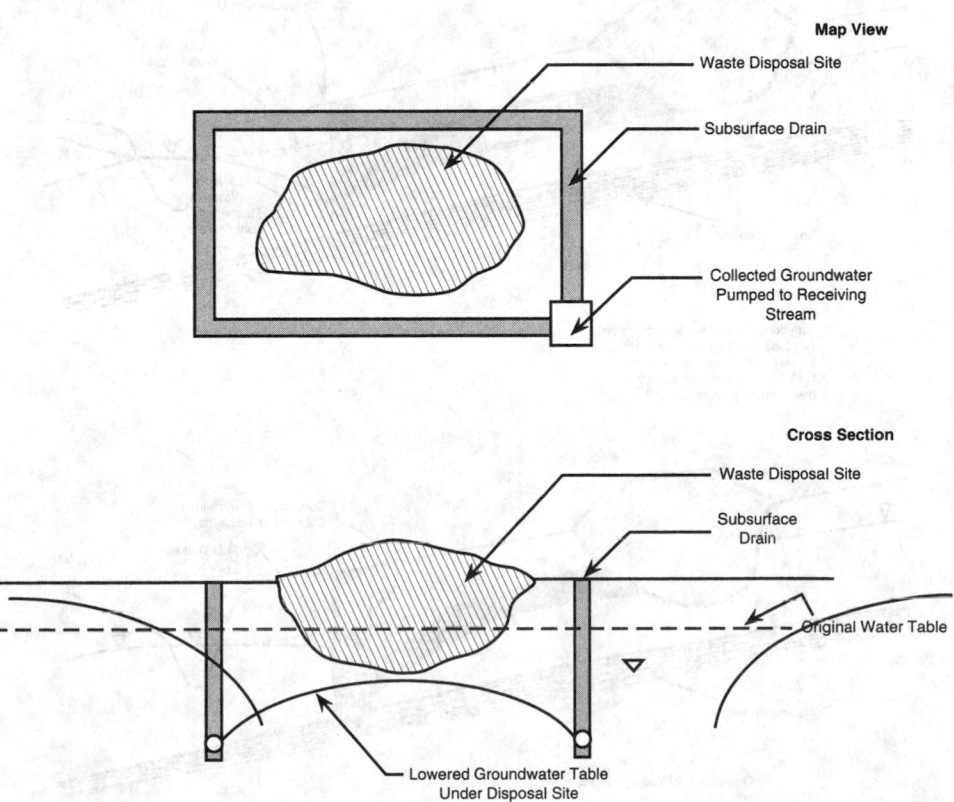

FIG. 9.17.4 Relief drains around the perimeter of a waste site. (Reprinted from U.S. Environmental Protection Agency, 1985, *Leachate plume management,* EPA/540/2-85/004, Washington, D.C.: U.S. EPA.)

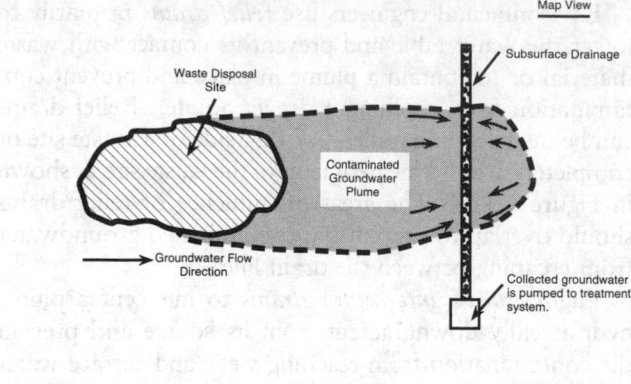

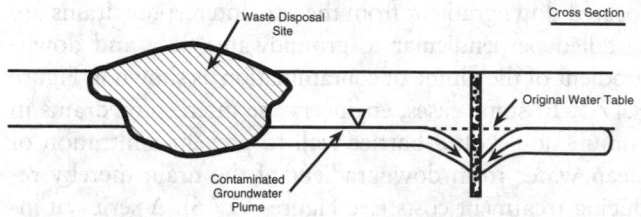

FIG. 9.17.5 Interceptor drains downgradient of the plume of contamination. (Reprinted from U.S. Environmental Protection Agency, 1985, *Leachate plume management*, EPA/540/2-85/004, Washington, D.C.: U.S. EPA.)

the drains, (3) the pipe diameter and slope, and (4) the envelope and filter materials around the pipe.

An interceptor drain should be installed perpendicular to the groundwater flow direction and downgradient from the plume of contamination. The drain should be installed on top of a layer of low hydraulic permeability to prevent underflow beneath the drain. The location of the drain should be selected so that the upgradient and downgradient influences of the drain completely capture the contamination plume. The upgradient and downgradient influences of an interceptor drain can be calculated using the following equations described by Van Hoorn and Vandemolen (1974) and Kuffs (1983):

$$D_u = 1.33m_sI \qquad 9.17(3)$$

$$D_d = \frac{KI}{Q} \cdot (d_e - h_d - D_2) \qquad 9.17(4)$$

where:

D_u = effective distance of drawdown upgradient, ft
m_s = saturated thickness of aquifer not affected by drainage, ft
I = hydraulic gradient
D_d = downgradient influence, ft
K = hydraulic conductivity, ft/day
Q = drainage coefficient, ft/day

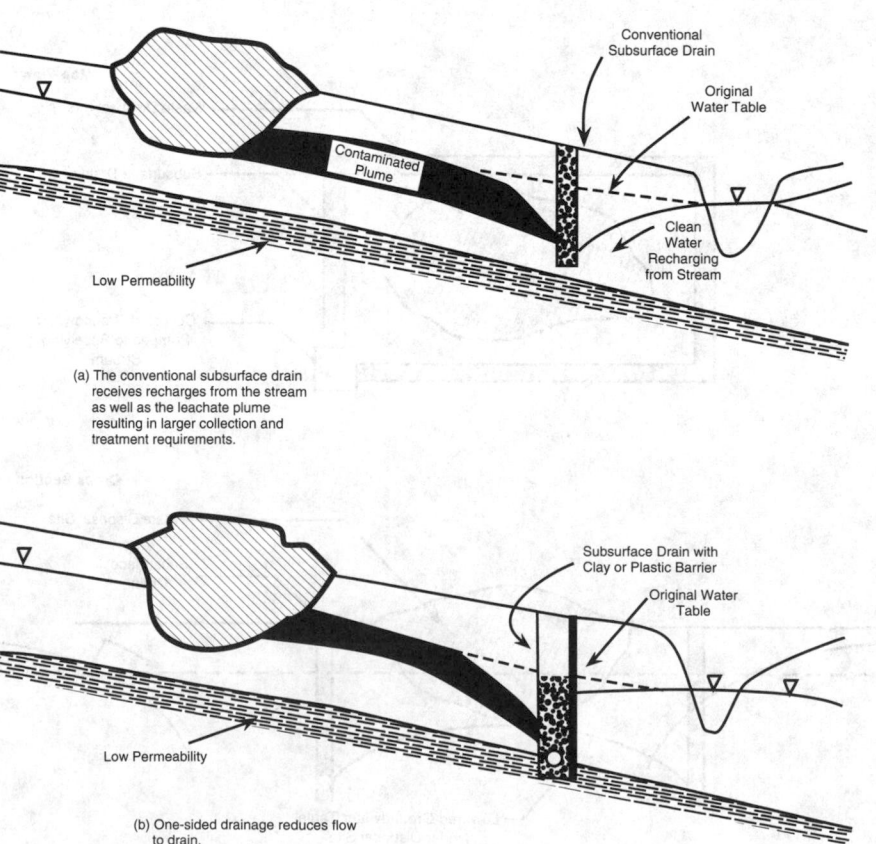

(a) The conventional subsurface drain receives recharges from the stream as well as the leachate plume resulting in larger collection and treatment requirements.

(b) One-sided drainage reduces flow to drain.

FIG. 9.17.6 Interceptor drains in conjunction with a barrier wall. (Reprinted from U.S. Environmental Protection Agency, 1985, *Leachate plume management*, EPA/540/2-85/004, Washington, D.C.: U.S. EPA.)

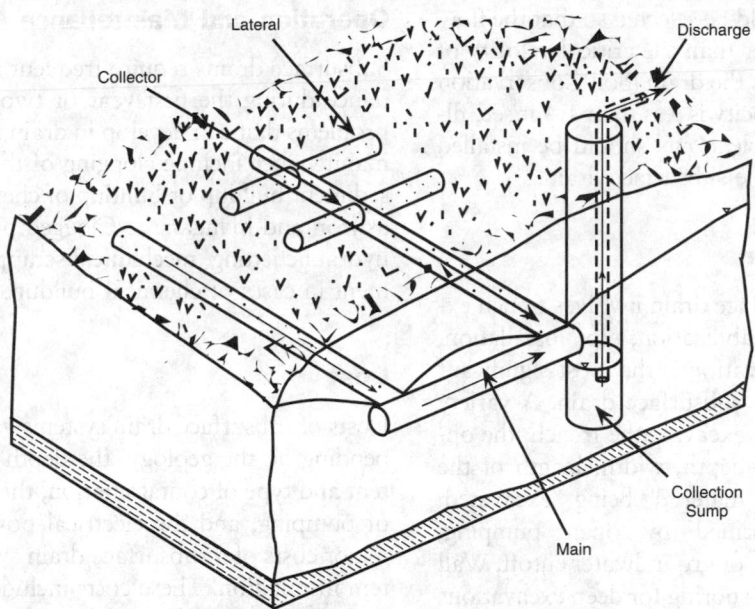

FIG. 9.17.7 Interceptor drains connected to a header. (Reprinted from U.S. Environmental Protection Agency, 1985, *Leachate plume management*, EPA/540/2-85/004, Washington, D.C.: U.S. EPA.)

d_e = depth of drain, ft

h_d = depth of drawdown, ft

D_2 = distance from ground surface to water table prior to drainage at the distance D_d downgradient from the drain, ft

The spacing between two parallel relief drains should be selected so that their combined drawdown is adequate to lower the water table beneath the waste. The minimum spacing, however, is often imposed by the boundaries of the waste material. The drain spacing depends on the hydraulic conductivity of the aquifer, the depth of the impermeable layer beneath the drain, the cross-sectional area of the drain, the water level in the drain, and precipitation and other sources of recharge. The spacing between two parallel drains resting on an impermeable barrier can be calculated with the use of the Wasseling (1973) equation as

$$L = \left(\frac{8KDH + 4KH^2}{Q}\right)^{0.5} \qquad 9.17(5)$$

where:

L = drain spacing, ft

K = hydraulic conductivity, ft/sec

D = distance between the water level in the drain and the impermeable layer, ft

H = height of the water table above the water level in the drain midway between the two drains, ft

Q = drain drainage rate per unit surface area, ft/sec

For a two-layered soil, Hooghoudt, as described by Wasseling (1973), developed a modified equation of

$$L = \left(\frac{8K_1DH + 4K_2H^2}{Q}\right)^{0.5} \qquad 9.17(6)$$

where K_1 and K_2 are the hydraulic conductivities above and below the drain, and d is the equivalent depth of the aquifer below the drain as illustrated in Figure 9.17.8. Using this equation involves either a trial-and-error procedure or the use of monographs which have been developed specifically for equivalent depth and drain spacing (U.S. EPA 1985b; Repa et al. 1982). Other equations for different subsurface configurations are available in Cohen and Miller (1983).

The diameter of the pipe can be calculated with the use of Manning's equation, assuming that the carrying capacity of the pipe is equal to the design seepage. The resulting equation (Luthin 1957) is

$$d = 0.892(q \cdot A)^{0.375} \cdot A^{-0.1875} \qquad 9.17(7)$$

where:

d = inside diameter of pipe, in

A = drainage area, acres

q = seepage coefficient, in/day

I = hydraulic gradient

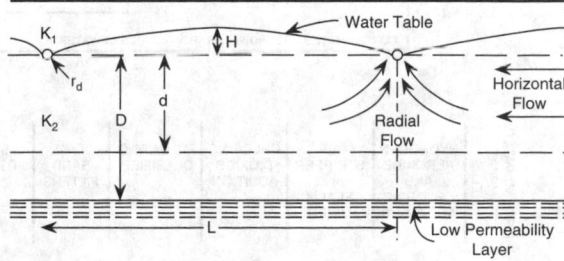

FIG. 9.17.8 Subsurface drain formulation. (Reprinted from U.S. Environmental Protection Agency, 1985, *Leachate plume management*, EPA/540/2-85/004, Washington, D.C.: U.S. EPA.)

The slope of the pipe should be selected so that the flow velocity in the drain is greater than the critical velocity of siltation of the soil that enters the drain (Soil Conservation Service 1973). When the velocity is less than 1.4 ft/sec, filter fabrics and silt traps or cleanouts should be installed around the pipe and along the subsurface drain.

Methods of Construction

The construction of a subsurface drain involves trench excavation, dewatering, wall stabilization, pipe installation, and backfilling. Trench excavation is the most significant step in the construction of a subsurface drain. A variety of excavation equipment can excavate the trench; the optimum is determined by the depth, width, length of the trench, and the type of material being excavated. Dewatering can be performed by open pumping, predrainage using well points, or groundwater cutoff. Wall stabilization methods include shoring for deep excavations or open cuts for shallow excavations. Continuous trenching machines can accomplish all excavation and pipe installation operations simultaneously (Oakley et al. 1994); however, this machinery is limited to small diameter and relatively shallow subsurface drains.

Another important aspect in subsurface drain installation is the placement of filter and envelope materials around the pipe to prevent soil particles from entering and clogging the pipe. Geotextiles and well-graded sand and gravel can be used as filter materials. The general requirement for envelopes is that their hydraulic conductivity is higher than that of the base material. Design procedures for filters and envelopes are in the Soil Conservation Service (1973).

Operation and Maintenance

Subsurface drains require frequent inspection and maintenance during the first year or two of operation. Typical problems that can develop in drainage systems and require maintenance include clogging of the drain or manhole by sediment buildup or buildup of chemical compounds such as iron and manganese. Clogged pipes can be cleaned by hydraulic jetting, mechanical scrapping, or chemical treatment in cases of chemical buildup.

Cost

Costs of subsurface drain systems vary from site to site depending on the geology, the depth of the aquifer, the extent and type of contamination, the periods and durations of pumping, and the electrical power requirements. The major costs of a subsurface drain system occur during system installation. These costs include excavation, dewatering, pipe bedding, filter and envelop materials, pipes, manholes, and pumps. Typical costs for subsurface drain installation are in Means (1994).

Advantages and Limitations

For shallow contaminations, subsurface drains are more cost-effective than wells particularly in aquifers with low or variable permeability. Construction methods for subsurface drains are simple, and operation costs are relatively low since flow to the underdrain is by gravity. In addition, subsurface drains provide considerable design flexibility since adjusting the depth or modifying the envelope material can alter the spacing to some extent. However, sub-

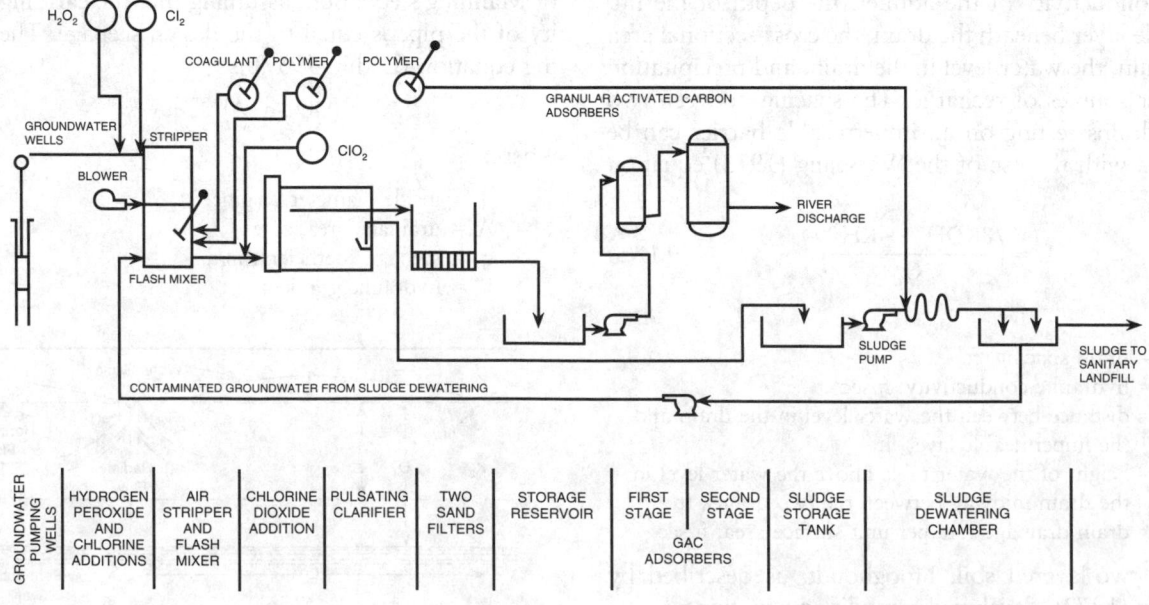

FIG. 9.17.9 Typical treatment trains. (Reprinted from North Atlantic Treaty Organization, 1993, *Demonstration of remedial action technologies for contaminated land and groundwater,* Vol. 2, pt. 2, no. 190, EPA/600/R-93/012, NATO Committee on the Challenges of Modern Society.)

surface drains are not suited to poorly permeable soils, to deep contaminant plumes, or beneath existing sites. In addition, subsurface drains require continuous and careful monitoring to assure adequate leachate collection and prevent pipe clogging.

Treatment Systems

The most commonly used treatment in pump-and-treat technologies is physical treatment. Physical treatment includes density separation, filtration, adsorption, air stripping, and reverse osmosis. Each of these processes can be used individually or in conjunction with others (e.g., treatment trains) as shown in Figure 9.17.9. In addition, most treatment systems include equalization and spill control to protect the treatment works from shock pollutants and hydraulic loadings. The selection of the process is usually based on the type of contaminant, influent concentration, effluent requirements, and cost.

DENSITY SEPARATION

Density separation is a process whereby the water and contaminant are separated based on their individual densities. This treatment is often used for the pretreatment of suspended solids or floating immiscible products that could be present in pumped groundwater. For suspended solids, the most commonly used equipment is clarifiers, settling chambers, and sedimentation basins as discussed in Chapter 8. For immiscible products, such as oil and grease, the most commonly used equipment is oil–water separators. Both suspended solids and oil and grease must generally be removed from contaminated groundwater prior to further treatment because these materials can foul instruments and interfere with other processes. Furthermore, oil and grease and suspended solids can damage the environment and cause a significant pollution problem to the receiving body of water.

The most common oil–water separators are the American Petroleum Institute (API) gravity separators and the parallel-plate separators. The design of an oil–water separator is based on the amount of oil present in the water, the oil droplet size distribution, the presence of surfactants, the specific gravity of the oil, and the water temperature. A step-by-step procedure for the design of an oil–water separator is in Corbitt (1990). Once the oil or floating product is at the surface, it can be removed from the water by slotted pipes, dip tubes, or belt or rope skimmers.

FILTRATION

Filtration is a process whereby suspended solids are removed from the influent by forcing the water through a filter of porous medium such as sand or sand with anthracite or coal. The purpose of filtration is to reduce the concentration of suspended solids, such as carbon columns, prior to certain treatment processes. The most common filter is a dual-media system with a layer of anthracite over a layer of sand. This filter provides better suspended solids removal with longer filter runs at higher flow rates than the more conventional single-medium filter (Corbitt 1990). The design of filters is based on the flow rate, flow scheme, and the type of medium used in the filter as discussed in Chapter 8. Up to 75% of suspended solids can be removed by dual-media filters operating at flow rates ranging from 2 to 8 gpm/ft (Oakley et al. 1994), bed depths of 24 to 48 inches, sand to anthracite ratios of 1:1 to 4:1, and a filter run of 8 to 148 hours (Corbitt 1990).

Filtration is a reliable and effective means of removing low levels of solids provided that the solid content does not vary greatly. Also, periodic filter backwashing is necessary to remove collected materials from the media. Typical backwash flow rates are 15 to 25 gpm/ft (Oakley et al. 1994) for eight to ten minutes (Corbitt 1990). The spent backwash water can be routed to the plant's headworks or to an intermediate process which provides settling.

CARBON ADSORPTION

In adsorption, the molecules of a dissolved contaminant become attached to the surface of a solid adsorbent. The most widely used adsorbent is granular activated carbon (GAC) because its porous structure provides a relatively large surface area per unit volume (1000–2000 m^2/g). Collection of the molecules on the surface of the adsorbent is due to chemical or physical forces. Chemical adsorption is due to actual chemical bonding at the solid's surface. Physical adsorption is due to van der Waals' forces, which are weak bonds compared to chemical adsorption. However, because of the weak nature of these bonds, adsorbed molecules can be easily removed with a change in the solute concentration or the addition of enough energy (regeneration) to overcome the bonds. This capacity to remove certain molecules adsorbed on carbon and, thus, the possibility for repeated carbon reuse is what allows activated carbon adsorption to be a cost-effective technology.

Environmental engineers commonly use carbon adsorption to remove organic contaminants from water or air; however, they also use it to remove a limited number of inorganic contaminants as shown in Table 9.17.1. The effectiveness of GAC depends on the molecular weight, structure, and solubility of the contaminant as well as the properties of the carbon, the water temperature, and the presence of impurities such as iron and manganese. The influence of each of these parameters on the absorbability of organic contaminants is shown in Table 9.17.2. As shown in this table, carbon adsorption is suitable for high molecular weight and low solubility and polarity compounds (U.S. EPA 1988), such as chlorinated hydrocarbons, organic phosphorous, carbonates, PCBs, phenols, and benzenes. GAC can also be used in conjunction with other treatment technologies. For example, GAC can be used to treat the effluent water or offgas from an air stripper (Crittenden 1988).

Design Considerations

The most important variables in designing carbon treatment systems are the contact time and the carbon usage rate, both of which depend on the flow rate, type of contaminant, and influent and effluent concentrations. The contact time is the time allowed for the pollutant to react with the carbon exterior and enter and react with the surface of the interior pores. The contact time is the result of dividing the volume of carbon by the flow rate. The carbon usage rate is the result of dividing the volume of carbon online by the volume of water treated when the required effluent concentration is exceeded (i.e., breakpoint).

TABLE 9.17.1 POTENTIAL FOR REMOVAL OF INORGANIC MATERIAL BY ACTIVATED CARBON

Constituents	Potential for Removal by Carbon
Metals of high sorption potential	
Antimony	Highly sorbable in some solutions
Arsenic	Good in higher oxidation states
Bismuth	Very good
Chromium	Good, easily reduced
Tin	Proven very high
Metals of good sorption potential	
Silver	Reduced on carbon surface
Mercury	CH_3HgCl sorbs easily Metal filtered out
Cobalt	Trace quantities readily sorbed, possibly as complex ions
Zirconium	Good at Low pH
Elements of fair-to-good sorption potential	
Lead	Good
Nickel	Fair
Titanium	Good
Vanadium	Variable
Iron	FE^{3+} good, FE^{2+} poor, but may oxidize
Elements of low or unknown sorption potential	
Cooper	Slight, possibly good if complexed
Cadmium	Slight
Zinc	Slight
Beryllium	Unknown
Barium	Very low
Selenium	Slight
Molybdenum	Slight at pH 6–8, good as complex ion
Manganese	Not likely, except as MnO_4
Tungsten	Slight
Free halogens	
F_2 fluorine	Will not exist in water
Cl_2 chlorine	Sorbed well and reduced
Br_2 bromine	Sorbed strongly and reduced
I_2 iodine	Sorbed very strongly, stable
Halides	
F, flouride	May sorb under special conditions
Cl^-, Br^-, I^-	Not appreciably sorbed

Source: U.S. Environmental Protection Agency, 1985, *Handbook, remedial action at waste disposal sites,* EPA/625/6-85/006 (Washington, D.C.: U.S. EPA).

TABLE 9.17.2 SUMMARY OF INFLUENCE OF CONTAMINANT PROPERTIES ON THE ABSORBABILITY OF ORGANICS

Parameter	Influence on Absorbability
Molecular weight	High molecular-weight compounds adsorb better than low molecular-weight compounds.
Solubility	Low-solubility compounds are adsorbed better than high-solubility compounds.
Structure	Nonpolar compounds adsorb better than polar compounds.
	Branched chains are usually more adsorbable than straight chains.
	Large molecules are more adsorbable than small molecules.
Substituent group	*Hydroxyl* generally reduces absorbability.
	Amino generally reduces absorbability.
	Carbonyl effect varies according to host molecule.
	Double bonds effect varies.
	Halogens effect varies.
	Sulfonic usually decreases absorbability.
	Nitro often increases absorbability.
Temperature	Adsorptive capacity decreases when the water temperature increases.
Properties of carbon	Adsorption is directly proportional to the surface area of the carbon used.
	Virgin carbon has more adsorptive capacity than regenerated carbon.
Other	Iron and manganese (if present at significant levels in the water) can precipitate onto the carbon, clog its pores, and cause rapid head loss.
	Biological growth on the surface of the carbon can enhance the removal efficiency and increase the carbon service life. If the growth is excessive, however, it can clog the carbon bed.
	Excessive amounts of suspended solids (above 50 ppm) or oil and grease (above 10 ppm) can affect the efficiency of the carbon.

Source: U.S. Environmental Protection Agency, 1985, *Handbook, remedial action at waste disposal sites*, EPA/625/6-85/006 (Washington, D.C.: U.S. EPA).

The goal of the design is to find the optimum contact time which provides the lowest carbon usage rate. Typical design parameters for carbon adsorption are shown in Table 9.17.3.

The contact time and carbon usage rate for a compound are usually determined through laboratory testing. A common test method is the bed depth service time (BDST) analysis, also called the dynamic column test study (Adams and Eckenfelder 1974). In this test method, three to four columns are connected in series as shown in Figure 9.17.10. Each column is filled with an amount of carbon which provides superficial contact times of fifteen to sixty minutes per column. Effluent from each column is ana-

lyzed for the chemicals of concern, and the effluent-to-influent concentration ratio is plotted against the volume of water treated by each column. Figure 9.17.11 shows an example of a dynamic test where four columns are used and each column represents fifteen minutes of contact time T_c. The curves obtained are called *breakthrough curves* since they represent the amount of contaminated water that has passed through the carbon bed before the maximum allowable concentration appears in the effluent.

Once the breakthrough curves are determined, the carbon usage rates can be calculated as:

$$q_c = \frac{V_c}{V_w} \qquad 9.17(8)$$

TABLE 9.17.3 TYPICAL DESIGN PARAMETERS FOR CARBON ADSORPTION

Parameters	Requirements
Contact time	Generally 10–50 min; may be as high as 2 hours for some industrial wastes
Hydraulic load	2–15 gpm/ft² depending on type of contact system; see Table 9.17.1
Backwash rate	Rates of 20–30 gpm/ft² usually produce 25–50% bed expansion
Carbon loss during regeneration	4–9% 2–10%
Weight of COD removed per weight of carbon	0.2–0.8
Carbon requirements PCT plant Tertiary plant	500–1800 lb/10⁶ gal 200–500 lb/10⁶ gal
Bed depth	10–30 ft

Source: U.S. Environmental Protection Agency, 1985, *Handbook, remedial action at waste disposal sites*, EPA/625/6-85/006 (Washington, D.C.: U.S. EPA).

where:

q_c = carbon usage rate, lb/gal
V_c = volume of carbon, lb
V_w = volume of water treated when the required effluent concentration is exceeded, gal

Carbon usage rates are then plotted for each contact time as shown in Figure 9.17.12. The optimum contact time t_{copt} is determined as the time which provides the lowest carbon usage rate. The optimum volume of carbon bed needed is calculated as

$$V_{copt} = T_{copt} \cdot Q \qquad 9.17(9)$$

where:

V_{copt} = optimum volume of carbon bed, lb
Q = flow rate, gal

The optimal tradeoff point between a lower carbon usage rate and a smaller carbon bed size can be found through analysis. A typical minimum contact time for gasoline contaminants is fifteen minutes. This contact time corresponds to a liquid loading rate of 2 gpm/ft² in a standard 20,000-lb and 10-ft-diameter carbon vessel (Noonan and Curtis 1990). Table 9.17.4 lists the contact times as well as carbon usage rates for several organics.

Methods of Construction

GAC is available from a number of suppliers in vessels of different sizes. The vessels are typically open-top, cylindrical steel tanks for gravity systems and closed-top, cylindrical steel tanks for pressure systems. Gravity systems are operated like sand filters and are generally used for high flows, such as at municipal wastewater treatment plants. Pressure systems are generally used for smaller flows and allow higher surface loading rates (5–7 gpm/ft² compared to 2–4 gpm/ft² for gravity systems) and pressure discharge to the distribution system, saving pumping costs (Noonan and Curtis 1990).

Activated carbon is commonly made from coal; other materials such as coconut shells, lignite, wood, tires, and pulp residues can also be used. In the formation of GAC, the material is subjected first to a high temperature to remove water and other vapors from it. Then, a superheated steam is released into the material (activation) to enlarge the pores and remove ashes from it (Noonan and Curtis 1990).

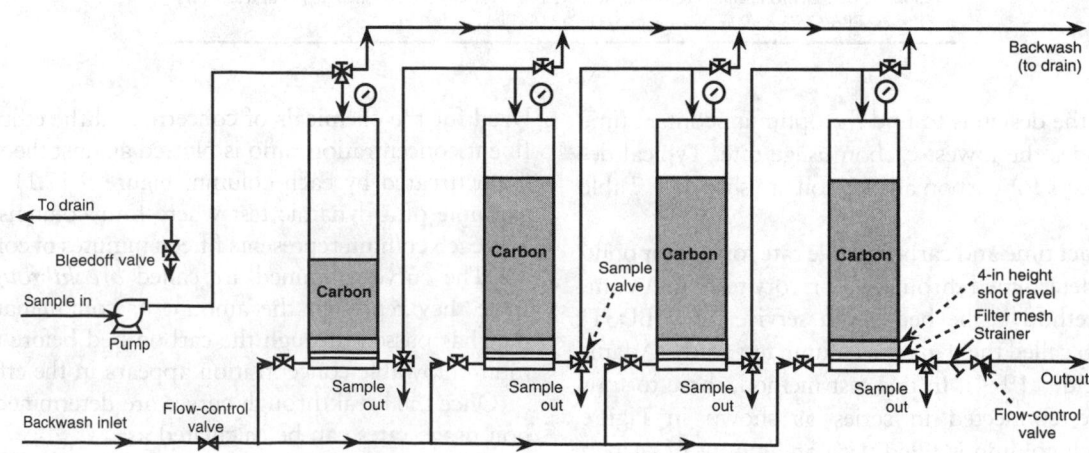

FIG. 9.17.10 Dynamic column test. (Reprinted from E.K. Nyer, 1992, *Groundwater treatment technology*, 2d ed., New York: Van Nostrand Reinhold.)

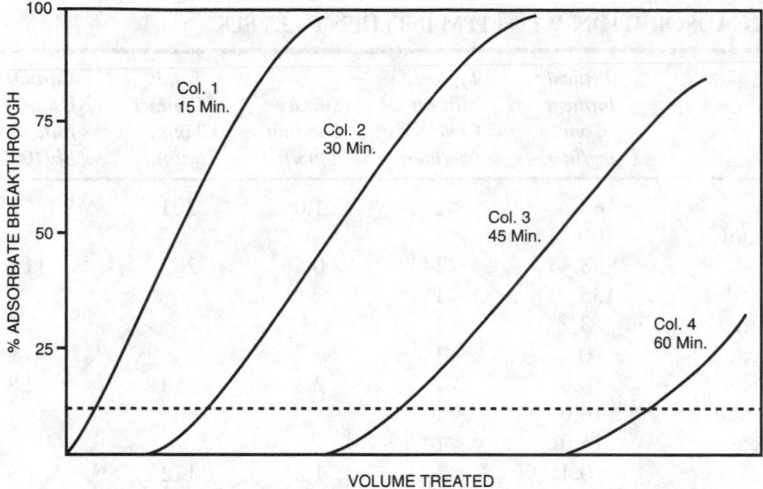

FIG. 9.17.11 Dynamic column test results breakthrough curves. (Reprinted from E.K. Nyer, 1992, *Groundwater treatment technology*, 2d ed., New York: Van Nostrand Reinhold.)

Operation and Maintenance

Activated carbon systems can be operated as upflow, expanded-bed columns or downflow, fixed-bed columns. Upflow expanded beds can tolerate higher suspended-solids loading than downflow beds. They also make efficient use of the carbon since fully exhausted carbon can be removed from the bottom of the bed while fresh carbon is added to the top. In addition, carbon beds can be operated in parallel (single-stage) or in series (multiple-stage) as shown in Figure 9.17.13. When operated in series, the leading contactor removes the majority of the contamination, while the second contactor removes any residual organics from the water. Furthermore, multiple-stage use allows a contactor to be completely exhausted before regeneration, while effluent quality remains protected by the subsequent contactor. When operated in parallel, contactors should stagger startup to permit bed-by-bed regeneration without reducing effluent quality.

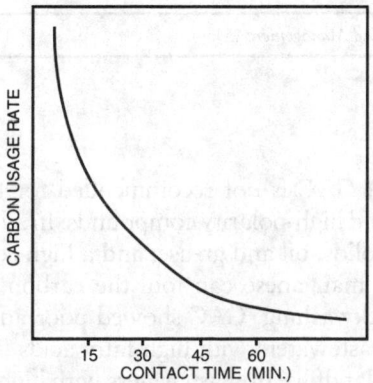

FIG. 9.17.12 Optimum carbon contact time. (Reprinted from E.K. Nyer, 1992, *Groundwater treatment technology*, 2d ed., New York: Van Nostrand Reinhold.)

When the adsorption capacity of the carbon is exhausted, the spent carbon can either be disposed of at a disposal site, regenerated, or reactivated for reuse. Offsite disposal at a landfill or an incinerator is the preferred method when the amount of carbon is small. For disposal at a landfill, testing and classifying the spent carbon are necessary to ensure that all regulations for disposal are being met. Spent carbon may be considered hazardous waste and may need to be disposed of at a hazardous waste landfill or burned at an incinerator where both the carbon and the hazardous waste are destroyed.

If the amount of spent carbon is large or the user has access to an offsite, multiuser facility, regeneration or reactivation for reuse may be the preferred solution. Regeneration exposes the spent carbon to steam to desorb the contaminants. Reactivation is conducted in electrical or multiple-heart furnaces where the temperature is high enough (up to 1800°F) to thermally destroy the contaminants and reactivate the carbon. Regeneration and reactivation can incur a 10 to 20% material loss and can change the adsorptive properties of the virgin grade material.

Cost

The capital costs of a GAC system include the costs of carbon, carbon vessels, pumps and piping, electrical equipment and controls, housing, design, and contingencies. The cost depends on the flow rates, type of contaminant, concentrations, and discharge requirements. Costs can vary from $0.10–1.50/1000 gal treated for flow rates of 100 mgd to $1.20–6.30/1000 gal treated for flow rates of 0.1 mgd (O'Brien 1983).

Operation and maintenance costs include labor, energy, carbon replacement, and sampling and monitoring. The major cost, however, is carbon replacement which is a function of the carbon usage rate. Typical carbon costs

TABLE 9.17.4   CARBON ADSORPTION WITH PPM INFLUENT LEVELS

System No.	Contaminants	Typical Influent Conc. (mg/liter)	Typical Effluent Conc. (mg/liter)	Surface Loading (gpm/ft²)	Total Contact Time (min)	Carbon Usage Rate (lb/1000 gal)	Operating Mode
1	Phenol	63	<1	1.0	201	5.8	Three fixed beds in series
	Orthorchlorophenol	100	<1				
2	Chloroform	3.4	<1	0.5	262	11.6	Two fixed beds in series
	Carbon tetrachloride	135	<1				
	Tetrachloroethylene	3	<1				
	Tetrachloroethylene	70	<1				
3	Chloroform	0.8	<1	2.3	58	2.8	Two fixed beds in series
	Carbon tetrachloride	10.0	<1				
	Tetrachloroethylene	15.0	<1				
4	Benzene	0.4	<1	1.21	112	1.9	Two fixed beds in series
	Tetrachloroethylene	4.5	<1				
5	Chloroform	1.4	<1	1.6	41	1.15	Two fixed beds in series
	Carbon tetrachloride	1.0	<1				
6	Trichloroethylene	3.8	<1	2.4	36	1.54	Two fixed beds in series
	Xylene	0.2–0.5	<1				
	Isopropyl Alcohol	0.2	<10				
	Acetone	0.1	<10				
7	Di-isoproply methyl phosphonate	1.25	<50	2.2	30	0.7	Single fixed bed
	Dichloropentadiene	0.45	<10				
1	1,1,1-Trichloroethane	143	<1	4.5	15	0.4	Single fixed bed in series
	Trichloroethylene	8.4	<1				
	Tetrachloethylene	26	<1				
2	Methyl T-butyl ether	30	<5	5.7	12	0.62	Two single fixed beds
	Di-isopropyl ether	35	<1				
3	Chloroform	400	<100	2.5	26	1.19	Four single fixed beds
	Trichloroethylene	10	<1				
4	Trichloroethylene	35	<1	3.3	21	0.21	Three single fixed series
	Tetrachloroethylene	170	<1				
5	1,1,1-Trichlorethane	70	<1	4.5	30	0.45	Two fixed beds in series
	1,1-Dichloroethylene	10	<1				
6	Trichlorethylene	25	<1	2.0	35	0.32	Single fixed bed
	Cis-1,2-dicloroethylene	15	<1				
7	Trichlorethylene	50	<1	1.6	42	0.38	Two single fixed beds
8	Cis-1,2-dichloroethylene	5	<1	1.91	70	0.25	Two fixed beds in series
	Trichloroethylene	5	<1				
	Tetrachloroethylene	10	<1				

Source: R.P. O'Brien, 1983, There is an answer to groundwater contamination, *Water/Engineering and Management* (May).

range from $0.60 per pound for regenerated carbon to $0.75 per pound for virgin, high-quality carbon (Noonan and Curtis 1990).

Advantages and Limitations

Carbon adsorption is an effective and simple treatment technology for volatile organic compounds. In addition, GAC can be used in conjunction with other treatment technologies.

However, GAC is not recommended for low-molecular-weight and high-polarity compounds. In addition, high-suspended solids, oil and grease, and a high concentration of iron and manganese can foul the carbon and require frequent backwashing. GAC showed poor adsorption capacity for wastewaters with high fatty acids (i.e., leachate from young landfills) or wastewaters with high BOD/COD and COD/TOC ratios (U.S. EPA 1977). Furthermore, the amount of carbon required, the frequency of regeneration and reactivation, and the potential need to handle the dis-

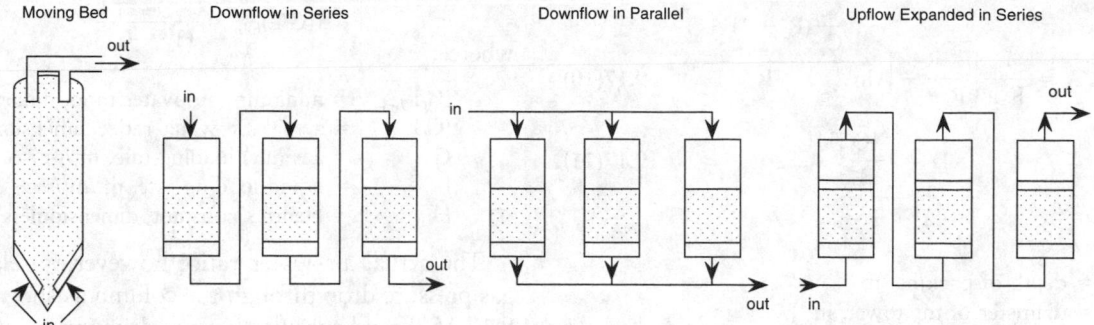

FIG. 9.17.13 Single-stage and multiple-stage contactors. (Reprinted from U.S. Environmental Protection Agency, 1985, *Handbook, remedial action at waste disposal sites,* EPA/625/6-85/006, Washington, D.C.: U.S. EPA.)

carded carbon as a hazardous waste make GAC a relatively expensive technology.

AIR STRIPPING

Air stripping is a mass-transfer process whereby volatile contaminants are stripped out of the aqueous solution and into the air. The process exposes the contaminated water to a fresh air supply which results in a net mass transfer of contaminants from the liquid phase to the gaseous phase. Contaminants are not destroyed by air stripping but rather are transferred into the air stream where they may need further treatment. Air stripping applies to volatile and semivolatile organic compounds. It does not apply to low volatility compounds, metals, or inorganic contaminants.

Several types of air stripping technologies are available including tray aeration, spray aeration, and packed towers. Among these technologies, packed tower aeration

(PTA) is the most commonly applied to remove volatile organics from groundwater. In a packed tower, the contaminated water comes in contact with a countercurrent flow of air. The packing material in the tower breaks the water into small droplets and thin films causing a large contact area where the mass transfer can take place. Figure 9.17.14 shows a typical treatment process using air stripping.

Design Consideration

The design of an air stripper is based on the flow rate, type of contaminant, concentration, temperature, and effluent requirements. The major design variables are the type of packing, gas pressure drop, and air-to-water ratio. Given those design variables, environmental engineers can determine the gas and liquid loading rates, tower diameter, and packing height by using the following mass-balance equation (Noonan and Curtis 1990):

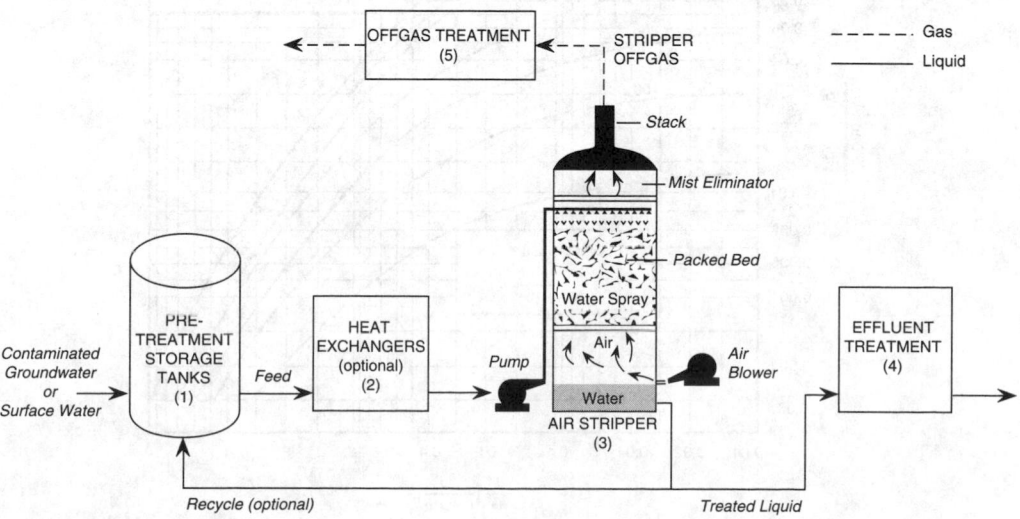

FIG. 9.17.14 Typical treatment process using air stripping. (Reprinted from U.S. Environmental Protection Agency, 1991, Air stripping of aqueous solutions, *Engineering Bulletin,* EPA/540/2-91/022, Washington, D.C.: U.S. EPA.)

$$Z_t = \frac{L}{K_1 a}\left[\frac{R}{R-1}\right] \cdot \ln \frac{\dfrac{C_i}{C_e}(R-1)+1}{R} \qquad 9.17(10)$$

$$D = \left[\frac{4Q}{\pi L}\right]^{0.5} \qquad 9.17(11)$$

where:

Z_t = depth of packing, m
D = diameter of the tower, m
L = liquid loading rate, m³/m²/sec
$K_1 a$ = overall liquid mass-transfer coefficient, sec⁻¹
R = stripping factor, dimensionless
C_i = influent concentration, mg/l
C_e = effluent concentration, mg/l
Q = flow rate m³/sec

The key variables to define in the preceding equations are the overall mass-transfer coefficient $K_1 a$ and the stripping factor R. The mass-transfer coefficient is a function of the type of packing, the liquid and gas flow rates, and the viscosity and density of the water. Therefore, the mass-transfer coefficient is usually determined from a pilot test on actual field data. When pilot testing is not feasible, theoretical correlations, such as those developed by Onda, Takeuchi, and Okumoto (1968), can be used.

The stripping factor R is related to the air–water ratio as follows (Noonan and Curtis 1990):

$$R = \frac{(G/L)_{act}}{(G/L)_{min}} \qquad 9.17(12)$$

$$(G/L)_{min} = \frac{1}{H}\frac{C_i - C_e}{C_i} \qquad 9.17(13)$$

where:

$(G/L)_{min}$ = minimum air–water ratio, dimensionless
$(G/L)_{actual}$ = actual air–water ratio, dimensionless
G = gas (air) loading rate, m³/m²/sec
L = liquid loading rate, m³/m²/sec
H = Henry's constant, dimensionless

The actual air–water ratio, however, is related to the gas pressure drop through the column as shown in Figure 9.17.15 (brand-specific pressure drop curves are available from packing vendors). Therefore, engineers should examine several combinations of air–water ratio and pressure drop to determine the most cost-effective design. A high pressure drop reduces the size of the tower and capital costs; however, it increases the size of the blower and operation costs. Studies have shown that the most cost-effective stripping factor R usually falls between 3 and 5 (Hand et al. 1986).

After a stripping factor is selected, the actual air–water ratio can be calculated with Equation 9.17(12), and the gas (air) loading rate can be obtained from Figure 9.17.15 for a given pressure drop. Then the tower height and diameter can be calculated with Equations 9.17(10) and 9.17(11), respectively. This procedure should be repeated for several combinations of stripping factor and pressure drop until the most cost-effective design is obtained. Several computer cost models can be used in this process (Nirmalakhandan, Lee, and Speece 1987; Cummins and Westrick 1983; Clark, Eilers, and Goodrick 1984).

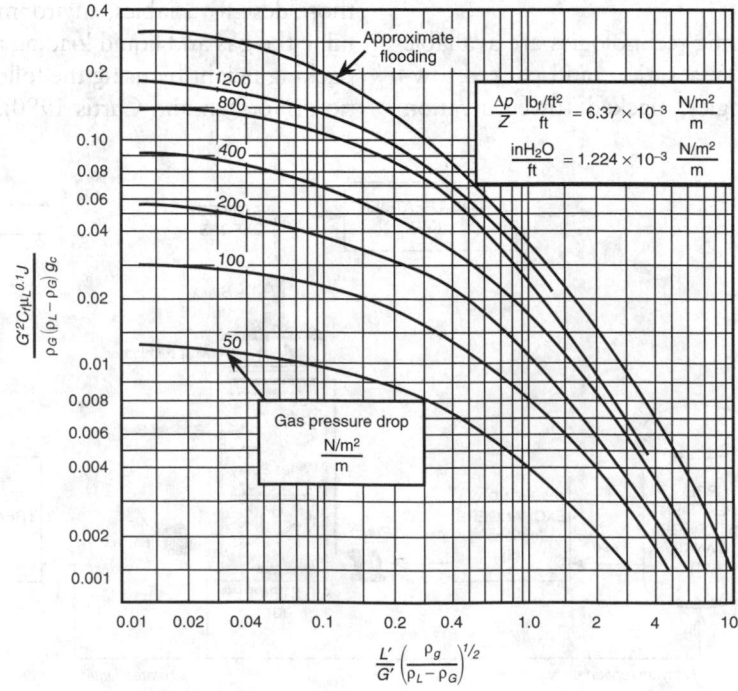

FIG. 9.17.15 Generalized pressure drop curves. (Reprinted from R.E. Treybal, 1980, *Mass transfer operations*, 3d ed., New York: McGraw-Hill.)

Methods of Construction

The components of a stripping tower include the tower shell, tower internals, packing, and air delivery systems. The tower shell can be made of aluminum, fiberglass, stainless steel, or coated carbon steel. Selecting the shell construction material is usually based on cost, structural strength, resistance to corrosion, and esthetics. Table 9.17.5 shows the advantages and disadvantages of several materials of construction.

Tower internals include the water distributor system inside the tower, the mist eliminator system, and the air exhaust ports. The environmental engineer should select the type of components that ensure optimal mass-transfer conditions at the most economical cost.

The packing material is an important component of the air stripping tower. Several types of packing materials are commercially available including plastic, metal, or ceramic (Perry and Green 1984). The selection is based on materials exhibiting a high mass-transfer rate and a low gas pressure drop. Plastic packings are often used because of their low price, corrosion resistance, and light weight. Table 9.17.6 shows the physical characteristics of common packing materials.

Other components of an air stripping system include the blower, noise control devices, and air filters. The blower is designed based on the air–water ratio and can be mounted on top of the tower or at the base. Sound mufflers control noise, and air filters prevent contact between the water and the air outside the tower.

Operation and Maintenance

In a packed air stripping tower, the water flows countercurrent to the air stream which is introduced at the bot-

TABLE 9.17.5 CONSTRUCTION MATERIALS FOR TOWER SHELLS WITH A PACKED AIR STRIPPER

Material	Advantages	Disadvantages
Aluminum	Lightweight; Low cost; Corrosion resistant; Excellent structural properties; Long life (> fifteen years); No special coating required	Poor resistance to water with pH less than 4.5 and greater than 8.6; Pitting corrosion occurs in the presence of heavy metals; Not well suited to high chloride water
Carbon Steel	Mid-range capital cost; Good structural properties; Long life if properly painted and maintained	Requires coating inside and outside to prevent corrosion, leading to increased maintenance; Heavier than aluminum or FRP
Fiberglass	Low cost; High chemical resistance to acidic and basic conditions, chlorides, and metals	Poorly defined structural properties; Short life (< ten years) unless more expensive resins used; Poor resistance to ultraviolet (UV) light (can be overcome with special coatings that must be maintained); Requires guy wires in most situations; Susceptible to extremes of temperature differential disturbing tower shape and interfering with distribution
Stainless Steel	Highly corrosion resistant; Excellent structural properties; Long life (> twenty years); No special coating required	Most expensive material for prefabricated towers; Susceptible to stress fracture corrosion in the presence of high chloride levels
Concrete	Aesthetics; Less prone to vandalism	Difficult to cast in one place leading to potential difficulties with cracks and leaks; More expensive than self-supporting prefab towers
Metal lined block and brick	Aesthetics; Less prone to vandalism; Prefab air stripper insert eliminates problems associated with cast in place towers	More expensive than self-supporting prefab towers

Source: K.E. Nyer, 1992, Groundwater treatment technology, 2d ed., New York: Van Nostrand Reinhold.

TABLE 9.17.6 PHYSICAL CHARACTERISTICS OF COMMON PACKING MATERIALS

Type	Size	Surface Area (ft^3/ft^3)	Void Space (%)	Packing Factor[a]
Dumped Packings				
Glitsch	0A	106	89	60
Mini-rings	1A	60.3	92	30
(Plastic)	1	44	94	28
	2A	41	94	28
	2	29.5	95	15
	3A	24	95.5	12
Tellerettes	1″(#1)	55	87	40
(Plastic)	2″(2-R)	38	93	18
	3″(3-R)	30	92	16
	3″(2-K)	28	95	12
Intalox	1″	63	91	33
Saddles	2″	33	93	21
(Plastic)	3″	27	94	16
Pall rings	⅝″	104	87	97
(Plastic)	1″	63	90	52
	1½″	39	91	40
	2½″	31	92	25
	3½″	26	92	16
Raschig rings	½″	111	63	580
(Ceramic)	¾″	80	63	255
	1″	58	73	155
	1½″	38	71	95
	2″	28	74	65
	3″	19	78	37
Jaegar	1″	85	90	28
Tri-Packs	2″	48	93	16
(Plastic)	3½″	38	95	12
Stacked Packing				
Delta	—	90	98	—
(PVC)				
Flexipac	Type 1	170	91	33
(Plastic)	Type 2	75	93	22
	Type 3	41	96	16
	Type 4	21	98	9

Source: R.E. Treybal, 1980, *Mass transfer operations,* 3d ed., New York: McGraw-Hill.

tom of the tower. In some configurations (e.g., induced draft systems), the air is drawn through the tower by the blower instead of being forced. The water supply pumps usually control the blowers to coordinate the air and water flows. The offgas from an air stripper may need to be treated, depending on air emission requirements, with the use of granular activated carbon, catalytic oxidation, or incineration (U.S. EPA 1985a). The liquid effluent from the air stripper may contain trace amounts of contaminants which can be treated by GAC.

Maintenance of air stripping systems is minimal and usually involves the blower. However, periodic inspection of the packing is required if the water contains high levels of iron, suspended solids, or microbial population.

During the aeration process, dissolved iron and manganese can be oxidized and deposited on the packing material. This deposit can build up and clog the packed bed and, therefore, reduce system efficiency. Pretreating the influent can control iron deposition. A high microbial population can lead to a biological build up within the packed bed and reduce system performance. This problem can also be prevented through pretreatment of the influent.

Cost

The capital costs of an air stripper include the costs of the tower shell, packing, tower internals, air delivery system, electrical equipment and controls, housing, design, and contingencies. The addition of an air treatment system roughly doubles the cost of an air stripping system (Lenzo and Sullivan 1989; U.S. EPA 1986a). The cost depends on the flow rate, volatility of the contaminant, concentration, and removal efficiency. Costs vary from $0.07–0.70/1000 gal for Henry's law coefficients of 0.01–1.0 to $7.00/1000 gal for Henry's law coefficients lower than 0.005 (Adams and Clark 1991).

Advantages and Limitations

Air stripping is a proven technology for treating water contaminated with volatile and semivolatile organic compounds. Removal efficiencies of greater than 98% for volatile organics and greater than or equal to 80% for semivolatile compounds have been achieved. Recent developments in this technology include high temperature air stripping and air rotary stripping to increase removal efficiencies (Bass and Sylvia 1992). The use of diffused air or bubble aeration air strippers for flows less than 50 gpm have also increased during the last five years.

The air stripping technology, however, is not effective in treating low volatility compounds, metals, or inorganics. Air emissions of volatile organics from the air stripper may need a separate treatment. In addition, the removal efficiency of air strippers is reduced for aqueous solutions with high levels of suspended solids, iron, manganese, or microbial population. Periodic cleaning of the packing material removes the deposits of these products.

OXIDATION AND REDUCTION

In chemical oxidation, the oxidation state of a contaminant is increased by the loss of electrons, while the oxidation state of the reactant is lowered. Conversely, in reduction, the oxidation state of a contaminant is decreased by the addition of electrons. Oxidizing or reducing agents can be added to contaminated water to destroy, detoxify, or convert the contaminants to less hazardous compounds. Many hazardous substances including various organics, sulfites, soluble cyanide- and arsenic-containing compounds, hydroxylamine, and chromates can be oxidized or reduced to forms which are more readily removed from groundwater (Huibregts and Kastman 1979).

Chemical Oxidation

Chemical oxidation involves adding oxidizing agents to the contaminated water and maintaining the pH at a proper level. The choice of an oxidizing agent depends on the substance or substances to be detoxified. Numerous oxidizing agents are available to detoxify a variety of compounds. The most commonly used agents are hydrogen peroxide, ozone, hypochlorite, chlorine, and chlorine dioxide because they tend not to form toxic compounds or residuals and are relatively inexpensive. Ozone and hydrogen peroxide have an advantage over oxidants containing chlorine because potentially hazardous chlorinated compounds are not formed (U.S. EPA 1986b).

Hydrogen peroxide is a stable and readily available substance that can oxidize many compounds. Industrial treatment plants have used hydrogen peroxide to detoxify cyanide and organic pollutants including formaldehyde, phenol, acetic acid, lignin sugars, surfactants, amines and glycol ethers, aldehydes, dialkyl sulfides, dithionate, and certain nitrogen and sulfur compounds (Envirosphere Company 1983).

Ozone is a strong oxidizing agent (gas) that is unstable and extremely reactive. Therefore, ozone cannot be shipped or stored but must be generated onsite immediately prior to application (U.S. EPA 1985b). Ozone rapidly decomposes to oxygen in solutions containing impurities. Ozone's half-life in distilled water at 68°F is twenty-five

minutes, while in groundwater it drops to eighteen minutes (Envirosphere Company 1983).

Hypochlorite is used in drinking water and municipal wastewater systems for the treatment and control of algae and biofouling organisms (U.S. EPA 1985b). In industrial waste treatments, hypochlorite is used for the oxidation of cyanide, ammonium sulfide, and ammonium sulfite (Huibregts and Kastman 1979). Sodium hypochlorite solutions at concentrations of 2500 mg/l are also used for the detoxification (by oxidation) of cyanide contamination from indiscriminate dumping (Farb 1978). However, because the principal products from chlorination of organic contaminants are chlorinated organics which can be as much of a problem as the original compound, hypochlorite treatment is limited.

Advanced Oxidation

Advanced oxidation uses UV radiation combined with ozone or hydrogen peroxide to enhance the oxidation rate of the compounds; reaction times can be 100 to 1000 times faster in the presence of UV light (U.S. EPA 1986b). UV light reacts with hydrogen peroxide molecules to form an hydroxyl radical, a powerful chemical oxidant. Specifically, hydrogen peroxide and UV light are used as shown in Figure 9.17.16 for the treatment of volatile organic compounds and other organic contaminants in contaminated groundwater (U.S. EPA 1993). In addition, hydrogen peroxide, ozone, and UV radiation are used as shown in Figure 9.17.17 for the oxidation of dissolved organic contaminants including chlorinated hydrocarbons

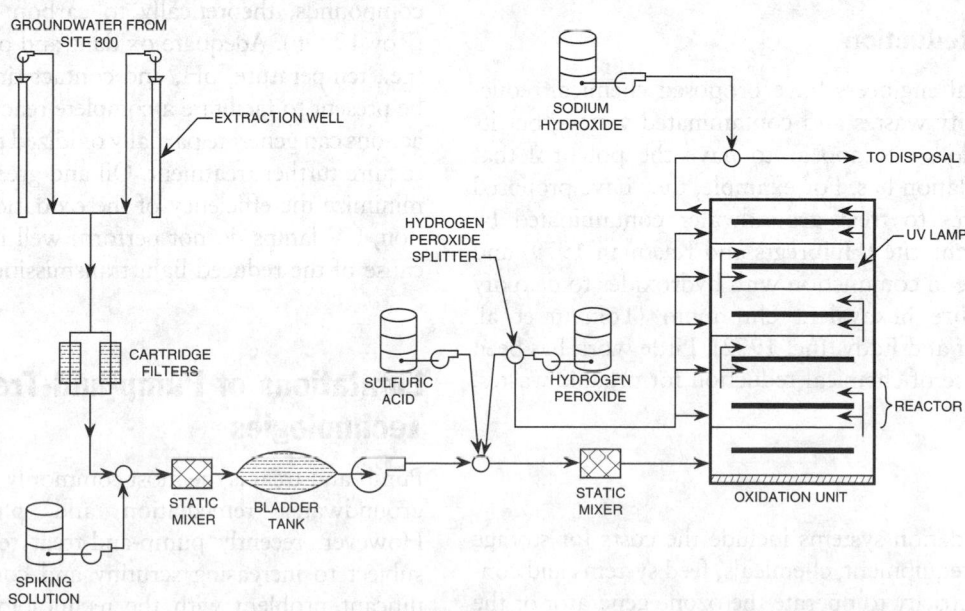

FIG. 9.17.16 Perox-pure chemical oxidation technology. (Reprinted from U.S. Environmental Protection Agency, 1993, *Perox-pure chemical oxidation technology—Perioxidation Systems, Inc.,* EPA/540/AR-93/501 Superfund Innovative Technology Evaluation, Washington, D.C.: U.S. EPA.)

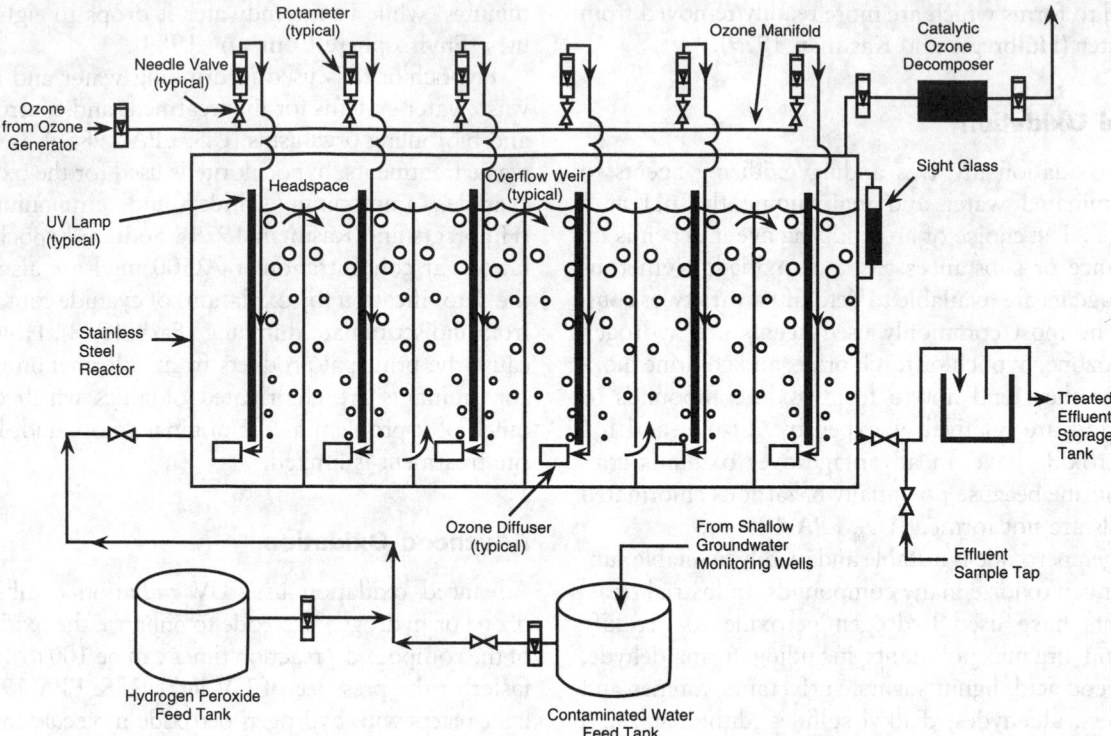

FIG. 9.17.17 Ultrox, UV/oxidation technology. (Reprinted from U.S. Environmental Protection Agency, 1990, *Ultraviolet radiation/oxidation technology—Ultrox International,* EPA/540/AS-89/012 Superfund Innovative Technology Evaluation, Washington, D.C.: U.S. EPA.)

and aromatic compounds in groundwater (U.S. EPA 1990).

Chemical Reduction

Environmental engineers have proposed chemical reduction to detoxify wastes and contaminated waters, but its application does not appear to have the potential that chemical oxidation has. For example, they have proposed sodium sulfites to treat groundwater contaminated by sodium hypochlorite (Huibregts and Kastman 1979) and ferrous sulfate in conjunction with hydroxides to detoxify and insolubilize hexavalent chromium (Tolman et al. 1978; Metcalf and Eddy, Inc. 1972). Little work has been done in the use of chemical reduction for organic wastes.

Cost

Costs for oxidation systems include the costs for storage and handling equipment, chemicals, feed systems and controls, and electricity to operate the ozone generator or the UV lamps. Costs for enhanced oxidation range from $0.15 to $70/1000 gal treated depending on the type of contaminants, their concentration, and the cleanup level (U.S. EPA 1993, 1990).

Advantages and Limitations

The principal advantage of chemical oxidation technology is the ability of oxidizing agents to degrade carbonaceous compounds, theoretically to carbon dioxide and water (Roy 1990b). Adequate oxidant and operating conditions (i.e., temperature, pH, and contact time), however, must be present to facilitate a complete reaction. Incomplete reactions can generate partially oxidized products which may require further treatment. Oil and grease in the water can minimize the efficiency of the oxidation process. In addition, UV lamps do not perform well in turbid waters because of the reduced light transmission (Roy 1990a).

Limitations of Pump-and-Treat Technologies

Pump-and-treat is the most commonly used technology for groundwater remediation and plume containment. However, recently pump-and-treat technology has been subject to increasing scrutiny and controversy. One significant problem with the technology is its inability to achieve cleanup goals within reasonable time frames (Galya 1994). At many sites where this technology is used, contaminant removal rates follow a relatively consistent pattern. After a period of initially steady reductions,

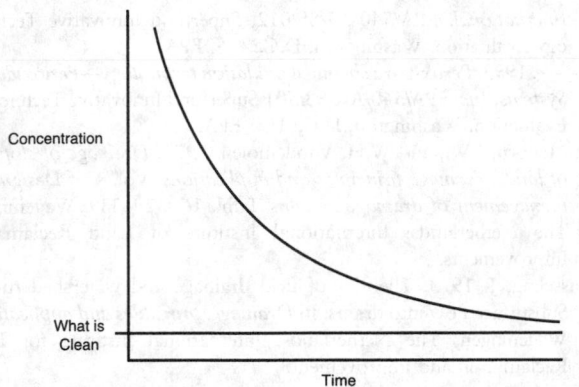

FIG. 9.17.18 Assumptotic behavior of pump-and-treat cleanup technologies. (Reprinted from K.E. Nyer, 1992, *Groundwater treatment technology,* 2d ed., New York: Van Nostrand Reinhold.)

groundwater contaminant concentrations tend to level off and remain fairly constant, with random fluctuations around an assumptotic limit (Tucker et al. 1989) as shown in Figure 9.17.18. The assumptotic concentration level may be higher than the specified cleanup target, and achieving cleanup goals within reasonable time frames may not be possible.

Therefore, pump-and-treat technology is not an effective approach by itself for the ultimate remediation of aquifers to health-based cleanup concentrations.

—Ahmed Hamidi

References

Adams, C.E., and W.W. Eckenfelder. 1974. *Process design techniques for industrial waste treatment.* Nashville, Tenn.: Enviro Press.

Adams, J.Q., and R.M. Clark. 1991. Evaluating the cost of packed-tower aeration and GAC for controlling selected organics. *Journal AWWA* 1:49–57.

Bass, D.H., and T.E. Sylvia. 1992. Heated air stripping for the removal of MTBE from recovered groundwater. Proceedings for the 1992 Petroleum Hydrocarbons and Organic Chemicals in Groundwater, 4–6 November, Houston, Tex.

Campbell, M.D., and J.H. Lehr. 1977. Well cost analysis. In *Water well technology.* 4th ed. McGraw-Hill.

Clark, R.M., R.G. Eilers, and J.A. Goodrick. 1984. VOCs in drinking water: Cost of removal. *Journal of Environmental Engineering* 110, no. 6:1146–1162.

Cohen, R.M., and W.J. Miller. 1983. Use of analytical models for evaluating corrective actions at hazardous waste disposal facilities. Proceedings of the Third National Symposium on Aquifer Restoration and Groundwater Monitoring. Worthington, Ohio: National Water Well Association.

Corbitt, R.A. 1990. Wastewater disposal. In *Standard handbook of environmental engineering,* edited by R.A. Corbitt. New York: McGraw-Hill.

Crittenden, J.C. et al. 1988. Using GAC to remove VOCs from air stripper off-gas. *Journal AWWA* 80, no. 5(May):73–84.

Cummins, M.D., and J.J. Westrick. 1983. Trichloroethylene removal by packed column air stripping: Field verified design procedure. In Proceedings ASCE Environmental Engineering Conference, 442–449. Boulder, Colo.: ASCE.

Envirosphere Company. 1983. *Evaluation of systems to accelerate stabilization of waste piles or deposits.* Cincinnati, Ohio: U.S. EPA.

Farb, D. 1978. *Upgrading hazardous waste disposal sites: Remedial approaches.* EPA-SW-677. Cincinnati, Ohio: U.S. EPA.

Galya, D. 1994. Evaluation of the effectiveness of pump and treat groundwater remediation system. In WEF Specialty Conference Series Proceedings on Innovative Solutions for Contaminated Site Management, 6–9 March, 323–342. WEF.

Hand, D.W. et al. 1986. Design and evaluation of an air stripping tower for removing VOCs from groundwater. *Journal of AWWA* 78, no. 9:87–97.

Huibregts, K.R., and K.H. Kastman. 1979. Development of a system to protect groundwater threatened by hazardous spills on land. Oil and Hazardous Material Spills Brands, Industrial Environmental Research Laboratory. Edison, N.J.: U.S. EPA.

Johnson Division, UOP, Inc. 1975. *Groundwater and wells.* Saint Paul, Minn.: Edward F. Johnson, Inc.

Kuffs, C. et al. 1983. Procedures and techniques for controlling the migration of leachate plumes. Ninth Annual Research Symposium on Land Disposal, Incineration and Treatment of Hazardous Waste.

Lenzo, F., and K. Sullivan. 1989. Groundwater treatment techniques, an overview of state-of-the-art in America. First US/USSR Conference on Hydrogeology, July. Moscow.

Luthin, J.N. 1957. *Drainage of agricultural lands.* Madison, Wis.: American Society of Agronomy.

Means. 1994. *Means site work and landscape cost data.* Means Southam Construction Information Network.

Metcalf and Eddy, Inc. 1972. *Wastewater engineering: Collection, treatment, and disposal.* New York: McGraw-Hill.

Nirmalakhandan, N., Y.H. Lee, and R.E. Speece. 1987. Designing a cost effective air stripping process. *Journal of AWWA* 79, no. 1:56–63.

Noonan, D.C., and J.T. Curtis. 1990. *Groundwater remediation and petroleum: A guide for underground storage tanks.* Chelsea, Mich.: Lewis Publishers.

Oakley, D. et al. 1994. The use of horizontal wells in remediating and containing a jet fuel plume—preliminary findings. WEF Specialty Conference Series Proceedings on Innovative Solutions for Contaminated Site Management, March, 331–342. Water Environment Federation.

O'Brien, R.P. 1983. There is an answer to groundwater contamination. *Water/Engineering and Management* (May).

Onda, K.H., Takeuchi, and Y. Okumoto. 1968. Mass transfer coefficients between gas and liquid phases in packed columns. *Journal of Chemical Engineering,* Japan 72, no. 12:684.

Perry, R.H., and D. Green. 1984. *Perry's chemical engineer's handbook.* 6th ed. New York: McGraw-Hill.

Powers, J.P. 1981. *Construction dewatering: A guide to theory and practice.* New York: John Wiley and Sons.

Repa, E. et al. 1982. *The establishment of guidelines for modeling groundwater contamination from hazardous waste facilities.* JRB Assoc. report prepared for the Office of Solid Waste, U.S. EPA.

Roy, K. 1990a. Researchers use UV light for VOC destruction. *Hazmat World* (May):82–93.

———. 1990b. UV-oxidation technology, shining star or flash in the pan? *Hazmat World* (June):35–50.

Soil Conservation Service. 1973. *Drainage of agricultural land.* Syosset, N.Y.: Water Information Center.

Tolman, A. et al. 1978. *Guidance manual for minimizing pollution from waste disposal sites.* EPA/600/2-78/142. Cincinnati, Ohio: U.S. EPA.

Tucker, W.A. et al. 1989. Technological limits of groundwater remediation: A statistical evaluation method. Proceedings of the Petroleum Hydrocarbons and Organic Chemicals in Groundwater, 15–17 Nov., Houston, Tex.

U.S. Environmental Protection Agency. 1977. *Wastewater treatment facilities for sewered small communities.* Washington, D.C.: U.S. EPA, Technology Transfer Division.

———. 1985a. *Handbook, remedial action at waste disposal sites.* EPA/625/6-85/006. Washington, D.C.: U.S. EPA.

———. 1985b. *Leachate plume management.* EPA/540/2-85/004. Washington, D.C.: U.S. EPA.

———. 1986a. *Mobile treatment technologies for superfund wastes.* EPA/540/2-86/003(f). Washington, D.C.: U.S. EPA.

———. 1986b. *Systems to accelerate in situ stabilization of waste deposits.* EPA/540/2-86/002. Cincinnati, Ohio: U.S. EPA.

———. 1988. A compendium of technologies used in the treatment of hazardous waste. EPA/540/2-88/1004. Washington, D.C.: U.S. EPA.

———. 1989. *Seminar publication on transport and fate of contaminants in the subsurface.* EPA/625/4-89/019. Cincinnati, Ohio: U.S. EPA.

———. 1990. *Ultraviolet radiation/oxidation technology—Ultrox International.* EPA/540/A5-89/012, Superfund Innovative Technology Evaluation. Washington, D.C.: U.S. EPA.

———. 1993. *Perox-pure chemical oxidation technology—Perioxidation Systems, Inc.* EPA/540/AR-93/501 Superfund Innovative Technology Evaluation. Washington, D.C.: U.S. EPA.

Van Hoorn, J.W., and W.H. Vandemolen. 1974. *Drainage of slopping of lands, drainage principles and applications.* Vol. 4 of *Design and management of drainage systems.* Publ. 16. 329–339. Wageningen, The Netherlands: International Institute of Land Reclamation Improvements.

Wasseling, J. 1973. Theories of field drainage and watershed runoff: Subsurface flow into drains. In *Drainage, principles and applications.* Wageningen, The Netherlands: International Institute for Land Reclamation and Improvement.

9.18
IN SITU TREATMENT TECHNOLOGIES

In situ treatment is an alternative to pump-and-treat technology and involves the underground destruction and neutralization of contaminants. The technology has the advantage of requiring minimal surface facilities and reducing public exposure to the contaminant. Theoretically, the technology could be applied to both organic and inorganic contaminants. However, in situ treatment is still relatively new and for the most part has been limited to organic compounds. The most commonly used in situ treatment methods include bioremediation, air sparging, and chemical detoxification.

Bioremediation

Bioremediation is a relatively new technology that has recently gained considerable attention. Bioremediation uses naturally occurring microorganisms to degrade and break down organic contaminants into harmless products consisting mainly of carbon dioxide and water. In situ bioremediation has two basic approaches. The first approach relies on the natural biological activities of indigenous microorganisms in the subsurface. The second approach is called *enhanced bioremediation* and involves stimulating the existing microorganisms by adding oxygen and nutrients. Most organic compounds are biodegradable, some faster than others. The rate of biodegradation, however, depends on the chemical structure of the compound as discussed in Section 9.12 and shown in Table 9.12.3. Figure 9.18.1 shows a simplified representation of a groundwater bioremediation system.

DESIGN CONSIDERATIONS

The design variables of bioremediation include the amount of bacteria, oxygen, and nutrients needed for the biodegradability of the contaminant as well as the characteristics of the subsurface environment. Given those variables, environmental engineers can determine an appropriate hydraulic design of the bioremediation system. Computer models such as BIOPLUME II (1986) can assist in the design of bioremediation systems.

The number of bacteria must be sufficient to consume all of the organic contaminants in a timely manner. Most sites have significant populations of indigenous microorganisms that can degrade a variety of organic contaminants. One gram of surface soil can contain from 0.1 to 1 billion cells of bacteria, 10 to 100 million cells of actinomycetes, and 0.1 to 1 million cells of fungi (Dockins 1980; Whitelaw and Edwards 1980). The microorganism population in soils is generally greatest in the surface horizons where the temperature, moisture, and energy supply is favorable for their growth. As the depth increases, the number of aerobic microorganisms decreases; however, anaerobic microorganisms can exist depending on the availability of nutrients and organic material. The type of microorganisms present on site and their optimal living conditions can be determined in the laboratory. If indigenous microorganisms are not present on site or if their number is not sufficient to consume all organic contaminants, appropriate exogenous microorganisms can be imported, or existing microorganisms can be stimulated with the addition of oxygen and nutrients.

In addition, aerobic bacteria require oxygen for their growth. Because the concentrations of dissolved oxygen in groundwater are generally low, adding oxygen supports the aerobic biodegradation of organic compounds in groundwater. The theoretical quantities of oxygen required to degrade an organic compound can be determined from stoichiometric analysis. For example, degradation of a simple organic acid, such as acetic acid, theoretically requires

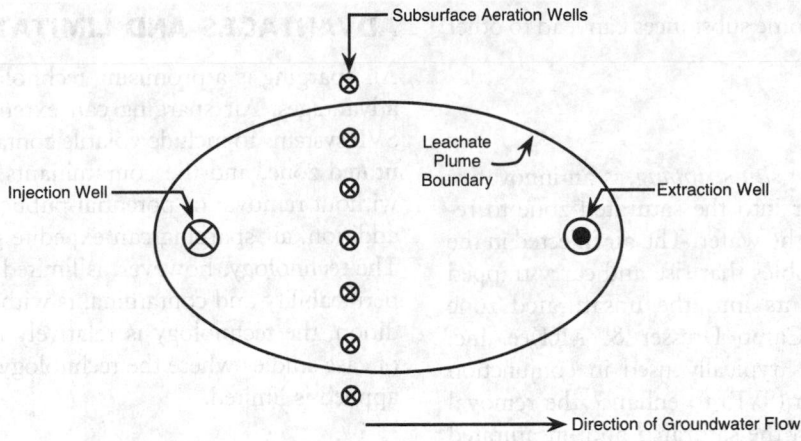

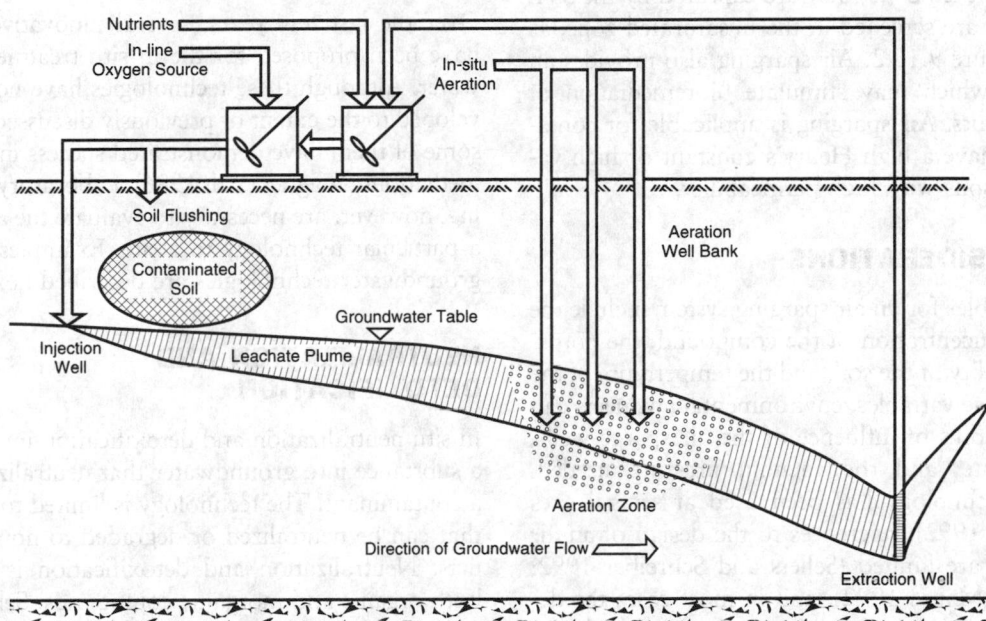

FIG. 9.18.1 Simplified representation of a groundwater bioremediation system. (Reprinted from U.S. Environmental Protection Agency, 1985, *Handbook, remedial action at waste disposal sites*, EPA/625/6-85/006, Washington, D.C.: U.S. EPA.)

1.1 mg of oxygen. Oxygen can be added in several ways, including aeration, oxygenation, and the use of hydrogen peroxide and other oxygen-containing compounds. Obviously, the use of these compounds requires careful control of the geochemistry and hydrology of the site.

Inorganic nutrients including nitrogen, phosphorous, and potassium are needed for proper bacterial growth and can limit cell growth if they are not present at sufficient levels. The groundwater may already contain levels of phosphorous and nitrogen, but these levels are probably insufficient for bacterial growth (Bouwer 1978; Doetsch and Cook 1973). The addition of nutrients, however, can contaminate the aquifer. Therefore, only the amount needed to sustain biological activity should be added.

Other factors limit the growth rate of bacteria and, therefore, the biodegradation of organic contaminants in groundwater. These factors include the pH, temperature, and toxicity of the contaminant. The appropriate range for these parameters should be determined in a treatability study.

ADVANTAGES AND LIMITATIONS

Bioremediation has several advantages over other cleanup technologies including cost, minimal surface facilities, and minimum public exposure to the contaminant. However, bioremediation suffers from several drawbacks (Lee et al. 1988). The technology is limited to aquifers with high permeability. Bacterial growth can be inhibited by one or more compounds at sites with mixed wastes. In addition, incomplete degradation of some substances can lead to other types of contamination.

Air Sparging

Air sparging, also called *in situ stripping*, is an innovative technology that injects air into the saturated zone to remove contaminants from the water. The air injected in the saturated area creates bubbles that rise and carry trapped and dissolved contaminants into the unsaturated zone above the water table (Camp Dresser & McKee, Inc. 1992). This technology is typically used in conjunction with soil vapor extraction (SVE) to enhance the removal rate of contaminants from the saturated and unsaturated zones (Bohler et al. 1990). As volatile organic compounds reach the unsaturated zone, they are captured by the SVE vapor wells that are screened in the unsaturated zone, as illustrated in Figure 9.18.2. Air sparging also provides an oxygen source which may stimulate bioremediation of some contaminants. Air sparging is applicable for contaminants which have a high Henry's constant or high vapor pressure in soils with high permeability.

DESIGN CONSIDERATIONS

The design variables for an air sparging system include the volatility and concentration of the compound, the porosity and permeability of the soil, and the temperature of the water. Given those variables, environmental engineers can determine the radius of influence of the air sparge wells, the air flow rate, and the vacuum pressure needed. Although the technology has been used at several sites (Loden and Fan 1992), references to the design of an air sparging system are limited (Sellets and Schreiber 1992; Marley, Li, and Magee 1992), and in most cases the design is based on empirical formulas or the results of pilot studies.

ADVANTAGES AND LIMITATIONS

Air sparging is a promising technology which has several advantages. Air sparging can extend the effectiveness of SVE systems to include volatile contaminants from the saturated zone, and the contaminants can be treated onsite without removal or potential public exposure to them. In addition, air sparging can expedite groundwater cleanup. The technology, however, is limited to aquifers with high permeability and contaminants with high volatility. In addition, the technology is relatively new, and the number of case studies where the technology has been successfully applied is limited.

Other Innovative Technologies

Over the last few years, several innovative technologies have been proposed for the in situ treatment of groundwater. Although these technologies have not yet been developed to the extent of previously discussed technologies, some of them have demonstrated success in actual site remediations (Wagner et al. 1986). Laboratory and pilot testing, however, are necessary to evaluate the applicability of a particular technology to a site. Examples of innovative groundwater technologies are described next.

NEUTRALIZATION AND DETOXIFICATION

In situ neutralization and detoxification involves injecting a substance into groundwater that neutralizes or destroys a contaminant. The technology is limited to contaminants that can be neutralized or degraded to nontoxic byproducts. Neutralization and detoxification is applicable to both organic and inorganic compounds. Selecting a treatment agent depends on the type of contaminant and the characteristics of the subsurface environment such as tem-

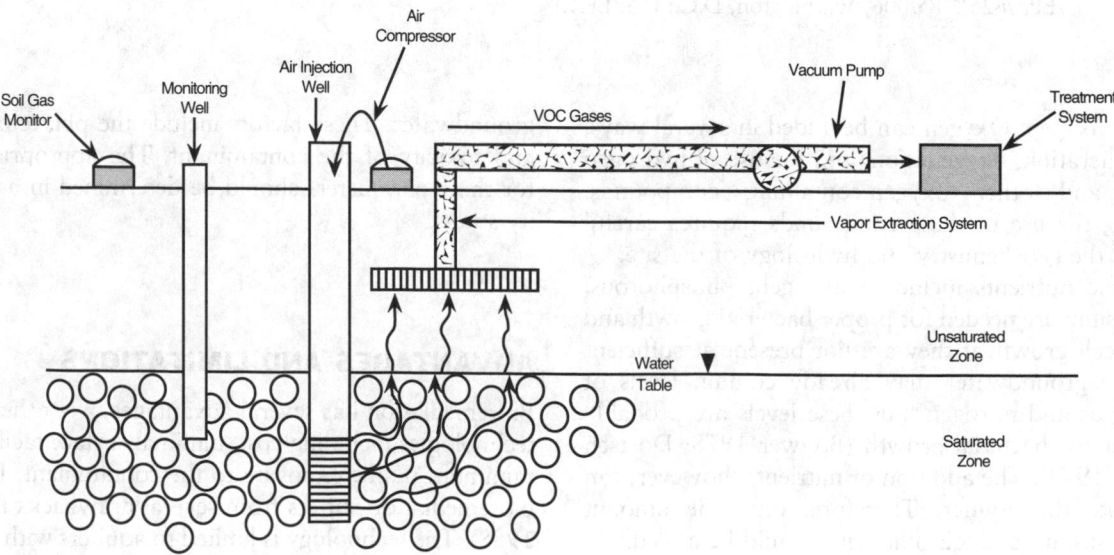

FIG. 9.18.2 Simplified representation of an air sparging system.

perature, permeability, pH, salinity, and conductivity. Examples of in situ treatment agents include hydrogen peroxide that can be injected directly into groundwater through existing monitoring wells or subsurface drains (Vigneri 1994). Hydrogen peroxide produces the hydrogen free radical OH, an extremely powerful oxidizer which progressively reacts with organic contaminants to produce carbon dioxide and water.

Other in situ neutralization and detoxification technologies include precipitation and polymerization. Precipitation involves injecting substances into the groundwater plume which form insoluble products with the contaminants, thereby reducing the potential for migration in groundwater (U.S. EPA 1985). This technique is mainly applicable to dissolved metals, such as lead, cadmium, zinc, and iron. Some forms of arsenic, chromium, and mercury and some organic fatty acids can also be treated by precipitation (Huibregts and Kastman 1979). The most common precipitation reagents include hydroxides, oxides, sulfides, and sulfates. As with other in situ techniques, precipitation is only applicable to sites with aquifers having high hydraulic conductivities. The major disadvantages of precipitation are that it can only be applied to a narrow, specific group of chemicals (mainly metals); that a potential groundwater pollutant may be injected; that toxic gases (as in sulfide treatment) may form; and that the precipate may resolubilize (U.S. EPA 1985).

In situ polymerization involves injecting a polymerization catalyst into the nonaqueous organic phase of a contaminant plume to cause polymerization (U.S. EPA 1985). The resulting polymer is gel-like and nonmobile in the groundwater flow regime. Polymerization is a specific technique that is applicable to organic monomers such as styrene, vinyl chloride isoprene, methyl methacrylate, and acrylonitrile (Huibregts and Kastman 1979). In a hazardous waste site where groundwater pollution has occurred over time, any organic monomers originally present would most likely have polymerized upon contact with the soil (U.S. EPA 1985). Therefore, in situ polymerization is a technique most suited for groundwater cleanup following land spills or underground leaks of a pure monomer. The major disadvantages of polymerization include its limited application and the difficulty of initiating sufficient contact of the catalyst with the dispersed monomer (Huibregts and Kastman 1979).

PERMEABLE TREATMENT BEDS

Permeable treatment beds are also in situ treatment techniques used at sites with relatively shallow groundwater tables. The concept of a permeable treatment bed involves excavating a trench, filling the trench with a permeable treatment material, and allowing the plume to flow through the bed thus physically removing or chemically altering the contaminants. The function of a permeable treatment bed is to reduce the quantities of contaminants

in the plume to acceptable levels. Potential problems with using a permeable treatment bed include saturation of the bed material, plugging of the bed with precipitates, and the short life of the treatment material (U.S. EPA 1985).

The selection of the appropriate bed material to treat the contaminants and the design of the bed are two elements that determine the effectiveness of a permeable treatment bed. The types of available treatment bed fill material include limestone, crushed shell, activated carbon, glauconitic greensands, and synthetically produced ion exchange resins. Ensuring proper physical design of the treatment bed requires a knowledge of the hydrogeology of the site (e.g., groundwater flow rate and direction, hydraulic conductivities) and the chemical characteristics of the plume (U.S. EPA 1985).

PNEUMATIC FRACTURING

Environmental engineers use pneumatic fracturing extraction and hot gas injection to treat in situ contamination located within low permeable formations (Accutech Remedial Systems, Inc. 1994). The process has been demonstrated at numerous sites and significantly increases subsurface permeability and contaminant mass removal (U.S. EPA 1993b). The process applies controlled bursts of high pressure air into a well through a proprietary injection and monitoring system. When the down-hole pressure exceeds the pressure of the formation, channels or fractures are created propagating from the fracture well. Once the permeability of the formation is increased, engineers inject hot gas air (250 to 300°F for pilot-scale and 300 to 600°F for full-scale design) under pressure to elevate the temperature of the fracture surface and volatilize contaminants located within the formation matrix. The extracted vapors are then treated by activated carbon during low-concentration process streams or by catalytic technology during high-concentration process streams.

The technology can be applied at depths to 50 feet and has a radius of influence of as much as 40 feet from the injection point (well). Subsurface air flow has been increased 150 times compared with the site's natural permeability. The technology, however, is not applicable for treating inorganic or nonvolatile organic compounds. In addition, applying the pneumatic fracturing process may be unnecessary at a site with a high natural permeability.

THERMALLY ENHANCED RECOVERY

The in situ steam enhanced extraction process, called thermally enhanced recovery (Praxis Environmental Services Inc. 1994), removes volatile and semivolatile organic compounds from an area of contaminated soil or groundwater without excavation. The process operates through the use of wells constructed in the contaminated soil. High-quality steam is added to the soil through some wells, called *injection wells*. Other wells, known as *extraction wells*, operate under vacuum to remove liquid and vapor contaminants and water from the soil. Injecting steam into the ground raises the temperature of the soil and causes the most volatile compounds to vaporize. In addition, pressure gradient is formed between the injection and extraction wells which drives the flow of steam and vaporized contaminants towards the extraction wells (U.S. EPA 1993a). Raising the temperature of the soil matrix also assists in removing less volatile compounds by increasing their in situ vapor pressure. After the entire soil mass being treated has reached the steam temperature, as determined by soil–temperature monitors, and steam breakthrough occurs at the extraction wells, the flow of steam continues only intermittently with a constant vacuum applied to the extraction wells. The vacuum extraction removes much of the remaining contamination. As the soil in the high permeability region cools, the steam remaining in the low permeability region evaporates the contaminants.

The technology is cost-effective for large and deep areas of contamination where technologies requiring excavation are difficult or impossible. The process can be applied in sections to treat an area of any size and depth. If the site, however, contains a high concentration (>200 ppm) of heavier-than-water organics, a possibility exists that these compounds might be mobilized downward into groundwater. In addition, treatment of shallow (<10 feet) contaminated areas is less cost-effective than deeper areas compared to other technologies.

—*Ahmed Hamidi*

References

Accutech Remedial Systems, Inc. 1994. *Pneumatic fracturing.* Keyport, N.J.

BIOPLUME II. 1986. *Computer model of two dimensional contaminant transport under the influence of oxygen limited biodegradation in groundwater.* Houston, Tex.: National Center for Groundwater Research, Rice University.

Bohler, U. et al. 1990. Air injection and soil air extraction as a combined method for cleaning contaminated sites: Observations from test sites in sediments and solid rocks. In *Contaminated Soil '90,* edited by F. Arench et al., 1039–1044. The Netherlands: Kluwser Academic Publ.

Bouwer, H. 1978. *Groundwater hydrology.* New York: McGraw-Hill.

Camp Dresser & McKee, Inc. 1992. *A technology assessment of soil vapor extraction and air sparging.* Risk Reduction Engineering Laboratory, Office of Research and Development. Cincinnati, Ohio: U.S. EPA.

Dockins, W.S. et al. 1980. Dissimilatory bacterial sulfate reduction in Montana groundwaters. *Geomicrobiology Journal* 2, no. 1:83–98.

Doetsch, and T.M. Cook. 1973. *Introduction to bacteria and their ecobiology.* Baltimore, Md.: University Park Press.

Huibregts, K.R., and K.H. Kastman. 1979. *Development of a system to protect groundwater threatened by hazardous spills on land.* Oil and Hazardous Material Spills Brands, Industrial Environmental Research Laboratory. Edison, N.J.: U.S. EPA.

Lee, M.D. et al. 1988. Biorestoration of aquifers contaminated with organic compounds. *CRC Crit. Rev. Environ. Control* 18:29–89.

Loden, M.E., and C.Y. Fan. 1992. Air sparging technology evaluation. Proceedings of 2nd National Research and Development Conference on the Control of Hazardous Materials, 328–334. San Francisco, Calif.

Marley, M.C., F. Li, and S. Magee. 1992. The application of a 3-D model in the design of air sparging systems. Proceedings of the Petroleum Hydrocarbons and Organic Chemicals in Groundwater 4–6 Nov., 377–392. Houston, Tex.: NGWA.

Praxis Environmental Services, Inc. 1994. *Thermally enhanced recovery in situ.* San Francisco, Calif.

Sellets, K.L., and R.P. Schreiber. 1992. Air sparging model for predicting groundwater cleanup rate. Proceedings of the Petroleum Hydrocarbons and Organic Chemicals in Groundwater, 4–6 Nov., 365–376. Houston, Tex.: NGWA.

U.S. Environmental Protection Agency. 1985. *Leachate plume management.* EPA/540/2-85/004. Washington, D.C.: U.S. EPA.

———. 1993a. In-situ steam enhanced extraction process. In *Superfund Innovative Technology Evaluation Program, Technology Profiles.* 6th ed. EPA/540/R-93/526. Washington, D.C.: U.S. EPA.

———. 1993b. Pneumatic fracturing extraction. In *Superfund Innovative Technology Evaluation Program, Technology Profiles.* 6th ed. EPA/540/R-93/526. Washington, D.C.: U.S. EPA.

Vigneri, R. 1994. *Groundwater remediation primer.* Wilmington, N.C.: Cleanox Environmental Services, Inc.

Wagner, K. et al. 1986. *Remedial action technology for waste disposal sites.* 2d ed. Park Ridge, N.J.: Noyes Data Corporation.

Whitelaw, K. and R.A. Edwards. 1980. Carbohydrates in the unsaturated zone of the chalk. *England Chemical Geology* 29, no. 314:281–291.

Storm Water Pollutant Management

9.19
INTEGRATED STORM WATER PROGRAM

Storm water is defined as storm water runoff, snowmelt runoff, and surface runoff and drainage. Storm water management is important in urban water systems, including water supply systems and wastewater systems. With increasing residential, commercial, and industrial development, stormwater has become an important issue.

Growing urbanization has a significant impact on the surrounding environment, creating problems such as nonpoint sources of water pollution. Because of changes in land-use patterns, pollutants in developed areas build up during dry periods and are washed off as runoff passes over land surfaces. Nonpoint sources account for about 45%, 76% and 65% of the degradation of estuaries, lakes, and rivers respectively (EPA 1989). In comparison, municipal and industrial point source discharges under National Pollution Discharge Elimination System (NPDES) control account for about 9–30% of the degradation of these water sources.

In contrast to our complex urban environment, the hydrological cycle shown in many hydrology textbooks is rather simplistic. Modification of natural drainage paths, damming of waterways, impoundment of water, reuse of stormwater, and implementation of new stormwater management processes result in highly intricate hydrological processes. The development of storm water runoff and its possible superimposition on dry weather flow in combined sewer systems are summarized in Figure 9.19.1. A detailed urban drainage subsystem is shown in Figure 9.19.2.

Integrated Management Approach

Storm water system components and functions interact with, and may also interfere with, each other. Integrated system management coordinates actions to achieve water quantity and quality control, focusing on issues such as floodplain management, erosion and sediment control, nonpoint source pollution, and preservation of wetlands and wildlife habitat. System management also facilitates cooperation among all levels of government, and helps to implement laws and regulations to control storm water pollution.

FEDERAL PROGRAMS

In the 1987 amendments to the Clean Water Act, Congress mandated development of a permit system for certain sources of storm water discharge, thus the EPA has established permit application requirements for industrial storm water discharges and municipal storm sewer system discharges. Pollutants entering storm water and surface water systems are now regulated as point sources under Section 402(p) and subject to the NPDES permit process.

The EPA also provides assistance and guidance to municipalities developing storm water management programs. Although there are several agencies with possible authority in this field, no federal agency has assumed general responsibility or control. Most actions taken to date have been local initiatives. Only the Soil Conservation Service has long-standing programs of storm water management.

However, many federal agencies are directly involved in flood hazard mitigation, flood control, and floodplain management. Although there is no federal agency directly mandated to plan and implement stormwater management programs, there are several agencies engaged in related activities.

The federal government exerts a broad influence via its many agencies. For example, in the Corps of Engineers' major structural flood control program, the federal agency consults with local agencies, but maintains field offices and staff for planning, construction, operation, and maintenance. In another approach, the Soil Conservation Service (SCS) has a nationwide network of conservation districts. The districts perform some functions autonomously, while other functions are carried out by the federal staff. In floodplain management, the Federal Emergency Management Agency (FEMA) has established fairly complete federal control, although actions affecting individuals are legally mandated by state laws and local ordinances. In this case, the financial incentives of the flood insurance program are the prime motivation for obtaining required state legislation and local ordinances.

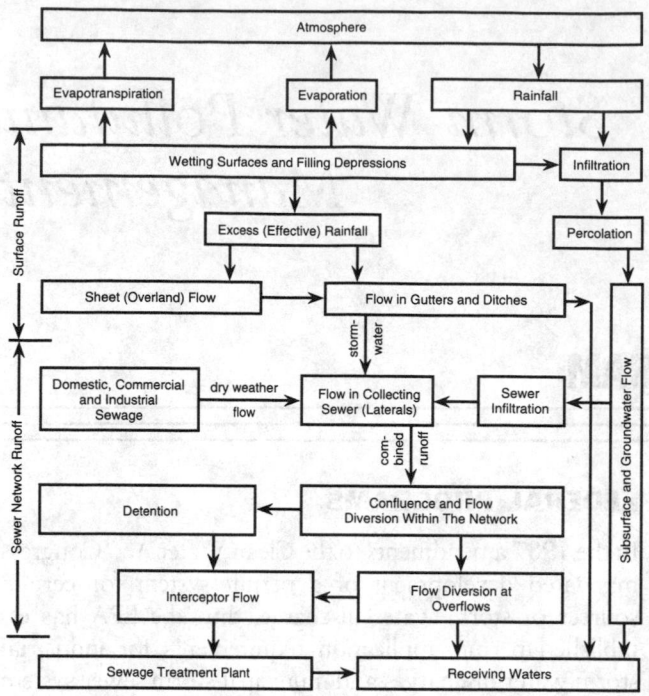

FIG. 9.19.1 Development of Stormwater Runoff and Flow in Combined Sewer Systems. (Reprinted, with permission, from W.F. Geiger, 1984, *Combined sewage quantity and quality—a contribution to urban drainage planning,* [Muenchen Technical Universitat, Muenchen].)

STATE PROGRAMS

State governments enable legislation providing for involvement in storm water management. For example, the Washington State Environmental Policy Act (SEPA) ensures that environmental values are considered by state and local government officials when making decisions. The Department of Ecology (State of Washington) recently completed a storm water rule, a highway storm water rule, and a model storm water ordinance for local governments. These rulings require the development of storm water management programs for cities and counties.

MUNICIPAL PROGRAMS

County-level involvement plays an important role in implementing comprehensive storm water management plans. The principal authority for storm water management is the government with jurisdiction, usually a municipality. Municipalities usually have legal control of:

- Erosion and sedimentation ordinances
- Floodplain ordinances
- Storm water drainage ordinances
- Zoning ordinances
- Building codes
- Grading ordinances

Storm water management is closely tied to future land use development and management. Existing and future

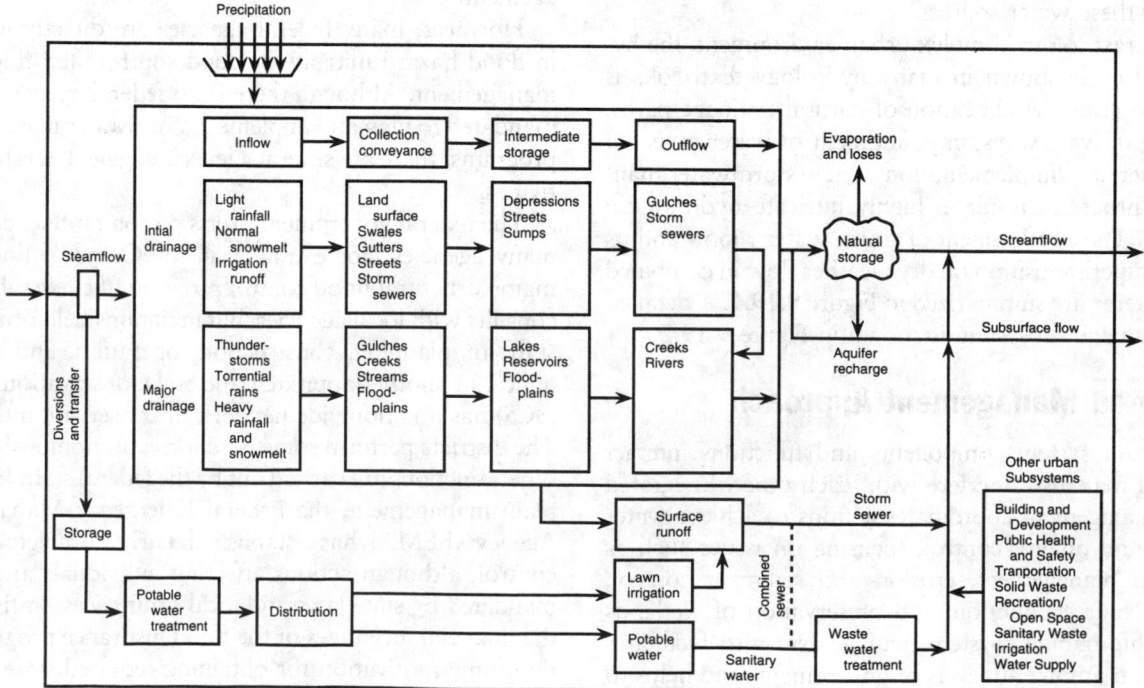

FIG. 9.19.2 The urban storm drainage subsystem.

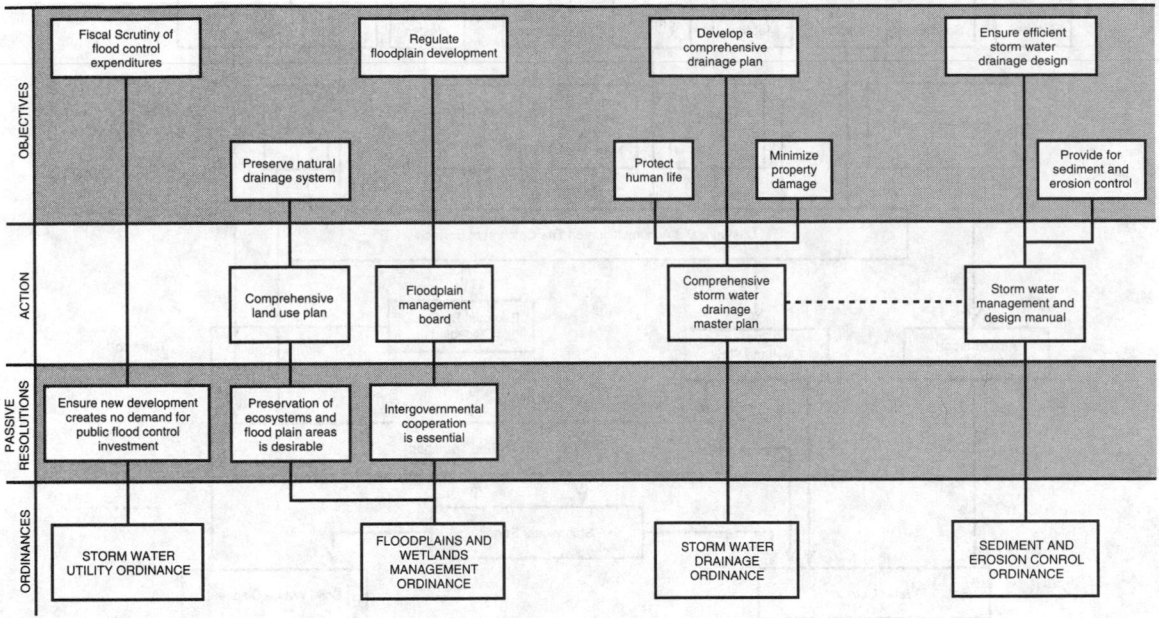

FIG. 9.19.3 Developing a comprehensive storm management program.

land use development are incorporated into an integrated storm water management program as presented in Figure 9.19.3.

Many municipalities now require developers to consider future development of watersheds when designing storm water drainage systems for new development. Detention facilities are frequently required in subdivision laws, zoning ordinances, building codes, and water pollution regulations.

—*Kent K. Mao*

9.20
NONPOINT SOURCE POLLUTION

Urban storm water pollution and most pollution in combined sewer overflows originates from nonpoint or diffuse sources. The processes controlling storm water quality are rather complex, as shown in Figure 9.20.1. In contrast to point source pollution, such as industrial and municipal treatment plant outfalls, these sources of pollution are numerous and their contributions are difficult to quantify. Diffuse pollution is a hydrologic process that closely follows the statistic character of rainfall, and must be evaluated similarly.

Urban nonpoint sources have been identified as a major cause of pollution of surface water bodies by the U.S. EPA (EPA 1984). In the 1988 Report to Congress (EPA 1990), the EPA stated that urban storm water runoff is the fourth most extensive cause of impaired water quality in the nation's rivers, and the third most extensive cause of impaired water quality in lakes. Combined sewer outflows (CSOs) are tenth on the list of significant sources of impairment for both surface-water bodies.

Major Types of Pollutants

Urban storm water runoff may transport many undesirable pollutants. The pollutants present, and their concentrations, are a function of the degree of urbanization, the type of land use, the densities of automobile traffic and animal population, and the degree of air pollution before rainfall. Major pollutant types are classified as follows:

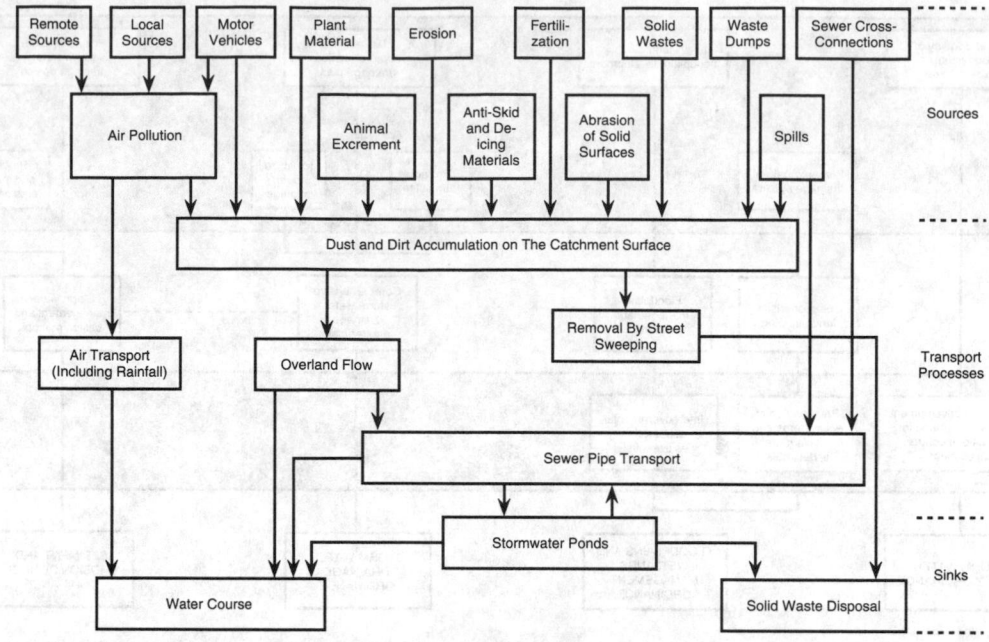

FIG. 9.20.1 Sources of pollutants in stormwater and pollutant pathways. (Reprinted from United Nations Educational, Scientific and Cultural Organization (UNESCO) 1987, Manual on drainage in urbanized areas, vol. 1, *Planning and design of drainage systems,* UNESCO Press.)

- Suspended sediments
- Oxygen-demanding substances
- Heavy metals
- Toxic organics—pesticides, PCBs
- Nutrients—nitrogen, phosphorous
- Bacteria and viruses
- Petroleum-based substances or hydrocarbons
- Acids and
- Humic substances—precursors for trihalomethane

Annual pollutant loadings for storm water and combined sewer overflows are given in Table 9.20.1 (UNESCO 1987).

Nonpoint Sources

Three basic processes generate pollutants during runoff:

ATMOSPHERIC DEPOSITION

Atmospheric deposition is generally divided into wet and dry deposition. Wet deposition is closely related to the levels of atmospheric pollution by traffic, industrial and domestic heating, and other sources. Urban rainfall is generally acidic, with below 5 pH units. The elevated acidity of urban precipitation damages pavements, sewers, and other

TABLE 9.20.1   ANNUAL UNIT POLLUTANT LOADINGS FOR STORMWATER AND COMBINED SEWER OVERFLOWS

| Source | Annual Pollutant Loadings (kg/ha/yr)* | | | | |
	Total Suspended Solids	BOD	COD	Total N	Total P
Runoff in storm sewers	100–6300	5–170	20–1000	2–12	0,2–2,2
Residential area runoff	600–2300	5–100	20–800	2–12	0,2–2,2
Commercial area runoff	100–800	40–90	100–1000	5–12	1,2–2,2
Industrial area runoff	400–1700	10–90	200–1000	5–10	1,0–2,1
Highway runoff	120–6300	90–170	180–3900	—	—
Combined sewer overflows	1200–5000	500–1300	500–3300	15–40	4–8

*1 kg/ha/yr = 0.89 lb/acre/yr

Source: Reprinted from United Nations Educational, Scientific and Cultural Organization (UNESCO), 1987, Manual on drainage in urbanized areas, vol. 1, *Planning and design of drainage systems,* (UNESCO Press).

building materials. Particles are then washed off the surface by stormwater.

Dry atmospheric deposits are fine particles originating from a distance (fugitive dust) or locally (traffic on unpaved roads, construction and industrial) sources. Dustfall rates vary from region to region. Rural dustfalls depend on soil condition; urban dustfalls are more related to local air pollution.

EROSION

Erosion of construction areas represents the largest source of sediments in urban runoff. Reported unit loads of sediment from urban construction sites ranged from 12 to 500 tons/half-yr (Novotny and Chesters 1981). Furthermore, building activities generate other pollutants such as chemicals from fertilizers and pesticides, petroleum products, construction chemicals (cleaning solvents, paints, acids and salts), and various solids. Grading exposes subsoil, increasing surface erosion due to stormwater runoff.

Erosion of urban lawns and park surfaces is usually low. Exceptions are open, unused lands, and construction sites.

Soil is a source of suspended solids, organics, and pesticide pollution. Despite the SCS's active promotion of erosion control, the U.S. Department of Agricultural estimates 57–76 million acres (21–31 Mha), about 15–25% of the nation's agricultural land, is in need of sediment control measures.

ACCUMULATION/WASHOFF

Most urban watersheds are dominated by accumulation and wash-off processes, depending on impervious areas. The accumulation of solids on impervious urban surface areas is described by Sartor and Boyd (1972), as shown in Figure 9.20.2.

The sources of urban diffused pollution are:

- Litter, including large-sized materials (greater than 3.2 mm) containing items such as cans, broken glass, vegetable residues and pet waste. Pet fecal deposits can reach alarming proportions in urban centers where large numbers of people reside in highly impervious zones.
- Medium size deposits (street dirt) represent the bulk of street surface-accumulated pollution. The sources are numerous and very difficult to identify and control. They may include traffic, road deterioration, vegetation resides, pets and other animal waste and residues, and decomposed litter.
- Traffic emissions are responsible for potentially toxic pollutants found in urban runoff, including lead, chromium, asbestos, copper, hydrocarbons, phosphorous, and zinc. Pollution also comes from particles of rubber abraded from tires.
- Road deicing salts applied in winter cause highly increased concentrations of salts in urban runoff.

Road salts are applied at rates of 100–300 kg/km of highway and contain sodium and calcium chloride.

- Pesticides and fertilizers applied onto grassed urban lands.

In fully developed urban areas, where most land surfaces are impervious because of paving and rooftops, washoff of deposited particles and their transport to the watercourse become the important mechanism. The relationship of imperviousness to the quantity of some pollutants are shown in Table 9.20.2.

Table 9.20.3 shows values and ranges of accumulation of street and surface pollutants estimated by Ellis (1986). A list of specific nonpoint sources is presented in Table 9.20.4. The list is not exhaustive. The importance of the sources varies with local conditions.

Direct Input from Pollutant Source

Nonpoint pollutants can also reach receiving waters by direct input from a pollutant source. Drainage systems include depressions, ditches, culverts, catch basins, wetlands, and creeks that collect water and transport pollutants to receiving waters. Pollutants may be directly introduced at specific sites in the system, independent of storm conditions. For example, substances may be poured into a catch basin, traveling directly into a creek or other receiving water.

In addition to cross-connections of sewage and industrial wastes from sanitary sewers, solid waste dumps, and failing septic tanks, solids accumulations and growth in sewers can also enter into storm sewers. Excess water from lawn watering and car washing is another example of direct input. Pollutant loadings from direct inputs are difficult to document and quantify.

—Kent K. Mao

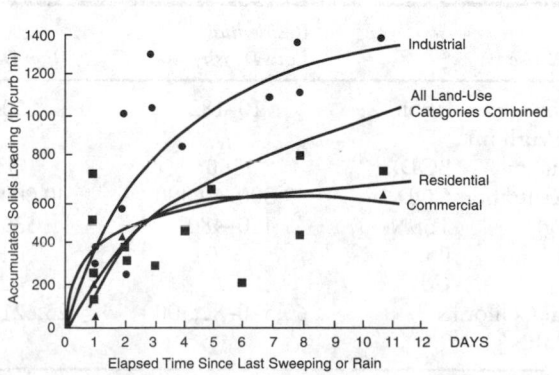

FIG. 9.20.2 Pollutant Accumulation for Different Urban Land Uses. (Reprinted, with permission from J.D. Sartor and S. Boyd, 1972, *Water pollution aspects of street surface contaminants,* U.S. Environmental Protection Agency (EPA), EPA Report R2–72–087, Washington, D.C.)

TABLE 9.20.2 ANNUAL STORM POLLUTANT EXPORT FOR VARIOUS LAND USE TYPES BY PERCENT IMPERVIOUS COVER (POUNDS/ACRE/YEAR)

General Land Use	Percent Imperviousness	Total Phosphorous	Total Nitrogen	BOD 5-Day	Zinc	Lead
Rural to	0	0.11	0.80	2.10	0.02	0.01
residential	5	0.20	1.60	4.00	0.03	0.01
	10	0.30	2.30	5.80	0.04	0.02
Large lot,	10	0.30	2.30	5.80	0.04	0.02
single family	15	0.39	3.00	7.70	0.06	0.03
	20	0.49	3.80	9.60	0.07	0.04
Medium density	20	0.49	3.80	9.60	0.07	0.04
single family	25	0.58	4.50	11.40	0.08	0.05
	30	0.68	5.20	13.30	0.10	0.05
	35	0.77	6.00	15.20	0.11	0.06
Townhouse	35	0.77	6.00	15.20	0.11	0.06
	40	0.86	6.70	17.10	0.12	0.07
	45	0.97	7.40	18.90	0.14	0.07
	50	1.06	8.20	20.80	0.15	0.08
Garden apartment	50	1.06	8.20	20.80	0.15	0.08
buildings	55	1.16	8.40	22.70	0.16	0.09
	60	1.25	9.60	24.60	0.18	0.09
High rise to light	60	1.25	9.60	24.60	0.18	0.09
commercial/industrial	65	1.35	10.40	26.40	0.19	0.10
	70	1.44	11.10	28.30	0.21	0.10
	75	1.54	11.80	30.20	0.22	0.11
	80	1.63	12.60	32.00	0.23	0.11
Heavy commercial to	80	1.63	12.60	32.00	0.23	0.11
shopping center	85	1.73	13.30	33.90	0.25	0.12
	90	1.82	14.00	35.80	0.26	0.13
	95	1.92	14.80	37.70	0.27	0.13
	100	2.00	15.40	39.20	0.28	0.14

NOTES: Assumed rainfall of 40 in/yr
Rural residential = 0.25–.5 dwelling units/acre
Large lot, single family = 1–1.5 dwelling units/acre
Medium density, single family = 2–10 dwelling units/acre
Townhouse and garden apartment = 10–20 dwelling units/acre
Pollutant loadings are for new developments only.

TABLE 9.20.3 SOLIDS ACCUMULATION AND ASSOCIATED POLLUTANT CONCENTRATIONS IN URBAN AREAS

Land Use		Residential Low Density	High Density	Commercial	Light Industrial	Highways
Solids accumulation (g/curb m)		10–182	30–210	13–180	80–288	13–1100
Pollutant concentration (μg/g)	BOD$_5$	5260	3370	7190	2920	2300–10,000
	COD	39,300–40,000	40,000–42,000	39,000–61,730	25,100	53,650–80,000
	Tot.N	460–480	530–610	410–420	430	223–1600
	Pb	1570	1980	2330	1390	450–2346
	Cd	3.2	2.7	2.9	3.6	2.1–10.2
Fecal Coliforms (MPN/g)		60,570–82,500	25,621–31,800	36,900	30,700	18,768–38,000

Source: Reprinted, with permission, from J.B. Ellis Pollutional aspects of urban runoff, *Urban runoff pollution,* ed. H.C. Torno, J. Marsalek, and M. Desbordes, 1–38. (New York, N.Y.: Springer, Verlag, Berlin).

TABLE 9.20.4   NONPOINT SOURCE POLLUTANTS

Source	N	O/G	T	S	O	M	P	H
Agricultural								
Nurseries	X		X	X	X			
Crop farms	X		X	X	X			
Livestock/hobby farms	X			X	X		X	X
Feed/seed/fertilizer supply	X		X		X			
Commercial/Retail								
Restaurants	X	X			X		X	
Dry cleaners			X					X
Garden centers	X		X	X	X	X		
Printing shops			X					
Urban Stormwater								
Roof washoff	X	X	X	X	X	X		X
Lawn/landscape washoff	X		X	X	X		X	
Yard debris	X			X	X		X	
Septic systems	X		X		X		X	
Household		X	X			X		
Miscellaneous								
Illicit dumps	X	X	X	X	X	X	X	
Cemeteries	X		X	X				
Warehouses		X	X		X	X		X
Fuel storage facilities		X	X			X		
Streambank erosion	X			X	X			
Ditch cleaning/defoliating	X		X	X	X			
Filling/diverting streams	X			X	X			
Loss of buffer zones				X	X			X
Boating and marinas	X	X	X		X		X	
Construction								
Clearing/grading	X	X		X	X			X
Building		X	X	X				
Transportation								
Roadways/parking lots	X	X	X	X	X	X		X
Service/repair stations		X	X	X	X	X		X
Car/truck washes	X	X	X	X	X	X		X
Oil change shops		X	X	X	X	X		X

N = nutrients; O/G = oils and greases; T = toxic chemicals; S = sediments; O = organics; M = metals; P = pathogens, bacteria; H = heat

References

Browne, F.X. and J.T. Grizzard. 1979. Nonpoint sources, J. Water Pol. Cont. Fed. 51, p. 1428.

Ellis, J.B. 1986. Pollutional aspects of urban runoff, in *Urban runoff pollution*, eds. H.C. Torno, J. Marsalek, and M. Desbordes, 1–38. New York, N.Y.: Springer Verlag, Berlin.

Novotny, V. and G. Chesters. 1981. *Handbook of nonpoint pollution: source and management*. New York, N.Y.: Van Nostrand Reinhold.

Rechow, K.H., M.N. Beaulac, and J.T. Simpon. 1980. *Modeling phosphorous loading and lake response under uncertainty: A manual and compilation of export coefficients*. U.S. Environmental Protection Agency (EPA) EPA 440–5–80–011. Washington, D.C.

Sartor, J.D. and G. Boyd. 1972. *Water pollution aspects of street surface contaminants*. U.S. Environmental Protection Agency (EPA) EPA R2–72–081. Washington, D.C.

United Nations Educational, Scientific and Cultural Organization (UNESCO). 1987. Manual on drainage in urbanized areas. vol. I. *Planning and design of drainage systems*. UNESCO Press.

U.S. Environmental Protection Agency (EPA). 1990. *National water quality inventory—1988 report to Congress*. U.S. EPA Office of Water. EPA Report 440–4–90–003. Washington, D.C.

9.21
BEST MANAGEMENT PRACTICES

Much emphasis is currently placed on controlling storm water pollution by attacking the problem at the source, instead of using more costly downstream treatment facilities. These source controls, termed "Best Management Practices" (BMPs), are judged most effective in reducing nonpoint source pollution to a level compatible with water quality goals.

Best Management Practices are classified into two groups:

- Planning, with efforts directed at future development and redevelopment of existing areas
- Maintenance and operational practices to reduce the impact of nonpoint source contamination from existing developed areas

Successful storm water pollution control depends on the effective implementation of proposed planning efforts and/or control practices. Legislation or ordinances encouraging or requiring conformance with intended BMPs has proven to be effective. Table 9.21.1 lists activities included in a typical source control program.

Planning

The first goal of planning is to develop a macroscopic management concept, preventing problems from short-sighted development of individual areas. The planner is interested in controlling storm water volume, rate, and pollutional characteristics of storm water runoff. Since the size of storm sewer networks and treatment plants relates directly to flow quantity, particularly the peak flowrate, reducing total volume or smoothing out peaks will result in lower construction costs.

TABLE 9.21.1 TYPICAL SOURCE CONTROLS

Activity or Area	Overview
S1.10—Fueling stations (both commercial and private)	Covered, concrete-paved pump island with separate drainage
S1.20—Vehicle/equipment wash and steam cleaning	Wash building or designated paved area with separate drainage containing oil-water separator
S1.30—Loading and unloading liquid materials	Conduct activities inside building or at dock with overhang or skirts to prevent drainage to storm drains; rail and tanker truck transfers require drip pans or paved areas, and operations and spill cleanup plans
S1.40—Above-ground tanks for liquid storage	Diked secondary containment area with stormwater drainage passing through oil-water separator
S1.50—Container storage of liquids, food wastes, or dangerous wastes	Containers kept indoors or under designated covered area with separate drainage and secondary containment for liquid wastes
S1.60—Outside storage of raw materials, by-products, or finished product (i.e., sand and gravel, lumber, concrete and metal)	Place materials under covered area, place temporary plastic sheeting over material, *or* pave the area and install treatment drainage system
S1.70—Outside manufacturing	Alter, enclose, cover, or segregate the activity; discharge runoff to sewer or process wastewater system; or use stormwater BMPs
S1.80—Emergency spill cleanup plans	Required for storing, processing, or refining oil products and producers of dangerous wastes
S1.90—Vegetation management /integrated pest management	Specific BMPs for seeding and planting, and pest management, including use of pesticides
S2.00—Maintenance of storm drainage facilities	Specific BMPs for maintenance (inspection, repair, and cleaning), disposal of contaminated water, and disposal of contaminated sediments

Source: Stormwater Management Manual for the Puget Sound Basin (Ecology 1992).

LAND USE PLANNING

Computer simulations are used to examine interacting pollutant sources in the watershed. By modeling the runoff process, a planner can predict the effects of proposed plans, and the ability of controls to solve potential problems. Several models are described in Section 10.9. Water quality criteria standards can be recommended after investigating pollution sources and the ability of receiving water to absorb loadings.

When watershed goals are set, the planning agency has two choices for achieving water quality standards. Individual sites can be forced to comply with the practices and performance standards set forth in the master plan, or the basin system must be designed and maintained as a public utility. Isolated development tracts can be controlled by requiring developers to follow specific source control practices, or a simple set of performance standards can be applied and the choice of practices can be left up to the developer. For example, the agency can require that runoff from developed sites must not exceed predevelopment intensity. The developer will have to minimize runoff-producing areas and provide detention facilities at the site.

Planners must also consider the effects of their actions on areas outside the watershed. For example, a system where storm flow is detained in a downstream watershed while it remains unregulated upstream can cause higher flood levels in a river than a completely unregulated system.

NATURAL DRAINAGE FEATURES

The key to preserving a natural drainage system for urbanizing areas is understanding the predevelopment water balance and designing to minimize interference with that system. The soil and hydrology of the site must be studied so that high-density, highly impervious locations, such as shopping centers and industrial complexes, are located in areas with low infiltration potential. Recharge areas should be preserved as open, undisturbed space in parks and woodlands. Runoff from developed areas should be directed to recharge areas and detained to use the full infiltration potential. Broad, grassy swales will slow runoff and maximize infiltration. The drainage plan can include variable depth detention ponds that rise during a runoff event and return to a base level during dry weather.

Realizing that the design goal is maximizing infiltration-recharge and minimizing runoff, the planner should incorporate the following techniques into a site plan:

- Roof leaders should discharge to pervious areas or seepage pits. Dry (French) wells, consisting of borings filled with gravel, can be used for infiltration of rooftop runoff.
- As much area as possible should be left in a natural, undisturbed state. Earthwork and construction traffic will compact soil and decrease infiltration.
- Steep slopes should be avoided. They contribute to erosion and lessen recharge.
- Large impervious areas should be avoided. Parking lots can be built in small units and drained to pervious areas.
- No development should be permitted in flood plains.

Porous pavement is an alternative to conventional pavement. (Thelen and Howe 1978; Dinitz 1980). It provides storage, enhancing soil infiltration to reduce surface and volume from an otherwise impervious area.

For parking lots and access roads, planners can use *modular pavement* systems. Pavers are placed on a prepared sand and gravel base, which overlays the subsoil. The voids of the pavers are filled with either sand, gravel, or sod. Frost problems are minimal.

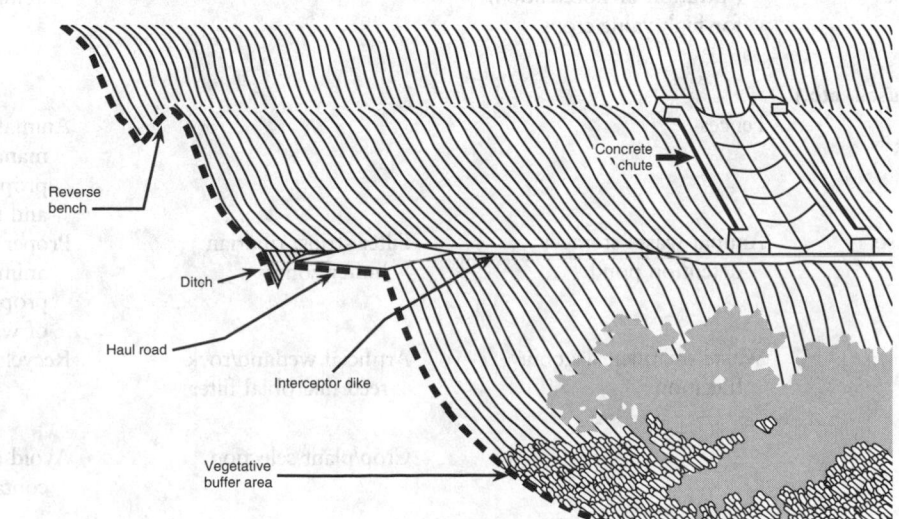

FIG. 9.21.1 Interception and diversion measures.

TABLE 9.21.2 SELECTING BMPs BY POLLUTANTS

Methods of Control	Structural	Vegetative	Management
Sediment (TSS, cobble embeddedness, turbidity)			
Control erosion on land and streambank	Terraces; diversions; grade stabilization structures; streambank protection and stabilization	Cover crops and rotations; conservation tillage; critical area planting	Contour farming; riparian area protection; proper grazing use and range management
Route runoff through BMPs that capture sediment	Sediment basins	Filter strip; grassed waterway; stripcropping; field borders	
Dispose of sediment properly			Beneficial use of sediment—wetland enhancement
Nutrients: N, P (nuisance algae, low dissolved oxygen, odor)			
Minimize sources	Animal waste system (lagoon, storage area); fences (livestock exclusion); diversions; terraces	Range management; crop rotations	Range and pasture management; proper stocking rate; waste composting; nutrient management
Uptake all that is applied to the land or contain and recycle/ reuse (dissolved form control—commercial nutrients)	Terrace; tailwater pit; runoff retention pond; wetland development	Cover crop; strip cropping; riparian buffer zone; change crop or grass species to one that is more nutrient demanding	Recycle/reuse irrigation return flow and runoff water; nutrient management; irrigation water management
Contain animal waste, process and land apply, or export to a different watershed (dissolved form control—animal waste)	Diversion; pit/pond/lagoon; compost facility	See 2(a)	Lagoon pump out; proper irrigation management
Minimize soil erosion and sediment delivery (adsorbed form control)	Terrace; diversion; stream-bank protection and stabilization; sediment pond; critical area treatment	Conservation tillage; filter strip; riparian buffer zone; cover crop	Nutrient management
Intercept, treat runoff before it reaches the water (suspended form control)	See 1–3; water treatment (filtration or flocculation) for high-value crops	Riparian buffer zone	See four preceding items, this column
Pathogens (bacteria, viruses, etc.)			
Minimize source	Fences		Animal waste management, especially proper application rate and timing
Minimize movement so bacteria dies	Animal waste storage; detention pond	Filter strips; riparian buffer zones	Proper site selection for animal feeding facility; proper application rate of waste
Treat water	Waste treatment lagoon; filtration	Artificial wetland/rock reed microbial filter	Recycle and reuse
Metals			
Control soil sources		Crop/plant selection	Avoid adding materials containing trace metals

Continued

TABLE 9.21.2 *Continued*

Methods of Control	Structural	Vegetative	Management
Control added sources	Tailwater pit; reuse/recycle system	Crop selection	Irrigation water management; integrated pest control
Treat water	Filtration	Artificial wetland/rock reed microbial filter system	
Salts/salinity			
Limit availability			Drip irrigation
Control loss	Evaporation basins; tailwater recovery pits; ditch lining; replace ditches with pipe	Crop selection; saline wetland buffer; land-use conversion	Irrigation water management
Pesticides and other toxins			
Minimize sources		Plant variety/crop selection	IPM; change planting dates; proper container disposal
Minimize movement and discharge	Terrace; sediment control basin; retention pond with water reuse/recycle system	Buffer zone; conservation tillage; filter strips (adsorbed control only); wetland enhancement	Irrigation water management; IPM
Treat discharge water	Carbon filter system (high-value crops)	Rock-reed microbial filter system/artificial wetland	
Physical habitat alteration			
Minimize disturbance within 100 feet of water	Road and turnrow realignment; fencing/livestock water crossing facility	Buffer strips; riparian buffer zones	Proper grazing management, including limiting livestock access
Control erosion on land	See sediment BMPs		
Maintain or restore natural riparian area vegetation and hydrology	Streambank stabilization; channel integrity repair	Wetland enhancement	Proper grazing use and range management; limit livestock access

Sources: U.S. EPA (1993); Brach (1990); Alexander (1993a); USDA, Soil Conservation Service (1988).

EROSION CONTROLS

Erosion control for construction and developing sites will have a major impact on the total pollution loads in receiving waters. Current estimates show that approximately 1500 sq mi of the United States is urbanized annually. All of this land area is exposed to accelerated erosion.

Following are basic guidelines and principles of erosion control. Reduce the area and duration of soil exposure. For example, various mining operation stages should be scheduled so that clearing, grubbing, scalping, grading and revegetation occur concurrently with extraction, so that a minimum area is exposed at one time.

Protect the soil with mulch and vegetable cover. For example, covering the soil surface with wood chips reduces construction site soil loss by 92%. Vegetation also has a marked effect on water quality. Temporary fast-growing grass can reduce erosion by an order of magnitude; sod-

ding can reduce erosion by two orders of magnitude. Straw mulch application can be combined with grass seeding for permanent surface protection.

Reduce the rate and volume of runoff by increasing the infiltration rate. A properly roughened and loosened soil surface will benefit plant growth, enhance water infiltration, and slow surface runoff.

Diminish runoff velocity with planned engineering works. A key concept in controlling soil erosion is to intercept runoff before it reaches a critical area and divert it to a safe disposal area. Interception and diversion are accomplished through various structures, including earth dikes, ditches, and combined ditch and dike structures (Figure 9.21.1).

Protect and modify drainage ways to withstand concentrated runoff from paved areas. To reduce the rate of flow and the resulting detachment and transport of soil particles in natural and manmade drainageways, grade can be controlled by the construction of flumes or other flow barriers across the channel. Bends in the channel, either natural or manmade, also impede flow.

Trap as much sediment as possible in temporary or permanent sedimentation basins.

Maintain completed works and assure frequent inspection for maintenance needs.

Principal cropland erosion control practices and BMPs for pollutants are summarized in Table 9.21.2.

Maintenance and Operational Practices

Proper maintenance and cleanliness of an urban area can have a significant impact on the quantity of pollutants washed from an area by storm water. Cleanliness of an urban area starts with control of litter, debris, deicing agents, and agricultural chemicals such as pesticides and fertilizers. Regular street repair and sweeping can further minimize pollutants in stormwater runoff. Proper drainage collection system use and maintenance can maximize control of pollutants by directing them to treatment or disposal.

URBAN POLLUTANT CONTROL

Litter Control

Used food containers, cigarettes, newspapers, sidewalk sweepings, lawn trimmings, and other materials carelessly discarded become street litter. Unless this material is prevented from reaching the street or is removed by street cleaning, it is often found in stormwater discharges. Enforcement of antilitter laws, convenient location of disposal containers, and public education programs are source control measures.

Chemical Use Control

Reducing the indiscriminate use and disposal of fertilizers, pesticides, oil and gasoline, and detergents is a frequently overlooked measure for reducing stormwater runoff pollution. Tree spraying, weed control, municipal fertilization of parks and parkways, and homeowner use of pesticides and fertilizers can be controlled by increasing public awareness of the potential hazards to receiving waters. Direct dumping of chemicals and debris into catch basins, inlets, and sewers is a significant problem that can only be addressed through educational programs, ordinances, and enforcement.

Street Sweeping

Street sweeping is used by most cities to remove accumulated dust, dirt and litter from street surfaces, but clean-

ing is usually done for aesthetic reasons. Street cleaning practices effectively attack the source of stormwater-related problems.

The type of cleaning equipment has an effect on the overall effectiveness of debris removal. Public awareness of street cleaning practice is essential for more efficient operations. Vehicles parked on the street during sweeping operations hamper efficiency and prevent cleaning of deposits.

Street Maintenance

Pavement conditions have an effect on the amount of street pollutants. Vehicles travelling over rough streets shake off more particulate matter. A large portion of solids also comes from cracks in the pavement.

Highway Deicing Management

Effective management of highway deicing practices can lessen receiving water impacts associated with chlorides, sodium, and suspended solids. Recommended alternatives for modifying deicing practices include: (1) judicious application of salt and abrasives; (2) reducing application rates using sodium and calcium salt premixers; (3) using better spreading and metering, and calibrating application rates; (4) prohibiting use of chemical additives; (5) providing improved salt storage areas; and (6) educating the public and operators about the effect of deicing technology and best management practices.

COLLECTION SYSTEM MAINTENANCE

The major objective of maintaining storm or combined sewer systems is to provide maximum transmission of flows to treatment and disposal, while minimizing overflows, bypasses, and local flooding conditions. This objective can be achieved by maintaining system facilities at peak capacity.

The significance of collection system maintenance as a best management practice is that when properly applied, extraneous solids and debris are removed in a controlled manner, not accumulated as pollutant sources to be flushed into receiving waters under storm conditions.

The basic part of a maintenance program is regular system inspection. Specific tasks include: (1) catchbasin maintenance; (2) cleaning (both deposits and root infestation) and flushing of pipes; (3) removal of excess shrubbery and debris from flood control channels and ditches; and (4) control of inflow and infiltration sources.

Sewer cleaning involves routine inspection of the sewer system. All plugged or restricted lines should be cleaned. Major problems in large-diameter sewers are siltation and accumulation of large debris like shopping carts and tree branches. In small-diameter sewers, siltation and penetration of tree roots are major problems. Benefits of sewer cleaning include reducing local flooding, emergency repairs, and pollutant loading. Increased carrying capacities and reduced blockages in interceptor/regulator works may directly reduce overflows.

Many types of sewer cleaning equipment are used, including hydraulic, mechanical, manual, and combination devices. The cleaning tool is pushed or pulled through the sewer to remove obstructions or cause them to be suspended in the flow and carried out of the system. However, large sewer and interceptor cleaning involves unique problems because several feet of sludge blanket can accumulate.

Regular flushing of sewers can ensure that sewer laterals and interceptors continue to carry their design capacity, as well alleviate solids buildup and reduce solid overflow.

Sewer flushing can be particularly beneficial in sewers with very flat slopes. If a modestly large quantity of water is periodically discharged through these flat sewers, small accumulations of solids can be washed from the system. This cleaning technique is effective only on freshly deposited solids.

Internally automatic flushing devices have been developed for sewer systems. An inflatable bag is used to stop flow in upstream reaches until a volume capable of generating a flush wave is accumulated. When the correct volume is reached, the bag is deflated with the assistance of a vacuum, releasing impounded water and cleaning the sewer segment.

INFLOW AND INFILTRATION

Extraneous flow entering a sewer is generally categorized as inflow or infiltration. Inflow generally occurs from surface runoff via roof leaders, yard and area away drains, and flooding of manhole covers. Infiltration usually occurs by water seeping into pipes or manholes from leaky joints, crushed or collapsed pipe segments, leaky lateral connections or other pipe failures. Extraneous flows may result in unnecessary pollution, as these reduce effective collection system and treatment plant facilities.

Details of principal methods of reducing both infiltration and inflow through rehabilitation are found in EPA 1977.

DRAINAGE CHANNEL MAINTENANCE

Maintenance of flood control channels covers a wide range of cleaning tasks. Debris to be removed ranges from trash, garbage, and yard trimmings to used tires and shopping carts.

—*Kent K. Mao*

References

Dinitz, E.V. 1980. *Porous pavement. Phase I. Design and operational criteria.* U.S. Environmental Protection Agency (EPA). EPA Report 600–2–80–135.

Stewart, B.A., D.A. Woodhiser, W.H. Wischmeier, J.H. Caro, and M.H. Frere. 1975. *Control of water pollution from cropland. Vol. I.* U.S. Environmental Protection Agency (EPA). EPA Report 600–2–75–026a. Washington, D.C.

Thelen, E. and L.F. Howe. 1978. *Porous pavement.* Philadelphia, Pa.: The Franklin Institute.

U.S. Environmental Protection Agency (EPA). 1977. *Sewer system evaluation, rehabilitation, and new construction: A manual of practice.* EPA Report 600–2–77–017d.

9.22
FIELD MONITORING PROGRAMS

The objectives of field monitoring water quality in drainage studies include:

- Analyzing the impact on receiving waters of (1) storm sewer discharges, (2) combined sewer overflows, (3) atmospheric fallout and urban activities, and (4) new facilities or treatment plants designed to reduce environment impacts.
- Identifying the contributions of various land uses to total pollution discharge, to optimize urban development and derive some regulations such as source control.

- Increasing existing treatment efficiency during wet weather in combined sewer systems.
- Analyzing of scour and deposit problems in sewers to define optimal cleaning sequences or to design facilities for better hydraulic conditions.

To fulfill these objectives, storm water discharges need to be sampled during dry-weather and wet-weather conditions. Water quality data gathered during dry weather provide a baseline and indicate point source impacts.

To trace contaminants and identify pollutant sources, a phased monitoring approach requires repeated investigation of land use activities in a basin. The program is expected to be an iterative process, as several rounds of sampling are generally required. Precise data are essential for calibrating and verifying nonpoint source models.

Experience proves that water quality data collection programs can be costly. Collection procedures have high manpower requirements, as frequent site visits are required. The cost of analyzing collected samples may increase rapidly with the number and types of pollutants studied. It is important that the parameters to be studied are carefully selected and limited to the essentials.

This section presents an outline of water quality parameters important in studies on urban stormwater discharges, and reveals the main difficulties in obtaining representative samples. Also included is a brief discussion on data analysis.

Selection of Water Quality Parameters

Water quality parameters included in urban hydrological studies may be divided into seven groups. Those parameters, relating to a specific drainage problem, are listed in Table 9.22.1, along with their detection limits, precision level of analysis, and study objectives. In most cases, only biochemical oxygen demand (BOD), chemical oxygen demand (COD), and total suspended solids (TSS) are initially studied, but if these parameters show high values, some other parameters can be taken into account (i.e. Kjeldahl nitrogen, total phosphorous, and volatile suspended solids [VSS]). As the program continues, some special investigation should be made on trace elements and other special parameters mentioned in Table 9.22.1.

Solids are good indicators of urban water quality, as they may contain pollutant materials. Suspended solids are closely related to other pollutant concentrations. In fact, sample uniformity is not easily achieved. Suspended solid concentrations are affected by the flow level, which is not taken into account by manual or automatic sampling. The sampler itself may also introduce effects that can modify the gradient profile of suspended solids. Conditions at the sampler intake cannot be adapted to the flow variations encountered in storm sewers or combined sewers during high flows.

If the sampler cannot be precisely measured in the collected samples, sample uniformity can be questionable. In most cases, suspended solids are regarded as a rough indicator of water quality, so this should be among the parameters selected.

Acquisition of Representative Samples

The number of sampling sites, the frequency of measurements, and the quality parameters to be measured should be chosen. This requires knowledge of the sewer network, significant building activities, street cleaning practices, atmospheric pollution sources affecting the experimental sites, erosion patterns in surrounding natural areas, industrial activities, seasonal or climatic changes, etc., in order to avoid erroneous judgements in understanding the phenomena studied.

The experimental design must be in agreement with the study objectives (Geiger 1981, Gideometeozdat 1984a, Wong and Marsalek 1981). However, trial and error procedures should be used at the beginning of the study for a few basic parameters (for example, BOD, COD, TSS) at a few sampling sites. This information should be used when determining the experimental design.

SAMPLING SITES AND LOCATION

Sampling sites must be chosen according to study objectives, but hydraulic conditions and constraints necessary to the adopted procedures should be given attention. The sampling site must be located at a section downstream of the study site, i.e. corresponding to well-known sewer systems, land use types, special activities, etc. It is recommended that highly turbulent sections with well mixed flow be sampled. However, for the study of sediment transport deposit, these conditions may not be suitable, as suspended solids in the highly turbulent section may be scattered.

For monitoring in-stream impacts, the area of interest should be bracketed by upstream and downstream stations. A control station on a hydrologically similar but undisturbed watershed can be used to determine baseline conditions.

Two types of monitoring stations are employed for nonpoint source surveys:

1. Small catchment stations ranging from 12 to 125 acres (5 to 50 ha) in size, are used to gather data on specific land uses or special areas. They are usually found on storm sewers, drainage ditches or small tributaries.

2. Another type of station is built to monitor larger basins of greater than 125 acres (50 ha), and measure nonpoint source pollution loads impacting a receiving body, such as stream channels or rivers.

There are cases where the final choice must be made from a group of catchments. In such cases, the technique of weighted suitability ratings, as developed for land use, is recommended (Alley 1977). Assignment of suitability values is perhaps the most subjective part of the schedule.

SAMPLING METHODS

Due to the transient nature of storm runoff phenomena, random collection of grab samples does not allow a true representation of pollutant transport. Even if grab samples are modified to concentrate on storm events, the error po-

TABLE 9.22.1 WATER QUALITY PARAMETERS SAMPLED IN URBAN DRAINAGE STUDIES

Parameters	Detection Limits	Precision Level (absolute or relative)	Study Objectives or Observations
Common Constituents and Indicators			
Chlorides		2.5–5%*	Impact of salts used for deicing
Water temperature		0.1°C–0.5°C	Cross-connections; parasites in waters
Conductivity (at 20°C)	5 μS/cm	5%	Changes during runoff, monitoring and control
pH		0.05–0.1 unit*	Rainfall quality analysis
Turbidity			Sediment transport
Nutrients			
Kjeldahl nitrogen		0.1 mg/l	Impact on receiving waters
Total phosphorus		5–15%*	Eutrophication process
Ammonia	0.001 mg/l	1–10%*	Impacts on detention basins with recreational purposes
Nitrites and nitrates	0.05 mg/l	4.5–18%	Cross-connections
Organic Indicators			
BOD$_5$ (5 day BOD)		2–25 mg/l	Impact on receiving waters by oxygen depletion
COD (with <1.5 g/l chlorides)		1–5% (if COD >50 mg/l)	Cross-connections
Trace Elements			
Lead			Impact on receiving waters; toxics accumulation in sediments
Zinc and other heavy metals	0.05–3 μg/l	2–10%†	
Solids			
TSS (at 105°C)	0.5 mg/l	2–5%	Turbidity, oxygen reduction, transport of toxics; increase of hydraulic roughness
VSS (at 550°C)	1 mg/l		Organic part, oxygen depletion
Settleable solids			Maintenance problems in sewers and detention basins in recreational areas
Bacterial Indicators			
Total coliforms			Impact on receiving waters with recreational use
Fecal coliforms			Detection of cross-connections
Special Parameters			
Persistent toxic substances (PTS) such as organochloride pesticides	0.00005 μg/l	0.005–0.05 g/l†	Impact on receiving waters Pollution of receiving waters sediments
Polyaromatic hydrocarbons	1–5 μg/l*		Bioaccumulation in food chains
Chlorinated benzenes	0.002–0.02 μg/l†		

*Depending on the instrument and/or analysis method.
†Depending on the substance analyzed.

tential remains quite high because of variations in pollutant concentrations during runoff events.

Two basic methods can provide estimates of pollutant loading during a storm event:

1. To determine total pollution loading during a storm event, a *flow-weighted composite method* is adequate. In these methods, either aliquot volume or time between aliquots is varied to construct a truly flow-weighted composite from many samples. Analyzing the composite sample and using synoptic flow data allow computation of an accurate estimate of runoff pollution loads, if the intervals between samples are short.

2. When, in addition to total pollution loading, it is necessary to investigate load variations during a storm event, the *sequential discrete procedure* must be used. A

series of samples is retrieved during a monitored runoff event. Following laboratory analysis of each sample and analysis of synoptic flow data, the runoff hydrograph and a curve of pollutant concentration or loading as a function of time may be plotted as shown in Figure 9.22.1. By determining the area under the curve, an accurate estimate of the total pollutant load for an event may be determined.

The interval between sample collection for the above procedures depends on the response time and duration of the storm. In general, at least four samples on the rising limbs and six samples on the recessing limbs should be collected for proper resolution of nonpoint source pollution loads in urban areas.

Samples may be collected either manually or by automatic samplers. Table 9.22.2 shows a matrix of advantages and disadvantages related to each sampling technique. A summary of methods used in urban stormwater sampling and comments on each was prepared by Shelley (Shelley and Kirkpatrick 1975).

Experimental results show sediment distribution in a stream cross section flowing at 5 ft/sec. An analysis of water quality constituents in the stream cross-section should be made to determine the distribution across the width and from top to the bottom of the stream. Samples should be tested for a suspended parameter (such as TSS) and a soluble parameter (such as orthophosphate). The testing should be carried out at a small runoff event and a moderate-to-high flow event if possible. Vertical sampling should be done using depth samplers (such as Kenmeyer bottles) or closeable bottles if the stream is more than 4 to 5 ft (1.2 to 1.5 m) deep. This factor should be considered in designing manual and automatic sampling procedures.

FLOW MEASUREMENT

Flow measurement is perhaps one of the most important aspects of designing an urban collections program. No data collecting task will be capable of achieving its goals if the precision and accuracy of the flow data required for load calculations are not considered.

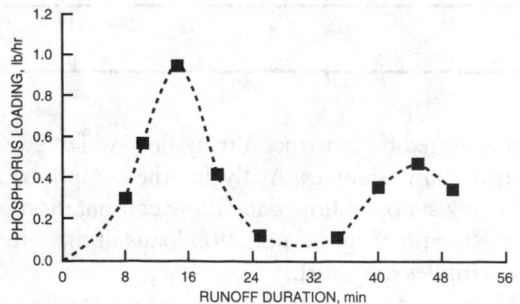

FIG. 9.22.1 Plot of total phosphorus loading at irongate catchment.

The flow measurement devices and methods can be classified according to the physical principles upon which their primary elements are based.

Channel Friction Coefficient Method

This indirect method, also referred to as the slope-area method, consists of measuring flow depth at a suitable cross-section and substituting the measured depth into an equation for uniform flow (such as the Manning equation) or critical flow. To complete the calculation, one must estimate the friction coefficient of the channel where the flow is to be measured, and know the channel slope and geometry.

The inference of flow rates from measured depths of flow is a rather inaccurate procedure. The main sources of error arise from the lack of uniformity and steadiness of flow, and the lack of certainty in estimating the friction coefficient.

Improved accuracy can be achieved by performing calibration in place, and developing an empirical rating curve for each measuring cross-section. In this case, the channel discharge (Q) is measured, generally by current meters, for various depths of flow, and the cross-section rating curve (Q vs depth of flow) is developed. This curve is then used to convert the observed stage to discharge.

Weirs

Measuring weirs are overflow structures built across a flow channel to measure discharge. For a given set of weir and channel geometry conditions, a single head value on the device may exist for each discharge under a free-flow, steady state regimen. The existence of such a relationship makes constructing a rating curve of head versus discharge a simple task. Such rating curves are available in the literature for most common configurations (such as rectangular weirs, V-notch weirs, vertical slot weirs, and trapezoidal weirs without the bottom part) (U.S. Department of Interior 1975).

One advantage of weirs is their large relative measurement range. However, weir installation in sewers reduces pipe capacity, may lead to solids accumulation (particularly in combined sewers), may distort flow hydrographs, and may limit operating range because of surcharging or submerging. These constraints will eliminate weirs from consideration for certain locations, but many of the above difficulties can be avoided in open-channel installations at outfalls. For these reasons, weirs should be used only under carefully controlled conditions, such as at detention basin outlets, where suspended solid concentrations are likely to be low.

Flumes

A measuring flume creates a constriction in the channel cross-section, causing a velocity change and, consequently,

TABLE 9.22.2 COMPARISON OF MANUAL AND AUTOMATIC SAMPLING
TECHNIQUES

Advantages	Disadvantages
Manual Grabs	
Appropriate for all pollutants	Labor-intensive
Minimum equipment required	Environment possibly dangerous to field personnel
	May be difficult to get personnel and equipment to the storm water outfall within the 30 min requirement
	Possible human error
Manual Flow-Weighted Composites (multiple grabs)	
Appropriate for all pollutants	Labor-intensive
Minimum equipment required	Environment possibly dangerous to field personnel
	Human error may have significant impact on sample representativeness
	Requires flow measurements taken during sampling
Automatic Grabs	
Minimizes labor requirements	Samples collected for O&G may not be representative
Low risk of human error	Automatic samplers cannot properly collect samples for VOC analysis
Reduced personnel exposure to unsafe conditions	Costly, numerous sampling sites require the purchase of equipment
Sampling may be triggered remotely or initiated according to present conditions	Requires equipment installation and maintenance
	Requires operator training
	May not be appropriate for pH and temperature
	May not be appropriate for parameters with short holding times (e.g., fecal streptococcus, fecal coliform, chlorine)
	Cross-contamination of aliquot if tubing/bottles not washed
Automatic Flow-Weighted Composites	
Minimizes labor requirements	Not acceptable for VOC sampling
Low risk of human error	Costly if numerous sampling sites require the purchase of equipment
Reduced personnel exposure to unsafe conditions	Requires equipment installation and maintenance, may malfunction
May eliminate the need for manual compositing of aliquots	Requires initial operator training
Sampling may be triggered remotely or initiated according to on-site conditions	Requires accurate flow measurement equipment tied to sampler
	Cross-contamination of aliquot if tubing/bottles not washed

a depth change. In critical flow flumes, the surface profile
in the constriction passes through the critical depth. The
flume discharge can then be directly related to the depth
immediately upstream of the throat.

Flumes are sometimes classified according to throat

shape. Common types include rectangular, trapezoidal,
semicircular, and composite throat flumes. Flumes with a
bottom contraction (a hump) are suitable for installation
in sewers. The Parshall flume, the cut-throat flume, and
the Palmer-Bowlus flume are also popular.

Rating curves for critical flume geometry may be constructed from solution of the Bernouilli Equation at points upstream of and in the flume throat. While they generally exhibit excellent characteristics of self-cleaning, flumes do not share the brand flow measurement characteristics of weirs.

There are a large number of other flume designs that can be used in drainage studies. For example, the Soil Conservation Service has HS, H, and HL flumes designed to measure small, moderate, and large runoff flows, respectively. These devices combine the best features of both flumes and weirs, with wide ranges of measurement, good self-cleaning characteristics, small head loss, and relative insensitivity to submergence.

Differential Pressure Methods

Traditionally, differential pressure flowmeters have been used to measure flows in full closed conduit. Two exceptions to this rule are the U.S. Geological Survey and University of Illinois sewer meters. Although these function as differential meters in the pressure flow region, they are also fully functional in the open-channel flow region, where they act as Venturi flumes. This dual mode of operation represents the main advantage of these flowmeters.

The U.S. Geological Survey (USGS) flowmeter is similar to flumes with a U-shaped throat. The flume does not obstruct the part of the pipe immediately below the crown, thus transition from open-channel flow to pressure flow is fairly smooth and head losses are reduced. Rating curves for the USGS flowmeter are available (Smoot 1975).

Dilution Method

In this method, a tracer is continuously injected at a constant rate into the flow, and tracer dilution by the metered flow is monitored at a downstream point. If a tracer absent in the meter flow is used, the following relationship applies

$$Q_D = q_T C_T / C_D \qquad 9.22(1)$$

where:

Q_D = flow upstream
q_T = tracer input flow
C_T = tracer input concentration
C_D = tracer concentration downstream

The dilution method has some definite advantages, because it is independent of flow characteristics, does not interfere with the flow and, consequently, does not cause any head loss. Using fluorescent dyes and ensuring complete tracer mixing, the method has a good range of measurement (1000 : 1), and can be fairly accurate (5%) (Alley 1977). Disadvantages are the discrete nature of measurement, as opposed to the preferred continuous measurements; the problem with automating the method; and the need for well-trained personnel. Consequently, the dilu-

tion method is mostly used for in-situ calibration of conventional flowmeters.

Basic characteristics of flow measurement methods discussed are summarized in Table 9.22.3.

Sampling Equipment
MANUAL SAMPLING

Certain manual techniques cannot be avoided in studies of urban runoff quality. Manual sampling is useful when setting up automatic equipment, selecting the sampling section, and the inlet location.

Manual sampling requires good logistic preparation. Field crews must be dispatched to sampling sites before the start of a runoff event, so that sampling can start at the beginning of runoff. This is particularly important in combined sewers which exhibit the first flush phenomenon with high pollutant loads occurring early during runoff events. Therefore, field crews may have to be stationed at sampling sites. Extensive field training is essential to ensure collection of adequate samples.

AUTOMATIC SAMPLING

To obtain necessary flow measurements along with storm water samples, two devices are required: one for flow metering and one for flow metering with an interconnection to insure synoptic collection of sample and flow data. Common characteristics of adequate devices are summarized below:

- Sample transport velocity of 3.0 fps or more to prevent sedimentation
- Minimum of 24 discrete sample bottles or ability to composite samples in one container
- 12 v dc supply option
- Constant sample size over different sampling lines for rising and falling streams
- Air purging of sampling intake line before and after sample collection
- Minimum $\frac{3}{8}$ in (or 1 cm) sample line
- No solids deposition in sample train
- Chemically inert surfaces in contact with sample

In general, the intake should point upstream, extended slightly upstream from any obstacles in the flow, and should not excessively obstruct flow to avoid clogging or damage. Locations are recommended along the pipe periphery at about one third of the average water depth above the bottom. The intake should be placed at a cross-section where the flow is highly turbulent and well mixed. At such locations, a single intake, instead of multiple intakes, may be acceptable.

Sample withdrawal is accomplished by a pump controlled by timers or flow meters. The best devices for urban pollution studies fall into the following categories of pumping methods: positive displacement, peristaltic, and

TABLE 9.22.3 CHARACTERISTICS OF SELECTED FLOW MEASUREMENT METHODS

| | Characteristics | | | | | | |
| | Suitable For | | Applicable At | | | Estimated | |
	Open Channel Flow	Pressure Flow	Outfall	Manhole	In Sewer Pipes	Accuracy* (%)	Relative Costs
Depth and channel friction coefficient	X	X	X	X		15–20	low
Depth and stage-discharge relationship	X		X	X		10–15	low
Weirs							
Rectangular	X		X			5†	low to
V-notch	X		X			5†	medium
Modified trapezoidal	X		X	X	X	5†	
Vertical slot	X		X	X	X	5†	
Flumes							
Cut-throat	X		X			5†	
Palmer-Bowlus	X		X	X	X	5†	medium
Parshall	X		X			5†	
USDA (H, HL and HS)	X		X			5†	
Differential pressure flowmeters							
U. of Illinois	X	X			X	5	medium
USGS	X	X			X	5	to high
Tracer dilution	X	X		X	X	5	medium

*Under favorable conditions.

†These relatively high accuracies correspond to well-designed, installed, and operated installations. Under less favorable circumstances, the accuracies would be somewhat lower, between 5 and 10%.

vacuum. Suction lift devices are the best means of sample withdrawal. Such devices have to operate near the flow sampled because the lift is limited to about 15 ft (5 m). Submersible positive displacement pumps are commonly used where equipment installation is restricted to locations too high above the water surface to operate in a suction lift mode. Although such pumps allow sampling at greater depth, they are susceptible to malfunction and clogging.

FLOWMETERING DEVICES

Selection of secondary devices for the continuous measurements necessary to convert from stage to discharge is an important facet of developing an automated monitoring program. Important criteria for these secondary devices include:

- Wide measurement range
- Accuracy and precision over the entire range
- Minimal calibration loss with time
- Insensitivity to suspended solids in flow
- Capacity to internally convert stage to discharge
- Capacity to trigger an associated sampler
- Unattended operations

Secondary devices are divided into four categories: float-operated devices; ultrasonic devices; bubbler devices (manometers and transducers); and combination bubbler-magnetic devices.

Bubbler-Operated Devices

In the simplest of designs, a float is connected to a strip chart or digital recorder via flexible steel tape. In most applications, float-type devices require a stilling well to damp out surges and rapid fluctuations in water surface elevation. In addition, most float-operated devices do not provide an internal stage-to-discharge conversion.

Ultrasonic Devices

These secondary devices rely upon the travel time of an ultrasonic signal from a transponder to the water surface and back. This type of meter functions in a noncontact mode, and is therefore free from clogging and freezing. However, ultrasonics are sometimes subject to spurious signals from floating matter and foam. Some devices have internally programmable read-only memories (PROMs) and microprocessor circuitry to provide stage-to-discharge conversion using the unique relationships of the primary device.

Bubbler Devices

In bubbler devices, gas is forced through a fixed orifice, oriented to assure that only static head is measured. The static pressure required to maintain a given bubble rate is proportional to the height of the water column above the

TABLE 9.22.4 SUMMARY OF DATA ANALYSIS METHODS

Level of Analysis and Methods	Examples	References
Design of Experiments		
Factor analysis	Choosing experimental catchments or measuring sites for a given experimental program: land uses catchments parameters water quality sampling Choosing number of experiments using physical models	Cochran & Cox, 1957; Kendall & Stuart, 1973; Snedecor & Cochran, 1957
Raw Data Criticism		
Double mass analysis	Testing for systematic errors in time data series such as cumulative rainfall or runoff depths at various points in the same climatic areas	
Parametric tests (Anderson test)	Testing the random aspect of a data series such as rainfall and runoff	Bennet & Franklin, 1967; Dagnelie, 1970; Haan, 1977; Pearson & Hartley, 1969
Nonparametric tests Variance ratio test, Bartlett's test, et al.	Testing of the hypothesis on equal variance of two populations: rainfall runoff, runoff quality data	Dagnelie, 1970; Kendall & Stuart, 1973; Kite, 1976; Pettitt, 1979
Wilcoxon, Mann-Whitney, Kruskal-Wallis, Wilks tests	Testing of the hypothesis on equal means and identical location of population: rainfall or runoff, runoff quality data from several catchments	
Statistical parameters Arithmetic mean (or geometric mean for data lognormally distributed) Variance or standard deviation	Comparison of samples and homogeneity testing Parameters can be time and/or flow weighted for runoff quality data	All books on statistical methods
Ranges Pearson's and Fisher's coefficients	Preliminary statistical analysis	
Point-Frequency Analysis		
Empirical frequency plotting Probability papers Plotting formulae	Analysis of a separate variable considered as a random variable: rainfall depths for various time intervals (I.D.F. curves) peak runoff risk analysis Choice of a theoretical probability distribution	Adamowski, 1981; Bennet & Franklin, 1967; Cunnane, 1973; Dagnelie, 1970; Haan, 1977; Kendall & Stuart, 1977a; Kite, 1976; Snedecor & Cochran, 1957; Yevjevich, 1972b
Theoretical probability (distributions discrete and continuous) Method of moments Method of maximum likelihood	Almost all hydrological variables (rainfall, runoff, quantity, quality) considered as a random variable	Chow, 1964; Dagnelie, 1970; Gumbel, 1960; Haan, 1977; Kendall & Stuart, 1977a; Kite, 1976; Linsley et al., 1975; Snedecor & Cochran, 1957; Viessman et al., 1977; Yevjevich, 1972b
Hypothesis testing and confidence intervals Tests of means and variances Goodness-of-fit tests	Testing the adequacy of a given probability distribution to a given sample	Chow, 1964; Dagnelie, 1970; Haan, 1977; Kendall & Stuart, 1977a; Kendall & Stuart, 1973; Kite, 1976; Snedecor & Cochran, 1957; Yevjevich, 1972b
Multivariate Analysis		
Simple Regression Analysis best fit procedure choice tests of fit spurious correlations	Applied to a pair of hydrological variables rainfall and runoff volumes runoff coefficients and imperviousness rainfall depths at two sites overland flow detention storage and discharge pollutants loads and peak runoff etc.	Chatfield & Collins, 1980; Haan, 1977; Morrison, 1976; Draper & Smith, 1966; Haan, 1977; Viessman et al., 1977

Continued

TABLE 9.22.4 *Continued*

Level of Analysis and Methods	Examples	References
Multivariate probability distributions	Applied to several independent variables considered to be purely random variables risk analysis in urban water management spatial rainfall depths distribution hydrological stochastic processes (discrete and continuous)	Adamowski, 1981; Dagnelie, 1970; Kendall & Stuart, 1977a; Kite, 1976; Yevjevich, 1972a; Yevjevich, 1972b
Multiple regressions analysis	Applied to one explained variable Y and to several explanatory variables Xi:	Chatfield & Collins, 1980; Draper & Smith, 1966; Haan, 1977; Pearson & Hartley, 1969; Robitaille and Bobbée, 1975; Stone, 1974; Yevjevich, 1972a
Simple matrix procedure of best fit	interpolation between a set of raingauges generation of data for incomplete data series	
Stepwise regression procedure	rainfall-runoff modeling at a given location versus rainfall and/or runoff at other locations	
Orthogonal regression procedure	runoff coefficients versus urban catchment parameters and rainfall characteristics	
Ridge regression procedure	lag times and times of concentration versus	
Cross validation procedure	catchments and rainfall parameters	
Better results when Xi variables are correlated	Urban runoff pollutant loads versus rainfall and runoff parameters, catchment characteristics such as land uses, imperviousness, slopes, etc.	
Interdependence analysis	Mostly for qualitative analysis. Not frequently applied in urban hydrology	Chatfield & Collins, 1980; Haan, 1977; Morrison, 1976
Correlation analysis	Just two variables	
Principal components analysis (P.C.A.)	More than two variables: reduction of dimensionality, preliminary analysis for regression procedures. Sometimes quantitative spatial distribution of rainfall urban runoff pollutants loads	
Factor analysis	Similar aims as P.C.A. but with assumption of a proper statistical model. Covariance analysis	
Cluster analysis	Grouping tests of individuals	
Discriminant analysis	Separation of individuals in two populations. Preliminary analysis for regression procedures	
Time Series Analysis	Testing the random aspect of a given variable for preliminary statistical analysis Stochastic modelling of hydrological processes (not very frequent in urban hydrology due to time and space intervals to be considered)	Bartlett, 1966; Box & Jenkins, 1970; Cox & Miller, 1968; Jenkins & Watts, 1968; Kendall & Stuart, 1977b; Yevjevich, 1972a
Trend analysis	Testing gradual natural or man-induced changes in data series	
Tests of randomness		
Least squares procedures	Changes in urban hydrological data due to	
Moving average methods	continuous urbanization	
Periodic analysis	Testing the existence of cycles: Seasonal aspects of rainfall, runoff, quality, quantity data Short cycles due to some industrial or domestic water uses	
Spectral analysis on the time domain (Autocorrelation function)	Testing the random aspect of a given process	
Spectral analysis on the frequency domain (Spectral density function)	Identifying Instantaneous Unit Hydrographs (IHU) for small urbanized catchments	

orifice. The static pressure may be measured either by the inclined manometer or pressure tranducer devices. Some devices are available with internal PROMs for flow data reduction. In fast flowing water, the dip tube may be protected by a simple still well: a concentrically placed perforated tube. Shortcomings include contaminant build-up on the dip tube in the vicinity of the measuring tube, and relatively low accuracy in the total part of the total pressure range.

Combination Bubbler-Magnetic Devices

These instruments rely on velocity-area measurement to compute instaneous flow rates. Stage measurements are made using conventional transducer bubblers. These data are converted to area measurements using a PROM that describes conduit geometry. Flow velocity measurements are made at the same time with an electromagnetic device located in a band attached to the conduit wall. Using the independent values of area and velocity, the device computes discharge.

Other Monitors

When planning any atmospheric precipitation measurement, contact the National Meteorological Institute or equivalent organization for expert assistance.

Various types of open containers and man-made natural surfaces are used to collect impurities deposited by atmospheric forces. Open containers are generally polyethylene, polypropylene, or glass funnels or cylinders. Various modified gauge types are designed for special purposes.

Precipitation intensity, volume, and duration data should be collected during qualified sample events, and for monitoring programs. Nonrecording gauges are used for measurement by most government hydrological and meteorological services. The ordinary rain gauge used for daily readings is a collector above a funnel leading to a receiver. Continuous registration has also been incorporated into rain gauges. Precipitation recorders in general use are the weighing, tipping-bucket, or float type. If the standard rain gauge is sited in the wind direction, it should be surrounded by an 0.4 m-high screen. A wind shield consisting of a frame of hanging strips is placed within 1 m of the recorder.

QA/QC Measures

A quality assurance/quality control program should be developed and implemented as part of a long-term monitoring program to provide assessment of techniques used during sample collection, storage, and analysis (EPA 1979b, 1980). EPA programs require QA/QC plans to be approved by the EPA prior to sample collection and analysis. The QA/QC plan should specify sample collection and preservation methods, maximum sample holding time, chain-of-custodian procedure, analytical techniques, accuracy and precision checks, detection limits, and data recording and documentation procedures.

SAMPLE STORAGE

The preceding steps will not guarantee a representative sample unless container selection and sample preservation methods meet the required standards. The choice of container and cap materials is very important due to the possibility of interference with constituents to be analyzed. Containers and all elements involved in sampling or compositing operations must be properly cleaned. More detailed information on container types and cleaning is found in EPA 1980b. Recommended operations are as follows:

Container Selection

Containers can introduce positive or negative errors in trace metal and inorganic measurements by contributing contaminants through leaching or surface desorption or, depleting concentrations through absorption. Samples to be analyzed for toxic metals can be stored in 1-l polyethylene or glass bottles with polypropylene caps. Teflon lid liners should be purchased or cut from sheet teflon and inserted in caps to prevent possible contamination from caps supplied with bottles.

Container Cleaning

Due to the sensitivity of tests examining waterborne trace metals, sample containers must be thoroughly cleaned. The following schedule must be followed for the preparation of all sample bottles and accessories, whether glass, polyethylene, polypropylene or Teflon:

- Wash with detergent and tap water
- Rinse with 1:1 nitric acid
- Rinse with tap water
- Rinse with 1:1 hydrochloric acid
- Rinse with tap water
- Triple rinse with distilled (or deionized) water

SAMPLE PRESERVATION

Water samples are susceptible to rapid physical or biological reactions that may take place between sampling and analysis. This time period can exceed 24 hr due to laboratory capacity needed to handle unpredictably varying amounts of samples resulting from aleatory rainfalls (Geiger 1981).

Preservation techniques are recommended to avoid sample changes resulting in large errors. Refrigeration of samples at 4°C is commonly used in fieldwork and helps to stabilize samples by reducing biological and chemical activity. All samples except metals must be refrigerated.

In addition to refrigeration, specific techniques are required for certain parameters. They consist of the addition of chemical compounds, biocides, etc. More detailed information can be found in EPA 1979a, 1980b.

The decision to eliminate a portion of the drainage system from further sampling must include a review of data QA/QC procedures. Review of contaminant data for drainage systems must be performed to ensure that analytical results are properly interpreted, and that detection of potential sources is not missed because of field or laboratory constraints.

Analysis of Pollution Data

When considering the significance of runoff pollutant contributions, both concentrations and total loadings must be examined. Receiving water concentrations are usually of prime concern. In principle, if pollutant concentrations do not exceed certain allowable maxima, detrimental effects will not occur. However, for many pollutants, allowable concentrations in water are not known. Because of sedimentation, accumulation of benthal deposits such as phosphorous, hydrocarbons, and heavy metals may be more significant than concentrations in the water. Receiving water column concentration and benthal accumulation depends more on mass loads of pollutants than of pollutant concentrations in the runoff.

Judging from the above, an estimate of mass loadings of the principal pollutants are of great importance for planning and management purposes. Priority should be placed on obtaining a reliable estimate of mass loadings entering a body of water because of urban runoff during a specific time, such as a year, and data collection plans should be so designed.

STORM LOADS

Data from nonpoint source monitoring studies are usually reduced to a pollutant load per storm basis. An event expected concentration (EMC) is multiplied by the value of runoff. EMCs are calculated by integrating the pollutograph (instantaneous load with time) with the hydrograph. After sampling several storms, these load per storm data are used to estimate annual load from the basin. It is assumed that the monitored storms are representative samples of storms usually occurring in the basin during the year (Whipple 1983).

ANNUAL LOADS

Regressions of total load versus total runoff from a family of storms give a slope in concentration units, which can be used to predict pollutant load for a specific quantity of runoff. To better represent the actual runoff process, base flows were abstracted from storm runoff or low-flow loads from storm load. Good correlations have been found using log-load versus log runoff volume, reflecting log-normal distribution of the concentration data. Nonpoint

source data are usually log-normal distributed, as are hydrologic events.

After mass loadings for a given land-use type are accumulated over a considerable period of time, results can be expressed in terms used to estimate loadings from that type of land use for the rest of the watershed(s) of interest. The approaches below are commonly used:

1. Annual loading/area of given land use, lbs/acre/yr
2. Annual loading/curb mi of given land use, lb/mi/yr
3. Annual loading/traffic volume, lbs/vehicle/yr
4. Annual loading/air pollution index, lbs/avg in/yr
5. Annual loading/runoff volume, lbs/million gal
6. Annual loading/precipitation amount, lbs/in (for specific area)

Number 1 assumes that pollution varies according to land use. This is the most commonly used method of predicting loading under future conditions. Number 2 assumes that pollution loading varies with the number of curb mi in various stages of development. Numbers 3 and 4 make similar assumptions regarding automobile traffic and air pollution. Numbers 5 and 6 are designed to convert loading data from specific storm events to annual average loadings, which are then converted to relationships with land use for predictive purposes.

SIMULATION MODEL CALIBRATION

Monitoring data can be used to calibrate sophisticated nonpoint source computer models. These models attempt to interpret the mechanisms involved in nonpoint source generation. Alternatives can then be evaluated using computer-simulated processes. Models exist for different types of basins, levels of complexity, and nonpoint source problems. They need good calibration data.

Statistical Analysis

Urban hydrological phenomena, especially those involving storms, give historical data that can be observed only once, and then will not occur again. Such collected data form an ever-growing sample of measurements. Even if some phenomena can be described by means of physical or rational theory, the input, rainfall, is commonly stochastic in nature and whole phenomena can be amenable to statistical interpretation and probability analysis.

Table 9.22.4 summarizes data analysis methods along with examples and references. Although it is impossible to summarize the many references in general or applied statistics, good basic knowledge is provided in books such as Kendall and Stuart's theory of statistics, or Haan's (1977) statistical methods in hydrology.

—Kent K. Mao
David H.F. Liu

References

Alley, W.M. 1977. *Guide for collection, analysis, and use of urban storm-water data,* conference report. American Society of Civil Engineers (ASCE) New York, N.Y.: American Society of Civil Engineers (ASCE).

Geiger, W.F. 1981. Continuous quality monitoring of storm runoff. *Water Science Technology.* Vol. 13, pp. 117–123. Munich: IAWRR/Pergamon Press Ltd.

Gideometeozdat. 1984. *Complex assessments of surface water quality.* (In Russian). p. 140. Leningrad.

Haan, C.T. 1977. *Statistical methods in hydrology.* Ames, Iowa: The Iowa State University Press.

Shelley, P.E., and G.A. Kirkpatrick. 1975. *An assessment of automatic flow samplers—1975.* U.S. Environmental Protection Agency (EPA). U.S. EPA 600-2-75-065. Washington, D.C.

Smoot, G.F. 1975. *A rain-runoff quantity—quality collection system.* Proceedings of a research conference on Urban Runoff Quantity and Quality. American Society of Civil Engineers (ASCE). New York, N.Y.

U.S. Department of Interior, Bureau of Reclamation. 1975. *Water Measurement Manual.*

U.S. Environmental Protection Agency. 1979a. *Methods for chemical analysis of water and waste.* Washington, D.C.: EPA.

U.S. Environmental Protection Agency. 1979b. *Monitoring requirements, methods, and costs for the nationwide urban runoff program.* EPA Report 600–9–76–014. Washington, D.C.

U.S. Environmental Protection Agency. 1980. *Monitoring toxic pollutants in urban runoff, a guidance manual.* U.S. Environmental Protection Agency (EPA), Office of Water Regulation and Standards.

Wong, J. and J. Marsalek. 1981. *Persistent toxic substances in urban runoff.* Proceedings of Storm Water Management Model Users Group Meeting, Niagara Falls, Ontario, Canada: U.S. Environmental Protection Agency (EPA). EPA Report, pp. 455–468.

Whipple, W., et al. 1983. *Stormwater management in urbanizing areas.* Englewood Cliffs, N.J.: Prentice-Hall.

9.23
DISCHARGE TREATMENT

Three types of treatment are used for wastewater discharges: biological, physical-chemical, and physical processes. Some systems use two or all types to achieve best water quality. The efficiency of various storm water and combined sewer overflow (CSO) treatment processes is given in Table 9.23.1 (Lager et al. 1977).

Biological Processes

The biological processes used for point sources are difficult to implement for stormwater discharges, which have low biochemical oxygen demand (BOD) nonpoint concentrations. These processes perform poorly or not at all when treating flows with irregular quantity or quality. It is very difficult to keep the biota alive between storm events. Wet weather reduces low organic concentrations, and the biomass is sensitive to toxic substances often present in urban stormwater runoff.

Physical-Chemical Processes

Physical-chemical processes show promise in overcoming shock loadings. Chemical coagulants enhance the separation of particles from liquid (see Chapter 7). Chemical addition is also effective in removing phosphorous, metals, and some organic colloids.

Physical Processes

Successfully demonstrated physical processes include fine-mesh screening, fine-mesh screening/high-rate filtration, sedimentation, sand and peat-sand filtration, fine-mesh screening/dissolved-air flotation, and swirl separation.

SWIRL-FLOW REGULATOR-CONCENTRATOR

The dual purpose swirl-flow regulator-solids-concentrator has shown a potential for simultaneous quality control (Field 1990). These devices have been applied to CSO; however, they can also be used for storm water runoff pollution control.

The swirl concentrator uses a swirl action to separate particles from liquids (Figure 9.23.1). Flow from combined sewers enters a diversion chamber and bar screen, removing the debris. The swirl facility is automatically activated when storm flows enter the lower portion of the circular chamber. Rotary motion causes liquids to follow a long spiral path, to be discharged from the chamber top through a downshaft. This overflow water can be disinfected and discharged or stored for later treatment. Because a flow deflector prevents chamber flow from completing its first revolution and merging with continuing inlet flow, there is a gently swirling rotational movement.

The settleable solids entering the chamber are spread over the full cross-section of the channel and settle quickly. Solids are entrained along the bottom around the chamber and are concentrated at the foul sewer outlet, where they are transported to the treatment plant.

The scum acts as a baffle, keeping floatables outside the overflow weir and preventing these from overflowing into

TABLE 9.23.1 EFFICIENCY OF VARIOUS STORM-WATER AND CSO TREATMENT PROCESSES

Process	Efficiency (%)				
	Suspended Solids	BOD$_5$	COD	Total P	TKN[a]
Physical—Chemical					
Sedimentation	20–60	50	34	20	
without chemicals					38
with chemicals	68	68	45		
Vortex separation	40–60	25–60	50–60		
Screening					
microstrainers	50–95	10–50	35	20	30
rotary screens	20–35	1–30	15	12	10
Sand–peat filters[b]	90	90	NA	70	50
Biological[c]					
Contact stabilization	75–95	70–90		50	50
Biodiscs	40–80	40–80	33		
Oxidation ponds	20–57	10–17		22–40	57
Aerated lagoons	92	91			
Facultative lagoons	50	50–90			

Source: Reprinted from J.A. Lager, W.G. Smith, W.G. Lynard, R.M. Finn, and E.J. Finnemore, 1977, *Urban stormwater management and technology: updates and users' guide* (U.S. Environmental Protection Agency (EPA), EPA Report 600–8–77–014, Municipal Environmental Research Laboratory, Office of Research and Development, Washington, D.C.).

[a]Total Kjeldahl nitrogen.

[b]After Galli (1990), peat-sand filters are similar to biological anaerobic-aerobic slow filters. They are applicable for treatment of urban runoff.

[c]Biological treatment is feasible only for CSOs.

the clean effluent. Floatables are directed by a floatable deflector to a floatable trap. The floatable trap is connected to a floatable storage area under the clear overflow weir plate. Floating material is drawn beneath the weir plate by the vortex and dispersed around the downshaft. Floating solids are retained here until after a storm event, when the water level recedes in the swirl chamber. As this occurs, trapped floatables are dropped and enter the foul sewer outlet, where they are transported to a sewage treatment plant.

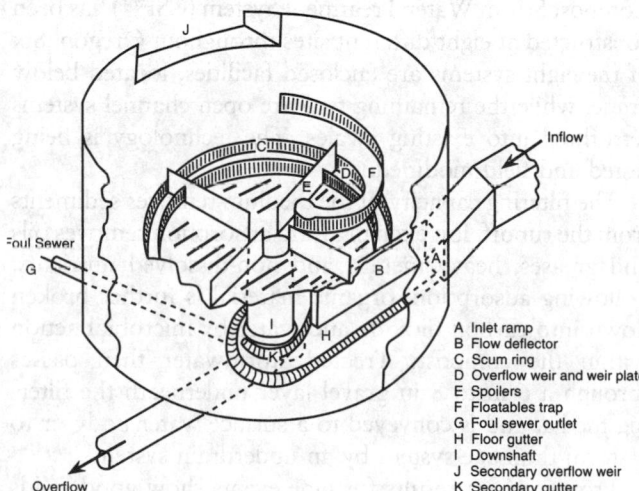

FIG. 9.23.1 An isometric view of a swirl regulator-concentrator.

A Inlet ramp
B Flow deflector
C Scum ring
D Overflow weir and weir plate
E Spoilers
F Floatables trap
G Foul sewer outlet
H Floor gutter
I Downshaft
J Secondary overflow weir
K Secondary gutter

A partial list of U.S. installations with experience in swirl-flow regulator-concentrator use was presented by Pisano (1989).

SAND FILTERS

Sand systems are usually off-line. A typical sand filtration system is comprised of an inlet structure with a presetting basin, a flow disperser, filtration media, an underdrain system, and a basin liner (Figure 9.23.2). For piped storm water systems, the inlet structure might be a manhole using a weir to divert low flows into the filtration system. For open-channel conveyance systems, the inlet might be a weir constructed within the flow path to divert low flows to the filtration system, while allowing higher flows to bypass the filtration system. Without a presettling basin, the filter medium may quickly become plugged with large sediments. This basin may not be necessary, if the sand filtration basin is used in place of an API oil/water separator, and if the contributing drainage area is small and completely impervious.

The primary pollutant removal mechanisms are filtration and sedimentation. Particulate matter such as sediments, oils and greases, and trace metals are removed by filtration as stormwater percolates through the sand filter. Sedimentation removes large particles, and filtration removes silt and clay-size particles.

Over time, sediment eventually penetrates the filter media surface, requiring replacement of the filter media.

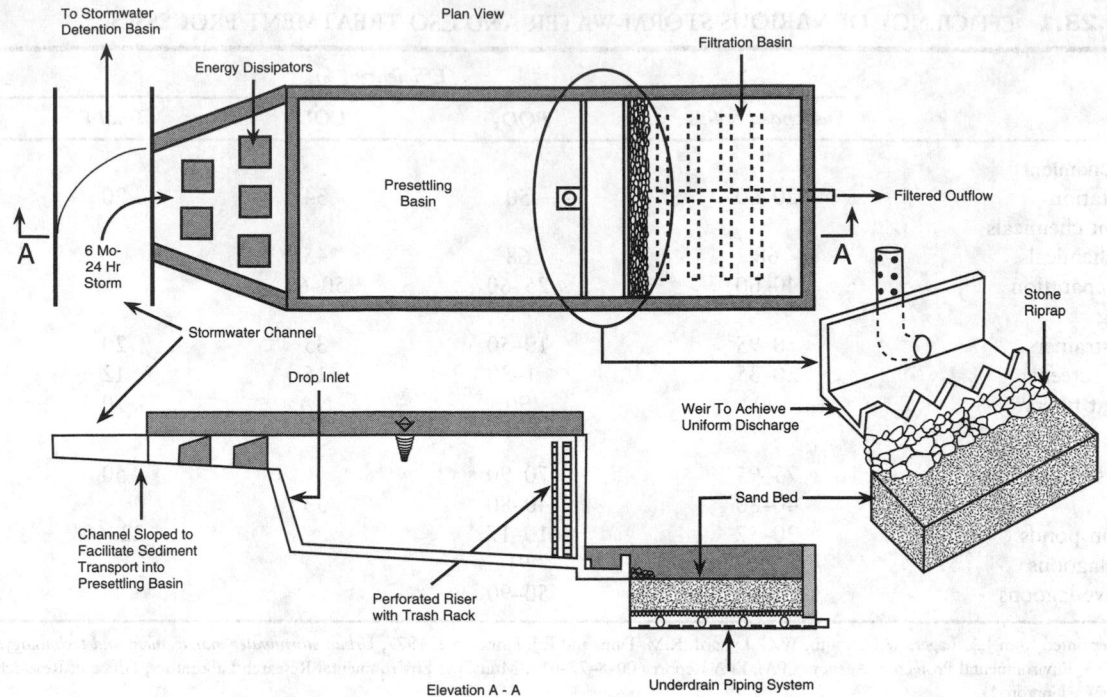

FIG. 9.23.2   Conceptual sand filtration basin system. (Reprinted from the City of Austin, 1988.)

Maintenance requirements can be intensive, depending upon sediment concentrations in surface runoff. Fifty acres is recommended as the maximum contributing drainage area for a sand filtration system (Schueler, Kumble and Heraty 1992).

ENHANCED FILTERS

Enhanced (peat-sand) filters use layers of peat, lime, and/or topsoil, and may also use a grass cover (Figure 9.23.3) to remove particulate pollutants. To minimize clogging, both sand and enhanced filters should be preceded by a solid-removing unit, such as a pond or a filter strip.

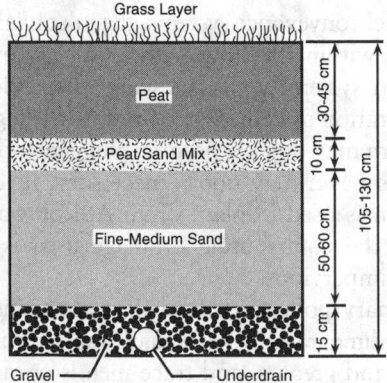

FIG. 9.23.3   Peat sand filter for storm water treatment. (Reprinted from J. Galli, 1990, *Peat sand filters: A proposed storm water management practice for urban areas* [Department of Environmental Programs, Metropolitan Washington Council of Governments, Washington, D.C.].)

Peat-sand filters provide high phosphorous, BAD, nitrogen and silt removal. Peat has a high removal affinity for adsorbing and removing toxic compounds (Novotny 1994), hence peat-containing filters are effective for removing priority pollutants.

COMPOST FILTERS

W&H Pacific conceived the idea of utilizing yard debris compost as a treatment and filtration medium for stormwater runoff. This medium removes organic and inorganic pollutants through adsorption, filtration, and biological processes (ion exchange and bioremediation). The Compost Storm Water Treatment System (CSF™) has been constructed at eight different sites throughout Oregon. Six of the eight systems are enclosed facilities, located below grade, while the remaining two are open channel systems retrofitted into existing swales. The technology is being tested and field modified.

The filtering capacity of the medium removes sediments from the runoff. Ion exchange and adsorption removes oils and greases, heavy metals, and non-dissolved nutrients. Following adsorption, organic material is further broken down into carbon dioxide and water by microbial action within the compost. Treated stormwater then passes through a 6 in to 8 in gravel layer underneath the filtering media, and is conveyed to a surface water body or to a storm drainage system by an underdrain system.

Prototype test results for nine events show good solids removal: 67% removal of COD, 40% removal of total phosphorous, 67% removal of copper, and better than

87% removal of zinc, aluminum and iron. The leaf compost has very good cation exchange capacity. However, like sand and enhanced filters, operating life depends on the frequency of preventive maintenance.

—*Kent K. Mao*

References

Field, R. 1990. Combined sewer overflows: Control and treatment. In *Control and treatment of combined-sewer overflows*. Moffa, P.E. ed. New York, N.Y.: Van Nostrand Reinhold.

Lager, J.A., W.G. Smith, W.G. Lynard, R.M. Finn, and E.J. Finnemore. 1977. *Urban stormwater management and technology: Updates and users' guide*. U.S. Environmental Protection Agency (EPA). EPA Report 600–8–77–014. Municipal Environmental Research Laboratory, Office of Research and Development.

Novotny, V., and H. Olem. 1994. *Water quality: Prevention, identification, and management of diffuse pollution*. New York, N.Y.: Van Nostrand Reinhold.

Pisano, W.C. 1989. Recent United States experience with designs and new German technology. In *Design of urban runoff quality controls*. L.A. Roesner, B. Urbonas, and M.B. Sonnen, eds. American Society of Civil Engineers (ASCE). New York, N.Y.

Schueler, T.R., P.A. Kumble, and M.A. Heraty. 1992. A current assessment of urban best management practices. Techniques for reducing non-point source pollution in the coastal zone. *Technical guidance manual*. Metropolitan Washington Council of Government. Office of Wetlands, Oceans, and Watersheds. Washington, D.C.

Bibliography

Aravin, V.L., and S.N. Numerov. 1965. *Theory of fluid flow in undeformable porous media*. New York: Daniel Davey.

Bear, J. 1972. *Dynamics of fluids in porous media*. Elsevier, Amsterdam.

Beasley, D.B. 1976. Simulation of the environmental impact of land use on water quality. In *Best management practices for non-point source pollution control*. U.S. Environmental Protection Agency (EPA). EPA Report 905-9-76-005. Washington, D.C.

Bennett, G.D. 1976. *Introduction to groundwater hydraulics*, Techniques of Water Resources Investigations, Chap. B2, Book 3, U.S. Geological Survey, Washington, D.C.

Bowen, R. 1980. *Groundwater*. Barking, Essex, England: Applied Science Publishers Ltd.

Cooper, H.H. 1966. The equation of groundwater flow in fixed and deforming coordinates. *J. Geophys. Res.* 71, no. 20:4785–4790.

Council on Environmental Quality. 1980. *Environmental quality—1978: The ninth annual report of the council of environmental quality*. Washington, D.C.: U.S. Government Printing Office.

Crawford, N.H., and R.K. Linsley. 1966. Digital simulation in hydrology: The Stanford Model IV. *Technical Report No. 39*. Stanford University, Department of Civil Engineering. Palo Alto, Calif.

Davis, S.N., and R.J.M. DeWiest. 1966. *Hydrogeology*. New York: John Wiley & Sons, Inc.

DeVries, J.J. 1975. *Groundwater hydraulics*. Aqua-Vu, Ser. A, no. 6, Communications of the Institute of Earth Sciences. Amsterdam: Free Reformed University.

DeWiest, R.J.M. 1965. *Geohydrology*. New York: John Wiley & Sons, Inc.

DeWiest, R.J.M., ed. 1969. *Flow through porous media*. New York: Academic Press, Inc.

Hantush, M.S. 1964. Hydraulics of wells. In *Advances in hydroscience*. Vol. 1, edited by V.T. Chow. New York: Academic Press, Inc.

Harr, M.E. 1962. *Groundwater and seepage*. New York: McGraw-Hill Book Company.

Heaney, J.P., and W.C. Huber. 1972. *Storm water management model; refinements, testing and decision making*. University of Florida, Department of Environmental Science. Gainesville, Fla.

Heath, R.C. 1983. *Basic groundwater hydrology*. Water Supply Paper 2220, U.S. Geological Survey. Washington, D.C.

Hubbert, M.K. 1940. The theory of groundwater motion. *J. Geol.* 48:785–944.

Jacob, C.E. Flow of groundwater. In *Engineering hydraulics*, edited by H. Rouse. John Wiley.

Javandel, I., C. Doughty, and C.F. Tsang. 1984. *Groundwater transport: Handbook of mathematical models*. Washington, D.C.: American Geophysical Union.

Lohman, S.W. 1972. *Groundwater hydraulics*. Professional paper 708, U.S. Geological Survey. Washington, D.C.

Marino, M.A., and J.N. Luthin. 1982. *Seepage and groundwater*. Developments in Water Science Series no. 13. New York: Elsevier Science Publishing Co., Inc.

McWhorter, D.B., and D.K. Sunada. 1977. *Groundwater hydrology and hydraulics*. Fort Collins, Colo.: Water Resources Publications.

Meinzer, O.E. [1923] 1960. *Outline of groundwater hydrology*. Water Supply Paper 494, U.S. Geological Survey. Washington, D.C.

Novotny, V., and G. Chesters. 1981. *Handbook of nonpoint pollution: Sources and management*. New York, N.Y.: Van Nostrand Reinhold.

Novotny, V., R. Imhoff, M. Olthoff, and P.A. Krenkel. 1989. *Karl Imhoff's handbook of urban drainage and wastewater disposal*. John Wiley.

Novotny, V., and H. Olem. 1994. *Water quality: Prevention, identification, and management of diffuse pollution*. New York, N.Y.: Van Nostrand Reinhold.

Petersen, D.F. 1957. Hydraulics of wells. *Trans. Am. Soc. Cir. Eng.* 122:502–517.

Pinneker, E.V., ed. 1983. *General hydrogeology*. Cambridge: Cambridge University Press.

Polubarinova-Kochina, P.Y. 1962. *Theory of groundwater movement*. Translated from Russian by R.J.M. DeWiest. Princeton, N.J.: Princeton University Press.

Todd, D.K. 1964. Groundwater. In *Handbook of applied hydrology*, edited by V.T. Chow. New York: McGraw-Hill Book Company.

United Nations Educational, Scientific and Cultural Organization (UNESCO). 1987. Manual on drainage in urbanized areas. *Vol. 1, Planning and design of drainage systems*. Paris, France: UNESCO Press.

U.S. Bureau of Reclamation. 1960. *Studies of groundwater movement*. Technical Memorandum 657. Denver, Colo.: U.S. Dept. of Interior.

U.S. Department of Interior, Water and Water Resources Service. [1977] 1981. *Groundwater manual*. Washington, D.C.: U.S. Government Printing Office.

U.S. Soil Conservation Service (SCS). 1975. Procedure for computing street and rill erosion on project areas. *SCS technical release no. 51.* Washington, D.C.

U.S. Soil Conservation Service (SCS). 1975. Urban hydrology for small watersheds. *SCS technical release no. 55*. Washington, D.C.

Whipple, W., Jr. et al. 1983. *Storm water management in urbanizing areas*. Englewood Cliffs, N.J.: Prentice Hall.

10

Solid Waste

R.C. Bailie | J.W. Everett | Béla G. Lipták | David H.F. Liu |
F. Mack Rugg | Michael S. Switzenbaum

Source and Effect

10.1
DEFINITION

For practical purposes, the term *waste* includes any material that enters the waste management system. In this chapter, the term *waste management system* includes organized programs and central facilities established not only for final disposal of waste but also for recycling, reuse, composting, and incineration. Materials enter a waste management system when no one who has the opportunity to retain them wishes to do so.

Generally, the term *solid waste* refers to all waste materials except hazardous waste, liquid waste, and atmospheric emissions. *CII waste* refers to wastes generated by commercial, industrial, and institutional sources. Although most solid waste regulations include hazardous waste within their definition of solid waste, *solid waste* has come to mean *nonhazardous solid waste* and generally excludes hazardous waste.

This section describes the types of waste that are detailed in this chapter.

Waste Types Included

This chapter focuses on two major types of solid waste: municipal solid waste (MSW) and bulky waste. MSW comprises small and moderately sized solid waste items from homes, businesses, and institutions. For the most part, this waste is picked up by general collection trucks, typically compactor trucks, on regular routes.

Bulky waste consists of larger items of solid waste, such as mattresses and appliances, as well as smaller items generated in large quantity in a short time, such as roofing shingles. In general, regular trash collection crews do not pick up bulky waste because of its size or weight.

Bulky waste is frequently referred to as C&D (construction and demolition) waste. The majority of bulky waste generated in a given area is likely to be C&D waste. In areas where regular trash collection crews take anything put out, the majority of bulky waste arriving separately at disposal facilities is C&D waste. In areas where the regular collection crews are less accommodating, however, substantial quantities of other types of bulky waste, such as furniture and appliances, arrive at disposal facilities in separate loads.

Waste Types Not Included

In a broad sense, the majority of nonhazardous solid waste consists of industrial processing wastes such as mine and mill tailings, agricultural and food processing waste, coal ash, cement kiln dust, and sludges. The waste management technologies described in this chapter can be used to manage these wastes; however, this chapter focuses on the management of MSW and the more common types of bulky waste in most local solid waste streams.

—*F. Mack Rugg*

10.2
SOURCES, QUANTITIES, AND EFFECTS

This section identifies the sources of solid waste, provides general information on the quantities of solid waste generated and disposed of in the United States, and identifies the potential effects of solid waste on daily life and the environment.

Sources

The primary source of solid waste is the production of commodities and byproducts from solid materials. Everything that is produced is eventually discarded. A secondary source of solid waste is the natural cycle of plant growth and decay, which is responsible for the portion of the waste stream referred to as yard waste or vegetative waste.

The amount a product contributes to the waste stream is proportional to two principal factors: the number of items produced and the size of each item. The number of items produced, in turn, is proportional to the useful life of the product and the number of items in use at any one time. Newspapers are the largest contributor to MSW because they are larger than most other items in MSW, they are used in large numbers, and they have a useful life of only one day. In contrast, pocket knives make up a negligible portion of MSW because relatively few people use them, they are small, and they are typically used for years before being discarded.

MSW is characterized by products that are relatively small, are produced in large numbers, and have short useful lives. Bulky waste is dominated by products that are large but are produced in relatively small numbers and have relatively long useful lives. Therefore, a given mass of MSW represents more discreet acts of discard than the same mass of bulky waste. For this reason, more data are required to characterize bulky waste to within a given level of statistical confidence than are required to characterize MSW.

Most MSW is generated by the routine activities of everyday life rather than by special or unusual activities or events. On the other hand, activities that deviate from routine, such as trying different food or a new recreational activity, generate waste at a higher rate than routine activities. Routinely purchased items tend to be used fully, while unusual items tend to be discarded without use or after only partial use.

In contrast to MSW, most bulky waste is generated by relatively infrequent events, such as the discard of a sofa or refrigerator, the replacement of a roof, the demolition of a building, or the resurfacing of a road. Therefore, the composition of bulky waste is more variable than the composition of MSW.

In terms of generation sites, the principal sources of MSW are homes, businesses, and institutions. Bulky waste is also generated at functioning homes, businesses, and institutions; but the majority of bulky waste is generated at construction and demolition sites. At each type of generation site, MSW and bulky waste are generated under four basic circumstances:

Packaging is removed or emptied and then discarded. This waste typically accounts for approximately 35 to 40% of MSW prior to recycling. Packaging is generally less abundant in bulky waste.

The unused portion of a product is discarded. In MSW, this waste accounts for all food waste, a substantial portion of wood waste, and smaller portions of other waste categories. In bulky waste, this waste accounts for the majority of construction waste (scraps of lumber, gypsum board, roofing materials, masonry, and other construction materials).

A product is discarded, or a structure demolished, after use. This waste typically accounts for 30 to 35% of MSW and the majority of bulky waste.

Unwanted plant material is discarded. This waste is the most variable source of MSW and is also a highly variable source of bulky waste. Yard wastes such as leaves, grass clippings, and shrub and garden trimmings commonly account for as little as 5% or as much as 20% of the MSW generated in a county-sized area on an annual basis. Plant material can be a large component of bulky waste where trees or woody shrubs are abundant, particularly when lots are cleared for new construction.

Packaging tends to be concentrated in MSW because many packages destined for discard as MSW contain products of which the majority is discarded in wastewater or enters the atmosphere as gas instead of being discarded as MSW. Such products include food and beverages, cleaning products, hair- and skin-care products, and paints and other finishes.

Quantities

The most important parameter in solid waste management is the quantity to be managed. The quantity determines the size and number of the facilities and equipment required to manage the waste. Also important, the fee collected for each unit quantity of waste delivered to the facility (the tipping fee) is based on the projected cost of operating a facility divided by the quantity of waste the facility receives.

The quantity of solid waste can be expressed in units of volume (typically cubic yards or cubic meters) or in units of weight (typically short, long, or metric tons). In this chapter, the word ton refers to a short ton (2000 lb). Although information about both volume and weight are important, using weight as the master parameter is generally preferable in record keeping and calculations.

The advantage of measuring quantity in terms of weight rather than volume is that weight is fairly constant for a given set of discarded objects, whereas volume is highly variable. Waste set out on the curb on a given day in a given neighborhood occupies different volumes on the curb, in the collection truck, on the tipping floor of a transfer station or composting facility, in the storage pit of a combustion facility, or in a landfill. In addition, the same waste can occupy different volumes in different trucks or landfills. Similarly, two identical demolished houses occupy different volumes if one is repeatedly run over with a bulldozer and the other is not. As these examples illustrate, the phrases "a cubic yard of MSW" and "a cubic yard of bulky waste" have little meaning by themselves; the phrases "a ton of MSW" and "a ton of bulky waste" are more meaningful.

Franklin Associates, Ltd., regularly estimates the quantity of MSW generated and disposed of in the United States under contract to the U.S. Environmental Protection Agency (EPA). Franklin Associates derives its estimates from industrial production data using the *material flows methodology,* based on the general assumption that what is produced is eventually discarded (see "Estimation of Waste Quantity" in Section 10.4). Franklin Associates estimates that 195.7 million tons of MSW were generated in the United States in 1990. Of this total, an estimated 33.4 million tons (17.1%) were recovered through recycling and composting, leaving 162.3 million tons for disposal (Franklin Associates, Ltd. 1992).

The quantity of solid waste is often expressed in pounds per capita per day (pcd) so that waste streams in different areas can be compared. This quantity is typically calculated with the following equation:

$$pcd = 2000T/365P \qquad 10.2(1)$$

where:

pcd = pounds per capita per day
T = number of tons of waste generated in a year
P = population of the area in which the waste is generated

Unless otherwise specified, the tonnage T includes both residential and commercial waste. With modification the equation can also calculate pounds per employee per day, residential waste per person per day, and so on.

Franklin Associates's (1992) estimate of MSW generated in the United States in 1990, previously noted, equates to 4.29 lb per person per day. This estimate is probably low for the following reasons:

Waste material is not included if Franklin Associates cannot document the original production of the material.
Franklin's material flows methodology generally does not account for moisture absorbed by materials after they are manufactured (see "Combustion Characteristics" in Section 10.3).

Table 10.2.1 shows waste quantities reported for various counties and cities in the United States. All quantities are given in pcd. Reports from the locations listed in the table indicate an average generation rate for MSW of 5.4 pcd, approximately 25% higher than the Franklin Associates estimate. Roughly 60% of this waste is generated in residences (residential waste) while the remaining 40% is generated in commercial, industrial, and institutional establishments (CII waste). The percentage of CII waste is usually lower in suburban areas without a major urban center and higher in urban regional centers.

Table 10.2.1 also shows generation rates for solid waste other than MSW. The quantity of other waste, most of which is bulky waste, is roughly half the quantity of MSW. The proportion of bulky and other waste varies, however, and is heavily influenced by the degree to which recycled bulky materials are counted as waste. The quantities of bulky waste shown for Atlantic and Cape May counties, New Jersey, include large amounts of recycled concrete, asphalt, and scrap metal. See also "Component Composition of Bulky Waste" in Section 10.3.

Franklin Associates (1992) projects that the total quantity of MSW generated in the United States will increase by 13.5% between 1990 and 2000 while the population will increase by only 7.3%. On a per capita basis, therefore, MSW generation is projected to grow 0.56% per year. No comparable projections have been developed for bulky waste. Table 10.2.2 shows the potential effect of this growth rate on MSW generation rates and quantities.

Effects

MSW has the following potential negative effects:

- Promotion of microorganisms that cause diseases
- Attraction and support of disease vectors (rodents and insects that carry and transmit disease-causing microorganisms)
- Generation of noxious odors
- Degradation of the esthetic quality of the environment
- Occupation of space that could be used for other purposes
- General pollution of the environment

TABLE 10.2.1 SOLID WASTE GENERATION RATES IN THE UNITED STATES

Location	Year	Residential Fraction of MSW (%)	Commercial/ Industrial Fraction of MSW (%)	Total MSW (pcd)	Bulky Waste (pcd)	Other Solid Waste (pcd)[a]	Total Solid Waste (pcd)
Atlantic County, NJ	1991	—	—	6.0	5.9	0.3	12.2
Bexar County, TX	1990	—	—	—	—	—	6.5
Cape May County, NJ	1990	—	—	6.6	6.0	0.6	13.2
Delaware (state)	1990	—	—	—	—	—	7.1
Fairfax County, VA	1991	55	45	4.8	1.3	0.0	6.1
Marion County, FL	1989	—	—	5.4	—	—	—
Middlesex County, NJ	1988	—	—	4.4	2.1	1.6	8.2
Minnesota Metro Area	1991	—	—	6.5	2.6	0.0	9.1
Monmouth County, NJ	1987	75	25	4.8	2.7	0.0	7.5
Monroe County, NY	1990	—	—	5.7	—	—	—
Rhode Island (state)	1985	52	48	4.9	—	—	—
San Diego, CA	1985	—	—	—	—	—	8.0
Sarasota County, FL	1989	—	—	—	—	—	9.2
Seattle, WA	1987	37	63	7.6	—	—	—
Somerset County, NJ	1989	—	—	4.2	1.5	0.6	6.3
Warren County, NJ	1989	—	—	3.2	0.4	0.9	4.5
Wichita, KA	1990	61	39	6.6	1.1	0.0	7.7
	Average[b]	56	44	5.4	2.6	0.5	8.1
	Minimum	37	25	3.2	0.4	0.0	4.5
	Maximum	75	63	7.6	6.0	1.6	13.2
USA (Franklin Associates)	1990	62	38	4.3	—	—	—

Sources: Data from references listed at the end of this section.

Note: pcd = pounds per capita per day

[a]Most waste in this category falls within the definition of either MSW or bulky waste. Specific characteristics vary from place to place.

[b]Because different information is available from different locations, the overall average is not the sum of the averages for the individual waste types.

Bulky waste also has the potential to degrade esthetic values, occupy valuable space, and pollute the environment. In addition, bulky waste may pose a fire hazard.

MSW is a potential source of the following useful materials:

- Raw materials to produce manufactured goods
- Feed stock for composting and mulching processes
- Fuel

Bulky waste has the same potential uses except for composting feed stock.

The fundamental challenge of solid waste management is to minimize the potential negative effects while maximizing the recovery of useful materials from the waste at a reasonable cost.

Conformance with simple, standard procedures for the storage and handling of MSW largely prevents the promotion of disease-causing microorganisms and the attrac-

TABLE 10.2.2 PROJECTED GENERATION OF MSW IN THE UNITED STATES IN THE YEAR 2000

Year	Population (in millions)	MSW Quantity Projected by Franklin Associates (millions of tons)	Per Capita Generation Based on Franklin Associates (lb/day)	Average Annual Growth of Per Capita Generation Represented (%)	Per Capita Generation Based on Average in Table 10.2.1 (lb/day)	MSW Quantity Based on Average in Table 10.2.1 (millions of tons)
1990	249.9	195.7	4.3	—	5.4	247.6
2000	268.3	222.1	4.5	0.56	5.7	281.0

Source: Data from Franklin Associates, Ltd., 1992, *Characterization of municipal solid waste in the United States: 1992 Update* (EPA/530-R-92-019, NTIS PB92-207-166, U.S. EPA).

Note: Derived from Table 10.2.1.

tion and support of disease vectors. Preventing the remaining potential negative effects of solid waste remains a substantial challenge.

Solid waste can degrade the esthetic quality of the environment in two fundamental ways. First, waste materials that are not properly isolated from the environment (e.g., street litter and debris on a vacant lot) are generally unsightly. Second, solid waste management facilities are often considered unattractive, especially when they stand out from surrounding physical features. This characteristic is particularly true of landfills on flat terrain and combustion facilities in nonindustrial areas.

Solid waste landfills occupy substantial quantities of space. Waste reduction, recycling, composting, and combustion all reduce the volume of landfill space required (see Sections 10.6 to 10.14).

Land on which solid waste has been deposited is difficult to use for other purposes. Landfills that receive unprocessed MSW typically remain spongy and continue to settle for decades. Such landfills generate methane, a combustible gas, and other gases for twenty years or more after they cease receiving waste. Whether the waste in a landfill is processed or unprocessed, the landfill generally cannot be reforested. Tree roots damage the impermeable cap applied to a closed landfill to reduce the production of leachate.

Solid waste generates odors as microorganisms metabolize organic matter in the waste, causing the organic matter to decompose. The most acute odor problems generally occur when waste decomposes rapidly, consuming available oxygen and inducing anaerobic (oxygen deficient) conditions. Bulky waste generally does not cause odor problems because it typically contains little material that decomposes rapidly. MSW, on the other hand, typically causes objectionable odors even when covered with dirt in a landfill (see Section 10.13).

Combustion facilities prevent odor problems by incinerating the odorous compounds and the microorganisms and organic matter from which the odorous compounds are derived (see Section 10.9). Composting preserves organic matter while reducing its potential to generate odors. However, the composting process requires careful engineering to minimize odor generation during composting (see Section 10.14).

In addition to odors, solid waste can cause other forms of pollution. Landfill leachate contains toxic substances that must be prevented from contaminating groundwater and surface water (see Section 10.13). Toxic and corrosive products of solid waste combustion must be prevented from entering the atmosphere (see Section 10.9). The use

of solid waste compost must be regulated so that the soil is not contaminated (see Section 10.14).

While avoiding the potential negative effects of solid waste, a solid waste management program should also seek to derive benefits from the waste. Methods for deriving benefits from solid waste include recycling (Section 10.7), composting (Section 10.14), direct combustion with energy recovery (Section 10.9), processing waste to produce fuel (Sections 10.8 and 10.12), and recovery of landfill gas for use as a fuel (Section 10.13).

— *F. Mack Rugg*

References

Cal Recovery Systems, Inc. 1990. *Waste characterization for San Antonio, Texas*. Richmond, Calif. (June).

Camp Dresser & McKee Inc. 1990a. *Marion County (FL) solid waste composition and recycling program evaluation*. Tampa, Fla. (April).

———. 1990b. *Sarasota County waste stream composition study*. Draft report (March).

———. 1991a. *Cape May County multi-seasonal solid waste composition study*. Edison, N.J. (August).

———. 1991b. *City of Wichita waste stream analysis*. Wichita, Kans. (August).

———. 1992. *Atlantic County (NJ) solid waste characterization program*. Edison, N.J. (May).

Cosulich, William F., Associates, P.C. 1988. *Solid waste management plan, County of Monroe, New York: Solid waste quantification and characterization*. Woodbury, N.Y. (July).

Delaware Solid Waste Authority. 1992. *Solid waste management plan*. (17 December).

Franklin Associates, Ltd. 1992. *Characterization of municipal solid waste in the United States: 1992 update*. U.S. EPA, EPA/530-R-92-019, NTIS no. PB92-207 166 (July).

HDR Engineering, Inc. 1989. *Report on solid waste quantities, composition and characteristics for Monmouth County (NJ) waste recovery system*. White Plains, N.Y. (March).

Killam Associates. 1989; 1991 update. *Middlesex County (NJ) solid waste weighing, source, and composition study*. Millburn, N.J. (February).

———. 1990. *Somerset County (NJ) solid waste generation and composition study*. Millburn, N.J. (May). Includes data for Warren County, N.J.

Minnesota Pollution Control Agency and Metropolitan Council. 1993. *Minnesota solid waste composition study, 1991–1992 part II*. Saint Paul, Minn. (April).

Rhode Island Solid Waste Management Corporation. 1987. *Statewide resource recovery system development plan*. Providence, R.I. (June).

San Diego, City of, Waste Management Department. 1988. *Request for proposal: Comprehensive solid waste management system*. (4 November).

SCS Engineers. 1991. *Waste characterization study—solid waste management plan, Fairfax County, Virginia*. Reston, Va. (October).

Seattle Engineering Department, Solid Waste Utility. 1988. *Waste reduction, recycling and disposal alternatives: Volume II—Recycling potential assessment and waste stream forecast*. Seattle (May).

Characterization

10.3
PHYSICAL AND CHEMICAL CHARACTERISTICS

This section addresses the characteristics of solid waste including fluctuations in quantity; composition, density, and other physical characteristics; combustion characteristics; bioavailability; and the presence of toxic substances.

Fluctuations in Solid Waste Quantities

Weakness in the economy generally reduces the quantity of solid waste generated. This reduction is particularly true for commercial and industrial MSW and construction and demolition debris. Data quantifying the effect of economic downturns on solid waste quantity are not readily available.

The generation of solid waste is usually greater in warm weather than in cold weather. Figure 10.3.1 shows two month-to-month patterns of MSW generation. The less variable pattern is a composite of data from eight locations with cold or moderately cold winters (Camp Dresser & McKee Inc. 1992, 1991; Child, Pollette, and Flosdorf 1986; Cosulich Associates 1988; HDR Engineering, Inc. 1989; Killam Associates 1990; North Hempstead 1986; Oyster Bay 1987). Waste generation is relatively low in the winter but rises with temperature in the spring. The surge of waste generation in the spring is caused both by increased human activity, including spring cleaning, and renewed plant growth and associated yard waste. Waste generation typically declines somewhat after June but remains above average until mid to late fall. In contrast, Figure 10.3.1 also shows the pattern of waste generation in Cape May County, New Jersey, a summer resort area (Camp Dresser & McKee Inc. 1991). The annual influx of tourists overwhelms all other influences of waste generation.

Areas with mild winters may display month-to-month patterns of waste generation similar to the cold-winter pattern shown in Figure 10.3.1 but with a smaller difference between the winter and spring/summer rates. On the other hand, local factors can create a distinctive pattern not generally seen in other areas, as in Sarasota, Florida (Camp Dresser & McKee Inc. 1990). The surge of activity and plant growth in the spring is less marked in mild climates,

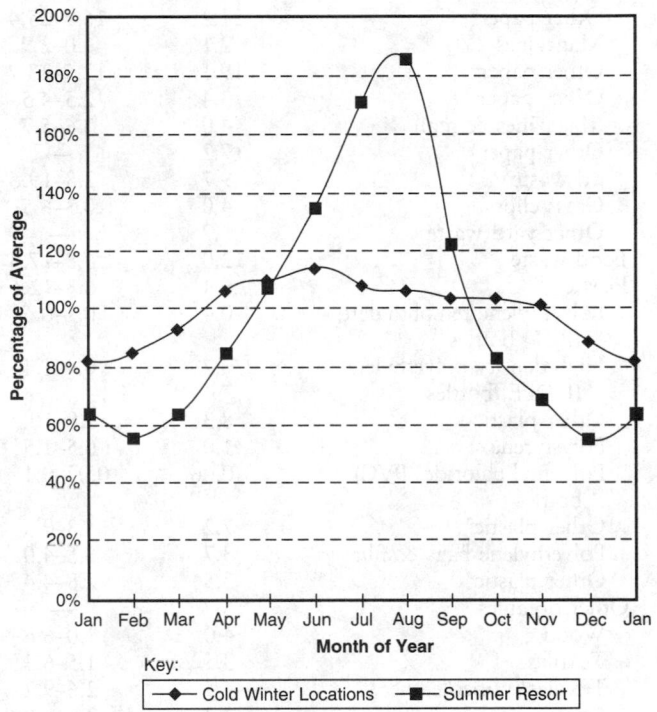

FIG. 10.3.1 Month-to-month variation in MSW generation rate.

and local factors can cause the peak of waste generation to occur in any season of the year.

Component Composition of MSW

Table 10.3.1 lists the representative component composition for MSW disposed in the United States and adjacent portions of Canada and shows ranges for individual components. Materials diverted from the waste stream for recycling or composting are not included. The table is based on the results of twenty-two field studies in eleven states plus the Canadian province of British Columbia. The ranges shown in the table are annual values for county-sized areas. Seasonal values may be outside these ranges, especially in individual municipalities.

TABLE 10.3.1   REPRESENTATIVE COMPONENT
COMPOSITION OF MSW

Waste Category	Representative Composition (%)[b]	Range of Reasonable Reported Values (%)[b]
Organics/Combustibles	86.6	—
Paper	39.8	—
Newspaper	6.8	4.0–13.1
Corrugated	8.6	3.5–14.8
Kraft	1.5	0.5–2.3
Corrugated & kraft	10.1	5.4–15.6
Other paper[a]	22.9	17.6–30.6
High-grade paper	1.7	0.6–3.2
Other paper[a]	21.2	16.9–25.4
Magazines	2.1	1.0–2.9
Other paper[a]	19.1	12.5–23.7
Office paper	3.4	2.5–4.5
Magazines & mail	4.0	3.6–5.7
Other paper[a]	17.2	—
Yard waste	9.7	2.8–19.6
Grass clippings	4.0	0.3–6.5
Other yard waste	5.7	—
Food waste	12.0	6.8–17.3
Plastic	9.4	6.3–12.6
Polyethylene terephthalate (PET) bottles	0.4	0.1–0.5
High-density polyethylene (HDPE) bottles	0.7	0.4–1.1
Other plastic	8.3	5.8–10.2
Polystyrene	1.0	0.5–1.5
Polyvinyl chloride (PVC) bottles	0.06	0.02–0.1
Other plastic[a]	7.2	5.3–9.5
Polyethylene bags & film	3.7	3.5–4.0
Other plastic[a]	3.5	2.8–4.4
Other organics	15.7	—
Wood	4.0	1.0–6.6
Textiles	3.5	1.5–6.3
Textiles/rubber/leather	4.5	2.6–9.2
Fines	3.3	2.8–4.0
Fines <½ inch	2.2	1.7–2.8
Disposable diapers	2.5	1.8–4.1
Other organics	1.4	—
Inorganics/Noncombustibles	13.4	—
Metal	5.8	—
Aluminum	1.0	0.6–1.2
Aluminum cans	0.6	0.3–1.2
Other aluminum	0.4	0.2–0.9
Tin & bimetal cans	1.5	0.9–2.7
Other metal[a]	3.3	1.1–6.9
Ferrous metal	4.5	2.8–5.5
Glass	4.8	2.3–9.7
Food & beverage containers	4.3	2.0–7.7
Other glass	0.5	—
Batteries	0.1	0.04–0.1
Other Inorganics		
With noncontainer glass	3.2	1.9–4.9
Without noncontainer glass	2.7	1.8–3.8

[a]Each "other" category contains all material of its type except material in the categories above it.
[b]Weight percentage

Residential MSW contains more newspaper; yard waste; disposable diapers; and textiles, rubber, and leather. Nonresidential MSW contains more corrugated cardboard, high-grade paper, wood, other plastics, and other metals.

The composition of MSW varies from one CII establishment to another. However, virtually all businesses and institutions generate a variety of waste materials. For example, offices do not generate only paper waste, and restaurants do not generate only food waste.

Component Composition of Bulky Waste

Fewer composition data are available for bulky waste than for MSW. Table 10.3.2 shows the potential range of compositions. The first column in the table shows the composition of all bulky waste generated in two adjacent counties in southern New Jersey, including bulky waste reported as recycled. The third column shows the composition of bulky waste disposed in the two counties, and the middle column shows the estimated recycling rate for each bulky waste component based on reported recycling and disposal. Note that the estimated overall recycling rate is almost 80%.

The composition prior to recycling is dramatically different from the composition after recycling. For example, inorganic materials account for roughly three quarters of the bulky waste before recycling but little more than one quarter after recycling. Depending on local recycling practices, the composition of bulky waste received at a disposal facility in the United States could be similar to the first column of Table 10.3.2, similar to the third column, or anywhere in between.

The composition of MSW does not change dramatically from season to season. Even the most variable component, yard waste, may be consistent in areas with mild climates. In areas with cold winters, generation of yard waste generally peaks in the late spring, declines gradually through the summer and fall, and is lowest in January and February. A surge in yard waste can occur in mid to late fall in areas where a large proportion of tree leaves enter the solid waste stream and are not diverted for composting or mulching.

Density

As discussed in Section 10.2, the density of MSW varies according to circumstance. Table 10.3.3 shows representative density ranges for MSW under different conditions. The density of mixed MSW is influenced by the degree of compaction, moisture content, and component composition. As shown in the table, individual components of MSW have different bulk densities, and a range of densities exists within most components.

TABLE 10.3.2 COMPONENT COMPOSITION OF BULKY WASTE AND THE POTENTIAL IMPACT OF RECYCLING

Waste Category	Composition of all Bulky Waste Generated (%)[a]	Composition of Bulky Waste Recycled (%)[a]	Composition of Bulky Waste Landfilled (%)[a]
Organics/Combustibles	24.7	37.9	73.4
Lumber	13.1	47.2	33.0
Corrugated cardboard	0.7	2.5	3.1
Plastic	1.0	18.8	3.7
Furniture	1.3	0.0	6.3
Vegetative materials	3.8	73.0	4.9
Carpet & padding	0.7	0.0	3.2
Bagged & miscellaneous	2.1	0.0	10.2
Roofing materials	1.2	0.4	5.9
Tires	0.3	100.0	0.0
Other	0.6	0.0	3.1
Inorganics/Noncombustibles	75.3	92.6	26.6
Gypsum board & plaster	1.8	3.9	8.3
Metal & appliances	15.4	92.5	5.5
Dirt & dust	1.2	0.0	5.8
Concrete	26.5	96.7	4.2
Asphalt	28.7	99.9	0.1
Bricks & blocks	1.3	81.8	1.1
Other	0.3	0.0	1.6
Overall	100.0	79.1	100.0

Sources: Data from Camp Dresser & McKee, 1992, *Atlantic County (NJ) Solid Waste Characterization Program* (Edison, N.J. [May]) and *Idem,* 1991, *Cape May County Multi-Seasonal Solid Waste Composition Study* (Edison, N.J. [August]).

[a]Weight percentage

TABLE 10.3.3 DENSITY OF MSW AND COMPONENTS

Material and Circumstance	Density (lb/cu yd)
Mixed MSW	
Loose	150–300
In compactor truck	400–800
Dumped from compactor truck	300–500
Baled	800–1600
In landfill	800–1400
Loose Bulk Densities	
Aluminum cans (uncrushed)	54–81
Corrugated cardboard	50–135
Dirt, sand, gravel, concrete	2000–3000
Food waste	800–1500
Glass bottles (whole)	400–600
Light ferrous, including cans	100–250
Miscellaneous paper	80–250
Stacked high-grade paper	400–600
Plastic	60–150
Rubber	200–400
Textiles	60–180
Wood	200–600
Yard waste	100–600

Within individual categories of MSW, bulk density increases as physical irregularity decreases. Compaction increases density primarily by reducing irregularity. Some compaction occurs in piles, so density tends to increase as the height of a pile increases. In most cases, shredding and other size reduction measures also increase density by reducing irregularity. The size reduction of regularly shaped materials such as office paper, however, can increase irregularity and decrease density.

Particle Size, Abrasiveness, and Other Physical Characteristics

Figure 10.3.2 shows a representative particle size distribution for MSW based on research by Hilton, Rigo, and Chandler (1992). Environmental engineers generally estimate size distribution by passing samples of MSW over a series of screens, beginning with a fine screen and working up to a coarse screen. As shown in the figure, MSW has no characteristic particle size, and most components of MSW have no characteristic particle size.

MSW does not flow, and piles of MSW have a tendency to hold their shape. Loads of MSW discharged from compactor trucks often retain the same shape they had in-

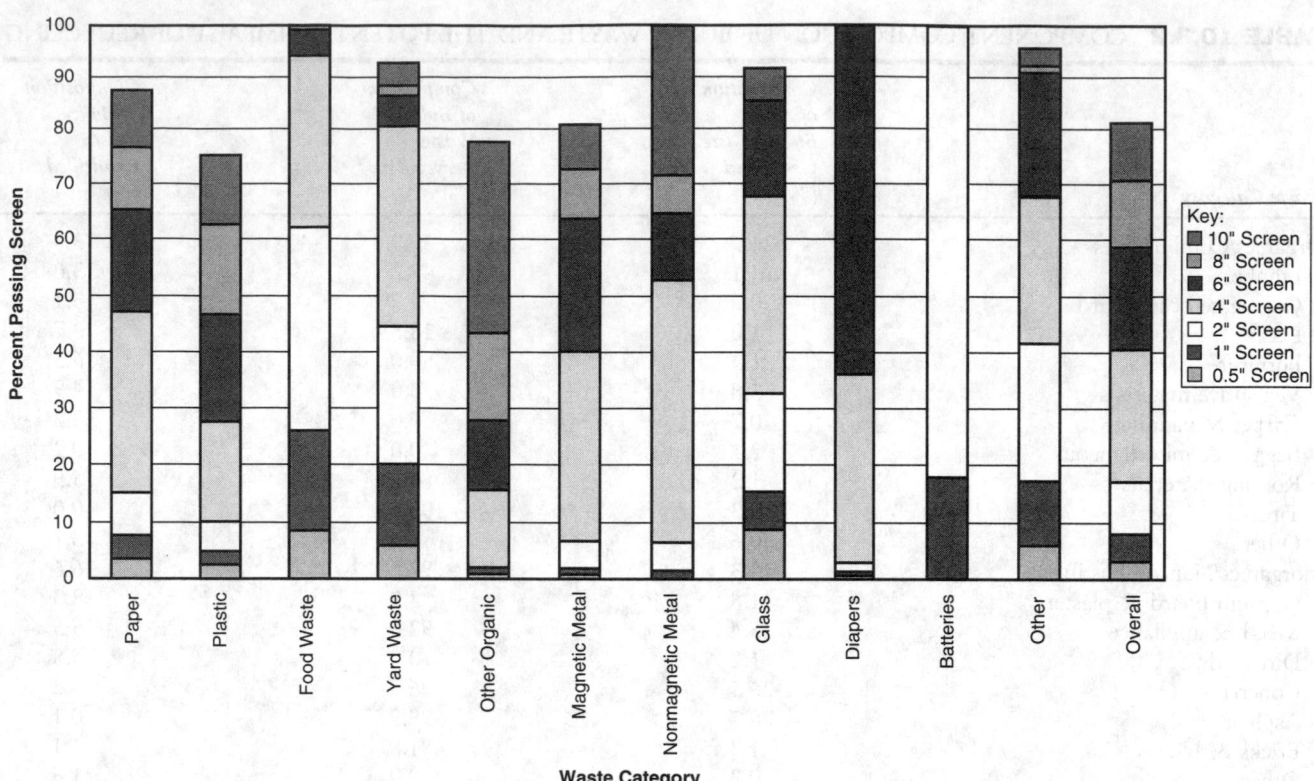

FIG. 10.3.2 Representative size distribution of MSW. (Adapted from D. Hilton, H.G. Rigo, and A.J. Chandler, 1992, Composition and size distribution of a blue-box separated waste stream, presented at *SWANA's Waste-to-Energy Symposium, Minneapolis, MN, January 1992.*)

side the truck. When MSW is removed from one side of a storage bunker at an MSW combustion facility, the waste on the other side generally does not fall into the vacated space. This characteristic allows the side on which trucks dump waste be kept relatively empty during the hours when the facility receives waste.

MSW tends to stratify vertically when mixed, with smaller and denser objects migrating toward the bottom and lighter and bulkier objects moving toward the top. However, MSW does not stratify much when merely vibrated.

Although MSW is considered soft and mushy, it contains substantial quantities of glass, metal, and other potentially abrasive materials.

Combustion Characteristics

Most laboratory work performed on samples of solid waste over the years has focused on parameters related to combustion and combustion products. The standard laboratory tests in this category are proximate composition, ultimate composition, and heat value.

PROXIMATE COMPOSITION

The elements of proximate composition are moisture, ash, volatile matter, and fixed carbon. The moisture content of solid waste is defined as the material lost during one hour at 105°C. Ash is the residue remaining after combustion. Together, moisture and ash represent the noncombustible fraction of the waste.

Volatile matter is the material driven off as gas or vapor when waste is subjected to a temperature of approximately 950°C for 7 min but is prevented from burning because oxygen is excluded. Volatile matter should not be confused with *volatile organic compounds* (VOCs). VOCs are a small component of typical solid waste. In proximate analysis, any VOCs present tend to be included in the result for moisture.

Conceptually, fixed carbon is the combustible material remaining after the volatile matter is driven off. Fixed carbon represents the portion of combustible waste that must be burned in the solid state rather than as gas or vapor. The value for fixed carbon reported by the laboratory is calculated as follows:

$$\% \text{ fixed carbon} = 100\% - \% \text{ moisture} - \% \text{ ash} - \% \text{ volatile matter} \quad \textbf{10.3(1)}$$

Table 10.3.4 shows a representative proximate composition for MSW. The values in the table are percentages based on dry (moisture-free) MSW. Representative moisture values are also provided. These moisture values are for MSW and components of MSW as they are received at a disposal facility. Because of a shortage of data for the proximate composition of noncombustible materials, these materials are presented as 100% ash.

The dry-basis values in Table 10.3.4 can be converted to as-received values by using the following equation:

$$A = D(100\% - M) \qquad 10.3(2)$$

where:

A = value for waste as received at the solid waste facility

D = dry-basis value

M = percent moisture for waste received at the solid waste facility

Between initial discard at the point of generation and delivery to a central facility, moisture moves from wet materials to dry and absorbent materials. The largest movement of moisture is from food waste to uncoated paper discarded with food waste. This paper includes newspaper, kraft paper, and a substantial portion of other paper from residential sources as well as corrugated cardboard from commercial sources.

Other sources of moisture in paper waste include water absorbed by paper towels, napkins, and tissues during use, and precipitation. Absorbent materials frequently exposed to precipitation include newspaper and corrugated cardboard. Many trash containers are left uncovered, and precipitation is absorbed by the waste. Standing water in dumpsters is often transferred to the collection vehicle.

The value of proximate analysis is limited because (1) it does not indicate the degree of oxidation of the combustible waste and (2) it gives little indication of the quantities of pollutants emitted during combustion of the waste. Ultimate analysis supplements the information provided by proximate analysis.

TABLE 10.3.4 REPRESENTATIVE PROXIMATE AND ULTIMATE COMPOSITION OF MSW

Waste Category	Proximate Composition—Dry Basis			Ultimate Composition—Dry Basis[a]						Moisture (%)
	Ash (%)	Volatile Matter (%)	Fixed Carbon (%)	Carbon (%)	Hydrogen (%)	Nitrogen (%)	Chlorine (%)	Sulfur (%)	Oxygen (%)	
Organics/Combustibles	7.7	82.6	9.6	48.6	6.8	0.94	0.69	0.22	35.0	32.5
Paper	6.3	83.5	10.1	43.0	6.0	0.36	0.17	0.17	43.8	24.0
Newspaper	5.2	83.8	11.1	43.8	5.9	0.29	0.14	0.24	44.4	23.2
Corrugated & kraft paper	2.2	85.8	12.1	46.0	6.4	0.28	0.14	0.22	44.8	21.2
High-grade paper	9.1	83.4	7.5	38.1	5.6	0.15	0.12	0.07	46.9	9.3
Magazines	20.4	71.8	7.9	35.0	5.0	0.05	0.07	0.08	39.4	8.6
Other paper	6.9	83.8	9.3	42.7	6.1	0.50	0.22	0.14	43.3	28.7
Yard waste	9.6	73.0	17.4	45.0	5.6	1.5	0.31	0.17	37.7	53.9
Grass clippings	9.7	75.6	14.7	43.3	5.9	2.6	0.60	0.30	37.6	63.9
Leaves	7.3	72.7	20.1	50.0	5.7	0.82	0.10	0.10	36.0	44.0
Other yard waste	12.5	70.5	17.0	40.7	5.0	1.3	0.26	0.10	40.0	50.1
Food waste	11.0	79.0	10.0	45.4	6.9	3.3	0.74	0.32	32.3	65.4
Plastic	5.3	93.0	1.3	76.3	11.5	0.26	2.4	0.20	4.4	13.3
PET bottles	1.3	95.0	3.6	68.5	8.0	0.16	0.08	0.08	21.9	3.6
HDPE bottles	2.4	97.4	0.2	81.6	13.6	0.10	0.18	0.20	1.9	7.0
Polystyrene	1.8	97.8	0.4	86.3	7.9	0.28	0.12	0.30	3.4	10.8
PVC bottles	0.6	46.2	3.2	44.2	5.9	0.26	40.1	0.89	7.6	3.2
Polyethylene bags & film	8.8	90.1	1.1	77.4	12.9	0.10	0.09	0.12	1.8	19.1
Other plastic	4.2	94.1	1.7	72.9	11.4	0.45	5.3	0.24	5.5	10.5
Other Organics	11.3	77.8	10.9	46.2	6.1	1.9	1.0	0.36	33.3	27.3
Wood	2.8	83.0	14.1	46.7	6.0	0.71	0.12	0.16	43.4	14.8
Textiles/rubber/leather	6.6	84.0	9.4	50.3	6.4	3.3	1.8	0.33	31.3	12.4
Fines	25.3	64.7	10.0	37.3	5.3	1.6	0.54	0.45	29.5	41.1
Disposable diapers	4.1	87.1	8.7	48.4	7.6	0.51	0.23	0.35	38.8	66.9
Other organics	31.3	58.8	9.9	44.2	5.3	1.8	2.2	0.81	14.4	8.0
Inorganics/Noncombustibles[b]	100	0	0	0	0	0	0	0	0	0
Overall	24.9	67.2	7.8	39.5	5.6	0.76	0.56	0.18	28.5	28.2

[a]Also includes ash values from first column of proximate analysis.
[b]Values assumed for the purpose of estimating overall values.

ULTIMATE COMPOSITION

Moisture and ash, as previously defined for proximate composition, are also elements of ultimate composition. In standard ultimate analysis, the combustible fraction is divided among carbon, hydrogen, nitrogen, sulfur, and oxygen. Ultimate analysis of solid waste should also include chlorine. The results are more useful if sulfur is broken down into organic sulfur, sulfide, and sulfate; and chlorine is broken down into organic (insoluble) and inorganic (soluble) chlorine (Niessen 1995).

Carbon, hydrogen, nitrogen, sulfur, and chlorine are measured directly; calculating oxygen requires subtracting the sum of the other components (including moisture and ash) from 100%. Table 10.3.4 shows a representative ultimate composition for MSW. The dry-basis values shown in the table can be converted to as-received values with use of Equation 10.3(2).

The ultimate composition of MSW on a dry basis reflects the dominance of six types of materials in MSW: cellulose, lignins, fats, proteins, hydrocarbon polymers, and inorganic materials. Cellulose is approximately 42.5% carbon, 5.6% hydrogen, and 51.9% oxygen and accounts for the majority of the dry weight of MSW. The cellulose content of paper ranges from approximately 75% for low grades to approximately 90% for high-grade paper. Wood is roughly 50% cellulose, and cellulose is a major ingredient of yard waste, food waste, and disposable diapers. Cotton, the largest ingredient of the textile component of MSW, is approximately 98% cellulose (Masterton, Slowinski, and Stanitski 1981).

Despite the abundance of cellulose, MSW contains more carbon than oxygen due to the following factors:

- Most of the plastic fraction of MSW is composed of polyethylene, polystyrene, and polypropylene, which contain little oxygen.
- Synthetic fibers (textiles category) contain more carbon than oxygen, and rubber contains little oxygen.
- The lower grades of paper contain significant quantities of lignins, which contain more carbon than oxygen.
- Fats contain more carbon than oxygen.

The nitrogen in solid waste is primarily in organic form. The largest contributors of nitrogen to MSW are food waste (proteins), grass clippings (proteins), and textiles (wool, nylon, and acrylic). Chlorine occurs in both organic and inorganic forms. The largest contributor of organic chlorine is PVC or vinyl. Most of the PVC is in the other plastic and textiles components. The largest source of inorganic chlorine is sodium chloride (table salt). Sulfur is not abundant in any category of combustible MSW but is a major component of gypsum board. The sulfur in gypsum is largely noncombustible but not entirely so. In Table 10.3.4, gypsum board is included in the Inorganics/Noncombustibles category, which is shown as 100% ash because of a lack of data on the ultimate composition.

The inorganic (noncombustible) waste categories contribute most of the ash in MSW. Additional ash is contributed by the inorganic components of combustible materials, including clay in glossy and high-grade paper, dirt in yard waste, bones and shells in food waste, asbestos in vinyl–asbestos floor coverings, fiberglass in reinforced plastic, and grit on roofing shingles.

HEAT VALUE

Table 10.3.5 shows the heat value of typical MSW based on the results of laboratory testing of MSW components. Calculations of the heat value based on energy output measurements at operating combustion facilities generally yield lower values (see Section 10.5).

The heat value shown for solid waste and conventional fuels in the United States, Canada, and the United Kingdom is typically the higher heating value (HHV). The HHV includes the latent heat of vaporization of the water created during combustion. When this heat is deducted, the result is called the lower heating value (LHV). For additional information see Niessen (1995).

The as-received heat value is roughly proportional to the percentage of waste that is combustible (i.e., neither moisture nor ash) and to the carbon content of the combustible fraction. The heat values of the plastics categories are highest because of their high carbon content, low ash content, and low-to-moderate moisture content. Paper categories have intermediate heat values because of their intermediate carbon content, moderate moisture content, and low-to-moderate ash content. Yard waste, food waste, and disposable diapers have low heat values because of their high moisture levels.

Bioavailability

Because microorganisms can metabolize paper, yard waste, food waste, and wood, this waste is classified as *biodegradable*. Disposable diapers and their contents are also largely biodegradable, as are cotton and wool textiles.

Some biodegradable waste materials are more readily metabolized than others. The most readily metabolized materials are those with high nitrogen and moisture content: food waste, grass clippings, and other green, pulpy yard wastes. These wastes are *putrescible* and have high *bioavailability*. Leaf waste generally has intermediate bioavailability. Wood, cotton and wool, although biodegradable, have relatively low bioavailability and are considered noncompostable within the context of solid waste management.

Toxic Substances in Solid Waste

Solid waste inevitably contains many of the toxic substances manufactured or extracted from the earth. Most

TABLE 10.3.5 REPRESENTATIVE HEAT VALUES OF MSW[a]

Waste Category	Dry-Basis Heat Value (HHV in Btu/lb)	Moisture Content (%)	As-Received Heat Value (HHV in Btu/lb)
Organics/Combustibles	9154	32.5	6175
Paper	7587	24.0	5767
Newspaper	7733	23.2	5936
Corrugated & kraft	8168	21.2	6435
High-grade paper	6550	9.3	5944
Magazines	5826	8.6	5326
Other paper	7558	28.7	5386
Yard waste	7731	53.9	3565
Grass clippings	7703	63.9	2782
Leaves	8030	44.0	4499
Other yard waste	7387	50.1	3689
Food waste	8993	65.4	3108
Plastic	16,499	13.3	14,301
PET bottles	13,761	3.6	13,261
HDPE bottles	18,828	7.0	17,504
Polystyrene	16,973	10.8	15,144
PVC bottles	10,160	3.2	9838
Polyethylene bags & film	17,102	19.1	13,835
Other plastic	15,762	10.5	14,108
Other organics	8698	27.3	6322
Wood	8430	14.8	7186
Textiles/rubber/ leather	9975	12.4	8733
Fines	6978	41.1	4114
Disposable diapers	9721	66.9	3222
Other organics	7438	8.0	6844
Inorganics/ Noncombustibles[b]	0	0.0	0
Overall	7446	28.2	5348

[a]Values shown are HHV. In HHV measurements, the energy required to drive off the moisture formed during combustion is not deducted.

[b]Values assumed for the purpose of estimating overall values.

toxic material in solid waste is in one of three categories:

- Toxic metals
- Toxic organic compounds, many of which are also flammable
- Asbestos

The results of studies of toxic metals in solid waste vary. Table 10.3.6 summarizes selected results of two comprehensive studies performed in Cape May County, New Jersey (Camp Dresser & McKee Inc. 1991a) and Burnaby, British Columbia (Chandler & Associates, Ltd. 1993; Rigo, Chandler, and Sawell 1993). Reports of both studies contain data for additional metals and materials, and the Burnaby reports contain results for numerous subcategories of the categories in the table. The Burnaby reports also analyze the behavior of specific metals from waste components during processing in an MSW incinerator.

Franklin Associates, Ltd. (1989) provided extensive information on sources of lead and cadmium in MSW, and Rugg and Hanna (1992) compiled detailed information on sources of lead in MSW in the United States.

Most MSW referred to as *household hazardous waste* is so classified because it contains toxic organic compounds. Large quantities of toxic organic materials from commercial and industrial sources were once disposed in MSW landfills in the United States, and many of these landfills are now officially designated as hazardous waste sites. The large-scale disposal of toxic organics in MSW landfills has been largely eliminated, but disposal of household hazardous waste remains a concern for many. Generally, household hazardous waste refers to whatever toxic materials remain in MSW, regardless of the source.

Estimates of the abundance of household hazardous waste vary. Reasons for the lack of consistency from one

TABLE 10.3.6 REPORTED METAL CONCENTRATIONS IN COMPONENTS OF MSW[a]

Waste Category	Arsenic CM	Arsenic BC	Cadmium CM	Cadmium BC	Chromium CM	Chromium BC	Copper CM	Copper BC	Lead CM	Lead BC	Mercury CM	Mercury BC	Nickel CM	Nickel BC	Zinc CM	Zinc BC
Organics/Combustibles																
Paper																
Newspaper	0.1	0.7	ND[b]	0.1	ND	49	17	18	ND	7	0.3	2	ND	28	58	21
Corrugated cardboard	0.2	0.6	ND	0.1	ND	2	13	3	19	4	0.2	0.1	6	4	56	10
Kraft paper	0.3	0.8	ND	0.1	5	5	11	11	15	9	0.1	0.5	ND	8	30	22
High-grade paper	0.7	1	ND	0.1	ND	3	7	8	ND	5	0.1	0.3	ND	8	28	208
Magazines	0.4	1	ND	0.2	4	11	46	32	ND	3	0.09	0.3	ND	13	88	27
Other	0.4	1	ND	1	4	27	52	25	9	182	0.07	0.3	ND	7	58	71
Yard waste	0.9	6	ND	5	4	87	10	571	14	137	0.1	1	3	21	89	321
Food waste	0.1	1	ND	2	ND	23	9	43	ND	72	0.02	0.3	2	5	20	186
Plastic																
PET	ND	0.8	ND	5	15	17	30	31	59	62	0.07	0.2	ND	8	21	97
HDPE	0.2	0.5	ND	3	52	15	14	24	211	61	0.1	0.2	ND	7	58	142
Film	0.5	0.6	ND	5	100	102	25	23	450	325	0.1	0.2	ND	7	120	658
Other	0.4	0.7	8	82	7	279	8	58	19	342	0.04	0.3	ND	40	69	231
Other organics																
Wood	34	24	ND	0.4	52	77	32	68	108	408	2	0.3	ND	3	205	174
Textiles & footwear	0.8	0.4	19	4	387	619	25	62	48	129	0.3	1	5	1	666	222
Fines	3	7	1	4	14	115	179	243	273	259	0.2	1	18	54	352	654
Disposable diapers	0.1	—	ND	—	1	—	2	—	ND	—	0.02	—	ND	—	28	—
Inorganics/Noncombustibles																
Metal																
Ferrous food & beverage cans	4	7	16	43	527	191	375	104	350	342	0.8	6	133	161	145	1552
Aluminum beverage cans	ND	0.4	ND	5	72	91	107	1105	30	41	0.7	0.4	54	21	80	229
Other metal	9	280	22	25	4702	768	6816	2082	1279	95	0.7	0.4	411	24	1675	199,000
Glass food & beverage containers	ND	2	ND	4	ND	91	ND	26	84	103	0.2	0.2	ND	15	ND	71
Household batteries																
Carbon-zinc & alkaline batteries[c]	7	2	53	1027	45	57	8400	6328	236	94	2900	136	—	512	180,000	103,000
Nickel-cadmium batteries	—	4	175,000	120,000	—	64	—	53	—	113	—	0.3	240,000	315	—	685
Other inorganics	1	12	ND	8	21	91	13	113	50	607	0.9	0.2	5	73	21	1997

Source: Data adapted from Camp Dresser & McKee Inc., 1991a, *Cape May County multi-seasonal solid waste composition study* (Edison, N.J. [August]); A.J. Chandler & Associates, Ltd. et al., 1993, *Waste analysis, sampling, testing and evaluation* (WASTE) *program: Effect of waste stream characteristics on MSW incineration: The fate and behaviour of metals. Final report of the mass burn MSW incineration study (Burnaby, B.C.), Vol. 1, Summary report* (Toronto [April]); and H.G. Rigo, A.J. Chandler, and S.E. Sawell, 1993, Debunking some myths about metals, in *Proceedings of the 1993 International Conference on Municipal Waste Combustion* (Williamsburg, Va. [30 March–2 April]).

[a]All values in mg/kg on an as-received basis. Values presented are based on reported results from studies in Cape May County, New Jersey and Burnaby, British Columbia. CM indicates Cape May, and BC indicates Burnaby.

[b]ND = Not detected.

[c]Current values for mercury are close to or below the Burnaby value.

study to another include the following:

Some quantity estimates include less toxic materials such as latex paint.

Most quantity estimates include the weight of containers, and many estimates include the containers even if they are empty.

Some quantity estimates include materials that were originally in liquid or paste form but have dried, such as dried paint and adhesives. Toxic substances can still leach from these dried materials, but drying reduces the potential leaching rate.

Strongly toxic organic materials, excluding their containers, appear to constitute well under 0.5% of MSW, and the toxic material is usually dispersed. Bulky waste typically contains no more toxic organic material than MSW, but bulky waste is more likely to contain concentrated pockets of toxic substances.

A statewide waste characterization study in Minnesota (Minnesota Pollution Control Agency 1992; Minnesota Pollution Control Agency and Metropolitan Council 1993) provides a detailed accounting of the household hazardous waste materials encountered.

Most of the asbestos in normal solid waste is in old vinyl–asbestos floor coverings and asbestos shingles. Asbestos in these forms is generally not a significant hazard.

—*F. Mack Rugg*

References

Camp Dresser & McKee Inc. 1990. *Sarasota County waste stream composition study.* Draft report (March).

———. 1991a. *Cape May County multi-seasonal solid waste composition study.* Edison, N.J. (August).

———. 1991b. *Cumberland County (NJ) waste weighing and composition analysis.* Edison, N.J. (January).

———. 1992. *Atlantic County (NJ) solid waste characterization program.* Edison, N.J. (May).

Chandler, A.J., & Associates, Ltd. et al. 1993. *Waste analysis, sampling, testing and evaluation (WASTE) program: Effect of waste stream characteristics on MSW incineration: The fate and behaviour of metals. Final report of the mass burn MSW incineration study (Burnaby, B.C.). Volume I, Summary report.* Toronto (April).

Child, D., G.A. Pollette, and H.W. Flosdorf. 1986. Waste stream analysis. *Waste Age* (November).

Cosulich, William F., Associates, P.C. 1988. *Solid waste management plan, County of Monroe, New York: Solid waste quantification and characterization.* Woodbury, N.Y. (July).

Franklin Associates, Ltd. 1989. *Characterization of products containing lead and cadmium in municipal solid waste in the United States, 1970 to 2000.* U.S. EPA (January).

HDR Engineering, Inc. 1989. *Report on solid waste quantities, composition and characteristics for Monmouth County (NJ) waste recovery system.* White Plains, N.Y. (March).

Killam Associates. 1990. *Somerset County (NJ) solid waste generation and composition study.* Millburn, N.J. (May).

Masterton, W.L., E.J. Slowinski, and C.L. Stanitski. 1981. *Chemical principles.* 5th ed. Philadelphia: Saunders College Publishing.

Minnesota Pollution Control Agency. 1992. *Minnesota solid waste composition study, 1990–1991 part I.* Saint Paul, Minn. (November).

Minnesota Pollution Control Agency and Metropolitan Council. 1993. *Minnesota solid waste composition study, 1991–1992 part II.* Saint Paul, Minn. (April).

Niessen, W.R. 1995. *Combustion and incineration processes: Applications in environmental engineering.* 2d ed. New York: Marcel Dekker, Inc.

North Hempstead, Town of (NY), transfer station scalehouse records, August 1985 through July 1986. 1986.

Oyster Bay, Town of (NY), transfer station scalehouse records, September 1986 through August 1987. 1987.

Rigo, H.G., A.J. Chandler, and S.E. Sawell. Debunking some myths about metals. In *Proceedings of the 1993 International Conference on Municipal Waste Combustion, Williamsburg, VA, March 30–April 2, 1993.*

Rugg, M. and N.K. Hanna. 1992. Metals concentrations in compostable and noncompostable components of municipal solid waste in Cape May County, New Jersey. *Proceedings of the Second United States Conference on Municipal Solid Waste Management, Arlington, VA, June 2–5, 1992.*

10.4
CHARACTERIZATION METHODS

This section describes and evaluates methods for estimating the characteristics of solid waste. The purposes of waste characterization are identified; and methods for estimating quantity, composition, combustion characteristics, and metals concentrations are addressed.

Purposes of Solid Waste Characterization

The general purpose of solid waste characterization is to promote sound management of solid waste. Specifically, characterization can determine the following:

The size, capacity, and design of facilities to manage the waste.

The potential for recycling or composting portions of the waste stream.

The effectiveness of waste reduction programs, recycling programs, or bans on the disposal of certain materials.

Potential sources of environmental pollution in the waste.

In practice, the immediate purpose of most waste characterization studies, including many extensive studies, is to comply with specific regulatory mandates and to provide information for use by vendors in preparing bids to design, construct, and operate solid waste management facilities.

The purposes of a waste characterization program determine the design of it. If all waste is to be landfilled, the characterization program should focus on the quantity of waste, its density, and its potential for compaction. The composition of the waste and its chemical characteristics are relatively unimportant. If all waste is to be incinerated, the critical parameters are quantity, heat value, and the percentage of combustible material in the waste. If recycling and composting are planned or underway, a composition study can identify the materials targeted for recovery, estimate their abundance in the waste, and monitor compliance with source separation requirements.

Basic Characterization Methods

Environmental engineers use one of two fundamental methods to characterize solid waste. One method is to collect and analyze data on the manufacture and sale of products that become solid waste after use. The method is called material flows methodology. The second method is a direct field study of the waste itself. Combining these two fundamental methods creates hybrid methodologies (for example, see Gay, Beam, and Mar [1993]).

The direct field study of waste is superior in concept, but statistically meaningful field studies are expensive. For example, a budget of $100,000 is typically required for a detailed estimate of the composition of MSW arriving at a single disposal facility, accurate to within 10% at 90% confidence. A skilled and experienced team can often provide additional information at little additional cost, including an estimated composition for bulky waste based on visual observation.

The principal advantage of the material flows methodology is that it draws on existing data that are updated regularly by business organizations and governments. This method has several positive effects. First, the entire waste stream is measured instead of samples of the waste, as in field studies. Therefore, the results of properly conducted material flows studies tend to be more consistent than the results of field studies. Second, updates of material flows studies are relatively inexpensive once the analytical structure is established. Third, material flows studies are suited to tracking economic trends that influence the solid waste stream.

The principal disadvantages of material flows methodology follow.

Obtaining complete production data for every item discarded as solid waste is difficult.

Although data on food sales are available, food sales bear little relation to the generation of food waste. Not only is most food not discarded, but significant quantities of water are added to or removed from many food items between purchase and discard. These factors vary from one area to another based on local food preferences and eating patterns.

Material flows methodology cannot measure the generation of yard waste.

Material flows methodology does not account for the addition of nonmanufactured materials to solid waste prior to discard, including water, soil, dust, pet droppings, and the contents of used disposable diapers.

Some of the material categories used in material flows studies do not match the categories of materials targeted for recycling. For example, advertising inserts in newspapers are typically recycled with the newsprint, but in material flows studies the inserts are part of a separate commercial printing category.

In performing material flows studies for the U.S. EPA, Franklin Associates bases its estimates of food waste, yard waste, and miscellaneous inorganic wastes on field studies in which samples of waste were sorted. Franklin Associates (1992) also adjusts its data for the production of disposable diapers to account for the materials added during use.

In general, the more local and the more detailed a waste characterization study is to be, the greater are the advantages of a direct field study of the waste.

Estimation of Waste Quantity

The best method for estimating waste quantity is to install permanent scales at disposal facilities and weigh every truck on the way in and again on the way out. An increasing number of solid waste disposal facilities are equipped with scales, but many landfills still are not.

In the United States, facilities without scales record incoming waste in cubic yards and charge tipping fees by the cubic yard. Since estimating the volume of waste in a closed or covered vehicle or container is difficult, the volume recorded is usually the capacity of the vehicle or container. Because this estimation creates an incentive to deliver waste in full vehicles, the recorded volumes tend to be close to the actual waste volumes.

For the reasons previously stated, expressing waste quantity in tons is preferable to cubic yards. This conversion is conceptually simple, as shown in the following equation:

$$M = VD/2000 \qquad 10.4(1)$$

where:

   M = mass of waste in tons
   V = volume of waste in cubic yards
   D = density of waste in pounds per cubic yard

If the density is expressed in tons per cubic yard, dividing by 2000 is unnecessary. In the United States, however, the density of solid waste is usually expressed in pounds per cubic yard.

Although simple conceptually, converting cubic yards to tons can be difficult in practice. The density of solid waste varies from one type of waste to another, from one type of vehicle to another, and even among collection crews. In small waste streams, local conditions can cause the overall density of MSW, as received at disposal facilities, to vary from 250 to 800 lb/cu yd. A conversion factor of 3.0 to 3.3 cu yd/tn (600 to 667 lb/cu yd) is reasonable for both MSW and bulky waste in many large waste streams; however, this conversion factor may not be reasonable for a particular waste stream.

At disposal facilities without permanent scales, environmental engineers can use portable scales to develop a better estimate of the tons of waste being delivered. Selected trucks are weighed, and environmental engineers use the results to estimate the overall weight of the waste stream.

Portable truck scales are available in three basic configurations: (1) platform scales designed to accommodate entire vehicles (or trailers), (2) axle scales designed to accommodate one axle or a pair of tandem axles at a time, and (3) wheel scales designed to be used in pairs to accommodate one axle or a pair of tandem axles at a time. Axle scales can be used singly or in pairs. Similarly, either one or two pairs of wheel scales can be used. When a single axle scale or a single pair of wheel scales is used, adding the results for individual axles yields the weight of the vehicle.

Platform scales are the easiest to use, but the cost can be prohibitive. The use of wheel scales tends to be difficult and time consuming. The cost of axle scales is similar to that of wheel scales, and axle scales are easier to use than wheel scales. The use of a pair of portable axle scales is recommended in the *Municipal solid waste survey protocol* prepared for the U.S. EPA by SCS Engineers (1979). Regardless of what type of scale is used, a solid base that does not become soft in wet weather is required.

Truck weighing surveys, like other waste characterization field studies, are typically conducted during all hours that a disposal facility is open during a full operating week. A full week is used because the variation in waste characteristics is greater among the hours of a day and among the days of a week than among the weeks of a month. Also, spreading the days of field work out over several weeks is substantially more expensive.

A truck weighing survey should be conducted during at least two weeks—one week during the period of minimum waste generation and one week during the period of maximum waste generation (see Section 10.3). One week during each season of the year is preferable. Holiday weeks should be avoided.

Weighing all trucks entering the disposal facility is rarely possible, so a method of truck selection must be chosen. A conceptually simple approach is to weigh every nth truck (for example, every 5th truck) that delivers waste to the facility. This approach assumes that the trucks weighed represent all trucks arriving at the facility. The total waste tonnage can be estimated with the following equation:

$$W = T(w/t) \qquad 10.4(2)$$

where:

   W = the total weight of the waste delivered to the facility
   T = the total number of trucks that delivered waste to the facility
   w = the total weight of the trucks that were weighed
   t = the number of trucks that were weighed

This approach is suited to a facility that receives a fairly constant flow of trucks. Unfortunately, the rate at which trucks arrive at most facilities fluctuates during the operating day. A weighing crew targeting every nth truck will miss trucks during the busy parts of the day and be idle at other times. Missing trucks during the busy parts of the day can bias the results; the trucks that arrive at these times tend to be curbside collection trucks, which have a distinctive range of weights. Also, having a crew and its equipment stand idle at slow times while waiting for the nth truck to arrive reduces the amount of data collected, which reduces the statistical value of the overall results.

A better approach is to weigh as many trucks as possible during the operating day, keeping track of the total number of trucks that deliver waste during each hour. A separate average truck weight and total weight is calculated for each hour, and the hourly totals are added to yield a total for the day. For this purpose, Equation 10.4(2) is modified as follows:

$$W = T_1(w_1/t_1) + T_2(w_2/t_2) \cdots + T_n(w_n/t_n) \quad 10.4(3)$$

where:

W = the total weight of the waste delivered to the facility

T_1 = the number of trucks that delivered waste to the facility in the first hour

T_2 = the number of trucks that delivered waste to the facility in the second hour

T_n = the number of trucks that delivered waste to the facility in the last hour of the operating day

w_1 = the total weight of the trucks that were weighed in the first hour

w_2 = the total weight of the trucks that were weighed in the second hour

w_n = the total weight of the trucks that were weighed in the last hour of the operating day

t_1 = the number of trucks that were weighed in the first hour

t_2 = the number of trucks that were weighed in the second hour

t_n = the number of trucks that were weighed in the last hour of the operating day

Estimating the statistical precision of the results is complex when the ratio of the weighed trucks to the unweighed trucks varies from hour to hour. (Klee [1991, 1993] provides a discussion of this statistical problem.)

Sampling MSW to Estimate Composition

As in all statistical exercises based on sampling, the acquisition of samples is a critical step in estimating the composition of MSW. The principal considerations in collecting samples are the following:

Each pound of waste in the waste stream to be characterized must have an equal opportunity to be represented in the final results.
The greater the number of samples, the more precise the results.

The greater the variation between samples, the more samples must be sorted to achieve a given level of precision.
The greater the time spent collecting the samples, the less time is available to sort the samples.
The more the waste is handled prior to sorting, the more difficult and less precise the sorting.

A fundamental question is the time period(s) over which to collect the samples. One-week periods are generally used because most human activity and most refuse collection schedules repeat on a weekly basis. Sampling during a week in each season of the year is preferable. Spring sampling is particularly important because generation of yard waste, the most variable waste category, is generally least in the winter and greatest in the spring.

Another fundamental question is whether to collect the samples at the places where the waste is generated or at the solid waste facilities where the waste is taken. Sampling at solid waste facilities is generally preferred. Collecting samples at the points of generation may be necessary under the following circumstances, however:

The primary objective is to characterize the waste generated by certain sources, such as specific types of businesses.
The identity of the facilities to which the waste is taken is not known or cannot be predicted with confidence for any given week.
The facilities are widely spaced, increasing the difficulty and cost of the sampling and sorting operation.
Access to the facilities cannot be obtained.
Sufficient space to set up a sorting operation is not available at the facilities.
Appropriate loads of waste (e.g., loads from the geographic area to be characterized) do not arrive at the facilities frequently enough to support an efficient sampling and sorting operation.

Sampling at the points of generation tends to be more expensive and less valid than sampling at solid waste facilities. The added expense results from the increased effort required to design the sampling protocol and the travel time involved in collecting the samples.

The decreased validity of sampling at the points of generation has two principal causes. First, a significant portion of the waste is typically inaccessible. Waste can be inaccessible because it is on private property to which access is denied or because it is in trash compactors. Some waste is inaccessible during the day because it is not placed in outdoor trash containers until after business hours and it is picked up early in the morning. The second major cause of inaccuracy is that the relative portion of the waste stream represented by each trash receptacle is unknown because the frequency of pickup and the average quantity in the receptacle at each pickup are unknown. Random selection of receptacles to be sampled results in under-sampling of the more active receptacles, which represent more waste.

These problems are generally less acute for residential MSW than for commercial or institutional MSW. Residential MSW is usually accessible for sampling from the curb on collection day or from dumpsters serving multifamily residences. Because households generate similar quantities of waste, random selection of households for sampling gives each pound of waste a similar probability of being included in a sample. In addition, because waste characteristics are more consistent from household to household than from business to business, flaws in a residential sampling program are generally less significant than flaws in a commercial sampling program.

A universal protocol for sampling solid waste from the points of generation is impossible to state because circumstances vary greatly from place to place and from study to study. The following are general principles to follow:

Collect samples from as many different sectors of the target area as possible without oversampling relatively insignificant sectors.

If possible, collect samples from commercial locations in proportion to the size of the waste receptacles used and the frequency of pickup.

Collect samples from single-family and multifamily residences in proportion to the number of people living in each type of residence (unless a more sophisticated basis is readily available). The required population information can be obtained from U.S. census publications.

Give field personnel no discretion in selecting locations at which to collect samples. For example, field personnel should not be told to collect a sample from Elm Street but rather to collect a sample from the east side of Elm Street, starting with the second house (or business) north from Park Street.

To the extent feasible, add all waste from each selected location to the sample before going on to the next location. This practice reduces the potential for sampling bias.

Collecting samples at solid waste facilities is less expensive than collecting them at the points of generation and is more likely to produce valid results. Sample collection at facilities is less expensive because no travel is required. Samples collected at facilities are more likely to represent the waste being characterized because they are typically selected from a single line of trucks of known size that contain the entire waste stream.

Collecting samples at solid waste facilities has two stages: selecting the truck from which to take the sample and collecting the sample from the load discharged from the selected truck.

SELECTING SAMPLES

Environmental engineers usually select individual trucks in the field to sample, but they can select trucks in advance to ensure that specific collection routes are represented in the samples. Possible methods for selecting trucks in the field include the following:

- Constant interval
- Progress of sorters
- Random number generator
- Allocation among waste sources

The American Society for Testing and Materials (1992) *Standard test method for determination of the composition of unprocessed municipal solid waste* (ASTM D 5231) states that any random method of vehicle selection that does not introduce a bias into the selection process is acceptable.

Possible constant sampling intervals include the following in which n is any set number:

- Every nth truck
- Every nth ton of waste
- Every nth cubic yard of waste
- A truck every n minutes

Collecting a sample from every nth truck is relatively simple but causes the waste in small trucks and partially full trucks to be overrepresented in the samples. Collecting a sample from the truck containing every nth ton of waste is ideal but is difficult in practice because the weight of each truck is not apparent from observation. Collecting a sample from the truck containing every nth cubic yard of waste is more feasible because the volumetric capacity of most trucks can be determined by observation. However, basing the sampling interval on volumetric capacity tends to cause uncompacted waste and waste in partially full trucks to be overrepresented in the samples.

Basing the sampling interval on either a set number of trucks or a set quantity of waste causes the pace of the sampling operation to fluctuate during each day of field work. This fluctuation can result in inefficient use of personnel and deviations from the protocol when targeted trucks are missed at times of peak activity.

Collecting a sample from a truck every n minutes is convenient for sampling personnel but causes the waste in small trucks and partially full trucks to be overrepresented and the waste in trucks that arrive at busy times to be underrepresented in the samples. This approach also causes overrepresentation of waste arriving late in the day because the time interval between trucks tends to lengthen toward the end of the day and because trucks arriving late tend to be partially full, especially if the facility charges by the ton rather than by the cubic yard.

Obtaining samples as they are needed for sorting is similar to collecting a sample every n minutes and has the same disadvantages. Regardless of the sampling protocol used, however, the sorters should be kept supplied with waste to sort even if the available loads do not fit the protocol. Having more data is better.

ASTM D 5231 specifically identifies the use of a random number generator as an acceptable method for random selection of vehicles to sample. A random number generator can provide random intervals corresponding to each of the predetermined intervals just discussed. For example, if a facility receives 120 trucks per day and 12 are to be sampled, one can either sample every 10th truck or use the random number generator to generate 12 random numbers from 1 to 120. Similarly, random intervals of waste tonnage, waste volume, or elapsed time can be generated.

Random sampling intervals have the same disadvantages as the corresponding constant sampling intervals plus the following additional disadvantages:

Random sampling intervals increase the probability that the field crew is idle from time to time.
Random sampling intervals increase the probability that the field crew has to work overtime.
Random sampling intervals increase the probability that targeted trucks are missed when too many randomly selected trucks arrive within too short a time period.

In many cases, sampling by waste source minimizes the problems associated with these types of interval sampling. Sources of waste from which samples can be selected include individual municipalities, individual waste haulers, specific collection routes, waste generation sectors such as the residential sector and the commercial sector, and specific sources such as restaurants or apartment buildings. In general, sampling by source makes sense if adequate information is available on the quantity of waste from each source to be sampled. Samples can be collected from each source in proportion to the quantity of waste from each source, or the composition results for the various sources can be weighted based on the quantity from each source.

In the best case, the solid waste facility has a scale and maintains a computer database containing the following information for each load of waste: net weight, type of waste, type of vehicle, municipality of origin, hauler, and a number identifying the individual truck that delivered the waste. This information, combined with information on the hauling contracts in effect in each municipality, is usually sufficient to estimate the quantity of household and commercial MSW from each municipality.

The municipality is often the hauler for household waste, and, in those municipalities, private haulers usually handle commercial waste. In other cases, the municipality has a contract with a private hauler to collect household waste and discourages the hauler from using the same vehicles to service private accounts. Household and commercial waste can also be distinguished by the types of vehicles in which they are delivered. Dominant vehicle types vary from one region to another.

If the solid waste facility has no scale, environmental engineers can use records of waste volumes in designing a sampling plan but must differentiate between compacted and uncompacted waste. Many facilities receive little uncompacted MSW, while others receive substantial quantities.

Because per capita generation of household waste is relatively consistent, environmental engineers can use population data to allocate samples of household waste among municipalities if the necessary quantity records are not available.

Field personnel must interview private haulers arriving at the solid waste facility to learn the origins of the load of waste. Information provided by the haulers is often incomplete. In some cases this information can be supplemented or corrected during sorting of the sample.

McCamic (1985) provides additional information.

COLLECTING SAMPLES

Most protocols, including ASTM D 5231, state that each selected truck should be directed to discharge its load in an area designated for sample collection. This provision is convenient for samplers but is not necessary if a quick and simple sampling method is used. ASTM D 5231 states that the surface on which the selected load is discharged should be clean, but in most studies preventing a sample from containing a few ounces of material from a different load of waste is unnecessary.

Understanding the issues involved in selecting a sampling method requires an appreciation of the nature of a load of MSW discharged from a standard compactor truck onto the surface of a landfill or a paved tipping floor. Rather than collapsing into a loose pile, the waste tends to retain the shape it had in the truck. The discharged load can be 7 or 8 ft high. In many loads, the trash bags are pressed together so tightly that pulling material for the sample out of the load is difficult. Some waste usually falls off the top or sides of the load, but this loose waste should not be used as the sample because it can have unrepresentative characteristics.

In general, one sample should be randomly selected from each selected truck, as specified in ASTM D 5231. If more than one sample must be taken from one load, the samples should be collected from different parts of the load.

A threshold question is the size of the sample collected from each truck. Various sample sizes have been used, ranging from 50 lb to the entire load. Large samples have the following advantages:

The variation (standard deviation) between samples is smaller, so fewer samples are required to achieve a given level of precision.
The distribution of the results of sorting the samples is closer to a normal distribution (bell-shaped curve).
The boundary area between the sample and the remainder of the load is smaller in proportion to the volume of the sample, making the sampler's decisions on

whether to include bulky items from the boundary area less significant.

Small samples have a single advantage: shorter collection and sorting time.

A consensus has developed (SCS Engineers 1979; Klee and Carruth 1970; Britton 1971) that the optimum sample size is 200 to 300 lb (91 to 136 kg). This size range is recommended in ASTM D 5231. The advantages of increasing the sample size beyond this range do not outweigh the reduced number of samples that can be sorted. If the sample size is less than 200 lb, the boundary area around the sample is too large compared to the volume of the sample, and the sampler must make too many decisions about whether to include boundary items in the sample.

Environmental engineers use several general procedures to obtain samples of 200 to 300 lb from loads of MSW, including the following:

Assembling a composite sample from material taken from predetermined points in the load (such as each corner and the middle of each side)
Coning and quartering
Collecting a grab sample from a randomly selected point using a front-end loader
Manually collecting a column of waste from a randomly selected location

Numerous variations and combinations of these general procedures can also be used.

The primary disadvantage of composite samples is the same as that for small samples: the large boundary area forces the sampler to make too many decisions about whether to include items of waste in the sample. A composite sample tends to be a judgement sample rather than a random sample. A secondary disadvantage of composite samples is that they take longer to collect than grab samples or column samples.

A variation of composite sampling is to assemble each sample from material from different loads of waste. This approach has the same disadvantages as composite sampling from a single load of waste and is even more time-consuming.

In coning and quartering, samplers mix a large quantity of waste to make its characteristics more uniform, arrange the mixed waste in a round pile (coning), and randomly select a portion—typically one quarter—of the mixed waste (quartering). The purpose is to combine the statistical advantages of large samples with the reduced sorting time of smaller samples. The coning and quartering process can begin with the entire load of waste or with a portion of the load and can be performed once or multiple times to obtain a single sample. ASTM D 5231 specifies one round of coning and quartering, beginning with approximately 1000 lb of waste, to obtain a sample of 200 to 300 lb.

Coning and quartering has the following disadvantages and potential difficulties compared to grab sampling or column sampling:

Substantially increases sampling time
Requires more space
Requires the use of a front-end loader for relatively long periods. Many solid waste facilities can make a front-end loader and an operator available for brief periods, but some cannot provide a front-end loader for the longer periods required for coning and quartering.
Tends to break trash bags, making the waste more difficult to handle
Increases sorting time by breaking up clusters of a category of waste
Reduces accuracy of sorting by increasing the percentage of food waste adhering to or absorbed into other waste items
Promotes loss of moisture from the sample
Promotes stratification of the waste by density and particle size. The biasing potential of stratification is minimized if the quarter used as the sample is a true pie slice, with its sides vertical and its point at the center of the cone. This shape is difficult to achieve in practice.

The advantage of coning and quartering is that it reduces the variation (the standard deviation) among the samples, thereby reducing the number of samples that must be sorted. Coning and quartering is justified if it reduces the standard deviation enough to make up for the disadvantages and potential difficulties. If coning and quartering is done perfectly and completely, sorting the final sample is equivalent to sorting the entire cone of waste, and the standard deviation is significantly reduced. Since the number of samples that must be sorted to achieve a given level of precision is proportional to the square of the standard deviation, coning and quartering can substantially reduce the required number of samples. Note, however, that the more thoroughly coning and quartering is performed, the more pronounced are each of the disadvantages and potential difficulties associated with this method.

A more common method of solid waste sampling is collecting a grab sample using a front-end loader. This method is relatively quick and can often be done by facility personnel without unduly disrupting normal facility operations. Sampling by front-end loader reduces the potential impact of the personal biases associated with manual sampling methods but introduces the potential for other types of bias, including the following:

Like shovel sampling, front-end loader sampling tends to favor small and dense objects over large and light objects. Large and light objects tend to be pushed away or to fall away as the front-end loader bucket is inserted, lifted, or withdrawn.

On the other hand, the breaking of trash bags as the front-end loader bucket penetrates the load of waste tends to release dense, fine material from the bags, reducing the representation of this material in the sample.

Front-end loader samples taken at ground level favor waste that falls off the top and sides of the load, which may not have the same characteristics as waste that stays in place. On dirt surfaces, front-end loader samples taken at ground level can be contaminated with dirt.

The impact of these biasing factors can be reduced if the sampling is done carefully and the sampling personnel correct clear sources of bias, such as bulky objects falling off the bucket as it is lifted.

In front-end loader sampling, sampling personnel can use different sampling points for different loads to ensure that the various horizontal and vertical strata of the loads are represented in the samples. They can vary the sampling point either randomly or in a repeating pattern. The extent of the bias that could result from using the same sampling point for each load is not known.

An inherent disadvantage of front-end loader sampling is the difficulty in estimating the weight of the samples. Weight can only be estimated based on volume, and samples of equal volume have different weights.

A less common method of solid waste sampling is manually collecting a narrow column of waste from a randomly selected location on the surface of the load, extending from the bottom to the top of the load. This method has the following advantages:

- No heavy equipment is required.
- Sampling time is relatively short.
- Because different horizontal strata of the load are sampled, the samples more broadly represent the load than grab samples collected using a front-end loader. Note, however, that loads are also stratified from front to back, and column samples do not represent different vertical strata.
- The narrowness of the target area within the load minimizes the discretion of the sampler in choosing waste to include in the sample.

The major disadvantage of column sampling is that manual extraction of waste from the side of a well-compacted load is difficult, and the risk of cuts and puncture wounds from pulling on the waste is substantial.

Of the many hybrid sampling procedures that combine features of these four general procedures, two are worthy of particular note. First, in the sampling procedure specified in ASTM D 5231, a front-end loader removes at least 1000 lb (454 kg) of material along one entire side of the load; and this waste is mixed, coned, and quartered to yield a sample of 200 to 300 lb (91 to 136 kg). Compared to grab sampling using a front-end loader, the ASTM method has the advantage of generating samples more

broadly representative of the load but has the disadvantage of increasing sampling time.

In a second hybrid sampling procedure, a front-end loader loosens a small quantity of waste from a randomly selected point or column on the load, and the sample is collected manually from the loosened waste. This method is safer than manual column sampling and provides more control over the weight of the sample than sampling by front-end loader. This method largely avoids the potential biases of front-end loader sampling but tends to introduce the personal biases of the sampler.

Number of Samples Required to Estimate Composition

The number of samples required to achieve a given level of statistical confidence in the overall results is a function of the variation among the results for individual samples (standard deviation) and the pattern of the distribution of the results. Neither of these factors can be known in advance, but both can be estimated based on the results of other studies.

ASTM D 5231 prescribes the following equation from classical statistics to estimate the number of samples required:

$$n = (t^*s/ex)^2 \qquad \textbf{10.4(4)}$$

where:

n = required number of samples
t^* = student t statistic corresponding to the level of confidence and a preliminary estimate of the required number of samples
s = estimated standard deviation
e = level of precision
x = estimated mean

Table 10.4.1 shows representative values of the coefficient of variation and mean for various solid waste components. The coefficient of variation is the ratio of the standard deviation to the mean, so multiplying the mean by the coefficient of variation calculates the standard deviation. Table 10.4.2 shows values of the student t statistic.

Table 10.4.1 shows the coefficients of variation rather than standard deviations because the standard deviation tends to increase as the mean increases, while the coefficient of variation tends to remain relatively constant. Therefore, the standard deviations for sets of means different from those in the table can be estimated from the coefficients of variation in the table.

The confidence level is the statistical probability that the true mean falls within a given interval above and below the mean, with the mean as the midpoint (the confidence interval or confidence range). A confidence level of 90% is generally used in solid waste studies. The confidence interval is calculated based on the results of the study (see Table 10.4.3 later in this section).

TABLE 10.4.1 REPRESENTATIVE MEANS AND COEFFICIENTS OF VARIATION FOR MSW COMPONENTS

Waste Category	Mean (%)	Coefficient of Variation[a] (%)
Organics/Combustibles	86.6	10
Paper	39.8	30
Newspaper	6.8	80
Corrugated	8.6	95
Kraft	1.5	120
Corrugated & kraft	10.1	85
Other paper[b]	22.9	40
High-grade paper	1.7	230
Other paper[b]	21.2	40
Magazines	2.1	160
Other paper[b]	19.1	40
Office paper	3.4	—
Magazines & mail	4.0	90
Other paper[b]	17.2	40
Yard waste	9.7	160
Grass clippings	4.0	300
Other yard waste	5.7	180
Food waste	12.0	70
Plastic	9.4	40
PET bottles	0.40	100
HDPE bottles	0.70	95
Other plastic	8.3	50
Polystyrene	1.0	95
PVC bottles	0.06	200
Other plastic[b]	7.2	50
Polyethylene bags & film	3.7	45
Other plastic[b]	3.5	80
Other organics	15.7	55
Wood	4.0	170
Textiles	3.5	—
Textiles/rubber/leather	4.5	110
Fines	3.3	70
Fines $<\frac{1}{2}$ inch	2.2	80
Disposable diapers	2.5	110
Other organics	1.4	160
Inorganics/Noncombustibles	13.4	60
Metal	5.8	70
Aluminum	1.0	70
Aluminum cans	0.6	95
Other aluminum	0.4	120
Tin & bimetal cans	1.5	70
Other metal[b]	3.3	130
Ferrous metal	4.5	85
Glass	4.8	70
Food & beverage containers	4.3	85
Batteries	0.1	160
Other inorganics		
With noncontainer glass	3.2	160
Without noncontainer glass	2.7	200

[a]Standard deviation divided by the mean, based on samples of 200 to 300 pounds.

[b]Each "other" category contains all material of the previous type except material in those categories.

TABLE 10.4.2 VALUES OF STUDENT t STATISTIC

Number of Samples (n)	Student t Statistic	
	90% Confidence	95% Confidence
2	6.314	12.706
3	2.920	4.303
4	2.353	3.182
5	2.132	2.776
6	2.015	2.571
7	1.943	2.447
8	1.895	2.365
9	1.860	2.306
10	1.833	2.262
12	1.796	2.201
14	1.771	2.160
17	1.746	2.120
20	1.729	2.093
25	1.711	2.064
30	1.699	2.045
41	1.684	2.021
51	1.676	2.009
61	1.671	2.000
81	1.664	1.990
101	1.660	1.984
141	1.656	1.977
201	1.653	1.972
Infinity	1.645	1.960

The desired level of precision is the maximum acceptable error, expressed as a percentage or decimal fraction of the estimated mean. Note that a lower precision level indicates greater precision. A precision level of 10% (0.1) is frequently set as a goal but is seldom achieved.

After a preliminary value for n based on a preliminary value for t* is calculated, the calculation is repeated with the value of t* corresponding to the preliminary value for n.

Equation 10.4(4) assumes that the values for each variable to be measured (in this case the percentages of each solid waste component in the different samples) are normally distributed (conform to the familiar bell-shaped distribution curve, with the most frequent value equaling the mean). In reality, solid waste composition data are not normally distributed but are moderately to severely skewed right, with numerous values several times higher than the mean. The most frequent value is invariably lower than the mean, and in some cases is close to zero. The greater the number of waste categories, the more skewed the distributions of individual categories are.

Klee (1991; 1993) and Klee and Carruth (1970) have suggested equations to account for the effect of this skewness phenomenon on the required number of samples. Use of these equations is problematic. Like Equation 10.4(4), they are designed for use with one waste category at a time. For waste categories for which the mean is large compared to the standard deviation, the equations yield higher numbers of samples than Equation 10.4(4). This result is intuitively satisfying because more data should be needed to quantify a parameter whose values do not follow a predefined, normal pattern of distribution. For waste categories for which the mean is less than twice as large as the standard deviation, however, these equations tend to yield numbers of samples smaller than Equation 10.4(4). This result is counterintuitive since no reason is apparent for why an assumption of nonnormal distribution should decrease the quantity of data required to characterize a highly variable parameter.

An alternative method of accounting for skewness is to select or develop an appropriate equation for each waste category based on analysis of existing data for that category. Hilton, Rigo, and Chandler (1992) provide the results of a statistical analysis of the skewness of individual waste categories.

Equation 10.4(4) gives divergent results for different solid waste components. Based on the component means and coefficients of variation shown in Table 10.4.1 and assuming a precision of 10% at 90% confidence, the number of samples given by Equation 10.4(4) is 45 for paper other than corrugated, kraft, and high-grade; almost 700 for all yard waste; and more than 2400 for just grass clippings. The value of Equation 10.4(4) alone as a guide in designing a sampling program is therefore limited.

An alternative method is to estimate the number of samples required to achieve a weighted-average precision level equal to the required level of precision. The weighted-average precision level is the average of the precision levels for individual waste categories weighted by the means for

the individual waste categories. The precision level for individual waste categories can be estimated with the following equation, which is Equation 10.4(4) solved for e:

$$e = t^*s/xn^{1/2} \qquad 10.4(5)$$

The precision level for each category is multiplied by the mean for that category, and the results are totaled to yield the weighted-average precision level. The number of samples (n) is adjusted by trial and error until the weighted-average precision level matches the required value.

Calculation of the weighted-average precision level is shown in Table 10.4.3 later in this section. Figure 10.4.1 shows the relationship of the weighted-average precision level to the number of samples and the number of waste categories based on the values in Table 10.4.1. Overall precision improves as the number of samples increases and as the number of waste categories decreases. This statement does not mean that studies involving greater number of categories are inferior; it simply means that determining a few things precisely is easier than determining many things precisely.

Sorting and Weighing Samples of MSW

In most cases, sorting solid waste should be viewed as an industrial operation, not as laboratory research. While accuracy is essential, the appropriate measure of accuracy is ounces rather than grams or milligrams. Insistence on an excessive level of accuracy slows down the sorting process, reducing the number of samples that can be sorted. This

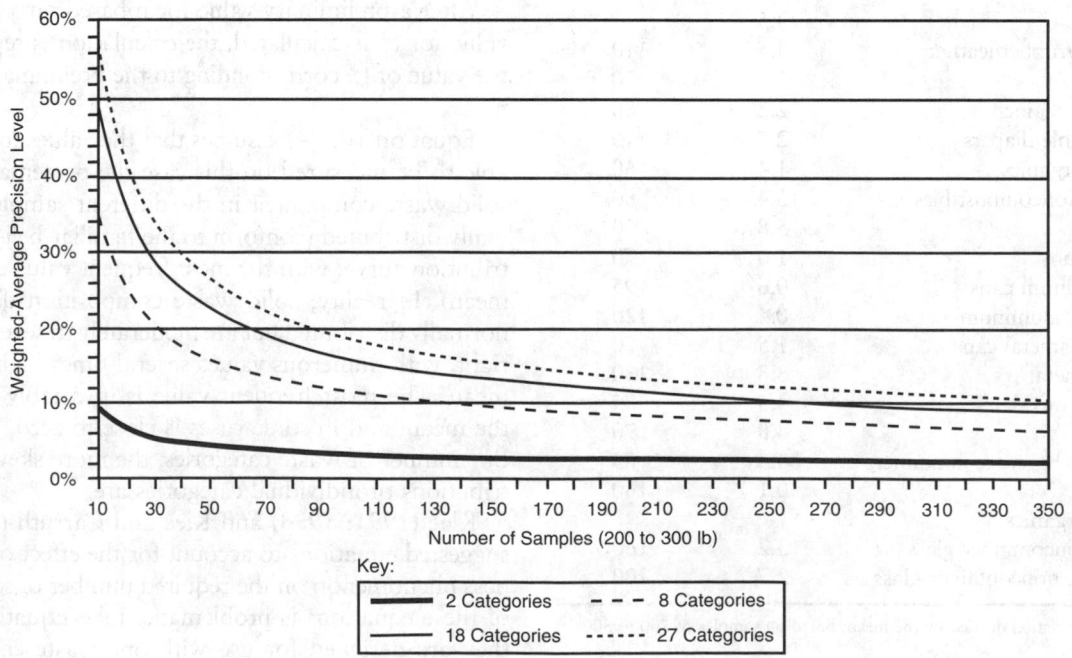

FIG. 10.4.1 Effect of the number of samples and the number of waste categories on weighted-average precision level (derived from Table 10.4.1).

reduction, in turn, reduces the statistical precision of the results. In the context of an operation in which a 10% precision level is a typical goal, inaccuracy of 1% is relatively unimportant.

The principles of industrial operations apply to solid waste sorting, including minimization of motion and maintenance of worker comfort and morale.

SORTING AREAS

A sorting area is established at the beginning of the field work and should have the following characteristics:

- A paved surface approximately 1000 sq ft in area and at least 16 ft wide
- Accessibility to vehicles
- Protection from precipitation and strong winds
- Heating in cold weather
- Separation from traffic lanes and areas where heavy equipment is used but within sight of arriving trucks

A typical sorting operation might use two sorting boxes and a crew of ten to twelve. The crew includes two sorting teams of four or five persons each, a supervisor, and a utility worker. The basic sorting sequence, starting when collection of the sample is complete, is as follows:

1. The sample is transported from the sampling point to the sorting area. A pickup truck or front-end loader can be used for this purpose.
2. The sampler gives the sorting supervisor a copy of a data form.
3. The sample is unloaded onto the surface of the sorting area.
4. Large items (e.g., corrugated cardboard and wood) and bags containing a single waste category (most often yard waste) are removed from the sample and set aside for weighing, bypassing the sorting box.
5. The remainder of the sample is transferred by increments into the sorting box, using broad-bladed shovels to transfer loose material.
6. The waste is sorted into the containers surrounding the sorting box.
7. The containers are brought to the scale, checked for accuracy of sorting, and weighed.
8. The gross weight of the waste and container and a letter symbol indicating the type of container are recorded on the data form.
9. If required, the waste in the containers is subsampled for laboratory analysis.
10. The containers are dumped in a designated receptacle or location.

The supervisor must ensure that each sample remains matched with the correct data form and that waste does not cross between samples.

SORTING CONTAINERS

Use of a counter-height sorting box speeds sorting, decreases worker fatigue, and encourages interaction among the sorters. All of these factors help build and sustain the morale of the sorters.

The following sorting box design has proven highly effective. The box is 4 ft wide, 6 ft long, 1 ft deep, and open at the top. It is constructed of $\frac{3}{8}$-in or $\frac{1}{2}$-in plywood with an internal frame of 2-by-3s or 2-by-4s. The long framing pieces extend 1 foot beyond the ends of the box at each bottom corner, like the poles of a stretcher. These framing pieces facilitate handling and extend the overall dimensions of the box to 4 ft by 8 ft by 1 ft. The box can lie flat within the bed of a full-sized pickup truck or standard cargo van.

A screen of $\frac{1}{2}$-in hardware cloth (wire mesh with $\frac{1}{2}$-in square openings) can be mounted in the bottom of the sorting box, $1\frac{1}{2}$ in from the bottom (the thickness of the internal framing pieces). If the screen is included, one end of the box must be open below the level of the screen to allow dumping of the fine material that falls through the screen. By allowing fine material to separate from the rest of the sample, the screen facilitates sorting of small items and makes dangerous items such as hypodermic needles easier to spot.

To facilitate dumping of the fines and to save space during transportation and storage, the sorting box is built without legs. During sorting, the sorting box is placed on a pair of heavy-duty sawhorses, 55-gal drums, or other supports. A support height of 32 in works well for a mixed group of male and female sorters. Fifty-five-gal drums are approximately 35 in high, approximately 3 in higher than optimum, and because of their size are inconvenient to store and transport.

The containers into which the waste is sorted should be a combination of 30-gal plastic trash containers and 5-gal plastic buckets. The 5-gal buckets are used for low-volume waste categories. Containers larger than 30 gal occupy too much space around the sorting box for efficient sorting and can be heavy when full. In a typical study with twenty-four to twenty-eight waste categories, each sorting crew should be equipped with approximately two dozen 30-gal containers and one dozen 5-gal buckets. In addition, each sorting crew should have several shallow plastic containers approximately 18 in wide, 24 in long, and 6 in deep.

For optimum use of space, the 30-gal containers should have rectangular rims. They should also have large handles to facilitate dumping. Recessed handholds in the bottom of the container are also helpful. In general, containers of heavy-duty HDPE are best. Because of their molded rims, these containers can be inverted and banged against pavement, the rim of a rolloff container, or the rim of a matching container to dislodge the material adhering to the inside of the container. The containers need not have wheels. Plastic containers slide easily across almost any flat surface.

Substantial field time can be saved when the containers of each type have fairly uniform weights so that each type of container can be assigned a tare weight rather than each container. When container weights are recorded on the data form after sorting, recording a letter code that refers to the type of container is faster than reading an individual tare weight on the container and recording it on the data form.

Assigning individual tare weights to containers weighing 2% more or less than the average weight for the container type is unnecessary. Batches of 5-gal buckets generally meet this standard, but many 30-gal containers do not. Ensuring that tare weights are consistent requires using portable scale when shopping for containers.

CONTAINER LABELING

Most sorting protocols, including ASTM D 5231, call for labeling each container to indicate which waste category is to be placed in it. When a sorting box is used, however, unlabeled containers have the following advantages:

The sorters are encouraged to establish a customary location for each waste category and sort by location, which is faster than sorting by labels.

When sorting is done by location rather than by labels, the containers can be placed closer to the sorters, which further speeds the sorting process.

Less time is required to arrange unlabeled containers around the sorting box after the sorted material from the previous sample has been weighed and dumped.

Keeping the containers unlabeled increases the flexibility of the sorting operation.

The flexibility gained by not labeling the containers has several aspects. First, different samples require multiple 30-gal containers for different waste categories. Second, many waste categories require a 30-gal container for some samples and only a 5-gal container for others. Third, the need for another empty container arises frequently in an active sorting operation, and grabbing the nearest empty container is quicker than searching for the container with the appropriate label.

Despite the advantages of unlabeled containers, the containers for food waste should be labeled. If individual containers are not designated for food waste, all containers will eventually be coated with food residue. This residue is unpleasant and changes the tare weights of the containers.

The tare weights of the food waste containers should be checked daily. Generally, checking the tare weights of other containers at the beginning of each week of field work is sufficient unless a visible buildup of residue indicates that more frequent checking is required.

SORTING PROCESS

The actual sorting of the sample should be organized in the following basic manner:

Each waste category is assigned a general location around the perimeter of the sorting box. In one effective arrangement, paper categories are sorted to one side of the sorting box, plastic categories are sorted to the other side, other organic categories are sorted to one end, and inorganic categories are sorted to the other end.

Each sorter is assigned a group of categories. With a typical sorting crew of four, each sorter is assigned the categories on one side or at one end of the box.

The sorters place their assigned materials in the appropriate containers and place other materials within reach of the sorters to which they are assigned.

Toward the end of sorting each sample, one of the shallow containers is placed in the middle of the sorting box, and all sorters place other paper in this container (see Table 10.4.1). This process can be repeated for food waste.

When only scattered or mixed bits of waste remain, sorting is suspended.

The material remaining above the screen in the sorting box, or on the bottom of a box without a screen, is scraped or brushed together and either (1) distributed among the categories represented in it in proportion to their abundance, (2) set aside as a separate category, or (3) set aside to be combined with the fine material from below the screen. ASTM D 5231 specifies the first alternative, but it should not be selected if the waste categories are to be subsampled for laboratory testing.

If the sorting box has a screen, the box is upended to allow the fine material from below the screen to fall through the slot at one end of the box. The material that falls out is swept together and shoveled into a container—preferably a wide, shallow container—for weighing.

WEIGHING SAMPLES

ASTM D 5231 specifies the use of a mechanical or electronic scale with a capacity of at least 200 lb (91 kg) and precision of 0.1 lb (0.045 kg) or better. When 30-gal containers are used in sorting samples of 200 to 300 lb, gross weights greater than 100 lb are unusual. Even if larger containers or sample sizes are used, sorting personnel should avoid creating containers with gross weights greater than 100 lb because they are difficult and dangerous to handle. For most sorting operations, a scale capacity of 100 lb is adequate. An electronic scale with a range of 0–100 lb is generally easier to read to within 0.1 lb than a mechanical scale with a range of 0–100 lb.

A platform-type scale is preferred. The platform should be 1 ft square or larger.

The digital displays on electronic scales make data recording easier and minimize recording errors by displaying the actual number to be recorded on the data form. When recording weights from a mechanical scale, interpolation between two values marked on the dial is often required. The advantages of mechanical scales are lower cost, reliability, and durability.

Ideally, one worker places containers on the scale, the supervisor checks the containers for accuracy of sorting and records the weights and container types, and two or more workers dump the weighed containers. If the containers are subsampled for laboratory analysis prior to being dumped, the process is much slower and fewer workers are required.

DUMPING SAMPLES

On landfills, the sorting containers are dumped near the sorting area for removal or in-place burial by facility personnel. In transfer stations and waste-to-energy facilities, the containers can be dumped on the edge of the tipping floor.

When the sorting area is separated from the disposal area, use of the sampling vehicle for disposal is difficult. Loads of waste that should be sampled can be missed, and sorting delays occur because the sampling vehicle is not available for dumping full containers from the previous sample. A better procedure is to dump the sorted waste in a rolloff container provided by the disposal facility. Facility personnel transport the rolloff container to the disposal area approximately once per day. The density of sorted waste is often as low as 150 lb/cu yd, so the rolloff tends to be filled more rapidly than expected. To facilitate dumping sorted waste over the sides, the rolloff container should not be larger than 20 cu yd (15.3 cu m).

Processing the Results of Sorting

After a sample is weighed and the gross weights and container types are recorded on the data form, the net weights are calculated and recorded on the data form. Total net weights are calculated for waste categories sorted into more than one container. Field personnel should calculate net category weights and total net sample weights after each day of sorting to monitor the size of the samples. Undersize samples decrease the accuracy and statistical precision of the results and can violate the contract under which the study is conducted. Oversize samples make sorting the required number of samples more difficult.

The net weights for each waste category in each sample are usually entered into a computer spreadsheet. For each waste category in each group of samples to be analyzed (for example, residential samples and commercial samples), the following should be calculated from the data in the spreadsheet:

- The percentage by weight in each sample
- The mean percentage within the group of samples
- The standard deviation of the percentages within the group of samples
- The confidence interval around the mean

Calculating the overall composition usually involves dividing the total weight of each waste category by the total weight of the samples rather than calculating the composition of each sample and averaging the compositions. If the samples have different weights, which is usually the case, these two methods yield different results. Calculating overall composition based on total weight creates a bias in favor of dense materials, which are more abundant in the heavier samples. Averaging the compositions of the individual samples is preferable because it gives each pound of waste an equal opportunity to influence the results. ASTM D 5231 specifies averaging of sample compositions.

Table 10.4.3 shows mean percentages, standard deviations, uncertainty values, precision levels, and confidence intervals for a group of 200 MSW samples with the characteristics shown in Table 10.4.1. The confidence intervals are based on the uncertainty values (sometimes called precision values). The uncertainty values are typically calculated with the following formula:

$$U_c = t^* s / n^{1/2} \qquad \textbf{10.4(6)}$$

where:

U_c = uncertainty value at a given level of confidence, typically 90%
t^* = student t statistic corresponding to the given level of confidence
s = sample standard deviation
n = number of samples

This equation is equivalent to the equation for calculating the precision level, Equation 10.4(5), with both sides multiplied by the mean, x. Dividing the uncertainty value by the mean yields the precision level. Adding the uncertainty values for all waste categories yields the weighted average precision level, weighted by the means for the individual waste categories.

Equation 10.4(6), like Equations 10.4(4) and 10.4(5), assumes that the percentage data are normally distributed. As previously discussed, this is not actually the case, and no reliable and reasonably simple method exists for estimating the effect of lack of normality on the statistical precision of the results.

Precision analysis can only be applied to groups of samples that are representative of the waste stream to be analyzed. For example, if 40% of the municipal waste stream is commercial waste but 60% of the samples sorted during a study are collected from commercial loads, statistical precision analysis of the entire body of composition data generated during the study is meaningless. Assuming that the commercial and residential samples represent the

TABLE 10.4.3 ILLUSTRATION OF WEIGHTED-AVERAGE PRECISION LEVEL AND CONFIDENCE INTERVALS[a]

Waste Category	Mean (%) (x)	Standard Deviation (%) (s)	Student t Statistic (t*) for 200 Samples (n) and 90% Confidence	Uncertainty Value (%) ($U_{90} = t^*s/n^{1/2}$)	Precision Level (%) (U_{90}/x)	90% Confidence Interval (%)
Newspaper	6.8	5.4	1.653	0.6	9.4	6.8 ± 0.6
Corrugated & kraft	10.1	8.6	1.653	1.0	9.9	10.1 ± 1.0
Other paper	22.9	9.2	1.653	1.1	4.7	22.9 ± 1.1
Yard waste	9.7	15.5	1.653	1.8	18.7	9.7 ± 1.8
Food waste	12.0	8.4	1.653	1.0	8.2	12.0 ± 1.0
PET bottles	0.4	0.4	1.653	0.05	11.7	0.4 ± 0.05
HDPE bottles	0.7	0.7	1.653	0.1	11.1	0.7 ± 0.1
Other plastic	8.3	4.1	1.653	0.5	5.8	8.3 ± 0.5
Wood	4.0	6.8	1.653	0.8	19.9	4.0 ± 0.8
Textiles/rubber/leather	4.5	5.0	1.653	0.6	12.9	4.5 ± 0.6
Fines	3.3	2.3	1.653	0.3	8.2	3.3 ± 0.3
Disposable diapers	2.5	2.8	1.653	0.3	12.9	2.5 ± 0.3
Other organics	1.4	2.2	1.653	0.3	18.7	1.4 ± 0.3
Aluminum	1.0	0.7	1.653	0.1	8.2	1.0 ± 0.1
Tin & bimetal cans	1.5	1.1	1.653	0.1	8.2	1.5 ± 0.1
Other metal	3.3	4.3	1.653	0.5	15.2	3.3 ± 0.5
Food & beverage containers	4.3	3.7	1.653	0.4	9.9	4.3 ± 0.4
Other inorganics	3.3	5.3	1.653	0.6	18.7	3.3 ± 0.6
Total or weighted average	100.0			10.1	10.1	100.0 ± 10.1

[a]Based on 200 samples, 90% confidence, and the eighteen waste categories listed in the table. Means and standard deviations are based on Table 10.4.1.

respective fractions of the waste stream from which they were collected, separate precision analysis of the commercial and residential results is valid. Representativeness is achieved by either random selection of loads to sample or systematic selection of loads based on preexisting data.

Visual Characterization of Bulky Waste

The composition of bulky waste is typically estimated by observation rather than by sorting samples. Visual characterization of bulky waste is feasible for several reasons: (1) most bulky waste is not hidden in bags, (2) most loads of bulky waste contain few categories of waste, and (3) the categories of waste present are usually not thoroughly dispersed within the load, as they are in loads of MSW. Conversely, sorting samples of bulky waste is problematic for several reasons: (1) because the variation among loads of bulky waste is large, a large number of trucks must be sampled, (2) because the waste categories are not thoroughly dispersed within the loads, the samples must be large, (3) sorting and weighing bulky waste is difficult and dangerous if not done with specialized mechanical equipment.

Estimating the composition of bulky waste based on observation has three phases. First, field personnel prepare field notes describing each load as the load is dumped, as the load sits on the tipping floor or landfill after dumping,

and as the heavy equipment operators move the load around the tipping floor or the working face of the landfill. Second, they determine or estimate the weight of each load. Third, they combine the field notes and load weights to develop an estimate of the composition of each load and of the bulky waste as a whole.

In general, the field notes should include the following elements for each load:

The date and exact time of day

The type of vehicle and its volumetric capacity (e.g., 30-cu-yd rolloff, 40-cu-yd trailer)

Any identifying markings that help match the field notes with the corresponding entry in the facility log for that day. Identifying markings that can be useful include the name of the hauler, the license plate number, and identifying numbers issued by regulatory agencies.

Either (1) a direct estimate of the by-weight composition of the load or (2) an estimate of the by-volume composition of the load combined with an indication of the amount of air space in each component.

If the facility does not have a scale, the facility log generally contains a volume for each load but no weight. If the volume of each load can be determined in the field, as it can when each truck or container is marked with its volumetric capacity, field notes do not have to be matched with log entries. Regardless of whether the facility log is used, the field notes should contain any information that can be helpful in estimating the weight of each load, including its total volume if different from the capacity of the vehicle in which it arrived.

Field personnel should visually characterize most if not all of the loads of bulky waste arriving at the solid waste facility during the period of field work. Because the composition of bulky waste varies from load to load, a large number of loads must be characterized.

Characterized loads of bulky waste should not be regarded as samples because they contain vastly different quantities of waste. The overall composition of bulky waste is not the mean of the results for individual loads, as with MSW. Rather, the overall composition is weighted in accordance with the weights of the individual loads. An estimate of the overall percentage of each component involves calculating the total quantity of the component in all observed loads and dividing it by the total weight of all observed loads, as illustrated by the following equation:

$$p_o = (p_1 w_1 + p_2 w_2 \cdots + p_n w_n)/w_o \qquad 10.4(7)$$

where:

p_o = the overall percentage of the component in the observed loads
p_1 = the percentage of the component in the first observed load
w_1 = the weight of the first observed load
p_2 = the percentage of the component in the second observed load
w_2 = the weight of the second observed load
p_n = the percentage of the component in the last observed load
w_n = the weight of the last observed load
w_o = the total weight of all observed loads

Before the overall composition can be calculated in this way, the weight of each load must be estimated. If the facility has a scale, environmental engineers can determine the actual weight of the observed loads by matching the field notes for each load with the corresponding entry in the facility log, based on the time of arrival and information about the truck and the load. The time of arrival recorded in the facility log is the time when the truck was logged in rather than the time when the load was discharged. Field personnel must determine the difference between the two times.

If the facility does not have a scale, environmental engineers must estimate the weight of each component and the total weight of the load by converting from cubic yards to tons. The following procedure is suggested:

The total volume of the load is distributed among the components of the load based on the field notes.
The weight of each component is estimated based on its volume and density. Table 10.3.3 shows density ranges for certain waste components.
The estimated component weights are added yielding the estimated total weight of the load.

The cost of a study can be reduced if the same person collects MSW samples for sorting and performs visual characterization of bulky waste during the same period of field work. This technique is feasible if loads of MSW and bulky waste are dumped in the same part of the facility and if a quick method is used for collecting MSW samples.

Sampling MSW for Laboratory Analysis

Obtaining meaningful laboratory results for MSW is difficult. The primary sources of difficulty are (1) the presence of many different types of objects in MSW and (2) the large size of these objects. Collecting small but representative samples from a homogeneous pile of small objects (e.g., a pile of rice) is easier than from a heterogeneous pile of large objects. Secondary sources of difficulty in sampling MSW include the uneven distribution of moisture and inconsistent laboratory procedures.

MIXED SAMPLE VERSUS COMPONENT SAMPLE TESTING

An initial choice to be made is whether to test mixed samples or individual waste components. Testing mixed samples is preferable when:

- The only purpose of the laboratory testing is to determine the characteristics of the mixed waste stream, such as heat value.
- The statistical precision of the laboratory results must be demonstrated.
- The study does not include sorting waste samples.
- No significant changes in the composition of the waste stream are anticipated.

Testing of individual waste components is necessary, of course, when the characteristics of individual waste components must be determined. In addition, component testing makes projecting the impact of changes in the component composition of the waste, such as changes caused by recycling and composting programs, possible. Component testing also enhances quality control because laboratory errors are easier to detect in the results for individual components than in those for mixed samples.

The procedures for collecting mixed samples for laboratory testing are essentially the same as those for collecting mixed samples for sorting. The preceding evaluation of these procedures also applies to the collection of mixed samples for laboratory testing, except for the comments concerning the impacts of various sampling procedures on the sorting process.

Laboratory samples of individual waste components are usually composite subsamples of samples sorted to estimate composition. In general, each component laboratory subsample includes material from each sorted sample. Material for the laboratory subsamples is collected from the sorting containers after the sorting and weighing are complete.

LABORATORY PROCEDURES

A fundamental question is how large should the samples sent to the laboratory be. The answer to this question depends on the procedures used by the laboratory. A state-of-the-art commercial laboratory procedure includes the following steps:

A portion of the sample material sent to the laboratory is weighed, dried, and reweighed to determine the moisture content. The limiting factor at this stage of the procedure is usually the size of the laboratory's drying oven.

A portion of the dried material is ground into particles of $\frac{1}{8}$ to $\frac{1}{4}$ in.

A portion of the $\frac{1}{8}$-to-$\frac{1}{4}$-in material is finely ground into as close to a powder as possible. For flexible plastic, dry ice must be added prior to fine grinding to make it more brittle.

The actual laboratory test is generally performed on 0.5 to 3 g of the finely ground material, depending on the type of test and the specific equipment and procedures.

Variations on this procedure include the following:

Most laboratories do not have equipment for grinding inorganic materials such as glass and metal. In combustion testing, this material is removed from the sample prior to grinding, then weighed and reported as ash. For metals testing, metal objects can be cut up by hand or drilled to create small pieces for testing. Glass and ceramics are typically crushed.

Many laboratories do not have fine grinding equipment, so they perform tests on relatively coarse material.

In addition to using different methods for preparing waste for testing, laboratories use different test methods.

The more sample material the laboratory receives, the more material they must exclude from the small quantity of material that is tested. The real question is not how large the samples should be but how field and laboratory personnel should share the task of reducing samples to a gram or two. For practical purposes, the maximum quantity sent to the laboratory should be the quantity the laboratory is prepared to spread out and mix in preparation for selecting the material to be dried. The minimum quantity should be the quantity the laboratory is prepared to dry and grind up.

Composite laboratory samples are typically accumulated in plastic trash bags, then boxed for shipment. An alternative is to accumulate the samples in 5-gal plastic buckets with lids. Plastic buckets are more expensive than plastic bags but have several advantages:

Plastic buckets (and their lids) are easier to label, and the labels are easier to read.

Adding material to plastic buckets is easier.

The lids, which are lifted only when material is added to the buckets, prevent moisture loss during the active sampling period.

Sample material can be compacted in plastic buckets if it is pushed down around the inside edge.

The buckets can be used as shipping containers.

The buckets can be reused if the laboratory ships them back.

COLLECTING MATERIAL FOR LABORATORY SUBSAMPLES

Three general methods for collecting material for laboratory subsamples from containers of sorted waste are blind grab sampling, cutting (or tearing) representative pieces from large objects, and selecting representative whole objects for inclusion in the sampling. Blind grab sampling is the preferred approach for waste that mainly consists of small objects. Cutting representative pieces is appropriate for waste consisting of large objects with potentially different characteristics. Selecting representative whole objects is appropriate for waste containing only a few different types of objects.

Blind grab samples should be collected by hand or with an analogous grasping tool. The objective is to extract the material from a randomly selected but defined volume within the container of sorted material. When scoops and shovels are used in sampling heterogeneous materials, they tend to create bias by capturing dense, small objects while pushing light, large objects away.

In collecting subsamples from containers of sorted waste, samplers must realize that because sorting progresses from larger objects to smaller, the objects at the top of the container tend to be smaller than those at the bottom. Objects of different sizes can have different characteristics, even within the same waste category. Therefore, the sampler must ensure that the objects at different levels of the containers are represented in the samples. Emptying the container onto a dry and reasonably clean surface prior to collecting the subsample may be necessary.

If the laboratory samples are tested for metals, objects with known metals content should not be represented in the samples. Instead, such objects should be weighed, and the laboratory results should be adjusted to reflect the quantities of metals they contain. For example, if 8 oz of lead weights are found in 10 tn of sorted waste, the weights represent 25 ppm of lead. The weights should be withheld from the laboratory sample, and 25 ppm should be added to the overall lead concentration indicated by the laboratory results. This procedure is more accurate than laboratory testing alone.

Review and Use of Laboratory Results

Laboratory procedures are imperfect, and errors in using the procedures and in calculating and reporting the results are common. Reviewing the results received from a laboratory to see if they make sense is important. This exercise is relatively straightforward for combustion characteristics because much is known about the combustion characteristics of solid waste and its component materials (see Section 10.3). Identification of erroneous laboratory results is more difficult for metals and toxic organic substances.

The following guidelines apply in an evaluation of reasonableness of laboratory results for combustion characteristics on a dry basis:

Dry-basis results for the paper, yard waste, plastics, wood, and disposable diapers categories should be close to those shown in Tables 10.3.4 and 10.3.5.

Greater variability must be accepted in individual results for food waste, textiles/rubber/leather, fines, and other combustibles because of the chemical variety of these categories.

The result for carbon must always be at least six times the result for hydrogen.

No oxygen result should be significantly higher than 50%.

For plant-based materials and mixed food waste, oxygen results should not be significantly less than 30% on an ash-free basis.

Among the paper categories, only those with high proportions of glossy paper, such as magazines and advertising mail, should have ash values significantly greater than 10%.

Nitrogen should be below 1% for all categories except grass clippings, other yard waste, food waste, textiles/rubber/leather, fines, and other organics (see Table 10.3.4).

Chlorine should be below 1% for all categories except for PVC bottles, other plastic, textiles/rubber/leather, and other organics.

Sulfur should be below 1% for all categories except other organics.

The laboratory should be willing to check its calculations and repeat the test if the calculations are not the source of the problem.

Estimating Combustion Characteristics Based on Limited Laboratory Testing

The combustion characteristics of individual waste categories on a dry basis are well documented and fairly con-

TABLE 10.4.4 HEAT VALUE ESTIMATES BASED ON BOIE, CHANG, AND DULONG EQUATIONS

Equation	Dry-Basis HHV (Btu/lb)	As-Received HHV (Btu/lb)
Boie	7395	5310
Chang	7479	5370
DuLong	7510	5392
Average	7461	5357
Laboratory values	7446	5348

sistent within categories. Moisture and component composition are more variable. One option, therefore, is to sort samples to estimate component composition and have subsamples tested for moisture only. Then, with the use of the documented values for the proximate and ultimate composition and heat value of each waste component, the overall combustion characteristics of the waste stream can be estimated.

Another potential cost-saving measure is to estimate heat value based on ultimate composition. Several equations have been proposed for this purpose (Niessen 1995):

BOIE EQUATION

$$HHV = 14,976C + 49,374H - 4644O + 2700N$$
$$+ 4500S + 1692Cl + 11,700P \quad 10.4(8)$$

CHANG EQUATION

$$HHV = 15,410 + 32,350H - 11,500S$$
$$- 20,010O - 16,200Cl - 12,050N \quad 10.4(9)$$

DULONG EQUATION

$$HHV = 14,095.8C + 64,678(H - O/8)$$
$$+ 3982S + 2136.6O + 1040.4N \quad 10.4(10)$$

where:

HHV = higher heating value in Btu/lb

Percentages for each element must be converted to decimals for use in these equations (i.e., 35% must be converted to 0.35). Using the values in Table 10.3.4 in the three equations yields the results shown in Table 10.4.4.

These values are close to the overall values in Table 10.3.5, which are based on laboratory testing of the same samples on which the ultimate composition in Table 10.3.4 is based. The laboratory-based values are closer to the average results for the three equations than to the results for any individual equation.

—*F. Mack Rugg*

References

American Society for Testing and Materials. 1992. *Standard test method for determination of the composition of unprocessed municipal solid waste.* ASTM Method D 5231-92 (September).

Britton, P.W. 1971. *Improving manual solid waste separation studies.* U.S. EPA (March).

Franklin Associates, Ltd. 1992. *Characterization of municipal solid waste in the United States: 1992 update.* U.S. EPA, EPA/530-R-92-019, NTIS no. PB92-207 166 (July).

Gay, A.E., T.G. Beam, and B.W. Mar. 1993. Cost-effective solid-waste characterization methodology. *J. of Envir. Eng.* (ASCE) 119, no. 4 (Jul/Aug).

Hilton, D., H.G. Rigo, and A.J. Chandler. 1992. Composition and size distribution of a blue-box separated waste stream. Presented at SWANA's Waste-to-Energy Symposium, Minneapolis, MN, January 1992.

Klee, A.J. 1991. *Protocol: A computerized solid waste quantity and composition estimation system.* Cincinnati: U.S. EPA Risk Reduction Engineering Laboratory.

———. 1993. New approaches to estimation of solid-waste quantity and composition. *J. of Envir. Eng.* (ASCE) 119, no. 2 (Mar/Apr).

Klee, A.J. and D. Carruth. 1970. Sample weights in solid waste composition studies. *J. of the Sanit. Eng. Div., Proc. of the ASCE* 96, no. SA4 (August).

McCamic, F.W. (Ferrand and Scheinberg Associates). 1985. *Waste composition studies: Literature review and protocol.* Mass. Dept. of Envir. Mgt. (October).

Niessen, W.R. 1995. *Combustion and incineration processes: Applications in environmental engineering.* 2d ed. New York: Marcel Dekker, Inc.

SCS Engineers. 1979. *Municipal solid waste survey protocol.* Cincinnati: U.S. EPA.

10.5
IMPLICATIONS FOR SOLID WASTE MANAGEMENT

This section addresses several aspects of the relationship between the characteristics of solid waste and the methods used to manage it. Implications for waste reduction, recycling, composting, incineration, and landfilling are included, as well as general implications for solid waste management as a whole.

MSW is abundant, unsightly, and potentially odorous; contains numerous potential pollutants; and supports both disease-causing organisms and disease-carrying organisms. Like MSW, bulky solid waste is abundant, unsightly and potentially polluting. In addition, the dry, combustible nature of some bulky waste components can pose a fire hazard. Because of these characteristics of MSW and bulky waste, a prompt, effective, and reliable system is required to isolate solid waste from people and the environment.

A beneficial use of solid waste is relatively difficult because it contains many different types of materials in a range of sizes. The only established use for unprocessed MSW is as fuel in mass-burn incinerators (see Section 10.9). Even mass-burn incinerators cannot handle unprocessed bulky waste. In the past, unprocessed bulky waste was used as fill material, but this practice is restricted today. In general, processing is required to recover useful materials from both MSW and bulky waste.

Implications for Waste Reduction

Waste reduction refers to reducing the quantity of material entering the solid waste management system. Waste reduction is distinguished from recycling, which reduces the quantity of waste requiring disposal but does not reduce the quantity of material to be managed.

Based on the composition of MSW (see Section 10.3), each of the following measures would have a significant impact on the quantity of MSW entering the solid waste management system:

- Leaving grass clippings on the lawn
- Increasing backyard composting and mulching of leaves and other yard wastes
- Selling products in bulk rather than in packages, with the consumer providing the containers
- Buying no more food than is eaten
- Substituting reusable glass containers for paper, plastic, and single-use glass containers
- Reusing shopping bags
- Placing refuse directly in refuse containers instead of using trash bags
- Using sponges and cloth hand towels in place of paper towels
- Continuing to use clothing and other products until they are worn out, rather than discarding them when they no longer look new
- Prohibiting distribution of unsolicited printed advertising

Leaving grass clippings on the lawn is becoming increasingly common because of disposal bans in some states and the development of mulching lawn mowers that cut the clippings into smaller pieces. Implementation of the other waste reduction measures on the list is unlikely in the United States because they do not conform to the prevailing standards of convenience, comfort, appearance, sanitation, and free enterprise.

Implications for Waste Processing

Fluctuations in waste generation must be considered when waste processing facilities are planned. If a facility must process the entire waste stream throughout the year, it must be sized to handle the peak generation rate. Storage of MSW for later processing is limited by concerns about odor and sanitation. Limitations on the storage of bulky waste are generally less severe, but long-term storage of combustible materials is usually restricted.

Processing systems for mixed solid waste must be capable of handling a variety of materials in a range of sizes.

Because solid waste does not flow, it must be hauled or moved by conveyor. Because objects in MSW do not readily stratify by size, screening of MSW generally requires a mixing action such as that produced by trommel screens. Abrasive materials in solid waste cause abrasive wear to handling and processing equipment. Heavy, resistant items can damage size reduction equipment. Size reduction is often required, however, because bulky items in solid waste tend to jam conveyors and other waste handling equipment.

Implications for Recovery of Useful Materials

Almost all solid waste materials can be recycled in some way if people are willing to devote enough time and money to the recycling effort. Because time and money are always limited, distinctions must be drawn between materials that are more and less difficult to recycle. Table 10.5.1 shows the compostable, combustible, and recyclable fractions of MSW. The materials listed as recyclable are those for which large-scale markets exist if the local recycling industry is well developed. The list of recyclable materials is different in different areas.

Approximately 75% of the MSW discarded in the United States is compostable or recyclable. No solid waste district of substantial size in the United States has documented a 75% rate of MSW recovery and reuse, however. Reasons for this include the following:

Some recyclable material becomes unmarketable through contamination during use.

A significant fraction of recyclable material cannot be recovered from the consumer.

A portion of both recyclable and compostable material is lost during processing (sorting recyclable materials or removing nonrecyclable and noncompostable materials from the waste stream).

Some compostable material does not decompose enough to be included in the finished compost product and is discarded with the process residue.

TABLE 10.5.1 COMBUSTIBLE, COMPOSTABLE, AND RECYCLABLE COMPONENTS OF MSW[a]

Waste Category	Percentage of Total[b]
Combustible, compostable, and recyclable	**22.6**
Newspaper	6.8
Corrugated cardboard	8.6
Kraft paper	1.5
High-grade paper	1.7
Magazines & mail	4.0
Recyclable and combustible but not compostable	**2.1**
PET bottles	0.4
HDPE bottles	0.7
Polyethylene film other than trash bags	1.0
Recyclable but not compostable or combustible	**7.9**
Aluminum cans	0.6
Tin & bimetal food & beverage cans	1.5
Other metal[c]	1.5
Glass food and beverage containers	4.3
Compostable and combustible but not recyclable	**44.7**
Other paper	17.2
Yard waste	9.7
Food waste	12.0
Disposable diapers	2.5
Fines	3.3
Combustible but not compostable or recyclable	**17.2**
Other plastic	7.3
Wood	4.0
Textiles/rubber/leather	4.5
Other organics	1.4
Not combustible or compostable or recyclable	**5.5**
Other aluminum	0.4
Other metal[c]	1.8
Batteries	0.1
Other inorganics	3.2
Total recyclable[a]	32.6
Total compostable	67.3
Total combustible	86.6

[a]Materials listed as recyclable are those for which large-scale markets exist in areas where the recycling industry is well developed.

[b]Derived from Table 10.3.1. Currently recycled materials are not included.

[c]A substantial portion of this category is readily recyclable, and a substantial portion is not. Some of the material listed here as nonrecyclable can be recovered in recyclable condition by an efficient ferrous recovery system at a combustion facility.

A portion of finished MSW compost cannot be marketed and must be landfilled.

In MSW discharged from compactor trucks, most glass containers are still in one piece, and most metal cans are uncrushed. Most glass and aluminum beverage containers are in recyclable condition. Many glass food containers and steel cans are heavily contaminated with food waste, however. Some of the recyclable paper in MSW received at disposal facilities is contaminated with other materials, but 50% or more is typically in recyclable condition.

The ratio of carbon to nitrogen (C/N ratio) is an indicator of the compostability of materials. To maximize the composting rate while minimizing odor generation, a C/N ratio of 25/1 to 30/1 is considered optimum. Higher ratios reduce the composting rate, while lower ratios invite odor problems.

Table 10.5.2 shows representative C/N ratios of compostable components of MSW. Controlled composting of food waste, with a C/N ratio of 14/1, is difficult unless large quantities of another material such as yard waste (other than grass clippings) are mixed in to raise the ratio. The C/N ratio moves above the optimum level as quantities of paper are added to the mixture, however.

Paper, leaves, and woody yard waste serve as effective *bulking agents* in composting MSW, so the addition of a bulking agent such as wood chips is generally unnecessary.

The metals content of MSW is a major concern in composting because repeated application of compost to land can raise the metals concentrations in the soil to harmful levels. Compost regulations usually set maximum metals concentrations for MSW compost applied to land. Most regulations do not distinguish between different forms of a metal. For example, the lead in printing ink on a plastic bag is treated the same as the lead in glass crystal even though the lead in printing ink is more likely to be released into the environment. Similarly, the hexavalent form of chromium found in lead chromate is treated the same as the elemental chromium used to plate steel even though the hexavalent form is more toxic than the elemental form.

Two extensive, recent studies of metals in individual components of MSW yielded contradictory results. A study in Cape May County, New Jersey found toxic metals concentrated in the noncompostable components of MSW (Camp Dresser & McKee Inc. 1991; Rugg and Hanna 1992). A study in Burnaby, British Columbia, however, found higher metals concentrations in the compostable components of MSW than were found in Cape May (see Table 10.3.6) (Rigo, Chandler, and Sawell 1993).

Disposable diapers are listed as compostable in Table 10.5.1 despite their plastic covers. The majority of the weight of disposable diapers is from the urine, feces, and treated cellulose inside the cover, all of which is compostable. Note, however, that most people wrap used diapers into a ball with the plastic cover on the outside, using the waist tapes to keep the ball from unraveling. Vigorous size reduction is required to prepare these diaper balls for composting.

Wood is biodegradable but does not degrade rapidly enough to be considered compostable. The same is true of cotton and wool fabrics, included in the textiles/rubber/leather category in Table 10.5.1.

Implications for Incineration and Energy Recovery

The heat value of MSW (4800–5400 Btu/lb) is lower than that of traditional fuels such as wood (5400–7200 Btu/lb), coal (7000–15,000 Btu/lb), and liquid or gaseous petroleum products (18,000–24,000 Btu/lb) (Camp Dresser & McKee 1991, 1992a,b; Niessen 1995). The heat value of MSW is sufficient, however, to sustain combustion without the use of supplementary fuel.

Heat value is an important parameter in the design or procurement of solid waste combustion facilities because each facility has the capacity to process heat at a certain rate. The greater the heat value of a unit mass of waste, the smaller the total mass of waste the facility can process.

The ash and moisture content of MSW is high compared to that of other fuels. Most of the ash is contained in relatively large objects that do not become suspended in the flue gas (Niessen 1995). Ash handling is a major consideration at MSW combustion facilities.

Because of its high ash and moisture content and low density, MSW has low *energy density* (heat content per unit volume) (Niessen 1995). Therefore, MSW combustion facilities must be designed to process large volumes of material.

The effect of recycling programs on the heat value of MSW is not well documented. Numerous attempts have been made to project the impact of recycling based on the

TABLE 10.5.2 REPRESENTATIVE C/N RATIOS OF COMPOSTABLE COMPONENTS OF MSW

Waste Category	C/N Ratio
Yard waste	29/1
Grass clippings	17/1
Leaves	61/1
Other yard waste	31/1
Food waste	14/1
Paper	119/1
Newspaper	149/1
Corrugated & kraft	165/1
High-grade paper	248/1
Magazines & mail	131/1
Other paper	85/1
Disposable diapers	95/1
Fines	23/1

Note: Derived from Table 10.3.4.

measured heat values of individual MSW components (for example, see Camp Dresser & McKee [1992a]). Little reliable data exist, however, that document the effect of known levels of recycling on the waste received at operating combustion facilities.

A reasonable assumption is that recycling materials with below-average heat values raises the heat value of the remaining waste, while recycling materials with above-average heat values reduces the heat value of the remaining waste. The removal of recyclable metal and glass containers increases heat value (and reduces ash content), while the recovery of plastics for recycling reduces heat value. The removal of paper for recycling also reduces heat value. Because recycled paper has a low moisture content, its heat value is 30% to 40% higher than that of MSW as a whole.

The increase in heat value caused by recycling glass and metal is probably greater than the reduction caused by recycling paper. Because plastics are generally recycled in small quantities, the reduction in heat value caused by their removal is relatively small. The most likely overall effect of recycling is a small increase in heat value and a decrease in ash content.

Sulfur in MSW is significant because sulfur oxides (SO_x) have negative effects and corrode natural and manmade materials. SO_x combines with oxygen and water to form sulfuric acid. A solid waste combustion facility must maintain stack temperatures above the dew point of sulfuric acid to prevent corrosion of the stack. Niessen (1995) provides additional information.

Like sulfur, chlorine has both health effects and corrosive effects. Combustion converts organic (insoluble) chlorine to hydrochloric acid (HCl). Because HCl is highly soluble in water, it contributes to corrosion of metal surfaces both inside and outside the facility (Niessen 1995).

Chlorine is a component of additional regulated compounds including dioxins and furans. Trace concentrations of dioxins and furans can be present in the waste or can be formed during combustion. Niessen (1995) provides additional discussion.

Oxides of nitrogen (NO_x) form during the combustion of solid waste, both from nitrogen in the waste and in the air. NO_x reacts with other substances in the atmosphere to form ozone and other compounds that reduce visibility and irritate the eyes (Niessen 1995).

Emissions of SO_x, NO_x, chlorine compounds, and hydrocarbons are regulated and must be controlled (see Section 10.9 and Niessen [1995]). Emissions of hydrocarbons and chlorine compounds other than HCl can generally be controlled by optimization of the combustion process. Maintaining complete control of the combustion of material as varied as MSW is difficult, however, so small quantities of hydrocarbons and complex chlorine compounds are emitted from time to time.

Combustion cannot destroy metals. Assuming that a combustion facility is designed with no discharge of the water used to quench the combustion ash, the toxic metals in the waste end up in the ash or are emitted into the air. Regulations limit the emission of toxic metals.

The tendency of a metal to be emitted from a combustion facility is a function of many factors such as:

- The volatility of the metal
- The chemical form of the metal
- The degree to which the metal is bound in other materials, especially noncombustible materials
- The degree to which the metal is captured by the air pollution control system

Emissions of a metal from a solid waste combustion facility cannot be predicted based on the abundance of the metal in the waste.

Mercury is the most volatile of the metals of concern, and a substantial portion of the mercury in MSW escapes capture by the air pollution control systems at MSW combustion facilities. The quantity of mercury in MSW has declined rapidly in recent years because battery manufacturers have eliminated most of the mercury in alkaline and carbon–zinc batteries. One cannot assume that a reduction in the quantity of mercury in batteries proportionately reduces the quantity emitted from MSW combustion facilities, however.

All but a small fraction of each metal other than mercury becomes part of the ash residue either because it never enters the facility stack or because it is captured by the air pollution control system. The environmental significance of a metal in combustion ash residue depends primarily on its leachability and the toxicity of its leachable forms. A portion of the ash residue from some MSW combustion facilities is regulated as hazardous waste because of the tendency of a toxic metal (usually lead or cadmium) to leach from the ash under the test conditions specified by the U.S. EPA.

Niessen (1995) and Chandler & Associates, Ltd. et al. (1993) provide additional information on the implications of solid waste characteristics with combustion as a disposal method. Niessen provides a comprehensive treatise on waste combustion from the perspective of an environmental engineer. The final report of Chandler & Associates, Ltd. et al. provides a detailed study of the relationships among metals concentrations in individual components of MSW, metals concentrations in stack emissions, and metals concentrations in various components of ash residue at a single MSW combustion facility.

Implications for Landfilling

The greater the density of the waste in a landfill, the more tons of waste can be disposed in the landfill. The density of waste in a landfill can be increased in a variety of ways, including the following:

- Using compacting equipment specifically designed for the purpose (Surprenant and Lemke 1994)

- Spreading the incoming waste in thinner layers prior to compaction (Surprenant and Lemke 1994)
- Shredding bulky, irregular materials such as lumber prior to landfilling

Because solid waste contains toxic materials (see Section 10.3), landfills must have impermeable liners and systems to collect water that has been in contact with the waste (leachate). The liner must be resistant to damage from any substance in the waste, including solvents. The first lift (layer) of waste placed on the liner must be free of large, sharp objects that could puncture the liner. For this reason, bulky waste is typically excluded from the first lift.

To some extent, the moisture content of waste placed in a landfill influences the quantity of the leachate generated. In most cases, however, a more important factor is the quantity of the precipitation that falls on the waste before an impermeable cap is placed over it.

For additional information, see Section 10.13.

—*F. Mack Rugg*

References

Camp Dresser & McKee Inc. 1991. *Cape May County multi-seasonal solid waste composition study.* Edison, N.J. (August).

——. 1992a. *Atlantic County (NJ) solid waste characterization program.* Edison, N.J. (May).

——. 1992b. *Prince William County (VA) solid waste supply analysis.* Annandale, Va. (October).

Chandler, A.J., & Associates, Ltd. et al. 1993. *Waste analysis, sampling, testing and evaluation (WASTE) program: Effect of waste stream characteristics on MSW incineration: The fate and behaviour of metals. Final Report of the Mass Burn MSW Incineration Study (Burnaby, B.C.).* Toronto (April).

Niessen, W.R. 1995. *Combustion and incineration processes: Applications in environmental engineering,* 2d ed. New York: Marcel Dekker, Inc.

Rigo, H.G., A.J. Chandler, and S.E. Sawell. 1993. Debunking some myths about metals. In *Proceedings of the 1993 International Conference on Municipal Waste Combustion, Williamsburg, VA, March 30–April 2, 1993.*

Rugg, M. and N.K. Hanna. 1992. Metals concentrations in compostable and noncompostable components of municipal solid waste in Cape May County, New Jersey. *Proceedings of the Second United States Conference on Municipal Solid Waste Management, Arlington, VA, June 2–5, 1992.*

Surprenant, G. and J. Lemke. 1994. Landfill compaction: Setting a density standard. *Waste Age* (August).

Resource Conservation and Recovery

10.6
REDUCTION, SEPARATION, AND RECYCLING

Municipal Waste Reduction

Waste reduction is the design, manufacture, purchase, or use of materials (such as products and packaging) which reduce the amount and toxicity of trash generated. Source reduction can reduce waste disposal and handling costs because it avoids the cost of recycling, municipal composting, landfilling, and combustion. It conserves resources and reduces pollution.

PRODUCT REUSE

Reusable products are used more than once and compete with disposable, or single-use, products. The waste reduction effect of a reusable product depends on the number of times it is used and thus the number of single-use products that are displaced.

Used household appliances, clothing, and similar durable goods can be reused. They can be donated as used products to charitable organizations. Such goods can also be resold through yard and garage sales, classified ads, and flea markets.

The following lists common source reduction activities in the private sectors (New Jersey Department of Environmental Protection and Energy 1992):

Office paper. Employees are encouraged to make two-sided copies, route memos and documents rather than making multiple copies, make use of the electronic bulletin board for general announcements rather than distributing memos, and limit distribution lists to essential employees.

Routing envelopes. After large routing envelopes are completely filled, employees can reuse them by simply pasting a blank routing form on the envelope face. Even large envelopes received in the mail can be converted to routing envelopes in this manner.

Paper towels. C-fold towels are replaced with roll towels.

Printers. Recharged laser printer toner cartridges are used.

Tableware. Nondisposable tableware (environmental mug program, china for conferences) is used.

Polystyrene containers. Reusable, glass containers are used, and all Styrofoam coffee cups in all office areas, shops, and the employee cafeteria are eliminated. Styrofoam peanuts are reused in offices or donated to local businesses.

Beverages and detergents. Some items are available in refillable containers. For example, some bottles and jugs for beverages and detergents are made to be refilled and reused by either the consumer or the manufacturer.

Cleaning rags. Reusable rags are used instead of throwaway rags.

Ringed note binder reuse. Employees take binders to one of several collection points at the facility where they are refurbished for reuse.

Laboratory chemicals. "Just-in-time" chemicals are delivered to labs to preclude stockpiling chemicals which eventually go bad. This method reduces hazardous waste disposal costs through source reduction.

Photocopy machines. New photocopying machines with energy-saving controls are used.

Batteries. Use of rechargeable batteries reduces garbage and keeps the toxic metals in batteries out of the waste stream. Using batteries with reduced toxic metals is another alternative.

INCREASED PRODUCT DURABILITY

When a consumer-durable product has a longer useful life, fewer units (such as refrigerators, washing machines, and tires) enter the waste stream. For instance, since 1973, the durability of the passenger tire has almost doubled as radial tires have replaced bias and bias-belted tires. Radial tires have an average life of 40,000 to 60,000 miles; the average life of bias tires is 15,000 miles, and bias-belted tires is 20,000 miles (Peterson 1989).

Other ways of reducing waste through increased product durability include:

Using low-energy fluorescent light bulbs rather than incandescent ones. These bulbs last longer, which means fewer bulbs are thrown out, and cost less to replace over time.

Keeping appliances in good working order by following the manufacturers' service suggestions for proper operation and maintenance

Whenever intended for use over a long period of time, choosing furniture, luggage, sporting goods, tools, and toys that standup to vigorous use

Mending clothes instead of throwing them away, and repairing worn shoes, boots, handbags, and brief cases

Using long-lasting appliances and electronic equipment with good warranties. Reports are available that list products with low breakdown rates and products that are easily repaired.

Refer to Section 3.2 for discussions on designing product line extension.

REDUCED MATERIAL USAGE PER PRODUCT UNIT

Reducing the amount of material used in a product means less waste is generated when the product is discarded. Consumers can apply this waste reduction approach in their shopping habits by purchasing packaged products in large container sizes. For example, the weight-to-volume ratio of a metal can for a sample food product declines from 5.96 with an 8-oz container (single serving size) to 3.17 with a 101-oz (institutional) size.

Other methods for reducing the material per product unit include:

Using wrenches, screwdrivers, nails, and other hardware available in loose bins. Purchasing grocery items, such as tomatoes, garlic, and mushrooms, unpackaged rather than prepackaged containers.

Using large or economy-size items of household products that are used frequently, such as laundry soap, shampoo, baking soda, pet foods, and cat litter. Choosing the largest size of food items that can be used before spoiling.

Using concentrated products. They often require less packaging and less energy to transport to the store, saving money as well as natural resources.

When appropriate, using products that are already on hand to do household chores. Using these products can save on the packaging associated with additional products.

DECREASED CONSUMPTION

Seldom-used items, like certain power tools and party goods, often collect dust and rust, take up valuable storage space, and ultimately end up in the trash. Renting or borrowing these items reduces consumption and waste. Infrequently used items can be shared among neighbors, friends, or family. Borrowing, renting, and sharing items save both money and natural resources.

Other ways to decrease consumption follow.

Renting or borrowing tools such as ladders, chain saws, floor buffers, rug cleaners, and garden tillers. In apartment buildings or co-ops, residents can pool resources and form banks to share tools and other equipment used infrequently. In addition, some communities have tool libraries, where residents can borrow equipment as needed.

Renting or borrowing seldom-used audiovisual equipment

Renting or borrowing party decorations and supplies such as tables, chairs, centerpieces, linens, dishes, and silverware

Sharing newspapers and magazines with others to extend the lives of these items and reduce the generation of waste paper

Before old tools, camera equipment, or other goods are discarded, asking friends, relatives, neighbors, or community groups if they can use them

REDUCING WASTE TOXICITY

In addition to reducing the amount of material in the solid waste stream, reducing waste toxicity is another component of source reduction. Some jobs around the home require the use of products containing hazardous components. Nevertheless, toxicity reduction can be achieved by following some simple guidelines.

Using nonhazardous or less hazardous components. Examples include choosing reduced mercury batteries and planting marigolds in the garden to ward off certain pests rather than using pesticides. In some cases, less toxic chemicals can be used to do a job; in others, some physical methods, such as sandpaper, scouring pads, or more physical exertion, can accomplish the same results as toxic chemicals.

When hazardous components are used, using only the amount needed. Used motor oil can be recycled at a participating service station. Leftover products with hazardous components should not be placed in food or beverage containers.

For products containing hazardous components, following all directions on the product labels. Containers must be labelled properly. For leftover products containing hazardous components, checking with the local environmental agency or chamber of commerce for any designated days for the collection of waste material such as leftover paints, pesticides, solvents, and batteries. Some communities have permanent household hazardous waste collection facilities that accept waste year around.

Separation at the Source

Kitchen designers and suppliers of kitchen equipment will need to become more sensitive to the needs of recycling.

Major manufacturers of kitchen equipment should make sorting drawers, lazy Susan sorting bins, and tilt-out bins as standard kitchen equipment. Kitchen designers should keep in mind small convenience items, such as automatic label scrapers, trash chutes, and can flatteners to make recycling more convenient.

The more finely household waste is separated, the greater its contribution to recycling. Figure 10.6.1 shows an approach where household waste is separated into four containers.

Container 1 would receive all organic or putrescible materials, including food-soiled paper and disposable diapers and excluding toxic substances and glass or plastic items. The contents of this container can be taken to a composting plant that also receives yard wastes and possibly sewage sludge and produces soil additives.

Container 2 would receive all clean paper, newspapers, cardboard, and cartons for paper processing, where contents are separated mechanically and sold to commercial markets.

Container 3 would receive clean glass bottles and jars and aluminum and tin cans free of scrap metals and plastics.

Container 4 would receive all other waste, including plastic, metal, ceramic, textile, and rubber items. (Later, a fifth container could be added for recyclable plastics.) The contents of this container can be considered nonrecyclable and sent to a landfill or a recycling plant for further separation. The contents of this container would represent about 12% of the total MSW.

Separate collections are required for trash items that are not generated on a daily basis, such as yard waste, brush

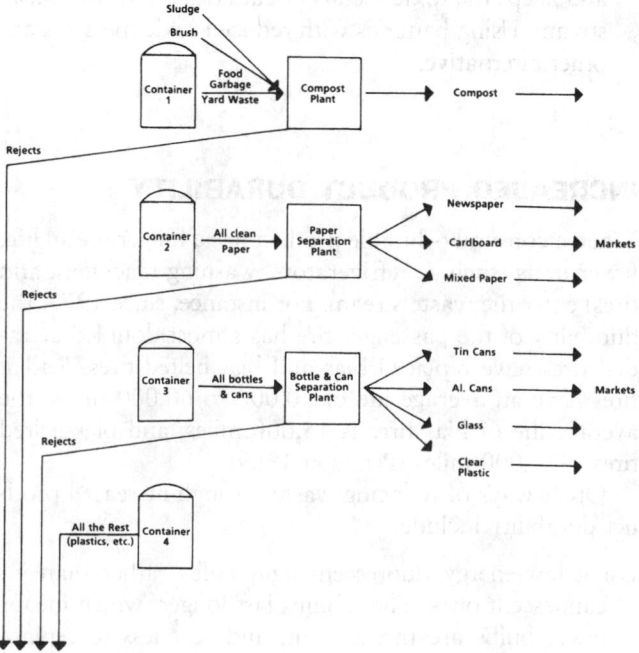

FIG. 10.6.1 Basic separation scheme.

and wood, discarded furniture and clothing, "white goods" such as kitchen appliances, toxic materials, car batteries, tires, used oil, and paint.

"BOTTLE BILLS"

In 1981 Suffolk County on Long Island outlawed nonreturnable soda bottles. By 1983 legislation had been passed in eight states requiring a 5-cent deposit on all soda bottles. The annual return rate on beer bottles in New York is nearly 90% and 80% of six billion soft drink and beer bottles. Further improvement was obtained by raising the deposit on nonrefillable containers to 10 cents and allowing the state to use part of the unredeemed deposits (at present kept by bottlers and totaling $64 million a year) to establish recycling stations.

Bottle bills, while having achieved partial success, should be integrated into overall recycling programs, which include office paper and newspaper recycling, cardboard collection from commercial establishments, curbside recycling, establishment of buy-back recycling centers, wood waste and metal recycling, glass and bottle collection from bars and restaurants, and composting programs. Advertising and public education are important elements in the overall recycling strategy. Street signs, door hangers, utility-bill inserts, and phone book, bus, and newspaper advertisements are all useful. The most effective long-range form of public education is to teach school-children the habits of recycling.

Recycling

PLASTIC

Plastics are strong, waterproof, lightweight, durable, microwavable, and more resilient than glass. For these reasons they have replaced wood, paper, and metallic materials in packaging and other applications. Plastics generate toxic by-products when burned and are nonbiodegradable when landfilled; they also take up 30% of landfill space even though their weight percentage is only 7% to 9%. Recent research has found that paper does not degrade in landfills either and because of compaction in the garbage truck and in the landfill, the original volume percentage of 30% in the kitchen waste basket is reduced 12% to 21% in the landfill. In addition, plastics foul the ocean and harm or kill marine mammals. Other problems include the toxic chemicals used in plastics manufacturing, the reliance on nonrenewable petroleum products as their raw material, and the blowing agents used in making polystyrene foam plastics, such as chlorofluorocarbons (CFCs), which cause ozone depletion. CFCs are now being replaced by HCFC-22 or pentane, which does not deplete the ozone layer but does contribute to smog. For these reasons, recycling appears to be the natural solution to the plastic disposal problem.

Unfortunately, recycling and reuse are not easily accomplished because each type of plastic must go through a different process before being reused. There are hundreds of different types of plastics, but 80% of plastic used in consumer products is either high-density polyethylene (milk bottles) or polyethylene terephthalate (large soda bottles). It is not yet possible to separate plastics by types because manufacturers do not indicate the type of plastic used. Plastic parts of automobiles are still uncoded, so salvagers cannot separate them by type. Even if recycled polystyrene were separated and could be used as a raw material for a plastics recycling plant, such plants are just beginning to be built and we do not know if they will be successful. For these reasons, environmentalists would prefer to stop using plastics altogether in certain applications.

TOXIC SUBSTANCES

The careless disposal of products containing toxic or hazardous substances can create health hazards if allowed to decompose and leach into the groundwater from landfills or if vaporized in incinerators. Since hazardous-waste landfills are limited, the available options are either to have manufacturers substitute toxic materials with nontoxic substances or recycle the products that contain toxic materials. Municipalities are just beginning to consider the requirements of toxic-waste recycling. Products that are toxic or contain toxic substances include paint, batteries, tires, some plastics, pesticides, cleaning and drain-cleaning agents, and PCBs found in white goods (appliances). Separate collections are also required for medical wastes.

Batteries play an important role in the recycling of toxic substances. Batteries represent a $2.5 billion-a-year market. At present, practically no batteries are being recycled in the United States. Battery manufacturers feel that recycling is neither practical nor necessary; instead, they feel that all that needs to be done is to lower the quantities of toxic materials in batteries. It is estimated that 28 million car batteries are landfilled or incinerated every year. This number contains 260,000 tons of lead, which can damage human neurological and immunological systems. The billions of household batteries disposed of yearly contain 170 tons of mercury and 200 tons of cadmium. The first can cause neurological and genetic disorders, the second, cancer. Some batteries also contain manganese dioxide, which causes pneumonia. When incinerated, some of these metals evaporate. The excessive emissions of mercury were the reason why Michigan temporarily suspended the operation of the incinerator in Detroit, the nation's largest.

Some states have recently initiated efforts to force manufacturers to collect and recycle or safely dispose of their batteries. The Battery Council International has prompted several states to pass laws requiring recycling of all used car batteries. In many European countries used batteries are returned to the place of purchase for disposal.

The disposal of white goods (appliances such as refrigerators, air conditioners, microwave ovens) is also a problem. Until 1979 appliance capacitors were allowed to contain PCBs (polychlorinated biphenyls). Even after the ban, some manufacturers were granted an extra year or two to deplete their inventories. When white goods are shredded, the "fluff" remaining after the separation of metals (consisting of rubber, glass, plastics, and dirt) is landfilled. When it was found that the "fluff" contains more than 50 ppm of PCBs, the Institute of Scrap Recycling Industries advised its 1,800 members not to handle white goods. The safe disposal of PCB-containing white goods would require scrap dealers to remove the capacitors before shredding. Similar toxic-waste disposal problems are likely to arise in connection with electronic and computing devices, the printed circuit boards of which contain heavy metals.

A long-range solution to toxic-waste disposal might be to require manufacturers of new products containing toxic substances to arrange for recycling *before* the product is allowed on the market, or at least to provide instruction labels describing the recommended steps in recycling.

PAPER

Paper used to be made of reclaimed materials such as linen rags. Rags were the raw materials used by the first paper mill built in the United States in 1690 in Philadelphia. Only in the nineteenth century did paper mills convert to wood-pulping technology. It takes seventeen trees to make a ton of paper. All Sunday newspapers in the United States, for example, require the equivalent of half a million trees every week. When paper is made from waste paper, it not only saves trees but also saves 4,100 kWh of energy per ton (the equivalent of a few months of electricity used by the average home), 7,000 gallons of water, 60 pounds of air-polluting emissions, and three cubic yards of landfill space and the associated tipping fees. The production of recycled paper also requires fewer chemicals and far less bleaching.

The paper output of the world has increased by 30% in the last decade. In 1990 the United States used more than 72 million tons of paper products, but only 25.5% of that (18.4 million tons) is made from recycled paper. This compares with 35% in Western Europe, almost 50% in Japan, and 70% in the Netherlands. There are some 2,000 waste-paper dealers in the United States who collect nearly 20 million tons of waste paper each year. In 1988, 20% of the collected waste paper was exported, mostly to Japan.

The waste-paper market is very volatile. In some locations the mixed office waste or mixed-paper waste (MPW) has no value at all and tipping fees must be paid to have them picked up. Therefore, what pays for collection and processing is not the prices paid for waste paper, but the savings represented by *not* landfilling them at $70/ton on the East Coast. A ton of old newspapers in California brings $25 to $35 because of the Japanese market demand. In the Northeast an oversupply in 1989 caused the waste-paper price to plummet from $15/ton to about −$10/ton. This oversupply also resulted in increased waste-paper exports to Europe, which in turn caused the collapse of the waste-paper market in Holland, where the value of a kilogram of waste paper dropped from eight cents to one cent.

Waste paper can be classified into "bulk" or "high" grade. The highest-grade papers are manila folders, hard manila cards, and similar computer-related paper products. High-grade waste paper is used as a pulp substitute, whereas bulk grades are used to make paper boards, construction paper, and other recycled paper products. The bulk grade consists of newspapers, corrugated paper, and MPW. MPW consists of unsorted waste from offices, commercial sources, or printing establishments. Heavy black ink used on newspaper reduces its value, however. The value of the paper is also reduced by the presence of other substances that interfere with a single-process conversion into pulp, such as the gum in the binding of telephone directories or the chemical coating of magazines.

The most effective way to create a waste-paper market is to attract a pulp and paper mill to the area. To keep such a plant in operation, however, requires a high-grade waste-paper supply of about 300 tons per day. In addition, facilities are also needed for wastewater treatment.

Newsprint Recycling

A large part of the waste-paper problem has to do with newsprint, which makes up 8% of the total MSW by weight. Some 13 million tons of newsprint are consumed yearly in the United States, 60% imported from Canada.

Connecticut requires the use of 20% recycled paper in the newspapers sold in the state today and 90% by 1998. Suffolk County on Long Island requires 40%. New York State reached a voluntary agreement with its publishers to achieve the 40% goal by the year 2000. Florida applies a ten-cent waste-recovery fee for every ton of virgin newsprint used and grants a ten-cent credit for every ton of recycled newsprint used.

The net effect of such legislation will be an increased and steady demand for waste paper, which is essential for the success of recycling. As the demand for waste paper products rises, paper manufacturers will also increase their capacity to produce recycled paper.

GLASS

About 13 million tons of glass are disposed of in the United States every year, representing more than 7% of the total MSW that is generated. But only about 12% of the total glass production is recycled. In comparison, Japan recycles about 50%.

Salvaged glass has been used in bricks and paving mixtures. "Glasphalt" can be made from a mixture of glass and asphalt or a mix of 20% ground glass, 10% blow sand, 30% gravel, and 40% limestone. In spite of all these other uses, the main purchasers of crushed glass are the glass companies themselves. The use of recycled crushed glass reduces both the energy cost and the pollutant emissions associated with glass making. Crushed glass is easily saleable, with a market almost as good as that for aluminum. Manufacturers use from 20% to as much as 80% of salvaged glass in their glass-making processes.

METALS

In the United States over 15 million tons of metals are discarded every year. This represents almost 9% of MSW by weight. We recycle 14% of our metallic wastes (nearly 64% of aluminum). During the last fifty years, more than half of the raw materials used in steel mills was recycled. At least one-third of the aluminum produced is from recycled sources.

Aluminum recycling is profitable and well established because it requires only 5% of the electric power to remelt aluminum as it does to extract it from bauxite ore. In 1990 the average price paid for crushed, baled aluminum cans was $1,050 per ton and some 55 billion aluminum cans (0.96 million tons) have been recycled. The recycling rate of aluminum increased from 61% in 1989 to 63.5% in 1990. Steel has also been recycled for generations, but the recycling of steel cans is relatively new. It was necessary to reduce the rust-preventing tin layer on the steel cans first, so that they might be added directly to steel furnaces. The recycling of steel cans has increased from 5 billion cans in 1988 to 9 billion in 1990 and its market value varied from $40 to $70 per ton in 1990 depending on location.

The main sources of scrap metals are cans, automobiles, kitchen appliances (white goods), structural steel, and farm equipment. The value of the noncombustibles in incinerator ash varies from area to area.

RUBBER

In the United States some two billion old tires have been discarded, and their number is growing by about 240 million a year. In the past tires were either piled, landfilled, burned, or ground up and mixed with asphalt for road surfacing. These "solutions" were expensive and often caused environmental problems because of the air pollution resulting from massive tire fires. Some newer rubber recycling processes have tried to overcome these limitations. The new processes do not pollute air or water because nothing is burned and no water is used. The tires are shredded and the polyester fibers removed by air classification. The steel from radial tires is removed magnetically. The remaining rubber powder is mixed with chemical agents that restore the ability of the "dead" rubber to bond with other rubber and plastic molecules. The vulcanized or "cured" tire rubber loses its ability to bond during the vulcanizing process.

Combining old rubber with "virgin" rubber or plastics results in an economically competitive product. The cost of virgin rubber is about 65 cents a pound and polypropylene costs about 68 cents, while the "reactivated" product is about 30 cents a pound ($600/ton).

INCINERATOR ASH

If all the MSW of New York City were incinerated, the residue would amount to 6,000 to 7,000 tons/day, representing a giant disposal problem. About 10% by weight of the incinerator residue is fly ash collected in electrostatic precipitators, scrubbers, or bag filters; the remaining 90% is bottom ash from the primary and secondary combustion chambers. This residue is a soaking-wet complex of metals, glass, slag, charred and unburned paper, and ash containing various mineral oxides. A Bureau of Mines test found that 1,000 pounds of incinerator residue yielded 166 pounds of larger-size ferrous metals, such as wire, iron items, and shredded cans. The total ferrous fraction was found to be 30.5% by weight; glass represented 50% of the total residue by weight.

Common practice in the U.S. is to recover some 75% of the ferrous metals through magnetic separation and to landfill the remaining residue. Incinerator residue has also been used as landfill cover, landfill road base, aggregate in cement and road building applications, and as aggregate substitute in paving materials.

Incinerator residue is processed to recover and reuse some of its constituents and thereby reduce the amount requiring disposal. Processing techniques include the recovery of ferrous materials through magnetic separation, screening the residue to produce aggregate for construction-related uses, stabilization through the addition of lime (which tends to minimize metal leaching), and solidification or encapsulation of the residue into asphaltic mixtures.

An incinerator-residue processing plant might consist of the following operations: (1) fly ash and bottom ash are collected separately, with lime mixed only with the fly ash; (2) ferrous materials are removed from the bottom ash; (3) the ferrous-free residue is screened to separate out the proper particle sizes for use as aggregate; and (4) the remaining oversized items and stabilized fly ash are landfilled. In a more sophisticated ash-processing plant, the ferrous removal and shredding (or oversize removal) are followed by melting of the ash (fusion), resulting in a glassy end-product. This high-tech process has some substantial advantages: It burns all the combustible materials, including dioxins and other trace organics, and encapsulates the metals, thereby preventing their leaching out. The resulting fused product is a glazed, nonabrasive, lightweight

TABLE 10.6.1 INCINERATOR ASH COMPOSITION

Component	Percentage
Ferrous metal	35
Glass	28
Minerals and Ash	16
Ceramics	8
Combustibles	9
Nonferrous Metal	4

TABLE 10.6.2 COMPOSITION OF INCINERATOR ASH (UNDER-2″ FRACTION)

Component	Percentage
Glass	37
Minerals and ash	21
Ferrous metals	19
Ceramics	9
Combustibles	8
Nonferrous metals	6

black aggregate. The fusion of combined incinerator ash and sewage sludge is currently practiced in Japan.

The first U.S. building to be built from recycled incinerator ash blocks is an 8,000-square-foot boathouse on the campus of the State University of New York at Stony Brook, Long Island. The ash comes from an incinerator in Peekskill and is mixed with sand and cement to form blocks that are as durable as standard cinder blocks. This technology has already been used in Europe. The blocks can be used to build seawalls, highway dividers, and sound barriers, in addition to regular buildings. It is the bottom ash (not the fly ash) portion that is considered safe for such applications.

The ash produced by one New York City incinerator has been extensively sampled and evaluated. Fly ash contains substantial quantities of organic materials. About 20% by weight is larger than 2″ (50.8 mm); the metal content of this fraction is over 80% by weight. The overall composition of all the incinerator residue (Table 10.6.1)

differed substantially from the composition of the under-2″ (50.8 mm) fraction (Table 10.6.2). The test also concluded that the New York State Department of Transportation specifications for Type 3 asphalt binder can be met if 10% combined incinerator ash is mixed in with 90% natural aggregate.

—*Béla G. Lipták*

References

New Jersey Department of Environmental Protection and Energy. 1992. *How to reduce waste and save money: Case studies from the private sector*. Division of Solid Waste Management, Office of Recycling and Planning, Bureau of Source Reduction and Market Development, Trenton, N.J. (July).

Peterson, Charles. 1989. What does "waste reduction" mean? *Waste Age* (January).

10.7
MATERIAL RECOVERY

The recycling of postconsumer material found in MSW involves (1) the recovery of material from the waste stream, (2) intermediate processing such as sorting and compacting, (3) transportation, and (4) final processing to provide a raw material for manufacturers. This section emphasizes separation and recovery, and applicable specifications for these materials. It focuses on those materials which are intended for short-term consumer usage, are discarded quickly, and are present in large quantities in the solid waste stream.

Role of MRFs and MRF/TFs

Material Recovery Facilities (MRFs) and Material Recovery/Transfer Facilities (MRF/TFs) are used as centralized facilities for the separation, cleaning, packaging, and shipping of large volumes of material recovered from MSW. These processes include:

Further processing of source-separated wastes from curbside collection programs. The type of source-separated material that is separated includes paper and cardboard from mixed paper and cardboard; aluminum from commingled aluminum and tin cans; plastics by class from commingled plastics; aluminum cans, tin cans, plastics, and glass from a mixture of these materials; and glass by color (clear, amber, and green).

Separating commingled MSW. All types of waste components can be separated from commingled MSW. Waste is typically separated both manually and mechanically. The sophistication of the MRF depends on (1) the number and types of components to be separated, (2) the waste diversion goals for the waste recovery program, and (3) the specifications to which the separated products must conform.

MRFs for Source-Separated Waste

MRFs for source-separated waste further separate paper and cardboard, aluminum and tin cans, and plastic and glass.

PAPER AND CARDBOARD

The principal types of paper recycled are old newspaper (ONP), old corrugated cardboard (OCC), high-grade paper, and mixed paper waste (MPW). These waste papers can be classified into *bulk* or *high-grade*. The highest grade of papers are manila folders, hard manila cards, and similar computer-related paper products. The bulk grade consists of newspapers, corrugated paper, and MPW. MPW consists of unsorted waste from offices, commercial sources, or printing establishments. High-grade waste paper is used as a pulp substitute, whereas bulk grades are used to make paper boards, construction paper, and other recycled paper products. The heavy black ink used on newspaper reduces its value. The value of paper is also reduced by the presence of other substances that interfere with the single-process conversion into pulp, such as gum in the binding of telephone directories or the chemical coating of magazines.

To ensure quality and minimize handling and processing, ONP should be separated from all other waste at or as close as possible to its source of generation. End users can reject an entire shipment of ONP where evidence exists that the paper was commingled with MSW. Care must also be taken to prevent contamination of the paper during collection, loading, transporting, unloading, processing, and storing.

In MRFs, mixed paper and cardboard are unloaded from the collection vehicle onto the tipping floor. There, cardboard and nonrecyclable paper items are removed. The mixed paper is then loaded onto a floor conveyor with a front-end loader. The floor conveyor discharges to an inclined conveyor that discharges into a horizontal conveyor. The horizontal conveyor transports the mixed paper past workers who remove any remaining cardboard from the mixed paper. The paper remaining on the conveyor is discharged to a conveyor located below the picking platform that is used to feed the baler. Once the paper has been baled, the cardboard is baled.

The Paper Stock Institute of America, which represents buyers and processors of waste paper, has listed thirty-three specialty grades whose specifications are agreed upon by buyers and sellers. Table 10.7.1 gives the specifications for the most common grades of postconsumer waste paper.

The four grades from lowest to highest quality are news (grade 6), special news (grade 7), special news de-ink quality (grade 8), and over-issue news (grade 9). Grades 6 and 7 are used primarily in the production of insulation and paperboard as well as in other applications where high quality (absence of contamination) is not of foremost importance. Grade 8 is used to make newspaper again, as is grade 9. Grade 9 is the grade that sellers find provides the most accessible market.

Paper shipped to a paper mill must meet mill specifications on outthrows and prohibited materials. *Outthrows* are defined as all papers that are so manufactured or treated or are in such form to be unsuitable for consumption as the grade specified. *Prohibitive materials* are defined as:

Any material in the packing of paper stock whose presence in excess of the amount allowed makes the packaging unsalable as the grade specified

Any material that may be damaging to equipment

The maximum amount of outthrows in grade specifications is the total of outthrows and prohibitive materials. Examples of prohibitive materials are sunburned newspaper, food containers, plastic or metal foils, waxed or treated paper, tissues or paper towels, bound catalogs or telephone directories, Post-its, and faxes or carbonless carbon paper. Other prohibitive materials are foreign materials such as dirt, metal, glass, food wastes, paper clips, and string.

ALUMINUM AND TIN CANS

Aluminum cans are one of the most common items recovered through municipal and commercial recycling programs because they are easily identified by residents and employers. They also provide higher revenues than other recyclable materials. The recycling of used beverage cans (UBCs) not only saves valuable landfill space but also minimizes energy consumption during the manufacturing of aluminum products. Manufacturing new aluminum cans from UBCs uses 95% less energy than producing them from virgin materials.

A successful aluminum recycling program must have interaction between various entities including those involved with collection, sorting and processing, reclamation, and reuse. Three generator sectors from which aluminum beverage containers can be recovered are residential house-

TABLE 10.7.1 SPECIFICATIONS FOR RECYCLED PAPER AND CARDBOARD

Grade Number	Class	Description	Prohibitive Materials, %	Total Outthrows, %
1	Mixed Paper	A mixture of various qualities of paper not limited to type of packing or fiber content	2	10
6	News	Baled newspapers containing less than 5% other papers	0.5	2.0
7	Special News	Baled, sorted, fresh, dry newspapers; not sunburned; free from paper other than news; containing not more than the normal percentage of rotogravure and colored sections	None permitted	2.0
8	Special News, De-ink Quality	Baled, sorted, fresh, dry newspapers; not sunburned; free from magazines, white blank, pressroom overissues, and paper other than news; containing not more than the normal percent of rotogravure and colored sections. This packaging must be free from tar.	None permitted	0.25
9	Overissue	Unused, overrun, regular newspaper printed on newsprint; baled or securely tied in bundles; containing not more than the normal percentage of rotogravure and colored sections	None permitted	None permitted
11	Corrugated	Baled, corrugated containers, Containers having liners of test liner, jute, or kraft	1.0	5.0
38	Sorted Colored Ledger	Printed or unprinted sheets, colored shavings, and cuttings of colored or ledger white sulfite or sulfate ledger, bond; and writing and other papers that have a similar fiber and filler content. This grade must be free of treated, coated, padded, or heavily printed stock.	None permitted	2.0
40	Sorted White Ledger	Printed or unprinted sheets, guillotined books, quire waste, and cuttings of white sulfite or sulfate ledger, bond, and writing and other papers that have a similar fiber and filler content. This grade must be free of treated, coated, padded, or heavily printed stock.	None permitted	2.0
42	Computer Printout	White sulfite or sulfate papers in forms manufactured for use in data processing machines. This grade can contain colored stripes and impact or nonimpact (e.g., laser) computer printing, and can contain not more than 5% of groundwood in the packing. All stock must be untreated and uncoated.	None permitted	2.0

Source: Paper Stock Institute, *Guidelines for paper stock* (Washington, D.C.: Institute of Scrap Recycling Inc.).

holds, commercial institutions, and manufacturing entities. Curbside collection programs recapture large quantities of recyclables. Aluminum UBCs can be separated as an individual commodity or commingled with other recyclables.

Steel food cans, which make up more than 90% of all food containers, are often called tin cans because of the thin tin coating used to protect the contents from corrosion. Some steel cans, such as tuna cans, are made with tin-free steel, while others have an aluminum lid and a steel body and are commonly called bimetal cans. All these empty cans are completely recyclable by the steel industry and should be included in any recycling program.

At the MRF, the collection vehicle discharges the commingled aluminum and tin cans into a hopper bin, which discharges to a conveyor belt. The conveyor transports the commingled cans past an overhead magnetic separator where the tin cans are removed. The belt continues past a pulley magnetic separator, where any tin cans not removed with the overhead magnet are taken out. The aluminum and tin cans, collected separately, are baled for shipment to markets.

At a reclamation plant, shredded aluminum cans are first heated in a delacquering process to remove coatings and moisture. Then they are charged into a remelting furnace. Molten metal is formed into ingots of 30,000 lb or more that are transferred to another mill and rolled into sheets. The sheets are sent to container manufacturing plants and cut into disks, from which cans are formed.

Aluminum markets have material specifications that regulate the extent of contamination allowed in each delivery as well as the method by which materials are prepared. For example, some markets prohibit aluminum foils and aluminum pans because they are usually contaminated. Noncontainer aluminum products purchased by dealers must simply be dry and free of contaminants.

PLASTIC AND GLASS

The recycling and reuse of plastics are not easily accomplished because each type of plastic must go through a different process before being reused. Hundreds of different types of plastics exist, but 80% of the plastics used in consumer products is either HDPE (milk and detergent bottles) or polyethylene terephthalate (PET) (large soda bottles). The most common items produced from postconsumer HDPE are detergent bottles and motor oil containers. Detergent bottles are usually made of three layers, with the center layer containing the recycled material.

Most plastic container manufacturers code their products. The code is a triangle with a number in the center and letters underneath. The number and letter indicate the resin from which the container is made:

1 = PET   (polyethylene terephthalate)
2 = HDPE (high-density polyethylene)
3 = V   (vinyl)

4 = LDPE (low-density polyethylene)
5 = PP   (polypropylene)
6 = PS   (polystyrene)
7 = Other (all other resins and multilayered material)

Still, keeping plastics separate is not easy. The most notorious look alikes are PET, the clear, shiny plastic that soda bottles are made from, and PVC, another clear plastic used mainly for packaging cooking oil. Because PVC starts to decompose at the temperature at which PET is just beginning to melt, one stray PVC bottle in a melt of 10,000 PET bottles can ruin the entire batch.

Container glass is the only glass being recycled today. Window panes, light bulbs, mirrors, ceramic dishes and pots, glassware, crystal, ovenware, and fiberglass are not recyclable with container glass and are considered contaminants in container glass recycling.

The consideration in container glass marketing is color separation. Permanent dyes are used to make different colored glass containers. The most common colors are green, brown, and clear (or colorless). In the industry, green glass is called emerald, brown glass is amber, and clear glass is flint. For bottles and jars to meet strict manufacturing specifications, only emerald or amber cullet (crushed glass) can be used for green and brown bottles, respectively.

At the MRF, the collection vehicle discharges the commingled plastic and glass into a hoppered bin, which discharges to a conveyor belt. The material is transported to a sorting area, where the plastic and glass are separated manually from the other materials. The remaining glass is color sorted and sent to a glass crusher. The waste is discharged to vibrating screens where broken glass falls through the openings in the screen. Any residual material is collected at the end of the vibrating screen. The crushed glass is loaded onto large trailers and transported to the vibrating screen. The residual material is disposed of in a landfill. The commingled plastic is separated further by visual inspection or according to the type (PET and HDPE) based on the imprinted code adopted by the plastic industry.

In a glass bottle manufacturing plant, specialized beneficiation equipment performs final cleaning to remove residual metals, plastic, and paper labels. The cullet is then mixed with the raw material used in the production of glass. After the batch is mixed, it is melted in a furnace at temperatures ranging from 2600 to 2800°F. The mix can burn at low temperatures if more cullets are used. The melted glass is dropped into a forming machine where it is blown or pressed into shape. The newly formed glass containers are slowly cooled in an annealing lehr. They are inspected for defects, packed, and shipped to the bottler.

At a reclamation facility, PET bottles and HDPE jugs are transformed into clean flakes. A resin reclamation facility chops and washes the chips to remove labels, adhesives, and dirt and separates the material from their com-

TABLE 10.7.2   SPECIFICATIONS FOR COLOR-
SORTED GLASS

| Color | Permissible Color Mix Levels, Percent | | | |
	Flint	Amber	Green	Other
Flint (clear)	97 to 100	0 to 3	0 to 1	0 to 3
Amber (brown)	0 to 5	95 to 100	0 to 5	0 to 5
Green	0 to 10	0 to 15	85 to 100	0 to 10

Source: American Society for Testing and Materials (ASTM), 1989, Standard specifications for waste glass as a raw material for the manufacture of glass containers, *1989 Annual book of standards,* Vol. 11.04 (Phila.: ASTM), 299–300.

ponents to produce a clean generic polymer. Clean PET is sold as flakes but most HDPE is made into pellets. The HDPE flakes are fed into an extruder and are compressed as they are carried toward the extrusion die. The combined heat from flow friction and supplemental heating bands causes the resin to melt, and volatile contaminants are vented from the mixture. Immediately before the die, the melted mixture passes through a fine screen that removes any remaining solid impurities. As the melt passes through the orifice, a rotating knife chops the strand into short segments, which fall into a water bath where they are cooled. The pellets are dried to a moisture content of about 0.5% and are packaged for shipment to the end user.

Glass used for new bottles and containers must be sorted by color and must not contain contaminants such as dirt, rocks, ceramics, and high-temperature glass cook-

ware. These materials, known as refractory materials, have higher melting temperatures than container glass and form a solid inclusion in the finished product. Table 10.7.2 gives the material specifications for color-sorted glass. The specifications in the Rotterdam glass processing plant limit the maximum amount of ceramics to 100 g per ton of crushed glass; the same limit for aluminum is only 6 g per ton.

Trade groups representing manufacturers and processors have established specifications for recycled plastics. These specifications are extensive and beyond the scope of this chapter. In general, buyers require postconsumer plastics to be well sorted, reasonably free of foreign material, and baled within a specified size and weight range.

MSW Processing

A solid waste processing plant in Rhode Island and the Sorain-Cechini MRF plant in Rome, Italy are two examples of MSW processing plants.

MRF PLANT FOR PARTIALLY SEPARATED MSW

In 1989, an 80 tons per day (tpd) MRF was started in Rhode Island (see Figure 10.7.1). Designed and operated by New England CR Inc. in conjunction with Maschinenfabrik Bezner of West Germany, this highly automated plant can sort and recover the recyclables from partially separated MSW containing metallic, glass, and plastic

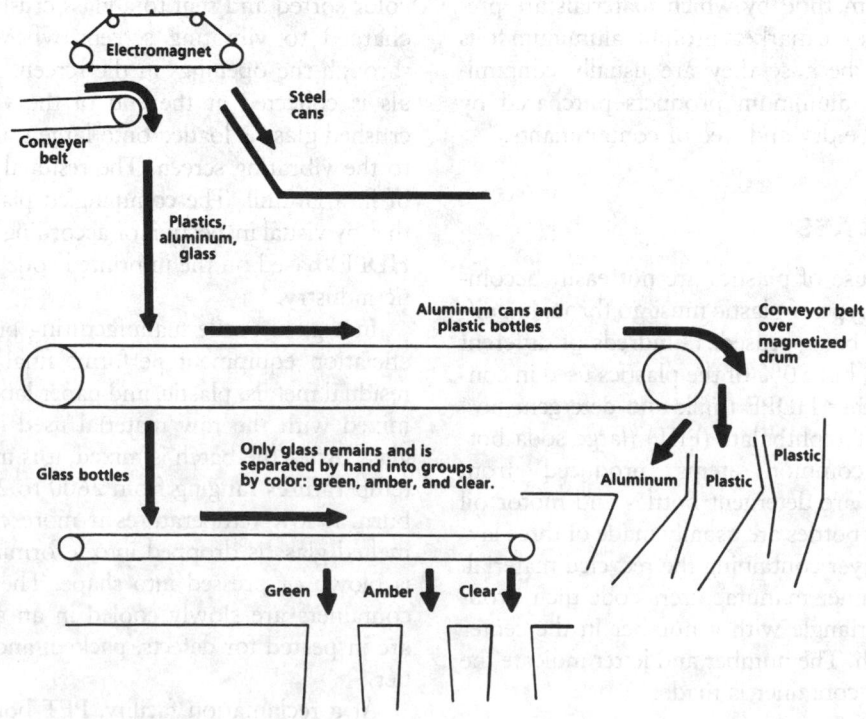

FIG. 10.7.1 The operation of a solid waste processing plant in Rhode Island. (Reprinted, with permission, from New England CRInc., 1989, *New York Times/Bohdan Osyczka,* [2 May], 160.)

cans, bottles, and other containers but not paper and organics. The partially separated MSW enters the plant on a conveyor belt, which first passes under an electromagnet that attracts the tin-plated steel cans and carries them off to be shredded. As the MSW falls, it encounters a rolling curtain of chains. The lighter objects (aluminum and plastic cans) cannot break through and are diverted toward a magnetized drum. The heavier (mostly glass) bottles pass through the curtain and arrive at a hand-separation belt, where they are separated manually by color.

As the plastic and aluminum containers reach the magnetic drum, the aluminum objects drop into a separate hopper. The plastic objects continue on the conveyor belt and are later sorted according to weight.

The plant design appears to be simple enough to guarantee reliability. The concept of this type of MRF plant is promising because it simplifies the process of source separation by allowing cans and containers of all types to be placed in the same bin.

MATERIAL RECOVERY PLANT

The Sorain-Cecchini MRF plant in Rome has been in operation for over twenty years. It recovers ferrous metals (6% by weight), aluminum (1%), organics (34%), and film

plastics (1%) while generating densified RDF (51%) and rejecting 7% oversized items. The total plant capacity is 1200 tn per day (Cachin and Carrera 1986).

The main processing steps involve magnetic separation for ferrous metal removal, eddy-current separation for aluminum recovery, rotary screens for separation by size, and air classifiers for separation by density. The overall process consists of eighty pieces of equipment, which are flexible and can be used in different combinations as market conditions change.

Figure 10.7.2 shows the resource recovery plant in Rome. The charging conveyor **1** is provided with a pickup device **2** that breaks the bags and removes bulky reject items. A leveling device **3** meters the waste-flow rate and removes rejects. The primary screen **4** separates the large (over 8 in) fraction from the smaller, heavier fraction. The approximately 55% large fraction of paper, wood, and plastic is fed to the 10–20 rpm large breaker **5**, which reduces particle size and breaks plastic bags. The large breaker automatically rejects any items it cannot break (about 2%). The output (53%) is sent to the large air classifier **6**, where the lighter (10%) sheet paper and plastic fraction is separated from the heavier (43%) cardboard, wood, and rags. The reject fraction consists mostly of white goods (appliances) but includes bulky items such as bedsprings. This fraction is hauled away by subcontractors.

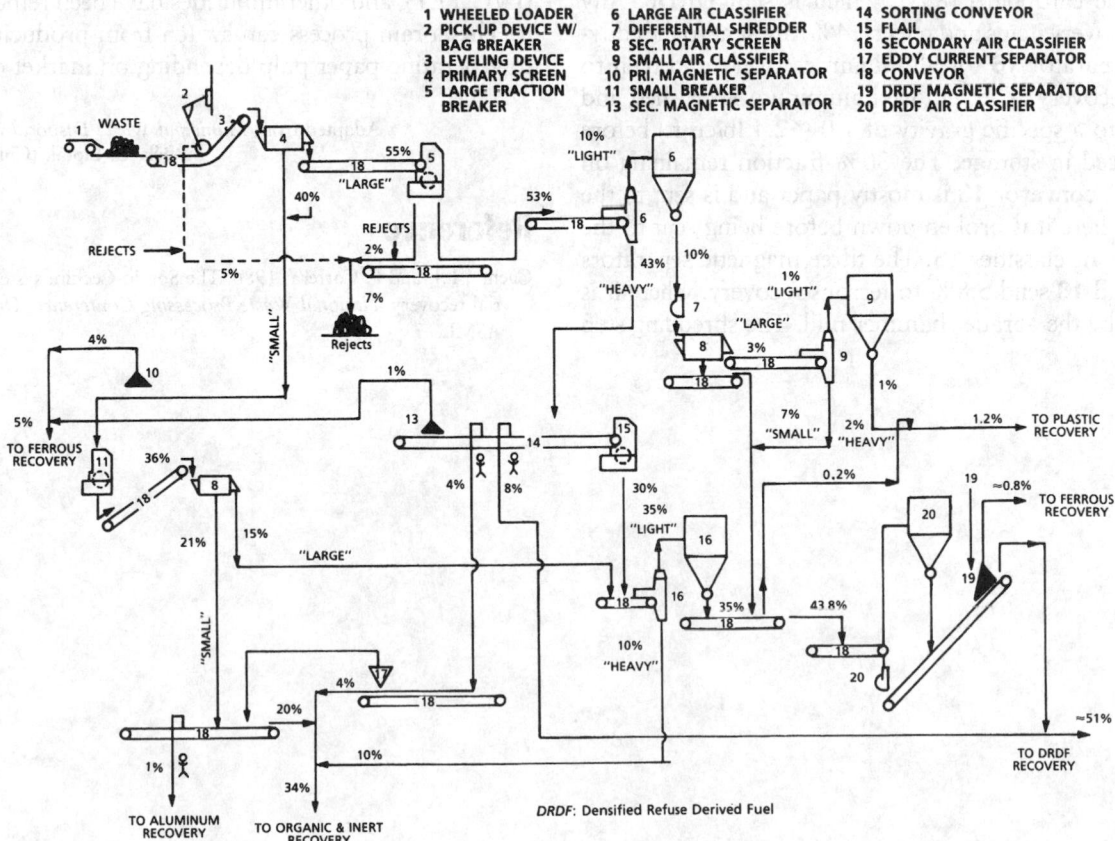

1 WHEELED LOADER	6 LARGE AIR CLASSIFIER	14 SORTING CONVEYOR	
2 PICK-UP DEVICE W/ BAG BREAKER	7 DIFFERENTIAL SHREDDER	15 FLAIL	
3 LEVELING DEVICE	8 SEC. ROTARY SCREEN	16 SECONDARY AIR CLASSIFIER	
4 PRIMARY SCREEN	9 SMALL AIR CLASSIFIER	17 EDDY CURRENT SEPARATOR	
5 LARGE FRACTION BREAKER	10 PRI. MAGNETIC SEPARATOR	18 CONVEYOR	
	11 SMALL BREAKER	19 DRDF MAGNETIC SEPARATOR	
	13 SEC. MAGNETIC SEPARATOR	20 DRDF AIR CLASSIFIER	

DRDF: Densified Refuse Derived Fuel

FIG. 10.7.2 Resource recovery plant in Rome, Italy. (Adapted from F.J. Cachin and F. Carrera, 1986, The Sorain-Cecchini system for material recovery, *National Waste Processing Conference, Denver, 1986* [ASME].)

The light fraction (10%) passes through a differential shredder **7**, which breaks up only the paper. It is followed by a rotary screen **8**, which separates the lighter 3% of the stream, containing the plastic film. This stream is sent to the small classifier **9**, where the 1% light fraction is taken to plastic recovery, while the remaining paper and rag fragments are included with the densified RDF (DRDF). The recovered plastic film (mostly polyethylene) is shredded into square-inch flakes, cleaned by washing, and air dried. The dry flakes are melted and fed to the extruder, and the pellets are shipped to plastic film manufacturers.

For metal separation, the 40% heavy fraction from the primary screen **4** passes through the primary magnetic separator **10**, which removes 4% and sends that fraction to ferrous recovery. The remaining fraction is further homogenized in the small breaker **11** and separated in the secondary rotary screen **8** into the 15% large fraction (over 4 in), consisting mainly of paper, wood, and plastics, and is sent to DRDF recovery. The 21% small fraction (under 4 in), consisting of organics, glass, ashes, and aluminum, is sent to a conveyor **17**; aluminum (1%) is removed by hand, and the rest (20%) is sent to organic recovery.

The 43% heavy fraction from the large air classifier **6** passes through the secondary magnetic separator **12**, which removes 1% and sends that fraction to ferrous recovery. The remaining fraction travels on the sorting conveyor **13**, where the semiautomatic devices and inspectors remove the cardboard (8%), which is sent DRDF. Any missed recovery items and rejects (4%) are sent to the eddy-current separator **16** for aluminum recovery and then to organic recovery. The removed aluminum is crushed and densified to a specific gravity of 1.0 (62.4 lb/cu ft) before being placed in storage. The 30% fraction remaining on the sorting conveyor **13** is mostly paper and is sent to the flail **14**, where it is broken down before being sent to secondary air classifier **15**. The three magnetic separators **10**, **12**, and **18** send 5.8% to ferrous recovery, where it is shredded by the abrader hammer mill. The shredding step is followed by cleaning through firing or washing and a final magnetic separation step to remove the nonmetals that were loosened by the abrader.

The organics fraction, left from the plant feed after the removal of metals, plastic film, and paper, is essentially a heavy fraction of small-sized particles containing organics, glass, ceramics, sand, ashes, hard plastics, and small pieces of wood. This fraction is placed into an aerobic digester and broken down into raw compost. After the removal of glass, ceramics, and other inorganic rejects, the raw compost is subcontracted for further processing. This processing splits the organic fraction into a feed fraction (a high-quality compost fraction) and a residue, which is usually landfilled.

The 15% large fraction from the secondary rotary screen **8** and the 30% from the flail **14** are sent to the secondary air classifier **15**, which removes all paper (35%) and sends it to the DRDF air classifier **19** together with the small fraction (7%) from the secondary rotary screen **8** and the heavy fraction (2%) from the small air classifier **9**. After the DRDF magnetic **18** removes the remaining ferrous metals, the DRDF is densified into flakes in the recovery line. The DRDF is stored in a specific gravity of 0.6 (38 lb/ft). The heavy fraction (10%) from the secondary air classifier **15** is sent to organic recovery.

The DRDF obtained is relatively clean, and its sulfur and chlorine content is low as most metals, hard plastics (PVC, PET), and other impurities have been removed from it. The Sorain process can switch from producing DRDF to generating paper pulp depending on market condition.

Adapted from *Municipal Waste Disposal in the 1990s* by Béla G. Lipták (Chilton, 1991).

Reference

Cachin, F.J. and F. Carrera. 1986. The Sorain-Cechini system for material recovery. *National Waste Processing Conference, Denver, 1986.* ASME.

10.8
REFUSE-DERIVED FUEL (RDF)

RDF is the combustible portion of MSW that has been separated from the noncombustible portion through processing such as shredding, screening, and air classifying. The RDF that remains after processing is highly combustible and can be used as is (fluffy material) or in pellet form.

RDF Preparation Plant

Figure 10.8.1 shows the process flow diagram of the RDF preparation plant in Haverhill, Massachusetts. This plant has been operating since 1984, feeding 100% RDF to a

250,000-lb/hr boiler designed for RDF service. In this facility, 1300 tpd of MSW are separated into 983 tpd of RDF fuel, 260 tpd of glassy residue which are landfilled, and 57 tpd of ferrous metals which are sold. The MSW passes through two parallel 70 tph Heil shredders producing an output particle size of 90% under 4 in (101.6 mm). The ferrous metals are removed by dings and head pulley magnets.

The shredded refuse passes through a two-stage, 12.5-ft diameter, 60-ft long (3.8 m × 18.24 m) trommel screen. The first stage has 1-in holes to remove the glassy residue. The second stage has 6-in (152.4 mm) holes that separate

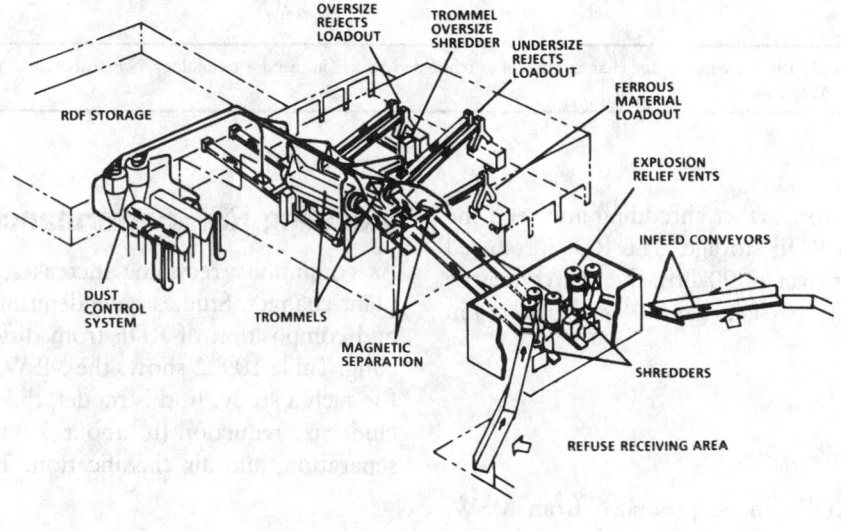

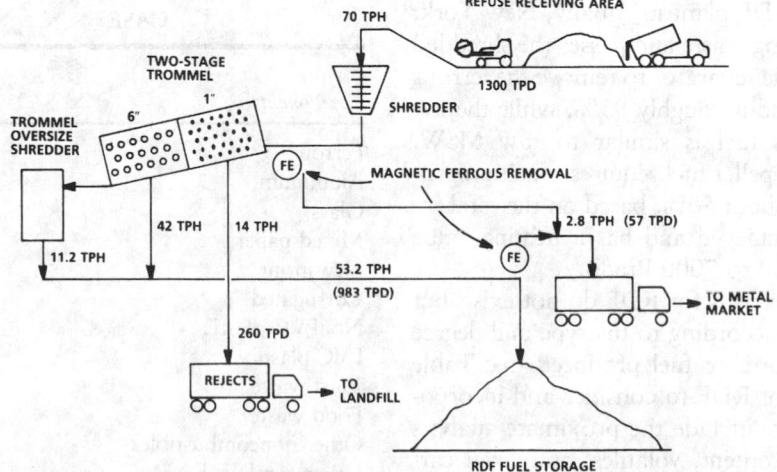

FIG. 10.8.1 RDF preparation plant in Haverhill, Massachusetts. (Reprinted, with permission, from D. Kaminski, 1986, Performance of the RDF delivery and boiler-fuel system at Lawrence, Massachusetts facility, *National Waste Processing Conference, Denver, 1986* [ASME].)

TABLE 10.8.1 ASTM CLASSIFICATION OF RDFS

Class	Form	Description
RDF-1 (MSW)	Raw	MSW with minimal processing to remove oversize bulky waste
RDF-2 (C-RDF)	Coarse	MSW processed to coarse particle size with or without ferrous metal separation such that 95% by weight passes through a 6-in square mesh screen
RDF-3 (f-RDF)	Fluff	Shredded fuel derived from MSW processed for the removal of metal, glass, and other entrained inorganics; particle size of this material is such that 95% by weight passes through a 2-in square mesh screen
RDF-4 (p-RDF)	Powder	Combustible waste fraction processed into powdered form such that 95% by weight passes through a 10 mesh screen (0.035 in. square)
RDF-5 (d-RDF)	Densified	Combustible waste fraction densified (compressed) into pellets, slugs, cubettes, briquettes, or similar forms
RDF-6	Liquid	Combustible waste fraction processed into a liquid fuel
RDF-7	Gas	Combustible waste fraction processed into a gaseous fuel

Source: R.E. Sommerland et al., 1988, Environmental characterization of refuse-derived-fuel incinerator technology, *National Waste Processing Conference, Philadelphia, 1988* (New York: ASME).

the oversized material for further shredding and send the under-6-in fraction to RDF storage. The RDF produced has a heating value of over 6000 Btu; the ash content is less than 15%, and its particle size is 97% under 4 in (101.2 mm).

Grades of RDF

Different grades of RDF can be produced from MSW. Generally the higher the fuel quality, the lower the fuel yield. For example, an RDF plant in Albany, New York, simply shreds the incoming waste and passes the shredded material across a magnetic separator to remove the ferrous component. The fuel yield is roughly 95%, while the average Btu value of this fuel is similar to raw MSW. Conversely, producing a pellet fuel requires much preprocessing. A field yield of about 50%, based on the total incoming waste, can be achieved and has a heating value which approximates 6500 to 7000 Btu/lb.

Industry-wide specifications for RDF do not exist, but RDF has been classified according to the type and degree of processing and the form of fuel produced (see Table 10.8.1). The properties of RDF to consider and incorporate into supply contracts include the proximate analysis (moisture content, ash content, volatiles, and fixed carbon); ultimate analysis (C, H, N, O, S, and ash percentage); higher heating value (HHV); and content of chlorine, fluorine, lead, cadmium, and mercury.

Modeling RDF Performance

As community recycling increases, the feed to an RDF plant changes. Studies have determined the heating value and composition of RDF from different degrees of recycling. Table 10.8.2 shows the MSW composition assumed for such a study. In this model, the pretreatment steps include size reduction (to about 5 cm), screening, magnetic separation, and air classification. Table 10.8.3 gives the

TABLE 10.8.2 COMPOSITION OF WASTE FOR BASE CASE

Component	Percent As-Received
Ferrous	5.5
Aluminum	0.9
Glass	9.5
Mixed paper	22.6
Newsprint	11.8
Corrugated	12.2
NonPVC plastic	2.9
PVC plastic	0.3
Yard waste	12.5
Food waste	2.5
Other noncombustible	9.5
Other combustible	9.8

Source: G.M. Savage and L.F. Diaz, 1986, Key issues concerning waste processing design, *National Waste Processing Conference, Denver, 1986* (ASME).

TABLE 10.8.3 CALCULATED MSW AND RDF PROPERTIES AND COMPOSITIONS RESULTING FROM DIFFERENT DEGREES OF RECYCLING

Scenario	Heating Value (Btu/lb Wet)	Percent Ash (Dry)	Ultimate Analysis (Percent)							Heavy Metal Analysis (mg/kg)									
			C	H	O	N	S	Cl	Sb	As	Ba	Cd	Cr	Cu	Pb	Hg	Ni	Zn
Baseline Case																		
MSW	3970	36.6	32.1	4.3	25.8	0.58	0.17	0.33	53	4.9	2160	14.4	210	720	630	18	220	290
RDF	5670	11.0	44.3	5.9	37.7	0.44	0.16	0.49	68	5.4	2620	14.0	200	170	500	23	40	160
30% Fe, Al, and Glass																		
MSW	4200	32.0	34.0	4.6	27.7	0.62	0.18	0.36	55	5.1	2220	15.0	200	570	600	18	160	270
RDF	5740	9.7	44.9	6.0	38.3	0.45	0.16	0.51	65	5.2	2510	13.4	190	140	470	22	30	130
30% Fe, Al, Glass, Newsprint, and Corrugated																		
MSW	4070	35.0	33.3	4.4	26.1	0.70	0.19	0.37	62	4.8	2500	16.5	210	600	550	21	170	300
RDF	5710	11.0	44.5	6.0	37.3	0.52	0.17	0.56	82	5.3	3200	16.2	210	160	440	28	30	160
30% Newsprint																		
MSW	3905	37.9	31.6	4.2	25.2	0.61	0.17	0.34	55	5.1	2250	15.0	210	750	590	18	230	300
RDF	5635	11.6	44.0	5.9	37.4	0.47	0.16	0.52	74	5.9	2860	15.1	200	180	450	25	40	170
50% PVC																		
MSW	3950	36.8	32.1	4.3	25.8	0.59	0.17	0.24	52	4.8	2190	14.4	210	730	640	18	220	290
RDF	5655	11.1	44.3	5.9	37.8	0.45	0.16	0.34	67	5.3	2680	14.0	200	170	500	23	40	160
50% Yard Waste																		
MSW	4055	37.6	31.7	4.2	25.5	0.50	0.16	0.33	55	5.1	2230	15.0	220	750	660	18	230	300
RDF	5720	11.0	44.2	5.9	37.8	0.41	0.16	0.49	69	5.5	2670	14.1	210	170	500	23	40	160
50% Food Waste																		
MSW	3990	36.7	32.2	4.3	25.8	0.57	0.17	0.32	53	5.0	2170	14.2	210	700	630	18	220	270
RDF	5680	11.0	44.3	5.9	37.8	0.44	0.16	0.48	68	5.4	2620	13.9	200	160	490	22	40	150

Source: Savage and Diaz, 1986.

properties of the MSW and the recovered RDF after various degrees of recycling.

The model shows that the ash content drops and the heating value rises as the MSW is processed into RDF, and the type and degree of recycling has only a limited effect on the ash content or heating value (Savage and Diaz 1986). The nitrogen content of the RDF is consistently lower than that of the MSW, and the sulfur content is relatively unaffected by processing; while PVC recycling has a substantial effect on the chlorine content of the RDF. The calculated heavy metal analysis shows that because of the magnetic separation of ferrous metals, the concentration of lead (Pb) and zinc (Zn) is lower in the RDF than in the MSW.

Modeling is a useful tool in the evaluation of RDF processes. One can estimate the effect of the degree of size reduction, the influence of the opening sizes in screening equipment, and the effect of placing shredders up or downstream of the screening or air-separation equipment. Some modeling calculations can also estimate the base/acid ratio, slagging index, and fouling index values, which can indicate likely maintenance and operating problems associated with a particular process.

Adapted from *Municipal Waste Disposal in the 1990s* by Béla G. Lipták (Chilton, 1991).

Reference

Savage, G.M. and L.F. Diaz. 1986. Key issues concerning waste processing design. *National Waste Processing Conference, Denver, 1986.* ASME.

Treatment and Disposal

10.9
WASTE-TO-ENERGY INCINERATORS

Incineration is the second oldest method for the disposal of waste—the oldest being landfill. By definition, incineration is the conversion of waste material to gas products and solid residues by the process of combustion. Combustion under optimal conditions can cut MSW 90% by volume and 75% by weight. Hot gases generated as a result of combustion exit the furnace and pass through boilers which recover energy in the form of steam. This steam can be sold directly or converted to electricity in a turbine. With dwindling landfill space, incineration reduces volume, but some scientists caution that incinerator residue is more dangerous and should not be disposed of in regular landfills.

The combustion process carries the risk of releasing air pollutants. Emissions from incinerators can include toxic metals and toxic organics. The primary goals of waste-to-energy incineration are to maximize combustion and min-imize pollution. Two other goals are high plant availability and low facility maintenance cost.

Mass-Burn and RDF Incinerators

Two main types of waste-to-energy incinerators are mass-burn incinerators and RDF incinerators. Figure 10.9.1 shows the typical structure of a waste-to-energy facility.

The more common *mass-burn incinerators* burn MSW as received with minimal onsite effort to separate objects that do not burn well or do not burn at all. (For example, bulky, oversized items such as tires, bedframes, fences, and logs are often separated by hand to avoid problems, but glass bottles and metals usually are not.)

RDF incinerators burn MSW that has been pre-processed and sorted (either on the site of the incinerator

① Tipping Hall	⑯ Stack
② Refuse Bunker	⑰ Control Room
③ Grapple and Refuse Crane	⑱ Deaerator Storage Tank
④ Crane Operator Control	and Heater
⑤ Charging Hopper	⑲ Motor Control Center
⑥ Overfire Air Fan	⑳ Maintenance Shop
⑦ Ram Feed	㉑ Heaters
⑧ Ignition Burner Fan	㉒ Condenser
⑨ Underfire Fan	㉓ Switchgear
⑩ Roller Grate	㉔ Id Fan
⑪ Ash Conveyors to	㉕ Turbine Crane
Materials Recovery	
⑫ Boiler	
⑬ Overfire Air Intake	
⑭ Turbine Generator	
⑮ Precipitator	

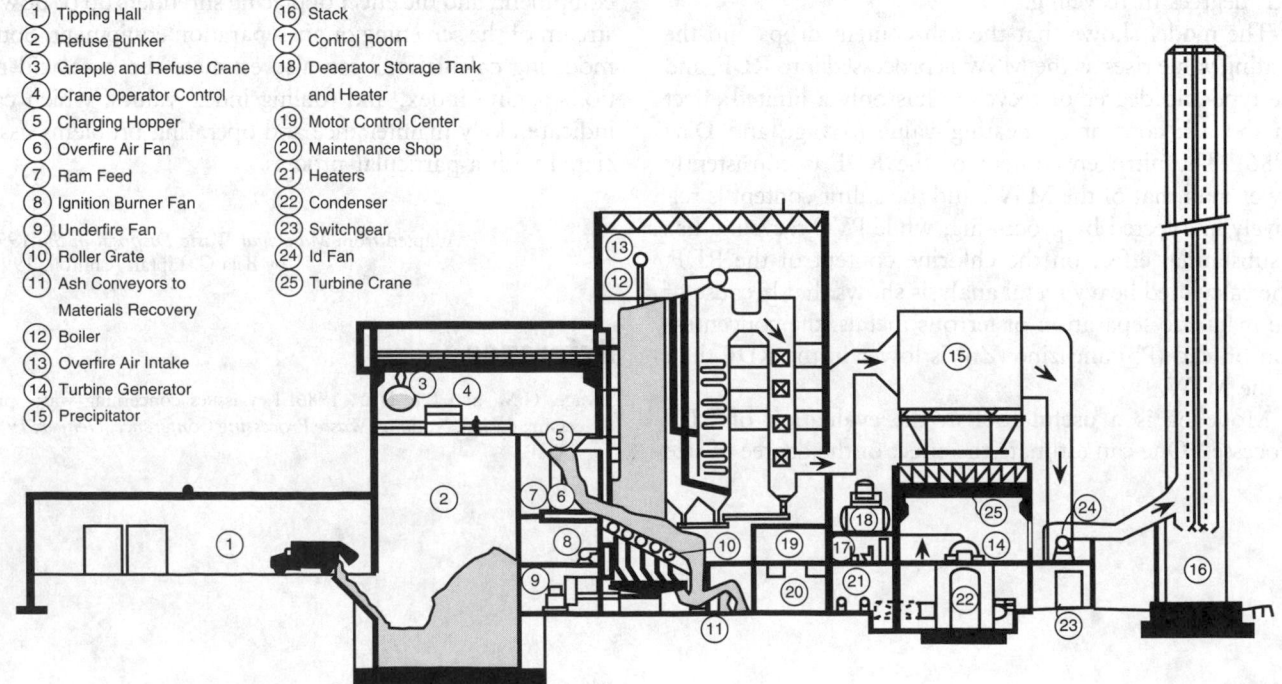

FIG. 10.9.1 Schematic of a typical waste-to-energy resource recovery facility.

or at separate processing facilities). Noncombustible and recyclable material such as ferrous metals, aluminum, and glass are separated mechanically and collected for processing and future sale or disposal. The combustible portion is converted to a more uniform, pellet fuel through particle reduction (usually 4- to 6-in pellets).

RDF technology is preferred by recycling-oriented users partly for economic reasons (e.g., income from the sale of aluminum), and partly because it cuts the incinerator residues to less than half and thereby reduces the amount of leftover material that must be landfilled. RDF-fired boilers can respond faster to load variations, require less excess air, and can operate at higher efficiencies. Comparisons of mass-burner performance on both raw MSW to simple prepared fuels show that prepared fuel plants have many advantages over the mass-burning technology (Sommer and Kenny 1984). However, RDF technology is still in the development stage. The majority of incinerators under construction are mass-burn. Part of the reason for this lack of development is the complexity of the RDF process, which remains an expensive and maintenance-intensive alternative to mass-burning.

Plant Design

The plant design for a waste-to-energy plant should consider state-of-the-art concepts as well as other design criteria.

CONCEPT OF STATE-OF-THE-ART

The term *state-of-the-art* for waste-to-energy plants refers to (1) the best technologies and operating practices for reducing the environmental impacts of these plants and (2) the best regularly attainable emission levels from them for certain air pollutants. The state-of-the-art in waste-to-energy plant design has been improving over time. Over the last decade, as landfill space has become scarce, interest in incineration has been renewed, environmental concerns have increased, and regulations have become more stringent.

The EPA's New Source Performance Standards (NSPS), proposed in 1989 and promulgated in February 1991, were the first regulations to broadly and specifically address the performance of MSW incinerators. The new regulations set standards in four basic areas: good combustion practice, emission levels for six pollutants, monitoring requirements, and operator training and certification. Table 10.9.1 summarizes these regulations.

DESIGN BASIS

In the design of waste-to-energy incinerators, the size of the plant is a critical factor. Planners need accurate information about the amount and type of waste the plant is to burn (see Sections 10.1 to 10.5) as well as projections for future solid waste management practices in the community.

Next, planners must determine what to burn. In keeping with the hierarchy of the Pollution Prevention Act (PPA), a state-of-the-art strategy provides for the maximum amount of source reduction and recycling, including composting, before incineration. Furthermore, materials that are not recyclable and are unsuitable for burning because they are noncombustible, explosive, or contain toxic substances or pollutant precursors, should be separated from the waste to be burned. These activities preserve natural resources, improve incinerator efficiency, and minimize pollutant emissions and ash quantity and toxicity.

A general, overriding principle in the design of a solid waste incinerator is to use the correct size incinerator for the amount of anticipated waste. Combustion is most efficient when an incinerator consistently burns the quantity and quality of MSW it was constructed to burn, as follows:

If the plant is oversized (i.e., if the amount of MSW available for burning is less than the plant was designed to take), it may operate less than full time. Each start up and shutdown causes unsteady burning conditions, resulting in reduced overall efficiency. Such unsteady state conditions increase the generation of incomplete combustion and particulates. More importantly, a plant that is oversized for the amount of waste available to burn has higher per ton disposal costs.

If an incinerator is undersized (that is, more MSW is available to be burned than originally planned), too much MSW may be loaded into the furnace. Overloading an incinerator can result in increased generation of incomplete combustion as well as an increased volume of unburned matter and ash. Also, an undersized incinerator that is not overloaded requires additional expenditures of alternative methods of waste disposal and recycling.

In determining the amount of MSW being generated, planners should collect actual waste data just prior to design and sizing. Waste composition studies should ideally sample waste from different neighborhoods at different times of the week and year, as shown in Figure 10.3.1. Some communities use average waste composition from other towns or cities to estimate their own waste composition. However, this method can be misleading since the composition of MSW changes not only from place to place but also over time.

Information about projected population growth and future trends in the volume and composition of waste is just as critical as current waste data, especially since waste management methods are changing. Incinerators are typically designed for at least a twenty-year lifetime, and incinerator arrangements often include long-term (fifteen- to thirty-

TABLE 10.9.1 KEY FEATURES OF NEW FEDERAL MSW INCINERATOR REGULATIONS (NSPS), COMPARED TO INFORM STATE-OF-THE-ART STANDARDS

New Source Performance Standards	INFORM State-of-the-Art Standard
Materials Separation	
None	Recyclables, noncombustibles, and wastes containing toxic materials or pollutant precursors removed
Good Combustion Practices	
Carbon monoxide emissions:	
50–150 ppm (depending on furnace type)	50 parts per million
Plant-specific maximum load level	
Plant-specific maximum flue gas temperature at inlet to final particulate control device	
Pollutant Emissions Levels	
(7% O_2, dry basis)	
PARTICULATES	
0.015 g per dry standard cu ft	0.010 grains per dry standard cubic foot
DIOXINS/FURANS	
30 nanograms per dry standard cu m— total dioxins and furans	0.10 nanograms per dry normal cubic meter— Eadon toxic equivalents
SULFUR DIOXIDE	
80% reduction, or 30 ppm (whichever is less stringent)	30 parts per million
HYDROGEN CHLORIDE	
95% reduction, or 25 ppm (whichever is less stringent)	25 parts per million
NITROGEN OXIDES	
180 ppm	100 parts per million
HEAVY METALS	
No individual standards; particulate emissions as surrogate	Not defined; further research needed to identify lowest regularly attainable emissions levels
Monitoring Requirements	
CONTINUOUS MONITORING	
Carbon monoxide, opacity, sulfur dioxide, nitrogen oxides	Furnace and flue gas temperature, steam pressure and flow, oxygen, carbon monoxide, opacity, sulfur dioxide, oxides of nitrogen
ANNUAL STACK TESTS	
Particulates, dioxins, furans, hydrogen chloride	Particulates, dioxins/furans, hydrogen chloride, metals
Operator Training and Certification	
American Society of Mechanical Engineers certification standards for chief facility operators and shift supervisors	Formal academic and practical education; supervised on-the-job training; formal testing; periodic reevaluation

Source: U.S. Environmental Protection Agency, 1991, Burning of hazardous waste in boilers and industrial furnaces, final ruling, *Federal Register 56*, no. 35 (21 February), 7134–7240.

year) contracts for the quantity of MSW to be delivered to the plant and for the quantity of energy to be sold. Knowing what potentially recyclable material is in the waste stream and in what quantities is essential (see Section 10.3).

Finally, plant designers need information about the composition of the waste stream to determine the optimal physical design of the plant. For instance, different materials generate different amounts of heat energy when burned, and knowing the anticipated overall Btu value is critical to planning boiler capacity and furnace structure. The variability of MSW (specifically density due to changes in composition) is another design consideration for volumetric material handling equipment for RDF incinerators.

Process Design

A typical incinerator system contains basic elements: a feed system, a combustion chamber, an exhaust system, and a residue disposal system. Ancillary equipment includes shredders and a material sorter in the front end and air pollution control devices and a heat recovery device at the back end of the incinerator. Modern incinerators in the United States use continuous-feed systems and moving grates in primary combustion chambers which are lined with refractory material (heat-resistant silica-based material). Secondary combustion chambers burn the gas or solids not burned in the primary combustion chambers before discharging to the air pollution control devices.

WASTE RECEIVING AND STORAGE

A state-of-the-art solid waste management system specifies exactly what waste can be burned (based on combustibility and content of the toxic materials and pollutant precursors). It ensures that prohibited materials are detected and removed from the waste. Table 10.9.2 compiles the materials that are prohibited at several MSW incinerators. In addition, stringy wire items, such as fencing and trolling wire, can become entangled in conveyors and should be removed from the MSW feed. Such specifications are stated in contracts between operators and municipalities. Plant operators should prevent prohibited materials from entering the plant or the furnace.

A preliminary view of the waste is recommended when incoming MSW trucks are weighed. Scales, preferably integrated into an automated recording system, should be provided to record the weight of the MSW entering the plant. Tipping floors, which resemble large warehouse floors, are better suited for visual inspection and the removal of unwanted items. State-of-the-art screening includes opening garbage bags on the tipping floor to identify unwanted items inside the bags. Radioactivity sensors are used as screening devices for hospital waste. The MSW is discharged from the tipping floor into the storage pit or directly into the furnace.

The storage provided depends on variations in the rate of truck delivery of MSW to the plant and the planned burning schedule. Storage permits MSW to be retained during peak loads and thus allows the combustion chambers to be sized for a lower average capacity. Large storage areas are generally required for MSW since it is quite bulky, with a bulk density between 250 and 350 lb/cu yd (180 and 240 kg/cu m). Provisions are often made for as much as one week's MSW at small incinerators to allow for downtime and other operating problems; two to three days of MSW storage is more common at larger plants (less than 500 tn/d). Planners should consider seasonable and cyclic variations and unplanned shutdowns in establishing plant storage requirements. The pit size is usually

TABLE 10.9.2 MATERIALS ROUTINELY PROHIBITED BY MASS-BURNING PLANTS

Bulky wastes (e.g., furniture) (may be acceptable if reduced in size)
Noncombustible wastes (not including glass bottles, cans, etc.)
Explosives
Tree stumps and large branches (may be acceptable if reduced in size)
Large household appliances (e.g., stoves, refrigerators, washing machines)
Vehicles and major parts (e.g., transmissions, rear ends, springs, fenders)
Marine vessels and major parts
Large machinery or equipment
Construction/demolition debris
Tires
Lead acid and other batteries
Ashes
Foundry sand
Cesspool and sewage sludge
Tannery waste
Water treatment residues
Cleaning fluids
Crank case and other mechanical oils
Automotive waste oil
Paints
Acids
Caustics
Poisons
Drugs
Regulated hospital and medical wastes
Infectious waste
Dead animals
Radioactive waste
Stringy wire (e.g., fencing and trolling wire)

Source: M.J. Clark, M. Kadt, and D. Saphire, 1991, *Burning garbage in the US*, edited by Sibyl R. Golden (New York: INFORM, Inc.).

calculated based on an MSW density of 350 lb/cu yd of pit volume.

Refuse tends to flow poorly and can maintain an angle of repose greater than 90°. Thus, plants commonly stack refuse in storage facilities to maximize storage capacity.

Storage pits are usually long, deep, and narrow. A pit can be located in front of the furnace or a pit can be situated on each side of the furnace. If the storage pit is over 25 ft in width, the refuse dumped from the trucks must generally be rehandled. The floor of storage pits is pitched to the facilities' drainage. Storage pits are constructed of reinforced concrete with steel plates or rails along the sides, which protect them against damage from the crane bucket. The pit is usually enclosed in the MSW storage building, in which combustion air for the furnace is drawn. This arrangement creates a slight vacuum inside the building which draws in atmospheric air and prevents the escape of odors and dust.

FEEDING SYSTEMS

The waste feed system introduces refuse into the incinerator from the tipping floor or pit (or, in case of an RDF fuel plant, from the preprocessing facilities). Of the two main types of refuse feed systems, a continuous loading system contributes to more efficient combustion than batch loading because it allows a more even flow of fuel.

In batch loading, the waste is introduced by a front-end loader that shoves the garbage, in discrete batches, into the furnace. The batch method adversely affects combustion since each load pushed into the incinerator causes a temporary overload, depleting available oxygen and creating poor combustion conditions. Variations in temperature due to air leaks into the furnace have an adverse impact on refractory material and increase air emissions. In small plants with floor dumps and stored MSW, feeding is accomplished on a semibatch basis by rams which push MSW directly to the furnace at 6- to 10-min cycles.

With continuous loading, a traveling bridge crane equipped with a grapple deposits waste, a few tons at a time, into the top of an inclined chute. The garbage moves down the chute onto the drying zone of a moving grate allowing for continuous introduction of waste into the furnace. RDF is typically continuously fed into the furnace.

A basic requirement of the continuous loading system is to keep the charging hopper to the furnace fired at all times and to protect against burnbacks of fire from the combustion area through the chute to the storage pit area.

Charging Cranes

Two types of cranes are widely used for handling refuse for municipal incinerators. The most versatile is the bridge crane with a clam-shell bucket. The bridge itself travels across the length of the storage pit while a trolley moves the bucket over the length of the bridge. With the bridge crane, the storage pit can be as wide as 30 ft. If the storage pit is wide, the crane has to travel to the far side of the pit to keep refuse from accumulating there. The time required to traverse the pit affects the carrying capacity of the system and wide pits with long bridges are not economical. Figure 10.9.2 shows a layout using a bridge crane. In large furnaces of more than 300 tpd capacity, bridge cranes are used.

The second type of crane, the monorail, can move in one direction only, along the rail at the center line of the pit. The range of the monorail is limited in regard to the pit. The pit width is limited to about 1 m wider than the width of the open bucket. If the storage pit is too wide, the bucket cannot move to the sides of the pit, and the material accumulates because of its tendency to cling together. The monorail system is normally designed to follow a straight path with the pit at one end and to lift MSW into charging hoppers at the other. In a medium 100- to 300-tpd plant, the monorail is often used.

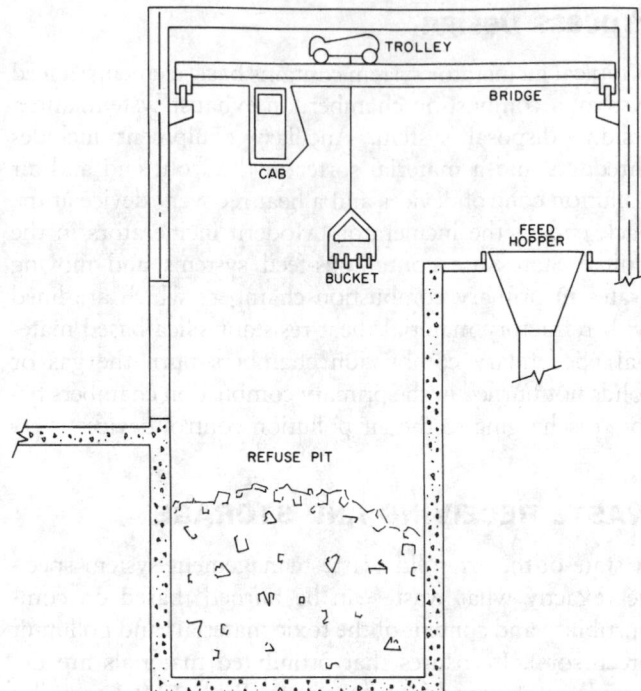

FIG. 10.9.2 Bridge crane installation.

Both crane types have either a clam-shell grapple or an abrasion-resistant steel bucket with a capacity between 30 and 150 cu ft (1–4 cu m). An automatic lubrication system for the crane is recommended, and a good preventive maintenance program is essential. Spare buckets are also recommended. Bridges, trolleys, and hoists travel at speeds of 6 ft per sec (100 m per min).

The traveling bridge is also used to mix the MSW. Mixing MSW facilitates combustion particularly if a large amount of one type of waste is discharged into one part of the storage pit. In the past, the crane was operated from an air-conditioned cab mounted on the bridge. With increasing frequency, crane operation is being centralized in a control room, usually located at the charging floor elevation and either over the tipping positions opposite the charging hoppers or close to the charging hoppers.

Charging Hoppers and Gates

Charging hoppers hold up some volume of refuse to guarantee a reasonably uniform waste flow into the incinerator. MSW enters charging hoppers in the following ways:

In larger plants where the hoppers are located above the storage pits, MSW is lifted by cranes.

In larger plants where the hoppers and storage area are at the same elevation, MSW is transferred into storage hoppers by ram feeders or by front-end loaders.

In plants under 100 tpd capacity, MSW is loaded directly from the trucks into the charging hoppers.

In multicell furnaces, each furnace cell usually has one charge hopper.

In a continuous-charging hopper, the outlet gate is kept open, and the air seal is maintained by the MSW and the movement of the mechanical grate charging the furnace. Most hoppers have an angle-of-slide surface of 30 to 60° from the vertical to prevent bridging. The feed chute is normally 4 ft (1.2 m) wide, to pass large objects with minimum bridging, and 12 to 14 ft (3.6 to 4.2 m) long from the hopper to the front end of the furnace. Because of its proximity to the combustion zone, the continuous-charging hopper is usually water cooled.

The continuous-charging hopper allows better furnace temperature control and thereby reduces the need for refractory maintenance. It also spreads the MSW more evenly across the grate, in a relatively uniform and thin layer, while sealing the furnace from cold air.

Lessons learned on RFD facilities suggest (1) using simple RDF floor storage, not bins that can become plugged; (2) using simple RDF transfer via a conveyor rather than pneumatic systems; (3) maintaining a uniform flow of RDF to boiler feeders and avoiding slug feeding, which results in unstable boiler control; and (4) using a proven RDF feeder, which maintains even grate distribution and is responsive to load change (Gibbs and Kreidler 1989).

THE FURNACE

The combustion zones in a refuse incinerator are commonly referred to as furnaces. Several common designs are currently in use: single-chamber furnaces, dual-chamber furnaces, multiple-chambered furnaces, rotary combustors, and fluidized combustors. The most common configuration includes the rectangular furnace, the multicell furnace, the vertical circular furnace, the combined rectangular furnace, and the rotary kiln. Furnaces can also be distinguished according to the type of grates used.

Because all large modern incinerators are continuous, this section discusses only continuous systems. Two classes of continuously feed furnaces are used today: refractory-lined and waterwall furnaces. Waterwall furnaces recover waste heat as well as reduce waste volume, while refractory furnaces are usually designed for volume reduction. Waterwall furnaces have water-filled tubes instead of refractory material lining the combustion chambers. As burning refuse transfers heat through the walls to the water in the tubes, these tubes form a cool wall which is in contact with the flame and hot gas. These cooler walls prevent the accumulation of slag on the side of the combustion chamber and produce steam.

Combustion Process

Efficient and even combustion is a key factor in minimizing the environmental impact of waste-to-energy incinerators, reducing both the amount of unburned material in the ash produced and the amount of air emissions. This reduction depends largely on the design of the furnace and the operating practices.

The following general guidelines foster good combustion (Licata 1986):

The grate should be covered with fuel (a uniform depth of garbage and trash) across its width. The depth at any location on the grate should be consistent with the air that can be delivered for combustion at that point.

The incinerator must include an air distribution system that apportions air according to the burning rate of waste along the entire length and width of the grate.

Underfire air should be introduced carefully. Depending on the technology, it can be concentrated in a small area or spread over a large area. Zones of high-pressure air and blowtorch effects should be eliminated. Bursts of air in one section of the fuel bed prevent even mixing of air in the burning refuse in other areas.

Air must be introduced into burning refuse both above and below the burning bed. Oxygen provided through the overfire system helps complete the combustion of any hydrocarbons (and particulates) not oxidized near the fuel bed.

Steps must be taken to prevent the buildup of slag within the furnace. Slag can damage the boiler system and also results in poor combustion by preventing proper air mixing in the fuel bed.

Gases generated in the incineration process should experience maximum mixing to facilitate oxygen reaching any unburned particles and to provide a maximum dwelling time for the gases before being released into the atmosphere.

The flue gas temperature should be at or above 1600°F for approximately 1 sec after leaving the fire bed. Figure 10.9.3 shows that these combustion conditions destroy

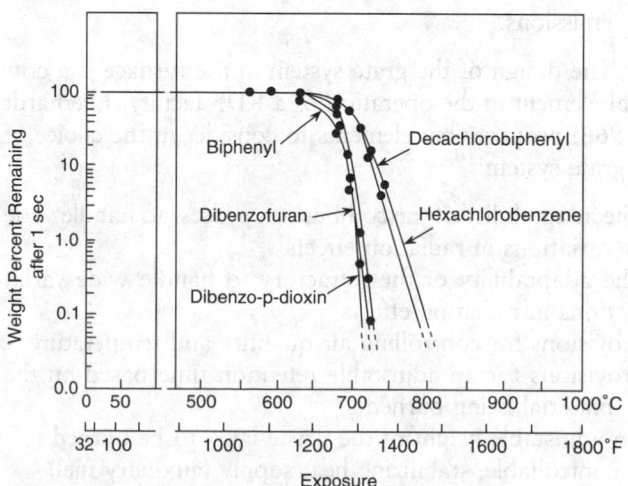

FIG. 10.9.3 Destruction efficiencies of various compounds as a function of temperature. (Reprinted, with permission, from *Air pollution control at resource recovery facilities*, 1984, California Air Resources Board [24 May].)

more than 99.9% of many effluent compounds, including dioxins and furans. Excessively high temperatures and extreme variations cause cracking and spalling, with rapid deterioration of refractories. The minimum burning temperature for carbonaceous waste to avoid the release of smoke is 1500°F (816°C). A temperature less than 1500°F permits the release of dioxins and furans.

Auxiliary burners can be added to maintain combustion efficiency. Reductions in combustion efficiency are usually due to one or more factors: start up and shutdown; large changes in moisture content, heat content, or the quantity of incoming refuse; and maladjustment of the air adjustment system. Auxiliary burners burn another, more uniform fuel (such as natural gas or oil). These burners are used when furnace temperature values fall below 1600°F, thereby stabilizing combustion by maintaining a minimum furnace temperature. Operators can increase residence time by reducing the amount of combustion air.

The design of the furnace interior affects combustion efficiency. Carefully placed protrusions from the furnace wall, called arches or bullnoses, can redirect the flow of air from the grate, guiding it into turbulent eddies within the furnace. Eddy currents maximize turbulence during the combustion of gases.

Grate Designs

The grate (stoker) serves dual functions:

1. Transports the solid waste and residue through the furnace to the point of residue discharge. The grate should be covered with a uniform depth of MSW across its width.
2. Promotes combustion by providing proper waste agitation and by permitting the passage of underfire air through the fuel bed. However, the agitation should not be so violent that it contributes to excessive particulate emissions.

The design of the grate system in the furnace is a critical element in the operation of a RDF facility. Eberhardt (1966) proposes ten elements to consider in the choice of a grate system:

The adaptability of the combustion process to handle wide variations in radiation effects
The adaptability of the refractory to handle wide variations in radiation effects
Provisions for controlling air quantity and temperature
Provisions for an adjustable retention time based on the material being burned
An adjustable height of the waste layer to be burned
A controllable, stabilizing heat supply (auxiliary fuel)
A controlled cooling of residue (by quenching)
A controlled flue gas temperature prior to impinging on the radiation heating surface
The capability of observing the fire layer and the fire gases

Technical design including:
—Prevention of reignition
—Positive conveyance of the refuse mass
—Serviceability and replaceability of worn-out parts
—Proper measuring and control systems

Grate Systems

The key factors for hot, uniform combustion are the constant mixing of air into the material being burned and the use of partially combusted material to heat and ignite new material introduced into the combustion chamber. Three major European grate designs have world-wide application:

Martin process (see Figure 10.9.4). In this design, the grate has a reverse reciprocal action; it moves alternately down and back to provide continuous motion of the refuse. The net motion of the refuse is downward toward the bottom of the furnace, but the agitation caused by oscillation of the grate causes considerable mixing of the burning refuse with the newly introduced material leading to rapid ignition and uniform burning.
Van Roll process (see Figure 10.9.5). This design has three sections: the first dries the newly introduced refuse and ignites it, the second serves as a primary combustion grate, and the last reduces the refuse to ash. Grate elements move so that at a given time for any pair of elements, one is moving and one is stationary. This process results in the refuse moving toward the bottom of the furnace, but the shuffling action of the grates agitates the fuel bed enhancing the combustion process.
VKW or Dusseldorf process (see Figure 10.9.6). This grate is comprised of several horizontal drums with a diameter of 1.5 m (5 ft). The shafts of the drums are parallel one after the other at a 30° slope. The drums are placed on 1.75-m (about 7-ft) centers. Each drum is built of bars (cast iron) in the form of arched segments

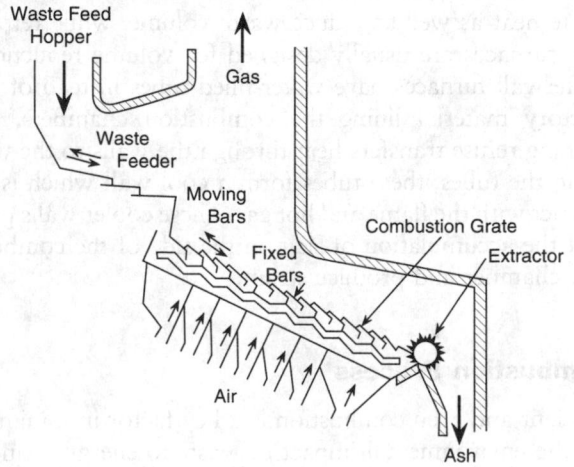

FIG. 10.9.4 Grate system for Martin resource recovery incinerator.

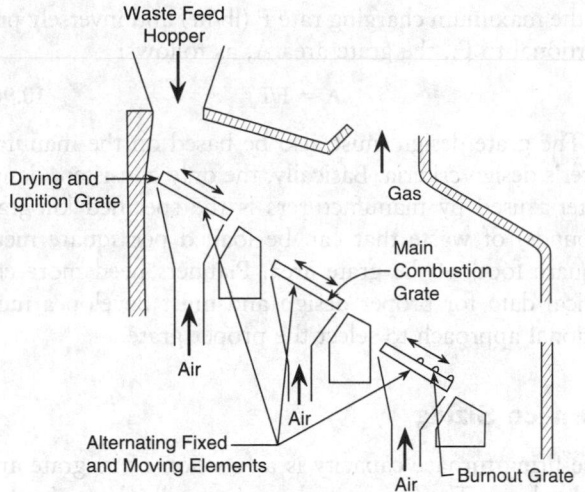

FIG. 10.9.5 Van Roll grate system for resource recovery incinerator.

which are keyed to a central element below. Each drum rests over a separate chamber to control underfire air. The unit rotates in the discharge direction at an adjustable peripheral speed which varies according to the constituents of the waste being burned. The drum shafts lie in the bearings placed in the outside walls of the unit, and each roller is fitted with a driving gear and can be regulated independently of the others. Ignition grates at the front end of the incinerator generally rotate at up to 15 m/hr (50 ft/hr). The burnout grates normally rotate at 5 m/hr (about 16 ft/hr) since they have little waste material to move. The room under the grate is divided into a zone for each roller, to which preheated or cooled flue gas (about 200 to 256°C) can be brought. A special feeding arrangement carries the refuse from the feeding chute to the grate.

Another aspect of grate design is the percentage of air openings provided in the grate. These air openings vary

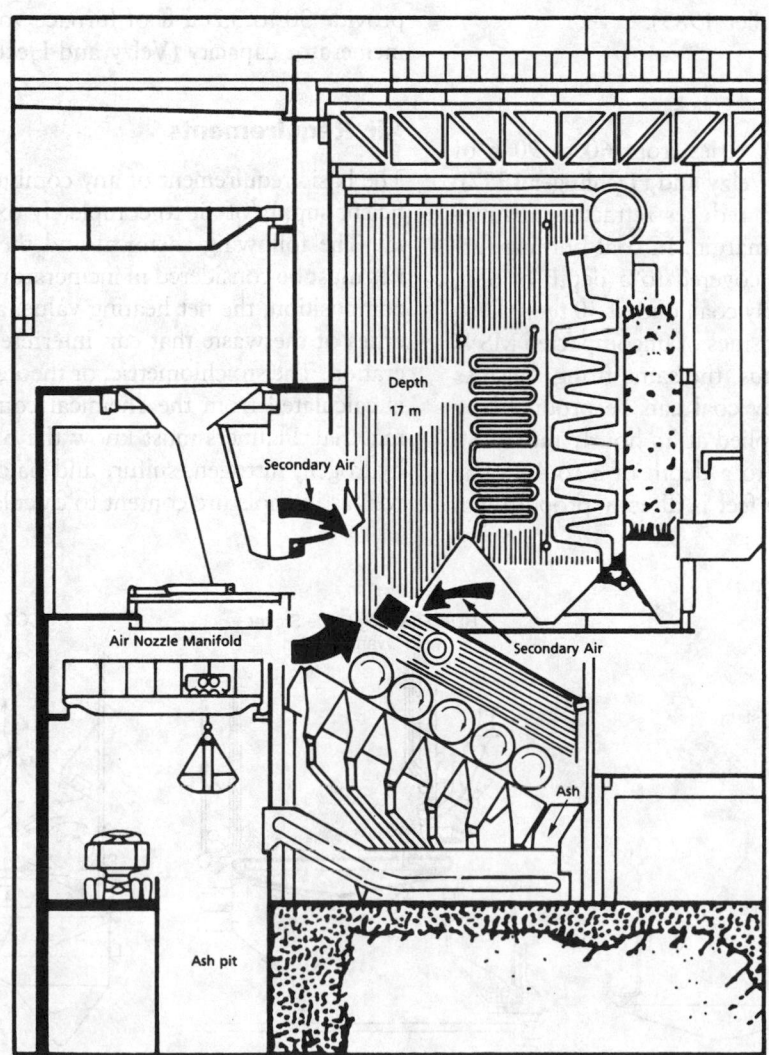

FIG. 10.9.6 VKM or Dusseldorf process (with installation of transverse manifold for 260 secondary air nozzles over roller grate at Wuppertal, Germany).

from 2 to over 30% of the grate area (Velzy 1968). Proponents of the larger openings feel that the siftings (the ash from the fuel bed) should be allowed to fall below the grate as soon as possible and large amounts of air should be permitted to pass through the bed to meet the combustion requirements of varying fuel characteristics. Proponents of the smaller openings cite advantages such as the small volume of siftings, the small amount of underfire air that is required, and the resulting shorter combustion flames, all of which reduce particle entrainment in the escaping gas (Velzy 1968).

In a technique employed at some U.S. facilities, waste is pneumatically injected into the furnace system and burned while suspended in the furnace chamber, rather than being burned on a grate (see Figure 10.9.7). With the removal of ferrous metals and other noncombustibles in typical RDF systems, a boiler system has evolved and has been in commercial operation at Biddeford, Maine, since 1987. The controlled combustion zone (CCZ) boiler design is a state-of-the-art boiler design for both wood and RDF boilers (Gibbs and Kreidler 1989).

Grate Sizing

The hourly burning rate (F_a) varies from 60 to 90 lb of MSW per sq ft of grate area (Velzy and Hechlinger 1987). An hourly rate of 60 lb/sq ft reduces refractory maintenance and provides a safety margin. In coal burning furnaces, the grates are usually covered to a depth of 6 in, which corresponds to an hourly coal load of 30 to 40 lb/sq ft. The heating values and densities of uncompacted MSW are less than half of that. Thus, the same firing densities (on a Btu basis) produced by coal can be produced by MSW when the MSW is supplied at an hourly rate of 60 lb/sq ft and covers the grate to a depth of 3 to 4 ft. The required grate area in square feet is directly proportional to the maximum charging rate F (lb/hr) and inversely proportional to F_a, the grate area A, as follows:

$$A = F/F_a \qquad 10.9(1)$$

The grate design must also be based on the manufacturer's design criteria. Basically, the only consistent design criteria used by manufacturers is the specified kilogram (pounds) of waste that can be loaded per square meter (square foot) of the grate area. Planners need more empirical data for proper design and must develop a more rational approach to select the proper grate.

Furnace Sizing

The firing furnace capacity is a function of its grate area and volume. The furnace volume is usually determined on the basis of an hourly heat release of 20,000 Btu/cu ft. If the hourly release rate is 20,000 Btu/cu ft and the heating value of the MSW is 5000 Btu/lb, the hourly firing rate is 4 lb/cu ft of furnace volume. A typical design basis is to provide 30 to 35 cu ft of furnace volume for each tpd of incinerator capacity (Velzy and Hechlinger 1987).

Air Requirements

The basic requirement of any combustion system is a sufficient supply of air to completely oxidize the feed material. The following chemical and thermodynamic properties must be considered in incinerator design: the elemental composition, the net heating value, and any special properties of the waste that can interfere with incinerator operation. The stoichiometric, or theoretical, air requirement is calculated from the chemical composition of the feed material. Planners must know the percentages of carbon, hydrogen, nitrogen, sulfur, and halogens in the waste as well as its moisture content to calculate the stoichiometric

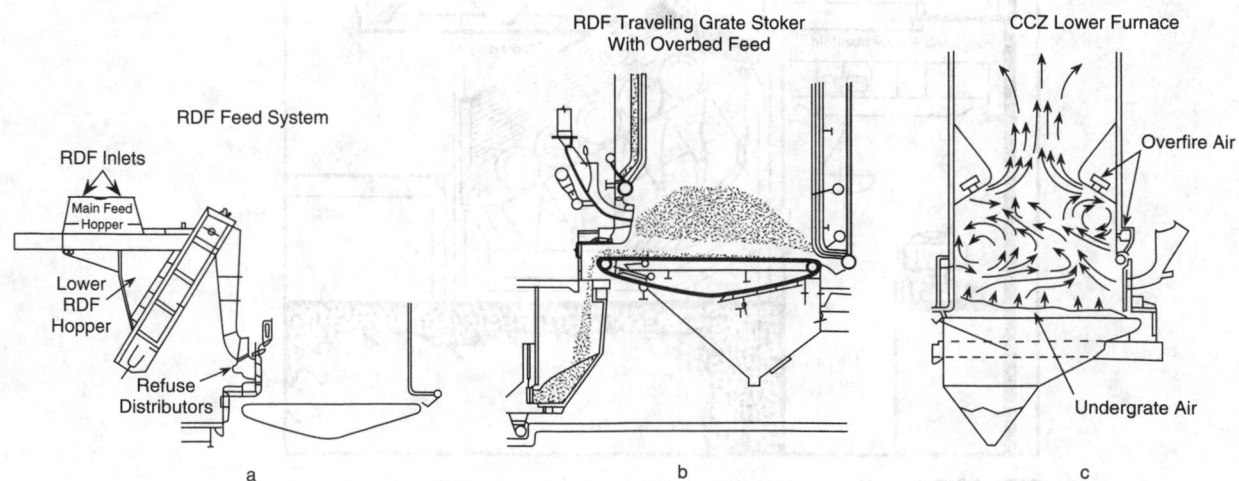

FIG. 10.9.7 RDF furnace. (Reprinted, with permission, from D.R. Gibbs and L.A. Kreidler, 1989, What RDF has evolved into, *Waste Age* [April].)

combustion air requirements and predict combustion air flow and flue gas composition.

Table 10.9.3 shows the stoichiometric oxygen requirements and combustion product yield for each waste component. The stoichiometric air requirement is determined directly from the stoichiometric oxygen requirement with use of the weight fraction of oxygen in air. Given temperature and pressure, the required volume of air can be calculated based on gas laws.

If perfect mixing could be obtained and waste burnout occurred instantaneously, only the stoichiometric requirement of air would be needed. However, neither of these phenomena occurs in real-world applications. Therefore, some excess air is required to ensure adequate waste–air contact. Excess air is usually expressed as a percentage of the stoichiometric air requirement. For example, 50% excess air implies that the total air supply to the incinerator is 50% higher than the stoichiometric requirement.

In general, the minimum excess air requirement for an incinerator depends on the degree of mixing achieved and waste-specific factors. Most incinerators require 80 to 100% excess air to burn all organics in the MSW (Wheless and Selna 1986). Incinerator operation is optimized when sufficient oxygen is provided to achieve complete combustion, but no more. Additional oxygen reduces thermal efficiency and increases nitrogen oxide generation.

The cold air volume required for proper combustion in the incinerator per unit weight of MSW can be calculated as follows (Essenhigh 1974):

$$\text{Total Cold Air Volume (cu ft/lb)} = B (1 - a - M)(S)(1 + e)$$
$$10.9(2)$$

where:

B = the dry and inert-free (DIF) heating value, in Btu/lb of MSW

TABLE 10.9.3 STOICHIOMETRIC OXYGEN REQUIREMENTS AND COMBUSTION PRODUCT YIELDS

Elemental Waste Component	Stoichiometric Oxygen Requirement	Combustion Product Yield
C	2.67 lb/lb C	3.67 lb CO_2/lb C
H_2	8.0 lb/lb H_2	9.0 lb H_2O/lb H_2
O_2	−1.0 lb/lb O_2	—
Cl_2	−0.23 lb/lb Cl_2	1.03 lb HCl/lb Cl_2 −0.25 lb H_2O/lb Cl_2
F_2	−0.42 lb/lb F_2	1.05 lb/HF/lb F_2 −0.47 lb H_2O/lb F_2
Br_2	—	1.0 lb Br_2/lb Br_2
I_2	—	1.0 lb I_2/lb I_2
S	1.0 lb/lb S	2.0 lb SO_2/lb S
P	1.29 lb/lb P	2.29 lb P_2O_2/lb P
Air N_2	—	3.31 lb N_2/lb $(O_2)_{stoich}$
Stoichiometric air requirement		4.31 lb Air/lb $O_{2(stoich)}$

a = the inert and ash fraction of the MSW
M = the moisture fraction of the MSW
S = the cubic feet of stoichiometric cold air required per Btu of heat release
e = the excess air fraction

In cases where metals are not burned and combustibles are predominantly organic, the value of S is approximately 0.01 (i.e., 1 cu ft of cold air per 100 Btu). This approximation is valid, generally to within 10 to 20%, for a wide range of organic fuel. Consequently, variations between wastes depend largely on their noncombustible content (a + M), particularly as the DIF calorific values lie within the narrow range of 8000 to 10,000 Btu/lb. The stoichiometric air requirements of most DIF waste are therefore 80 to 100 cu ft/lb or 6.4 to 8 lb of air per pound of waste. If the waste contains 50% ash and moisture, the calorific values drop to 4000 to 5000 Btu/lb. If this waste is fired at 150% excess, the air requirements are 8 to 10 lb of air per pound of waste as fired.

In modern, mechanical, grate furnace chambers, the underfire and overfire air are usually provided by separate blower systems. Underfire air is admitted to the furnace under the grates and through the fuel bed. It supplies primary air for the combustion process and also cools the grates. Underfire air is usually more than half of the total air (50 to 70%). Particulate emissions from incinerators tend to increase with heat release and underfire air flow, while they tend to decrease with increasing fuel particle size (see Figure 10.9.8).

Overfire air can be introduced at two levels:

Immediately above the fuel bed to promote turbulence and mixing and to complete the combustion of volatile gases driven off the bed of burning solid waste.

From rows of nozzles placed high on the furnace wall. These nozzles allow secondary overfire air to be introduced into the furnace to promote additional turbulence of gases and control temperature. The number, size, and location of the overfire inlet ports determine the amount of turbulence and backmixing in the stirred reaction region above the burning waste. See Figure 10.9.6. For good combustion, the overfire air system must have broad flexibility to accommodate changes in fuel moisture, ash content, and Btu value.

Operators control flue temperature and smoking by modulating the total air flow and the underfire-to-overfire air ratio. For most U.S. grate designs, the required underfire air pressure is about 3 in of water. The overfire air pressure is adjusted so that entrance velocities at the nozzle are high enough to guarantee high turbulence without impinging on the opposite wall and residence times are long enough to assure complete combustion.

Influent air is usually at an ambient temperature, normally 27°C (80°F). It can get as high as 1650°C (2100 to 2500°F) in the immediate proximity of the flame. When the gas leaves the combustion chamber, the temperature

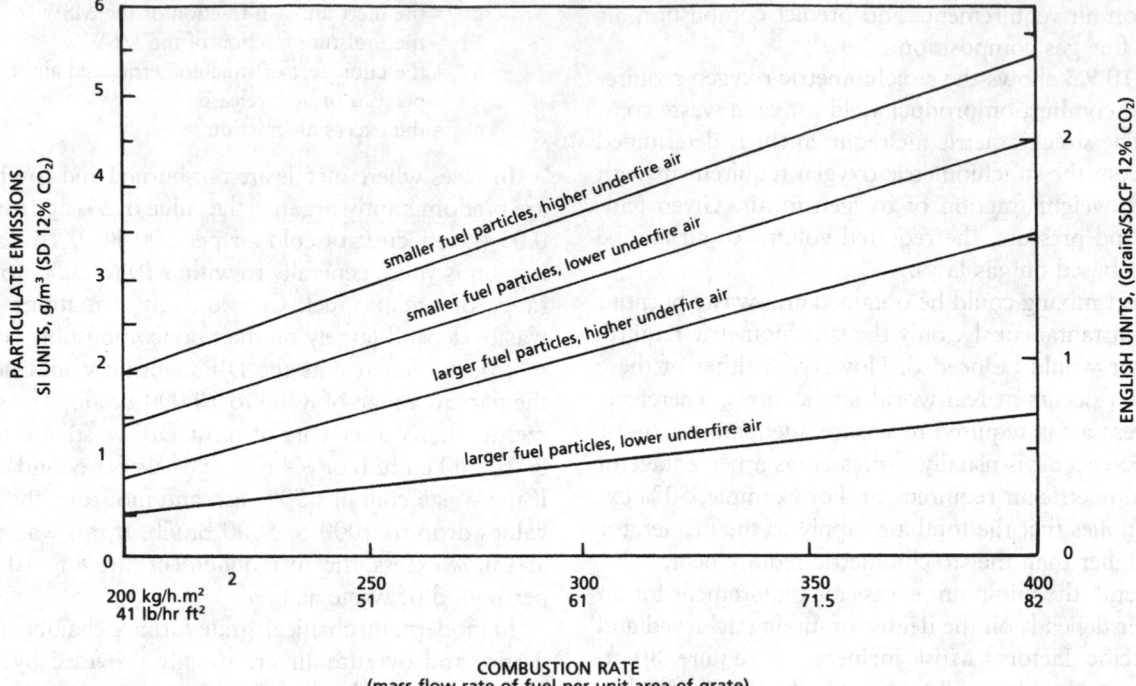

FIG. 10.9.8 Effects of combustion rate, underfire air, and fuel particle size on particulate emissions generated by combustion of wood waste. (Reprinted, with permission, from K.L. Tuttle, 1986, Combustion generated particulate emissions, *National Waste Processing Conference, Denver, 1986* [ASME].)

should be reduced to between 760 and 1000°C (1400 to 1800°F). If air pollution control devices are installed, induced draft fans must be installed, and the temperature should probably not exceed 260 to 370°C (500 to 700°F).

The mathematical modeling of the incinerator presented by Essenhigh (1974) provides a better understanding of the combustion processes taking place in incinerators. Figure 10.9.9 describes the gas-phase (II) and solid-phase (I) zones in a top-charged incinerator (overbed feed).

Calculating Heat Generation

Calculating the amount of heat generated through the incineration of MSW is necessary to determine how much auxiliary fuel is needed for combustion. The moisture content of MSW ranges from 20 to 50% by weight, and the combustible content is 25 to 70% by weight. The heating value of MSW depends on its composition.

Assuming that the average heating value of the combustible is 8500 Btu/lb and the moisture and inert concentration of the MSW is known, environmental engineers can estimate the heat content of MSW using Figure 10.9.10. If the heating value of the combustibles is less than 8500 Btu/lb, the number in Figure 10.9.10 must be multiplied by the ratio of the actual heating value divided by 8500.

Table 10.9.4 gives a material balance of burning 100 lb of MSW. The table assumes the MSW to have a heat content of 5000 Btu/lb, a moisture content of 22.4%, and a noncombustible content of 19% and that it contains 28

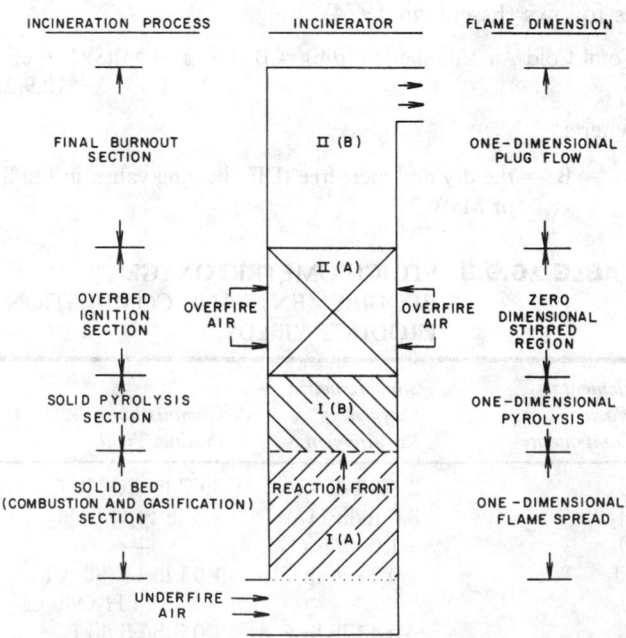

FIG. 10.9.9 Incinerator for continuous overbed feed of waste. Schematic represents solid bed, zone I, with overbed, zone II. For overbed feed, zone I has subzones, including I(A), the combustion and gasification section on the grate, and I(B), the drying and pyrolysis above I(A). Zone II (overbed combustion) has a backmix (stirred) region called subzone II(A), followed by a plug flow burnout region, subzone II(B). With underfeed, zone I is inverted, with drying and pyrolysis below the combustion subzone and the reaction front moving down instead of up as shown.

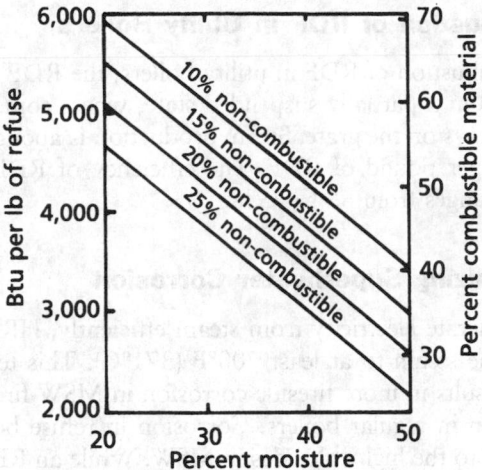

FIG. 10.9.10 Moisture–heat content relation with 8500 Btu/lb combustible material. (Reprinted, with permission, from Velzy and Hechlinger 1987.)

lb of carbon and 0.6 lb of hydrogen. It also assumes that 1–3 lb of combustibles escape unburned, and 140% excess air is needed to cool the refractories. Therefore, the total air required is 2.4 times the stoichiometric requirement, or 8.24 lb of air per pound of MSW.

TABLE 10.9.4 MATERIAL BALANCE FOR FURNACE (IN LB/100 LB OF REFUSE)

Input:		
Refuse		
Combustible material		
Cellulose	52.74	
Oils, fats, etc.	5.86	58.6
Moisture		22.4
Noncombustible		19.0	100.0
Total air, at 140% excess air		
Oxygen	191.0	
Nitrogen	633.0	824.0
Moisture in air		11.0
Residue quench water		5.0
Total		940.0
Output:		
CO$_2$ (28 × 3.667)		102.7
Air—Oxygen (191–80)	111.0	
Nitrogen	633.0	744.0
Moisture		
In refuse	22.4	
From burning cellulose	29.3	
From burning hydrogen	5.4	
In air	11.0	
In residue quench water	5.0	73.1
Noncombustible material		19.0
Unaccounted for		1.2
Total		940.0

Source: C.O. Velzy and R.S. Hechlinger, 1987, Incineration, Section 7.4 in *Mark's standard handbook for mechanical engineers*, 9th ed., edited by T. Baumeister and E.A. Avallone (New York: McGraw-Hill).

An enthalpy-balance calculation for this example shows that the enthalpy of each pound of existing gas is 455 Btu higher than the enthalpy of the 80°F air that enters (Velzy and Hechlinger 1987). Based on this enthalpy rise, the expected flue gas temperature of 1680°F can be read from Figure 10.9.11. This flue gas temperature is low enough to protect the refractory.

HEAT RECOVERY INCINERATORS (HRIs)

Three types of HRI designs are used to burn MSW: mass-burning in refractory-walled furnaces, mass-burning in waterwall furnaces, and combustion of RDF in utility boilers.

Mass-Burning in Refractory-Walled Furnaces

In mass-burning in refractory-walled furnaces, the waste-heat boiler, located downstream of the furnace, receives heat from the flue gases. Older HRIs tend to be refractory-walled designs, and their steam production is usually limited to 1.5 to 1.8 lb of steam per pound of MSW burned, assuming that the heating value of MSW is 4400 Btu/lb. In older furnaces, the larger the furnace, the lower the surface-to-volume ratio, because less surface exists to cool the flame. These units need higher quantities of combustion air to prevent overheating the wall, which results in slagging and deterioration. Approximately 50 to 60% of the heat generated in the combustion process can be recovered from such systems.

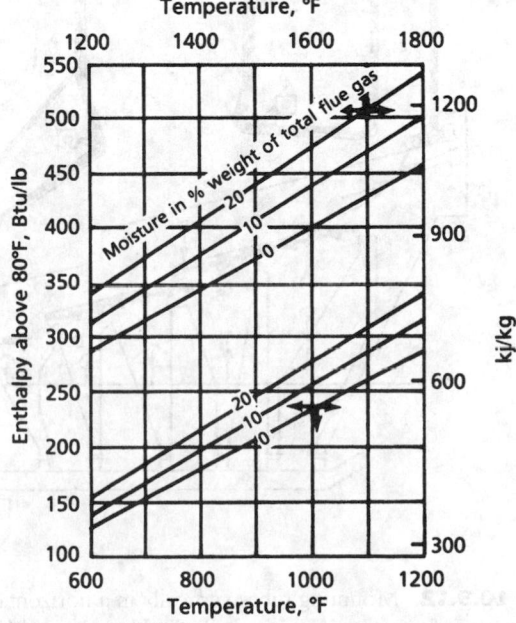

FIG. 10.9.11 Enthalpy of flue gas above 80°F. (Reprinted, with permission, from Velzy and Hechlinger 1987.)

Mass-Burning in Waterwall Furnaces

In mass-burning in waterwall furnaces, most or part of the refractory in the furnace chamber is replaced by waterwalls made of closely spaced steel tubes welded together to form a continuous wall. Water is continuously circulated through these tubes. In newer waterwall designs, the steam production is around 3 lb of steam per pound of MSW. The increase in thermal efficiency is mostly due to a reduction in the excess air (from about 150% for refractory-walled furnaces to about 80% for waterwall furnaces).

Coating a substantial height of the primary combustion chamber, which is subject to higher temperatures and flame impingement, with a thin coat of silicon carbide refractory material and limiting the average gas velocities to under 15 ft/sec (4.5 m/sec) is recommended. Gas velocities entering the boiler convection bank should be less than 30 ft/sec (9.0 m/sec) (Velzy 1986). The efficiency of heat recovery in such units ranges from 65 to 70%.

Combustion of RDF in Utility Boilers

In combustion of RDF in utility boilers, the RDF is often burned in a partially suspended state, where some of the RFD stays on the grate. Steam production is about 3 lb of steam per pound of RDF. The efficiency of RDF boiler units ranges from 65 to 75%.

Minimizing Superheater Corrosion

To generate electricity from steam efficiently, HRIs must heat the steam to at least 700°F (371°C). This temperature results in more fireside corrosion in MSW-fired boilers than in regular boilers. Corrosion in refuse boilers is related to the high chlorides in MSW. While an RDF processing system can remove some of the material containing chlorides, removing chloride-containing material in the RDF processing system is not a realistic means to prevent boiler corrosion. High-nickel-alloy superheater tubes (e.g., Inconel 825) minimize superheater corrosion in addition to protecting the furnace from overloading and providing

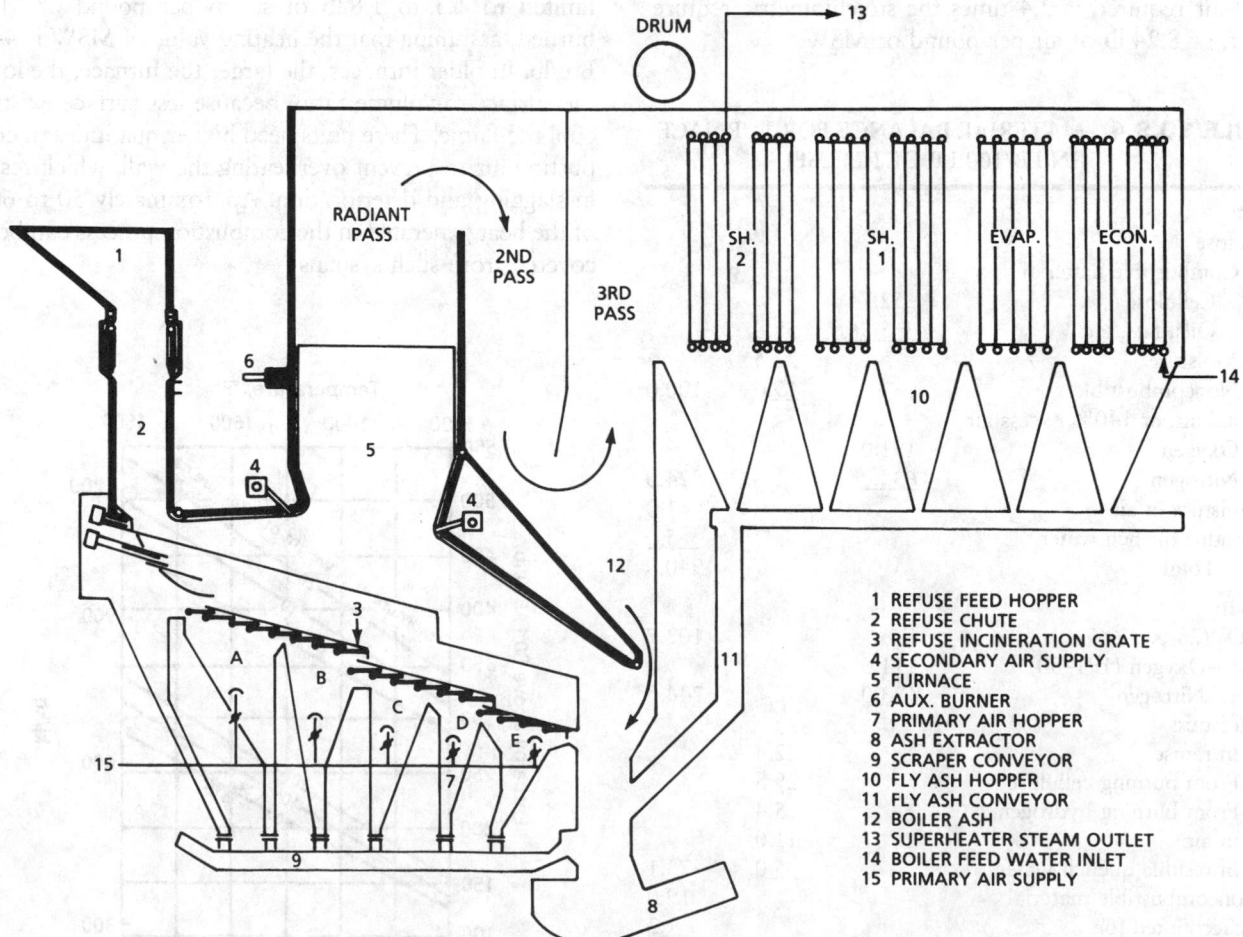

1 REFUSE FEED HOPPER
2 REFUSE CHUTE
3 REFUSE INCINERATION GRATE
4 SECONDARY AIR SUPPLY
5 FURNACE
6 AUX. BURNER
7 PRIMARY AIR HOPPER
8 ASH EXTRACTOR
9 SCRAPER CONVEYOR
10 FLY ASH HOPPER
11 FLY ASH CONVEYOR
12 BOILER ASH
13 SUPERHEATER STEAM OUTLET
14 BOILER FEED WATER INLET
15 PRIMARY AIR SUPPLY

FIG. 10.9.12 Mounting tubes vertically in a horizontal superheater section to prevent particle velocity increases. (Reprinted, with permission, from A.J. Licata, R.W. Herbert, and U. Kaiser, 1988, Design concepts to minimize superheater corrosion in municipal waste combustors, *National Waste Processing Conference, Philadelphia, 1988* [New York: ASME].)

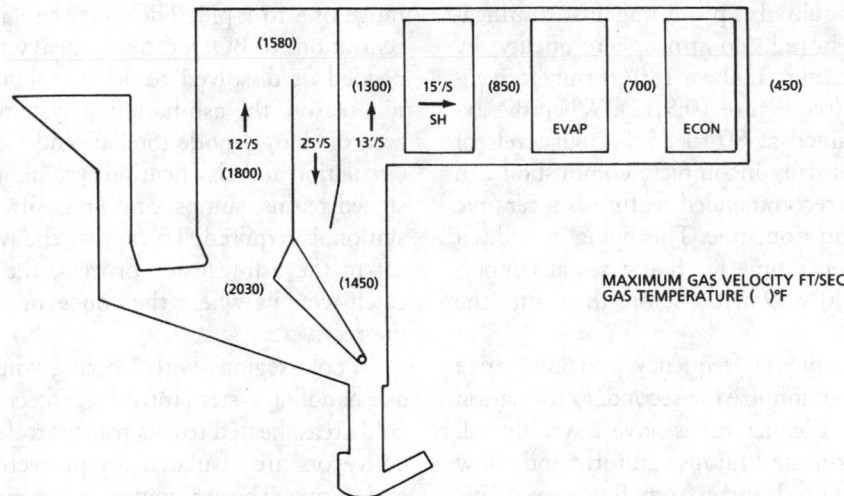

FIG. 10.9.13 Boiler design criteria for corrosion and erosion control. (Reprinted, with permission, from Licata, Herbert, and Kaiser 1988.)

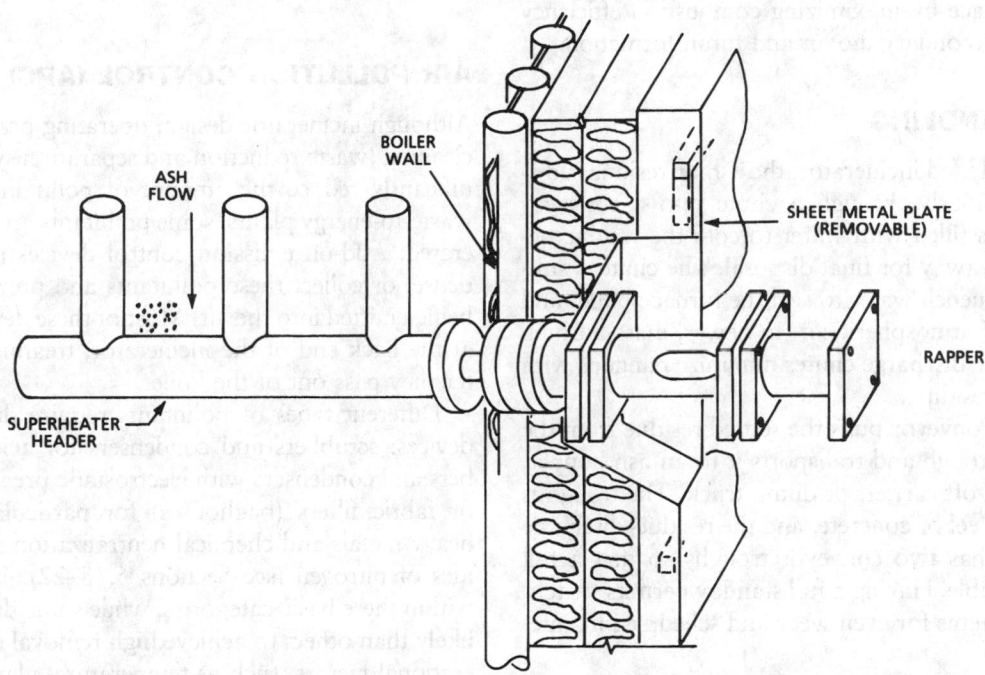

FIG. 10.9.14 Rapper boiler superheater headers. (Reprinted, with permission, from Licata, Herbert, and Kaiser 1988.)

rugged furnace walls. Hydrogen chloride corrosion begins by penetrating a slag layer on the superheater tubes. The tubes must be kept clean by soot blowers or mechanical rapping. Chlorides in hot gases become corrosive and can destroy a superheater.

With improved superheater designs, the operating superheater temperature can be increased from 750 to 825 or 900°F (Licata, Herbert, and Kaiser 1988). This temperature can be achieved when gas velocities are kept between 15 and 18 ft/sec to minimize the erosion caused by the impact of the particles. In addition, tubes should be liberally spaced to mitigate the increase in velocity as ash

buildup occurs. Figures 10.9.12 and 10.9.13 show the recommended superheater design criteria for velocities and temperatures.

Another design improvement is the elimination of the harmful effects of soot-blowing by steam or air which damages the protective oxide film, creates hot spots from nonuniform cleaning, and reentrains ash into the flue gas. Rapping rather than blowing can eliminate these effects (Licata, Herbert, and Kaiser 1988). Figure 10.9.14 shows pneumatically actuated mechanical rappers that allow deposits to slide down the tube surfaces into the hoppers below.

The boiler design should also protect against stratification (which can result in reduced atmosphere quality) by forcing the flue-gas stream to make a 180° turn before entering the superheater (see Figure 10.9.12). When the excess air level is maintained at 80 to 85%, high levels of CO concentration caused by incomplete combustion can be prevented. Another recommended feature is a ceramic lining for the postcombustion zone. This lining provides a 1-sec (minimum) residence time for flue gases at temperatures in excess of 1800°F (980°C) before they enter the superheater section.

An increased soot removal frequency and innovative cleaning techniques can minimize the secondary formation of dioxins and furans. Cleaner tubes have fewer fly ash particles on which dioxins and furans can form and allow more heat to be transferred away from flue gases. This heat transfer further cools the gases below the 450°F (250°C), which is conducive to dioxin and furan formation. Additionally, minimizing the production of precursors in the furnace by maximizing combustion efficiency helps decrease secondary dioxin and furan formation.

RESIDUE HANDLING

In a continuously fed incinerator, the ash or residue is discharged continuously through a chute into a conveyor trough, which is filled with water to cool the residue before it is hauled away for final disposal. The chute is submerged under quench water to seal the furnace outlet and prevent entry of atmospheric air. In newer, mass-burning facilities, full-size discharge chutes minimize hangups with large pieces of residue.

The residue conveyor pulls the settled residue from the bottom of the trough and transports it to an ash hopper, storage bin, roll-off carrier, or dump truck. The trough is constructed of steel or concrete, and the residue–discharge system usually has two conveyor troughs so that a full standby is available. Having a full standby permits switching between systems for even wear and scheduled maintenance.

As the conveyor carries the residue, most of the quench water runs off and returns to the trough. The conveyor should run at velocities not exceeding 5 to 10 ft/min (1.5 to 3 m/min) for good dewatering and minimal wear (Stelian and Greene 1986). The moisture content of the ash is usually 25 to 40% or more by weight. Reducing the water content of the ash minimizes transportation costs and water pollution. By reducing the speed of variable speed conveyors, operators can achieve this reduction by maximizing the residence time of residue on the wet-drag conveyor. Wet-drag conveyors can operate at slopes up to 45°, but some operators prefer lower slopes to protect bulky items from rolling back.

The design of a residue-handling system should minimize the discharge of water pollutants. Ash can be acid or alkaline; therefore, the water pH must be controlled in the range of 6 to 9 pH. The water can also contain high concentrations of BOD, dioxins, heavy metals, and other suspended or dissolved toxic or polluting constituents. For this reason, the ash-handling system must operate in the zero discharge mode (Stelian and Greene 1986). A water circulation and clarification system, including properly designed basins, sumps, and an easily maintained pumping station, is required. To capture the water that might drain off in the ash-transfer process, the system should have catch troughs where the conveyor transfers the ash into the receiver.

In cold regions with freezing winter temperatures, the ash-handling system must be protected against freezing. In cold areas, heated trucks transport the ash, and the fly-ash conveyors are insulated for protection against corrosion and caking. The ash conveyor can unload wet residue into a temporary container or directly into a transport vehicle for removal from the site. In mass-burn systems, directly discharging into dump trucks is best and simplest.

AIR POLLUTION CONTROL (APC)

Although incinerator design, operating practices, and fuel cleaning (waste reduction and separation systems) can significantly reduce the amount of pollutants produced in waste-to-energy plants, some pollutants are inevitably generated. Add-on emission control devices neutralize, condense, or collect these pollutants and prevent them from being emitted into the air. Most of these devices are placed at the back end of the incinerator, treating flue gases after they pass out of the boiler.

Different types of pollutants require different control devices: scrubbers and condensers for acid gases, scrubbers and condensers with electrostatic precipitators (ESPs) or fabric filters (baghouses) for particulates and other heavy metals and chemical neutralization systems for oxides of nitrogen (see Sections 5.18–22). Variations exist within these basic categories; while some devices are more likely than others to achieve high removal efficiencies, operational factors, such as temperature, play a key role.

Acids, Mercury, Dioxin, and Furan Emissions

Scrubbers, followed by an efficient particulate control device, are the state-of-the-art equipment for controlling emissions of acids such as hydrogen chloride, sulfur dioxide, and sulfuric acid. Scrubbers generally use impaction, condensation, and acid–base reactions to capture acid gases in flue gas. Since greater removal efficiencies usually accompany greater condensation, devices that lower gas temperatures and thus increase condensation can enhance scrubber effectiveness. The lower temperatures also allow mercury, dioxins, and furans to condense so that they can subsequently be captured by a particulate device.

Three types of scrubbers are in use: wet scrubbers, spray-dry scrubbers, and dry injection scrubbers. The first two scrubbers are condensers, while dry injection scrubbers require a separate condenser (either a humidifier or a heat exchanger). In all cases, temperature and, for dry and spray-dry scrubbers, the amount of lime (an alkaline substance that neutralizes acids) are the key factors affecting scrubber effectiveness. In general, to maximize emission control, the scrubber should be adequately sized, operate at temperatures below 270°F, and allow flue gas circulation through the scrubber for at least 10–15 sec.

WET SCRUBBERS

Wet scrubbers capture acid gas molecules onto water droplets; sometimes alkaline agents are added in small amounts to aid in the reaction (see Section 5.21). New designs report on removing over 99% of the hydrogen chloride and, in some cases, sulfur dioxide and over 80% of the dioxin, the lead, and mercury (Hershkowitz 1986). The disadvantages include the added cost to treat the wastewater produced, corrosion of the metal parts, and incompatibility with the fabric type of the particulate control device. However, wet scrubbers collect gases as well as particulates, especially sticky ones.

SPRAY-DRY OR SEMI-DRY SCRUBBERS

With these scrubbers, acid gases are captured by impaction of the acid gas molecules onto an alkaline slurry, such as lime. Here, the evaporation water from the scrubbing liquid is carefully controlled so that when the material reaches the bottom of the tower, it is a dry powder (a dry fly ash and lime mixture). This method eliminates the scrubber water that must be treated or disposed; additionally, the power requirements and corrosion potential are reduced. Emission tests have demonstrated control efficiencies of 99% or better for hydrogen chloride and sulfur dioxide removal under optimal conditions (temperatures below 300°F, sufficiently high lime/acid ratios, and sufficiently long gas residence time in the scrubber). Dioxins were also considerably reduced (Hershkowitz 1986).

DRY INJECTION SCRUBBERS

Dry injection scrubbers inject dry powdered lime or another agent that reacts with the acid gases in flue gas. In one research test, removal efficiencies of 99% for hydrogen chloride and 96% for sulfur dioxide were measured under optimal temperature conditions (230°F); dioxins were also considerably reduced (Platt et al. 1988).

TRENDS

A report by the German equivalent to the U.S. EPA predicts a trend toward wet scrubbing because of better elimination of sulfur dioxide, heavy metals, and other toxic substances (McIlvaine 1989). In addition, spray drying and other dry processes have the disadvantage of increased residue production. The report does cite potential problems with wet scrubbers; however, water treatment and heavy metal precipitation and evaporation are promising solutions. Typically, a German APC system uses a packed tower to remove hydrogen chloride, sulfur dioxide, and condensed heavy metals and a high-efficiency ESP to remove dust.

Particulate and Heavy Metal Emissions

The emissions of particulates and heavy metals are best reduced by collecting them in one of two basic types of add-on particulate control devices: fabric filters and ESPs (see Section 5.20). Heavy metals are captured because they are condensed out of flue gas onto the particles. These devices are designed to operate at temperatures lower than 450°F for flue gas leaving the boiler; some operate at temperatures as low as 250°F, which is beneficial for condensing and collecting acids, volatile metals, and organics. The state-of-the-art level for particulate emission is 0.010 g per dry cu ft.

FABRIC FILTERS

Fabric filters (also called baghouses) are a state-of-the-art particulate control technology with a consistent 99% removal efficiency over the range of particulate sizes. Figure 10.9.15 shows a schematic diagram of a scrubber followed by a baghouse for particulate control. Particulates as small as 0.1 microns can be captured. The accumulated particulates or fly ash fall into a hopper when the fabric filters are cleaned, and this ash must be disposed of appropriately. Table 10.9.5 lists the advantages and disadvantages of fabric filter systems.

ESPs

ESPs consist of one or more pairs of electric charge plates or fields. The particulates in flue gases are given an electric charge, forcing them to stick to the oppositely charged plate. ESPs with four or more fields are state-of-the-art. Table 10.9.6 lists the advantages and disadvantages of ESPs.

CYCLONES

A third type of particulate control device is the cyclone, a mechanical device that funnels flue gases into a spiral, creating a centrifugal force that removes large particles. When combined with baghouses and ESPs cyclones improve their efficiency by removing larger particles before they reach these other more efficient devices.

TRENDS

When they are placed after a scrubber, particulate control devices also collect heavy metals and other pollutants that have condensed out of flue gas onto particle surfaces. Placing a scrubber first helps lower the temperature of gases entering a fabric filter. However, wet scrubbers cannot precede fabric filters because the wet particles in flue

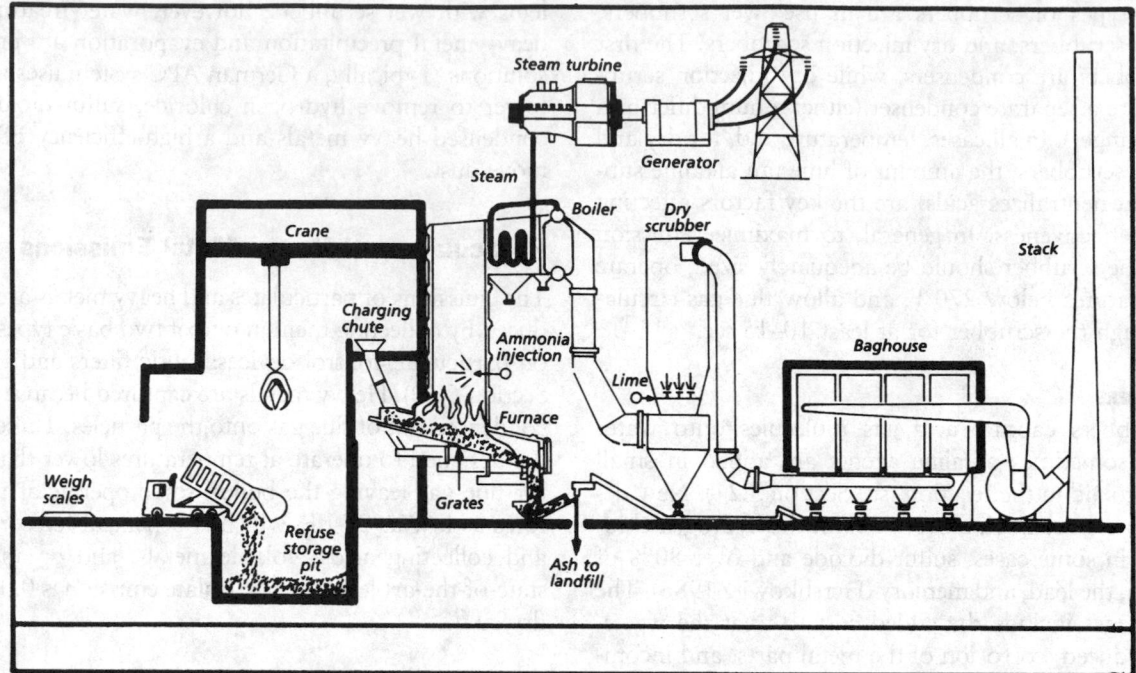

FIG. 10.9.15 Schematic of Commerce waste-to-energy plant in southern California. (Reprinted, with permission, from Commerce Refuse-to-Energy Authority.)

gases clog the filters. Thus, facilities with wet scrubbers place their scrubbers after the particulate control device.

The smallest particles are the most potentially damaging when inhaled into the lungs, and dioxins, furans, acid gases, and heavy metals are adsorbed in the largest quantities on these smaller particles. Thus, a state-of-the-art particulate control device should achieve even lower emission levels for particulates below 2 microns in diameter.

Since many heavy metals condense at temperatures of 450°F (230°C), both ESPs and fabric filters collect heavy

TABLE 10.9.5 ADVANTAGES AND DISADVANTAGES OF FABRIC FILTER SYSTEMS

Advantages:

High particulate (coarse to submicron) collection efficiencies

Dry collection and solids disposal

Relatively insensitive to gas stream fluctuations. Efficiency and pressure drop are unaffected by large changes in inlet dust loading for continually cleaned filters

Corrosion and rusting of components usually not a problem

No hazard of high voltage, simplifying maintenance and repair and permitting the collection of flammable dust

Use of selected fibrous or granular filter aids (precoating) which permits the high-efficiency collection of submicron smokes and gaseous contaminants

Filter collectors available in a number of configurations, resulting in a range of dimensions and inlet and outlet flange locations to suit a range of installation requirements

Simple operation

Disadvantages:

Special refractory mineral or metallic fabrics (that are still in the developmental stages and can be expensive) required for temperatures in excess of 550°F

Fabric treatments to reduce dust seeping or to assist in the removal of the collected dust required for certain particulates

A fire or explosion hazard due to concentrations of some dusts in the collector ($\approx$ 50 g/cu m) when a spark or flame is accidently admitted. Fabrics can burn if readily oxidizable dust is being collected.

High maintenance requirements (bag replacements, etc.)

Fabric life shortened at elevated temperatures and in the presence of acid or alkaline particulate or gas components

Crusty caking or plugging of the fabric caused by hydroscopic materials, condensation of moisture (or tarry), and adhesive components which may require special additives

Respiratory protection for maintenance personnel required in replacing the fabric

Medium pressure-drop requirements, typically in the range of 4 to 10 in of water

TABLE 10.9.6 ADVANTAGES AND DISADVANTAGES OF ESPs

Advantages:
  High particulate (coarse and fine) collection efficiencies with a relatively low expenditure of energy
  Dry collection and solids disposal
  Low pressure drop (typically less than 0.5 in of water)
  Designed for continuous operation with minimum maintenance requirements
  Low operating costs
  Capable of operation under high pressure (to 150 psi) or vacuum conditions
  Capable of operation at high temperatures (to 1300°F)
  Capable of handling large gas flow rates effectively

Disadvantages:
  High capital costs
  Sensitive to fluctuations in gas stream conditions (flow, temperature, particulate and gas composition, and particulate loading)
  Difficulty in collecting certain particulates due to extremely high or low resistivity characteristics
  Relatively large space requirements for installation
  Explosion hazard when treating combustible gases and collecting combustible particulates
  Special precautions required to safeguard personnel from high voltage equipment
  Ozone produced by the negatively charged discharge electrodes during gas ionization
  Sophisticated maintenance personnel required

metals that condense onto particulate matter. Effective mercury emission control technology, while evolving, has not been implemented in MSW incinerators. A volatile metal, mercury vaporizes under the high temperatures of combustion although recent research suggests that mercury can also be present as mercuric chloride, mercuric oxides, and mercury solids. Whereas most vaporized metals return to a solid state when combustion gases cool, mercury remains in the vapor state. Wet scrubbing, activated carbon and sodium sulfide technologies show promising results.

Mercury control requires that the vapor be adsorbed onto particulates or absorbed into a liquid which is evaporated to leave the solids. The mercury-laden solids are collected in traditional collection devices. Some technologies, used in conjunction with other pollution control systems, can simultaneously remove dioxins, furans, mercury, and other metals as well as acid gases and particulates (Seigies and Trichon 1993).

Nitrogen Oxide Emissions

State-of-the-art control of nitrogen oxides requires both minimizing the formation of nitrogen oxides in the furnace and transforming them into nitrogen and water. Strategies for minimizing formation include using appropriate furnace designs (such as flue gas recirculation and dual-chambered furnaces) and operating practices (such as optimal temperatures and amount of excess air). See Section 5.22. Techniques for destroying nitrogen oxides involve injecting chemicals that neutralize them.

Chemical injection devices use ammonia, urea, or other compounds to react with nitrogen oxides to form nitrogen and water. The technologies for neutralizing and re-

moving nitrogen oxides from flue gases are called selective noncatalytic reduction (SNCR) and selective catalytic reduction (SCR). (See Section 5.23.) Both technologies have been successfully demonstrated on MSW incinerators. Wet scrubbing and condensation also have the capacity to control nitrogen oxides.

Emission Control Devices

The arrangement of emission control devices other than the devices for nitrogen oxides is usually standard: a scrubber and condenser, followed by a particulate collector, followed by an induction fan that sucks flue gases up to the stack. Two reasons for this arrangement are:

Fabric filters cannot operate at the high temperatures at which gases exit the boiler without risk of fire. Thus, placing the scrubber between the boiler and the fabric filter or ESP permits cooling and often humidification that prevent fire. Cooling the gases also plays a role in reducing acid gas, mercury, and dioxin emissions.

Dioxins and heavy metals are trapped more effectively by particulate control devices when they are first condensed out of the flue gas and adsorbed onto the surface of particulate matter, as happens in a scrubber–condenser system.

An alternate arrangement, common in European plants, involves an ESP followed by a wet scrubber. The ESP is not damaged by high temperatures, and the wet scrubber cools and condenses gases and captures particulates.

The location of control devices for nitrogen oxides depends on the type of technology used. These devices can be in the furnaces or the boiler as well as at the back end of the plant.

Ash Management

The first priority in state-of-the-art ash management is to reduce both the volume and toxicity of the residue left after burning MSW. Removing noncombustibles and material containing toxic substances from the MSW before incineration followed by efficient combustion accomplishes this reduction. The amount of toxic material in ash has been increasing as more effective air pollution control devices capture more pollutants in the fly ash.

State-of-the-art ash management practices are designed to minimize worker and citizen exposure to potentially toxic substances in ash during handling, treatment, and storage, long-term storage, or reuse. Safe ash management has several components:

The bottom ash or residue (noncombustible and partially burned solids left in the incinerator) and fly ash (material captured by emission control devices) is kept separate for rigorous handling of the potentially more toxic fly ash.

The ash is contained while still in the plant. A closed system of conveyors is preferable to handling ash in the open.

The ash is transported wet in leakproof, covered trucks to disposal sites.

The ash is treated to minimize its potential toxic impact.

The ash is disposed in ash-only monofills because codisposal of ash with MSW increases the leachability of the ash when it is exposed to acid.

Fly ash from APC is fine-grained, not unlike soot from fireplaces. For every ton of MSW burned, approximately $\frac{1}{4}$ tn becomes some form of ash. Fly ash accounts for about 10 to 15% of the total ash residue; the remaining 85 to 90% is bottom ash.

Operational data from resource recovery incineration facilities throughout the world indicate certain heavy metals, such as lead and cadmium, tend to concentrate in the fly ash, scrubber residue, and small particles (less than $\frac{3}{8}$ in) in the bottom ash. Heavy metals, including lead, cadmium, and total soluble salts (including chlorides and sulfates), are potentially leachable components which can impact the environment. Leachable components are those chemical species which dissolve in water and are transported with water. The toxicity characteristic leaching procedure (TCLP) and numerous other techniques exist to estimate the potential environmental impact resulting from ash generation, handling, and disposal.

Two main categories of ash treatment, both recently developed and being improved, are fixation or cementation and vitrification. Both techniques minimize the environmental impact of ash and enable its reuse in situations such as cinderblocks, reefs, and roads. A few incinerators have onsite vitrification facilities.

Another new treatment technology involves washing the toxic materials out of the ash with hot water and then treating the water to remove soluble toxic materials. The system has been used in Europe, particularly in incinerators with wet scrubbers (Clark, Kadt, and Saphire 1991).

Instrumentation

Continuous process monitors (CPMs) and *continuous emission monitors (CEMs)* track the performance of incinerators so that when combustion upsets or high emissions of one or more pollutants occur, timely corrective measures can be implemented (see Section 5.15). These monitors are usually connected to alarms that warn plant operators of any combustion, emission, or other operating condition that requires attention. Table 10.9.7 lists the typical instrumentation on continuous-feed incinerators for closed-loop control of the temperature and draft controllers (Shah 1974).

State-of-the-art CPMs and CEMs measure nine operating and emission factors: furnace and flue gas temperatures, steam pressure and flow, oxygen, carbon monoxide, sulfur dioxide, nitrogen oxides, and opacity (a crude measure of particulates). Continuous monitoring of hydrogen chloride is possible and may soon be a state-of-the-art requirement. By monitoring parameters that indicate combustion efficiency (carbon monoxide, oxygen, and furnace temperature), plant operators also obtain indications of levels of incomplete combustion. Operators must perform frequent maintenance, including periodic calibration, on continuous monitors to ensure their accuracy.

TABLE 10.9.7   INSTRUMENT LIST FOR CONTINUOUS-FEED INCINERATOR

Temperature Recorders

Furnace temperature at furnace sidewall near outlet, range 38 to 1250°C

Stoker compartment temperature, range 38 to 1250°C

Dust collector inlet temperature, range 38 to 500°C

Temperature Controllers

Furnace outlet temperature controlled by regulating total air from forced draft fan; set point in 800 to 1000°C range

Dust collector inlet temperature controlled by regulating water spray into flue gas; set point in 300 to 400°C range

Draft Gauges

Forced-draft-fan outlet duct

Induced-draft-fan inlet duct

Furnace outlet

Stoker compartments

Differential gauge across dust collector

Draft Controller

Furnace draft control by regulating damper opening

Oxygen Analyzer

Furnace outlet

Adapted from *Municipal Waste Disposal in the 1990s* by Béla G. Lipták (Chilton, 1991).

References

Clark, M.J., M. Kadt, and D. Saphire. 1991. *Burning garbage in the US* Edited by Sibyl R. Golden. New York: INFORM, Inc.

Eberhardt, H. 1966. European practices in refuse and sewage sludge disposal by incineration. *ASME National Incinerator Conference, New York, 1966.*

Essenhigh, R.H. 1974. Incinerators—the incineration process. Vol. 2 in *Environmental engineers' handbook,* edited by B.G. Lipták. Radnor, Pa.: Chilton Book Company.

Gibbs, D.R. and L.A. Kreidler. 1989. What RDF has evolved into. *Waste Age* (April).

Hershkowitz. 1986. Garbage burning: Lessons from Europe: Consensus and controversy in four European states. New York: INFORM, Inc.

Licata, A.J. 1986. Design for good combustion. *24 January, 1986, Municipal Solid Waste Forum, Marine Sciences Research Center, State University of New York, 1986.*

Licata, A.J., R.W. Herbert, and U. Kaiser. 1988. Design concepts to minimize superheater corrosion in municipal waste combustors. *National Waste Processing Conference, Philadelphia, 1988.* New York: ASME.

McIlvaine, R.W. 1989. Incineration and APC trends in Europe. *Waste Age* (January).

Platt, Brenda et al. 1988. *Garbage in Europe technologies, economics, and trends.* Institute for Local Self Reliance (May).

Seigies, J. and M. Trichon. 1993. Waste to burn. *Pollution Engineering* (15 February).

Shah, I.S. 1974. Scrubbers. Sec. 5.12–5.21 in *Environmental engineers handbook,* edited by B.G. Lipták. Radnor, Pa.: Chilton Book Company.

Sommer, Jr., E.J. and G. Kenny. 1984. Effects of materials recovery on waste-to-energy conversion at Gallatin, Tennessee mass fired facility. *Proc. Waste Processing Conf.* New York: ASME.

Stelian, J. and H.L. Greene. 1986. Operating experience and performance of two ash handling systems. *National Waste Processing Conference, Denver, 1986.* ASME.

U.S. Environmental Protection Agency. 1991. Burning of hazardous waste in boilers and industrial furnaces, final ruling. *Federal Register* 56, no. 35 (21 February): 7134–7240.

Velzy, C.O. 1968. The enigma of incinerator design. *ASME Winter Annual Meeting, New York, 1968.*

Velzy, C.O. and R.S. Hechlinger. 1987. Incineration. Sec. 7.4 in *Mark's standard handbook for mechanical engineers,* 9th ed. Edited by T. Baumeister and E.A. Avallone. New York: McGraw-Hill.

Wheless, E. and M. Selna. 1986. Commerce refuse-to-energy facility: An alternative to landfilling. *National Waste Processing Conference, Denver, 1986.* ASME.

10.10 SEWAGE SLUDGE INCINERATION

Sewage sludge, the stabilized and digested solid waste product of the wastewater treatment process, can be disposed of by landfilling, incineration, composting, or ocean dumping. Nature returns organic material to the soil as fertilizer. Organic material becomes waste when it is not returned to the soil but instead is burned, buried, or dumped in the ocean. These unhealthy practices began when chemical fertilizers took the market away from sludge-based compost and when industrial waste began to contaminate sewage sludge with toxic metals (lead and cadmium), making it unusable for agricultural purposes. Until recently, the bulk of the sewage sludge generated by metropolitan areas has been either landfilled or dumped in the ocean. These options are gradually disappearing and as a result municipalities will have to make some hard decisions. (See Sections 7.31 to 7.56).

Sludge Incineration Economics

Incinerating sewage sludge has been practiced for the last sixty years. Early designs were either flash-drying or multiple-hearth types, while in recent years fluidized-bed incinerators have also been used. The flash-drying process has a low capital cost and is flexible in that it can produce the amount of dried sludge that markets need; the remainder can be incinerated. Its limitations are the added cost of pay fuel and the associated odor and pollution problems. The multiple-hearth design, the most widely used for sludge incineration, reduces odor and pollution but provides less operating flexibility because it cannot dry the sludge without incinerating it. The most recent and advanced design is the fluidized-bed sludge incinerator, which can operate in either the combustion or pyrolysis mode. The exhaust temperature from a fluidized-bed incinerator is higher than from a multiple-hearth furnace so afterburners are less likely to be needed to control odor.

The auxiliary fuel cost of sludge incineration is higher with fluidized-bed incinerators than with multiple hearths. The cost varies according to the moisture content of the sludge and the degree of heat recovery (Sebastian 1974a). Eliminating the need for auxiliary fuel requires that the dry–solid content exceed 25% for multiple-hearth and 32% for fluidized-bed incinerators (Sebastian 1974a). In some fluidized-bed installations in Japan, operating costs have been cut in half through heat recovery (Henmi, Okazawa, and Sota 1986).

In a multiple-hearth incinerator with a feed containing 10% solids, the ash is about 10% of the feed. Table 10.10.1 gives the composition of incinerator ashes. The ash is either landfilled or marketed as a soil conditioner. Table 10.10.2 gives the composition of Vitalin, the ash from Tokyo's Odai plant. (The Japanese word lin means

TABLE 10.10.1 TYPICAL ANALYSIS OF ASH FROM TERTIARY QUALITY-ADVANCED WASTE TREATMENT SYSTEMS

Content	Percent of Total Sample			
	Lake Tahoe 11/19/69	Lake Tahoe 11/25/69	Minn.–St. Paul 9/30/69	Cleveland 3/2/70
Silica (SiO_2)	23.85	23.72	24.87	28.85
Alumina (Al_2O_3)	16.34	22.10	13.48	10.20
Iron oxide (Fe_2O_3)	3.44	2.65	10.81	14.37
Magnesium oxide (MgO)	2.12	2.17	2.61	2.13
Total calcium oxide (CaO)	29.76	24.47	33.35	27.37
Available (free) calcium oxide (CaO)	1.16	1.37	1.06	0.29
Sodium (Na)	0.73	0.35	0.26	0.18
Potassium (K)	0.14	0.11	0.12	0.25
Boron (B)	0.02	0.02	0.006	0.01
Phosphorus pentoxide (P_2O_5)	6.87	15.35	9.88	9.22
Sulfate ion (SO_4)	2.79	2.84	2.71	5.04
Loss on ignition	2.59	2.24	1.62	1.94

phosphorus). The term soil conditioner is used instead of fertilizer because the phosphate content is less than 12%, the nitrogen content is under 6%, and the total NPK content is less than 20%.

Incineration Processes

A description of the flash-dryer, multiple-hearth, fluidized-bed and fluidized-bed with heat recovery incineration processes follows.

FLASH-DRYER INCINERATION

The flash-dryer incinerator process was first introduced in the 1930s as a low-capital-cost, space-saving alternative to air drying sludge on sand beds. This method of drying is advantageous because the resulting heat-dried sludge is virtually free of pathogens and weed seeds and the process is flexible enough to produce only the amount of dried sludge that could be marketed. The disadvantages of this process are dust and odor. These problems, while manageable through the use of dust collectors and afterburners, make the flash-dryer less popular than multiple-hearth and fluidized-bed incinerators.

TABLE 10.10.2 COMPOSITION OF MULTIPLE-HEARTH INCINERATION ASH FROM THE ODAI PLANT IN TOKYO

Silica oxide	30.00	Potassium	1.00
Magnesium oxide	3.30	Nitrogen	0.20
Calcium oxide	30.00	Manganese	0.06
Phosphoric oxide	6.20	Copper	0.61
Ferric oxide	18.20	Boron	200.00 ppm

Note: Ash is marketed under the trade name Vitalin.

In the flash-drying process (see Figure 10.10.1), the wet, dewatered sludge is mixed with dry sludge from the dryer cyclone. This preconditioned mixture contacts a gas stream of 1000 to 1200°F, which moves it at a velocity of several thousand feet per minute. In this turbulent, high-temperature zone, the moisture content of the sludge is reduced to 10% or less in only a few seconds. As the mixture enters the dryer cyclone, the hot gases are separated from the fine, fluffy heat-dried sludge. Depending on the mode of operation, the flash-dried sludge is either sent to the sludge burner and incinerated at 1400°F, or it is sent to the fertilizer cyclone and recovered as a saleable fertilizer product. When the incoming wet sludge contains about 18% solids, about 6500 Btu ($\frac{1}{2}$ lb of coal plus 1 cu ft of natural gas) are required to produce 1 lb of dry sludge (15,000 kJ/kg) (Shell 1979).

MULTIPLE-HEARTH INCINERATION

Multiple-hearth incineration was developed in 1889 and was first applied to sludge incineration in the 1930s. It is the most widely used method of sludge incineration (Sebastian 1974b). The multiple-hearth furnace consists of a steel shell lined with a refractory (see Figure 10.10.2). Horizontal brick arches separate the interior into compartments. The sludge is fed through the roof by a screw feeder or a belt feeder and flapgate at a rate of about 7 to 12 lb per sq ft. Rotating rabble arms push the sludge across the hearth to drop holes, where it falls to the next hearth. As the sludge travels downward through the furnace, it turns into a phosphate-laden ash (see Tables 10.10.1 and 10.10.2).

The sludge is dried in the upper, or first, operating zone of the incinerator. In the second zone, it is incinerated at a temperature of 1400 to 1800°F (760 to 982°C) and deodorized. In the third zone, the ash is cooled by the in-

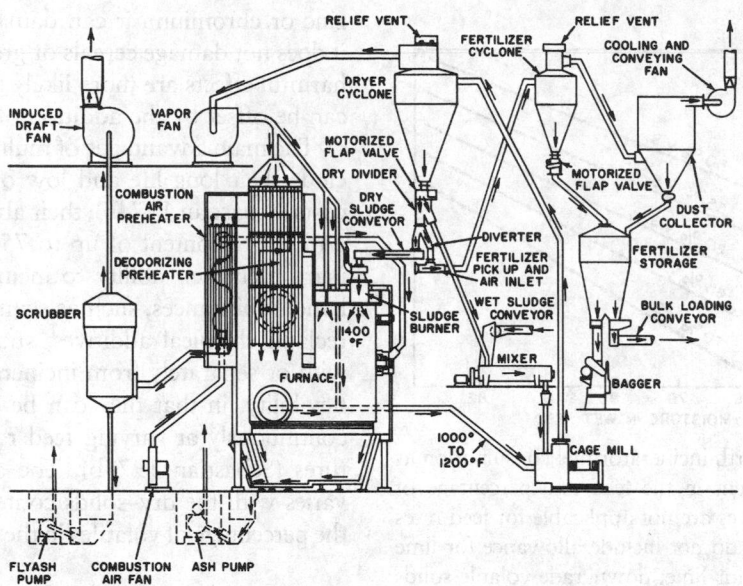

FIG. 10.10.1 Sludge drying and incineration using a deodorized flash-drying process.

coming combustion air. The air, which travels in counterflow with the sludge, is first preheated by the ash, then participates in the combustion, and finally sweeps over the cold incoming sludge drying it until the moisture content is about 48%. At this percentage of moisture content, a phenomenon called *thermal jump* occurs as the sludge enters the combustion zone. The thermal jump allows the sludge to bypass the temperature zone where the odor is distilled. The exhaust gases are 500 to 1100°F (260 to 593°C) and are usually odor-free. The sludge temperature profile across the furnace is shown in Table 10.10.3.

The pollution control equipment usually includes three-stage impingement-type scrubbers for particulate and sulfur dioxide removal and standby after-burners, which de-

stroy malodorous substances such as butyric and caproic acids. The need for afterburners is a function of the exhaust gas temperature. Usually at temperatures above 700 to 800°F (371 to 427°C) in a well-controlled incinerator where the combustion process is complete, afterburners are not necessary for odor-free operation. If combustion is not complete, however, the exhaust gas temperature might have to rise to 1350°F (732°C) before the odor is distilled. In such cases, installing an afterburner is less expensive than using auxiliary fuel to achieve such high exhaust temperatures.

If the incoming sludge contains 75% moisture and if 70% of the sludge solids are volatile, the incineration process produces about 10% ash. The ash can be used as a soil conditioner and as the raw material for bricks, concrete blocks, and road fills, or it can be landfilled. In the United States the supply of phosphates is sufficient for less than a century (Sebastian 1974b), so the phosphate content of sludge ash is important. If the ash also contains

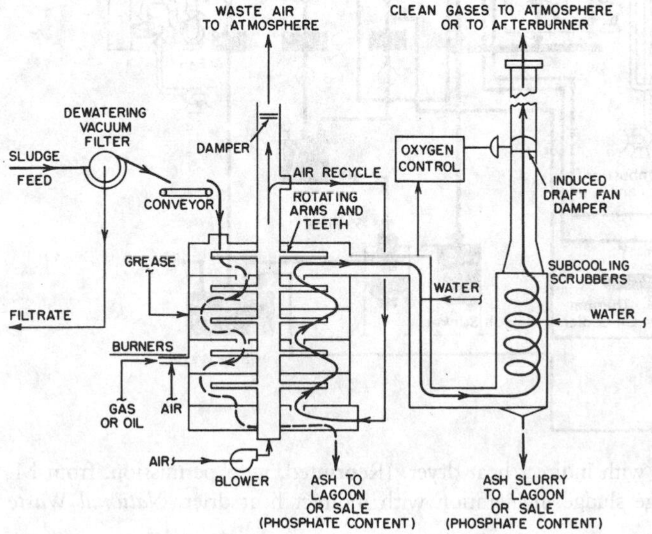

TABLE 10.10.3 SLUDGE FURNACE TEMPERATURE PROFILE

Hearth No.	Approximately at Half Capacity (°F)	Nominal Design Capacity (°F)
1	670	800
2	1380	1200
3	1560	1650
4	1450	1450
5	1200	1200
6	325	300

FIG. 10.10.2 Multiple-hearth incineration of sludge.

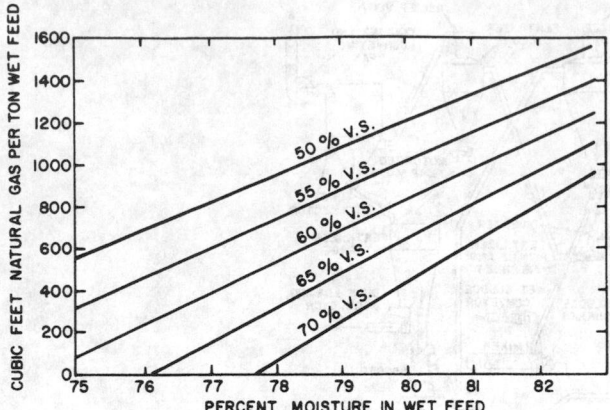

FIG. 10.10.3 Multiple-hearth incinerator fuel consumption as a function of moisture content in the feed and percentage of volatile solids. *Notes.* 1. Curves are not applicable for feed rates below 4 tn per hr. 2. Curves do not include allowance for lime as a filter aid. 3. To correct for lime, downgrade volatile solids according to lime dosage. Assuming lime forms calcium hydroxide, each pound of CaO forms 1.32 lb calcium hydroxide. 4. Natural gas calorific value is assumed to be 100 Btu per cu ft. 5. Heat content of sludge is based on 10,000 Btu per lb of volatile solids. 6. %V.S. represents percentage of volatile solids in the feed.

zinc or chromium, it can damage certain crops although it does not damage cereals or grass (Sebastian 1974b). The harmful effects are more likely to occur in acidic soils and can be offset by the addition of lime to the sludge.

The main advantages of multiple-hearth incinerators include their long life and low operating and maintenance costs (Sebastian 1974b); their ability to handle sludges with a moisture content of up to 75% without requiring auxiliary fuel; their ability to incinerate or pyrolize hard-to-handle substances, such as scum or grease; their ability to reclaim chemical additives, such as lime in combination with or separately from incinerating the sludge; and their flexibility, in that they can be operated intermittently or continuously at varying feed rates and exit-gas temperatures (Sebastian 1974b). The auxiliary fuel requirement varies with the dry–solids content of the sludge and with the percentage of volatiles in the solids (see Figure 10.10.3).

FLUIDIZED-BED INCINERATION

Fluidized-bed incineration can handle sewage sludge containing as much as 35% solids. The sludge is injected into a fluidized bed of heated sand. The incinerator is a vertical cylinder with an air distribution plate near the bottom, which allows the air to enter the sand bed while also sup-

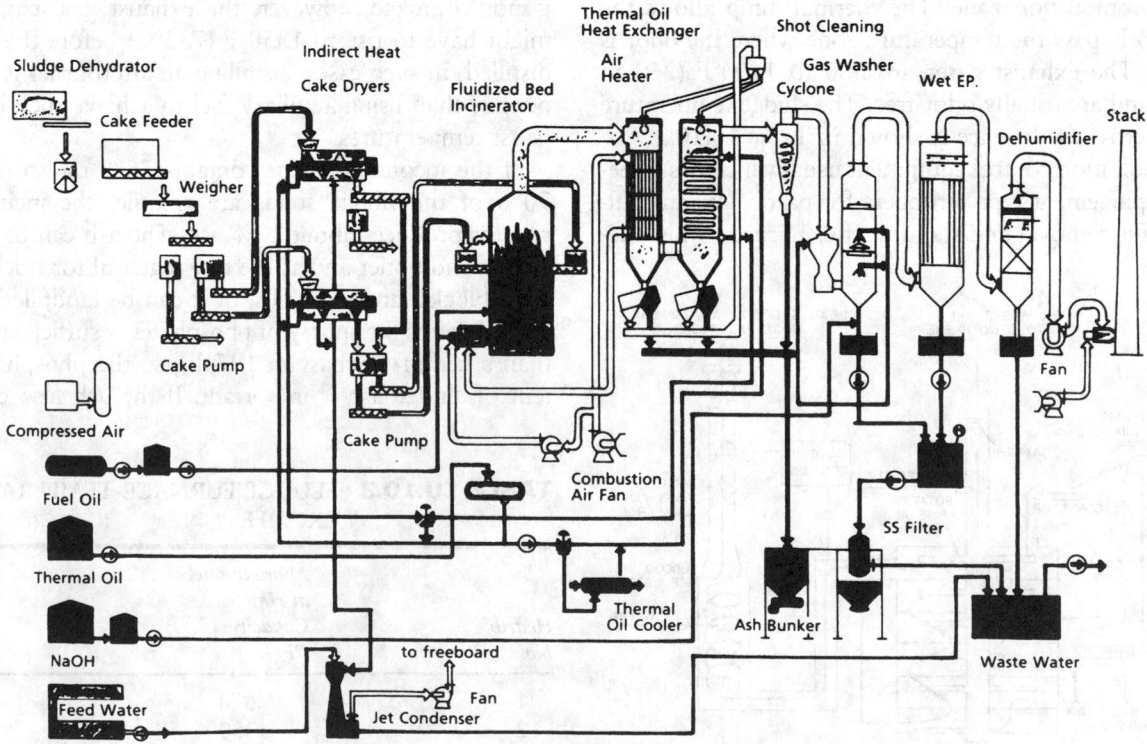

FIG. 10.10.4 Flow diagram of sewage sludge incineration plant with indirect heat dryer. (Reprinted, with permission, from M. Henmi, K. Okazawa, and K. Sota, 1986, Energy saving in sewage sludge incineration with indirect heat drier, *National Waste Processing Conference, Denver, 1986* [ASMER].)

TABLE 10.10.4   OPERATIONAL DATA OF FLUIDIZED-BED SLUDGE INCINERATOR WITH AND WITHOUT INDIRECT HEATING

	Direct Incineration (Without Drying)			Incineration with Dried Cake				
	Design	Run 1	Run 2	Design	Run 3	Run 4	Run 5	Run 6
Input Cake (tpd)	80	60	80	96	60	80	96	70
Cake Property								
Moisture (%)	78	83–86 (84.7)	← ←	75–80 (78)	83–86 (84.8)	←	←	78–80 (79)
Combustibles (%ds)	56–64 (60)	80.2–80.7 (80.5)	← ←	56–64 (60)	80.1–80.2 (80.2)	79.1–79.4 (79.3)	79.3–80.4 (84.7)	(80)
Lower heating value (MJ/kg ds)	12.6	18.5–18.9 (18.7)	←	12.6	18.7	18.8	18.5	18.8
Dried Cake								
Moisture (%)	same	same	same	66–73 (70)	80.1–80.9 (80.6)	79.8–80.6 (80.2)	79.2–80.2 (79.7)	70–72 (71)
Weight (tpd)	80	60	80	70	46.7	61.8	72.4	43
Supporting Fuel								
(l/h)	312	238	265	47	92	94	96	0
(l/tn cake)	94	95	80	12	37	28	24	0
Furnace Temperature								
Sand bed (°C)	800	790	780–820 (800)	800	770–810 (780)	770–810 (780)	760–810 (780)	(780)
Outlet (°C)	850	860	850–950 (900)	800	760–800 (780)	760–800 (780)	760–810 (780)	(780)
Fluidizing air (cu m Normal/h)	7500	6400	7500	5500	5500	5500	5500	5500
Excess air ratio	1.40	1.62	1.52	1.56	2.02	1.61	1.41	(1.8)

Source: M. Hemmi, K. Okazawa, and K. Sota, 1986, Energy saving in sewage sludge incineration with indirect heat drier, *National Waste Processing Conference, Denver,* 1986 (ASMER).

Note: Values in parentheses are average values. All units are in metric (SI).

porting it. As the air flow increases, the bed expands and becomes fluidized. The solid waste in the sludge can be destructed by either combustion or pyrolysis.

During combustion the organic material is turned into carbon dioxide:

$$\text{Sewage Sludge} + \text{Oxygen} \longrightarrow CO_2 + H_2O + \text{Ash} + \text{Heat} \quad 10.10(1)$$

The heat of combustion helps maintain the fluidized bed at a temperature of about 1400°F (760°C).

In the pyrolysis process, the sludge is decomposed in the presence of inert gases at 1400°F (760°C), which yields hydrogen, methane, carbon monoxide, and carbon dioxide. For pyrolysis, auxiliary heat is required to maintain the fluidized bed at the high reaction temperature:

$$\text{Sewage Sludge} + \text{Inert Gas} \longrightarrow$$
$$H_2 + CO + CO_2 + CH_4 + \text{Auxiliary Heat} \quad 10.10(2)$$

The operation of the fluidized-bed incinerator is optimized by control of the airflow rate and the bed temper-

ature. Since reaction rates are related to bed mixing and the source of agitation is the fluidizing air, operators can adjust reaction rates by changing the airflow supply. The bed temperature is usually maintained between 1300 and 1500°F (704 to 815°C). For complete combustion (odor-free operation), about 25% excess air is needed (Rabosky 1974).

Organic material can be deposited on the sand particles (agglomerative mode) and removed by continuous or intermittent withdrawal of excess bed material. An alternative mode of operation (nonagglomerative) combines the organic ashes with exhaust gases and collects them downstream with dust collectors.

The main advantages of fluidized-bed incinerators include the uniformity of the bed, the elimination of stratification and hot or cold spots, the high rate of heat transfer for rapid combustion, the elimination of odor and the need for afterburners, and the low maintenance requirements of the process. The disadvantages include the high operating-power requirement, the need for auxiliary fuel

TABLE 10.10.5 OPERATING COSTS OF FLUIDIZED-BED SLUDGE INCINERATION WITH AND WITHOUT HEAT RECOVERY

Items	Utility Unit Cost	Normal Undried Cake Incineration		Newly Developed Indirectly Dried Cake Incineration	
		Amount	Cost	Amount	Cost
Plant Capacity		80 tn/d		96 tn/d	
Operation Cost			US$/d		US$/d
1. Supporting fuel	0.35 US$/l	6360 l/d	2226	2304 l/d	806
2. Electricity	0.1 US$/kWh	10800 kWh/d	1080	10300 kWh/d	1030
3. Chemical (NaOH)	0.35 US$/kg	220 kg/d	77	151 kg/d	53
4. Lubrications	2.6 US$/l	0.8 l/d	2	0.8 l/d	2
Total per Day	—	—	3385 US$/d	—	1891 US$/d
Unit Treatment Cost per Input Cake Vol.	—	—	42.3 US$/tn	—	19.7 US$/tn

Source: Henmi, Okazawa, and Sota 1986.
Note: All units are in metric (SI).

if the dry solid concentration is less than 32% and the need for dust-collection devices.

FLUIDIZED-BED INCINERATION WITH HEAT RECOVERY

The addition of heat-recovery equipment can increase the capacity of fluidized-bed incinerators by about 20% (Henmi, Okazawa, and Sota 1986). The plant shown in Figure 10.10.4 has been in operation in Tokyo since 1984 (Henmi, Okazawa, and Sota 1986). Heat recovery involves inserting a heat exchanger into the stream of the hot gases, which generates a supply of hot thermal oil. The oil is then used as the energy source to heat the sludge cake dryers.

The hot oil passes through the hollow inside of the motor-driven screws, while the sludge cake is both moved and

TABLE 10.10.6 CONCENTRATION OF POLLUTANTS IN ATMOSPHERIC EMISSIONS AND IN ASH PRODUCED BY A FLUIDIZED-BED TYPE SLUDGE INCINERATOR

Items	Regulations	Run 1	Run 4
Exhaust Gas			
Sulfur oxides (SO_x)	292 ppm	4 ppm	25 ppm
Nitrogen oxides (NO_x)	250 ppm	—	41 ppm
Hydrogen chloride (HCl)	93 ppm	3 ppm	—
Dust density	0.05 g/m³N	0.006 g/m³N	0.002 g/m³N
Residual Ash			
Amount	7.0 tn/d	2.4 tn/d	2.9 tn/d
Ignition loss	<15%	0.6%	0.6%
Dissolution to water			
Alkyl mercury (Hg)	—	<0.0005	—
Total mercury (Hg)	0.05	<0.0005	—
Cadmium (Cd)	0.1	<0.05	—
Lead (Pb)	1	<0.2	—
Phosphorus (P)	0.2	<0.01	—
Chromium 6+ (Cr[VI])	0.5	<0.05	—
Arsenic (As)	0.5	<0.19	—
Cyanide (—CN)	1	<0.05	—
Polychlorinated biphenyls (PCB)	0.003	<0.0005	—

Source: Henmi, Okazawa, and Sota 1986.
Notes: Values marked (—) were not measured. Suffix "N" means the value converted at normal condition of 273 K, 1 atm.

TABLE 10.10.7   CONCENTRATION OF WASTEWATER GENERATED BY FLUIDIZED-BED INCINERATOR

Content (mg/l)	Feed Water	Dryer Condensate	Wet Electrostatic Precipitation Effluent	Dehumidifier Effluent
Tkj—N	24.71	29.67	24.97	24.52
NH4$^+$—N	21.14	24.40	21.24	20.75
Org—N	3.57	5.27	3.73	3.77
NO2$^-$—N	0.18	0.16	N.D.	0.11
NO3$^-$—N	0.63	0.60	0.98	0.62
T—N	25.52	30.43	25.95	25.25
BOD	6.86	41.5	2.24	3.44
CODmn	12.2	21.4	38.8	12.5
SS	3.3	15.7	84.0	2.3
Cl$^-$	74.3	74.5	58.3	76.1

Source: Henmi, Okazawa, and Sota 1986.

heated by these screws. As shown in Table 10.10.4, the cake is dried to a substantial degree (some 20% of the inlet flow is evaporated) as the sludge cake passes through the cake dryer. The heating oil circulates in a closed cycle and is maintained at about 480°F (250°C) inside the screw-conveyor dryers by the throttling of two three-way valves. One valve can increase the outlet oil temperature from the cake dryers by blending in warmer inlet oil; the other can lower the outlet temperature by sending some of the oil through an oil cooler.

The operators of the Tokyo incinerator feel that the total capital cost of the plant is unaffected by the addition of the heat-recovery feature because the cost of the heat-transfer equipment is balanced by the reduced capacity requirement. The operating costs, on the other hand, are cut in half with the heat-recovery system (Table 10.10.5).

Another interesting feature of this system is the method of cleaning the accumulation of ash and soot from the heat-transfer surfaces. This automated system uses 3- to 5-mm-diameter steel-shot balls that are dropped every three to six hours from the top of the hot-air heaters. The random movement of the balls removes the dust from the heater tubes. The dust is removed at the bottom, while the balls are collected and returned to the top.

Table 10.10.6 gives the composition of the ash residue and the stack gases (after they have been cleaned by wet electrostatic precipitation); both meet Japanese regulations. Table 10.10.7 gives the composition of the wastewater produced by this process. According to the operators, the process produces almost no odor.

Adapted from *Municipal Waste Disposal in the 1990s* by Béla G. Lipták (Chilton, 1991).

References

Henmi, M., K. Okazawa, and K. Sota. 1986. Energy saving in sewage sludge incineration with indirect heat drier. *National Waste Processing Conference, Denver, 1986.* ASMER.

Rabosky, J.G. 1974. Incineration—fluidized bed incineration. Vol. 3, Sec. 2.23 in *Environmental engineers' handbook,* edited by B.G. Lipták. Radnor, Pa.: Chilton Book Company.

Sebastian, F. 1974a. Incinerator economics. Vol. 3, Sec. 2.17 in *Environmental engineers' handbook,* edited by B.G. Lipták, Radnor, Pa.: Chilton Book Company.

———. 1974b. Multiple hearth incineration. Vol. 3, Sec. 2.22 in *Environmental engineers' handbook,* edited by B.G. Lipták. Radnor, Pa.: Chilton Book Company.

Snell, J.R. 1974. Flash drying or incineration. Vol. 1, Sec. 8.6 in *Environmental engineers' handbook,* edited by B.G. Lipták. Radnor, Pa.: Chilton Book Company.

10.11
ONSITE INCINERATORS

Air Preheater Co., Inc.; American Schack Co., Inc.; Aqua-Chem, Inc.; BSP Corp., Div. of Envirotech Systems, Inc.; Bartlett-Snow; Beloit-Passavant Corp.; Best Combustion Engrg. Co.; Bethlehem Corp.; Brule Pollution Control Systems; C-E Raymond; Carborundum Co.; Pollution Control Div., Carver-Greenfield Corp.; Coen Co.; Combustion Equipment Assoc. Inc.; Copeland Systems, Inc.; Dally Engineering-Valve Co.; Dorr-Oliver Inc.; Dravo Corp.; Environmental Services Inc.; Envirotech; First Machinery Corp.; Foster Wheeler Corp.; Fuller Co.; Garver-Davis, Inc.; Haveg Industries Inc.; Holden, A. F., Co.; Hubbell, Roth & Clark, Inc.; Intercontek, Inc.; International Pollution Control, Inc.; Ishikawajima-Harima Heavy Industries Co., Ltd.; Kennedy Van Saun Corp.; Klenz-Aire, Inc.; Koch & Sons, Inc.; Koch Engrg. Co., Inc.; Kubota, Ltd., Chuo-Ku, Tokyo, Japan; Lawler Co., Leavesley Industries; Lurgi Gesellschaft fuer Waerm & ChemotecHnik mbH, 6 Frankfurt (Main) Germany; Maxon Premix Burner Co., Inc.; Melsheimer, T., Co., Inc.; Midland-Ross Corp., RPC Div.; Mid-South Mfg. Corp.; Mine & Smelter Supply Co.; MSI Industries; Mitsubishi Heavy Industries, Ltd., Tokyo, Japan; Monsanto Biodize Systems, Inc.; Monsanto Enviro-Chem Systems Inc.; Nichols Engrg. & Research Corp.; North American Mfg. Co.; Oxy-Catalyst, Inc.; P.D. Proces Engrg. Ltd., Hayes, Middlesex, England; Peabody Engrg. Corp.; Picklands Mather & Co., Prenco Div.; Plibrico Co.; Prenco Mfg. Co.; Pyro Industries, Inc.; Recon Systems Inc.; Renneburg & Sons Co.; Rollins-Purle, Inc.; Ross Engrg. Div., Midland-Ross Corp.; Rotodyne Mfg. Corp.; Rust Engrg. Co., Div. of Litton Industries; Sargent, Inc.; Surface Combustion Div., Midland-Ross Corp.; Swenson, Div. of Whiting Corp.; Tailor & Co., Inc.; Takuma Boiler Mfg. Co., Ltd., Osaka, Japan; Thermal Research & Engrg. Co.; Torrax Systems, Inc.; Vulcan Iron Works, Inc.; Walker Process Equip., Div. of Chicago Bridge & Iron Co.; Westinghouse Water Quality Control Div., Infilco; Zink, John, Co.; Zurn Industries, Inc.

This section describes some of the smaller incinerator units used onsite in domestic, commercial, and industrial applications. Onsite incineration is a simple and convenient means of handling the waste transportation problem since it reduces the volume of disposable waste. Onsite incinerators are smaller and their fuel more predictable in composition than the MSW burned in municipal incinerators. Therefore, this section discusses separately considerations that affect the location, selection, and operating practices of these onsite units.

Location

The onsite incinerator should be located close to larger sources of waste and expected waste collection routes. Onsite incinerators are constructed of 12-gauge steel casing with high-temperature (over 1000° F) insulation and high-quality refractory lining. Indoor installations are preferred, but even when the incinerator is situated outdoors, the charging and cleanout operations should be shielded from the weather. Incinerator rooms should be designed for two-hour fire resistance and should comply with the National Fire Protection Association (NFPA) recommendations contained in bulletin NFPA No. 82.

The Incinerator Institute of America (IIA) separates incinerators into nine classes according to their use and size (see Table 10.11.1) and provides minimum construction and performance standards for each class. The NFPA has

TABLE 10.11.1   CLASSIFICATION OF INCINERATORS

Class I—Portable, packaged, completely assembled, direct-feed incinerators having not over 5 cu ft storage capacity or 25 lb/hr burning rate, suitable for type 2 waste.

Class IA—Portable, packaged or job assembled, direct-feed incinerators having a primary chamber volume of 5 to 15 cu ft or a burning rate of 25 lb/hr up to, but not including, 100 lb/hr of type 0, 1, or 2 waste; or a burning rate of 25 lb/hr up to, but not including, 75 lb/hr of type 3 waste.

Class II—Flue-fed, single chamber incinerators with more than 2 cu ft burning area for type 2 waste. This incinerator type is served by one vertical flue functioning as a chute for both charging waste and carrying the products of combustion to the atmosphere. This incinerator type is installed in apartment or multiple dwellings.

Class IIA—Chute-fed, multiple chamber incinerators for apartment buildings with more than 2 cu ft burning area, suitable for type 1 or 2 waste. (Not recommended for industrial installations). This incinerator type is served by a vertical chute for charging waste from two or more floors above the incinerator and a separate flue for carrying the products of combustion to atmosphere.

Class III—Direct-feed incinerators with a burning rate of 100 lb/hr and more suitable for burning type 0, 1, or 2 waste.

Class IV—Direct-feed incinerators with a burning rate of 75 lb/hr or more suitable for burning a type 3 waste.

Class V—Municipal incinerators suitable for type 0, 1, 2, or 3 waste or a combination of all four wastes and are rated in tons per hour or tons per twenty-four hours.

Class VI—Crematory and pathological incinerators suitable for burning type 4 waste.

Class VII—Incinerators designed for specific by-product waste, type 5 or 6.

Note: For waste type numbers, see Tables 10.11.2 and 10.11.3.

TABLE 10.11.2 CLASSIFICATION OF WASTES TO BE INCINERATED

Classification of Waste Type and Description	Principal Components	Approximate Composition, % by Weight	Moisture Content, %	Incombustible Solids, %	Btu Value/lb of Refuse as Fired	Requirement for Auxiliary Fuel Btu per lb of Waste	Recommended Minimum Btu Burner Input per lb Waste	Density lb/cu ft
0 Trash	Highly combustible waste, paper, wood, cardboard cartons, and up to 10% treated papers, plastic or rubber scraps; commercial and industrial sources	Trash 100	10	5	8500	0	0	8–10
1 Rubbish	Combustible waste, paper, cartons, rags, wood scraps, combustible floor sweepings; domestic, commercial, and industrial sources	Rubbish 80 Garbage 20	25	10	6500	0	0	8–10
2 Refuse	Rubbish and garbage; residential sources	Rubbish 50 Garbage 50	50	7	4300	0	1500	15–20
3 Garbage	Animal and vegetable wastes, restaurants, hotels, markets; institutional commercial, and club sources	Garbage 65 Rubbish 3	70	5	2500	1500	3000	30–35
4 Animal solids and organics	Carcasses, organs, solid organic wastes; hospital, laboratory, abattoirs, animal pounds, and similar sources	Animal and human tissue 100	85	5	1000	3000	8000 (5000 primary) (3000 secondary)	45–55
5 Gaseous, liquid, or semiliquid	Industrial process wastes	Variable	Dependent on major components		Variable	Variable	Variable	Variable
6 Semisolid and solid	Combustibles requiring hearth, retort, or grate equipment	Variable	Dependent on major components		Variable	Variable	Variable	Variable

TABLE 10.11.3 INDUSTRIAL WASTE TYPES

Type of Wastes	Description
1	Mixed solid combustible materials, such as paper and wood
2	Pumpable, high heating value, moderately low ash, such as heavy ends, tank bottoms
3	Wet, semisolids, such as refuse and water treatment sludge
4	Uniform, solid burnables, such as off-spec or waste polymers
5	Pumpable, high ash, low heating value materials, such as acid or caustic sludges, or sulfonates
6	Difficult or hazardous materials, such as explosive, pyrophoric, toxic, radioactive, or pesticide residues
7	Other materials to be described in detail

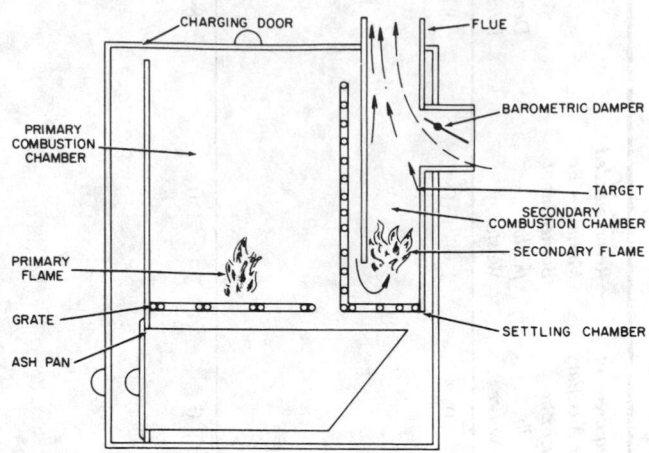

FIG. 10.11.1 Features of a domestic incinerator.

also instituted similar classifications and construction standards in its standard *Incinerators and rubbish handling*. The IIA also classifies waste into seven types (see Tables 10.11.2 and 10.11.3). Planners must also comply with local and state codes when selecting an incinerator.

Selection

The first step in incinerator selection is to record the volume, weight, and classes of waste collected for a period of at least two weeks. The survey should be checked against typical waste-production rates. The maximum daily operation can be estimated as three hours for apartment buildings; four hours for schools; six hours for commerical buildings, hotels, and other institutions; and seven hours per shift for industrial installations.

The results of the waste survey help to determine whether a continuous or batch-type incinerator should be installed. Batch-type units consist of a single combustion chamber (see Figure 10.11.1). If the batch furnace has no grate, the ash accumulation reduces the rate of burning. The batch incinerator is sized according to the weight of each type of waste per batch at the number of batches per day. The continuous incinerator consists of two chambers:

one for charge storage and the other an evacuated chamber for combustion. The charge chamber can be loaded at any time. Sizing is based on the pounds-per-hour burning rate required.

The nature and characteristics of the waste are usually summarized in a form such as that in Table 10.11.4. Most incinerator manufacturers offer standard, pre-engineered packages for waste types 0, 1, 2, 3, and 4 (see Tables 10.11.2 and 10.11.3). Waste types 5 and 6 usually require unique designs because the physical, chemical, and thermal characteristics of these wastes are variable. Type 6 waste tends to have low heating values but contains material that can cause intense combustion. Plastics and synthetic rubber decompose at high temperatures and form complex organic molecules that require auxiliary heat and high turbulence before they are fully oxidized. In extreme cases, three combustion chambers are necessary; operators must recycle flue gases from the secondary combustion chamber back into the primary chamber to complete the combustion process.

Charging

Incinerators can be charged manually or automatically; they can also be charged directly (see Figure 10.11.2) or

TABLE 10.11.4 WASTE ANALYSIS SHEET

% Ash _____ % Sediment _____ % Water _____

Waste material soluble in water? _____ Water content well mixed, emulsified? _____

If there are solids in the liquid, what is their size range? _____

Conradson carbon _____ Corrosion (copper strip) _____

Is the material corrosive to carbon steel? _____ Corrosive to brass? _____

What alloy is recommended for carrying the fluid? _____

Distillation data (if applicable) 10% at _____ °F; 90% at _____ °F; end point _____ °F.

Flash point _____ °F; fire point _____ °F; pour point _____ °F.

Viscosity _____ SSF at 122°F or _____ SSU at 100°F.

pH _____ ; acid number _____ ; base number _____

Heating value _____ Btu/gal   Specific gravity (H_2O = 1.0) _____

Will the material burn readily? _____

Toxic? (explain) _____

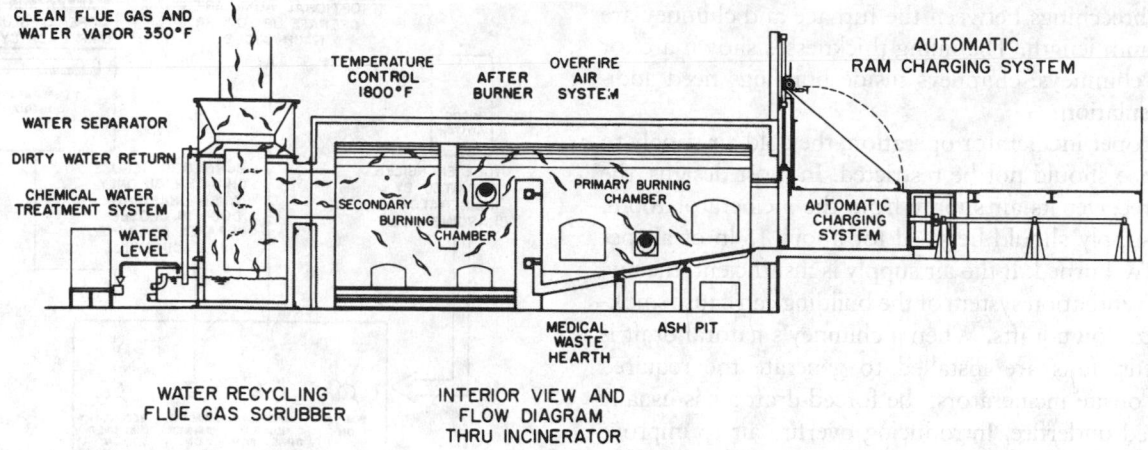

FIG. 10.11.2 Incinerator with a ram-feed system.

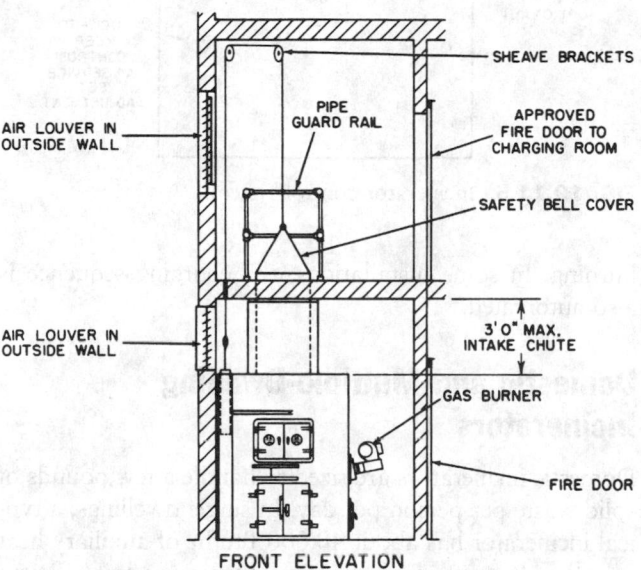

FIG. 10.11.3 Top-charging incinerator.

from charging rooms (see Figure 10.11.3). Direct incinerators are the least expensive but are limited in their hourly capacities to 500 lb, while indirect incinerators operate at capacities up to 1000 lb/hr. A manually charged incinerator (see Figure 10.11.3) is fed through a bell-covered chute from the floor above the furnace. This labor-saving design also guarantees good combustion efficiency and protection against flashbacks. The separate charging room is also convenient for sorting waste for recycling. Incinerators can also be fed from the same floor where the furnace is located. This arrangement also permits sorting and is labor-efficient although the radiant heat can be uncomfortable for the operator.

In high-rise buildings, the installation of a waste chute eliminates the labor involved in charging the incinerator (see Figure 10.11.4). The chute automatically directs the solid waste into a top-charged, mechanical incinerator. The charging rate can be regulated by rotary star feeders or by charging gates that open at 15- to 30-min intervals. Both offer protection against momentary overloading.

Hydraulic plungers or rams offer a more controlled method of automatic charging. The movement of the reciprocating plunger forces the refuse from the bottom of the charge hopper into the furnace (see Figure 10.11.2). This method is the most common for automatic charging for capacities exceeding 500 lb per hour.

Incinerators that burn sawdust or shredded waste are frequently charged by screw feeders or pneumatic conveyors. Screw feeders are at least 6 in (15 cm) in diameter and are designed with variable pitch to minimize the compression (and therefore blocking) of the shredded waste. Container charging, which is being used in a few isolated cases, has the advantage of protecting against exposure to flashback caused by aerosol cans or the sudden combustion of highly flammable substances.

Accessories

For smaller incinerators, chimneys provide sufficient draft to discharge flue gases at a high enough point where no nuisance is caused by the emissions. A fully loaded chemistry should provide at least 0.25 in of water draft (-62 Pa). Table 10.11.5 lists the diameters and heights of chimneys according to the weight rate of waste burned in a continuously charged multiple-chamber incinerator. The table assumes that the incinerator uses no dilution air and

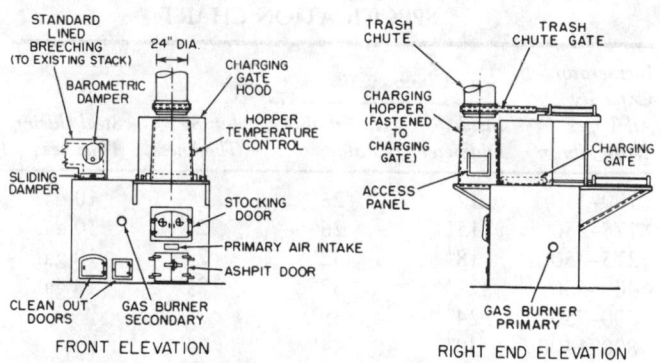

FIG. 10.11.4 Incinerator with automatic charging system.

that the breechings between the furnace and chimney are of minimum length. The lining thicknesses shown are for outdoor chimneys; chimneys inside buildings need additional insulation.

For proper incinerator operation, the cold air supply to the furnace should not be restricted. In most designs, the furnace receives its air supply from the incinerator room. The air supply should be sized for about 15 lb of air per lb of MSW burned. If the air supply is insufficient, the mechanical ventilation system of the building can cause smoking due to downdrafts. When a chimney's natural draft is insufficient, fans are installed to generate the required draft. In onsite incinerators, the forced-draft air is usually introduced underfire. Introducing overfire air to improve combustion efficiency is not widely used in onsite units.

When the waste is wet or its heating value is low, auxiliary fuels are needed to support combustion. In continuously charged incinerators, the primary burner is sized for 1500 Btu per lb of type 3 waste or for 3000 Btu per lb of type 4 waste (see Table 10.11.2). The heat capacity of the secondary burner is also 3000 Btu per lb of waste. When the incinerator is fully loaded, the secondary burner runs for only short periods at a time.

Controls

Onsite incinerators are frequently operated automatically from ignition to burndown (see Figure 10.11.5). The cycle is started by the microswitch on the charging door, which automatically starts the secondary burner, the water flow to the scrubber, and the induced-draft fan. When the door is closed, the primary burner is started and stays on for an adjustable time of up to an hour or until the door is reopened. Unless interrupted by a high-temperature switch, the secondary burner stays on for up to five hours. To provide each charge with the same preset burndown protection, the secondary burner timer is reset every time the charging door opens. Both burners have overtemperature and flame-failure safety controls. Also, a separate cycle timer controls the induced-draft fan and scrubber water flow to guarantee airflow and scrubbing action during

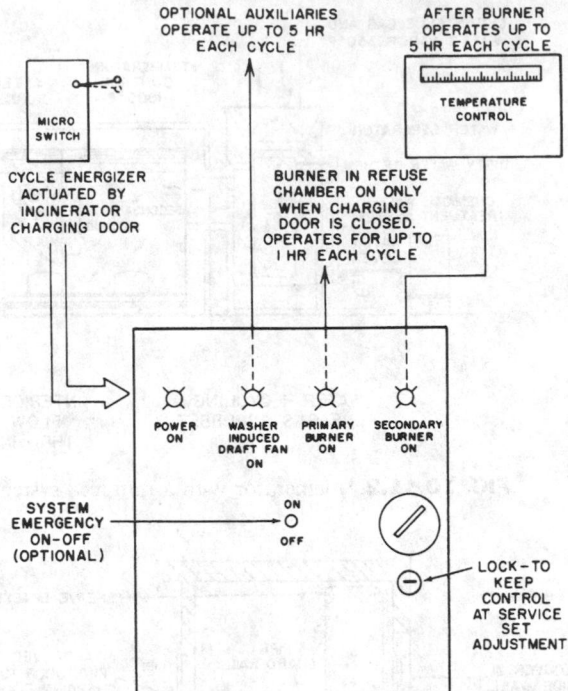

FIG. 10.11.5 Incinerator control system.

burning. In some installations, the charging sequence is also automated.

Domestic and Multiple-Dwelling Incinerators

Domestic incinerators are sized to handle a few pounds of solid waste per person per day. In single dwellings, a typical incinerator has about 40,000 Btu/hr of auxiliary heat capacity. Because domestic incinerators are less efficient than their municipal counterparts, the amount of auxiliary fuel used is high. The domestic incinerator in Figure 10.11.1 has two combustion chambers. The main purpose of the secondary chamber is to eliminate smoke and odor. As a result, the pollutant emissions from domestic incinerators are not excessive (see Table 10.11.6).

In multiple dwellings, the main purpose of incineration is to reduce the volume of the MSW prior to disposal. The refuse from a dwelling of 500 residents producing 2000 lb/day of MSW at a density of 4 lb/cu ft fills 100 trash cans. If incinerated onsite, the residue fits into 10 trash cans.

Incinerators in multiple dwellings can either be chute fed or flue fed. In the chute-fed design, waste is discharged into the chute and then into the incinerator feed hopper in the basement (see Figure 10.11.4). In the flue-fed design (see Figure 10.11.6), the chimney also serves as the charging chute for the waste, which falls onto grates above an ash pit inside a boxlike furnace. The main purpose of the charging door is to ignite the waste, while the purpose of the underfire and overfire air ports is to manually set the airflow for smokeless burning. The walls of the incinera-

TABLE 10.11.5 CHIMNEY SELECTION AND SPECIFICATION CHART

| Incinerator Capacity All Types Waste, lb/hr | Chimney Size | | | |
	Inside Diameter	Height above Grate	Lining Thickness	Steel Casing Thickness
100–150	12″	26′	2″	10 ga.
175–250	15″	26′	2″	10 ga.
275–350	18″	32′	2″	10 ga.
400–550	21″	37′	2½″	10 ga.
600–750	24″	39′	3″	¼″
800–1400	30″	44′	3″	¼″
1500–2000	36″	49′	3½″	¼″

TABLE 10.11.6 INCINERATOR EMISSIONS—
TYPICAL VALUES

Pollutant	New Domestic Wastes Incinerator	Municipal Incinerator with Scrubber
Particulates, grain/SCF	0.01–0.20	0.03–0.40
Carbon monoxide, ppm	200–1000	<1000
Ammonia, ppm	<5	—
Nitrogen oxides, ppm	2–5	24–58
Aldehydes, ppm	25–40	1–9

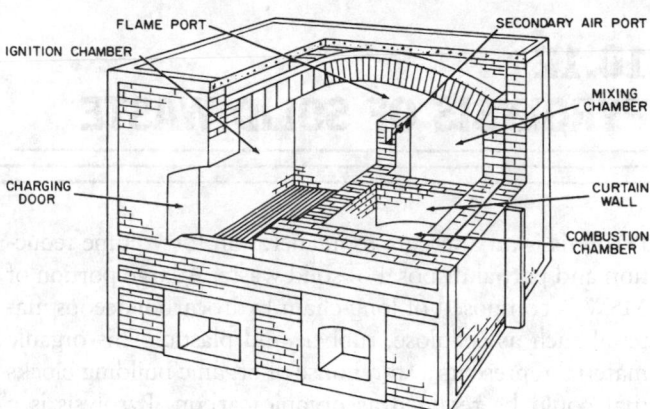

FIG. 10.11.7 Retort-type, multiple-chamber incinerator.

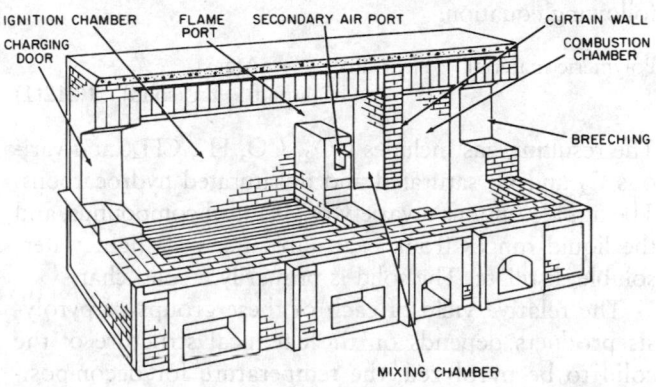

FIG. 10.11.8 Inline multiple-chamber incinerator.

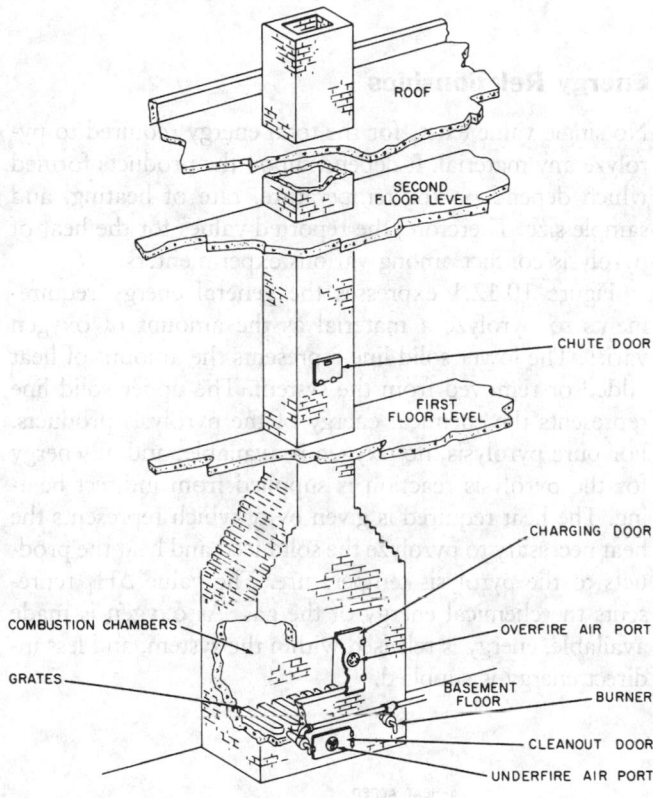

FIG. 10.11.6 Flue-fed incinerator.

tor consist of two brick layers with an air space between them. The inner layer is made of 4.5 in of firebrick and the 9-in outer layer is made of regular brick.

Flue-fed apartment incinerators have a draft-control damper in the stack, right above the furnace. This damper is pivoted and counterweighted to close when a chute door opens to charge refuse into the furnace. As a result, draft at the furnace remains relatively constant. To withstand flame impingement, the draft-control damper should be made of 20-gauge 302 stainless steel.

Miscellaneous Onsite Incinerators

Some incinerator designs are specially developed for on-site industrial applications. The outstanding design fea-

ture of the retort incinerator (see Figure 10.11.7) is the multiple chambers connected by lateral and vertical breechings; the combustion gases must pass through several U-turns for maximum mixing. The inline design (see Figure 10.11.8) also emphasizes good flue-gas mixing. Here, the combustion gases are mixed by passing through 90° turns in the vertical plane only. Both designs are available in mobile styles for use in temporary applications such as land clearance or housing construction. The retort design is for smaller waste-burning capacities (under 800 lb/hr), while the inline design is for higher burning rates.

Rotary incinerators for burning solid or liquid wastes can be continuous or batch and can be charged manually or by automatic rams. Their capacities range from 100 to 4000 lb/hr. For burning waste that contains chlorinated organics, the incinerator chamber must be lined with acid-resistant brick, and the combustion gases must be sent through absorption towers to remove the acidic gases from the flue gas.

Adapted from *Municipal Waste Disposal in the 1990s* by Béla G. Lipták (Chilton, 1991).

10.12
PYROLYSIS OF SOLID WASTE

Pyrolysis is an alternate to incineration for volume reduction and partial disposal of solid waste. A large portion of MSW is composed of long-chain hydrocarbonaceous material such as cellulose, rubber, and plastic. This organic material represents a storehouse of organic building blocks that could be retained as organic carbon. Pyrolysis is a process that is less regressive than incineration and recovers much of the chemical energy.

Long-chain organic material disintegrates when exposed to a high-temperature thermal flux according to the following equation:

$$\text{Polymeric material} + \text{Heat flux} \longrightarrow a\text{A(gas)} \\ + b\text{B(liquid)} + c\text{C(solid)} \quad \textbf{10.12(1)}$$

The resulting gas includes CO_2, CO, H_2, CH_4, and various C_2 and C_3 saturated and unsaturated hydrocarbons. The liquid contains a variety of chemical compounds, and the liquid ranges from a tar substance to a light water-soluble distillate. The solid is primarily a solid char.

The relative yield of each of these groups of pyrolysis products depends on the chemical structure of the solid to be pyrolyzed, the temperature for decomposition, the heating rate, and the size and shape of the material.

If the products of pyrolysis react with oxygen, they react according to the following equations:

$$\text{A(gas)} + O_2 \longrightarrow CO_2 + H_2O + \text{heat} \quad \textbf{10.12(2)}$$

$$\text{B(liquid)} + O_2 \longrightarrow CO_2 + H_2O + \text{heat} \quad \textbf{10.12(3)}$$

$$\text{C(solid)} + O_2 \longrightarrow CO_2 + H_2O + \text{heat} \quad \textbf{10.12(4)}$$

Pure pyrolysis involves only the reaction in Equation 10.12(1), the destructive distillation in an oxygen-free atmosphere. This definition can be expanded to include systems in which a limited amount of oxygen is made available to the process to release enough chemical energy for the pyrolysis reaction.

Comparing the results of various experimental investigations on pyrolysis is difficult because of the many variables influencing the results. No reliable design methods have been developed that allow for the scale-up of the experimental results. However, certain guiding principles underlying all pyrolysis systems can help in the selection of a process that most likely satisfy a particular need.

The process and operating conditions vary depending upon the relative demand for the char, liquid, and gas from the process.

Pyrolysis Principles

An understanding of the energy relationships, the effect of thermal flux, solid size, and the types of equipment is requisite to an understanding of pyrolysis principles.

Energy Relationships

No single value exists for the total energy required to pyrolyze any material. It depends upon the products formed which depend on the temperature, rate of heating, and sample size. Therefore, the reported values for the heat of pyrolysis conflict among various experimenters.

Figure 10.12.1 expresses the general energy requirements to pyrolyze a material as the amount of oxygen varies. The lower solid line represents the amount of heat added or removed from the system. The upper solid line represents the chemical energy of the pyrolysis products. For pure pyrolysis, no oxygen is available, and all energy for the pyrolysis reaction is supplied from indirect heating. The heat required is given by q, which represents the heat necessary to pyrolyze the solid feed and heat the products to the pyrolysis temperature. The value ΔH_1 represents the chemical energy of the gas. As oxygen is made available, energy is released within the system, and less indirect energy is supplied.

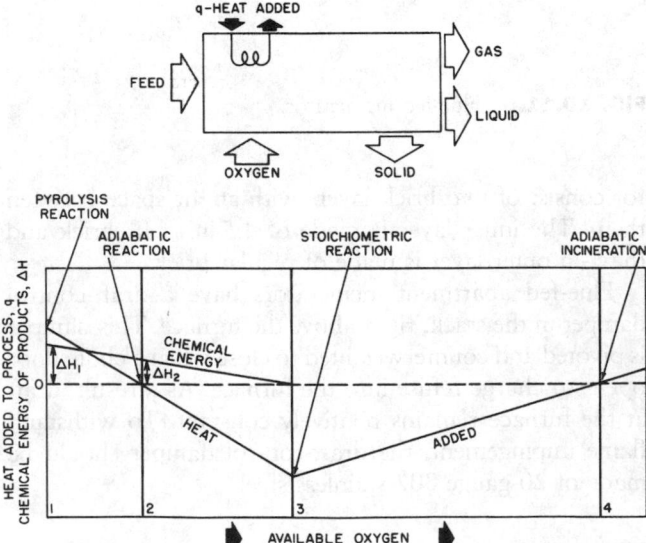

FIG. 10.12.1 General energy requirements for solid–gas reactions as a function of oxygen availability.

At point 2, an adiabatic condition is reached where the heat released from the oxidation of a portion of the pyrolysis products can furnish the energy required for the pyrolysis reaction as well as the energy necessary to heat the pyrolysis products, oxidation products, and nitrogen to the pyrolysis temperature. The value ΔH_2 represents the total chemical energy of the gas under these conditions. The larger fraction of the energy goes to sensible heat if nitrogen is present and ΔH_2 is smaller.

As the available oxygen increases, heat must be removed to maintain a constant reaction temperature. At point 3, the stoichiometric oxygen for complete combustion is reached, and the reaction products contain no chemical energy. Additional oxygen acts only as a coolant; therefore, less energy must be removed until point 4 is reached. This point is where the feed is being incinerated adiabatically, and no heat recovery is possible. This figure shows that the combined energy of the pyrolysis products is higher when the available oxygen is reduced. An advantage of the oxygen dependency is that it eliminates the limitation of pyrolysis systems on the rate of external heat demand. When enriched oxygen is used rather than air, the fraction of energy tied up in sensible heat is less, leaving more chemical energy in the pyrolysis products (the greater the fraction of chemical energy).

Effect of Thermal Flux

The products resulting from the thermal destruction of hydrocarbonaceous solids depend upon the maximum temperature of pyrolysis and the time needed to bring the feed to this temperature. The products formed during slow heating are far different than the products obtained during rapid heating. At very slow heating rates to low temperatures, the molecule has sufficient time to break at the weakest level and reorganize itself into a more thermally stable solid that becomes increasingly hard to destroy. On rapid heating to a high temperature, the molecule explodes and forms a range of smaller organic molecules.

For the cellulose molecule, slow heating forms high char yields and low gas and liquid yields. The gas is composed primarily of CO, H_2O, and CO_2 and has a low heating value. For rapid heating rates and high temperature, the gas yield increases and the liquid is smaller. The gas is composed primarily of CO_2, CO, H_2, CH_4, C_2H_2, and C_2H_4 and has a reasonable heating value. For intermediate heating rates and temperatures, high liquid yields are obtained. The gas produced is composed of many C_1, C_2, C_3, and C_4 compounds and has a high heating value.

Table 10.12.1 shows some values obtained for pyrolysis gas obtained at 1300 and 1600°F at a slow heating rate in a retort. For comparison the table also shows 1450°F pyrolysis at a high heating rate in a fluidized bed. Both systems pyrolyzed MSW.

Solid Size

For a large retort requiring indirect heating, the time required to pyrolyze a batch often exceeds twenty-four hours. The products change drastically as the reaction proceeds because a long time is needed for the center of the batch to reach the pyrolysis temperature. The mass near the center goes through a much slower heating cycle than the material near the walls. For the produced gas and liq-

TABLE 10.12.1 PYROLYSIS PRODUCTS OF MSW

Product Data	Composition of Pyrolysis Products		
Speed of Pyrolysis	Slow	Slow	Fast
Pyrolysis Temperature, °F	1382	1652	1450
Weight %			
Residue	11.59	7.7	3.0
Gas	23.7	39.5	61.0
Tar	1.2	0.2	26.0
Light Oil	0.9		
Liquor	55.0	47.8	4.0
Gas (Volume %)			
H_2	30.9	51.9	37.16
CO	15.6	18.2	35.50
CH_4	22.6	12.7	11.10
C_2H_6	2.05	0.14	not
C_2H_4	7.56	4.68	measured
CO_2	18.4	11.4	16.3
Btu/ft^3	563	447	366
10^6 Btu/ton	5.42	7.93	6.36
Gas Volume, cu ft/tn	9620	17,300	17,400

uids to be collected, they must pass through thick layers of pyrolysis char, and numerous secondary reactions result. For these reasons, the pyrolysis products of wood from a large retort can contain more than 120 products.

The same conditions are true for individual particles. For a material having the thermal properties of wood, the time required for the center temperature of a sphere to reach the surface temperature can be given by the following equation:

$$t = 0.5r^2 \qquad\qquad 10.12(5)$$

where r is the radius in inches and t is in hours. This equation indicates that about one hour is needed for the center of a 3-in particle to approach the surface temperature.

Types of Equipment

Several types of equipment are available for the pyrolysis of waste. The general types include retorts, rotary kilns, shaft kilns, and fluidized beds. The type of equipment and the manner of contacting have a significant effect on the pyrolysis product yield.

The retort has the longest application history and has been used extensively to make wood charcoal and naval stores. It is a batch system where the retort is charged, sealed, and heated externally. The heating cycle is long (often over twenty-four hours). The products are complex. They are normally solid char and a pyroligneous acid plus the gas produced which is used as the energy source for indirect heating. The process is limited by the rate of heat addition; a typical analysis for demolition lumber shows a yield of 35% char, 30% water, 12% wood tar, 5% acetic acid, 3% methanol, and 15% gas with a heating value of 300 Btu/cu ft.

Rotary Kiln

The rotary kiln is more flexible and provides increased heat transfer. The kiln can be heated indirectly, or the heat can be furnished by partially burning the pyrolysis products. The gas flow can be either parallel or countercurrent to the waste flow. The gas and liquid products do not have to escape through thick layers of char as in the batch retort; therefore, fewer complex solid–gas reactions occur. The heat cycle is much faster than in the retort, and the gas yield is higher and the liquid yield lower. The size of the indirectly fired kiln, because of the high temperatures involved and the need to transfer energy through the walls, is severely limited. The maximum capacity is in the range of 2 tn per hour for wood waste and is similar for solid waste.

If a limited amount of oxygen is used as the energy needed for pyrolysis, refractory lined kilns can be used, and large systems become feasible. If the oxygen and the feed are introduced countercurrently, the oxygen contacts the pyrolysis char first and tends to burn this char to furnish the heat for pyrolysis, which reduces the char yield. If air is introduced parallel to the feed, the oxygen reacts with the raw feed and the pyrolysis gas and gives a lower gas yield.

Rotary equipment, however, is more expensive to build, is more difficult to design with positive seals, and requires more maintenance.

Shaft Kiln

In the shaft kiln, the solids descend through a gas stream. The oxygen enters the bottom countercurrently to the feed and burns the char product reaching the bottom of the shaft. The combustion gases produced flow past the solid feed causing pyrolysis. The char is used to furnish the energy for pyrolysis. This use is undesirable if the char is a valuable product and, in that case, the pyrolysis gas can be used to preheat the air.

Fluidized-Bed Reactor

The fluidized-bed reactor is a system where the heat transfer rate is rapid. This design gives low liquid yields and high gas yields. In the fluidized bed, the feed is injected into a hot bed of agitated solids. To keep the bed in the fluid state, the system passes gas upward through the bed. If air is introduced into the bed, the oxygen contacts and reacts with the pyrolysis gas more readily than the char does, reducing the gas yield. To assure that the produced pyrolysis gas does not react with oxygen, operators can remove the char produced and burn it in a separate unit and return the hot gas used to pyrolyze the feed into the fluidized bed. The capacity is limited by the sensible heat available from this gas. In a fluidized bed, circulating the solids in the bed adds heat. The heat source necessary to pyrolyze the feed can be heated solids which can be easily added and removed in the fluidized bed.

Experimental Data

Little data is published on the pyrolysis of MSW, and no data is published for full-scale operating units. The data in Tables 10.12.1 and 10.12.2 are based on solid waste from a small pilot plant scale or bench scale lab experiments.

Table 10.12.2 presents the yield of various materials exposed to a 1500°F temperature. The yields of gas, liquid, and char vary widely between materials. The feed in all cases is newspaper, and the final pyrolysis temperature is 1500°F. An increase in the gas yield and a decrease in liquid organics due to an increase of the heating rate is evident from these data. The total amount of energy available from the gas also increases with the heating rate.

TABLE 10.12.2 PYROLYSIS YIELDS, IN WEIGHT PERCENT OF REFUSE FEED

Type of Waste Feed	Gas	Water	CnHmOx	Char C + S	Ash
Ford Hardwood	17.30	31.93	20.80	29.54	0.43
Rubber	17.29	3.91	42.45	27.50	8.85
White Pine Sawdust	20.41	32.78	24.50	22.17	0.14
Balsam Spruce	29.98	21.03	28.61	17.31	3.07
Hardwood Leaf Mixture	22.29	31.87	12.27	29.75	3.82
Newspaper I	25.82	33.92	10.15	28.68	1.43
II	29.30	31.36	10.80	27.11	1.43
Corrugated Box Paper	26.32	35.93	5.79	26.90	5.06
Brown Paper	20.89	43.10	2.88	32.12	1.01
Magazine Paper I	19.53	25.94	10.84	21.22	22.47
II	21.96	25.91	10.17	19.49	22.47
Lawn Grass	26.15	24.73	11.46	31.47	6.19
Citrus Fruit Waste	31.21	29.99	17.50	18.12	3.18
Vegetable Food Waste	27.55	27.15	20.24	20.17	4.89

Table 10.12.1 gives the products for a slow pyrolysis process at 1382 and 1652°F along with data for fast pyrolysis at 1450°F. The increase in gas yield and decrease in organic liquid with an increase in reactor temperature are evident. The hydrocarbon fraction of the gas decreases from 32.2 to 17.5% with an increase in temperature from 1382 to 1652°F, while the H_2 and CO portion increases from 46.5 to 70.1%. This data shows that higher temperature pyrolysis gives significantly higher yields to lower Btu gas. The higher temperature apparently results in the destruction of hydrocarbons in the gas. Comparing values for solid waste is difficult because of the variability between feed stocks of MSW. Table 10.12.1 shows the data for a run at 1450°F where pyrolysis is rapid along with data for slow pyrolysis. These data indicate that the fast pyrolysis at 1450°F gives results which are closer to those of the slow, high-temperature pyrolysis process. The total CO and H_2 is 72.6%, the total hydrocarbon 10.1%+ (undoubtedly higher but only CH_4 is evaluated), and the gas volume 17,400 cu ft/tn. Unfortunately, the data for fast pyrolysis is not complete, and a full comparison on the yield is not possible.

Table 10.12.3 presents data for a process to convert MSW to fuel oil. The temperature for pyrolysis is low (932°F), and the reaction rate rapid. This low-temperature pyrolysis gives higher yields of organic liquids, and the gas has significant quantities of C_2–C_7 hydrocarbons not present at higher temperatures. Reducing the temperature for this same process by several hundred degrees results in an increase in the gas yield of 80%.

The data available substantiate the guiding principles previously outlined and explain the composition and quantities of the products and how they are affected by changes in composition, temperature, and heating rate. They do

TABLE 10.12.3 FUEL OIL PRODUCTION FROM MSW

Char fraction, 35 wt%, heating value 9000 Btu/lb

CO	48.8 wt%
H	3.9
N	1.1
S	0.03
Ash	31.8
Cl	0.2
O (by difference)	12.7

Oil fraction, 40 wt%, heating value 12,000 Btu/lb

C	60%
H	8
N	1
S	0.2
Ash	0.4
Cl	0.3
O_2 (by difference)	20.0

Gas fraction, 10 wt%, heating value 600 Btu/cu ft

H_2O	0.1 mol%
CO	42.0
CO_2	27.0
H_2	10.5
CH_3Cl	<0.1
CH_4	5.9
C_2H_6	4.5
C_3–C_7	8.9

Water fraction contains:
Acetaldehyde
Acetone
Formic acid
Furfural
Methanal
Methylfurfural
Phenol

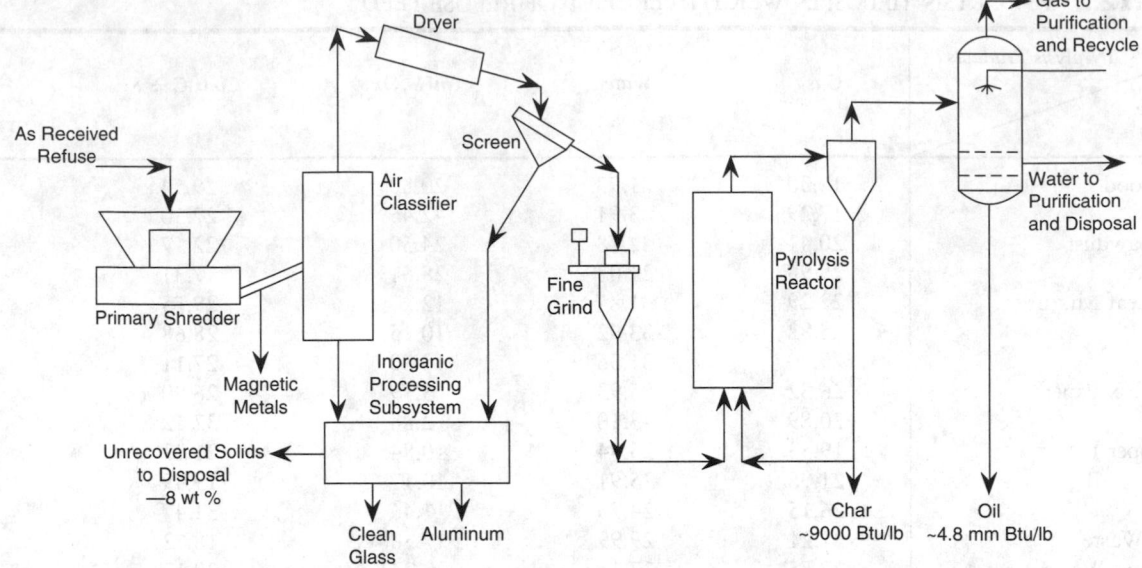

FIG. 10.12.2 Schematic diagram of Occidental Flash Pyrolysis System for the organic portion of MSW. (Reprinted, with permission, from G.T. Preston, 1976, Resource recovery and flash pyrolysis of municipal refuse, presented at Inst. Gas Technol. Symp., Orlando, FL, January.)

not accurately predict the products from a particular process or waste. However, the data do furnish sufficient evidence for environmental engineers to suggest the type of process for a given application.

Status of Pyrolysis

Pyrolysis is widely used as an industrial process to produce charcoal from wood, coke and coke gas from wood, coke and coke gas from coal, and fuel gas and pitch from heavy petroleum fractions. In spite of these industrial uses, the pyrolysis of MSW has not been as successful. No large-scale pyrolysis units are used for MSW operation in the United States as of April 30, 1995.

Only one full-scale MSW pyrolysis system was built in the United States. A simplified flowsheet of the Occidental Flash Pyrolysis System is shown in Figure 10.12.2. The front-end system consists of two stages of shredding, air classification, trommeling, and drying to produce finely divided RDF. Because of the short residence time of RDF in the reactor, this process is described as flash pyrolysis. The heat required for the pyrolysis reaction in the reactor is supplied from recirculation of the hot char. The hot char is removed from the reactor, passed through an external fluidized bed in which some air is added to partially oxidize the char, and recirculated to furnish energy for the endothermic pyrolysis reaction which yields the liquid by-products.

The end products were gases, pyrolytic oil, char, and residues. The liquid product had several noxious qualities making it a poor substitute for Bunker C fuel oil. It was corrosive, requiring special storage and fuel nozzles, and

was more difficult to pump and smelled poorly. These qualities resulted largely from highly oxygenated organics (including acids). Furthermore, the oil produced had a moisture content of 52%, not the 14% predicted from the pilot plant results. The increase in moisture in the oil decreased the energy content to 3600 Btu/lb, compared to the 9100 Btu/lb predicted by the pilot plant tests. The 100 tpd plant was built in El Cajun, California but never ran successfully and was shut down after only two years of operation.

The principal causes for the failure of pyrolysis technology appear to be the inherent complexity of the system and a lack of appreciation by system designers of the difficulties of producing a consistent feedstock from MSW (Tchobanoglous, Theisen, and Vigil 1993).

Although systems such as the Occidental Flash Pyrolysis System were not commercial successes, they produced valuable design and operational data that can be used by future designers. If the economics associated with the production of synthetic fuels change, pyrolysis may again be an economical, viable process for the thermal processing of solid waste. However, if gaseous fuels are required, gasification is a simpler, more cost-effective technology.

—R.C. Bailie (1974)
and David H.F. Liu (1996)

Reference

Tchobanoglous, G., H. Theisen, and S. Vigil. 1993. *Integrated solid waste management.* McGraw-Hill, Inc.

10.13
SANITARY LANDFILLS

The landfill is the most popular disposal option for MSW in the United States. Not only has it traditionally been the least-cost disposal option, it is also a solid waste management necessity because no combination of reduction, recycling, composting, or incineration can currently manage the entire solid waste stream. Barring unforeseen technological advances, landfills will always be needed to handle residual waste material.

Landfill Regulations

Solid waste landfills are federally regulated under Subtitle D of the Resource Conservation and Recovery Act of 1976 (RCRA). In the past, landfill regulation was left to the discretion of the individual states. The Solid Waste Disposal and Facility Criteria, promulgated by the U.S. EPA, specify how MSW landfills are to be designed, constructed, operated, and closed and were implemented in 1993 and 1994. The criteria were developed to ensure that municipal landfills do not endanger human health and are based on the assumption that municipal landfills receive household hazardous waste and hazardous waste from small generators. States are required to adopt regulations at least as strict as the EPA criteria. Although some states had some acceptable regulations in place, many did not. The EPA is currently considering criteria for non-hazardous industrial waste landfills.

In most states, landfills constructed under the new regulations are more expensive to construct and operate than past landfills because of requirements concerning daily cover, liners, leachate collection, gas collection, monitoring, hazardous waste exclusion, closure and postclosure requirements, and financial assurances (to cover anticipated closure and postclosure costs). Some of the major aspects of the landfill criteria are briefly described below next (40 *CFR* Parts 257–258).

LOCATION RESTRICTIONS

Location restrictions exclude landfills from being near or within certain areas to minimize environmental and health impacts. Table 10.13.1 summarizes location restrictions. Other location restrictions not mentioned in the federal disposal criteria but found in other federal state regulations include public water supplies, endangered or threatened species, scenic rivers, recreation or preservation areas, and utility or transmission lines.

EMISSIONS, LEACHATE, AND MONITORING

Gaseous Emissions

Landfills produce gases comprised primarily of methane and carbon dioxide. Emissions are controlled to avoid explosive concentrations of methane or a build-up of landfill gases that can rupture the cover liner or kill cover vegetation. Landfill design and monitoring must ensure that the concentration of CH_4 is less than 25% of the lower explosion limit in structures at or near the landfill and less than the lower explosion limit at the landfill property boundary.

A final rule announced by EPA in March, 1996 requires large landfills that emit volatile organic compounds in excess of 50 megagrams (Mg) per year to control emissions by drilling collection wells into the landfill and routing the gas to a suitable energy recovery or combustion device. It also requires a landfill's surface methane concentration to be monitored on a quarterly basis. If the concentration is greater than 500 parts per million, the control system must be modified or expanded to insure that the landfill gas is collected. The rule is expected to effect only the largest 4% of landfills in the United States.

Leachate

Leachate is water that contacts the waste material. It can contain high concentrations of COD, BOD, nutrients, heavy metals, and trace organics. Regulations require leachate to be collected and treated to avoid ground or surface water contamination. Composite bottom liners are required, consisting of an HDPE geomembrane at least 60 mil over 2 ft of compacted soil with a hydraulic conductivity of less than 1×10^{-7} cm/sec. However, equivalent liner systems can be used, subject to approval. The composite liner is covered with a drainage layer and leachate collection pipes to remove leachate for treatment and maintain a hydraulic head of less than 1 ft. Leachate is generally sent directly to a municipal wastewater treatment plant but can be pretreated, recirculated, or treated on-site.

Surface Water

Leachate generation can be reduced when water is kept from entering the landfill, especially the working face. Surface water control also reduces erosion of the final

TABLE 10.13.1   SITING LIMITATIONS CONTAINED IN SUBTITLE D OF THE RCRA AS ADOPTED BY THE EPA

Location	Siting Limitation
Airports	Landfills must be located 10,000 ft from an airport used by turbojet aircraft, 5000 ft from an airport used by piston-type aircraft. Any landfills closer must demonstrate that they do not pose a bird hazard to aircraft.
Flood plains	Landfills located within the 100-year floodplain must be designed to not restrict flood flow, reduce the temporary water storage capacity of the floodplain, or result in washout of solid waste, which would pose a hazard to human health and the environment.
Wetlands	New landfills cannot locate in wetlands unless the following conditions have been demonstrated: (1) no practical alternative with less environmental risk exists, (2) violations of other state and local laws do not exist, (3) the unit does not cause or contribute to significant degradation of the wetland, (4) appropriate and practicable steps have been taken to minimize potential adverse impacts, and (5) sufficient information to make a determination is available.
Fault areas	New landfill units cannot be sited within 200 ft of a fault line that has had a displacement in Holocene time (past 10,000 years).
Seismic impact zone	New landfill units located within a seismic impact zone must demonstrate that all contaminant structures (liners, leachate collection systems, and surface water control structures) are designed to resist the maximum horizontal acceleration in lithified material (liquid or loose material consolidated into solid rock) for the site.
Unstable areas	Landfill units located in unstable areas must demonstrate that the design ensures stability of structural components. The unstable areas include areas that are landslide prone, are in karst geology susceptible to sinkhole formation, and are undermined by subsurface mines. Existing facilities that cannot demonstrate the stability of the structural components must close within five years of the regulation's effective date.

Source: Data from G. Tchobanoglous, H. Theissen, and S. Vigil, 1993, *Integrated solid waste management: Engineering principles and management issues* (New York: McGraw-Hill).

cover. Regulations require preventing flow onto the active portion of the landfill (i.e., the working face) during peak discharge from the twenty-five-year storm of twenty-four-hour duration. Collection and control of water running off the active area during the twenty-five-year storm of twenty-four-hour duration is also required. Landfills should have no discharges that violate the Clean Water Act.

Daily Cover

Exposed waste must be covered with at least 6 in of soil at the end of each operating day. Alternative covers, such as foam or temporary blankets, can be approved for use.

Hazardous Waste

Hazardous waste should be kept out of the landfill so that the quality of leachate and gaseous emissions is improved. Landfill operations must have a program for detecting and preventing the disposal of regulated hazardous wastes.

Monitoring

Monitoring is done to identify, quantify, and track contaminants and to determine where and when corrective action should take place. At least three wells, one upgradi-

ent and two downgradient, should be maintained, and the well water should be tested at specified intervals. Most sites have more than three wells; the applicable regulations vary by state. Monitoring is conducted before, during, and after the landfill operating period. Remedial action is required when downgradient water quality is significantly worse than upgradient water quality.

Closure and Postclosure

To reduce, control, or retain leachate, gaseous emissions, and surface water, landfill operators must close the landfill properly and maintain it until waste material stabilizes. They must install and maintain a final cover to keep rainwater out of the landfill and establish vegetation to reduce erosion. The postclosure period is thirty years, during which all previously mentioned regulations must be followed, erosion must be controlled, and site security must be maintained.

Financial Assurance

In case of bankruptcy or other circumstances, facilities must be closed in a proper manner. The financial capability to safely close the facility at any time during its operational life must be maintained by the operator in a manner acceptable to the governing agency.

Siting New Landfills

Proper siting of sanitary landfills is crucial to providing economic disposal while protecting human health and the environment. The siting process consists of the following tasks (Walsh and O'Leary 1991a):

- Establishing goals and gathering political support
- Identifying facility design basis and need
- Identifying potential sites within the region
- Selecting and evaluating in detail superior sites
- Selecting the best site
- Obtaining regulatory approval

Goals include delineating the region to be served, facility lifetime, target tipping fees, maximum hauling distance, potential users, and landfill services. Political support is crucial to successful siting. Because opposition to a new landfill is almost always present, strong political support for a new landfill must exist from the start of the siting process. A solid waste advisory council—made up of interested independent citizens and representatives of interested groups—should be formed early in the process, if one does not already exist.

The design basis and needs of a landfill depend on the applicable regulations and the required landfill area (which in turn depend on the amount of waste to be handled and the required lifetime). The amount of waste to be handled depends on the present and future population served by the landfill, the projected per capita waste generation rate, and the projected recycling, composting, and reduction rates.

Developing a new landfill involves finding the most suitable available location. The main criteria involved in siting a new landfill are:

SITE SIZE—The site should have the capacity to handle the service area's MSW for a reasonable period of time.

SITE ACCESS—All-weather access roads with sufficient capacity to handle the number and weight of waste transport vehicles must be available.

HAUL DISTANCE—This distance should be the minimum distance that does not conflict with social impact criteria.

LOCATION RESTRICTIONS—These restrictions are summarized in Table 10.13.1. Additional or stricter constraints can also be imposed.

PHYSICAL PRACTICALITY—Sites with, for example, surface water or steep slopes should be avoided.

LINER AND COVER SOIL AVAILABILITY—This soil should be available onsite; Offsite sources increase construction and operating costs.

SOCIAL IMPACT—Siting landfills far from residences and avoiding significant traffic impacts minimizes this impact.

ENVIRONMENTAL IMPACT—The effect on environmentally sensitive resources, such as groundwater, surface water, wetlands, and endangered or threatened species should be minimized. Siting landfills on impermeable soils with a deep water table avoids groundwater impacts.

LAND USE—The land around the potential site should be compatible with a landfill.

LAND PRICE AND EASE OF PURCHASE—A potential site is easier to purchase if it is owned by one or a few parties.

ESTIMATING REQUIRED SITE AREA

Before attempting to identify potential landfill sites, planners must estimate the area requirement of the landfill. Landfill sizing is a function of:

- Landfill life (typically five to twenty-five years)
- Population served
- Waste production per person per day
- Extent of waste diversity
- Shape and height of the landfill
- Landfill area used for buffer zone, offices, roads, scalehouse, and optional facilities such as MRF, tire disposal and storage, composting, and convenience center

A number of formulas can help determine the acreage required for waste disposal (Tchobanoglous, Theissen, and Vigil 1993; Noble 1992). The total annual waste produced by the population to be served by the landfill for each year of the expected landfill's life is estimated as:

$$V_{ip} = \frac{(365 \text{ d/yr})PW_g(1 - f)}{C_d} \qquad \textbf{10.13(1)}$$

where:

V_{ip} = annual in-place waste volume (cu yd/yr)
P = population served by landfill in a given year
W_g = waste generation in a given year (lb/person/d)
f = fraction of waste stream diverted in a given year
C_d = specific density of the waste (lb/cu yd)

Population predictions for the years of expected landfill operation can usually be obtained from local government agencies. The total amount of waste generated in a community per person can be developed from waste characterization studies. State or national data can be used if no other data are available. Data on recycling trends should be gathered locally.

As landfilling costs increase, larger and heavier compactors are becoming more common, resulting in higher compaction. Compaction densities achieved in landfills vary from around 800 to as high as 1400 lb/cu yd depending on the type of compaction equipment used. Values of 1000 to 1200 lb/cu yd are often used as estimates.

Cover material adds to the amount of material placed in the landfill, reducing the landfill's effective volume. Typical waste-to-cover-soil-volume ratios are in the range of 4:1 to 10:1. A value of 5:1 or 4:1 is often assumed, indicating that for every 4 or 5 cu yd of waste, 1 cu yd of cover soil is deposited. In a 5:1 ratio, the cover-soil-to-waste ratio is 1 divided by 5, or 0.2. Incorporating waste and cover soil, the annual in-place waste and soil volume is:

$$V_{ap} = V_{ip}(1 + CR) \qquad 10.13(2)$$

where:

V_{ap} = annual in-place waste and soil volume, including waste and cover soil (cu yd/yr)
CR = cover-soil-to-waste ratio

Sometimes planners assume that all of the cover soil will come from the landfill excavation. In this case, all of the soil material excavated from the landfill ends up in the landfill. With this assumption, planners can estimate the area using the assumed shape and height of the landfill above ground level and the sum of the annual in-place waste volume the landfill expects to receive. Height regulations are generally included in state or local landfill regulations and vary from place to place.

The simplest shape that can be assumed is the cube. A more realistic shape is the flat-topped pyramid. Both shapes are shown in Figure 10.13.1. The volume of the cubic landfill is $V = (H)(B^2)$, where H = the height and B = the length of the base. Thus, area = B^2 = V/H. The volume of the flat-topped pyramid landfill, with a square base and 3:1 side slopes, is:

$$V = HB_{3:1}^2 - 6H^2B_{3:1} + 12H^3 \qquad 10.13(3)$$

where a 3:1 side slope means that for every 3 ft horizontal run, the slope rises 1 ft. Solving for B with the quadratic equation gives:

$$B_{3:1} = \frac{6H^2 \pm \sqrt{4HV_{ip} - 12H^4}}{2H} \qquad 10.13(4)$$

Planners can determine the area by squaring $B_{3:1}$. They can take the buffer zone and area needed for roads, facilities, and lagoons into account by increasing B or B^2. If 4:1 side slopes are used:

$$B_{4:1} = \frac{8H^2 + \sqrt{4HV_{ip} - 21.33H^4}}{2H} \qquad 10.13(5)$$

If planners do not assume that all of the excavated soil comes from the landfill excavation, then their solution must incorporate the annual in-place waste and soil volume, the excavated landfill volume, the aboveground landfill volume, and the cover-soil-to-waste ratio. Excavation side slopes are often assumed to be 1:1 but may be more gradual. Excavation bottom slopes are slight and can be assumed level for the purpose of initial size estimates.

EXCLUSIVE AND NONEXCLUSIVE SITING CRITERIA

Landfill siting criteria can be divided into two main groups: exclusive and nonexclusive criteria. If a site fails an exclusive criterion, it is excluded from consideration. Exclusive criteria include federal, state, or local location restrictions or physical restrictions. Exclusive criteria can be applied with maps and transparent overlays. For example, a U.S. Geological Survey (USGS) quadrangle map can be used as the base map. Transparent overlays with darkened restricted areas can be placed over the base map, as shown in Figure 10.13.2. Areas that remain clear are

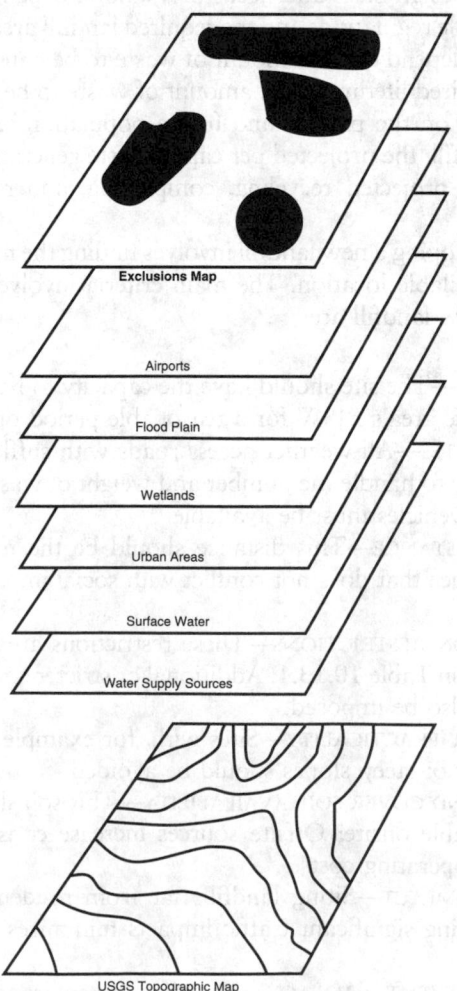

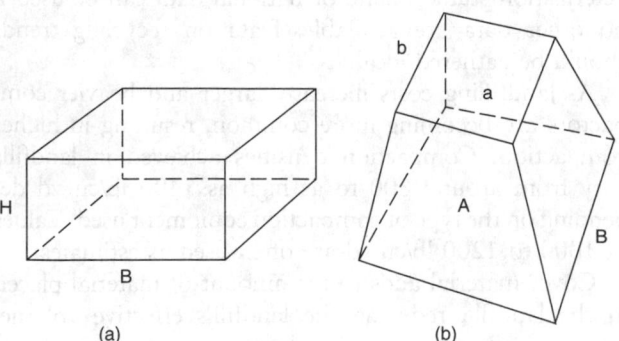

FIG. 10.13.1 Cubic (a) and pyramid (b) landfill shapes.

FIG. 10.13.2 Exclusive criteria mapping with overlays.

considered potential landfill sites. If information is available in digitized format, geographical information systems (GIS) can be used to complete overlay analyses (Siddiqui 1994).

A small number of potential landfill sites are selected from the areas that remain after the exclusive criteria are applied. The final selection process uses nonexclusive criteria, such as hydrogeological conditions, hauling distance, site accessibility, and land use. This process can be done in one or two steps.

If digitized data exist for the entire region under investigation, planners can rank the remaining areas using GIS and an appropriate decision making model, such as the analytical hierarchy process (Siddiqui 1994; Erkut and Moran 1991). For example, USGS soil maps are available in digitized format and include depth to water table, depth to bedrock, soil type, and slope although information is available only down to 5 ft. Planners can also use USGS digitized maps to identify urban areas, rivers and streams, and land use.

If areas are ranked, planners use this information, along with nondigitized information and field inspection, to select a number of sites from the best areas. Otherwise, planners select a number of sites based only on nondigitized information and simple field assessments without the aid of area rankings.

Once a number of sites have been identified, planners should rank the sites in a scientifically justifiable manner using established decision making models such as the analytical hierarchy procedure, interaction matrices, or multiattribute utility models (Camp Dresser & McKee 1984; Morrison 1974; Anandalingham and Westfall 1988–89). This process identifies a small number of sites, usually less than four, to undergo detailed investigations regarding hydrogeologic characteristics such as drainage patterns, geologic formations, groundwater depth, flow directions, and natural quality and construction characteristics of site soils. In addition, detailed information about existing land use, available utilities, access, political jurisdictions, and land cost is gathered (Walsh and O'Leary 1991a). Planners use this information to select the site for which regulatory approval will be sought.

Hydrogeologic information is crucial to the final site selection and has many uses. The main consideration is the proximity of groundwater, groundwater movement, and the potential for attenuation of leachate should it escape from the landfill. The proximity of groundwater is simply the depth to the groundwater table. Measuring the piezometric elevation of the water table in a number of wells on and around the potential site determines groundwater movement. The direction of flow is perpendicular to the lines of constant piezometric elevation. Groundwater movement is important (1) to assess the potential for landfill contamination to impact human health, for example if nearby drinking water wells are downgradient of the landfill site, and (2) to determine the placement of monitoring

wells should a site be selected. Leachate attenuation is a function of mechanical filtration; precipitation and coprecipitation; adsorption, dilution, and dispersion; microbial activity; and volatilization, most of which can be assessed via a subsurface investigation (O'Leary and Walsh 1991c).

The approval process can be demanding. Application writers should work closely with state permitting personnel, who can offer guidance on what is acceptable. To keep costs low, planners should start the siting process with the consideration of large areas based on limited and readily available information and end it with the selection of one site based on detailed information on a small number of sites.

At some point during the selection and permitting of a new landfill, planners hold at least one, and perhaps several, public meetings to ensure that public input is obtained concerning the selection of the landfill site. Generally all landfill sites inconvenience some portion of the local population, and thus most sites generate some public opposition. The siting process must be clear, logical, and equitable. The site selected must be the best available site. However, even if the best site is selected, equity considerations may necessitate offering compensation to residents near the site.

Design

Landfill design is a complex process involving disciplines such as geomechanics, hydrology, hydraulics, wastewater treatment, and microbiology. Design goals can include the following (Walsh and O'Leary 1991c):

- Protection of groundwater quality
- Protection of air quality
- Production of energy
- Minimization of environmental impact
- Minimization of disposal costs
- Minimization of dumping time for site users
- Extension of site lifetime
- Maximum use of land upon site closure

Planners must consider the final use of the landfill site during the design process to ensure that landfill design and operation are compatible with the end use.

Table 10.13.2 summarizes the sanitary landfill design steps. Table 10.13.3 summarizes the landfill design factors. Procurement of the requisite permits can take several years, thus the design process, an integral part of any application, should be started long before the current disposal option is scheduled to close.

The design package includes plans, specifications, a design report, an operator's manual, and cost estimates (Walsh and O'Leary 1991b). An important part of landfill design is the development of maps and plans which describe the landfill's construction and operation, including the location map, base map, site preparation map, devel-

TABLE 10.13.2 SANITARY LANDFILL DESIGN STEPS

1. Determination of solid waste quantities and characteristics
   a. Existing
   b. Projected
2. Design of filling area
   a. Selection of landfilling method based on site topography, bedrock, and groundwater
   b. Specification of design dimensions: cell width, length, and depth; fill depth; liner thickness; interim cover thickness; and final cover thickness
   c. Specification of operational features: method of cover application, need for imported soil for cover or liner, equipment requirements, and personnel requirements
3. Design features
   a. Leachate controls
   b. Gas controls
   c. Surface water controls
   d. Access roads
   e. Special working areas
   f. Special waste handling
   g. Structures
   h. Utilities
   i. Convenience center
   j. Fencing
   k. Lighting
   l. Washracks
   m. Monitoring facilities
   n. Landscaping
4. Preparation of design package
   a. Development of preliminary site plan of fill areas
   b. Development of landfill contour plans: excavation plans; sequential fill plans; completed fill plans; fire, litter, vector, odor, surface water, and noise controls
   c. Computation of solid waste storage volumes, cover soil requirement volumes, and site life
   d. Development of final site plan showing normal fill areas; special working areas (i.e., wet weather areas), leachate controls, gas controls, surface water controls, access roads, structures, utilities, fencing, lighting, washracks, monitoring facilities, and landscaping
   e. Preparation of elevation plans with cross sections of excavated fill, completed fill, and phase development of fill at interim points
   f. Preparation of construction details: leachate controls, gas controls, surface water controls, access roads, structures, and monitoring facilities
   g. Preparation of ultimate landuse plan
   h. Preparation of cost estimate
   i. Preparation of design report
   j. Preparation of environmental impact assessment
   k. Submission of application and obtaining required permits
   l. Preparation of operator's manual

Source: Data from P. Walsh and P. O'Leary, 1991, Landfill site plan preparation, *Waste Age* 22, no. 9: 97–105 and E. Conrad et al., 1981, *Solid waste landfill design and operation practices,* EPA draft report, Contract no. 68-01-3915.

opment plans, cross sections, phase plans, and the completed site map (Walsh and O'Leary 1991c).

The location map is a topographic map which shows the relationship of the landfill to surrounding communities, roads, etc. The base map usually has a scale of 1 in to 200 ft and contour lines at 2 to 5 ft intervals. It includes the property line, easements, right-of-ways, utility corridors, buildings, wells, control structures, roads, drainage ways, neighboring properties, and land use. The site preparation map shows fill and stockpile areas and site facilities. The landfill should be designed so that the excavated material is used quickly as cover. Development plans show the landfill base and top elevations and slopes. Cross sections at various places and times during the landfill lifetime should also be developed. Phase plans show the order in which the landfill is constructed, filled, and closed. The completed site map shows the elements of the proposed end use and includes the final landscaping. Construction details should be available detailing leachate controls, gas controls, surface water controls, access roads, structures, and monitoring facilities.

Equally important as the design maps is the site design report, which describes the development of the landfill in sequence (Walsh and O'Leary 1991b). The four major elements of the design report are:

- Site description
- Design criteria
- Operational procedures
- Environmental safeguards

Landfill Types

Two types of lined landfills are excavated and area. At excavated landfills, soil is excavated from the area where waste is to be deposited and saved for use as daily, intermediate, or final cover. Excavated landfills are constructed on sites where excavation is economical and the water table is sufficiently below the ground surface. Area landfills do not involve soil excavation and are built where excavation is difficult or the water table is near the surface. All cover soil is imported to area landfills. Both types of landfills are lined; the excavated landfill on the bottom of the excavation, the area landfill on the ground surface.

If the entire available area is lined at the beginning of a landfill's life, the large lined area collects rainwater for the life of the landfill, generating a large quantity of unnecessary leachate. For this reason, landfill liners and leachate collection systems are constructed in phases. Each phase consists of construcing a liner and leachate collection system on a portion of the available area, depositing waste in the lined area, and installing intermediate cover. Construction of the next phase begins before the current

TABLE 10.13.3 LANDFILL DESIGN FACTORS

Factors	Remarks
Access	Paved all-weather access roads to landfill site; temporary roads to unloading areas
Land area	Area large enough to hold all community waste for a minimum of five years, but preferably ten to twenty-five years; area for buffer strips or zones also
Landfilling method	Based on terrain and available cover; most common methods are excavated and area landfills
Completed landfill characteristics	Finished slopes of landfill, 3 or 4 to 1; height to bench, if used, 50 to 75 ft; slope of final landfill cover, 3 to 6%
Surface drainage	Drainage ditches installed to divert surface water runoff; 3 to 6% grade maintained on finished landfill cover to prevent ponding; plan to divert storm water from lined but unused portions of landfill
Intermediate cover material	Use of onsite soil material maximized; other materials such as compost produced from yard waste and MSW also used to maximize the landfill capacity; typical waste-to-cover ratios from 5 to 1 to 10 to 1
Final cover	Multilayer design; slope of final landfill cover, 3 to 6%
Landfill liner, leachate collection	Multilayer design incorporating the use of a geomembrane and soil liners. Cross slope for leachate collection systems, 1 to 5%; slope of drainage channels, 0.5 to 1.0%. Size of perforated pipe, 4 in; pipe spacing, 20 ft
Cell design and construction	Each day's waste forms one cell; cover at end of day with 6 in of earth or other suitable material; typical cell width, 10 to 30 ft; typical lift height including intermediate cover, 10 to 14 ft; slope of working faces, 2:1 to 3:1
Groundwater protection	Any underground springs diverted; if required, perimeter drains, well point system, or other control measures installed. If leachate leakage occurs, control with impermeable barriers, pump and treat, or active or passive bioremediation
Landfill gas management	Landfill gas management plan developed including extraction or venting wells, manifold collection system, condensate collection facilities, vacuum blower facilities, flaring facilities, and energy production facilities; operating vacuum located at well head, 10 in of water
Leachate collection	Maximum leachate flow rates determined and leachate collection pipe and trenches sized; leachate pumping facilities sized; collection pipe materials selected to withstand static pressures corresponding to the maximum height of the landfill
Leachate treatment	Pretreatment determined based on expected quantities of leachate and local environmental and political conditions
Environmental requirements	Vadose zone gas and liquid monitoring facilities installed; up- and downgradient groundwater monitoring facilities installed; ambient air monitoring stations located
Equipment requirements	Number and type of equipment based on the type of landfill and the capacity of the landfill
Fire prevention	Onsite water available; if nonpotable, outlets must be marked clearly; proper cell separation prevents continuous burn-through if combustion occurs

Source: Data from G. Tchobanoglous, H. Theissen, and S. Vigil, 1993, *Integrated solid waste management: Engineering principles and management issues* (New York: McGraw-Hill).

phase is filled so that it is ready to receive waste as soon as the current phase is filled. The liners and leachate collection systems of adjacent phases are usually tied together.

The size of landfill phases depends on the rate at which waste is deposited in the landfill, local precipitation rates, state permitting practice, and site topography. At landfills receiving large amounts of waste per day, phase size can be chosen so that phase construction equipment is always in use. As soon as the construction of one phase ends, construction of the next phase begins. Smaller landfills cannot operate this way.

Figure 10.13.3 shows several points in a normal, excavated, landfill lifetime, simplified because the landfill has only one phase. Part (a) shows the landfill just before waste is deposited. The liner is installed at grades that cause leachate to flow toward leachate collection pipes. Groundwater monitoring wells are also installed. Part (b) shows the second waste lift of an operating landfill cell being created. Each lift consists of a layer of daily waste cells. Each daily cell consists of the waste deposited during a single operating day. Daily cells are separated by the cover soil applied at the end of each day. To keep the daily cover and litter to a minimum, operators should keep the working face as small as possible.

Temporary roads on the landfill allow truck traffic easy access to the working face of the landfill. During wet weather, use of a special easy access area for waste disposal may be necessary.

Part (c) of Figure 10.13.3 shows the completed landfill. Five lifts are created, the final cover is installed, vegetation is established, and gas collection wells are installed.

Landfill excavations have sloping bottoms and sides. Excavated side slopes are generally not more than a ratio of 1:1. Their stability must be checked, typically with rotational or sliding-block methods. Bottom slopes are generally 1 to 5%. However, when landfills are built on sloping terrain, bottom slopes can be steeper, requiring stability analysis as well. Operators must also check the stability of the synthetic liner on steeper slopes to ensure that it does not slip or tear. This analysis is based on the friction force between the liner and the material just below the liner. Planners should estimate the bearing capacity of the soil below the landfill and future settlement to ensure that problems associated with differential settling do not ensue after waste is deposited in the landfill. Finally, the pipes used in the leachate collection system must be able to bear the weight of the waste placed on top.

The side slopes of the top of the landfill are generally a ratio of 3:1 or 4:1. Large landfills have benches, or terraces, on the side slope to help reduce erosion by slowing down water as it flows down the sides. The central portion of the top of the landfill is relatively flat because height limitations keep landfills from being pointed cones. However, a slight slope (3 to 6%) is maintained to encourage run-off.

Leachate Control

Water brought in with the waste, precipitation, and surface run-on can increase the amount of water in the landfill, called leachate. Leachate, especially from new landfills, can have high concentrations of COD, BOD, nutrients, heavy metals, and trace organics (Tchobanoglous, Theissen, and Vigil 1993). Leachate that contacts drinking water supplies can result in contamination. For this reason, liners and collection systems are used to minimize the leachate that escapes from landfills. Unless testing indicates that it is not a pollutant, collected leachate is treated before being released in a controlled manner into the environment.

The factors affecting leachate generation are climate, site topography, the final landfill cover material, the vegetative cover, site phasing and operating procedures, and the type of waste material in the fill (O'Leary and Walsh 1991c). Obviously, with all else equal, the more rainfall, the more infiltration into the landfill and the more leachate. Topography can affect the amount of water entering or leaving the landfill site. One purpose of the final cover is to keep water from entering the fill. Current federal regulations require the final cover to have a hydraulic conductivity at least as low as the bottom composite liner. Unless exemptions are made, this requirement means that the final cover must include a geosynthetic layer. If a drainage layer is included in the final cover, this layer further reduces the amount of water infiltrating the fill. Vegetative cover on the final cover reduces infiltration by intercepting precipitation and encouraging transpiration. As already mentioned, proper site phasing keeps the amount of exposed liner area small, thus reducing the collection of rainwater. Finally, the waste deposited in the landfill contains some water, and the resulting moisture content varies with location and waste type. For example, wastewater treatment plant sludges contain significant amounts of moisture. Planners can estimate the amount of leachate generated by a landfill using water balance equations or the EPA's HELP model (Tchobanoglous, Theissen, and Vigil 1993; O'Leary and Walsh 1991c).

Leachate controls are the final cover, the surface water controls that keep water from running onto the landfill, the liner, the leachate collection system, the leak detection system, and the leachate disposal system.

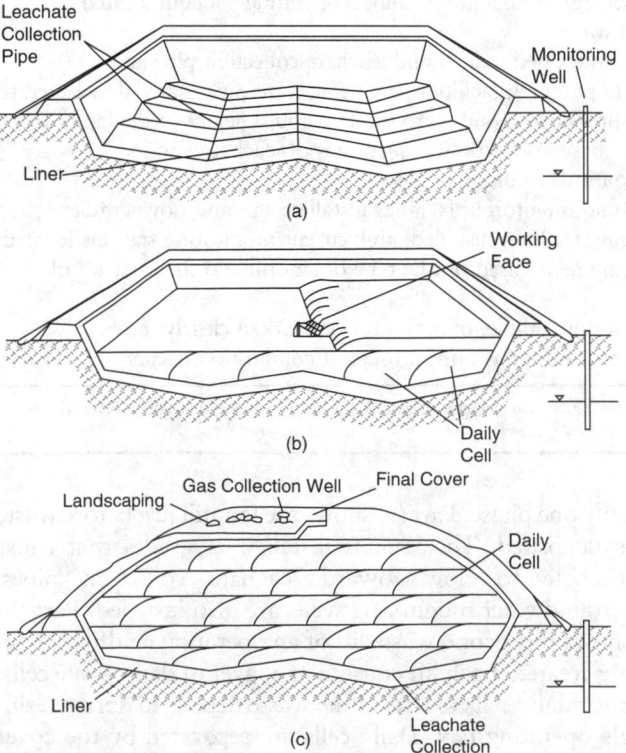

FIG. 10.13.3 Development of a landfill: (a) excavation and installation of landfill liner, (b) placement of solid waste in landfill, (c) cutaway through completed landfill. (Adapted from G. Tchobanoglous, H. Theissen, and S. Vigil, 1993, *Integrated solid waste management: Engineering principles and management issues* [New York: McGraw-Hill].)

FINAL COVER AND SURFACE WATER CONTROLS

The final cover creates a relatively impermeable barrier over the fill area which keeps rainwater from entering. The slope, soil type, and vegetation determine the surface water run-off characteristics of the site. Planners can determine run-off quantities and peak flows using standard hydrologic run-off techniques, such as the rational method or TR 55. The control of surface run-off generally requires berms to be constructed around the fill area, but drainage ditches can also be used. A detention pond is generally required as well.

LINERS

A landfill can be thought of as a bathtub. Liners make the bottom and sides of the landfill less permeable to the movement of water. Figure 10.13.4 shows a typical liner system. Federal regulations call for a composite liner, consisting of an HDPE geomembrane at least 60 mil thick (1000 mil equals one inch) over 2 ft of compacted soil (clay) with a hydraulic conductivity of less than 1×10^{-7} cm/sec. Equivalent or better alternative liner systems are approved in some cases.

Constructing the soil liner requires spreading and compacting impermeable soil in several lifts, ensuring that the soil contains near optimum moisture content and compaction for minimum permeability. Compaction is usually done with large vehicles with sheepsfoot wheels. Synthetic membranes used in composite landfill liners must be at least 60 mil thick. These membranes can be damaged by heavy equipment and are generally protected with a carefully applied layer of sand, soil, or MSW. Geotextiles can also be used to protect geosynthetic liners.

COLLECTION AND LEAK DETECTION SYSTEMS

Just as a bathtub has a drain to remove bath water, the landfill has a mechanism to remove leachate. The liner in Figure 10.13.3(a) is graded to direct any leachate reaching the liner surface into a leachate collection system. Liner systems are not leak proof. Collecting leachate and removing it from the landfill reduces the hydraulic head on the liner, thus reducing fluid flow through the liner. To speed the lateral flow of leachate once it reaches the bottom of the landfill, a drainage layer is placed over the composite liner (see Figure 10.13.4). The drainage layer can be made of coarse media such as sand or shredded tires, though geonets (high-strength geosynthetic grids less than $\frac{1}{2}$ in thick capable of transmitting high quantities of water) are also common. Geotextiles minimize clogging of the drainage layers by excluding particles. Drainage layers slope toward collection pipes, which direct leachate toward a sump or directly out of the landfill.

Figure 10.13.5 shows a typical leachate collection pipe cross section. The leachate collection pipe is laid in a gravel trench wrapped with a geotextile which allows water to enter the leachate trench but keeps out small particles that could clog the gravel or pipe. Leachate collection trenches lay on top of the liner and travel along local hydraulic low points. The leachate collection system carries leachate out of the landfill cell through the liner or dumps leachate into a sump which is pumped over the side of the liner.

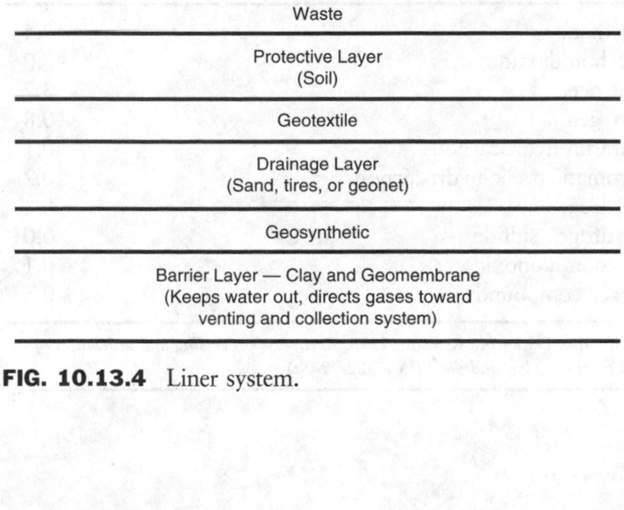

LEACHATE DISPOSAL SYSTEMS

Leachate can be treated by recycling, onsite treatment, or discharge to a municipal wastewater treatment plant. Recycling leachate involves reapplying collected leachate at or near the top of the landfill surface, thus providing additional contact between leachate and landfill microbes. Recycling can reduce BOD and COD and increase pH—with subsequent reduction in heavy metals concentrations. Furthermore, leachate recycling evens the flow of leachate that is removed from the landfill and can enhance the stabilization of the landfill (O'Leary and Walsh, 1991c). Onsite treatment can involve physical, chemical, or biological treatment processes. However, leachate from recently deposited waste is a high-strength wastewater. Furthermore, leachate characteristics change dramatically

Waste
Protective Layer (Soil)
Geotextile
Drainage Layer (Sand, tires, or geonet)
Geosynthetic
Barrier Layer — Clay and Geomembrane (Keeps water out, directs gases toward venting and collection system)

FIG. 10.13.4 Liner system.

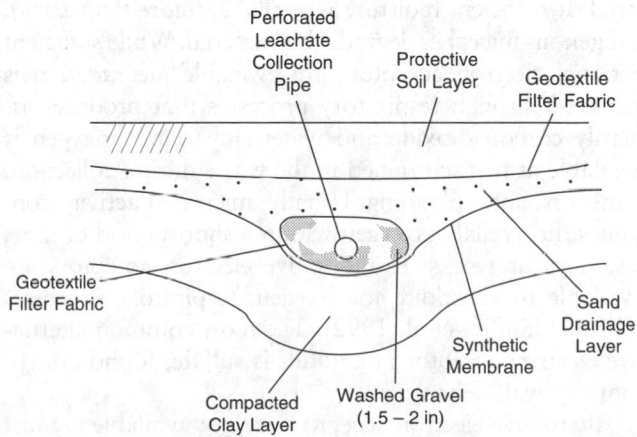

FIG. 10.13.5 Leachate collection pipe and trench. (Adapted from G. Tchobanoglous, H. Theissen, and S. Vigil, 1993, *Integrated solid waste management: Engineering principles and management issues* [New York: McGraw-Hill].)

over the course of the landfill's life. Consequently, treatment processes should be carefully designed and constructed. The most common option is to use a nearby municipal wastewater treatment plant. Leachate is usually transported to the facility by tanker truck, but a pipeline is economic in some cases. Using a municipal wastewater treatment plant to treat a high-strength wastewater may involve extra charges or pretreatment requirements.

LEACHATE MONITORING

The last leachate control element is monitoring. Some landfills use lysimeters, geosynthetic membranes placed in the ground, to detect and collect material directly under the landfill. However, monitoring is most commonly accomplished by collecting groundwater from wells located around the landfill, both upgradient and downgradient of the landfill. Upgradient wells are important in determining whether downgradient contamination is caused by the landfill or some upgradient event. Groundwater is monitored regularly for a number of inorganic and organic constituents (*CFR* 40 Parts 257–258). Detection of a contaminant at a statistically significant higher concentration than background levels results in increased monitoring requirements. Detection of contaminants at concentrations above groundwater protection levels requires the operator to assess corrective measures. Based on this assessment, a corrective measure is selected that protects human health and the environment, attains the applicable groundwater protection standards, controls the source(s) of release to the maximum extent possible, and complies with the applicable standards for managing any waste produced by the corrective measures. Corrective measures may involve pump and treat, impermeable barriers, or bioremediation.

Gas Control

Waste material deposited in landfills contains organic material. If sufficient moisture is available (more than 20%), indigenous microbes degrade this material. While sufficient external electron acceptors are available, degradation is achieved through respiratory processes that produce primarily carbon dioxide and water. Invariably, oxygen is available at first, entrained in the waste during collection, transport, and unloading. Usually, microbial activity consumes the available oxygen within a short period of time, i.e., days or weeks. If alternative electron acceptors are available to substitute for oxygen, respiratory processes continue (Suflita et al. 1992). The most common alternative electron acceptor in landfills is sulfate, found in gypsum dry wall debris.

Alternative electron acceptors are not available in most of the volume of a typical landfill; subsequently, fermentative processes predominate, ultimately producing landfill gas that is primarily carbon dioxide and methane (see Table 10.13.4). Observed gas yields are less than the theoretical maximum based on stoichiometry and are generally in the range of 4000 cu ft per tn of waste (O'Leary and Walsh 1991b).

Landfill gas must be removed from landfills. The final cover, used to keep out water and support vegetation for erosion control, can trap landfill gases. A build-up of gas in a landfill can rupture the final cover. In addition, the vegetative cover can be killed if the pore space in the final cover topsoil becomes saturated with landfill gases. Finally, methane is explosive if present in sufficient concentration, above 5%. Methane traveling through the landfill or surrounding soils can collect at explosive concentrations in nearby buildings. Migration distances greater than 1500 ft have been observed (O'Leary and Walsh 1991a). For these reasons, landfill gases must be vented or collected.

Gas control can be accomplished in a passive or active manner. Passive landfill gas control relies on natural pressure and convection to vent gas to the atmosphere or flares. Passive systems consist of gas venting trenches or wells, either in the landfill or around it. However, passive systems are not always successful because the pressure generated by gas production in the landfill may not be enough to push landfill gas out.

Active gas control removes landfill gases by applying a vacuum to the landfill. In other words, the landfill gases are pumped out. However, overpumping draws air into the landfill, slowing the production of more methane. After expensive landfill gas extraction equipment is installed, slowing methane production is not desirable. If migration control is the primary purpose of active gas control, recovery wells can be placed near the perimeter of the landfill. However, landfill gas can be an energy source, in which case vertical or horizontal recovery wells are typically placed in the landfill. Landfill gas with 50% methane has a heating value of 505 Btu/standard cu ft, about half that of natural gas (O'Leary and Walsh 1991a).

Collected landfill gas can be vented, burned without energy recovery, or directed to an energy recovery system.

TABLE 11.13.4 TYPICAL LANDFILL GAS COMPONENTS

Component	Percent
Methane	47.4
Carbon dioxide	47.0
Nitrogen	3.7
Oxygen	0.8
Paraffin hydrocarbons	0.1
Aromatic-cyclic hydrocarbons	0.2
Hydrogen	0.1
Hydrogen sulfide	0.01
Carbon monoxide	0.1
Trace compounds	0.5

Source: Data from R. Ham, 1979, *Recovery, processing and utilization of gas from sanitary landfills*, EPA 600/2-79-001.

When energy is to be recovered, the gas can be piped directly into a boiler, upgraded to pipeline quality, or cleaned and directed to an onsite electricity engine-generator. The first two options are feasible only if a boiler or gas pipeline is located near the landfill, which is not common.

Because of the explosive and suffocative properties of landfill gases, special safety precautions are recommended (O'Leary and Walsh 1991a):

- No person should enter a vault or trench on a landfill without checking for methane gas or wearing a safety harness with a second person standing by to pull him to safety.
- Anyone installing wells should wear a safety rope to prevent falling into the borehole.
- No smoking is allowed while gas wells or collection systems are being drilled or installed or when gas is venting from the landfill.
- Collected gas from an active system should be cleared to minimize air pollution and a potential explosion and fire hazard.

Personnel entering the landfill through gas collection manholes must carry an air supply.

Gas monitoring wells should be placed around the landfill if methane migration could threaten nearby buildings. Gas wells are used to measure gas pressure and to recover gas from soil pore space. The explosive potential of gases can be measured with portable equipment.

Site Preparation and Landfill Operation

Site preparation involves making a site ready to receive MSW and can include (O'Leary and Walsh 1991d):

- Clearing the site
- Removing and stockpiling the soil
- Constructing berms around the landfill for aesthetic purposes and surface water control. Berms are usually constructed around each landfill phase.
- Installing drainage improvement, if necessary. These improvements can include drainage channels and a lagoon.
- Excavating fill areas as phases are built (only for excavated landfills)
- Installing environmental protection facilities, including a liner, leachate collection system, gas control equipment, groundwater monitoring equipment, and gas monitoring equipment
- Preparing access roads
- Constructing support facilities, including a service building, employee facilities, weigh scale, and fueling facilities
- Installing utilities, including electricity, water, sewage, and telephone

- Constructing fencing around the perimeter of the landfill
- Constructing a gate and entrance sign as well as landscaping
- Constructing a convenience center, either for small vehicles to unload waste (to minimize traffic at the working face) or for the collection of recyclables
- Installing litter control fences
- Preparing construction documentation

An efficient landfill is operated so that vectors, litter, and environmental impacts are minimized, compaction is maximized, worker safety is ensured, and regulations are met or exceeded. Regulations control or influence much of the daily landfill operation. For example, regulations require some or all of the following:

- Traffic control
- An operating plan
- Control of public access, unauthorized traffic, litter, dust, disease vectors, and uncontrolled waste dumping
- Measurement of all refuse
- Control of fires
- Minimization of the working face area
- Minimization of litter scatter from the working area
- Frequent cleaning of the site and site approaches
- 6 in of soil cover on exposed waste at the end of the operating day
- Special provisions to handle bulky wastes
- Separation of salvage or recycling operations from the working face
- Exclusion of domestic animals
- Safety training for employees
- Annual reports and daily record keeping

Landfill equipment falls into four groups: site construction; waste movement and compaction; cover movement, placement, and compaction; and support functions (O'Leary and Walsh 1991d). Conventional earth moving equipment is usually used in landfill construction. However, specialized equipment is required for liner installation. The vehicles that bring waste to the landfill dump on the working face. Therefore, operators accomplish waste movement and compaction at the landfill by moving and spreading the waste around the working face and traveling over it several times with heavy equipment, usually compactors or dozers. If soil is used as cover material, it is transported using scrapers or trucks. If trucks are used, additional equipment is needed for loading. Cover soil compaction is done by the same equipment that compacts the waste. The use of an alternative cover material, such as foam or blankets, may require special equipment. A common support vehicle is the water truck, which reduces road dust and controls fires. The selection of landfill equipment depends on budget and the daily capacity of the site.

Closure, Postclosure, and End Use

Both the design and operation must consider the closure and postclosure periods, as well as the end use. Typical end uses include green areas, parks, and golf courses. As phases are closed, the final or intermediate cover may be applied depending on whether the top elevation of the landfill has been reached. Vertical gas vents or recovery wells can be installed as the final elevations are reached. Horizontal gas recovery wells are installed at specified height intervals as the phases are filled. As the side slopes of the landfill are completed, many aspects of final closure can also be completed, including final cover installation and revegetation.

Figure 10.13.6 shows a typical final cover cross section. The surface layer consists of top soil and is used to support vegetation. The vegetation reduces erosion and aesthetically improves the landfill. Grasses are the most common vegetation used, but other plants are used, including trees. Just below the surface layer is the optional drainage layer, used to minimize the hydraulic head on the barrier layer. The drainage layer can be sand or a geonet and is protected from clogging by a geotextile. The next layer is the hydraulic barrier. Current regulations require it to have a hydraulic conductivity at least as low as the bottom liner. Therefore, the barrier layer usually includes a geomembrane. A subbase layer may be necessary to protect the barrier layer.

Final closure involves installing the remaining final cover, planting the remaining vegetation, and adding any fencing required to maintain site security. Revegetation depends on a number of factors (O'Leary and Walsh 1992b). First, the cover soil must be deep enough to sustain the planted species. Grasses require at least 60 cm, while trees require at least 90 cm. The final cover topsoil should be

stabilized with vegetation as soon as possible to avoid erosion. Operators should determine the soil characteristics before planting and add lime, fertilizer, or organic matter as required. The bulk density should be measured, and, if too high, amended. Species should be chosen that are landfill tolerant (Gilman, Leone, and Flower 1981; Gilman, Flower, and Leone 1983). Grasses and ground covers should be planted first. If possible, seeds should be embedded in the soil. Trees or shrubs, if used, should be planted only one or two years after grasses are planted. If grasses cannot survive on the landfill, the same will be true of trees and shrubs. The most common problems encountered with revegetation of landfill surfaces are poor soil, root toxicity, low oxygen concentration in the soil pore space, low nutrient value, low moisture content, and high soil temperature. Operators should develop a landfill closure plan which addresses control of leachate and gases, drainage and cover design, and environmental monitoring systems. The postclosure period is currently specified by regulation to be at least thirty years after closure. During this time, surface water drainage control, gas control, leachate control, and monitoring continue. The general problems that must be addressed during this period are the maintenance of required equipment and facilities, the control and repair of erosion, and the repair of problems associated with differential settlement of the landfill surface.

Special Landfills

The distinction between the modern sanitary landfill and hazardous waste landfill is blurred, except the latter usually has two or three liner systems and multiple leachate collection systems (O'Leary and Walsh 1992a). Landfills similar to the sanitary landfill are sometimes built to handle special waste. Special waste is high-volume waste that is not hazardous and can be easily handled separate from the municipal waste stream.

Separate disposal is advantageous if a dedicated disposal facility is required, the waste is perceived to have special associated risks, or the waste carries a lower risk than MSW. An example of a waste with special risks is infectious waste which, though relatively innocuous in the ground, must be handled with special care so that disposal facility workers are not infected. In this case, a dedicated facility may be required for worker safety. An example of a low-risk waste is construction and demolition waste. In this case, using a disposal facility with lower performance standards can reduce disposal cost. Thus, a special landfill is dedicated to one or a few classes of special waste material. Examples of special material include coal-fired electric power plant ash, MSW incinerator ash, construction and demolition debris, infectious waste, asbestos, or any nonhazardous industrial waste subject to subtitle D regulations.

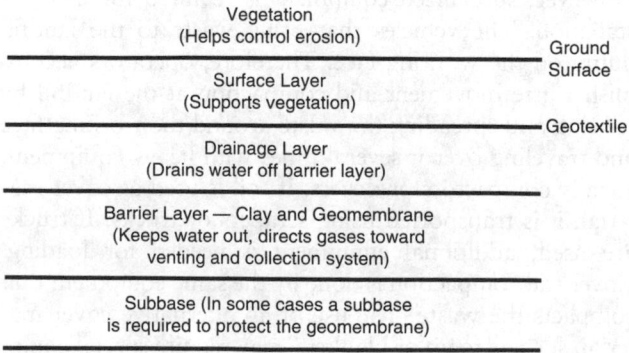

FIG. 10.13.6 Typical final cover.

Conclusion

In the United States, the landfill is the most popular disposal option for MSW. Traditionally, it has been the least-cost disposal option, and it is also a solid waste management necessity because no combination of reduction, recycling, composting, or incineration can currently manage the entire solid waste stream. Developing a new landfill involves site location, landfill design, site preparation, and landfill construction. Locating a new landfill can involve significant public participation. Federal regulations specify many location, design, operation, monitoring, and closure criteria. These regulations reduce the incidence of unacceptable pollution caused by landfills.

—J.W. Everett

References

Anandalingham, G. and M. Westfall. 1988–1989. Selection of hazardous waste disposal alternative using multi-attribute theory and fuzzy set analysis. *Journal of Environmental Systems* 18, no. 1: 69–85.

Camp Dresser & McKee Inc. 1984. *Cumberland County landfill siting report.* Edison, N.J.

CFR 40 Parts 257 and 258. *Federal Register* 56, no. 196: 50978–51119.

Erkut, E. and S. Moran. 1991. Locating obnoxious facilities in the public sector: An application of the analytic hierarchy process to the municipal landfill siting decision. *Socio-Economic Planning Sciences* 25, no. 2: 89–102.

Gilman, E., F. Flower, and I. Leone. 1983. Standardized procedures for planting vegetation of completed sanitary landfill. EPA 600/2-83-055.

Gilman, E., I. Leone, and F. Flower. 1981. The adaptability of 19 woody species in vegetating a former sanitary landfill. *Forest Science* 27, no. 1: 13–18.

Morrison, T.H. 1974. Sanitary landfill site selection by the weighted rankings method. Masters thesis, University of Oklahoma, Norman, Okla.

Noble, G. 1992. *Siting landfills and other LULUs.* Lancaster, Pa.: Technomic Publishing Company, Inc.

O'Leary, P. and P. Walsh. 1991a. Landfill gas: Movement, control, and uses. *Waste Age* 22, no. 6: 114–122.

———. 1991b. Landfilling principles. *Waste Age* 22, no. 4: 109–114.

———. 1991c. Leachate control and treatment. *Waste Age* 22, no. 7: 103–118.

———. 1991d. Sanitary landfill operation. *Waste Age* 22, no. 11: 99–106.

———. 1992a. Disposal of hazardous and special waste. *Waste Age* 23, no. 3: 87–94.

———. 1992b. Landfill closure and long-term care. *Waste Age* 23, no. 2: 81–88.

Siddiqui, M. 1994. Municipal solid waste landfill site selection using geographical information systems. Masters thesis, University of Oklahoma, Norman, Okla.

Suflita, J., C. Gerba, R. Ham, A. Palmisano, W. Rathje, and J. Robinson. 1992. The world's largest landfill: A multidisciplinary investigation. *Environmental Science and Technology* 26, no. 8: 1486–1495.

Tchobanoglous, G., H. Theissen, and S. Vigil. 1993. *Integrated solid waste management: Engineering principles and management issues.* New York: McGraw-Hill.

Walsh, P. and P. O'Leary. 1991a. Evaluating a potential sanitary landfill site. *Waste Age* 22, no. 8: 121–134.

———. 1991b. Landfill site plan preparation. *Waste Age* 22, no. 10: 87–92.

———. 1991c. Sanitary landfill design procedures. *Waste Age* 22, no. 9: 97–105.

10.14
COMPOSTING OF MSW

In the United States, 180 million tn, or 4.0 lb per person per day of MSW were generated in 1988 (U.S. EPA 1990). The rate of generation has increased steadily between 1960 and 1988, from 88 million to 180 million tn per day (U.S. EPA 1990). Furthermore, the rate continues to increase (Steuteville and Goldstein 1993). In 1988, 72% of the MSW was landfilled. At the same time, due to strict federal regulations, mainly the RCRA, the number of landfills has decreased (U.S. Congress 1989). For the protection of human health and the environment, old landfills are being closed and new ones must be carefully constructed, operated, and monitored even when the landfill is closed. Thus the cost of disposing MSW by landfilling has greatly increased.

The increasing rate of generation, decreasing landfill capacity, increasing cost of solid waste management, public opposition to all types of management facilities, and concerns for the risks associated with waste management has led to the concept of integrated solid waste management (U.S. EPA 1988). Integrated solid waste management refers to the complementary use of a variety of waste management practices to safely and effectively handle MSW with minimal impact on human health and the environment. An integrated system contains some or all of the following components:

- Source reduction
- Recycling of materials

- Incineration
- Landfilling

The U.S. EPA recommends a hierarchical approach to solve the MSW generation and management problems. Using the four components of integrated solid waste management, the hierarchy favors source reduction, which is aimed at reducing the volume and toxicity of waste. Recycling is the second favored component. Recycling diverts waste from landfills and incinerators and recovers valuable resources. Landfills and incinerators are lower in the hierarchy but are recognized as necessary in the foreseeable future to handle some waste.

Essentially, the goal of integrated solid waste management is to promote source reduction, reuse, and recycling while minimizing the amount of waste going to incinerators and landfills. Composting is included in the recycling component of the hierarchy. This section discusses the composting of MSW.

Aerobic Composting in MSW Management

The organic fraction of MSW includes food waste, paper, cardboard, plastics, textiles, rubber, leather, and yard waste. Organic material makes up about half of the solid waste stream (Henry 1991) (see Section 10.5). Almost all organic components can be biologically converted although the rate at which these components degrade varies. Composting is the biological transformation of the organic fraction of MSW to reduce the volume and weight of the material and produce compost, a humus-like material that can be used as a soil conditioner (Tchobanoglous, Theissen, and Vigil 1993).

Composting is gaining favor for MSW management (Goldstein and Steuteville 1992). It diverts organic matter from landfills, reduces some of the risks associated with landfilling and incineration, and produces a valuable byproduct (compost). At the present time, twenty-one MSW composting plants are operating in the United States (Goldstein and Steuteville 1992). Most of these plants compost a mixed MSW waste stream. This number does not include a larger number of operations which deal solely with organic material, primarily from commercial establishments (grocery stores, restaurants, and institutions) and those facilities composting yard waste. Finstein (1992) states that over 200 such yard waste facilities are in New Jersey alone.

Applications of aerobic composting for MSW management include yard waste, separated MSW, commingled MSW, and cocomposting with sludge.

SEPARATED AND COMMINGLED WASTE

Yard waste composting includes leaves, grass clippings, bush clippings, and brush. This waste is usually collected separately in special containers. Yard waste composting is increasing especially since some states, as a part of their waste diversion goals, are banning yard waste from landfills (Glenn 1992). The U.S. EPA (1989); Strom and Finstein (1985); and Richard, Dickson, and Rowland (1990) provide detailed descriptions of yard waste composting.

Separated MSW refers to the use of mechanical and manual means to separate noncompostable material from compostable material in the MSW stream before composting. The mechanical separation processes involve a series of operations including shredders, magnetic separators, and air classification systems. The sequence is often referred to as front-end processing. Front-end processing prepares the feedstock for efficient composting in terms of homogeneity and particle size. Front-end processing also removes the recyclable components and thus insures a higher-quality compost product since the material which causes product contamination is removed. Still, significant amounts of metals and trace amounts of household hazardous waste are often found after mechanical separation. For this reason, source-separated material is the preferred feedstock to produce the highest quality compost product.

On the other hand, composting partially processed, commingled MSW can divert waste from landfills when the product quality is not too demanding. The compost can also be used as intermediate landfill cover (Tchobanoglous, Theissen, and Vigil 1993). Recently, a planning guide was published for mixed organic composting (Solid Waste Composting Council 1991).

The organic fraction of MSW can be mixed with wastewater treatment plant sludge for composting. This process is commonly known as cocomposting. In general, a 2:1 mixture of compostable MSW to sludge is used as the starting point. Sludge dewatering may not be necessary.

While MSW contains a high percentage of biodegradable material (yard waste, food waste, and paper), one must decide prior to composting whether to keep the organic material separate from the other components of MSW or to begin with mixed MSW and extract the organic material later for composting. For example, yard waste (particularly leaves) is often kept separate from the rest of MSW and composted. This separation allows easier composting (than with mixed MSW) and yields a product with low levels of contamination. The disadvantage is that separate collection of yard waste is necessary.

COCOMPOSTING RETRIEVED ORGANICS WITH SLUDGE

The principles of composting and a description of the process technology are presented in Section 7.43 for sludge composting. While the fundamentals of sludge composting are applicable to MSW composting, several significant differences exist. The major difference involves preprocessing when MSW is composted. As shown in Figure

10.14.1, receiving, the removal of recoverable material, size reduction, and the adjustment of waste properties (e.g., the C:N ratio and the addition of moisture and nutrients) are essential steps in preparing MSW for composting. Obviously, different preprocessing strategies are needed for source-separated organic MSW and yard waste. Also, the degree of preprocessing depends on the type of composting process used and the specifications for the final compost product (Tchobanoglous, Theissen, and Vigil 1993).

MSW composting employs the same techniques as sludge composting: windrow, aerated static pile, and in-vessel systems. Tchobanoglous, Theissen, and Vigil (1993)

note that over the past fifty years, more than fifty types of proprietary commercial systems have been developed and applied worldwide. In general, they are variations of these three basic techniques.

Municipal Composting Strategies

Today, a large degree of public opposition to all types of waste management facilities and concerns for the risks associated with waste management exists. Composting, however, is often perceived as a safer alternative to either landfilling or incineration (Hyatt 1991) and is ranked higher

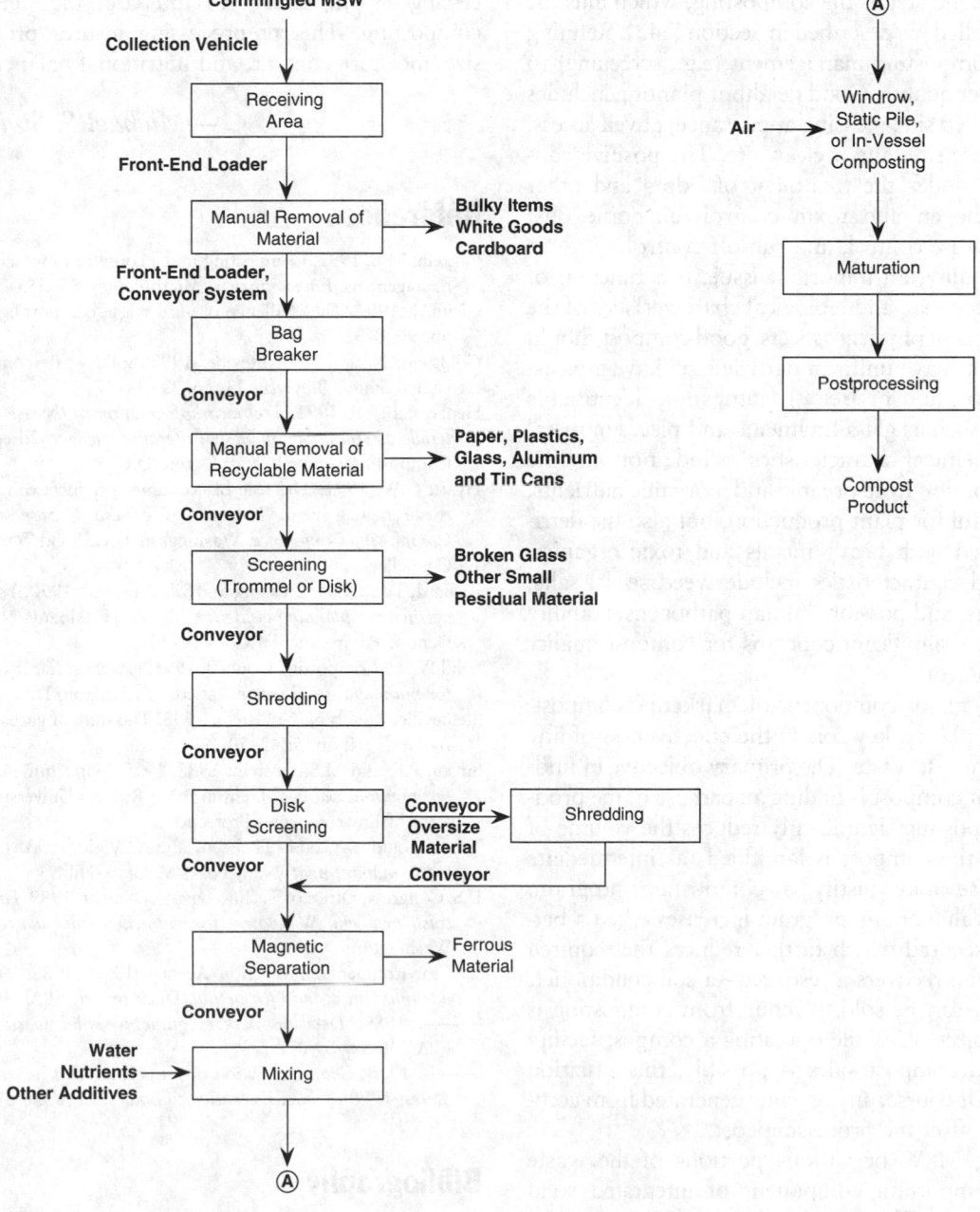

FIG. 10.14.1 Generalized flow diagram for the composting process. (Reprinted, with permission, from G. Tchobanoglous, H. Theissen, and S. Vigil, 1993, *Integrated solid waste management* [New York: McGraw-Hill].)

in the integrated solid waste management hierarchy. Nonetheless, composting facilities must be carefully planned and managed for successful operation. The key elements are elucidated by the Solid Waste Composting Council (1991) and include:

1. Recovery and preparation of compostables
2. Composting
3. Refining
4. Good neighbor planting
5. Positive control of litter, dust, odors, noise, and runoff

The first step involves preprocessing (as previously described). This processing results in the preparation of a good feed stock for composting and the recovery of recyclables. The second step is the composting, which must be properly controlled (as described in Section 7.43). Refining involves postcomposting management (e.g., screening) to improve product quality. Good neighbor planting includes a carefully selected site, pleasing appearance, paved access, parking, a secure site, and a clean site. The positive control element includes the treatment of odors and other emissions, pathogen and toxin control, air-borne dust management, noise control, and run-off control.

Compost quality, an important issue, is a function of the physical, chemical, and biological characteristics of the product. In terms of physical aspects, good compost should be dark in color; have uniform particle size; have a pleasant, earthy odor; and be free of clumps and identifiable contaminants, such as glass fragments and pieces of metal and plastic. Chemical characteristics include not only the positive contribution from organic and inorganic nutrients, which are helpful for plant production, but also the detriments associated with heavy metals and toxic organics. Other chemical characteristics include weed seeds, salts, plant pathogens, and possibly human pathogens. Stability and maturity are significant concerns for compost quality and process control.

Quality is a major component of marketing compost, and marketing plays a key role in the effectiveness of any program to compost waste. The primary objective in finding a market for compost is finding an end use of the product. Since composting significantly reduces the volume of MSW, even if the compost is landfilled (as intermediate cover) that use may justify a composting program. However, the value of any program increases when a better end use is secured which further reduces the required landfill space and recovers a resource—a soil conditioner. While compost can be sold, revenue from composting is a secondary objective. While operating a compost facility for profit from compost sales is possible, this situation rarely occurs. Of course, any revenue generated from compost sales can offset the processing cost.

Composting MSW or various portions of the waste stream is an important component of integrated solid waste management. The use of composting is part of the strategy of minimizing incineration and landfilling while promoting source reduction and recycling. At least 50% of the MSW stream is compostable. Composting diverts these materials from less beneficial disposal methods and provides a more environmentally sound MSW program.

A central issue is the tradeoff between collection ease and management concerns. Source separated organics are easier to compost and yield a compost product of higher quality but require separate collection. The use of composting processes and the type of waste to be composted (mixed MSW versus source separation) must be integrated into the overall waste management plan for a given region. In terms of mixed MSW, preprocessing is important to obtain a high-quality product. Regardless of the final compost use or source of feedstock, some degree of preprocessing is necessary to prepare the feedstock for composting. This preprocessing insures proper particle size, moisture content, and nutritional balance.

—*Michael S. Switzenbaum*

References

Finstein, M.S. 1992. Composting in the contest of municipal solid waste management. *Environmental Microbiology* 58: 355–374.

Glenn, J. 1992. The challenge of yard waste composting. *BioCycle* 33, no. 9: 30–32.

Goldstein, N. and R. Steuteville. 1992. Solid waste composting in the United States. *BioCycle* 33, no. 11: 44–47.

Henry, C.L., ed. 1991. *Technical information of the use of organic materials as soil amendments: A literature review.* 2d ed. Solid Waste Composting Council. Washington, D.C.

Hyatt, G.W. 1991. The role of consumer products companies in solid waste management. *Proceedings of the Northeast Solid Waste Composting Conference.* Washington, D.C.: Solid Waste Composting Council.

Richard, T.L., N.M. Dickson, and S.J. Rowland. 1990. *Yard waste management: A planning guide for New York.* Albany, N.Y.: N.Y. State Dept of Environmental Conservation.

Solid Waste Composting Council. 1991. *Compost facility planning guide for municipal solid waste.* 1st ed. Washington, D.C.

Steuteville, R. and N. Goldstein. 1993. The state of garbage in America. *BioCycle* 34, no. 5: 42–50.

Strom, P.F. and M.S. Finstein. 1985. *Leaf composting manual for New Jersey municipalities.* Trenton, N.J.: Rutgers University and the N.J. Dept of Environmental Protection.

Tchobanoglous, G., H. Theissen, and S. Vigil. 1993. *Integrated solid waste management.* New York: McGraw-Hill.

U.S. Congress, Office of Technology Assessment. 1989. *Facing America's trash problem. What next for municipal solid waste?* OTA-0-424. Washington, D.C.

U.S. Environmental Protection Agency (EPA). 1988. *The solid waste dilemma: an agenda for action.* Draft report. EPA/530/SW-88-052.

———. 1989. *Decision makers guide to solid waste management.* EPA/530-SW-89-072.

———. 1990. *Characterization of municipal solid waste in the United States: 1990 update. Executive summary.* EPA/530-SW-90-042A.

Bibliography

American Society for Testing and Materials (ASTM). 1989. Standard specifications for waste glass as a raw material for the manufacture of glass containers. E708-79 (Reapproved 1988). Vol. 11.04 in *1989 Annual book of standards*, 299–300, Philadelphia: ASTM.

Bagchi, A. 1990. Design, construction and monitoring of sanitary landfill. New York: John Wiley & Sons.

Baillie, R.C. and M. Ishida. 1971. Gasification of solid waste materials in fluidized beds. *69th National A.I.Ch.E. Meeting, Cincinnati, Ohio, May 1971*.

Bergvall, G. and J. Hult. 1985. *Technology, economics, and environmental effects of solid waste treatment*. Final report #3033, DRAV Project 85:11. Sweden (July).

Cal Recovery Systems, Inc. 1990. *Waste characterization for San Antonio, Texas*. Richmond, Calif. (June).

California Integrated Waste Management Board. 1991. Unpublished preliminary data from a waste characterization study in Downey and Commerce, CA. Study by CalRecovery, Inc., Hercules, Calif. (Samples collected July 1988; data dated 1991.)

CalRecovery, Inc. 1989. *Waste characterization study for Berkley, California*. (December).

———. 1992. *Conversion factor study—In-vehicle and in-place waste densities*. (March).

Camp Dresser & McKee, Inc. Unpublished data developed by field personnel during waste characterization studies.

———. 1989. *Polk County (FL) waste composition analysis*. (September).

———. 1990. *Cumberland County (NJ) waste weighing and composition analysis*. Edison, N.J. (January).

———. 1990. *Sarasota County (FL) waste stream composition study*. Draft report (March).

———. 1991. *Cape May County (NJ) multi-seasonal solid waste composition study*. Edison, N.J. (August).

———. 1991. *City of Ontario (CA) source reduction and recycling evaluation*. Ontario, Calif. (March).

———. 1991. *City of Wichita integrated solid waste management plan*. Wichita, Kans. (December).

———. 1992. *Atlantic County (NJ) solid waste characterization program*. Edison, N.J. (May).

———. 1992. *Bay County (FL) waste composition analysis report*. (September).

———. 1992. *Frederick County (VA) solid waste composition analysis*. Annandale, Va. (June).

———. 1992. *Jacksonville (FL) waste composition study*. Tallahassee, Fla.

———. 1992. *Prince William County (VA) solid waste supply analysis*. Annandale, Va. (October).

———. 1993. *Berkeley and Dorchester Counties (NC) waste characterization study*. Raleigh, N.C. (April).

———. 1993. *Lake County municipal solid waste characterization study*. Chicago (November).

———. 1993. *Scott Area (IA) municipal solid waste characterization study*. Chicago (February).

———. 1993. *Wake County/City of Raleigh (NC) commercial, institutional, and industrial solid waste characterization study*. Raleigh, N.C. (February).

Cashin Associates, P.C. 1990. *Town of Oyster Bay commercial waste stream analysis*. Plainview, N.Y. (July).

CFR 40 Parts 257 and 258. *Federal Register 56*, no. 196: 50978–51119.

CH2M Hill Engineering, Ltd. 1993. *Waste flow and recycling audit, Greater Vancouver Regional District*. Vancouver (January).

Conrad, E., J. Walsh, J. Atcheson, and R. Gardner. 1981. *Solid waste landfill design and operation practices*. EPA draft report, Contract no. 68-01-3915.

Diaz, L.F. et al. 1993. Composting and recycling municipal solid waste. Chap. 3 in *Waste characterization*. Boca Raton, Fla.: Lewis Publishers.

Glysson, E.A. 1989. Solid waste. In *Standard handbook of environmental engineering*. McGraw-Hill.

Goff, J.A. 1993. Waste from airports. *Waste Age* (January).

———. 1993. Waste from malls. *Waste Age* (February).

Graham, B. 1993. Collection equipment and vehicles. Chap. 27 in *The McGraw-Hill recycling handbook*, edited by H.F. Lund. McGraw-Hill, Inc.

Ham, R. 1979. *Recovery, processing and utilization of gas from sanitary landfills*. EPA 600/2-79-001.

Harrison, B. and P.A. Vesilind. 1980. Design and management for resource recovery. Vol. 2 of *High technology—A failure analysis*. Ann Arbor, Mich.: Ann Arbor Science.

Hill, R.M. 1986. Three types of low-speed shredder designs. *National Waste Processing Conference, Denver, 1986*. ASME.

Hilton, D., H.G. Rigo, and A.J. Chandler. 1992. Composition and size distribution of a blue-box separated waste stream. Presented at SWANA's Waste-to-Energy Symposium, Minneapolis, MN, January 1992.

Holmes, J.R. 1983. Waste management options and decisions. In *Practical waste management*, edited by J.R. Holmes. Chichester, England: John Wiley & Sons.

Institute for Solid Wastes, American Public Works Association. 1975. *Solid waste collection practice*. 4th ed. Chicago.

Kaiser, E.R. and S.B. Friedman. 1968. *Pyrolysis of refuse component combustion*. (May): 31–36.

Kaminski, D. 1986. Performance of the RDF delivery and boiler-fuel system at Lawrence, Massachusetts facility. *National Waste Processing Conference, Denver, 1986*. ASME.

Killam Associates. 1989; 1991. *Middlesex County (NJ) solid waste weighing, source, and composition study*. Millburn, N.J. (February).

Lipták, B.G. 1991. *Municipal waste disposal in the 1990s*. Radnor, Pa.: Chilton Book Company.

Liu, David H.F. 1974. Solid waste characterization. In *Environmental engineers handbook*, edited by B.G. Lipták. Radnor, Pa.: Chilton Book Company.

Lund, Herbert F. 1993. *The McGraw-Hill recycling handbook*. New York: McGraw-Hill, Inc.

Mallan, G.M. 1971. A total recycling process for municipal solid wastes. Paper 46C, *Nat. A.I.Ch.E., Atlantic City, August 29–September 1, 1971*.

Malloy, M.G. 1993. Waste from hospitals. *Waste Age* (July).

National Solid Wastes Management Association. 1985. Technical Bulletin 85-6. Washington, D.C. (October).

Non-Burn system for total waste stream. 1987. *BioCycle* (April): 30–31.

O'Leary, P. and P. Walsh. 1991. Landfill gas: Movement, control and uses. *Waste Age* 22, no. 6: 114–122.

———. 1991. Landfilling principles. *Waste Age* 22, no. 4: 109–114.

———. 1991. Leachate control and treatment. *Waste Age* 22, no. 7: 103–118.

———. 1991. Sanitary landfill operation. *Waste Age* 22, no. 11: 99–106.

———. 1992. Disposal of hazardous and special waste. *Waste Age* 23, no. 3: 87–94.

———. 1992. Landfill closure and long-term care. *Waste Age* 23, no. 2: 81–88.

Paper Stock Institute. *Guidelines for paper stock*. Washington, D.C.: Institute of Scrap Recycling Institute Inc.

Pfeffer, J. 1992. *Solid waste management engineering*. Englewood Cliffs, N.J.: Prentice-Hall.

Portland Metropolitan Service District. 1993. *Waste stream characterization study*. Results for fall 1993.

Preston, G.T. 1976. Resource recovery and flash pyrolysis of municipal refuse. Presented at Inst. Gas Technol. Symp., Orlando, FL, January 1976.

Rabasca, L. 1993. Waste from restaurants. *Waste Age* (March).

Robinson, W., ed. 1986. *The solid waste handbook*. New York: John Wiley & Sons.

Rugg, M. 1992. *Lead in municipal solid waste in the United States: Sources and forms*. Edison, N.J.: Camp Dresser & McKee Inc. (June).

San Diego, City of, Waste Management Department. 1988. *Request for proposal: Comprehensive solid waste management system*. (4 November).

Sanner, W.S., C. Crtuglio, J.G. Walters, and D.E. Wolfson. 1970. *Conversion of municipal and industrial refuse into useful materials by pyrolysis.* RI 7428. U.S. Dept. of Interior, Bureau of Mines (August).

Santhanam, C.J. 1974. Flotation techniques. Vol. 3 of *Environmental engineers handbook,* edited by B.G. Lipták. Radnor, Pa.: Chilton Book Company.

Savage, G.M., L.F. Diaz, and C.G. Golueke. 1985. Solid waste characterization. Results of waste composition study in Santa Cruz County, Calif. *BioCycle* (November/December).

Schaper, L.T. and R.C. Brockway. 1993. Transfer stations. In *The McGraw-Hill recycling handbook,* edited by H.F. Lund. McGraw-Hill, Inc.

Scher, J.A. 1971. Solid waste characterization techniques. *Chem. Eng. Prog.* 67 (March).

SCS Engineers. 1991. *Waste characterization study, solid waste management plan, Fairfax County, Virginia.* Reston, Va. (October).

Seattle Engineering Department, Solid Waste Utility. 1988. *Waste reduction, recycling and disposal alternatives: Volume II—Recycling potential assessment and waste stream forecast.* Seattle (May).

Snell, J.R. 1974. Size reduction and compaction equipment. Vol. 3 of *Environmental engineers handbook,* edited by B.G. Lipták. Radnor, Pa.: Chilton Book Company.

Solid waste management: Technology assessment. 1975. Schenectady, N.Y.: General Electric.

Sommerland, R.E., W.R. Seeker, A. Finkelstein, and J.D. Kilgroe. 1988. Environmental characterization of refuse-derived-fuel incinerator technology. *National Waste Processing Conference, Philadelphia, 1988.* New York: ASME.

Steven, W.K. 1989. When the trash leaves the curb: New methods improve recycling. *New York Times,* 2 May.

Surprenant, G. and J. Lemke. 1994. Landfill compaction: Setting a density standard. *Waste Age* (August).

Tchobanoglous, G., H. Theisen, and R. Eliassen. 1977. *Solid wastes: Engineering principles and management issues.* New York: McGraw-Hill.

Tchobanoglous, G., H. Theisen, and S. Vigil. 1993. *Integrated solid waste management.* McGraw-Hill, Inc.

———. 1993. *Integrated solid waste management: Engineering principles and management issues.* New York: McGraw-Hill.

Tuttle, K.L. 1986. Combustion generated particulate emissions. *National Waste Processing Conference, Denver, 1986.* ASME.

U.S. Environmental Protection Agency (EPA). 1976. *Decision makers' guide in solid waste management.* 2d ed. Washington, D.C.: U.S. EPA.

———. 1992. *The consumer's handbook for reducing solid waste.* EPA 530-K-92-003. U.S. EPA (August).

Vesilind, P.A. and A.E. Reimer. 1980. *Unit operations in resource recovery engineering.* Englewood Cliffs, N.J.: Prentice-Hall.

Walsh, P. and P. O'Leary. 1991. Evaluating a potential sanitary landfill site. *Waste Age* 22, no. 8: 121–134.

———. 1991. Landfill site plan preparation. *Waste Age* 22, no. 10: 87–92.

———. 1991. Sanitary landfill design procedures. *Waste Age* 22, no. 9: 97–105.

Hazardous Waste

Paul A. Bouis | Mary A. Evans | Lloyd H. Ketchum, Jr. | David H.F. Liu | William C. Zegel

Sources and Effects

11.1
HAZARDOUS WASTE DEFINED

Purpose and Scope

Hazardous waste is often defined as waste material that everyone wants picked up but no one wants put down. The legal and scientific definitions have become more complex as more compounds are found and more is learned about the toxicity of compounds and elements. The Resource Conservation and Recovery Act (RCRA) hazardous waste regulations (40 CFR §261 1987) provide the legal definition of hazardous waste. This definition is not always clear because the regulations are written in language general enough to apply to all possible situations, including unusual terminology, several exemptions, and exclusions.

The purpose of this section is to present the various definitions of hazardous waste in a manner useful to the environmental engineer. To be a hazardous waste, material must first conform to the definition of *waste*; second, it must fit the definition of *solid waste*; and third, it must fit the definition of *hazardous waste*. The environmental engineer must test the material against each of these definitions. This section assumes that the generator can demonstrate whether the material is indeed a waste.

Definition of Solid Waste

Solid waste need not literally be a solid. It may be a solid, a semisolid, a liquid, or a contained gaseous material. In accordance with RCRA regulations, a solid waste is any discarded material that is not specifically excluded by the regulation or excluded by granting of a special variance by the regulatory agency. Discarded material is considered abandoned, recycled, or inherently wastelike. Materials are considered abandoned if they are disposed of, burned or incinerated, or accumulated, stored, or treated (but not recycled) before being abandoned.

Materials are considered recycled if they are recycled or accumulated, stored, or treated before recycling. However, materials are considered solid waste if they are used in a manner constituting disposal, burned for energy recovery, reclaimed, or accumulated speculatively. Table 11.1.1 presents various classes of materials and general situations in which they would be considered solid wastes.

Inherently wastelike materials are solid wastes when they are recycled in any manner. This includes:

- Certain wastes associated with the manufacturing of tri-, tetra-, or pentachlorophenols or tetra-, penta-, or hexachlorobenzenes (for listed wastes F020, F021, F022, F023, F026, and F028, see the following section for an explanation of F designations
- Secondary materials that, when fed to a halogen acid furnace, exhibit characteristics of hazardous waste or are listed as hazardous waste (see section 2.2)
- Other wastes that are ordinarily disposed of, burned, or incinerated
- Materials posing a substantial hazard to human health and the environment when they are recycled.

For a material to be considered recycled and not a solid waste, the material must be used or reused in making a product without reclamation. The material is also considered recycled if it is used as an effective substitute for commercial products or returned to the process from which it was generated without reclamation. In this latter case, the material must be a substitute for raw material feedstock, and the process must use raw materials as its principal feedstocks.

The process for determining whether a waste is a solid waste is summarized in Figure 11.1.1.

Definition of Hazardous Waste

A solid waste is classified as a hazardous waste and is subject to regulation if it meets any of the following four conditions:

The waste is a characteristic hazardous waste, exhibiting any of the four characteristics of a hazardous waste: ignitability, corrosivity, reactivity, or toxicity (see Section 11.4 Hazardous Waste Characterization).

The waste is specifically listed as hazardous in one of the four tables in Part 261, Subpart D of the RCRA regulations: Hazardous Wastes From Nonspecific Sources,

TABLE 11.1.1   CONDITIONS UNDER WHICH COMMON MATERIALS ARE SOLID WASTES

Material	Use Constituting Disposal*	Energy Recovery Fuel†	Reclamation‡	Speculative Accumulation§
Spent Materials	Solid Waste	Solid Waste	Solid Waste	Solid Waste
Sludge	Solid Waste	Solid Waste	Solid Waste	Solid Waste
Sludge Exhibiting Characteristics of Hazardous Waste	Solid Waste	Solid Waste	NOT a Solid Waste	Solid Waste
By-products	Solid Waste	Solid Waste	Solid Waste	Solid Waste
By-products Exhibiting Characteristics of Hazardous Waste	Solid Waste	Solid Waste	NOT a Solid Waste	Solid Waste
Commercial Chemical Products	Solid Waste	Solid Waste	NOT a Solid Waste	NOT a Solid Waste
Scrap Metal	Solid Waste	Solid Waste	Solid Waste	Solid Waste

*Use constituting disposal includes application to or placement on the land, and use in the production of (or incorporation in) products that are applied to or placed on the land. Exceptions are made for materials that are applied to the land in ordinary use.

†Energy recovery fuel includes direct burning, use in producing a fuel, and incorporation in a fuel. However, selected commercial chemical products are not solid wastes if their common use is fuel.

‡Reclamation includes materials processed to recover useable products, or regenerated. Examples are recovery of lead from old automobile batteries or used wheel weights and regeneration of spent catalysts or spent solvents.

§Speculative accumulation refers to materials accumulated before the precise mechanism for recycle is known. This designation can be avoided if: the material is potentially recyclable; a feasible means for recycle is available; and during each calendar year the amount of material recycled or transferred to another site for recycling equals at least 75% of the material accumulated at the beginning of the period.

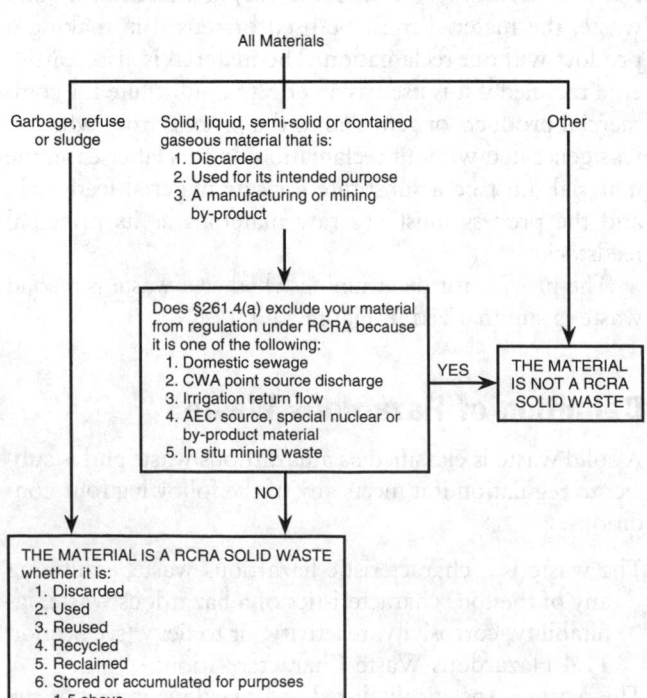

FIG. 11.1.1 Definition of a solid waste.

Hazardous Wastes From Specific Sources, Acute Hazardous Wastes, or Toxic Wastes.

The waste is a mixture of a listed hazardous waste and a nonhazardous waste.

The waste is declared hazardous by the generator of the waste. This is true even if the waste is not hazardous by any other definition and was declared hazardous in error.

The environmental engineer is referred to Section 261.3 of the RCRA regulations (40 CFR §261.3) for more information on exceptions to these criteria. A hazardous waste must be a solid waste and thus may be in the form of a solid, semisolid, liquid, or contained gas.

The EPA developed *listed wastes* by examining different types of wastes and chemical products to see if they exhibited one of the characteristics of a hazardous waste, then determining whether these met the statutory definition of hazardous waste, were acutely toxic or acutely hazardous, or were otherwise toxic. The following series letters denote the origins of such wastes.

F Series includes hazardous wastes from nonspecific sources (e.g., halogenated solvents, nonhalogenated solvents), electroplating sludges, cyanide solutions from plating batches). These are generic wastes commonly produced by manufacturing and industrial processes.

K Series is composed of hazardous waste from specific sources (e.g., brine purification muds from the mercury cell process in chlorine production where separated, purified brine is not used and API separator sludges). These are wastes from specifically identified industries, such as wood preserving, petroleum refining and organic chemical manufacturing.

P Series denotes acutely hazardous waste of specific commercial chemical products (e.g., potassium silver cyanide, toxaphene, or arsenic oxide) including discarded and off-specification products, containers, and spill residuals.

U Series includes toxic hazardous wastes that are chemical products, (e.g., xylene, DDT, and carbon tetrachloride) including discarded products, off-specification products, containers, and spill residuals.

Acute hazardous wastes are defined as fatal to humans in low doses, or capable of causing or contributing to serious irreversible, or incapacitating reversible illness. They are subject to more rigorous controls than other listed hazardous wastes.

Toxic hazardous wastes are defined as containing chemicals posing substantial hazards to human health or the environment when improperly treated, stored, transported, or disposed of. Scientific studies show that they have toxic, carcinogenic, mutagenic, or teratogenic effects on humans or other life forms.

The environmental engineer needs to understand when a waste becomes a hazardous waste, since this change initiates the regulatory process. A solid waste that is not excluded from regulation (see previous sections) becomes a hazardous waste when any of the following events occur:

- For listed wastes—when the waste first meets the listing description
- For mixtures of solid waste and one or more listed wastes—when a listed waste is first added to the mixture
- For other wastes—when the waste first exhibits any of the four characteristics of a hazardous waste

After a waste is labeled hazardous, it generally remains a hazardous waste forever. Some characteristic hazardous wastes may be declared no longer hazardous if they cease to exhibit any characteristics of a hazardous waste. However, wastes that exhibit a characteristic at the point of generation may still be considered hazardous even if they no longer exhibit the characteristic at the point of land disposal.

Figures 11.1.2 and 11.1.3 summarize the process used to determine whether a solid waste is a hazardous waste and whether it is subject to special provisions for certain hazardous wastes.

EXCLUSIONS

The regulations allow several exemptions and exclusions when determining whether a waste is hazardous. These exclusions center on recycled wastes and several large-vol-

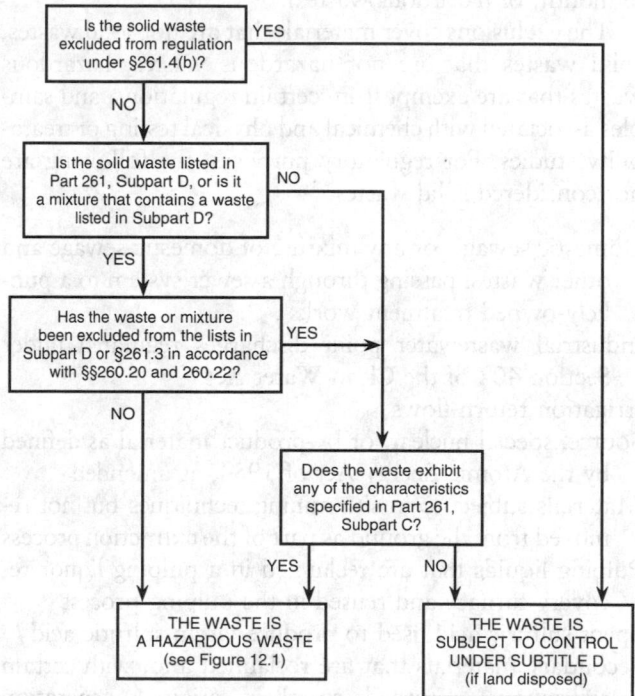

FIG. 11.1.2 Definition of a hazardous waste.

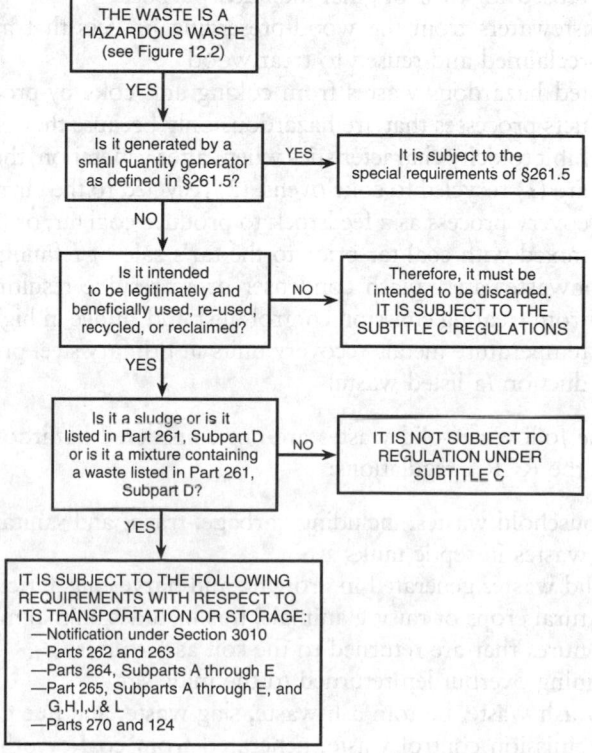

FIG. 11.1.3 Special provisions for certain hazardous waste.

ume or special-interest wastes. Wastes specifically excluded from regulation include industrial wastewater discharges, nuclear materials, fly ash, mining overburden, drilling fluids, and ore processing wastes. A major exemption is also granted to small-quantity generators of hazardous wastes (i.e., those generating less than 100 kg/month [220 lb/month] of hazardous wastes).

The exclusions cover materials that are not solid wastes, solid wastes that are not hazardous wastes, hazardous wastes that are exempt from certain regulations, and samples associated with chemical and physical testing or treatability studies. For regulatory purposes, the following are not considered solid wastes:

Domestic sewage, or any mixture of domestic sewage and other wastes, passing through a sewer system to a publicly-owned treatment works

Industrial wastewater point discharges regulated under Section 402 of the Clean Water Act

Irrigation return flows

Source, special nuclear, or by-product material as defined by the Atomic Energy Act of 1954, as amended

Materials subject to in situ mining techniques but not removed from the ground as part of the extraction process

Pulping liquids that are reclaimed in a pulping liquor recovery furnace and reused in the pulping process

Spent sulfuric acid used to produce virgin sulfuric acid

Secondary materials that are reclaimed and, with certain restrictions, returned to their original generation process(es) and reused in the production process

Spent wood-preserving solutions that are reclaimed and reused for their original intended purpose

Wastewaters from the wood-preserving process that are reclaimed and reused to treat wood

Listed hazardous wastes from coking and coke by-products processes that are hazardous only because they exhibit toxicity characteristics when, after generation, they are (1) recycled to coke ovens, (2) recycled to the tar recovery process as a feedstock to produce coal tar, or (3) mixed with coal tar prior to the tar's sale or refining

Nonwastewater splash condenser dross residue resulting from treating emission control dust and sludge in high-temperature metals-recovery units in primary steel production (a listed waste)

The following solid wastes are not considered hazardous by the RCRA regulations:

Household wastes, including garbage, trash, and sanitary wastes in septic tanks

Solid wastes generated in growing and harvesting agricultural crops or raising animals; this includes animal manures that are returned to the soil as fertilizers

Mining overburden returned to the mine site

Fly ash waste, bottom ash waste, slag waste, and flue gas emission control waste, generated from coal or other fossil fuels combustion

Drilling fluids, produced waters, and other wastes associated with the exploration, development, or production of crude oil, natural gas, or geothermal energy

Waste that could be considered hazardous based on the presence of chromium *if* it can be demonstrated that the chromium is not in the hexavalent state. Such a demonstration is based on information showing only trivalent chromium in the processing and handling of the waste in a non-oxidizing environment, or a specific list of waste sources known to contain only trivalent chromium.

Solid waste from extracting, beneficiating, and processing of ores and minerals

Cement kiln dust waste, unless the kiln is used to burn or process hazardous waste

Before an environmental engineer concludes a company or concern is not subject to regulation under RCRA, the engineer should confirm this conclusion via the RCRA Hotline (1-800-424-9346). Preferably, the decision should also be confirmed by an attorney or other qualified professional familiar with RCRA regulations.

SMALL-QUANTITY GENERATORS (40 CFR §261.5)

A small-quantity generator is conditionally exempt if it generates no more than 100 kg of hazardous waste in a calendar month. In determining the quantity of hazardous waste generated in a month, the generator does not need to include hazardous waste removed from on-site storage, only waste generated that month. Also excluded is waste that is counted more than once. This includes hazardous waste produced by on-site treatment of already-counted hazardous waste, and spent materials that are generated, reclaimed, and subsequently reused on site, so long as such spent materials have been counted once.

The limits on generated quantities of hazardous waste are different for acute hazardous waste (P list). The limit is equal to the total of one kg of acute hazardous waste or a total of 100 kg of any residue or contaminated soil, waste, or other debris resulting from the clean-up of any spilled acute hazardous wastes.

With exceptions, wastes generated by conditionally exempt small-quantity generators are not subject to regulation under several parts of RCRA (Parts 262 through 266, 268, and Parts 270 and 124 of Chapter 2, and the notification requirements of section 3010). The primary exception is compliance with section 262.11, hazardous waste determination. Hazardous wastes subject to these reduced requirements may be mixed with nonhazardous wastes and remain conditionally exempt, even though the mixture exceeds quantity limits. However, if solid waste is mixed with a hazardous waste that exceeds the quantity exclusion level, the mixture is subject to full regulation. If hazardous wastes are mixed with used oil and this mixture is to be burned for energy recovery, the mixture is subject to used oil management standards (Part 279 of RCRA).

RECYCLABLE MATERIALS (40 CFR §261.6)

Recycled hazardous wastes are known as recyclable materials. These materials remain hazardous, and their identification as recyclable materials does not exempt them from regulation. With certain exceptions, recyclable materials are subject to the requirements for generators, transporters, and storage facilities. The exceptions are wastes regulated by other sections of the regulations and wastes that are exempt, including: waste recycled in a manner constituting disposal; waste burned for energy recovery in boilers and industrial furnaces; waste from which precious metals are reclaimed; or spent lead-acid batteries being reclaimed. Wastes generally exempt from regulation are reclaimed industrial ethyl alcohol, used batteries or cells returned to a battery manufacturer for regeneration, scrap metal, and materials generated in a petroleum refining facility. Recycled used oil is subject to used oil management standards (Part 279 of RCRA).

CONTAINER RESIDUE (40 CFR §261.7)

Any hazardous waste remaining in a container or an inner liner removed from an empty container is not subject to regulation. The problem is determining whether a container is empty or not. RCRA regulations consider a container empty when all possible wastes are removed using common methods for that type of container, and no more than an inch (2.5 cm) of residue remains on the bottom of the container or liner. Alternately, a container with a volume of 110 gal or less can be considered empty if no more than 3% of the capacity, by weight, remains in the container or liner. Larger containers are considered empty when no more than 0.3% of capacity, by weight, remains in the container or liner. If the material in the container was a compressed gas, the container is considered empty when its pressure is reduced to atmospheric pressure.

Regarding acute hazardous waste (P list), the test for an empty container is much more stringent. The container or inner liner must be triple-rinsed using a solvent capable of removing the commercial chemical product or manufacturing chemical intermediate. Alternative cleaning methods can be used if they are demonstrated to be equivalent to or better than triple rinsing. Of course, a container can also be considered empty if a contaminated liner is removed.

—*Mary A. Evans*
William C. Zegel

References

Code of Federal Regulations. (1 July 1987): Title 40, sec. 261.
U.S. Environmental Protection Agency (EPA). 1986. *RCRA orientation manual.*" Office of Solid Waste, Washington, D.C.

11.2
HAZARDOUS WASTE SOURCES

The *reported* quantities of hazardous waste generated in the U.S. remained in the range of 250–270 million metric tn per year through most of the 1980s. Figure 11.2.1 indicates which industrial sectors generate these wastes. The majority of hazardous waste is generated by the chemical manufacturing, petroleum, and coal processing industries. As Figure 11.2.2 shows, waste generation is not broadly distributed throughout these industries; instead, a few dozen facilities account for most waste generation. While it is striking that a few dozen manufacturing facilities generate most of the country's hazardous wastes, these waste generation rates must be viewed in context. Figure 11.2.3 shows that 250–270 million tn of hazardous waste generated annually are over 90% wastewater. Thus, the rate of generation of hazardous constituents in the waste is prob-

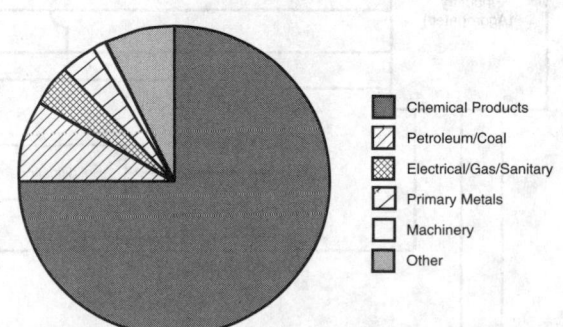

FIG. 11.2.1. Hazardous waste generation in 1986, classified by industry sector. (Reprinted from U.S. Environmental Protection Agency (EPA), 1988, *1986 national survey of hazardous waste treatment, storage, disposal and recycle facilities,* EPA/530-SW-88/035.)

ably on the order of 10 to 100 million tons per year. In relation to the 300+ million tons of commodity chemicals produced annually and the 1000 million tons of petroleum refined annually (C&E News 1991), the mass of hazardous constituents in waste is probably less than 5% of all chemical production.

Examples of basic industries and types of hazardous wastes produced are listed in Table 11.2.1, illustrating the

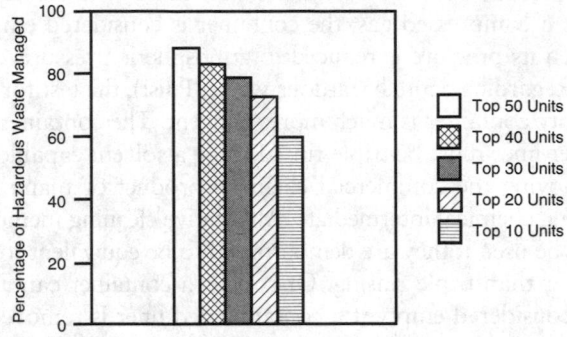

FIG. 11.2.2 Percentages of hazardous waste managed in the 50 largest facilities in 1986. (Reprinted from U.S. EPA, 1988.)

wide range and complexity of the wastes. However, these few examples do not adequately suggest the numbers and kinds of hazardous chemical constituents in hazardous wastes to be managed. There are approximately 750 listed wastes in 40 CFR Part 261, and countless more characteristic wastes. The intensity of industrial competition constantly engenders the introduction of new products, thus wastes are generated at an awesome pace.

Hazardous Waste from Specific Sources (40 CFR §261.32)

The following solid wastes are listed as hazardous wastes from a specific source unless they meet an exclusion. Except for K044, K045, and K047, which are reactive wastes, they are toxic wastes.

WOOD PRESERVATION

Bottom sediment sludge from wastewater treatment in wood-preserving processes using creosote or pentachlorophenol (K001) is a hazardous waste.

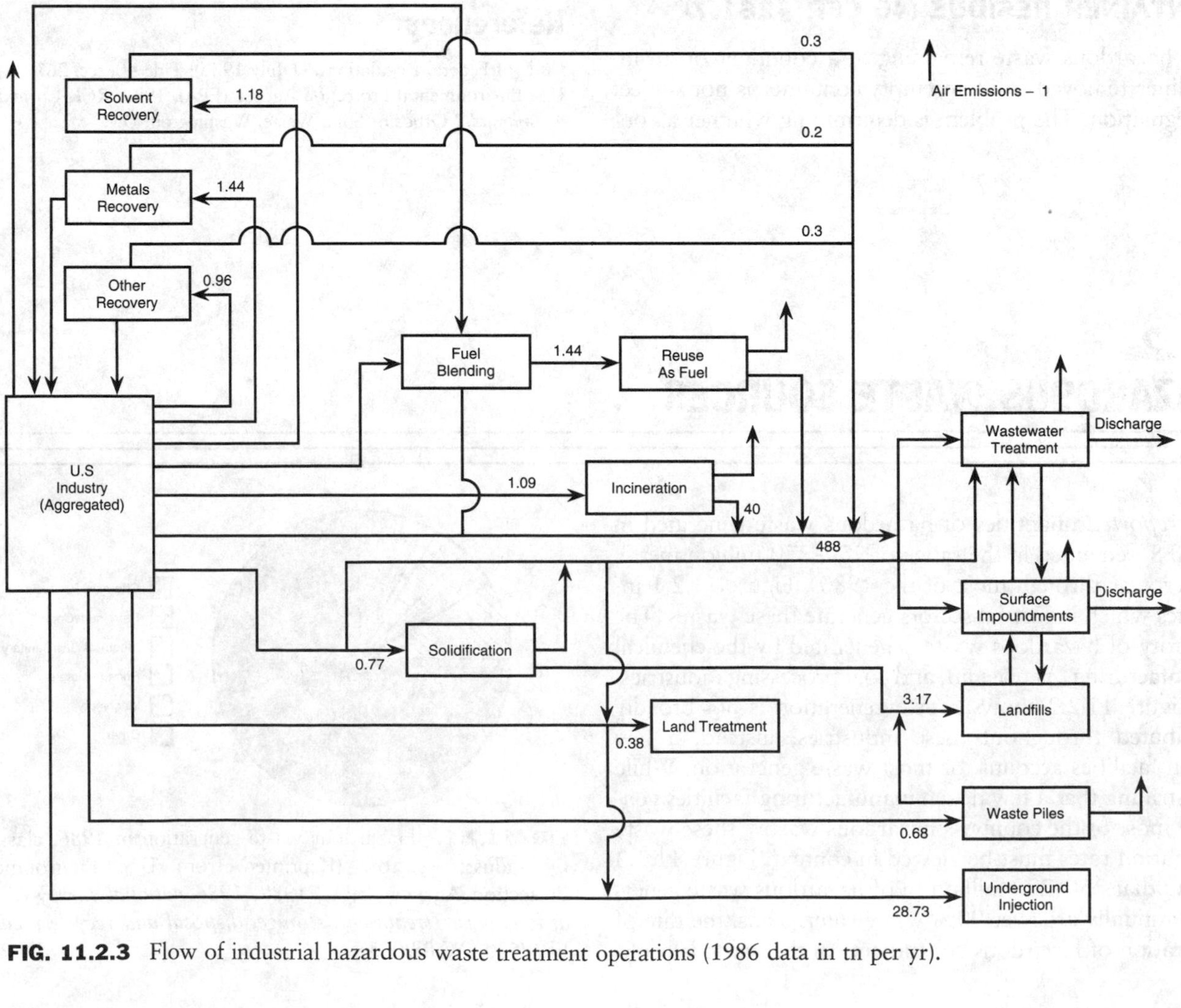

FIG. 11.2.3 Flow of industrial hazardous waste treatment operations (1986 data in tn per yr).

TABLE 11.2.1   TYPES OF HAZARDOUS WASTE

Industry	Wastes Produced

Chemical Manufacturing	• Spent solvents and still bottoms
	    White spirits, kerosene, benzene, xylene, ethyl benzene, toluene, isopropanol, toluene diisocyanate, ethanol, acetone, methyl ethyl ketone, tetrahydrofuran, methylene chloride, 1,1,1-trichloroethane, trichloroethylene
	• Ignitable wastes not otherwise specified (NOS)
	• Strong acid/alkaline wastes
	    Ammonium hydroxide, hydrobromic acid, hydrochloric acid, potassium hydroxide, nitric acid, sulfuric acid, chromic acid, phosphoric acid
	• Other reactive wastes
	    Sodium permanganate, organic peroxides, sodium perchlorate, potassium perchlorate, potassium permanganate, hypochlorite, potassium sulfide, sodium sulfide
	• Emission control dusts and sludges
	• Spent catalysts

Construction	• Ignitable paint wastes
	    Ethylene dichloride, benzene, toluene, ethyl benzene, methyl isobutyl ketone, methyl ethyl ketone, chlorobenzene
	• Ignitable wastes not otherwise specified (NOS)
	• Spent solvents
	    Methyl chloride, carbon tetrachloride, trichlorotrifluoroethane, toluene, xylene, kerosene, mineral spirits, acetone
	• Strong acid/alkaline wastes
	    Ammonium hydroxide, hydrobromic acid, hydrochloric acid, hydrofluoric acid, nitric acid, phosphoric acid, potassium hydroxide, sodium hydroxide, sulfuric acid

Metal Manufacturing	• Spent solvents and solvent still bottoms
	    Tetrachloroethylene, trichloroethylene, methylene chloride, 1,1,1-trichloroethane, carbon tetrachloride, toluene, benzene, trichlorofluoroethane, chloroform, trichlorofluoromethane, acetone, dichlorobenze, xylene, kerosene, white spirits, butyl alcohol
	• Strong acid/alkaline wastes
	    Ammonium hydroxide, hydrobromic acid, hydrochloric acid, hydrofluoric acid, nitric acid, phosphoric acid, nitrates, potassium hydroxide, sodium hydroxide, sulfuric acid, perchloric acid, acetic acid
	• Spent plating wastes
	• Heavy metal wastewater sludges
	• Cyanide wastes
	• Ignitable wastes not otherwise specified (NOS)
	• Other reactive wastes
	    Acetyl chloride, chromic acid, sulfides, hypochlorites, organic peroxides, perchlorates, permanganates
	• Used oils

Paper Industry	• Halogenated solvents
	    Carbon tetrachloride, methylene chloride, tetrachloroethylene, trichloroethylene, 1,1,1-trichloroethane, mixed spent halogenated solvents
	• Corrosive wastes
	    Corrosive liquids, corrosive solids, ammonium hydroxide, hydrobromic acid, hydrochloric acid, hydrofluoric acid, nitric acid, phosphoric acid, potassium hydroxide, sodium hydroxide, sulfuric acid
	• Paint wastes
	    Combustible liquid, flammable liquid, ethylene dichloride, chlorobenzene, methyl ethyl ketone, paint waste with heavy metals
	• Solvents
	    Petroleum distillates

Source: Reprinted from U.S. Environmental Protection Agency (EPA), *Does your business produce hazardous wastes?* (Office of Solid Waste and Emergency Response, (EPA/530-SW-010, Washington, D.C.)

INORGANIC PIGMENTS

Hazardous wastes include wastewater treatment sludge from the production of various metal-based pigments: chrome yellow and orange (K002), molybdate orange (K003), zinc yellow (K004), chrome green from the solvent recovery column in the production of toluene diiosocyanate via phosgenation of toluenediamine (K005), anhydrous and hydrated chrome-oxide green (K006), iron blue (K008), and oven residue from the production of chrome-oxide green (K008).

ORGANIC CHEMICALS

Numerous hazardous wastes occur in organic chemical production facilities. In the production of acetaldehyde from ethylene, distillation bottoms (K009) and distillation side cuts (K010) are hazardous wastes. In acrylonitrile production, the bottom streams from the wastewater stripper (K011), the acetonitrile column (K013), and the acetonitrile purification column (K014) are hazardous wastes. In 1,1,1-trichlorethane production, hazardous wastes include spent catalyst from the hydrochlorinator reactor (K028), waste from the product steam stripper (K029), distillation bottoms (K095), and heavy ends from the heavy end column (K096).

In the production of toluenediamine via hydrogenation of dinitrotoluene, hazardous wastes are generated in reaction by-product water from the drying column (K112) and condensed liquid light ends (K113), vicinals (K114), and heavy ends (K115) from the purification of toluenediamine.

In the production of ethylene dibromide via bromination of ethylene, hazardous wastes result from reactor vent gas scrubber wastewater (K117), spent adsorbent solids (K118), and still bottoms (K136) from purification.

Hazardous wastes are found in heavy ends or still bottoms from benzyl chloride distillation (K015), ethylene dichloride in ethylene dichloride production (K019), and vinyl chloride in vinyl chloride monomer production (K020). Heavy ends or distillation residues from carbon tetrachloride production (K016); the purification column in the production of epichlorohydrin (K017); the fractionation column in ethyl chloride production (K018); the production of phenol/acetone from cumene (K022); the production of phthalic anhydride from naphthalene (K024); the production of phthalic anhydride from ortho-xylene (K094); the production of nitro-benzene by the nitration of benzene (K025); the combined production of trichloroethylene and perchloroethylene (K030); the production of aniline (K083); and the production of chlorobenzenes (K085) are also hazardous wastes.

Other sources of hazardous wastes include distillation light ends from the production of phthalic anhydride from ortho-xylene (K093) or naphthalene (K024); aqueous spent antimony catalyst waste from fluoromethanes production (K021); stripping still tails from the production of methyl ethyl pyridines (K026); centrifuge and distillation residues from toluene diisocyanate production (K027); process residues from aniline extraction in aniline production (K103); combined wastewater streams generated from nitrobenzene/aniline production (K104); the separated aqueous stream from the reactor product washing step in the production of chlorobenzenes (K105); and the organic condensate from the solvent recovery column in the production of toluene diisocyanate via phosgenation of toluenediamine.

INORGANIC CHEMICALS

Chlorinated hydrocarbon waste from the purification step of the diaphragm cell process using graphite anodes (K073); wastewater treatment sludge from the mercury cell process (K106); and brine purification muds from the mercury cell process where separately prepurified brine is not used (K071) are hazardous wastes related to the production of chlorine.

PESTICIDES

Hazardous wastes are generated in the production of nine pesticides: MSMA and cacodylic acid, chlordane, creosote, disulfoton, phorate, toxaphene, 2,4,5–T, 2,4–D, and ethylenebisdithiocarbamic acid and its salts. In MSMA and cacodylic acid production, hazardous waste is generated as by-product salts (K031). In chlordane production, hazardous wastes include: wastewater treatment sludge (K032); wastewater and scrub water from the chlorination of cyclopentadiene (K033); filter solids from the filtration of hexachlorocyclopentadiene (K034); and vacuum stripper discharge from the chlordane chlorinator (K097). Wastewater treatment sludges generated in creosote production (K035) are also defined as hazardous waste. Hazardous wastes from the production of disulfoton are still bottoms from toluene reclamation distillation (K036), and wastewater treatment sludges (K037). Phorate production generates hazardous wastes from washing and stripping wastewater (K038), wastewater treatment sludge (K040), and filter cake from filtration of diethylphosphorodithioic acid (K039).

Wastewater treatment sludge (K041) and untreated process wastewater (K098) from toxaphene production and heavy ends, or distillation residues from tetrachlorobenzene in 2,4,5–T production (K042) are hazardous wastes. Similarly, 2,6–dichlorophenol waste (K043) and untreated wastewater (K099) from 2,4–D production are hazardous wastes.

Hazardous wastes from the production of ethylenebisdithiocarbamic acid and its salts are: process wastewaters (including supernates, filtrates, and washwaters) (K123); reactor vent scrubber water (K124); filtration, evaporation, and centrifugation solids (K125); and baghouse dust and floor sweepings in milling and packaging operations (K126).

EXPLOSIVES

Hazardous wastes from explosives production include: wastewater treatment sludges from manufacturing and processing explosives (K044) and manufacturing, formulation, and loading lead-based initiating compounds (K046); pink or red water from TNT operations (K047); and spent carbon from the treatment of wastewater-containing explosives (K045).

PETROLEUM REFINING

Dissolved air flotation (DAF) float (K048), slop oil emulsion solids (K049), heat exchanger bundle cleaning sludge (K050), API separator sludge (K051), and tank bottoms from storage of leaded fuel (K052) are hazardous wastes.

IRON AND STEEL

Emission control dust and sludges from primary steel production in electric furnaces (K061) and spent pickle liquor generated in steel finishing operations (K062) are hazardous wastes.

SECONDARY LEAD

Emission control dust and sludge (K069) and waste solution from acid leaching of emission control dust and sludge (K100) are hazardous wastes.

VETERINARY PHARMACEUTICALS

Wastewater treatment sludges generated in the production of veterinary pharmaceuticals from arsenic or organo-ar-

senic compounds (K084), distillation tar residues from the distillation of aniline-based compounds (K101), and residue from the use of activated carbon for decolorization (K102) are hazardous wastes.

INK FORMULATION

Solvent washes and sludges, caustic washes and sludges, or water washes and sludges from cleaning tubs and equipment used in ink formulation from pigments, driers, soaps, and stabilizers containing chromium and lead are hazardous wastes (K086).

COKING

Ammonia still lime sludge (K060) and decanter tank tar sludge (K087) are hazardous wastes.

Hazardous Wastes from Nonspecific Sources (40 CFR §261.31)

Hazardous wastes are also generated from nonspecific sources, depending upon the type of waste. Table 11.2.1 lists a number of these categories, although it is by no means an exhaustive listing.

—*Mary A. Evans*
William C. Zegel

Reference

Code of Federal Regulations. (1 July 1981): Title 40, sec. 261.3.

11.3
EFFECTS OF HAZARDOUS WASTE

It is virtually impossible to describe a "typical" hazardous waste site, as they are extremely diverse. Many are municipal or industrial landfills. Others are manufacturing plants where operators improperly disposed of wastes. Some are large federal facilities dotted with contamination from various high-tech or military activities.

While many sites are now abandoned, some sites are partially closed down or still in active operation. Sites range dramatically in size, from quarter-acre metal plating shops to 250-sq mi mining areas. The wastes they contain vary widely, too. Chief constituents of wastes in solid, liquid, and sludge forms include heavy metal, a common by-product of electroplating operations, and solvents or degreasing agents.

Human Health Hazards

Possible effects on human and environmental health also span a broad spectrum. The nearly uninhibited movement, activity, and reactivity of hazardous chemicals in the atmosphere are well established, and movement from one medium to another is evident. Hazardous wastes may enter the body through ingestion, inhalation, dermal absorption, or puncture wounds.

Human health hazards occur because of the chemical and physical nature of the waste, and its concentration and quantity; the impact also depends on the duration of exposure. Adverse effects on humans range from minor temporary physical irritation, dizziness, headaches, and nausea to long-term disorders, cancer or death. For example, the organic solvent carbon tetrachloride (CCl_4) is a central nerve system depressant as well as an irritant and can cause irreversible liver or kidney damage. Table 11.3.1 shows the potential effects of selected hazardous substances.

Site Safety

Transportation spills and other industrial process or storage accidents account for some hazardous waste releases. Such releases can result in fires, explosions, toxic vapors, and contamination of groundwater used for drinking.

Danger arises from improper handling, storage, and disposal practices (refer to Section 11.11 on Treatment, Storage, and Disposal Requirements). At hazardous waste sites, fires and explosions may result from investigative or remedial activities such as mixing incompatible contents of drums or from introduction of an ignition source, such as a spark from equipment.

A site safety plan is needed to establish policies and procedures for protecting workers and personnel during clean-up and day-to-day waste-handling activities. The minimum contents of a site safety plan are listed in Table 11.3.2.

TABLE 11.3.1 HEALTH EFFECTS OF SELECTED HAZARDOUS SUBSTANCES

Chemical	Source	Health Effects
Pesticides		
DDT	Insecticides	Cancer; damage to liver, embryos, bird eggs
BHC	Insecticides	Cancer, embryo damage
Petrochemicals		
BENZENE	Solvents, pharmaceuticals and detergents	Headaches, nausea, loss of muscle coordination, leukemia, damage to bone marrow
VINYL CHLORIDE	Plastics	Lung and liver cancer, depression of central nervous system, suspected embryotoxin
Other Organic Chemicals		
DIOXIN	Herbicides, waste incineration	Cancer, birth defects, skin disease
PCBs	Electronics, hydraulic fluid, fluorescent lights	Skin damage, possible gastro-intestinal damage, possibly cancer-causing
Heavy Metals		
LEAD	Paint, gasoline	Neurotoxic; causes headaches, irritability, mental impairment in children; brain, liver, and kidney damage
CADMIUM	Zinc, batteries, fertilizer	Cancer in animals, damage to liver and kidneys

Source: World Resources Institute and International Institute for Environment and Development, 1987; *World Resources 1987,* (New York, N.Y.: Basic Books, pp. 205–06.

TABLE 11.3.2 SITE SAFETY PLANS

• Name key personnel and alternates responsible for site safety.
• Describe the risks associated with each operation conducted.
• Confirm that personnel are adequately trained to perform their job responsibilities and to handle the specific hazardous situations they may encounter.
• Describe the protective clothing and equipment to be worn by personnel during various site operations.
• Describe any site-specific medical surveillance requirements.
• Describe the program for periodic air monitoring, personnel monitoring, and environmental sampling, if needed.
• Describe the actions to be taken to mitigate existing hazards (e.g., containment of contaminated materials) to make the work environment less hazardous.
• Define site control measures and include a site map.
• Establish decontamination procedures for personnel and equipment.
• Set forth the site's standard operating procedures for those activities that can be standardized, and where a checklist can be used.
• Set forth a contingency plan for safe and effective response to emergencies.

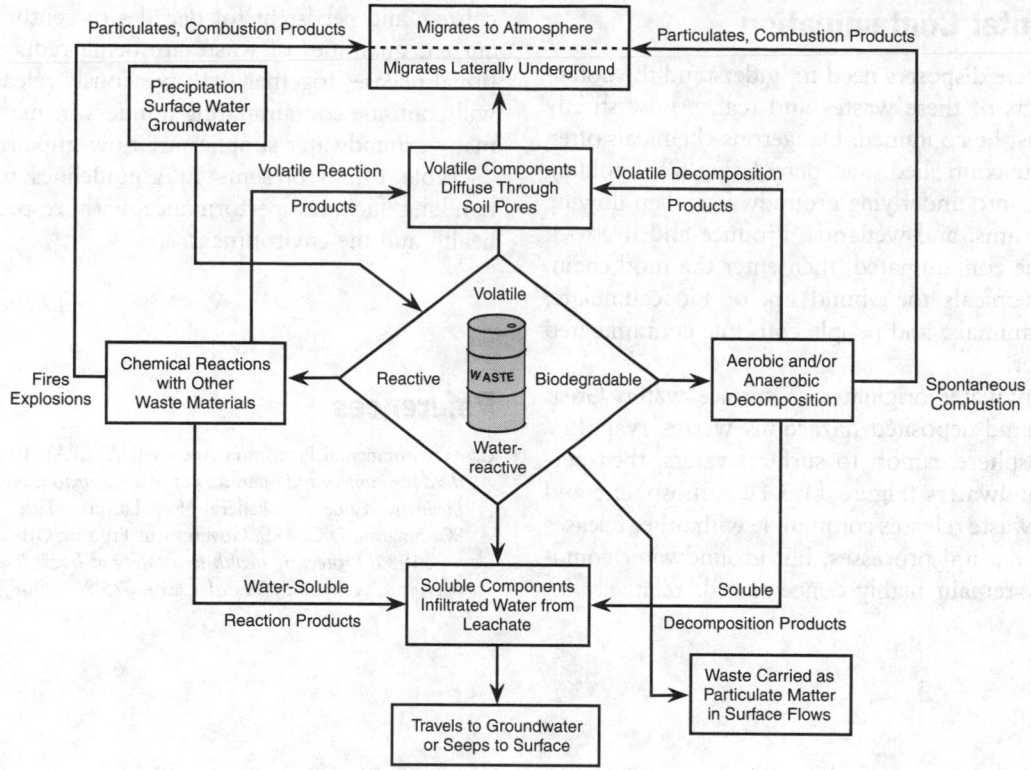

FIG. 11.3.1 Initial transport processes at waste disposal sites (EPA).

TABLE 11.3.3 ENVIRONMENTAL PERFORMANCE GUIDELINES

Prevention of adverse effects on air quality considering
1. Volume and physical and chemical characteristics of facility waste, including potential for volatilization and wind dispersal
2. Existing quality of the air, including other sources of contamination and their cumulative impact on the air
3. Potential for health risks caused by human exposure to waste constituents
4. Potential damage to wildlife, crops, vegetation, and physical structures caused by exposure to waste constituents
5. Persistence and permanence of the potential adverse effects

Prevention of adverse effects on surface water quality considering
1. Volume and physical and chemical characteristics of facility waste
2. Hydrogeological characteristics of the facility and surrounding land, including topography of the area around the facility
3. Quantity, quality, and directions of groundwater flow
4. Patterns of rainfall in the region
5. Proximity of facility to surface waters
6. Uses of nearby surface waters and any water quality standards established for those surface waters
7. Existing quality of surface water, including other sources of contamination and their cumulative impact on surface water
8. Potential for health risks caused by human exposure to waste constituents
9. Potential damage to wildlife, crops, vegetation, and physical structures caused by exposure to waste constituents
10. Persistence and permanence of the potential adverse effects

Prevention of adverse effects on groundwater quality considering
1. Volume and physical and chemical characteristics of the waste in the facility, including its potential for migration through soil or through synthetic liner materials
2. Geologic characteristics of the facility and surrounding land
3. Patterns of land use in the region
4. Potential for migration of waste constituents into subsurface physical structures
5. Potential for migration of waste constituents into the root zone of food-chain crops and other vegetation
6. Potential for health risks through human exposure to waste constituents
7. Potential damage to wildlife, crops, vegetation, and physical structures through exposure to waste constituents
8. Persistence and permanence of potential adverse effects

Environmental Contamination

Hazardous waste disposers need to understand the potential toxic effects of these wastes and realize how strictly the wastes must be contained. Dangerous chemicals often migrate from uncontrolled sites, percolating from holding ponds and pits into underlying groundwater, then flowing into lakes, streams, and wetlands. Produce and livestock in turn become contaminated, then enter the food chain. Hazardous chemicals then build up, or bioaccumulate, when plants, animals, and people consume contaminated food and water.

Most groundwater originates as surface water. Great quantities of land-deposited hazardous wastes evaporate into the atmosphere, runoff to surface waters, then percolate to groundwaters (Figure 11.3.1). Atmospheric and surface water waste releases commingle with other releases or are lost to natural processes, but groundwater contamination may remain highly concentrated, relatively localized, and persistent for decades or centuries. Although current quantities of waste are being reduced, any additional releases together with previously released materials will continue contaminating aquifers in many areas, and many groundwater supplies are now impaired.

Table 11.3.3 presents EPA guidelines for hazardous handling facilities performance with respect to human health and the environment.

—*David H.F. Liu*

References

U.S. Environmental Protection Agency (U.S. EPA). 1981. *Interim standard for owners and operators of new hazardous waste land disposal facilities.* Code of Federal Regulations. Title 40, Part 267. Washington, D.C.: U.S. Government Printing Office.

———. 1985. *Protecting health and safety at hazardous waste sites: an overview.* Technology Transfer, EPA 625/9–25/006, Cincinnati, OH.

Characterization, Sampling, and Analysis

11.4
HAZARDOUS WASTE CHARACTERIZATION

Criteria

The EPA applies two criteria in selecting four characteristics as inherently hazardous in any substance:

The characteristics must be listed in terms of physical, chemical, or other properties causing the waste to meet the definition of a hazardous waste in the act; and
The properties defining the characteristics must be measurable by standardized, available testing protocols.

The second criterion was adopted because generators have the primary responsibility for determining whether a solid waste exhibits any of the characteristics. EPA regulation writers believed that unless generators were provided with widely available and uncomplicated methods for determining whether their wastes exhibited the characteristics, the identification system would not work (U.S. EPA 1990).

Because of this second criterion, the EPA did not add carcinogenicity, mutagenicity, bioaccumulation potential, or phytotoxicity to the characteristics. The EPA considered the available protocols for measuring these characteristics either insufficiently developed, too complex, or too highly dependent on skilled personnel and professional equipment. In addition, given the current knowledge of such characteristics, the EPA could not confidently define the numerical threshold levels where characteristic wastes would present a substantial hazard (U.S. EPA 1990).

Characteristics

As testing protocols become accepted and confidence in setting minimum thresholds increases, more characteristics may be added. To date, waste properties exhibiting any or all of the existing characteristics are defined in 40 CFR §261.20–261.24.

CHARACTERISTIC OF IGNITABILITY

Ignitability is the characteristic used to define as hazardous those wastes that could cause a fire during transport, storage, or disposal. Examples of ignitable wastes include waste oils and used solvents.

A waste exhibits the characteristics of ignitability if a representative sample of the waste has any of the following properties:

1. It is a liquid, other than an aqueous solution containing less than 24% alcohol by volume, and has flash point less than 60°C (140°F), as determined by a Pensky-Martens Closed Cup Tester (using the test method specified in ASTM Standard D-93-79 or D-93-80) or by a Setaflash Closed Cup Tester (using the test method specified in ASTM Standard D-3278-78).
2. It is not a liquid and is capable, under standard temperature and pressure, of causing fire through friction, absorption of moisture, or spontaneous chemical changes and, when ignited, burns so vigorously and persistently that it creates a hazard.
3. It is an ignitable compressed gas as defined in the 49 Code of Federal Regulations 173.300 DOT regulations.
4. It is an oxidizer as defined in the 49 Code of Federal Regulations 173.151 DOT regulations.

A waste that exhibits the characteristic of ignitability but is not listed as a hazardous waste in Subpart D of RCRA has the EPA hazardous waste number of D001.

CHARACTERISTIC OF CORROSIVITY

Corrosivity, as indicated by pH, was chosen as an identifying characteristic of a hazardous waste because wastes with high or low pH can react dangerously with other wastes or cause toxic contaminants to migrate from certain wastes. Examples of corrosive wastes include acidic wastes and used pickle liquor from steel manufacture. Steel corrosion is a prime indicator of a hazardous waste since wastes capable of corroding steel can escape from drums and liberate other wastes.

A waste exhibits the characteristic of corrosivity if a representative sample of the waste has either of the following properties:

1. It is aqueous and has a pH less than or equal to 2 or greater than or equal to 11.5, as determined by a pH meter using an EPA test method. The EPA test method for pH is specified as Method 5.2 in "Test Methods for the Evaluation of Solid Waste, Physical/Chemical Methods."
2. It is a liquid and corrodes steel (SAE 1020) at a rate greater than 6.35 mm (0.250 inch) per year at a test temperature of 55°C (130°F), as determined by the test method specified in NACE (National Association of Corrosion Engineers) Standard TM-01-69 and standardized in "Test Methods for the Evaluation of Solid Waste, Physical/Chemical Methods."

A waste that exhibits the characteristic of corrosivity but is not listed as a hazardous waste in Subpart D has the EPA hazardous waste number of D002.

CHARACTERISTIC OF REACTIVITY

Reactivity was chosen as an identifying characteristic of a hazardous waste because unstable wastes can pose an explosive problem at any stage of the waste management cycle. Examples of reactive wastes include water from TNT operations and used cyanide solvents.

A waste exhibits the characteristic of reactivity if a representative sample of the waste has any of the following properties:

1. It is normally unstable and readily undergoes violent change without detonating.
2. It reacts violently with water.
3. It forms potentially explosive mixtures with water.
4. When mixed with water, it generates toxic gases, vapors, or fumes in a quantity sufficient to present a danger to human health or the environment.
5. It is a cyanide- or sulfide-bearing waste which, when exposed to pH conditions between 2 and 11.5, can generate toxic gases, vapors, or fumes in a quantity sufficient to present a danger to human health or the environment.
6. It is capable of detonation or explosive reaction if subjected to a strong initiating source or if heated under confinement.
7. It is readily capable of detonation or explosive decomposition or reaction at standard temperature and pressure.
8. It is a forbidden explosive as defined in the 49 Code of Federal Regulations 173.51, or a Class A explosive as defined in the 49 Code of Federal Regulations 173.53, or a Class B explosive as defined in the 49 Code of Federal Regulations 173.88 DOT regulations.

A waste that exhibits the characteristic of reactivity but is not listed as a hazardous waste in Subpart D has the EPA hazardous waste number of D003.

CHARACTERISTIC OF TOXICITY

The test, toxicity characteristic leaching procedure (TCLP), is designed to identify wastes likely to leach hazardous concentrations of particular toxic constituents into the groundwater as a result of improper management. During the TCLP, constituents are extracted from the waste to stimulate the leaching actions that occur in landfills. If the concentration of the toxic constituent exceeds the regulatory limit, the waste is classified as hazardous.

TABLE 11.4.1 MAXIMUM CONCENTRATION OF CONTAMINANTS FOR RCRA TOXICITY CHARACTERISTICS

EPA Hazardous Waste Number	Contaminant	Maximum Concentration (mg/L)	EPA Hazardous Waste Number	Contaminant	Maximum Concentration (mg/L)
D004	Arsenic[a]	5.0	D036	Hexachloro-1,3-butadiene	0.5
D005	Barium[a]	100.0	D037	Hexachloroethane	3.0
D019	Benzene	0.5	D008	Lead[a]	5.0
D006	Cadmium[a]	1.0	D013	Lidane[a]	0.4
D022	Carbon tetrachloride	0.5	D009	Mercury[a]	0.2
D023	Chlordane	0.03	D014	Methoxychlor[a]	10.0
D024	Chlorobenzene	100.0	D040	Methyl ethyl ketone	200.0
D025	Chloroform	6.0	D041	Nitrobenzene	2.0
D007	Chromium	5.0	D042	Pentachlorophenol	100.0
D026	o-Cresol	200.0	D044	Pyridine	5.0
D027	m-Cresol	200.0	D010	Selenium	1.0
D028	p-Cresol	200.0	D011	Silver[a]	5.0
D016	2,4-D[a]	10.0	D047	Tetrachloroethylene	0.7
D030	1,4-Dichloroben-zene	7.5	D015	Toxaphene[a]	0.5
D031	1,2-Dichloroethane	0.5	D052	Trichloroethylene	0.5
D032	1,1-Dichloroethy-lene	0.7	D053	2,4,5-Trichloro-phenol	400.0
D033	2,4-Dinitrotoluene	0.13	D054	2,4,6-Trichloro-phenol	2.0
D012	Endrin[a]	0.02	D017	2,4,5-TP (Silvex)[a]	1.0
D034	Heptachlor (and its hydroxide)	0.008	D055	Vinyl chloride	0.2
D035	Hexachlorobenzene	0.13			

[a]Formerly EP Toxicity Contaminants.
Source: Code of Federal Regulations, Title 40, sec. 261.24.

If the extract from a representative waste sample contains any of the contaminants listed in Table 11.4.1 at a concentration equal to or greater than the respective value given, the waste exhibits the toxicity characteristic. Where the waste contains less than 0.5 percent filterable solids, the waste itself is considered to be the extract. A waste that exhibits the toxicity characteristic but is not a listed hazardous waste has the EPA hazardous waste number specified in Table 11.4.1. The TCLP test replaced the EP toxicity test in September 1990 and added 25 organic compounds to the eight metals and six pesticides that were subject to the EP toxicity test.

Specific Compounds

Information about waste is needed to evaluate the health effects, determine the best method of handling, and evaluate methods of storage, treatment or disposal. Items of interest include:

- Physical properties such as density or viscosity
- Toxicity in water
- Permissible exposure limits (PELs) in the air
- Health hazards
- Precautions
- Controls
- Emergency and first aid procedures
- Disposal methods

There are a number of references that define the properties of specific compounds (Sax 1984, Sittig 1985, Weiss 1986), however, no current source defines the impact of hazardous mixtures.

—*David H.F. Liu*

References

Sax, N. 1984. *Dangerous properties of hazardous materials*. 6th ed. New York, N.Y.: Van Nostrand Reinhold.
Sittig, M. 1985. *Handbook of toxic and hazardous chemicals and carcinogens*. 2d ed. Park Ridge, N.J.: Noyes Publications.
U.S. Environmental Protection Agency (EPA). 1990. *RCRA orientation manual*. Office of Solid Waste. Washington, D.C.
Weiss, G. 1986. *Hazardous chemical data book*. 2d ed. Park Ridge, N.J.: Noyes Publications.

11.5
SAMPLING AND ANALYSIS

Safety and data quality are the two major concerns when sampling hazardous waste. Where environmental data are collected, quality assurance provides the means to determine data quality. This entails planning, documentation and records, audits, and inspections. Data quality is known when there are verifiable and defensible documentation and records associated with sample collection, transportation, sample preservation and analysis, and other management activities.

Sampling Equipment and Procedures

SAFETY

Samples must be secured in a manner ensuring the safety of the sampler, all others working in the area, and the surroundings.

If the source and nature of the hazardous waste are known, the sampler should study the properties of the material to determine the necessary safety precautions, including protective clothing and special handling precautions.

If the nature of the hazardous waste is unknown, such as at an abandoned waste disposal site, then the sampler should take additional precautions to prevent direct contact with the hazardous waste. Stored, abandoned, or suspect waste will often be containerized in drums and tanks. Such containers and materials buried under abandoned waste sites pose special safety problems (De Vera, Simmons, Stephens, Storn, 1980; EPA 1985). Care must be exercised in opening drums or tanks to prevent sudden releases of pressurized materials, fire, explosions, or spillage.

SAMPLING EQUIPMENT

Drums should be opened using a spark-proof brass bung wrench. Drums with bulged heads are particularly dangerous. The bulge indicates that the contents are under extreme pressure. To sample a bulged drum, a remotely operated drum opening device should be used, enabling the sampler to open the drum from a safe distance. Such operations should be carried out only by fully trained technicians in full personnel protective gear.

Liquid waste in tanks must be sampled in a manner that represents the contents of the tank. The EPA specifies that the *colawassa* sampler is used for such sampling. The colawassa is a long tube with a stopper at the bottom that opens or closes using the handle at the top. This device enables the sampler to retrieve representative material at any depth within the tank. The colawassa has many shortcomings, including the need for completely cleaning it and removal of all residues between each sampling. This is difficult, and it also creates another batch of hazardous waste to be managed.

A glass colawassa, which eliminates sample contamination by metals and stopper materials, is available through technical and scientific supply houses. In most situations, ordinary glass tubing can be used to obtain a representative sample, and can be discarded after use.

Bomb samplers that are lowered into a liquid waste container, then opened at the selected depth, are also useful in special situations.

Long-handled dippers can be used to sample ponds, impoundments, large open tanks, or sumps: however these devices cannot cope with stratified materials. Makeshift devices using tape or other porous or organic materials introduce the likelihood of sample contamination.

Dry solid samples may be obtained using a thief or trier, or an augur or dipper. Sampling of process units, liquid discharges, and atmospheric emissions all require specialized equipment training.

The EPA has published several guidance documents detailing hazardous waste, soil, surface water and groundwater and waste stream sampling (EPA 1985a, 1985b; De Vera et al. 1980; Evans and Schweitzer 1984).

Procedures used or materials contacting the sample should not cause gain or loss of pollutants. Sampling equipment and sample containers must be fabricated from inert materials and must be thoroughly cleaned before use. Equipment that comes into contact with samples to be analyzed for organic compounds should be fabricated of (in order of preference):

- Glass (amber glass for organics; clear glass for metals, oil, cyanide, BOD, TOC, COD, sludges, soil, and solids, and others)
- Teflon (Teflon lid liners should be inserted in caps to prevent contamination normally supplied with bottles)
- Stainless steel
- High-grade carbon steel
- Polypropylene
- Polyethylene (for common ions, such as fluoride, chloride, and sulfate)

Classic commercial analytic schedules require a sample of more than 1,500 ml. Commercial field samplers collect samples of 500 to 1,000 ml. If such volumes are insufficient, multibottle samples can be collected. Special containers may be designed to prolong sample duration.

PROCEDURES

Representative samples should be obtained to determine the nature of wastes.

If the waste is in liquid form in drums, it should be completely mixed (if this is safe) before sampling, and an aliquot should be taken from each container. Within a group of drums containing similar waste, random sampling of 20% of the drums is sufficient to characterize the wastes. If the sampler is unsure of the drum contents, each must be sampled and analyzed.

If the waste source is a manufacturing or waste treatment process solid, composite sampling and analysis are recommended. In such cases, an aliquot is periodically collected, composited, and analyzed.

If the solid waste is in a lagoon, abandoned disposal facility, tank, or similar facility, three-dimensional sampling is recommended. Although samples collected three-dimensionally are sometimes composited, they are usually analyzed individually. This process characterizes the solid waste and aids in determining whether the entire quantity of material is hazardous.

CHAIN OF CUSTODY RECORD

PROJECT										SAMPLERS: (Signed)	
LAB #	STATION	DATE	TIME	SAMPLE TYPE						NUMBER OF CONTAINERS	REMARKS
				WATER	SEDIMENT	TISSUE	AIR	OIL	OTHER		

RELINQUISHED BY: (Signed)	RECEIVED BY: (Signed)		DATE/TIME
RELINQUISHED BY: (Signed)	RECEIVED BY: (Signed)		DATE/TIME
RELINQUISHED BY: (Signed)	RECEIVED BY: (Signed)		DATE/TIME
RELINQUISHED BY: (Signed)	RECV'D BY MOBILE LAB FOR FIELD ANAL.: (Signed)		DATE/TIME
DISPATCHED BY: (Signed)	DATE/TIME	RECEIVED FOR LAB BY: (Signed)	DATE/TIME
METHOD OF SHIPMENT:			

FIG. 11.5.1 Example chain of custody record. Distribution: Original—accompany shipment; One copy—survey coordinator-field files.

If the source and nature of the material is known, sampling and analysis are limited to the parameters of concern. When the waste is unknown, a full analysis for 129 priority pollutants is often required.

SAMPLE PRESERVATION

Aqueous samples are susceptible to rapid chemical and physical reactions between the sampling time and analysis. Since the time between sampling and analysis could be greater than 24 hours, the following preservation techniques are recommended to avoid sample changes resulting in errors: all samples except metals must be refrigerated. Refrigeration of samples to 4°C is common in fieldwork, and helps stabilize samples by reducing biological and chemical activity (EPA 1979).

In addition to refrigeration, specific techniques are required for certain parameters (see section 10.9). The preservation technique for metals is the addition of nitric acid (diluted 1:1) to adjust the pH to less than 2, which will stabilize the sample up to 6 months; for cyanide, the addition of 6N caustic will adjust the pH to greater than 12, and refrigeration to 4°C, which will stabilize the sample for up to 14 days. Little other preservation can be performed on solid samples.

Quality Assurance and Quality Control

Quality assurance has emerged significantly during the past decade. Permit compliance monitoring, enforcement, and litigation are now prevalent in the environmental arena. Only documented data of known quality will be sustained under litigation. This section focuses on two areas.

SAMPLE CUSTODY

Proper chain-of-custody procedures allow sample processing and handling to be traced and identified from the time containers are initially prepared for sampling to the final disposition of the sample. A chain-of-custody record (Figure 11.5.1) should accompany each group of samples from the time of collection to their destination at the analytical laboratory. Each person with custody of the samples must sign the chain-of-custody form, ensuring that the samples are not left unattended unless properly secured.

Within the laboratory, security and confidentiality of all stored material should always be maintained. Analysts should sign for any sample removed from a storage area for performing analyses and note the time and date of returning a sample to storage. Before releasing analytical results, all information on sample labels, data sheets, tracking logs, and custody records should be cross-checked to ensure that data are consistent throughout the record. Gummed paper custody seals or custody tape should be used to ensure that the seal must be broken when opening the container.

PRECISION AND ACCURACY

One of the objectives of the QA or QC plan is to ensure that there is no contamination from initial sampling through final analysis. For this reason, duplicate, field blank, and travel blank samples should be prepared and analyzed.

Duplicate sampling requires splitting one field sample into two aliquots for laboratory analysis. Typically, 10% of the samples should be collected in duplicate. Duplicates demonstrate the reproducibility of the sampling procedure.

A travel blank is a contaminant-free sample prepared in the laboratory that travels with empty sample bottles to the sampling site and returns to the laboratory with the samples. Typically, two travel blanks are prepared and shipped. Travel blanks identify contamination in the preparation of sample containers and shipping procedures.

Field blanks are empty sampling bottles prepared using contaminant-free water following general field sampling procedures for collection of waste samples. These are returned to the laboratory for analysis. Field blanks identify contamination associated with field sampling procedures.

For liquid samples, all three types of the above QA/QC samples are prepared. For soils, semi-soils, sludges, and solids, only duplicate samples are typically prepared.

The field supervisor of sample collection should maintain a bound logbook so that field activity can be completely reconstructed without relying on the memory of the field crew. Items noted in the logbook should include:

- Date and time of activity
- Names of field supervisor and team members
- Purpose of sampling effort

TABLE 11.5.1   CATEGORIZATION OF PRIORITY POLLUTANTS

Volatile Organics			
acrolein	2-nitrophenol	dimethyl phthalate	endosulfan sulfate
acrylonitrile	4-nitrophenol	2,4-dinitrotoluene	endrin
benzene	parachlorometacresol	2,6-dinitrotoluene	endrin aldehyde
bis(chloromethyl)ether	1,2,4-trichlorobenzene	1,2-diphenylhyrazine	heptachlor
bromoform	phenol	fluoranthene	heptachlor epoxide
carbon tetrachloride	2,4,6-trichlorophenol	fluorene	PCB-1016
chlorobenzene		hexachlorobenzene	PCB-1221
chlorodibromomethane	**Base and Neutral Organics**	hexachlorobutadiene	PCB-1232
pentachlorophenol	acenaphthene	hexachlorocyclo-	PCB-1242
2-chloroethyl vinyl ether	acenaphtylene	pentadiene	PCB-1248
chloroform	anthracene	hexachloroethane	PCB-1254
dichlorobromomethane	benzidine	indeno(1,2,3-cd)-pyrene	PCB-1260
1,2-dichloroethane	benzo(a)anthracene	isophorone	toxaphene
1,1-dichloroethane	benzo(a)pyrene	naphthalene	
1,1,-dichloroethylene	benzo(ghi)perylene	nitrobenzene	**Metals**
1,2-dichloropropane	benzo(k)fluoranthene	N-nitrosodi-n-	antimony
1,2-dichloropropylene	3,4-benzo-fluoranthene	propylamine	arsenic
ethylbenzene	bis(2-chloroethoxy) methane	N-nitrosodimethylamine	beryllium
methyl bromide	bis(2-chloroethyl)ether	N-nitrosodiphenylamine	cadmium
methyl chloride	bis(2-chloroisopropyl)-	phenanthrene	chromium
methylene chloride	ether	pyrene	copper
1,1,2,3-tetrachloroethane	bis(2-ethylhexyl)phthalate	2,3,7,8-tetrachloro-	lead
tetrachloroethylene	4-bromophenyl phenyl	dibenso-p-dioxin	mercury
toluene	ether		nickel
1,2-trans-dichloroethylene	butyl benzyl phthalate	**Pesticides and PCBs**	selenium
1,1,1-trichloroethane	2-chloro-naphthalene	aldrin	silver
1,1,2-trichloroethane	4-chlorophenyl phenyl	alpha-BHC	thallium
trichloroethylene	ether	beta-BHC	zinc
vinyl chloride	chrysene	gamma-BHC	
	di-n-butyl phthalate	delta-BHC	**Cyanides**
Acid-Extractable Organics	di-n-octyl phthalate	chlordane	
2-chlorophenol	dibenzo(a,h)anthracene	4,4'-DDD	**Asbestos**
2,4-dichlorophenol	1,2-dichlorobenzene	4,4'-DD chloroethane	
2,4-dimethylphenol	4,4'-DDT	dieldrin	
4,6-dinitro-o-cresol	1,4-dichlorobenzene	alpha-endosulfan	
	diethyl phthalate	beta-endosulfan	

Source: Reprinted from U.S. Environmental Protection Agency (EPA), 1980–1988, *National Pollutant Discharge Elimination System,* Code of Federal Regulations, Title 40, Part 122. (Washington, D.C.: U.S. Government Printing Office).

- Description of sampling site
- Location of sampling site
- Sampling equipment used
- Deviation(s) from standard operating procedures
- Reason for deviations
- Field observations
- Field measurements
- Results of any field measurements
- Sample identification
- Type and number of samples collected
- Sample handling, packaging, labeling, and shipping information

The logbook should be kept in a secure place until the project activity is completed, when the logbook should be kept in a secured project file.

Analysis

If the source and nature of the waste is known, sampling and analysis are limited to the parameters of concern. If the waste is unknown, a full spectrum analysis is often required, including analysis for the 129 priority pollutants. Table 11.5.1 divides priority pollutants into seven categories (EPA 1980–1988).

Table 11.5.2 presents the recommended analytical procedures for the following categories: volatile organics, acid-extractable organics, base and neutral organics, pesticides and PCBs, metals, cyanides, asbestos, and others. Typically, organic analysis is performed using gas chromatography and mass spectrometry (GC/MS). Typical sensitivity is on the order of 1–100 parts per billion (ppb), depending on the specific organic compound and the concentration of compounds that may interfere with the analysis. This technique gives good quantification and excellent qualification about the organics in the waste.

A number of references should be consulted before determining the analytical protocols for the waste sample (EPA 1979; EPA 1977; EPA 1985a; EPA 1979a; APHA 1980).

Because analysis of hazardous waste samples is costly, it is beneficial to prepare several samples and subject them to one of several screening procedures. Depending on the data obtained, the analytical program can then focus on the major constituents of concern, resulting in cost savings. Recommended screening tests include: pH; conductivity; total organic carbon (TOC); total phenols; organic scan (via GC with flame ionization detector); halogenated (via GC with electron capture detector); volatile organic

TABLE 11.5.2 RECOMMENDED METHOD FOR ANALYSIS

Analytical Category	Recommended Method for Analysis*
Volatile organics	GC/MS (USEPA Method 624)
Acid-extractable organics	GC/MS (USEPA Method 625)
Base and neutral organics	GC/MS (USEPA Method 625)
TCDD (dioxin)	GC/MS (USEPA Method 608)
Pesticides and PCBs	GC/MS (USEPA Method 625)
Metals	Atomic absorption (flame or graphite)†
Mercury	Cold vapor atomic absorption spectroscopy
Cyanide	EPA colorimetric method
Asbestos	Fibrous asbestos method
Anions (SO_4^{2-}, F^-, Cl^-)	Ion chromatography
Oil and grease	Freon extraction and gravimetric measurement
Purgeable halocarbons	GC (USEPA Method 601)
Purgeable aromatics	GC (USEPA Method 602)
Acrolein and acrylonitrile	GC (USEPA Method 603)
Phenols	GC (USEPA Method 604)
Benzidine	GC (USEPA Method 605)
Pthalate esters	GC (USEPA Method 606)
Nitrosamines	GC (USEPA Method 607)
Pesticides and PCBs	GC (USEPA Method 608)
Nitroaromatics and isophorone	GC (USEPA Method 609)
Polynuclear aromatic hydrocarbons	GC (USEPA Method 610)
Chlorinated hydrocarbons	GC (USEPA Method 611)
TCDD (dioxin screening)	GC (USEPA Method 612)

*GC/MS = gas chromatography/mass spectrometry; GC = gas chromatography.
†Graphite furnace is a more sensitive technique.
Source: Reprinted from U.S. EPA, 1980–1988.

scan; nitrogen-phosphorous organic scan; and metals (via inductively coupled plasma or atomic emission spectroscopy).

—*David H.F. Liu*

References

American Public Health Association (APHA). 1980. *Standard methods for the examination of water and wastewater*. 15th ed. APFA. New York, N.Y.

De Vera, E.R., B.P. Simmons, R.D. Stephens, and D.L. Storn, 1980. *Samplers and sampling procedures in hazardous waste streams*. EPA 600–2–80–018, Cincinnati, Oh.

Evans, R.B., and G.E. Schweitzer. 1984. Assessing hazardous waste problems, *Environmental science and technology*. 18(11).

U.S. Environmental Protection Agency (EPA). 1977. *Sampling and analysis procedure for screening of industrial effluent for priority pollutants*. Effluent Guideline Division. Washington, D.C.

———. 1979. *Method for chemical analysis of water and waste*. EPA 600–4–79–020. Washington, D.C.

———. 1979a. *Guidelines establishing procedures for analysis of pollutants*. Code of Federal Regulations, Title 40, Part 136. Washington, D.C.: U.S. Government Printing Office.

———. 1980–1988. *National pollutant discharge elimination system*. Code of Federal Regulations, Title 40, Part 122. Washington, D.C.: U.S. Government Printing Office.

———. 1985. *Protecting health and safety at hazardous waste sites: an overview*, Technology Transfer EPA, 625–9–85–006. Cincinnati, Oh.

———. 1985a. *Characterization of hazardous waste sites—a methods manual; vol II, available sampling methods*. EPA 600–4–84–075. Washington, D.C.

———. 1985b. *Test methods for evaluating solid waste, physical/chemical methods*. 2d ed. SW-846. Washington, D.C.

11.6
COMPATIBILITY

Wasteloads are frequently consolidated before transport from point of generation to point of treatment or disposal. Accurate waste identification and characterization is necessary to:

- Determine whether wastes are hazardous as defined by regulations
- Establish compatibility grouping to prevent mixing incompatible wastes
- Identify waste hazard classes as defined by the Department of Transportation (DOT) to enable waste labeling and shipping in accordance with DOT regulations
- Provide identification to enable transporters or disposal operators to operate as prescribed by regulations.

Most wastes are unwanted products of processes involving known reactants. Thus, the approximate compositions of these wastes are known. Wastes of unknown origin must undergo laboratory analysis to assess their RCRA status, including testing for the hazardous properties of ignitability, reactivity, corrosivity, or toxicity in accordance with methods specified in the regulations (See Section 11.4).

Once a waste is identified, it is assigned to a compatibility group. One extensive reference for assigning groups is a study of hazardous wastes performed for the EPA by Hatayama et al (1980). A waste can usually be placed easily in one of the groups shown in Figure 11.6.1, based on its chemical or physical properties. The compatibility of various wastes is shown in Figure 11.6.1, which indicates the consequences of mixing incompatible wastes. Complete compatibility analysis should be carried out by qualified professionals to ascertain whether any waste can be stored safely in proximity to another waste.

—*William C. Zegel*

Reference

Hatayama et al. 1980. *A method for determining the compatibility of hazardous wastes*. U.S. Environmental Protection Agency (EPA). Office of Research and Development. EPA 600–2–80–076. Cincinnati, Oh.

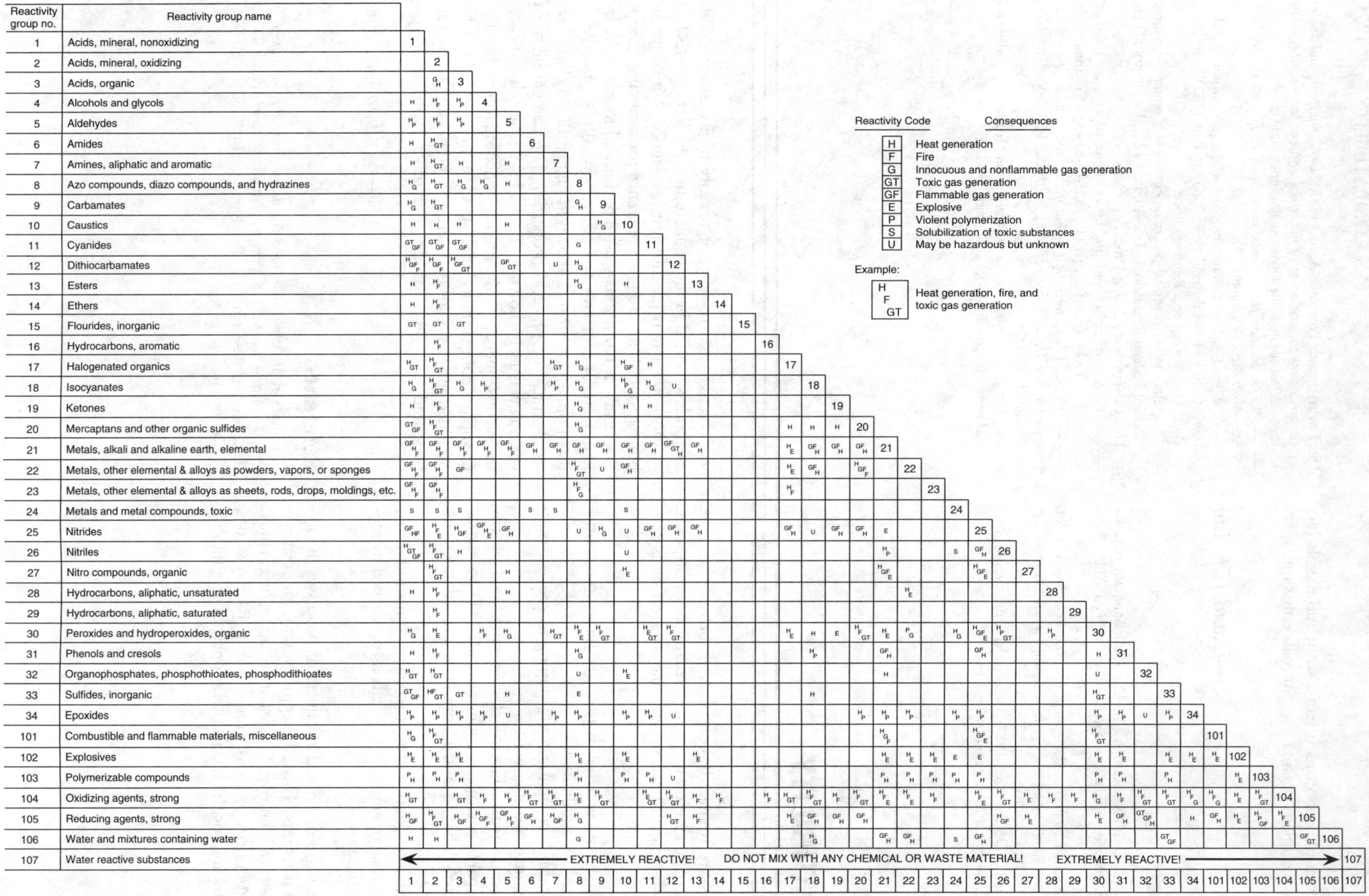

FIG. 11.6.1 Hazardous waste compatibility chart. (Reprinted from Hatayama et al. 1980, *A method for determining the compatibility of hazardous wastes*, U.S. Environmental Protection Agency [EPA] [Office of Research and Development. EPA 600-2-80-076, Cincinnati, Oh].)

Risk Assessment and Waste Management

11.7
THE HAZARD RANKING SYSTEM AND THE NATIONAL PRIORITY LIST

The Comprehensive Environmental Response, Compensation, and Liability Act (CERCLA) of 1980, better known as Superfund, became law "to provide for liability, compensation, cleanup and emergency response for hazardous substances released into the environment and the cleanup of inactive hazardous waste disposal sites." CERCLA was intended to give the EPA authority and funds to clean up abandoned waste sites and to respond to emergencies related to hazardous waste.

If a site poses a significant threat, the EPA uses its Hazard Ranking System (HRS) to measure the relative risk. Based upon this ranking system, sites warranting the highest priority for remedial action become part of the National Priority List (NPL).

The HRS ranks the potential threat posed by facilities based upon containment of hazardous substances, route of release, characteristics and amount of substances, and likely targets. HRS methodology provides a quantitative estimate of the relative hazards posed by a site, taking into account the potential for human and environmental exposure to hazardous substances. The HRS score is based on the probability of contamination from three sources—groundwater, surface water, and air—on the site in question. The HRS score assigned to a hazardous site reflects the potential hazards relative to other sites (Hallstedt, Puskar & Levine 1986).

S_M is the potential for harm to humans or the environment from migration of a hazardous substance to groundwater, surface water, or air; it is a composite of scores of each of the three routes

S_{FE} is the potential for harm from flammable or explosive substances

S_{DC} is the potential for harm from direct contact with hazardous substances at the site

The score for each of these hazard modes is obtained from a set of factors characterizing the facility's potential to cause harm as shown in Table 11.7.1. Each factor is assigned a numerical value according to the prescribed criteria. This value is then multiplied by a weight factor, yielding the factor score.

The factor scores are then combined: scores within a factor category are added together, then the total scores for each factor category are multiplied together. S_M is a composite of the scores of three possible migration routes:

$$S_M = \frac{1}{1.73} \sqrt{S_{gw}^2 + S_{sw}^2 + S_a^2} \qquad 11.7(1)$$

Figure 11.7.1 shows a typical worksheet for calculating the score for groundwater. Other worksheets are included in 40 CFR Part 300, Appendix A (1987).

Use of the HRS requires considerable information about the site, its surroundings, the hazardous substances present, and the geology in relation to the aquifers. If the data are missing for more than one factor in connection with the evaluation of a route, then that route score becomes 0, and there is no need to assign scores to factors in a route set at 0.

The factors that most affect an HRS site score are the proximity to a densely populated area or source of drinking water, the quantity of hazardous substances present, and toxicity of those hazardous substances. The HRS methodology has been criticized for the following reasons:

There is a strong bias toward human health effects, with only slight chance of a site in question receiving a high score if it represents only a threat or hazard to the environment.

Because of the human health bias, there is an even stronger bias in favor of highly populated affected areas.

The air emission migration route must be documented by actual release, while groundwater and surface water routes have no such documentation requirement.

The scoring for toxicity and persistence of chemicals may be based on site containment, which is not necessarily related to a known or potential release of toxic chemicals.

TABLE 11.7.1   RATING FACTORS FOR HAZARD RANKING SYSTEM

Hazard Mode	Category	Groundwater Route	Surface Water Route	Air Route
Migration	Route charcteristics	Depth to aquifer of concern	Facility slope and intervening terrain	
		Net precipitation	1-year 24-hour rainfall	
		Permeability of unsaturated zone	Distance to nearest surface water	
		Physical state	Physical state	
	Containment	Containment	Containment	
	Waste characteristics	Toxicity/persistence, Quantity	Toxicity/persistence, Quantity	Reactivity/incompatibility, Toxicity, Quantity
	Targets	Groundwater use	Surface water use	Land use
		Distance to nearest well/population served	Distance to sensitive environment	Population within 4-mile radius
			Population served/distance to water intake downstream	Distance to sensitive environment
Fire and explosion	Containment	Containment		
	Waste characteristics	Direct evidence		
		Ignitability		
		Reactivity		
		Incompatibility		
		Quantity		
	Targets	Distance to nearest population		
		Distance to nearest building		
		Distance to nearest sensitive environment		
		Land use		
		Population within 2-mile radius		
		Number of buildings within 2-mile radius		
Direct contact	Observed incident	Observed incident		
	Accessibility	Accessibility of hazardous substances		
Direct contact	Observed incident	Observed incident		
	Accessiblity	Accessibility of hazardous substances		
	Containment	Containment		
	Toxicity	Toxicity		
	Targets	Population within 1-mile radius		
		Distance to critical habitat		

Source: U.S. Environmental Protection Agency.

A high score for one migration route can be more than offset by low scores for other migration routes.

Averaging the route scores creates a bias against sites with only one hazard, even though that hazard may pose an extreme threat to human health and the environment.

The EPA provides quality assurance and quality control for each HRS score to ensure that site evaluations are performed on a consistent basis. HRS scores range from 0 to 100, with a score of 100 representing the most hazardous sites. Generally, HRS scores of 28.5 or higher will place a site on the NPL. Occasional exceptions have been made in this priority ranking to meet the CERCLA requirement that a site designated as top priority by a state be included on the NPL.

When the EPA places a hazardous waste site on the NPL, it also issues a summary description of the site and its threat to human health and the environment. Some typical examples are in EPA files, and in Wentz's book (1989).

(This discussion follows C.A. Wentz, *Hazardous Waste Management*, McGraw-Hill, pp 392–403, 1989.)

References

Code of Federal Regulations, Title 40, Part 300, Appendix A, 1987.

Hallstedt, G.W., M.A. Puskar, and S.P. Levine, 1986. Application of hazard ranking system to the prioritization of organic compounds identified at hazardous waste remedial action site. *Hazardous waste and hazardous materials,* Vol. 3, No. 2.

Wentz, C.A. 1989. *Hazardous waste management.* McGraw-Hill, Inc.

Facility Name: _____ Date: _____

Surface Water Route Work Sheet						
Rating Factor	Assigned Value (Circle One)		Multi-plier	Score	Max. Score	Ref. (Section)
☐1 Observed Release	0 45		1		45	4.1
If observed release is given a value of 45, proceed to line ☐4.						
If observed release is given a value of 0, proceed to line ☐2.						
☐2 Route Characteristics						4.2
Facility Slope and Intervening Terrain	0 1 2 3		1		3	
1-yr 24-hr Rainfall	0 1 2 3		1		3	
Distance to Nearest Surface Water	0 1 2 3		2		6	
Physical State	0 1 2 3		1		3	
Total Route Characteristics Score					15	
☐3 Containment	0 1 2 3		1		3	4.3
☐4 Waste Characteristics						4.4
Toxicity/Persistence	0 3 6 9 12 15 18		1		18	
Hazardous Waste Quantity	0 1 2 3 4 5 6 7 8		1		8	
Total Waste Characteristics Score					26	
☐5 Targets						4.5
Surface Water Use	0 1 2 3		3		9	
Distance to a Sensitive Environment	0 1 2 3		2		6	
Population Served/	0 4 6 8 10		1		40	
Distance to Water	12 16 18 20					
Intake Downstream	24 30 32 35 40					
Total Targets Score					55	
☐6 If line ☐1 is 45, multiply ☐1 ∞ ☐4 ∞ ☐5						
If line ☐1 is 0, multiply ☐2 ∞ ☐3 ∞ ☐4 ∞ ☐5					64,350	
☐7 Divide line ☐6 by 64,350 and multiply by 100			$S_{SW} =$			

FIG. 11.7.1 Surface water route worksheet.

11.8
RISK ASSESSMENT

The term "risk" refers to the probability that an event will have an adverse effect, indirectly or directly, on human health or welfare. Risk is expressed in time or unit activity, e.g., cancer cases per pack of cigarettes smoked. Risk assessment takes into account the cumulative effects of all exposure. For example, in assessing the risk that a person will suffer from air pollution, both indoor and outdoor pollution must be taken into account.

The function of an effective hazardous materials management program is to identify and reduce major risks. This involves both risk assessment and risk management. The flowchart in Figure 11.8.1 shows the factors affecting the hazardous waste risk assessment procedure. This procedure begins with identification of the waste and the laws and regulations pertaining to that waste. When the waste is identified, its toxicity and persistence must be determined

to evaluate the risk of human and the environmental exposure. The risk management process involves selecting a course of action based on the risk assessment.

One way to highlight differences between risk assessment and risk management is to look at differences in the information content of the two processes. Data on tech-

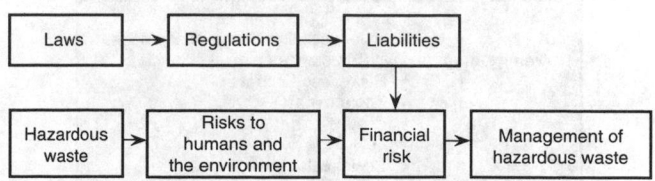

FIG. 11.8.1 Factors affecting the risk assessment of hazardous waste.

nological feasibility, on costs, and on the economic and social consequences of possible regulatory decisions are of critical importance to risk management but not to risk assessment. As statutes require, risk managers consider this information with risk assessment outcomes to evaluate risk management options and make environmental decisions (Figure 11.8.2).

Environmental risk assessment is a multi-disciplinary process. The risk assessment procedure is an iterative loop that the assessor may travel several times. It draws on data, information, and principles from many scientific disciplines, including biology, chemistry, physics, medicine, geology, epidemiology, and statistics. After evaluating individual studies for conformity with standard practices within each discipline, the most relevant information from each is combined and examined to determine the risk. Although studies from single disciplines are used to develop risk assessment, such studies alone are not regarded as risk assessment or used to generate risk assessments.

Review of Basic Chemical Properties

Before exploring the major components of risk assessment, some basic chemical properties and their relationships to biological processes must be reviewed.

Chemical Structure.   The chemical structure of a substance is the arrangement of its atoms. This structure determines the chemical's properties, including how a chem-

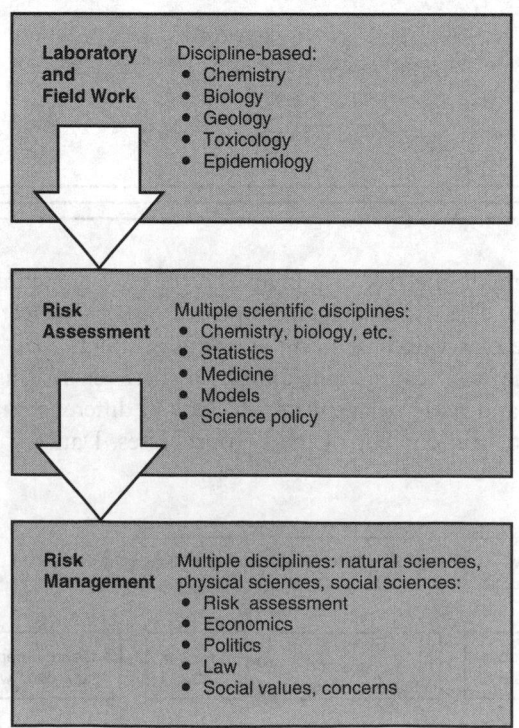

FIG. 11.8.2 Disciplines contributing to environmental decisions.

ical will combine with another substance. Because different structural forms of a chemical may exhibit different degrees of toxicity, the chemical structure of the substance being assessed is critical. For example, the free cyanide ion dissolved in water is highly toxic to many organisms (including humans); the same cyanide combined with iron is much less toxic (blue pigment). Cyanide combined with an organic molecule may have completely different toxic properties.

Solubility.   Solubility is a substance's ability to blend uniformly with another. The degree of water and lipid (fat) solubility of a chemical is important in risk assessment. Solubility has significant implications for activities as diverse as cooking or chemical spill cleanup. To estimate the degree of potential water contamination from a chemical spill, it is necessary to know the chemical makeup of the material spilled to judge the extent that chemical contamination will be dispersed by dissolving in water. Likewise, the degree of lipid solubility has important implications, particularly in such processes as bioaccumulation.

Bioaccumulation.   The process of chemical absorption and retention within organisms is called bioaccumulation. For example, a fat-soluble organic compound ingested by a microorganism is passed along the food chain when an organism eats the microorganism, then another predator eats the organism. The organic compound, because it is fat-soluble, will concentrate in the fat tissue of each animal in the food chain. The pesticide DDT is an example of a chemical that bioaccumulates in fish, and then in humans and birds eating those fish.

Transformation.   Biotransformation and transformation caused by physical factors exemplify how chemical compounds are changed into other compounds. Biotransformation is the change of one compound to another by the metabolic action of a living organism. Sometimes such a transformation results in a less toxic substance, other times in a more toxic substance.

Chemical transformation is prompted by physical agents such as sunlight or water. A pesticide that is converted into a less toxic component by water in a few days following application (e.g., malathion) carries a different long-term risk than a pesticide that withstands natural degradation or is biotransformed into a toxic compound or a metabolite (e.g., DDT). The ability to withstand transformation by natural processes is called *persistence*.

Understanding basic chemical and physical properties helps to determine how toxic a chemical can be in drinking water or in the food chain, and whether the substance can be transported through the air and into the lungs. For example, when assessing the risk of polychlorinated biphenyls (PCBs), it must be recognized that they biodegrade very slowly and that they are strongly fat-soluble, so they readily bioaccumulate. When monitoring their presence, it must also be recognized that they are negligibly soluble in water: concentrations will always be much higher in the fat tissue of a fish, cow, or human than in the blood, which has a higher water content.

RA Paradigms

The risk assessment paradigm published in *Risk Assessment in the Federal Government: Managing the Process* (National Academy of Science [NAS] 1983), provides a useful system for organizing risk science information from these many different sources. In the last decade, the EPA has used the basic NAS paradigm as a foundation for its published risk assessment guidance and as an organizing system for many individual assessments. The paradigm defines four fields of analysis describing the use and flow of scientific information in the risk assessment process (Figure 11.8.3).

The following paragraphs detail those four fields of analysis. Each phase employs different parts of the information database. For example, hazard identification relies primarily on data from biological and medical sciences. Dose-response analysis uses these data in combination with statistical and mathematical modeling techniques, so that the second phase of the risk analysis builds on the first.

HAZARD IDENTIFICATION

The objective of hazard identification is to determine whether available scientific data describes a causal relationship between an environmental agent and demonstrated injury to human health or the environment. In humans, observed injuries may include birth defects,

neurological damage, or cancer. Ecological hazards might result in fish kills, habitat destruction, or other environmental effects. If a potential hazard is identified, three other analyses become important for the overall risk assessment.

Chemical toxicities are categorized according to the various health effects resulting from exposure. The health effects, often referred to as *endpoints,* are classified as acute (short-term) and chronic (long-term). *Acute* toxic effects occur over a short period of time (from seconds to days), for example: skin burns from strong acids and poisonings from cyanide. *Chronic* toxic effects last longer and develop over a much longer period of time, and include cancer, birth defects, genetic damage, and degenerative illnesses.

A wide variety of reference materials provide basic toxicity data on specific chemicals, including:

Registry of Toxic Effects of Chemical Substances (RTECS), (U.S. Department of Health and Human Services)
Health Assessment Guidance Manual (U.S. Department of Health and Human Services 1990)
The Handbook of Toxic and Hazardous Chemicals and Carcinogens (Sittig 1985)
Threshold Limit Values (TLVs) for Chemical Substances and Physical Agents and Biological Exposure Indices (BEIs) American Conference of Governmental Industrial Hygienists (ACGIH 1990)
Integrated Risk Information System (IRIS), a database supported by EPA Office of Research and Development, Environmental Criteria and Assessment Office (MS-190), Cincinnati, Oh. 45268, Telephone: 513-569-7916

The U.S. Environmental Protection Agency has classified some 35,000 chemicals as definitely or potentially harmful to human health. However, the risk resulting from ex-

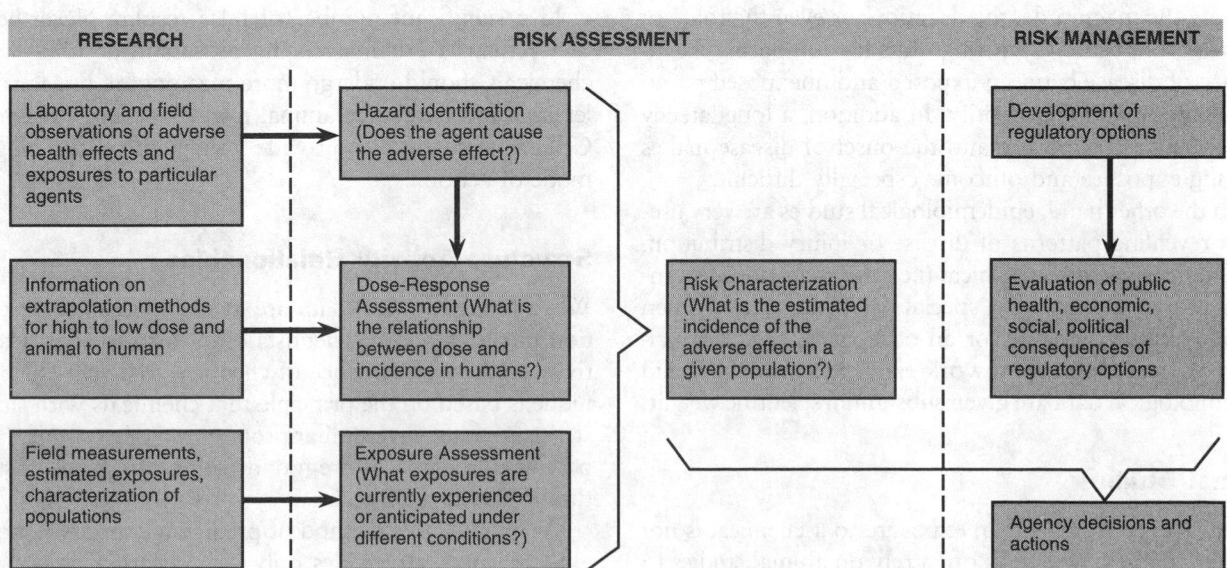

FIG. 11.8.3 Elements of risk assessment and risk management. (Reproduced with permission from *Risk Assessment in the Federal Government: Managing the Process,* 1983, The National Academy of Science [NAS], Washington, D.C.: The National Academy Press.

posure to more than one of these substances at the same time is not known (Enger, Kormelink, Smith & Smith 1989).

The following estimation techniques are commonly used to learn about human toxicity (Nally 1984).

Clinical Studies

The strongest evidence of chemical toxicity to humans comes from observing individuals exposed to the chemical in clinical studies. Scientists can determine direct cause and effect relationships by comparing the control groups (individuals not exposed to the chemical) to the exposed individuals. For obvious moral and ethical reasons, there is a limit to testing toxicity directly on humans. For example, tests for acute toxicity, such as allergic skin reactions, might be permissible, but tests for chronic toxicity, such as cancer, would be unacceptable.

Epidemiological Studies

As clinical studies frequently cannot be performed, scientists gather data on the incidence of disease or other ill effects associated with human exposure to chemicals in real-life settings. The field of *epidemiology* studies the incidence and distribution of disease in a population. This type of information is after the fact and in the case of cancer, comes many years after the exposure. Nevertheless, while epidemiological studies cannot unequivocally demonstrate direct cause and effect, they often can establish convincing and statistically significant associations. Evidence of a positive association carries the most weight in risk assessment.

Many factors limit the number of chemicals examined in epidemiological studies. Often there is no mechanism to verify the magnitude, the duration, or even the route of individual exposure. Control groups for comparing the incidence of disease between exposed and unexposed populations are difficult to identify. In addition, a long latency period between exposure and the onset of disease makes tracking exposure and outcome especially difficult.

On the other hand, epidemiological studies are very useful in revealing patterns of disease or injury distribution, whether these are geographical (i.e., the incidence of stomach cancer in Japan), for a special risk group (i.e., women of child-bearing age), or for an occupation (i.e., the incidence of cancer in asbestos workers). When available, valid epidemiological data are given substantial scientific weight.

Animal Studies

Since evidence from human exposure to a chemical is not usually available, scientists often rely on animal studies to determine the toxicity of a chemical. The objective of animal studies is to determine, under controlled laboratory conditions, the chemical dose that will produce toxic effects in an animal. This information is used to predict what may occur in humans under normal exposure conditions. Toxic effects that occur in laboratory animals often occur in humans exposed to the same agents. Scientists recognize, however, that animal tests may not be conclusive for humans.

Routes of exposure in animal studies are designed to mimic the routes of possible human exposure. Ideally, a suspected food contaminant would be tested in a feeding study, a suspected skin surface irritant in a dermal irritation study, and a potential air contaminant in an inhalation study. However, it is not always possible to administer a test dose of the chemical to an animal via the expected route of exposure in humans (for instance, if it alters the color or odor of feed) so other methods must be devised.

Test-Tube Studies

Test-tube or in vitro studies involving living cells are particularly useful in testing whether a chemical is a potential carcinogen. Some of these tests are for mutagenicity or the ability to alter genetic material. Mutagenicity is believed to be one way in which carcinogens initiate cancer. These are often referred to as short-term tests because they require only a few hours or days, as opposed to several years required for long-term carcinogenicity studies in laboratory animals. The Ames mutagenicity test, which uses bacteria strains that reproduce only in the presence of a mutagen, is the best-known short-term test.

One of the major drawbacks of these cellular tests is that even with the addition of enzyme mixtures and other useful modifications, they are far simpler than the complex human organism. The human body's sensitive biological systems and remarkable defense mechanisms protect against chemical attack. The cellular tests lack the complexities of whole, integrated organisms, thus, they yield a significant number of false results. Nevertheless, they remain a useful screening process in deciding which chemicals should undergo more meaningful, but far more lengthy and expensive animal testing for carcinogenicity. Cellular tests can also provide insight into a carcinogen's mode of action.

Structure-Activity Relationships

When limited (or no) data are available from the estimation methods above, scientists often turn to structure-activity studies for evidence of chemical toxicity. This technique is based on the principle that chemicals with similar structures may have similar properties. For example, many potential carcinogens are found within categories of structurally similar chemicals.

At present, this method of predicting toxicity is not an exact science; it provides only an indication of potential hazard. However, as the technique develops along with the understanding of biological mechanisms, structure-activity relationships will evolve into a more precise predictive tool.

Animal studies are currently the preferred method for determining chemical toxicity. Although they are less convincing than human studies, animal studies are more convincing than test-tube and structure-activity studies. They are also easier to schedule, an industry has evolved around performing them.

The uncertainties associated with animal toxicity studies are discussed below.

The Testing Scheme

The selection of toxicological tests is crucial to any experimental program. Similarly, decisions regarding: the amount of chemical to be tested; the route of exposure; the test animal species; the composition of the test population (homogeneous or heterogeneous); the effects to be observed; and the duration of the study affect the usefulness and reliability of the resulting data. Although based on scientific judgment, all such decisions introduce elements of subjectivity into the testing scheme. The outcome of the test may be shaped by the specific nature of the test itself. For example, the decision to conduct an inhalation study might preclude discovering toxic effects via a different route of exposure. For this reason, a route of exposure is selected to approximate real-life conditions.

Demonstration of carcinogenicity requires strict observance of analytical protocols. NCI (IRLG 1979) presents criteria for evaluating experimental designs (see Table 11.8.1). Laboratory data not developed in compliance with these protocols are questionable.

Synergism/Antagonism

In vivo animal experiments are controlled studies that allow the isolation of individual factors to determine a specific cause and effect relationship. However, critics point out that such tests, although useful, are not absolute indicators of toxicity. As such tests are specific, *synergistic effects* from human exposure to more than one chemical are not detected. These tests may also overlook *antagonistic effects* where one chemical reduces the adverse effect of another.

DOSE-RESPONSE RELATIONSHIP

When toxicological evaluation indicates that a chemical may cause an adverse effect, the next step is to determine the potency of the chemical. The dose-response analysis determines the relationship between the degree of chemical exposure (or dose) and the magnitude of the effect (response) in the exposed organism. Scientists use this analysis to determine the amount of a chemical that causes tumor development in skin irritation, animals, or death in animals.

Dose-response curves are generated from various acute and chronic toxicity tests. Depending on chemical action, the curve may rise with or without a threshold. As Figure 11.8.4 shows, the TD_{50} and TD_{100} points indicate the doses associated with 50% and 100% occurrence of the measured toxic effect; also shown are the No Observable Adverse Effect Level (NOAEL) and Lowest Observable Adverse Effect Level (LOAEL). The NOAEL is assumed to be the basis for the Acceptable Daily Intake (ADI).

Figure 11.8.5 illustrates the threshold and no-threshold dose-response curve. In both cases, the response normally reaches a maximum, after which the dose-response curve becomes flat.

To estimate the effects of low doses, scientists extrapolate from the observed dose-response curve. Extrapolation models extend laboratory results into ranges where observations are not yet available or possible. Most current models are not based exclusively upon known biol-

TABLE 11.8.1   CRITERIA FOR EVALUATING CARCINOGEN EXPERIMENTS ON ANIMALS

Criteria	*Recommendations*
Experimental design	Two species of rats, and both sexes of each; adequate controls; sufficient animals to resolve any carcinogenic effect; treatment and observation throughout animal lifetimes at range of doses likely to yield maximum cancer rates; detailed pathological examination; statistical analyses of results for significance
Choice of animal model	Genetic homogeneity in test animals, especially between exposed and controls; selection of species with low natural-tumor incidence when testing that type of tumor
Number of animals	Sufficient to allow for normal irrelevant attrition along the way and to demonstrate an effect beyond the level of cancer in the control group
Route of administration	Were tumors found remote from the site of administration? No tumors observed should demonstrate that absorption occurred
Identity of the substance tested	Exposure to chemicals frequently involves mixtures of impurities. What effect did this have on results? What is the significance of pure compound results? Also, consider the carrier used in administration
Dose levels	Sufficient to evoke maximum tumor incidence
Age of treatment	Should be started early
Conduct and duration of bioassays	Refer to NCI's "Guidelines for Carcinogen in Small Rodents"

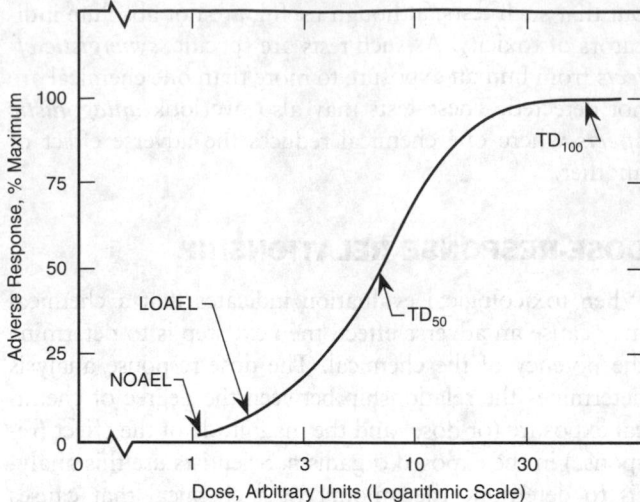

FIG. 11.8.4 Hypothetical dose-response curve. (Adapted from ICAIR Life Systems, Inc., 1985, *Toxicology Handbook*, prepared for EPA Office of Waste Programs Enforcement, Washington, D.C.)

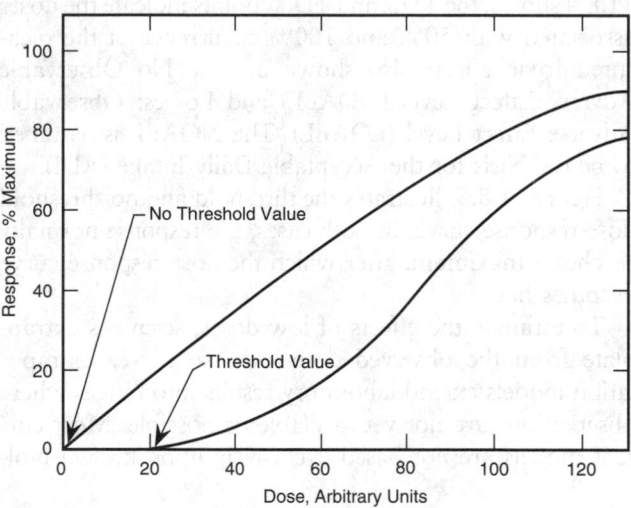

FIG. 11.8.5 Hypothetical dose-response curves. (Adapted from ICAIR Life Systems, Inc., 1985.)

ogy or toxicology, but are largely mathematical constructs based upon assumptions carrying varying degrees of uncertainty. How accurately these extrapolated low-dose responses correspond to true human risk remains a scientific debate.

Animal to Human Extrapolation

In extrapolating from animal data to potential human toxicity, a number of conversion factors are used to account for the differences between humans and animals. Factors must consider individual differences within a species; account for different sensitivities (Table 11.8.2); and note the variations between the two species, such as differences in weight, surface area, metabolism, and absorption.

Extrapolation from animal species to humans has elements of both science and art. It serves as a best estimate, neither invalid nor absolute truth. Although each step of assessment is laden with controversy, animal testing is the generally accepted approach for predicting human toxicity.

EXPOSURE ANALYSIS

Determining toxicity and exposure is necessary in a chemical risk assessment. Exposure to a chemical can occur through direct or indirect routes. Direct exposure is easier to identify, for example, exposure to nicotine and carbon monoxide from smoking cigarettes, or exposure to a pesticide from swimming in a contaminated lake. Indirect exposure can be somewhat more elusive, for example, mercury exposure by eating fish from mercury-contaminated waters. Whether direct or indirect, human exposure to chemicals will be dermal (skin contact), oral (contact by ingestion), and/or inhaled (contact by breathing).

Assessing human exposure to a chemical involves first determining the magnitude, duration, frequency, and route of exposure; and second, estimating the size and nature of the exposed population. Questions include to what concentration of chemical is a person exposed? How often does exposure occur—is it long-term or short-term, continuous or varied? What is the route of exposure—is the chemical in foods, consumer products, or in the workplace? Is the chemical bioactive, or is it purged from the human system without causing any harmful effects? Are special risk groups, such as pregnant women, children, or the elderly, exposed?

Sources of Uncertainty

Exposure measurements and estimates are difficult to obtain, and full of uncertainty. Often, less data exists about human exposure to chemicals than about chemicals' inherent toxicities. When estimating chemical exposure, it is important to be aware that exposure can come from different sources at varying rates—some intermittent, others continuous. Frequently, people assume that exposure comes from only one source and that they only need to monitor levels from that source. However, people may be exposed to different sources at various times and in various quantities. These considerations make estimating exposure very difficult. Below are some sources of uncertainty in estimating chemical exposure.

Monitoring Techniques

When chemical contamination is suspected, it is necessary to identify the baseline, or background concentration, of the chemical before onset of contamination. Subtracting the background concentration from the total concentration detected provides an accurate measure of the exposure resulting from contamination.

TABLE 11.8.2 FACTORS INFLUENCING HUMAN RESPONSE TO TOXIC COMPOUNDS

Factor	Effect
Dose	Larger doses correspond to more immediate effects
Method of administration	Some compounds nontoxic by one route and lethal by another (e.g., phosgene)
Rate of administration	Metabolism and excretion keep pollutant concentrations below toxic levels
Age	Elderly and children more susceptible
Sex	Each sex has hormonally controlled hypersensitivities
Body weight	Inversely proportional to effect
Body fat	Fat bioconcentrates some compounds (large doses can occur in dieters due to stored pollutants)
Psychological status	Stress increases vulnerability
Immunological status	Influences metabolism
Genetic	Different metabolic rates
Presence of other diseases	Similar to immunological status; could be a factor in cancer recurrence
Pollutant pH and ionic states	Interferes or facilitates absorption into the body
Pollutant physical state	Compounds absorbed on particulates may be retained at higher rate
Chemical milieu	Synergisms, antagonisms, cancer "promoters," enhanced absorption
Weather conditions	Temperature, humidity, barometric pressure, and season enhance absorption

It should be noted that scientists cannot measure zero concentration of a chemical—zero concentration is not a scientifically verifiable number. Instead, the terms "nothing detected" or "below the limit of measurement techniques" are used.

Sampling Techniques

In determining the location and number of samples for analysis, samples must accurately represent exposure levels at the place and time of exposure. There may be a difference between soil sampled at the surface or at a depth of six inches. Proper scientific methods must be observed, as an error in sampling will be propagated throughout the entire analysis. Furthermore, enough samples must be taken to allow statistical analysis crucial to ensuring data reliability.

Past Exposure

It is often difficult to determine the past exposure to a chemical. Most epidemiological studies are initiated after symptoms associated with exposure have occurred and after the amount or duration of exposure has changed. Unless detailed records are maintained, as in some workplace environments, the exact amount of exposure must be estimated.

Extrapolation to Lifetime Exposure

When initial exposure is measured (e.g., an industrial worker exposed to about 50 ppm of ethylene oxide for 8 hr per day), an extrapolation is made to determine the exposure over a lifetime of such activity. Information gathered from a small population segment must be extrapolated to the entire population. Such extrapolation often does not account for individual variability in exposures within the population.

RISK CHARACTERIZATION

Although the preceding analyses examine all relevant data to describe hazards, dose-response, or exposure, no conclusions are drawn about the overall risk. The final analysis addresses overall risk by examining the preceding analyses to *characterize* the risk. This process fully describes the expected risk through examining exposure predictions for real-world conditions, in light of the dose-response information from animals, people, and special test systems.

Risk is usually identified as a number. When the risk concern is cancer, the risk number represents the probability of additional cancer cases. For example, an estimate for pollutant X might be expressed as 1×10^{-6} or simply 10^{-6}. This means one additional case of cancer projected in a population of one million people exposed to a certain level of Pollutant X over their lifetimes.

A numerical estimate is only as good as the data it is based on. Scientific uncertainty is a customary and expected factor in all environmental risk assessment. Measurement uncertainty refers to the usual variances accompanying scientific measurement, such as the range (10 ± 1). Sometimes the data gap exists because specific measurements or studies are missing. Sometimes the data gap is more broad, revealing a fundamental lack of understanding about a scientific phenomenon.

The 1983 paradigm and EPA risk assessment guidelines stress the importance of identifying uncertainties and presenting them as part of risk characterization.

The major sources of uncertainty are: (1) difficulty in estimating the amount of chemical exposure to an individual or group; (2) limited understanding of the mechanisms determining chemical absorption and distribution within the body; and (3) reliance on animal experiment data for estimating the effects of chemicals on human organs. All of these areas rely upon scientific judgments even though judgments may vary significantly among experts.

Despite differing views within the scientific community on certain issues, a process has emerged for dealing with these differences. Beginning with peer reviews of each scientific study, this process assures accurate data interpretation by qualified specialists. The next step involves an interdisciplinary review of studies relevant to the risk assessment, where differences of interpretation are fully aired. This structured peer-review process is the best means available to resolve differences among experts.

In summary, despite the limitations of risk assessment, quantifying the best estimate of risk is important in preventing harmful chemical exposure. However, understanding the limits of such estimates and indicating the degree of uncertainty is equally important for sound decision-making.

PUBLIC PERCEPTION OF RISK

The nine criteria in Table 11.8.3 are identified as influencing public perceptions of risk. The characteristics of the criteria on the left contribute to perceptions of low risk, while the criteria on the right contribute to perceptions of higher risk.

Several general observations about perceptions of risk have been made. People tend to judge exposure to involuntary activities or technologies as riskier than voluntary ones. The obvious reason for this perception is that voluntary risk can be avoided whereas involuntary risk cannot. The amount of pesticide residues in food or the concentration of contaminants in drinking water is an involuntary decision for the public. Therefore, the public must turn to the government to regulate these activities and technologies.

Catastrophic events are perceived as riskier than ordinary events. For example, the chance of a plane crash

TABLE 11.8.3   CRITERIA INFLUENCING PUBLIC
PERCEPTION OF RISK

Criteria	Characteristics Perceived as Lower Risk	Characteristics Perceived as Higher Risk
origin	natural	manmade
volition	voluntary	involuntary
effect manifestation	immediate	delayed
severity (number of people affected per incident)	ordinary	catastrophic
controllability	controllable	uncontrollable
benefit	clear	unclear
familiarity of risk	familiar	unfamiliar
exposure	continuous	occasional
necessity	necessary	luxury

killing many individuals is perceived as riskier than the chance of an auto accident killing one or two people. Although the severity of a plane crash is higher, the probability of occurrence is much lower, thus the risk may be lower. In addition, delayed effects, such as cancer, are dreaded more than immediate effects such as poisoning.

Determining the acceptability of a risk to society is a social, not scientific, decision. This determination is influenced greatly by public perception of the risk, and is often reflected in legislation. The variation in the perception of risk can be related to the determination of an acceptable level of risk with various value judgments superimposed upon these perceptions. For example, laboratory tests identified saccharin as an animal carcinogen, requiring the FDA to ban it. However, the U.S. Congress determined that using saccharin was an acceptable risk, and prevented a ban due to perceived public benefits. No absolute answer can be provided to the question, "How safe is safe enough?" Determining acceptable levels of risk and making those value judgments is a very difficult and complex task.

To determine the acceptable risk for noncarcinogens, a *safety factor* is applied. Although it is rooted in science, selection of a safety factor is more of a rule of thumb, or an art. This factor is used when determining the safe dose to humans to compensate for uncertainties in the extrapolation process. This safe dose is known as the *acceptable daily intake* (ADI). The ADI amount of a chemical should not cause any adverse effects to the general human population even after long-term, usually lifetime, exposure. An ADI is calculated by dividing the NOAEL by a safety factor.

Risk Management

Risk assessment estimates the magnitude and type of risk from exposure to a potentially hazardous chemical. The government frequently decides to manage the risk. Public decision-makers are called upon to make the judgments: to synthesize the scientific, social, economical, and political factors and determine the acceptable risk for society. They need to reexamine the issues raised in risk assessment and address the following questions:

- Is the chemical economically important or essential?
- Is there a safer alternative?
- Is the risk of chemical exposure voluntary or involuntary?
- Can exposure be reduced?
- What are the benefits associated with use of this chemical?
- Are those individuals or societies subjected to risks the ones receiving the benefits?
- What are the costs of avoidance?
- What are the public perceptions of the risk?
- What level of risk is acceptable?
- Are some risks perceived as unacceptable no matter what the benefits?

Over the years, many laws have been enacted to protect human health, safety, and the environment, providing a basic framework for risk management decisions. Each law reflects state-of-the-art understanding at the time of its enactment, as well as the political concerns and the public perceptions at that time. Regulators must make their decisions within the constraints of the applicable laws. These laws generally do not prescribe risk assessment methodologies. However, many environmental laws do provide very specific risk management directives.

Statutory risk management mandates can be divided into roughly three categories: pure-risk; technology-based standards; and reasonableness of risks balanced with benefits.

PURE-RISK STANDARDS

Pure-risk standards, sometimes termed zero-risk, are mandated or implied by only a few statutory provisions. Following are two examples of such standards:

The "Delaney clause" of the Federal Food, Drug, and Cosmetic Act prohibits the approval of any food additive that has been found to induce cancer in humans or animals.

The provisions of the Clean Air Act pertaining to national ambient air quality standards require standards for listed pollutants that "protect the public health allowing an adequate margin of safety," i.e., that assure protection of public health without regard to technology or cost factors.

TECHNOLOGY-BASED STANDARDS

Technology-based laws, such as parts of the Clean Air Act and the Clean Water Act, impose pollution controls based on the best economically available or practical technology. Such laws tacitly assume that benefits accrue from the use of the medium (water or air) into which toxic or hazardous substances are discharged, and that complete elimination of discharge of some human and industrial wastes into such media currently is not feasible. The basis for imposing these controls is to reduce human exposure, which indirectly benefits health and environment. The goal is to provide an ample margin of safety to protect public health and safety.

NO UNREASONABLE RISK

A number of statutes require balancing risks against benefits in making risk management decisions. Two examples include:

The Federal Insecticide, Fungicide, and Rodenticide Act requires the EPA to register pesticides that will not cause "unreasonable adverse effects on environment." The phrase refers to "any unreasonable risks to man or the environment taking into account the economic, social, and environmental costs and benefits of the use of any pesticide."

The Toxic Substances Control Act mandates that the EPA is to take action if a chemical substance "presents or will present an unreasonable risk of injury to health or the environment." This includes considering the substance's effects on human health and the environment; the magnitude of human and environmental exposure; the benefits and availability of such substances for various uses; and the reasonably ascertainable economic consequences of the rule.

The RCRA embodies both technology-based and pure-risk-based standards. Congress and the EPA have attempted to craft RCRA regulations in pure-risk-based rationales, but the large numbers of mixtures and the variety of generator/source operations have made that approach exceedingly difficult. As a result, the RCRA focuses on the following regulatory mechanisms:

- Identifying wastes that are hazardous to human health and the environment, and capturing them in a cradle-to-grave management system
- Creating physical barriers to isolate the public from contact with identified hazardous wastes
- Minimizing generation of hazardous wastes
- Encouraging reuse, recycling, and treatment of hazardous wastes
- Providing secure disposal for wastes that cannot otherwise be safely managed

—*David H.F. Liu*

References

American Conference of Governmental Industrial Hygienists. 1990. *Threshold limit values for chemical substances and physical agents and biological exposure indices (BEIs)*. Cincinnati, Oh.

Enger, E.D., R. Kormelink, B.F. Smith, and Smith, R.J. 1989. *Environmental science: the study of Interrelationships*, Dubuque, Iowa: Wm. C. Brown Publishers.

Interagency Regulatory Liaison Group (IRLG). 1979. Scientific bases for identification of potential carcinogens and estimation of risk. *Journal of The Cancer Institute* 63(1).

Nally, T.L. 1984. *Chemical risk: a primer*, Department of Government Relations and Science Policy, American Chemical Society, Washington, D.C.

National Academy of Science (NAS). 1983. *Risk Assessment in Federal Government: Managing the Process*, Washington, D.C.: National Academy Press.

Patton, D. 1993. The ABC of risk assessment, *EPA Journal*, February/March.

Sittig, M. 1985. *Handbook of toxic and hazardous chemicals and carcinogens*, 2nd ed. Park Ridge, N.J.: Noyes Publications.

U.S. Department of Health and Human Services. 1990. *Health assessment guidance manual*. Published by the Agency for Toxic Substances and Disease Registry. Atlanta, Ga.

U.S. Department of Health and Human Services, *Registry of toxic effects of chemical substances*. Superintendent of Documents. Washington, D.C.: U.S. Government Printing Office.

11.9
WASTE MINIMIZATION AND REDUCTION

The first step in establishing a waste minimization strategy is to conduct a waste audit. The key question at the onset of a waste audit is "why is this waste present?" The environmental engineer must establish the primary cause(s) of waste generation before seeking solutions. Understanding the primary cause is critical to the success of the entire investigation. The audit should be waste-stream oriented, producing specific options for additional information or implementation. Once the causes are understood, solution options can be formulated. An efficient materials and waste trucking system that allows computation of mass balances is useful in establishing priorities. Knowing how much raw material is going into a plant and how much is ending up as waste allows the engineer to decide which plant and which waste to address first.

The first four steps of a waste audit allow the engineer to generate a comprehensive set of waste management options. These should follow the hierarchy of source reduction first, waste exchange second, recycling third, and treatment last.

In the end, production may be abandoned because the product or resulting by-product poses an economic hazard that the corporation is not willing to underwrite. These include cases where extensive testing to meet the TSCA (Toxic Substance Control Act) is required. Other such cases include the withdrawal of pre-manufacturing notice applications for some phthalate esters processes, and the discontinuation of herbicide and pesticide production where dioxin is a by-product.

Source Reduction and Control

INPUT MATERIALS

Source control investigations should focus on changes to input materials, process technology, and the human aspect of production. Input material changes can be classified into three elements: purification, substitution, and dilution.

Purification of input materials prevents inert or impure materials from entering the production process. Such impurities cause waste because the process must be purged to prevent undesirable accumulation. Examples of purified input materials include diionized water in electroplating and oxygen instead of air in oxychlorination reactors for ethylene dichloride production.

Substitution involves replacing a toxic material with a less toxic or more environmentally desirable material. Industrial applications of substitution include: using phosphates instead of dichromates as cooling water corrosion inhibitors; using alkaline cleaners instead of chlorinated solvents for degreasing; using solvent-based inks instead of water-based inks; and replacing cyanide cadmium plating bath with noncyanide bath.

Dilution is a minor component of input material changes. An example of dilution is the use of a more dilute solution to minimize dragouts in metal parts cleaning.

TECHNOLOGY CHANGES

Technology changes are made to the physical plant. Examples include process changes; equipment, piping or layout changes; changes to operating settings; additional automation; energy conservation; and water conservation.

Process Change

Innovative technology is often used to develop new processes to achieve the same products, while reducing waste. Process redesign includes alteration of existing processes by adding new unit operations or implementation of new technology to replace outmoded operations. For example, a metal manufacturer modified a process to use a two-stage abrasive cleaner and eliminated the need for a chemical cleaning bath.

A classic example of a process change is the staged use of solvent. An electronics firm switched from using three different solvents—mineral spirits for machine parts, perchloroethylene for computer housings, and a fluorocarbon-mineral blend for printed circuit boards—to a single solvent system. Currently, fresh solvent is used for the printed circuit boards, then reused to degrease the computer housings, and finally, to degrease the machine parts. This practice not only reduces solvent consumption and waste, it eliminates potential cross-contamination of solvents, regenerates a single stream for recycling, and simplifies safety and operating procedures (U.S. EPA 1989).

Equipment, Piping, or Layout Changes

Equipment changes can reduce waste generation by reducing equipment–related inefficiencies. The capital required for more efficient equipment is justified by higher productivity, reduced raw material costs, and reduced waste materials costs. Modifications to certain types of equipment can require a detailed evaluation of process characteristics. In this case, equipment vendors should be consulted for information on the applicability of equipment for a process. Many equipment changes can be very simple and inexpensive.

Examples include installing better seals to eliminate leakage or simply putting drip pans under equipment to collect leaking material for reuse. Another minor modification is to increase agitation and alter temperatures to prevent formation of deposits resulting from crystallization, sedimentation, corrosion, or chemical reactions during formulating and blending procedures.

Operational Setting Changes

Changes to operational settings involve adjustments to temperature, pressure, flow rate, and residence time parameters. These changes often represent the easiest and least expensive ways to reduce waste generation. Process equipment is designed to operate most efficiently at optimum parameter settings. Less waste will be generated when equipment operates efficiently and at optimum settings. Trial runs can be used to determine the optimum settings. For example, a plating company can change the flow rate of chromium in the plating bath to the optimum setting and reduce the chromium concentrations used, resulting in less chromium waste requiring treatment.

Additional Control/Automation

Additional controls or automation can result in improved monitoring and adjustment of operating parameters to ensure maximum efficiency. Simple steps involving on-stream set-point controls or advanced statistical process control systems can be used. Automation can reduce human error, preventing spills and costly downtime. The resulting increase in efficiency can increase product yields.

PROCEDURAL CHANGES

Procedural changes are improvements in the ways people affect the production process. All referred to as *good operating practices* or *good housekeeping*, these include operating procedures, loss prevention, waste segregation, and material handling improvement.

Material Loss Prevention

Loss prevention programs are designed to reduce the chances of spilling a product. A hazardous *material* becomes an RCRA hazardous *waste* when it is spilled. A long-term, slow-release spill is often hard to find, and can create a large amount of hazardous waste. A material loss prevention program may include the following directives:

- Use properly designed tanks and vessels for their intended purpose only
- Maintain physical integrity of all tanks and vessels
- Install overflow alarms for all tanks and vessels

- Set up written procedures for all loading, unloading, and transfer operations
- Install sufficient secondary containment areas
- Forbid operators to bypass interlocks or alarms, or to alter setpoints without authorization
- Isolate equipment or process lines that are leaking or out of service
- Install interlock devices to stop flow to leaking sections
- Use seal-less pumps
- Use bellow seal valves and a proper valve layout
- Document all spillage
- Perform overall material balances and estimate the quantity and dollar value of all losses
- Install leak detection systems for underground storage tanks in accordance to RCRA Subtitle I
- Use floating-roof tanks for VOC control
- Use conservation vents on fixed-roof tanks
- Use vapor recovery systems (Metcalf 1989)

Segregating Waste Streams

Disposed hazardous waste often includes two or more different wastes. Segregating materials and wastes can decrease the amount of waste to be disposed. Recyclers and waste exchangers are more receptive to wastes not contaminated by other substances. The following are good operating practices for waste segregation:

- Prevent hazardous wastes from mixing with non-hazardous wastes
- Isolate hazardous wastes by contaminant
- Isolate liquid wastes from solid wastes

Materials Tracking and Inventory Control

These procedures should be used to track waste minimization efforts and target areas for improvement.

- Avoid over-purchasing
- Accept raw materials only after inspection
- Ensure that no containers stay in inventory longer than the specified period
- Review raw material procurement specifications
- Return expired materials to the supplier
- Validate shelf-life expiration dates
- Test outdated materials for effectiveness
- Conduct frequent inventory checks
- Label all containers properly
- Set up manned stations for dispensing chemicals and collecting wastes

Production Scheduling

The following alterations in production scheduling can have a major impact on waste minimization:

- Maximize batch size

- Dedicate equipment to a single product
- Alter batch sequencing to minimize cleaning frequency (light-to-dark batch sequence, for example)
- Schedule production to reduce cleaning frequency

Preventive Maintenance

These programs cut production costs and decrease equipment downtime, in addition to preventing waste release due to equipment failure.

- Use equipment data cards on equipment location, characteristics, and maintenance
- Maintain a master preventive maintenance (PM) schedule
- Maintain equipment history cards
- Maintain equipment breakdown reports
- Keep vendor maintenance manuals handy
- Maintain a manual or computerized repair history file

Cling

The options for minimizing wastes that cling to containers include:

- Use large containers instead of small condensers whenever possible
- Use containers with a one-to-one height-to-diameter ratio to minimize wetted area
- Empty drums and containers thoroughly before cleaning or disposal

PRODUCT CHANGES

Product Substitution

Changing the design, composition, or specifications of end products allows fundamental change in the manufacturing process or in the end use of raw materials. This can lead directly to waste reduction. For example, the manufacture of water-based paints instead of solvent-based paints involves no hazardous toxic solvents. In addition, the use of water-based paints reduces volatile organic emissions to the atmosphere.

Product Reformulation

Product reformulation or composition changes involves reducing the concentration of hazardous substances or changing the composition so that no hazardous substances are present. Reformulating a product to contain less hazardous material reduces the amount of hazardous waste generated throughout the product's lifespan. Using a less hazardous material within a process reduces the overall amount of hazardous waste produced. For example, a

company can use nonhazardous solvents in place of chlorinated solvents.

Dow Chemical Company achieved waste reduction through changes in product packaging. A wettable powder insecticide, widely used in landscape maintenance and horticulture, was originally sold in 2-lb metal cans. The cans had to be decontaminated before disposal, creating a hazardous waste. Dow now packages the product in 4-oz water-soluble packages which dissolve when the product is mixed with water for use (U.S. Congress 1986).

Product Conservation

One of the most successful methods of product conservation is the effective management of inventory with specific shelf-lives. The Holston Army Ammonium Plant reduced waste pesticide disposal from 440 to 0 kg in one year by better management of stocks (Mill 1988).

WASTE EXCHANGE

Waste exchange is a reuse function involving more than one facility. An exchange matches one *industry's* output to the input requirement of another. Waste exchange organizations act as brokers of hazardous materials by purchasing and transporting them as resources to another client. Waste exchanges commonly deal in solvents, oils, concentrated acids and alkalis, and catalysts. Limitations include transport distance, purity of the exchange product, and reliability of supply and demand.

Waste exchanges were first implemented and are now fairly common in Europe; there are few in the U.S. Although more exchanges have recently been set up in this country, they are not widely accepted because of liability concerns. Even when potential users of waste are found, they must be located fairly close to the generator. Waste transportation requires permits and special handling, increasing the cost.

Recycling and Reuse

Recycling techniques allow reuse of waste materials for beneficial purposes. A recycled material is used, reused, or reclaimed [40 CFR §261.1 (c)(7)]. Recycling through use or reuse involves returning waste material to the original process as a substitute for an input material, or to another process as an input material. Recycling through reclamation involves processing a waste for recovery of a valuable material or for regeneration. Recycling can help eliminate waste disposal costs, reduce raw material costs, and provide income from saleable waste.

Recycling is the second option in the pollution prevention hierarchy and should be considered only when all source reduction options have been investigated and implemented. Recycling options are listed in the following order:

- Direct reuse on-site
- Additional recovery on-site
- Recovery off-site
- Sale for reuse off-site (waste exchange)

It is important to note that recycling can increase a generator's risk or liability as a result of the associated material handling and management. Recycling effectiveness depends upon the ability to separate recoverable waste from other process waste.

DIRECT ON-SITE REUSE

Reuse involves finding a beneficial purpose for a recovered waste. Three factors to consider when determining the potential for reuse are:

- The chemical composition of the waste and its effect on the reuse process
- The economic value of the reuse waste and whether this justifies modifying a process to accommodate it
- The availability and consistency of the waste to be reused
- Energy recovery

For example, a newspaper advertising printer purchased a recycling unit to produce black ink from various waste inks. Blending different colors of ink with fresh black ink and black toner, the unit creates black ink. This mixture is filtered to remove flakes of dried ink, and is used in lieu of fresh black ink. The need to ship waste ink for offsite disposal is eliminated. The price of the recycling unit was recovered in nine months, based on savings in fresh ink purchases and costs of waste ink disposal (U.S. EPA 1989).

In another example, an oil skimmer in a holding tank enables annual capture and recycling of 3000 gallons of waste oil from 30,000 gallons of oily waste water disposed to waste landfills. (Metcalf 1989).

ADDITIONAL ON-SITE RECOVERY

Recycling alternatives can be accomplished either on-site or off-site and may depend on a company's staffing or economic constraints. On-site recycling alternatives result in less waste leaving a facility. The disadvantages of on-site recycling lie in the capital outlay for recycling equipment, the need for operator training, and additional operating costs. In some cases, the waste generated does not warrant the installation costs for in-plant recycling systems. However, since on-site alternatives do not involve transportation of waste materials and the resulting liabilities, they are preferred over off-site alternatives.

For instance, sand used in casting processes at foundries contains heavy metal residues such as copper, lead, and zinc. If these concentrations exceed Toxicity Characteris-

tics Leaching Procedure (TCLP) standards, the sand is a hazardous waste. Recent experiments demonstrated that 95% of the copper could be precipitated and recovered (McCoy and Associates 1989). In another example, a photoprocessing company uses an electrolytic deposition cell to recover silver from rinse water used in film processing equipment. By removing the silver from the wastewater, the wastewater can be discharged to the sewer without additional pretreatment.

OFFSITE RECOVERY

If the amount of waste generated on-site is insufficient for a cost-effective recovery system, or if the recovered material cannot be reused on-site, off-site recovery is preferable. Materials commonly reprocessed off-site are oils, solvents, electroplating sludges and process baths, scrap metal, and lead-acid batteries. The cost of off-site recycling depends upon the purity of the waste and the market for the recovered materials.

The photoprocessing company mentioned above also collects used film and sells it to a recycler. The recycler burns the film and collects the silver from residual ash. By removing the silver from the ash, the fly ash becomes nonhazardous (EPA 1989).

SALE FOR REUSE OFF-SITE

See the preceding discussions on waste exchange. The most common reuses of hazardous waste include wastewater used for irrigation and oil field pressurization; sludges used as fertilizers or soil matrix; and sulfuric acid from smelters.

Recycling methods, including numerous physical, chemical and biological technologies will be discussed in Section(s) 11.15 and 11.18.

—David H.F. Liu

References

Code of Federal Regulations, Title 40, sec. 261.1.

McCoy and Associates, Inc. 1989. *The hazardous waste consultant.* (March-April).

Metcalf, C., ed. 1989. *Waste reduction assessment and technology transfer (WRATT) training manual.* The University of Tennessee Center for Industrial Services. Knoxville, Tenn.

Mill, M.B. 1988. Hazardous waste minimization in the manufacture of explosives, in *Hazardous waste minimization in the department of defense.* Edited by J.A. Kaminski. Office of the Deputy Assistant Secretary of Defense (Environment). Washington, D.C.

U.S. Congress, Office of Technology Assessment. 1986. *Serious reduction of hazardous waste.* Superintendent of Documents. Washington, D.C.: Government Printing Office.

U.S. Environmental Protection Agency (EPA). 1989. *Waste minimization in metal parts cleaning.* Office of Solid Waste and Emergency Response, Report No. EPA/530-SW-89-049. Washington, D.C.

11.10
HAZARDOUS WASTE TRANSPORTATION

The EPA's cradle-to-grave hazardous waste management system attempts to track hazardous waste from generation to ultimate disposal. The system requires generators to establish a manifest or itemized list form for hazardous waste shipments. This procedure is designed to ensure that wastes are direct to, and actually reach, permitted disposal sites.

Generator Requirements

The *generator* is the first element of the RCRA cradle-to-grave concept, which includes generators, transporters,

treatment plants, storage facilities, and disposal sites. Generators of more than 100 kg of hazardous waste or 1 kg of acutely hazardous waste per month must, with a few exceptions, comply with all generator regulations.

Hazardous waste generators must comply with all DOT legislation regulating transport of hazardous materials, as well as other hazardous waste regulations promulgated by both DOT and the EPA. Table 11.10.1 summarizes the requirements, indicates the agency responsible for compliance, and provides a reference to the Code of Federal Regulations.

TABLE 11.10.1   EPA AND DOT HAZARDOUS WASTE TRANSPORTATION REGULATIONS

Required of	Agency	Code of Federal Regulations
Generator/Shipper		
1. Determine if waste is hazardous according to EPA listing criteria	EPA	40 CFR 261 and 262.11
2. Notify EPA and obtain I.D. number; determine that transporter and designated treatment, storage, or disposal facility have I.D. numbers	EPA	40 CFR 262.12
3. Identify and classify waste according to DOT Hazardous Materials Table and determine if waste is prohibited from certain modes of transport	DOT	49 CFR 172.101
4. Comply with all packaging, marking, and labeling requirements	EPA	40 CFR 262.32 (b),
	DOT	49 CFR 173, 49 CFR 172, subpart D, and 49 CFR 172, subpart E
5. Determine whether additional shipping requirements must be met for the mode of transport used.	DOT	49 CFR 174–177
6. Complete a hazardous waste manifest	EPA	40 CFR 262, subpart B
7. Provide appropriate placards to transporter	DOT	49 CFR 172, subpart F
8. Comply with record-keeping and reporting requirements	EPA	40 CFR 262, subpart D
Transporter/Carrier		
1. Notify EPA and obtain I.D. number	EPA	40 CFR 263.11
2. Verify that shipment is properly identified, packaged, marked, and labeled and is not leaking or damaged	DOT	49 CFR 174–177
3. Apply appropriate placards	DOT	49 CFR 172.506
4. Comply with all manifest requirements (e.g., sign the manifest, carry the manifest, and obtain signature from next transporter or owner/operator of designated facility)	DOT	49 CFR 174–177
	EPA	40 CFR 263.20
5. Comply with record-keeping and reporting requirements	EPA	50 CFR 263.22
6. Take appropriate action (including cleanup) in the event of a discharge and comply with the DOT incident reporting requirements	EPA	40 CFR 263.30–31
	DOT	49 CFR 171.15–17

Source: Reprinted from U.S. Environmental Protection Agency.

The regulatory requirements for hazardous waste generators contained in 40 CFR Part 262 include:

- Obtaining an EPA ID number
- Proper handling of hazardous waste before transport
- Establishing a manifest of hazardous waste
- Recordkeeping and reporting

EPA ID NUMBER

The EPA and primacy states monitor and track generator activity by an identification number to each generator. Without this number, the generator is barred from treating, storing, disposing of, transporting, or offering for transportation any hazardous waste. Furthermore, the generator is forbidden from offering the hazardous waste to any transporter, or treatment, storage, or disposal (TSD) facility that does not also have an EPA ID number. Generators obtain ID numbers by notifying the EPA of hazardous waste activity, using the standard EPA notification form.

PRETRANSPORT REGULATIONS

Pretransport regulations are designed to ensure safe transportation of a hazardous waste from origin to ultimate disposal; to minimize the environmental and safety impacts of accidental releases; and to facilitate control of any releases that may occur during transportation. In developing these regulations, the EPA adopted those used by the Department of Transportation (DOT) for transporting hazardous materials (49 CFR 172, 173, 178 and 179). These DOT regulations require:

Proper packaging to prevent hazardous waste leakage under normal or potentially dangerous transport conditions such as when a drum of waste falls from a truck or loading dock;

Labeling, marking, or placarding of the package to identify characteristics and dangers associated with the waste.

These pretransport regulations apply only to generators shipping waste off-site.

Briefly, individual containers are required to display "Hazardous Waste" markings, including the proper DOT shipping name, using the standardized language of 49 CFR Sections 172.101 and .102. The labels on individual containers must display the correct hazard class as prescribed by Subpart E of Part 172. Examples of DOT labels and placards are shown in Figures 11.10.1 and 11.10.2. Placards are important in case of accidents because they are highly visible. Efforts are now in progress for international adoption of hazardous marking, labeling and placarding conventions.

WASTE ACCUMULATION

A generator may accumulate hazardous waste on-site for 90 days or less, provided the following requirements are met:

Proper Storage. The waste must be properly stored in containers or tanks marked "Hazardous Waste" with the date accumulation began.

Emergency Plan. A contingency plan and emergency procedures are developed. Generators must have a written emergency plan.

Personnel Training. Facility personnel must be trained in the proper handling of hazardous waste.

The 90-day period allows more cost effective transportation. Instead of paying to haul several small shipments of waste, the generator can accumulate enough for one big shipment.

FIG. 11.10.1 DOT labels for hazardous materials packages. Source: Reprinted from U.S. Department of Transportation.

FIG. 11.10.2 DOT placards for hazardous substances. (*Source:* Reprinted from U.S. Department of Transportation.)

If hazardous waste is accumulated on-site for more than 90 days, the generator is considered an operator of a storage facility and becomes subject to Subtitle C requirements including permitting. Under temporary, unforeseen, or uncontrollable circumstances, the 90-day period may be extended for up to 30 days by the EPA Regional Administrator on a case-by-case basis.

Small quantity generators (SQGs), defined as those producing 100–1000 kg of hazardous waste per month, are accorded an exception to this 90-day accumulation period. The Hazardous and Solid Waste Amendments (HSWA) require, and the EPA developed, regulations allowing such generators to accumulate waste for 180 days, or 270 days if waste must be shipped over 200 miles, before SQGs are considered to be operating a storage facility.

THE MANIFEST

The Uniform Hazardous Waste Manifest is the key to the cradle-to-grave management system (Figure 11.10.3). Using the manifest, generators and regulators can track the movement of hazardous waste from the point of generation to the point of ultimate treatment, storage, or disposal (TSD).

The HSWA requires manifests to certify that generators have programs in place to reduce waste volume and toxicity to the degree the generator determines economically practicable. In addition, the treatment, storage, or disposal method chosen by the generator must be the best method currently available to minimize risks to human health and the environment.

Generators must prepare manifests properly since they are responsible for the production and ultimate disposition of hazardous wastes. Some common mistakes found on manifests are (Turner 1992):

Omission of the 24-hr emergency response telephone number. As of December 31, 1990, the DOT required inclusion of a 24-hr telephone number for use if an incident should occur during transportation. Shippers and carriers should look closely at this section to ensure its proper completion.

Omission of the manifest document number. Many generators use this control number to indicate the number of shipments made during a specified period. Others use it to indicate shipments from a specific section of their facility.

Misunderstanding of the generator name and mailing address. The address listed should be the location managing the return manifest form. The 12-digit EPA identification number is site specific in that it is assigned to the physical location where the hazardous waste is generated.

Improper entry of shipping name, hazard class or UN/NA numbers. 49 CFR Sec. 172.202 specifies the proper order for entering a basic description on a shipping document. The technical or chemical group names may be entered in parentheses between the proper shipping name and hazard class.

The manifest is part of a controlled tracking system. Each time waste is transferred from a transporter to a designated facility or to another transporter, the manifest must be signed to acknowledge receipt of the waste. A copy of the manifest is retained by each link in the transportation chain. Once the waste is delivered to the designated facility, the owner or operator of the facility must send a copy of the manifest back to the generator. This system ensures that the generator has documentation that the hazardous waste has reached its destination.

The multiple-copy manifest is initially completed and signed by the hazardous waste generator. The generator retains Part 6 of the manifest, sends Part 5 to the EPA or the appropriate state agency, and provides the remainder to the transporter. The transporter retains Part 4 of the manifest and gives the remaining parts to the TSD facility upon arrival. The TSD facility retains Part 3 and sends Parts 1 and 2 to the generator and the regulatory agency, or agencies, respectively. Throughout this transition, the hazardous waste shipment is gener-

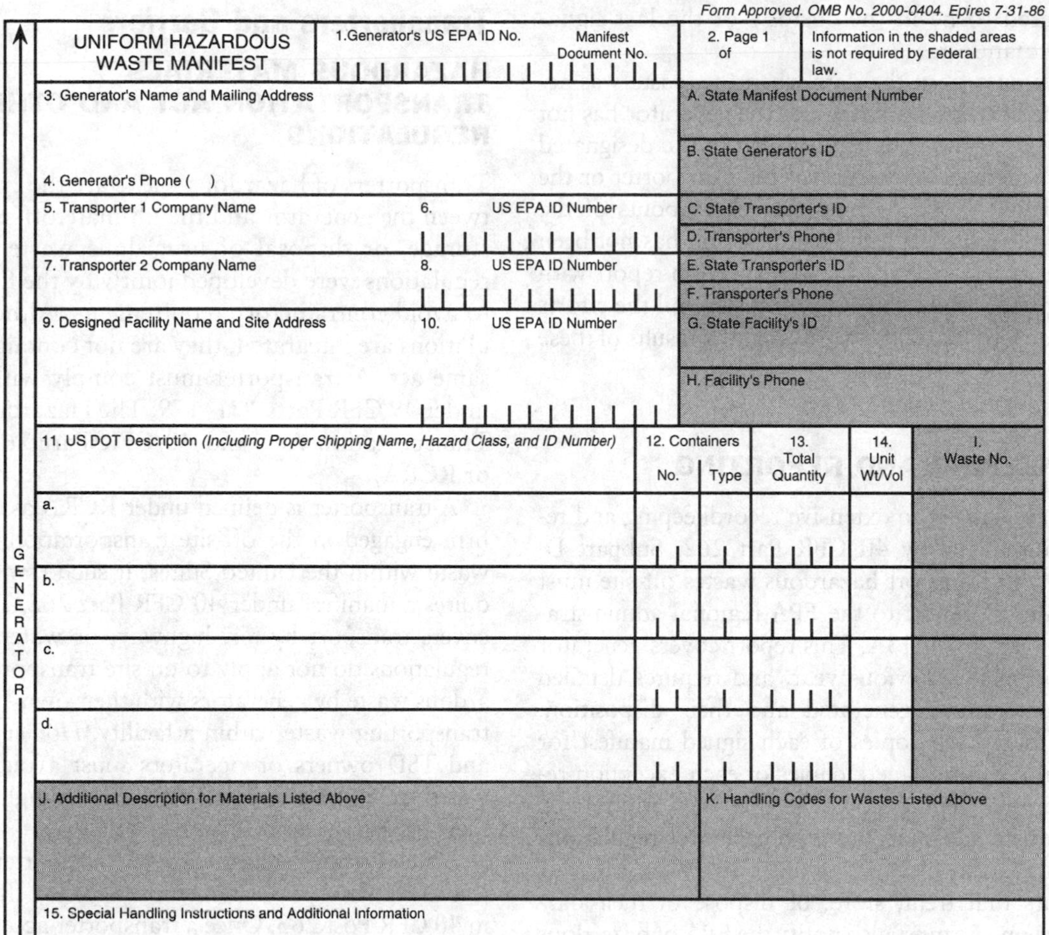

Form Approved. OMB No. 2000-0404. Epires 7-31-86

UNIFORM HAZARDOUS WASTE MANIFEST

1. Generator's US EPA ID No.

Manifest Document No.

2. Page 1 of

Information in the shaded areas is not required by Federal law.

3. Generator's Name and Mailing Address

A. State Manifest Document Number

B. State Generator's ID

4. Generator's Phone ( )

5. Transporter 1 Company Name

6. US EPA ID Number

C. State Transporter's ID

D. Transporter's Phone

7. Transporter 2 Company Name

8. US EPA ID Number

E. State Transporter's ID

F. Transporter's Phone

9. Designed Facility Name and Site Address

10. US EPA ID Number

G. State Facility's ID

H. Facility's Phone

11. US DOT Description *(Including Proper Shipping Name, Hazard Class, and ID Number)*	12. Containers No.	12. Containers Type	13. Total Quantity	14. Unit Wt/Vol	I. Waste No.
a.					
b.					
c.					
d.					

J. Additional Description for Materials Listed Above

K. Handling Codes for Wastes Listed Above

15. Special Handling Instructions and Additional Information

16. GENERATOR'S CERTIFICATION: I hereby declare that the contents of this consignment are fully and accurately described above by proper shipping name and are classed, packed, marked, and labeled, and are in all respects in proper condition for transport by highway according to applicable international and national government regulations.

Unless I am a small quantity generator who has been exempted by statute or regulation from the duty to make a waste minimization certification under Section 3002(b) of RCRA. I also certify that I have a program in place to reduce the volume and toxicity of waste generated to the degree I have determined to be economically practicable and I have selected the method of treatment, storage, or disposal currently available to me which minimizes the present and future threat to human health and the environment.

Printed/Typed Name | Signature | Month Day Year

GENERATOR

17. Transporter 1 Acknowledgement of Receipt of Materials

Printed/Typed Name | Signature | Month Day Year

18. Transporter 2 Acknowledgement of Receipt of Materials

Printed/Typed Name | Signature | Month Day Year

TRANSPORTER

19. Discrepancy Indication Space

20. Facility Owner or Operator: Certification of receipt of hazardous materials covered by this manifest except as noted in item 19.

Printed/Typed Name | Signature | Month Day Year

FACILITY

EPA Form 8700-22 (Rev. 4-85) Previous edition is obsolete.

FIG. 11.10.3 Uniform hazardous waste manifest.

ally considered to be in the custody of the last signatory on the manifest.

If 35 days pass from the date when the waste was accepted by the initial transporter and the generator has not received Part 1 of the manifest form from the designated facility, the generator must contact the transporter or the designated facility to determine the whereabouts of the waste. If 45 days pass and the manifest still has not been received, the generator must file an exception report with the EPA regional office. The report must detail the efforts of the generator to locate the waste, and the results of these efforts.

RECORDKEEPING AND REPORTING

Generators are subject to extensive recordkeeping and reporting requirements by 40 CFR Part 262, Subpart D. Generators who transport hazardous wastes off site must submit an annual report to the EPA regional administrator on EPA form 8700-13A. This report covers generator activities during the previous year, and requires detailed accounting of wastes generated and their disposition. Generators must keep copies of each signed manifest for 3 years from the date signed, copies of each exception report, each annual report, copies of analyses, and related determinations made in accord with generator regulations (40 CFR Part 262).

Generators that treat, store, or dispose of their hazardous waste on-site must also notify the EPA of hazardous waste activity, obtain an EPA ID, apply for a permit, and comply with permit conditions. They too must submit an annual report containing a description of the type and quantity of hazardous waste handled during the year, and the method(s) of treatment, storage, or disposal used.

EXPORT AND IMPORT OF HAZARDOUS WASTE

Export of hazardous waste from the U.S. to another country is prohibited unless:

- Notification of intent to export has been provided to the EPA at least 60 days in advance of shipment;
- The receiving country has consented to accept the waste;
- A copy of the EPA "Acknowledgment of Consent" accompanies the shipments; and
- The hazardous waste shipment conforms to the terms of the receiving country's consent (40 CFR §262.52).

Any import of hazardous waste from another country into the U.S. must comply with the requirements of 40 CFR Part 262, i.e., *the importer becomes the generator*, for RCRA regulatory purposes.

Transporters and Carriers

HAZARDOUS MATERIALS TRANSPORTATION ACT AND OTHER REGULATIONS

Transporters of hazardous waste are the critical link between the generator and the ultimate off-site treatment, storage, or disposal of hazardous waste. Transporter regulations were developed jointly by the EPA and DOT to avoid contradictory requirements. Although the regulations are integrated, they are not contained under the same act. A transporter must comply with regulations under 49 CFR Parts 171–179, The Hazardous Materials Transportation Act, and 40 CFR Part 263 (Subtitle C of RCRA).

A transporter is defined under RCRA as any person or firm engaged in the off-site transportation of hazardous waste within the United States, if such transportation requires a manifest under 40 CFR Part 262. This definition covers transport by air, highway, or water. Transporter regulations do not apply to on-site transportation of hazardous waste by generators with their own TSDs, or TSDs transporting waste within a facility. *However,* generators and TSD owners or operators must avoid transporting waste over public roads that pass through or alongside their facilities (Figure 11.10.4).

Under certain circumstances a transporter may be subject to regulatory requirements other than those contained in 40 CFR Post 263. Once a transporter accepts hazardous waste from a generator or another transporter, the transporter can store it for up to 10 days without being subject to any new regulations. However, if storage time exceeds 10 days, the transporter is considered to be operating a storage facility and must comply with the regulations for such a facility. In addition, transporters who bring hazardous waste into the United States or mix hazardous wastes of different DOT shipping descriptions by placing

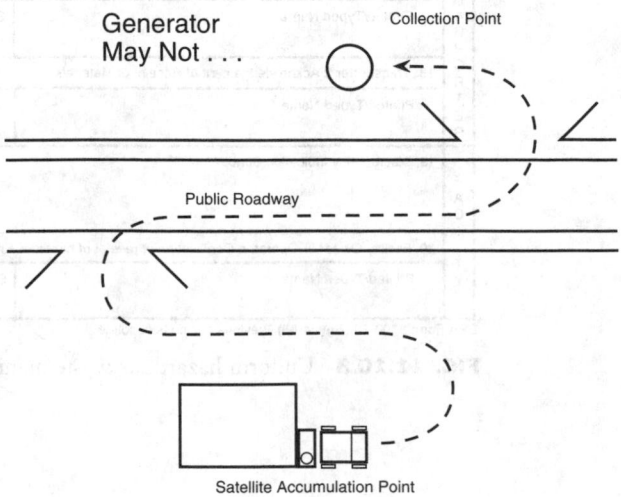

FIG. 11.10.4 Off-site transportation of hazardous waste.

them in the same container are classified as generators and must comply with the generator regulations.

A transporter is subject to regulations including obtaining an EPA ID number, complying with the manifest system, and dealing with hazardous waste discharges.

The transporter is required to deliver the entire quantity of waste accepted from either the generator or another transporter to the facility designated on the manifest. If the waste cannot be delivered as the manifest directs, the transporter must inform the generator and receive further instructions, such as returning the waste or taking it to another facility. Before handing the waste over to a TSD, the transporter must have the TSD facility operator sign and date the manifest. One copy of the manifest remains at the TSD facility while the other stays with the transporter. The transporter must retain a copy of the manifest for three years from the date the hazardous waste was accepted.

Even if generators and transporters of hazardous waste comply with all appropriate regulations, transporting hazardous waste can still be dangerous. There is always the possibility of an accident. To deal with this possibility, regulations require transporters to take immediate action to protect health and the environment if a release occurs by notifying local authorities and/or closing off the discharge area.

The regulations also give officials special authority to deal with transportation accidents. Specifically, if a federal, state, or local official, with appropriate authority, determines that immediate removal of the waste is necessary to protect human health or the environment, the official can authorize waste removal by a transporter without an EPA ID or a manifest.

MODES OF TRANSPORT

A 1981 report, prepared for the EPA, estimated that 96% of the 264 million tn of hazardous wastes generated each year were disposed at the site where they were generated. By 1989, the National Solid Wastes Management Association (NSWMA 1989) stated that trucks traveling over public highways move over 98% of hazardous wastes that are treated off-site. Another perspective can be gained from statistics for hazardous materials transportation. Rail transportation moves about 8% of the hazardous materials shipped, but 57% of the *ton-miles* of hazardous materials shipped (U.S. Office of Technology Assessment 1986).

The highway transport mode is regarded as the most versatile, and is the most widely used. Tank trucks can access most industrial sites and TSD facilities. Rail shipping requires expensive sidings, and is suitable for very large quantity shipments. Cargo tanks are the main carriers of bulk hazardous materials; however, large quantities of hazardous wastes are shipped in 55-gal drums.

Cargo tanks are the main carriers of bulk hazardous materials over roads. Cargo tanks are usually made of steel or aluminum alloy, or other materials such as titanium, nickel, or stainless steel. They range in capacity from 4,000 to 12,000 gal. Federal road weight laws usually limit motor vehicle weights to 80,000 lb gross. Table 11.10.2 lists DOT cargo tank specifications for bulk shipment of common hazardous materials and example cargos.

As stated above, rail shipments account for about 8% of the hazardous materials transported annually, with about 3,000 loads each day. However, the proportion of hazardous waste shipments is unknown.

The major classifications of rail tank cars are pressure and nonpressure (for transporting both gases and liquids). Both categories have several subclasses, which differ in test pressure, presence or absence of bottom discharge valves, type of pressure relief system, and type of thermal shielding. Ninety percent of tank cars are steel; aluminum is also common.

DOT tank car design specifications are detailed in 49 CFR Part 179. Rail car specification numbers for trans-

TABLE 11.10.2 CARGO TANK TABLE

Cargo Tank Specification Number	Types of Commodities Carried	Examples
MC-306 (MC-300, 301, 302, 303, 305)[a]	Combustible and flammable liquids of low vapor pressure	Fuel oil, gasoline
MC-307 (MC-304)	Flammable liquids, Poison B materials with moderate vapor pressure	Toluene, diisocyanate
MC-312 (MC-310, 311)	Corrosives	Hydrochloric acid, caustic solution
MC-331 (MC-330)	Liquified compressed gases	Chlorine, anhydrous ammonia
MC-338	Refrigerated liquified gases	Oxygen, methane

[a]The numbers in parentheses designate older versions of the specifications; the older versions may continue in service but all newly constructed cargo tanks must meet current specifications. (*Source: Code of Federal Regulations*, Title 49, sections 172.101 and 178.315–178.343).

TABLE 11.10.3 PRESSURE RAIL CARS

Class	Material	Insulation	Test pressure	Relief valve setting	Notes
DOT 105	Steel, aluminum	Required	100	75	No bottom outlet or washout; only one opening in tank; chlorine
			200	150	
			300	225	
			400	300	
			500	375	
			600	450	
DOT 112	Steel	None	200	150	No bottom outlet or washout; anhydrous ammonia
			340	225	
				280	
			400	300	
				330	
			500	375	
DOT 114	Steel	None	340	255	Similar to DOT 105; optional bottom outlet; liquefied petroleum gas
			400	300	

Source: Reprinted from Office of Technology Assessment, 1986, *Transportation of Hazardous Materials* (U.S. Congress, Washington, D.C.).

TABLE 11.10.4 NONPRESSURE RAIL CARS

Class	Material	Insulation	Test pressure	Relief valve setting	Notes
DOT 103	Steel, aluminium, stainless steel, nickel	Optional	60	35	Optional bottom outlet; whiskey
DOT 104	Steel	Required	60	35	Similar to DOT 103
DOT 111	Steel, aluminum	Optional	60	35	Optional bottom outlet and bottom washout
DOT 111A	Steel, aluminum	Optional	100	75	Hydrochloric acid

Source: Reprinted from Office of Technology Assessment, 1986.

porting pressurized hazardous materials are DOT 105, 112, and 114 (Table 11.10.3); for unpressurized shipments the numbers are DOT 103, 104, and 111 (Table 11.10.4). Specifications call for steel jacket plate and thickness ranging from 11 ga (approximately $\frac{1}{8}$ in.) to $\frac{3}{4}$ in and aluminum jacket plate thickness of $\frac{1}{2}$ to $\frac{5}{8}$ in. Capacities for tank cars carrying hazardous materials are limited to 34,500 gal or 263,000 lb gr wt. It is proposed that the gross rail load (GRL) limits on 100-tn trucks be increased to 286,000 lb gr.

Because of regulations and industry initiatives, the tank car of the future may be only three to five years away (Snelgrove 1995). This tank car design will most likely be based on non-accident release (NAR) products. Changes will probably include safety valves or surge devices to replace the safety vent; elimination of bottom loading; improved versions of today's manway design; and the equivalent of pressure heads for non-pressure DOT 111A-specification tank cars.

In recent years, the DOT has significantly revised hazardous materials classifications, hazard communications, and packaging requirements to agree with other national and international United Nations (UN) codes. In the coming years, these regulations will cover every shipping container, from drums and intermediate bulk containers (IBCs), to tractor trailers, and rail tank cars.

—*David H.F. Liu*

References

Code of Federal Regulations, Title 40, Part 262.
Code of Federal Regulations, Title 49, Parts 172, 173, 178 and 179.
National Solid Waste Management Association (NSWMA). 1989. *Managing hazardous waste: fulfilling the public trust.* Washington, D.C.
Turner, P.L. 1992. Preparing hazwaste transport manifests. *Environmental protection*, December 1992.
U.S. Environmental Protection Agency (EPA) (1986). *RCA Orientation Manual.* Office of Solid Waste. Publication No. EPA 530-SW-86-001. Washington, D.C.: U.S. Government Printing Office.

Treatment and Disposal

11.11
TREATMENT, STORAGE, AND DISPOSAL REQUIREMENTS

Treatment, storage, and disposal facilities (TSDs) are the last link in the cradle-to-grave hazardous waste management system. All TSDs handling hazardous waste must obtain operating permits and abide by treatment, storage, and disposal regulations. TSD regulations establish performance standards for owners and operators to minimize the release of hazardous waste into the environment.

The original RCRA establishes two categories of TSDs based on permit status. Section 3005(a) of the act specifies that TSDs must obtain a permit to operate. The first category consists of *interim status* facilities that have not yet obtained permits. Congress recognized that it would take many years for the EPA to issue all permits, therefore, the interim status was established. This allows those who own or operate facilities existing as of November 19, 1980, and who are able to meet certain conditions, to continue operating as if they have a permit until their permit application is issued or denied. The second category consists of facilities with permits.

Under Section 3004(a) of the act, the EPA was required to develop regulations for all TSDs. Although only one set was required, the EPA developed two sets of regulations, one for interim status TSDs and the other for permitted TSDs. While developing TSD regulations, the EPA decided that owners and operators of interim status facilities should meet only a portion of the requirements for permitted facilities.

General Facility Standards

Both interim status and permit standards consist of administrative and nontechnical requirements, and technical and non-specific requirements. The interim status standards, found in 40 CFR Part 265, are primarily good housekeeping practices that owners and operators must follow to properly manage hazardous wastes. The permit standards found in 40 CFR Part 264 are design and operating criteria for facility-specific permits.

As detailed in Section 11.10, all facilities handling hazardous wastes must obtain an EPA ID number. Owners and operators must ensure that wastes are correctly identified and managed according to the regulations. They must also ensure that facilities are secure and operating properly. Personnel must be trained to perform their duties correctly, safely, and in compliance with all applicable laws, regulations, and codes. Owners and operators must:

Conduct waste analyses before starting treatment, storage, or disposal in accord with a written waste analysis plan. The plan must specify tests and test frequencies providing sufficient information on the waste to allow management in accordance with the laws, regulations, and codes.

Install security measures to prevent inadvertent entry of people or animals into active portions of the TSDF. The facility must be surrounded by a barrier with control entry systems or 24-hr surveillance. Signs carrying the warning "Danger—Unauthorized Personnel Keep Out" must be posted at all entrances. Precautions must be taken to avoid fires, explosions, toxic gases, or any other events threatening human health, safety, and the environment.

Conduct inspections according to a written schedule to assess facility compliance status and detect potential problem areas. Observations made during inspections must be recorded in the facility's operating log and kept on file for 3 years. All problems noted must be remedied.

Conduct training to reduce the potential for mistakes that might threaten human health and the environment. In addition, the Occupational Safety and Health Administration (OSHA) now requires each TSD to implement a hazard communication plan, a medical surveillance program, and a health and safety plan. Decontamination procedures must be in place and employees must receive a minimum of 24 hr of health and safety training.

Properly manage ignitable, reactive, or incompatible wastes. Ignitable or reactive wastes must be protected from sources of ignition or reaction, or be treated to eliminate the possibility. Owners and operators must ensure that treatment, storage, or disposal of ignitable, reactive, or incompatible waste does not result in damage to the containment structure, or threaten human health or the environment. Separation of incompatible wastes must be maintained.

Comply with local standards to avoid siting new facilities in locations where floods or seismic events could affect waste management units. Bulk liquid wastes are prohibited from placement in salt domes, salt beds, or underground mines or caves.

PREPAREDNESS AND PREVENTION

Facilities must be designed, constructed, maintained, and operated to prevent fire, explosion, or any unplanned release of hazardous wastes that could threaten human health and the environment. Facilities must be equipped with:

An internal communication or alarm system for immediate emergency instructions to facility personnel
Telephone or two-way radio capable of summoning emergency assistance from local police, fire, and emergency response units
Portable fire extinguishers, along with fire, spill control, and contamination equipment
Water at adequate volume and pressure to supply water hoses, foam-producing equipment, automatic sprinklers, or water spray systems

All communication and emergency equipment must be tested regularly to ensure proper emergency operation. All personnel must have immediate access to the internal alarm or emergency communication system. Aisle space must allow unobstructed movement of personnel and equipment during an emergency.

Owners and operators of TSDs must make arrangements to:

Familiarize police, fire, and emergency response teams with the facility, wastes handled and their properties, work stations, and access and evacuation routes
Designate primary and alternate emergency response teams where more than one jurisdiction might respond
Familiarize local hospitals with the properties of hazardous wastes handled at the facility, and the type of injuries or illnesses that could result from events at the facility

CONTINGENCY PLAN AND EMERGENCY PROCEDURE

A contingency plan must be in effect at each TSDF. The plan must minimize hazards from fires, explosions, or any release of hazardous waste constituents. The plan must be implemented immediately whenever there is a fire, explosion or release that would threaten human health or the environment.

The contingency plan must

Describe personnel actions to implement the plan
Describe arrangements with local police, fire, and hospital authorities, as well as contracts with emergency response teams to coordinate emergency services

List names, addresses, and phone numbers of all persons qualified to act as emergency coordinators for the facility
List all emergency equipment, communication, and alarm systems, and the location of each item
Include an evacuation plan for facility personnel

The contingency plan must be maintained at the facility and at all emergency response facilities that might provide services. It must be reviewed and updated whenever any item affecting the plan is changed. A key requirement is designating an emergency coordinator to direct response measures and reduce the adverse impacts of hazardous waste releases.

General Technical Standards for Interim Status Facilities

The objective of the RCRA interim status technical requirements is to minimize the potential for environmental and public health threats resulting from hazardous waste treatment, storage, and disposal at existing facilities waiting for an operating permit. The general standards cover three areas:

- Groundwater monitoring requirements (Subpart F)
- Closure, postclosure requirements (Subpart G)
- Financial requirements (Subpart H)

GROUNDWATER MONITORING

Groundwater monitoring is required for owners or operators of surface impoundments, landfills, land treatment facilities, or waste piles used to manage hazardous wastes. These requirements assess the impact of a facility on the groundwater beneath it. Monitoring must be conducted for the life of the facility, except at land disposal facilities, which must monitor for up to 30 years after closing.

The groundwater monitoring program requires installing a system of four monitoring wells: one up-gradient from the waste management unit and three down-gradient. The down-gradient wells must be placed to intercept any waste from the unit, should a release occur. The up-gradient wells must provide data on groundwater that is not influenced by waste coming from the waste management unit (called background data). If the wells are properly located, data comparisons from up-gradient and down-gradient wells should indicate if contamination is occurring.

After the wells are installed, the owner or operator monitors them for one year to establish background concentrations for selected chemicals. These data form the basis for all future comparisons. There are three sets of parameters for background concentrations: drinking water, groundwater quality, and groundwater contamination.

If a significant increase or decrease in pH is detected for any of the indicator parameters, the owner or operator must implement a groundwater assessment program to determine the nature of the problem. If the assessment shows contamination by hazardous wastes, the owner or operator must continue assessing the extent of groundwater contamination until the problem is ameliorated, or until the facility is closed.

CLOSURE

Closure is the period when wastes are no longer accepted, during which owners or operators of TSD facilities complete treatment, storage, and disposal operations, apply final covers to or cap landfills, and dispose of or decontaminate equipment, structures, and soil.

Following the closure, a 30-yr postclosure period is established for facilities that do not *close clean* as described below. Postclosure care consists of the following at minimum:

- Groundwater monitoring and reporting
- Maintenance and monitoring of waste containment systems
- Continued site security

Clean closure may be accomplished by removing all contaminants from impoundments and waste piles. At a minimum, owners and operators of surface impoundments and waste piles that wish to close clean must conduct soil analyses and groundwater monitoring to confirm that all wastes have been removed from the unit. The EPA or state agency may establish additional clean closure requirements on a case-by-case basis. A successful demonstration of clean closure eliminates the requirement for postclosure care of the site.

FINANCIAL REQUIREMENTS

Financial requirements were established to ensure funds are available to pay for closing a facility, for rendering postclosure care at disposal facilities, and to compensate third parties for bodily injury and property damage caused by accidents related to the facility's operation. There are two kinds of financial requirements:

- Financial assurance for closure and postclosure
- Liability coverage for injury and property damage

To meet financial assurance requirements, owners and operators must first prepare written cost estimates for closing their facilities. If postclosure care is required, a cost estimate for providing this care must also be prepared. These cost estimates must reflect the actual cost of the activities outlined in the closure and postclosure plans, and are adjusted annually for inflation. The cost estimate for closure is based on the point in the facility's operating life when closure would be most expensive. Cost estimates for postclosure monitoring and maintenance are based on projected costs for the entire postclosure period.

The owner or operator must demonstrate to the EPA or state agency an ability to pay the estimated amounts. This is known as financial assurance. The owner/operator may use one or a combination of the following six mechanisms to comply with financial assurance requirements: trust fund, surety bond, letter of credit, closure/postclosure insurance, corporate guarantee for closure, and financial test. All six mechanisms are adjusted annually for inflation or more frequently if cost estimates change.

The Subpart H requirements for these mechanisms are extensive. Readers with particular interest in the details should examine 40 CFR Parts 264 and 265, Subpart H. Liability insurance requirements include coverage of at least $1 million (annual aggregate of at least $2 million) per sudden accidental occurrence, such as fire or explosion. Owners and operators must also maintain coverage of at least $3 million per occurrence (annual aggregate of at least $6 million), exclusive of legal defense costs, for nonsudden occurrences such as groundwater contamination. Liability coverage may be demonstrated using any of the six mechanisms allowed for assurance of closure or postclosure funds.

—David H.F. Liu

11.12
STORAGE

Many early disasters and current Superfund sites grew from uncontrolled accumulation of hazardous wastes. Congress and the EPA sought to impose rigorous controls and accountability on all who accumulate and store hazardous wastes through the RCRA statutes and EPA regulations.

The RCRA defines *storage* as holding hazardous waste for a temporary period, after which the hazardous waste is treated, disposed of, or stored elsewhere. The accumulation of hazardous waste beyond a prescribed period, usually 90 days, is considered storage. The owner or operator of a facility where waste is held for more than 90 days must apply for a permit before starting accumulation, and must comply with regulations pertaining to storage facilities.

Since the primary function of containers and tanks is storage, this overview of 40 CFR Parts 264/265 Subparts I and J includes *permitted* and *interim status* standards for container and tank use. The four general types of land disposal or long-term storage facilities—surface impoundments, waste piles, landfills, and underground injection, are discussed briefly.

Many concerns about storage facilities can be addressed by following proper procedures for storage of materials. Table 11.12.1 lists fundamental storage and handling procedures.

TABLE 11.12.1 PROPER PROCEDURES FOR STORAGE OF HAZARDOUS MATERIALS

Use personal protective equipment
Be familiar with specific hazards of material being handled
Obey all safety rules
Do not smoke while handling materials
Store chemicals according to manufacturer's instructions, away from other chemicals or environmental conditions that could cause reactions
Face labels on containers out
Keep stacks straight and aligned
Check for location accuracy
Do not stack containers too high
Check for loose closures
Place into proper locations as soon as possible
Do not block exits or emergency equipment
Report all spills or leaks immediately

Containers

The 55-gal drum remains a standard container for hazardous waste. The several DOT-specified 55-gal drums are the most frequently used container for collection, storage, shipment, and disposal of liquid hazardous wastes (EPA 1990).

Selecting the proper drum or container requires consulting DOT regulations. The process begins with the 49 CFR Sec. 172.101 Hazardous Materials Transportation Act (MHTA), which is frequently referred to as the heart of MHTA. About 16,000 materials and substances are listed, followed by twelve columns with transport, packaging, and identification requirements.

Drums used in hazardous waste management must be in good condition, clean, free of rust, dents, and creases. In addition, the regulations require:

Containers holding hazardous wastes must always be closed, except when wastes are added or removed
Wastes must be compatible with containers (i.e., corrosive wastes should not be stored in metal containers)
Wastes in leaking or damaged containers must be recontainerized
Containers must be handled properly to prevent ruptures and leaks
Incompatible wastes must be prevented from mixing
Inspections must be conducted to assess container condition

Containers holding ignitable or reactive wastes must be located at least 15 m (50 ft) from the facility's property line.

Permit requirements for containers are similar to the interim status requirements with the following exceptions:

Liquid hazardous waste containers must be placed in a containment system capable of containing leaks and spills. This system must have sufficient capacity to contain 10% of the volume of all containers, or the volume of the largest container, whichever is greater.
When closing a container, all hazardous waste and hazardous waste residues must be removed, unless the container is to be disposed of as hazardous waste.

Tanks

Subpart J regulations apply to stationary tanks storing wastes that are hazardous under Subtitle C of the RCRA. General operating requirements fall into five basic areas:

Tank assessment must be completed to evaluate structural integrity and compatibility with the wastes that the tank system is expected to hold. The assessment covers design standards, corrosion protection, tank tests, waste characteristics, and tankage.

Secondary containment and release detection is required unless the tank does not contain free liquids and is located in a building with impermeable floors. A secondary containment system must be designed, installed, and operated to prevent liquid migration out of the tank system, and to detect and collect any releases that occur. Containment systems include liners, vaults, and double-walled tanks.

Operating and maintenance requirements require the management of tanks to avoid leaks, ruptures, spills, and corrosion. This includes a freeboard or containment structure to prevent and contain escaping wastes. A shut-off or bypass system must be installed to prevent liquid from flowing into a leaking tank.

Response to releases must include immediate removal of the leaking tank contents. The areas surrounding the tank must be visually inspected for leaks and spills. Based on the inspection, further migration of spilled waste must be stopped, and contaminated soils and surface water must be disposed of in accordance with RCRA requirements. All major leaks must be reported to the EPA or state agency.

Closure and postclosure requirements include removing all contaminated soils and other hazardous waste residues from the tank storage area at the time of closure. If decontamination is impossible, the storage area must be closed following the requirements for landfill (EPA 1990).

Surface Impoundments

Surface impoundments are used to reduce waste volume through evaporation, while containing and concentrating residue within liners. Wastes are added directly to lined depressions in the ground known as pits, lagoons, treatment basins, or ponds. Long-term storage more accurately describes the process.

All surface impoundments are required to have at least one liner and be located on an impermeable base. New surface impoundments, replacements, or lateral expansions must include:

- Two or more liners
- A leachate collection system between the liners
- Groundwater monitoring as prescribed in Subpart F

Requirements include preventing liquid from escaping due to overfilling or runoff, and preventing erosion of dikes and dams. Liners must meet permit specifications for materials and thickness. During construction and installation, liners must be inspected for uniformity, damage, and imperfection.

The double-liner system for an impoundment facility is shown in Figure 11.12.1.

Waste Piles

Hazardous waste piles now exist on many industrial sites. Volatile components in such waste piles are available for evaporation and subject to wind and water erosion. They may be leached by percolation of rainfall and runoff. Piles containing minerals or metal values may be leached with weak acid or caustic to recover the value. Unless carefully constructed over an impervious base, leachate escapes to the subsurface, contaminating groundwater or emerging as base flow in streams.

The waste pile has become accepted practice, followed by landfilling, when pile size becomes a problem. Recent determinations of EP or TCLP toxicity will bring many dross and fluff piles within RCRA control. RCRA regulations for waste piles are similar to those for landfills.

Owners or operators of waste piles used for treatment or storage of noncontainerized accumulations of solid, nonflowing hazardous wastes may choose to comply with

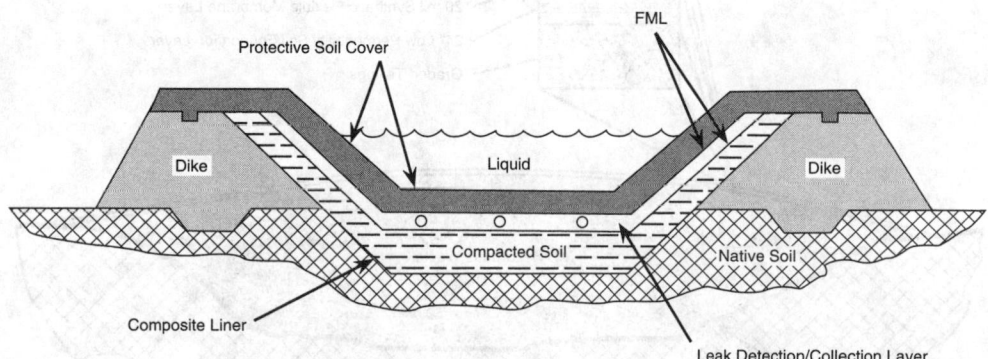

FIG. 11.12.1 Schematic—cross section of a liquid waste impoundment double liner system. (Reprinted, with permission, from W.C. Blackman, 1992, *Basic Hazardous Waste Management* [Boca Raton, Fla.: Lewis Publishers].)

waste pile or landfill requirements. Waste piles used for disposal must comply with landfill requirements. The requirements for managing storage and treatment waste piles include protecting the pile from wind dispersion. The pile must also be placed on an impermeable base compatible with the waste being stored. If hazardous leachate or runoff is generated, control systems must be imposed.

Landfills

Sanitary landfills were developed for municipal refuse disposal to replace open dumps (see Section 10.13). New secure landfills are used to bury non-liquid hazardous wastes in synthetically lined depressions. Secure landfills for hazardous waste disposal are now equipped with double liners, leak detection, leachate monitoring and collection, and groundwater monitoring systems. Synthetic liners are a minimum of 30 mil thickness.

Liner technology has improved greatly and continues to do so. Very large sections of liner fabric now minimize the number of joints. Adjacent sections are welded together to form leak-proof joints with a high degree of integrity. Liners are protected by sand bedding or finer materials free of sharp edges or points which might penetrate the inner fabric. Another layer of bedding protects the inner layer from damage by machinery working the waste. Some states allow one of the liners to be natural clay. The completed liner must demonstrate low permeability and must include a leachate collection system.

Leachate detection and collection systems are equipped with access galleys or other means of leachate removal. Double liner, leakage detection, and leachate collection systems are shown in Figure 11.12.2.

Leachate caps are detailed by the EPA. Figure 11.12.3 is a cross-section of a typical cap design. The objectives of cap design are to protect the cells from erosion, to route

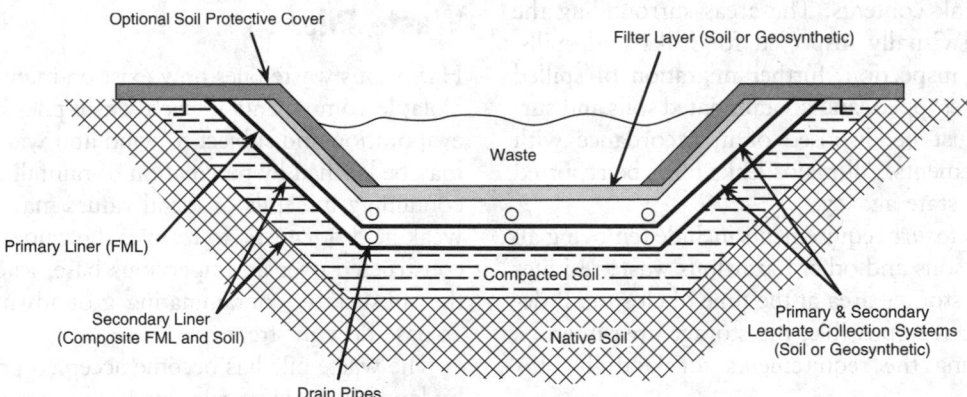

FIG. 11.12.2 Schematic—cross section of a secure landfill double liner system. *Credit:* (Reprinted with permission, from W.C. Blackman, 1992.)

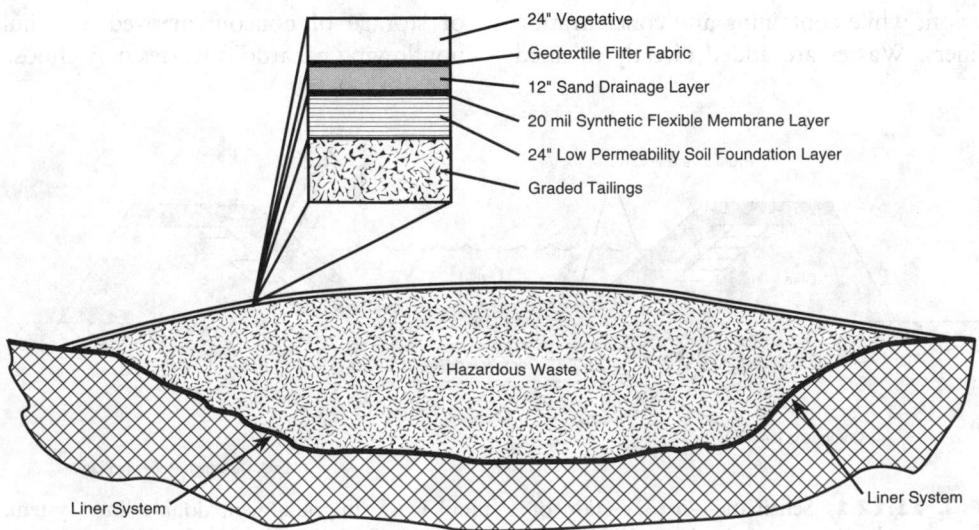

FIG. 11.12.3 Land disposal site cap designed for maximum resistance to infiltration (CH2M-Hill, Denver, CO). (Reprinted, with permission, from W.C. Blackman, 1992.)

potential runoff around and away from the cap, and to prevent buildup of gases generated within the landfill.

Groundwater monitoring schemes are designed to provide up-gradient (background) water quality data, and to detect down-gradient differences in critical water quality parameters. The RCRA requires a minimum of one up-gradient and three down-gradient monitoring wells to detect leakage from landfills (EPA 1981, 1987).

Landfills present two general classes of problems. The first class includes fires, explosions, production of toxic fumes, and related problems from the improper management of ignitable, reactive, or incompatible wastes. Thus, owners and operators are required to analyze wastes to provide enough information for proper management. They must control the mixing of incompatible wastes in landfill cells, and place ignitable and reactive wastes in landfills only when the waste has been rendered unignitable or nonreactive (EPA 1990).

The second class of landfill problems concerns the contamination of surface and groundwater. To deal with problems, interim status regulations require diversion of runoff away from the active face of the landfill; treatment of liquid or semisolid wastes so they do not contain free liquids; proper closure and preclosure care to control erosion and the infiltration of rainfall; and crushing or shredding landfill containers so they cannot collapse later leading to subsidence and breaching of the cover. Groundwater monitoring as described in Subpart F is required, as is collection of rainwater and other runoff from other active faces of the landfill. Segregation of waste such as acids, that would mobilize, make soluble, or dissolve other wastes or waste constituents is required (EPA 1990).

In the HSWA, Congress prohibited disposal of non-containerized liquid hazardous waste, and hazardous waste containing free liquids, in landfills.

Such landfills should be situated away from groundwater sources. These safeguards should be followed because there is no guarantee that engineering solutions will be able to contain the wastes in perpetuity. A well-built facility may allow sufficient leadtime for remedial action before environmental damage occurs.

Secure landfills meeting new RCRA standards may, under temporary variances, be able to accept a few hazardous wastes for which alternative disposal methods have not been developed. Secure landfills may also accept hazardous wastes that are treated to the best demonstrated available technology.

Underground Injection

Underground injection involves using specially designed wells to inject liquid hazardous waste into deep earth strata containing non-potable water. Through this method, a wide variety of waste liquids are pumped underground into deep permeable rocks that are separated from fresh water aquifers by impermeable layers of rock above, below, and

lateral to the waste layer. The depth of an injection ranges from 1,000 to 8,000 ft and varies according to the geographical factors of the area. HSWA prohibits the disposal of hazardous waste within $\frac{1}{4}$ mi of an underground source of drinking water.

Figure 11.12.4 is a cross-section of a typical injection well. To prevent plugging of the injection equipment, wastes are usually pretreated to remove solids greater than one micron. The well must be constructed to assure that potable water zones are isolated and protected. At minimum, well casings must be cemented and must extend through all potable water zones.

Deep-well disposal uses limited formation space, is expensive in construction and operation, and the subject of ever-tightening regulations. For hazardous liquid waste to be deep-well injected, the following criteria must apply: the hazardous liquid waste must have a low volume and a high concentration of waste, cannot cause an unfavorable reaction with material in the disposal zone, must be biologically inactive, must be noncorrosive, and must be

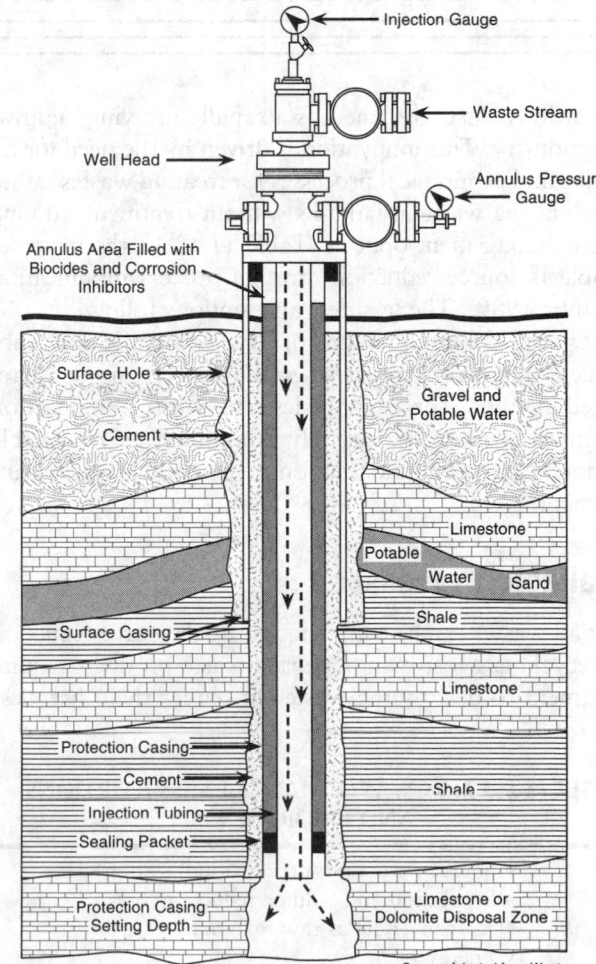

FIG. 11.12.4 Schematic cross section design of a hazardous waste injection well. (Reprinted and adapted, with permission, from C.A. Wentz, 1989, *Hazardous Waste Management* [McGraw-Hill, Inc.].)

difficult to treat by other methods. Thus, the method should be used only for those wastes with no other feasible management options.

Due to faulty construction or deterioration, there is a potential for leakage from some old wells. Detection of a leak and remedial action may not be feasible because of the nature and location of the leakage. Because of the difficulty associated with monitoring subsurface migration of liquid waste, the potential for geographical disturbances to the underground injection system, or the geographical nature of the land, underground injection wells are severely restricted in most states.

—*David H.F. Liu*

11.13
TREATMENT AND DISPOSAL ALTERNATIVES

Hazardous waste treatment is a rapidly growing, innovative industry. This innovation is driven by the need for effective and economical processes for treating wastes rather than placing waste in landfills without treatment. Among waste management options (Table 11.13.1), the most desirable is source reduction through process modification (Combs 1989). The less desirable options follow.

If waste can be eliminated or reduced significantly, subsequent treatment processes become unnecessary or are reduced in scope. These highly desirable waste minimization alternatives must be carefully considered as reasonable technical and economic solutions to hazardous waste management.

Available Processes

Not all wastes can be eliminated through source reduction or recycling. Most manufacturing waste products require treatment to destroy the wastes or render them harmless to the environment. Technological options for waste handling depend upon waste type, amount, and operating cost. Figure 11.13.1 aligns categories of industrial wastes with the treatment and disposal processes usually applied.

Numerous chemical, physical, and biological treatments are applicable to hazardous wastes. Many such treatment processes are used in by-product recovery and volume reduction processes. All wastes should first be surveyed and characterized to determine which treatment or destruction process should be used.

Hazardous wastes may be organic or inorganic. Water will dissolve many of these substances, while others have limited solubility. Sodium, potassium, and ammonium salts are water soluble, as are mineral acids. Most halogenated inorganics, except fluoride, are soluble; while many carbonates, hydroxides, and phosphates are only slightly soluble. Alcohols are highly soluble, but aromatics and long-chained petroleum-based organics are of low solubility. Solubility is critical in chemical treatment processes.

The following treatment alternatives are detailed in Figure 11.13.1.

Low-concentration effluents and other wastewaters usually require modest capital and operating costs to treat before discharging into municipal sewers.

Strong acids and alkalis can be neutralized to prevent characterization as hazardous wastes under the RCRA corrosivity criteria. Frequently, industrial water may be acid or basic, requiring neutralization before any other treatment. It may be feasible to mix an acidic waste stream with a basic stream to change the pH to a more neutral level of 6 to 8.

References

U.S. Environmental Protection Agency (EPA). 1981. *Guidance document for subpart F air emission monitoring—land disposal toxic air emissions evaluation guidelines*. Report No. PB87-155578. National Technical Information Service. Springfield, Va.

———. 1987. *Background document on bottom liner performance in double-lined landfills and surface impoundments*. Report PB87-182291. National Technical Information Service. Springfield, Va.

——— 1987. *Background information on proposed liner and leak detection rule*. Report No. PB87-191383. National Technical Information Service. Springfield, Va.

——— 1990. *RCRA orientation manual*, 1990 Edition, Superintendent of Documents. Washington, D.C.: Government Printing Office.

TABLE 11.13.1 WASTE MANAGEMENT OPTIONS AND PRIORITIES

- Source reduction (process modification)
- Separation and volume reduction
- Exchange/sale as raw material
- Energy recovery
- Treatment
- Secure ultimate disposal (landfill)

Source: Reprinted with permisstion from G.D. Combs, 1989, *Emerging treatment technologies for hazardous waste*, Section XV, Environmental Systems Company (Little Rock, AR).

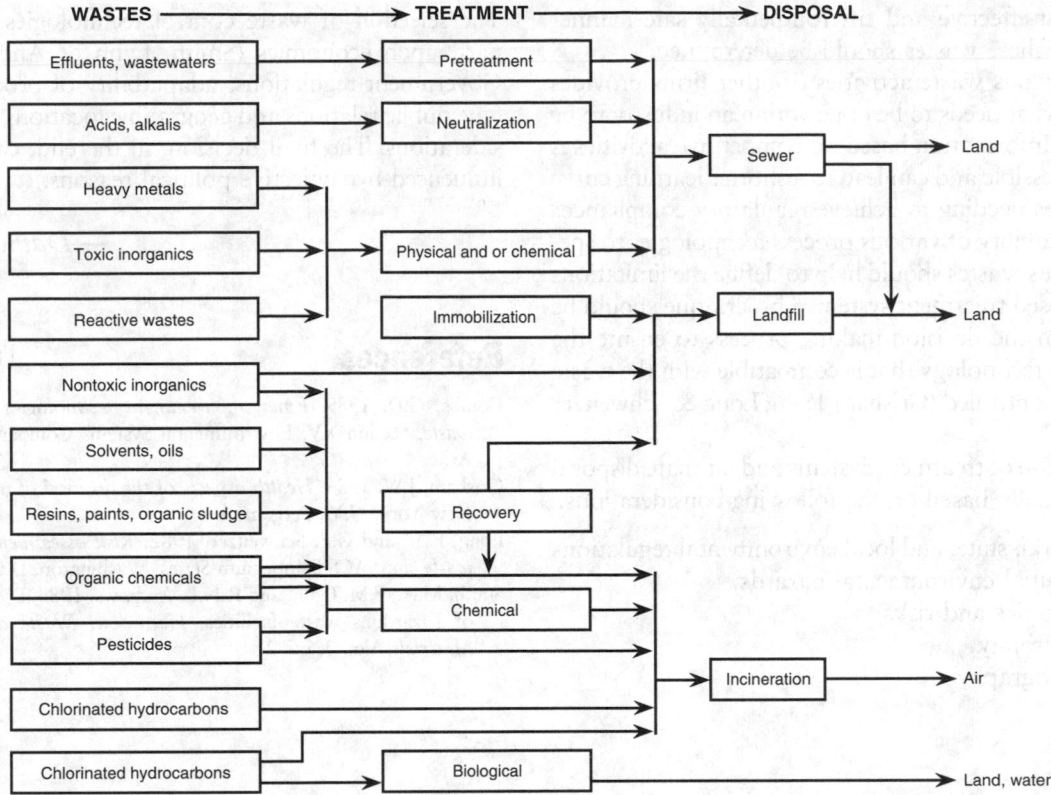

FIG. 11.13.1 Treatment and disposal alternatives for industrial wastes. (Reprinted and adapted with permission from Charles A. Wentz, 1989, *Hazardous Waste Management,* New York, N.Y.: McGraw-Hill, Inc.)

As heavy metals are virtually impossible to destroy, they must be managed by immobilization techniques. After heavy metals have undergone fixation processes and are nonleachable, they can be placed in landfills.

Reactive wastes and toxic inorganics, such as hexavalent chromium and aqueous cyanide-bearing wastes, must be handled carefully prior to the chemical treatments and separation processes that will make them environmentally acceptable. Hexavalent chromium is highly toxic. When it is reduced to trivalent chromium it can be precipitated as chromium hydroxide, which is much less toxic and more acceptable for subsequent recovery or disposal. A common method for treating aqueous cyanide waste is alkaline chlorination.

Should inorganic waste streams contain sufficient amounts of metals or other potentially valuable resources, recovery via physical and chemical processes is highly desirable. Recovery potential must be studied on a case-by-case basis, considering the estimated value of the quantities available, the market acceptance of the recovered materials, and the public perception of recycling and reusing such waste products.

Organic wastes such as solvents, resins, paints, sludges, and chemicals offer considerable recovery potential. Separation techniques such as distillation or extraction can recover valuable hydrocarbon streams for energy or chemical process industry use. However, organic recovery processes still produce a concentrated but significant volume of hazardous waste that eventually must be destroyed or landfilled.

The destruction of hazardous wastes, such as chlorinated hydrocarbons and pesticides, that cannot be eliminated or recovered involves incineration or biological treatment. Incineration is the third alternative in the EPA's preventive hierarchy, after source reduction and recycling. It is preferred because it eliminates potential problems in landfill disposal or other interim waste management processes.

Biological treatment also offers the potential for complete destruction of biodegradable hazardous wastes. The development of specialized microbes for efficient destruction systems eliminates the need for landfill disposal.

Ultimate disposal of products from hazardous waste management facilities will affect the air, water, and land. There is simply no way to avoid placing the waste by-products of our society and technology into our air, water, and land.

Process Selection

The various waste streams managed in a facility should be surveyed. The waste streams should then be characterized using sampling and analytical techniques to quantify potential threats to human health and the environment. Then the most cost-effective and environmentally safe manner of managing these wastes should be determined.

The hazardous waste activities of other firms provides insight into what needs to be done within an industry to be competitive. Information based on competitive activities is generally accessible and can lead to a shorter learning curve for companies needing to achieve regulatory compliance.

The adaptability of various process technologies to specific hazardous wastes should help to define the limitations of any proposed treatment system. This critique should be made early in the decision-making process to ensure the selection of a technology that is compatible with the waste stream to be controlled (Grisham 1986; Long & Schweitzer 1982).

The selection of treatment systems and ultimate disposal options is usually based on the following considerations.

- Federal, state, and local environmental regulations
- Potential environmental hazards
- Liabilities and risks
- Geography
- Demography

The selection of waste control technologies is based, in part, upon economics (Smith, Lynn & Andrews 1986). Government regulations, adaptability of process technology, public relations and geographic locations are also considerations. The final decision, in the end, can be largely influenced by subjective political reasons.

—*David H.F. Liu*

References

Combs, G.D. 1989. *Emerging treatment technologies for hazardous waste.* Section XV. Environmental Systems Company. Little Rock, Ar.

Grisham, J.W. 1986. *Health aspects of the disposal of waste chemicals.* New York, N.Y.: Pergamon.

Long, F.A., and G.E. Schweitzer. 1982. *Risk assessment at hazardous waste sites.* ACS Symposium Series. Washington, D.C.

Smith, M.A., F.M. Lynn, and R.N.L. Andrews. 1986. Economic impacts of hazardous waste facilities. *Hazardous Waste and Hazardous Materials,* Vol. 3, no. 2.

11.14
WASTE DESTRUCTION TECHNOLOGY

PARTIAL LIST OF SUPPLIERS

Liquid Injection Incinerators
Ensco Environmental Services: TRANE Thermal Co.; Coen Co. Inc.; John Zink Co.; Vent-o-Matic Incinerator Corp.; Lotepro Co.

Rotary Kiln Incinerators
S.D. Myers, Inc.; American Industrial Waste of ENCSO, Inc. (mobile); Exceltech, Inc.; Coen Co.; International Waste Energy Systems; Thermal, Inc.; Lurgi Corp.; Komline Sanderson; Winston Technology, Inc. (mobile); Volland, U.S.A.; Von Roll: DETOXCO Inc.

Fluidized Bed Incinerators
Lurgi Corp.; G.A. Technologies; Waste-Tech Services, Inc.: Dorr-Oliver; Combustion Power; Niro Atomizer

Wet Air Oxidation
Zimpro Inc.; Modar Inc.; Vertox Treatment Systems

Supercritical Water Oxidation
Vertox Corporation; Modar Inc.

Incineration offers advantages over other hazardous waste treatment technologies, and certainly over landfill operations. Incineration is an excellent disposal technology for all substances with high heat release potentials. Liquid and solid hydrocarbons are well adapted to incineration. Incineration of bulk materials greatly reduces the volume of wastes. Any significant reduction in waste volume makes management simpler and less subject to uncertainty.

If wastewater is too dilute to incinerate, yet too toxic to deepwell or biotreat, it is a good candidate for Wet Air Oxidation. Unlike other thermal processes, Wet Oxidation produces no smoke, fly ash or oxides. Spent air from the system passes through an adsorption unit to meet local air quality standards. Operating results show destruction approaching or exceeding 99% for many substances, including cyanides, phenols, sulfides, chlorinated compounds, pesticides, and other organics.

This section focuses on the various types of incineration, wet oxidation, and supercritical water oxidation processes.

Incineration

Incineration is a versatile process. Organic materials are detoxified by destroying the organic molecular structure through oxidation or thermal degradation. Incineration provides the highest degree of destruction and control for a broad range of hazardous substances (Table 11.14.1). Design and operating experience exists and a wide variety of commercial incineration systems are available.

TABLE 11.14.1 SUMMARY OF INCINERATOR DESTRUCTION TEST WORK

Waste	Incineration[a]	Destruction Efficiency of Principal Component (%)
Shell aldrin (20% granules)	MC	99.99
Shell aldrite	MC	99.99
Atrazine (liquid)	MC	99.99
Atrazine (solid)	MC	99.99
Para-arsanilic acid	MS	99.999
Captan (solid)	MC	99.99
Chlordane 5% dust	LI	99.99
Chlordane, 72% emulsifiable concentrate and no. 2 fuel oil	LI	99.999
Chlorinated hydrocarbon, trichloropropane, trichlorethane, and dichloroethane predominating	HT	99.92 99.98
Chloroform	MS	99.999
DDT 5% oil solution	LI	99.99
DDT (solid)	MM	99.970 to 99.98
DDT 10% dust	MC	99.99
20% DDT oil solution	LI	99.98
DDT 25% emulsifiable concentration	LI	99.98
DDT 25% emulsifiable concentrate	MC	99.98 to 99.99
DDT oil 20% emulsified DDT waste oil—1.7% PCB	TO	99.9999
DDT powder	MS	99.998
Dieldrin—15% emulsifiable concentrate	LI	99.999
Dieldrin—15% emulsifiable concentrates and 72% chlordane emulsifiable concentrates (mixed 1:3 ratio)	LI	99.98
Diphenylamine-HCl	MS	99.999
Ethylene manufacturing waste	LI	99.999
GB ($C_4H_{10}O_2PP$)	MS	99.99999969
Herbicide orange	RL	99.999 to 99.985
Hexachlorocyclopentadiene	LI	99.999
Acetic acid, solution or kepone	RKP	99.9999
Toledo sludge and kepone coincineration	RKP	99.9999
Lindane 12% emulsifiable concentrate	LI	99.999
Malathion	MS	99.999 to 99.9998
Malathion 25% wet powder	MC	99.99
Malathion 57% emulsifiable concentrate	MC	99.99
Methyl mathacrylate (MMA)	FB	99.999
0.3% Mirex bait	MC	98.21 to 99.98
Mustard	MS	99.999982
Nitrochlorobenzene	LI	99.99 to 99.999
Nitroethane	MS	99.993
Phenol waste	FB	99.99
Picloram	MC	99.99
Picloram, (tordon 10K pellets)	MC	99.99
PCBs	RK	99.999964
PCB capacitors	RK	99.5 to 99.999

(Continued on next page)

TABLE 11.14.1 *(Continued)*

Waste	Incineration[a]	Destruction Efficiency of Principal Component (%)
PCB	CK	99.9998
Polyvinyl chloride waste	RK	99.99
Toxaphene 20% dust	MC	99.99
Toxaphene 60% emulsifiable concentrate	MC	99.99
Trichlorethane	MS	99.99
2,4-D low-volatile liquid ester	LI	99.99
2,4,5-T (Weedon™)	MM	99.990 to 99.996
2,4,5-T	SH	99.995
2,4,5-T	SH	99.995
2,4,5-T	SH	92
2,4,5-T	SH	99.995
VX ($C_{11}H_{26}O_2PSN$)	MS	99.999989 to 99.9999945
Zineb	MC	99.99

Source: J. Corini, C. Day, and E. Temrowski, 1980 (Sept. 2). Trial Burn Data (unpublished draft) Office of Solid Waste, U.S. Environmental Protection Agency, Washington, D.C.

[a]MC = multiple chamber; MS = molten salt combustion; LI = liquid injection; HT = 2 high-temp. incinerators; MM = municipal multiple-hearth sewage sludge incinerator; TO = thermal oxidizer waste incinerator; RL = 2 identical refractory-lined furnaces; RKP = rotary kiln pyrolyzer; FB = fluidized bed; CK = cement kiln; SH = single-hearth furnace.

Detoxified hazardous wastes include combustible carcinogens, mutagens, teratogens, and pathological wastes. Another advantage of incineration is the reduction of leachable wastes from landfills. Incineration of contaminated soils is increasing. The EPA, for example, employed a mobile incinerator to decontaminate 40 tn of Missouri soil that was contaminated with 4 lb of dioxin compounds.

Different incineration technologies are used to handle various types of hazardous waste. The four most common incinerator designs are liquid injection (sometimes combined with fume incineration), rotary kiln, fixed hearth and fluidized bed incinerators.

The four major subsystems of hazardous waste incineration are: (1) waste preparation and feeding, (2) combustion chamber(s), (3) air pollution control, and (4) residue and ash handling. The normal orientation of these subsystems is shown in Figure 11.14.1, along with typical process component options.

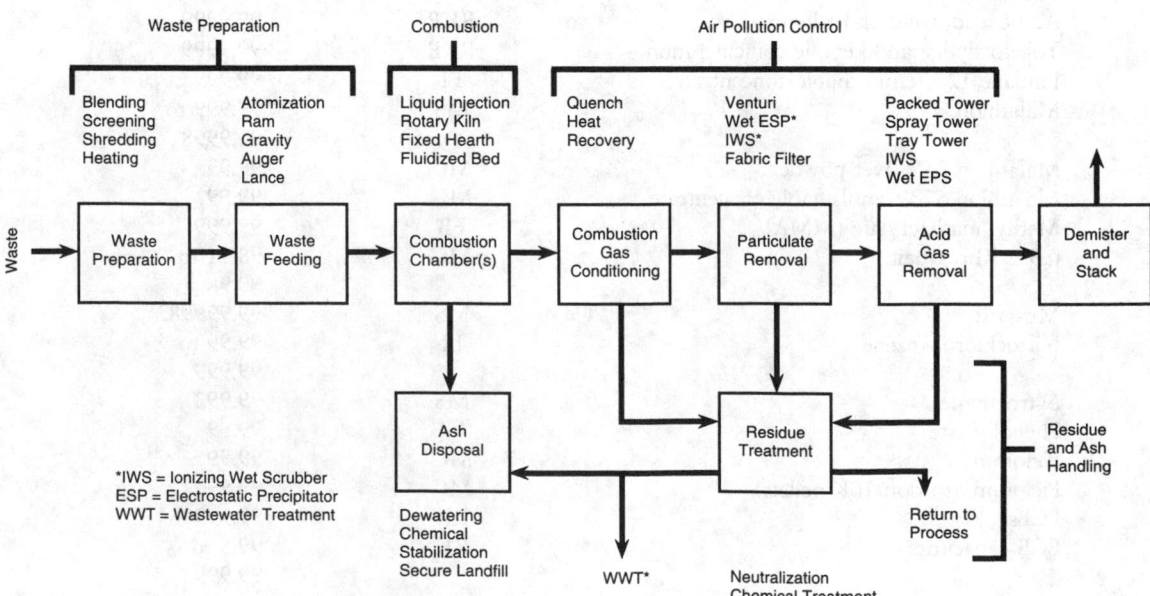

FIG. 11.14.1 General orientation of incineration subsystems and typical process component options. (Reprinted, with permission, from Dempsey and Oppelt 1993).

INCINERATOR SYSTEM DESIGN

Incinerator design plays a key role in ensuring adequate destruction of waste. Important data on waste characteristics needed to design an effective incineration system are listed in Table 11.14.2.

The major incinerator design factors significantly affecting thermal destruction of hazardous waste include:

Temperature

Temperature is probably the most significant factor in ensuring proper destruction of hazardous waste in incinerators. The *threshold temperature* is defined as the operating temperature to initiate thermal destruction of hazardous waste. The threshold temperature ensures waste destruction and allows cost-effective operation.

Residence Time

Incinerator volume determines the residence time for a given flow rate. Residence time, combined with thermal destruction temperature, ensures compliance with destruction and removal efficiency (DRE) regulations. Sufficient residence time must be allowed to achieve DREs, as well as to ensure conversion to desirable incinerator products.

Turbulence

Turbulence is used to attain desirable DREs and to cut operating temperature and residence time requirements. The incinerator configuration affects the degree of turbulence. Pumps, blowers, and baffles should be selected based upon the type of waste to be incinerated and the desired DREs. Heat transfer and fluid flow should be considered in the turbulence requirements.

TABLE 11.14.2 IMPORTANT THERMAL
TREATMENT DATA NEEDS

Need	Purpose
Heat Content (HHV and LHV)	Combustiblity
Volatile Matter Content	Furnace Design
Ash Content	Furnace Design, Ash Handling
Ash Characteristics	Furnace Design
Halogen Content	Refractory Design, Flue Gas Ductwork Specification, APC Requirements
Moisture Content	Auxiliary Fuel Requirements
Heavy Metal Content	Air Pollution Control

NOTE: Generally, the data needs for evaluating thermal processes include the data needed for physical treatment for the purpose of feed mechanism design.

Pressure

Thermal destruction systems, which operate at slightly positive elevated pressures, require nonleaking incinerators. Pressurized systems require high-temperature seals for trouble-free operation.

Air Supply

Incomplete combustion products result from insufficient residence time, temperature, or air. The thermal destruction unit must be supplied with amounts of oxygen or air higher than the stoichiometric amount required, to ensure that products of hydrocarbon combustion ultimately result in carbon dioxide and water.

Construction Materials

Most incinerators are constructed with materials selected for continuous trouble-free operation with many hazardous wastes and under many destructive conditions. Materials of construction range from ordinary steel to exotic alloys. The chemical and physical properties of the wastes to be incinerated must be well-defined for selection of materials to ensure a longer operating life and fewer maintenance problems for the incinerator.

Auxiliaries

Numerous additional features must be considered:

Feed systems must be designed to incorporate the hazardous wastes identified by market surveys.

Afterburners may be needed to ensure proper DRE capability.

Downstream treatment is usually necessary to neutralize and remove undesirable destruction products such as mineral acids.

Ash removal may play a key role in the thermal destruction of solid and semi-solid wastes.

Combustion Chambers

Many hazardous wastes are incinerated in industrial boilers and furnaces. However, hazardous waste combustion in boilers is limited by the amount of chlorine in the waste stream, because most industrial boilers do not use scrubbers for hydrogen chloride.

The physical form of the waste and its ash content determine the type of combustion chamber selected. Table 11.14.3 provides selection considerations for the four major combustion chamber designs as a function of different forms of waste (EPA 1981; Dempsey & Oppelt 1993). Incinerator systems derive their names from the types of combustion chambers used.

TABLE 11.14.3 APPLICABILITY OF MAJOR INCINERATOR TYPES TO WASTES OF VARIOUS PHYSICAL FORM

	Liquid Injection	Rotary Kiln	Fixed Hearth	Fluidized Bed
Solids:				
Granular, homogeneous		X	X	X
Irregular, bulky (pallets, etc.)		X	X	
Low melting point (tars, etc)	X	X	X	X
Organic compounds with fusible ash constituents		X	X	X
Unprepared, large, bulky material		X	X	
Gases:				
Organic vapor laden	X	X	X	X
Liquids:				
High organic strength aqueous wastes	X	X	X	X
Organic liquids	X	X	X	X
Solids/liquids:				
Waste contains halogenated aromatic compounds (2,200°F minimum)	X	X	X	
Aqueous organic sludge		X		X

Source: Reprinted with permission from C.R. Dempsey and E.T. Oppelt, 1993, Incineration of hazardous waste: a critical review update, *Air & Waste,* Vol. 43, 1993.

LIQUID INJECTION INCINERATORS

Liquid injection incinerators are applicable for pumpable liquid waste. These units (Figure 11.14.2) are usually simple, refractory-lined cylinders (either horizontally or vertically aligned) equipped with one or more waste burners. Liquid wastes are injected through the burner(s), atomized to fine droplets and burned in suspension. Burners, as well as separate injection nozzles, may be oriented for axial, radial or tangential firing. Improved use of combustion space and higher heat release rates can be achieved by using swirl or vortex burners, or designs involving tangential entry. A forced draft must be supplied to the combustion chamber for the necessary mixing and turbulence.

Good atomization is critical to achieving high destruction efficiency in liquid combustors. Nozzles have been developed to produce mists with mean particle diameters as low as 1 micron (μm), as compared to oil burners, which yield oil droplets in the 10 to 50 μm range. Atomization may be obtained by low pressure air or steam (25 to 100 psig), or mechanical (hydraulic) means using specially designed orifices (25 to 250 psig).

Vertical, downward-oriented liquid injection incinerators are preferred when wastes are high in inorganic salts and fusible ash content; horizontal units may be used with low ash waste. In the past, the typical capacity of liquid injection incinerators was 30 MM Btu/hr heat release. However, units as high as 210 MM Btu/hr are in operation.

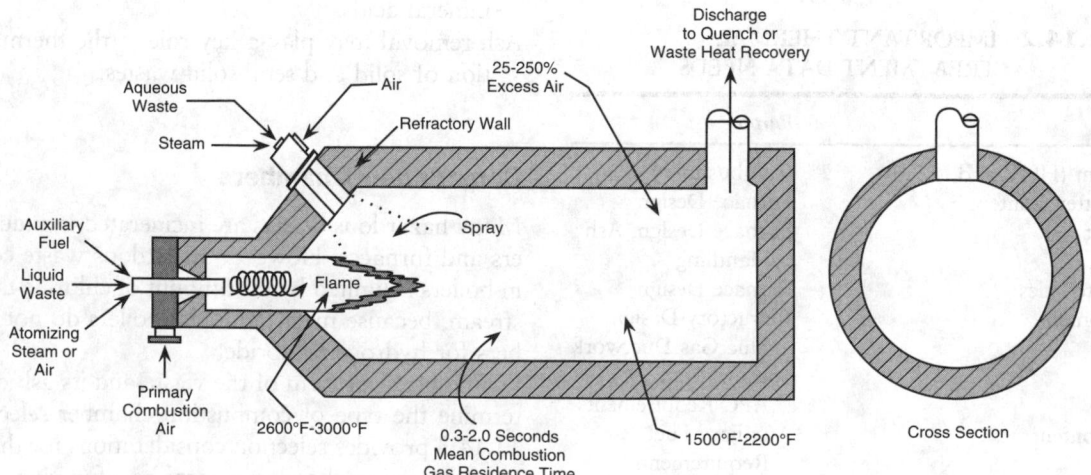

FIG. 11.14.2 Typical liquid injection combustion chamber. (Reprinted, with permission, from Dempsey and Oppelt 1993.)

ADVANTAGES

- Incinerates a wide range of liquid wastes
- Requires no continuous ash removal system, other than for air pollution control
- Capable of a high turndown ratio
- Provides fast temperature response to changes in the waste-fuel flow rate
- Includes virtually no moving parts
- Allows low maintenance costs
- Is a proven technology

DISADVANTAGES

- Must be able to atomize liquids through a burner nozzle except for certain limited applications
- Must provide for complete combustion and prevent flame impingement on the refractory
- Susceptible to plugging. High percent solids can cause problems
- No bulk solids capability

ROTARY KILN INCINERATORS

Rotary kiln incinerators (Figure 11.14.3) are more versatile, as they are used to destroy solid wastes, slurries, containerized wastes, and liquids. Because of this, these units are frequently incorporated into commercial off-site incineration facilities and used for Superfund remediation.

The rotary kiln is a horizontal, cylindrical, refractory-lined shell mounted on a slight slope. Rotation of the shell transports waste through the kiln and mixes the burning solid waste. The waste moves concurrently or countercurrently to the gas flow. The residence of waste solids in the kiln is generally 0.5 to 1.5 hrs. This is controlled by kiln rotation speed (typically 0.5 to 1 rpm), waste feed rate, and in some instances, internal dams to retard waste movement through the kiln. The feed rate is regulated, limiting the waste processed to 20% or less of kiln volume.

Rotary kilns are typically 5–12 ft in diameter and 10–30 ft in length. Rotary kiln incinerators generally have a length-to-diameter ratio (L/D) of 2:8. Smaller L/D ratios result in less particulate carryover. Higher L/D ratios and slower rotational speeds are used when waste materials require longer residence time.

The primary function of the kiln is converting solid wastes to gases through a series of volatilization, destructive distillation, and partial combustion reactions. An afterburner, connected directly to the discharge end of the kiln, completes gas-phase combustion reactions. Gases exiting the kiln are directed to the afterburner chamber.

Some recent systems have a "hot cyclone" installed between the kiln and afterburner to remove solid particles that might create slagging problems in the afterburner. The afterburner may be horizontally or vertically aligned, and functions on the same principles as a liquid injection in-

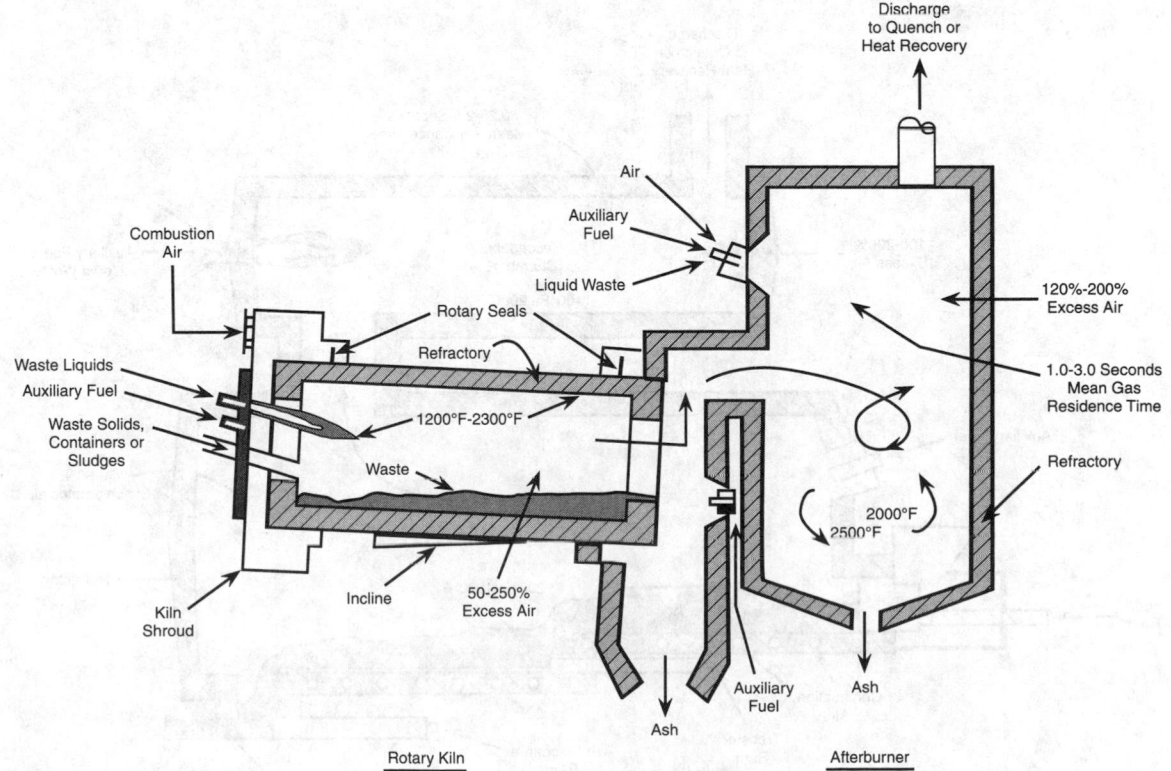

FIG. 11.14.3 Typical rotary kiln/afterburner combustion chamber. (Reprinted, with permission, from Dempsey and Oppelt 1993.)

cinerator. In fact, many facilities also fire liquid hazardous waste through separate waste burners in the afterburner. Afterburners and kilns are usually equipped with auxiliary fuel-firing systems to bring the units up to temperature and to maintain the desired operating temperatures. Some operators fire aqueous waste streams into afterburners as a temperature control measure. Rotary kilns are designed with a heat release capacity of up to 150 MM Btu/hr in the United States; Average units are typically around 60 MM Btu/hr.

ADVANTAGES

- Incinerates a wide variety of liquid and solid wastes
- Receives liquids and solids separately or in combination
- Not hampered by materials passing through a melt phase
- Includes feed capability for drums and bulk containers
- Permits wide flexibility in feed mechanism design
- Provides high turbulence and air exposure of solid wastes
- Continuous ash removal does not interfere with waste burning
- Adapts for use with wet-gas-scrubbing system
- Permits residence time of waste to be controlled by adjusting rotational speed of the kiln

- Allows many wastes to be fed directly into the incinerator without preparations such as preheating or mixing
- Operates at temperatures in excess of 2500°F (1400°C), destroying toxic compounds that are difficult to degrade thermally
- Uses proven technology

DISADVANTAGES

- Requires high capital installation costs, especially for low feed rates
- Necessitates operating care to prevent refractory damage from bulk solids
- Permits airborne particles to be carried out of the kiln before complete combustion
- Frequently requires large excess air intakes due to air leakage into the kiln by the kiln end seals and feed chute. This affects supplementary air efficiency
- Causes high particulate loadings into the air-pollution control system
- Allows relatively low thermal efficiency

FIXED HEARTH INCINERATORS

Fixed hearth incinerators, also called controlled air, starved air, or pyrolytic incinerators, are the third technology for hazardous waste incineration. These units employ a two-stage combustion process, much like rotary kilns.

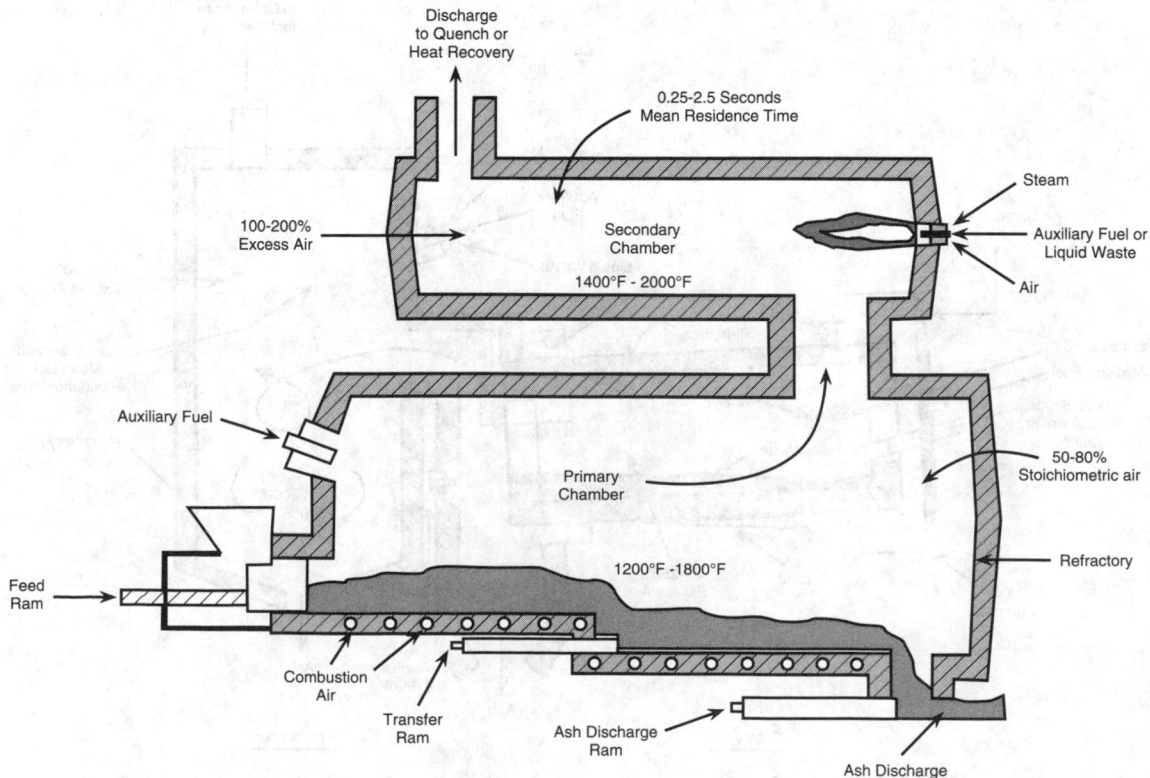

FIG. 11.14.4   Typical fixed hearth combustion chamber. (Reprinted, with permission, from Dempsey and Oppelt 1993.)

As shown in Figure 11.14.4, waste is ram-fed or pumped into the primary chamber, and burned at roughly 50–80% of stoichiometric air requirements. This starved air condition causes the volatile waste to be vaporized by the endothermic heat provided in oxidation of the fixed carbon fraction. The resulting smoke and pyrolytic products consist primarily of methane, ethane, and other hydrocarbons; carbon monoxide and combustion products pass to the secondary chamber. Here additional air is injected to complete combustion, which occurs spontaneously or through the addition of supplementary fuels. Primary chamber combustion reactions and turbulent velocities are maintained at low levels by the starved-air conditions, minimizing particulate entrainment and carryover. With the addition of secondary air, the total excess air for fixed hearth incinerators is 100–200%.

Fixed hearth units tend to be smaller in capacity than liquid injection or rotary kiln incinerators because of the physical limitations in ram-feeding and transporting large amounts of waste material through the combustion chamber. Lower capital costs and reduced particulate control requirements make them more attractive than rotary kilns for smaller on-site installations.

ADVANTAGE

- Represents proven technology

DISADVANTAGES

- Requires more labor
- Operates at a temperature lower than necessary for acceptable waste destruction

FLUIDIZED BED INCINERATORS

Fluidized bed combustion systems have only recently been applied in hazardous waste incineration. Fluidized bed incinerators may be either circulating or bubbling bed designs (Chang et al. 1987). Both types consist of single refractory-lined vessels partially filled with particles of sand, aluminum, calcium carbonate or other such materials. Combustion air is supplied through a distributor plate at the bottom of the combustor (Figure 11.14.5) at a rate sufficient to fluidize (bubbling) or entrain part of the bed material (recirculating bed). In the recirculating bed design, air velocities are higher and the solids are blown overhead, separated in a cyclone, then returned to the combustion chamber (Figure 11.14.6). Operating temperatures are normally in the 1400–1600°F range. Excess air requirements range from 25–150%.

Fluidized bed incinerators are used primarily for liquids, sludges, or shredded solid materials, including soil. To allow good circulation of waste materials and removal of solid residues within the bed, all solids require pre-screening or crushing to a size less than 2 in in dia.

Fluidized bed incinerators offer: high gas-to-solids ratios, high heat transfer efficiencies, high turbulence in both

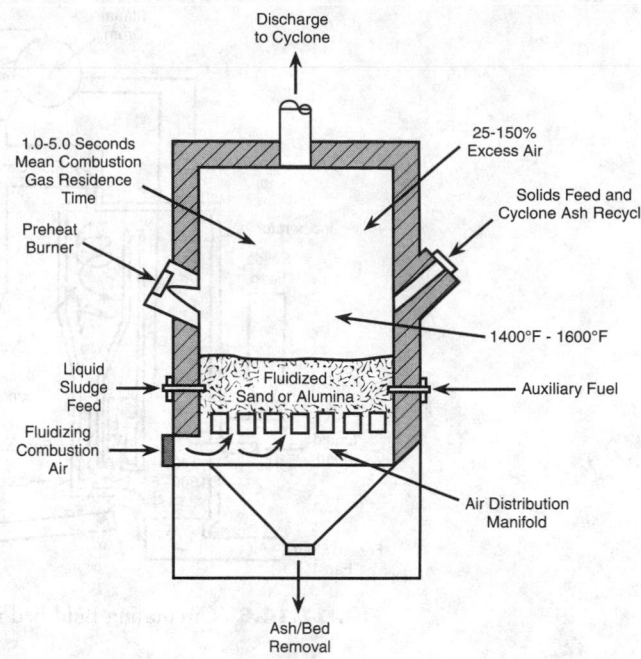

FIG. 11.4.5 Typical fluidized bed combustion chamber. (Reprinted, with permission, from Dempsey and Oppelt 1993).

gas and solid phases, uniform temperatures throughout the bed, and the potential for in-situ gas neutralization by lime, limestone, or carbonate addition. Fluidized beds also have the potential for solid agglomeration in the bed, especially if salts are present in waste feeds.

ADVANTAGES

- Burns solid, liquid, and gaseous wastes
- Simple design has no moving parts
- Compact design due to high heating rate per volume
- Low gas temperatures and excess air requirements minimize nitrogen oxide formation
- Large active surface area enhances combustion efficiency
- Fluctuations in feed rate and composition are easily tolerated due to large heat capacity

DISADVANTAGES

- Residual materials are difficult to remove
- Fluid bed must be prepared and maintained
- Feed selection must prevent bed damage
- Incineration temperatures limited to 1500°F max to avoid fusing bed material
- Little experience on hazardous waste combustion

A wide range of innovative technologies such as high- and low-temperature plasmas, molten salt, molten glass and molten metals baths have merged since the passage of RCRA (Freeman 1990). Many such techniques are now in development.

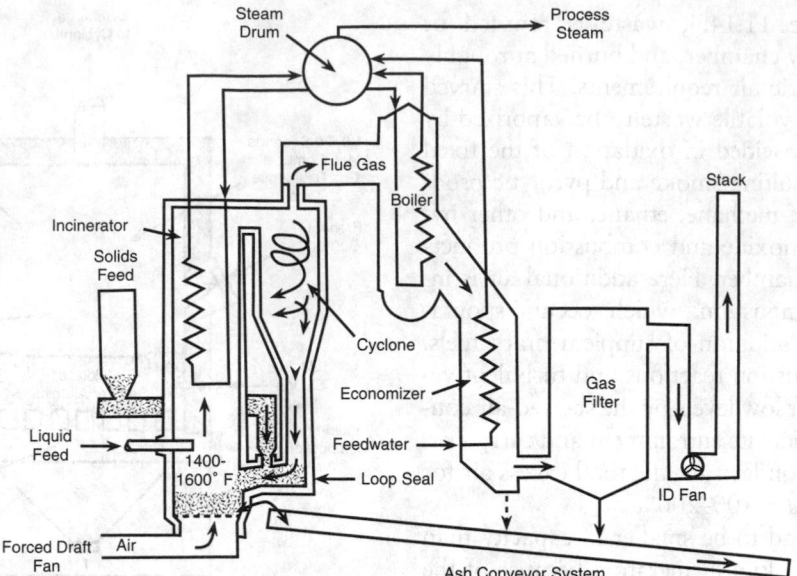

FIG. 11.14.6 Circulating fluid-bed incinerator for hazardous waste.

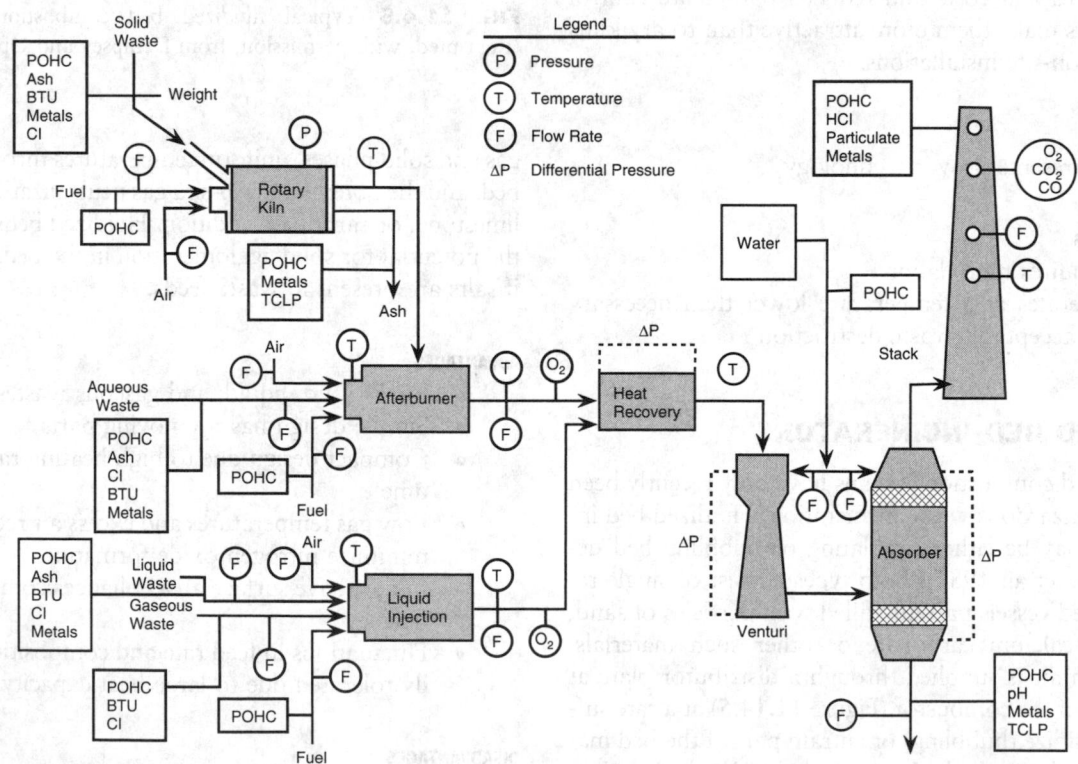

FIG. 11.14.7 Potential sampling points for assessing incinerator performance. (Reprinted, with permission, from Dempsey and Oppelt 1993).

PROCESS PERFORMANCE

Performance measurement is undertaken for any of the following three purposes:

- Establishing initial or periodic compliance with performance standards (e.g., trial burns)
- Routine monitoring of process performance and direct process control (e.g., continuous monitoring)

- Conducting performance measurements for research and equipment development

Figure 11.14.7 illustrates sampling points for assessing incinerator performance. In trial burn activities, sampling activities focus on collecting of waste feed and stack emission samples. Ash and air pollution control system residues are also sampled and analyzed. Sampling of input and output around individual unit components, e.g. scrubbers, may also be conducted in research or equipment evaluation studies.

Trial Burns

Trial burns provide regulatory agencies with the data to issue operating permits. Consequently, trial runs are directed to show that plants achieve the RCRA limits under the desired operating conditions. These RCRA limits are:

A destruction and removal efficiency (DRE) of greater than 99.99% for each of subject principal organic hazardous constituents (POHCs). (Note: the Toxic Substance Control Act [TSCA] requires that incinerators burning PCB and dioxin-containing waste achieve 99.9999% DRE.)

A particulate emission of less than 180 mg per dry standard cu meter (0.08 grains/dry ft) of stack gas (corrected to 7% O_2)

Hydrogen chloride (HCl) emissions less than 4 lb/hr (2.4 kg/hr) or greater than 99% removal efficiency

Trial burns test the plant's operating conditions and ability to meet the three RCRA limits. The EPA recommends three or more runs under any one set of conditions, with varying conditions, or with different waste feed characteristics.

Operating Permits

Permits should allow plants to incinerate the expected types and quantities of waste, at the necessary feed rates and within an acceptable range of operating conditions. Permit conditions must be flexible, with limits that are reasonably achievable. Based on trial burn results, operating permits may specify certain criteria such as:

- Maximum concentration of certain POHCs in waste feed
- Maximum waste feed rate or maximum total heat input rate
- Maximum air feed rate or maximum flue gas velocity
- Minimium combustion temperature
- Maximum carbon monoxide content of stack gas
- Maximum chloride and ash content of waste feed

Sampling and Analysis

The EPA provides guidance on sampling and analysis methods for trial burns designed to measure facility compliance with the RCRA incinerator standards (EPA 1981, 1983, 1989, 1990c; Gorman, et al. 1985).

Table 11.14.4 outlines sampling methods typically involved in RCRA trial burns. Sampling method numbers refer to methods identified in EPA guidance documents

and reports (Harris et al. 1984; EPA 1990). The EPA has a computerized data base including a reference directory on the availability and reliability of sampling and analysis methods for designated POHCs.

Assuring Performance

Key control parameters used to trigger fail-safe controls are presented in Table 11.14.5. The parameters are divided into three groups:

Group A parameters are continuously monitored and interlocked to the automatic waste feed cutoff.

Group B parameters are set to ensure that worst-case conditions demonstrated in the trial run are not exceeded during continuous operation. They are not linked with the automatic waste cutoff.

Group C parameters are based on equipment manufacturers' design and operating specifications. They are set independently of trial-run results and are not linked with automatic waste feed cutoff.

No individual real-time monitoring performance indicators appear to correlate with actual organic DRE. No correlation between indicator emissions of CO or HC and DRE has been demonstrated in field-scale incinerator operations, although CO is a conservative indicator of organic emissions. It may be that combinations with other potential real-time indicators, such as surrogate destruction, may be desirable.

Wet Air Oxidation
PROCESS DESCRIPTION

The patented Zimmermann process involves flameless or wet combustion in aqueous solution or dispersions (Zimmermann 1954). Unlike other thermal processes, wet air oxidation does not require dewatering before combustion and creates no air pollution. In aqueous dispersion, a wide range of organic and hazardous industrial wastes can be oxidized to carbon dioxide and water by the addition of air or oxygen. Water, the bulk of the aqueous phase, catalyzes oxidation reactions so they proceed at relatively low temperatures (350–650°F). At the same time, water moderates the oxidation rates by evaporation.

Figure 11.14.8 shows a simplified flow scheme of a continuous air oxidation system. The waste liquor is mixed with air and is preheated by steam during process startup and by hot reactor effluent during operation to 300°–400°F. At this reactor inlet temperature oxidation starts, with the associated heat release further increasing the temperature as the liquid air mixture moves through the reactor. The higher the operating temperature, the greater the destruction of organic pollutants for the same residence time period. The operating temperature cannot

TABLE 11.14.4 SAMPLING METHODS AND ANALYSIS PARAMETERS

Sample	Sampling frequency for each run	Sampling method[a]	Analysis parameter[b]
1. Liquid waste feed	Grab sample every 15 min	S004	V&SV-POHCs, Cl, ash, ult, anal., viscosity, HHV, metals
2. Solid waste feed	Grab sample from each drum	S006, S007	V&SV-POHCs, Cl, ash, HHV, metals
3. Chamber ash	Grab one sample after all runs are completed	S006	V&SV-POHCs, TCLP[d], HHV, TOC, metals
4. Stack gas	Composite	Method 0010 (3h) (MM5)	SV-POHCs
	Composite	Method 5[f]	Particulate, H_2O
	Composite	Method 0011	Formaldehyde
	Composite	Method 0050	HCl, Cl_2
	Composite	Method 0030 (2h) (VOST)	V-POHCs
	Three pairs of traps		
	Composite in Tedlar gas bag	Method 0040	V-POHCs[c]
	Composite	Method 3 (1-2 h)	CO_2 and O_2 by Orsat
	Continuous	CEM	CO, CO_2, O_2, SO_2
	Composite	Method 0012	Trace metals[e]
5. APCD Effluent (liquid)	Grab sample every ½ h	S004	V&SV-POHCs, Cl, pH, metals
6. APCD Residue (solid)	Grab sample every ½ h	S006	V&SV-POHCs, metals

[a]VOST denotes volatile organic sampling train; MM5 denotes EPA Modified Method 5; SXXX denotes sampling methods found in "Sampling and Analysis Methods for Hazardous Waste Combustion"[83]; CEM denotes Continuous Emission Monitor (usually nondispersive infrared).

[b]V-POHCs denotes volatile principal organic hazardous constituents (POHCs); SV-POHCs denotes semivolatile POHCs; HHV denotes higher heating value; TOC denotes Total Organic Carbon.

[c]Gas bag samples may be analyzed for V-POHCs only if VOST samples are saturated and not quantifiable or if the target POHC is too volatile for VOST.

[d]TCLP = toxicity characteristic leaching procedure[192].

[e]Metals captured by the Multiple Metals Sampling Train.

[f]Method 5 can be combined with Method 0050 or Method 0011.

Source: Reprinted, with permission, from Dempsey and Oppelt, 1993.

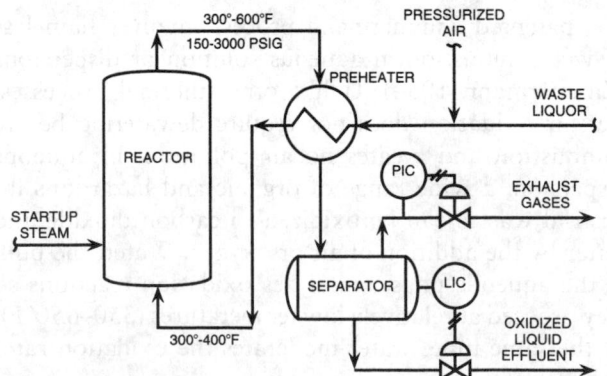

FIG. 11.14.8 Wet oxidation process.

exceed the critical temperature of water (705°F), because the continuous presence of a liquid water phase is essential (Liptak 1974).

A consequence of high operating temperature is the need to run the process at high pressure (300–3000 psig) to keep water from vaporizing. The static pressure energy of the exhaust gases can drive an air compressor or gen-

erate electric power, while the thermal energy of the reactor effluent can be used for steam generation.

Detoxified priority pollutants and products stay in the aqueous phase. Materials such as sulfur compounds, chlorinated hydrocarbons, or heavy metals end up in their highest oxidation state, i.e., sulfates, hydrochloric acid, or salt. Air pollutants are controlled because oxidation takes place in water at low temperatures and no fly ash, dust, sulfur dioxide or nitrogen oxide is formed.

Typically, 80% of the organic substances will be completely oxidized. The system can accommodate some partially halogenated compounds, but highly chlorinated species such as PCBs are too stable for complete destruction without adding a catalyst or very high pressure and temperature (Kiang & Metry 1982).

Control of a wet air oxidation system is relatively simple, as the system is self-regulating. Oxidation occurs in a massive amount of water, which provides an effective heat sink and prevents the reaction from running away. Should a surge of organic material enter the reactor, the air would be depleted, or the heat liberated by additional oxidation would form more steam.

TABLE 11.14.5 CONTROL PARAMETERS

Group	Parameter[a]
Group A Continuously monitored parameters are interlocked with the automatic waste feed cutoff. Interruption of waste feed is automatic when specified limits are exceeded. The parameters are applicable to all facilities.	1. Minimum temperature measured at each combustion chamber exit 2. Maximum CO emissions measured at the stack or other appropriate location 3. Maximum flue gas flowrate or velocity measured at the stack or other appropriate location 4. Maximum pressure in PCC and SCC 5. Maximum feed rate of *each* waste type to *each* combustion chamber[b] 6. The following as applicable to the facility: • Minimum differential pressure across particulate venturi scrubber • Minimum liquid-to-gas ratio (L/G) and pH to wet scrubber • Minimum caustic feed to dry scrubber • Minimum kVA settings to ESP (wet/dry) and kV for ionized wet scrubber (IWS) • Minimum pressure differential across baghouse • Minimum liquid flowrate to IWS
Group B Parameters do *not* require continuous monitoring and are thus *not* interlocked with the waste feed cutoff systems. Operating records are required to ensure that trial burn worst-case conditions are not exceeded.	7. POHC incinerability limits 8. Maximum total halides and ash feed rate to the incinerator system 9. Maximum size of batches or containerized waste[b] 10. Minimum particulate scrubber blowdown or total solids content of the scrubber liquid
Group C Limits on these parameters are set independently of trial burn test conditions. Instead, limits are based on equipment manufacturer's design and operating specifications and are thus considered good operating practices. Selected parameters do *not* require continuous monitoring and are *not* interlocked with the waste feed cutoff.	11. Minimum/maximum nozzle pressure to scrubber 11. Maximum total heat input capacity for each chamber 13. Liquid injection chamber burner settings: • Maximum viscosity of pumped waste • Maximum burner turndown • Minimum atomization fluid pressure • Minimum waste heating value (only applicable when a given waste provides 100% heat input to a given combustion chamber) 14. APCD inlet gas temperature[c]

[a]PCC denotes primary combustion chamber; SCC denotes secondary combustion chamber; APCD denotes air pollution control device; kVA denotes kilovolt-amperes; ESP denotes electrostatic precipitator.
[b]Items 5 and 9 are closely related.
[c]Item 14 can be a group B or C parameter.
Source: Reprinted, with permission, from Dempsey and Oppelt, 1993.

PROCESS CHARACTERISTICS

The wet air oxidation process has three basic reaction mechanisms: hydrolysis, mass transfer, and chemical kinetics. Table 11.14.6 gives brief explanations of the mechanisms and their major influences. The four basic steps encountered in the oxidation of hydrocarbon pollutants are:

$$\text{Hydrocarbon} + \text{oxygen} \longrightarrow \text{alcohol}$$
$$\text{Alcohol} + \text{oxygen} \longrightarrow \text{aldehyde}$$
$$\text{Aldehyde} + \text{oxygen} \longrightarrow \text{acid}$$
$$\text{Acid} + \text{oxygen} \longrightarrow \text{carbon dioxide} + \text{water}$$

Nearly all organic materials in industrial waste break down into several intermediate compounds before complete ox-

TABLE 11.14.6 WET AIR OXIDATION PROCESS REACTION MECHANISMS[a]

Reaction Mechanism	Typical Effects	Strongest Influences
Hydrolysis	Dissolves solids	pH
	Splits long-chain hydrocarbons	Temperature
Mass Transfer	Dissolves, absorbs oxygen	Pressure
		Presence of liquid-gas interface
Chemical Kinetics	Oxidizes organic chemicals	Temperature
		Catalysts
		Oxygen activity

[a]Courtesy of Plant Engineering, Barrington, IL.

idation occurs. The process is efficient in total organic carbon reduction for most compounds, but not for acetates or benzoic acid.

APPLICABILITY/LIMITATIONS

This process is used to treat aqueous waste streams containing less than 5% organics, pesticides, phenolics, organic sulfur, and cyanide wastewaters. At ethylene plants, Zimpro Passavant provides wet air oxidation units for converting caustic liquors into nonhazardous effluents that can be treated biologically (Zimpro Environmental, Inc. 1993). These liquors are produced during scrubbing of ethylene gases. Table 11.14.7 lists hazardous wastes that are good candidates for the wet air oxidation process.

Skid-mounted units can be situated at disposal sites for pretreatment of hazardous liquid before deep welling, or for carbon regeneration and sludge oxidation.

TABLE 11.14.7   THE EPA HAZARDOUS WASTE LIST—GOOD CANDIDATES FOR WAO

F Classification:

F004, F005	Spent non-halogenated solvents and still bottoms.
F006	Sludges from electroplating operations.
F007, F011, F015	Spent cyanide bath solutions.
F016	Coke oven, blast furnace gas scrubber sludges.

K Classification:

K009-K015	Bottoms, bottom streams, and side cuts from production of acetaldehyde and acrylonitrile.
K017-K020	Heavy ends or still bottoms from epichlorohydrin, ethyl chloride, ethylene dichloride or vinyl chloride operations.
K024-K025	Distillation bottoms from production of phthalic anhydride and nitrobenzene.
K026	Stripping still tails from production of methyl ethyl pyridines.
K027	Residues from toluene diisocyanate production.
K029-K030	Bottoms, ends, stripper wastes from tricholorethylene, perchloroethylene production.
K035	Creosote sludges.
K045	Spent carbon from explosives wastewater.
K052	Leaded petroleum tank bottoms.
K058-K059	Leather tanning, finishing sludges.

P Classification (discarded commercial chemical products):

P024	p-chloroaniline
P029-P030	Copper cyanide, cyanides
P048	2,4-dinitrophenol
P052-P054	Ethylcyanide, ethylenediamine, ethyleneimine.
P063-P064	Hydrocyanic acid, isocyanic acid
P077	p-nitroaniline
P081	Nitroglycerine
P090	Pentachlorophenol
P098	Potassium cyanide
P101	Propionitrile
P106	Sodium cyanide

U Classification:

U007 & U009	Acrylamide, acrylonitrile
U130	Hexachlorocyclopentadiene
U135	Hydrogen sulfide
U152-U153	Methacrylonitrile, methanethiol
U159	Methyl ethyl ketone

Other Hazardous Wastes (SIC Code Numbers):

2865	Vacuum still bottoms from maleic anhydride production.
	Fractionating residues, benzene and chlorobenzene recovery.
	Residues from distillation of 1-chloro-4-nitrobenzene.
	Methanol recovery bottoms, heavy ends, methyl methacrylate production.
2869	Ends, distillation from carbaryl production.
	Ethylene dichloride distillation ends in vinyl chloride production.
	Quench column bottoms, acrylonitrile production.
	Aniline production still bottoms.
3312	Cyanide-bearing wastes from steel finishing.

Source: Reprinted, with permission, from Zimpro Environmental, Inc., 1993, wet air oxidation—solving today's hazardous wastewater problems. *Bulletin WAO-100.*

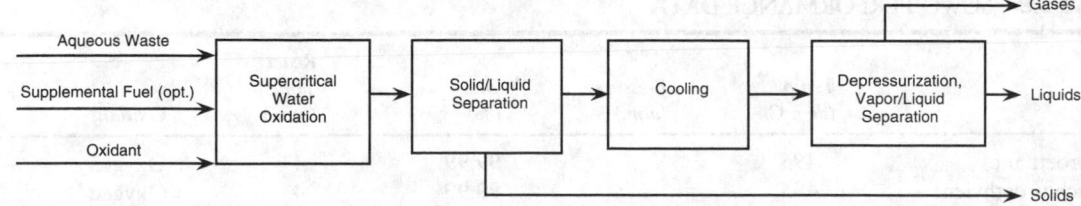

FIG. 11.14.9 SCWO Schematic. (Reprinted from U.S. EPA, 1992).

This technology is not recommended for aromatic halogenated organics, inorganics, or for large volumes of waste. It is not appropriate for solids or viscous liquids.

Status. Available at commercial scale.

Supercritical Water Oxidation

Supercritical water oxidation (SCWO) is an emerging waste treatment technology. There are no full scale SCWO systems in operation, but large bench- and pilot-scale data are available.

PROCESS DESCRIPTION

SCWO is basically a high-temperature, high-pressure process. In SCWO, decomposition occurs in the aqueous phase above the critical point of water ($374°C/221$ atm or $705°F/3248$ psi). A schematic of a generic SCWO process is shown in Figure 11.14.9. The feed is typically an aqueous waste. An oxidant such as air, oxygen, or hydrogen peroxide must be provided unless the waste itself is an oxidant.

Many of the properties of water change drastically near its critical point ($374°C/221$ atm): the hydrogen bonds disappear and water becomes similar to a moderately polar solvent; oxygen and all hydrocarbons become completely miscible with water; mass transfer occurs almost instantaneously; and solubility of inorganic salts drops to ppm range. Thus, inorganic salt removal must be considered in the design of a SCWO reactor (Thomason, Hong, Swallow & Killilea 1990).

Two process approaches have been evaluated: an above-ground pressure vessel reactor (Modar), and the use of an 8000–1000–ft deep well as a reactor vessel (Vertox). Figure 11.14.10 is a schematic of a subsurface SCWO reactor. Subsurface reactors consist of aqueous liquid waste columns deep enough that the material near the bottom is subject to a pressure of at least 221 atm (Gene Syst, 1990). To achieve this pressure solely through hydrostatic head, a water column depth of approximately 12,000 ft is required. The influent and effluent will flow in opposite directions in concentric vertical tubes. In surface SCWO systems, the pressure is provided by a source other than gravity, and the reactor is on or above the earth's surface.

The supercritical water process is best suited for large volume (200 to 1000 gpm), dilute (in the range of 1–10,000 mg/l COD), aqueous wastes that are volatile and have a sufficiently high heat content to sustain the process. In many applications, high Btu, nonhazardous waste can be mixed with low Btu hazardous waste to provide the heat energy needed to make the process self-sustaining. Emissions or residues include gaseous effluents (nitrogen and carbon dioxide), precipitates of inorganic salts, and liquids containing only soluble inorganic acids and salts. The advantages are rapid oxidation rates, complete oxidation of organics, efficient removal of inorganics, and no off-gas processing required (EPA 1992).

Significant bench- and pilot-scale SCWO performance data are available. Typical destruction efficiencies (DEs) for a number of compounds are summarized in Table 11.14.8. Although several low DEs are included in this table to illustrate that DE is proportional to both temperatures and time, DEs in excess of 99% can be achieved for nearly all pollutants (EPA 1992). Table 11.14.8 shows that using hydrogen peroxide as an oxidant in SCWO systems produces DEs significantly higher than those obtained using of air and oxygen.

APPLICABILITY/LIMITATIONS

Supercritical water oxidation is used to treat a wide variety of pumpable aqueous organic solutions, slurries, and mixed organic and inorganic waste (EPA 1992). Sophisticated equipment and long-term continuous operations have not been demonstrated, thereby limiting its use. Demonstra-

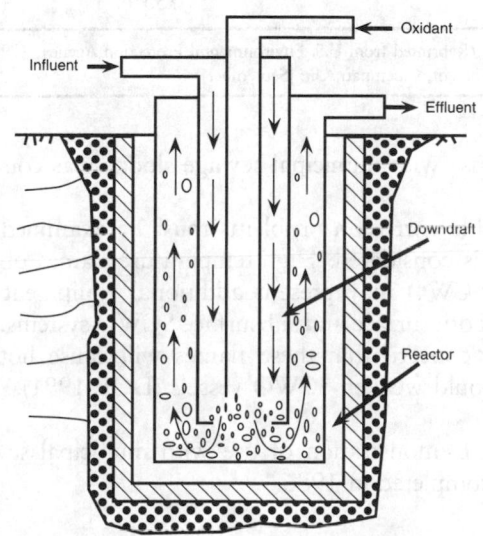

FIG. 11.14.10 Subsurface SCWO Reactor

TABLE 11.14.8 SCWO PERFORMANCE DATA

Pollutant	Temp. (deg. C)	Pressure (atm.)	DE (%)	React Time (min.)	Oxidant	Feed Conc. (mg/L)
1,1,1-Trichloroethane	495		99.99	4	Oxygen	
1,1,2,2-Tetrachloroethylene	495		99.99	4	Oxygen	
1,2-Ethylene dichloride	495		99.99	4	Oxygen	
2,4-Dichlorophenol	400		33.7	2	Oxygen	2,000
2,4-Dichlorophenol	400		99.440	1	H_2O_2	2,000
2,4-Dichlorophenol	450		63.3	2	Oxygen	2,000
2,4-Dichlorophenol	450		99.950	1	H_2O_2	2,000
2,4-Dichlorophenol	500		78.2	2	Oxygen	2,000
2,4-Dichlorophenol	500		>99.995	1	H_2O_2	2,000
2,4-Dimethylphenol	580	443	>99	10	$H_2O_2 + O_2$	135
2,4-Dinitrotoluene	410	443	83	3	Oxygen	84
2,4-Dinitrotoluene	528	287	>99	3	Oxygen	180
2-Nitrophenol	515	443	90	10	Oxygen	104
2-Nitrophenol	530	430	>99	15	$H_2O_2 + O_2$	104
Acetic acid	400		3.10	5	Oxygen	2,000
Acetic acid	400		61.8	5	H_2O_2	2,000
Acetic acid	450		34.3	5	Oxygen	2,000
Acetic acid	450		92.0	5	H_2O_2	2,000
Acetic acid	500		47.4	5	Oxygen	2,000
Acetic acid	500		90.9	5	H_2O_2	2,000
Activated sludge (COD)	400	272	90.1	2		62,000
Activated sludge (COD)	400	306	94.1	15		62,000
Ammonium perchlorate	500	374	99.85	0.2	None	12,000
Biphenyl	450		99.97	7	Oxygen	
Cyclohexane	445		99.97	7	Oxygen	
DDT	505		99.997	4	Oxygen	
Dextrose	440		99.6	7	Oxygen	
Industrial sludge (TCOD)	425		>99.8	20	Oxygen	
Methyl ethyl ketone	505		99.993	4	Oxygen	
Nitromethane	400	374	84	3	None	10,000
Nitromethane	500	374	>99	0.5	None	10,000
Nitromethane	580	374	>99	0.2	None	10,000
o-Chlorotoluene	495		99.99	4	Oxygen	
o-Xylene	495		99.93	4	Oxygen	
PCB 1234	510		99.99	4	Oxygen	
PCB 1254	510		99.99	4	Oxygen	
Phenol	490	389	92	1	Oxygen	1,650
Phenol	535	416	>99	10	Oxygen	150

Source: (Reprinted from U.S. Environmental Protection Agency, 1992, *Engineering Bulletin: Supercritical water oxidation,* [EPA 540–S–92–006], Office of Research and Development, Cincinnati, Oh. [September]).

tion of use with municipal sewage sludge was completed in 1985.

Possible corrosion problems must be examined when SCWO is considered. High-temperature flames observed during SCWO may present additional equipment problems in both surface and subsurface SCWO systems. There is some concern that these flames will cause hot spots which could weaken SCWO vessels (DOE 1991).

Status. Demonstration of use with municipal sewerage sludge completed in 1986.

—*David H.F. Liu (1995)*

References

Chang, D.P.Y., et al. 1987. Evaluation of a pilot-scale circulating bed combustor as a potential hazardous waste incinerator. *JAPCA,* Vol. 37, no. 3.

Dempsey, C.R., and E.T. Oppelt. 1993. Incineration of hazardous waste: a critical review update. *Air & Waste,* Vol. 43.

Freeman, H. 1990. *Thermal processes: innovative hazardous waste treatment technology series,* Lancaster, PA: Technomic Publishing Company.

Gene Syst International Inc. 1990. *The gravity pressure vessel* (June).

Gorman, P., et al. 1985. *Practical Guide to Trial Burns for Hazardous Waste Incineration.* U.S. EPA, EPA 600–R2-86–050. NTIS PB 86-190246 (November).

Harris, J.C., D.J. Larsen, and C.E. Rechsteiner. 1984. *Sampling and analysis methods for hazardous waste combustion.* EPA 600–8-84–002. PB 84-155545 (February).

Kiang, Y.H., and A.A. Metry. 1982. *Hazardous waste processing technology.* Ann Arbor, Mi: Ann Arbor Science.

Liptak, B.G. 1974. *Environmental engineers' handbook.* Vol. 3.2.15. Radnor, Pa: Chilton Book Company.

Thomason, Terry B., G.T. Hong, K. Swallow, and W.R. Killilea. 1990. The Modar supercritical water oxidation process. *Innovative hazardous waste treatment technology series.* Vol. 1. Technomic Publishing Company, Inc.

U.S. Department of Energy (DOE). 1991. *Supercritical Oxidation Destroys Toxic Wastes.* NTIS Technical Note (February).

U.S. Environmental Protection Agency (EPA). 1981. *Engineering handbook on hazardous waste incineration.* SW-889. NTIS PB 81-248163 (September).

———. 1983. *Guidance Manual for Hazardous Waste Incinerator Permits.* EPA-SW-966. NTIS PB 84-100577 (March).

———. 1989. *Guidance on setting permit conditions and reporting trial burn results.* EPA 625–6-89–019 (January).

———. 1990. *Handbook: quality assurance/quality control (QA/QC) procedures for hazardous waste incinerators.* EPA 625–6-89–023. NTIS PB 91-145979 (January).

———. 1990c. *Methods manual for compliance with the BIF regulations.* EPA 530-SW-91-010. NTIS PB 90-120-006 (December).

———. 1992. *Engineering bulletin: Supercritical water oxidation.* EPA 540–S–92–006. Office of Research & Development. Cincinnati, Oh. (September).

Zimmerman, F.J., U.S. Patent No. 2,665,249 (Jan. 5, 1954).

Zimpro Environmental Inc. 1993. *Wet air oxidation–solving today's hazardous wastewater problems.* Bulletin #WAO-100.

11.15
WASTE CONCENTRATION TECHNOLOGY

PARTIAL LIST OF SUPPLIERS

Sedimentation: Chemical Waste Management Inc.; Dorr-Oliver Inc.; Eimco Process Equipment Co.; Wyo Ben Inc.; National Hydro Systems Inc.; Sharples Stokes Div., Pennwalt; Water Tech Inc.; AFL Industries

Centrifugation: Clinton Centrifuge Inc.; ALFA Laval Inc.; Tetra Recovery Systems; Dorr-Oliver Inc.; Bird Environmental Systems; Western States Machine; Fletcher; Astro Metallurgical; Barrett Centrifugals; Donaldson Industrial Group; GCI Centrifuges; General Production Services Inc.; IT Corp.; Ingersoll Rand Environmental; Master Chemical Corp. System Equipment; Sartorius Balance Div., Brinkman; Sharples Stokes Div., Pennwalt; Tekmar Co.; Thomas Scientific

Evaporation: Resources Conservation Company (mobile brine concentration systems); Kipin Industries; APV Equipment Inc.; Ambient Technical Div., Ameribrom Inc.; Analytical Bio Chem Labs; Aqua Chem Water Technologies; Capital Control Co., Inc.; Dedert Corp.; HPD Inc.; Industrial Filter & Pump Manufacturing; Kimre Inc.; Fontro Co., Inc.; Lancy International Inc.; Luwa Corp.; Licon Inc.; Rosenmund Inc.; Sasakura International American Corp.; Spraying Systems Co.; Votator Anco Votator Div.; Wallace & Tiernan Div., Pennwalt; Wastesaver Corp.; Weathermeasure Weathertronics; Wheaton Instruments

Air Stripping: OH Materials; Carbon Air Services; Detox Inc.; IT Corporation; Oil Recovery Systems Inc.; Resource Conservation Company; Terra Vac Inc.; Advanced Industrial Technology; Baron Blakeslee Inc.; Beco Engineering Co.; Calgon Carbon Corp.; Chem Pro Corp; D.R. Technology Inc.; Delta Cooling Towers; Detox Inc.; Hydro Group Inc.; IPC Systems; Kimre Inc.; Munters Corp.; NEPCCO; North East Environmental Products; Oil Recovery System Inc.; Tri-Mer Corp.; Wright R.E. Associates Inc.

Distillation: Exceltech, Inc.; Kipin Industries; Mobil Solvent Reclaimers, Inc.; APV Equipment Inc.; Ace Glass Inc.; Artisan Industries Inc.; Gilmont Instruments Inc.; Glitsch Inc.; Hoyt Corp.; Licon Inc.; Progressive Recovery Inc.; Rosenmund Inc.; Sutcliffe Croftshaw; Tekmar Co.; Thomas Scientific; Vera International Inc.; Vic Manufacturing Co.; Industrial Div.; Wheaton Instruments; York Otto H. Co., Inc.

Soil Flushing: Critical Fluid Systems; IT Corp.

Liquid/Liquid Extraction: Resources Conservation Co.

Filtration: Calgon Carbon Corp.; Carbon Air Services Inc; Chemical Waste Management; Industrial Innovations Inc., Krauss-Maffei; Komline Sanderson; Bird Machine Co.; D.R. Sperry, Inc., Dorr-Oliver

Carbon Adsorption: Calgon Carbon Corp.; Carbon Air Services Inc.; Zimpro Inc.; Chemical Waste Management

Reverse Osmosis: Osmonics, Inc.; Artisan Industries Inc.

Ion Exchange: Calgon Carbon; Dionex; DeVoe-Holbein; Davis Instrument Mfg Co., Inc.; Ecology Protection Systems, Inc.; Envirex Inc.; Industrial Filter & Pump Mfg.; Lancy International Inc.; McCormack Corp.; Osmonic Membrane Sys. Div.; Pace International Corp.; Permutit Co., Inc.; Serfilco LTD.; Techni Chem., Inc.; Thomas Scientific; Treatment Technologies; Water Management Inc.; Western Filter Co.

Chemical Precipitation: Mobile Systems-Rexnord Craig; Ecolochem Inc., Dravo Corp.; Detox Inc.; Envirochem Waste Management Services; Chemical Waste Management Inc.; Andco Environmental Processes Inc.; Ensotech Inc.; Tetra Recovery Systems

Chemical and physical waste treatment processes are used for removal rather than destruction. A more appropriate term for non-destructive processes is *concentration technologies* (Martin & Johnson 1987). Physical treatment processes use physical characteristics to separate or concentrate constituents in a waste stream. Residues then require further treatment and ultimate disposal. Chemical treatment processes alter the chemical structure of wastes, producing residuals that are less hazardous than the original waste.

In this section, physical treatment processes are organized into four groupings: gravity; phase change; dissolution; and size, adsorptivity, or ionic characteristics (Table 11.15.1). Important physical treatment data needs are presented in Table 11.15.2.

TABLE 11.15.1 PHYSICAL TREATMENT PROCESS

Gravity Separation:
- Sedimentation
- Centrifugation
- Flocculation
- Oil/Water Separation
- Dissolved Air Flotation
- Heavy Media Separation

Phase Change:
- Evaporation
- Air Stripping
- Steam Stripping
- Distillation

Dissolution:
- Soil Washing/Flushing
- Chelation
- Liquid/Liquid Extraction
- Supercritical Solvent Extraction

Size/Adsorptivity/Ionic Characteristics:
- Filtration
- Carbon Adsorption
- Reverse Osmosis
- Ion Exchange
- Electrodialysis

The following chemical treatment processes discussed in this section are commonly used for waste treatment applications. These include

- pH adjustment (for neutralization or precipitation)
- Oxidation and reduction
- Hydrolysis and photolysis
- Chemical oxidation (ozonation, electrolytic oxidation, hydrogen peroxide)
- Chemical dehalogenation (alkaline metal dechlorination, alkaline metal/polyethylene glycol, based-catalyzed dechlorination)

Important chemical treatment data needs are presented in Table 11.15.3.

Gravity Separation

SEDIMENTATION

Description

Sedimentation is a settling process in which gravity causes heavier solids to collect at the bottom of a containment vessel, separated from the suspending fluid. Sedimentation

TABLE 11.15.2 PHYSICAL TREATMENT DATA NEEDS

Data Need	Purpose
For Solids	
Absolute Density	Density Separation
Bulk Density	Storage Volume Required
Size Distribution	Size Modification or Separation
Friability	Size Reduction
Solubility	Dissolution
(in H$_2$O, organic solvents, oils, etc.)	
For Liquids	
Specific Gravity	Density Separation
Viscosity	Pumping & Handling
Water Content (or oil content, etc.)	Separation
Dissolved Solids	Separation
Boiling Pt/Freezing Point	Phase Change Separation, Handling and Storage
For Liquids/Solid Mixtures	
Bulk Density	Storage & Transportation
Total Solids Content	Separation
Solids Size Distribution	Separation
Suspended Solids Content	Separation
Suspended Solids Settling Rate	Separation
Dissolved Solids Content	Separation
Free Water Content	Storage & Transport
Oil and Grease Content	Separation
Viscosity	Pumping and Handling
For Gases	
Density	Separation
Boiling (condensing) Temp.	Phase Change Separation
Solubility (in H$_2$O, etc.)	Dissolution

TABLE 11.15.3 IMPORTANT CHEMICAL TREATMENT DATA NEEDS

Data Need	Purpose
pH	pH Adjustment Needs, Corrosivity
Turbidity/Opacity	Photolysis
Constituent analysis	Treatment Need
Halogen Content	Dehalogenation

Note: Generally, the data needs for evaluating and comparing chemical treatment technologies include the data needs identified for physical treatment technologies.

TABLE 11.15.4 IMPORTANT SEDIMENTATION DATA NEEDS

Data Need	Purpose
Viscosity of aqueous waste	High viscosity hinders sedimentation
Oil and grease content of waste stream	Not applicable to wastes containing emulsified oils
Specific gravity of suspended solids	Must by greater than 1 for sedimentation to occur

can be accomplished using a batch process or a continuous removal process. Several physical arrangements where the sedimentation process is applied are shown in Figure 11.15.1.

The top diagram illustrates a settling pond. Aqueous waste flows through while suspended solids are permitted to gravitate and settle out. Occasionally the settling particles (sludges) are removed, so this system is considered a semibatch process.

The middle diagram shows a circular clarifier equipped with a solids-removal device. This facilitates continuous clarification, resulting in a lower solid content outlet fluid.

The sedimentation basin is shown in the bottom diagram. It uses a belt-type solids collector mechanism to force solids to the bottom of the basin's sloped edge, where they are removed.

The efficiency of sedimentation treatment depends upon the depth and surface area of the basin, settling time (based on the holding time), solid particle size, and the flow rate of the fluid.

Applicability/Limitations

Sedimentation is considered a separation process only. Typically, some type of treatment process for aqueous liquids and sludges will follow. Use is restricted to solids that are more dense than water. It is not suitable for wastes consisting of emulsified oils. Important sedimentation data are summarized in Table 11.15.4.

Status. This is a conventional process.

CENTRIFUGATION

Description

Centrifuge involves physical separation of fluid mixture components based on their relative density. A rapidly rotating fluid mixture within a rigid vessel deposits the more dense solid particles farthest from the axis of rotation, while liquid supernatant lies separated near the axis. Centripetal forces in centrifugation are similar to gravitational forces in sedimentation, except the centripetal forces are thousands of times stronger than gravitational forces, depending upon centrifuge diameter and rotational speed.

Applicability/Limitations

This treatment is limited to dewatering sludges (including metal-bearing sludges), separating oils from water, and clarification of viscous gums and resins. Centrifuges are generally better suited than vacuum filters for dewatering sticky or gelatinous sludges. Disc-type centrifuges (Figure 11.15.2) can be used to separate three component mixtures (e.g. oil, water, and solids). Centrifuges often cannot be used for clarification since they may fail to remove less dense solids and those small enough to remain in suspension. Recovery and removal efficiencies may be improved if paper or cloth filters are used.

Status. This process is commercially available.

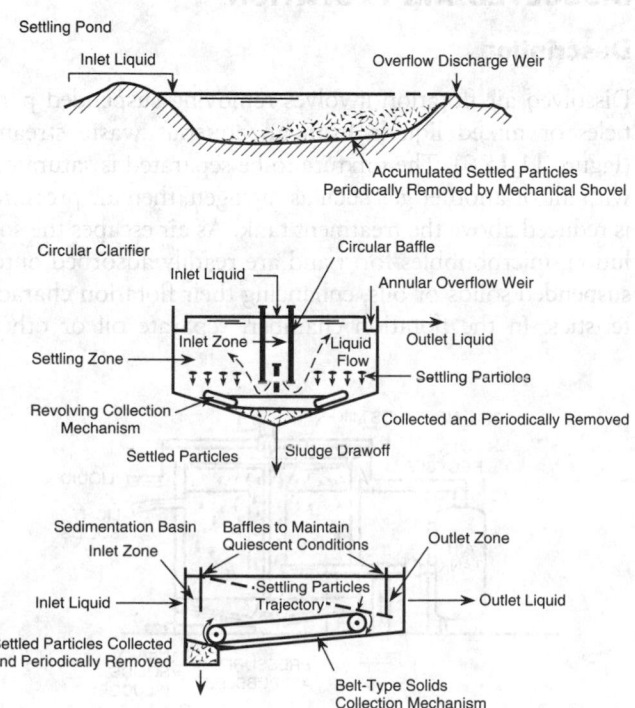

FIG. 11.15.1 Representative types of sedimentation.

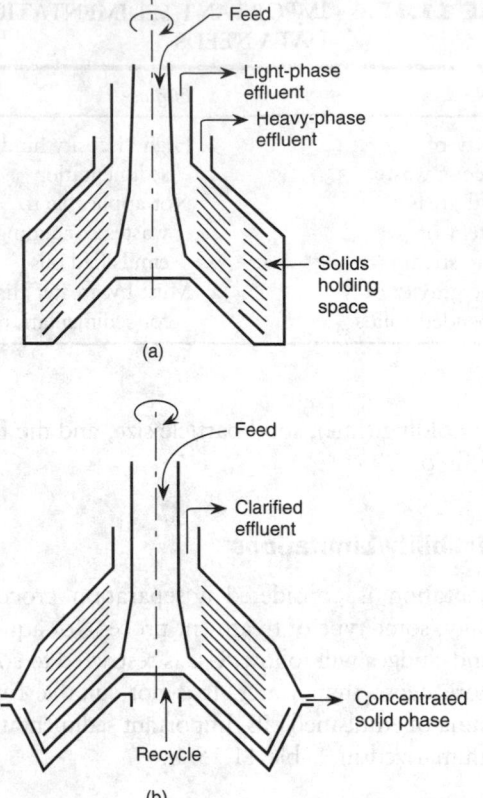

FIG. 11.15.2 Disk-centrifuge bowls. (*a*) separator, solid wall; (*b*) recycle clarifier, nozzle discharge.

FLOCCULATION

Description

Flocculation is used to enhance sedimentation or centrifugation. The waste stream is mixed while a flocculating chemical is added. Flocculants adhere readily to suspended solids and to each other (agglomeration), and the resultant particles are too large to remain in suspension. Flocculation is used primarily for the precipitation of inorganics.

Availability/Limitations

The extent of flocculation depends upon waste stream flow rate, composition, and pH. This process is not recommended for a highly viscous waste stream. Table 11.15.5 presents the important flocculation data needs.

Status. Flocculation is a conventional, demonstrated treatment technique.

OIL/WATER SEPARATION

Description

As in sedimentation, the force of gravity can be used to separate two or more immiscible liquids with sufficiently different densities, such as oil and water. Liquid/liquid sep-

TABLE 11.15.5 IMPORTANT FLOCCULATION DATA NEEDS

Data Need	Purpose
pH of waste	Selection of flocculating agent
Viscosity of waste system	Affects settling of agglomerated solids; high viscosity not suitable
Settling rate of suspended solids	Selection of flocculating agent

aration occurs when the liquid mix settles. Thus, flow rates in continuous processes must be kept low. The waste flows into a chamber, where it is kept quiescent, and permitted to settle. The floating oil is skimmed off the top using an oil skimmer while the water or effluent flows out of the lower portion of the chamber. Acids may be used to break oil/water emulsion and to enhance this process for efficient oil removal.

Availability/Limitations

Effectiveness can be influenced by waste stream flow rate, temperature, and pH. Separation is a pretreatment process if the skimmed oil requires further treatment.

Status. Mobile phase separators are commercially available.

DISSOLVED AIR FLOTATION

Description

Dissolved air flotation involves removing suspended particles or mixed liquids from an aqueous waste stream (Figure 11.15.3). The mixture to be separated is saturated with air or another gas such as nitrogen, then air pressure is reduced above the treatment tank. As air escapes the solution, microbubbles form and are readily adsorbed onto suspended solids or oils, enhancing their flotation characteristics. In the flotation chamber, separate oil or other

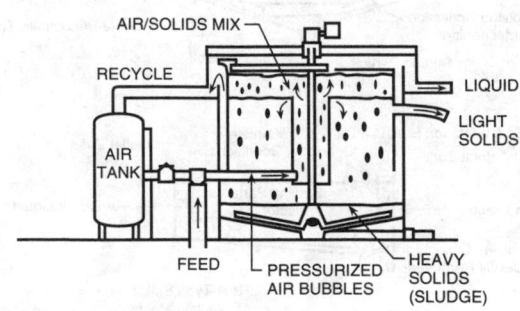

FIG. 11.15.3 Recycle flow dissolved air flotation system. Source: Peabody-Welles, Roscoe, Il.

floats are skimmed off the top while aqueous liquids flow off the bottom.

Applicability/Limitations

This technology is only applicable for waste with densities close to water. Air emission controls may be necessary if hazardous volatile organics are present.

Status.   This is a conventional treatment process.

HEAVY MEDIA SEPARATION

Description

Heavy media separation is used to process two solid materials with significantly different absolute densities. Mixed solids are placed in a fluid with a specific gravity adjusted to allow lighter solids to float while heavier solids sink. Usually, the separating fluid or heavy medium is a suspension of magnetite in water. The specific gravity is adjusted by varying the amount of magnetite powder used. Magnetite is easily recovered magnetically from rinsewaters and spills, then reused.

Availability/Limitations

This type of separation is used to separate two insoluble solids with different densities. Limitations include the possibility of dissolving solids and ruining the heavy media; the presence of solids with densities similar to those solids requiring separation; and the inability to cost-effectively separate magnetic materials, because of the need to recover magnetite.

Status.   Commonly used in the mining industry to separate ores from tailings.

Phase Change

EVAPORATION

Description

Evaporation is the physical separation of a liquid from a dissolved or suspended solid by applying energy to make the liquid volatile. In hazardous waste treatment, evaporation may be used to isolate the hazardous material in one of the two phases, simplifying subsequent treatment. If the hazardous waste is volatilized, the process is usually called *stripping*.

Availability/Limitations

Evaporation can be applied to any mixture of liquids and volatile solids provided the liquid is volatile enough to evaporate under reasonable heating or vacuum conditions (both the liquid and the solid should be stable under those conditions). If the liquid is water, evaporation can be carried out in large ponds using solar energy. Aqueous waste can also be evaporated in closed process vessels using steam energy. The resulting water vapor can be condensed for reuse. Energy requirements are minimized by techniques such as vapor recompression or multiple effect evaporators. Evaporation is applied to solvent waste contaminated with nonvolatile impurities such as oil, grease, paint solids or polymeric resins. Mechanically agitated or wipe-thin-film evaporators (Figure 11.15.4) are used. Solvent is evaporated and recovered for reuse. The residue is the bottom stream, typically containing 30 to 50% solids.

Status.   This process is commercially available.

AIR STRIPPING

Description

Air stripping is a mass transfer process in which volatile contaminants in water or soils are evaporated into the air. Organics removal from wastewater via air stripping depends upon temperature, pressure, air-to-water ratio, and surface area available for mass transfer. Air-to-water volumetric ratios may range from 10 : 1 up to 300 : 1. Contaminated off-gas and stripped effluent are the resulting residuals. Volatile hazardous materials must be recaptured for subsequent treatment to preclude air pollution.

Availability/Limitations

This process is used to treat aqueous wastes that are more volatile, less soluble (e.g., chlorinated hydrocarbons such as tetrachloroethylene) and aromatic (e.g., toluene). Limitations include temperature dependency, as stripping efficiency is impacted by changes in ambient temperature. In addition, the presence of suspended solids may reduce efficiency. If the concentration of volatile organic contaminants (VOCs) exceeds about 100 ppm, another separation process, e.g. steam stripping, is usually preferred.

Status.   This process is commercially available.

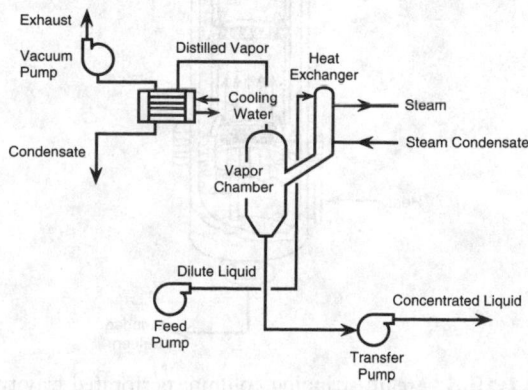

FIG. 11.15.4 Typical single effect evaporator, falling film type.

STEAM STRIPPING

Description

Steam stripping uses steam to evaporate volatile organics from aqueous wastes. Steam stripping is essentially a continuous fractional distillation process carried out in a packed or tray tower. Clean steam, rather than reboiled bottoms, provide direct heat to the column, and gas flows from the bottom to the top of the tower (Figure 11.15.5). The resulting residuals are contaminated steam condensate, recovered solvent and stripped effluent. The organic vapors and the raffinate are sent through a condenser in preparation for further purification treatment. The bottom requires further consideration as well. Possible post-treatment includes incineration, carbon adsorption, or land disposal.

Availability/Limitations

Steam stripping is used to treat aqueous wastes contaminated with chlorinated hydrocarbons, aromatics such as xylenes, ketones such as acetone or MEK, alcohols such as methanol, and high-boiling-point chlorinated aromatics such as pentachlorophenol. Steam stripping will treat less volatile and more soluble wastes than will air stripping and can handle a wide concentration range (e.g., from less than 100 ppm to about 10% organics). Steam stripping requires an air pollution control (APC) mechanism to eliminate toxic emissions.

Status.   Conventional, well documented.

DISTILLATION

Description

Distillation is simply evaporation followed by condensation. The separation of volatile materials is optimized by controlling the evaporation-stage temperature and pressure, and the condenser temperature. Distillation separates miscible organic liquids for solvent reclamation and waste volume reduction. Two types of distillation processes are batch distillation and continuous fractional distillation.

Availability/Limitations

Distillation is used to separate liquid organic wastes, primarily spent solvents, for full or partial recovery and reuse. Both halogenated and nonhalogenated solvents can be recovered via distillation. Liquids to be separated must have different volatilities. Distillation for recovery is limited by the presence of volatile or thermally reactive suspended solids. If constituents in the input waste stream form an *azeotrope* (a specific mixture of liquids exhibiting maximum or minimum boiling point with the individual constituents), the energy cost to break the azeotrope can be prohibitive.

Batch distillation in a heated still pot with condensation of overhead vapors is easily controlled and flexible, but cannot achieve the high product quality typical of continuous fractional distillation. Small, packaged-batch stills treating one drum or less per day are becoming popular for on-site recovery of solvents. Continuous fractional distillation is accomplished in tray columns or packed columns ranging up to 40 ft in diameter and 200 ft high. Each is equipped with a reboiler, a condenser, and an accumulator. Unit capacity is a function of the processed waste, purity requirements, reflux ratios, and heat input. Fractional distillation is not applicable to liquids with high viscosity at high temperature, liquids with high solid concentrations, polyurethanes, or inorganics.

Status.   Commercially available.

Dissolution

SOIL FLUSHING/SOIL WASHING

Soil is comprised of fine-grained (e.g., silt and clay) and coarse-grained (e.g., sand and gravel) particles, organic materials (e.g., decayed plant and animal matter), water, and air. Contaminants bind readily, chemically or physically, to silt, clay, and organic matter. Silt, clay, and organic mat-

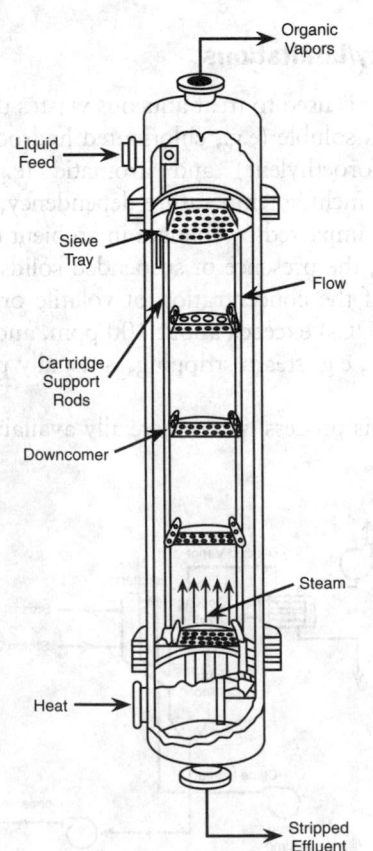

FIG. 11.15.5 Steam stripping column, perforated tray type.

ter bind physically to sand and gravel. When soil contains large amounts of clay and organic materials, contaminants attach more easily to the soil and are more difficult to remove.

Description

Soil flushing is an in-situ extraction of inorganic and organic compounds from soil, and is accomplished by passing extractant solvents through the soils using an injection and recirculation process. Solvents may include: water, water-surfactant mixtures, acids, bases (for inorganics), chelating agents, oxidizing agents, or reducing agents. Soil washing consists of similar treatments, but the soil is excavated and treated at the surface in a soil washer.

A simplified drawing of the soil washing process is illustrated in Figure 11.15.6. The contaminated soil is removed to a staging area, then sifted to remove debris and large objects such as rocks. The remaining material enters a soil scrubbing unit, is mixed with a washing solution, and agitated. The washing solution may be water, or may contain some additives like detergent to remove contaminants. Then the washwater is drained and the soil is rinsed with clean water. The heavier sand and gravel particles in the processed soil settle out and are tested for contaminants. If clean, these materials can be used on site or taken elsewhere for backfill. If contaminated, these materials may undergo soil washing again.

The contaminated silt and clay in the washwater settle out and are then separated from the washwater. The washwater, which also contains contaminants, undergoes wastewater treatment processes for future recycling use. This wastewater may contain additives that interfere with the wastewater treatment process. If so, the additives must be removed or neutralized by pretreatment methods before wastewater treatment. The silts and clays are then tested for contaminants. If clean, these materials can be used on the site or taken elsewhere for backfill. If contaminated, these materials may undergo soil washing again, or be collected for alternate treatment or off-site disposal in a permitted RCRA landfill.

Availability/Limitations

Soil flushing and washing fluids must have: good extraction coefficients; low volatility and toxicity; capability for safe and easy handling, and, most important, be recoverable and recyclable. This technology is very promising in extracting heavy metals from soil, although problems are likely in dry or organically-rich soils. Surfactants can be used to extract hydrophobic organisms. Soil type and uniformity are important. Certain surfactants, when tested for in-situ extraction, clogged soil pores and precluded further flushing.

Status.   The U.S. EPA in Edison, New Jersey, has a mobile soil washer; other systems are under development.

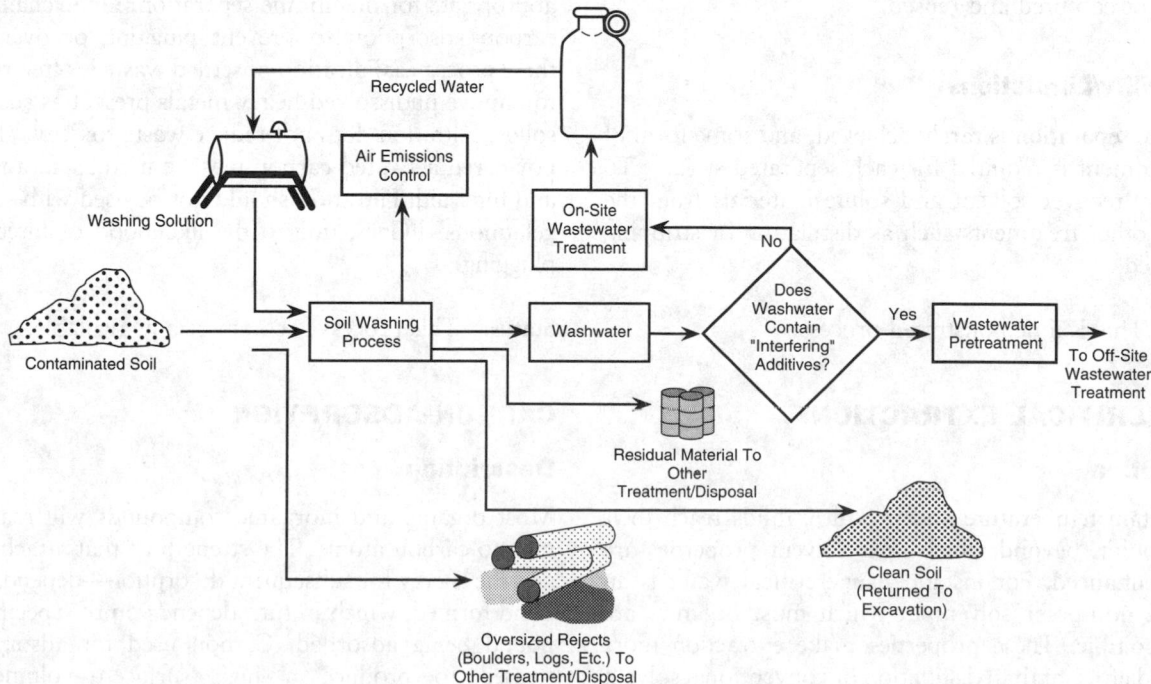

FIG. 11.15.6   Simplified soil washing process flow. (Reprinted from U.S. Environmental Protection Agency (EPA), 1992, *A citizen's guide to glycolate dehalogenation* [EPA 524–F–92–005], Office of Solid Waste and Emergency Response [March]).

CHELATION

Description

A chelating molecule contains atoms that form ligands with metal ions. If the number of such atoms in the molecule is sufficient, and if the molecular shape is such that the final atom is essentially surrounded, then the metal will be unable to form ionic salts which can precipitate out. Thus chelation is used to keep metals in solution and to aid in dissolution for subsequent transport and removal (e.g., soil washing).

Applicability/Limitations

Chelating chemicals are chosen for their affinity to particular metals (e.g., EDTA and calcium). The presence of fats and oils can interfere with the process.

Status. Chelating chemicals are commercially available.

LIQUID/LIQUID EXTRACTION

Description

Two liquids that are well mixed or mutually soluble may be separated by liquid/liquid extraction. The process requires that a third liquid be added to the original mix. This third liquid must be a solvent for one of the original components, but must be insoluble in and immiscible with the other. The final solvent and solute stream can be separated by distillation or other chemical means, and the extracting solvent captured and reused.

Availability/Limitations

Complete separation is rarely achieved, and some form of post–treatment is required for each separated stream. To effectively recover solvent and solute materials from the process, other treatments such as distillation or stripping are needed.

Status. This is a demonstrated process.

SUPERCRITICAL EXTRACTION

Description

At a certain temperature and pressure, fluids reach their critical point, beyond which their solvent properties are greatly enhanced. For instance, supercritical water is an excellent non-polar solvent in which most organics are readily soluble. These properties make extraction more rapid and efficient than distillation or conventional solvent extraction methods. Presently, the use of supercritical carbon dioxide to extract hazardous organics is being investigated.

Availability/Limitations

This technology may be useful in extracting hazardous waste from aqueous streams. Specific applicability and limitations are not yet known.

Status. This process has been demonstrated on a laboratory scale.

Size/Adsorptivity/Ionic Characteristics

FILTRATION

Description

Filtration is the separation and removal of suspended solids from a liquid by passing the liquid through a porous medium. The porous medium may be a fibrous fabric (paper or cloth), a screen, or a bed of granular material. The filter medium may be precoated with a filtration aid such as ground cellulose or diatomaceous earth. Fluid flow through the filter medium may be accomplished by gravity, by inducing a partial vacuum on one side of the medium, or by exerting mechanical pressure on a dewaterable sludge enclosed by filter medium.

Availability/Limitations

Filtration is used to dewater sludges and slurries as pretreatment for other processes. It is also a polishing step for treated waste, reducing suspended solids and associated contaminants to low levels. Pretreatment by filtration is appropriate for membrane separation, ion exchange, and carbon adsorption to prevent plugging or overloading these processes. Filtration of settled waste is often required to remove undissolved heavy metals present as suspended solids. Filtration does not reduce waste toxicity, although powdered activated carbon may be used as an adsorbent and filter aid. Filtration should not be used with sticky or gelatinous sludges, due to the likelihood of filter media plugging.

Status. This process is commercially available.

CARBON ADSORPTION

Description

Most organic and inorganic compounds will readily attach to carbon atoms. The strength of that attachment—and the energy for subsequent desorption—depends on the bond formed, which in turn depends on the specific compound being adsorbed. Carbon used for adsorption is treated to produce a high surface-to-volume ratio (900 : 1,300 sq.m/g), exposing a practical maximum number of carbon atoms for active adsorbtion. This treated carbon is said to be *activated* for adsorption. When activated carbon has adsorbed so much contaminant that its adsorptive capacity is severely depleted, it is said to be *spent*. Spent carbon can be regenerated, but for strongly adsorbed contaminants, the cost of such regeneration is higher than simple replacement with new carbon.

Availability/Limitations

This process is used to treat single-phase aqueous organic wastes with high molecular weight and boiling point, and low solubility and polarity; chlorinated hydrocarbons such as tetrachloroethylene; and aromatics such as phenol. It is also used to capture volatile organics in gaseous mixtures. Limitations are economic, relating to how rapidly the carbon becomes spent. As an informal guide, concentrations should be less than 10,000 ppm; suspended solids less than 50 ppm; and dissolved inorganics, oil, and grease less than 10 ppm.

Status.   Conventional, demonstrated.

REVERSE OSMOSIS

Description

In normal osmotic processes, solvent flows across a semi-permeable membrane from a dilute solution to a more concentrated solution until equilibrium is reached. Applying high pressure to the concentrated side causes the process to reverse. Solvent flows from the concentrated solution, leaving an even higher concentration of solute. The semi-permeable membrane can be flat or tubular, and acts like a filter due to the pressure driving force. The waste stream flows through the membrane, while the solvent is pulled through the membrane's pores. The remaining solutes, such as organic or inorganic components, do not pass through, but become more and more concentrated on the influent side of the membrane.

Availability/Limitations

For efficient reverse osmosis, the semi-permeable membrane's chemical and physical properties must be compatible with the waste stream's chemical and physical characteristics. Some membranes will be dissolved by some wastes. Suspended solids and some organics will clog the membrane material. Low-solubility salts may precipitate onto the membrane surface.

Status.   Commercial units are available.

ION EXCHANGE

Description

Although some ion exchange media occur naturally, this process normally uses specially formulated resins with an exchangeable ion bonded to the resin with a weak ionic bond. Ion exchange depends upon the electrochemical potential of the ion to be recovered versus that of the exchange ion; it also depends upon the concentration of the ions in the solution. After a critical relative concentration of recoverable ion to exchanged ion in the solution is exceeded, the exchanged resin is said to be spent. Spent resin is usually recharged by exposure to a concentrated solution of the original exchange ion, causing a reverse exchange. This results in regenerated resin and a concentrated solution of the removed ion, which can be further processed for recovery and reuse. This process is used to remove toxic metal ions from solution to recover concentrated metal for recycling. The residuals include spent resins and spent regenerants such as acid, caustic, or brine.

Availability/Limitations

This technology is used to treat metal wastes including cations (e.g., Ni^{2+}, Cd^{2+}, Hg^{2+}) and anions (e.g., CrO_4^{2-}, SeO_4^{2-}, $HAsO_4^{2-}$). Limitations are selectivity and competition, pH, and suspended solids. Concentrated waste streams with greater than 25,000 mg/L contaminants can be more cost-effectively separated by other means. Solid concentrations greater than 50 mg/L should be avoided to prevent resin blinding.

Status.   This is a commercially available process.

ELECTRODIALYSIS

Description

Electrodialysis concentrates or separates ionic species contained in a water solution. In electrodialysis, a water solution is passed through alternately placed cation-permeable and anion-permeable membranes (Figure 11.15.7). An

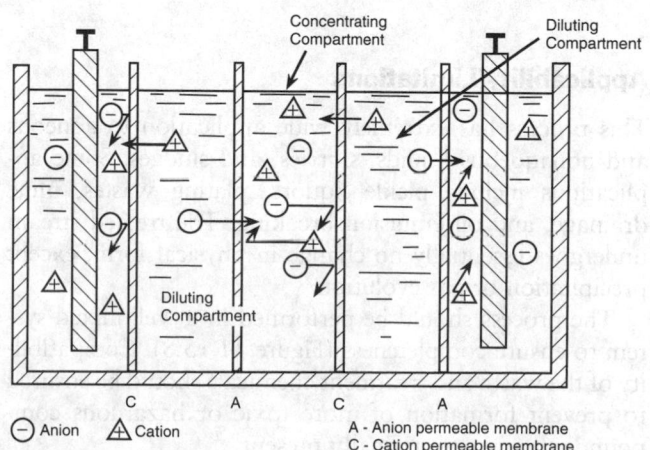

FIG. 11.15.7 Electrodialysis. An electric current concentrates the dissolved ions in compartments adjacent to those between the electrodes.

electrical potential is applied across the membrane to provide the motive force for ion migration. The ion-selective membranes are thin sheets of ion exchange resins reinforced by a synthetic fiber backing.

Availability/Limitations

The process is well established for purifying brackish water, and was recently demonstrated for recovery of metal salts from plating rinse.

Status. Units are being marketed to reclaim metals of value from rinse streams. Such units can be skid mounted and require only piping and electrical connections.

Chemical Treatment Processes

NEUTRALIZATION

Description

When an ionic salt is dissolved in water, several water molecules break into their ionic constituents of H^+ and OH^-. Neutralization is the process of changing the constituents in an ionic solution until the number of hydrogen ions (H^+) is balanced by the hydroxyl (OH^-) ions. Imbalance is measured in terms of the hydrogen ion (H^+) concentration, and is described as the solution's pH. Neutrality, on the pH scale, is 7; an excess of H^+ ions (acidity) is listed at between 0 and 7; and an excess of hydroxy or OH^- ions (alkalinity) is indicated as between 7 and 14. Neutralization is used to treat waste acids and alkalis (bases) to eliminate or reduce reactivity and corrosivity. Neutralization is an inexpensive treatment, especially if waste alkalis can be used to treat waste acid and vice/versa. Residuals include neutral effluents containing dissolved salts, and any precipitated salts.

Applicability/Limitations

This process has extremely wide application to aqueous and nonaqueous liquids, slurries, and sludges. Some applications include pickle liquors, plating wastes, mine drainage, and oil emulsion breaking. The treated stream undergoes essentially no change in physical form, except precipitation or gas evolution.

The process should be performed in a well-mixed system to ensure completeness (Figure 11.15.8). Compatibility of the waste and treatment chemicals should be ensured to prevent formation of more toxic or hazardous compounds than were originally present.

Status. This is a common industrial process.

CHEMICAL PRECIPITATION

Description

Like neutralization, chemical precipitation is a pH adjustment process. To achieve precipitation, an acid or base is added to a solution to adjust the pH to a point where the constituents to be removed reach their lowest solubility. Chemical precipitation facilitates the removal of dissolved metals from aqueous wastes. Metals may be precipitated from solutions by the following methods.

Alkaline agents, such as lime or caustic soda, are added to waste streams to raise the pH. The solubility of metals decreases as pH increases, and the metal ions precipitate out of the solution as hydroxide (Figure 11.15.9).

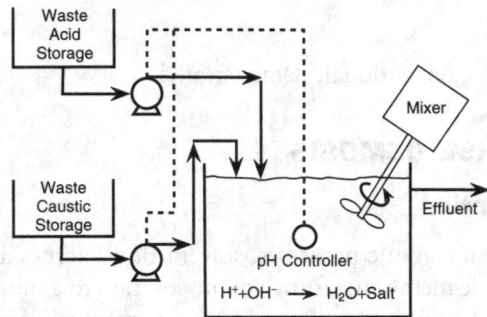

FIG. 11.15.8 Simultaneous neutralization of acid and caustic waste.

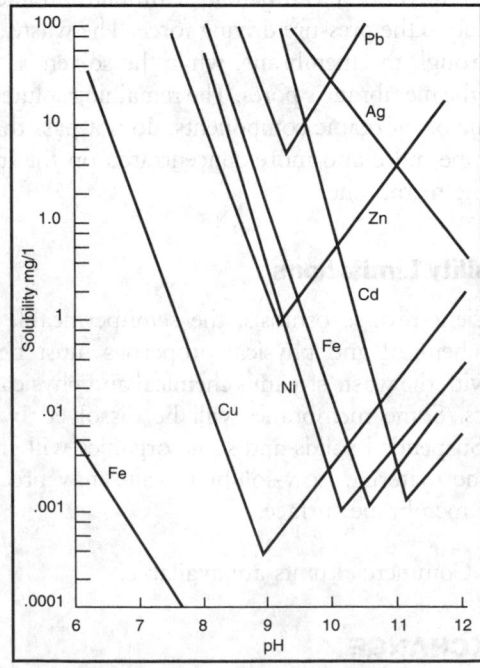

Experimentally determined solubilities of metal hydroxides.

FIG. 11.15.9 Solubilities of metal hydroxides at various pH's. (Reprinted, with permission from Graver Water.)

Soluble sulfides, such as hydrogen or sodium sulfide, and insoluble sulfides, such as ferrous sulfide, are used for precipitation of heavy metals. Sodium bisulfide is commonly used for precipitating chromium out of solution. Sulfates, including zinc sulfate or ferrous sulfate, are used for precipitation of cyanide complexes.

Carbonates, especially calcium carbonate, are used directly for precipitation of metals. In addition, hydroxides can be converted into carbonates with carbon dioxide, and easily filtered out.

Hydroxide precipitation with lime is most common; however, sodium sulfide is sometimes used to achieve lower effluent metal concentrations. Solid separation is effected by standard flocculation/coagulation techniques. The residuals are metal sludge and treated effluent with an elevated pH and, in the case of sulfide precipitation, excess sulfide.

The metal's valence state is important in the process of precipitation. For example, ferrous iron is considerably more soluble than ferric iron, making oxidizing agent treatment to convert ferrous iron to ferric iron an essential part of the iron-removal process. Another example is hexavalent chromium, Cr^{+6}, which is more soluble than the less hazardous trivalent form. Chromates must be reduced before removal of trivalent chromium in a precipitation process. Also, the engineer must consider the possibility of complex ion formation when dealing with waste water containing ammonia, fluoride, cyanide, or heavy metals. For example, an iron complex may be the ferrocyanide ion, which is soluble, and remains in solution unless the complex is broken by chemical treatment.

Applicability/Limitations

This technology is used to treat aqueous wastes containing metals. Limitations include the fact that metals have different optimum pH levels for precipitation. Chelating and complexing agents can interfere with the process. Organics are not removed except through adsorptive carryover. The resulting sludge may be hazardous by definition, but often may be taken off the list by special petition.

Precipitation has many useful applications to hazardous waste treatment, but laboratory jar tests should be made to verify the treatment. The jar test is used to select the appropriate chemical; determine dosage rates; assess mixing, flocculation and settling characteristics; and estimate sludge production and handling requirements.

Status. Commercially available.

OXIDATION AND REDUCTION

Description

Oxidation and reduction must take place in any such reaction. In any oxidation reaction, the oxidation state of one compound is raised, while the oxidation state of another compound is reduced. Oxidation and reduction change the chemical form of a hazardous material: rendering it less toxic; changing its solubility, stability, or separability; or otherwise changing it for handling or disposal purposes. In the reaction, the compound supplying oxygen, chlorine or another negative ion, is called the oxidizing agent while the compound supplying the positive ion and accepting the oxygen is called the reducing agent. The reaction can be enhanced by catalysis, electrolysis or irradiation.

Reduction lowers the oxidation state of a compound. Reducing agents include: iron, aluminum, zinc, and sodium compounds. For efficient reduction, waste pH should be adjusted to an appropriate level. After this is accomplished, the reducing agent is added and the resulting solution is mixed until the reaction is complete. This treatment can be applied to chemicals such as hexavalent chromium, mercury, and lead. Other treatment processes may be used in conjunction with chemical reduction.

Cyanide-bearing wastewater generated by the metal-finishing industry, is typically oxidized with alkaline chlorine or hypochlorite solutions. In this process, the cyanide is initially oxidized to a less toxic cyanate and then to carbon dioxide and and nitrogen in the following reactions:

$$NaCN + Cl_2 + 2\,NaOH \longrightarrow NaCNO + 2\,NaCl + H_2O$$
$$11.15(1)$$

$$2\,NaCNO + 3\,Cl_2 + 4\,NaOH \longrightarrow$$
$$2\,CO_2 + N_2 + 6\,NaCl + 2\,H_2O \quad 11.15(2)$$

In the first step, the pH is maintained at above 10, then the reaction proceeds in a matter of minutes. In this step great care must be taken to maintain relatively high pH values, because at lower pHs there is a potential for the evolution of highly toxic hydrogen cyanide gas. The second reaction step proceeds most rapidly around a pH of 8, but not as rapidly as the first step. Higher pH values may be selected for the second step to reduce chemical consumption in the following precipitation steps. However, cyanide complexes of metals, particularly iron and to some extent nickel, cannot be decomposed easily by the cyanide oxidation method.

Cyanide oxidation can also be accomplished with hydrogen peroxide, ozone, and electrolysis.

Applicability/Limitations

The process is nonspecific. Solids must be in solution. Reaction can be explosive. Waste composition must be well known to prevent the inadvertent production of a more toxic or more hazardous end product.

Status. This is a common industrial process.

HYDROLYSIS

Description

Hydrolysis is the breaking of a bond in a non-water-soluble molecule so that it will go into ionic solution with water.

$$XY + H_2O \longrightarrow HY + XOH \qquad 11.15(3)$$

Hydrolysis can be achieved by: adding chemicals, e.g., acid hydrolysis; irradiation, e.g., photolysis; or biological means, e.g., enzymatic bond cleavage. The cloven molecule can then be further treated by other means to reduce toxicity.

Applicability/Limitations

Chemical hydrolysis applies to a wide range of otherwise refractory organics. Hydrolysis is used to detoxify waste streams of carbamates, organophosphorous compounds and other pesticides. Acid hydrolysis as an in-situ treatment must be performed carefully due to potential mobilization of heavy metals. In addition, depending on the waste stream, products may be unpredictable and the mass of toxic discharge may be greater than the waste originally input for treatment.

Status.   Common industrial process.

CHEMICAL OXIDATION

Oxidation destroys hazardous contaminants by chemically converting them to nonhazardous or less toxic compounds that are stable, less mobile, or inert. Common oxidizing agents are ozone, hydrogen peroxide, hypochlorites, chlorine, and chlorine dioxide. Current research shows that combining these reagents, or combining ultraviolet (UV) light and oxidizing agent(s) makes the process more effective.

The effectiveness of chemical oxidation on general contaminant groups is shown in Table 11.15.6 (U.S. EPA 1991). Chemical oxidation depends on the chemistry of the oxidizing agents and the chemical contaminants. Table 11.15.7 lists selected organic compounds by relative oxidization ability. The oxidation process is nonselective; any oxidizable material reacts. Chemical oxidation is also a part of the treatment process for cyanide-bearing wastes and metals such as arsenic, iron, and manganese. Metal oxides formed in the oxidation process precipitate more readily out of the solution.

Some compounds require a combination of oxidizing agents or the use of UV light with an oxidizing agent.

TABLE 11.15.6   EFFECTIVENESS OF CHEMICAL OXIDATION ON GENERAL CONTAMINANT GROUPS FOR LIQUIDS, SOILS, AND SLUDGES[a]

Contaminant Groups	Liquids	Soils, Sludges
Organic		
Halogenated volatiles	■	▼
Halogenated semivolatiles	■	▼
Nonhalogenated volatiles	■	▼
Nonhalogenated semivolatiles	■	▼
PCBs	■	□
Pesticides	■	▼
Dioxins/Furans	▼	□
Organic cyanides	■	■
Organic corrosives	▼	▼
Inorganic		
Volatile metals	■	▼
Nonvolatile metals	■	▼
Asbestos	□	□
Radioactive materials	□	□
Inorganic corrosives	□	□
Inorganic cyanides	■	■
Reactive		
Oxidizers	□	□
Reducers	■	▼

■ Demonstrated Effectiveness: Successful treatability test at some scale completed
▼ Potential Effectiveness: Expert opinion that technology will work
□ No Expected Effectiveness: Expert opinion that technology will not work
[a] Enhancement of the chemical oxidation process is required for the less easily oxidizable compounds for some contaminant groups.

Source: Reprinted, from U.S. Environmental Protection Agency (EPA), 1991, *Engineering Bulletin: chemical oxidation treatment,* (EPA 540–2–91–025, Office of Research and Development, Cincinnati, Oh. [September]).

TABLE 11.15.7 SELECTED ORGANIC COMPOUNDS
BY RELATIVE ABILITY TO BE
OXIDIZED

Ability to be Oxidized	Examples
High	phenols, aldehydes, amines, some sulfur compounds
Medium	alcohols, ketones, organic acids, esters, alkyl-substituted aromatics, nitro-substituted aromatics, carbohydrates
Low	halogenated hydrocarbons, saturated aliphatics, benzene

Source: Reprinted, from U.S. EPA, 1991.

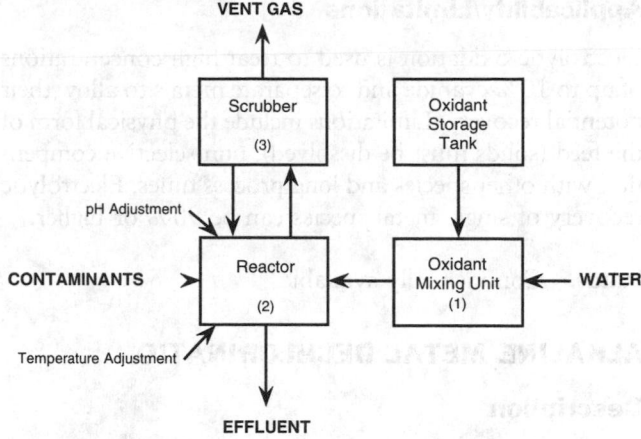

FIG. 11.15.10 Process flow diagram for chemical oxidation system. (Reprinted from U.S. EPA, 1991.)

Polychlorinated biphenyls (PCBs) do not react with ozone alone, but have been destroyed by combined UV and ozone treatment. Enhanced chemical oxidation has been used at several Superfund sites (U.S. EPA 1990a).

Description

Chemical oxidation increases the oxidation state of a contaminant and decreases the oxidation state of the reactant. The electrons lost by the contaminant are gained by the oxidizing agent. The following equation is an example of oxidation reaction:

$$NaCN + H_2O_2 \longrightarrow NaCNO + H_2O \quad 11.15(4)$$

Figure 11.15.10 details the process flow for a chemical oxidation system. The main component is the process reactor. Oxidant is fed into the mixing unit (1), then the reactor (2). Reaction products and excess oxidant are scrubbed before venting to the ambient air. Reactor pH and temperature are controlled to ensure completion at the reaction. The reaction can be enhanced by adding UV light.

Systems that combine ozone with hydrogen peroxide or UV radiation are catalytic ozonation processes. They accelerate ozone decomposition, increasing hydroxyl radical concentration, and promoting oxidation of the compounds. Specifically, hydrogen peroxide, hydrogen ion, and UV radiation have been found to initiate ozone decomposition and accelerate oxidation of refractory organics via free radical reaction. Reaction times can be 100 to 1000 times faster in the presence of UV light. Minimal emissions result from the UV-enhanced system.

Applicability/Limitations

This process is nonspecific. Solids must be in solution. It may be exothermic or explosive or require addition

of heat. Waste composition must be well-known to prevent producing a more toxic or hazardous end product. Oxidation by hydrogen peroxide is not applicable for in situ treatment. However, it may be used for surface treatment of contaminated groundwater sludge. Oxidation is not cost-effective for highly concentrated waste because of the large amount of oxidizing agent required.

Ozone can be used to pretreat wastes to break down refractory organics or to oxidize untreated organics after biological or other treatment processes. Ozone is currently used to destroy cyanide and phenolic compounds. Rapid oxidation offers advantages over the slower alkaline chlorination method. Limitations include the physical form of the waste (i.e., sludges and solids are not readily treated) and non-selective competition with other species. Ozonation systems have higher capital costs because ozone generators must be used.

The cost of generating UV lights and the problems of scaling or coating on the lamps are two of the major drawbacks to UV-enhanced chemical oxidation systems. They do not perform well in turbid waters or slurries because reduced light transmission lowers the effectiveness.

Status. Commercially available.

ELECTROLYTIC OXIDATION

Description

In this process, cathodes and anodes are immersed in a tank containing waste to be oxidized, and a direct current is imposed on the system. This process is particularly applicable to cyanide-bearing wastes. Reaction products are ammonia, urea, and carbon dioxide. During decomposition, metals are plated out on the cathodes.

Applicability/Limitations

Electrolytic oxidation is used to treat high concentrations of up to 10% cyanide and to separate metals to allow their potential recovery. Limitations include the physical form of the feed (solids must be dissolved), non-selective competition with other species and long process times. Electrolytic recovery of single metal species can be 90% or higher.

Status.   Commercially available.

ALKALINE METAL DECHLORINATION

Description

This process of chemical dechlorination displaces chlorine from chlorinated organic compounds contained in oils and liquid wastes. Typically, wastes are filtered before entering the reactor system and encountering the dechlorinating reagent. The great affinity of alkali metals for chlorine (or any halide) is the chemical basis of this process.

Successive treatment includes additional centrifugation and filtration. By-products include chloride salts, polymers, and heavy metals. Several chemical dechlorination processes are based on a method developed by the Goodyear Tire and Rubber Company in 1980. The original method uses sodium naphthalene and tetrahydrofuran to strip chlorine atoms from PCBs, polymerizing the biphenols into an inert condensible sludge. The reactor is blanketed with nitrogen because the reagents are sensitive to air and water, and an excess of reagent to chlorine is required. The Goodyear Company has not commercially developed this technology; however, several companies have modified the method by substituting their own proprietary reagents for the naphthalene. The equipment is mobile and can be transported on semi-trailers.

Applicability/Limitations

Such processes are used to treat PCBs, other chlorinated hydrocarbons, acids, thiols, and dioxins. Moisture content

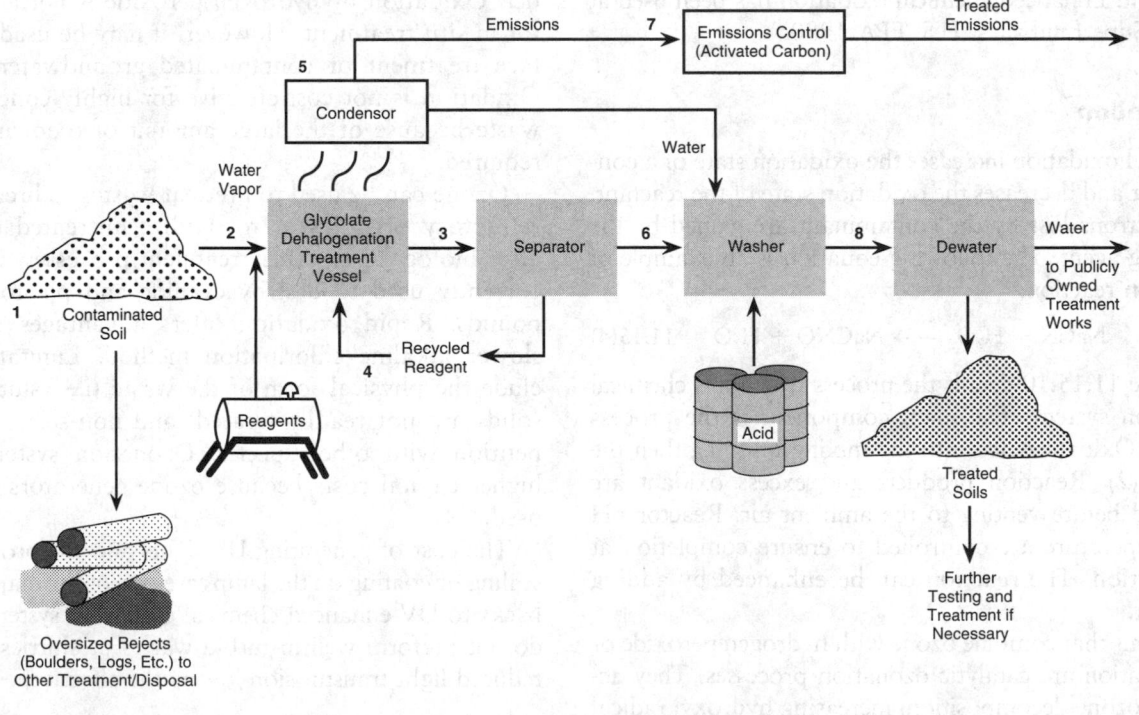

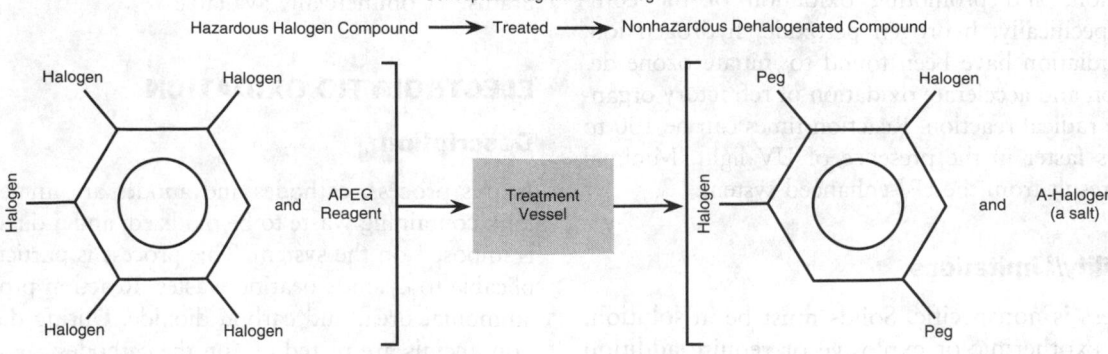

FIG. 11.15.11 Glycolate dehalogenation process flow. (Reprinted, from U.S. EPA, 1992 [March].)

adversely affects the rate of reaction, therefore dewatering should be a pretreatment step. Waste stream concentrations are also important.

Status. Commercially available.

ALKALINE METAL/POLYETHYLENE GLYCOL (APEG)

Description

In 1978, EPA-sponsored research led to the development of the first in a series of APEG reagents, which effectively dechlorinate PCBs and oils. These reagents were alkali metal/polyethylene glycols which react rapidly to dehalogenate halo-organic compounds of all types (Figure 11.15.11).

In the APEG reagents, alkali metal is held in solution by large polyethylene anions. PCBs and halogenated molecules are soluble in APEG reagents. These qualities combine in a single-phase system where the anions readily displace the halogen atoms. Halogenated aromatics react with PEGs resulting in the substitution of halogenated aromatics for chlorine atoms to form a PEG ether. The PEG ether decomposes to a phenol.

The effectiveness of APEG on general contaminant groups for various matrices is shown in Table 11.15.8 (U.S. EPA, 1990b).

A variation of APEG, referred to as ATEG, uses potassium hydroxide or sodium hydroxide/tetraethylene glycol, and is more effective on halogenated aliphatic compounds.

Figure 11.15.11 is a schematic of the APEG treatment process. Waste preparation includes excavating and/or moving the soil to the process where it is normally screened (1) removing debris and large objects and producing particles small enough to allow treatment in the reactor without binding the mixer blades.

Typically, reagent components are mixed with contaminated soil in the reactor (2). Treatment proceeds inefficiently without mixing. The mixture is heated to between 100°C and 180°C. The reaction proceeds for 1–5 hrs. depending upon the type, quantity, and concentration of the contaminants. The treated material goes from the reactor to a separator (3), where the reagent is removed and can be recycled (4).

During the reaction, water is evaporated in the reactor, condensed (5), and collected for further treatment or recycled through the washing process, if required. Carbon filters (7) are used to trap any volatile organics that are

TABLE 11.15.8 EFFECTIVENESS OF APEG TREATMENT ON GENERAL CONTAMINANT GROUPS FOR VARIOUS MATRICES

Contaminant Groups	Effectiveness			
	Sediments	Oils	Soil	Sludge
Organic				
Halogenated volatiles	▼	▼	▼	▼
Halogenated semivolatiles	▼	▼	▼	▼
Nonhalogenated volatiles	□	□	□	□
Nonhalogenated semivolatiles	□	□	□	□
PCBs	■	■	■	■
Pesticides (halogenated)	▼	■	■	▼
Dioxins/Furans	■	■	■	■
Organic cyanides	□	□	□	□
Organic corrosives	□	□	□	□
Inorganic				
Volatile metals	□	□	□	□
Nonvolatile metals	□	□	□	□
Asbestos	□	□	□	□
Radioactive materials	□	□	□	□
Inorganic corrosives	□	□	□	□
Inorganic cyanides	□	□	□	□
Reactive				
Oxidizers	□	□	□	□
Reducers	□	□	□	□

■ Demonstrated Effectiveness: Successful treatability test at some scale completed
▼ Potential Effectiveness: Expert opinion that technology will work
□ No Expected Effectiveness: Expert opinion that technology will not work

Source: Reprinted, from U.S. Environmental Protection Agency (EPA), 1990, *Engineering Bulletin: Chemical dehalogenation treatment: APEG treatment* (EPA 540-2-90-015, Office of Research and Development, Cincinnati, Oh. [September]).

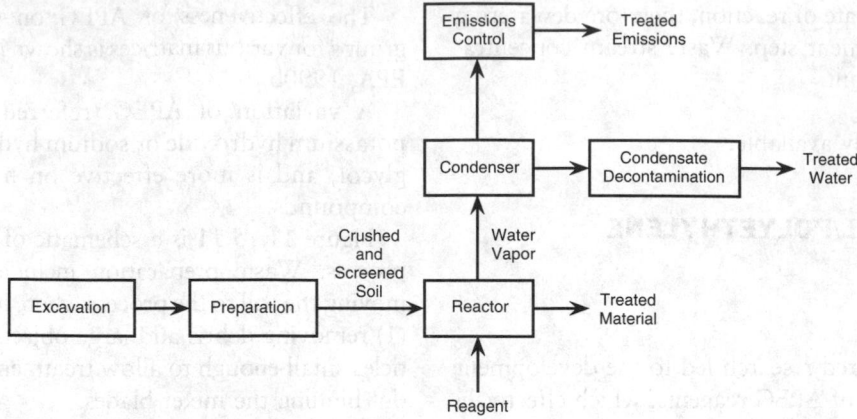

FIG. 11.15.12 BCD process flow schematic. (Reprinted, from U.S. Environmental Protection Agency, 1992, *BCD detoxification of chlorinated wastes.* Office of Research and Development, Cincinnati, Oh. [September].)

not condensed. In the washer (6), soil is neutralized by the addition of acid. It is then dewatered (8) before disposal.

Applicability/Limitations

Dehalogenation is effective in removing halogens from hazardous organic compounds such as dioxins, furans, PCBs, and chlorinated pesticides; rendering them non-toxic. APEG will dehalogenate aliphatic compounds if the mixture reacts longer and at temperatures significantly higher than for aromatics. This technology usually costs less than incineration.

Treatability tests should be conducted before the final selection of the APEG technology. Operating factors such as quantity of reagents, temperature, and treatment time should be defined. Treated soil may contain residual reagents and treatment by-products that should be removed by washing the soil with water. The soil should also be neutralized by lowering the pH before final disposal.

Specific safety aspects must be considered. Treatment of certain chlorinated aliphatics in high concentrations with APEG may produce potentially explosive compounds (e.g., chloroacetylenes) or cause a fire hazard.

Status. This process has been field tested.

BASED-CATALYZED DECOMPOSITION (BCD)

Description

Based-catalyzed decomposition is another technology for removing chlorine molecules from organic substances.

$$\text{Chlorinated Products} \xrightarrow[\text{Catalyst, base, } \Delta]{\text{Hydrogen donors}}$$

$$\text{Dechlorinated products}$$
$$+$$
$$\text{sodium chloride} \qquad 11.15(5)$$

The BCD process (Figure 11.15.12) embodies the following steps: mixing the chemicals with the contaminated matrix (such as excavated soil or sediment or liquids, containing these toxic compounds), and heating the mixture at 320–340°C for 1–3 hr. The off-gases are treated before releasing to the atmosphere. The treated receptor remains are nonhazardous, and can be either disposed of according to standard methods, or further processed to separate components for reuse.

Applicability/Limitations

Laboratory and bench-scale tests demonstrated this technology's ability to reduce PCBs from 4,000 ppm to less than 1 ppm. The BCD process requires only 1–5% reagent by weight. The reagent is also much less expensive than the APEG reagent. BCD also is regarded as effective for pentachlorophenol (PCP), PCBs, pesticides (halogenated), herbicides (halogenated), dioxins and furans. Again, BCD is not intended as an in situ treatment. Treatability studies should be conducted before the final selection.

Status. This process has been field tested.

—*David H.F. Liu*

References

Martin, E.J., and J.H. Johnson, Jr. 1987. *Hazardous waste management engineering.* Chapter 3. New York, N.Y.: Van Nostrand Reinhold.

U.S. Environmental Protection Agency (EPA). 1990. *Technology evaluation report: SITE program demonstration of the Ultrox R international ultraviolet radiation/oxidation technology.* EPA 540–5–89–012 (January).

———. 1990b. *Engineering Bulletin: Chemical dehalogenation treatment: APEG treatment.* EPA 540–2–90–015. Office of Research and Development. Cincinnati, Oh. (September).

———. 1991. *Engineering Bulletin: Chemical oxidation treatment.* EPA 540–2–91–025. Office of Research and Development. Cincinnati, Oh. (October).

11.16
SOLIDIFICATION AND STABILIZATION TECHNOLOGIES

PARTIAL LIST OF SUPPLIERS

Portland Cement Pozzolan Process: Aerojet Energy Conversion Co.: ATCOR, Inc.; Chem-Nuclear System, Inc.; Delaware Custom Materials; Energy, Inc.; General Electric Co.; Hittman Nuclear and Development Co.; Stock Equipment Co.; Todd Research and Technical Div., United Nuclear Industries; Westinghouse Electric Co.

Asphalt-Based (Thermoplastic); Microencapsulation (Thermoplastic); Microencapsulation: Werner A. Pfleidler; Aerojet Energy Conversion Co.; Newport News Industrial Corp.

Solidification techniques encapsulate hazardous waste into a solid material of high structural integrity. Encapsulation involves either fine waste particles, microencapsulation, or a large block or container of wastes, macroencapsulation (Conner 1990). Stabilization techniques treat hazardous waste by converting it into a less soluble, mobile, or toxic form. Solidification/stabilization (S/S) processes utilize one or both of these techniques.

The goal of S/S processes is the safe ultimate disposal of hazardous waste. Four primary reasons for treating the waste are to:

- Improve handling characteristics for transport on-site or to an off-site TSD facility
- Limit the mobility or solubility of pollutants contained in the waste
- Reduce the exposed area allowing transfer or loss of contained pollutants
- Detoxify contained pollutants

Applications

Table 11.16.1 summarizes the effectiveness of S/S on general contaminant groups for soils and sludges. The fixing and binding agents for S/S immobilize many heavy metals and solidify a wide variety of wastes including spent pickle liquor, contaminated soils, incinerator ash, wastewater treatment filter cake, waste sludge, and many radionuclides (EPA 1990). In general, S/S is considered as an established full-scale technology for nonvolatile heavy metals, although the long-term performance of S/S in Superfund applications has yet to be demonstrated (EPA 1991).

Technology Description

S/S processes can be divided into the following broad categories: inorganic processes (cement and pozzolanic) and organic processes (thermoplastic and thermosetting).

CEMENT-BASED PROCESSES

These processes generally use Portland cement and sludge along with certain other additives (some proprietary) including fly ash or other aggregate to form a monolithic, rock-like mass. Type I Portland cement, the cement normally used in construction, is generally used for waste fixation. Type II is used in the presence of moderate sulfate concentrations (150–1500 mg/kg), and Type V, for high sulfate concentrations (greater than 1500 mg/kg).

These processes are successful on many sludges generated by the precipitation of heavy metals. The high pH of the cement mixture keeps the metals as insoluble hydrox-

TABLE 11.16.1 EFFECTIVENESS OF S/S ON GENERAL CONTAMINANT GROUPS FOR SOIL AND SLUDGES

Contaminant Groups	Effectiveness Soil/Sludge
Organic	
Halogenated volatiles	□
Nonhalogenated volatiles	□
Halogenated semivolatiles	■
Nonhalogenated semivolatiles and nonvolatiles	■
PCBs	▼
Pesticides	▼
Dioxins/Furans	▼
Organic cyanides	▼
Organic corrosives	▼
Inorganic	
Volatile metals	■
Nonvolatile metals	■
Asbestos	■
Radioactive materials	■
Inorganic corrosives	■
Inorganic cyanides	■
Reactive	
Oxidizers	■
Reducers	■

KEY: ■ Demonstrated Effectiveness: Successful treatability test at some scale completed.
▼ Potential Effectiveness: Expert opinion that technology will work.
□ No Expected Effectiveness: Expert opinion that technology will/does not work.

Source: Reprinted, from U.S. Environmental Protection Agency, 1993, *Engineering bulletin: solidification/stabilization of organics and inorganics,* (EPA 540–5–92–015. Office of Research and Development, Cincinnati, Oh.

ides or carbonate salts. Metal ions may also be taken up into the cement matrix.

Additives such as clay, vermiculite, and soluble silicate improve the physical characteristics and decrease leaching losses from the resulting solidified sludge. Many additives are proprietary.

POZZOLANIC PROCESSES

These lime-based stabilization processes depend on the reaction of lime with a fine-grained siliceous material and water to produce a hardened material. The most common pozzolanic materials used in waste treatment are fly ash, ground blast-furnace slag, and cement-kiln dust. As all these materials are waste products to be disposed of, the fixation process can reduce contamination from several wastes. Other additives, generally proprietary, are often added to the sludge to enhance material strength or to help limit migration of problem contaminants from the sludge mass.

Lime and Portland cement are the setting agents, but gypsum, calcium carbonate, and other compounds may also be used. Lime-based and cement-based processes are better suited for stabilizing inorganic wastes rather than organic wastes. Decomposition of organic material in sludge after curing can result in increased permeability and decreased strength of the material.

Certain processes fall in the category of cement-pozzolanic processes. In this case both cement and lime-siliceous materials are combined to give the best and most economical containment for waste.

THERMOPLASTIC PROCESSES

Bitumen stabilization techniques (including bitumen, paraffin and polyethylene) were developed for use in radioactive waste disposal and later adapted for handling industrial wastes. In a bitumen process, the waste is dried and then mixed with bitumen, paraffin or polyethylene (usually at temperatures greater than 100°C). The mixture solidifies as it cools, then is placed in a container, such as a steel drum or a thermoplastic coating, before disposal.

A variation of the bitumen process uses an asphalt emulsion that is miscible with the wet sludge. This process can be conducted at a lower temperature than a bitumen process. The emulsion-waste mixture must be dried before disposal.

The type of waste sludges that can be fixed with bitumen techniques is limited. Organic chemicals that act as solvents with bitumen cannot be stabilized. High concentrations of strong oxidizing salts such as nitrates, chlorates, or perchlorates will react with bitumen and cause slow deterioration.

ORGANIC POLYMER PROCESSES

The major organic polymer process (including urea-formaldehyde, unsaturated polyesters) currently in use is the urea-formaldehyde process. In the process, a monomer is added to the waste or sludge and thoroughly mixed. Next, a catalyst is added to the mixture and mixing continues until the catalyst is dispersed. The mixture is transferred to another container and allowed to harden. The polymerized material does not chemically combine with the waste. Instead, a spongy mass forms, trapping the solid particles while allowing some liquid to escape. The polymer mass can be dried before disposal.

Table 11.16.2 compares the advantages and disadvantages of the above S/S processes. Table 11.16.3 illustrates the compatibility of selected waste categories with S/S processes.

Figures 11.16.1 and 11.16.2 depict generic elements of typical ex situ and in situ S/S processes for soils and sludges. Ex situ processing involves: (1) excavation to remove the contaminated waste from the subsurface; (2) classification to remove oversize debris; (3) mixing; and (4) off-gas treatment. In situ processing has only two steps: (1) mixing; and (2) off-gas treatment. Both processes require a system for delivering water, waste, and S/S agents in proper proportions; and a mixing device (e.g., rotary drum paddle or auger). Ex situ processing requires a system for delivering treated waste to molds, surface trenches, or subsurface injection. The need for off-gas treatment using vapor col-

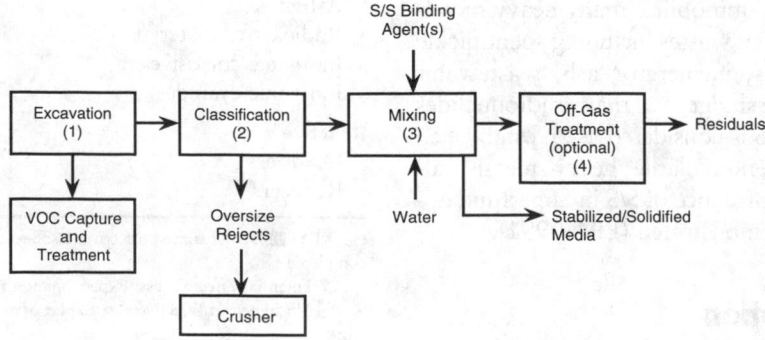

FIG. 11.16.1 Generic elements of a typical ex situ S/S process. (Reprinted, from U.S. EPA, 1993.)

TABLE 11.16.2 COMPARISON FOR SOLIDIFICATION AND STABILIZATION PROCESSES

Process	Description	Advantages	Disadvantages
Cement	Slurry of wastes and water is mixed with portland cement to form a solid	Low costs; readily available mixing equipment; relatively simple process; suitable for use with metals	Solids are suspended, not chemically bound; therefore are subject to leaching; doubles waste volume; requires secondary containment; incompatible with many wastes (organics, some sodium salts, silts, clays, and coal or lignite)
Pozzolanic	Waste is reacted with lime and a fine-grained siliceous material (fly ash, ground blast furnace slag, cement kiln dust) to form a solid	Low cost; readily available mixing equipment; suitable for power-plant wastes (FGD sludges, etc.) as well as a wide range of industrial wastes, including metals, waste oil, and solvents	Increases waste volume; may be subject to leaching; requires secondary containment
Thermoplastic	Waste is dried, heated, and dispensed through a heated plastic matrix of asphalt bitumen, paraffin or polyethylene	Less increase in volume than with cement- or lime-based processes; reduced leaching relative to cement- or lime-based processes; suitable for radioactive wastes and some industrial wastes	Wastes must be dried before use; high equipment costs; high energy costs; requires trained operators; incompatible with oxidizers, some solents and greases, some salt, and chelating/complexing agents; requires secondary containment
Organic polymers	Waste is mixed with a prepolymer and a catalyst that causes solidification through formation of a spongelike polymer matrix; urea-formaldehyde or vinyl ester-styrene polymers are used	Suitable for insoluble solids; very successful in limited limited applications	Pollutants arc not chemically bound, subject to leaching; strongly acidic leach water may be produced; requires special equipment and operators; some of the catalysts used are corrosive; harmful vapors may be produced; incompatible with oxidizers and some organics; some resins are biodegradable and decompose with time

Source: Reprinted, from U.S. Environmental Protection Agency, 1980, *Guide to the disposal of chemically stabilized and solidified waste,* (EPA SW–872, Cincinnati, Oh. [September]).

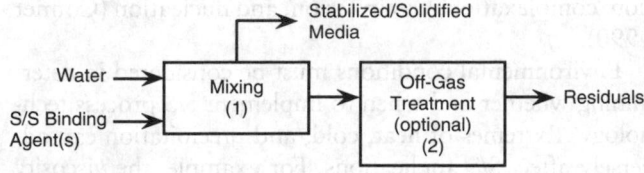

FIG. 11.16.2 Generic elements of a typical in situ S/S process. (Reprinted, from U.S. EPA, 1993.)

lection and treatment modules is specific to the S/S process. Also, hazardous residuals from some pretreatment technologies must be disposed of using appropriate procedures.

Technology Limitations

Tables 11.16.4 and 11.16.5 summarize factors that interfere with stabilization and solidification processes.

TABLE 11.16.3 COMPATIBILITY OF WASTE CATEGORIES WITH SOLIDIFICATION/STABILIZATION TECHNIQUES

Waste Component	Treatment Type						
	Cement-Based	Lime-Based	Thermoplastic	Organic Polymer (UF)[a]	Surface Encapsulation	Self-Cementing	Classification and Synthetic Mineral Formation
Organics							
1. Organic solvents and oils	Many impede setting, may escape as vapor	Many impede setting, may escape as vapor	Organics may vaporize on heating	May retard set of polymers	Must first be absorbed on solid matrix	Fire danger on heating	Wastes decompose at high temperatures
2. Solid organics (e.g., plastics, resins, tars)	Good–often increases durability	Good–often increases durability	Possible use as binding agent	May retard set of polymers	Compatible– many encapsulation materials are plastic	Fire danger on heating	Wastes decompose at high temperatures
Inorganics							
1. Acid wastes	Cement will neutralize acids	Compatible	Can be neutralized before incorporation	Compatible	Can be neutralized before incorporation	May be neutralized to form sulfate salts	Can be neutralized and incorporated
2. Oxidizers	Compatible	Compatible	May cause matrix breakdown, fire	May cause matrix breakdown	May cause deterioration of encapsulating materials	Compatible if sulfates are present	High temperatures may cause undesirable reactions
3. Sulfates	May retard setting and cause spalling unless special cement is used	Compatible	May dehydrate and rehydrate, causing splitting	Compatible	Compatible	Compatible	Compatible in many cases
4. Halides	Easily leached from cement, may retard setting	May retard setting, most are easily leached	May dehydrate	Compatible	Compatible	Compatible if sulfates are also present	Compatible in many cases
5. Heavy metals	Compatible	Compatible	Compatible	Acid pH solubilizes metal hydroxides	Compatible	Compatible if sulfates are present	Compatible in many cases
6. Radioactive materials	Compatible	Compatible	Compatible	Compatible	Compatible	Compatible if sulfates are present	Compatible

Source: Reprinted, from U.S. EPA, 1980.
[a]Urea-formaldehyde resin.

Physical mechanisms that interfere with the S/S process include incomplete mixing due to high moisture or organic chemical content. This results in only partial wetting or coating of particles with stabilizing and binding agents, and the aggregation of untreated waste into lumps (EPA 1986). Chemical mechanisms that interfere with S/S of cement-based systems include chemical adsorp-

tion, complexation, precipitation, and nucleation (Conner 1990).

Environmental conditions must be considered in determining whether and when to implement S/S process technology. Extremes of heat, cold, and precipitation can adversely affect S/S applications. For example, the viscosity of one or more of the materials may increase rapidly with

TABLE 11.16.4 SUMMARY OF FACTORS THAT MAY INTERFERE WITH STABILIZATION PROCESSES

Characteristics Affecting Processing Feasibility	Potential Interference
VOCs	Volatiles not effectively immobilized; driven off by heat of reaction. Sludges and soils containing volatile organics can be treated using a heated extruder evaporator or other means to evaporate free water and VOCs prior to mixing with stabilizing agents.
Use of acidic sorbent with metal hydroxide wastes	Solubilizes metal.
Use of acidic sorbent with cyanide wastes	Releases hydrogen cyanide.
Use of acidic sorbent with waste containing ammonium compounds	Releases ammonia gas.
Use of acidic sorbent with sulfide wastes	Releases hydrogen sulfide.
Use of alkaline sorbent (containing carbonates such as calcite or dolomite) with acid waste	May create pyrophoric waste.
Use of siliceous sorbent (soil, fly ash) with hydrofluoric acid waste	May produce soluble fluorosilicates.
Presence of anions in acidic solutions that form soluble calcium salts (e.g., calcium chloride acetate, and bicarbonate)	Cation exchange reactions—leach calcium from S/S product increases permeability of concrete, increases rate of exchange reactions.
Presence of halides	Easily leached from cement and lime.

Source: Reprinted, from United States Environmental Protection Agency, 1991, *Technical resources document on solidification/stabilization and its application to waste material (Draft),* (Contract No. 68–CO–003, Office of Research and Development, Cincinnati, Oh.).

falling temperature, or the cure rate may be unacceptably slowed.

Depending on the waste and binding agents involved, S/S processes can produce hot gases, including vapors that are potentially toxic, irritating, noxious to workers or communities downwind from the processes. In addition, if volatile substances with low flash points are involved, and the fuel-air ratio is favorable, there is a potential for fire and explosion.

Taking S/S processes from bench-scale to full-scale operation involves inherent uncertainties. Variables such as ingredient flow-rate control, material balance, mixing, materials handling and storage, and weather may all affect a field operation. These potential engineering difficulties emphasize the need for field demonstration before full-scale implementation.

Performance Testing

Treated wastes are subjected to physical tests to (1) determine particle size and distribution, porosity, permeability, and dry and wet density; (2) evaluate bulk-handling properties; (3) predict the reaction of a material to applied stresses in embankments and landfills; and (4) evaluate durability under freeze/thaw and wet/dry weathering cycles.

Chemical leach testing determines the chemical stability of treated wastes when in contact with aqueous solutions encountered in landfills. The procedures demonstrate the immobilization of contaminants by the S/S processes. Many techniques for leach testing are available. The major variables encountered in different leaching procedures are: the nature of the leaching solution; waste to leaching solution ratios; number of elutions of leaching solutions used; contact time of waste and leaching solution; surface area of waste exposed; and agitation technique used. Treated wastes must meet certain maximum leachate concentrations when subject to the Toxicity Characteristic Leaching Procedure (TCLP) determination (see Section 11.4).

—David H.F. Liu

References

Connor, J.R. 1990. *Chemical fixation and solidification of hazardous waste.* New York, N.Y.: Van Nostrand Reinhold.

U.S. Environmental Protection Agency (EPA). 1987. *Handbook—remedial action at waste disposal sites (rev.).* EPA 625–6–85–006. Washington, D.C. (January).

———. 1990. *Stabilization/solidification of CERCLA and RCRA wastes: physical test, chemical testing procedures, technology, and field activities.* EPA 625–6–89–022. Cincinnati, Oh. (May).

———. 1991. *Technical resources document on solidification/stabilization and its application to waste material (draft).* Contract No. 68–C0–0003. Office of Research and Development. Cincinnati, Oh.

TABLE 11.16.5 SUMMARY OF FACTORS THAT MAY INTERFERE WITH SOLIDIFICATION PROCESSES

Characteristics Affecting Processing Feasibility	Potential Interference
Organic compounds	Organics may interfere with bonding of waste materials with inorganic binders.
Semivolatile organics or poly-aromatic hydrocarbons (PAHs)	Organics may interfere with bonding of waste materials.
Oil and grease	Weaken bonds between waste particles and cement by coating the particles. Decrease in unconfined compressive strength with increased concentrations of oil and grease.
Fine particle size	Insoluble material passing through a No. 200 mesh sieve can delay setting and curing. Small particles can also coat larger particles, weakening bonds between particles and cement or other reagents. Particle size $>\frac{1}{4}$ inch in diameter not suitable.
Halides	May retard setting, easily leached for cement and pozzolan S/S. May dehydrate thermoplastic solidification.
Soluble salts of manganese, tin, zinc, copper, and lead	Reduced physical strength of final product caused by large variations in setting time and reduced dimensional stability of the cured matrix, thereby increasing leachability potential.
Cyanides	Cyanides interfere with bonding of waste materials.
Sodium arsenate, borates, phosphates, iodates, sulfides, and carbohydrates	Retard setting and curing and weaken strength of final product.
Sulfates	Retard setting and cause swelling and spalling in cement S/S. With thermoplastic solidification may dehydrate and rehydrate, causing splitting.
Phenols	Marked decreases in compressive strength for high phenol levels.
Presence of coal or lignite	Coals and lignites can cause problems with setting, curing, and strength of the end product.
Sodium borate, calcium sulfate, potassium dichromate, and carbohydrates	Interferes with pozzolanic reactions that depend on formation of calcium silicate and aluminate hydrates.
Nonpolar organics (oil, grease, aromatic hydrocarbons, PCBs)	May impede setting of cement, pozzolan, or organic-polymer S/S. May decrease long-term durability and allow escape of volatiles during mixing. With thermoplastic S/S, organics may vaporize from heat.
Polar organics (alcohols, phenols, organic acids, glycols)	With cement or pozzolan S/S, high concentrations of phenol may retard setting and may decrease short-term durability; all may decrease long-term durability. With thermoplastic S/S, organics may vaporize. Alcohols may retard setting of pozzolans.
Solid organics (plastics, tars, resins)	Ineffective with urea formaldehyde polymers; may retard setting of other polymers.
Oxidizers (sodium hypochlorite, potassium permanganate, nitric acid, or potassium dichromate)	May cause matrix breakdown or fire with thermoplastic or organic polymer S/S.
Metals (lead, chromium, cadmium, arsenic, mercury)	May increase setting time of cements if concentration is high.
Nitrates, cyanides	Increase setting time, decrease durability for cement-based S/S.
Soluble salts of magnesium, tin, zinc, copper and lead	May cause swelling and cracking within inorganic matrix exposing more surface area to leaching.
Environmental/waste conditions that lower the pH of matrix	Eventual matrix deterioration.
Flocculants (e.g., ferric chloride)	Interference with setting of cements and pozzolans.
Soluble sulfates >0.01% in soil or T50 mg/L in water	Endangerment of cement products due to sulfur attack.
Soluble sulfates >0.5% in soil or 2000 mg/L in water	Serious effects on cement products from sulfur attacks.
Oil, grease, lead, copper, zinc, and phenol	Deleterious to strength and durability of cement, lime/fly ash, fly ash/cement binders.
Aliphatic and aromatic hydrocarbons	Increase set time for cement.
Chlorinated organics	May increase set time and decrease durability of cement if concentration is high.
Metal salts and complexes	Increase set time and decrease durability for cement or clay/cement.
Inorganic acids	Decrease durability for cement (Portland Type I) or clay/cement.
Inorganic bases	Decrease durability for clay/cement; KOH and NaOH decrease durability for Portland cement Type III and IV.

Source: Reprinted from U.S. EPA, 1991.

11.17
BIOLOGICAL TREATMENT

PARTIAL LIST OF SUPPLIERS

Activated Sludge: Polybac Corp.; Detox Inc.; Ground Decontaminaton Systems

Rotating Biological Contactors: Polybac Corp.; Detox Inc.; Ground Decontaminaton Systems

Bioreclamation: FMC

Biological degradation of hazardous organic substances is a viable approach to waste management. Common processes are those originally utilized in treating municipal wastewaters, based on aerobic or anaerobic bacteria. In-situ treatment of contaminated soils can be performed biologically. Cultures used in biological degradation processes can be native (indigenous) microbes, selectively adapted microbes, or genetically altered microorganisms.

Table 11.17.1 shows that every class of anthropogenic compound can be degraded by some microorganism. Anthropogenic compounds such as halogenated organics are relatively resistant to biodegradation. One reason for this is the naturally present organisms often cannot produce the enzymes necessary to transform the original compound to a point where resultant intermediates can enter common metabolic pathways and be completely mineralized.

Several of the most persistent chlorinated compounds, such as TCE, appear to be biodegradable only through co-metabolism. Co-metabolism involves using another substance as a source of carbon and energy to sustain microbial growth. The contaminant is metabolized gratuitously due to a lack of enzyme specificity. To stimulate co-metabolism in bioremediation, a co-substrate is added to the

TABLE 11.17.1 EXAMPLES OF ANTHROPOGENIC COMPOUNDS AND MICROORGANISMS THAT CAN DEGRADE THEM

Compound	Organism
Aliphatic (nonhalogenated)	
Acrylonitrile	Mixed culture of yeast mold, protozoan bacteria
Aliphatic (halogenated)	
Trichloroethane, trichloroethylene, methyl chloride, methylene chloride	Marine bacteria, soil bacteria, sewage sludge
Aromatic compounds (nonhalogenated)	
Benzene, 2,6-dinitrotoluene, creosol, phenol	*Pseudomonas* sp, sewage sludge
Aromatic compounds (halogenated)	
1,2-; 2,3-; 1,4-Dichlorobenzene, hexachlorobenzene, trichlorobenzene	Sewage sludge
Pentachlorophenol	Soil microbes
Polycyclic aromatics (nonhalogenated)	
Benzo(a)pyrene, naphthalene	*Cunninghamells elogans*
Benzo(a)anthracene	*Pseudomonas*
Polycyclic aromatics (halogenated)	
PCBs	*Pseudomonas, Flavobacterium*
4-Chlorobiphenyl	Fungi
Pesticides	
Toxaphene	*Corynebacterium pyrogenes*
Dieldrin	Anacystic nidulans
DDT	Sewage sludge, soil bacteria
Kepone	Treatment lagoon sludge
Nitrosamines	
Dimethylnitrosamine	*Rhodopseudomonas*
Phthalate esters	Micrococcus 12B

Source: Reprinted, with permission from Table 1 of H. Kobayashi and B.E. Rittmann, 1982, Microbial removal of hazardous organic compounds, *Environmental Science and Technology,* (vol. 16, p. 173A).

contaminated site, to induce growth of microorganisms whose enzymes can degrade both the co-substrate and the original pollutant. Even inherently toxic inducers, such as phenol or toluene, are sometimes added to stimulate bacterial production of enzymes to degrade polyaromatic hydrocarbons and chlorinated aliphatics.

Table 11.17.2 lists important treatment data needs for biological treatments.

This section describes biological processes applicable to hazardous waste.

Aerobic Biological Treatment

DESCRIPTION

Hydrocarbons are catabolized or broken down into simpler substances by microorganisms using aerobic respiration, anaerobic respiration, and fermentation. In general, aerobic degradation processes are more often used for biodegradation because the degradation process is more rapid and more complete, and problematic products such as methane and hydrogen sulfide are not produced. However, anaerobic degradation is important for dehalogenation.

In aerobic respiration, organic molecules are oxidized to carbon dioxide (CO_2), water, and other end products using molecular oxygen as the terminal electron acceptor. Oxygen is also incorporated into the intermediate products of microbial catabolism through oxidase enzyme action, making these products more susceptible to further biodegradation. Microorganisms metabolize hydrocarbons by anaerobic respiration in the absence of molecular oxygen using inorganic substrates as terminal electron acceptors. Naturally occurring aerobic bacteria can decompose natural and synthetic organic materials to harmless or stable forms by mineralizing them to CO_2 and water. Some anthropogenic compounds appear refractory to biodegradation by naturally occurring microbial populations because of environmental influences, lack of solubility, and the absence of required enzymes, nutrients or other factors. However, properly selected or engineered micro-

bial populations, maintained under environmental conditions conducive to their metabolic activity are an important means of biological transformation or degrading these otherwise refractory wastes.

All microorganisms require adequate levels of inorganic and organic nutrients, growth factors (vitamins, magnesium, copper, manganese, sulfur, potassium, etc.), water, oxygen, carbon dioxide and sufficient biological space for survival and growth. One or more of these factors is usually in limited supply. In addition, various microbial competitors adversely affect each other in struggling for these limited resources. Other factors influencing microbial degradation rates include microbial inhibition by chemicals in the waste to be treated, the number and physiological state of the organisms as a function of available nutrients, the seasonal state of microbial development, predators, pH, and temperature. Interaction between these and other potential factors can cause wide variations in degradation kinetics.

For these and other reasons, aerobic biodegradation is usually carried out in processes where many of the requisite conditions can be controlled. Such processes include conventional activated-sludge processes, with modifications such as sequencing batch reactors, and aerobic-attached growth biological processes such as rotating biological contactors and trickling filters. Recently developed genetically engineered bacteria are reported to be effective for biological treatment of specific, relatively uniform, hazardous wastes.

APPLICABILITY/LIMITATIONS

Used to treat aqueous waste contaminated with low levels (e.g., BOD less than 10,000 mg/L) of nonhalogenated organic and certain halogenated organics. Treatment requires consistent, stable operating conditions.

Status.   Conventional, broadly used technology

Activated Sludge

DESCRIPTION

Activated sludge treatment breaks down organic contaminants in aqueous waste streams through aerobic microorganisms' activity. These microorganisms metabolize biodegradable organics. This treatment includes conventional activated sludge processes and modifications such as sequencing batch reactors. The aeration process includes pumping the waste to an aeration tank where biological treatment occurs. Following this, the stream is sent to a clarifier where the liquid effluent (the treated aqueous waste) is separated from the sludge biomass (Figure 11.17.1). Aerobic processes can significantly reduce a wide range of organic, toxic and hazardous compounds. However, only dilute aqueous wastes (less than 1%) are normally treatable.

TABLE 11.17.2   IMPORTANT BIOLOGICAL TREATMENT DATA NEEDS

Data Need	Purpose
Gross Organic Component (BOD, TOC)	Treatability
Priority Pollutant Analysis	Toxicity to Process Microbes
Dissolved Oxygen	Aerobic Reaction Rates/ Interference with Anaerobic System
Nutrient Analysis (NH_3, NO_3, PO_4, etc.)	Nutrient Requirements
pH	pH Adjustment
ORP	Chemical Competition

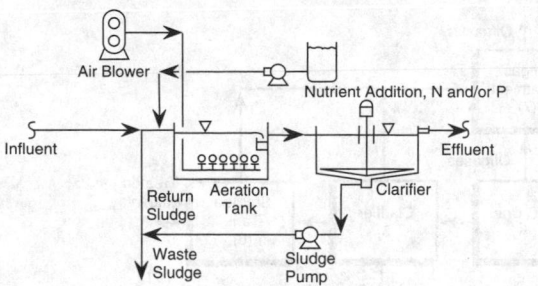

FIG. 11.17.1 Activated sludge process.

APPLICABILITY/LIMITATIONS

The treatment requires consistent, stable operating conditions. Activated sludge processes are not suitable for removing highly chlorinated organics, aliphatics, amines, and aromatic compounds from waste streams. Some heavy metals and organic chemicals are harmful to the organisms. When using conventional open aeration tanks and clarifiers, volatile hazardous materials may escape.

Status. Conventional, well developed

Rotating Biological Contactors

DESCRIPTION

Rotating biological contactors (RBCs) aerobically treat aqueous waste streams, especially those containing alcohols, phenols, phthalates, cyanides, and ammonia. Primary treatment (e.g., clarifiers or screens) to remove materials that could settle in RBC tanks or plug the discs, is often essential for good operation. Influents containing high concentrations of floatables (e.g., grease) require treatment with a primary clarifier or an alternate removal system (EPA 1984; EPA 1992).

A typical RBC unit consists of 12-ft-dia plastic discs mounted along a 25-ft horizontal shaft. The disc surface is normally 100,000 sq ft for a standard unit and 150,000

sq ft for a high density unit. Figure 11.17.2 details a typical RBC system.

As the discs rotate through leachate at 1.5 rpm, a microbial slime forms on the discs. These microorganisms degrade the organic and nitrogenous contaminants present in the waste stream. During rotation, about 40% of the discs' surface area is in contact with the aqueous waste, while the remaining area is exposed to the atmosphere. The rotation of the media through the atmosphere causes oxygenation of the attached organisms. When operated properly, the shearing motion of the discs through the aqueous waste causes excess biomass to shear off at a steady rate. Suspended biological solids are carried through the successive stages before entering the secondary clarifier.

The RBC treatment process involves a number of steps as indicated in Figure 11.17.3. Typically, aqueous waste is transferred from a storage or equalization tank (1) to a mixing tank (2) where chemicals are added for metal precipitation, nutrient adjustment, and pH control. The waste stream then enters a clarifier (3) where solids are separated from the liquid. The clarifier effluent enters the RBC (4) where the organics and/or ammonia are converted to innocuous products. The treated waste is then pumped into a second clarifier (5) for removal of biological solids. After secondary clarification, the effluent enters a storage tank (6) where, depending upon remaining contamination, the waste may be stored pending additional treatment or discharged to a sewer system or surface stream. Throughout this process, offgases should be collected for treatment (7).

In addition to maximizing the system's efficiency, staging can improve the system's ability to handle shock loads by absorbing the impact in stages. Staging, which employs a number of RBCs in series, enhances biochemical kinetics and establishes selective biological cultures acclimated to successively decreasing organic loading. As the waste stream passes from stage to stage, progressively increasing levels of treatment occur.

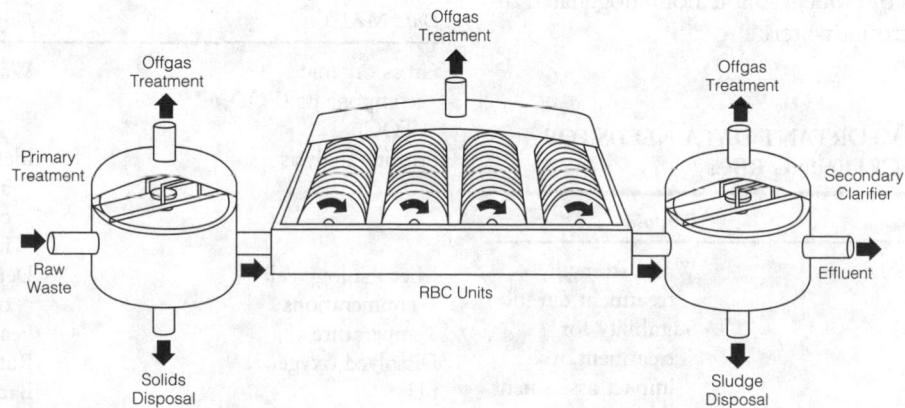

FIG. 11.17.2 Typical RBC plant schematic. (Reprinted, from U.S. Environmental Protection Agency (EPA), 1992, *Engineering Bulletin: rotating biological contactors*, [EPA 540–5–92–007, October].)

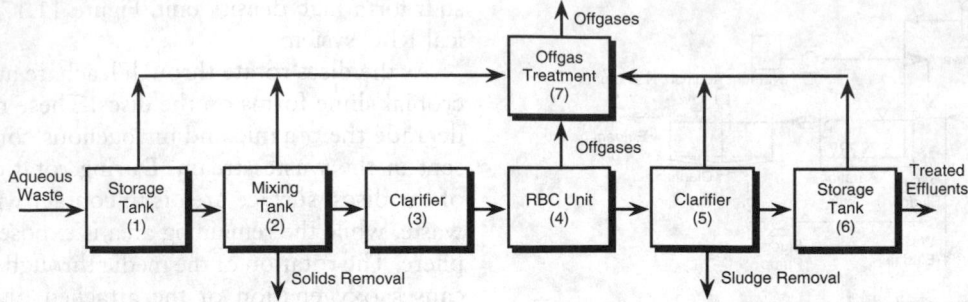

FIG. 11.17.3 Block diagram of the RBC treatment process. (Reprinted, from U.S. EPA, 1992.)

Factors effecting the removal efficiency of RBC systems include the type and concentration of organics present, hydraulic residence time, rotational speed, media surface area exposed and submerged, and pre- and post-treatment activities. See Section 7.24 on the design of RBCs.

APPLICABILITY/LIMITATIONS

Rotating biological contacts are not sufficient to remove highly chlorinated organics, aliphatics, amines or aromatic compounds. Some heavy metals and organic chemicals are harmful to the organisms.

Table 11.17.3 lists the important data needed for screening RBCs.

Status.   Conventional process

Bioreclamation
DESCRIPTION

Bioreclamation uses aerobic microbial degradation in treating contaminated areas. It is used for in-situ treatment using injection/extraction wells or excavation processes. Extracted water, leachates or wastes are oxygenated, nutrients and bacteria are added, and the liquids are reinjected into the ground. Bacteria can then degrade wastes still in the soil. This treatment has successfully reduced the contamination levels of biodegradable nonhalogenated organics in soils and groundwater.

TABLE 11.17.3   IMPORTANT DATA NEEDS FOR SCREENING RBCs

Data Need	Purpose
Gross organic components (BOD, TOC)	Waste strength, treatment duration
Priority pollutant analyses (organics, metals, pesticides, CN, phenols)	Suitability for treatment, toxic impact assessment
Influent temperature	Feasibility in climate

APPLICABILITY/LIMITATIONS

For in-situ treatment, limitations include site geology and hydrogeology restricting waste pumping and extraction, along with reinjection and recirculation. Ideal soil conditions are neutral pH, high permeability and a moisture content of 50–75%. Biological treatment systems are used to treat soils contaminated with pentachlorophenol, creosote, oils, gasoline, and pesticides.

Table 11.17.4 lists important bioreclamation data needs.

Status.   Demonstrated process

Anaerobic Digestion
DESCRIPTION

All anaerobic biological treatment processes reduce organic matter, in an oxygen-free environment, to methane and carbon dioxide. This is accomplished using bacteria cultures, including facultative and obligate anaerobes. Anaerobic bacterial systems include:

- Hydrolytic bacteria (catabolized saccharide, proteins, lipids)

TABLE 11.17.4   IMPORTANT BIORECLAMATION DATA NEEDS

Data Need	Purpose
Gross organic components (BOD, TOC)	Waste strength, treatment duration
Priority analysis	Identify refractory and biodegradable compounds, toxic impact
Microbiology cell enumerations	Determine existence of dominant bacteria
Temperature	Feasibility in climate
Dissolved oxygen	Rate of reaction
pH	Bacteria preference
Nutrient analysis NH$_3$, NO$_3$, PO$_4$, etc.	Nutrient requirements

- Hydrogen-producing acetogenic bacteria (catabolized products of hydrolytic bacteria, e.g., fatty acids and neutral end products)
- Homolactic bacteria (catabolized multicarbon compounds to acetic acid)
- Methanogenic bacteria (metabolized acetic acid and higher fatty acids to methane and carbon dioxide)

Figure 11.17.4 is a schematic diagram of biological reaction in an anaerobic system.

Strict anaerobics require totally oxygen-free environments and an oxidation reduction potential of less than −0.2 V. Microorganisms in this group are known as methanogenic consortia and are found in anaerobic sediments or sewage sludge digesters. These organisms play an important role in reductive dehalogenation reactions, nitrosamine degradation, reduction of epoxides to olefins, reduction of nitro-groups, and ring fission of aromatic structures.

Available anaerobic treatment concepts are based on approaches such as the classic well-mixed system, the two-stage system and the fixed bed system.

In a well-mixed digester system, a single vessel is used to contain the wastes being treated and all bacteria must function in that common environment. Such systems typically require long retention times, and the balance between acetogenic and methanogenic populations is easily upset.

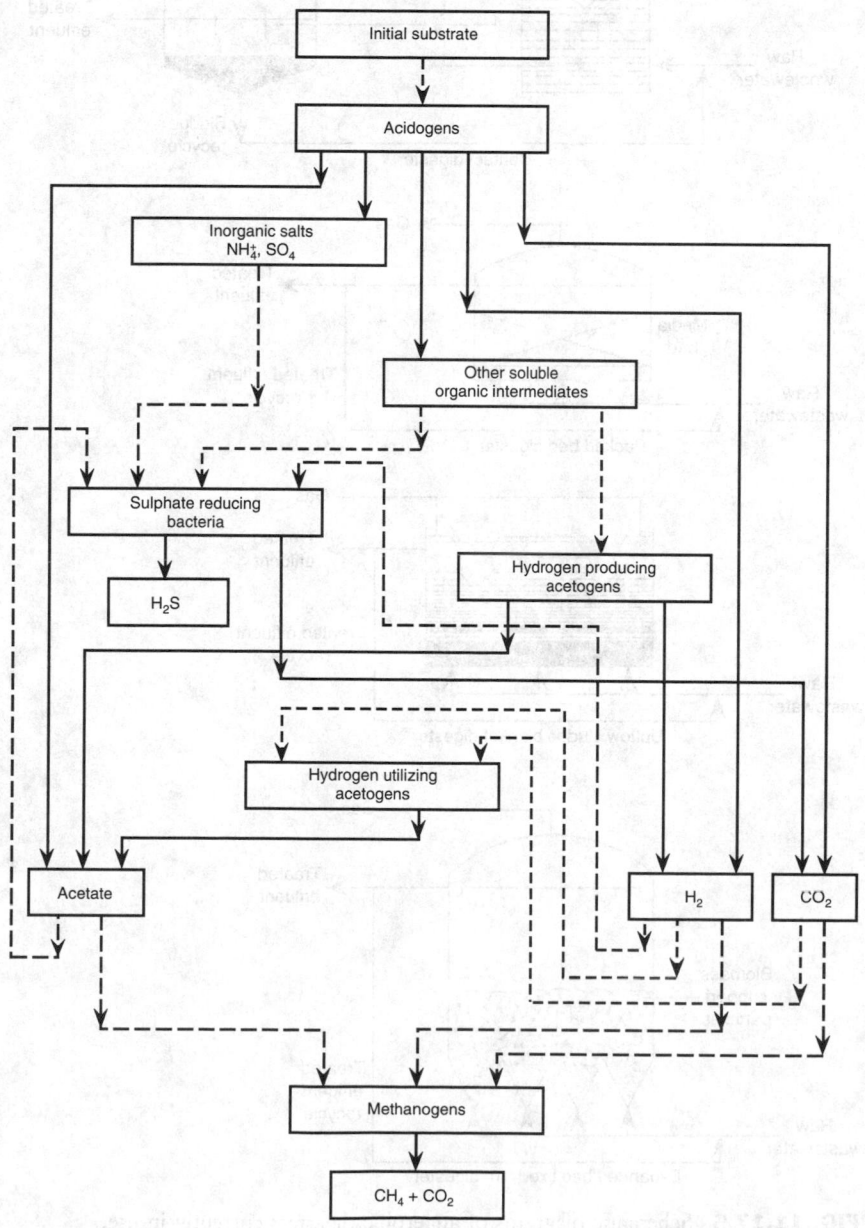

FIG. 11.17.4 Schematic diagram of biological reactors in anaerobic systems.

In the two-stage approach, two vessels are used to maintain separate environments, one optimized for acetogenic bacteria (pH 5.0), and the other optimized for methanogenic bacteria (pH 7.0). Retention times are significantly lower and upsets are uncommon in this approach.

The fixed bed approach (for single- or 2-staged systems) utilizes an inert solid media to which the bacteria attach. Low solids wastes are pumped through columns of the bacteria-rich media. Use of such supported cultures allows reduced retention times since bacterial loss through washout is minimized. Organic degradation efficiencies can be quite high.

A number of proprietary engineered processes based on these types of systems are being actively marketed. Each has distinct features, but all utilize the fundamental anaerobic conversion to methane and carbon dioxide (Figure 11.17.5).

APPLICABILITY/LIMITATIONS

This process is used to treat aqueous waste with low to moderate levels of organics. Anaerobic digestion can handle certain halogenated organics better than aerobic treatment. Stable, consistent operating conditions must be

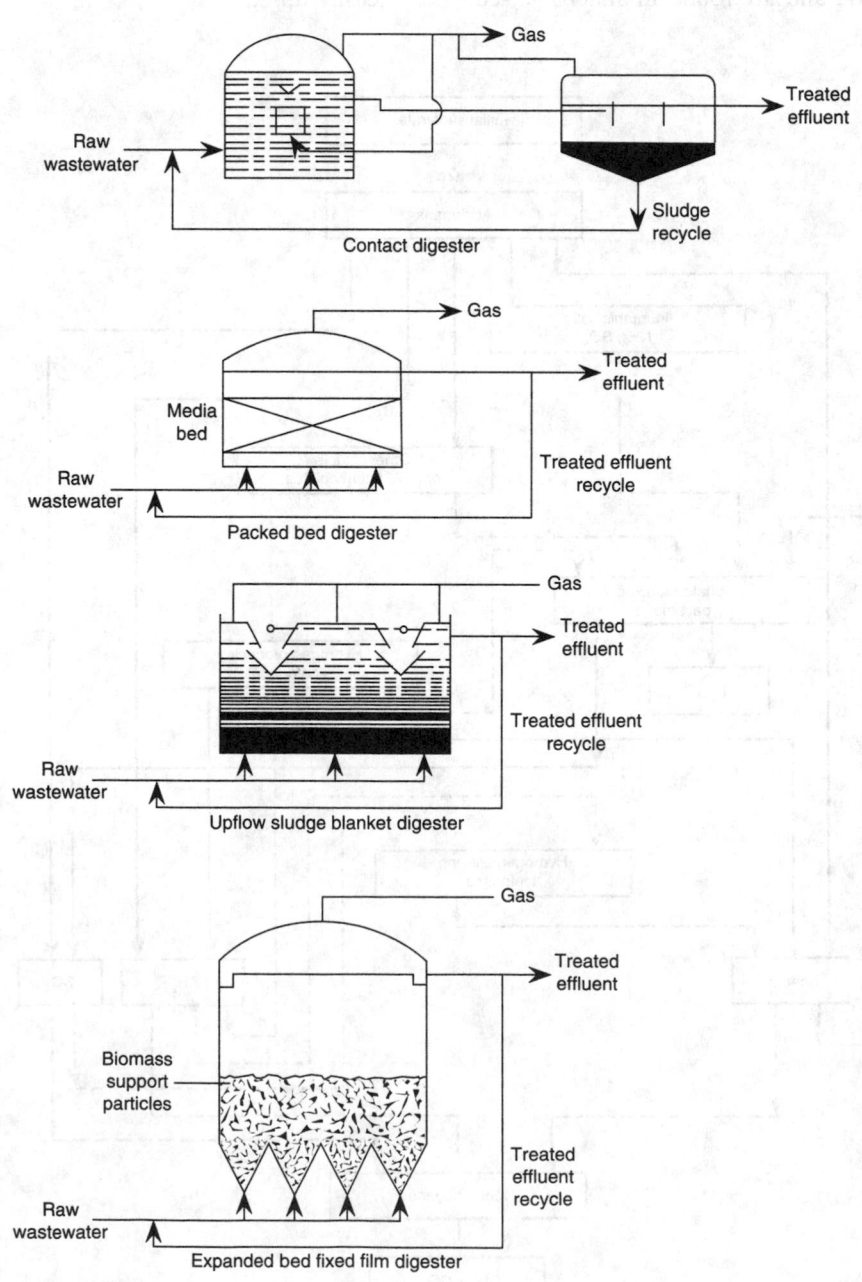

FIG. 11.17.5   Schematic diagrams of anaerobic digesters currently in use.

maintained. Anaerobic degradation can take place in native soils, but in a controlled treatment process, an air-tight reactor is required. Since methane and carbon dioxide gases are formed, it is common to vent the gases or burn them in flare systems. However, volatile hazardous materials could readily escape via such gas venting and flare systems. Thus controlled off-gas burning could be required. Depending upon the nature of the waste to be treated, the off-gas could be used as a source of energy.

Status.   Available and widely used in POTWs.

White-Rot Fungus

DESCRIPTION

Lignin-degrading white-rot fungus (phanerochaete chrysosporium) degrades a broad spectrum of organopollutants, including chlorinated, lignin-derived by-products of the Kraft pulping process. White-rot fungus degrades aliphatic, aromatic, and heterocyclic compounds. Specifically, white-rot fungus has been shown to degrade indane, benzo(a)-pyrene, DDT, TCDD, and PCBs to innocuous end products. The studies performed, to date, suggest that the white-rot fungus may prove to be an extremely useful microorganism in the biological treatment of hazardous organic waste.

Note: Certain plants, such as specific strains of *Brassica* (mustards), accumulate heavy metals when growing in metal contaminated soils, forming the basis for a process called phytoremediation. These plants can accumulate up to 40% of their biomass as heavy metals, including lead [Atlas 1995].

APPLICABILITY/LIMITATIONS

This technology is in the development phase and has been applied only in laboratory environs.

Status.   Demonstrated on laboratory scale

—*David H.F. Liu*

References

Atlas, R.M. 1995. Bioremediation. *C&E News* (3 April).
U.S. Environmental Protection Agency (EPA). 1984. *Design information on rotating biological contactors.* EPA 600–2–84–106 (June).
———. 1992. *Engineering Bulletin: rotating biological contactors.* EPA 540–S–92–007 (October).

11.18
BIOTREATMENT BY SEQUENCING BATCH REACTORS

FEATURE SUMMARY

Type of Process: Biological treatment of liquid hazardous wastewaters.

Type of Reactor: Sequencing Batch Reactor (SBR), a fill-and-draw, activated sludge-type system where aeration and settling occur in the same tank.

Type of Aeration and Mixing Systems: Jet-aeration systems are common and allow mixing either with or without aeration; however, other aeration and mixing systems are used.

Type of Decanters: Most decanters, including some which are patented, float or otherwise maintain inlet orifices slightly below the water surface to avoid removal of both settled and floating solids.

Type of Tanks: Steel tanks, appropriately coated for corrosion control, are most common; however, concrete and other materials may be used. Concrete tanks are favored for municipal treatment of domestic wastewaters.

Partial List of Suppliers of SBR Equipment: Aqua-Aerobics Systems; Austgen-Biojet; Bioclear Technology; Envirodyne Systems; Fluidyne; Jet Tech; Mass Transfer; Purestream; Transenviro

The Sequencing Batch Reactor (SBR) is a periodically operated, activated sludge-type, dispersed-growth, biological wastewater treatment system. Both biological reactions and solids separation are accomplished in a single reactor, but at different times during a cyclic operation. In comparison, continuous flow activated sludge systems use two reactors, one dedicated to biological reactions and the other dedicated to solids separation. Once constructed, these two-tank systems offer little flexibility because changing reactor size is difficult. However, the SBR is flexible, as the time dedicated to each function can be adjusted. For example, reducing the time dedicated to solids separation provides additional time for the completion of biological reactions. Other advantages of the SBR system are described below, along with a description of the SBR operating cycle. The SBR originally was developed to treat domestic wastewater and is now used to effectively treat industrial and other organic wastewaters containing hazardous substances.

Process Description

Influent wastewater is added to a partially filled reactor. The partially filled reactor contains biomass acclimated to the wastewater constituents during preceding cycles. Once the reactor is full, it behaves like a conventional activated sludge aeration basin, but without inflow and outflow. After biological reactions are completed, and aeration and mixing is discontinued, the biomass settles and the treated supernatant is removed. Excess biomass is wasted at any convenient time during the cycle. Frequent wasting results in holding the mass ratio of influent substrate to biomass nearly constant from cycle to cycle. In contrast, continuous flow systems hold the mass ratio of influent substrate to biomass constant by adjusting return sludge flow rates continually as influent flow rates and characteristics and settling tank underflow concentrations vary.

No specific SBR reactor shape is required. The width-to-length ratio is unimportant, although this is a concern with conventional continuous flow systems. Deep reactors improve oxygen transfer efficiency and occupy less land area. The SBR shown in Figure 11.18.1 uses an egg-shaped reactor that offers most of the advantages of a spherical reactor, and provides a deeper reactor. Along with improved oxygen transfer efficiency, deep reactors allow a higher fraction of treated effluent removal during decanting. Similar to a spherical reactor, the egg-shaped reactor has a minimum surface area to volume ratio resulting in lower heat loss, less material needed in reactor construction, and less energy required for mixing.

The small reactor top is easily enclosed to contain volatile organics, or to direct exhaust gases for removal in an absorber. During filling, floating materials are forced together towards the top center for easy removal. The con-verging bottom improves thickening of the settled solids, reduces sediment accumulation, and allows easy solids removal from a single center point.

The egg-shaped reactor is constructed as a single piece, eliminating the need for special reinforcement at static tension points or seams. The bottom section of the egg is buried in the ground, supporting the reactor without a special foundation.

Modes of Operation

The illustration of an SBR in Figure 11.18.1 includes five discrete periods of time, with more than one operating strategy possible during any time period. The five discrete periods are defined as:

1. *Idle*: waiting period
2. *Fill*: influent is added
3. *React*: biological reactions are completed
4. *Settle*: solids separate from treated effluent
5. *Draw*: treated effluent is removed

Each discrete period is detailed in the following sections.

IDLE

Idle is considered the beginning of the cycle, although there exists no true beginning after the initial start-up. Idle occurs between draw, the removal of treated effluent, and fill, the beginning of influent addition. Idle time may be short or long depending on flow rate variation and operating strategy. When the influent flow rates are constant and predictable, and when flow equalization is provided upstream, idle is nearly eliminated. Idle time is long during periods of low influent flow, and short during periods of high influent flow. During operations with variable idle times, the SBR also provides flow equalization. Idle can also be used to accomplish other functions, such as sludge wasting, and mixing to condition the biomass to a low substrate concentration.

STATIC, MIXED AND AERATED FILL

Developing an operating strategy for the fill period is a complex problem for designers of hazardous wastewater SBR systems. Domestic wastewater treatment systems seldom require laboratory treatability studies, because these systems follow conservative design approaches and municipal wastewater flow rates and characteristic variations are predictable. Laboratory treatability studies are almost always needed to design SBRs and to select the appropriate fill policy for hazardous wastes. The following describes alternative fill policies that must be developed during treatability studies.

SBR influent may require pretreatment. The decision to provide screening and degritting is made on the same basis used by designers of conventional continuous flow

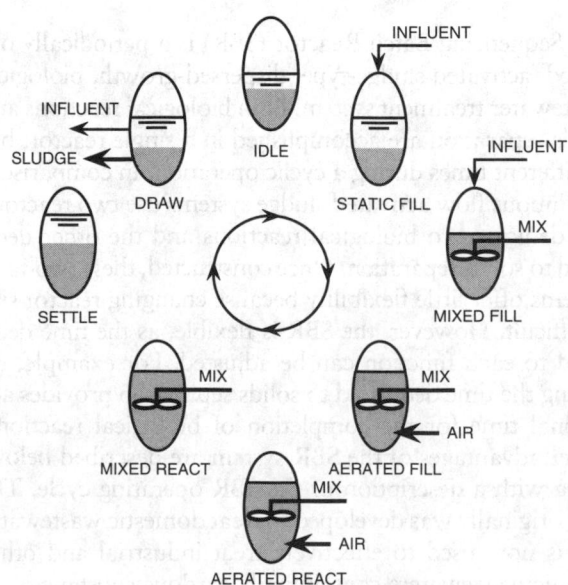

FIG. 11.18.1 Illustration of one cycle in a single SBR.

plants. Upstream flow equalization allows for rapid fill rates (i.e., higher than influent flow rates). This results in reduced cycle times and reduced reactor size. The use of rapid fill periods also results in the accumulation of high substrate concentrations. Upstream flow equalization is not necessary because idle is a normal SBR function. However, upstream flow equalization may be selected, allowing wastewaters with highly variable characteristics, or from more than one process to be blended for more uniformity. In addition, high concentrations resulting from spill events can be caught and kept from interfering with the biological process. Flow equalization basins are often included in SBR systems because they are inexpensive, provide added system flexibility, and reduce or eliminate idle time.

Static Fill

Static fill introduces influent wastewater into the SBR with little or no mixing and contact with the settled biomass, resulting in a high substrate concentration when mixing first begins. High substrate concentrations result in high reaction rates. In addition, such concentrates control sludge bulking because they favor organisms that form more dense floc particles over filament forming organisms. Bulking sludge is a common problem in continuous flow systems where substrate concentrations are always low. Finally, static fill favors organisms that produce internal storage products during high substrate conditions, a requirement for biological phosphorous removal. However, static fill time should be limited if an influent biodegradable constituent is present at concentrations toxic to the organisms.

Mixed Fill

Mixed fill begins the biological reactions by bringing influent organics into contact with the biomass. Mixing without aeration reduces the energy needed for aeration, because some organics are biologically degraded using residual oxygen or alternative electron acceptors, such as nitrate-nitrogen. If nitrate-nitrogen is the electron acceptor, a desirable denitrification reaction occurs. The period when alternative electron acceptors are present and oxygen is absent is called *anoxic*. Anaerobic conditions develop after all electron acceptors are consumed, and fermentation reactions may occur. Thus, in a single reactor, aerobic, anoxic and anaerobic treatment conditions exist by correctly varying mixing and aeration policies during fill.

Aerated Fill

Aerated fill begins the aerobic reactions that are later completed during react. Aerated fill reduces cycle time because the aerobic reactions occur during the fill period. In some cases, a biodegradable influent constituent may be present in concentrations that are toxic to the organisms. When that condition exists, aeration during fill begins early to limit concentration of that constituent. For example, if the wastewater constituent is toxic at 10 mg/L, and present in the influent at a concentration of 30 mg/L, aeration should begin prior to the reactor becoming three-quarters full, if the SBR liquid volume at the end of draw was one-half the volume at the end of fill. This assumes toxic constituent degradation is at a rate greater than or equal to its rate of addition. If the rate of degradation is low, then aeration should begin earlier, or a higher volume of treated effluent should be held in the SBR for more dilution.

REACT

Aeration is often provided during the fill react period to complete the aerobic reactions. However, aerated react alternated with mixed react will provide alternating periods of aerobic and anoxic, or even anaerobic, conditions. This is a normal procedure for nitrification and denitrification. During periods of aeration, nitrate concentration increases as organic nitrogen and ammonia are converted to nitrites and nitrates. The mixed react results in anoxic conditions needed for denitrification and the conversion of nitrates to nitrogen gas. Anaerobic conditions are necessary if some waste constituents are degraded only anaerobically, or partially degraded anaerobically followed by a complete degradation under aerobic conditions.

For mixed wastes, the easily degraded constituents are removed first, and the more difficult to degrade constituents are removed later during extended periods of aeration. Long periods of aerated react, after removal of soluble substrates, may be necessary to condition the biomass, to remove internal storage products, or to aerobically digest the biomass. Aerated react may also be stopped soon after the soluble substrate is removed. This saves energy and maximizes sludge production, which is desirable when separate anaerobic sludge digestion is used to stabilize these waste solids and to produce methane, an energy-rich and useful by-product.

SETTLE

Settle is normally provided under quiescent conditions in the SBR; however, gentle mixing during the early stages of settle may result in both a clearer effluent and a more concentrated settled biomass. Unlike continuous flow systems, settle occurs without inflow or outflow, and the accompanying currents that interfere with settling.

DRAW

The use of a floating decanter, or a decanter that moves downward during draw, offers several advantages. Draw is initiated earlier because the effluent is removed from near the surface while the biomass continues to settle at lower depths. The effluent is removed from a selected depth below the surface by maintaining outlet orifices or slots at a fixed depth below the variable water surface. This avoids removal of floating materials and results in effluent removal from high above the settled biomass. Floating decanters allow maximum flexibility, because fill and draw volumes can be varied from time to time, or even from cycle to cycle. However, lower-cost fixed-level decanters can be used if the settle period is extended, to assure that the biomass has settled below the decanter orifices. Fixed-level decanters can be made somewhat more flexible, if they are designed to allow operators to occasionally lower or raise the location of the decanter.

Rapid draw rates allow use of smaller reactors, but cause high surges of flow in downstream units and in receiving waters. Effluent flow equalization tanks or reduced draw rates will reduce peak flow discharges.

Figure 11.18.2 illustrates the hydraulic conditions over two days in a three-tank SBR system under design flow conditions. The illustration shows influent flow into Tank 1 beginning at 6:00 A.M. The treatment strategy provides a static fill (Fs) for 1.67 hr followed by an aerated fill for 1.0 hr. The influent is diverted to Tank 2 at 8:40 A.M., and to Tank 3 at 11:20 A.M. In Tank 1, a 2.33 hr react is followed by a 1.0 hr settle and a 1.0 hr draw. Tank 1 idle occurs from 1:00 P.M. to 2:00 P.M. At 2:00 P.M. the influent is again diverted to Tank 1, and the cycle is repeated. As shown in Figure 11.18.2, each tank cycle is the same, with an 8-hr cycle divided as follows: fill 2.67 hr (i.e., static fill, 1.67 hr and aerated fill, 1.00 hr), react 2.33 hr, settle 1.0 hr, draw 1.0 hr, and idle 1.0 hr. Total aeration time is 3.33 hr in both aerated fill and react.

The shaded areas of the illustration show that influent flows continuously into one of the three tanks. Under design conditions, flow occurs at a constant rate and every cycle is identical. Effluent is not continuous, and is illustrated by the cross-hatched areas. In this design flow example, each cycle fill time is 2.67 hr and draw time is 1.0

hr, resulting in an effluent flow rate equal to 2.67 times the influent flow rate.

Under actual flow conditions, diurnal flow rates vary. Typically, flow rates increase throughout the morning, peak in the early afternoon, and decrease later in the day with minimum flow in the early morning hours. Figure 11.18.3 illustrates a three-tank SBR system with a typical diurnal flow variation. Note the short fill and idle periods during high flow rate times, and the long fill and idle periods during low flow rate times. During peak flow, no idle period exists (e.g., about 2:00 P.M. of Day 2 in Tank 1).

Laboratory Treatability Studies

Most SBR systems for treatment of industrial wastewaters, especially those containing hazardous wastes, must be designed based on treatability studies. This section outlines the procedures and equipment needed to perform these studies. Figure 11.18.4 illustrates a laboratory SBR system.

The reactors are 1-L to 4-L *reaction kettles*, typically covered to control volatile organics. The small reactors require less influent wastewater and effluent disposal, and the larger reactors allow collection of larger sample volumes. A gas collection tube, containing activated carbon or other organic absorbent, prevents the escape of volatile organics and measures the extent of organic volatilization. Two gas collection tubes are connected in series to prevent volatile organics in laboratory air from contributing to those released from the reactor. The reactor can be mixed with or without aeration using a magnetic stirrer and a star-head stir bar. Air is supplied through a diffuser from a laboratory air supply. A gas collection tube containing an organic absorbent may be used to prevent organics in the supply air from entering the SBR (not illustrated). To minimize evaporation during SBR aeration, a water humidifier is included in the air stream. Peristaltic tubing pumps add influent and remove effluent from the

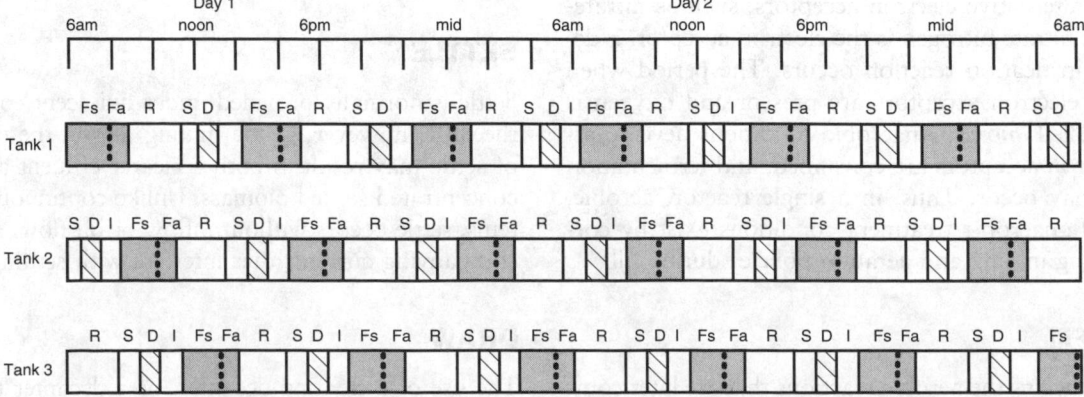

FIG. 11.18.2 Design flow conditions (i.e., constant flow rate). Illustrations show static fill (Fs), aerated fill (Fa), react (R), settle (S), draw (D), and idle (I) for a three-tank system. Fill periods are shaded to demonstrate that inflow continuously occurs into one of the three tanks, and draw periods are cross-hatched to show the periodic nature of effluent flow.

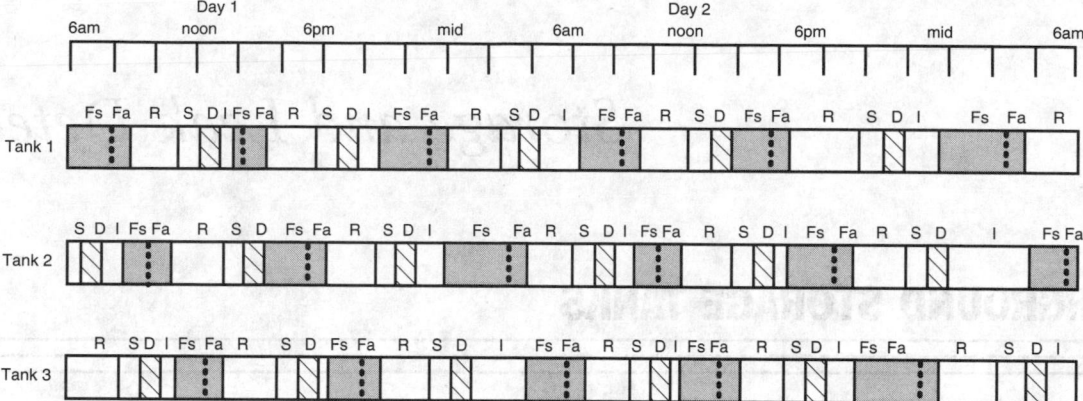

FIG. 11.18.3 Typical flow conditions (i.e., diurnal flow rate variation). Illustrations show static fill (Fs), aerated fill (Fa), react (R), settle (S), draw (D), and idle (I) for a three-tank system. Fill periods are shaded to demonstrate that inflow continuously occurs into one of the three tanks, and draw periods are crosshatched to show the periodic nature of effluent flow.

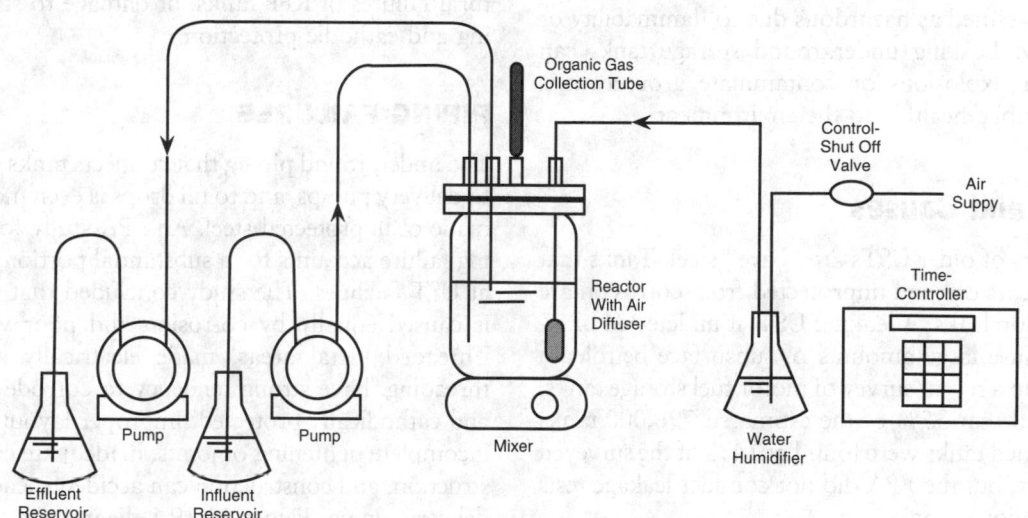

FIG. 11.18.4 Illustration of a laboratory SBR system.

SBR. Liquid level switches can be used to control the fill and draw volume, or, as an alternative, suction tubes can be placed at an elevation to prevent pumping excessive volumes. The suction tube alternative will result in too little addition or removal if minor changes occur in the pumping rates, but this is minimized by frequent monitoring and occasional manual over-ride at the end of a pumping cycle, or simply recorded to reflect a slightly different loading rate. System control is provided with a microprocessor-based timer and controller that turns pumps on and off, opens and closes the air shut-off valve, and turns the mixer on and off at appropriate times.

Laboratory studies include a start-up period to develop a biomass enriched for organisms acclimated to the wastewater constituents. Start-up time is minimized by including these organisms in the initial seed biomass. Aeration of some wastewaters alone will result in an accumulation of a suitable biomass. At other times, activated sludge collected from nearby municipal or industrial wastewater treatment plants may be needed. Finally, for difficult to degrade wastes, organisms are taken from: soil samples collected from waste spill sites; sediments collected from nearby receiving waters; or from residues in contact with waste constituents for long periods of time. During start-up, fill and draw periods are controlled manually after substrate reduction is observed. Once an enriched and acclimated biomass is developed, automatic SBR operation is used to determine the appropriate operating strategy, as described above under the explanation of the different phases of the SBR cycle.

—Lloyd H. Ketchum, Jr.

Storage and Leak Detection

11.19
UNDERGROUND STORAGE TANKS

The terms *underground storage tanks, USTs,* and *UST systems* include underground storage tank vessels, and the connected underground piping. Thus, above–ground tanks with extensive underground piping may also be regulated under RCRA Subtitle I. Usually associated with gasoline service stations, these tanks are also used to store materials that are classified as hazardous due to flammability or combustibility. Leaking underground storage tanks can cause fires or explosions or contaminate groundwater, threatening public health and the environment.

Problems and Causes

Large numbers of older USTs are "bare" steel. Tanks that are over 10 years old and unprotected from corrosion are likely to develop leaks. A leaking UST, if undetected or ignored, can cause large amounts of subsurface petroleum product loss. In a recent survey of motor fuel storage tanks, the EPA found that 35% of the estimated 796,000 tanks leak. Abandoned tanks were found at 14% of the surveyed establishments, but the EPA did not conduct leakage tests on these abandoned tanks.

Underground storage tanks release contaminants into the subsurface environment because of one or more of the following factors:

GALVANIC CORROSION

The most common failure of underground tank systems is galvanic corrosion of the tank or piping. Corrosion can be traced to failure of corrosion protection systems due to pinholes in the coating or taping, depletion of sacrificial anodes, corrosion from the inside due to the stored product, and various other reasons. Many corrosion–related leaks are found in systems that have no corrosion protection at all. However, tanks with corrosion protection can also corrode if the protective coating is damaged during installation; the sacrificial anodes are not replaced when required; the current is switched off in impressed current systems; or the protection system is not designed properly for the soil condition and stored liquids.

FAULTY INSTALLATION

Installation failures include inadequate backfill, allowing movement of the tank, and separation of pipe joints. These tanks receive a substantial portion of structural support from backfill and bedding. Mishandling can cause structural failures of RFP tanks, or damage to steel tank coating and cathodic protection.

PIPING FAILURES

The underground piping that connects tanks to each other, to delivery pumps, and to fill drops is even more frequently made of unprotected steel. An EPA study found that piping failure accounts for a substantial portion of large spills at UST facilities. The study concluded that piping failure is caused equally by corrosion and poor workmanship. Threaded metal areas, made electrically active by the threading, have strong tendency to corrode if not coated and cathodically protected. Improper layout of pipe runs, incomplete tightening of joints, inadequate cover pad construction, and construction can accidents lead to failure of delivery piping. Figure 11.19.1 diagrams a typical service station tank and piping layout.

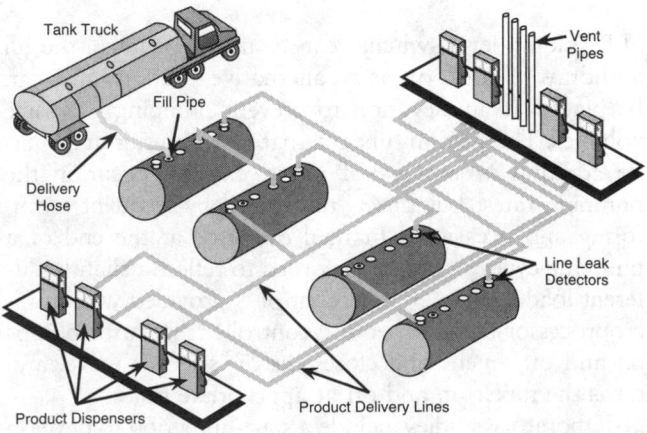

FIG. 11.19.1 Typical service station tank and piping layout (EPA).

SPILLS AND OVERFILLS

Spills and overfills, usually the result of human error, aggravate release problems at UST facilities. In addition to direct contamination effects, repeated spills of petroleum or hazardous waste can intensity the corrosiveness of the soils. These mistakes can be corrected by following the correct tank filling practices required by 40 CFR 280. Other causes include: delivery source failure, shutoff valve failure, and tank level indicator failure.

COMPATIBILITY OF UST AND CONTENTS

Compatibility between tanks and contents means that fuel components will not change the physical or mechanical properties of the tank. Compatibility for liners requires that fuel components not cause blistering, underfilm corrosion or internal stress or cracking. There are concerns that some FRP tanks or liners may not be compatible with some methanol-blended (or possible ethanol-blended) fuels. A fuel tank should be designed to handle fuel compositions of the future.

Owners and operators of businesses with FRP-constructed or lined tanks should consult appropriate standards of the American Petroleum Institute.

UST Regulations

EPA regulations establishing controls for new and existing underground storage tanks became effective in December 1988. Following is a list of important requirements:

DESIGN, CONSTRUCTION, AND INSTALLATION

Tanks and piping may be constructed of fiberglass, coated steel (asphalt or paint), or metal without additional corrosion protection. If constructed of coated steel, tanks and piping must also contain corrosion-protection devices such as cathodic protection systems. Such corrosion-protection devices must be regularly inspected. If constructed of metal without corrosion protection, records must be maintained showing that a corrosion expert has determined that the site is not corrosive enough to cause leaks during the tank's operating life. Tanks must be installed properly and precautions must be taken to prevent damage.

Information on the design (including corrosion protection), construction, and installation of tanks and piping is available on notification forms filed in the designated state office, usually the state UST office. Reports of inspections, monitoring, and testing of corrosion protection devices are on file at the UST site or must be made available to the implementing agency upon request.

Recent developments in preventing leakage include using asphalt coated steel tanks, double-walled fiberglass tanks, double-walled steel tanks, epoxy-coated steel tanks, fiberglass-coated steel tanks, fiberglass-coated double-walled tanks, synthetic underground containment liners, and tanks placed in subgrade vaults (see Figure 11.19.2). Some suppliers offer double-walled pipes for added safety. Lined trenches offer secondary containment for single-walled pipes.

SPILLS AND OVERFILLS CONTROL

Except for systems filled by transfer of no more than 25 gallons at one time, UST systems must use one or more overfill prevention devices. These devices include sensors to detect tank capacity level, automatic flow shutoff valves, and spill catchment basins.

Owners and operators must report spills and overfills to the implementing agency within a reasonable time period.

The standards apply to new UST systems and some existing tanks.

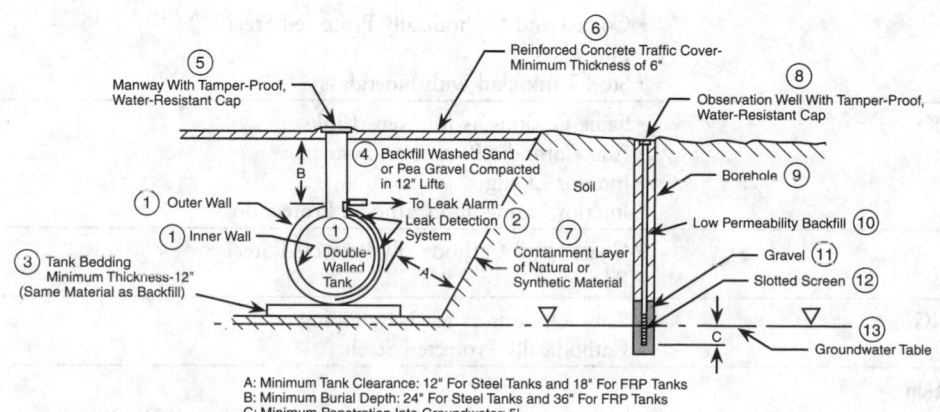

FIG. 11.19.2 Double wall tank and leak detection system. (Reprinted from New York Department of Environmental Conservation, 1982, *Siting manual for storing hazardous substances: a practical guide for local officials.* New York, N.Y.)

REPAIRS

Tank repairs must be conducted in accordance with a code of practice developed by a nationally recognized association or an independent testing laboratory. Following repairs, tightness tests may be required.

Owners must maintain repair records on-site or must make them available to the implementing agency upon request.

These repair standards apply to new and existing UST systems.

LEAK DETECTION

Release detection requirements differ between petroleum USTs and hazardous waste USTs. Petroleum UST systems may choose from among five primary release detection methods, for example: 1) automatic tank gauging that tests for product loss and conducts inventory control; 2) testing or monitoring for vapor within the soil gas of the tank area; and 3) testing or monitoring for liquids in the groundwater. Hazardous substance tanks must use secondary containment systems such as double-walled tanks or external liners, unless the owner has obtained a variance.

Owners must maintain records pertaining to: system leak detection methods; recent test results to detect possible leaks; and maintenance of release detection equipment; and must make these available to the agency on request. Information on leak detection method(s) used by UST systems is also on the notification form on file in each state's UST office.

These leak detection standards apply to new UST systems.

TABLE 11.19.1   MINIMUM REQUIREMENTS FOR COMPLIANCE WITH UST REGULATIONS (EPA)

Leak Detection

NEW TANKS *2 Choices*	• Monthly Monitoring* • Monthly Inventory Control and Annual Tank Tightness Every 5 Years (You can only use this choice for 10 years after installation.)

EXISTING TANKS
3 Choices
	• Monthly Monitoring*
	• Monthly Inventory Control and Annual Tank Tightness Testing
	  (This choice can be used until December 1998.)
	• Monthly Inventory Control and Tank Tightness Testing Every 5 Years
	  (This choice can only be used for 10 years after adding corrosion protection and
	  spill/overfill prevention or until December 1998, whichever date is later.)

NEW & EXISTING PRESSURIZED PIPING *Choice of one from each set*	• Automatic Flow Restrictor • Annual Line Testing • Automatic Shutoff Device -and- • Monthly Monitoring* • Continuous Alarm System (except automatic tank gauging)
NEW & EXISTING SUCTION PIPING *3 Choices*	• Monthly Monitoring* (except automatic tank gauging) • Line Testing Every 3 Years • No Requirements—if slope and check valve conditions are met

Corrosion Protection

NEW TANKS *3 Choices*	• Coated and Cathodically Protected Steel • Fiberglass • Steel Tank clad with Fiberglass
EXISTING TANKS *4 Choices*	• Same Options as for New Tanks • Add Cathodic Protection System • Interior Lining • Interior Lining and Cathodic Protection
NEW PIPING *2 Choices*	• Coated and Cathodically Protected Steel • Fiberglass
EXISTING PIPING *2 Choices*	• Same Options as for New Piping • Cathodically Protected Steel

Spill/Overfill Prevention

ALL TANKS	Catchment Basins -and- • Automatic Shutoff Devices -or-
	                               • Overfill Alarms -or-
	                               • Ball Float Valves

*Monthly Monitoring includes: Automatic Tank Gauging, Ground-water Monitoring, Vapor Monitoring, Other Approved Methods, Interstitial Monitoring

OUT-OF-SERVICE SYSTEMS AND CLOSURE

During temporary closures, owners must continue all usual system operation, maintenance, and leak detection procedures, and must comply with release reporting and cleanup regulations if a leak is suspected or confirmed. Release detection, however, is not required if the UST system is empty.

If a UST system is taken out of service for more than 12 months, it must be permanently closed unless it meets certain performance standards and upgrading requirements. Before a tank is permanently closed, the owner or operator must test for system leaks: if a leak is found, the owner or operator must comply with corrective-action regulations. Once the tank is permanently out of service, it must be emptied, cleaned, and either removed from the ground or filled with an inert solid material.

Closure procedures are available at each tank site or must be made available to the implementing agency upon request.

These closure standards apply to all new and existing UST systems.

Financial Responsibility

These regulations specify the amount and scope of coverage required for corrective action and for compensating third parties for bodily injury and property damage from leaking tanks. The minimum coverage required varies depending on the number of tanks owned, from $1,000,000 for up to 100 USTs to $2,000,000 for more than 100 USTs.

Various mechanisms may be used to fulfill coverage requirements, such as self-insurance, indemnity contracts, insurance, standby trust funds, or state funds. Quick action may ensure that available funds are directed toward coverage. Particular attention should be paid to self-insured tank owners. However, even large companies may go bankrupt, leaving the contracting engineer unprotected.

Information on financial responsibility for new tanks must be filed with the EPA regional office. Owners must also maintain evidence of financial responsibility at tank sites or places of business, or make such evidence available upon request of the implementing agency.

These financial responsibility standards apply to most new and existing UST systems.

The technical requirements for UST rules are summarized in Table 11.19.1.

—*David H.F. Liu*

11.20
LEAK DETECTION AND REMEDIATION

Contamination caused by leaky USTs often may not be detected until it is widespread, and difficult and expensive to correct. Regular tests and inspections of tanks and piping are necessary to ensure that leaks are detected early and prevented promptly. The extent of releases and their migration are characterized to plan corrective actions.

If a tank is leaking more than 0.05 gph, it is a leaker. Less than 0.05 gph is beyond the scope of measurement ability and the tank is considered tight. Present technology is imprecise in detecting leaks smaller than 0.05 gph. This standard is listed in NFPA 329, Final Test, now renamed Precision Test.

Tank Monitoring

There are four general methods for detecting leaks in USTs:

- Volumetric (quantitative) leak testing and leak rate measurement
- Non-volumetric (qualitative) leak testing

- Inventory monitoring
- Environmental monitoring

These methods can be used independently or in combination.

Figure 11.20.1 illustrates some of these leak detecting alternatives.

Regardless of the UST monitoring techniques used, the effects of major variables must be compensated for (Table 11.20.1). It is important for USTs to be tested under conditions close to normal operating conditions without uncovering the tanks. Uncovering tanks or piping is expensive and time-consuming, and can cause new leaks. Inadvertent pressurization during testing may rupture the tank and piping.

VOLUMETRIC LEAK TESTING

Table 11.20.2 summarizes common volumetric tank testing systems. These systems may be used to detect leaks and ascertain tightness of tanks and associated piping.

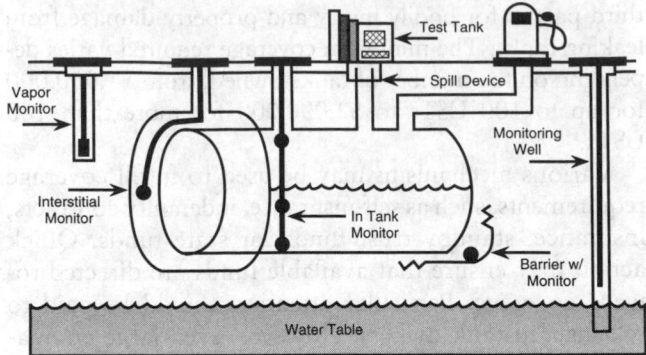

FIG. 11.20.1 Leak detection alternatives (EPA).

An experimental device that detects leaks based on laser interferometry is currently being developed by SRI International under contract to API (Figure 11.20.2). The device aims a laser beam at the underground tank and the beam is reflected to a detector that computes the liquid level in the tank. Test results to date indicate that this de-

vice can detect liquid level changes in micro-inches. The API has specified that the device must instantly detect leaks as small as 0.05 gph.

NONVOLUMETRIC LEAK TESTING

Table 11.20.3 presents a number of nonvolumetric (qualitative) leak testing methods. If a leak is occurring, volumetric testing is used to determine the rate of the leak.

INVENTORY MONITORING

This involves thorough record keeping of product purchases and consumption, regular inspections, and recognition of conditions indicating leaks. This simple, low-cost leak-detecting method is applicable to any product stored or transported in pipelines. In addition, it does not require interruption of tank service or a set degree of tank fullness. However, this method requires good bookkeeping and will not detect small leaks. Table 11.20.4 summarizes three common inventory monitoring techniques.

TABLE 11.20.1 MAJOR VARIABLES AFFECTING LEAK DETECTION

Variable	Impact
Temperature change	Expansion or contraction of a tank and its contents can mask leak and/or leak rate.
Water table	Hydrostatic head and surface tension forces caused by groundwater may mask tank leaks partially or completely.
Tank deformation	Changes or distortions of the tank due to changes in pressure or temperature can cause an apparent volume change when none exists.
Vapor pockets	Vapor pockets formed when the tank must be overfilled for testing can be released during a test or expand or contract from temperature and pressure changes and cause an apparent change in volume.
Product evaporation	Product evaporation can cause a decrease in volume that must be accounted for during a test.
Piping leaks	Leaks in piping can cause misleading results during a tank test because many test methods cannot differentiate between piping leaks and tank leaks.
Tank geometry	Differences between the actual tank specifications and nominal manufacturer's specification can affect the accuracy of change in liquid volume calculations.
Wind	When fill pipes or vents are left open, wind can cause an irregular fluctuation of pressure on the surface of the liquid and/or a wave on the liquid-free surface that may affect test results.
Vibration	Vibration can cause waves on the free surface of the liquid that can cause inaccurate test results.
Noise	Some nonvolumetric test methods are sound-sensitive, and sound vibrations can cause waves to affect volumetric test results.
Equipment accuracy	Equipment accuracy can change with the environment (e.g., temperature and pressure).
Operator error	The more complicated a test method, the greater the chance for operator error, such as not adequately sealing the tanks if required by the test method in use.
Type of liquid stored	The physical properties of the liquid (including effects of possible contaminants) can affect the applicability or repeatability of a detection method (e.g., viscosity can affect the sound characteristics of leaks in acoustical leak-detection methods).
Power vibration	Power vibration can affect instrument readings.
Instrumentation limitation	Instruments must be operated within their design range or accuracy will decrease.
Atmospheric pressure	A change in this parameter has the greatest effect when vapor pockets are in the tank, particularly for leak-rate determination.
Tank inclination	The volume change per unit of level change is different in an inclined tank than in a level one.

Source: Reprinted, from U.S. Environmental Protection Agency (EPA), 1986, *Underground tank leak, detection methods: a state-of-the-art review* (EPA, EPA 600–2–80–001, Washington, D.C.).

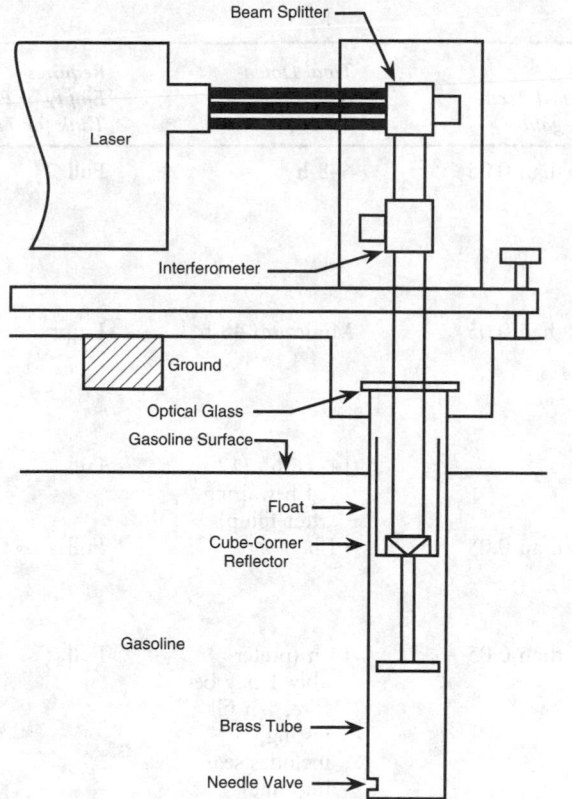

FIG. 11.20.2 Laser interferometer used to measure level changes. (Source: U.S. EPA.)

ENVIRONMENTAL MONITORING

Monitoring wells are the most prevalent form of environmental monitoring for USTs. With environmental effects monitoring, it is difficult to determine which tank is leaking when there is more than one tank. These methods do not provide information on leakage rates or the size of the leak; however, once installed, a leak effects monitoring system enables more frequent checking for leaking tanks than the other methods. Table 11.20.5 presents the principal environmental monitoring methods.

An early warning monitoring technique of double-walled tanks involves monitoring the space between the inner and outer walls of the tank, using either fluid sensors or pressure sensors. This is accurate and applicable with any double-walled tank.

Vapor wells may be used to detect hazardous gases or vapors released into the soil surrounding the UST. Gas detectors or portable gas sampling devices can be used to monitor for gaseous contaminants.

Groundwater monitoring wells may be used to detect or define the movement of leaked substances in a groundwater table. This typically entails drilling monitoring wells, installing monitoring casings, and performing chemical analyses. Table 11.20.6 presents a generalized groundwater sampling protocol.

Table 11.20.7 summarizes the advantages and disadvantages of the various leak monitoring techniques, including those presented in this section.

TABLE 11.20.2 VOLUMETRIC LEAK TESTING METHODS

Method	Principle	Claimed Accuracy, gal/h	Total Downtime for Testing	Requires Empty or Full Tank for Test
Ainlay tank integrity testing	Pressure measurement by a coil-type manometer to determine product level change in a propane bubbling system	0.02	10–12 h (filled a night before 1.5-h testing)	Full
ARCO HTC underground tank detector	Level change measurement by float and light-sensing system	0.05	4–6 h	No
Certi-Tec testing	Monitoring of pressure changes resulting from product level changes	0.05	4–6 h	Full
"Ethyl" tank sentry	Level change magnification by a J tube manometer	Sensitive to 0.02-in level change	Typically 10 h	No
EZY-CHEK leak detector	Pressure measurement to determine product level change in an air bubbling system	Less than 0.01	4–6 h (2 h waiting after fillup, 1-h test)	Full
Fluid-static (standpipe) testing	Pressurizing of system by a standpipe; keeping the level constant by product addition or removal; measuring rate of volume change	Gross	Several days	Full

(Continued on next page)

TABLE 11.20.2 *(Continued)*

Method	Principle	Claimed Accuracy, gal/h	Total Downtime for Testing	Requires Empty or Full Tank for Test
Heath Petro Tite tank and line testing (Kent-Moore)	Pressurizing of system by a standpipe; keeping the level constant by product addition or removal; measuring volume change; product circulation by pump	Less than 0.05	6–8 h	Full
Helium differential pressure testing	Leak detection by differential pressure change in an empty tank; leak rate estimation by Bernoulli's equation	Less than 0.05	Minimum 48 h	Empty
Mooney tank test detector	Measuring level change with a dip stick	0.02	14–16 h* (12 to 14 h waiting after fillup)	Full
PACE tank tester	Magnification of pressure change in a sealed tank by using a tube (based on manometer principle)	Less than 0.05	14 h	Full
PALD-2 leak detector	Pressurizing system with nitrogen at three different pressures; level measurement by an electrooptical device; estimate of leak rate based on the size of leak and pressure difference across the leak	Less than 0.05	14 h (preferably 1 day before, 1-h fill testing, includes sealing time)	Full
Pneumatic testing	Pressurizing system with air or other gas; leak rate measurement by change in pressure	Gross	Several hours	No
Tank auditor	Principle of buoyancy	0.00001 in the fill pipe; 0.03 at the center of a 10.5-ft-diameter tank	1.5–3 h	Typically full
Two-tube laser interferometer system	Measuring level change by laser beam and its reflection	Less than 0.05	4–5 h†	No (at existing level)

*Including the time for tank end stabilization when testing with standpipe.
†Including 1 to 2 h for reference tube temperature equilibrium.
Source: Reprinted from U.S. EPA, 1986.

Corrective Technologies

The most important considerations are the volume and type of substance released and constraining site features that can hinder or prevent effective implementation of a technology. To lesser extent, the financial ability of responsible parties to implement certain technologies and the impact on facility production or service operations also should be considered.

Table 11.20.8 summarizes potential applicable corrective action technologies commensurate with release volume and chemical characteristics. Applicable site data needs for potential technologies are presented in Table 11.20.9.

Initial Response Action. The first response action must minimize immediate risk to human health and the environment; all remaining product must be removed from leaky tanks. Table 11.20.10 lists potential situations and their associated initial corrective actions.

Permanent Response Action. Table 11.20.11 gives examples of permanent corrective actions for a variety of site-specific problems.

—David H.F. Liu

TABLE 11.20.3 NONVOLUMETRIC LEAK TESTING METHODS

Method	Principle	Claimed Accuracy, gal/h	Total Downtime for Testing	Requires Empty or Full Tank for Test
Acoustical Monitoring System (AMS)	Sound detection of vibration and elastic waves generated by a leak in a nitrogen-pressurized system; triangulation techniques to detect leak location	Does not provide leak rate; detects leaks as low as 0.01 gal/h	1–2 h	No
Leybold-Heraeus helium detector, Ultratest M2	Rapid diffusivity of helium; mixing of a tracer gas with products at the bottom of the tank; helium detected by a sniffer mass spectrometer	Does not provide leak rate; helium could leak through 0.005-in leak size	None	No
Smith & Denison helium test	Rapid diffusivity of helium; differential pressure measurement; helium detection outside a tank	Provides the maximum possible leak detection based on the size of the leak (does not provide leak rates); helium could leak through 0.05-in leak size	Few—24 h (excludes sealing time)	Empty
TRC rapid leak detector for underground tanks and pipes	Rapid diffusion of tracer gas; mixing of a tracer gas with product; tracer gas detected by a sniffer mass spectrometer with a vacuum pump	Does not provide leak rate; tracer gas could leak through 0.005-in leak size	None	No
Ultrasonic leak detector (Ultrasound)	Vacuuming the system (5 lb/in^2); scanning entire tank wall by ultrasound device; noting the sound of the leak by headphones and registering it on a meter	Does not provide leak rate; a leak as small as 0.001 gal/h of air could be detected; a leak through 0.005-in could be detected	Few hours (includes tank preparation and 20-min test)	Empty
VacuTect (Tanknology)	Applying vacuum at higher than product static head; detecting bubbling noise by hydrophone; estimating approximate leak rate by experience	Provides approximate leak rate	1 h	No
Varian leak detector (SPY2000 or 938–41)	Similar to Smith & Denison	Similar to Smith & Denison	Few—24 h (excludes sealing time)	Empty

Source: Reprinted from U.S. EPA, 1986.

TABLE 11.20.4 INVENTORY MONITORING METHODS

Method	Principle	Claimed Accuracy
Gauge stick	Measuring project level with dip stick when station is closed	Gross
MFP-414 TLG leak detector	Monitoring product weight by measuring pressure and density at the top, middle, and bottom of tank	Sensitive to 0.1% of product height change
TLS-150	Using electronic level measurement device or programmed microprocessor inventory system	Sensitive to 0.1-in. level change

Source: Reprinted from U.S. EPA, 1986.

TABLE 11.20.5 ENVIRONMENTAL MONITORING METHODS

Method	Principle
Collection sumps	Using collection mechanism of product in collection sump through sloped floor under the storage tank
Dye method	Hydrocarbon detection by use of soluble dye through perforated pipe
Groundwater and soil sampling	Water and soil sampling
Interstitial monitoring in double-walled tanks	Monitoring in interstitial space between the walls of double-walled tanks with vacuum or fluid sensors
LASP	Diffusion of gas and vapor to a plastic material
Observation wells	Product sensing in liquid through monitoring wells at areas with high groundwater
Pollulert and Leak-X	Difference in thermal conductivity of water and hydrocarbons through monitoring wells
Remote infrared sensing	Determining soil temperature characteristic change due to the presence of hydrocarbons
Surface geophysical methods	Hydrocarbon detection by ground-penetrating radar, electromagnetic induction, or resistivity techniques
U-tubes	Product sensing in liquid; collection sump for product directed through a horizontal pipe installed under a tank
Vapor wells	Monitoring of vapor through monitoring well

Source: Reprinted from U.S. EPA, 1986.

TABLE 11.20.6 GENERALIZED GROUNDWATER SAMPLING PROTOCOL

Step	Goal	Recommendations
Hydrologic measurements	Establish nonpumping water level.	Measure the water level to ± 0.01 ft (± 0.3 cm)
Well purging	Remove or isolate stagnant H_2O which would otherwise bias representative sample.	Pump water until well purging parameters (e.g., pH, T, Ω^{-1}, Eh) stabilize to $\pm 10\%$ over at least two successive well volumes pumped.
Sample collection	Collect samples at land surface or in well-bore with minimal disturbance of sample chemistry.	Pumping rates should be limited to -100 mL/min for volatile organics and gas-sensitive parameters.
Filtration and preservation	Filtration permits determination of soluble constituents and is a form of preservation. It should be done in the field as soon as possible after collection.	Filter: Trace metals, inorganic anions and cations, alkalinity. Do not filter: TOC, TOX, volatile organic compound samples; other organic compound samples only when required.
Field determinations	Field analyses of samples will effectively avoid bias in determining parameters and constituents which do not store well; e.g., gases, alkalinity, pH.	Samples for determining gases, alkalinity, and pH should be analyzed in the field if at all possible.
Field blanks and standards	These blanks and standards will permit the correction of analytical results for changes which may occur after sample collection: preservation, storage, and transport.	At least one blank and one standard for each sensitive parameter should be made up in the field on each day of sampling. Spiked samples are also recommended for good QA/QC.
Sample storage and transport	Refrigerate and protect samples to minimize their chemical alteration prior to analysis.	Observe maximum sample holding or storage periods recommended by the Agency. Documentation of actual holding periods should be carefully performed.

TABLE 11.20.7 COMPARISON OF VARIOUS LEAK MONITORING TECHNIQUES

Approach	Description	Applications	Substances Detected	Relative Cost	Advantages/Disadvantages
Inventory Control	A system based on product record keeping, regular inspections, and recognition of the conditions which indicate leaks.	Any storage tanks and buried pipelines.	Any product stored or transported.	Low	The technique is widely applicable to any product stored or transported in pipelines. However, it requires good bookkeeping, and will not detect small leaks.
Thermal Conductivity Sensors	Uses a probe that detects the presence of stored product by measuring thermal conductivity.	Can monitor groundwater or normally dry areas.	Any liquid.	Medium	Primary advantage is early detection which makes it possible for leaks and spills to be corrected before large volumes of material are discharged. Typically requires ¼ inch of product on ground/water interface in wet (groundwater) applications.
Electric Resistivity Sensors	Consists of one or series of sensor cables that deteriorate in the presence of the stored product, thereby indicating a leak.	Can monitor groundwater or normally dry areas.	Any liquid.	Medium	Primary advantage is the early detection of spills. Once a leak or spill is detected the sensors must be replaced. Can detect small as well as large leaks.
Gas Detectors	Used to monitor the presence of hazardous gases in vapors in the soil.	Areas of highly permeable, dry soil, such as excavation backfill or other permeable soils, above ground-water table.	Highly volatile liquids, such as gasoline.	Medium	Once the contaminant is present and detected, gas detectors are no longer of use until contamination has been cleaned up.
Sampling	Grabbing soil or water samples from area for analysis.	Universal; primarily used to collect groundwater samples, as would be the case with tanks stored in high ground-water area.	Any substance	High	Highly accurate intermittent evaluation tool. However, does not provide continuous monitoring.
Interstitial Monitoring in Double-Walled Tanks	Monitors pressure level or vacuum in space between walls of a double-walled tank.	Double-walled tanks.	Pressure sensors monitor tank integrity and are applicable with any stored liquid. Fluid sensors monitor presence of any liquid in a normally dry area and are also applicable with any stored liquid.	High	Accurate technique which is applicable with any double-walled tanks.

(Continued on next page)

TABLE 11.20.7 (Continued)

Approach	Description	Applications	Substances Detected	Relative Cost	Advantages/Disadvantages
Groundwater Monitoring Wells (wet wells)	Wet wells are used to detect and determine the extent of contamination in ground-water tables.	Area-wide or local monitoring for groundwater contamination from underground storage tanks and pipelines. May be used for periodic sampling or may employ one of the sensors described above to detect leaks or spills.	Any hazardous liquids which can be detected by on-site instruments or laboratory analysis.	Medium to High	The type, number and location of wet wells depends upon the site's hydrogeology, the direction of groundwater flow, and the type of spill containment and spill collection systems used.
Vapor (sniff) Wells	Vapor wells are used to detect and monitor the presence of hazardous gases and vapors in the soil.	Area-wide or local monitoring of the soil surrounding underground storage tanks and pipe-lines.	Many different combustible and non-combustible gases and vapors.	Low	The type, number and location of vapor wells depend upon the extent of the spill, the volatility of the product, and the soil characteristics. Vapor wells are subject to contamination from surface spills and cannot be used at contaminated sites.
Dyes and Tracers	Substances with a characteristic color or other characteristics (e.g., radioactive tracers) that can be used to trace the origin of a spill.	Area-wide monitoring of underground tanks and buried pipelines.	Dye itself is detected visually or with the use of instruments.	Low Medium	Dye or tracer could be low in cost, but the time required to perform a study could be great. Also may require the drilling of observation wells to trace the dye of other material. Radioactive tracers require a license and approval from the Nuclear Regulatory Commission of the U.S. Department of Labor. Therefore they are generally discouraged.

TABLE 11.20.8 APPLICATIONS OF TYPICAL CORRECTIVE ACTION TECHNOLOGIES

Technology	Small- to moderate-volume recent gasoline or petroleum release (gas station or tank farms)	Large-volume or long-term chronic gasoline or petroleum release (gas station or tank farms)	Release from tanks containing hazardous substances (organic)	Release from tanks containing hazardous substances (inorganic)
Removal and excavation of tank, soil, and sediment				
Tank removal	●	●	●	●
Soil excavation	●	●	●	●
Sediment removal		●	●	●
On-site and off-site treatment and disposal of contaminants				
Solidification or stabilization		●	●	
Landfilling				
Landfarming	●	●	●	
Soil washing				
Thermal destruction		●	●	
Aqueous waste treatment	●	●	●	●
Deep well injection				
Free product recovery				
Dual pump systems	●	●	●	
Floating filter pumps	●	●	●	
Surface oil and water separators	●	●	●	
Groundwater recovery systems				
Groundwater pumping	●	●	●	●
Subsurface drains	●			●
Subsurface barriers				
Slurry walls		●	●	●
Grouting		●	●	●
Sheet piles				
Hydraulic barriers	●	●	●	●
In situ treatment				
Chemical treatment				
Physical treatment			●	●
Soil flushing	●	●	●	●
Biostimulation	●	●	●	●
Groundwater treatment				
Air stripping	●	●	●	
Carbon adsorption	●	●	●	
Biological treatment	●	●	●	
Precipitation, flocculation, sedimentation				●
Dissolved air flotation				●
Groundwater treatment				
Granular media filtration				●
Ion-exchange resin adsorption				●
Oxidation-reduction				●
Neutralization				●
Steam stripping			●	
Reverse osmosis				●
Sludge dewatering				●

(Continued on next page)

TABLE 11.20.8 *(Continued)*

Technology	Small- to moderate-volume recent gasoline or petroleum release (gas station or tank farms)	Large-volume or long-term chronic gasoline or petroleum release (gas station or tank farms)	Release from tanks containing hazardous substances (organic)	Release from tanks containing hazardous substances (inorganic)
Vapor migration control, collection, and treatment				
Passive collection systems		•	•	
Active collection systems		•	•	
Ventilation of structures	•	•	•	
Adsorption			•	
Flaring				
Surface water and drainage controls				
Diversion and collection systems	•	•	•	•
Grading	•	•	•	•
Capping		•	•	•
Revegetation	•	•	•	•
Restoration of contaminated water supplies and sewer lines				
Alternative central water supplies		•	•	•
Alternative point-of-use water supplies	•	•	•	•
Treatment of central water supplies		•	•	•
Treatment of point-of-use water supplies		•	•	•
Replacement of water and sewer lines		•	•	•
Cleaning and restoration of water and sewer lines	•	•	•	•

TABLE 11.20.9 SITE INFORMATION FOR USE IN EVALUATING CORRECTIVE ACTION ALTERNATIVES

Technology	Geographic and Topographic Characteristics								Land and Water Use Patterns				Hydrogeologic Characteristics											
	Precipitation	Temperature	Evapotranspiration	Topography	Accessibility	Site Size	Proximity to Surface Water	Proximity to Human Interfaces	Current Water-use Patterns	Future Water-use Patterns	Current Land-use Patterns	Growth Projections	Soil Profiles	Soil Physical Properties	Soil Chemical Properties	Depth to Bedrock	Depth to Groundwater	Aquifer Physical Properties	Groundwater Flow Rate (Volume)	Groundwater Flow Direction	Recharge Areas	Recharge Rates	Aquifer Characteristics	Natural Groundwater Quality
Removal/excavation of tank, soil and sediment																								
Tank removal		•	•	•	•								•											
Soil excavation		•	•	•	•								•	•	•			•						
Sediment removal		•	•	•	•								•											
On-site and off-site treatment and disposal of contaminants																								
Solidification and stabilization	•		•	•	•	•	•	•	•	•	•		•	•	•	•	•	•	•	•	•	•	•	•
Landfilling	•	•	•	•	•	•	•	•	•	•	•		•	•	•	•	•	•	•	•	•	•	•	•
Landfarming	•		•	•	•	•	•	•	•	•	•		•	•	•									
Soil washing	•	•	•	•	•	•	•	•	•	•	•		•	•	•	•	•	•	•	•	•	•	•	•
Thermal destruction		•		•	•	•	•	•	•	•	•		•	•	•									
Aqueous waste treatment	•	•		•	•	•	•	•					•	•	•	•	•	•	•	•	•	•	•	•
Deep well injection		•		•	•	•	•	•					•	•	•	•	•	•	•	•	•	•	•	•
Free product recovery																								
Dual pump systems				•	•								•			•	•	•	•	•	•	•	•	•
Floating filter pumps					•								•			•	•	•	•	•	•	•	•	•
Surface oil and water separators					•			•					•			•	•	•	•	•	•	•	•	•
Groundwater recovery systems																								
Groundwater pumping	•	•			•				•	•	•	•	•			•	•	•	•	•	•	•	•	•
Subsurface drains	•				•				•	•	•	•	•			•	•	•	•	•	•	•	•	•
Subsurface barriers																								
Slurry walls	•			•	•	•			•	•	•	•	•	•	•	•	•	•	•	•	•	•	•	•
Grouting	•	•		•	•	•			•	•	•	•	•	•	•	•	•	•	•	•	•	•	•	•
Sheet piles				•	•	•							•	•	•	•	•	•	•	•	•	•	•	•
Hydraulic barriers		•		•	•	•							•	•	•	•	•	•	•	•	•	•	•	•
In situ treatment																								
Soil flushing	•	•		•	•	•		•	•	•	•	•	•	•	•	•	•	•	•	•	•	•	•	•
Biostimulation	•	•		•	•	•		•	•	•	•	•	•	•	•	•	•	•	•	•	•	•	•	•
Chemical treatment	•	•		•	•	•		•	•	•	•	•	•	•	•	•	•	•	•	•	•	•	•	•
Physical treatment	•	•		•	•	•		•					•	•	•									
Groundwater treatment																								
Air stripping	•	•	•	•	•	•																	•	•
Carbon adsorption	•		•	•	•	•																	•	•

(Continued on next page)

1365

TABLE 11.20.9 (Continued)

Technology	Hydrogeologic Characteristics												Land and Water Use Patterns				Geographic and Topographic Characteristics							
	Natural Groundwater Quality	Aquifer Characteristics	Recharge Rates	Recharge Areas	Groundwater Flow Direction	Groundwater Flow Rate (Volume)	Aquifer Physical Properties	Depth to Groundwater	Depth to Bedrock	Soil Chemical Properties	Soil Physical Properties	Soil Profiles	Growth Projections	Current Land-use Patterns	Future Water-use Patterns	Current Water-use Patterns	Proximity to Human Interfaces	Proximity to Surface Water	Site Size	Accessibility	Topography	Evapotranspiration	Temperature	Precipitation
Groundwater treatment *cont'd.*																								
Biological treatment	●	●	●	●		●							●	●	●	●	●		●	●	●	●		●
Precipitation, flocculation, sedimentation	●	●	●	●		●							●	●	●	●	●		●	●	●	●		●
Dissolved air flotation	●	●	●	●		●							●	●	●	●	●		●	●	●	●		●
Granular media filtration	●	●	●	●		●							●	●	●	●	●		●	●	●	●		●
Ion-exchange resin adsorption			●	●		●							●	●	●	●	●		●	●	●	●		●
Oxidation-reduction	●	●	●	●		●							●	●	●	●	●		●	●	●	●		●
Neutralization	●	●	●	●		●													●	●	●	●		●
Steam stripping	●	●	●	●		●													●	●	●	●		●
Reverse osmosis	●	●																	●	●	●			
Sludge dewatering	●	●																	●	●	●			
Vapor migration control, collection, and treatment																								
Passive collection systems										●	●	●	●	●			●		●	●	●			●
Active collection systems										●	●	●	●	●			●		●	●	●			●
Ventilation of structures										●	●	●	●	●			●		●	●	●			●
Adsorption										●	●	●	●	●			●		●	●	●			●
Flaring										●	●	●	●	●			●		●	●	●			
Surface water and drainage controls																								
Diversion and collection systems											●	●	●	●	●	●		●	●	●	●	●		●
Grading											●	●	●	●	●	●		●	●	●	●	●		●
Capping											●	●	●		●	●		●	●	●	●	●		●
Revegetation											●	●	●		●	●		●	●	●	●	●		●
Restoration of contaminated water supplies and sewer lines																								
Alternative central water supplies	●	●		●									●	●	●	●	●	●	●	●	●	●	●	●
Alternative point-of-use water supplies	●	●		●									●	●	●	●	●	●	●	●	●	●	●	●
Treatment of central water supplies	●	●				●							●	●	●	●	●	●	●	●	●	●	●	●
Treatment of point-of-use water supplies	●	●				●							●	●	●	●	●	●	●	●	●	●	●	●
Replacement of water and sewer lines						●							●	●	●	●	●	●	●	●	●	●		
Cleaning and restoration of water and sewer lines						●												●	●	●	●	●		

Note: Technologies in italics are likely to be used in response to UST releases at gasoline stations.

TABLE 11.20.10 POTENTIAL INITIAL RESPONSE SITUATIONS AND ASSOCIATED CORRECTIVE ACTIONS

Situation	Tank repair or removal	Free product recovery	Groundwater recovery and treatment	Subsurface barriers	Soil excavation	Vapor migration control and collection	Sediment removal	Surface water diversion drainage	Alternative or treatment central water supply	Alternative or treatment point-of-use water supply	Restoration of utility, water, and sewer lines	Evacuation of nearby residents	Restricted egress or ingress
Groundwater contamination													
Existing public or private wells	•	•	•	•									
Potential future source of water supply	•	•	•	•	•								
Hydrologic connection to surface water	•	•	•	•									
Soil contamination													
Potential for direct human contact: nuisance or health hazard	•				•		•	•					
Agricultural use	•				•		•	•					
Potential source of future releases to ground water	•	•	•	•	•	•							
Surface water contamination													
Drinking water supply	•	•	•	•					•	•	•		
Source or irrigation water	•	•	•	•					•	•	•		
Water-contact recreation	•							•					•
Commercial or sport fishing	•							•					•
Ecological habitat	•							•					
Other hazards													
Danger of fire or explosion	•					•						•	•
Property damage to nearby dwellings						•						•	•
Vapors in dwellings	•					•						•	•

Source: Reprinted from U.S. Environmental Protection Agency (EPA), 1987.

TABLE 11.20.11 POTENTIAL SITE-SPECIFIC PROBLEMS AND ASSOCIATED PERMANENT CORRECTIVE ACTIONS

	Removal or excavation of soil and sediments	On-site and off-site treatment and disposal of contaminants	Free product recovery	Groundwater recovery systems	Subsurface barriers	In situ treatment	Groundwater treatment	Vapor migration control, collection, and treatment	Surface water drainage controls	Restoration of contaminated water supplies and sewerlines
Volatilization of chemicals into air								•		
Hazardous particulates released to atmosphere	•							•		
Dust generation by heavy construction or other site activities										
Contaminated site runoff	•								•	
Erosion of surface by water									•	
Surface seepage of released substance	•		•						•	
Flood hazard or contact of surface water body with released substance	•								•	
Released substance migrating vertically or horizontally			•	•	•					
High water table which may result in groundwater contamination or interfere with other corrective action				•	•					
Precipitation infiltrating site and accelerating released substance migration								•		•
Explosive or toxic vapors migrating laterally underground					•			•		

(Continued on next page)

TABLE 11.20.11 *(Continued)*

	Removal or excavation of soil and sediments	On-site and off-site treatment and disposal of contaminants	Free product recovery	Groundwater recovery systems	Subsurface barriers	In situ treatment	Groundwater treatment	Vapor migration control, collection, and treatment	Surface water drainage controls	Restoration of contaminated water supplies and sewerlines
Contaminated surface water, groundwater, or other aqueous or liquid waste	●	●		●	●	●	●			
Contaminated soils	●	●				●	●			
Toxic and/or explosive vapors that have been collected		●								
Contaminated stream banks and sediments	●	●								
Contaminated drinking water distribution system		●			●	●	●			
Contaminated utilities								●		●
Free product in groundwater and soils	●	●	●	●	●	●	●	●	●	●

Source: Reprinted from U.S. EPA, 1987.

Radioactive Waste

11.21
PRINCIPLES OF RADIOACTIVITY

Radioactivity in the environment provokes public reaction faster than any other environmental occurrence. The mere word *radioactivity* evokes fear in most people, even trained and skilled workers in the field. This fear has been etched in the public mind by such names as Hiroshima, Three Mile Island, and Chernobyl. This legacy of fear has made it difficult for proponents of the usefulness of radioactivity to gain the public trust.

Handling radioactivity in the environment was formerly the territory of the nuclear engineer. This has changed dramatically during the past twenty-five years. Today, individuals working in the environmental arena may be required to deal with radiological issues as one part of a broad environmental program. The aim of this section, and the remainder of the chapter, is to help such individuals grasp the principles of radioactivity as they pertain to environmental engineering.

The following questions must always be answered when one encounters or suspects the presence of radiation:

- What type and how much radioactive material is present?
- How can it be handled safely?
- How can it be contained and/or disposed, including classification and transportation?

Radioactivity enters the environment from natural and man-made sources. Radioactivity can exist as gaseous, liquid or solid materials. Radon is a well-known example of a radioactive gas. Water often contains dissolved amounts of radium and uranium. Solid radioactive waste is produced from many sources, including the uranium and rare earth mining industries, laboratory and medical facilities, and the nuclear power industry.

The Environmental Protection Agency (EPA) has the authority to develop federal radiation protection guidelines for release of radioactivity into the general environment and for exposure of workers and the public. The Nuclear Regulatory Commission (NRC) and individual states authorized by the NRC, called agreement states, see Figure 11.21.1, implement the EPA's general environmental standards through regulations and licensing actions. These standards are usually based on recommendations developed by the International Atomic Energy Agency (IAEA).

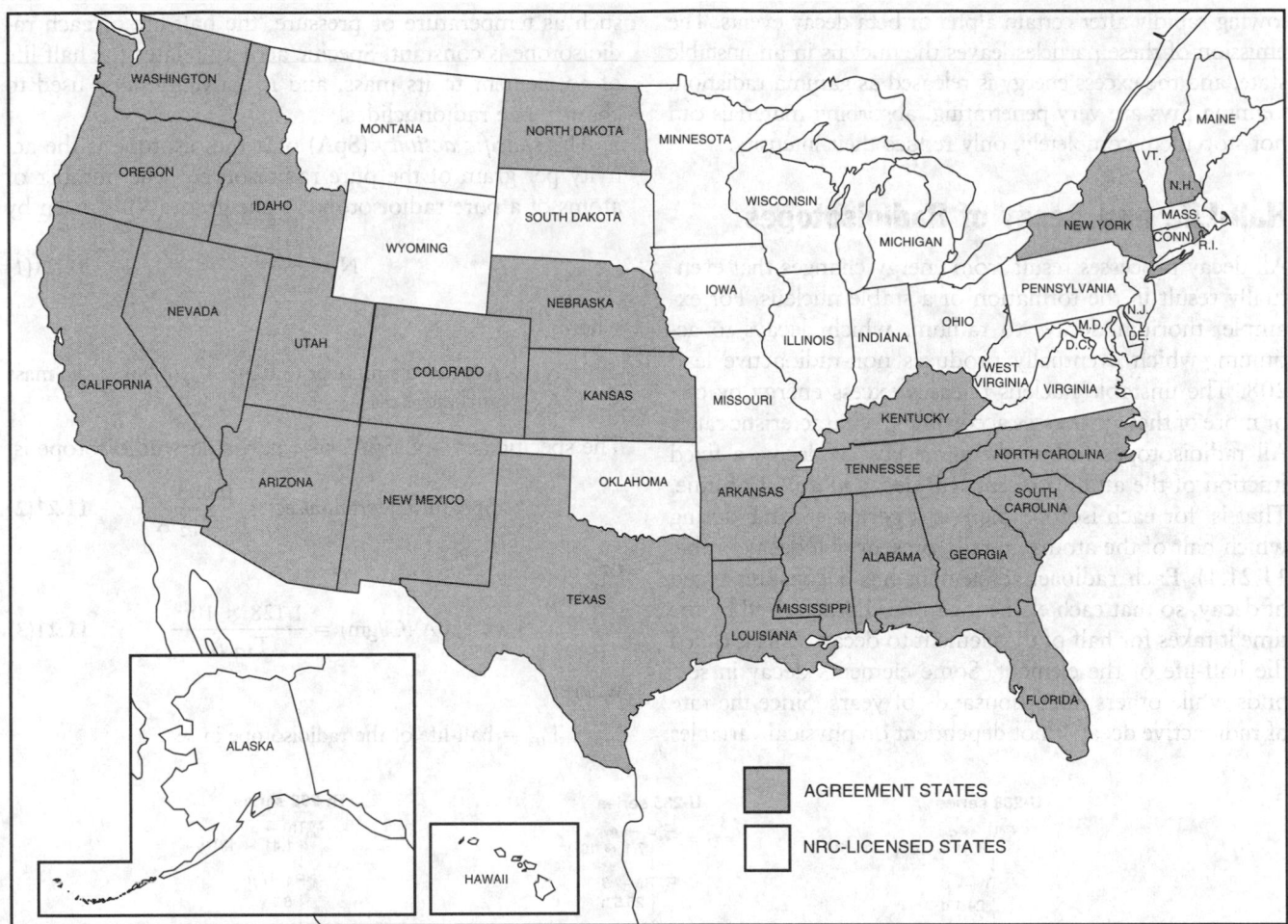

FIG. 11.21.1 States with NRC-licenses or agreements for possession of radioactive materials.

Types of Radioactivity

Radioactivity is defined as the property possessed by some elements with spontaneously emitting alpha particles (α), beta particles (β), or sometimes gamma rays (γ) by the disintegration of the nuclei of atoms. It is a naturally occurring phenomenon, it can not be stopped, and it has been taking place since the beginning of time. The process of unstable nuclei giving off energy to reach a stable condition is called radioactive decay. This process produces nuclear radiation, and the emitting isotopes are called radionuclides (radio isotopes). All isotopes of elements with atomic numbers larger than 83 (Bismuth) are radioactive. A few elements with lower atomic numbers, such as potassium and rubidium, have naturally occurring isotopes that are also radioactive. The kind of ionizing radiation emitted, the amount of energy, and the period of time it takes to become stable differs for each radioactive isotope. The following three types of radiation can be emitted.

ALPHA PARTICLES

Emitted by many high-atomic-number natural and man-made radioactive elements such as thorium, uranium and plutonium, alpha decay does not lead directly to stable nuclei. Intermediate isotopes are first produced, then these undergo further decay. The relatively high mass of the alpha particle means that for a given energy, the velocity is relatively low. The heavy, slow-moving, highly charged particles are completely absorbed by a few centimeters of air. After absorption, they are released to the atmosphere as harmless helium gas.

BETA PARTICLES

Emitted by both high and low atomic weight radioactive elements, the beta particle is an electron possessing kinetic energy due to the speed with which it is emitted from the nucleus. The velocities of the more energetic betas approach the speed of light. Beta decay is the most common mode of radioactive decay among artificial and natural radioisotopes. The range of beta particles in air may be more than a meter for energetic betas, and such particles will penetrate several meters of aluminum.

GAMMA RAYS

This type of emission consists not of particles but quanta of energy, similar to radio-waves, but containing much higher levels of energy. This emission is a secondary process following rapidly after certain alpha or beta decay events. The emission of these particles leaves the nucleus in an unstable state, and the excess energy is released as gamma radiation. Gamma rays are very penetrating: absorbing materials can not stop them completely, only reduce their intensity.

Half-Life and Decay of Radioisotopes

All decay processes result from energy changes that eventually result in the formation of a stable nucleus. For example, thorium decays to radium, which decays to actinium, which eventually produces non-radioactive lead 208. The unstable nucleus releases excess energy by one or more of these processes according to characteristic rates. All radioisotopes follow the same law of decay: a fixed fraction of the atoms present will decay in a unit of time. That is, for each isotope there is a period of time during which half of the atoms initially present will decay (Table 11.21.1). Each radioactive element has a constant speed of decay, so that each element can be characterized by the time it takes for half of the element to decay. This is called the half-life of the element. Some elements decay in seconds while others take thousands of years. Since the rate of radioactive decay is not dependent on physical variables

such as temperature or pressure, the half-life of each radioisotope is constant. Specific activity relates the half-life of an element to its mass, and is conventionally used to characterize radionuclides.

The *specific activity* (SpA) of a radioisotope is the activity per gram of the pure radioisotope. The number of atoms of a pure radioisotope in one gram (N) is given by

$$N = \frac{N_A}{A} \qquad 11.21(1)$$

where:

N_A = Avogadro's number (6.0248×10^{23})/nuclidic mass
A = nuclidic mass

The specific activity (SpA) of a particular radioisotope is:

$$SpA \text{ (disintegration/sec)} = \frac{0.693 \, N_A}{T_{1/2} \, A} \qquad 11.21(2)$$

or

$$SpA \text{ (Ci/gm)} = \frac{1.128 \times 10^{13}}{T_{1/2} \, A} \qquad 11.21(3)$$

where:

$T_{1/2}$ = half-life of the radioisotope in sec.

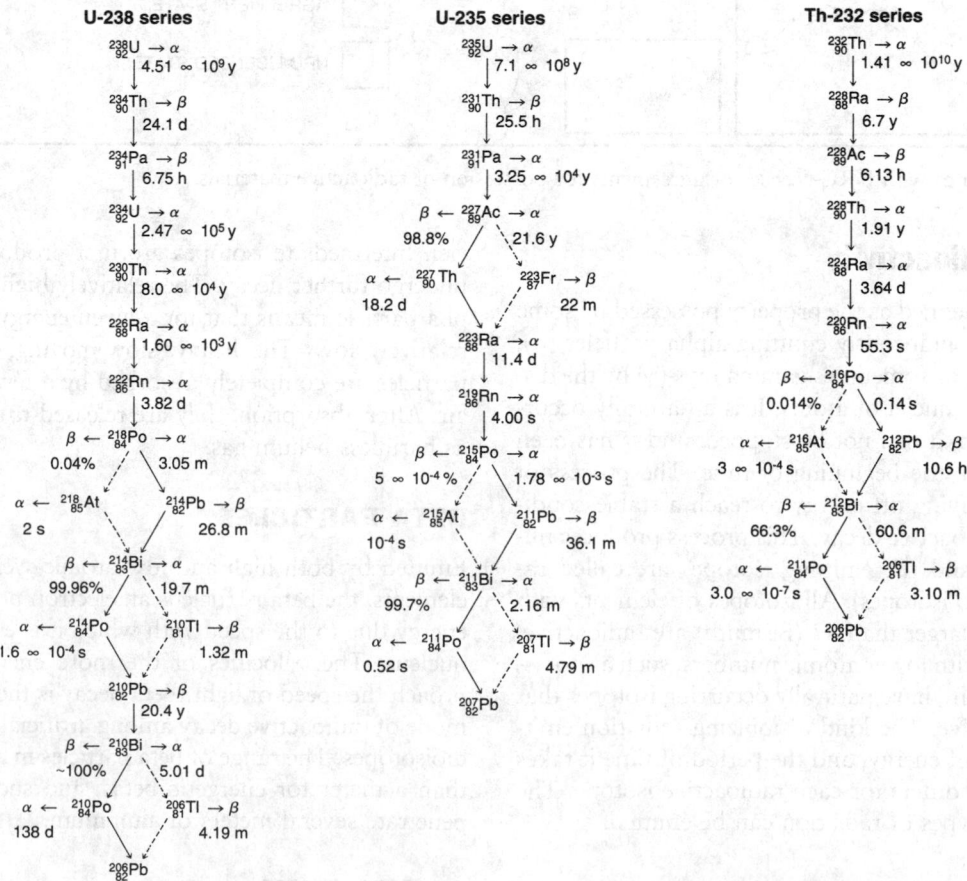

FIG. 11.21.2 Emissions and half-lives of members of radioactive series. (Reprinted, with permission, from R.C. Weast, ed. 1978. *Handbook of chemistry and physics.* CRC Press, Inc.)
[a]The abbreviations are y, year; d, day; m, minute; and s, second.

TABLE 11.21.1 HALF-LIFE AND DECAY MODE OF SELECTED RADIOISOTOPES

Element	Half-Life Duration	Type of Emission	Element	Half-Life Duration	Type of Emission
$^{14}_{6}C$	5770 y	(β^-)	$^{226}_{88}Ra$	1590 y	(α)
$^{13}_{7}N$	10.0 m	(β^+)	$^{228}_{88}Ra$	6.7 y	(β^-)
$^{24}_{11}Na$	15.0 h	(β^-)	$^{228}_{89}Ac$	6.13 h	(β^-)
$^{32}_{15}P$	14.3 d	(β^-)	$^{228}_{90}Th$	1.90 y	(α)
$^{40}_{19}K$	$1.3 \times 10^9 \, y$	$(\beta^-$ or $E.C.)$	$^{232}_{90}Th$	$1.39 \times 10^{10} \, y$	$(\alpha, \beta^-,$ or $S.F.)$
$^{60}_{27}Co$	5.2 y	(β^-)	$^{233}_{90}Th$	23 m	(β^-)
$^{87}_{37}Rb$	$4.7 \times 10^{10} \, y$	(β^-)	$^{234}_{90}Th$	24.1 d	(β^-)
$^{90}_{38}Sr$	28 y	(β^-)	$^{223}_{91}Pa$	27 d	(β^-)
$^{115}_{49}In$	$6 \times 10^{14} \, y$	(β^-)	$^{233}_{92}U$	$1.62 \times 10^5 \, y$	(α)
$^{131}_{53}I$	8.05 d	(β^-)	$^{234}_{92}U$	$2.4 \times 10^5 \, y$	$(\alpha$ or $S.F.)$
$^{142}_{58}Ce$	$5 \times 10^{15} \, y$	(α)	$^{235}_{92}U$	$7.3 \times 10^8 \, y$	$(\alpha$ or $S.F.)$
$^{198}_{79}Au$	64.8 h	(β^-)	$^{238}_{92}U$	$4.5 \times 10^9 \, y$	$(\alpha$ or $S.F.)$
$^{208}_{81}Tl$	3.1 m	(β^-)	$^{239}_{92}U$	23 m	(β^-)
$^{210}_{82}Pb$	21 y	(β^-)	$^{239}_{93}Np$	2.3 d	(β^-)
$^{212}_{82}Pb$	10.6 h	(β^-)	$^{239}_{94}Pu$	24,360 y	$(\alpha$ or $S.F.)$
$^{214}_{82}Pb$	26.8 m	(β^-)	$^{240}_{94}Pu$	$6.58 \times 10^3 \, y$	$(\alpha$ or $S.F.)$
$^{206}_{83}Bi$	6.3 d	$(\beta^+$ or $E.C.)$	$^{241}_{94}Pu$	13 y	$(\alpha$ or $\beta^-)$
$^{210}_{83}Bi$	5.0 d	(β^-)	$^{241}_{95}Am$	458 y	(α)
$^{212}_{83}Bi$	60.5 m	$(\alpha$ or $\beta^-)$	$^{242}_{96}Cm$	163 d	$(\alpha$ or $S.F.)$
$^{207}_{84}Po$	5.7 h	$(\alpha, \beta^+,$ or $E.C.)$	$^{243}_{97}Bk$	4.5 h	$(\alpha$ or $E.C.)$
$^{210}_{84}Po$	138.4 d	(α)	$^{245}_{98}Cf$	350 d	$(\alpha$ or $E.C.)$
$^{212}_{84}Po$	$3 \times 10^{-7} \, s$	(α)	$^{253}_{99}Es$	20.0 d	$(\alpha$ or $S.F.)$
$^{216}_{84}Po$	0.16 s	(α)	$^{254}_{100}Fm$	3.24 h	$(S.F.)$
$^{218}_{84}Po$	3.0 m	$(\alpha$ or $\beta^-)$	$^{255}_{100}Fm$	22 h	(α)
$^{215}_{85}At$	$10^{-4} \, s$	(α)	$^{256}_{101}Md$	1.5 h	$(E.C.)$
$^{218}_{85}At$	1.3 s	(α)	$^{254}_{102}No$	3 s	(α)
$^{220}_{86}Rn$	54.5 s	(α)	$^{257}_{103}Lr$	8 s	(α)
$^{222}_{86}Rn$	3.82 d	(α)	$^{263}_{106}(106)$	0.9 s	(α)
$^{224}_{88}Ra$	3.64 d	(α)			

Symbol in parentheses indicates type of emission; *E.C.* = K-electron capture, *S.F.* = spontaneous fission; y = years, d = days, h = hours, m = minutes, s = seconds.

The curie (Ci) is thus the quantity of any radioactive material in which the number of disintegrations is 3.7×10^{10} per second. This is a rather large amount of radioactivity, and smaller quantities are expressed in such units as *millicuries* (1 mCi = 10^{-3} Ci), *microcuries* (1 μCi = 10^{-6} Ci), and *picocuries* (1 pCi = 10^{-12} Ci). Since the curie is a measure of the emission rate, it is not a satisfactory unit for setting safety standards for handling radioactive materials.

Radioactivity originates from natural and man-made sources. Man-made radioactive materials produce *artificial radioactivity*. The radioactivity produced from nuclear fission in a nuclear reactor is a classic example of artificial radioactivity.

Naturally occurring radioisotopes of higher atomic number elements belong to chains of successive disintegrations. The original element, which starts the whole decay series, is called the parent. The new elements formed are called daughters, and the whole chain is called a family. The parent of a natural radioactive series undergoes a series of disintegrations before reaching its stable form. When a series is in secular equilibrium, one Ci of the parent will coexist with one Ci of each of the daughters. Three series, the uranium, actinium, and thorium, make up most of the naturally radioactive elements found in the periodic table (Figure 11.21.2).

—Paul A. Bouis

11.22
SOURCES OF RADIOACTIVITY IN THE ENVIRONMENT

Radioactivity in the environment comes from natural and man-made sources (Figure 11.22.1). Although natural radioactivity is the most likely to be encountered in the environment due to its widespread dispersal, man-made radioactivity poses the greatest environmental risk. Natural radioactivity harnessed by man and not properly disposed of is also a potential threat to the environment. There are five basic sources of radioactivity in the environment: the nuclear fuel cycle, mining activities, medical and laboratory facilities, nuclear weapons testing and seepage from natural deposits.

Nuclear Fuel Cycle

The nuclear fuel cycle is defined as the activities carried out to produce energy from nuclear fuel. These activities include, but are not limited to, mining of uranium-containing ores, enrichment of uranium to fuel grade specifications, fabrication and use of fuel rods, and isolation and storage of waste produced from power plants. The nuclear

fuel cycle is shown in Figure 11.22.2. Note that commercial fuel reprocessing, currently practiced in Europe, was discontinued in the United States in 1972 for safety and security reasons. The Department of Energy (DOE), however, does reprocess most of its spent fuels (U.S. DOE 1988).

Mining Activities

Mining, processing, and the use of coal, natural gas, phosphate rock, and rare earth deposits result in the concentration and release or disposal of large amounts of low–level radioactive material (UNSCEAR 1977). Coal–fired power plants release as much radioactivity (radon) to the environment as nuclear facilities, and the fly ash residue contains low levels of several natural radioisotopes. Natural gas is one of many radon sources in the environment. Phosphate rock always has associated natural radioisotopes; in many cases, tailings from phosphate operations have levels above those allowed by the

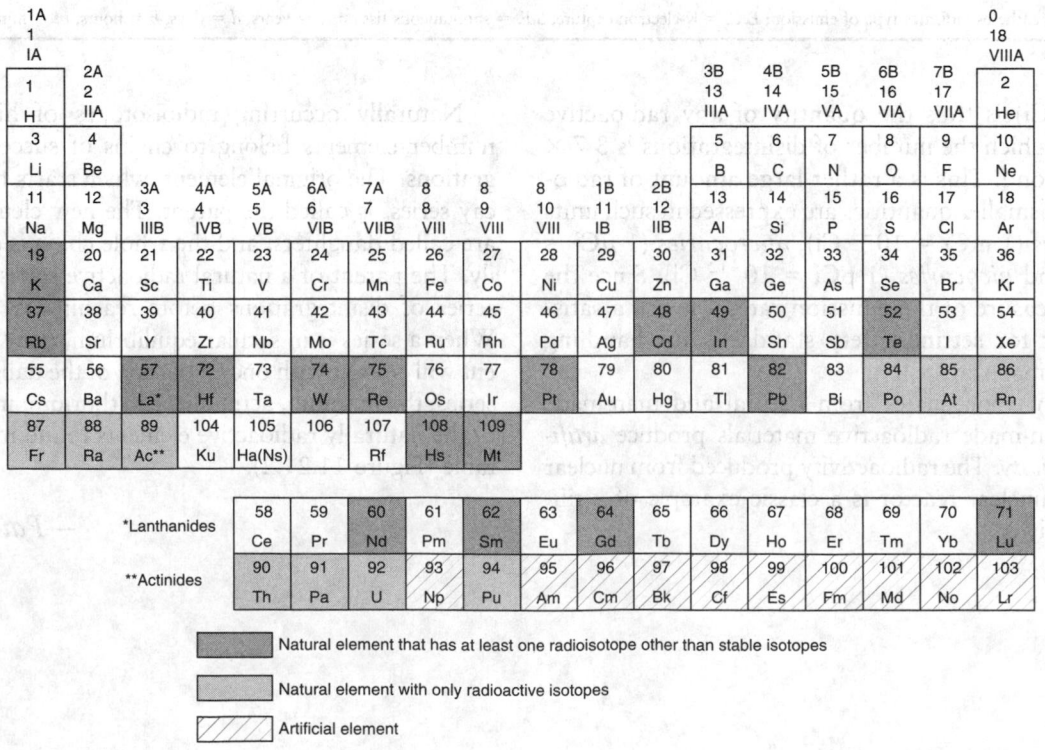

FIG. 11.22.1 Periodic table showing different types of radioisotopes.

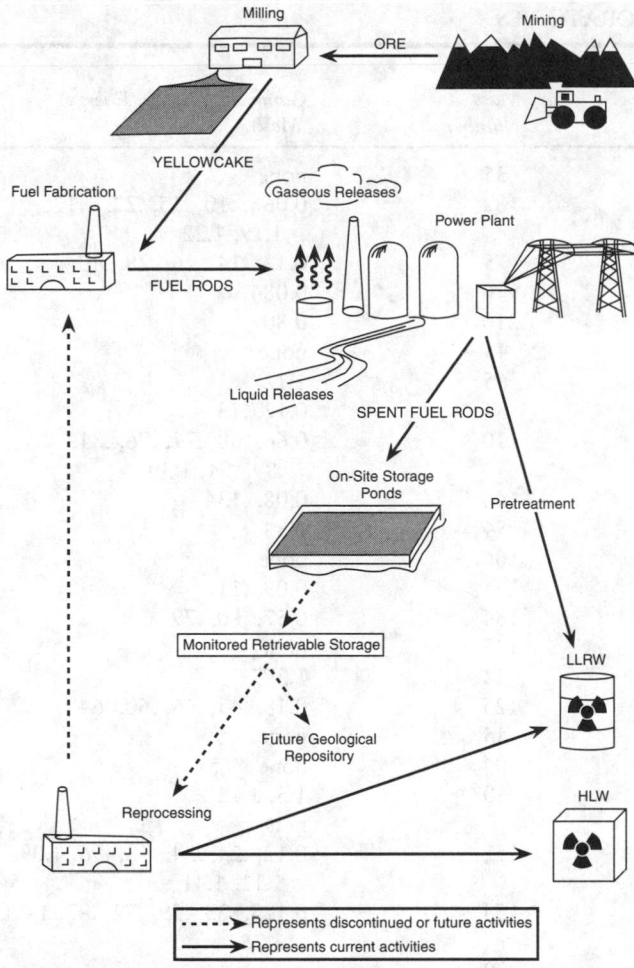

FIG. 11.22.2 The nuclear fuel cycle.

Medical and Laboratory Facilities

Radioisotopes are used extensively in medical facilities, biomedical research laboratories, and to a lesser extent in other types of laboratories (Table 11.22.2). Clinical use of radioisotopes is expanding rapidly in such areas as cancer treatment and diagnostic testing. The lack of waste management plans at many of these facilities results in frequent misclassification of materials as radioactive (Party & Gershey 1989). Relatively large amounts of radioisotopes are used in clinical procedures. Although most of these isotopes are strong emitters of gamma radiation, they have short half lives (Table 11.22.3).

Nuclear Weapons Testing

The use of nuclear devices in weapons is the primary cause of radioactive fallout, although the nuclear accident at Chernobyl and various volcanic eruptions have also contributed. Tritium (^3H) and several isotopes of iodine, cesium and strontium are found in the environment largely because of nuclear testing. In the United States, most radioactive waste is a by-product of nuclear weapons production. It is estimated that 70% of U.S. radioactive waste results from defense department activities (Eisenbud 1987). The DOE is currently trying to remedy many defense sites due to past poor waste management practices.

Natural Deposits

The majority of radioactivity in groundwater is due to seepage from natural deposits of uranium and thorium (Table 11.22.4). Strict guidelines for acceptable levels of radionuclides in drinking water exist (U.S. EPA 1986). The EPA has established maximum levels for radium (U.S. EPA 1976, 1980) to monitor for the presence of natural radionuclides (Table 11.22.5). Radon, a colorless, odorless, inert, radioactive gas that seeps out of the earth has been found at dangerously high levels in inadequately ven-

NRC for release to the environment. Processing rare-earth containing ores produces concentrated waste high enough in radioactivity to be disposed of as low-level radioactive waste. If monazite ore is the rare-earth source, nearly one ton in ten must be disposed of in this manner. Disposal costs have dramatically reduced imports of monazite (Table 11.22.1).

TABLE 11.22.1 U.S. IMPORTS FOR CONSUMPTION OF MONAZITE, BY COUNTRY

Country	1985 Quantity (Metric Tons)	1986 Quantity (Metric Tons)	1987 Quantity (Metric Tons)	1988 Quantity (Metric Tons)	1989 Quantity (Metric Tons)
Australia	5,694	2,660	—	382	180
India	—	300	—	—	—
Indonesia	—	—	—	1,144	794
Malaysia	—	—	527	197	—
Thailand	—	—	594	201	—
Total	5,694	2,960	1,121	1,924	974
RBO content[a]	3,132	1,628	617	1,058	536

[a]Estimated.
Source: Reprinted and adopted from U.S. Bureau of the Census.

TABLE 11.22.2 RADIOISOTOPES ENCOUNTERED IN LABORATORIES

Half-Life	Element and Symbol	Atomic Number	Mass Number	Gamma Radiation Energy (MeV)
88 days	Sulfur (S)	16	35	none
115 days	Tantalum (Ta)	73	182	0.068, .10, .15, .22, 1.12, 1.19, 1.22
120 days	Selenium (Se)	34	75	0.12, .14, .26, .28, .40
130 days	Thulium (Tm)	69	170	0.084
138 days	Polonium (Po)	84	210	0.80
165 days	Calcium (Ca)	20	45	none
245 days	Zinc (Zn)	30	65	1.12
270 days	Cobalt (Co)	27	57	0.12, .13
253 days	Silver (Ag)	47	110	0.66, .68, .71, .76, .81, .89, .94, 1.39
284 days	Cerium (Ce)	58	144	0.08, .134
303 days	Manganese (Mn)	25	54	0.84
367 days	Ruthenium (Ru)	44	106	none
1.81 yr	Europium (Eu)	63	155	0.09, .11
2.05 yr	Cesium (Cs)	55	134	0.57, .60, .79
2.6 yr	Promethium (Pm)	61	147	none
2.6 yr	Sodium (Na)	11	22	1.277
2.7 yr	Antimony (Sb)	51	125	0.18, .43, .46, .60, .64
2.6 yr	Iron (Fe)	26	55	none
3.8 yr	Thallium (Tl)	81	204	none
5.27 yr	Cobalt (Co)	27	60	1.3, 1.12
11.46 yr	Hydrogen (H)	1	3	none
12 yr	Europium (Eu)	63	152	0.12, .24, .34, .78, .96, 1.09, 1.11, 1.41
16 yr	Europium (Eu)	63	154	0.123, .23, .59, .72, .87, 1.00, 1.28
28.1 yr	Strontium (Sr)	38	90	none
21 yr	Lead (Pb)	82	210	0.047
30 yr	Cesium (Cs)	55	137	0.661
92 yr	Nickel (Ni)	28	63	none
1602 yr	Radium (Ra)	88	226	0.186
5730 yr	Carbon (C)	6	14	none
2.12×10^5 yr	Technetium (Tc)	43	99	none
3.1×10^5 yr	Chlorine (Cl)	17	36	none

Source: Reprinted, from U.S. Department of Health, Education and Welfare (HEW), 1970, *Radiological health handbook,* rev. ed. (HEW, Rockville, Md., [January]).

TABLE 11.22.3 PRINCIPAL CLINICALLY ADMINISTERED RADIOISOTOPES

Radionuclide	Principal Uses	Half-life	Typical Dose (mCi)	Number of Procedure[a]	Total Curies	% of Total
^{99m}Tc	Organ imaging	6.0 hr	4–25	8,040,000	116,580	96
^{309}Tl	Myocardial & parathyroid imaging	74.0 hr	2	960,000	1,920	2
Ga	Tumor/infection diagnosis	78.1 hr	5	600,000	3,000	2
I	Thyroid imaging	8.1 days	0.1	960,000	96	<0.1
Total				10,560,000	121,596	100

[a]Annual number of procedures in the United States.

TABLE 11.22.4 AVERAGED SOIL AND ROCK CONCENTRATION OF URANIUM AND THORIUM

| | U | Th |
	(mg kg^{-1})	
Rocks		
Igneous		
Silica (granites)	4.7	20.0
Intermediate (diorites)	1.8	8.0
Mafic (basalt)	0.9	2.7
Ultramafic (dunites)	0.03	6.0
Sedimentary		
Limestones	2.2	1.7
Carbonates	2.1	1.9
Sandstones	1.5	3.0
Shales	3.5	11.0
(Mean value in earth's crust)	3.0	11.4
Soils		
Typical range	1–4	2–12
World average	2	6.7
Average specific activity (pCi/kg^{-1})	670	650

Source: Reprinted, with permission, from M. Boyle, 1988, Radon testing of soils, *Environmental Science Technology* [22(12):1397–1399].

TABLE 11.22.5 CONCENTRATIONS OF RA^{226} AND DAUGHTERS IN CONTINENTAL WATERS (pCi per liter)

	Ra^{226}	Rn^{222}	Pb^{210}	Po^{210}
Deep wells	1–10	10^4–10^5	<0.1*	~0.02
Ground water	0.1*–1	10^2–10^3	<0.1*	~0.01
Surface water	<1	10	<0.5	—
Rainwater	—	10^3–10^5†	0.5–3	~0.5

*Below detection limits.
†As determined through presence of short-lived Rn^{222} daughters.

tilated buildings. Radon originates from the radioactive decay of uranium, thorium and/or radium (Figure 11.22.3). The EPA states that levels above four pCi/L should be reviewed for possible corrective actions (Boyle 1988). (See Sections 5.31 and 5.32 for further information on this subject.)

—Paul A. Bouis

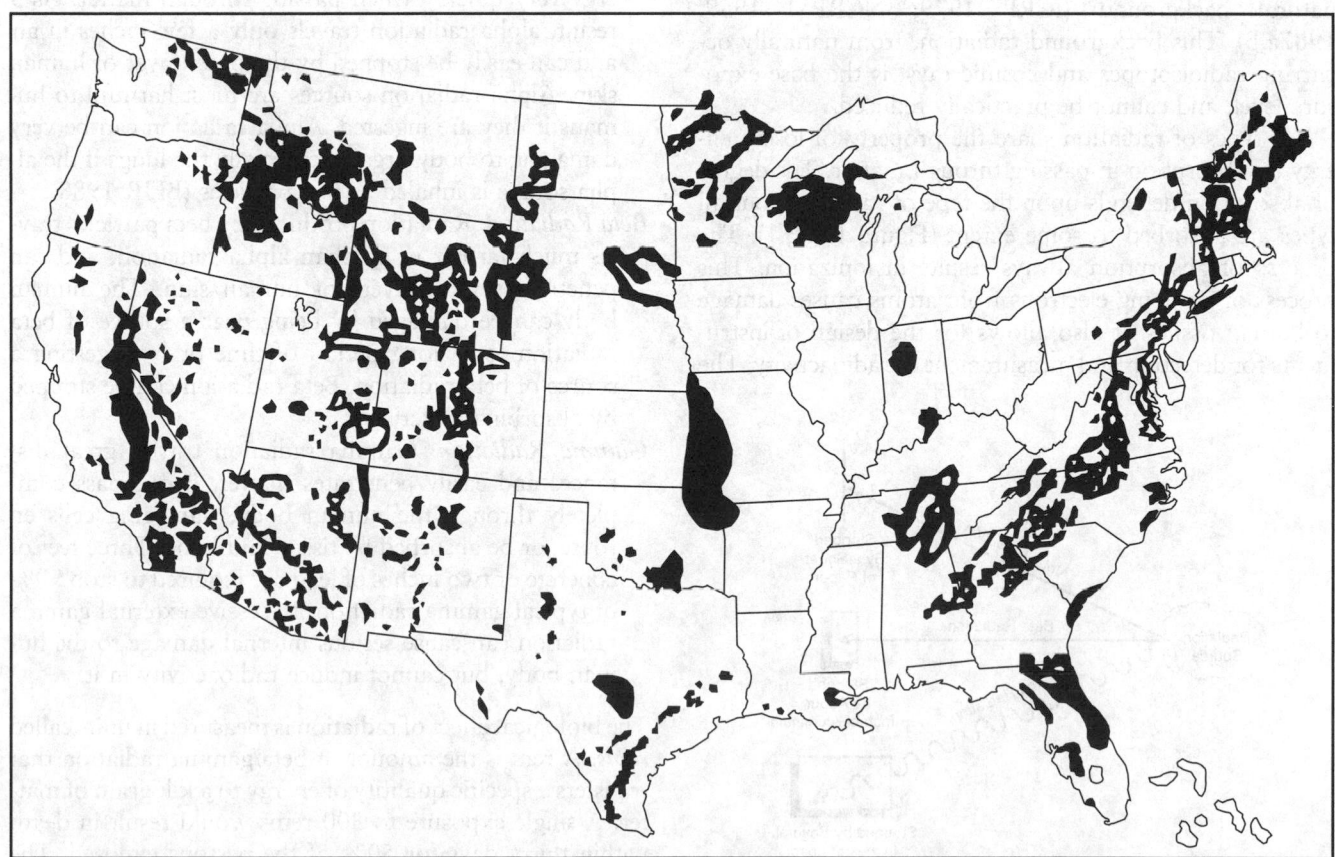

FIG. 11.22.3 Areas with potentially high radon levels. (Reprinted, with permission, from M. Boyle, 1988, Radon testing of soils, *Environmental Science Technology* [22(12):1397–1399].)

References

Boyle, M. 1988. Radon testing of soils, *Environmental Science Technology*, 22(12):1397–1399.

Eisenbud, M.E. 1987. *Environmental radioactivity from natural, industrial, and military sources.* 3rd ed. Orlando, Fla.: Academic Press.

Party, E.A., and E.L. Gershey. 1989. Recommendations for radioactive waste reduction in biomedical/academic institutions. *Health Physics* 56(4):571–572.

U.S. Department of Energy (DOE). 1988. *Database for 1988: spent fuel and radioactive waste inventories, projections, and characteristics.* DOE/RW–0006, Rev. 4. Washington, D.C.

United Nations Scientific Committee on the Effects of Atomic Radiation (UNSCEAR). 1977. *Sources and effects of ionizing radiation.* New York, N.Y.

U.S. Environmental Protection Agency (EPA). 1976. Drinking water regulations, radionuclides. *Federal Register* 41:28402.

———. 1980. *Prescribed procedures of measurement of radioactivity in drinking water.* EPA 600–4–80–032.

———. 1986. Water pollution control: radionuclides: advance notice of proposed rulemaking. 40 CFR Part 141, 34836: *Federal Register* 51:189.

11.23
SAFETY STANDARDS

Radioactivity presents special hazards because it cannot be detected by the normal human senses. Strict safety standards have been established by international organizations to ensure that exposure to workers is minimized and that the public is not exposed to radiation from other than the natural background (ICRP 1979; NCRPM 1959, 1987a,b). This background radiation, from naturally occurring radioisotopes and cosmic rays, is the base exposure level, and cannot be practically reduced.

All types of radiation share the property of losing energy by absorption in passing through matter. The degree of absorption depends upon the type of radiation, but all types are absorbed to some extent (Figure 11.23.1). The process of absorption always results in ionization. This process of stripping electrons from atoms causes damage to human tissues. It also allows for the design of instruments for detection and measurement of radioactivity. The properties of the various radiations determine the protective measures needed and the methods of measurement. Three types of radiation exist.

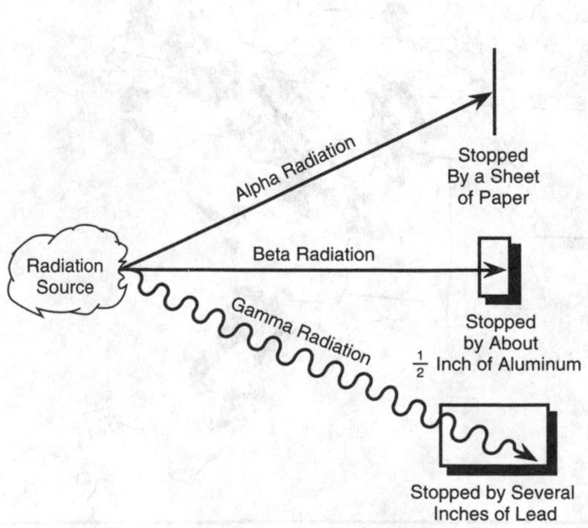

Alpha Radiation: Radiation from alpha particles loses energy very quickly when passing through matter. As a result, alpha radiation travels only a few inches in air and can easily be stopped by the outer layer of human skin. Alpha radiation sources are most harmful to humans if they are ingested. Alpha radiation can be very damaging to body organs, especially the lungs if the alpha source is inhaled as fine particles (BEIR 1988).

Beta Radiation: Radiation produced by beta particles travels much farther in air than alpha radiation, and can penetrate several layers of human skin. The human body can be damaged by being near a source of beta radiation for a long period of time or by ingesting a source of beta radiation. Beta radiation can be stopped by absorbing materials.

Gamma Radiation: Gamma radiation travels great distances and easily penetrates matter. It can pass completely through the human body, damaging cells en route, or be absorbed by tissue and bone. Three feet of concrete or two inches of lead are required to stop 90% of typical gamma radiation. Excessive external gamma radiation can cause serious internal damage to the human body, but cannot induce radioactivity in it.

The biological effect of radiation is measured in units called *rems*. A rem is the amount of beta/gamma radiation that transfers a specific quantity of energy to a kilogram of matter. A single exposure to 300 rems would result in death within thirty days for 50% of the persons exposed. The unit of dose is difficult to put into perspective, however, a comparison of the allowable doses helps. The permissi-

FIG. 11.23.1 Relative penetrating power of alpha, beta, and gamma radiation.

ble level for occupational radiation exposure is five rems per year to the whole body. It is believed that this level can be absorbed for a working lifetime without any sign of biological damage. Background radiation is measured in millirems (0.001 rem).

The average person is exposed to ionizing radiation from many sources. The environment, and even the human body, contains naturally occurring radioactive materials. Cosmic radiation contributes additional exposure. The use of x-rays and radioisotopes in medicine and dentistry adds to the public exposure. Table 11.23.1 shows the estimated average individual exposure in millirems from natural background and other sources.

Protection from Exposure

Maximum permissible levels of external and internal radiation (Table 11.23.2) have been set by the National Council on Radiation Protection (NCRP) and by the International Commission on Radiological Protection (ICRP). In addition, the practice of keeping exposures *As Low As Reasonably Achievable* (NRC 1976) is recommended by these and many other organizations. This means that every activity involving exposure to radiation should be planned to minimize unnecessary exposure to workers and the public. For further explanations of radiation protection the reader is referred to the many references available (Henry 1969, Olishifski 1981, Schapiro 1981).

Basic Radiation Safety

Safety practices for handling radioactive materials are aimed at protecting individuals from external and internal hazards.

TABLE 11.23.1 U.S. GENERAL POPULATION EXPOSURE ESTIMATES

Source	Average Individual Dose mrem/yr
Natural Background	100
Mining Releases	5
Medical	90
Nuclear Fallout	7
Nuclear Energy	0.3
Consumer Products	0.03
Total	≈200

TABLE 11.23.2 MAXIMUM PERMISSIBLE DOSE EQUIVALENT FOR OCCUPATIONAL EXPOSURE

Combined whole body occupational exposure	
Prospective annual limit	5 rems in any one year
Retrospective annual limit	10–15 rems in any one year
Long-term accumulation	$(N-18) \times 5$ rems, where N is age in years
Skin	15 rems in any one year
Hands	75 rems in any one year (25/qtr)
Forearms	30 rems in any one year (10/qtr)
Other organs, tissues and organ systems	15 rems in any one year (5/qtr)
Fertile women (with respect to fetus)	0.5 rem in gestation period
Dose limits for the public, or	
occasionally exposed individuals:	
Individual or occasional	0.5 rem in any one year
Students	0.1 rem in any one year
Population dose limits	
Genetic	0.17 rem average per year
Somatic	0.17 rem average per year
Emergency dose limits—Life saving:	
Individual (older than 45, if possible)	100 rems
Hands and forearms	200 rems, additional (300 rems, total)
Emergency dose limits—Less urgent:	
Individual	25 rems
Hands and forearms	100 rems, total
Family of radioactive patients:	
Individual (under age 45)	0.5 rems in any one year
Individual (over age 45)	5 rems in any one year

Reprinted from National Committee on Radiation Protection (NCRP), 1975, *Review of the current state of radiation protection philosophy.* NCRP Publication No. 43.

TABLE 11.23.3   CLASSIFICATION OF ISOTOPES ACCORDING TO RELATIVE RADIOTOXICITY PER UNIT ACTIVITY (THE ISOTOPES IN EACH CLASS ARE LISTED IN ORDER OF INCREASING ATOMIC NUMBER)

CLASS 1 (very high toxicity)

Sr-90 + Y-90, *Pb-210 + Bi-210 (RaD + E), Po-210, At-211, Ra-226 + percent *daughter products, Ac-227, *U-233, Pu-239, *Am-241, Cm-242.

CLASS 2 (high toxicity)

Ca-45, *Fe-59, Sr-89, Y-91, Ru-106 + *Rh-106, *I-131, *Ba-140 + La-140, Ce-144 + *Pr-144, Sm-151, *Eu-154; *Tm-170, *Th-234 + *Pa-234, *natural uranium.

CLASS 3 (moderate toxicity)

*Na-22, *Na-24, P-32, S-35, Cl-36, *K-42, *Sc-46, Sc-47, *Sc-48, *V-48, *Mn-52, *Mn-54, *Mn-56, Fe-55, *Co-58, *Co-60, Ni-59, *Cu-64, *Zn-65, *Ga-72, *As-74, *As-76, *Br-82, *Rb-86, *Zr-95 − *Nb-95, *Nb-95, *Mo-99, Tc-98, *Rh-105, Pd-103 − Rh-103, *Ag-105, Ag-11, Cd-109 − *Ag-109, *Sn-113, *Te-127, *Te-129, *I-132, Cs-137 − *Ba-137, *La-140, Pr-143, Pm-147, *Ho-166, *Lu-177, *Ta-182, *W-181, *Re-183, *Ir-190, *Ir-192, Pt-191, *Pt-193, *Au-198, *Au-199, Tl-200, Tl-202, Tl-204, *Pb-203.

CLASS 4 (slight toxicity)

H-3, *Be-7, C-14, F-18, *Cr-51, Ge-71, *Tl-201.

*Gamma-emitters.

Source: International Atomic Energy Agency (IAEA), *Safe handling of radionuclides,* Safety Series No. 1 (Vienna, Austria: IAEA).

EXTERNAL RADIATION

Protection from external radiation is accomplished by adhering to the principles of maximizing distance, minimizing time, and shielding individuals from the radioactive source. Exposure levels are readily measured using conventional radiation-measuring devices (IAEA 1976). This allows the distance, time, and the necessary amount of shielding to be determined to minimize exposure.

INTERNAL RADIATION

The most frequent routes for radioactive materials intake are through inhalation or open wounds. Air monitoring in areas where radioactive materials are handled is always recommended. Simple methods and measuring devices exist. Periodic testing of urine, body fluids, and excrement is also recommended as a secondary means of determining if radioisotopes have been ingested. The toxicity of various isotopes is shown in Table 11.23.3.

—Paul A. Bouis

References

Committee on the Biological Effects of Ionizing Radiation (BEIR). 1988. *Health risks from radon and other internally deposited alpha emitters.* BEIR IV Report. Washington, D.C.: National Academy Press.

Henry, H.F. 1969. *Fundamentals of radiation protection.* New York, N.Y.: Wiley Interscience.

International Atomic Energy Agency (IAEA). 1976. Manual on radiological safety in uranium and thorium mines and mills. *Safety Series No. 43,* Vienna, Austria.

International Commission on Radiation Protection (ICRP). 1979. Limits for intake of radionuclides by workers. *ICRP Publication 30.* New York, N.Y.: Pergamon Press.

National Committee on Radiation Protection and Measurements (NCRP). 1959. Maximum permissible body burdens and maximum permissible concentrations of radionuclides in air and water for occupational exposure. *NBS Handbook No. 69.*

———. 1987a. Recommendations on limits for exposure to ionizing radiation. *NCRP Report No. 91.* Bethesda, Md.

———. 1987b. Ionizing radiation exposure of the population of the United States. *NCRP Report No. 93.* Bethesda, Md.

U.S. Nuclear Regulatory Commission (NRC). 1976. Operating philosophy for maintaining occupational radiation exposures as low as is reasonably achievable. *NRC Regulatory Guide 8.10.*

Olishifski, J.B. 1981. *Fundamentals of industrial hygiene.* Chicago, Il.: National Safety Council.

Schapiro, J. 1981. *Radiation protection: A guide for scientists and physicians,* 2d ed. Cambridge, Mass.: Harvard University Press.

11.24
DETECTION AND ANALYSIS

FEATURE SUMMARY

Types of Measurement: A. Radiation surveys; B. Personnel monitoring; C. Radiological analysis

Types of Detection: A. Film and thermoluminescent; B. Gas filled devices; C. Scintillation counters; D. Gamma-ray spectrometers

Range: A. Personnel monitoring, millirems to rems; B. Radiological analysis, picocuries to curies

Approximate Cost: A. Personnel monitoring, $500 to $100,000, Typical—$1000 to $10,000 for basic monitoring capabilities; B. Radiological analysis, $3,000 to $100,000, Typical—$15,000 for basic analytical capabilities, $50,000 with gamma ray capabilities

PARTIAL LIST OF SUPPLIERS

A. Survey and personnel monitoring: Amersham Corp.; Dosimeter Corp.; Eberline Instrument Corp.; Edmund Scientific Co.; EG&G Ortec/EG&G Berthold; Health Physics Instruments; Keithley Instruments, Inc.; Lab Safety Supply, Inc.; Ludlum Measurements, Inc.; Oxford Instruments, Inc.; Radiation Monitoring Devices, Inc.; TN Technologies, Inc.

B. Radiological Analysis: Canberra Industries, Inc.; EG&G Ortec/EG&G Berthold; Health Physics Instruments; Oxford Instruments, Inc.; PGC Scientific Corp.; Premier American Technologies, Corp.; Princeton Gamma-Tech, Inc.; Teledyne Isotopes

Detection and quantification of radioactivity is critical to personnel and environmental safety, since radioactivity cannot be detected by the human senses. Devices to measure the dose of radiation received, and the amount of the radionuclide present are available. In spite of this, there are few qualified commercial enterprises that handle radiological monitoring or analysis. Most radiological expertise is confined to government or industrial institutions whose primary responsibility or business involves radioactive materials. Some general-purpose radiological services are provided by individual state laboratories.

Radiation Monitoring

Radiation measurement for the purpose of protection from exposure requires proper measuring devices and proper measurement techniques. Equipment and units of measure for health protection are designed in terms of absorbed radiation dose. This absorbed dose is defined as the energy imparted to matter by ionizing radiation per unit mass of irradiated material at a given location. This unit of absorbed dose is called a *rad*. The dose equivalent is a quantity used in radiation protection to express all radiation exposure on a common scale. The unit of dose equivalent is the *rem*. Rads of γ and β radiation are normally equivalent to rems, and are used interchangeably. The *sievert* (Sv) is the equivalent SI unit and is equal to 100 rems.

Radiation doses can be measured using survey meters when instantaneous dose readings of a particular area are required, or by badges or dosimeters for longer-term measurements. Radiation surveys will often measure doses from the source, surface contamination (using wipe samples), and airborne contaminants.

SURVEY INSTRUMENTS

Geiger Mueller Counter

This meter is able to measure all ionizing radiation and is therefore used for area surveys. However, it uses a gas-filled detector; it cannot identify the specific source of radioactivity; it has a poor efficiency for gamma radiation; and it can easily be overloaded.

Ionization Chamber

Portable versions of this chamber are used routinely in radiation protection surveys. These detectors give a direct indication of the exposure rate to gamma rays. In many cases, the detectors have a thin window and a removable shield allowing measurement of high energy alpha and beta particles.

Scintillation Counter

This flat probe is used extensively for surface contamination surveys (IAEA 1976). It uses a ZnS (Ag) scintillator which only responds to alpha particles. NOTE: the aluminized plastic film window is susceptible to pinhole leaks.

Film Badges

Used to monitor worker exposure to ionizing radiation, the film badge is a photographic film packet used to measure approximate long-term exposure. It is usually read monthly or quarterly. The badge may contain two or more films of differing sensitivity, and it may contain filters that shield part of the film from certain types of radiation.

Thermoluminescent Dosimeter (TLD)

This is the primary method used for long term radiation monitoring. These devices can be configured as finger or body badges. The TLD badge is usually made of lithium fluoride, which is capable of storing absorbed ionizing radiation, then releasing this energy in the form of visible light when heated. The amount of light released is then related to the amount of radiation exposure. The TLD is rugged in design, and is independent of absorption energy for gamma radiation.

DEVICE CALIBRATION

Periodic calibration of survey instruments and dosimeter readers must be done by qualified experts. Check standards, consisting of encapsulated known quantities of radioisotopes, are commercially available and should be used daily to check that these devices are in proper working order.

Radioactivity Analysis

Detection, identification, and quantification of the specific radionuclide responsible for the presence of ionizing radiation is very often required. Identification of the radionuclide facilitates the design of systems to safely and economically handle radioactive materials. Many different types of instruments, ranging in price from several thousand to several hundred thousand dollars, are available. Field measuring devices are often used to detect and approximate the amount of radiation present. Identification and quantification is usually performed in a controlled laboratory area, equipped to minimize background interference and gross contamination, and to ensure personnel safety.

Radiological analysis must be done using good laboratory practices and sound scientific judgement (Knoll 1989). Many measurements are done at extremely low levels (NCRP 1976), and poor laboratory technique often leads to erroneous conclusions (NCRP 1978). The laboratory area dedicated to radiological analysis is often called the counting room. The gross alpha-beta measurement is the most frequently run analysis. This relatively inexpensive analysis screens samples to determine if more elaborate analysis for specific radionuclides is required. Specific radionuclide analyses are usually done if dose estimates are desired, if long-term trends are being established, or if regulatory requirements are mandated. Analyses of specific radionuclides are usually more expensive and require greater expertise to interpret the results. These analyses are based on accurately measuring the half-life or the energy of the radiation emitted. Chemicals are often separated prior to instrumental analysis to eliminate potential interferences or to concentrate the radionuclide of interest. Gamma-emitting materials can be identified and quantified quickly with little or no sample preparation by using gamma ray spectroscopy.

A sample representing the area or situation under analysis must be obtained. Survey instruments can be effective in obtaining a proper sample. Liquid and solid samples are very amenable to analysis by gamma ray spectroscopy. Air samples can be collected using continuous samplers. Air-grab samples can be taken directly into evacuated scintillation cells if alpha emitters are being monitored.

ANALYTICAL COUNTING INSTRUMENTS

Proportional Counters

These instruments, mid-range in price, are most frequently used for gross alpha/beta testing. They are gas-filled detectors that operate on the principle that gas ionization by radiation can, with the proper circuitry, produce an electrical signal proportional to the amount of radiation that caused the ionization. P-10 gas, a methane-containing mixture, is used as the counting gas. Three types of counters are routinely used: internal, end window, and thin window. Internal counters are especially suited for low-level environmental samples (Greenberg, Clesceri & Eaton 1992). End window and thin window counters are more suitable for moderate to high level radioactivity samples. These are easier to maintain and less prone to contamination than internal counters. All proportional counters require a calibration curve be run to quantify results (Figure 11.24.1).

Scintillation Counter

In this counter, light flashes produced in a scintillator (ZnS [Ag]) by ionizing radiation are converted into electrical pulses by a photo-multiplier tube. The scintillation counter is especially suited for the analysis of alpha-emitting

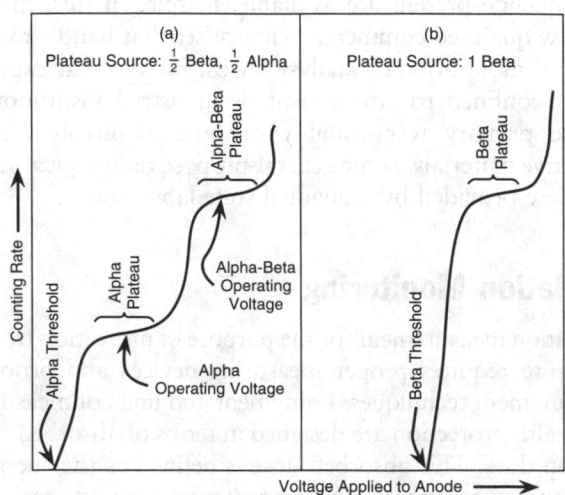

FIG. 11.24.1 Calibration curve-counting rate vs. anode voltage curve for internal proportional counter with P-10 gas.

gaseous radioisotopes such as radon. Radium 226 analysis by radon de-emanation into a Lucas cell is a classic use of the scintillation counter (Greenberg, Clesceri & Eaton 1992) (Figure 11.24.2). Calibration of each cell is required to obtain quantitative results.

Gamma Ray Spectroscopy

Simultaneous analysis of multiple specific radionuclides can be done using gamma ray spectroscopy (Heath 1974). This method is applicable to the analysis of gamma-emitting radionuclides with gamma energies ranging from 80 keV to approximately 2000 keV. The technique minimizes the sample preparation required to do radiochemical analysis. Using a NaI detector, it is possible to routinely analyze four to eight gamma-emitting radionuclides. Personal computer-based, high-resolution intrinsic germanium detector systems are now used almost exclusively in gamma ray spectroscopy. These systems can analyze an almost unlimited number of radionuclides, and are especially suited for low-level analysis of environmental samples. A comparison of the superior resolution of an intrinsic germanium detector is shown in Figure 11.24.3. The most frequently used photo-energy peaks for common radionuclides are shown in Table 11.24.1. Liquid and solid samples can be placed directly in a Marineli beaker for analysis. An efficiency calibration is required for quantitative analysis. The result of this calibration is an efficiency versus energy curve (Figure 11.24.4).

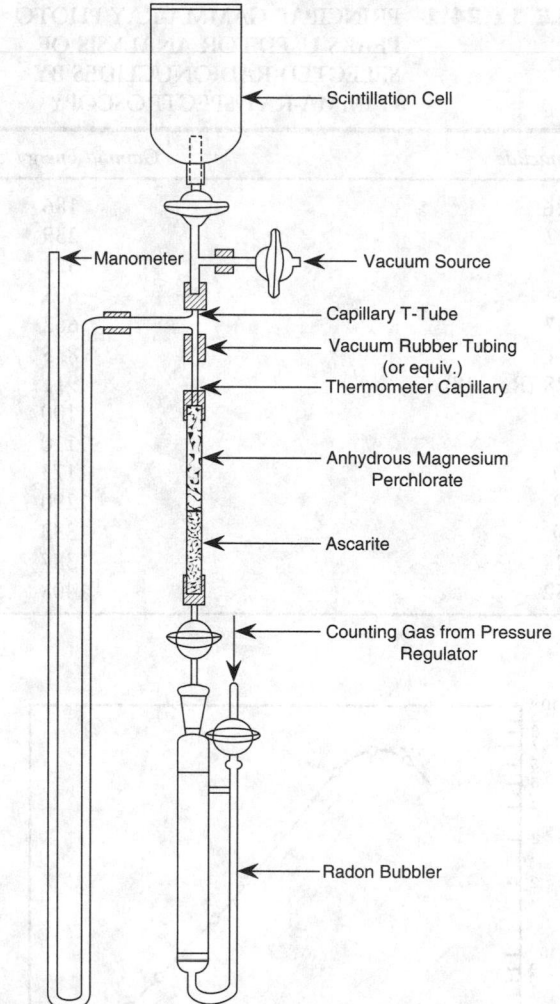

FIG. 11.24.2 De-emanation assembly.

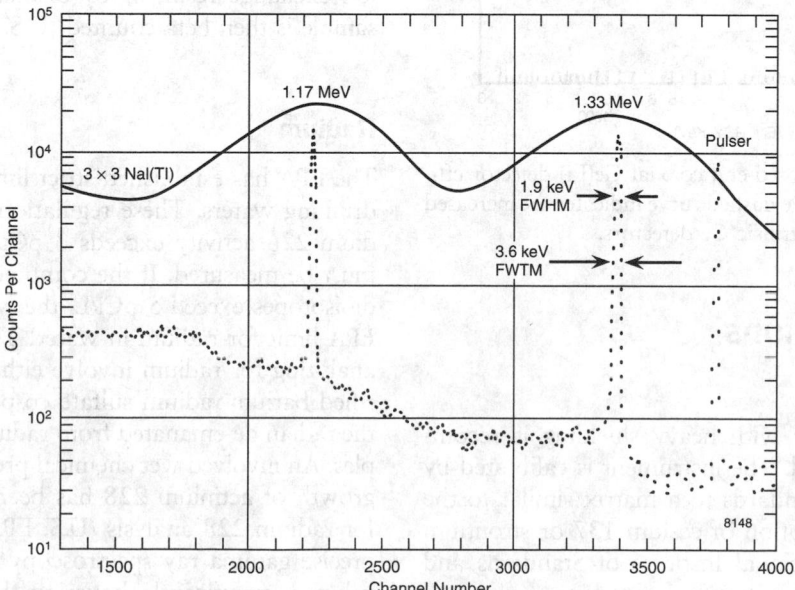

FIG. 11.24.3 ^{60}Co spectrum showing resolutions and peak-to-compton ratios for an intrinsic Ge detector and a NaI(Tl) detector.

TABLE 11.24.1 PRINCIPAL GAMMA-RAY PHOTO PEAKS USED FOR ANALYSIS OF SELECTED RADIONUCLIDES BY GAMMA-RAY SPECTROSCOPY

Radionuclide	Gamma Energy keV
Ra 226	186
Pb 212	239
Cr 51	321
Cs 134	605
Cs 137	662
Mn 54	835
Ac 228 (Ra 228)	911
Co 58	1100
Zn 65	1110
Co 60	1173
Co 58	1290
Co 60	1333
Ce 144	1387
Eu 152	1408

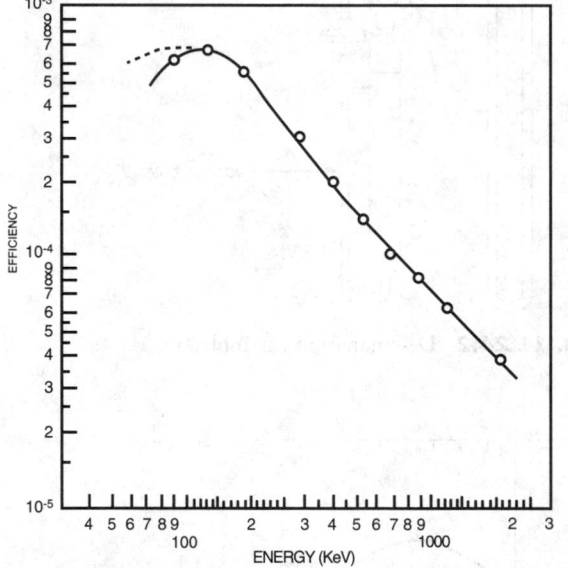

FIG. 11.24.4 Typical closed-end coaxial Ge(Li) detector efficiency calibration curve. The dashed curve indicates the increased low-energy efficiency of intrinsic Ge detectors.

ANALYTICAL METHODS

Gross Alpha-Beta

A proportional counter with heavy shielding is recommended for this method. The instrument is calibrated by adding radionuclide standards to a matrix similar to the sample. A standard solution of cesium 137 or strontium 90 certified by the National Institute of Standards and Technology (NIST) is suitable for gross beta analysis. A solution of natural uranium, thorium, plutonium 239 or americium 241 is recommended for gross alpha analysis. Gross alpha-beta results are always reported in comparison to a specific standard.

The sample, usually a liquid, is evaporated onto a *planchette,* to a thin film. The standard is prepared in the same manner. Counts from the sample are then compared to the standard. Samples suspected of containing fission or artificial radionuclides can be tested for gross beta using either the cesium or strontium standard. Environmental samples suspected of containing natural radionuclides can be tested for gross alpha using any of the alpha standards previously listed. Careful attention must be paid to self absorption of alpha and beta particles due to sample thickness on the planchette whenever test results are evaluated.

Radioactive Cesium

An extremely hazardous fission product, the interim EPA drinking water regulations limit cesium 134 to 80 pCi/L and cesium 137 to 200 pCi/L. Samples suspected of containing moderate or high levels of cesium can be tested by gamma ray spectroscopy. Low-level environmental samples can be purified and concentrated by co-precipitation with ammonium phosphomolyodate and analyzed either by gamma ray analysis or by beta counting (Kreiger 1976).

Radioactive Iodine

Radioiodine originates from nuclear weapons testing and from the nuclear fuel cycle. Fission products may contain iodines 129 through 135. The EPA drinking water maximum for iodine 135 is 3 pCi/L. Samples are preconcentrated either by precipitation as PdI_2, absorption on an anion exchange resin, or by distillation. The concentrated sample is then beta counted (U.S. EPA 1980).

Radium

The EPA has established strict limits on radium in public drinking waters. These regulations require that if the radium 226 activity exceeds 3 pCi/L, radium 228 activity must be measured. If the combined activities of these radioisotopes exceed 5 pCi/L, the water supply exceeds the EPA limit for radium in water. The standard methods of analyzing for radium involve either alpha counting a purified barium-radium sulfate co-precipitate, or measuring the radian de-emanated from radium 226-containing samples. An involved wet chemical procedure based on the ingrowth of actinium 228 has been published by the EPA for radium 228 analysis (U.S. EPA 1980). Simpler, more precise gamma ray spectroscopy methods have been developed, significantly lowering the detection limits (U.S. EPA 1980).

Strontium

Nuclear fission produces radioactive strontium isotopes. Strontium 90 is an extremely hazardous isotope. Upon ingestion it tends to concentrate in bone. Analysis of strontium involves tedious and complicated wet procedures of large samples. It is impossible to separate the isotopes of strontium, therefore strontium 90 is actually determined by measuring the amount of its daughter, yttrium 90. The final purified concentrate is beta counted using cesium 137 as the calibration standard.

Tritium

Tritium is found in the environment as a result of natural cosmic rays, nuclear weapons testing, and the nuclear fuel cycle. Tritium eventually decays by beta emission to helium. Analysis consists of an alkaline permanganate distillation, mixing with a liquid scintillator, and beta counting with a liquid scintillation spectrometer.

Uranium

Uranium is found in most drinking water supplies as a soluble carbonate. Uranium 238 is the primary isotope found in these waters. Standard uranium methods involve complicated wet procedures combined with ion exchange purification prior to alpha counting with a proportional counter (Barker 1965). A direct fluorescence analyzer is now commercially available, considerably simplifying this analysis.

—Paul A. Bouis

References

Analytical chemistry, lab guide edition, Vol. 65, No. 16. American Chemical Society.

American Laboratory. *Buyers' Guide Edition.* Vol. 26, No. 4. International Scientific Communications, yearly publication.

Barker, F.B., et al. 1965. *Determination of uranium in natural waters.* U.S. Geological Survey, Water Supply Paper 1696-C. Washington, D.C.: U.S. Government Printing Office.

Greenberg, A.E., L.S. Clesceri, and A.D. Eaton. 1992. *Standard methods for the examination of water and wastewater,* 18th ed. APHA. Washington, D.C.

Heath, R.L. 1974. *Gamma ray spectrum catalogue, Ge(Li) and Si(Li) spectrometry.* ANCR-1000-2. National Technical Information Service. Springfield, Va.

International Atomic Energy Agency (IAEA). 1976. Manual on radiological safety in uranium and thorium mines and mills. *Safety Series No. 43.* Vienna, Austria: IAEA.

Knoll, G.F. 1989. *Radiation detection and measurements.* New York, N.Y.: J. Wiley & Sons.

Kreiger, H.L. 1976. *Interim radiochemical methodology for drinking water.* EPA 600-4-75-008 (Revised). U.S. Environmental Protection Agency, Environmental Monitoring and Support Lab. Cincinnati, Oh.

National Council on Radiation Protection and Measurements (NCRP). 1976. Environmental radiation measurements. *NCRP Report No. 50.* Washington, D.C.

———. 1978. A handbook of radioactivity measurement procedures. *NCRP Report No. 58.* Washington, D.C.

U.S. Environmental Protection Agency (EPA). 1980. *Prescribed procedures for measurements of radioactivity in drinking water.* EPA 600-4-80-032. Environmental Monitoring and Support Lab. Cincinnati, Oh.

11.25
MINING AND RECOVERY OF RADIOACTIVE MATERIALS

The nuclear fuel cycle begins with the exploration and mining of uranium-containing ores. Although few active mining sites are currently in operation, there are numerous closed mines throughout the world. Many of the facilities in the United States, built and operated under contract to the DOE, have not been properly remediated. The proliferation of sites occurred during a period in the 1970s when uranium prices skyrocketed. Ores containing as little as 0.2% U_3O_8 were processed during this period.

A simplified schematic of a typical uranium mill is shown in Figure 11.25.1. Closure of mining and mill sites is strictly regulated by the NRC and its agreement states. Detailed criteria for closure can be found in the Code of Federal Regulations (10 CFR Part 40). The mining and milling of uranium ores produces large quantities of contaminated rock, sludge, gases, and liquids. These materials contain varying concentrations of the radioactive ore and its daughters such as radon, radium, lead, thorium,

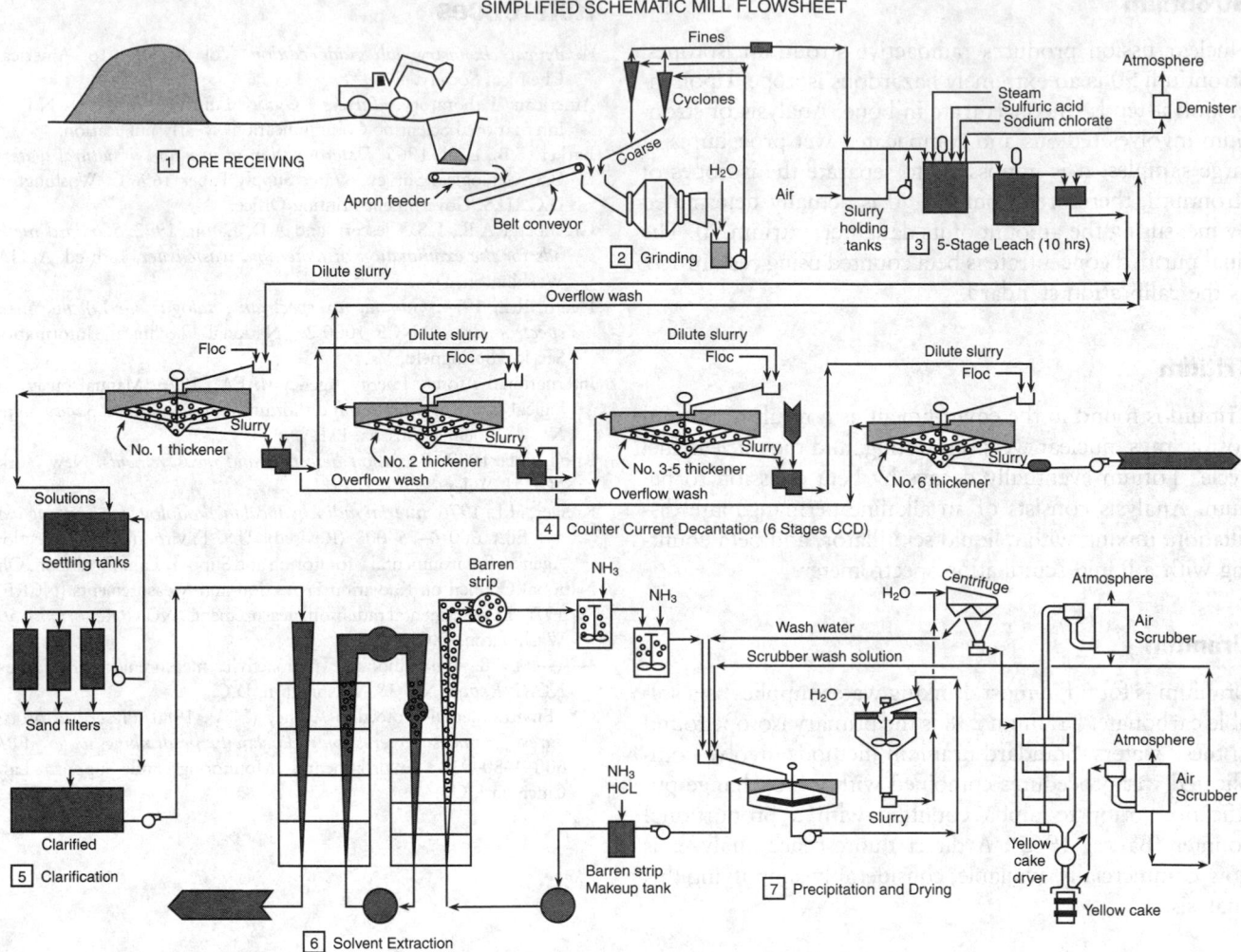

FIG. 11.25.1 Process schematic for a typical uranium mill.

and bismuth. Tailings and waste produced from the extraction or concentration of uranium or thorium from any ore processed primarily to recover *source material* is called by-product material. Source material is defined as any material containing more than 0.05% uranium and/or thorium by weight. Exemptions from most NRC regulations exist for many commercial uses of source material. By-product material is regulated as radioactive waste, including surface wastes from uranium solution extraction. Underground ore bodies depleted by such techniques are not considered by-product material.

Proper handling and disposal of waste classified as by-product material is a large part of any mining and milling operation (IAEA 1976). The typical process produces waste at almost every step. Most wastes are put into tailings ponds where uranium is periodically recovered. These ponds are normally highly acidic due to the large quantities of acid used in the ore leaching step. They may also be contaminated with organic solvents or ion exchange resins if these are used in the recovery step. Many of these

impoundments have contaminated local groundwaters (UNSCEAR 1977). Treatments for runoff from uranium mills have been developed; a typical treatment process is shown in Figure 11.25.2. This process is designed to produce an effluent that comes close to meeting the drinking water limits of 5 pCi/L total radium and 3 pCi/L of radium 226. The concentrated radioactive (radium) sludge (Tsivoglou & O'Connell 1965) is then handled as a low-level radioactive waste.

Tailings ponds, even after closure, are a constant source of radon from the decay of radium. Tailings are sometimes used as building materials, posing a potential health hazard from radon seepage.

Non-radioactive mining such as phosphate rock operations can produce tailings containing uranium, thorium, and radium at levels above those permissible for release to the environment. These tailings also often find their way into commerce as building materials.

—*Paul A. Bouis*

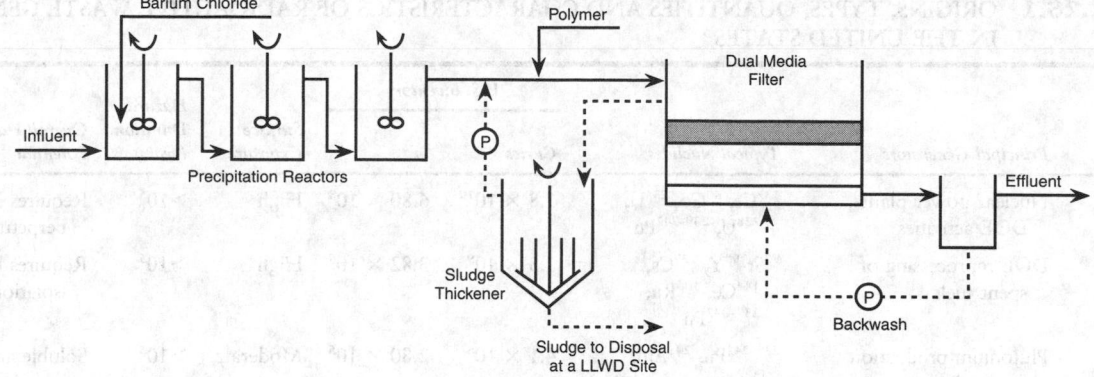

FIG. 11.25.2 Process schematic of a typical radium removal operation.

References

International Atomic Energy Agency (IAEA). 1976. *Manual on radiological safety in uranium and thorium mines and mills*. Safety Series No. 43. IAEA. Vienna, Austria.

Tsivoglou, E.C., and R.S. O'Connell. 1965. Nature, volume and activity of uranium mill wastes. *Radiological Health and Safety in Mining and Milling of Nuclear Materials*. IAEA. Vienna, Austria.

United Nations Scientific Committee on the Effects of Atomic Radiation (UNSCEAR). 1977. *Sources and effects of ionizing radiation*. New York, N.Y.

11.26
LOW-LEVEL RADIOACTIVE WASTE

Low-level radioactive waste is a general term for a wide range of materials contaminated with radioisotopes (Gershey, Klein, Party & Wilkerson 1990; Burns 1988). Industries and hospitals, medical, educational and research institutions, private and government laboratories, and nuclear fuel cycle facilities using radioactive materials generate low-level radioactive wastes as part of normal operations. These wastes are generated in many physical and chemical forms, and at many levels of contamination. Low-level radioactive waste (LLRW) accounts for only one percent of the activity (curies, bequerels) but eighty-five percent of the volume of radioactive waste generated in the United States. The NRC defines LLRW as "radioactive material subject to NRC regulations that is *not* high-level waste, spent nuclear fuel, or mill tailings and which NRC classifies in 10 CFR Part 61 as low-level radioactive waste."

Table 11.26.1 shows the origins of most radioactive wastes. Figure 11.26.1 shows general classifications for all radioactive wastes. Low-level wastes fall under four categories:

1. Below regulatory concern
2. Generator disposed
3. Class A, B, or C
4. Greater than class C

Approximately two million cubic feet of LLRW are disposed of annually at currently operating commercial disposal sites. The nuclear fuel cycle accounts for over fifty percent of this volume, and more than eighty percent of the activity.

Although contact with radioactive waste in the environment should be minimal, due to the highly regulated nature of the waste handling protocols, the ongoing design, operation, and maintenance of the numerous sites are ongoing activities requiring the expertise of environmental engineers and scientists.

Waste Classification

No worldwide agreement has been reached for classification of radioactive wastes. This is contrary to the rules established for release of radioactive materials to the environment and for protection of the general public and workers from radiological exposure. However, most countries agree that waste is best classified from the point of view of disposal. The NRC, in 10 CFR Part 61, classifies low-level radioactive waste based on its suitability for near surface disposal. According to the NRC, classifying radioactive waste involves two factors:

TABLE 11.26.1 ORIGINS, TYPES, QUANTITIES AND CHARACTERISTICS OF RADIOACTIVE WASTE GENERATED IN THE UNITED STATES

Waste	Principal Generators	Typical Nuclides	U.S. Inventory Curies	U.S. Inventory m	Surface Exposure	Hazard Duration (years)	Overall Hazard Potential
Spent fuel	Nuclear power plants, DOE activities	^{137}Cs, ^{60}Co, ^{235}U, ^{238}U, $^{239-242}$Pu	1.8×10^{10}	6.80×10^3	High	$>10^5$	Requires isolation in perpetuity
High-level	DOE reprocessing of spent fuels	^{90}Sr-^{90}Y, ^{137}Cs, ^{144}Ce, ^{106}Ru, $^{239-242}$Pu	1.3×10^9	3.82×10^5	High	$>10^5$	Requires long-term isolation
Transuranic	Plutonium production for nuclear weapons	$^{239-242}$Pu, ^{241}Am, ^{244}Cm	4.1×10^6	2.80×10^5	Moderate	$>10^5$	Soluble and respirable
Mill tailings	Mining and milling of uranium/thorium ores	^{235}U, ^{230}Th, ^{226}Ra	1.4×10^5	1.20×10^8	Low	$>10^4$	Hazard to worker
Greater than Class C	Nuclear power plants, users and manufacturers of sealed-source devices	^{60}Co, ^{137}Cs, ^{90}Sr, ^{241}Am	2.40×10^6	1.30×10^2	High	500	High
Low-level DOE	Various processes, including decontamination and remedial action cleanup projects	Fission products, ^{235}U, ^{230}Th, α-bearing waste, ^{3}H	1.4×10^7	2.40×10^6	Unknown	$>10^3$	High, poorly managed in the past
Low-level commercial							
Class A	Fuel cycle, power plants, industry, institutions		3.6×10^5	1.3×10^6	Low	200	Low
Class B	Principally power plants and industry		9.5×10^5	2.7×10^4	Moderate	$\sim10^3$	Moderate
Class C	Power plants, some industry		2.5×10^6	6.5×10^3	High	$>10^5$	High

TABLE 11.26.2 OVERVIEW OF CLASSES A, B, AND C, WASTE CHARACTERISTICS

Characteristic	Class A Waste	Class B Waste	Class C Waste
Concentration	low concentrations of radionuclides	higher concentrations of radionuclides	highest concentration of radionuclides
Waste Form	must meet minimum waste form requirements / does not require stabilization (but may be stabilized)	must meet minimum waste form requirements / requires stabilization for 300 years	must meet minimum waste form requirements / requires stabilization for 300 years
Examples	typically contaminated protective clothing, paper, laboratory trash	typically resins and filters from nuclear power plants	typically nuclear reactor components, sealed sources, high activity industrial waste
Intruder Protection	after 100 years, decays to acceptable levels to an intruder / requires no additional measures to protect intruder	after 100 years, decays to acceptable levels to an intruder, provided waste is recognizable / requires stabilization to protect intruder	after 500 years, decays to acceptable levels to an intruder / requires stabilization and deeper disposal (or barriers) to protect intruder
Segregation	unstable Class A must be segregated from Classes B and C	need not be segregated from Class C	need not be segregated from Class B

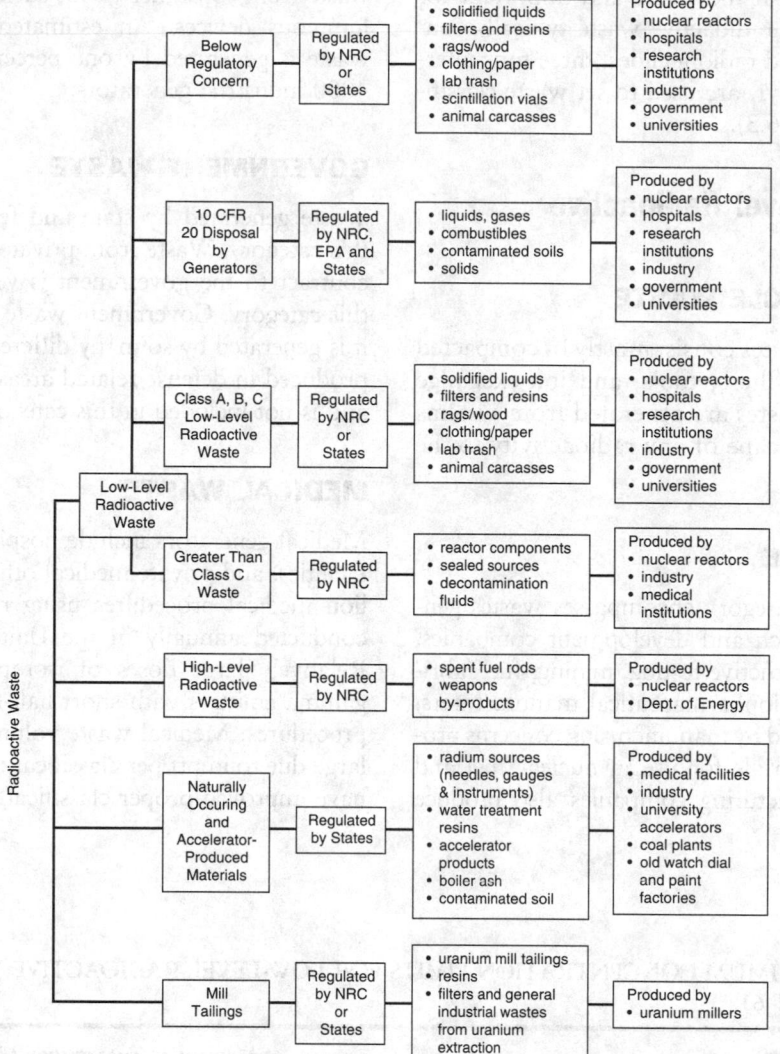

FIG. 11.26.1 General classifications of radioactive waste.

1. Long-lived radionuclide concentrations posing potential hazards that will persist long after such precautions as institutional controls, improved waste forms and deeper disposal have ceased to be effective
2. Shorter-lived radionuclide concentrations for which institutional controls, waste forms, and disposal methods are effective

Low-level radioactive waste is classified as Class A, B, and C waste. An overview of the characteristics of wastes in these classes is shown in Table 11.26.2. 10 CFR §61.54 defines these classes as follows:

1. Class A wastes are usually segregated from other waste classes at the disposal site. The physical form and characteristics must meet the minimum requirements set forth in these regulations (10 CFR §61.56[a]), e.g., contains less than 1% liquid by volume, etc. If Class A waste also meets the stability requirements set forth in 10 CFR §61.56(b), it is not necessary to segregate the waste for disposal.

2. Class B wastes must meet more rigorous waste form requirements to ensure stability after disposal.
3. Class C wastes must meet more rigorous waste form requirements, and also require additional measures at the disposal facility to protect against inadvertent intrusion.

Wastes with form and disposal methods more stringent than Class C are not acceptable for near surface disposal. These wastes must be disposed of in geological repositories.

Classification by specific long- and short-lived radionuclide concentrations is also given in 10 CFR §61.54. The reader is referred to this section for details.

The 10 CFR Part 61 radioactive waste classification is a systematic attempt to control the potential dose to man from disposed waste. System components include site characteristics, site design and operation, institutional controls, waste forms, and intruder barriers. The quantity and type of radionuclides permitted in each class are based on these various disposal components and on radioactive material concentrations expected in the waste and important for disposal. Since low-level radioactive waste typically contains short- and long-lived radionuclides, three time intervals, 100, 300, and 500 yr, are used to set waste classification limits (Table 11.26.3).

Sources of Low-Level Radioactive Waste

NUCLEAR FUEL CYCLE WASTE

Fuel cycle and utility wastes consist mostly of compacted trash and dry wastes, filters, tools, and ion-exchange resin. Many of these wastes are generated from systems designed to minimize escape of any radioactivity to the environment.

INDUSTRIAL WASTE

The industrial LLRW category encompasses wastes generated by private research and development companies, manufacturers, non-destructive testing, mining, fuel fabrication facilities, and radiopharmaceutical manufacturers. Most wastes are generated by manufacturing concerns producing radioactive materials for use in nuclear fuel and non-fuel cycles. Manufacturing companies also produce

waste from consumer goods such as smoke detectors and luminous devices. An estimated ninety-five percent of waste is generated by one percent of the approximately 4000 industrial generators.

GOVERNMENT WASTE

Waste generated by state and federal agencies falls into this category. Waste from private facilities working under contract to the government is very often excluded from this category. Government waste is the most diverse since it is generated by so many different organizations. LLRW produced in defense-related areas is handled by the DOE and is not included in this category.

MEDICAL WASTE

Medical generators include hospitals and clinics, research facilities, and private medical offices. More than 120 million medical procedures using radioactive materials are conducted annually in the United States (SNM 1988). Relatively large doses of isotopes, frequently powerful gamma emitters with short half-lives, are used in clinical procedures. Medical waste volumes were historically too large due to improper classifications. Rising disposal costs have improved proper classification.

TABLE 11.26.3 MAXIMUM CONCENTRATION LIMITS FOR LOW-LEVEL RADIOACTIVE WASTE FROM 10 CFR PART 61

Radionuclide	Half-life (years)	Maximum Concentration Limits $(Ci/m^3)^a$ Class A	Class B	Class C
Nuclides with half-lives				
<5 years^a	<5.0	700.000	NL^b	···
^{60}Co	5.3	700.000	NL	···
^{3}H	11.3	40.000	NL	···
^{90}Sr	28.0	0.040	150.0	7000.00
^{137}Cs	30.0	1.000	44.0	4600.00
^{63}Ni	92.0	3.500	70.0	700.00
^{63}Ni in activated metal	92.0	35.000	700.0	7000.00
^{14}C	5,730.0	0.800	···	8.00
^{14}C in activated metal	5,730.0	8.000	···	80.00
^{94}Nb in activated metal	20,000.0	0.020	···	0.20
^{59}Ni in activated metal	80,000.0	22.000	···	220.00
^{99}Tc	212,000.0	0.300	···	3.00
^{129}I	17,000,000.0	0.008	···	0.08
α-emitting transuranic nuclides with half-lives <5 years	<5.0	10.000 nCi/g	···	100.00 nCi/g
^{242}Cm	0.45	2,000.000 nCi/g	···	20,000.00 nCi/g
^{241}Pu	13.2	350.000 nCi/g	···	3,500.00 nCi/g

^{a}Including, but not limited to: ^{32}P, ^{35}S, ^{51}Cr, ^{54}Mn, ^{55}Fe, ^{58}Co, ^{59}Fe, ^{65}Zn, ^{67}Ga, ^{125}I, ^{131}I, ^{134}Cs, ^{144}Ce, and ^{192}Ir.
^{b}No upper limit on concentration.

ACADEMIC WASTE

Academic waste includes university hospitals and university medical and nonmedical research facilities. It tends to be low in activity and relatively high in volume, often due to improper classification of some materials as radioactive waste.

GREATER THAN CLASS C WASTE

Greater than Class C (GTCC) wastes contain concentrations of radionuclides greater than Class C limits established in 10 CFR Part 61. These wastes, as mentioned earlier, cannot be disposed of as LLRW but must go to a geological repository. GTCC waste comes primarily from decontamination and decommissioning of nuclear power plants. Nonutility generators include manufacturers of sealed sources used as measuring devices. GTCC waste volume is projected to expand during the next twenty years as more nuclear plants are decommissioned.

BELOW REGULATORY CONCERN WASTE

Below regulatory concern (BRC) wastes have radioactive content so low that unregulated release does not pose an unacceptable risk to public health or safety (Table 11.26.4). This class was established to make practical, timely determinations of when wastes need to go to a licensed LLRW site. The low-level radioactive waste policy amendments act of 1985 established procedures for acting expeditiously on petitions to exempt specific radioactive waste streams from NRC regulations (NRC 1986). Petitions already filed could dramatically reduce the total LLRW needing disposal.

TABLE 11.26.4 EXAMPLES OF MATERIALS EXEMPT FROM LICENSING REQUIREMENTS UNDER 10 CFR PART 31

Product[a]	Permissible Activity ($\leq$)
Static-elimination devices	500 μCi ^{210}Po
Ion-generating tubes	500 μCi ^{210}Po or 50 mCi ^{3}H
Luminous devices in aircraft	10 Ci ^{3}H or 300 mCi ^{147}Pm
Calibration sources	5 μCi ^{241}Am
Ice-detection devices	50 μCi ^{90}Sr
Prepackaged in vitro/clinical testing kits	10 μCi ^{125}I/test
	10 μCi ^{131}I/test
	10 μCi ^{14}C
	50 μCi ^{3}H
	20 μCi ^{59}Fe
	10 μCi ^{55}Fe

[a]The use of thorium in gas mantles, vacuum tubes, welding rods, incandescent lamps, photographic films, and finished optical lenses is also not regulated. Naturally occurring radioactive materials (NORM) present in geologic specimens, petroleum drilling wastes, and rare earth minerals processing wastes (with the exception of uranium and thorium) are also not regulated.

MIXED WASTE

Mixed low-level radioactive waste contains both radioactive and hazardous components and meets, respectively, NRC's definition of low-level radioactive waste in 10 CFR Part 61, and the Environmental Protection Agency's definition of hazardous material in 40 CFR Part 261. Although any type of low-level waste may be "mixed," surveys of waste generators indicate that less than five percent of the wastes to be sent to commercial sites would be classified as mixed (Bowerman, Davis & Siskind 1986). An example of a mixed waste would be a contaminated flammable extraction solvent used in radioisotope recovery. NRC deregulation of scintillation fluids containing minimal quantities of ^{3}H and ^{14}C has eliminated the largest source of mixed waste from disposal as LLRW.

Quantities of LLRW Generated

Each year, the DOE national low-level waste management program publishes data on both national and state-specific LLRW commercially disposed of in the United States (Fuchs & McDonald 1993). Data are categorized by disposal site, generator category, waste class, volume, and radionuclide activity. A distinction is made between LLRW shipped directly for disposal by generators, and waste handled by an intermediary. Wastes are subdivided into five categories:

- Academic
- Government
- Industrial
- Medical
- Utility

The volume of LLRW disposed of at commercial sites exceeded 3,500,000 ft^3 in 1980 (LLWMP 1982). The volume of LLRW disposed of at these sites since that time has steadily declined. In 1992, commercial LLRW disposal facilities received a total volume of 1,743,279 ft^3 of waste containing an activity of 1,000,102 curies. Waste distribution by disposal site is presented in Table 11.26.5. Tables 11.26.6 and 11.26.7 provide typical radionuclide and waste forms associated with commercial LLRW. Table

TABLE 11.26.5 DISTRIBUTION OF LOW-LEVEL RADIOACTIVE WASTE RECEIVED AT DISPOSAL SITES IN 1992

Site	Volume (ft^3)	Percent of Total	Activity (curies)	Percent of Total
Barnwell	830,512	48	815,974	82
Beatty	514,726	29	90,205	9
Richland	398,041	23	93,923	9
Total	1,743,279	100	1,000,102	100

TABLE 11.26.6 REPORTED LOW-LEVEL RADIOACTIVE WASTE RADIONUCLIDES RECEIVED AT DISPOSAL SITES IN 1992 FOR DIRECT AND NONDIRECT SHIPMENTS IN ORDER OF HIGHEST TO LOWEST ACTIVITY LEVELS

Nondirect	Reactors	Academic	Medical	Industrial	Government
H-3	Fe-55	Pm-147	Cs-137	H-3	Sr-90
Cs-137	Co-60	H-3	Sr-90	Co-60	Co-60
Fe-55	Ni-63	Co-60	Ni-63	Cs-137	Fe-55
Co-60	Mn-54	I-129	Co-57	Fe-55	U-238
S-35	Ag-110m	S-35	Ba-133	S-35	Mn-54
Ni-63	Cs-137	Cr-51	Ra-226	Ir-192	Ni-63
Co-58	Co-58	P-32	Rn-222	U-238	Co-58
C-14	Cr-51	C-14		Sr-90	Ra-226
Mn-54	Cs-134	I-131		Th-232	C-14
Kr-85	H-3	Ca-45		Th-228	H-3
Cs-134	Cd-109	Ni-63		Ce-144	Eu-152
Sr-90	Sb-125	Ra-226		P-32	U-235
P-32	Sr-90	Co-57		Ni-63	Ni-59
Cr-51	Ni-59	I-123		Ag-110m	Co-57
Zn-65	Nb-95	Re-186		Sb-125	Tc-99
I-125	Zr-95	Cu-67		Ra-228	Am-241
Am-241	Fe-59	Cu-64			Eu-154
Ni-59	C-14	K-40			I-125
Sb-125		Sr-85			Cs-137
Fe-59		Zn-65			
Ra-226		Ag-108			

TABLE 11.26.7 TYPICAL WASTE FORMS BY GENERATOR CATEGORIES

Academic
Compacted trash or solids
Institutional laboratory or biological waste
Absorbed liquids
Animal carcasses

Government
Compacted trash or solids
Contaminated plant hardware
Absorbed liquids

Industrial
Depleted uranium
Compacted trash or solids
Contaminated plant hardware
Absorbed liquids
Sealed sources

Medical
Compacted trash or solids
Institutional laboratory or biological waste
Absorbed liquids
Sealed sources

Utilities
Spent resins
Evaporator bottoms and concentrated waste
Filter sludge
Dry compressible waste
Irradiated components
Contaminated plant hardware

11.26.8 shows volume and activity according to generator category.

LLRW Commercial Disposal Sites

There were only two low-level radioactive waste disposal sites operating in the United States in 1994. Located in Barnwell, South Carolina and Richland, Washington, these facilities handle all low-level waste generated in the United States. Beginning in 1993, federal law allowed these states to refuse to accept any low-level waste generated outside their borders. The low-level radioactive waste policy act of 1980 made each of the 50 states responsible for dis-

TABLE 11.26.8 LOW-LEVEL RADIOACTIVE WASTES RECEIVED AT COMMERCIAL DISPOSAL SITES IN 1992

Generator Category	Volume (ft³)	Activity (curies)
Academic	44,322.34	1,724.39
Government	158,186.17	40,780.08
Industrial	908,451.86	100,089.80
Medical	26,251.32	397.80
Utility	606,066.85	857,110.38
Total	1,743,278.54	1,000,102.45

posal of its own low-level waste. Amendments passed in 1985 strengthened the act and established a firm decision-making timeline for the states (Table 11.26.9). Federal law gives each state the option of establishing a disposal site within its borders, or forming a partnership or compact, with other states to dispose of low-level waste on a regional basis (Figure 11.26.2). Each state must take title to all waste generated within its borders by January 1, 1996, whether or not a disposal facility is operating at that time.

Beginning in 1993, the Washington site restricted LLRW received for disposal to the Northwest and Rocky Mountain compact states. The South Carolina facility is to remain open until 1996, but stopped accepting waste from outside the southeast compact in 1994. At this time, states without access to an operational disposal site will have to store their LLRW until a location is identified for disposal.

LLRW Reduction Processes

Volume reduction is the single most-used technique to minimize the cost and environmental impact of low-level radioactive waste disposal. NRC-mandated volume-reduction measures have been very effective in minimizing nuclear power industry waste for LLRW disposal sites (Table 11.26.10). It is estimated that an 80% volume reduction could be achieved by many institutions and industries through segregation, decay of short-lived isotopes, compaction, regulated sewer disposal, exclusion of scintillation fluids, and incineration.

WASTE MINIMIZATION

Waste minimization is the simplest, most economical, and often most overlooked method to achieve significant volume reduction. Careful preplanning of activities, minimizing the use of clean materials and disposable protective equipment, and proper maintenance in radioactive areas are simple, effective strategies for minimizing LLRW generation.

SEGREGATION

Segregation of waste according to physical form, chemical composition half-life, and NRC classifications (A, B, C) should be an integral part of any volume reduction plan. Labeling all waste according to International Commission on Radiation Protection (ICRP) guidelines facilitates their eventual shipment to the disposal site.

DECAY

Medical and academic LLRW is commonly stored, allowing decay of short-lived radionuclides to innocuous levels, so that wastes can be disposed of according to their non-radiological properties (termed *hold-for-decay disposal*).

Storage is regulated as an operational matter, subject to the same public health and environmental protection requirements. The hold-for-decay practice is best suited for small volumes of waste containing discrete radionuclides

TABLE 11.26.9 DEADLINES DEFINED BY THE LOW-LEVEL RADIOACTIVE WASTE POLICY AMENDMENTS ACT OF 1985

Date	Legislated Action(s)
January 1, 1986	Each state to have joined compact or to have enacted legislation indicating intention to develop its own site; surcharge not to exceed $10 per cubic foot.
July 1, 1986	Generators in states that did not meet the January 1, 1986 deadline are subject to doubled surcharges until December 31, 1986.
January 1, 1987	Generators in states that did not meet January 1, 1986 deadline may be denied access to operating disposal sites.
January 1, 1988	Compacts to have named host states; unaligned states to have developed siting plan and schedule and to have delegated authority for development; surcharge not to exceed $20 per cubic foot; noncompliance states subject to doubled surcharges.
July 1, 1988	Noncompliance states subject to quadrupled surcharges.
January 1, 1989	Generators in states and compacts that did not meet the January 1, 1988 deadline to be denied access to operating disposal sites.
January 1, 1990	Compacts and unaligned states to file a complete operating license application; letter from governor stating that the unaligned state will have provisions for LLRW disposal in place by December 31, 1992, may be submitted in lieu of application; surcharge not to exceed $40 per cubic foot; failure to comply may result in denial of access to operating disposal sites.
January 1, 1992	All compacts and unaligned states to file operating license applications; letter from governor no longer sufficient for compliance status.
January 1, 1993	Sited compacts to be empowered to restrict import of non-compact LLRW.
January 1, 1996	Surcharge rebates cease.

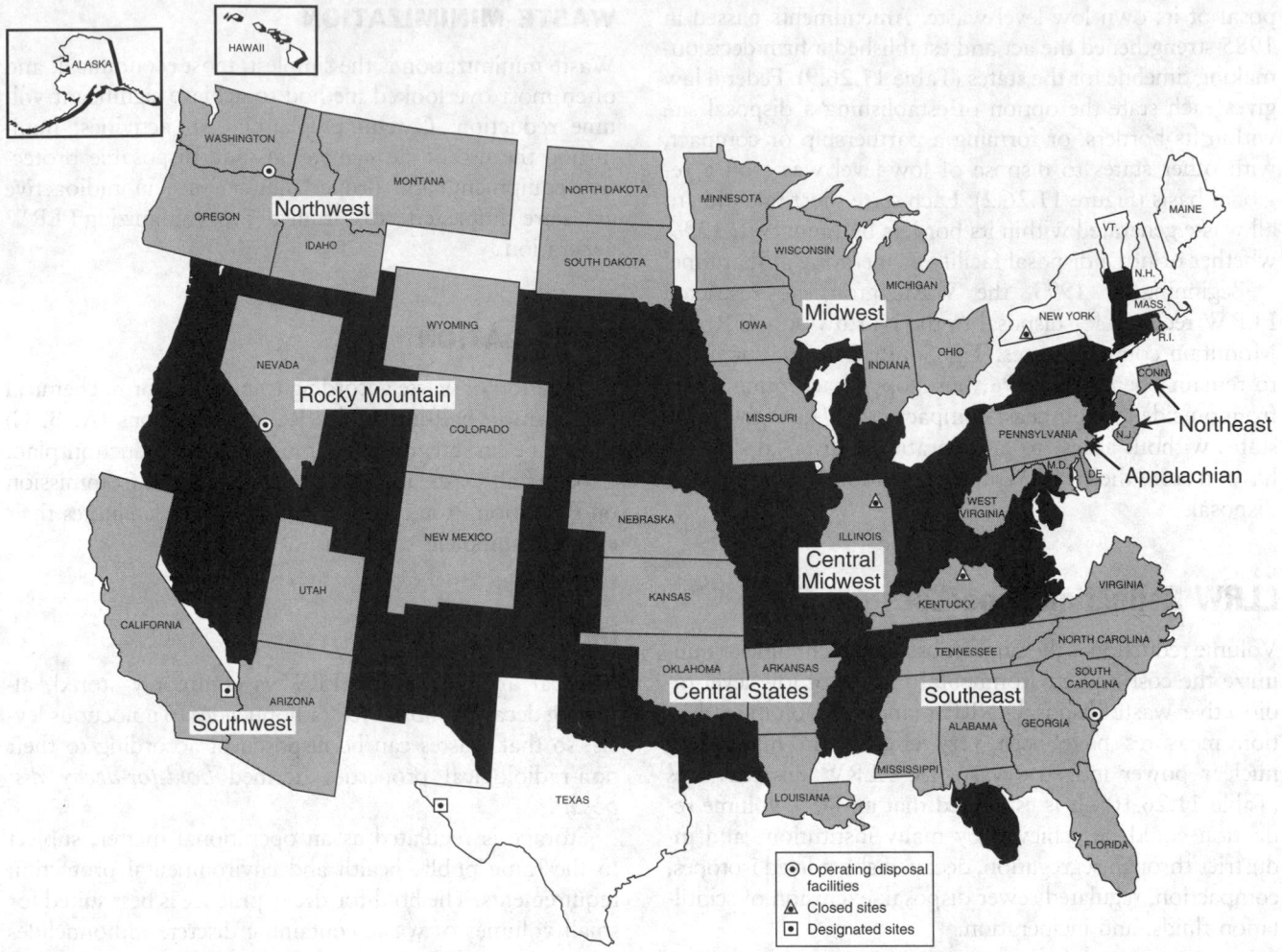

FIG. 11.26.2 U.S. map showing compact alignments, the operating and closed LLRW disposal facilities, and the designated proposed sites in California and Texas. Unaligned states are shown in white. The existing sites are shown by circles; closed sites by triangles; and proposed sites by squares. The Northeast and Southwest compacts are not contiguous and will require travel outside their respective regions in order to transport LLRW from generators to the disposal facilities.

TABLE 11.26.10 COMPARISON OF VOLUMES OF VARIOUS NRC CLASSES OF LLRW SHIPPED TO COMMERCIAL DISPOSAL SITES IN 1987 AND 1992

Class	1987		1992	
	ft^3	%	ft^3	%
A	1,796,695	97.4	1,676,007	96.1
B	39,128	2.1	41,599	2.4
C	8,687	0.5	25,673	1.5

with very short half-lives. Wastes containing long-lived radionuclides such as fission products are not amenable to this practice due to: larger volumes; wide variety of physical and chemical form and radionuclide content; and long storage times needed for decay.

SEWAGE DISPOSAL

Sewage disposal of radionuclides is authorized by the NRC under 10 CFR §20.303. The reader is referred to this section and the limits in Appendix B, Table I, Column 2 for details. The total quantity of licensed and other radioactive material, excluding ^3H and ^{14}C, disposed in sewers cannot exceed one curie per year. The quantity of ^3H may not exceed five curies per year, and the quantity of ^{14}C may not exceed one curie per year.

DEREGULATION

Deregulation of LLRW poses little hazard to the public or the environment and could significantly decrease the volume of LLRW. The NRC, as previously discussed, has established a petition mechanism for the deregulation of generic wastes. An example of the effect of such a petition comes from an exemption for biomedical institutions

(Fortom and Goode 1986). This petition proposes on-site incineration of solid biomedical waste containing a maximum of one curie of ^3H and one hundred millicuries of ^{14}C per year. The resulting ash would be disposed of as sanitary waste. This petition estimates a 90% reduction in institutional waste presently sent to LLRW disposal sites. It has not been enacted, in part due to concern for clean air requirements not related to radioactivity.

DEWATERING

Radioactive waste dewatering is an effective and efficient method for volume reduction. In addition, radioactive waste must not contain more than 0.5% freestanding water to be accepted at LLRW disposal sites. Centrifugation, filtration, and evaporation are standard techniques used to dewater wastes.

COMPACTION

Compaction is the primary volume-reduction method. Uncompacted waste has a typical density of approximately 130 kg/m^3 and can be increased three- to fourfold using a standard (20,000 psi) compactor. Super compactors can increase the density by a factor of ten. Shredding waste prior to compaction can also reduce the final volume. Compaction methods cannot be applied to hard and dense waste items for which volume reduction would be minimal. During compacting, potentially contaminated gases, liquids, and particulates are expelled from the waste and must be trapped by an off-gas (scrubber) treatment system.

INCINERATION

A large portion of LLRW is combustible and suitable for incineration. Used in combination with compacting, one-hundred fold volume reductions can be achieved. Radioactive waste incineration is an expensive and potentially troublesome treatment technique. Most European countries incinerate combustible radioactive waste prior to disposal. In the United States, incineration is reserved for cases where maximum volume reduction is required, and/or sophisticated off-gas treatment is not necessary. Clean air requirements make it increasingly difficult to build commercial incinerators.

Several waste characteristics are important in relation to incinerator performance. With very compact materials, combustion may be incomplete. Certain materials such as plastics (PVC) produce corrosive (HCl) gases that can damage the incinerator and must be scrubbed prior to release to the environment. The correct temperature must be maintained to ensure complete combustion. Since furnace temperature is controlled by the calorific value of the waste, the moisture content, and the combustion rate, it is clear that the feed rate is critical to successful incineration. The use of supplemental fuel to control combustion is dis-

couraged unless it is already contaminated with radioactive materials.

Liquid and Gaseous Effluent Treatment

LIQUID EFFLUENTS

LLRW is produced from the clean-up of drainings and cooling water at nuclear power plants, manufacturing sites, and R&D laboratories where radioactive materials are handled. These low-activity wastes are usually treated to remove most radionuclides, then discharged to the environment. Low-activity wastes can be collected and mixed for a more uniform effluent or segregated to utilize specific treatments for the individual components. If the first approach is utilized, the usual wastewater treatments of flocculation, precipitation, absorption, filtration, and ion exchange can be adapted to radioactive wastes (Table 11.26.11). Provisions must be made for water discharging and for drying, compacting, and disposing of the solids produced. Presently, solids are sent to a LLRW disposal site. Radium removal, covered in the section on mining and milling, is a good example of a specific treatment process.

If the total solids content of the contaminated water is low, if the volume is small, or if a final polishing of effluents is necessary, ion exchange may be a suitable treatment method. At nuclear power plants, ion exchange, filtration, evaporation, and reverse osmosis are the major processes used for contaminated water treatment (Figure 11.26.3).

GASEOUS EFFLUENTS

The primary source of radioactive gaseous effluents to the environment is from nuclear power plants. Coal-fired power plants also emit particulate radionuclides and are treated by conventional stack gas technology. Effluents from nuclear reactors include noble gas isotopes, radioiodines, tritium and some fission products (heavy water reactors). Typical treatment processes are shown schematically in Figure 11.26.3.

Conditioning Techniques

Proper LLRW disposal is closely regulated by the NRC and its agreement states. The application for a license to handle radioactive materials requires a sound disposal plan for any radioactive waste produced. The NRC permits LLRW disposal via six methods outlined in 10 CFR Part 20.

1. Transfer of waste to an authorized recipient.
2. Disposal by release into a sanitary sewerage system (meets limits in 10 CFR §20.303).

TABLE 11.26.11 TREATMENT PROCESSES FOR REMOVAL OF RADIOACTIVE WASTES

Process	Decontamination Factor[a] Individual Radionuclides	Mixed Fission Products[b]
Conventional		
Coagulation and settling	0–100+	2–9.1
Clay addition, coagulation and settling	0–100	1.1–6.2
Sand filtration	1–100	
Coagulation, settling and filtration	1–50	1.4–13.3
Lime-soda ash softening	2–100	
Ion exchange, cation	1.1–500	2.0–6.1
Ion exchange, anion	0–125	
Ion exchange, mixed bed	11–3300	50–100
Solids-contact clarifier	1.9–15	2.0–6.1
Evaporation	1.00–10,000	
Nonconventional		
Phosphate	1.2–1000	125–250
Metallic dusts	1.1–1000	1.1–8.6
Clay treatment	0–100+	
Diatomaceous earth	1.1–∞	
Sedimentation	<1.05	
Activated sludge	1.03–8.2	4.8–9.8
Trickling filter	1.05–37	3.5–6.1
Sand filter	8.3–100	1.9–50
Oxidation ponds	<1.1–20	

[a]Decontamination factor = $\dfrac{\text{initial concentration}}{\text{final concentration}}$

[b]Where no data are listed it implies lack of information and not the unsuitability of the process.

3. Release to the environment, if material is below the maximum permissible concentration (MPC) in 10 CFR Part 20 appendix B, Table II.
4. Disposal by incineration according to 10 CFR §20.305, especially for waste oils and scintillation fluids.
5. Disposal of certain specific waste without regard to its radioactivity 10 CFR §20.306 (e.g., 0.05 mCi ^{3}H or ^{14}C).
6. Specific procedures approved as part of licensing to handle radioactive materials.

Radioactive waste is normally disposed of as a solid, except for liquids released to sanitary sewers or other water systems when radioactivity levels are below the maximum permissible concentration (MPC). In contrast to other types of waste, where pollutants can be eliminated by treatment, radioactivity can only be reduced by decay time. Thus the disposal methods used at NRC-authorized disposal sites are for solids and are based on the decay time required to make them non-radioactive. The correct preparation of radioactive waste is the first step to ensure the waste is disposed of economically and according to all applicable regulations.

Conditioning of radioactive wastes can include segregation, pretreatment, processing, and packaging. These techniques are covered in other sections of this chapter. Here, conditioning refers only to the various immobilization techniques used to prevent radioisotopes leaching into the environment. Immobilization is often used to help meet the NRC stability requirements for Class B and C waste and even for some forms of Class A and mixed wastes. The principle immobilization techniques are cementation, bituminization, polymerization, and vitrification. All of these techniques will increase the volume of radioactive waste.

CEMENTATION

Cement is used to solidify liquid waste. Cementation is relatively inexpensive but prone to leaching. The radioactive waste reacts with the cement and is bound to it. Waste compatibility must be verified, and special cement formulations are sometimes required to insure the product sets. This technique is sometimes used to dry a solid waste so that it contains less than 0.5% freestanding liquid.

BITUMINIZATION

The use of bitumen or asphalt is a classic immobilization technique. The process, carried out at the relatively high

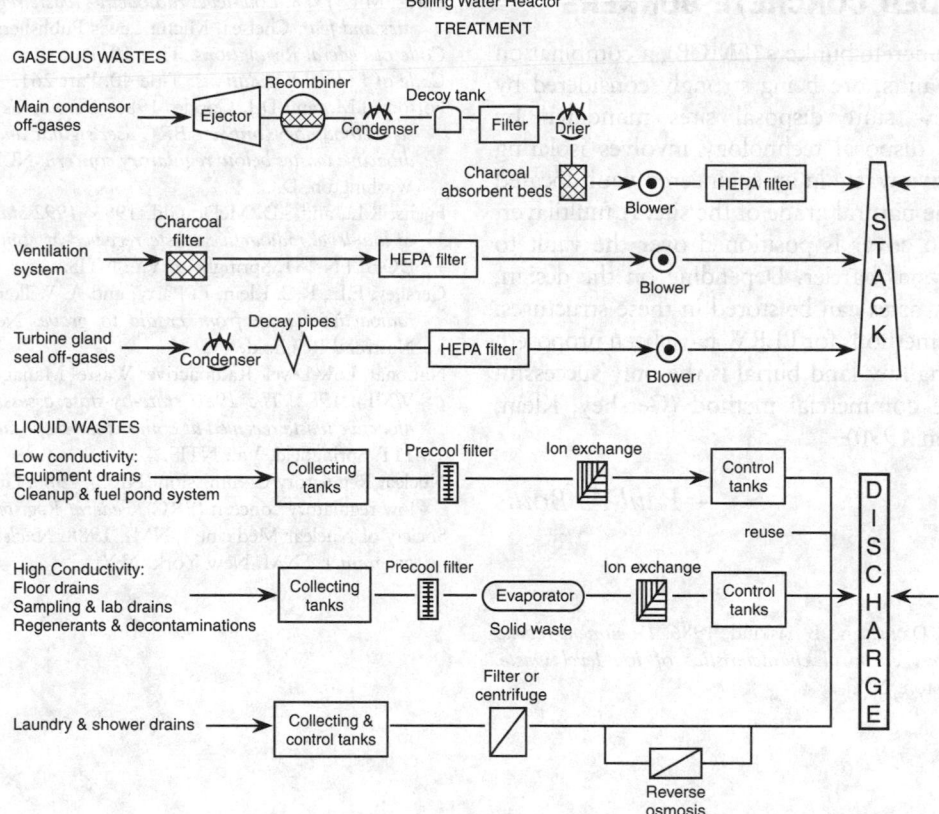

FIG. 11.26.3 Specific treatment and volume-reduction methods for nuclear plants. (Reprinted from International Atomic Energy Agency, 1986.)

temperature of $\geq 150°C$, is dangerous and requires specialized equipment. The product is less subject to normal leaching, but is susceptible to fire damage. The product also has a tendency to swell from the release of gases.

POLYMERIZATION

Polymerization of liquid and semi-liquid LLRW by in situ addition of monomers and initators is a relatively new technique. The process must be carefully adopted to the type of waste being immobilized. The product has shortcomings similar to bitumen waste.

VITRIFICATION

Vitrification in borosilicate waste is an expensive technique very rarely used in the immobilization of LLRW.

Disposal Techniques

Disposal of LLRW in the United States has been based on some form of land burial since ocean dumping was banned in the 1960s. The facilities must be on a site designed, operated, closed, and controlled after closure to meet all criteria in 10 CFR Part 61. Releases to the environment must be as low as reasonably achievable (ALARA), and waste containment systems must be effective until the radioactivity has decayed to MPC levels.

SHALLOW LAND BURIAL

Shallow land burial (SLB) in trenches, often plastic lined, is the most economical disposal method. Prepackaged or preconditioned waste is carefully stacked into the trench, then covered with the excavated earth. Radioactivity can be successfully confined in the burial area if leaching of the waste by groundwater or rainwater can be reduced to negligible levels. Thus, careful geological, geochemical and hydrological studies must be made for burial site location.

DISPOSAL VAULTS

Below-ground vaults (BGV) and above ground vaults (AGV) are enclosed, engineered structures built to hold the most hazardous low-level radioactive wastes, such as Class C or greater than Class C (GTCC). The long-term effectiveness of this expensive solution has been questioned by proponents of SLB disposal (Gershey, Klein, Party & Wilkerson 1990).

EARTH-MOUNDED CONCRETE BUNKERS

Earth-mounded concrete bunkers (EMCB), a combination of trenches and vaults, are being strongly considered by many of the new state disposal sites mandated by Congress. EMCB disposal technology involves isolating low-level radioactive waste in an engineered vault located above or below the natural grade of the site. A multilayer, engineered earthen cover is positioned over the vault to provide an additional barrier. Depending on the design, Class A, B, or C wastes can be stored in these structures.

Other disposal methods for LLRW have been proposed, but at this time shallow land burial is the only successful and cost effective commercial method (Gershey, Klein, Party & Wilkerson 1990).

—*Paul A. Bouis*

References

Bowerman, B.S., R.E. Davis, and B. Siskind. 1986. *Document review regarding hazardous chemical characteristics of low-level waste.* NUREG, BNL. Upton, N.Y.

Burns, M.E. 1988. *Low-level radioactive waste regulations: science, politics and fear.* Chelsea, Mich.: Lewis Publishers.

Code of Federal Regulations. Title 10, Part 61.

Code of Federal Regulations. Title 40, Part 261.

Fortom, J.M., and D.J. Goode. 1986. *Deminimis waste impacts analysis methodology: impacts-BRC user's guide and methodology for radioactive wastes below regulatory concern.* NUREG/CR-3585, NRC, Washington, D.C.

Fuchs, R.L., and S.D. McDonald. 1993. *1992 state-by-state assessment of low-level radioactive waste received at commercial disposal sites.* DOE/LLN-181. Springfield, Va.: NTIS.

Gershey, E.L., R.C. Klein, E. Party, and A. Wilkerson. 1990. *Low-level radioactive waste: from cradle to grave.* New York, N.Y.: Van Nostrand Reinhold.

National Low-Level Radioactive Waste Management Program (LLWMP). 1982. *The 1980 state-by-state assessment of low-level radioactive waste received at commercial disposal sites.* DOE/LLWMP-11T. Springfield, Va.: NTIS.

Nuclear Regulatory Commission (NRC). 1986. Guideline for wastes below regulatory concern (BRC). *Federal Register 51,* 30839.

Society of Nuclear Medicine (SNM). 1988. *Nuclear medicine self-study program 1.* SNM, New York, N.Y.

11.27
HIGH-LEVEL RADIOACTIVE WASTE

High-level radioactive waste consists of spent fuel elements from nuclear reactors, waste produced from reprocessing, and waste generated from the manufacture of nuclear weapons. All these wastes are highly regulated and controlled due to the dangerously high levels of radiation and the security issues caused by their plutonium content. Strict licensing requirements for the storage of spent nuclear fuel and high-level radioactive waste are specified in 10 CFR Part 72.

Spent nuclear fuel has been withdrawn from a reactor, has undergone at least one year of decay since being used as an energy source in a power reactor, and has not undergone chemical reprocessing. Spent fuel is normally stored on-site at nuclear power plants in an independent spent fuel storage installation (ISFSI). An ISFSI is defined in 10 CFR Part 72 as a complex designed and constructed for the interim storage of spent nuclear fuel and other radioactive materials associated with spent fuel storage.

Spent fuel reprocessing was discontinued in the United States in 1972, except for the DOE, which continues to reprocess most of its spent fuel. France, Germany and several other major nuclear power producers also reprocess their spent fuel. Reprocessing improves the cost effectiveness of nuclear power by recycling recovered uranium and plutonium. The reprocessing of spent fuel, using the PUREX process developed in the United States, involves dissolution in large volumes of acid, liquid/liquid extraction, chemical reduction, and precipitation (Lanham & Runiou 1949, Flagg 1961, Koch 1979). The highly radioactive waste produced from reprocessing is classified by the NRC as a high-level radioactive waste or HLW in 10 CFR Part 72.

Spent fuel elements, HLW, and other highly radioactive wastes, such as transuranic wastes, require permanent containment. The disposal method must be designed to allow decay of the longest-lived radionuclides present in significant amounts in the waste. This means a time period of several hundreds of thousands of years.

Burial in engineered geological repositories is the only current option being seriously considered on a worldwide basis. Except for TRU waste, no site has been selected in the U.S., making it necessary for power plants and the DOE to continue storing waste on site. TRU waste generated by the DOE from various weapons programs is being disposed of at the waste isolation pilot plant (WIPP), a geological repository constructed in a bedded salt dome in New Mexico (Kohn 1987).

Many books and publications are available on the subject of HLW and the reader is referred to these for further details (Delange 1987, IAE 1981, Gertz 1989).

—*Paul A. Bouis*

References

Code of Federal Regulations. Title 10, Sec. 72.

Delange, M. 1987. LWR spent fuel reprocessing at La Hague: ten years on. Proc. Int. Conf. Nucl. Fuel Reprocc. *Waste Management.* Vol. 1, Societe Francaise d'Energie Nucleaire. Paris.

Flagg, J.F. 1961. *Chemical processing of reactor fuels.* London: Academic Press.

Gertz, C.P. 1989. Yucca Mountain, Nevada: is it a safe place for isolation of high-level radioactive waste? *Waste Management,* Vol. 1:9–11.

International Association of Energy. 1981. Underground disposal of radioactive wastes—basic guidance. *Safety Series No. 54.*

Koch, G. 1979. *Existing and projected reprocessing plants: a general review.* Atomkernenerg/Kerntech, Vol. 33:241.

Kohn, K. 1987. Kerntech, Vol. 51:157–160.

Lanham, W.B., and T.C. Runiou. 1949. *Purex process for plutonium and uranium recovery.* U.S. Atomic Energy Commission (USAEC) Report ORNZ-479.

Ullmann's encyclopedia of industrial chemistry. 1993. 5th ed. Vol. A22, pp. 499–591. Weinheim, Germany: VCH.

11.28
TRANSPORT OF RADIOACTIVE MATERIALS

Approximately 2,500,000 packages of radioactive materials are shipped per year in the United States. The vast majority of these shipments involves small or intermediate quantities of material in relatively small packages. The U.S. Department of Transportation (DOT) has regulatory responsibility for safety in the transportation of radioactive materials. The DOT updates transport regulations to keep pace with the changing transportation scene. The NRC has promulgated requirements, in 10 CFR Part 71, for licensees delivering radioactive materials for transport. The principle sources of federal regulations pertaining to transport of radioactive materials are listed in Table 11.28.1. An excellent review of DOT regulations is available from the U.S. Government Printing Office (DOT 1983).

Materials Subject to DOT Regulations

For transportation purposes, radioactive materials arc defined as materials that emit ionizing radiation and have a specific activity greater than 0.002 mci/g are not regulated by the DOT or IAEA. The International Atomic Energy Agency (IAEA) has established international regulations

TABLE 11.28.1 SOURCES OF FEDERAL REGULATIONS

Title 49: U.S. Department of Transportation's Hazardous Materials Regulations, Parts 100–177 and 178–199
Main Headings
49 CFR 106—Rulemaking Procedures
49 CFR 107—Hazardous Materials Program Procedures
49 CFR 171—General Information, Regulations and Definitions
49 CFR 172—Hazardous Materials Tables and Hazardous Materials Communications Regulations
49 CFR 173—Shippers—General Requirements for Shipments and Packagings
49 CFR 174—Carriage by Rail
49 CFR 175—Carriage by Aircraft
49 CFR 176—Carriage by Vessel
49 CFR 177—Carriage by Public Highway
49 CFR 178—Shipping Container Specifications
49 CFR 179—Specifications for Tank Cars

Title 10: U.S. Nuclear Regulatory Commission
10 CFR 71—Packaging of Radioactive Materials for Transport and Transportation of Radioactive Materials Under Certain Conditions

Title 39: U.S. Postal Service
Domestic Mail Manual, U.S. Postal Service Regulations, Part 124. (Postal Regulations for Transport of Radioactive Matter are published in U.S. Postal Service Publication 6, and in the U.S. Postal Manual.)

and requirements (IAEA 1978). Materials not subject to DOT regulations may be subject to use or transfer regulations issued by the NRC or even the EPA.

REGULATIONS FOR SAFE TRANSPORT

A primary consideration in safe transportation of radioactive materials is the use of proper packaging for the specific radioactive material to be transported. In order to determine the packaging requirements, the following questions must be answered.

1. What radionuclides are being shipped? 49 CFR §173.435 contains a listing of over 250 specific radionuclides. Certain ground rules for dealing with unlisted or unknown radionuclides, or with mixtures of radionuclides, appear in 49 CFR §173.433.
2. What quantity of the radionuclides is being shipped? Packaging requirements are related to the activity of the material.
3. Is the radionuclide material *normal* or *special* form? Special form refers to materials that, if released from a package, would present a direct external radiation hazard, but not from contamination (Figure 11.28.1). Figure 11.28.2 details normal form materials that are, therefore, any radioactive materials that do not qualify as special form.

QUANTITY LIMITS AND PACKAGING

The quantity or specific activity of a radioactive material determines the packaging requirements. The regulations use A_1 and A_2 values as points of reference for quantity limitations for every radionuclide. Every radionuclide is assigned an A_1 and an A_2 value. These two values, in curies, are the maximum activity of that radionuclide that may be transported in a Type A package (Figure 11.28.3). Table 11.28.2 gives examples of A_1 and A_2 values for some typical radionuclides. Type B

quantities (Figure 11.28.4) are defined as exceeding the appropriate A_1 or A_2 value. Type B packages, highway route controlled quantities, and fissile radioactive materials are additionally controlled by the NRC regulations in 10 CFR Part 71.

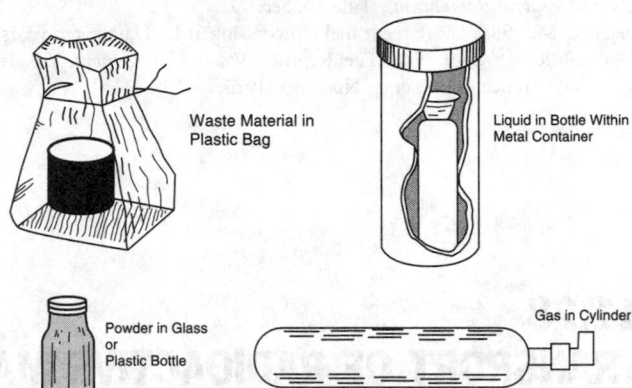

FIG. 11.28.2 Normal Forms of Radioactive Materials 49 CFR §173.403(s). Normal form materials may be solid, liquid or gaseous and include material that has not been qualified as special form. Type A Package Limits are A_2 Values.

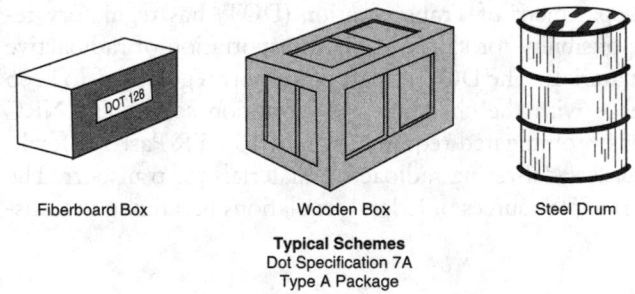

Typical Schemes
Dot Specification 7A
Type A Package

FIG. 11.28.3 Typical Type A Packaging. Package must withstand normal conditions (49 CFR §173.465) of transport, without loss or dispersal of radioactive contents.

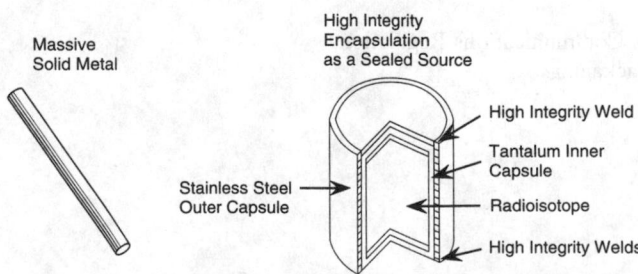

FIG. 11.28.1 *Special Form* R.A.M. (49 CFR §§173.403[z] and 173.469[a]). May present a direct radiation hazard if released from package, but presents little hazard due to contamination. Special form R.A.M. may be a *natural* characteristic, i.e., massive solid metal, or *acquired* through high integrity encapsulation.

TABLE 11.28.2 TYPE A PACKAGE QUANTITY LIMITS FOR SELECTED RADIONUCLIDES (ADDITIONAL RADIONUCLIDES ARE LISTED IN 49 CFR §173.435)

Symbol of Radionuclide	Element and Atomic Number	A_1 (Ci) (Special Form)	A_2 (Ci) (Normal Form)
^{14}C	Carbon (6)	1000	60
^{137}Cs	Cesium (55)	30	10
^{99}Mo	Molybdenum (42)	100	20
^{235}U	Uranium (92)	100	0.2
^{226}Ra	Radium (88)	10	0.05
^{201}Pb	Lead (82)	20	20

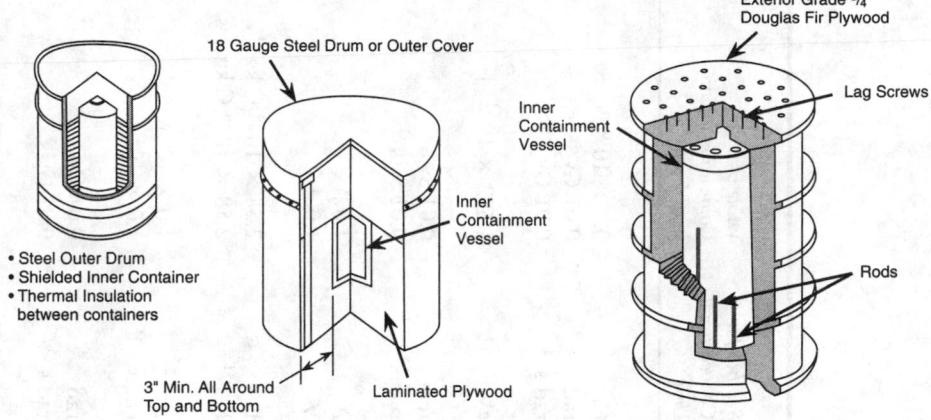

FIG. 11.28.4 Typical Type B Packagings. Package must stand both normal (49 CFR §173.465) and accident (10 CFR Part 71) test conditions without loss of contents.

TABLE 11.28.3 REMOVABLE EXTERNAL RADIOACTIVE CONTAMINATION: WIPE LIMITS

Contaminant	Maximum Permissible Limits	
	uCi/cm²	dpm/cm²
Beta/gamma-emitting radionuclides: all radionuclides with half-lives less than ten days; natural uranium; natural thorium; uranium-235; uranium-238; thorium-232; thorium-228 and thorium-230 when contained in ores or physical concentrates	10^{-5}	22
All other alpha-emitting radionuclides	10^{-6}	2.2

uCi/cm² = microcuries per square centimeter.
dpm/cm² = disintegrations per minute per square centimeter.

EXTERNAL RADIATION AND CONTAMINATION LEVELS

Radiation levels may not exceed certain dose rates at any point from the package's external surface.

A. 200 millirems per hour at the surface
B. 10 millirems per hour at one meter from the surface.

If the package is transported in an "exclusive use" closed transport vehicle, the maximum radiation levels may be:

A. 1000 millirems per hr on the accessible surface of the package
B. 200 millirems per hr at the external surface of the transport vehicle
C. 10 millirems per hr at two meters from external surface of the vehicle
D. 2 millirems per hr in any position in the vehicle occupied by a person.

TABLE 11.28.4 RADIOACTIVE MATERIALS PACKAGES MAXIMUM RADIATION LEVEL LIMITATIONS (SEE SECTIONS 173.441(A) AND (B)

Radiation level (dose) rate at any point on external surface of any package of R.A.M. may not exceed:
  A. 200 millirem per hr.
  B. 10 millirem per hr at one meter (*transport index* may not exceed 10).
Unless the packages are transported in an *exclusive use* closed transport vehicle (aircraft prohibited), then the maximum radiation levels may be:
  A. 1000 millirem per hr on the accessible external package surface.
  B. 200 millirem per hr at external surface of the vehicle.
  C. 10 millirem per hr at two meters from external surface of the vehicle.
  D. 2 millirem per hr in any position of the vehicle which is occupied by a person.

DOT regulations also prescribe limits for control of removable (non-fixed) radioactive contamination as shown in Table 11.28.3. Maximum levels for materials packages are covered in Table 11.28.4. A conversion chart (Table 11.28.5) and a list of NRC contacts are also provided for reference purposes.

—*Paul A. Bouis*

References

International Atomic Energy Agency (IAEA). 1978. Regulations for the safe transportation of radioactive materials. *Safety Series No. 6.* IAEA.
U.S. Department of Transportation (DOT). 1983. A review of the department of transportation regulations for transportation of radioactive materials. Washington, D.C.: U.S. Government Printing Office.

TABLE 11.28.5 CONVERSION FACTORS FOR IONIZING RADIATION

Quantity	Symbol for Quantity	Expression in SI Units	Special Name for SI Units	Expression in Symbols for SI Units	Symbols Using Special Names and Other Units	Conventional Units	Symbol for Conventional Unit	Value of Conventional Unit in SI Units
					Conversion Between SI and Other Units			
Activity	A	1 per second	becquerel	s^{-1}	Bq	curie	Ci	3.7×10^{10} Bq
Absorbed dose	D	joule per kilogram	gray	$J\,kg^{-1}$	Gy	rad	rad	0.01 Gy
Absorbed dose rate	$\dot{D}$	joule per kilogram second		$J\,kg^{-1}\,s^{-1}$	Gy s⁻¹	rad	rad s⁻¹	0.01 Gy s⁻¹
Average energy per ion pair	W	joule		J		electronvolt	eV	1.602×10^{-19} J
Dose equivalent	H	joule per kilogram	sievert	$J\,kg^{-1}$	Sv	rem	rem	0.01 Sv
Dose equivalent rate	$\dot{H}$	joule per kilogram second		$J\,kg^{-1}\,s^{-1}$	Sv s⁻¹	rem per second	rem s⁻¹	0.01 Sv s⁻¹
Electric current	I	ampere		A		ampere	A	1.0 A
Electric potential difference	U, V	watts per ampere	volt	$W\,a^{-1}$	V	volt	V	1.0 A
Exposure	X	coulomb per kilogram		$C\,kg^{-1}$		roentgen	R	2.58×10^{-4} C kg⁻¹
Exposure rate	$\dot{X}$	coulomb per kilogram second		$C\,kg^{-1}\,s^{-1}$		roentgen	R s⁻¹	2.58×10^{-4} C kg⁻¹ s⁻¹
Fluence	ϕ	1 per meter squared		m^{-2}		1 per centimeter squared	cm⁻²	1.0×10^{4} n⁻²
Fluence rate	Φ	1 per meter squared second		$m^{-2}\,s^{-1}$		1 per centimeter squared second	cm⁻² s⁻¹	1.0×10^{4} m⁻² s⁻¹
Kerma	K	joule per kilogram	gray	$J\,kg^{-1}$	Gy	rad	rad	0.01 Gy
Kerma rate	$\dot{K}$	joule per kilogram second		$J\,kg^{-1}\,s^{-1}$	Gy s⁻¹	rad per second	rad s⁻¹	0.01 Gy s⁻¹
Lineal energy	y	joule per meter		$j\,m^{-1}$		kiloelectron volt per micrometer	keV μm⁻¹	1.602×10^{-10} J m⁻¹
Linear energy transfer	L	joule per meter		$j\,m^{-1}$		kiloelectron volt per micrometer	keV μm⁻¹	1.602×10^{-10} J m⁻¹
Mass attenuation coefficient	μ/p	meter squared per kilogram		$m^{2}\,kg^{-1}$		centimeter squared per gram	cm² g⁻¹	0.1 m² kg⁻¹

(Continued on next page)

TABLE 11.28.5 (Continued)

Quantity	Symbol for Quantity	Expression in SI Units	Special Name for SI Units	Expression in Symbols for SI Units	Symbols Using Special Names	Conventional Units	Symbol for Conventional Unit	Value of Conventional Unit in SI Units
Mass energy transfer coefficient	μ_t/ρ	meter squared per kilogram		$m^2\ kg^{-1}$		centimeter squared per gram	$cm^2\ g^{-1}$	$0.1\ m^2\ kg^{-1}$
Mass energy absorption coefficient	μ_{en}/ρ	meter squared per kilogram		$m^2\ kg^{-1}$		centimeter squared per gram	$cm^2\ g^{-1}$	$0.1\ m^2\ kg^{-1}$
Mass stopping power	S/ρ	joule meter squared per kilogram		$J\ m^2\ kg^{-1}$		MeV centimeter squared per gram	$MeV\ cm^2\ g^{-1}$	$1.602 \times 10^{-14}\ J\ m^2\ kg^{-1}$
Power	P	joule per second	watt	$J\ s^{-1}$	W	watt	W	1.0W
Pressure	P	newton per meter squared	pascal	$N\ m^{-2}$	Pa	torr	torr	(101325/760)Pa

Conversion Between SI and Other Units

Quantity	Symbol for Quantity	Expression in SI Units	Special Name for SI Units	Expression in Symbols for SI Units	Symbols Using Special Names	Conventional Units	Symbol for Conventional Unit	Value of Conventional Unit in SI Units
Radiation chemical yield	G	mole per joule		$mol\ J^{-1}$		molecules per 100 electron volts	molecules $(100\ eV)^{-1}$	1.04×10^{-7} mole J^{-1}
Specific energy	z	joule per kilogram	gray	$J\ kg^{-1}$	Gy	rad	rad	0.01 Gy

Converting SI Units/Non-SI Units

To Convert:

From	To	Multiply By
becquerel (Bq)	curie	2.7×10^{-11}
curie (Ci)	becquerel	3.7×10^{10}
gray (Gy)	rad	100
rad (rad)	gray	0.01
sievert (Sv)	rem	100
rem (rem)	sievert	0.010

Taken from the National Council on Radiation Protection and Measurements Report No. 82. "SI Units in Radiation Protection and Measurements". Reproduced by permission of the copyright owner. Information regarding data in these tables is presented in the publication "NCRP Report No. 82" and is available from NCRP, 7910, Woodmont Avenue, Suite 1016, Bethesda, Maryland 20814.

NRC Contacts for Further Information

Division of Low-Level Waste Management and Decommissioning NMSS
U.S. Nuclear Regulatory Commission
1555 Rockville Pike
Rockville, MD 20852
(301) 415-7000

Public Affairs
U.S. Nuclear Regulatory Commission
1555 Rockville Pike
Rockville, MD 20852
(301) 415-7715

State Liaison Officer
Region I
475 Allendale Road
King of Prussia, PA 19406
(610) 337-5246

State & Government Affairs Staff Director
Region II
101 Marietta Street NW, Suite 2900
Atlanta, GA 30323
(404) 331-5597

State and Government Affairs Director
Region III
801 Warnerville Road
LaSalle, IL 60532-4351
(630) 829-9500

State Liaison Officer
Region IV
Parkway Central Plaza Building
611 Ryan Plaza Drive, Suite 400
Arlington, TX 76011-8064
(817) 860-8100

State Liaison Officer
Region V
1450 Maria Lane, Suite 300
Walnut Creek, CA 94596-5368
(510) 975-0200

Bibliography

American Society for Testing and Materials (ASTM). 1983. *A standard guide for examining the incompatibility of selected hazardous waste based on binary chemical mixtures.* Philadelphia, Pa.

Berlin, R.E., and C.C. Stanton. 1989. *Radioactive waste management.* New York, N.Y.: Wiley.

Blackman, W.C., Jr. 1992. *Basic hazardous waste management.* Boca Raton, Fla.: Lewis Publishers.

Chapman, N.A., and I.G. McKinley. 1987. *The geological disposal of nuclear waste.* Chichester, Great Britain: John Wiley & Sons.

Chenoweth, D. 1995. DOT penalties for shipping container violations. *Chemical Processing* (April).

Cheremisinoff, P. 1990. Biological treatment of hazardous wastes, sludges, and wastewater. *Pollution Engineering* (May).

Cheremisinoff, P. 1992. *A guide to underground storage tanks: evolution, site assessment, and remediation.* Englewood Cliffs, N.J.: Prentice-Hall, Inc.

David, M.L. and D.A. Cornwell. 1991. *Introduction to environmental engineering.* New York, N.Y.: McGraw-Hill, Inc.

Gershey, E.L., R.C. Klein, E. Party, and A. Wilkerson. 1990. *Low-level radioactive waste: from cradle to grave.* New York, N.Y.: Van Nostrand Reinhold.

Gollnick, D.A. 1988. *Basic radiation protection technology.* 2nd ed. Pacific Radiation Corp. Altadena, Calif.

Greenberg, A.E., L.S. Clesceri, and A.D. Eaton. 1992. *Standard methods for the examination of water and wastewater.* 18th ed. APHA. Washington, D.C.

Harris, M. 1988. Inhouse solvent reclamation efforts in Air Force maintenance operations, of *Hazardous waste minimization within the Department of Defense,* edited by J.A. Kaminsky. Office of the Deputy Assistant Secretary of Defense (Environment) Washington, D.C.

International Atomic Energy Agency (IAEA). 1986. Assessment of the radiological impact of the transport of radioactive materials. *Technical Document 398.* Vienna, Austria.

International Atomic Energy Agency (IAEA). 1987. Safe management of wastes from the mining and milling of uranium and thorium ores. *Safety Series No. 85.* Vienna, Austria.

International Atomic Energy Agency (IAEA). 1988. Immobilization of low intermediate level radioactive wastes and polymers. *Technical Report 289.* Vienna, Austria.

International Atomic Energy Agency (IAEA). 1989. *Nuclear power and fuel cycle: status and trends.* Vienna, Austria.

Irvine, R.L. and L.H. Ketchum, Jr. 1988. Sequencing batch reactors for biological waste water treatment. *Critical Reviews in Environmental Control,* 18(4):255–294.

Leiter, J.L., ed. 1989. *Underground storage tank guide.* Salisbury, Md.: Thomas Publishing Group.

Maillet, J. and C. Sombret. 1988. High-level waste vitrification: the state-of-the-art in France. *Waste Management* 88(2):165–172.

Mattus, A.J., R.D. Doyle, and D.P. Swindlehurst. 1988. Asphalt solidification of mixed wastes. *Waste Management* 89(1):229–234.

Murray, R.L. 1989. *Understanding radioactive waste.* 3rd ed. Columbus, Oh.: Battelle Press.

National Council on Radiation Protection and Measurements (NCRP). 1987. Ionizing radiation exposure of the population of the United States. *NCRP Report No. 93.* Bethesda, Md.

National Low-Level Waste Management Program. 1993. *The 1992 state-by-state assessment of low-level radioactive wastes received at commercial disposal sites.* DOE/LLW-181. Washington, D.C.

Party, E.P., and E.L. Gershey. 1989. Recommendations for radioactive waste reduction in biomedical/academic institutions. *Health Physics* 56(4):571–572.

Snelgrove, W.L. and B.O. Paul. 1995. The chemical industry and tank car development. Chemical Processing (April).

Theodore, L. and Y.C. McGuinn. 1992. *Pollution Prevention.* New York, N.Y.: Van Nostrand Reinhold.

Ullmann's encyclopedia of industrial chemistry, 5th ed. 1993. Vol. A22:499–591. Weinheim, Germany: VCH.

United Nations Scientific Committee on the Effects of Atomic Radiation (UNSCEAR). 1977. *Sources and effects of ionizing radiation.* New York, N.Y.

U.S. Department of Transportation (DOT). 1983. *A review of the Department of Transportation regulations for transportation of radioactive materials.* Washington, D.C.

U.S. Environmental Protection Agency (EPA). 1985. Minimum technology guidance on double liner systems for landfills and surface impoundments, design, construction, and operation. *Report No. PB87-151072*. National Technical Information Service. Springfield, Va.

———. 1986. *RCRA orientation manual*. Office of Solid Waste. Washington, D.C.

———. 1987. Underground storage tank corrective action technologies. EPA 625–6–87–015. Washington, D.C.

———. 1987. Handbook—Groundwater. EPA 625–6–87–016. Office of Research and Development. Cincinnati, Oh.

———. 1987. *A compendium of technologies used in the treatment of hazardous wastes*. EPA 625–8–87–014 (September).

———. 1988. Musts for USTs. *Report No. 530–UST–88–008*. Office of Underground Storage. Washington, D.C. (September).

———. 1989. *Hazardous waste incineration measurement guidance manual*. EPA 625–6–89–021 (June).

———. 1990. *Guidance on PIC controls for hazardous waste incinerators*. EPA 625–530–SW–90–040 (April).

———. 1990. *Engineering Bulletin: Soil washing treatment*. Office of Research and Development. Cincinnati, Oh. (September).

———. 1990. *RCRA orientation manual*. 1990 edition. Superintendent of Documents. Washington, D.C.: U.S. Government Printing Office.

———. 1992. *A citizen's guide to soil washing*. EPA 524–F–92–003. Office of Solid Waste and Emergency Response. Washington, D.C. (March).

Wagner, H.N., and L.E. Ketchem. 1989. *Living with radiation*. Baltimore, Md.: The Johns Hopkins University Press.

Wentz, C. 1989. *Hazardous waste management*. McGraw-Hill, Inc.

Index

A²/O process, 829–830
Absorbent materials, 496–497
Absorption
  and air quality monitoring, 305–306
  in gaseous emission control, 377–384
    commercial applications, 382–384
    operations, 377–381
  and qualitative and quantitative chemical
      analysis, 842–843
Absorptive silencers, 503
Academic waste as source of low-level ra-
      dioactive, 1389
Accelerators, 973
Acceptable daily intake (ADI), 1282
Accumulation/washoff, 1117
Acidic reagents, role of, in neutralization,
      839
Acidification, 956
Acid mine drainage, 1052
Acid rain
  definition of, 232, 370
  effects of, 265
    on forests, 265
    on groundwater, 265–266
    on health, 266
    on materials, 266
    on soil, 265
    on surface water, 266
Acids
  fermentation of, 715
  in groundwater contamination, 1060–1061
  as proton donor, 582
  recovery of values, 975
  regeneration of, 976
Acid sewer, 617
Acoustical linings, 497
Acoustical separation, 496–497
Acoustic impedance, 451
Acoustic power, 472
  of source, 453
Acoustic privacy, 470
Acoustic trauma, 467
  damage-risk criteria, 467–468
Activated carbon, 388, 932
Activated carbon adsorption, 434, 946–947
Activated carbon for mercury, adsorption on,
      966
Activated carbon process, 397
Activated sludge, 669–670, 1342–1343

Activated-sludge processes, 698
  in sewage treatment plants, 677–678, 698,
      773
  description, 698
  design, 706–707
    artificial aeration, 703–704
    contractor retention time, 702–703
    costs, 704
    effluent quality, 704
    liquid-solids separation, 704
    microorganism concentration, 702
    organic loading, 701–702
  flow diagrams, 699
  kinetic models
    CFSTR, 704–705
    CFSTR-in-series, 705–706
    contact-stabilization activated-sludge,
        706
    PF, 705
    step-aeration activated-sludge, 706
  microbiology, 698–699
  operational problems, 707–708
  and secondary clarification, 722–723
  in soluble organics removal, 773
Acute hazardous wastes, 1255
Adaptability in product life extension, 112
Adhesions, 347
Adiabatic elapse rate, 273
Administration laws
  finding regulations, 6
  government agencies
    executive, 4
    judicial, 4
    legislative, 4
  judicial review of agency actions, 5
    deference to, 5–6
    exhaustion, 5
    standards of review, 5
  limitations on agencies, 4–5
Administrative agencies, 4
  judicial review of actions, 5
  limitations on, 4–5
  reference to, 5–6
Administrative laws, 3
Adsorbents, 388
  selection of, 388
Adsorber design, 388
Adsorption
  activated carbon, 434

in air quality monitoring, 306–307
  with chemical reaction, 434–435
  in gaseous emission control
    absorber design in, 388–392
    equipment for, 386–387
    isotherm, 385–386
  in groundwater, 1057–1058, 1061–1062
Adsorption isotherm, 385–386
Adsorption operation, 388–391
Advanced oxidation, 1105–1106
Advection, 1064–1065
Aerated grit chamber, 651
Aerated lagoons, 712
Aerated skimming tanks, 654
Aeration
  and composting, 921
  extended, 677–678
  mixing by, 638–639
  step, 677
  tapered, 677
Aerobes, 686
  obligate, 686
Aerobic bacteria, 687
Aerobic biological treatment, 1342
Aerobic composting for municipal solid waste
      (MSW) management, 1244–1245
Aerobic digestion, 862–864
  advantages of, 863
  disadvantages of, 863–864
Aerobic fluidized-bed treatment
  of industrial wastewater, 780–781
  of municipal wastewater, 778–780
Aerobic lagoons, 939
Aerobic ponds, 711
Aerobic sludge digestion, 682–683
Aerodynamic cut diameter, 368
Aerotolerant anaerobes, 686
Agitators, 425–426
Agricultural operations in groundwater conta-
      mination, 1052–1053
Agriculture, impact of global warming on,
      268–269
Air
  condition of, xvii–xix
  and fuel flow patterns, 376
Air cleaners, 443–444
Air drying
  area needed for, 917
  mechanical sludge removal, 917